997
W9-BBZ-380

REFERENCE BOOK
DOES NOT CIRCULATE

UMBC LIBRARY

The Wiley

ENCYCLOPEDIA *of* ENERGY AND THE ENVIRONMENT

Volume 1

The Wiley ENCYCLOPEDIA *of* ENERGY AND THE ENVIRONMENT

Volume 1

Attilio Bisio
Sharon Boots
Editors

A Wiley-Interscience Publication
John Wiley & Sons, Inc.
New York / Chichester / Brisbane / Toronto / Singapore

REF
TJ
163.235
W55
1997
v. 1

This text is printed on acid-free paper.

Copyright© 1997 by John Wiley & Sons, Inc.

All rights reserved. Published simultaneously in Canada.

Reproduction or translation of any part of this work beyond that permitted by Section 107 or 108 of the 1976 United States Copyright Act without the permission of the copyright owner is unlawful. Requests for permission or further information should be addressed to the Permissions Department, John Wiley & Sons, Inc., 605 Third Avenue, New York, NY 10158-0012.

Library of Congress Cataloging in Publication Data:

The Wiley Encyclopedia of energy and the environment / Attilio Bisio, Sharon Boots, editors.
p. cm.
Abridged ed. of: Encyclopedia of energy technology and the environment. c1995.
"A Wiley-Interscience publication."
Includes bibliographical references and index.
ISBN 0-471-14827-X (cloth : alk. paper)
1. Power resources—Handbooks, manuals, etc. 2. Environmental protection—Handbooks, manuals, etc. I. Bisio, Attilio. II. Boots, Sharon. III. Title: Encyclopedia of energy technology and the environment.
TJ163.235.W55 1996
333.79'03—dc20 96-2734

Printed in the United States of America

10 9 8 7 6 5 4 3 2

EDITORIAL STAFF

Editors: **Attilio Bisio**
Sharon Boots
Paula Siegel

Associate Editor: **Hannah Ben-Zvi**

Editorial Board: **David Gushee, Congressional Research Service**
Peter Lederman, New Jersey Institute of Technology
Ben Luberoff, Consultant
Jim Mathis, New Jersey Commission of Science and Technology
David Richman, Department of Energy
Jefferson Tester, Massachusetts Institute of Technology

CONTRIBUTORS

A. Warren Adam, *Consultant, Rockford, Illinois,* Organic Rankine Engines

K. A. G. Amankwah, *Syracuse University, Syracuse, New York,* Hydrogen Storage Systems

John Appleby, *Texas A&M University, College Station, Texas,* Fuel Cells

Emil Attanasi, *United States Geological Survey, Reston, Virginia,* Petroleum Reserve (Oil and Gas Reserve)

William Banta, *American University, Washington, D.C.,* Oceanography

Franco Barbir, *University of Miami, Coral Gables, Florida,* Transportation Fuels, Hydrogen

Larry Baxter, *Sandia National Laboratories, Livermore, California,* Ash

Larry Beaumont, *R. W. Beck, Denver, Colorado,* Waste-to-Energy, Economics

Richard Bechtold, *EA Engineering, Silverspring, Maryland,* Methanol Vehicles

Martin Bender, *The Land Institute, Salina, Kansas,* Agriculture and Energy

Ron Berglund, *TRC, Houston, Texas,* Exhaust Control, Industrial

J. Bernaden, *Johnson Control Inc., Milwaukee, Wisconsin,* Energy Management, Principles

Richard A. Birdsley, *USDA Forest Service, Northeastern Forest Station, Radnor, Pennsylvania,* Carbon Storage in Forests

Attilio Bisio, *Atro Associates, Mountainside, New Jersey,* District Heating; Life Cycle Analysis; Nuclear Power; OPEC; Renewable Resources

Frank Black, *Atmospheric Research and Exposures Assessment, EPA, Research Triangle Park, North Carolina,* Automobile Emissions

George Blomgren, *Eveready Battery Co., Westlake, Ohio,* Batteries

Karl Boer, *University of Delaware, Newark, Delaware,* Global Health Index

Gary R. Boss, *General Accounting Office, Washington, D.C.,* Energy Policy Planning

Ugo Bozzano, *UOP, Des Plains, Illinois,* Reformulated Gasoline

Pamela J. Brodowicz, *EPA, Ann Arbor, Michigan,* Air Pollution, Automobile, Toxic Emissions

F. J. Brooks, *GE Industrial and Power Systems, Schenectady, New York,* Gas Turbines

Bill Browning, *Rocky Mountain Institute, Snowmass, Colorado,* Lighting, Energy Efficient

Gregory Bryant, *Consultant, Eugene, Oregon,* Human Powered Vehicles

Bruce Bunker, *Pacific Northwest Laboratories, Richland, Washington,* Nuclear Materials, Radioactive Tank Wastes
Andrew F. Burke, *Consultant, Idaho Falls, Idaho,* Hybrid Vehicles
Joan Bursey, *Radian Corp., Research Triangle Park, North Carolina,* Environmental Analysis, Mass Spectrometry
Richard J. Camm, *University of Manchester, Institute of Science and Technology, Manchester, United Kingdom,* Computer Applications for Energy Efficient Systems
Penny M. Carey, *EPA, Warren, Michigan,* Air Pollution: Automobile, Toxic Emissions
Lon Carlson, *Illinois State University, Normal, Illinois.* Environmental Economics
John A. Casazza, *CSA Energy Consultants, Arlington, Virginia,* Electric Power Systems and Transmission
Clarence D. Chang, *Mobil Research and Development, Princeton, New Jersey,* Fuels, Synthetic, Liquid
Scott Chaplin, *Rocky Mountain Institute, Snowmass, Colorado,* Recycling and Reuse, Energy savings; Water Efficiency
Norman Chigier, *Carnegie Mellon University, Pittsburgh, Pennsylvania,* Combustion Systems, Measurements; Liquid Fuel Spray Combustion
Robert H. Clark, *Consultant, Canada,* Tidal Power
Jerry L. Clayton, *U.S. Geological Survey, Reston, Virginia,* Coalbed Gas
Stephen Cleary, *Medical College of Virginia, Richmond, Virginia,* Electromagnetic Fields, Health Effects
Jeffrey F. Clunie, *R. W. Beck and Associates, Denver, Colorado,* Waste-to-Energy Economics
Art Cohen, *University of South Carolina, Columbia, South Carolina,* Peat: Environment and Energy Uses
Richard N. Cooper, *Harvard University, Cambridge, Massachusetts,* Global Climate Change, Mitigation
Robert M. Counce, *University of Tennessee, Knoxville, Tennessee,* Gas Cleanup Absorption
John Counsil, *Santa Rosa, California,* Geothermal Energy
Burton B. Crocker, *Chesterfield, Missouri,* Air Pollution Control Methods
Daniel A. Crowl, *Michigan Technological University, Houghton, Michigan,* Hazard Analysis of Energy Facilities
Charles Coutant, *Oak Ridge National Laboratory, Oak Ridge, Tennessee,* Thermal Pollution, Power Plants
Vincent Covello, *Columbia University, New York,* Risk Assessment; Risk Communication
T. A. Czuppon, *The M. W. Kellog Company, Houston, Texas,* Hydrogen
Bukin Danley, *University of Wollongong, Wollongong, NSW, Australia,* Coal Availability
Douglas Decker, *Johnson Control, Inc., Milwaukee, Wisconsin,* Energy Management, Principles
V. R. Dhole, *University of Manchester Institute of Science and Technology, Manchester, UK,* Computer Applications for Energy Efficient Systems
William G. Dukek, *Summit, New Jersey,* Aircraft Fuels
K. G. Duleep, *EEA, Arlington, Virginia,* Automotive Engines, Efficiency
Jim Dyer, *Rocky Mountain Institute, Water Program, Snowmass, Colorado,* Water Efficiency
Alan D. Eastman, *Phillips Petroleum Research Center, Bartlesville, Oklahoma,* Hydrocarbons
W. W. Eckenfelder, *Eckenfelder, Inc., Nashville, Tennessee,* Water Quality Management
Eugene Ecklund, *EA Engineering, Silverspring, Maryland,* Methanol Vehicles
James A. Edmonds, *Pacific Northwest Laboratory, Washington, D.C.,* Carbond Dioxide Emissions, Fossil Fuel
William D. Ehmann, *University of Kentucky, Lexington, Kentucky,* Radiation Hazards, Health Physics; Radiation Monitoring
Raymond J. Ehrig, *ARISTECH Chemical Corporation, Monroeville, Pennsylvania,* Recycling, Tertiary, Plastics
Thomas D. Ellis, *DuPont Engineering, Newark, Delaware,* Incineration
William R. Ellis, *Raytheon Constructor, Ebasco Division, New York, New York,* Fusion Energy
Ramon Espino, *Exxon Research and Engineering, Annandale, New Jersey,* API Engine Service Categories; Afterburning; Carbon Residue; Exhaust and Gas Recirculation; Heat Balance; Knock; Middle Distillate; Octane Number; Vapor Lock
D. B. Firth, *University of Manchester Institute of Science and Technology, Manchester, United Kingdom, Computer Applications for Energy Efficient Systems*
Jack Fishman, *NASA, Hampton, Virginia,* Ozone, Tropospheric
Robert A. Frosch, *General Motors Research Laboratories, Warren, Michigan,* Global Climate Change, Mitigation
Mark Gattuso, *UOP, Des Plaines, Illinois,* Gum in Gasoline
N. J. Gernert, *Thermacore, Inc., Lancaster, Pennsylvania,* Cold Fusion
Lewis M. Gibbs, *Chevron Research & Technology Company, Richmond, California,* Transportation Fuels, Automotive
Jennifer Gilitz, *Consultant, Worcester, Massachusetts,* Aluminum
J. Duncan Glover, *Failure Analysis Associates, Inc., Framingham, Massachusetts,* Electric Power Distribution
Diane R. Gold, *Channing Laboratory, Boston, Massachusetts,* Air Pollution, Indoor
Henry Gong, Jr., *Rancho Los Amigos Medical Center, Downey, California, and University of Southern California, School of Medicine,* Air Pollution, Health Effects
Michael Grub, *Royal Institute of Internal Affairs, London, UK,* Carbon Dioxide Emissions, Fossil Fuels
David Gushee, *Congressional Research Service, Library of Congress, Washington, D.C.,* Energy Taxation, Automobile Fuels
George Hagerman, *SEASUN Power Systems, Alexandria, Virginia,* Wave Power
Daniel Halacy, *Consultant, Lakewood, Colorado,* Solar Cells, Solar Cooking, Solar Heating, Solar Thermal Electric
Scott Han, *Mobil Research and Development Corp., Princeton, New Jersey,* Fuels, Synthetic, Liquid Fuels
John Hem, *United States Geologial Survey, Menlo Park, California,* Water Quality Issues
Byron Y. Hill, *Union Carbide Corporation, South Charleston, West Virginia,* Gas Cleanup Absorption
Ronald Hites, *Indiana University, Bloomington,* Environmental Analysis, Mass Spectrometry Advances
Richard Houghton, *Woods Hole Research Center, Woods Hole, Massachusetts,* Carbon Cycle
James Hower, *University of Kentucky, Lexington,* Coal Availability
Thomas Hunt, *Advance Modular Power Systems, Ann Arbor, Michigan,* Sodium Heat Engines
Adrian C. Hutton, *University of Wollongong, Wollongong, NSW, Australia,* Coal Availability
Jiri Janata, *Pacific Northwest Laboratory, Richland, Washington,* Environmental Analysis, Chemical Sensor Applications
Peter A. Johnson, *United States Congress, Washington, D.C.,* Nuclear Power, Managing Nuclear Materials
Andrew Jones, *Rocky Mountain Institute, Water Program, Snowmass, Colorado,* Water Efficiency
Jack Kaye, *NASA, Washington, D.C.,* Ozone, Stratospheric
James P. Kelly, *Pacific Northwest Laboratory, Richland, Washington,* Environmental Analysis, Optical Spectroscopy
William Kenney, *Consultant, Florham Park, New Jersey,* Energy Efficiency, Calculations; Thermodynamics, Process Analysis
Naresh Khosla, *Enviro Management Research, Inc., Springfield, Virginia,* Energy Auditing
Donald L. Klass, *Entech International, Inc., Barrington, Illinois,* Fuels from Biomass
Annette Koklauner, *Gas Research Institute, Washington, D.C.,* Petroleum Markets
Paul Komor, *Office of Technology Assessment, United States Congress, Washington, D.C.,* Energy Consumption in the United States
S. A. Kruz, *The M. W. Kellog Company, Houston, Texas,* Hydrogen
Bill Kuhn, *Pacific Northwest Laboratories, Richland, Washington,* Nuclear Materials, Radioactive Tank Wastes
Karl S. Kunz, *Pennsylvania State University, University Park, Pennsylvania,* Electric Heating

James Langman, *Earth Island Institute, San Francisco, California,* Environmental and Conservation Organizations
William Lanouette, *General Accounting Office, Washington, D.C.,* Energy Policy Planning
Ben E. Law, *U.S. Geological Survey, Reston, Virginia,* Coalbed Gas
Salvatore Lazzari, *Congressional Research Service, Library of Congress, Washington, D.C.,* Energy Taxation, Automobile Fuels; Energy Taxation, Subsidies for Biomass; Energy Taxation; Biomass
Thomas H. Lee, *Massachusetts Institute of Technology, Cambridge, Massachusetts,* Global Climate Change, Mitigation
D. J. LeKang, *CSA Energy Consultants, Arlington, Virginia,* Electric Power Systems and Transmission
William S. Linn, *Rancho Los Amigos Medical Center, Downey, California, and University of Southern California, School of Medicine,* Air Pollution, Health Effects
Amory Lovins, *Rocky Mountain Institute, Snowmass, Colorado,* Agriculture and Energy; Lighting, Energy Efficient; Supercars
L. Hunter Lovins, *Rocky Mountain Institute, Snowmass, Colorado,* Agriculture and Energy
Greg Marland, Oak Ridge National Laboratory, Oak Ridge, Tennessee, Global Climate Change, Mitigation
Amy R. Marrow, *General Accounting Office, Washington, D.C.,* Energy Policy Planning
Susan L. Mayer, *Congressional Research Service, Library of Congress, Washington, D.C.,* Clean Air Act
Loch McCabe, *Resource Recycling Systems, Inc., Ann Arbor, Michigan,* Waste Management Planning
Robin S. McDowell, *Pacific Northwest Laboratory, Richland, Virginia.* Environmental Analysis, Optical spectroscopy.
Jon McGowen, *University of Massachusetts, Amherst,* Wind Power
David E. Mears, *Unocal,* Hydrocarbons
Richard Meyer, *U.S. Geological Survey, Reston, Virginia,* Bitumen
Mark Mills, *Mills, McCarthy & Associates, Inc., Chevy Chase, Maryland,* Transportation Fuels, Electricity
John Mooney, *Englehard Corporation, Iselin, New Jersey,* Exhaust Control, Automotive
J. H. Moore, *GE Industrial and Powers Systems, Schenectady, New York,* Steam Turbines
Michael Morrison, *Caminus Energy Limited, Cambridge, United Kingdom,* Carbon Dioxide
Armand Moscovici, *Kerrite Co., Seymour, California,* Insulation, Electric; Wire and Cable Coverings
Andrew Moyad, *Environmental Protection Agency, Washington, D.C.,* Energy Consumption in the United States; Nuclear Power, Decommissioning power plants
Carl Moyer, *ACUREX Environmental Corp., Mountain View, California,* Alcohol Fuels
D. S. Newsome, *The M. W. Kellog Company, Houston, Texas,* Hydrogen
Les Norford, *Massachusetts Institute of Technology, Cambridge, Massachusetts,* Building Systems
B. K. Parekh, *University of Kentucky Lexington, Kentucky,* Coal Availability
Joseph J. Perona, *University of Tennessee, Knoxville,* Gas Cleanup, Absorption
Anthony Pietsch, *Chandler, Arizona,* Closed Brayton Cycle
Richard Pinkham, *Rocky Mountain Institue, Water Program, Snowmass, Colorado,* Water Efficiency
Thomas Perdy, *Johnson Control, Inc., Milwaukee, Wisconsin,* Energy Management, Principles
Thomas Polk, *Annandale, Virginia,* Recycling
David E. Prinzing, *Foster Wheeler Environmental Corp., Sacramento, California,* Fuel Resources
Rod Quinn, *Pacific Northwest Laboratory, Richland, Washington,* Nuclear Materials, Radioactive tank wastes
Sam Raskin, *California Energy Commission, Sacramento, California,* Commercial Availability of Energy Technology
Raymond Regan, *Pennsylvania State University, University Park,* Activated sludge
Danny Reible, *Louisiana State University, Baton Rouge,* Chemodynamics
Dudley Rice, *U.S. Geological Survey, Denver, Colorado,* Coalbed Gas
James Robinson, *Trevose, Pennsylvania,* Water Conditioning
D. H. Root, *United States Geological Survey, Reston, Virginia,* Petroleum Reserve (Oil and Gas Reserve)
Arthur H. Rosenfeld, *Lawrence Berkeley Laboratory, University of California, Berkeley, California,* Global Climate Change, Mitigation
Marc Ross, *University of Michigan, Ann Arbor,* Energy Efficiency
Robin Roy, *United States Congress, Washington, D.C.,* Nuclear Power, Decomissioning Power Plants; Nuclear Power; Safety of Aging Power Plants
Edward S. Rubin, *Carnegie Mellon University, Pittsburgh, Pennsylvania,* Global Climate Change, Mitigation
Ted Russell, *Carnegie Mellon University, Pittsburgh, Pennsylvania,* Air Quality Modeling
Michael G. Ryan, *USDA Forest Service, Fort Collins, Colorado,* Carbon Balance Modeling
Sheppard Salon, *Rensselaer Polytechnic Institute, Troy, New York,* Electric Power Generation
Harry J. Sauer, Jr., *University of Missouri, Rolla, Missouri,* Air Conditioning
Barbara Schaefer-Pederson, *UOP, Des Plaines, Illinois,* Hydrocracking
Donald Scherer, *Bowling Green State University, Bowling Green, Ohio,* Sustainable Resources, Ethics
Ron Schmitt, *Amoco Corporation, Chicago, Illinois,* Petroleum Refining, Emissions and Wastes
James A. Schwartz, *Syracuse University, Syracuse, New York,* Hydrogen Storage Systems
Andrew C. Scott, *University of London, Surrey, United Kingdom,* Coal Availability
Neil Seldman, *Institute for Local Self Reliance, Washington, D.C.,* Recycling, History in the United States
Richard J. Seymour, *Texas A&M University, College Station,* Renewable Resources from the Ocean
Robert M. Shaubach, *Thermacore, Inc., Lancaster, Pennsylvania,* Cold Fusion
Eric Silberhorn, *Technology Sciences Group, Washington, D.C.,* PCB's
E. A. Skrabek, *Fairchild Space and Defense Company, Germantown, Maryland,* Thermoelectric Energy Conversion
B. Smith, *University of Manchester Institute of Science and Technology, Manchester, United Kingdom,* Computer Applications for Energy Efficient Systems
William Smith, *Yale University, New Haven, Connecticut,* Acid Rain
L. Douglas Smoot, *Brigham Young University, Provo, Utah,* Coal Combustion
James Speight, *Laramie, Wyoming,* Extra Heavy Oils; Fuels, Synthetic, Gaseous Fuels; Hydroprocessing; Kerosene; Liquefied Petroleum Gas; Natural Gas; Pipelines; Tar Sands; Underground Gasification; Visbreaking
John S. Spencer, Jr.,*USDA, St. Paul, Minnesota,* Forest Resources
Edward M. Stack, *University of South Carolina, Columbia,* Peat: Environment and Energy Use
Andrew Steer, *The World Bank, Washington, D.C.,* Global Environmental Change, Population effect
Donald H. Stedman, *University of Denver, Denver, Colorado,* Automobile Emissions, Control
Andrew Steer, *The World Bank, Washington, D.C.,* Global Environmental Change, Population and Economic Growth
Dan Steinmeyer, *Monsanto, St. Louis, Missouri,* Energy Conservation: Energy Management, Process
J. Hugo Steven, *ICI Chemicals & Polymers Ltd., United Kingdom,* Refrigerant Alternatives
Deborah D. Stine, *National Academies of Science and Engineering, Washington, D.C.,* Global Climate Change, Mitigation
Frank Stodolsky, *Argonne National Laboratory, Washington, D.C.,* Transportation Fuels, Natural Gas
Robert Szaro, *USDA Forest Service, Washington, D.C.,* Biodiversity Maintenance
Patrick Ten Brink, *Caminus Energy Ltd., Cambridge, UK,* Carbon Dioxide Emissions, Fossil Fuels

Gregory Thompson, *UOP, Des Plaines, Illinois,* Hydrocracking
David Tillman, *Foster Wheeler Environmental Corporation, Sacramento, California,* Fuel Resources; Fuels from Waste
Thomas W. Tippett, *UOP, Des Plaines, Illinois,* Hydrocracking
Sergio C. Trindade, *SET International Ltd., Scarsdale, New York,* Transportation Fuels, Ethanol Fuels in Brazil
Diane E. Vance, *Analytical Services Organization, Oak Ridge, Tennessee,* Radiation Hazards, Health Physics Radiation Monitoring
John Vandermeulen, *Bedford Institute of Oceanography, Nova Scotia, Canada,* Oil Spills
Carl Vansant, *HCI Publications, Kansas City, Missouri,* Hydropower
Luis A. Vega, *Pacific International Center for High Technology Research, Honolulu, Hawaii,* Ocean Thermal Energy Conversion
Walter Vergara, *The World Bank, Washington, D.C.,* Transportation Fuels, Ethanol Fuels in Brazil
T. Nejat Veziroglu, *University of Miami, Coral Gables, Florida,* Transportation Fuels, Hydrogen
Jud Virden, *Pacific Northwest Laboratories, Richland Washington,* Nuclear Materials, Radioactive Tank Wastes
Karl Vorres, *Argonne National Laboratory, Argonne, Illinois,* Lignite and Brown Coal
Michael P. Walsh, *Arlington, Virginia,* Air Pollution, Automobile; Clean Air Act, Mobile Sources
Richard Waring, *Oregon State University, Corvallis, Oregon,* Carbon Balance Modeling
Jeffrey Warshauer, *Foster Wheeler Environmental Corporation, Sacramento, California,* Fuel Resources
David Wear, *Forestry Sciences Laboratory, Research Triangle Park, North Carolina,* Forestry, Sustainable
Mary Wees, *HDR Engineering, Omaha, Nebraska,* Waste to Energy
Kenneth Wilund, *Argonne National Laboratory, Washington, D.C.,* Transportation Fuels, Natural Gas
Barry M. Wise, *Pacific Northwest Laboratory, Richland, Washington,* Environmental Analysis, Chemical Sensor Applications
George Wolff, *GM Research and Environmental Staff, Warren, Michigan,* Air Pollution
Charles Wood, *Southwest Research Institute, San Antonio, Texas,* Automotive Engines
Markus Zahn, *MIT, Cambridge, Massachusetts,* Electromagnetic Fields
John Zerbe, *Forest Products Laboratory, Madison, Wisconsin,* Wood for Heating and Cooking

PREFACE

The Wiley Encyclopedia of Energy and the Environment brings to its readers an accessible introduction to topics in energy production, energy use, and on how these technologies affect the environment. Issues of technology are presented along with legal and policy issues, providing an integrated and fully comprehensive source.

As the editors Attilio Bisio and Sharon Boots said in their introduction to the full edition, "Energy touches all aspects of life on this planet." Because these topics permeate our lives, it is important for information on these issues to be available to readers at all levels of expertise. In this Encyclopedia, you will find a basic treatment of energy related topics and their impact on the environment as well as bibliographic sources for further reading. Extensive cross-referencing, the self-contained format of the articles, and the index allow for readers approaching the information from varied starting points to be lead directly to the information sought.

The 1995 Association of American Publishers, Inc. Award for Excellence in Professional/Scholarly Publishing, Chemistry, was awarded to the **Encyclopedia of Energy Technology and the Environment.** The work presented here is based on that award-winning, technical work and aims to bring the same quality in treating issues of energy production and the environment to a broader audience.

Reviews of the technical Encyclopedia cited it as the "definitive encyclopedia on energy technology" and stated that the "publication lives up to every expectation that one might have of an encyclopedia dealing with energy technologies and their environmental impacts." This condensed version retains the quality, the scope and the clarity, while presenting the material in a manner more accessible to readers without scientific training. Additional photographs, labeled as plates, have been added to emphasize and illustrate concepts fully.

We hope that you will come to rely on **The Wiley Encyclopedia of Energy and the Environment** as a useful, accurate and informative reference.

This work would not have been possible without the considerable effort and talent of the contributors; we take this opportunity to thank them.

Hannah Ben-Zvi
Associate Editor

CONVERSION FACTORS

Often in dealing with questions about energy, it is useful to be able to convert from one set of units to another. Most often, energy units have their origin in some historic event such as the development of steam power. Therefore, a variety of units are used throughout the world.

The following tables are provided:

1. Common Energy Conversion Factors
2. Aggregate Energy Equivalents
3. Weights of Typical Petroleum Products
4. Equivalence of Mass and Energy
5. Energy Factors
6. Power Factors
7. Fluid Flow Factors
8. Useful Factors for Global Climate Change

Table 1. Common Energy Conversion Factors

	Joule	Quadrillion BTU	Kilogram Calorie	Metric Ton of Coal Equivalent
1 Joule	1	947.9×10^{-21}	239×10^{-6}	34.1×10^{-12}
1 Quadrillion BTU	1055×10^{15}	1	252×10^{12}	36.0×10^{6}
1 Kilogram Calorie	4184	3966×10^{-18}	1	142.9×10^{-9}
1 Metric Ton of Coal Equivalent	29.3×10^{9}	27.8×10^{-9}	7×10^{6}	1
1 Barrel of Oil Equivalent	6119×10^{6}	5.8×10^{-9}	1462×10^{3}	0.21
1 Metric Ton of Oil Equivalent	44.8×10^{9}	42.4×10^{-9}	10.7×10^{6}	1.53
1 Cubic Meter of Natural Gas	37.3×10^{6}	35.3×10^{-12}	8905	1272×10^{-6}
1 Terawatt Year	31.5×10^{18}	29.9	7537×10^{12}	1076×10^{6}

Table 2. Aggregate Energy Equivalents

1 MBDOE = 1 million barrels per day of oil equivalent
= 50 million tons of oil equivalent per year
= 76 million metric tons of oil equivalent per year
= 57 billion cubic meters of natural gas per year
= 2.2 10^{18} joules per year
= 530 × 10^{12} kilocalories per year
= 2.1 × 10^{15} Btus per year = 2.1 quads
= 620 10^{9} kwh per year

1 QUAD = 1 quadrillion Btus = 10^{15} Btus
= 500,000 petroleum barrels a day
182,500,000 barrels per year
= 40,000,000 short tons of bituminous coal = 36,363,636 metric tons
= 1 trillion (10^{12}) cubic feet of natural gas
= 100 billion (10^{11}) kwh (based on 10,000 Btu/kwh heat rate)

1 kilowatt-hour of hydropower
= 10 × 10^{3} Btus
= 0.88 lb coal
= 0.076 gallon crude oil
= 10.4 cubic feet of natural gas

1 MTCE = one million short tons of coal equivalent
= 4.48 × 10^{6} barrels of crude oil
= 67 tons of crude oil
= 25.19 × 10^{12} cubic feet of natural gas

Table 3. Weights of Typical Petroleum Products[a]

	Pounds per U.S. Gallon	Pounds per 55-gal Drum	Kilograms per Cubic Meter	Barrels (42-gal) per Short Ton	Barrels (42-gal) per Metric Ton
LP-Gas	4.52	248	541.6	10.5	11.6
Aviation gasoline	5.90	325	707.0	8.2	8.9
Motor gasoline	6.17	339	739.3	7.7	8.5
Kerosene	6.76	372	810.0	7.0	7.8
Distillate fuel oils	7.05	388	845.8	6.8	7.5
Lubricating oils	7.50	413	898.7	6.3	7.0
Residual fuel oils	7.88	434	944.2	6.0	6.7
Paraffin Wax		367	800.1	7.1	7.9
Grease		458	998.8	5.7	6.3
Asphalt		477	1039.2	5.5	6.1

[a] Source: U.S. Energy Information Administration, *Monthly Energy Review.*

Table 4. Equivalence of Mass and Energy

	1 Electron Mass	1 Atomic Mass Unit	1 Gram
Million electron volts	0.511	931.5	5.61 × 10^{26}
Joules	8.19 × 10^{-14}	1.49 × 10^{-10}	8.99 × 10^{13}
Btu	7.76 × 10^{-17}	1.42 × 10^{-13}	8.52 × 10^{10}
Kilowatt hours	2.27 × 10^{-20}	4.15 × 10^{-17}	2.50 × 10^{7}
Quads	7.76 × 10^{-35}	1.42 × 10^{-31}	8.52 × 10^{-8}

Table 5. Energy Factors

	1 Electron-Volt	1 Joule	1 British Thermal Unit	1 Kilocalorie	1 Kilowatt-Hour
Electron volt	1	6.24 × 10^{18}	6.58 × 10^{21}	2.61 × 10^{22}	2.25 × 10^{25}
Joule	1.60 × 10^{-19}	1	1054	4184	3.6 × 10^{6}
Calorie	3.83 × 10^{-20}	0.24	252	1000	860.4 × 10^{5}
Btu	1.52 × 10^{-22}	9.48 × 10^{-4}	1	3.97	3413
Kilocalorie	3.83 × 10^{-23}	2.39 × 10^{-4}	0.252	1	860.4
Kilowatt-hour	4.45 × 10^{-26}	2.78 × 10^{-7}	2.93 × 10^{-4}	1.16 × 10^{-3}	1
Megawatt-day	1.85 × 10^{-30}	1.16 × 10^{-11}	1.22 × 10^{-8}	4.84 × 10^{-8}	4.17 × 10^{-5}
Quad	1.52 × 10^{-40}	9.48 × 10^{-22}	10^{-18}	3.97 × 10^{-18}	3.41 × 10^{-15}

Table 6. Power Factors

	1 Btu per Day	1 Kilowatt-Hour per Year	1 Watt (W)	1 Kilowatt	1 Megawatt	1 Gigawatt	1 Quad per Year
Btu/day	1	9.35	81.95	8.2×10^{4}	8.2×10^{7}	8.2×10^{10}	2.74×10^{15}
Kilowatt/year	0.11	1	8.77	8766	8.8×10^{6}	8.77×10^{9}	2.93×10^{14}
Watts	0.012	0.114	1	1000	10^{6}	10^{9}	3.34×10^{13}
Kilowatts	1.22×10^{-5}	1.14×10^{-4}	0.001	1	1000	10^{6}	3.34×10^{10}
Megawatts	1.22×10^{-8}	1.14×10^{-7}	10^{-6}	0.001	1	1000	3.34×10^{7}
Gigawatts	1.22×10^{-11}	1.14×10^{-10}	10^{-9}	10^{-6}	0.001	1	3.34×10^{4}
Quad/yr	3.65×10^{-16}	3.41×10^{-15}	2.99×10^{-14}	2.99×10^{-11}	2.99×10^{-8}	2.99×10^{-5}	1

Table 7. Fluid Flow Factors

	1 Gallon per Day	1 Acre-Foot per Year	1 Cubic Foot per Minute	1 Cubic Meter per Second	1 Billion Gallons per Day
Gallon/day	1	892	1.077×10^{4}	2.28×10^{7}	10^{9}
Acre-cubic feet/yr	1.12×10^{-3}	1	12.07	2.56×10^{4}	1.12×10^{6}
Cubic feet/min	9.28×10^{-5}	8.28×10^{-2}	1	2119	9.28×10^{4}
Cubic feet/sec	1.55×10^{-6}	1.38×10^{-3}	1.67×10^{-2}	35.31	1547
Cubic meters/sec	4.38×10^{-8}	3.91×10^{-5}	4.72×10^{-4}	1	43.8
Billion gallons/day	10^{-9}	8.92×10^{-7}	1.08×10^{-5}	2.28×10^{-2}	1

Table 8. Useful Quantities for Global Climate Change

Quantity	Value	Ref.[a]
Solar constant	1.375 kilowatts square meters	1
Earth mass	5.976×10^{24} kilogram	2
Equatorial radius	6.378×10^{6} meters	2
Polar radius	6.357×10^{6} meters	2
Mean radius	6.371×10^{6} meters	
Surface area	5.101×10^{14} square meters	
Land area	1.481×10^{14} square meters	3
Ocean area	3.620×10^{14} square meters	4
Ice sheets and glaciers area	0.14×10^{14} square meters	5
Mean land elevation	840 meters	4
Mean ocean depth	3730 meters	
Mean ocean volume	1.350×10^{18} cubic meters	4
Ocean mass	1.384×10^{21} kilograms	
Mass of atmosphere	5.137×10^{18} kilograms	6
Equatorial surface gravity	9.780 meters/second	2

[a] Sources

1. D. V. Hoyt, "The Smithsonian Astrophysical Observatory Solar Constant Program," *Rev. Geophys. Space Physics* **17,** 427–458 (1979).
2. F. Press and R. Siever, *Earth,* W. H. Freeman and Company, San Francisco, 1974.
3. B. K. Ridley, *The Physical Environment,* Ellis Horwood, Ltd., West Sussex, England, 1979.
4. H. W. Menard and S. M. Smith, "Hypsometry of Ocean Basin Provinces," *J. Geophys. Res.* **71,** 4305–4325 (1966), adopted as reference standard by Bolin (10).
5. M. F. Meier, ed. *Glaciers, Ice Sheets, and Sea Level: Effect of a CO_2-Induced Climatic Change.* DOE/ER-60235-1, U.S. Department of Energy, Carbon Dioxide Research Division, Office of Basic Energy Sciences, Washington, D.C., 1985.
6. K. E. Trenberth, "Seasonal Variations in Global Sea-Level Pressure and the Total Mass of the Atmosphere." *J. Geophys. Res.* **86,** 5238–5246 (1981); this supercedes value adopted as reference standard by Bolin (10).

The Wiley ENCYCLOPEDIA *of* ENERGY AND THE ENVIRONMENT

Volume 1

API ENGINE SERVICE CATEGORIES

RAMON ESPINO
Exxon Research and Engineering
Annandale, New Jersey

Engine Service Categories are intended to provide the consumer with recommendations concerning which engine oil should be used in a given vehicle and given climate conditions. The essence of the Service Categories is a set of tests that the engine oil must pass in order to claim a performance level. Since 1947 the tests have become increasingly more severe as well as numerous. For the gasoline engines the categories have gone from SA, alphabetically to SG, with Category SH being introduced in mid-1993. For diesel engines the highest classification level is presently CF and CG level performance expected in 1994. In 1947 the Lubrication Committee of the America Petroleum Institute (API) published the first classification and since that time, numerous revisions have been made in both the categories that define gasoline as well as diesel engines. An important concern of auto makers, which is also shared by engine oil marketers as well, has been the increasing proliferation of service categories.

In 1990 the American and Japanese automobile manufacturers had requested that the system be significantly revamped. In 1992 an agreement was reached between the API and Japanese and American auto makers. The engine oil additive manufacturers also joined in the agreement under the auspices of the Chemical Manufacturers Association. The agreement has a number of critical elements:

- Definitions of base stocks and additive changes that circumscribe the testing protocol required when changes in base stocks or additives are needed to pass an engine test. The changes in question include the concentration as well as the chemical composition of the additive or base stock.
- Testing procedures are based on sound statistical concepts for the engine tests that must be passed to claim a given category. The procedures include protocols for engine stand selection and for monitoring engine test repeatability.
- A strengthen system is used to monitor the quality claims of engine oils in the market place.

A number of antiquated Service Categories will be abolished by the API, but a new label will appear in many containers: this one sponsored by the engine manufacturers in the U.S. and Japan.

BIBLIOGRAPHY

Reading List

Marketing Technical Bulletin 93-3, Society of Automotive Engineers Inc. (SAE).

ACID RAIN

WILLIAM SMITH
Yale University
New Haven, Connecticut

Acid rain is appropriately described as an old environmental problem with a new image. Acid rain, more than any other environmental contaminant, has focused societal concern on ecosystem toxicology. Robert Angus Smith (1) is credited with first using the term *acid rain* in 1872. During the first half of the 20th century numerous European investigators added insight to this emerging environmental challenge. The longest continuous record of rain chemistry in the United States has been maintained at the Hubbard Brook Experimental Forest (U.S. Department of Agriculture, Forest Service) in central New Hampshire since 1963 (2). The federal governments of both Canada and the United States initiated systematic rain chemistry monitoring in 1976 (Canadian Network for Sampling Precipitation) and 1978 (National Atmospheric Deposition Program), respectively. Results of the latter program reveal that the acidity of precipitation in the eastern portion of the United States is approximately 10 times more acid than the western portion (Fig. 1). (See also AIR QUALITY MODELING; FOREST RESOURCES; FORESTRY SUSTAINABLE.)

Over the 150-yr history of acid rain research, concerns have been raised in regard to the ability of acid rain to erode and corrode buildings, sculpture, monuments, and other structures (see Plate I); to reduce visibility; to impair human health; and to impact adversely agricultural as well as stream, lake, and forest ecosystems.

In recognition of the expansive size of forest ecosystems, the multiple values of forests to people, and the potential for adverse interaction of regional-scale air pollutants on forest health, this article describes the current understanding of the relationship between forests and acid rain (4).

FORESTS

Forests cover approximately 33% of the terrestrial surface of the earth. Forests dominate the landscape in the United States by covering roughly 40% of the land area (approximately 4×10^{12} m^2, or 1 billion acres) and occurring in all 50 states. Native tree species that grow in the United States number 850. Foresters group forest stands of similar species composition and ecological development into forest types. A total of 250 distinct forest types are recognized in Canada and the United States (excluding Hawaii).

Forest ecosystems are characterized by enormous variability. Forest systems may differ in soil type, climate, aspect, elevation, age, and health as well as species composition and development stage. Forests may be uneven aged, even aged, all aged, mature, and overmature. Forests may

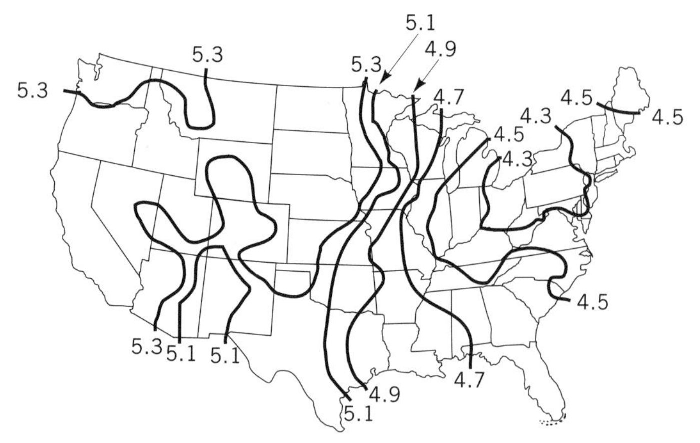

Figure 1. Average annual hydrogen ion concentrations in precipitation, expressed as pH and adjusted for amount of precipitation, for 1990. From Ref. 3.

be reproduced by seed, sprouting, and planting. Some forests have their structure completely shaped by natural forces, some may be influenced by human forces along with natural forces, and others may be completely artificial in design and establishment.

The human values associated with forest systems are almost as varied as forest types. Some values are traditional (eg, wood), long and widely appreciated, and quantifiable in standard economic terms. Other values of more recent appreciation (eg, drugs and pesticides) are developing in acceptance and sophistication and are not quantifiable in standard economic terms. For convenience, forest values may be thought of as products or services.

ACID RAIN

At room temperature, pure water dissociates to produce equal amounts (0.0000001 g/L) of hydrogen (H^+) and hydroxyl (OH^-) ions. When these ionic concentrations are equal, the solution is said to be neutral. The concentration of hydrogen ions in a solution is represented by the pH scale. The pH number reflects the negative logarithm of the hydrogen ion concentration. At neutrality, when hydrogen and hydroxyl concentrations are equal, pH is 7.0. Acid solutions in water are defined as those solutions that

Plate I. These two images of an outdoor statue show the effects of extended exposure to acid rain. The photograph on the right shows the statue prior to decay; the image on the left clearly shows the erosion due to acid rain. Courtesy of the National Office for the Preservation of Landmarks, Westfälisches Germany.

have hydrogen ion concentrations greater than the hydroxyl concentration (pH < 7). Keep in mind that the pH scale is logarithmic and that a change from pH 7.0 to pH 6.0 represents a 10-fold increase in hydrogen ion concentration and, therefore, a 10-fold increase in acidity.

Natural rain, including precipitation in relatively clean or unpolluted regions, is naturally acidic, with a pH in the range of 5.0 to 6.0. This natural acidity results from the oxidation of carbon oxides and the subsequent formation of carbonic acid. Formic and acetic acids, originating primarily from natural sources, may also contribute minor amounts of acidity to precipitation.

In regions downwind from electric-generating power stations employing fossil fuels, industrial regions, or major urban centers, precipitation can be acidified below pH 5.0. Precipitation with a pH less than 5.0 is designated acid rain. This human-caused acidification of precipitation results primarily from the release of sulfur dioxide (SO_2) and nitrogen oxides (NO and NO_2) from smokestacks and tailpipes. The sulfur and nitrogen oxides are subsequently oxidized to sulfate (SO_4^{2-}) and nitrate (NO_3^-), hydrolyzed, and returned to earth as sulfuric (H_2SO_4) and nitric (HNO_3) acids.

The atmosphere deposits acidity onto the landscape both during and in between precipitation events. In the latter case, termed dry deposition in contrast to wet deposition, the acids are delivered in the gas phase or in association with fine particles (aerosols). *Acid deposition* is a term that includes acid delivery in the form of precipitation (rain, snow, fog, and cloud moisture) plus dry deposition. In view of the importance of both wet and dry deposition in acid transfer from the atmosphere to the biosphere, acid deposition is a much more appropriate descriptor than acid rain.

The Hubbard Brook Experimental Forest, which is part of the White Mountain National Forest in central New Hampshire, (45°56′ N, 71°45′ W), has been a focal point for North American study of acid deposition. Despite the fact that Hubbard Brook is more than 100 km from any large urban-industrial area, the average annual pH since 1963 has generally fallen in the 4.0–4.4 range.

It is important to recognize that both the positive and negative effects of acid deposition on ecosystems involve the consideration of hydrogen ions, sulfate ions, nitrate ions, and heavy metal ions in acid deposition. Selected forests are judged to be at special risk to the adverse effects of acid deposition. These forests receive especially high exposure to hydrogen, sulfate, or nitrate ions due to their proximity to primary sources of sulfuric and nitric acids or because of their high elevation.

ADVERSE IMPACT OF ACID DEPOSITION ON FOREST SYSTEMS

Assessment of air pollution impact on forest systems is extremely challenging for a variety of reasons; three of the most important are forest system variability, deficiency of understanding of ecosystem- and landscape-scale phenomena, and large variation in system exposure to acid deposition. In general, forest disturbance from air pollutants is exposure related, and dose–response thresholds for a specific pollutant are different among the various organisms of the ecosystem. Ecosystem response is, therefore, a complex process. In response to low exposure to air pollution, the vegetation and soils of an ecosystem function as a sink or receptor. When exposed to intermediate loads, individual plant species or individual members of a given species may be subtly and harmfully affected by nutrient stress, impaired metabolism, predisposition to entomological or pathological stress, or direct induction of disease.

For North American forests, it is generally concluded that acid deposition influences on forest systems are neutral, ie, no adverse effects can be discriminated from natural forest dynamics. Actual effects may be slightly stimulatory (eg, nitrogen fertilization via nitrate input) or contributory to multiple-factor forest stress (eg, in high elevation, high risk (montane) forest ecosystems). In the latter case, acid deposition is presumed to be highly interactive with other stresses, subtle in manifestation, and long-term (several decades) in development. The primary hypotheses for these subtle effects are listed below. Not all of these hypotheses are supported by equal scientific evidence. The first five hypotheses are the ones that are best understood and supported by the greatest evidence; they are summarized in the following sections.

Tree Population Interaction	Forest Ecosystem Perturbation
Increased rate of soil acidification causes altered nutrient availability and root disease.	Population dynamics, tree competition, and species composition.
Cation nutrients are leached from foliage to throughfall and stem flow.	Biogeochemical cycle rates.
Cation nutrients are leached below soil horizons of active root uptake.	Biogeochemical cycle rates.
Increased available soil aluminum results in fine-root morbidity.	Population dynamics, tree competition, and species composition.
Increased available heavy metal and hydrogen ion concentrations in soil result in enhanced root uptake or impact on soil microbiota.	Decomposer impact, biogeochemical cycle rates, and species composition.
Deposition causes alteration of carbon allocation to maintenance respiration or repair or to above-ground instead of below-ground tissues.	Productivity and energy storage.
Deposition increases or decreases phytophagous arthropod activity.	Consumer impact and insect population dynamics.

Tree Population Interaction	Forest Ecosystem Perturbation
Deposition increases or decreases microbial pathogen activity.	Consumer impact and pathogen population dynamics.
Deposition increases or decreases abiotic stress influence (temperature, moisture, wind, and nutrient stresses).	Population dynamics, tree competition, and species composition.
Increased soil weathering alters soil cation availability.	Biogeochemical cycle rates.
Increased nitrogen (sulfur) deposition alters nitrogen (sulfur) cycle dynamics.	Biogeochemical cycle rates.
Deposition increases or decreases microbial symbioses.	Productivity and energy storage.
Deposition impacts one or more processes of reproductive or seedling metabolism.	Population dynamics, tree competition, and species composition.
Deposition impacts a critical metabolic process, eg, photosynthesis, respiration, water uptake, translocation, or evapotranspiration.	Population dynamics, tree competition, and species composition.

Soil Acidification

Processes that acidify forest soils include those that increase the number of negative charges, such as organic matter accumulation or clay formation, and those that remove basic cations, such as leaching of bases in association with an acid anion. Weathering by carbonic acid, organic acids (podzolization), humification, and cation uptake by roots all increase the negative charge of forest soils.

The evidence for acidification (ie, lowering of pH) of forest soils at the present rates of acid deposition in North America is not great. Forest soils at greatest risk to pH reduction from acid deposition are restricted to those limited soil types characterized by no renewal by fresh soil deposits, low cation exchange capacity, low clay and organic matter content, low sulfate adsorption capacity, high input of acidic deposition without significant base cation deposition, high present pH (5.5–6.5), and deficiency of easily weatherable materials to a 1-m depth. In addition, these high risk forest systems would need to be exposed to significant levels of acid deposition for decades.

Cation Depletion from Foliage

Vegetative leaching refers to the removal of substances from plants by the action of aqueous solutions such as rain, dew, mist, and fog. Inorganic chemicals leached from plants include all the essential macroelements. Potassium, calcium, magnesium, and manganese are typically leached in the greatest quantities. A variety of organic compounds, including sugars, amino acids, organic acids, hormones, vitamins, and pectic and phenolic substances, are also leached from vegetation. As the maturity of leaves increases, susceptibility to nutrient loss via leaching also increases and peaks at senescence. Leaves from healthy plants are more resistant to leaching than leaves that are injured, infected with microbes, infested with insects, or otherwise under stress.

Deciduous trees lose more nutrients from foliage than do coniferous species during the growing season. Conifers, however, continue to lose nutrients throughout the dormant season. The stems and branches of all woody plants lose nutrients during both the growing and dormant seasons.

The mechanism of leaching is presumed to be primarily a passive process. Cations are lost from free space areas within the plant. Under uncontaminated, natural environmental conditions, little if any cations are thought to be lost from within cells or cell walls. Pollutant exposure may predispose foliage to leaching loss by cuticular erosion, membrane dysfunction, or metabolic abnormality (Fig. 2).

Field investigations of foliar leaching have typically compared throughfall and stem flow chemistry to direct deposition chemistry to evaluate leaching. If cation enrichment is detected in stem flow or throughfall, relative to precipitation collected in the open, foliar leaching is pre-

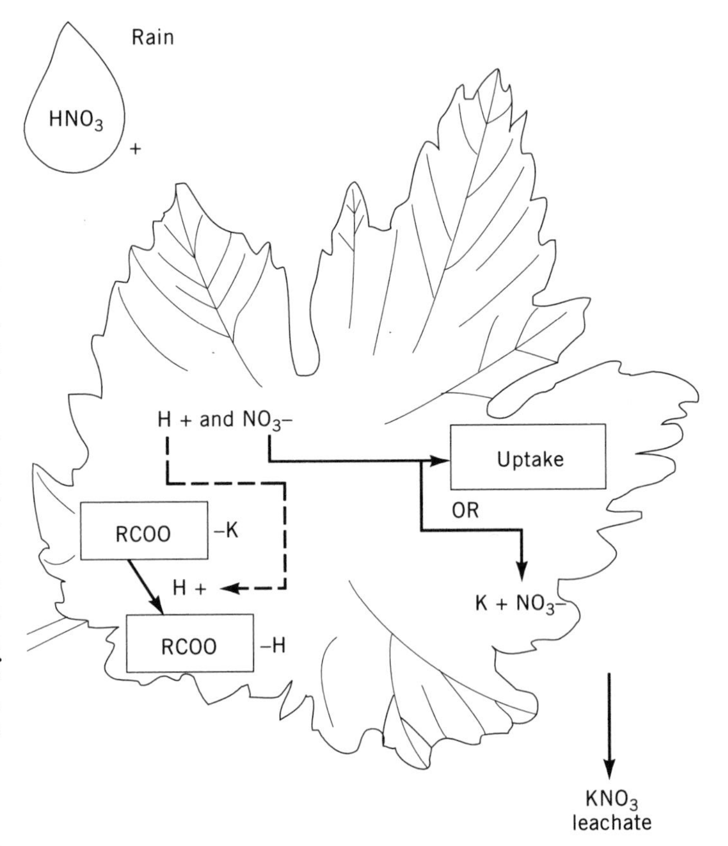

Figure 2. Foliar leaching can result from the deposition of nitric acid. Hydrogen ions from nitric acid may displace nutrient cations held on leaf cell wall exchange sites. Nitrate ions from the acid, if not taken up by the leaf or leaf microorganisms, may be available to combine with potassium and to remove this nutrient from the leaf. If potassium ions are not resupplied by the roots, forest trees may suffer from nutrient stress. From Ref. 5.

sumed. Numerous studies have provided evidence for cation enrichment in throughfall collected under forest canopies. The significance of foliar cation depletion, however, in overall forest health is generally judged relatively unimportant.

Cation Depletion from Upper Soil Horizons

Most metallic nutrient elements taken up by trees are absorbed as cations and exist in three forms in the soil: (*1*) slightly soluble components of mineral or organic material, (*2*) adsorbed into the cation exchange complex of clay and organic matter, and (*3*) small quantities in the soil solution. As water migrates through the soil profile, the movement of cations from any of the three compartments is known as cation depletion leaching.

The relative mobility of cations leached from decomposing forest litter is typically sodium > potassium > calcium > magnesium. The leaching rate largely depends on the generation of a supply of mobile anions. In forest soils, the anions are produced along with hydrogen cations and occur as acids. The hydrogen cations have a powerful substituting capability and can readily replace other cations adsorbed in the soil.

It has been proposed that in forest soils that are subject to the deposition of air pollutants sulfuric and nitric acids may provide the primary, or a significant, source of H^+ for cation displacement and mobile anions for cation transport. Soil cation leaching requires the mobility of the anion associated with the leaching acid. This results from the requirement for charge balance in soil solutions, a requirement that disallows cation leaching without associated mobile anions. Nitric and sulfuric acid deposited from the atmosphere supply mobile anions in the form of nitrate (NO_3^-) and sulfate (SO_4^{2-}), respectively. In most soils, rapid biological uptake may immobilize the NO_3^- anion. As a result, risk of cation loss associated with this anion may be restricted to ecosystems rich in nitrogen where biological immobilization of NO_3^- is minimal. In a similar manner, SO_4^{2-} may be immobilized in weathered soils by adsorption to free iron and aluminum oxides. In other soils, however (especially those low in free iron or aluminum or high in organic matter, which appears to block SO_4^{2-} adsorption sites), SO_4^{2-} may readily combine with nutrient cations and leach the elements beyond the zone of fine-root uptake by forest trees.

The movement of nutrient cations via leaching in forest soil profiles is an extremely important component of forest nutrient cycling. The evidence that has been provided by numerous experiments subjecting soil lysimeters to natural or artificially acidified precipitation indicates a potential for a meaningful acid precipitation influence on the soil leaching process. The threshold for significant increases in the rate of movement of calcium, potassium, and magnesium appears to require precipitation in the pH range of 3–4 for most systems examined. For certain forests, eg, subalpine balsam fir in New England, the threshold of increased leaching may be higher and in the range of pH 4.0–4.5.

Despite the efficacy of SO_4^{2-} leaching in selected soils, it must be realized that forest soils vary greatly in their susceptibility to it.

Forest soils have been judged more vulnerable to leaching influence by acid precipitation than are agricultural soils. However, unless the acidity of precipitation increases substantially or the buffering capacity of forest soils declines significantly, acid rain influence will not quickly or dramatically alter the productivity of most temperate forest soils via cation depletion.

Aluminum Toxicity

Aluminum toxicity to forest trees may be expressed indirectly via soluble aluminum interference with nutrient (cation) uptake by roots or directly via soluble aluminum interference with one or more tree metabolic processes. Aluminum toxicity may influence forest tree-root health; acid deposition plus natural acidifying processes may increase soil acidity enough to increase the amount of soluble aluminum available for root uptake.

Much evidence has been gathered concerning the mechanisms of aluminum toxicity to plants. Direct root effects include reduced cell division associated with aluminum binding of DNA, reduced root growth caused by inhibition of cell elongation, and destruction of epidermal and cortical cells. Indirect effects of aluminum stress, especially in forest trees, may involve interference with the uptake, translocation, or use of required nutrients such as calcium, magnesium, phosphorus, potassium, and other essential elements.

Reductions in calcium uptake by roots have been associated with increases in aluminum uptake. As a result, an important toxic effect of aluminum on forest tree roots could be calcium deficiency. Mature forest trees have a higher calcium requirement than agricultural crops. An integrated hypothesis concerning aluminum interference with calcium uptake in red spruce in the northeastern United States has been proposed (6). It has been stressed that calcium is incorporated at a constant rate per unit volume of wood produced and is not recovered from sapwood as it matures into heartwood. As a result, when aluminum and calcium are present in approximately equimolar concentrations within the soil solution or when the aluminum: calcium ratio exceeds 1, aluminum will reduce calcium uptake by competition for binding sites in the cortical apoplast of fine roots. Reduced calcium uptake will suppress cambial growth (annual ring widths) and predispose trees to disease and injury by biotic stress agents when the functional sapwood becomes less than 25% of the cross-sectional area of the stem.

Heavy Metal Toxicity

Human activities cause increases in heavy metal deposition from the atmosphere for antimony, cadmium, chromium, copper, lead, molybdenum, nickel, silver, tin, vanadium, and zinc. Small increases also have been recorded in selected environments for cobalt, manganese, and mercury. Upper soil horizons, especially forest floors with persistent organic matter accumulation, represent important sinks for heavy metals from any source, including the atmosphere. At sufficient concentration, all heavy metals are toxic to biological systems.

Direct toxicity to trees and direct and indirect toxicity to other biotic components of the forest soil by heavy metals are exposure dependent. Exposure is a function of both deposition from the atmosphere and chemical availability. More than 90% of certain heavy metals deposited from the atmosphere may be biologically unavailable. Heavy metals may be absorbed or chelated by organic matter (humic and fulvic acids); clays; or hydrous oxides of aluminum, iron, or manganese. Heavy metals also may be complexed with soluble low molecular weight compounds. Soluble cadmium, copper, and zinc may be chelated in excess of 99%. Adsorbed heavy metals remain in equilibrium with chelated metals. Heavy metals also may be precipitated as inorganic compounds of low solubility such as oxides, phosphates, and sulfates.

Adsorption, chelation, and precipitation are strongly regulated by soil pH. As pH decreases and soils become more acid, heavy metals generally become more available for biological uptake. Natural forest soils generally become more acid as they mature. Acidification in excess of natural processes is possible especially in soils with a pH greater than 5. Under this circumstance, soil acidification, associated with acid deposition may result in increased biological availability of heavy metals in the forest floor.

While components of the forest floor soil biota may be impacted by trace metals, decomposition rates or other soil processes may remain unchanged due to adaptation or shifts in species composition unreflected in gross soil process activities.

In regard to direct toxicity of heavy metal ions to plant roots, copper, nickel, and zinc toxicities have occurred frequently in grossly polluted environments. Cadmium, cobalt, and lead toxicities have occurred less frequently. Acute direct toxicity caused by heavy metal deposition or mobilization by acidification probably does not occur in forest trees located outside of urban, roadside, or selected point source industrial and electric-generating environments.

PROGNOSIS

At the regional scale, the two most important air pollutants influencing terrestrial ecosystems are acid rain and oxidant (ozone) pollutants (6). Research indicates that regional air pollution is one of the significant contemporary stresses imposed on some temperate forest ecosystems. Global air pollution, with its associated capability to cause rapid climate change, has the potential to alter dramatically forest ecosystems in the next century. The integrity, productivity, and value of forest systems is intimately linked to air quality. Failure to give careful consideration to forest resources in societal considerations of energy technologies and in management and regulation of air resources is unthinkable.

BIBLIOGRAPHY

1. E. B. Cowling, *Environ. Sci. Technol.* **16,** 110A–123A (1982).
2. G. E. Likens, F. H. Bormann, R. S. Pierce, J. S. Eaton, and N. M. Johnson, *Biogeochemistry of a Forested Ecosystem,* Springer-Verlag, New York, 1977.
3. National Acid Precipitation Assessment Program, *1990 Integrated Assessment Report,* Washington, D.C., 1991.
4. W. H. Smith in D. C. Adriano and A. H. Johnson, eds., *Acidic Precipitation, Vol. II, Biological and Ecological Effects,* Springer-Verlag, New York, 1989, pp. 165–188.
5. W. H. Smith, *Air Pollution and Forests,* Springer-Verlag, New York, 1990.
6. W. C. Shortle and K. T. Smith, Aluminum-induced calcium deficiency syndrome. *Science* **240,** 1017–1018 (1988).

ACTIVATED SLUDGE

RAYMOND REGAN
Pennsylvania State University
University Park, Pennsylvania

Each day wastewater (sewage is the old term) is produced as a result of domestic, municipal, and industrial activities. Wastewater includes various inorganic and organic contaminants that must be removed so that the water may be safely returned to the environment. Episodes of serious pollution damage have occurred in the past when wastewaters were improperly or incompletely treated. This article will summarize some of the important features of the activated sludge process, a widely used method of wastewater treatment.

BASIC PROCESS

This is how the activated sludge process works.

Wastewater collected from domestic and municipal sources is generally allowed to pass through a series of preliminary treatment steps basically to remove large debris (floating pieces of wood, rags, etc), materials that might damage mechanical parts of the tanks, pipes, and pumps (sand and gritty solids), and to provide conditions so that organic solids will be more readily removed from the wastewater (size reduction, odor reduction). Preliminary is followed by primary treatment, a processing step which allows organic and inorganic solids to be settled and thereby removed from the wastewater. Also during primary treatment, floating materials, such as food preparation discards and greases, are removed. When used for domestic and municipal wastewater, the treatment provided by the primary process is suitable for removing approximately 35% of the BOD_5 (biochemical oxygen demand measured after 5 days when incubated at 20°C) and 50% of the total SS (suspended solids) which are settable by gravity under the provisions provided for in the treatment plant. The discharge after primary treatment is then followed by further BOD and SS removal using the secondary treatment.

Based on the characteristics of the industrial wastewater, preliminary and primary treatment may not be needed. In this situation treatment may be initiated immediately with activated sludge.

Activated sludge is the principal process used to provide secondary levels of treatment of all types of wastewater containing biodegradable organics, which may be present in either soluble, colloidal or settable forms. The activated sludge process is suitable for removing the remainder of the BOD_5 and total SS. This process involves two main components, namely (*1*) the aeration/mixing tank and (*2*) the secondary clarifier/settling tank. Organics and other inorganic nutrients (food sources) present in the wastewater are introduced into the aeration/mixing tank where they are mixed with a suspension of microorganisms (referred to as mixed liquor suspended solids, MLSS) with atmospheric air. The air provides for the dual purposes of contacting the food sources with the MLSS by mixing, and supplying the oxygen needed by the microorganisms to metabolize the food. Microbial protoplasm is generated as part of the life cycle of the activated sludge microorganisms. The microbial population includes various species of bacteria, protozoa and rotifers.

After a period of aeration, the treated wastewater and MLSS are introduced into a clarifier. The clarifier provides the conditions needed for gravity separating of the MLSS from the treated wastewater. The clarified effluent is commonly disinfected and then discharged to a surface waterbody. The gravity settled MLSS is divided into two parts, namely (*1*) return flow to the aeration tank and (*2*) waste flow to remove the excess microbial protoplasm (waste sludge) produced during the aeration step. The activated sludge process is operated continuously. Therefore, the plant operators must understand how the growth of activated sludge microorganisms is effected by changes in the wastewater and other environmental factors controlling the life cycle of the microlife involved.

PROCESS DEVELOPMENTS

Activated sludge has changed little from the original report of Arden and Lockett in 1914 (1). Basically process changes have been made as a means of solving specific operating problems. Three general problem areas were (*1*) oxygen supply to meet the needs of microorganisms, (*2*) wastewater characteristics and (*3*) nutrient removal.

CONCLUSION

An expanding number of biotreatment processes are becoming available. Our knowledge of biochemical systems continues to improve and new ways are being developed to take advantage of the biologically medicated reactions. For activated sludge microorganisms, the sky seems to be the limit.

BIBLIOGRAPHY

1. E. Arden, and W. F. Lockett, "Experiments on the Oxidation of Sewage Without the Aid of Filters," *J. Soc. Chem. Indust. London* **33,** 523–539 (1914).

AFTERBURNING

Ramon Espino
Exxon Research and Engineering
Annandale, New Jersey

In a conventional turbojet or turbofan engine a significant portion of the air entering the engine bypasses the combustion zone. The temperature in the combustion chamber and therefore the amount of fuel that can be burned in the chamber is limited by the materials available today to make turbine blades. As a result, engineers must design engines with increasing amount of air bypassing the combustor. The thrust provided by this bypass air results solely from its increase in pressure and temperature provided by compression and heat exchange with the combustor zone. The bypass air exchanges heat with the combustor zone and cools it. Modern turbofan engines have air bypass ratios of 10 and greater.

Afterburning is a concept developed during the 1960s and applied in engine design since the 1970s. It takes advantage of this hot air that bypasses the combustion and burns additional fuel to provide additional engine thrust. The design of engines that have afterburners is quite complex and their operation at subsonic speeds is not very fuel efficient. Afterburners are used mainly in aircraft that operate at supersonic flow. These are mainly military aircraft and presently the Concorde supersonic transport. In the case of military aircraft, afterburners are also used at subsonic speeds since fuel economy is not a critical factor. Military jets use their afterburners to achieve better take off, higher rates of climb, and improved maneuverability.

The basic design of an afterburning turbojet or turbofan aircraft consists of a diffuser section to slow down the very hot gases exiting the combustion/turbine section and a combustion zone where additional fuel is sprayed into the air and burned. Ignition is provided by either spark ignition or by a small flame (hot streak) from the main combustor. The design of this combustion chamber is complex since one needs to maintain combustion while reducing drag. The size of a jet engine with an afterburner is double the size of a conventional engine. Finally, the afterburner unit must have the capability of expanding the gas exit area (nozzle) when the afterburners are operating in order to accommodate the increase in volumetric gas flow. The thrust of all jet engines arises from the difference in gas velocity at the exit of the engine exhaust and the velocity of the inlet air or the aircraft speed (see Aircraft engines).

The superiority of an afterburning engine over a conventional turbojet engine is demonstrated only at supersonic speeds.

BIBLIOGRAPHY

Reading List

W. J. Hesse and N. V. S. Mumford, Jr., *Jet Propulsion for Aerospace Applications,* 2nd ed., Pitman Publishing Corp., 1964, Chapt. 12.

J. L. Kerrebrock, *Aircraft Engines and Gas Turbines,* The MIT Press, Cambridge, Mass., 1978, Chapt. 4.

AGRICULTURE AND ENERGY

Amory B. Lovins
L. Hunter Lovins
Rocky Mountain Institute
Snowmass, Colorado

Marty Bender
The Land Institute
Salina, Kansas

The sustainability of an agriculture depends not only on the balance between the energy it uses and produces but also on its ability to preserve its soil and water resources. In the United States, during the past 200 years, at least one-third of the cropland topsoil has been lost. Studies within the past decade indicate that from 25 to 50% more soil per acre is being lost now than in the 1930s.

Water has also become a serious resource problem since a full two-thirds of the groundwater pumped in the United States is used to irrigate crops. About one-fourth of this withdrawal is overdraft, ie, water drawn at a greater rate than it refills (1). (See WATER QUALITY ISSUES; WATER QUALITY MANAGEMENT).

Concerns over a possible future oil shortage has encouraged many people to look to agriculture to provide substantial fuels. Alcohol from biomass offers the hope of a renewable domestic source of liquid fuel to provide the mobility on which modern society depends. However, a significant biomass fuels program holds a large risk; if fuels are regarded as more important than soil and water, it would only contribute to the collapse it was meant to help forestall.

PETROFARMING

Farms in the United States have provided an ever-higher yield from the land while requiring fewer and fewer people to work on the land. Supermarkets offer a staggering array of produce, available almost regardless of season or weather. Exports have earned thus a reputation as the breadbasket of the world. However, more food is possible only because of a temporary overabundance of subsidized fossil fuels. Thoughtful analysts for years have pointed out that highly mechanized, chemicalized, and capitalized farming cannot be sustained without cheap energy. As energy prices rose in the late 1970s, various studies were conducted to ascertain the dependence of U.S. agriculture on fossil fuels (Table 1). Energy consumption in U.S. agriculture has not changed much since that time. As of 1974, on-farm energy use was 3.1 percent of the total U.S. energy budget, and food processing, distribution, and preparation accounted for 13.5 percent.

The early uses of oil-derived fertilizers and pesticides yielded remarkable results. The profits made from the use of deceptively inexpensive fossil-fuel feedstocks were irresistible. Fossil fuels allowed the farmer to overwork the land while still increasing the yield. As the soil mining accelerated erosion and diminished soil quality, more and more fertilizer was needed to maintain crop production. The primary reasons for the reduced yields on eroded soils are low nitrogen content, impaired soil structure, deficient organic matter, and reduced availability of moisture.

Use of pesticides also shows similarly diminishing returns. Although the chemicals were at first highly successful, many insects have quickly evolved resistance to insecticides, and similarly, a number of weeds have evolved resistance to some herbicides. Around 1948, at the outset of the synthetic insecticide era, when the U.S. used 50 million pounds of insecticides, the insects destroyed 7 percent of preharvest crops. Today, using 600 million pounds, the U.S. loses 13 percent of crops before harvest (3).

The addiction to petrofarming came on so gradually and insidiously that it required serious price shocks, as happened in 1974 and 1979, to bring this problem to public attention. Today, American agriculture uses more petroleum products than any other industry in the nation (4). The largest direct agricultural user of energy is farm machinery, including trucks and automobiles, which accounts for almost three-fifths of all directly used energy. Energy is also used to dry crops to prevent spoilage and to maintain livestock, dairy, and poultry.

Another important direct use of energy is irrigation. Pumping up groundwater from the San Joaquin and Imperial valleys, for example, has made the water table recede so that increasing amounts of electricity are needed to pump it; agriculture is California's biggest single users of electricity. In the Pacific Northwest, part of the justification for building several nuclear power plants under the Washington Public Power Supply System was for irrigation of the western slope desert of Washington state. WPPSS entered the largest municipal bankruptcy in history.

The energy embodied in chemicals and equipment, ie, indirect energy use, is not much less than direct use. About forty million tons of fertilizer are applied to America's fields each year, approximately 330 pounds for each person in the country (see Plate I). Similarly, pesticides are a fossil-fuel based system of control. A full 80 percent of the one billion pounds spread annually comes out of oil wells. The energy required to purify pesticide-contaminated water supplies should also be included. The energy

Table 1. Energy Consumption with the U.S. Food System[a]

Sector of Food System	Energy Consumed (quadrillion kJ)	Percent U.S. Total Energy Use	Percent of U.S. Food System Energy Use
On farm			
Direct energy	1.4	1.7	10
Indirect energy	1.1	1.4	8
Food processing	3.8	4.9	30
Distribution system	1.3	1.7	10
Commercial food service	2.1	2.7	17
Home food preparation	3.3	4.2	25
Total	12.9	16.6	100

[a] Ref. 2.

Plate I. Tractors pulling field implements and the use of anhydrous ammonia fertilizer tanks are both energy intensive technologies that are based on fossil fuels. Courtesy of Marty Bender.

embodied in manufacturing equipment is estimated to be 0.3 quads (5).

Farmers continue to depend on fossil fuels for a number of basic reasons, ie, many agronomic methods are capital intensive; farmers lack financial flexibility; perceived risk; and lack of research support on alternatives (6).

Most importantly, petroleum-powered machines have been substituted for human and animal labor. This makes it difficult for farmers to switch from machines back to human and animal labor.

Both ecology and economics are conspiring to put an end to petrofarming. The inflation caused by the escalating price of fossil fuel is giving farmers little choice.

ENERGY FOR FOOD PROCESSING

On-farm energy use accounts for only 18% of the energy consumption within the U.S. food system. Food processing and distribution account for 40%, and preparation accounts for the remaining 42%. Since consumer demands dictate how food is processed, distributed, and prepared, we should look at what happens beyond the farm.

More than three-quarters of the food grown on farms is processed before shipment to the consumer. The food-processing industry is the fourth largest energy consumer of the Standard Industrial Classification groupings, with only primary metals, chemicals, and petroleum ahead of it.

The eleven most energy-intensive food-processing industries as a whole derive about 48% of their energy from natural gas, 28% from electricity, 9% from coal, 7% from residual fuel oils, and 8% from other fuels.

Finally, food-processing waste represents another energy loss. The energy cost of the processed food includes the energy invested in the processing and embedded in the waste. Additionally, as a consequence of the material losses, more raw food is required to obtain a given amount of processed food. The food-processing industries annually generate 14.4 million tons of solid waste, which is 9.6% of that generated by manufacturing industries in the United States every year. The 262 billion gallons of wastewater produced by the food-processing industry represent a twelve-day supply of water for U.S. urban domestic use (7).

SEEDS OF CHANGE

Most farmers go deeper into debt because they are advised only on how to raise production, not on how to cut the costs of production in water, energy, chemicals, and machines (8). The heavy use of nitrogen fertilizer has resulted in dangerously high nitrate levels in the groundwater of some areas. Drinking water contaminated by nitrogen can cause severe health problems. Pesticides are also contaminating groundwater in some areas. Heavy irrigation, made economically possible by cheap oil and gas, is eroding the soil and rapidly salinizing what is left. Soil erosion threatens farm productivity, and eroded soil is filling streams and rivers with silt and the air with dust.

Either the economic or the ecological failure alone should be enough to bring about a change in agricultural methods. Together, they make an urgent need for change.

Conservation and efficiency improvements can range from weather-stripping buildings to changing the timing of irrigation systems to making better use of farm machinery. The last opportunity alone represents a saving of 0.1 quad or 21 percent over present use. Almost half of this figure could be saved solely by switching over from gasoline to diesel engines. Another potential area for immense energy saving is in irrigation. Some experts report that half of the energy used in irrigation could be eliminated by improved techniques and by better pumping equipment. If farmers were to use low-temperature grain drying where climate permits, solar drying could save half the supplemental heat required (9).

The energy consumed in fertilizer applications can in many cases be much reduced, either because present usage is more than is necessary or because at least part of the nutrient additives can come from organic, rather than inorganic, sources. Only one soil test is performed for every 162 acres planted, so the nutrient levels of most cropland are not known.

Livestock manure can also be substituted for inorganic fertilizers with considerable energy savings. U.S. livestock manure production is estimated at 1.7 billion tons per year. More than half of it is produced in feedlots and confinement rearing. If one-fifth of the manure from confinement rearing and feedlots were used as fertilizer, it could serve seventeen million acres at ten tons per acre and save 0.07 quads. At the same time, manure would add organic matter to the soil, which increases beneficial bacteria and fungi, makes plowing easier, improves soil texture, and reduces erosion.

Rotating crops with legumes can supply nitrogen to cropland in considerable quantities. If this procedure were performed on fifteen million acres of corn, which is about one-fourth of U.S. corn acreage, about 0.07 quads would be saved. Using pesticides only where and when they are necessary would reduce pesticide consumption by up to one-half. This would amount to a saving of 0.03 quad, but the major benefit from decreased pesticide use is not the energy saved; rather it is the reduced contamination of the countryside.

Because of the soil erosion and energy costs associated with moldboard plowing, various forms of conservation tillage are being rapidly adopted, such as reduced-till, minimum-till, no-till. Conservation tillage can reduce erosion on many soils from 50 to 90 percent. Additional benefits of conservation tillage are lower costs for equipment, labor, and fuel, and increased soil moisture retention.

Conservation tillage has inherent problems that include increased pest populations (insects, nematodes, rodents, fungi), increased susceptibility to plant disease, herbicide carry-over that locks farmers into continuous planting of corn, evolution of weed resistance to herbicides, and shifts in weed species, to perennial weeds become a problem for which there are no fully effective chemical solutions. In addition, the farmer takes a greater economic risk.

No-till can reduce fuel use in field operations by as much as 90%; in some instances it has actually increased crop yields. However, this energy saving is partly offset because no-till requires 30 to 50% more pesticide than conventional tillage needs. In addition, no-till soil sometimes needs extra nitrogen. The result, as numerous studies have shown, is that net energy savings for no-till on individual fields range from zero to about 10%; occasionally, savings run as high as 30% (10).

Conservation tillage has become too reliant on herbicides to control weeds. Compared to research done on herbicides, little effort has been made to combine conservation tillage with integrated pest management, the system for managing insects, diseases, and weeds by combining resistant crop varieties, beneficial organisms, and crop rotations, plus other techniques. Such management uses pesticides only where necessary, decreasing the impact on target organisms including humans.

A sustainable agriculture has to result in significant energy savings. It has been estimated that the direct energy savings from such agriculture could be 0.8 to 0.9 quads. Combined with the indirect savings of oil formerly used to replace soil fertility, the total would be 1.1 to 2.2 quads, or a financial savings of $13 to $26 billion (1992 dollars) each year (11).

It is imperative for farmers to implement techniques that remove them from the fossil-fuel treadmill (see Plate II). Thus, interest has turned to biomass fuels.

BIOMASS FUELS

A renewable liquid fuel program based on biomass feedstocks must adhere to four principles if it is to be truly sustainable.

1. *The land comes first.* All operations must be based on a concern for soil fertility and long-term environmental compatibility.
2. *Efficiency is vital.* Both the vehicle for which the fuel is intended and the means of converting the biomass into fuel must be efficient.
3. *Wastes are the source.* Use farming and forestry wastes as the principal feedstocks; no crop should be grown just to make fuels.
4. *Sustainability is a goal.* The program should be a vehicle for the reform of currently unsustainable farming and forestry practices.

PUT LAND MAINTENANCE FIRST

Renewable must mean sustainable in the very long run. No biomass program can long endure unless it is based on the preservation and enhancement of soil fertility, water quality, and the biotic community on which agriculture de-

Plate II. The use of draft horses and of a photovoltaic array are both examples of renewable energy technologies. Courtesy of Marty Bender.

pends. The subsidized, corn-based ethanol/Gasohol program—largely ignores these concerns. Thus the first requirement of a sustainable biomass fuels program is to choose feedstocks whose production is not energy intensive and does not cause intolerable soil erosion and degradation.

Of critical importance is the proper scale of the biomass conversion techniques and their efficiency. Proper scale is essential to minimize the costs of collecting and transporting the feedstock. No definitive study has yet shown what scale is most cost effective, and there is probably no universal answer to this question, but the same diseconomies of large scale that have lately been described for electrical generating plants will probably dictate that biomass systems be relatively small in most circumstances (12).

Biomass conversion plants could be as small as mobile pyrolyzers; these devices heat a feedstock—usually a woody one—with little or no ash, producing, among other possible products, char, a low-grade fuel and gas, and a heavy oil akin to condensed woodsmoke. They could be hauled on the back of a truck to go wherever there are small collections of feedstocks. A conversion plant the size of a milk-bottling plant could serve a half-dozen towns (13).

Initially, integrated systems that use both the fuel product and its byproducts at the conversion site—the farm, the food-processing plant, or the pulp mill where the feedstock is available—make the most economic sense.

It is sometimes argued that many dispersed biomass fuel plants cannot produce enough total fuel to be nationally important. Even if this were true, a biofuels program meeting most of the needs of farms themselves would be important. Farms now get some 93 percent of their fuel from oil and gas and are often on the end of long and precarious supply lines. If the world oil trade is disrupted, farmers will suffer a double problem, ie, curtailed supplies boost fuel prices, while curtailed crop exports depress crop prices.

Conversion efficiency will decide whether a biomass fuels program will result in a net gain or a net loss of energy. The conversion technology assumed in most official studies of biomass fuels is borrowed from other processes that are inappropriate for biomass conversion. But it is also easy to find more efficient methods. Ethanol stills, for example, have been, and can still be, enormously improved. Today some commercial processes need only 25,000 Btus to go all the way to 200-proof ethanol, and the best demonstrated processes have reduced this to a mere 8,000–10,000 Btus. Stills can also use solar or, in some regions, geothermal heat.

PUT WASTES TO USE

Most biomass studies assume the use of special crops, notably grains, grown specifically for conversion to alcohol. Even better, however, is to run the biomass program on wastes. Many attractive feedstocks are currently a disposal problem. In California, for instance, rice straw is now burned in open fires. Used as a biomass feedstock, it would, coincidentally, solve a major air pollution problem. Each region has its example of a potential biomass feedstock, from apple pumice in Washington state to energy studies in Washington, D.C.

There are several possible exceptions to the proscription against special crops. Among the most attractive potential biomass crops are plants of the family Euphorbiaceae, including spurges, cassavas, and poinsettias, for example. They come in hundreds of varieties, adapted to conditions ranging from deserts to rain forests. They share a sap rich in various resins, terpenes, and other fuels or fuel feedstocks. One study concluded that *Euphorbia* planted on a land area equal to that of Arizona could meet one-fourth of current U.S. petroleum consumption. *Euphorbia* grows well on otherwise marginal lands and has minimal water requirements. With careful attention to the impact of such species on the soil structure of sensitive lands, this could provide an attractive second crop in many regions (14). Other dryland crops might offer similar potential. Another possibility for special cropping is cattails. Douglas Pratt at the University of Minnesota has shown that sustainable, ecologically sensitive cropping of cattails could yield just under one quad of liquid fuels per day—up to one-fifth of ultimate national needs for vehicular liquid fuels.

This article is an updated version of "Energy and Agriculture," in W. Jackson, W. Berry, and B. Coleman, eds., *Meeting the Expectations of the Land: Essays in Sustainable Agriculture and Stewardship,* North Point Press, San Francisco, 1984, pp. 68–86.

BIBLIOGRAPHY

1. U.S. Water Resources Council, *The Nation's Water Resources: 1975–2000,* U.S. Government Printing Office, Washington, D.C., 1978, p. 18.
2. Booz-Allen and Hamilton, Inc., *Energy Use in the Food System,* Office of Industrial Programs, Federal Energy Administration, FEA/D-76/083 U.S. Government Printing Office, Washington, D.C. 1976); U.S. Department of Agriculture, *Energy and U.S. Agriculture: 1974 and 1978,* Economics, Statistics, and Cooperatives Service Statistical Bulletin no. 632 (U.S. Government Printing Office, Washington, D.C. 1979), p. 64; J. S. Steinhart and C. F. Steinhart, *Science* **184** (4134), 33–42 (1974); R. A. Friedrich, *Energy Conservation for American Agriculture,* Ballinger, Cambridge, Mass, 1978, pp. 55, 69, 104.
3. R. den Bosch, *The Pesticide Conspiracy,* Anchor Press/Doubleday, New York: 1980, p. 24.
4. R. A. Friedrich, *Energy Conservation for American Agriculture* Cambridge, Mass. 1978, p. 55.
5. W. Jackson, *New Roots for Agriculture,* Friends of the Earth, San Francisco 1980, pp. 26, 28.
6. H. F. Breimyer, "Outreach Programs of the Land Grant Universities: Which Publics Should They Serve?" Keynote address at Kansas State University for conference with same name and title, 1976.
7. U.S. Bureau of the Census, *Statistical Abstracts,* 206, see W. Vegara, ref. 1, p. 1864.
8. S. Tifft, "Farmers Are Taking Their PIK," *Time,* 14 (July 25, 1983).
9. Ref. 9, pp. 56, 69.
10. M. Hinkle, "Problems with Conservation Tillage," *J. Soil and Water Conserv.* **38**(3), 201–206 (1983). E. Axell, "The Toll of No-Till," *Soft Energy Notes* 3 (6), 14–16, 1981; W. Lockeretz,

"Energy Implications of Conservation Tillage," *Journal of Soil and Water Conservation* 38(3), 207–21 (1983); D. Locker, "Problems Remain with No-Till Technology," *Prairie Sentinel* **1**(3), 8–10 (1982).

11. W. Jackson and M. Bender, "Saving Energy and Soil," *Soft Energy Notes* **3**(6), 7 (1981).
12. A. B. Lovins and L. Hunter Lovins, *Brittle Power: Energy Strategy for National Security,* Brick House, Andover, Mass. 1982, p. 335.
13. Solar Energy Research Institute, *A Survey of Biomass Gasification,* 3 Vols. National Technical Information Service, Springfield, Va., 1979; National Research Council, *Energy for Rural Development: Renewable Resources and Alternative Technologies for Developing Countries,* National Academy Press, Washington, D.C., 1981; National Research Council, *Producer Gas: Another Fuel for Motor Transport,* National Academy Press, Washington, D.C. 1983; and I. E. Cruz, *Producer-Gas Technology for Rural Applications,* Food and Agriculture Organization, Rome, 1985.
14. C. E. Wyman, R. L. Bain, N. D. Hinman and D. J. Stevens, "Ethanol and Methanol from Cellulosic Biomass," in T. B. Johansson, H. Kelly, A. K. N. Reddy, & R. H. Williams, *Renewable Energy: Sources for Fuels and Electricity,* Island Press, Washington, D.C., 1993, pp. 865–923.

AIR CONDITIONING

Harry J. Sauer, Jr.
University of Missouri
Rolla, Missouri

The term *air conditioning* in its broadest sense implies control of any or all of the physical and chemical qualities of the air. *Heating, ventilating, and air conditioning* (HVAC) systems have as their primary function either (*1*) the generation and maintenance of comfort for occupants in a conditioned space, or (*2*) the supplying of a set of environmental conditions (high temperature and high humidity; low temperature and high humidity, etc) for a process or product within a space.

Today there is a professional engineering society devoted to promoting the art and science of air conditioning and related fields—the American Society of Heating, Refrigerating, and Air-Conditioning Engineers (ASHRAE). ASHRAE techniques and methodology are widely accepted throughout the industry and as a result will be often cited herein.

HISTORICAL NOTES

In order to understand the current design criteria and trends it is helpful to know something of the past. As in other fields of technology, the accomplishments and failures of the past affect the current and future design concepts.

Towering above his contemporaries was the "Father of Air Conditioning," Willis H. Carrier (1876–1950), who through his brilliant analytical and practical accomplishments contributed more to the advancement of the developing industry than any other individual. In 1902, faced with a challenge to solve the lithographic industry's perennial problem of poor color register with every change in the weather, Carrier designed and installed for the Sackett-Wilhelms Lithographing & Printing Company in Brooklyn the first engineered year-round air-conditioning system, providing the four main functions of heating, *cooling, humidifying,* and *dehumidifying.* In December 1911, Carrier presented at the Annual Meeting of ASME his epoch-making paper, Rational Psychrometric Formulae, which related *dry-bulb temperature, wet-bulb temperature* and *dew-point temperatures* of the air as well as its sensible, latent and total heat, and enunciated the theory of adiabatic saturation. These formulas, together with the accompanying psychrometric chart, became the authoritative basis for all fundamental calculations in the air-conditioning industry.

Comfort air conditioning made its major breakthrough in motion picture theaters in Chicago in 1919–1920 employing CO_2 machines, and in Los Angeles in 1922 using an ammonia compressor. New centrifugal type of compressors provided the impetus for several theaters on Times Square. Towards the end of the decade, General Electric introduced the first self-contained room air conditioner. Product technology then advanced by leaps and bounds. The air-source heat pump and the large lithium bromide water-chiller were important innovative breakthroughs right after the war. In the 1950s came automobile air conditioners and rooftop heating and cooling units. The rooftop units contained a gas- or oil-fired heating surface, a mechanical cooling system complete with air-cooled condenser, supply/exhaust fans, a filter, and an outdoor and return damper section, all assembled with the necessary controls. Only the room thermostat had to be mounted separately.

People have now reached the moon and are seeking further space conquests. For environmental control of astronaut and spacecraft, air conditioning in its most sophisticated form has permitted people to surmount the hostile climates of moon and space, and made possible all of the significant manned missions beyond the planet Earth.

Beginning with the oil embargo of 1973, the HVAC field underwent tremendous change. As detailed in a 1989 Department of Energy Report (15), more than half of the nation's four million commercial–industrial–institutional buildings made use of energy conservation measures for heating and cooling.

Another report (6) by the Energy Information Administration details how residential heating has changed over the past 40 years. The share of electricity as the main space heating fuel in new homes has grown from a 15% level in the 1960s to more than 55% today. Much of this growth is due to the *electric heat pump.* This growth has been at the expense of natural gas, whose share of the market has declined from 70% for homes built in the sixties to 29% today.

The breakdown of residential primary fuel use and the type of heating equipment is as follows:

Primary Fuel Type

NATURAL GAS (50×10^6 HOMES)	
Central warm-air furnaces	31.6 Use, $\times 10^6$
Steam/hot water system	9.2
Floor, wall pipeless furnace	5.1
Room heater/other	4.0

Primary Fuel Type	
ELECTRICITY (17.9×10^6 HOMES)	
Central warm air furnace	6.9
Built-in electric units	5.1
Heat pumps	4.5
Other	1.1
FUEL OIL (10.9×10^6 HOMES)	
Steam/hot water system	6.3
Central warm air furnace	4.0
Other	0.5
LPG (4.1×10^6 HOMES)	
Central warm air furnace	2.4
Room heater	0.9
Other	0.8
KEROSENE (1.3×10^6 HOMES)	

Table 1 provides very rough estimates of the relative size of heating and cooling units required for various applications.

BUILDING DESIGN LOADS

The first step in sizing and/or selecting the components for an air-conditioning system is the calculation of the design *heating load* (winter) and the design *cooling load* (summer). The design heat loss in winter is determined for the winter inside and outside design conditions of temperature and humidity; the design heat gain in summer at the summer design conditions.

For the estimation of the space heat and moisture loads, the size, construction, and use of the space must be known. The schedule of occupancy and activity of occupants must be estimated. All heat and/or moisture emitting equipment in the space and schedules of operation must be determined. The space conditions to be maintained as well as the environment external to the space must be specified.

Standard procedures for calculating design heat losses and gains are described in the *ASHRAE Handbook 1993 Fundamental, ASHRAE Cooling and Heating Load Calculation Manual* (7), and the Air Conditioning Contractors of America *Load Calculation for Residential Winter and Summer Air Conditioning, Manual J* (8).

Table 1. Design Heating and Cooling Load

Building Type	Air Conditioning (ft^2/ton)	Heating [(Btu/h)/ft^3]
Apartment	450	4.9
Bank	250	3.2
Department store	250	1.1
Dormitory	450	4.9
House	700	3.2
Medical center	300	4.9
Night club	250	3.2
Office		
Interior	350	3.2
Exterior	275	3.2
Post office	250	3.2
Restaurant	250	3.2
School	275	3.2
Shopping center	250	3.2

Indoor Design Conditions

The comfort environment is the result of simultaneous control of temperature, humidity, cleanliness and air distribution within the occupant's vicinity. This set of factors includes mean radiant temperature as well as the air temperature, odor control, and control of the proper acoustic level within the occupant's vicinity. The necessary criteria, indexes and standards for use where human occupancy is concerned are available in the ASHRAE literature (1,9) and elsewhere. Direct indexes of the sensation of comfort for the human body include the following: dry-bulb temperature; dew-point temperature; wet-bulb temperature; relative humidity; and air movement.

Physiologists recognize that sensations of comfort and of temperature may have different physiological and physical bases, and each type should be considered separately. This dichotomy was recognized in the ANSI/ASHRAE Standard 55 (9), where *thermal comfort* is defined as "that state of mind which expresses satisfaction with the thermal environment." Unfortunately, the majority of our current predictive charts are based on comfort defined as a sensation "that is neither slightly warm or slightly cool." Figure 1 shows the winter and summer comfort zones specified in ANSI/ASHRAE Standard 55. Offices, homes, schools, shops, theaters, and many others can be approximated well with these specifications.

Outdoor Design Conditions: Weather Data

Design outdoor conditions of temperature and humidity for most locations in the United States and many locations throughout the world may be found in the refs. 1 and 10. A sample of the ASHRAE weather data is given in Table 2. The winter design temperature usually selected is the outdoor air temperature that is exceeded 97.5% of the time. The temperature that is exceeded 99% of the time may be chosen when a more conservative design is desired. The summer design temperature to be used is the outdoor air temperature that is exceeded 2.5% of the time. Because humidity is also important in sizing air-conditioning equipment, the mean coincident wet-bulb temperature is tabulated along with the design (dry-bulb) temperature values.

Winter design temperatures are presented in Column 5.

For summer design, the recommended design dry-bulb and wet-bulb temperatures and mean daily range are presented in Columns 6, 7, and 8.

Infiltration and Ventilation

Infiltration is the air leakage through cracks and interstices, around windows and doors, and through floors and walls of a building of any type from a low one story house or commercial building to a multistory skyscraper. The magnitude of infiltration depends on the type of construction, workmanship, and condition of the building.

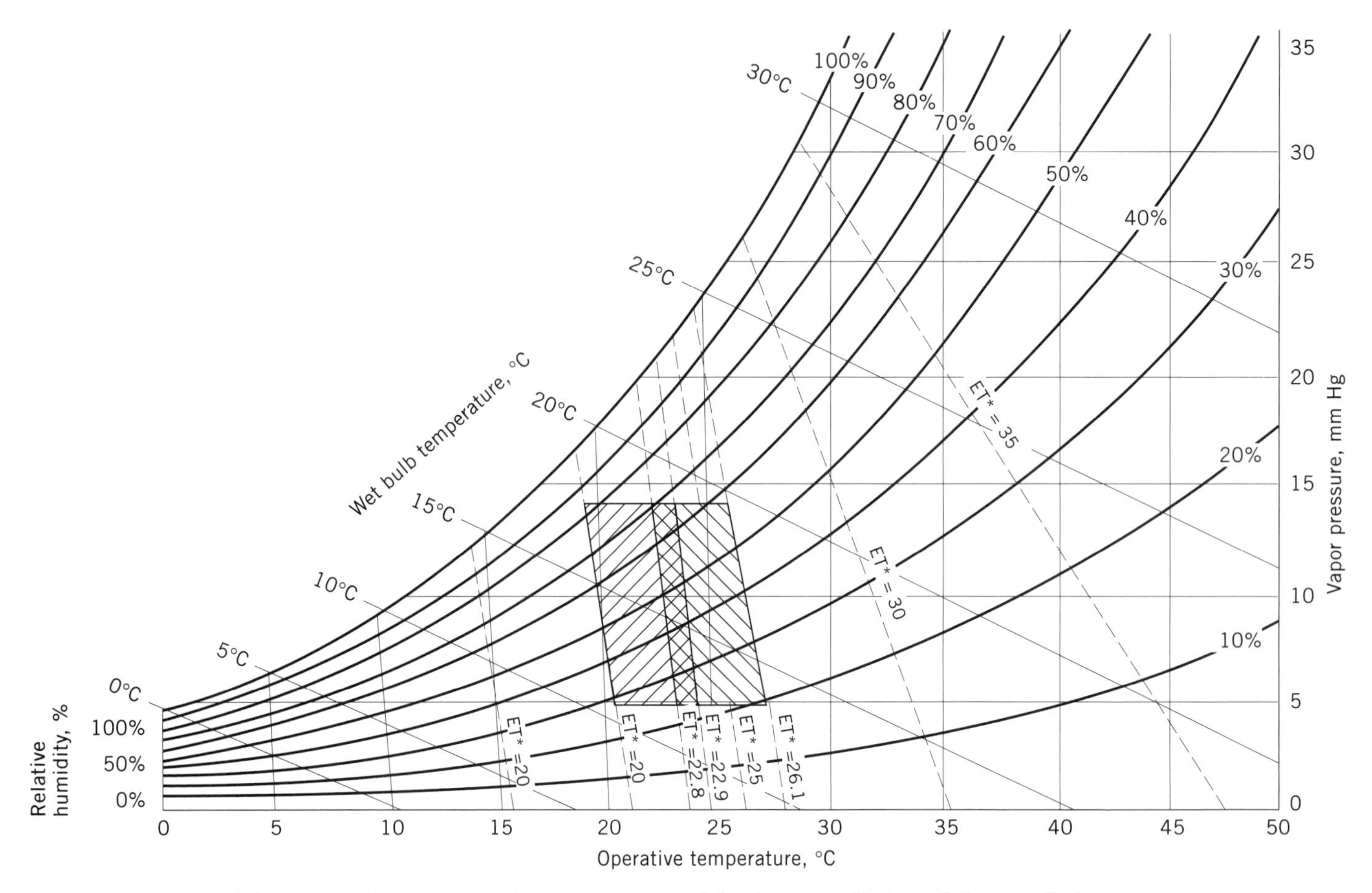

Figure 1. ASHRAE comfort zones. Courtesy of the American Society of Heating, Refrigerating, and Air-Conditioning Engineers.

Natural ventilation, on the other hand, is the intentional displacement of air through specified openings, such as windows, doors, and by ventilators.

Outside *air infiltration* may account for a significant proportion of the heating or cooling requirements for buildings. It is therefore important to be able to make an adequate estimate of its contribution with respect to both design loads and seasonal energy requirements. Air infiltration is also an important factor in determining the relative humidity that will occur in buildings or, conversely, the amount of humidification or dehumidification required to maintain given humidities.

Heat gain due to infiltration must be included whenever the outdoor air introduced mechanically by the system is not sufficient to maintain a positive pressure within the enclosure to prevent any infiltration.

There are two methods of estimating air infiltration in buildings. In one case the estimate is based on measured leakage characteristics of the building components and selected pressure differences. This is known as the *crack method,* since cracks around windows and doors are usually the major source of air leakage. The other method is known as the *air change method* and consists of estimating a certain number of air changes per hour for each room, the number of changes assumed being dependent upon the type, use, and location of the room. The crack method is generally regarded as being more accurate, provided that leakage characteristics and pressure differences can be properly evaluated. Otherwise the air change method may be justified. The accuracy of estimating infiltration for design load calculations by the crack or component method is restricted both by the limitations in information on air leakage characteristics of components and by the difficulty of estimating the pressure differences under appropriate design conditions of temperature and wind.

The introduction of some outdoor air into conditioned spaces generally is necessary. Local codes and ordinances frequently specify outdoor air ventilation requirements for public places and for industrial installations.

ASHRAE Standard 62, Ventilation for Acceptable Indoor-Air Quality, (11) is aimed at establishing better control of indoor contaminants without incurring a large energy penalty. Studies have linked increased ventilation rates with fewer respiratory infections. Other studies have shown the need for 7.2 L/s per person (15 cfm/person) to remove occupant generated odors effectively.

Table 3 provides examples of the outdoor air requirements for ventilation specified by ASHRAE Standard 62.

Typical infiltration values in housing in North America vary by about a factor of ten, from tight housing with seasonal air change rates at 0.2 per hour to housing with infiltration rates as great as 2.0 per hour. In general, the minimum outside air requirement is 0.5 air changes per hour to overcome infiltration by producing a slight positive pressure within the structure. The minimum outside air standard for ventilation is (7.2 L/s) 15 cfm per person.

Table 2. Excerpt from ASHRAE Weather Data Table[a]

				Winter,[b] °C		Summer,[c] °C							Prevailing Wind		Temp. °C	
Col. 1	Col. 2	Col. 3	Col. 4	Col. 5		Col.6			Col. 7	Col. 8			Col. 9		Col. 10	
	Lat.	Long.	Elev.	Design Dry Bulb		Design Dry-Bulb and Mean Coincident Wet-Bulb			Mean daily Range	Design Wet-Bulb			Winter	Summer	Median of Annual Extr.	
State and Station[d]	° ′	° ′	m	99%	97.5%	1%	2.5%	5%		1%	2.5%	5%	m/s		Max.	Min.
Arizona																
Douglas AP	31 27	109 36	1249	−3	−1	37/17	35/17	34/17	17	21	21	20			40.2	−10.0
Flagstaff AP	35 08	111 40	2136	−19	−16	29/13	28/13	27/12	17	16	16	15	NE 3	SW	32.2	−24.2
Fort Huachuca AP (S)	31 35	110 20	1422	−4	−2	35/17	33/17	32/17	15	21	20	19	SW 3	W		
Kingman AP	35 12	114 01	1079	−8	−4	39/18	38/18	36/18	17	21	21	21				
Nogales	31 21	110 55	1159	−2	0	37/18	36/18	34/18	17	22	21	21	SW 3	W		
Phoenix AP (S)	33 26	112 01	339	−1	1	43/22	42/22	41/22	15	24	24	24	E 2	W	44.9	−2.9
Prescott AP	34 39	112 26	1528	−16	−13	36/16	34/16	33/16	17	19	18	18				
Tucson AP (S)	32 07	110 56	780	−2	0	40/19	39/19	38/19	14	22	22	22	SE 3	WNW	42.7	−7.1
Winslow AP	35 01	110 44	1492	−15	−12	36/16	35/16	34/16	18	19	18	18	SW 3	WSW	39.3	−18.0
Yuma AP	32 39	114 37	65	2	4	44/22	43/22	42/22	15	26	26	25	NNE 3	WSW	46.0	−.7
District of Columbia																
Andrews AFB	38 5	76 5	85	−12	−10	33/24	32/23	31/23	10	26	24	24				
Washington, National AP	38 51	77 02	4	−10	−8	34/24	33/23	32/23	10	26	25	24	WNW 6	S	36.4	−13.7
Florida																
Belle Glade	26 39	80 39	5	5	7	33/24	33/24	32/24	9	26	26	26			34.8	−.6
Cape Kennedy AP	28 29	80 34	5	2	3	32/26	31/26	31/26	8	27	26	26				
Daytona Beach AP	29 11	81 03	9	0	2	33/26	32/25	31/25	8	27	26	26	NW 4	E		
Fort Lauderdale	26 04	80 09	3	6	8	33/26	33/26	32/26	8	27	26	26	NW 5	ESE		
Fort Myers AP	26 35	81 52	5	5	7	34/26	33/26	33/25	10	27	26	26	NNE 4	W	34.9	1.6
Minnesota																
Albert Lea	43 39	93 21	372	−27	−24	32/23	31/22	29/22	13	25	24	23				
Alexandria AP	45 52	95 23	436	−30	−27	33/22	31/22	29/21	13	24	23	22			35.1	−33.3
Bemidji AP	47 31	94 56	424	−35	−32	31/21	29/21	27/19	13	23	22	21	N 4	S	34.7	−38.3
Brainerd	46 24	94 08	374	−29	−27	32/23	31/22	29/21	13	24	23	22				
Duluth AP	46 50	92 11	435	−29	−27	29/21	28/20	26/19	12	22	21	20	WNW 6	WSW	32.7	−33.0

[a] Reprinted with permission of the American Society of Heating, Refrigerating and Air-Conditioning Engineers.
[b] Winter design data are based on the 3-month period, December through February.
[c] Summer design data are based on the 4-month period, June through September.
[d] AP or AFB following the station name designates airport or Airforce base temperature observations. Co designates office locations within an urban area that are affected by the surrounding area. Undesignated stations are semirural and may be compared to airport data.

However, local ventilation ordinances must be checked and may require greater quantities of outdoor air.

Heat Transfer Coefficients

The design of a heating, refrigerating, or air-conditioning system, including selection of building *insulation* or sizing of piping and ducts, or the evaluation of the thermal performance of parts of the system such as chillers, heat exchangers, fans, etc, is based on the principles of heat transfer. Two terms are used to indicate the relative insulating value of materials and sections of walls, floors, and ceilings. One is called U-factor and the other, R number. The U factor indicates the rate at which heat flows through a specific material or a building section, such as the one shown in Figure 2. The smaller the U factor, the better the insulating value of the material or group of materials making up the wall, ceiling, or floor.

The R number indicates the ability of one specific material, or a group of materials in a building section, to resist heat flow through them. Many of the insulating materials now have their R number stamped on the outside of the package, batt, or blanket.

Heating Load Methodology

Prior to designing a heating system, an estimate must be made of the maximum probable heat loss of each room or space to be heated, based on maintaining a selected indoor air temperature during periods of design outdoor weather conditions. The heat losses may be divided into two groups: (*1*) the transmission losses or heat transmitted through the confining walls, floor, ceiling, glass, or other surfaces, and (*2*) the infiltration losses or heat required to warm outdoor air which leaks in through cracks and crevices, around doors and windows, or through open doors and windows, or heat required to warm outdoor air used for ventilation.

The ideal solution to the basic problem which confronts the designer of a heating system is to design a plant that has a capacity at maximum output just equal to the heating load which develops when the most severe weather conditions for the locality occur. However, where night set-

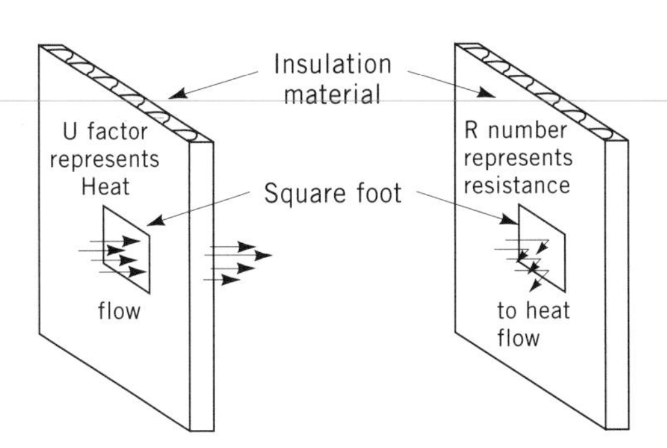

Figure 2. U factor and R number.

Table 3. Ventilation Requirements (ASHRAE Standard 62)[a]

Application	Estimated Maximum[b] Occupancy P/1000 ft^2 or 100 m^2	Outdoor Air Requirements cfm/ person	L/s · person	cfm/ft^2	L/s · m^2	Comments
Offices						
Office space	7	20	10			Some office equipment
Reception areas	60	15	8			may require local exhaust.
Telecommunication centers and data entry areas	60	20	10			
Conference rooms	50	20	10			Supplementary smoke-removal equipment may be required.
Public Spaces						
				cfm/ft^2	L/s · m^2	
Corridors and utilities				0.05	0.25	
Public restrooms, cfm/wc or urinal		50	25			Mechanical exhaust
Locker and dressing rooms				0.5	2.5	with no recirculation is recommended.
Smoking lounge	70	60	30			Normally supplied by transfer air, local mechanical exhaust; with no recirculation recommended.
Elevators				1.00	5.0	Normally supplied by transfer air.
Photo studios	10	15	8			
Darkrooms	10			0.50	2.50	
Pharmacy	20	15	8			
Bank vaults	5	15	8			
Duplicating, printing				0.50	2.50	Installed equipment must incorporate positive exhaust and control (as required) of undesirable contaminants (toxic or otherwise).
Education						
Classroom	50	15	8			
Laboratories	30	20	10			Special contaminant
Training shop	30	20	10			control systems may
Music rooms	50	15	8			be required for
Libraries	20	15	8			processes or functions including laboratory animal occupancy.
Locker rooms				0.50	2.50	
Corridors				0.10	0.50	
Auditoriums	150	15	8			
Smoking lounges	70	60	30			Normally supplied by transfer air. Local mechanical exhaust with no recirculation recommended.
Hospitals, Nursing and Convalescent Homes						
Patient rooms	10	25	13			Special requirements or
Medical procedure	20	15	8			codes and pressure
Operating rooms	20	30	15			relationships may
Recovery and ICU	20	15	8			determine minimum ventilation rates and filter efficiency. Procedures generating contaminants may require higher rates.

Private Spaces		
Living areas	0.35 air changes per hour but not less than 15 cfm (7.5 L/s) per person	For calculating the air changes per hour, the volume of the living spaces shall include all areas within the conditioned space. The ventilation is normally satisfied by infiltration and natural ventilation. Dwellings with tight enclosures may require supplemental ventilation supply for fuel-burning appliances, including fireplaces and mechanically exhausted appliances. Occupant loading shall be based on the number of bedrooms as follows: first bedroom, two persons; each additional bedroom, one person. Where higher occupant loadings are known, they shall be used.
Kitchens	100 cfm (50 L/s) intermittent or 25 cfm (12 L/s) continuous or openable windows	Installed mechanical exhaust capacity. Climatic conditions may affect choice of the ventilation system.
Baths, Toilets	50 cfm (25 L/s) intermittent or 20 cfm (10 L/s) continuous or openable windows	Installed mechanical exhaust capacity.

[a] Reprinted with permission of the American Society of Heating, Refrigerating and Air-Conditioning Engineers

[b] Net occupiable space.

back is used, some excess capacity may be needed unless the owner is aware that under some conditions of operation he/she may not have the ability to elevate the temperature.

In most cases, economics interferes with the attainment of this ideal. Studies of weather records show that the most severe weather conditions do not repeat themselves every year. If heating systems were designed with adequate capacity for the maximum weather conditions on record, there would be considerable excess capacity during most of the operating life of the system.

In many cases, occasional failure of a heating plant to maintain a preselected indoor design temperature during brief periods of severe weather is not critical. However, the successful completion of some industrial or commercial processes may depend upon close regulation of indoor temperatures. These are special cases and require extra study before assigning design temperatures.

Normally the heating load is estimated for the winter design temperature usually occurring at night; therefore, no credit is taken for the heat given off by internal sources (people, lights, etc.) In industrial plants, quite a different condition may exist, and heat sources, if always available during occupancy, may be substituted for a portion of the heating requirements. In no case should the actual heating installation (exclusive of heat sources) be reduced below that required to maintain at least 4.4°C in the building.

Heat transfer through basement walls and floors to the ground depends on (*1*) the difference between the air temperature within the room and that of the ground and outside air, (*2*) the material of the walls or floor, and (*3*) conductivity of the surrounding earth.

Although the practice of drastically lowering the temperature when the building is unoccupied may be effective in reducing fuel consumption, additional equipment capacity is required for pickup.

Cooling Load Principles

The design cooling load must take into account the heat gain into the space from outdoors as well as the heat being generated within the space. In calculating the heating load, the effects of the sun were ignored. In making heat gain calculations, the effect of solar radiation on glass, walls, and roofs must be included. In addition, internal loads caused by lights, people, cooking, motors, and anything that produces heat must be considered.

Peak cooling loads are more transient than heating loads. Radiative heat transfer within a space and thermal storage cause the thermal loads to lag behind the instantaneous heat gains and losses. This lag is important, especially with cooling loads, in that the peak is both reduced in magnitude and delayed in time compared to the heat gains that cause it. Today various calculation methods are used to account for the transient nature of cooling loads. Some widely used methods of performing load analysis for building elements include the following: Transfer Function Method (TFM); Cooling Load Temperature Difference/Cooling Load Factor (CLTD/CLF) Method; Total Equivalent Temperature Differential/Time Averaging (TETD/TA) Method; Response Factor Method; and Finite Difference Method. The primary procedure for sizing cooling equipment is the Transfer Function Method (TFM).

Space *heat gain* is the rate at which heat enters into and/or is generated within a space at a given instant of time. Heat gain is classified by (*1*) the mode in which it enters the space, and (*2*) whether it is a sensible or latent gain. The first classification is necessary because different fundamental principles and equations are used to calculate different modes of energy transfer. The heat gain occurs in the form of (*1*) solar radiation through transparent surfaces; (*2*) heat conduction through exterior walls and roofs; (*3*) heat conduction through interior partitions, ceilings, and floors; (*4*) heat generated within the space by occupants, lights, and appliances; (*5*) energy transfer as a result of ventilation and infiltration of outdoor air; or (*6*) miscellaneous heat gains. The second classification, sensible or latent, is important for proper selection of cooling equipment. The heat gain is sensible when there is a direct addition of heat to the conditioned space by any or all mechanisms of conduction, convection, and radiation. The heat gain is latent when moisture is added to the space (eg, by vapor emitted by occupants). To maintain a constant humidity ratio in the enclosure, water vapor in the cooling apparatus must condense out at a rate equal to its rate of addition into the space. The amount of energy required to do this, the latent heat gain, essentially equals the product of the rate of condensation and latent heat of condensation. The distinction between sensible and latent heat gain is necessary for cooling apparatus selection. Any cooling apparatus has a maximum sensible heat removal capacity and a maximum latent heat removal capacity for particular operating conditions.

Space cooling load is the rate at which heat must be removed from the space to maintain room air temperature at a constant value. The sum of all space instantaneous heat gains at any given time does not necessarily equal the cooling load for the space at that same time. The space heat gain by radiation is partially absorbed by the surfaces and contents of the space and does not affect the room air until some time later. The thermal storage effect can be important in determining an economical cooling equipment capacity.

In general, the design cooling load determination for a building requires data for most or all of the following:

- Full building description, including the construction of the walls, roof, windows, etc, and the geometry of the rooms, zones, and building. Shading geometries may also be required.
- Sensible and latent internal loads because of people, lights, and equipment and their corresponding operating schedules.
- Indoor and outdoor design conditions.
- Geographical data such as latitude and elevation.
- Ventilation requirements and amount of infiltration.
- Number of zones per system and number of systems.

Commercially available computer loads programs are available for both mainframe and microcomputer which, when input with the above data, will calculate both the heating and cooling loads as well as perform a psychrometric analysis of the system. Output typically includes peak room and zone loads, supply air quantities, and total system (coil) loads.

ENERGY ESTIMATING

After the peak loads have been evaluated, the equipment must be selected with capacity sufficient to offset these loads. The air supplied to the space must be at the proper conditions to satisfy both the sensible and latent loads which were estimated. However, peak load occurs but a few times each year and partial load operation exists most of the time. Since operation is predominantly at partial load, partial load analysis is at least as important as the selection procedure.

It is often necessary to estimate the energy requirements and fuel consumption of environmental control systems for either short or long terms of operation. These quantities can be much more difficult to calculate than design loads or required system capacity.

The simplest procedures assume that the energy required to maintain comfort is a function of a single parameter, the outdoor dry-bulb temperature. The more accurate methods include consideration of solar effects, internal gains, heat storage in the walls and interiors, and the effects of wind on both the building envelope heat transfer and infiltration. The most sophisticated procedures are based on hourly profiles for climatic conditions and operational characteristics for a number of typical days of the year, or better still, a full 8760 hours of operation.

Calculation of the fuel and energy required by the primary equipment to meet these loads gives consideration to efficiencies and part-load characteristics of the equipment. Often it will be necessary to keep track of the different forms of such energy, such as electrical, natural gas, or oil. In some cases where calculations are done to assure compliance with codes or standards, it is necessary to convert these energies to source energy or resource consumed, as opposed to that delivered to the building boundary.

For air conditioners and heat pumps, the American Refrigeration Institute (ARI) has established performance ratings and testing conditions and publishes directories of certified products (12,13). The *Coefficient of Performance* (COP) is the scientifically accepted measure of heating or cooling performance. The COP is defined as

$$\text{COP} = \frac{\text{Heating (or cooling) provided by system}}{\text{Energy consumed by the system}}$$

The total heating output of a heat pump excludes supplemental resistance heat. The COP is a dimensionless number and thus the heating or cooling output and the energy input must be expressed in the same units. The *energy efficiency ratio (EER)* is defined in exactly the same way as the COP; however, for the EER, the heating or cooling must be expressed in Btu and the energy consumed by the system in watt-hours. The EER is numerically equal to the $3.413 \times$ COP.

Federal regulations specify two additional ratings for heat pumps and air conditioners: the *Heating Seasonal Performance Factor* (*HSPF*) and the *Seasonal Energy Efficiency Ratio* (*SEER*). The HSPF combines the effects of heat pump heating under a range of weather conditions applicable for a particular geographical region with performance losses due to coil frost, defrost, cycling under part-load conditions, and use of supplemental resistance heating during defrost. Standard HSPFs are determined by manufacturers for each of the six Department of Energy (DOE) regions and are provided on equipment labels. The SEER is a measure of seasonal cooling efficiency under a range of weather conditions assumed to be typical of the DOE region in question, as well as performance losses due to cycling under part-load operation. The SEER is defined as the total cooling provided during the cooling season (Btu) divided by the total energy consumed by the system (watthours). Currently, SEERS are reported for conventional heat pumps and air conditioners according to a simplified procedure that does not take weather into account explicitly.

The sophistication of the energy calculation procedure used can often be inferred from the number of separate ambient conditions and/or time increments used in the calculations. Thus, a simple procedure may use only one measure, such as annual degree days, and will be appropriate only for simple systems and applications. Here such methods will be called single-method measures. Improved accuracy may be obtained by the use of more information, such as the number of hours anticipated under particular conditions of operation. These methods, of which the bin method is the most well known, are referred to as simplified multiple-measure methods. The most elaborate methods currently in use perform energy balance calculations at each hour over some period of analysis, typically one year. These are called the detailed simulation methods, of which there are a number of variations. These methods require hourly weather data, as well as hourly estimates of internal loads, such as lighting and occupants.

Single-measure methods for estimating cooling energy are less well-established than those for heating. This is because the indoor–outdoor temperature difference under cooling conditions is typically much smaller than under heating conditions.

Even when used with applications apparently as simple as residential cooling, single-measure estimates can be seriously inaccurate if they do not consider all factors that significantly affect energy use.

In preparing energy estimates, it should be realized that any estimating method will produce a more reliable result over a long period of operation than over a short period.

Degree Day Method (Heating)

The traditional degree day procedure for estimating heating energy requirements is based on the assumption that, on a long-term average, solar and internal gains will offset heat loss when the mean daily outdoor temperature is 18.3°C, and that fuel consumption will be proportional to the difference between the mean daily temperature and 18.3°C. This basic concept can be represented in an equation stating that energy consumption is directly proportional to the number of degree days in the estimation period.

Table 4. Annual Degree Days, 18.3°C (65°F) Base

City, State	Heating-Degree Days	Cooling-Degree Days
Albuquerque, N.M.	4348	1345
Atlanta, Ga.	2961	1469
Bismark, N.D.	8851	528
Boston, Mass.	5634	674
Cheyenne, Wy.	7381	308
Chicago, Ill.	6639	713
Fort Worth, Texas	2405	2500
Kansas City, Kansas	4711	1475
Los Angeles, Calif.	2061	357
Miami, Fla.	214	4189
New Orleans, La.	1385	2653
New York, N.Y.	5219	1027
Omaha, Neb.	6612	1007
Phoenix, Ariz.	1765	3334
Pittsburgh, Pa.	5897	732
Portland, Maine	7511	292
Salt Lake City, Utah	6052	958
San Francisco, Calif.	3015	98
Seattle-Tacoma, Wash.	5145	134
Washington, D.C.	4224	1491

The *degree days* equation has undergone several stages of refinement in an attempt to make it agree as closely as possible with the available measured data on an average basis. The currently recommended form is

$$E = \frac{qL \times DD \times 24}{\Delta t \times k \times V} C_D$$

where E = fuel or energy consumption for the estimate period, kW · h or Btu; qL = design heat loss, W or Btu · h; DD = number of 18.3°C (65°F) degree days for the estimate period; Δt = design temperature difference, °C; k = overall efficiency factor; V = heating value of fuel, units consistent with q_L and E; and C_D = empirical correction factor.

Table 4 lists the average number of heating and cooling degree days 18.3°C (65°F base) for various cities in the United States. Some typical heating values are natural gas, 1050 Btu/ft^3; propane, 90,000 Btu/gal; and No. 2 fuel oil, 140,000 Btu/gal.

Table 5 gives approximate values for k. In the case of heat pumps, q must be in Btuh and 1000 × HSPF is to be used for the product of k and V. The errors inherent in the established 18.3°C (65°F) based method may be adjusted by the use of an empirical factor, C_D. Table 6 gives the best presently available values for C_D.

The variable base degree day (VBDD) method is a generalization of the widely used degree day method. It retains the familiar degree day concept, but counts degree days based on the balance point temperature, defined as the average outdoor temperature at which the building requires neither heating nor cooling.

The degree-day procedure is intended to recognize that, although the energy transferred out of a building is proportional to the difference between interior space temperature and the outside temperature, the heating equipment needs to meet only the part not covered by free heat from internal sources such as lights, equipment, occupants, and solar gain.

Cooling Degree Day Method

Cooling degree days are widely available to a base temperature of 18.3°C, but tabulations of degree days from base 7.2°C to 18.3°C are also available from the National Climatic Center, Asheville, N.C. (*Degree Days to Selected Bases,* in microfiche or published form). Table 4 provides cooling degree day values for several U.S. cities.

Cooling energy is predicted in a similar fashion as described for heating earlier, ie,

$$E_C = \frac{q_g \times CDD \times 24}{\Delta t_d \times SEER \times 1000}$$

where E_C = energy consumed for cooling, kW · h; q_g = design cooling load, Btu · h; CDD = Fahrenheit cooling degree days; Δt = design temeprature difference, °F; $SEER$ = seasonal energy efficiency ratio, Btu · h/W; and 1000 = W/kW.

Detailed Energy and Systems Stimulation

Energy analysis programs have come to be fundamental tools used by engineers in making decisions regarding building energy use. Energy programs, along with life-cycle costing routine, are used to quantify the impact of proposed energy conservation measures in existing buildings. In new building design, energy programs are used to aid in determining the type and size of building systems and components as well as to explore the effects of design tradeoffs.

Table 5. Efficiency Factor, k^a

Heating System	k Value
Conventional gas-fired forced-air furnace or boiler	0.60–0.70
Conventional gas-fired gravity-air furnace	0.57–0.67
Gas-fired forced-air furnace or boiler with typical energy conservation devices, intermittent ignition device, automatic vent damper	0.65–0.75
Gas-fired forced-air furnace or boiler with sealed combustion chamber, intermittent ignition device and automatic vent damper	0.70–0.78
Gas-fired pulse combustion furnace or boiler	0.80–0.90
Oil-fired furnace or boiler	0.50–0.75

[a] Reprinted with permission of the American Society of Heating, Refrigerating and Air-Conditioning Engineers.

Table 6. Correction Factor, C_D[a]

Quality of Construction and Relative Use of Electrical Appliances	Number of Degree Days 18.3°C (65°F)								
	1000	2000	3000	4000	5000	6000	7000	8000	9000
Well-constructed house Large quantities of insulation, tight fit on doors and windows, well sealed openings. Large use of electrical appliances. Large availability of solar energy at the house.	0.48	0.45	0.42	0.39	0.36	0.37	0.38	0.39	0.40
House of average construction Average quantities of insulation, average fit on doors and windows, partially sealed openings. Average availability of solar energy at the house. Average use of electrical appliances.	0.80	0.76	0.70	0.65	0.60	0.61	0.62	0.69	0.67
Poorly constructed house Small quantities of insulation, poor fit on doors and windows, unsealed openings. Small use of electrical appliances. Small availability of solar energy at the house.	1.12	1.04	0.98	0.90	0.82	0.85	0.88	0.90	0.92

[a] Reprinted with permission of the Electric Power Research Institute from *Heat Pump Manual,* 1985.

BASIC HVAC SYSTEM CALCULATIONS

Application of Thermodynamics to HVAC Processes

It is important to note that the energy (q_c) and moisture (m_c) transfers at the conditioner cannot be determined from the space heat and moisture transfers alone. The effect of the outdoor ventilation air must also be included as well as other system load components. The designer must recognize that items such as fan energy, duct transmission, roof and ceiling transmissions, heat of lights, bypass and leakage, type of return air system, location of main fans, and actual vs. design room conditions are all related to one another, to component sizing, and to system arrangement.

The most powerful analytical tools of the air-conditioning design engineer are the first law of thermodynamics or energy balance, and the conservation of mass or mass balance. These conservation laws are the basis for the analysis of moist air processes. The following sections demonstrate the application of these laws to specific HVAC processes.

In many air-conditioning systems, air is taken from the room and returned to the air-conditioning apparatus where it is reconditioned and supplied again in the room. In most systems, the return air from the room is mixed with outdoor air required for ventilation.

Figure 3 shows a typical air-conditioning system. Outdoor air (o) is mixed with return air (r) from the room and enters the apparatus (m). Air flows through the conditioner and is supplied to the space (s). The air supplied to the space picks up heat q_s and moisture m_w and the cycle is repeated.

The problem of air conditioning a space usually reduces to the determination of the quantity of moist air that must be supplied and the necessary condition which it must have in order to remove given amounts of energy and water from the space and be withdrawn at a specified condition.

The quantity q_s denotes the net sum of all rates of heat gain upon the space, arising from transfers through boundaries and from sources within the space. This heat gain involves addition of energy alone and does not include energy contributions due to addition of water (or water vapor). It is usually called the sensible heat gain. The quantity m_w denotes the net sum of all rates of moisture gain upon the space arising from transfers through boundaries and from sources within the space. Each pound of moisture injected into the space adds an amount of energy equal to its specific enthalphy h. Assuming steady-state conditions, the governing equations for the process across the space are

$$m_a h_s + m_w h_w - m_a h_r + q_s = 0$$

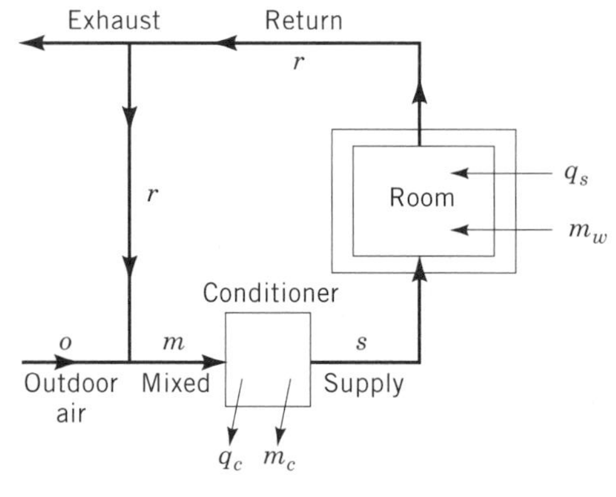

Figure 3. Typical air-conditioning system processes for cooling (14). Reprinted with permission.

and

$$m_a W_s + m_w = m_a W_r$$

where m, q, h, and W are the appropriate mass flow rate, heat-transfer rate, enthalpy, and humidity ratio, respectively.

When moist air is cooled to a temperature below its dew point as it is conditioned, some of the water vapor will condense and leave the air stream. Although the actual process path will vary considerably depending on the type surface, surface temperature, and flow conditions, the heat and mass transfer can be expressed in terms of the initial and final states.

For the system of Figure 3, the energy and mass balance equations for the conditioner are

$$m_c = m_a(W_m - W_S)$$
$$q_c = m_a[(h_m - h_s) - (W_m - W_s)h_c]$$

This process of heating and humidifying, generally required during the cold months of the year, follows the same mass and energy equations as given directly above.

Since the specific volume of air varies appreciably with temperature, all calculations should be made with the mass of air instead of volume. However, volume values are required for selection of coils, fans, ducts, etc. A practical method of using volume, but still including mass (weight) so that accurate results are obtained, is the use of volume values based on measurement at ASHRAE standard conditions. The value considered standard is 1.204 kg of dry air/m^3 (0.83 m^3/kg of dry air) [0.075 lb of dry air/ft^3 (13.33 ft^3/lb of dry air)], which corresponds to approximately 15.5°C at saturation and 20.6°C, dry [at 101.4 kPa (14.7 psia)].

Air-conditioning design often requires calculation of the following:

1. Sensible heat gain corresponding to the change of dry-bulb temperature for a given air flow (standard conditions). The sensible heat gain q_s, as a result of a difference in temperature Δt between the incoming air and leaving air flowing at ASHRAE standard conditions, is approximated by

$$q_s = 1.232(l/\text{s})\Delta t$$

and in I-P units

$$q_s = 1.10(cfm)\Delta t$$

2. Latent heat gain corresponding to the change of humidity ratio (W) for given air flow (standard conditions). The latent heat gain in watts (Btu h), as a result of a difference in humidity ratio (ΔW) between the incoming and leaving air flowing at ASHRAE standard conditions, is

$$q_l = 3012(l/s)\Delta W$$

In I-P units

$$q_l = 4840\ (cfm)\Delta W$$

AIR CONDITIONING SYSTEMS

System Categories

The central system is the basic system of air conditioning, originating with the forced-warm-air heating and ventilating systems, which has centrally located equipment and distributed tempered air through ducts. It was found that cooling and dehumidification equipment added to these systems produced satisfactory air conditioning in spaces where heat gains were relatively uniform throughout the conditioned area (Fig. 4). Complication in the use of this system appeared when heat sources varied within the space served. When this occurred, the area had to be divided into sections or zones, and the central system supplemented by additional equipment and controls.

As the science of air conditioning progressed, variations in the basic design were required to meet the functional and economic demands of individual buildings. Now air-conditioning systems are categorized according to the means by which the controllable cooling is accomplished in the conditioned area. They are further segregated to accomplish specific purposes by special equipment arrangements.

There are four basic types of HVAC systems used in commercial and institutional buildings: all-air systems; air-and-water systems; all-water systems; and unitary equipment systems.

The first three types generally fall into the central system category if there is a single heating and/or cooling source for multiple areas of the building.

An *all-air* system provides through a system of ductwork all of the heating and/or cooling needed by the conditioned space. The basic all-air systems are single-zone duct systems, multizone systems, dual duct systems, terminal reheat systems, and variable air volume (VAV) systems.

The basic *air–water* system provides cooled primary air to a conditioned space from a central duct system with heating and/or additional cooling provided by a central hydronic (hot or chilled water) system to terminal units in the conditioned space. There are many variations of this system with the airside being dual duct or VAV systems and the hydronic side either two or four pipe units. Fan-coil units supplied with primary ventilation air also fall into this category. The widest use of air–water systems has been in office buildings where heating and cooling are provided simultaneously in different zones of the building.

All-water systems accomplish both sensible and latent space cooling by circulating chilled water from a central refrigeration system through cooling coils in terminal units located in the space being conditioned. Fan-coil units and unit ventilators are the most common terminal units. Heating is provided by circulating hot water through the same coil and piping system (2-pipe), through a separate coil and piping system (4-pipe), or through a separate coil but with a common return piping system (3-pipe). Alternatively, electric resistance or steam coils may be used for heating. Ventilation must be provided either through wall apertures, by an interior zone system, or by a separate ventillation air system, possibly capable of humidifying ventilation air during the winter.

Unitary air-conditioning equipment is an assembly of factory-matched components for inclusion in field assem-

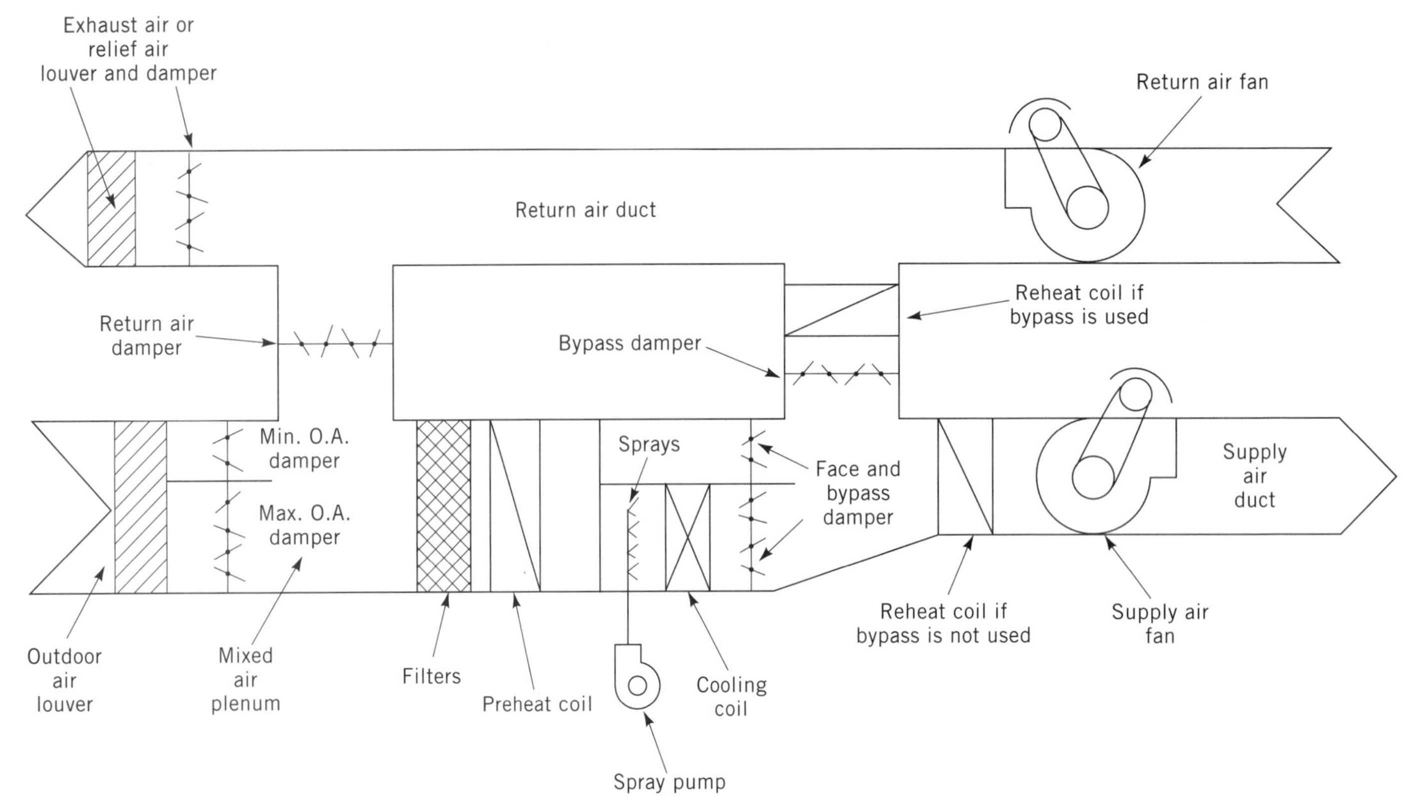

Figure 4. Single-zone central system. Courtesy of SMACNA.

bled air-conditioning systems. Unitary equipment tends to serve zoned systems, with each zone served by its own unit.

VEHICULAR AIR-CONDITIONING SYSTEMS

Automotive Cooling

All passenger cars sold in the United States must meet federal defroster requirements and thus ventilation systems and heaters are included in the basic vehicle design. Even though trucks are excluded from federal regulations, all U.S. manufacturers included heater/defrosters as standard equipment. Air conditioning remains an extra-cost option on nearly all vehicles. Thus, the environmental control system of modern automobiles consists of one or more of the following: heater–defroster, ventilation, and cooling and dehumidifying (air-conditioning) systems. The integration of the heater–defroster and ventilation systems is common.

Aircraft Cooling

Aircraft air-conditioning equipment must meet additional requirements beyond those of building air-conditioning systems. The equipment must be more compact, lightweight, highly reliable, and unaffected by aircraft vibration and landing impact. The aircraft air-conditioning and pressurization system must maintain an aircraft environment for safety and comfort of passengers and crew and for the proper operation of onboard electronic equipment, through the wide and often rapidly changing range of ambient temperature and pressure.

Air conditioning a modern wide-body aircraft actually entails the use of a total environment control system consisting of the following subsystems:

- Bleed-air control
- Air-conditioning package
- Cabin temperature control
- Cabin pressure control

The bleed-air control system regulates the pressure, temperature, and flow of engine compressor bleed air before it is supplied to the air-conditioning system. The power source for air-cycle refrigeration is the same as that used for pressurizing the cabin. The amount of bleed air extracted from either the main engines or the auxiliary power unit (APU) by an aircraft air conditioning system depends on the cooling loads, ventilation requirements, and aircraft pressurization demands. Traditionally, the cooling load establishes the minimum requirement.

LIFE-CYCLE COSTS

A properly engineered HVAC system must also be economical. The selection of any system is often based on a compromise between its performance and economic merits. The system selected is usually determined by the user's needs, the designer's experience, local building codes, first costs, most efficient use of source energy, and the projected operating costs. When the choice is between alternatives which give apparent equal results, the system with the lowest life-cycle costs should be chosen. This directs a decision toward a system with the lowest long term costs. This overall cost may be divided into two main categories, owning costs and operating costs.

FUTURE DEVELOPMENTS

There will certainly continue to be concern about the quality of both the interior and exterior environment, about energy consumption and conservation, and, of course, about costs. Indoor air quality (IAQ) and the sick building syndrome (SBS) will continue to be of concern. Alternative refrigerants to the CFCs and HCFCs and new cooling techniques will remain under development. Microelectronics will continue to provide for better monitoring and control of HVAC systems. The new scroll compressor will experience market growth as will multi-speed compressors for year-round heat pump air-conditioning systems. The market for variable speed motors for both fans and compressors will expand. The commissioning of HVAC systems will become commonplace.

BIBLIOGRAPHY

1. *ASHRAE Handbook 1993 Fundamentals,* American Society of Heating, Refrigerating, and Air-Conditioning Engineers, Inc., Atlanta, Ga., 1993.
2. *ASHRAE Handbook 1992 HVAC Systems and Equipment,* American Society of Heating, Refrigerating, and Air-Conditioning Engineers, Inc., Atlanta, Ga., 1992.
3. *ASHRAE Handbook 1991 Refrigeration,* American Society of Heating, Refrigerating and Air-Conditioning Engineers, Inc., Atlanta, Ga., 1991.
4. *ASHRAE Handbook 1990 Applications,* American Society of Heating, Refrigerating, and Air-Conditioning Engineers, Inc., Atlanta, Ga., 1990.
5. *Commercial Buildings Consumption and Expenditures, 1986,* DOE, Superintendent of Documents, U.S. Printing Office, Washington, D. C., 1989.
6. *Housing Characteristics 1987,* DOE/EIA, Superintendent of Documents, U.S. Printing Office, Washington, D.C., 1989.
7. F. C. McQuiston and J. D. Spitler, *Cooling and Heating Load Calculation Manual,* 2nd ed., American Society of Heating, Refrigerating, and Air-Conditioning Engineers, Atlanta, Ga., 1992.
8. *Load Calculation for Residential Winter and Summer Air Conditioning, Manual J,* Air Conditioning Contractors of America, Washington, D.C., 1975.
9. *ASHRAE Standard 55-1981,* American Society of Heating, Refrigerating and Air-Conditioning Engineers, Inc., Atlanta, Ga., 1981.
10. Departments of the Air Force, Army, and Navy, *Engineering Weather Data,* AFM 88-29, #008-070-00420-8, U.S. Government Printing Office, Washington, D.C., 1978.
11. *ASHRAE Standard 62-1989* American Society of Heating, Refrigerating and Air-Conditioning Engineers, Inc., Atlanta, Ga., 1989.
12. *Directory of Certified Unitary Air Conditioners, Unitary Air Source Heat Pumps, Sound-Rated Outdoor Unitary Equipment, Solar Collectors,* Air-Conditioning and Refrigeration Institute, Arlington, Va., semi-annually.
13. *Directory of Certified Applied Air Conditioning Products, Air Cooling and Air Heating Coils, Packaged Terminal Air Conditioners, Central Station Air-Handling Units, Packaged Terminal Heat Pumps, Room Fan-Coil Air Conditioners, Water Source Heat Pumps,* Air-Conditioning and Refrigeration Institue, Arlington, Va., semi-annually.
14. H. J. Sauer, Jr., and R. H. Howell, *Heat Pump Systems,* John Wiley & Sons, Inc., New York, 1983.

AIR POLLUTION

George Wolff
GM Environmental and Energy Staff
Detroit, Michigan

Air pollution is the presence of any substance in the atmosphere at a concentration high enough to produce an objectionable effect on humans, animals, vegetation or materials, or to alter the natural balance of any ecosystem significantly. These substances can be solids, liquids, or gases, and can be produced by anthropogenic activities or natural sources. In this article, however, only nonbiological material will be considered. Airborne pathogens and pollens, molds, and spores will not be discussed. Airborne radioactive contaminants will not be discussed either, except for radon, which will be discussed in the context of an indoor air pollutant. (See also GLOBAL CLIMATE CHANGE; GLOBAL ENVIRONMENTAL CHANGE, POPULATION EFFECT; GLOBAL HEALTH INDEX.)

Initial perceptions of objectionable effects of air pollutants were limited to those easily detected: odors, soiling of surfaces (see Plate I), and smoke-belching stacks. Later, it was the concern over longer term, chronic effects that led to the initiation of National Ambient Air Quality Standards (NAAQS) for six "criteria" pollutants (so named because EPA is required to summarize published information on each pollutant—these summaries are called Criteria Documents) in the United States in the early 1970s. The six criteria pollutants were: sulfur dioxide (SO_2), carbon monoxide (CO), nitrogen dioxide (NO_2), ozone (O_3), suspended particulates, and nonmethane hydrocarbons (NMHC, now referred to as volatile organic compounds or VOC). These Criteria Pollutants captured the attention of regulators because they were ubiquitous, there was substantial evidence linking them to health effects at high concentrations, three of them (O_3, SO_2, and NO_2) were also known phytotoxicants (toxic to vegetation), and they were fairly easy to measure. The NMHC were listed as Criteria Pollutants because NMHC are precursors to O_3. However, shortly after NMHC were so designated, it became obvious that the simple empirical relationship that the U.S. Environmental Protection Agency (EPA) used to relate NMHC concentrations to O_3 concentrations was not valid; consequently, NMHC were dropped from the criteria list. In the late 1970s, EPA added lead (Pb) to the list. Particulate matter with an aerodynamic

Plate I. Cincinnati's City Hall Building undergoes a cleaning after accumulating 34 years worth of dirt. The striking difference between the cleaned side (right) and the not-yet-cleaned side (left) illustrates the effects of air pollution. Courtesy of the National Air Pollution Control Administration.

diameter of less than or equal to 10 μm, PM_{10}, was added to the list in 1987.

Until the 1970s, one air pollution axiom was "dilution is the solution to pollution," and the result was that tall smoke stacks were built (see Plate II). When stratospheric O_3 depletion and global warming became issues, air pollution became viewed in a global context. At the same time that the geographic scale of the issue was expanding, the number of pollutants of concern also increased. In the 1970s, it was realized that hundreds of potentially toxic chemicals were being released to the atmosphere. As detection capabilities improved, these chemicals were, indeed, measured in the air. This led to the establishment of a "hazardous air pollutant" category which included any potentially toxic substance in the air that was not a criteria pollutant.

Plate II. It has been shown that high concentrations of O_3 and sulfate haze from tall smoke stacks can be transported hundreds of miles across state and international borders. Courtesy of the United States Environmental Protection Agency.

AIR POLLUTION COMPONENTS

Air pollution can be considered to have three components: sources, transport and transformations in the atmosphere, and receptors. The sources are any process, device, or activity that emits airborne substances. When the substances are released, they are transported through the atmosphere. Some of the substances interact with sunlight, or other substances in the atmosphere, and are transformed into different substances. Air pollutants that are emitted directly to the atmosphere are called primary pollutants. Pollutants that are formed in the atmosphere as a result of transformations are called secondary pollutants. The reactants that undergo the transformation are referred to as precursors. An example of a secondary pollutant is O_3, and its precursors are nitrogen oxides (NO_x = nitric oxide [NO] + NO_2) and NMHC. The receptor is the person, animal, plant, material, or ecosystem affected by the emissions.

Sources

There are three types of air pollution sources: point, area, and line sources. A point source is a single facility that has one or more emissions points. An area source is a collection of smaller sources within a particular geographic area. For example, the emissions from residential heating would be treated collectively as an area source. A line source is a one-dimensional, horizontal configuration. Roadways are an example. Most, but not all, emissions emanate from a specific stack or vent. Emissions emanating from sources other than stacks, such as storage piles or unpaved lots, are classified as fugitive emissions.

EPA requires that individual states develop emissions inventories for all primary pollutants and precursors to secondary pollutants that are classified as criteria or hazardous air pollutants. Details on the construction of emissions inventories have been presented by EPA (1).

Emissions rates for a specific source can be measured directly by inserting sampling probes into the stack or vent. While this has been done for most large point sources, it would be an impossible task to do for every individual source included in an area inventory. Instead, emission factors (based on measurements from similar sources or engineering mass-balance calculations) are applied to the multitude of sources. An emission factor is a statistical average or quantitative estimate of the amount of a pollutant emitted from a specific source-type as a function of the amount of raw material processed, the amount of product produced, or the amount of fuel consumed. Emission factors for most sources have been compiled (2).

Each year, EPA publishes a summary of air pollution emissions and air quality trends for the Criteria Pollutants (3).

Transport and Transformation

Once emitted into the atmosphere, air pollutants are transported and may be transformed. The fate of a particular pollutant depends upon the stability of the atmosphere and the stability of the pollutant in the atmosphere. The former will determine the concentration of the species, initially in the atmosphere, while the latter will determine the persistence of the substance in the atmosphere. The stability of the atmosphere depends on the ventilation. The stability of a pollutant depends on the presence or absence of clouds or fog, the presence or absence of precipitation, the pollutant's solubility in water, the pollutant's reactivity with other atmospheric constituents (which may be a function of temperature), the concentrations of other atmospheric constituents, the pollutant's stability in the presence of sunlight, and the deposition velocity of the pollutant.

In order to illustrate some fundamental principles of atmospheric stability, it is useful to examine a simple model which ignores transformations. The form of the Gaussian Plume Model is

$$X(x,y,z=0) = \frac{Q}{\pi\sigma_y\sigma_z u} \exp - \{(H^2/2\sigma_z^2) + (y^2/2\sigma_y^2)\}$$

where X is the concentration at the receptor located on the ground (z = 0); Q is the pollutant release rate; σ_y and σ_z are the crosswind and vertical plume standard deviations, which are functions of the atmospheric stability and the distance downwind (x); u is the mean wind speed; H is the effective stack height (which is equal to the height of release only if the plume is not buoyant); and x and y are the downwind and crosswind distances. As the ventilation increases (ie, u, σ_y and σ_z), the concentration of the pollutants decrease for a given emission rate Q. The atmospheric stability is determined by comparing the actual lapse-rate to the dry adiabatic lapse-rate. An air parcel warmer than the surrounding air will rise and cool at the dry adiabatic lapse-rate of 9.8°C/1000 m, the atmosphere is unstable, the σs become larger, and the concentrations of pollutants are lower. As the lapse rate becomes smaller, the dispersive capacity of the atmosphere declines and reaches a minimum when the lapse rate becomes positive. When the lapse rate becomes positive, a temperature inversion exists. Temperature inversions form every evening in most places as the heat from the earth's surface is radiated upward, and the air in the lower layers of the atmosphere is cooled. However, these inversions are usually destroyed the next morning as the sun heats the earth's surface, which in turn heats the air adjacent to the surface, and convective activity (mixing) is initiated. Most episodes of high pollution concentrations are associated with multiday inversions.

The stability or persistence of a pollutant in the atmosphere depends on the pollutants atmospheric residence time, which depends on the pollutant's reactivity with other atmospheric constituents, surfaces, or water. Mean atmospheric residence times and principal atmospheric sinks for a variety of species are given in Table 1. Species like SO_2, NO_x (NO and NO_2), and coarse particles have lifetimes less than a day; thus important environmental impacts from these pollutants are usually within close proximity to the emissions sources. Secondary reaction products may have a larger zone of influence, however, depending on the residence times of the products.

Particles with diameters less than 2.5 μm have negligible settling velocities and, therefore, have residence times which are considerably longer than those of larger particles. As a result, observations have shown multiday transport of haze produced by fine particles over distances of more than a thousand kilometers (7). The longer lifetimes of the greenhouse gases, those listed below CO in Table 1,

Table 1. Mean Residence Time (τ) of Species in the Atmosphere

Species	CAS Registry Number	τ	Dominant Sink[a]	(Location)[b]	Refs.
SO_2	*[7446-09-5]*	0.5 days	OH	T	4
NO_x	*[10102-44-0] [10102-43-9]*	0.5 days	OH	T	4
Coarse particles (dia. > 2.5 μm)		<1 day	S, P	T	5
O_3 (tropospheric)	*[10028-15-6]*	90 days[c]	NO, uv, Sr, O	T	4
Fine particles (dia. < 2.5 μm)		5 days[d]	P	T	5
CO	*[630-08-0]*	100 days	OH	T	4
CO_2	*[124-38-9]*	120 yr[e]	O	T	6
CH_4	*[74-82-8]*	7–10 yr	OH	T	4
CFC-11	*[75-69-4]*	65–75 yr	uv	St	4,6
CFC-12	*[75-71-8]*	110–130 yr	uv	St	4,6
N_2O	*[10024-97-2]*	120–150 yr	uv	St	4,6
CFC-113	*[76-13-1]*	90 yr	uv	St	4

[a] Sinks: (OH), reaction with OH; S, sedimentation; P, precipitation scavenging; NO, reaction with NO; uv, photolysis by ultraviolet radiation; Sr, destruction at surfaces; O, adsorption or destruction at oceanic surface.
[b] Location of sink: *T,* troposphere; St, stratosphere.
[c] Tropospheric residence time only; shorter lifetime applies to urban areas where NO quickly destroys O_3; upper limit applies to the remote troposphere.
[d] Applies to particles released in the lower troposphere only; the most important sink is scavenging by precipitation, so in the absence of precipitation, these particles will remain suspended longer.
[e] Combined lifetime for atmosphere, biosphere, and upper ocean.

result in the accumulation and relatively even distribution of these gases around the globe.

To determine the fate of a pollutant after it is released, two approaches, monitoring and modeling, are available. Monitoring of the criteria pollutants is done routinely by state and local air-pollution agencies in most large urban areas and in some other areas as well. Monitoring is expensive and time consuming, however, and even the most extensive urban networks are insufficient to assess the geographic distribution of pollutants accurately. Consequently, various air pollution models are employed to estimate the three-dimensional distributions of pollutants. The types of models used vary in both the scale of the area covered and in the number of processes treated. The smallest scale is the microscale which extends from the emission source to less than 10-km downwind. The Gaussian Plume Model, described above, is an example of a microscale model. From 10-km to about 100-km downwind, mesoscale or urban-scale models are applied. Such models are used to describe the pollutant patterns within and downwind of urban areas. From 100 km to about 1000 km, synoptic- or regional-scale models are employed. These models are used to estimate pollution patterns in areas the size of the eastern U.S. Above 1000 km, global-scale or general-circulation models are used, and these are the models used to calculate the distributions of species with long atmospheric residence times like the greenhouse gases. With each increase in scale, the complexity of the meteorological processes increases greatly. In addition, the complexity of the model depends on the number and kinds of transformation processes which are included (see AIR POLLUTION MODELING).

Receptors

The receptor is the person, animal, plant, material, or ecosystem affected by the pollutant emission (see Plate III). With respect to acid deposition, the receptors which have generated the most concern are certain aquatic ecosystems and certain forest ecosystems, although there is also some concern that acid deposition adversely affects some materials. For visibility-reducing air pollutants, CFCs, and greenhouse gases, the receptor is the atmosphere.

AIR QUALITY MANAGEMENT

In the United States, the framework for air quality management was established by the 1965 Clean Air Act (CAA) and subsequent amendments in 1970 and 1977, and the comprehensive amendments of 1990. The CCA defined two categories of pollutants: criteria pollutants and hazardous air pollutants. For the criteria pollutants, the CAA required that EPA establish NAAQS and emissions standards for some large new sources and for motor vehicles, and gave the primary responsibility for designing and implementing air quality improvement programs to the states. For the hazardous air pollutants, only emissions standards for some sources are required. The NAAQS apply uniformly across the United States whereas emissions standards for criteria pollutants can vary somewhat, depending on the severity of the local air-pollution problem and whether an affected source already exists or is proposed as a new source. In addition, individual states have the right to set their own ambient air quality and emissions standards (which must be at least as stringent as the Federal standards) for all pollutants and all sources except motor vehicles. With respect to motor vehicles, the CAA allows the states to choose between two sets of emissions standards: the Federal standards or the more stringent California standards.

The two levels of NAAQS, primary and secondary, are listed in Table 2. Primary pollutant standards were set to protect public health with an adequate margin of safety. Secondary standards, where applicable, were chosen to protect public welfare, including vegetation. The pollutant

Plate III. In the instance depicted above, white birch leaves act as the receptor for sulfur dioxide pollutant emissions. The leaf on the right is healthy. The discolored leaf on the left shows the damaging effect of air pollution. Courtesy of the United States Department of Agriculture.

PM_{10} refers to particulate matter with an aerodynamic diameter less than or equal to 10 μm. According to the CAA, the scientific bases for the NAAQS are to be reviewed every 5 years so that the NAAQS levels reflect current knowledge. In practice, however, the review cycle takes considerably longer. In order to analyze trends in Criteria Pollutants nationwide, the EPA has established three types of monitoring systems. The first is a network of 98

Table 2. National Ambient Air Quality Standards

	Primary[a]		Secondary[a]		
Pollutant	μg/m^3	ppm	μg/m^3	ppm	Averaging Time
PM_{10}	50		50		Annual arithmetic mean
	150		150		24-h[b]
SO_2	80	(0.03)			Annual arithmetic mean
	365	(0.14)			24-h[b]
			1300	(0.50)	3-h[b]
CO	(10)	9			8-h[b]
	(40)	35			1-h[b]
NO_2	(100)	0.053	(100)	0.053	Annual arithmetic mean
Pb	1.5		1.5		Maximum quarterly average
O_3	(235)	0.12	(235)	0.12	Maximum daily[c] 1-h average

[a] Parenthetical value is an approximately equivalent concentration.
[b] Not to be exceeded more than once per year.
[c] Not to be exceeded on more than three days in three years.

National Air Monitoring sites (NAMS), located in areas with high pollutant concentrations and high population exposures. In addition, EPA also regularly evaluates data from the State and Local Monitoring system (SLAMS) and from Specific Purpose Monitors (SPM). These three types of stations comprise the 274-site national monitoring system, which is required to meet rigid quality-assurance criteria. To determine if an area meets the NAAQS, the states are required to monitor the concentrations of the Criteria Pollutants in areas that are likely to be near or exceed the NAAQS. If an area exceeds a NAAQS for a given pollutant, it is designated as a nonattainment area for that pollutant, and the state is required to develop and implement a State Implementation Plan (SIP). The SIP is a strategy designed to achieve emissions reductions sufficient to meet the NAAQS within a specific deadline. The deadline is determined by the severity of the local pollution problem. Areas that receive long deadlines (six years or more) must show continuous progress by reducing emissions by a specified percentage each year. For SO_2 and NO_2, the initial SIPs were very successful in achieving the NAAQS in most areas. However, for other criteria pollutants, particularly O_3 and to a lesser extent CO, many areas are starting a third round of SIP preparations with little hope of meeting the NAAQS in the near future. If a state misses an attainment deadline, fails to revise an inadequate SIP, or fails to implement SIP requirements, EPA has the authority to enforce sanctions such as banning construction of new stationary sources and the withholding of federal grants for highways.

In nonattainment areas, the degree of control on small sources is left to the discretion of the state, and it is largely determined by the degree of required emissions reductions. Large existing sources must be retrofitted with "reasonable available control technology" (RACT) to minimize emissions. All large new sources and existing sources that undergo major modifications must meet EPA's new source performance standards at a minimum; and in nonattainment areas they must be designed with "lowest achievable emission rate" (LAER) technology, and emissions offsets must be obtained. Offsets require that emissions from existing sources within the area must be reduced below legally allowable levels so that the amount of the reduction is greater than or equal to the emissions expected from the new source.

In attainment areas, new large facilities must be designed to incorporate the "best available control technology" (BACT). In no situation can the facility cause a new violation in the NAAQS.

Large sources of SO_2 and NO_x may also require additional emission reductions because of the 1990 Clean Air Act Amendments. To reduce acid deposition, the amendments require that nationwide emissions of SO_2 and NO_x be reduced on an annual basis by 10 million and 2 million tons, respectively, by the year 2000.

Once a substance is designated by EPA as a Hazardous Air Pollutant (HAP), EPA has to promulgate a NESHAP (National Emission Standard for Hazardous Air Pollutants), which is designed to protect public health with an ample margin of safety. The 1990 Clean Air Act Amendments identify 189 HAPs. These are further discussed in the Air Toxics section.

AIR POLLUTION ISSUES

Photochemical Smog

Photochemical smog is a complex mixture of constituents formed when VOCs and NO_x are irradiated by sunlight. From an effects perspective, O_3 is the primary concern and it is the most abundant species formed in photochemical smog (See Plate IV). Extensive studies have shown that O_3 is a lung irritant and a phytotoxicant.

Photochemical smog is a summertime phenomenon for most parts of the United States because temperatures are too low and sunlight is insufficient during the other seasons. In the warmer parts of the country, especially in Southern California, the smog season begins earlier and lasts into the fall.

There is a significant clean air background O_3 concentration that varies with season and latitude. The clean air background is defined as the concentrations measured at pristine areas of the globe. It consists of natural sources of O_3, but it undoubtedly contains some anthropogenic contribution because it may have increased since the last century (8). In the summertime in the United States, the average background is about 0.04 ppm (9). This background

Plate IV. Photochemical smog envelops New York City. Metropolitan areas consistently have the highest O_3 concentrations in the United States, with levels frequently above the National Ambient Air Quality Standards. Courtesy of AERO SERVICE.

O_3 has four sources: intrusions of O_3-rich stratospheric air, *in situ* O_3 production from methane (CH_4) oxidation, the photooxidation of naturally emitted VOCs from vegetation, and the long-range transport of O_3 formed from the photooxidation of anthropogenic VOCs and NO_x emissions. Although there are several mechanisms which will transport O_3-rich air from the stratosphere into the lower troposphere, the most important appears to be associated with large-scale eddy transport that occurs in the vicinity of upper air troughs of low pressure associated with the jet stream (10). This is an intermittent mechanism, so the contribution of stratospheric O_3 to surface O_3 will have a considerable temporal variation. On very rare occasions, this mechanism has produced brief ground level concentrations exceeding 0.12 ppm (11). The other three mechanisms will be described below.

In the presence of sunlight ($h\nu$), NO_2 photolyzes and produces O_3

$$NO_2 + h\nu \rightarrow NO + O \quad (1)$$

$$O + O_2 + M \rightarrow O_3 + M \quad (2)$$

$$NO + O_3 \rightarrow NO_2 + O_2 \quad (3)$$

where M is any third body molecule (most likely N_2 or O_2 in the atmosphere) that remains unchanged in the reaction. This process produces some steady-state concentration of O_3 that is a function of the initial concentrations of NO and NO_2, the solar intensity, and the temperature. Although these reactions are extremely important in the atmosphere, the steady-state O_3 produced is much lower than the concentrations usually observed, even in clean air. In order for O_3 to accumulate, there must be a mechanism that converts NO to NO_2 without consuming a molecule of O_3, as does reaction 3. Reactions among hydroxyl radicals (OH) and hydrocarbons or VOC constitute such a mechanism. In clean air, OH may be generated by:

$$O_3 + h\nu \rightarrow O_2 + O(^1D) \quad (4)$$

$$O(^1D) + H_2O \rightarrow 2\,OH \quad (5)$$

where $O(^1D)$ is an excited form of an O atom that is produced from a photon at a wavelength between 280 and 310 nm. This produces a "seed" OH which can produce the following chain reactions:

$$OH + CH_4 \rightarrow H_2O + CH_3 \quad (6)$$

$$CH_3 + O_2 + M \rightarrow CH_3O_2 + M \quad (7)$$

$$CH_3O_2 + NO \rightarrow CH_3O + NO_2 \quad (8)$$

The NO_2 will then photolyze and produce O_3 (equations 1 and 2). The CH_3O radical continues to react:

$$CH_3O + O_2 \rightarrow HCHO + HO_2 \quad (9)$$

and the HO_2 radical also forms more NO_2:

$$HO_2 + NO \rightarrow NO_2 + OH \quad (10)$$

which will result in more O_3. In addition, OH is regenerated, and it can begin the cycle again by reacting with another CH_4 molecule. Further, the formaldehyde photodissociates

$$HCHO + h\nu \xrightarrow{a} H_2 + CO \quad (11)$$

$$\xrightarrow{b} HCO + H \quad (12)$$

$$HCO + O_2 \rightarrow HO_2 + CO \quad (13)$$

$$H + O_2 \rightarrow HO_2 \quad (14)$$

and the HO_2 from both equations 13 and 14 will form additional NO_2. Furthermore, the CO is oxidized:

$$CO + OH \rightarrow CO_2 + H \quad (15)$$

and the H radical can form another NO_2 (eqs. 14 and 10). Thus, the oxidation of one CH_4 molecule is capable of producing three O_3 molecules and two OH radicals by this reaction sequence. However, this chain reaction is less than 100% efficient because there are many competing chain-terminating reactions. Two examples are

$$HO_2 + HO_2 \rightarrow H_2O_2 + O_2 \quad (16)$$

$$OH + NO_2 \xrightarrow{M} HNO_3 \quad (17)$$

Both of these reactions terminate the chain by scavenging a free radical. On the average, however, the chain results in a net production of O_3 and OH.

In a polluted or urban atmosphere, O_3 formation by the CH_4 oxidation mechanism is overshadowed by the oxidation of other VOCs. The "seed" OH can be produced from equations 4 and 5, but the photodisassociation of carbonyls and nitrous acid (HNO_2, formed from the reaction of OH + NO and other reactions) are also important sources of OH in polluted environments. An imperfect, but nevertheless useful, measure of the rate of O_3 formation by VOC oxidation is the rate of the initial OH–VOC reaction. The rate of the OH-CH_4 reaction is much slower than any other OH–VOC reaction. Table 3 contains the reaction rates with OH relative to the OH–CH_4 rate for some commonly occurring VOCs and their median concentrations from 39 cities. Also shown for comparison are the relative reaction rates between OH and two VOC species emitted by vegetation: isoprene and α-pinene. It is obvious from the data in Table 3 that there is a wide range of reactivities. The reaction mechanisms by which the VOCs are oxidized are analogous to, but much more complex than, the CH_4 oxidation mechanism. The fastest reacting species are the natural VOCs emitted from vegetation. However, natural VOCs also react rapidly with O_3, so they are a source as well as a sink of O_3. Whether the natural VOCs are a net source or sink is determined by the natural VOCs to NO_x ratio and the sunlight intensity. At high VOC/NO_x ratios, there will be insufficient NO_2 formed to offset the O_3 loss. However, when O_3 reacts with these internally bonded olefinic compounds, carbonyls are formed and, the greater the sunshine, the better the chance the carbonyls will photolyze and produce OH, which will initiate the O_3-forming chain reactions.

Once the sun sets, O_3 formation ceases and, in an urban area, it is rapidly scavenged by freshly emitted NO by

Table 3. Median Concentration of the Ten Most Abundant Ambient Air Hydrocarbons in 39 U.S. Cities and their Relative Reactivity with OH

Compound	CAS Registry Number	Median Concentration[a] ppbC	Relative Reactivity with OH[b,c]
Isopentane	[*463-82-1*]	45.3	494
n-Butane	[*106-97-8*]	40.3	351
Toluene	[*108-88-3*]	33.8	831
Propane	[*74-98-6*]	23.5	143
Ethane	[*74-84-0*]	23.3	36
n-Pentane	[*109-66-0*]	22.0	480
Ethylene	[*74-85-1*]	21.4	1013
m-, *p*-xylene	[*108-38-3*]	18.1	3117
p-xylene	[*106-42-3*]	18.1	1818
2-Methylpentane	[*107-83-5*]	14.9	
Isobutane	[*75-28-5*]	14.8	325
	Biogenic species		
α-Pinene	[*80-56-8*]		7792
Isoprene	[*78-79-5*]		12078

[a] Ref. 12.
[b] Ref. 13.
[c] Relative to CH_4 + OH reaction at 298°C.

equation 3. On a typical summer night, however, a nocturnal inversion begins to form around sunset, usually below a few hundred meters. Consequently, the surface-based NO emissions are trapped below the top of the inversion. Above the inversion to the top of the mixed layer (usually about 1500 m), O_3 is depleted at a much slower rate. The next morning, the inversion dissipates and the O_3-rich air aloft is mixed down into the O_3-depleted air near the surface. This process, in combination with the onset of photochemistry as the sun rises, produces the sharp increase in surface O_3.

Although photochemical smog is a complex mixture of many primary and secondary pollutants and involves a myriad of atmospheric reactions, there are characteristic pollutant concentration versus time profiles that are generally observed within and downwind of an urban area during a photochemical smog episode. In particular, the highest O_3 concentrations are generally found 10–100 km downwind of the urban emissions areas, unless the air is completely stagnant. This fact, in conjunction with the long lifetime of O_3 in the absence of high concentrations of NO, means that O_3 is a regional problem. In the Los Angeles basin, high concentrations of O_3 are transported throughout the basin, and multiday episodes are exacerbated by the accumulation of O_3 aloft, which is mixed to the surface daily. On the East Coast, a typical O_3 episode is associated with a high pressure system anchored offshore, producing a southwesterly flow across the region. As a result, emissions from Washington, D.C., travel and mix with emissions from Baltimore and over a period of a few days will continue traveling northeastward through Philadelphia, New York City, and Boston. Under these conditions, the highest O_3 concentrations typically occur in central Connecticut (14).

It is obvious from the above discussion that in order to reduce O_3 in a polluted atmosphere, reductions in the precursors (VOC and NO_x) are required. However, the choice of whether to control VOC or NO_x or both as the optimum control strategy depends on the local VOC/NO_x ratio. This is illustrated in the O_3-isopleth diagrams in Figure 1. Although the four different chemical mechanisms used to obtain Figure 1 give somewhat different results, the shape of the isopleths are quite similar and the key features are summarized in Figure 2. The first region in the upper left is the NO_x-inhibition region. In this region, a decrease in NO_x alone results in an increase in O_3, but a decrease in VOC decreases O_3. The region at the bottom right is the VOC or HC saturation region where reducing VOCs has no effect on the O_3. On the other hand, a reduction in NO_x in this region results in lower O_3. In the middle is the knee region, where reductions in either reduce O_3. The upper boundary of this region will vary from day to day and from place to place as its location is a function of the reactivity of the VOC mix and the sunlight intensity. As a guideline, VOC controls alone are generally the most efficient way to reduce O_3 in areas with a median 6 A.M. to 9 A.M. VOC/NO_x ratio of 10:1 or less; areas with a higher ratio may need to consider NO_x reductions as well (17). The 1990 Clean Air Act Amendments require that O_3 nonattainment areas reduce both VOC and NO_x from big stationary sources unless the air quality benefits are greater in the absence of NO_x reductions. Determining a workable control strategy is further complicated by the transport issue. For example, on high O_3 days in the Northeast, the upwind air entering Philadelphia and New York City frequently contains O_3 already near or over the NAAQS as a result of emissions from areas to the west and south (18). Consequently, control strategies must be developed on a coordinated, multistate regional basis.

Because of the complex mixture of VOCs in the atmosphere, the composition of the reaction products and intermediate species is even more complex. Some of the more important species produced in the smog process include hydrogen peroxide (H_2O_2), peroxyacetyl nitrate (PAN), aldehydes (particularly formaldehyde, HCHO [*50-00-0*]), nitric acid (HNO_3), and particles. The H_2O_2 is formed and dissolved in cloud droplets; is an important oxidant, responsible for oxidizing SO_2 to H_2SO_4, the primary cause of acid precipitation.

At high enough concentrations, PAN is a potent eye irritant and a phytotoxicant. Aldehydes are important because they are temporary reservoirs of free radicals and HCHO is a known carcinogen. Nitric acid is important because, next to H_2SO_4, it is the second most abundant acid in precipitation. In addition, in Southern California it is the main cause of acid fog.

Particles are the principal cause of the haze and the associated brown color that is often apparent with smog. The three most important types of particles produced in smog are composed of organics, sulfates, and nitrates.

Volatile Organic Compounds (VOC). VOCs include any organic-carbon compound that exists in the gaseous state in the ambient air. Sources of VOCs include any process or activity that uses organic solvents, coatings, or fuel. Emissions of VOCs are important because some are toxic

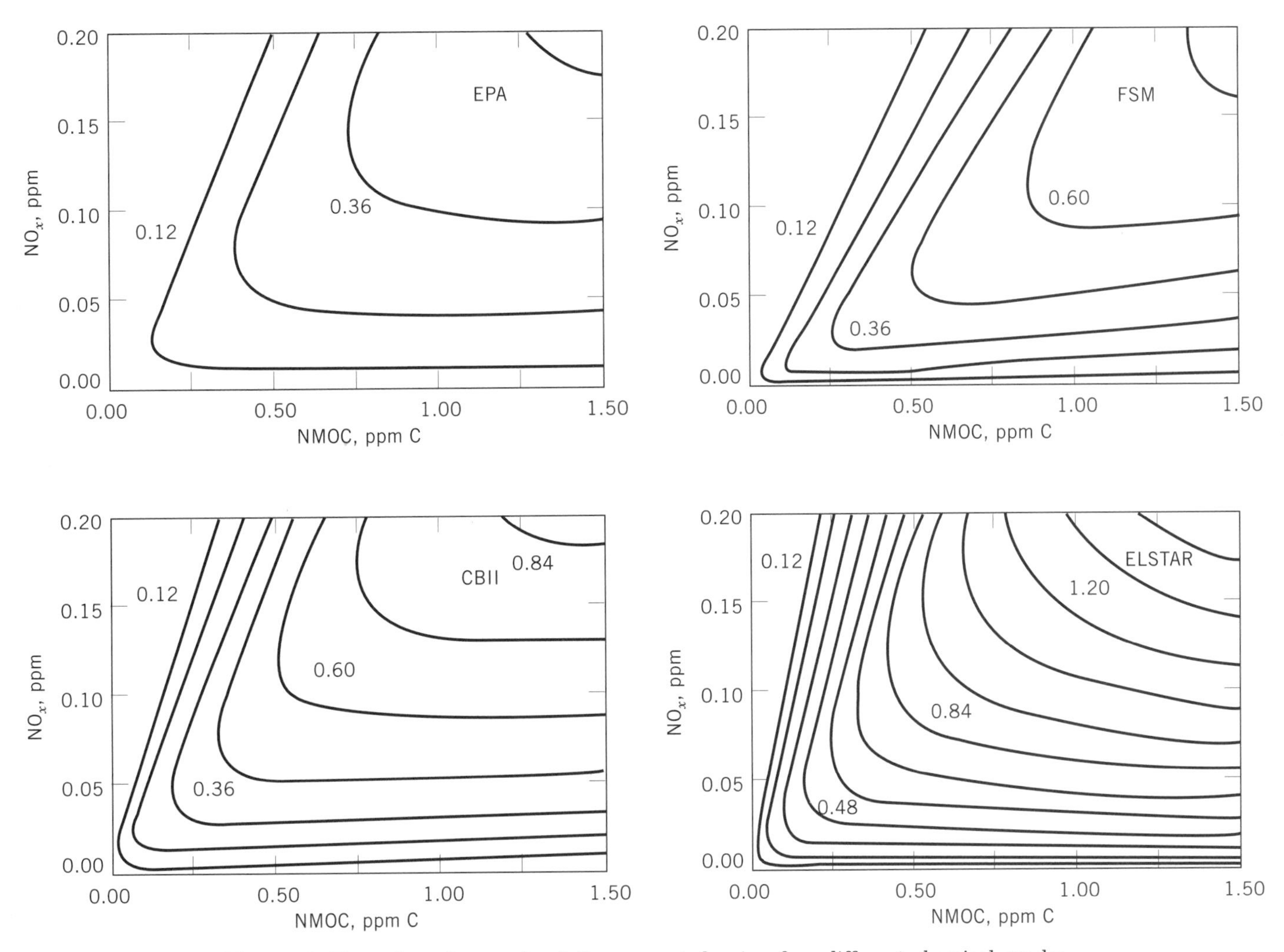

Figure 1. Examples of ozone isopleths generated using four different chemical mechanisms (15). Courtesy of Pergamon Press.

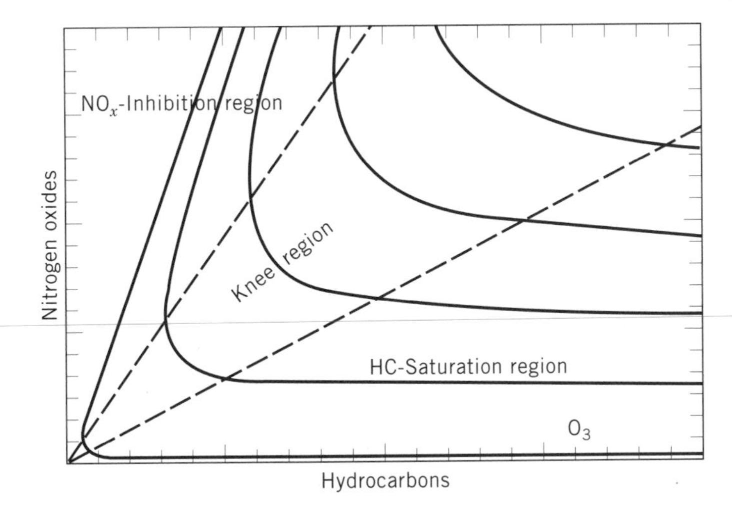

Figure 2. Typical O_3 isopleth diagram showing the three chemical regimes (16). Courtesy of Pergamon Press.

by themselves and most are precursors of O_3 and other species associated with photochemical smog. As a result of control measures designed to reduce O_3, VOC emissions are declining in the United States.

Nitrogen Oxides (NO_x). In air-pollution terminology, nitrogen oxides include the gases NO and NO_2. Most of the NO_x is emitted as NO, which is oxidized to NO_2 in the atmosphere (see eqs. 3 and 8). All combustion processes are sources of NO_x. At the high temperatures generated in the combustion process, some N_2 will be converted to NO in the presence of O_2. In general, the higher the combustion temperature, the more NO_x produced. Since NO_2 is one of the original Criteria Pollutants and it is a precursor to O_3, it has been the target of successful emissions reduction strategies for two decades in the United States. As a result, in 1987, all areas of the United States, with the exception of the Los Angeles/Long Beach area, were in compliance with the NAAQS for NO_2.

Throughout the United States, however, NO_x remains an important issue because it is an essential ingredient of photochemical smog and some of the NO_x is oxidized to

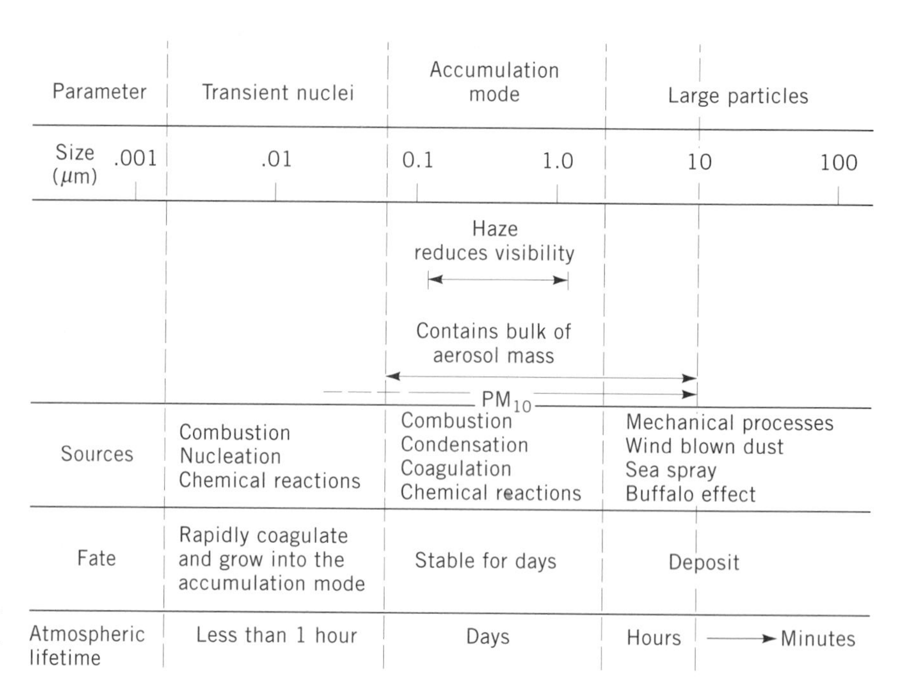

Figure 3. Some important aerosol characteristics (modified from ref. 20, which was adopted in part from ref. 21).

HNO_3, an essential ingredient of acid precipitation and fog. In addition, NO_2 is the only important gaseous species in the atmosphere that absorbs visible light, and in high enough concentrations can contribute to a brownish discoloration of the atmosphere.

Sulfur Oxides (SO_x). The combustion of sulfur-containing fossil fuels, especially coal, is the primary source of SO_x. Between 97 and 99% of the SO_x emitted from combustion sources is in the form of SO_2, which is a Criteria Pollutant. The remainder is mostly SO_3, which in the presence of atmospheric water vapor is immediately transformed into H_2SO_4 [*7664-93-9*], a liquid particulate. Both SO_2 and H_2SO_4 at sufficient concentrations produce deleterious effects on the respiratory system. In addition, SO_2 is a phytotoxicant. As with NO_2, control strategies designed to reduce the ambient levels of SO_2 have been highly successful. However, the 1990 Clean Air Act Amendments require additional SO_2 reductions because of the role that SO_2 plays in acid deposition. In addition, there is some concern over the health effects of H_2SO_4 particles, which are not only emitted directly from some sources, but are also formed in the atmosphere from the oxidation of SO_2 (19).

Carbon Monoxide (CO). Carbon monoxide is a colorless, odorless gas emitted during the incomplete combustion of fuels. CO is emitted during any combustion process, and transportation sources account for about two-thirds of the CO emissions nationally. However, in certain areas, woodburning fireplaces and stoves contribute most of the observed CO. CO is absorbed through the lungs into the blood stream and reacts with hemoglobin to form carboxyhemoglobin, which reduces the oxygen carrying capacity of the blood.

Emissions of CO in the United States peaked in the late 1960s, but have decreased consistently since that time as transportation sector emissions significantly decreased.

Particulate Matter

In the air pollution field, the terms particulate matter, particulates, particles, and aerosols are used interchangeably and all refer to finely divided solids and liquids dispersed in the air. In 1987, the EPA promulgated new standards for ambient particulate matter using a new indicator, PM_{10}. PM_{10} is particulate matter with an aerodynamic diameter of 10 μm or less. The 10 μm diameter was chosen because 50% of the 10 μm particles deposit in the respiratory tract below the larynx during oral breathing. The fraction deposited decreases with increasing particle diameter above 10 μm. The primary and secondary PM_{10} NAAQS are given in Table 2.

Atmospheric aerosols can be classified into three size modes: the nuclei-, accumulation-, and large- or coarse-particle modes. Some important characteristics of these modes are illustrated in Figure 3.

Figure 4 shows the mass size distribution of typical ambient aerosols. Note the mass peaks in the accumulation

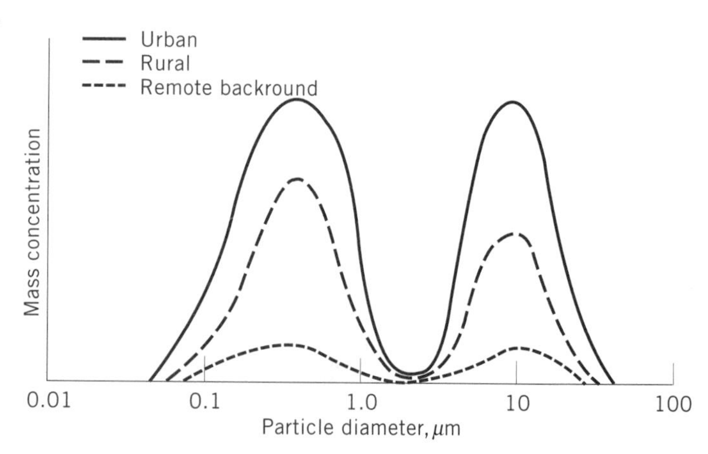

Figure 4. Typical size distributions of atmospheric particles.

mode and between 5 and 10 μm. The minimum in the curve at about 1–2.5 μm is due to a lack of sources for these particles. Coagulation is not significant for the accumulation-mode particles, and particles produced by mechanical process are larger than 2.5 μm. Consequently, particles less than about 2.5 μm have different sources than particles greater than 2.5 μm. Because of this, it is convenient to classify PM_{10} into a coarse-particulate-mass mode (CPM, diameter ≥ 2.5–10 μm) and a fine-particulate-mass mode (FPM < 2.5 μm). By knowing the relative amounts of CPM and FPM as well as the chemical composition of the principal species, information on the sources of the PM_{10} can be deduced. In urban areas, the CPM and FPM modes are usually comparable in mass. In rural areas, the FPM is generally lower than in urban areas, but higher than the CPM mass. The reason for this is that a significant fraction of the rural FPM is generally transported from upwind sources, whereas most of the CPM is generated locally.

The most abundant FPM species are sulfate (SO_4^{2-}), carbon (organic carbon, OC, and elemental carbon, EC), and nitrate (NO_3^-) compounds which generally account for 70–80% of the FPM. In the eastern United States, SO_4^{2-} compounds are the dominant species, but very little SO_4^{2-} is emitted directly into the atmosphere. Most of the SO_4^{2-} is a secondary aerosol formed from the oxidation of SO_2, and in the eastern United States the source of most of the SO_2 is coal-burning emissions. In the atmosphere, the principal oxidation routes include: homogeneous oxidation of SO_4^{2-} by OH, and the heterogeneous oxidation in water droplets by hydrogen peroxide (H_2O_2), O_3, or, in the presence of a catalyst, O_2. Atmospheric particles which have been identified as catalysts include many metal oxides and soot. The water droplets include cloud and dew droplets as well as aerosols which contain sufficient water. Under high relative humidity conditions, hygroscopic salts deliquesce and form liquid aerosols. Sulfate initially forms as sulfuric acid (H_2SO_4), which rapidly reacts with available ammonia (NH_3) to form ammonium bisulfate (NH_4HSO_4). If sufficient NH_3 is present, the final product is ammonium sulfate [$(NH_4)_2SO_4$]. In some urban areas in the western United States, NO_3^- is more abundant than SO_4^{2-}. The NO_3^- in the FPM exists primarily as NH_4NO_3. However, acidic SO_4^{2-} (H_2SO_4 or NH_4HSO_4) readily reacts with NH_4NO_3 and abstracts the NH_3, leaving behind gaseous HNO_3. Consequently, unless there is sufficient NH_3 to completely convert all of the SO_4^{2-} to $(NH_4)_2SO_4$, NH_4NO_3 will not accumulate in the atmosphere. The western U.S. cities have sufficient NH_3, which is due in a large part to the presence of animal feedlots, to neutralize the SO_4^{2-} and to allow the NH_4NO_3 to accumulate. In general, in the eastern United States, there is insufficient NH_3 to neutralize the SO_4^{2-}, so NH_4NO_3 does not accumulate. OC is a principal constituent of the FPM at all sites. The main sources of OC are combustion and atmospheric reactions involving gaseous VOCs. As is the case with VOCs, there are hundreds of different OC compounds in the atmosphere. A minor but ubiquitous aerosol constituent is elemental carbon (EC). EC is the nonorganic, black constituent of soot. Its structure is similar to graphite crystals with many imperfections. Combustion and pyrolysis are the only processes that produce EC, and diesel engines and wood burning are the most significant sources. Crustal dust and water make up most of the remaining FPM mass.

Lead

Lead (Pb) [*7439-92-1*] is of concern because of its tendency to be retained and accumulated by living organisms. When excessive amounts accumulate in the human body, lead can inhibit the formation of hemoglobin and produce life-threatening lead poisoning. In smaller doses, lead is also suspected of causing learning disabilities in children. From 1982 to 1991, nationwide lead emissions decreased 90%, with the primary source, transportation, showing a 97% reduction (3). The most important reason for this dramatic reduction was the removal of lead compounds from fuels, primarily gasoline.

Air Toxics

There are thousands of commercial chemicals used in the United States. Hundreds of these substances are emitted into the atmosphere and have some potential to affect human health adversely at certain concentrations. Some are known or suspected carcinogens. Identifying all of these substances and promulgating emissions standards is beyond the present capabilities of existing air-quality management programs. Consequently, toxic air pollutants (TAPs) need to be prioritized based on risk analysis, so that those posing the greatest threats to health can be regulated. Although the criteria pollutants were so designated because they can have significant public-health impacts, the criteria pollutants are not considered TAPs because the criteria pollutants are regulated elsewhere in the CAA. A distinguishing feature between TAPs and criteria pollutants is that criteria pollutants are considered national issues while TAPs, on the other hand, are most often isolated issues, localized near the source of the TAP emissions.

There are three types of TAP emissions: continuous, intermittent, and accidental. Intermittent sources can be routine emissions associated with a batch process or a continuous process operated only occasionally. An accidental release is an inadvertent emission. A dramatic example of this type was the release of methyl isocyanate in Bhopal, which was responsible for over 2000 deaths. As a result of this accident, the U.S. Congress created Title III, a free-standing statute included in the Superfund Amendments and Reauthorization Act (SARA) of 1986. Title III provides a mechanism by which the public can be informed of the existence, quantities, and releases of toxic substances, and requires the states to develop plans to respond to accidental releases of these substances. Further, it requires anyone releasing specific toxic chemicals above a certain threshold amount to annually submit a toxic-chemical-release form to EPA. At present, there are 308 specific chemicals subject to Title III regulation (22).

A valuable resource that contains a listing of many potential TAPs is the American Conference of Governmental Industrial Hygienists' *Threshold Limit Values (TLV)* (23).

The 1970 Clean Air Act required that EPA provide an ample margin of safety to protect against Hazardous Air

Pollutants (HAPs) by establishing national emissions standards (NESHAPs) for certain sources. From 1970 to 1990, over 50 chemicals were being considered for designation as HAPs, but EPA's review process was completed for only 28 chemicals. Of the 28, NESHAPs were promulgated for only eight substances: beryllium [*7440-41-7*], mercury [*7436-97-6*], vinyl chloride [*75-01-4*], asbestos [*1332-21-4*], benzene [*71-43-2*], radionuclides, inorganic arsenic [*7440-38-2*], and coke-oven emissions. EPA decided not to list ten of the substances and intended to list the other ten substances as HAPs (24). However, in the 1990 Clean Air Act Amendments, 189 substances are listed (Table 4) that EPA must regulate by enforcing "maximum achievable control technology (MACT)." The Amendments mandate that EPA issue MACT standards for all sources of the 189 substances by the year 2000. In addition, EPA must determine the risk remaining after MACT is in place and develop health-based standards that would limit the cancer risk to one case in one million exposures. EPA may add or delete substances from the list.

Table 4. Substances Listed as Hazardous Air Pollutants in the 1990 Clean Air Act Amendments

Substance	(CAS Registry Number)	Substance	(CAS Registry Number)	Substance	(CAS Registry Number)
Acetaldehyde	[*75-07-0*]	Dimethyl carbamoyl chloride	[*79-44-7*]	Parathion	[*56-38-2*]
Acetamide	[*60-35-5*]	Dimethyl formamide	[*68-12-2*]	Pentachloronitrobenzene	[*82-68-8*]
Acetonitrile	[*75-05-8*]	1,1-Dimethyl hydrazine	[*54-14-7*]	Pentachlorophenol	[*87-86-5*]
Acetophenone	[*98-86-2*]	Dimethyl phthalate	[*131-11-3*]	Phenol	[*108-95-2*]
2-Acetylaminofluorene	[*53-96-3*]	Dimethyl sulfate	[*77-78-1*]	*p*-Phenylenediamine	[*106-50-3*]
Acrolein	[*107-02-8*]	4,6-Dinitro-o-cresol, and salts	[*534-52-1*]	Phosgene	[*75-44-5*]
Acrylamide	[*79-06-1*]	2,4-Dinitrophenol	[*51-28-5*]	Phosphine	[*7803-51-2*]
Acrylic acid	[*79-10-7*]	2,4-Dinitrotoluene	[*121-14-2*]	Phosphorus	[*7723-14-0*]
Acrylonitrile	[*107-13-1*]	1,4-Dioxane	[*123-91-1*]	Phthalic anhydride	[*85-44-9*]
Allyl chloride	[*107-05-1*]	1,2-Diphenylhydrazine	[*122-66-7*]	Polychlorinated biphenyls	[*1336-36-3*]
4-Aminobiphenyl	[*92-67-1*]	Epichlorohydrin	[*106-89-8*]	1,3-Propane sultone	[*1120-74-4*]
Aniline	[*62-53-3*]	1,2-Epoxybutane	[*106-88-7*]	beta-Propiolactone	[*57-57-8*]
o-Anisidine	[*90-04-0*]	Ethyl acrylate	[*140-88-5*]	Propionaldehyde	[*123-38-6*]
Asbestos	[*1332-21-4*]	Ethyl benzene	[*100-41-4*]	Propoxur (Baygon)	[*114-26-1*]
Benzene	[*71-43-2*]	Ethyl carbamate	[*51-79-6*]	Propylene dichloride	[*78-87-5*]
Benzidine	[*92-87-5*]	Ethyl chloride	[*75-00-3*]	Propylene oxide	[*75-56-9*]
Benzotrichloride	[*98-07-7*]	Ethylene dibromide	[*106-93-4*]	1,2-Propylenimine	[*75-55-8*]
Benzyl chloride	[*100-44-7*]	Ethylene dichloride	[*107-06-2*]	Quinoline	[*91-22-5*]
Biphenyl	[*92-52-4*]	Ethylene glycol	[*107-21-1*]	Quinone	[*106-51-4*]
Bis(2-ethylhexyl)phthalate	[*117-81-7*]	Ethylene imine	[*151-56-4*]	Styrene	[*100-42-5*]
Bis(chloromethyl)ether	[*542-88-1*]	Ethylene oxide	[*75-21-8*]	Styrene oxide	[*96-09-3*]
Bromoform	[*75-25-2*]	Ethylene thiourea	[*96-45-7*]	2,3,7,8-Tetrachlorodibenzo-*p*-dioxin	[*1746-01-6*]
1,3-Butadiene	[*106-99-0*]	Ethylidene dichloride	[*75-34-3*]	1,1,2,2-Tetrachloroethane	[*79-34-5*]
Calcium cyanamide	[*156-62-7*]	Formaldehyde	[*50-00-0*]	Tetrachloroethylene	[*127-18-4*]
Caprolactam	[*105-60-2*]	Heptachlor	[*76-44-8*]	Titanium tetrachloride	[*7550-45-0*]
Captan	[*133-06-2*]	Hexachlorobenzene	[*118-74-1*]	Toluene	[*108-88-3*]
Carbaryl	[*63-25-2*]	Hexachlorobutadiene	[*87-68-3*]	2,4-Toluene diamine	[*95-80-7*]
Carbon disulfide	[*75-15-0*]	Hexachlorocyclopentadiene	[*77-47-4*]	2,4-Toluene diisocyanate	[*584-84-9*]
Carbon tetrachloride	[*56-23-5*]	Hexachloroethane	[*67-72-1*]	*o*-Toluidine	[*95-53-4*]
Carbonyl sulfide	[*463-58-1*]	Hexamethyl-1,6-diisocyanate	[*822-06-0*]	Toxaphene	[*8001-35-2*]
Catechol	[*120-80-9*]	Hexamethylphosphoroamide	[*680-31-9*]	1,2,4-trichlorobenzene	[*120-82-1*]
Chloramben	[*133-90-4*]	Hexane	[*110-54-3*]	1,1,2-Trichloroethane	[*79-00-5*]
Chlordane	[*57-74-9*]	Hydrazine	[*302-01-2*]	Trichloroethylene	[*79-01-6*]
Chlorine	[*7782-50-5*]	Hydrochloric acid	[*7647-01-0*]	2,4,5-Trichlorophenol	[*95-95-4*]
Chloroacetic acid	[*79-11-8*]	Hydrogen fluoride	[*7664-39-3*]	2,4,6-Trichlorophenol	[*88-06-2*]
2-Chloroacetophenone	[*532-27-4*]	Hydroquinone	[*123-31-9*]	Trimethylamine	[*121-44-8*]
Chlorobenzene	[*108-90-7*]	Isophorone	[*78-59-1*]	Trifluralin	[*1582-09-8*]
Chlorobenzilate	[*510-15-6*]	Lindane (all isomers)	[*58-89-9*]	2,2,4-Trimethylpentane	[*540-84-1*]
Chloroform	[*67-66-3*]	Maleic anhydride	[*108-31-6*]	Vinyl acetate	[*108-05-4*]
Chloromethyl methyl ether	[*107-30-2*]	Methanol	[*67-56-1*]	Vinyl bromide	[*593-60-2*]
Chloroprene	[*126-99-8*]	Methoxychlor	[*72-43-5*]	Vinyl chloride	[*75-01-4*]
Cresols/cresylic acid	[*1319-77-3*]	Methyl bromide	[*74-83-9*]	Vinylidene chloride	[*75-35-4*]
o-Cresol	[*95-48-7*]	Methyl chloride	[*74-87-3*]	Xylenes (isomers and mixture)	[*1330-20-7*]
m-Cresol	[*108-39-4*]	Methyl chloroform	[*71-55-6*]	*o*-Xylenes	[*95-47-6*]
p-Cresol	[*106-44-5*]	Methyl ethyl ketone	[*78-93-3*]	*m*-Xylenes	[*108-38-3*]
Cumene	[*98-82-8*]	Methyl hydrazine	[*60-34-4*]	*p*-Xylenes	[*106-42-3*]
2,4-D, salts and esters	[*94-75-7*]	Methyl iodide	[*74-88-4*]	Antimony compounds	
DDE	[*3547-04-4*]	Methyl isobutyl ketone	[*108-10-1*]	Arsenic compounds	
Diazomethane	[*334-88-3*]	Methyl isocyanate	[*624-83-9*]	Beryllium compounds	
Dibenzofurans	[*132-64-9*]	Methyl methacrylate	[*80-62-6*]	Cadmium compounds	
1,2-Dibromo-3-chloropropane	[*96-12-8*]	Methyl-*tert*-butyl ether	[*1634-04-4*]	Chromium compounds	
Dibutylphthalate	[*84-74-2*]	4,4-Methylene bis(2-chloroaniline)	[*101-14-4*]	Cobalt compounds	
1,4-Dichlorobenzene(p)	[*106-46-7*]	Methylene chloride	[*75-09-2*]	Coke oven emissions	
3,3-Dichlorobenzidene	[*91-94-1*]	Methylene diphenyl diisocyanate	[*101-68-8*]	Cyanide compounds	
Dichloroethyl ether	[*111-44-4*]	4,4′-Methylenedianiline	[*101-77-9*]	Glycol ethers	
1,3-Dichloropropene	[*542-75-6*]	Naphthalene	[*91-20-3*]	Lead compounds	
Dichlorvos	[*62-73-7*]	Nitrobenzene	[*98-95-3*]	Manganese compounds	
Diethanolamine	[*111-42-2*]	4-Nitrobiphenyl	[*92-93-3*]	Mercury compounds	
N,N-Diethyl aniline	[*121-69-7*]	4-Nitrophenol	[*100-02-7*]	Fine mineral fibers	
Diethyl sulfate	[*64-67-5*]	2-Nitropropane	[*79-46-9*]	Nickel compounds	
3,3-Dimethoxybenzidine	[*119-90-4*]	*N*-Nitroso-*N*-methylurea	[*684-93-5*]	Polycyclic organic matter	
Dimethyl aminoazobenzene	[*60-11-7*]	*N*-Nitrosodimethylamine	[*62-75-9*]	Radionuclides (including radon)	
3,3′-Dimethyl benzidine	[*119-93-7*]	*N*-Nitrosomorpholine	[*59-89-2*]	Selenium compounds	

Because EPA was so slow in promulgating standards for HAPs prior to the 1990 Amendments, most states developed and implemented their own TAP control programs.

Odors

The 1977 Clean Air Act Amendments directed EPA to study the effects, sources, and control feasibility of odors. Although no Federal legislation has been established to regulate odors, individual states have responded to odor complaints by enforcing common nuisance laws. Because about 50% of all citizen air pollution complaints concern odors, it is clear that a disagreeable odor is perceived as an indication of air pollution. However, many substances can be detected by the human olfactory system at concentrations well below those considered harmful.

Visibility

Although there is no NAAQS designed to protect visual air quality, the 1977 Clean Air Act Amendments set as a national goal "the remedying of existing and prevention of future impairment of visibility in mandatory Class I Federal areas which impairment results from man-made pollution." Class I areas are certain national parks and wildernesses that were in existence in 1977. The 1977 Amendments also directed EPA to promulgate appropriate regulations to protect against visibility impairment in these areas. In 1981, EPA directed 36 states to amend their State Implementation Plans to develop control programs for visual impairment that could be traced to particular sources. This type of impairment is called plume blight, and it was the initial focus of EPA's effort because it involved easily identifiable sources. The 1990 Clean Air Act Amendments direct EPA to promulgate appropriate regulations to address regional haze in affected Class I areas. EPA has not dealt with a third type of visibility impairment, urban-scale haze, because the source-receptor relationships are extremely complex (25).

Visibility or visual range is the maximum distance at which a black object can be distinguished from the horizon. b_{ext} is a useful indicator of the inverse of visual range, and it is widely used as an indicator of visual air quality. The total extinction can be written as the sum of a number of components:

$$b_{ext} = b_{sp} + b_R + b_{ap} + b_{ag} \quad (18)$$

where b_{sp} is the light extinction due to light scattered by particles; b_R is the light scattered by air molecules (Rayleigh scattering) and is a function of the atmospheric pressure; b_{ap} is the light absorbed by particles; and b_{ag} is the light absorbed by gas molecules.

Rayleigh scattering accounts for only a minor part of the extinction, except on the clearest days. In general, light scattering is dominated by particles, primarily fine particles. The most efficient light-scattering particles are those that are the same size as the wavelengths of visible light (0.4–0.7 μm). As shown in Figure 4, a peak in the mass distribution occurs in the size range comparable to the wavelength range of visible light. The particles in this size range, therefore, almost always dominate b_{sp}. Exceptions to this occur during fog, precipitation, and dust storms. On a per mass basis, the most efficient light-scattering fine particles are hygroscopic particles, such as sulfate, nitrate, and ammonium particles, which will sorb significant amounts of water at moderate to high relative humidities. As the particles sorb water, they become more efficient light scatterers. Light absorption by particles in the atmosphere is almost exclusively due to elemental carbon which also scatters light. The only common light-absorbing gaseous pollutant is NO_2, which usually accounts for a few percent or less of the total extinction.

As mentioned above, there are three scales of visual impairment: plume blight, urban-scale haze, and regional-scale haze. Plume blight occurs when a plume from a large point source travels into an otherwise clean area and interferes with the viewing of a particular vista. Such events can occur anywhere, but are usually most noticeable in the western U.S. and have been observed in many scenic areas. Most frequently, plume blight is associated with sulfates from a sulfur dioxide-emitting point source. Most large urban areas occasionally experience urban haze, but the public perception of the haze is highest in those cities with scenic mountain vistas like Los Angeles and Denver. Most of the light extinction in urban haze can be accounted for by sulfates (sulfuric acid and the ammonium salts), nitrate (as ammonium nitrate), organic carbon, and elemental carbon. Regional haze refers to the situation where the haze extends for hundreds of miles. Regional haze is usually dominated by sulfates. In the Southwest, occasional haze obscures scenic vistas over large portions of the area, and is attributed to a combination of sulfates from coal-fired power plant and smelter emissions, and transport of urban, Southern California haze which is composed mainly of carbonaceous particles, nitrates and, to a lesser extent, sulfates from southern California. In the East, a denser sulfate-dominated haze (ie, in the rural West, mean sulfate concentrations are $\sim 1\ \mu g/m^3$, while in the rural East they average $\sim 8\ \mu g/m^3$; also, the relative humidity is generally much higher in the East) frequently extends over much of the area during the summer. The primary source of the Eastern haze is coal-burning emissions. Natural haze, caused mainly by aerosols generated from biogenic VOC emissions from vegetation, was historically cited as the regional haze in areas such as the Blue Ridge and Smoky Mountain Regions. Except on the cleanest days, however, sulfate haze now dominates natural haze in these regions (26).

Air pollutants can also cause discolorations of the atmosphere. The most common are brownish discolorations (eg, the "brown Los Angeles haze" and the "brown clouds" observed in Denver and elsewhere). Three factors can contribute to the brown tint. The first is nitrogen dioxide which is a brownish gas. This is most commonly viewed in a plume of NO_2. In the urban hazes, the effect of NO_2 is usually overwhelmed by the effects caused by particles. Since fine particles preferentially scatter blue light in the forward direction, the light viewed through an optically thin cloud with the sun behind the observer is deficient in the blue wavelengths and appears brown. In dense haze clouds, the preferential scattering is masked by multiple-scattering effects and the haze is seen as white.

Acid Deposition

Acid deposition is the deposition of acids from the atmosphere to the surface of the earth. The deposition can be dry or wet. Dry deposition refers to the process whereby acid gases or their precursors or acid particles come in contact with the earth's surface and are retained by the surface. The principal species associated with dry acid-deposition are SO_2(g), acid sulfate particles (H_2SO_4 and NH_4HSO_4), and HNO_3(g). In general, the dry acid deposition is estimated to be a small fraction of total acid deposition because most of the dry deposited SO_4^{2-} and NO_3^- has been neutralized by basic gases and particle in the atmosphere. However, the sulfate and nitrate deposited from dry deposition is estimated to be a significant fraction of the total sulfate and nitrate deposition. More specific estimates are not possible because current spatial and temporal dry deposition data are insufficient. On the other hand, there are abundant data on wet-acid deposition. Wet-acid deposition or acid precipitation is the process by which acids are deposited by the rain or snow. The principal dissolved acids are H_2SO_4 and HNO_3. Other acids, such as HCl and organic acids, usually account for only a minor part of the acidity, although organic acids can be significant contributors in remote areas.

Both acid particles and gases can be incorporated into cloud droplets. Particles are incorporated into droplets by nucleation, Brownian diffusion, impaction, diffusiophoresis (transport into the droplet induced by the flux of water vapor to the same surface), thermophoresis (thermally induced transport to a cooler surface), and electrostatic transport. Advective and diffusive attachment dominate all other mechanisms for pollutant gas uptake by cloud droplets. Modeling and experimental evidence suggest that most of the H_2SO_4 is formed in cloud water droplets. SO_2 diffuses into the droplet and is oxidized to H_2SO_4 by one of several mechanisms. At pHs greater than about 5.5, oxidation of SO_2 by dissolved O_3 is the dominant reaction. At lower pHs, SO_2 oxidation is dominated by the reaction with H_2O_2. Under some conditions, oxidation by O_2, catalyzed by metals or soot, may contribute to the formation of H_2SO_4. Most of the HNO_3 in precipitation is due to the diffusion of HNO_3 into the droplet. However, there is observational evidence that some HNO_3 is formed in the droplets, but the mechanism has not been identified.

The pH of rainwater in equilibrium with atmospheric CO_2 is 5.6, a value that was frequently cited as the natural background pH of rainwater. However, in the presence of other naturally occurring species such as SO_2, SO_4^{2-}, NH_3, organic acids, sea salt, and alkaline crustal dust, the "natural" values of unpolluted rainwater will vary between 4.9 and 6.5 depending upon time and location. Across the United States, the mean annual average pH varies from 4.2 in western Pennsylvania to 5.7 in the West. Precipitation pH is generally lowest in the eastern United States within and downwind of the largest SO_2 and NO_x emissions areas. In general, the lowest pH precipitation occurs in the summer.

Since SO_2 and NO_2 are criteria pollutants, their emissions are regulated. In addition, for the purposes of abating acid deposition in the United States, the 1990 Clean Air Act Amendments require that nationwide SO_2 and NO_x emissions be reduced by 10 million tons/year and 2 million tons/year, respectively, by the year 2000. The reasons for these reductions are based on concerns which include acidification of lakes and streams, acidification of poorly buffered soils, and acid damage to materials. An additional concern was that acid deposition was causing the die-back of forests at high elevations in the eastern United States and in Europe. Although a contributing role of acid deposition cannot be dismissed, the primary pollutant suspected in the forest decline issue is now O_3.

Global Warming (The Greenhouse Effect)

Solar energy, mostly in the form of visible light, is absorbed by the earth's surface and reemitted as longer wavelength infrared (ir) radiation. Certain gases in the atmosphere, primarily water vapor and to a lesser degree CO_2, have the ability to absorb the outgoing ir radiation which is translated to heat. The result is a higher atmospheric equilibrium temperature than would occur in the absence of water vapor and CO_2. This temperature enhancement is called the greenhouse effect, and gases that have the ability to absorb ir and produce this effect are called greenhouse gases. Without the naturally occurring concentrations of water vapor and CO_2, the earth's mean surface temperature would be $-18°C$ instead of the present 17°C. There is concern that increasing concentrations of CO_2 and other trace greenhouse gases due to human activities will enhance the greenhouse effect and cause global warming. Speculated scenarios based on global warming include the following: an alteration in existing precipitation patterns, an increase in the severity of storms, the dislocation of suitable land for agriculture, the dislocation and possible extinction of certain biological species and ecosystems, and the flooding of many coastal areas due to rising sea levels resulting from the thermal expansion of the oceans, the melting of glaciers, and, probably less so, from the melting of polar ice caps.

A list of the greenhouse gases (not including water), present concentrations, current rates of increase in the atmosphere, and estimates of relative greenhouse efficiencies and atmospheric residence times are presented in Table 5. Ozone is not included in this table because there is insufficient information available for quantifying the global radiative influence of O_3 (28). CFCs are included in the table, but their warming potentials should be used with caution because there presently is a controversy over whether the warming due to the presence of CFCs is partially or completely offset by the cooling resulting from O_3 losses in the lower stratosphere (28). The principal sources of greenhouse gases, as they are understood today, are summarized in Table 6.

From the analyses of air trapped in Antarctic and Greenland ice, the concentrations of greenhouse gases (except for O_3) in the preindustrial atmosphere (averaged over $\sim$ 1000 years) can be estimated quite accurately. The enhancement of the greenhouse effect due to current concentrations of greenhouse gases relative to preindustrial concentrations is called the "enhanced greenhouse effect" or radiative forcing. Using a radiative convective model, the contributions from the various greenhouse gases to the radiative forcing in the 1980s can be estimated (6). Such

Table 5. Summary of Important Greenhouse Gases

Gas	CAS Registry Number	Present Concentrations[a]	Concentration Increase,[a] %/Year	Warming Potential[b,c]	Atmospheric Residence Times,[b,d] Years
CO_2		350 ppm	0.3	1	120[e]
CH_4		1.68 ppm	0.8–1	11	10.5
N_2O		340 ppb	0.2	270	132
$CFCl_3$ (CFC-11)	[*75-69-4*]	226 ppt	4	3400	55
CF_2Cl_2 (CFC-12)	[*75-71-8*]	392 ppt	4	7100	116
$CHClF_2$ (HCFC-22)	[*75-46-6*]	100 ppt	7	1600	16
$C_2H_2F_4$ (HFC-134a)	[*811-97-2*]			1200	16
$C_2Cl_3F_3$ (CFC-113)	[*76-14-2*]	30–70 ppt	11	4500	110
CH_3CCl_3 (Methychloroform)	[*71-55-6*]	125 ppt	7	100	6
CCl_4 (Carbon tetrachloride)	[*56-23-5*]	75–100 ppt	1	1300	47

[a] From Ref. 27.
[b] From Ref. 28.
[c] Relative to CO_2. This is based on a 100 year time horizon (28).
[d] Residence times may be slightly different from those reported in Table 1 because the primary source is different.
[e] From Ref. 6.

estimates are shown in Figure 5. At present, CO_2 accounts for about half of the radiative forcing. The relative contribution from CO_2 has been shrinking and will continue to do so because other species that are much more efficient ir absorbers are increasing in concentration at a faster relative rate than CO_2.

Although there is strong evidence linking temperature and CO_2 changes, the cause and effect has not only not been demonstrated, but it is not clear which is the cause and which is the effect. The lack of a definitive relationship may also be obscured by changes in other factors that affect the earth's heat budget, such as increased atmospheric aerosols or cloud cover as well as natural climatic cycles.

Predictions of future temperature changes due to increased greenhouse gas forcing are made using global circulation models (GCMs). The GCMs are sophisticated, but incomplete models that incorporate expressions for the basic physical processes that govern the dynamics of the atmosphere and allow for some atmosphere–ocean interactions.

Table 6. Principal Sources of Greenhouse Gases[a]

Gases	Principal Sources
CO_2	Fossil fuel combustion, deforestation oceans, respiration
CH_4	Wetlands, rice paddies, enteric fermentation (animals), biomass burning, termites
N_2O	Natural soils, cultivated and fertilized soils, oceans, fossil fuel combustion
O_3	Photochemical reactions in the troposphere, transport from stratosphere
CFC-11	Manufacturing of foam, aerosol propellant
CFC-12	Refrigerant, aerosol propellant, manufacturing of foams
HCFC-22	Refrigerant, production of fluoropolymers
CFC-113	Electronics solvent
CH_3CCl_3	Industrial degreasing solvent
CCl_4	Intermediate in production of CFC-11, -12, solvent

[a] Ref. 27.

Stratospheric O_3 Depletion. In the stratosphere, O_3 is formed naturally when O_2 is dissociated by ultraviolet (uv) solar radiation in the wavelength (λ) region 180–240 nm:

$$O_2 + uv \rightarrow O + O \quad (19)$$

$$O + O_2 + M \rightarrow O_3 + M \quad (20)$$

where M is any third body molecule (most likely N_2 or O_2 in the atmosphere) that remains unchanged in the reaction. Uv radiation in the 200–300 nm λ region can also dissociate the O_3:

$$O_3 + uv \rightarrow O_2 + O \quad (21)$$

In this last reaction, O_3 is responsible for the removal of uv-B radiation (λ = 280–330 nm) that would otherwise reach the earth's surface. The concern is that any process that depletes stratospheric ozone will increase the uv-B (in the 293–320 nm region) reaching the surface. Increased uv-B will lead to increased incidence of skin cancer and could have deleterious effects on certain ecosystems. The

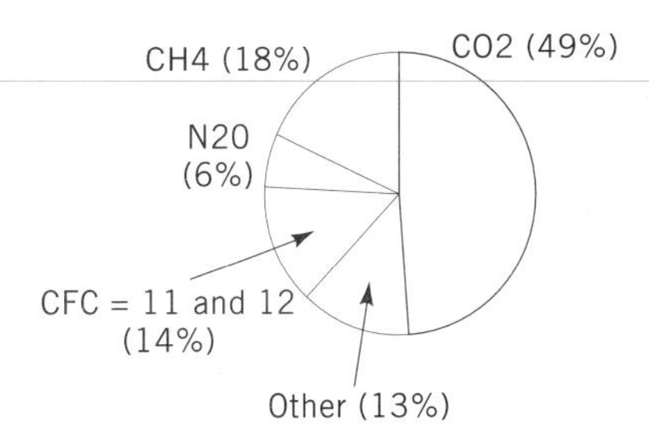

Figure 5. Estimates of greenhouse-gas contributions to global warming in the 1980s.

first concern over O_3 depletion was from NO_x emissions from a fleet of supersonic transport aircraft that would fly through the stratosphere and cause (29):

$$NO + O_3 \rightarrow NO_2 + O_2 \quad (22)$$

$$NO_2 + O \rightarrow NO + O_2 \quad (23)$$

The net effect of this sequence is the destruction of 2 molecules of O_3 since the O would have combined with O_2 to form O_3. In addition, the NO acts as a catalyst because it is not consumed, and therefore can participate in the reaction sequence many times.

In the mid-1970s, it was realized that the chlorofluorocarbons (CFCs) in widespread use because of their chemical inertness, would diffuse unaltered through the troposphere into the mid-stratosphere where they would be photolyzed by uv ($\lambda < 240$ nm) radiation. For example, CFC-12 would photolyze forming Cl and ClO radicals:

$$CF_2Cl_2 + uv \rightarrow CF_2Cl + Cl \quad (24)$$

$$CF_2Cl + O_2 \rightarrow CF_2O + ClO \quad (25)$$

The following reactions would then occur:

$$Cl + O_3 \rightarrow ClO + O_2 \quad (26)$$

$$ClO + O \rightarrow Cl + O_2 \quad (27)$$

In this sequence the Cl also acts as a catalyst, and two O_3 molecules are destroyed. Before the Cl is finally removed from the atmosphere in 1–2 years by precipitation, each Cl atom will have destroyed approximately 100,000 O_3 molecules (30). Another class of compounds, halons, are also ozone-depleting compounds. The halons are bromochlorofluorocarbons or bromofluorocarbons that are widely used in fire extinguishers. Although their emissions and thus their atmospheric concentrations are much lower than the most common CFCs, they are of concern because they are 3 to 10 times more destructive to O_3 than the CFCs.

Evidence that stratospheric O_3 depletion is occurring comes from the discovery of the Antarctic O_3 hole. Other data show that globally, stratospheric O_3 concentrations have declined during the winter, spring, and summer in both the northern and southern hemispheres at middle and high latitudes (28). Declines were most evident during winter months (31,32).

In 1976, the United States banned the use of CFCs as aerosol propellants. However, because of the Antarctic ozone hole and the observed global decreases in stratospheric ozone, there has been increased support for a faster phaseout. In 1990, an agreement was reached among 93 nations to accelerate the phaseout and completely eliminate the production of CFCs by the year 2000. The 1990 Clean Air Act Amendments contain a phaseout schedule for CFCs, halons, carbon tetrachloride, and methyl chloroform. Such steps will stop the increase of CFCs in the atmosphere but, because of their long lifetimes, they will remain in the atmosphere for centuries.

Indoor Air Pollution

Indoor air pollution is simply the presence of air pollutants in indoor air. The focus of this section is on air in residential buildings as opposed to the industrial environment which would be covered under industrial hygiene. The concentrations of indoor pollutants depend upon the strength of the indoor sources of the pollutants as well as the ventilation rate of the building and the outdoor concentrations of the pollutants. In response to the energy crisis in the early 1970s, new buildings were constructed more airtight. Unfortunately, airtight structures created a setting conducive to the accumulation of indoor air pollutants. Numerous sources and types of pollutants found indoor can be classified into seven categories: tobacco smoke, radon, emissions from building materials, combustion products from inside the building, pollutants which infiltrate from outside the building, emissions from products used within the home, and biological pollutants.

Tobacco smoke contains a variety of air pollutants. In addition to TSP, burning tobacco emits CO, NO_x, formaldehyde, benzopyrene, nicotine, phenols, and some metals such as cadmium and arsenic (33).

Radon-222 (Rn) is a naturally occurring, inert, radioactive gas formed from the decay of radium-226 (Ra). Because Ra is a ubiquitous, water-soluble component of the earth's crust, its daughter product, Rn, is found everywhere. Although Rn receives all of the notoriety, the principal health concern is not with Rn itself, but with its alpha (α) particle-emitting daughters (radioactive decay products). Since Rn is an inert gas, inhaled Rn will not be retained in the lungs. With a half-life of 4 days, Rn decays to polonium-218 (Po-218) with the emission of an α particle. It is Po-218, an α emitter with a half-life of 3 minutes, and Po-214, also an α-emitter with a half-life of 1.6×10^{-4} seconds, that are of most concern. Po-218 decays to lead-214 (a β-emitter with a $t_{1/2} = 27$ minutes), which decays to bismuth-214 (a β-emitter with a $t_{1/2} = 20$ minutes), which decays to Po-214. When inhaled, the Rn daughters, either by themselves or attached to an airborne particle, are retained in the lung and the subsequent α emission irradiate the surrounding lung tissue. Rn can enter buildings through emissions from soil, water, or construction materials. By far, the soil route is the most common, and construction material the least common source of Rn contamination (there have been isolated incidents where construction materials contained high levels of Ra). In the ambient air, Rn concentrations are typically 0.25–1.0 picoCurries per liter (pC/L), while the mean concentration in U.S. residences is about 1.2 pC/L (62). However, it is estimated that there are 1 million residences that have concentrations exceeding 8 pC/L, which is the action level for remedial action recommended by the National Council on Radiation Protection and Measurements (34). Remedial action consists of (*1*) reducing the transport of Rn into the building by sealing cracks with impervious fillers and installing plastic or other barriers that have been proven effective; (*2*) removing the daughters from the air by filtration; and (*3*) increasing the infiltration of outside air with an air-exchanger system.

Of the pollutants emitted from construction materials

within the home, asbestos [*1332-21-4*] has received the most attention. Asbestos is a generic term for a number of naturally occurring fibrous, hydrated silicates. Asbestos was incorporated into many common products including roofing materials, wallboard, insulation, spray-on fireproofing and insulating material, floor tiles, pipes, filters, draperies, pot holders, brake linings, etc (35). In the 1940s and 1950s however, evidence accumulated linking exposure to the airborne fibers with asbestosis (pulmonary interstitial fibrosis), lung cancer, and mesothelioma (a rare form of cancer of the lung or abdomen). Although all forms of asbestos were implicated in the early studies, very recent studies indicate that most of the asbestos-related diseases are due to exposure to airborne amphiboles rather than the most common type, chrysotile, and to fibers greater than or equal to 5 μm in length (36). In the 1970s, the spray-on application of asbestos was banned and substitutes were found for many products. Nevertheless, asbestos was used liberally in buildings for several decades, and many of them are still standing. Asbestos in building materials does not spontaneously shed fibers, but when the materials become damaged by normal decay, renovation, or demolition, the fibers can become airborne. When such situations arise, specific procedures should be followed to contain and remove the damaged materials.

Another important pollutant emanating from building material is formaldehyde (HCHO). Formaldehyde is important because of its irritant effects and its suspected carcinogenicity. Although traces of formaldehyde can be found in the air in virtually every modern home, mobile homes and homes insulated with urea–formaldehyde foam have the highest concentrations. Plywood is also a source of formaldehyde as the layers of wood are held together in a similar urea–formaldehyde resin adhesive. In general, however, particle board contains more adhesive per unit mass, so the emissions are greater. Urea foam is an efficient insulation material that can be injected into the sidewalls of conventional homes. Production of the foam peaked in 1977, when about 170,000 homes were insulated. When improperly formulated or installed, the foam can emit significant amounts of formaldehyde. In 1982, the use of the foam was banned in the United States. Other sources of formaldehyde indoors are paper products, carpet backing, and some fabrics.

Whenever unvented combustion occurs indoors or when venting systems attached to combustion units malfunction, a variety of combustion products will be released to the indoor environment. Products of combustion include CO, NO, NO_2, fine particles, aldehydes, polynuclear aromatics, and other organic compounds. Especially dangerous sources are unvented gas and kerosene space heaters which discharge pollutants directly into the living space. The best way to prevent the accumulation of combustion products indoors is to make sure all units are properly vented and properly maintained.

Pollutants from outdoors can also be drawn inside under certain circumstances. Incorrectly locating an air intake vent downwind of a combustion exhaust stack can cause this condition. High outdoor pollutant concentrations can infiltrate buildings. Unreactive pollutants such as CO will diffuse through any openings in the building and pass unaltered through any air-intake system.

Air contaminants are emitted to the indoor air from a wide variety of activities and consumer products. Some of these are summarized in Table 7. It is obvious from this list that most indoor activities produce some types of pollutants. When working with these products or engaging in these activities, care should be exercised to minimize exposures by proper use of the products and by providing adequate ventilation.

Biological air pollutants found indoors include airborne bacteria, viruses, fungi, spores, molds, algae, actinomycetes, and insect and plant parts. Many of the microorganisms multiply in the presence of high humidity. The microorganisms can produce infections, disease, or allergic reactions, while the nonviable biological pollutants can produce allergic reactions. The most notable episode was the 1976 outbreak of Legionella (Legionnaires') disease in Philadelphia, where American Legion convention attendees inhaled Legionella virus from a contaminated central air-conditioning system. A similar incident in an industrial environment occurred in 1981 when more than 300

Table 7. Types of Emissions of Indoor Air Pollutants Associated with Various Activities and Consumer Products

Activity or Product	Intentional Aerosol Production	Unintentional Aerosol Production	Evaporation or Sublimation	Unintentional Outgassing
Cleaning	X	X	X	X
Painting	X		X	X
Polishing	X		X	X
Stripping	X	X	X	
Refinishing	X		X	X
Hobbies, Crafts	X	X	X	X
Deodorizer	X		X	
Insecticide	X		X	
Disinfectant	X		X	
Personal grooming product	X		X	X
Plastic				X

From Ref. 37.

workers came down with "Pontiac fever" as a result of inhalation exposure to a similar aerosolized virus from contaminated machining fluids (38). Better preventative maintenance of air management systems and increased ventilation rates reduce the concentrations of all species, and this should reduce the incidence of adverse affects.

BIBLIOGRAPHY

1. *Procedures for Emission Inventory Preparation,* Volumes I–IV, Publication No. EPA 450/4-81-026A-E, U.S. Environmental Protection Agency, Research Triangle Park, N.C., 1981.
2. *Compilation of Air Pollution Emission Factors,* Publication No. AP-42, 5th ed., U.S. Environmental Protection Agency, Research Triangle Park, N.C., 1989.
3. *National Air Quality & Emissions Trends Report, 1991,* Publication No. EPA-450-R-92-001, U.S. Environmental Protection Agency, Research Triangle Park, N.C., 1992.
4. V. Ramanathan and co-workers, *Review Geophys.* **25,** 1441 (1987).
5. P. Warneck, *Chemistry of Natural Atmospheres,* Academic Press, New York, 1988, p. 367.
6. Intergovernmental Panel on Climate Change, *Scientific Assessment of Climate Change, Section 2, Radiative Forcing of Climate,* United Nations, New York, 1990, p. 14.
7. G. T. Wolff, N. A. Kelly, and M. A. Ferman, *Science* **311,** 703 (1981).
8. A. M. Hough and R. G. Derwent, *Nature* **344,** 645 (1990).
9. N. A. Kelly, G. T. Wolff, M. A. Ferman, *Atmos. Environ.* **16,** 1077 (1978).
10. W. Johnson and W. Viezee, *Atmos. Environ.* **15,** 1309 (1981).
11. W. Attmannspacher and R. Hartmannsgruber, *Pure Appl. Geophysics.* **106–108,** 1091 (1973).
12. R. L. Seila, W. A. Lonneman, and S. A. Meeks, *Determination of C_2 to C_{12} Ambient Air Hydrocarbons in 39 U.S. Cities from 1984 through 1986,* Publication No. EPA/600/3-89/058, U.S. Environmental Protection Agency, Research Triangle Park, N.C., (1989).
13. P. Warneck, *Chemistry of Natural Atmospheres,* Academic Press, New York, 1988, pp. 721–729.
14. G. T. Wolff and co-workers, *Environ. Sci. Technol.* **11,** 506 (1977).
15. A. M. Dunker, S. Kumar, and P. H. Berzins, *Atmos. Environ.* **18,** 311 (1984).
16. N. A. Kelly and R. G. Gunst, *Atmos. Environ.* **24A,** 2991 (1990).
17. *Catching Our Breath. Next Steps for Reducing Urban Ozone,* U.S. Office of Technology Assessment, Washington, D.C., 1989, pp. 101–102.
18. G. T. Wolff and co-workers, *J. Air Pollut. Control Assoc.* **27,** 460 (1977).
19. *An Acid Aerosols Issue Paper,* Publication No. EPA/600/8-88/005F, U.S. Environmental Protection Agency, Washington, D.C., 1989.
20. G. T. Wolff, *Annals NY Acad. Sci.* **338,** 379 (1980).
21. K. Willeke and K. T. Whitby, *J. Air Pollut. Control Assoc.* **25,** 529 (1975).
22. P. W. Fisher, R. M. Currie, and R. J. Churchill, *J. Air Pollut. Control Assoc.* **38,** 1376 (1988).
23. *Threshold Limit Values and Biological Exposure Indices,* American Conference of Governmental and Industrial Hygienists, Cincinnati, Ohio, 1989, p. 124.
24. J. A. Cannon, *J. Air Pollut. Control Assoc.* **36,** 562 (1986).
25. J. C. Mesta, in P. S. Bhardwaja, ed., *Visibility Protection Research and Policy Aspects,* Air and Waste Management Association, Pittsburgh, Pa. 1987, pp. 1–8.
26. M. A. Ferman, G. T. Wolff, and N. A. Kelly, *J. Air Pollut. Control Assoc.,* **31,** 1074 (1981).
27. *Policy Options for Stabilizing Global Climate,* U.S. Environmental Protection Agency, Washington, D.C., 1990.
28. *Scientific Assessment of Ozone Depletion: 1991,* World Meteorological Organization/United Nations Environment Programme, 1991.
29. P. J. Crutzen, *Quart J. Royal Meteorol Soc.* **96,** 320 (1970).
30. M. J. Molina and F. S. Rowland, *Nature* **249,** 810 (1974).
31. M. B. McElroy and R. J. Salawitch, *Science* **243,** 763 (1989).
32. F. S. Rowland, *Amer. Scientist* **77,** 36 (1989).
33. California Department of Consumer Affairs, *Clean Your Room, Compendium on Indoor Air Pollution,* Sacramento, Calif. 1982, pp. III.Ei–III.E.II.
34. Mueller Associates, Inc., Syscon Corporation and Brookhaven National Laboratory, *Handbook of Radon in Buildings,* Hemisphere Publishing Corporation, New York, 1988, p. 95.
35. P. Brodeur, *New Yorker* **44,** 117 (1968).
36. B. T. Mossman and co-workers, *Science* **247,** 294 (1990).
37. *Indoor Air Pollutants,* National Academy Press, Washington, D.C., 1981, p. 101.
38. L. A. Herwaldt and co-workers, *Ann. Intern. Med.* **100,** 333 (1984).

General References

References 3, 4, 5, 6, 27, 28, 31, 33 and 37 and the following books and reports constitute an excellent list for additional study. Reference 5 is an especially useful resource for global atmospheric chemistry.

J. H. Seinfeld, *Atmospheric Chemistry and Physics of Air Pollution,* John Wiley & Sons, Inc., New York, 1986.

B. J. Finlayson-Pitts and J. N. Pitts, Jr., *Atmospheric Chemistry Fundamentals and Experimental Techniques,* John Wiley & Sons, Inc., New York, 1986.

T. E. Graedel, D. T. Hawkins, and L. D. Claxton, *Atmospheric Chemical Compounds Sources, Occurrence and Bioassay,* Academic Press, New York, 1986.

Air Quality Criteria for Ozone and Other Photochemical Oxidants, Publication No. EPA/600/8-84-020F (5 Volumes), U.S. Environmental Protection Agency, Research Triangle Park, N.C., 1986.

EPA publishes separate Criteria Documents for all the Criteria Pollutants, and they are updated about every five years.

Atmospheric Ozone 1985, World Meteorological Organization, Geneva, Switzerland (3 volumes)_an excellent compendium on tropospheric and stratospheric processes.

G. T. Wolff, J. L. Hanisch, and K. Schere, eds., *The Scientific and Technical Issues Facing Post-1987 Ozone Control Strategies.* Air and Waste Management Association, Pittsburgh, Pa., 1988.

J. H. Seinfeld, "Urban Air Pollution: State of the Sciences," *Science* **243,** 745 (1989).

S. H. Schneider, "The Greenhouse Effect: Science and Policy," *Science,* **243,** 771 (1989).

AIR POLLUTION: AUTOMOBILE

MICHAEL P. WALSH
Arlington, Virginia

Across the entire globe, motor vehicle use has increased tremendously. In 1950, there were about 53 million cars on the world's roads; only four decades later, the global automobile fleet is over 430 million, more than an eightfold increase. On average, the fleet has grown by about 9.5 million automobiles per year over this period. Simultaneously, the truck and bus fleet has been growing by about 3.6 million vehicles per year (1). While the growth rate has slowed in the highly industrialized countries, population growth and increased urbanization and industrialization are accelerating the use of motor vehicles elsewhere. If the approximately 100 million two-wheeled vehicles around the world is included (growing at about 4 million vehicles per year over the last decade), the global motor vehicle fleet is now approximately 675 million.

One result is that most of the major industrialized areas of the world have been experiencing serious motor vehicle pollution problems. To deal with these problems North America, Europe, and Japan have developed significant motor vehicle pollution control programs that have led to tremendous advances in petrol car control technologies. At present, similar technologies are under intensive development for diesel cars and trucks and significant breakthroughs are starting to appear with production diesel vehicles. See also AIR POLLUTION: AUTOMOBILE, TOXIC EMISSIONS.

Motor vehicle-related air pollution problems are not limited to the OECD countries. Areas of rapid industrialization are now starting to note similar air pollution problems to those of the industrialized world. Cities such as Mexico, Delhi, Seoul, Singapore, Hong Kong, Sao Paulo, Manila, Santiago, Bangkok, Taipei, and Beijing already experience unacceptable air quality or are projecting that they will in the relatively near future. See also AIR QUALITY MODELING.

The purpose of this article is to survey what is presently known about transportation related air pollution problems, to summarize the adverse impacts which result, to review actions under way or planned to address these problems, and to estimate future trends. Recent technological developments will also be summarized.

THE PROBLEM

Motor vehicles emit large quantities of carbon monoxide, hydrocarbons, nitrogen oxides, and such toxic substances as fine particles and lead. Each of these can cause adverse effects on health and the environment. Because of the growing vehicle population and the high emission rates, serious air pollution problems have been an increasingly common phenomena in modern life. Initially, these problems were most apparent in center cities but recently lakes and streams and even remote forests have experienced significant degradation as well. As more and more evidence of human impacts on the upper atmosphere accumulates, concerns are increasing that motor vehicles are contributing to global changes that could alter the climate of the planet.

Pollutants from vehicles are not an academic concern; they impair health, destroy vegetation, and inhibit the general quality of life. In regard to health especially, their effects are most pronounced in the very old, the very young, and those least able to cope.

Many sources contribute to air pollution, but the motor vehicle has been singled out as especially serious. In large part, this is due to the great number of vehicles in congested urban areas and the overall volume of pollutants which they emit. The problem is compounded by the fact that vehicles emit pollutants in close proximity to the breathing zones of people.

Until recently most of the air pollution problems were considered local in nature, but over the last two decades the evidence has been increasing that some of the most severe impacts may occur over large distances and over long periods of time with the effect far removed from the source. A comprehensive look at air pollution today must consider localized adverse impacts, regional and continental effects, and even global changes.

Urban

Carbon Monoxide. Over 90% of the carbon monoxide emitted in cities generally comes from motor vehicles. Numerous studies in humans and animals have now demonstrated that those individuals with weak hearts are placed under additional strain by the presence of excess CO in the blood. In addition, fetuses, sickle cell anemics, and young children are also especially susceptible to exposure to low levels of CO. Also, evidence indicates that CO may contribute to elevated levels of tropospheric ozone.

Oxides of Nitrogen. NO_x emissions from vehicles and other sources produce a variety of adverse health and environmental effects. They react chemically with hydrocarbons to form ozone and other highly toxic pollutants. Next to sulfur dioxide, NO_x emissions are the most prominent pollutant contributing to acidic deposition. Direct exposure to nitrogen dioxide (NO_2) leads to increased susceptibility to respiratory infection, increased airway resistance in asthmatics, and decreased pulmonary function. Short-term exposures to NO_2 have resulted in a wide ranging group of respiratory problems in school children (cough, runny nose, and sore throat are among the most common) as well as increased sensitivity to urban dust and pollen by asthmatics. Some scientists believe that NO_x is a significant contributor to the dying forests throughout central Europe.

Lead. Because lead is added to petrol, motor vehicles have been the major source of lead in the air of most cities. Several studies have now shown that children with high levels of lead accumulated in their baby teeth experience more behavioral problems, lower IQs, and decreased ability to concentrate. Most recently, in a study of 249 children from birth to 2 yr, it was found that those with prenatal umbilical cord blood lead levels at or above 10 μg/dL consistently scored lower on standard intelligence tests than those at lower levels.

Diesel Particulate. Uncontrollable diesels emit approximately 30–70 times more particulate than gasoline-fueled engines equipped with catalytic converters and burning unleaded fuel. These particles are small and respirable (less than 2.5 μm) and consist of a solid carbonaceous core on which myriad compounds adsorb. These include

- Unburned hydrocarbons.
- Oxygenated hydrocarbons.
- Polynuclear aromatic hydrocarbons.
- Inorganic species such as sulfur dioxide, nitrogen dioxide, and sulfuric acid.

It is now well established that diesel particulate represents a serious health hazard in many urban areas around the world. Cities as diverse as Santiago, Taipei, Mexico City, Manila, Bangkok, Seoul, and Jakarta stand out as experiencing particularly high levels. The World Health Organization has concluded that diesel particulate is a probable human carcinogen. Several human epidemiological studies seemed to point to this conclusion. "In the two most informative cohort studies (of railroad workers), one in the USA and one in Canada, the risk for lung cancer in those exposed to diesel engine exhaust increased significantly with duration of exposure in the first study and with increased likelihood of exposure in the second."

Other studies focusing on noncancer health effects have raised equally alarming concerns. By correlating daily weather, air pollutants, and mortality in five U.S. cities, scientists have discovered that nonaccidental death rates tend to rise and fall in near lockstep with daily levels of particulates—more than with other pollutants. Because the correlation held up even for low levels (in one city, with just 23% of the U.S. limit on particulates), these analyses suggest that as many as 60,000 U.S. residents per year may die from breathing particulates at or below legally allowed levels. Confirmation of the new findings by other researchers would make airborne particulate levels the largest known "involuntary environmental insult" to which Americans are exposed (2).

In Germany, the diesel engine has enjoyed some of its greatest success. Based on its conclusion that diesel particulate reflects an occupational health hazard, the German government has issued guidelines intended to discourage the use of diesel engines in occupational settings when possible (eg, substitute electric fork lifts for diesels) or to reduce emissions to the greatest extent feasible (eg, by using particulate traps or filters).

Other Toxics. The 1990 Clean Air Act (CAA) directed the EPA to complete a study of emissions of toxic air pollutants associated with motor vehicles and motor vehicle fuels. The study found that in 1990, the aggregate risk is 720 cancer cases. For all years, 1,3-butadiene is responsible for the majority of the cancer incidence, ranging from 58 to 72% of the total. This is due to the high unit risk of 1,3-butadiene. Gasoline and diesel particulate matter, which are considered to represent motor vehicle polycyclic organic matter (POM) in this report, are roughly equal contributors to the risk. The combined risk from gasoline and diesel particulate matter ranges from 16 to 28% of the total, depending on the year examined. Benzene is responsible for roughly 10% of the total for all years. The aldehydes, predominately formaldehyde, are responsible for roughly 4% of the total for all years.

A variety of studies have found that in individual metropolitan areas, mobile sources are one of the most important and possibly the most important source category in terms of contributions to health risks associated with air toxics. For example, according to the EPA, mobile sources may be responsible for almost 60% of the air pollution–related cancer cases in the United States per year.

Regional

Tropospheric Ozone. The most widespread air pollution problem in areas with temperate climates is ozone, one of the photochemical oxidants that results from the reaction of nitrogen oxides and hydrocarbons in the presence of sunlight. Many individuals exposed to ozone suffer eye irritation, cough and chest discomfort, headaches, upper respiratory illness, increased asthma attacks, and reduced pulmonary function. Numerous studies have also demonstrated that photochemical pollutants inflict damage on forest ecosystems and seriously impact the growth of certain crops.

Acidification. Acid deposition results from the chemical transformation and transport of sulfur dioxide and nitrogen oxides. Evidence indicates that the role of NO_x may be of increasing significance with regard to this problem.

Global

Climate Modification. A significant environmental development during the 1980s was the emergence of global warming or the greenhouse effect as a major international concern. Pollutants associated with motor vehicle use (eg, CO_2, CH_4, N_2O, etc) can increase global warming by changing the chemistry in the atmosphere to reduce the ability of the sun's reflected rays to escape. These greenhouse gases have been shown to be accumulating in recent years.

Stratospheric Ozone Depletion. The release of chlorofluorocarbons (CFCs) used in vehicle air conditioning equipment can destroy the earth's protective ozone shield. During the Antarctic spring, the ozone hole spans an area the size of North America. At certain altitudes over the Antarctic, the ozone is destroyed almost completely. By allowing more ultraviolet radiation to penetrate to the earth's surface, such a loss is expected to increase the incidence of skin cancer and impair the human immune system. Increased ultraviolet radiation might also harm the life-supporting plankton that dwell in the ocean's upper levels, thus jeopardizing marine food chains that depend on the tiny plankton.

Carbon Dioxide. Virtually the entire global motor vehicle fleet runs on fossil fuels, primarily oil. For every gallon of oil consumed by a motor vehicle, about 8.6 kg of carbon dioxide (containing about 2.4 kg of carbon) go directly into the atmosphere. In other words, for a typical fill-up at the service station (estimated at 57 L of gas), about 136 kg of carbon dioxide are eventually released into the atmosphere. Globally, motor vehicles account for about 33% of world oil consumption and about 14% of the world's carbon dioxide emissions from fossil fuel burning.

CFCs. A principal source of CFCs in the atmosphere is motor vehicle air conditioning, and in 1987 approximately 48% of all new cars, trucks, and buses manufactured worldwide were equipped with air-conditioners. (CFCs also are used as a blowing agent in the production of seating and other foam products, but this is a considerably smaller vehicular use.) Annually, about 120,000 t of CFCs are used in new vehicles and in servicing air-conditioners in older vehicles. In all, these uses account for about 28% of global demand for CFC-12. As agreed under the Montreal Protocol, CFCs are to be completely phased out of new vehicles by the turn of the century, a welcome but long overdue step. CFCs emitted over the next decade will cause damage to the planet for the next two centuries.

Motor Vehicles are a Dominant Pollution Source

Because the growth has been so great, motor vehicles are now generally recognized as responsible for more air pollution than any other single human activity. The primary pollutants of concern from motor vehicles include the precursors to ground level ozone, hydrocarbons (HC) and nitrogen oxides (NO_x), and carbon monoxide (CO).

HC, NO_x, and CO in Industrialized Countries. Motor vehicles are the dominant source of these air pollutants in Europe.

> The primary source category responsible for most NO_x emissions is road transportation roughly between 50 and 70 per cent. . . . Mobile sources, mainly road traffic, produce around 50 per cent of anthropogenic VOC emissions, therefore constituting the largest man-made VOC source category in all European OECD countries.

In calendar year 1990, the U.S. EPA estimated that transportation sources were responsible for 63% of the carbon monoxide, 38% of the NO_x, and 34% of the hydrocarbons (HC). Based on data regarding evaporative "running losses," the HC contribution from vehicles may actually be substantially higher.

Beyond the U.S. and Europe, for OECD countries as a whole, motor vehicles are the dominant source of carbon monoxide (66%), oxides of nitrogen (47%), and hydrocarbons (39%).

Selected Developing Countries. While not as well documented, it is increasingly clear that motor vehicles are also the principal source of many of the pollution problems that are plaguing the developing world. By way of examples, the air pollution problems of a few countries are summarized below.

India. The motor vehicle related air pollution problem in India is already severe and worsening. While the problem of diesel smoke and particulate is the most apparent, carbon monoxide, nitrogen dioxide, hydrocarbons, and ozone levels also exceed internationally accepted levels. High leaded petrol levels also exist. The problem continues to worsen as the vehicle population continues to grow and age; vehicles once introduced into use remain active much longer than in highly industrialized areas.

Thailand. The motor vehicle–related air pollution problem in Thailand, especially in regard to diesel particulate, is already severe and worsening. While the problems of carbon monoxide and ambient lead levels also exceed internationally accepted levels, they are apparently stabilizing. The overall vehicle population continues to increase at a rate of about 10% per year. The number of light pickup trucks, vans, and motorcycles are also growing rapidly. The adverse consequences of the pollution are especially severe because the lifestyle and climate is such that public exposure to high pollution levels is great.

Indonesia. The motor vehicle related air pollution problem in Jakarta is especially severe; a combination of densely congested traffic, poor vehicle maintenance, and large numbers of diesels and two-stroke engined (smoky) motorcycles is the cause. While air quality monitoring data are not well documented, the problems of carbon monoxide and ambient lead levels also exceed internationally accepted levels.

Philippines. A large proportion of the vehicles are diesel fueled and most of these, especially the "Jeepneys" emit excessive smoke. Due to a lack of instrumentation, actual air quality data do not exist at present. However, it is generally recognized that particulate, lead, carbon monoxide, and possibly ozone levels exceed internationally accepted levels. Based on analysis of fuel consumption, motor vehicles are estimated to be responsible for approximately 50% of the particulate, 99% of the carbon monoxide, 90% of the hydrocarbons, and 5% of the sulfur dioxide. Fuel quality is poor as reflected by the sulfur content of as much as 1.0 wt % (compared with 0.3 wt % in the United States) and a lead content of up to 1.16 g/L (compared with 0.15 g/L in the European Community).

Mexico City. The number of automobiles in Mexico City has grown dramatically in the past several decades; there were approximately 48,000 cars in 1940, 680,000 in 1970, 1.1 million in 1975, and 3 million in 1985. The Mexican government estimates that fewer than half of these cars are fitted with even modest pollution control devices. In addition, more than 40% of the cars are over 12 years old, and of these, most have engines in need of large repairs. The degree to which the existing vehicles are in need of maintenance is reflected by the results of the "voluntary" inspection and maintenance (I/M) program run by the DF during 1986 through 1988. Of the over 600,000 vehicles tested (209,638 in 1986; 313,720 in 1987; and 80,405 in the first four months of 1988), about 70% failed the gasoline vehicle standards and 85% failed the diesel standards (65HSU).

International Agreements Are Increasing Control Pressure

Historically, the thrust toward vehicle pollution controls has been directed by individual countries acting to address their local air pollution concerns, but there has evolved over the past few years an international focus. This is most evident in the Common Market, of course, where countries have joined together to prevent pollution controls from becoming a barrier to trade. But most recently, there have been a series of international protocols designed to address pollution problems of a global nature. The Montreal Protocol designed to reduce emissions of CFCs has received the most attention in this regard. However, two others are of special interest to the subject of this article.

NO_x Protocol Increases Focus on Vehicle Controls. Following the eighth session of the Working Group on Nitrogen Oxides (February 1988) and a further informal consultation of heads of delegations (April 1988), the draft *Protocol Concerning the Control of Emissions of Nitrogen Oxides or Their Transboundary Fluxes* was developed. The protocol will, as a first step, commit signatories to a "freeze" of their NO_x emissions or transboundary fluxes by 1994 on the basis of 1987 levels. At the same time, parties to the protocol must apply emission standards "based on the best available technologies which are economically feasible" to the new stationary and mobile emission sources and to introduce pollution control measures for principal existing stationary sources. They will also have to make unleaded fuel available in sufficient quantities, at least along main international transit routes, to facilitate the circulation of vehicles equipped with catalytic converters.

The second step to combat NO_x pollution, which should start not later than January 1, 1996, consists of agreed emission reductions on the basis of "critical loads" of nitrogen compounds in the environment: vegetation, soils, groundwater, and surface waters. At meetings in Sofia, Bulgaria, in November 1988 countries signed these agreements to control NO_x emissions over the next decade and presumably beyond which could also have a significant impact on both mobile and stationary source requirements in Europe. A total of 12 European countries, including France, Germany, Italy, and Spain, agreed to reduce NO_x by 30% over the next decade, an ambitious program. In summary, the protocol should increase the pressure to reduce NO_x from all sources.

VOC Protocol. The United States has joined with other European countries to form an "ad hoc workgroup," the first step toward developing an international protocol to control emissions of volatile organic compounds. After the signing of the NO_x agreement in Sofia, Bulgaria, a workgroup was formed to discuss methods for researching VOC emissions and control strategies. According to a draft report, the executive body for the Convention on Long Range Transboundary Air Pollution, recognizing damage from VOCs, which react with NO_x to form ozone, agreed to prepare "necessary substantiation for appropriate internationally agreed measures and proposals for a draft protocol to the Convention" aimed at controlling VOCs. As part of the development of the U.N. ECE protocol, a technical annex dealing with mobile sources was drafted at a meeting in Switzerland (April 6, 1990). A principal conclusion is that closed-loop three-way catalyst technology is cleaner and more efficient than either engine modifications or lean settings with open-loop catalysts.

Technology Option	Emission Level	Cost	Fuel Consumption
Uncontrolled	400		<100
Engine modifications	100	base	100
Lean setting w/oxidation catalyst	50	150–200	100
Closed-loop, three way catalyst	10	250–400	95
Advanced closed-loop catalyst	6	350–600	90

Global Warming Protocols. ***CFCs.*** As noted earlier, the U.N. IPCC process has concluded as a matter of scientific consensus that the global warming concern is real and currently under way. As a result, pressure is building at an international level to develop a collective approach. The first manifestation of this was the Montreal Protocol and its subsequent amendment, which is designed to eliminate the use of CFCs before the end of the century. Beyond this, however, and spurred in part by recent events in the Middle East, more and more countries are calling for substantial cuts in so-called CO_2 equivalent emissions. In addition to pressures for improving vehicle fuel efficiency (and therefore CO_2 emissions), this includes potential reductions in CO, HC and NO_x, which as noted earlier can increase global accumulations of methane, carbon dioxide, nitrous oxide, and ozone in direct and indirect ways; these gases are all greenhouse gases.

Carbon Dioxide. A total of 155 nations, including the European Community, signed the climate change treaty by the close of the United Nations Conference on Environment and Development, according to the U.N. Treaty Office. The treaty set guidelines for cutting emissions of greenhouse gases but stopped short of mandating firm targets and timetables for the cuts. During earlier negotiations at U.N. headquarters, the United States blocked nations from including firm targets in the treaty.

The treaty said the "return by the end of the present decade to earlier levels of anthropogenic emissions of carbon dioxide and other greenhouse gases not controlled by the Montreal Protocol would contribute" to industrialized countries "modifying longer-term trends" of emissions. It said also the aim for developed countries is to return "individually or jointly to their 1990 levels these emissions of carbon dioxide and other greenhouse gases." The EC, in signing the treaty, reaffirmed its goal of stabilizing carbon dioxide emissions by 2000.

Conclusions Regarding the Forces at Work

Air pollution problems worldwide remain serious and growing. As the principal source of HC, CO and NO_x emissions in most areas of the world, motor vehicles continue to receive priority attention from national governments and international bodies. Increasingly diesel particulate and other toxic compounds are also receiving attention.

GOVERNMENT RESPONSES TO THE PROBLEM

In 1959, California adopted legislation that called for the installation of pollution control devices as soon as three workable control devices were developed. At that time, the auto manufacturers repeatedly asserted that the technology to reduce emissions did not exist.

In 1964, the state of California was able to certify that three independent manufacturers had developed workable, add-on devices. This triggered the legal requirement that new automobiles comply with California's standards beginning with the 1966 model year. Soon afterward the large domestic manufacturers announced that they too could and would clean up their cars with technology that they had developed. Thus independently developed devices were unnecessary.

Subsequent to California's pioneering efforts, and as a reuslt of recognition of the national nature of the auto pollution problem in 1964, Congress initiated federal motor vehicle pollution control legislation. As a result of the 1965 Clean Air Act Amendments, the 1966 California auto emission standards were applied nationally in 1968.

In December 1970, the Clean Air Act was amended by Congress, "to protect and enhance the quality of the nation's air resources." Congress took particular notice of the significant role of the automobile in the nation's effort to reduce ambient pollution levels by requiring a 90% reduction in emissions models from the level previously prescribed in emissions standards for 1970 (for carbon monoxide and hydrocarbons) and 1971 (for nitrogen oxides). Congress clearly intended to aid the cause of clean air by mandating levels of automotive emissions that would essentially remove the automobile from the pollution picture.

In many ways the serious effort to control motor vehicle pollution can be considered to have begun with the passage of the landmark 1970 law. In 1977, the act was "fine tuned" by Congress, delaying and slightly relaxing the auto standards, imposing similar requirements on trucks, and specifically mandating in use directed vehicle inspection and maintenance programs in the areas with the most severe air pollution problems. Just recently, Congress passed the 1990 Clean Air Act Amendments and they were signed into law by President Bush on November 15, 1990. (A summary of these provisions is contained in Appendix A.)

Gasoline-Fueled Vehicles

Gasoline vehicle pollution control efforts reflect an approximately 30-yr effort to date. Initial crankcase HC controls were first introduced in the early 1960s followed by exhaust CO and HC standards later that decade. By the early to mid 1970s, most major industrial countries had initiated some level of vehicle pollution control program.

During the mid to late 1970s, advanced technologies were introduced on most new cars in the United States and Japan. These technologies resulted from a conscious decision to "force" the development of new approaches and were able dramatically to reduce CO, HC, and NO_x emissions beyond previous systems. As knowledge of these technological developments on cars spread, and as the adverse effects of motor vehicle pollution became more widely recognized, more and more people across the globe began demanding the use of these systems in their countries. During the mid 1980s, Austria, the Netherlands, and the FRG adopted innovative economic incentive approaches to encourage purchase of low pollution vehicles. Australia, Canada, Finland, Austria, Norway, Sweden, Denmark, and Switzerland have all decided to adopt mandatory requirements. Even rapidly industrializing, developing countries such as Brazil, Chile, Taiwan, Hong Kong, Mexico, Singapore, and South Korea have adopted stringent emissions regulations.

After years of delay, the European Economic Community has also made significant strides. As 1990 came to a close, the European Council of Environmental Ministers reached unanimous agreement to require all new models of light-duty vehicles by 1992–1993 to meet emission standards roughly equivalent to U.S. 1987 levels. Further they voted to require the commission to develop a proposal before December 31, 1992 which, taking account of technical progress, will require a further reduction in limit values (presumably the proposal should be roughly equivalent to the recently adopted U.S. standards of 0.25 NMHC, 0.4 NO and 3.4 CO).

Just as Europe was moving toward parity with U.S. standards, the United States, and to a much greater extent California, embarked on a course that could prove just as momentous to the 1990s as the 1970 Clean Air Act was to the 1970s and 1980s. Many significant changes to the federal motor vehicle emission control program will result from the passage of the Clean Air Act Amendments of 1990, including the following: the design, manufacture, and certification of new vehicles to lower tailpipe and evaporative emission standards; enhanced inspection and maintenance programs for seriously polluted areas; more durable pollution control systems; and the use of less polluting gasoline and alternative fuels.

The state of California has responded to the ozone nonattainment problem in the South Coast Air Basin with a long-term plan of new VOC and oxides of nitrogen (NO_x) control initiatives designed to achieve significant reductions in emissions of these and other criteria pollutants from current levels. To further reduce motor vehicle emissions, the California Air Resources Board (CARB) has established stringent new vehicle exhaust emission standards. Compliance with these standards will be achieved through a combination of advanced vehicle emission control technology and clean burning fuels.

The California motor vehicle emission control program is distinguished from the federal program in its general approach and philosophy as well as its more stringent emission standards and program requirements. The California Air Resources Board characterizes its program as a technology-forcing approach, which is more stringent and flexible than the federal program. Because the California program is more flexible and responsive than the federal program, in-use compliance concerns can be addressed more rapidly. The California Air Resources Board has indicated that it is committed to taking all steps necessary to ensure that California-certified vehicles meet certification standards in-use for the useful lifetime of the vehicle.

To control the evaporative emissions from vehicles in use, both EPA and CARB have developed a sequence of events designed to test vehicles for compliance with evaporative emission regulations. The California Air Resources Board was first to adopt their new procedures but EPA tightened them even more early in 1994. Although both testing programs include the events of preconditioning, diurnal heat builds and exhaust, running loss, and hot soak tests, differences exist that complicate a comparison of the emission control differences that might result from the two programs. It appears likely, however, that the EPA and CARB procedures will eventually be closely matched.

Diesel Vehicles and Engines

Not surprisingly, in view of the already serious health concerns and the continued growth in the use of diesel vehicles, many countries are pushing equally hard to reduce diesel emissions as gasoline vehicle emissions.

North America. ***United States.*** U.S. emission control requirements for smoke from engines used in heavy-duty trucks and buses were first implemented for the 1970 model year. These opacity standards were specified in terms of percent of light allowed to be blocked by the smoke in the diesel exhaust (as determined by a light extinction meter). Heavy-duty diesel engines produced during model years 1970 through 1973 were allowed a light extinction of 40% during the acceleration phase of the certification test and 20% during the lugging portion; 1974 and later model years are subject to smoke opacity standards of 20% during acceleration, 15% during lugging, and 50% at maximum power.

The first diesel exhaust particulate standards in the world were established for cars and light trucks in an EPA rule making published on March 5, 1980. Standards of 0.6 grams per mile (gpm) were set for all cars and light trucks starting with the 1982 model year dropping to 0.2 and 0.26 gpm for 1985 model year cars and light trucks, respectively. In early 1984, EPA delayed the second phase of the standards from 1985 to 1987 model year. Almost simultaneously, California decided to adopt its own diesel particulate standards: 0.4 gpm in 1985, 0.2 gpm in 1986 and 1987, and 0.08 gpm in 1989.

Subsequently, EPA revised the 0.26 gpm diesel particulate standard for certain light-duty trucks. Light-duty diesel trucks with a loaded vehicle weight of 1700 kg or greater, otherwise known as LDDT2s, were required to meet a 0.50 gpm standard for 1987 and 0.45 gpm level for 1988–1990. For the 1991 and later model years the standard was tightened to 0.13 gpm.

Particulate standards for heavy-duty diesel engines were promulgated by the EPA in March 1985. Standards of 0.60 grams per brake-horsepower-hour (G/BHP-h) (0.80 g/kW · h) were adopted for 1988 through 1990 model years, 0.25 (0.34) for 1991 through 1993 model years and 0.10 (0.13) for 1994 and later model years. Because of the special need for bus control in urban areas, the 0.10 (0.13) standard for these vehicles was to go into effect in 1991, three years earlier than for heavy-duty trucks.

Table 1 shows the standards adopted for heavy-duty diesel engines and vehicles as a result of the Clean Air Act Amendments of 1977. The U.S. Clean Air Act of 1977 required that the regulations for NO_x be based on "the maximum degree of emission reduction which can be achieved by means reasonable expected to be available." The PM standard was to require "the greatest degree of emission reduction available" according to the EPA's determination, considering cost, noise, energy, and safety. The limit was intended to be technology forcing. These regulations have an important influence on the rate at which new, less polluting technologies penetrate heavy-duty engine markets.

From 1970 to 1983, the U.S. regulations required demonstration of compliance on the U.S. 13-mode steady-state test procedure. In 1984, either the steady-state or the new U.S. transient test procedure could be used, but the latter has been required for 1985 and later engines. In addition, U.S. regulations permit no crankcase emissions of HC, except for turbo-charged engines. Compliance is required over the full useful life of 1985 and later engines, defined as 8 yr or 110,000—290,000 m, depending on the size of engine.

Changes in the 1990 Clean Air Act Amendments. As noted above, on November 15, 1990, the U.S. Clean Air Act Amendments of 1990 were signed into law. This legislation crafted a new challenge for diesel manufacturers. In summary, for cars, it requires all vehicles to meet the California particulate standard of 0.08 gpm. It initially postponed the 0.1 g/BHP-h urban bus particulate standard from 1991 until 1993. Beginning in 1994, however, the new law requires buses operating more than 70% of the time in large urban areas to cut particulate by 50% compared with conventional heavy duty vehicles (i.e., to 0.05 g/BHP-h). Furthermore, it mandates EPA to conduct testing of vehicles to determine whether urban buses are meeting the standard in use over their full useful lives. If EPA determines that 40% or more of the buses are not, it must establish a low pollution fuel requirement. Essentially, this provision allows the use of exhaust after treatment devices to reduce diesel particulate to a low level, provided they work in the field; if they fail, EPA is to mandate alternative fuels. Table 2 summarizes the heavy-duty diesel particulate standards required by current law. See also CLEAN AIR ACT, MOBILE SOURCES.

Table 1. U.S. Emission Standards for Heavy-Duty Diesel Engines

Model Year	NO_x, g/BHP-h	HC, g/BHP-h	CO, g/BHP-h	PM, g/BHP-h	Smoke,[a]% A	B	C
1988	10.7	1.3	15.5	0.60	20	15	50
1990	6.0	1.3	15.5	0.60	20	15	50
1991	5.0	1.3	15.5	0.25[b]	20	15	50
1994	5.0	1.3	15.5	0.10	20	15	50

[a] *A,* average acceleration; *B,* lug down peak; *C,* acceleration average.
[b] For urban buses, the standard is 0.10 g/BHP-h.

Table 2. Heavy-duty Vehicle Standards

Heavy-duty Trucks (3856 kg GVWR[a] or more)	
1988–1990 federal and California	0.6 g/BHP-h
1991–1993 federal and California	0.25 g/BHP-h
1994 and later model year federal and California	0.1 g/BHP-h
Urban Buses	
1988–1990 federal and California	0.6 g/BHP-h
1991–1992	
federal	0.25 g/BHP-h
California	0.1 g/BHP-h
1993 federal and California	0.1 g/BHP-h
1994 and later model year	
federal[b]	0.05 g/BHP-h
California	0.1 g/BHP-h

[a] Gross vehicle weight rating.
[b] EPA may relax to 0.07 g/BHP-h, based on technical considerations.

Beyond the vehicles themselves, EPA has issued a rule limiting the sulfur content of diesel fuel to 0.05 wt. % after October 1, 1993. This decision not only directly lowers particulate (sulfur) emissions but also paves the way for the use of catalytic control technology for diesel particulate since it reduces the concern over excessive sulfate emissions. The rule also allows vehicle manufacturers to certify their engines for compliance with the 1991 to 1993 emission limits with fuel of 0.10 sulfur content.

The statutory standards for HC, CO, and NO_x for heavy-duty vehicles and engines created by the 1977 Clean Air Act Amendments were eliminated in the 1990 Amendments. In their place, a general requirement that standards applicable to emissions of HC, CO, NO_x, and particulate reflect

> the greatest degree of emission reduction achievable through the application of technology which the Administrator determines will be available for the model year to which such standards apply, giving appropriate consideration to cost, energy, and safety factors associated with the application of such technology.

Standards adopted by EPA that are currently in effect will remain in effect unless modified by EPA. EPA may revise such standards, including standards already adopted, on the basis of information concerning the effects of air pollutants from heavy-duty vehicles and other mobile sources, or the public health and welfare, taking into consideration costs.

A 4.0 g/BHP-h NO_x standard is established for 1998 and later model year gasoline- and diesel-fueled heavy-duty trucks. The useful life provisions of California apply for light and medium trucks and EPA is authorized to delay the 1998 NO_x standard for heavy trucks for 2 yr on the basis of technological feasibility. EPA is also mandated to promulgate regulations that require that urban buses in large metropolitan areas that have their engines rebuilt after January 1, 1996, shall be retrofitted to meet emissions standards that "reflect the best retrofit technology and maintenance practices."

California. Under the U.S. system, California has been allowed to adopt its own vehicle emission standards. As noted in Table 2, California retained its 0.1 particulate standard for urban buses starting wtih the 1991 model year. It also adopted new emission standards for medium-duty vehicles following a public hearing on June 14, 1990. The regulation expands the definition of medium-duty vehicles to include vehicles weighing between 6000 and 14,000 lb GVWR, establishes a chassis test procedure for those vehicles (which should greatly facilitate in use recall testing), and expands the useful life requirements to 120,000 mi.

The new emission standards are set out in Table 3. Manufacturers will be required to meet 50% certification compliance in 1995 and 100% compliance in 1996. Less stringent in-use compliance will be permitted through the 1997 model year with full 100% certification and in-use compliance required for the 1988 model year.

Demonstrating the same degree of leadership with regard to diesels as it has with gasoline fueled vehicles, the California Air Resources Board (CARB) initiated a process to explore tighter emission standards for 1996 and later model year urban buses. The staff initially was considering a 2.5 g/BHP-h NO_x standard and a 0.05 g/BHP-h particulate standard. For other classes of heavy-duty engines, the CARB staff is examining the feasibility of a 2.0 g/BHP-h NO_x standard and a 0.05 g/BHP-h particulate standard to be phased in beginning possibly as early as the 1998 model year.

Transit Buses. In light of the strong opposition of the transit industry to the proposed 1996 standards for transit buses (2.5 NO_x and 0.05 particulate) that would require the use of alternative fueled engines, the CARB staff modified its proposal as summarized below.

The staff had originally proposed that all transit buses with a gross vehicle weight rating (GVWR) of 14,001 lb and above be subject to the proposed standards. The staff has modified its proposed definition to include only transit buses that would typically use a heavy-duty engine of >33,000 lb GVWR. This change would align California's definition of transit buses with EPA's definition. Staff has

Table 3. California Emission Standards

	50,000-mi Standards, g/mi			120,000-mi Standards, g/mi			
Test Weight, lb	NMHC[a]	CO	NO_x	NMHC[a]	CO	NO_x	PM
0–3750	0.25	3.4	0.4	0.36	5.0	0.55	0.08
3751–5750	0.32	4.4	0.7	0.46	6.4	0.98	0.10
5751–8500	0.39	5.0	1.1	0.56	7.3	1.53	0.12
8501–10,000	0.46	5.5	1.3	0.66	8.1	1.81	0.15
10,001–14,000	0.60	7.0	2.0	0.86	10.3	2.77	0.18

[a] Nonmethane hydrocarbons.

determined that this revision would be beneficial to small transit agencies as they tend to operate a greater proportion of smaller buses (<33,000 lb GVWR), which would not be exempt under this new proposed definition.

The staff proposed two options for transit buses: base emission standards and optional low-emission vehicle (LEV) standards: The following base emission standards are proposed for transit buses beginning with the 1996 model year: THC, 1.3 g/BHP-h; NMHC, 1.2 g/BHP-h; CO, 15.5 g/BHP-h; NO_x, 4.0 g/BHP-h; and PM, 0.05 g/BHP-h. CARB staff believe that it is possible to meet a 4.0 g/BHP-h NO_x standard and below with heavy-duty (HD) diesel engines. In addition, with the revised proposed transit bus definition of >33,000 lb GVWR, engine manufacturers will be able to concentrate their efforts on improving emissions for just a few engine families. Furthermore, in 1993, diesel fuel quality regulations will limit the sulfur content to 0.05% and the aromatic HC content to 10%, which will also lower diesel engine emissions.

In addition to the base emission standards, the ARB would encourage the air quality districts to develop a program for NO_x emission credits for those transit agencies that would like to have the option of purchasing buses that are certified to a standard of 2.5 g/BHP-h NO_x. The ARB is currently developing technical guidelines for such an emission credit program, which will assist those districts interested in implementing a NO_x emission credit system. Staff proposes, beginning with the 1994 model year, that transit buses must be certified to the following optional LEV standards to be eligible for such an emission credit program: THC, 1.3 g/BHP-h; NMHC, 1.2 g/BHP-h; CO, 15.5 g/BHP-h; NO_x, 2.5 g/BHP-h; and 0.05 g/BHP-h.

Staff believes that alternative fuels will be the primary way of meeting the proposed optional LEV standards. The Detroit Diesel Corporation 1M model 6V-92TA methanol engine has been certified to levels of 1.7 NO_x and 0.03 PM and the 3M model has been certified to 2.3 NO_x and 0.06 PM. The Cummins L10 compressed natural gas engine has recently been certified to levels of 2.0 NO_x and 0.02 PM.

Currently, all HD vehicles are required to use closed crankcase emission control systems or positive crankcase ventilation (PCV) systems except for petroleum-fueled diesel-cycle engines that use turbochargers, pumps, blowers, or superchargers for air induction.

With the advent of low emission standards, crankcase emissions are becoming a significant percentage of the exhaust emissions and need to be controlled as well. In fact, EPA has indicated that their 1979 analysis, which led to their decision to allow nonnaturally aspirated petroleum-fueled diesel engines to emit crankcase gases, is outdated and represents a worst-case cost-effectiveness situation for diesels. In addition, several manufacturers have voluntarily controlled crankcase emissions of diesel engines by routing the emissions through an oil separator, and into the turbocharger. Therefore, EPA will now require that all transit bus engines have PCV systems beginning in 1996.

Heavy-duty Engines. Based on the technology feasibility assessment carried out by a contractor, CARB staff has recommended that sales-weighted average emission standards for NO_x and PM emissions be set for heavy-duty vehicles. The sales-weighted averaging scheme would allow manufacturers some time to develop low emission engines without overly impacting all engine lines. With this scheme, low emission engines can be averaged with base engines to produce a sales-weighted average.

To implement the averaged emission standards, four low emission truck classifications and three low emission bus classifications were defined. The four low emission truck classifications are the transitional low emission truck (TLET), the low emission truck (LET), the ultra low emission truck (ULET), and the zero emission truck (ZET). The three low emission bus classifications are the low emission bus (LEB), the ultra low emission bus (ULEB), and the zero emission bus (ZEB). The transitional category for buses has been eliminated as the bus standard proposed by the ARB for 1996 is lower than TLET levels.

The various low emission classifications also include a total hydrocarbon standard, a carbon monoxide standard, a reactivity-adjusted nonmethane organic gas (RANMOG) standard, a formaldehyde (HCHO) standard, and a total aldehyde standard. It was recommended that all standards be applied to all fuels in a fuel-neutral manner.

Retrofit Requirements. The rule making to establish control requirements for existing diesel heavy-duty engines is on hold pending resolution of the requirements for transit buses.

Off-the-road Vehicles or Engines. California has also taken the lead in regard to off-road emissions. At its January 10, 1992 meeting, the California Air Resources Board adopted emission standards for off-road diesel engines above 175 HP. New 1996 through 2000 model year engines must meet standards equivalent in stringency to the 1990 on-road heavy-duty diesel engine standards. Actual standards would be 6.9 g/BHP-h for NO_x and 0.4 g/BHP-h particulate, based on an eight-mode steady-state test largely culled from the ISO 8178 procedure. Beginning in the year 2001, the proposal is intended to be equivalent to 1991 on-road heavy-duty engine standards, 5.8 g/BHP-h NO_x and 0.16 g/BHP-h particulate. Beginning in 2000, CARB includes requirements for engines over 750 HP for the first time, requiring that they meet the 1996–1999 proposed levels.

CARB estimates that the 1996 standards will likely require modifications to fuel injection timing, fuel injectors, combustion chambers and jacket water after coolers. The 2000 standards will require further improvements in after cooling, improved oil control, and possibly electronic fuel injection systems. CARB noted in its proposal that alternative fuels and particulate after treatment devices may be used but no 1991 on highway engine needed them to meet similar standards.

Canada. In March 1985, in parallel with a significant tightening of gaseous emission standards, Canada adopted the U.S. particulate standards for cars and light trucks (0.2 and 0.26 gpm, respectively) to go into effect in the 1988 model year. Subsequently, Canada also decided to adopt U.S. standards for heavy-duty engines for 1988 as well. U.S. manufacturers have committed themselves to marketing 1991 and subsequent technology heavy-duty engines in Canada in the absence of specific regulations.

Mexico. In a short period of time, Mexican officials have made dramatic progress in putting an aggressive motor vehicle pollution control program in place. They have not yet adequately addressed their heavy-duty diesel particulate problem, however. As part of the refinery upgrade currently under way, diesel fuel sulfur levels will be lowered to 0.1 wt %. In addition, adoption of U.S. 1994 heavy truck standards is under consideration but has been delayed because the U.S. requirements do not address the high altitude conditions that exist in Mexico City.

Western Europe. ***Common Market.*** *Light-duty Vehicles.* The European Community Environmental ministers decided in December 1987 to adopt a particulate standard for light duty diesels of 1.1 g per test (1.4 for conformity of production). However, in response to pressure from Germany, which has a large light-duty diesel population, and thus was pushing for much tighter standards, the ministers ordered the commission to develop a second-step proposal by the end of 1989.

As part of the consolidated directive released on February 2, 1990, the commission proposed new standards of 0.19 g/km for type approval and 0.24 g/km for conformity of production. Subsequently, following critical comments from the European Parliament, the proposed levels were reduced to 0.14 and 0.18, respectively. At the Council of Ministers meeting in October 1990 there was still some debate over the conformity of production (COP) standard, with Germany and Greece pushing for a level of 0.14 to 0.15 rather than the commission proposal of 0.18.

At the European Council of Environmental Ministers meeting on December 21, 1990, final, unanimous agreement was reached. Specifically the ministers decided to:

1. Adopt the commission proposal as amended at the October 29 meeting, to require all light-duty vehicles to meet emission standards of 2.72 g/km CO, 0.97 g/km of HC plus NO_x, 0.14 g/km of particulates for type approval and 3.16 g/km CO, 1.13 g/km for jf23HC plus NO_x, 0.18 g/km particulates for conformity of production.
2. To require the commission to develop a proposal before December 31, 1992, which, taking account of technical progress, will require a further reduction in limit values.
3. To have the council decide before December 31, 1993, on the standards proposed by the commission.
4. To allow countries before 1996 to encourage introduction of vehicles meeting the proposed requirements through "tax systems" that include pollutants and other substances in the basis for calculating motor vehicle circulation taxes.

At the local level, the German parliament agreed to include diesel vehicles in its low pollution vehicle tax incentive program. To qualify, cars must achieve a particulate standard of 0.08 g/km. (This standard is approximately halfway between the current U.S. particulate standard of 0.2 g/m and the California particulate standard of 0.08 g/m.)

Heavy-duty Vehicles. European nations have for many years negotiated harmonized standards within the U.N. Economic Commission for Europe (U.N. ECE). Once adopted in this international organization, those standards may be promulgated in law by individual countries. Regulation ECE 24 governed smoke emissions and a similar standard was required in the European Communities through EC Directive 72/306. The amount of light passing through the exhaust gas is measured with the Bosch scale and method at full load for various engine sizes.

The standards for gaseous pollutants of Regulation ECE 49, which were adopted in 1983, were no where legally required until 1986. In a first step toward reducing emissions from heavy-duty vehicles, the EEC adopted Directive 88/77/EEC, which applied as of October 1, 1990. In early 1990, after months of debate, and delayed slightly by the tremendous effort associated with the consolidated directive for light-duty vehicles, the European Commission released its proposal for cleaning up diesel trucks. The draft directive proposed compulsory EC norms to be introduced in two stages:

1. After July 1, 1992, new types of diesel engines and diesel-fuel vehicles will be required to meet the following standards:
   - carbon monoxide (CO g/kW·h = 4.5)
   - hydrocarbons (HC g/kW·h = 1.1)
   - nitrogen oxides (NO_x g/kW·h = 8)
   - particulates according to engine power: less than 85 kW (PT g/kW·h = 0.63) and more than 85 kW (PT g/kW·h = 0.36)
2. After October 1, 1996, the new diesel engines and vehicles will be required to comply with the following standards:
   - carbon monoxide (CO g/kW·h = 4)
   - hydrocarbons (HC g/kW·h = 1.1)
   - nitrogen oxide (NO_x g/kW·h = 7)
   - particulate emissions regardless of the power of the engine (PT g/kW·h = 0.3 or 0.2)

For particulate emissions, the proposal provided that the council would decide by September 30, 1994, at the latest, between the figures indicated in the proposal and the results of a report prepared by the commission before the end of 1993. The report was to indicate progress made in regard to (*1*) techniques for monitoring atmospheric pollution from diesel engines, (*2*) the availability of improved diesel fuel (sulfur content, aromatics content, cetane number) and a corresponding reference fuel to test emissions, and (*3*) a new statistical method for monitoring production compliance, which must still be adopted by the commission, needed to make a system based on a single standard operational.

In presenting the commission's proposal, Commissioner Ripa di Meana recalled the conclusions of the Task Force on Environment/Single Market of December 1989, which underlined the fact that road transport's impact on the environment could increase after 1992. From 1985 to 1992, he estimated that road transport will increase its share of goods transport from 50 to 70%. After 1992, he indicated that the opening of frontiers should increase road transport even more.

Shortly after the commission proposal, in a dramatic, surprise decision, the UK government announced on

March 22, 1990 its decision to support the full U.S. heavy-duty vehicle emissions control program. This reflects British conclusions that a transient test procedure is necessary actually to achieve low in use vehicle emissions rates of nitrogen oxides and particulates.

After receiving comments from the Parliament, at the October 1, 1991, meeting of the Council of Environmental Ministers, final agreement was reached on the directive. The unanimous decision by EC environment ministers applies to both particulate and gaseous emissions (Table 4).

The directive takes effect in two stages. New vehicle models were required to conform by July 1, 1992; all new vehicles by October 1, 1993. The second stage begins October 1, 1995, for new vehicle models, and October 1, 1996, for all vehicles. The dates represent a compromise from the original proposal by the EC Commission, in that the first stage was pushed back by nine months and the second stage was brought forward by a year.

Fuel with 0.05 sulfur content will have to be available for the second stage. Manufacturers will be allowed to certify their engines using this new fuel immediately. The sulfur content in diesel fuel will be reduced to no more than 0.2% by weight beginning October 1, 1994, and to 0.05% by weight starting October 1, 1996. As of October 1, 1995, there must be a balanced distribution of available diesel fuel, so that at least 25% of diesel fuel in each member country has a sulfur content of not more than 0.05%.

Royal Commission Report. Just as the EC was about to reach its heavy-duty vehicle decision, the Royal Commission on Environmental Pollution released a report on heavy-duty diesel particulate. Introducing the report, the chairman noted that diesel vehicles accounted for about 25% of the national emissions of oxides of nitrogen and were the principal source of smoke in urban areas. Significant points of the report are as follows:

1. Acceptable air quality levels are being exceeded regularly in London and other British cities as a result of nitrogen oxides emissions, to which diesel vehicles are major contributors. Further diesel exhaust has been classified by the World Health Organization as a probable carcinogen.
2. As proposed by the UK government, the European Community's test for new heavy-duty diesel engines should be made more demanding, along the lines of the test now used in the United States.
3. Financial incentives should be developed for lower polluting engines to accelerate the replacement of older, higher polluting engines with newer, cleaner engines.

Table 4. EC Heavy-duty Diesel Requirements

EC Step	CO	HC	NO_x	PART
1 Type Approval	4.5	1.1	8.0	0.63 (<85 kW) 0.36 (>85 kW)
COP	4.9	1.23	9.0	0.7 (<85 kW) 0.4 (>85 kW)
Step 2	4.0	1.1	7.0	0.15
Step 3 (2000) or later		new test possible		

4. The government should proceed urgently with trials of traps fitted to exhausts to catch particulates; a grant should be offered for fitting these to buses.
5. A heavy-duty engine diagnostic test should be developed to improve the actual condition of engines; new engines should be required to maintain good emissions performance over prolonged periods of operation.
6. Tighter limits for emissions from urban buses may be needed to reduce emissions in urban areas.
7. Incentives should be created for the early introduction and use of low sulfur fuel. Furthermore, subject to the outcome of a study, the government should seek to encourage bus operators to make use of alternative fuels such as petrol, LPG, CNG, or electricity.

German Field Studies Continuing. Germany has not been waiting for EC regulations to deal with the urban bus problem. For some time it has had an extensive program under way in cooperation with industry with 1500 buses distributed in urban areas across the country fitted with exhaust aftertreatment systems. Preliminary findings indicate that the after treatment technology has advanced significantly.

Non–Common Market Countries of Europe. Several other European countries have been moving toward more stringent diesel particulate requirements than those of the EC. Sweden adopted the U.S. passenger car standard in 1989 and other EFTA countries took similar actions. These countries have also been seriously considering more stringent requirements for trucks and buses. Sweden announced its intention to adopt requirements that will bring about the same degree of control as the United States by 1995. Specifically, U.S. 1990 requirements (including diesel particulate) were adopted by Sweden on a voluntary basis for 1990 light duty trucks, and on a mandatory basis in 1992. Regarding heavy trucks, the EFTA countries have been striving for U.S.-type engine technology, trying to derive the equivalent environmental benefits.

The regulatory situation in Europe is changing rapidly, however. In early 1992, the EC and EFTA countries finally reached agreement on the European Economic Area (EEA), the world's biggest (with 380 million people) and wealthiest single market. This agreement significantly alters the shape of motor vehicle pollution control in Europe. Several of the EFTA countries have also applied for membership in the EC. One can foresee that most if not all future vehicle emissions regulation in Europe will take place under the auspices of the EC. At the Edinburgh summit, the EC ministers agreed to accelerate the consideration of full community membership for Sweden, Norway, Austria, and Finland.

Asia Pacific Region

Japan. Japanese smoke standards have applied to both new and in-use vehicles since 1972 and 1975, respectively. The maximum permissible limits for both are 50% opacity; however, the new vehicle standard is the more stringent because smoke is measured at full load, while in-use vehicles are required to meet standards under the less severe no-load acceleration test.

Japan also established emission limits for heavy-duty engines as summarized in Table 5. All engines must meet the maximum value during certification testing. Also, Japanese limits are expressed in terms of parts per million (ppm), or concentrations in the exhaust gas. This differs from other countries' standards relative to engine power outputs. Without detailed information on engine power outputs and air flows in Japan, then, it is not possible to compare the Japanese with other OECD member countries' standards. In addition, the emission levels allowed in Japan for direct injection engines are higher than for indirect injection engines. Gasoline vehicles are measured for compliance on a six-mode steady-state test, while diesel vehicles are measured on a different six-mode steady-state test.

On December 22, 1989, the Central Council for Environmental Pollution Control submitted the *Approach to Motor Vehicle Exhaust Emission Control Measures in Years Ahead* to the director-general of the Environment Agency. It includes a diesel particulate standard for the first time. In accordance with the recommendation, the Ministry of Transport revised the related laws, ordinances and regulations during 1990. Principal provisions include the following.

The emission levels of nitrogen oxides from diesel-powered motor vehicles are reduced by 38% in the case of large-size diesel-powered trucks (with direct injection-type engines) and by 56% in the case of medium-size diesel-powered passenger motor vehicles.

The NO_x emission level for the direct injection type diesel-powered motor vehicles is required to be the same as that for indirect injection type diesel-powered motor vehicles.

The emission level of particulate matter from diesel-powered motor vehicles is reduced by more than 60%. Besides the diesel smoke standards hitherto enforced, a new particulate standard was introduced, reducing the level by more than 60% from the present level by providing two-phase target values.

The emission level of diesel smoke will also be cut in half by providing two-phase target values.

The sulfur content in diesel (light) oil will be reduced from the present level of 0.4 to 0.5 wt % to 0.2 wt % and eventually to 0.05 wt %.

In Tokyo alone, the number of diesel-powered vehicles has risen from 467,000 in 1985 to 721,000. Such high growth has raised concerns about whether the government will succeed in reducing NO_x levels by the year 2000 as it has pledged to do. As a result efforts have been under way to develop additional control measures for this pollutant. Japan's Environment Agency (JEA) on June 26, 1990 announced new regulations calling for phased reductions of nitrogen oxide emissions from gasoline-powered vehicles, beginning in 1994, and from diesel-powered vehicles, beginning in 1997. Previously, the agency had planned to enforce these requirements from 1999. NO_x emissions from diesel-powered vehicles of 2.5 L or less must be reduced by 65% from 1997, the announcement said. However, it stopped short of setting targets for diesel-powered vehicles with larger engines.

An intense debate is under way within the government regarding the schedule for implementation of new truck emissions requirements, and it appears that the Environment Ministry is losing. The net result would be a significant delay. It appears that MITI and the MOT have succeeded in pushing through a grace period that would delay the implementation schedule by providing a grace period of up to 8 yr for trucks under 2.5 t and 9 yr for those over.

Also on June 26, the JEA decided to reduce the sulfur contents in light fuel oil, which is used as fuel for diesel-powered vehicles, to 0.05% from the current 0.4%. The government enacted a bill to regulate diesel powered NO_x vehicle emissions in urban areas during the Diet (parliament) session that ended June 19, 1990. The law requires vehicle operators to change vehicles that fail to achieve 1990 emission requirements to less-polluting ones. Those that failed to meet the new standards will not be issued vehicle inspection certificates and be kept off the road.

South Korea. The vehicle pollution control program in Korea continues to progress, but can still barely keep pace with the high rate of growth in the domestic vehicle population. From a current total of approximately 3 million vehicles throughout South Korea, the domestic population is expected to grow to about 15 million by the year 2000. In addition, while gasoline vehicles are gradually being controlled (with 50,000-m durability requirements for U.S. 83 standards introduced in 1990) diesels remain a serious unsolved problem. Approximately 45% of the annual kilometers in the country are accumulated by diesels, mainly trucks and buses. As reported previously, a several-stage strategy is under way to address this problem (6).

Reduce the Number of Diesel Vehicles. A program is under way to gradually discourage diesel engines in light trucks and buses. For the first stage during 1990 and 1991, 15-person mini buses and jeeps have been designated; for the second stage, 1992–1993, 35-person buses and 3-t cargo trucks have been designated as targets.

Diesel engine modifications to fumigate the diesel fuel with 30% LPG have been under investigation for several years. While this concept worked well in the laboratory, difficulties have been experienced in the field due to im-

Table 5. Heavy-duty Vehicle Emission Limits in Japan, ppm

Engine	NO_x	HC	CO	Smoke	Effective Date
Diesel					
direct injection	520	670	980	50% rate of contamination	Oct. 1, 1989/Sept. 1, 1990[a]
indirect injection	350	670	980	50% rate of contamination	Oct. 1, 1989/Sept. 1, 1990[a]
Gasoline	850	520	1.6%		April 1, 1991

[a] Applies to new models/all models.

proper mixing. A new test is under way with electronically controlled mixing. In addition, the government is exploring the feasibility of natural gas buses. At present, about 2 million t of LNG are imported into Seoul each year from Indonesia; about 80% is used in utilities and 20% in space heaters. It would be possible to make fuel available for buses if the government concludes natural gas buses are feasible.

Increase Horsepower. The excessive smoke levels from diesel buses are at least partly caused by overloading underpowered engines. Therefore, the government has decided that new bus engines should increase their power from 185 to 235 HP.

Tighten In-use Smoke Standards. The in-use diesel smoke standard was tightened from 50% opacity to 40% in 1990. The government hopes this action will encourage good vehicle maintenance.

Develop Diesel Particulate Trap Systems. The National Environmental Protection Institute is conducting a 3-yr research program to evaluate particulate controls. Starting in August 1990, an effort was undertaken to identify the most promising systems. Laboratory evaluations are still under way. Initial results with a trap system using cerium fuel additives have been encouraging.

Emission Standards for Light and Heavy Trucks. Particulate emission standards have also been promulgated for new light and heavy trucks. By January 1, 1996 light trucks should be 0.31 g/km and heavy trucks, 0.9 g/kW·h. By January 1, 2000, light trucks should be 0.05–0.16 g/km, heavy trucks, 0.25 g/kW·h, and city buses 0.10 g/kW·h.

Singapore. The government of Singapore is moving rapidly to introduce state of the art pollution controls. As of January 1, 1991, all new diesel vehicles have been required to comply with ECE R 24.03.

Hong Kong. The government has decided to impose new car standards from January 1992 as follows:

Vehicle Type	Petrol	Standards for Diesel
Private car, taxi	U.S. 1987, FTP75; Japan 1987, 10-mode	U.S. 1987, FTP75; Japan 1990, 10-mode
Light goods vehicle, light bus not more than 1.7 t	U.S. 1990, FTP75; Japan 1988–1990, 10-mode	U.S. 1990, FTP75; Japan 1988, 10-mode
Light goods vehicle, light bus more than 1.7 t but not more than 2.5	U.S. 1990, FTP75; Japan 1988–1990, 10-mode	U.S. 1988–1990, FTP75; Japan 1988–1990, 6-mode

Taiwan (ROC). The Taiwan EPA continues to move forward with a comprehensive approach to motor vehicle pollution control. Building on its previous adoption of U.S. 1983 standards for light-duty vehicles (starting July 1, 1990) it has decided to move to U.S. 1987 requirements, which include the 0.2 gpm particulate standard as of July 1, 1995. Heavy-duty diesel particulate standards almost as stringent as U.S. 1990, 6.0 g/BHP·h NO_x and 0.7 g/BHP·h particulate, using the U.S. transient test procedure, will go into effect on July 1, 1993. While 90% of the current heavy-duty vehicles in Taiwan are of Japanese origin, none of these have yet been able to obtain approval at these levels; only Cummins and Navistar currently have been approved. It is intended that U.S. 1994 standards, 5.0 NO_x and 0.25 particulate, will be adopted soon, probably for introduction by July 1, 1997. Diesel fuel currently contains 0.5 wt % sulfur. A proposal to reduce levels to 0.3 by 1993 and 0.05 by 1997 is currently under consideration.

Latin America. *Brazil.* A move is under way to convert diesel buses in Sao Paulo to natural gas (CNG). Petrobras has a large surplus of natural gas. The have stopped flaring and constructed a pipeline from Rio to Sao Paulo. The municipal government in Sao Paulo has developed a 10-yr plan to convert 10,000 buses to CNG. The first CNG station went into operation in 1993, for 200 buses. Also Daimler Benz has developed a CNG bus with low HC, CO, particulate, and noise leels. NO_x levels will increase by 15% initially, because the buses will not be equipped with catalysts. DB has indicated that it intends to add catalysts at some later date.

CETESB is investigating whether or not it makes sense for the natural gas to be used in vehicles as CNG or whether it should be converted to methanol and then used. The principal advantage in their eyes is that the use of a liquid fuel would be less disruptive to the existing infrastructure. In addition, when equipped with a catalyst, the overall emissions should be lower.

Diesel fuel sulfur levels in Brazil average about 1%. CETESB is trying to get Petrobras to introduce a lower sulfur fuel, initially in the cities, starting with buses. Petrobras says the best they can do is to introduce 0.5% in the cities, several years from now. To get to 0.05% would require about $1 billion in investment, according to Petrobras, and they are expanding all available capital on increased exploration.

The motor industry is pushing the federal government to authorize the use of light-duty diesel vehicles, which is presently not allowed in Brazil. CETESB supports the present prohibition on the grounds that diesel fuel is highly subsidized and, therefore, should be used only for transport of passengers (buses) and goods. In addition, diesel oil represents the bottle neck of PETROBRAS (the consumption of diesel in Brazil defines the volume of oil which has to be refined) and, therefore, its quality has been affected over the years with the addition of both heavy and light refinery projects to meet demand. Some motor industry sources claim that Brazilian diesel use results in an increase of about one Bosch unit if compared with the European diesel.

Finally, CETESB believes that particulate and other emissions from these cars will offset any advantage due to the low CO. Besides, the introduction of these vehicles into the market means that a number of less polluting vehicles, such as the alcohol fueled, will be substituted by the diesel cars. In regard to heavy-duty diesel particulates CETESB agreed with industry to introduce EURO 1 and EURO 2 standards; however this agreement must be approved by CONAMA before it becomes effective.

Chile. Santiago, Chile, has a serious diesel particulate problem caused in large part by urban buses. To address this problem it has introduced a stringent smoke inspection program. In addition, it introduced a one day a week ban on driving with exemptions granted only to diesel buses equipped with catalysts or traps. In October 1992, the exemption program was replaced by an auction system designed to reduce the number of buses. Essentially, only 6000 buses have been granted a license to operate in the center of the city, down from approximately 9000. While the criteria for granting such licenses did not explicitly include emissions, it is intended to include particulate or smoke levels in a follow up program.

Conclusions

Diesel particulate controls are getting increasingly stringent around the world. As it has previously with gasoline-fueled vehicles, California is taking the lead in advancing diesel particulate control. CARB believes that the TLET, LET, and LEB emission levels are achievable by using current technology methanol or natural gas engines. For diesel-fueled engines to meet base 1996 bus and base 1998 truck standards, new or radically redesigned versions of current engines and a significant advancement in after-treatment technology will be required. According to its contractor, Acurex, by using a combination of high pressure fuel injection, varaible geometry turbocharger, air-to-air aftercooler, optimized combustion chamber, electronic unit injections with minimized sac volumes, rate shaping, exhaust gas recirculation, and sophisticated electronic control of all engine systems, diesel engines could meet a 2.5 g/BPH-h NO_x standard at 0.15 g/BHP-h PM with a 5% penalty in fuel economy. Particulate traps could be used to reduce the particulate emissions to 0.05 g/BHP-h. Durability of the engine may be reduced to 80% of the 1994 counterpart. Several manufacturers and research organizations are studying these refinements in diesel technology and predict such engines may be available as early as 1998. Concentrated research and development, tied with demonstration programs, will be needed to bring these engines into reality.

Motorcycles

On February 2, 1988 the Swiss government decided to set new emission standards for motorcycles to go into effect by October 1990. These prescriptions will focus mainly on VOC control for two-stroke engines, whereas the standards for four-stroke motorcycles remain unchanged.

	Standards by Oct. 1, 1990	Present Standards	
	Two-stroke Engines, g/km	Two-stroke Engines, g/km	Four-stroke Engines, g/km
CO	8.0	8.0	13.0
VOC	3.0	7.5	3.0
NO_x	0.1	0.1	0.3

On July 1, 1991, the motorcycle standards in Taiwan were tightened to 4.5 g/km for CO (from 8.8), and 3.0 for HC and NO_x combined (from 6.5), based on the ECE R40 test procedure. With introduction of these requirements, Taiwan has the most stringent motorcycle control in the world, likely requiring use of catalytic converter controls.

GLOBAL EMISSIONS PROJECTIONS INDICATE MORE MUST BE DONE

Global Vehicle Emissions Trends by Region

Based on a continuation of the strong motor vehicle control programs in the United States and Japan and the recent tightening of requirements in EC Europe, these estimates indicate that global CO, HC, and NO_x emissions will remain fairly stable throughout the next decade. Beyond that point, however, emissions of all three pollutants will start to increase due to the projected continued growth in vehicle populations both in OECD countries but especially in other areas of the world where emissions controls are frequently minimal. This upturn in hydrocarbon and NO_x emissions will occur shortly after the turn of the century; for carbon monoxide, the downward trend is expected until about 2020 when it will also turn up.

Global Vehicle Emissions Trends, by Vehicle Type

An analysis of the trends in global emissions of CO, HC, and NO_x by vehicle type provides some startling insights. First of all, not surprisingly, cars remain the dominant source of CO for the foreseeable future. However, motorcycles, most of which are two stroke, are seen to be a significant contributor to HC emissions around the world, a fact that is largely ignored in the West due to their small contribution in that region. In regard to NO_x emissions, heavy-duty trucks are a large and rapidly growing contributor, due to the minimal NO_x control of these vehicles in most regions of the world.

TECHNOLOGICAL DEVELOPMENTS

Gasoline-Fueled Vehicles

California has demonstrated a commitment to achieving the lowest possible in use emissions levels through a comprehensive program of tighter standards, extended durability requirements, and technological (eg, onboard diagnostics) and regulatory (eg, defect reporting) innovation. Some important elements of the CARB program as it existed in early 1990 have in broad terms been mandated by the Clean Air Act Amendments of 1990 to apply to federal vehicles across the country. In regard to vehicle technology, the only comparable period of emissions and fuel technological push was following the 1970 Clean Air Act Amendments when within 5 yr unleaded fuel went from virtually zero to being the norm for all new cars, catalytic converters, which previously had virtually no real-world experience, were introduced in four out of five new cars, and the electronics revolution transformed the control mechanisms of such critical emission control components as spark and air fuel management systems. Other than controls that have already emerged from the laboratory such as preheated converters, it is difficult to anticipate fundamental technological breakthroughs that may emerge over the course of the coming decade. Therefore, in a context in which strong regulatory requirements create the most favorable environment for technological breakthroughs, another element of conservatism in the analysis that follows is that it can only be based on currently known technologies.

Technology Overview. The level of tailpipe hydrocarbon emissions from modern vehicles is primarily a function of the engine-out emissions and the overall conversion efficiency of the catalyst, both of which depend highly on proper function of the fuel and ignition systems. A fairly comprehensive system has evolved. A significant portion of the HC and CO emissions are generated during cold start, when the fuel system is operating in a rich mode and the catalyst has not yet reached its light-off temperature. There are many technological improvements, which are currently becoming widespread or are on the horizon, that make more stringent control of HC and CO feasible. These advances are expected not only to reduce the emission levels that can be achieved in the certification of new vehicles but also to reduce the deterioration of vehicle emissions in customer service.

First is the trend toward increased use of fuel injection. Fuel injection has several distinct advantages over carburetion as a fuel control system: more precise control of fuel metering, better compatibility with digital electronics, better fuel economy, and better cold-start function. Fuel metering precision is important in maintaining a stoichiometric air: fuel ratio for efficient three-way catalyst operation. Efficient catalyst operation, in turn, can reduce the need for dual-bed catalysts, air injection, and EGR. Better driveability from fuel injection has been a motivating force for the trend to convert engines from carburetion to fuel injection. In fact, it has been projected that the percentage of new California light-duty vehicles with fuel-injection will reach 95% by the early 1990s, with 70% being multipoint. Because of the inherently better fuel control provided by fuel-injection systems, this trend is highly consistent with more stringent emissions standards.

Fuel injection's compatibility with onboard electronic controls enhances fuel metering precision and also gives manufacturers the ability to integrate fuel control and emissions control systems into an overall engine management system. This permits early detection and diagnosis of malfunctions, automatic compensation for altitude, and to some degree, adjustments for normal wear. Carburetor choke valves, long considered a target for maladjustment and tampering, are replaced by more reliable cold-start enrichment systems in fuel-injected vehicles.

Closed-loop feedback systems are critical to maintain good fuel control, although when they fail, emissions can increase significantly. In fact, the CARB in-use surveillance data show that failure of components in the closed-loop system frequently has been associated with high emissions. The CARB's new requirement for onboard diagnostics will enable the system to alert the driver when something is wrong with the emission control system and will help the mechanic to identify the malfunctioning component.

Second, improvements to the fuel control and ignition systems, such as increasing the ability to maintain a stoichiometric air: fuel ratio under all operating conditions and minimizing the occurrence of spark plug misfire, will result in better overall catalyst conversion efficiency and less opportunity for catastrophic failure. These improvements, therefore, have a twofold effect: (*1*) limiting the extra engine-out emissions that would be generated by malfunctions and (*2*) helping keep the catalyst in good working condition.

Third, a trend that bodes well for catalyst deterioration rates (and therefore in-use emissions) is EPA's lead phaseout, which reduced the lead content of leaded gasoline by about 90% (to 0.1 g/gal) beginning in January 1986. The new Clean Air Act will actually lead to a complete ban on lead in gasoline in a few years. The phaseout will also reduce the small lead content of unleaded gasoline (since these amounts are due to contact with leaded gasoline facilities), which will reduce gradual low-level catalyst poisoning. Both catalyst and O_2 sensor durability will benefit from these lower gasoline lead contents.

Finally, there are alternative catalyst configurations that could and likely will be used in the future to meet lower emission standards. It is likely that dual-bed catalysts will be phased out over time, but a warmup catalyst (preceding the TWC) could be used for cold-start hydrocarbon control. To avert thermal damage and lower the catalyst deterioration rate, this small catalyst could be bypassed at all times other than during cold-start. Warm-up air injection could also be used with a single-bed TWC for cold-start hydrocarbon control. As hydrocarbon standards are lowered, preheated catalysts will likely become a more important element of the pollution control system of many cars.

Other variables include engine-out emissions, which depend highly on the conditions in the combustion chamber. Over the past few years, combustion chamber geometry and turbulence levels have been optimized in an effort to minimize emissions and maximize fuel economy. Fast-burn combustion, which is being used more and more, involves changes to chamber geometry, turbulence, and the location of the spark plug. It allows greater use of EGR for NO_x control without hurting efficiency, making simultaneous control of hydrocarbons and NO_x easier.

Ignition misfire is often due to fouled or faulty spark plugs, deteriorated spark plug wires, or other ignition component malfunctions. Greater durability of the ignition system, especially spark plugs, limits misfires and resulting thermal damage to catalysts. New ignition systems currently under development (and being used experimentally by Saab and Nissan) may virtually eliminate misfires and the need for high voltage spark plug wires.

There is inevitable uncertainty associated with predicting the specific technology that manufacturers will apply to future vehicles to comply with the mandates of the LEV program. Historical evidence demonstrates that strong regulatory requirements create a favorable environment for technological breakthroughs. For example, in the 5-yr period following the adoption of the 1970 Clean Air Act Amendments, unleaded fuel use, catalytic converter technology, and electronic control mechanisms quickly became the norm. It is likely that compliance with the LEV emission standards will result in the development of new technology that is not fully accounted for in this analysis.

Virtually all vehicles used the following emission control technology to meet California 1991 emission standards:

Closed-loop, single-bed, three-way catalysts.

Fuel injection, usually multipoint.

No secondary air.

To meet the 1994 federal standards as well as to respond to competitive pressures manufacturers are likely to use the following:

Improved catalyst formulations with higher noble metal loadings.

Sequential multipoint fuel injection.

Direct-fire (distributorless) ignition systems.

The latter two technologies allow for more precise control of air: fuel ratio and spark timing, relative to current technology. Improvements to catalyst formulations, coupled with reduction in gasoline contaminants such as sulfur and lead, and reduction in oil additive-based contaminants such as phosphorus will lead to reductions in catalyst deterioration, the principal cause of emissions deterioration of well-maintained cars. In addition, the electronic control and OBD systems may reduce production line variability and assist in improving in-use durability.

Evolutionary improvements to engines will also aid in meeting the 1994 federal standards. These improvements will include tuned inlet manifolds designed to reduce cylinder-to-cylinder variability, revised pistons to reduce crevice volume and oil consumption, and the incorporation of fast-burn combustion chambers. A number of engines in production today already feature such improvements.

Transitional low emission vehicles must meet a standard of 0.125 g/mi for NMHC, 3.4 g/mi for CO, and 0.4 g/mi NO_x. The California Air Resources Board projected in 1990 that for small- and medium-displacement engines, only heated fuel preparation systems and/or close-coupled catalyst completed systems will be required to meet TLEV standards. Many current gasoline engine families have certified to TLEV standards with calibration changes and evolutionary changes to hardware. The calibration changes required to meet TLEV standards will include spark retard during warmup, more careful control of air: fuel ratio after cold start, and (possibly) electronic control strategies to control precisely individual cylinder air : fuel ratio.

Low emission vehicles must meet a standard of 0.075 g/mi for CH, 3.4 g/mi for CO, and 0.2 g/mi for NO_x. Based on the most recent data, it is clear that the pace of technological progress has exceeded expectations and that the low emission vehicle standards will require less advanced technology than previously expected. Less than a year ago, to produce LEVs, manufacturers were expected to use advanced fuel control strategies, greater catalytic loading for better NO_x control for some models, and electrically heated catalysts (EHCs). In the spring of 1992, however, Ford submitted to CARB a certification application containing data that strongly indicated that LEV levels can be achieved without using EHCs. In addition to these certification materials, CARB became aware of other significant advances being made by the automotive industry to reduce emission levels. Because the rate of progress has exceeded original expectations, CARB has revised its assessment of the technologies needed in each low emission vehicle category.

At an Air Resources Board public meeting held June 11, 1992, to consider the status of implementation of the low emission vehicle program, the CARB staff outlined its current technology assessment. In summary, for TLEVs, it continues to project that only modest fuel control and catalyst improvements will be necessary. For LEVs, small- to medium-displacement engines should be able to achieve the standards with fuel control improvements and improved conventional catalysts; larger eight-cylinder engines may still require the use of poststart heated EHCs or other similarly effective technologies. For ULEVs, in addition to fuel control improvements and greater catalyst loading, CARB expects that poststart heated EHCs will be sufficient for most vehicles, although prestart heated units may be needed for some applications.

To support its new assessment of LEV technology, the CARB staff presented emission data that showed that the emission levels from a 1993 Ford Escort TLEV and a 1992 Oldsmobile Achieva were at or below the LEV standards. Just after the June 11 meeting, the CARB certified the 1993 model of the Achieva and the 1993 Honda Civic to TLEV standards. In Table 6 the certification emission levels of the Escort, Achieva, and Civic TLEVs are summarized. The Escort and Civic were certified on Indolene while the Achieva was certified on California Phase 2 gasoline.

Although certified as TLEVs, exhaust emissions of the Escort were only 77% of the LEV standard, while the emissions from the Achieva were even lower, only 65% of the LEV standard for NMOG at 50,000 miles. At 100,000 miles, the margins for compliance are even greater: 69% and 54% for the Escort and Achieva, respectively.

Advanced Engine Modifications to Achieve Low In-use Levels. The emission levels of the two vehicles were surpris-

Table 6. Certification Emission Levels (g/mi) of Three 1993 TLEV Engine Families

Mileage	Model	Displacement, L	NMOG	CO	NO_x
50,000	Ford Escort	1.9	0.058	1.0	0.2
50,000	Olds Achieva	2.3	0.049	0.4	0.1
50,000	Honda Civic	1.5	0.076	0.3	0.2
50,000	LEV Standard		0.075	3.4	0.2
100,000	Ford Escort	1.9	0.062	1.3	0.3
100,000	Olds Achieva	2.3	0.049	0.4	0.1
100,000	Honda Civic	1.5	0.086	0.3	0.2
100,000	LEV Standard		0.090	4.2	0.3

ingly low in view of the fact that these vehicles used little or none of the advanced technologies available to significantly reduce emissions even further. In Table 7, the advanced technologies incorporated by the Escort and Achieva are listed.

As is clearly seen from Table 7, the Escort incorporates only two of the identified technologies, while the Achieva does not use any. Because both vehicle types have already shown certification emission levels (which includes a 100,000-mi durability demonstration) well below the LEV standards, even without the addition of some of these available technologies, the prospects for these and similar vehicles attaining compliance with the LEV standards in-use are excellent. With the additional technologies, virtually all but the largest vehicles should be able to be certified as LEVs without the use of EHCs.

Dual Oxygen Sensors. Toyota has been using dual oxygen sensor systems since 1988. It makes good engineering sense that dual oxygen sensors will help maintain low emission levels in-use. As oxygen sensors age, their warmup response slows considerably, and the air : fuel ratio at which the sensor switches from low to high voltage (and vice versa) can shift significantly from stoichiometric, thereby increasing emissions. Because the second sensor is placed downstream of the catalyst, it operates in a relatively low temperature environment and is better protected from poisons. It, therefore, can be used to compensate for slow response and to adjust for changes in the switch point of the front sensor. In this way, a dual oxygen sensor system helps maintain good fuel control, and consequently good emissions control, as vehicles age.

Sequential Fuel Control. Precise injection timing may be helpful in minimizing hydrocarbon emissions under steady-state conditions. An effective way to use sequential fuel injection is to optimize injection timing to occur while the intake valve is open. This can be accomplished by using aerated fuel injectors to eliminate the need to rely on evaporation, thereby allowing direct injection to significantly less fuel into the combustion chamber.

Aerated Fuel Injectors. Toyota's aerated fuel injection system demonstrates emission benefits not only over the full range of steady-state conditions but under transient and cold conditions as well. Use of aerated fuel injectors ensures good atomization during cold starts and, therefore, permits injection directly into the combustion chamber when the intake valve is open. This strategy reduces the amount of excess fuel needed to avoid driveability problems due to wall wetting, thereby minimizing emission increases and fuel economy degradation. This is especially important during transient engine operation (eg, during rapid accelerations) when excess fuel is normally added to prevent engine stumbles and sags.

Table 7. Advanced Technologies Used on the Escort and Achieva

Technology	Escort	Achieva
Heated oxygen sensor	X	
Air pump		
Sequential multiport fuel injection	X	
Aerated fuel injectors		
Heated fuel preparation		
Cylinder fueling torque control		
Dual oxygen sensor		
Adaptive transient fuel control		

Adaptive Fuel Control. CARB believes that while adaptive control of steady-state engine operating conditions has been used for some time by industry, only a few manufacturers have been successful in developing adaptive strategies for transient operating conditions. In fact, to the knowledge of the CARB staff, the first application of this technology is on a few 1992 Toyota models, and it is aware of only one other major manufacturer that is planning to employ adaptive transient fuel control in the near future.

Individual Cylinder Torque Control. CARB staff discussions with numerous manufacturers indicate that measurement of cylinder torque is being examined as a means for controlling the entire fueling strategy for their engines, enabling them to reduce the dependence on oxygen sensors for primary fuel altogether. This same technique is being used by the automotive industry for misfire detection as part of the CARB's On-Board Diagnostic II (OBD II) misfire detection requirements. It may not be necessary to monitor all engine operating conditions to determine the level of fuel compensation needed over the full range of engine operating conditions if a particular injector is underfueling or overfueling (eg, due to a partially restricted or leaking injector). In CARB's view, torque changes become more pronounced as load increases. Under more moderate load conditions and more representative vehicle driving modes, as opposed to idle conditions, torque fluctuation should be readily detectable well within a one air : fuel ratio variation.

Heated Fuel Injectors and Heated Fuel Preparation Systems. Fuel heating strategies have been used on production vehicles for some time now. Mercedes Benz has used an intake manifold heater on the 2.3-L engines in its 190 series since 1991. As with aerated fuel injectors, improved fuel vaporization allows the use of leaner air : fuel mixtures during cold engine operation, thereby providing an even greater HC and CO emission reduction.

Air Injection. To the extent that air–fuel enrichment is needed during cold engine operation to provide acceptable driveability, air injection can be utilized to provide the additional air needed fully to oxidize the excess HC and CO emissions in the catalytic converter. The need for air injection can be lessened by incorporating aerated fuel injectors, sequential fuel injection, and adaptive transient fuel compensation, either singly or in combination, to reduce the level of cold engine operating enrichment needed to achieve acceptable driveability. Air injection is a highly effective means of reducing NMOG and CO emissions but may not be needed to meet the relatively less stringent Tier I or TLEV emission levels. In fact, none of the 1993 TLEVs certified by the CARB utilize air injection. However, air pumps have been an important element in achieving reduced NMOG and CO emissions on the CARB's EHC-equipped vehicles and would also be effective in achieving LEV emission levels for vehicles without EHCs.

Heated Oxygen Sensors. Heated oxygen sensors, a technology commonly used on many of today's vehicles, will also help reduce emissions. Heating oxygen sensors is im-

Table 8. In-use Compliance Test Results (g/mi) from Some 1989 Production Vehicles[a]

Manufacturer	Engine Family	NMHC	CO	NO_x
Volvo	KVV2.3V5FE8X	0.21	2.43	0.19
Ford	FKM2.2V5FXC4	0.23	4.88	0.14
Mitsubishi	KMT2.0V5FC18	0.22	2.19	0.20
Applicable standards		0.39	7.0	0.4

[a] Results shown are the average of 10 tests; mileage on the test vehicles ranged from 31,000 to 49,000.

portant because as oxygen sensors age, they require higher operating temperatures to maintain adequate responsiveness to changes in air : fuel ratio, particularly during cold start operation. Slow response sensors can prolong the time required to switch from open-loop to closed-loop operation or provide poor fuel control during the initial closed-loop operating period, thereby resulting in increased emissions. Oxygen sensor deterioration of this nature is most likely to occur after about 50,000 mi of driving. The deterioration can be masked, however, by electrically heating the sensor to operating temperature, thereby reducing the time needed to initiate closed-loop operation and minimizing emission increases due to improper air : fuel ratios caused by slow oxygen sensor response rates.

Additional Rhodium Loading for NO_x Control. The rhodium levels of many current vehicles are sufficient to enable achievement and maintenance of a 0.2 g/mi NO_x standard in-use. CARB in-use compliance records reveal that several 1989 models were able to achieve 0.2 g/mi NO_x levels in-use (even without the application of advanced fuel controls). The emission results of these vehicles are listed in Table 8.

Some manufacturers may need to add rhodium to match the levels used by these better performing vehicles. Adding rhodium to the catalyst is only one option for achieving low NO_x emissions. Improved control of the air : fuel ratio at stoichiometric can also improve NO_x emissions. Another option was proposed in the July 1992, issue of *Automotive Engineering,* which concluded that less expensive palladium could be a viable replacement for some of the rhodium in catalytic converters. For these reasons, the CARB considers increased rhodium loading to be an effective strategy for reducing NO_x emissions and for maintaining NO_x standards in use.

Ultra-low emission vehicles must meet a standard of 0.04 g/mi for HC, 1.7 g/mi for CO, and 0.2 g/mi for NO_x. At the very low emission levels of 0.040 HC (NMOG), it appears today that it will be necessary to have some form of additional exhaust aftertreatment. Several types of after-treatment are being investigated:

Close-coupled start catalysts.
Exhaust port catalysts.
Electrically heated catalysts.
Hydrocarbon molecular sieves.

Electrically Heated Catalysts. The electrically heated catalyst (EHC) has received the most attention and has demonstrated the potential to meet ULEV standards even in large, heavy cars. The ARB, EPA, and a catalyst manufacturer (CAMET) have collaborated in extensive testing of these types of catalysts. The main drawback of these catalysts had been that they impose a large current drain on the battery for 15–30 s before cold start. While the battery problems are not large at ambient temperatures of 21°C they may be more significant at winter temperatures of −7°C, when heating requirements are increased while battery capacity is decreased.

Innovative solutions to these problems are being researched. To support its upcoming rule making to establish reactivity adjustment factors for Phase 2 gasoline-fueled low emission vehicles, the CARB staff recently tested several late model vehicles that were retrofitted with EHCs. The warmed-up emissions performance of these vehicles are considered representative of the emission levels CARB expects to see from low emission vehicles (Table 9).

For the purpose of obtaining reactivity data to support its November rulemaking, the CARB staff has continued to conduct tests on EHC-equipped vehicles. Its most recent data using Camet's latest EHC on the Buick indicate that energy requirements range from 29 to 42 W · h (Table 10).

Table 9. Per Bag Emission Results (g/mi) for Late Model Year Vehicles[a]

Vehicle	Test	Bag 1	Bag 2	Bag 3	FTP Composite
1992 Lexus LS400	NMHC	0.204	0.000	0.000	0.041
	CO	1.766	0.078	0.079	0.429
	NO_x	0.313	0.131	0.248	0.201
1988 Chevrolet Corsica	NMHC	0.230	0.003	0.019	0.054
	CO	1.692	0.142	0.794	0.643
	NO_x	0.284	0.125	0.366	0.224
1990 Toyota Celica	NMHC	0.192	0.000	0.009	0.042
	CO	2.003	0.076	0.194	0.508
	NO_x	0.666	0.154	0.084	0.241
1991 VW Jetta	NMHC	0.125	0.002	0.023	0.034
	CO	2.645	0.852	1.470	1.392
	NO_x	0.357	0.038	0.184	0.144

[a] Equipped with EHCs by the CARB; vehicles operated on 1989 industry average gasoline.

Table 10. EHC Energy Requirements of Recent CARB Tests[a]

Test Date	Vehicle	Test Number	Watt · Hours	NMHC
7/10/92	Buick	28-10	33.9	0.027
7/14/92	Buick	28-11	37.2	0.025
7/16/92	Buick	28-13	38.0	0.032
7/24/92	Buick	28-15	28.8	0.027
7/29/92	Buick	28-16	33.8	0.044
7/31/92	Buick	28-17	41.7	0.022

[a] The July 1992 CARB test data involved various EHC heating and air injection strategies that contributed to the variability in the results.

Even then, the EHCs being used on CARB test vehicles are stand-alone units to facilitate fabrication of the new exhaust systems. Ideally, the EHCs should be close mounted to the main catalyst and contained in the same housing to minimize heat loss, thereby reducing energy requirements even further. The energy requirements of other new EHC designs arranged in this manner are much lower, as shown in Table 11. These results on a Honda Accord using an EHC developed by Corning, Inc. show NMHC emissions of 0.022 g/mi for 25 s of preheat while using only 14.8 W · h of electrical energy. For 25 s of postheat, NMHC emissions were 0.046 g/mi with an energy requirement of 14.8 W · h. The test results also indicate that heating time, electrical energy demand, and emissions should decrease even further if a 24-V energy source is used.

Energy Sources for EHCs. Recent advances made by developers of advanced power sources such as ultracapacitors and sealed bipolar lead acid (SBLA) batteries have been encouraging. Idaho National Engineering Laboratory (INEL) has provided the CARB with test data from a Panasonic carbon-based ultracapacitor and a mixed-metal oxide ultracapacitor. The data indicate that Panasonic's ultracapacitor is capable of surviving 350,000 EHC charge-discharge cycles with minimal performance losses. Tests conducted with a Camet EHC indicated that the Panasonic ultracapacitor could easily provide the energy needed to achieve a catalyst temperature well above light-off within 10 s. The ultracapacitor assembly weighs less than 4.536 kg and is similar in size to a conventional automotive battery. The mixed-metal oxide ultracapacitor appears to have the potential for even better performance than the Panasonic device and should be much smaller.

Other organizations have also recognized the potential of the EHC market and are conducting research and development on advanced energy sources for EHCs. A large battery manufacturer has been developing a deep-discharge lead-acid battery intended specifically for EHC applications. This battery has been successfully life cycle tested and is claimed to be lighter and smaller than conventional lead–acid batteries. The CARB has also been working with two other organizations to develop an SBLA battery; one of these organizations is also developing the mixed metal oxide ultracapacitor referenced above for installation on one of the CARB's test vehicles.

CARB believes that current leakage from ultracapacitors is not a problem. Improvements in contamination control have curtailed current leakage to insignificant levels. INEL has tested ultracapacitors with current leakage in the microampere to milliampere range. Also, it is well recognized in this industry that ultracapacitor current leakage can be virtually eliminated by simply maintaining a potential difference across the ultracapacitor, eg, by using the vehicle battery.

Other Strategies. While CARB has identified several technologies to achieve the low emission vehicle standards, it emphasizes that the regulations contain sufficient flexibility to accommodate nearly any strategy that can be used to reduce emissions. One viable strategy for achieving LEV and ULEV emission levels is the use of bypass start catalysts. In such a system, a small, quick light-off catalyst is mounted close to the exhaust manifold upstream of the main catalyst. During cold starts, exhaust emissions are routed to the start catalyst, thus bypassing the main catalyst. Because the start catalyst is rapidly heated to light-off temperature, it treats the exhaust gases almost immediately. The exhaust gases are routed back to the main catalyst once it has reached a sufficient operating temperature.

Bypass start catalysts should have quick light-off characteristics similar to EHCs but without the power requirements of EHCs or any fuel economy penalty. Also, because the start catalyst is used only for a short period of time during cold starts, it should be fairly durable despite its close-coupled position. Therefore, although the CARB staff believes that EHCs will continue to be more effective in the long-term, bypass start catalysts appear to be a viable strategy for manufacturers that consider EHCs to be undesirable.

In addition, there are a number of other strategies that may be used to produce a LEV or ULEV. Some of the options being explored by the automotive industry include direct injection two-stroke engines, hydrocarbon vapor traps, exhaust gas ignition, and hybrid electric vehicles. EPA has completed its initial evaluation of the direct injection two-stroke engine with promising results. One vehicle, a Ford Fiesta, had average composite FTP emission levels for all testing conducted at EPA of 0.05/g/mi nonmethane hydrocarbons, 0.2 g/mi CO, and 0.2 g/mi NO_x. In addition, average combined fuel economy was 50.2 mpg. The vehicle is equipped with a single close coupled oxidation catalyst. The CO levels and fuel economy values espe-

Table 11. Corning EHC Test Results (June 8, 1992)

Description	Vehicle	Watt · Hours	NMHC
12-V battery, 25 s preheat	Honda Accord	14.8	0.022
24-V battery, 5 s preheat	Honda Accord	12.0	0.019
24-V battery, 25 s postheat	Honda Accord	11.6	0.027
12-V battery, 25 s postheat	Honda Accord	14.8	0.046

cially stand out relative to other clean, efficient production vehicles tested by EPA.

The hydrocarbon trap or molecular sieve is a zeolite-type material that can adsorb hydrocarbons at low temperatures, and then release them when heated to higher temperatures. It has been suggested that such materials can be used to adsorb cold-start HC before catalyst light off, and then release the HC after the catalyst is operating at high efficiency. Such a material is more energy efficient than an EHC system, and is potentially cheaper as it may not require an electrical heating and control system. Unfortunately, zeolites are temperature sensitive and little is known about their durability. In addition, research on these materials has been kept highly confidential, and no data are publicly available to gauge their efficiency or durability.

The California Air Resources Board has projected that alternative fuel (CNG and methanol) vehicles will be able to meet the LEV standards with less additional technology than gasoline-fueled vehicles. In fact, the low vapor pressure of methanol requires more cold-start enrichment and presents greater difficulty in controlling cold-start related emissions. Cold-start related formaldehyde can also be an obstacle (as it has a higher reactivity index) in meeting the reactivity weighted HC standards. However, recent tests on M85 vehicles with an electrically heated catalyst have been encouraging, as aldehyde emissions on the FTP were less than 5 mg/mi. CARB staff also stated that recent tests on a M85 vehicle at low mileage showed the potential to meet even the ULEV standards with a start catalyst–main catalyst system without electrical heat. Methanol-fueled vehicles emit a range of oxygenated compounds that are improperly measured by current systems used to measure HC emissions from gasoline vehicles. It is not clear if the measurements of HC from the M85 vehicles cited by CARB staff accounted for all of oxygenated HC emissions.

Conclusions Regarding Gasoline-fueled Technology

Starting in 1975, when HC and CO standards were tightened as a result of the 1970 Clean Air Act Amendments, the first generation oxidation catalysts were introduced on a vast majority of all new cars in the United States. By 1981, when the NO_x standard was tightened, the second-generation systems, three-way catalysts took over much of the market. As a result of the Clean Air Act Amendments of 1990 as well as the California decisions of 1989, the third-generation systems, with greater durability will begin to be introduced in 1993. As a result of the California decisions of September 1990 (coupled with Section 177 of the Clean Air Act, which allows other states to adopt the California program, a fourth generation will likely begin to be introduced by the mid to late 1990s, one based on either very rapid light off or even preheating.

FACTORS AFFECTING DIESEL EMISSIONS

Diesel engine emissions are determined by the combustion process. This process is central to the operation of the diesel engine. As opposed to Otto-cycle engines (which use a more or less homogeneous charge) all diesel engines rely on heterogeneous combustion. During the compression stroke, a diesel engine compresses only air. The process of compression heats the air to about 700° to 900°C, which is well above the self-ignition temperature of diesel fuel. Near the top of the compression stroke, liquid fuel is injected into the combustion chamber under tremendous pressure, through a number of small orifices in the tip of the injection nozzle. The quantity of fuel injected with each stroke determines the engine power output.

The high pressure injection atomizes the fuel. As the atomized fuel is injected into the chamber, the periphery of each jet mixes with the hot air already present. After a brief period known as the ignition delay, this fuel–air mixture ignites. In the premixed burning phase, the fuel–air mixture formed during the ignition delay period burns very rapidly, causing a rapid rise in cylinder pressure. The subsequent rate of burning is controlled by the rate of mixing between the remaining fuel and air, with combustion always occurring at the interface between the two. Most of the fuel injected is burned in this diffusion burning phase, except under light loads.

A mixture of fuel and exactly as much air as is required to burn the fuel completely is called a "stoichiometric mixture." The air : fuel ratio is defined as the ratio of the actual amount of air present per unit of fuel to the stoichiometric amount. In diesel engines, the fact that fuel and air must mix before burning means that a substantial amount of excess air is needed to ensure complete combustion of the fuel within the limited time allowed by the power stroke. Diesel engines, therefore, always operate with overall air : fuel ratios which are considerably lean of stoichiometric.

The air : fuel ratio in the cylinder during a given combustion cycle is determined by the engine power requirements, which govern the amount of fuel injected. Diesels operate without throttling, so that the amount of air present in the cylinder is essentially independent of power output, except in turbocharged engines. The minimum equivalence ratio for complete combustion is about 1.5. This ratio is known as the smoke limit, because smoke increases dramatically at equivalence ratios lower than this. The smoke limit establishes the maximum amount of fuel that can be burned per stroke and thus the maximum power output of the engine. This minimum air : fuel ratio explains why NO_x reduction catalysts of the type used on gasoline automobiles, which rely on stoichiometric air : fuel ratios, are not effective for diesel engines.

Pollutant Formation

The principal pollutants emitted by diesel engines are oxides of nitrogen, (NO_x), sulfur oxides (SO_x), particulate matter, and unburned hydrocarbons. Diesels are also responsible for a small amount of CO as well as visible smoke, unpleasant odors, and noise. In addition, like all engines using hydrocarbon fuel, diesels emit significant amounts of CO_2, which has been implicated in the so-called greenhouse effect. With thermal efficiencies typically in excess of 40%, however, diesels are the most fuel efficient of all common types of combustion engines. As a result, they emit less CO_2 to the atmosphere than any other type of engine doing the same work.

The NO_x, HC, and most of the particulate emissions from diesels are formed during the combustion process and can be reduced by appropriate modifications to that process, as can most of the unregulated pollutants. The sulfur oxides, in contrast, are derived directly from sulfur in the fuel, and the only feasible control technology is to reduce fuel sulfur content. Most SO_x is emitted as gaseous SO_2, but a small fraction (typically 2–4%) occurs in the form of particulate sulfates.

Diesel particulate matter consists mostly of three components: soot formed during combustion, heavy hydrocarbons condensed or adsorbed on the soot, and sulfates. In older diesels, soot was typically 40–80% of the total particulate mass. Developments in in-cylinder emissions control have reduced the soot contribution to particulate emissions from modern emission controlled engines considerably, however. Most of the remaining particulate mass consists of heavy hydrocarbons adsorbed or condensed on the soot. This is referred to as the soluble organic fraction (SOF) of the particulate matter. The SOF is derived partly from the lubricating oil, partly from unburned fuel, and partly from compounds formed during combustion. The relative importance of each of these sources varies from engine to engine.

In-cylinder emission control techniques have been most effective in reducing the soot and fuel derived SOF components of the particulate matter. As a result, the relative importance of the lube oil and sulfate components has increased. In the emission-controlled engines under development, the lubricating oil accounts for as much as 40% of the particulate matter, and the sulfates may account for another 25%. Lube oil emissions can be reduced by reducing oil consumption, but this may adversely affect engine durability. The only known way to reduce sulfate emissions is to reduce the sulfur content of diesel fuel.

The gaseous hydrocarbons and the SOF component of the particulate matter emitted by diesel engines include many known or suspected carcinogens and other toxic air contaminants. These include polynuclear aromatic compounds (PNA) and nitro-PNA, formaldehyde and other aldehydes, and other oxygenated hydrocarbons. The oxygenated hydrocarbons are also responsible for much of the characteristic diesel odor.

Oxides of nitrogen (NO_x) from diesels is primarily NO_2. This gas forms from nitrogen and free oxygen at high temperatures close to the flame front. The rate of NO_x formation in diesels is a function of oxygen availability, and is exponentially dependent on the flame temperature. In the diffusion burning phase, flame temperature depends only on the heating value of the fuel, the heat capacity of the reaction products and any inert gases presenting temperature of the initial mixture. In the premixed burning stage, the local fuel : air ratio also affects the flame temperature, but this ratio varies from place to place in the cylinder and is very hard to control.

In diesel engines, most of the NO_x emitted is formed early in the combustion process, when the piston is still near top dead center (TDC). This is when the temperature and pressure of the charge are greatest. Recent work by several researchers (3,4) indicates that most NO_x is actually formed during the premixed burning phase. It has been found that reducing the amount of fuel burned in this phase can significantly reduce NO_x emissions.

NO_x can also be reduced by actions that reduce the flame temperature during combustion. These actions include delaying combustion past TDC, cooling the air charge going into the cylinder, reducing the air : fuel mixing rate near TDC, and exhaust gas recirculation (EGR). Because combustion always occurs under near stoichiometric conditions, reducing the flame temperature by "lean burn" techniques, as in spark-ignition engines, is impractical.

Particulate matter (or soot) is formed only during the diffusion burning phase of combustion. Primary soot particles are small spheres of graphitic carbon, approximately 0.01 mm in diameter. These are formed by the rapid polymerization of acetylene at moderately high temperatures under oxygen deficient conditions. The primary particles then agglomerate to form chains and clusters of linked particles, giving the soot its characteristic fluffy appearance. During the diffusion burning phase, the local gas composition at the flame front is close to stoichiometric, with an oxygen rich region on one side and a fuel rich region on the other. The moderately high temperatures and excess fuel required for soot formation are thus always present.

Most of the soot formed during combustion is subsequently burned during the later portions of the expansion stroke. Typically, less than 10% of the soot formed in the cylinder survives to be emitted into the atmosphere. Soot oxidation is much slower than soot formation, however, and the amount of soot oxidized is heavily dependent on the availability of high temperatures and adequate oxygen during the later stages of combustion. Conditions which reduce the availability of oxygen (such as poor mixing, or operation at low air : fuel ratios), or which reduce the time available for soot oxidation for soot oxidation (such as retarding the combustion timing) tend to increase soot emissions.

The SOF component of diesel particulate matter consists of heavy hydrocarbons condensed or adsorbed on the soot. A significant part of this material is unburned lubricating oil, which is vaporized from the cylinder walls by the hot gases during the power stroke. Some of the heavier hydrocarbons in the fuel may also come through unburned, and condense on the soot particles. Finally, heavier hydrocarbons may be synthesized during combustion, possibly by the same types of processes that produce soot. Pyrosynthesis of polynuclear aromatic hydrocarbons during diesel combustion has been demonstrated in Ref. 5.

Diesel hydrocarbon emissions (as well as the unburned fuel portions of the particulate SOF) occur primarily at light loads. They are caused by excessive fuel : air mixing, which results in some volumes of air : fuel mixture that are too lean to burn. Other HC sources include fuel deposited on the combustion chamber walls or in combustion chamber crevices by the injection process, fuel retained in the orifices of the injector that vaporizes late in the combustion cycle, and partly reacted mixture that is subjected to bulk quenching by too rapid mixing with air. Aldehydes (as partially reacted hydrocarbons) and the small amount of CO produced by diesels are probably formed in the same processes as the HC emissions.

The presence of polynuclear aromatic hydrocarbons and their nitro derivatives in diesel exhaust is of special concern, because these compounds include many known mutagens and suspected carcinogens. A significant portion of these compounds (especially the smaller two- and three-ring compounds) are apparently derived directly from the fuel. Typical diesel fuel contains several percent PNA by volume. Most of the larger and more dangerous PNAs, on the other hand, appear to form during the combustion process, possibly via the same acetylene polymerization reaction that produces soot (6). Indeed, the soot particle itself can be viewed as essentially a large PNA molecule.

Visible smoke is caused primarily by the soot component of diesel particulate matter. Under most operating conditions, the exhaust plume from a properly adjusted diesel engine is normally invisible, with a total opacity (absorbance and reflectance) of 2% or less. Visible smoke emissions from heavy-duty diesels are generally the result of operating at air : fuel ratios at or below the smoke limit or to poor fuel–air mixing in the cylinder. These conditions can be prevented by proper design. The particulate reductions required to comply with the U.S. 1991 emissions standards essentially eliminate visible smoke emissions from properly functioning engines.

Noise from diesel engines is due principally to the rapid combustion (and resulting rapid pressure rise) in the cylinder during the premixed burning phase. The greater the ignition delay, and the more fuel is premixed with the air, the greater this pressure rise and the resulting noise emissions will be. Noise emissions and NO_x emissions thus tend to be related reducing the amount of fuel burned in the premixed burning phase will tend to reduce both. Other noise sources include those common to all engines, such as mechanical vibration, fan noise, and so forth. These can be minimized by appropriate mechanical design.

Odor characteristic of diesel engines is believed to be due primarily to partially oxygenated hydrocarbons (aldehydes and similar species) in the exhaust. These are believed to be due primarily to slow oxidation reactions in volumes of air : fuel mixture too lean to burn normally. Unburned aromatic hydrocarbons may also play a significant role. The most significant aldehyde species are benzaldehyde, acetaldehyde, and formaldehyde, but other aldehydes such as acrolein (a powerful irritant) are significant as well. Aldehyde and odor emissions are closely linked to total HC emissions experience has shown that modifications which reduce total HC tend to reduce aldehydes and odor as well.

Influence of Engine Variables

The engine variables having the greatest effects on diesel emission rates are the air : fuel ratio, rate of air–fuel mixing, fuel injection timing, compression ratio, and the temperature and composition of the charge in the cylinder. Most techniques for in-cylinder emission control involve manipulating one or more of these variables.

Air : Fuel Ratio. The ratio of air to fuel in the combustion chamber has an extremely important effect on emission rates for hydrocarbons and particulate matter. As discussed above, the power output of the engine is determined by the amount of fuel injected at the beginning of each power stroke. At high air : fuel ratios (corresponding to light load), the temperature in the cylinder after combustion is too low to burn out residual hydrocarbons, so emissions of gaseous HC and particulate SOF are high. At lower air : fuel ratios, less oxygen is available for soot oxidation, so soot emissions increase. As long as the equivalence ratio remains above about 1.6, this increase is relatively gradual. Soot and visible smoke emissions in a direct injection diesel engine show a strong nonlinear increase below the smoke limit at an air : fuel ratio of about 1.5, however.

In naturally aspirated engines (those without a turbocharger), the amount of air in the cylinder is independent of the power output. Maximum power output for these engines is normally smoke limited, ie, limited by the amount of fuel that can be injected without exceeding the smoke limit. Maximum fuel settings on these engines represent a compromise between smoke emissions and power output. Where diesel smoke is regulated, this compromise must result in smoke opacity below the regulated limit. Otherwise, opacity is limited by the manufacturer's judgment of commercially acceptable smoke emissions.

In turbo-charged engines, increasing the fuel injected per stroke increases the energy in the exhaust gas, causing the turbo charger to spin more rapidly and pump more air into the combustion chamber. For this reason, power output from turbo charged engines is not usually smoke limited, although this depends on the rated power of the engine. Instead, it is limited by design limits on variables such as turbocharger speed and engine mechanical stresses. However, turbo-charged engines are smoke limited at low speeds.

Turbo-charged engines do not normally experience low air : fuel ratios during steady-state operation. Low air : fuel ratios can occur during transient accelerations, because the inertia of the turbo charger rotor keeps it from responding instantly to an increase in fuel input. Thus the air supply during the first few seconds of a full power acceleration is less than the air supply in steady-state operation. To overcome this problem, turbo-charged engines in highway vehicles commonly incorporate an acceleration smoke limiter. This device limits the fuel flow to the engine until the turbocharger has time to respond. The setting on this device must compromise between acceleration performance driveability and low smoke emissions. In the United States acceleration smoke emissions are limited by regulation; elsewhere, they are limited by the manufacturer's judgment of commercial acceptability.

Air–Fuel Mixing. The rate of mixing between the compressed charge in the cylinder and the injected fuel is among the most important factors in determining diesel performance and emissions. During the ignition delay period, the mixing rate determines the amount of fuel that mixes with the air and is thus available for combustion during the premixed burning phase. The higher the mixing rate, the greater the amount of fuel burning in premixed mode, and the higher the noise and NO_x emissions will tend to be.

In the diffusion burning phase, the rate of combustion is limited by the rate at which air and fuel can mix. The more rapid and complete this mixing, the greater the amount of fuel that burns near piston top dead center, the higher the fuel efficiency, and the lower the soot emissions. Too rapid mixing, however, can increase hydrocarbon emissions, especially at light loads as small volumes of air : fuel mixture are diluted below the combustible level before they have a chance to burn. Unnecessarily intense mixing also dissipates energy through turbulence, increasing fuel consumption.

In engine design practice, it is necessary to strike a balance between the rapid and complete mixing required for low soot emissions and best fuel economy, and too rapid mixing leading to high NO_x and HC emissions. The primary factors affecting the mixing rate are the fuel injection pressure, the number and size of injection orifices, any swirling motion impared to the air as it enters the cylinder during the intake stroke, and air motions generated by combustion chamber geometry during compression. Much of the progress in in-cylinder emissions control over the last decade has come through improved understanding of the interactions between these different variables and emissions, leading to improved designs.

Air–fuel mixing rates in present emission controlled engines are the product of extensive optimization to assure rapid and complete mixing under nearly all operating conditions. Poor mixing may still occur during "lug down" high torque operation at low engine speeds. Turbocharger boost, air swirl level, and fuel injection pressure are typically poorer in these "off design" conditions. Maintenance problems such as injector tip deposits can also degrade air–fuel mixing, and result in greatly increased emissions.

Injection Timing. The timing relationship between the beginning of fuel injection and the top of the compression stroke has an important effect on diesel engine emissions and fuel economy. For best fuel economy, it is preferable that combustion begin at or somewhat before top dead center. Because there is a finite delay between the beginning of injection and the start of combustion, it is necessary to inject the fuel somewhat before this point (generally 5 to 15° of crankshaft rotation before, although this could be 10 to 25° for engines in Europe).

Because fuel is injected before the piston reaches top dead center, the charge temperature is still increasing as the charge is compressed. The earlier fuel is injected, the cooler the charge will be, and the longer the ignition delay. The longer ignition delay provides more time for air and fuel to mix, increasing the amount of fuel that burns in the premixed combustion phase. In addition, more fuel burning at or just before top dead center increases the maximum temperature and pressure attained in the cylinder. Both of these effects tend to increase NO_x emissions.

On the other hand, earlier injection timing tends to reduce particulate and light load HC emissions. Fuel burning in premixed combustion forms little soot, while the soot formed in diffusion combustion near TDC experiences a relatively long period of high temperatures and intense mixing and is thus mostly oxidized. The end of injection timing also has a major effect on soot emissions; fuel injected more than a few degrees after TDC burns more slowly, and at a lower temperature, so that less of the resulting soot has time to oxidize during the power stroke. For a fixed injection pressure, orifice size, and fuel quantity, the end of injection timing is determined by the timing of the beginning of injection.

The result of these effects is that injection timing must compromise between PM emissions and fuel economy on the one hand and noise, NO_x emissions, and maximum cylinder pressure on the other. The terms of the compromise can be improved to a considerable extent by increasing injection pressure, which increases mixing and advances the end of injection timing. Another approach under development is split injection, in which a small amount of fuel is injected early to ignite the main fuel quantity, which is injected near TDC.

Compared with uncontrolled engines, modern emission controlled engines generally exhibit moderately retarded timing to reduce NO_x in conjunction with high injection pressures to limit the effects of retarded timing of PM emissions and fuel economy. The response of fuel economy and PM emissions to retarded timing is not linear; up to a point, the effects are relatively small, but beyond that point deterioration is rapid. Great precision in injection timing is necessary; a change of one degree crank angle can have a significant impact on emissions. The optimal injection timing is a complex function of engine design, engine speed and load, and the relative stringency of emissions standards for the different pollutants. To attain the required flexibility and precision of injection timing has posed a principal challenge to fuel injection manufacturers.

Compression Ratio. Diesel engines rely on compression heating to ignite the fuel, so the engine's compression ratio has an important effect on combustion. A higher compression ratio results in a higher temperature for the compressed charge and thus in a shorter ignition delay is to reduce NO_x emissions, while the higher flame temperature would be expected to increase them. In practice, these two effects nearly cancel, so that changes in compression ratio have little effect on NO_x.

Emissions of gaseous HC and of the SOF fraction of the particulate matter are reduced at higher compression ratios, as the higher cylinder temperature increases the burnout of hydrocarbons. Soot emissions may increase at higher compression ratios, however. Because the higher compression is achieved by reducing the volume of the combustion chamber, this results in a larger fraction of the air charge being sequestered in "crevice volumes" such as the top and sides of the piston, where it is not available for combustion early in the power stroke. Thus the effective air : fuel ratio in the combustion chamber decreases, and soot emissions go up. This effect can be limited (and overall air utilization and power output improved) by reducing crevice volumes to the maximum extent possible.

Engine fuel economy, cold starting, and maximum cylinder pressures are also affected by the compression ratio. For an idealized diesel cycle, the thermodynamic efficiency is an increasing function of compression ratio. In a real engine, however, the increased thermodynamic efficiency is offset after some point by increasing friction, so that a point of maximum efficiency is reached. With most heavy-duty diesel engine designs, this optimal compression ratio is about 12 to 15. To ensure adequate starting ability under cold conditions, however, most diesel engine designs require a somewhat higher compression ratio; in the range of 15 to 20 or more. Generally, higher speed engines with smaller cylinders require higher compression ratios for adequate cold starting.

Charge Temperature. Reducing the temperature of the air charge going into the cylinder has benefits for both PM and NO_x emissions. Reducing the charge temperature directly reduces the flame temperature during combustion and thus helps to reduce NO_x emissions. In addition, the colder air is denser, so that (at the same pressure) a greater mass of air can be contained in the same fixed cylinder volume. This increases the air:fuel ratio in the cylinder and thus helps to reduce soot emissions. By increasing the air available while decreasing piston temperatures, charge air cooling can also make possible a significant increase in power output. Excessively cold charge air can reduce the burnout of hydrocarbons, and thus increase light load HC emissions, however. This can be counteracted by advancing injection timing, or by reducing charge air cooling at light loads.

Charge Composition. NO_x emissions depend heavily on flame temperature. By altering the composition of the air charge to increase its specific heat and the concentration of inert gases, it is possible to decrease the flame temperature significantly. The most common way of accomplishing this is through exhaust gas recirculation (EGR). At moderate loads, EGR has been shown to be capable of reducing NO_x emissions by a factor of two or more, with little effect on particulate emissions. Although soot emissions are increased by the reduced oxygen concentration, particulate SOF and gaseous HC emissions are reduced, due to the higher in-cylinder temperature caused by the hot exhaust gas. EGR cannot be used at high loads, however, because the displacement of air by exhaust gas would result in an air:fuel ratio below the smoke limit and thus high soot and PM emissions.

Emissions Trade-offs. It is apparent from the foregoing discussion that there is an inherent conflict between some of the most powerful diesel NO_x control techniques and particulate emissions. This is the basis for the much discussed trade-off relationship between diesel NO_x and particulate emissions. This trade-off is not absolute, various NO_x control techniques have varying effects on soot and HC emissions, and the importance of these effects varies as a function of engine speed and load. These tradeoffs do place limits on the extent to which any one of these pollutants can be reduced, however. To minimize emissions of all three pollutants simultaneously requires careful optimization of the fuel injection, fuel–air mixing, and combustion processes over the full range of engine operating conditions.

Influence of Fuel Qualities

The quality and composition of diesel fuel can have important effects on pollutant emissions. The area of fuel effects on diesel emissions has seen a great deal of study in the last few years, and a large amount of new information has become available. These data indicate that the fuel variables having the most important effects on emissions are the sulfur content and the fraction of aromatic hydrocarbons contained in the fuel. Other fuel properties may also affect emissions, but generally to a much lesser extent. A recent extensive study carried out for the Dutch Ministry of Environment for the EEC's Motor Vehicle Emissions Group (MVEG) indicates that the volatility of the diesel fuel (85 or 90% distilled temperatures) is the most important factor followed in importance by sulfur and aromatics content, in that order. The apparent discrepancy with the U.S. findings might be the result of lower volatility of European diesel fuels, compared with those in the United States. The subject is currently being studied. Finally, the use of fuel additives may have a significant impact on emissions.

Sulfur Content. Diesel fuel for highway use normally contains between 0.1 and 0.5% by weight sulfur, although some (mostly less developed) nations permit 1% or even higher sulfur concentrations. Sulfur in diesel fuel contributes to environmental deterioration both directly and indirectly. Most of the sulfur in the fuel burns to SO_2, which is emitted to the atmosphere in the diesel exhaust. Because of this, diesels are significant contributors to ambient SO_2 levels in some areas. This makes them indirect contributors to ambient particulate levels and acid deposition as well. In the United States, diesel fuel accounted for about 629,220 t of SO_2 in 1984, or about 3% of all SO_2 emissions during the same period. For Europe, diesel fuel for road traffic adds approximately 1.6% of total SO_2 emissions, a share that is growing.

Most of the fuel sulfur that is not emitted as SO_2 is converted to various metal sulfates and to sulfuric acid during the combustion process or immediately afterward. Both of these materials are emitted in particulate form. The typical rate of conversion in a heavy-duty diesel engine is about 2 to 3% of the fuel sulfur and about 3 to 5% in a light-duty engine. Even at this rate, sulfate particles typically account for 0.05 to 0.10 g/BHP-h of particulates in heavy-duty engines. In a Dutch program, a sulfur change from 0.28 to 0.07% resulted in a 10 to 15% decrease in particulate emissions. The effect of the sulfate particles is increased by their hygroscopic nature; they tend to absorb significant quantities of water from the air.

Certain precious metal catalysts can oxidize SO_2 to SO_3, which combines with water in the exhaust to form sulfuric acid. The rate of conversion with the catalyst depends on the temperature, space velocity, and oxygen content of the exhaust and on the activity of the catalyst; generally, catalyst formulations that are most effective in oxidizing hydrocarbons and CO are also most effective at oxidizing SO_2. The presence of significant quantities of sulfur in diesel fuel thus limits the potential for catalytic converters or catalytic trap-oxidizers for use in controlling PM and HC emissions.

Sulfur dioxide in the atmosphere oxidizes to form sulfate particles, in a reaction similar to that which occurs with the precious metal catalyst. Viewed in another way, the presence of the catalyst merely speeds up a reaction that would occur anyway (although this can have a significant effect on human exposure to the reaction products). According to analysis by the California Air Resources Board staff (6) roughly 0.54 kg of secondary particulate is formed per kg of SO_2 emitted in the South Coast Air Basin. For a diesel engine burning fuel of 0.29 wt% sulfur at 0.19 kg of fuel per horsepower per hour, this is equivalent to 0.85 g/BHP-h. For comparison, the average rate of primary or directly emitted particulate emissions from heavy-duty engines in use was about 0.8 g/BHP-h (7).

Quite aside from its particulate-forming tendencies, sulfur dioxide is recognized as a hazardous pollutant in its own right. The health and welfare effects of SO_2 emissions from diesel vehicles are probably much greater than those of an equivalent quantity emitted from a utility stack or industrial boiler, because diesel exhaust is emitted close to the ground level in the vicinity of roads, buildings, and concentrations of people.

Volatility. Diesel fuel consists of a mixture of hydrocarbons having different molecular weights and boiling points. As a result, as some of it boils away on heating, the boiling point of the remainder increases. This fact is used to characterize the range of hydrocarbons in the fuel in the form of a "distillation curve" specifying the temperature at which 10%, 20%, etc of the hydrocarbons have boiled away. A low 10% boiling point is associated with a significant content of relatively volatile hydrocarbons. Fuels with this characteristic tend to exhibit somewhat higher HC emissions than others. Formerly, a relatively high 90% boiling point was considered to be associated with higher particulate emissions. More recent studies (8) have shown that this effect is spurious; the apparent statistical linkage was due to the higher sulfur content of these high boiling fuels.

In a Dutch study for the EC MVEG, however, the test fuels were composed of two sets at clearly different 85 or 90% boiling points, among which sulfur content varied independently. A highly significant effect of 85 or 90% boiling point temperatures was found, in addition to a significant effect of sulfur and probably significant effect of aromatics contents. A typical effect of a 20°C change in 85% boiling point is 0.05 g/kW·h at present particulate levels. As mentioned earlier, this may be related to generally higher, 85 to 90 percentage points, that in the test fuels went up to 350 or 360°C. Commercial diesel fuels in Europe show values up to about 370°C.

Aromatic Hydrocarbon Content. Aromatic hydrocarbons are hydrocarbon compounds containing one or more "benzene-like" ring structures. They are distinguished from paraffins and naphthenes, the other principal hydrocarbon constituents of diesel fuel, which lack such structures. Compared to these other components, aromatic hydrocarbons are denser, have poorer self ignition qualities, and produce more soot in burning. Ordinarily, "straight run" diesel fuel produced by simple distillation of crude oil is fairly low in aromatic hydrocarbons. Catalytic cracking of residual oil to increase gasoline and diesel production results in increased aromatic content, however. A typical straight run diesel might contain 20 to 25% aromatics by volume, while a diesel blended from catalytically cracked stocks could have 40–50% aromatics.

Aromatic hydrocarbons have poor self ignition qualities, so that diesel fuels containing a high fraction of aromatics tend to have low cetane numbers. Typical cetane values for straight run diesel are in the range of 50–55; those for highly aromatic diesel fuels are typically 40 to 45, and may be even lower. This produces more difficulty in cold starting, and increased combustion noise, HC, and NO_x due to the increased ignition delay.

Increased aromatic content is also correlated with higher particulate emissions. Aromatic hydrocarbons have a greater tendency to form soot in burning, and the poorer combustion quality also appears to increase particulate SOF emissions. Increased aromatic content may also be correlated with increased SOF mutagenicity, possibly due to increased PNA and nitro-PNA emissions. There is also some evidence that more highly aromatic fuels have a greater tendency to form deposits on fuel injectors and other critical components. Such deposits can interfere with proper fuel/air mixing, greatly increasing PM and HC emissions.

Other Fuel Properties. Other fuel properties may also have an effect on emissions. Fuel density, for instance, may affect the mass of fuel injected into the combustion chamber at full load, and thus the air/fuel ratio. This is because fuel injection pumps meter fuel by volume, not by mass, and the denser fuel contains a greater mass in the same volume. Fuel viscosity can also affect the fuel injection characteristics, and thus the mixing rate. The corrosiveness, cleanliness, and lubricating properties of the fuel can all affect the service life of the fuel injection equipment—possibly contributing to excessive in-use emissions if the equipment is worn out prematurely.

Fuel Additives. Several generic types of diesel fuel additives can have a significant effect on emissions. These include cetane enhancers, smoke suppressants, and detergent additives. In addition, some additive research has been directed specifically at emissions reduction in recent years. Although some moderate emission reductions have been demonstrated, there is yet no consensus on the widespread applicability or desirability of such products.

Cetane enhancers are used to enhance the self ignition qualities of diesel fuel. These compounds (generally organic nitrates) are generally added to reduce the adverse impact of high aromatic fuels on cold starting and combustion noise. These compounds also appear to reduce the aromatic hydrocarbons' adverse impacts on HC and PM emissions, although PM emissions with the cetane improver are generally still somewhat higher than those from a higher quality fuel able to attain the same cetane rating without the additive. In the MVEG study cited earlier, no significant effect of ashless cetane improving additives could be detected on NO_x or particulates.

Smoke suppressing additives are organic compounds of calcium, barium, or (sometimes) magnesium. Added to diesel fuel, these compounds inhibit soot formation during the combustion process, and thus greatly reduce emissions of visible smoke. Their effects on the particulate SOF are not fully documented, but one study (9) has shown a significant increase in the PAH content and mutagenicity of the SOF with a barium additive. Particulate sulfate emissions are greatly increased with these additives, since all of them readily form stable solid metal sulfates, which are emitted in the exhaust. The overall effect of reducing soot and increasing metal sulfate emissions may be either an increase or decrease in the total particulate mass, depending on the soot emissions level at the beginning and the amount of additive used.

Detergent additives (often packaged in combination with a cetane enhancer) help to prevent and remove coke deposits on fuel injector tips and other vulnerable locations. By thus maintaining new engine injection and mixing characteristics, these deposits can help to decrease in-use PM and HC emissions. A study for the California Air Resources Board (10) estimated the increase in PM emissions due to fuel injector problems from trucks in use as being more than 50% of new-vehicle emissions levels. A significant fraction of this excess is unquestionably due to fuel injector deposits.

CONTROL TECHNOLOGIES FOR DIESEL FUELED VEHICLES

Diesel engine emissions are determined by the characteristics of the combustion process within each cylinder. Primary engine parameters affecting diesel emissions are the fuel injection system, the engine control system, the air intake port and combustion chamber design, and the air charging system. Actions to reduce lubricating oil consumption can also impact HC and PM emissions. Further, beyond the engine itself, exhaust aftertreatment systems such as trap oxidizers and catalytic converters can play a significant role. Finally, modifications to conventional fuels as well as alternative fuels can impact emissions. The following sections will review the status of each of these areas.

Engine Modifications

Air Motion and Combustion Chamber Design. The geometries of the combustion chamber and the air intake port control the air motion in the diesel combustion chamber, and thus play an important role in air/fuel mixing and emissions. A number of different combustion chamber designs, corresponding to different basic combustion systems, are in use in heavy duty diesel engines at present. This section outlines the basic consumption systems in use, their advantages and disadvantages, and the effects of changes in combustion chamber design and air motion on emissions.

Combustion Systems. Diesel engines used in heavy duty vehicles use several different types of combustion systems. The most fundamental difference is between direct injection (DI) engines and indirect injection (IDI) engines. In an indirect injection engine, fuel is injected into a separate "prechamber," where it mixes and partly burns before jetting into the main combustion chamber above the piston. In the more common direct injection engine, fuel is injected directly into a combustion chamber formed out of the top of the piston. DI engines can be further divided into high swirl and low swirl.

Fuel/air mixing in the direct injection engine is limited by the fuel injection pressure and any motion imparted to the air in the chamber as it enters. In high swirl DI engines, a strong swirling motion is imparted to the air entering the combustion chamber by the design of the intake port. These engines typically use moderate to high injection pressures, and three to five spray holes per nozzle. Low swirl engines rely primarily on the fuel injection process to supply the mixing. They typically have very high fuel injection pressures and six to nine spray holes per nozzle.

In the indirect injection engine, much of the fuel/air mixing is due to the air swirl induced in the prechamber as air is forced into it during compression, and to the turbulence induced by the expansion out of the prechamber during combustion. These engines typically have better high speed performance than direct injected engines, and can use cheaper fuel injection systems. Historically, IDI diesel engines have also exhibited lower emission levels, especially NO_x, than DI engines but, with recent developments in DI engine emission controls, this is no longer the case. The main disadvantage of the IDI engine is that the extra heat and frictional losses due to the prechamber result in a 5-10 percent reduction in fuel efficiency compared to a DI engine.

A number of advanced, low emitting fuel efficient high swirl DI engines have recently been introduced and it appears that these engines will completely displace the existing IDI designs.

DI Combustion Chamber Design. Changes in the engine combustion chamber and related areas have demonstrated a major potential for emission control. Design changes to reduce the crevice volume for DI diesel cylinders increase the amount of air available in the combustion chamber. Changes in combustion chamber geometry—such as the use of a reentrant lip on the piston bowl—can markedly reduce emissions by improving air/fuel mixing and minimizing wall impingement by the fuel jet. Optimizing the intake port shape for best swirl characteristics has also yielded significant benefits. Several manufacturers are considering variable swirl intake ports, to optimize swirl characteristics across a broader range of engine speeds.

Fuel Injection. The fuel injection system, one of the most important components in a diesel engine, includes the process by which the fuel is transferred from the fuel tank to the engine, and the mechanism by which it is injected into the cylinders. The precision, characteristics and timing of the fuel injection determine the engine's power, fuel economy, and emissions characteristics.

The important areas of concentration in fuel injection system development have been on increased injection pressure, increasingly flexible control of injection timing, and more precise governing of the fuel quantity injected. Some manufacturers are also pursuing technology to vary the rate of fuel injection over the injection period, in order to reduce the amount of fuel burning in the premixed combustion phase. Reductions in NO_x and noise emissions and maximum cylinder pressures have been demonstrated using this approach. Systems offering electronic control of these quantities, as well as fuel injection rate, have been introduced. Other changes have been made to the injection nozzles themselves, to reduce or eliminate sac volume and to optimize the nozzle hole size and shape, number of holes, and spray angle for minimum emissions.

High fuel injection pressures are desirable in order to improve fuel atomization and fuel/air mixing, and to offset the effects of retarded injection timing by increasing the injection rate. It is well established that higher injection pressures reduce PM and/or smoke emissions. High injection pressures are most important in low swirl, direct injection engines, since the fuel injection system is responsible for most of the fuel/air mixing in these systems. For this reason, low swirl engines tend to use unit injector systems, which can achieve peak injection pressures in excess of 1,500 bar.

Engine Control Systems. Traditionally, diesel engine control systems have been closely integrated with the fuel injection system, and the two systems are often discussed together. These earlier control systems (still in use on most engines) are entirely mechanical. The last few years have seen the introduction of an increasing number of computerized electronic control systems for diesel engines. With the introduction of these systems, the scope of the engine control system has been greatly expanded.

The advent of computerized electronic engine control systems has greatly increased the potential flexibility and precision of fuel metering and injection timing controls. In addition, it has made possible whole new classes of control functions, such as road speed governing, alterations in control strategy during transients, synchronous idle speed control, and adaptive learning—including strategies to identify and compensate for the effects of wear and component variation in the fuel injection system.

By continuously adjusting the fuel injection timing to match a stored "map" of optimal timing vs. speed and load, an electronic timing control system can significantly improve on the NO_x/particulate and NO_x/fuel economy tradeoffs possible with static or mechanically variable injection timing. Most electronic control systems also incorporate the functions of the engine governor and the transient smoke limiter. This helps to reduce excess particulate emissions due to mechanical friction and lag time during engine transients, while simultaneously improving engine performance. Potential reductions in PM emissions of up to 40% have been documented with this approach.

Turbocharging and Intercooling. A turbocharger consists of a centrifugal air compressor feeding the intake manifold, mounted on the same shaft as an exhaust gas turbine in the exhaust stream. By increasing the mass of air in the cylinder prior to compression, turbocharging correspondingly increases the amount of fuel that can be burned without excessive smoke, and thus increases the potential maximum power output. The fuel efficiency of the engine is improved as well. The process of compressing the air, however, increases its temperature, increasing the thermal load on critical engine components. By cooling the compressed air in an intercooler before it enters the cylinder, the adverse thermal effects can be reduced. This also increases the density of the air, allowing an even greater mass of air to be confined within the cylinder, and thus further increasing the maximum power potential.

Increasing the air mass in the cylinder and reducing its temperature can reduce both NO_x and particulate emissions as well as increase fuel economy and power output from a given engine displacement. Most heavy duty diesel engines are presently equipped with turbochargers, and most of these have intercoolers. In the U.S., virtually all engines will be equipped with these systems by 1991. Recent developments in air changing systems for diesel engines have been primarily concerned with increasing the turbocharger efficiency, operating range, and transient response characteristics; and with improved intercoolers to reduce the temperature of the intake charge further. Tuned intake air manifolds (including some with variable tuning) have also been developed, to maximize air intake efficiency in a given speed range.

Lubricating Oil Control. A significant fraction of diesel particulate matter consists of oil derived hydrocarbons and related solid matter; estimates range from 10 to 50%. Reduced oil consumption has been a design goal of heavy duty diesel engine manufacturers for some time, and the current generation of diesel engines already uses fairly little oil compared to their predecessors. Further reductions in oil consumption are possible through careful attention to cylinder bore roundess and surface finish, optimization of piston ring tension and shape, and attention to valve stem seals, turbocharger oil seals, and other possible sources of oil loss. Some oil consumption in the cylinder is required with present technology, however, in order for the oil to perform its lubricating and corrosion protective functions.

Aftertreatment Systems

Trap Oxidizers. A trap oxidizer system consists of a durable particulate filter (the "trap") positioned in the engine exhaust stream, along with some means for cleaning the filter by burning off ("oxidizing") the collected particulate matter. The construction of a filter capable of collecting diesel soot and other particulate matter from the exhaust stream is a straightforward task, and a number of effective trapping media have been developed and demonstrated. The most challenging problem of trap oxidizer system development has been with the process of "regenerating" the filter by burning off the accumulated particulate matter.

Diesel particulate matter consists primarily of a mixture of solid carbon coated with heavy hydrocarbons. The ignition temperature of this mixture is about 500–600°C, which is above the normal range of diesel engine exhaust temperatures. Thus, special means are needed to assure regeneration. Once ignited, however, this material burns to produce very high temperatures, which can easily melt or crack the particulate filter. Initiating and controlling the regeneration process to ensure reliable regeneration without damage to the trap is the central engineering problem of trap oxidizer development.

Traps. Presently, most of the trap oxidizer sytems under development are based on the cellular cordierite ceramic monolith trap. These traps can be formulated to be highly efficient (collecting essentially all of the soot, and a large fraction of the particulate SOF), and they are relatively compact, having a large surface area per unit of volume. They can also be coated or impregnated with catalyst material to assist regeneration.

The high concentration of soot per unit of volume with the ceramic monolith, however, makes these traps sensitive to regeneration conditions. Trap loading, temperature, and gas flow rates must be maintained within a fairly narrow window. Otherwise, the trap fails to regenerate fully, or cracks or melts due to overheating.

An alternative trap technology is provided by ceramic fiber coils. These traps are composed of a number of individual filtering elements, each of which consists of a number of thicknesses of silica fiber yarn wound on a punched metal support. A number of these filtering elements are suspended inside a large metal can to make up a trap.

Numerous other trapping media have been tested or proposed, including ceramic foams, corrugated mullite fiber felts, and catalyst coated stainless steel wire mesh.

Regeneration. Numerous techniques for regenerating particulate trap oxidizers have been proposed, and a great deal of development work has been invested in many of these. These approaches can generally be divided into two groups: passive systems and active systems. Passive systems must attain the conditions required for regeneration during normal operation of the vehicle. The most promising approaches require the use of a catalyst (either as a coating on the trap or as a fuel additive) in order to reduce the ignition temperature of the collected particulate matter. Regeneration temperatures as low as 420°C have been reported with catalytic coatings, and even lower temperatures are achievable with fuel additives.

Active systems, on the other hand, monitor the buildup of particulate matter in the trap and trigger specific actions intended to regenerate it when needed. A wide variety of approaches to triggering regeneration have been proposed, from diesel fuel burners and electric heaters to catalyst injection systems.

Passive regeneration systems face special problems on heavy duty vehicles. Exhaust temperatures from heavy duty diesel engines are normally low, and recent developments such as charge air cooling and increased turbo charger efficiency are reducing them still further. Under some conditions, therefore, it would be possible for a truck to drive for many hours without exceeding the exhaust temperature (around 400–450°C) required to trigger regeneration.

Engine and catalyst manufacturers have experimented with a wide variety of catalytic material and treatments to assist in trap regeneration. Good results have been obtained both with precious metals (platinum, palladium, rhodium, silver) and with base metal catalysts such as vanadium and copper. Precious metal catalysts are effective in oxidizing gaseous HC and CO, as well as the particulate SOF, but are relatively ineffective at promoting soot oxidation. Unfortunately, these metals also promote the oxidation of SO_2 to particulate sulfates such as sulfuric acid (H_2SO_4). The base metal catalysts, in contrast, are effective in promoting soot oxidation, but have little effect on HC, CO, NO_x or SO_2. Many experts believe that ultimately precious metal catalysis must be an important element of an effective particulate control system because it specifically attacks the "bad actors."

Catalyst coatings also have a number of advantages in active systems, however. The reduced ignition temperature and increased combustion rate due to the catalyst mean that less energy is needed from the regeneration system. Regeneration will also occur spontaneously under most duty cycles, greatly reducing the number of times the regeneration system must operate. The spontaneous regeneration capability also provides some insurance against a regeneration system failure. Finally, the use of a catalyst may make possible a simpler regeneration system.

Although normal heavy duty diesel exhaust temperatures are not high enough under all operating conditions to provide reliable regeneration for a catalyst coated trap, the exhaust temperature can readily be increased by changes in engine operating parameters. Retarding the injection timing, bypassing the intercooler, throttling the intake air (or cutting back on a variable geometry turbo charger), and/or increasing the EGR rate all markedly increase the exhaust temperature. Applying these measures all the time would seriously degrade fuel economy, engine durability, and performance. The presence of an electronic control system, however, makes it possible to apply them very selectively to regenerate the trap. Since they would be normally needed only at light loads, the effects on durability and performance would be imperceptible.

Fuel additives may play a key role in trap based systems although concerns have been raised about possible toxicity if metallic additives were widely used. Cerium based additives which do not appear to raise these concerns have been found especially promising in recent fleet studies in Athens buses; they were able to lower engine out particulate emissions as well as facilitate regeneration.

Catalytic Converters. Like a catalytic trap, a diesel catalytic converter oxidizes a large part of the hydrocarbon constituents of the SOF, as well as gaseous HC, CO, odor creating compounds, and mutagenic emissions. Unlike a catalytic trap, however, a flow through catalytic converter does not collect any of the solid particulate matter, which simply passes through in the exhaust. This eliminates the need for a regeneration system, with its attendant technical difficulties and costs. The particulate control efficiency of the catalytic converter is, of course, much less than that of a trap. However, a particulate control efficiency of even 25–35% is enough to bring many current development engines within the target range for the U.S. 1994 emissions standard.

Diesel catalytic converters have a number of advantages. First, in addition to reducing particulate emissions, the oxidation catalyst greatly reduces HC, CO, and odor emissions. The catalyst is also very efficient in reducing emissions of gaseous and particle bound toxic air contaminants such as aldehydes, PNA, and nitro-PNA. While a precious metal catalyzed particulate trap would have the same advantages, the catalytic converter is much less complex, bulky, and expensive. Unlike the trap, the catalytic converter has little impact on fuel economy or safety, and it will probably not require replacement as often. Also, unlike the trap oxidizer, the catalytic converter is a relatively mature technology—millions of catalytic converters are in use on gasoline vehicles, and diesel catalytic converters have been used in underground mining applications for more than 20 years.

The disadvantage of the catalytic converter is the same as with the precious metal catalyzed particulate trap: sulfate emissions. The tendency of the precious metal catalyst to convert SO_2 to particulate sulfates requires the use of low sulfur fuel: otherwise, the increase in sulfate emissions would more than counterbalance the decrease in SOF. Also on the road durability has not yet been demonstrated in heavy duty applications for the required 270,000 mile useful life.

NO_x Reduction Techniques

Under appropriate conditions, NO_x can be chemically reduced to form oxygen and nitrogen gases. This process is used in modern closed-loop, three-way catalyst equipped gasoline vehicles to control NO_x emissions. However, the process of catalytic NO_x reduction used on gasoline vehicles is inapplicable to diesels. Because of their heterogeneous combustion process, diesel engines require substantial excess air, and their exhaust thus inherently contains significant excess oxygen. The three-way catalysts used on automobiles require a precise stoichiometric mixture in the exhaust in order to function, in the presence of excess oxygen, their NO_x conversion efficiency rapidly approaches zero.

A number of aftertreatment NO_x reduction techniques which will work in an oxidizing exhaust stream are currently available or under development for stationary pollution sources. These include selective catalytic reduction (SCR), selective noncatalytic reduction (Thermal Denox (tm)), and reaction with cyanuric acid (RapReNox). However, each of these systems requires a continuous supply of some reducing agent such as ammonia or cyanuric acid to react with the NO_x. Because of the need for frequent replenishment of this agent, and the difficulty of ensuring that the replenishment is performed when needed, such systems are considered impractical for vehicular use. Even if the replenishment problems could be resolved, these systems would raise serious questions about crash safety and possible emissions of toxic air contaminants.

Fuel Modifications

Modifications to diesel fuel composition have drawn considerable attention as quick and cost effective means of reducing emissions from existing vehicles. Regulations mandating low sulfur fuel (0.05 wt %) have been promulgated in both Europe and the U.S. In addition to a direct reduction in emissions of SO_2 and sulfate particles, reducing the sulfur content of diesel fuel reduces the indirect formation of sulfate particles from SO_2 in the atmosphere.

A number of well-controlled studies have demonstrated the ability of detergent additives in diesel fuel to prevent and remove injector tip deposits, thus reducing smoke levels. The reduced smoke probably results in reduced PM emissions as well, but this has not been demonstrated as clearly, due to the great expense of PM emissions tests on in use vehicles. Cetane improving additives are also likely to result in some reduction in HC and PM emissions in marginal fuels.

Alternative Fuels

The possibility of substituting cleaner burning alternative fuels for diesel fuel has drawn increasing attention during the last decade. Motivations advanced for this substitution include conservation of oil products and energy security, as well as the reduction or elimination of particulate emissions and visible smoke. Care is necessary in evaluating the air quality claims for alternative fuels, however. While many alternative fuel engines do display greatly reduced particulate and SO_2 emissions, emissions of other gaseous pollutants such as unburned hydrocarbons, CO, and in some cases NO_x and aldehydes may be much higher than from diesels.

The principal alternative fuels presently under consideration are natural gas and methanol made from natural gas, and in limited applications, LPG.

Natural Gas. Natural gas has many desirable qualities as an alternative to diesel fuel in heavy duty vehicles. Clean burning, cheap, and abundant in many parts of the world, it already plays a significant vehicular role in a number of countries. The major disadvantage of natural gas as a motor fuel is its gaseous form at normal temperatures.

Pipeline quality natural gas is a mixture of several different gases but the primary constituent is methane, which typically makes up 90–95% of the total volume. Methane is a nearly ideal fuel for Otto cycle (spark ignition) engines. As a gas under normal conditions, it mixes readily with air in any proportion, eliminating cold start problems and the need for cold start enrichment. In its purest form, it is flammable over a fairly wide range of air fuel ratios. With a research octane number of 130 (the highest of any commonly used fuel), it can be used with engine compression ratios as high as 15:1 (compared to 8–9:1 for gasoline), thus giving greater efficiency and power output. The low lean flammability limit permits operation with extremely lean air fuel ratios having as much as 60 percent excess air. On the other hand, its high flame temperature tends to result in high NO_x emissions, unless very lean mixtures are used.

Because of its gaseous form and poor self ignition qualities, methane is a poor fuel for diesel engines. Since diesels are generally somewhat more efficient than Otto cycle engines, natural gas engines are likely to use somewhat more energy than the diesels they replace. The high compression ratios achievable with natural gas limit this efficiency penalty to about 10 percent of the diesel fuel consumption, however.

Options for using natural gas in heavy duty vehicle engines are limited to the following:

Fumigation, or mixing the gas with the diesel intake air, to be ignited by diesel fuel injected in the conventional way;

Conversion of the existing diesel engine to Otto cycle operation; or

Replacement of the diesel engine with a conventional spark ignition engine.

Fumigation. Fumigation is the easiest way to use natural gas in a diesel engine. Injection and combustion of the diesel fuel ignites and burns the alternative fuel as well. Since the gas supplies much of the energy for combustion, the diesel fuel delivery for a given power level is reduced resulting in reduced smoke and particulate (PM) emissions at high load. However, incomplete combustion (especially at light loads) usually increases CO and HC emissions considerably. The increased HC emissions are of less concern with natural gas than with other hydrocarbon fuels, since the principal component, methane, is non toxic and has very low photochemical reactivity. On the other hand, methane is a very active greenhouse gas, and the other, minor components of the HC emissions include some formaldehyde as well as higher molecular weight hydrocarbons.

Spark Ignition Engines. The modifications required to convert a diesel engine to Otto cycle operation are machining the cylinder head to accept a spark plug instead of a fuel injector; redesign of the pistons to reduce the compression ratio; replacement of the fuel injection pump with an ignition system and distributor; replacement of exhaust valves and valve seats with wear resistant materials; and addition of a carburetor (or fuel injection system) and throttle assembly.

CO emissions are governed by oxygen availability, while NO_x emissions are primarily a function of flame temperature. For natural gas engines, typical NO_x emissions at stoichiometry are about 10 to 15 g/BHp–h. This can be reduced to less than 2 g/BHp–h, however, through the use of a three way catalyst and closed loop mixture controls like those on light duty passenger cars. The durability of such catalysts under the high temperatures experienced in heavy duty operation has not yet been demonstrated, however. The durability of electronic control systems under heavy duty conditions is also unproven.

An alternative approach to NO_x control is to operate very lean. Through the use of high energy ignition systems and careful optimization, homogeneous mixtures can be lean enough to lower NO_x emissions to the 2 g/BHp-h range. The driveability of such engines under transient conditions has not yet been demonstrated, however. The lean combustion limit can be extended even further by using a stratified charge strategy. Stationary engines using this technique have demonstrated NO_x emissions less than one g/BHp-h.

Liquefied Petroleum Gas (LPG). Liquefied petroleum gas is already widely used as a vehicle fuel in the U.S., Canada, the Netherlands, and elsewhere. As a fuel for spark ignition engines, it has many of the same advantages as natural gas, with the additional advantage of being easier to carry aboard the vehicle. Its major disadvantage is the limited supply, which would rule out any large scale conversion to LPG fuel.

The technologies available for LPG are the same as those available for natural gas: fumigation, or spark ignition using either stoichiometric or very lean mixtures. Due to the lower octane value of LPG, the compression ratio (and thus the thermal efficiency) possible with this fuel in spark ignition operation is lower than with natural gas, although still considerably higher than with gasoline. Aside from this, the engine technologies involved are very similar. Due to the lower octane value (and higher photochemical reactivity) of LPG, however, it is not as good a candidate for use in fumigation as natural gas.

Like natural gas, LPG in spark ignition engines is expected to produce essentially no particulate emissions (except for a small amount of lubricating oil), very little CO, and moderate HC emissions. NO_x emissions are a function of the air fuel ratio and other engine variables such as spark timing and compression ratio. LPG does not burn as well under very lean conditions as natural gas, so the NO_x levels achievable through lean burn technology are expected to be somewhat higher, probably in the range of 3 to 5 g/BHp-h. For stoichiometric LPG engines, the use of a three way catalyst and closed loop air fuel mixture control results in very low NO_x emissions, assuming that such systems can be made sufficiently durable.

Methanol. Methanol has many desirable combustion and emissions characteristics, including good lean combustion characteristics, low flame temperature (leading to low NO_x emissions) and low photochemical reactivity.

As a liquid, methanol can either be burned in an Otto cycle engine or injected into the cylinder as in a diesel. With a fairly high octane number of 112, and excellent lean combustion properties, methanol is a good fuel for lean burn spark ignition (SI). Its lean combustion limits are similar to those of natural gas, while its low energy density results in a low flame temperature compared to hydrocarbon fuels, and thus lower NO_x emissions. Methanol burns with a sootless flame and contains no heavy hydrocarbons. As a result, particulate emissions from methanol engines are very low—consisting essentially of a small amount of unburned lubricating oil.

Methanol's high octane number results in a very low cetane number, so that it is more difficult to use methanol in a diesel engine without some supplemental ignition source. Investigations to date have focused on the use of ignition improving activities, spark ignition, grow plug ignition, or dual injection with diesel fuel. Converted heavy duty diesel engines using each of these approaches have been developed and demonstrated.

The low energy density of methanol means that a large amount (about three times the mass of diesel fuel) is required to achieve for the same power output. Therefore, the diesel injection pump supplied with the engine would not be suitable in most cases; a larger volumetric capacity is required. In addition, diesel injection pumps are fuel lubricated. Since methanol is a poor lubricant, a separate oil supply to the pump is required. Other changes to the high pressure lines, injector nozzels, and so forth are required to prevent cavitation and premature wear. All of these changes are straightforward, however, and injection pumps suitable for use with methanol have been produced.

Emissions. Methanol combustion does not produce soot, so particulate emissions from methanol engines are limited to a small amount of lubricating oil. Methanol's flame temperature is also lower than that for hydrocarbon fuels, resulting in NO_x emissions which are typically 50 percent lower. CO emissions are generally comparable to or somewhat greater than those from a diesel engine; these emissions can be controlled with a catalytic converter, however.

The principal pollution problems with methanol come from emissions of unburned fuel and formaldehyde. Methanol (at least in moderate amounts) is relatively innocuous. It has low photochemical reactivity, and while acutely toxic in large doses, displays no significant chronic toxicity effects. Formaldehyde, the first oxidation product of methanol, is much less benign, however. A powerful irritant and suspected carcinogen, it also displays very high photochemical reactivity. While all combustion engines produce some formaldehyde, some early generation methanol engines exhibited greatly increased emissions compared to diesels. The potential for large increases in formaldehyde emissions with the widespread use of methanol vehicles has raised considerable concern about what would otherwise be a very benign fuel from an environmental standpoint. DDC has made major advances in formaldehyde control and levels are currently at or below the diesel engines they replace.

Formaldehyde emissions can be reduced through changes in combustion chamber and injection system design, and are also readily controllable though the use of catalytic converters, at least under warmed up conditions. In fact, the DDC engine equipped with a catalyst has attained emission levels as low as 0.06 formaldehyde, along with 0.2 HC, 0.8 CO, 2.2 NO_x and 0.04 particulate. This system is now available commercially.

ALTERNATIVE FUTURES: THE TECHNOLOGICAL STATE OF THE ART

Continued growth in emissions from the transportation sector is not inevitable. Even relatively modest steps can significantly lower emissions in the near term, and combined with slight reductions in future vehicle growth patterns, overall stability in emissions from the transport sector is possible.

In the state of the art case, emissions were estimated as if all vehicles in the world were to adopt the most stringent set of requirements considered technologically feasible today. These include currently adopted California low emission vehicle (LEV) standards for cars and light trucks, heavy truck requirements scheduled to be introduced during the 1990s in the U.S., volatility controls on petrol to reduce evaporative and refueling hydrocarbon emissions, enhanced inspection and maintenance programs to maximize the actual effectiveness of emissions standards and refueling controls.

If all vehicles in the world were to adopt these requirements, CO, HC, and NO_x emissions would go down significantly over the next twenty years. Two points are worth emphasizing:

1. Using today's state of the art technology on all vehicles, emissions are dramatically lower than they would be in the base case. For example, hydrocarbons are only a third of what they would otherwise be; in the case of NO_x, they are less than half.
2. Because of their high growth rates, even with application of state of the art controls, the non OECD countries of the world will be much more important in the future.

CONCLUSIONS

1. Motor vehicles are the largest single source of man made VOC, NO_x, and CO in the OECD as well as many rapidly industrializing countries. Controls which reduce both VOC (as well as CO) and NO_x emissions from this source to the maximum extent technologically feasible are therefore most effective at reducing ozone concentrations. In addition, motor vehicles are probably the major source of toxic pollutants as well and a significant contributor to potential climate altering emissions.
2. Depending on the ambient concentration, these emissions can cause or contribute to a wide range of adverse health and environmental effects including eye irritation, cough and chest discomfort, headaches, heart disease, upper respiratory illness, increased asthma attacks and reduced pulmonary function. The most recent studies indicate that these emissions can cause cancer and exacerbate mortality and morbidity from respiratory disease. In addition, studies indicate that air pollution seriously impairs the growth of certain crops, reduces visibility, and in sensitive aquatic systems such as small lakes and streams can destroy fish and other forms of life and damages forests.

   Whether it be localized urban problems, regional smog or global changes, it is clear that motor vehicles are a dominant source of air pollution.
3. Technologies such as closed loop three way catalysts have been developed which have the potential to reduce substantially vehicle emissions in a cost effective manner. Application of these state of the art technologies can improve vehicle performance and driveability, reduce maintenance and is consistent with improved fuel economy. Evaporative controls are also readily available and cost effective. The effectiveness of state of the art emissions controls can be improved by in use vehicle directed programs such as inspection and maintenance, recall and warranty.
4. As a result of catalysts and other controls introduced to date, CO, HC and NO_x levels are substantially below what they would otherwise be.
5. If today's state of the art emissions controls were introduced on all new vehicles around the world, it would be possible to continue to reduce vehicle emissions of CO, HC and NO_x while simultaneously absorbing the expected vehicle growth, at least until early in the next century.
6. Advances in the state of the art for vehicle technology are emerging and coupled with more modest ve-

hicle growth, they can constrain global vehicle emissions if they are widely utilized and effectively implemented.

Looking ahead to the future, it is clear that several technological challenges remain for reduced vehicle emissions. Particular areas of intense activity include:

1. Preheated or very quick light off systems which will enable manufacturers to comply with the stringent California LEV requirements.
2. Diesel flow through particulate catalysts which can lower particulate and the organics associated with diesel combustion without converting too much sulfur dioxide to sulfate.
3. Diesel trap oxidizers which can virtually eliminate diesel particulate and the associated organics.
4. Lean NO_x catalysts which would enable such technologies with inherent lean operating advantages, ie, diesels or two strokes, to take advantage of these capabilities without increasing NO_x emissions.

BIBLIOGRAPHY

1. *World Motor Vehicle Data,* Various Editions, Motor Vehicle Manufacturers Association of the United States, Inc.
2. Gardiner, Foley and Shapiro, "Priority Revision of the PM-10 NAAQS" Memo to the US EPA Administrator, July 19, 1993.
3. W. P. Cartellieri and W. F. Wachter, *Status Report on a Preliminary Survey of Strategies to Meet US-1991 HD Diesel Emission Standards Without Exhaust Gas Aftertreatment,* SAE Paper No. 870342, Society of Automotive Engineers, Warrendale, Pa., 1987.
4. W. R. Wade, C. E. Hunter, F. H. Trinker, and H. A. Cikanek, *The Reduction of NO_x and Particulate Emissions in the Diesel Combustion Process,* ASME Paper No. 87-ICE-37, American Society of Mechanical Engineers, New York, 1987.
5. P. T. Williams, G. E. Andrews, and K. D. Bartle *Diesel Particulate Emissions: The Role of Unburnt Fuel in the Organic Fraction Composition,* SAE Paper No. 870554, Society of Automotive Engineers, Warrendale, Pa., 1987.
6. California Air Resources Board staff, *Status Report: Diesel Engine Emissions Reductions through Modification of Motor Vehicle Diesel Fuel Specifications,* CARB, Sacramento, Calif., 1984.
7. Engine Manufacturer's Association "Heavy Duty Diesel Engine In-Use Emission Testing Meeting," briefing package for a presentation to the California Air Resources Board Staff, El Monte, Calif., 1985.
8. J. C. Wall and S. K. Hoekman. *Fuel Composition Effects on Heavy Duty Diesel Particulate Emissions,* SAE Paper No. 841364, Society of Automotive Engineers, Warrendale, Pa., 1984.
9. W. M. Draper, J. Phillips, and H. W. Zeller, *Impact of a Barium Fuel Additive on the Mutagenicity and Polycyclic Aromatic Hydrocarbon Content of Diesel Exhaust Particulate Emissions,* SAE paper, 1988.
10. C. S. Weaver, and R. F. Klausmeier. *Heavy Duty Diesel Vehicle Inspection and Maintenance Study: Final Report,* 4 Vol., report under ARB Contract No. A4-151-32, Radian Corporation, Sacramento, Calif., 1988.

AIR POLLUTION: AUTOMOBILE, TOXIC EMISSIONS

Pamela J. Brodowicz
Penny M. Carey
EPA
Ann Arbor, Michigan

Motor vehicle emissions are extremely complex and hundreds of compounds have been identified. This article attempts to summarize what is known about air toxic emissions from gasoline and diesel fueled motor vehicles. Specific pollutants or pollutant categories discussed include benzene, formaldehyde, 1,3-butadiene, acetaldehyde, and diesel particulate matter, all of which have been considered in previous analyses of air toxics and are also currently or in the process of being classified by the U.S. Environmental Protection Agency (EPA) as known or probable human carcinogens. For each pollutant, information is provided on emissions, atmospheric transformation, ambient levels, and health effects.

In order for a compound to be assigned an EPA classification, the experimental data must be evaluated through a process outlined in the *EPA Guidelines for Carcinogenic Risk Assessment* (1). The data from carcinogenic studies (both human and animal) are divided into five separate categories by weight of evidence collected. These five categories are defined as sufficient evidence, limited evidence, inadequate evidence, no data, and no evidence for carcinogenicity. Each compound is evaluated, giving the human and animal study criteria each its own category designation.

Human studies depend on case reports of individual cancer patients and/or epidemiological studies. These studies must be without bias, logically approached, organized, and analyzed, and the probability of chance must be ruled out in order to infer a causal relationship. Animal studies must be long-term studies, conducted in more than one species, strain, or experiment, with adequate dosage, exposure, follow-up and reporting, adequate numbers of animals, good survival rates, and the increase of tumors of the malignant type, or a combination of malignant and benign.

The categories above characterizing human and animal weight of evidence classifications can now be utilized to allow a tentative assignment to one of five categories. The tumor studies are combined with all other relative information to determine if the categorization of the weight of evidence is appropriate for a particular agent. The five categories are described below:

Group A: Human Carcinogen. Requires *sufficient evidence* from human epidemiological studies to support a causal relationship between exposure and the agent.

Group B: Probable Human Carcinogen. B1–Human evidence is *limited* no matter what is found in the animal studies. B2–If animal studies are sufficient and in humans the evidence is *inadequate, no data exists, or no evidence* is found.

Group C: Possible Human Carcinogen. There is *limited* evidence in animals in the absence of human data.

Group D: Not Classifiable as to Human Carcinogenicity. There are *inadequate* human and animal data or no data are available.

Group E: Evidence of Noncarcinogenicity For Humans. *No evidence* has been found in at least two animal studies or in both adequate human epidemiologic and animal studies. This is based on available evidence and does not rule out the possibility that this agent may be carcinogenic under other conditions.

BENZENE

Benzene is a clear, colorless, aromatic hydrocarbon which is both volatile and flammable. Benzene is present as a gas in both exhaust and evaporative emissions from motor vehicles. Benzene in the exhaust, expressed as a percentage of total organic gases (TOG), varies depending on control technology (eg, type of catalyst) and the levels of benzene and aromatics in the fuel, but is generally about 3 to 5%. Percent TOG of benzene in evaporative emissions depends on control technology (ie, fuel injector or carburetor) and fuel composition (eg, benzene level and Reid Vapor Pressure, or RVP), and is generally about 1%.

Benzene is quite stable in the atmosphere, with the only benzene reaction which is important in the lower atmosphere being the one with OH radicals (2). Yet even this reaction is relatively slow. Benzene itself will not be incorporated into clouds or rain to any large degree because of its low solubility. Benzene is not produced by atmospheric reactions. The most important source of a person's daily exposure to benzene is active tobacco smoking, accounting for roughly half of the total populations' exposure (3). Outdoor concentrations of benzene, due mainly to motor vehicles, account for roughly one-quarter of the total exposure. Of the mobile source contribution, the majority comes from the exhaust component. The annual average ambient level of benzene in urban areas ranges from 4 to 7 $\mu g/m^3$, based on 1987–1990 urban air monitoring data (4,5). Based on available data, maximum exposure levels to benzene in environments heavily impacted by motor vehicles range from 40 $\mu g/m^3$ from in-vehicle exposure (6), to 288 $\mu g/m^3$ from exposure during refueling (7).

Long-term exposure to high levels of benzene in air has been shown to cause cancer of the tissues that form white blood cells (leukemia), based on epidemiology studies with workers (8,9). Leukemias and lymphomas (lymphoma is a general term for growth of new tissue in the lymphatic system of the body), as well as other tumor types, have been observed in experimental animals that have been exposed to benzene by inhalation or oral administration (8,9). Inhalation exposure of mice to benzene (10) has also provided an animal model for the type of cancer identified most closely with occupational exposure, acute myelogenous leukemia. Exposure to benzene and/or its metabolites has also been linked with genetic changes in humans and animals (11), and increased proliferation of mouse bone marrow cells (12). Furthermore, the occurrence of certain chromosomal aberrations in individuals with known exposure to benzene may serve as a marker for those at risk for contracting leukemia (13). EPA has classified benzene as a Group A human carcinogen. The International Agency for Research on Cancer (IARC) also defines benzene as an agent that is carcinogenic to humans.

Research has been conducted on the fate of benzene in experimental animals (9,14). These studies demonstrate that species differ with respect to their ability to metabolize benzene. These differences may be important when choosing an animal model for human exposures and when extrapolating high dose exposures in animals to the low levels of exposure typically encountered in occupational situations. The recent development of a benzene physiologically based pharmacokinetic model should help in performing interspecies and route-to-route extrapolations of cancer data.

A number of adverse noncancer health effects have been associated with exposure to benzene (8,9,14). Benzene is known to cause disorders of the blood. People with long-term exposure to benzene at levels that generally exceed 50 ppm (162,500 $\mu g/m^3$) may experience harmful effects on blood-forming tissues, especially bone marrow. These effects can disrupt normal blood production and cause a decrease in important blood components, such as red blood cells and blood platelets, leading to anemia and a reduced ability to clot. Exposure to benzene at comparable or even lower levels can be harmful to the immune system, increasing the chance for infection and perhaps lowering the body's defense against tumors by altering the number and function of the body's white blood cells. Studies in humans and experimental animals indicate that benzene may be a reproductive and developmental toxicant (14).

FORMALDEHYDE

Formaldehyde, a colorless gas at normal temperatures, is the simplest member of the family of aldehydes. Formaldehyde gas is soluble in water, alcohols, and other polar solvents. It is the most prevalent aldehyde in vehicle exhaust and is formed from incomplete combustion of the fuel. Formaldehyde is emitted in the exhaust of both gasoline and diesel-fueled vehicles. It is not a component of evaporative emissions. The TOG percentage of formaldehyde in exhaust varies from roughly 1 to 4%, depending on control technology and fuel composition.

Formaldehyde exhibits extremely complex atmospheric behavior (2). It is present in emissions, but is also formed by the atmospheric oxidation of virtually all organic species. It is ubiquitous in the atmosphere because it is formed in the atmospheric oxidations of methane and biogenic (produced by a living organism) hydrocarbons. Formaldehyde is photolyzed readily, and its photolysis is an important source of photochemical radicals in urban areas. It is also destroyed by reaction with OH and NO_3. The major carbon-containing product of all gas-phase formaldehyde reactions is carbon monoxide. Because formaldehyde is often the dominant source of radicals in urban atmospheres, formaldehyde concentrations have a feedback effect on the chemical residence time of other atmospheric species. Formaldehyde is highly water soluble and participates in a complex set of chemical reactions within clouds. The product of the aqueous-phase oxidation of formaldehyde is formic acid. The mobile source contribution to ambient formaldehyde levels contains both primary (ie, direct emissions from motor vehicles) and secondary formaldehyde (ie, formed from photooxidation of volatile organic compounds, or VOCs from vehicles). The mobile source contribution is difficult to quantify, but it appears that at least 30% of formaldehyde in the ambient air may be attributable to motor vehicles (14). The annual average ambient level of formaldehyde in urban areas is roughly 5 $\mu g/m^3$, based on 1990 air monitoring data (5) which accounted for ozone interference. Based on available data, maximum exposure levels to formaldehyde in environments heavily impacted by motor vehicles range from 4.9 $\mu g/m^3$ from exhaust exposure at a service station (15), to 41.8 $\mu g/m^3$ from parking garage exposure (7).

Studies in experimental animals provide sufficient evidence that long-term inhalation exposure to formaldehyde causes an increase in the incidence of squamous cell carcinomas of the nasal cavity (9,14,16). In addition, the distribution of nasal tumors in rats has been better defined; the findings suggest that not only regional exposure, but also local tissue susceptibility may be important for the distribution of formaldehyde-induced tumors (9,14). Epidemiological studies in occupationally exposed workers suggest that long-term inhalation of formaldehyde may be associated with tumors of the nasopharyngeal cavity, nasal cavity, and sinus (9,14,16). The evidence for an association between lung cancer and occupational formaldehyde is tenuous, and collectively, the recent studies do not conclusively demonstrate a causal relationship between cancer and exposure to formaldehyde in humans. EPA has classified formaldehyde as a Group B1 probable human carcinogen. IARC also defines formaldehyde as an agent that is probably carcinogenic to humans. Both classifications are based on limited evidence for carcinogenicity in humans and sufficient evidence for carcinogenicity in animals.

Recent work on the pharmacokinetics of formaldehyde has focused on the validation of measurement of DNA-protein adducts, or cross-links (DPX), as an internal measure of dose for formaldehyde exposure (9,14). DPX is the binding of DNA to a protein to which formaldehyde is bound, thus forming a separate entity that can be quantified. This is considered a more accurate way to measure the amount of formaldehyde that is present inside a tissue. An internal dosimeter for formaldehyde exposure is desirable because the inhaled concentration of formaldehyde may not reflect actual tissue exposure levels. The difference in inhaled concentration and actual tissue exposure level is due to the action of multiple defense mechanisms that act to limit the amount of formaldehyde that reaches cellular DNA. These studies have provided more accurate data with which to quantify the level of formaldehyde in the cell.

Noncancer adverse health effects associated with exposure to formaldehyde in humans and experimental animals include irritation of the eyes, nose, throat, and lower airway at low levels (0.05–10 ppm or 123–12,300 $\mu g/m^3$) (9,14,16). There is also suggestive, but not conclusive, evidence in humans that formaldehyde can affect immune function. Studies in humans and experimental animals indicate that formaldehyde may be a reproductive and developmental toxicant (14). Adverse effects on the liver and kidney have also been noted in experimental animals exposed to higher levels of formaldehyde.

1,3-BUTADIENE

1,3-Butadiene is a colorless, flammable gas at room temperature, is insoluble in water, and its two conjugated double bonds make it highly reactive. 1,3-Butadiene is formed in vehicle exhaust by the incomplete combustion of the fuel and is not present in vehicle evaporative and refueling emissions. The TOG percentage of 1,3-butadiene in exhaust varies from roughly 0.4 to 1.0%, depending on control technology and fuel composition.

1,3-Butadiene is transformed rapidly in the atmosphere (2). There are three chemical reactions of 1,3-butadiene which are important in the ambient atmosphere: reaction with hydroxyl radical (OH), reaction with ozone (O_3), and reaction with nitrogen trioxide radical (NO_3). All three of these reactions are relatively rapid, and all produce formaldehyde and acrolein, species which are themselves toxic and/or irritants. The oxidation of 1,3-butadiene by NO_3 produces organic nitrates as well. Incorporation of 1,3-butadiene into clouds and rain will not be an important process due to the low solubility of 1,3-butadiene. 1,3-Butadiene is probably not produced by atmospheric reactions. Current estimates indicate that mobile sources account for approximately 94% of the total 1,3-butadiene emissions (17). The remaining 1,3-butadiene emissions (6%) come from stationary sources mainly related to industries producing 1,3-butadiene and those industries that use 1,3-butadiene to produce other compounds. The annual average ambient level of 1,3-butadiene in urban areas (excluding high point source areas) ranges from roughly 0.2 to 1.0 $\mu g/m^3$, based on 1987–1990 urban air monitoring data (4,5). Based on a single study (18), in-vehicle exposure to 1,3-butadiene was found to average 3.0 $\mu g/m^3$.

Long-term inhalation exposure to 1,3-butadiene has been shown to cause tumors in several organs in experimental animals (9,14,19). One recent study (9,14) demonstrates the occurrence of cancer in mice at additional sites at lower concentrations of 1,3-butadiene than those of earlier studies. Studies in humans exposed to 1,3-butadiene suggest that this chemical may cause cancer (9,14,19). These epidemiological studies of occupationally exposed workers are inconclusive with respect to the carcinogenicity of 1,3-butadiene in humans, however, because of a lack of adequate exposure information and concurrent exposure to other potentially carcinogenic substances. Studies in animals also indicate that 1,3-butadiene can alter the genetic material (9,14,19). EPA has classified 1,3-butadiene as a Group B2 probable human carcinogen. IARC has classified 1,3-butadiene as an agent that is possibly carcinogenic to humans. Both classifications are based on inadequate evidence for carcinogenicity in humans and sufficient evidence for carcinogenicity in animals.

Studies on the fate of 1,3-butadiene in the body have focused on the mechanism behind the differences in carcinogenic responses seen between species (20–21). Recent pharmacokinetic research has found marked differences among mice, rats, and human tissue preparations in their ability to metabolize 1,3-butadiene and its metabolites (22). The results suggest that the effective internal dose of DNA-reactive metabolites may be less in humans than in mice for a given level of exposure.

Exposure to 1,3-butadiene is also associated with adverse noncancer health effects (9,14,19). Exposure to high levels (on the order of hundreds to thousands ppm) of this chemical for short periods of time can cause irritation of the eyes, nose, and throat, and exposure to very high levels can cause effects on the brain leading to respiratory paralysis and death. Studies of rubber industry workers who are chronically exposed to 1,3-butadiene suggest other possible harmful effects, including heart disease, blood disease, and lung disease. Studies in animals indicate that 1,3-butadiene at exposure levels of greater than 1,000 ppm (2.2×10^6 μg/m^3) may adversely affect the blood-forming organs. Reproductive and developmental toxicity has also been demonstrated in experimental animals exposed to 1,3-butadiene.

ACETALDEHYDE

Acetaldehyde is a saturated aldehyde that is a colorless liquid and volatile at room temperature. Acetaldehyde is found in vehicle exhaust and is formed as a result of incomplete combustion of the fuel. Acetaldehyde is emitted in the exhaust of both gasoline and diesel-fueled vehicles. It is not a component of evaporative emissions. Percent TOG of acetaldehyde in exhaust varies from roughly 0.4 to 1.0 percent, depending on control technology and fuel composition.

The atmospheric chemistry of acetaldehyde is similar in many respects to that of formaldehyde (2,23). Like formaldehyde, it can be both produced and destroyed by atmospheric chemical transformation. However, there are important differences between the two. Acetaldehyde photolyzes, but much more slowly than formaldehyde. Whereas formaldehyde produces CO upon reaction or photolysis, acetaldehyde produces organic radicals that ultimately form peroxyacetyl nitrate (PAN) and formaldehyde. Acetaldehyde is also significantly less water soluble than formaldehyde. Acetaldehyde is ubiquitous in the environment and is naturally released (24). The mobile source contribution to ambient acetaldehyde levels contains both primary and secondary acetaldehyde. Data from emission inventories and atmospheric modeling indicate that roughly 40% of the acetaldehyde in ambient air may be attributable to mobile sources. The annual average ambient level of acetaldehyde in urban areas is roughly 3 μg/m^3, based on available 1990 urban air monitoring data (5) which accounted for ozone interference. Based on a single study (15), the in-vehicle exposure level of acetaldehyde was found to average 13.7 μg/m^3.

There is sufficient evidence that acetaldehyde produces cytogenetic damage in cultured mammalian cells (24,25). Although there are not many studies done with whole animals (*in vivo*) (24–26), they suggest that acetaldehyde produces similar effects *in vivo*. Thus, the available evidence indicates that acetaldehyde is mutagenic and may pose a risk for somatic cells (all body cells excluding the reproductive cells). Current knowledge, however, is inadequate with regard to germ cell (reproductive cell) mutagenicity because the available information is insufficient to support any conclusions about the ability of acetaldehyde to reach mammalian gonads and produce heritable genetic damage.

Studies in experimental animals provide sufficient evidence that long-term inhalation exposure to acetaldehyde causes an increase in the incidence of squamous cell carcinomas of the nasal cavity (24–26). One epidemiological study, in occupationally exposed workers, was insufficient to suggest that long-term inhalation of acetaldehyde may be associated with an increase in total cancers (24–26). EPA has classified acetaldehyde as a Group B2 probable human carcinogen.

Non-cancer effects in studies with rats and mice showed acetaldehyde to be moderately toxic by the inhalation, oral, and intravenous routes (24–26). Acetaldehyde is a sensory irritant that causes a depressed respiration rate in mice. In rats, acetaldehyde increased blood pressure and heart rate after exposure by inhalation. The primary acute effect of exposure to acetaldehyde vapors is irritation of the eyes, skin, and respiratory tract. At high concentrations, irritation and ciliastatic effects can occur, which could facilitate the uptake of other contaminants. Clinical effects include reddening of the skin, coughing, swelling of the pulmonary tissue, and localized tissue death. Respiratory paralysis and death have occurred at extremely high concentrations. It has been suggested that voluntary inhalation of toxic levels of acetaldehyde would be prevented by its irritant properties, since irritation occurs at levels below 200 ppm (360,000 μg/m^3). The new genotoxicity studies, which utilize lower concentrations of acetaldehyde, do not produce chromosomal aberration and/or cellular mutations.

The research into reproductive and developmental effects of acetaldehyde is based on intraperitoneal injection, intravenous, or oral administration of acetaldehyde to rats and mice, and also *in vitro* (cell culture) studies (24–26). Little research exists that addresses the effects of inhalation of acetaldehyde on reproductive and developmental effects. The *in vitro* and *in vivo* studies provide evidence to support the fact that acetaldehyde may be the causative factor in birth defects observed in fetal alcohol syndrome.

DIESEL PARTICULATE MATTER

Diesel exhaust particulate matter consists of a solid core composed mainly of carbon, a soluble organic fraction, sulfates, and trace elements (27). Light-duty diesel engines emit from 30 to 100 times more particles than comparable catalyst-equipped gasoline vehicles (27). Diesel particulate matter is mainly attributable to the incomplete combustion of fuel hydrocarbons, though some may be due to engine oil or other fuel components. The particles may also become coated with adsorbed and condensed high molecular weight organic compounds.

Diesel particulate matter has not been explicitly modeled to determine its atmospheric transformation and residence times because of its inherent complexity. In order to accomplish this modeling, consideration needs to be given to the relative abundance of the various polycyclic organic matter (POM) species in the atmosphere, the availability of emissions data, and determining a location's specific area, mobile, and point sources. Using the ambient national average total suspended particle (TSP) concentration (28,29), and the percent contribution of diesel particulate matter to TSP, the concentration of diesel particulate matter in ambient air can be estimated. For 1990, the estimated resultant concentration of diesel particulate matter is 2.5 $\mu g/m^3$.

Studies in experimental animals provide sufficient evidence that long-term inhalation exposure to high levels of diesel exhaust causes an increase in the induction of lung tumors in two strains of rats and two strains of mice (30). Studies have concentrated on the hypothesis that the carbon core of diesel particulate matter is the causative agent in the genesis of lung cancer (31,32). By exposing rats to carbon black and diesel soot and comparing the results to diesel exhaust itself, the tumor response to diesel exhaust and carbon black is qualitatively similar. Also, as a result of extensive studies, the direct-acting mutagenic activity of both particle and gaseous fractions of diesel exhaust has been shown (30). In two key epidemiological studies on railroad workers occupationally exposed to diesel exhaust (30), it was observed that long-term inhalation of diesel exhaust produced an excess risk of lung cancer. Collectively, the epidemiological studies show a positive, though limited, association between diesel exhaust exposure and lung cancer. EPA currently considers diesel particulate matter to be a Group B1 probable human carcinogen; however, a formal risk assessment is still in progress.

An understanding of the pharmacokinetics associated with pulmonary deposition of diesel exhaust particles and their adsorbed organics is critical in understanding the carcinogenic potential of diesel engine emissions. The pulmonary clearance of diesel exhaust particles has multiple phases and involves several processes, including a relatively rapid transport system and slow macrophage-mediated (a white blood cell that engulfs and digests foreign particles) processes (30). The observed dose-dependent increase in the particle burden of the lungs is due, in part, to an overloading of alveolar macrophage function. The resulting increase in particle retention has been shown to increase the bioavailability of particle adsorbed mutagenic and carcinogenic components, such as benzo[a]pyrene and 1-nitropyrene. Experimental data also indicate the ability of the alveolar macrophage to metabolize and solubilize the particle-adsorbed components. Although macromolecular binding of diesel exhaust particle-derived POM and the formation of DNA adducts following exposure to diesel exhaust have been reported, a quantitative relationship between these and increased carcinogenicity is not available.

A number of adverse noncancer health effects have also been associated with exposure to acute, subchronic, and chronic diesel exhaust at levels found in the ambient air (30). Most of the effects observed through acute and subchronic exposure are respiratory tract irritation and diminished resistance to infection. Increased cough and phlegm and slight impairments in lung function have also been documented. Animal data indicate that chronic respiratory diseases can result from long-term (chronic) exposure to diesel exhaust. It appears that normal, healthy adults are not at high risk to serious noncancer effects of diesel exhaust at levels found in the ambient air. The data base is inadequate to form conclusions about sensitive subpopulations.

BIBLIOGRAPHY

1. *Fed. Reg.*, **51**(185), 33992–34003 (Sept. 24, 1986).
2. M. P. Ligocki and co-workers, *Atmospheric Transformation of Air Toxics: Benzene, 1,3-Butadiene, and Formaldehyde,* Systems Applications International, San Rafael, Calif., SYSAPP-91/106, 1991.
3. L. A. Wallace, *Environ. Health Perspect.* **82,** 165–169, (1989).
4. Environmental Protection Agency (EPA), *AIRS User's Guide,* vol. 1-7, Office of Air Quality Planning and Standards, Research Triangle Park, N.C., 1989.
5. R. A. McAllister and co-workers, *Urban Air Toxics Monitoring Program,* EPA-450/4-91-001, 1990.
6. EPA, *The Total Exposure Assessment Methodology (TEAM) Study: Summary and Analysis:* vol. 1, EPA/600/6-87/002a, Office of Research and Development, Washington, D.C., June 1987.
7. A. Wilson and co-workers, *Air Toxics Microenvironment Exposure and Monitoring Study,* South Coast Air Quality Management District, El Monte, Calif., 1991.
8. EPA, *Interim Quantitative Cancer Unit Risk Estimates Due to Inhalation of Benzene,* Office of Health and Environmental Assessment, Carcinogen Assessment Group, Washington, D.C., 1985.
9. Clement Associates, Inc., *Motor Vehicle Air Toxics Health Information,* September 1991.
10. E. P. Cronkite and co-workers, *Environ. Health Perspect.* **82**(97-108), 1989.
11. *IARC monographs* **29,** 345–389 (1982).
12. R. D. Irons and co-workers, *Proc. Natl. Acad. Sci.* **89,** 3691–3695 (1992).
13. M. Lumley, H. Barker, and J. A. Murray, *Lancet* **336,** 1318–1319 (1990).
14. EPA, *Motor Vehicle-Related Air Toxics Study,* EPA 420-R-93-005, Office of Mobile Sources, Ann Arbor, Mich., April 1993.
15. D. C. Shikiya and co-workers, *In-Vehicle Air Toxics Characterization Study in the South Coast Air Basin,* South Coast Air Quality Management District, El Monte, Calif., May 1989.
16. EPA, *Assessment of Health Risks to Garment Workers and Certain Home Residents from Exposure to Formaldehyde,* Office of Pesticides and Toxic Substances, April 1987.
17. EPA, *Locating and Estimating Air Emissions from Sources of 1,3-Butadiene,* EPA-450/2-89-021, Office of Air Quality Planning and Standards, Research Triangle Park, N.C., 1989.
18. C. C. Chan and co-workers, *Commuter's Exposure to Volatile Organic Compounds, Ozone, Carbon Monoxide, and Nitrogen Dioxide,* Air and Waste Management Association Paper 89-34A.4, 1989.
19. EPA, *Mutagenicity and Carcinogenicity Assessment of 1,3-Butadiene,* EPA/600/8-85/004F, Office of Health and Environmental Assessment, Washington, D.C., 1985.
20. G. A. Csanády and J. A. Bond, *CIIT Activities* **11**(2), 1–8 (1991).
21. G. A. Csanády and J. A. Bond, *Toxicologist* **11,** 47, 1991.
22. L. Recio and co-workers, "Biotransformation of butadiene by hepatic and pulmonary tissues from mice, rats, and humans: relationship to butadiene carcinogenicity and *in vivo* mutage-

nicity," proceedings from the *Air and Waste Management Association's specialty conference "Air Toxics Pollutants from Mobile Sources: Emissions and Health Effects,"* October 16–18, 1991.

23. M. P. Ligocki and G. Z. Whitten, *Atmospheric Transformation of Air Toxics: Acetaldehyde and Polycyclic Organic Matter,* Systems Applications International, San Rafael, Calif., (SYSAPP-91/113), 1991.
24. EPA, *Health Assessment Document for Acetaldehyde,* EPA-600/8-86/015A (External Review Draft), Office of Health and Environmental Assessment, Environmental Criteria and Assessment Office, Research Triangle Park, N.C., 1987.
25. *Proposed Identification of Acetaldehyde as a Toxic Air Contaminant,* Part B Health assessment, California Air Resources Board, Stationary Source Division, August, 1992.
26. EPA, *Integrated Risk Information System (IRIS),* Office of Health and Environmental Assessment, Environmental Criteria and Assessment Office, Cincinnati, Ohio, 1992.
27. National Research Council, *Impacts of Diesel-Powered Light-duty Vehicles: Diesel Technology,* National Academy Press, Washington, D.C., 1982.
28. EPA, *National Air Pollutant Emission Estimates 1940–1990,* EPA-450/4-91-026, Office of Air Quality Planning and Standards, Research Triangle Park, NC, 1991.
29. EPA, *National Air Quality and Emissions Trends Report, 1990,* EPA-450/4-91-023, Research Triangle Park, NC, Office of Air Quality Planning and Standards, 1991.
30. EPA, *Health Assessment Document for Diesel Emissions; Workshop Review Draft,* EPA-600/8-90/057A, Office of Health and Environmental Assessment, Washington, D.C., 1990.
31. U. Heinrich and co-workers, "The carcinogenic effects of carbon black particles and tar/pitch condensation aerosol after inhalation exposure of rats," *Seventh International Symposium on Inhaled Particles,* Edinburgh, Sept. 16–20, 1991.
32. J. L. Mauderly and co-workers, "Influence of particle-associated organic compounds on carcinogenicity of diesel exhaust," *Seventh International Symposium on Inhaled Particles,* Edinburgh, Sept. 16–20, 1991.

AIR POLLUTION CONTROL METHODS

BURTON B. CROCKER
Chesterfield, Missouri

Air pollution has been defined as the presence in ambient air of one or more contaminants of such quantity and time duration as to be injurious to human, plant, or animal life; property; or the conduct of business (1,2) or so as to alter significantly the natural balance of an ecosystem (3). The effect of a time–dosage relationship has been clearly considered by the U.S. EPA in the establishment of the National Ambient Air Quality Standards (NAAQS) for Criteria Pollutants (Table 1). Because of more recent concerns about the effect of release of greenhouse gases and of stratospheric O_3-depleting gases to the atmosphere, the last phrase has been added to the definition above (3).

Selection of pollution-control methods is generally based on the need to control ambient air quality and to achieve compliance with NAAQS standards for criteria pollutants or, in the case of nonregulated contaminants, to protect human health and vegetation. There are three elements to a pollution problem: a source, a receptor affected by the pollutants, and the transport of the pollutants from source to receptor. Modification or elimination of any one of these elements can change the nature of a pollution problem. For instance, tall stacks that disperse the effluent, modifying the transport of pollutants, can reduce nearby pollution levels. Although better dispersion aloft can solve a local problem, if done from numerous sources, it can cause a regional problem, such as the acid rain problems mentioned above. Atmospheric dilution as a control measure has been discussed (4–16). A better approach is to control emissions at the source.

There are three main classes of pollutants: gases, particulates (which may be either liquid or solid or a combination), and odors (which may originate as gases or particulates). Although odors are controlled as are other pollutants, they are often discussed separately because of the different methods used for their sensing and measurement. Many effluents contain several contaminants: one or two may be present as gases; the others often exist as liquid or solid particulates of various sizes. The possibility that effluent pollutants may be present in more than one physical state must be taken into account in sampling, analysis, and control.

PRINCIPAL ENERGY SOURCES OF AIR POLLUTANTS

Principal energy sources producing and emitting air pollutants to the atmosphere are fossil fuel combustion for electric power generation, steam production, and space heating. Another large source is fuel combustion for transportation. On a mass-emission basis, U.S. EPA estimates show the following percentage of total U.S. emissions by source: transportation, 43%; stationary fuel combustion, 29%; industrial processes, 16%; solid waste disposal (incineration), 4%; and miscellaneous, 8%. This article discusses air pollution control techniques for stationary sources only (see AUTOMOBILE EMISSIONS, AIRCRAFT ENGINES, AUTOMOTIVE; AIR POLLUTION: AUTOMOBILE TOXIC EMISSIONS).

Stationary consumption of fossil fuels takes place in boilers (about 33% of the total U.S. fossil fuel consumption), residential use, and direct and indirect heating of processes such as steel production and rolling, nonferrous metallurgical processes as well as many types of process industry operations.

On a combustion heat release capacity basis, the total U.S. utility boiler capacity is about 80% of the combined U.S. capacity for all other classes of boilers. More than 50% of the U.S. utility boiler capacity burns coal or residual oil. These two fuels present the most significant air pollution problems, because both contain ash and often sulfur. In the past, these boilers have also been designed for high flame temperatures, which maximizes the fixation of NO_x by reaction between N_2 and O_2 in combustion air. It is estimated that there are 4,000–5,000 utility boilers in the United States, 700,000–800,000 industrial boilers, and >1.5 million commercial or institutional boilers.

Industrial boilers average about 6.3×10^9 J/h heat release, and commercial boilers are even smaller. Approximately 33% of these boilers are natural gas or distillate oil fired. The remainder present complicated pollution-control problems because of their large number, their small size, and their proximity to population centers. In addition,

Table 1. National Ambient Air Quality Standards

	Maximum Permissible Concentration[a]			
Pollutant	Averaging Time	Primary Standard	Secondary Standard	Measurement Method
Sulfur oxides	annual arithmetic mean	80 $\mu g/m^3$ (0.3 ppmv)		West-Gaeke Pararosaniline
	24 h max	365 $\mu g/m^3$ (0.14 ppmv)		
	3 h max		1300 $\mu g/m^3$ (0.5 ppmv)	
Particulates (PM_{10})	annual arithmetic mean	50 $\mu g/m^3$	same	gravimetric 24-h high volume sample with PM_{10} classifying head
	24 h max	150 $\mu g/m^3$	same	
Carbon monoxide	8 h max	10 mg/m^3 (9 ppmv)	same	nondispersive infrared analyzer
	1 h max	40 mg/m^3 (35 ppmv)	same	
Ozone	1 h max	235 $\mu g/m^3$	same	gas-phase chemiluminescence analyzer
Hydrocarbons	3 h max (6–9 am)	160 $\mu g/m^3$ (0.24 ppmv as CH_4)	same	flame ionization detector
Nitrogen oxides	annual arithmetic mean	100 $\mu g/m^3$ (0.05 ppmv as NO_2)	same	chemiluminescence analyzer
Lead	calender quarter arithmetic mean	1.5 $\mu g/m^3$	same	lead analysis by atomic absorption spectrometry on extract from high volume sample catch

[a] Standards for periods shorter than annual average may be exceeded once per year.

many of these boilers have fluctuating load swings and unsteady operating conditions, which influence the rate and type of emissions released.

For boilers burning coal, the largest specific release, uncontrolled, is typically SO_2 (in tons released per ton of fuel fired). This can be appreciably reduced by burning low sulfur coal. NO_x is the second highest pollutant released, with particulate emissions close behind. Hydrocarbon emissions are generally quite low but important in the overall atmospheric pollution situation. Although even lower, there are small quantities of aldehydes released that are important as precursers of photochemical irritants.

MEASUREMENT OF AIR POLLUTION

Measurement techniques are divided into two categories: ambient and source measurement. Ambient air samples often require detection and measurement in the ppmv to ppbv (parts by volume) range, whereas source concentrations can range from tenths of a volume percent to a few hundred ppmv. Federal regulations (17,18) require periodic ambient air monitoring at strategic locations in a designated air quality control region. The number of required locations and complexity of monitoring increases with region population and with the normal concentration level of pollutants. Continuous monitoring is preferable, but for particulates one 24-h sample every 6 days may be permitted. In some extensive metropolitan sampling networks, averaged results from continuous monitors are telemetered to a single data-processing center. Special problems have been investigated using portable, vehicle-carried, or airborne ambient sampling equipment. The use of remote-guided miniature aircraft has been reported as a practical, cost-effective ambient sampling method (19). Ambient sampling may fulfill one or more of the following objectives: (*1*) establishing and operating a pollution alert network, (*2*) monitoring the effect of an emission source, (*3*) predicting the effect of a proposed installation (compliance with prevention of significant deterioration (PSD) regulations requires 1 yr of background ambient air monitoring at the proposed installation site before filing an application for a construction permit for a new installation), (*4*) establishing seasonal or yearly trends, (*5*) locating the source of an undesirable pollutant, (*6*) obtaining permanent sampling records for legal action or for modifying regulations, and (*7*) correlating pollutant dispersion with meteorological, climatological, or topographic data and with changes in societal activities.

The problems of source sampling are distinct from those of ambient sampling. Source gas may have high temperature or contain high concentrations of water vapor or entrained mist, dust, or other interfering substances so that particulates or gases may be deposited on or absorbed into the grain structure of the gas-extractive sampling probes. Depending on the objective or regulations, source sampling may be infrequent, occasional, intermittent, or continuous. Typical objectives are (*1*) demonstrating compliance with regulations; (*2*) obtaining emission data; (*3*) measuring product loss or optimizing process operating variables; (*4*) obtaining data for engineering design, such as for control equipment; (*5*) determining collector efficiency or acceptance testing of purchased equipment; and (*6*) determining need for maintenance of process or control equipment.

AIR POLLUTION AND CONTROL REGULATIONS

There has been considerable improvement, especially in industrial areas, in U.S. air quality since the adoption of the Clean Air Act of 1972. Appreciable reductions in particulate emissions and in SO_2 levels are especially evident. In 1990, however, almost every metropolitan area was in nonattainment status on ozone air quality standards; 50 metropolitan areas exceeded the CO standard and between 50 and 100 exceeded the PM_{10} standard for particulate level (20).

The U.S. Congress adopted a new clean air act in 1990 that has three areas of emphasis: acid rain reduction in the northeastern United States; severe limitation on atmospheric emissions of 189 chemicals on the EPA hazardous or toxic substance list; and tightened regulations on vehicular exhaust, reformulated vehicular fuels, and vehicles capable of using alternative fuels (ozone compliance and smog reduction). Regulations associated with acid rain prevention emphasize reductions in sulfur oxide and NO_x emissions from combustion processes, especially coal-fired power boilers in the Midwest. The chemical process industry and their customers will be increasingly under pressure to eliminate atmospheric releases of VOCs and carcinogenic-suspect compounds.

Minimizing Pollution Control Cost

Although the first impulse for emission reduction is often to add a control device, this may not be the environmentally best or least costly approach. Process examination may reveal changes or alternatives that can eliminate or reduce pollutants, decrease the gas quantity to be treated, or render pollutants more amenable to collection. Following are principles to consider for controlling pollutants without the addition of specific treatment devices (21):

1. Eliminate the source of the pollutant.
    - Seal the system to prevent interchanges between system and atmosphere.
        - Use pressure vessels.
        - Interconnect vents on receiving and discharging containers.
        - Provide seals on rotating shafts and other necessary openings.
    - Change raw materials, fuels, etc to eliminate the pollutant from the process.
    - Change the manner of process operation to prevent or reduce formation of, or air entrainment of, a pollutant.
    - Change the type of process step to eliminate the pollutant.
    - Use a recycle gas or recycle the pollutants rather than using fresh air or venting.
2. Reduce the quantity of pollutant released or the quantity of carrier gas to be treated.
    - Minimize entrainment of pollutants into a gas stream.
    - Reduce number of points in system in which materials can become airborne.
    - Recycle a portion of process gas.
    - Design hoods to exhaust the minimum quantity of air necessary to ensure pollutant capture.
3. Use equipment for dual purposes, such as a fuel combustion furnace to serve as a pollutant incinerator.

Steps such as the substitution of low sulfur fuels or nonvolatile solvents, change of raw materials, lowering of operation temperatures to reduce NO_x formation or volatilization of process material, and installation of well-designed hoods (22–29) at emission points to reduce effectively the air quantity needed for pollutant capture are illustrations of the above principles.

Selection of Control Equipment

Engineering information (30) needed for the design and selection of pollution control equipment include knowledge of the properties of pollutants (chemical species, physical state, particle size, concentration, and quantity of conveying gas) and effects of pollutant on surrounding environment. The design must consider likely future collection requirements. Advantages of alternative collection techniques must be determined:

1. Collection efficiency.
2. Ease of reuse or disposal of recovered material.
3. Ability of collector to handle variations in gas flow and loads at required collection efficiencies.
4. Equipment reliability and freedom from operational and maintenance attention.
5. Initial investment and operating cost.
6. Possibility of recovery or conversion of contaminant into a salable product.

Known engineering principles must be applied even in areas of extremely dilute concentration.

Emission standards may set collection efficiency, but specific regulations do not exist for many trace emissions. In such cases emission targets must be set by dose–exposure time relationships obtained from effects on vegetation, animals, and humans. With such information, a list of possible treatment methods can be made (Table 2).

Control devices that are too inefficient for a particular pollutant or too expensive can then be stricken from the list. Grade-efficiency curves should be consulted in evaluating particulate collection devices, and the desirability of dry or wet particulate collection should be considered, especially with respect to material recycle or disposal.

Other factors to be evaluated are capital investment and operating cost, material reuse or alternative disposal economics, relative ruggedness and reliability of alternative control devices, and the ability to retain desired efficiency under all probable operating conditions. Control equipment needs to be both rugged and reliable. Efficiency of control devices varies with processing conditions, flow rate, temperature, emission concentration, and particle size. Control devices should be designed to handle these

Table 2. Checklist of Applicable Devices for Control of Pollutants

	Pollutant Classification			
			Particulate	
Equipment Type	Gas	Odor	Liquid	Solid
Absorption	•	•		
aqueous solution				
nonaqueous				
Adsorption	•	•		
throw-away canisters				
regenerable stationary beds				
regenerable traveling beds				
chromatographic adsorption				
Air dispersion (stacks)	•	•	•	•
Condensation	•	•		
Centrifugal separation (dry)			•	•
Chemical reaction	•	•		
Coagulation and particle growth			•	•
Filtration				
fabric and felt bags				•
granular beds			•	•
fine fibers			•	•
Gravitational settling				•
Impingement (dry)				•
Incineration	•	•	•	•
Precipitation, electrical				
dry			•	•
wet	•	•	•	•
Precipitation, thermal			•	•
Wet collection[a]	•	•	•	•

[a] Includes cyclonic, dynamic, filtration, inertial impaction (wetted targets, packed towers, turbulent targets), spray chambers, and venturi.

variations. Combinations of gaseous and particulate pollutants can be especially troublesome as gaseous removal devices are often unsuitable for heavy loadings of insoluble solids. Concentrations of soluble particulates up to 11 g/m^3 have been handled in gas absorption equipment with some success. However, to ensure rapid particle solution, special consideration must be given to wet–dry interfaces and adequate liquid quantities.

CONTROL OF GASEOUS EMISSIONS

Five methods are available for controlling gaseous emissions: absorption, adsorption, condensation, chemical reaction, and incineration. Atmospheric dispersion from a tall stack, considered as an alternative in the past, is now less viable. Absorption is particularly attractive for pollutants in appreciable concentration; it is also applicable to dilute concentrations of gases having high solvent solubility. Adsorption is desireable for contaminant removal down to extremely low levels (less than 1 ppmv) and for handling large gas volumes that have quite dilute contaminant levels. Condensation is best for substances having rather high vapor pressures. Where refrigeration is needed for the final step, elimination of noncondensible diluents is beneficial. Incineration, suitable only for combustibles, is used to remove organic pollutants and small quantities of H_2S, CO, and NH_3. Specific problem gases such as sulfur and nitrogen oxides require combinations of methods and are discussed separately.

Absorption

Absorption is a diffusional mass-transfer operation by which a soluble (or semisoluble) gaseous component can be removed from a gas stream by causing the absorbable component to dissolve in a solvent liquid through gas–liquid contact. The driving force for absorption is the difference between the partial pressure of the soluble gas in the gas mixture and the vapor pressure of the solute gas in the liquid film in contact with the gas. If the driving force is not positive, no absorption will occur. If it is negative, desorption or stripping will occur and pollution of the gas being treated will actually be enhanced.

Absorption systems can be divided into those that use water as the primary absorbing liquid and those that use a low volatility organic liquid. The system can be a simple absorption in which the liquid (usually water) is used in a single pass and then disposed of while still containing the adsorbed pollutant. Alternatively, the pollutant can be separated from the absorbing liquid and recovered in a pure, concentrated form by stripping or desorption. The absorbing liquid is then used in a closed circuit and is continuously regenerated and recycled. Regeneration alternatives to stripping are pollutant removal through precipitation and settling; chemical destruction through neutralization, oxidation, or reduction; hydrolysis; solvent extraction; pollutant liquid adsorption; and so on.

Absorption is one of the most frequently used methods for removal of water-soluble gases. Acidic gases such as HCl, HF, and SiF_4 can be absorbed in water efficiently and readily, especially if the last contact is made with water that has an alkaline pH. Less soluble acidic gases such as SO_2, Cl_2, and H_2S can be absorbed more readily in a dilute caustic solution. The scrubbing liquid may be made alkaline with dissolved soda ash or sodium bicarbonate, or with NaOH (usually no higher a concentration in the scrubbing liquid than 5–10%). Lime is a cheaper and more plentiful alkali, but its use directly in the absorber may lead to plugging or coating problems if the calcium salts produced have only limited solubility. A technique often used, such as in the two-step flue gas desulfurization process, is to have a NaOH solution inside the absorption tower, and then to lime the tower effluent externally, precipitating the absorbed component as a slightly soluble calcium salt. The precipitate may be removed by thickening, and the regenerated sodium alkali solution is recycled to the absorber. Scrubbing with an ammonium salt solution is also employed. In such cases, the gas is often first contacted with the more alkaline solution and then with the neutral or slightly acid contact to prevent stripping losses of NH_3 to the atmosphere.

Nonaqueous Systems. Although water is the most common liquid used for absorbing acidic gases, amines (monoethanol, diethanol, and triethanolamine; methyldiethanolamine; and dimethylanaline) have been used for absorbing SO_2 and H_2S from hydrocarbon gas streams. Such absorbents are generally limited to solid particulate-free systems, because solids can produce difficult-to-handle sludges as well as use up valuable organic absorbents. Furthermore, because of absorbent cost, absorbent regeneration must be practiced in almost all cases.

Types and Arrangements of Absorption Equipment. Absorption requires intimate contact between a gas and a liquid. Usually means are provided to break the liquid up into small droplets or thin films (which should be constantly renewed through turbulence) to provide high liquid surface area for mass transfer and a fresh, unsaturated surface film for high driving force. The most commonly used devices are packed and plate columns, open spray chambers and towers, cyclonic spray chambers, and combinations of sprayed and packed chambers.

Adsorption

The attractive forces in a solid that exist between atoms, molecules, and ions, holding the solid together, are unsatisfied at the surface and thus are available for holding other materials such as liquids and gases. This phenomenon is known as adsorption. If the solid is produced in a highly porous form with extensive pores and microstructure, its adsorptive capacity can be greatly enhanced. Sorption can be used as a pollution-control measure to remove pollutant gases from an otherwise harmless gas desired to be released to the atmosphere. Adsorption is quite adaptable for removing such contaminants, especially VOCs, down to extremely low levels (less than 1 ppmv). Adsorption's best applications are (*1*) handling large volumes (hundred thousands of CFM) with dilute pollution levels and (*2*) removing the contamination level, regardless of gas quantity, down to only trace pollutant levels. Removal of solvent losses from large quantities of ventilation air is an example of the former, and recovery of toxic and hazardous vapors to extremely low concentrations exemplifies the latter. In any case, absorption may be used alone for the entire control requirement or in combination with other removal methods. In the latter case, adsorption typically becomes the final cleanup step because of its capability of achieving low emission concentrations.

In being adsorbed, a gas molecule travels to an adsorption site on the surface of the solid, where it is held by attractive forces and loses much of its molecular motion. This loss of kinetic energy is released as heat. The heat of adsorption is often close in magnitude to the heat of condensation for the species being adsorbed. Thus adsorption is always exothermic. Desorption is a reversal of the adsorption process and heat must be supplied to cause desorption to take place. Hence temperature rise tends to reverse the process or cause a loss in capacity of the sorbent. Cooling of the sorption bed (which is often difficult because of poor heat transfer within the bed) or precooling of the gas stream to be treated is desirable to provide a sink for the heat of adsorption being released. Some sorption processes can occur so strongly that they are irreversible, ie, the adsorbed material can only be desorbed by removal of some of the solid substrate. Such a process is referred to as chemisorption. For example, oxygen, under certain circumstances can be adsorbed so strongly on activated carbon that it can be removed from the solid only in the form of CO or CO_2.

The adsorbing solid is called the adsorbent or sorbent; the adsorbed material, the adsorbate or sorbate. A thorough discussion of adsorption processes for air pollution and design equations have been given (31), and other more general references are also available (32–41).

Types of Adsorbents. Commercially important adsorbents are activated carbon, other simple or complex metallic oxides, and impregnated sorbents. Activated carbon is a general adsorbent. It is composed primarily of a single species of neutral atoms with no electrical gradients between molecules. Thus there are no significant potential gradients to attract or orient polar molecules in preference to nonpolar molecules. For this reason, carbon has less selectivity than other sorbents and is one of the few that will work in absorbing organics from a humid gas stream. Because the polar water molecules attract each other as strongly as the neutral carbon, the latter tends to be slightly selective for organic molecules. However, some water is adsorbed, especially if its partial pressure is greater than that of the organic molecules. The water being adsorbed must be taken into consideration in selecting the sorptive capacity to be provided in the design. Typical sources of activated carbon are coconut and other nut shells, fruit pits, bituminous coal, hardwoods, and petroleum coke and residues.

Simple and complex metal oxides are polar and have a much greater degree of selectivity than carbon and a great preference for polar molecules. They can be useful for removal of a particular species from the gas stream but are ineffective when moisture is present, because most of these adsorbents are excellent desiccants. Siliceous adsorbents include materials such as silica gel, Fuller's and diatomaceous earth, synthetic zeolites, and molecular sieves. They are available in a wide range of capacities, with the best equaling the capacity of the best activated carbons.

Impregnated sorbents fall into three general classes: (*1*) those impregnated with a chemical reactant or reagent, (*2*) those in which the impregnant acts as a continuous catalyst for pollutant oxidation or decomposition, and (*3*) those in which the impregnant acts only intermittantly as a catalyst. Reagent impregnants chemically convert the pollutant into a harmless or adsorbable pollutant. Carbon may be impregnated with 10–20% of its mass, with bromine being used to react with olefins. Thus ethylene, which is poorly adsorbed from an air stream because of its low molecular weight, is converted at the brominated surface to 1,2-dibromoethane, which is readily adsorbed. Other impregnant reagents are iodine for collecting mercury vapor, lead acetate for collecting H_2S, and sodium silicate for collecting HF. Other applications of impregnated sorbents, continuous and intermittent catalytic sorbents have been discussed (42).

Desorption or Disposal. After the pollutants have been adsorbed, they may be disposed of by discarding the saturated adsorbent or, alternatively, the sorbent may be regenerated. Disposal may be attractive when the quantity of material to be adsorbed is small, occurs infrequently, or must be recovered in an inconvenient location, such as breathing losses from a tank vent in a remote area. In these cases, the cost of fresh adsorbent is insignificant compared with the cost or inconvenience of attempting regeneration. However, before disposal can be considered, one must ascertain the nature of the chemical species that has been adsorbed. If toxic or hazardous materials have been adsorbed, the entire adsorbent cartridge and sorbent must now be considered as hazardous or toxic, which will preclude its disposal except in ways approved by applicable regulations. Even if the adsorbate is not hazardous, one must consider whether it would be leachable from the sorbent, if disposed in an ordinary landfill. When disposal is desired, often the adsorbent is contained in a paper carrier or disposable cartridge. If so, and the sorbent is carbon, incineration may be considered the best and safest method of ultimate disposal. Also, sometimes it is possible to return the spent sorbent to the manufacturer for regeneration.

When economics dictate regeneration of the sorbent, desorption may be carried out *in situ* by a number of methods: (*1*) heating the bed, (*2*) evacuating the bed, (*3*) stripping with an inert gas, (*4*) replacement of the adsorbate with a more readily adsorbed material, or (*5*) a combination of two or more of these methods.

Types of Adsorption Equipment. Five distinct types of gaseous adsorption devices are available: (1) disposable and rechargeable canisters, (2) fixed regenerable beds, (3) traveling bed adsorbers, (4) fluid bed adsorbers, and (5) chromatographic bag houses.

For small flows, those of an intermittent or infrequent nature, and those with low sorbate concentration, disposable charges of carbon may be used.

Regeneratable beds are used when economics so dictates: the volume of gas treated or sorbate concentration is high enough to make recovery attractive, or the cost of fresh sorbate is higher than regeneration. For handling larger gas flows or higher pollutant concentrations for which a static bed would be too rapidly exhausted, a fluid bed adsorber might be used.

A newer development, described as chromatographic adsorption, consists of injecting a cloud of adsorbent particles into the gas stream. Adsorption occurs during concurrent flow of effluent and suspended particles in pneumatic transport. The sorbent is removed from the gas stream along with the adsorbed pollutant in a conventional bag filter. Some final adsorption takes place as the gas flows through the layers of adsorbent on the surface of the filter bags.

In another new technology, called pressure-swing adsorption (43), the adsorption bed is subjected to relatively short pulses of higher pressure gas containing the species to be adsorbed. The bed is then partially regenerated by reducing the pressure and allowing the adsorbed material to vaporize. By controlling the gas flow and the pressure, the pollutant can be transferred from effluent to another gas stream. This same result can be achieved by temperature fluctuation, but bed temperature changes cannot be as rapidly controlled as bed pressures.

Adsorption Applications. Typical pollution applications are

1. Odor control in food processing.
2. Solvent and odor control in chemical and manufacturing processes, including cleaning and degreasing operations; paint, coating, and printing operations; pulp and paper manufacture; and in tanneries.
3. Odor control from foundries and animal laboratories.
4. Radioactive gas control in the nuclear industry.

Small carbon-filled canisters have been mandated on U.S. automobiles in recent years to control evaporative fuel emissions from gasoline engines. Similar adsorption equipment is being used to control vapor emissions from fuel-tank filling in areas in noncompliance with NAAQS standards for ozone. Implementation of the 1990 Clean Air Reauthorization Act provisions for control of toxic and hazardous organic vapor emissions should result in increased use of adsorption pollution control applications (44).

Condensation

Condensation has been discussed as a pollutant control method (44–47). It is most applicable for vapors fairly close to their dew points. As such, it is suitable for hydrocarbons and organic compounds that have reasonably higher boiling points than ambient conditions and are present in the effluent gas in appreciable concentrations. Pollutants having reasonably low vapor pressures at ambient temperatures may be controlled satisfactorily in water-cooled or even air-cooled condensers. For more volatile solvents, two-stage condensation may be required, using cooling water in the first stage, and refrigeration in the second. Condensation is not a practical method of total control for reasonably volatile toxic or hazardous organics in appreciable concentrations in streams of noncondensables if the effluent concentration must be reduced to a few ppmv. Condensation may still be useful as a preliminary treatment method to recover valuable solvents or to reduce the capacity required for the final treatment method. Partial condensation can be useful when the stripped gas can be recycled to the process (rather than vented) or when it can be used as primary combustion air. (The remaining pollutants are incinerated.) Condensation can be an attractive pretreatment method when it can serve to cool the gas before final control by adsorption.

Type of Equipment. Condensation cooling can occur either by direct contact or by indirect cooling. In the former, the vapor is brought into intimate contact with a cooled or refrigerated liquid. With indirect cooling, a surface condenser having metal tubes is commonly employed. The tubes are cooled with another fluid on the other side of the wall. When appreciable noncondensibles are present, compression of the gas stream before treatment can reduce the temperature required to achieve the desired pollutant partial pressure. However, gas compression is seldom economical for pollution control unless higher pressures are needed for other process reasons. When low temperatures are required, consideration must be given to the possible presence of other materials that could solidify at the required temperature. Icing on condenser surfaces will quickly foul the heat-transfer capability of the condenser.

In direct contact condensers, intimate mixing and contact is brought about between the cold liquid and the gas to effect as close an approach to thermal and mass-transfer equilibrium as possible. The cold liquid may be sprayed into the gas in a spray tower or a jet eductor, or a tower with gas–liquid contacting internals can be used. The contacting tower can be a packed tower, sieve-plate tower, disk and doughnut or segmental baffle plate tower, or even a slat-packed chamber. Because the liquid becomes heated by both the gas and the condensing vapor, lower dewpoints can be reached with counterflow of gas and liquid, but many parallel flow devices are used. The cooling liquid is frequently recirculated through an external heat exchanger and recycled.

The recirculated cool liquid is often water, if temperatures near its freezing point are not required. (Even below the freezing point of water, a water–antifreeze mixture can be used if water itself is not being condensed from the gas, which would, of course, dilute the antifreeze mixture.) However, an appropriate low vapor pressure liquid can be used as the recirculated liquid. It may even be the substance being condensed from the gas stream. This has the advantage that no further steps are required to separate the material being condensed from the liquid being recirculated. Whatever the recirculated liquid, it should be remembered that the treated gas is going to be close to vapor–liquid equilibrium with the cooling liquid. Thus if the liquid has appreciable vapor pressure, the gas could become polluted with vapors from the cooling liquid.

When the recirculated cool liquid is a different substance from the vapor being condensed, consideration must be given to how the two materials will be separated. Use of a cool liquid with a low solubility or miscibility with the condensed vapor is often helpful, because a simple phase separator can be used.

Surface condensers are most often used when the vapors to be condensed constitute the major portion of the gas stream with only a small amount of noncondensibles present to be vented. Under such conditions, tubular condenser type heat exchangers may be used. When noncondensibles predominate, tubes finned on the gas side will give better heat transfer unless the condensing vapor will tend to scale the heat-transfer surface. In such cases, tubular condensers designed for ease of gas-side tube cleaning are used. Condensers may be either vertical or horizontal. Horizontal units are frequently pitched to provide for good drainage. When dewpoints no lower than 10°C above atmospheric air temperatures are satisfactory, air-cooled heat exchangers of the fin-fan type can be employed.

Design of either type of condenser follows heat-transfer methods used for gas dehumidifying design. Applicable heat-transfer methods for this purpose have been briefly reviewed and summarized (47).

Chemical Reaction

Removal of an objectional or hazardous gaseous pollutant by chemical reaction has interesting possibilities. It is difficult to generalize about such means because they are so specific to the chemistry of the species of concern. In addition, the process suitability can also vary with the pollutant concentration in the gas stream as well as the temperature and composition of the carrier gas. The unit operations of absorption and wet scrubbing provide opportunities to carry on chemical reactions by adding a chemical reactant to the absorbing or scrubbing liquid, such as an alkali to enhance the absorption of an acidic gas (discussed above.) Furthermore, the chemical nature of the contacting liquid could change or destroy the pollutant vapor (if reactive in nature) by the presence of an oxidizing agent, such as potassium permanganate ($KMnO_4$), hydrogen peroxide, ozone, strong (and hot) nitric acid, hypochlorites, and chlorates. In addition, agents can be added to remove the absorbed vapors from the liquid by precipitation, forming insoluble compounds with the gaseous pollutant. Likewise, in adsorption, comment has already been made about impregnated adsorbents, such as with bromine, iodine, lead acetate, and sodium silicate. Another category is that of catalytic adsorbents, which are oxcar catalysts resulting in the oxidation of organic pollutants. More process development research has been carried out on chemical methods for removal of SO_x and NO_x, probably because of their wide-spread occurrence in flue gases. Gas–solid reactions are feasible for removal of specific gaseous pollutants by injection of dry solids into the conveying gas steam. Hydrated lime injection to remove SO_2 is a prime example of such applications.

Some fundamental objectives to consider for chemical reactions are (*1*) convert the pollutant into a different material with a lower vapor pressure (ideally into a liquid or solid particle) or into another chemical species that is more easily collectible, (*2*) convert the pollutant molecule into a harmless or at least less harmful molecule, (*3*) destroy the pollutant.

Incineration

Incineration of gaseous contaminants is also known as thermal oxidation and fume incineration. It is primarily applicable to gaseous impurities that can be oxidized or burned to decompose the original molecules to simpler nonhazardous compounds such as CO_2 and H_2O. In a word, the process is simply controlled combustion. As such, it may be used to destroy airborne or air-mixed hydrocarbon gases, other organic vapors, and similar blowdown gases; mercaptans; and undesirable inorganic gases such as H_2S, HCN, CO, H_2, and NH_3. VOC emissions are frequently destroyed by incineration (48,49) as are gases evolving from landfills.

Substituted organic vapors can also be decomposed by incineration, but it is important to consider the complications imposed when passing the substituted radical through the combustion process. For instance, halogen-substituted organics will result in the release of the corresponding acidic hydrogen halide gas. Sulfur-containing groups will generally be oxidized to SO_2. These acidic gases generally must be scrubbed or otherwise removed from the incinerator effluent before the combustion products are released to the atmosphere.

Incineration also may be used to destroy odors in those cases in which the odor substance can be decomposed by combustion. With proper design, aerosol incineration could destroy combustible airborne liquid or solid particles using a burner much like one used to burn pulverized coal or activated sludge from sewerage treatment. However, literature references for aerosol incineration are essentially nonexistent. In such an incinerator, rapid ignition of particles by quickly heating them above their kindling temperature is required as well as providing adequate residence time and flame space for complete combustion. Particulates have a considerably slower burning rate than combustible gases. (Means to reduce the particulate size to subsieve size would greatly enhance the combustion speed.) The possible presence of noncombustible residue that can produce a solid or molten ash and its subsequent handling must also be considered.

Thermal oxidizers can be designed to yield from 95 to >99% destruction of all combustible compounds. They can be designed with a capacity to handle from 0.5 to 236 m^3/s and for inlet combustible pollutant concentrations ranging from 100 to 2000 ppmv.

Incinerator Precautions. Waste gases to be incinerated must be appreciably below the lower explosive limit for mixtures of the combustibles with air to prevent explosions and to protect equipment and personnel. This means only gases that are dilute in combustibles may be incinerated. The usual safety practice is to limit operation to those gases that are no more concentrated than 25% of the lower explosive limit.

Upstream processes must be protected against the possibility of flame flashback. If the waste gas is hydrogen rich, ordinary flame arrestors may not provide adequate flashback protection.

As an alternative to incineration, consideration should be given to collecting valuable hydrocarbons and organic solvents by other means for reuse, such as condensation, rather than destruction by combustion. If the quantities are appreciable, recovery for fuel value alone may be worthwhile. Gases containing sufficient combustibles to support combustion are not burned in an incinerator. Rather they are burned in flares, or waste heat recovery boilers, or used for process heat.

Incinerator Types and Operation. Two general types of gaseous incinerators are in use: direct flame and catalytic. Gases also can be incinerated by indirect heating through a heat-conducting wall, but a higher temperature is generally required for their ignition and combustion than when a flame is present.

In the direct flame type incinerator, gases are heated in a fuel-fired refractory chamber to their autoignition temperature where oxidation occurs with or without a visible flame. Autoignition temperatures vary with chemical structure, but are generally in the range of 540–760°C. Use of a fuel flame aids in both combustion and mixing. The presence of the flame itself does not change the oxidation process, but it does influence the time, temperature, and turbulence factors. The fuel flame may be either short and intense (as occurs with a premix fuel burner) or luminous and diffuse. The latter accelerates ignition of entrained particles because of flame radiation. Nevertheless, the entire waste stream must be heated to the autoignition temperature to oxidize the gaseous components.

A short intense fuel flame promotes turbulence and offers an opportunity to increase the effectiveness of the holding time provided at incineration temperature. It is believed to result in more economical use of burner fuel.

Figure 1 illustrates a simple stand-alone refractory incinerator and one built into the base of a stack. These are adequate for occasional oxidation of process blow-down gases preparatory to process shutdown for annual maintenance. However, they would have high fuel operating costs if used continuously. For such use, heat exchange or recovery is added to preheat the incoming waste gas with the hot exhaust.

Two types of thermal energy recovery systems are in common use: regenerative and recuperative. Regenerative systems use ceramic (or other dense, inert material) in stationary beds to capture heat from the combustion exhaust gases. As the bed approaches the combustion temperature, the incinerator exhaust gas is switched to a lower temperature bed. The waste gas to be incinerated is passed through the hot bed to preheat the waste gas before its entry into the combustion process. When the gas flows between various beds are switched, the process is reminiscent of the ceramic-filled stoves of the steel blast furnace industry. However, by using multiple beds, regenerative systems have achieved up to 95% recovery of the energy released in the combustion process.

Thermal efficiency depends on the process operating characteristics. With relatively constant waste gas flow and combustible concentration, heat regeneration can

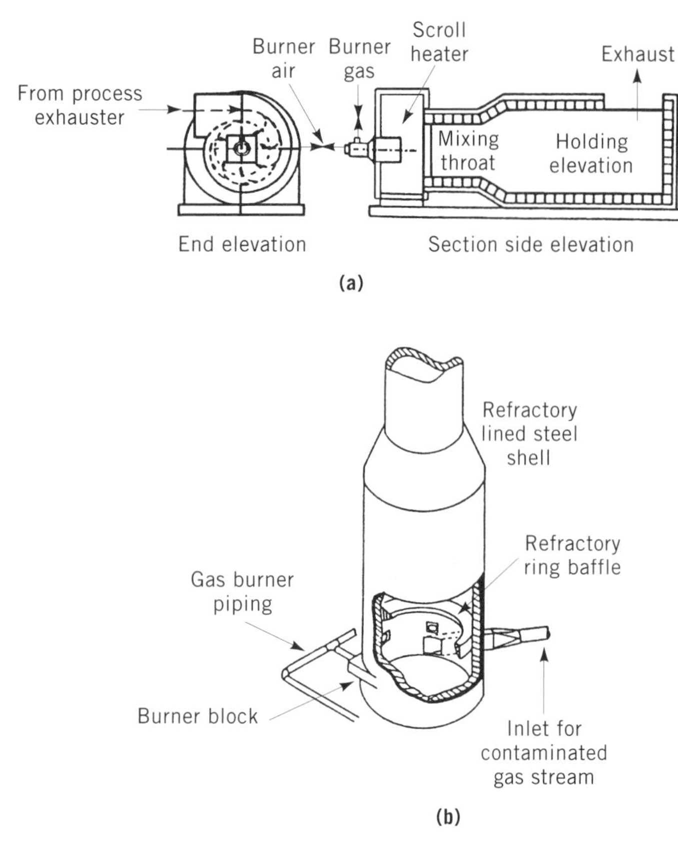

Figure 1. Simple direct flame incinerator types without heat interchange. (**a**) A modified scroll-type refractory air heater modified with a holding chamber to keep the gases at the temperature and time required for their complete oxidation. Reproduced from Ref. 50 with permission of Academic Press. (**b**) A vertical combustion chamber built into the base of a stack. The refractory ring baffle divides the chamber into a turbulent inlet section for gas and flame mixing, heat transfer, and ignition, followed by a temperature–time holding section to ensure complete destruction.

come close to a "no external fuel requirement" (except for start-up conditions and normal safety fuel requirements for pilot flames). However, processes with cyclic operation, or highly variable waste gas total flow, and highly variable combustible content are not as compatible for heat regeneration application.

Recuperation. Recuperative systems exchange heat directly from one gas stream to the other with a heat transfer surface. Typically, the interchanger may be a metallic shell and tube heat exchanger (perhaps finned to improve effective coefficients). Other alternatives are a rotating ceramic heat wheel, heated by one stream and cooled by the other, or the use of a metallic Lungstrum air preheater from steam boiler practice.

The maximum thermal recovery from a recuperative system is on the order of 70%. The advantage of the recuperative systems is the relatively short time needed for them to reach thermal equilibrium with changing conditions in incinerator operation. Their versatility yield themselves to responding well to cyclic operating conditions.

Catalytic Incineration. Catalytic incinerators oxidize substances at temperatures below those at which they would burn in air. Typical oxidation temperatures are around 260°C, but vary with chemical species being incinerated. Naphtha requires a gas and catalyst temperature of 230°C to initiate catalytic oxidation. Methane requires a temperature of 400°C, and catalytic oxidation of hydrogen can be initiated at room temperature. In general, the initial oxidation temperature decreases with increase in molecular size. Chemical structure also affects the initiation temperature. Within compounds with the same number of carbon atoms, the initiation temperature is lowest for highly unsaturated aliphatic compounds. Initiation temperature increases with structure in the following order (lowest to highest): acetylenic, olefinic, normal paraffin, branched-chain paraffin, aromatic.

Fuel must be used initially to start a cold catalytic incinerator, but once the catalyst is heated, frequently no further fuel is needed to keep the oxidation going. Thus the advantages of catalytic oxidation are lower fuel requirements, no NO_x formation (because of the low oxidation temperature), and compactness. Heat exchange is often needed to preheat the incoming waste gas with the oxidized exhaust gases. However, if the application is not carefully selected, many problems can arise. The catalyst can be poisoned by the presence of heavy metals, phosphates, and arsenic, destroying its activity. Its activity can be decreased temporarily by the presence of halogen and sulfur compounds, necessitating its off-line rejuvenation, often in a high temperature hydrogen environment. Furthermore, the surface of the catalyst may be rendered inactive with coatings of soot or fused inorganic dusts, requiring cleaning of the catalyst surfaces. Temperature control of a catalytic incinerator is critical, because the catalyst is easily damaged by overheating. Thus these units are often restricted to applications in which gas flow rate and combustible concentration is relatively constant.

Flares. Flares burning externally in the atmosphere, without containment or shielding, are unlikely to destroy hazardous and toxic combustible gases totally. The causes include too quick flame temperature quenching by the atmosphere before combustion is complete; variations in wind turbulence and direction from minute to minute; and excessive heat losses from the flame to the surroundings by radiation. To produce efficient oxidation, the flare should be enclosed on the sides of the flame with a lightweight housing and insulated internally with a high temperature, low heat capacity, fiberous material. This shields the flame from changes in external turbulence and reduces flame heat losses to the atmosphere.

Checking Incinerator Combustion Efficiency. If the combustion in a gaseous incinerator is incomplete, compounds such as aldehydes, organic acids, carbon, and carbon monoxide will usually be present.

Therefore, one method of testing for combustion efficiency is to collect a sample of the effluent and have it laboratory screened for the presence of these incomplete combustion products. Because many of these products have a distinctive and sometimes irritating odor, another accepted test method is to form an odor panel of five or six individuals with sensitive noses and have them cautiously sniff the effluent. If all agree that they can detect no odor in the effluent, combustion is considered to be acceptable. More detailed information on design and application of gaseous incinerators is available (50–56).

Specific Problem Gases

Sulfur dioxide, nitrogen oxides, and vehicular exhaust gases are widespread gas pollutants that present specific problems. The U.S. Clean Air Reauthorization Act of 1990 requires greater control of emission of these gases. Germany and Denmark adopted acid rain regulations on sulfur and nitrogen oxides earlier. Another widely emitted gas, carbon dioxide, is predicted to become a problem emission in the not too distant future because of its greenhouse effect. A few exploratory research studies are beginning to be published on control techniques for CO_2, apart from avoiding its emission by ceasing to burn fossil fuels (57,58).

Major sources of SO_x are the combustion of sulfur-containing fossil fuels, the manufacture of sulfuric acid and its recycled sludge acid purification, sulfur recovery or disposal from petroleum processing, nonferrous smelting, and pulp and paper manufacture. Combustion emissions are controlled by substituting a low sulfur fuel source, by fuel desulfurization and refining, and by sulfur removal either in the combustion process or from the flue gas. Many methods of sulfur removal from flue gas have been developed and voluminous literature is available (59–83). Utility experience with more than 100 flue gas desulfurization (FGD) plants in commercial operation in the United States has been developed. Removal of 90–95% of the SO_2 in the flue gas is practical for high sulfur coal, and 70–80% removal can be achieved for low sulfur coal.

Table 3 lists a number of FGD processes.

Nitrogen Oxides. Principal sources of nitrogen oxide emission are nitrogen fixation during high temperature combustion, nitric acid manufacture and concentration, organic nitrations, and vehicular emissions. During combustion in the presence of air, N_2 and O_2 react in the high temperature flame to produce NO. This reversible reaction favors NO formation as the temperature increases. Unfortunately, the kinetic rate for the reverse (decomposition) reaction drops essentially to 0 at temperatures below 870°C. Thus, as the flue gas cools, complete reversion of the NO to N_2 and O_2 does not take place. Nitrogen present in fuel compounds tends to oxidize much more readily to NO_x than does air. This NO_x from fuel nitrogen is essentially additive in concentration to that which is produced thermally from the excess air in high temperature combustion.

The reaction kinetics are usually too slow in most high temperature furnaces to produce equilibrium amounts of NO at the flame temperature. Nevertheless, in coal and oil combustion, it is fairly easy to produce NO_x concentrations of 1000–2000 ppmv with no more than 5–10% excess air.

NO reacts with O_2 in the atmosphere at a slow but steady rate such that all NO produced ends up in the atmosphere as NO_2. Combustion chemistry and NO_x formation have been reviewed (147). NO formation during combustion can be reduced by five methods: (*1*) maintaining low excess air (0.5% O_2 or less in the flue gas), (*2*) two-stage combustion, (*3*) flue gas recirculation, (*4*) lowering flame and combustion temperatures, and (*5*) burner and combustion chamber modifications. Combinations of two or more methods are also beneficial. Another method to reduce NO formation is to reduce firing capacity. This reduces the NO produced by lowering flame and gas temperatures, since the gases are chilled faster due to a larger ratio of heat sink area to heat input.

Selective Catalytic Reduction. A number of chemical processes have been developed for the reduction of NO_x which, while costly, are capable of reducing NO_x emission levels to 80–100 ppmv (or even lower). These have generally been grouped in the literature as selective catalytic reduction (SCR) and selective noncatalytic reduction (SNCR) processes. SCR (83–88) is the most commercially developed and applied method, using postcombustion technology. It was developed in Japan and applied primarily to oil-fired boilers with capacity totaling 10,000 MW. Subsequently, it has been applied in Germany to large coal-fired boilers. The original Japanese catalysts were platinum based, alumina supported in the shape of plates, or with rectangularly cored gas passages. Alternatively, the honeycomb design was available. The honeycomb catalyst was more compact per contact surface area and preferred when space limitation was important. The plate and cored–block design had lower pressure drop and could be cleaned when used in dusty boiler areas such as before the precipitator.

Recent literature shows that catalysts that are more resistant to poisoning or in other physical shapes, such as granular catalyst, are being developed. The American power industry has been exploring the use of SCR technology along with FGD. The SCR process can be installed similarly to the European installations, in a hot and dusty atmosphere, except that the catalyst would be subjected to SO_2. (Present catalysts appear to be degraded structurally in acidic environments.) An alternative is to install the SCR unit after FGD scrubbing.

However, the American power industry has recently been giving more attention to dry scrubbing with alkaline adsorbents. This would allow use of SCR after sulfur removal and filter collection of all particulates without dropping the gas temperature too greatly. Perhaps an even better alternative is to remove SO_2 and NO_x simultaneously with dry scrubbing, using a sorbent that can remove both simultaneously.

Selective Noncatalytic Reduction Processes. A small number of competing processes have been developed that depend solely on the thermal energy to supply the activating energy for the reaction. Thermal DeNox is the oldest of these; others are urea injection and urea–methanol injection.

Thermal DeNox. The Exxon Thermal DeNox process introduces ammonia alone into the gas stream for NO_x reduction. The process has been used primarily on gas-fired boilers and on petroleum industry still heaters. There have been some installations on oil and coal-fired utility boilers, on glass furnaces, and on some municipal solid waste incinerators. On the latter applications, removal efficiencies of 40–60% have been attained.

Urea Injection. Urea injection, developed and patented by EPRI and marketed by Fuel Tech, Inc., under the name

Table 3. Commercialized Flue Gas Desulfurization Processes

Process Name	Process Description	References
	Wet Throw-away Processes	
Dual alkali system	SO_2 absorbed in tower with $NaOH–Na_2SO_3$ recycle solution. CaOH or $CaCO_3$ added externally to precipitate $CaSO_4$, regenerate NaOH; make up NaOH or Na_2CO_3 added; process attempts to eliminate scaling and plugging problems of limestone slurry scrubbing	73,77
Limestone slurry scrubbing	Limestone slurry scrubs flue gas. SO_2 absorbed, reacted to $CaSO_3$; further air oxidized to $CaSO_4$, settled, and removed as sludge; lower cost and simpler than other processes; disadvantages: abrasive and corrosive, plugging and scaling, poor dewatering of $CaSO_4$	77–79,82
Dowa process	Similar to dual alkali except $Al_2(SO_4)_3$ solution used in scrubber; limestone addition regenerates reactant, precipitating $CaSO_4 \cdot 2H_2O$ crystals that dewater more readily; reduces plugging and scaling	
CHIYODA thoroughbred 121 process	Single vessel used to absorb SO_2 with limestone slurry and oxidize product to gypsum	
Forced oxidation	Limestone scrubbing, products air oxidized to gypsum in separate tank	
Lime spray dying	Wet–dry process; lime slurry absorbs SO_2 in vertical spray dryer forming $CaSO_3–CaS$, H_2O evaporated before droplets reach bottom or wall; dry solids collected in bag house along with fly ash	77
	Dry Throw-away Processes	
Direct injection	Pulverized lime or limestone injected into flue gas (often through burner); SO_2 absorbed on solid particles; high excess alkali required for fairly low SO_2 absorption; finer grinding, lime preheat, and flue gas humidification benefit removal; particulate collected in bag house	74,75,81
Trona sorption	Trona (natural Na_2CO_3) or Nacolite (natural $NaHCO_3$) injected into boiler; SO_2 absorbed to higher extent than with dry lime; product collected in bag house; also can capture high quantity of NO_x	74,75,80,81
	Wet Regenerative Processes	
Wellman-Lord	After flue gas pretreatment, SO_2 absorbed into Na_2SO_3 solution; solids and chloride purged, SO_2 stripped, regenerating Na_2SO_3, and SO_2 processed to S	76,77
Magnesium oxide process	SO_2 absorbed from gas with $Mg(OH)_2$ slurry, giving $MgSO_3–MgSO_4$ solids that are calcined with coke or other reducing agent, regenerating MgO and releasing SO_2	
Citrate-scrubbing	SO_2 absorbed with buffered citric acid solution; SO_2 reduced with H_2S to S; H_2S produced on site by reduction of S with steam and methane	73
Flakt-Boliden process	SO_2 absorbed with buffered citric acid solution; SO_2 stripped from solution with steam	
Aqueous carbonate	SO_2 absorbed into Na_2CO_3 solution in spray dryer, producing dry Na_2SO_3 particles	
SULF-X process	FeS slurry absorbs SO_2; product calcined producing S vapors that are condensed	
Conosox process	K_2CO_3 and K salt solutions absorb SO_2, forming K_2SO_4, which is converted to thiosulfate and $KHSO_3$, which is converted to H_2S; regenerates K_2CO_3	
	Dry Regenerative Processes	
Westvaco	SO_2 adsorbed in activated carbon fluid bed. SO_2, H_2O, and $\frac{1}{2}O_2$ react at 65–150°C forming H_2SO_4; in next vessel, $H_2SO_4 + 3H_2S$ at 150°C gives $4S + 4H_2O$; bed temperature is increased to vaporize some S; remaining S reacts with H_2 to H_2S	
Copper oxide adsorption	SO_2 adsorbed on copper oxide bed forming $CuSO_4$; bed is regenerated with H_2 or $H_2–CO$ mixture, giving concentrated SO_2 stream; bed is reduced to Cu but reoxidized for SO_2 adsorption	74

NOxOUT process (89), has been commercially installed on two boilers in Europe. It is generally applied by making a 50% solution of urea in water, which is sprayed into the boiler flue gas to react with NO_x.

Urea–Methanol Injection. Emcotek has patented a two-stage process in which urea solution is injected into the gas stream, followed somewhat further downstream with the injection of methanol. The principal purpose of the methanol injection is to reduce ammonia slip and to reduce deposits in the air preheater. NO_x reduction of 65 to 80% is reported with ammonia slip below 5 ppmv in gas temperature ranges of 815–1040°C.

Other SNCR Research. Other SNCR research processes have involved the injection of ammonium sulfate and volatilized cyanuric acid to reduce NO_x (90–91). The cyanuric acid process appears to have good prospects for NO_x reduction from diesel engine exhaust (both large stationary engines and over-the-road engines), but cyanuric acid is a high cost reagent for large-scale use. Recent SNCR possibilities for NO_x control have been discussed (92).

Other NO_x Emission Source Controls. Control of NO_x emissions from nitric acid and nitration operations is usually achieved by NO_2 reduction to N_2 and water, using natural gas in a catalytic decomposer (93–95). NO_2 from nitric acid–nitration operations is also controlled by absorption in water to regenerate nitric acid.

CONTROL OF PARTICULATE EMISSIONS

The removal of particles (liquids, solids, or mixtures) from a gas stream requires deposition and attachment to a surface. The surface may be continuous, such as the wall and cone of a cyclone or the collecting plates of an electrostatic precipitator, or it may be discontinuous, such as spray droplets in a scrubbing tower. Once deposited on a surface, the collected particles must be removed at intervals without appreciable reentrainment in the gas stream. One or more of seven physical principles (Table 4) are frequently employed to move particles from the bulk gas stream to the collecting surface. In some instances, a few other principles such as diffusiophoresis and methods of particle growth and agglomeration have also been used.

Gravity Settling

The gravity settling chamber is one of the oldest forms of gas–solid separation. It may be nothing more than a large room where the well-distributed gas enters at one end and leaves at the other. Such chambers were used at the turn of the century for collecting products such as lamp-black. Although mechanical conveyors might minimize labor costs, gravity settlers have largely disappeared because of bulky size and low collection efficiency. They are generally impractical for particles smaller than 40–50 μm.

Centrifugal and Cyclonic Collection

A cyclonic collector is a stationary device that uses gas in vortex flow, produced either by tangential entry of the gas or by spin baffles with axial gas inlet, to collect particles. Centrifugal force acting on the particles in the gas stream causes them to migrate to the cylindrical containing wall where they are collected by inertial impaction. Because the centrifugal force developed can be many times that of gravity, small particles can be collected in a cyclone, especially in a cyclone of small diameter.

Cyclone Efficiency. Most cyclone manufacturers provide grade-efficiency curves to predict overall collection efficiency of a dust stream in a particular cyclone. Many investigators have attempted to develop a generalized grade-efficiency curve for cyclones (108). One problem is that a cyclone's efficiency is affected by its geometric design.

Cyclone Problems. Problems may be encountered in cyclone application because of fouling and caking or from erosion or when using multiple cyclones. Multiple cyclones are designed so that each cyclone handles a prorated share of gas and dust and the overall efficiency of the system is the same as that calculated for an individual unit. This is the case, however, only when each cyclone receives identical dust fractions (size and loading) and gas flow. Because cyclone efficiency increases with flow and dust loading and is affected by particle-size distribution, the design of the inlet gas distribution system must accomplish the proper distribution. Otherwise, those cyclones with lower gas flow and dust concentration (and perhaps finer dust) will have much poorer efficiency. When multiple cyclones share a common dust hopper, it is important that all cyclones have essentially uniform pressures at the cone apex. Wall caking, unequal gas flow or dust distribution resulting from pressure drop decreases that occur with increases in dust loading, or partial plugging of cone or cyclone inlets can all cause unequal apex pressures. Unequal pressures will cause gas from higher pressure cyclones to flow into the dust hopper and back into the cyclones having lower apex pressure. This short-circuiting can result in heavy dust reentrainment and decreased efficiency.

Fouling of cyclone walls is usually caused by sticky or hydroscopic particulates, or by moisture or other vapor condensation. To prevent particle sticking, a highly polished finish, a graniteware glass coating, or fluorocarbon plastic lining may be used on the walls or alternatively, revolving wall scrapers. Cables or chains are sometimes suspended from the center of the vortex outlet. Vortex flow causes the cable or chain to rotate and thus scrape the wall, sometimes freeing the buildup. Condensation must be prevented by decreasing the dewpoint of the gas or by heating or insulating the cyclone wall. In extreme situations, the cyclones may be enclosed in a heated chamber.

Erosion can be a severe problem even in well-designed cyclone installations when handling a heavy loading of coarse and abrasive angular particles. One answer is lining the cyclone, using protective materials. The cyclone may be made of wear-resistant plate. Linings may be hard and thick sacrificial castings; wear-resistant applied welded coatings; or of rubber, ceramic shapes, reinforced castable, or brick. Alternatively, cyclones may be used in series, increasing the velocity as dust loading is decreased. In this method, first-stage efficiency will be low, but the bulk of the coarser and more abrasive particles will be collected, permitting higher velocities in second- and even third-stage cyclones. Dust particles smaller than 5–10 μm do not cause appreciable erosion.

Other problems affecting cyclone efficiency are usually caused by abuse or poor maintenance. Problems may arise from temperature warpage, rough interior surfaces, overlapping plates and rough welds, or misalignment of parts, such as an uncentered (or cocked) vortex outlet in the barrel.

Other Centrifugal Collectors. Cyclones and modified centrifugal collectors are often used to remove entrained liquids from a gas stream. Cyclones for this purpose have been described (109,110). The rotary stream dust separator (111,112), a newer dry centrifugal collector with improved collection efficiency on particles down to 1–2 μm, is considered more expensive and hence has been found

Table 4. Physical Principles Affecting Particle Movement and Collection

Control Principle	Related Equations	Conditions and Assumptions	References
Gravity settling	$U_t = \frac{D_p^2(\rho_t - \rho_g)g}{18\,\mu g} = \left[\frac{4\,D_p(\rho_t - \rho_g)g}{3\rho_g C_D}\right]^{\frac{1}{2}}$	free falling, rigid sphere; fluid continuum; for $N_{RE} < 0.1$, viscous, streamline flow, no wake formation, $C_D = 24/N_{RE}$; for $N_{RE} > 0.1$, other functions of N_{RE} must be used to calculate C_D	96–98
	$K_m \cong 1 + \frac{A\lambda}{D_p}$	Cunningham-Stokes correction factor for small particles when fluid does not behave as continuum; in air, for $D_p = 1\ \mu m$, $K_m = 1.17$; for $D_p = 0.1\ \mu m$, $K_m = 2.7$	
Centrifugal deposition	$\frac{dr}{dt} = U_{t,n} = \frac{D_p^2 \rho_t V_{cT}^2}{18\,\mu_g r}$	the tangential velocity, V_{cT}, thus the particle velocity, $U_{t,n}$, is a function of the radius r, usually $V_{cT} \propto 1/r^n$; for free gas rotation and conservation of momentum, $n = 1$; for cyclones, n varies between 0.5 and 0.7	99,100
Flow line interception	$\eta = \frac{1}{2.00 - \ln N_{RE}}\left[(1 + N_{SF})\ln(1 + N_{SF}) - \frac{N_{SF}}{2}\frac{(2 + N_{SF})}{(1 + N_{SF})}\right]$	for cylindrical target; N_{SF} is the ratio of particle diameter, D_p, to the diameter of the collector; assumes that particle will be collected if it approaches within $D_p/2$ from collector; none of the particles are reentrained	101,102
Inertial impaction	$N_{SI} = \frac{U_t U_o}{D_b} = \frac{K_m(\rho_t - \rho_g)D_p^2 U_o g}{18\,\mu_g D_b}$	viscous flow; Stokes' law region; physically, N_{SI} is the stopping distance in a quiescent fluid of a particle with initial velocity U_0/D_p; Figure 21 relates N_{SI} to collection efficiency	102,103
Diffusional deposition	$D_v = \frac{K_m k T}{3\pi\,\mu_g D_g}$	particle diffusivity from Stokes-Einstein equation; assumes Brownian motion; D_p same order of magnitude or greater than mean free path of gas molecules (0.1 μm at NTP)	
	$\eta_D = \frac{4}{N_{PE}}(2 + 0.557\,N_{SC}^{\frac{3}{8}} N_{RE}^{\frac{1}{2}})$	efficiency for spherical collector; $N_{SC} < 10^5$ or $D_p \le 0.5\ \mu m$ in ambient air	104
	$\eta_D = \frac{\pi}{N_{PE}}\left(\frac{1}{\pi} + 0.55\,N_{SC}^{\frac{1}{3}} N_{RE}^{\frac{1}{3}}\right)$	efficiency for cylindrical collector; $1 < N_{RE} < 10^4$ and $N_{SC} < 100$	
Electrostatic precipitation	$U_e = \frac{E_o E_p D_p}{4\pi\,\mu_g K_v}$	conductive spherical particles; streamline gas flow; gas behaves as continuum; negligible particle acceleration	105
	$\eta = 1 - e^{-(U_e A_e/q_e)}$	Deutsch-Anderson equation; assumes no reentrainment from collector; well-mixed turbulent flow, turbulent eddies small compared to precipitator dimensions	
Thermal precipitation	$U_r = -\frac{1}{5(1 + \pi a/8)}\frac{k_{gTR}}{P}\frac{dT}{dX}$	D_p less than mean free path of gas; free molecular or Knudsen flow regime; a = fraction of inelastic collisions, usually taken as 0.81	106,107
(Thermophoresis)	$U_r = -\frac{3\,K_m \mu_g}{2\rho_g T(2 + k_p/k_g)}\frac{dT}{dX}$	D_p same order of magnitude or greater than mean free path of gas molecules	

less attractive than cyclones unless improved collection in the 2–10-μm particle range is a necessity. A number of inertial centrifugal force devices as well as some others termed dynamic collectors have been described in the literature (111).

Electrostatic Precipitation

An electrical precipitator can collect either solid or liquid particles efficiently. Using a special design, it can also collect solids and liquids in combination and adsorb gases. Particles entering a precipitator are charged in an electric field and then move to a surface of opposite polarity where deposition occurs. For particles $\le 2\ \mu m$, electrical forces are stronger than any other collectional force. Thus precipitators have the highest energy use efficiency. Advantages are low pressure drops (often 250–500 Pa), low electric power consumption, and low operating costs (mostly capital-related charges). In addition, particulates are recovered in an agglomerated form, rendering them more easily collectible in case of reentrainment. Unfortunately, precipitators are also the most capital-intensive of all control devices, and mechanical, electrical, and process problems can cause poor on-stream time and reliability.

Precipitators are currently used for high collection efficiency on fine particles.

Precipitators can be classed as single or two stage. In single-stage units, used for industrial gases, the particles are charged and collected in the same electrical field. Negative discharge (gas-ionizing) polarity is practically always used in single-stage precipitators, because higher voltages can be achieved without sparkover. However, negative polarity also produces O_3 from O_2-containing gases. In two-stage precipitation, the particles are charged in an ionizing section and collected or precipitated in a separate section. Two-stage units are used for air purification, air-conditioning (qv), or ventilation. They are operated at lower voltage so there is less electrical hazard, less sparking, and fewer fires. These units have also received some consideration for the collection of condensed hydrocarbon mists. A positive polarity is always used for room air-conditioning to avoid ozone formation.

Electrically Augmented Particle Collection. In the 1970s considerable developmental research was devoted to improved fine particle collection by using a combination of electrostatic forces and other collection mechanisms such as inertial impaction or Brownian diffusion. A number of devices were developed and some successfully applied. Little subsequent progress has been made, presumably because of the reduced availability of governmental funding. As mentioned earlier, the electrostatic attractive force is the strongest collecting force available for particles finer than 2–2.5 μm. It is, therefore, logical to couple it with other collecting mechanisms in an attempt to build more highly efficient and cheaper fine-particle collectors. The forces operating between charged and uncharged bodies decrease in magnitude in the order (113): (*1*) coulomb force, the attraction of a charged particle to an oppositely charged particle or surface; (*2*) charged particle–uncharged conducting collector; (*3*) uncharged particle–charged collector; and (*4*) charged particle in a space-charged repulsive field. These relationships have been quantified (114). The coulomb force, or migration of a particle in a gradient electric field, is, of course, employed in both single-stage and two-stage precipitators. Electrostatics is also operative for charged particles and oppositely charged water droplets. A collection efficiency of 85–95% for industrial gases in such a single-stage vertical spray tower and up to 99% collection efficiency utilizing two towers in series have been demonstrated (115,116). A number of retrofit installations have been made on existing spray towers to enhance collection of submicrometer particles.

An investigation of the collection of charged submicrometer particulates on an oppositely charged 0.8–2.0-mm fluidized bed of sand particles found up to 90% efficiency (117). The charged particle–uncharged collecting surface principle has been used and marketed in a device where particles are initially charged in a negative polarity ionizer before entering a grounded, irrigated, cross-flow, packed bed of Tellerette packing (118). The charged particles are brought close to wetted surfaces in the confines of tortuous paths through the bed, inducing a "mirror image" charge of opposite polarity in the water film that causes the particles to impact the water for collection. This device can be used for simultaneous particulate collection and gas absorption. Submicrometer particles have been collected at 85–90% efficiency in single-stage units and at 98% efficiency using two units in series. The principle of charging particles in an ionizer and collecting them dry on uncharged filter media such as cellulose fibers has been practiced for many years in some types of electronic air cleaners. This principle has been applied (119) to improved collection of fine particles in a bag dust filter. Another application of the charged particle–uncharged collector principle was its application to fine particle collection in venturi (wet) scrubbers. Precharging of particles in an ionizer permitted efficient collection of submicrometer particles in the venturi with up to a 50% reduction in venturi pressure drop, a large savings in energy. Unfortunately, a device marketed by Chemical Construction Co. disappeared after the company's bankruptcy, but a similar adaptation is available in France. A charged-droplet scrubber whose commercial design was installed on steel furnace fumes has been described (120). Severe corrosion and electrical loss problems resulted in withdrawal of the scrubber.

Research devoted to optimizing design of various electric augmentation hardware items includes exploration (121) of parameters for preionizer design and discussion (122) of detailed mathematical relations for types of charged-droplet scrubbing and means of charging spray particles. Wet scrubbing appears to have a valid place in air-pollution control because of its relatively low capital investment compared with other control techniques for fine particles. However, wet scrubbing is a marginal control method because of its poor efficiency–energy relationship on fine particles, and thus electric augmentation should be an attractive means of overcoming the weakness of wet scrubbing at moderate cost.

Particle Filtration

Filtration devices for particle collection can be divided into three categories: cloth filtration using either woven or felted fabrics in a bag or envelope, paper and mat filters, and in-depth aggregate bed filtration. The first type is used for dry particle removal from gases but cannot be employed when liquid particulates are present or condensation is imminent. Subclasses of cloth filters are dust filters using cloth in the form of a single envelope (pocket filter) and housings containing rows of stacked cannister filters. The pocket filter has low dust-handling capacity, and when pressure drop becomes too high, the element must be removed and either discarded or manually cleaned. It is used primarily for low dust loads, occasional emissions, or as a safeguard against broken bags after a normal bag-house filter. Likewise, the multiple canister filters cannot be cleaned and must be replaced once high pressure drop occurs. Filters in the form of fiber pads or pleated paper in frames are used for preparing clean air for process use or ventilation. They have limited dust-holding capacity and are seldom used for air pollution control.

Several types of aggregate-bed filters are available that provide in-depth filtration. Both gravel and particle-bed filters have been developed for removal of dry particulates but have not been used extensively. Filters have also been developed using a porous ceramic or porous metal filter surface. Mesh beds of knitted wire mesh, plastic, or glass fibers are used for the removal of liquid particulates and mist. They will also remove solid particles, but will plug rapidly unless irrigated or flushed with a particle-dissolving solvent.

Wet Scrubbing

Scrubbers can be highly effective for both particulate collection and gas absorption. Costs can be quite reasonable for the required efficiency, but the addition of water treatment for recycle or for waste disposal may make the total cost as great as any other collection method, depending on water treatment complications. Although scrubbers automatically provide cooling of hot gases, the water-saturated effluent may produce offensive plume condensation in cold weather. Many moist effluents become more corrosive than dry ones. Solids accumulation may occur at wet–dry interfaces and icing problems may occur around the stack in winter. For efficient fine-particle collection, energy consumption may also exceed that of dry collectors by an appreciable amount.

Scrubbers make use of a combination of particulate collection mechanisms. It is difficult to classify scrubbers predominantly by any one mechanism, but for some systems, inertial impaction and direct interception predominate. Semrau (123) proposed a contacting power principle for correlation of dust-scrubber efficiency: the efficiency of collection is proportional to power expended, and more energy is required to capture finer particles. This principle is applicable only when inertial impaction and direct interception are the mechanisms employed. Furthermore, the correlation is not general because different parameters are obtained for differing emissions collected by different devices. However, in many wet scrubber situations for constant particle-size distribution, Semrau's power law principle, roughly applies:

$$N_t = \alpha P_T^{\gamma}$$

The constants α and γ depend on the physical and chemical properties of the system, the scrubbing device, and the particle-size distribution in the entering gas stream.

Table 5 can be used as a rough guide for scrubber collection in regard to minimum particle size collected at 85% efficiency. In some cases, a higher collection efficiency can be achieved on finer particles under a higher pressure drop.

Wet scrubber collection efficiency may be unexpectedly enhanced by particle growth. Vapor condensation, high turbulence, and thermal forces acting within the confines of narrow passages can all lead to particle growth or agglomeration. Of these, vapor condensation produced by cooling is the most common. Condensation will occur preferentially on existing particles, making them larger, rather than producing new nuclei. Careful experimentation (125) has shown that the addition of small quantities of nonfoaming surfactants to the scrubbing water can enhance the collection of hydrophobic dust particles without further energy expenditure.

Venturi and High Energy Scrubbers. The venturi scrubber has been studied more intensively than any other wet scrubber, perhaps because of its ability to efficiently scrub any size particle by changing the pressure drop. The design readily lends itself to mathematical modeling. Gas is accelerated in the throat to velocities of 60–150 m/s where water is added either as a spray, as solid jets, or as a wall-flowing sheet. The water is atomized into small droplets by the high speed gas. Aerosol particles, moving at close to the gas speed, collide with the accelerating liquid droplets and are captured. A cyclonic collector is needed to remove the water mist produced by the venturi. Collection mechanisms have been studied (104,126–131). The liquid drop size produced has been investigated (132); the effect of water-injection method on venturi performance, examined (133); and a generalized pressure-drop prediction method has been developed (134).

Venturi scrubbers can be operated at 2.5 kPa to collect many particles coarser than 1 μm efficiently. Smaller particles often require a pressure drop of 7.5–10 kPa. When most of the particulates are smaller than 0.5 μm and are hydrophobic, venturis has been operated at pressure drops from 25 to 32.5 kPa. Water injection rate is typically 0.67–1.4 m^3 of liquid per 1000 m^3 of gas, although rates as high as 2.7 are used. Increasing water rates improves collection efficiency.

Wet Scrubber Entrainment Separation. Fiber pads and beds to collect fine liquid entrainment have been discussed. Unless fog is present from condensation, scrubber entrainment will be coarse and the high efficiency of fiber beds is not needed. Entrainment separators for scrubbers

Table 5. Particle Size Collection Capabilities of Various Wet Scrubbers[a]

Type of Scrubber	Pressure Drop, Pa[b]	Minimum Collectible Particle Dia, μm[c]
Gravity spray towers	125–375	10
Cyclonic spray towers	500–2,500	2–6
Impingement scrubbers	500–4,000	1–5
Packed and moving bed scrubbers	500–4,000	1–5
Plate and slot scrubbers	1,200–4,000	1–3
Fiber bed scrubbers	1,000–4,000	0.8–1
Water jet scrubbers		0.8–2
Dynamic		1–3
Venturi	2,500–18,000	0.5–1

[a] Refs. 114, 124.
[b] To convert Pa to psi, multiply by 1.450×10^{-4}.
[c] Minimum particle size collectible with approximately 85% efficiency.

are usually centrifugal swirl vanes, hook and zigzag eliminators, or momentum separators using a change in direction. Control of recycle water cleanliness is important to good wet scrubber performance and poor scrubber performance has been traced (135) to entrainment of dirty scrubbing liquid and also to temperature flashing (evaporation) of fine particulate contained in recycle spray water.

Developing Particulate Control Technology

Present control methods for particulates are least efficient in the size range from 0.2 to 2.5 μm; this range is the most costly to collect and energy intensive. Health studies indicate that particles in this size range are also those that penetrate most deeply into, and often become deposited in, the human respiratory system. This is the principal reason for the U.S. EPA change in the ambient air quality standard from total suspended particulate (TSP) concentration to a PM_{10} standard (ambient air particles equal to or smaller than 10-μm aerodynamic diameter). The new standard will undoubtedly place even more emphasis on the need to collect particulates in this difficult-to-control fine-particle range, and therefore, collection of this size range is most in need of improvements in technology. Improved collection requires the use of a separating force that is independent of gas velocity or of the growth of particles that can be more readily collected. Particle growth can be accomplished through coagulation (agglomeration), chemical reaction, condensation, and electrostatic attraction. Promising separation forces are the "flux forces" involving diffusiophoresis, thermophoresis, electrophoresis, and Stefan flow. Although particle growth techniques and flux-force collection theoretically can be considered independently, both phenomena are applied in many practical devices.

ODOR CONTROL

Odor is a subjective preception of the sense of smell. Its study is still in a developmental stage: information including a patent index has been compiled (136), 124 rules of odor preferences have been listed (137), detection and recognition threshold values have been given (138), and odor technology as of 1975 has been assessed (139). Odor control involves any process that gives a more acceptable perception of smell, whether as a result of dilution, removal of the offending substance, or counteraction or masking.

Odor Measurement

Both static and dynamic measurement techniques exist for odor. The objective is to measure odor intensity by determining the dilution necessary so that the odor is imperceptible or doubtful to a human test panel, ie, to reach the detection threshold, the lowest concentration at which an odor stimulus may be detected. The recognition threshold is a higher value at which the chemical entity is recognized. An odor unit, o.u., has been widely defined in terms of 0.0283 m^3 of air at the odor detection threshold. It is a dimensionless unit representing the quantity of odor that when dispersed in 28.3 L of odor-free air produces a positive response by 50% of panel members. Odor concentration is the number of cubic meters that one cubic meter of odorous gas will occupy when diluted to the odor threshold. Selection of people to participate in an odor panel should reflect the type of information or measurement required, eg, for evaluation of an alleged neighborhood odor nuisance, the test subjects should be representative of the entire neighborhood. However, threshold determinations may be done with a carefully screened panel of two or three people (140).

Static Dilution Methods. A known volume of odorous sample is diluted with a known amount of nonodorous air, mixed, and presented statically or quiescently to the test panel. The ASTM D1391 syringe dilution technique is the best known of these methods and involves preparation of a 100-mL glass syringe of diluted odorous air, which is allowed to stand 15 min to ensure uniformity. The test panel judge suspends breathing for a few seconds and slowly expels the 100-mL sample into one nostril. The test is made in an odor-free room with a minimum of 15 min between tests to avoid olfactory fatigue. The syringe dilution method is reviewed from time to time by the ASTM E18 Sensory Evaluation Committee, which suggests and evaluates changes. Instead of a syringe, a test chamber may be used, which can be as large as a room (141,142).

Dynamic Dilution Methods. In this method, odor dilution is achieved by continuous flow. Advantages are accurate results, simplicity, reproducibility, and speed. Devices known as dynamic olfactometers control the flow of both odorous and pure diluent air, provide for ratio adjustment to give desired dilutions, and present multiple, continuous samples for test panel observers at ports beneath odor hoods.

Odor Control Methods

Absorption, adsorption, and incineration are all typical control methods for gaseous odors; odorous particulates are controlled by the usual particulate control methods. However, carrier gas, odorized by particulates, may require gaseous odor control treatment even after the particulates have been removed. For oxidizable odors, treatment with oxidants such as hydrogen peroxide, ozone, and $KMnO_4$ may sometimes be practiced; catalytic oxidation has also been employed. Odor control as used in rendering plants (143), spent grain dryers (144), pharmaceutical plants (145–147), and cellulose pulping (148) has been reviewed (149–157); some reviews are presented in two symposium volumes (152,153) from APCA specialty conferences. The odor-control performances of activated carbon and permanganate–alumina for reducing odor level of air streams containing olefins, esters, aldehydes, ketones, amines, sulfide, mercaptan, vapor from decomposed crustacean shells, and stale tobacco smoke have been compared (154). Activated carbon produced faster deodorization in all cases. Activated carbon adsorbers have been used to concentrate odors and organic compounds from emission streams, producing fuels suitable for incineration (155). Both air pollution control and energy recovery were accomplished. A summary of odor control technology is available (156).

Acknowledgment

Much of this article has been taken from J. Kroschwitz and M. Howe-Grant, eds., *Kirk-Othmer Encyclopedia of Chemical Technology,* 4th ed. (*ECT*4), Vol. 1, John Wiley & Sons, Inc., New York, 1991 and M. Grayson, ed., *Kirk-Othmer Encyclopedia of Chemical Technology,* 3rd ed. (*ECT*3), John Wiley & Sons, Inc., New York, 1978.

BIBLIOGRAPHY

1. *Guiding Principles of State Air Pollution Legislation,* U.S. Department of Health, Education, and Welfare, Washington, D.C., 1965.
2. Sect. 1420, Chapt. 111, *General Laws,* Chapt. 836, *Acts of 1969,* the Commonwealth of Massachusetts, Department of Public Health, Division of Environment, Health, Bureau of Air Use Management.
3. G. T. Wolff, in "Air Pollution," *ECT* 4, Vol. 1, p 711.
4. D. H. Slade, ed., *Meteorology and Atomic Energy 1968,* U.S. Atomic Energy Commission, July 1968; available as *TID-24190,* Clearinghouse for Federal Scientific and Technical Information National Bureau of Standards, U.S. Department of Commerce, Springfield, Va.
5. D. B. Turner, *Workbook of Atmospheric Dispersion Estimates, US EPA, OAP, Pub. AP26,* Research Triangle Park, N.C., revised 1970, U.S. Deparatment Printing Office Stock No. 5503-0015.
6. A. D. Busse and J. R. Zimmerman, *User's Guide for the Climatological Dispersion Model, US EPA Pub. No. EPA-R4-73-024,* Research Triangle Park, N.C., Dec. 1973.
7. M. Smith, ed., *Recommended Guide for the Prediction of the Dispersion of Airborne Effluents,* American Society of Mechanical Engineers, New York, 1968.
8. G. A. Briggs, "Plume Rise," *USAEC Critical Review Series TID-25075,* NTIS, Springfield, Va., 1969.
9. *Effective Stack Height: Plume Rise, US EPA Air Pollution Training Institute Pub. SI:406,* with Chapts. D, E, and G by G. A. Briggs and Chapt. H by D. B. Turner, 1974.
10. J. E. Carson and H. Moses, *J. APCA* **19,** 862 (Nov. 1969).
11. H. Moses and M. R. Kraimer, *J. APCA* **22,** 621 (Aug. 1972).
12. G. A. Briggs, *Plume Rise Predictions, Lectures on Air Pollution and Environmental Impact Analyses,* American Meteorological Society, Boston, Mass., 1975.
13. *Guideline on Air Quality Models, OAQPS Guideline Series,* U.S. Environmental Protection Agency, Research Triangle Park, N.C., 1980.
14. G. A. Schmel, *Atmos. Environ.* **14,** 983–1011 (1980).
15. N. E. Bowne, R. J. Londergan, R. J. Minott, D. R. Murray, "Preliminary Results from the EPRI Plume Model Validation Project—Plains Site," Report EPRI EA-1788, Electric Power Research Institute, Palo Alto, Calif., 1981.
16. N. E. Bowne, "Atmospheric Dispersion," in S. Calvert and H. M. Englund, eds., *Handbook of Air Pollution and Technology,* John Wiley & Sons, Inc., New York, 1984, 859–891.
17. *Code of Federal Regulations* 40 (CFR 40), *Fed. Reg.,* C-50–99.
18. Ref. 17, part 58.
19. W. W. Lund and R. Starkey, *J. Air Waste Manage Assn.* **40**(6), 896–897 (June 1990).
20. "Hard Realities: Air and Waste Issues of the 90s" (Report of 18th Government Affairs Seminar), *J. Air and Waste Management Assn.* **40**(6), 855–860 (June 1990); also *Proceedings of the 18th Air and Waste Management Assn. Govt. Affairs Seminar,* Air and Waste Management Assn., Pittsburgh, Pa., 1990.
21. R. H. Perry, ed., *Engineering Manual,* 3rd ed., McGraw-Hill, Inc., New York, 1976. Text used with permission.
22. *Industrial Ventilation,* 15th ed., Sect. 4, American Conference of Governmental Industrial Hygienists, Committee on Industrial Ventilation, Lansing, Mich., 1978.
23. R. Jorgensen, *Fan Engineering,* 7th ed., Buffalo Forge Company, Buffalo, N.Y., 1970, pp. 471–480.
24. W. E. L. Hemeon, *Plant and Process Ventilation,* Industrial Press, Inc., New York, 1954.
25. J. M. Dalla Valle, *Exhaust Hoods,* Industrial Press Inc., New York, 1952.
26. J. L. Alden, *Design of Industrial Exhaust Systems for Dust and Fume Removal,* 3rd ed., Industrial Press Inc., New York, 1959.
27. J. A. Danielson, ed., *Air Pollution Engineering Manual, Pub. No. 999-AP-40,* U.S. Department of Health, Education, and Welfare, Cincinnati, Ohio, 1973, Chapt. 3.
28. H. D. Goodfellow, in J. A. Buonicore and T. Davis, eds., *Air Pollution Engineering Manual,* Von Nostrand Reinhard, New York, 1992, Chapt. 6, pp. 155–206.
29. B. B. Crocker, "Capture of Hazardous Emissions," in *Proceedings, Control of Specific (Toxic) Pollutants* (Conference, Feb. 1979, Gainsville, Fla.), Air Pollution Control Assn., Pittsburgh, Pa., 1979, pp. 415–433.
30. B. B. Crocker, *Chem. Eng. Prog.* **64,** 79 (Apr. 1968).
31. B. B. Crocker and K. B. Schnelle, Jr., in Ref. 16, chapt. 7, pp. 135–192.
32. D. M. Ruthven, "Adsorption" in *ECT* 4, Vol. 1, pp. 493–528.
33. J. D. Sherman and C. M. Yon, "Adsorption, Gas Separation," in *Ect* 4, Vol. 1, pp. 529–573.
34. M. Suzuki, *Adsorption Engineering,* Kodansba-Elsevier, Tokyo, 1990.
35. A. E. Rodrigues, M. D. LeVan, and D. Tondeur, *Adsorption, Science and Technology,* NATO ASI E158, Kluwer, Amsterdam, 1989.
36. R. T. Yang, *Gas Separation by Adsorption Processes,* Butterworths, Stoneham, Mass., 1987.
37. P. Wankat, *Large Scale Adsorption and Chromatography,* CRC Press, Boca Raton, Fla., 1986.
38. T. Vermeulen, M. D. LeVan, N. K. Hiester, and G. Klein, in Ref. 57, Sec. 16.
39. D. M. Ruthven, *Principles of Adsorption and Adsorption Processes,* Wiley-Interscience, New York, 1984.
40. A. Turk, in A. C. Stern, ed., *Air Pollution,* 3rd ed., Vol. 5, Academic Press, New York, 1977, pp. 329–363.
41. R. J. Buonicore, in Ref. 45, pp. 31–52.
42. Ref. 40, pp. 337–339.
43. W. M. Edwards, in R. H. Perry and D. Green, eds., *Perry's Chemical Engineers' Handbook,* 6th ed., McGraw Hill Book Co., New York, 1984, pp. 16–36.
44. J. J. Spivey, *Environ. Progress* **7**(1), 31–40 (Feb. 1988).
45. S. M. Hall, *J. Air Waste Manage Assn.* **40**(3), 404–407 (Mar. 1990).
46. Ref. 31, pp. 185–189.
47. A. J. Buonicore, in Ref. 28, pp. 52–58.
48. V. S. Katari, W. M. Vatavuk, and A. H. Wehe, Part I, *J. APCA* **37**(1), 91 (Jan. 1987); Part II, *J. APCA* **37**(2), 198–201 (Feb. 1987); M. Kosusko and C. M. Nunez, *J. Air Waste Manage Assn.* **40**(2), 254–255 (Feb. 1990); M. A. Palazzolo and B. A. Tichenor, *Environ. Progress* **6,** 172–176 (Aug. 1987).

49. E. N. Ruddy and L. A. Carroll, *Chem. Eng. Progress* **89**(7), 28–35 (July 1993).
50. H. J. Paulus, in A. C. Stern, ed., *Air Pollution,* 2nd ed., Vol. 3, Academic Press, New York, 1968, p. 528.
51. J. Hirt, in Ref. 16, Chapt. 8, pp. 193–201.
52. A. J. Buonicore, in Ref. 28, pp. 58–70.
53. D. R. van der Vaart, M. W. Vatavuk, and A. H. Wehe, *J. Air Waste Manage. Assoc.* **41,** 92–98 (Jan. 1991); 497–501 (April 1991).
54. P. Acharya, S. G. DeCicco, and R. G. Novak, *J. Air Waste Manage. Assoc.,* **41,** 1605–1615 (Dec. 1991).
55. L. Takacs and G. L. Moilanen, *J. Air Waste Manage. Assoc.* **41,** 716–722 (May 1991).
56. M. A. Palazzolo and B. A. Tichenor, *Environ. Prog.* **6,** 172–176 (Aug. 1987).
57. T. B. Simpson, *Environ. Prog.* **10,** 248–250 (Nov. 1991).
58. H. Herzog, D. Goglomb, and S. Zemba, *Environ. Prog.* **10,** 64–74 (Feb. 1991).
59. R. W. Coughlin, R. D. Siegel, and C. Rai, eds., *AIChE Symp. Ser.* **70,** (137) (1974). Contains four papers on flue gas desulfurization, four papers on coal desulfurization, and three papers on petroleum desulfurization.
60. C. Rai and R. D. Siegel, eds., *AIChE Symp. Ser.* **71,** (148) (1975). Contains seven papers on flue gas desulfurization, two on petroleum desulfurization, one on coal desulfurization, and fifteen on NO_x control.
61. J. A. Cavallaro, A. W. Dearbrouck, and A. F. Baher, *AIChE Symp. Ser.* **70,** (137); 114–122 (1974).
62. K. S. Murthy, H. S. Rosenberg, and R. B. Engdahl, *J. APCA* **26,** 851–855 (Sept. 1976).
63. A. V. Slack, *Chem. Eng. Prog.* **72,** 94–97 (Aug. 1976).
64. R. M. Jimeson and R. R. Maddocks, *Chem. Eng. Prog.* **72,** 80–88 (Aug. 1976).
65. *AIChE Symp. Ser.* **68,** (126) (1972). Contains four papers on flue gas desulfurization and two on NO_x control.
66. *Control Technology: Gases and Odors,* APCA Reprint Series, Air Pollution Control Assn., Pittsburgh (Aug. 1973). A reprint of 1970–1973 APCA journal articles: six papers on flue gas desulfurization, three on NO_x control, and two on SO_2 control from pulp and paper mills.
67. *Sulfur Dioxide Processing,* Reprints of 1972–1974 *Chem. Eng. Prog.* articles, AIChE, New York (1975). Contains thirteen papers on flue gas desulfurization, two on SO_2 control in pulp and paper, one on sulfuric acid tail gas, one on SO_2 from ore roasting, and two on NO_x from nitric acid.
68. E. L. Plyler and M. A. Maxwell, eds., *Proceedings, Flue Gas Desulfurization Symposium,* (New Orleans, La., Dec. 1973). *Pub. No. EPA-650/2-73-038,* U.S. EPA, Research Triangle Park, N.C., 1973. Contains 34 papers on flue gas desulfurization.
69. *Proceedings: Symposium on Flue Gas Desulfurization,* (Atlanta, Ga., Dec. 1974), U.S. EPA, Research Triangle Park, N.C.
70. *Proceedings: Symposium on Flue Gas Desulfurization* (New Orleans, Mar. 1976), U.S. EPA, Research Triangle Park, N.C. Contains 36 papers dealing with flue gas desulfurization.
71. *Proceedings of the Third Stationary Source Combustion Symposium, EPA 600/7-79-050a (NTIS PB 292 539)* U.S. EPA, Industrial Environmental Research Laboratory, Research Triangle Park, N.C., Feb. 1979.
72. G. M. Blythe and co-workers, *Survey of Dry SO_2 Control Systems, EPA-600/7-80-030 (NTIS PB 80 166853),* U.S. EPA, Industrial Research Laboratory, Research Triangle Park, N.C., Feb. 1980.
73. *Definitive SO_2 Control Process Evaluation: Limestone Double Alkali and Citrate FGD Process,* EPA Pub. EPA-600/7-79-177, Aug. 1979.
74. J. D. Mobley and K. J. Lim, Ref. 16, chapt. 9, pp. 193–213.
75. G. M. Blythe and co-workers, *Survey of Dry SO_2 Control Systems, EPA Pub. EPA-600/7-80-030 (NTIS PB 80-166853),* U.S. EPA, Research Triangle Park, N.C., Feb. 1980.
76. R. Pedroso, *An Update of the Wellman-Lord Flue Gas Desulfurization Process, EPA Pub. EPA-600/2-76-136a,* May 1976.
77. T. W. Devitt, in Ref. 16, pp. 375–417.
78. *Flue Gas Desulfurization Systems and SO_2 Control, Pub. GS-6121,* Electric Power Research Institute, Palo Alto, Calif., October 1988. An abstracted bibliography of EPRI Reports and Projects.
79. *Introduction to Limestone Flue Gas Desulfurization, Pub. CS-5849,* Electric Power Research Institute, Palo Alto, Calif., 1988.
80. *SO_2 Removal by Injection of Dry Sodium Compounds, Pub. RP-1682,* Electric Power Research Institute, Palo Alto, Calif., 1987.
81. C. S. Chang and C. Jorgensen, *Environ. Progress* **6**(1), 26 (Feb. 1987).
82. M. T. Melia and co-workers, "Trends in Commercial Applications of FGD," in *Proceedings: Tenth Symposium on FGD, EPRI Report CS-5167,* Electric Power Research Institute, Palo Alto, Calif. (May 1987).
83. *Proceedings of the EPA/EPRI First Combined Flue Gas Desulfurization and Dry SO_2 Control Symposium,* Oct. 25–28, 1988; St. Louis, Mo., Electric Power Research Institute, Palo Alto, Calif.
84. Ref. 113, pp. 203–213.
85. D. Eskinazi, J. E. Cichanowicz. W. P. Linak, and R. E. Hall, *JAPCA* **39,** 1131–1139 (Aug. 1989).
86. A. Kokkinos, J. E. Cichanowicz, R. E. Hall, and C. B. Sedman, *J. Air Waste Manage. Assoc.* **41,** 1252–1259 (Sept. 1991).
87. D. Cobb, L. Glatch, J. Ruud, and S. Snyder, *Environ. Prog.* **10,** 49–59 (Feb. 1991).
88. L. C. Hardison, G. J. Nagl, and G. E. Addison, *Environ. Prog.* **10,** 314–318 (Nov. 1991).
89. *C&EN* 22 (April 18, 1988).
90. R. A. Perry and D. L. Siebers, *Nature* **324,** 657–658 (1986).
91. R. K. Lyon, *Environ. Sci. Technol.* **21**(3), 231–236 (1987).
92. S. L. Chen, R. K. Lyon, and W. R. Seeker, *Environ. Prog.* **10,** 182–185 (Aug. 1991).
93. O. J. Adlhart, S. G. Hindin, and R. E. Kenson, *Chem. Eng. Prog.* **67,** 73 (Feb. 1971); D. J. Newman, *Chem. Eng. Prog.* **67,** 79 (Feb. 1971).
94. G. R. Gillespie, A. A. Boyum, and M. F. Collins, *Chem. Eng. Prog.* **68,** 72 (Apr. 1972).
95. R. M. Reed and R. L. Harvin, *Chem. Eng. Prog.* **68,** 78 (Apr. 1972).
96. R. Clift and W. H. Gauvin, *Can. J. Chem. Eng.* **49,** 439 (1971).
97. F. A. Zenz and D. F. Othmer, *Fluidization and Fluid Particle Systems,* Reinhold, New York, 1960, pp. 206–207.
98. C. A. Lapple and co-workers, *Fluid and Particle Mechanics,* University of Delaware, Newark, Del., 1956, p. 292.
99. C. B. Shepherd and C. E. Lapple, *Ind. Eng. Chem.* **31,** 972 (1939).

100. C. B. Shepherd and C. E. Lapple, *Ind. Eng. Chem.* **32,** 1246 (1940).
101. K. E. Lunde and C. E. Lapple, *Chem. Eng. Prog.* **53,** 385 (Aug. 1957).
102. C. Y. Shen, *Chem. Rev.* **55,** 595 (1955).
103. W. E. Ranz and J. B. Wong, *Ind. Eng. Chem.* **44,** 1371 (1952).
104. H. F. Johnstone and M. H. Roberts, *Ind. Eng. Chem.* **41,** 2417 (1949).
105. H. J. White, *Industrial Electrostatic Precipitation,* Addison-Wesley Publishing Co., Reading, Mass., 1963.
106. B. Singh and R. L. Byers, *Ind. Eng. Chem. Fund.* **11,** 127 (1972).
107. L. Waldmann, *Z. Naturforsch,* **14A,** 589 (1959).
108. D. Leith and W. Licht, *AIChE Symp. Ser.* **68**(126), 196–206 (1972).
109. A. C. Stern, K. J. Caplan, and P. D. Bush, *Cyclone Dust Collectors,* American Petroleum Institute, New York, 1955; H. J. Tengbergen in K. Rietema and C. G. Verner, eds., *Cyclones in Industry,* Elsevier Publishing Co., Amsterdam, 1961 (in English).
110. B. B. Crocker in "Phase Separation" in Ref. 43, pp. 18-70–18-88.
111. W. Strauss, *Industrial Gas Cleaning,* 2nd ed., Pergamon Press, Inc., Oxford, 1975.
112. K. J. Caplan in Ref. 50, Chapt. 43.
113. D. W. Cooper, "Fine Particle Control by Electrostatic Augmentation of Existing Methods," *Preprint 75-02.1, 68th APCA Annual Meeting, Boston, Mass., June 15–20, 1975.*
114. K. A. Nielsen and J. C. Hill, *Ind. Eng. Chem. Fundam.* **15,** 149, 157 (1976).
115. M. J. Pilot, *J. APCA* **25,** 176 (Feb. 1975).
116. M. J. Pilot and D. F. Meyer, *University of Washington Electrostatic Spray Scrubber Evaluation, NTIS PB 252653,* Apr. 1976.
117. K. Zahedi and J. R. Melcher, "Electrofluidized Beds in the Filtration of Submicron Particulate," *Preprint 75-57.8, 68th APCA Annual Meeting, Boston, Mass., June 15–20, 1975.*
118. W. L. Klugman and S. V. Sheppard, "The Ceilcote Ionizing Wet Scrubber," *Preprint 75-30.3, 68th APCA Annual Meeting, Boston, Mass., June 15–20, 1975.*
119. Helfritch, *Chem. Eng. Progress* **73,** 54–57 (Aug. 1977).
120. C. W. Lear, W. F. Krieve, and E. Cohen, *J. APCA* **25,** 184 (Feb. 1975).
121. D. H. Pontius, L. G. Felix, and W. B. Smith, "Performance Characteristics of a Pilot Scale Particle Charging Device," *Preprint 76-42.6, 69th APCA Annual Meeting, Portland, Oregon, June 27–July 1, 1976.*
122. J. R. Melcher and K. S. Sachar, "Charged Droplet Scrubbing of Submicron Particulate," *NTIS Pub. PB-241262,* Aug. 1974.
123. K. T. Semrau, C. W. Margnowski, K. E. Lunde, and C. E. Lapple, *Ind. Eng. Chem.* **50,** 1615 (1958).
124. R. L. Lucas in R. H. Perry and C. H. Chilton, eds., *Chemical Engineers Handbook,* 5th ed., McGraw-Hill, Inc., New York, 1973, p. 20–98. G. J. Celenza, *Chem. Eng. Prog.* **66,** 31 (Nov. 1970).
125. H. E. Hesketh, "Atomization and Cloud Behavior in Wet Scrubbers," *U.S.-USSR Symposium on Control of Fine Particulate Emissions,* Jan. 15–18, 1974.
126. H. F. Johnstone, R. B. Field, and M. C. Tassler, *Ind. Eng. Chem.* **46,** 1601 (1954).
127. W. Barth, *Staub* **19,** 175 (1959).
128. S. Calvert and D. Lundgren, *J. APCA* **18,** 677 (Oct. 1968).
129. S. Calvert, D. Lundgren, and D. S. Mehta, *J. APCA* **22,** 529 (July 1972).
130. R. H. Boll, *Ind. Eng. Chem. Fundam.* **12,** 40 (Jan. 1973).
131. H. E. Hesketh, *J. APCA* **24,** 939 (Oct. 1974).
132. R. H. Boll, L. R. Flais, P. W. Maurer, and W. L. Thompson, *J. APCA* **24,** 934 (Oct. 1974).
133. S. W. Behie and J. M. Beeckmans, *J. APCA* **24,** 943 (Oct. 1974).
134. K. G. T. Hollands and K. C. Goel, *Ind. Eng. Chem. Fundam.* **14,** 16 (Jan. 1975).
135. T. R. Blackwood, *Environ. Progress* **7**(1), 71–75 (Feb. 1988).
136. J. P. Cox, *Odor Control and Olfaction,* Pollution Sciences Publishing Company, Lynden, Washington, 1975.
137. R. W. Moncrieff, *Odour Preferences,* John Wiley & Sons, Inc., New York, 1966.
138. W. H. Stahl, ed., *Compilation of Odor and Taste Threshold Values Data,* American Society for Testing and Materials Data Series 48, Philadelphia, Pa., 1973.
139. P. N. Cheremisinoff and R. A. Young, eds., *Industrial Odor Technology Assessment,* Ann Arbor Science Publishers, Inc., Ann Arbor, Mich., 1975.
140. J. Wittes and A. Turk, *Correlation of Subjective–Objective Methods in the Study of Odors and Taste,* American Society for Testing and Materials Special Technical Publication 440, Philadelphia, Pa., 1968, pp. 49–70.
141. G. Leonardos, D. Kendall, and N. Barnard, *J. APCA* **19,** 91 (1969).
142. A. Turk, *Basic Principles of Sensory Evaluation,* American Society for Testing and Materials Special Technical Publication #433, 1968, pp. 79–83.
143. R. M. Bethea and co-workers, *Environ. Sci. Technol.* **7,** 504 (1973).
144. M. W. First and co-workers, *J. APCA* **24,** 653 (1974).
145. G. A. Herr, *Chem. Eng. Prog.* **70,** 65 (1974).
146. D. E. Quane, *Chem. Eng. Prog.* **70,** 51 (1974).
147. D. J. Eisenfelder and J. W. Dolen, *Chem. Eng. Prog.* **70,** 48 (1974).
148. J. E. Paul, *J. APCA* **25,** 158 (1975).
149. J. E. Yocom and R. A. Duffee, *Chem. Eng.* **77**(13), 160 (1970).
150. M. Beltran, *Chem. Eng. Prog.* **70,** 57 (1974).
151. R. M. Bethea, *Engineering Analysis and Odor Control,* Chapt. 13, pp. 203–214, ref. 371.
152. *Proceedings, State of the Art of Odor Control Technology Specialty Conference,* March 1974, Air Pollution Control Association, Pittsburgh, Pa., 1974.
153. *Proceedings, State of the Art of Odor Control Technology Specialty Conference,* March 1977, Air Pollution Control Association, Pittsburgh, Pa., 1977.
154. A. Turk, S. Mehlman, and E. Levine, *Atmos. Environ.* **7,** 1139 (1973).
155. W. D. Lovett and F. T. Cunniff, *Chem. Eng. Prog.* **70,** 43 (May 1974).
156. W. H. Prokop, in Ref. 28, Chapt. 5, pp. 147–154.

Reading List

R. G. Bond and C. P. Straub, eds., *Handbook of Environmental Control,* Vol. 1, CRC Press, Cleveland, Ohio, 1972.

A. J. Buonicore and W. T. Davis, eds., *Air Pollution Engineering Manual,* Van Nostrand Reinhold, New York, 1992.

S. Calvert and H. M. Englund, eds., *Handbook of Air Pollution Technology,* John Wiley & Sons, Inc., New York, 1984.

P. N. Cheremisinoff, ed., *Encyclopedia of Environmental Control Technology,* Vol. 2, Gulf Publishing Co., Houston, 1989.
P. N. Cheremisinoff and R. A. Young, *Air Pollution Control and Design Handbook,* Parts 1 and 2, Marcel Dekker, New York, 1977.
R. A. Corbett, *Standard Handbook of Environmental Engineering,* McGraw-Hill Publishing Co., New York, 1989.
C. N. Davies, ed., *Aerosol Science,* Academic Press, Inc., New York, 1966.
H. E. Hesketh, *Air Pollution Control,* Ann Arbor Science Publishers, Ann Arbor, Mich., 1979.
L. H. Keith, *Compilation of EPA's Sampling and Analysis Methods,* Lewis Publishers, Inc. (CRC Press Inc, Boca Raton, FL), 1991.
W. Licht, *Air Pollution Control Engineering: Basic Calculations for Particulate Collection,* Marcel Dekker, Inc., New York, 1980.
B. Y. H. Liu, ed., *Fine Particles—Aerosol Generation, Measurement, Sampling and Analysis,* Academic Press, Inc., New York, 1976.
J. J. McKetta, *Unit Operations Handbook,* Vols. 1 and 2, Marcel Dekker, Inc., New York, 1992.
H. C. Perkins, *Air Pollution,* McGraw-Hill, Inc., New York, 1974.
R. H. Perry and D. Green, eds., *Perry's Chemical Engineers' Handbook,* 6th ed., McGraw-Hill Book Co., New York, 1984.
W. Ruch, ed., *Chemical Detection of Gaseous Pollutants,* Ann Arbor Science Publishers, Inc., Ann Arbor, Mich., 1966.
J. H. Seinfeld, *Air Pollution: Physical and Chemical Fundamentals,* McGraw-Hill, Inc., New York, 1975.
A. C. Stern, ed., *Air Pollution,* 3rd ed., Vols. 1–5, Academic Press, Inc., New York, (Vols. 1–3, 1976; Vols. 4–5, 1977).
W. Strauss, *Industrial Gas Cleaning,* 2nd ed., Pergamon Press, Oxford, 1975.
L. Theodore and A. J. Buonicore, *Air Pollution Control Equipment—Selection, Design, Operation and Maintenance,* Prentice-Hall, Englewood Cliffs, N.J., 1982.
L. K. Wang and N. C. Pereira, eds., *Handbook of Environmental Engineering,* Vol. 1—Air and Noise Pollution Control, The Humana Press, Clifton, N.J., 1979.
K. Wark and C. F. Warner, *Air Pollution: Its Origin and Control,* 2nd ed., Harper & Row, Publishers, New York, 1981.
P. O. Warner, *Analysis of Air Pollutants,* John Wiley & Sons, Inc., New York, 1976.

AIR POLLUTION: HEALTH EFFECTS

HENRY GONG, JR.
WILLIAM S. LINN
Rancho Los Amigos Medical Center
Downey, California and
University of Southern California School of Medicine

The recognition that air pollution can impair health is not new. Galen (born 130 AD) recognized the dangers from exposures to air contaminants in certain occupations. Treatises on occupational lung diseases and a recognition of community air pollution appeared during the Middle Ages. Governmental regulations intended to control air pollution have existed since the early 1300s. For example, harsh punishment (including death) was in store for people who violated regulations for burning coal in 14th-century England. Air pollution itself was subsequently identified as a direct cause of death during episodes in which pollutant levels were elevated well above those that occur on a daily basis, usually in association with certain meteorologic conditions. Thus, sharp increases in death rates were documented during severe pollution episodes in Belgium (Meuse River Valley) in 1930, the United States (Donora, Pennsylvania) in 1948, and England (London) in 1952 and 1962. The most notable recent example of a major episode was the London fog of December 1952 when approximately 4,000 people, out of an exposed population of 8 million, died prematurely from high levels of coal smoke. This disaster prompted serious efforts to control traditional air pollution, ie, coal smoke containing sulfur dioxide (SO_2) and a wide variety of particulate matter. The first quantitative analyses of public health statistics in Britain, along with the first report specifically associating air pollution with ill health and arguing for mitigation measures, were published during the 1660s (1).

In the United States the current system of air pollution regulations and related health research dates only from the late 1960s. Substantial progress has been made in understanding and preventing the health effects of environmental pollutants. Episodes of pollution still occur in the United States but not to the extent as in past decades. Internationally, ambient concentrations of pollutants have fallen to small fractions of their pre-1950 levels throughout the more affluent industrialized nations.

Although historical episodes of air pollution have been definitely associated with health effects, the major concern today has shifted to more subtle issues. The current emphasis is on the extent of health hazard posed by chronically low levels and intermittently high levels of outdoor and indoor pollution to which large populations are routinely exposed (2). The protection of at-risk or unusually susceptible groups is also important. Nitrogen dioxide (NO_2), ozone (O_3), particulates (including acidic aerosols), environmental tobacco smoke (ETS), asbestos, radon, and volatile organic compounds (VOCs) have become prominent health issues in the lay news media and biomedical research. Causal relationships of current air pollution to public health, such as diverse pollutant mixtures to which people are exposed, different exposure risks and heterogeneous responses within populations, and the multiplicity of environmental and lifestyle factors that can also contribute to increased morbidity and mortality, are not clear for many reasons. On the other hand, air pollution is a suspected contributor to exacerbations and premature deaths from asthma, as well as other chronic lung and heart diseases, on the basis of many statistical studies. Thus, the emphasis has shifted from avoiding unusually severe episodes with clinical diseases among highly exposed individuals to protecting large segments of the population against unacceptable risk from lower levels of pollution.

This article reviews scientific evidence concerning health effects of contemporary outdoor and indoor air pollution. Although many suggested health effects of environmental pollution remain controversial, epidemiologic, clinical, and animal toxicologic research has clearly identified changes at the molecular, cellular, organ, individual, and population levels that merit attention. This article initially reviews general concepts: common components of air pollution and methods used to measure them, health effects and their determinants, and the primary scientific approaches of assessing health effects. Respiratory effects are emphasized since the airways and lungs are the target sites for many inhaled pollutants. Known and uncertain aspects of the current database, as well as strengths and limitations of different research methodologies, will be discussed. Specific pollutants and their health effects will then be reviewed, along with possible control strategies. This review emphasizes information pertinent to the United States (U.S.), where many air pollutants have been systematically monitored, regulated, and subjected to formal health risk assessment. Substantive air quality problems in other developed countries are similar, although priorities and regulatory approaches may differ from those in the United States. The health effects of contemporary air pollution in other countries are reported elsewhere (2). It should be recognized that this review is necessarily superficial and that knowledge about health effects and risks from air pollution is continually evolving. Also, this article provides a current foundation and perspective for further reading, understanding, and questioning about this complex, multifaceted topic. See also AIR POLLUTION INDOOR; AIR POLLUTION: AUTOMOBILE; AIR POLLUTION CONTROL METHODS.

AIR POLLUTION

Definitions

Air pollution may be defined as the contamination of outdoor or indoor air by a natural or man-made agent in such a way that the air becomes less acceptable for intended uses, which, in this context, is the maintenance of human health. Although natural pollution sources are sometimes important, most pollution-related health problems result from man-made (anthropogenic) pollution involving mobile sources (eg, automobiles), outdoor stationary sources (eg, power plants, smelters, and factories), and various indoor sources (eg, building materials and combustion). Physically, air pollutants are dispersed into the atmosphere as gases, fibers, or as suspensions of liquid or solid particles in air (aerosols). Gases, in the form of discrete molecules, form true solutions within air. Fibers are arbitrarily defined as particles having a length at least three times their width. Aerosols may contain particles either of uniform size (monodisperse) or of different sizes (polydisperse or heterodisperse). The practicalities of particle monitoring dictate a focus on aerodynamic properties rather than the physical size of particles. Particles that move in an air stream, similar to one-micrometer spherical water droplets, have an aerodynamic diameter of one micrometer (μm), regardless of their physical size, shape, and density. Thus, aerosols are characterized by a mass median aerodynamic diameter (MMAD), a measure of central tendency, and a geometric standard deviation (GSD), a measure of dispersion.

Outdoor Air Pollutants

Clean air is a relative term. Chemically pristine or "pure" air is never found in nature because numerous natural sources continually contribute various agents, while meteorologic conditions mix and disperse them. For example, volcanic action and erosion inject particulate matter into the air, while decomposition of organic matter releases organic and inorganic particulates and gases. Globally, natural sources exceed man-made emissions and account for most of the atmospheric particulates and trace gases, except for SO_2 and carbon monoxide (CO). Levels of specific air pollutants at any given location depend on complex interactions of natural and anthropogenic sources which change over time (daily, seasonally, or annually) (3). Local anthropogenic sources often predominate. Thus, pollution patterns and associated health risks vary widely at different times and places because sources and meteorologic conditions continually change.

The major ambient pollutants with known or suspected adverse health effects are listed below.

Primary Pollutants	Secondary Pollutants
Carbon monoxide	Ozone
Sulfur dioxide	Nitrogen dioxide
Nitrogen oxides	Peroxyacetyl nitrate
Nitric oxide	Nitric and nitrous acid
Nitrogen dioxide	Suspended particulates
Volatile organic compounds	Sulfuric acid and sulfate salts
Suspended particulates	Nitrate salts
Metal compounds	Organic aerosols
Dusts	
Soots	

Some community air pollutants are directly released into the air. These are called primary pollutants. Those air pollutants not directly released into the atmosphere but are formed within it by chemical reactions among primary pollutants and normal constituents of air are called secondary pollutants. Secondary pollutants have been increasingly emphasized in health research. Classic urban air pollution from the combustion of coal and oil fuel, as observed in the northeastern U.S. and Europe, is characterized by primary pollutants with reducing chemical properties, ie, carbon particulates and SO_2. On the other hand, Southern California and other dry sunny regions have a characteristic oxidizing photochemical "smog" resulting from a complicated series of photochemical reactions involving primary pollutants (nitrogen oxides and hydrocarbons). Episodes of acidic sulfate particles (in the northeastern U.S.) and nitric acid vapors (in California) result from the oxidation of primary pollutants SO_2 and nitrogen oxides, respectively. Nonetheless, the distinction between primary and secondary pollutants is not always clear or practical. Both primary and secondary pollutants frequently contribute to the ambient pollutant mix in a given locale so that one cannot generally ascribe a health outcome to either alone. Certain pollutants may be primary or secondary, such as NO_2.

Sources and emission burdens of individual pollutants will not be discussed in detail here but are covered in other articles and cited references (3,4). In general, coal combustion emits more SO_2, CO, and particulates than oil combustion, whereas emission of nitrogen oxides (NO_x) is similar for both fuels. The fate of effluents from fossil fuel combustion depends on the manner in which they are released. Large modern stationary sources, such as electric power plants and smelters, typically discharge through tall stacks. This type of discharge diminishes local pollutant concentrations at ground level (except during episodic downdrafts) and enhances pollutant dispersion but increases residence and reaction times in the atmosphere, favoring the production of secondary pollutants and regional pollution problems. On the other hand, home furnaces discharge their effluents near roof level, resulting in higher, more localized ground level concentrations. The greatest source of air pollutants in the U.S. is vehicular exhaust. Fuel combustion in conventional gasoline and diesel engines is generally incomplete and results in exhaust gases containing varying amounts of CO, SO_2, NO_x, VOCs, and particulates. Industrial processes constitute a major and extremely varied and complex source of effluents. Specific industries vary greatly in the types and amounts of emissions during manufacturing, shipping, or storage of raw or processed products. The three main sources in terms of total tons emitted per year, are the chemical, primary metal, and paper industries.

Atmospheric particulates are characterized by numerous chemical species and multimodal size patterns (5). Ambient particles can be separated into two size modes–coarse and fine. Coarse particles are larger than 2.5 μm and are formed by mechanical, grinding, and other dispersive processes. Their major chemical constituents in urban areas are calcium, silicon, and aluminum. Fine particles (median size 2.5 μm or less) are generally secondary pollutants, formed by condensation processes from the vapor or gaseous state, followed by agglomeration. Sulfate and nitrate are the two largest constituents in terms of mass in outdoor urban air. From the biomedical standpoint, the aerodynamic size of particulates in air determines the potential for adverse health effects. Many coarse particles are too large to be inhaled beyond the nose and mouth (upper airways) and are not usually considered health hazards. The current federal standard regulates only particles 10 μm or less in aerodynamic diameter (PM_{10}) since these sizes are in the respirable range and potentially responsible for health effects. Thus, the upper size limit for collection of particles in contemporary air pollution studies is typically 10 μm MMAD.

Table 1. Common Indoor Air Pollutants[a] with Known or Possible Adverse Health Effects

Pollutant	Sources
Tobacco smoke	Cigarettes, cigars, pipes
Volatile organic compounds	Solvents, cleaning chemicals, paints, glues, polishes, waxes, aerosol sprays, pesticides
Carbon monoxide	Gas and wood stoves, kerosene heaters
Nitrogen dioxide	Gas and wood stoves, kerosene heaters
Formaldehyde	Furniture stuffing, paneling, particle board, foam insulation, veneer furniture, carpets, draperies
Ozone	Outdoor ozone, photocopying machines, laser printers, electrical air cleaners
Particulates	Multiple sources, eg, wood and coal stoves, fireplaces, tobacco smoke
Asbestos	Insulation, vinyl tiles
Radon	Soil gas, water, building materials, utility natural gas

[a] Though not always considered air pollutants, bioaerosols occur both indoors and outdoors as potential infectious or antigenic agents; they include spores, pollens, molds, bacteria, viruses, animal proteins.

Indoor Air Pollutants

Indoor air pollutants originate from both outdoor and indoor sources (Table 1) (6–9). Generally, outdoor pollutants which enter indoor environments are modified or lowered in concentration by various physical and chemical processes. For example, outdoor ozone reacts rapidly with surfaces and water vapor ("sinks") inside buildings so that its concentration is usually reduced by half or more. The concentrations of indoor air contaminants depend on many factors such as the source strength and ventilation within the indoor space. However, when indoor sources or human activities emit pollutants at rates exceeding their removal rates by ventilation or surface reactions, the indoor pollutants can accumulate to levels much higher than outdoor concentrations. Exposure to a wide range of indoor pollutants in different microenvironments (eg, residences, workplaces, schools or public buildings) is universal and virtually unavoidable. The most common exposures are to combustion emissions from appliances such as cooking ranges and furnaces and to hydrocarbon emissions from furnishing materials, hobbies, or cleaning activities. The principal pollutants generated by combustion are CO, NO_2, nitrous acid, SO_2, formaldehyde, VOCs, and particulate matter. Environmental tobacco smoke (ETS) also contributes high levels of gases and particulates. Asbestos-containing materials in thermal insulation, wallboard, and flooring release fibers into the air if disturbed or removed. Radon decay products vary in indoor concentrations and depend on the radium concentration and permeability of the soil, presence of openings through the foundations, and building ventilation. Allergens and other biological agents are present in indoor air in the form of fecal material from house dust mites and other insects, fungal spores, and dander from pets. Although biological agents are not always considered pollutants, they may impose a greater aggregate burden of illness than non-biological agents in some circumstances.

Other Environments with Air Pollutants

Microenvironments associated with transportation have not been well characterized but available evidence suggests that they can transiently expose occupants to high levels of different agents. For example, increased levels of CO, NO_2 and NO_x, particulates, and VOCs from automobile exhaust have been measured in automobile passenger compartments. Biologic aerosols (molds) from automobile air conditioners have also been documented. Localized high pollution levels have been measured in heavy traffic, at intersections, "street canyons" of cities, underground parking garages, parking lots, and drive-up facilities where automobile emissions are not substantially diluted before contact with people. Interestingly, some pollutants may be removed via local chemical interactions, eg, ozone is scavenged by nitric oxide in automobile exhaust, while acidic aerosols are neutralized by oral ammonia (10).

Occupational exposures to inorganic and organic dusts, VOCs, and particulates are well-documented health risks (11,12). Historically, major concerns and research have focused on inordinately high levels in the industrial setting. However, exposures with health consequences can also occur in "white-collar" offices and commercial settings where combustion products, VOCs, formaldehyde, biologic aerosols, or accumulated CO_2 are prevalent.

Air Quality Standards

Outdoor Pollutant Standards. The protection of public health has been the major driving force in air pollution control, an effort that includes legislated standards based upon judgments of scientific knowledge about health effects. The Clean Air Act of 1970, with subsequent amendments, is the basic law governing protection of air quality in the United States. The U.S. Environmental Protection Agency (EPA) is responsible for formulating and implementing the regulations governing two broad categories of air pollutants: "criteria" and "hazardous" air pollutants. The criteria pollutants are those agents identified by the U.S. EPA as airborne contaminants that "cause or contribute air pollution which may reasonably be anticipated to endanger public health" (42 USC 7408). The criteria pollutants have less alarming health effects at low exposure levels but occur widely and affect a large proportion of the population, causing a potentially great overall health impact. The Clean Air Act assumes that some level of exposure to criteria pollutants can be permitted without harm to health.

The EPA Administrator is required to 1) determine the lowest exposure level for each criteria pollutant which causes adverse health effects in the most susceptible segments of the population, and 2) set primary National Ambient Air Quality Standard (NAAQS) on the basis of this information. The NAAQS must prevent known adverse health effects and include a margin of safety to protect against health effects yet to be discovered. The Administrator must decide what constitutes an adverse health effect and how large a margin of safety is needed. Primary standards must be set without regard to cost. Cost may be considered in deciding what specific pollution control measures are required and how soon they must be implemented. Implementation is usually the responsibility of state and local governments. Thus, specific pollution control strategies and regulations vary from place to place, even though air quality standards apply nationwide.

The EPA Scientific Staff has primary responsibility for compiling and evaluating appropriate scientific evidence, and then recommending a standard to the Administrator. The staff's work is reviewed by a Clean Air Scientific Advisory Committee (CASAC) composed of independent scientists who advise the EPA on health-related issues (13,14). Other independent scientists are frequently consulted by the EPA staff or the CASAC to provide additional review. Virtually all relevant, peer-reviewed information from the scientific literature is summarized in an Air Quality Criteria Document. A shorter staff paper is then prepared to advise the EPA Administrator as to which information is most important and how it should be applied in the choice of a standard. A period of public comment is allowed when a document is issued or a standard is proposed. The Clean Air Act requires that this entire process operate on a five-year cycle. In practice, the process may take considerably longer. At the end of each cycle the EPA Administrator will either reaffirm or modify the existing Standard or set a new NAAQS on the basis of current scientific information.

Six pollutants that commonly occur in outdoor air are regulated by the U.S. EPA as criteria pollutants with individual NAAQS (Table 2): ozone, nitrogen dioxide, sulfur dioxide, carbon monoxide, particulate matter which is 10 μg or less in aerodynamic diameter (PM_{10}), and lead. The averaging times for the standards are chosen to provide a biologically relevant measure of exposure and, at the same time, a margin of safety. For example, the current NAAQS for CO is set at 9 parts per million (ppm), by volume, for an 8-hour average and 35 ppm for a one-hour average so that the blood concentrations of carboxyhemoglobin (COHb) are less than 2%, based on pharmacokinetic data about CO uptake and elimination in the body. State and local governments are required by federal law to monitor air quality in most urban areas of the U.S. In 1990, over 4,000 monitoring sites reported data to the EPA's Aerometric Information Retrieval System. Areas that do not meet the air quality standards are called "nonattainment areas." The number of nonattainment areas has decreased dramatically since the 1970s. However, many heavily pop-

Table 2. National Ambient Air Quality Standards (NAAQS)[a] in United States, 1993

Pollutant	Maximum Allowable Concentration	Averaging Time
Ozone (O_3)	0.12 ppm (235 $\mu g/m^3$)	One hour
Nitrogen dioxide (NO_2)	0.053 ppm (100 $\mu g/m^3$)	Annual arithmetic mean
Sulfur dioxide (SO_2)	0.14 ppm (365 $\mu g/m^3$)	24 hours
	0.03 ppm (80 $\mu g/m^3$)	Annual arithmetic mean
Carbon monoxide (CO)	35 ppm (40 $\mu g/m^3$)	One hour
	9 ppm (10 $\mu g/m^3$)	8 hours
Particulates (PM_{10})[b]	150 $\mu g/m^3$	24 hours
	50 $\mu g/m^3$	Annual arithmetic mean
Lead (Pb)	1.5 $\mu g/m^3$	Quarterly arithmetic mean

[a] Primary standards designed to protect public health.
[b] Particles with aerodynamic diameter of 10 μm or less.

ulated urban and rural areas in the U.S. remain in nonattainment for one or more of the criteria pollutants. For example, 98 counties with 135 million people are presently classified as nonattainment areas for ozone (15). Seventy counties (mostly in western U.S.) are nonattainment areas for inhalable particulates (PM_{10}). Fifty counties in the Ohio River Valley and northeastern U.S. are in nonattainment for sulfur dioxide. Only metropolitan Los Angeles is in nonattainment for nitrogen dioxide. Outdoor levels of airborne lead have decreased significantly since the removal of lead from gasoline in the mid-1970s, and all monitored areas meet the lead standard.

The Clean Air Act also aims at controlling levels of other airborne chemicals for which no ambient air quality standard can be applied but "which may reasonably be anticipated to result in an increase in mortality or an increase in serious irreversible, or incapacitating reversible illness" (42 USC 7412). Title III of the 1990 Amendments (16) lists 189 substances or groups of substances as Hazardous Air Pollutants (HAP) or "air toxics" (see examples below).

	Compounds with:
Benzo[a]pyrene	Arsenic
Benzene	Beryllium
Carbon disulfide	Cadmium
Carbon tetrachloride	Chromium
Chlorine	Cobalt
Chloroform	Lead
Formaldehyde	Manganese
Hydrochloric acid	Mercury
Phenol	Nickel
Toluene	

The toxic air pollutants, which include organic compounds such as benzene and mineral compounds such as cadmium, are associated with cancer or other catastrophic health effects. Technology-based emission standards are imposed on industrial sources of HAP to protect public health. Subsequent to implementation and compliance with these standards, an analysis of the remaining "residual risks" is also required.

The EPA Administrator may also establish secondary air quality standards to prevent unfavorable effects of air pollutants on public welfare, eg, damage to natural ecosystems, crop loss, or minor effects on humans that are considered nuisances rather than adverse health effects.

Indoor Pollutant Guidelines. There are no federal regulations of indoor air quality per se. Rather, the standards most widely applied to indoor air are those used to regulate worker exposures in industrial environments (17). Workplaces are regulated by the Occupational Safety and Health Administration (OSHA). The American Conference of Governmental Industrial Hygienists (ACGIH) annually publishes a list of recommended maximum acceptable workplace exposure concentrations for more than 500 toxic chemicals and dusts. These Threshold Limit Values (TLV), are defined as air concentrations of chemicals to which nearly all workers may be exposed for 40-hour work weeks over a working lifetime without ill effects. The TLV are not legal regulations but many of them have been adopted by OSHA as Permissible Exposure Limits (PEL). The TLV apply to healthy workers and should not be extended to the general population, which includes potentially more susceptible groups such as young children, the elderly, and the sick who are usually exposed for periods longer than 40 hours a week to contaminants in the home environment.

Standards for ventilation in nonresidential buildings are set by the American Society of Heating, Refrigeration, and Air Conditioning Engineers (ASHRAE). For example, ASHRAE (18) recommends a minimum indoor ventilation of 0.57 m^3 (20 ft^3) per minute per person. Indoor air quality guidelines or standards have also been recommended by other professional and governmental groups (eg, Consumer Product Safety Commission). These guidelines usually specify either ventilation requirements or acceptable indoor concentrations of specific air contaminants. They must not be interpreted as absolute values below which adverse health effects will not occur. Guidelines are developed on a pollutant-by-pollutant basis and do not consider toxic interactions that may occur in multi-contaminant indoor environments. These interactions may be antagonistic, additive, or more than additive (synergistic). Synergistically interacting pollutants might cause health effects even when exposure levels are very low. Standards or even rough guidelines are currently unavailable for most indoor pollutants derived from biologic sources (eg, allergens, microbes). When guidelines are available, one can measure contaminant levels and compare them with guidelines to judge the degrees of health risk. However, results, particularly from a single test, must be interpreted cautiously. Concentrations of indoor air contaminants vary widely depending on such factors as: the type, condition, and use of the source; indoor temperature and relative humidity; ventilation; air mixing; and the season. Also, guidelines reflect the limited state of knowledge at a given time and are subject to change; for example, the TLV for ozone has been reduced from 1.0 ppm to 0.1 ppm.

HEALTH EFFECTS

Concern about the health risks of air pollution reflects the frequent exposures to numerous pollutants in various environments, the diverse mechanisms by which these pollutants might cause health effects or diseases, and the wide range of susceptibility to pollutants in the population. Humans typically inhale 10,000 to 20,000 L (or approximately 11 to 22 kilograms or 25 to 50 pounds) of air daily, so that doses of pollutants inhaled at even low concentrations may become biologically significant with sustained exposure. The respiratory tract of humans possesses important physical, chemical, and immunologic defense mechanisms for clearing and detoxifying inhaled agents. However, the defense systems may be impaired by disease, overwhelmed by large pollutant doses, or may not be fully effective during long-term exposures.

Definition of Adverse Health Effect

There is no broadly accepted objective definition of adverse health effects caused by air pollution (19), despite widespread recognition of air pollution as a potentially large public health problem. Adversity is perceived differently by patients, practicing physicians, and scientists and depends greatly on individual circumstances. It can be defined generally as a biological change that reduces the level of well-being or functional capacity, with the implication of long-term consequences (20). Adverse respiratory health effects can be defined as medically significant changes as indicated by one or more of the following: (*1*) interference with normal activity of the affected person or persons; (*2*) episodic respiratory illness; (*3*) incapacitating illness; (*4*) permanent respiratory injury; and/or (*5*) accelerated or premature respiratory dysfunction. Examples include: (*1*) transient shortness of breath due to asthma, requiring interruption of work or school; (*2*) increased frequency of full-blown asthma attacks, requiring medical attention; (*3*) acute bronchitis or shortness of breath requiring bed rest; (*4*) pulmonary fibrosis (lung scarring) resulting from repeated lung inflammation; and (*5*) statistically lower average lung function in population groups with higher exposure levels. The final situation might not be of immediate medical significance for typical members of the affected population but portends an increased risk of disability and early death from lung disease.

Many pollution-associated effects are nonspecific (ie, are vague or transient or have possible causes other than air pollution) and go unrecognized. As investigative techniques become more sophisticated, subtle effects will be detected that may be equivocal in medical significance. For example, a small, temporary measured change in lung function may show statistical significance, eg, difference from baseline measurement with less than 5% probability of being a chance variation ($p < 0.05$). However, the change may or may not be biologically or medically relevant. Thus, not all measurable biologic, physiologic, or clinical changes are necessarily adverse for the purposes of the Clean Air Act. "Trivial" or "nuisance" effects might include low carboxyhemoglobin concentrations (<2.5%) from CO exposure or eye, nose and throat irritation during exposures to photochemical air pollution. On the other hand, some subtle, temporary physiologic changes may have medically significant long-term medical consequences. Exposure to some air pollutants may produce small but statistically significant changes in the responsiveness of airways (bronchial passages) to inhaled nonspecific bronchoconstrictors (ie, drugs that constrict the airways, such as methacholine) (21). This physiologic finding suggests not only a biological reaction to an air pollutant stimulus (ie, alteration of airway responsiveness) but also inflammation of airways, raising the possibility of acute exacerbation of asthma or even the induction of chronic airways hyperresponsiveness. The latter is believed to occur in occupational asthma caused by exposures to, for example, toluene diisocyanate, a synthetic chemical, or plicatic acid, a natural product found in cedar sawdust.

A theoretical continuum of increasing effects from a given air pollution exposure in various segments of the population is presented in Figure 1. The more intense the biological effects, the smaller is the proportion of people who experience the effects. The largest proportion of the population has some physiological or subclinical effects from air pollution, while smaller numbers of individuals develop pathophysiologic changes, morbidity (clinical illness), and death. The threshold cited above in which responses may be considered adverse and thus requires regulatory action, lies between physiologic and pathophysiologic changes. This boundary is not always well defined and varies with specific circumstances of exposure and host susceptibility. An increase in exposure concentration or duration tends to shift the middle or right portions of the curve upward, ie, to move more people into the range of adverse effects.

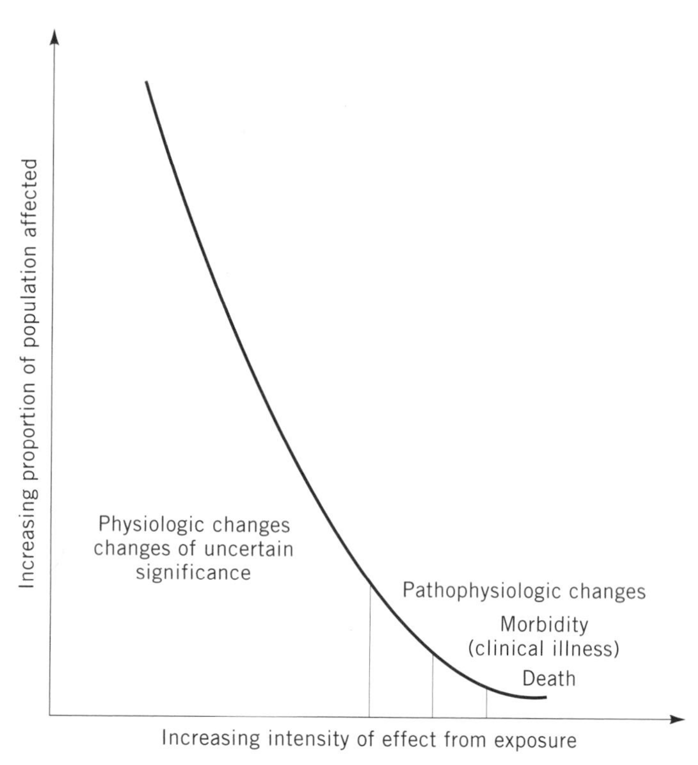

Figure 1. Relationships between increasing intensity of effect from exposure to air pollution and the proportions of the population which are affected. See text for details.

Determinants of Health Effect

The complex pathway leading to health effects begins with the generation and discharge of air pollutants from indoor and outdoor sources, continues with personal exposure to the pollutants, and, finally, may reach biologic effects of varying intensity (Fig. 2) (22). The intermediate stages in this process influence whether an exposure does or does not result in a measurable effect. Factors that affect the health outcome include the pollutant's concentration, its physical and chemical characteristics, the exposed individual's location (microenvironment) and activity pattern, the duration (contact time), and the susceptibility of the exposed individual. Exposure is defined as contact between the body and the external environment containing the agent of concern. Exposure is one (but not the only) important determinant of the dose of an agent at target sites in the human body. Important components of exposure are the concentration of the agent in inhaled air, the duration of inhalation of the contaminated air, and the ventilation rate (volume of air inhaled per unit of time) during the period. The same exposure may be achieved by various

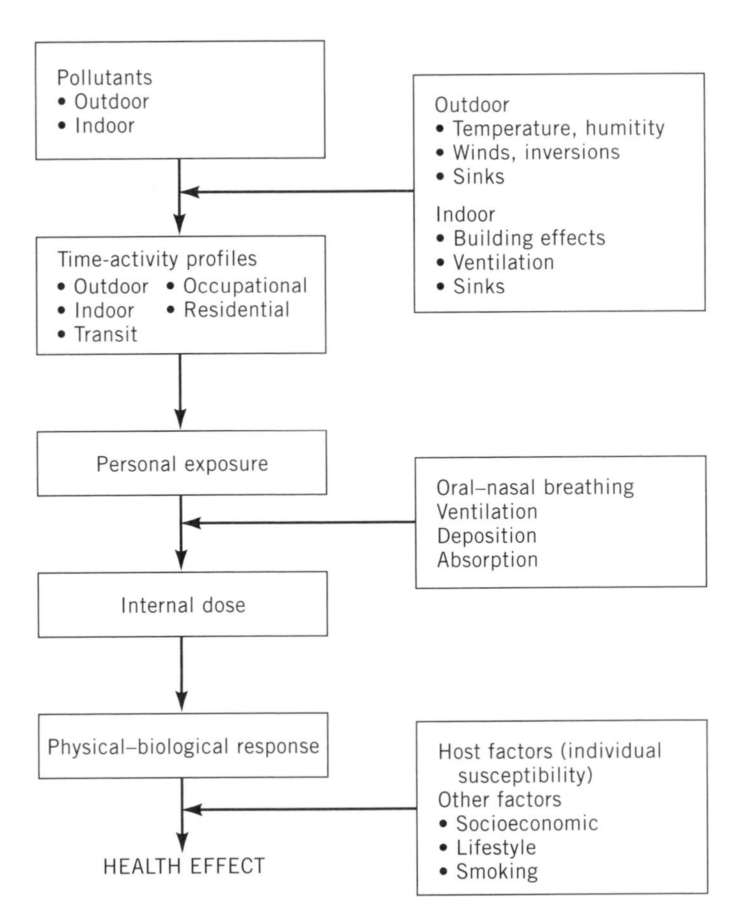

Figure 2. Factors that affect the development of adverse health effects.

combinations of concentration, duration, and ventilation rates. However, short-term peak exposures may elicit biologic effects that are different from that observed in longer-term, lower-level exposures with the same product of concentration, time, and ventilation rate. The respiratory tract is the most important route of exposure and thus the critical target organ system for most common inhaled pollutants, other than carbon monoxide.

Exposures can be estimated (modeled) most accurately by evaluating a person's time-activity pattern, ie, determining how much time a person spends in various locations (microenvironments) with measured or estimated air pollutant levels and how active the person is (reflecting how much air he/she breathes in) in each microenvironment (22–25). Most people are exposed to a wide range of pollutants at low levels during their normal daily activities; some individuals can be exposed to certain pollutants at levels much higher than average. The indoor environment is an important location for exposure because most people spend 80–90% of their time indoors. However, the more critical index is total exposure, which is the sum of exposures received indoors, outdoors, and in transit. The relative contributions of each setting to total exposure will vary according to age, gender, and socioeconomic, occupational, and health status. It is frequently too costly or technically unfeasible to measure air pollution exposures directly for a given individual, subgroup, or population. Thus, exposure surrogates or indices are often used to estimate exposures. These indirect measures may be pollutant concentrations in air at different but relevant times or places; measurable concentrations of the agent of interest or a related substance within the body (eg, carboxyhemoglobin [COHb] from CO inhalation); or mathematical models of exposure based on a combination of indirect measures (22–26). For example, ambient levels of O_3, NO_2, CO, and other outdoor pollutants are often used to estimate exposures for the residents of a particular community. These estimates may be reasonable for the few people outdoors near the monitoring stations. However, the accuracy of exposure estimates usually decreases with increasing distance from the stations and increasing time spent indoors and in transit. The best estimates are those based on direct measurements of personal exposure, eg, by real-time recording of CO concentrations with portable monitors, or biochemical measurements closely related to internal dose, eg, COHb in blood or nicotine/cotinine in saliva. The selection of an exposure estimator depends on the goals of the study, required level of accuracy, and costs. Generally, the cost of estimating exposure is directly proportional to the certainty of the estimate (accuracy).

The other important link in the pathway from pollutant to health effect is the internal dose, which is the quantity of agent deposited at a target site within the body where toxic effects occur. Dose and exposure are not necessarily equivalent for inhaled particles and gases. Penetration into the lung and retention at potential sites of injury depend on the physical and chemical properties of the pollutant (eg, particle size, solubility, reactivity), the exposed person's ventilation rate, the proportion of air breathed through the nose versus the mouth, airway anatomy, and efficiency of clearance mechanisms (27). Highly water-soluble gases, such as sulfur dioxide and formaldehyde, are almost completely extracted by the upper airway (nasopharynx) of a resting subject during a brief exposure, whereas less soluble gases, such as NO_2 and O_3, penetrate to the small airways and alveoli. The penetration of particles into the lung and sites of deposition within the lung depend on the aerodynamic size of the particle. Particles greater than 10 μm MMAD are effectively removed by the upper airway, whereas smaller particles penetrate and deposit in the lower airways and alveoli where local clearance mechanisms (mucociliary clearance, macrophages, permeation into blood or lymphatic vessels) are active. The response of the respiratory tract to inhaled fibers depends on fiber width, length, and susceptibility to dissolution. Regardless of the type of pollutant, the heightened ventilation during exercise increases the amount of inhaled air and the proportion of oral-to-nasal breathing, thus enhancing the dose of inhaled pollutants.

Dose measurement is more difficult and uncertain than exposure measurement. The dose of many air pollutants cannot be accurately measured in most human exposure situations. Thus, exposure–response, rather than dose–response, relationships are measured or modeled for most pollutants. In general, the likelihood of a health effect shows a strong positive correlation with exposure as well as dose. The relationship between exposure and response may have different forms, depending on the mechanisms of action (Fig. 3) (2). The shape and slope of the exposure–response relationship have important biologic and clinical

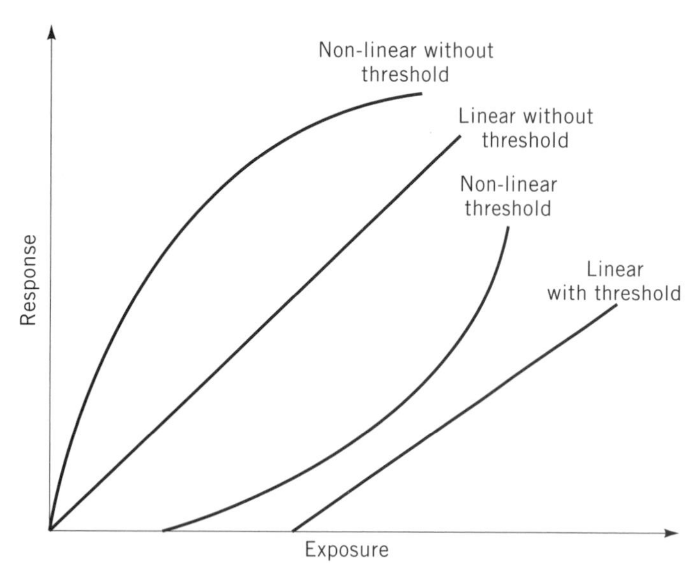

Figure 3. Theoretical exposure–response relationships for inhaled pollutants. Response indicates a biological response, eg, a change in symptoms, lung function, biochemistry, or activity of target cells, tissues, or organs.

implications. Response curves having a threshold that must be exceeded to produce a response or disease indicate that levels below the threshold are without risk. Thus, only individuals who exceed the threshold are at risk and a regulation could permit a safe or acceptable level of exposure. In contrast, curves without a threshold imply that any level of exposure conveys some risk. For example, the linear no-threshold relationship is widely used to assess risks of carcinogenesis for regulatory purposes. However, the assumption that a linear no-threshold model applies for carcinogenesis remains very controversial. In practice, distinguishing among the different exposure–response curves may be difficult using either animal or human studies. Even when a threshold exists, it may be very low in some susceptible subgroups so that, for practical purposes, there is no safe exposure level.

The site of deposition of the inhaled material largely determines the type of clinical response (2). For example, nasal deposition of pollen may result in rhinitis in individuals with hay fever. Sulfur dioxide, aeroallergens, radon, or asbestos may enter the conducting airways (trachea, bronchi, and terminal bronchioles) where bronchoconstriction or bronchogenic carcinoma may develop. Certain fungi and inorganic dusts can affect the lung parenchyma or alveoli (where gas exchange occurs) and result in hypersensitivity pneumonitis and pneumoconiosis, respectively. The mechanisms by which deposited inhaled agents injure the airways and lungs are not fully understood but can be broadly categorized as acute irritation and inflammation, chronic inflammation accompanied by a fibrotic response (scarring), immediate and cell-mediated immune responses, and carcinogenesis (see below).

Mechanism	Pollutant
Bronchoconstriction	Sulfur dioxide, acidic aerosols
Inflammation	Ozone, environmental tobacco smoke
Fibrosis	Asbestos, silica
Carcinogenesis	Radon, asbestos, formaldehyde, active smoking, environmental tobacco smoke

The likelihood of an adverse response depends on the mass of pollutant deposited, the site of deposition, the rate of clearance or detoxification, and characteristics of the exposed person that determine susceptibility.

Certain individuals and groups may have one or more characteristics that place them above the norm or usual risk for adverse health effects (28). The Clean Air Act specifically requires protection of populations at risk, ie, groups having a significantly higher probability of developing illness or other abnormal health status than the general population (see below).

Group	Air Pollutant
Healthy young adults who exercise or work outdoors	Ozone
Children and adolescents	Ozone, lead
Pregnant women and fetuses	Carbon monoxide
Patients with asthma	Sulfur dioxide, ozone, nitrogen dioxide, particulates (sulfuric acid, sulfates)
Patients with chronic obstructive pulmonary disease (COPD)	Carbon monoxide, particulates, sulfur dioxide
Patients with ischemic heart disease	Carbon monoxide, particulates, sulfur dioxide

Inherent individual characteristics (eg, age, gender, psychologic factors, or possibly race/ethnicity) and interactions with other environmental factors (eg, nutrition, preexisting disease, medications, lifestyle or socioeconomic factors, access to health care) or agents (eg, tobacco smoking and asbestos exposure) determine a group's biological responsiveness to a specific pollutant. Particular groups may differ in their sensitivity to different pollutants. For example, most asthmatics are very sensitive to SO_2 in comparison with healthy people. On the other hand, healthy athletes develop noticeable ozone-related responses after heavy exercise, whereas asthmatics generally have little excess sensitivity to ozone in controlled exposures.

Another way to understand the potential health impact of air pollution on susceptible groups is to review the size of at-risk populations. In 1991, a total of 514 counties and 20 cities in the U.S. were designated as nonattainment areas for one or more of the six criteria pollutants. Age-specific national prevalence rates for medical conditions (National Health Interview Survey, Centers for Disease Control) and the 1990 county-specific U.S. census data can be used to derive the prevalence of populations at risk (29) (Table 3). The nonattainment sites represent an estimated 164 million persons (66% of the resident U.S. population), including 63% (approximately 31 million) of preadolescent

Table 3. Estimated Populations at Risk[a] Residing in Communities That Have Not Attained One or More National Ambient Air Quality Standard[b], United States, 1991[c,d]

Population Subgroup at Risk	Pollutant[e]	At-risk Population Living in Nonattainment Areas No.	% [f]
Preadolescent children (aged ≤13 yrs)	PM-10, SO_2, O_3, NO_2	31,528,939	63
Elderly (aged ≥65 yrs)	PM-10, SO_2, O_3	18,846,666	60
Persons with pediatric asthma[g]	PM-10, SO_2, O_3, NO_2	2,285,061	61
Adults (aged ≥18 yrs) with asthma	PM-10, SO_2, O_3, NO_2	4,279,413	66
Persons with chronic obstructive pulmonary disease[h]	PM-10, SO_2, O_3, NO_2	8,831,970	64
Persons with coronary heart disease[i]	CO	3,493,847	33
Pregnant women[j]	CO, Pb	1,602,045	38
Children aged ≤5 yrs	Pb	74,312	3

[a] Populations at-risk estimates should be quoted individually and not added to form totals. These categories are not mutually exclusive.
[b] Only a portion of some communities are designated as nonattainment areas for some pollutants. The totals in this document are based on entire county populations and may, therefore, reflect an overrepresentation of the true populations at risk.
[c] Estimated total U.S. population living in the 514 counties and 20 cities designated as nonattainment areas = 164 million (66% of the U.S. population).
[d] Ref. 29.
[e] PM-10 = particulate matter with a diameter ≤10 μm; SO_2 = sulfur-dioxide; O_3 = ozone; NO_2 = nitrogen dioxide; CO = carbon monoxide; and Pb = lead.
[f] The proportion of each population subgroup at risk of the total population in the category.
[g] Asthma in persons aged <18 years.
[h] Includes chronic bronchitis and emphysema.
[i] Includes ischemic and cerebrovascular heart disease.
[j] The estimated number of pregnant women in each county is derived from the number of live births. Fetal losses and multiple births may have an impact on the accuracy of these estimates.

children (13 years or less in age), 60% (19 million) of persons aged 65 years or greater, and 64% (9 million) of persons with chronic obstructive pulmonary disease (COPD). In addition, 61% of children with asthma and 65% of adults with asthma, and 33% of pregnant women and patients with coronary artery disease reside in nonattainment areas. The data do not include estimates for healthy persons who exercise extensively outdoors, an important group at risk for health effects from ozone and other outdoor pollutants. The increased ventilation associated with exercise enhances oral breathing (bypassing nasal defenses) and penetration of inhaled air into the deep lung, resulting in increased total dose. Thus, the estimates of at-risk populations in nonattainment areas may substantially underestimate the number of persons potentially exposed to unhealthy air quality.

Clinical Effects

Table 4 lists common symptoms, signs, and diagnoses associated with air pollutants. Effects in individuals can vary from trivial events to interference with normal activity to death. As discussed previously, some mild and temporary symptoms (eg, eye irritation) may be considered nuisances, while more persistent, activity-limiting symptoms such as wheezing from bronchospasm or chest pains from ischemic heart disease (ie, condition with inadequate oxygen supply to the heart muscle) would be considered adverse health effects. The table also indicates that pollutant-related effects can involve the respiratory tract at all levels (upper airways, lower airways, and alveoli) and may involve target organs outside the lungs. For example, inhaled carbon monoxide can inordinately increase myocardial oxygen needs during exercise, resulting in decreased exercise performance in healthy athletes and the early onset of chest pains (angina pectoris) in patients with coronary artery disease. Similarly, volatile organic compounds (VOCs) do not generally impair lung function but can produce neuropsychologic effects (30). Temporal patterns of

Table 4. Symptoms, Signs, and Diagnoses Associated with Exposure to Air Pollutants

Upper Respiratory Tract Effects	
Rhinitis	
Sinusitis	
Pharyngitis	
Lower Respiratory Tract Effects	
Cough, wet or dry	Worsening of asthma or COPD[a]
Chest tightness or pain	Bronchitis
Shortness of breath	Pneumonia
Wheezing	Lung cancer
Nonrespiratory Effects	
Eye irritation	Decreased exercise performance
Fatigue	Worsening of angina pectoris (cardiac chest pains)
Malaise	Malignancy
Dizziness	
Headache	
Nausea	
Skin irritation	
Fever	

[a] Chronic obstructive pulmonary disease.

clinical effects vary widely. Whereas asthmatics develop respiratory exacerbations within minutes of inhaling SO_2 and susceptible individuals develop hypersensitivity symptoms within hours after fungal exposure, nonsmokers or smokers exposed to ETS or asbestos may develop a lung malignancy only after a latent period of several decades. Not explicitly shown in Table 7 are subclinical effects, ie, physiologic, biochemical, or cellular measures that indicate pollutant-induced exposure or effect. There is frequently controversy as to the clinical significance of these objective but subclinical alterations, some of which are discussed later in relation to specific pollutants.

The above-described responses refer to individuals. On a group or population (epidemiologic) basis, adverse health outcomes from air pollution exposure take on greater public health significance since larger numbers of persons are potentially affected. Epidemiologic evidence of adverse health outcomes can be classified according to whether the outcomes are related to acute (episodic) or chronic (long-term) exposures to air pollution (Table 5). Epidemiologic findings will be reviewed in relation to specific pollutants.

Risk Assessment

Air quality regulatory officials must compare risks and benefits of possible air pollution conditions and their control in deciding how best to protect public health. The discipline of risk assessment gives a rational structure to that process (31–33). Risk assessment is a set of decision rules widely applied in the United States for identifying and quantifying risks of chemicals and other events for adverse health effects. They are

1. Hazard identification: Is an agent causally linked to the health effect of concern?
2. Exposure assessment: What is the extent of exposure?
3. Dose-response assessment: What is the relationship between level of exposure and risk of the health effect?
4. Risk characterization: What is the risk to human health, including uncertainties?

Table 5. Sources of Evidence for Adverse Health Effects of Air Pollution in Populations[a]

Acute or Episodic Effects at More Versus Less Polluted Times
Increases in general indices of ill-health, eg, school or work absences, reduced activity days
Small temporary declines in lung function
Increased incidence of respiratory symptoms or infections
Increased physician visits or emergency room visits
Increased hospital admission rates for respiratory diseases
Increased rates of respiratory or total mortality
Chronic or Long-Term Effects in More Versus Less Polluted Regions
Increased prevalence of symptoms or respiratory disease indicators
Cross-sectional differences in lung function, eg, lower average function and/or increased percentage of population with abnormal function
Longitudinal differences in lung function, eg, slower increases in children and/or faster declines in adults
Increased mortality from respiratory disease, including long-term changes in mortality within same region coinciding with long-term changes in pollution

[a] Modified from Ref. 55.

To assure safety or a level of risk judged to be acceptable, the analysis depends on accurate and valid measurements of pollutants (34) and accurate and valid characterization of risks. In other words, the scope of risk assessment extends beyond testing for an exposure–response association to the quantification of risk at various levels of exposure and the assessment of factors that modify the exposure–response relationship. The analysis consists of evaluating the distribution of individual exposures in the population (eg, mean, variance, upper 10th percentile), the causes of high exposures (eg, sources, environmental pathways, exposure locations and settings, and time–activity patterns), and the relationship between exposure and dose. The results of risk assessment can be used to identify gaps in knowledge requiring further research, to assign priorities among environmental hazards, and to select approaches for managing risks. The strengths, limitations or costs, and methodological basis of risk assessment are controversial in both scientific and political areas (35). Risk assessment of Hazardous Air Pollutants (HAP) is especially difficult. Many non-criteria agents are classified as HAP because they are known or suspected to cause adverse, often catastrophic, health effects (36,37). Quantitative estimation of health risks is difficult for this group because of inadequate data on exposures, doses, and effects. Preliminary estimates suggest that up to 2,000 cancer cases per year may result from outdoor exposures to 45 of the 189 HAP (36). Noncancer health effects may be widespread and include nonmalignant respiratory disease, hematopoietic abnormalities, neurotoxicity, renal toxicity, and reproductive and developmental abnormalities. Approximately 50 million people live near emission sources where estimated concentrations of one or more HAP exceed "levels of concern" for noncancer health effects (37). However, a paucity of data is available to estimate actual exposures for either the general population or communities potentially at risk. Careful and extensive epidemiologic evaluations will be required to support quantitative risk assessment in these situations.

METHODS OF ASSESSING HEALTH EFFECTS

General

Research on the health effects of air pollution falls into three complementary disciplines: animal toxicologic studies, clinical (controlled exposure) studies of human subjects, and epidemiologic investigations. Research may be classified according to its approach (laboratory experiments, field observations, or review of public health statistics) or its subjects (human beings or laboratory animals). Each discipline has its particular strengths and limitations (discussed below). Results from all three disciplines are necessary to the deliberative process of setting air quality standards. Ideally, research findings should be confirmed by independent studies from other disciplines.

However, generally consistent findings from different studies within one discipline are usually considered sufficient to guide most public-policy decisions.

The simplest laboratory studies of either animal or human subjects compare one defined pollutant exposure against a control (reference) condition in which the subjects breathe clean air but are otherwise treated the same way as in the pollutant exposure. In other words, the only essential difference in the exposures is the test atmosphere. The "exposed" and "control" subjects may be the same individuals studied at different times (cross-over study design) or may be different but comparable individuals (parallel-group study design). More definitive laboratory studies include a control condition (clean air) and multiple pollutant exposures at different concentrations to determine an exposure–response relationship. A statistically significant response that increases as the pollutant exposure increases gives added confidence that the pollutant is responsible and supports a causal role of the agent.

Animal (Toxicology) Studies

Animal studies provide maximal control of experimental conditions as well as a broad range of possible exposures and endpoints. Animals are typically exposed to known levels of one or more pollutants generated in laboratory exposure chambers and their responses observed (38–40). Many potential interferences such as air environment, nutrition, activity, and exposure to infection or other disease can be tightly controlled or eliminated throughout the lifespan of the animals. Results from carefully designed and performed studies can be confirmed by replication in different laboratories. Animal studies provide the important advantages of testing possible mechanisms of action of pollutants and evaluating possible interventions to prevent or limit adverse responses. On the other hand, even clear-cut findings from animal toxicology may be of uncertain relevance to humans.

Clinical Studies

Clinical studies involve temporarily exposing human volunteers to controlled concentrations of specific pollutants in a laboratory either by inhalation via a mouthpiece or mask or by unencumbered breathing in an environmentally controlled exposure chamber. Effects of acute exposure can be objectively measured with tests similar to those used by physicians in clinical practice and can be related to simultaneous subjective effects (symptoms). Lung function tests are typically used in studies of inhaled pollutants. Two very important measures of lung function are forced vital capacity (FVC), the volume of a maximum breath forced out; and forced expired volume in one second (FEV_1), the volume of the maximum breath forcibly exhaled in the first second (41). These rapid, simple, precise tests provide extensive exposure–response information that can be compared among different laboratories. Airway resistance (Raw) or specific airway resistance (SRaw = Raw/lung volume) is another common test of lung function. Increases in either Raw or SRaw indicate abnormally increased airflow obstruction because of bronchoconstriction (excessive shortening of airway smooth muscles). Other measures of pollutant–induced effects include bronchial challenges with inhaled bronchoconstricting substances (eg, methacholine, histamine, carbachol) (21) or exercise, and tests for cellular and biochemical changes in the lung, ie, by analysis of bronchoalveolar lavage (BAL) fluid obtained with a flexible fiberoptic bronchoscope (26). The wide range of research goals, study designs, and procedures in controlled clinical studies are described in more detail elsewhere (42–47). To a considerable extent, clinical studies resemble animal studies and clinical trials of new therapy in terms of their experimental designs and scientific rigor (tight control of experiments; clear demonstration of cause and effect) and relevance to human health. However, the subjects' diverse experiences and exposures away from the laboratory may introduce interferences (confounders and effect modifiers). Small groups of volunteers may not be representative of large populations exposed to community air pollution. Researchers often recruit subjects expected to be sensitive to pollutants to observe a "worst-case" response. Ethically, clinical studies are limited to exposures that cause only mild and temporary disturbances in health. Practically, the laboratory exposures are limited to small numbers of subjects and relatively low-level, short-term exposures that resemble natural conditions and do not interfere unduly with normal activities. Persistent health effects or effects that require a long time to develop (long latent period) must be investigated by animal toxicology or epidemiology.

Epidemiologic Studies

Epidemiology is the study of the distribution and determinants of disease in human populations. The primary aims of air pollution epidemiology are to identify and quantify health effects of environmental agents under ordinary exposure conditions (22). As such, epidemiologic studies potentially have the highest relevance to public health. Field studies of well-defined groups of people in their usual environment measure their health status by interviews and/or clinical tests and attempt to determine whether health changes are associated with pollution exposures. Some studies do not involve actual field observations but involve review of data routinely collected in communities with different air qualities by health or demographic agencies. Conventional epidemiologic approaches used to study the adverse effects of inhaled pollutants are the cross-sectional study (where a large group is tested at only one point in time), the cohort study (where a group of people is followed closely for a period of time), and the case–control study (where similar persons with and without a response or disease are matched and their exposures are estimated retrospectively). Each design has advantages and disadvantages for examining the effects of environmental exposures (48,49).

Epidemiologic studies are usually observational rather than experimental or interventional in nature, ie, the investigator does not usually control the exposures of the study subjects. Epidemiology has been an effective tool for investigating pollutants with either very strong or very specific effects. However, epidemiologic studies are more limited when investigating agents with weak effects (but still of public health concern). As a result, the relationship between exposure to an inhaled agent and health effect of concern may be altered by biases: selection, misclassification, and confounding (48). Bias may increase or decrease the strength of statistical association. Selection bias refers to differential patterns of subject participation depending on exposure and disease status. Misclassification bias refers to error in measuring either pollutant exposure or the health outcome. Random misclassification tends to show a lack of effect of exposure and is most frequent in epidemiologic studies of air pollution in which exposures are estimated using limited measurements or surrogates. Statistical power (ability of a study to detect exposure–effect associations) declines as the degree of random misclassification increases. Differential misclassification refers to nonvalid measures, such as from interview data in a case–control study. Confounding bias refers to another risk factor that alters the effect. Specifically, a confounder (eg, temperature) affects an outcome measure (eg, lung function) and may be correlated with the risk factor (eg, ambient concentration of ozone). An effect modifier (eg, exercise) is a variable that modifies, by antagonism or potentiation, the effect of an air pollutant (eg, ozone) on an outcome measure (eg, lung function) but does not directly affect the outcome measure. In summary, these factors can produce misleading statistical associations or mask true associations between air pollution and health outcomes. These factors may not be controllable but they should be accounted for before one can make causal inferences about the relationship between air pollution and health effects.

Accurate ascertainment of exposures and health status of the study population is usually very difficult, particularly in longitudinal studies where the cumulative health effects of pollutants are sought. For these reasons, epidemiologic studies are time-consuming, expensive, and difficult to conduct. Many older epidemiologic studies have reported results that are controversial or difficult to interpret quantitatively, in part because the studies relied on a few centrally located outdoor air monitoring stations to represent the exposure of large populations. The often invalid assumption that centralized exposure estimates for population groups will hold for individuals within those groups who have health effects is known as the "ecological fallacy." Inaccuracies in exposure assessment (misclassification) are being reduced by technological advances and methodological improvements in epidemiologic studies. Carefully developed questionnaires about indoor air quality and time-activity patterns may help to establish relative rankings of exposure for different individuals or groups (50). Questionnaires are important when monitoring is not possible and may indicate when and where monitoring is necessary. As previously discussed, individual exposures vary greatly within large populations depending on individual time–activity patterns and pollutant exposures within each microenvironment. The accuracy of individual or group exposure estimates can be improved by use of portable monitors that can record pollutant concentration when placed in various locations where individuals spend time, or when personally carried on clothing by an individual. Other epidemiologic issues remain such as the relationship between exposure and dose to the target tissue.

Interpretation and Application of Research Results

In the aggregate, a large amount of information about health effects from air pollution is available. However, much still remains unknown or uncertain, and new research results often raise more questions than they answer. Research priorities should be established based on existing knowledge and judgments about the greatest risks, since resources will never be sufficient to investigate all biomedical questions. Even when a particular health effect appears well understood in scientific terms, it may be difficult to decide on a policy to control it. Any policy decision will influence many public risks and benefits, and its overall effect may be difficult to predict (51). The limits of science and its regulatory applications are discussed elsewhere (52).

No single scientific study is completely reliable, all-inclusive, and relevant to public health. Any research data must be considered in relation to those in other studies. All reported studies require critical evaluation by an independent party (eg, journal editorial board, regulatory agency), regardless of whether the final results are alarming or reassuring, or are consistent with one's personal beliefs or not. Thus, an informed evaluation of the quality and validity of research reports can be based on the following questions:

1. What question is the research attempting to answer? All research should ideally add new, meaningful information to the scientific database. Studies should have a specific, identifiable question posed by the experimenter, ie, a hypothesis which will be tested. However, the hypothesis may or may not be highly relevant to the specific question of whether air pollution affects health. Relevance is a key but not always a direct or predictable criterion. Although some basic science research provides indirect data about subclinical effects, this research may still have important applications in the future. For example, one study may expose rats to low levels of sulfur dioxide to find the smallest dose capable of producing a certain effect. On the other hand, another study might deliberately use massive exposure to induce injury in order to investigate repair processes. The former is more immediately relevant to air pollution health risks than the latter example, although both involve the same experimental animal and pollutant. However, the latter example has scientific relevance in terms of better mechanistic understanding of pollutant injury.
2. How well does an experiment simulate real life? No experiment is perfect in this respect, but some are more realistic than others. Ideally, the concentrations of pollutants, duration of exposure, secondary stresses (temperature, humidity, exercise), the health status of the subjects, and exposure to other pollutants should closely resemble the circumstances naturally encountered by people exposed to community air pollution. The subjects should be comparable to other populations of interest when controlled exposures or epidemiologic studies are designed. For example, a study showing no effects in indoor desk

workers does not rule out effects in heavily breathing outdoor laborers.

3. How do observed effects relate to pollutant exposure/dose? The 16th-century alchemist–physician, Paracelsus, stated "There is nothing which is not a poison. The dose makes the poison." This basic principle of toxicology has been consistently verified (53). Even essential substances such as oxygen become injurious at abnormally high exposure levels. Some common toxins such as CO and formaldehyde are products of normal metabolism and are present at low concentrations in the body without harmful effects. Ideally, an experiment should cover a range of concentrations to understand the health risks of air pollutants. A response pattern that consistently increases as exposure–concentration increases, generally indicates two important findings: A cause-and-effect relationship is present and a threshold (minimum dose necessary to produce a detectable effect) may exist. One can then estimate whether ambient pollution levels produce exposures exceeding a threshold.
4. Are the results significant? Variability in behavior is a characteristic of all living organisms and must be taken into account when measuring the effects of stresses such as air pollutants. Biostatistical analyses are necessary to provide confidence that the biological results are either random (reflecting chance variability) or nonrandom (pattern related to the experimental factors, eg, air pollutant exposure). Statistical analyses use laws of probability to test the (null) hypothesis, typically with an arbitrarily chosen probability (p) value of <0.05, indicating less than a 5% chance that the distribution was obtained by chance alone and was not influenced by the experimental factor. In other words, the statistically significant data have only one chance in 20 of being a "false-positive" finding (type I or alpha error). A "false-negative" finding (type II or beta error) is also a major problem. A real effect that occurs inconsistently, only in a sensitive subgroup, or is small compared to the usual variability between tests may be missed if too few subjects and/or measurements are available for analysis. In general, a "positive" finding (an effect of exposure) should be supported by evidence of statistical significance as well as precise, unbiased measurements. A "negative" finding (no effect of exposure) should be supported by evidence that the study had adequate statistical power since many studies do not provide sufficient precision to completely exclude the possibility of some increased risk of effect from the experimental agent. Interval estimation to derive confidence intervals is another useful approach to statistical hypothesis testing (48).

   The other important dimension of "significance" is the biological or clinical relevance of the findings. Statistical analyses relate to numerical probabilities and not necessarily to underlying biological realities or plausibility. Test results that reflect biology can validate statistical results. For example, small, statistically significant changes in lung function or cellular function should also be considered in relation to human structure and function to decide if the alteration has possible clinical relevance. Thus, precise and unbiased experimental measurements are important, and the results must be carefully interpreted if a direct biomedical implication is possibly present.

   Epidemiologic results cannot prove causality but only imply causality. Statistical methods cannot establish proof of cause and effect but can define an association with a certain probability. Careful analysis is especially important, and causal inference is especially difficult, when one attempts to differentiate the effects of a single pollutant (eg, ozone) from a complex pollutant mixture. The causal significance of an association is a matter of judgment that goes beyond any statement of statistical probability. To judge the causal significance of the association between an air pollutant and health effect, nine criteria (48,54) must be used, none of which is pathognomonic by itself: strength of association (high incidence rates); consistency (replication of findings in different populations); specificity (a cause leads to an effect); temporality of exposure and effect; biologic gradient (exposure–response relationship); biological plausibility; coherence (systematic or methodologic connectedness or interrelatedness especially when governed by logical principles); experimental evidence; and analogy. As argued by Bates (55), the epidemiologic database regarding health effects from contemporary air pollution must be actively evaluated using the central question of coherence. Coherence may exist at three different levels: within epidemiological data; between epidemiological and animal data; and between epidemiologic, controlled human exposure, and animal data. For some pollutant mixes (SO_2–particulate complex) epidemiologic coherence appears to exist over a wide range of air quality levels and human populations, but findings from other disciplines are not necessarily coherent with epidemiologic findings.
5. Has the work been peer reviewed? A peer review process for research papers submitted for publication improves the quality of scientific research and published reports. This process is the best means to assure adequate quality control of new information, but is still imperfect. Poor quality work is occasionally published and good work may have delays in publication. Preliminary data in the form of abstracts or oral presentations or work presented in news media are not peer-reviewed and cannot be assumed to be scientifically valid or accurate. Some agencies, eg, the U.S. EPA, require only peer-reviewed references to be cited and discussed in its criteria documents.

SELECTED POLLUTANTS AND THEIR HEALTH EFFECTS

Ozone, nitrogen dioxide, sulfur dioxide, particulates (including sulfuric acid), carbon monoxide, and environmental tobacco smoke have been selected for more detailed review because they represent ubiquitous pollutants with documented or strongly suspected health effects that potentially affect large proportions of the population and/or susceptible groups. Sources, ambient concentrations, biological mechanisms, and health findings from animal, clinical, and epidemiologic studies will be addressed for each pollutant. The reader is referred to other reviews (7–9,45,56–58) and EPA criteria documents for more details. It should be remembered that this selection is somewhat arbitrary and that these pollutants do not exist alone in nature but in mixtures with other compounds. Other environmental agents may have similar health effects, and still other agents may present health risks beyond those mentioned here. Lessons to be learned from the example pollutants may be generalized to other agents.

Ozone

Sources and Ambient Levels. Ozone (O_3) is the most important of the class of pollutants known as photochemical oxidants. Oxidants alter biological molecules by adding oxygen atoms with greater facility than atmospheric oxygen (O_2). Nitrogen dioxide and peroxyacetyl nitrate (PAN) are other common oxidant pollutants which appear to present less health risk than ozone.

Ozone is a secondary pollutant which forms in the troposphere by photochemical reactions of primary pollutants, ie, nitrogen oxides and nonmethane hydrocarbons (or volatile organic compounds) (59–62). These precursors originate from essentially all common combustion sources, particularly motor vehicles. The VOCs also enter the atmosphere by evaporation of volatile materials, such as gasoline, paint thinners, and cleaning solvents, and as natural emissions from vegetation.

Unlike most pollutants, ozone is a natural and beneficial constituent of the upper atmosphere (stratosphere). Stratospheric ozone absorbs potentially harmful solar radiation that would otherwise reach the earth's surface. Normal atmospheric mixing brings modest amounts of this ozone to ground levels, resulting in concentrations of 0.03–0.05 ppm in the clean air in remote areas. At one time, some medical authorities believed that ozone was a health-promoting substance because of its presence in clean air and its disinfectant properties. However, scientific observations of its substantial health risks date back as far as the 1850s. It is ironic that ozone today is spatially maldistributed in the atmosphere, ie, less than desired amounts are present in the stratosphere and greater than desired amounts are present in the troposphere. The maldistribution can be attributed to man-made ozone-destroying compounds (eg, chlorofluorocarbons) in the stratosphere and combustion emissions in the troposphere.

Significant efforts have been made to decrease photochemical oxidant air pollution, of which ozone is the most potent and common constituent. Nevertheless, the control of tropospheric ozone is a very difficult, as well as important, regulatory task. The most severe ozone pollution is found in situations with little wind (eg, a temperature inversion over a mountain-lined valley), abundant sunlight, high temperatures, and traffic congestion with large numbers of motor vehicles. Ambient ozone is characteristically highest during summer and fall months ("smog season"), although episodes can occur at other times. Ozone generally reaches its peak outdoor levels during mid-day, eg, frequently greater than 0.2 ppm and occasionally 0.30–0.35 ppm for one or more hours in Los Angeles. Multiple hours exceeding 0.12 ppm ozone can occur throughout the day and persist as episodes for 3–10 days in many urban and rural locations (59,61). Long-range transport of precursor pollutants may raise ambient ozone concentrations in areas hundreds of miles downwind from the sources. This condition commonly occurs in the northeastern U.S. during the summer. Ozone is the only known outdoor air pollutant for which permissible ambient exposures can exceed workplace permissible exposure limits (0.10 ppm, 8-h average). To some extent, nitric oxide (NO) from motor vehicular exhaust scavenges ozone in urban traffic and during the night.

Indoor ozone levels are usually much lower than concurrent outdoor concentrations since indoor sources (eg, photocopiers, laser printers) are relatively few and ozone reacts rapidly with air conditioning and building materials. Recent measurements indicate that indoor ozone values closely track outdoor levels, and, depending on the air exchange rate, are 20–80% of outdoor concentrations (63). Significant indoor/outdoor ozone ratios (>0.5) exist in homes without air conditioning and with open windows. Most people probably receive the greatest part of their personal ozone exposure (concentration $\times$ duration) while indoors because indoor levels are usually not negligible and people spend most of their time indoors.

Mechanisms. The health effects of ozone originate from its very reactive oxidizing nature and its active inflammatory reactions at all levels of the respiratory tract. The mechanisms of action of ozone in humans have been evaluated at tissue and cellular levels with various techniques. Ozone is efficiently removed (approximately 95%) by the entire respiratory tract in resting people (64). The extrathoracic airway (nose, mouth, pharynx, and larynx) removes about 40% of inhaled ozone, while the intrathoracic airways remove about 90% of the remaining ozone. Mode of breathing (oral versus nasal) has a relatively small effect on removal efficiency.

Ozone-induced pulmonary dysfunction consists of an involuntary inhibition of full inspiration, reduction of total lung capacity and vital capacity, and a concomitant decrease in maximal expiratory flow rates (65). Ozone may directly stimulate airway sensory fibers (possibly nonmyelineated C-fibers) which inhibit inspiratory muscle contraction. Either direct or indirect stimulation via biochemical mediators may explain the effects of an inhaled anticholinergic medication (atropine), which blocks ozone-induced increase in airway resistance (Raw) but only partially prevents the decrease in forced expiratory flow rates and does not prevent respiratory symptoms or decreases in vital capacity. Inhaled beta agonists (bronchodilator medications) are ineffective in preventing ozone-induced responses. Other potentially pertinent acute airway alterations include increased permeability of respiratory epithelium (ie, decreased integrity of the air–blood barrier) and increased activity of the mucociliary particle clearance system in the trachea and bronchi in nonsmokers following 2-hour exposures to 0.4 ppm ozone (64). Of note is that exposure to 0.2 ppm ozone accelerates peripheral airway clearance without significant abnormalities in routine breathing tests.

The inflammatory injury produced by inhaled ozone is evident not only from respiratory symptoms and increased methacholine responsiveness but also from results of bronchoalveolar lavage (BAL) and nasal lavage (NAL) in healthy adults undergoing exposures to 0.08 to 0.6 ppm ozone for 2.0–6.6 hours. As compared to air-exposure BALs, BAL fluid following exposure to 0.4 ppm ozone shows significantly increased polymorphonuclear leukocytes (PMN) whether sampled at 3 or 18 hour following exposure (66,67). In addition, BAL fluid shows significant ozone-induced increases in vascular permeability (total protein, albumin, immunoglobuin G), enzymes (immunoreactive neutrophil elastase, PMN elastase activity), and inflammatory markers (fibronectin, complement 3a, prostaglandin E2 and $F_{2\alpha}$, thromboxane B_2, urokinase plasminogen activator). Many of these markers, as well as other substances (lactate dehydrogenase, interleukin-6, and α_1-antitrypsin), are detected in BAL fluid when healthy adults are exposed for 6.6 hours to 0.08 and 0.10 ppm ozone (levels below the federal standard) (68). Although the BAL fluid following 0.08 ppm-ozone exposure shows smaller increases in inflammatory mediators and no major increase in protein and fibronectin, the overall results indicate that even ozone levels as low as 0.08 ppm are sufficient to initiate an inflammatory reaction in the lung. Similar increases in PMN and albumin are found in nasal lavage (NAL) performed 18 hours following exposure to 0.4 or 0.5 ppm ozone (69). Significant increases in plasma prostaglandin $F_{2\alpha}$ occurs in ozone-responsive subjects during and after exposures to ozone. Indomethacin, an inhibitor of the cyclooxygenase pathway, significantly reduces ozone-induced decrements in FVC and FEV_1 but not the increase in nonspecific bronchial reactivity, suggesting that the latter occurs via a separate, noncyclooxygenase mechanism.

Effects

Animal Studies. The effects of ozone are most pronounced in the lungs although alterations in other organs have not been ruled out. Numerous studies in different animal species have investigated various changes in the lungs after ozone exposure (14,39,40,59,62,70–74). Damage typically occurs at sites where the smallest airways (respiratory bronchioles) merge into alveoli, resulting in a respiratory bronchiolitis. The involved airways show evidence of injury and repair, ie, inflammation, edema, increased permeability (leakage) of proteins in the airways, cellular hyperplasia, replacement of normal alveolar lining cells (Type I epithelial cells) by primitive, more ozone-resistant cells (Type II epithelial cells), and scarring (fibrosis). Larger, more proximal bronchi are relatively less affected, perhaps because their surfaces are better protected by a mucus layer. Similarly, the alveoli are relatively spared perhaps because most ozone reacts with airway epithelium before entering the alveoli, which are also lined by a possibly protective surfactant layer. The nasal passages, however, can be vulnerable to inhaled ozone. The structural effects in the lower airways are most pronounced at ozone exposures of 0.5 ppm or higher (above the ambient range), but they can also occur in exposures to as little as 0.2 ppm for two hours. Exposures lasting weeks or months, either continuously or intermittently, to concentrations as low as 0.12 ppm produce persistent damage. Some studies at higher than ambient concentrations have shown airway recovery after a few days or weeks in clean air, whereas other studies indicate that ozone-induced damage persists.

Lung function changes in ozone-exposed animals typically include rapid and shallow breathing and increased resistance to air flow (Raw). These effects occur with 0.20–0.30 ppm for as little as two hours. At higher concentrations (above 0.5 ppm), the ozone-injured airways become hyperreactive and constrict in response to inhalation of non-specific bronchoconstrictors such as methacholine. Hyperreactivity presumably reflects underlying inflammation (21).

A variety of cellular and biochemical changes also occurs in the lungs at ozone concentrations of 0.2 ppm or lower. These alterations are complex and probably reflect ozone-induced inflammation. An influx of PMN and increased production of antioxidant substances (eg, vitamin E) occur in most species. Increased collagen deposition may occur with resulting fibrosis, making the lungs less compliant, ie, abnormally stiff. The implications of these findings for humans are unclear.

Ozone exposure may disturb natural defenses against microbes, as demonstrated by studies in which rodents are exposed to ozone and then to bacteria. These studies assess the proportion of ozone-exposed animals that die of respiratory infection (pneumonia) or the amount of viable bacteria remaining in the lungs. Control animals are kept in ozone-free air before exposure to bacteria. Massive amounts of bacteria are used, so that even some control animals die. Exposure concentrations as low as 0.1–0.2 ppm ozone may be sufficient to overwhelm defense mechanisms and increase death rates. The defenses which are altered by ozone include the immune system, mucociliary clearance, and alveolar macrophages.

Clinical Studies. Physiologic and symptomatic responses at current ambient concentrations of ozone have been extensively evaluated in healthy nonsmoking young adults and, to a lesser extent, in children and the elderly (14,45,59,62,64,72,75). Group responses in controlled exposures with intermittent exercise have largely correlated to the "effective dose" of ozone, which is the product of concentration × ventilation rate × duration. Most healthy adults have little response with 2–3 hours of resting exposure to ambient ozone concentrations (less than 0.3 ppm). Healthy young adults show variable but definite decreases in vital capacity and flow rates (eg, FVC, FEV_1) and increases in airway resistance (Raw) following 2-hour exposures to ≤0.20 ppm ozone during intermittent moderate exercise (76). As the ozone concentration increases, a strong sigmoid-shaped dose–response relationship begins at 0.12–0.16 ppm and reaches a plateau near 0.4 ppm. Exposures with heavier or more continuous exercise levels (with corresponding increases in ventilation) elicit significant respiratory responses at lower ozone levels. For example, young adults continuously exercising for 7 hours (approximately 40 L/min ventilation) in 0.08, 0.10, and 0.12 ppm ozone, showed significant progressive concentration- and duration-related reductions of FEV_1 and symptoms (especially cough and pain on deep inspiration), as well as increased sensitivity to inhaled methacholine, indicating ozone-induced airways hyperresponsiveness (77,78). Exercise performance by athletes is adversely affected by ozone exposure (79,80). Ozone-induced respiratory symptoms and lung dysfunction generally parallel each other in adults but not necessarily in children, who tend to report fewer respiratory symptoms despite comparable physiologic responses.

Several studies have evaluated whether photochemical pollutant mixture (including ambient ozone) is more toxic than ozone alone. This question can be addressed in two ways: Volunteers can be exposed to mixed pollutants of known concentrations generated in a chamber or exposed to actual oxidant air pollution in a laboratory setting. The latter has been performed in different areas in Los Angeles using a mobile exposure facility. In both cases, the most irritant responses are apparently due to ozone, and there is no clear evidence of substantial extra toxicity from other pollutants that typically accompany ozone in ambient pollution in Los Angeles.

Several potentially ozone-sensitive populations have been evaluated, but no group has yet been confirmed to show markedly greater responsiveness to inhaled ozone than observed in healthy young adults. Children, adolescents, and women appear as or slightly more sensitive than adult males in terms of lung function changes. Children show small but statistically significant decreases in FEV_1 following exposure to 0.12 ppm ozone or ambient Los Angeles air (with 0.14 ppm ozone). However, comparisons by age or gender are complicated by inherent differences in body size, lung volume, and exercise capability. Older subjects (>50 years) are less responsive to 0.20–0.45 ppm ozone and an oxidant mixture of ozone, NO_2, and peroxyacetyl nitrate (PAN) than young adults in acute 1- or 2-hour exposures. Surprisingly, healthy smokers and subjects with asthma, allergic rhinitis, COPD, or ischemic heart disease have not been conclusively shown to be more responsive than their respective controls in controlled ozone exposures. For example, adolescent asthmatics are not more responsive to 0.12 or 0.18 ppm ozone or combined ozone (0.12 ppm) and NO_2 (0.18 ppm) than healthy adolescents. Only following a 2-hour exposure to 0.4 ppm ozone with intermittent heavy exercise (ventilation 52 L/min) do adult asthmatics demonstrate greater airflow obstruction than normal subjects, but they still have similar induction of symptoms and changes in lung volumes and bronchial reactivity to methacholine. However, exposure to ozone (0.12 ppm) can increase reactivity to inhaled allergens in asthmatic subjects (81). Observed group differences (or lack thereof) may relate to multiple factors, eg, sample size, subject selection, medication usage, underlying bronchial reactivity, effective dose of ozone, and ambient exposures to inhaled irritants.

Although group differences tend to be small, individuals may show large differences in response to ozone despite being physically similar and receiving the same exposure conditions. Individual lung function response is fairly reproducible for at least 3 months in healthy young adults exposed to >0.18 ppm, although older subjects have a less consistent pattern. Individual ozone responses may be modified by repeated ambient (seasonal) exposures or by other factors which change underlying airway responsiveness. At present, an individual's sensitivity to ozone can be determined only by actual exposure, although asthma or asymptomatic nonspecific airway hyperreactivity may imply increased risk of ozone-induced lung dysfunction. Such airway-hyperreactive, ozone-responsive individuals may have difficulty developing tolerance to ozone.

Tolerance (or adaptation) results when humans (and animals) are exposed repeatedly to ozone in the laboratory. Most volunteers exposed to high concentrations (eg, 0.4 ppm) for several hours on five consecutive days show a characteristic pattern in which the worst symptoms and lung dysfunction occur during the first two exposure days, followed by less and less response on subsequent days. There is minimal response by the fourth or fifth day. The attenuated response to ozone is lost in a week or two if exposure ceases. The long-term health consequences of this phenomenon are unknown, in part, because it remains unproven that tolerance occurs naturally or is widespread in the community.

Epidemiologic Studies. Epidemiologists evaluate the short-term (acute) and long-term (chronic) health effects in people who reside in ozone-polluted areas (14,59,62,72). Short-term effects are typically evaluated as changes in reported symptoms, lung function, athletic performance, or rate of hospital admissions, in various cites such as Los Angeles and less polluted areas (Houston, Tucson, and locales in northeastern U.S. and southeastern Canada). Long-term effects are typically evaluated in terms of the respiratory health of residents in a high-pollution community, as judged by questionnaires and/or lung function tests, in comparison with the health of a matched population in a cleaner environment. Some of these studies will be reviewed.

Summer camp studies at diverse times and places have consistently shown oxidant-related changes in lung function in children. Declines in FVC, FEV_1, and peak expiratory flow (PEF) have been strongly associated with increasing ozone levels. In a 1982 study in Mendham, New Jersey, an air pollution episode occurred that involved four days of hazy weather and a peak one-hour ozone level of 0.186 ppm. Decrements in lung function were present not only during the episode but also one week following the episode.

Children participating in the Harvard Six-City Study of respiratory effects of air pollution performed lung function tests at each of six weekly sessions. Ambient ozone was monitored hourly at nearby monitoring stations. The results were in good agreement with those in camp studies despite relatively low ozone levels, ie, 0.007–0.078 ppm. A study in Tucson also showed a relationship between ambient exposure to ozone (0.12 ppm or less) and reductions in lung function.

A large-scale study in the Los Angeles area estimated the influence of pollution exposure on the rate of asthma attacks, as recorded daily by panels of asthmatic volunteers in six different communities. The best predictor of an attack on a given day was an attack on the previous day, regardless of air quality. However, increasing oxidants (primarily ozone) were associated with an increased likelihood of an attack. Similar relationships were found in Houston, Texas.

Regional ambient levels of ozone and other pollutants were compared against total hospital admissions throughout southern Ontario in Canada, for several years during the 1970s. Ambient ozone did not exceed 0.12 ppm on an hourly average. Admissions for respiratory illness during the summer tended to increase with increasing ozone, sulfur dioxide, sulfate, and temperature levels a day or two earlier. Admissions for illnesses other than respiratory illness did not show this relationship with air pollution. It remains difficult to prove which environmental factors were responsible for the increased hospital admissions, although the combination of ozone and acidic aerosols has been suggested as the most likely cause.

One large-scale study of chronic effects compared the results of lung function tests and symptom interviews among four middle-class communities in Los Angeles County with different types and levels of pollution. Some measures were worse on the average in high-ozone areas, but there was no clear overall pattern of ill health attributable to higher pollution levels. A similar association was found in another study which compared nonsmoking Seventh-Day Adventist populations between Los Angeles and other less polluted California communities.

Preliminary results from an autopsy study (82) in Los Angeles County have implicated ambient air pollution as a cause of pathologic lung effects. The lungs of 107 ostensibly healthy youths (age 15–25 years) who died by accident or violence were evaluated. Approximately 80% of the youths demonstrated some degree of respiratory bronchiolitis or other anatomic changes and 27% of the changes were considered severe and extensive. The cause of the pathologic lesions have remained unknown since the smoking and drug use histories of the youths were not determined. The causative pollutant(s) also were unknown, if, indeed, the changes are related to ambient photochemical pollution. Nonetheless, the results have suggested possible pollution-related changes in young persons. Further studies of this nature with epidemiologic data (eg, comparisons of similarly selected autopsy specimens in cities with different air qualities) will be necessary to better understand the significance of the lung changes.

Nitrogen Dioxide

Sources and Ambient Levels. Atmospheric nitrogen will thermally combine with oxygen to form nitric oxide (NO) whenever any common fuel is burned in air. Although nitric oxide is comparatively nontoxic at ambient concentrations, it can be oxidized rapidly to form nitrogen dioxide (NO_2), which has greater toxic potential. Nitrogen dioxide is the most important and ubiquitous nitrogen oxide (NO_x) from the viewpoint of human exposure. Nitrogen dioxide is a fairly water-insoluble, reddish–brown gas with moderate oxidizing capability. The major source of this pollutant in the atmosphere is the combustion of fossil fuels in stationary points (industrial heating and power generation) and motor vehicles (internal combustion engines). Indoor sources of NO_2 are unvented combustion appliances such as gas stoves and gas-fired water heaters.

Like other outdoor pollutants, urban levels of NO_2 vary according to time of day, season, meteorological conditions, and human activities (83–85). The most severe outdoor NO_2 pollution is typically found at places and times of heavy motor vehicular traffic. Typically, morning and afternoon peak NO_2 levels (following traffic emissions of NO) are superimposed on a low ambient background level. Maximum 30-minute and 24-hour outdoor values of NO_2 can be 0.45 ppm and 0.21 ppm, respectively, in metropolitan Los Angeles, which has the highest measured NO_2 concentrations in the U.S.

Indoor exposures to NO_2 may be more important and widespread than outdoor exposures for most people, ie, indoor NO_2 is the major contributor to personal exposure (7,83,85–87). Low ventilation rates ("tight" buildings), use of gas-fueled appliances, and tobacco smoking tend to produce higher indoor than outdoor concentrations of NO_2. Short-term peak concentrations in poorly ventilated kitchens with gas stoves may reach a maximum 1-hour level of 0.25–1.0 ppm and, during cooking, peak levels as high as 4 ppm in the kitchen.

Mechanisms. Potential mechanisms of airway injury by NO_2 exposure in humans have been evaluated with bronchoalveolar lavage (BAL) (45). Several laboratories have found no evidence of decreased cellular viability or increased cell numbers or inflammatory cells in BAL fluid from healthy subjects exposed continuously or intermittently for 3–3.5 hours to 0.6 to 4.0 ppm NO_2 with intermittent exercise. Alveolar macrophages collected after a 3-hour *in vivo* continuous exposure to 0.6 ppm NO_2 tend to inactivate influenza virus *in vitro* less effectively than cells collected after air exposure. However, the effect of NO_2 exposure on controlled influenza infection in humans is inconclusive. On the other hand, increased numbers of BAL lymphocytes and mast cells are found 4, 8, and 24 hours after exposure to 4 ppm NO_2 ($\times$ 20 minutes), with a return to baseline (pre-exposure) BAL cell values at 72 hours. The clinical significance of these changes is unclear since significant changes in symptoms, lung function, or airway reactivity did not occur in these subjects. The lack of a significant neutrophil response in the BAL fluid is noteworthy since NO_2 has been associated with alterations in the activity of alpha-1-proteinase inhibitor (α1-PI) and alpha-2-macroglobulin in BAL fluid. A functional or immunologic decrease in these lung antiproteases could potentially lead to a protease–antiprotease imbalance which would favor degradation of alveolar elastin and potentially cause emphysema if the condition persisted. Acute exposures to NO_2 do not significantly increase total protein or albumin in BAL fluid, suggesting that low-level NO_2 exposure (under experimental conditions investigated so far) does not alter respiratory epithelial permeability in normal subjects. Continuous exposure to NO_2 in humans evokes greater physiologic and cellular responses than intermittent peak exposures.

Effects

Animal Studies. Rats, mice, guinea pigs, and monkeys have been studied in both short- and long-term experiments with NO_2 (70,85,87–89). Nitrogen dioxide acts similarly to ozone in animals in that it injures the respiratory bronchioles, stimulates production of natural antioxidant substances, and increases susceptibility to respiratory infection. Usually a higher concentration of NO_2 than ozone is necessary to produce similar effects, and the toxic NO_2 concentration is larger than commonly occurs during ambient pollution episodes. However, enlarged air spaces due to alveolar destruction (emphysema-like changes) have been reported after six months of continuous exposure to 0.1 ppm NO_2 in conjunction with daily two-hour exposures to 1 ppm. Changes in lung collagen have been reported after less than a month of daily four-hour exposures to 0.25 ppm NO_2.

Biochemical changes may occur at NO_2 concentrations within or slightly above the ambient range. Changes in red blood cell metabolism, liver function, and kidney function in animals have been reported in exposures to 0.5 ppm or less lasting a few hours to a few days. Increased death rates after bacterial inhalation challenges have occurred in long-term (several months) exposures to as little as 0.5 ppm and short-term (three-hour) exposures to 2 ppm. The health significance of these findings for humans remains unclear.

Clinical Studies. Clinical studies using a wide range of exposure conditions have demonstrated mixed results regarding symptoms and lung function, suggesting that NO_2 exerts weak respiratory effects in the laboratory setting (45,84,85,87–91). Healthy adolescents, young adults, and older individuals have failed to consistently demonstrate acute effects of NO_2 on pulmonary function at levels ranging from 0.1 to 4.0 ppm. One study reported increased airway resistance with 0.24 ppm NO_2 but not with 0.48 ppm in resting subjects, suggesting a nonuniform dose–response pattern. Similarly, adult and adolescent subjects with primarily mild asthma have generally not responded to exposures of 0.12–4.0 ppm for 20–120 minutes. However, some asthmatics cooking on gas ranges in the home environment have demonstrated statistically significant decreases in FVC and PEF when the average NO_2 level was >0.30 ppm, although other environmental exposures could not be excluded (92). Patients with moderately severe COPD have not shown consistent lung function changes with exposures to as high as 2.0 ppm NO_2 for 1 hour. In general, subjects in the above protocols lacked significant respiratory symptoms during and following NO_2 exposure. Exposure of healthy young adults to a combination of NO_2 (0.6 ppm) and ozone (0.3 ppm) during one hour of sustained heavy exercise has not augmented symptoms or lung dysfunction above that observed with ozone alone.

Effects of NO_2 on airway reactivity have been inconsistent (45,93). Normal subjects do not change airway reactivity after exposures to as high as 0.6 ppm NO_2. However, enhanced cholinergic bronchial reactivity has been reported following exposures to 2.0 ppm ($\times$ 1 hour) at rest and 1.5 ppm ($\times$ 3 hours) during intermittent exercise. Some asthmatic subjects have not shown enhanced reactivity to inhaled methacholine, histamine, cold air, or SO_2 following exposures to 0.1 to 3.0 ppm NO_2 for two hours or less. On the other hand, NO_2 exposures have potentiated subsequent responses to the same stimuli in other asthmatic subjects. Small numbers of subjects (resulting in lack of statistical power to adequately test a hypothesis), variation in exposure protocols, and different protocols of airway challenges, have been possible sources of difficulty when comparing responses between studies. A simplified "meta-analysis" (93) combining results from 20 studies of asthmatics and 5 studies of healthy subjects exposed to NO_2 showed that there was an overall trend among both groups for airway responsiveness to increase, especially in resting exposures in asthmatics and with concentrations >1 ppm in healthy subjects.

Overall, nitrogen dioxide, at typical ambient indoor and outdoor concentrations, appears to be at least an order of magnitude weaker than ozone as an acute oxidizing agent in the laboratory setting. To date, the acute effects of NO_2 on lung health appear small, and clinically significant effects are unlikely or infrequent. However, this contaminant is probably the least understood of those discussed in this review. The inconsistent exposure–response data are currently difficult to reconcile.

Epidemiologic Studies. Epidemiologic studies of NO_2 effects are difficult because of its ubiquity indoors and outdoors and its coexistence (colinearity) with other ambient pollutants of equal or greater toxic potential (eg, ozone) (84,85,87,90). For example, studies of long-term oxidant-induced health effects in California may have some relevance to NO_2 since ambient levels of NO_2 are usually high where other oxidant levels (primarily due to ozone) are also high. Unfortunately, it is very difficult to specify effects of NO_2 apart from those of ozone, which would probably be considered the most likely cause of most health effects in high-oxidant communities.

One epidemiologic study that specifically addressed NO_2 effects was conducted near an ammunition plant in Chattanooga, Tennessee, in the late 1960s and early 1970s. Nitrogen dioxide was the predominant, although not the only, air pollutant emitted from the plant. In high-pollution areas near the plant, both schoolchildren and their parents tended to have high rates of lower-respiratory illness, when compared with residents of cleaner areas. When pollution levels improved, the illness rate appeared to decline although it remained highest in the most polluted area. Unfortunately, the standard air monitoring method then in use was eventually found to be inaccurate, leaving doubts about the reliability of NO_2 concentrations reported for the study areas.

Subsequent epidemiologic information on NO_2 has come from surveys comparing the health status of adults and children living in homes with electric stoves (with low NO_2 levels) and in homes with gas stoves (with comparatively high NO_2 levels, often exceeding current EPA air quality standards). Some studies have associated small increases in illness rates or decreases in lung function with gas stoves; other studies have reported no association. The inconsistencies may be related to numerous factors. For example, NO_2 is not the only pollutant emitted by gas stoves. Nitrogen dioxide, which was frequently not directly measured, might also be emitted by other indoor sources. Thus far, it is difficult to confirm that gas stoves in general and indoor NO_2 in particular present a meaningful health risk since most epidemiologic studies did not carry out precise exposure assessments on their study populations (86). However, a meta-analysis (94) of results from epidemiologic studies has suggested an increase of at least 20% in the odds of respiratory illness in children exposed to a long-term increase of 30 $\mu g/m^3$ NO_2.

A prospective 18-month cohort study (95), involving 1,205 healthy infants of nonsmoking parents in Albuquerque, New Mexico, evaluated whether exposure to NO_2 increases the incidence or duration, or both, of respiratory illness in infants. No association was found between NO_2 exposure and the incidence rates for any illness category or between the presence of a gas stove and illness incidence. The results cannot be generalized to older children, adults, or to more susceptible groups such as infants of smoking parents. Over 75% of the measured NO_2 concentrations in the homes were less than 0.02 ppm, whereas 5% were greater than 0.04 ppm. These levels have been similar to those in many locations in the U.S. but are much lower than those in heavily polluted cities and in poorly ventilated inner-city apartments.

Thus, the study results cannot be generalized to infants exposed to greater than 0.04 ppm NO_2. This study is considered to be one of the most carefully designed and executed assessments of the effects of NO_2 on respiratory illness conducted to date. The findings are reassuring for the segment of the very young population without uncommonly heavy exposures. Available epidemiologic evidence on the acute health effects from NO_2 exposure is insufficient to support the establishment of a short-term standard (NAAQS). However, the EPA has established a long-term standard using a maximum permissible annual average concentration (see Table 2). Apart from any direct health risk, NO_x pollution is important to health because it contributes to ozone formation.

Sulfur Dioxide

Sources and Ambient Levels. Natural fuels such as coal and oil usually contain sulfur. During combustion, the sulfur combines with oxygen to form sulfur dioxide (SO_2) gas, which is emitted into the atmosphere with other products of combustion. Additional sulfur-containing products form as small particles (eg, sulfuric acid aerosols), either during the combustion process itself or by further chemical reactions in the atmosphere. Additional particles arise from incompletely burned fuel and noncombustible mineral impurities in the fuel. Because they typically occur together in the ambient environment, SO_2 and particulate pollution must often be considered together in scientific and regulatory efforts (49,96,97). Particulate pollutants (including sulfuric acid) are discussed in the next section.

The primary outdoor source of SO_2 is the domestic, commercial, and industrial combustion of sulfur-containing fossil fuels (coal and oil). Metal smelters tend to be the largest point sources of SO_2. Power plants and other large industrial establishments are smaller point sources but constitute a larger cumulative source. Oil- or coal-heated homes and commercial buildings may also contribute substantially to SO_2 levels in urban areas. The highest ambient concentrations occur where the "plume" of emissions from a large point source reaches the ground. Many modern industries employ tall smokestacks, which reduce local ground-level pollution because the plumes are either greatly diluted before reaching ground level or dispersed elsewhere by winds. However, the long-range transport of the pollutants leads to high ambient concentrations and acid rain in downwind areas unless additional emission controls are used.

Apart from large point sources, urban SO_2 concentrations in the U.S. do not often exceed 0.2–0.3 ppm as a one-hour average. This range reflects a marked improvement relative to the 1950's, and earlier, when uncontrolled burning of coal and high-sulfur oil led to severe SO_2-particulate pollution episodes and high morbidity and mortality rates. However, down drafts near power plants may produce 5- to 10-minute peaks exceeding 1 ppm.

Indoor levels of SO_2 are dependent on outdoor concentration and the air-exchange rate. Indoor levels are generally much lower than outdoor values, owing to absorption of the gas on furniture, walls, clothing, and in ventilation systems, as well as infrequent indoor sources.

Mechanisms. Various drug pretreatment studies have provided insights into the mechanisms of SO_2-induced bronchoconstriction in asthma (45,98). Reflex parasympathetic pathways are involved since atropine inhibits the airway response to inhaled SO_2 in normal and in asthmatic subjects. However, anticholinergic drugs have not been completely protective in all asthmatics. Cromolyn sodium and nedocromil sodium provide greater protection, which is dose-dependent, than either atropine or ipratropium bromide. The combination of cromolyn and ipratropium more effectively inhibits SO_2-induced bronchoconstriction than either drug alone. The mode of action of cromolyn and nedocromil is unknown but may be related to mast cell stabilization or inhibition of mediator release. Inhaled H_1-antihistamine (clemastine) and bronchodilators (albuterol and metaproterenol) effectively prevent SO_2-induced bronchoconstriction. Pretreatment with an inhaled corticosteroid (beclomethasone dipropionate) fails to block the acute airway response to 0.75 ppm SO_2 in asthmatics.

Little information is available about the effects of inhaled SO_2 on cellular and mediator responses in the lung (45). However, dose-dependent increases in BAL lymphocytes, mast cells, macrophages, and total cells are present in healthy volunteers 24 hours following 20-minute exposures to 4, 5, and 8 ppm SO_2. BAL cell responses do not increase further after exposure to 11 ppm SO_2. Significant increases of these BAL cells occur at varying times (4, 8, 24, and 72 hours after exposure to 8 ppm SO_2) and generally peak by 24 hours. All cell counts return to pre-exposure baseline values by 72 hours after exposure. Neither albumin nor the ratio of helper/cytotoxic suppressor lymphocytes are changed at any time after exposure. No significant changes in lung function or increases in symptoms occur under these conditions. Although the above SO_2 exposures exceed the U.S. TLV of 5 ppm, peak indoor industrial levels of 8 ppm are not uncommon. These results indicate that acute SO_2 exposures induce cellular responses in otherwise asymptomatic (nonasthmatic) individuals with little change in lung function.

Effects

Animal Studies. Exposures to SO_2 of several minutes to hours in guinea pigs and other species have shown decreases in respiratory rate and increases in Raw and bronchial reactivity (97). These effects occur only at concentrations above 1 ppm. Airway resistance increases within a few minutes after exposure begins and recovers within a few minutes to hours after exposure ends, depending on the severity of the initial response.

Longer-term studies (ranging from a week to more than a year) have evaluated effects of SO_2 on lung injury and repair and on respiratory defense mechanisms. Evidence of airway epithelial injury has been found with light microscopy at 5 ppm or above. Chronic bronchitis (similar to that in humans) can be induced in dogs by exposure to 200 ppm SO_2 for 8 months (2 hours a day, 4–5 days a week) (99). Within the first 2–4 weeks of exposure, the dogs develop cough, mucus hypersecretion, airway epithelial thickening, an increase in the number (hyperplasia) of goblet cells, hypertrophy and hyperplasia of mucous glands, chronic airflow obstruction, and persistent lung inflammation as demonstrated by an increase in PMN in BAL fluid. On the other hand, airway responsiveness to inhaled methacholine decreases 2- to 3-fold within 8 weeks of SO_2 exposure. Daily exposures of dogs to 50 ppm for 10–11 months (with the same schedule as previously described) results in similar histologic and physiologic changes but no inflammatory cell infiltration of the airways or changes in histamine or methacholine responsiveness (100). Dogs exposed to 15 ppm SO_2 for 5 months do not demonstrate the histologic or physiologic changes noted above. These findings suggest that chronic airway inflammation induced by SO_2 may be associated with decreased responsiveness (attenuation) to inhaled bronchoconstricting agents.

Functional alterations of the immune system occur at 2 ppm SO_2 while respiratory defenses against inhaled bacteria appear unimpaired at exposure levels up to 5 ppm. Sulfur dioxide alters mucociliary clearance at ambient and higher concentrations, but the specific findings differ among species. In some species, SO_2, appears to delay clearance of inhaled tracer particles, suggesting greater retention and potential for infection or toxicity from inhaled substances. In other species, SO_2 accelerates clearance, suggesting a stress response. At present, too little is known about normal clearance mechanisms to judge the health significance of the changes attributed to SO_2.

Clinical Studies. In healthy adolescents and adults, acute exposures to high concentrations (>5 ppm) of SO_2 uniformly result in bronchoconstriction, whereas lower levels (0.15–2.0 ppm) do not elicit significant symptoms or impair lung function even with intermittent exercise and different inhalation routes (45,97,98,101,102). Older subjects (55–73 years) are not greatly more reactive to inhaled SO_2 than younger subjects. The most frequently reported symptoms are confined primarily to the upper airways, eg, an unpleasant taste or odor. The exposure combination of SO_2 (0.4 and 1.0 ppm) and ozone (0.3 and 0.4 ppm) does not produce an additive or synergistic effect.

Individuals with asthma or allergy with exercise-induced bronchospasm (EIB) represent a well-documented sensitive population, although there is some inter-individual variability. Markedly symptomatic bronchoconstriction occurs at <1 ppm SO_2 at rest and ≥0.5 ppm during moderate exercise for 10–30 minutes. At 0.25 ppm or less, some asthmatics have asymptomatic increases in specific airway resistance (SRaw). Healthy nonasthmatic individuals are usually not affected at these low concentrations. Dose-response relationships have been consistently found. Symptomatic bronchoconstriction in asthmatics can occur within 3 minutes of inhaling 0.5 or 1.0 ppm. Lengthening the exposure beyond 10 minutes does not appreciably influence the magnitude of response. Bronchoconstriction reverses spontaneously at rest within 30 minutes in clean air or within an hour with continued exposure.

Atopic (allergic) adolescents with EIB and asthmatic adolescents decrease their FEV_1 > 15% following 30-minute rest and 10-minute moderate exercise with 0.5 and 1 ppm SO_2. Atopic adolescents without EIB and healthy adolescents have similar responses to inhaled SO_2, indicating that atopy alone is insufficient for SO_2 sensitivity and that hyperreactive airways is a critical factor. Patients with moderately severe COPD do not have significant symptoms or changes in lung function or oxygen saturation with 1-hour exposures to 0.4 and 0.8 ppm SO_2 during intermittent light exercise, although it is uncertain whether low dose rates and/or airway reactivity and continued medication use, influence the results.

The bronchoconstrictor effect of SO_2 (as low as 0.10 ppm) can be potentiated in asthmatics by oral breathing, increased ventilation, subfreezing, dry air (−10° to −20°C) and prior exposure to 0.12 ppm ozone. Hot air (38°C) does not enhance the SO_2 response. The relationship between the responses to methacholine and SO_2 is unclear. Freely breathing asthmatics frequently develop significant bronchoconstriction at ≥0.5 ppm SO_2. Although the threshold is higher with unencumbered breathing than with oral breathing alone, the former does not prevent bronchoconstriction, particularly with heavy exercise. As with many inhaled pollutants, ventilation during exercise increases the effective dose to the airways and facilitates EIB. Oral and, to a lesser extent, oronasal breathing during exercise also decreases the protective SO_2-removal capability of the nose and enhances EIB (103).

Tolerance to SO_2 develops in asthmatic subjects during repeated exposures. Lung function responses are significantly attenuated in asthmatics recurrently exposed to 0.2–1.0 ppm SO_2 for ≤1 hour during exercise. The mechanism of the tolerance is unknown. Unfortunately, the attenuation is not practically protective since the response is generally modest, short-lived (<24 hours), and the initial exposure can result in a severe bronchospastic attack that halts further exercise or other activity.

Inhaled SO_2 may also affect the upper airway since as much as 99% of this water-soluble gas may be absorbed from the inspired air by the nasopharynx. Inhalation of

1.0 ppm SO_2 through a face mask has been reported to significantly increase nasal work of breathing in adolescent asthmatics. However, other investigators could not confirm either increased nasal symptoms or resistance with brief exposures to 1–4 ppm SO_2 in nonasthmatic subjects with rhinitis or in subjects with asthma and allergic rhinitis, suggesting that SO_2 responsiveness in the upper airways is not consistent among subjects and is not uniform throughout the respiratory tract.

Epidemiologic Studies. The majority of epidemiologic studies reviewed here deal with the effects of combined sulfur dioxide and particulate pollution because of their strong tendency to coexist. When health effects are found, scientists cannot be certain whether they are caused by sulfur dioxide alone, particulate pollutants alone, their combination, or an unidentified factor that tends to accompany air pollution.

Three separate international incidents between 1930 and 1952 demonstrated that several days of fog and intense sulfur dioxide-particulate pollution were associated with large increases in morbidity (illness rates) and respiratory mortality (death rates) (49,104,105). All occurred in areas with a heavy concentration of industry and uncontrolled coal burning. In the Meuse Valley of Belgium in December, 1930, about 60 premature deaths occurred in older persons with preexisting cardiac or lung disease, and complaints of respiratory irritation were widespread among all age groups. In Donora, Pennsylvania, in October, 1948, about 20 premature deaths from cardiopulmonary causes occurred in a population of 14,000. 40% of the population had some respiratory symptoms during the episode, and the incidence was higher among people with preexisting chronic respiratory illness. Although air monitoring data were not available, retrospective estimations of ambient sulfur dioxide concentrations ranged from 0.5 to 2 ppm. In London during December, 1952, 3,500–4,000 premature deaths were recorded. The maximum weekly death rate was more than twice its usual value. Sulfur dioxide levels exceeded 1 ppm as a 24-hour average. Peak particulate concentrations exceeded 4 milligrams per cubic meter (mg/m^3) or about 10 times the level considered normal at that time (but now considered as severe pollution).

Since 1952, numerous reviews have examined the relationship between London death rates and more moderate pollution (49,97,104,105). The U.S. EPA (97) drew four conclusions: (*1*) markedly increased death rates occurred, mainly among the elderly and chronically ill, at particulate concentrations above 1 mg/m^3 and SO_2 concentrations above 0.4 ppm as 24-hour averages, particularly if these concentrations persisted for several days; (*2*) during the episodes high humidity or fog was important, perhaps because it facilitated the formation of sulfuric acid or other acidic aerosols; (*3*) increased death rates were associated with particulate levels of 500 micrograms per cubic meter ($\mu g/m^3$) to 1 mg/m^3 and SO_2 levels of 0.2 to 0.4 ppm; and (*4*) small but probably real increases in death rates are likely at particulate levels below 500 mg/m^3, but not at SO_2 levels below 0.2 ppm.

A number of studies have attempted to explain the variation in death rates among different U.S. metropolitan areas in terms of their different pollution levels. Sulfur dioxide, total particulate matter, and particulate sulfate all have been suggested to cause increased mortality rates (49). The results obtained in such analyses are quite sensitive to the choice of metropolitan areas, air pollution measurements, statistical analysis, and adjustments for confounding factors. Different studies tend to yield different conclusions as to which pollutants were associated with high death rates. The U.S. EPA has concluded that full reliance cannot be placed on these analyses to assess the increased risk of mortality associated with SO_2–particulate air pollution.

Field epidemiologic studies of school-age children in North America and Europe suggest that temporary losses of lung function occur during SO_2–particulate pollution episodes. The 24-hour average concentrations measured in these episodes are 0.1 to 0.2 ppm for SO_2 and 200–400 mg/m^3 for particulate matter. Most children show only very small decreases in FVC or FEV_1. However, a few individuals have larger, perhaps medically significant losses in lung function. Test results appear to return to their pre-episode levels after several weeks.

Additional field studies have examined respiratory health in western U.S. communities where much of the SO_2 comes from smelters. Short-term exposures in these areas may be quite severe (with concentrations near 1 ppm for an hour or more), despite relatively low annual average concentrations. A comparison of four Utah towns with similar particulate levels but different SO_2 levels (annual averages ranging from near zero to above 0.4 ppm) showed that nonsmoking young adults in the most polluted community were more than twice as likely to report persistent cough, as compared to similar people in cleaner communities. An excess cough without lung function decrement was reported in Arizona school children living in smelter communities, as compared to children in cleaner areas.

Long-term studies are fewer in number. One large-scale investigation compared six northeastern and midwestern U.S. cities of differing pollution levels. More than 10,000 children were examined. Across different cities, the frequency of chronic cough was related to annual average levels of SO_2, total suspended particulates (TSP), and particulate sulfate. The frequency of lower respiratory illness was associated with total particulate levels but not SO_2. However, within cities, regional differences in pollution have not appeared to affect health measures. Most lung function differences between cities have not related significantly to pollution levels. The annual average levels of SO_2 ranged from zero to 0.07 ppm across the six cities. The annual average levels of particulate matter ranged between 30 to 160 mg/m^3 for TSP and 4 to 20 mg/m^3 for the sulfate fraction.

Particulates and Acidic Aerosols

Sources and Ambient Levels. Airborne particulate matter is a complex mixture of organic and inorganic solids-in-gas and liquid-in-gas aerosols with heterogeneous physicochemical composition, size, and biologic activity (3,5,102). Unlike other criteria pollutants, "particulates" is not a precise chemical or physical term. Particulates may be solid or liquid, chemically active or inert, soluble or insoluble in water, or may combine all these properties in different components of a particle mixture. "Smoke," "soot," "fumes," "dust," and "haze" are common terms applied to some of the diverse forms of particulate pollution. They may be emitted directly from a source (primary) or formed in the atmosphere from pollutant gases (secondary). Particulates are emitted from a wide range of natural (eg, dust storms, volcanoes, pollens, bacteria, fungi) and man-made sources (eg, coal burning, automobiles, incinerators, industrial processes, and power plants). The more widespread and clinically relevant sources are man-made and are usually concentrated in heavily populated areas. Air monitoring practices in the U.S. originally determined the concentration of total suspended particulates (TSP) by drawing a known volume of air through a highly efficient filter over a 24-hour period and measuring the gain in filter weight. Concentrations determined by this method may range from 30 $\mu g/m^3$ in clean rural air to 300 $\mu g/m^3$ or more during an urban pollution episode. Urban concentrations have been much higher in the past.

Chemical analysis of the filter samples is complex and expensive. Air quality standards relate only to the mass of particles and not to their chemical composition, except when they contain lead. A complex mixture of particles will be generated in areas with extensive coal and oil burning. Some particles have a sooty or oily nature, containing carbon or hydrocarbons from incomplete fuel combustion. Other particles will be "fly ash," ie, minerals (especially metal oxides) from noncombustible impurities in the fuel. Other particles (especially fine particles, <2.5 μm in diameter) may arise from photochemical reactions between emitted sulfur and nitrogen oxides in the atmosphere to form strong acids such as sulfuric acid (H_2SO_4), nitric acid (HNO_3). These acids can exist as airborne or aerosolized solution droplets (fog or "acid rain") with extremes in acidity that vary with meteorological factors and *in situ* interactions with other chemicals. Fog is an important factor in particulate pollution (see discussion of historical pollution episodes in "Sulfur Dioxide"). Natural fog consists of water droplets approximately 5–20 μm in diameter, which may dissolve substantial amounts of water-soluble pollutants. Clean fog droplets are slightly acidic because they dissolve some atmospheric CO_2. However, polluted fog water may contain 1,000 times the normal amount of acid. Sulfuric acid (H_2SO_4) will predominate in coal- and oil-burning areas with high levels of ambient SO_2. Most of the H_2SO_4 in ambient air results from combustion-related SO_2, which is oxidized to sulfur trioxide and hydrated to H_2SO_4 (107,108). The distribution of acidic aerosols is complex and dependent on the rates of oxidation of SO_2 and neutralization of H_2SO_4 by ammonia (NH_3) from human, animal, and fertilizer sources. With time, ammonia gas in the atmosphere or alkaline minerals in particles will neutralize the acid, forming sulfate salts. However, the partially neutralized forms of H_2SO_4, ie, ammonium bisulfate (NH_4HSO_4), letovicite ($(CNH_4)_3$ $H(SO_4)_2$) ammonium sulfate ($(NH_4)_2$ SO_4), remain strongly acidic. The acid sulfates are primarily in the fine particle size range (<2.5 μm in diameter) and can persist in the atmosphere for days and be transported downwind for long distances.

In areas with extensive motor vehicular contribution to photochemical oxidant smog, particles are likely to contain hydrocarbons and related organic substances, sulfate salts from SO_2, or nitrate salts derived from NO_2. Nitric acid (HNO_3) vapor will predominate in these areas.

Health risk from particulates depends on the effective inhaled dose (as with any pollutant) and the dose depends on both particle size and concentration. Size is expressed as aerodynamic diameter which depends on physical size, shape, and density. An airborne particle of 1 micron (micrometer, μm) in aerodynamic diameter, will settle to the ground under the influence of gravity at the same speed as a sphere that is 1 μm in physical diameter and as dense as water. Particles from any source will vary in size. Ambient air typically contains "fine" particles (largely derived from man-made pollution, with aerodynamic diameters ranging from 2.5 to less than 0.1 μm) and "coarse" particles (largely derived from soil, plants, or sea salt, with aerodynamic diameters ranging from 2.5 to 50 μm). Particles larger than 40–50 μm will not remain airborne for long and settle rapidly in windless conditions. Most particles larger than 10 μm are deposited in the upper airway during normal nasal breathing at rest and present minimal health risk.

A more relevant measurement of ambient particles, PM_{10} (particulates ≤ 10 μm in aerodynamic diameter), has been implemented to replace TSP measurements in order to focus on respirable particulates and their potential health effects. Particles ranging from several microns to less than one micron have presented the most health risk since they are most likely to be inhaled and deposited in the lungs and are chemically the most bioreactive. Most man-made pollutant particles fall within this respirable size range. Respirable particles that reach the tracheobronchial region tend to deposit at bifurcations of the airways; these branch points receive much higher doses of potentially toxic particles than other areas. Unless they are dissolved in the airway mucus lining, clearance of deposited insoluble particles is effected within minutes by the mucociliary defense mechanism (mucus and ciliary motion). With mouth breathing and increased ventilation (usually with exercise), more particles will penetrate deeper into the respiratory tract. During heavy exercise with mouth breathing, an estimated 10–20% of particles of 10 μm will enter the tracheobronchial tree. Very fine particles (less than 1 μm in diameter) may be exhaled or deposit in the peripheral small airways and alveoli where clearance by macrophages and lymphatics may take hours, months, or years. Some fine particles are hygroscopic, and thus attract water vapor and increase in mass during inhalation, causing them to deposit in larger, more central airways. Clearance may be impaired by some lung diseases and by some types of air pollution exposures.

The spatial and temporal distributions of ambient acidic aerosols in North America have not been monitored until recently, owing to the complex sampling systems required. Sulfate (SO_4^{2-}) has been considered to be an adequate surrogate for acidic species since it represents the predominant ambient acidic aerosol (H_2SO_4) and is easy to measure. However, the sulfate ion probably is not the toxicologically active portion of particulate sulfates, and high sulfate levels do not necessarily indicate the presence of high acid levels (107). Although the hydrogen ion (H^+) appears to be a critical marker of biological potency, most contemporary studies have expressed the concentration of

H^+ as $\mu g/m^3$ of H_2SO_4. Where measured, the most frequent values of H_2SO_4 are <5 $\mu g/m^3$, and the highest levels of H_2SO_4, occurring most frequently in the summer, are 20–40 $\mu g/m^3$ (over a range of 1–12 hours) in various parts of North America. These concentrations are below the current EPA standard of 150 $\mu g/m^3$ (24-hour average) for PM_{10}. However, ambient acid aerosols are being strongly considered as a new EPA-regulated criteria pollutant owing to their distinct health risk as compared to other particulates (106). Almost 700 $\mu g/m^3$ (1-hour average) has been estimated to have occurred in London during 1962, and even higher levels were likely present in previous years. The 8-hour threshold level of H_2SO_4 for occupational exposure is 1000 $\mu g/m^3$, reflecting the observation that H_2SO_4 is not especially irritating to healthy people when inhaled.

Data for indoor levels of acidic aerosols and gases are currently scarce. A sampling of residences in the Boston area have revealed low indoor/outdoor ratios for HNO_3 and strong (free or titratable) acid (H^+) during winter and summer. However, nitrous acid (HONO), sulfate (SO_4^{2-}), nitrite (NO_2^-), nitrate ($NO3^-$), NH_3, and ammonium (NH_4^+) have showed high indoor/outdoor ratios during both seasons. The significance of neutralization of acidic aerosols by indoor NH_3 remains to be determined.

Mechanisms. Research in acid-related mechanisms of action in humans is largely limited to characterization of the types of bronchoconstrictive stimuli present in acidic aerosols or fog (45). Titratable acidity (total available H^+) appears to be more important than pH (free H^+) alone, since oral NH_3 neutralizes inhaled H^+. The bronchoconstrictor potency of sulfate aerosols is proportional to their acidity, and inhaled buffered acids produce greater bronchoconstriction in asthmatics than unbuffered acids at the same pH, presumably by transiently decreasing airway surface pH. However, the quantitative relationship between airway response and aerosol H^+ content (dose) and the importance of differentiating the H^+ of specific acidic gases and aerosols remain uncertain. Bisulfite (HSO_3^-) without SO_2 or hypoosmolar acid aerosol can independently produce bronchoconstriction in asthmatics. Interactions of acid aerosols with co-pollutants such as oxidants may also play a role in toxicity, although combined ozone and H_2SO_4 exposures in the laboratory have not shown excess respiratory effects.

The pulmonary effect of a two-hour exposure to 1000 $\mu g/m^3$ H_2SO_4 (0.9 μm MMAD and 1.9 GSD) during intermittent exercise in healthy young adults was evaluated by BAL 18 hours following exposure (109). When compared to results from a sodium chloride exposure (control), the acid exposure did not show significant differences in alveolar cell subpopulations or alveolar macrophage function. The subjects did not have lower respiratory symptoms or changes in lung function during and after acid exposure.

Effects

Although many ambient acid aerosols and other particulate matter have potential health significance, sulfuric acid will be emphasized in this section because of its respirable nature and highest irritant potential of the particulates. Sulfate and nitrate salts have shown few effects in humans, except possibly at concentrations much higher than ambient.

Animal Studies. A substantial fraction of animal studies have dealt with particulate derivatives of SO_2, especially H_2SO_4 and certain salts that might be formed from atmospheric H_2SO_4. The latter include ammonium sulfate and ammonium bisulfate (the products of complete or partial neutralization of the acid by NH_3 gas), zinc ammonium sulfate, and sulfate salts of zinc, iron, copper, and manganese. Most of these studies have used particles of respirable sizes. Sulfuric acid appears to be the most potent respiratory irritant (102,108,110–112). In the guinea pig, dose-related increases in Raw have been observed after one-hour exposures to 100 $\mu g/m^3$ and higher. These concentrations are within the possible ambient range for total particulate matter, but probably exceed maximum contemporary ambient H_2SO_4 levels, since only a small fraction of all particulate pollution is likely to be H_2SO_4. However, ambient levels of H_2SO_4 are not well documented since only a few measurements are available and acidic particles make up a significant proportion of the fine particles. Changes in mucociliary clearance have been reported in dogs, rabbits, and donkeys during short-term and long-term exposures to H_2SO_4. Although results from different studies have been inconsistent, acute mucociliary clearance changes in humans and rabbits are similar. Continuous exposures of monkeys, dogs, and guinea pigs for a year or longer have not shown obvious damage to the lungs or blood at concentrations from less than 100 $\mu g/m^3$ to 1 mg/m^3. Thickening of airway walls occurs in monkeys at concentrations above 2 mg/m^3. Sulfuric acid and ammonium salts appear to have little effect on animal lung defenses against inhaled bacteria. Although lung defenses against infection are impaired by some metal salts, the concentrations required are much higher than ambient levels.

Sulfuric acid exposures have been studied in combination with other gaseous and particulate pollutants to compare the toxicity of mixtures and single pollutants. In some studies, ozone and H_2SO_4 together have shown excess toxic effects when ozone concentrations are near ambient levels and H_2SO_4 levels are much higher than ambient (eg, 1 mg/m^3). Coexisting sulfates and other salt aerosols appear to increase bronchoconstriction from SO_2 in guinea pigs. Exposures to mixed carbon and H_2SO_4 aerosols, both at concentrations much higher than ambient, suggest excess effects although the results have been inconsistent.

Particulate substances other than sulfates have been studied extensively to assess occupational hazards. In general, no effects have been found except at concentrations well above the ambient range.

Clinical Studies. The health effects of acidic aerosols in controlled human exposures have been summarized elsewhere (45,97,98,102,106,108,113,114). Normal subjects generally show no significant change in symptoms or lung function following 10- to 240-minute exposures to H_2SO_4 alone (10–2000 $\mu g/m^3$) or in combination with 0.3 ppm ozone with varying exercise. Nonspecific airway reactivity following brief exposures to 1000 μg/m^3 (1.1 μm diameter) of H_2SO_4, NH_4HSO_4, $(NH_4)_2SO_4$, and sodium bisulfate ($NaHSO_4$) increases according to particle acidity, with H_2SO_4 and NH_4HSO_4 significantly enhancing reactivity even though they cause only marginal changes in post-exposure lung function. However, one study has indicated that bronchial reactivity to methacholine is not significantly enhanced with exposure to 2000 μg/m^3 H_2SO_4 of droplet sizes ranging between 1–20 μm.

Results of acute acid aerosol exposures in asthmatics have been mixed and less consistent than with SO_2, although it appears that many individuals with hyperreactive airways may be sensitive to inhaled H_2SO_4. In some studies, significant bronchoconstriction (adjusting for exercise effect) has not been detected in adult asthmatics exposed to as much as 2000 μg/m^3 H_2SO_4 for 10–120 minutes with intermittent exercise. However, other studies demonstrate acutely symptomatic, dose-dependent bronchoconstriction in adult asthmatics following exposures to H_2SO_4, beginning at 450 μg/m^3. A relationship between the acidity of different sulfate aerosols and degree of induced bronchoconstriction is observed. Airway reactivity to the cholinergic drug carbachol is enhanced following H_2SO_4 exposure, but reactivity to cold air inhalation is not affected. Adolescent asthmatics have been reported to be more sensitive than adult asthmatics with a 40-minute exposure to 100 μg/m^3 H_2SO_4 but this observation has not been confirmed by other investigations, suggesting that acid-sensitive and acid-insensitive subgroups of young asthmatics may exist. Most of the above studies have used fine H_2SO_4 particles less than 1 μm in aerodynamic diameter.

Inhaled H_2SO_4 acutely affects mucociliary clearance of tracer particles inhaled concurrently or after the acid exposure in healthy and asthmatic subjects. Mucociliary clearance may be more sensitive than lung function tests to detect the effect of acid. Reported clearance effects vary with dose, acid concentration, tracer particle size, and exposure duration in a complex manner. One-hour exposures to inhaled H_2SO_4 (100 and 1000 μg/m^3) show faster clearance at the low exposure concentration and slower clearance at the higher level when the inhaled tagged tracer aerosol is 7.6 μm in aerodynamic diameter. This experience suggests that lower doses stimulate clearance, whereas higher doses damage the mucociliary apparatus. However, clearance is slowed during both concentrations with a smaller tagged tracer (0.5 μm), reflecting more similar deposition sites of H_2SO_4 and tracer. Mucociliary clearance is transiently slowed in nonmedicated asthmatics following inhalation of 3.9-μm aerosol tracer and 1-hour exposure to 1000 μg/m^3 H_2SO_4. Bronchial clearance is significantly greater after 2-hour (compared to 1-hour) exposure to H_2SO_4 at 100 μg/m^3, and the reduced clearance persists as long as 3 hours following exposure. Pulmonary function does not significantly change following the above exposures. Thus, the duration of acute H_2SO_4 exposure affects mucociliary clearance both in magnitude and in persistence. The latter may correlate with some reports of delayed symptoms and bronchial hyperreactivity following H_2SO_4 exposure. Sulfuric acid fog (10.3 μm MMAD; pH 2.0; osmolarity 30 mOsm) significantly increases tracheal and outer lung zone clearance in healthy adults without affecting symptoms, lung function, or airways responsiveness to methacholine.

The inconsistent physiologic results with ambient levels of sulfuric acid warrant further clarification. The explanation may relate to biological variability of the subjects, with more sensitive subjects selected by one investigator and comparatively less sensitive volunteers by another, either by chance or by using different selection criteria. Other experimental differences may be important such as exercise intensity, times at which responses are measured, and use of mask or unencumbered breathing. Ammonia gas occurs naturally in the upper airways and can neutralize inhaled acid aerosols, presumably making them less toxic. The effectiveness of acid neutralization may vary according to the size of the acid particles, manner of breathing, amount of NH_3 present (dependent on both biological and microenvironmental characteristics), and use of acidic beverages to purposely reduce endogenous NH_3 (115–117).

Epidemiologic Studies. By necessity, most epidemiologic studies have assessed the SO_2–particulate complex rather than particulate pollution itself. Combined SO_2–particulate studies are discussed in the section on sulfur dioxide. Reviews of the possible health effects of particulate pollution have drawn different conclusions about contemporary particulate levels, pointing out the difficulties and uncertainties faced by researchers in this field (49,97,98,102,105,106,108,118). Most early epidemiologic studies have not included measurements of acid aerosols since they were not (and still are not) part of routine aerometric monitoring networks. However, subsequent studies have included a determination of acid aerosols and have suggested that acid aerosols may aggravate symptoms in subjects with preexisting lung disease by irritation of the airways.

A potentially important finding is the association between contemporary respirable particulate matter (or a more complex pollution mix associated with respirable matter and sulfur oxides or metal ions) and daily mortality across diverse U.S. cities (119). Small but statistically significant associations of particulate pollution with all-cause daily mortality rates have been demonstrated in a number of U.S. cities or communities (Detroit, Philadelphia, Los Angeles, Steubenville, Ohio, and Santa Clara County in California) and in the Utah Valley, the site of a steel mill. This association has occurred when PM_{10} levels were below the present NAAQS (see Table 2). The increments of pollution-associated mortality have been strongest for cardiopulmonary deaths and respiratory mortality in Philadelphia. Respiratory morbidity in healthy persons and asthmatics (120) and PM_{10} pollution have shown positive correlations in the Utah Valley and Seattle, Washington. PM_{10} concentrations in Salt Lake City, Utah, have accounted for a 2–3% reduction in lung function in adult smokers with mild to moderate airflow limitation, after statistical adjustment for smoking habits (121).

All studies have used centrally located fixed-site monitors to estimate population exposures. Personal exposures to particulates have also been measured in some studies. Most studies have not been able to adequately control for cigarette smoking, temperature, season, winter-time infections, and other potential ecologic differences between communities. The particulate–mortality association has been difficult to explain mechanistically and awaits further animal and human toxicologic studies to provide an adequate interpretation. Until then, the epidemiologic data suggest that the current NAAQS for PM_{10} may not be protecting against adverse health effects with an adequate margin of safety as mandated by the Clean Air Act.

Carbon Monoxide

Sources and Ambient Levels. Carbon monoxide (CO) is formed during combustion of any carbon-containing fuel. Carbon monoxide occurs naturally as a trace constituent of the troposphere. The major source of outdoor carbon monoxide in urban areas is incompletely burned fuels (122–125). Motor vehicular exhaust accounts for two-thirds of outdoor carbon monoxide pollution, although industrial and domestic sources also contribute. In cities like Los Angeles the ambient concentrations occasionally exceed 20 ppm as a one-hour average. Before the use of catalytic converters in automobiles, community levels were substantially higher. Average background levels are higher in winter and lower in summer. Local concentrations may exceed 40 ppm on streets with very heavy traffic or in tunnels and parking garages. Thus, people who spend considerable time in such areas may receive much greater doses than would be predicted from outdoor fixed-site air monitoring stations.

Indoor microenvironments present the most important CO exposures for the majority of individuals (122–125). Indoor concentrations of carbon monoxide are a function of outdoor levels, indoor sources, infiltration, ventilation, and air mixing between and within rooms. In residences without indoor sources, average indoor CO concentrations are approximately equal to average outdoor levels. Homes with attached garages as well as homes adjacent to heavily traveled roadways, or parking garages may have high indoor CO concentrations. Personal carbon monoxide exposure studies have shown that the highest CO exposures (9–35 ppm) occur in indoor microenvironments associated with such transportation sources. Unvented gas and kerosene space heaters used indoors for prolonged periods may cause residential CO concentrations to exceed the NAAQS of an 8-hour average of 9 ppm or a one-hour average of 35 ppm. Intermittent indoor sources such as gas cooking ranges can result in peak CO levels in excess of 9 ppm and 24-hour average concentrations on the order of 1 ppm.

Of all the criteria pollutants, carbon monoxide (along with lead) is most widely recognized as a poison (126,127). Hundreds of people die every year by accident or suicide from grossly elevated indoor CO concentrations, often in the thousands of ppm. Combustion in a poorly ventilated space (eg, with a defective or improperly vented stove or heater or a car engine running in a closed garage) causes the pollution in these cases. With proper care, such serious risks can be avoided, and indoor CO levels need not be much higher than outdoor levels. Normal cooking and heating and tobacco smoking still produce some indoor CO.

Mechanism. The toxicity of carbon monoxide is due to its binding affinity for hemoglobin to form carboxyhemoglobin (COHb), a central feature of CO toxicology (123–127). The relative affinity of CO for hemoglobin is approximately 200-fold greater than that of oxygen, so that only a small concentration of CO in air can form a significant amount of COHb in blood. The COHb not only reduces the amount of hemoglobin available to carry oxygen, but also impairs the release of oxygen to body tissues. The result is reduced delivery of oxygen to tissues that require oxygen. The risk of significantly impaired oxygenation due to CO exposure increases with altitude. Oxygen requirements are most critical in the heart and brain, so most health research on CO focuses on either cardiovascular or neuropsychologic function.

A unique feature of CO exposure is that there is a biological marker of the dose the individual has received. The body burden of CO at any given time can be objectively measured by determining the percentage of COHb in a sample of blood or by measuring carbon monoxide in exhaled breath, after holding the breath for 20–30 seconds to equalize the CO between the circulating blood and alveoli. At low levels, each additional 1% of COHb in blood adds about 5 or 6 ppm of CO in the exhaled breath. Conversely, each additional 5 or 6 ppm of CO in polluted air can raise the blood COHb concentration by another 1%. The approximate COHb level resulting from any exposure can be calculated, although the actual situation is usually more complicated. In toxicologic studies, the exposure is frequently described in terms of percent COHb rather than CO concentration and exposure time. At any given concentration of CO in air, up to 10 hours of exposure will be required to "load" the hemoglobin with CO to the calculated maximum COHb concentration. Loading time is shortened by exercise, which increases ventilation and cardiac output. The body requires a slightly longer time in clean air to fully eliminate previously acquired CO because of the tight binding of CO to hemoglobin. (The conversion of COHb to oxyhemoglobin is less thermodynamically favored than the opposite reaction.) A healthy nonsmoker residing in clean air will have about 0.5% COHb which is produced by normal metabolism. Light smokers typically have 2–3% COHb, whereas heavy smokers typically have 5% or more. Cigarette smoking represents a special case since, for the smoker, smoking almost always dominates over personal exposure from other sources of carbon monoxide. Ambient CO pollution exerts little effect on smokers' preexisting COHb elevation except to prolong it during periods of smoking abstinence. Thus, the environmental exposure of nonsmokers to CO is the primary concern in terms of public health.

Animal Studies. The 1979 EPA criteria document (123) concluded that altered behavior and impaired learning might occur in laboratory animals at CO levels as low as 100 to 150 ppm. Some effects were observed in single short-term exposures (eg, 2 hours), as well as long-term exposures. However, behavioral studies in laboratory animals did not demonstrate significant effects in schedule-controlled behavior below 20–30% COHb.

Carbon monoxide exposures producing COHb concentrations of up to 39% do not produce consistent effects on lung parenchyma and vasculature or on alveolar macrophages (124). Although alveolar epithelial permeability increases during exposures to high concentrations of CO, no accumulation of edema (lung water) or significant changes in diffusing capacity of the lung have been found.

Inhaled CO (producing 2–12% COHb) can cause disturbances in cardiac rhythm and conduction in healthy as well as cardiac-impaired animals. The lowest-observable-effect level varies, depending upon the exposure regimen and species. Cardiac arrhythmias have been observed consistently after long-term intermittent exposures to CO at 100 ppm or more, producing 8–13% COHb. In a few cases, effects are reported at CO near 50 ppm, producing 4–7% COHb. Small increases in hemoglobin and hematocrit with COHb levels of 9% probably represent a compensation for reduced oxygen delivery caused by CO. Scarring or degeneration of heart muscle has been observed in some cases at the higher COHb levels. Increased myocardial ischemia was produced by COHb as low as 5% in dogs with experimentally obstructed coronary arteries. Evidence is conflicting about CO-induced enhancement of atherosclerosis and platelet aggregation.

Clinical Studies. Considerable information has been obtained on the systemic toxicity of CO, its direct effects on blood and other tissues, and the functional effects in various organs (122–129). Most laboratory-based human health studies relevant to ambient CO exposures have focused on maximal exercise performance, mental tasks, and cardiac function. A few studies have exposed volunteers to CO combined with another pollutant such as ozone; they have failed to show any additive or synergistic interaction from the combined exposure.

When the oxygen-carrying capacity of blood is reduced by the formation of COHb, the maximum rate of oxygen delivery to body tissues must decrease, potentially limiting maximal oxygen uptake by exercising muscles. Thus, CO exposure must reduce maximum exercise capability. Numerous investigators have attempted to quantify the amount of CO needed to affect exercise, ie, to determine a dose-response relationship. Decreases in maximum work rate and oxygen consumption have been consistently observed at COHb concentrations of 5% and higher and occasionally at 2–4% COHb in young, healthy nonsmoking volunteers. The desired COHb has been achieved by using higher-than-ambient CO concentrations and brief exposure times or by using several-hour exposures to 30–40 ppm CO. Submaximum exercise is not significantly affected by small increases in COHb for most sedentary healthy people.

Patients with coronary artery disease are vulnerable to myocardial ischemia which is manifested clinically by characteristic symptoms of angina pectoris (pressure–pain in chest) and by electrocardiographic (ECG) abnormalities (S-T depression). These ischemic events can be precipitated by exercise which increases myocardial oxygen requirements. The effect of CO exposure on cardiac patients can be assessed during a carefully controlled exercise test on a treadmill. The length of exercise time before the onset of angina or ECG changes should significantly decrease after CO exposure as compared to a control study with clean air, if CO (as COHb) impairs myocardial oxygenation. Carbon monoxide exposures that raise COHb to 2–4% have decreased the time to angina in some studies. A multicenter study (128) assessed a large subject population (to increase statistical power) and found that the lowest observable adverse effect level in patients with stable angina is between 3 and 4% COHb, representing a 1.5–2.2% increase in COHb from baseline values. The effect appears small compared to typical day-to-day and person-to-person variability of the test results; it is not surprising that earlier, smaller-scale studies did not show consistent results. Thus, patients with exercise-induced angina clearly are a sensitive group within the general population that is at increased risk for experiencing health effects at ambient or near-ambient CO exposure concentrations that result in COHb levels of 5% or less. Exposure sufficient to achieve 6% COHb has been shown to adversely affect exercise-related arrhythmia in patients with coronary artery disease (129).

Carbon monoxide has several effects on the nervous system. Cerebral blood flow increases, presumably due to CO-induced hypoxia, even at very low exposure levels. Studies of neurological function in CO-exposed subjects have employed numerous endpoints, including vigilance (eg, ability to respond appropriately to random signals), sensory or time discrimination (eg, visual sensitivity in the dark), complex sensory-motor performance tasks (eg, driving simulation), sleep activity, and electroencephalograms (EEG). Behaviors requiring sustained attention and performance are most sensitive to CO although the results across studies are not totally consistent. The lowest COHb level at which changes have been reliably demonstrated is 5%.

Epidemiologic Studies. Community population studies of CO in ambient air have not found significant relationships with pulmonary function, symptoms, and disease (124). Some studies have specifically addressed the relationship between CO and death rates from myocardial infarctions, cardiopulmonary symptoms, or deaths in newborn infants. Confounding factors have been abundant and results have been inconclusive. One important factor is that individual personal exposure to CO does not directly correlate with ambient CO concentrations as measured by fixed-site monitors alone because of the mobility of people, highly localized nature of CO sources, and spatial and temporal variability of CO measurements. Personal CO measurements are required to augment fixed-site monitoring data when total exposure is to be evaluated.

Environmental Tobacco Smoke

Sources, Ambient Levels, and Mechanisms. Environmental tobacco smoke (ETS) consists of a wide array of gas and particulate constituents in sidestream smoke released from the cigarette's burning end and mainstream smoke exhaled by the smoker. The inhalation of ETS is referred to as passive, involuntary, or second-hand smoking, whereas the inhalation of smoke that is drawn directly from the cigarette into the mouth or lungs is referred to as active smoking. The lower temperature in the burning cone of the smoldering cigarette releases more partial pyrolysis products in sidestream smoke as compared to mainstream smoke. As a result, sidestream smoke contains higher concentrations of some toxic and carcinogenic substances than mainstream smoke. (Table 6) Only a minority of measured substances released from a single non-filtered cigarette have dramatically higher concentrations in sidestream than in mainstream smoke, but some known carcinogens have sidestream levels greater than 3 times

Table 6. Selected Constituents of Cigarette Smoke[a] and Their Concentrations in Sidestream Smoke (SS) Relative to Mainstream Smoke (MS)

Constituents	MS	SS/MS Ratio
Gas Phase		
Carbon dioxide	20–60 mg	8.1
Carbon monoxide	10–20	2.5
Methane	1.3 mg	3.1
Acetylene	27 μg	0.8
Ammonia	80 μg	73.0
Hydrogen cyanide	430 μg	0.25
Methylfuran	20 μg	3.4
Acetonitrile	120 μg	3.9
Pyridine	32 μg	10.0
Dimethylnitrosamine	10–65 μg	62.0
Particulate Phase		
Tar	1–40 mg	1.3
Water	1–4 mg	2.4
Toluene	108 μg	5.6
Phenol	20–150 μg	2.6
Methylnaphthalene	2.2 μg	28.0
Pyrene	50–200 μg	3.6
Benzo[a]pyrene	20–40 μg	3.4
Aniline	360 μg	30.0
Nicotine	1.0–2.5 mg	2.7
2-Naphthylamine	2 ng	39.0

[a] Principal gas and particulate components present in mainstream smoke from a single nonfiltered cigarette.

that in mainstream smoke, eg, dimethylnitrosamine, pyrene, and benzo(a)pyrene. Dilution by room air can markedly reduce the concentrations inhaled by passive smokers in comparison to those inhaled by the active smoker. Thus, passive smokers are exposed to quantitatively smaller and potentially qualitatively different smoke than mainstream smokers (130–133). Substantial environmental exposure to the numerous toxic (irritant and carcinogenic) agents from combustion of tobacco can potentially result in significant health effects.

Tobacco smoke can fill an indoor environment with high levels of respirable particles, nicotine, polycyclic aromatic hydrocarbons, CO, acrolein, and NO_2, to name a few. Their indoor concentrations depend on multiple factors such as the number of smokers, intensity of smoking, type and number of burning cigarettes, size of the room, ventilation, residence times of the individual and smoke, and use of air-cleaning devices. Monitoring studies have been conducted in public buildings and, to a lesser extent, in homes and offices. Several components of tobacco smoke have been measured as markers of the contribution of tobacco combustion to indoor air pollution. Respirable particles have been measured most frequently because they are present in large concentrations in both mainstream and sidestream smoke. Over 80% of the ETS in a room results from sidestream smoke. However, particles are a nonspecific marker of ETS because numerous other sources also discharge particles. Other components (eg, nicotine, NO_2, benzene) have also been measured as indoor markers.

Time–activity patterns strongly influence personal exposure to ETS (133). Certain groups may have greater exposures in certain locations. For example, residential exposure can be greater for infants of smoking parents than for other individuals such as nonsmoking adolescents. Nonsmoking adults may be exposed heavily at home with a smoking spouse and/or at the workplace. Concentrations of respirable particles and nicotine (contained in the vapor phase of ETS) vary widely in homes. A smoker of one pack of cigarettes daily can contribute 20 $\mu g/m^3$ to 24-hour indoor particle concentrations, and two heavy smokers can contribute more than 300 $\mu g/m^3$. Average nicotine levels in ETS at home can be 4 $\mu g/m^3$ and as high as 20 $\mu g/m^3$ in some 24-hour measurements. Indoor benzene, xylenes, ethylbenzene, and sytrene are increased in homes with smokers as compared to homes without smokers. Bars, restaurants, and other public places where smoking may be intense, show particulate concentrations as high as 700 $\mu g/m^3$. Transportation environments (trains, buses, automobiles, and airplanes) also have been documented to be contaminated by ETS. For example, NO_2, respirable particles, and nicotine in commercial aircraft are much higher with smoking than without smoking and higher in smoking than in nonsmoking sections (134). However, all passengers are exposed to nicotine (according to markers of personal exposure) regardless of seating in smoking or nonsmoking sections of the airplane cabin, especially with recirculation of cabin air (135).

Accurate exposure assessment is critical to evaluate the effects of indirect or low-dose exposure, especially when small increases in risk are present, such as with ETS. The major methodologic problem of misclassification of smoking status (ie, former smokers categorized as nonsmokers; exposed subjects as nonexposed and *vice versa*) can bias observed health effects of passive smoking. However, this problem can be mitigated with valid biological markers. A reliable and accurate biologic marker of exposure to ETS improves exposure and dose estimates, as well as validates questionnaire responses about exposure (26,133). At present, the most sensitive and specific biological markers for tobacco smoke exposure are nicotine and its metabolite, cotinine. Exposure to tobacco smoke (and perhaps ingestion of some raw vegetables) is usually the only source of nicotine or cotinine in body fluids. Nicotine concentrations in body fluids reflect recent exposures since its circulating half-life is generally less than two hours. Cotinine has a blood or plasma half-life from 10 to 40 hours, rendering it detectable for 3–6 days and capable of providing information about more subacute exposures to tobacco smoke. Cotinine measurements on the urine or saliva of passive smokers are fairly accurate and inexpensive, making them suitable for epidemiologic surveys. Cotinine levels in passive smokers range from less than 1% to 8% of the cotinine levels measured in active smokers. Thiocyanate (a metabolite of hydrogen cyanide, which is a component of tobacco smoke), exhaled CO, and COHb can distinguish active smokers from nonsmokers but are not as sensitive and specific as cotinine for assessing passive exposure to tobacco smoke. However, all the above markers are of minimal value in assessing the risk of lung cancer in children or adults since they indicate only current and not cumulative exposure. Nonetheless, the data on biological markers provide ample evidence that ETS exposure leads to absorption, circulation, and excretion of tobacco smoke components and add to the biological plausibility of associations between passive smoking and disease as documented in epidemiologic studies.

Effects

The health impact of passive smoking has yet to be fully characterized but is clearly an important public health issue since 30% of American adults are active smokers and 60% of all homes have at least one smoker. The information on the health effects of passive smoking has been largely derived from observational epidemiologic studies (130–139). In these studies ETS exposure has been primarily estimated by questionnaires and biological markers. The reliability of subjects in indicating the smoking status of spouses or parents appears high, but quantification of smoking and time spent with a smoker is less reliable. At present, both biological markers and carefully formulated questionnaires are necessary to assess acute and cumulative ETS exposures in epidemiologic studies.

Respiratory Effects in Children. In general, the association between exposure to ETS and increased respiratory disease is stronger in young children than in adults. Epidemiologic studies have linked passive smoking in children to increased occurrence of lower respiratory tract illness (presumably infectious in etiology) during infancy and childhood. Multiple investigations throughout the world have demonstrated an increased risk of bronchitis and pneumonia during the first year of life in infants with smoking parents. Maternal (rather than paternal) smoking alone can increase the incidence of lower respiratory illness in a dose-related fashion. A similar effect of passive smoking has not been readily identified after the first year of life, perhaps reflecting the change in time–activity patterns as the infant matures.

Numerous large surveys have demonstrated increased frequency of cough, phlegm, and wheeze in school-aged children of smoking parents. The frequency of reported respiratory symptoms in children increases with the number of smokers in the home, suggesting a dose-response relationship. For example, smoking by parents (largely maternal) have increased the frequency of persistent cough in their children by 30% in a Harvard study of 10,000 school children in six U.S. communities. However, the association of passive smoking with childhood asthma is less consistent despite the association with wheeze. Nonetheless, maternal smoking can adversely affect children (especially boys) with asthma, as supported by adverse effects on lung function, symptom frequency, and airway responsiveness to inhaled agonists (eg, histamine), and increased medication use and emergency room visits. Urinary cotinine concentrations are directly related to increased frequency of asthma exacerbations and reduced lung function in asthmatic children of smoking parents (140). On the other hand, the severity of children's asthma will decrease with less smoking exposure from smoking parents (141). Some studies have also found associations between passive smoking and middle-ear disease, sore throat, and school absenteeism.

Longitudinal studies indicate that passive smoking is associated with small reductions in the rate of lung function growth during childhood in children of smoking parents (especially mothers) in comparison with those of nonsmokers.

Respiratory Effects in Adults. Studies have generally evaluated the effects of ETS exposure from a smoking spouse and have found inconsistent evidence of increased chronic respiratory symptoms in adults. Similarly, epidemiologic and experimental exposure studies of asthmatic adults have yet to demonstrate consistent effects of ETS in exacerbating asthma in adults, although these studies have largely involved small numbers of subjects. Some studies, however, have indicated that a subset of ETS-sensitive adult asthmatics develop significant reductions in lung function as well as increased symptoms (142).

Lung function in adults has not been found conclusively to be adversely affected by passive smoking in cross-sectional studies. Some studies suggest that chronic exposure to ETS impairs airflow in the small peripheral airways. No data indicate that emphysema can be caused by passive smoking alone. However, further research is warranted because of the widespread exposure to ETS in workplaces and homes.

The association of passive smoking and lung cancer has generated considerable attention and controversy. The association has biologic plausibility because of the presence of respirable carcinogens in sidestream smoke and the lack of a documented threshold dose for respiratory carcinogenesis in active smokers. On the other hand, many epidemiologic studies have been affected by misclassifications of smoking or exposure status, inaccurate quantification of smoking habits, failure to assess and control for potential confounding factors (eg, diet, other exposures), and misdiagnosis of lung metastases (from cancers of other organs) as primary lung cancers. The assumption that spouse smoking alone represents ETS exposure may miss exposures outside the home or within the home. Gender-related differences may be present since women smokers generally smoke less and generate less ETS than male smokers. Inconsistencies among studies should be anticipated in view of the relatively small populations in many studies and the difficulties in estimating exposure. Thus, evidence from case-control and cohort studies does not uniformly indicate increased risk of lung cancer in persons exposed to ETS. Despite these methodologic issues, the collective weight of evidence indicates a substantial increase in the risk of lung cancer in subjects involuntarily exposed to cigarette smoke, especially in female nonsmokers married to smoking spouses. Risk appears to increase with increasing dose of ETS. Expert reviews by the National Research Council (132), U.S. Surgeon General (136), U.S. EPA (138), and the International Agency for Research on Cancer of the World Health Organization (139) have concluded that passive smoking causally increases the risk of lung cancer in nonsmokers. The U.S. EPA (138) has officially classified ETS as a known carcinogen. The magnitude of lung cancer risk remains uncertain. Simplifying assumptions about the extent and degree of exposure to ETS, exposure–response relationships, etc, must be made in risk assessments. With these limitations, involuntary smoking is estimated to be associated with 3,000 to 8,000 lung cancer deaths per year, with relative risks from 1.3

to over 2.0 versus the non-exposed population (132,143), which are small in relation to risks of typical, active smoking. In other words, assuming a mean risk of 35% (132), an average lifetime passive smoke exposure from a smoking spouse increases a nonsmoker's risk by about 35%, as compared with the risk of 1000% (or tenfold) for a lifetime of active smoking. Although the risks are small (if estimates are accurate) for individual passive smokers, their public health significance lies in the number of excess lung cancers caused by passive smoking (approximately 3000–5000 per year with this risk estimate). Regardless of the exact numbers, the calculations demonstrate that passive smoking must be considered an important cause of lung cancer deaths from the public health perspective since ETS exposure is involuntary and subject to control (ie, is avoidable).

Cardiovascular Effects. Although active smoking has been extensively established as a cause of cardiovascular disease and causes (along with lung cancer) 300,000 to 400,000 deaths annually, relatively few studies have examined passive smoking as a risk factor (133). There is no consistent data that passive smoking poses an increased risk for accelerating atherogenesis, angina pectoris/coronary artery disease, or cardiac mortality. Some epidemiologic studies indicate a two-to-threefold increase in the risk of ischemic cardiac death in passive smokers married to a male smoker, after adjustment for other known risk factors. However, cigarette smoking increases ambient CO levels and even increments of 1–2% COHb can decrease the time to the occurrence of angina during exercise. Similarly, passive smoking may adversely affect athletic performance in healthy subjects during controlled exposures to ETS. The postulated components of ETS which might affect cardiovascular function are CO and nicotine. The former (as COHb) decreases oxygen delivery to the heart muscle, whereas the latter releases epinephrine and increases oxygen demand. These studies are unlikely to be truly blinded because of the unmistakable odor of tobacco smoke. However, there is no strong evidence that cardiopulmonary responses have a psychogenic basis (144). Further research is necessary to clarify any link between cardiovascular disease and ETS.

Other Effects. Some small studies suggest that exposure to ETS is associated with health effects outside the lungs, although consistent evidence is lacking. Excess risk of malignancies at sites other than the lungs (eg, breast, cervix, nasal sinus) in nonsmoking women with smoking spouses has been reported. However, these associations show weak biologic plausibility since the cancers in question are not increased in active smokers. The associations may arise by chance or by the effect of bias (confounding factors). Several cohort studies have shown increased risk of all-cause mortality (15–40%) in passive smokers. This trend parallels cardiovascular mortality of passive smokers in some studies. Annoyance and irritation effects of ETS on the eyes, upper and lower airways are common complaints from passive smokers. ETS-sensitive individuals who are experimentally exposed to sidestream tobacco smoke report more nasal congestion, runny nose (rhinorrhea), throat irritation, headache, chest discomfort, and cough, as well as statistically significant increases in nasal resistance to airflow without changes in lung function (145).

CONTROL STRATEGIES

General

Because of the large number of potential sources of air pollutants found in outdoor and indoor locations and the wide range of health effects, different levels of preventive strategy may be considered: (*1*) Primary prevention refers to elimination or reduction of exposure to the pollutant before the onset of an adverse health effect or disease; (*2*) Secondary prevention refers to detection of asymptomatic adverse health effects, thus permitting early intervention to eliminate or reduce exposure and possibly prevent clinically evident disease; and (*3*) Tertiary prevention refers to management of a clinical effect or disease caused by the pollutant to reduce morbidity and mortality. For all pollutants, primary prevention is the most desirable and will be emphasized here. This prevention is frequently accomplished by public education and regulatory policies to limit emissions and exposures, although there are also important individual actions which may be effective.

Steps can be taken to reduce risks from environmental pollutants despite persistent controversies and points of uncertainty about their health effects. Table 7 summarizes approaches to control the health effects of air pollutants. These approaches essentially attempt to control pollutant emissions, reduce human exposure, or mitigate illness caused or exacerbated by exposure. Often multiple strategies must be pursued concomitantly to achieve substantial benefits. Each approach has inherent risk–benefit considerations which may involve scientific data, clinical experience, ethical judgments, economics, or other factors. The practicality and effectiveness of any given approach may vary markedly in different localities, depending on their physical and socioeconomic characteristics.

Outdoor Pollution

An ambitious national health objective for the year 2000 in the U.S. is to increase the proportion of persons who live in counties that have not exceeded any outdoor air quality standard during the previous 12 months from 49.7% in 1988 to 85% (objective 11.5) (146). The large population exposed to poor air quality and the substantial proportion who belong to high-risk subgroups underscore the importance of improved air quality for health promotion and disease prevention. How can this objective be achieved?

For some pollutants, only national policy and regulation (eg, under the Clean Air Act) can reduce risk. Pollution abatement (action intended to reduce or eliminate pollution hazards at their sources) is usually undertaken by a state or local regulatory agency, often with statutory authority delegated by the federal government (EPA). Public and private sectors contribute to the regulatory decision-making process. Community-wide or region-wide pollution abatement is a long-term process. The abatement of more limited or specific local problems may have faster progress since the sources, costs of control, and benefits of control

Table 7. Approaches to Control the Health Effects of Air Pollutants

A. *Political*

1. "Command and control" method that legally limits or forbids certain emissions; a common approach of state or federal implementation plans under the Clean Air Act
2. "Emissions trading" method that limits the quantity of emission permits distributed among sources through market mechanism; favored by economists to maximize efficiency of pollution controls
3. Direct taxes on measurable emissions
4. Taxes on goods or services associated with pollution

B. *Technological*

1. Source modification
   a. Discontinue or isolate processes that emit pollutants
   b. Substitution of less-polluting for more-polluting processes
   c. Emission controls on sources
2. Ventilation/dispersion
   a. Dilution, eg, by tall stacks (trade-off of high regional versus low local pollution)
   b. Exhaust from work areas with internal sources
3. Removal from human-occupied space
   a. Physical filtration
   b. Adsorption
   c. Catalytic destruction
   d. Electrostatic precipitation

C. *Personal*

1. Avoidance by changing activity patterns, job location
2. Medical management, eg, asthmatics inhale bronchodilator prior to exercising in SO_2
3. Education of others to practice 1 and 2

are more easily understood or managed. For example, motor vehicular emissions are being strictly controlled in the State of California with diverse strategies, eg, catalytic converters, reformulated gasoline, "alternative" or "clean" fuels, "zero-emission" electric vehicles. Some strategies may have greater impacts and cost-effectiveness than others (51). On the other hand, reduction of acidic aerosols or ozone can be affected only by multifaceted, broadly applied regulatory strategies since the pollutants have multiple sources and spread widely in the atmosphere. In this process, there is the continual need for additional education, communication, and coordination among industries, health agencies, environmental groups, and the general public.

When outdoor pollutant exposure presents a clear risk to health and immediately effective abatement is unlikely, individual protective actions are necessary. These actions are often satisfactory for the involved individuals but are unlikely to be fully effective throughout a community. Individual protection should be considered only a temporary alternative to abatement of outdoor pollution.

For some pollutants, the individual can directly reduce risks (147). Removing the hazard from the environment is feasible in some instances (eg, hypersensitivity to a nonessential household product) but not easy with community air pollution. Conversely, removing oneself from an environment perceived or known to be hazardous (ie, personal avoidance of exposure) is frequently more effective although not necessarily as practical. Modifying time–activity patterns to limit time outside during pollution episodes often represents the most effective strategy. Staying indoors and avoiding unnecessary outdoor activity during community pollution episodes are commonly recommended by physicians and public agencies, especially for susceptible individuals or patients with lung disorders. This recommendation assumes that there are no indoor pollutant sources or hazards and that outdoor pollutant levels fall when air migrates indoors. The latter is usually true for reactive pollutants such as ozone and SO_2 but not for others such as particulates. Even cleaning shag carpeting with portable vacuum cleaners can significantly increase the number of respirable airborne particles, which may remain suspended in air for as long as one hour (148). Central vacuum cleaners generate fewer airborne particles during carpet cleaning than portable vacuum cleaners. Some medications (eg, albuterol, metaproterenol, cromolyn, nedocromil) commonly used by asthmatics and other respiratory patients can prevent acute respiratory effects (wheezing or asthma attack) induced by SO_2.

More drastic individual protective measures include changes of job and residence. These choices may involve substantial costs or risks in moving the affected person from the polluted environment, and there is no assurance of overriding benefits. They require careful evaluation of the environments in question in order to verify, insofar as possible, that the existing environment is indeed contributing to the problem and the proposed alternative is truly preferable and helpful.

Indoor Pollution

Personal modification of indoor sources is potentially effective in reducing pollutant concentrations (17,147). The available methods cover a broad range, from the control of the indoor environment where the sources are located, to modifying the sources themselves. Controlling indoor temperature and relative humidity may affect the rate at which a pollutant is generated and transported within a residence or building. For example, central, forced-air heating and cooling systems can mix pollutants within homes. Avoiding high relative humidity is important for preventing the growth of house dust mites, fungi, or various infectious pathogens which may become airborne. Low humidity may reduce emissions of formaldehyde from building materials and furnishings, as well as inhibit the growth of the aforementioned biological agents. Levels of other pollutants are not greatly affected by climate control. Some sources may be removed from the living space or a source may be replaced by a functionally equivalent product that emits less pollution. Occupant exposure may be reduced by isolating the source in a separate area and limiting access to it. Behavioral adjustments are the simplest, least expensive, and most effective means of improving indoor air quality, eg, prohibiting smoking indoors or completely segregating smoking areas; proper use and maintenance of gas appliances; proper use of humidifiers/

dehumidifiers and pesticides; proper storage of cleaning materials, paints, and solvents; proper ventilation while painting or pursuing hobbies. Radon concentrations in homes can be measured inexpensively, and techniques are available for mitigating and avoiding unacceptable radon concentrations. Prevention and cessation of smoking control the hazards of both active and passive smoking, such as lung cancer, with or without concomitant asbestos or radon exposure.

Respiratory protective equipment such as masks has been developed for use in the workplace to minimize exposure to toxic gases and particles. Somewhat more sophisticated masks may include an efficient particulate filter and/or a chemical sorption agent (eg, activated carbon) to remove some pollutant gases. Many simple masks, even when effective against some particulate pollution, present a number of problems. They are uncomfortable, interfere with oral activities, do not always fit snugly, are ineffective against gases and small particles, and add to the work of breathing, which cannot be tolerated by persons with respiratory disease. For these reasons, masks are considered a last resort in air-quality protection.

Reduction of indoor air contaminants may be accomplished by natural ventilation, air purification (or more accurately, air cleaning), and mechanical ventilation systems. Personal protection by air cleaning implies remaining in a contaminated environment but cleaning the air before it is breathed. Air-cleaning devices vary greatly in size, portability, cost, and capabilities for maximum rate of cleaning and type of pollutants removed (17,147,149,150). The typical domestic air cleaner consists of a portable box containing an electrically operated blower and cleaning device. Polluted air is drawn in through the rear of the unit, passed through the purification system, and discharged through the front of the unit. The user must breathe directly in front of the clean air outlet to receive maximal benefit because of dilution by untreated room air. If the unit is to clean the air of an entire room, its flow rate (volume of clean air discharged per minute) must be a substantial fraction of the room's total volume and larger than the rate of air exchange between the room and its surroundings. Achieving these standards usually requires that the room be tightly closed, in which case there must be no significant indoor source of contaminants that the purifier cannot remove completely.

The basic mechanisms of commercially available household air-cleaning devices are as follows: high-efficiency particulate arresting (HEPA) filters, electrostatic precipitators, electret filters, negative ionizers, and ozonators. Each type has advantages and disadvantages which are reviewed elsewhere (17,147,149,150); only the most commonly used types are briefly reviewed.

Most particulate filters consist of a fibrous mat in which particles must contact and adhere to the fibers, thus removing the particles from the airstream. Common glass–fiber furnace filters are effective only against relatively large dust particles. Particles less than 1 μm are the most difficult to filter but make up a substantial fraction of the particulate matter in polluted air. Paper HEPA filters can efficiently remove particles of a wide size range and retain the same size particles with up to 99.97% efficiency. They do not control gases or sequestered allergens such as house dust mites. Increasing the filtration efficiency increases cost, both for the filter material itself and for the energy needed to move air against increased resistance of the filter. Particulate filters of any type must be discarded and replaced regularly since heavily loaded filters will obstruct airflow and may release previously captured particles.

Electrostatic precipitators use a high-voltage electric field that imparts charges to passing particles. The air then passes between pairs of charged plates to which oppositely charged particles are attracted and adhere. Although they are generally less efficient than HEPA filters, electrostatic precipitators can achieve moderate to high removal efficiencies of small particles (eg, ETS, house dust mite, and animal allergens) and larger particles (eg, pollens, mold spores) with optimal design and operating conditions. They remove radon decay products attached to particles but not unattached radon decay products. Small home systems must be deactivated regularly for cleaning the collection plates; otherwise, collection efficiency declines and some ozone may be produced by electrical arcing to the dirty collector surfaces. Large systems use automatic cleaning mechanisms. Electrostatic precipitators do not require periodic replacement and offer relatively little resistance to airflow, which can be increased to improve effectiveness.

Chemical filters are solid substances, usually in granular or pellet form. They do not actually filter but adsorb pollutant gas molecules or react with them to form more innocuous substances. An effective adsorbent must have a large surface area containing properly-sized sub-microscopic pores to capture and retain pollutant molecules. Activated carbon (charcoal) is the most common general-purpose chemical filter medium. It effectively removes organic gases of moderate to large molecular weight, including many irritating and malodorous VOCs (except tobacco smoke). Activated carbon effectively removes ozone, is only moderately effective against NO_2 and SO_2, and is ineffective against NO, CO, and organic gases of low molecular weight, such as formaldehyde. The charcoal or airstream must be chemically analyzed to determine the saturation status of the carbon. Used carbon can be either replaced or reactivated by treatment with hot air or steam to desorb (revolatize) the contaminants. Finally, activated alumina (aluminum oxide) impregnated with potassium permanganate may be more effective than activated carbon against nitrogen and sulfur oxides. Alumina-permanganate is effective against organic compounds that are readily oxidized but not against CO. Unlike activated carbon, alumina-permanganate is noncombustible but its dust is irritating and corrosive. Its functional condition may be monitored by observing the color change from purple to brown as the permanganate is consumed.

In principle, the most effective general-purpose recirculating air cleaners are those that incorporate a chemical filter and a HEPA filter. This combination should be capable of removing most particulate pollutants and most gases other than those that are relatively unreactive or of low molecular weight, such as carbon monoxide. However, the strong odor of tobacco smoke from its gas phase can be difficult to remove with either HEPA filters or electrostatic precipitators and requires separate control measures for removal (eg, ventilation).

Regular maintenance (cleaning and/or filter replacement) is absolutely essential to maintain the effectiveness of any device. In buildings equipped with central air conditioners, air-cleaning devices may be installed in the ductwork to allow the whole building's air to be treated. Particulate filtration, chemical filtration, electrostatic precipitation, or a combination may be used. The choice usually depends on economics and the nature of the pollutants to be controlled. Air-conditioning systems typically recirculate about 90% of the indoor air and draw only about 10% from outside air to minimize energy requirements for heating and cooling. Cleaning devices may be placed in the outside air inlet if there are no significant contaminant sources inside the building and maximum economy is desired. More typically, both internal and external sources of pollution are important, in which case both recirculated air and outside air should be cleaned by installing the cleaning devices in the main air-supply duct. HEPA filters can effectively remove most microbes and nonviable pollutant particles. In large buildings an engineering study is necessary to design or evaluate a system that will be satisfactory in terms of ventilation, temperature and humidity control, contaminant removal, energy use, and maintenance costs (151). Fire safety is another critically important design consideration: nonflammable materials should be used to avoid ignition and spreading smoke throughout a buildling.

The clinical perspective of the above air cleaning devices, however, is important to consider. The lack of a standard test program for these air cleaners has resulted in a scarcity of objective information regarding their performance beyond the general claims of the manufacturers. The few published evaluations of these devices indicate a wide range of performance. Similarly, the clinical utility of personal air cleaning is largely untested and anecdotal; air cleaners have not been objectively shown to improve health by reducing indoor pollutants, except in a very few small-scale studies. For example, although allergic symptoms in some patients have been reported to decrease during the domestic use of HEPA filters (152), a consensus group (153) has concluded that the lack of objective results about the health-promoting action of indoor air-cleaning devices in double-blind studies precludes any firm medical recommendations about their use by patients. The use of room air-cleaning devices in the absence of other forms of environmental control is not considered reasonable.

Acknowledgments

Parts of this chapter are based on the following sources: American Lung Association, *Health Effects of Ambient Air Pollution*, American Lung Association, New York, 1989. H. Gong, Jr., "Health Effects of Air Pollution. A Review of Clinical Studies," *Clin Chest Med* **13,** 201–230 (June 1992). American Lung Association, *Personal Protection Against Air Pollution: The Physician's Role*, American Lung Association, New York, 1981.

This publication is partly based on the authors' research which is supported by the California Air Resources Board, Electric Power Research Institute, Health Effects Institute, Southern California Edison Co., and the U.S. Environmental Protection Agency.

BIBLIOGRAPHY

1. P. Brimblecombe, *J. Air Pollut. Control Assoc.* **26,** 941–45 (Oct. 1976).
2. J. M. Samet and M. J. Utell, *J. Am. Med. Assoc.* **266,** 670–675 (Aug. 7, 1991); T. Schneider, S. D. Lee, G. J. R. Wolters, and L. D. Grant, eds., "Atmospheric Ozone Research and Its Policy Implications." *Proceedings of the 3rd US-Dutch International Symposium, Nijmegen, The Netherlands, May 9–13, 1988*, Elsevier Science Publishers B.V., Amsterdam, 1989; G. Hoek, *Acute Effects of Ambient Air Pollution Episodes on Respiratory Health of Children*, Thesis, Departments of Epidemiology and Public Health and Air Pollution, Agricultural University Wageningen, The Netherlands, Sept. 14, 1992; U. Mohr, D. V. Bates, H. Fabel, and M. J. Utell, eds., *Advances in Controlled Clinical Inhalation Studies*, Springer-Verlag, Berlin, 1993.
3. R. B. Schlesinger, "Atmospheric Pollution," *Otolaryngol Head Neck Surg.* **106,** 642–649 (June, 1992).
4. U.S. Environmental Protection Agency, Office of Air Quality Planning and Standards, *National Air Quality and Emissions Trends Report, 1989*, EPA-450/4-91-003, U.S. EPA, Research Triangle Park, N.C., 1991.
5. R. W. Shaw, *Sci. Amer.* **257,** 96–103 (Aug. 1987).
6. National Research Council, *Indoor Pollutants*, National Academy Press, Washington, D.C., 1981.
7. J. M. Samet, M. C. Marbury, and J. D. Spengler, *Am. Rev. Respir. Dis.* **136,** 1486–1508 (Dec. 1987).
8. J. M. Samet, M. C. Marbury, and J. D. Spengler, *Am. Rev. Respir. Dis.* **137,** 221–242 (Jan. 1988).
9. D. R. Gold, *Clin. Chest Med.* **13,** 215–229 (June 1992).
10. T. V. Larson, D. S. Covert, R. Frank, and R. J. Charlson, *Science* **197,** 161–163 (July 8, 1977).
11. J. B. L. Gee, W. K. C. Morgan, and S. M. Brooks, eds., *Occupational Lung Disease*, Raven Press, New York, 1984.
12. N. A. Esmen and M. A. Mehlman, eds., *Occupational and Industrial Hygiene: Concepts and Methods*, Volume 8, "Advances in Modern Environmental Toxicology," Princeton Scientific Publishers, Inc., Princeton, N.J., 1984.
13. M. Lippman, *Aerosol Sci. Technol.* **6,** 93–114 (1987).
14. M. Lippman, *J. Exposure Analysis Environ. Epidemiol.*, 103–129 (Jan.–Mar. 1993).
15. Office of Technology Assessment, *Catching Our Breath: Next Steps to Reducing Urban Ozone* (OTA-0-412), U.S. Government Printing Office, Washington D.C., July 1989.
16. Clean Air Act (CAA) Amendments, P.L. 101–549 (Nov. 15, 1990).
17. J. Samet, "Environmental Controls and Lung Disease," *Am. Rev. Respir. Dis.* **142,** 915–939 (Oct. 1990).
18. ASHRAE standard 55–1981, *The Thermal Environment Conditions for Human Occupancy*, American Society of Heating, Refrigerating, and Air Conditioning Engineers, 1989.
19. American Thoracic Society, *Am. Rev. Respir. Dis.* **131,** 666–668 (Apr. 1985).
20. I. T. T. Higgins, *J. Air Pollut. Control Assoc* **33,** 661–663 (July 1983).
21. H. A. Boushey, M. J. Holtzman, J. R. Sheller, and J. A. Nadel, *Am. Rev. Respir. Dis.* **121,** 389–413 (Feb. 1980).

22. National Research Council, Committee on the Epidemiology of Air Pollutants, *Epidemiology and Air Pollution*, National Academy Press, Washington D.C., 1985.
23. National Research Council, *Human Exposure Assessment of Airborne Pollutants: Advances and Opportunities*, National Academy Press, Washington D.C., 1991.
24. K. Sexton and P. B. Ryan, "Assessment of Human Exposure to Air Pollution: Methods, Measurements, and Models," in A. Y. Watson, R. R. Bates, and D. Kennedy, eds., *Air Pollution, the Automobile, and Public Health*, National Academy Press, Washington D.C., 1988.
25. K. Sexton, S. G. Selevan, D. K. Wagener, and J. A. Lybarger, *Arch. Environ. Health* **47,** 398–407 (Nov./Dec. 1992).
26. National Research Coouncil, Subcommittee on Pulmonary Toxicology, *Biologic Markers in Pulmonary Toxicology*, National Academy Press, Washington D.C., 1989.
27. M. Lippmann, "Introduction and Background," in M. Lippmann, ed., *Environmental Toxicants. Human Exposures and Their Health Effects*, Van Nostrand Rheinhold Co., Inc., New York, 1992, pp. 1–29.
28. J. D. Brain, B. D. Beck, A. J. Warren, and R. A. Shaikh, *Variations in Susceptibility to Inhaled Pollutants. Identification, Mechanisms, and Policy Implications*, The Johns Hopkins University Press, Baltimore, Md., 1988.
29. *MMWR Morb. Mortal Wkly. Rep.* **42,** 301–304 (Apr. 30, 1993).
30. L. Mølhave, "Volatile Organic Compounds and The Sick Building Syndrome," in M. Lippmann, ed., *Environmental Toxicants. Human Exposures and Their Health Effects*, Van Nostrand Rheinhold Co., Inc., New York, 1992, pp. 633–646.
31. National Research Council, Committee on the Institutional Means for Assessment of Risks to Public Health, *Risk Assessment in the Federal Government: Managing the Process*, National Academy Press, Washington D.C., 1983.
32. U.S. Environmental Protection Agency, *Indoor Air-Assessment. A Review of Indoor Air Quality Risk Characterization Studies,* Environmental Criteria and Assessment Office, Research Triangle Park, N.C., Mar., 1991, EPA/600/8-90/044.
33. A. C. Upton, "Perspectives on Individual and Community Risks," in M. Lippmann, ed., *Environmental Toxicants. Human Exposures and Their Health Effects*, Van Nostrand Rheinhold Co., Inc., New York, 1992, pp. 647–661.
34. C. M. Spooner, *Clin. Chest Med.* **13,** 179–192 (June 1992).
35. E. K. Silbergeld, *Environ. Health Perspect. 1993* **101,** 100–104 (June 1993).
36. U.S. Environmental Protection Agency, *Cancer Risk from Outdoor Exposures to Air Toxics*, Washington D.C., 1990, EPA-450/1-90-004A.
37. B. Hasset-Sipple and I. L. Cote, *MMWR Morb. Mortal. Wkly. Rep.* **40,** 278–279 (May 13, 1990).
38. R. F. Phalen, *Inhalation Studies: Foundations and Techniques*, CRC Press, Boca Raton, Fla., 1984.
39. D. E. Gardner, J. D. Crapo, and E. J. Massaro, eds., *Toxicology of the Lung*, Raven Press, New York, 1988.
40. J. Crapo, F. J. Miller, B. Mossman, W. A. Pryor, and J. P. Kiley, *Am. Rev. Respir. Dis.* **145,** 1506–1512 (June 1992).
41. D. V. Bates, *Respiratory Function in Disease*, 3rd ed., W.B. Saunders, Philadelphia, Pa., 1989.
42. J. D. Hackney, W. S. Linn, and E. L. Avol, *Environ. Sci. Technol.* **18,** 115A–122A (April 1984).
43. R. Frank, J. J. O'Neil, M. J. Utell, J. D. Hackney, J. Van Ryzin, and P. E. Brubaker, eds., *Inhalation Toxicology of Air Pollution: Clinical Research Considerations*, American Society for Testing and Materials, Philadelphia, Pa., 1985, ASTM no. 872.
44. L. J. Folinsbee, "Human Clinical Inhalation Exposures. Experimental Design, Methodology, and Physiological Responses," in D. E. Gardner, J. D. Crapo, E. J. Massaro, eds., *Toxicology of the Lung*, Raven Press, New York, 1988, pp. 175–199.
45. H. Gong, Jr., *Clin. Chest. Med.* **13,** 201–230 (June 1992).
46. U. Mohr, D. V. Bates, H. Fabel, M. J. Utell, eds., *Advances in Controlled Clinical Inhalation Studies*, Springer-Verlag, Berlin, 1993.
47. H. Gong, Jr., "Air Pollutants As a Provocative Challenge in Humans," in S. Spector, ed., *Provocation Testing in Clinical Allergy*, Marcel Dekker, New York, in press.
48. K. J. Rothman, *Modern Epidemiology*, Little, Brown and Company, Boston, Mass., 1986.
49. F. W. Lipfert, *Environ. Sci. Technol.* **19,** 764–770 (Sept. 1985).
50. M. D. Lebowitz, J. J. Quackenboss, M. L. Soczek, M. Kollander, and S. Colome, *J. Air Pollut. Control Assoc.* **39,** 1411–1419 (Nov. 1989).
51. J. G. Calvert, J. B. Heywood, R. F. Sawyer, and J. H. Seinfeld, *Science* **261,** 37–45 (July 2, 1993).
52. W. W. Lowrance, *Modern Science and Human Values*, Oxford University Press, New York, 1985.
53. M. A. Ottoboni, *The Dose Makes the Poison: A Plain-Language Guide to Toxicology*, 2nd ed., Van Nostrand Reinhold, New York, 1991.
54. A. B. Hill, *Proc. Royal Soc. Med.* **58,** 295–300 (May 1965).
55. D. V. Bates, *Environ. Res.* **59,** 336–349 (Dec. 1992).
56. R. A. Bethel, *Semin. Respir. Med.* **8,** 253–258 (Jan. 1987).
57. A. J. Wardlaw, *Clin. Exp. Allergy* **23,** 81–96 (Feb. 1993).
58. M. Lippmann, ed., *Environmental Toxicants. Human Exposures and Their Health Effects*, Van Nostrand Rheinhold Co., Inc., New York, 1992.
59. U.S. Environmental Protection Agency, *Air Quality Criteria for Ozone and Other Photochemical Oxidants*, Environmental Criteria and Assessment Office, Research Triangle Park, N.C., Aug., 1986, EPA/600/8-84/020eF.
60. J. H. Seinfeld, *Science* **243,** 745–752 (Feb. 10, 1989).
61. P. J. Lioy and R. V. Dyba, *Toxicol. Indust. Health* **5,** 493–505 (Mar. 1989).
62. M. Lippman, "Ozone," in M. Lippman, ed., *Environmental Toxicants, Human Exposures and Their Health Effects*, Van Nostrand Rheinhold Co., Inc., New York, 1992, pp. 465–519.
63. C. J. Weschler, C. J. Shields, and D. V. Naik, *J. Air Pollut. Control Assoc.* **39,** 1562–1568 (Dec. 1989).
64. U.S. Environmental Protection Agency, *Summary of Selected New Information on Effects of Ozone on Health and Vegetation*, Environmental Criteria and Assessment Office, Research Triangle Park, N.C., Jan., 1992, EPA/600/8-88/105F.
65. M. J. Hazucha, D. V. Bates, and P. A. Bromberg, *J. Appl. Physiol.* **67,** 1535–1541 (Apr. 1989).
66. J. Seltzer and co-workers, *J. Appl. Physiol.* **60,** 1321–1326 (Apr. 1986).
67. H. S. Koren and co-workers, *Am. Rev. Respir. Dis.* **139,** 407–415 (Feb. 1989).
68. R. B. Devlin and co-workers, *Am. J. Respir. Cell Mol. Biol.* **4,** 72–81 (Jan. 1991).

69. H. S. Koren, G. E. Hatch, and D. E. Graham, *Toxicology* **60,** 15–25 (Jan./Feb. 1990).
70. M. G. Mustafa and D. F. Tierney, *Am. Rev. Respir. Dis.* **118,** 1061–1090 (Dec. 1978).
71. M. A. Mehlman and C. Borek, *Environ. Res.* **42,** 36–53 (Feb. 1987).
72. M. Lippmann, *J. Air Pollut. Control Assoc.* **39,** 672–694 (May 1989).
73. E. S. Wright, D. M. Dziedzic, and C. S. Wheeler, *Toxicol. Lett.* **51,** 125–145 (Apr. 1990).
74. W. S. Tyler, M. D. Julian, and D. M. Hyde, *Semin. Respir. Med.* **13,** 94–113 (Mar. 1992).
75. B. E. Tilton, *Environ Sci. Technol.* **23,** 257–263 (Mar. 1989).
76. W. F. McDonnell and co-workers, *J. Appl. Physiol.* **54,** 1345–1352 (May 1983).
77. L. J. Folinsbee, W. F. McDonnell, and D. H. Horstman, *J. Air Pollut. Control Assoc.* **38,** 28–35 (Jan. 1988).
78. D. H. Horstman, L. Folinsbee, P. J. Ives, S. Abdul-Salaam, and W. F. McDonnell, *Am. Rev. Respir. Dis.* **142,** 1158–1163 (Nov. 1990).
79. W. C. Adams, *Sports Med.* **4,** 395–424 (Mar. 1987).
80. H. Gong, Jr., *J. Sports Med. Physical Fitness* **27,** 21–29 (Mar. 1987).
81. N. A. Molfino and co-workers, *Subjects Lancet* **338,** 199–203 (July 27, 1991).
82. R. P. Sherwin, *Clin. Toxicol.* **29,** 385–400 (Sept. 1991).
83. National Academy of Sciences, *Nitrogen Oxides*, National Academy Press, Washington D.C., 1977.
84. S. V. Dawson, and M. B. Schenker, *Am. Rev. Respir. Dis.* **120,** 281–292 (Aug. 1979).
85. U.S. Environmental Protection Agency, *Air Quality Criteria for Oxides of Nitrogen*, Environmental Criteria and Assessment Office, Research Triangle Park, N.C., 1982.
86. J. J. Quackenboss, J. D. Spengler, M. S. Kanarek, R. Letz, and C. P. Duffy, *Environ Sci Technol* **20,** 775–789 (Aug. 1986).
87. U.S. Environmental Protection Agency, *Air Quality Criteria for Oxides of Nitrogen*, Environmental Criteria and Assessment Office, Research Triangle Park, N.C., Aug. 1991, EPA600/8-8/049c.
88. P.E. Morrow, *J. Toxicol. Environ. Health* **13**(2–3), 205–227 (1984).
89. R. B. Schlesinger, "Nitrogen Oxides," in M. Lippmann, ed., *Environmental Toxicants. Human Exposures and Their Health Effects*, Van Nostrand Rheinhold Co., Inc., New York, 1992, pp. 412–453.
90. J. M. Samet and M. J. Utell, *Toxicol. Indust. Health* **6,** 247–262 (Mar. 1990).
91. H. Magnussen, *Eur. Respir. J.* **5,** 1040–1042 (May 1992).
92. I. F. Goldstein and co-workers, *Arch. Environ. Health* **43,** 138–142 (Mar./Apr. 1988).
93. L. J. Folinsbee, *Toxicol. Indust. Health* **8,** 273–283 (Sept./Oct. 1992).
94. V. Hasselblad, D. M. Eddy, and D. J. Kotchmar, *J. Air Waste Manag. Assoc.* **42,** 662–671 (May 1992).
95. J. M. Samet and co-workers, "Nitrogen Dioxide and Respiratory Illness in Children, *Health Effects Institute Research Report No. 58*, June, 1993.
96. National Academy of Sciences, *Sulfur Oxides*, National Academy Press, Washington D.C., 1978.
97. U.S. Environmental Protection Agency, *Air Quality Criteria for Particulate Matter and Sulfur Oxides*, Environmental Criteria and Assessment Office, Research Triangle Park, N.C., Dec., 1982, EPA-600/8-82-029aF-cF.
98. U.S. Environmental Protection Agency, *Second Addendum to Air Quality Criteria for Particulate Matter and Sulfur Oxides (1982): Assessment of Newly Available Health Effects Information*, Environmental Criteria and Assessment Office, Research Triangle Park, N.C., December 1986, EPA/600/8-86/020F.
99. S. A. Shore and co-workers, *Am. Rev. Respir. Dis.* **135,** 840–847 (Apr. 1987).
100. P. D. Scanlon, J. Seltzer, R. H. Ingram, Jr., L. Reid, and J. Drazen, *Am. Rev. Respir. Dis.* **135,** 831–839 (Apr. 1987).
101. Subcommittee on Public Health Aspects of the Energy Committee on Public Health, New York Academy of Medicine, "Environmental Effects of Sulfur Oxides and Related Particulates," *Bulletin New York Acad Med* **54,** 983–1278 (Dec. 1978).
102. M. Lippmann, "Sulfur Oxides-Acidic Aerosols and SO_2," in M. Lippmann, ed., *Environmental Toxicants. Human Exposures and Their Health Effects*, Van Nostrand Rheinhold Co., Inc., New York, 1992, pp. 543–574.
103. M. T. Kleinman, *J. Air Pollut. Control Assoc.* **34,** 32–37 (Jan. 1984).
104. J. R. Goldsmith and L. T. Friberg, "Effects of Air Pollution on Human Health," in A. C. Stern, ed., *Air Pollution*, 3rd. ed., Academic Press, New York, 1977.
105. J. R. Withey, *Toxicol Indust Health* **5,** 519–554 (May 1989).
106. F. W. Lipfert, S. C. Morris, and R. E. Wyzga, *Environ. Sci. Technol.* **23,** 1316–1322 (Nov. 1989).
107. P. J. Lioy and J. M. Waldman, *Environ Health Perspect.* **79,** 15–34 (Feb. 1989).
108. American Thoracic Society, *Am. Rev. Respir. Dis.* **144,** 464–467 (Aug. 1991).
109. M. W. Frampton, K. Z. Voter, P. E. Morrow, N. J. Roberts, Jr., D. J. Culp, C. Cox, and M. J. Utell, *Am. Rev. Respir. Dis.* **146,** 626–632 (Sept. 1992).
110. R. B. Schlesinger, *Environ. Health Perspect.* **63,** 25–38 (Nov. 1985).
111. R. B. Schlesinger, *Environ. Health Perspect.* **79,** 1212–126 (Feb. 1989).
112. J. M. Gearhart and R. B. Schlesinger, *Environ. Health Perspect.* **79,** 127–137 (Feb. 1989).
113. M. J. Utell, *Environ. Health Perspect.* **63,** 39–44 (Nov. 1985).
114. L. J. Folinsbee, *Environ. Health Perspect.* **79,** 195–199 (Feb. 1989).
115. T. V. Larson, *Environ. Health Perspect.* **79,** 7–13 (Feb. 1989).
116. M. J. Utell, J. A. Mariglio, P. E. Morrow, F. R. Gibb, and D. M. Speers, *J. Aerosol Med.* **2,** 141–147 (1989).
117. D. M. Norwood, T. Wainman, P. J. Lioy, and J. M. Waldman, *Arch. Environ. Health* **47,** 309–313 (July/Aug. 1992).
118. J. D. Spengler, M. Brauer, and P. Koutrakis, *Environ Sci. Technol.* **24,** 946–955 (July 1990).
119. M. J. Utell and J. M. Samet, *Am. Rev. Respir. Dis.* **147,** 1334–1335 (June 1993).
120. J. Schwartz, D. Slater, T. V. Larson, W. E. Pierson, and J. Q. Koenig, *Am. Rev. Respir. Dis.* **147,** 826–831 (Apr. 1993).
121. C. A. Pope III and R. E. Kanner, *Am. Rev. Respir. Dis.* **147,** 1336–1340 (June 1993).
122. National Academy of Sciences, *Carbon Monoxide*, National Academy Press, Washington D.C., 1977.
123. U.S. Environmental Protection Agency, *Air Quality Criteria for Carbon Monoxide*, Environmental Criteria and Assess-

ment Office, Research Triangle Park, N.C., 1979, EPA-600/8-79-022.
124. U.S. Environmental Protection Agency, *Air Quality Criteria for Carbon Monoxide*, Environmental Criteria and Assessment Office, Research Triangle Park, N.C., Mar. 1990, EPA/600/8-90/045A.
125. M. T. Kleinman, "Health Effects of Carbon Monoxide," in M. Lippmann, ed., *Environmental Toxicants. Human Exposures and Their Health Effects*, Van Nostrand Rheinhold Co., Inc., New York, 1992, pp. 98–118.
126. R. J. Shepherd, *Carbon Monoxide: The Silent Killer*, Charles C. Thomas, Springfield, Ill., 1983.
127. S. R. Thom, "Carbon Monoxide Poisoning," in D. H. Simmons and D. F. Tierney, eds., *Current Pulmonology, Volume 13*, Mosby Year Book, St. Louis, Mo., 1992, pp. 289–309.
128. E. N. Allred and co-workers, *N. Engl. J. Med.* **321,** 1426–1432 (Nov. 23, 1989).
129. D. S. Sheps and co-workers, *Ann. Intern. Med.* **113,** 343–351 (Sept. 1, 1990).
130. S. T. Weiss, I. B. Tager, M. Schenker, and F. E. Speizer, *Am. Rev. Respir. Dis.* **128,** 932–942 (Nov. 1983).
131. S. T. Weiss, *J. Respir. Dis.* **9,** 46–62 (Jan. 1988).
132. National Research Council, *Environmental Tobacco Smoke: Measuring Exposures and Assessing Health Effects*, National Academy Press, Washington D.C., 1986.
133. J. M. Samet, "Environmental Tobacco Smoke," in M. Lippmann, ed., *Environmental Toxicants. Human Exposures and Their Health Effects*, Van Nostrand Rheinhold Co., Inc., New York, 1992, pp. 231–265.
134. National Research Council, Committee on Airliner Cabin Air Quality, *The Airliner Cabin Environment: Air Quality and Safety*, National Academy Press, Washington, D.C., 1986.
135. M. E. Mattson and co-workers, *J. Am. Med. Assoc.* **261,** 867–872 (Feb. 10, 1989).
136. U.S. Department of Health and Human Services, *The Health Consequences of Involuntary Smoking: A Report of the Surgeon General*, U.S. Government Printing Office, Wash., D.C., 1986, DHHS publication No. (CDC) 87-8398.
137. J. E. Fielding and K. J. Phenow, *New Engl. J. Med.* **319,** 1452–1460 (Dec. 1, 1988).
138. U.S. Environmental Protection Agency, *Respiratory Health Effects of Passive Smoking: Lung Cancer and Other Disorders*, Washington D.C., May 1992, EPA/600/6-90/006B.
139. International Agency for Research on Cancer, *IARC Monographs on the Evaluation of the Carcinogenic Risk of Chemicals to Humans: Tobacco Smoking*, Vol. 38, World Health Organization, IARC, Lyon, France, 1986.
140. B. A. Chilmonczyk and co-workers, *New Engl. J. Med.* **328,** 1665–1669 (June 10, 1993).
141. A. B. Murray and B. J. Morrison, *J. Allergy Clin. Immunol.* **91,** 102–110 (Jan. 1993).
142. B. Danuser, A. Weber, A. L. Hartmann, and H. Krueger, *Chest* **103,** 353–358 (Feb. 1993).
143. J. L. Repace and A. H. Lowrey, *Risk Analysis* **10,** 27–37 (Mar. 1990).
144. R. J. Shephard, *Arch. Environ. Health* **47,** 123–130 (Mar./Apr. 1992).
145. R. Bascom, T. Kulle, A. Kagey-Sobotka, and D. Proud, *Am. Rev. Respir. Dis.* **143,** 1304–1311 (June 1991).
146. U.S. Public Health Service, *Healthy People 2000: National Health Promotion and Disease Prevention Objectives*, U.S. Department of Health and Human Services, Public Health Service, Wash., D.C., 1991, DHHS no. (PHS) 91-50213.
147. American Lung Association, *Personal Protection Against Air Pollution: The Physician's Role*, American Lung Association, New York, 1981.
148. R. M. Sly, S. H. Josephs, and D. M. Eby, *J. Allergy* **54,** 209–212 (Mar. 1985).
149. "Air Purifiers," *Consumer Reports* **54,** 88–93 (Feb. 1989).
150. "Household Air Cleaners," *Consumer Reports* **57,** 657–662 (Oct. 1992).
151. M. J. Ellenbecker, *Clin. Chest Med.* **13,** 193–199 (June 1992).
152. R. E. Reisman, P. M. Mauriello, G. B. Davis, J. W. Georgitis, and J. M. DeMasi, *J. Allergy Clin. Immunol.* **85,** 1050–1057 (June 1990).
153. H. S. Nelson, S. R. Hirsch, J. L. Ohman, Jr., T. A. E. Platts-Mills, C. E. Reed, and W. R. Solomon, *J. Allergy Clin. Immunol.* **82,** 661–669 (Oct. 1988).

AIR POLLUTION: INDOOR

Diane R. Gold
Channing Laboratory
Department of Medicine
Brigham and Women's Hospital
Harvard Medical School
Boston, Massachusetts

Occupational pulmonary medicine began as the study of the effects of dusts and gases on the respiratory health of industrial workers. In the more economically developed countries, the proportion of adults working in the primary industries has decreased since World War II, while the proportion of adults working in offices, schools, and hospitals has increased. Investigation into indoor air pollution in these nonsmoke-stack-industrial settings was stimulated in the 1970s by an increase in health complaints that occurred after the sealing of buildings and the decreased provision of fresh air in response to the call for energy conservation during the "oil crisis." Research interest in air pollution in homes as well as offices also has reflected a growing awareness of the importance of the indoor environment, because residents of more developed countries spend most of their lives indoors (1). In poorer countries interest in indoor air pollution centers around concern for the effects of the combustion products of unvented cooking and heating fuels, and for the increasing morbidity and mortality associated with the increase of cigarette smoking.

Since the review "Health Effects and Sources of Indoor Air Pollution (2,3) was presented, additional monographs and reviews have been published (4,5). Proceedings of meetings on indoor air quality are also available (6). The American Thoracic Society has published a practical report on environmental controls and lung disease (7). This article presents an update on the health effects of each of the main indoor air pollutants and the modalities available to control them. Health effects of indoor exposure to radon and asbestos are not discussed, since they are presented elsewhere in this encyclopedia. The diagnosis and control of building-related illness that cannot be ascribed to a specific pollutant will be discussed. Finally, air quality issues related to hospitals will be reviewed. See also Air Pollution Control Methods; Air Pollution: Health

EFFECTS; ENERGY MANAGEMENT, PRINCIPLES; RISK ASSESSMENT.

TOBACCO SMOKE

Active Smoking

Tobacco smoke is an aerosol containing several thousand substances that are distributed as gases (vapors) or particles. Components of the particulate phase include human carcinogens such as nickel, 2-naphthylamine and 4-aminobiphenyl, and suspected human carcinogens such as benzopyrene. Carbon monoxide, sulfur dioxide, nitrogen oxides, benzene, toluene, formaldehyde, ammonia, formic acid, acrolein, and acetic acid are some of the components of the vapor phase (8). Nicotine is partitioned between the vapor and particulate phase.

Active cigarette smoking has been well-established as a main preventable cause of bronchitis, emphysema, and lung cancer; the association with an increased risk for cardiovascular disease in men and women is also well recognized (9–11). Recent studies have focused on the association between active smoking, nonspecific bronchial reactivity, atopy, and elevated IgE levels (12).

Smoking Cessation

The beneficial effects of smoking cessation were presented in the 1990 Surgeon General's report (13). The Surgeon General's message has been generally accepted in the United States; 80% of smokers surveyed wanted to stop smoking (14). By 1987 nearly one-half of all living adults who ever smoked had quit smoking (13). The techniques that most of these individuals used in order to quit are unknown.

A doctor's advice to quit smoking has been demonstrated to increase patient cessation rates (13). Smoking cessation interventions recently reviewed by Fisher and associates include temptation management, cue extinction, aversive techniques, contingency management, and pharmacological treatment of tobacco dependence (15). By integrating these individual tactics, the American Lung Association Freedom from Smoking program has achieved a 29% smoking cessation rate at the end of 12 months of follow-up (15).

Attempts to stop initiation of smoking have met with mixed results; smoking initiation rates in the United States have actually been increasing among less educated young women (14). Since the overall decrease in smoking in the United States, cigarette companies have successfully targeted advertising to minority communities and to citizens of developing countries (14).

Passive Smoking

Environmental tobacco smoke exposes nonsmokers to many of the same toxins inhaled by active smokers (8). Salivary, plasma, and urinary cotinine have been established as markers for environmental tobacco smoke inhalation (16,2). Studies on the health effects of environmental tobacco smoke have recently been reviewed by the U.S. Environmental Protection Agency (EPA) (17), which concluded that environmental tobacco smoke is responsible for inducing new asthma cases and for increasing the number and severity of episodes among asthmatic children in the United States. In addition, environmental tobacco smoke exposure predicts acute respiratory infections (18), wheezing (18,19) and bronchial responsiveness (20) in children without a diagnosis of asthma. Additional epidemiologic studies of infants have found environmental tobacco smoke exposure associated with an increased incidence of middle-ear effusions (21) and hospitalization for chest illness (72). Published cross-sectional and longitudinal data add to the evidence for a significant effect of passive smoking on level and growth of lung function in children. Kauffman and associates studied 1160 French children between 6 and 10 years of age and found an association between maternal smoking and a decrease in FEV_1 and FEF_{25-75} (23). In a cohort of 362 children followed for an average of 9 years, Lebowitz and co-workers found slower growth of lung function in children with parents who smoke (24). A study of 80 healthy infants from East Boston suggested that maternal smoking during pregnancy may impair *in utero* airway development or alter lung elastic properties (25). Researchers are currently investigating whether the greatest effects of environmental tobacco smoke on the lung architecture and susceptibility are already established *in utero*, or whether disease associated with such exposures results from cumulative exposures during infancy and childhood.

In adults there is strong evidence for an association between environmental tobacco smoke and acute symptoms of irritation measured subjectively as eye, nose, or throat irritation, or more objectively, with eye-blink tests (8). The threshold for perception of tobacco smoke odor is tenfold lower than that for the development of irritation (4). The data for adverse effects of passive smoke on chronic respiratory symptoms or lung function in adults are less conclusive.

In 1986 the National Research Council and the Surgeon General's report concluded that involuntary smoking can cause lung cancer in nonsmokers, but that more data were necessary to estimate the magnitude of the risk in the U.S. population. Case-control studies published since 1986 have added supporting evidence for a relationship between environmental tobacco smoke and lung cancer. In a case-control study of 191 patients from the United States, Janerick and associates concluded that approximately 17% of lung cancers among nonsmokers could be attributed to high levels of exposure to cigarette smoke during childhood and adolescence (26). Their study did not find that exposure to spouses' smoke was an independent predictor of lung cancer risk. In contrast, Japanese (27,28) contribute strong evidence for an excess risk of lung cancer related to spousal smoking. This may reflect a higher exposure to spousal smoke in Japanese as compared to subjects in the Janerick study, where higher cumulative exposures may have occurred either in childhood or at work. A recent case-control study of 618 primary lung cancer patients from Missouri showed an increased lung cancer risk for lifetime nonsmokers with exposure of more than 49 pack-years from all household members or from spouses only (29).

Control of Environmental Tobacco Smoke

A totally smoke-free environment cannot be produced through presently available air-cleaning devices or through policies restricting smoking to limited areas. The commercially available modular cleaning devices can reduce the particulate load from environmental tobacco smoke but will not eliminate the constituents of tobacco smoke in the gaseous phase (7,30). Many of these gaseous elements may be irritants or carcinogens. The high ventilation rate required to eliminate the odor of the smoke is often costly and impractical from an engineering point of view. Consequently, many private and public institutions have established policies restricting or banning indoor smoking. Recent recommendations of the Joint Commission on Hospital Accreditation have specified that all hospitals are to become smoke free, and that this recommendation is to be part of the Review of Hospitals.

COMBUSTION PRODUCTS

Carbon Monoxide

Carbon monoxide results from any incomplete combustion of carbonaceous material. Water heaters, gas or coal heaters, and gas stoves are all indoor sources of carbon monoxide. Camping stoves and lanterns are sources of carbon monoxide that in the unvented space of a tent have led to accidental carbon monoxide poisoning and death. High indoor levels of carbon monoxide can also result from the entrainment of outdoor car, bus, or truck exhaust into the ventilation system of a building (31). This is a risk particularly in buildings that have delivery areas where vehicles are parked with the motors running.

Hourly carbon monoxide concentrations in homes during cooking with gas stoves range from 2 to 6 parts per million (ppm); one-hour averages may exceed 12 ppm (2). During city commuting, carbon monoxide levels in cars (6–15 ppm) may average 2 to 5 times the concentrations measured in homes (2). Cigarette smoke contains over 2% carbon monoxide; the average concentration in smoke that reaches the lungs is about 400 ppm (31).

Carbon monoxide produces toxic effects through several mechanisms (32). Because its affinity for hemoglobin is approximately 200 times greater than that of oxygen, it displaces oxygen, lowering the oxygen-carrying capacity of blood. It interferes with delivery of oxygen to the tissues by causing a leftward shift of the oxyhemoglobin dissociation curve. Carbon monoxide can also directly impair oxygen diffusion into mitochondria, interfere with intracellular oxidation, and increase platelet adhesiveness.

In a clean environment, for a nonsmoker, the normal carboxyhemoglobin saturation is 0.6%. Smoking one pack of cigarettes per day can lead to a daytime carboxyhemoglobin level of 5% to 6%; two to three packs per day can result in a carbon monoxide saturation of 7% to 9% (31). Methylene chloride, used as a paint stripper, is metabolized to carbon monoxide and after 3-hour exposure, even in a well-ventilated room, can result in a carboxyhemoglobin saturation of 16% (31).

Carbon monoxide poisoning has its most toxic acute effects on the organs with high oxygen requirements, the heart and the brain (31). The carboxyhemoglobin level is only an approximate predictor of the effect of carbon monoxide on the brain. The effects of very high levels of carbon monoxide have been well documented. Significant deterioration of judgment, calculation, and manual dexterity have been detected at levels above 20%; levels of 50% or greater can lead to coma, convulsions, or death (31). The low level and chronic effects of carbon monoxide are less well understood. Bunnell and Horvath found performance of complex tasks was impaired with increasing carboxyhemoglobin level (0–10%) at high work loads (60% maximum oxygen consumption) (32). It was recently demonstrated that exposure to inhaled carbon monoxide (200 ppm resulting in a carboxyhemoglobin of 6%) produced direct adverse effects on the heart by facilitating ventricular ectopy in patients with coronary artery disease (33,34). A multicenter study (35) suggested that exposure of patients with stable coronary disease to carbon monoxide can induce earlier subjective and objective evidence of myocardial ischemia at carboxyhemoglobin levels as low as 2% to 4%. At low levels carbon monoxide can lead to nonspecific flulike symptoms that often go unrecognized. Symptoms of headache, lethargy, nausea, or dizziness in children with carboxyhemoglobin levels between 2% and 10% have been noted (36).

Oxygen is the treatment for carbon monoxide poisoning. For an adult breathing air at 1 atmosphere, blood carboxyhemoglobin levels are reduced by half in 320 minutes; with an FIO_2 of 100%, levels are reduced by half in 80 minutes; and in a hyperbaric chamber at 3 atmospheres, levels are reduced by half in 23 minutes (37).

Nitrogen Dioxide

The National Ambient Air Quality Standard for an annual mean of NO_2 in ambient air is 53 parts per billion (ppb); on average, gas stoves add 15–25 ppb to the background concentration in a home (2), with peak levels in the kitchen of 200 to 1000 ppb. About 50% of homes in the United States have gas cooking appliances (2).

NO_2 is an oxidant that not only causes direct lung injury, but may increase susceptibility to respiratory infection through its effect on ciliary clearance and macrophage function. At high concentrations, NO_2 is known to cause diffuse pulmonary damage, not only in the animal model but in humans. Silo-fillers may develop on occupational lung disease resulting from high-dose NO_2 exposure. Freeman and co-workers demonstrated that with chronic exposure to a low level of 800 ppb of NO_2, occasional reduction or absence of bronchiolar cilia occurred in rats, but the reduction was not present in nonexposed lungs (38).

Meta-analysis utilizing the epidemiologic literature on indoor exposure to NO_2 at levels produced by gas stoves and heating in households in developed countries suggests a significant but small association (for an exposure increment of 15 ppb, odds ratio = 1:2; 95% Confidence Interval 1.14 to 1.28) (Lucas Neas, Doctoral Dissertation, 1991) between that pollutant and respiratory symptoms in children. In their study of 6,273 children aged 7 to 11 from six cities in the United States, Neas (39), and Dockery and colleagues (40) measured indoor exposure averaging 15 ppb higher in homes with gas stoves and pilot lights. They found increases in individual respiratory symptoms of 5–29%, with a combined symptom odds ratio of 1:40. In contrast, the study of 6 to 9-year-old children in the southeast region of the Netherlands did not find an effect of NO_2 on respiratory symptoms (41). A recent study found no association between nitrogen dioxide exposure and the incidence rates for respiratory illnesses in a population of healthy infants and toddlers exposed to a range of 0 to 40 ppb. They pointed out that these levels were lower than nitrogen dioxide concentrations in heavily polluted cities and in poorly ventilated inner-city apartments, but were equivalent to levels observed in many U.S. locations (42). Although recent studies of ambient NO_2 suggest an effect on pulmonary function (43), environmental chamber studies and indoor studies on NO_2 and pulmonary function are inconclusive (44). Certain asthmatics may be sensitive to acute exposure to NO_2; in a study of 11 asthmatics observed while cooking on a gas range, FVC and peak flow rates tended to drop with exposure to more than 300 ppb (45).

Biofuel

Burning of biofuel can result in production of particulates, carbon monoxide, nitrogen dioxide, sulfur dioxide (SO_2), formaldehyde, volatile organic compounds, and numerous other substances, depending on the type of fuel. In the 1970s, with the oil embargo, the shipment of woodstoves to the United States increased 10-fold (7). The newer "airtight" residential woodstoves should not leak combustion by-products into the home except during startup, stoking, and reloading. Traynor and co-workers reported indoor carbon monoxide concentrations of 0.4 to 2.8 ppm with particle concentrations only slightly above background (0 to 30 $\mu g/m^3$) with "airtight" stoves. In contrast, "nonairtight" stoves produced average levels of 1.8 to 14 ppm of carbon monoxide and particle levels of 200 to 1900 $\mu g/m^3$ (7). In developing countries, burning of wood, charcoal, crop residues, or animal dung for cooking or heating is often done without a flue or chimney and with poor ventilation. In 390 households in India, Menon found that the levels of particulates, measured during cooking, ranged between 4000 and 21,000 $\mu g/m^3$ (46). These particulate levels from biofuel combustion contrast with average particulate levels of about 40 $\mu g/m^3$ and peak concentrations of 120 $\mu g/m^3$ in homes where two packs per day of cigarettes are smoked (46).

In Shenyang, China, after adjustment for smoking, the risk for lung cancer increased in proportion to years of sleeping on brick beds heated by coal-burning stoves that were located directly under the bed (47). In Shenyang and Harbin, the number of meals cooked by deep frying and the frequency of smokiness during cooling were also associated with increased risk of lung cancer (48). In Nepal, Pandey et al. showed an association between chronic bronchitis in nonsmoking women and domestic smoke exposure as measured by the number of hours spent daily near the stove (49). Cooking in poor countries is often performed by women carrying their infants on their backs. In The Gambia, after adjustment for parental (paternal) smoking, an association between carriage on the mother's back and a child's development of an episode of difficulty in breathing was found (odds ratio of 2:80; 95% confidence interval: 1.29, 6.09) (50). Infant mortality rates due to pneumonia are higher in less-developed countries; exposure to smoke from biomass fuel may increase the vulnerability of intants to severe respiratory infection (51). In a group of 30 nonsmoking patients from Mexico, Sandoval and co-workers observed a syndrome of interstitial lung disease, with a mixed restrictive-obstructive pattern on pulmonary function testing and signs of pulmonary hypertension, which they attributed to prolonged woodsmoke inhalation (52). Studies of effects of indoor air pollution in developing countries are difficult to perform because of lack of appropriate control populations in morbidity studies, and because of the large numbers of subjects required to study mortality as an end point. The existing literature on indoor air pollution in developing countries was recently reviewed (53).

Few data are available on woodsmoke and respiratory illness in developed countries. In a study of 63 children in Michigan, an association between woodsmoke exposure and an increased prevalence of moderate to severe respiratory symptoms during the winter was found (54), yet in a similar study in Massachusetts, did not find such an association (55). In the study of 6 U.S. cities, an odds ratio of 1:32 (95% confidence interval (CI) of 0.99 to 1.76) for respiratory illness was found in households with wood-burning stoves as compared to households with other sources of heating (55).

Control of Combustion-Related Pollutants

Control options for combustion-related indoor air pollution in developed countries are outlined in the report by the American Thoracic Society (ATS) workshop on environmental controls and lung disease (7). Appliances should be ventilated, and should receive periodic service and maintenance. A visual inspection may not detect the source of a carbon monoxide leak detectable with a carbon monoxide monitor. Pilot lights should be turned off or eliminated. The cooking range should not be used for heat. Optimally, gas stoves should be exhausted to the outdoors. Woodstoves should be airtight; it should not be possible to smell the smoke indoors. The presence of soot on the inside wall above a fireplace suggests that pollutants are probably entering the house. More research is necessary on control options and interventions appropriate to the specific cultures and resources available in less-developed countries.

BIOLOGIC AGENTS

Biologic agents in buildings can cause disease through immune mechanisms, through infectious processes, or through direct toxicity (56). Of the immune-mediated respiratory diseases, allergic rhinitis and asthma are primarily associated with IgE antibody to indoor allergens, while humidifier fever and hypersensitivity pneumonitis are associated with IgG, IgA, and IgM antibody (Type-III immune complex–induced injury) and cell-mediated Type-IV hypersensitivity.

Asthma and Biologic Agents

In the United States, asthma prevalence and severity appear to be increasing. Rates of asthma and wheeze are particularly high among black peoples and inner-city inhabitants (57,58). These epidemiologic patterns of asthma prevalence may, in part, relate to changing patterns of indoor exposures to aeroallergens.

Mites, cockroaches, fungi, cats, dogs, and rodents are common sources of indoor allergens (7). Outdoor allergens such as pollens can also be entrained into the indoor ventilation system. Socioeconomic status, cultural factors, and housing conditions may be determinants of sensitization (59). In the emergency-room study of asthmatics from Charlottesville, Virginia, it was found that the prevalence of IgE antibody to one of five common inhalant allergens (dust mites, cats, cockroaches, grass pollen, and ragweed pollen) was significantly higher in asthmatics (75%) than controls (14%) (60). Whereas most asthmatics were allergic to dust mite, antibody to cats was a risk factor only among white patients, and IgE antibody to cockroaches was a risk factor only among black patients. These racial differences in allergy correlated with socioeconomic and housing conditions. The house dust mite (*Der p* I, *Der f* I) is one of the most important of the indoor aeroallergens (61). Possibly because of low humidity and temperature, homes at high altitudes in the Alps have been demonstrated to have a paucity of mites in dust (62). The use of fitted carpets, detergents effective in cool water, and tighter insulation resulting in a narrow range of temperatures may have resulted in conditions "more congenial to mite and man" (63). In a cohort of British children with a family history of allergy, they demonstrated that exposure in early childhood to house-dust-mite allergens was an important determinant of subsequent development of asthma. All but one of the children with asthma at the age of 11 had been exposed at 1 year of age to more than 10 μg of house dust mites per gram of dust. Several studies have demonstrated that levels of antigen exposure to house dust mites is associated with the degree of sensitization (64,65). Emergency-room visits and hospitalization for asthma have been related to house-dust-mite sensitization (59,66).

Molds or fungi, which thrive in conditions of home dampness and in humidifiers and vaporizers, have been associated with sensitization, wheeze symptoms, and asthma (67–69) in adults and children. Home dampness itself has been associated with increased risk of cough, phlegm, and acute respiratory illness (70,71). The high prevalence of *Alternaria* sensitivity among U.S. asthmatic children suggests the importance of this fungal antigen (72). Mold exposure may cause immediate-type hypersensitivity or may promote respiratory disease through toxic nonallergic reactions.

Sampling of Aeroallergens and Diagnosis of Type I Hypersensitivity

Refs. 62, 73–75 provide details on the sampling of aeroallergens. Ref. 76 provides a perspective on interpretation of mold/fungal measurements. While fungi and antigens from house dust mite, cockroach, cat, and dog are quantifiable in house dust, more information is needed to increase our understanding of the relationships between levels of antigen or fungi in dust and the development or exacerbation of wheeze and asthma (7). Particularly with fungi, it is not certain whether air samples rather than dust samples better represent personal exposure.

Diagnosis of sensitivity to known indoor allergens may be established by history and, for some allergens, by skin testing. Standard skin testing is not available for most fungi (7), but is available for *Alternaria*. For dust mite, cat, dog, cockroach, and few fungal allergens, specific IgE antibody can be measured through radioallergosorbent tests (RAST) (7).

Without a specific identifiable allergen, diagnosis of sensitivity to an indoor exposure can often be made only by history and by demonstration of a worsening of peak flow or peak flow variability while indoors and an improvement of symptoms and peak flows or spirometry when away from the suspected indoor setting. If the asthmatic symptoms are severe, interpretation of improvement may be confounded by the administration of more intensive asthma therapy at the same time as removal from the suspected indoor setting.

Control Measures for Aeroallergens

Control measures for indoor allergens have recently been reviewed (75). The beneficial effects of portable air filtration devices over and above those associated with central air conditioning are thought to be minimal (75,77). Although pollen entrained into a room may be effectively filtered by household air cleaners, the beneficial effect of air cleaners is considered small in removing allergens such as those from mites and cats, which exist in surface reservoirs (75). The measures can be time-consuming and costly; sensitivity to house dust mite should be documented before implementing them. Some of these measures may not be practical for people with poor housing and little money.

Mold abatement can be accomplished by prevention of the intrusion of basement air into living space, dehumidification of the basement, cleaning of moldy surfaces with 10% chlorine bleach, and cleaning of humidifiers and vaporizers with chlorine bleach before each use and by the keeping of relative humidity below 50% (7). The American Thoracic Society recommends that, because of their high rate of fungal contamination, water-spray humidifying devices should not be used in areas with asthmatics or immunocompromised persons (7).

Platts-Mills and co-workers demonstrated that over a period of months, rigorous avoidance of domestic allergens in a hospital setting substantially decreased bronchial reactivity in a group of dust mite–sensitive asthmatic patients (62). Although the majority of allergic patients lose their symptoms within days or weeks after elimination of the exposure, repeated exposure to certain allergens may lead to chronic, irreversible airway hyperresponsiveness. This has been most clearly demonstrated in occupational settings, with exposure to toluene diisocyanide (which can cause either asthma or hypersensitivity pneumonitis) and western red cedar (7).

Hypersensitivity Pneumonitis and Humidifier Fever

In hypersensitivity pneumonitis, sensitizing antigens can be microorganisms such as actinomycetes, bacteria, fungi, amoebae, animal and plant proteins including pigeon or parakeet antigen, rat urine, low-molecular-weight chemicals such as detergent enzymes, toluene diisocyanate, or pharmaceutical products (78).

Humidifier lung, a hypersensitivity pneumonitis in response to antigens such as thermophilic actinomycetes, results from indoor exposure to contaminated cooling, heating, or humidifying units. Clinically, the acute disease consists of fever, chills, cough, and dyspnea, usually with interstitial and alveolar nodular infiltrates on chest X rays. Restrictive changes are seen on pulmonary function testing. Chronic exposure can lead to progressive dyspnea and interstitial fibrosis, with decreasing lung volumes and diffusing capacity. Acutely, on lung biopsy the tissue reaction typically consists of both an alveolar and interstitial inflammation, with prominence of lymphocytes, plasma cells, macrophages with foamy cytoplasm, granulomata, and giant cells with birefringent material. Humidifier fever is a milder form of the disease, with fewer pulmonary manifestations.

In the process of differentiating hypersensitivity pneumonitis from illnesses not due to a hypersensitivity reaction, but causing similar symptoms (eg, gram-negative toxin, *Legionella*), identification of the inciting antigen can be difficult. Three episodes of an acute, flulike illness associated with manipulations on the central air handling systems of an office building have been described (79). Symptoms were consistent with hypersensitivity pneumonitis, but no single etiologic agent could be identified. As in the Hodgson study, a search for the inciting antigen can be performed through volumetric culture plate sampling for microorganisms, though the organisms that are isolated may not be representative of personal exposure. Large volumes of air can be collected for immunologic analyses, using sera from affected patients. One disadvantage to this method is that many of the immunoassays are sensitive only to far larger amounts of antigen than would be necessary to cause symptoms in an already sensitized patient (7). Precipitating antibody testing has a use in diagnosis, but limitations include improper antigen-antibody combining ratios, the relative insensitivity of the test, and the possibility of false-positive reactions (78). Exposed subjects can have positive precipitins without the presence of disease. Skin testing is not clinically useful because many crude allergen preparations cause nonspecific irritation.

Reexposure of the patient to the suspected environment can establish the diagnosis of hypersensitivity pneumonitis, with limited risk to the patient, though not necessarily with identification of the inciting antigen. Laboratory inhalation challenge has been used clinically to identify specific etiologic agents; acute bronchospasm or severe pneumonitis are risks of the procedure (78), which should be performed in centers familiar with diagnostic challenge testing to the antigen of interest.

INFECTION

Ventilation-Related Transmission: Legionella

Legionella is a common bacterial contaminant of air conditioning and potable water systems. Since the 1976 epidemic in Philadelphia, there have been a number of reports of sporadic and nosocomial cases of legionellosis in buildings whose water systems contained the organism (3). Seven cases of nosocomial legionellosis were reported that occurred over a 7-month period in a community hospital in upstate New York (80). A case-control study demonstrated that length of hospital stay and proximity of patients' rooms to the ward shower were significant risk factors for acquiring legionellosis. The ward showers and hospital hot-water system were contaminated with *L. pneumophilia* serogroup 1. A large outbreak of Legionnaires' disease associated with Stafford District Hospital in England was reported (81). Among 68 hospitalized patients with confirmed legionellosis, 22 died. An additional 35 patients had suspected legionellosis and nearly one-third of the hospital staff had legionella antibodies. In a case-control study of the first 53 inpatient cases to be recognized, people who had visited the outpatient department were almost 100 times more likely to have developed the disease than those who had not. The cases were more likely to be smokers and more likely to have chronic illness.

Legionnella pneumophilia was identified in a chiller unit that cooled the air entering the outpatient department through the air-conditioning system. The staff with *Legionella* antibodies had not necessarily worked in the outpatient department, but were more likely to have worked in areas of the hospital ventilated by contaminated air from the cooling tower. Prior to the Stafford outbreak the water treatment company contracted to maintain the air-conditioning plants had twice isolated the epidemic strain of *Legionella* from water in the cooling tower. The outbreak ended before control measures were instituted.

Control of Legionella

The control measures utilized in Stafford included chlorinating the cooling-tower water to 50 ppm followed by continuous chlorination to 5 ppm free residual chlorine, raising the hot-water temperature in the calorifiers to 60–63°C to provide temperatures of 55–63°C at the tap, and continuous chlorination of the cold-water systems to 1–2 ppm free residual chlorine. The chilling unit for the outpatient department was replaced, the air conditioning to the outpatient department and floors above was turned off, and all spray attachments to wash-basin taps were removed.

The finding of *Legionella* in the cooling waters of air-conditioning systems is not uncommon. Better design and maintenance of hospital ventilation systems can result in disease prevention if an aerosol cannot be generated from contaminated cooling water and that bacterial counts of *Legionella* in the water are kept low. Chlorination alone may not entirely eradicate the organism; control of water temperature may also be needed. In the potable water systems, Farrell and co-workers recommend that the hot-water feed be kept at temperatures of 60°C or above and the cold-water feed temperatures do not exceed 20°C (82).

Ventilation-Related Transmission of Other Infections

Ventilation-related transmission of other infections is well reviewed in Refs. 2 and 83. Cases of nosocomial infection with *Aspergillus* have been reported primarily in immunocompromised patients. Viral infections with measles or chicken pox are easily spread to nonimmune or immunocompromised subjects. Nosocomial acquisition of tuberculosis usually requires prolonged or intimate exposure to droplet nuclei. A study of the spread of tuberculosis on a U.S. naval vessel in 1965 revealed that 140 members of the crew converted their tuberculin skin tests and 7 had evidence of active pulmonary tuberculosis after living on the same ship as a man with active pulmonary tuberculosis (84). The minimal contact between the source case and many of the tuberculin converters strongly suggested that the infection was spread by the ventilation system.

Humidity and Infection Transmission

Arundel and Sterling reviewed the literature on the effects of relative humidity on the transmission of infectious diseases for the Air Pollution Control Association (85). Of 8 epidemiologic studies, 7 suggested that the incidence of acute respiratory illness was lower in buildings with relative humidity levels between 40% and 60%, compared with buildings with lower levels of relative humidity. Very low humidity may dry protective mucous membranes and may also increase airborne or fomite transmission of organisms. Limited experimental data on the survival of viruses suggest maximum survival at low and high relative humidity and minimum survival at midrange humidity (84). More studies are needed on the risks and benefits of building and room humidification.

FORMALDEHYDE

There are numerous sources of formaldehyde in homes and offices. Because of the many health complaints by residents of homes insulated with urea–formaldehyde foam, use of this insulation has virtually ceased in the United States (3). But plywood, particle board, and pressed-wood products remain a main source of formaldehyde exposure, particularly in mobile homes. Formaldehyde is present in grocery bags, waxed paper, facial tissues, paper towels, colored newsprint, disposable sanitary products, floor coverings, carpet backings, adhesive binders, fire retardants, permanent-press cloths, household cleaning agents, cosmetics, deodorants, shampoos, fabric dyes, inks, and disinfectants. It is also present in natural gas, kerosene, and tobacco (86).

In a study of 28 residences, the range of formaldehyde levels in residences with urea–formaldehyde foam was 0.02–0.13 ppm; the range in control residences was 0.02–0.07 ppm (85). An "energy-efficient" nonurea-formaldehyde foam-insulated home evaluated by Gammage was found to have levels in the range of 0.13–0.17 ppm (85). The median level in 65 mobile homes where complaints had been noted was 0.47 ppm, with a range of less than 0.01–3.68 ppm (85).

Formaldehyde is a volatile gas that is highly water soluble, with irritant effects on the mucous membranes of the eyes and upper respiratory tract. Formaldehyde exposure can result in neuropsychologic effects that have been well-documented. Its acute human health effects are reviewed in Ref. 3. Sterling estimates that 10–20% of the general population may be sensitive to the irritant effects of formaldehyde; for these people the odor threshold can be as low as 0.01 ppm (85). Although wheeze has occasionally been reported in response to formaldehyde, immediate-type hypersensitivity to the substance has not been conclusively demonstrated because formaldehyde is a strong irritant and skin testing for hypersensitivity is not useful. An unpublished study of immunologic responses in several hundred formaldehyde-exposed individuals has demonstrated only one individual with formaldehyde-specific IgE antibody. IgM- and IgG-mediated responses were more common (G. Pier, 1991). Specific bronchial provocation with formaldehyde, though not widely available, is useful for diagnosis of formaldehyde sensitivity. Reactivity to very high concentrations of 2 ppm of formaldehyde in 12 of 230 Finnish workers referred for possible sensitivity have been reported (3). Their decrease in pulmonary function may have been effected through irritant mechanisms. Asthma and Type I reactivity related to formaldehyde exposure is probably very unusual.

The regulatory agencies recognize formaldehyde as a "probable human carcinogen" (87). In animals formaldehyde causes squamous-cell carcinoma; the dose response is nonlinear (3). This has led the Risk Estimation Panel of the Consensus Workshop to propose that formaldehyde should not be considered to have a threshold for cancer induction; a consensus has not been reached regarding a risk estimation model (3).

Evidence for carcinogenicity in humans is limited. Evidence is strongest for an association between formaldehyde and nasal and nasopharyngeal cancer (86,88). Excess risk of lung cancer in a study of 26,561 industrial workers with exposure to formaldehyde was found, but also with exposure to a variety of other substances that they believe could have been carcinogens and cocarcinogens (89,90). A clear dose–response relationship was not found. Reported also, was an excess of nasopharyngeal and oropharyngeal cancer. Significant excesses of brain cancer have been found in embalmers, anatomists, and pathologists, but not in formaldehyde-exposed industrial workers (3).

VOLATILE ORGANIC COMPOUNDS AND BUILDING-RELATED ILLNESS

In addition to formaldehyde, hundreds of other volatile organic compounds have been found in indoor air environments. Common sources of organic compounds include solvents, gasoline, printed materials, photocopying machines, chlorinated water, dry-cleaned clothes, pesticides, silicone caulk, floor adhesive, particle board, moth crystals, floor wax, wood stain, paint, furniture polish, floor finish, carpet shampoo, room freshener, and vinyl tiles (3,85,91). Even in a survey of two New Jersey communities with strong outdoor sources of volatile organic compounds, investigators from the Environmental Protection Agency found levels of most volatile organic compounds were 5 to 10 times higher indoors than outdoors (7). There is consistenty among surveys in detection of certain specific organics; 22 of the same compounds were found both in a United States building survey and in a West German survey (90).

Many of the volatile organic compounds identified in studies of indoor environments are known to be mucous irritants. Chronic exposure to high levels of some of the solvents can have effects on the central nervous system and on other organ systems such as the liver. Some known and suspected carcinogens are among these organic compounds: benzene, tetrachloroethylene, trichloroethane, and trichloroethylene. Limonene and toluene are on the EPA's priority list for carcinogenicity testing (85).

In the studies of nonindustrial indoor environments, levels of the compounds have been an order of magnitude less than the Occupational Safety and Health maximum permissible exposure levels. The individual and combined effects of low levels of volatile organic compounds have been difficult to assess in laboratory and in epidemiologic studies. In a longitudinal study of personnel from 14 Swedish primary schools, Norback, Torgen, and Edling assessed the relationship between volatile organic compounds and symptoms of eye, skin, and upper airway irritation, headache, and fatigue (92). They found a relationship between levels of VOCs and the repeated reporting of the same symptom both in 1982 and in 1986. This relationship remained after adjusting for other significant illness predictors, including "hyperreactivity," psychosocial dissatisfaction index, and wall-to-wall carpeting in the workplace. The mean indoor concentration of volatile organic compounds ranged from 70 to 180 $\mu g/m^3$.

Sterling and Sterling suggest that fluorescent lighting may generate reactions between low levels of volatile organic compounds (93). In some settings, the resultant photochemical smog, rather than the chemical antecedents, may be the most potent irritant. In their study of office workers, self-reported eye irritation decreased after replacement of fluorescent lamps with lamps emitting less ultraviolet light. No objective measurement of chemical exposure was available; the symptom improvement could have been related to a direct effect of improved lighting rather than to a reduction of photochemical smog.

Studies of health effects of volatile organic compounds sometimes have some objective measure of exposure, but they often rely solely on self-reporting of symptoms as a measure of effect. Laboratory researchers have been searching for objective markers of mucous irritation. Eye-blinking as a measure of irritation can be measured on an objective scale. Nasal lavage may be useful in evaluation of the upper respiratory tract for signs of inflammation (94). Statistically significant increases in neutrophils in nasal lavage fluid of 14 subjects exposed to a mixture of volatile organic compound has been reported (95). The mixture and quantity used (25 mg/m^3) was felt to be representative of what is found in new homes and office buildings. Although it may be impractical to use nasal lavage to evaluate large populations, the technique could be applied to subgroups among a population complaining of irritant symptoms.

Nasal lavage has not been found to be a useful tool for assessment of effects of volatile organic compounds in atopic asthmatics, who often have a high percentage of neutrophils in their lavage fluid, even without volatile organic compound exposure. With asthmatics, measurements of variability in peak flow on work days and on days off, and serial spirometric measurements may be of more use. In a chamber study of 11 subjects with bronchial reactivity to histamine and a history of asthma, Harving, Dahl, and Molhave found a decline in FEV1 with an 85-minute exposure to 25 mg/m^3 volatile organic compounds (96). The mean decline of 90.7% was statistically different from the initial value but not statistically significant from the decline to 97.4% of baseline found after sham exposure.

TIGHT BUILDING SYNDROME

Since the 1970s, there have been many outbreaks of work-related illnesses in buildings not contaminated by industrial processes. Typical building-related symptoms have included runny or stuffy nose, eye irritation, cough, chest tightness, fatigue, headache, and malaise.

Of 356 cases of building-related illnesses, the National Institute for Occupational Safety and Health found that 39% involved identifiable indoor or outdoor contaminants, and 50% were associated with inadequate provision of fresh air with no identifiable contaminant (3). The etiology of the building-related illnesses was unknown in 11% of cases. The terminology has not been uniform in the description of outbreaks of building-related illness of unknown etiology. The terms "tight building syndrome" and "sick building syndrome" have been used variably. In this article the term "tight building syndrome" is restricted to epidemics of illness in buildings with inadequate supply of fresh air and without specific contaminants associated with symptoms.

Provision and Distribution of Fresh Air

The adequacy of the supply of fresh air is often assessed by the number of cubic feet per minute provided per person, or by the level of CO_2 in a room. In the 1970s, with the pressure on energy conservation, the American Society for Heating, Refrigerating, and Air Conditioning Engineers revised downward its recommended ventilation standard for fresh air supply in the absence of smoking, recommending 5–10 cubic feet per minute (cfm) per person (3). Their guidelines recommended between 20–30 cfm for spaces where smoking was permitted. Buildings constructed according to these guidelines might not supply sufficient fresh air for the comfort of the occupants. On average, a supply of 15 cfm/occupant corresponds with a CO_2 level of approximately 1000 ppm (J. McCarthy, 1989). The newly adopted American Society for Heating, Refrigerating, and Air Conditioning Engineers standard 62-1989 (Ventilation for Indoor Air Quality) calls for maintaining CO_2 levels below 1000 ppm. Depending on other conditions in a building, its occupants may already begin to experience discomfrt and irritant symptoms at CO_2 levels of 800 to 1000 ppm. CO_2 at this level is probably a marker for poor air exchange, not a cause of symptoms.

The economic incentives for reducing the fresh air supply are evident. The annual cost of supplying and conditioning a cubic foot of air per minute may range from $2.00 to more than $4.00. Schools, office buildings, and arenas often require fresh air supply rates greater than 100,000 cfm resulting in an annual cost to the institution of more than $200,000 per year (3).

Even in a building with adequate overall supply of fresh air, the distribution of the air may be uneven, resulting in poor supply of air to some employees. This often occurs when partitions are placed in large spaces, to create small cubicle workspaces.

Researchers have not established what it is about poor

supply of fresh air that causes the symptoms of tight building syndrome. The illnesses are often probably multifactorial, with retension of odors and low levels of many of the previously mentioned pollutants. A review of the literature suggests that in poorly ventilated buildings the presence of fewer negative ions may lead to decreased concentration and reaction time. Field studies are inconclusive as to whether the generation of negative ions reduces the prevalence of symptoms common in tight buildings (97).

The effect of varying levels of outdoor air supply on the symptoms of sick building syndrome was studied (98). It concluded that increasing ventilation with outdoor air above the current standard did not eliminate symptoms of sick building syndrome in the office buildings studied. An accompanying editorial points out that "this study . . . does not address the more interesting question of whether increased ventilation with outdoor air can reduce the rate of symptoms in buildings with low rates of ventilation" (99).

Temperature and Humidity

In addition to provision of adequate supply of fresh air, the actual flow of air, temperature, and humidity can effect comfort in buildings. The 1981 American Society for Heating, Refrigerating, and Air Conditioning Engineers standards define thermal environmental parameters that most—80% or more—sedentary occupants in indoor environments will find satisfactory. They recommend operative temperatures of 22.8–26°C at 50% relative humidity in summertime conditions and 20–23.6°C at 30% relative humidity in wintertime conditions. Indoor humidities below 20% are considered unacceptable because of the effect of dry air on both the mucous membranes and the skin. Cleaning and maintenance of internal ducts and filters, particularly in the setting of increased humidity, is crucial, and it is often done inadequately. Notably, most filters visible to the office workers are present for the protection of engineering structures and not for the health of the employee; they are not necessarily the filters to be concerned about. Dirty, wet, blocked systems can be the source of organisms and the reason for inadequate provision of fresh air.

Other Factors

In buildings where temperature and humidity, provision of fresh air, and distribution of fresh air are considered adequate, and where no specific contaminants are identified, building-related complaints of fatigue and malaise may, in part, relate to poor lighting and the physical and psychological effects of working in windowless environments. Psychological factors such as job dissatisfaction have been associated with physical symptoms. "Hysteria," or bias as a source of building-related illness, is a diagnosis of exclusion after what can be a long and potentially expensive investigation.

Evaluation of Building-Related Complaints

The pulmonologist/epidemiologist who wants to evaluate the relationship between conditions in a building and the health of employees as a group and as individuals needs to work hand in hand with engineers and industrial hygienists. Engineers and industrial hygienists are needed to evaluate the following:

1. Is the building "tight"? That is to say, what is the air turnover, distribution of supply of fresh air, temperature, and humidity in the work spaces in the building?
2. Are there any "significant" levels of indoor air pollutants in the building, for example, volatile organic substances from new carpets or photocopying machines, carbon monoxide from entrained exhaust produced by standing vehicles, mold from damp carpets, glass fiber or asbestos insulation?

More work is necessary to standardize questionnaires regarding building-related complaints. The pulmonologist/epidemiologist evaluates the following conditions:

1. What is the overall prevalence of irritant symptoms and symptoms of change in mental status in the building? Ideally, symptom rates are evaluated before and after changes are made to the building.
2. Are the symptoms clustered in any particular work space?
3. What is the significance of the measured levels of pollutants, and the markers of air turnover, temperature, and humidity for acute and chronic symptoms:
   a. In the entire group of building occupants?
   b. In individuals why may be "sensitive"?

If a building with a cluster of building-related illnesses is demonstrated to have poor supply of fresh air, the solution to the problem may be to provide more fresh air and adequate distribution of that air. This can be a difficult and expensive engineering task, involving extensive building maintenance as well as redesign of the building ventilation system.

Continued illness after improvement of the ventilation system may mean:

1. Continued inadequate ventilation;
2. Continued presence of a contaminant;
3. Continued disease and nonspecific reactivity following specific sensitization. It would be very unusual for this to happen in more than a few "sensitive" individuals.
4. Continued psychological concern and reporting bias.

The individual clinician is often limited to investigating whether an individual's illness symptoms are caused by building-related exposures. Decision trees for the clinician are available (4). The individual clinician can take several measures:

1. Take a careful, detailed work exposure history. Request data regarding products and materials with which the patient works. In many states Right to Know laws require that this information be provided.

2. Document disease. If the patient has respiratory complaints, provide the patient with a peak flow meter and diary to assess the level, diurnal pattern and variability of peak flow at work and on days off. For workers on the day shift, peak flows in the morning are usually lower than peak flows in the afternoon; an afternoon drop in peak flow may provide evidence of a relationship between work- or building-related exposure and lower pulmonary function.
3. Document change in disease with change in environment. If possible, try not to change medication at the same time as a change in work environment is being implemented; this may make it difficult to assess whether a change in the environment is related to improvement in symptoms. It may be appropriate to recommend that the patient stay away from the building; changes in symptoms and pulmonary function before and after leaving the building should be documented.
4. Perform specific diagnostic procedures when they are available and appropriate (eg, skin or RAST testing for specific allergens; transbronchial biopsy for suspected hypersensitivity pneumonitis).

The clinician faced with an individual patient who has building-related complaints may not be able to perform the function of either the epidemiologist or the industrial hygienist. In practice, issues that relate to job security, politics, finances, and the law often heavily influence the decision whether or not to pursue a full investigation of a building. Employers should be encouraged to recognize the importance of demonstrating an honest attempt to identify causes and institute corrective actions in terms of employer-employee relationships, employee productivity, and the avoidance of legal or regulatory action.

HOSPITAL-RELATED ILLNESS

Certain forms of indoor air pollution are more common in hospitals. Formaldehyde and ventilation-related transmission of infection have already been discussed. This section will briefly summarize recent literature regarding exposure of health personnel to antineoplastic drugs, anesthetic agents, and ethylene oxide. It will conclude with a review of literature regarding asthma in hospital workers.

Antineoplastic Agents

As Selevan points out in her study of occupational exposure of nurses to antineoplastic drugs, many antineoplastic agents have been reported to be carcinogenic, mutagenic, and teratogenic (100). In the past it has been common for nurses to prepare these agents in areas without hoods. Detectable amounts of these agents have been measured when hoods have not been used. Only vertical laminar-flow hoods reduce workers' exposure sufficiently; mutagenic agents have been found in the urine of workers using horizontal laminar-flow hoods (101,102). Not all gloves are equally protective; surgical latex gloves are less permeable to many chemotoxic drugs than the polyvinyl chloride gloves previously recommended (101).

Sotaniemi et al. reported liver damage in three consecutive head nurses after years of handling cytostatic drugs. In the first case, the 34-year-old nurse had prepared cytostatic solutions for infusion with bare hands and without a face mask in an open place (103). In her case-control study of nurses in 17 Finnish hospitals, Selevan found an association between fetal loss and occupational exposure to antineoplastic drugs during the first trimester of pregnancy (odds ratio = 2:30 [95% confidence interval 1.20 to 4.39]) (99). Although the study has been criticized for possible reporting bias in the use of self-reported fetal loss, the Selevan study does lend support to the argument that caution should be exercised in the handling of antineoplastic drugs. Work practice guidelines available for personnel dealing with cytotoxic agents include recommendations that antineoplastic agents be mixed in environmentally controlled settings by trained pharmacists. Strict protocols are also established for the administration and disposal of antineoplastic drugs, and for dealing with spills.

Anesthetic Agents

At anesthetic doses, many anesthetic agents have been demonstrated to have toxic effects on organ systems in sensitive individuals. At anesthetic doses most agents have been found to have adverse effects on reproductive events in rodents. Less well understood are the effects on hospital personnel of anesthetic gases at subanesthetic doses. In a review of the literature (104), it was concluded that there is insufficient evidence to conclude that occupational exposure to anesthetic agents causes increased rates of fetal loss or congenital anomalies. However, Rowland et al. recently reported reduced fertility among women employed as dental assistants exposed to high levels of nitrous oxide (105). Data regarding the effects of large exposures to anesthetic agents suggest the importance of maintaining an adequate anesthetic scavenging system. Studies from the 1970s have reported levels of nitrous oxide ranging from 139 to 7000 ppm, and halothane levels of 10 to 85 ppm in the breathing zone of anesthesiologists. Currently United States hospital regulations limit contamination levels in the operating room to 25 ppm of nitrous oxide and 1 to 2 ppm of halogenated agents. This is accomplished by "removal of the waste gases with a gas-scavenging trap over the escape valve of the tank and a length of tubing to provide a conduit to the outside" (103).

Ethylene Oxide

Ethylene oxide is widely used for sterilization of medical supplies and equipment within hospitals. For certain types of sterilization safer alternatives have not been identified.

In industrial accidents, short exposure to high concentrations of this agent has led to nausea, headache, respiratory irritation, bronchitis, and pulmonary edema. Direct contact with small amounts of liquid ethylene oxide may cause eye and skin irritation (106). Delayed sensitization has been reported from residues of the agent in renal dialysis equipment. With chronic exposure in hospital workers, neurological deficits have been reported, including deficits in smell, sural nerve velocity, cognition, memory, and coordination (107).

In animal models, ethylene oxide is mutagenic (108). A case-control study by Ehrenbery suggested that chronic exposure led to lymphocytosis and anemia (107). In 1986 8 cases of leukemia among 733 ethylene oxide exposed workers were reported compared with an expected 0.8 cases, and 6 cases of stomach cancer compared with 0.65 cases expected (107).

NIOSH has concluded that ethylene oxide should be treated as a cancer and reproductive hazard. In the 1977 assessment of aeration procedures in hospitals the National Institute for Occupational Safety and Health, airborne ethylene oxide concentrations of 300 to 500 ppm were noted behind some aeration cabinets (105). The survey noted a number of conditions resulting in "unnecessary, preventable exposure of hospital staff" to ethylene oxide. Federal Regulations from 1984 require an 8-hour, time-weighted, average permissible exposure limit of no more than 1 ppm (108). Short-term exposure over a 15-minute period may not exceed 10 ppm. Frequent health monitoring is required if employees are exposed above the action level of 0.5 ppm. Recommendations for modifications of workplace design and practice in hospitals should be detailed and specific.

Asthma in Hospitals

In a cross-sectional questionnaire study comparing respiratory therapists with controls, it was found that respiratory therapists in Rhode Island reported higher asthma rates (odds ratio = 3:2; 95% confidence interval, 1.9 to 5.5) (109). A similar excess of asthma was found in Massachusetts respiratory therapists (110). The mechanism for this excess of asthma has not been delineated. In the past, asthma in hospital personnel has been associated with exposure to hexachlorophene, formaldehyde, psyllium laxatives, and laboratory animals (109). Sensitization can also occur during the preparation of pharmaceutical products, particularly antibiotics.

Because of recent episodes of fatal and life-threatening anaphylaxis, recent attention has been focused on sensitization and allergy to latex, dust, and powder in the gloves of health care workers (75). A study of 28 medical center employees diagnosed with rhinitis or asthma was shown to be caused by exposure to dust from latex gloves (110). While the allergy is believed to be IgE mediated, the principal allergenic protein has not been identified.

SUMMARY

This article summarizes the health effects of indoor air pollutants and the modalities available to control them. The pollutants discussed include active and passive exposure to tobacco smoke, combustion products of carbon monoxide, nitrogen dioxide, products of biofuels including wood and coal, biologic agents leading to immune responses, such as house dust mites, cockroaches, fungi, animal dander and urine, biologic agents associated with infection such as *Legionella* and tuberculosis, formaldehyde, and volatile organic compounds. An approach to assessing building-related illness and "tight-building" syndrome is presented. Finally, the article reviews recent data on hospital-related asthma and exposures to potential respiratory hazards such as antineoplastic agents, anesthetic gases, and ethylene oxide.

Acknowledgments

This article has been reprinted from D. R. Gold, *Clinics in Chest Medicine* **13**(2), 215–229 (June 1992) Courtesy of W.B. Saunders Company.

BIBLIOGRAPHY

1. A. Szalai, ed., *The Use of Time: Daily Activities of Urban and Surburban Populations in Twelve Countries*, Mouton, The Hague, 1972.
2. J. M. Samet, M. C. Marbury, and J. D. Spengler, *Am. Rev. Respir. Dis.* **136**(6), 1486–1508 (1987).
3. J. M. Samet, M. C. Marbury, and J. D. Spengler, Part II, *Am. Rev. Respir. Dis.* **137**(1), 221–242 (1988).
4. J. E. Cone and M. J. Hodgson, ed., "Problem buildings: Building-associated Illness and the Sick Building Syndrome," *Occupational Medicine: State of the Art Reviews* **4**(4) (Oct.–Dec., 1989).
5. J. M. Samet and J. D. Spengler, ed., *Indoor Air Pollution. A Health Perspective*, The Johns Hopkins Press, Baltimore, 1991.
6. "Indoor Air '90." *Proc. of the 5th International Conference on Indoor Air Quality and Climate*, Vol. 1–4, Ottawa 1990.
7. J. Samet (chairman), *Am. Rev. Respir. Dis.* **142**(4), 915–939 (1990).
8. National Research Council (NRC), *Environmental Tobacco Smoke: Measuring Exposures and Assessing Health Effects*, National Academy Press, Washington, D.C., 1986.
9. G. A. Colditz, R. Bonita, M. J. Stampfer, and co-workers, *NEJM* **318**(15), 937–941 (1988).
10. U.S. Department of Health and Human Services. Public Health Service, Office on Smoking and Health, *Smoking and Health: A Report of the Surgeon General.* U.S. Department of Health and Human Services. U.S. Government Printing Office, Washington, D.C., 1979. DHHS 79-50066.
11. W. C. Willett, A. Green, M. J. Stampfer, and co-workers, *N. Eng. J. Med.* **317**(21), 1303–1309 (1987).
12. S. T. Weiss and D. Sparrow, *Airway Responsiveness and Atopy in the Development of Chronic Lung Disease*, Raven Press, New York, 1989.
13. U.S. Department of Health and Human Services. Public Health Service, Office on Smoking and Health. *Health Benefits of Smoking Cessation: A Report of the Surgeon General*, U.S. Department of Health and Human Services. U.S. Government Printing Office, Washington, D.C., 1990.
14. U.S. Department of Health and Human Services. Public Health Service, Office on Smoking and Health, *Reducing the Health Consequences of Smoking: Twenty-five Years of Progress. A Report of the Surgeon General* U.S. Department of Health and Human Services. U.S. Government Printing Office, Washington, D.C. 1989, DHHS (CDC) 89-8411.
15. E. B. Fisher, D. Haire-Joshu, G. D. Morgan, and co-workers, *Am. Rev. Respir. Dis.* **142**(3), 702–720 (1990).
16. R. A. Etzel, "A Review of the Use of Saliva Cotinine as a Marker of Tobacco Smoke Exposure," *Prev. Med.* **19**(2), 190–197 (1990).
17. United States Environmental Protection Agency, *Respiratory Health Effects of Passive Smoking: Lung Cancer and Other Disorders*, United States Environmental Protection

Agency. Office of Research and Development, Washington, D.C., 1992.

18. S. M. Somerville, R. J. Rona, and S. Chinn, *J. Epidemiol. Community Hlth.* **42**(2), 105–110 (1988).
19. D. R. Neuspiel, D. Rush, N. R. Butler, and co-workers, "Parental Smoking and Post-infancy Wheezing in Children: A Prospective Cohort Study," *Am. J. Public Health* **79**(2), 168–171 (1989).
20. S. M. Somerville, R. J. Rona, and S. Chinn, "Passive Smoking and Respiratory Conditions in Primary School Children," *J. Epidemiol. Community Hlth.* **42**(2), 105–110 (1988).
21. B. D. Reed and L. J. Lutz, *Fam. Med.* **20**(6), 426–430 (1988).
22. Y. Chen, W-X. Li, S. Yu, and W. Quian, *Inter. J. Epidemiol.* **17**(2), 348–355 (1988).
23. F. Kauffmann, I. B. Tager, A. Munoz, and co-workers, *Am. J. Epidemiol.* **129**(6), 1289–1299 (1989).
24. M. D. Lebowitz, C. J. Holberg, R. J. Knudson, and co-workers, *Am. Rev. Respir. Dis.* **136**(1), 69–75 (1987).
25. J. P. Hanrahan, I. B. Tager, M. R. Segal, T. D. Tosteson, R. G. Castile, H. Van Vunakis, S. T. Weiss, and F. E. Speizer, *Am. Rev. Respir. Dis.* **145**(5), 1129–1135 (1992).
26. D. T. Janerich, W. D. Thompson, L. R. Varela, and co-workers, *N. Eng. J. Med.* **323**(10), 632–636 (1990).
27. S. Akiba, H. Kato, W. J. Blot, and co-workers, *Cancer Res.* **46**(9), 4804–4807 (1986).
28. T. Hiriyama, *Prev. Med.* **13**(6), 680–690 (1984).
29. R. C. Brownson, M. C. R. Alavanja, E. T. Hock, and T. S. Loy, *Am. J. Public Health* **82**(11), 1525–1530 (1992).
30. "Household Air Cleaners," *Consumer Reports*, Consumers Union, Yonkers, NY., 1992.
31. R. R. Beard, "Inorganic compounds of oxygen nitrogen and carbon." In: Patty's *Industrial Hygiene and Toxicology, 2C*, John Wiley & Sons., Inc., New York, 1982.
32. D. E. Bunnell and S. M. Horvath, *Aviat. Space Environ. Med.* **59**(12), 1133–1138 (1988).
33. D. S. Sheps, M. C. Herbst, A. L. Hinderliter, and co-workers, *Ann. Intern. Med.* **113**(5), 343–351 (1990).
34. S. M. Walden and S. O. Gottlieb, *Ann. Intern. Med.* **113**(5), 337–338 (1990).
35. E. N. Allred, E. R. Bleecher, B. R. Chaitman, and co-workers, *N. Engl. J. Med.* **321**(21), 1426–1432 (1989).
36. M. D. Baker, F. M. Henretig, and S. Ludwig, *J. Pediatrics* **113**(3), 501–504 (1988).
37. R. P. Smith, "Toxic Responses of the Blood," in Casarett and Doull's *Toxicology*, Macmillan Publishing Company, New York, 1986, p. 231.
38. G. Freeman, S. C. Crane, N. J. Furiosi, and co-workers, *Am. Rev. Respir. Dis.* **106**(4), 563–579 (1972).
39. L. M. Neas, D. W. Dockery, J. H. Ware, J. D. Spengler, F. E. Speizer, and B. G. Ferris, Jr., *Am. J. Epidemiol.* **134**(2), 204–219 (1991).
40. D. W. Dockery, J. D. Spengler, L. M. Neas, and co-workers "An Epidemiologic Study of Respiratory Health Status and Indicators of Indoor Air Pollution from Combustion Sources. In Combustion Processes and the Quality of the Indoor Environment," *Proceedings of the AWMA Specialty Conference*, September 1988. Pittsburgh, Pa., Air and Waste Management Association, 1989, p. 262 (AWMA specialty conference no. TR-15).
41. B. Brunekreef, P. Fischer, D. Houthuijs, et al., "Health effects of indoor NO_2 pollution," In: B. Selfer, H. Esdorn, M. Rischer, and co-eds., *Indoor Air '87: Proceedings of the 4th International Conference on Indoor Air Quality and Climate; Vol. 1, volatile organic compounds, combustion gases, Particles and Fibres, Microbiological Agents*, Berlin, Federal Republic of Germany, Institute for Water, Soil and Air Hygiene, 1987, p. 304.
42. J. M. Samet, W. E. Lambert, B. J. Skipper, A. H. Cushing, W. C. Hunt, S. A. Young, L. C. McLaren, M. Schwab, and J. D. Spengler, "Nitrogen Dioxide and Respiratory Illness in Children. Part I: Health Outcomes." In: *Nitrogen Dioxide and Respiratory Illness in Children. Health Effects Institute.* Capital City Press, Montpelier, VT, 1993.
43. J. Schwartz and S. Zeger, *Am. Rev. Respir. Dis.* **141**(1), 62–67 (1990).
44. J. M. Samet and M. J. Utell, *Toxicol. Indust. Hlth.* **6**(2), 247–262 (1990).
45. I. F. Goldstein, K. Lieber, L. R. Andrews, G. Koutrakis, E. Kazembe, P. Huang, and C. Hayes, *Arch. Environ. Health* **43**(2), 138–142 (1988).
46. M. R. Pandey, K. R. Smith, J. S. M. Boleij, and co-workers, *Lancet* 427–29, February 25, 1989.
47. Z-Y. Xu, W. J. Blot, H-P. Xiao, and co-workers, *J. Natl. Cancer Inst.* **81**(23), 1800–1806 (1989).
48. A. H. Wu-Williams, X. D. Dai, W. Blot, Z. Y. Xu, X. W. Sun, H. P. Xiao, B. J. Stone, S. F. Yu, Y. P. Feng, A. G. Ershow, J. Sun, J. F. Faumeni, and B. E. Henderson, *Br. J. Cancer* **62**(6), 982–987 (1990).
49. M. R. Pandey, R. P. Neupane, and A. G. Gautam, "Domestic Smoke Pollution and Acute Respiratory Infection in Nepal." In: B. Sefert, H. Esdorn, M. Fischer, et al., eds., *Fourth International Conference on Indoor Air Quality and Climate*, Vol. 4. Berlin, Institute for Water, Soil, and Air Hygiene. 1987, p. 25.
50. H. Campbell, J. R. M. Armstrong, and P. Byass, *Lancet* 1012 (May 6, 1989).
51. N. M. H. Graham, *Epidemiol. Rev.* **12,** 149–178 (1990).
52. J. Sandoval, J. Salas, M. L. Martinez-Guerra, A. Gomez, C. Martinez, A. Portales, A. Palomar, M. Villegas, and R. Barrios, *Chest* **103**(1), 12–20 (1993).
53. B. H. Chen, C. J. Hong, M. R. Pandey, and K. R. Smith, *World Health Statistics Quart.* **43**(3), 127–138 (1990).
54. R. E. Honicky, J. S. Osborne, and C. A. Akpom, *Pediatrics* **75**(3), 587–593 (1985).
55. W. E. Pierson, J. Q. Koenig, and E. J. Bardana, *West. J. Med.* **151**(3), 339–342 (1989).
56. H. Burge, *J. Allergy Clin. Immunol.* **86**(5), 687–701 (1990).
57. J. Schwartz, D. Gold, D. W. Dockery, and co-workers, *Am. Rev. Respir. Dis.* **142**(3), 555–562 (1990).
58. D. R. Gold, A. Rotnitzky, A. I. Damokosh, J. H. Ware, F. E. Speizer, B. G. Ferris, Jr., and D. W. Dockery, *Am. Rev. Respir. Dis.* **148,** 10–18 (1993).
59. L. E. Gelber, L. H. Seltzer, J. K. Bouzoukis, S. M. Pollart, M. D. Chapman, and T. A. E. Platts-Mills, *Am. Rev. Respir. Dis.* **147**(3), 557–578 (1993).
60. T. A. E. Platts-Mills, S. Pollart, G. Fiocco, and co-workers, *J. Allergy Clin. Immunol.* **79,** 184 (abstract) (1987).
61. T. A. E. Platts-Mills, *J. Allergy Clin. Immunol.* **83**(2 Pt. 1), 416–427 (1989).
62. W. R. Solomon and K. P. Mathews, "Aerobiology and Inhalant Allergens," in *Allergy Principles and Practice.* Vol. I., CV Mosby Company, St. Louis, 1988, p. 312.
63. R. Sporik, S. T. Holgate, T. A. E. Platts-Mills, and co-workers, *N. Engl. J. Med.* **323**(8), 502–507 (1990).
64. S. Lau, G. Falkenhorst, A. Weber, I. Wethmann, and co-workers, *J. Allergy Clin. Immunol.* **84**(5 Pt. 1), 718–725 (1989).

65. M. Wickman, S. L. Nordvall, G. Pershagen, and co-workers, *J. Allergy Clin. Immunol.* **88**(1), 89–95 (1991).
66. S. M. Pollard, M. S. Chapman, G. P. Fiocco, and co-workers, *J. Allergy Clin. Immunol.* **83**(5), 875–882 (1989).
67. S. B. Lehrer, L. Aukrust, and J. E. Salvaggio, *Clin. Chest Med.* **4**(1), 23–41 (1983).
68. M. Lopez and J. E. Salvaggio, *Clin. Rev. Allergy* **3**(2), 183–196 (1985).
69. J. Salvaggio and L. Aukrust, *J. Allergy Clin. Immunol.* **68,** 327–346 (1981).
70. B. Brunekreef, D. W. Dockery, F. E. Speizer, J. H. Ware, J. D. Spengler, and B. G. Ferris, *Am. Rev. Respir. Dis.* **140,** 1363–1367 (1989).
71. R. E. Dales, R. Burnett, and H. Zwanenburg, *Am. Rev. Respir. Dis.* **143,** 505–509 (1991).
72. P. J. Gergen, P. C. Turkeltaub, and M. G. Kova, *J. Allergy Clin. Immunol.* **80,** 669–679 (1987).
73. H. Burge, *J. Allergy Clin. Immunol.* **86**(5), 687–701 (1990).
74. H. Burge, "Indoor Sources for Airborne Microbes," in R. B. Gammage and S. V. Kaye, *Indoor Air and Human Health*, Lewis, Chelsea, Michigan, 1984, p. 139.
75. A. M. Pope, R. Patterson, and H. Burge, eds., *Indoor Allergens*, National Academy Press, Washington, D.C., 1993.
76. P. P. Kozak and J. Gallup, "Endogenous Mold Exposure: Environmental Risk to Atopic and Nonatopic Patients," In R. B. Gammage and S. V. Kaye, *Indoor Air and Human Health*, Lewis, Chelsea, Michigan, 1984, p. 149.
77. H. S. Nelson, S. R. Hirsch, J. L. Ohman, T. E. Platts-Mills, C. E. Reed, and W. R. Solomon, *J. Allergy Clin. Immunol.* **82,** 661–669 (1988).
78. M. Lopez and J. E. Salvaggio, "Hypersensitivity Pneumonitis," in Murray/Nadal, *Textbook of Respiratory Medicine*, W.B. Saunders, Philadelphia, Pa., 1988, p. 1606.
79. M. J. Hodgson, P. R. Morey, J. S. Simon, et al., *Am. J. Epidemiol.* **125,** 631–637 (1987).
80. J. P. Hanrahan, D. L. Morse, V. B. Scharf, et al., *Am. J. Epidemiol.* **125**(4), 639–649 (1987).
81. M. C. O'Mahony, R. E. Stanwell-Smith, H. E. Tillett, and co-workers, *Epidemiol. infect.* **104,** 361–380 (1990).
82. I. D. Farrell, J. E. Barker, E. P. Miles, et al., *Epidemiol. Infect.* **104,** 381–387 (1990).
83. M. J. Finnegan and C. A. C. Pickering, *Clin. Allergy* **16,** 389–405 (1986).
84. V. N. Honk, D. C. Kent, J. H. Baker, and co-workers, *Arch. Environ. Hlth.* **16,** 26–35 (1968).
85. A. V. Arundel and E. M. Sterling, "Review of the Effect of Relative Humidity on the Transmission of Infectious Diseases," For presentation at the 80th Annual Meeting of APCA. Vancouver, British Columbia, Theodur D. Sterling, Ltd., 1987.
86. D. A. Sterling, "Volatile Organic Compounds in Indoor Air: An Overview of Sources, Concentrations, and Health Effects," in R. B. Gammage and S. V. Kaye, *Indoor Air and Human Health*, Lewis, Chelsea, Michigan, 1984.
87. International Agency for Research on Cancer, "IARC Monographs on the Evaluation of Carcinogenic Risks to Humans," *Overall evaluations of Carcinogenicity: an Updating of IARC Monographs Volumes 1 to 42. Supplement 7.* Lyon, France, World Health Organization, 1987.
88. D. Luce, M. Gerin, A. Leclerc, J. Morcet, J. Brugere, and M. Goldberg, *Int. J. Cancer* **53,** 224–231 (1993).
89. A. Blair, P. A. Stewart, M. O'Berg, et al., *JNCI* **76,** 1071–1084 (1986).
90. C. R. Buncher, *J. Occup. Med.* **31**(11), 885 (1989).
91. B. A. Tichenor and M. A. Mason, *JAPCA* **38,** 264–268 (1988).
92. D. Norback, M. Torgen, and C. Edling, *Br. J. Indust. Med.* **47,** 733–741 (1990).
93. E. Sterling and T. Sterling, *Canadian J. Public Hlth.* **74,** 385–392 (1983).
94. H. S. Koren and R. B. Devlin, *Ann. N.Y. Acad. Sci.* **641,** 215–224 (1992).
95. H. S. Koren, D. E. Graham, and R. B. Devlin, *Arch. of Environ. Health* **47,** 39–44 (1992).
96. H. Harving, R. Dahl, and L. Molhave, *Am. Rev. Respir. Dis.* **143,** 751–754 (1991).
97. M. J. Finnegan and C. A. C. Pickering, *Clin. Allergy* **16,** 389–405 (1986).
98. R. Menzies, R. Tamblyn, J. Farant, J. I. Hanley, F. Nunes, and R. Tamblyn, *N. Engl. J. Med.* **328,** 821–827 (1993).
99. K. Kreiss, *N. Engl. J. Med.* **328,** 877–878 (1993).
100. S. G. Selevan, M-L. Lindbohm, R. W. Hornung, and co-workers, *N. Engl. J. Med.* **313,** 1173–1178 (1985).
101. Office of Occupational Medicine, "Work Practice Guidelines for Personnel Dealing with Cytotoxic (Antineoplastic) Drugs," In: *Occupational Safety and Health Administration Instruction Publication 8-1.1*, Office of Occupational Medicine, January 29, 1986.
102. R. W. Anderson, W. H. Puckett, W. J. Dana, and co-workers, *Am. Soc. Hosp. Pharm.* **39,** 1881–1887 (1982).
103. E. A. Sotaniemi, S. Sutinen, A. J. Arranto, and co-workers, *Acto. Med. Scand.* **214,** 181–189 (1983).
104. T. N. Tannenbaum and R. J. Goldberg, *J. Occup. Med.* **27,** 659–668 (1985).
105. A. S. Rowland, D. D. Baird, C. R. Weinberg, D. L. Shore, C. M. Shy, and A. J. Wilcox, *N. Engl. J. Med.* **327,** 993–997 (1992).
106. A. R. Glaser, *Special Occupational Hazard Review with Control Recommendations for the Use of Ethylene Oxide as a Sterilant in Medical Facilities*, U.S. Department of Health, Education, and Welfare, Center for Disease Control, National Institute for Occupational Safety and Health, Division of Criteria Documentation and Standards Development, Priorities and Research Analysis Branch, Rockville, Maryland, 1977.
107. W. J. Estrin, S. A. Cavalieri, P. Wald, and co-workers, *Arch. Neurol.* **44**(12), 1283–1286 (1987).
108. V. L. Dellarco, W. M. Generoso, G. A. Sega, and co-workers, *Environ. Mol. Mutagen* **16**(2), 85–103 (1990).
109. D. G. Kern and H. Frumkin, *Ann. Intern. Med.* **110**(10), 767–773 (1989).
110. D. C. Christiani and D. G. Kern, *Am. Rev. Respir. Dis.* **143,** A441 (1991).

AIR QUALITY MODELING

Armistead G. Russell
Carnegie Mellon University
Pittsburgh, Pennsylvania
Jana B. Milford
University of Colorado
Boulder, Colorado

To better understand how particular emissions are affecting the atmosphere and thus to devise strategies to limit harmful outcomes, it is necessary to be able to describe quantitatively the source–impact relationship. This is the role of atmospheric models. Such mathematical air quality models have been developed to provide a powerful framework for understanding the dynamics of pollutants in the atmosphere and for assessing the impact emission sources have on pollutant concentrations.

Two commonly used classes of models have been developed. Empirical models provide an understanding of source impacts by statistically analyzing historical air quality data. Diagnostic models provide a comprehensive description of the detailed physics and chemistry of compounds in the atmosphere, following the evolution of pollutant emissions to their ultimate fate. At the heart of the model is a system of mathematical routines that integrate the effects of individual processes. The complexity and computational intensity of modern models have necessitated the development of algorithms for fast and accurate mathematical solution techniques. Air quality models are being applied to solving such problems as urban smog, acid deposition, regional ozone, haze in scenic regions, and the destruction of the protective stratospheric ozone layer.

Mathematical models have grown increasingly detailed in descriptions of air pollutant dynamics and are thus key tools for gaining scientific understanding of atmospheric processes. These models also are the most practical and scientifically defensible means of relating pollutant emissions to air quality. They are widely used in regulatory planning and analysis, as indicated in Figure 1. A wide variety of models is used to address problems ranging from indoor air pollution to regional acid deposition and global climate change. The models used are remarkably similar.

Development of a mathematical air quality model proceeds through four stages. In the conceptual stage, a working set of relationships approximating the physical system is derived. Next, these relationships are expressed as mathematical equations, giving a formal description of the idealized system. The third step is the computational implementation of the model, including development of the algorithms and computer code needed to solve the equations given various inputs. The final step is the application of the model, including acquisition and processing of the necessary input data, and evaluation of model results. See also Air pollution; Carbon balance modeling; Life cycle analysis; Risk assessment.

AIR QUALITY MODELS

Models used in air pollution analysis fall into two classes: empirical–statistical and deterministic, as shown in Figure 2. In the former, the model statistically relates observed air quality data to the accompanying emission patterns, and chemistry and meteorology are included only implicitly. In the latter, analytical or numerical expressions describe the complex transport and chemical processes that affect air pollutants. Pollutant concentrations are determined as explicit functions of meteorology, topography, chemical transformation, surface deposition, and source characteristics (1). A listing of many of the air quality models, including status, applications, and the model formulations, is available (2). Each formulation involves approximations and has certain strengths and limitations.

Empirical–Statistical Models

Empirical–statistical models are based on establishing a relationship between historically observed air quality and the corresponding emissions and other explanatory factors. The linear rollback model is simple to use and re-

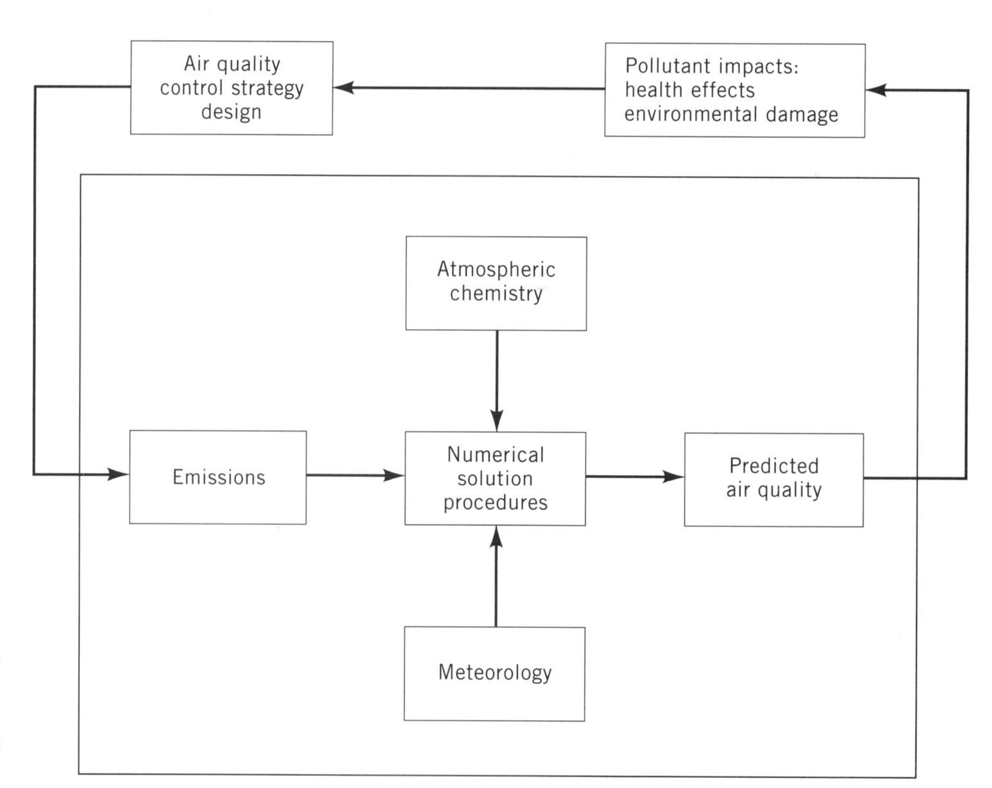

Figure 1. Schematic of the role of an air quality model, in the air quality control planning process. Some studies, eg, the prediction of indoor air pollutant concentrations, require more than one model.

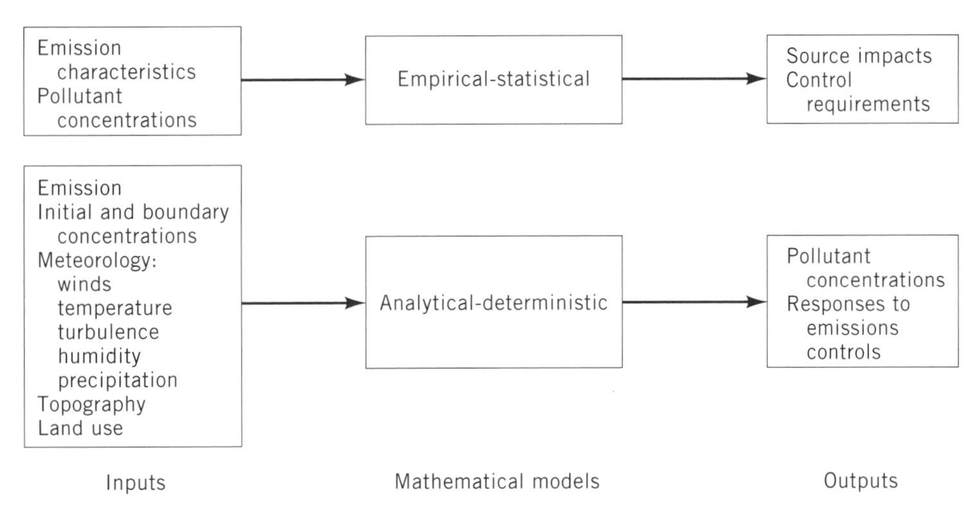

Figure 2. Inputs, types of models, and outputs used in air quality modeling studies.

quires few data, and for these reasons has been widely applied (3,4). Linear rollback models assume that the highest measured pollutant concentration is proportional to the basinwide emission rate, plus the background value, ie,

$$c_{\max} = aE + c_b \tag{1}$$

where $c_{\max}$ is the maximum measured pollutant concentration, E is the emission rate, c_b is the background concentration resulting from sources outside the modeling region, and a is a constant of proportionality. The constant implicitly accounts for the dispersion, transport, deposition, and chemical reactions of the pollutant. The allowable emission rate E_a necessary to reach a desired ambient air quality goal c_d can be calculated from

$$\frac{E_a}{E_o} = \frac{c_d - c_b}{c_{\max} - c_b} \tag{2}$$

where E_o is the emission rate that prevailed at the time that $c_{\max}$ was observed. Presumably, pollutant concentrations at other times decrease toward background levels as emissions are reduced. The linear rollback model is a simplified approach, and its application is limited. Nonlinear processes, such as chemical reactions and spatial or temporal changes in emission patterns, are not accounted for in the rollback model.

Time Series and Regression Models

As in many other fields, relationships between pollutant levels and explanatory factors such as meteorological variables have been analyzed using regression models. Day-to-day variations in peak ozone concentrations, eg, can be closely reproduced by simple linear models that use temperature and relative humidity as independent variables (5). Time series models for pollutants also have been developed, accounting for temporal correlations between pollutant levels, with or without incorporation of "exogenous" explanatory variables such as temperature or emissions rates (6). Over short time horizons, time series and regression models have been used successfully to forecast pollutant levels (7,8). Such models have been used, eg, to predict whether field study crews should be mobilized in expectation of smog episodes. A clear limitation of time series and regression models is that in general they cannot provide information about how pollutant levels would respond to emissions controls.

Receptor-Oriented Model

A third class of empirical–statistical models is the receptor-oriented model, which has been used extensively for estimating the contributions that distinguishable sources such as automobiles or municipal incinerators make to particulate matter concentrations (9–18). Attempts have also been made to track nonreacting gases (19), and under special conditions, reactive organic compounds (20–22), to their sources using receptor modeling methods. Receptor models compare the measured chemical composition of particulate matter at a receptor site with the chemical composition of emissions from the primary sources to identify the source contributions at the monitoring location. There are three principal categories of receptor models: chemical mass balance, multivariate, and microscopic. Hybrid analytical and receptor (or combined source–receptor) models have been used (23), but further investigation into their capabilities is required.

Deterministic Models

Deterministic air quality models describe in a fundamental manner the individual processes that affect the evolution of pollutant concentrations. These models are based on solving the atmospheric diffusion–reaction equation, which is in essence the conservation-of-mass principle for each pollutant species (24):

$$\frac{\delta c_i}{\delta t} + \overline{U} \cdot \nabla c_i = \nabla \cdot D_i \nabla c_i + R_i(c_1, c_2, c_3, \ldots c_n) + S_i(\bar{x}, t)$$
$$i = 1, 2, 3, \ldots n \tag{3}$$

where c_i is the concentration of species i, $\overline{U}$ is the wind velocity vector, D_i is the molecular diffusivity of species i, R_i is the net production (depletion if negative) of species i

by chemical reaction, S_i is the emission rate of i, and n is the number of species. R can also be a function of the meteorological variables. Equation 3 states that the time rate of change of a pollutant depends on convective transport (term 2), diffusion (term 3), chemical reactions (term 4), and emissions (term 5), together with initial and boundary conditions. Deposition and surface level emissions enter as boundary conditions at the ground:

$$E - v_d c = -K_{zz}\frac{\delta c}{\delta z}$$

where E is the ground level emissions, v_d is the deposition velocity, and K_{zz} is the vertical diffusivity. The turbulence closure problem makes it necessary to approximate the atmospheric diffusion equation, usually by K-theory (25,26):

$$\frac{\delta c_i}{\delta t} + \langle \overline{U} \rangle \cdot \nabla \langle c_i \rangle = \nabla \cdot K \nabla \langle c_i \rangle + R_i(\langle c_1 \rangle, \langle c_2 \rangle, \ldots \langle c_n \rangle) + \langle S_i(x, t) \rangle \qquad i = 1, 2, 3, \ldots n \tag{4}$$

where the angle brackets $\langle \quad \rangle$ indicate an ensemble average, and K is the turbulent (eddy) diffusivity tensor. Known analytical solutions exist only for the simplest source distributions and chemical reaction mechanisms, $\langle S_i \rangle$ and R, in equation 4.

Examination of equation 5 shows that if there are no chemical reactions, ($R = 0$) or if R is linear in $\langle c_i \rangle$ and uncoupled, then a set of linear, uncoupled differential equations are formed for determining pollutant concentrations. This is the basis of transport models, which may be transport only or transport with linear chemistry. Transport models are suitable for studying the effects of sources of CO and primary particulates on air quality, but not for studying reactive pollutants such as O_3, NO_2, HNO_3, and secondary organic species.

Lagrangian Models. There are two distinct reference frames from which to view pollutant dynamics. The most natural is the Eulerian coordinate system, which is fixed at the earth's surface and in which a succession of different air parcels are viewed as being carried by the wind past a stationary observer. The second is the Lagrangian reference frame, which moves with the flow of air, in effect maintaining the observer in contact with the same air parcel over extended periods of time. Because pollutants are carried by the wind, it is often convenient to follow pollutant evolution in a Lagrangian reference frame, and this perspective forms the basis of Lagrangian trajectory and Lagrangian marked-particle or particle-in-cell models. In a Lagrangian marked-particle model, the center of mass of parcels of emissions are followed, traveling at the local wind velocity, while diffusion about that center of mass is simulated by an additional random translation corresponding to the atmospheric diffusion rate (27,28).

Gaussian Plume Model. One of the most basic and widely used transport models based on equation 4 is the Gaussian plume model. Gaussian plume models for continuous sources can be obtained from statistical arguments or can be derived by solving:

$$\overline{U}\frac{\delta c}{\delta x} = K_{yy}\frac{\delta^2 c}{\delta y^2} + K_{zz}\frac{\delta^2 c}{\delta z^2} \tag{5}$$

where $\overline{U}$ is the temporally and vertically averaged wind velocity; x, y, and z are the distances in the downwind, crosswind, and vertical directions, respectively; and K_{yy} and K_{zz} are the horizontal and vertical turbulent diffusivities, respectively. For a source with an effective height H, emission rate Q, and a reflecting (nonabsorbing) boundary at the ground, the solution is

$$c(x, y, z) = \frac{Q}{2\pi \overline{U}\sigma_y(x)\sigma_z(x)} \exp\left[\frac{-y^2}{2\sigma_y^2(x)}\right]\left[\exp\frac{-(z-H)^2}{2\sigma_z^2(x)} + \exp\frac{-(z+H)^2}{2\sigma_z^2(x)}\right]$$

This solution describes a plume with a Gaussian distribution of pollutant concentrations, such as that in Figure 3, where $\sigma_y(x)$ and $\sigma_z(x)$ are the standard deviations of the mean concentration in the y and z directions. The standard deviations are the directional diffusion parameters, and are assumed to be related simply to the turbulent diffusivities, K_{yy} and K_{zz}. In practice, $\sigma_y(x)$ and $\sigma_z(x)$ are functions of x, $\overline{U}$, and atmospheric stability (2,29–31).

Gaussian plume models are easy to use and require relatively few input data. Multiple sources are treated by superimposing the calculated contributions of individual sources. It is possible to include the first-order chemical decay of pollutant species within the Gaussian plume

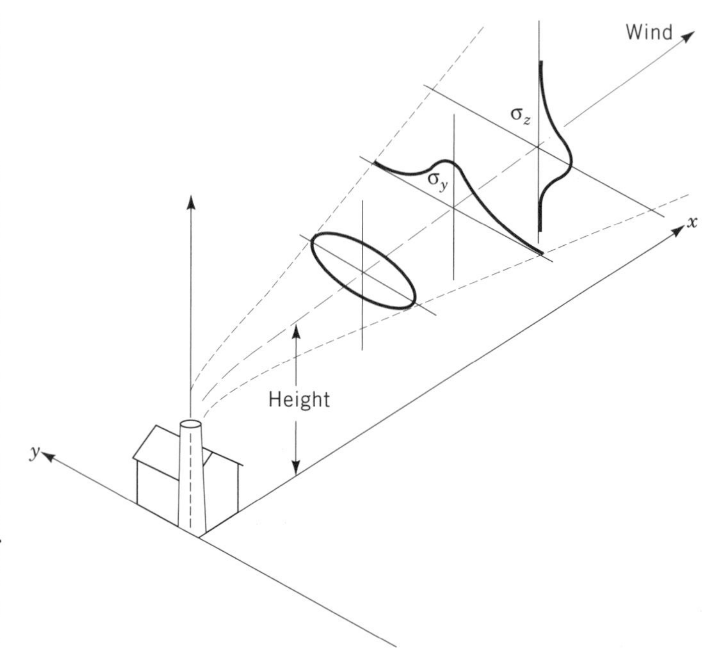

Figure 3. Diffusion of pollutants from a point source. Pollutant concentrations have separate Gaussian distributions in both the horizontal (y) and vertical (z) directions. The spread is parameterized by the standard deviations (σ), which are related to the diffusivity (K).

framework. For chemically, meteorologically, or geographically complex situations, however, the Gaussian plume model fails to provide an acceptable solution.

Eulerian Models. Of the Eulerian models, the box model is the easiest to conceptualize. The atmosphere over the modeling region is envisioned as a well-mixed box, and the evolution of pollutants in the box is calculated following conservation-of-mass principles including emissions, deposition, chemical reactions, and atmospheric mixing.

Eulerian "grid" air quality models are the most complex, but potentially the most powerful, involving the least-restrictive assumptions. They are also the most computationally intensive. Grid models solve a finite approximation to equation 4, including temporal and spatial variation of the meteorological parameters, emission sources, and surface characteristics. Grid models divide the modeling region into a large number of cells, horizontally and vertically, which interact with each other by simulating diffusion, advection, and sedimentation (for particles) of pollutant species (Fig. 4). Input data requirements for grid models are similar to those for Lagrangian trajectory models, but in addition, data on background concentrations (boundary conditions) at the edges of the grid system are required. Eulerian grid models predict pollutant concentrations throughout the entire airshed. Over successive time periods the evolution of pollutant concentrations and how they are affected by transport and chemical reaction can be tracked.

Modeling Chemically Reactive Compounds. A number of compounds are formed or destroyed in the atmosphere by a series of complex, nonlinear chemical reactions, eg, stratospheric ozone. Models that not only describe pollutant transport but also account for complex chemical transformations, $R(\bar{c},t)$ in equation 5, are necessary for these systems. Such models are also required to study the dynamics of chemically reactive primary pollutants such as nitric oxide, NO, and pollutants that are primary as well as secondary in origin, eg, nitrogen dioxide [*10102-44-0*], NO_2, and formaldehyde. Addition of the capability to describe a series of interconnected chemical reactions greatly increases requirements for computer storage as well as computing time and input data requirements. Increased computational demands arise because the evolution of the interacting species must be followed simultaneously, leading to a system of coupled, nonlinear differential equations.

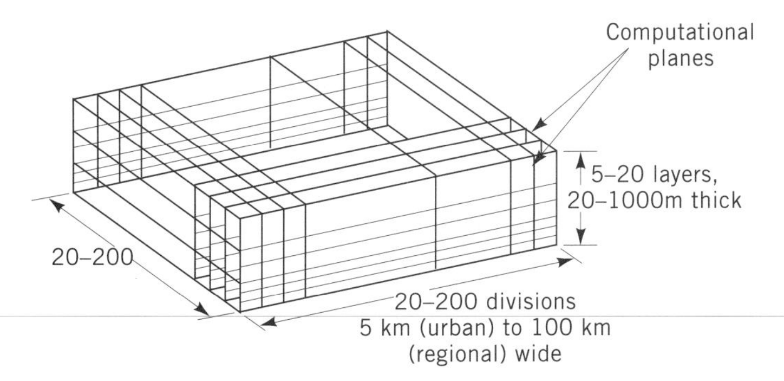

Figure 4. A typical model application would include a domain of about 60 × 100 horizontal cells, and 5 to 20 layers vertically, for a total of about 60,000 computational cells. Vertical cell spacing is usually nonuniform with layers 20- to 1000-mm thick (with the finest resolution near the ground). Horizontal resolution depends on the size of the modeling domain. Urban applications use grids of about 5 × 5 km, while regional models use grids of about 20 to 100 km per side.

Box, Lagrangian trajectory, and Eulerian grid models have all been developed to include nonlinear chemical reactions. Box models assume that the pollutants are mixed homogeneously within the modeling region, an assumption that is often inappropriate. Trajectory and grid models resolve pollutant dynamics spatially and have been used widely and with success particularly for studying photochemical smog and acid deposition problems (30,32–43).

Temporal and Spatial Resolution

The temporal and spatial resolutions of models can vary from minutes to a year and from meters to hundreds of kilometers. The minimum meaningful resolution of a model is determined by the input data resolution and the structure of the model. Statistical models generally rely on several years worth of measurements of hourly or daily pollutant concentrations. The resolution of the input data represents the minimum resolution of a statistical model. Resolution of analytical models is limited by the spatial and temporal resolution of the emissions inventory, the meteorological fields, and the grid size chosen for model implementation. For modeling urban air basins, the size of individual grid cells is on the order of a few kilometers per side, whereas for modeling street canyons, the cell size must be reduced to a few meters on each edge. At the other extreme, regional models have horizontal resolutions varying from 20 to 100 km, and global models may have a resolution as coarse as a few thousand kilometers. The temporal resolution of models ranges from about 15 min to a few hours or days.

The information desired from modeling studies often depends on processes that occur on spatial scales much smaller than the resolution of most air quality models. The modeling of NO_x air quality in street canyons involves small-scale processes of this sort. Introduction of point-source emissions into grid-based air quality models likewise involves a mismatch between the high concentrations that exist near the source versus the lower concentrations computed by a model that immediately mixes those emissions throughout a grid cell of several kilometers on each side. Because of computational time constraints, it has often been considered impractical to describe fully the processes that take place on a scale smaller than the main model grid, ie, subgrid scale. Nevertheless, values obtained from large-scale calculations should be accurate over the spatial averaging scale adopted by the model.

MODEL COMPONENTS

A model's ability to correctly predict pollutant dynamics and to apportion source contributions depends on the accuracy of the individual process descriptions and input data, and the fidelity with which the framework reflects the interactions of the processes.

Turbulent Transport and Diffusion

There are two pollutant transport terms in equation 5: an advection term, in which pollutants are carried along with the time-averaged mean wind flow, and a dispersion term, representing transport resulting from local turbulence. The averaging time that determines the mean winds is related to the spatial scale of the system being modeled. Minutes may be appropriate for urban-scale simulations, multihour averages for the regional scale, and daily to weekly averages for determining long-term concentrations of nonreactive pollutants.

Turbulent transport is determined by complex interactions between meteorological conditions and topography.

Removal Processes

Pollutant removal processes, particularly dry deposition and scavenging by rain and clouds, are a primary factor in determining the dynamics and ultimate fate of pollutants in the atmosphere.

Representation of Atmospheric Chemistry Through Chemical Mechanisms

A complete description of atmospheric chemistry within an air quality model would require tracking the kinetics of many hundreds of compounds through thousands of chemical reactions. Fortunately, in modeling the dynamics of reactive compounds such as peroxyacetyl nitrate [*2278-22-0*] (PAN), $C_2H_3NO_5$, O_3, and NO_2, it is not necessary to follow every compound. Instead, a compact representation of the atmospheric chemistry is used. Chemical mechanisms represent a compromise between an exhaustive description of the chemistry and computational tractability. The level of chemical detail is balanced against computational time, which increases as the number of species and reactions increases. Instead of the hundreds of species present in the atmosphere, chemical mechanisms include on the order of 50 species and 100 reactions.

Three different types of chemical mechanisms have evolved as attempts to simplify organic atmospheric chemistry: surrogate (44,45), lumped (46–49), and carbon bond (50–52). These mechanisms were developed primarily to study the formation of O_3 and NO_2 in photochemical smog but can be extended to compute the concentrations of other pollutants, such as those leading to acid deposition (38,40).

Surrogate mechanisms use the chemistry of one or two compounds in each class of organics to represent the chemistry of all the species in that class.

Lumped mechanisms are based on the grouping of chemical compounds into classes of similar structure and reactivity.

The carbon bond mechanisms (50–52), a variation of a lumped mechanism, splits each organic molecule into functional groups using the assumption that the reactivity of the molecule is dominated by the chemistry of each functional group.

Aerosol Dynamics

Inclusion of a description of aerosol dynamics within air quality models is of primary importance because of the health effects associated with fine particles in the atmosphere, visibility deterioration, and the acid deposition problem. Aerosol dynamics differ markedly from gaseous pollutant dynamics in that particles come in a continuous distribution of sizes and can coagulate, evaporate, grow in size by condensation, be formed by nucleation, or be deposited by sedimentation. Furthermore, the species mass concentration alone does not fully characterize the aerosol. The particle size distribution, which changes as a function of time, and size-dependent composition determine the fate of particulate air pollutants and their environmental and health effects. Particles of about 1 μm in diameter or smaller penetrate the lung most deeply and represent a substantial fraction of the total aerosol mass, as shown in Figure 5. The origin of these fine particles is difficult to identify because much of the fine particle mass is formed by gas-phase reaction and condensation in the atmosphere.

Simulation of aerosol processes within an air quality model begins with the fundamental equation of aerosol dynamics that describes aerosol transport (term 2), growth (term 3), coagulation (terms 4 and 5), and sedimentation (term 6):

$$\frac{\delta n}{\delta t} + \nabla \cdot \overline{U} n + \frac{\delta I}{\delta v} = \frac{1}{2}\int_0^v \beta(\bar{v}, v - \bar{v}) n(\bar{v}) n(v - \bar{v}) d\bar{v} - \int_0^\infty \beta(\bar{v}, v) n(\bar{v}) n(v) d\bar{v} - \nabla \cdot Cn \quad (6)$$

where n is the particle size distribution function, $\overline{U}$ is the fluid velocity, I is the droplet current that describes particle growth and nucleation resulting from gas-to-particle conversion, v is the particle volume, β is the rate of particle coagulation, and C is the sedimentation velocity. Modeling the formation and growth of aerosols is done by sectioning the size distribution n into discrete ranges. Then the size and chemical composition of an aerosol is followed as it evolves by condensation, coagulation, sedimentation, and nucleation.

Air Quality Model Inputs

Inputs to analytical air quality models can be broadly grouped as those dealing with meteorology, emissions, topography, and atmospheric concentrations. Meteorological inputs generally control the transport rate of pollutants and are used to determine reaction rates and the depositional flux of compounds. Topography influences transport and deposition. Observed compound concentrations are used to specify both initial and boundary conditions for model simulations. Especially for pollution problems involving organic compounds, emissions are a key input subject to considerable uncertainty. Although emissions from primary industrial facilities or utilities may be reasonably well known, emissions from residential or commercial facilities, mobile sources, and natural sources are often roughly estimated and difficult to verify.

The data requirements for applying models differ greatly among model types. For a Gaussian plume model, the required data could include as little as the mean wind velocity, source emission rate, atmospheric stability (and hence diffusivity), effective source height, and air quality

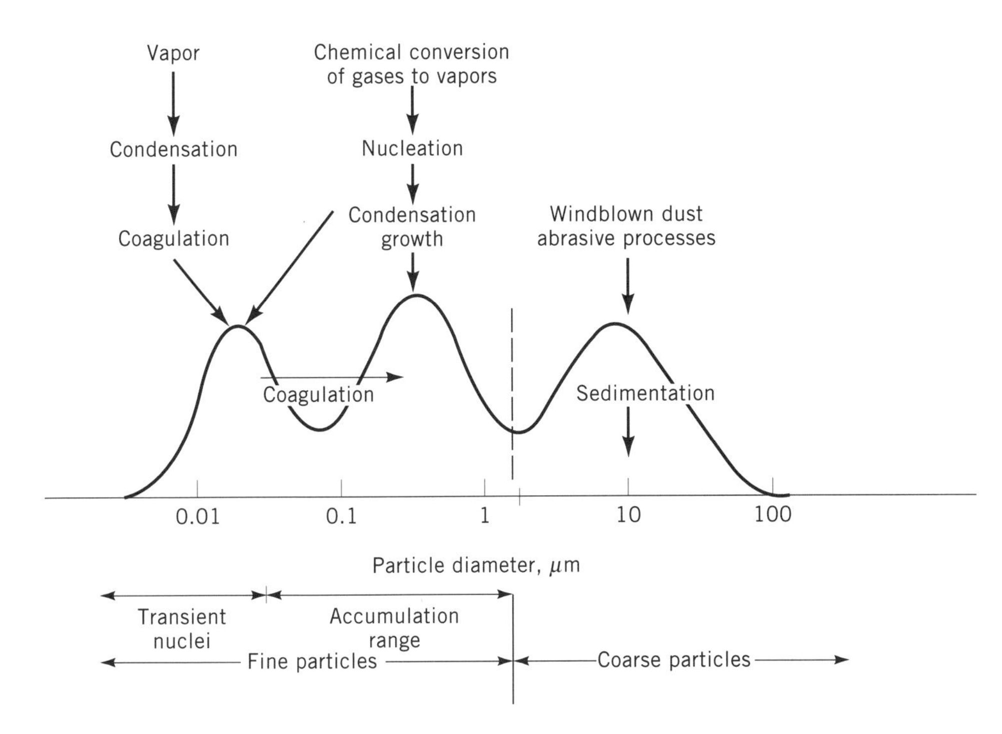

Figure 5. Size distribution of an urban aerosol showing the three modes containing much of the aerosol mass. The fine mode contains particles produced by condensation of low volatility gases. The mid-range, or accumulation mode, results from coagulation of smaller aerosols and condensation of gases on preexisting particles. Coarse particulates, the largest aerosols, are usually generated mechanically.

data for comparison with model predictions (53). At the other extreme, a large grid model that incorporates chemical kinetics requires considerably more information. Ideally a comprehensive model evaluation study would incorporate spatially and temporally resolved meteorological data; eg, winds, temperature, and solar insolation; temporally varying emissions for every species in each cell of the modeling region; topographical data, eg, land use and elevation; initial species concentrations; boundary conditions for each species; and concentration data for comparison against model predictions (42).

Mathematical and Computational Implementation

Solution of the complex systems of partial differential equations governing both the evolution of pollutant concentrations and meteorological variables, eg, winds, requires specialized mathematical techniques.

Advances in more efficient numerical algorithms as well as in computational capabilities are facilitating increased chemical and physical detail of atmospheric models and making their use and the interpretation of results more direct. One advance is the use of multiple scales in modeling urban and regional pollutant transport and chemistry simultaneously (Fig. 6). Fine scales are used over cities, and coarse scales over less populated domains. Condensed phase chemistry and dynamics are computationally intensive, though are now being included in model implementation. Visualization of air quality dynamics and synthetic construction of smog allows for quicker interpretation of emission impacts (54,55).

Model Evaluation

One question that always needs to be asked is how well an air quality model is performing, and whether the performance is adequate for the purposes of a particular application. Central to model evaluation is assessment of how closely simulated pollutant concentrations match those observed for some historical episode. As an example of guidelines for this type of performance evaluation, the California Air Resources Board (56) developed a series of tests and performance standards for evaluating the ability of photochemical models to predict ozone concentrations. Beyond model performance, however, is the question of whether the predictions are accurate for the right reason. To address this issue, separate model components such as chemical mechanisms and intermediate modeling steps such as interpolation of windfield data are tested by comparison with results from laboratory experiments or detailed field studies.

Sensitivity and uncertainty analysis are valuable tools for gaining insight into the performance of complex models such as those used for ozone and acid deposition. Analyses in which uncertain parameters or inputs are varied one-at-a-time to observe the effect on the outputs have been applied to most models, if only informally during model development or in diagnosing poor performance. Taking advantage of the simple form of the differential equations that describe reaction kinetics, comprehensive functional analysis techniques that calculate local sensitivities to many parameters simultaneously have been applied to chemical mechanisms used in air quality (57–60) models and have highlighted influential reactions for which additional research might provide substantial benefits. Methods for global uncertainty analysis such as FAST (61,62) and Monte Carlo, which are essentially schemes for sampling from probability distribution functions for uncertain model parameters, have also been applied to chemical mechanisms and simplified models (63–66). Rapid expansion in computing power has lowered one of the barriers to conducting global uncertainty analyses for complex air quality models. However, development of probability distributions for the many (and potentially correlated) parameters in these models is still a formidable task.

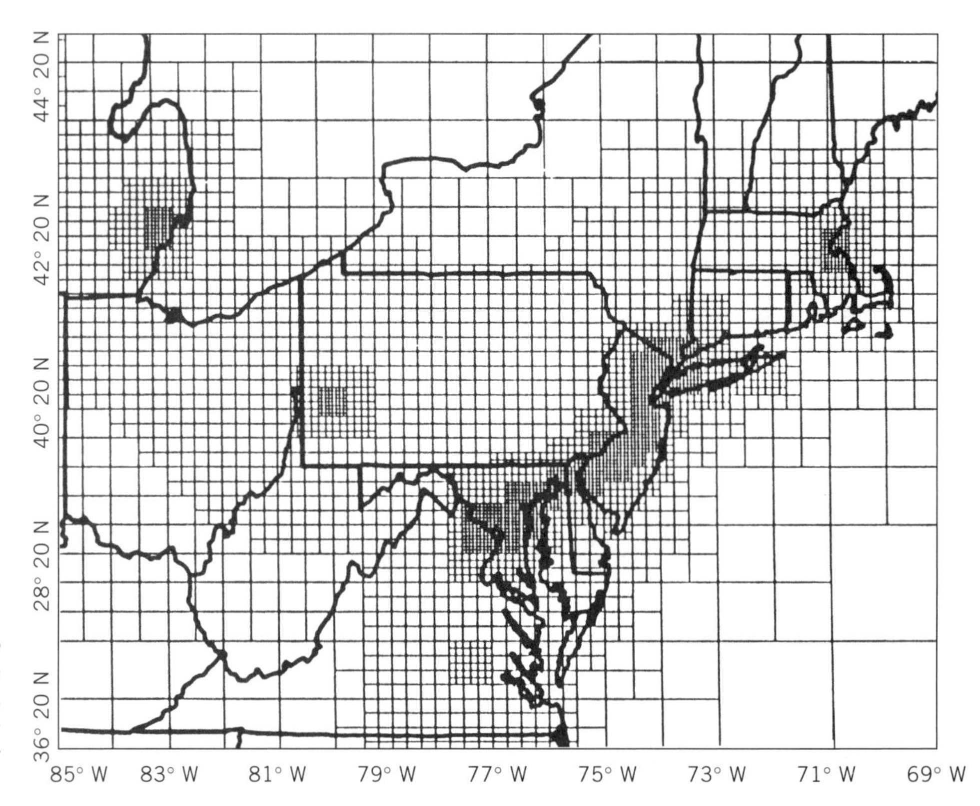

Figure 6. Multiscale grid applied to the northeastern United States. Fine grids are used in dense source regions, larger grids where less activity is present. This allows for more efficient computational implementation.

APPLICATION OF AIR QUALITY MODELS

Both receptor and analytical air quality models have proven to be powerful tools for understanding atmospheric pollutant dynamics and for determining the impact of sources on air quality.

Receptor Models

Receptor models, by their formulation, are effective in determining the contributions of various sources to particulate matter concentrations.

Receptor models are suitable for predicting the outcome of perturbations in some sources but not others. They are, however, good for determining the sources of particulate matter when an accurate emissions inventory is not available. Dispersion models, on the other hand, are well-suited for modeling the impact of a wide variety of emissions changes that would result from changed emission control regulations but rely totally on an input emissions inventory, which may be uncertain or difficult to obtain.

Analytical Modeling Studies

Analytical air quality models have been used the most in modeling the dynamics of pollutants at local and urban scales.

Of great interest is the use of chemically active air quality models for studying and controlling urban air pollution. This is because of the high costs of controls, the complexity of the system, and the historic lack of success in reducing ozone levels in the urban areas. Interest in extending the application of chemically active models to the regional scale has heightened because the ozone problem has been recognized as extending well beyond urban areas.

The lack of success in reducing ozone levels can be partially ascribed to the nonlinearity of the system. An incremental change in the emissions of precursors to a secondary pollutant such as ozone need not lead to a proportional change in the pollutant concentration, or any change at all.

Regional oxidant and acid deposition models came into use later than urban photochemical models because of the increased computational intensity, the need to describe more physical and chemical processes, and the later regulatory mandate for development and use. Regional models are similar to urban-scale photochemical models, differing primarily in their horizontal resolution, 20–100 km versus 4–5 km, and in their treatment of cloud processes and liquid-phase chemistry. Applications of regional models also cover longer periods because of the increased residence time of polluted air masses within their domain.

Regional ozone modeling of the Northeast has shown that air pollution problems across cities within this region are linked (68). Air masses starting in Washington, D.C., can travel up the coast, impacting downwind cities such as Baltimore and New York. Emissions in New York further impact Connecticut and Massachusetts. These studies have shown that the type of controls that are most effective in one city may be counterproductive in another. For example, modeling studies have found NO_x control to be effective for controlling ozone in Boston, Washington, D.C., and Philadelphia, but lead to increased concentrations in New York (68). The horizontal resoluton of the model used for this study was about 20 km, giving some detail in urban areas but not as much as urban-scale models provide.

A variety of models have been developed to study acid deposition. Sulfuric acid is formed relatively slowly in the atmosphere, so its concentrations are believed to be more uniform than ozone, especially in and around cities. Also, the impacts are viewed as more regional in nature. This allows an even coarser horizontal resolution, on the order of 80 to 100 km, to be used in acid deposition models. Atmospheric models of acid deposition have been used to determine where reductions in sulfur dioxide emissions would be most effective. Many of the ecosystems that are most sensitive to damage from acid deposition are located in the northeastern United States and southeastern Canada. Early acid deposition models helped to establish that sulfuric acid and its precursors are transported over long distances, eg, from the Ohio River Valley to New England (69–71). Models have also been used to show that sulfuric acid deposition is nearly linear in response to changing levels of emissions of sulfur dioxide (72).

In the mid-1980s, the destruction of the ozone layer above the Antarctic was recognized to be a potential environmental disaster, and chemically active transport models were used to identify the most important processes leading to depleted stratospheric ozone levels. As with ground-level chemistry, the gas-phase chemistry that occurs in the stratosphere is relatively well understood, but the incorporation of heterogeneous chemistry is an ongoing challenge. In addition, the models that have incorporated stratospheric chemistry in the greatest detail have oversimplified the transport and dynamics of stratospheric circulations. Conversely, three-dimensional models that focus on the dynamics of the atmosphere have used limited treatments of stratospheric chemistry.

Zero- through three-dimensional models have been used to simulate the stratospheric ozone problem. Zero-dimensional models focus on radiative transfer and chemistry at a single point. One-dimensional models incorporate vertical diffusive transport as well. Two-dimensional models currently represent the best compromise between computational tractability and chemical detail (73) and are resolved by latitude as well as in the vertical dimension. Stronger flows and greater homogeneity over changes in longitude make two-dimensional treatments reasonable, except in the polar regions. Three-dimensional models are needed to study the dynamics of polar regions, including exchanges with air at lower latitudes.

Heterogeneous chemistry occurring on polar stratospheric cloud particles of ice and nitric acid trihydrate has been established as a dominant factor in the aggravated seasonal depletion of ozone observed to occur over Antarctica. Preliminary attempts have been made to parameterize this chemistry and incorporate it in models to study ozone depletion over the poles (74) as well as the potential role of sulfate particles throughout the stratosphere (75).

Models can be used to study human exposure to air pollutants and to identify cost-effective control strategies. In many instances, the primary limitation on the accuracy of model results is not the model formulation but the accuracy of the available input data (76). Another limitation is the inability of models to account for the alterations in the spatial distribution of emissions that occurs when controls are applied. The more detailed models are currently able to describe the dynamics of unreactive pollutants in urban areas.

Because of the expanded scale and need to describe additional physical and chemical processes, the development of acid deposition and regional oxidant models has lagged behind that of urban-scale photochemical models. An additional step up in scale and complexity, the development of analytical models of pollutant dynamics in the stratosphere is also behind that of ground-level oxidant models, in part because of the central role of heterogeneous chemistry in the stratospheric ozone depletion problem. In general, atmospheric liquid-phase chemistry and especially heterogeneous chemistry are less well understood than gas-phase reactions such as those that dominate the formation of ozone in urban areas. Development of three-dimensional models that treat both the dynamics and chemistry of the stratosphere in detail is an ongoing research problem.

BIBLIOGRAPHY

1. K. L. Demerjian, "Photochemical Diffusion Models for Air Quality Simulation: Current Status," in *Proceedings of the Conference on the State of the Art of Assessing Transportation-Related Air Quality Impacts*, Oct. 22–24, 1975, Transportation Research Board, Washington, D.C., 1975.
2. P. Zanetti, *Air Pollution Modeling*, Van Nostrand Reinhold, New York, 1990.
3. D. S. Barth, *J. Air Pollut. Control Assoc.* **20,** 519–523 (1970).
4. *Final Air Quality Management Plan*, 1982 rev., South Coast Air Quality Management District and Southern California Association of Governments, El Monte, Calif., 1982.
5. M. Das, *Multivariate Analysis of Atlanta Air Quality Data*, University of Connecticut, Mar. 1993.
6. D. P. Chock, T. R. Terrell, and S. B. Levitt, *Atmos. Environ.* **9,** 978–989 (1975).
7. G. Fronza, A. Spirito, and A. Tonielli, *Appl. Math. Model.* **3,** 409–415 (1979).
8. P. Bacci, P. Bolzern, and G. Fronza, *J. Appl. Meterol.* **20,** 121–129 (1981).
9. S. K. Friedlander, *Environ. Sci. Technol.* **7,** 235–240 (1973).
10. S. L. Heisler, S. K. Friedlander, and R. B. Husar, *Atmos. Environ.* **7,** 633–649 (1973).
11. G. Gartrell and S. K. Friedlander, *Atmos. Environ.* **9,** 279–299 (1975).
12. D. F. Gatz, *Atmos. Environ.* **9,** 1–18 (1975).
13. D. F. Gatz, *J. Appl. Meteorol.* **17,** 600–608 (1978).
14. G. E. Gordon, *Environ. Sci. Technol.* **14,** 792–800 (1980).
15. J. G. Watson, R. C. Henry, J. A. Cooper, and E. S. Macias in E. S. Macias and P. K. Hopke, eds., *Atmospheric Aerosol: Source/Air Quality Relationships*, Symp. Ser. No. 167, American Chemical Society, Washington, D.C., 1981, pp. 89–106.
16. G. R. Cass and G. J. McRae, *Environ. Sci. Technol.* **17,** 129–139 (1983).
17. J. G. Watson, *J. Air Pollut. Control Assoc.* **34,** 619–623 (1984).
18. P. K. Hopke, *Receptor Modeling in Environmental Chemistry*, John Wiley & Sons, Inc., New York, 1985.
19. R. J. Yamartino in S. C. Dattner and P. K. Hopke, eds., *Receptor Models Applied to Contemporary Pollution Problems*, Air Pollution Control Association, Pittsburgh, Pa., 1983, pp. 285–295.
20. P. F. Nelson, S. M. Quigley, and M. Y. Smith, *Atmos. Environ.* **17,** 439–449 (1983).
21. W. J. O'Shea and P. A. Scheff, *JAPCA* **38,** 1020–1026 (1988).
22. P. F. Aronian, P. A. Scheff, and R. A. Wadden, *Atmos. Environ.* **23,** 911–920 (1989).

23. J. E. Core, P. L. Hanrahan, and J. A. Cooper, in Ref. 11, pp. 107–123.
24. R. G. Lamb and J. H. Seinfeld, *Environ. Sci. Technol.* **7,** 253–261 (1973).
25. R. G. Lamb, *Atmos. Environ.* **7,** 257–263 (1973).
26. J. H. Seinfeld, *Atmospheric Chemistry and Physics of Air Pollution*, Wiley-Interscience, New York, 1986.
27. R. G. Lamb and M. Neiburger, *Atmos. Environ.* **5,** 239–264 (1971).
28. G. R. Cass, *Atmos. Environ.* **15,** 1227–1249 (1981).
29. F. A. Gifford, *Nucl. Saf.* **2,** 47–57 (1961).
30. D. B. Turner, *J. Appl. Meteorol.* **3,** 83–91 (1964).
31. D. B. Turner, *A Workbook of Atmospheric Dispersion Estimates*, Public Health Service Publication No. 999-AP-26, U.S. Government Printing Office, Washington, D.C., 1967.
32. S. D. Reynolds, P. W. Roth, and J. H. Seinfeld, *Atmos. Environ.* **7,** 1033–1061 (1973).
33. A. C. Lloyd and co-workers, *Development of the ELSTAR Photochemical Air Quality Simulation Model and Its Evaluation Relative to the LARPP Data Base*, Environmental Research and Technology Report, No. P-5287-500, West Lake Village, Calif., 1979.
34. S. D. Reynolds, T. W. Tesche, and L. E. Reid, *An Introduction to the SAI Airshed Model and Its Usage*, Report No. SAI-EF79-31, Systems Applications, Inc., San Rafael, Calif., 1979.
35. J. H. Seinfeld, *J. Air Waste Management* **38,** 616–645 (1988).
36. D. P. Chock, A. M. Dunker, S. Kumar, and C. S. Sloane, *Environ. Sci. Technol.* **15,** 933–939 (1981).
37. G. R. Carmichael, T. Kitada, and L. K. Peters, *Atmos. Environ.* **20,** 173–188 (1986).
38. G. R. Carmichael, L. K. Peters, and R. D. Saylor, *Atmos. Environ.* **25A,** 2077–2090 (1991).
39. A. G. Russell and G. R. Cass, *Atmos. Environ.* **20,** 2011–2025 (1986).
40. J. G. Chang and co-workers, *J. Geophys. Res.* **92,** 14,681–14,700 (1987).
41. S. A. McKeen and co-workers, *J. Geophys. Res.* **96,** 10,809–10,845 (1991).
42. G. J. McRae and J. H. Seinfeld, *Atmos. Environ.* **17,** 501–522 (1983).
43. A. G. Russell, K. McCue, and G. R. Cass, *Environ. Sci. Technol.* **22,** 1336–1347 (1988).
44. T. E. Graedel, L. A. Farrow, and T. A. Weber, *Atmos. Environ.* **10,** 1095–1116 (1976).
45. M. C. Dodge, *Combined Use of Modeling Techniques and Smog Chamber Data to Derive Ozone-Precursor Relationships*, Report No. EPA-600/3-77-001a, U.S. Environmental Protection Agency, Research Triangle Park, N.C., 1977.
46. A. H. Falls and J. H. Seinfeld, *Environ. Sci. Technol.* **12,** 1398–1406 (1978).
47. R. Atkinson, A. C. Lloyd, and L. Winges, *Atmos. Environ.* **16,** 1341–1355 (1982).
48. W. R. Stockwell, *Atmos. Environ.* **20,** 1615–1632 (1986).
49. F. W. Lurmann, W. P. L. Carter, and L. A. Coyner, *A Surrogate Species Chemical Reaction Mechanism for Urban-Scale Air Quality Simulation Models*, Report No. EPA/600/3-87-014, U.S. Environmental Protection Agency, Research Triangle Park, N.C., 1987.
50. J. P. Killus and G. Z. Whitten, *A New Carbon Bond Mechanism for Air Quality Modeling*, Report No. EPA 60013-82-041. U.S. Environmental Protection Agency, Research Triangle Park, N.C., 1982.
51. G. Z. Whitten and co-workers, *Modeling of Simulated Photochemical Smog with Kinetic Mechanism*, Vols. I and II, Report No. EPA-600/3-79-001a, U.S. Environmental Protection Agency, Research Triangle Park, N.C., 1979.
52. M. W. Gery, G. Z. Whitten, J. P. Killus, and M. C. Dodge, *J. Geophys. Res.* **94,** 12,925–12,956 (1989).
53. E. Weber, *Air Pollution, Assessment Methodology and Modeling*, Plenum Press, New York, 1982.
54. G. J. McRae and A. G. Russell, *J. Comp. Phys.* **4,** 227–233 (1990).
55. Eldering and co-workers, *Environ. Sci. Technol.* 626–636 (1993).
56. *Technical Guidance Document: Photochemical Modeling*, Sacramento, Calif., California Air Resources Board, Aug. 1990.
57. A. M. Dunker, *J. Chem. Phys.* **81,** 2385–2393 (1984).
58. M. Dermilap and H. Rabitz, *J. Chem. Phys.* **74,** 3362–3375 (1981).
59. S. N. Pandis and J. H. Seinfeld, *J. Geophys. Res.* **94,** 1105–1126 (1989).
60. J. B. Milford, D. Gao, A. G. Russell, and G. J. McRae, *Environ. Sci. Technol.* **26,** 1179–1189 (1992).
61. A. H. Falls, G. J. McRae, and J. H. Seinfeld, *Int. J. Chem. Kin.* **11,** 1137–1162 (1979).
62. G. J. McRae, J. W. Tilden, and J. H. Seinfeld, *Comput. Chem. Eng.* **6,** 15–25 (1982).
63. R. S. Stolarski, D. M. Butler, and R. D. Rundel, *J. Geophys. Res.* **83,** 3074–3078 (1978).
64. D. H. Ehhalt, J. S. Chang, and D. M. Butler, *J. Geophys. Res.* **84,** 7889–7894 (1979).
65. R. Derwent and O. Hov, *J. Geophys. Res.* **93,** 5185–5199 (1988).
66. A. E. Thompson and R. W. Stewart, *J. Geophys. Res.* **96,** 13089–13108 (1991).
67. T. Y. Chang, R. H. Hammerle, S. M. Japan, and I. T. Salmeen, *Environ. Sci. Technol.* **25,** 1190–1197 (1991).
68. N. C. Possiel, L. B. Milich, and B. R. Goodrich, eds., *Regional Ozone Modeling for Northeast Transport (ROMNET)*, Report No. EPA-450/4-91-002a, U.S. Environmental Protection Agency, Research Triangle Park, N.C., 1991.
69. H. Rodhe, P. Crutzen, and A. Vanderpol, *Tellus* **33,** 132–141 (1981).
70. A. Eliassen, *J. Appl. Meteor.* **19,** 231–240 (1980).
71. *U.S.–Canada Memorandum of Intent on Transboundary Air Pollution, Atmospheric Sciences and Analysis Work Group 2 Phase III Final Report*, U.S. Environmental Protection Agency, Washington, D.C., 1983.
72. J. S. Chang and co-workers, *The Regional Acid Deposition Model and Engineering Model*, State-of-Science–Technology Report 4, National Acid Precipitation Assessment Program, Washington, D.C., 1989.
73. *Scientific Assessment of Stratospheric Ozone: 1989*, United Nations Environment Program and World Meteorological Organization, New York, 1989.
74. W. B. DeMore and co-workers, *Chemical Kinetics and Photochemical Data for Use in Stratospheric Ozone Modeling*, Evaluation No. 8, JPL Publ. 87–41, Jet Propulsion Laboratory, Pasadena, Calif., 1987.
75. D. J. Hoffmann and S. Solomon, *J. Geophys. Res.* **94,** 5029 (1989).
76. W. R. Pierson, A. W. Gertler, and R. L. Bradow, *J. Air Waste Manage. Assoc.* **40,** 1495–1504 (1990).

AIRCRAFT FUELS

William G. Dukek
Consultant
Summit, New Jersey

Aircraft fuels have evolved since World War II from aviation gasoline for piston engines to kerosene for gas turbine jet engines. Today aviation gasoline is a shrinking specialty product used by general aviation while commercial airlines, business aircraft, and military aviation have turned jet fuel into a principal product of the petroleum industry ranking with automotive fuels and industrial oils in importance. Petroleum products and uses; Kerosene.

The gas turbine powerplant which has revolutionized aviation derives basically from the steam turbine adapted to a different working fluid. The difference is crucial with respect to fuel because steam can be generated by any heat source whereas the gas turbine requires a fuel that efficiently produces a very hot gas stream and is also compatible with the turbine itself. The hot gas stream results from converting chemical energy in fuel directly and continuously by combustion in compressed air. It is expanded in a turbine to produce useful work in the form of jet thrust or shaft power.

The jet engine combines three basic steps (Fig. 1). Air is compressed in a series of stages, fuel is burned continuously and intensively in the compressed air, and the hot gas is then expanded through a turbine which extracts energy to run the compressor and the fan and also provide shaft power. The residual exhaust mixes with fan air and exits through a nozzle as a high velocity jet which provides forward thrust to the system. The gas turbine concept was adapted to aircraft in 1930. At high speed and altitude this type of engine improved in efficiency and made jet propulsion competitive with propellor drive (1). The W-1 engine of 1937 has since proliferated into much more efficient aircraft powerplants ranging in size from small auxiliary power units to the giant turbofans which propel wide-bodied airliners. In its most advanced form, the aviation gas turbine for supersonic aircraft, a second combustion step is added to the three basic steps of compression, combustion, and turbine work extraction. Additional fuel is burned in the turbine exhaust to increase the velocity of the jet from the nozzle and gain more thrust.

Ground turbines carry out the same three basic steps of the jet engine but residual energy of the turbine exhaust is extracted by a waste heat boiler or regenerator. Industrial turbines have heavy compressors, external combustion systems, and massive turbine sections. Lightweight design for aircraft required new high temperature metals and advanced concepts for cooling compression, and combustion. Today, these lightweight types have been adapted for many ground applications in gas compression, pumping, electrical power generation, and general purpose utility units. Because the gas turbine is fundamentally a heat conversion device the combustor or gas generator is the heart of the engine. Combustion of fuel with a portion of compressed air is continuous. Early developers of gas turbines depended upon the experience with oil-fired furnaces to design fuel nozzles and combustion chambers (2).

The first fuels burned in a turbine combustor were petroleum liquids: kerosene in the prototype jet engine and diesel fuel in the industrial turbine, selected because of their availability and convenience. After fifty years of experience the gas turbine has been adapted to a wide variety of combustible gases and liquids. Although such a wide-ranging appetite for fuels capable of releasing heat energy would appear to classify the gas turbine as fuel-insensitive, clearly defined limits on fuel properties have developed to optimize its performance. The aircraft propulsion unit is most demanding, accepting only certain liquid distillates; the industrial turbine least demanding. The

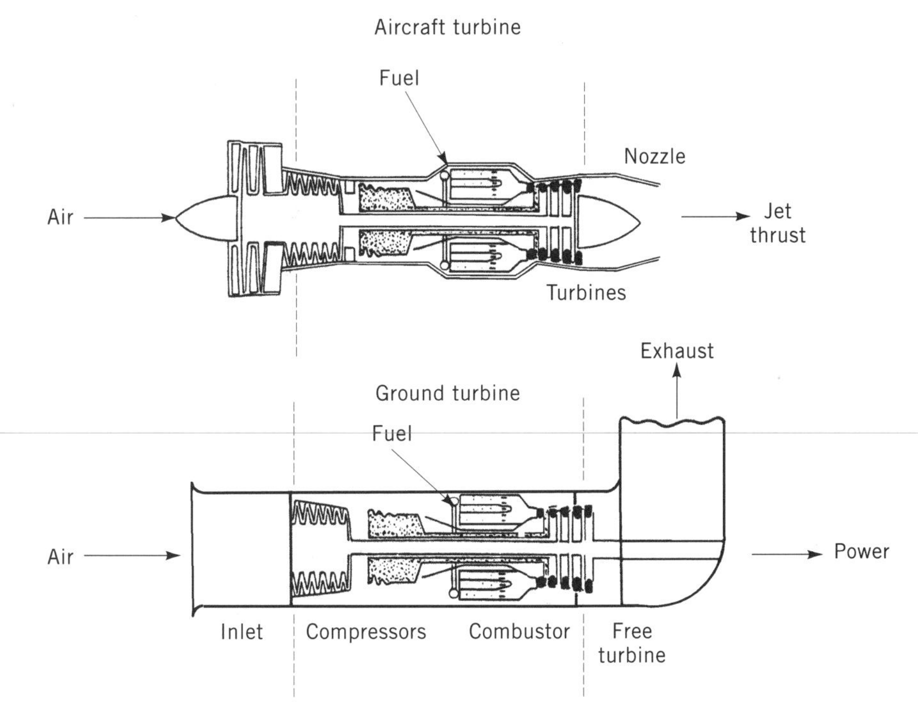

Figure 1. Schematic of aircraft and ground gas turbines.

special requirements of gas turbine fuels that have developed relate to the individual processes that take place in the engine itself and in the support systems for handling fuels. It is evident that aviation gas turbines place more severe demands on liquid fuels than ground turbines and that gaseous fuels are much easier to utilize since they need to supply only heat energy.

A kerosene for turbine research to converse gasoline for the impending war effort was used because it had better low temperature properties than a diesel fuel. This decision tended to establish a dividing line between air and ground turbines; subsequent developments led to lighter fuels for jet engines and heavier fuels for industrial turbines. Aircraft have continued to fly on kerosene fuel or a wide-cut blend of heavy naphtha and kerosene. Supersonic transports and advanced military vehicles are also operated on kerosene fuel. Most ground turbines utilize liquid fuel systems based on standard No. 2 diesel or home burner fuel.

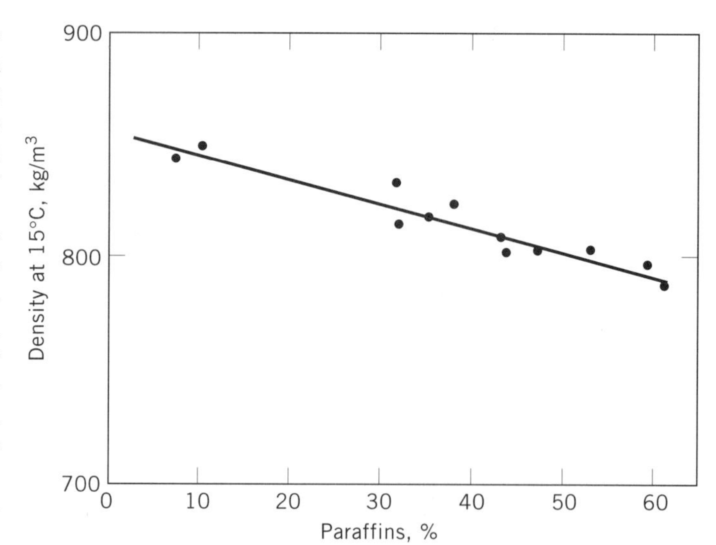

Figure 2. Density vs paraffin content of gas turbine fuels (150–288°C fractions).

GAS TURBINE PRODUCTS COMPOSITION

Because the jet aircraft is a weight-limited vehicle, a high premium is assigned to hydrocarbon fuels with a maximum gravimetric heat content or hydrogen-to-carbon ratio. Paraffinic fuels have high mass heat contents (about 45 MJ/kg or 10.8×10^3 kcal/kg) which decrease very slowly as chain length (molecular weight) increases. However, because the density of these fuels is low, the heat content on a volumetric basis is less than that of nonparaffinic fuels. Since fuels are measured and sold on a volumetric basis, the aircraft user obtains less energy for each cubic meter of the preferred paraffinic fuel but more energy per unit mass. Ground turbine users are not concerned with weight limitations and seek maximum energy per unit of cost which means that heavier fuels are more economical for ground application.

Fuel properties are largely determined by the nature of the crude oil from which it is derived. This is illustrated by the typical density–paraffin content curve for the kerosene fractions (150–288°C) of different crudes (Fig. 2). Aromatics in this cut are of special concern because of their effects on combustion and elastomers. The mass spectrometric data in Table 1 illustrate that a wide range of single and multiring aromatics is possible even within a total aromatic concentration of 25%, the maximum level permitted by most specifications.

Specifications

The global nature of jet aircraft operations has mandated that aviation fuel quality be closely controlled in every part of the world. Specifications tend to be industry standards issued by a consensus organization like ASTM or by a government body rather than manufacturer's requirements. In Table 2 are listed some of the requirements of the principal grades of civil and military aviation turbine fuels in use throughout the world. Domestic and international airlines use fuel modeled on ASTM specification grades. Specifications issued by the British Ministry of Defence, the recommendations of the International Air Transport Association, and ASTM specifications are coordinated as a Joint System Check List to control quality at many worldwide airports where jointly operated fueling systems prevail. Specifications issued by former Soviet governments are similar to Free World aircraft fuels but not identical in all respects. International airlines that fly to East Bloc countries can utilize both types of fuels without difficulty. Except for severe Arctic areas where Jet B fuel is used, essentially all civil aviation uses kerosene fuel. About half of the demand is Jet A fuel marketed in the United States and half is Jet A1, the international fuel marketed outside the United States similar to Jet A except

Table 1. Composition[a] of 150–288°C Kerosenes

Crude Source	Middle East			North Africa	United States				South America			
Saturates, wt %	78.8	82.4	80.3	83.1	85.5	81.0	76.8	81.7	78.8	76.8	76.1	74.5
Paraffins	63.0	61.7	54.0	47.4	44.4	37.9	35.3	44.0	34.9	31.3	9.0	6.0
Cycloparaffins, single ring	10.6	12.0	14.5	23.8	24.1	22.9	27.3	26.9	27.3	29.0	33.3	31.7
Cycloparaffins, two rings	4.7	7.6	8.7	9.8	12.0	17.9	11.3	9.6	12.8	12.6	24.9	28.1
Cycloparaffins, 2 + rings	0.9	1.1	3.2	3.1	5.0	2.3	2.9	1.2	3.8	3.9	8.9	8.7
Aromatics, wt %	18.4	16.3	18.2	14.7	14.0	17.7	21.9	17.0	19.9	22.0	23.7	25.5
Single ring	16.9	14.1	14.8	11.3	11.8	13.9	19.0	15.1	14.8	16.6	16.4	16.8
Two rings	1.4	2.0	3.0	3.2	2.1		2.6	1.9	4.7	5.1	3.6	7.7
2 + rings	0.1	0.1	0.2	0.2	0.1	3.8	0.3		0.5	0.4	0.7	1.0

[a] By mass spectrometric analysis.

Table 2. Selected Specification Properties of Civil and Military Aviation Gas Turbine Fuels

	ASTM D1655[a]		Mil-T-5624-N[a]		Mil-T-83133C[a]
Characteristic	Jet A kerosene A/L and	Jet B wide-cut Gen Avn	JP-4 wide-cut USAF	JP-5 kerosene USN	JP-8 kerosene USAF
Composition					
Aromatics, vol % max	25[b]	25[b]	25	25	25
Sulfur, wt % max	0.3	0.3	0.4	0.4	0.3
Volatility					
Dist. Temp Max °C — 10% rec'd	205			205	205
Dist. Temp Max °C — 50% rec'd		190	190		
Dist. Temp Max °C — End pt	300		270	300	300
Flash pt, °C min	38			60	38
Vapor pressure at 38°C, kPa Max (psi)		21(3)	14–21 (2–3)		
Density at 15°C, kg/m³	775–840	751–802	751–802	788–845	775–840
Fluidity					
Freezing pt, °C max	−40[c]	−50	−58	−46	−47
Viscosity at −20°C, mm²/s (= cSt) max	8.0			8.5	8.0
Combustion					
Heat content, MJ/kg, min[d]	42.8	42.8	42.8	42.6	42.8
Smoke pt, mm, min	18[e]	18[e]	20	19	19
H_2 Content, wt % min			13.5	13.4	13.4
Stability					
Test temp[f], °C min	245	245	260	260	260

[a] Full specification requires other tests.
[b] For aromatics above 20% (22% for Jet A1) users must be notified.
[c] International airlines use Jet A1 with −47°C freeze point.
[d] To convert MJ to kcal, multiply by 239.
[e] Plus naphthalenes 3 vol % max unless smoke point exceeds 25.
[f] Thermal stability test by D3241 to meet 3.3 kPa (25 mm Hg) pressure drop and Code 3 Deposit Rating.

for freezing point. A lower temperature limit of −47°C for the (wax) freezing point of Jet A1 is required for long duration international flights. The same fuel is labeled JP-8 by the United States Air Force and used at most bases outside the United States to replace JP-4, the wide-cut fuel originally developed after World War II to assure maximum availability. JP-4 is still the primary Air Force fuel used at U.S. bases but is being replaced by JP-8 in a continuing conversion program. The other military fuel of importance is JP-5, the high flash point kerosene used by the Navy to ensure handling safety aboard carriers.

Manufacture

Aviation turbine fuels are primarily blended from straight-run distillates rather than cracked stocks because of specification limitations on olefins and aromatics. However, in refineries where heavy gas oil is hydrocracked, ie, a process that involves catalytic cracking in a hydrogen atmosphere, aviation fuels can include hydrocracked components since they are free of olefins, low in sulfur, and stable. Hydrocracking provides a means to extend availability of gasolines and distillates by converting most of the heavy ends in crudes to lighter products.

A schematic of a typical refinery processing sequence appears in Figure 3. In this example, the basic processes after distillation are catalytic reforming for gasoline and hydrogen processing of naphtha, kerosene, and gas oil using the hydrogen from reforming. A hydrocracking unit usually requires a separate source of hydrogen. The approximate boiling ranges of the sidestream cuts and finished gas turbine fuels are shown.

The distillation cut-points must be closely controlled to yield a product that meets the requirements of flash point on one hand and freezing point on the other. In practical terms, a jet fuel requires a fraction of about 60°C initial boiling point and a final boiling point not exceeding 300°C. The lighter portion of this fraction contains gasoline components and meets the specification for JP-4. The heavier portion above 160°C must be tailored to either the Jet A or Jet A1 freezing point, but the final boiling point is dependent on crude composition. With higher aromatic crudes, undercutting is required to meet compositional limits or a test such as smoke point. Lower boiling components will compensate for aromatic or freezing point limitations, but sometimes it is impossible to take advantage of them because a dual-purpose kerosene usually requires a higher flash point than jet fuel. The distilled fractions from the crude are apt to contain mercaptans or organic acids in excess of specification limits. A caustic wash is normally used to control acidity and remove traces of hydrogen sulfide. Removal or conversion of odorous mercaptans is carried out in a sweetening process. The most widely used chemical treatment today, Merox sweetening, utilizes dissolved air to oxidize mercaptans to disulfides over a fixed bed cobalt chelate catalyst (3). It has the advantage of simplicity and minimum waste disposal problems but does not lower the sulfur content. A modern

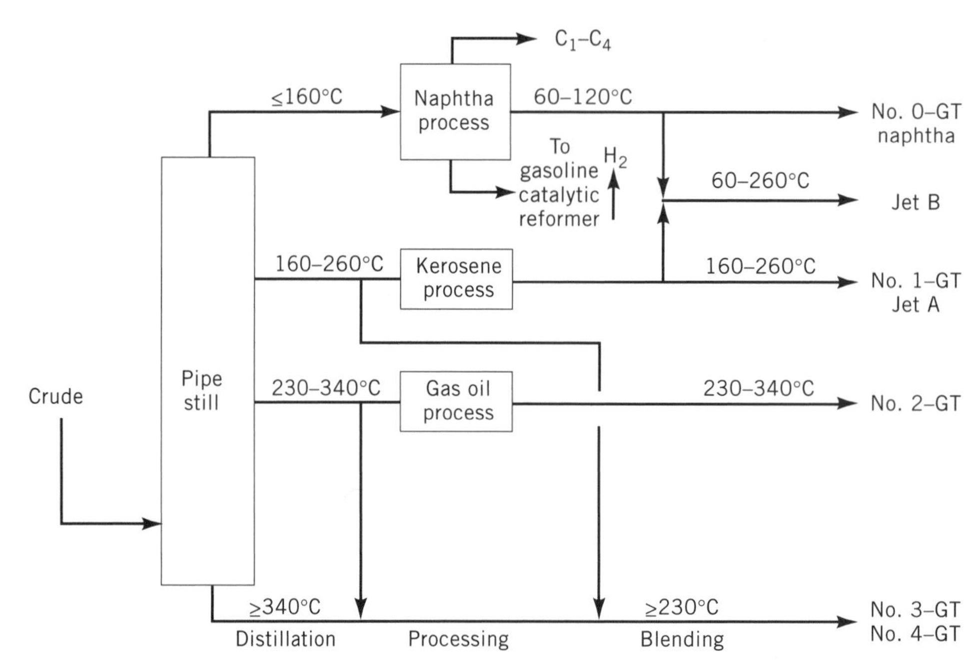

Figure 3. Refinery process for gas turbine fuels. Processing may include hydrogen treating.

version of the old doctor process, Bender sweetening, uses lead plumbite deposited on a fixed bed to carry out the mercaptan oxidation and spent doctor regeneration processes simultaneously, but product quality is more difficult to control and waste disposal problems are greater.

Most of the crudes available in greatest quantity today are high in sulfur and yield products that must be desulfurized to meet specifications or environmental standards. The process most widely used is mild catalytic hydrogenation. Hydrofining uses a fixed bed catalyst, usually cobalt molybdate on alumina, to convert all but ring sulfur compounds to H_2S which is then removed (and recovered) from recycle gas. At reactor temperatures below 300°C only mercaptans react. This technique is called hydrosweetening. At reactor temperatures of 400–500°C desulfurizations levels are 80% or better, and other reactive species including some multiring sulfur compounds are hydrogenated. The product of the hydrofiner is caustic washed to remove traces of dissolved H_2S and passed through sand or clay to blend tanks.

Hydrogen treating removes oxygen containing species such as phenols and naphthenic acids that are found in some crudes. The former tend to perform as natural inhibitors of hydrocarbon oxidation by trapping peroxy radicals while the latter act as natural lubricating agents. Recognition of this side-effect of hydrogen treatment led refiners to add antioxidants to hydrotreated components; this procedure is now a specification requirement.

Blending of two or more components is carried out to match as closely as practical the various specification constraints. At this point, additives are usually introduced, eg, an antioxidant to inhibit gum formation in storage or a metal deactivator to deactivate any dissolved metal ions (see Table 2). Military fuels have made the addition of anti-icing agents and corrosion inhibitors mandatory, the former because military aircraft fuel systems do not include fuel filter heaters to prevent ice formation, the latter to reduce the tendency of fuels to pick up rust particles from pipelines and tanks. The Air Force also requires that antistatic additive be included in JP-4 to reduce the hazard of static generation and discharge in handling fuel and in operating aircraft with foam-filled tanks. It was discovered that the anti-icing additives inhibit growth of microorganisms in aircraft and storage tanks and that certain corrosion inhibitors are excellent lubricity agents for prolonging the life of engine pumps. With civil fuels, anti-icing agents are not required since air transports are equipped with fuel filter heaters to prevent ice formation. Corrosion inhibitors are not normally used in civil fuels since they tend to degrade the efficiency of filter-coalescers and would be removed in the clay filters that are commonly installed at receiving terminals and airports. However, antistatic additives are widely used in civil fuels, particularly Jet A1 at international airports. While antistatic additive is introduced into some Jet A to safeguard against static discharge in truck filling, the widespread use of clay filters in the United States to clean up fuel before airport delivery makes refinery addition impractical. Table 3 also lists a boron containing biocide used as a shock treatment additive to inhibit the tendency of organisms to form fungal mats in aircraft fuel tanks.

Distribution and Handling

Preserving the quality of gas turbine fuels between the refinery and the point of use is an important but difficult task. The difficulty arises because of the complicated distribution systems of multiproduct pipelines which move fuel and sometimes introduce contaminants. The importance is reflected by the sensitivity of gas turbine engines and fuel systems to water, corrosion products, metal salts, microorganisms, and other extraneous materials that can be introduced by the distribution system. Most products moved through pipelines contain additives, eg, detergents, deicers, antifouling dispersants, and rust inhibitors in gasoline, corrosion inhibitors and dispersants in heating oil; or ignition promotors in diesel fuel. Traces of these additives left on pipeline walls are picked up by nonadditive jet fuel which makes it difficult for pipeline operators to deliver gas turbine fuel of the same quality as that introduced into the line.

The principal means of removing contaminants are tank settling and filtration. With aviation fuels it is common practice to install several stages of cartridge type filter–coalescers between the storage tank and the aircraft delivery point. Figure 4 is a schematic of a typical airport fueling system which transfers fuel by underground hydrant lines. Filter elements of fiberglass covered with synthetic fabric are designed to coalesce water and particulates at high flow rates. Coalesced water is prevented from passing with fuel by a hydrophobic barrier filter.

Tank settling as a means of contaminant removal is not very efficient with fuels having the viscosity of kerosene. It is common practice to design tanks with cone down drains and floating suctions to facilitate water and solids removal. Contamination control by filtration would be relatively easy if it were not for surface-active materials produced during fuel manufacture or picked up in distribution systems from pipe walls and tank bottoms. These surfactants stabilize water-in-oil emulsions, disarm filter elements, coat solid particles, and promote growth of organisms at fuel–water interfaces. When they reach the aircraft tank, the surfactants tend to trap water containing dissolved salts and particulates, causing fungal growths, coating attack, and metal corrosion.

Gas turbine fuels can contain natural surfactants if the crude fraction is high in organic acids, eg, naphthenic (cycloparaffinic) acids of 200–400 mol wt. These acids readily form salts that are water soluble and surface-active. Older treating processes for sulfur removal can leave sulfonate residues which are even more powerful surfactants. Refineries have installed processes for surfactant removal. Clay beds to adsorb these trace materials are widely used and salt towers to reduce water levels will also remove water soluble surfactants. In the field clay filters designed as cartridges mounted in vertical vessels are also used extensively to remove surfactants picked up in fuel pipelines or contaminated tankers or barges.

Specifications for gas turbine fuels prescribe test limits that must be met by the refiner who manufactures fuel; however, it is customary for fuel users to define quality control limits for fuel at the point of delivery or custody transfer. These limits must be met by third parties who distribute and handle fuels on or near the airport. Tests on receipt at airport depots include appearance, distillation, flash point (or vapor pressure), density, freezing point, smoke point, corrosion, existing gum, water reaction, and water separation. Tests on delivery to the aircraft include appearance, particulates, membrane color, free water, and electrical conductivity.

The extensive processing and cleanup steps carried out on gas turbine fuels produce a purified liquid dielectric of high resistivity which is capable of retaining electrical charges long enough for buildup of large surface voltages. Because fuel is filtered at high flow rates through filter media of extensive surface area, the ionic species that adsorb on the filter surface generate a static charge. The current carried by fuel increases sharply as fuel flows through a filter and then decreases at a rate determined by the fuel conductivity. The charge remaining in the fuel at any given point is determined by the residence time available for charges to recombine. In a hose delivering fuel to an aircraft, only a few seconds are available for charge relaxation. The result is a possible hazard: a tank filled with charged fuel that under some circumstances can discharge its energy to ground in a spark capable of igniting fuels mists or vapors (4).

This aspect of fuel handling has received much attention because of a number of accidents that have resulted in tank explosions, most often in filling tank trucks but also in fueling aircraft. Several approaches have been taken to reduce the risk of static discharge. The most common method requires introduction of an additive to increase the electrical conductivity of the fuel to a safe level, ie, to speed up charge relaxation to a fraction of a second. Charge generation can be decreased by using low-charging screens instead of filters or by reducing flow rates. The conductivity additives used in fuel are readily removed in clay filters, or lost in pipeline movement and must be monitored by conductivity tests. The additives can also degrade water separation properties and lower filter efficiency. Nevertheless the safety record since antistatic

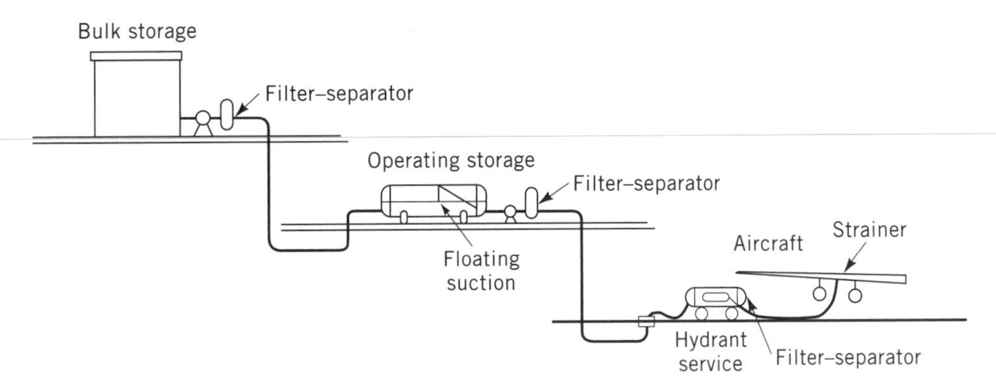

Figure 4. Typical airport hydrant system.

additives were introduced far outweighs the negative effects on other fuel properties.

PERFORMANCE COMBUSTION

The primary reaction carried out in the gas turbine combustion chamber is oxidation of a fuel to release its heat content at constant pressure. Atomized fuel mixed with enough air to form a close-to-stoichiometric mixture is continuously fed into a primary zone. There, its heat of formation is released at flame temperatures determined by the pressure. The heat content of the fuel is therefore a primary measure of the attainable efficiency of the overall system in terms of fuel consumed per unit of work output. Net rather than gross heat content is a more significant measure because heat of vaporization of the water formed in combustion cannot be recovered in aircraft exhaust.

The most desirable gas turbine fuels for use in aircraft after hydrogen are hydrocarbons. Fuels that are liquid at normal atmospheric pressure and temperature are the most practical and widely used aircraft fuels; kerosene with a distillation range from 150 to 300°C is the best compromise to combine maximum mass heat content with other desirable properties.

Liquid fuel is injected through a pressure-atomizing or an airblast nozzle. This spray is sheared by air streams into laminae and droplets that vaporize and burn. Because the atomization process is so important for subsequent mixing and burning, fuel injector design is as critical as fuel properties. Figure 5 is a schematic of the processes occurring in a typical combustor.

Droplet size, particularly at high velocities, is controlled primarily by the relative velocity between liquid and air and in part by fuel viscosity and density (5). Surface tension has a minor effect. Minimum droplet size is achieved when the nozzle is designed to provide maximum physical contact between air and fuel. Hence primary air is introduced within the nozzle to provide both swirl and shearing forces. Vaporization time is characteristically related to the square of droplet diameter and is inversely proportional to pressure drop across the atomizer (5).

The vapor cloud of evaporated droplets burns like a diffusion flame in the turbulent state rather than as individual droplets. In the core of the spray, where droplets are evaporating, a rich mixture exists and soot formation occurs. Surrounding this core are a rich mixture zone where CO production is high and a flame front exists. Air entrainment completes the combustion, oxidizing CO to CO_2 and burning the soot. Soot burnup releases radiant energy and controls flame emissivity. The relatively slow rate of soot burning compared with the rate of oxidation of CO and unburned hydrocarbons leads to smoke formation. This model of a diffusion-controlled primary flame zone makes it possible to relate fuel chemistry to the behavior of fuels in combustors (5).

Aromatics readily form soot in the fuel-rich spray core as their hydrogens are stripped, leaving a carbon-rich benzenoid structure to condense into large molecular aggregates. Multiring aromatics form soot more readily than single-ring aromatics and exhibit greater smoking tendency and greater flame luminosity. At the other extreme, *n*-paraffins undergo little cyclization in the fuel-rich spray core, holding soot formation and flame radiation to a minimum. Other hydrocarbon structures exhibit smoke and radiation characteristics between the extremes of multiring aromatics and *n*-paraffins (6). Low luminosity fuels are synonymous with minimum flame radiation and clean (soot-free) combustion. In mixed fuels, two-ring aromatics have been shown to have a ten-fold greater effect luminosity rating than single-ring aromatics (7). Luminous flames in a gas turbine raise temperatures of the metal in the combustor liner and turbine inlet by the direct transmission by radiation of the heat energy in the flame core.

Air in the gas turbine combustor is carefully divided between primary and secondary uses. Primary air promotes the atomization process by adding swirl to the fuel exiting the nozzle; it must be limited to ensure that the flame core remains fuel-rich to avoid flame blow-out and facilitate reignition. Most of the air in the gas turbine combustor is directed either toward secondary reactions beyond the primary flame core or toward dilution of the hot exhaust products so that a uniform temperature gas is presented to the turbine section. The secondary reactions involve intensely turbulent mixing under stoichiometric or fuel-lean mixtures to complete the combustion to CO_2 and burn up the soot particles formed in the fuel-rich core. A combustor design that causes proper mixing is critical for ensuring smoke-free exhaust products. Nevertheless, in any given combustor there is a relationship between smoke output and the hydrogen content of fuel (8).

A gas turbine used in aircraft must be capable of handling a very wide span of fuel and air flows because the thrust output, ie, pressure, covers the range from idle to

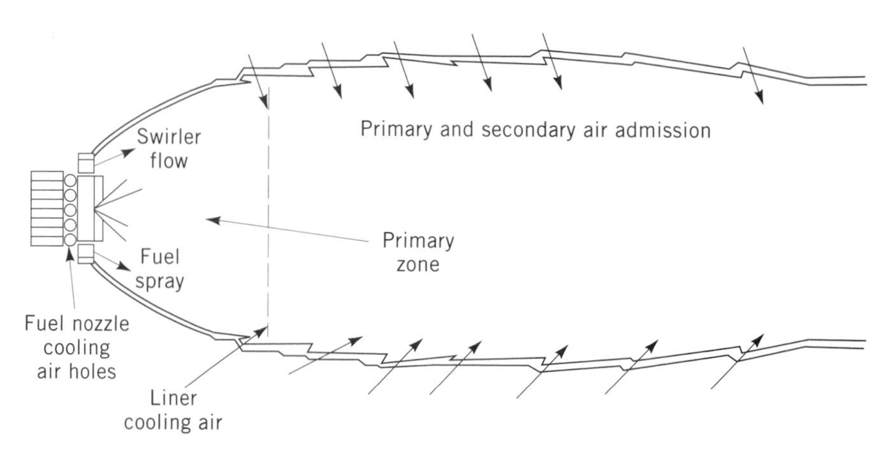

Figure 5. Gas turbine combustion chamber.

full-powered take-off. To accommodate this degree of flexibility in the combustor, fuel nozzles usually are designed with two streams, primary and secondary flow, or with alternate rows of nozzles that turn on only when secondary flow (or full thrust power) is needed. It is more difficult to vary the air streams to match the different fuel flows and as a consequence a combustor optimized for cruise conditions, ie, most of the aircraft's operation, operates less efficiently at idle and full thrust.

Exhaust emissions of CO, unburned hydrocarbons, and nitrogen oxides reflect combustion conditions rather than fuel properties. The only fuel component that degrades exhaust is sulfur; the SO_2 levels in emissions are directly proportional to the content of bound sulfur in the fuel. Sulfur levels in fuel are determined by crude type and desulfurization processes. Specifications for aircraft fuels impose limits of 3000–4000 ppm total sulfur but average levels are half of these values. Sulfur levels in heavier fuels are determined by legal limits on stack emissions.

Control of nitrogen oxides in aircraft exhaust is of increasing concern because nitrogen oxides react with ozone in the protective layer of atmosphere which exists in the altitude region where supersonic aircraft operate. Research is under way to produce a new type of combustor which minimizes NO_x formation. It is an essential component of the advanced propulsion unit needed for a successful supersonic transport fleet.

Another class of pollutant in exhaust does serious damage to the turbine itself. In this case the damage is hot corrosion of the metal blades and vanes by alkali metal (primarily sodium) salts which remove blade coatings and promote intergranular corrosion. The source of the salts can be fuel (as metal salts dissolved or entrained in water) or the air that enters the gas turbine in huge quantities. Sulfur oxides in the exhaust react with alkali to form sulfates which produce a low melting flux on the oxide coatings. The problem is especially serious with gas turbines used in ground (or marine) locations because heavier fuels are apt to contain metal contaminants. For example, some crudes contain soluble vanadium which forms V205 leading to hot corrosion.

Stability

Aviation fuel is exposed to a wide range of thermal environments and these greatly influence required fuel properties; eg, in the tank of a long-range subsonic aircraft, fuel temperatures can drop so low that ice crystals, wax formation, or viscosity increase may affect the fuel system performance. On its way to the engine the fuel then absorbs heat from airframe or engine components and in fact is used as a coolant for engine lubricant. Current high thrust engines expose fuel in the main engine control, pump, and manifold to an intense thermal environment, and fuel temperatures steadily rise as flow reaches the nozzle. Fuel stability assumes primary importance since freedom from deposits within the fuel system is essential for both performance and life.

The fuel systems of ground based turbines are far less critical since coolants other than fuel can be used and fuel lines can be well insulated. The tendency for deposit formation in fuel is not a concern in ground systems.

Deposits sometimes block fuel nozzles and distort fuel spray patterns, leading to skewed temperature distribution with the possibility of burnout of turbine parts by a hot streak exhaust. These deposits are sometimes associated with metal containing particulates but are, in general, another manifestation of fuel instability.

Oxidation of hydrocarbons by the air dissolved in fuel is catalyzed by metals and leads to polymer formation, ie, varnish and sludge deposits, by a chain reaction mechanism involving free radicals. Since it is impossible to exclude air dissolved in fuel, oxidation stability is controlled by eliminating species prone to form free radicals and by introducing antioxidants. An accelerated test that simulates the critical temperature regimes of the gas turbine engine fuel manifold and nozzle, ASTM D3241, the Jet Fuel Thermal Oxidation Test (JFTOT) has been developed to measure the oxidation stability of aircraft gas turbine fuels. In the JFTOT test deposits formed in the heated test section in 2.5 h at a specified temperature are assessed. This relative rating rates fuels that would normally be expected to operate thousands of hours in an engine without deposit formation. In those few instances when deposits have been observed in service, the fuel has been shown to fail JFTOT tests at the specification temperature.

The reactive species that initiate free radical oxidation are present in trace amounts. Extensive studies (9) of the autooxidation mechanism have clearly established that the most reactive materials are thiols and disulfides, heterocyclic nitrogen compounds, diolefins, furans, and certain aromatic–olefin compounds. Because free radical formation is accelerated by metal ions of copper, cobalt, and even iron (10) the presence of metals further complicates the control of oxidation. It is difficult to avoid some metals, particularly iron, in fuel systems.

It is possible to deactivate a metal ion by adding a compound such as disalicyclidene alkyl diamine which readily forms a chelate with most metal atoms to render them ineffective. Metal deactivator has been shown to reduce dramatically oxidation deposits in the JFTOT test and in single tube heat exchanger rigs. The role of metal deactivator in improving fuel stability is complex since quantities beyond those needed to chelate metal atoms act as passivators of metal surfaces and as antioxidants (11).

Oxidation deposits in aircraft engines are related to the thermal stresses imposed by heat soakback which results from the off/on cycles associated with many landings and takeoffs. In contrast, ground turbines tend to operate longer at constant flow and do not subject fuel to the same degree of thermal stress. Anticipating the introduction of higher performance aircraft which will place greater thermal stress on fuel, the USAF has undertaken a program to upgrade the stability of JP-8 by additive.

Volatility

The volatility of aircraft gas turbine fuel is controlled primarily by the aircraft itself and its operating environment. For example, limits of the vapor pressure of aviation gasoline were dictated by the vapor and liquid entrainment losses that could occur in a piston aircraft capable of climbing to an altitude (about 6000 m) where the vapor pressure of the warm fuel exceeds atmospheric pressure (48 kPa or 7 psi). Since early military jet aircraft could climb even higher (12000 m) and faster, it was necessary to limit further the vapor pressure of military gas turbine fuels to 21 kPa (3 psi). Originally, the United States followed the British lead and selected a kerosene of very low vapor pressure for military jet engines. Because projected production of this fuel was limited, fuel availability was the eventual determining factor for the selection of the wide-cut gasoline type fuel JP-4 as the fuel for military jet aircraft. The volatility of kerosene fuel is measured not by its vapor pressure but by its temperature at the point where its vapors just prove to be flammable, ie, the flash point.

Commercial aviation utilizes low volatility kerosene defined by a flash point minimum of 38°C. The flammability temperature has been invoked as a safety factor for handling fuels aboard aircraft carriers; Navy JP-5 is a low volatility kerosene of minimum flash point of 60°C, similar to other Navy fuels.

The dependence of vapor pressure on temperature for the fuels most commonly used in gas turbines appears in Figure 6 (12). The points on the abscissa reflect the flash point temperatures used to define the volatility of higher molecular weight fuels. When vapor pressure itself is limited, as with JP-4 or Jet B, a test temperature of 38°C is specified.

A minimum volatility is frequently specified to assure adequate vaporization under low temperature conditions. It can be defined either by a vapor pressure measurement or by initial distillation temperature limits. Vaporization promotes engine start-up. Fuel vapor pressure assumes an important role particularly at low temperature. For example, if fuel has cooled to −40°C, as at Arctic bases, the amount of vapor produced is well below the lean flammability limit.

In this case, an energetic spark igniter must vaporize enough fuel droplets to initiate combustion. Start-up under the extreme temperature conditions of the Arctic is a main constraint in converting the Air Force from volatile JP-4 to kerosene type JP-8, the military counterpart of commercial Jet A1.

The lower volatility of JP-8 is a primary factor in the U.S. Air Force conversion from JP-4 since fires and explosions under both combat and ordinary handling conditions have been attributed to the use of JP-4. In examining the safety aspects of fuel usage in aircraft, a definitive study (13) of the accident record of commercial and military jet transports concluded that kerosene type fuel is safer than wide-cut fuel with respect to survival in crashes, inflight fires, and ground fueling accidents. However, the difference in the overall accident record is small because most accidents are not fuel-related.

Low Temperature Fuel Flow

The decrease in the temperature of fuel in the tank of an aircraft during a long duration flight produces a number of effects which can influence flight performance. Fuel viscosity increases, necessitating more pumping energy in the tank boost pump and in the engine pump. Fuel becomes saturated with water and droplets of free water form and settle. Those carried with fuel may form ice on the cold filter which protects the fuel control. For this reason, filter heaters are used in civil aircraft to avoid ice blockage and fuel starvation. Military aircraft avoid the complication of a filter heater and depend instead on an anti-icing additive in the fuel to depress the freezing point of water. At still lower temperatures, crystals of wax form in fuel. The temperature at which these wax crystals disappear is called the freezing point and distinguishes Jet A1 used by long-range international airlines from Jet A used by domestic airlines for relatively short duration flights.

The increase in fuel viscosity with temperature decrease is shown for several fuels in Figure 7. The departure from linearity as temperatures approach the pour point illustrates the non-Newtonian behavior created by wax matrices. The freezing point appears before the curves depart from linearity. It is apparent that the low tempera-

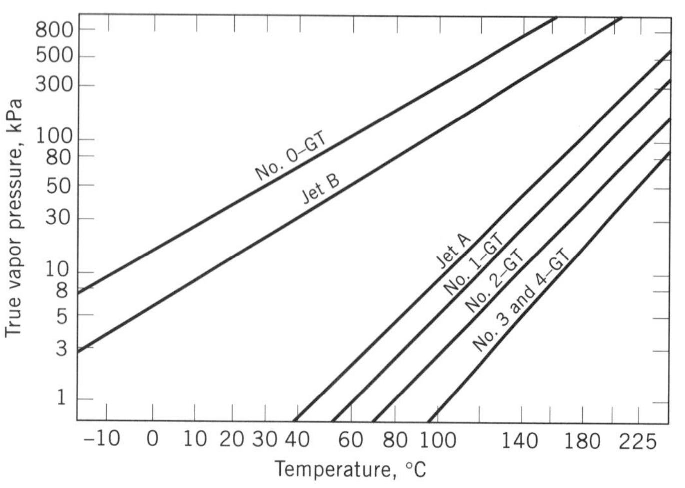

Figure 6. Variation of vapor pressure with temperature for gas turbine fuels (13). To convert kPa to psi, divide by 6.9.

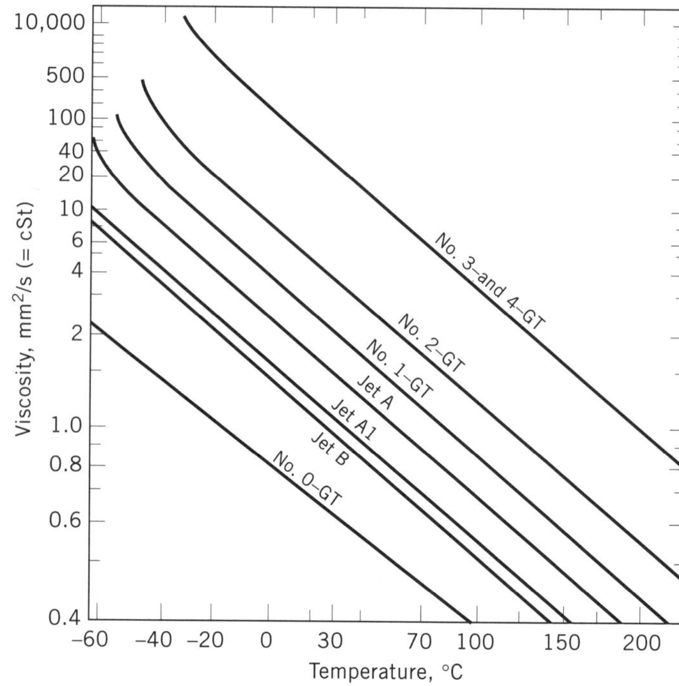

Figure 7. Variation of viscosity with temperature for gas turbine fuels (13).

ture properties of fuel are closely related to its distillation range as well as hydrocarbon composition. Wide-cut fuels have lower viscosities and freezing points than kerosenes while heavier fuels used in ground turbines exhibit much higher viscosities and freezing points.

Low temperature viscosities have an important influence on fuel atomization and affect engine starting. Cycloparaffinic and aromatic fuels reach unacceptably high levels of viscosity at low temperatures. A kinematic viscosity of 35 mm2/s represents the practical upper limit for pumps on aircraft while much higher limits are acceptable for ground installations.

Water in Fuel

The solubility of water in a hydrocarbon is related to the molecular structure of the hydrocarbon and its temperature. Aromatics dissolve about five times more water than corresponding paraffins and in turn lower molecular weight paraffins dissolve more water than longer chain compounds. Figure 8 is a plot of water solubility for certain pure hydrocarbons and typical jet fuels. The water saturation values in fuel vary considerably because of composition effects. For example, wide-cut fuel normally dissolves more water than does kerosene, but because the latter may contain twice as many aromatics, the higher molecular weight fuel may in fact exhibit an equivalent water saturation curve. At a temperature of 23°C, a Coordinating Research Council program on Jet A fuel showed a range of 56–120 ppm (14).

The slope of the water solubility curves for fuels is about the same, and is constant over the 20–40°C temperature range. Each decrease of 1°C decreases water solubility about 3 ppm. The sensitivity of dissolved water to fuel temperature change is important. For example, the temperature of fuel generally drops as it is pumped into an airport underground hydrant system because subsurface temperatures are about 10°C lower than typical storage temperatures. This difference produces free water droplets, but these are removed by pumping fuel through a filter–coalescer and hydrophobic barrier before delivery into aircraft.

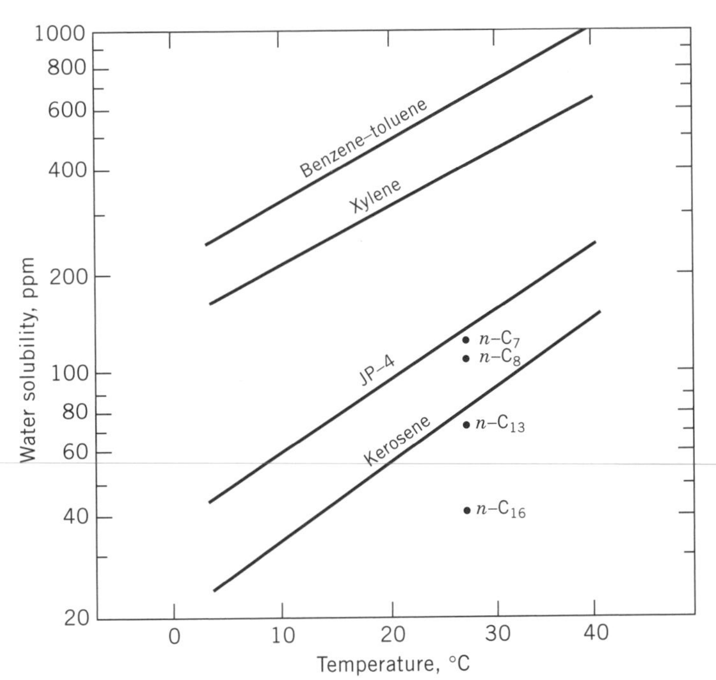

Figure 8. Variation of water solubility with temperature of gas turbine fuels (13).

The amount of water actually dissolved in fuel at any given temperature is determined by equilibration with the water in the atmosphere above the fuel. If a tank is vented to atmosphere, fuel may enter saturated with water but lose half its dissolved water when the relative humidity is 50%. Conversely, a tank of fuel in a humid environment will rapidly pick up water. Many storage tanks contain floating roofs which effectively eliminate any opportunity for dissolved water to be removed as vapor. The tank in the aircraft behaves like a storage tank on the ground. In the upper atmosphere, the partial pressure of water is very low and fuel tends to dry out; dissolved water leaves fuel and exits from the vent. An aircraft descending through clouds will pick up water and saturate its fuel. At the end of flight, fuel in a tank is apt to be both cold and cloudy with free water.

A stable cloud of water in fuel usually means that a surfactant is present to form a stable water-in-oil emulsion. Smaller droplets resist natural coalescence processes. A surfactant that is potent as an emulsifying agent is apt to disarm the coalescing filters allowing excess water to be delivered with fuel.

The effects of water in fuel inside a tank, particularly an aircraft tank, are important because of the demonstrated proclivity of free water to form undrainable pools where microorganisms can flourish. In ground storage tanks, an attempt is made to exclude water bottoms by proper design and regular draining practices. This is more difficult in aircraft because pools of water freeze in flight and ice may not melt when fueling turnarounds are rapid. In a large storage tank, organisms (usually fungi and yeasts) develop at fuel–water interfaces forming a cuff of growth which can plug filters. In the aircraft, the growth usually takes place on tank surfaces, forming a fungal mat under which metabolic products such as organic acids penetrate polymeric coatings to attack the aluminum skin itself. The growth may also affect the capacitance gage used to read liquid level.

It is common practice to curb growth of organisms by biocidal treatment. In storage tanks, a water soluble agent is used. Aircraft tanks are opened periodically for hand cleaning and subsequent treatment with a fuel-soluble boron-containing biocide. The anti-icing additive used by the military to lower the freezing point of water, 2 methoxyethanol, happens also to inhibit growth of organisms effectively. Therefore, aircraft tank treatment to remove fungal mats is needed only for commercial transports.

Water plays a primary role in corrosion of the metal walls of tanks (15) and pipes and increases the tendency for high speed pumps to produce wear particles and exhibit shortened life. Formation of corrosion products can be controlled by addition of corrosion inhibitors, a mandatory additive in military fuels. However, corrosion inhibitors may also degrade other fuel properties and adversely affect ground filtration equipment. Thus, they are not generally acceptable in commercial fuels where rigorous attention is given to clean and dry fuels upon aircraft fueling.

Compatibility and Corrosion

Gas turbine fuels must be compatible with the elastomeric materials and metals used in fuel systems. Elastomers are used for O-rings, seals, and hoses as well as pump parts and tank coatings. Polymers tend to swell when in contact with aromatics and improve their sealing ability, but degree of swell is a function of both elastomer type and aromatic molecular weight. Rubbers can also be attacked by peroxides that form in fuels that are not properly inhibited.

Attack on metals can be a function of fuel components as well as of water and oxygen. Organic acids react with cadmium plating and zinc coatings. Traces of H_2S and free sulfur react with silver used in older piston pumps and with copper used in bearings and brass fittings. Specification limits by copper and silver strip corrosion tests are required for fuels to forestall these reactions.

Boundary lubrication of rubbing surfaces such as those found in high speed pumps and fuel controls has been found to be related to the presence in fuel of species capable of forming a chemisorbed film that reduces friction and wear (16). The lubricity of fuel is attributed to polar materials (17), multiring aromatics (17), and a thiohydroindane species (18). Oxygen containing compounds, particularly long chain acids, tend to adsorb on metal surfaces and act as lubricity agents. Corrosion inhibitors which tend to form tenacious films on metal surfaces are generally excellent lubricity agents. Refining processes to reduce sulfur and control acidity tend to degrade a fuel's lubrication propensity. Corrosion inhibitors are mandatory in military fuels since they also improve the lubricity of fuels by extending the life of pumps and controls.

Both friction and wear measurements have been used to study boundary lubrication of fuel because sticking fuel controls and pump failures are principal field problems in gas turbine operation. An extensive research program of the Coordinating Research Council has produced a ball-on-cylinder lubricity test (BOCLE), standardized as ASTM D5001, which is used to qualify additives, investigate fuels, and assist pump manufacturers (19).

ECONOMIC ASPECTS

The exacting list of specification requirements for aviation gas turbine fuels and the constraints imposed by delivering clean fuel safely from refinery to aircraft are the factors that affect the economics. Compared with other distillates such as diesel and burner fuels, kerosene jet fuels are narrow-cut specialized product and usually command a premium price over other distillates. The prices charged for jet fuels tend to escalate with the basic price of crude, a factor which seriously undermined airline profits during the Iran revolution as crude prices increased sharply.

Availability and cost of jet fuels are also affected by the interrelationships of the market for other petroleum products. An excellent example of these interrelationships was developed for NASA in a study of how jet fuel property changes affected producibility and cost in the U.S., Canada, and Europe (20). For example the western United States is a unique market compared with the rest of the country because demand for gasoline and jet fuel is high while demand for burner and boiler fuels is low. This product distribution encourages high conversion refineries with particular emphasis on hydrocracking to make lighter products. At the same time environmental concerns over urban air quality are pressuring refineries to market reformulated gasolines and low sulfur diesel fuels. Refining these products and also producing specification jet fuel has tended to limit availability and increase the cost of aviation fuels in the western United States.

The majority of the hundreds of refineries operated by petroleum companies in different parts of the world to make local products such as gasoline and burner fuels also produce jet fuels. Even a small refinery with simple equipment can make suitable jet fuel if it has access to the right crude. However, the principal supply of both civil and military jet fuels is produced in large refineries. Many are located near large cities and airports and are frequently connected by pipeline directly to the airport. Modern airports have extensive storage and handling facilities operated by local authorities, by petroleum companies or consortia, or by the airlines themselves.

Worldwide demand for the jet fuels specified in Table 2 amounted to about 3 million B/D (477 k cubic meters per day) in 1990. About one half of this demand was kerosene Jet A sold in the United States. One third represented kerosene Jet A1 for delivery to international airlines outside the United States, while the balance comprised various military fuels used by air forces around the world.

AVIATION GASOLINE

The specifications for aviation gasoline are even more exacting than specifications for jet fuel (see selected specifications, Table 3). Antiknock requirements dictate that high grade fuel be blended mostly with iso-octane and tetraethyl lead. Table 4 summarizes the key requirements of ASTM D910 the specification used by both civil and military aircraft. Antiknock ratings, determined in a single-cylinder laboratory engine, represent the detonation limit of a spark-ignited fuel–air mixture under the same conditions as the ASTM D2700 Motor Method used to rate automotive gasoline. A second rating under rich conditions of a fuel–air mixture is required using ASTM D909 Supercharge Method to ensure adequate take-off power. Both the lean and rich ratings are related to iso-octane rated at 100 octane or performance number. Use of too low an octane grade can wreck an engine by detonation; hence, each grade is identified by a color dye. Higher powered engines require Grade 100/130 produced as either green fuel with 4mL/USG of TEL or blue fuel with half this amount of lead. Low compression engines tolerate less lead and require Grade 80/87 with only 0.5mL/USG of TEL.

The volatility requirements for all grades are determined by the need to vaporize fuel for cold starting and to control the rate of engine warm-up. Controls on distilla-

Table 3. Selected Specification Properties of Aviation Gasoline[a]

Characteristic	Grade 80	Grade 100	Grade 100LL
Knock rating, lean min O.N.	80	100	100
Knock rating, rich min P.N.	87	130	130
Color	Red	Green	Blue
Tetraethyl lead, g Pb/Lmax	0.19	1.12	0.56
Distillation T, °C, 10% max		75	
°C, FBP max		170	
Vapor pressure, kPa[b]		38–49	
Net heat content, MJ/kg[c]		43.54	
Sulfur, % mass, max		0.05	
Freezing point, max, °C		−58	

[a] Detailed requirements appear in ASTM D910.
[b] To convert kPa to psi, multiply by 0.145.
[c] To convert MJ/kg to Btu/lb, multiply by 430.2.

tion are also needed to exclude undesirable heavy ends while vapor pressure limits are specified to avoid excessive vapor formation at high altitude. Other exacting requirements for aviation gasoline include long-term storage stability, noncorrosiveness, low sulfur, good low temperature performance, and minimum water reaction. The specific energy minimum tends to limit the aromatics used for blending. Antioxidants are mandatory.

The shrinking market for all grades of aviation gasoline has reduced the numbers of producers, fostered exchange arrangements, and increased the cost of distribution and handling. The resulting high price of aviation gasoline has encouraged operators of low compression engines to replace Grade 80 with unleaded premium motor gasoline, a practice supported by the Federal Aviation Administration which has issued supplemental type certificates for certain models of aircraft and engines.

FUTURE TRENDS

Aircraft Fuels

Demand for aviation gas turbine fuels has been growing more rapidly than other petroleum products since 1960, about 3–5% per year compared with 1% for all oil products. This strong demand reflects a current and predicted growth in worldwide air traffic of 4–7% annually until the end of the century. Total world oil demand will be up by 15% by the year 2000, but aviation fuel demand will increase by 50–125%. However, the fraction of the oil barrel devoted to aviation, now about 8%, will increase only slightly.

Changes in the crude supply outlook are in fact decreasing the potential yield of the straight-run kerosene fraction needed not only for aviation but also for blending into diesel and heating fuels. For example, Arabian Gulf crudes, the most abundant in the world, yield less kerosene than the primary U.S. crude. This situation will require more conversion processes at refineries such as hydrocracking of gas oil, a process that yields high quality but more expensive aviation kerosene.

The vulnerability of the aviation market was exposed in the winter of 1973–1974 when the Arab oil embargo caused actual shortages of jet fuel (21). Coupled with the dramatic rise in crude and product prices, the supply crisis precipitated a new generation of fuel-efficient aircraft and a relaxation of specification limits such as higher freezing point to −47°C and higher aromatics to 25%. It has been

Table 4. Additives Used in Aviation Gas Turbine Fuels

Class of Additive	Chemical Types	Purpose	Specification Status[a] Civil	Specification Status[a] Military
Antioxidant	Alkyl phenylene diamines, hindered alkylphenols	Improve oxidation stability–storage and high temperature	O	O
Metal deactivator	Disalicylidene alkyl diamine	Remove metal ions that catalyze oxidation	O	O
Antiicing	Glycol ether	Prevent low temperature filter icing		R
Corrosion inhibitor	Dimer acid, phosphate ester	Reduce development of corrosion products, improve lubrication		R
Antistatic	Alkyl chromium salt and calcium sulfosuccinate, olefin–sulfur dioxides and polyamines	Increase conductivity	OR	R
Biocide	Boron ester	Inhibit organic growth in fuel tanks	A	

[a] O = optional, R = required, A = airline use in maintenance.

estimated that 18% of the world's crudes produce a 150–250°C fraction containing more than 20% aromatics. Two of these, Heavy Arabian and North Slope, represent the important new sources of marginal jet fuel.

Operation of aircraft gas turbines on a wider cut than the 160–260°C fraction of crude to expand availability was considered in a 1977 ASTM Symposium (22). In the case of Jet A, a reduction of flash point from 38°C to 32°C would increase yield by 17% while an increase in Jet A1 freezing point to the −40°C Jet A level used domestically would add about 25% to the kerosene pool.

Synthetic fuels derived from shale or coal will have to supplement domestic supplies from petroleum some day. Aircraft gas turbine fuels producible from these sources were assessed during the 1970s. Shale-derived fuels can meet current specifications if steps are taken to reduce the nitrogen levels. However, extracting kerogen from shale rock and denitrogenating the jet fuel are energy-intensive steps compared with petroleum refining; it has been estimated that shale jet fuel could be produced at about 70% thermal efficiency compared with 95% efficiency for petroleum (23). Such a difference would represent much higher cost for a shale product.

Synthetic jet fuel derived from coal is even more difficult and expensive since the best of the conversion processes produces a fuel very high in aromatics. With hydrogenation, overall thermal efficiency is only 50%. Without additional hydrogenation, the gas turbine fuels would contain 60–70% aromatics.

A third alternative, production of liquid jet fuel from processing of abundant natural gas, is a more promising and cheaper source of high quality product than shale or coal.

Fuels for Advanced Aircraft

The area of greatest traffic growth is predicted to be the Pacific basin where distances are great and the greatest incentive exists for an advanced aircraft that could greatly reduce travel time. Research at the National Aeronautic and Space Administration is proceeding on a high speed commercial transport (HSCT) that would satisfy this market by the year 2010. The aircraft would carry about 300 passengers nonstop over 8000 miles at three times the speed of sound and reduce travel time from 14 to 6 hours. To be economically and environmentally viable, the HSCT would have to produce less nitrogen oxides in exhaust to cause lower sonic boom and to operate on the same kerosene as subsonic aircraft.

The first commercial supersonic transport, the Concorde, operates on Jet A1 kerosene but produces unacceptable noise and exhaust emissions. Moreover, it is limited in capacity to 100 passengers and to about a 3000 mile range. At supersonic speed of Mach 2, the surfaces of the aircraft are heated by ram air. The surfaces heated by ram air can raise the temperature of fuel held in the tanks to 80°C. Since fuel is the coolant for airframe and engine subsystems, fuel to the engine can reach 150°C (24). An HSCT operated at Mach 3 would place much greater thermal stress on fuel. To minimize the formation of thermal oxidation deposits, it is likely fuel delivered to the HSCT would have to be deoxygenated.

NASA is also considering a more advanced aircraft such as Mach 5 to cut Pacific travel time to about 3 h, but in this case kerosene fuel is no longer acceptable and liquefied natural gas or liquefied hydrogen would be needed to provide the necessary cooling and stability. However, a completely new fueling system would be required at every international airport to handle these cryogenic fuels.

In the speed regime of Mach 6 and beyond, hypersonic vehicles, hydrogen is the fuel of choice because of its high energy content and combustion kinetics. For example, a first stage to orbit reusable booster for space vehicles would employ a supersonic combustor ramjet engine burning hydrogen, the only fuel capable of maintaining combustion in the shock wave. A booster that replaced 80% of the present weight of oxidizer would revolutionize space transportation by replacing the current space shuttle as a delivery vehicle.

Use of kerosene fuel rather than air as a coolant focuses attention on another fuel property (specific heat) which is a measure of its efficiency to absorb thermal energy. Heat capacity increases with temperature but decreases with density which favors a low density paraffinic fuel from petroleum or natural gas sources.

Because of tank heating, fuel volatility is also more critical in supersonic aircraft. For example, the Concorde tank is pressurized to prevent vapor losses which could be significant at high altitude where fuel vapor pressure may equal atmospheric pressure. The tank can reach 6.9 kPa at the end of flight. The need to deoxygenate fuel for thermal stability in the HSCT will doubtless require a similar pressurized system.

BIBLIOGRAPHY

1. W. G. Dukek, A. R. Ogston, and D. R. Winans, *Milestones in Aviation Fuels,* AIAA No. 69-779, American Institute of Aeronautics and Astronautics, New York, July 1969.
2. R. S. Schlaefer and S. D. Heron, *Development of Aircraft Engines and Fuels,* Harvard Business School, Boston, Mass., 1950.
3. K. M. Brown, *Commercial Results with UOP MEROX Process for Mercaptan Extraction in U.S. and Canada,* UOP Booklet 267, UOP, Inc., Des Plaines, Ill., 1960.
4. W. M. Bustin and W. G. Dukek, *Electrostatic Hazards in the Petroleum Industry,* John Wiley & Sons, Inc., Chichester, U.K., 1983.
5. N. A. Chigler, "Atomization and Burning of Liquid Fuel Sprays," in *Progress in Energy and Combustion Science,* Vol. 21 No. 4, Pergamon Press, Inc., Elmsford, N.Y., 1976.
6. *Microburner Studies of Flame Radiation as Related to Hydrocarbon Structure,* Report 3752-64R, Phillips Petroleum Co., Navy Buweps Contract NOw 63-0406d May 1964.
7. M. Smith, *Aviation Fuels,* G. T. Foulis & Co. Ltd., Henley-on-Thames, U.K., 1970.
8. H. F. Butze and R. C. Ehlers, *Effect of Fuel Properties on Performance of a Single Aircraft Turbojet Combustor,* NASA TM X-71789, National Aeronautics and Space Administration, Lewis Research Center, Cleveland, Ohio, Oct. 1975.
9. W. F. Taylor, *Development of High Stability Fuels,* contract N00140-74-C-0618, Naval Air Propulsion Center, Trenton, N.J., Dec. 1975.
10. W. F. Taylor, *J. Appl. Chem.* **18,** 25 (1968).

11. R. H. Clark, "The Role of Metal Deactivator in Improving the Thermal Stability of Aviation Kerosenes," *3rd Intl. Conference on Stability and Handling of Liquid Fuels,* London, Sept. 1988.
12. H. C. Barnett and R. R. Hibbard, *Properties of Aircraft Fuels,* TN 3276, NASA, Lewis Research Center, Cleveland, Ohio, Aug. 1956.
13. *Aviation Fuel Safety–1975,* CRC Report no. 482, Coordinating Research Council, Inc., Atlanta, Ga., Nov. 1975.
14. *Survey of Electrical Conductivity and Charging Tendency Characteristics of Aircraft Turbine Fuels,* CRC Report no. 478, Coordinating Research Council, Inc., Atlanta, Ga., Apr. 1975.
15. H. R. Porter and co-workers, *Salt Driers Assure Really Dry Fuel to Aircraft,* SAE 710440, Society of Automotive Engineers, Warrendale, Pa.
16. J. J. Appeldoorn and W. G. Dukek, *Lubricity of Jet Fuels,* SAE 660712, Society of Automotive Engineers, Warrendale, Pa.
17. J. K. Appeldoorn and F. F. Tao, *Wear,* 12 (1968).
18. R. A. Vere, *SAE Trans.* **78,** 2237 (1969).
19. W. G. Dukek, *Ball-on-Cylinder Testing for Aviation Fuel Lubricity,* SAE 881537, Society of Automotive Engineers, Warrendale, Pa.
20. G. M. Varga and A. J. Avella, *Jet Fuel Property Changes and Their Effect on Producibility and Cost in U.S., Canada and Europe,* NASA Contract Report 174840, NASA, Cleveland, Ohio, Feb. 1985.
21. A. G. Robertson and R. E. Williams "Jet Fuel Specifications: The Need for Change," *Shell Aviation News,* Shell Oil Co. Houston, Tex., 1976, p. 435.
22. *Factors in Using Kerosine Jet Fuel of Reduced Flash Point,* ASTM STP 688, American Society for Testing and Materials, Philadelphia, Pa., 1979.
23. W. Dukek and J. P. Longwell, "Alternative Hydrocarbon Fuels for Aviation," *Esso Air World,* No. 4, Exxon International Co., Florham Park, N.J.
24. H. Strawson and A. Lewis, *Predicting Fuel Requirements for the Concorde,* SAE 689734, Society of Automotive Engineers, Warrendale, Pa., Oct. 1988.

ALCOHOL FUELS

MICHAEL D. JACKSON
CARL B. MOYER
Acurex Corporation

The use of alcohols as motor fuels gained considerable interest in the 1970s as substitutes for gasoline and diesel fuels, or, in the form of blend additives, as extenders of oil supplies. In the United States, most applications involved the use of low level ethanol [*64-17-5*] blends in gasoline [*8006-61-9*]. Brazil, however, launched a major program to substitute ethanol for gasoline in 1976, beginning with a 22% alcohol ethanol–gasoline blend and adding dedicated ethanol cars in the 1980s. By 1985 ethanol cars accounted for 95% of new car sales in Brazil. By 1988 Brazil had 3,000,000 automobiles, about 30% of the total automobile population, dedicated to ethanol. The United States has demonstration vehicles using alcohols (mostly methanol [*67-56-1*]), but otherwise has not yet passed beyond the use of limited amounts of alcohols and of ethers produced from alcohols as gasoline components. However, proposals continue to be made to implement alcohol programs similar to that of Brazil (1). Other nations such as New Zealand, Germany, and Sweden also investigated the use of alcohols as a transportation fuel.

The benefits of alcohol fuels include increased energy diversification in the transportation sector, accompanied by some energy security and balance of payments benefits, and potential air quality improvements as a result of the reduced emissions of photochemically reactive products. The Clean Air Act of 1990 and emission standards set out by the State of California may serve to encourage the substantial use of alcohol fuels, unless gasoline and diesel technologies can be developed that offer comparable advantages.

See also AIR POLLUTION: AUTOMOBILE; AIR POLLUTION: AUTOMOBILE, TOXIC EMISSIONS; AUTOMOBILE EMISSIONS, CONTROL; CLEAN AIR ACT, MOBILE SOURCES; EXHAUST CONTROL, AUTOMOTIVE; METHANOL VEHICLES; TRANSPORTATION FUELS, AUTOMOTIVE FUELS; TRANSPORTATION FUELS, ETHANOL FUELS IN BRAZIL.

PROPERTIES OF ALCOHOL FUELS

Table 1 summarizes key properties of ethanol and methanol as compared to other fuels. Both alcohols make excellent motor fuels, although the high latent heats of vaporization and the low volatilities can make cold-starting difficult in vehicles having carburetors or fuel injectors in the intake manifold where the fuel must be vaporized prior to being introduced into the combustion chamber. This is not the case for direct injection diesel-type engines using methanol or ethanol. Both methanol and ethanol have high octane values and allow high compression ratios having increased efficiency and improved power output per cylinder. Both have wider combustion envelopes than gasoline and can be run at lean air–fuel ratios with better energy efficiency. However, ethanol and methanol have very low cetane numbers and cannot be used in compression–ignition diesel-type engines unless gas temperatures are high at the time of injection. Manufacturers of heavy-duty engines have developed several types of systems to assist autoignition of directly injected alcohols. These include glow plugs or spark plugs, reduced engine cooling, and increased amounts of exhaust gas recirculation or, in the case of two-stroke engines, reduced scavenging. Additives to improve cetane number have been effective as have dual-fuel approaches, in which a small amount of diesel fuel is used as an ignitor for the alcohol.

There are particular alcohol fuel safety considerations. Unlike gasoline or diesel fuel, the vapor of methanol or ethanol above the liquid fuel in a fuel tank is usually combustible at ambient temperatures. This poses the risk of an explosion should a spark or flame find its way to the tank such as during refueling. Additionally, a neat methanol fire has very little luminosity and, consequently, fire fighting efforts can be difficult in daylight. However, low luminosity also implies low radiative fluxes from the fire. This, combined with the high latent heat of vaporization, means that the heat release of a methanol fire is low relative to one of gasoline or diesel fuel. Because methanol or ethanol are both water-soluble, fires can be successfully controlled by dilution with large amounts of water, a tactic

Table 1. Properties of Fuels[a]

Properties	Methanol	Ethanol	Propane	Methane	Isoctane	Unleaded gasoline	Diesel Fuel #2
Constituents	CH_3OH	CH_3CH_2OH	C_3H_8	CH_4	C_8H_{18}	$C_nH_{1.87n}$ (C_4 to C_{12})	$C_nH_{1.8n}$ (C_8 to C_{20})
CAS Registry Number	*[67-65-1]*	*[64-17-5]*	*[74-98-6]*	*[74-82-8]*	*[540-84-1]*	*[8006-6-9]*	
Molecular weight	32.04	46.07	44.10	16.04	114.23	≈110	≈170
Element composition, wt %							
C	37.49	52.14	81.71	74.87	84.12	86.44	86.88
H	12.58	13.13	18.29	25.13	15.88	13.56	13.12
O	49.93	34.73	0	0	0	0	0
Density at 16°C and 101.3 kPa[b], kg/m³	794.6	789.8	505.9	0.6776	684.5	721–785	833–881
Boiling point at 101.3 kPa[b], °C	64.5	78.3	6.5	−161.5	99.2	38–204	163–399
Freezing point, °C	−97.7	−114.1	−188.7	−182.5	−107.4		<−7
Vapor pressure at 38°C, kPa[b]	31.9	16.0	1.297	0.5094	11.8	48–108	negligible
Heat of vaporization, ΔH_v, MJ/kg[c]	1.075	0.8399	0.4253	0.5094	0.2712	0.3044	0.270
Gross heating value, MJ/kg[c]	22.7	29.7	50.4	55.5	47.9	47.2	44.9
Net heating value							
MJ/kg[c]	20.0	27.0	46.2	50.0	44.2	43.9	42.5
MJ/m³[d]	15,800	21,200	23,400		30,600	32,000	35,600
Stoichiometric mixture net heating value, MJ/kg[c]	2.68	2.69	2.75	2.72	2.75	2.83	2.74
Autoignition temperature, °C	464	363	450	537	418	260–460	257
Flame temperature, °C	1,871	1,916	1,988	1,949	1,982	2,027	1,993
Flash point, °C	11	13			4	−43 to −39	52–96
Flame speed at stoichiometry, m/s	0.43		0.40	0.37	0.31	0.34	
Octane ratings							
Research	106	107	112	120	100	92–98	
Motor	92	89	97	120	100	80–90	
Cetane rating	0–5						>40
Flammability limits, vol % in air	6.72–36.5	3.28–18.95	2.1–9.5	5.0–15.0	1.0–6.0	1.4–7.6	1.0–5.0
Stoichiometric air–fuel mass ratio	6.46	8.98	15.65	17.21	15.10	14.6	14.5
Stoichiometric air–fuel volumetric ratio	7.15	14.29	23.82	9.53	59.55	55	85
Water solubility	complete	complete	no	no	no	no	no
Sulfur content, wt %	0	0	0	0	0	<0.06	<0.5

[a] Refs. 2–7.
[b] To convert kPa to psi, multiply by 0.145.
[c] To convert MJ/kg to Btu/lb, multiply by 430.3
[d] To convert MJ/m³ to Btu/gal, multiply by 3.59.

that simply spreads gasoline fires. Nevertheless, fire-extinguishing foams are the preferred alcohol fire-fighting method.

Some potential problems of alcohol fuels have been addressed by adding small amounts of gasoline or specific hydrocarbons to the fuel, reducing the flammability envelope and providing luminosity in case of fire.

USES OF ALCOHOL FUELS

Early applications of internal combustion engines featured a variety of fuels, including alcohols and alcohol–hydrocarbon blends. In 1907 the U.S. Department of Agriculture investigated the use of alcohol as a motor fuel. A subsequent study by the U.S. Bureau of Mines concluded that engines could provide up to 10% higher power on alcohol fuels than on gasoline (8). Mixtures of alcohol and gasoline were used on farms in France and in the United States in the early 1900s (9). Moreover, the first Ford Model A automobiles could be run on either gasoline or ethanol using a manually adjustable carburetor (1). However, the development of low cost gasoline pushed other automobile fuels into very minor roles and the diesel engine further solidified the hold of petroleum fuels on the transportation sector. Ethanol was occasionally used, particularly in rural regions, when gasoline supplies were short or when farm prices were low.

Methanol has been used as a motor racing fuel for many decades. Its high latent heat of vaporization cools the incoming charge of air to each cylinder. Increasing the mass of air taken into each cylinder increases the power developed by each stroke, providing a turbocharger effect. Furthermore, methanol has a higher octane value than gasoline allowing higher compression ratios, greater efficiency, and higher output per unit of piston displacement volume. These power increases are advantages in racing as is the simple means by which methanol fuel quality and uniformity can be verified.

The transparency of methanol flames is usually a safety advantage in racing. In the event of fires, drivers have some visibility and the lower heat release rate of methanol provides less danger for drivers, pit crews, and spectators.

Partly for these reasons, methanol has been the required fuel of the Indianapolis 500 since 1965. Methanol is also used in many other professional and amateur races. However, transparency of the methanol flames has also been a disadvantage in some race track fires. The invisibility of the flame has confused pit crews, delayed fire detection, and caused even trained firefighters problems in locating and extinguishing fires.

Low level blends of ethanol and gasoline enjoyed some

popularity in the United States in the 1970s. The interest persists into the 1990s, encouraged by the exemption of low level ethanol–gasoline blends from the Federal excise tax as well as from state excise taxes in many states.

ENERGY DIVERSIFICATION AND ENERGY SECURITY

The ethanol program in Brazil addressed that country's oil supply problems in the 1970s, at times improved the balance of payments, and served to strengthen the economy of the sugar production portion of the agricultural sector. Although the benefits are difficult to quantify because of the very high inflation rate (10) the Brazilian ethanol program generally met its goals. Ironically the inability of ethanol to substitute successfully for diesel fuel in the heavy-duty truck sector necessitated keeping refinery outputs up to provide sufficient available diesel fuel. Thus gasoline was in surplus in Brazil and oil imports were not as much affected as originally hoped. The Brazilian program demonstrated that petroleum substitution strategies need to address the "whole barrel" product slate of oil refineries.

In the 1990s world events precipitated renewed interest in energy diversification strategies for the U.S. transportation sector. However, few measures are in place to encourage fuel alternatives outside of the exemption from the Federal excise tax on motor fuel granted to ethanol blends. The Alternative Motor Fuels Act of 1988 did extend credits to automobile manufacturers in the calculation of corporate average fuel economy (CAFE) for vehicles that use methanol or ethanol or natural gas; electric vehicles had previously been granted such a credit. Under the Act's provisions, neither ethanol nor methanol is counted as fuel consumed in the calculation of fuel economy. Thus vehicles that use alcohol have very high fuel economy ratings, reflecting the value of these vehicles in reducing oil imports. The credits take effect in model-year 1993. The fuel economy calculation assumes that methanol and ethanol fuels in commerce contain 15% by volume gasoline. Vehicles that can use either alcohol fuels or gasoline receive a reduced CAFE credit. The maximum credit that can be earned by a manufacturer for selling vehicles capable of using petroleum fuels is capped because alcohol fuels usage by these vehicles is not assured.

ALCOHOL AVAILABILITY

Methanol

If methanol is to compete with conventional gasoline and diesel fuel it must be readily available and inexpensively produced. Thus methanol production from a low-cost feed stock such as natural gas [*8006-14-2*] or coal is essential. There is an abundance of natural gas (see NATURAL GAS) worldwide and reserves of coal are even greater than those of natural gas.

Natural Gas Reserves. U.S. natural gas reserves could support a significant methanol fuel program. 1990 proved, ie, well characterized amounts with access to markets and producible at current market conditions U.S. resources are 4.8 trillion cubic meters (168 trillion cubic feet = 168 TCF). U.S. consumption is about one-half trillion m^3 (18 TCF) or 18 MJ equivalents per year, but half of that is imported. Estimates of undiscovered U.S. natural gas reserves range from 14 to 16 trillion m^3 (492 to 576 TCF) or roughly a 30-year supply at current U.S. consumption rates. Additional amounts of natural gas may become available from advanced technologies.

If 10% of the U.S. gasoline consumption were replaced by methanol for a twenty year period, the required reserves of natural gas to support that methanol consumption would amount to about one trillion m^3 (36 TCF) or twice the 1990 annual consumption. Thus the United States could easily support a substantial methanol program from domestic reserves. However, the value of domestic natural gas is quite high. Almost all of the gas has access through the extensive pipeline distribution system to industrial, commercial, and domestic markets and the value of gas in these markets makes methanol produced from domestic natural gas uncompetitive with gasoline and diesel fuel, unless oil prices are very high.

It is therefore more relevant to examine world resources of natural gas in judging the supply potential for methanol. World proved reserves amount to approximately 1.1×10^{15} m^3 (40,000 TCF) (11). As seen in Figure 1, these reserves are distributed more widely than oil reserves.

Using estimates of proven reserves and commitments to energy and chemical uses of gas resources, the net surplus of natural gas in a number of different countries that might be available for major fuel methanol projects has been determined. These are more than adequate to support methanol as a motor fuel.

Coal Reserves. As indicated in Table 2, coal is more abundant than oil and gas worldwide. Moreover, the U.S. has more coal than other nations: U.S. reserves amount to about 270 billion metric tons, equivalent to about 11×10^{16} MJ (1×10^{20} BTU = 6600 quads), a large number compared to the total transportation energy use of about 3.5×10^{14} MJ (21 quads) per year (11). Methanol produced from U.S. coal would obviously provide better energy security benefits than methanol produced from imported natural gas. At present however, the costs of producing methanol from coal are far higher than the costs of producing methanol from natural gas.

Biomass. Methanol can be produced from wood and other types of biomass. The prospects for biomass reserves are noted below.

Ethanol

From the point of view of availability, ethanol is extremely attractive because it can be produced from renewable biomass. Estimates of the amount of ethanol that could be produced from biomass on a sustained basis range from about 3×10^{13} to more than 8×10^{14} MJ/yr (2–50 quad/yr), about half of which would derive from wood crops (11).

ECONOMIC ASPECTS

Alcohol Production

Studies to assess the costs of alcohol fuels and to compare the costs to those of conventional fuels contain significant uncertainties. In general, the low cost estimates indicate

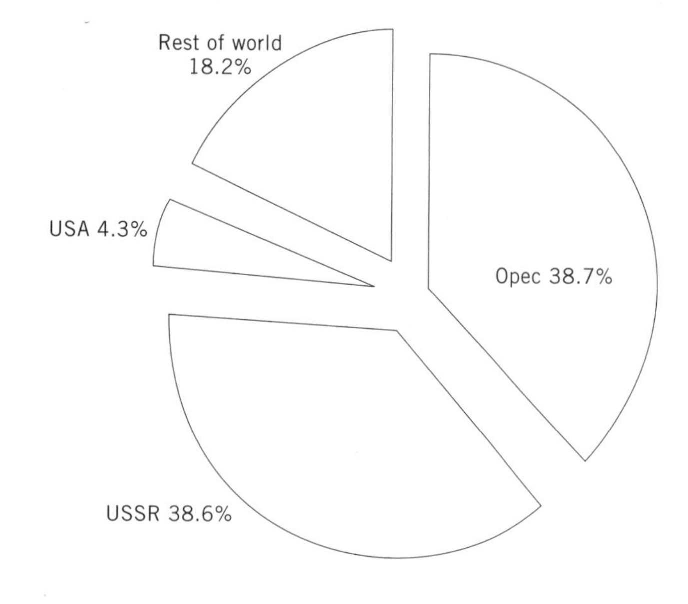

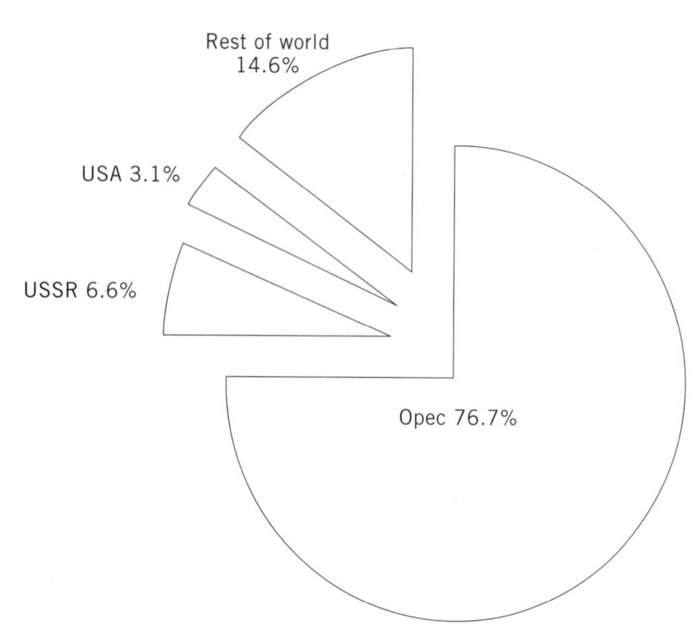

Figure 1. Distribution of the world proven gas and oil reserves as of January 1, 1988. (**a**) 107.5 × 10^{12} m^3 (3,800 TCF) natural gas; (**b**) 140 m^3 (890 barrels) oil.

that methanol produced on a large scale from low cost natural gas could compete with gasoline when oil prices are around 14¢/L ($27/bbl). This comparison does not give methanol any credits for environmental or energy diversification benefits. Ethanol does not become competitive until petroleum prices are much higher.

Methanol

Produced from Natural Gas. Cost assessments of methanol produced from natural gas have been performed (13–18). Projections depend on such factors as the estimated costs of the methanol production facility, the value of the feedstock, and operating, maintenance, and shipping costs. Estimates vary for each of these factors. Costs also depend on the value of oil. Oil price not only affects the value of natural gas, it also affects the costs of plant components, labor, and shipping.

Estimates of the landed costs of methanol (the costs of methanol delivered by ship to a bulk terminal in Los Angeles), vary between 8.9 and 15.6 cents per liter (33.6 and 59.2 cents per gallon). Estimates range from 7.9 to 11.1 cents per liter (30 to 42 cents per gallon) in a large-scale established methanol market and 11.9 to 19.3 cents per liter (45 to 73 cents per gallon) during a small volume transition phase (18).

Estimated pump prices must take terminaling, distribution, and retailing costs into account as well as any differences in vehicle efficiency that methanol might offer. Estimates range from no efficiency advantage to about 30% improvement in fuel efficiency for dedicated and optimized methanol light-duty vehicles. Finally, the cost comparison must take into account the fuel specification for methanol. Most fuel methanol for light-duty vehicles is in the form of 85% methanol, 15% gasoline by volume often termed M85.

The sum of the downstream costs adds roughly 7.9 cents per liter (30¢/gal) and the adjustment of the final cost for an amount of methanol fuel equivalent in distance driven to an equal volume gasoline involves a multiplier ranging from 1.6 to 2.0, depending on fuel specification and the assumed efficiency for methanol light-duty vehicles as compared to gasoline vehicles. The California Advisory Board has undertaken such cost assessment (11).

A cost assessment must also take into account the volume of methanol use ultimately contemplated. A huge methanol program designed to replace most of the U.S. gasoline use would be likely to increase the value of even remotely located natural gas. A big program would also tend to decrease the price of oil, making it more difficult for methanol to compete as a motor fuel. Nevertheless, a balanced assessment of all the studies appears to indicate that a large scale methanol project could provide a motor fuel that competes with gasoline when oil prices are not less than about $0.17/L (1988 U.S. $27/bbl).

In small volume transition phases, methanol cannot compete directly in price with gasoline unless oil prices become very high, with the possible exception of a few scenarios in which low cost methanol is available from expansions to existing methanol plants currently serving the chemical markets for methanol. Energy diversification benefits have not been quantified but the potential air quality benefits have been studied in the work of the California Advisory Board on Air Quality and Fuels. The investment required to introduce methanol might well be justified on air quality grounds, at least in those areas such as California where air pollution programs involve substantial costs. However, the relative advantage of methanol depends on the emissions levels from future vehicles and on the costs of cleaner gasolines that might be able to offer environmental benefits that compete at least to some extent with the environmental benefits of methanol.

Produced from Coal. Estimates of the cost of producing methanol from coal have been made by the U.S. Depart-

Table 2. World Estimated Recoverable Reserves of Coal in Billions of Metric Tons[a]

Location	Anthracite and Bituminous Coal[b]	Lignite[c]	Total
North America			
Canada	4.43	2.42	6.85
Mexico	1.91		1.91
United States	231.13	32.71	263.84
total	*237.47*	*35.13*	*272.60*
Central and South America			
total	*5.13*	*0.02*	*5.15*
Western Europe			
total	*32.20*	*58.23*	*90.43*
Eastern Europe and USSR			
USSR	150.19	94.53	244.72
total	*182.82*	*139.17*	*321.99*
Middle East			
total	*0.18*		*0.18*
Africa			
total	*64.67*		*64.67*
Far East and Oceania			
Australia	29.51	36.20	65.71
China	98.79		98.79
total	*130.39*	*38.61*	*169.00*
World total	*652.86*	*271.16*	*924.02*

[a] Ref. 12.
[b] Includes subanthracite and subbituminous.
[c] Includes brown coal.

ment of Energy (DOE) (12,17) and they are more uncertain than those using natural gas. Experience in coal-to-methanol facilities of the type and size that would offer the most competitive product is limited. The projected costs of coal-derived methanol are considerably higher than those of methanol produced from natural gas. The cost of the production facility accounts for most of the increase (11). Coal-derived methanol is not expected to compete with gasoline unless oil prices exceed $0.31/L ($50/bbl). Successful development of lower cost entrained gasification technologies could reduce the cost so as to make coal-derived methanol competitive at oil prices as low as $0.25/L ($40/bbl) (17).

These cost comparisons do not assign any credit to methanol for environmental improvements or energy security. Energy security benefits could be large if methanol were produced from domestic coal.

Produced from Biomass. Estimates for methanol produced from biomass indicate (11) that these costs are higher than those of methanol produced from coal. Barring substantial technological improvements, methanol produced from biomass does not appear to be competitive.

Ethanol

Accurate projections of ethanol costs are much more difficult to make than are those for methanol. Large scale ethanol production would impact upon food costs and have important environmental consequences that are rarely cost-analyzed because of the complexity. Furthermore, for corn, the most likely large-scale feedstock, ethanol costs are strongly influenced by the credit assigned to the protein by-product remaining after the starch has been removed and converted to ethanol.

Cost estimates of producing ethanol from corn have many uncertainties (11). Most estimates fall into the range of $0.26 to 0.40 per liter ($1 to 1.50/gal), after taking credits for protein by-products, although some estimates are lower. These estimates do not make ethanol competitive with oil until oil prices are above $0.38/L ($60/bbl) (17).

For these reasons, ethanol is most likely to find use as a motor fuel in the form of a gasoline additive, either as ethanol or ethanol-based ethers. In these blend uses, ethanol can capture the high market value of gasoline components that provide high octane and reduced vapor pressure.

Impact of Incremental Vehicle Costs

The costs of alcohol fuel usage may include other costs associated with vehicles. Incremental vehicle costs have been estimated by the Ford Motor Company for the fuel-flexible vehicle that can use gasoline, methanol, ethanol, or any mixture of these, to be in the range of $300 per vehicle assuming substantial production (14). This cost may or may not be passed along to the consumer, because of incentives provided to the manufacturer by the Alternative Motor Fuels Act of 1988. There also may be incremental vehicle operating costs resulting from increased lubrication or increased frequency of oil changes.

Infrastructure Requirements

In general, infrastructure requirements resulting from the expanded use of alcohol fuels are not especially greater than those involved in the production and refining of oil. However, for a corresponding delivery of energy, the capital costs of alcohol infrastructure would presumably be larger, as capital costs of production facilities appear to be larger than corresponding oil refinery costs for an equivalent amount of energy output. Moreover, infrastructure costs for storage and distribution facilities would be higher. Facilities for hydrocarbon fuels are generally not compatible with methanol and ethanol. Thus a program to introduce substantial amounts of alcohol fuel might well require existing infrastructure modifications. These changes can be especially difficult and costly for underground pipelines and tanks, which need to be replaced or modified.

In California, the South Coast Air Quality Management District has implemented a local rule requiring that one new or replacement underground tank at each gasoline retail facility must be suitable for methanol. Replacement of an existing tank can cost $50,000 and perhaps more. But many small retail outlets are being replaced by larger more efficient ones, and many older underground tanks are being replaced to prevent possible leaks and hence underground contamination. Therefore the compatibility rule allows for the gradual development of a methanol-compatible infrastructure at low costs. The extra costs for methanol-compatible storage ranges from negligible to about $4000 per tank and dispenser, depending on the technical choices that would otherwise be made for gasoline-only facilities.

VEHICLE TECHNOLOGY AND VEHICLE EMISSIONS

One of the reasons that U.S. automobile manufacturers showed more interest in alcohols as alternative fuels in the late 1970s and early 1980s is because alcohol's energy density is closer to gasoline and diesel than other alternatives such as compressed natural gas. They reasoned that consumers would be more comfortable with liquid fuels, envisioning little change in the fuel distribution of alcohols. Most of the research in the 1970s focused on converting light-duty vehicles, to alcohol fuels. Towards the late 1970s, researchers also began to turn their attention to heavy-duty applications. In heavy-duty engines the emissions benefits of alcohols are far clearer than in light-duty vehicles. However, it is also much harder to design heavy-duty engines to use the low cetane number alcohols.

It was not until the early 1980s that the potential air quality benefits of alcohol fuels started to be investigated. It was about five years later that proponents argued that alcohols could provide significant air quality benefits in addition to energy security benefits. Low level blends of ethanol and gasoline were argued to provide lower CO emissions. The exhaust from light-duty methanol vehicles was thought to be less reactive in the formation of ozone. Uses of alcohols fuels in heavy-duty engines showed substantially reduced mass emissions in contrast to light-duty experience which showed about the same mass emissions but a reduced reactivity of the exhaust components.

LIGHT-DUTY VEHICLES

Use of Low Level Blends. The first significant U.S. use of alcohols as fuels since the 1930s was the low level 10% splash blending of ethanol in gasoline, which started after the oil crisis of the 1970s. This blend, called gasohol, is still sold in commerce although mostly by independent marketers and distributors instead of the major oil companies. EPA provided a waiver for this fuel allowing for a 6.9 kPa (1 psi) increase in the vapor pressure of gasohol over that of gasoline.

In the first years of gasohol use some starting and driveability problems were reported (19). Not all vehicles experienced these problems, however, and better fuel economy was often indicated even though the energy content of the fuel was reduced. Gasohol was exempted from the federal excise tax amounting to a $0.16/L ($0.60/gal) subsidy. Without this subsidy, ethanol would be too expensive for use even as a fuel additive.

Nearly four billion L/yr of ethanol are added to gasoline and sold as gasohol (18). The starting or driveability difficulties have been solved, in part, by the advances in vehicle technology employing fuel feedback controls.

Methanol was also considered as a gasoline additive. Table 3 summarizes some of the oxygenated compounds approved by EPA for use in unleaded gasoline. EPA waivers were granted to: Sun Oil Company in 1979 for 2.75% by volume methanol with an equal volume of tertiary butyl alcohol [*75-65-0*] (TBA) up to a blend oxygen total of 2% by weight oxygen; ARCO in 1979 for up to 7% by volume TBA; and ARCO in 1981 for the use of the blends containing a maximum of 3.5% by weight oxygen. These last blends are gasoline-grade TBA (GTBA) and OXINOL having up to 1:1 volume ratio of methanol to GTBA. Petrocoal was also granted a waiver to market up to 10% by volume methanol and cosolvents in gasoline but this waiver was revoked in 1986 after automobile manufactures complained of significant material compatibility problems and openly warned consumers against gasolines containing methanol. A waiver was issued to Du Pont in 1985 which allowed addition of up to 5% methanol to gasoline having a mixture of 2.5% cosolvents. None of these additives became very popular (20).

Vehicle Emissions. Gasohol has some automotive exhaust emissions benefits because adding oxygen to a fuel leans out the fuel mixture, producing less carbon monoxide [*630-08-2*] (CO). This is true both for carbureted vehicles and for those having electronic fuel injection.

Urban areas such as Denver, Phoenix, and others at high altitudes have problems complying with health-based carbon monoxide standards in part because of automobile emissions. Vehicles calibrated for operation at sea level that are operated at high altitudes run rich, producing more CO. Blends such as gasohol cause the engine to operate leaner because of the oxygen in the fuel. There are larger CO reductions using oxygenated blends in older, carbureted engines. But even the newer technology vehicles have lower CO emissions using gasohol and other oxygenated fuel because of periods of open loop operation, especially during cold starts.

Blended fuels increase the vapor pressure of the resulting mixture so that more hydrocarbons are evaporated into the atmosphere during operation, refueling, or periods of extended parking. Although these hydrocarbons can react with NO_x emissions in sunlight to form ozone, atmospheric modeling has indicated that ozone is probably not

Table 3. EPA Approved Oxygenated Compounds for Use in Unleaded Gasoline[a]

Compound[b]	Broadest EPA Waiver	Date	Maximum Oxygen, wt %	Maximum Oxygenate, vol %
Methanol	substantially similar	1981		0.3
Propyl alcohols	substantially similar	1981	2.0	(7.1)[c]
Butyl alcohols	substantially similar	1981	2.0	(8.7)[c]
Methyl *tert*-butyl ether (MTBE)	substantially similar	1981	2.0	(11.0)[c]
Tert-amyl methyl ether (TAME)	substantially similar	1981	2.0	(12.7)[c]
Isopropyl ether	substantially similar	1981	2.0	(12.8)[c]
Methanol and butyl alcohol or higher mol wt alcohols in equal vol	substantially similar	1981	2.0	5.5
Ethanol	gasohol	1979, 1982	(3.5)[c,d]	10.0
Gasoline grade *tert*-butyl alcohol (GTBA)	ARCO	1981	3.5	(15.7)[c]
Methanol + GTBA (1:1 max ratio)	ARCO (OXINOL)	1981	3.5	(9.4)[c]
Methanol at 5 vol % max + 2.5 vol % min cosolvent	Du Pont[f]	1985	3.7	[e]
	Texas methanol (OCTAMIX)[g]	1988	3.7	[e]

[a] Ref. 21.
[b] All blends of these oxygenated compounds are subject to ASTM D 439 volatility limits except ethanol. Contact the EPA for current waivers and detailed requirements, U.S. Environmental Protection Agency, Field Operations and Support Division (EN-397F), 401 M Street, S. W., Washington, D. C. 20460.
[c] Calculated equivalent for average specific gravity gasoline (0.737 specific gravity at 16°C, NIPER Gasoline Report). Calculated equivalent depends on the specific gravity of the gasoline.
[d] Value shown is for denatured ethanol. Neat ethanol blended at 10.0 vol % produces 3.7 wt % oxygen.
[e] Varies with type of cosolvent.
[f] The cosolvents are any one or a mixture of ethanol, propyl, and butyl alcohols. Corrosion inhibitor is also required.
[g] The cosolvents are a mixture of ethanol, propyl, butyl and higher alcohols up to octyl alcohol. Corrosion inhibitor is also required.

increased as a result of the higher fuel volatility for two reasons (see AIR QUALITY MODELING). First, CO is also an ozone precursor so reducing CO reduces ozone. Second, the hydrocarbon species are somewhat less reactive because of the lower reactivity of ethanol. Furthermore, programs for oxygenated fuel use are focused at high CO occurrences during the year. These usually occur in the wintertime, whereas most areas violate ozone standards in the summer months. Therefore, oxygenated fuel programs, as a CO control strategy, do not generally interfere with ozone attainment strategies. However, programs should be individually evaluated (20).

Ethanol blends can also have an effect on NO_x emissions. Scattered data indicate that NO_x may increase as oxygenates are added to the fuel.

Other countries have also investigated the use of low level alcohol blends as an energy substitution strategy as well as to reduce exhaust emissions of lead (22,23). Brazil implemented low level ethanol–gasoline blends throughout the twentieth century during times of oil shortages or as a hedge against international fluctuations in sugar prices. Blends ranged from 15 to 42%. In 1975 Proalcool, Brazil's ethanol fuel program, was initiated and required the blending of 20% by volume of ethanol in gasoline. This was not totally achieved throughout Brazil until about 1986 when a 22% ethanol–gasoline blend was standardized. Once the fuel was standardized, engine modifications for new vehicles were made, including higher compression and adjustments to the carburetor and timing (22).

Germany also evaluated the gasoline–alcohol blends using methanol. Early programs used 15% methanol added to gasoline (M15). This program required vehicles to be designed for this fuel. Modifications included changes to the fueling system for air–fuel control and vehicle material changes to be compatible with the higher methanol concentrations. The program ended in 1982. M15 was concluded to be feasible if higher vehicle costs could be offset by the possibility of lower fuel costs (24). Lower level blends were also investigated using up to 3% methanol with 2 or more percent of a suitable cosolvent. Unlike M15, gasoline vehicles could use this blend without any modifications (25). Germany has for several years now used low level blends of methanol in their gasoline.

Retrofits. Retrofits are vehicles designed for conventional fuels modified so that the vehicles can operate on alcohol. Generally, because both ethanol and methanol have lower energy densities, the quantity of fuel entering the engine must be increased to get the same power. Also, because the alcohols have slightly different combustion characteristics, engine parameters such as ignition timing need to be adjusted. To optimize performance and fuel economy the compression ratio of the engine should be increased. However, the economics of these conversions are such that the least amount of changes are made and adjustments to engine compression ratio is typically not done.

Retrofits were popular at the beginning of alcohol fuel programs. Kits were introduced that modified only the fuel flow rate into the engine, but material changes were also necessary because both ethanol and methanol are more corrosive to metals than gasoline. Retrofitting allowed maximum market penetration without having to wait for fleet roll over or for manufacturers to market new vehicles. There was some success in converting light-duty vehicles to methanol. Bank of America operated a fleet of 292 converted Ford and Chevrolet vehicles in the late 1970s and into the 1980s before oil prices collapsed. A conversion kit for these vehicles included hardware, material changes, and a fuel additive to help minimize corrosion (26,27).

The California Energy Commission (CEC) evaluated the conversion of 1980 Ford Pintos equipped with 2.3-L four-cylinder engines, feedback-controlled carburetors, and three-way catalysts. Four vehicles were left unmodified, four converted to methanol, and four converted to ethanol. The basic changes required to use the alcohols were: the terneplate coating in the fuel tanks was stripped; fuel level sending units and carburetors were chromated to inhibit corrosion; and the air–fuel ratio, timing, and fuel vaporization mechanisms were recalibrated for proper combustion of alcohol fuels and to comply with emission standards. Two methanol- and two ethanol-fueled engines had special pistons installed to raise compression ratios from 9:1 to 12:1 for better efficiency. These vehicles were operated for 18 months accumulating 272,000 km (169,300 miles). The methanol conversions averaged 25,000 km; ethanol conversions, 21,700 km; and gasoline controls, 22,100 km. Although both the methanol and ethanol vehicles were designed to operate on 100% alcohol, they utilized M94.5 (94.5 vol % methanol and 5.5% isopentane [*78-78-4*] added to improve cold starts and engine warmup) and CDA-20 (ethanol denatured using 2 to 5% unleaded gasoline) (28,29).

This conversion program indicated that vehicles could be converted to alcohols. Good fuel economy was obtained; methanol vehicles averaged 4.7 km/L (11.0 mpg) or 9.1 km/L (21.3 mpg) on an equivalent energy basis compared to 8.3 km/L (19.5 mpg) for the gasoline control vehicles. No driveability problems were reported and vehicles had no problems starting (lowest temperature was −1.1°C). Both the ethanol and methanol Pintos showed increased upper cylinder wear over gasoline engines. Poor lubrication from using alcohols and excess fueling because of carburetor float problems contributed to the higher wear rates. Hydrocarbon, carbon monoxide, and NO_x emissions were less for methanol, 0.14, 3.2, and 0.3 g/km, respectively, than for gasoline, 0.25, 5.6, and 0.6 g/km.

The biggest problems of the CEC Pinto fleet were that vehicle conversions were expensive and alcohol fuels were more expensive than gasoline. Changes to the fuel tank, fuel lines, and the carburetor were too labor-intensive to be done cheaply. However, these changes if designed, could be made during the vehicle manufacturing at little additional cost (30). Brazil priced ethanol at 65% the cost of gasoline (10) so that conversions could be cost-effective because of the savings on the fuel costs.

Other significant disadvantage of retrofits were the quality of the conversion kits and the ability of the conversions to meet emission regulations over the useful vehicle life. The initial phases of the Brazilian ethanol program also suffered because of poor quality vehicle retrofits. The quality was so poor that the program almost failed after a fairly substantial number of vehicles were converted and the ethanol fuel infrastructure was in place. Further incentives and automobile manufacturers introducing new vehicles designed for ethanol stabilized the program (31).

For these reasons, CEC and DOE concluded that the only cost-effective method of getting alcohol fueled vehicles would be from original equipment manufacturers (OEM). Vehicles produced on the assembly line would have lower unit costs. The OEM could design and ensure the success and durability of the emission control equipment.

Dedicated Vehicles. Only Brazil and California have continued implementing alcohols in the transportation sector. The Brazilian program, the largest alternative fuel program in the world, used about 7.5% of oil equivalent of ethanol in 1987 (equivalent to 150,000 bbl of crude oil per day). In 1987 about 4 million vehicles operated on 100% ethanol and 94% of all new vehicles purchased that year were ethanol-fueled. About 25% of Brazil's light-duty vehicle fleet (10) operate on alcohol. The leading Brazilian OEMs are Autolatina (a joint venture of Volkswagen and Ford), GM, and Fiat. Vehicles are manufactured and marketed in Brazil.

In contrast the California program has some 600 demonstration vehicles (32). Both Ford and Volkswagen participated in the dedicated vehicle phase of the California program. In 1981 Volkswagen provided the first alcohol vehicles produced on an assembly line, forty (19 methanol, 20 ethanol, and 1 gasoline) VW Rabbits and light-duty trucks were manufactured. Design incorporated continuous port fuel injection 12.5:1 compression ratio, a new ignition system calibration, and a heat exchanger for faster oil warmup. The entire fuel system was designed using materials compatible with methanol and ethanol. These vehicles operated until 1983 and logged 728,000 km of service. This fleet used the same fuel as the ethanol and methanol Pinto retrofits.

In 1981 Ford also provided CEC with 40 Escorts designed to operate on M94.5 and 15 gasoline vehicles to serve as controls. These vehicles had accumulated over 3.4 million km of service as of March 1986. The 1981 Escorts were modified to use methanol. The 1.6-L gasoline engine had a production piston used in European 1.6-L engines to raise the compression ratio to 11.4:1. Other field modifications included spark plug change, a carburetor throttleshaft material change, carburetor float redesign, and the replacement of tin-plated fuel tanks with ones of stainless-steel.

In 1983 the methanol fleet was expanded with the purchase of 506 Ford Escorts. These vehicles are equipped with engines and fuel systems redesigned from Ford's standard 1.6-L gasoline-fueled Escort. Ford also produced five advanced technology vehicles equipped with electronic fuel injection and microprocessor control, with the goal of improving fuel economy and reducing NO_x emissions to 0.25 g/km. The emissions control on these vehicles used the same technology as on the carbureted gasoline versions: standard three-way catalyst, exhaust gas recirculation, and air injection. The 1981 and 1983 Escorts have logged over 48 million km in service and some vehicles have reached gasoline equivalent fuel usage per kilometer over the lifetime of the vehicle.

The methanol fuel specification was changed for the Escorts, at the end of 1983 from M94.5 to a blend of 90% methanol and 10% unleaded gasoline (M90). In the summer of 1984 the fuel was further modified to include 15% gasoline (M85). This change from isopentane was made because gasoline was cheaper. In addition, M85 has a gasoline odor and taste, and in daylight the flame is more visible than either M94.5 or M90. Another safety benefit of the added gasoline is the increased volatility creating a richer air–vapor mixture much less likely to burn or explode in closed containers than neat or 100 percent methanol.

The results of the California fleet demonstrations indicated that fuel economy on an energy basis was equal to or better than gasoline, especially using vehicles having higher compression. None of these engines, however, was fully optimized for methanol so additional improvements are possible. Driveability was also good for the methanol vehicles. An acceleration test of two 1983 methanol-fueled vehicles and a similar 1984 model resulted in: 1983 fuel injection (EFI) methanol, 14.53 s; 1983 carbureted methanol, 15.51 s; and 1984 carbureted gasoline, 19.10 s.

Tests demonstrate that methanol vehicles can meet stringent emission standards for HC, CO, and NO_x as indicated in Figure 2. The primary benefit of methanol, however, is not the amount of hydrocarbons emitted but rather that methanol-fueled vehicles emit mainly methanol which is less reactive in the formation of ozone than the variety of complex organic molecules in gasoline exhaust. Formaldehyde [*50-00-0*] emissions from methanol vehicles are increased in comparison to gasoline vehicles. Tests of 1983 Escorts showed tailpipe levels as high as 62 mg/km, well above typical gasoline levels of 2 to 7 mg/km. The 1981 Rabbits ranged from about 6 to 14 mg/km and the 1981 Escorts had levels less than 7 mg/km. All results were obtained on relatively low mileage vehicles. Deterioration of catalyst effectiveness could increase these emissions.

The vehicles investigated were mostly adaptions of gasoline technology. For example, automobile manufacturers recommended that catalytic converters designed for gasoline automobiles be installed on vehicles in Brazil to control acetaldehyde [*75-07-0*] emissions. Brazil decided against catalysts because gasoline vehicles at that time did not have these systems. Similarly, California adopted M85 to aid in cold starting and to provide some measures of perceived safety. Research to find other additives that would assist in cold starting and provide safety characteristics at a reasonable price have been relatively unsuccessful (33). But another way to overcome the issue of cold starting is to design and optimize engines to operate on 100% methanol (M100). EPA has been pursuing M100 for several years with good success (34). Cold starting is not a problem for direct injected engines where high pressure is used to atomize the fuel and results indicate that a light-duty, direct injection engine can attain very low emissions having good fuel economy and driveability. However, safety concerns of using M100 in general commerce need to be addressed (35).

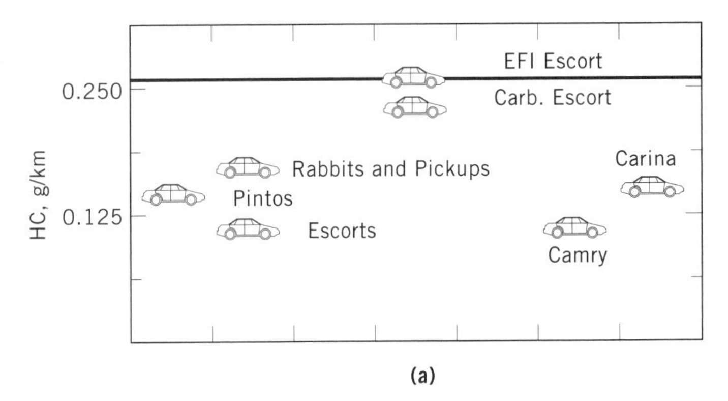

(a)

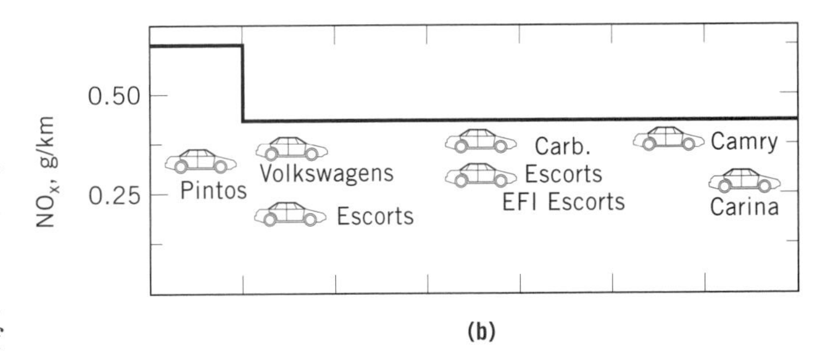

(b)

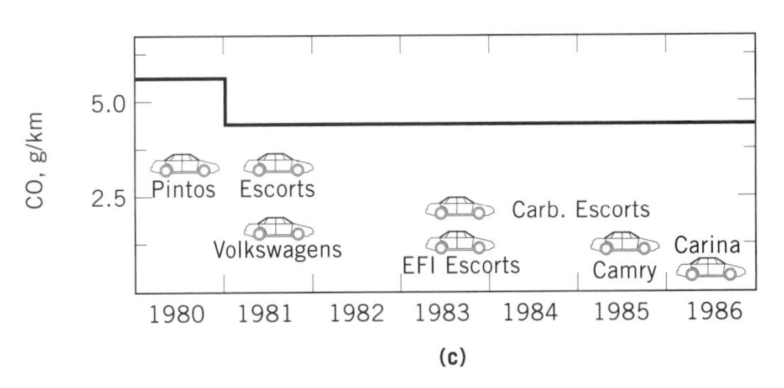

(c)

Figure 2. CEC methanol-fueled vehicle exhaust profile for (**a**) HC, hydrocarbons; (**b**) NO_x; and (**c**) CO. Solid line represents State of California standard maximum emissions. Methanol HC emissions are calculated as $CH_{1.85}$ and not corrected for flame-ionization detector response.

Fuel Flexible Vehicles. Using dedicated alcohol fuel vehicles pointed to the importance of a wide distribution of fueling stations. Methanol-fueled vehicles require refueling more often than gasoline vehicles.

In 1981, the Dutch company TNO in cooperation with the New Zealand government converted a gasoline engine to a flexible fuel vehicle by adding a fuel sensor. The sensor determined the amount of oxygen in the fuel and then used this information to mechanically adjust the carburetor jets. The initial mechanical system was crude, but the advancement of engine and emission controls, in particular the use of electronics and computers, has brought about substantial refinement (36).

Ford first tried the flexible fuel system on an Escort and called it the "flexible fueled vehicle" or FFV. As seen in Figure 3, the system included building into the electronics any necessary calibrations for gasoline and methanol fuels, adding a sensor to determine the amount of methanol in the fuel, and making necessary material changes to fuel wetted components. The sensor is one of the most critical parts of this system: its output determines parameters such as the amount of fuel to be injected and engine timing. Fuel injectors must also have a wider response range. The engine compression ratio was not changed because the

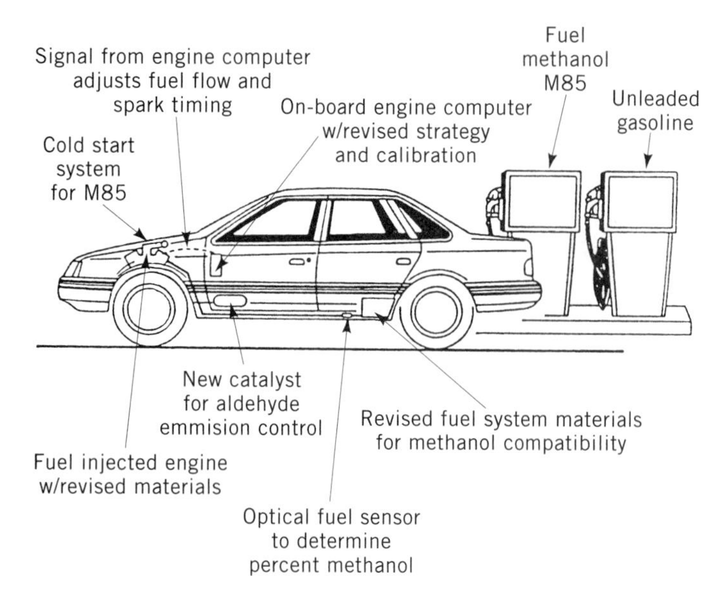

Figure 3. Components of a Ford flexible fuel vehicle (FFV).

vehicle is designed to operate as well on gasoline as on methanol.

Of course, FFV drivers do not have to use methanol. Emissions benefits are not obtained if methanol is not used, and fuel economy is not optimized for methanol nor are emissions. However the State of California has concluded that advantages offered by the flexibility of the FFV far outweigh the disadvantages (37).

Many U.S. and foreign automobile manufacturers are developing a fuel flexible vehicle in the 1990s. Ford has developed FFVs for 5-L engines used in Crown Victorias and for 3.0-L engines used in their Taurus car line. GM followed Ford with a variable fueled vehicle (VFV) and applied this technology first to the 2.8-L engine family used in the Corsica, and more recently to the 3.1-L engine family in the Lumina (see Fig. 4). Prototype flexible fuel vehicles are also being developed by Volkswagen, Chrysler, Toyota, Nissan, and Mitsubishi. California is in the process of obtaining an additional 5,000 of these vehicles in the next several model years (MY92 nd MY93) to be used by government and private fleets.

The experience using fuel flexible vehicles has been surprisingly successful. California is operating about 200 vehicles and driveability is excellent on whatever fuel is used. Tests performed on a Ford Crown Victoria showed slightly better fuel economy (4%) and better acceleration (6%) on methanol than gasoline (38). The fuel flexible technology is not limited to methanol but with electronic calibration changes also works for ethanol and gasoline combinations. Changes can be made by adjusting the engine maps in the computer.

EPA, the Air Resources Board (ARB), and others are investigating the possible emission benefits of alcohol fueled vehicles. EPA and ARB adopted regulations for hydrocarbon mass emissions which accounted for the oxygen components in the exhaust, so called organic mass hydrocarbon equivalent (OM-HCE). The regulations required methanol vehicles to meet the same OMHCE value as gasoline hydrocarbons, which in California is 0.155 g/km in 1993. The trend is to account also for the total mass and the reactivity of individual hydrocarbon species and the measure being proposed for total mass is nonmethane organic gases or NMOG. Reactivity of the individual species that make up NMOG are estimated (39) to give a value of ozone/km.

Vehicle emissions have been monitored over the last several years on fuel flexible vehicles and depending on when the tests were performed, reported in total hydrocarbons, OMHCE, or NMOG. The vehicle testing performed to date has shown that methanol FFVs can provide emissions benefits. Figure 5 shows the emissions from a GM Corsica for various gasoline methanol mixtures (40). This vehicle was designed to meet the California standards of 0.155 g/km hydrocarbon, 2.1 g/km CO, and 0.25 g/km NO_x. In addition California requires methanol vehicles to meet a formaldehyde standard of 9.3 mg/km. The total organic emissions decrease with increasing methanol, whereas formaldehyde increases. NO_x and CO vary but appear unaffected by methanol content. The Corsica data were taken on a green catalyst (low vehicle mileage) and some deterioration of these emissions levels can be expected with age.

Figure 6 shows data for four vehicles operated on gasoline and M85, two having an electrically heated catalyst (EHC). The two vehicles equipped with EHC both showed low values of NMOG and estimated ozone production. These data seem to indicate that methanol vehicles result in less ozone than comparable gasoline vehicles. However, the data only include exhaust emissions and not evaporative or running losses. These later sources of emissions should be lower using methanol because of the lower reactivity of the alcohols.

HEAVY-DUTY VEHICLES

The use of alcohols in heavy-duty engines developed more slowly than in light-duty engines primarily because the majority of heavy-duty engines are diesels. Diesels are unthrottled, stratified charge engines which autoignite fuel by heat generated during compression. Engine speed and load are modulated by varying the quantity of fuel injected into the cylinder rather than by throttling the fuel–air mixture as done in the Otto cycle or spark-ignition engine. Unthrottled air aspiration reduces pumping losses which in turn increases the engine's thermal efficiency. Diesel engines also are designed for higher compression ratios resulting in further efficiency improvements. The higher efficiency and excellent reliability and durability of these engines make them attractive for heavy-duty applications in trucks, buses, and off road equipment. Unfortunately, their low cetane number limited the compability of alcohol fuels with diesel engines without modifications to assist ignition.

Not until the early 1980s were prototype heavy-duty engines developed to operate on methanol and ethanol, having efficiencies equal to or better than corresponding diesel engines. These engines were quieter and had considerably cleaner exhaust emissions. Mass emissions of NO_x and particulates were substantially lower when the cleaner alcohol fuels were used, overcoming the inherent NO_x-particulate tradeoff of diesel engines. The biggest challenge fac-

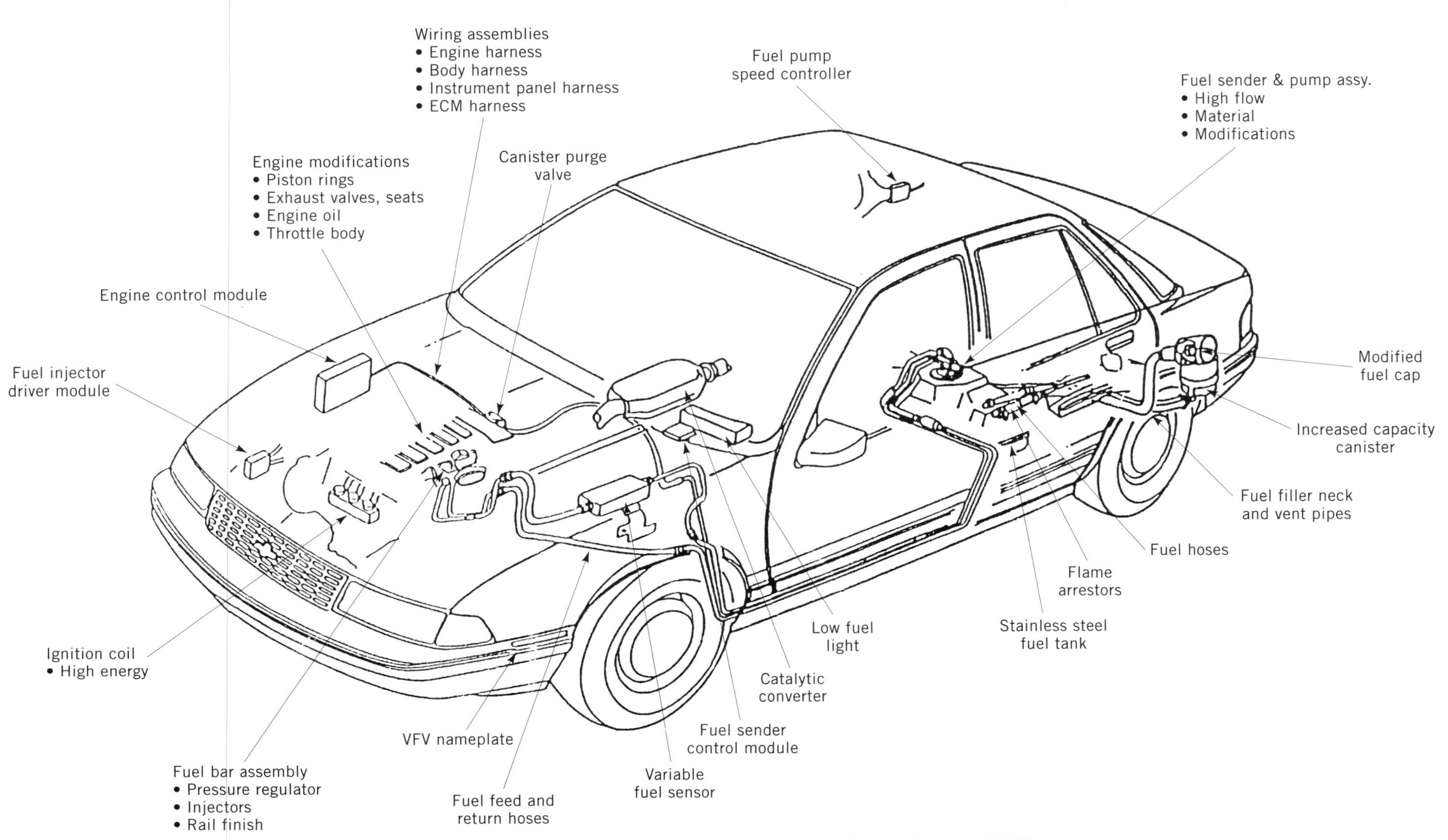

Figure 4. Components of a Lumina methanol variable fueled vehicle (VFV).

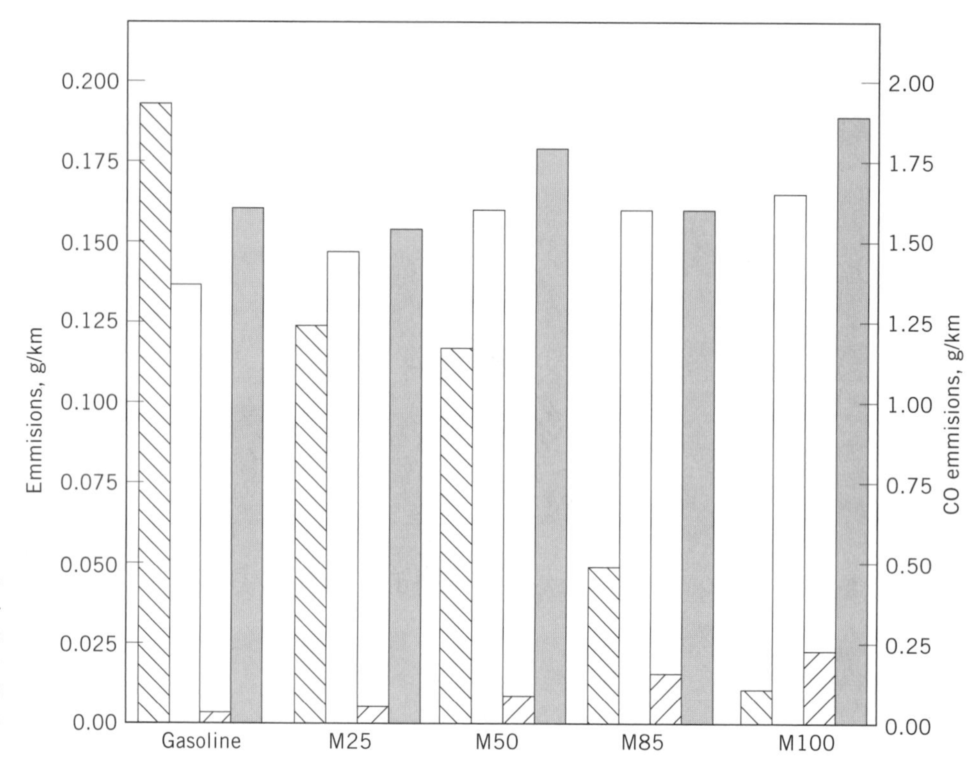

Figure 5. Emissions from a GM Corsica VFV for gasoline and gasoline–methanol mixtures where ▧ represents total organic material, including hydrocarbons, methanol, and formaldehyde; □ represents NO_x; ▨ formaldehyde; and ■ carbon monoxide.

ing engine manufacturers in the 1980s and 1990s was to make the alcohol engines as reliable and durable as heavy-duty diesel engines which operate in the range of 0.3 to 0.5 million kilometers without major engine maintenance and repair.

Technology Options. Because alcohols are not easily ignited in diesel-type engines changes in engine hardware or modifications to the fuel are needed. Engine modifications can include the addition of a spark ignition system or an additional fuel injection system to provide dual fuel capabilities. Fuel modifications involve the addition of cetane improvers. Low level blends do not work because methanol and diesel fuels are not miscible. Some research to emulsify diesel and alcohols was carried out, but was never successful (42). Other investigations involved adding

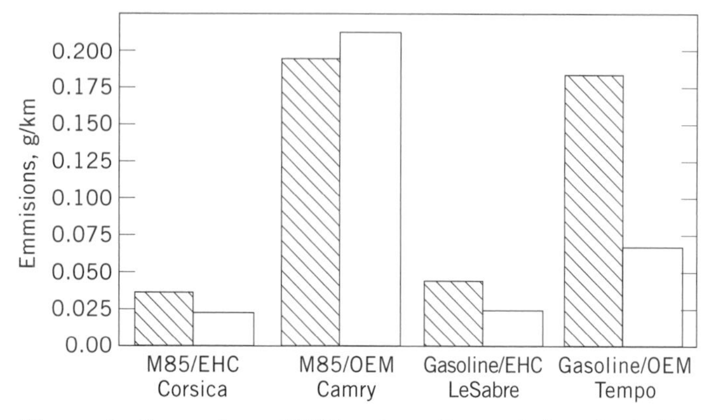

Figure 6. Comparison of M85 and gasoline emissions (including ozone) corresponding to ▧ estimated ozone (ref. 40) and □ NMOG for vehicles having an electrically heated catalyst (EHC) or not (OEM) (ref. 41).

a separate fuel system including two fuel tanks, fuel lines, injection pumps, and injectors. In this approach diesel was used to ignite the fuel mixtures, and at low speeds–low loads diesel was the primary fuel. At high speeds–high loads alcohol was the primary fuel. This dual fuel approach was both cumbersome and expensive (43).

One successful method for using alcohols was fumigation. In this technique alcohol is atomized in the engine's intake air either by carburetion or injection. Diesel is directly injected into the cylinder and the combined air–alcohol and diesel mixture is autoignited. Diesel consumption is reduced by the energy of the alcohol in the intake air. This approach, although technically feasible, also requires separate fuel systems for the diesel and alcohol fuels. Additionally, the amount of alcohol used is limited by the amount that can be vaporized into the intake air. This approach is more appropriate as an engine retrofit where total energy substitution is not the primary objective (44).

Other possible technologies involve either assisted ignition or cetane improvers. Assisted ignition approaches can be divided between direct injection, stratified charge type engines, and engines converted back to Otto cycles, by throttling and lowering engine compression.

Dedicated Vehicles. As late as 1982, researchers were still arguing the worthiness of alcohols as fuels for heavy-duty engines (45). Pioneer work on multifuel engines led to modifications in diesel engines to burn neat or 100% alcohol. The German manufacturers were the first to provide prototype methanol engines. Daimler-Benz modified their four stroke M 407 series diesel engine to operate on 100% methanol, by converting the diesel version to a spark-ignited, Otto cycle engine. This required lowering the compression, adding a spark ignition system, and carburetion (throttling). To get back some of the efficiency loss caused by going to a throttling and lower compression, the Daimler-Benz design incorporated a heat exchanger to vaporize the methanol using engine cooling water. Vaporized methanol was introduced into the engine using a standard gaseous carburetor–mixing device.

The M 407 hGO methanol engine is a horizontal, water-cooled, inline six-cylinder configuration (46). Basic combustion is similar to the conventional spark-ignition Otto cycle with one significant exception. Lean combustion at part load is possible for two reasons: because of methanol's favorable flammability limits and because methanol is vaporized and introduced as a gas. Equivalence ratios (air–fuel ratio relative to stoichiometric) greater than 2 are possible without misfire, and minimum fuel consumption is obtained at an equivalence ratio of about 1.8. In the higher load range, the engine is controlled by the air–fuel ratio, rather than by intake throttling, so efficiency is increased relative to the conventional spark-ignition engine. Intake throttling is used for control in the lower load range.

The first methanol bus in the world was placed in revenue service in Auckland, New Zealand in June 1981. It was a Mercedes O 305 city bus using the M 407 hGO methanol engine. This vehicle operated in revenue service for several years with mixed results. Fuel economy on an equivalent energy basis ranged from 6 to 17% more than diesel fuel economy. Power and torque matched the diesel engine and drivers could not detect a difference. Reliability and durability of components was a problem. Additional demonstrations took place in Berlin, Germany and in Pretoria, South Africa, both in 1982.

The world's second methanol bus was introduced in Auckland shortly after the first. This was a M.A.N. bus with a M.A.N. FM multifuel combustion system utilizing 100% methanol. The FM system, more similar to a diesel engine, is a direct injection, high compression engine using a spark ignition. Fuel is injected into an open chamber combustion configuration in close proximity to the spark plugs which ignite the air–fuel mixture. Near the spark plugs the air–fuel mixture is rich and combustion proceeds to the lean fuel air mixtures in the rest of the cylinder. The air–fuel charge is thus stratified in the cylinder and these types of engines are often called lean burn, stratified charge. Engine hardware is similar to the diesel version including a high pressure injection pump and a compression ratio comparable to diesel (19:1). This technology was applied to M.A.N.'s 2566 series engines, an inline 6-cylinder engine, and for buses is configured horizontally (47). Like the Mercedes, it is a four stroke engine.

This technology was tested using diesel fuel, gasoline, methanol, and ethanol. A.M.A.N. SL 200 bus having the M.A.N. D2566 FMUH methanol engine was also demonstrated in Berlin. Results of these tests were somewhat mixed. Fuel economy was 12% less than a comparable diesel bus, but driveability was very good. Because the methanol fueled bus was not smoke limited at low speeds, higher torque was possible and bus drivers used this advantage to accelerate faster from starts. Emissions results indicated a considerable advantage in using fuels such as methanol. CO and NO_x were reduced compared to diesel engines and particulates were virtually eliminated.

The success of the New Zealand and German programs were instrumental in implementing a similar bus demonstration in California in 1982 (48). The primary objective was to assess the viability of using methanol in heavy-duty engines. The project focused on evaluating engine durability, fuel economy, driveability, and emissions characteristics. CEC also initiated a demonstration project for off-road heavy-duty vehicles using a multifuel tractor capable of operating on either neat methanol or ethanol (30).

M.A.N. and Detroit Diesel Allison, now Detroit Diesel Corporation (DDC), agreed to participate in the California bus program. DDC provided a methanol version of their 6V-92TA engine, which along with the DDC 71 series, is the most commonly used bus engine in the United States. The engine is a compression-ignited, two-stroke design having a displacement of 9.1 liters and power rating of 20,700 W (277 hp). Several design changes were incorporated for operation on methanol, including electronic unit injectors (EUI) for more precise fuel control, an increased compression ratio, a bypass blower, and glow plugs. Compression ignition is achieved by maintaining the cylinder temperatures above the autoignition temperature of methanol. Air is diverted around the blower, reducing the amount of air entering the cylinders. Glow plugs are used as a starting aid and also at low speeds and low loads to maintain the cylinder temperatures necessary for autoignition. The methanol-fueled engine is turbocharged and equipped with a blower (supercharged).

The engine was the first to incorporate compression ignition of alcohols (7). Low cetane fuels can autoignite, however, provided the in-cylinder temperature is high enough and fuel injection correctly timed. This compression ignition works for methanol as well as ethanol and gasoline.

The California bus program was run at Golden Gate Transit District (GGTD) and continued through late 1990 (49). M.A.N. supplied two European SU 240 coaches for this project, one diesel powered and one methanol powered. DDC provided a GM RTS coach powered by methanol. GGTD already had a RTS diesel powered coach. Results indicate that methanol is a viable fuel for heavy-duty engines in general and transit in particular. Driveability including starting, full and partial throttle acceleration, and deceleration was as good or better using the methanol buses as compared to their diesel counterparts. Figure 7 illustrates the comparison of full throttle acceleration. Detailed fuel economy tests were also performed. Figures 8 and 9 compare steady-state and transient fuel consumption tests, respectively. The transient tests were performed using the Fuel Consumption Test Procedure, Type II (50). Methanol is comparable to diesel in steady-state fuel usage tests, but methanol consumption is higher at idle and during accelerations. The idle fuel consumption is higher because methanol cannot burn as lean as diesel fuel. Higher transient fuel consumption is a result of poor combustion factors resulting from poor fuel atomization, air control, and over fuelling. The methanol engine should not inherently be worse than the diesel engine during accelerations, if good combustion can be maintained.

The biggest problem of the GGTD program was engine and vehicle reliability and durability (see Fig. 10). Components needing the most frequent replacement were electronic unit fuel injectors and glow plugs, followed by the electronic control system (controlled power to the glow plugs), throttle position sensor, fuel pump, and fuel cooler fans. Other problems included increased engine deposits

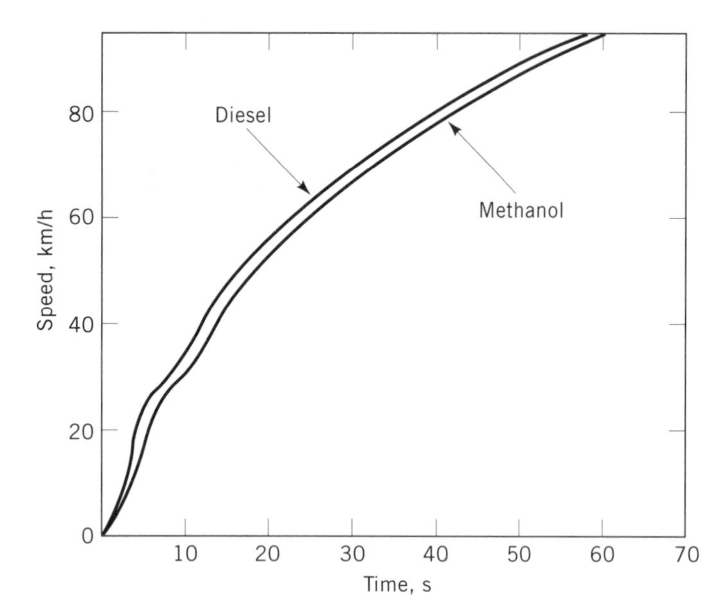

Figure 7. Full-throttle acceleration for diesel and methanol-powered GM RTS coaches having simulated full-seated passenger loads of 43 passengers.

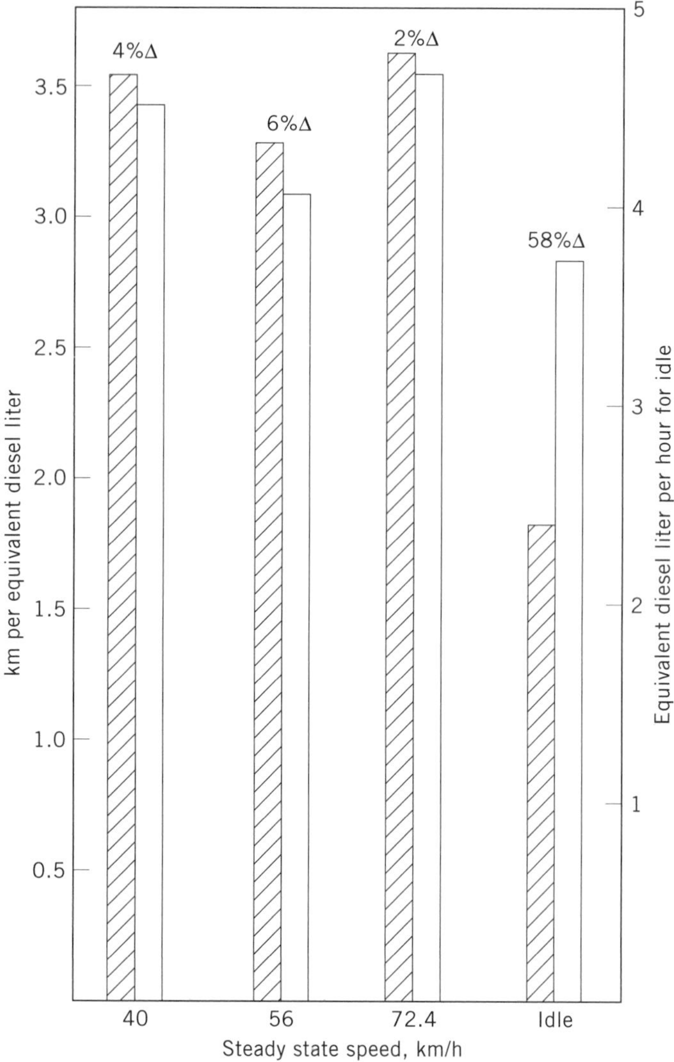

Figure 8. Fuel consumption for GM RTS coaches using ▨ diesel and □ methanol as fuel. Differences in fuel consumption are indicated by the symbol Δ. To convert km/L to mpg, multiply by 2.35.

and ring and liner wear (51). The M.A.N. engine had similar but fewer problems; the components having the lowest lifetime were spark-plugs.

California continued the development of methanol powered vehicles primarily because of the substantial emission benefits (52). Then in 1986, the U.S. EPA promulgated technology-forcing standards for on-road, heavy-duty diesel engines which had been basically uncontrolled (53). These standards were also adopted in California by the ARB for buses in 1991 and all heavy-duty engines in 1993. Diesel engine manufacturers have made significant improvements in technology and new diesel engines are projected to meet standards without a particulate trap. These improvements have made the diesel engine more competitive with alcohol fueled engines.

In 1987 Seattle Metro purchased 10 new American built M.A.N. coaches powered by methanol. Six GM buses powered by DDC methanol engines entered revenue service at Triboro Coach in Jackson Heights, New York, 2 GM buses in Medicine Hat Transit in Medicine Hat, Manitoba, and 2 Flyer coaches in Winnipeg Transit, Winnipeg, Manitoba, Canada. An additional 45 DDC powered methanol buses were introduced in California as indicated by Table 4. Figure 11 shows the distance accumulation of alternate-fueled buses in the four California transit properties.

Many of the development problems identified in the first bus programs were carried into the more recent demonstration projects. Spark-plug life continued to be an issue at Seattle Metro and the project was terminated in 1990 because of costs of replacement parts. Costs were compounded when M.A.N. decided to discontinue manufacturing buses in the United States. DDC engines also continued to have problems with fuel injectors and glow plugs. Unit injectors were failing for a variety of reasons but the biggest problem was plugging injector tips. Injectors on some buses had to be changed at mileages as low as 1600 to 3200 km compared to diesel injectors which last up to 100 times as long. DDC and Lubrizol have since developed a fuel additive that when added to methanol at 0.06% by volume substantially reduces injector failures (see Fig. 12).

Many improvements have been made in both combustion and emissions control from the first experimental engine operating at GGTD (54). The new DDC preproduction engines have increased compression, 23:1 compared to 19:1, allowing the glow plugs to function only during starting. The rest of the time the cylinder temperatures are high enough to autoignite methanol. This revision increased the life of glow plugs from an average of 11,900 to 22,100 km between failures. Fuel economy has also improved as have exhaust emissions. Tests performed on an engine dynamometer following the Federal test procedure are many times better than California's 1991 bus standard as shown in Table 5.

Because of the success of these various alcohol fuel programs, heavy-duty demonstrations have been extended to other applications as shown in Table 4. The majority of

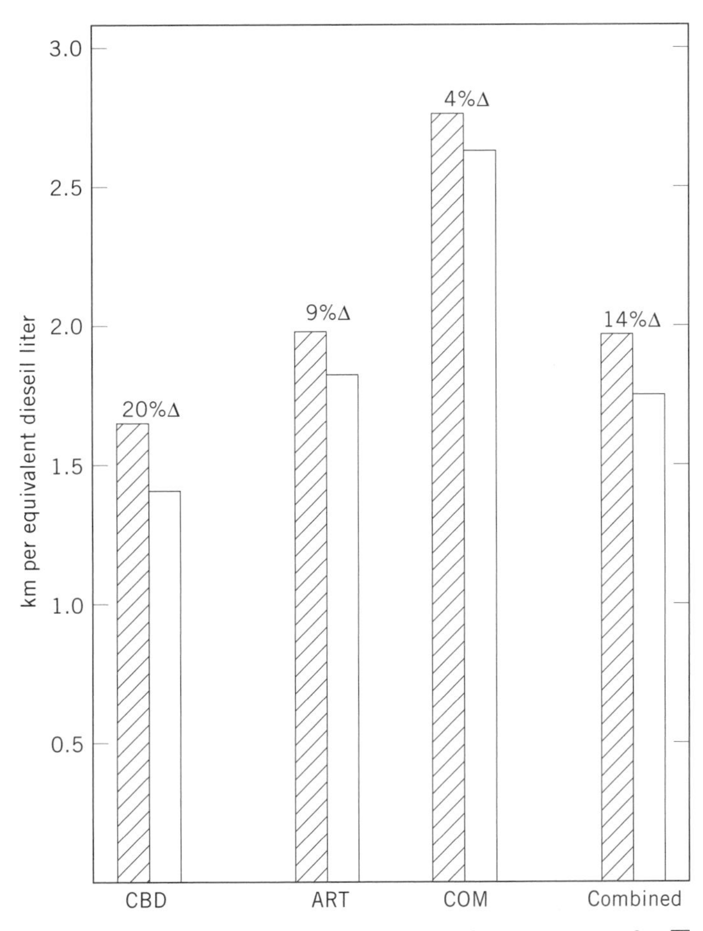

Figure 9. GM RTS coaches transient fuel consumption for ▨ diesel and methanol. SAE type road test, UMTA ADB duty cycle. Differences in fuel consumption are indicated by the symbol Δ. CBD (Central Business District), 4.3 stops per km, maximum speed 32 km/h, 18 m deceleration, 7 s dwell time, 19.3 km; ART (arterial), 1.24 stops per km, maximum speed 64 km/h, 61 m deceleration, 7 s dwell time, 13 km; COM (commuter), one stop every 6.4 km, maximum speed of 88.5 km/h, 150 m deceleration, 20 s dwell time, 13 km. To convert km/L to mpg, multiply by 2.35.

applications are still either in transit or school buses, but California has also begun a program to utilize methanol engines in heavy-duty trucking applications (55). Domestic engine manufacturers participating in this program include Caterpillar, Cummins, DDC, Ford, and Navistar. The Caterpillar and Navistar engines are four stroke engines having glow plugs to assist in igniting methanol. These engines are very different from the two stroke DDC engine and from each other. The Navistar (56) uses a shield glow plug which neither Caterpillar (57) nor DDC do. Navistar claims these glow plugs provide both good combustion and long life. The combustion chambers and fuel control schemes differ from manufacturer to manufacturer. Ford converted a diesel engine to an Otto cycle and added electronically controlled port fuel injection, throttling, and a distributorless spark-ignition system.

Methanol has been shown to be a viable fuel for a variety of trucking applications in a local yet fairly large geographical area (58). And although the results focus primarily on dedicated methanol engines and vehicles in the U.S., the conclusions are nearly transferrable to ethanol fuels. DDC's 6V-92TA methanol engine operates on ethanol using a different engine calibration optimized for good performance and emissions. Additional hardware modifications are not anticipated.

Cetane Improvers. Compared to dedicated alcohol engines, fewer hardware changes need to be made to heavy-duty engines using cetane improvers. The early research on using cetane improvers and alcohol fuels for heavy-duty diesel engines was performed by the Germans in the late 1970s-early 1980s (59). The possibilities of using cetane improvers with ethanol for heavy-duty vehicles operating in Brazil were investigated (60). The work indicated that nitrates are the most effective cetane improvers for alcohols (61) and the Brazilian program focused on nitrates that could be manufactured from sugar cane, the feedstock for ethanol production. Additives considered included: butyl nitrate, isoamyl nitrates, 2-ethoxyethyl nitrate, and

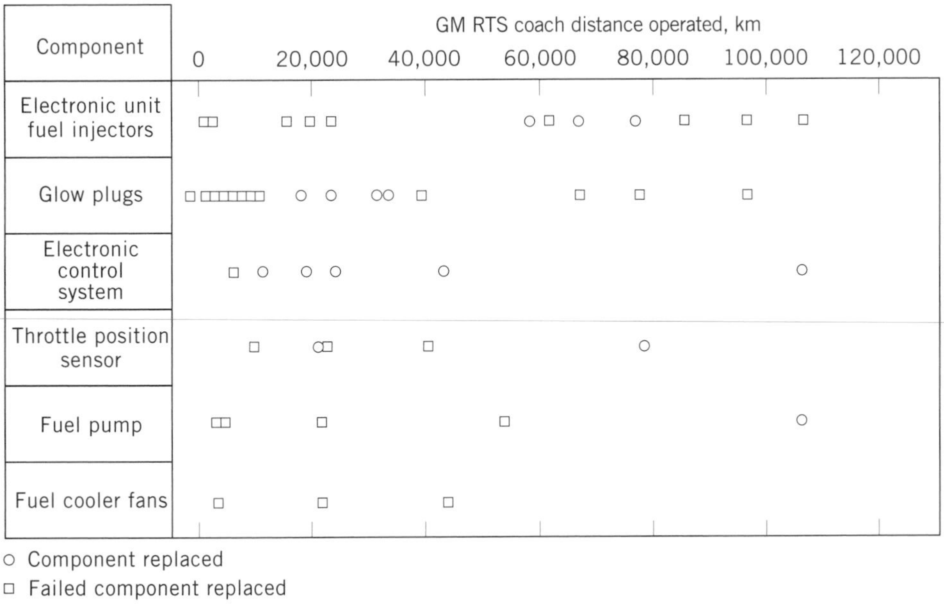

Component
GM RTS coach distance operated, km
0 20,000 40,000 60,000 80,000 100,000 120,000
Electronic unit fuel injectors
Glow plugs
Electronic control system
Throttle position sensor
Fuel pump
Fuel cooler fans
○ Component replaced
□ Failed component replaced

Figure 10. Components of GM RTS methanol-powered coach replaced, ○, and replaced because of component failure, □, as a function of distance operated.

Table 4. Distribution of California's Heavy-Duty Alternative Fuel Demonstration Vehicles

Transit District	No. of Vehicles	Fuel	OEM/Engine
South Coast Area Transit	1	methanol	DDC 6V-92TA
Riverside Transit District	3	methanol	DDC 6V-92TA
Southern California RTD	30	methanol	DDC 6V-92TA
Southern California RTD	12	methanol/Avocet[a]	DDC 6V-92TA
Southern California RTD	1	methanol	MAN D2566 MUH
Orange County Transit District	2	methanol/Avocet[a]	Cummins L10
Orange County Transit District	2	CNG[b]	Cummins L10
Orange County Transit District	2	LPG[b]	Cummins L10
		School bus demo	
Various fleets	50	methanol	DDC 6V-92TA
Various fleets	10	CNG	Bluebird/Teogen GM 454
		Trucking applications	
City of Los Angeles	1	methanol	GMC DDC 6V-92TA
City of Los Angeles	1	methanol/Avocet[a]	Peterbuilt Cummins L10
City of Glendale	1	methanol	Peterbuilt Caterpillar 3306
Golden State Foods	1	methanol	Freightliner DDC 6V-92TA
Federal Express	1	methanol	Freightliner DDC 6-71 TA
Arrowhead	1	methanol	Ford/Ford 6.61
SCE	1	methanol	Volvo/DDC 6V-92TA
South Lake Tahoe	1	methanol	International/Navistar DTG-460
Waste Management	2	methanol	Volvo/DDC 6-71TA

[a] Avocet is a cetane improver.
[b] CNG, compressed natural gas, and LPG, liquefied petroluem gas, are also used as alternative fuels.

ethylene glycol nitrates. The selection of the improver depends on the method of the additive modifying the ignition delay over the entire speed and load range. Ideally ethanol should match the same ignition delay behavior as diesel.

Four buses converted to ethanol started operation in 1979 and the engine used was an OM 352, 6 cylinder, direct injection engine rated at 96,200 watt (129 hp). The ignition improver was n-hexyl nitrate [*20633-11-8*] which was later changed to triethylene glycol dinitrate [*111-22-8*] (TEGDN). TEGDN was mixed with ethanol at 5% or less by volume (60). The test fleet was further expanded to a variety of trucks manufactured and marketed by Mercedes-Benz in Brazil. 1,700 heavy-duty trucks were converted to ethanol for use in the more gruelling sugar cane industry. Engines converted included the OM 352 O (5.7 L, 96,200 watt) and OM 355/5 O (9.7 L, 141,000 watt). Modifications to the engines to use ignition improved ethanol included increasing the fuel delivery capacity of the fuel injection pump, changing the fuel injection nozzles, and making material compatible with the TEGDN/ethanol fuel. The engines provided equal or better power and torque than the unmodified models. Some durability problems, which arose because of lack of lubrication or fuel material incompatibilities, were solved by adding lubricants to the injection pump plungers or by changing materials. Emissions were also generally lower than those from the equivalent diesel engine. Even NO_x was lower because of the lower flame temperatures compared to diesel (62).

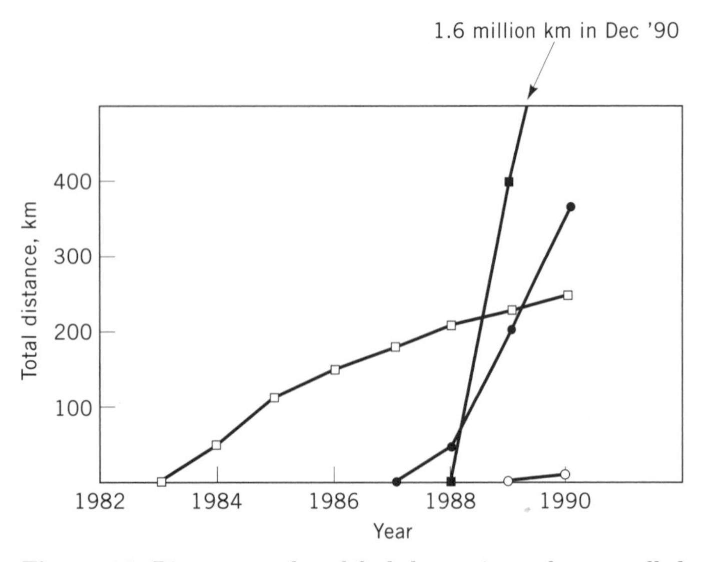

Figure 11. Distance methanol-fueled transit coaches travelled per year of operation in the □ Golden Gate Transit District (GGTD); ● Riverside Transit Agency (RTA), ■ Southern California Rapid Transit District (SCRTD); and ○ Orange County Transit District (OCTD).

The biggest drawback was the cost of ethanol compared to diesel fuel and the cost of the TEGDN and ethanol mixture compared to diesel. Unlike in the United States, diesel fuel in Brazil is considerably less expensive on an energy basis than gasoline, because gasoline is taxed at a higher rate. This is generally the case in Europe as well. So, although technically feasible, the costs were too high for Brazil to convert many heavy-duty vehicles to ethanol.

Additional research for both ethanol and methanol showed that the amount of ignition improver could be reduced by systems increasing engine compression (63). Go-

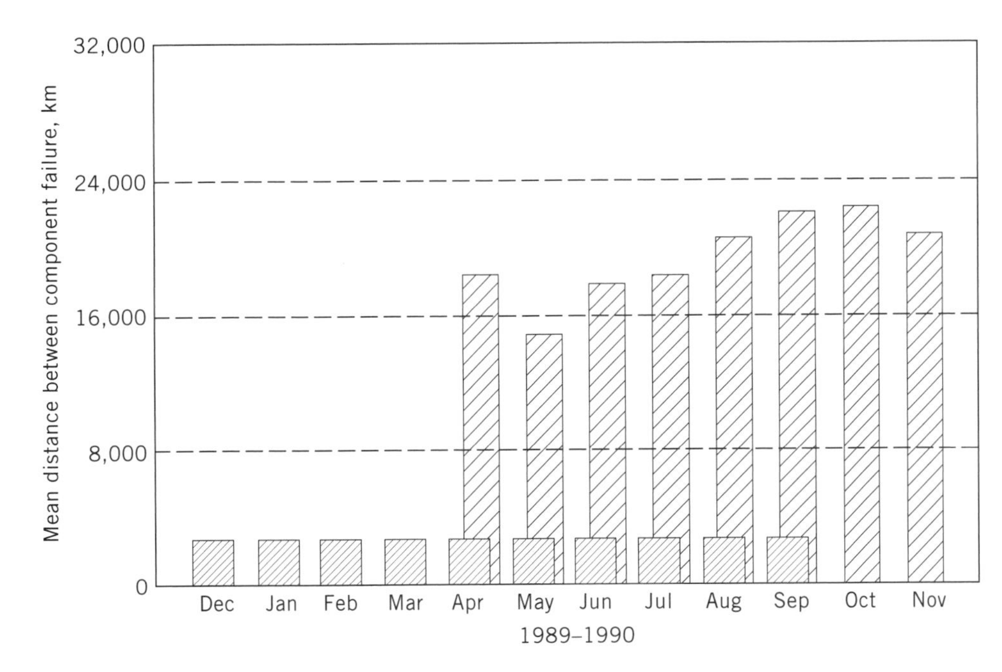

Figure 12. Methanol and coach fleet cumulative distance driven before fuel injector failure, where ▨ represents operation without, and ▨ represents operation with fuel additive.

ing from 17:1 to 21:1 reduced the amount of TEGDN required for methanol from 5% by volume to 3%. Ignition-improved methanol exhibited very low exhaust emissions compared to diesels: particulate emissions were eliminated except for small amounts associated with engine oil, NO_x was even lower with increased compression, and CO and hydrocarbons were also below diesel levels.

Auckland Regional Authority converted two M.A.N. buses to use a cetane improver and methanol and South Africa investigated the use of methanol with a proprietary cetane improver. Four Renault buses were converted in Tours, France to operate on ethanol and a cetane improver, Avocet, manufactured by Imperial Chemical Industries (ICI). The results of these demonstrations were also technically successful; slightly better fuel economy was obtained on an energy basis and durability issues were much less than the earlier tests using dedicated engines.

Cetane improvers were first investigated in the U.S. as part of a demonstration project to retrofit DDC 6V-71 two stroke engines to methanol for three transit buses operating in Jacksonville, Florida. This project started out using hardware conversion where the engine was modified to control air flow and a glow plug was added to the system. Although this was basically the same approach as used in the dedicated 6V-92TA methanol engine, the 6V-71 employed mechanical injectors as opposed to electronic injectors. These mechanical injectors were failing even with 1% castor oil [*8001-79-4*] added to the methanol. Jacksonville thus decided to use a nitrate based cetane improver containing a lubrication additive and a corrosion inhibitor and the project proved the viability of cetane-improved methanol.

The first U.S. engine manufacturer to evaluate the use of cetane improvers was Cummins Engine Company. Their methanol designed L10 engines were converted based on the knowledge gained in Brazil with ethanol and 4.5% TEGDN in their 14-L engine (64). They increased the fuel capacity of their injection pumps, for instance, and modified the combustion process to match the diesel start of combustion by increasing the compression ratio from 15.8 to 16.1:1 and adding turbocharging (both increased in-cylinder temperatures). For the methanol L10 development Avocet at 5% by volume mixed with 100% methanol was selected (65). Engine modifications were only made to the fuel system. These included fuel pump changes for increased capacity, a different camshaft to change injection timing, and larger injectors. Changes were also made

Table 5. Exhaust Emissions and California 1991 Bus Standards, g/kW · h[a]

Exhaust Component	1991 California Standards g/(bhp · h)[a]	1991 California Standards g/(kW · h)	DC 6V-92TA (Catalyst) g/(bhp · h)[a]	DC 6V-92TA (Catalyst) g/(kW · h)
OMHCE	1.3	1.8	0.10	0.14
CO	15.5	21.1	0.22	0.3
NO_x	5.0	6.8	2.3	3.1
particulates	0.10	0.14	0.05	0.07
aldehydes	0.10/0.05[b]	0.14/0.07[b]	0.04	0.05

[a] bhp = brake horsepower; 1 bhp = 0.735 kW.
[b] The 0.10 and 0.14 values apply from 1993 to 1995; the 0.05 and 0.07 values apply beginning in 1996.

in various materials to make them compatible with methanol. The resultant L10 matched diesel power and torque, but emissions results were mixed. Data showed low emissions of particulates, but higher HC, CO, and NO_x at low speed, low load operation. These data suggest that additional optmization may be necessary for the L10 methanol engine.

The L10 methanol engines are currently being used in several demonstration fleets in California. The first use was in the City of Los Angeles Peterbilt dump truck (58), a part of the CEC truck demonstration project. Methanol L10s are also being used in the OCTD comparative alternative fuel demonstration project. In this project methanol, CNG, and LPG L10 engines are being compared to each other in a transit bus application. In the Canadian methanol in large engines (MILE) project, two L10 methanol engines were operated for several years in refuse haulers in Vancouver, British Columbia.

The largest retrofit demonstration project in the U.S. is underway at SCRTD where 12 vehicles using DDC 6V-92 with mechanical injectors are being modified to use methanol with Avocet (66). This project has evaluated the changes necessary to optimize the conversions for both fuel economy and emissions. Using the two stroke engine, changes had to be made to reduce the air into the engine, increase compression from 17 to 23:1, and use better ring packages. These changes gave both good fuel economy and reduced emissions of NO_x by 50%, considerably lowered particulates, and maintained levels of CO and hydrocarbon. New York has also modified a 8V-71 for use on Avocet-improved methanol and had similar results (67).

The success of these tests indicate that cetane-improved alcohol is technically feasible. Engines can be designed to provide equal diesel power and torque characteristics having lower NO_x and particulate emissions than diesel. However, if it is not necessary to achieve lowest possible emissions, only changes in fuel rate are required, rather than engine changes, and the commercial application of this approach depends mostly on the cost of the cetane improvers. The price of Avocet is about $4/L in small quantities and adding 3.0% in methanol nearly doubles the cost of methanol from $0.13 to $0.22/L. If Avocet were produced in larger quantities its price would drop considerably.

The biggest potential use of the cetane-improver approach may be in vehicle retrofits where for environmental reasons bus and truck fleets may be required to convert to cleaner burning fuels.

AIR QUALITY BENEFITS OF ALCOHOL FUELS

In the 1970s evaluations of alcohol fuel programs always considered environmental impacts and objectives even though the main thrust of the programs was toward energy security and diversification benefits. Assessments of performance identified these fuels as consistent with environmental goals and by the mid 1980s, the environmental benefits of the alcohol fuels had become the chief driving force for their further consideration. Detailed assessments were made of photochemical smog and air toxics reductions that might be obtained from the wide use of alcohol fuels in light-duty vehicles. Methanol received the most evaluation, because it appeared to be far more cost competitive than ethanol. The potential benefits of alcohols used in heavy-duty diesel-type engines were also studied.

The most comprehensive air quality study, supported by the California ARB and the South Coast Air Quality Management District (68), showed that if gasoline and methanol cars emitted the same amounts of carbon, an assumption that seemed reasonable based on emissions test data taken throughout the 1980s, and if methanol cars had formaldehyde emissions controlled to 9.3 mg/km (equal to the current California formaldehyde emissions standard for methanol automobiles), then substituting M85 for gasoline would produce a 9% reduction in the peak summer-day afternoon ozone level and a 19% reduction in exposure to ozone levels above the Federal standard of 0.12 ppm. These reductions constituted a substantial fraction of the reductions that would be obtained by eliminating all the emissions from vehicles. Additional assumptions were that exhaust carbon emissions were at the level of 0.15 g/km, equal to the planned certification standard for new vehicles in California beginning in 1993 (but not really expected to be characteristic of in-use vehicles) and that the distribution of hydrocarbon species in the exhaust resembled that of cars tested in the 1980s.

The results of the study, conducted at Carnegie Mellon University, are now generally accepted as the best available guides for the smog-reducing benefits of a methanol substitution strategy, at least for the conditions prevailing in the Los Angeles basin. Overall, replacement of a conventional gasoline vehicle by an equivalent M85 vehicle should provide about a 30% reduction in smog-forming potential. A 100% result would be earned by eliminating the vehicle entirely. Vehicles using M100 would provide substantially greater benefits than M85 vehicles.

Benefits depend upon location. There is reason to believe that the ratio of hydrocarbon emissions to NO_x has an influence on the degree of benefit from methanol substitution in reducing the formation of photochemical smog (69). Additionally, continued testing on methanol vehicles, particularly on vehicles which have accumulated a considerable number of miles, may show that some of the assumptions made in the Carnegie Mellon assessment are not valid. Air quality benefits of methanol also depend on good catalyst performance, especially in controlling formaldehyde, over the entire useful life of the vehicle.

Methanol substitution strategies do not appear to cause an increase in exposure to ambient formaldehyde even though the direct emissions of formaldehyde have been somewhat higher than those of comparable gasoline cars. Most ambient formaldehyde is in fact secondary formaldehyde formed by photochemical reactions of hydrocarbons emitted from gasoline vehicles and other sources. The effects of slightly higher direct formaldehyde emissions from methanol cars are offset by reduced hydrocarbon emissions (68).

Methanol use would also reduce public exposure to toxic hydrocarbons associated with gasoline and diesel fuel, including benzene, 1,3-butadiene, diesel particulates, and polynuclear aromatic hydrocarbons. Although public formaldehyde exposures might increase from methanol use in garages and tunnels, methanol use is expected to reduce overall public exposure to toxic air contaminants.

ALCOHOL FUELS USAGE AS AN AIR QUALITY STRATEGY

The cost-effectiveness of methanol substitution as an air quality strategy has been studied in some detail. Air quality planners usually rate cost-effectiveness in terms of dollars per ton of reactive hydrocarbons controlled (or removed from the inventory of emissions). Typical costs for controlling reactive hydrocarbon emissions in the U.S. are in the range of several hundreds of dollars per ton. In the Los Angeles area, the average costs of future hydrocarbon control measures average about $500 per metric ton, although some individual measures have cost-effectiveness figures above $10,000 per ton. Methanol substitution appears to be a viable and competitive control strategy. Cost-effectiveness is linked to the price of oil. Methanol appears to have a cost-effectiveness ranging from a few thousand dollars to several tens of thousands of dollars per ton (18,70–73). In heavy-duty engines, the alcohols may offer cost-effective reductions of particulates and NO_x emissions. A recent study indicates that the use of methanol was competitive with the use of cleaner diesel fuels and diesel particulate traps under some circumstances (74).

The potential air quality impacts of ethanol use have not yet been studied in detail.

Global Warming Impacts

Several studies have been made of the global warming impacts of alcohol fuels. The most useful assessments cover the entire life cycle from raw material feedstock production through processing, distribution, and fuel usage. They also consider global gases in addition to carbon dioxide (75,76). Results reflect the influence of assumptions but methanol is expected to provide slight reductions in global warming impacts compared to gasoline. Ethanol evaluations are less certain.

PUBLIC SAFETY ISSUES

Several investigators have assessed the comparative safety of methanol and conventional hydrocarbon fuels (14,77–79). The ingestion toxicity of methanol has been of some concern because of the number of gasoline ingestions associated with siphoning and in-home accidental ingestions. The use of gasoline in small engines such as those used for lawnmowers, leafblowers, and other small utility applications results in most of the siphoning ingestions and in-home accidental ingestions of gasoline. This potential problem is addressed by discouraging methanol use in small engines, by labeling and public education, and by positive siphoning prevention screens in the fill pipes of vehicles. These screens have been required in recent purchases of methanol vehicles by California agencies.

Skin contact with methanol may present a greater health threat than skin contact with gasoline and diesel fuel and is being evaluated.

The fire hazard of methanol appears to be substantially smaller than the fire hazard of gasoline, although considerably greater than the fire hazard of diesel fuel. The lack of luminosity of a methanol flame is still a concern to some, and M85 (or some other methanol fuel with an additive for flame luminosity) may become the standard fuel for this reason.

In reviewing the full range of health and safety issues associated with all alternative fuels, the California Advisory Board determined that there were no roadblocks that would prevent the near term deployment of either methanol or ethanol, assuming that adequate safety practices were followed appropriate to the specific nature of each fuel (14).

THE FUTURE OF ALCOHOL FUELS

In the late 1980s attempts were made in California to shift fuel use to methanol in order to capture the air quality benefits of the reduced photochemical reactivity of the emissions from methanol-fueled vehicles. Proposed legislation would mandate that some fraction of the sales of each vehicle manufacturer be capable of using methanol, and that fuel suppliers ensure that methanol was used in these vehicles. The legislation became a study of the California Advisory Board on Air Quality and Fuels. The report of the study recommended a broader approach to fuel quality and fuel choice that would define environmental objectives and allow the marketplace to determine which vehicle and fuel technologies were adequate to meet environmental objectives at lowest cost and maximum value to consumers. The report directed the California ARB to develop a regulatory approach that would preserve environmental objectives by using emissions standards that reflected the best potential of the cleanest fuels.

The ARB adopted a regulatory package for light-duty vehicles in 1990 that modifies the historically uniform approach to vehicle emissions, in which each and every vehicle in a regulated class must meet the same emissions standard. The new approach adopts emissions standards that apply on the average to the entire sales mix of vehicles sold by each manufacturer in each of several broad weight classes of vehicles. Thus vehicles that use fuels such as methanol and ethanol having air quality benefits in the form of lower levels of photochemical reactivity have the emissions adjusted to reflect the lower smog forming tendency of these fuels. This regulatory approach provides a powerful incentive for vehicle manufacturers to certify at least some of the sales mix of vehicles on fuels such as methanol and ethanol.

The future market response to the new form of emissions regulation is unknown. For the purpose of meeting new vehicle emissions standards, however, it is still not clear whether some combination of new emissions control approaches and reformulated gasolines can provide benefits equal to those of methanol and ethanol. It is possible that the new emissions standards will simply result in improved gasoline technologies, and that, despite the prospective air quality advantages of the alcohol fuels, the market result of the new standards will simply be cleaner gasolines. However, in 1990 the U.S. Alternative Fuels Council agreed on a goal of a 25% share of nonpetroleum transportation fuels by 2005. Although this goal may not become part of a national energy plan, it represents the first official statement of a specific goal to substitute for the use of petroleum in transportation. Alcohol fuels could capture a large part of this 25% share of nonpetroleum fuels, although vehicles powered by natural gas and electric energy will no doubt win some acceptance. For the ordinary passenger car, alcohol fuels may offer the most gasolinelike alternative in terms of range, comparable costs, and compatibility with the current gasoline/diesel storage and distribution infrastructure.

Acknowledgment

Reprinted from *Kirk-Othmer Encyclopedia of Chemical Technology*, 4th ed., Vol. 1, John Wiley & Sons, Inc., New York, 1991.

BIBLIOGRAPHY

1. J. W. Shiller, "The Automobile and the Atmosphere" in *Energy: Production, Consumption and Consequences,* National Academy Press, Washington, D.C., 1990, pp. 111–142.
2. *Engineering Data Book,* Gas Processors Suppliers Association, 1972.
3. *Reference Data for Hydrocarbons and Petro-Sulfur Compounds,* Phillips Petroleum Company, 1962.
4. J. B. Heywood, *Internal Combustion Engine Fundamentals,* McGraw-Hill, New York, 1988.
5. *Fire Hazard Properties of Flammable Liquids, Gases, and Volatile Solids,* Pub. # NFPA 325M, National Fire Protection Association, 1984.
6. P. F. Schmidt, *Fuel Oil Manual,* Industrial Press, 1969.
7. R. R. Toepel, J. E. Bennethum, R. E. Heruth, "Development of a Detroit Diesel Allison 6V-92TA Methanol-Fueled Coach Engine," *SAE Paper 831744, SAE Fuels and Lubricants Meeting* (San Francisco, Calif., Oct. 31–Nov. 3, 1983), Society of Automotive Engineers, Warrendale, Pa., 1983.
8. T. Powell, "Racing Experiences with Methanol and Ethanol-based Motor Fuel Blends," *SAE Paper 750124,* Society of Automotive Engineers, Warrendale, Pa., Feb., 1975.
9. Sypher-Mueller International, Inc., *Future Transportation Fuels: Alcohol Fuels, Energy, Mines and Resources—Canada,* Project Mile Report, A Report on the Use of Methanol in Large Engines in Canada, May 1990.
10. S. C. Trindade and A. V. de Carvalho, "Transportation Fuels Policy Issues and Options: The Case of Ethanol Fuels in Brazil" in D. Sperling, ed., *Alternative Transportation Fuels,* Quorum Books, New York, 1989, pp. 163–185.
11. *First Interim Report,* United States Alternative Fuels Council, Washington, D.C., Sept. 30, 1990.
12. Office of Policy, Planning, and Analysis, *Assessment of Costs and Benefits of Flexible and Alternative Fuel Use in the U.S. Transportation Sector,* Technical Report #3 (Methanol Production and Transportation Costs) Pub. # DOE/P/E–0093, U.S. Department of Energy, Washington, D.C., Nov. 1989.
13. Bechtel Corporation, *California Fuel Methanol Cost Study: Chevron Corporation, U.S.,* Vol. 1 (Executive Summary, Jan. 1989); Vol. 2, (Final Report, Dec. 1988), San Francisco, Calif., 1988–1989.
14. California Advisory Board on Air Quality and Fuels, Vol. 1, *Executive Summary;* Vol. 2, *Energy Security Report;* Vol. 3, *Environmental Health and Safety Report;* Vol. 4, *Economics Report;* Vol. 5, *Mandates and Incentives Report;* San Francisco, Calif., June 13, 1990.
15. California Energy Commission, *Methanol as a Motor Fuel: Review of the Issues Related to Air Quality, Demand, Supply, Cost, Consumer Acceptance and Health and Safety,* Pub. # P500–89–002, Sacramento, Calif., April 1989.
16. California Energy Commission, *Cost and Availability of Low Emission Motor Vehicles and Fuels,* Sacramento, Calif., Aug. 1989.
17. National Research Council, *Fuels to Drive Our Future,* National Academy Press, Washington, D.C., 1990.
18. Office of Technology Assessment, *Replacing Gasoline: Alternative Fuels for Light-Duty Vehicles,* Pub. # OTA–E–364, U.S. Congress, Washington, D.C., Sept. 1990.
19. J. L. Keller, *Hydrocarbon Process.* (May 1979).
20. Office of Mobile Sources, *Analysis of the Economic and Environmental Effects of Methanol as an Automotive Fuel,* U.S. Environmental Protection Agency, Ann Arbor, Mich., Sept. 1989.
21. *Alcohols and Ethers Blended with Gasoline,* Pub. # 4261, American Petroleum Institute, Washington, D.C., Chapt. 4, pp. 23–27.
22. *7th Int. Symp. on Alcohol Fuels,* Institut Francais du Petrole, Editions Technip, Paris, France, 1986.
23. *8th Int. Symp. on Alcohol Fuels,* New Energy and Industrial Technology Development Organization, Sanbi Insatsu Co., Ltd. Tokyo, Japan, Nov. 13–16, 1988.
24. H. C. Wolff, "German Field Test Results on Methanol Fuels M100 and M15," *American Petroleum Institute 48th Midyear Refining Meeting* (Session on Oxygenates and Oxygenate-Gasoline Blends as Motor Fuels) May 11, 1983.
25. H. Menrad, "Possibilities to Introduce Methanol as a Fuel, An Example from Germany," *6th Int. Symp. on Alcohol Fuels Technology,* (Ottawa, Canada, May 21–25, 1984), Vol. 2.
26. L. Schieler, M. Fischer, D. Dennler, and R. Nettell, "Bank of America's Methanol Fuel Program: An Insurance Policy that is Now a Viable Fuel," *6th Int. Symp. on Alcohol Fuels Technology,* (Ottawa, Canada, May 21–25, 1984), Vol. 2.
27. R. N. McGill and R. L. Graves, "Results from the Federal Methanol Fleet, A Progress Report," *8th Int. Symp. on Alcohol Fuels* (Tokyo, Japan, Nov. 13–16, 1988).
28. *Alcohol Energy Systems: Alcohol Fleet One Test Report,* Pub. # 500–82–058, California Energy Commission, Sacramento, Calif., Aug. 1983.
29. F. J. Wiens and co-workers, "California's Alcohol Fleet Test Program: Final Results" *6th Int. Symp. on Alcohols Fuels Technology* (Ottawa, Canada, May 21–25, 1984), Vol. 3.
30. Acurex Corporation, *California's Methanol Program: Evaluation Report,* Vol. 2 (Technical Analyses), Pub. # P500–86–012A, California Energy Commission, Sacramento, Calif., June 1987.
31. S. C. Trindade and A. V. de Carvalho, "Utilization of Alcohol Fuels in Brazil: Early Experience, Current Situation, and Future Prospects," *Int. Symp. on Introduction of Methanol-Powered Vehicles* (Tokyo, Japan, Feb. 19, 1987).
32. Acurex Corporation, *California's Methanol Program: Evaluation Report,* Vol. 1 (Executive Summary), Pub. #P500–86–012, California Energy Commission, Sacramento, Calif., Nov. 1986.
33. E. R. Fanick, L. R. Smith, J. A. Russell and W. E. Likos, "Laboratory Evaluation of Safety-Related Additives for Neat Methanol Fuel," SAE Paper 902156, (SP840), Society of Automotive Engineers, Warrendale, Pa., Oct. 1990.
34. U. Hilger, G. Jain, E. Scheid, and F. Pischinger, "Development of a Direct Injected Neat Methanol Engine for Passenger Car Applications," SAE Paper 901521, *SAE Future Transportation Technology Conf. and Expo.* (San Diego, Calif., Aug. 13–16, 1990).

35. P. A. Machiele, "A Perspective on the Flammability, Toxicity, and Environmental Safety Distinctions Between Methanol and Conventional Fuels," *AIChE 1989 Summer National Meeting* (Philadelphia, Pa., Aug. 22, 1989). American Institute of Chemical Engineers.
36. J. V. D. Weide and R. J. Wineland, "Vehicle Operation with Variable Methanol/Gasoline Mixtures," *6th Int. Symp. on Alcohol Fuels Technology* (Ottawa, Canada, May 21–25, 1984), Vol. 3.
37. T. B. Blaisdell, M. D. Jackson, and K. D. Smith, "Potential of Light-Duty Methanol Vehicles," SAE Paper 891667, *SAE Future Transportation Technology Conf. and Expo.* (Vancouver, Canada, Aug. 7–10, 1989), Society of Automotive Engineers, Warrendale, Pa.
38. Mobile Source Division, *Alcohol Fueled Vehicle Fleet Test Program: 9th Interim Report,* Pub. ARB/MS–89–09, California Air Resources Board, El Monte, Calif., Nov. 1989.
39. W. Carter, *Ozone Reactivity Analysis of Emissions from Motor Vehicles,* (Draft Report for the Western Liquid Gas Association), Statewide Air Pollution Research Center, University of California at Riverside, July 11, 1989.
40. P. A. Gabele, *J. of Air Waste Management* **40**(3), 296–304 (Mar. 1990).
41. S. Albu, "California's Regulatory Perspective on Alternate Fuels," *13th North American Motor Vehicle Emissions Control Conf.* (Tampa, Fla., Dec. 11–14, 1990), Mobile Source Division, California Air Resources Board, El Monte, Calif.
42. A. Lawson, A. J. Last, A. S. Desphande, and E. W. Simmons, "Heavy-Duty Truck Diesel Engine Operation on Unstabilized Methanol/Diesel Fuel Emulsion," SAE Paper 810346, (SP–480) *Int. Congress and Expo* (Detroit, Mich., Feb. 23–27, 1981) Society of Automotive Engineers, Warrendale, Pa.
43. B. M. Bertilsson, "Regulated and Unregulated Emissions from an Alcohol-Fueled Diesel Engine with Two Separate Fuel Injection Systems," *5th Int. Symp. on Alcohol Fuel Technology* (Auckland, New Zealand, May 13–18, 1982) Vol. 3.
44. R. A. Baranescu, "Fumigation of Alcohols in a Multicylinder Diesel Engine: Evaluation of Potential," SAE Paper 860308, *SAE Int. Congress and Expo.* (Detroit, Mich., Feb. 24–28, 1986) Society of Automotive Engineers, Warrendale, Pa.
45. R. G. Jackson, "Workshop on Diesel Fuel Substitution," *5th Int. Symp. on Alcohol Fuel Technology* (Auckland, New Zealand, May 13–18, 1982), Vol. 4.
46. H. K. Bergmann and K. D. Holloh, "Field Experience with Mercedes-Benz Methanol City Buses," *6th Int. Symp. on Alcohol Fuels Technology* (Ottawa, Canada, May 21–25, 1984), Vol. 1.
47. A. Nietz and F. Chmela, "Results of Further Development in the M.A.N. Methanol Engine," *6th Int. Symp. on Alcohol Fuels Technology* (Ottawa, Canada, May 21–25, 1984), Vol. 1.
48. M. D. Jackson, C. A. Powars, K. D. Smith, and D. W. Fong, "Methanol-Fueled Transit Bus Demonstration," *Paper 83–DGP–2,* American Society of Mechanical Engineers.
49. M. D. Jackson, S. Unnasch, C. Sullivan, and R. A. Renner, "Transit Bus Operation with Methanol Fuel," *SAE Paper 850216, SAE Int. Congress and Expo.* (Detroit, Mich., Feb. 25–Mar. 1, 1986) Society of Automotive Engineers, Warrendale, Pa.
50. *Joint TMC/SAE Fuel Consumption Test Procedure, Type 2,* SAE J1321, SAE Recommended Practice Approved October 1981, Society of Automotive Engineers, Warrendale, Pa., 1981.
51. M. D. Jackson, S. Unnasch, and D. D. Lowell, "Heavy-Duty Methanol Engines: Wear and Emissions," *8th Int. Symp. on Alcohol Fuels* (Tokyo, Japan, Nov. 13–16, 1988).
52. K. D. Smith, "California's Methanol Program," *7th Int. Symp. on Alcohol Fuels,* (Paris, France, Oct. 20–23, 1986).
53. *EPA Emissions Standards, Code of Federal Regulations,* Vol. 40, Chapt. 1, Sect. 86.088–11, 86.091–11, and 86.094–11, U.S. Environmental Protection Agency, U.S.G.P.O., Washington, D.C., 1987.
54. J. Jaye, S. Miller, and J. Bennethum, "Development of the Detroit Diesel Corporation Methanol Engine," *8th Int. Symp. Alcohol Fuels* (Tokyo, Japan, Nov. 13–16, 1988).
55. R. A. Brown, J. A. Nicholson, M. D. Jackson, and C. Sullivan, "Methanol-Fueled Heavy-Duty Truck Engine Applications: The CEC Program." *SAE Paper 890972, SAE 40th Annual Earthmoving Industry Conf.* (Peoria, Ill., April 11–13, 1989) Society of Automotive Engineers, Warrendale, Pa.
56. R. Baranescu and co-workers, "Prototype Development of a Methanol Engine for Heavy-Duty Application-Performance and Emissions," *SAE Paper 891653, SAE Future Transportation Technology Conf.* (Aug. 7–20, 1989), Society of Automotive Engineers, Warrendale, Pa.
57. R. Richards, "Methanol-Fueled Caterpillar 3406 Engine Experience in On-Highway Trucks," *SAE Paper 902160, (SP–840),* Society of Automotive Engineers, Warrendale, Pa., Oct., 1990.
58. M. D. Jackson, C. Sullivan, and P. Wuebben, "California's Demonstration of Heavy-Duty Methanol Engines in Trucking Applications," *9th Int. Symp. on Alcohol Fuels* (Milan, Italy, April 9–12, 1991).
59. W. Bandel, "Problems in the Application of Ethanol as Fuel for Utility Vehicles," *2nd Int. Symp. on Alcohol Fuel Technology* (Wolfsburg, Germany, Nov. 21–23, 1977).
60. W. Bandel and L. M. Ventura, "Problems in Adapting Ethanol Fuels to the Requirements of Diesel Engines," *40th Int. Symp. on Alcohol Fuels* (Guaruja, Brazil, Oct. 1980).
61. A. J. Schaefer and H. O. Hardenburg, "Ignition Improvers for Ethanol Fuels," SAE Paper 810249, SP–840, *SAE Int. Congress. and Expo.* (Detroit, Mich., Feb. 23–27, 1981).
62. E. P. Fontanello, L. M. Ventura, and W. Bandel, "The Use of Ethanol with Ignition Improver as a CI-Engine Fuel in Brazilian Trucks," *7th Int. Symp. on Alcohol Fuels* (Paris, France, Oct. 20–23, 1986).
63. H. O. Hardenburg, "Comparative Study of Heavy-Duty Engine Operation with Diesel Fuel and Ignition-Improved Methanol," *SAE Paper 872093, SAE Int. Fuels and Lubricants Meeting and Expo.* (Toronto, Canada, Nov. 2–5, 1987), Society of Automotive Engineers, Warrendale, Pa.
64. E. J. Lyford-Pike, F. C. Neves, A. C. Zulino, and V. K. Duggal, "Development of a Commercial Cummins NT Series to Burn Ethanol Alcohol," *7th Int. Symp. on Alcohol Fuels* (Paris, France, Oct. 20–23, 1986).
65. A. B. Welch, W. A. Goetz, D. Elliott, J. R. MacDonald, and V. K. Duggal, "Development of the Cummins Methanol L10 Engine with an Ignition Improver," *8th Int. Symp. on Alcohol Fuels* (Tokyo, Japan, Nov. 13–16, 1988).
66. S. Unnasch and co-workers, "Transit Bus Operation with a DDC 6V–92TAC Engine Operating on Ignition-Improved Methanol," *SAE Paper 902161,* (SP–840), *SAE Int. Fuels and Lubricants Meeting and Expo.* (Tulsa, Oklahoma, Oct. 22–23, 1990).
67. C. M. Urban, T. J. Timbario, and R. L. Bechtold, "Performance and Emissions of a DDC 8V–71 Engine Fueled with Cetane Improved Methanol," *SAE Paper 892064, SAE Int. Fuels and Lubricants Meeting and Expo.* (Baltimore, Md., Sept. 25–28, 1989) Society of Automotive Engineers, Warrendale, Pa.

68. J. N. Harris, A. R. Russell, and J. B. Milford, *Air Quality Implications of Methanol Fuel Utilization, SAE Paper 881198,* Society of Automotive Engineers, Warrendale, Pa., 1988.
69. T. Y. Chang and S. Rudy, Urban Air Quality Impact of Methanol-Fueled Compared to Gasoline-Fueled Vehicles, in W. Kohl, ed., *Methanol as an Alternative Fuel Choice: An Assessment,* Johns Hopkins Foreign Policy Institute, Washington, D.C., 1990, pp. 97–120.
70. C. B. Moyer, S. Unnasch, and M. D. Jackson, *Air Quality Programs as Driving Forces for Methanol Use,* Transportation Research, Vol. 23A, No. 3, 1989, pp. 209–216.
71. T. Lareau, *The Economics of Alternative Fuel Use: Substituting Methanol for Gasoline,* American Petroleum Institute, Washington, D.C., June 1989.
72. Mobile Source Division, *Alcohol-Fueled Vehicle Fleet Test, 9th Interim Report* ARB/MS–89–09, California Air Resources Board, El Monte, Calif., Nov. 1989.
73. *Proposed Regulations for Low Emission Vehicles and Clean Fuels,* Staff Report, California Air Resources Board, Sacramento, Calif., Aug. 13, 1990.
74. S. Unnasch, C. B. Moyer, M. D. Jackson, and K. D. Smith, *Emissions Control Options for Heavy-Duty Engines, SAE 861111,* Society of Automotive Engineers, Warrendale, Pa., 1986.
75. M. A. DeLuchi, "Emissions of Greenhouse Gases from the Use of Gasoline, Methanol, and Other Alternative Transportation Fuels," in W. Kohl, ed., *Methanol as an Alternative Fuel Choice: An Assessment,* Johns Hopkins Foreign Policy Institute, Washington, D.C., 1990, pp. 167–199.
76. S. C. Unnasch, B. Moyer, D. D. Lowell, and M. D. Jackson, *Comparing the Impacts of Different Transportation Fuels on the Greenhouse Effect, Pub. # 500–89–001,* California Energy Commission, Sacramento, Calif., April, 1989.
77. *Methanol Health and Safety Workshop* (Los Angeles, Calif., Nov. 1–2, 1989) South Coast Air Quality Management District, El Monte, Calif., 1989.
78. P. A. Machiele, "A Perspective on the Flammability, Toxicity, and Environmental Safety Distinctions Between Methanol and Conventional Fuels," *AIChE 1989 Summer National Meeting* (Philadelphia, Pa., Aug. 22, 1989), American Association of Chemical Engineers.
79. P. A. Machiele, "A Health and Safety Assessment of Methanol as an Alternative Fuel," in W. Kohl, ed., *Methanol as an Alternative Fuel: An Assessment,* Johns Hopkins Foreign Policy Institute, Washington, D.C., 1990, pp. 217–239.

General References

K. Boekhaus and co-workers, "Reformulated Gasoline for Clean Air: An ARCO Assessment," *2nd Biennial U.C. Davis Conf. on Alternative Fuels* (July 12, 1990).

California's Methanol Program, Evaluation Report, Vol. 2 (Technical Analyses), Pub. # P500–86–012A, California Energy Commission, Sacramento, Calif., June 1987.

A. V. de Carvalho, "Future Scenarios of Alcohols as Fuels in Brazil," *3rd Int. Symp. on Alcohol Fuels Technology* (Asilomar, Calif., May 29–31, 1979).

C. L. Gray, Jr. and J. A. Alson, *Moving American to Methanol,* University of Michigan Press, Ann Arbor, Mich., 1985.

P. A. Lorang, Emissions from Gasoline-Fueled and Methanol Vehicles, in W. L. Kohl, ed., *Methanol as an Alternative Fuel Choice: An Assessment* Johns Hopkins Foreign Policy Institute, Washington, D.C., 1990, pp. 21–48.

J. H. Perry and C. P. Perry, *Methanol: Bridge to a Renewable Energy Future* University Press of American, Lanham, Md., 1990.

D. Sperling, *New Transportation Fuels,* University of California Press, Berkeley, Calif., 1988.

D. Sperling, *Brazil, Ethanol, and the Process of Change, Energy,* Vol. 12, No. 1, pp. 11–23, 1987.

E. Supp, *How to Produce Methanol from Coal,* Springer-Verlag, Berlin, 1990.

J. V. D. Weide and W. A. Ramackers, "Development of Methanol and Petrol Carburation Systems in the Netherlands," *2nd Int. Symp. on Alcohol Fuel Technology,* (Wolfsburg, Germany, Nov. 21–23, 1977).

1987 World Methanol Conf. (San Francisco, Calif., Dec. 1–3, 1987) Crocco and Associates, Houston, Tex.

1989 World Methanol Conf. (Houston, Tex., Dec. 5–7, 1989), Crocco and Associates, Houston, Tex.

ALUMINUM

Jennifer S. Gitlitz
Clark University
Worcester, Massachusetts

Aluminum has become one of the most important metals on the global market in terms of tonnage produced and dollar value: with an annual production of 721 Mt, only crude steel surpasses it, while copper lags behind at 9.3 Mt (1). The success of the industry has been accompanied by a wide range of environmental impacts, most of which are unknown to the average consumer of aluminum products. Direct impacts of the primary aluminum production process include emissions of fluorides, sulfur dioxide, and several important greenhouse gases, and the generation of large quantities of solid wastes, some of which may be hazardous. Indirect effects include the consumption of a tremendous amount of electrical energy with all of its associated environmental impacts: notably emissions of greenhouse gases and those contributing to acid rain, and the large-scale degradation and displacement of wildlife habitat and human communities by hydroelectric dams. See also Energy efficiency; Super cars; Life cycle analysis; Recycling.

PROPERTIES

Aluminum has often been called a wonder-metal. It is lightweight, and when alloyed with small quantities of other metals, is extremely strong. It is nonmagnetic, odorless and tasteless, and can be embossed, painted and printed on with ease. Because it resists corrosion by weather and by strong acids, it is ideal for use in the transportation and storage of many chemicals. It can be shaped into a wide variety of forms using heat and mechanical means. These properties combine to make it desirable in many applications, including food and drink packaging, household furniture and appliances, and cars, planes and trains. Its flexibility and resistance to corrosion give it an advantage over steel in these same applications. A good electrical conductor, aluminum is replacing heavier

Plate I. Aluminum transmission cable is rolled onto a reel. Overhead conductor cable is a common use for aluminum. Courtesy of Reynolds Metals Company.

steel and more expensive copper in wires and cables throughout the world (see Plate I). Aluminum can be fabricated in many ways. Much like steel, molten aluminum can be cast into parts, and solid ingots can be rolled into sheets of various thicknesses, including extremely thin

Plate II. The most common use of aluminum is for containers and packaging. Within this area, the largest component is beverage-can manufacturing. Beverage cans, as seen here, are closely tracked from production to consumption and through recycling. Courtesy of Reynolds Metals Company.

foil. It can be extruded through shaped holes into various rods and bars, drawn (or punched) into desired shapes, or blown in powder form into complex molds and then heated, or sintered, to get its final form.

COMMON END USES AND PRODUCTS

Figure 1 shows the end use breakdown of net aluminum shipments in the United States in 1992. This includes aluminum that is fabricated in the United States and imported. Containers and packaging is the single largest principal end use of aluminum, constituting almost a third of U.S. net aluminum shipments (see Plate II). Containers and packaging is the only sector of the aluminum market that is closely tracked from production through consumption through recycling, because its largest component is used beverage cans (UBCs), which are recycled in a "closed-loop," or into cans again. This sector of the market

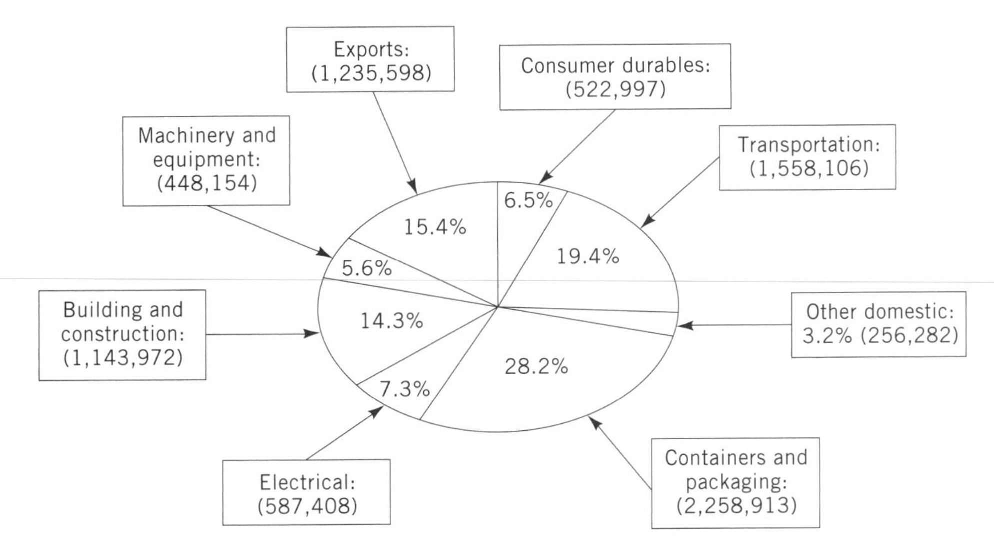

Figure 1. Distribution of net aluminum shipments by principal market in the United States, 1992 (t). From Ref. 2.

is also important because it is the only one that is specifically designed to have a short product lifetime, in many cases a few minutes between consumer purchase and disposal, as in the case of beer or pop cans, or the aluminum linings of aseptic containers (drink boxes). Aluminum foil is usually only used once, as are TV-dinner trays, pie tins, and closures for glass bottles. It is up to the consumer to recycle these items, and in areas where curbside recycling does not yet exist, the vast majority of non-can items will end up in the garbage.

Unlike packaging, other aluminum products have lifetimes of several years, as in the case of cars, to over fifty years, in the case of buildings and bridges. After containers, the transportation sector is the next largest market. This includes parts and bodies for passenger automobiles, commercial and military planes and rockets, railroad cars, trucks, trailers, busses, etc. If recent trends continue, these uses will show more growth than all other market sectors in the coming decade. To improve gas mileage, car manufacturers in Japan, Europe, and the United States are "lightweighting" cars by substituting aluminum for steel in many applications.

Building and construction materials include window and door frames, screens, awnings, aluminum siding, girders and supports for bridges and buildings, mobile homes, and highway signs and guardrails. Electrical uses primarily consist of wires and cables. Consumer durables include aluminum furniture, appliances, cooking utensils, etc.

BAUXITE MINING AND ALUMINA PRODUCTION

Mining and Processing Bauxite Ore

The third most abundant element on Earth, aluminum comprises 8.3% of the Earth's crust by weight on average, or 15.7% alumina. Higher than average concentrations of aluminum oxide, or alumina (Al_2O_3), can be found in a wide range of rocks, including granite and basalt. Economically-recoverable alumina is extracted from bauxite, a reddish-brown ore containing 40–60% hydrated alumina by weight. The bauxite deposits mined today contain alumina which has been concentrated by surface weathering of permeable rocks under tropical conditions spanning geologic time. Although alumina can be recovered from materials other than bauxite, including clays, alunite, anorthosite, coal wastes and oil shales, these production processes are less economically feasible, and have only been employed in Eastern Bloc countries (3,4).

In some areas including parts of Australia and Jamaica, bauxite ore lies beneath a few feet of vegetation and topsoil, and can be scooped up with front-loaders and draglines with relative ease. In other areas, the deposits lie beneath up to 200 feet of overburden (unusable topsoil and rock), which is frequently vegetated. Thicknesses of ore bodies range widely: from a few feet in Arkansas, an average of 16 feet in Brazil, and 30 feet in Australia (6). Strip mining or open pit mining accounts for almost all production, because underground mining is prohibitively expensive. After vegetation is removed, the overlying rock is blasted to expose and loosen the ore. Large pieces are broken up, and are hauled to the processing site in dump trucks, rail cars or aerial trams. Strip mining has traditionally led to severe erosion and loss of topsoil. Millions of tons of overburden are generated in producing the world's aluminum. According to the U.S. Bureau of Mines, ratios of overburden to recovered ore have been as high as 13:1 in the state of Arkansas (5). Overburden has historically been left as open mounds, easily washing into lakes, streams, and the ocean, degrading the health of freshwater and marine ecosystems. However, there is an increasing industry trend in strip-mining to place the overburden back in the excavated holes, and to attempt to revegetate the site.

After mining, sizing and washing removes impurities such as sand and clay, and in the process creates about 0.7 tons of bauxite tailings in slurry form per ton alumina produced. These wet residues are usually disposed of on land, and have created huge mud lakes which have contaminated local water sources worldwide. New technologies now can reduce the moisture content of the mud to permit a more benign "dry stacking" method of disposal. The washed bauxite is then sun-dried or "calcined" (baked) in kilns to reduce its moisture content, and is then shipped to domestic alumina plants or is exported by ship for refining. Some alumina refineries are located near bauxite mines and far from smelters; some are near smelters, utilizing bauxite that is shipped from afar. In other cases, production facilities for bauxite, alumina and aluminum are integrated. Transportation is the largest single cost factor in producing alumina (6).

Alumina Production by the Bayer Process

Figure 2 is a simplified diagram of the Bayer Process, which is used in making virtually all of the world's alumina. Because the composition of bauxite ores around the world varies, alumina plants are designed to treat specific bauxites (8). First, the dried bauxite is size-reduced in a rod mill (stages 1 and 2) to particles in the 1 mm range. A recycled, concentrated solution of sodium hydroxide (2NaOH) or sodium carbonate (Na_2CO_3) is added to the ground bauxite, producing a slurry of 40–50% solids, which is fed to a series of digesters (stage 3), where more sodium hydroxide solution ("liquor") is added, and heated under pressure. The basic reaction in the digesters (stage 1) is:

$$Al_2O_3 \cdot x\ H_2O + \text{impurities} + 2\ NaOH \rightarrow 2NaAlO_2 + (x + 1)\ H_2O + \text{red mud}$$

The heat, pressure and caustic solution cause the alumina to dissolve into a sodium aluminate solution ($2NaAlO_2$). The mixture is brought through a heat-exchange "flash vessel" (stage 4) where its temperature and pressure are lowered, and the heat is recovered to pre-heat the recycled sodium hydroxide liquor. Excess water is removed, and in (stage 5), spent liquor (NaOH) is added to dilute the solution, to prevent the $NaAlO_2$ from precipitating prematurely. Stages 6–9 involve solid-liquid separation, wherein the sodium aluminate solution undergoes three types of filtration to remove insoluble impurities (red sand and mud). Gravity separators or wet cyclones are employed to remove the coarse sand. The red mud residues from the filtration are discharged into a waste lake, after being filtered to remove as much caustic liquor as possible. The resulting fluid is a clear brown sodium aluminate solution

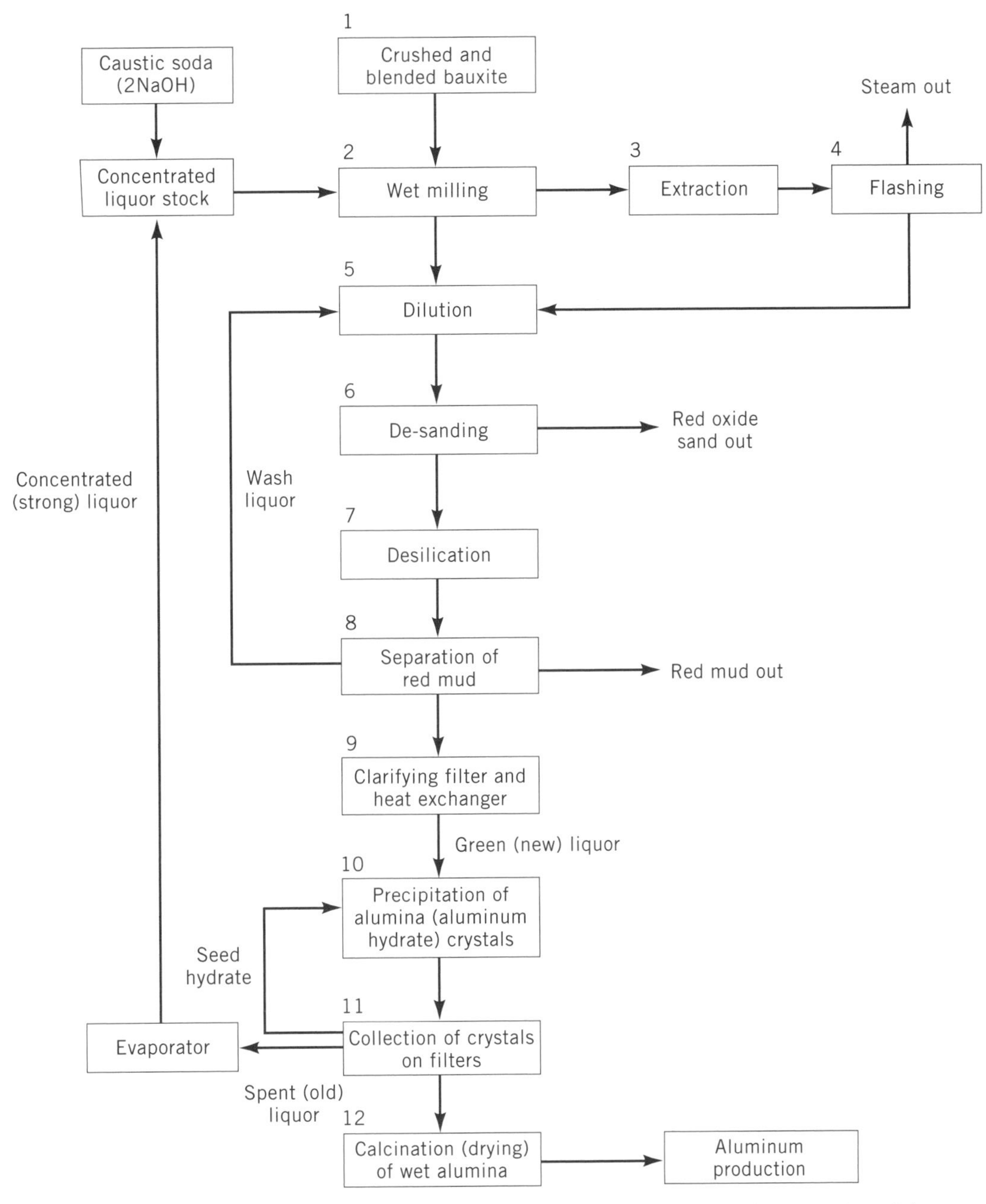

Figure 2. Simplified flow chart of the Bayer process (7). Reprinted with permission of the Controller of Her Britannic Majesty's Stationery Office.

known as "new liquor" or "green liquor." It is subjected to another set of heat exchangers before it is precipitated (7,9). The equation:

$$NaAlO_2 + 2\,H_2O \xrightarrow{\text{seeding with } Al_2O_3 \cdot 3H_2O \text{ crystals}} NaOH + Al(OH)_3$$

describes the reaction wherein alumina is precipitated out from the supersaturated sodium aluminate solution by seeding the solution with aluminum trihydrate crystals (stage 10), which serve as nucleation points for the dissolved alumina. The chemical reaction is essentially the reverse of the earlier digestion reaction. The crystals grow as the reaction proceeds for up to three days in large agitated tanks. During this stage, the spent liquor is regenerated to yield caustic solution which is then recycled, along with any unprecipitated alumina, to the earlier digestion stage of the process (stage 11). The precipitated alumina is then washed and filtered for the last time to remove any residues. Flash evaporators remove excess water to attain a free moisture content of 8–10%. Some alumina is saved as "seed" for the precipitation process; the remainder is calcined in a rotary kiln at 1,200–1,300°C to further reduce the moisture content and produce anhydrous alumina powder (stage 12). Industry trends are to move from producing fine powders to producing a coarser grain product that requires less calcination. Coarse, sand-like alumina powder has the advantage of large surface area, and can be used to trap fluoride gas emissions in dry scrubbers in primary aluminum smelting operations, and can also be recycled as feedstock for the cell. Finished metallurgical grade alumina contains 0.3–0.8% soda, less than 0.1% iron oxide and silica, and trace amounts of other impurities (10).

Energy and Materials Requirements for Bauxite and Alumina Production

From 2.2–3.5 tons of bauxite are required for every ton of alumina produced, depending on the grade of the ore, and about 2 tons of alumina are required for each ton of aluminum. Using higher grade ores and the best technologies, the resulting ratio of bauxite ore to primary aluminum is approximately 4.5:1 (11). According to the Argonne National Laboratory, the range of energy required to mine and process bauxite is $0.26–1.06 \times 10^6$ Btu/short ton of bauxite, and the amount of fuel oil needed to ship the bauxite from producing countries to ports in the United States ranged from $0.75–2.0 \times 10^6$ Btu/short ton of bauxite. The alumina production process is much more energy-intensive, requiring 18.6×10^6 Btu/short ton of alumina (12). Added together, about 35×10^6 Btu are required per short ton of primary aluminum produced.

Waste Products from Bauxite and Alumina Production

In the usual alumina production process described above, there are between 1.3 and 1.5 tons of red mud produced per ton of aluminum. Although these muds pose a significant disposal problem, they are still exempt from hazardous waste regulation under the Resource Conservation and Recovery Act (RCRA) in the United States. Disposal is largely unregulated in major bauxite and alumina-producing countries. Alcoa has developed a "combination process" whereby lower grade ores (with higher silica contents and thus more aluminum silicate lost in the muds) can be exploited. Their process enables the recovery of alumina from the red mud by heating the mud with calcium carbonate (limestone) and sodium carbonate (soda ash) to produce a sodium aluminate solution, which is returned to the digesters to resume the process described. The waste is called "brown mud." (13).

THE HALL-HÉROULT SMELTING PROCESS

Primary aluminum is not produced in massive fiery furnaces like those used to smelt iron; it is made electrolytically in rectangular vats (called "cells" or "pots") measuring several meters across. The smelting operation is a continuous (rather than a batch) process. The individual cells are constructed out of rectangular steel shells lined with a refractory (non-burnable) material such as insulating brick. The refractory material on the floor of each cell is overlain with a carbon lining and a carbon cathode. This in turn is overlain with a layer of molten aluminum. The molten aluminum lies beneath a "bath" of molten synthetic cryolite, or Na_3AlF_6, at a temperature of 960–970°C. The cryolite serves as a medium to dissolve the alumina and minor additives, and to conduct electric current from the anode to the aluminum. Alumina powder is lowered from overhead hoppers onto a "crust" of cooler cryolite which forms on top of the cell. The crust serves to thermally insulate the contents of the pot, and to drive off any remaining moisture in the alumina. Periodically, the crust is mechanically broken and the alumina is stirred into the bath.

Large carbon anodes are suspended by steel bars into each cell. When a strong electric current (ranging from 50–225 kilo-Amperes) is passed through the anode into the bath, it causes the alumina to split into oxygen and molten aluminum. The oxygen combines with carbon on the anode to produce carbon dioxide (CO_2) and carbon monoxide (CO), which are vented to the atmosphere, along with fluoride compounds that are created as a result of partial evaporation of the molten cryolite.

There are two main methods of smelting used today, each relying on a different type of anode. The older and more inefficient "Søderberg" process employs only one large anode per cell. The anode is comprised of a moist carbon paste that is continually baked as it descends into the cell, while fresh paste is added from above. All modern smelters use the more efficient "prebake" method, using anodes which are usually produced in on-site facilities from petroleum coke and coal tar pitch, by-products of the petroleum, coal, and steel manufacturing industries, which if not used in aluminum production, might have to be disposed of alternatively. Each prebake cell has 10 to 20 anodes. As the bottoms of the anodes are slowly consumed, they are lowered further into the bath to maintain a constant distance between the anode and the cathode (about 1.5 inches). Each anode must be replaced every few weeks, but this can be done in a staggered fashion without disrupting production. In contrast, spent potlinings need only be replaced once every five years, which is fortunate because replacement entails suspending production. Figure 3 is an illustration of a prebake cell.

The molten aluminum is periodically siphoned off through the tops of the cells into a holding furnace, and is then poured into "ingots" or "billets," which are large blocks or bars of solid aluminum (see Plate III).

Under normal operating conditions, potlines run 24 hours a day, 365 days a year. If the electricity supply to the cells is cut off for more than three hours at a time, the molten aluminum will "freeze," or solidify in the pots, a mess that can take months and millions of dollars to clean up. This production restriction sets aluminum apart from some other large energy-intensive industries: aluminum smelters cannot purchase "interruptible power" from utilities. Their contracts usually have clauses which specify that they must take first priority in the event of power shortages.

DIRECT BY-PRODUCTS OF THE SMELTING PROCESS

The primary aluminum production process generates waste products that can impact the environment at local, regional and global levels. The following sections are concerned with solid wastes and airborne emissions from the smelting process, including fluoride compounds that can harm vegetation and animal life in the smelter locale; sulfur dioxide, a contributor to acid rain formation on a regional level; and greenhouse gases including carbon dioxide, carbon monoxide, and compounds containing carbon

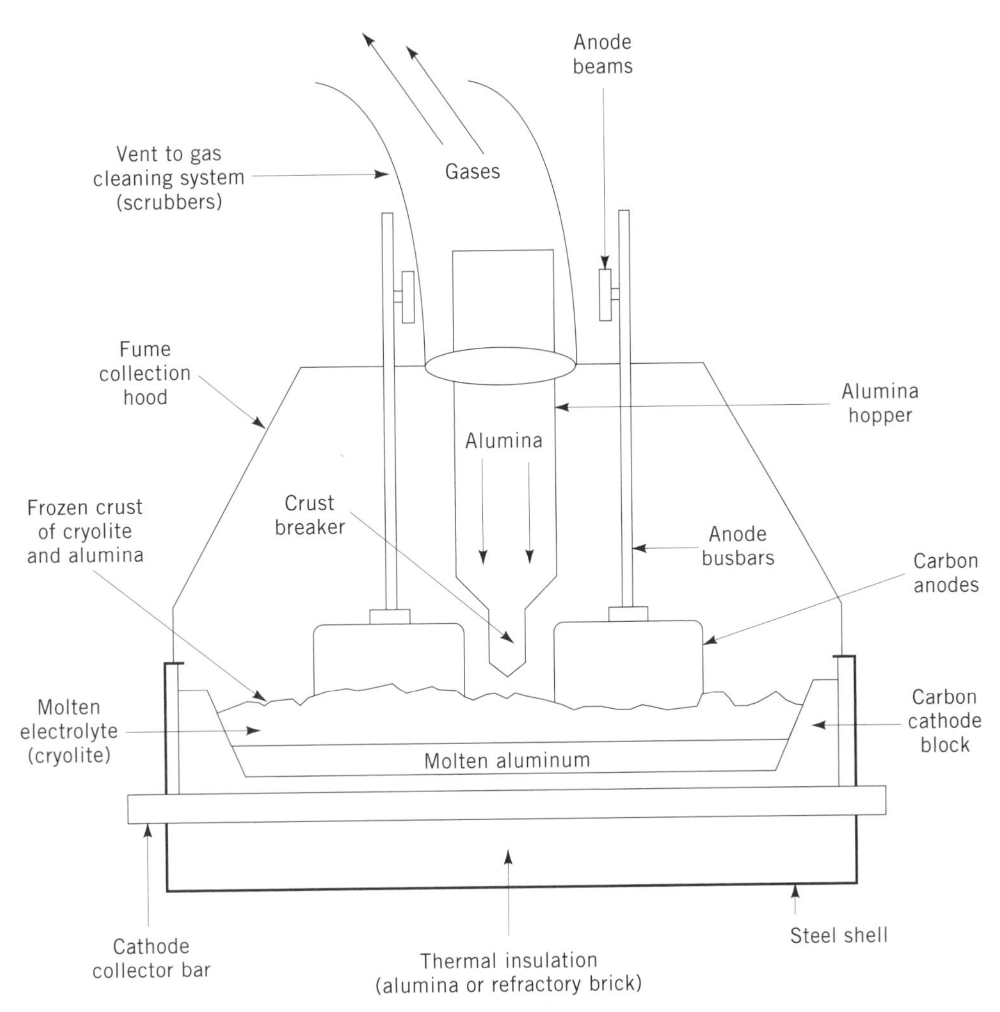

Figure 3. Cross section of a center-worked prebake anode cell.

and fluorine. These gases trap long-wave radiation in the atmosphere, possibly leading to global warming. Solid wastes include spent carbon potlinings which can be leached by rain, contaminating groundwater. Other by-products from anode manufacture, smelting, and product forming which will not be treated here include wastewater, solid wastes other than potlinings, dusts generated by alumina and coke handling, and emissions from anode forming and baking including carbon dust, particulate and gaseous fluorides, NO_x, SO_2, and tar vapor.

Local Damages from Fluoride Emissions

During normal operations, a variety of fluorides in gaseous and particulate form evaporate from the hot cryolite bath, and are released from the anodes and the bath when the frozen crust is broken to add alumina. Particulates include cryolite (Na_3AlF_6), aluminum fluoride (AlF_3), calcium fluoride (CaF_2) and chiolite ($Na_5Al_3F_{14}$), while gases include hydrogen fluoride (HF) and silicon tetrafluoride (SiF_4) (14). The particulates are collected by bag filters. Since the mid-1970s, all new smelters have installed hoods over the cells to capture fluoride fumes and process them using dry and wet scrubbers. Dry scrubbers employ alumina which combines with the fluoride gases and is then recycled into the cell bath as fresh cryolite, thus reducing total fluorides emissions per ton aluminum (F_t/t) to 0.75–1.0 kg. The U.S. emissions standard set by the Environmental Protection Agency is 1 kg F_t/t. Some smelters, including older Søderberg plants which have not been able to afford costly equipment retrofitting, use wet scrubbers and electrostatic precipitators, enabling them to emit a net of 3–6 kg F_t/t (15). The generation of toxic wash liquors which must then be disposed of is a disadvantage of wet scrubbers. In addition to or instead of these primary gas cleaning systems, some smelters use secondary cleaning systems wherein potroom gases are processed by wet scrubbers after collection through roof vents. Due to the high cost of cleaning large volumes of potroom air with very dilute fluoride concentrations, these measures are the exception rather than the norm.

Before any control measures were taken, smelters released almost all fluorides to the atmosphere (15–25 kg F_t/t), and have been blamed for widespread environmental damage in localized areas. For example, in parts of Norway, wild and domesticated animals have contracted fluorosis of the teeth and bones from ingestion of contaminated vegetation in the vicinity of aluminum smelters (16). In Japan, fluoride emissions from the Kanbara aluminum smelter reportedly killed all vegetation within a 10-kilo-

Plate III. Molten aluminum is gradually hardened into ingot form. This 30,000-pound ingot (13.6 metric tons) will be made into as many as 750,000 aluminum beverage cans. Courtesy Reynolds Aluminum Recycling Company.

meter radius around the factory, leading the Japanese people to reject a proposed aluminum smelter in Okinawa in 1973, and to force the domestic aluminum industry to invest in new projects abroad, notably a large smelter in Indonesia which uses power from new dams built on the Asahan River in Sumatra (17).

When emissions lead to fluoride concentrations exceeding 40 parts per million on the surface of leaves or in plant tissue, effects can include foliar lesions, impaired growth, and reduced yield. The plants most sensitive to fluoride damage are conifers, certain berries and fruits, and grasses such as corn and sorghum. Forage plants are moderately sensitive, while most field crops, vegetables, and deciduous trees are resistant to fluoride damage. When grazing animals ingest sufficient quantities of contaminated forage, they are susceptible to bone lesions, lameness, reduced appetite, abnormal dental development, and diminished lactation. The fluoride remains in their bones, however; it is not absorbed by their flesh, milk, and eggs, and thus poses no threat to consumers. Despite these risks to animals and different types of vegetation, many smelters around the world (including most U.S. smelters) are located in agricultural areas that produce grains and vegetables for human consumption, feed and livestock, orchard products, and timber. Inhalation of ambient air around smelters is not reported to pose a further threat to animals nor humans in the vicinity, with the exception of workers in smelters with substandard fume collection units, where fluoride emissions contaminate potroom air, posing an occupational health risk (18).

Greenhouse Gases: Carbon Dioxide, Carbon Monoxide, and Fluorides

During electrolysis, 400–500 kilograms of carbon anode are consumed per ton of aluminum produced. This comes from the basic production reaction:

$$2\ Al_2O_3 + 3C \rightarrow 4Al + 3\ CO_2$$

At 100% current efficiency, the production of one ton of aluminum would thus generate 1.22 tons of carbon dioxide (CO_2), but because cells are not perfectly efficient, actual emissions are higher: between 1.5–1.8 tons CO_2 per ton of aluminum produced (16). Additional emissions from anode production bring total CO_2 generation to about 2 tons per ton of aluminum produced, yielding a global total of approximately 38 Mt of CO_2 emissions, based on 1992 world production (19.2 Mt).

When the amount of alumina in a cell falls below a certain level, an "anode effect" occurs. In addition to reducing the amount of molten aluminum produced, an anode effect causes several other greenhouse gases to be emitted, including carbon monoxide (CO), and the halocarbons dicarbon hexafluoride (C_2F_6, also known as CFC-14) and carbon tetrafluoride (CF_4, also known as CFC-116). In various texts, CF_4 is referred to as tetrafluoromethane and tetraflurocarbon, C_2F_6 is referred to as hexafluoroethane. The carbon in these compounds comes from the anodes, while the fluorine atoms come from the cryolite bath (Na_3AlF_6). Although CO is produced in significant amounts during anode effects, much of it is converted to CO_2 in burners prior to leaving plant stacks. The emission of halocarbons containing fluoride, however, may present a far greater threat to global warming than both CO and CO_2 emissions.

Global Implications of Halocarbon Emissions

At present, the primary aluminum industry is the only known generator of CF_4 and C_2F_6 emissions. Due to the difficulty in measuring these gases at the point of generation, and to a very limited number of reliable measurements of current atmospheric concentrations, a scientific consensus has not yet emerged as to the exact amount of CF_4 and C_2F_6 generated per ton of aluminum produced, or as to whether there are other industrial or natural sources of these compounds (19).

Primary aluminum producers can only reduce current emissions rates of CF_4 and C_2F_6 by reducing the frequency and severity of anode effects. Even if aluminum production were to cease today, present atmospheric stocks of CF_4 and C_2F_6 would not be measurably depleted for many thousands of years.

The formation of CF_4 and C_2F_6 is determined by a range of factors, including the type of cell used, operating temperature, and the frequency, duration, and severity of anode effects, all of which are highly variable among smelters. An anode effect can last one to several minutes, depending on how quickly an operator responds by adding alumina. Because they may occur only 0.1 times per cell per day in modern smelters using the best technology, and 2–3 times in older prebake and Søderberg plants, it is difficult to infer a worldwide average output. Multiplying a conservative average emission value of 1 kg CF_4/ton aluminum produced by the 100-year GWP of 5,000 yields a result of 5 tons of CO_2 equivalent per ton, while an average emission value of 0.1 kg C_2F_6/tons times 10,000 yields 1 ton of CO_2 equivalent, for a total of 6 tons of CO_2 equivalent per ton aluminum produced. This yields an estimated world total of 115 million tons of CO_2 equivalent per year. Due to scientific uncertainties, however, this estimate may be in error by several hundred percent (Isaksen, 20).

Sulfur Dioxide

During electrolysis, organic sulfur present in the anodes reacts with alumina to form sulfur dioxide (SO_2). Since the coke and pitch used in anodes contain an average of 3% sulfur, 25–30 kilograms of SO_2 are generated per ton of aluminum produced, for a worldwide total of 480–575 thousand tons per year. Due to environmental regulations in many countries, however, some SO_2 is removed by wet scrubbers before being released to the atmosphere, so actual global emissions may be closer to 250,000 TPY, or about 1.5% of the total anthropogenic flow of SO_2 to the atmosphere as a result of metals smelting and fossil fuel combustion (15). Once airborne, SO_2 emissions can drift hundreds of miles before combining with other compounds to form acid rain. Acid rain has caused much damage to forest vegetation and lakes in temperate zones throughout the Northern Hemisphere.

Solid Wastes: Spent Potlinings

As previously stated, the life of carbon cathodes is 4–5 years, generating 40–70 kg of spent carbon potlinings and refractory materials per ton aluminum produced. Over the years, hundreds of thousands of tons of these wastes have been created, much of which is stored in open piles on the grounds of aluminum production facilities worldwide. Atmospheric nitrogen combines with the carbon in the tailings to produce cyanide, which if not properly monitored and controlled by linings and leachate recovery systems, can pose a serious groundwater contamination hazard. Fluorides accumulated from the bath are also present in the wastes, accounting for 15–30% of total mass. The remainder is comprised of carbon (30–45%), alumina (15–20%), and trace amounts of calcium, silicon, iron, sulfur, and nitrogen compounds. Research is underway to process spent potlinings by incineration, intentional leaching for cryolite recovery, by reuse as fuel in other industries, and by conversion to inert substances for use in manufacturing bricks, cement, mineral wool, and steel (21).

Improvements in Smelting Technology

The aluminum industry continually strives to reduce energy consumption and improve emissions from the smelting process. Computer programs have been implemented to improve current efficiency in the cell from 70% at the turn of the century to 95% by 1990, and to reduce overall power consumption from 40 to under 13 kWh/kg aluminum during the same period (22). Design improvements have been made to lengthen anode life and thus reduce CO_2 emissions, and to reduce the frequency of anode effects and thus lower CF_4 and C_2F_6 emissions (23). Improved gas cleaning equipment has been adopted in many smelters, especially in countries with environmental regulations.

ELECTRICITY USE IN PRIMARY ALUMINUM SMELTING

In addition to the direct emissions and effluents enumerated above, aluminum production is responsible for significant indirect environmental impacts that stem from its heavy use of electricity. In all likelihood, the environmental damages created by electricity generation for aluminum exceed those of process-related waste products. The damages occur at all spatial scales: global effects include greenhouse gas emissions from power plants' combustion of fossil fuels; regional effects of fossil fuel combustion include acid rain; and local effects include smog, and a myriad of impacts associated with hydroelectric dams.

The Impacts of Electric Generation Using Fossil Fuels and Nuclear Energy

The range of environmental costs associated with nuclear energy is generally well-known, as is the possibility of global climate change resulting from fossil fuel combustion. While policymakers, regulators, and the public often call on electric utilities to improve plant efficiency and institute conservation measures to reduce their emissions, they are less diligent in targeting major power consumers to determine the portion of responsibility they might bear in this regard. It is a useful exercise to look at the emissions from electric generation consumed by the world aluminum industry.

The combustion of fossil fuels is also responsible for generating oxides of nitrogen and sulfur which contribute to acid rain, especially in regions where environmental regulation is lax, and pollution control equipment is inadequate or nonexistent. Information about a variety of other environmental damages associated with coal mining and oil extraction is widely available.

Although the amount of nuclear energy consumed by smelters is slightly greater than the amount consumed by oil-generated electricity, it does not have the same emissions consequences. Because the aluminum industry consumes nuclear energy at a rate which only slightly exceeds the output of one large nuclear plant operating at full capacity (about 1,400 MW), its impacts are less significant relative to those of electric generation using fossil fuels.

The Impacts of Hydroelectric Dams Associated with Smelters

This section briefly examines some specific damages wrought by hydro projects associated with smelters in various parts of the world.

Over the last fifty years, aluminum smelters have been the sole impetus for constructing new large dams in many states and countries, including Brazil, British Columbia,

Ghana, and Indonesia. Without the electric demand provided by smelters in these countries, the Tucuruí, Kenney, Volta, and Akosombo dams might never have been built. In other nations, including Chile, China, Egypt, Norway, Russia, Quebec, the United States and Venezuela, the aluminum industry is at the very least an important contributing factor in the development of existing and proposed dams. By providing large blocks of steady, long-term demand, one or more aluminum smelters can help assure an electric utility of the revenues needed to pay for dam construction.

The flooding of large tracts of land has had deleterious consequences for both human communities and wildlife habitat, especially among indigenous people who practice subsistence hunting, fishing and farming. Worldwide, significant numbers of people have been relocated to new villages and towns, often against their wills and without more than a few weeks of warning. Many that live in proximity to the dams or smelters have had their lives disrupted by the loss of access to and degradation of hunting or farming lands: as in James Bay, Brazil and British Columbia, and by the dams' exacerbation of diseases such as schistosomiasis, malaria, and river blindness: as in Egypt and Ghana. There are also severe secondary effects on the communities where displaced peoples are relocated, including overcrowding, competition for land and employment, incompatible cultural practices of different groups in the new community, and a variety of other urban problems caused by the large influx of construction workers and their families, as in Sumatra in Indonesia, Egyptian Nubia, and the Volta River Valley in Ghana.

The areas encompassing and surrounding these reservoirs contain unique wildlife habitats and agricultural areas which suffer irreversible changes as a result of the extensive flooding and alteration of water flow regimes. The diverse wildlife habitats include a large tract of Amazonian rainforest in the Tocantins river basin which almost definitely contained undiscovered species of plants and insects before being inundated by the mammoth Tucuruí reservoir, huge swaths of temperate forest and wetland along the La Grande, Opinaca, Caniapiscau and Eastmain rivers in James Bay, the largest remaining contiguous wilderness area in North America, and fertile riparian zones along the Nile, a river which cuts through thousands of miles of desert, and whose annual alluvial flood ecosystem has enabled the Nubians to farm for millennia. Aside from contributing to possible species extinction in tropical areas, reservoirs are a lower quality habitat than free-flowing rivers. Their water chemistry is lower in dissolved oxygen, a condition harmful to many fish. The decomposition of drowned vegetation often enables the release of toxic methylmercury which is absorbed by aquatic plants and fish, and bioaccumulates as it rises in the food chain, creating potential hazards for the fish, waterfowl and humans that feed at higher trophic levels. Reservoirs in relatively level terrain also create many square kilometers of shallow waters and a fluctuating shoreline, which leads to eutrophication and can also cause large migratory animals to get stuck in the mud. Dams also present a physical impediment to the migration of many fish species, some endangered. At this writing, few dams built in association with aluminum smelters contain fish ladders. Because large hydroelectric dams alter the seasonal patterns of freshwater flow from large rivers into estuaries, and therefore the balance of nutrient influx, ice formation, and water salinity, they may also cause long-term damage to the breeding grounds of migratory waterfowl. Adequate scientific research has yet to be conducted to determine the long-term effects of dams in various ecosystems.

Finally, two mechanisms operating in large reservoirs may also contribute to global warming. Flooding destroys forests, thus precluding the continual uptake of atmospheric CO_2 by trees and other terrestrial plants in undisturbed ecosystems. Anaerobically decaying vegetation in the inundated area also releases methane, a potent greenhouse gas. One researcher has speculated that smelters that rely on hydroelectric reservoirs with low ratios of power capacity to flooded area, such as the Akosombo in Ghana, may contribute more to global warming through methane emissions than comparable smelters using coal-fired electricity (24).

Tertiary effects of dams include damage to agricultural lands, for example the salinization of newly-irrigated lands in Egypt which had formerly relied on rainfall and annual floods. The availability of irrigation, electric power, and roads (with the concomitant access to goods and services they provide), enables the commercialization of agriculture, including the use of gasoline and pesticides in areas that had formerly used traditional farming methods without significant environmental consequences, and the possible acceleration of erosion. These changes are often accompanied by a transition from a diverse crop base for local or domestic consumption to a monocropping scheme focused on animal feeds or luxury export crops. The presence of dam and smelter access roads in previously roadless areas also opens up new opportunities for mining, logging, oil and gas drilling, illegal hunting, and human settlement in remote areas hosting sensitive wildlife communities and indigenous people.

COST OF PRODUCTION

A 1988 study by the U.S. Bureau of Mines found that material inputs, such as alumina, coal, petroleum coke, soda, and lime, are the most expensive components in producing primary aluminum, accounting for about 33% of worldwide average total costs. The price of these commodities does not fluctuate dramatically by region or over short periods of time. Overhead, recovery of capital, transportation, profits, and taxes together accounted for 31.6% of total costs. Labor only accounted for 14% of production costs; the degree of training required limits how low wages can go, regardless of location. Energy is the second most expensive component of primary aluminum production, accounting for about 21% of total costs, although other studies have estimated this fraction to be as high as 30% (25).

The lion's share of energy costs are spent on the electricity used in smelting; a much smaller portion is fossil fuel expenditures used in mining and processing raw materials. By choosing where to build new smelters, aluminum corporations can exert some control over their electricity costs. Corporations seek regions where electricity sources are abundant and relatively cheap to develop, and where host governments have favorable industrial policies. The availability of local bauxite and other materials is of secondary concern.

PRODUCTION AND TRADE OF BAUXITE, ALUMINA AND ALUMINUM

Aluminum and its precursors, bauxite and alumina, are international commodities. A given ton of aluminum can be mined, refined, smelted, fabricated, sold, recycled and then refabricated in several different parts of the world. Because electricity expenditures comprise one fifth to one third of total production costs, while transportation costs from mine to smelter comprise less than 1% of costs, it is standard practice to ship bauxite or alumina halfway around the world to take advantage of cheap electricity contracts, many of which are based on hydroelectricity. A Toyota Corolla, for example, may contain aluminum atoms that originated in bauxite mines on the west coast of Australia, were processed into aluminum ingot using hydroelectricity generated in the mountains of British Columbia, and were manufactured into auto parts in Japan before being shipped to the United States or Europe for sale.

The World's Bauxite Reserves

Energy rather than materials availability is the limiting factor in the development of primary aluminum smelters. Unlike many other nonrenewable resources, the source material bauxite is not at all in short supply. In 1989, world reserves of bauxite were estimated to be 22 billion tons, up from an estimated 1 billion tons in 1945. This growth was due to increases in exploration leading to major bauxite discoveries, and to technological improvements which now make it possible to exploit lower-grade deposits previously classified as "subeconomic" reserves. Current estimates of the world's total bauxite resource base (which include subeconomic, hypothetical and speculative reserves), range from 50–70 billion tons. The world's largest known bauxite reserves are located in tropical and subtropical zones, with four countries accounting for 67% of world reserves: Guinea (5.6 billion tons), Australia (4.4 billion tons), Brazil (2.8 billion tons), and Jamaica (2 billion tons). The United States has a very small economic reserve base in comparison: about 40 million tons, all in Arkansas. Deposits in Hawaii, Washington and Oregon amount to an additional 100 Mt, but they are considered subeconomic at this time (26).

The Main Producing Countries

Bauxite is mined in 28 countries, and alumina is processed in 25 countries. About 90% of world production is refined to alumina; the rest is used in refractories and abrasives. As Table 1 shows, world bauxite production in 1992 was 103.6 million tons. The top five primary aluminum producers are the United States, Russia, Canada, Brazil and Australia, together producing 11 Mt, or 58% of the world total. In 1992, there were fifty aluminum-producing countries, up from forty a few years prior (the increase is due to the breakup of the Soviet Union and Yugoslavia, and the establishment of new independent republics with existing smelter capacity). The number of producing countries will grow should several smelter proposals come to fruition in countries with no existing capacity.

By studying Table 1, it becomes apparent that countries with vast reserves of bauxite *and* cheap energy (such as Australia, Brazil and the former USSR) tend to dominate in all three production stages: bauxite, alumina and aluminum. Because these countries do not have substantial domestic markets, they are also the most prolific exporters of aluminum. Nations that possess large supplies of bauxite but have no energy sources are also major exporters, notably Guinea and Jamaica. Countries without substantial bauxite but with cheap energy and/or access to the major consuming markets tend to be in the top echelon of alumina and aluminum producers, clearly relying on im-

Table 1. Principal Bauxite, Alumina, and Primary Aluminum Producing Countries[a]

Bauxite (1992)[b]	Mt	%	Alumina (1989)[c]	Mt	%	Aluminum (1992)[d]	Mt	%
1. Australia	39.95	38.6	1. Australia	10.80	27.9	1. United States	4.04	21.0
2. Guinea	13.77	13.3	2. United States	4.67	12.1	2. Russia	2.70	14.0
3. Jamaica	11.30	10.9	3. U.S.S.R.	3.50	9.0	3. Canada	1.95	10.1
4. Brazil	10.80	10.4	4. Jamaica	2.22	5.7	4. Brazil	1.22	6.3
5. India	4.48	4.3	5. Brazil	1.60	4.1	5. Australia	1.20	6.2
6. Russian Fed.[e]	4.00	3.9	6. Suriname	1.57	4.1	6. China[f]	0.95	4.9
7. Suriname	3.25	3.1	7. China[f]	1.44	3.7	7. Norway	0.81	4.2
8. China	3.00	2.9	8. India	1.42	3.7	8. Germany	0.60	3.1
9. Guyana	2.30	2.2	9. Canada	1.40	3.6	9. Venezeula	0.60	3.1
10. Greece	2.10	2.0	10. Venezuela	1.24	3.2	10. India	0.50	2.6
Subtotal, top 10	94.95	91.6	Subtotal, top 10	29.86	77.2	Subtotal, top 10	14.57	75.8
18 Other Countries	8.68	8.4	15 Other Countries	8.84	22.8	40 Other Countries	4.65	24.2
World total	103.63	100.0	World total	38.70	100.0	World total	19.2	100.0

[a] Mt = production in millions of tons; % = percent of total world production.
[b] *World Resources 1994–1995,* World Resources Institute, Oxford University Press, 1994.
[c] E. Sehnke and P. Plunkert, Ref. 3.
[d] Aluminum Statistical Review for 1992, Ref. 2.
[e] Data for the newly independent republics of the former U.S.S.R. and Yugoslavia were first collected separately in 1992.
[f] Estimate.

ported bauxite and/or alumina. Canada and the United States are examples of the latter category, but the long-term future of the U.S. industry is threatened by rising energy costs, in part due to environmental considerations. Because the major producers of bauxite are not necessarily the major aluminum producers, it is evident that the nations with the highest per capita consumption rates, such as Japan, the U.S., and several European nations, reap the economic benefits of adding value by processing domestic and imported ingot into semifabricated and finished products. In effect, the transnationals whose ownership and management are based in the U.S., Europe and Japan are importing low-cost energy in the form of aluminum ingot, and exporting the environmental costs of primary aluminum smelting to the newly producing countries.

Production by Private and State-Owned Corporations

As Table 2 illustrates, ten principal transnational and state-run corporations account for almost 60% of world capacity. In 1990, the top five alone accounted for 43% of world capacity: the (former) Soviet Government, Alcoa, Alcan, the Chinese Government, and Reynolds. During the last decade, a number of minor producers have made significant inroads into the market, eroding the virtual supremacy of what used to be called "the Big Six," or Alcan, Alcoa, Kaiser, Reynolds, Pechiney and Alusuisse.

THE FATE OF DISCARDED ALUMINUM

After aluminum consumer products have become obsolete, they are discarded, then either landfilled, burned, or recycled as "scrap" into new products. Because about 95% of the total energy used to make primary aluminum is not needed to produce secondary aluminum, including *all* the electrical energy, recycling saves the producer a tremendous amount of money. Scrap also provides a guaranteed supply of domestic aluminum which can serve as a partial buffer against periodically uncertain international market conditions. These two benefits have made scrap aluminum play an important role in total aluminum production. In 1992, for example, a full third of the total U.S. aluminum supply (primary and secondary production plus imports) was comprised of scrap—an amount equivalent to 68% of the primary aluminum produced. In Europe, about 35% of the supply was comprised of scrap in 1990. A 39-country survey conducted by the Aluminum Association, a U.S. trade organization, showed that 5.75 million tons of scrap were recovered for recycling that year, or 43.2% of the amount of primary aluminum produced in those countries. If the ratio of scrap recovery to primary aluminum production is similar in the rest of the world, then approximately 7.7 Mt of scrap may have been recovered worldwide in 1990 (1,28).

Table 2. Top Ten Aluminum Corporations (Private and State-owned), 1991[a]

Corporation	Ingot Capacity, t	% of World Capacity
1. Soviet government	2,740,000	13.7
2. Alcoa	1,867,560	9.3
3. Alcan	1,702,000	8.5
4. Chinese government	1,315,000	6.6
5. Reynolds Metals	1,023,500	5.1
6. Pechiney (French government)	851,000	4.2
7. Norsk-Hydro	623,000	3.1
8. Amax	541,630	2.7
9. Maxxam (includes Kaiser)	528,710	2.6
10. Venezuela	498,500	2.5
Total	11,690,900	58.3
World ingot capacity, 1991	20,042,000	100.0

[a] Ref. 27. World capacity figures do not correlate with production figures in Table 5 because smelters which rely on expensive fuels are used as "swing capacity," only operating when world demand is high.

Sources of Scrap for Recycling

There are two sources of secondary aluminum: old scrap and new scrap. There are also two types of new scrap: "home scrap" is generated by the plant that produced the aluminum; it is remelted in-house without ever having entered the market. "Prompt industrial scrap" consists of cuttings, end pieces, mistakes and product overruns. It is generated in the plant of a fabricator or manufacturer of aluminum products, and is uncontaminated by debris or unwanted metals. New scrap has always been recycled. "Old scrap" or "post-consumer scrap," on the other hand, consists of products that have been purchased by consumers, and have later become old or obsolete, then discarded (29) (see Plate IV). In the United States, the ratio of old scrap to new was 1.4 to 1 in 1992, and has been rising by about 4% per year since 1982. The 39-country survey mentioned earlier did not break the scrap down into the categories of old and new scrap, which is unfortunate because only the old scrap number reveals how much aluminum is being kept out of landfills worldwide.

Landfill Rates and Replacement Production

The amount of primary aluminum produced annually worldwide has increased by 6 million tons since 1974 (2). This is roughly equivalent to the amount of aluminum landfilled every year. There is reason to believe that more than five Mt of aluminum are being landfilled annually, due to inadequate recycling opportunities, insufficient public education, and low market prices that discourage more aggressive recycling.

Precise data on the amount of aluminum landfilled worldwide do not exist, but one can try to extrapolate based on United States data. The U.S. Environmental Protection Agency has estimated that 1.54 Mt of aluminum were landfilled in 1988 in the United States alone—a quantity that surely would have astounded Karl Bayer a century earlier (30). This was equivalent to 72.6% of all the scrap recovered that year, and to 39% of the primary aluminum produced. If it is assumed that similar percentages apply to the rest of the world, then between 5.5 and 7 Mt of aluminum may be doomed to landfill burial each year, an amount equivalent to the production of about twenty-nine 250,000-ton primary smelters. Were this estimate five times too large, it would be no less appalling.

These numbers are backed up in part by data on used beverage container (UBC) recycling rates in selected countries including the United States. In 1989, beverage cans accounted for 11% of Western Europe's total aluminum

Plate IV. Aluminum cans are purchased from the public, and then shredded into small chips so that they can be used for reclamation. The cans pictured at a recycling facility here, are about to be shredded. Courtesy of Reynolds Metals Company.

consumption of about 4 Mt. In some industrialized nations with high per capita consumption rates, such as the United Kingdom, recycling programs have been widely developed only in the last five years, leaving much aluminum in the garbage. In 1989, the rate of UBC recycling was only 5% in the United Kingdom, 20% in Greece, and 22% in Switzerland. Where recycling programs have been in place longer, the 1990 UBC recovery rates are higher: 61% in the United States, 63% in Canada, and 87% in Sweden, where recycling is mandatory. In 1990 Japan initiated a Y100 million campaign to boost its UBC recycling rate from a dismal 40%. European aluminum producers kicked off similar campaigns in 1989. These "green" moves have been calculated to capitalize on the public's environmental sentiments, in an attempt to win market share from steel cans (28,31). Such a shift took place in the 1970s and early 1980s in the United States.

The used beverage container is the only aluminum product that is tracked from production through recycling, due to the distinct "closed-loop" nature of the recycling process: UBC is usually collected separately from other aluminum scrap, and is remelted, rolled into can sheet, and made into brand new cans which may be back on the shelves within a few weeks of being collected at curbside or deposited at a recycling center. Despite the fact that nine U.S. states have some form of beverage container deposit legislation, and that convenient curbside recycling programs are now available to about 100 million people in thousands of American cities, a great deal of cans still go to waste. As Table 3 illustrates, 534 thousand tons of beverage cans were landfilled in 1993. The financial losses from not recycling this aluminum exceed half a billion dollars. Because the rate of aluminum consumption has continued to grow each year, all landfilled aluminum must be replaced by a similar quantity of primary aluminum: in this case, a quantity equivalent to the maximum annual production capacity of four smelters in New York, Oregon, and Montana. The energy requirements to do this are prodigious, including approximately 8 TWh of electricity, an amount which could power more than 800,000 average U.S. households for a year. As earlier sections of this article pointed out, there are also airborne emissions and solid and liquid effluents associated with this "replacement" production.

The recycling rates of other aluminum products have not been tracked because they are mixed together at the point of collection, and as previously stated, because their

Table 3. Landfilled Beverage Cans and Wasted Production, 1993[a]

	# of Cans	Million Pounds	Metric Tons
U.S. beverage can production (29.53 cans/lb)[a]	94.3 billion	3,193	1,448,491
Cans recycled (63.1% of total production)[a]	59.5 billion	2,015	913,998
Cans landfilled (36.9% of total production)	34.8 billion	1,178	543,493
Annual production capacity of four U.S. smelters[b]			
Alcoa Aluminum, Massena, New York			127,000
Reynolds Metals, Massena, New York			123,000
Reynolds Metals, Troutdale, Oregon			121,000
Columbia Falls Aluminum, Columbia Falls Montana			163,000
Total capacity			534,000
Foregone revenues by not recycling 534,493 tons aluminum (at $1,014/ton):[a]			$542 million
Electricity required to replace this quantity with primary aluminum (at 15 kWh/kg):			8.0 TWh
World consumption of electricity from natural gas for primary aluminum:[c]			9.5 TWh
Approximate # of U.S. homes 8.0 TWh could power for a year (at 1.127 kW/home):			812,092

[a] Ref 32.
[b] Ref. 33.
[c] International Primary Aluminum Institute, Ref. 34.

useful lifetimes range from several years to many decades. It is difficult to be more precise about the amount of aluminum landfilled. Because aluminum comprises only 2.3% of all U.S. municipal solid waste by volume, and only 1.1% by weight, there is little incentive for the governmental agencies who commission waste composition studies to be more accurate in examining the contribution of aluminum.

The Role of Aluminum Companies in Initiating Recycling

Without the efforts of the aluminum industry, recycling of all sorts would not have progressed as far as it has today. After World II, the intense recycling efforts that had been sustained in the national interest were abandoned. Disposal became equated with peacetime and prosperity. It was not until the growing environmental consciousness of the 1960s, coupled with a slow rise in energy prices, that aluminum recycling experienced a renaissance, at least in the United States.

Because the energy savings from recycling aluminum are so great, the aluminum industry encouraged recycling before it was environmentally popular, and long before the current landfill crisis was felt. The industry was motivated not only by energy savings, but by a scarcity in supply. As the all-aluminum beverage can gained in popularity during the 1960s and 1970s, it took over market share from glass bottles and steel or bi-metal cans. Although the shift was promoted by primary aluminum producers to keep wartime capacity in production, the growth was so rapid that they could not keep pace with the demand. Recycling was a way of ensuring adequate supply for continued can production. It became an insurance policy against an interruption in primary supplies due to geopolitical conflicts, uncertain smelter contracts, and rising energy prices.

In the United States, Reynolds, Coors and Alcoa were especially instrumental in establishing buyback centers for cans in the late 1960's and early 1970's, when no other recycling options were available. They distributed pamphlets touting the benefits of recycling, and developed programs that encouraged the participation of schools and community groups. These early recycling programs helped to re-establish recycling in the public consciousness, and paved the way for the collection of other materials, such as newspapers, steel cans and glass bottles. At first the collection efforts were spurred by volunteers with minimal cooperation from producers, but as the shortage of landfill space has intensified, local and state governments have taken on the operation of many large-scale recycling programs. This trend is also evident in Europe and Japan.

Obstacles to Recycling: Artificially Low Prices and Inadequate Public Education

It is clear from the foregoing discussion that aggressive recycling efforts are still needed to keep this valuable metal out of the trash. A 1991 survey of secondary smelter operators in the United States revealed that the single most important obstacle to increased secondary recovery is the low price of aluminum, especially that of non-UBC scrap. As the primary price rises, the competitive position of secondary aluminum improves; as the primary price drops, more aluminum is landfilled. Three quarters of survey respondents thought that the U.S. EPA estimate of 1.5 million tonnes of aluminum landfilled annually was fairly accurate. Some respondents said that market volatility discourages a number of would-be scrap collectors, and that the absence of consistent operating margins discourages scrap processing. The lack of mandatory recycling legislation, and the fact that many recycling opportunities are too far away, too inconvenient, or under-publicized, were also cited as lesser obstacles to higher aluminum recovery rates (35).

There may be yet another obstacle to increased secondary recovery: inadequate public awareness of the environmental effects of primary aluminum production. In recent years, the aluminum industry has aggressively used images of recycling to present itself as a "green" industry, without telling the public the full range of environmental costs of producing primary aluminum. Advertisements in major newspapers such as *USA Today* and London's *Financial Times,* and in numerous environmentally-oriented journals, employ abstract and oversimplified "energy savings" arguments to encourage aluminum can recycling, and to promote the use of aluminum packaging based solely on its recyclability. Using full-color pictures of butterflies and smiling little girls, they also relay anecdotes about how the revenues from various aluminum can recycling programs have been used to bolster civic organizations in small towns, to fund school tree-planting projects, and to provide medical care for marine mammals. While it is commendable for the aluminum industry to encourage the public to recycle beverage cans for all of these reasons, it may not be enough. For one, information about recycling non-UBC scrap is not widespread. These "feel-good" advertisements also convey the overall impression that aluminum consumption does not have serious environmental impacts. Even if beverage cans were recycled at rates approaching 90%, there would still be a need for primary aluminum production to replace the amount landfilled, and to contribute to many other non-beverage can end uses.

CONCLUSION

If primary aluminum production continues to grow at rates even approaching those of the late 20th century, the environmental repercussions will be severe. More rivers will be dammed, and more power plants constructed to burn fossil fuels that threaten an already-burdened global atmosphere. The problem of solid waste disposal may grow ahead of the industry's ability to control it, despite promising research and development. The extent of these problems will be determined by the rate of demand growth, which in turn will be determined by at least four factors operating at a global spatial scale:

1. Absolute population growth,
2. Growth in per capita consumption of existing products using aluminum,
3. The increasing proportion of aluminum to other materials in various uses (such as cars), and
4. The worldwide rate of aluminum recycling.

Because each of these factors is complex, it is impossible to predict future demand growth with any precision. What is known, however, is that the world population is expected to double by the year 2020, and that as historically closed socialist markets such as Russia and China open their doors to capitalist goods, their per capita consumption will grow substantially. In the interest of minimizing environmental damage in light of the first and second factors, it is useful to ask if the third and fourth factors can be controlled.

Factor 3 entails questioning the uses of aluminum. Comparative lifecycle analyses are being used to evaluate which materials are most environmentally benign for certain end uses. For example, many investigators are convinced that refillable glass or plastic bottles have lower overall environmental impacts for delivering single-serving drinks than aluminum beverage cans, even taking aggressive recycling into account. On the other hand, it is probably desirable to increase the proportion of aluminum to steel in transportation vehicles because its lighter weight enables a significant savings of fossil fuels over the life of a vehicle. These questions cannot be answered simply, but they do deserve further inquiry.

Changing factor 4 will demand that we re-examine the institutional framework in which people in society manufacture, consume and dispose of products. As previously stated, only a handful of states have deposit legislation on the books, and recycling rates are higher in these states than in those lacking such legislation. Rather than fight for new deposit laws on a state-by-state basis, many activists and policymakers are calling for a federal "bottle bill" to encourage recycling nationwide. Others are advocating "advance disposal fees" (ADF's), or taxes on the producers of manufactured goods which are tied to the amount of their products which end up in the wastestream. The present system removes manufacturers' responsibility for disposal, making it the sole fiscal burden of city and county governments. ADF's would pass the cost of disposal from the general taxpayer on to the consumer of products. Still others want to see externalities incorporated into the price of energy and primary materials production; carbon taxes and the removal of subsidies for the acquisition of land, water, and raw materials are examples. Again, these measures would pass the responsibility of paying for resource extraction from the public on to the consumer of finished goods.

Finally, many activists believe that certain areas should be declared off-limits to resource development for environmental and social reasons, including various wild rivers, coastal environments serving as wildlife refuges, and forests with importance for global biodiversity and the preservation of indigenous cultures. If more areas were protected from energy and mineral development, the price of energy and materials would increase due to competition for a more limited resource base. This in turn would hasten the transition from a global economy based on primary commodities to one that relies more heavily on secondary materials.

BIBLIOGRAPHY

1. World Resources Institute, "World Resources 1993–94." Oxford University Press, 1994, Table 21.4, pp. 338–339.
2. "Aluminum Statistical Review: Historical Supplement," The Aluminum Association, Washington, D.C., 1982; "Aluminum Statistical Review for 1992," The Aluminum Association, Washington, D.C., 1993.
3. "The Next Phase for Aluminum," *The Mining Journal (London)* **312**(8002) (Jan. 13, 1989); E. Sehnke and P. Plunkert, "Bauxite, Alumina, and Aluminum," *Minerals Yearbook 1989,* U.S. Department of the Interior Bureau of Mines, Washington, D.C., Nov. 1990.
4. S. Y. Shen, *Energy and Materials Flows in the Production of Primary Aluminum,* Prepared by Argonne National Laboratory for the Department of Energy, (ANL-CNSV-21), Oct., 1981, pp. 2–5; I. J. Polmear, ed., *"Light Alloys: Metallurgy of the Light Metals,"* 2nd ed., Edward Arnold, London, 1989, p. 8.
5. E. Sehnke and P. Plunkert, Ref. 3, pp. 23–26.
6. B. Richardson, "Super-Companies," National Film Board of Canada, 1990; S. Y. Shen, Ref. 4, p. 2.
7. R. N. Crockett, "Bauxite, Alumina and Aluminum," Mineral Resources Consultative Committee, Mineral Dossier No. 20, Institute of Geological Sciences, London, 1978, p. 17.
8. H. F. Kurtz and L. H. Baumgardner. "Aluminum," in *Mineral Facts and Problems,* U.S. Bureau of Mines, U.S. Government Printing Office, Washington, D.C., 1980, p. 18.
9. Ref. 8, p. 19; N. Jarrett, "Aluminum Production," *Encyclopedia of Materials Science and Engineering,* p. 166.
10. Ref. 7, p. 17; Ref. 8, p. 19; N. Jarrett, Ref. 9, p. 166; E. Sehnke and P. Plunkert, Ref. 3, p. 27; A. S. Russell, "Aluminum," in *McGraw-Hill Encyclopedia of Science and Technology,* 1987, p. 417.
11. S. Y. Shen, Ref. 4, p. 3; S. H. Patterson and J. R. Dyni, "Aluminum and Bauxite," in *United States Mineral Resources,* U.S. Geological Survey Professional Paper 820, U.S. Government Printing Office, Washington, D.C., 1973, p. 36.
12. S. Y. Shen, Ref. 4, p. 12.
13. I. J. Polmear, Ref. 4, p. 11; E. Sehnke and P. Plunkert, Ref. 3, p. 1; S. Y. Shen, Ref. 4, p. 14; B. Richardson, Ref. 6.
14. *Review of New Source Performance Standards for Primary Aluminum Reduction Plants,* U.S. Environmental Protection Agency, Office of Air Quality Planning and Standards, Research Triangle Park, N.C. (EPA-450/3-86-010) Sept. 1986, pp. 3–29.
15. Y. Lallement and P. Martyn, "Industrial Sustainable Development in the Context of the Primary Aluminum Industry," *UNEP Industry and Environment,* July–Dec. 1989.
16. R. Huglen and H. Kvande, "Global Considerations of Aluminum Electrolysis on Energy and the Environment," Hydro-Aluminum a.s., Stabekk, Norway. (TMS 1994 conference paper, location unknown).
17. R. Pardy, and co-workers, "Purari Overpowering Papua New Guinea?" The International Development Action for Purari Group, IDA, Fitzroy, Victoria 3065, Australia, 1978, p. 181; "Asahan Dam: Energy for Whom?" *Environesia,* **3**(2), (June 1989), Published by WALHI, The Indonesian Environmental Forum, Jakarta, 1989.
18. *Primary Aluminum Draft Guidelines for Control of Fluoride Emissions from Existing Primary Aluminum Plants,* U.S. Environmental Protection Agency, Office of Air Quality Planning and Standards, Research Triangle Park, NC 27711 (EPA-450/2-78-049a), Feb. 1979.
19. H. I. Schiff, "The Aluminum Industry's Contribution to Atmospheric Fluorocarbons," *Proceedings, Workshop on Atmospheric Effects,* in Ref. 18.
20. *Proceedings, Workshop on Atmospheric Effects,* in Ref. 18; and I. S. A. Isaksen and co-workers, "An Assessment of the Role of CF_4 and C_2F_6 as Greenhouse Gases," Center for International Climate and Energy Research, Oslo, September 1992.

21. The Aluminum Association, *Proceedings of the Workshop on Storage, Disposal and Recovery of Spent Potlining,* Washington, D.C., December 3–4, 1981; "Draft: Pollution Prevention and Abatement Guidelines for the Aluminum Industry," Prepared by ICF Kaiser Engineers International, USA, for United Nations Industrial Development Organization, (US/GLO/91/202-UNIDO/World Bank Industrial Pollution Guidelines), Jan. 1993.
22. H. A. Øye and R. Huglen, "Managing Aluminum Reduction Technology: Extracting the Most from Hall-Héroult," *JOM,* November 1990.
23. *Energy Efficiency in U.S. Industry: Accomplishments and Outlook,* A Report for the Global Climate Coalition by the EOP Group, Inc., Oct. 1993, p. 49.
24. D. Abrahamson, "Greenhouse Gas Production From Aluminum Production: Some Policy Implications." *Proceedings, Workshop on Atmospheric Effects, Origins, and Options for Control of Two Potent Greenhouse Gases: CF_4 and C_2F_6,* sponsored by the Global Change Division of the U.S. Environmental Protection Agency Office of Air and Radiation, April 21–22, 1993, Washington, D.C.
25. D. R. Wilburn and D. A. Buckingham, "Assessing the Availability of Bauxite, Alumina and Aluminum through the 1990's," *JOM* (Nov 1989).
26. E. Sehnke and P. Plunkert, Ref. 3, p. 25; Patterson Ref. 12, p. 39.
27. *Primary Aluminum Plants Worldwide, Part 1—Detail, U.S.* Bureau of Mines, Dec. 1990.
28. K. Gooding, "Aluminum Producers Woo the Green Vote," *Financial Times,* London, June 16, 1989. Calculation: 43.2%. * 17.8 Mt (world production, 1990) = 7.7 Mt.
29. P. Plunkert, "Aluminum Recycling in the United States—A Historical Perspective," Presented at *TMS 2nd International Symposium-Recycling of Metals and Engineered Materials,* Williamsburg, Va., Oct. 28–31, 1990, p. 2.
30. *Characterization of Municipal Solid Waste in the United States, 1990 Update,* U.S. Environmental Protection Agency, Office of Solid Waste and Emergency Response, Washington, DC, (EPA/530-SW-90-042), June 1990.
31. "Japanese Aluminum Battle for Can Market." *Mining Journal* **314**(8074), 460 (June 8, 1990).
32. *Resource Recycling,* June 1994, Table 4, p. 26.
33. *Primary Aluminum Plants Worldwide, Part II-Summary,* U.S. Bureau of Mines, Washington, D.C., 1990.
34. *World Energy Outlook to the Year 2010,* International Energy Agency, Paris 1993; *Electrical Power Utilization Annual Report for 1990,* International Primary Aluminum Institute, London, 1991.
35. J. S. Gitlitz, unpublished survey, 1991.

ASH

Larry L. Baxter
Sandia National Laboratories
Livermore, California

The principal means of energy production from solid and heavy liquid fuels is combustion. During combustion, inorganic components of most solid and many liquid fuels lead to the formation of ash. Inorganic material typically accounts for about 10% of the mass of solid fuels, varying from over 50% to less than 1%. Coal represents the single most significant solid fuel in worldwide energy production, and statistics from coal-based electric power production in the United States accurately illustrate the magnitude of ash-related issues during combustion. About 90.7 million t of ash per year, or 363 kg per person per year, are generated in the United States from coal combustion. By comparison, Western Europe generates about 36.3 million t annually (1). Studies indicate that ash deposition, commonly referred to as fouling and slagging, causes unnecessary lost production that costs utilities between $\$5.2 \times 10^{-5}$ and $\$1.3 \times 10^{-4}$ d per kilowatt hour (kW·h) of coal-based electricity production (1985 U.S. dollars). Financial losses occurring due to other ash-related issues, such as tube leaks due to erosion or corrosion by ash, can be as much as a factor of 10 higher (2). In the United States, where coal accounts for 56% of the 2.8×10^{12} kW·h of utility power production (1992 statistics) (3), preventable ash-related production losses for U.S. utilities can be estimated to range from many hundreds of millions of dollars a year to a few billion dollars a year. These represent the bulk of unanticipated costs associated with ash-related problems. In addition, ash-related issues play a dominate role in the anticipated costs associated with scheduled shutdowns and maintenance. In general, ash-related problems can conservatively be estimated to represent a $1 to $2 billion (1990 U.S. dollars) a year problem in the U.S. electric power industry. The U.S. experience with ash during coal combustion does not differ from that of other major coal-consuming countries, nor does coal present unusually difficult ash-related problems. Energy production by combustion of essentially all solid fuels in all regions of the world is accompanied by significant ash management problems. See also Coal combustion; Combustion modeling; Commercial availability of energy technology.

Boiler designers are keenly aware of the pronounced impact of ash on boiler performance and maintenance. Considerations of ash behavior dominate both the design and operation of combustors that burn solid and some liquid fuels. An appreciation of these dominant effects of ash-related issues on combustor design can be gained by considering designs of typical coal combustors. Figure 1 illustrates two boiler designs, both intended to produce the same amount and quality of steam electricity from coal. The first section of the boiler is called the furnace and consists of a tall narrow chamber with heat absorbing walls referred to as waterwalls. These walls are composed of steam- and water-containing tubes welded together by membranes, which are small pieces of steel between them. The second section of the boiler is referred to as the convection pass and consists of heat-exchanging tube bundles hanging in the bulk gas flow, usually with no membranes bonding them together. Coal is pneumatically transported to the burners where it is blended with combustion air as it is introduced into the bottom of the furnace. A utility power station producing 600 MW of electricity burns coal at a rate of about 181 t/h and generates about 18 t/h of ash. The combustion of this coal forms a fireball in the bottom of the furnace. Char and fly ash are carried from the fireball to the top of the furnace, through the convection pass, and eventually into cleanup systems. Dimensions for the boiler depend on its generating capacity. They are denoted as height (H), width (W), and depth (D) for a boiler designed for a typical U.S. bituminous steam coal in

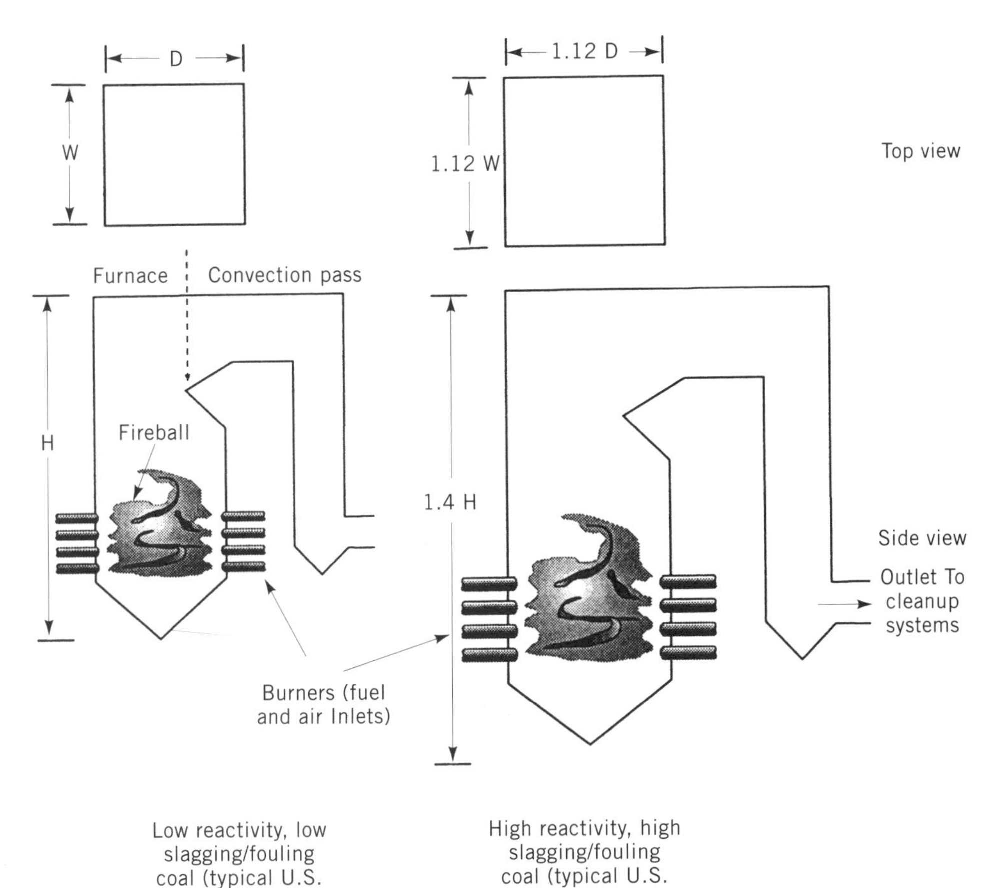

Figure 1. The effect of ash deposition (fouling and slagging) on boiler dimensions.

the design shown on the left. Utility boilers range in electrical generating capacity from about 100 to 1200 MW, with a typical 600 MW bituminous coal boiler being roughly 61 m tall and 12–15 m wide and deep. Bituminous coals typically present relatively small fouling and slagging problems relative to the lower rank subbituminous coals and lignites. The dimensions for the boiler on the right are shown relative to those for the boiler on the left. The boiler illustrated on the right has the same electrical generating capacity, but is designed for a typical U.S. lignite which poses significantly greater ash deposition (fouling and slagging) problems than bituminous coals.

The slowest step of coal (and most solid fuel) combustion is the reaction of residual char with oxygen. This reaction occurs much faster with chars generated from lignites than with those generated from bituminous coals (4). On this basis, one might expect that boilers designed for bituminous coals would be larger than those designed for lignites. The opposite trend, as is illustrated in Figure 1, attests to the dominate influence of ash on boiler design and operation. As can be seen, the lignite boiler illustrated on the right typically has 25% greater cross-sectional area and 40% greater height, for a total of 75% greater volume, than the bituminous coal boiler on the left (5). Other design features, such as tube spacing in the convection pass and maximum allowable furnace exit gas temperature, are also strongly affected by ash deposition considerations. The large change in boiler dimensions and design occur specifically to address concerns about the behavior of inorganic material in the two types of coals. This specific example applies to pulverized-coal-fired systems but is also representative of the influence ash considerations have on other technologies, such as gasifiers and fluid-bed combustors, as well as with other fuels, such as biomass and refuse. Inorganic material, from which ash is derived, has some effect on mining or collection, transportation, and preparation of fuels. As important as these effects are, the properties of inorganic material become especially important during combustion.

Principal changes in technology during the history of coal combustion are attributable in large measure to the influence of ash on combustor design and performance (6,7). These changes led to the evolution of boiler designs from moving or fixed grates to the entrained-flow, pulverized-coal boiler design currently used by most utility power stations such as those indicated in Figure 1. Advanced technology may lead to still other designs, ranging from gasifiers to fluid beds. In the opinion of many practitioners, the behavior of inorganic material in the fuel will continue to dominate design considerations, whatever technology is employed. Liquid fuels contain far less ash than most solid fuels, the former containing rarely more than 0.2% ash. Yet even this small quantity of ash is capable of causing severe ash deposition and corrosion problems in combustors. This article defines the terminology used in discussing inorganic materials and briefly describes some of the analytical techniques used to quantify its properties. The effects of inorganic material on energy and environmental issues are then traced from the origin of the inorganics in fuel to their disposal. The discussion is intended to be general and is primarily based on ashes found in coal, biomass and refuse, and heavy oils.

DEFINITIONS

The term *ash* is often used to refer to inorganic material in a general sense. Practitioners refer to ash in the fuel, fly ash, ash deposits, bottom ash, etc. In more precise terms, ash refers to material that does not combust under a prescribed set of conditions. In this precise sense, fuels contain inorganic material, not ash. Ash is a product of combustion or gasification of the fuel. Inorganic material comprises both mineral matter and atomically distributed inorganic components. In common usage, the term *mineral matter* is sometimes used to refer to all inorganic material. More precise discussions refer to mineral matter as the portion of the inorganic material that is mineral in nature, ie, granular material that is identifiable and classifiable by common mineralogical techniques. This contrasts with atomically distributed material, which is either an integral part of the organic structure or in some ionic form absorbed on surfaces or dissolved in moisture inherent in the fuel. Common classes and examples of mineral matter found in fuels include oxides (silica, rutile), silicates (kaolinite, illite), sulfides (pyrite), carbonates (calcite), and sulfates (iron sulfate, gypsum). This material occurs in the fuel as grains that may be either intimately associated with (inherent) or extraneous to (adventitious) the fuel matrix. Minerals intimately associated with the fuel commonly occur as bands of inorganic material within a coal seam or concentrated in certain sections of a plant. This reflects their origin as detrital material (see below) or material with specific biological function. Grains vary in size from nanometers to millimeters in the raw fuel. Fuels such as coal often undergo washing, pulverization, or other fuel pretreatments, reducing maximum grain sizes to tens of microns. The variety of forms of mineral matter in fuels can be appreciated by considering the forms of silica and silicates. Grains may be dense, as is common with silica in coal, or porous, as is common for silicates in coal. Porous, imbedded grains of mineral matter are often impregnated with organic material. Grains may also be either crystalline, as is common for silica in coal, or amorphous, as is common for silica in biomass. Mineral matter may be hydrated, as is common for silicates in coal and silica in biomass, or anhydrous, as is common for silica in coal. The specific forms of minerals in most fuels can be traced to their biological and geological origin and history (see below).

Atomically distributed material includes alkali and alkaline earth elements, usually in ionic forms and often bound as cations to the organic structure. Other forms of atomically distributed inorganics include chemisorbed, complexed, and solvated material. While alkali and alkaline earth materials often dominate the atomically distributed portion of inorganic material, at least small fractions of many other elements can be found in atomically distributed forms (8). The distinctions between organic and inorganic portions of the fuel are not always clear when discussing atomically distributed material. For this discussion, only carbon, hydrogen, nitrogen, sulfur, and oxygen that is covalently bound to a hydrocarbon structure is considered organic. Thus potassium chloride in biomass, vanadium in oils, calcium cations in coal, and other similar materials with clear biological functions in living compounds are considered inorganic in this discussion. Similarly, sulfur, oxygen, and carbon in the fuel are considered inorganic when they occur in the forms of minerals.

Ash deposition in combustors is commonly referred to as fouling and slagging. There is little agreement in the precise distinction between fouling and slagging deposits. The term *slag* has connotations of molten material, whereas *foul* has connotations of solid material. The distinction between molten and solid is, in practice, one of degrees. Only a small fraction of modern combustors operate with running, fluid slag on the walls. Most operate with ash deposits that have some plastic character when stressed but do not appreciably flow. The distinction between fouling and slagging used in this discussion is that fouling deposits are confined to the convection pass and slagging deposits are confined to the furnace. This distinction is less arbitrary than it may appear. Furnace deposits occur predominantly on flat or reasonably flat surfaces such as waterwalls. Gas temperatures in the furnace are in the range of 871–1649°C. Deposits in the convection pass occur predominately on tube surfaces. Gas temperatures in the convection pass range from 149 to 1093°C. Differences in temperatures and geometries typically lead to distinct differences in deposit characteristics.

ORIGIN AND COMPOSITION OF INORGANIC MATERIAL

Figure 2 presents a useful framework for considering the relationships among the various common fuels. The lower left corner of the figure comprises a coalification diagram,

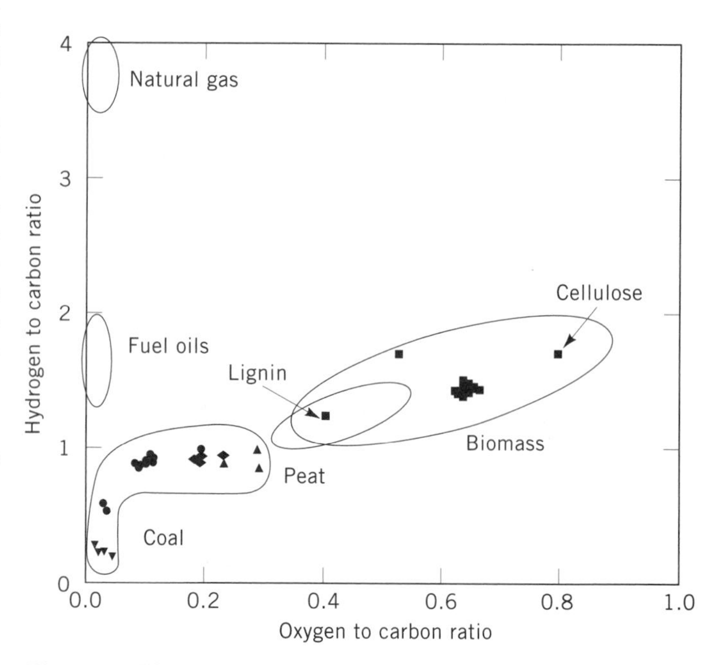

Figure 2. The relationships between various fuels based on their major atomic compositions.

which illustrates the changes in coal structure as it matures from lignite through subbituminous and bituminous coal, eventually to anthracite (9,10). Figure 2 represents an expansion of the coalification diagram and shows, eg, that biomass is a logical extension of the coalification process. Premium fuels, such as fuel oils and natural gas, are also illustrated. Liquid fuels derived from solids (pyrolysis oils, tars, etc) are slightly enriched in hydrogen and depleted in oxygen relative to the solid from which they are derived, with specific compositions dependent on processing temperature and similar parameters (11).

Figure 2 illustrates that one distinguishing feature of solid fuels is their oxygen content. Oxygen in such fuels is commonly found in the form of functional groups such as carbonyls and hydroxyls. These forms of oxygen represent potential sites were inorganic materials, in particular alkali and alkaline earth ions, bond with the organic structure. As could be inferred from Figure 2, the amount of this organically bound inorganic material often increases as the oxygen content of the fuel increases. The other principal forms of inorganic material in fuels are not related to the fuel structure. They include salts and other loosely bound materials, often dissolved in moisture in the fuel, inherent particulate inorganic material that results from either biological or geological processes, and adventitious inorganic material that typically results from mining or fuel handling processes.

The most prevalent inorganic elements found in solid fuels are generally oxygen, chlorine, silicon, aluminum, titanium, iron, calcium, magnesium, sodium, potassium, sulfur, and phosphorus. Nitrogen is almost exclusively found as part of the organic structure. Sulfur is found in both organic and inorganic forms. The amount of the latter 10 elements in the ash is commonly reported on an oxide basis for reasons discussed below. They only occasionally occur in the forms of oxides in the fuels. In liquid fuels, an additional element, vanadium, is of considerable importance. The proportion of the various components can vary significantly from fuel to fuel, as discussed below for each principal fuel type.

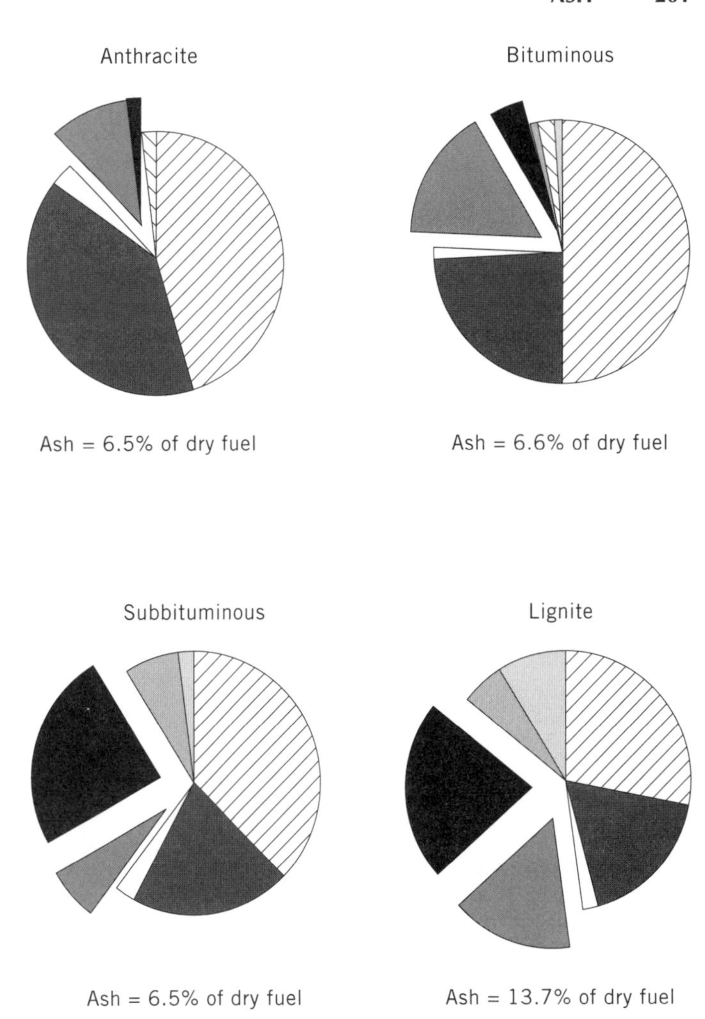

Figure 3. Representative compositions of ashes from four coals of various rank, expressed as oxides. These data are on a sulfur- and phosphorus-free basis.

Coal

Figure 3 presents typical ash compositions for several different ranks of coal in terms of the metal oxides. A primary component of all coal ashes is silicon, with significant amounts of aluminum. Subbituminous coals typically have higher calcium and lower iron contents than either bituminous coals or anthracites. Lignite ash composition varies more widely than ash for the other ranks of coal. The data shown in Figure 3 are all derived from U.S. coals. These U.S. coals are representative of a great many coals worldwide, although analyses of coal ashes from other parts of the world differ somewhat from those shown in Figure 3.

Ash compositions are usually reported on a normalized, sulfur- and phosphorus-free basis (Fig. 3). The amount of sulfur and phosphorus retained in the ash generally increases with increasing amount of alkali or alkaline earth material. Alkali and alkaline earth material that is not incorporated in silicates reacts with sulfur to form sulfates that are stable at low temperatures (below 800 to 900°C). The amount of sulfate in high rank (anthracite and bituminous coals) fuels is usually small (less than 5%). Subbituminous coals and lignites often contain as much as 25% of their ash in the form of sulfates.

The data in Figure 3 represent typical results from standard analyses of coals. They indicate little about the forms or modes of occurrence of inorganic material in coal. Geological and biological histories affect the amount and mode of occurrence in inorganic material in fossil fuels. For example, atomically distributed calcium in coals arises in part from dissolution of detrital calcite and incorporation of the dissolved material in the organic matrix. Pyrites develop to some extent by combination of organic sulfur with available iron by bacterial processes (6). The mode of occurrence of inorganic material in fuels follows reasonably consistent geographic patterns. Figures 4 and 5 indicate the partitioning of the alkali metals sodium and potassium as a function of coal rank. The total amount of sodium (Fig. 4) and potassium (Fig. 5) is divided into four forms: water soluble, ion exchangeable, acid (HCl) soluble, and residual. These four fractions are determined by a pro-

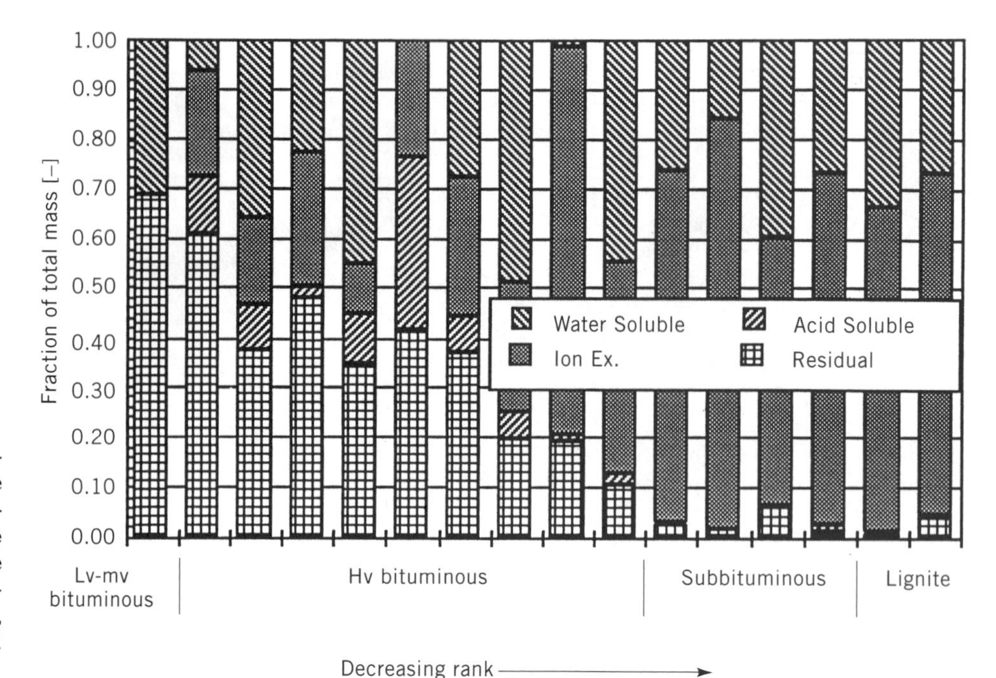

Figure 4. Mode of occurrence of sodium as a function of rank for a range of coals. The atomically dispersed sodium, represented by the sum of the water-soluble and ion-exchangeable material, increases steadily with decreasing coal rank or, equivalently, with increasing coal oxygen content (cf. Fig. 5).

cedure known as chemical fractionation, the repeatability and precision of which have been determined, but it has not been standardized.

Water-soluble inorganic material generally includes ions that dissolve readily in water (eg, chlorides) or chemisorbed material that is bound in the coal by weak bonds. Ion-exchangeable material includes organically associated inorganic species that complex, eg, with carboxyl or other functional groups. The sum of the water-soluble and ion-exchangeable fractions comprise the atomically dispersed (as opposed to mineralogical) fraction of the inorganic material. Certain minerals (carbonates, sulfates, etc) are soluble in hydrochloric acid and can be distinguished from other materials (many oxides, sulfides, etc) that are not.

The data indicate clear trends with respect to both rank and geography. High rank coals generally tend to have less atomically dispersed alkali material as a fraction of total alkali material. Chemically similar materials, such as sodium and potassium, occur in significantly different modes in the fuels. For example, about half of the sodium in high rank coals and nearly all of the sodium in low rank coals is atomically dispersed. By contrast, essentially none of the potassium in high rank coals and only about half of the potassium in low rank coals is atomically dispersed. Alkali material that is not atomically dispersed is typically found in clays and other impurities in the coal.

The other components show less variation as a function of rank. Aluminum occurs primarily in the form of the clays such as illite, kaolinite, and montmorillonite in all ranks of coal. Silicon also occurs in clays, with a lesser but

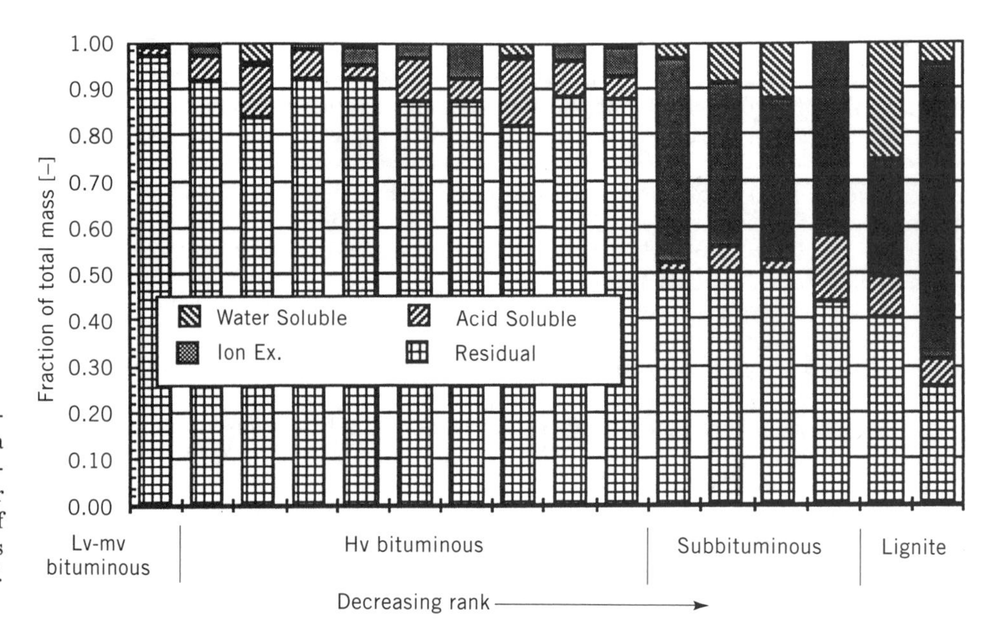

Figure 5. Mode of occurrence of potassium as a function of rank for a range of coals. The atomically dispersed potassium follows a similar trend as sodium, but the amount of atomically dispersed potassium differs significantly from that of sodium (cf. Fig. 4).

significant fraction (1–30% of the silicon) in the form of quartzlike material (free silica). Titanium occurs primarily in the form of rutile, with a small fraction in the organic matrix. Calcium and magnesium occur as carbonates, oxides, and silicates.

The total amount of ash in coals usually varies from 5 to 15% of the dry mass of the fuel. Lignites are the most common exceptions, with ash contents sometimes reaching the 50% range. Total coal ash in some coals is reduced by washing, which is usually performed at or near the mine. Further reductions are achieved by beneficiation of several varieties. Physical beneficiation processes typically selectively remove dense, adventitious inorganic components such as pyrite. Chemical beneficiation processs can be designed either to remove some ash components selectively or to remove all components of the ash. The ash content of coals can be reduced to arbitrarily low levels by combinations of physical and chemical coal beneficiation and processing, but economics generally favor minimal processing for ash removal.

Biomass and Refuse

There is greater diversity in the types of ash found in various forms of biomass and refuse-derived fuels than in coal. Figure 6 illustrates typical results of ash analyses for four typical biomass fuels. In each case, the analyses are based on uncontaminated samples of biomass. Refuse-derived fuels exhibit even wider variations in their ash compositions. Handling or processing of material often increases the total amount of ash and its composition by addition of soil or other extraneous material. The figure illustrates how ash from heartwood or other woody, slowly growing, large-diameter biomass is often dominated by calcium and potassium. Leaves, stems, and other rapidly growing biomass material of small diameter have both more total ash and less calcium relative to other components. The biological activity of rapidly growing material requires greater amounts of inorganic material for both structural and metabolic reasons. Straws and grasses incorporate relatively large amounts of typically hydrated silica in the form of structural components (so-called silica cells) that give the plant rigidity to prevent lodging. Complete discussions of forms and biological functions of inorganic material in biomass are available (12).

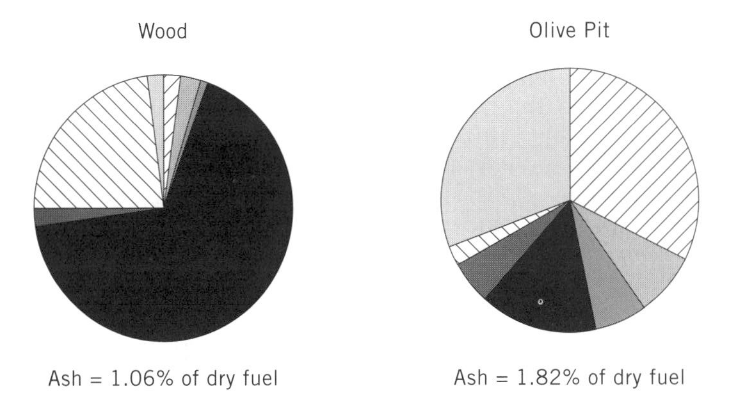

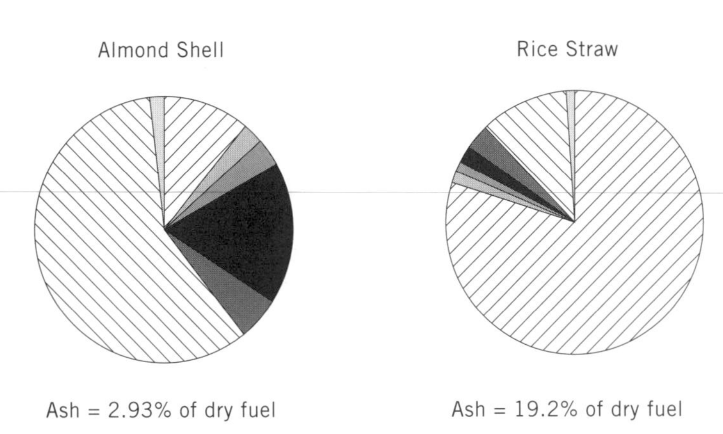

Figure 6. Typical ash analyses from four forms of biomass.

The mode of occurrence of inherent inorganic material in biomass can usually be traced directly to its biological function. For example, essentially all of the potassium in essentially all forms of biomass is atomically dispersed, in strong contrast to the forms of potassium in coal. Nearly all of the silica in biomass is hydrated silica (opal), again in strong contrast to coal. Extraneous mineral matter in fuels reflects the local environment. For example, aluminum is a plant toxin and has low concentration in the inherent inorganic material of biomass fuels. Its ubiquitous nature in soils makes it useful as an indicator of soil contamination of biomass fuels. By contrast, aluminum is a major component in the inherent minerals in coal, where it occurs almost exclusively in clays (hydrated aluminosilicates).

Refuse fired directly in waste to energy plants is commonly referred to as municipal solid waste (MSW). Processed refuse is referred to as refuse-derived fuel (RDF). The amount and composition of both MSW and RDF vary to such a large extent that representative results are of little value. Typically, the amount of ash from these fuels is high, its composition is related directly to its source, and the content of noxious materials such as heavy metals and chlorine is higher than coal, biomass, or fuel oils.

Liquid Fuels

The quantity of inorganic material in fuel oils generally increases with decreasing fuel volatility. The type of inorganic material differs from that in solid fuels. The primary components of concern in liquid fuels include alkali material (primarily sodium and potassium), vanadium, and nickel. Concentrations of these inorganic materials range from a few to a few hundred ppm by weight in the fuel. Some of the material has its origin in the animal and plant matter from which the oils are formed. For example, potassium and sodium in pyrolysis oils is commonly derived from organically associated potassium or sodium in plants. Some vanadium and nickel in petroleum-based oils possibly derives from its proposed role as a porphyrin in ancient animal life. Additional inorganic material in liquid fuels derives from environmental and process sources. Inorganic material tends to concentrate in the heaviest fractions of any distillation or pyrolysis process. Therefore, most of the inorganic material is found in chars or asphalts resulting from the processes, with increasingly smaller amounts found in increasingly light fractions.

ANALYTICAL METHODS

There are both standard and developing analytical methods for determining the amount and properties of ash in fuels. The most active U.S. institute involved in developing standards for analytical procedures is the American Society for Testing and Materials (ASTM). The D and E series of ASTM standards include most of the ash-related information for coal and biomass–refuse-derived fuel, respectively. Similar standards are commonly developed on a worldwide basis by the International Organization for Standardization (ISO). Standards are also promulgated under the direction of other governments, trade and professional organizations, and research societies. These institutions publish and update their standards periodically, and reviews with more detailed discussions of the procedures are available in several reference books (13). The most relevant and common procedures are briefly reviewed below, with some indication of potential biases in the standardized procedures and, where applicable, means of avoiding them. Also described are several promising developing procedures for fuel characterization.

Standard Procedures

Total Ash. Standard methods to determine total ash in fuels involve maintaining a sample in an oxidizing environment while heating it to some prescribed temperature until its weight no longer changes. Different standards are developed for different fuels. For example, coals and cokes are typically heated to 750°C, whereas biomass is heated to 600°C. The details of the standardized techniques attempt to result in a sample that is fully oxidized, with particular emphasis on avoiding formation of carbonates and sulfates. The total ash values thus determined are typically at least 10% lower than the corresponding total inorganic material in the fuel. The difference arises because the forms of the species in the fuel (pyrites, carbonates, sulfates, chlorides, and silicates) typically weigh more than the corresponding, fully oxidized material. Furthermore, much of the inorganic material is hydrated in its pristine form. Standard methods for determining mineral matter (as opposed to ash) involve partially or completely digesting samples in acids or other strong reagents.

The principal objection to these standard methods is that they underestimate the total amount of inorganic material. In addition to converting the inorganics into different and usually lighter forms, volatile inorganics such as sodium and potassium are partially vaporized during sample heating. Alternative procedures are sometimes used, particularly by research labs, to avoid these problems. The alternative procedures have not been formally standardized, although most have been at least documented and are de facto standards for a particular lab.

Ash Elemental Composition. The elemental composition of ash is typically determined by digesting the ash sample and analyzing the resulting fluid by atomic emission or absorption spectroscopy. The high temperature source for the analysis is most commonly an inductively coupled plasma. Other common techniques for the analysis include neutron activation and x-ray fluorescence spectroscopy. The former is typically the most accurate and has the lowest detection limits, but requires equipment and facilities not widely available. The latter is rapid but is the most subject to analysis error than the other techniques. The set of 10 elements included in the analysis of most solid fuels are silicon, aluminum, titanium, iron, calcium, magnesium, sodium, potassium, sulfur, and phosphorus. Chlorine in ash is sometimes included. It is of particular importance in biomass ashes, where chlorine contents are usually higher than for most coals.

The common techniques for determining the elemental composition of ash are based on atomic composition. Results of the analysis are reported on an atomic or elemental basis or, more commonly, converted to an oxide basis. The sum of the oxides can be compared with the measured total ash content of the fuel as a consistency check. If the measured ash exceeds the sum of the oxides, as is usually the case, the difference is attributed to the presence of unmeasured species in the ash, incomplete oxidation of the ash (carbonates, sulfates, etc weigh more than their respective oxides), or experimental error. Normal procedures yield ash and sum-of-oxide numbers that agree within about 5% for many fuels.

The elemental composition of fuels ash varies in predictable ways with geographic or biological origin of the fuel. For example, coals from most western U.S. mines exhibit higher calcium concentrations and lower iron concentrations than those mined in the Midwest or eastern U.S. Herbaceous and annual plants exhibit higher concentrations of silicon and potassium and lower concentrations of calcium than the bulk of most trees and other perennial plants. However, actively growing portions of perennial plants, such as leaves and twigs, exhibit high potassium concentrations. Oils often contain soluble forms of inorganic material, including vanadium, potassium, and sodium.

Principal objections to procedures used to determine ash composition follow directly from those to total ash determination. During high temperature ashing, volatile inorganic components of the fuel are often vaporized, leading to errors in the determination of ash composition. These problems can be avoided by using lower temperature procedures in preparing the ash for analysis, as discussed above.

Fusion Temperatures. The tendency of ash to sinter on combustor walls, and the desire to predict the temperature ranges in which this occurs, has led to the development of ash fusion temperature measurements. These measurements can be performed in a completely automated fashion once the ash is prepared. Ash samples are pressed into the shape of a pyramid 19 mm high with the base forming an equilateral triangle 6.4 mm on a side and placed in a furnace. The samples are shaped like pyramids, but they are almost universally referred to as cones. Furnace temperature is increased at a rate of 8 ± 3°C/min. Changes in the shape of the cone are monitored as the furnace temperature increases. The procedure yields four critical temperatures, defined as follows: (*1*) the initial deformation temperature (tip of cone first becomes rounded), (*2*) the softening temperature (height equals width of the base),

(*3*) the hemispherical temperature (height equals one-half of the width of the base), and (*4*) the fluid temperature (mass has spread to a flat layer no greater than 1.6 mm thick). These measurements are performed under both oxidizing (air or vitiated flow from a lean flame) and reducing (nominal 60% CO, 40% CO_2 gas mixture) environments.

Results of this test are sometimes several hundred degrees lower when conducted under reducing conditions than under oxidizing conditions. A primary reason for the difference is the presence of iron in the ash. Under oxidizing conditions, iron is predominately present as ferric ion (Fe^{3+}) whereas reducing conditions favor ferrous (Fe^{2+}) or, in some cases, metallic iron. Ferric iron is less likely to form eutectics or pure phases that melt at low temperatures than is ferrous iron. As the amount of iron in ash decreases, the differences between fusion temperatures measured in a reducing temperature and those measured on an oxidizing basis decrease.

Objections to ash fusion temperature measurements derive from both the technique and the interpretation of the results. The technique involves heating samples at a prescribed rate to a reasonably high temperature. The reactions that lead to the formation of liquids are often transport or kinetically controlled and occur over time periods much longer than are allowed for in the fusion temperature technique. Ash deposits in combustors have minimum lifetimes (time between soot blowing cycles, eg) of about 8 h. For this reason, the experimental technique is not too representative of actual combustor conditions.

A second objection is that the ash fusion temperature measurements are performed with fuel ash that may have a markedly different composition than the ash in deposits in a combustor fired with that fuel. Ash deposition can be selective in regard to elemental composition, sometimes highly selective. This selectivity depends on location within the combustor, combustor operating conditions, and combustor design in addition to the type of fuel fired. The reasons for this selectivity are discussed and illustrated later.

Ash Viscosity. Combustors designed to produce and remove molten slag are sensitive to ash viscosity. The most common specification of ash viscosity is the T_{250} temperature, or the temperature at which the slag viscosity is 25 Pa·s. This temperature is usually measured using a controlled-environment, high temperature, rotating-bob viscometer. Molten slag is introduced into the viscometer at temperatures of 1430°C or higher. Sample viscosity is monitored as sample temperature is reduced until the viscosity reaches a value of 25 Pa·s. T_{250} values for slags from eastern coals typically range from 1150° to more than 1500°C, with higher iron content leading to lower T_{250} values. Increasing alkali and alkaline earth concentrations often also lead to decreasing T_{250} values.

Other Analyses. The analyses discussed thus far are used primarily to address the issue of ash deposit formation during combustion. Inorganic material in fuels also affects fuel handling and preparation, potential sulfur and particulate emissions, corrosion and metal wastage, and ash salability. Analyses that give insight into these issues include the concentration of pyrite in the fuel, fly ash resistivity, concentrations of chlorine and other corrosive agents, and carbon or toxic element concentrations in the ash. Pyrite (FeS_2) is a common component of high rank coals. Pyrite is a hard mineral that is difficult to pulverize and that contributes to the overall sulfur content of the fuel. It is often the dominant form of iron in unweathered coal. Combustors using electrostatic precipitators (ESPs) as fly ash clean-up devices are sensitive to the electrical conductivity of fly ash. Specific resistivities on the order of 10^9 to 10^{10} ohm-cm yield optimal ESP performance. Higher resistivities tend to excessively insulate the collection plates, decreasing the effective strength of the electric field. High resistivities typically occur with low sulfur, low carbon, high alkali fly ash. Lower resistivities can lead to arcing in the precipitator and excessive reentrainment of collected fly ash. Corrosion and metal wastage are related to the formation of vanadium-, alkali-, sulfur-, or chlorine-laden deposits on surfaces and the abrasion of silica particles as they strike surfaces. Carbon content and other properties of ash influence its potential resale value. The impacts of ash properties on ash marketability are discussed below.

Indices of Ash Behavior. Industrial practice has led to a large number of indices of ash behavior based on the standard analyses discussed above. These are reviewed in detail elsewhere (14,15). The most commonly used are briefly indicated here.

Oxides are usually divided into acidic and basic classes. The most prevalent acidic oxides are silica, alumina, and titania (alumina and titania are actually amphoteric, but are classified as acidic in the application of ash behavior indices). Basic oxides are formed from iron and the alkali and alkaline earth elements. The most commonly cited index of ash deposition behavior is the base:acid ratio. While this is largely an empirically derived number, it has some basis in fundamental concepts. Various components of ash slag are viewed as chain formers, modifiers, or combinations. Acidic oxides are chain formers, whereas basic components are modifiers. Iron can be either a modifier or a combination, depending on its oxidation state. Aluminum is combination of former and modifier. Chain formers typically increase the length of silica-based polymers in molten silicates, increasing the viscosity of the material. Modifiers tend to destabilize the polymers, leading to lower viscosities. The empirically developed ratio of basic to acidic components of the ash relates in this way to the structure of coal slags.

There are many similar empirical indices of ash behavior. A large number of them are perturbations of the base:acid-ratio approach, but some are completely independent of it. They are, in large measure, the tools that have been used to design and operate boilers in the past. However, their successful application generally depends on some field experience with similar situations and, even then, they are perhaps only accurate in anticipating problems slightly more than half of the time. They commonly fail when extrapolated to situations new to the person using them or to situations different from those from which they were developed.

Developing Procedures

The standardized procedures discussed above are the basis for most design, operation, and fuel selection decisions in industry. There usefulness in understanding and mitigating the effects of ash on combustors and the environment is beyond question. There is general agreement, however, that these analyses and their interpretation do not provide sufficient information to address practical issues reliably, eg, to ensure trouble-free operation of combustors. The standardized techniques are even less adequate to develop detailed understanding of ash transformations and deposition. Some of the more obvious shortcomings of these analyses include (*1*) a lack of information about the mode of occurrence of the various ash components, (*2*) neglect of selective deposition of ash components, (*3*) implicit assumptions that ash is a homogeneous entity, neglecting the particle-to-particle variations, and (*4*) neglect of the influence of operating conditions and boiler design on ash deposition tendencies. The potential importance of these issues is large. For example, potassium or sodium incorporated in a clay is relatively benign with regard to ash deposition and corrosion, whereas the same amount of material in an atomically dispersed form greatly increases the deleterious effects of ash deposition and corrosion. There are several developing procedures that address the shortcomings of the standardized procedures. Some of the more significant ones are discussed here and illustrated below.

Chemical Fractionation. Figure 7 schematically illustrates a procedure called chemical fractionation which is used to determine the modes of occurrence of many inorganic components (8,16–18). A series of leaching agents selectively remove various components of fuels, with the leaching agents arranged in order of increasing aggressiveness. Interest focuses primarily on 11 inorganic components (silican, aluminum, titanium, iron, calcium, magnesium, sodium, potassium, sulfer, phosphorus, and chlorine). The modes of occurrence of each element are deduced from the fraction of material removed by each of the leaching agents. The leaching agents chosen for use most commonly include water, ammonium acetate, and hydrochloric acid and sometimes nitric acid.

Dissolved salts such as chlorides often comprise the majority of water-soluble material (stage 1). Other atomically dispersed materials such as alkali and alkaline earth elements comprise much of the material removed by ammonium acetate (stage 2). These two categories of material are broadly classified as atomically dispersed and include the elements most likely to vaporize or otherwise be released to the gas phase during combustion. Their subsequent condensation or chemical reactions underlie many of the most severe deposition and corrosion problems in combustion systems. Hydrochloric acid dissolves carbonates, but not sulfides, and is useful for determining the fraction of, eg, calcium carbonate, in the sample. The residual material comprises primarily oxides, sulfides, or other relatively stable materials. Principal among these are oxides (silica and titania), aluminosilicates typically derived from clays found in soils (kaolinite, montmorillonite, and illite) and sulfides (pyrite). While some of these compounds are largely chemically inert at room temperature, their high temperature chemistry has profound influence on ash deposit growth and property development in combustors. Chemical fractionation results, combined with standardized coal analyses, are the essential inputs to recently developed ash deposition models (19,20).

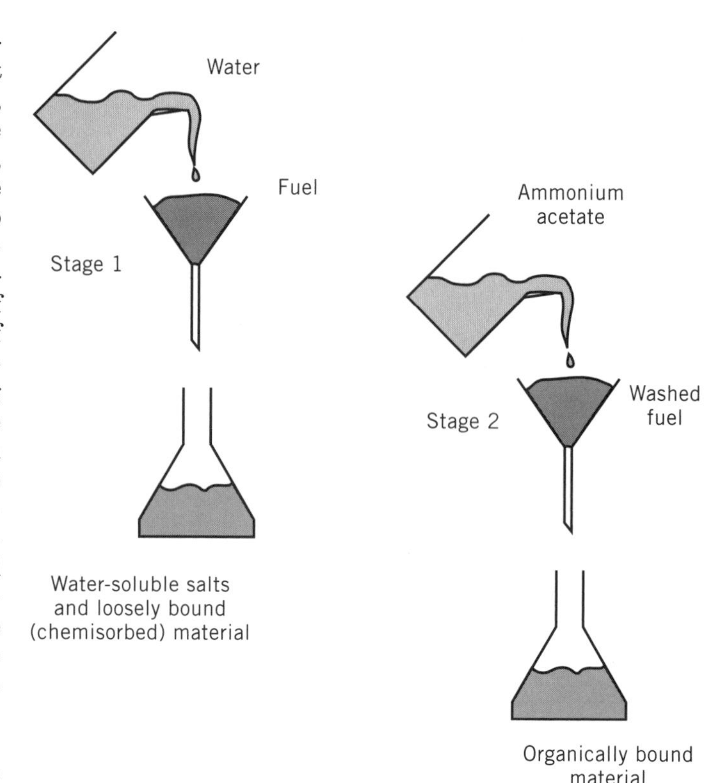

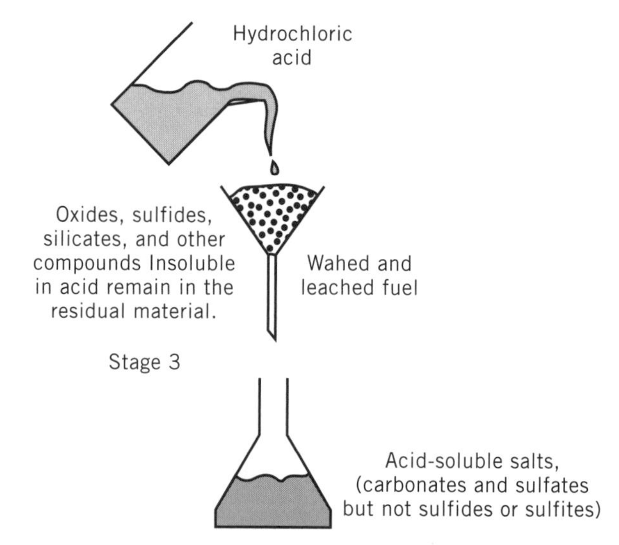

Figure 7. Chemical fractionation procedure illustrating each of the major extractions performed and examples of chemical species removed from the sample at each leaching step.

Scanning Electron Microscope Techniques. Techniques based on scanning electron microscopy (sem) capitalize on this instruments ability to resolve the size and elemental composition of individual mineral grains in fuels. A technique that has come to be called computer-controlled scanning electron spectroscopy (ccsem) combines the automation available from computers with the elemental analysis capabilities of modern electron microscopes to produce size- and composition-classified descriptions of mineral matter in fuels. Several publications discuss and illustrate the technique in detail (21–24). The technique is not standardized, but in all of its incarnations mineral grains are identified in an sem image of the fuel and their size and elemental composition are determined. The probable chemical species or the grain is inferred from the elemental composition combined with look-up tables or logic processes. The accuracy and precision of the technique have been investigated, and recommendations for avoiding common errors are available (25). The results of ccsem analyses form the basis for some of the most recent descriptions of transformations and deposition of inorganic material (26–28). Ccsem provides information about the mineral portion of the inorganic material that is not available by alternative means.

TRANSFORMATIONS OF INORGANIC MATERIAL DURING COMBUSTION

During combustion, inorganic material transforms into ash through several mechanisms. A portion, rarely more than 30%, of the inorganic material is released from the coal particle in the form of either vapors or fumes. This occurs through vaporization, with atomically dispersed species exhibiting high vapor pressures (usually alkali metal-containing inorganic species) participating most actively (29). The partitioning of the alkali species between clays and atomically dispersed forms (see Figs. 4 and 5) is particularly important because the latter are strongly affected by vaporization whereas the former are essentially unaffected. Even when the overall combustion stoichiometry is fuel lean, reducing conditions often prevail inside individual fuel particles, significantly enhancing vaporization rates of many materials. Organic and inorganic reactions can also lead to mass release under many combustion conditions (30,31). The residual inorganic material forms a residual particle which, in turn, goes through several transformations. Mineral grains may fragment during rapid heating (23). The residual char may fragment, with the number of fragments formed increasing with increasing initial char particle size and increasing char porosity (32). Individual mineral grains and adventitious inorganic material go through a series of chemical reactions. For example, sulfides and carbonates decompose, silicates absorb available alkali and alkaline earth material, many species change phase, and grains often agglomerate to form more complex chemical mixtures (33,34).

Chlorine plays a particularly important role in the vaporization of inorganic material. Alkali chlorides are among the most stable, high temperature, alkali-containing species that form in combustion environments. Alkali chlorides also represent a favored form of chlorine in these environments (35,36). Therefore, the extent of alkali vaporization for chlorinated fuels often depends as strongly on the chlorine concentration as on the alkali concentration. The amount and effects of chlorine in coal have been carefully studied (37), but most U.S. coals and many international coals have negligible amounts of chlorine. Exceptions include some Illinois basin coals in the United States and coals mined from seaboard regions. Biomass and refuse derived fuels, on the other hand, may contain significant amounts of chlorine. Chlorine is also a significant factor in corrosion of both metallic and ceramic combustor components.

The result of these transformations is a stream of inorganic vapors and fly ash with a wide range of sizes and a wide variety of compositions. Because alkali material is preferentially vaporized, submicron-size fume particles in the fly ash that form by nucleate recondensation and coagulation are usually enriched in alkali. These alkali-laden particles and vapors generally represent a small fraction of the total fly ash, but they play an inordinate role in deposit formation and property development (38–40). Condensates on surfaces enhance the probability of particles adhering to the surface when they impact. Alkali reacts with silica and, more slowly, with silicates to form low melting point compounds that are also more likely to adhere to surfaces when they impact. Iron also plays an important role in the collection of ash particles on boiler surfaces (41). Even a thin layer of rigid particles enhances the probability of larger particles adhering to a surface when they impact (42,43). The super-micron material generally originates from agglomeration of nonvolatilized, atomically dispersed material and inorganic grains. Fly ash less than 10 μm in size (PM10) is of particular concern as a respiratory health hazard. PM10 particles are also more difficult to remove from a flue gas than larger particles. Detailed studies of fly ash formation document the development of fly ash properties and their dependence on the combustion environment and fuel properties (28,44,45).

DEPOSIT FORMATION

The most significant aspect of inorganic material during combustion is its propensity to form ash deposits on combustor surfaces (see Plate I). The processes that contribute to deposit formation include inertial impaction (and particle capture), thermophoresis, condensation, and heterogeneous reaction. The extent to which each of these mechanisms contributes to deposit formation depends on combustor design, combustor operating conditions, and

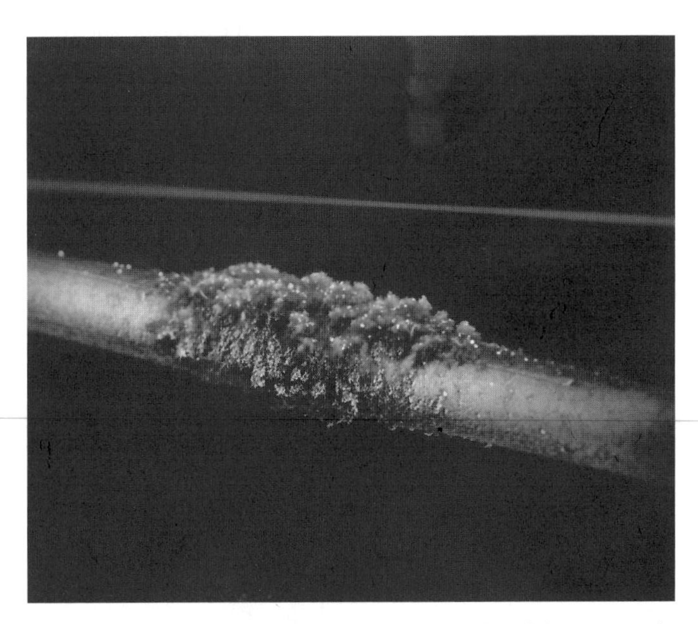

Figure 12. Ash deposit forming on a simulated heat transfer tube. The beam above the ash deposit originates from a laser used to measure fly ash particle size and velocity. Courtesy of Sandia National Laboratories.

fuel properties (46). Each of these processes is briefly reviewed below.

Inertial Impaction

Inertial impaction (Figure 8) is most often the process by which the bulk of the ash deposits. Particles depositing on a surface by inertial impaction have sufficient inertia to traverse the gas stream lines and impact on the surface. The particle capture efficiency describes the propensity of these particles to stay on the surface once they impact. The rate of inertial impaction depends almost exclusively on target geometry, particle size and density, and gas flow properties. The capture efficiency depends strongly on these parameters and on particle composition and viscosity (27,47). It also depends on deposit surface composition, morphology, and viscosity (42,43). The relative magnitudes of the characteristic times and dimensions of particle and fluid relaxation processes control the rate of inertial impaction. Specifically, inertial impaction occurs when the distance a particle travels before it fully adjusts to changes in the fluid velocity is larger than the length scale of an object, or target, submerged in the fluid. The particle Stokes number is defined as the ratio of these length scales.

Inertial impaction is illustrated schematically in Figure 8 for the case of a cylinder in cross-flow. Two particles are illustrated as they approach the cylinder. Both respond to the gas flow field around the cylinder by beginning to move around the cylinder on approach. The inertia of both particles overwhelms the aerodynamic drag forces, and they impact on the cylinder. One is shown rebounding and the other sticking to the surface. Gas stream lines, including recirculation zones, are shown in light gray.

This process is most important for large particles (σ10 μm) and results in a coarse-grained deposit. The impaction rates are highest at the cylinder stagnation point, decreasing rather rapidly with angular position along the surface as measured from this stagnation point. At angular displacements larger than about 50° (as measured from the forward stagnation point), the rate of inertial impaction drops to essentially zero under conditions typical of combustor operation. Rates of inertial impaction can be correlated to particle, fluid, and target properties (20,48). The efficiency with which impacting particles adhere to surfaces has more complex dependencies on particle, surface, and gas properties (49).

Thermophoresis

Thermophoresis is a process of particle transport in a gas due to local temperature gradients. Under some circumstances, thermophoretic deposition accounts for a dominant fraction of the submicron particulate on a surface. Under most conditions relevant to coal combustion, however, the other mechanisms of deposition contribute a larger fraction of the total deposit mass than thermophoresis.

Thermophoretic forces on a particle may be induced either by the temperature gradient in the gas in which the particle is suspended or as a consequence of a temperature gradient in the particle itself. The origin of thermophoretic forces on a particle can be appreciated from the following, overly simplified argument. A particle suspended in a fluid with a strong temperature gradient interacts with molecules that have higher average kinetic energies on the side with the hot fluid than on the side with the cold fluid. The energetic collisions of the high energy molecules on the hot side of the particle create a stronger force than those of the low energy molecules on the cold side. This gives rise to a net force on the particle. In general, these forces act in the direction opposite to that of the temperature gradient, although they can act in the direction of the gradient under certain conditions of particle surface temperature.

An illustration of thermophoretic deposition is presented in Figure 9. Thermophoretic deposits are finer

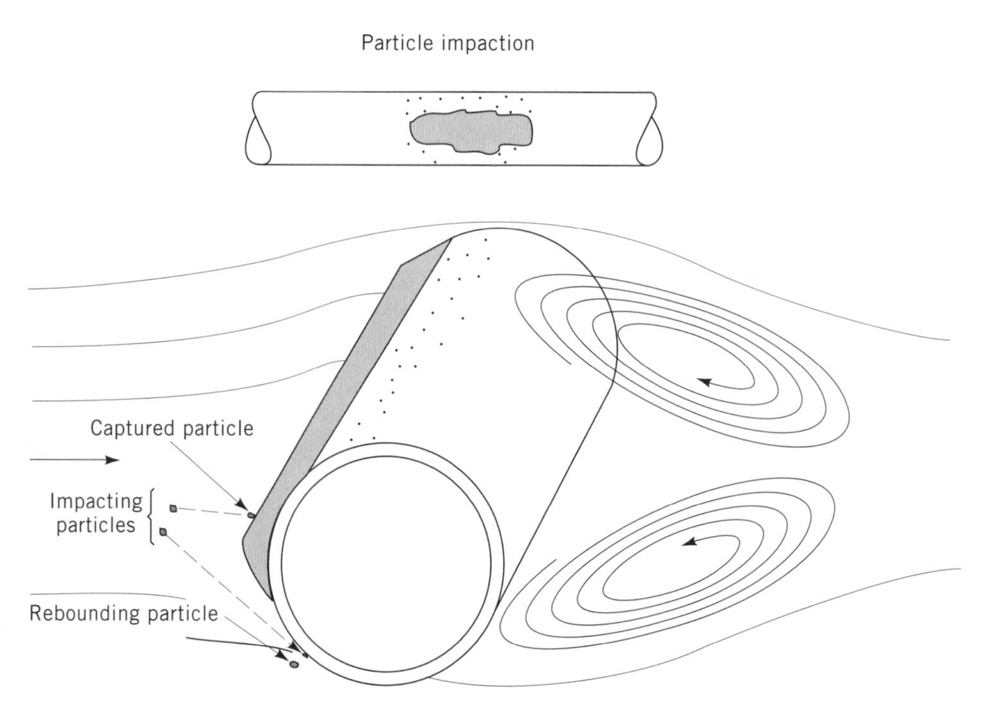

Figure 8. Inertial impaction mechanism on a cylinder in cross flow. One rebounding and one sticking particle are also illustrated.

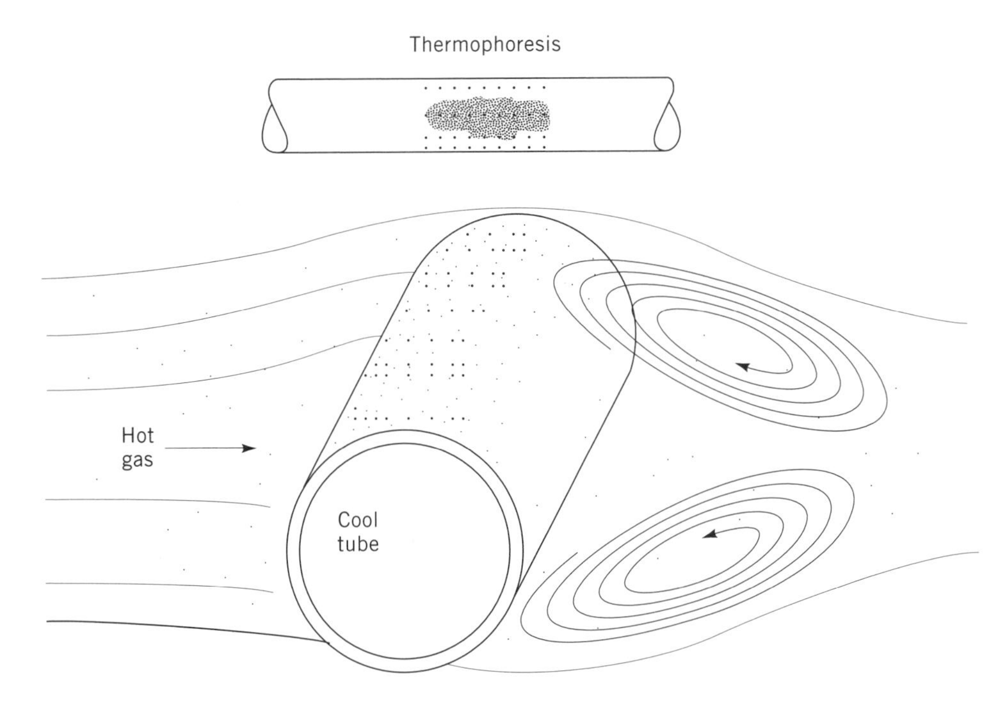

Figure 9. Thermophoretic deposition on a tube in cross flow.

grained and more evenly distributed around the tube surface than deposits formed by inertial impaction, as indicated. With increasing deposit accumulation on the tube surface, there is a decrease in the temperature gradient in the thermal boundary layer, decreasing the rate of thermophoresis. Mathematical descriptions of thermophoresis are available at several levels of complexity (17,50–52).

Condensation

Condensation is the mechanism by which vapors are collected on surfaces cooler than the local gas. An illustration of deposition by condensation on a tube in cross-flow is presented in Figure 10. The amount of condensate in a deposit depends strongly on the mode of occurrence of the inorganic material in the coal. Low rank (subbituminous) coals, lignites, and other similar fuels have the potential of producing large quantities of condensable material. Furthermore, the role of condensate in determining deposit properties can be substantially greater than the mass fraction of the condensate in the deposit might suggest. For example, condensate increases the contacting area between an otherwise granular deposit and a surface by several orders of magnitude. This increases by the difficulty of removing the deposit from the surface by a similar amount. Condensate can also increase the contacting area between particles by many orders of magnitude, having profound influences in the bulk strength, thermal conduc-

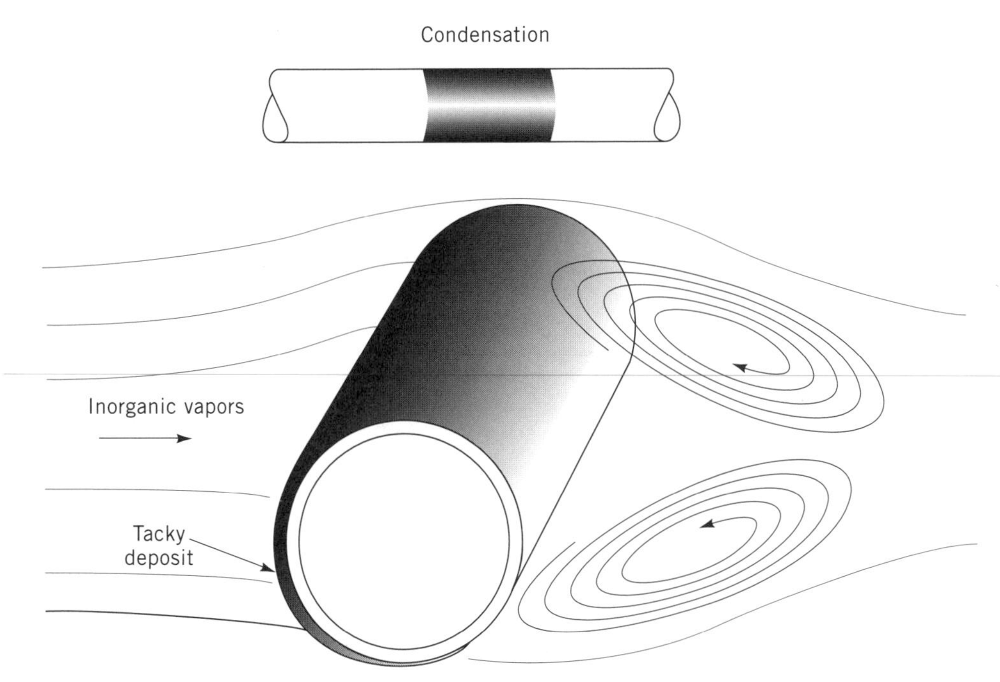

Figure 10. Condensation on a tube in cross flow.

tivity, mass diffusivity, etc. of the deposit. Condensation is a relatively minor contributor to the development of deposits and their properties for most high rank coals. However, in lower grade fuels condensation becomes a significant contributor.

All vapors that enter the thermal boundary layer around a cool surface and subsequently are deposited on the surface can be thought of as condensate. Condensation occurs by at least three mechanisms: (*1*) vapors may traverse the boundary layer and heterogeneously condense on the surface or within the porous deposit, (*2*) vapors may homogeneously nucleate to form a fume and subsequently deposit by thermophoresis on the surface, and (*3*) vapors may heterogeneously condense on other particles in the boundary layer and arrive at the surface by thermophoresis (39,52,53).

Condensation deposits have no granularity at length scales larger than 0.5 μm and are more uniformly deposited on the tube than either thermophoretically or inertially deposited material. The deposits are tacky and have a strong influence on the surface capture efficiency.

Chemical Reaction

Inertial impaction and thermophoresis describe the two most significant mechanisms for transporting particulate material to a surface. Condensation involves transportation of vapors to a surface by means of a physical reaction and phase change. Chemical reactions (Fig. 11) complete the mechanisms by which mass can be accumulated in a deposit. These involve the heterogeneous reaction of gases with materials in the deposit or, less commonly, with the deposition surface itself. Some of the chemical species found in deposits are not stable at gas temperatures, alkali sulfates being typical examples. The sole source of these species is heterogeneous reactions between gas phase constituents and constituents of the lower temperature deposits.

Among the most important chemical reactions with respect to ash deposition are sulfation, alkali absorption, and oxidation. The principal sulfating species of concern are compounds containing the alkali metals, sodium and potassium. Sodium and potassium in the forms of condensed hydroxides and possibly chlorides are susceptible to sulfation (35). The formation of these materials often leads to highly reflective deposits, substantially reducing heat transfer rates in the combustor (54,55). Silica absorbs alkali material to form silicates. Silicates are less rigid and melt at lower temperatures than silica. The transformations of silica to silicates in deposits can induce sintering and significant changes in deposit properties (56,57). These reactions are slower than sulfation. Residual char often deposits with the inorganic material on combustor surfaces. However, the char oxidizes with locally available oxygen to produce deposits with little residual carbon. In coal combustion, carbon typically accounts for less than 2% of the overall deposit mass. Chemical reactions, such as sulfation of alkali species and combustion of residual carbon in the ash, are similar to condensation in their mathematical treatment (40). Both condensation and chemical reactions are strongly temperature dependent and give rise to spatial variation in ash deposit composition. Sulfation can become particularly important when sulfur-laden fuels are blended with alkali-laden fuels (58).

EXPERIMENTAL ILLUSTRATIONS OF ASH DEPOSITION

The morphological and chemical differences in ash deposits produced by different mechanisms, as discussed above,

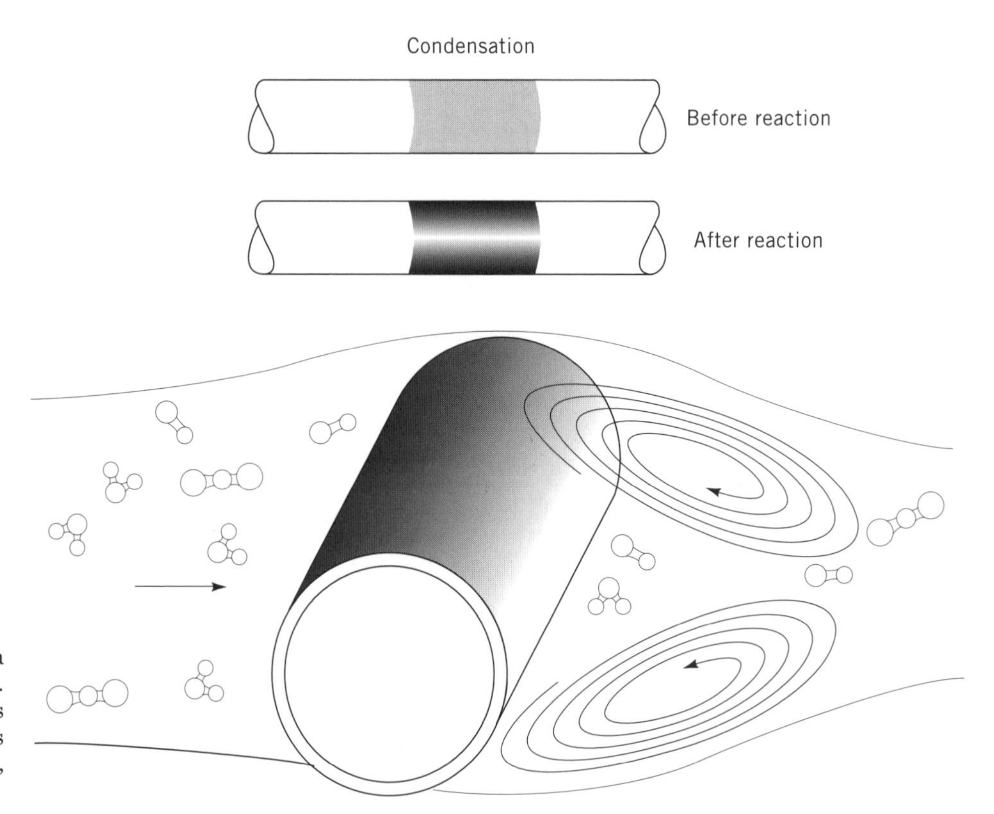

Figure 11. Chemical reaction as a mechanism of ash deposit formation. Gas-phase species such as alkali vapors and oxides of sulfur react with deposits and lead to formation of alkali silicates, alkali sulfates, and other species.

can be demonstrated through experiment at both pilot and commercial scale (59). A deposit generated during experimental combustion of wheat straw illustrates several of the principal mechanisms of ash deposition in a single deposit. The white material on the surface of the tube in cross flow to either side of the bulk of the deposit and on the surfaces of many of the particles is composed of potassium salts that, after condensation on the surface, reacted with sulfur to form sulfates. The particles that comprise the bulk of the deposit are primarily silica, the single most prevalent element in wheat straw ash. In a commercial furnace, the deposit of this nature would extend the length of the probe, rather than form a localized mass as occurs in the experimental facility. Deposit morphologies also depend on geometry and fuel type. For example, bituminous coals in the United States typically have little atomically dispersed inorganic material and large quantities of mineral grains, whereas U.S. lignites have larger quantities of atomically dispersed material. The particulate material in the coal usually forms large fly ash particles, whereas significant fractions of the atomically dispersed material forms vapors. Therefore, the dominant mechanism for ash deposition for bituminous coals is commonly inertial impaction, whereas lignite deposits are typically initiated by condensation. This will be illustrated qualitatively by examples of deposits formed in the Sandia Multifuel Combustor. The combustor itself is described in more detail later.

The composition of deposits found in commercial systems may vary substantially from that of the fuel used to fire the system. This selective deposition occurs because different chemical species are involved in different mechanisms of ash deposit formation. For example, Figure 12 illustrates the elemental composition of a deposit collected during commercial operation of a biomass combustor burning a blend of fuels. The compositions of ash from the fuel blend and deposits found on the upper furnace walls are illustrated in the figure. While the fuel ash was composed primarily of silicon and other material, the deposit was primarily composed of alkali and sulfur. The primary mechanism of deposition in this case is condensation followed by chemical reaction. Condensation involves primarily the atomically dispersed volatile materials such as alkali metals.

ASH DISPOSAL AND ENVIRONMENTAL CONCERNS

In the United States, approximately 363 kg of ash is generated per person per year from coal-fired power plants alone. The bulk of this ash is fly ash, material collected in the flue gas clean-up systems of power plants. A growing economic and environmental issue is the ultimate disposition of this material.

Coal, like any material extracted from the earth, contains low levels of inorganic contaminants that cause some concern. Landfills represent one of the most common traditional means of ash disposal. Some concern has been expressed that harmful ash components may be leached from the landfill and find their way into sensitive ecosystems or present a public health problem. If all harmful components of ash were assumed to find their way into ground water, most fly ash would be considered hazardous waste. However, most of the potentially harmful components of ash are not soluble or are only slightly soluble (60). After studying this issue, the regulatory and legal institutions in the United States determined that ash from coal-fired power plants need not be treated as hazardous waste. In several other countries, land filling of ash is not allowed, although the regulations are based more on lack of landfill space than concern about environmental or health hazards associated with the ash. In all countries, alternative markets for this ash are sought.

Two of the primary uses for coal ash and similar ashes from other fuels are as a component in Portland cement and in concrete. Some coal fly ash exhibits cementlike properties in its natural state, while others do so only when mixed with water and lime. The latter are referred

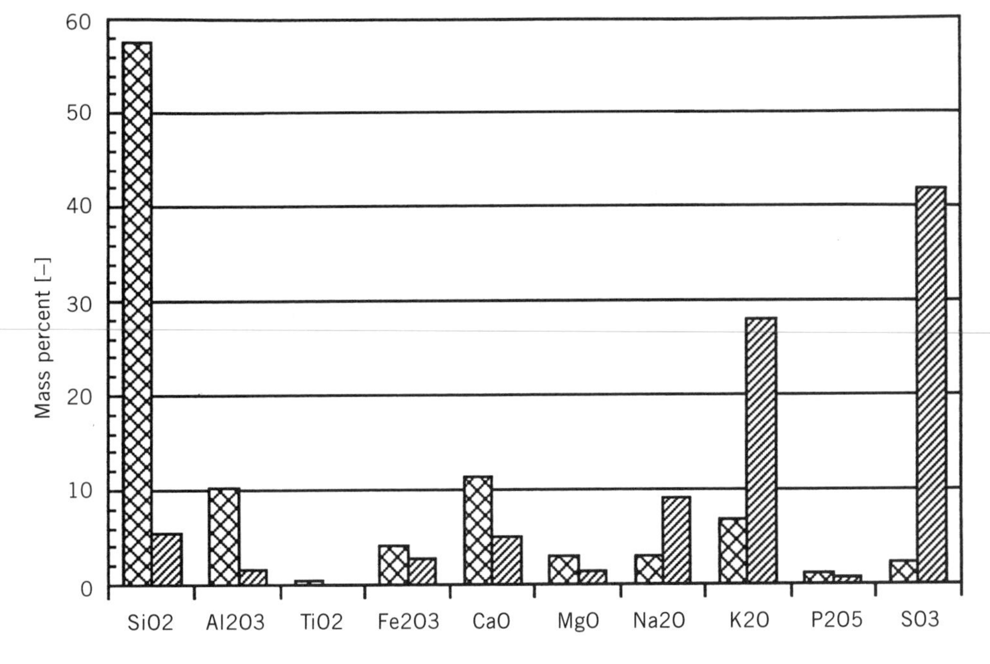

Figure 12. Elemental composition of biomass fuel and ash deposits formed on the upper walls and ceiling of the furnace. The significant differences in composition arise due to mechanisms of deposition that are selective in which elements they involve.

to as pozzolans. Both types of ashes participate chemically in reactions that occur as concrete cures and have value as an additive during cement production. Possibly the most important measure of fly ash suitability for use in the cement industry is its carbon content. Concrete used in applications subjected to temperature changes (eg, freezing temperatures) contains small bubbles to provide relief for expansion and contraction stresses. Carbon in fly ash behaves much like activated carbon and absorbs these bubbles, producing an inferior material. Concrete also becomes increasingly discolored if the fly ash contains high carbon content or other colored components. This presents problems that are largely cosmetic but nonetheless present marketing problems for the cement industry. Commonly, the loss on ignition (weight loss during prescribed heating conditions) of fly ash for use in cement is usually limited to less than a few percent. The amount of carbon in fly ash from low rank coal is typically low (1% or less), with increasing amounts of carbon as coal rank increases and as inorganic material in the original fuel decreases. In all cases, large fly ash particles contain far greater concentrations of carbon than small fly ash particles. Carbon in fly ash also depends on the type of combustor used to generate the ash. Typically, ashes from fixed or slowly moving grates burning either coal or biomass contain higher amounts of carbon than ashes from bubbling or, especially, circulating fluid beds. Besides carbon content, specifications for fly ash to be used in cement manufacture sometimes include a combined amount of SiO_2, Al_2O_3, and Fe_2O_3 of not less than 70%, no more than 5% MgO, and no more than 4% SO_3. Its moisture content should be less than 10% and its available alkali should be less than 1.5% as Na_2O. Fly ash used in concrete (ie, not as cement but as aggregate in concrete) is less tightly constrained. Ash from low rank coals typically contains high concentrations of CaO (25–35%) and, if sulfur contents can be kept low, represents an exceptionally good cement additive. Fly ash used in concrete reduces the amount of water required, improves the strength, improves workability, lowers permeability, decreases heat generation during curing, and can lead to lower bulk densities to the extent it contains cenospheric particles. It normally should have a surface area of around 3000 cm^2/g and 88% of it should pass a 325 mesh (44 μm) (7).

Other uses for fly ash include backfill, road beds, foundry powder, fire brick (especially for cenospheric particles), and sandblasting grit. In addition to these applications, bottom ash is used as an aggregate for block manufacturing. Typically, 20% or less of the fly ash generated in the United States finds practical use due to costs associated with shipping and handling material and historically low fees for landfilling. In Europe and other densely populated areas with less available landfill volume, larger fractions of fly ash are generally used (>30% of bituminous fly ash in 1974, eg) (61). The United States will likely follow this trend as landfill volume becomes increasingly scarce.

Passage of the Clean Air Act Amendments in the United States in 1990 (62) ushered in major new legislation with regard to inorganic components of fuels in the United States. In these amendments, the EPA is directed to promulgate emission regulations for any compound that contains arsenic, beryllium, cadmium, chromium, cobalt, cyanide, lead, manganese, mercury, nickel, selenium, any radionuclide (including radon), or fine mineral fibers. Combustion systems potentially emit large quantities of some of these materials due less to their concentrations in the flue gas than to the volume of the flue gas produced. Preliminary studies indicate that mercury and possibly arsenic and selenium will be the compounds of greatest concern for most power generation facilities.

BIBLIOGRAPHY

1. J. F. Unsworth, D. J. Barratt, and P. T. Roberts in L. L. Anderson, ed., *Coal Science and Technology,* Elsevier, Amsterdam, The Netherlands, 1991.
2. R. P. Johnson, R. G. Plangemann, and M. D. Rehm, *Coal Quality Effects on Boiler Performance: An Eastern Perspective,* EPRI Report, RP2256-8, Electric Power Research Institute, 1991.
3. Energy Information Office, *Monthly Energy Review,* DOE/EIA-0035(93/09), U.S. Department of Energy, Washington, D.C., Sept. 1993.
4. R. E. Mitchell, R. H. Hurt, L. L. Baxter, and D. R. Hardesty, *Compilation of Sandia Coal Char Combustion Data and Kinetic Analyses: Milestone Report,* SAND92-8208, Sandia, 1992.
5. S. C. Stultz, and J. B. Kitto, eds., *Steam—Its Generation and Use,* Babcock & Wilcox Co., Barberton, Ohio, 1992.
6. E. Raask, *J. Inst. Energy* **57,** 231 (1984).
7. W. T. Reid in M. A. Elliot, ed., *Chemistry of Coal Utilization: Second Supplementary Volume,* John Wiley & Sons, Inc., New York, 1981, pp. 1389–1446.
8. R. B. Finkelman, Ph.D. Dissertation, University of Maryland, 1980.
9. N. Berkowitz, *The Chemistry of Coal,* Elsevier, Amsterdam, The Netherlands, 1985.
10. D. W. van Krevelen, *Coal: Typology—Chemistry—Physics—Constitution,* Elsevier, Amsterdam, The Netherlands, 1981.
11. D. C. Elliot in E. J. Soltes, T. A. Milne, eds., *Pyrolysis Oils from Biomass: Producing, Analyzing, and Upgrading,* American Chemical Society, Washington D.C., 1988, pp. 55–65.
12. H. Marschner, *Mineral Nutrition of Higher Plants,* Harcourt Brace Jovanovich, London, 1986.
13. C. Karr Jr., eds., *Analytical Methods for Coal and Coal Products,* Vol. 1 Academic Press, Inc., New York, 1978.
14. E. C. Winegartner, *Coal Fouling and Slagging,* ASME, New York, 1974.
15. R. W. Bryers, *Symposium on Slagging and Fouling in Steam Generators,* ASME, New York, 1987.
16. L. L. Baxter and D. R. Hardesty, *The Fate of Mineral Matter During Pulverized Coal Combustion: January–March,* SAND92-8210, Sandia National Laboratories, 1992.
17. L. L. Baxter and D. R. Hardesty, *The Fate of Mineral Matter During Pulverized Coal Combustion: April—June,* SAND92-8227, Sandia National Laboratories, 1992.
18. S. A. Benson and P. L. Holm, *Ind. Eng. Chem. Prod. Res. Dev.* **24,** 145–149 (1985).
19. M. F. Abbott, R. E. Douglas, C. E. Fink, N. J. DeIuliis, and L. L. Baxter, paper presented at the Engineering Foundation Conference on the Impact of Ash Deposition on Coal-Fired Plants, 1993.
20. L. L. Baxter and D. R. Hardesty, *The Fate of Mineral Matter During Pulverized Coal Combustion,* SAND91-8217, Sandia National Laboratories, 1991.

21. A. Carpenter and N. Skorupska, *Coal Combustion—Analysis and Testing,* IEA Coal Research, 1993.
22. F. R. Karner, C. J. Zygarlicke, D. W. Brekke, E. N. Steadman, and S. A. Benson, *Power Eng.,* 35–38 (1994).
23. L. L. Baxter, *Prog. Energy Combustion Sci.* **16,** 261–266 (1990).
24. H. Yu, J. E. Marchek, N. L. Adair, and J. N. Harb, *The Impact of Ash Deposition on Coal Fired Power Plants.* Taylor & Francis, 1993.
25. N. Y. C. Yang and L. L. Baxter, paper presented at the Conference on Inorganic Transformations and Ash Deposition During Combustion, 1991.
26. J. M. Beér, L. S. Monroe, L. E. Barta, and A. F. Sarofim, paper presented at the Seventh International Pittsburgh Coal Conference, 1990.
27. S. Srinivasachar, C. L. Senior, J. J. Helble, and J. W. Moore, paper presented at the Twenty-Fourth Symposium (International) on Combustion. 1992.
28. C. J. Zygarlicke, M. Ramanathan, and T. A. Erickson in S. A. Benson, ed., *Inorganic Transformations and Ash Deposition During Combustion,* American Society of Mechanical Engineers, New York, 1992, pp. 525–544.
29. J. J. Helble and A. F. Sarofim, *J. Colloid Interface Sci.* **128,** 348–362 (1989).
30. L. L. Baxter and R. E. Mitchell, *Combustion Flame* **88,** 1–14 (1992).
31. L. L. Baxter, R. E. Mitchell, and T. H. Fletcher, *Combustion Flame,* in press.
32. L. L. Baxter, *Combustion Flame* **90,** 174–184 (1992).
33. S. Srinivasachar, J. J. Helble, and A. A. Boni, *Prog. Energy Combustion Sci.* **16,** 281–292 (1990).
34. S. Srinivasachar and co-workers, *Prog. Energy Combustion Sci.* **16,** 293–302 (1990).
35. S. Srinivasachar, J. J. Helble, D. O. Ham, and G. Domazetis, *Prog. Energy Combustion Sci.* **16,** 303–309 (1990).
36. J. J. Helble, S. Srinivasachar, and A. A. Boni, *Combustion Sci. Technol.* **81,** 193–205 (1992).
37. J. Stringer and D. D. Banerjee, eds., *Chlorine in Coal,* Elsevier, Amsterdam, The Netherlands, 1991.
38. L. J. Wibberly and T. F. Wall, *Fuel* **61,** 93–99 (1982).
39. J. J. Helble, M. Neville, and A. F. Sarofim, paper presented at the Twenty-First Symposium (International) on Combustion, 1986.
40. L. L. Baxter, *Biomass Bioenergy* **4,** 85–102 (1993).
41. A. T. S. Cunningham, W. H. Gibb, A. R. Jones, F. Wigley, and J. Williamson, paper presented at the Conference on Inorganic Transformations and Ash Deposition During Combustion, 1991.
42. C. L. Wagoner and X.-X. Yan, paper presented at the Conference on Inorganic Transformations and Ash Deposition During Combustion, 1991.
43. S. M. Smouse and C. L. Wagoner, paper presented at the Conference on Inorganic Transformations and Ash Deposition During Combustion, 1991.
44. R. J. Quann, M. Neville, and A. F. Sarofim, *Combustion Sci. Technol.* **74,** 245–265 (1990).
45. S.-G. Kang, J. J. Helble, A. F. Sarofim, and J. M. Beér, paper presented at the Twenty-Second Symposium (International) on Combustion, 1988.
46. L. L. Baxter and R. W. DeSollar, *Fuel* **72,** 1411–1418 (1993).
47. J. Srinivasachar, J. J. Helble, and A. A. Boni, paper presented at the Twenty-Third Symposium (International) on Combustion, 1990.
48. R. Israel and D. E. Rosner, *Aerosol Sci. Technol.* **2,** 45–51 (1983).
49. L. L. Baxter, K. R. Hencken, and N. S. Harding, paper presented at the Twenty-Third Symposium (International) on Combustion, 1990.
50. S. A. Gökoglu and D. E. Rosner, *Int. J. Heat Mass Transfer* **27,** 639–646 (1984).
51. S. A. Gökoglu and D. E. Rosner, *AIAA J.* **24,** 172–179 (1986).
52. D. E. Rosner, and M. Tassopoulos, *AIChE J.* **35,** 1497–1508 (1989).
53. J. L. Castillo and D. E. Rosner, *Int. J. Multiphase Flow* **14,** 99–120 (1988).
54. L. L. Baxter, G. H. Richards, D. K. Ottesen, and J. N. Harb, *Energy Fuels* **7,** 755–760 (1993).
55. G. H. Richards and co-workers, paper presented at the Twenty-Fifth Symposium (International) on Combustion, 1994.
56. L. L. Baxter and co-workers, paper presented at the Second International Conference on Combustion Technologies for a Clean Environment, 1993.
57. L. L. Baxter, paper presented at the EPRI Conference on The Effects of Coal Quality on Power Plants, 1992.
58. L. L. Baxter and L. Dora, Paper No. 92-JPGC-FACT-14, ASME, New York, 1992.
59. N. S. Harding and M. C. Mai, in R. W. Bryers and K. S. Vorres, eds., *Mineral Matter and Ash Deposition from Coal* Engineering Trustees, Inc., 1990, pp. 375–399.
60. C. W. Gehrs, D. S. Shriner, S. E. Herbes, E. J. Salmon, and H. Perry in M. A. Elliot, ed., *Chemistry of Coal Utilization: Second Supplementary Volume,* John Wiley & Sons, Inc., New York, 1981, pp. 2159–2223.
61. H. Jüntgen, J. Klein, K. Knoblauch, H.-J. Schröter, and J. Schulze in Ref. 60.
62. *Federal Register: Part III; Air Contaminants,* Final Rule, 29 CFR Part 1910 (1989).

AUTOMOBILE EMISSIONS

F. M. Black
Mobile Source Emissions Research Branch
U.S. Environmental Protection Agency
Research Triangle Park, North Carolina

A relationship between automobile emissions and air pollution was first suggested in the United States by studies that began during the 1940s in California. Air pollution characterized by plant damage, eye and throat irritation, stressed rubber cracking, and decreased visibility was detected in Los Angeles as early as 1943. By 1948, the California legislature had established air pollution control districts empowered to curb emission sources. Initial efforts were aimed at reducing industrial particulate emissions and resulted in improved visibility, but eye irritation and other smog symptoms remained.

See also Clean air act, mobile sources; Air pollution; automobile, toxic emissions; Exhaust control, automotive; Automobile engines; Methanol vehicles; Transportation fuels—automotive fuels; Reformulated gasoline; Alcohol fuels; Transportation fuels—ethanol fuels in Brazil.

Research directed by A. J. Haagen-Smit at the California Institute of Technology, published in 1952, demonstrated that in the presence of sunlight and nitrogen dioxide (NO_2), hydrocarbon (HC) compounds react to form a variety of oxidants, including ozone, that could account for many of the observed environmental effects (1). Because the automobile was thought to be a significant source of oxidant-precursors, the Automobile Manufacturers Association established the Vehicle Combustion Products Committee in 1953 to further define the problem and to search for solutions. Simultaneously, California began an effort to reduce stationary-source HC emissions in the Los Angeles area. However, smog remained a serious problem, and in 1958, the decision was made to require some form of automotive HC emissions control. In 1959, the California legislature enacted air quality standards for oxidants and carbon monoxide (CO), and in 1960, the California Motor Vehicle Pollution Control Board was created to implement these standards. Its function was to set specifications for vehicle exhaust and evaporative emissions and to certify that vehicles sold in California met the specifications. It was estimated that achieving the air quality goals for 1970 would require an 80% reduction of motor vehicle HC emissions and a 60% reduction of motor vehicle CO emissions (2). As illustrated in Figure 1 for cars without emission controls, engine exhaust accounted for about 60% of polluting emissions, crankcase vapors for about 20%, and fuel evaporation for about 20% (3). California's first emission control requirement addressed crankcase blowby emissions. In 1963, devices to eliminate crankcase emissions were required on all cars sold in California and were installed on all new automobiles nationally. Tailpipe HC and CO emissions were first regulated by California in 1966.

The federal government became involved with air pollution in 1955 by empowering the Department of Health, Education, and Welfare to provide technical assistance for resolution of problems related to air pollution. This activity was enhanced by the Clean Air Act of 1963, which was structured to stimulate state and local air pollution control efforts. A 1965 amendment to the Clean Air Act specifically authorized the writing of national standards for emissions from all new motor vehicles sold in the United States Control of motor vehicle exhaust emissions was required nationally beginning in 1968 (4). Gasoline vehicles were required to be equipped with closed crankcase ventilation systems, and tailpipe concentrations of HC and CO emissions were lmited to prescribed concentrations (275 ppm HC, 1.5% CO) over a defined dynamometer driving schedule. Beginning in 1970, a preproduction certification practice controlled the mass emission rates (eg, grams emitted per vehicle mile traveled, g/mi) of regulated compounds, as summarized in Table 1 (5).

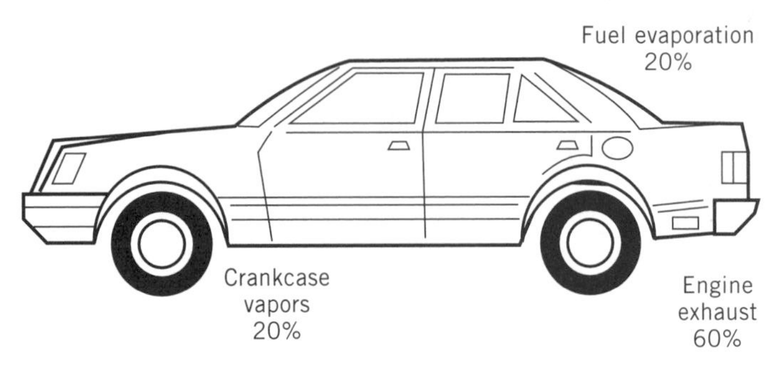

Figure 1. Sources of automobile pollutants prior to emissions control (3).

Since motor vehicle emissions control was instituted in the United States, standards have been progressively reduced and test procedures have been improved. For example, 1971 light-duty vehicle tailpipe emissions, certified at 2.2 g/mi total hydrocarbon (THC) and 23 g/mi CO, were determined using a seven-mode steady-state driving schedule. Since 1972, transient driving schedules have been used. Current tailpipe emission standards are 0.41 g/mi THC, 3.4 g/mi CO, 1.0 g/mi oxides of nitrogen [NO_x, the sum of nitric oxide (NO) and NO_2], and 0.2 g/mi particulate matter. In 1971, motor vehicle evaporative emissions, certified at 6 g per test, were determined with a gravimetric carbon trap technique by collecting emissions from selected point sources on the vehicle. Beginning in 1978, evaporative emissions were determined with a sealed housing for evaporative determination (SHED) technique, in which the entire vehicle is enclosed so that emissions are collected from all evaporative sources. The current evaporative emissions standard is 2 g per test. Starting in 1974, tailpipe emissions from heavy-duty vehicles [gross vehicle weight (GVW) 3856 kg (8500 lb) or more] were regulated using a 13-mode steady-state driving schedule. Since 1985, transient driving schedules have been used. The tailpipe emission standards have ranged from 40 g per brake-horsepower-hour (g/bhph) CO in 1974 to current gasoline truck and bus standards of 1.1 g/bhph THC, 14.4 g/bhph CO, and 5.0 g/bhph NO_x [standards are higher for GVW greater than 6350 kg (14,000 lb)]. The evaporative emission standard for heavy-duty gasoline trucks and buses is 3 g per test. Current diesel truck and bus standards are 1.3 g/bhph THC, 15.5 g/bhph CO, 5.0 g/bhph NO_x, and 0.1 g/bhph particulate (0.05 g/bhph for urban buses). Summarizing, tailpipe HC, CO, NO_x, and particulate matter emissions, evaporative HC emissions, and fuel economy currently are regulated with light-duty passenger cars and trucks; similarly, tailpipe and evaporative emissions (from vehicles using gasoline fuel) are regulated with heavy-duty trucks and buses, but not fuel economy.

The emission standards are enforced initially by a program of preproduction design review, testing, and certification. Production prototypes are certified to emit at or below standard levels for 80,450 km (50,000 miles). The federal government also administers assembly-line audits and surveillance-recall programs designed to ensure that production motor vehicles perform as well as the certification prototypes. Thus, emission controls are examined at three stages: initially, to certify that cars are designed to meet federal emission standards before they are manufactured (preproduction); subsequently, to ensure that assembly-line cars actually meet the emission standards they were designed and certified to meet (production); and finally, to ensure that the emission control systems on cars that have been in normal everyday consumer use for several years, having accumulated substantial mileage, perform at an acceptable level (postproduction). If a substantial number of properly maintained and operated vehicles of a specific "family" fail an emission standard during in-

Table 1. U.S. Motor Vehicle Emissions and Fuel Economy Standards

Model Year	Vehicle Category	Tailpipe Hydrocarbon	Carbon Monoxide	Oxides of Nitrogen	Evaporative Hydrocarbon	Particulate	Fuel Economy
1970[a]	LDV, LDT	2.2 g/mi	23 g/mi	—	—		
1971	LDV, LDT	2.2 g/mi	23 g/mi	—	6 g/test[b]		
1972[a]	LDV, LDT	3.4 g/mi	39 g/mi	—	2 g/test		
1973	LDV, LDT	3.4 g/mi	39 g/mi	3.0 g/mi	2 g/test		
1974[d]	HDGV, HDDV	—	40 g/bhphr	—	—		
1975[e]	LDV	1.5 g/mi	15 g/mi	3.1 g/mi	2 g/test		
	LDT	2.0 g/mi	20 g/mi	3.1 g/mi	2 g/test		
1977	LDV	1.5 g/mi	15 g/mi	2.0 g/mi	2 g/test		
1978	LDV	1.5 g/mi	15 g/mi	2.0 g/mi	6 g/test[f]	—	18 mpg
	LDT	2.0 g/mi	20 g/mi	3.1 g/mi	6 g/test		
1979	LDV	1.5 g/mi	15 g/mi	2.0 g/mi	6 g/test	—	19 mpg
	LDT	1.7 g/mi	18 g/mi	2.3 g/mi	6 g/test		
	HDGV, HDDV	1.5 g/bhphr	25 g/bhphr	—	—		
1980	LDV	0.41 g/mi	7.0 g/mi	2.0 g/mi	6 g/test	—	20 mpg
1981	LDV	0.41 g/mi	3.4 g/mi	1.0 g/mi	2 g/test	—	22 mpg
	LDT	1.7 g/mi	18 g/mi	2.0 g/mi	2 g/test		
1982	LDV	0.41 g/mi	3.4 g/mi	1.0 g/mi	2 g/test	—	24 mpg
1983	LDV	0.41 g/mi	3.4 g/mi	1.0 g/mi	2 g/test	—	26 mpg
1984	LDV	0.41 g/mi	3.4 g/mi	1.0 g/mi	2 g/test	—	27 mpg
	LDT	0.8 g/mi	10 g/mi	2.3 g/mi	2 g/test		
1985	LDV	0.41 g/mi	3.4 g/mi	1.0 g/mi	2 g/test	0.6 g/mi	27.5 mpg
	LDT	0.8 g/mi	10 g/mi	2.3 g/mi	2 g/test	0.6 g/mi	
[g]	HDGV (<14k)	2.5 g/bhphr	40.1 g/bhphr	—	3 g/test		
[g]	(> 14K)				4 g/test		
[g]	HDDV	1.3 g/bhphr	15.5 g/bhphr	—	—		
1987	LDV	0.41 g/mi	3.4 g/mi	1.0 g/mi	2 g/test	0.2 g/mi	27.5 mpg
	LDT	0.8 g/mi	10 g/mi	1.2 g/mi	2 g/test	0.26 g/mi	
	HDGV (<14K)	1.3 g/bhphr	15.5 g/bhphr	6.0 g/bhphr	3 g/test		
	(>14K)	2.5 g/bhphr	40 g/bhphr	6.0 g/bhphr	4 g/test		
	HDDV	1.3 g/bhphr	15.5 g/bhphr	6.0 g/bhphr	—		
1988	HDGV (<14k)	1.1 g/bhphr	14.4 g/bhphr	6.0 g/bhphr	3 g/test		
	(>14K)	1.9 g/bhphr	37.1 g/bhphr	6.0 g/bhphr	4 g/test		
	HDDV	1.3 g/bhphr	15.5 g/bhphr	6.0 g/bhphr	—	0.6 g/bhphr	
1991	HDGV (<14K)	1.1 g/bhphr	14.4 g/bhphr	5.0 g/bhphr	3 g/test		
	HDDV	1.3 g/bhphr	15.5 g/bhphr	5.0 g/bhphr	—	0.25 g/bhphr	
1993	HDDV (bus)	1.3 g/bhphr	15.5 g/bhphr	5.0 g/bhphr	—	0.1 g/bhphr	
1994[h]	LDV	0.25 g/mi	3.4 g/mi	0.4 g/mi	2 g/test	0.08 g/mi	27.5 mpg
[h]	LDT	0.32 g/mi	4.4 g/mi	0.7 g/mi	2 g/test	0.26 g/mi	
	HDDV	1.3 g/bhphr	15.5 g/bhphr	5.0 g/bhphr	—	0.1 g/bhphr	
	urban bus					0.05 g/bhphr	
1995[i]	LDT	0.32 g/mi	4.4 g/mi	0.7 g/mi	2 g/test	0.08 g/mi	
1998	HDDV	1.3 g/bhphr	15.5 g/bhphr	4.0 g/bhphr	—	0.1 g/bhphr	

Notes: LDV = light-duty passenger car, LDT = light-duty truck, HDGV = heavy-duty gasoline truck/bus, HDDV = heavy-duty diesel truck/bus, g/mi = grams per mile, g/bhphr = grams per brake horsepower hour, mpg = miles per gallon.

[a] Seven-mode steady-state driving procedure.
[b] Carbon trap technique.
[c] CVS-72 transient driving procedure.
[d] Thirteen-mode steady-state driving procedure.
[e] CVS-75 transient driving procedure.
[f] SHED technique.
[g] Transient driving procedure.
[h] Phased in from 1994 to 1996.
[i] PM phased in from 1995 to 1997.

use surveillance, the manufacturer can be required to recall the entire vehicle family and restore each vehicle to satisfactory operating condition at no cost to the owner.

Studies during and since the 1970s have revealed that large percentages of vehicles on roadways are exceeding the emission standards for which they were designed. This results primarily from inadequate maintenance by vehicle owners and, in some instances, from owners or their mechanics intentionally disabling emission control systems. Inspection and maintenance (I&M) and antitampering (ATP) programs have been instituted to identify such vehicles and to require their repair at the owners' expense. The 1977 Clean Air Act Amendments required state or local governments to administer I&M programs in all cities exceeding carbon monoxide or ozone air quality standards. A U.S. Environmental Protection Agency (EPA) 1988 tampering survey (7259 vehicles, 15 cities) suggested that 23% of vehicles not covered by I&M and ATP programs had emissions control system tampering, compared with 16% in areas with I&M and ATP (6). The 1990 Clean Air Act amendments require enhancements of programs to identify and repair inoperative vehicles (7).

The U.S. EPA periodically publishes National Air Pollutant Emission Estimates that assess the relative importance of transportation and other anthropogenic sources of air pollution (8). As suggested in California during the 1940s, transportation sources do contribute significantly to air pollution; in 1985, transportation sources contributed about 36% of volatile organic carbon (VOC) emissions, 72% of CO emissions, and 45% of NO_x emissions, according to estimates based on national annual average emission rates [calculated at the vehicle-miles-traveled (VMT) weighted annual average temperature of 14°C]. Figure 2 illustrates the trends in these emissions from 1940 to 1985. Mobile source emission rates are sensitive to ambient temperature, as are air pollution problems. As indicated in Figure 3, ozone air quality violations occur most frequently during the hot summer months, when automobile HC emission rates are elevated (relative to emissions at 14°C); and CO air quality violations occur most frequently during the cold winter months, when automobile CO emission rates are elevated (relative to emissions at 14°C) (9,10). Thus, the contribution of transportation sources to the atmospheric VOC burden is greater than 36% during periods when ozone is an air quality problem, and their contribution to the atmospheric CO burden is greater than 72% during periods when CO is an air quality problem.

Motor vehicles emission estimates generally are based on estimates of fleet average HC, CO, and NO_x emission rates (g/mi) obtained with the U.S. EPA computer model MOBILE and estimates of fleet VMT (11). MOBILE calculates emission rates for eight individual vehicle types in two altitude ranges of the United States. The emission estimates depend on various conditions such as ambient temperature, speed, and vehicle model year mileage accrual rates. The computer model permits user specification of many of the variables affecting average vehicle emission rates. The model can estimate emission rates for any calendar year between 1960 and 2020. The 25 most recent vehicle model years are considered to be in operation in each calendar year.

Recent roadway studies have suggested that MOBILE may underestimate the HC and CO emission rates of actual roadway motor vehicles (12). The computer model estimates are based on motor vehicle emissions data obtained with laboratory dynamometer simulations of roadway driving conditions. The dynamometer simulations may not represent actual roadway conditions, and the vehicles tested may not represent actual roadway motor vehicles. The laboratory dynamometer driving simulations used to estimate emission rates were developed during the late 1960s and early 1970s and may not adequately represent today's driving patterns (13), and poorly maintained or tampered-with vehicles may not be appropriately represented. Current research efforts are examining both the characteristics of roadway vehicles (eg, their emissions control performance) and the adequacy of laboratory simulations for determining both tailpipe and evaporative emission rates.

MOTOR VEHICLE EMISSIONS CONTROL

Beginning in the early 1950s, Americans tended to move from cities to surrounding suburbs. Because the majority of the suburbs were not served by mass transit, this population shift created an attractive market for automobile manufacturers. The number of vehicles produced and sold increased dramatically, as did construction of highways between cities and suburbs. Families began to own more than one vehicle, and it became more common for each family member to have a vehicle. As vehicle ownership and use increased, so did the pollutant levels in and around cities, as the suburbanites commuted daily between their businesses and employment in the city and

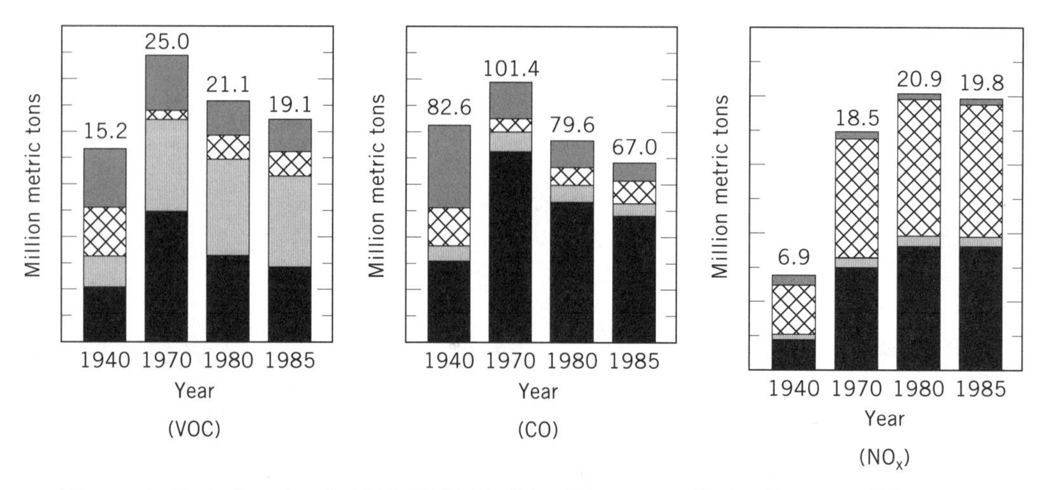

Figure 2. Emissions trend, 1940–1985 (8); (■) solid waste and miscellaneous; (⊠) stationary combustion; (⊡) industrial processes; (■) transportation.

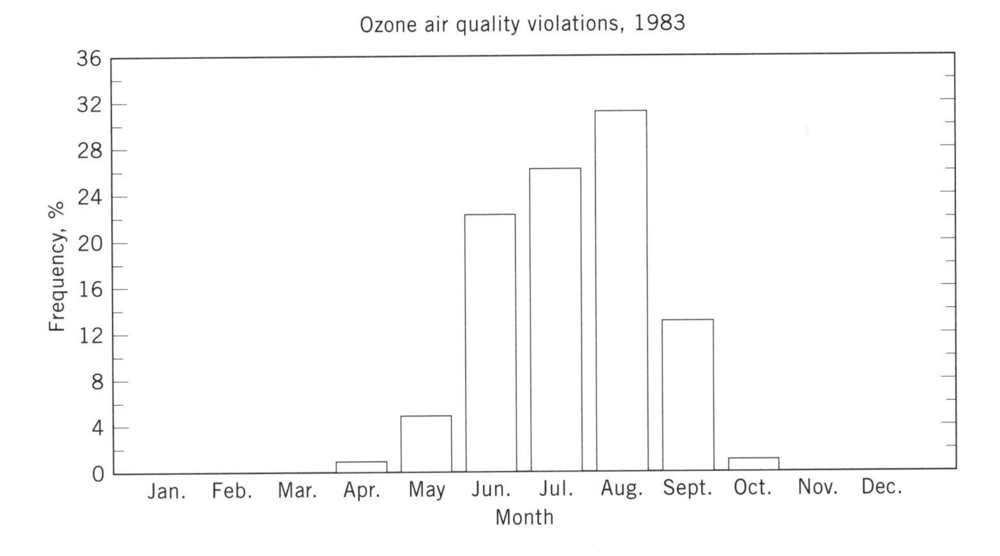

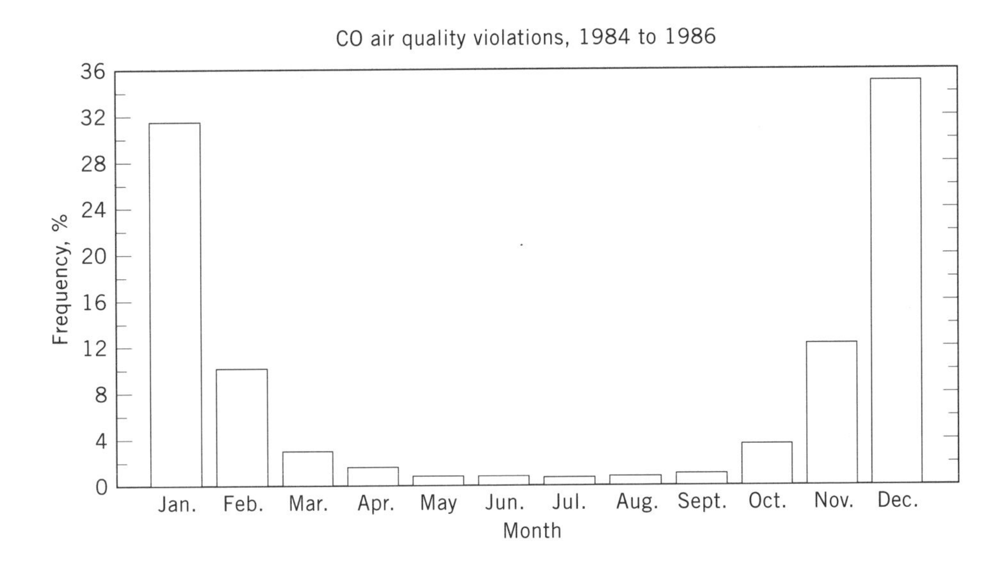

Figure 3. Monthly frequency of ozone and carbon monoxide air quality violations.

their homes in the suburbs (3). As indicated in Figure 4, the automobile remains the favored form of transportation in the United States (14). If the current rate of VMT growth continues, the contribution of motor vehicle emissions to air pollution is projected to bottom out in the mid to late 1990s and begin to grow in magnitude, unless further emission control improvements are realized (15). Changes in both vehicle technology and fuel formulations are expected to be used to achieve further emissions reductions.

Efforts by the U.S. government and motor vehicle industries to improve air quality in U.S. cities have resulted in a dramatic reduction of emission rates from roadway motor vehicles. Estimates obtained with MOBILE 4 suggest that from 1975 to 1990, fleet average nonmethane hydrocarbon (NMHC) and CO emission rates were reduced by about 70% and NO_x emission rates by about 55% (11), as shown in Figure 5. During that same period, however, VMT increased about 45%, substantially offsetting the emission rate reductions. Most of the emission reductions achieved to date have resulted from the changes in vehicle technology responding to new federal motor vehicle emission standards. The emissions certification practice involves measurement of the emission rates of regulated compounds (or groups of compounds) with laboratory simulations of roadway driving conditions. Figure 6 schematically illustrates a typical light-duty chassis dynamometer test cell and a speed-versus-time display of the CVS-75 driving schedule used for laboratory simulation of roadway driving conditions. Figure 7 provides a flowchart overview of light-duty motor vehicle tailpipe and evaporative emissions certification test procedures. The heavy-duty vehicle certification procedure involves a transient engine dynamometer test wherein engine speed and load are varied according to a prescribed schedule. The engine dynamometer test is used for heavy-duty vehicles because in the United States heavy-duty truck engines and chassis often are manufactured by different companies. The adequacy of

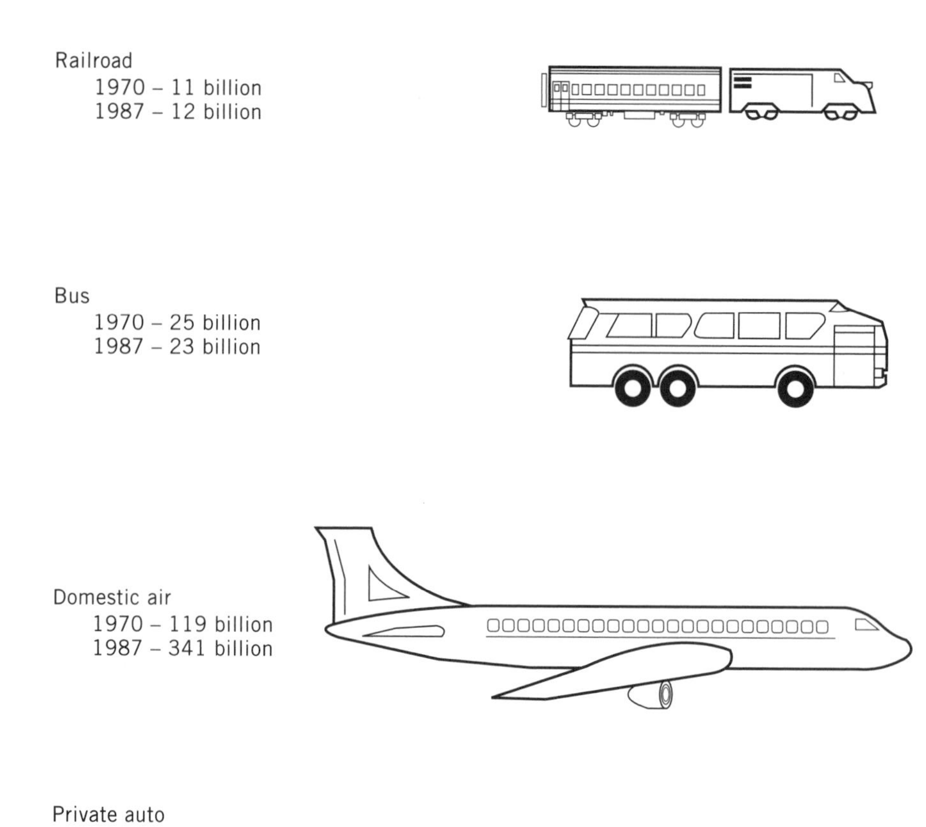

Figure 4. How Americans travel (passenger miles).

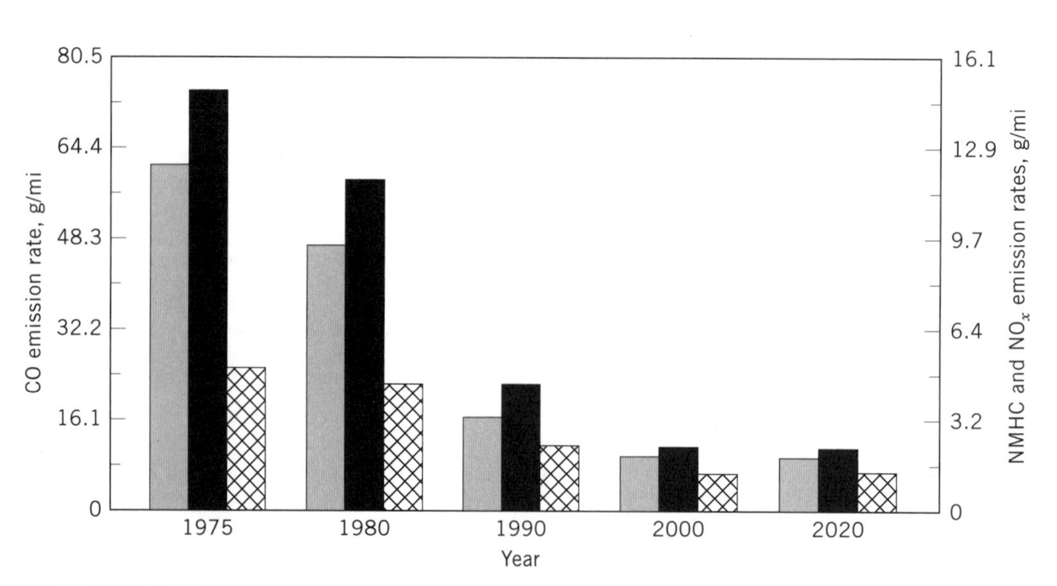

Figure 5. Fleet average NMHC, CO, NO_x, emission rates—1975–2020 (11); (⚃) NMHC; (■) CO; (⊠) NO_x. 1975–1990 reductions: NMHC, 73%; CO, 71%; NO_x, 55%. Note: 16–29°C diurnal range; 79.3 × 10^3 Pa (11.5 psi), 72.4 × 10^3 Pa (10.5 psi) (89) RVP fuel; 31.5 km/h (19.6 mph); I&M, antitampering.

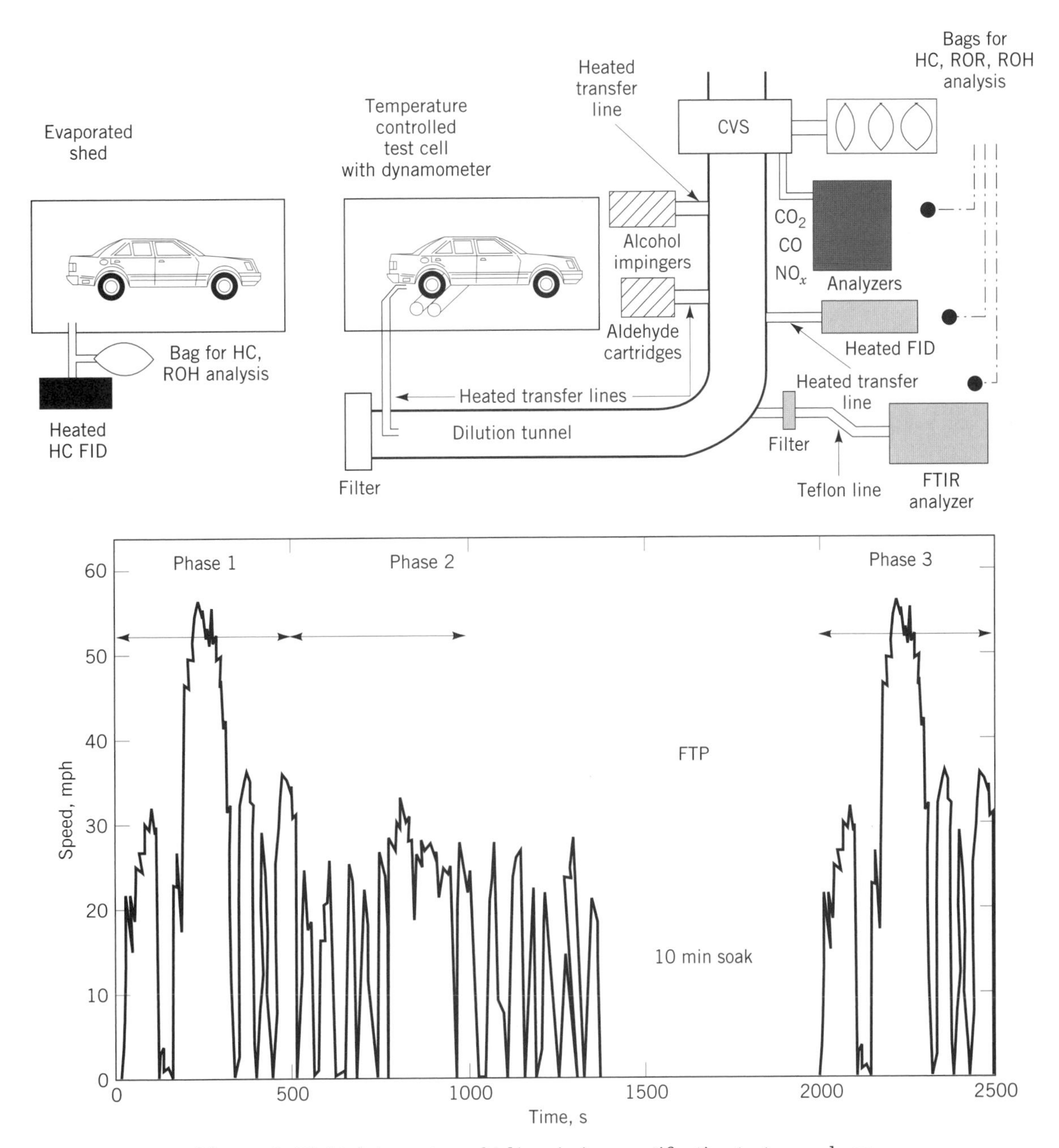

Figure 6. Light-duty motor vehicle emissions certification test procedures.

current practice for laboratory simulation of roadway driving conditions is being examined, and both tailpipe and evaporative emission test procedures may be modified in the near future.

The three major categories of emissions from motor vehicles without emissions control are crankcase ventilation, fuel evaporation, and engine exhaust. Control technologies have been developed to reduce emission rates from each of these sources.

Crankcase Emissions

Crankcase emissions, responsible for approximately 20% of all harmful automotive pollutants before emission controls, result from compression gases being forced past the piston rings on both the compression and power strokes, resulting in an accumulation of "blowby" gases in the crankcase. These gases mix with vapors from the agitated lubricating oil and must be vented from the crankcase area to prevent damaging pressures from developing. Before the 1960s, a road draft tube was used to ventilate the crankcase, emitting the pollutants into the atmosphere. By 1968, all vehicles manufactured in the United States were equipped with a closed crankcase ventilation system that prevented the escape of blowby gases and oil vapor to the atmosphere. Figure 8 is a schematic diagram of a typical positive crankcase ventilation (PCV) system. During idle and cruise periods, the emissions are vented to the intake manifold where they are mixed with the cylinder fuel–air charge for subsequent combustion. In the closed crankcase ventilation system, the fresh air intake, located in the air cleaner, has a dual role. It not only provides

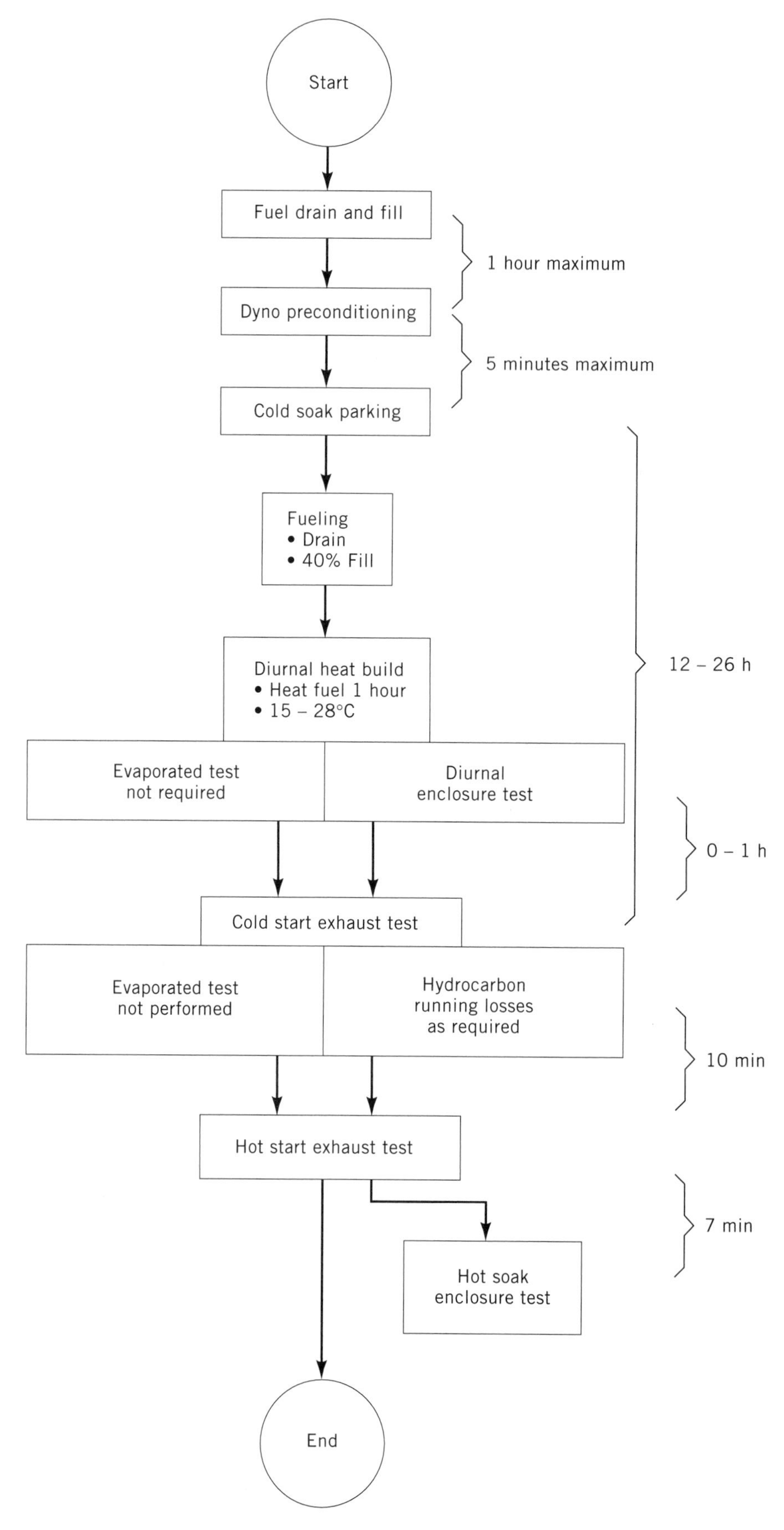

Figure 7. Light-duty motor vehicle emissions certification test sequence.

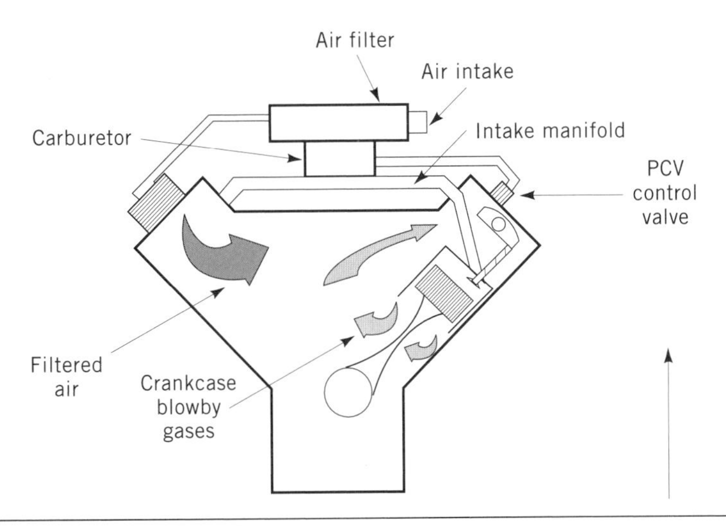

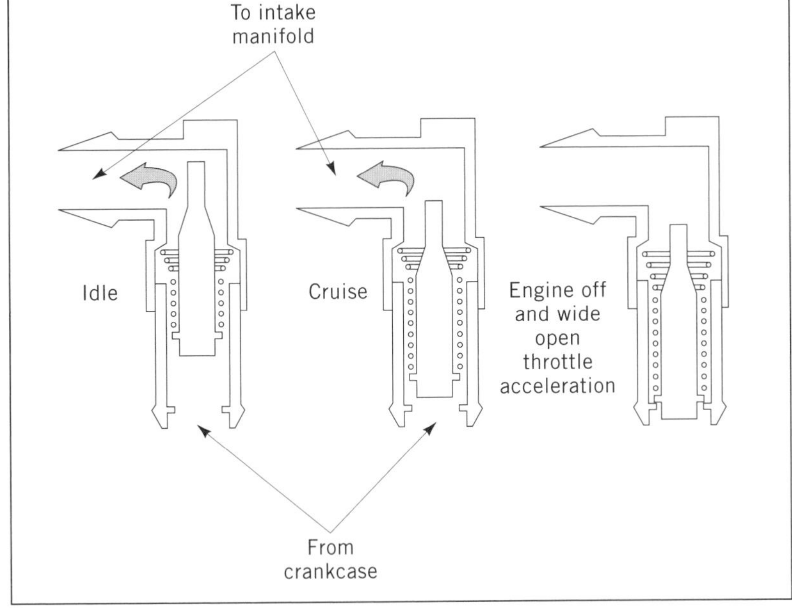

Figure 8. Crankcase emissions control (3).

fresh air for crankcase ventilation, but it also allows overload release of blowby gases into the carburetor airstream should the PCV valve fail to control the buildup of blowby gases; this prevents excess gases from escaping into the atmosphere. With this system functioning properly, crankcase emissions are essentially eliminated as a source of air pollution from motor vehicles (3).

Evaporative Emissions

Annual average fuel evaporation was responsible for about 20% of harmful automobile pollutants prior to emissions control, the severity increasing with elevated ambient temperature and fuel volatility. The primary sources of evaporative hydrocarbon emissions were the fuel tank and carburetor bowl, both of which were vented to the atmosphere. Another problem was gasoline spillage when fuel tanks were overfilled. The evaporative emissions control system is designed to prevent gasoline leakage and to trap vapors. Figure 9 is a schematic diagram of a typical system. The key design features include a sealed fuel tank cap, a fuel tank fill limiter, a liquid–vapor separator, and a carbon canister. The vented fuel tank filler cap is replaced by a sealed cap with the fuel tank vented through a carbon canister designed to store the fuel vapors. A pressure/vacuum relief valve is incorporated into the sealed cap to prevent damage to the tank should excessive internal or external pressures exist. The fuel tank has been redesigned to provide approximately 10% of the total tank volume for expansion space should the fuel expand because of temperature changes. Overfill protection is provided by a filler neck that assures the expansion space is maintained when the tank is filled. Tank vapor ventilation is controlled by having a vent tap at or near the expansion chamber dome. A foam-type filter or a vapor separator allows the vapors to pass to a storage point, while preventing the passage of liquid which is returned to the fuel tank. For most vehicles, the liquid–vapor separator assembly is located within the vent line. Vehicles may be equipped with separate expansion chambers, with sepa-

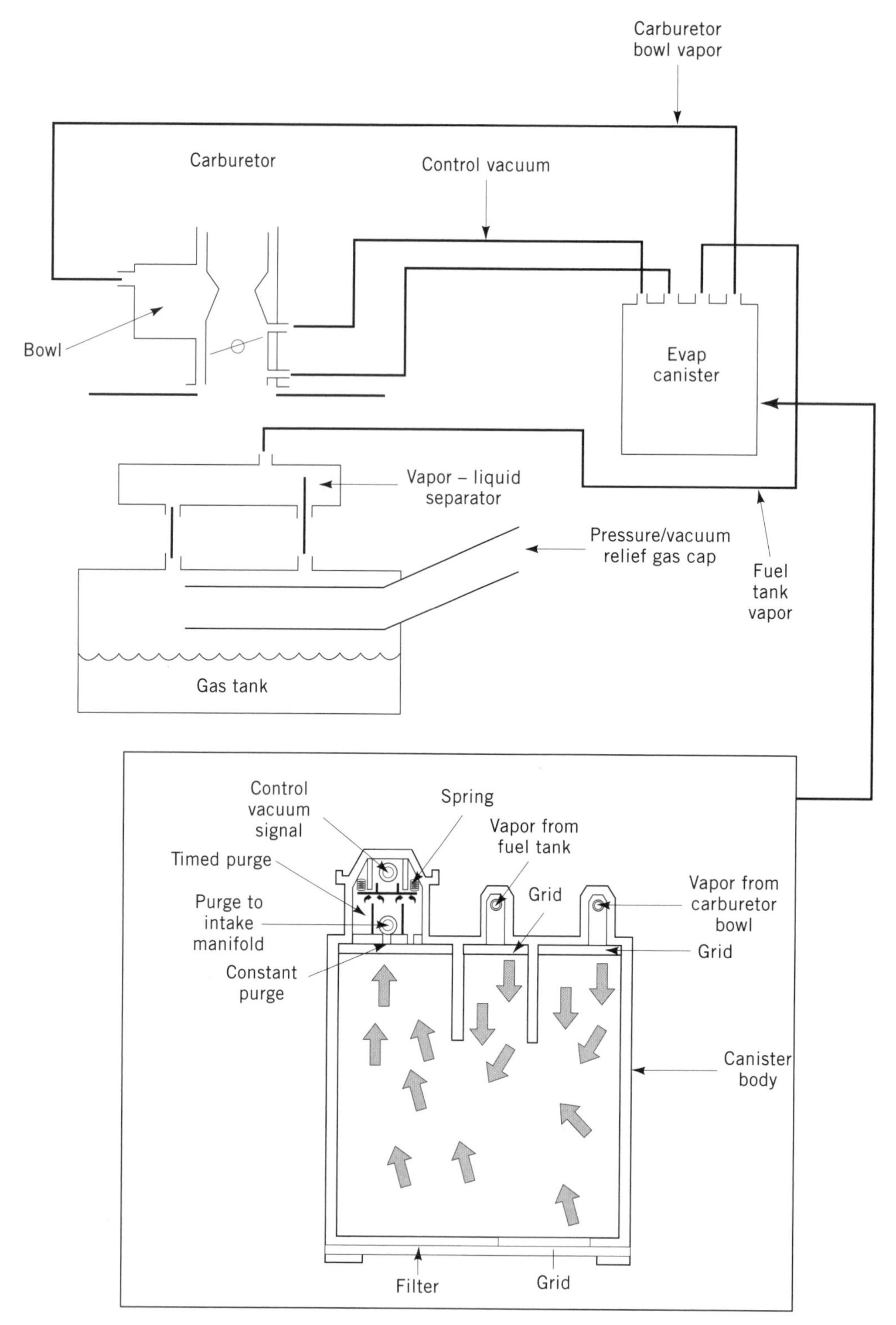

Figure 9. Evaporative emissions control (3).

rate evaporation chambers, and with one or more check valves in the lines. These components usually are located near the fuel tank.

An important component of the fuel evaporative control system is the canister of activated charcoal (porous carbon). Activated charcoal can absorb and store fuel vapors, and it can be cleaned by purging with fresh air, allowing it to be reused many times. When the engine is not running, the fuel vapors are routed to the canister, where they are absorbed and stored on the activated charcoal. When the engine is started, either engine vacuum or the airflow through the carburetor air cleaner causes a metered amount of air to pass over and through the charcoal, purging the vapors. The purge air is routed into the engine's induction system. Various types of purge control systems have been used. The evaporative emissions control system schematically illustrated in Figure 9 uses a combination of constant and demand purging. Constant purging is provided by a continuous restricted flow through the canister to the intake manifold, whereas the demand purging is provided by flow only when the throttle plate is open (not during idle operation) (3).

Another category of evaporative emissions receiving recent attention is "running-loss" emissions, which occur when the vapor generated while the vehicle is being operated on the roadway exceeds the system's canister storage and purging capacity. This category of emissions can be significant during hot summer days, when ozone (O_3) problems prevail.

Fuel volatility, often expressed in terms of Reid vapor pressure (vapor pressure at 38°C; RVP), strongly affects the magnitude of motor vehicle evaporative emissions. Higher evaporative emission rates are associated with higher gasoline volatilities. Historically, the volatility of marketed gasolines has increased as vehicle designs have tolerated more volatile fuels (i.e., have become more resistant to fuel delivery system vapor lock). National average summer gasoline volatilities increased 14% from 1974 [65.5×10^3 Pa (9.5 psi) RVP] to 1985 [74.5×10^3 Pa (10.8 psi) RVP]; the increase varied regionally from 10 to 20% (16). The volatility of federal emissions certification gasoline is 60×10^3–63.4×10^3 Pa (8.7–9.2 psi) RVP. Because of the disparity between certification and marketed gasoline volatilities, certified evaporative emissions control systems performed inadequately under roadway conditions. Since 1989, the U.S. EPA has regulated marketed gasoline volatility (9). Summer (May to September) gasoline volatilities currently vary regionally from 53.8×10^3 to 62.1×10^3 Pa (7.8–9.0 psi) RVP (17).

Engine Exhaust Emissions

Engine exhaust was estimated to be responsible for about 60% of pollution from automobiles prior to emissions control. The technology that has evolved to reduce engine exhaust emissions is complex and varies from vehicle to vehicle. The control system may include many different elements or subsystems, such as the following:

- thermostatically controlled air cleaner,
- air injection systems,
- ignition timing controls,
- increased temperature control,
- transmission or speed-controlled spark system,
- exhaust gas recirculation system,
- catalytic converter system,
- engine feedback systems,
- carburetor and choke modifications,
- fuel injection (throttle body and port) systems, and
- computer-based engine controls.

Figure 10 is a schematic diagram of several elements of a typical exhaust emissions control system. A key factor in simultaneous reduction of emissions of HC, CO, and NO_x is accurate control of the air–fuel ratio; many of the control system components serve that purpose. A common practice is to operate the engine slightly fuel rich (ie, at an air–fuel ratio lower than the stoichiometric ratio), optimizing NO_x conversion in the front half of the catalyst, and then to inject air midway, optimizing conversion of excess HC and CO in the rear half of the catalyst, as illustrated in Figure 11. Figure 12 shows the impact of the air–fuel ratio on engine exhaust emissions and on catalyst conversion efficiency. This system is commonly referred to as a

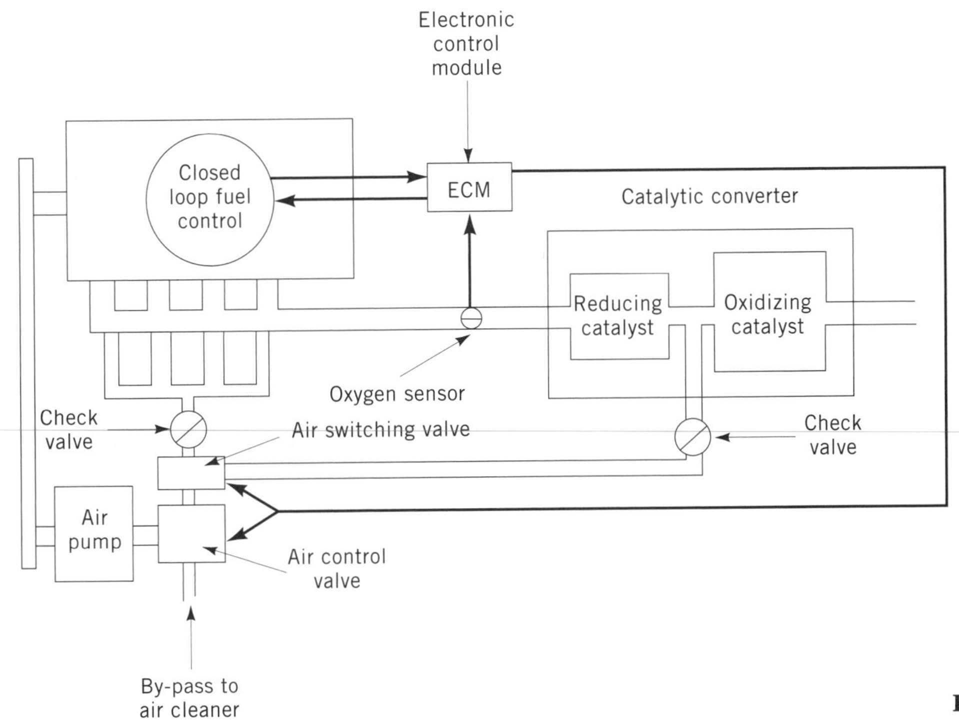

Figure 10. Tailpipe emissions control (3).

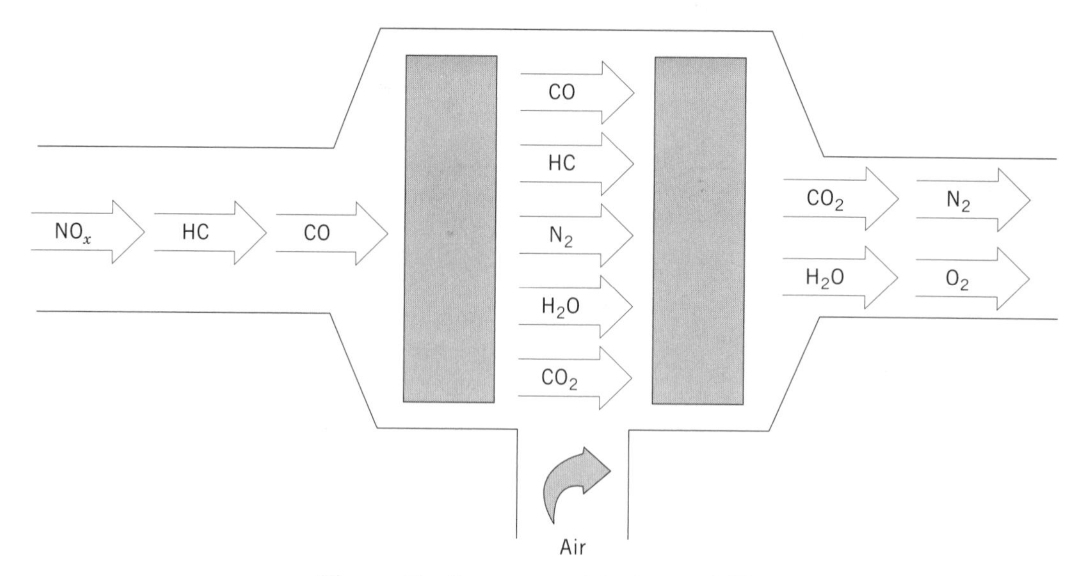

Figure 11. Three-way catalyst control (3).

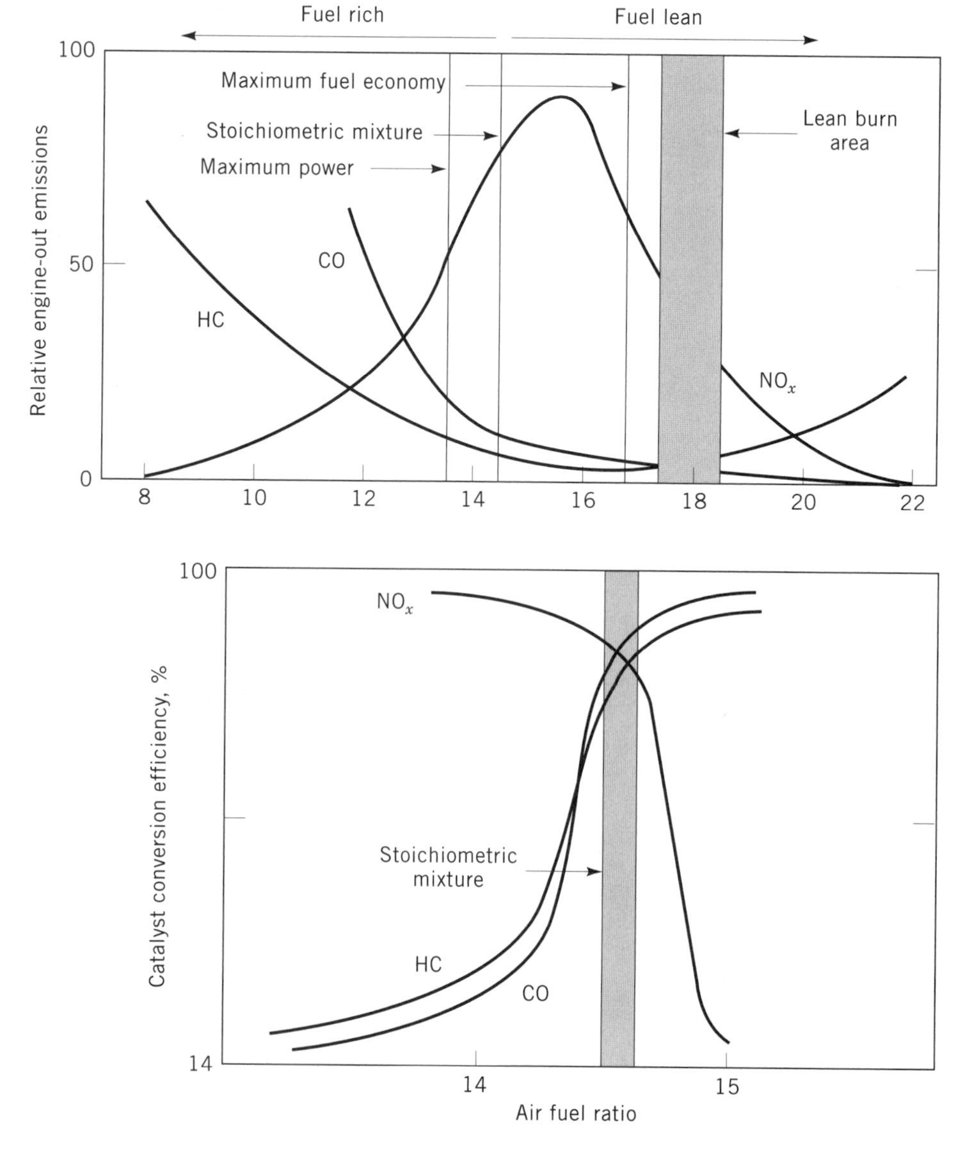

Figure 12. Emissions as a function of air–fuel ratio (18,19).

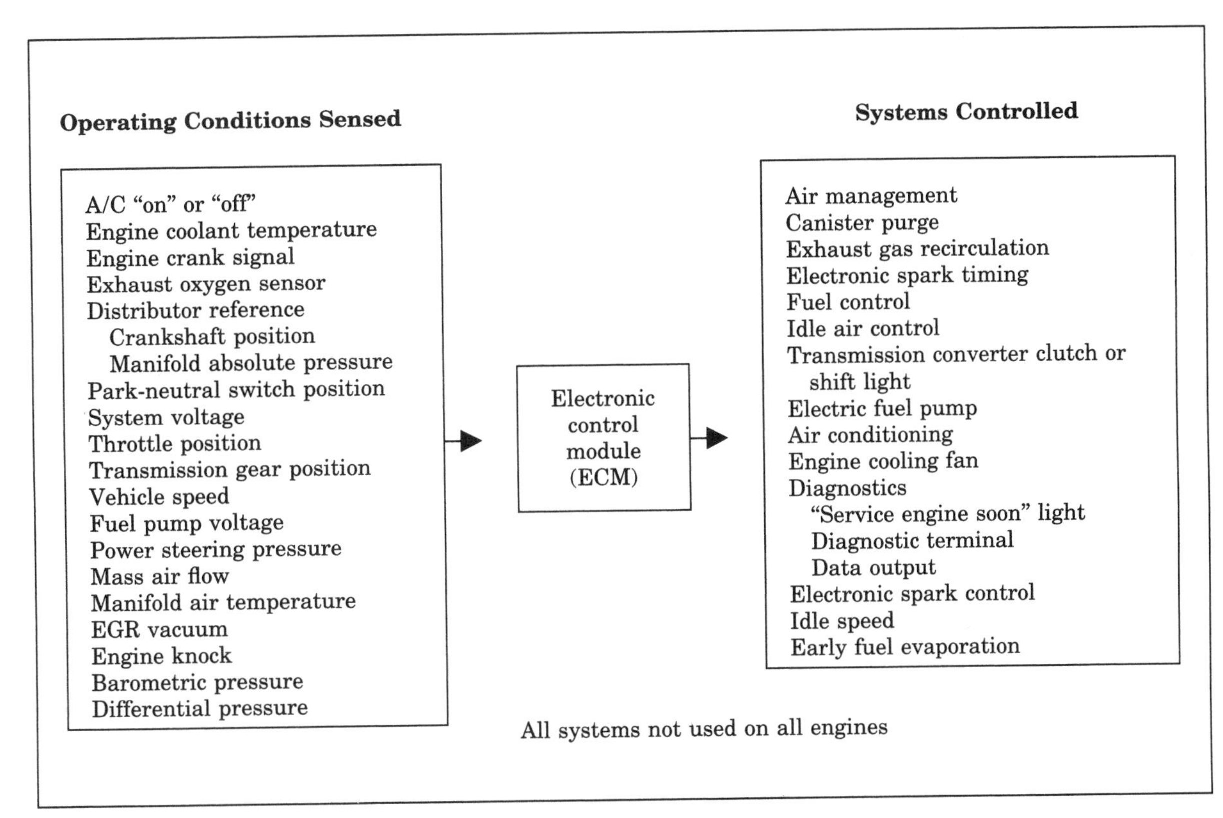

Figure 13. Electronic emission control (3).

"three-way" catalyst, because of its combined reduction of HC, CO, and NO_x. Fuel injection and air injection are controlled by an electronic control module (ECM) in association with an exhaust oxygen sensor. The ECM is the keystone of the entire emissions control system, acquiring information from numerous sensors and controlling numerous elements and devices of the engine and the emissions control system, as summarized in Figure 13 (3).

In addition to catalytic reduction, exhaust gas recirculation (EGR) often is used to reduce NO_x emissions. Because formation of NO_x is sensitive to combustion-zone peak temperature, systems that lower the peak temperature will reduce NO_x emissions. Peak temperature can be reduced by recirculating some exhaust gas into the intake air–fuel mixture. Figure 14 schematically illustrates a typical EGR system. The amount of recirculated exhaust gas must be

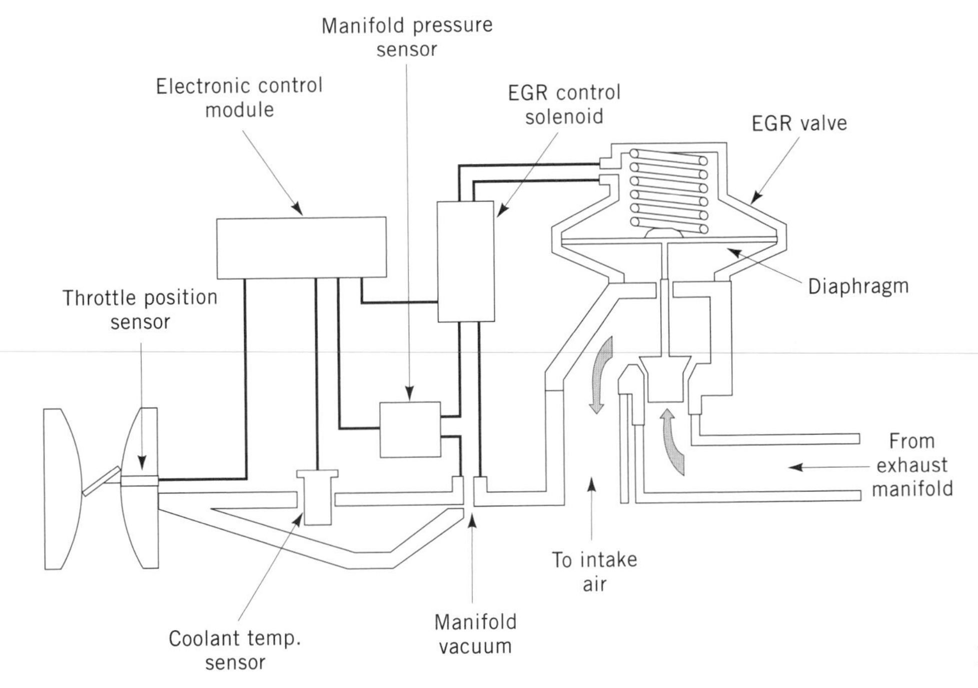

Figure 14. Exhaust gas recirculation (3).

carefully controlled. If too much exhaust gas is supplied at the wrong time (which may occur if the EGR valve sticks in the open position), the engine may stall or idle rough. If not enough exhaust gas is recirculated (which may occur if the vacuum activation system fails), NO_x will not be reduced (3).

Tampering

Emission control technology can significantly reduce the contribution of motor vehicles to atmospheric pollution, but only if the equipment is properly maintained. During 1988, the emissions control systems of 23% of U.S. vehicles not covered by I&M and ATP were purposely rendered inoperative, compared with 16% of vehicles in areas with I&M and ATP (6). As shown in Figure 15, of 7259 vehicles surveyed, 19% were clearly tampered with and an additional 12% may have been tampered with (i.e., components became inoperative either through tampering or as a result of poor maintenance). Table 2 shows the frequency of tampering with various components. Tampering has serious implications for emissions because a disconnected air pump can increase HC emissions 200% and CO emissions 800%, a disconnected EGR system can increase NO_x emissions 175 %, a missing or damaged catalytic converter can increase HC emissions 475% and CO emissions 425%, and a disabled oxygen sensor can increase HC emissions 445% and CO emissions 1242% (20).

Inspection and maintenance and ATP reduce tampering somewhat, but rates continue to be excessive, and the U.S. Congress has amended the Clean Air Act to address this problem. Measures for improved, more effective identification of vehicles requiring repair include enhanced I&M programs providing more comprehensive and accurate evaluation of emission control system performance, onboard diagnostic equipment providing the operator with continuous information on the performance of the vehicle's emission control systems, and enforcement programs using remote monitoring of emissions from vehicles on roadways.

Table 2. Control System Tampering

Component/System	Tampering Rate (%)
Catalytic converter	5
Filler neck restrictor	6
Air pump system	11
Air pump belt	8
Air pump/valve	7
Aspirator	2
Positive crankcase ventilation	6
Evaporative control system	6
Exhaust gas recirculation	7
Heated air intake	3
Oxygen sensor and computer system	1

Note: From ref. 6. Vehicles with aspirated air systems are not equipped with other listed air injection components and conventional systems do not include aspirators.

MOTOR VEHICLE EMISSIONS CHARACTERISTICS

Certification procedures, emissions regulations, and emissions characteristics are different for light-duty [less than 3856 kg (8500 lb) GVW] and heavy-duty (3856 kg GVW or more) motor vehicles. Light-duty gasoline passenger cars and trucks are responsible for more than 90% of the motor vehicle NMHC and CO emissions in the United States and about 67% of motor vehicle NO_x emissions, and heavy-duty trucks and buses are responsible for more than 90% of fine particulate emissions (8). This distribution is due in part to differences in emissions characteristics (most light-duty vehicles have catalyst-equipped gasoline engines, and most heavy-duty vehicles have diesel engines) and in part to the distribution of VMT. In 1989, light-duty gasoline-fueled vehicles accounted for about 92% motor vehicle miles traveled and heavy-duty diesel-fueled vehicles for about 4% (11). Therefore, efforts to characterize motor vehicle NMHC and CO emissions have focused on light-duty gasoline vehicles, particulate matter emissions on heavy-duty diesel vehicles, and NO_x emissions on both categories of vehicles.

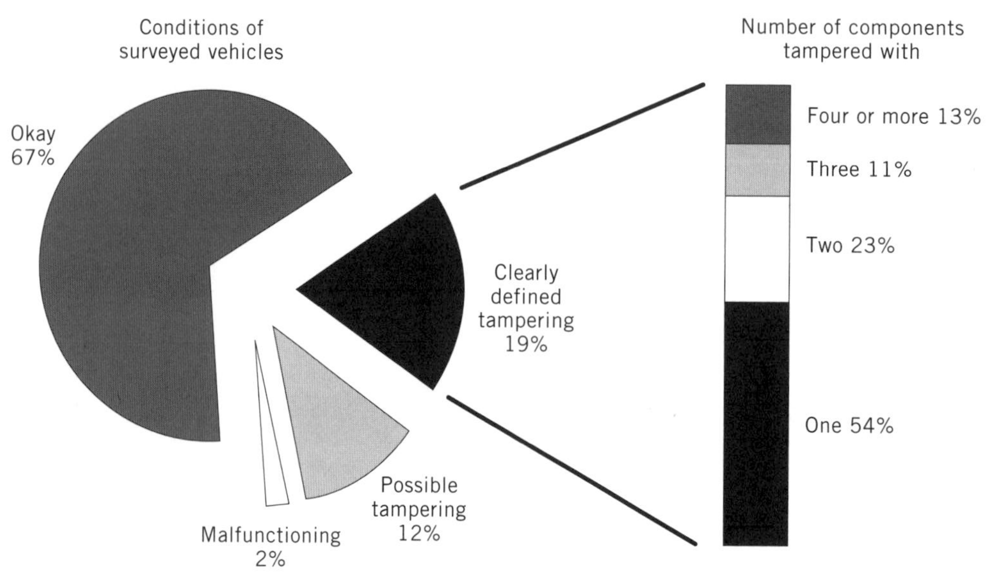

Figure 15. Frequency of emissions control system tampering (6).

Motor vehicle emissions (especially those of gasoline-fueled vehicles) are sensitive to changes in operating conditions such as ambient temperature, average speed, altitude, and age (because of malfunctions). Figure 16 shows how NMHC, CO, and NO_x emissions vary with ambient temperature and speed, as estimated by MOBILE 4 (11). Federal emissions certification practice involves a narrow window within this range of conditions: 31.5 km/h (19.6 mi/h) average speed and 21.1°C average temperature, with a diurnal cycle of 15.6°–28.9°C. Hydrocarbons are emitted

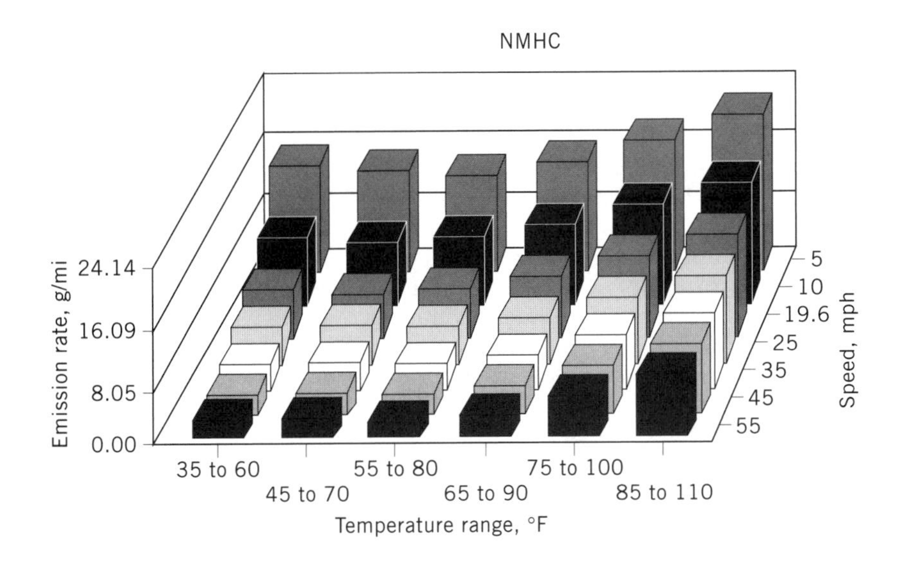

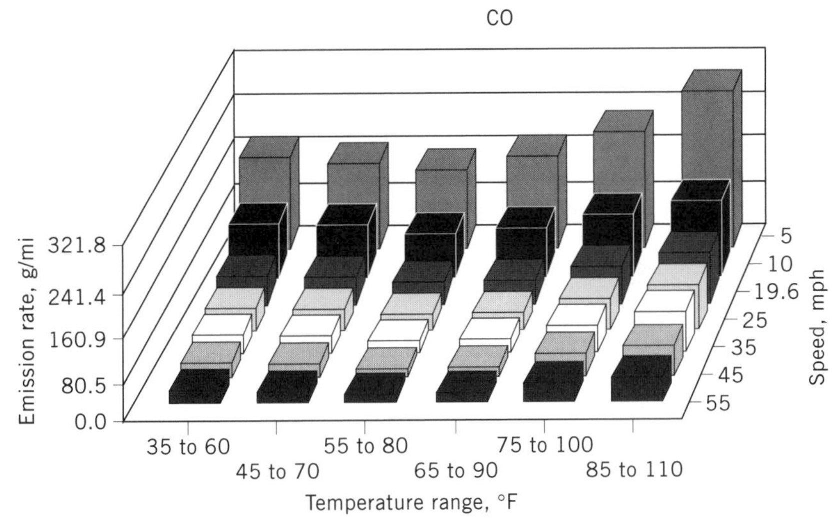

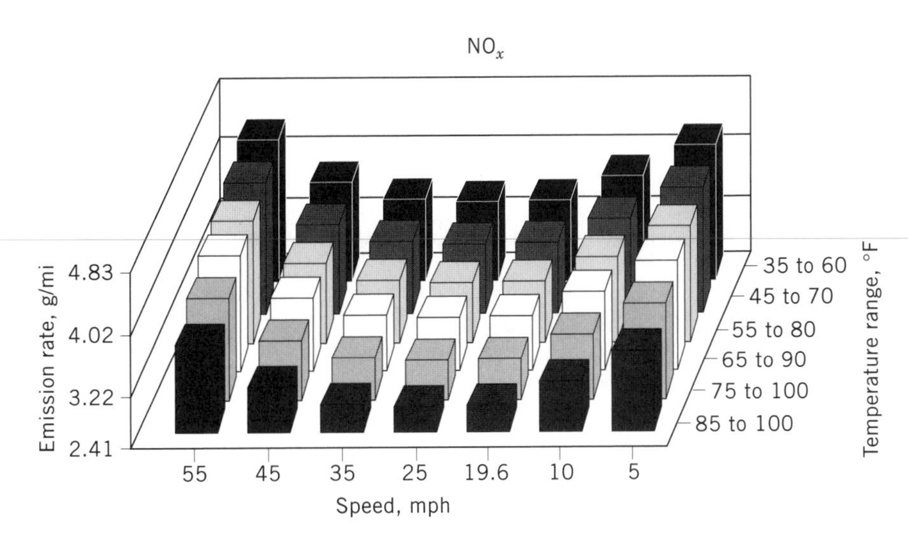

Figure 16. Motor vehicle emissions as a function of speed and ambient temperature (11).

from the tailpipe and from evaporative sources (running losses) when vehicles are operated on roadways, from evaporative sources (diurnal and hot soak) when vehicles are parked, and from the gasoline tanks when vehicles are refueled. Figure 17 illustrates the relative contribution of each of the sources at various average vehicle speeds and ambient temperatures for 11.5- and 9.0-RVP gasoline volatilities. The relative contribution of evaporative sources is much greater at higher temperatures. It should be noted that in the MOBILE 4 model, vehicle evaporative emissions are not affected by changes in vehicle average speed, nor are refueling emissions affected by fuel volatility or changes in vehicle average speed. Available data are not adequate to define the mathematical algorithms necessary to incorporate these variables in the model. Since the control technology for evaporative emissions involves a regenerative carbon canister that generally is purged more rapidly at higher average speeds and engine loads, one would expect the emission rates to vary with speed.

Historically, the primary incentive for regulation of motor vehicle emissions has been to improve air quality by reducing O_3, CO, and particulate matter concentrations. The impact of motor vehicle emissions on atmospheric O_3 is sensitive to many variables, including the relative importance of other O_3 precursor sources, the organic emissions composition (and thus their reactivity), and local HC–NO_x ratios. The relative contributions of motor vehicles and other sources (anthropogenic and biogenic) vary from one region to another, depending on ambient temperature, traffic congestion (average speed), and many other manageable and unmanageable variables. The emissions composition depends on many of these same variables, the composition of the fuel, and the characteristics of the vehicle fleet (e.g., the age of the vehicles, distribution of trucks, buses, and cars). Table 3 illustrates how U.S. commercial gasoline and diesel fuels can vary in composition. As indicated in Table 4, emissions from tailpipe, evaporative, and refueling sources are expected to vary in composition; the aggregate HC emissions composition for a vehicle depends on the relative contribution of each category, which varies with speed, temperature, and other factors (22). Since HC and NO_x emission rates differ in sensitivity to operating variables (as shown in Fig. 16), the HC–NO_x ratio of emissions also varies with operating conditions, as shown in Figure 18.

In addition to O_3 precursor, CO, and particulate matter emissions, recent attention has been directed to toxic automobile emissions such as benzene, 1,3-butadiene, formaldehyde, and gasoline vapors. The emission rates of these hazardous substances are sensitive to many of the variables discussed above. For example, benzene emission rates are sensitive to fuel benzene level and to higher molecular weight fuel aromatics (eg, toluene and xylenes), which can be a source of tailpipe benzene through the dealkylation processes that occur during combustion (23–25). The limited data available suggest that tailpipe benzene emission rates increase as temperature decreases but that the benzene fraction of THC emissions remains relatively constant (26). Few data are available relating the composition of evaporative emissions to temperature; however, distillation theory suggests that aromatics constitute a larger fraction of the emissions at higher temperatures.

Estimates of cancer risk from motor vehicle emissions are presented in Table 5 (27). The largest potential risk is associated with diesel particulate emissions, which have been the subject of much study. The U.S. EPA is scheduled to publish a comprehensive diesel emissions hazard assessment during 1994. Diesel particles include elemental and organic carbon, sulfate, and small amounts of trace metals (28,29). The cancer risk from these particles is

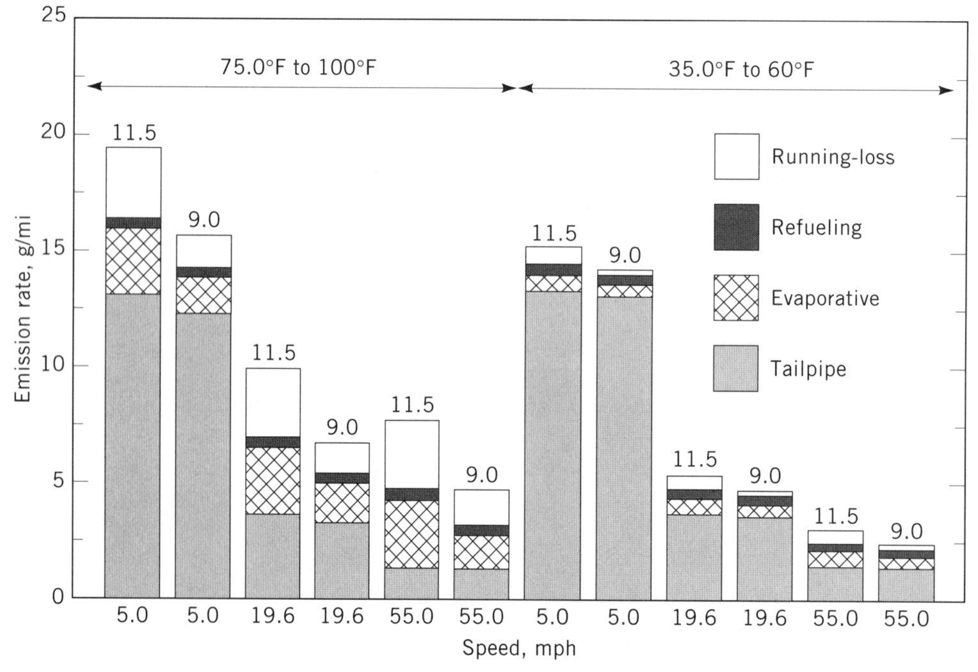

Figure 17. NMHC motor vehicle emissions distribution (1985): tailpipe, evaporative, refueling, and running losses (11).

Table 3. National Fuel Composition Survey: Summer 1989[a]

	Aromatic (%)	Olefins (%)	Saturates (%)	MTBE (%)	RVP (psi)	Sulfur (wt %)	Octane, (R + M)/2	Cetane No.
Premium unleaded								
Maximum	67.0	17.8	82.5	10.3	10.9	0.071	97.3	
Minimum	11.0	1.1	30.6	<0.1	5.9	<0.001	87.0	
Average	37.1	6.2	56.6	<1.9	8.9	<0.016	92.3	
Intermediate								
Maximum	48.6	23.1	66.6	8.3	10.3	0.111	91.1	
Minimum	25.1	2.9	37.6	0.1	8.2	0.005	87.2	
Average	32.5	11.4	55.9	<1.6	9.1	0.036	89.3	
Regular unleaded								
Maximum	42.2	42.8	73.2	5.4	10.9	0.187	94.9	
Minimum	17.7	1.0	33.9	<0.1	7.7	<0.001	84.4	
Average	29.9	11.8	58.1	<0.2	8.9	<0.037	87.3	
No. 2 diesel								
Maximum	43.3	2.3	86.1	—	—	0.60	—	56.4
Minimum	13.3	0.4	55.3	—	—	0.03	—	39.2
Average	32.7	0.9	66.2	—	—	0.32	—	45.4

[a] From ref. 21. MTBE = methyltertiarybutyl ether; RVP = Reid vapor pressure.
[b] Research octane + motor octane)/2.

associated primarily with adsorbed and condensed organic substances; nitropolycyclic aromatic hydrocarbon compounds are an important subgroup (30). Technology-forcing particulate emissions standards have resulted in improved engine designs and fuel formulations (31–33). Diesel fuel sulfur content was reduced to 0.05 wt % in 1993 (the average was 0.32 wt % in the 1989 national fuel composition survey).

Motor vehicles also contribute to the atmospheric burden of the radiatively important trace gases that influence global climate (34,35). Motor vehicle engine emissions of greatest importance in this regard are CO_2, methane (CH_4), and nitrous oxide (N_2O). When emissions associated with fuel production, distribution, and use in motor vehicles are considered, CO_2 is responsible for about 77% of the global warming effect from motor vehicle fuel combustion with conventional gasoline fuels and about 91% of the global warming effect with conventional diesel fuels (34). Carbon dioxide emission rates are directly related to fuel economy, and fleet average fuel economy has improved in response to federal standards and consumer market demands, as shown in Figure 19. Table 6 presents CO_2 emission rates for 1975, 1985, and 1995 (22). From 1975 to 1985, fleet average CO_2 emission rates decreased about 22%, but VMT increased about 30%, thus increasing the atmospheric CO_2 burden from motor vehicle emissions.

Table 4. Tailpipe, Evaporative, and Refueling Emissions Hydrocarbon Composition

Organic Classification	Tailpipe Emissions	Evaporative Emissions	Refueling Emissions
Paraffinic, %	55	72	85
n-Butane, %	5	23	32
Isopentane, %	4	15	19
Olefinic, %	18	10	11
Aromatic, %	25	18	4
Acetylenic, %	2	0	0

Chlorofluorocarbon (CFC) emissions from automobile air conditioners also are important greenhouse gases. Amann estimated that vehicle lifetime difluordichloromethane (CFC-12) emissions could have a greater global climate impact than exhaust CO_2 emissions with gasoline fuel, assuming 27.5 mpg fuel economy and a 10,000 CFC-12 to CO_2 mass-based relative global warming effect (36). Replacements for CFC-12 that have less greenhouse effect are being evaluated, with 1,1,1,2-tetrafluoroethane (HFC-134a) a current leading candidate. The greenhouse effect of HFC-134a is estimated to be about 10% that of CFC-12 (36). Further, air conditioner servicing practices are changing to include capture and reuse of CFCs, rather than venting to the atmosphere.

Fleet average methane emission rates are estimated at about 0.1 g/mi by MOBILE 4 at 23°C, 9.0-RVP fuel, for 1989. Catalyst emission control systems do not reduce methane as effectively as other hydrocarbon compounds,

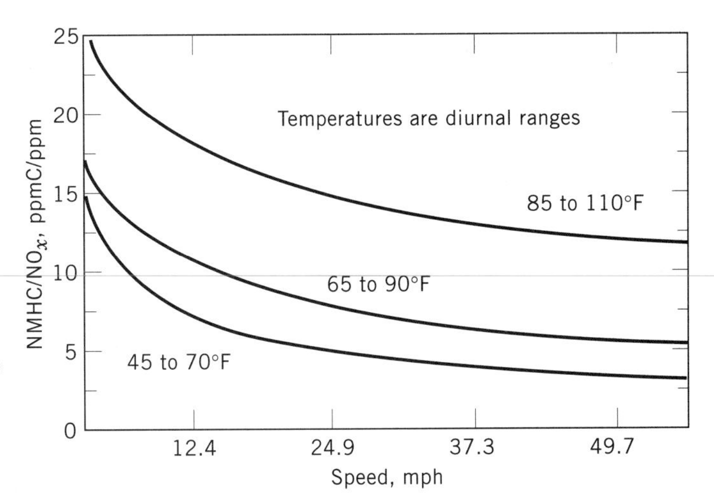

Figure 18. The variation of NMHC–NO_x ratio with speed and ambient temperature (11).

Table 5. Motor Vehicle Emissions Risk Estimates, Cancer Incidents Per Year[a]

	1986	1995
Diesel particles	178–860	106–662
Formaldehyde	46–86	24–43
Benzene	100–155	60–107
Gasoline vapors	17–68	24–95
1,3-Butadiene	236–269	139–172
Gasoline particles	1–176	1–156
Asbestos	5–33	Not determined
Cadmium	<1	<1
Ethylene dibromide	1	<1
Acetaldehyde	2	1
Total	586–1650	355–1236

[a] From ref. 27.

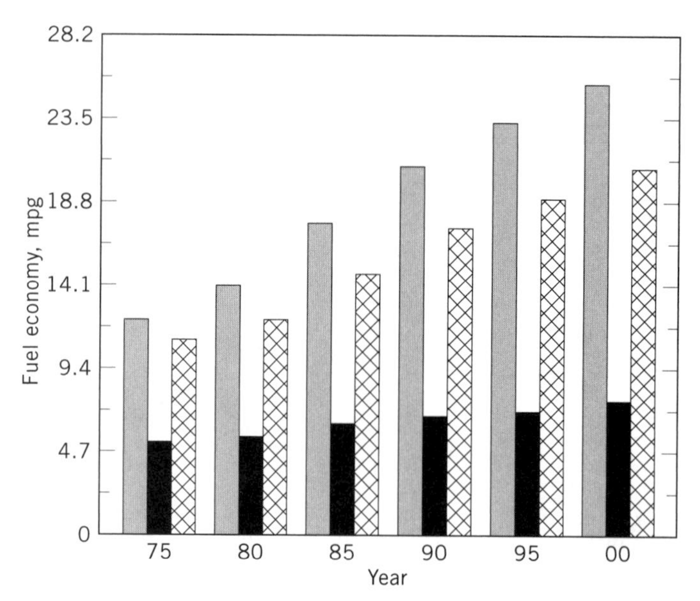

Figure 19. Fuel economy trend: (▤) light duty; (■) heavy duty; (⊠) all vehicles.

so methane emission rates have not been reduced as much as NMHC emission rates. Typically, methane constitutes as increasingly larger fraction of the THC emission rate as THC is reduced. In an examination of consumer passenger cars, methane accounted for 7.2% (0.33 g/mi) of 4.58 g/mi THC for 1975 model year cars and 24% (0.14 g/mi) of 0.57 g/mi THC for 1982 model year cars (37).

Few data on motor vehicle N_2O emissions are available. Table 7 provides an analysis of available data (22). The catalysts used to reduce other emissions may increase N_2O emission rates by about an order of magnitude.

FUTURE DIRECTIONS

Although significant progress has been made in reducing motor vehicle emission rates, more than 80 million Americans still live in areas exceeding U.S. national ambient air quality standards (38). As indicated in Figure 20, excessive levels of O_3, CO, and fine particulate matter are the prevalent air quality problems, each with significant contribution from motor vehicle emissions. A growing data base also suggests that motor vehicles contribute significantly to the population's exposure to other hazardous compounds such as benzene, formaldehyde, gasoline vapors, and 1,3-butadiene. For these reasons and others, the U.S. Congress has amended the Clean Air Act to reduce the impact of motor vehicles on risk to public health and welfare. In an effort to render the total transportation system as environmentally benign as possible, both motor vehicle emissions and fuel composition are being regulated.

In the 1990 Clean Air Act amendments, Congress required a number of changes to current mobile source emissions regulations. Target dates have been established for a two-phase reduction of tailpipe emission standards. Phase I will reduce light-duty motor vehicle emission rates to 0.25 g/mi for NMHC, to 0.40 g/mi for NO_x, and to 0.08 g/mi for particulate matter (from current standards of 0.4, 1.0, and 0.2 g/mi, respectively) by 1994. Phase II, if enacted (depends on air quality in 1997), will require further reductions of emission rates to 0.125 g/mi for NMHC, to 0.2 g/mi for NO_x, and to 1.7 g/mi for CO (from the current standard of 3.4 g/mi) by about 2003.

The amendments also require the use of "clean fuel" motor vehicles in centrally fueled fleets in the nation's most severe nonattainment areas. Additionally, a demonstration program in California will provide for the introduction of 150,000 clean fuel cars per year beginning in 1996 and 300,000 per year beginning in 1999. Management of the emission characteristics of the clean fuel vehicles will be similar to the California Low-Emission Vehicle and Clean Fuel Program (discussed below).

The amended Clean Air Act also calls for regulation of the composition of gasoline in specified nonattainment areas by limiting the volumetric fractions of benzene, requiring a specific amount of oxygen in the fuel (2% during ozone nonattainment periods and 2.7% during CO nonattainment periods, probably by addition of ethanol, methyl-*tert*-butyl ether, or ethyl-*tert*-butyl ether), and requiring reduction of volatility. By 1995, ozone-forming volatile organic compounds and toxic compounds (the aggregate of

Table 6. Motor Vehicle CO_2 Emission Rates (g/mi)

Year	LDGV	LDGT	HDGV	LDDV	LDDT	HDDV	Types
1975	567.0	636.8	988.6	455.8	328.6	2081.6	685.8
1985	434.4	532.3	856.8	379.9	436.0	1731.4	537.6
1995	336.0	458.2	893.5	321.0	405.5	1461.1	431.5

Note: LDGV = light-duty gasoline vehicle; LDGT = light-duty gasoline truck; HDGV = heavy-duty gasoline vehicle; LDDV = light-duty diesel vehicle; LDDT = light-duty diesel truck; HDDV = heavy-duty diesel truck.

Table 7. Motor Vehicle N_2O Emission Rates

Vehicle Category[a]	Emission Factor (mg/mi)	Emission Factor Range (mg/mi)
LDGV (no-catalyst)	6.0	3.1–15.9
LDGV (catalyst)	61.0	3.1–234.0
LDDV	—	10.9–48.0
HDGV	73.1	48.0–97.0
HDDV	47.0	31.1–74.0

Note: LDGV = light-duty gasoline vehicle; LDDV = light-duty diesel vehicle; HDGV = heavy-duty gasoline vehicle; HDDV = heavy-duty diesel vehicle.

benzene, 1,3-butadiene, formaldehyde, acetaldehyde, and polycyclic organic matter) will be reduced 15% by gasoline reformulation from the emissions characteristic of 1990 model year vehicles using national average unleaded gasoline, without concurrent increase in NO_x emissions. By 2000, the required reductions in volatile organic compounds and toxic compounds will be increased to 25%. The Auto/Oil Air Quality Improvement Research Program has demonstrated that fuel aromatic and olefinic hydrocarbon fractions, oxygen level, sulfur level, and distillation T_{90} are among the fuel specifications that can be used to manipulate motor vehicle emissions characteristics (39).

Additional standards reducing CO emissions at low ambient temperatures (−6.7°C) and controlling refueling emissions are required by the amended Clean Air Act. The amendments also include a variety of requirements addressing emissions from malfunctioning in-use vehicles, including on-board emission control system diagnostics, enhanced inspection and maintenance programs, and remote sensing of emissions from vehicles on roadways as an enforcement tool.

In addition to these federal programs, California approved new state vehicle and fuel regulations in September 1990, requiring even greater reduction of motor vehicle emissions (40). Compliance with California's more stringent emission standards can be achieved with advanced vehicle emission control technology, cleaner burning fuels, or a combination of the two (encouraging cooperation between the vehicle and fuel industries). The program requires a phased introduction of low-emission vehicles and the clean fuels used by those vehicles. The low-emission vehicles are defined according to four sets of emissions criteria, described in Table 8. The vehicle–fuel categories include transitional-low-emissions vehicles (TLEVs), low-emissions vehicles (LEVs), ultra-low-emissions vehicles (ULEVs), and zero-emissions vehicles (ZEVs). Beginning in the mid-1990s, vehicle manufacturers will be required to certify portions of their new motor vehicles to TLEV, LEV, ULEV, or ZEV standards. New California motor vehicles under 1701 kg (3750 lb) must certify to fleet average nonmethane organic gas (NMOG) emission standards of 0.25 g/mi by 1994 and 0.062 g/mi by 2003 (other standards apply to larger vehicles). The clean-fuel element of the regulation requires that the fuels used to certify the low-emissions vehicles be readily available for consumer use. Other states may opt to require California motor vehicles if considered necessary to achieve the air quality required by the 1990 Clean Air Act amendments.

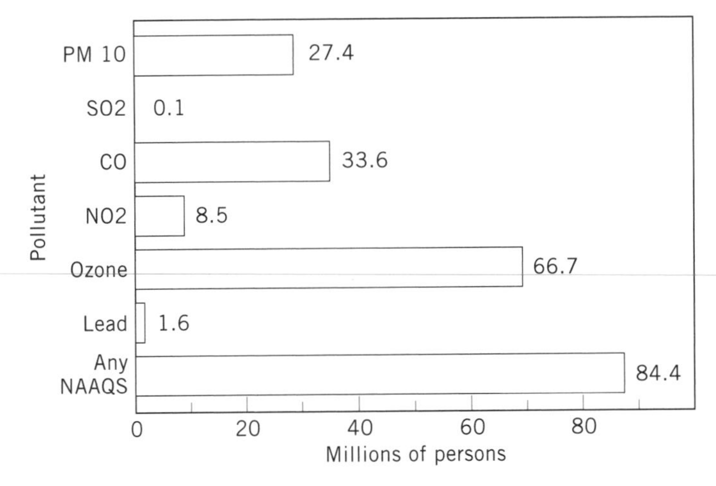

Figure 20. Persons living in counties with air quality violating NAAQS, 1989 (38).

Table 8. California Low-Emissions Motor Vehicle Standards

Category	NMOG[a] (g/mi)	CO (g/mi)	NO_x (g/mi)
Transition low-emission vehicle	0.125	3.4	0.4
Low-emission vehicle	0.075	3.4	0.2
Ultra-low-emission vehicle	0.040	1.7	0.2
Zero-emission vehicle	0.000	0.0	0.0

[a] Nonmethane organic gases before reactivity adjustment.

Clearly, alternatives to conventional petroleum-based fuels will be important to the United States in the future, for both environmental and economical–political reasons. The nation needs to improve its energy security by decreasing its dependence on imported oil. That is, it needs to diversify its sources of transportation energy and develop motor vehicle technologies compatible with fuels produced from natural gas, coal, and renewable biomass such as agricultural crops (corn, wheat, and sugar cane). About 63% of U.S. petroleum consumption is associated with transportation (41). In addition to the Clean Air Act

Table 9. Alternative Fuel Possibilities

Replacements for gasoline and/or diesel fuel
- Methanol
- Ethanol
- Compressed natural gas
- Liquefied petroleum gas
- Electricity

Gasoline and diesel fuel reformulation
- Varied gasoline volatility, T90, aromatic and olefinic hydrocarbon, sulfur, and oxygen content
- Varied diesel sulfur and aromatic hydrocarbon content, natural cetane number, and cetane improver

amendments requiring regulation of fuel composition, the Alternative Motor Fuels Act of 1988 encourages development and widespread use of such alternative fuels as methanol, ethanol, and compressed or liquified natural gas (CNG or LNG) (42).

Alternative fuels often are classified as fuels to replace conventional gasoline and diesel fuels or fuels to extend conventional fuels with environmentally favorable reformulations. Table 9 lists several replacement fuels and fuel reformulations currently under consideration, and Figure 21 illustrates some of the feedstocks and production processes being evaluated (41). Table 10 lists several of the emissions impacts associated with the replacement fuels being studied. Potential for reduced O_3, CO, particulate matter, and other toxic compounds in urban air has been demonstrated, but several important uncertainties have been identified requiring further study, such as the significance of formaldehyde emissions from methanol-fueled motor vehicles.

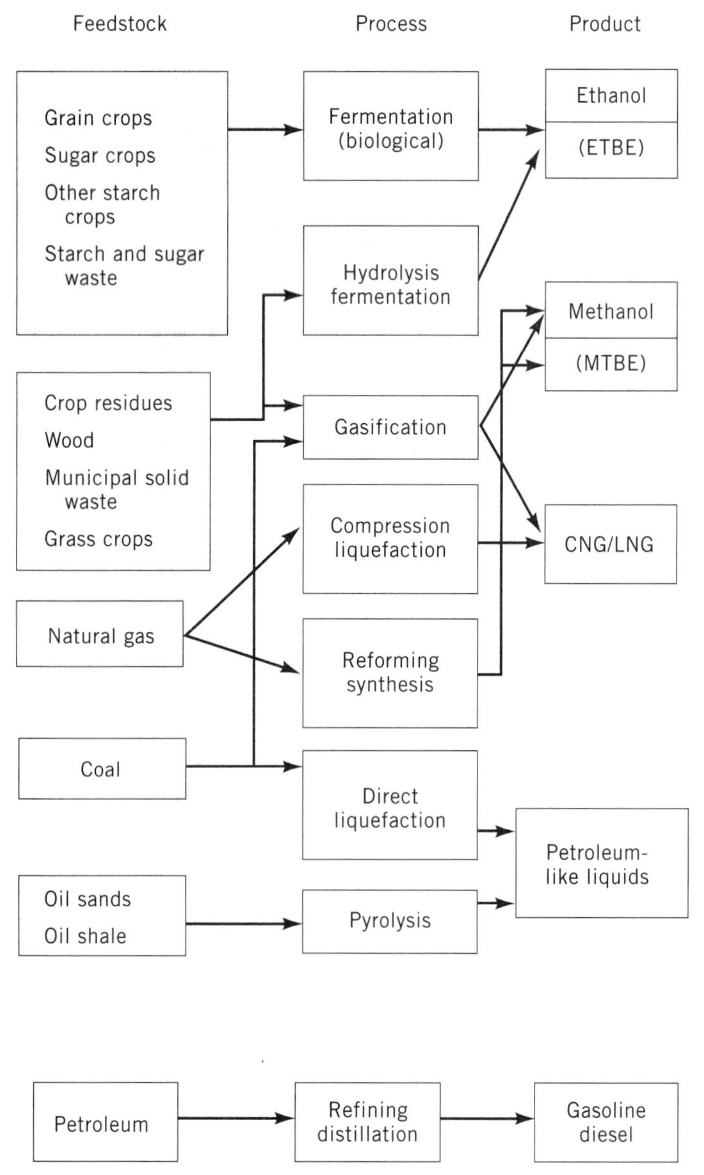

Figure 21. Alternative fuel feedstocks and processes (41).

Table 10. Potential Emissions Impact of Replacement Fuels

Ozone precursor organic emissions reactivity reduced: methanol, ethanol, CNG (methane) less photochemically reactive than gasoline hydrocarbon mixtures
Evaporative hydrocarbon emission rate reduced
Aldehyde emissions elevated
  Methanol: formaldehyde
  Ethanol: acetaldehyde
Ozone benefit sensitive to many variables
  Formaldehyde (and possibly methyl nitrite) emission rate
  Local VOC/NO_x ratio
  Relative significance of motor vehicle and other sources of local VOC and NO_x burden
CO and NO_x impact depends on engine design
  Those designed for stoichiometric combustion emit less NO_x but provide little CO improvement (lower flame temperature with methanol, ethanol)
  Those designed for fuel lean combustion emit less CO but provide little NO_x improvement (versus closed-loop 3-way catalyst gasoline engines)
Particulate emissions from heavy-duty diesel engines significantly reduced with methanol and CNG fuels
NO_x emissions from heavy-duty diesel engines reduced with methanol fuels
Toxic organic (benzene and other gasoline hydrocarbons) emissions reduced

The primary gasoline specifications that are being studied for reformulation are volume percentages of aromatics and olefinics, sulfur level, T_{90}, volatility, and oxygen level. Some of the potential changes in emissions associated with changes in these specifications are given in Table 11.

SUMMARY

Many emissions improvements have been realized with U.S. motor vehicle fleets since the late 1960s as a result of cooperation between government and industry. But further improvements will be required for many major cities to achieve the air quality required by the Clean Air Act. Miles traveled by the motoring public continue to increase,

Table 11. Potential Emissions Impact of Fuel Reformulations

Increased fuel oxygen reduces CO and HC emissions (technology sensitive, less impact with advanced "adaptive learning" closed-loop 3-way catalysts) and increases NO_x.
Reduced fuel volatility reduces evaporative (diurnal, hot soak, running-loss) emissions, and summer CO.
Reduced fuel aromatic and olefinic hydrocarbons reduce ozone precursor reactivity (important at lower vocal VOC/NO_x ratios).
Reduced fuel aromatic hydrocarbons reduce toxic benzene emissions.
Increased fuel MTBE (ETBE) increases isobutylene emissions.
Increased fuel oxygenates (ethanol, MTBE, ETBE) increase aldehyde emissions.
Reduced fuel T90 reduces HC emissions with current technology vehicles and increases CO with old technology vehicles.
Reduced fuel sulfur reduces HC, CO, and NO_x with current technology vehicles.

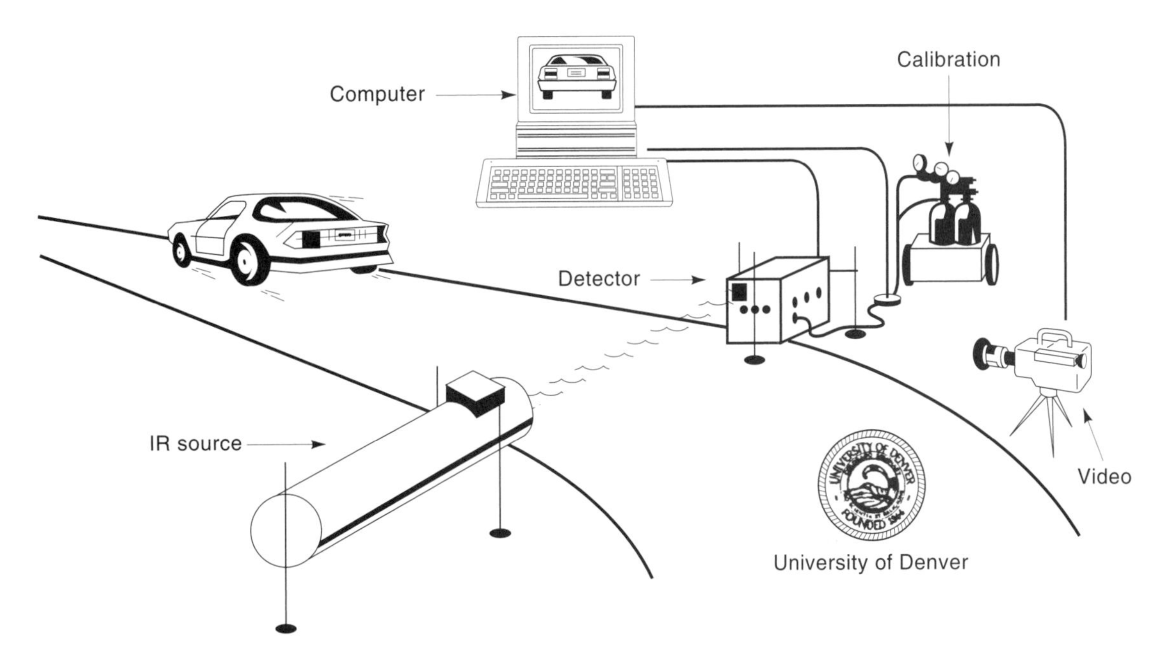

Figure 1. A schematic diagram of the University of Denver on-road emissions monitor. It is capable of monitoring emissions at vehicle speeds betwee 4.0 and 241 km/h in under 1 s/vehicle.

ity. Reduction in the signal caused by absorption of light by the molecules of interest reduces the voltage output. One way of conceptualizing the instrument is to imagine a typical garage-type NDIR instrument in which the separation of the ir source and detector is increased from 0.08 to 6–12 m. Instead of pumping exhaust gas through a flow cell, a car now drives between the source and the detector.

Because the effective plume path length and amount of plume seen depends on turbulence and wind, the FEAT can only directly measure ratios of $CO:CO_2$ or $HC:CO_2$. These ratios are constant for a given exhaust plume and by themselves are useful parameters to describe the combustion system. However, with a fundamental knowledge of combustion chemistry, it is possible to determine many other parameters of the vehicle's operating characteristics from these ratios, including the instantaneous air:fuel ratio, grams of CO or HC emitted per liter of gasoline (g CO/L or g HC/L) burned, and the percent CO or percent HC in the exhaust gas. Most vehicles show ratios of zero, because they emit little to no CO or HC. To observe a $CO:CO_2$ ratio greater than zero, the engine must have a fuel-rich air:fuel ratio and the emission control system, if present, must not be fully operational. A high $HC:CO_2$ ratio can be associated with either fuel-rich or fuel-lean air:fuel ratios coupled with a missing or malfunctioning emission-control system. A lean air:fuel ratio, while impairing driveability, does not produce CO in the engine. If the air:fuel ratio is lean enough to induce misfire then a large amount of unburned fuel is present in the exhaust manifold. If the catalyst is absent or nonfunctional, then high HC will be observed in the exhaust without the presence of high CO. To the extent that the exhaust system of this misfiring vehicle contains some residual catalytic activity, the HC may be partially or totally converted to a $CO:CO_2$ mixture.

INSTRUMENT DETAILS

The present design of University of Denver FEAT instruments incorporates CO (4.6 μm), CO_2 (4.3 μm), HC (3.4 μm), and background (3.9 μm) channels using interference filters built into Peltier-cooled lead selenide detectors. The instrument uses a mirror to collect the light and focus it onto a spinning 12-faceted polygon mirror that provides a periodic detector illumination of 2400 Hz. The reflected light from each facet of the rotating mirror sweeps across a series of four focusing mirrors that in turn direct the light onto the four detectors as shown in Figure 2. Each detector thus gets a burst of full signal from the source in a sequential fashion for each measurement mode.

Each detector provides a pulse train at 2400 Hz equivalent to the intensity of the ir radiation detected at its specific wavelength. Electronic circuitry averages 24 of these pulses, subtracts any background signal, and provides the averaged DC level to four signal ports. These are connected to a computer through an analogue-to-digital converter. Figure 3 shows 0.5 s of zero-corrected exhaust data voltages from the HC, reference, CO, and CO_2 detectors. Voltage levels are monitored in front of and behind each passing vehicle to eliminate effects of variable background concentrations.

All data from the CO, CO_2, and HC channels (I) are corrected by ratio to the reference channel (I_0). This procedure eliminates other sources of opacity such as soot, turbulence, spray, license plates, etc from providing data that could be incorrectly identified as CO or HC. These reference-corrected values are then apportioned according to the detector-specific response factors and the resulting values for CO and HC are compared with CO_2 as shown in Figure 4. The resulting slope for $CO:CO_2$ (and $HC:CO_2$) is the path-independent measurement for that vehicle.

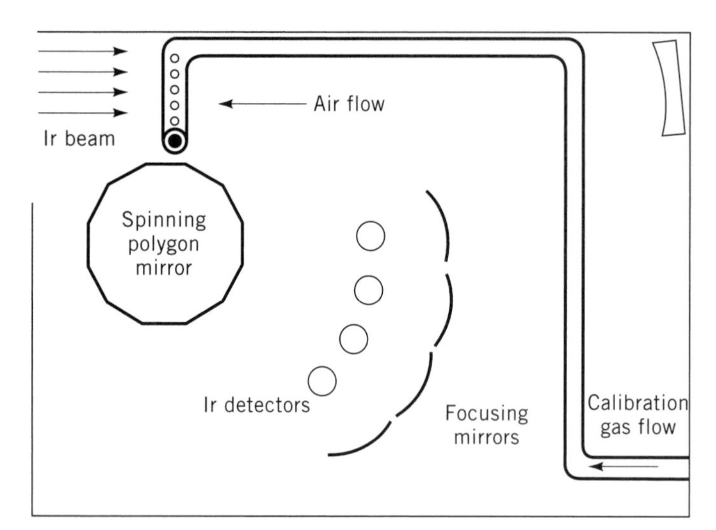

Figure 2. Schematic view (not to scale) of the optical bench layout for a University of Denver remote sensor. In-coming radiation is received by a pickup mirror in the upper right and focused off of the spinning polygon mirror and a detector-focusing mirror onto each of the four detectors for the reference, CO_2, HC, and CO. The daily gas calibrations are performed by releasing a small amount of gas directly into the beam path via the delivery tube shown. Note that the delivery tube continues up toward the viewer for an additional 6 cm before ending.

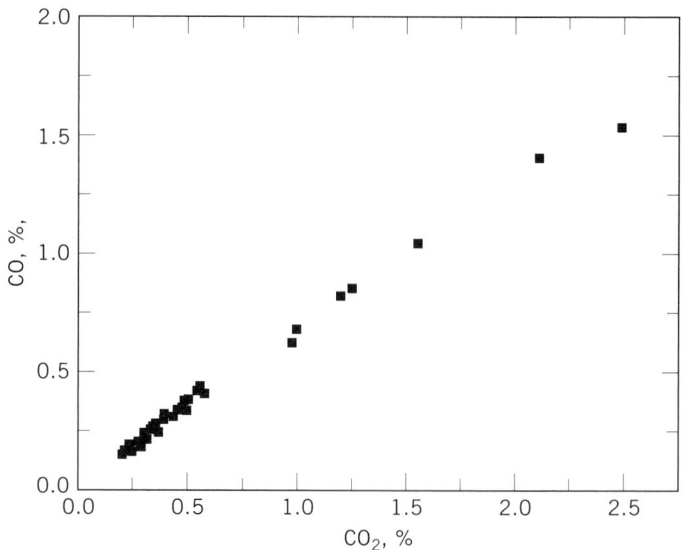

Figure 4. The final $CO:CO_2$ correlation graph of the data in Figure 3 used to obtain a unitless slope. The $HC:CO_2$ ratio is determined via the same process.

Software written for these instruments computes percent CO, percent CO_2, and percent HC on a dry, excess air-corrected basis from the measured $CO:CO_2$ and $HC:CO_2$ ratios. These reported values are designed to be equivalent to the percentages a standard garage analyzer would measure if one could run along behind the vehicle with a tailpipe probe. The percent HC is reported as an equivalent concentration of propane. This procedure is different from the reported HC measurements in most analyzers used in I/M programs. Most I/M instruments are tested for a single propane : hexane response ratio. All subsequent calibrations are performed with propane, yet the data are reported as "hexane equivalent" by dividing the measured number by the propane : hexane response ratio (a divisor usually close to two). The response factor for the FEAT is also close to two. Nevertheless, HC data are reported in propane units because the device is, in fact, calibrated daily with propane not hexane.

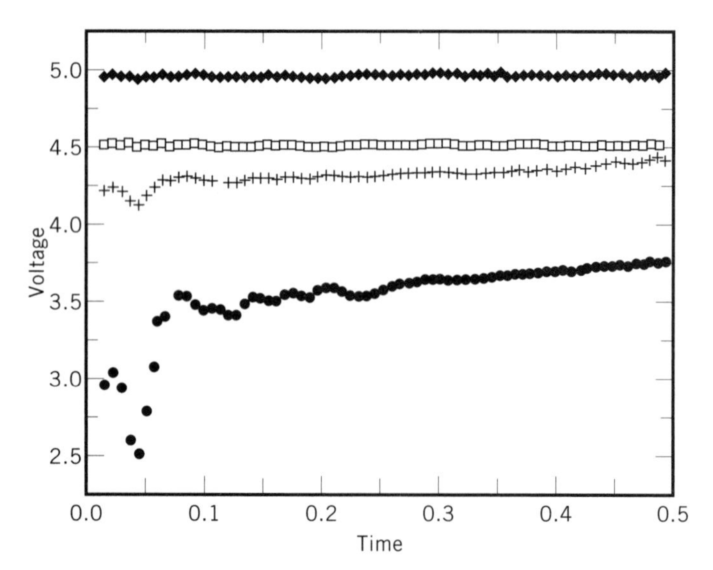

Figure 3. For each of the four detectors, 0.5 s of data. HC, ◆; reference, □; CO, +; and CO_2, ●. As the tailpipe exhaust rolls through the beam, the CO_2 voltages rise and fall until the plume finally dissipates.

The FEAT can be accompanied by a video system to record license plates. The video camera is coupled directly into the data analysis computer so that the image of the back of each passing vehicle is frozen onto the video screen. The computer writes the date, time, and the calculated exhaust CO, HC, and CO_2 percentage concentrations at the bottom of the image. These images are stored on videotape or digital storage media.

Before each day's operation in the field, a quality assurance calibration is performed on the instrument with the system set up in the field. A puff of gas designed to simulate all measured components of the exhaust is released into the instrument's path from a cylinder containing industry-certified amounts of CO, CO_2, and propane. The ratio readings from the instrument are compared with those certified by the cylinder manufacturer and used as a daily adjustment factor.

The FEAT is effective across traffic lanes of up to 15 m in width. It can be operated across double lanes of traffic with additional video hardware; however, the normal operating mode is on single-lane traffic (8). The FEAT operates most effectively on dry pavement, as rain, snow, and tire spray from a wet pavement scatter the ir beam. At suitable locations exhaust from more than 2,000 vehicles per hour can be monitored. The unit has been tested successfully for vehicle speeds exceeding 241 km/h. The FEAT has been used to measure the emissions of more than 750,000 vehicles worldwide. Table 1 compares the results for many of the locations that have been sampled (9–16).

Table 1. Worldwide Data Summary

Location	Date	Number of Measurements	Mean Percent CO	Median Percent CO	Mean Percent HC	Median Percent HC
United States						
Denver	1991–1992	35,945	0.74	0.11	0.057	0.033
Denver	1993	58,894	0.58	0.13	0.022	0.013
Chicago	1990	13,640	1.10	0.37	0.139	0.087
Chicago	1992	8,733	1.04	0.25	0.088	0.064
California	1991	91,679	0.82	0.14	0.076	0.042
Provo, Utah	1991–1992	12,066	1.17	0.45	0.220	0.127
El Paso, Tex.	1993	15,986	1.22	0.37	0.073	0.044
Mexico						
Juarez	1993	7,640	2.96	2.18	0.170	0.091
Mexico City	1991	31,838	4.30	3.81	0.214	0.113
Europe						
Göteborg, Sweden	1991	10,285	0.71	0.14	0.058	0.046
Denmark	1992	9,038	1.71	0.67	0.177	0.058
Thessaloniki, Greece	1992	10,536	1.40	0.55	0.155	0.082
London, UK	1992	11,666	0.96	0.17	0.136	0.071
Leicester, UK	1992	4,992	2.32	1.61	0.212	0.131
Edinburgh, UK	1992	4,524	1.48	0.69	0.129	0.084
Asia						
Bangkok, Thailand	1993	5,260	3.04	2.54	0.948	0.567
Hong Kong	1993	5,891	0.96	0.18	0.054	0.037
Kathmandu, Nepal	1993	11,227	3.85	3.69	0.757	0.363
Seoul, Korea	1993	3,104	0.82	0.26	0.044	0.019
Taipei, Taiwan	1993	12,062	1.49	0.88	0.062	0.050
Australia						
Melbourne	1992	15,908	1.42	0.57	0.107	0.058

The FEAT has been shown to give accurate readings for CO and HC in double-blind studies of vehicles both on the road and on dynamometers (17–19). A fully instrumented vehicle with tailpipe emissions controllable by the driver or passenger has been used in a series of drive-by experiments with the vehicles emissions set for CO between 0% and 10% and between 0% and 0.35% (propane) for HC to confirm the accuracy of the on-road readings (20). The comparison between the on-board measurements and the FEAT measurements is shown in Figure 5. The results

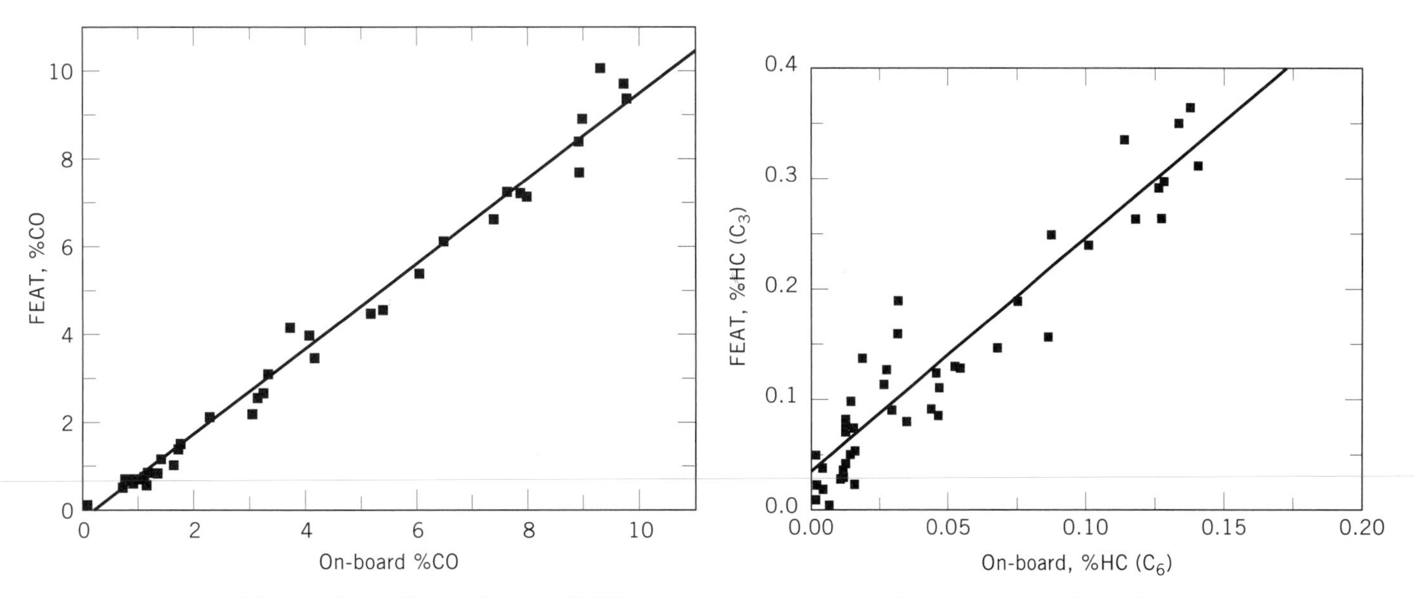

Figure 5. (**a**) Comparison of FEAT measurements to on-board measurements for carbon monoxide. The equation for the line is FEAT percent CO = 0.98 · on-board percent CO − 0.23 with an r^2 = 0.99. (**b**) Comparison of FEAT measurements to on-board measurements for hydrocarbons. The equation for the line is FEAT percent HC (C_3) = 2.16 · on-board percent HC (C_6) + 0.028 with an r^2 = 0.85. Note that because the FEAT reports HC as propane, the correct slope for the comparison with the on-board instrument will be 2.

have an accuracy of ± 5% for CO and ± 15% for HC. Recently, the abilities to measure nitric oxide and exhaust opacity have been added. NO is measured by ultraviolet absorption, whereas opacity is determined by means of ir absorption at 3.9 μm. Diesel soot (black) and so called "steam" from cold cars both cause observable opacity. Current units measure visible steam as HC. A system to eliminate this interference is under development.

REAL-WORLD VEHICLE EMISSION CHARACTERISTICS

Through the use of FEAT it is now possible to collect data quickly and easily on a large number of vehicles. This enables the study of important questions concerning the automobile's contribution to urban air quality and what mitigation actions might be taken. Following is a brief overview of general characteristics about automobile emissions that have been determined so far through the use of FEAT. Several interesting data sets have been collected that highlight the usefulness of large real-world databases.

Not all cars have equal emissions. Data show that a small fraction of the passing vehicles is responsible for half or more of the emissions in any given area. In Denver half the emissions come from only 7% of the vehicles. In Kathmandu half the emissions come from 25% of the vehicles. Figure 6 shows data collected at a single site in the Los Angeles area. At this location half of the CO is emitted by 7% of the vehicles and half of the HC is emitted by 11% of the vehicles. The few vehicles emitting half of the CO and HC are referred to as *gross polluters*. For automobile emissions the old adage the "tail wags the dog" holds true.

The overall characteristics of these fleets are similar regardless of age, location, or the presence or absence of I/M programs and can be mathematically described by a γ distribution (21). Most vehicles show mean emissions of 1% CO and 0.1% HC (as propane) or less in the exhaust. The newer the fleet the more skewed the emissions. This is because more of the vehicles have near zero emissions, and thus a smaller number of gross polluters dominates the total emissions.

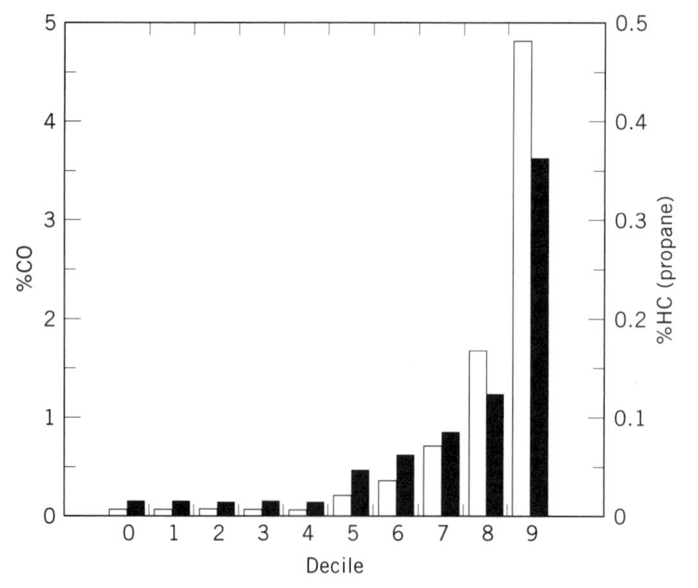

Figure 6. FEAT data collected in 1991 in El Monte, California (Rosemead Boulevard site). The fleet is rank ordered and divided into deciles; the average percent CO and percent HC for each decile is plotted. The solid bars denote CO while the empty bars are HC data. The first five deciles are displayed as an average of all five. The graph illustrates that automobile emissions are not normally distributed.

The good news is that for the U.S. fleet 50% of the vehicles produce only 4% of the CO emissions and 16% of HC, using current gasoline formulas as fuel. This shows that alternative and reformulated fuels are not likely to solve a problem that apparently arises due to a lack of maintenance.

Not all gross polluters are old vehicles (only about 25% of pre-1975 vehicles in the United States). In fact, even the majority of precatalyst vehicles emit relatively low levels. There is a strong correlation between fleet age and fleet emissions (Fig. 7**a**). However, this correlation has less to do with emissions-control technology than it does with vehicle maintenance. Any well-maintained vehicle regardless of age can be relatively low emitting, as seen by the mean emissions for all model years in the lowest-emitting quintile. It can also be seen that the most rapid deterioration in emissions occurs during the first 11 yr of ownership. It is interesting to note that the vehicles with the most rapidly deteriorating emissions are all newer technology vehicles with computer-controlled fuel delivery systems and three-way catalysts (installed in the United States after 1982). All of the studies note that these vehicles have negligible emissions when first purchased, emphasizing the need for proper maintenance.

If the problem is not the old clunker, then which vehicles are responsible? The product of the average emissions (Fig. 7a) and the fleet distribution (Fig. 7b) shown in Figure 7c, represents the overall contribution for each model year. What is obvious is that although older vehicles (pre-1980) have higher average emissions, it is the newer and more prevalent (post-1980) vehicles that actually contribute the most to the total. This has important implications for regulatory agencies that design programs around vehicle age only.

For HC, the fleet emissions tend to be less skewed than for CO, with a larger percentage of vehicles responsible for the majority of the fleet emissions. Only four cities (Bangkok, Hong Kong, Kathmandu, and Taipei) have half of the HC emissions produced by more than 15% of the fleet. Most of the same conclusions that are drawn regarding CO emissions and fleet characteristics hold true for HC emissions, because HC emissions increase as engine combustion gets richer and produces more CO. Two cities in Asia (Bangkok and Kathmandu) stand out for HC emissions, in large part due to the high percentage of two-cycle engines (necessarily high HC emitters by design), many of which also appear to be poorly tuned and maintained.

REMOTE SENSING APPLICATIONS

The ability quickly and unobtrusively to monitor real vehicles under real conditions opens up the possibility for many new applications. The ideas have run the gamut

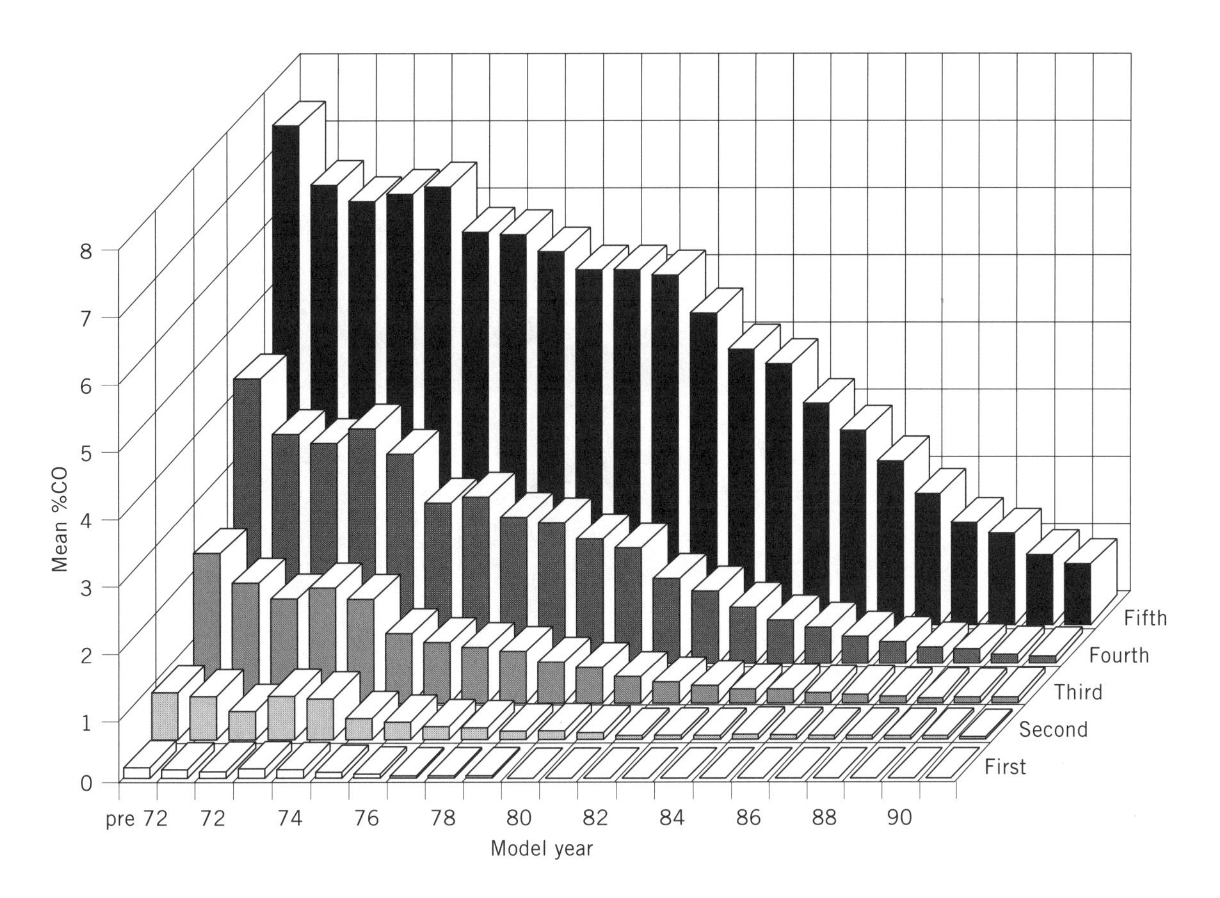
Mean %CO
8
7
6
5
4
3
2
1
0
pre 72
72
74
76
78
80
82
84
86
88
90
Model year
Fifth
Fourth
Third
Second
First

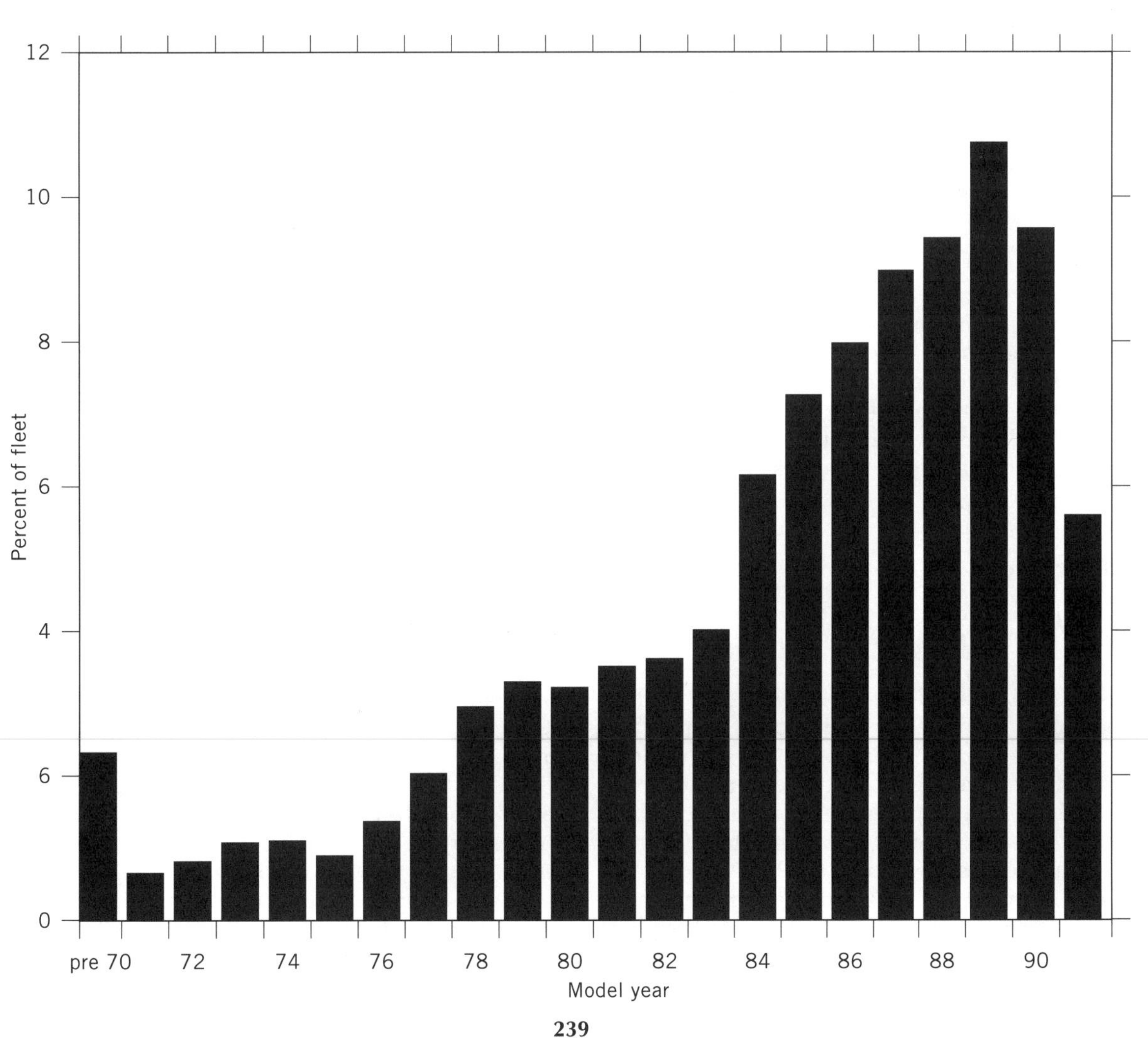
Percent of fleet
12
10
8
6
4
6
0
pre 70
72
74
76
78
80
82
84
86
88
90
Model year

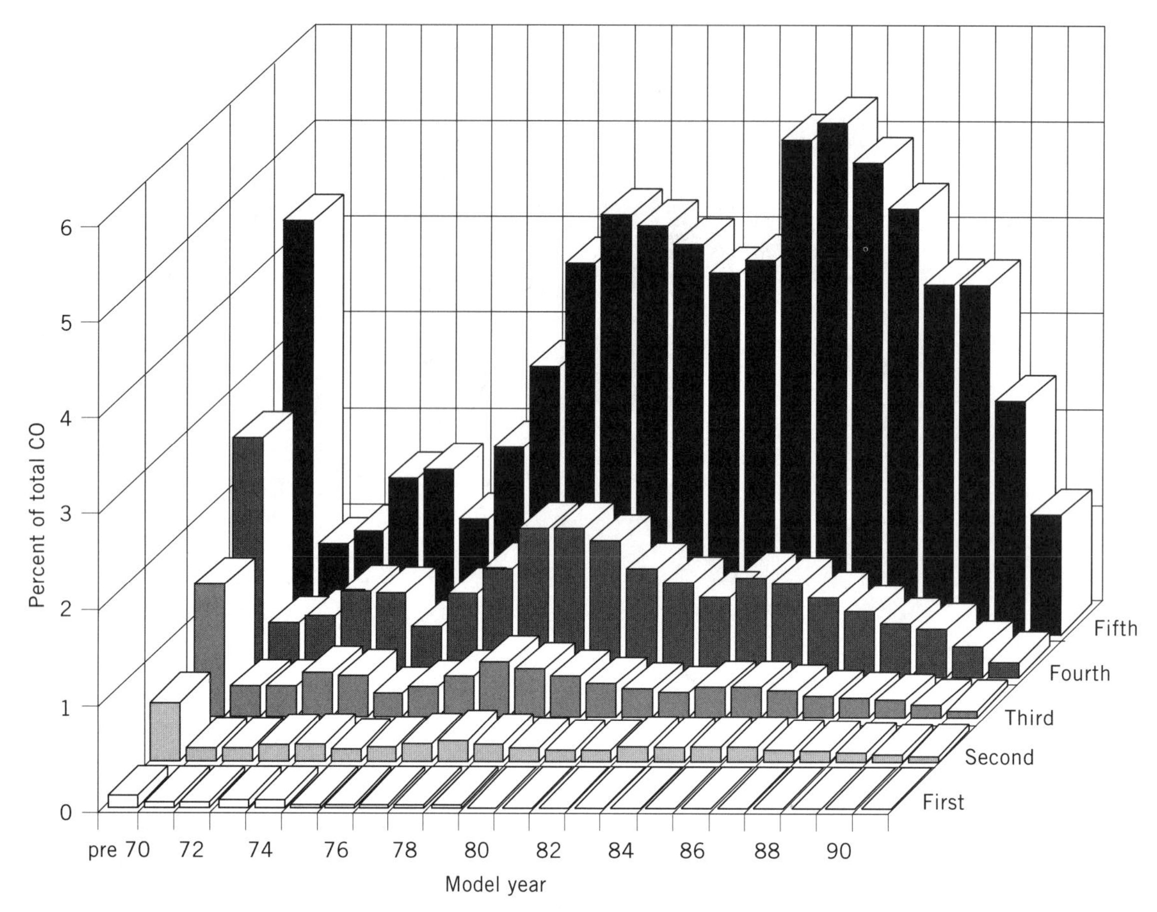

Figure 7. (**a**) The 1991 California data for CO presented as emission factors by model year for each of the data's quintiles. (**b**) The distribution of vehicles by model year for the data shown in part **a**. (**c**) The product of parts **b** and **c**, showing the average contribution of CO emissions by model year. Note that the majority of emissions is contributed by vehicles in the fifth quintile that are post-1980 models.

from pure research on emissions and control equipment performance to using the measurements as a basis for emission pricing schemes (by which the toll or vehicle registration is prorated according to the emission level). In the following section, data from several research programs that have used FEAT and the potential implications and logical extensions of that data to emission-control programs are addressed.

Roadside Pullover Studies

The ability of FEAT to identify gross polluters has been tested in several pilot studies. Two such programs in California have sought out explanations for vehicles with excessive emissions. In both studies, FEAT was used to identify gross polluting vehicles. A police officer stops the vehicle and performs a roadside inspection of the vehicle's tailpipe emissions and emission-control equipment status. In California, this type of inspection is called a smog check and is required once every 2 yr for gasoline-powered vehicles registered in most parts of the state.

The first California study involved the nonrandom inspection of 60 vehicles, 50 of which were stopped for having FEAT readings of greater than 2% CO (17). What was discovered was that of the 50 vehicles stopped for elevated CO emissions 45 failed the roadside inspection. One of the chief reasons for the high failure rate was that 12 of the 45 vehicles were found to have tampered emissions-control equipment. This means that emission-control equipment that was originally on the vehicle had been removed or otherwise disabled. The most glaring example was a 1984 GMC pickup that was originally purchaed equipped with a diesel engine but was found to have a gasoline engine without any required emissions-control equipment. Because this vehicle was still registered as a diesel-powered vehicle, it was exempt from the biannual state inspection program.

To verify these results on a larger sampling of vehicles, an expanded pullover program was conducted during June 1991 in a suburb of Los Angeles (13). Following a similar protocol, vehicles were selected nonrandomly based on a 4% CO or 0.4% HC on-road emissions level. Two inspection teams conducted roadside inspections of the vehicles that were stopped. In 2 weeks of measurements 60,487 measurements on 58,063 vehicles were performed. In all 3,271 gross polluting vehicles were identified, from which 307 vehicles were stopped and inspected. In the roadside inspection, 92% failed, with tampered emissions control equipment again a leading reason. A total of 41% of the vehicles were found out of compliance and an additional 25% had defective emissions-control equipment that may not have been caused by intentional tampering. The tailpipe portion of the test had an 85% failure rate.

These data strongly suggest that excessive tailpipe emissions are often the result of vehicles with broken or defective emissions-control equipment. This undoubtly has been influenced by the willful tampering that has taken place in a large number of the examples. It is also apparent that the scheduled testing program that California has depended on to find these tampered vehicles is being circumvented by the public. The classic example of this is the case of the diesel to gasoline engine switch mentioned previously (an additional case was found in the second pullover study). Aggressive antitampering programs using FEAT as probable cause for a pullover inspection could begin quickly and effectively to address another segment of the excessive mobile source tailpipe emissions.

Provo Pollution Prevention Program

In a study conducted in Provo, Utah, gross polluting vehicles were recruited from the public through FEAT measurements to investigate repair costs and effectiveness (22). Provo was chosen because it is an area that regularly violates the national ambient CO standard during the winter months. Measurements were made at two off-ramps (one northbound and one southbound) from I15 at the entrance to Provo during November 1991 and January 1992 in search of vehicles that exceeded a 4% CO emission level on at least two occasions. Two ramps were used to impose a control on the study, ie, vehicles were only recruited from one of the two ramps. In this way it was possible convincingly to compare emission changes at one ramp with those at the companion ramp as a judge of overall repair effectiveness. At these locations it was observed that half of the CO emissions were being produced by only 9% of the vehicles. For the entire program 17,000 measurements of more than 10,000 individual vehicles were made.

With the help of the city and a local community college, 47 out of 114 identified vehicles were recruited with the promise of a rental vehicle and free emission-related repairs. All of the work was carried out by local repair shops with the community college overseeing that the work paid for was in fact performed. Repairs ranged from the simple, such as fixing automatic chokes, to the extensive repair of replacing a cam shaft, lifters, and timing chain. In all, repairs for the 47 vehicles averaged $195 with an additional $43/vehicle spent on providing rental cars. In April 1992, through more on-road emission measurements, the effectiveness of the diagnosis and repairs at reducing CO emissions from these 47 vehicles was evaluated.

Table 2 shows, by model year, the 28 vehicles that were successfully remeasured. On average, the repairs resulted in a 50% reduction in the observed on-road emissions, with 25 of the remeasured vehicles showing statistically significant CO emission reductions. The remaining three vehicles were measured to have statistically the same or increased CO emission when compared with the before-repair measurements. These repair reductions were also found to be statistically significant when compared with the control group of vehicles that were identified from the control ramp but not included in the repair program.

In terms of the cost effectiveness of the repairs, it was estimated that CO was reduced from the 47 vehicles for approximately $181/t. This compares favorably with an estimate for I/M programs of $708/t (23), oxygenated fuel programs of $0 to $1147/t (9,23,24), and of $930/t for an old vehicle scrappage program (25). A major advantage of this approach is the ability to measure directly the program's effectiveness, allowing alterations and adjustments to be made to improve the results further.

Swedish Vehicle Study

In the 1991 data from Los Angeles, the European nameplate vehicles tend to have the lowest emissions. It was speculated that this is because they are well maintained. Data from the California Air Resources Board (CARB) listing manufacturer-specific failure rates for smog check reinforces this perception (26). The CARB data show Saab and Volvo with the lowest and third lowest smog check failure rates, respectively.

In September 1991, a study was conducted in Göteborg, Sweden (14). The location was a freeway interchange ramp (Gullbergsmotet) just across the river from the Volvo factory and downriver from the Saab manufacturing facility. In the Swedish study, emissions from 4011 Saabs and Volvos were measured. Sweden has a stringent I/M program (fail badly and the vehicle is towed to a repair shop). Sweden mandated closed-loop catalytically controlled systems in 1988. They were phased in during the 1987 model year, with about 50% of the vehicles. The 1986 and older Saabs and Volvos in Sweden are not equipped with any type of catalytic convertor.

The data from Sweden (4011 vehicles) and Los Angeles (536 vehicles) were used to examine the effects of technology and maintenance on vehicle emissions. By comparing

Table 2. Repair Data Summary According to Vehicle Emissions Technology Grouping for the 28 Vehicles that Were Successfully Remeasured after Repairs

Model Year Grouping	Number of Vehicles	Average g CO/L before Repairs	Average g CO/L after Repairs	Average g CO/L Reduction	Average g CO/L for Provo Fleet
post-1982	6	5931	2540	3391	867
1981–1982	1	5821	2305	3516	1911
1975–1980	16	5314	2702	2612	2589
pre-1975	5	6662	4285	2377	4012

these presumably well-maintained high technology vehicles to the well-maintained lower technology Swedish vehicles, the effects of technology ought to be readily observable. Figure 8 shows the emission data for CO and HC. For 1978–1986 model years, the CO and HC emissions of the Los Angeles vehicles average about 0.4% and 0.04%, respectively. For the Swedish vehicles of the same model years, the CO and HC emissions average about 1.5% and 0.08%, respectively. The improved technology of the Los Angeles fleet of Saabs and Volvos has clearly resulted in lower emissions, even for older vehicles. For 1988 model years and newer, when both fleets incorporated the same technology, the Swedish vehicles in Los Angeles and Göteborg are indistinguishable.

The dramatic drop in average vehicle emissions in Sweden following the 1987–1988 introduction of catalysts is only barely decernable in the first three quintiles of the California database (Fig. 7**a**), because catalysts were introduced longer ago. In Melbourne, Australia, catalysts were introduced in 1986. The dramatic improvement shown in Swedish vehicles also is not observed in the 15,908 vehicle Australian database. It was suspected that Australian maintenance is more like California and less like Sweden.

To examine further the effect of maintenance on emissions, emissions from the Los Angeles fleet of 1978–1986 model year U.S. vehicles were compared with the same model year noncatalyst vehicles in Sweden. The Swedish vehicles averaged 1.5% CO and 0.08% HC. The emissions of U.S. vehicles were slightly lower for CO and comparable for HC. In other words, the well-maintained Swedish noncatalyst vehicles emit nearly the same CO and HC as the overall (less well-maintained) U.S. fleet in Los Angeles. This demonstrates that a high level of maintenance is as important as technology to the higher emitting (on average) older model year vehicles.

Finally, emissions from Saabs and Volvos in Los Angeles are higher in the pre-1976 fleet than in the Swedish fleet. Because vehicles rust faster in Sweden, the pre-1976 fleet is much older, on average, in Los Angeles. The older Saabs have two-stroke engines, which are notorious for HC emissions and often tuned to produce high CO; thus it is not surprising that the older fleet in Los Angeles has higher average emissions.

Swedish-manufactured vehicles appear to be well maintained in both Sweden and Los Angeles. In both locations, they have used computer-controlled port fuel injection for more than 20 yr. In Los Angeles, these vehicles also have used catalysts since 1980, whereas in Sweden catalysts were not introduced until 1987. These data have been used to conduct two thought experiments in which the citizens of Los Angeles are imagined to all drive Swedish nameplate vehicles. The first assumes that all vehicles are constructed, operated, and maintained as in Los Angeles (ie, their emissions match the entire California fleet for all makes). The second assumes they are constructed, operated, and maintained as in Sweden. The overall emissions of the vehicle fleet measured in California in the 1991 study averaged 0.79% CO and 0.076% HC. The same age distribution as this overall fleet but with the emissions distribution of the Swedish-manufactured vehicles currently in use in Los Angeles gives average CO and HC emissions of 0.49% and 0.056%, respectively. The same age distribution but with the emissions of the Swedish-manufactured vehicles currently in use in Sweden gives average CO and HC emissions of 0.9% and 0.066%, respectively. The better maintenance with catalytic control provides a reduction of 38% and 26% for CO and HC, respectively. The better maintenance alone provides an increase of 14% for CO and a reduction of 13% for HC. It was concluded that better maintenance of the current fleet in Los

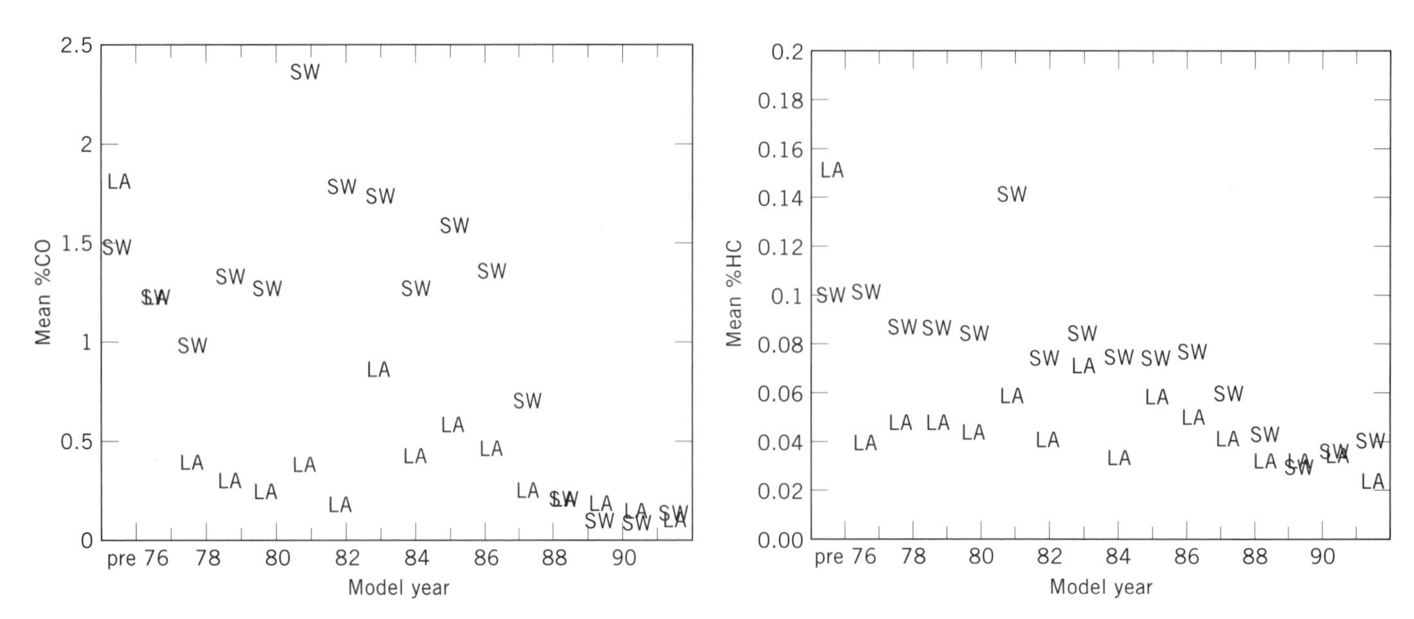

Figure 8. (**a**) Average percent CO for Saabs and Volvos measured in Los Angeles (LA) during the summer of 1991 compared with the same model year vehicles measured in Göteborg, Sweden (SW), during September of the same year. (**b**) Average percent HC for Saabs and Volvos measured in the same study.

Angeles could provide on-road emissions reductions greater than 25% for both CO and HC. To the extent that remote sensing offers lower cost emissions testing, the money saved could be spent on more important diagnosis and repair functions.

CONCLUSIONS

On-road remote sensing can now be used to analyze the emissions of a large fleet of vehicles in a cost-effective manner without inconveniencing the driving public. The statistics of the data can be used to plan and evaluate emission-control programs. The identification of gross polluters can be used as a component of a program designed to ensure that those vehicles receive effective repair (an inspection and maintenance program). Excessive on-road readings have been used as evidence for pulling vehicles over to the side of the road to check for tampering with the emissions-control system finding a high incidence of such behavior.

Remote sensing can be applied at the same site under the same conditions periodically to determine how well efforts are working to reduce on-road emissions. Because the correlation between low emissions and proper maintenance is high, most people operate vehicles that do not contribute significantly to pollution. These people could be rewarded for their socially responsible behavior based on low on-road emissions readings. On-road remote sensing of motor vehicle emissions is a new tool that has progressed from a university prototype to a commercially available system, which is either portable or mounted in a mobile van. Everywhere the system has been tried, local air pollution officials have suggested new ways in which this tool could be applied to solving their mobile source emissions problems.

BIBLIOGRAPHY

1. *Federal Register*, Part II, **21**(60) (1966).
2. *Federal Register*, Part II, **33**(108) (1968).
3. *Federal Register*, Part II, **35**(214) (1970).
4. *Federal Register*, Part II, **36**(128) (1971).
5. M. P. Walsh, *Critical Analysis of the Federal Motor Vehicle Control Program*, NESCAUM, Albany, N.Y., 1988.
6. G. A. Bishop, J. R. Starkey, A. Ihlenfeldt, W. J. Williams, and D. H. Stedman, *Anal. Chem.* **61,** 671A–676A (1989).
7. U.S. Pat. 5,210,702 (May 11, 1993), G. Bishop and D. H. Stedman (to Colorado Seminary).
8. G. A. Bishop, Y. Zhang, S. E. Mclaren, P. L. Guenther, S. P. Beaton, J. E. Peterson, D. H. Stedman, W. R. Pierson, K. T. Knapp, R. B. Zweidinger, J. W. Duncan, A. Q. McArver, P. J. Groblicki, and J. F. Day, *J. Air Waste Manage. Assoc.* **44,** 169–175 (1994).
9. PRC Environmental Management, Inc., *Performance Audit of Colorado's Oxygenated Fuels Program, Final Report to the Colorado State Auditor Legislative Services*, Denver, Dec. 1992.
10. G. A. Bishop and D. H. Stedman, *Environ. Sci. Technol.* **24,** 843–847 (1990).
11. D. H. Stedman, G. A. Bishop, J. E. Peterson, P. L. Guenther, I. F. McVey, and S. P. Beaton, *On-Road Carbon Monoxide and Hydrocarbon Remote Sensing in the Chicago Area*, ILENR/RE-AQ-91/14, Illinois Department of Energy and Natural Resources, Springfield, Ill., 1991.
12. D. H. Stedman, G. Bishop, J. E. Peterson, and P. L. Guenther, *On-Road CO Remote Sensing in the Los Angeles Basin*, Contract No. A932-189, California Air Resources Board, Sacramento, Calif., 1991.
13. D. H. Stedman, G. A. Bishop, S. P. Beaton, J. E. Peterson, P. L. Guenther, I. F. McVey, and Y. Zhang, *On-Road Remote Sensing of CO and HC Emissions in California*, Contract No. A032-093, California Air Resources Board, Sacramento, Calif., 1994.
14. J. E. Peterson, D. H. Stedman, and G. A. Bishop, *J. Air Waste Manage. Assoc.* (1991).
15. Å. Sjödin, *Rena Och Smutsiga Bilar: En pilotstudie av avgasutsläpp från svenska fordon i verklig trafik*, Swedish Environmental Research Institute, Göteberg, 1991.
16. S. P. Beaton, G. A. Bishop, and D. H. Stedman, *J. Air Waste Manage. Assoc.* **42,** 1424–1429 (1992).
17. D. R. Lawson, P. J. Groblicki, D. H. Stedman, G. A. Bishop, and P. L. Guenther, *J. Air Waste Manage. Assoc.* **40,**1096–1105 (1990).
18. D. H. Stedman and G. A. Bishop, *Remote Sensing for Mobile Source CO Emission Reduction,* EPA 600/4-90/032, U.S. Environmental Protection Agency, Las Vegas, Nev., 1991.
19. D. Elliott, C. Kaskavaltizis, and T. Topaloglu, *Soc. Automotive Eng.* **922314** (1992).
20. L. L. Ashbaugh, D. R. Lawson, G. A. Bishop, P. L. Guenther, D. H. Stedman, R. D. Stephens, P. J. Groblicki, B. J. Johnson, and S. C. Huang, paper presented at the A&WMA International Specialty Conference on PM10 Standards and Nontraditional Source Controls, Phoenix, 1992.
21. Y. Zhang, G. A. Bishop, and D. H. Stedman, *Environ. Sci. Tech.*, in press.
22. G. A. Bishop, D. H. Stedman, J. E. Peterson, T. J. Hosick, and P. L. Guenther, *J. Air Waste Manage. Assoc.* **43,** 978–990 (1993).
23. A. M. Michelsen, J. B. Epel and R. D. Rowe, *Economic Evaluation of Colorado's 1988–1989 Oxygenated Fuels Program,* RCG/Hagler, Bailly, Inc., Boulder, Colo., 1989.
24. J. B. Epel, M. Skumanich and R. Rowe, *The Colorado Oxygenated Fuels Program: Evaluation of Program Costs for 1989–1990 and for Alternative Program Designs,* RCG/Hagler, Bailly, Inc., Boulder, Colo., 1990.
25. *SCRAP A Clean-Air Initiative from UNOCAL,* UNOCAL Corp., Los Angeles, 1991.
26. *CVS News,* Jan. 1992.

AUTOMOTIVE ENGINES–EFFICIENCY

K. G. Duleep
EEA
Arlington, Virginia

Automotive engine efficiency can be defined in terms of the percent of fuel energy that is converted to useful work, but should not be confused with vehicle fuel efficiency, or fuel economy. Automotive fuel economy is broadly understood by the public as a measure of the distance traveled per unit volume of fuel consumed (because most automobiles use liquid fuel), and is usually expressed in the United States in miles per gallon, or MPG. In other developed countries, fuel consumption is preferred to fuel economy and is usually expressed in liters per 100 km or the volume of fuel consumed per unit distance. Fuel economy (or consumption) of vehicles usually varies by the size, as defined by the physical exterior dimensions, and the weight. The larger or heavier the vehicle, the lower the MPG values typically obtained. However, this need not imply that the engines in larger cars have proportionally lower efficiency. Because larger cars demand more power to be driven a specific distance, fuel economy will be lower than for a smaller car with an engine of the same efficiency as that in the larger car.

See also THERMODYNAMICS; ENERGY EFFICIENCY; ENERGY EFFICIENCY, CALCULATIONS; KNOCK; OCTANE NUMBER.

Almost all currently produced automobiles and trucks worldwide use the Otto cycle engine or the Diesel cycle engine for motive power. Most of the Otto cycle engines use gasoline and are of the "four-stroke" type, because the "two-stroke" type has been too inefficient and polluting to be widely used in automobiles (see Plate I). The Otto cycle two-stroke engine is still used in small motorcycles and other two-wheelers, especially in developing countries, and new versions in the prototype stage hold the promise of low emisisons and high efficiency. Diesel cycle engines use diesel fuel and are of both the two- and four-stroke types, but their primary use has been in commercial heavy trucks because they are significantly more fuel efficient and much heavier than Otto cycle engines of the same power. Other engines, such as the gas turbine, have been installed in prototypes or research vehicles but have never progressed to production status because of various real and perceived drawbacks. Electrical batteries and fuel cells, which are not heat engines, have also been tried for automotive use and could replace heat engines in the future, but are not considered in the following discussion on engine efficiency of conventional heat engines.

Heat engine efficiency can be stated in several ways. One intuitively appealing method is to express the useful energy produced by an engine as a percent of the total heat energy that is theoretically liberated by combusting the fuel. This is sometimes referred to as "the first law" efficiency, implying that its basis is the first law of thermodynamics, the law of conservation of energy. Another potential but less widely used measure is based on the second law of thermodynamics, which governs how much of that heat can be converted to work. Given a maximum combustion temperature (usually limited by engine material considerations and by emission considerations), the second law postulates a maximum efficiency based on an idealized heat engine cycle called the Carnot cycle. The ratio of the "first law efficiency" to the Carnot cycle efficiency can be used as a measure of how efficiently a particular engine is operating with reference to the theoretical maximum based on the second law of thermodynamics. However, the most common measure of efficiency used by automotive engineers is termed *brake specific fuel consumption* (bsfc), which is the amount of fuel consumed per unit time per unit of power produced. In the United States, the bsfc of engines is usually stated in pounds of fuel per brake horsepower-hour, whereas the more common metric system measurement units are in grams per kilowatt-hour (g/kwh). The term 'brake' here refers to a common method historically used to measure engine shaft power. Of course, all three measures of efficiency are related to each other.

Plate I. Four-stroke engines are the most common for automobile use. Courtesy of Chrysler Corporation.

The efficiency of Otto and Diesel cycle engines are not constant but depend on the operating point of the engine as specified by its torque output and shaft speed (revolutions per minute, or RPM). Engine design considerations, frictional losses, and heat losses result in a single operating point where efficiency is highest. This maximum efficiency usually occurs at relatively high torque and at low to mid-RPM within the operating RPM range of the engine. At idle, the efficiency is zero because the engine is consuming fuel but not producing any useful work.

When considering automotive engine efficiency, the maximum efficiency need not, by itself, be an indicator of the average efficiency under normal driving conditions, because engine speed and torque vary widely under normal driving. The maximum efficiency of an engine is of interest to automotive designers, but a more practical measure of efficiency is its average efficiency during "normal" driving. In most developed countries, vehicle fuel economy figures provided to consumers are measured over a government-prescribed driving cycle, or cycles, which represent "normal" driving. In the United States, two cycles are used, one for city driving, at an average speed of about 32 km/h (20 mph) and one for high-way driving, at an average speed of about 80 km/h (50 mph). These driving cycles provide convenient reference measures over which engine efficiency can be evaluated. A composite fuel economy, which is a weighted sum of city and highway fuel economy, is often used, and fuel efficiency benefits of technology improvements are with reference to the composite, unless otherwise stated. The engine operating points when the vehicle is driven on these reference cycles depends on the weight and size of the car, and how the engine is geared to the wheels. Hence, the same engine can display different average efficiencies over the driving cycle depending on the gear ratios employed and the power demanded by the vehicle. These considerations are important to the discussion on engine efficiencies that follows.

THEORETICAL ENGINE EFFICIENCY

The characteristics features common to all piston internal combustion engines are (a) intake and compression of the air or air-fuel mixture; (b) raising the temperature (and hence, the pressure) of the compressed air by combustion of fuel; (c) the extraction of work from the high-pressure products of combustion by expansion: (d) exhaust of the products of combustion. The four-stroke cycle requires two complete revolutions of the crankshaft or four up-and-down motions of the piston, to complete the entire cycle from intake to exhaust. The first engine to use this approach successfully was built by N.A. Otto in 1876. The two-stroke cycle, which requires only one revolution of the crankshaft to complete the cycle, was developed in 1878 by Sir Dugald Clerk. The Otto cycle engine is sometimes called the spark ignition (s.i.) because the combustion is initiated by an externally powered spark. It is also called the gasoline engine, because most spark ignition engines use gasoline, although many other liquid and gaseous hydrocarbon fuels can be used in this engine. Combustion of the homogenous air-fuel mixture takes place very quickly relative to piston motion, and is represented in idealized calculations as an event occurring at constant volume.

According to classical thermodynamic theory (1), the thermal efficiency (η) of an idealized Otto cycle, starting with intake air-fuel mixture drawn in at atmospheric pressure, is given by

$$\eta = 1 - 1/r^{n-1}$$

where r is the compression (and expansion) ratio and n is the ratio of specific heat at constant pressure to that at constant volume for the mixture. The equation shows that efficiency increases with increasing compression ratio.

Compression ratios are limited by the octane number of gasoline, which is a measure of its resistance to preignition, or "knock." At high compression ratios, the heat of compression of the air-fuel mixture becomes high enough to induce spontaneous combustion of small pockets of the mixture, usually those in contact with the hottest parts of the combustion chamber. These spontaneous combustion events are like small explosions that can damage the engine and reduce efficiency depending on when they occur during the cycle. Higher-octane-number gasolines prevent these events, but also cost more and require greater energy expenditure for manufacture at the refinery. The octane number is measured using two different procedures, resulting in two different ratings for a given fuel, called "motor octane" and "research octane" number. Octane numbers displayed at the pump are an average of research and motor octane numbers, and most engines sold in the United States require regular gasoline with a pump octane number of 87.

Using an n value of 1.4 for air, the equation predicts an efficiency of 58.47% at a compression ratio of 9 : 1, common in today's engines. A value of $n = 1.26$ is more correct for products of combustion of a stoichiometric mixture of air and gasoline. A stoichiometric mixture corresponds to an air-fuel ratio of 14.7 : 1, and this air-fuel ratio is commonly used in most cars sold in the United States today. At this air-fuel ratio, calculated efficiency is about 43.5%. Actual engines yield still lower efficiencies even in the absence of mechanical friction, because of heat transfer to the walls of the cylinder and the inaccuracy associated with assuming combustion to be instantaneous. Figure 1 shows the pressure-volume cycle of a typical spark ignition engine (1) and its departure from the ideal relationships.

The Diesel engine was developed by Rudolf Diesel about 20 years after Otto cycle engine was invented. The Diesel engine differs from the spark ignition engine in that only air, rather than the air-fuel mixture, is compressed. The diesel fuel is sprayed into the combustion chamber at the end of compression in a fine mist of droplets, and the diesel fuel ignites spontaneously upon contact with the compressed air because of the heat of compression. As a result, this engine is also referred to as a compression ignition (c.i.) engine. The sequence of processes (ie, intake, compression, combustion, expansion, and exhaust) are similar to those of a Otto cycle. However, the combustion process occurs over a relatively long period and is represented in idealized calculations as an event occurring at constant pressure, that is, combustion occurs as the piston moves downward to increase volume and decrease pressure at a rate offsetting the pressure rise due to heat release. Figure 2 shows the pressure-volume cycles for a typical diesel engine and its relationship to the ideal Diesel cycle (2). If the ratio of volume at the end of the combustion period to the volume at the beginning of the period is r_c, or the "cut-off ratio," the thermodynamic efficiency of the idealized constant-pressure combustion cycle (3) is given by

$$\eta = \frac{1}{r^{n-1}}\left[\frac{r_c^n - 1}{n(r_c - 1)}\right].$$

It can be seen that for $r_c = 1$, the combustion occurs at constant volume and the efficiency of the Diesel and Otto cycle are equivalent.

The term r_c also measures the interval during which fuel is injected, and increases as the power output is increased. The efficiency equation shows that as r_c is increased, efficiency falls so that the idealized Diesel cycle is less efficient at high loads. The combustion process also is responsible for a major difference between Diesel and Otto cycle engines: in an Otto cycle engine, intake air is throttled to control power while maintaining near-constant air-fuel ratio; in a Diesel engine, power control is achieved by varying the amount of fuel injected while keeping the air mass inducted per cycle at near-constant levels. In most operating modes, combustion occurs with considerable excess air in a c.i. engine, whereas combustion occurs at or near stoichiometric air-fuel ratios in modern s.i. engines.

At the same compression ratio, the Otto cycle has the higher efficiency. However, Diesel cycle engines normally operate at much higher compression ratios, because there are no octane limitations associated with this cycle. In fact, spontaneous combustion of the fuel is required in such engines, and the ease of spontaneous combustion is measured by a fuel properly called cetane number. Most current c.i. engines require diesel fuels with a cetane number over 40.

In practice, there are two kinds of c.i. engine, the direct injection type (DI) and the indirect injection type (IDI). The DI type uses a system in which fuel is sprayed directly

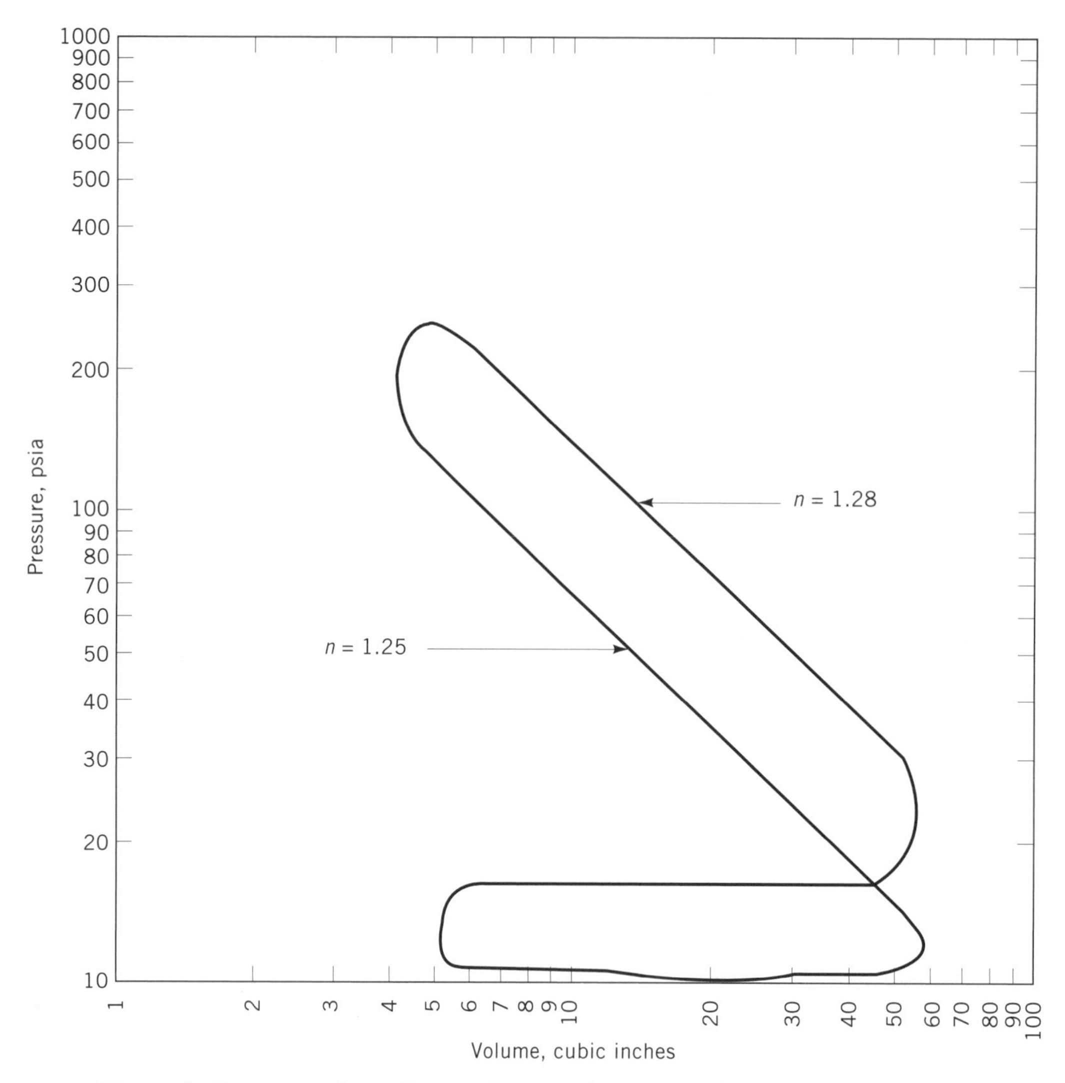

Figure 1. Pressure–volume diagram for a gasoline engine. Compression ratio = 8.7.

into the combustion chamber. The fuel spray is premixed and partially combusted with air in a prechamber in the IDI engine, before the complete burning of the fuel in the main combustion chamber occurs. DI engines generally operate at compression ratios of 15 to 20 : 1, whereas IDI engines operate at 18 to 23 : 1. For illustration, the theoretical efficiency of a c.i. engine with a compression ratio of 20 : 1, operating at a cutoff ratio of 2, is about 54% (for combustion with excess air, n is approximately 1.3). In practice, these high efficiencies are not attained, for reasons similar to those outlined for s.i. engines.

ACTUAL AND ON-ROAD EFFICIENCY

Four major factors affect the efficiency of s.i. and c.i. engines. First, the ideal cycle cannot be replicated because of thermodynamic and kinetic limitations of the combustion process, and the heat transfer that occurs from the cylinder walls and combustion chamber. Second, mechanical friction associated with the motion of the piston, crankshaft, and valves consumes a significant fraction of total power. Because friction is a stronger function of engine speed than torque, efficiency is degraded considerably at light-load and high-RPM conditions. Third, aerodynamic frictional losses associated with air flow through the air cleaner, intake manifold and valves, exhaust manifold, silencer, and catalyst are significant, especially at high air-flow rates through the engine. Fourth, pumping losses associated with throttling the air flow to achieve part-load conditions in spark ignition engines are very high at light loads. C.i. engines do not usually have throttling loss, and their part-load efficiencies are superior to those of s.i. engines. Efficiency varies with both speed and load for both engine types.

Hence, production spark ignition or compression ignition engines do not attain the theoretical values of efficiency, even at their most efficient operating point. In general, for both types of engines, the maximum efficiency point occurs at an RPM that is intermediate to idle and maximum RPM, and at a level that is 70% to 85% of maximum torque. "On-road" average efficiencies of engines used in cars and light trucks are much lower than peak efficiency, because the engines generally operate at very

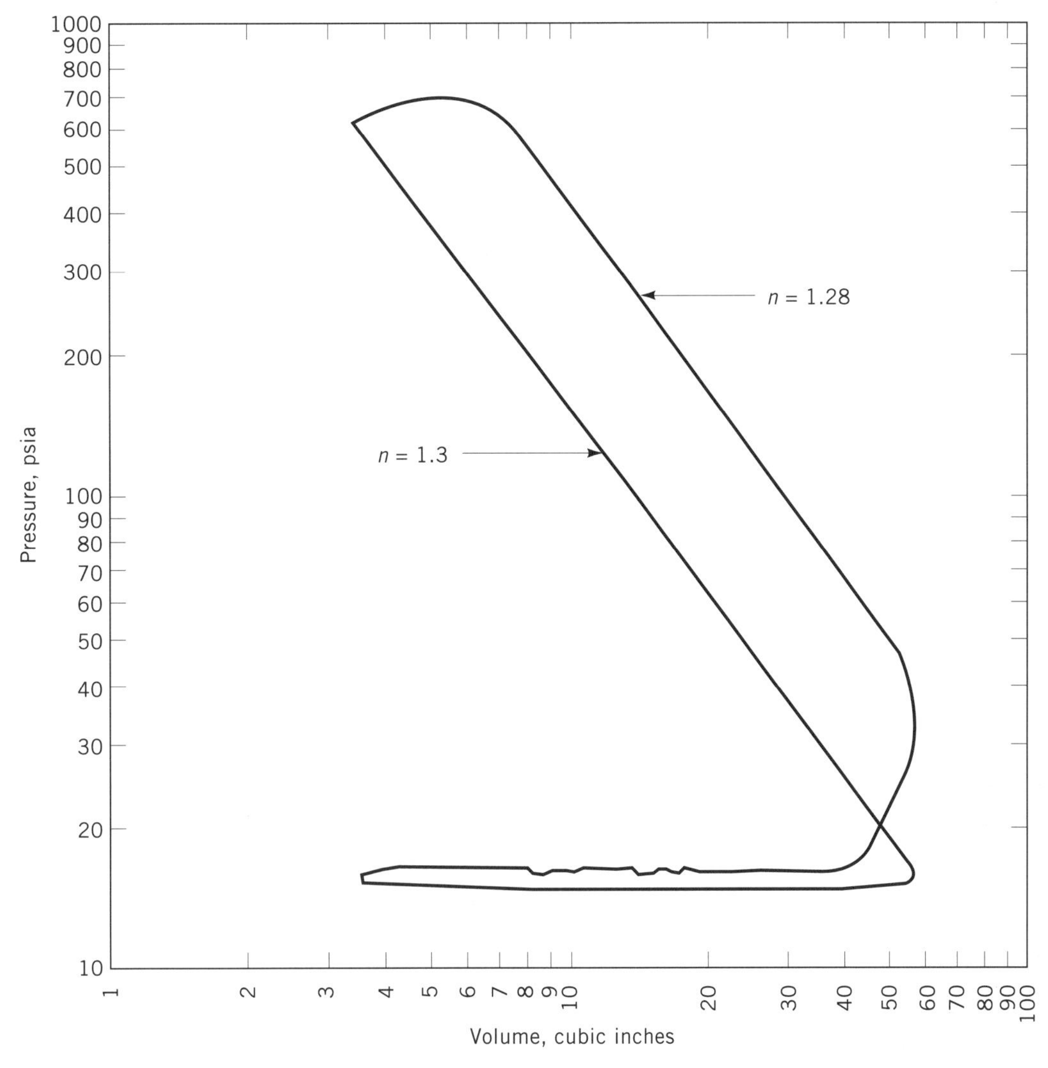

Figure 2. Pressure–volume diagram for a diesel engine. Compression ratio = 17.0.

light loads during city driving and steady-state cruise on the highway. High power is used only during strong accelerations, at very high speeds or when climbing steep gradients. The high-load conditions are relatively infrequent, and the engine operates at light loads much of the time during normal driving.

Figures **3a** and **b** illustrate the brake specific fuel consumption (bsfc) maps for an s.i. and an IDI c.i. engine of late-1970s vintage (4). Lines of constant bsfc resemble "islands," with the lowest bsfc indicated in the central island. The spark ignition engine using gasoline is of 1.6 liters' displacement, and the diesel is of 2.1 liters' displacement. Both engines produce approximately the same power, 48 kw, and also have very similar torque ratings over the 1000 to 4500 rpm range. The c.i. produces less specific power (power per unit displacement) because it always operates with excess air; the brake specific fuel consumption of 260 g/kw-h corresponds to an efficiency of about 33%. In contrast, the lowest bsfc of 270 g/kw-h for the s.i. engine corresponds to an efficiency of about 32%. Although the peak efficiency is comparable, the part-load efficiency of the s.i. engine is significantly worse than that of the c.i. engine of equal power. The difference at part load is largely due to the effect of pumping loss associated with throttling in an s.i. engine, which more than compensates for the higher frictional loss in a c.i. engine.

THE SOURCES OF ENERGY LOSS

During normal driving, the heat of fuel combustion is lost to a variety of sources and only a small fraction is converted to useful output, resulting in the low values for on-road efficiency. Figure 4 provides an overview of the heat balance for a typical modern small car with a spark ignition engine under a low-speed (40 km/h) and a high-speed (100 km/h) condition (7). At very low driving speeds typical of city driving, most of the heat energy is lost to the engine coolant. Losses associated with "other waste heat" include radiant and convection losses from the hot engine block, and heat losses to the engine oil. A similar heat loss diagram for a Diesel c.i. would indicate lower heat loss to the exhaust and coolant and an increased fraction of heat converted to work, especially at the low-speed condition.

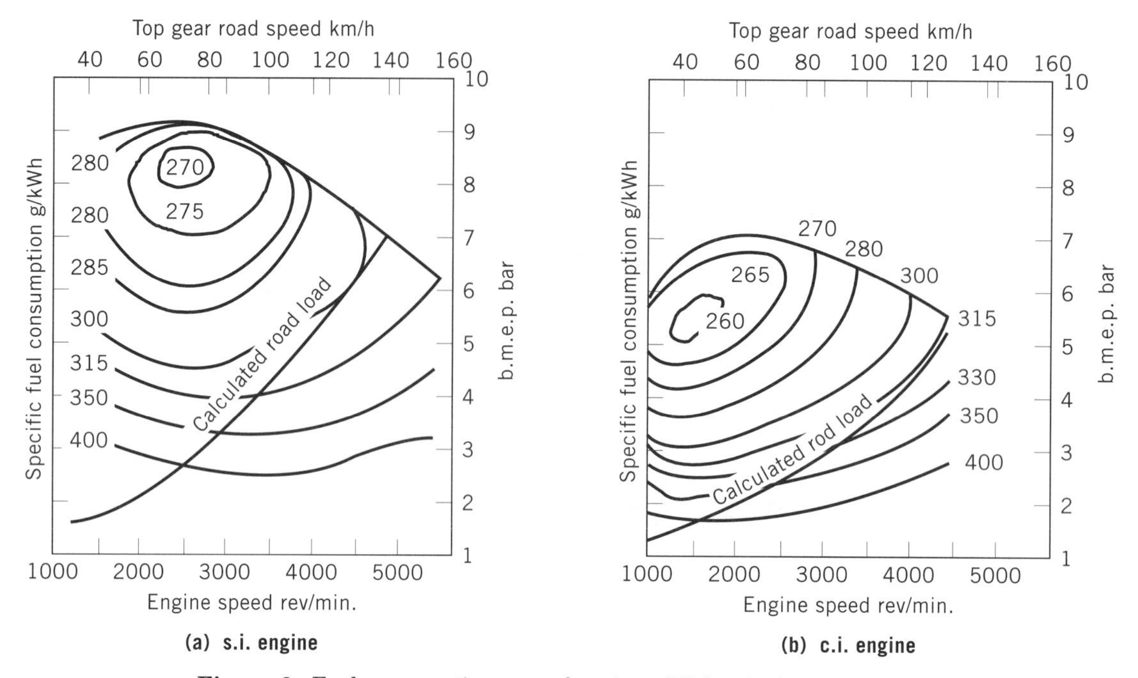

Figure 3. Fuel consumption map of engines. With calculated road load.

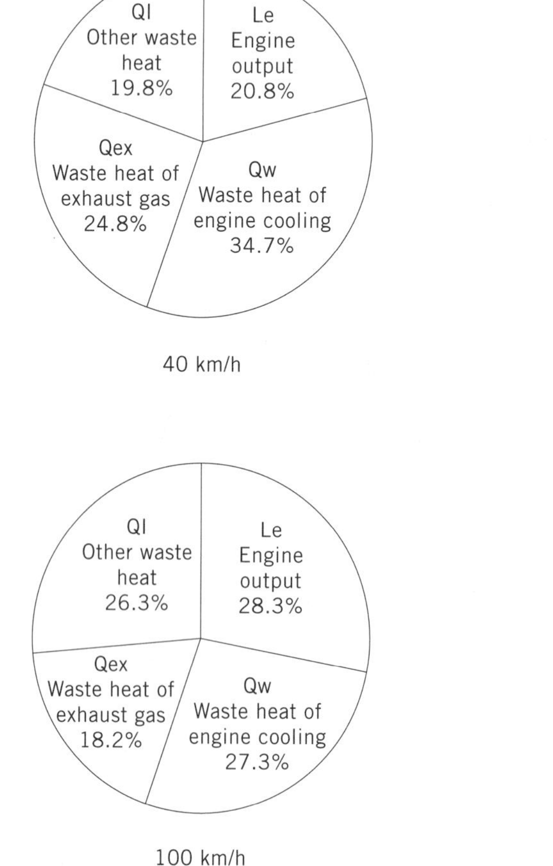

Figure 4. Heat balance of a passenger car equipped with 1500cc engine.

During stop-and-go driving conditions typical of city driving, efficiencies are even lower than those indicated in Figure 4, because of the time spent at idle, where efficiency is zero. Under the prescribed U.S. city cycle conditions, typical modern spark ignition engines have an efficiency of about 18%, and modern IDI c.i. engines have an efficiency of about 21%.

Another method of examining the energy losses is by allocating the power losses starting from the power developed within the cylinder. The useful work corresponds to the area that falls between the compression and expansion curve depicted in Figures 1 and 2.

The pumping work that is subtracted from this useful work, referred to as indicated work, is a function of how widely the throttle is open, and to a lesser extent, the speed of the engine. Figure 5 shows the dependence of specific fuel consumption (or fuel consumption per unit of work) with load, at constant (low) engine RPM (6). Pumping work represents only 5% of indicated work at full-load, low-RPM conditions, but increases to over 50% at light loads of less than two-tenths of maximum power.

Mechanical friction and accessory drive power on the other hand, increase non-linearly with engine speed, but do not change much with the throttle setting. Figure 6 shows the contribution of the various engine components as well as the alternator, water pump and oil pump to total friction, expressed in terms of mean effective pressure, as a function of RPM (7). The mean effective pressure is a measure of specific torque, or torque per unit of displacement; typical engine brake mean effective pressure (bmep) of spark ignition engines that are not supercharged range from 8.5 to 10 bar. Hence, friction accounts for about 25% of total indicated power at high RPM (~6000), but only for about 10% of indicated power at low RPM (~2000) in spark ignition engines. Friction in c.i. engines is higher

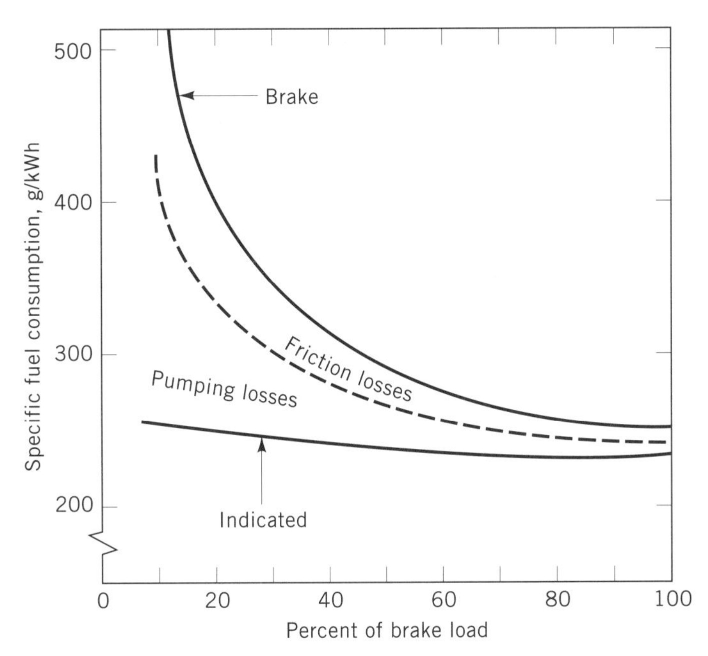

Figure 5. Specific fuel consumption vs engine load.

because of the need to maintain an effective pressure seal at high compression ratios, and the friction mean effective pressure is 30% to 40% higher than that for a dimensionally similar s.i. engine at the same RPM. Because the brake mean effective pressure of a diesel is also lower than that of a gasoline engine, friction accounts for 15% to 16%

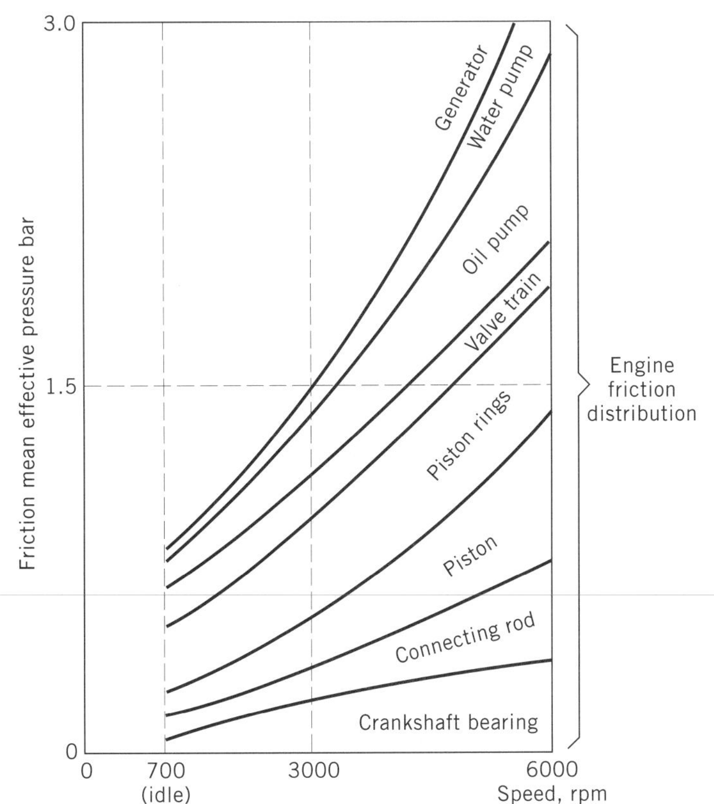

Figure 6. Friction distribution as a function of engine speed.

of indicated maximum power even at 2000 RPM. Typical bmep of naturally aspirated c.i. engines range from 6.5 to 7.5 bar.

EVOLUTION OF SPARK IGNITION ENGINES

During the 1980s, most automotive engine manufacturers have improved engine technology to improve thermodynamic efficiency, reduce pumping loss, and decrease mechanical friction and accessory drive losses.

Design Parameters

Engine valvetrain design is a widely used method to classify spark ignition engines. The first spark ignition engines were of the side-valve type, but such engines have not been used in automobiles for several decades, although some engines used in off-highway applications, such as lawn mowers or forklifts, continue to use this design. The overhead valve (OHV) design supplanted the side-valve engine by the early 1950s, and continues to be used in many U.S. engines in a much-improved form. The overhead cam engine (OHC) is the dominant design used in the rest of the developed world. The placement of the camshaft in the cylinder heads allows the use of a simple, lighter valvetrain, and valves can be opened and closed more quickly as a result of the reduced inertia. This permits better flow of intake and exhaust gases, especially at high RPM, with the result that an OHC design can produce greater power at high RPM than an OHV design of the same displacement.

A more sophisticated version of the OHC engine is the double overhead cam (DOHC) engine, in which two separate camshafts are used to activate the intake and exhaust valves. The DOHC design permits a very light valvetrain, as the camshaft can actuate the valves directly without any intervening mechanical linkages. The DOHC design also allows some layout simplification, especially in engines that feature two intake valves and two exhaust valves (four-valve). The four-valve engine has become popular since the mid-1980s, Japanese manufacturers, in particular, have embraced the DOHC four-valve design. The DOHC design permits higher specific output than an OHC design, with the four-valve DOHC design achieving the highest ratings.

Typical specific output values for the different designs (8) in the early 1990s are as follows:

- OHV 40 to 45 BHP/liter
- OHC 50 to 55 BHP/liter
- DOHC 55 to 60 BHP/liter
- DOHC 4-valve 60 to 70 BHP/liter

A DOHC four-valve engine (9) has a maximum BMEP of up to 12 bar, and is expected to be the dominant engine design in the near future.

Thermodynamic Efficiency

Increases in thermodynamic efficiency within the limitations of the Otto cycle are obviously possible by increasing the compression ratio. However, compression ratio is also fuel octane limited, and increases in compression ratio depend on how the characteristics of the combustion chamber and the timing of the spark can be tailored to prevent knock while maximizing efficiency.

Spark timing is associated with the delay in initiating and propagating combustion of the air-fuel mixture. To complete combustion before the piston starts its expansion stroke, the spark must be initiated a few crank angle degrees ("advance") before the piston reaches top dead center. For a particular combustion chamber, compression ratio, and air-fuel mixture, there is an optimum level of spark advance for maximizing combustion chamber pressure and hence, fuel efficiency. This level of spark advance is called MBT for "maximum for best torque." However, MBT spark advance can result in knock if fuel octane is insufficient to resist preignition at the high pressures achieved with this timing. Hence, there is an interplay between spark timing and compression ratio in determining the onset of knock. Retarding timing from MBT reduces the tendency to knock but decreases fuel efficiency. Emissions of hydrocarbons and oxides of nitrogen (NO_x) are also dependent on spark timing and compression ratio, so that emission-constrained engines require careful analysis of the knock, fuel efficiency, and emission tradeoffs before the appropriate value of compression ratio and spark advance can be selected. In conventional two-valve spark ignition engines with a compression ratio of 8.5 to 9:1, spark timing close to MBT appears to be possible with regular gasoline having an octane rating of 87, even under the stringent 1993 emission regulations in force in the United States (8).

Electronic control of spark timing has made it possible to set spark timing closer to MBT relative to engines with mechanical controls. Because of production variability and inherent timing errors in a mechanical ignition timing system, the average value of timing in mechanically controlled engines had to be retarded significantly from the MBT timing so that the fraction of engines with higher-than-average advance due to production variability would be protected from knock. The use of electronic controls coupled with magnetic or optical sensors of crankshaft position has reduced the variability of timing between production engines and has also allowed better control during transient engine operation. More recently, engines have been equipped with knock sensors which are essentially vibration sensors tuned to the frequency of knock. These sensors allow for advancing ignition timing to the point where trace knock occurs, so that timing is optimal for each engine produced regardless of production variability. Figure 7 shows the ignition timing advance possible and the increase in torque with a trace knock controller in an engine of recent design. Electronic spark timing control has improved engine efficiency and fuel economy by 2% to 3% in the 1980–1990 period (10).

High-swirl, fast-burn combustion chambers have been developed during the 1980s to reduce the time taken for the air-fuel mixture to be fully combusted. The shorter the burn time, the more closely the cycle approximates the theoretical Otto cycle with constant volume combustion, and the greater the thermodynamic efficiency. Reduction in burn time can be achieved by having a turbulent vortex within the combustion chamber that promotes flame propagation and mixing. The circular motion of the air-fuel mixture is known as swirl, and turbulence is also enhanced by shaping the piston so that gases near the cylinder wall are pushed rapidly toward the center in a motion known as "squish." Recent improvements in flow visualization and computational fluid dynamics have allowed the optimization of intake valve, inlet port, and combustion chamber geometry to achieve desired flow characteristics. Typically, these designs have resulted in a 2% to 3% improvement in thermodynamic efficiency and fuel economy. The high-swirl chambers also allow higher compression ratios and reduced "spark advance" at the same fuel octane number. The use of these types of combustion chambers has allowed the compression ratio for two-valve engines to increase from 8:1 in the early 1980s to 9:1 in the early 1990s, and further improvements are likely.

Compression ratios in the future are likely to increase further with improved electronic control of spark timing and improvements in combustion chamber design. In

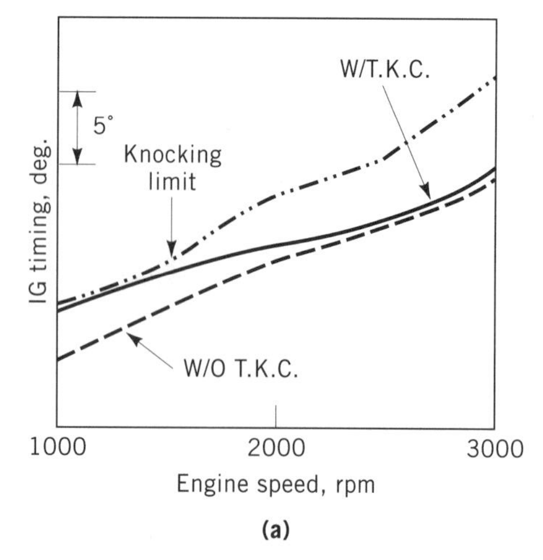

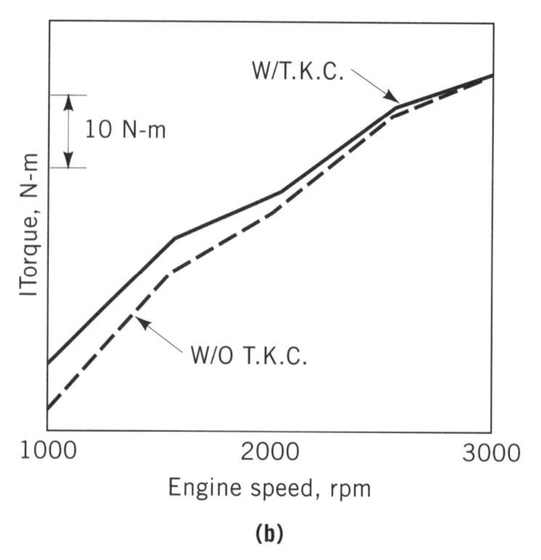

Figure 7. Trace knock control. (**a**) Ignition timing with/without trace knock control. (**b**) Effect of trace knock control on torque.

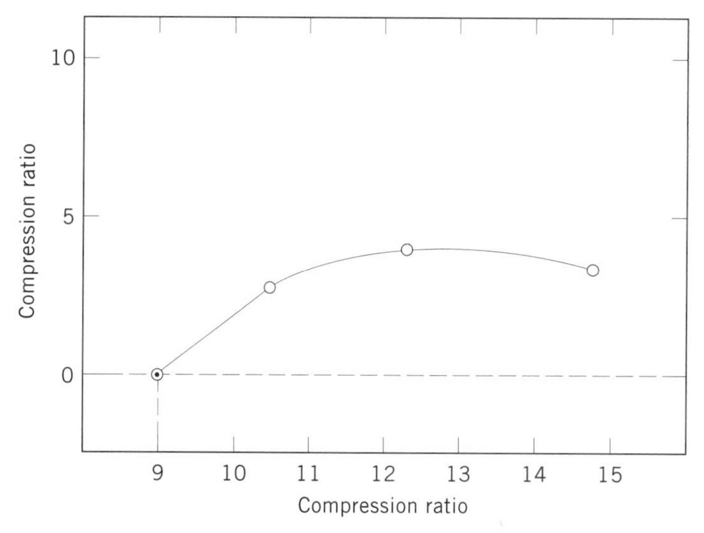

Figure 8. Effect of compression ratio on fuel economy.

newer engines of the four-valve DOHC type, the spark plug is placed at the center of the combustion chamber, and the chamber can be made very compact by having a nearly hemispherical shape. Engines incorporating these designs have compression ratios up to 10:1 while still allowing the use of regular 87 octane gasoline. Increases beyond 10:1 are expected to have diminishing benefits in efficiency and fuel economy (13) as shown in Figure 8, and compression ratios beyond 12:1 are not beneficial unless fuel octane is raised simultaneously.

Reduction in Mechanical Friction

Mechanical friction losses are being reduced by converting sliding metal contacts to rolling contacts, reducing the weight of moving parts, reducing production tolerances to improve the fit between pistons and bore, and improving the lubrication between sliding or rolling parts. Friction reduction has focused on the valvetrain, pistons, rings, crankshaft, crankpin bearings, and oil pump.

Valvetrain friction accounts for a larger fraction of total friction losses at low engine RPM than at high RPM. The sliding contract between the cam that activates the valve mechanism through a pushrod in an OHV design, or a rocker arm in an OHV design, can be substituted with a rolling contact by means of a roller cam follower, as illustrated in Figure 9. Roller cam followers have been found to reduce fuel consumption by 3% to 4% during city driving and 1.5% to 2.5% in highway driving (12). The use of lightweight valves made of ceramics or titanium is another possibility for the future. The lightweight valves reduce valve train inertia and also permit the use of lighter springs with lower tension. Titanium alloys are also being considered for valve springs that operate under heavy loads. There alloys have only half the shear modules of steel, and fewer coils are needed to obtain the same spring constant. A secondary benefit associated with lighter valves and springs is that the erratic valve motion at high RPM is reduced, allowing increased engine RPM range and power output (13).

The pistons and rings contribute to approximately half of total friction. The primary function of the rings is to minimize leakage of the air-fuel mixture from the combustion chamber to the crankcase, and oil leakage from the crankcase to the combustion chamber. The ring pack for most current engines is composed of two compression rings and an oil ring. The rings have been shown to operate hydrodynamically over the cycle, but metal-to-metal contact occurs often at the top and bottom of the stroke (14). The outward radial force of the rings are a result of installed ring tension, and contribute to effective sealing and friction. A wide variety of low-tension ring designs have been introduced in the 1980s especially since the need to conform to axial diameter variations or bore distortions has been reduced by improved cylinder manufacturing techniques. Reduced-tension rings have yielded friction reduc-

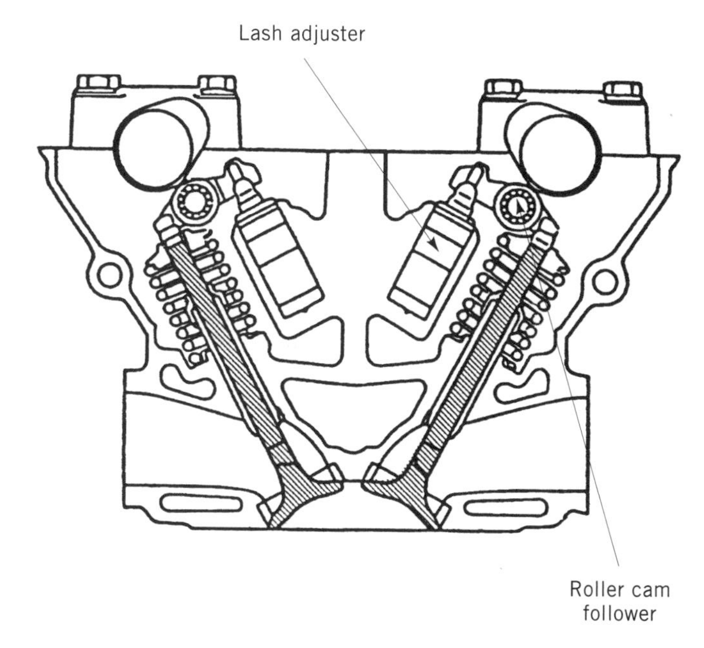

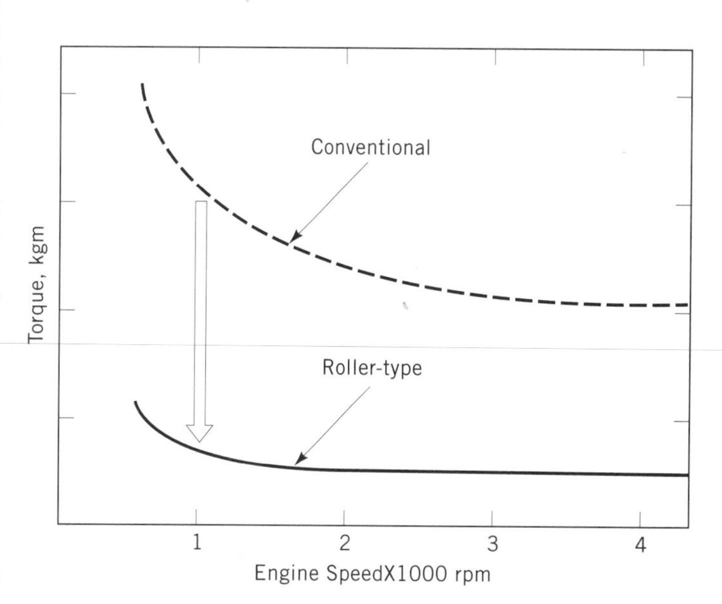

Figure 9. Roller cam follower.

tion in the range of 5% to 10%, with fuel economy improvements of 1% to 2%. Elimination of one of the two compression rings has also been tried on some engines, and two-ring pistons may be the low-friction concept for the 1990s (15).

Pistons have also been redesigned to decrease friction. Before the 1980s, piston had large "skirts" to absorb side forces associated with side-to-side piston motion due to engine manufacturing inaccuracies. Pistons with reduced skirts diminish friction by having lower surface area in contact with the cylinder wall, but this effect is quite small. A larger effect is obtained for the mass reduction of a piston with smaller skirts, and piston skirt size has seen continuous reduction in the 1980s (Fig. 10). Reducing the reciprocating mass reduces the piston-to-bore loading; analytical results indicate that 25% mass reduction reduces friction mean effective pressure by 0.7 kPa at 1500 RPM (6). Secondary benefits include reduced engine weight and reduced vibration. Use of advanced materials also result in piston weight reduction. Current lightweight pistons use hypereutectic aluminum alloys, and future pistons could use composite materials such as fiber-reinforced plastics. Advanced materials can also reduce the weight of the connecting rod, which also contributes to the side force on a piston.

The crankshaft bearings include the main bearings that support the crankshaft and the crankpin bearings, and are of the journal bearing type. These bearings contribute to about 25% of total friction while supporting the stresses transferred from the piston. The bearings run on a film of oil, and detailed studies of lubrication requirements have led to optimization of bearing width and clearances to minimize engine friction. Studies on the use of roller bearings rather than journal bearings in this application have shown that further reduction in friction is possible (16). Crankshaft roller bearings are currently used only in some two-stroke engines, such as outboard motors for boat propulsion, but their durability in automotive applications has not been established. The use of roller bearings may contribute to a 2% to 3% improvement in fuel economy.

Coatings of the piston and ring surfaces with materials to reduce wear also contribute to friction reduction. The top ring, for example, is normally coated with molybdenum, and new proprietary coating materials with lower friction are being introduced. Piston coatings of advanced high-temperature plastics or resin have recently entered the market and are claimed to reduce friction by 5% and fuel consumption by 1% (17).

The oil pumps generally used in most engines are of the gear pump type. Optimization of oil-flow rates and reduction of the tolerances for the axial georotor clearance have led to improved efficiency, which translates to reduced drive power. Friction can be reduced by 2% to 3% with improved oil pump designs for a gain in fuel economy of about 0.5% (14).

Improvements to lubricants used in the engine also contribute to reduced friction and improved fuel economy. There is a relationship between oil viscosity, oil volatility, and engine oil consumption. Reduced viscosity oils traditionally resulted in increased oil consumption, but the development of viscosity index (VI) improvers had made it possible to tailor the viscosity with temperatures to formulate "multigrade" oils such as 10W-40 (these numbers refer to the range of viscosity covered by a multigrade oil). These multigrade oils act like low-viscosity oils during cold starts of the engine, reducing fuel consumption, but retain the lubricating properties of higher-viscosity oils after the engine warms up to normal operating temperature. The recent development of 5W-30 oils and 5W-40 oils can contribute to a fuel economy improvement by further viscosity reduction of 0.5% to 1% relative to the higher-viscosity 10W-40 oil. Friction modifiers containing molybdenum compounds have also reduced friction without affecting wear or oil consumption. Future synthetic oils combining reduced viscosity and friction modifiers can offer good wear protection, low oil consumption, and extended drain capability along with small improvements to fuel economy in the range of 1% to 1.5% over current 10W-40 oils (18).

Reduction in Pumping Loss

Reductions in flow pressure loss can be achieved by reducing the pressure drop that occurs in the flow of air (air-fuel mixture) into the cylinder, and the combusted mixture through the exhaust system. However, the largest part of pumping loss during normal driving is due to throttling, and strategies to reduce throttling loss have included variable valve timing and "lean-burn" systems.

The pressure losses associated with the intake system and exhaust system have been typically defined in terms of volumetric efficiency, which is a ratio of the actual airflow through an engine to the air flow associated with fill-

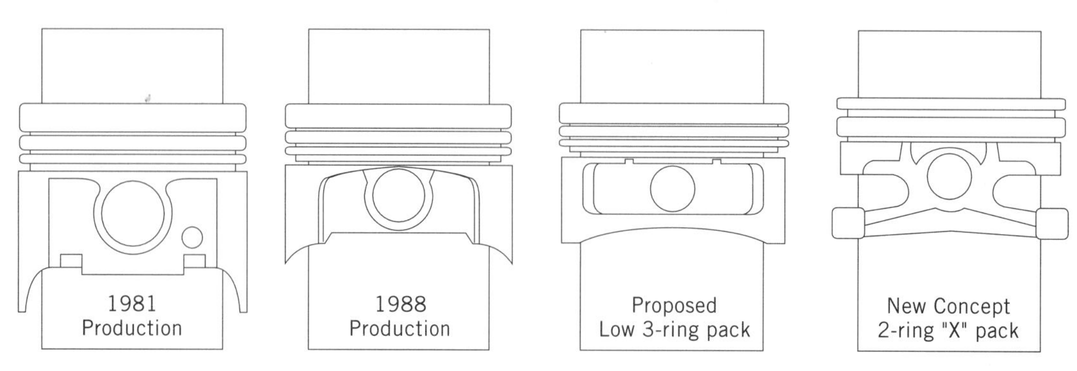

Figure 10. Evolution of piston design.

ing the cylinder completely. The volumetric efficiency can be improved by making the intake air-flow path as free of flow restrictions as possible through the air filters, intake manifolds, and valve ports. The shaping of valve ports to increase swirl in the combustion chamber can lead to reduced volumetric efficiency, leading to a tradeoff between combustion and volumetric efficiency.

More important, the intake and exhaust processes are transient, as they occur only over approximately half a revolution of the crankshaft. The momentum effects of these flow oscillations can be exploited by keeping the valves open for durations greater than half a crankshaft revolution. During the intake stroke, the intake valve can be kept open beyond the end of the intake stroke, because the momentum of the intake flow results in a dynamic pressure that sustains the intake flow even when the piston begins the compression stroke. A similar effect is observed in the exhaust process, and the exhaust valve can be held open during the initial part of the intake stroke. These flow momentum effects depend on the velocity of the flow, which is directly proportional to engine RPM. Increasing the valve opening duration helps volumetric efficiency at high RPM, but hurts it at low RPM. Increasing the exhaust valve opening time during part of the intake stroke when the intake valve is also open ("valve overlap") results in good high-RPM performance, but poor low-RPM performance when the exhaust gases lack sufficient momentum and are drawn back into the cylinder, diluting the intake change with burned gases, which can lead to rough combustion. Valve timing and overlap are selected to optimize the tradeoff between high and low RPM performance characteristics.

Improving Volumetric Efficiency

The oscillatory intake and exhaust flows can allow volumetric efficiency to be increased by exploiting resonance effects associated with pressure waves similar to those in organ pipes. The intake manifolds can be designed with pipe lengths that resonate, so that a high-pressure wave is generated at the intake valve as it is about to close, to cause a supercharging effect. Exhaust manifolds can be designed to resonate to achieve the opposite pressure effect to purge exhaust gases from this cylinder. For a given pipe length, resonance occurs only at a certain specific frequency, and its integer multiples so that, historically, "tuned" intake and exhaust manifolds could help performance only in certain narrow RPM ranges. The incorporation of resonance tanks using the Helmholtz resonator principle, in addition to tuned length intake pipes, has led to improved intake manifold design that provide benefits over broader RPM ranges. More recently, variable resonance systems have been introduced, in which the intake tube lengths are changed at different RPM by opening and closing switching valves to realize smooth and high torque across virtually the entire engine speed range. Typically, the volumetric efficiency improvement is in the range of 4% to 5% over fixed resonance systems (10).

Another method to increase efficiency is by increasing valve area. A two-valve design is limited in valve size by the need to accommodate the valves and spark plug in the circle defined by the cylinder bore. The active flow area is defined by the product of valve circumference and lift. Increasing the number of valves is an obvious way to increase total valve area and flow area, and the four-valve system which increases flow area by 25% to 30% over two-valve layouts, has gained broad acceptance. The valves can be arranged around the cylinder bore and the spark plug placed in the center of the bore to improve combustion. While the peak efficiency or bsfc of a four-valve engine may not be significantly different from a two-valve engine, there is a broader range of operating conditions where low bsfc values are realized, as shown in Figure 11 (13). Analysis of additional valve layout designs that take into account the minimum required clearance between valve seats and the spark plug location suggests that five valve designs (three intake, two exhaust) can provide an additional 20% increase in flow area, at the expense of increased valvetrain complexity (19). Additional valves do not provide further increases in flow area either be-

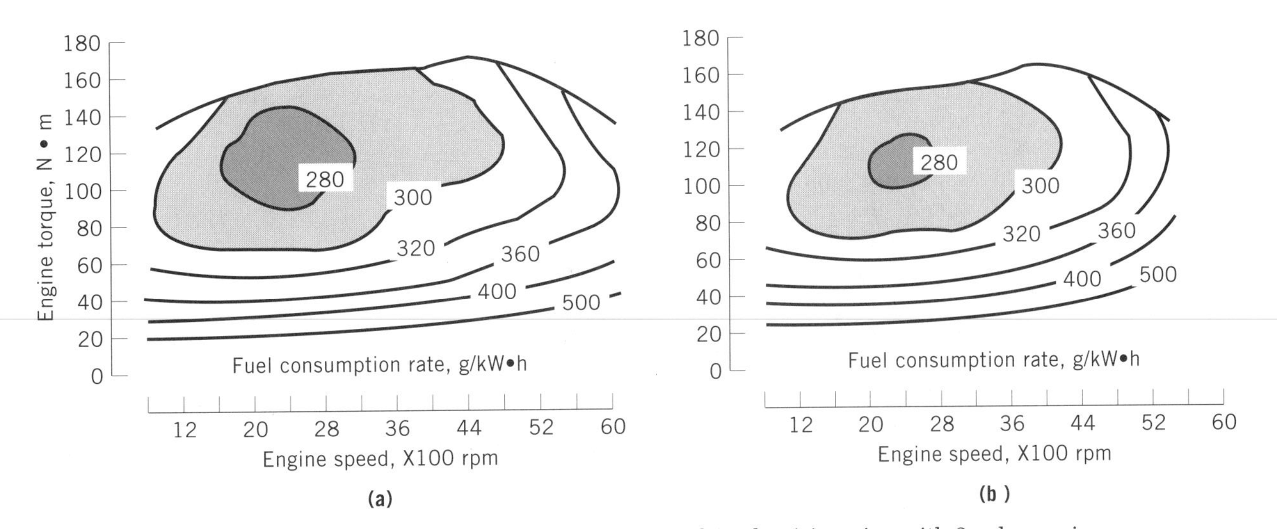

Figure 11. Fuel consumption map: Comparison of 4-valve **(a)** engine with 2-valve engine **(b)**.

cause of noncentral plug locations or valve-to-valve interference.

Efficiency improvements can be realized by changing the valve overlap period to provide less overlap at idle and low engine speeds, and greater overlap at high RPM. In DOHC engine, where separate crankshafts actuate the intake and exhaust valves, the valve overlap period can be changed by rotating the camshafts relative to each other. Such mechanisms have been commercialized in 1990–1991 by Nissan and Mercedes; these engines show low-RPM torque improvements of 7% to 10% with no sacrifice in maximum horsepower attained in the 5500 to 6000 RPM range. Variable valve overlap period is just one aspect of a more comprehensive variable valve timing system, as described in the following section.

Reduction in Throttling Loss

Under most normal driving conditions the throttling loss is the single largest contributor to reduction in engine efficiency. In s.i. engines, the air is throttled ahead of the intake manifold by means of a butterfly valve that is connected to the accelerator pedal. The vehicle's driver demands a power level by depressing or releasing the accelerator pedal, which in turn opens or closes the butterfly valve. The presence of the butterfly valve in the intake air stream creates a vacuum in the intake manifold at part throttle conditions, and the intake stroke draws in air at reduced pressure, resulting in pumping losses. These losses are proportional to the intake vacuum and disappear at wide-open throttle.

Measures to reduce throttling loss are varied. The horsepower demand by the driver can be satisfied by any combination of torque and RPM because

$$Power = Torque \times RPM$$

The higher the torque, the lower the RPM to satisfy a given power demand. Higher torque implies less throttling, and the lower RPM also reduces friction loss so that the optimum theoretical fuel efficiency at a given level of horsepower demand occurs at the highest torque level the engine is capable of. In practice, the highest level is never chosen because of the need to maintain a large reserve of torque for immediate acceleration, and also because engine vibrations are a problem at low RPM, especially near or below engine speeds refererd to as "lugging" RPM. Nevertheless, this simple concept can be exploited to the maximum by using a small displacement, high specific output engine in combination with a multispeed transmission with five or more forward gears. The larger number of gears allows selection of the highest torque/lowest RPM combination for fuel economy at any speed and load, while maintaining sufficient reserve torque for instantaneous changes in power demand. A specific torque increase of 10% can be used to provide a fuel economy benefit of 3% to 3.5% if the engine is downsized by 8% to 10%.

"Lean-burn" is another method to reduce pumping loss. Rather than throttling the air, the fuel flow is reduced so that the air-fuel ratio increases, or becomes "leaner." (In this context, the c.i. engine is a lean-burn engine.) Most s.i. engines, however, do not run well at air-fuel ratios leaner than 18:1, as the combustion quality deteriorates under lean conditions. Engines constructed with high swirl and turbulence in the intake charge can run well at air-fuel ratios up to 21:1. In a vehicle, lean-burn engines are calibrated lean only at light loads to reduce throttling loss, but run at stoichiometric or rich air-fuel ratios at high loads to maximize power. The excess air combustion at light loads has the added advantage of having a favorable effect on the polytropic coefficient (n) in the efficiency equation. Modern lean-burn engines do not eliminate throttling loss, but the reduction is sufficient to improve vehicle fuel economy by 8% to 10% (21). The disadvantage of lean-burn is that such engines cannot yet use catalytic controls to reduce emissions of oxide of nitrogen (NO_x), and the in-cylinder NO_x emission control from running lean is sometimes insufficient to meet stringent NO_x emissions standards. However, these are developments in "lean NO_x catalysts" that could allow lean-burn engines to meet the most stringent NO_x standards proposed in the future.

Another type of lean-burn s.i. engine is the stratified charge engine. Current research is focused on direct-injection stratified charge (DISC) engines in which the fuel is sprayed into the combustion chamber, rather than into or ahead of the intake valve. Typically, this enables the air-fuel ratio to vary axially or radially in the cylinder, with the richest air-fuel ratios present near the spark plug or at the top of the cylinder. Stratification requires very careful design of the combustion chamber shape and intake swirl, and of the fuel injection system. Laboratory tests have indicated the potential to maintain stable combustion at total air-fuel ratios as high as 40:1 (21). Maintaining stratification over a wide range of loads and speeds has been problematic in practice. As a result, DISC engines have not been commercialized, but remain an interesting possibility for the future.

Variable valve timing is another method of reducing throttling loss. By closing the intake valve early, the intake process occurs over a smaller fraction of the cycle, resulting in a lower vacuum in the intake manifold. It is possible to completely eliminate the butterfly valve that throttles air and achieve all part-load settings by varying the intake-valve opening duration. However, at very light load, the intake valve is open for a very short duration, and this leads to weaker in-cylinder gas motion and reduced combustion stability. At high RPM, the throttling loss benefits are not realized fully. Throttling occurs at the valve when the valve closing time increases raltive to the intake stroke duration at high speeds, because of the valvetrain inertia. Hence, throttling losses can be decreased by 80% at light-load, low-RPM conditions, but by only 40% to 50% at high RPM, even with fully variable valve timing (22).

Variable valve timing can also provide a number of other benefits, such as reduced valve overlap at light loads/low speeds (discussed earlier) and maximized output over the entire range of engine RPM. Fully variable valve timing can result in engine output levels of up to 100 BHP/liter at high RPM with little or no effect on low speed torque. In comparison to an engine with fixed valve timing that offers equal performance, fuel efficiency improvements of 7% to 10% are possible. The principal drawback has historically been the lack of a durable and low-cost mechanism to implement valve timing changes. Recently, Honda has commercialized a two-stage system in its four-valve/cylinder engines in which, depending on engine speed and load, one of two valve timing and lift schedules are realized for the intake valves. This type of engine has been combined with lean-burn to achieve remarkable efficiency in a small car (23). Fuel economy benefits of up to 15% are possible with such strategies.

Another simpler version of the system of variable valve timing simply shuts off individual cylinders by deactivating the valves. For example, an eight-cylinder engine can operate at light load as a four-cylinder engine (by deactivating the valves for four of the cylinders) and as a six-cylinder engine at moderate load. Such systems have also been tried on four-cylinder engines in Japan with up to two cylinders deactivated at light load. At idle, such systems have shown a 40% to 45% decrease in fuel consumption, and composite fuel economy has improved by 10% to 12% because both pumping and frictional losses are reduced by cylinder deactivation (24). However, the system has problems associated with noise, vibration, and emissions that have resulted in reduced acceptance in the marketplace.

EVOLUTION OF THE COMPRESSION IGNITION (DIESEL) ENGINE

Fuel Efficiency Relative to S.I. Engines

Compression ignition engines, commonly referred to as Diesel engines, are in widespread use. Most c.i. engines in light-duty vehicle applications are of the indirect-injection type (IDI), whereas most c.i. engines in heavy-duty vehicles are of the direct-injection type. In comparison to s.i. engines, c.i. engines operate at much lower brake mean effective pressures of (typically) about 7 to 8 bar at full load. Maximum power output of a c.i. engine is limited by the rate of mixing between the injected fuel spray and hot air. At high fueling levels, inadequate mixing leads to high black smoke, and the maximum horsepower is usually smoke limited for most current c.i. engines. Naturally aspirated diesel engines for light-duty vehicle use have specific power outputs of 25 to 35 BHP per liter, which is about half the specific output of a modern s.i. engine. However, fuel consumption is significantly better, and c.i. engines are preferred over s.i. engines where fuel economy is important.

Because of the combustion process and the high internal friction of a c.i. engine, maximum speed is typically limited to less than 4500 RPM, which partially explains the lower specific output of c.i. engines. In light-duty vehicle use, an IDI engine can display between 20% and 40% better fuel economy depending on whether the comparison is based on engines of equal displacement or of equal power output in the same RPM range (4). The improvement is largely due to the superior part-load efficiency of the c.i. engine, as there is no throttling loss. At high vehicle speeds (>120 km/h), the higher internal friction of the c.i. engine offsets the reduced throttling loss, and the fuel efficiency difference between s.i. and c.i. engines narrows considerably.

Most of the evolutionary improvements for compression ignition engines in friction and pumping loss reduction are conceptually similar to those described for s.i. engines, and this section focuses on the unique aspects of c.i. engine improvements.

Design Parameters

C.i. engines have also adopted some of the same valvetrain designs as those found in s.i. engines. While most c.i. engines are of the OHV type, many recent European c.i. engines for passenger car use are of the OHC type. The c.i. engine is not normally run at high RPM, so the difference in specific output between an OHV and an OHC design is small. The OHC design does permit a simpler and lighter cylinder block casting, which is beneficial for overcoming some of the inherent weight liabilities. OHC designs also permits the camshaft to activate the fuel injector directly in "unit injector" designs, which are capable of high injection pressure and very fine atomization of the fuel spray. DOHC designs are not used as of 1993 although they are possible in future four-valve engines of the DI type.

Thermodynamic Efficiency

The peak efficiency of an IDI engine is comparable to or only slightly better than the peak efficiency of an s.i. engine, based on average values for engines in production. The contrast between theoretical and actual efficiency is notable; part of the reason is that the prechamber in the IDI diesel is a source of energy loss. The design of the prechamber is optimized to promote swirl and mixing of the fuel spray with air, but the prechamber increases total combustion time. Its extra surface area also results in more heat transfer into the cylinder head. As can be seen from Figure 12, the prechamber is connected to the main combustion chamber by a small passage (or passages), and the flow of hot, partially combustsed gas through these passages at high velocity promotes further mixing but also results in pressure losses. The main advantage of the prechamber is that it promotes smoother and more complete combustion and does not require a very-high-pressure fuel-injection system (4).

Direct-injection (DI) systems avoid the heat and flow losses from the prechamber by injecting the fuel into the combustion chamber. The combustion process in DI diesels occurs in two phases. The first phase consists of an ignition delay period followed by spontaneous ignition of the fuel droplets. The second phase is characterized by diffusion burning of the droplets. The fuel-injection system must be capable of injecting very little fuel during the first phase, and providing highly atomized fuel and promoting intensive mixing during the second. Historically, the mixing process has been aided by creating high swirl in the combustion chamber to promote turbulence. However, high swirl and turbulence also lead to flow losses and heat losses, thus reducing efficiency. The newest concept is the "quiescent" chamber in which all the mixing is achieved by injecting fuel at very high pressures to promote fine atomization and complete penetration of the air in the

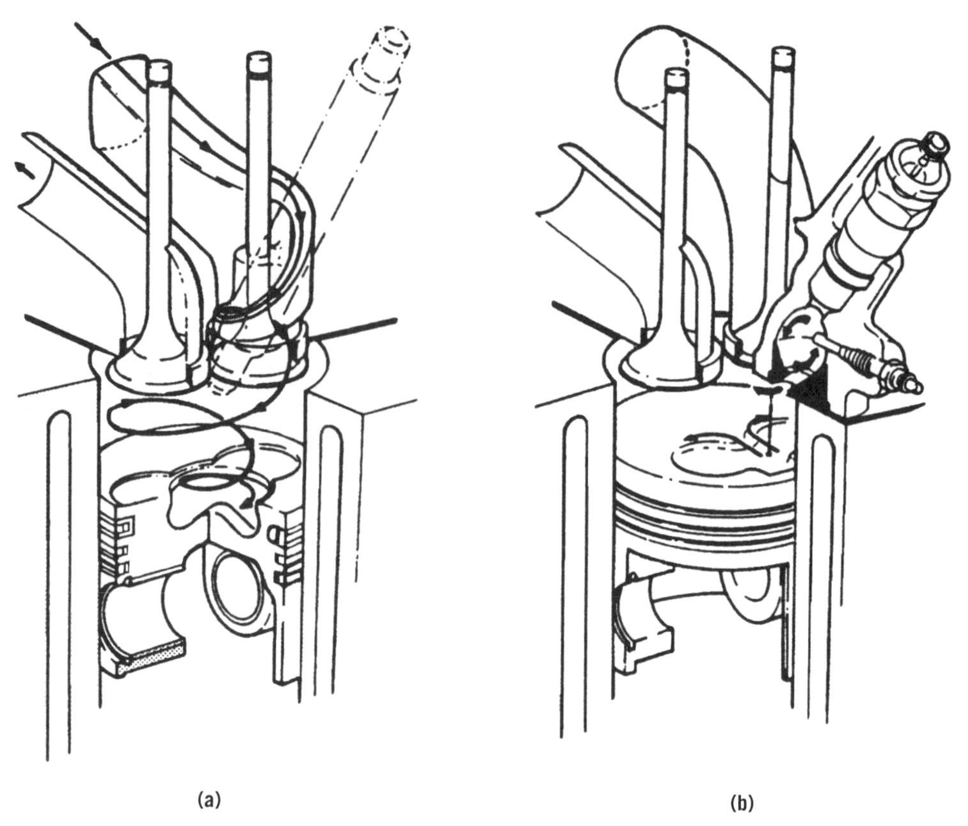

Figure 12. Comparison of combustion chambers of a DI **(a)** and IDI **(b)** engine.

combustion chamber. New fuel injection systems using "unit injectors" can achieve pressures in excess of 1000 bar, almost twice as high as injection pressures used previously. Quiescent combustion chamber designs with high-pressure fuel-injection systems have proved to be very fuel efficient and are coming into widespread use in heavy-duty truck engines. These systems have the added advantage of reducing particulate and smoke emissions (2).

DI engines have only recently entered the light-duty vehicle market, but these engines still use swirl-type combustion chambers. In combination with turbocharging, the new DI engines have attained peak efficiencies of 43%. Fuel economy improvements in the composite cycle relative to IDI engines are in the 12% to 15% range, and are up to 40% higher than naturally aspirated s.i. engines with similar torque characteristics (25). It is not clear if quiescent combustion chambers will ever be used in DI engines for cars, because the size of the chamber is quite small and fuel impingement on cylinder walls is a concern. The bsfc map of a modern high-speed automotive DI engine is shown in Figure 13.

Although the efficiency equation shows that increasing compression ratio has a positive effect on efficiency, practical limitations preclude any significant efficiency gain through this method. At high compression ratios, the size of the combustion chamber is reduced, and the regions of "dead" air trapped between the cylinder and piston edges and crevices became relatively large, leading to poor air utilization, reduced specific output, and, potentially, more smoke. Moreover, the stresses on the engine increase with increasing compression ratio, making the engine heavy and bulky. Currently, the compression ratios are already somewhat higher than optimal to provide enough heat of compression so that a cold start at low temperature is possible.

The "adiabatic diesel" has also received much attention. In this c.i. engine, the combustion chamber walls are insulated to prevent heat loss; the availability of low-thermal-conductivity ceramic materials has provided the impetus for this research. In such an insulated engine, the reduction in heat lost to the walls simply increases heat lost to exhaust gas, because the expansion ratio of a piston engine is fixed. Recovery of the waste heat of exhaust by a separate power turbine can provide significant gains in total efficiency. Unfortunately, the characteristics of turbomachinery are such that they recover energy only over a restricted range of operating conditions, typically conditions close to the maximum torque and RPM of the engine. Engines incorporating a waste heat recovery turbine, known as turbocompound engines, are best suited to heavy-duty vehicles where full power operation is common, but are not well suited to light-duty vehicles (26).

Insulated engines have other disadvantages that may prevent their commercialization. The hot cylinder walls heat the intake gases, reducing volumetric efficiency and specific power output from the piston engine. The high temperatures also lead to NO_x emission problems and problems with fuel ignition upon wall contact. Hence, theoretical efficiency advantages have not been attained in practice. However, some manufacturers have reported modest successes in insulating the prechamber in IDI diesels, although efficiency gains have been small (27).

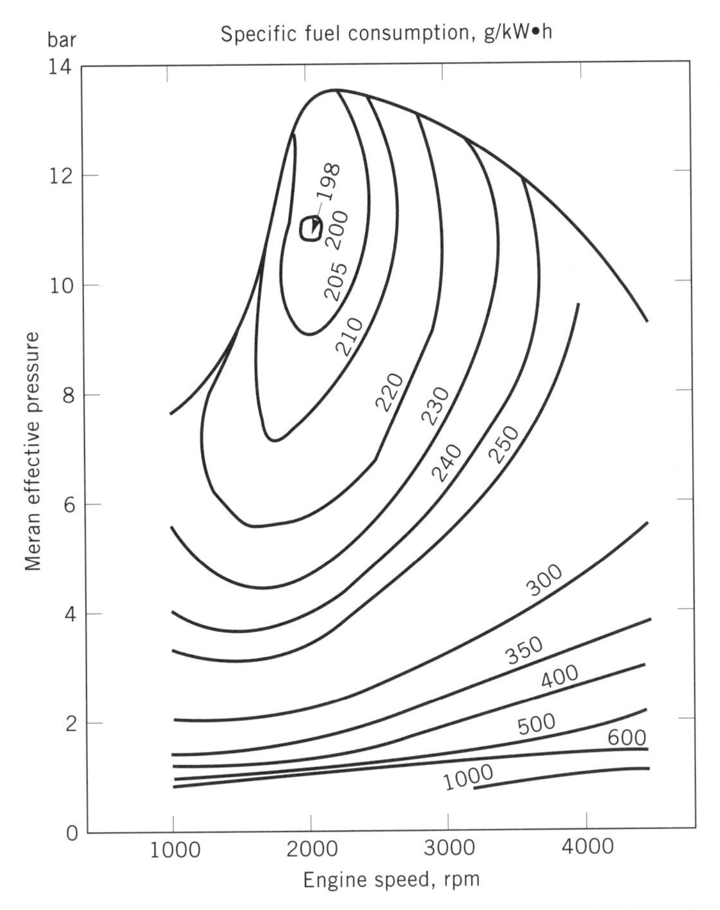

Figure 13. Fuel consumption map for a modern turbocharged DI diesel passenger car engine.

Friction and Pumping Loss

Most of the friction-reducing technologies that can be adopted in s.i. engines are conceptually similar to those that can be adopted for diesels. There are limitations to the extent of reduction of ring tension and piston size owing to the high compression ratio of c.i. engines, but roller cam followers, optimized crankshaft bearings, and multigrade lubricants have also been adopted for c.i. engine use. Because friction is a larger fraction of total loss, a 10% reduction in fraction in a c.i. engine can lead to a 3% to 4% improvement in fuel economy.

Pumping losses are not as significant a contributor to overall energy loss in a c.i. engine, but tuned intake manifolds and improved valve port shapes and valve designs have also improved the volumetric efficiency of modern c.i. engines. Four-valve designs, in widespread use in the heavy truck market, have only recently appeared in passenger cars, but their benefits are smaller in c.i. engine use because of the low maximum RPM relative to s.i. engines. Nevertheless, the four-valve head with a centrally mounted injector is particularly useful in DI engines because it allows for symmetry in the fuel spray with resultant good air utilization.

Variable valve timing or any form of valve control holds little benefit for c.i. engines because of the lack of throttling loss or the need for high RPM performance. Valve timing can be varied to reduce the effective compression ratio so that a very high ratio can be used for cold starts, and a lower, more optimal, ratio can be used for fully warmed up operation. In future, improvements to volumetric efficiency and reduction in throttling loss are expected to contribute a 3% to 5% improvement in fuel economy.

INTAKE CHARGE BOOSTING

Most c.i. and s.i. engines for light vehicles use intake air at atmospheric pressure. One method to increase maximum power at wide-open throttle is to increase the density of air supplied to the intake by precompression. This permits a smaller displacement engine to be substituted without loss of power and acceleration performance. The use of a smaller displacement engine reduces pumping loss at part load and friction loss. However, intake charge compression has its own drawbacks. Its effect is similar in some ways to raising the compression ratio as regards peak cylinder pressure, but maximum cylinder pressure is limited in s.i. engines by the fuel octane. Charge-boosted engines generally require premium gasolines with higher octane number if the charge-boost levels are more than 0.2 to 0.3 bar over atmospheric in vehicles for street use. Racing cars use boost levels up to 1.5 bar in conjunction with a very high octane fuel such as methanol. This limitation is not present in a c.i. engine, and charge boosting is much more common in c.i. engine applications.

Intake charge boosting is normally achieved by the use of turbochargers or superchargers. Turbochargers recover the wasted heat and pressure in the exhaust through a turbine, which in turn drives a compressor to boost intake pressure (see Plate II). Superchargers are generally driven

Plate II. Turbocharged engines have boosted intake charge caused by the recapturing of heat and pressure in the exhaust through a turbine. This engine utilizes air-to-air aftercooling. Courtesy of Ford.

by the engine itself and are theoretically less efficient than a turbocharger. Many engines that use either device also use an aftercooler that cools the compressed air as it exits from the supercharger or turbocharger before it enters the s.i. engine. The aftercooler increases engine specific power output by providing the engine with a denser intake charge, and the lower temperature also helps in preventing detonation, or knock. Charge boosting is useful only under wide-open throttle conditions in s.i. engines, which occur rarely in normal driving, so that such devices are usually used in high-performance vehicles. In c.i. engines, charge boosting is effective at all speeds and levels.

Turbochargers

Turbochargers in automotive applications are of the radial flow turbine type. The turbine extracts pressure energy from the exhaust stream and drives a compressor that increases the pressure of the intake air (27). The process is shown schematically in Figure 14. There are a number of issues that affect the performance of turbomachinery, some of which are a result of natural laws governing the interrelationship between pressure, airflow, and turbocharger speed. Turbochargers do not function at light load because there is very little energy in the exhaust stream. At high load, the turbocharger's ability to provide boost is a nonlinear function of exhaust flow. At low engine speed and high load, the turbocharger provides little boost, but boost increases very rapidly beyond a certain flow rate that is dependent on the turbocharger size. The turbocharger also has a maximum flow rate, and the matching of a turbochargers' flow characteristics to a piston engines' flow requirements involves a number of tradeoffs (28). If the turbocharger is sized to provide adequate charge boost at moderate engine speeds of 2000 to 3000 RPM, high-RPM boost is limited and there is a sacrifice in maximum power. A larger turbocharger capable of maximizing power at high RPM sacrifices the ability to provide boost at normal driving conditions. At very low RPM (for example, when accelerating from a stopped condition), no practical design

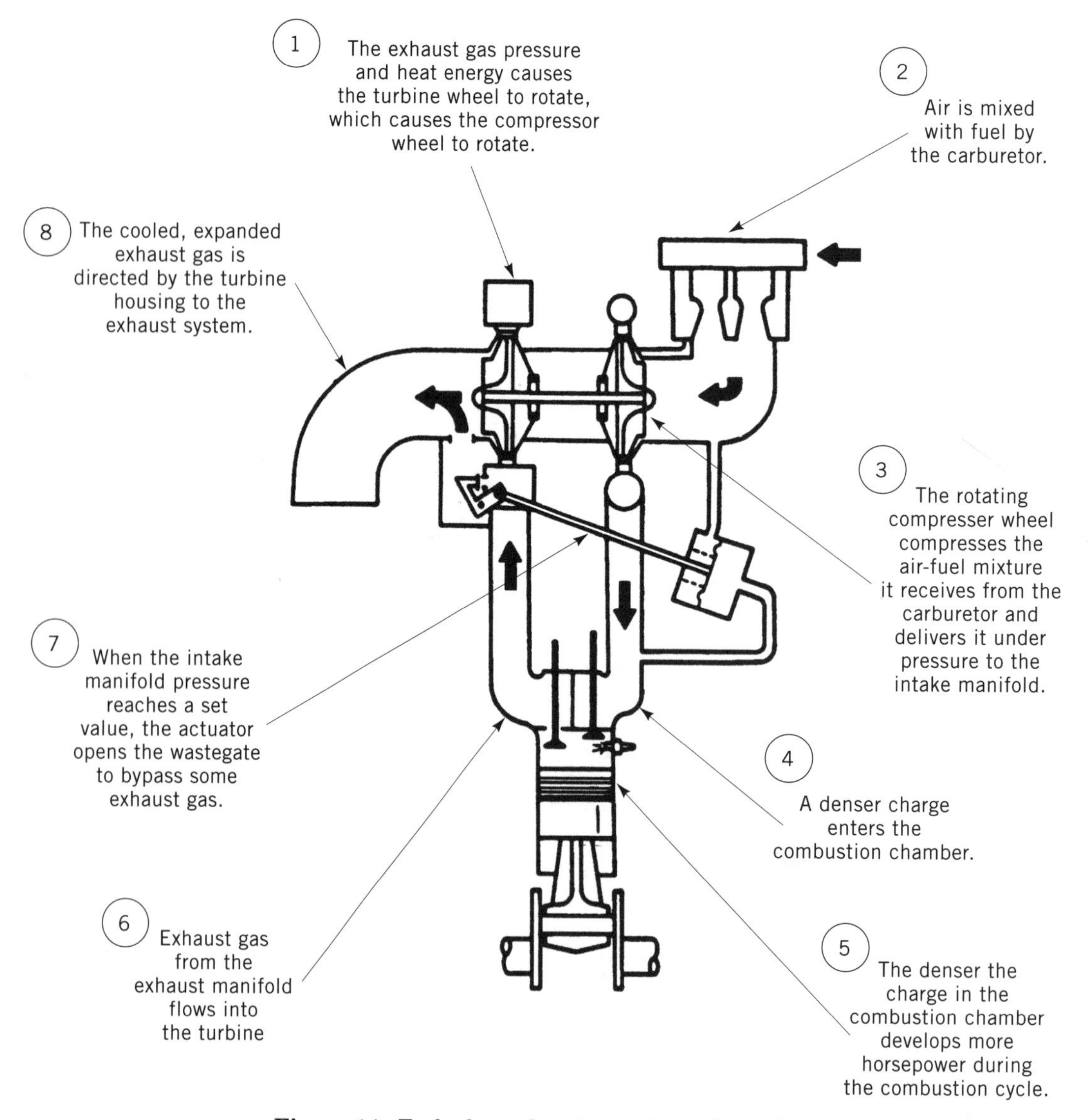

Figure 14. Turbocharged engine system schematic.

provides boost immediately. Moreover, the addition of a turbocharger requires the engine compression ratio to be decreased by 1.5 to 2 points (or 1 to 1.5 with an aftercooler) to prevent detonation. The net result is that turbocharged engines have *lower* brake specific fuel efficiencies than engines of equal size, but can provide some efficiency benefit when compared to engines of equal midrange or top-end power. During sudden acceleration, the turbocharger does not provide boost instantaneously because of its inertia, and turbocharged vehicles have noticeably different acceleration characteristics than naturally aspirated vehicles. This factor, coupled with the lack of low speed boost, has limited consumer acceptance of turbocharged s.i. engines to those primarily interested in very high performance.

Maximum boost pressure in an s.i. engine is limited by the fuel octane and engine compression ratio, and boost is capped at this level by bleeding the exhaust gas energy before it reaches the turbine. A turbocharged engine that emphasizes midrange boost can be substituted for a naturally aspirated engine that is 30% to 35% larger, and a net fuel economy gain of about 7% to 10% is possible because of pumping loss reduction. Turbocharged engines that maximize high-RPM power are not considered to be fuel efficiency enhancing because the lack of low-speed boost inherently limits that potential to downsize the engine without a large sacrifice in low-speed performance. Such designs can increase specific power by 40% to 50% at high RPM.

Turbochargers are much better suited to c.i. engines because these engines are unthrottled and the combustion process is not knock limited. Airflow at a given engine load/speed setting is always higher for a c.i. engine relative to an s.i. engine, and this provides a less restricted operating regime for the turbocharger. The lack of a knock limit also allows increased boost and removes the need to cap boost pressure under most operating conditions. In general, turbocharged c.i. engines offer up to 50% higher specific power and torque, and about 10% better fuel economy than naturally aspirated c.i. engines of approximately equal torque capability (26).

Superchargers

Most s.i. engine superchargers are driven off the crankshaft and are of the Roots blower or positive displacement pump type. In comparison to turbochargers, these superchargers are bulky and weigh considerably more. In addition, the superchargers are driven off the crankshaft, absorbing 3% to 5% of the engine power output depending on pressure boost and engine speed.

The supercharger, however, does not have the low-RPM boost problems associated with turbochargers, and also can be designed to nearly eliminate any time lag in delivering the full boost level. As a result, the superchargers are more acceptable to consumers from a drivability viewpoint. The need to reduce engine compression ratio and the supercharger's drive power requirement detract from overall efficiency. In automotive applications, a supercharged engine can replace a naturally aspirated engine that is 30% to 35% larger in displacement, with a net pumping loss reduction (30). Overall, fuel economy improves by about 8% or less, if the added weight effects are included.

Superchargers are less efficient in combination with c.i. engines, because these engines run lean even at full load, and the power required for compressing air is proportionally greater. Supercharged c.i. engines in passenger car applications are not commercially available, because the turbocharger appears far more suitable in these applications.

ALTERNATIVE HEAT ENGINES

A number of alternative engines types have been researched for use in passenger cars but have not yet proved successful in the marketplace. A brief discussion of the suitability of four engines for automotive power plants follows.

Wankel Engines

The Wankel engine is the most successful of the four engines in that it has been in commercial production in limited volume for two decades. The thermodynamic cycle is identical to that of a four-stroke engine, but the engine does not use a reciprocating piston in a cylinder. Rather, a triangular rotor spins eccentrically inside a figure eight–shaped casing. The volume trapped between the two rotor edges and the casing varies with rotor position, so that the intake, comparison, expansion, and exhaust stroke occur as the rotor spins through one revolution. The engine is very compact relative to a piston s.i. engine of equal power, and the lack of reciprocating parts provides very smooth operation. However, the friction associated with the rotor seals is high, and the engine also suffers from more heat losses than an s.i. engine. For these reasons, the Wankel engine's efficiency has always been below that of a modern s.i. piston engine (31).

Two-Stroke Engines

The two-stroke engine is widely used in small motorcycles but was considered too inefficient and polluting for use in passenger cars. A more recent development is the use of direct-injection stratified charge (DISC) combustion with this type of engine. One of the major problems with the two-stroke engine is that the intake stroke overlaps with the exhaust stroke, resulting in some intake mixture passing uncombusted into the exhaust. The use of a DISC design avoids this problem because only air is inducted during intake. Advanced fuel-injection systems have been developed to provide a finely atomized mist of fuel just before spark initiation and sustain combustion at light loads. The two-stroke engines of this type are thermodynamically less efficient than four-stroke DISC engines, but the internal friction loss and weight of a two-stroke engine are much lower than those of a four-stroke engine of equal power (32). As a result, the engine may provide fuel economy equal or superior to that of a DISC (four-stroke) engine when installed in a vehicle. Experimental prototypes have achieved good results, but the durability and emissions performance of advanced two-strokes is still not established.

Gas Turbine Engines

These engines are widely used to power aircraft, and considerable research has been completed to assess their use in automobiles. Such engines use continuous combustion of fuel, which holds the potential for low emissions and multifuel capability. The efficiency of the engine is directly proportional to the combustion temperature of the fuel, which has been constrained to 1200°C by the metals used to fabricate turbine blades. The use of high-temperature ceramic materials for turbine blades and the use of regenerative exhaust waste heat recovery were expected to increase the efficiency of gas turbine engines to levels significantly higher than the efficiency of s.i. engines (33).

In reality, such goals have not yet been attained, partly because the gas turbine components become less aerodynamically efficient at the small engine sizes suitable for passenger car use. Part-load efficiency is a major problem for gas turbines because of the nonlinear efficiency changes with air-flow rates in turbomachinery. In addition, the inertia of the gas turbine makes it poorly suited to passenger car applications, where speed and load fluctuations are rapid in city driving. As a result, there is little optimism that the gas turbine–powered car will be a reality in the foreseeable future.

Stirling Engines

These engines have held a particular fascination for researchers because the cycle closely approximates the ideal Carnot cycle, which extracts the maximum amount of work theoretically possible from a heat source. This engine is also a continuous combustion engine like the gas turbine engine. While the engine uses a piston to convert heat energy to work, the working fluid is enclosed and heat is conducted in and out of the working fluid by heat exchangers. To maximize efficiency, the working fluid is a gas of low molecular weight, such as hydrogen or helium. Prototye designs of Stirling engine have not yet attained efficiency goals and have had other problems, such as the containment of the working fluid (34). The Stirling engine is, like the gas turbine, not well suited to applications where the load and speed change rapidly, and much of the interest in this engine has faded in recent years.

BIBLIOGRAPHY

1. C. Lichty, *Combustion Engine Processes*, 7th ed., 1967.
2. M. J. Hower and co-workers, *The New Navistar T444E Direct Injector Turbocharged Diesel Engine*, SAE Paper 930269, March 1993.
3. H. R. Ricardo, *The High Speed Internal Combustion Engine*, 4th ed., 1953.
4. H. W. Barnes Moss, and W. M. Scott, *The High Speed Engine for Passenger Cars*, Institute of Mechanical Engineers Paper C15/75, 1975.
5. H. Omori and S. Ogino, *Waste Heat Recovery of Passenger Car Using a Combination of Rankine Bottoming Cycle and Evaporative Engine Cooling System*, SAE Paper 930880, March 1993.
6. J. H. Tuttle, *Controlling Engine Load by Means of Early Intake Valve Closing*, SAE Paper 820408, 1982.
7. FEV of America, *Spark Ignition Engine Development*, 1992.
8. Martin Marietta Energy Systems, *Documentation of the Attributes of Technologies to Improve Automotive Fuel Economy*, Contractor Report, 1993.
9. J. Abthoss and co-workers, *Diamler Benz 2.3 Litre, 16-valve High Performance Engine*, SAE Paper 841226, 1984.
10. T. Sakono and co-workers, *Mazda New Lightweight and Compact V6 Engines*, SAE Paper 920677, 1992.
11. K. Suzuki and M. Takimoto, "Fuel Economy Potential and Prospects," Presentation to the National Academy of Sciences, July 1991.
12. Automotive Engineering, *Technical Highlights of the 1987 Automobiles*, **94**(10), Volume (Oct. 1987).
13. Toyota Motors, *Toyota Engine Technology*, 1990.
14. J. T. Kovach, E. A. Tsakiris, and L. T. Wong, *Engine Friction Reduction for Improved Fuel Economy*, SAE Paper 820085, 1982.
15. Nissan Research and Development, *The SR18DI Engine*, Nissan Press Release, 1990.
16. Automotive Engineering, *Roller Bearings for I.C. Engines?* **95**(4) (1987).
17. A. Tanake, T. Sugiyama, and A. Kotani, *Development of Toyota JZ Type Engine*, SAE Paper 930881, 1993.
18. T. J. Cousineau, T. F. McDonnell, and D. G. Witt, *Second-Generation SAW 5W-30 Passenger Car Oils*, SAE Paper 8615, 1986.
19. K. Aoi, K. Nomura, and H. Matsuzaka, *Optimization of Multi-Valve, Four Cycle Engine Design*, SAE Paper 860032, 1986.
20. Honda Motors, *The VTEC-E Engine*, Honda Press Information, July 30, 1991.
21. M. Misumi, R. Thring, and S. Ariga, *An Experimental Study of a Low Pressure DISC Engine Concept*, SAE Paper 900653, 1990.
22. Y. Urata and co-workers, *A Study of Vehicle Equipped with Non-Throttling SI Engine with Early Intake Valve Closing Mechanism*, SAE Paper 930820, 1993.
23. *EPA Test Car List*, Honda Civic VX, 1994.
24. K. Hatano and co-workers, *Development of a New Multi-Mode Variable Valve Timing Engine*, SAE Paper 93078, 1993.
25. D. Stock and R. Bauder, *The New Audi 5-Cylinder Turbo Diesel Engine*, SAE Paper 900648, 1990.
26. C. Amman, "Promises and Challenges of the Low-Heat Rejection Diesel," GM Research Publication GMR-6188, 1988.
27. H. Kawamura, *Development Status of Isuzu Ceramic Engine*, SAE Paper 880011, 1988.
28. H. H. Dertian and co-workers, *Turbochanging Ford's 2.3 Liter Spark Ignition Engine*, SAE Paper 790312, 1979.
29. H. Hiereth and G. Withelm, *Some Special Features of the Turbocharged Gasoline Engine*, SAE Paper 790207, 1979.
30. *EPA Test Car List*, Vehicle Specifications, 1992.
31. H. A. Kuck, V. Fleischer, and W. Schnorbus, *VW's New 1.3L High Performance Supercharged Engine*, SAE Paper 860102, 1986.
32. C. Amman, *The Automotive Engine—A Future Perspective*, GM Research Publication GMR-6653, 1989.
33. K. Hellman and co-workers, *Evaluation of Research Prototype Vehicle Equipped with DISC Two Stroke Engines*, EPA Technical Report EPA/AA/CTAB 92-01, 1992.
34. C. Amman, *Why Not A New Engine?* SAE Paper 801428, 1980.

B

BATTERIES

George E. Blomgren
Eveready Battery Co., Inc.

Batteries are storehouses for electrical energy on demand. Common sizes range from large house-sized batteries for utility storage to several liter-sized batteries for starting, lighting, and ignition of vehicles to tiny coin- and button-sized cells for electronic applications that require only small amounts of capacity (see Plate I). The most important aspect of battery technology is that the chemicals involved in the battery reaction are converted directly into electrical energy in contrast to heat engines, which involve an intermediate thermal or combustion process to convert the chemical energy into electrical energy. The result of this direct conversion is that all the free energy of the chemical system is available for the conversion rather than being limited by the Carnot cycle, as are all heat engines. Furthermore, the conversion of electrical to mechanical energy is quite efficient because of the inherent simplicity and low friction of electrical motors. This has led to many applications for batteries, although the practical large-scale use of batteries in vehicle motive power has remained elusive. Conversion of electricity to light or sound is also efficient and easily controlled. As a result, many applications for batteries in lighting and the creation and reproduction of sound have also been developed.

There are, of course, losses in batteries as in any other energy conversion device. As a result, much of the history of battery technology has been concerned with finding ways to reduce these losses to as low a level as possible. All types of losses increase with current, so they become particularly important as the application demands higher current (see also FUEL CELLS).

The three main types of batteries are primary, secondary, and reserve. A primary battery is designed to be discharged only one time and then discarded. A secondary battery is rechargeable and can be used like the primary battery, then recharged and used again, repeating the cycle until the capacity fades or sudden loss of capacity occurs, due usually to an internal short circuit. A reserve battery involves a special construction that normally keeps active materials well separated until the time for use occurs; then an activation device readies the battery for use. This type of battery is designed for long storage before use.

The term *battery* is usually used to mean one or more electrically connected galvanic cells, although some authors restrict the definition to more than one cell. Some other useful definitions follow.

Anode is the negative electrode of a primary cell and is always associated with oxidation or the release of electrons into the external circuit. In a rechargeable cell, the anode is the negative pole during discharge and the positive pole during charge.

Cathode is the positive electrode of a primary cell and is always associated with reduction or taking of electrons from the external circuit. In a rechargeable cell, the cathode is the positive pole during discharge and the negative pole during charge.

Electrolyte is a material that provides ionic conductivity between the positive and negative electrodes of the cell.

Separator is a physical barrier between the positive and negative electrodes to prevent direct shorting of the electrodes. Separators must be permeable to ions, but must not conduct electrons. They must be inert in the total environment.

Open-circuit voltage is the voltage measured across the terminals of the cell or battery when no external current is flowing. When measured on a single cell, it is usually close to the thermodynamic electromotive force (emf).

Closed-circuit voltage is the voltage measured across the terminals of the cell or battery when current is flowing into the external circuit.

Discharge is the operation of a cell when current flows spontaneously from the battery into an external circuit.

Charge is the operation of a cell when an external source of current reverses the electrochemical reactions of the cell to restore the battery to its original state of charge.

Internal resistance or *impedance* of the battery is the resistance or impedance of the battery to the flow of current. This resistance operates in addition to the resistance of the external load.

Electrochemical couple is the combination of the electrode reactions of the anode and cathode to form the complete galvanic cell. The number of electrons given up by the anode to the external circuit must be identical with the number of electrons withdrawn from the external circuit by the cathode.

Batteries have had both a positive and a negative impact on the environment. Lead, cadmium, and mercury are highly toxic elements that have frequently been discarded to land fills or incinerators. This practice has changed greatly, and the use of some of these elements has also declined sharply. The positive aspect of their use has been as well-contained energy conversion devices that contribute much to the needs of mankind. This aspect is made emphatic by the requirement in the near future that zero emission vehicles be produced in the United States and the fact that only battery-powered electric vehicles qualify for this application. Also, many nonpolluting energy conversion devices such as photovoltaic systems require the concomitant use of rechargeable batteries for energy storage so the devices can be used at night or in cloudy conditions.

PRIMARY CELLS

As noted previously, primary cells are galvanic cells designed to be discharged only once. Manufacturers of primary cells strongly recommend that consumers should not attempt to recharge them because of possible safety hazards, such as leakage or gas generation causing venting.

Plate I. Batteries range in size from those that power hearing-aids to those that start cars. All batteries are electrically connected galvanic cells. Courtesy of Eveready Battery.

The cells are designed to have the maximum possible energy in each cell size because of the single discharge, so that comparisons between battery types on the basis of the energy density in Wh/L, and the specific energy, in Wh/kg, are especially apt.

The main categories of primary cells are carbon-zinc cells (heavy duty and general purpose), alkaline cells (cylindrical and miniature), and lithium cells.

The growth of sales has been tied mostly to growth in the electronics industry, although the sales in battery-operated toys has also shown substantial growth. The market in other parts of the world reflects the historical dominance of carbon-zinc cells. The high proportion of lithium cells reflects the important photographic market in Japan. Western European sales are similar to the Japanese sales, whereas the sales in less developed countries are almost totally dominated by carbon-zinc batteries.

A number of multinational suppliers of batteries have manufacturing facilities in many countries and a broad line of products. These companies account for a major share of the batteries manufactured in the world. Table 1 lists the companies that have sales of primary batteries of greater than $100 million, the regions of their headquarters, and their product lines.

Table 1. Battery Companies Manufacturing Primary Cells

	Cell Type		
Company	Carbon–Zinc	Alkaline	Lithium
North America			
Duracell International		x	x
Eastman Kodak		x	
Eveready Battery Co.	x	x	x
Rayovac Corporation	x	x	x
European Economic Community			
SAFT			x
Varta Batterie AG	x	x	x
Far East			
Gold Peak	x		
Matsushita	x	x	x
Sanyo	x	x	x
Toshiba	x	x	x
Yuasa	x	x	x

Carbon–Zinc Cells

Carbon–zinc batteries are the most commonly found primary cells worldwide and are produced in every major country. They traditionally have a carbon rod (for cylindrical cells) or a carbon-coated plate (for flat cells) to collect the current at the cathode with a zinc anode, which is also used as the primary container of cylindrical cells, and it is this combination that has led to their name. There are two basic verions, the Leclanché cell and the zinc chloride or heavy-duty cell. Both types have zinc anodes and manganese dioxide cathodes and include zinc chloride in the electrolyte. The Leclanché cell also has an electrolyte saturated with ammonium chloride (additional undissolved ammonium chloride is usually added to the cathode) whereas the zinc chloride cell has at most a small amount of ammonium chloride added to the electrolyte. Both types are dry cells in the sense that there is no excess liquid electrolyte in the system. The zinc chloride cell is often made with synthetic manganese dioxide and gives higher capacity than the Leclanché cell, which uses inexpensive natural manganese dioxide for the active cathode material. Because MnO_2 is only a modest conductor, the cathodes in both types of cell contain 10% to 30% carbon black in order to distribute the current. Because of the ease of manufacture and the long history of the cell, the system can be found in many sizes and shapes.

As noted previously, carbon-zinc cells perform best under intermittent use. Thus many standardized tests have been devised that are appropriate to these applications, such as light and heavy flashlight tests, radio tests, cassette tests, and motor (toy) tests. The most frequently used tests are detailed in reference (2), the American National Standards Institute (ANSI) tests. The tests are carried out at constant resistance and the results reported in minutes or hours of service. Figure 1 shows typical results under a light load for different size cells, and Figure 2 shows results for different types of R20 (D) size cells under heavy intermittent load. Typical computed values of the energy density and specific energy for R20 (D) and R6 (AA) size cells of the Leclanché and zinc chloride battery types are given in Table 2.

Note that, for heavy-duty cells, the specific energies (Wh/kg) are higher for the larger R20 cells than for the R6 cells, whereas the energy densities (Wh/L) are higher for the R6 cells. This is due to the fact that the can weight makes a much greater contribution to the weight of the system for small cells, whereas the smaller cathode of the R6 cell allows a more efficient design of the cell. These relationships are not exactly preserved in the general-pur-

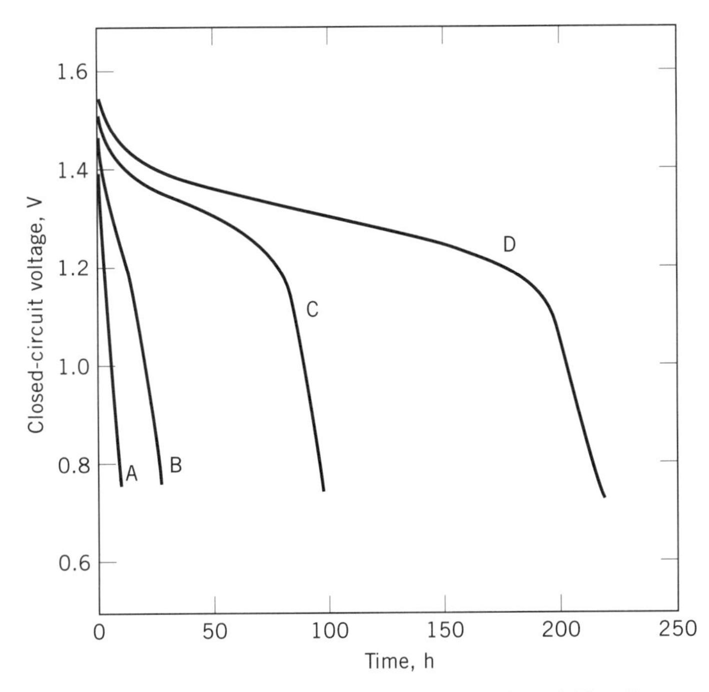

Figure 1. Hours of service on 40-Ω discharge for 4 h/d radio test at 21°C for A, RO3 "AAA"; B, R6 "AA"; C, R14 "C"; and D, R20 "D" paper-lined, heavy-duty zinc chloride cells.

pose cells in Table 2, because the R6 cell shown here is now made in the more efficient zinc chloride cell design. The effect of the discharge rate is especially pronounced for the general-purpose cells. On intermittent tests, the heavy-duty cell operates at high efficiency even at high rate. On continuous test at high rate, heavy duty cells provide 60% to 70% of the intermittent service, whereas general purpose cells give only 30% to 50% of the intermittent service values.

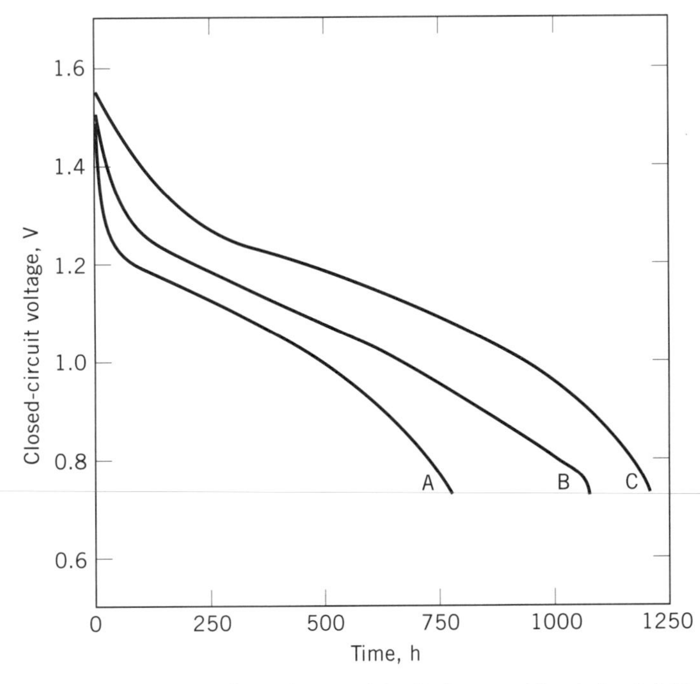

Figure 2. Hours of service on 4-Ω discharge, 15 min/h, 8 h/d, LIF test (see Table 1) at 21°C for A, general purpose (pasted Leclanché) "D"-size; B, premium Leclanché (paper-lined Leclanché); and C, heavy-duty (paper-lined $ZnCl_2$) cells.

Table 2. Specific Energies and Energy Densities of Carbon Zinc Cells[a]

Battery Designation and Type	Test, Ω[b,c]	Wh/L	Wh/kg
R20 (D) size cells:			
General purpose Leclanché	40Ω Radio	120	73
	LIF	70	41
Heavy duty zinc chloride	40Ω Radio	160	100
	LIF	150	90
R6 (AA) Size Cells:			
General purpose zinc chloride	75Ω Radio	130	72
	LIF	100	51
Heavy duty zinc chloride	75Ω Radio	170	81
	LIF	150	68

[a] Calculations based on data from reference 1.
[b] The radio test is one 4 h continuous discharge daily.
[c] Each discharge performed at 2.25Ω on the light industrial flashlight (LIF) test: one 4 min discharge at 1 h intervals for eight consecutive hours each day, with 16-h rest periods; ie, 32 min of discharge per day.

Alkaline Cells

Primary alkaline cells use sodium hydroxide or potassium hydroxide as the electrolyte. The alkaline cells of the 1990s are mostly of the limited electrolyte, dry cell type. Most primary alkaline cells are made using zinc as the anode material; a variety of cathode materials can be used. Primary alkaline cells are commonly divided into two classes, based on type of construction: the larger, cylindrical shaped batteries, and the miniature, button-type cells. Cylindrical alkaline batteries are mainly produced using zinc-manganese dioxide chemistry (manufacture of cylindrical zinc-mercury oxide cells has now been discontinued, because of environmental concerns). Miniature cells are produced with a much larger number of chemical systems, to meet the needs of particular applications.

Cylindrical alkaline cells are zinc-manganese dioxide cells having an alkaline electrolyte, which are constructed in the standard cylindrical sizes, R20 "D," R14 "C," R6 "AA," R03 "AAA," and a few other less common sizes. They can be used in the same types of devices as ordinary Leclanché and zinc chloride cells. Moreover, their high level of performance makes them ideally suited for applications such as toys, audio devices, and cameras.

Alkaline batteries having high output capacity and high current-carrying ability are now made by many manufacturers throughout the world. There is continuing competition among manufacturers to improve the performance of cylindrical alkaline batteries.

The alkaline cell derives its power from the reduction of the manganese dioxide cathode and the oxidation of the zinc anode.

Certain impurities on zinc can act as catalysts for the generation of hydrogen, thereby greatly increasing the corrosion rates. For this reason, zinc in alkaline cells must be of high purity and careful control must be exercised over the level of the harmful impurities. Moreover, other components of the cell must not contain harmful levels of impurities that might dissolve and migrate to the zinc anode.

Alternatively, there are also inhibitors that can decrease the rate of hydrogen generation and thus decrease the corrosion. Mercury, effective at inhibiting zinc corrosion, has long been used as an additive to zinc anodes. More recently, however, because of increased interest in environmental issues, the amount of mercury in alkaline cells has been sharply reduced.

Alkaline manganese-dioxide batteries have relatively high energy density, as can be seen from Table 3. This is due in part to the use of highly pure materials, formed into electrodes of near optimum density. Moreover, the cells are able to function well with a rather small amount of electrolyte.

The performance of the cells is influenced not only by the relatively high theoretical capacity, but also by the fact that the cells can provide good efficiency at various currents over a wide variety of conditions. The high conductivity and low polarization of the alkaline electrolyte, combined with the good electronic and ionic conductivity in the electrodes, leads to high performance even under heavy drain conditions.

Batteries tend to perform more poorly as the operating temperature is decreased because of decreased conductivity of the electrolyte and slower electrode kinetics. Batteries tend to perform better at higher temperature only up to the point that loss of performance occurs because of cell venting and drying out or parasitic reactions in the cell. Overall, alkaline cells have less performance loss at low and high temperatures than do Leclanché cells.

Lithium Cells

Cells with lithium anodes are generally called lithium cells regardless of the cathode. They can be conveniently separated into two different types: a) cells with solid cathodes, and b) cells with liquid cathodes. Cells with liquid cathodes have liquid electrolytes. At least one component of the electrolyte solvent and the cathode active material are the same. Cells with solid cathodes may have liquid or solid electrolytes but, except for the lithium-iodine system (see the following), solid electrolyte systems have not yet matured to commercial status.

All the cells take advantage of the inherently high energy of lithium metal and its unusual film-forming property. This property allows for lithium compatibility with solvents and other materials with which it is thermodynamically unstable, yet permits high electrochemical activity when the external circuit is closed. This is possible because the film formed is conductive to lithium ions, but not to electrons. Thus, corrosion reactions occur only to the extent of forming thin, electronically insulating films on lithium that are both coherent and adherent to the base metal. The thinness of the film is important in order to allow reasonable rates of transport of lithium ions through the film when the circuit is closed. Many materials, such as water and alcohols, which are thermodynamically unstable with lithium, do not form this kind of passivating film.

Much detailed analytical study has been required to establish materials for use as solvents and solutes in lithium batteries. Among the best organic solvents that have been found are the cyclic esters, such as propylene carbonate (PC), ethylene carbonate (EC), and butyrolactone, and the ethers, such as dimethoxyethane (DME), the glymes, tetrahydrofuran (THF), and dioxolane. Among the most useful electrolyte salts are lithium perchlorate, lithium trifluoromethanesulfonate, lithium tetrafluoroborate, and, more recently, lithium hexafluorophoshate. Lithium hexafluoroarsenate has good electrolyte properties, but because of possible environmental problems with arsenic-containing materials, it is now used mostly in military applications.

A limitation of these so-called organic electrolytes is their relatively low conductivity, compared to aqueous electrolytes. This limitation, combined with the generally slow kinetics of the cathode reactions, has forced the use of certain designs such as thin electrodes and very thin separators for all lithium batteries. This usage led to the development of coin cells rather than button cells for miniature batteries and jelly-roll or spiral-wound designs rather than bobbin designs for cylindrical cells. Many of the cylindrical cells have glass-to-metal hermetic seals, although this is becoming less common because of the high cost associated with this type of seal. Alternatively, cylindrical cells have compression seals carefully designed to minimize the ingress of water and oxygen and the egress of volatile solvent. These construction changes are costly, and the high price of the lithium cell has limited its use thus far. However, the energy densities are superior and, in some applications, there is a definite economic benefit compared to aqueous systems.

The Li/MnO_2 cell is becoming the most widely used 3-volt solid cathode lithium primary battery. The critical step in obtaining good performance is the heat treatment

Table 3. Characteristics of Aqueous Primary Batteries

Parameter	Carbon Zinc (Zn/MnO_2)	Alkaline Manganese Dioxide (Zn/MnO_2)	Mercuric Oxide (Zn/HgO)	Silver Oxide (Zn/Ag_2O)	Zinc Air (Zn/O_2)
Nominal voltage, V	1.5	1.5	1.35	1.5	1.25
Working voltage, V	1.2	1.2	1.3	1.55	1.25
Specific energy, Wh/kg	40–100	80–95	100	130	230–400
Energy density, Wh/L	70–170	150–250	400–600	490–520	700–800
Temperature range, °C	−40 to 50	−40 to 50	−40 to 60	−40 to 60	−40 to 50
Storage Operating	−5 to 55	−20 to 55	−10 to 55	−10 to 55	−10 to 55

of the MnO_2 to at least 300°C before incorporation in the cathode (3). This removes from the EMD any excess water, which can otherwise be dissolved in the electrolyte during storage or discharge and affect the operation of the anode, as well as activate the cathode for the discharge reaction.

The energy density of the system depends on the type of cell and the current drain. Table 4 gives the specification of lithium batteries and shows that Li-MnO_2 is very competitive with the other systems. The coin cells are widely used in electronic devices such as calculators, watches, and so on whereas the cylindrical cells have found their main application in fully automatic cameras.

The lithium–iron disulfide battery is unusual among lithium cells in that the voltage is a nominal 1.5 V. The cells were first manufactured in button-cell sizes, and because of the voltage similarity, they may be used as direct replacements for Zn/Ag_2O cells (4). Due to the lower conductivity of the organic electrolyte compared to KOH electrolyte in the Zn/Ag_2O cell, they cannot be used in the highest drain applications, but such uses are relatively few. Recently, AA-size, cylindrical lithium–iron disulfide batteries were introduced (5).

The performance of the cylindrical cell is superior to alkaline cells at all rates of discharge, but because of the high-area electrode construction, the performance advantage is magnified during high rate discharge. Table 4 shows the energy density and comparison to Tables 2 and 3 shows the superiority of the cells to aqueous alkaline and Leclanché cells and the parity for Zn/Ag_2O button cells. The cell is widely used as a power source for electronic camera flash guns; three to five times more flashes are obtained than with use of alkaline cells. Automatic cameras, camcorders, and many other constant power electronic applications show similar gains over the alkaline cell, as a range of four to eight times the energy from alkaline cells is obtained (5).

Lithium-iodine cells are widely used in medical applications (6). Iodine forms a charge transfer complex with poly(2-vinylpyridine) (PVP), which makes up the cathode for these cells. In one process for making the cell, the two materials (with excess iodine) are heated together to form the molten charge transfer complex. Then, the liquid is poured into the waiting cell, which already contains the lithium anode and a cathode collector screen. After cooling and solidification of the cathode, the cell is hermetically sealed. In another process, the iodine-PVP mixture is reacted and then pelletized. The pellets are then pressed into the lithium to form a sandwich structure, which is then inserted into the waiting cell and hermetic sealing is done. The direct contact of the excess iodine and the lithium results in the immediate reaction of the two to form an *in situ* LiI film, which then protects the lithium from further reaction.

The Li-I_2 battery system has allowed simple heart pacers to operate for as long as 10 years compared to the 1.5 to 2 years of operation for alkaline batteries. This is mostly because of the very high stability of the lithium batteries allowing for trouble-free operation. Many other lithium batteries have been tried in this application, but the Li-I_2 system is now almost universally used, and the gain in patient health from fewer surgeries is substantial.

The limitation to low current densities seems intrinsic with the Li-I_2 system, however, so medical electronics developers are seeking equally stable batteries with higher current capabilities. The lithium–silver vanadium oxide battery shows promise for the defibrillator application (6).

Less expensive coin-type cells have also been developed for consumer electronics applications, but the severe current limitations have restricted the use of the cells.

The second class of lithium batteries is the liquid cathode group. The first successful lithium cell of the category was the lithium–sulfur dioxide cell (7). Lithium–sulfur dioxide cells (8) generally use either acetonitrile (AN) or PC or a mixture of the two as cosolvents with the SO_2, usually in amounts as high as 50 vol. %. This has the advantage of lowering the vapor pressure of the sulfur dioxide which at 25°C is about 300 kPa (3 atm).

The liquids are usually chilled, however, to make management of the toxic SO_2 gas easier during cell filling and sealing, which are the last steps in the cell construction. AN can be used as a cosolvent only because of the excellent film-forming properties of SO_2. Lithium is known to react extensively with AN in the absence of sulfur dioxide. The other properties of AN, relatively high dielectric constant and very high fluidity, enhance the electrolyte conductivity and permit construction of high rate cells. The cell reaction proceeds according to

$$2Li + SO_2 \rightarrow Li_2S_2O_4 \tag{1}$$

where the cathode reaction, reduction of SO_2, takes place on a highly porous, >80%, carbon electrode. The lithium dithionite product is insoluble in the electrolyte and precipitates in place in the cathode as it is formed. The reac-

Table 4. Characteristics of Lithium Primary Batteries

Cathode: Cell System	Nominal Voltage, V	Working Voltage, V	Specific Energy, Wh/kg	Energy Density, Wh/L	Operating Temperature, Range, °C
Solid Cathodes					
Carbon monofluoride (Li/CF)	3	2.5–2.7	200	400	−20 to 60
Manganese dioxide (Li/MnO_2)	3	2.7–2.9	200	400	−20 to 55
Iron disulfide (Li/FeS_2)	1.5	1.3–1.7	250	460	−20 to 60
Iodine (Li/I_2)	3	2.4–2.8	250	920	37
Liquid Cathodes					
Sulfur dioxide (Li/SO_2)	3	2.7–2.9	250	440	−55 to 70
Thionyl chloride (Li/$SOCl_2$)	3.6	3.3–3.5	300	950	−50 to 70

tion involves only one electron per SO_2 molecule, which, along with the high volume of cosolvent in the cell, limits the capacity and energy density. Lithium bromide is the most commonly used salt, although sometimes $LiAsF_6$ is used, especially in reserve cell configurations. Because of the high rate capability, the cell has been widely used for military applications.

For both this system and the thionyl chloride system, safety has been a major focus, partly because of the high toxicity of the materials. Also, the potentially high rate of reaction during an accidental short circuit or other incident has led to some runaway reactions of high energy. However, tight control has allowed usage and they have become important batteries for military communications. In addition, the careful engineering of electrodes and vent mechanisms has controlled incidents. The cells have been subjected to extensive abuse testing.

All sulfur dioxide batteries have hermetic glass-to-metal seals to prevent loss of the highly volatile sulfur dioxide. They also have pressure-operated vents to accomplish emergency evacuation of the cell under abuse to prevent explosions. Safety has also been improved by using a balanced cell design (ie, an equal capacity of lithium and sulfur dioxide), so that low-rate discharge uses up all the lithium. The performance of these cells is generally optimized for high-current-density operation, and the battery is usually limited by blockage of the porous cathode structure with the insoluble lithium dithionite reaction product.

Both types of batteries have high energy density and specific energy (Table 4), because of the efficient use of space characteristic of a liquid cathode cell. Many cell types and sizes have been manufactured and many of these cells have been combined in series-parallel networks in batteries. They have been mostly made under contract for government use or, in some cases, for original equipment use for electronic device manufacturers. It is difficult for a consumer to purchase the cells separately.

Lithium-thionyl chloride cells, discovered shortly after $Li\text{-}SO_2$ batteries (9), have very high energy density. One of the main reasons is the nature of the cell reaction

$$4Li + 2SOCl_2 \rightarrow 4LiCl + S + SO_2. \qquad (2)$$

The reaction involves two electrons per thionyl chloride molecule. Also, one of the products, SO_2, is a liquid under the internal pressure of the cell, facilitating a more complete use of the reactant. Finally, no cosolvent is required for the solution, because thionyl chloride is a liquid having only a modest vapor pressure at room temperature. The electrolyte salt most commonly used in $LiAlCl_4$. Initially, the sulfur product is also soluble in the electrolyte, but as the composition changes to higher SO_2 composition and the sulfur concentration builds up, a saturation point is reached and the sulfur precipitates. The end of discharge occurs as the anode is used up because the cells are usually anode limited. However, if very high discharge rates are used, the collector, which is usually made of a porous carbon such as that in the SO_2 cell, becomes blocked with the insoluble LiCl. In either case, the discharge curve is flat over time, then declines abruptly at the end of life. The system has an open-circuit voltage of 3.6 V. Because of the high activity of the cathode and the good mass transfer of the liquid catholyte, the cell is capable of very-high-rate discharges.

Several applications are developing for the lithium–thionyl chloride system. Because of the excellent voltage control, the battery is finding increasing use on electronic circuit boards to supply a fixed voltage for memory protection and other standby functions. These cells are designed in low-rate configuration, which maximizes the energy density and cell stability. Military and space applications are also developing because of the wide range of temperature/performance capability along with the high specific energy. They are also used in medical devices such as neurostimulators and drug delivery systems (6).

SECONDARY BATTERIES

Secondary or rechargeable batteries can be conveniently divided into the categories of lead acid, alkaline, and others including lithium/lithium-ion batteries. Lead acid and iron air batteries are treated in another section; only the actively produced cells among the others are discussed in this section.

However, as energy densities of rechargeable cells have improved with the introduction of nickel metal hydride cells and lithium ion cells, as well as improvements in nickel-cadmium cells, the markets for these batteries have also improved—and at a faster rate of growth than the primary cell market.

Several other systems are in various stages of research and development, but have not yet reached production status. These systems are summarized in Table 5. The lithium and lithium ion cells are now on the market, but are in limited production and sales figures are not yet available.

Rechargeable Alkaline Batteries

The different types of alkaline cells are summarized in Table 6 along with their principal applications and a list of advantages and disadvantages. Table 7 gives some of the electrochemical properties, the open-circuit voltages, the capacities of typical cells, the energy densities and specific energies at the usual rates of discharge for the cells.

Other Secondary Batteries

As in the field of rechargeable alkaline batteries, a great many other electrochemical systems have been proposed, developed and produced for some period of time. This section deals with currently produced batteries and briefly discusses the systems that are under development.

A group of battery systems called lithium-ion, rocking chair, or swing batteries is of great current interest. The concept of the batteries is deceptively simple. A lithiated transition metal oxide such as $LiCoO_2$, $LiNiO_2$, or $LiMn_2O_4$ is mixed with a conductor and binder and coated on a metal foil for use as a positive electrode. Only the $LiCoO_2$ is used at present in commercial cells. The material is in the discharged state. The negative electrode is composed of one of several different kinds of carbon or graphite and a binder and coated on another metal foil and is also in the discharged state. The cell is charged after production.

Table 5. Rechargeable Battery Systems in Research and Development

Common Name	Nominal Voltage	Comments
Nickel–zinc	1.65	Limited cycle life, high rate capability
Zinc–bromine	1.85	Bromine complex in circulating liquid phase, containment problems
Lithium aluminum–FeS	1.33	High temperature fused salt electrolyte, corrosion problems
Sodium–sulfur	2.1	High temperature liquid electrodes, solid β-Al_2O_3 separator, containment problems
Sodium–nickel chloride (or iron chloride)	2.5 (or 2.3)	Solid β-Al_2O_3 separator, solid positive electrode, with high temperature molten salt as catholyte, molten sodium negative electrode
Aluminum–air	1.6	Replaceable negative electrode, circulating electrolyte, low efficiency negative electrode
Lithium–TiS_2	2.3	Room temperature operation, air sensitive positive electrode material, lithium plating problems
Lithium–MnO_2	3.0	Room temperature operation, somewhat air sensitive positive, lithium plating problems
Lithium–V_6O_{13}	2.5	Solid polymer electrolyte based on radiation cross linking, must maintain electrode-polymer interfaces through long term cycling

The nature of the cell reaction on charge is that the positive electrode is oxidized in a homogeneous reaction to form a lithium-deficient transition metal oxide with the transition metal in a higher oxidation state (10):

$$LiCoO_2 \leftrightharpoons Li_xCoO_2 + (1-x)Li^+ + (1-x)e^- \quad (3)$$

The negative electrode reaction depends to a certain extent on the nature of the carbon (11). Batteries of this type have been on the market for only about two years in special applications such as portable telephones, camcorders, and computers (12). The midlife voltage is about 3.6 V and the charged open-circuit voltage is about 4.1 V. The unit cells are cylindrical, jelly-roll designs, often of the same size as nickel-cadmium batteries. Because of the substantial voltage difference from aqueous systems, however, they are sold only in battery packs. The energy density of the cells is as high as 250 Wh/l and the specific energy 115

Table 6. Alkaline Rechargeable Batteries

System	Battery Types	Applications	Advantages	Disadvantages
Nickel–Cadmium	1. Prismatic cells with pocket electrodes 2. Vented prismatic cells with sintered electrodes 3. Sealed cylindrical cells with jelly roll electrodes and prismatic cells with stacked electrodes. 4. Button cells with pressed powder electrodes	1. Stationary power-Remote power uses, load leveling, etc. 2. Aircraft batteries for starting and standby applications. 3. Portable high energy, high power applications. 4. Portable low power applications.	1. Easy to manufacture; long history 2. High rate capability, high electrode efficiencies, high cycle life. 3. Capable of mass production, no maintenance design. 4. Mass production capable, small size.	1. Low energy per volume and weight, relatively low efficiency. 2. Suitable for large size only, high maintenance. 3. Relatively low electrode efficiencies. 4. Relatively low current capability. All-cadmium toxicity.
Nickel–Hydrogen	Pressure vessel-thin electrode stacks.	Satellite power systems.	Very high cycle life and high electrode efficiency. Easily charged and abuse resistant.	Poor energy density, difficult to manufacture, poor charge retention, very expensive.
Nickel–metal hydride	Sealed cylindrical, jelly roll electrodes	Portable high energy, high power applications.	High energy density, high negative electrode efficiency. High cycle life capability. No cadmium or lead.	Poor charge retention compared to Ni-Cd. Cost and availability of rare earth and transition metal elements.
Manganese dioxide–zinc	Sealed cylindrical cells with concentric cylinder electrodes	Portable low to moderate power applications.	Inexpensive, direct replacement for primary alkaline batteries, environmental advantages.	Very poor cycle life, rapid capacity fade. Mercury contamination in some designs.
Silver–zinc	Prismatic cells with flat plate electrodes	Military aerospace, submarine and communications.	Very high rate and energy density.	Very expensive due to silver component, poor cycle life.

Table 7. Electrochemical Properties of Rechargeable Alkaline Batteries

Cell Type	Open Circuit Voltage Cell Capacity	Energy Density, Wh/L	Specific Energy, Wh/kg
Nickel–cadmium			
1. Stationary	[1.3 Volts]		
2. Aircraft	1. 315 Ah	1. 36	1. 21
3. Sealed cylindrical or prismatic	2. 40 Ah	2. 72	2. 31
	3. 4 Ah	3. 90	3. 34
4. Button	4. 60 mAh	4. 63	4. 21
Nickel–hydrogen	[1.3 Volts] 55 Ah	60	50
Nickel–metal hydride	[1.3 Volts] 1.5 Ah	184	55
Manganese dioxide–zinc	[1.5 Volts] 1.5 Ah	240 (cycle 1)	80 (Cycle 1)
		125 (cycle 10)	42 (Cycle 10)
Silver–zinc	[1.6 Volts] 100 Ah	190	90

Wh/kg. Sony Energytech quotes a cycle life at 1200 cycles, reasonably high current capability and a self-discharge rate of 10% per month (12). A & T Battery Corp. of Japan claims similar performance with their battery (13) as does Japan Storage Battery (14). In order to ensure safety, a complicated and expensive charging system that monitors the voltage of each cell in the battery has been developed. Considerable work continues on rechargeable lithium batteries with liquid electrolytes because of the high energy densities and specific energies possible with this technology (15). Typical positive electrode materials under study include MnO_2, V_2O_5, and TiS_2. Many of the problems with the liquid electrolytes for rechargeable lithium cells have been discussed (16).

Another approach to the rechargeable lithium and lithium-ion cells is to use an immobilized polymer electrolyte as the ionically conductive medium rather than a liquid electrolyte with a polymer separator. Plasticized polymers are currently under intensive study and show promise of enhanced battery safety and high energy at moderate to low current drains (16). Some positive electrode materials under study for polymer electrolytes include V_6O_{13}, MnO_2, TiS_2, and $LiMn_2O_4$.

Two elevated-temperature batteries have received extensive development efforts because of the possibility that they could become the favored power source for electric vehicles (15). The U.S. Department of Energy and many private companies and other governments have contributed greatly to the effort, although the batteries are not yet available for market testing. The two systems are the iron sulfide–lithium aluminum battery with a molten chloride electrolyte and the sodium polysulfide-sodium battery with a solid beta alumina separator.

RESERVE BATTERIES

This type of battery has been developed for applications that require an intense discharge of high energy and power, and sometimes operation at low ambient temperature. The batteries must also have the capability of a long inactive shelf period and rapid activation to the ready state. These batteries are usually classified by the mechanism of activation employed. The types are a) water-activated batteries, using fresh or sea water, b) electrolyte-activated batteries, using the complete electrolyte or only the solvent, c) gas-activated batteries, using the gas as an active cathode material or as part of the electrolyte, and d) heat-activated batteries or thermal batteries, using a solid salt electrolyte that is activated by melting on application of heat. Activation involves adding the missing component just before use, which can be done in a very simple way such as pouring the water into an opening in the cell for water-activated cells or in a more complicated way by using pistons, valves, or heat pellets activated by gravitational or electric signals for the electrolyte- or thermal-activation types. Such batteries may be stored for 10 to 20 years while awaiting use. Reserve batteries are usually manufactured under contract for various government agencies (mostly defense departments) although occasional industrial or safety uses have been found.

Many battery types are used in reserve cells, including several discussed under Primary Batteries. For example, the lithium–thionyl chloride and the lithium–sulfur dioxide systems are often used in reserve configurations, in which electrolyte, stored in a sealed compartment, is released on activation and may be forced by a piston or inertial forces into the interelectrode space. Another new reserve battery is the lithium–vanadium pentoxide battery, which performs at high rates and high energy. Most applications for such batteries are in mines and fuzes used in military ordnance.

An interesting variant of the liquid cathode reserve battery is the lithium water battery in which water is the liquid cathode and is also used as the electrolyte. A certain amount of corrosion of the lithium occurs, but sufficient lithium to compensate for the loss is provided. The reaction product is soluble lithium hydroxide. In some cases, a solid cathode material, such as silver oxide, or another liquid reactant, such as hydrogen peroxide, is used in combination with the lithium anode and aqueous electrolyte to improve the rate or decrease the amount of gas given off by the system. These cells are mostly used in the marine environment where water is available or compatible with the cell reaction product. Common applications are for torpedo propulsion and powering sonobuoys and submersibles. An older system still used for these applications is the magnesium–silver chloride seawater-activated battery. This cell is much heavier than the corresponding lithium cell, but the buoyancy of the sea makes this less of a detriment. The magnesium-silver chloride cell is also useful for powering emergency communication devices for airplane crews whose planes have come down in the sea.

The last type of reserve cell is the thermally activated cell. The older types use calcium or magnesium anodes; newer types use lithium alloys as anodes. The cathodes for the older calcium anode cells are typically metal chromates such as calcium chromate. The anode reaction product is calcium chloride, and the cathode product is a mixed calcium chromium oxide of uncertain composition. One of the best cathodes for lithium alloy cells is FeS_2, which forms a system very similar in reaction mechanism to the lithium–iron disulfide cell.

The heat pellet used for activation is usually a mixture of a reactive metal such as iron or zirconium and an oxidant such as potassium perchlorate. Many millions of these batteries have been manufactured for military ordnance—rockets, bombs, missiles, and so on.

BIBLIOGRAPHY

1. G. W. Heise and N. C. Cahoon, eds., *The Primary Battery*, Vol. 1, John Wiley & Sons, New York, 1971.
2. *American National Standard Specification for Dry Cells and Batteries*, ANSI C18.1-1979, American National Standards Institute, New York, May, 1979.
3. H. Ikeda, T. Saito, and H. Tamura, in J. P. Gabano, ed., *Lithium Batteries,* Academic Press, New York, 1983, p. 384.
4. M. B. Clark in Ref. 3, p. 115.
5. R. A. Langan, *Wescon/92, Conference Record*, Western Periodicals Co., Ventura, CA, 1992, p. 551.
6. C. F. Holmes in *Lithium Batteries: New materials, Developments and Perspectives*, Elsevier, New York, 1994, p. 377.
7. U.S. Pat. 3,567,515 (1971), D. L. Maricle and J. P. Mohns (to American Cyanamid).
8. C. R. Walk in Ref. 3, p. 281.
9. Ger. Pat. 2,262,256 (1972), G. E. Blomgren and M. L. Kronenberg (to Union Carbide Corp.).
10. C. Delmas in G. Pistoia, ed., *Lithium Batteries: New Materials, Developments and Perspectives*, Elsevier, New York, 1994, p. 457.
11. J. R. Dahn, A. K. Sleigh, H. Shi, B. M. Way, W. J. Weydanz, J. N. Reimers, Q. Zhong, and U. von Sacken, in Ref. 10, p. 1.
12. *J. Electronics Industry*, **40**(2), Series No. 462, 28 (Feb. 1993).
13. I. Kuribayashi, *JEE*, **30:**322, 51 (Oct. 1993).
14. *J. Electronics Industry*, **41**(2) Series No. 474, 18 (Feb. 1994).
15. P. R. Gifford, "Batteries (Secondary Cells)" in *Kirk-Othmer Encyclopedia of Chemical Technology,* 4th ed., John Wiley & Sons, Inc., New York, 1992, pp. 1107–1178.
16. L. A. Dominey, in Ref. 15, p. 137.

BIODIVERSITY MAINTENANCE

ROBERT C. SZARO
USDA Forest Service, Research
Washington, D.C.

One of today's most pressing environmental issues is the conservation of biodiversity. Many factors threaten the world's biological heritage. The challenge is for nations, government agencies, organizations, and individuals to protect and enhance biodiversity while continuing to meet people's needs for natural resources. This challenge exists from local to global scales. If not met, future generations will live in a biologically impoverished world and perhaps one that is less capable of producing desired resources as well. (See also ENVIRONMENTAL AND CONSERVATION ORGANIZATION; FOREST RESOURCES; FORESTRY, SUSTAINABLE.)

Biodiversity is a multi-faceted issue that crosses all traditional resource boundaries and requires a true interdisciplinary approach to develop potential solutions. Many of the concepts and theories on biodiversity are still in the formulation stages but this situation does not prevent the implementation of a variety of projects designed for its conservation. Many pieces of the puzzle have been under study for years and provide an adequate base for conservation decisions. Threatened, endangered, and sensitive species programs have long been the focal point of biodiversity concerns on the species level but they represent only the tip of the iceberg. The magnitude of the problem (more than 500 threatened and endangered species in the U.S. alone with at least another 1000 proposed for listing) makes it clear that we do not have the resources for intense management programs for all species if only because of the enormity of the problem. There is even less hope in tropical environments where the potential numbers of species, particularly insect species, number in the millions (1). Perhaps because organic diversity is so much larger than previously imagined, it has proved difficult to express as a coherent subject of scientific inquiry (2).

Clearly every effort should be made to conserve biodiversity (3–5). The conservation of biodiversity encompasses genetic diversity of species populations, richness of species in biological communities, processes whereby species interact with one another and with physical attributes within ecological systems, and the abundance of species, communities, and ecosystems at large geographic scales (6). In recent years, traditional uses of forested lands in North America have become increasingly controversial (7). The demands and expectations placed on these resources are high and widely varied, calling for new approaches that go beyond merely reacting to resource crises and concerns (8).

The greatest challenge for managing biodiversity lies in preparing for the disturbances and changes that are a fundamental feature of natural ecosystems. Many of these disturbances are the direct result of human activities. Expanses of pristine forest that support an enormous diversity of wildlife and plants and a richness of human cultures are being rapidly converted into vast wastelands that support a few tough, fire-resistant weeds and perhaps some cattle, while people scrounge for food and fuelwood from the newly–degraded soils and sparse shrubbery (9). Yet, we cannot conserve biodiversity simply by preserving areas and trying to prevent all changes, whether naturally occurring or human-caused. Nor can we conserve biodiversity by trying to maximize diversity on any particular site.

How can land managers react to the oftentimes painful dilemmas they face on an almost daily basis when making management decisions that can have potentially devastating impacts on ecosystem stability? The discipline of Conservation Biology has been described as a "crisis discipline, where limited information is applied in an uncertain environment to make urgent decisions with sometimes irrevocable consequences" (10). This really speaks to the heart of all land managers. They find themselves trying to find the balance between maintaining and sustaining forest systems while still providing the forest products needed by people. Trade-offs will be inevitable and will necessitate formulating and using alternative land management strategies to provide an acceptable mix of commodity production, amenity use, protection of environmental and ecological values, and biodiversity. Conserving biodiversity now is likely to alter immediate access to resources currently in demand in exchange for increasing the likelihood that long-term productivity, availability, and access are assured.

But is this dilemma something new? Are we the first to wrestle with these kinds of decisions? With massive simplification of landscapes? Plato in approximately 2350 B.C. describes an area in ancient Greece that was stripped of its soil following clearing and grazing (11). In fact, since the development of agriculture, there have been extensive modifications to the natural vegetation cover of every continent except Antarctica (12). Yet, never before have there been so many humans on earth taking advantage of its resources.

It is hardly surprising then, that global awareness and concerns for conserving biodiversity are continually increasing. When we have concerns for biodiversity we are saying we have a concern for all life and its relationships (4). As arguably the most intelligent species on earth we have a responsibility to try as much as possible for the continuance of all forms of life. But how can we go about this? The first step of trying to determine the amount, variety, and distribution of species, ecosystems, and landscapes will require more comprehensive inventories which must be followed by monitoring efforts to determine the impacts of management activities. Next, strategies must be developed and implemented for the preservation, maintenance, and restoration of forest ecosystems. These efforts should also incorporate strategies for the sustainable use of forest resources including more efficient utilization, recycling programs, and forest plantations in order to meet human needs.

DEFINITION OF BIODIVERSITY

Simply stated, biodiversity is the variety of life on earth and its myriad of processes (13). It includes all lifeforms, from the unicellular fungi, protozoa, and bacteria to complex multicellular organisms such as plants, fishes, and mammals. The recently negotiated Global Convention on Biological Diversity (1992) defined it as follows:

> "Biological diversity" means the variability among living organisms from all sources including, inter alia, terrestrial, marine and other aquatic ecosystems and the ecological complexes of which they are a part; this includes diversity within species and of ecosystems (14,15).

Biodiversity can be divided into four levels: genetic, species, ecosystem, and landscape diversity. Most attention is often given to *Species Diversity*, the number of different kinds of organisms found at a particular locale, and how it varies from place to place and even seasonally at the same place. A less obvious aspect of biodiversity, *Genetic Diversity*, is the variety of building blocks found within individuals of a species. It allows populations of a species to adapt to environmental changes. *Ecosystem Diversity* is the distinctive assemblage of species that live together in the same area and interact with the physical environment in unique ways. The interaction of organisms within the ecosystem is often more than the simple sum of its parts. On a broad regional scale, *Landscape Diversity* refers to the placement and size of various ecosystems across the land surface and the linkages between them.

IMPORTANCE OF BIODIVERSITY

Why should people care about all this variety of life, about sustaining and enhancing genetic resources, recovering endangered species, restoring riparian areas, maintaining old-growth forests, or conserving trees, insects, and marshes? The answer is both aesthetic and practical. A diversity of living things provides subtle needs. People enjoy picnicking, visiting seashores, and a variety of other recreational activities. Our homes, air, livestock, vegetables, fruits, and grains all derive from the products of diverse and healthy ecosystems. Diverse communities of plants, animals, and microorganisms also provide indispensable ecological services. They recycle wastes, maintain the chemical composition of the atmosphere, and play a major role in determining the world's climate. Moreover, in many cultures, diversity and the maintenance of mountains or other landforms are important because of their religious significance. Many people also feel that we must maintain biodiversity because our role as the dominant species on earth confers upon us the responsibility for the wise and careful stewardship of life.

But the value of biodiversity goes far beyond the aesthetic aspects (16). We hardly know a fraction of the species on this planet, especially in tropical ecosystems, despite years of intensive scientific effort. Every year species are lost before we have a chance to know anything about them. Who can guess what potentially valuable foods, medicines, and commercial products are forever lost with each extinction? Wild plants, animals, and microorganisms have provided essential products since humans first walked the earth, including virtually everything we eat. They continue to provide a basis for human society to respond to future changes.

Waiting in the wings, are tens of thousands of species that are edible, and many are demonstrably superior to those already in use. The vast insect faunas contain large numbers of species that are potentially superior crop pollinators, control agents for weeds, and parasites of insect pests (17). A case in point is in the area of natural sweeteners where a plant found in West Africa, the katemfe (*Thaumatococcus danielli*), produces proteins that are 1,600 times sweeter than sucrose (17). One in ten plant species contains anticancer substances of variable potency, but relatively few have been bioassayed. Similarly, few species have been screened for other potential medicinal values even though about a third of all modern prescription drugs contain compounds that plants have evolved to defend themselves against their enemies (18).

Probably most important, however, is that interacting communities of plants, animals, and microorganisms functioning as ecosystems also provide us with many indispensable services. Balanced systems recycle wastes, maintain the chemical composition of the atmosphere, and play a major role in determining the world's climate. The "ecosystem services" provided to us gratis also include supplying us with fresh water, generating soils, supplying plant pollinators, and maintaining a giant genetic "library" (18).

LOSSES OF BIODIVERSITY

Over the past few decades, the rate of global biotic impoverishment has increased dramatically. Exponential growth in human populations and even faster growth in consumption of the world's natural resources, have led to high rates of loss of species and habitats. Current rates of species loss exceed anything found in the past 65 million years (19). If the trend continues, by 2050 we may see the loss of up to one quarter of the world's species (20) and potentially dramatic changes in the climate and hydrology of entire regions such as Amazonia (21). The biotic resources we stand to lose are of immediate future value to humanity and essential for the maintenance of productive ecosystems.

Many of our most serious problems are centered in the tropics, where biodiversity is highest and losses of species and whole ecosystems is being lost most rapidly (22). In developing countries, the issues are most intense, because for hundreds of millions of people, the struggle is for survival (23). The destruction of Third World forests amounts to more than 11 million hectares annually (7.5 million ha closed forest and 3.8 million ha open forest) (23). Between 1950 and 1983, forest and woodland areas dropped 38% in Central America and 24% in Africa (23). Four factors are of special importance: (1) the explosive growth of human populations; (2) widespread and extreme poverty; (3) an ignorance of the ways in which to carry out productive agriculture and forestry; and (4) government policies that encourage the wasteful uses of forest resources (22,23).

Archaeological evidence points to the interaction between humans and tropical forests that extends far into the past when population densities were actually higher than they are today (24,25). In Mexico, studies clearly document the existence of ancient civilizations with high population densities integrated within tropical forest ecosystems. Examples are both the Olmec and Maya civilizations of southeastern Mexico that existed in that region for a combined period of at least 3000 years (26). Population densities in the rural Mayan area today is only about 5 people per km^2 compared to the peak of 400–500 people per km^2 during the height of the Olmec and Maya civilizations (26). These findings indicate the current extensive areas of tropical forests in Mexico that have been cut over the last 50 years were not untouched primeval forest but the result of regeneration since the last cycle of abandonment (24).

Recent tropical deforestation is associated with a pervasive cycle of initial timber extraction followed by shifting cultivation, land acquisition, and subsequent conversion to pasture (27) which leads to loss of forest resources, reduction of biodiversity, and impoverishment of rural people (24). The effect of past civilizations on the structure and composition of today's forests is more than just an intriguing question but is important in determining those practices used by those civilizations to maintain the tropical biodiversity left by previous generations. In fact, one of the primary causes of tropical deforestation in Mexico is due to the neglect of traditional people's vast experience with resource management. The persistence of forest resources and ecosystems following widespread human intervention indicates that a knowledge of management techniques practiced by ancient civilizations, such as the Olmec and Maya, could help in reverting current processes of landscape degradation in the tropics (24).

INVENTORY AND ASSESSMENT

Implementing biodiversity goals will require resources and knowledge. The role of science in the conservation of biodiversity is critical. More research to improve methodologies, distributional and status information, and strategies based on sound information will ultimately provide the basis for all sound policy and management decisions. Current scientific understanding of ecological processes is far from perfect. Existing resources could be reallocated, but additional resources will be necessary for improving efforts in inventory, monitoring, and basic research. Better inventories and assessments are needed of current conditions, abundances, distributions, and management direction for genetic resources, species populations, biological communities, and ecological systems. Even in the United States with the intensity of scientific effort focused on natural resources, it is extremely difficult to assess the overall status of biodiversity. Rarely have studies been done examining all vascular plants and vertebrates and their relationships in any given ecosystem, let alone the thousands of other species found within it: Particularly as Maini (1992) point out that "forests are a rich repository of planet earths' genetic heritage" (28). Forests are usually delineated by the presence of a few dominant species but this delineation barely touches the surface of their species richness. For example, in two pine systems in the southeastern United States, tree species make up less than 10% of the plant and vertebrate animal species. And this arrangement does not take into account all the thousands of other species likely to be found. R. M. May (1992) states "If we speak of total number of species, then to a good approximation everything is a terrestrial insect" (29). In fact, if one uses an estimate of 3–5 million (29) as total species on earth, then vascular plants (5.4–13.3%) and vertebrates (0.9–1.5%) only represent 6.3 to 14.8%. These percentages may be off by an order of magnitude if Erwin's (1988) hypothesis of 30 to 50 million species is closer to the true number of species (30). This example illustrates the extent of the problem as the traditional approaches to forest management have not incorporated a consideration of the vast majority of species (30). Even on an area the size and geographic scope of a typical National Forest in the United States, native biodiversity can easily encompass thousands of species of plants and animals, dozens to hundreds of identifiable biological communities, and an incomprehensible number of pathways, processes, and cycles through which all that life is interconnected. Obviously, it is not possible to address each and every aspect of this complexity. Therefore, identification is of specific aspects of diversity, such as distinct species, biological communities, or ecological processes that warrant special consideration (31).

Biodiversity Inventories

Biodiversity at the species, community and ecosystem levels can be assessed by surveys (analyses of the geographic distribution of the biota) and inventories (catalogues of the elements of biodiversity present at a site or in a region). Surveys and inventories are critical in helping determine which areas are in greatest need of preservation or management, and in identifying threatened species and habitats. A wide range of such activities needs to be supported, including:

- Rapid Assessment Programs, which give a "snapshot" of species richness in an area.
- Surveys, to map out the distribution of the Earth's ecosystems. Such analyses would be greatly aided by the development of a common and consistent scheme to classify communities and ecosystems. There is currently no widely accepted vegetation/ecological classification system.
- Broad-based inventory efforts, focused on poorly known habitats (soil communities, tropical and marine systems) that take advantage of existing protected sites, such as UNESCO-MAB's Biosphere Reserves and analogous national networks, wherever possible. However, degraded and multiple-use habitats should also be included, as well as better-understood taxa, such as vertebrates and vascular plants, that can be used as "benchmarks" for habitat quality.
- Intensive inventories, which attempt to determine all of the species present, from microbes to vertebrates, should be undertaken at two or three sites around the world. Such analyses will require intense effort and a long time-frame (a decade or more), but they are key to answering such questions as how biodiversity at the species level affects diversity at the ecosystem and community levels) (32).

It is a challenging task to formulate an integrated, concise and relevant approach to inventory. However, there are some recommended broad principles outlined in the Keystone Biodiversity Report (1991) on the components of an inventory program (13). These include:

- The inventory should be hierarchical and "top-down" in the sense that landscape level assessments such as the Fish and Wildlife Service's "gap analysis" are used to identify priorities for inventory at the local level, and local assessments are used to identify priorities at the site level.
- The inventory should make maximum use of existing data management systems.
- The inventory should be landscape based in the sense that abundance and distribution of plant and animal species are correlated with soils, vegetation, plant and animal community characteristics, and landscape features.
- The inventory at a minimum should include natural vegetation, all vertebrate and vascular plant species and at least some indicator species of non-vascular plants and invertebrates, and some indicators of other elements of biological diversity, such as sensitive communities or human-influenced processes and elements of structural diversity.
- Provision should be made for systematic inventories of all candidate, threatened, endangered, and sensitive species and for all other elements that are imperiled due to human activities or natural events.
- Inventories should be guided by an inter-agency master plan that coordinates acquisition of aerial photography, soil survey, vegetation survey, and vertebrate inventory that ensures compatibility of data within and among agencies.
- The above mentioned master plan should be implemented for all regional ecosystems and vegetation mapping and inventory of vertebrates should be completed within the next 10 years.
- The inventory should be compatible with, and feed information directly into, development and implementation of Geographic Information System (GIS) methodology, monitoring programs, and research activities.
- The inventory should provide the basis for determining species (including genetic level assessment), species groups, population guilds, habitats, landscapes or processes that require more intensive studies.
- Inventories should be coordinated with and make maximum use of the fifty state Heritage Program databases, procedures and technology.
- The inventory process should identify levels or intensities of inventory that are appropriate for each level of planning, type of management activity or impact, type of land classification or degree of rarity or sensitivity of the element being inventoried.
- The inventory should have a strong element of quality control and assurance, including setting specific standards of accuracy and precision, timing the inventory to encompass the life-cycles of the target elements, standardizing methods and databases to the extent possible, and using trained personnel to conduct the inventories.

The need for more specific data and more efficient ways for collecting and managing data will lead to significant changes in inventory processes including the use of methods and technology that will: (1) provide resource estimates for specific geographic units and evaluate the reliability of such estimates; (2) display estimates and units spatially; (3) make maximum use of existing information and new technology, such as remote sensing and geographic information systems; (4) provide a baseline for monitoring changes in the extent and condition of the resource; (5) eliminate redundant data collection, develop common terminology, and promote data sharing through corporate data bases; (6) utilize information management systems to provide maximum flexibility for data integration, manipulation, sharing, and responding to routine and special requests; and (7) provide up-to-date data bases using modeling techniques, accounting procedures, and reinventories.

Monitoring Effects of Land Management and Conservation Programs

Present concepts of monitoring vary depending on who is expressing them, their background, and the objectives of the monitoring being considered. There is a need for greater coordination, with considerable direction and standardization set at both National and Regional levels. Monitoring means different things to different people. Just what it is depends solely on monitoring objectives. This definition should emphasize the need to exercise great care in defining the objectives for monitoring. Monitoring should provide sufficient information about the abundance of animals or plants targeted for monitoring to assure that current management practices are not threatening the long-term viability of their populations (33). But viability concerns have added an additional layer of complexity to monitoring programs particularly when we try to derive a number for minimum viable population size. This theoretical concept espoused by Michael Soule (34) is useful from the standpoint that there probably is some minimum size population threshold that, when crossed, will lead to the demise of a species' population. The level of this threshold clearly varies with the species concerned; for passenger pigeons it probably was in the hundreds of thousands while for desert pupfish the number may be as low as 200. But as useful a concept this may seem, in the real world its close to impossible to determine minimum viable population size with any degree of certainty. There are too many variables involved; demographic variables such as immigration emigration, recruitment, birth rates, survivorship, dispersal mechanisms, etc, catastrophic events, habitat loss and even changing climatic conditions. It is difficult to envision that we can determine minimum viable population size for every species. Perhaps a more subjective approach, that being that larger numbers of a species are obviously better then fewer numbers of the same species, should be considered. Maybe we should try to maintain species with as large a population size as practical.

Monitoring efforts are often severely hampered by the lack of prior planning and thought given to the desired results from any given monitoring effort. It is not enough to select a management indicator species, guild, or other monitoring target with the idea that this will allow us to assess the impact of any given management activity (35,36). Ideally, the results from monitoring should feedback into the system to correct or fine tune management activities.

Effort should not be wasted on monitoring systems that fail to give the level of confidence needed to deduce the most likely effects of management activities on wildlife, fish or sensitive plant resources. In an era when humankind's activities are the dominant force influencing biological communities, proper management requires understanding of pattern and process in biological systems and the development of assessment and evaluation procedures that assure protection of biological resources (37,38). It is essential that appraisals of these resources give us the ability to forecast the consequences of human-induced environmental changes accurately (39). But we have a long way to go in this process.

We must first have a clear understanding of our goals and objectives. The next step is to assess risk and assign priorities. We need to perform a kind of environmental triage. We must be able to say when enough is enough. With limited financial and physical resources at our disposal we may have to make the highly undesirable decision that we no longer will try to prevent the extinction of a particular species. Once those species have been determined and not forgetting about the importance of ecological processes, we need to formulate the types of questions that need to be answered to determine that we are meeting our goals and objectives. It is absolutely critical to ask the right question in the first place. Why monitor shade cover over perennial streams in order to maintain water temperature for trout when directly monitoring stream temperature is a more appropriate measure. Thus our efforts will be geared to maintaining as much shade cover as needed to maintain water temperature.

Monitoring should also have a strong element of quality control and assurance, including setting specific levels of accuracy and precision, timing the inventory to encompass the life-cycles of the target species, as much as possible, and standardizing methods and databases for all management units, especially when monitoring the same species. However, whatever is done must be as cost-effective as possible. Some possibilities are risk analysis, increasing the scope of monitoring efforts, and determining the needs of monitoring objectives. We might want to limit direct monitoring only on high risk species, on a priority basis, while relying on habitat relationships for most other species. Monitoring efforts should be spread over as large a geographical base as possible (and feasible) to reduce the cost per forest and to increase the scope of applicability. Whenever possible, monitoring should only be asked to detect declining trends because of the potential cost savings (almost 90%) (33).

Along with this monitoring is the need to develop a quality control and assurance program that ensures: (1) objectives are measurable (and thus monitorable); (2) appropriate measurement techniques and procedures are being used; and (3) management thresholds are clearly identified and incorporated into the planning process so that, if crossed, they automatically trigger a reanalysis of the planned activities.

The advent of geographic information systems holds future promise for the development of a comprehensive biological information system (40,41). Much work and coordination will be involved to bring this information system to pass including building in mechanisms to continually check and update the information in the database. It is not enough to set up the system and then use it without regard to the dynamics of ecological systems.

MAINTAINING BIODIVERSITY

The responsibility for biodiversity belongs to all people and institutions: both public and private (31). The repeated association of biodiversity with preservationist approaches leads to the perception that biodiversity requires wilderness and preserves and cannot be sustained where human activities are prevalent. This view has disastrous consequences. The World Resources Institute in 1986 reported that only 3% of the surface of the earth had major protection status as a national park, nature preserve, wilderness, or sanctuary. How much diversity can be sustained on that land base, even if it were doubled or tripled, politically unlikely events (31)? It is clear that the major accomplishments on behalf of biodiversity must occur in conjunction with human activities.

To maintain biodiversity, we must ensure that a sufficient amount of each ecosystem is conserved and managed through a variety of actions that address different and related concerns. And because these actions must occur on lands under a variety of ownerships, goals, and uses, considerations for biodiversity must be blended into a myriad of management approaches (31). We must strive to understand the functions and processes of natural ecosystems, and make the wise, tough decisions that are necessary to maintain and enhance the productivity of those systems for all purposes and uses. This means that biodiversity, and an understanding of ecosystems, should be the underlying basis for the management of all lands.

Restoration of Biodiversity

While research and management are urgently needed to slow continuing losses of biodiversity, the remediation of past losses can help offset unavoidable future losses. Restoration of ecosystems and biological communities is one important means of maintaining biodiversity, or at least of slowing its net loss. Biodiversity is threatened not only by a reduction of habitat acreage and of connections between habitats, but also a degradation of quality of the remaining habitats. Restoration activities respond to these problems by restoring eliminated habitat types (eg, native prairies and wetlands) and enhancing the condition of remaining habitat fragments. By restoring both the extent and quality of important habitats, restoration programs provide refuges for species and genetic resources that might otherwise be lost. Moreover, surrounding landscapes are habitats that disperse into these disturbed areas, and so restoration programs can also affect the recovery and renewed diversity of their biota.

Restoration is not a substitute for the preservation or good management and is both time-consuming and expensive. In tropical forests, the incredible diversity and complexity of the ecosystem make restoration of the original vegetation and ecosystems particularly difficult (9,43). But even though it addresses the symptom of deforestation rather than the causes, restoration ecology is worth serious consideration. It can speed regeneration in managed systems, make non-productive land productive again, relieve pressure on natural forest resources, and protect closed-canopy forest. It is a strategy that focuses on areas of severe erosion and soil compaction, where quick action is desperately needed (9).

Many techniques are used to restore ecosystems, depending on the ecosystem and impact type being addressed. These techniques include vegetation planting to control erosion, fertilization of existing vegetation to encourage growth, removal of contaminated soils, fencing to exclude cattle, reintroduction of extirpated species, restoration of hydrologic connections to wetlands, and others. However, not all of these restorations strategies are compatible with the goal of maintaining biodiversity. For example, eroding lands can be rapidly restored by introducing some types of aggressively spreading plants, but the same plants can spread beyond the site and imperil the diversity of flora in adjacent areas. Thus, restoration actions can be either a savior or a nemesis for regional biodiversity, depending on their design, application methods, and existing conditions of the landscape.

Just as there are many restoration techniques, there are a very large number of species that exist in habitats that are candidates for restoration. Each has particular environmental requirements, minimum viable population size, and expected recovery rate and pattern–knowledge which is essential to evaluating restoration potential. Although responses of some species to specific restoration techniques is known, the theoretical basis is weak for grouping species so that results can be extrapolated to other combinations of techniques and species. Ecosystem restoration does not always require intervention. Left to natural processes, many ecosystems will return to something like their pre-disturbance condition if populations of original species still exist nearby (44). For example, a temperate climate and productive soils promote natural reestablishment of forests in most regions of the United States. However, restoration technologies can speed the recovery of communities and ecosystems after disturbance and can enhance in-situ conservation (44).

Ecosystem Management

What exactly does "ecosystem management" mean? An ecosystem is a community of organisms and its environment that functions as an integrated unit. Forests are ecosystems. So are ponds, rivers, rotting logs, rangelands, whole mountain ranges, and the planet. Ecosystems occur at many different scales, from micro sites to the biosphere. They constantly change in their species composition, structure, and function over time. And they do not have natural boundaries. All ecosystems grade into others; all are nested within a matrix of larger ecosystems. We arbitrarily describe the boundaries of ecosystems for specific purposes.

Management means using skill or care in treating or handling something. Thus, "ecosystem management" means the use of skill and care in handling integrated units of organisms and their environments (45). It implies that the system, or integrated ecological unit, is the context for management rather than just its individual parts. The pivotal question is, what should we manage ecosystems for? Goals are not obvious from definitions or from merely stating that one manages ecosystems. Ecosystem management is the means to an end. It is not an end in itself. Ecosystems are usually managed for specific purposes such as: producing, restoring, or sustaining certain ecological conditions; desired resource uses and products; vital environmental services; and aesthetic, cultural, or spiritual values. This idea is an important point. There is much discussion these days about whether ecosystem management means to preserve the intrinsic values or natural conditions of ecosystems, with resource uses and products as secondary byproducts, or whether it means to produce desired resource values, uses, products, and services in ways that do not impair the long-term sustainability, diversity, and productivity of the ecosystems. It is unlikely that a consensus will ever be reached on this point; however, this debate does not mean that the conservation of biodiversity should not be a common concern of all humankind whatever our individual desires or objectives. In some places the emphasis will be on ecological conditions and environmental services. In others it will be on resource products and uses. Overall, the mandate should be to protect environmental quality while also producing resources that people need. Therefore, ecosystem management cannot be reduced to a simple matter of choosing to have either resource products or natural conditions. It must chart a prudent course to attain both of these goals together. This policy can only happen in areas large enough to allow for compatible patterns of different uses and values in a landscape or region.

A thorough understanding of vegetation dynamics is critical for sound vegetation management and the maintenance of landscape diversity. The role of landscape ecology in this regard is to focus attention on hierarchy theory that considers vegetation patterns at different scales (46). Natural areas should also be a part of any long-term landscape management strategy designed to conserve landscape diversity (46). The degree of predictability decreases as the scale decreases to specific sites where most management strategies are formulated (47). Successful management strategies will therefore be more likely when planning is done on a broader scale. Planning also needs to consider the dynamics of landscape scale forest patterns, both natural and managed, and their effects on hydrology, wildlife, and other resources. The results need to be geared toward long-term design of forest management to minimize cumulative effects while maintaining biodiversity. Geographic information systems (GIS) can be used to effectively integrate land uses for forestry, agriculture, wildlife, tourism and other non-commodity values. Graphic displays of resource information allow for an easier conceptualization of the land management patterns and their interrelationships. GIS allows many different scenarios to be tried and evaluated for the possible long term effects of proposed management activities.

Future Strategies for Conserving Biodiversity

Biodiversity provides a focus for a global conservation program (48–50). But even the best-conceived conservation program could not have anticipated the enormous changes of the 20th century, with its resultant pressures on natural ecosystems. Current global concerns about worldwide losses of massive expanses of forest ecosystems and their associated species provide opportunities for resource management agencies to build on past accomplishments and chart new courses in conservation for multiple benefits. Conserving biodiversity involves restoring, protecting, conserving, or enhancing the variety of life in an area so that the abundances and distributions of species and communities provide for continued existence and normal ecological functioning, including adaptation and extinction. This definition does not mean all things must occur in all areas, but that all things must be cared for at some appropriate geographic scale.

In recent years, traditional uses of public lands in the United States and around the world have become increasingly controversial (7). The demands and expectations placed on these resources are high and widely varied, a situation calling for new approaches that go beyond merely reacting to resource crises and concerns (3,8). These new approaches must incorporate fundamental shifts in the scale and scope of conservation practice (49). These approaches include the shift of focus from the more traditional single-species and stand level management approach, to management of communities and ecosystems (50). The greatest challenge for conserving biodiversity is developing an integrative ecosystem approach to land management on a broader landscape scale. Spatial scale, be it local, regional, or global, greatly influences our perceptions of biodiversity (51). An understanding of the importance of scale is critical to accurately assess the impact of land management practices on biodiversity since many significant biological responses and cumulative management effects develop at the landscape level. Planners and managers are increasingly aware that adequate decisions cannot be made solely at the stand level. If context is ignored in conservation decisions and the surrounding landscape changes radically in pattern and structure, patch content also will be altered by edge effects and other external influences (52). Particularly when, land-use patterns characteristic of a human-dominated landscape are ones in which large, continuous tracts of natural habitat become increasingly fragmented and isolated by a network of developed lands.

The protection and maintenance of biodiversity is a long-term issue which will create problems in political systems that deal primarily with the short-term goals and objectives. The most obvious conflicts will be political and financial. There are inherent biases in a market economy that tend to result in environmental degradation. The environmental costs are generally passed to the public at large. Attempts to internalize such costs are resisted strongly by private industry. For example, installation of scrubbers in smokestacks and catalytic converters on cars in the United States were not welcomed by industry. Similar results are likely in actions to protect biodiversity. The Tellico Dam in the United States was chosen over possible extinction of a small fish, the snail darter. Today the timber industry is against setting aside old growth forests on the basis of the impacts on local jobs and economies. The actual monetary costs may not be as large as the political costs. For every action taken, there will be winners and losers. The loss of diversity will have long-term effects as we lose what are essentially building blocks for human survival. Using a financial analogy, we should manage our biological resources so that we can live off the interest. Living off the principal eventually leads to bankruptcy.

Moreover, ecosystems by their very nature are in a constant state of flux, always shifting and changing from one condition to the next (52). Reproductive strategies and other ecological characteristics of many ecosystems are strikingly adjusted to disturbance regimes resulting from catastrophes such as fire, insect outbreaks, and flooding. Change per se is not necessarily something to be avoided. It ultimately may be the underlying motivating factor in management decisions. Land-managers may wish to alter vegetation structure or composition to emphasize rarer or endangered species. Old management paradigms are difficult to shed, but only new, dynamic efforts are likely to succeed in conserving the biodiversity of all our ecosystems (3). Ecosystem-level management is going to require new approaches in planning, monitoring, coordinating, and administering. We are being challenged by shifts in public attitudes to manage our public as well as private lands for uses other than commodity production. A new paradigm is needed, one that balances all uses in the management process and looks beyond the immediate benefits. Maintaining biodiversity requires attention to a wider array of components in determining management options as well as the management of larger landscape units. There will be trade-offs, commodity production may decline in the short-term, but in the long-term these trade-offs will result in gains in sustained productivity while maintaining biodiversity with its complete range of ecological processes.

Threatened and endangered species have long been the focus of biodiversity concerns at the level of plant and animal species, but they represent only one aspect of a larger issue: conservation of the full variety of life, from genetic variation in species populations to the richness of ecosystems in the biosphere (31). And for certain species and biological communities, the pressing concern is perpetuation or enhancement of the genetic variation that provides for long-term productivity, resistance to stress, and adaptability to change. A biologically diverse forest holds a greater variety of potential resource options for a longer period of time than a less diverse forest. It is more likely to be able to respond to environmental stresses and adapt to a rapidly changing climate. And it may be far less costly in the long run to sustain a rich variety of species and biological communities operating under largely natural ecological processes than to resort to the heroic efforts now being employed to recover California condors (*Gymnogyps californianus*), peregrine falcons (*Falco peregrinus*), and grizzly bears (*Ursus horribilis*). Resource managers know from experience that access to resources is greater and less costly when forests and rangelands are sufficiently healthy and diverse.

It is easy to understand why endangered species have received the focus of attention. Many are large, easily observable, and oftentimes aesthetically pleasing. Thereby, most efforts at restoration and rehabilitation have been directed towards endangered species and harvested species. However, endangered species are fundamental indicators of environmental disturbance. Since extinction is a process, not a simple event, the recognition that a species is endangered is little more than a snapshot of a moving vehicle. Attempts at therapy most often address symptoms rather than causes. We have failed to communicate successfully why rehabilitation and restoration beyond the narrow focus of the endangered and harvested, are essential. Processes from which endangered species arise are primarily degradation at the landscape scale: fragmentation, physiographic alteration, vegetation removal or replacement, pollution, etc (54). The environmental variables which affect the health and welfare of all the flora and fauna also affect people: water and air quality, recycling of organic and inorganic substances, microclimate, etc. Loss of biodiversity means loss of ecological services and options for the future. The cost of replacing ecological services, already great, will increase to staggering proportions. The real and potential wealth represented by conserved biodiversity cannot be replaced (54). So by shifting the focus of conservation activities from the species and population level to the landscape level we may start to make progress (54).

The tough choices posed in the spotted owl (*Strix occidentalis*) case in the Pacific Northwest of the United States typify many future issues as the conservation of biodiversity becomes a higher social priority. Regardless of the eventual outcome of this issue, there is an important lesson to be learned: Conserving biodiversity will not be cheap or noncontroversial. Federal land management agencies in the United States have increasingly come under fire over management decisions that appear to decrease biodiversity. The USDA Forest Service faces numerous appeals and lawsuits on forest plans for insufficient and sometimes conflicting consideration of forest biodiversity in management decisions. The dispute over the spotted owl and old growth forests is the most visible example of how tough it is to blend the conservation of biodiversity with other uses and values of public resources. It illustrates the reality of "no free lunch" in resource allocations. Even though parks, reserves, set-asides, and easements are critical components in the mix for the conservation of biodiversity, they will become more difficult to come by and ultimately will require an expansion beyond the "reserve mentality" (55). Multiple-use of public lands is deeply ingrained. Somehow we have to come up with management prescriptions for our public lands that will allow both consumptive and nonconsumptive uses but will do so in such a way that no net loss of native species will occur. Such a prescription will require that the livestock, timber, and mining industries take their turns at the trough instead of always going to the head of the line. This prescription will require encouraging resource conservation and recycling programs that reduce the need for raw materials from public lands (55). The scope of conservation has been too constrained and steps must be taken to incorporate the benefits of biodiversity and the use of biological resources into local, regional, national and international economies (42,56).

Future conservation at larger scales will always be confounded by the potentially large number of political authorities that conduct land management practices on watershed, basin, or even landscape scales (57). Imposing coarse scale structure on the landscape, like protecting large blocks or habitat or the creation of regional-scale corridors, can preserve connectivity in the face of some acceptable amount of habitat destruction (58). Management of larger parcels for permanent and temporary openings has the potential to reduce edge impacts and allow for the regeneration of larger stands of mature forest which can benefit area-sensitive species. The juxtaposition of habitat patches within a landscape context and its importance to the maintenance of wildlife diversity is essential, but only beginning to be appreciated (51). Unfortunately, many management practices used today often ignore the natural dynamics of ecosystems (3). Managers need to be open-minded about the use of a suite of disturbance-creating tools to manipulate flora and fauna, particularly in systems that have evolved under a disturbance regime.

The scale and scope of conservation has been too restricted (42). Spatial scale be it local, regional, or global, greatly influences our perceptions of biodiversity (51). Understanding the importance of scale is critical to accurately assessing the impact of land management practices on biodiversity. The scale of a conservation endeavor affects the strategy involved, the determination of realistic goals, and the probability of success (57). For example, a strategy to maximize species diversity at the local level does not necessarily add to regional diversity. In fact, oftentimes in our hast to "enhance" habitats for wildlife we have emphasized "edge" preferring species at the expense of "area" sensitive ones and consequently may have even decreased regional diversity. It is important to realize that principles that apply at smaller scales of time and space do not necessarily apply to longer time periods and larger spatial scales (51). Long-term maintenance of species and their genetic variation will require cooperative efforts across entire landscapes (42). This maintenance is consistent with the growing scientific sentiment that biodiversity should be dealt with at the scale of habitats or ecosystems rather than species (58). However, how to deal with this technically is still in the developmental stages.

The key to maintaining global biodiversity is ecosystem and landscape protection and management, not crisis management of an increasing number of endangered species. The time to save a species is when it is still common. To meet the challenge we must continually develop new programs as information becomes available. Adaptive management which depends on monitoring the consequences of various management treatments, should be used to adjust and refine management practices (52). Much remains to be learned about managing wildlands for biodiversity, and only through extensive cooperative efforts can we provide and protect essential diversity for ourselves and future generations.

BIBLIOGRAPHY

1. T. L. Erwin, "The Tropical Forest Canopy: The Heart of Biodiversity," in E. O. Wilson and E. M. Peter, eds., *Biodiversity*. National Academy Press, Washington, D.C., 1988, pp. 123–129.
2. R. C. Szaro, "Biodiversity and Biological Realities," in D. Murphy, ed., *Proceedings of the National Silvicultural Workshop, Cedar City, Utah*, May, 1991, USDA Forest Service, Washington, D.C., 1993.
3. R. C. Szaro, "Conserving Biodiversity in the United States: Opportunities and Challenges," in D. Bandu, H. Singh and A. K. Maitra, eds., *Environmental Education and Sustainable Development (Proceedings of the 3rd International Conference on Environmental Education*, Panjim, Goa, India, October 3–7, 1989, Indian Environmental Society, New Delhi, India, 1990, pp. 173–180.
4. R. C. Szaro, "Biodiversity Inventory and Assessment," in A. Evendon, Compiler, *Proceedings—Northern Region Biodiversity Workshop*, USDA Forest Service, Northern Region, Missoula, Mont., 1992, pp. 57–62.
5. K. E. Evans and R. C. Szaro, "Biodiversity: Challenges and Opportunities for Hardwood Utilization, *Diversity* **6,** 27–29 (1990).
6. C. Harrington, D. Debell, M. Raphael, K. Aubry, A. Carey, R. Curtis, J. Lehmkuhl and R. Miller, *Stand-level Information Needs Related to New Perspectives*, Pacific Forest and Range Experiment Station, Olympia, Wash., 1990.
7. R. Szaro and B. Shapiro, *Conserving Our Heritage: America's Biodiversity*, The Nature Conservancy, Arlington, Va., 1990.
8. R. C. Szaro and H. Salwasser, "The Management Context for Conserving Biological Diversity," 10th World Forestry Congress, Paris, France, Sept., 1991, *Revue Forestiere Francaise, Actes Proceedings* **2,** 530–535 (1991).
9. J. Gradwohl and R. Greenberg, *Saving the Tropical Forests*, Island Press, Washington, D.C., 1988.
10. L. A. Maquire, "Risk Analysis for Conservation Biologists," *Conservation Biology* **5,** 123–125 (1991).
11. R. T. T. Formann, "The Ethics of Isolation, the Spread of Disturbance, and Landscape Ecology," in M. G. Turner, ed., *Landscape Heterogeneity and Disturbance*, Springer-Verlag, New York, 1987, pp. 213–229.
12. D. A. Saunders, R. J. Hobbs, and C. R. Margules, "Biological Consequences of Ecosystem Fragmentation: A Review," *Conservation Biology* **5,** 18–32 (1991).
13. Keystone Center, *Final Consensus Report of the Keystone Policy Dialogue on Biological Diversity on Federal Lands*, The Keystone Center, Keystone, Colorado, 1991.
14. United Nations Environment Programme, *Convention on Biological Diversity*. Na. 92-7807, Nairobi, Kenya, 1992.
15. United Nations Environment Programme, *Conference for The Adoption of The Agreed Text of the Convention on Biological Diversity*, Na. 92-7825. Nairobi, Kenya, 1992.
16. D. Ehrenfeld, "Why Put a Value on Biodiversity?" in E. O. Wilson and F. M. Peter, eds., *Biodiversity*, National Academy Press, Washington, D.C., 1988, pp. 212–216.
17. E. O. Wilson, "The Biological Diversity Crisis," *Bioscience* **35,** 703–704 (1985).
18. P. R. Ehrlich, "Extinctions and Ecosystem Functions: Implication for Humankind," in R. J. Hoague, ed., *Animal Extinction: What Everyone Should Know*, Smithsonian Institution Press, Washington, D.C., 1985, p. 162.
19. E. O. Wilson, "The Current State of Biological Diversity," in E. O. Wilson and F. M. Peter, eds., *Biodiversity*, National Academy Press, Washington, D.C., 1988, pp. 3–18.
20. J. A. McNeely, K. R. Miller, W. V. Reid, R. A. Mittermeier, and T. B. Werner, *Conserving the World's Biological Diversity*, World Conservation Union (IUCN), Gland, Switzerland; World Resources Institute, Conservation Internationa, World Wildlife Fund–U.S. and the World Bank, Washington, D.C., 1990, p. 193.
21. E. Salati and P. B. Vose, "Depletion of Tropical Rain Forests," *Ambio* **12,** 67–71 (1983).
22. P. H. Raven, "We're Killing Our World: Preservation of Biological Diversity," *Vital Speeches of the Day*, May 15, 1987, pp. 472–478.

23. R. Repetto, *The Forest for the Trees? Government Policies and the Misuse of Forest Resources*, World Resources Institute, Washington, D.C, 1988.
24. A. Gomez–Pompa and A. Kaus, "Traditional Management of Tropical Forests in Mexico," in A. B. Anderson, ed., *Alternatives to Deforestation: Steps Toward Sustainable Use of the Amazon Rain Forest*, Columbia University Press, New York, 1990, pp. 45–64.
25. J. R. Parsons, "The Changing Nature of New World Tropical Forests Since European Civilization," in *The Use of Ecological Guidelines for Development in the American Humid Tropics*, 1975, IUCN Publication New Series No. 31, pp. 22–38.
26. B. L. Turner, II, "Population Density in the Classic Maya lowlands: New Evidence for Old Approaches," *Geographical Review* **66,** 73–82 (1976).
27. W. L. Partridge, "The Humid Tropics Cattle Ranching Complex: Cases from Panama Reviewed," *Human Organization* **43,** 76–80 (1984).
28. J. S. Maini, "Sustainable Development of Forests," *Unasylva* **43**(169), 3–8 (1992).
29. R. M. May, "Past Efforts and Future Prospects Towards Understanding How Many Species There Are," in O. T. Solbrig, H. M. van Emden and P. G. W. J. van Oordt, eds., *Biodiversity and Global Change*, International Union of Biological Sciences, Paris, Monograph No. 8, 1992.
30. T. L. Erwin, "The Tropical Forest Canopy: the Heart of Biotic Diversity," in E. O. Wilson and F. M. Peter, eds., *Biodiversity*, National Academy Press, Washington, D.C, 1988, pp. 123–129.
31. H. Salwasser, "Conserving Biological Diversity: A Perspective on Scope and Approaches," *Forest Ecology and Management* **35,** 79–90 (1990).
32. O. T. Solbrig, ed., *From Genes to Ecosystems: A Research Agenda for Biodiversity*, International Union of Biological Scientists, Cambridge, Mass., 1991.
33. J. Verner, "Future Trends in Management of Nongame Wildlife: A Researchers Viewpoint," in J. B. Hale, L. B. Best and R. L. Clawson, eds., *Management of Nongame Birds in the Midwest—A Developing Art*, North Central Section, Wildlife Society, Madison, Wis., 1986, pp. 149–171.
34. M. E. Soule, *Viable Populations for Conservation*, Cambridge University Press, N.Y., 1987.
35. R. C. Szaro, "Guild Management: An Evaluation of Avian Guilds as a Predictive Tool," *Environmental Management* **10,** 681–688 (1986).
36. R. C. Szaro and R. P. Balda, *Selection and Monitoring of Avian Indicator Species: An Example from a Ponerosa Pine Forest in the Southwest*, Rocky Mountain Forest and Range Experiment Station, Fort Collins, Colo., 1982, USDA Forest Service General Technical Report RM-89.
37. J. R. Karr, "Biological Monitoring and Environmental Assessment: A Conceptual Framework," *Environmental Management* **11,** 249–256 (1987).
38. F. B. Goldsmith, ed., *Monitoring for Conservation and Ecology*, Chapman and Hall, New York, 1991.
39. T. W. Hoekstra and C. H. Flather, "Theoretical Basis for Integrating Wildlife in Renewable Resource Inventories," *Journal of Environmental Management* **24,** 95–110 (1986).
40. Council of Environmental Quality, "Linking Ecosystems and Biodiversity," in *Environmental Quality: 21st Annual Report*, Washington, D.C., 1990, pp. 135–187.
41. F. W. Davis, D. M. Stoms, J. E. Estes, J. Scepan, and J. M. Scott, "An Information Systems Approach to the Preservation of Biological Diversity," *International J. of Geographic Information Systems* **4,** 55–78 (1990).
42. Global Biodiversity Strategy, *Guidelines for Action to Save, Study, and Use Earth's Biotic Wealth Sustainably and Equitably*, World Resources Institute, Washington, D.C.; The World Conservation Union (IUCN), Gland, Switzerland; United Nations Environment Programme (UNEP), Nairobi, Kenya, 1992.
43. R. L. Peters and T. E. Lovejoy, eds., *Global Warming and Biological Diversity*, Yale University Press, New Haven, Conn., 1992.
44. W. V. Reid and K. R. Miller, *Keeping Options Alive: The Scientific Basis for Conserving Biodiversity*, World Resources Institute, Washington, D.C., 1989.
45. S. Woodley, J. Kay, and G. Francis, *Ecological Integrity and the Management of Ecosystems*, St. Lucie Press, Ottawa, Ont., 1993.
46. W. A. Niering, "Vegetation Dynamics (Succession and Climax) in Relation to Plant Community Management," *Conservation Biology* **1,** 287–295 (1987).
47. D. L. Urban, R. V. O'Neill, and H. H. Shugart, "Landscape Ecology," *Bioscience* **37,** 119–127 (1987).
48. O. T. Solbrig, H. M. van Emden, and P. G. W. J. van Oordt, eds., *Biodiversity and Global Change, Monograph No. 8*, International Union of Biological Sciences, Paris, 1992.
49. IUCN/UNEP/WWF, *Caring for the Earth. A Strategy for Sustainable Living*, The World Conservation Union (IUCN), Gland, Switzerland; United Nations Environment Programme (UNEP); World Wide Fund for Nature (WWF), 1991.
50. J. Lubchenco, A. M. Olson, L. B. Brubaker, S. R. Carpenter, M. M. Holland, S. P. Hubbell, S. A. Levin, J. A. MacMahon, P. A. Matson, J. M. Melillo, H. A. Mooney, C. H. Peterson, H. R. Pulliam, I. A. Real, P. J. Regal, and P. G. Risser, "The Sustainable Biosphere Initiative: An Ecological Research Agenda," *Ecology* **72,** 371–412 (1991).
51. T. R. Crow, "Biological Diversity and Silvicultural Systems," in *Proceedings of the National Silvicultural Workshop: Silvicultural Challenges and Opportunities in the 1990's*, USDA Forest Service, Timber Management, Washington, D.C. (1991).
52. R. F. Noss, "Biodiversity Conservation at the Landscape Scale," in R. C. Szaro, ed., *Biodiversity in Managed Landscapes: Theory and Practice*, Oxford University Press (In Press).
53. E. D. Schulze and H. A. Mooney, *Biodiversity and Ecosystem Function*, Springer–Verlag, New York, 1993.
54. P. Bridgewater, D. W. Walton, and J. R. Busby, "Creating Policy on Landscape Diversity," in R. C. Szaro, ed., *Biodiversity in Managed Landscapes: Theory and Practice*, Oxford University Press (In Press).
55. P. F. Brussard, D. D. Murphy, and R. F. Noss, "Strategy and Tactics for Conserving Biological Diversity in the United States," *Conservation Biology* **6,** 157–159 (1992).
56. K. R. Miller, "Conserving Biodiversity in Managed Landscapes," in R. C. Szaro, ed., *Biodiversity in Managed Landscapes: Theory and Practice*, Oxford University Press (In Press).
57. S. M. Pearson, M. G. Turner, R. H. Gardner, and R. V. O'Neill, "Scaling Issues for Biodiversity Protection," in R. C. Szaro, ed., *Biodiversity in Managed Landscapes: Theory and Practice*, Oxford University Press, in press.
58. M. L. Hunter, *Wildlife, Forests, and Forestry: Principles of Managing Forests for Biodiversity*, Prentice-Hall, Inc., Englewood Cliffs, N.J., 1990.

BITUMEN

RICHARD MEYER
U.S. Geological Survey
Reston, Virginia

Occurrences of natural bitumen are now known throughout the world, in a number of varieties. All have been exploited at various times, most notably in the 19th and early 20th centuries, when the most abundant variety, natural asphalt, was widely used as paving material. This application has since been mostly displaced by manufactured asphalt, which makes possible a much more accurately controlled mixture of the bitumen and stone. However, the quantities of natural asphalt worldwide are exceedingly large, representing nearly three trillion barrels of oil in place, and the technology for its recovery and conversion to petroleum products is available. The present constraints on the exploitation of natural bitumen are economic and environmental, not technological.

DEFINITION

Natural asphalt is commonly referred to as natural bitumen or as bitumen, even though the term bitumen technically includes the other, minor species. Natural asphalt describes one species of natural bitumen. The term has been used for any altered crude oil (1). A brief description of this terminology is found in ref. 2.

Such terms as tar sand, oil sand, hydrocarbonaceous rock, and oil-impregnated rock refer to natural asphalt plus the rock which contains it. This rock is most frequently sandstone or a friable sand cemented by the natural asphalt and contains a significant proportion of fine material, predominantly or entirely clay. The rock may also be limestone or dolomite. In this discussion, such rock is simply referred to as asphalt rock, suitably modified according to the lithology. In the subsurface, the asphalt-containing rock constitutes a reservoir.

Natural bitumen is a black, highly complex mixture of molecules, some of which, the asphaltenes in particular, are extremely large and, therefore, heavy (high molecular weight). These large molecules act like colloids and thus affect the ability of the fluid to flow.

The choice of criteria for distinguishing bitumen from crude oil must be arbitrary. A gas-free dynamic viscosity measured at reservoir temperature of 10,000 centipoises (cP) in customary units or its metric equivalent, 10,000 millipascal seconds (mPa·s), appears to satisfy most requirements for such a standard: (*1*) Viscosity can be determined by any of several widely applied laboratory methodologies, such as the rotational viscometer. (*2*) 10,000 cP closely approximates the maximum viscosity at which most oils can be recovered from a reservoir without thermally enhanced oil recovery methods. (*3*) 10,000 mPa·s is a value reached by many scientists and engineers (2) or falls within the range of values suggested by others (3,4).

NATURAL BITUMEN RESOURCES

Reserves of pyrobitumens in countries of the former Soviet Union (FSU) are as follows (millions of tons): Tajikistan, Uzbekistan, and Kirgizstan together, greater than 1,000; Georgia, 2.1; European Russia, 7; and Asian Russia, 141. Ozocerite reserves are reported to be as follows (millions of tons): Tajikistan, Uzbekistan, and Kirgizstan together, 3; Georgia, 0.5; Turkmenistan, 1.5; Ukraine, 2; and Asian Russia, 0.5. Similar deposits are present in many other countries, including the United States but quantities have not been reported or else, as is frequently the case, the deposits have been exhausted.

Estimated natural asphalt resources for the world are given in Table 1.

Table 1. World Natural Asphalt Resources, by Area, in Measured Deposits (million barrels)[a]

Area	Demonstrated	Inferred
North America		
Canada		
Alberta	1,697,359	831,000
Melville Island		100
Canada total	1,697,359	831,100
United States		
Alaska		11,000
Alabama		7,970
California	6,210	2,540
Kentucky	1,720	1,690
Missouri–Kansas–Oklahoma	220	2,730
New Mexico	91	
Oklahoma	42	22
Texas	2,670	300
Utah	16,856	6,429–12,362
Wyoming	120	70
United States total	27,929	32,751–38,684
Area total	*1,725,288*	*863,851–869,784*
South America		
Trinidad	60	
Venezuela	62	
Area total	*122*	
Europe		
Albania	371	
Azerbaijan	82	
Georgia	636	
Germany	315	
Italy	1,260	
Romania	25	
Russia	94,516	
Ukraine	599	
Area total	*97,804*	
Asia		
Kazakstan	2,112	
Kirgizstan, Tajikistan, Uzbekistan	48	
People's Republic of China	10,050	
Russia	91,779	
Turkmenistan	3	
Area total	*103,992*	

Table 1. *(continued)*

Area	Demonstrated	Inferred
Africa		
Madagascar	1,600	3,400
Nigeria	42,740	42,000
Zaire	30	
Area total	*44,370*	*45,400*
Middle East		
Syria	13	
Area total	*13*	
Southeast Asia		
Indonesia	8	
Phillippines	3	
Area total	*11*	
World Totals	1,971,600	909,251–915,184
World Demonstrated Plus Inferred	2,880,851–2,886,784	

[a] Ref. 5, modified from ref. 6 and ref. 7.

NATURAL BITUMEN PRODUCTION

Today, almost all rock asphalt mining is for the purpose of bitumen extraction, the bitumen then being upgraded to synthetic crude oil.

RECOVERY

Three methods are employed for the recovery of natural asphalt from the earth for the purpose of converting it to petroleum products. Each of these methods either is in use or else has been tested for use in Alberta. These methods are (*1*) surface mining, (*2*) in situ thermal, and (*3*) combined mining and thermal.

ENVIRONMENTAL CONSIDERATIONS

Many of the environmental considerations required in the process of converting a natural asphalt deposit to a suite of useful products are no different from those involved with conventional crude oil. The objective is to minimize each step that leads to environmentally undesirable consequences and to maximize an upgraded product as rich as possible in hydrogen and as lean as possible in carbon and nonhydrocarbon contaminants. A major difference between natural asphalt and conventional crude oil is that much of it is, and in the future probably will be produced by surface mining rather from oil wells.

Exploration

The natural bitumens, except for natural asphalt and reservoir bitumen, occur in small deposits and those that have been found occur on the surface of the earth. The reservoir bitumens are disseminated in the subsurface and hold no economic interest. Therefore, these species of bitumen have no environmental impacts in the exploration process or in their later exploitation and need not be considered further.

Natural asphalt is found in the subsurface but is discovered only by serendipity in the search for crude oil. Surface deposits of natural asphalt are known and did not incur particular environmental impacts in their discovery.

Recovery and Extraction of Bitumen

Subsurface in situ production of asphalt requires heat to reduce the viscosity and fresh water steam as the heat carrier has proved to be most effective. Usually, 3–7 barrels of steam as water are needed to produce a barrel of bitumen. Recycling of produced water reduces the requirement for fresh water and eases the problem of produced water disposal. Produced water may contain some or all of dispersed or soluble oil, heavy metals, radionuclides, treatment chemicals, salt, and dissolved oxygen (8). Many of the most toxic compounds in the dispersed oil droplets are volatile and escape, unless trapped in sediment somewhere in the production system. Of the soluble compounds, phenols are biodegradable. Levels of polynuclear aromatic compounds are below toxic levels in produced water. Heavy metals in produced water, such as cadmium and lead, may be at toxic levels and must be treated for removal or else properly disposed. The problem of radionuclides is strongly dependent on geographic location and thus is site specific. Risk-assessment studies suggest that predicted excess human cancers from this source would be comparable to background concentrations of radium. Treatment chemicals are generally present below toxic limits, with the exception of those injected as slugs. Slugs containing such chemicals, including scale and corrosion inhibitors, may be captured separately to permit proper disposal of the concentrate. If the produced water is saline it can be treated at the surface facility. Usually, the produced water is low in oxygen since that is the nature of formation water; its low oxygen content, inimical to life, must be considered if the water is discharged at the surface.

Recycled produced water can be used directly as makeup water for steam if total dissolved solids (TDS) are less than 8000 ppm, as is the case at Peace River. At higher TDS levels, some treatment is required.

Use of the hot water bitumen-extraction process leads to very large tailings in the form of sludge; at Syncrude, this amounts to about 3.5×10^5 million cubic m per day (9) in a pond covering 4.25 square miles. If left alone, this material may take centuries to consolidate. Clay minerals become thoroughly dispersed in the sludge, which also contains residual bitumen and fine-grained non-clay minerals (10). A number of solutions to the sludge problem have been proposed but at present the material is still segregated in ponds; it cannot be used or recycled.

Until recently oily-sand waste was simply used for surfacing country roads (11). This procedure must now give way to methods yielding clean sand and separated oil. The Taciuk processor, previously described as a recovery process to replace hot water extraction and the attendant tailings ponds, also may be used to clean oily wastes. The result is clean, disposable sand, hydrocarbon liquids acceptable to a refinery, and fuel gas for plant use (12). The effluent solids from the ATP are comprised of 93 wt % bulk tailings, 4.9 wt % flue-gas cyclone fines, 1.7 wt % baghouse fines, and 0.4 wt % hydrocarbon cyclone fines (13). The solids mix is a dry silty sand, free of hydrocarbons, with only 21 ppm organic carbon, the latter being a function of the degree of coke burn-off in the combustion zone. Chlorides and sulfates are contained primarily in the flue gas particulates and could be separately disposed. Analysis of the leachate (13) found the presence of phenols and heavy metals low or not detectable, oil and grease not detectable, total carbon low, and concentrations of chlorides and sulfate below permissible levels for potable water. The effluent water from the preheat zone and the sour water from the oil fractionator showed high phenol concentrations, biological and chemical oxygen demand, and sulfur, and are thus similar to refinery process water generated in thermal cracking of high-sulfur heavy oil and residue. These waters would have to be treated or disposed in deep wells. Off-gas could be flared or used as fuel. The flue gas from combustion of the coke on the spent solids is low in SO_x, NO_x, H_2S, and solids and can be safely discharged from the stacks.

Upgrading

With respect to upgrading, the goal is to minimize the production of residue and, particularly, sulfur. This can be done most effectively by use of high-conversion, hydrogen-addition processes to replace the current carbon-rejection reactors. High-pressure reactors will sharply reduce residue and in fact increase yield. Ref. 14 compares examples of different upgrading schemes with respect to their environmental impacts. For example, Flexicoking for carbon-rejection is used, giving moderate conversion; LC-Fining plus delayed coking for hydrogen addition plus catalysis, giving medium conversion; and VEBA-COMBI-Cracking (VCC) for hydrogen addition giving high conversion. Assuming each is treating 60,000 bbl/day of Cold Lake undiluted bitumen, the inputs to, and outputs from each process are presented in Table 2. Flexicoking yields

Table 2. Summary of Net Upgrader Inputs and Outputs[a]

	VCC	LC-Fining/ Delayed Coker	Flexicoker
Inputs			
Crude, bbl/day	60,000	60,000	60,000
Natural gas, 10^9 J/day[b]	49,899	36,725	0
Electric power, MW·h	25	33	14
Additive, t/day	42	0	0
Catalyst kg/day	0	5443	0
Outputs			
Synthetic crude, bbl/day	63,910	60,920	51,950
Sulfur, t/day	361	352	359
Coke, t/day	0	576	29
Low btu gas (10^9 J/day)[b]	0	0	18,502
Residue, t/day	279	0	0
Spent catalyst, kg/day	0	5443	0

[a] Ref. 14.
[b] To convert J to Btu, divide by 1.054×10^3.

Table 3. Emissions from Selected Upgraders[a]

	VCC	LC-Fining/ Delayed Coker	Flexicoker
SO_2, (kg/h)	8.00	8.20	9.90
NO_x, (kg/h)	1.66	1.60	9.80
CO_2, (t/h)	0.88	0.84	5.20

[a] Ref. 14

a large quantity of low-Btu gas at the expense of liquids product, the gas being used for fuel, reducing the electric power requirement. The two hydrogen-addition processes increase yield by converting gas to liquids with hydrogen. The synthetic crude oil from the three processes is very similar in quality. Table 3 shows the emissions from the upgraders, based on production of 1000 bbl/day of synthetic crude. All meet legal emission limits but the two hydrogen-addition processes are much superior to Flexicoking carbon rejection. The carbon-rejection yield of SO_2 is higher because of the very large amount of coke gasified; as well, the NO_x and CO_2 are much higher due to the combustion of the low-Btu gas, which also adds SO_2. Particulate emissions depend upon the degree of stack gas protection but will be highest for the delayed coker, less for the Flexicoker, and none for the VCC, which produces no coke.

The LC-Finer/delayed coker will produce about three times as much gypsum as the VCC, the gypsum being a by-product of flue-gas desulfurization. Such gypsum is generally disposed in landfills.

VCC catalyst residue may be used as binder in coke manufacturing but the amount of catalyst residue from LC-Fining is greater and must be disposed in landfill after leaching for vanadium. Significant amounts of vanadium may be produced from the spent catalysts. Processes which yield such useful byproducts as vanadium are to be preferred because they lead to higher synthetic oil production without increasing emissions.

Ref. 14 concludes that hydrogen-addition processes, especially those of high conversion like VCC, have the least impact because of gaseous and solid emissions. They also have the greatest liquid yields.

Upgrading is the final step in natural asphalt utilization prior to transportation to the refinery. The high-conversion hydrogen-addition upgrading processes such as VCC are most advantageous with respect to product output and environmental consequences in terms of emission of gases and particulates. Such advantages will increase in merit as demand for the products increases and environmental constraints become more stringent. At present, however, the process used will be the one that is most cost effective. However, it also must meet demand for high-quality products that can themselves meet stringent regulatory requirements and for processes that will lead to the capture and productive use of carbon dioxide from combustion and process sources (15).

Land Use

The sludge waste resulting from use of the hot-water extraction process impacts the environment in several ways. (*1*) Areally, it removes land from other uses. (*2*) It may trap wildlife, most notably wildfowl, although protective steps have been instituted. (*3*) It would despoil any stream into which it leaked, which has resulted in stringent disposal regulations.

The land required for surface mining operations cannot be utilized for other purposes until mining stops and reclamation is completed.

With respect to in situ recovery, a number of environmental advantages accrue to the drilling of slant wells from central pads rather than from pattern spacing. In terms of surface disturbance, this means fewer roads and location sites. The pads concentrate oily and produced water wastes in one place. Drilling pads make it feasible to collect vent gases and utilize them at the site, rather than flare them. Both economically and environmentally, the use of pads is essential in soft ground areas of tidelands, swamp, muskeg, and permafrost.

Acknowledgments

The technical assistance and critical reviews of C. J. Schenk and Kathy Varnes, U.S. Geological Survey, Denver, Colorado, R. W. Stanton, U.S. Geological Survey, Reston, Virginia, and D. E. Towson, Consulting Engineer, Calgary, Alberta, are gratefully acknowledged.

BIBLIOGRAPHY

1. J. Connan and B. M. van der Weide, in G. V. Chilingarian and T. F. Yen, eds., *Bitumens, Asphalts and Tar Sands,* Elsevier, Amsterdam, 1978, pp. 27–55.
2. R. F. Meyer and Wallace de Witt, *U.S. Geological Survey Bulletin 1944,* 14 pp., 1990.
3. A. M. Danyluk, B. Galbraith, and R. A. Omana, "Toward Definitions of Heavy Crude and Tar Sands," in R. F. Meyer, J. C. Wynn, and J. C. Olson, eds., *The Future of Heavy Crude and Tar Sands, Second International Conference,* McGraw-Hill, New York, 1984, pp. 3–6.
4. E. M. Khalimov, and co-workers, "Geological Problems of Natural Bitumen," *World Petroleum Congress, 11th, Proceedings,* vol. 2, pp. 57–70, 1983.
5. R. F. Meyer, and J. M. Duford, "Resources of Heavy Oil and Natural Bitumen Worldwide," in R. F. Meyer, and E. J. Wiggins, eds., *The Fourth UNITAR/UNDP International Conference on Heavy Crude and Tar Sands Proceedings,* vol. 2, 1988, pp. 277–307.
6. I. S. Goldberg, ed., "Prirodnye Bitumy SSSR [National Bitumens of the USSR]," *Nedra, Leningrad,* **195,** [223] (1981).
7. Energy Resources Conservation Board [Alberta], "Alberta's Reserves of Crude Oil, Oil Sands, Gas, Natural Gas Liquids and Sulphur," *ERCB ST 92-18,* pp. 3-1–3-6, 1992.
8. M. T. Stephenson, *JPT,* **44**(5), 548–550, 602–603 (1992).
9. J. A. Franklin, and Dusseault, *Rock Engineering Applications,* New York, McGraw-Hill, 1991, 431 pp.
10. M. B. Dusseault, D. W. Scafe, and J. D. Scott, *AOSTRA Journal of Research,* **5**(4), 303–320 (1989).
11. R. N. Houlihan, and R. G. Evans, "Development of Alberta's Oil Sands," in R. F. Meyer, and E. J. Wiggins, eds., *The Fourth UNITAR/UNDP International Conference on Heavy Crude and Tar Sands Proceedings,* vol. 1, pp. 95–110, 1989.
12. Alberta Oil Sands Technology and Research Authority, *AOSTRA, a 15 Year Portfolio of Achievement,* Edmonton, Alberta, Canada, 176 pp., [1990].
13. L. R. Turner, "Treatment of Oil Sands and Heavy Oil Production Wastes Using the AOSTRA Taciuk Process," in R. F. Meyer, ed., *Heavy Crude and Tar Sands—Hydrocarbons for the 21st Century, UNITAR International Conference on Heavy Crude and Tar Sands,* 5th, Alberta Oil Sands Technology and Research Authority, Edmonton, Canada, v. 4, pp. 347–356, 1991.
14. F. W. Wenzel, "Comparison of Different Upgrading Schemes with Respect to their Environmental Impacts," in R. F. Meyer, ed., *Heavy Crude and Tar Sands—Hydrocarbons for the 21st Century, UNITAR International Conference on Heavy Crude and Tar Sands, 5th,* Alberta Oil Sands Technology and Research Authority, Edmonton, Canada, **4,** 357–370 (1991).
15. Alberta Oil Sands Technology and Research Authority, *1992: Annual Report,* 99 p.

BRAYTON CYCLE (CLOSED)

Anthony Pietsch
Chandler, Arizona

The closed Brayton cycle bears the name of George Brayton (1839–1892), a New England inventor who produced a continuous flow internal combustion reciprocating engine that incorporated many unique design features to control combustion and burn a variety of fuel (1). It was supplanted by the Otto engine that at that time had a higher efficiency.

Today, the cycle utilizes high speed gas turbine machinery and perhaps should have been called the Closed Gas Turbine Cycle.

Cycle Description

The closed Brayton gas turbine cycle, like the open cycle gas turbine used extensively for aircraft propulsion and utility power, is a continuous flow process. The open cycle gas turbine uses ambient air as the working fluid. The closed Brayton gas turbine is modified to use heat exchangers to add and reject the heat in the cycle. The benefits of closing the cycle are many, including:

- A wide selection of working fluids can be chosen
- A wide variety of energy sources can be utilized such as coal, oil, gas, solar and nuclear
- The system can be pressurized which results in dramatic size and weight reductions as well as performance gains
- The system can be hermetically sealed to operate in hostile environments such as undersea and in space

Figure 1 is a schematic of the closed Brayton cycle. The compressor raises the pressure and temperature of the gas. The gas temperature is raised further by heat transfer from the waste heat of the turbine exhaust via the recuperator. The gas temperature is further increased in the heat source heat exchanger. The gas is then expanded through the turbine which provides power to drive the compressor and the load. The remaining waste heat is rejected by means of a low temperature heat exchanger to complete the cycle.

The addition of the input heat exchanger, recuperator and waste heat exchanger significantly increases the size

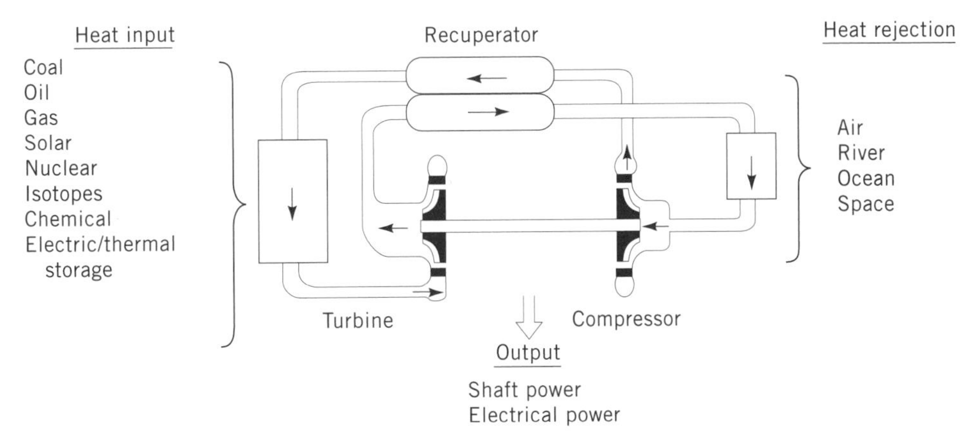

Figure 1. Closed Brayton cycle schematic.

and weight of the closed Brayton system when it is compared to the open cycle gas turbine. These disadvantages exclude it only from being used for aircraft propulsion. For terrestrial, marine, undersea and space applications, it is very competitive.

The closed Brayton cycle can achieve efficiencies from mid 30% values for solar powered space systems to in excess of 50% for nuclear powered terrestrial systems. This high range of efficiencies can be achieved at power levels as low as a few kilowatts to more than 350 megawatts using only one turbomachinery set at each power level.

The recuperator plays a large role in the accomplishment of these efficiencies because it recycles as much as two-thirds of the heat in the cycle, thus minimizing the heat added and rejected. When a high degree of recuperation is utilized, the cycle optimizes at low pressure ratios (less than 3:1 for diatomic gases such as air or nitrogen and 2:1 for monatomic gases such as helium, neon, argon, krypton and xenon and mixtures of these). The reason is to minimize the addition of heat during compression and the drop in temperature when expanded through the turbine. This optimization applies whether the gas turbine is closed or open cycle. This system is counter to the optimization of unrecuperated gas turbines where high pressure and expansion ratios are utilized to improve efficiency.

With the loop closed it is possible to raise the pressure level in the system to 50 to 75 atmospheres (5.07 to 7.6 MPa). This level dramatically increases the mass flow per unit of volume through the system resulting in significant improvement in the power density of the components (ie, the size and weight of the turbomachinery and the heat exchangers). This result is true for the megawatt size system outputs. At low power levels such as a few kilowatts, the pressure may be subatmospheric in the loop to keep the size of the turbomachinery from becoming minuscule in size and inefficient due to clearance losses.

The part load efficiency of the system is excellent. Power output is adjusted by pumping gas into and out of the loop to change the mass flow. Power output can be reduced 85 to 90% with no loss in efficiency. The explanation for this is that the heat exchangers, which are sized for the full load condition, become effectively oversized for part load. This circumstance results in lower fractional pressure losses and improved heat transfer performance. At the low power end, the parasitic losses in the system cause the efficiency to rapidly fall off.

The use of helium as the working fluid, greatly increases the heat transfer in the heat exchangers, particularly the recuperator. Helium is the ideal working fluid for systems that are directly coupled with a gas cooled nuclear reactor.

In some systems it is optimum to tailor the working fluid such as mixing xenon with helium to achieve a desired molecular weight of the gas. This tailoring is usually done because the sonic velocity and low molecular weight of helium adversely impacts the design of the turbomachinery (ie, too many turbomachinery stages are needed or their size becomes too large). Fortunately, there is a synergistic relationship when helium is mixed with other inert gases that results in higher heat transfer coefficients.

Another feature of using a closed loop is the avoidance of foreign object damage or having combustion products foul the loop, both of which cause a loss of performance as well as maintenance costs.

In the United States a large number of companies have investigated the use of the closed Brayton system for various applications (1,2). A partial listing includes Westinghouse, Allied Signal, Rockwell International, United Technologies, General Atomics, LaFleur Corp., Babcock and Wilcox, Grumman and General Electric.

The government agencies that have provided funding are the National Aeronautics and Space Agency (NASA), Department of Energy (DOE), Department of Defense (DOD), and the Maritime Administration (MARAD). Some of these agencies (NASA, DOE, DOD) have assembled systems and performed tests at their respective sites. Research and study funding has also been provided by a consortia of utility power companies, the Electric Power Research Institute (EPRI).

Environmental Aspects

The Clean Air Act of 1970 was a strong stimulus to conduct many of the studies that were made to adapt the closed Brayton cycle to vehicular, rail, ship, refrigeration, and utility applications. This attempt was due to the continuous combustion process that permitted better emission control augmented by the high cycle efficiencies that resulted in less fuel being burned. The oil embargo of 1973 further accelerated these studies to find ways to reduce fuel consumption and use indigenous fuels such as coal.

The possibility of using the high grade waste heat rejected from the closed Brayton cycle for heating, cogeneration, or industrial processes further decreased the fuel consumption with the result of further reducing unwanted pollution. The higher heat rejection temperature also made dry cooling very effective. This ability eliminates the heating of river and lake water normally used to cool power plants.

BIBLIOGRAPHY

1. C. F. McDonald, "Large Closed-Cycle Gas Turbine Power Plants," in *Sawyers Gas Turbine Engineering Handbook* 3rd ed., vol. 2, Chapter 9, 1985.
2. A. Pietsch, "Closed-Cycle Gas Turbines 50MW and Smaller," in *Sawyers Gas Turbine Engineering Handbook,* 3rd ed., vol. 2, Chapter 8, 1985.

BUILDING SYSTEMS

LES NORFORD
Massachusetts Institute of Technology
Cambridge, Massachusetts

Buildings are lighted and conditioned primarily for the benefit of their occupants. Conditioning requirements also exist for many industrial buildings, ranging from very stringent regulation of temperature, humidity and air quality in semiconductor manufacturing plants to relaxed criteria for warehouses. This article will focus on the needs of people and not processes, because processes are difficult to consider in general and because the indoor environment is becoming increasingly important as people spend more time within buildings, now as much as 90% (1). After occupant needs are established, this article will review how these needs can be met with low-energy, so-called passive techniques, and finally consider active mechanical and electrical systems.

See also AIR CONDITIONING; DISTRICT HEATING; AIR POLLUTION: INDOOR; LIGHTING, ENERGY EFFICIENT.

Occupant Needs

Thermal. The body is a source of heat, converting approximately 20% of the energy released by the metabolism of food into work and the remaining 80% into heat. Heat production depends on activity level, which ranges from 40 W per m^2 of body surface area while sleeping to 10 times this level, or 400 W/m^2, for strenuous sports activities (2). To maintain the near-constant deep-body temperature necessary for survival, heat associated with the body's metabolic rate must be transferred to the environment.

The body exchanges heat with its surroundings, via conduction, convection, radiation and evaporation of water. In common situations, conduction of heat through direct contact with hot or cold surfaces is typically very small, convection and radiation each account for about 40% of the net heat transfer and evaporation of water vapor providing the remaining 20% (3). Convective heat transfer depends on the surrounding air temperature, the surface temperature of exposed parts of the body and of clothing, which in turn is governed by the thermal resistance of the clothing, and the speed at which the air moves past the body. Radiation depends on the surface temperature of clothing and the body and of surrounding surfaces. Evaporation of water vapor, from the lungs or as sweat from the skin, is influenced by the amount of water in the air. The heat transfer processes have been modeled in detail (4).

People regulate heat flow from the body in a number of involuntary and voluntary ways. Sweating, shivering, and dilation and contraction of the blood vessels near the skin to control skin temperature are involuntary mechanisms (2,5). Choice of clothing, activity level, curling up or stretching out, and migration within rooms, buildings or across continents are voluntary means. Voluntary activities are motivated not only by survival but by more stringent thermal comfort criteria. That is, a feeling of satisfaction with the thermal environment occurs within a narrow range of conditions that permit the body to regulate its temperature.

Buildings are designed and operated to ensure the thermal comfort of the occupants. Temperatures that heating and cooling systems are sized and controlled to maintain are based on tests of large numbers of people who are subjected to different conditions and asked to identify what is comfortable. These results can be adjusted for wall surface temperatures that differ from the room air temperature, as would be the case for people working near large, poorly insulated glass in winter. Relative humidity limits and bounds on airspeed have been similarly derived.

Thermal comfort tests have shown that a single indoor thermal environment will probably not satisfy all of the occupants. People in buildings dress in a variety of styles. Tests have shown different comfort preferences in different geographical regions and between men and women (3,4). Even for a uniform population there is a range of preferences. The abundant data on the subject have been distilled into equations that predict the mean vote of occupants concerning their thermal comfort, on a seven-point scale, and the percent of occupants who will be dissatisfied for a given predicted mean vote (4). Guidelines for dry-bulb temperature, mean-radiant temperature, humidity and air-speed control in buildings typically provide a range of conditions within which 80% of test subjects were satisfied (2,6).

Most space-conditioning systems in commercial buildings provide little or no choice to individuals. Choice can come from local control within an individual work space or, for a small but growing percentage of the work force, from freedom to move within a building or outside it, to the point of staying home and telecommuting. Workers in buildings may be more productive with local controls that allow individually tuned conditions. Energy-consumption simulations of one local-control system, for which productivity changes have not been documented, showed the system capable of saving 7% of the energy of a more typical system or requiring 15% more energy, depending on configuration (7). Small changes in productivity are difficult to reliably measure and attribute to single factors but, if established, can justify considerable expense in capital and operating costs of conditioning systems. Labor costs about 100–200 times more than the energy to condition a building, although strictly economic arguments fail to capture environmental externalities that have not been embedded in fuel costs, a societal dilemma that confronts industry and transportation as well as buildings.

Air Quality. Recently surveyed occupants of large office buildings have claimed that air quality and air movement affect their comfort (8). Comfort is influenced by pollutants that are offensive but not harmful, such as body odors, as well as those pollutants that have been shown to have adverse health effects, such as many volatile organic compounds. In addition, there are pollutants such as radon and carbon monoxide that cannot be detected by the human senses but can be present in quantities considered to be dangerous. Table 1 lists several pollutants and, where specified, maximum concentration limits.

Indoor air is considered to be acceptable if outdoor air brought into the building meets the ambient limits shown in Table 1 and if there is sufficient outdoor air or active filtration to keep into pollutant levels below the mandated maximum levels. For CO_2, the maximum level of 0.1% does not represent a threshold for health effects but instead is a surrogate for body odors, which influence comfort.

Lighting. The amount of light admitted to or generated in a building should meet the needs of the occupants, which vary with tasks that range from walking down stairs to precision machining. Light can be characterized by an objective measurement of the luminance of such luminous surfaces as lamps, reflectors, a sheet of paper, windows, walls, and the sky, or by the illuminance, the amount of luminance flux falling onto a unit surface area. The eye detects luminance, which can be related to illuminance through the reflectivity of a surface. The ability to discern the detail of an object, for example the difference between the letters 'C'and 'O,' depends on the visual or radial size of the object, the amount of allowed time, the contrast between the object and its background, the luminance of the background, and the required accuracy of the visual task. Visibility studies have attempted to quantify the importance of each of these factors, with the goal of identifying the required background luminance given values for the other factors. That is, for a specified time and accuracy to identify an object of given size and contrast with its background, background luminance can be selected. However, the calculated luminance varied over orders of magnitude for modest changes in contrast, a sensitivity that raised doubts about the accuracy and applicability of the supporting and placed no reasonable bound on luminance and the associated electric lighting capable of producing the desired luminance (13).

Visual performance studies have attempted to directly show how speed and accuracy in performing a task vary with lighting conditions and therefore answer questions about the impact of lighting levels on productivity. Early studies required participants to match split rings with similar orientation (14) while more recent work scored the ability of participants to identify mistakes in lists of similar numbers (15). The relative visual performance from the latter tests shows a sharp increase as contrast or luminance increase from very low values, followed quickly by very small increase in performance as lighting conditions further improve.

As a potential basis for lighting standards, visual performance studies suffer from being tied to particular tasks that may not accurately represent a given business. Also, lighting must be provided in circulation and recreation spaces where visual performance criteria may not be appropriate. Accordingly, visual satisfaction studies have placed participants in controlled environments and asked them to express whether or not they are satisfied. These studies are similar to thermal comfort studies that have the same goal and have similar results: a Gaussian curve of percent satisfied as a function of horizontal illuminance. In one study the curve peaked at 80 percent satisfaction at 2000 lumens per square meter, with 10% finding that light level too dark and the remaining 10 percent considering it too bright (14). For circulation areas, the appropriate measure is vertical illuminance required to reveal facial features. Visual preference studies have also identified desirable luminances for room surfaces. Surfaces which have excessively high luminances or represent a sharp change in luminance from neighboring surfaces may be a source of direct or reflected glare, which can reduce visual performance in a direct, measurable way or cause discomfort that can indirectly affect well being and productivity.

Quantitative lighting criteria attempt to strike a balance between visual performance, visual preferences and prevailing economic conditions. In 1979 the Illuminating Engineering Society provided a simplified set of illuminance levels for lighting designers (16). Illuminance varied from 22 to 22,000 lumens/m^2 (2–2000 footcandles), covering eight tasks that ranged from walking in public areas

Table 1. Maximum Acceptable Concentrations of Pollutants Found in Buildings

Pollutant	Limit	Ref.
Radon	4 picocuries/liter (EPA guideline)	9
	0.027 working level	10
Asbestos	0.1 fibers longer than 5 $\mu m/cm^3$ abatement level (NIOSH)	
Tobacco smoke	dilution with smoke-free air	10
Formaldehyde	0.4 ppm (state standard)	10
Other volatile organics	occupational standards	11
	0.1 occupational standards for general public	10
Carbon dioxide	0.10%	10
Microorganisms and allergens	No standards	
Nitrogen dioxide	100 $\mu g/m^3$ ambient air long-term limit	12
Sulfur dioxide	80 $\mu g/m^3$ ambient long-term average	12
Carbon monoxide	40 mg/m^3 ambient one-hour limit	12
Total particulates	75 $\mu g/m^3$ ambient long-term average	12
Oxidants	235 $\mu g/m^3$ ambient one-hour limit	12

to sewing and surgery. Within each category are a triplet of slightly different illuminances, with guidelines for selecting a single value on the basis of the age of the occupants and the required speed and accuracy of the work to be performed.

Illuminance levels as the sole criteria for lighting design can lead to undue preoccupation with providing a uniform, wall-to-wall lighting that does not recognize spatial and temporal variability in occupants' needs. Surface luminance guidelines, based on absolute luminance levels or ratios of luminances from different surfaces within a room, may prohibit bright light sources that cause glare, such as bare incandescent bulbs, but do not account for luminance as a source of both visual variety and stimulation, as would be produced by a brilliant chandelier. In response to the failure of quantitative standards to discriminate desirable and uncomfortable lighting with similar numerical values, another approach to lighting design concentrates on the information content of light. Occupants of buildings require visual information to satisfy biological needs as well as task needs (17). Biological needs include security, orientation, definition of personal space, and relation to nature. Features of a room that help to fulfill these needs should be emphasized while visual distractions should be minimized.

Understanding the process by which humans interpret visual information has led to fresh insights about lighting. Light levels should be higher during the day, to prevent the inside of a building from appearing unduly gloomy compared to outdoors, and lower at night. Humans are insensitive to constant spatial gradients in luminances, relieving the need for uniform lighting levels and making possible the extensive use of daylighting. Changes in the gradient are acceptable if justified by the building structure, making coffered ceilings acceptable but raising cautions about scallops of light and shadow on walls for no apparent reason. Luminances can be high if occupants understand and appreciate the reason, which might be the brilliance of a chandelier or a patch of sunshine. Illuminance levels can be very low in such circulation areas as subway stations, with sufficient light to guide riders to the platforms.

Bioclimatic Chart

Thermal comfort preferences can be summarized graphically in charts that use temperature and moisture content as axes. One of the most useful graphs is a bioclimatic chart, Figure 1, which defines a comfort region as a function of dry-bulb temperature and relative humidity (3).

The thermal resistance of clothing and the body's metabolic rate are implicit variables. Radiation, from the sun or from surfaces that affect the mean radiant temperature, and air speed alter the boundaries of the comfort region, as indicated. That is, wind can compensate for temperatures above the comfort region, while solar radiation or proximity to heated surfaces that increase the mean radiant temperature can balance temperatures below the comfort region.

The chart also shows evaporation as a cooling mechanism. Just as the body loses heat by evaporation of water vapor, the air surrounding a person will be cooled if it gives up energy to evaporate water. Evaporative cooling clearly works better in dry climates, as indicated on the chart, and requires that air flow over the water to be evaporated. If the air is stagnant, it will reach saturation and evaporative cooling will cease.

It is important to note that the chart is defined for outdoor conditions. For use within buildings, it must be modified to account for the sources of heat that raise the indoor temperature above outside conditions—lights, appliances, solar radiation that is absorbed by structure and furnishings which in turn heat the air. The impact of these heat sources can be quantified with the aid of an energy balance for the building.

BUILDING ENERGY BALANCE

A thermal balance equation for a building can be formed on the basis of conservation of energy. This equation relates energy stored within the building structure to heat flows into or out of the building:

$$\left(\sum_i m_i \cdot C_i\right) \cdot \frac{dT_{in}}{dt} = q_{\text{int}} + q_{solar} + q_{aux} + q_{cond} + q_{airflow}$$

Here, the mass of materials within the building is combined with the specific heats. The rate at which energy is stored is proportional to the time rate of change of the inside-air temperature, in the simple approximation that the air and the masses within the building are all at nearly the same temperature, T_{in}. The inside temperature rises in response to internal heat flows, from lights or office equipment, and from solar energy that enters through windows and to a lesser extent is absorbed by outside walls. Auxiliary heating or cooling comes from mechanical equipment. Conduction through the building walls includes surface heat transfer due to both convection and radiation while the airflow term embraces both forced ventilation and infiltration. Both conduction and airflow are proportional to the indoor-outdoor temperature difference (18) and the sum of the two terms can be expressed as $K(T_{in} - T_{out})$.

At steady state, there is no net flow of energy into or out of the building mass. With no mechanical heating or cooling, the equilibrium indoor-air temperature can be related to the outside temperature and sources of heat gain; its value can be used instead of the outdoor temperature to apply the bioclimatic chart to indoor conditions:

$$T_{in\text{-}equilibrium} = \frac{q_{\text{int}} + q_{solar}}{K} + T_{out}$$

The steady-state equation can also be used to determine the magnitude of ventilative cooling required to maintain the indoor temperature at or below a specified upper bound, or the amount of auxiliary heating or cooling needed when ventilation alone is insufficient:

$$q_{aux} = K \cdot (T_{in\text{-}setpo\,\text{int}} - T_{out}) - q_{\text{int}} - q_{solar}$$

The building mass can absorb heat during the day, reducing the need for cooling during the day and heating at night, if its temperature is allowed to increase. The air

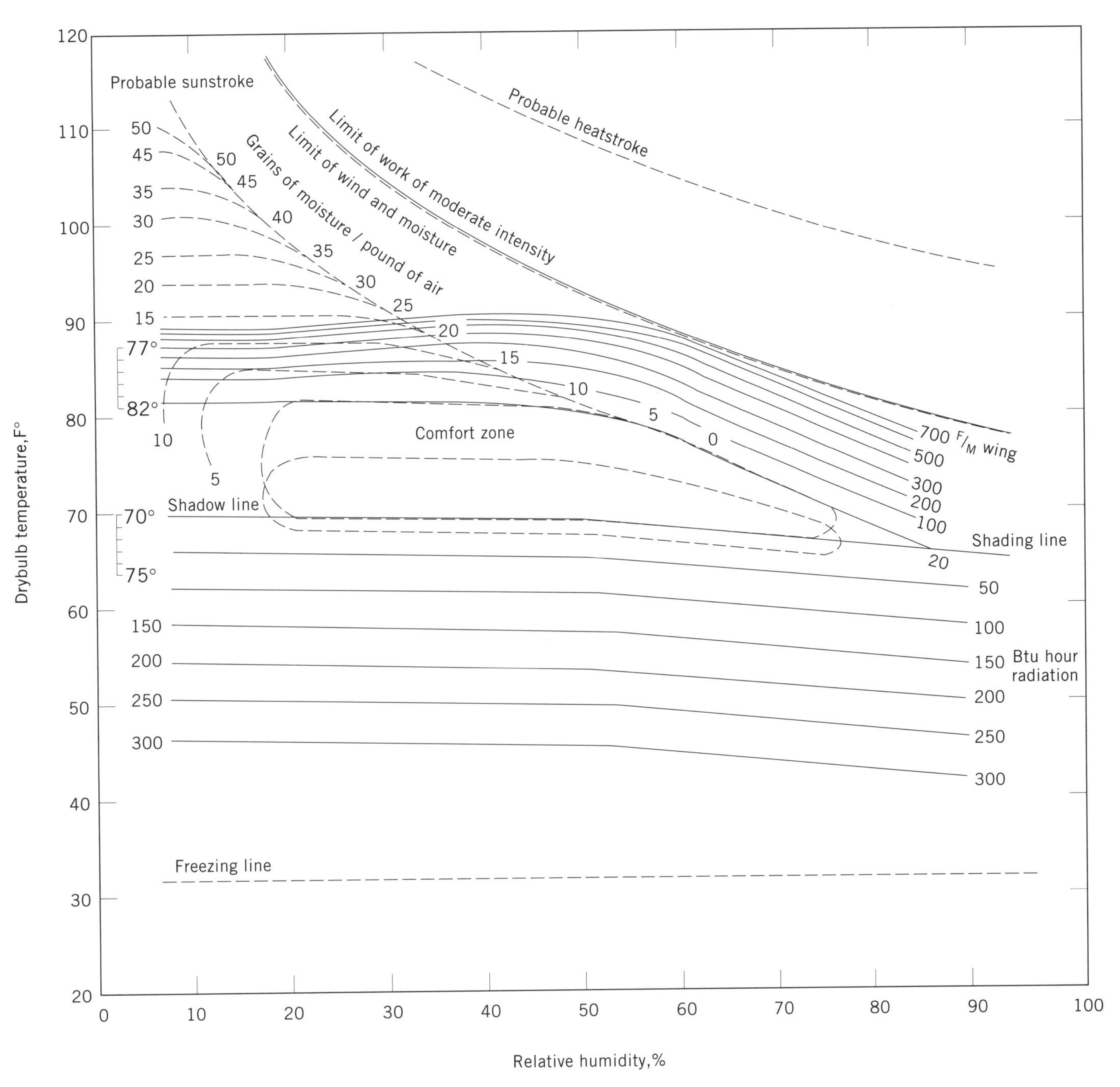

Figure 1. Bioclimatic chart for U.S. moderate zone inhabitants (3).

temperature must increase as well. The more mass that can effectively store heat, the lower the rise in temperature.

PASSIVE CONDITIONING AND LIGHTING

The bioclimatic chart identifies a number of strategies for providing comfortable conditions when the dry-bulb temperature and relative humidity fall outside the thermal comfort region. The building structure, its windows and openings can effectively stretch the comfort region, by promoting ventilation and controlling solar radiation, or shrink it, by passing heat and air inside when not needed. Solar heat can be used to increase the radiant temperature felt by an occupant or to increase the equilibrium temperature, via absorption of radiant energy by building mass and subsequent convective transfer to room air. Passive conditioning and lighting can reduce energy bills and environmental degradation, save space within a building otherwise required by mechanical systems, and reduce building capital costs.

Ventilation

Air flows across openings in a building when there is a pressure drop created by wind or by differences between indoor and outdoor temperatures. The pressure due to air movement can be derived from conservation of momentum, expressed as Bernoulli's equation for steady, irrotational flow of incompressible frictionless fluids (19):

$$p_{stagnation} = \frac{\rho \cdot v^2}{2}$$

where ρ is the density of air and v is the mean air velocity.

What follows are a number of empirical relationships. The mean air velocity, increases from zero at ground level up to its free-stream value, varying with height raised to a power of 0.1–0.3, depending on terrain. When wind strikes a building, the increase in pressure p_w on the windward surface is about 80% of what would be achieved if the wind completely stagnated at an obstruction of infinite size. On the leeward side, the pressure drop has a slightly lower magnitude and the opposite sign (20). Finally, the turbulent volumetric flow through an opening varies with area, pressure drop across the opening and an empirical discharge coefficient, according to an orifice equation (19):

$$\dot{V} = C_d \cdot A \cdot \sqrt{\frac{2 \cdot \Delta p}{\rho}}$$

With a number of openings in the building, the indoor pressure will adjust to conserve mass: the mass flow into the building will, in steady state, equal the flow going out. Wind-driven ventilation can be enhanced by building designs that provide ample operable windows or other openings on facades that are windward and leeward relative to prevailing winds in warmer months. The surface-to-volume ratio of the building should be large, which may lead to floor areas less than what building codes might permit as the maximum on a given site. Older office buildings, lacking mechanical cooling, often had larger interior courtyards and double-loaded corridors, rather than expansive open-plan office bays and fully used core spaces, with no intent of using natural ventilation. In relatively benign climates, concerns about energy consumption have prompted a renewed interest in natural ventilation, particularly in smaller commercial buildings. In very hot climates, vernacular architecture has made use of roof-mounted wind scoops to capture a breeze and direct it into a dwelling.

Buoyancy-driven airflow within buildings works exactly as the flow of hot air within a chimney. The pressure of a column of air decreases with height, according to the equation of hydrostatics (19):

$$p = p_o - \rho \cdot g \cdot h$$

where p_0 is the pressure where the height is taken to be zero and g is the gravitational constant. Density is inversely related to temperature through the ideal gas law. When indoor temperatures are warmer than outside, the pressure inside will decrease more gradually with height. Outside air will be drawn into the building at low levels, where inside pressure is lower than out, and expelled at higher levels, where inside pressure exceeds outside pressure. The relation between volumetric flow and pressure drop across an open window governs buoyancy-driven as well as wind-driven airflows. In climates and building types where buoyancy-driven flows are desirable, window openings should be of generous size and at maximum vertical spacing. Within the building there should be ample paths for air to flow upward and out. Examples of architectural features that promote buoyancy-driven ventilation and examples of calculation of airflows are found in references 21, 22 and 23.

The cooling effect of buoyancy-driven flows comes from the velocity of the air and the enhanced removal of body heat by convection. While the volumetric flow can remove unwanted heat from the building, indoor temperatures must exceed outdoor levels for airflow to be generated. Wind-driven flows can provide additional cooling by reducing indoor temperatures to the outside value, as shown in the steady-state energy equation when the conductivity term K becomes very large.

Both wind-driven and buoyancy-driven airflows can create unwanted heat flow across the building envelope. These unintended flows can be drawn through small cracks around window and door frames and through other openings in the building structure. Passive heating strategies work better with well-sealed buildings that minimize heat flow due to infiltration. However, even when airflow is not needed on the basis of thermal criteria, fresh air is still required to maintain satisfactory air quality.

Heating, Cooling and Lighting with the Sun

Solar energy can serve as the energy source for a passive heating system, particularly on the scale of a house. Solar radiation can be brought directly into living spaces, can be confined to a sunspace attached to the house, or can be absorbed in a glass-covered, dark colored wall named after its original developer, Felix Trombe of France (23,24). The design of a few low-rise office buildings has also made use of solar energy, particularly by incorporating atria to collect heat that can maintain the temperature of the atria at near-comfortable conditions and potentially provide additional heat for storage and distribution through the building (25).

Solar energy enters the building at the time of day when it is least needed; outside temperatures are at their warmest and, for commercial buildings, internal loads often at their largest. To offset the need for auxiliary heat at night, solar energy must be stored and later released. Storage can be provided by mass inherent in the structure or by mass provided specifically for the solar heating system. For example, drums of water or masonry walls exposed to the sun store heat and release it naturally, when the temperature of the mass exceeds the indoor-air temperature. Less frequently, the thermal mass may be a phase-change material that can store substantial heat without large increases in temperature. Isolating the storage medium from the remainder of the house, as with a sunspace, allows the diurnal temperature cycle in the storage area to exceed that which full-time occupants would likely label as comfortable.

To complete the system, stored heat must be transferred to the building. A Trombe wall can transfer heat by conduction through the wall and then by convection and radiation to the indoor air, or by convection from the side of the wall facing the sun to air that flows to occupied spaces through holes in the upper part of the wall, as

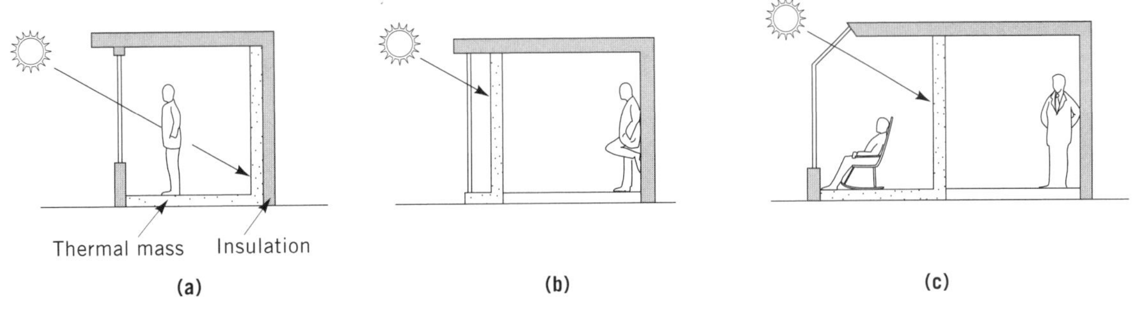

Figure 2. Three main types of passive solar space heating systems. (**a**) Direct gain; (**b**) Trombe wall; (**c**), sunspace (23).

shown in Figure 2. Alternatively, heat collected in a sunspace or atrium can be drawn by a fan through ducting to underground storage in a rock bed, then withdrawn by fan to heat occupied portions of the building.

Although solar heating systems are often mechanically quite simple, they nevertheless require careful design to specify the size of windows and type of glazing, the size and configuration of the storage area and the means of heat transfer to the building. Design goals may include maximum reduction of the annual heating needs, where the benchmark is a building of comparable size and design that does not include a passive-heating system, as well as minimal amplitude of the diurnal temperature oscillations. System design is complicated by variations in the design of the building itself, which is often an integral part of the system: passive heating systems are not a mass-produced item. Design techniques include rational analysis of individual buildings using heat-transfer calculations and correlations derived from previous simulations (21–23,25).

In warmer weather, use of solar energy depends on whether there is a need for heat at night even though daytime temperatures are within or exceed the comfort region of the bioclimatic chart. The mass of the building structure can reduce the thermal impact of solar radiation that penetrates opaque walls by conduction or is transmitted and conducted through windows. Heat stored in adobe walls or tile floors can be vented through open windows at night or be used to keep indoor temperatures above cool outdoor conditions. Windows in occupied spaces should be shaded, either with structure or vegetation, to reduce solar heat gain at times of day and year when the bioclimatic chart indicates that no solar radiation is needed, day or, through storage, night, to ensure thermal comfort. Fixed shading devices typically are a compromise design, because the position of the Sun is symmetric about the summer solstice while outdoor temperatures are higher after the solstice, due to the thermal mass of the Earth. Vegetation, which grows in response to light and warmth, can in some cases more closely match admitted solar radiation to seasonal changes in outdoor temperatures. The sun can be used to draw air through a building by the buoyancy effect, without directly heating the building interior to create the driving force, by means of a solar chimney admits solar radiation above the occupied regions (24). Design strategies for shading and passive cooling are found in references 21, 22, 23 and 25.

The Sun is also the source of natural lighting. Daylighting is difficult to control due to its intensity, variability, and attendant heat. The Sun is a more efficient light source than many man-made devices, producing a luminous flux at the Earth of 90 lumens for each Watt of radiant energy (13). Incandescent lights typically produce 15 lm/W, the addition of a halogen gas improves the efficiency to about 23 lm/W, fluorescent lamps vary from 55–90 lm/W and high-intensity discharge lamps with metal halides also operate at about 90 lm/W (26). Light in excess of that required for visual tasks can be distracting or fatiguing, as the eye muscles adjust the size of the pupil to regulate the luminous energy admitted to the retina. Illuminating Engineering Society guidelines (16) specify a 3:1 maximum allowable ratio between the luminance, or brightness, of a visual task and the luminance of the adjacent background and a 10:1 variation between the task and distant backgrounds within a room.

Daylighting design in buildings centers on bringing light in above eye level and diffusing it, to evenly illuminate large areas. Windows are a good source of daylight when separated by a horizontal lightshelf into a lower portion for a view of the outdoors and an upper, lighting portion for reflecting light off the shelf, onto the ceiling, and then down into the room. The lightshelf may be solely on the outside of the glass or on the inside as well, and are sized to reduce glare and shade the lower or vision part of the window. On the roof, vertically mounted glass will admit light into "scoops" that direct the light downward into occupied spaces, corridors or atria, as shown in Figure 3. The light can be directed to a wall, producing a luminous surface, or to work surfaces, through diffusing baffles as needed to control glare. Skylights, mounted horizontally, also admit light but are not as easily shaded at times of year when the heat associated with the light is not desirable. Light scoops are similarly designed although glare is usually less of a problem due to their position in a building. Light scoops benefit from indirect radiation bouncing off the roof, particularly if the roof is covered with a highly reflective material such as marble chips in lieu of the gravel typically used to protect the waterproof roofing membrane (27).

Methods for calculating horizontal illuminance in daylit spaces (13) start with estimation of solar illuminance and sky luminance under clear and cloudy conditions. The lumen method then estimates horizontal illuminance from the luminance at windows or skylights and correlations

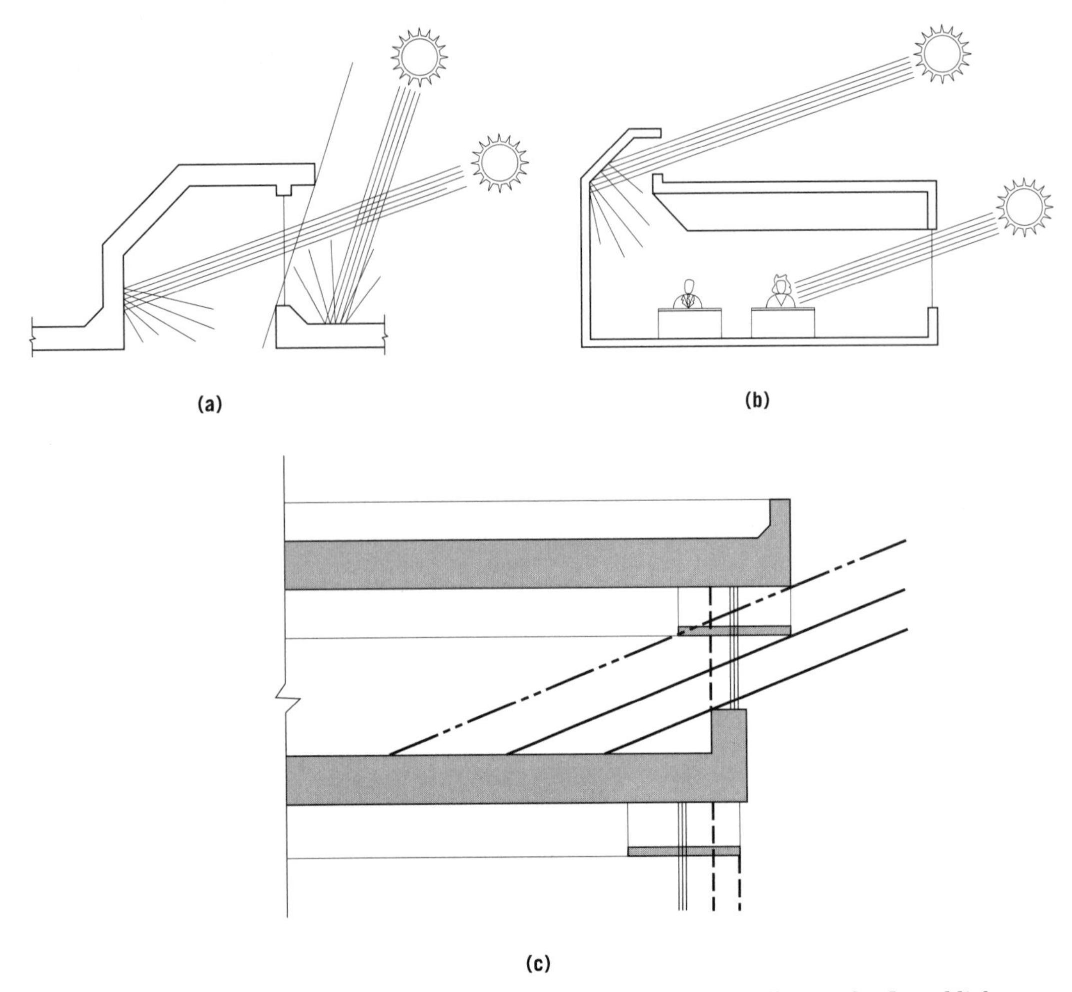

Figure 3. (**a**) Sunscoops receive direct sunlight and sky light as well as roof-reflected light. (**b**) Sunscoops are better than windows for controlling potential glare of low-angle sunlight. (**c**) South wall section diagram at 2PM on December 15 shows how interior lightshelf as designed controls glare at eye level. From ref. 17.

based on the geometry and surface properties of the room, which acts as an optical cavity. The daylight-factor method, an alternative procedure appropriate for predominantly cloudy conditions, excludes direct sunlight and defines analytic expressions for the contribution of luminance directly from the sky and also reflected light from the ground and external surfaces. Designers frequently use scale models for complex geometries that can be tested outdoors (27) and simulation programs to reduce computation time.

Evaporative Cooling

The temperature of a stream of air will be reduced if the air transfers some of its energy to evaporate water. Passive evaporative cooling strategies require wind or buoyancy forces to generate the flow of air. While the water can be in a pool, spraying the water into the air increases the surface area between air and water and therefore the evaporation rate. Evaporation can also cool wetted building surfaces. Examples of evaporative cooling include courtyard fountains, pools upwind of plazas, wood scoops where air flows over dampened material, roof sprays, and water walls bounding occupied areas (23,28). The visual and acoustic appeal of splashing water augment the cooling that can be achieved by evaporation. Evaporative cooling also plays a role in active conditioning systems, where it can be more precisely designed. In passive applications, uncertainty about its efficacy, due to lack of experience or inadequate modeling methods, has inhibited its use by architects as a prominent feature in high-profile projects where there is an expectation of substantial energy benefit.

HEATING, VENTILATING AND AIR CONDITIONING SYSTEMS

The Need for Mechanical Systems

For many of the world's people, living in tropical and subtropical areas, the shells of residential and smaller commercial buildings have been designed to provide shelter from rain and modest thermal protection. Entrances to hotel lobbies in Hawaii may have large overhangs and supporting structure but no completely sealed walls. The airport terminal in Jeddah, Saudi Arabia used by Muslim pilgrims (29) has a fabric, tent-like roof but open sides, to screen the sun but bring in breezes. Homes in India require ample airflow for inefficient wood stoves used for cooking.

But there are many other situations where a combination of internal sources of heat, occupant expectations, and harsh climate make mechanical heating and cooling a necessity. Heating is required in homes in latitudes well removed from the Equator. Cooling brings indoor temperatures and humidity within the comfort region identified on the bioclimatic chart. While one person's luxury may often be another's necessity, mechanical systems are increasingly prevalent.

When climatic conditions become a foe rather than a friend, at least during a portion of the year, the building shell becomes much more a barrier than a sieve. Houses in Scandinavian countries are tightly sealed to minimize unwanted air leakage that increases winter heat loss. Wind-driven and buoyancy-driven ventilation is minimized. Even if the thermal balance in the house can be maintained with little or no heat from a mechanical system, equipment is still required to bring fresh air into the house and to exhaust this air. The complexity of the mechanical system is balanced by the control it offers: an even amount of airflow, regardless of changes in weather, that can be carefully directed through a house. Fresh air typically is brought into all rooms of the house and is exhausted in bathrooms and the kitchen (30).

In commercial buildings, mechanical ventilation systems make it possible to increase the floor area of a structure on a given site. No longer must office areas be sufficiently shallow for everyone to be close to an operable window. But the advantages of carefully controlled ventilation in harsh weather are often countered by occupants' inability to open up the building during mild weather.

Most buildings incorporate some passive heating, cooling and lighting features: a simple window admits solar energy and light. Most commercial buildings worldwide and most residential buildings in industrialized countries require active thermal conditioning and lighting as well. Active systems that work well in conjunction with passive strategies are preferable from an environmental perspective and offer occupants more contact with the outdoors. Also preferable are active systems that can efficiently and simultaneously serve regions of a building that may have very different thermal and lighting requirements.

Thermal Zones

Most space-conditioning systems must provide heating and cooling that varies both spatially and temporally. Late-afternoon solar radiation pours through windows on the West facade of a building on a chilly day, while the East side needs heat. Rooms in the core of a building, buffered by conditioned spaces on all sides, have no heat loss. With heat gain throughout the year due to lights and office equipment, these spaces require constant cooling, even when perimeter spaces need heating.

The space-conditioning system is quite simple for a building with a single thermal zone. Because the space is thermally uniform, a single thermostat is sufficient to measure slight deviations in temperature with respect to the set point and trigger either heating or cooling, which can be delivered with either conditioned air transported to the thermal zone via ducts or with hot or cold water. Single-zone systems typically have a constant flow of air. Whether air or water is the transport media, the thermostat controls the flow of chilled or heating water through heat exchangers in the duct or in the room itself. From the thermal balance relationship, it is expected that the amount of heating will increase as outside temperatures drop, while cooling increases with outside temperature above the cooling balance point. In most houses, energy consumption strongly correlates with outside temperature (31).

Larger buildings offer behave as multiple thermal zones. Zonal boundaries may match physical partitions or may be fuzzy borders between perimeter and core areas in an open office plan. Each thermal region requires its own thermostat. Different types of heating and cooling systems can be judged in part on the basis of how efficiently they can condition numerous thermal zones. Some older systems cool all the air centrally and then reheat it locally in those zones that require heat, a wasteful practice similar to opening all the windows in a house when the kitchen is warm and expecting the heating system to compensate. With such systems, the relationship between heating or cooling and outside temperature is different than in a single-zone building. More heat is required at mild temperatures, for example, because some of that heat is tempering air that has been deliberately cooled.

Thermal loads in buildings vary over time as well as from zone to zone at a single moment. Passive heating and cooling strategies seek to smooth out the diurnal or even seasonal variation in cooling and heating loads. Some mechanical systems use water or other substances for energy storage to accomplish the same task. Thermal storage can also serve as repository of heat from zones that require cooling and a source of heat for other zones, eliminating the need for simultaneous heating and cooling.

Ventilation

A standard issued by the American Society of Heating, Refrigerating and Air-Conditioning Engineers (ASHRAE) governs building ventilation rates for acceptable indoor air quality (32). Ventilation systems may also serve a number of other purposes, distributing heating or cooling and maintaining a specified pressure within a building, but all systems must maintain pollutant levels below the maximum values shown in Table 1. Typically, most pollutants in an office environment are controlled by careful selection of building materials and the ventilation system removes CO_2 and other bioeffluents. An amount of outdoor air proportional to the maximum expected number of occupants is usually set as a fixed minimum air intake; the current standard prescribes 20 L/s-occupant. An alternative, performance-based methodology is also permitted in the standard, whereby the ventilation system controls airflow to maintain indoor CO_2 concentrations below 1000 ppm. This alternative will reduce outdoor-air intakes when occupancy is below the design level and when CO_2 levels are building up to equilibrium values after periods of reduced occupancy (33), saving the energy required to condition the outdoor air at times of year when that is needed. Control of airflow to regulate CO_2 requires CO_2 sensors and additional control logic.

Air Movement with Fans. Consider a system that provides heated or cooled air, as shown in Figure 4. First, there are one or more central fans to circulate the air. This circulation process provides for filtration and tempering via heating or cooling, to maintain the building in thermal equilibrium. The supply fan also draws into the building the required amount of outdoor air for occupants. The supply fan raises the pressure of the air an amount sufficient to push the air through the ducts and into the occupied areas of the building. Often there is a return fan to bring most of the air the back. The remaining air leaves the building via exhausts in toilet or electrical rooms, and by exfiltrating through cracks in the building shell. Laboratory and industrial buildings may require large amounts of outside air, in excess of the needs of occupants, to remove pollutants and have large exhaust fans to remove all of the air, with no recirculation. All of this air must be tempered, for the benefit of occupants and the experimental or industrial process.

The mass flow of air out of the building must equal, in steady state, the amount of air coming in. Designers typically assume that the outdoor air intake is the only source of fresh air when the ventilation system is operating, because the ventilation system is designed to pressurize the building, but recent measurements have indicated that there may be locally depressurized areas and some infiltration as well. While pressurizing a building offers enhanced comfort to occupants working near outside walls by eliminating cold drafts of infiltrating winter air, it also drives moist interior air into the walls, where it can condense without proper attention to vapor barriers.

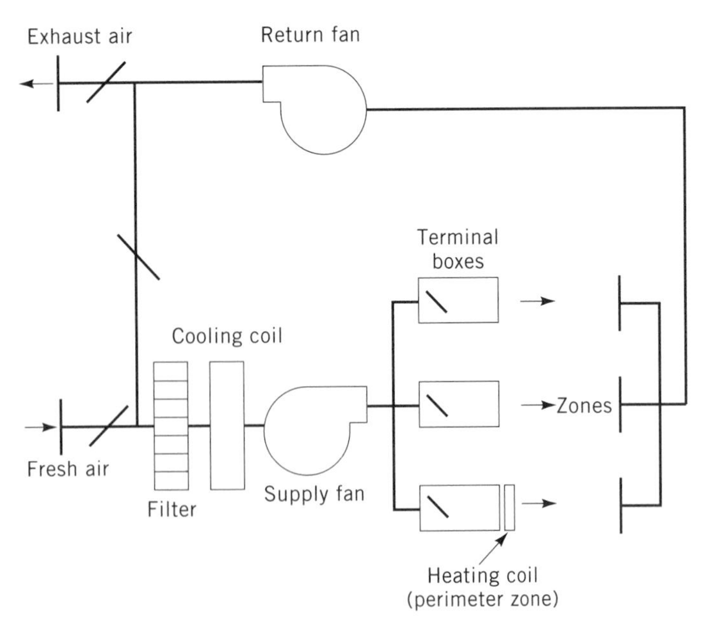

Figure 4. Schematic diagram of basic VAV system. Cool air is supplied to the terminal boxes at a constant temperature; terminal boxes vary the flow as necessary to cool individual spaces, or zones. During the heating season, the cooling coil is off, and terminal boxes add heat locally as required. Terminal boxes may use built-in fans to recirculate room air and provide a constant outlet volume, or the outlet volume may vary, with no recirculation. The return fan tracks the flow of the supply fan to maintain a fixed fresh air flow rate. A large building may have one or more VAV systems serving many zones.

Displacement Ventilation. Typically, ventilation is provided to occupied spaces via diffusers that mix the incoming air with air already in the room. This is intended to ensure that outdoor air is distributed throughout the room and to prevent occupants from directly feeling a strong stream of heated or cooled air that may be thermally uncomfortable. An alternative approach, implemented in some European buildings in recent years, introduces ventilation air near the floor and avoids mixing. The goal is to produce a nearly uniform, upward-moving flow that sweeps heat and pollutants, particularly bioeffluents, to the ceiling, where they are exhausted. This approach, known as displacement ventilation (34,35), offers a higher ventilation efficiency than well-mixed ventilation, by reducing pollutant concentrations in the breathing zone. It can be successful if the thermal gradient is not so large as to be uncomfortable and if stratification is not unduly compromised by thermal plumes by local heat sources. Cooled ceilings are the subject of current investigation as a means of providing additional cooling to zones with large heat sources.

Air and Water Systems for Heating and Cooling

Systems for conditioning buildings can be classified by the transport medium used for cooling: air or water (36). All-air systems employ central cooling equipment and deliver cooled air to end users but may transport hot water for local heating; mixed air-water systems provide a portion of the required cooling centrally and the remainder locally, via cooling coils; and all-water systems provide all cooling locally but still need to bring in outdoor air, either centrally or locally.

All-Air Systems. ***Single-Duct Systems.*** Removing heat from a building involves both sensible and latent heat transfer. Air passing over a cooling coil will leave at a lower temperature and, if the exit temperature is below the dew point, water vapor will condense on the surface of the coils. Latent cooling is often about one third of the total, seasonal cooling load in commercial buildings, a proportion that varies with local climate.

Cooling coils can be located in a central mechanical room or locally, in individual occupied spaces. If air is cooled centrally, as shown in Figure 4, it not only provides the amount of outdoor air required to maintain indoor air quality but also preserves the thermal balance at the required temperature. Centrally located cooling offers the advantages of centralized maintenance and removal of condensed water vapor, at the cost of constraints on efficiently serving multiple thermal zones, as will be discussed, and relatively inefficient transport of thermal energy. Cooling water can be more cheaply pumped than can cool air of the same energy content be pushed through ducts by fans. In addition, cool air requires ducts larger than water pipes, occupying valuable space within the building shell (36).

The simplest system maintains a constant flow of air to a building that behaves as a single thermal zone. For multiple zones, the most basic extension of this system still keeps the airflow constant to each zone. The temperature of the air can be kept constant or can be controlled to maintain thermal equilibrium in the zone with the largest heat gains. In the former case, all zones need a source of heat at all times except during the hottest weather for which the system was designed. In the latter, the controlling zone alone requires no source of heat. Such systems are holdovers from an era of cheap energy and seemingly infinite, secure energy supplies.

Single-Duct Systems with Variable Airflow. The waste associated with simultaneous heating and cooling can be partially eliminated by varying the flow of air to each thermal zone, while still maintaining a fixed temperature. Thermostats in each zone throttle the airflow to maintain the desired space temperature, down to a minimal airflow required for air quality. At this point the thermostat controls a source of heat in the zone.

Varying the airflow has the secondary advantage of reducing the energy consumed by fans. Energy savings depend on how much energy is dissipated across dampers in the system. Zone-level thermostats drive local dampers. As they close, these dampers tend to raise the air pressure in the supply-air duct, causing dampers at the fan itself to throttle. The combined action of the dampers results in modest energy savings if the central dampers are located immediately downstream of the fan and higher savings if the dampers are placed at the inlet to centrifugal fans and impart a swirl to the incoming air that reduces fan power. Further, if the fan blades are inclined at a forward angle, savings due to flow control approach those due to replacing the central dampers with controllers that vary the speed of the fan motor. These variable-speed motor drives (VSDs) have been measured to save 35–60% of the energy used by fans with inlet dampers (37). Larger savings are typically achieved when a fan is not constrained to maintain pressure in the duct at a specified level, as is true for single-zone ventilation systems and most return or exhaust fans. In this case, fan power varies with the third power of airflow. Supply fans which serve multiple zones by maintaining the duct at a constant pressure show a weaker dependence of power on flow and reduced savings. For multi-zone supply fans, further savings can be achieved by connecting the *local* dampers and thermostats in each zone of a building to a central supervisory controller which slows the central fan until at least one local damper is fully open. This has been shown (38) to save 20–40% of the energy needed when a supply fan has a VSD but local thermostat action cannot be centrally monitored.

Dual-Duct Systems. Single-duct systems serving multiple zones provide heat locally, through hot water or electric resistance coils. Dual-duct systems, in contrast, provide warm as well as cool air and are truly all-air systems, as seen from the occupied spaces. In its simplest form, air from both ducts is tapped in each thermal zone and regulated by the thermostat, which mixes the two streams in the proportion needed to maintain the desired temperature. This offers effective control with no possibility for water leakage in occupied spaces but involves simultaneous production of both heated and cooled airstreams. Improvements focus on controlling the temperature of the air in the two ducts to reduce the energy penalty of coincident heating and cooling.

Combined Air and Water Systems. ***Induction Systems.*** Reducing the volume of cooled air distributed to occupied spaces requires that air temperature be lowered or that some cooling be provided locally. So-called induction cooling systems cool the air centrally to remove latent heat, confining condensate to a central location. The dry air is distributed via fans to office areas, where it is forced out through nozzles at high speed, inducing room air into the airstream. The warmer, higher volume of mixed air flows over secondary coils and into the room. Less space and smaller fans are needed for air distribution, but chilled water must be piped and pumped to office areas, where it is a potential source of leakage, and the systems are considered to be noisier than all-air systems by some mechanical engineers.

All-Water Systems. Performing all cooling locally will reduce airflow back to the minimal level required for air quality. Local cooling can be accomplished with cooling coils, now equipped with drip pans and drains for the condensate, or local heat pumps. Chilled water lines must be carefully insulated to prevent drippage in occupied areas. Heat pumps can be coupled with large water storage tanks and are particularly effective in serving buildings where there is simultaneous need for heating and cooling, or where there is substantial cooling needed during the day and heating at night.

Use of Outdoor Air for Cooling. In houses, the indoor equilibrium temperature, defined earlier, is typically about 3–5°C above the outside temperature. Passive-solar houses will have a much larger temperature difference, with indoor temperatures in the comfort range even when outdoor temperatures are much lower. This is achieved with large solar gains and small thermal conductivity. The indoor equilibrium temperature is also well above outdoor temperatures, by 10–15°C, for many commercial buildings, where internal heat gains are large. This temperature difference reduces the need for heat in the winter but necessarily yields a longer cooling season. With outdoor air intakes at the minimum amount needed to meet ventilation standards, this leads to situations when cooling equipment is running even though the outdoor-air temperature is lower than the indoor temperature.

All-air systems are often designed with outdoor-air intakes large enough to bring in more outdoor air than in needed for ventilation, to provide “free” cooling in mild weather. Such a so-called economizer strategy (39) requires an exhaust-air duct to remove the extra airflow from the building, and additional controls to vary the amount of outdoor air required to maintain the supply-air temperature at a set point. Only if outdoor air alone is not sufficient will the cooling equipment be turned on. Outdoor airflow is returned to its minimum value when its energy content exceeds that of the indoor air returned to the supply fan, as measured by enthalpy or approximated by temperature alone.

Economizers that rely on additional outdoor air require more space within a building for air-intake ducting. This is acceptable in small buildings, where the fans are in a basement or roof-top equipment room, but may be undesirable in high rise office buildings, where there are separate fan systems for each floor, outdoor air is ducted to each floor from a small number of intakes, and larger ducts reduce space available for occupants. In addition, mixed air-water and all-water systems cannot take advantage of free cooling from outdoor air. In these cases, it is still possible to take advantage of outdoor air in mild weather by water-side economizers that remove heat from the cooling water not with chillers but with outdoor air, via a cooling tower. The tower is normally used to reject heat from a chiller: the chiller removes heat from the water, transfers it to a separate condenser-water loop, and the condenser water is cooled in the tower by evaporation and convection. For free cooling, the condenser water loop is coupled to the building chilled-water loop via a heat exchanger, rather than a chiller. Less frequently, the cooling water is sent straight to the tower and then strained to remove impurities before use within the building. Water-side economizers are effective as long as the outdoor wetbulb temperature (the equilibrium temperature of a wetted thermometer in a moving stream of air) is below the desired chilled-water temperature.

Motive Power. As has been noted, the electrical power required by fan motors in variable-air-volume systems can be reduced when airflows are below design values by employing VSDs in lieu of dampers or vanes at the fan. The VSDs regulate pressure or airflow. VSDs, which are based on electronic components similar in principle to the transistor so ubiquitous in home and office electronics but capable of handling large amounts of current, are also used in a variety of other applications. In buildings, these applications include chilled-water and hot-water pumping. Here, the control is very similar to that for supply fans: the pump speed is regulated to maintain a constant pressure in the piping loop and individual throttle valves vary in position to control the flow of water to such end uses as radiators and convectors. VSDs are also used in condenser-water piping loops, to vary the flow of condenser water to cooling towers, in cooling-tower fans, and in chiller compressor motors. The energy savings depends on the distribution of required air or water flows throughout a year and on the type of control that would have been provided without VSDs. For example, cooling towers with multiple cells and multiple fans for each cell can closely match fan power to the thermal load from the condenser water and VSDs may offer little improvement.

High-efficiency motors also offer energy savings. Motors which convert electrical power to mechanical power at an efficiency higher than those at the low end offered by manufacturers, and at higher cost, are effective when operating hours are long, loads are large and the added investment is quickly paid back. While efficiency improvements are just a few percentage points, from 90% to 94% as an example, the savings can be substantial for a large motor in constant use.

Selection and Design of Systems. System selection depends on the individual experience of the design engineer and a host of issues that concern the building owner and occupants, ranging from first cost and operating cost to space for equipment, maintenance, noise, accuracy of control of temperature and humidity and cleanliness. Packaged, roof-top units that provide ventilation and conditioning are overwhelmingly popular in small commercial buildings. These units may be constant volume or, for multiple zones, variable volume. All-air, variable airflow systems are a common choice in larger buildings, where fans, filters and cooling equipment are not packaged but are assembled into a system on-site. Once system selection is made, an engineer employs design procedures, available in the form of manuals (39) or computerized algorithms for component sizing and selection available from equipment manufacturers.

Sources of Heating and Cooling

Equipment. Heat released from the combustion of fossil fuels can be used to directly condition buildings, either individually or via a district heating system. Fossil or nuclear fuels that are burned to generate electricity can also be a source of heat for buildings, using a heat-delivery network. Power plants typically operate at efficiencies less than 40%, due to mechanical losses and limitations embodied in the Second Law of Thermodynamics. The Second Law requires that some portion of the heat extracted from a high-temperature source be delivered to a low-temperature sink. Typically, power plants use a body or water or a cooling tower to dispose of the unusable heat. Ideally, buildings can serve as a heat sink, making use of waste heat via a districting heating system. Practically, the cost and thermal losses of distribution pipe reduce the thermal advantages. In addition, a combined heat-and-power plant must extract steam at the end of the turbine before it reaches as low a temperature as can be achieved with an electricity-only plant, a further loss in efficiency.

Heat in buildings can also be provided by the drop of electrical potential across resistive wire. The primary energy required at the power plant is larger by a factor of 2.5–3.0 (the reciprocal of the overall efficiency of generating, transmitting and distributing electricity). Hydroelectric power eliminates the thermodynamic inefficiency of generating electricity via a heat source. The overall inefficiency of electric resistance heat, computed at the power plant, is usually reflected in the higher cost of electricity compared to fossil fuels, when normalized by delivered energy. However, electric-resistance heat has lower capital and installation costs, requires less space, and is more easily controlled to match thermal loads in different thermal zones.

Heat pumps are a popular type of heating equipment in some areas not served by natural gas utilities and where cooling loads are high enough that mechanical cooling is needed in summer. Like mechanical cooling equipment, heat pumps are heat engines essentially run in reverse (40). Instead of taking heat from a high temperature source, producing work, and rejecting the remaining heat to a low temperature sink, a heat pump or chiller takes heat out of a low temperature source by means of mechanical work and adds this heat to a high-temperature sink. For a heat pump, the low-temperature source is outdoors; for the chiller, it is inside the building.

Both a heat pump and a chiller use the same four basic components, shown in Figure 5. Heat flows from the low-temperature source across the heat-exchange surfaces of the evaporator to the colder low-pressure liquid refrigerant, which is evaporated. The slightly superheated gas is compressed to a higher pressure and temperature. The hot gas then transfers heat, primarily due to a phase change back to the liquid state, across the condenser to outdoor air or condenser water. Finally, the pressure is dropped across an expansion valve, which is regulated to maintain a constant superheat temperature at the evaporator outlet (41).

The heat pump can be run as a chiller in the summer by directing the compressor discharge to the outdoor rather than indoor heat exchanger and uses the same distribution system. Like the chiller, its efficiency is highest when the indoor-outdoor temperature difference or thermal hill is small. Water-cooled condensers offer higher efficiencies than air-cooled units, because the water can be cooled via evaporation to a temperature that approaches the outdoor wet-bulb temperature. The cost of this improved performance is consumption of water to replace what is evaporated. However, higher efficiency translates to less make-up water at those power plants that use cooling towers.

Thermal Storage. Thermal storage is applied in commercial buildings for two major and different reasons: to make the building as a whole behave as a single thermal zone, thereby minimizing simultaneous heating and cooling during occupied periods as well as daytime cooling followed by nighttime heating; and to shift cooling and less frequently heating to nighttime periods, when electricity rates are lower. In the first case, thermal storage is provided by large water tanks, which serve as a heat source or sink for individual heat pumps. One example of this technology, a 150,000 square foot office building in Boston, Massachusetts, uses 750,000 gallons of water.

In the second case, chillers are operated at night to lower the temperature of tanks of water or to freeze the water. Ice storage is a more complex technology but has a higher energy storage density. Ice storage requires chillers to operate at lower evaporator temperatures than chilled-water storage. For ice-buildup systems, the chillers cool a glycol solution to below-freezing temperatures and the glycol, passing through pipes or coiled tubes in a storage tank, freezes the stored water. During the day, the glycol solution is circulated to cooling coils in the ventilation system, extracting heat from the building that is then transferred to the storage tanks, melting the ice. In ice-harvesting systems, the refrigerant in the chiller directly freezes a thin layer of ice on plates above the storage tank. Hot refrigerant gas is periodically pulsed to the plates to melt the boundary layer of ice, allowing the remaining sheet to fall into the tank. Recent experimental work has produced a "slippery" ice slurry that falls off the plates without need for the hot-gas cycle, which, even when properly tuned, reduces the efficiency of the ice-making system by about 10%.

The overall efficiency of ice-storage systems, measured in delivered cooling per unit of fuel consumed at the power plant, suffers from low evaporator temperatures in the chiller, a penalty partially eliminated with chilled-water storage. Both types of systems benefit from lower nighttime outdoor temperatures for the condenser. Ventilation systems that take advantage of melted ice or glycol solu-

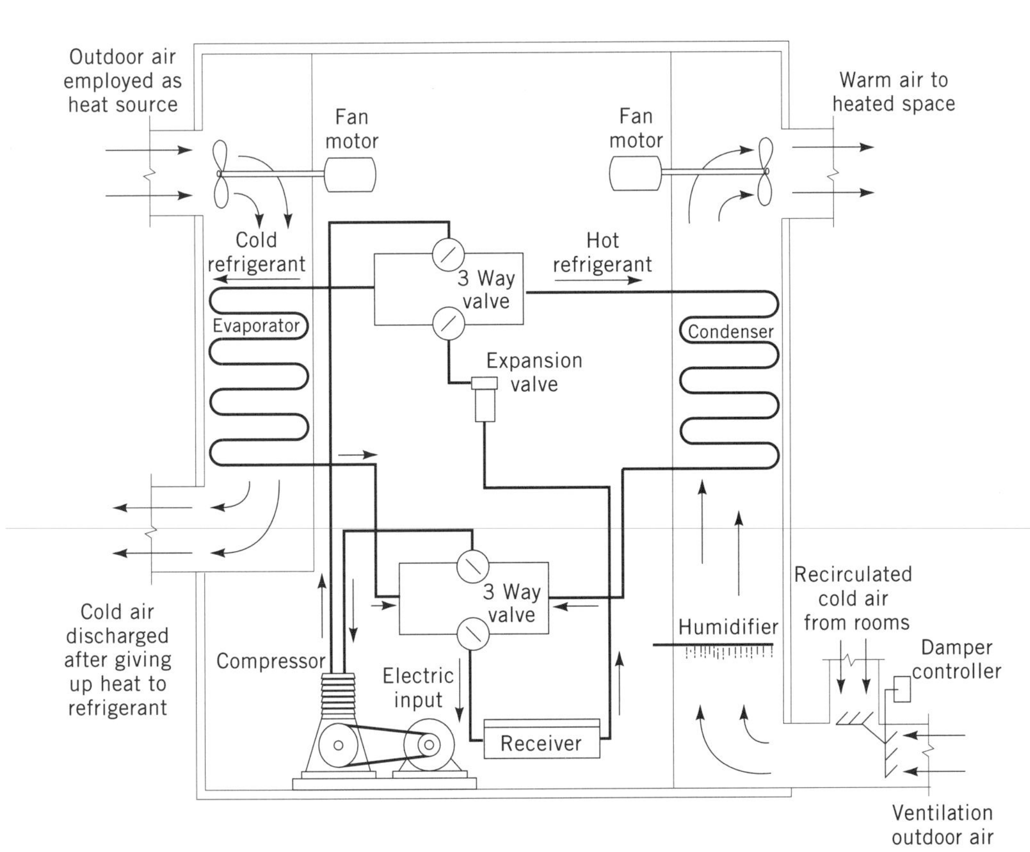

Figure 5. Diagram of heat pump used to heat residence (41).

tions near the freezing point can provide the same cooling to the building with lower airflows by reducing the temperature of the supply air. The ventilation system can use smaller fans and ducts, which take up less space and are less expensive to purchase and operate. Reduce space for ducting yields more usable space within a given building volume, a substantial economic advantage. Less appreciated but of major importance is the higher marginal efficiency of electricity production during off-peak periods, which contributes to lower electricity prices.

The mass inherent in the building structure may also be used for thermal storage, to shift loads to off-peak utility periods. This strategy has been rarely practiced because it requires more a more capable control system than is often installed in buildings. Before a building is occupied in the morning, the air temperature in a building is cooled to a point at or near the lower bound of the comfort region established by the building tenants. This sub-cooling requires more energy than conventional cooling strategies or dedicated thermal storage, because the building mass at lower temperature absorbs more heat from its environment. However, it is done at times of lower energy costs. In the day, the building air temperature is allowed to rise, letting heat flow into the cooler mass. If the thermostat is simply set back up to its normal position after the sub-cooling period, the benefits of sub-cooling will last for a relatively short period of time and the required chiller power will then be as high as if no sub-cooling were provided. Many electric utility rates for commercial buildings include a charge for demand as well as energy consumption, and smoothing out electricity demand is a primary goal of thermal storage. Electronic controls make it possible to carefully and gradually raise the thermostat set point over the day, to minimize the cost of running the chiller.

CFCs. Modern air-conditioning and refrigeration equipment uses chlorofluorcarbon (CFC) compounds as refrigerants; as a class these compounds are non-flammable, low in toxicity, and very stable. Fully halogenated CFCs have all hydrogen atoms replaced with chlorine or fluorine atoms. Examples of fully halogenated CFCs (42) include R-11, trichlorofluormethane (CCl_3F), and R-12, dichlorodifluoromethane (CCl_2F_2), as well as those derived from hexachloroethane (C_2Cl_6): R-113 (CCl_2FCClF_2), R-114 ($CClF_2CClF_2$), and R-115 ($CClF_2CF_3$). Examples of CFCs that are not fully halogenated include R-21, dichlorofluoromethane ($CHCl_2F$), and R-22, chlorodifluoromethane ($CHClF_2$). In many cases, CFCs that are used as refrigerants eventually leak or are discharged to the atmosphere. Those that are fully halogenated are more stable and a higher percentage of chlorine atoms reach the stratosphere and deplete, via catalytic reaction, the concentration of ozone molecules that screens the Earth's surface from ultraviolet radiation from the Sun. The relative ozone-depletion potential for the fully halogenated compounds varies inversely with the order listed above and ranges from 1.0 for R-11 to 0.32 for R-115. By contrast, R-22 has an ozone depletion potential of 0.05 (43).

R-11 and R-12 are used in refrigeration equipment and in foam insulation materials. Both are used in centrifugal chillers in commercial buildings, while R-11 is used in refrigerators, freezers, automobile air conditioners, water coolers and dehumidifiers. R-114 is used in shipboard chillers. The non-fully halogenated refrigerant R-22 is used in larger centrifugal compressors and in reciprocating and screw compressors. R-22 dominated 1985 refrigerant purchases by companies that are members of the American Refrigeration Institute, with 77% of the total, followed by R-11 with 11% and R-12 with 10% (43). The fully halogenated CFCs are being phased out as part of an international agreement known as the Montreal Protocol. Replacement refrigerants may be near drop-in substitutes (R-123 or R-141b for R-11 and R-134a for R-12) or may require substantial equipment modification to accommodate higher operating pressures and temperatures (R-22 for R-12). In the future, R-22 and all other CFCs containing any chlorine may be banned, leading to a need for entirely new refrigerants or such alternatives as ammonia or propane. National energy impacts of changes in the use of fully halogenated refrigerants range from a penalty of about 3% of national energy consumption to a gain of about 1%, as a depending on refrigerant substitutes and advanced technologies for insulations and refrigeration cycles (44). In all cases, the dominant component of the overall energy impact is refrigerators and freezers, not centrifugal compressors.

Improved Control Strategies

Building systems rarely if ever operate at optimal efficiency. Mathematical techniques for optimization are well known but can only be applied when these criteria are met: there is a model for the dynamic system being controlled, which in this case is the thermal environment within the building, as affected by the thermal dynamics of the building structure and the space conditioning system; the model captures all of the dynamics that affect the variable to be minimized, which might be cost to maintain a given environment; and the parameters of the model can be accurately determined via prior knowledge or suitable experiments. An accurate model makes the most demands on parameter estimation and most models have therefore considered a limited number of variables, in some cases at time intervals as long as an hour. Improved computational speeds and more use of distributed intelligence in building control systems have made it possible to gather and process more information, encouraging recent efforts to perform on-line optimization calculations (45).

Optimization of equipment operation within an individual building is more reasonably based on cost rather than energy consumption, because the cost of electricity often varies with time of day. Thermal storage systems may be financially attractive to building owners and customers for this reason. More efficient base-load power plants are one factor that may contribute to lower electricity rates at night: during the day, these plants may be augmented with less-efficient plants to serve peak loads. In this case, thermal storage may lead to less fuel consumption at the power plant than would be the case with day-time operation of chillers even though, at the building, the chiller used to make ice at the building may operate at lower efficiency than conventional equipment due to lower evaporator temperatures.

Electricity rates that show a simple step change between on-peak and off-peak periods do not capture hourly fluctuations in the marginal costs that a utility incurs in generating and distributing electricity to its customers. Several electric utilities have computed and transmitted hourly prices to customers capable or responding to them, and tests have shown that careful choice of the least costly hours to charge a thermal storage system can nearly double the benefit of storage in reducing utility costs, relative to no storage (46).

Optimization serves as one basis for fault detection, where a fault is defined as monitored equipment performance below the calculated optimum. Faults can also be detected by identifying departures from historic data and abnormal hours of equipment operation. The same trend toward electronic control of building systems that is enhancing optimization also promotes on-line fault detection.

LIGHTING

Light Sources

Lighting systems can be designed on the basis of illuminance levels, luminance ratios, and direct attention to biological needs, or some combinations of all three. In any case, lighting starts with lamps and fixtures. The principal types of lamps can be classified as incandescent, in which a filament is heated to the point where it emits light in the visible as well as infrared spectra and discharge lamps, in which gas under low or high pressure emits light in response to electrical excitation. Low-pressure discharge lamps include those based on sodium vapor, which can also operate at high pressure, and fluorescent lamps. High-pressure discharge lamps, which offer a concentrated source of light from a small discharge tube (14), include sodium, mercury and metal halide lamps, in which halide salts improve the color rendition of ordinary mercury lamps. Metal halide lamps, long used outdoors in such applications as sports fields and parking lots, are also available for circulation spaces and indirect lighting in office areas, where the lamp points toward the ceiling. Discharge lamps require a device to ignite the arc and a device to limit the current of the arc; high-pressure discharge lamps need an outer glass tube to protect the very hot discharge tube.

Lamps vary in their efficiency, as already noted, and in color rendition. The most efficient low-pressure sodium lamps are more often used in outdoor applications because the visible output is dominated by the yellow part of the spectrum. Fluorescent lamps come with a wide variety of visible spectra, some more toward the blue and others the red. Color output can be controlled to some extent by varying the mix of phosphors that coat the inside of the fluorescent tube and are excited by colliding ultraviolet photons from the mercury vapor produced by the electric arc between the two ends of the tube.

The efficiency at which light is provided to work spaces or circulation depends on the light fixture as well as the lamp itself. The amount of light delivered per Watt of electrical power will be higher for fixtures that direct light downward rather than toward the ceiling. Direct light has a higher potential for glare than indirect and may not be acceptable in rooms with computer displays unless the light fixture includes a carefully designed diffuser.

Small light sources are more easily controlled to distribute light where needed than large ones. Incandescent spot lights are often used in retail stores to spotlight merchandise. Task lighting in the home and office is directed to illuminate small areas. Compact fluorescent lamps adapted to incandescent light electrical sockets bring increased efficiency to local lighting. Fixture manufacturers have recognized the benefits of small fluorescent lamps and have increased the variety of fixtures that are sized to accept such lamps, which are larger than most incandescent lamps.

In addition to size, color and efficiency, lamps can be classified on the basis of luminous output, luminance of the lamp surface, and lifetime. Incandescent lamps, because they are smaller than most discharge lamps, will have higher luminance for a given luminous flux. They also have lifetimes less than 10% that of discharge lamps. Lamps, regardless of type, deteriorate over time and the luminous output decreases; as a result, illuminance in a room when lamps are new should exceed the minimum acceptable value, which will be attained just prior to relamping.

Office Lighting

Lighting for open-plan offices usually involves a regular array of ceiling-mounted lamps, designed to provide a uniform illumination at desk height. The number and spacing of a given lamp fixture can be determined on the basis of room geometry and surface reflectances, for standard lamp designs and configurations. A uniform illumination is most immune to changes in office layout and necessarily the most wasteful, with light provided to unused or infrequently used areas. However, office furnishings can sufficiently reduce desk-top illumination to the point where task lighting is required. In this case, reduced general-area lighting combined with user-controlled task lighting offers a system that is potentially very efficient and that gives office workers the opportunity to vary the lighting in their individual work environment. In office areas near a source of natural light, photosensors that dim or switch the general-area lighting can achieve further energy savings. The control must be sufficiently smooth to maintain occupant visual comfort, which favors the relatively new and expensive dimmable ballasts over switchable ballasts for fluorescent lamps. Case studies of newly constructed or renovated buildings have shown that lighting power of about 15–20 W/m^2, as specified by standards, can be reduced to about 5-7 W/m^2 by use of daylighting sensors, reduced general-area lighting and task lights (47).

Electric lighting design, especially for simple systems with uniform levels of light, can be done with procedures based on manuals that describe a room as three optical cavities: from the ceiling to the height of the lamps, from the lamps to the work plane, and from the work plane to the floor. The size and surface properties of each cavity alone with the amount and distribution of luminance from the lamps affects the illuminance at the work plane (16). More complex designs, including those involving task lighting and complex wall and ceiling surfaces, may be designed with computerized methods or mock-ups.

PERFORMANCE CRITERIA

Performance criteria include such factors as lifetime, noise, required service and pollutant emission as well as peak-power demand and annual energy consumption. Energy performance of building systems as well as individual pieces of equipment is subject to labeling as well as regulation. Labels with energy consumption and operating costs (common on many residential appliances), manufacturers' specification sheets, and information-dissemination campaigns conducted by government agencies, environmental groups and energy utilities give consumers opportunity to make rational purchase decisions. Regulation is intended to remove from the market the most inefficient equipment, protecting bill-payers from purchase decisions made in ignorance or by those not paying for operating costs. Many regulations originate in non-binding, building-energy standards (32) that are incorporated into state building codes and therefore made into enforceable limits. The standards are formed in a consensus-building process that involves manufacturers as well as environmental advocates. For example, the standard for energy-efficient design of new buildings except small houses published by the American Society of Heating, Refrigerating and Air-Conditioning Engineers and the Illuminating Engineering Society of North America specifies a number of performance criteria, some of which are summarized in Table 2.

The criteria prescribe minimal acceptable performance. The standard also includes a path to compliance based on performance; for example, a building envelope will meet the standard if the *overall* thermal conductivity does not exceed a specified limit, even if some individual elements do not meet the prescriptive criteria. Further, the standard includes a third path to compliance, the energy-cost budget approach, which provides designers the most flexibility but requires that a design energy cost be computed and compared with a target. Such an approach allows for extensive use of glazing, for example, if it can be shown that the solar heat gain and daylighting benefits offset the increased thermal conductivity.

The National Energy Policy Act adopted in the U.S. in 1992 mandates that states review commercial building codes and certify that energy performance criteria are at least as stringent as those included in the ASHRAE standard. Similarly, state-level residential codes must meet or exceed the model energy code published by the Council of American Building Officials (48). The Energy Policy Act also requires energy-efficiency information programs for lights, office equipment and windows, mandates minimum efficiency criteria for heating and cooling equipment manufactured after January 1, 1994, and specifies minimum efficiencies for electric motors.

Table 2. Performance Criteria for Energy-Efficient Building Design

Component	Performance Criteria
Lighting	Electrical power per floor area
Building envelope	Thermal conductivity of individual elements
Systems	Limited oversizing
	Zone controls to limit simultaneous heating and cooling
	Economizer controls to reduce cooling costs in mild weather
	Fan power per unit flow rate
	Frictional losses in pumping systems
Equipment	Efficiencies of air conditioners, chillers, heat pumps, boilers and furnaces

BIBLIOGRAPHY

1. P. E. McNall, "Indoor Air Quality Status Report," in *CLIMA 2000,* VVS Kongres-VVS Messe, Copenhagen, 1985, Vol. 1, pp. 13–26.
2. American Society of Heating, Refrigerating and Air-Conditioning Engineers (ASHRAE), *Handbook of Fundamentals,* ASHRAE, Atlanta, 1989, pp. 8.1–8.32.
3. V. Olgyay and A. Olgyay, *Design with Climate,* Princeton University Press, Princeton, New Jersey, 1963, pp. 16.
4. P. O. Fanger, *Thermal Comfort: Analysis and Applications in Environmental Engineering,* McGraw-Hill, New York, pp. 19–64.
5. B. Givoni, *Man, Climate & Architecture,* Von Nostrand Reinhold, New York, 1981, pp. 30–59.
6. *Thermal Environmental Conditions for Human Occupancy: ANSI/ASHRAE Standard 55-1981,* American Society of Heating, Refrigerating and Air-Conditioning Engineers, Atlanta, 1981.
7. J. E. Seem and J. E. Braun, "The Impact of Personal Environmental Control on Building Energy Use," in *ASHRAE Transactions,* American Society of Heating, Refrigerating and Air-Conditioning Engineers, Atlanta, Vol. 98, Part 1, 1992.
8. F. S. Baumann, R. S. Helms, D. Faulkner, E. A. Arens, and W. J. Fisk, *ASHRAE Journal,* **35**(3), (1993).
9. *Radon Reduction Methods, a Homeowner's Guide,* U.S. Environmental Protection Agency, Washington, D.C. 1986.
10. *Ventilation for Acceptable Indoor Air Quality: ASHRAE Standard 62-1989,* American Society of Heating, Refrigerating and Air-Conditioning Engineers, Atlanta, 1989.
11. National Institute of Occupational Safety and Health, "Criteria for Recommended Standards," U.S. Department of Health and Human Services, various dates.
12. U.S. Environmental Protection Agency, *National Primary and Secondary Ambient Air Quality Standards,* Code of Federal Regulations, Title 40, Part 50.
13. J. P. Murdoch, *Illuminating Engineering from Edison to the Laser,* MacMillan, New York, 1985.
14. J. B. deBoer and D. Fischer. *Interior Lighting.* Philips Technical Library, Kluwer, Antwerp. 1981, p. 19.
15. M. S. Rea, *Journal of the Illuminating Engineering Society,* **15,** (2) 41–57 (1986).
16. *IES Lighting Handbook Reference Volume,* Illuminating Engineering Society of North America, New York, 1981.
17. W. M. C. Lam, *Perception and Lighting as Formgivers for Architecture,* Von Nostrand Reinhold, New York, 1992.
18. F. C. McQuiston and J. D. Parker, *Heating, Ventilating, and Air Conditioning Analysis and Design,* John Wiley & Sons, Inc., New York, 1988.
19. W-H Li and S-H Lam, *Principles of Fluid Mechanics,* Addison-Wesley, Reading, Mass. 1976.

20. E. Simiu and R. H. Scanlan, *Wind Effects on Structures: An Introduction to Wind Engineering,* John Wiley & Sons, Inc., New York, 1986.
21. D. Watson and K. Labs, *Climatic Building Design,* McGraw-Hill, New York, 1983.
22. G. Z. Brown and co-workers, *Insideout: Design Procedures for Passive Environmental Technologies,* John Wiley & Sons, Inc., New York, 1992.
23. N. Lechner, *Heating, Cooling, Lighting Design Methods for Architects,* Wiley Interscience, New York, 1991.
24. F. Moore, *Environmental Control Systems Heating Cooling Lighting,* McGraw-Hill, New York, 1993.
25. "Enerplex: Office Complex Exploring Sophisticated Energy Solutions," *Architectural Record,* **170**(6) (May).
26. *Sylvania GTE Light Sources Product Information/Engineering Bulletins,* Sylvania Lighting Center, Danvers, Mass.
27. W. M. C. Lam, *Sunlighting as Formgiver for Architecture,* Von Nostrand Reinhold, New York, 1986.
28. *Structure, Space, and Skin: The Works of Nicholas Grimshaw & Partners,* Phaidon Press, London, 1993.
29. *Skidmore, Owings & Merrill: Architecture and Urbanism 1973–1983,* Van Nostrand Reinhold, New York, 1983, p. 382.
30. A. Blomsterberg, *Ventilation and Airtightness in Low-Rise Residential Buildings: Analysis and Full-Scale Measurements,* Swedish Council for Building Research D10:1990.
31. M. F. Fels, "PRISM: an Introduction," in *Energy and Buildings,* Elsevier Sequoia S. A., Lausanne, Switzerland, Volume 9, Numbers 1&2, 1986.
32. American Society of Heating, Refrigerating and Air-Conditioning Engineers (ASHRAE), *ASHRAE Standard 90.1-1989,* "Energy Efficient Design of New Buildings Except New Low-Rise Residential Buildings," Atlanta, 1989.
33. A. K. Persily and W. S. Dols, "The Relation of CO_2 Concentration to Office Building Ventilation."
34. M. Sandberg and C. Blonquist, *ASHRAE Transactions,* **95,** Pt. 2 (1989).
35. A. K. Melikov, G. Langkilde, and B. Derbiszewski, *ASHRAE Transactions,* **96,** Pt. 1 (1990).
36. American Society of Heating, Refrigerating and Air-Conditioning Engineers (ASHRAE), *Handbook of HVAC Systems and Equipment,* Atlanta, 1989, pp. 1.1–4.4.
37. D. Lorenzetti, and L. Norford, *ASHRAE Transactions,* **98,** Pt. 2 (1992).
38. D. M. Lorenzetti and L. K. Norford, "Pressure Reset Control of Variable Air Volume Ventilation Systems," *Proceedings of the ASME International Solar Energy Conference, Washington, D.C.,* 1993.
39. *Air-Conditioning Systems Design Manual,* American Society of Heating, Refrigerating and Air-Conditioning Engineers, Atlanta, 1993.
40. J. B. Fenn, *Engines, Energy, and Entropy,* W. H. Freeman, New York, 1982.
41. B. H. Jennings, *Environmental Engineering Analysis and Practice,* International Textbook Co., New York, 1970.
42. R. Downing, "Development of Chlorofluorocarbon Refrigerants," in *CFCs: Time of Transition,* American Society of Heating, Refrigerating and Air-Conditioning Engineers, Atlanta, 1989.
43. R. J. Denny, 1989. "The CFC Footprint," in *CFCs: Time of Transition,* American Society of Heating, Refrigerating and Air-Conditioning Engineers, Atlanta, 1989.
44. S. K. Fischer and F. A. Creswick, 1989. "How Will CFC Bans Affect Energy Use?" in *CFCs: Time of Transition,* American Society of Heating, Refrigerating and Air-Conditioning Engineers, Atlanta.
45. Z. Cumali, *ASHRAE Transactions,* **94,** Pt. 1, (1988).
46. B. Daryanian, R. E. Bohn, and R. D. Tabors, *IEEE Transactions on Power Systems* **6**(4), 1356–1365 (1991).
47. R. K. Watson and co-workers, "Office Building Retrofit to Produce Energy Savings of 75%," in *Proceedings of the ACEEE 1992 Summer Study on Energy Efficiency in Buildings,* American Council for an Energy-Efficient Economy, Washington, D.C., Vol. 1, p. 251.
48. Energy Policy Act of 1992, *Public Law 102-486-October 24, 1992.*

C

CARBON BALANCE MODELING

RICHARD H. WARING
Oregon State University
Corvallis, Oregon
MICHAEL G. RYAN
U.S.D.A. Forest Service
Fort Collins, Colorado

Carbon is a constituent of all terrestrial life; the energy locked in its chemical bonds in living and dead plant material provides sustenance for animals and microbes. Plants extract energy from sunlight to combine carbon dioxide (CO_2) from the atmosphere and water to form carbohydrates, which serve as a source of energy and the substrate from which leaves, wood, and roots are built (see Fig. 1). Dead plant material is consumed by soil arthropods and microbes, and in the process releases nutrients. Most of the carbon is quickly returned as CO_2 to the atmosphere; however, a small fraction of the organic debris is converted by microbes into complex organic molecules that remain in the soil for many years.

Modeling the carbon balance requires the development of mathematical expressions to describe the rate that carbon flows through the cycle (from the atmosphere into biota and the soil, and out again). There is a plethora of models that describe some or all of the carbon cycle. Most are designed to answer specific management or research questions; a few are more general. On a small scale, models may estimate the effects of management on crop, range, or forest yields (2). At larger scales, models forecast commodity production, examine interactions between biota and the atmosphere, and predict the effects of global change.

This review of carbon balance modeling concentrates on the principles underlying the equations in quantitative treatments and processes described by those equations, in the context of global change. The mathematical expressions, which are not presented here, vary, depending on the purpose and form of particular models, and can be obtained from the literature cited. In addition, the focus is on forests rather than crops or grasslands, as forests represent the most complicated types of vegetation and store more than 70% of the total terrestrial carbon in their biomass and soil organic matter. Forests demonstrate the principles by which most vegetation captures and exchanges carbon with the atmosphere.

See also AIR QUALITY MODELING; CARBON CYCLE; CARBON STORAGE IN FORESTS; FOREST RESOURCES; GLOBAL CLIMATE CHANGE, MITIGATION.

CARBON DIOXIDE AND THE CARBON CYCLE

The sum total of carbon on the Earth and in its atmosphere is constant. What varies is the fraction present in the atmosphere, in terrestrial vegetation and soils, in the oceans, and locked in sediments of limestone and dolomite or in fossil fuels (see Fig. 2). Carbon moves between the atmosphere and the biota in the form of carbon dioxide, a gas that in the atmosphere is transparent to short-wave radiation from the sun but traps long-wave radiation from the relatively cool surface of the Earth. As a result, atmospheric CO_2 has the potential, along with other gases, to cause atmospheric heating and affect weather patterns globally, ie, the "greenhouse effect."

As industrial societies have recaptured and utilized the energy in fossil fuels during the last 150 years, the atmospheric pool of carbon has increased by more than 160 billion tons. In the last 30 years, the increase in CO_2 has exceeded the rate of increase at any time previously. Deforestation has played a part in increasing the atmospheric CO_2 level, but the main contributor is combustion of fossil fuel. Because the amount of carbon released in the combustion of fossil fuel is now about 5.4×10^9 annually, whereas the increase in atmospheric content is only about 3.4×10^9 annually (Fig. 2), there is an imbalance in the carbon source–sink relationships between the Earth and atmosphere. The difference between carbon released in fossil-fuel consumption and the increase in atmospheric carbon content indicates that either the oceans or the land, or both, must be increasing their efficiency in capturing CO_2.

At the scale of a forest or agricultural field, increased carbon dioxide concentrations can affect growth and decay processes, and consequently the rate of CO_2 accumulation in the atmosphere. To understand any given response, the processes involved must be analyzed. Although understanding of many key processes is still limited, as well as the ability to make accurate measurements of some variables, considerable progress can be made if the carbon budget for each system can be closed, ie, accounting for the fluxes in and out of the system, and hence the net gains, losses, and changes in carbon storage.

TYPES OF MECHANISTIC MODELS

An array of models has been developed to estimate components of the forest carbon balance. In a recent review (4), models have been classified into four types: physiological, ecosystem, plant succession, and regional or global. Physiological models simulate carbon, water, and occasionally nutrient transfers (fluxes) for single plants or small areas of perhaps a hectare (10^4 m^2) with a homogeneous cover of a single species. The time-step applied in physiological models generally ranges from an hour to a year. These models usually describe leaf-level processes, such as photosynthesis, in great detail because they are well-understood. Less understood processes, such as carbon allocation within plants, are usually modelled in much less detail (5–7).

Ecosystem models determine the same fluxes for vegetation as physiological models but generalize responses for groups of species with similar characteristics. These models may apply to areas of several hectares and resolve time at daily, monthly, and yearly intervals for periods up to

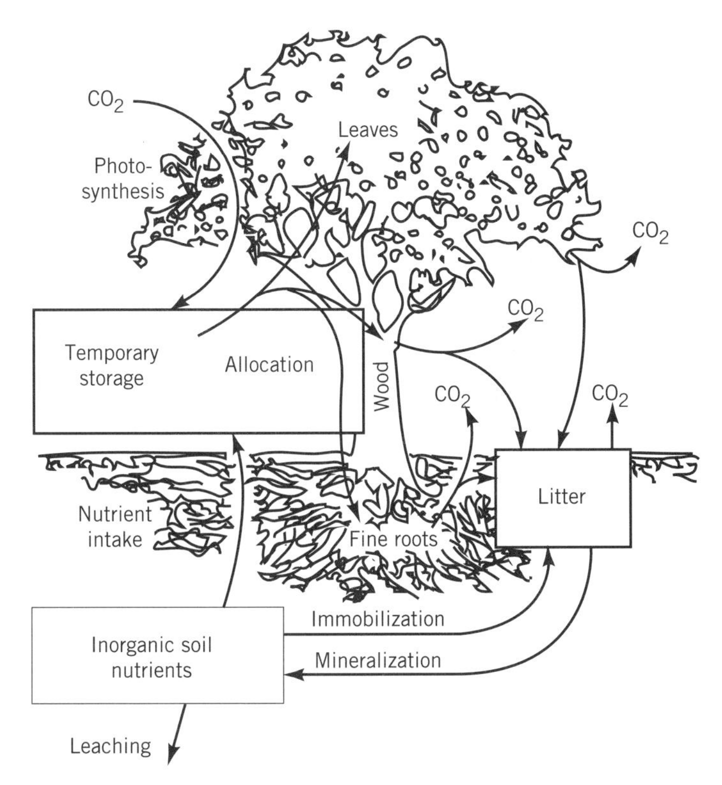

Figure 1. The carbon balance of an ecosystem represents a set of counterbalancing processes (1).

several decades. Most ecosystem models assume that the vegetative cover does not significantly change over the selected time period. They simplify the leaf-level detail of many physiological models and tend to focus on larger-scale processes such as carbon storage and decomposition in the soil and the interactions of nitrogen and other essential nutrient with the carbon cycle (8,9).

Modeling Compositional Changes in Vegetation Over Time

Most carbon balance models that predict exchanges between the land and the atmosphere do not incorporate changes in the composition of vegetation. Yet the composition of vegetation will change as those species that may have established on disturbed areas grow and create environments favorable for other species adapted to more shaded conditions. Disturbances, which include insect and disease outbreaks, fire, wind damage, and logging, may occur at different scales. They may kill individual plants, selectively exclude certain species, or, in the case of fire, consume all aboveground biomass. The consequences of such disturbances may be modelled and predicted in stochiastic terms (ie, as probabilities), as a function of climatic conditions, the presence of native and introduced pests and pathogens, and soil conditions associated with fertility, aeration, and slope stability.

A series of "plant succession" and forest "gap" models predict compositional changes and the growth and death of species for a wide range of vegetation. After disturbances, these succession models use simple equations to estimate tree growth and probability functions to predict when trees die and new seedlings take their place (see Fig. 3). Because of the probability functions, model predictions vary widely from simulation to simulation. Mean estimates must be derived, therefore, from the average of many simulations. Examples of plant succession models have been described (10,11). Later versions of these models incorporate nutrient cycling (12,13) and considerable physiological detail (14).

Predictions from plant succession models have favorably compared with the record determined from pollen cores for patterns of vegetation change that followed the last full-glacial period in the eastern United States (10). With the rapid changes that are projected in regional climates and land use, however, rates of disturbances are

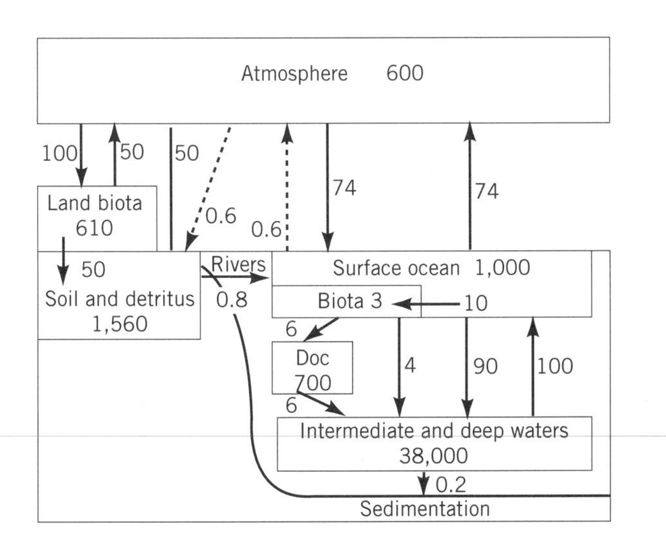

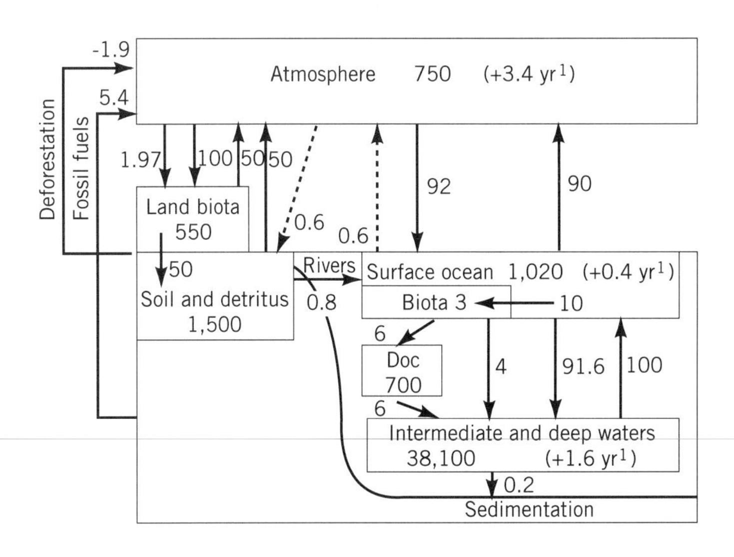

Figure 2. The global carbon cycle reservoirs and fluxes are contrasted between (**a**) the pre-industrial carbon cycle and (**b**) the carbon cycle averaged for 1980–1989. Differences caused by human activities are indicated in bold numbers in (**b**). Arrows indicate estimated annual fluxes in or out of various reservoirs; units are in billions of tons (Gt) (3).

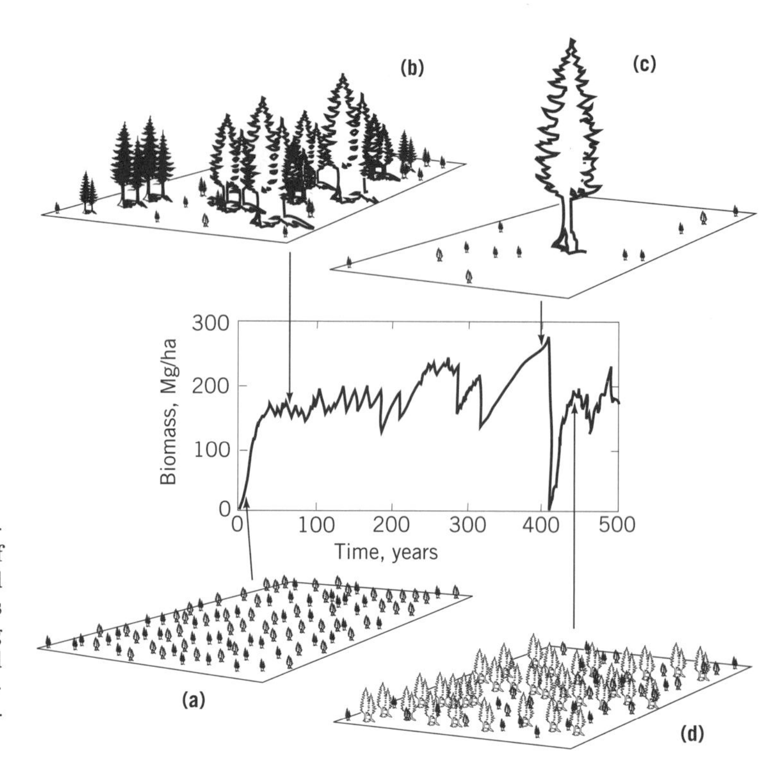

Figure 3. The changing structure of a forest can be modelled to take into account natural succession, probability of disturbance from fire, insects, disease, windstorms, and harvesting activities by humans. In this diagram, changes are depicted for a cycle that starts with (**a**), regenerating forest with many small trees, progresses to (**b**), a mixed forests with fewer trees, (**c**), harvesting of all but one dominant tree, and (**d**) regeneration of a mixture of species following death of the dominant tree (10).

likely to increase; some components of regional flora may begin to migrate and others may become extinct. These types of changes are difficult to model, even with greater knowledge of the behavior of individual species and populations. Simplification of succession models, not greater complexity, is needed to extend the models to regional or global scales.

Several approaches to scaling have proven useful. The simplest approach to estimating the effects of climatic change on primary production is to extrapolate correlations obtained between current climate and production rates associated with various types of vegetation (15,16). More sophisticated approaches combine ecosystem models with succession models (13,17); at the global scale, the precise distribution of the main vegetation types becomes critical. Some imaginative approaches have been outlined but await implementation and critical testing (18,19).

COMPONENTS OF TERRESTRIAL CARBON MODELS

Most terrestrial carbon models contain terms that represent reservoirs or pools where carbon accumulates, and include a set of equations that describe the rates at which carbon is transferred from one structural component to another. The next section introduces the components of a typical ecosystem level model.

Structural Components

Carbon balance models of terrestrial ecosystems that calculate fluxes of CO_2 to the atmosphere contain at least three structural components: living plant material, dead plant material, and carbon stored in the soil. Because humans remove a significant fraction of carbon from forests by making wood products that last for several to many decades, some models track the type and amount of materials harvested each year and the rates at which various products breakdown and are recycled into the atmosphere (20). Although a simple balance sheet may be constructed with only three structural components, reasonable predictions of the rates at which carbon cycles through a system demand a more detailed knowledge of structure. Most ecosystem models recognize leaves and woody material in stems and roots (as in Fig. 1); as these structures die, the carbon in them is released to the atmosphere or incorporated into soils at different rates. Many ecosystem models, therefore, recognize a number of separate soil reservoirs (4,8).

ECOSYSTEM PROCESSES

In undisturbed ecosystems, the rates of carbon accumulation and loss over long periods are mostly in balance. Through photosynthesis, CO_2 is transformed into simple carbohydrates that provide energy and building blocks for the construction of all structural elements. Respiration metabolically degrades large organic molecules into CO_2, which is released back to the atmosphere. The basic functioning of ecosystems, however, depends not only on the cycling of carbon, but also on water and nutrients. Currently, there are just a handful of ecosystem models that combine carbon, water, and nutrients cycles (4,5,8,21,22).

The following section outlines the primary processes involved in plant community and ecosystem carbon balances, and discusses the state and stage of development of the models being used to simulate those processes.

Photosynthesis

Photosynthesis is the process by which green plants convert CO_2 into carbohydrates, using the energy of sunlight. About 50% of the sun's radiant radiation is in the visible part of the spectrum (400-700 nm), which is the range that provides the energy for photosynthesis. The intensity and duration of sunlight varies with climate, season, and latitude, and can be modelled, but is more accurately measured directly. In addition to sunlight, chlorophyll and associated photosynthetic enzymes in leaves require nitrogen for their synthesis, and limitations in the availability of nitrogen and certain other elements reduce the capacity of leaves to photosynthesize. The strong correlation between photosynthesis and leaf nitrogen content illustrates how carbon and nutrient cycles are linked.

The light energy absorbed by chlorophyll and other leaf pigments is used by enzymes embedded in the chloroplasts of leaf tissue to split water into hydrogen and oxygen, and to combine CO_2 into simple sugars. The rate of CO_2 diffusion into leaves is restricted by pores, called stomata, that open and close, depending on environmental conditions. Stomata are especially sensitive to air humidity and dry soil, but also respond to light, ambient CO_2 concentrations, and temperature. When the rate of water loss by diffusion through stomata is greater than the rate of the movement through the plant into the leaves, the leaf water content falls and stomata close. The wet mesophyll cells behind stomatal pores are essential to trap CO_2, but necessarily involve the loss of the water, a process called transpiration.

Transpiration is a physical process driven by the energy balance of leaves. Radiant energy falling on the leaves must be disposed of as sensible heat, driven by temperature gradients, or latent heat, ie, the energy required to evaporate water. If the rate of evaporation is too slow, leaf temperatures rise and the humidity deficit at the leaf surfaces increases; stomata then tend to close. The opening and closing of stomata represent a compromise between limiting the loss of water while maximizing the rate that CO_2 enters the leaf for photosynthesis. As a result, stomatal opening is closely related to net photosynthesis.

The horizontal distribution of leaves in plant canopies affects the efficiency with which light is intercepted. These effects are recognized in most models, some of which use complex procedures to account for differences in leaf arrangement (23). Large, flat leaves that form umbrella-like canopies around individual trees capture light more effectively than small clumps of dispersed needles. In general, the amount of light intercepted by plant communities is calculated in terms of amounts intercepted by cumulative layers of leaves. Forests composed of broad-leaved trees intercept nearly all solar radiation with the equivalent of about six layers of leaves projected onto an equal area of ground surface (referred to as leaf area index). In coniferous forests, made up of trees with narrow crowns, light penetrates more easily through the canopy. For this reason, coniferous forests may display up to 12 layers of projected leaf area (24).

Transpiration and Evaporation

Because the movement of water through ecosystems affects the carbon cycle in many ways, water and carbon balance models share many features, as illustrated in Figure 4. Canopy leaf area is critical in both cases; in addition to absorbing solar energy for photosynthesis, leaves intercept precipitation and evaporate water from their surface or, as noted earlier, through their stomata in the process of transpiration. Radiant energy can raise leaf temperatures and increase the humidity deficit of the air at leaf surfaces.

The size of vegetation affects the transfer of heat, water vapor, and CO_2. The foliage in tall vegetation is usually near ambient air temperatures, because air flow is turbulent around such vegetation and transfer processes are more effective. Shorter vegetation, particularly that with broad leaves, can be substantially heated above air temperature if wind speeds are low and transpiration is constrained, preventing evaporative cooling (25). This interdependence of carbon and water fluxes requires, in most cases, that additional environmental variables be measured or modelled.

Consequently, one of the variables required for water balance models is total radiant energy; clearly, another is precipitation. Water that is not evaporated from the surfaces of vegetation enters the litter or soil, or in some cases runs off the surface. The water storage capacity of the rooting zone must be known in order to model the water balance of plant communities. When between two-thirds to three-quarters of the available water has been extracted from the soil, the imbalance between the rate of loss from leaves and the rate of uptake by the generally sparse network of roots in the deeper part of the soil profile causes water stress (drought) that may induce partial or complete stomatal closure. Extreme or prolonged drought may also lead to the shedding of leaves. Given values of the appropriate variables, ie, leaf area index, atmospheric factors, and a stomatal response function, models can provide good estimates of changes in root zone water storage (see Fig. 5).

Respiration by Plants

Green plants expend metabolic energy and respire CO_2 to the atmosphere during photosynthesis (photorespiration), in producing new tissue (construction respiration), and in maintaining live tissue (maintenance respiration). Photorespiration only occurs when enough light exists for photosynthesis; maintenance and construction respiration can occur in the light and dark. All three types of respiration respond positively to increasing temperature for different reasons.

Photorespiration is a product of the same enzyme that converts CO_2 into simple sugars. This enzyme will bind with O_2 as well as with CO_2; when it does, there is a breakdown of organic molecules and production of CO_2. As temperature increases, the CO_2 concentration within leaves normally decreases, because photosynthesis continues while stomata begin to close and restrict CO_2 diffusion into the leaf.

Currently, few ecosystem models treat photorespiration explicitly. With projected increases in atmospheric CO_2 and temperature, however, the process of photorespiration becomes more significant: the compensation point where photosynthesis and respiration are in balance should be at a lower light level than today's environment permits (26). Photorespiration does not occur in some plants, such as corn and other related tropical grasses, so CO_2 rise may affect competitive interactions between species regardless of a temperature change.

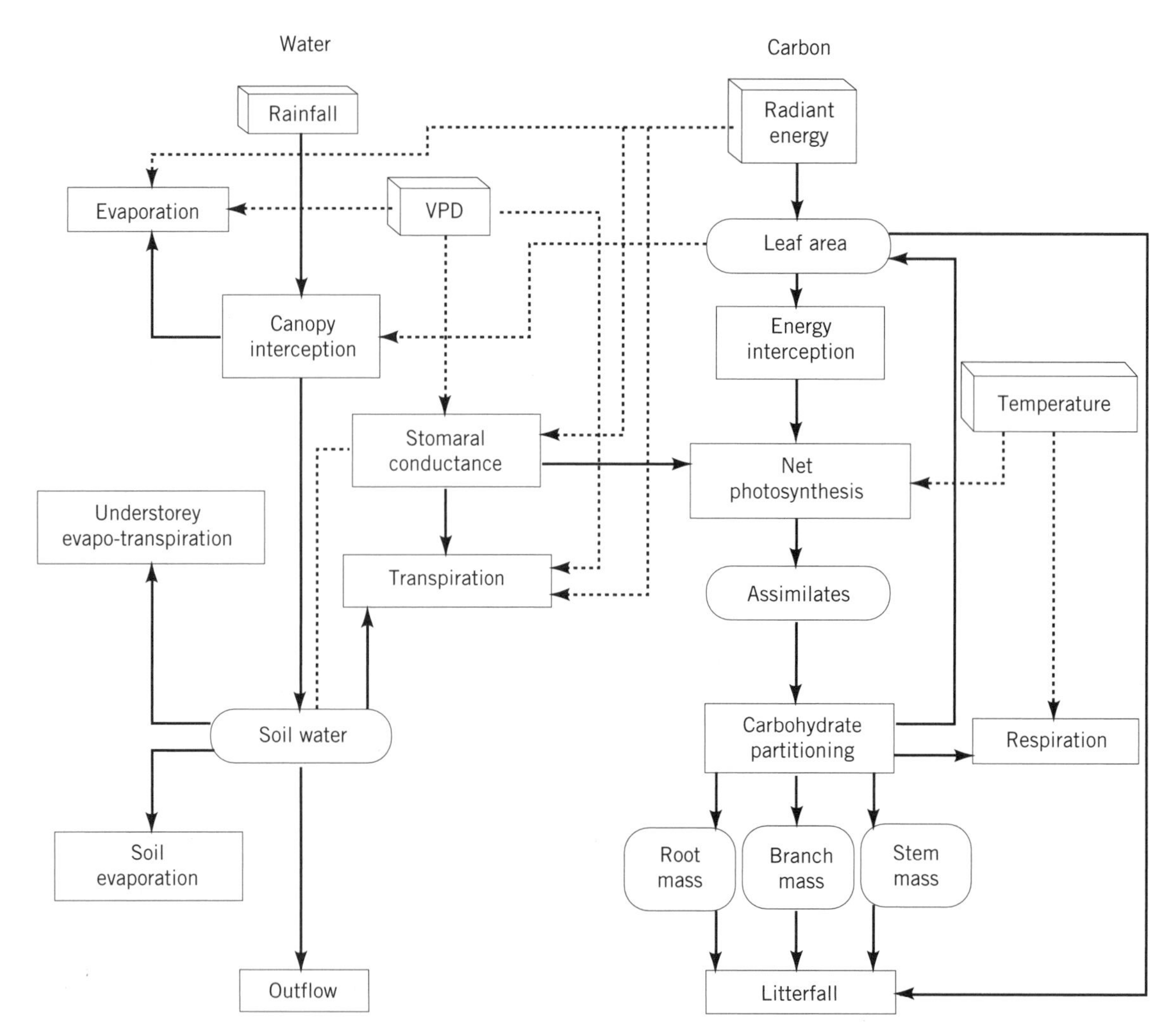

Figure 4. Coupled plant carbon and hydrologic models share many features. Transpiration and photosynthesis are both limited when stomatal conductance and leaf area are reduced. Increasing temperature potentially allows greater evaporation and transpiration, and also affects photosynthesis and respiration. As soils dry, carbon allocated to root growth usually increases until water stores in the soil are fully exhausted. Reprinted with permission of R. E. McMurtrie, University of New South Wales, Australia.

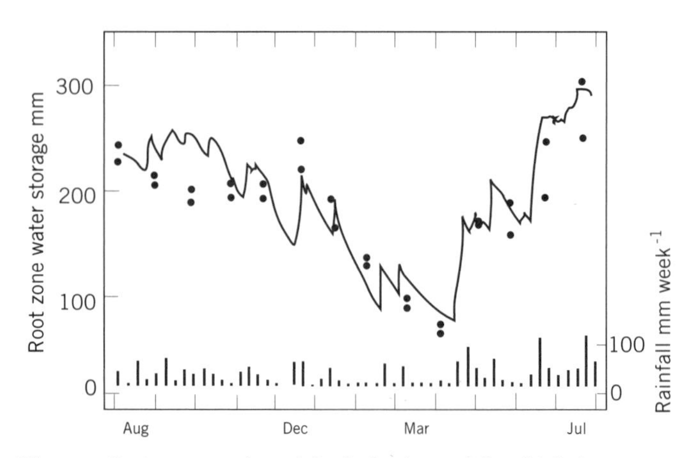

Figure 5. A process-based hydrologic model, which integrates leaf area and controls on stomatal conductance, can closely match predicted changes in root zone water storage with those observed (●). Water storage was measured in the top 3.25 m of the root zone of a 9-year-old plantation of pine growing in Australia (5).

Construction of plant tissue is metabolically expensive. During the production of organic compounds and their assembly into structure, construction respiration costs are incurred, equivalent to about 15–50% of the carbon in the material produced. Carbon incorporated into organic tissues comprises about 50% of the dry weight of plants. If growth increases with temperature, respiration associated with construction cost must therefore increase proportionally. The growth of aboveground plant material is relatively easy to measure but root production is exceedingly difficult, particularly the growth of small-diameter roots that are produced and die within a year or less. Increase in knowledge about root growth will occur upon the development of reliable models of photosynthesis, aboveground growth, and respiration; the carbon not accounted for will then be allocated belowground.

Maintenance respiration occurs in all living cells in the process of restoring enzymes, maintaining gradients of ions, and converting inorganic nutrients into biochemical compounds. Leaves are usually the most active respiratory organs, with the highest concentration of enzymatic machinery and living cells. Small-diameter roots are also active in acquiring inorganic nutrients and in converting these into biochemical compounds. In ecosystems with tall vegetation, the stems and large-diameter roots also significantly contribute to the total maintenance respiration. In general, maintenance respiration can be modelled as a linear function of the increasing nitrogen content in the living cells of plants and as an exponential function of increasing temperature (27). Because annual net photosynthesis also tends to increase with rising temperatures and an extended growing season, the relative cost of maintenance respiration as a fraction of net annual CO_2 uptake increases linearly, not exponentially, as shown in Figure 6. Although the maintenance respiration required to support the vascular system is small compared with leaves, it increases substantially with increasing temperature.

Microbial Respiration

Most of the organic matter produced by plants, whether eaten by animals or directly transferred into the litter and soil, is eventually consumed by microbes. Microbial respiration produces CO_2 while releasing nutrients from the organic matter. Decomposition of organic material depends on the local environment, where microbial activity occurs. Thus, the decomposition rates differ substantially between dead standing trees, leaf litter, and soil.

Drying and freezing conditions essentially halt microbial activity everywhere. In soils, however, conditions are often sufficiently favorable, even under snow cover and during extended drought, to allow some decomposition to proceed at reduced rates. Flooding reduces the amount of O_2 available and decreases the rate of decomposition, but it can stimulate anaerobic production of methane (CH_4) by specialized microbes. Methane production has been estimated at about 3% of the total net primary production in rice paddies, sedge meadows, and bogs (28). The environments supporting types of vegetation that vent large amounts of methane to the atmosphere are easily recognized and become important when considering whole landscapes.

In general, microbial respiration associated with the decay of surface leaf litter is closely correlated with surface moisture and temperature conditions. Evaporation is also a function of these same variables, so it is not surprising that the litter decay rates (turnover) and annual evaporation from soil and transpiration from plants (evapotranspiration) are related. The relation with evapotranspiration provides a means of estimating decomposition rates across regions (see Fig. 7). The chemical composition of litter also affects its rate of decay. Deciduous leaves generally decay more rapidly than evergreen leaves, because they contain a more favorable balance of nitrogen-rich proteins to hard-to-decay organic molecules, such as the lignin, that binds together cell walls in plants. In modeling decomposition, therefore, the lignin–nitrogen ratio, or some other index of organic matter quality, is often taken into account to predict the annual weight loss of leaf litter (see Fig. 8).

Soil Carbon Accumulation

A small fraction of the total carbon entering the detrital pool of dead organic material, in the form of leaves, stems, roots, and root exudates, is converted to organic acids and other products of decomposition that become part of the soil, or are dissolved in water and eventually enter streams. Soils carbon decays more slowly than that in litter because some is bound to clay particles in a form that resists further decomposition. In some cold, anaerobic environments, pure organic material may also accumulate and form peat. Even following drastic disturbance such as clear-cutting of forests, little change occurs in the total storage of soil carbon if forest cover is allowed to return.

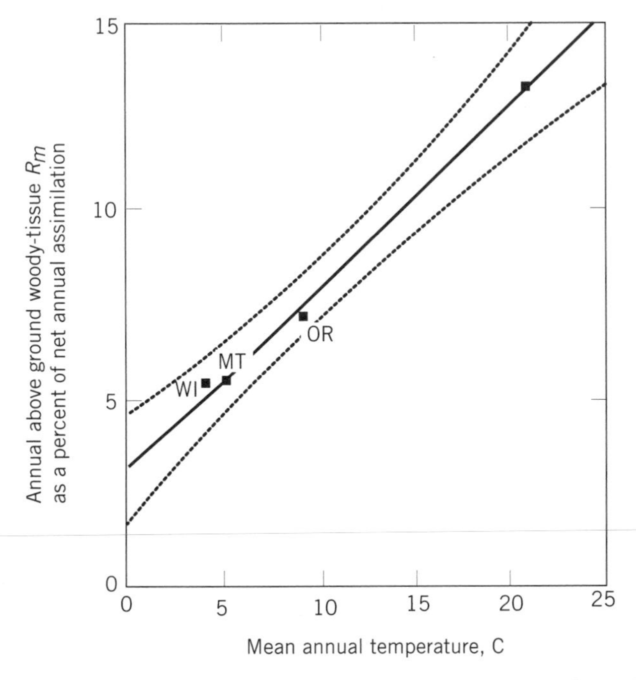

Figure 6. Maintenance respiration (R_m) of aboveground woody tissue in evergreen trees growing in Wisconsin (WI), Montana (MT), Oregon (OR), and Florida (FL) increases linearly as a fraction of the total estimated net annual assimilation of carbon. Reprinted with permission of M. G. Ryan, Rocky Mountain Forest and Range Experiment Station, Fort Collins, Colorado.

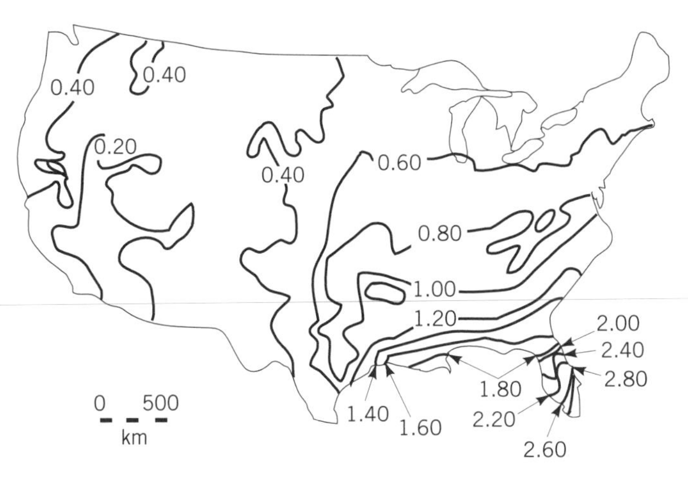

Figure 7. Rates of decomposition of fresh litter in the United States predicted by a simulation model using actual evapotranspiration as a predictive variable. Isopleth values are the fractional loss rate of mass from fresh litter during the first year of decay (29).

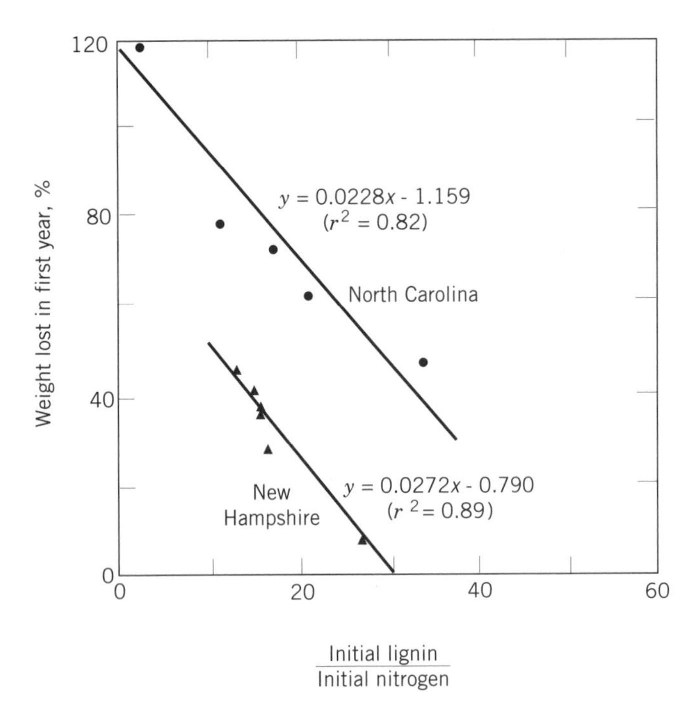

Figure 8. Decomposition of fresh leaves as a function of the lignin–nitrogen ratio of various species in New Hampshire and North Carolina forests shows parallel slopes associated with differences in mean annual temperature. Reprinted with permission of the authors and Academic Press (30).

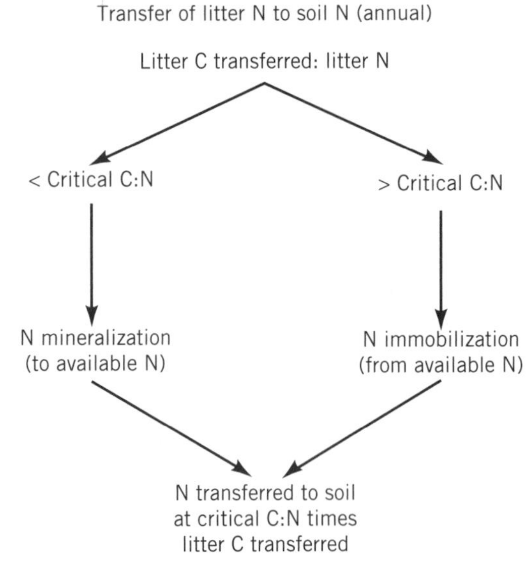

Figure 9. The fraction of nitrogen (N) and carbon (C) transferred annually to soil is partly a function of critical C–N ratios in most ecosystem models. The total amount of litter entering the soil depends on the expected death of leaves, woody material, and roots. When the system is disturbed, additional material may be added or removed. Reprinted with permission of Raymond Hunt, University of Montana, Missoula, Montana.

Conversion to permanent pasture or agricultural use will, however, commonly reduce soil carbon levels. With climate change and changing land-use patterns, a new equilibrium in soil carbon content is likely to result as vegetation patterns shift.

The fraction of carbon transferred from the litter to the soil is modelled in relation to decomposition rates of a single large reservoir, or of several reservoirs consisting of different materials that decay at different rates. In the soil, separate pools may receive the breakdown products. A rapidly decaying pool of organic material provides nutrients to plants, whereas a more recalcitrant carbon pool turns over carbon and releases nutrients much more slowly.

Ecosystem models generally couple the transfer of carbon to soil with the release of organically bound nitrogen and other essential minerals to plant roots, as illustrated in Figure 9. Decomposition rates are usually calculated at monthly time steps based on soil temperature, water content, and the carbon–nitrogen ratio of the decaying organic matter. When conditions are favorable, nutrients in their elemental form are released and made available to plants.

Models usually assume that all soil carbon is confined to the upper meter of soil or less, although this is not the case. One of the most difficult problems in modeling soil carbon is obtaining accurate estimates of carbon distribution throughout the entire rooting zone that may, in some ecosystems, extend to a depth of 10 m or more.

Nutrient Uptake and Carbon Allocation by Plants

The pool of nutrients released in decomposition of organic matter by microbial activity is supplemented from other sources, which include inputs from precipitation and chemical exchanges with minerals on the exposed surfaces of clays and organic residues. Most fertile soils hold calcium, magnesium, and potassium ions readily available on mineral exchange sites. Soil acidity, texture, organic matter content, and the type of clay minerals present strongly affect the availability of these essential bases. Uptake of all elements essential for plant growth must annually balance what is incorporated into new tissues, minus that lost in the shedding of leaves and other nutrient-rich organic residues.

Carbon allocation to roots generally increases over shoot growth when nutrients, particularly nitrogen and phosphorus, are not readily available. Including other nutrients in models would be desirable because the total complex affects carbon allocation and, through competition, the composition of plant communities (21). A more mechanistic approach to modeling interactions between carbon, water, and nutrients has recently been published (31). This model links transport of water, carbohydrates, and nutrients to movement through a plant's vascular system, which includes xylem, which transfers water and nutrients from roots to shoots, and phloem, which transfers carbohydrates and amino acids from leaves to roots and other organs.

A close linkage exists between carbon allocation in plants and the decomposition process because reduced nutrient availability lowers foliar nutrient levels, which in turn limits decomposition. Some carbon balance models adjust the carbon–nitrogen ratio in foliage to mirror that in litter, and use the ratio as a control on modeling carbon allocation (22).

Some checks on the reliability of how models allocate carbon can be obtained by comparing independent estimates of the amount of carbon required to grow and main-

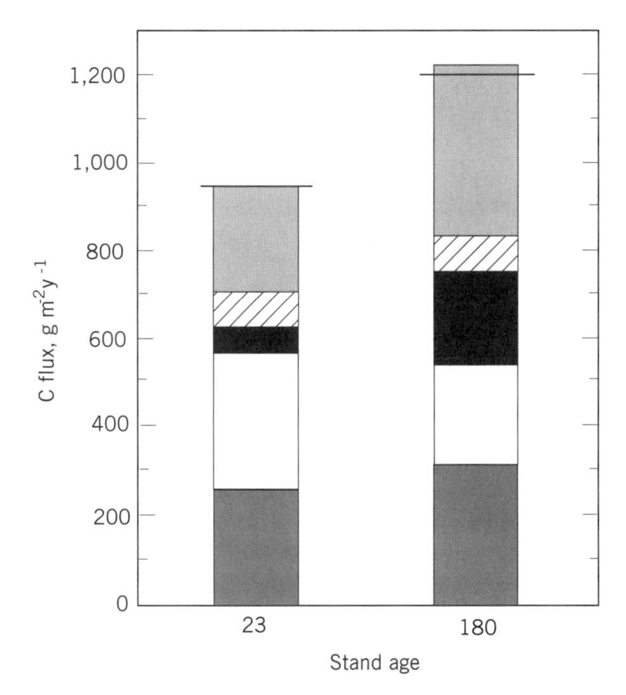

Figure 10. Annual carbon flux estimates for a 23- and 180-year-old forest of fir trees show good agreement between the amount of photosynthate available estimated by a simulation model (_ _) and the sum of production (p), construction (c), and maintenance respiration (m) by foliage (F); stems (S); and roots (R) estimated independently. Reprinted with permission of M. G. Ryan and *Tree Physiology* (32).
F_m S_{p+c}
F_{p+c} R_{m+p+c}
S_m — Photosynthesis

tain foliage, stems, and roots against that provided through photosynthesis (e.g., see Fig. 10). Where the forest is in equilibrium, an independent estimate of carbon allocation to roots for growth and respiration is obtained by measuring the annual CO_2 flux from the soil, minus the carbon in aboveground litterfall (33). Under varying conditions, changes in how the carbon is allocated to above- and belowground components can thus be estimated and empirically modelled.

Most models assume plant roots are evenly distributed throughout the soil or that all plants growing in a particular environment have similar rooting characteristics. Water and nutrients are not equally available to plant species; the roots of some are restricted to surface horizons, whereas others extend more deeply. When plants with deep roots are absent from a community, mineral cycling from the lower profile stops. As a result, the surface organic and mineral layers may become impoverished. Such vertically defined interactions in the soil are absent in most current ecosystem models, although not all (34).

TESTING AND SIMPLIFYING CARBON BALANCE MODELS

Testing Models

The reliability of carbon balance models may be assessed by:

- Direct comparisons with field measurements
- Projecting simulations over long periods
- Predicting changes in hydrologic and nutrient components
- Applying sensitivity analysis to identify critical relationships

A variety of approaches is desirable because some questions are not subject to direct experimentation (eg, changing climate), whereas others require simultaneous measurements of many variables for long periods.

Comparison of model predictions with field measurements is valuable but may be difficult to accomplish. The amount of carbon in aboveground biomass is easily measured, whereas that in roots is more difficult to estimate accurately. Carbon stocks in soil are large and vary from site to site, particularly where rooting depth is affected by rocks. Net fluxes of CO_2 from the soil upward through the vegetation can be measured, and various transfer equations tested on a short-term basis (35). Although improvements in micrometeorological instrumentation can provide accurate net exchange measurements of carbon dioxide from forest ecosystems over short periods (see Fig. 11), difficulties in making accurate measurements of growth, respiration, and photosynthesis over extended periods limit the data sets available to compare model predictions.

Using model simulations to make predictions over many years can illustrate whether the transfer rates estimated for a single year are reasonable. If such simulations resulted in litter accumulations on the ground to unrealistic depths, for example, clearly the estimated rates of annual litterfall must be too high or the rate of decomposition too low. Because of the close link between the carbon, water, and nutrient cycles, comparing simulations of nutrient and water export from monitored streams draining closed basins can provide another type of model validation (37).

Sensitivity analysis can identify critical assumptions and interactions. For example, in simulating the response of a pine forest to a doubling of CO_2 in the atmosphere, the net annual growth should increase less than 8% because decomposition rates should limit nitrogen availability (38). Sensitivity analyses can also indicate the relative need for more accurate measurements of certain variables. This might be accomplished, for example, by comparing changes in predicted growth, litter accumulation, transpiration, and photosynthesis in response to a 10% change in precipitation, carbon–nitrogen ratios, soil water storage, and leaf area index.

Simplifying Models

Carbon balance models must be simplified to make them applicable to larger spatial and temporal scales. Sensitivity analysis, discussed earlier, is one way of identifying those variables that can be estimated less precisely without losing much accuracy. An alternative approach is to compare predictions made by refined models with those from simpler models. Thus, hourly predictions of canopy photosynthesis derived from a model with separate layers of leaves can be compared with a whole canopy model that generates estimates of photosynthesis from meteorological data averaged for the day (38,39). If the comparisons indicate that the simpler models provide acceptable results,

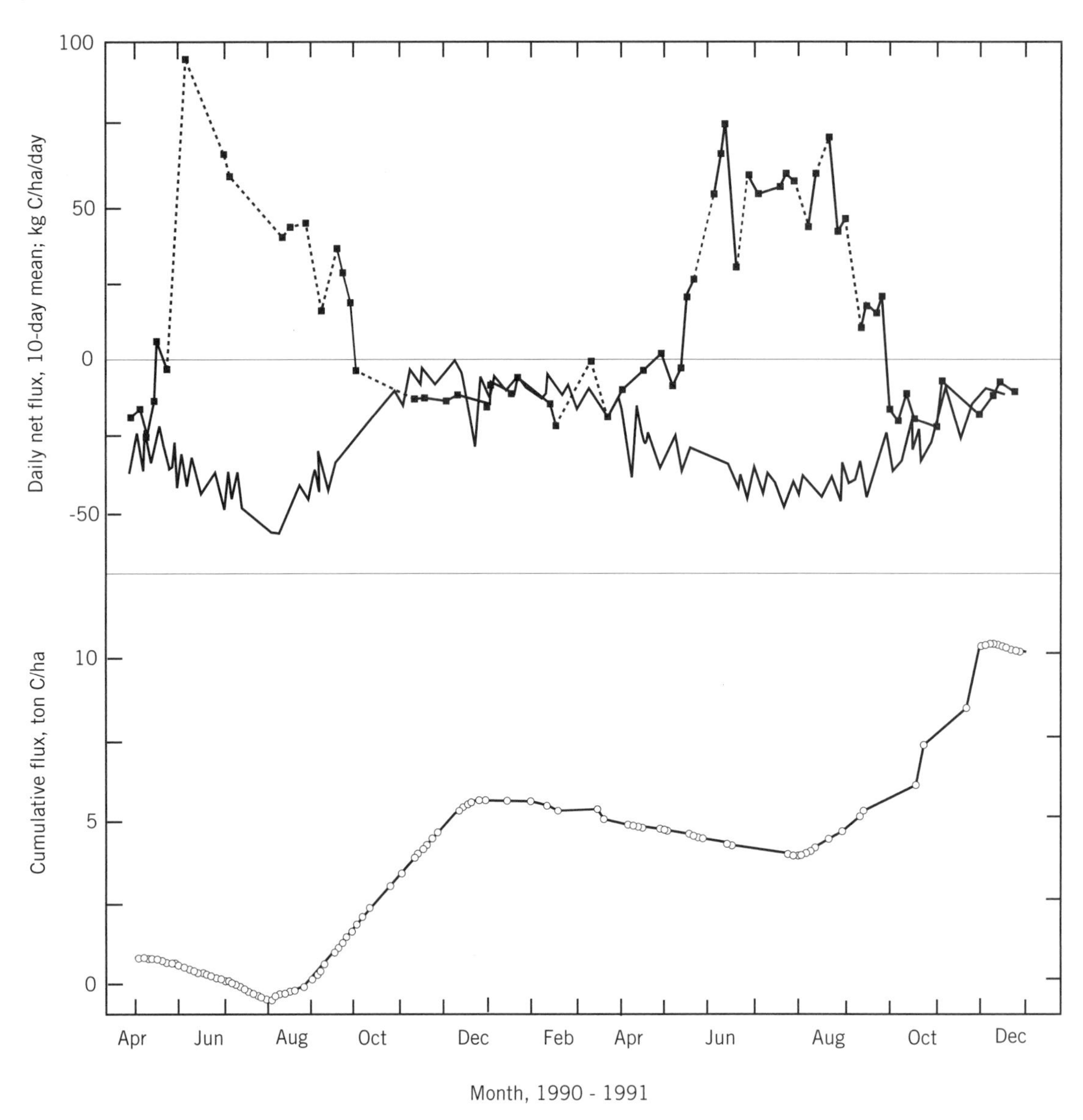

Figure 11. Net ecosystem exchanges measured with eddy correlation methods from a tower over a northern deciduous hardwood forest in Massachusetts show large seasonal differences. In the upper diagram are 10-day means for daily net CO_2 flux from the atmosphere to the ecosystem, the middle diagram presents respiration on a similar time scale, and the lower graph shows cumulative flux (36).

they can serve as the basis for further simplifications that predict photosynthesis, respiration, evapotranspiration, decomposition, and growth on a monthly basis. Finally, monthly balance sheets can be compared with annual runoff from gauged watersheds, annual litterfall, and annual aboveground production. Sequential comparisons, from detailed to less complex models, help identify the extent to which simplifications are warranted and the minimum detail needed to provide a certain reliability of estimates from coarser-scale models.

SCALING CARBON BALANCE MODELS SPATIALLY

Although the composition of vegetation differs over time and space, some ecosystem models have been simplified to assume that generalizations of process rates can be made across landscapes by recognizing only the broadest categories of vegetation (19,40).

An even simpler model of energy utilization efficiency (41) has proven applicable to a wide range of forests in the western United States, as illustrated in Figure 12. In this model, the upper limits to production are set by the ability of vegetation to absorb solar radiation. At extremes in latitude and where leaf area development is restricted, the growth of all vegetation is limited. Even at mid-latitudes where forests with dense canopies dominate, production is severely constrained if frost, soil drought, or excessive atmospheric humidity deficits occur over extended periods (see Fig. 13).

Satellite coverage of the Earth now provides a means of estimating seasonal changes in leaf area index (43) and

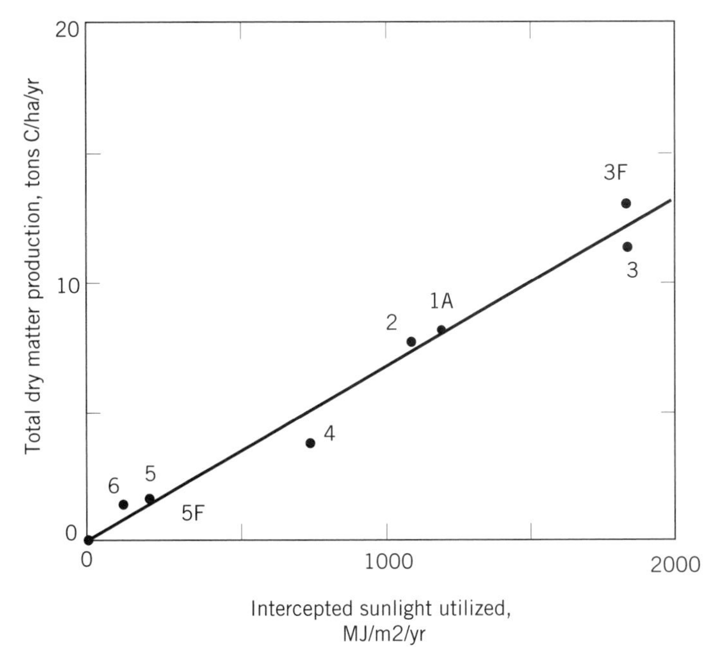

Figure 12. The net amount of carbon that is transferred into dry matter production increases proportionally with the amount of visible radiation that can be intercepted and converted by photosynthesis into carbohydrates when stomata are open. The numbered sites refer to ecosystems dominated by (1A) deciduous alder, (2) mixed forests of conifers with deciduous oak, (3) fertilized (F) and unfertilized fast-growing conifers on the western slopes of the Cascade Mountains, (4) subalpine forests, (5) fertilized (F) and unfertilized open pine forest, and (6) juniper woodland (42).

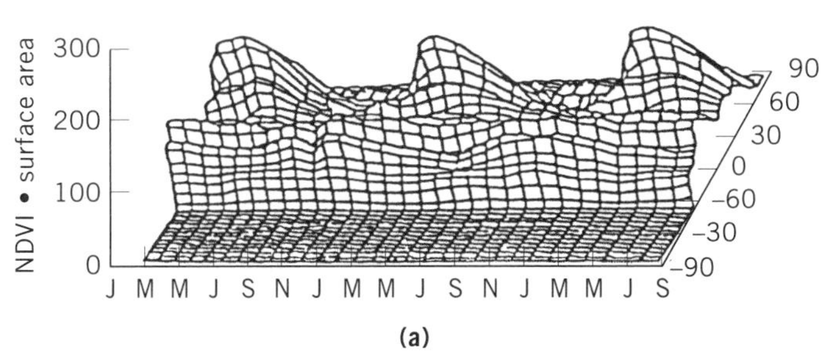

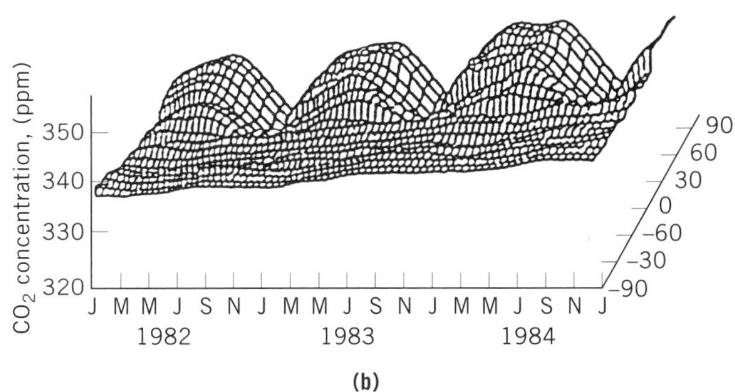

Figure 14. (**a**) A satellite-derived index of greenness (NDVI), related to reflectance in the near-ir and red parts of the spectrum, varies most at mid- and far northern latitudes and least at the equator. (**b**) Seasonal variation in the concentration of CO_2 in the atmosphere parallels changes in vegetation greenness (47).

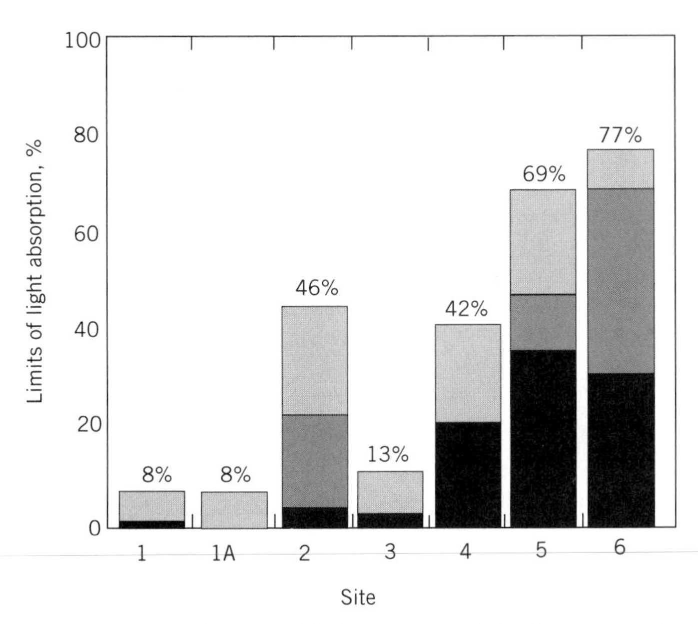

Figure 13. The fraction of light intercepted by forest canopies varies from about 20 to nearly 100% across the environmental gradient, corresponding to the forest types described in Figure 12. As environments become harsher, the percentage of light that is intercepted and can be used in photosynthesis is progressively reduced from > 90% by the two coastal forests (1 and 1A) to < 25% in the juniper woodland (42). ▢, humidity deficit; ▩, drought, ■, freezing

cloud cover (44). This capability, combined with a network of climatic stations, provides the essential information for modeling seasonal changes in photosynthesis, transpiration, growth, and other ecosystem processes (45). The first general global-scale ecosystem model using satellite-derived estimates of monthly vegetative cover has recently appeared (46). Previously, as illustrated in Figure 14, seasonal variations in atmospheric CO_2 concentrations were shown to vary with this same satellite-derived index of vegetative greenness associated with photosynthetic activity (47).

These models are being improved to incorporate annual variations in climate and atmospheric CO_2. A few global models include atmospheric circulation and provide a basis of comparison with measured variations in atmospheric CO_2 (46,48–50). These models will be improved further as more is learned about the coupling between land, atmosphere, and oceans (51,52), and satellite-derived data is extended to estimate climatic variation as well as vegetative cover (53).

CONCLUSION

Carbon balance modeling is still in its infancy, but a sufficient number of models is now available to allow comparisons. Leaf-level processes are well-understood and have been the first to be generalized and scaled in time and space. The implications of changes in species composition are recognized as important but continue to present difficulties to modellers. Increased understanding of how nutrients and water interact with carbon cycling is leading to more realistic, but more complicated, models. At the same time, simplified models are providing new insights at regional and global scales. Improved monitoring techniques in micrometeorology, atmospheric chemistry, and remote sensing are available to help assess the reliability of larger-scale predictions. Field experiments, model comparisons, and sensitivity analyses will continue to advance understanding and improve model predictive capacities.

Acknowledgments
We appreciate comments made on earlier drafts of this article by Joe Landsberg, Warwick Silvester, Beverley Law, Jeanne Panek, and Lise Waring.

BIBLIOGRAPHY

1. *The Ecosystems Center Annual Report,* Marine Biological Laboratory, Woods Hole, Mass., 1987.
2. L. A. Joyce and R. N. Kickert, "Applied Plant Growth Models for Grazing Lands, Forests, and Crops," in *Plant Growth Modelling for Resource Management,* Vol. 1, CRC Press, Boca Raton, Fl., 1987, pp. 17–55.
3. U. Siegenthaler and J. L. Sarmiento, *Nature* **365,** 119–125 (1993).
4. G. I. Ågren, R. E. McMurtrie, W. J. Parton, J. Pator, and H. H. Shugart, *Ecological Applications* **1,** 118–138 (1991).
5. R. E. McMurtrie, D. A. Rook, and F. M. Kelliher, *Forest Ecology and Management* **39,** 381–413 (1990).
6. Y.-P. Wang and P. G. Jarvis, *Agricultural and Forest Meteorology* **51,** 257–280 (1990).
7. G. B. Bonan, *Journal of Geophysiocal Research* **96,** 7301–7312 (1991).
8. E. B. Rastetter, M. G. Ryan, G. R. Shaver, J. M. Melillo, K. J. Nadelhoffer, J. E. Hobbie, and J. D. Aber, *Tree Physiology* **9,** 101–126 (1991).
9. R. L. Sanford, Jr., W. J. Parton, D. S. Ojima, and D. J. Lodge, *Biotropica* **23,** 364–372 (1991).
10. H. H. Shugart, *A Theory of Forest Dynamics,* Springer-Verlag, New York, 1984.
11. D. B. Botkin, J. F. Janak, and J. R. Wallis, *Journal of Ecology* **60,** 849–872 (1972).
12. J. D. Aber and J. M. Melillo, FORTNITE: A Computer Model of Organic Nitrogen Dynamics in Forest Ecosystems," *University of Wisconsin Research Bulletin R3130,* 1982.
13. J. Pastor and W. M. Post, *Nature* **334,** 55–58 (1988).
14. A. D. Friend, H. H. Shugart, and S. W. Running, *Ecology* **74,** 792–797 (1993).
15. H. Lieth, *Nature and Resources* **8,** 5–10 (1972).
16. E. Box, *Macroclimate and Plant Forms: An Introduction to Predictive Modeling in Phytogeography,* Dr. W. Junk Publishers, The Hague, the Netherlands. 1981.
17. I. C. Burke, T. G. F. Kittel, W. K. Lauenroth, P. Snook, C. M. Yonker, and W. J. Parton, *Bioscience* **41,** 685–692 (1991).
18. F. I. Woodward, *Climate and Plant Distribution,* Cambridge University Press, Cambridge, U.K., 1987.
19. I. C. Prentice, W. Cramer, S. P. Harrison, R. Leemans, R. A. Monserud, and A. M. Solomon, *Journal of Biogeography* **19,** 117–134 (1992).
20. M. E. Harmon, W. K. Ferrell, and J. F. Franklin, *Science* **247,** 699–702 (1990).
21. W. J. Parton, J. W. B. Stewart, and C. V. Cole, *Biogeochemistry* **5,** 109–131 (1988).
22. S. W. Running and S. T. Gower, *Tree Physiology* **9,** 147–160 (1991).
23. J. M. Norman, "Scaling Processes Between Leaf and Canopy Levels," in *Scaling Physiological Processes: Leaf to Globe,* Academic Press, San Diego, Calif., 1993, pp. 41–76.
24. J. D. Marshall and R. H. Waring, *Ecology* **67,** 975–979 (1986).
25. J. Grace, "Some Effects of Wind on Plants," in *Plants and Their Atmospheric Environment,* Blackwell Scientific Publications, London, 1981, pp. 31–56.
26. S. P. Long and P. R. Hutchin, *Ecological Applications* **1,** 139–156 (1991).
27. M. G. Ryan, *Ecological Applications* **1,** 157–167 (1991).
28. G. J. Whiting and J. P. Chanton, *Nature* **364,** 794–795 (1993).
29. V. Meentemeyer, "Climatic Regulation on Decomposition Rates of Organic Matter in Terrestrial Ecosystems," in *Environmental Chemistry and Cycling Processes,* National Technical Information Service, Springfield, Va., 1978, pp. 779–789.
30. J. M. Melillo, A. D. McGuire, D. W. Kicklighter, B. Moore, III, C. J. Vorosmarty, and A. L. Schloss, *Nature* **363,** 234–240 (1993).
31. R. C. Dewar, *Functional Ecology* **7,** 356–368 (1993).
32. M. G. Ryan, *Tree Physiology* **9,** 255–266 (1991).
33. J. W. Raich and K. J. Nadelhoffer, *Ecology* **70,** 1346–1354 (1989).
34. P. J. H. Sharpe, J. Walker, L. K. Penridge, and H.-I. Wu, *Ecological Modelling* **29,** 189–213 (1985).
35. D. D. Baldocchi, "Scaling Water Vapor and Carbon Dioxide Exchange from Leaves to Canopy: Rules and Tools," in *Scaling Physiological Processes: Leaf to Globe,* Academic Press, San Diego, Calif., 1993, pp. 77–114.
36. S. C. Wofsy, M. L. Gouldon, J. W. Munger, S.-M. Fan, P. S. Bakwin, B. C. Duabe, S. L. Bassow, and F. A. Bazzaz, *Science* **260,** 1314–1317 (1993).
37. F. H. Bormann, G. E. Likens, T. G. Siccama, R. S. Pierce, and J. S. Eaton, *Ecological Monographs* **44,** 255–277 (1974).
38. R. E. McMurtrie, H. N. Comins, M. U. F. Kirschbaum, and Y.-P. Wang, *Australian Journal of Botany* **40,** 657–677 (1992).
39. Y.-P. Wang, R. E. McMurtrie, and J. J. Landsberg, "Modelling Canopy Photosynthetic Productivity," in *Crop Photosynthesis: Spatial and Temporal Determinants,* Elsvier Science Publishers, Amsterdam, the Netherlands, 1992, pp. 43–67.
40. J. M. Melillo, J. D. Aber, and J. F. Muratore, *Ecology* **63,** 621–626 (1982).
41. J. L. Monteith, *Philosophical Transactions of the Royal Society, Series B* **281,** 277–294 (1977).
42. J. Runyon, R. H. Waring, S. N. Goward, and J. M. Welles, *Ecological Applications* **4,** 226–237 (1994).
43. M. Spanner, L. Johnson, J. Miller, R. McCreight, J. Runyon, P. Gong, and R. Pu, *Ecological Applications* **4,** 258–271 (1994).
44. T. Eck and D. Dye, *Remote Sensing of the Environment* **38,** 135–146 (1991).
45. P. J. Sellers, J. A. Berry, G. J. Collatz, C. B. Field, and F. G. Hall, *Remote Sensing of the Environment* **42,** 187–216 (1992).
46. C. S. Potter, J. T. Randerson, C. B. Field, P. A. Matson, P. M. Vitousek, H. A. Mooney, and S. A. Klooster, *Global Biogeochemical Cycles* **7,** 811–841 (1993).
47. C. J. Tucker, I. Y. Fung, C. D. Keeling, and R. H. Gammon, *Nature* **319,** 195–199 (1986).

48. P. P. Tans, T. J. Conway, and T. Nakazawa, *Journal of Geophysical Research* **94,** 5151–5172 (1989).
49. P. P. Tans, I. Y. Fung, and T. Takahashi, *Science* **247,** 1431–1438 (1990).
50. P. J. Sellers, S. O. Los, C. J. Tucker, C. O. Justice, D. A. Dazlich, G. J. Collatz, and D. A. Randall, *Journal of Climate* (in press).
51. R. F. Keeling, *Nature* **363,** 399–400 (1993).
52. G. D. Farquhar, J. Lloyd, J. A. Taylor, L. B. Flanagan, J. P. Syvertsen, K. T. Hubick, S. C. Wong, and J. R. Ehleringer, *Nature* **363,** 439–443 (1993).
53. S. N. Goward, R. H. Waring, D. G. Dye, and J. Yang, *Ecological Applications* **4,** 322–343 (1994).

CARBON CYCLE

Richard Houghton
Woods Hole Research Center
Woods Hole, Massachusetts

Concern about the effects of a rapid global warming has focused attention on the global carbon cycle, first because carbon dioxide (CO_2) has been the most important human-induced greenhouse gas in the past and is expected to remain so in the future (1) and second because an understanding of the carbon cycle is required for predicting how future emissions will affect the rate of warming. The rate and extent of a global warming depend on the concentration of CO_2 and other greenhouse gases in the atmosphere. The concentration of CO_2 in the atmosphere, in turn, is a function of the emissions of CO_2 to the atmosphere from the combustion of fossil fuels and deforestation (sources) and the removal of CO_2 from the atmosphere by the oceans and terrestrial ecosystems (sinks). To constrain the rate of warming within acceptable limits, the concentration of CO_2 must be stabilized. Stabilization of the concentration requires a reduction in current emission rates. The decision of how much to reduce emissions requires an understanding of the relationship between emissions and atmospheric concentrations. If scientists do not know the relationship, there is no accurate way to determine policies regarding energy use and land use, the major determinants of CO_2 emissions. The distinction between emissions and concentration is an important one. Slowing the rate of warming will require stabilization of atmospheric concentrations, and stabilization of concentrations will require a reduction in emissions.

See also Air quality modeling; Carbon balance modeling; Carbon storage in forests; Forest resources; Global climate change, mitigation.

Stabilization of emissions at present rates will not prevent climatic change. At present rates of emission, concentrations of greenhouse gases in the atmosphere will increase to the point at which global warming will occur many times more quickly than it did 10,000 yr ago, with the retreat of the glaciers. Carbon dioxide is estimated to have accounted for more than half of the increased radiative forcing in the 1980s and is projected to account for more than half in the future. Because the concentration of CO_2 in the atmosphere is determined not only by its human-induced emissions but also by natural processes that remove the gas from the atmosphere, understanding the global carbon cycle is essential for predicting future concentrations of CO_2 and calculating the relationship between emissions and the rate of warming.

The processes important in the global carbon cycle vary depending on the time frame of interest. Over millions of years, the processes that govern the distribution of carbon are largely geological: weathering of the land surface, sedimentation of organic and inorganic material on the ocean floor, vulcanism, and sea floor spreading. At the other end of the spectrum are processes such as photosynthesis and hydration of a molecule of CO_2 at the sea surface that occur over seconds. The time scale of immediate interest to the current inhabitants of the earth and to the next few generations is years to a century. The geological processes are too slow to be important in the next decades; the rapid processes are integrated into other processes more appropriately studied over years. This discussion emphasizes processes that are important over years to a century.

THE NATURAL CYCLE

In its most simple abstraction, the global carbon cycle has only four major compartments or reservoirs: atmosphere, oceans, land, and fossil fuels. The reservoirs and the exchanges of carbon shown in Figure 1 pertain to the natural world before human interference. The actual date is probably 6000 yr ago, although human disturbance affected the storage of carbon in vegetation and soils with the development of settled agriculture about 10,000 yr ago. Well before that, fire may have been used for hunting. But 10,000 yr ago the earth's climate was different enough from today's to have affected the distribution of forests and hence the amount of carbon held on land. Thus the natural cycle shown in Figure 1 is an attempt to represent a world with a climate similar to today's but before major human effects. The important point is that the amount of carbon in the atmosphere (600 Pg) was about equal to the amounts in the surface layers of the ocean (1000 Pg) and in the

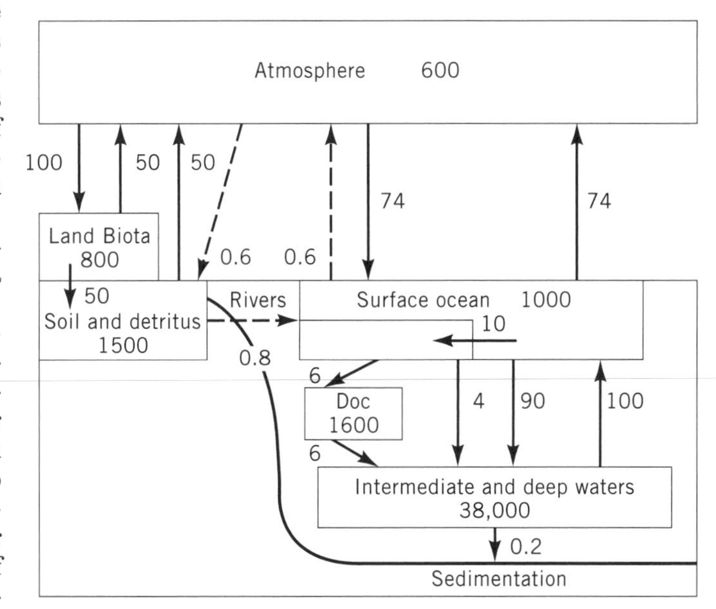

Figure 1. The global carbon cycle before human disturbance (about 6000 BP). Units are petagrams carbon and petagrams of carbon per year. Modified from Ref. 2.

terrestrial biota (800 Pg), a condition still true today. Because the exchanges of carbon between the atmosphere and these two reservoirs are rapid, changes in either reservoir has the potential to cause large and rapid changes in the concentration of atmospheric CO_2.

Most of the carbon in the atmosphere occurs as CO_2, but its concentration is low. Nitrogen and oxygen gases comprise more than 99% of the atmosphere; CO_2 accounts for about 0.03%. If 0.03% were the accuracy of measurement today, there would be no evidence for its increase; however CO_2 is presently measured within a few tenths of a part per million by volume (ppmv), or 0.00003%. In 1990, the concentration of CO_2 was 353 ppmv. For the 1000 yr before the Industrial Revolution, the concentration was about 280 ppmv and varied by less than 10 ppmv (Fig. 2). This preindustrial concentration of CO_2 was equivalent to about 600 Pg carbon (1 Pg = 1 billion t).

The concentration of CO_2 in the atmosphere has varied over most of the earth's history. Analyses of CO_2 in the bubbles of air trapped in glacial ice from Vostok, Antarctica, show that the concentrations varied over the last 160,000 yr, in phase with the advance and retreat of glaciers (7). The concentration was about 180 ppmv during glacial periods and about 280 ppmv during interglacial periods (Fig. 3). The concentration of CO_2 increased as global temperature increased and decreased as temperature fell. The coupling is evidence for a positive feedback between the carbon cycle and global climate. It suggests not only a greenhouse effect, in which atmospheric CO_2 warms the earth, but another effect by which warming raises the CO_2 concentration of the atmosphere. The greenhouse effect was advanced almost a century ago (8). Only recently have scientists been concerned about the positive feedback to a warming. The glacial–interglacial difference of 100 ppmv corresponds to a temperature difference of about 5°C, globally (10°C at Vostok). For comparison, the warming projected for the year 2100 under the business-as-usual scenario of the Intergovernmental Panel on Climatic Change (IPCC) is 2° to 5°C (1).

A small amount of the carbon in the atmosphere exists

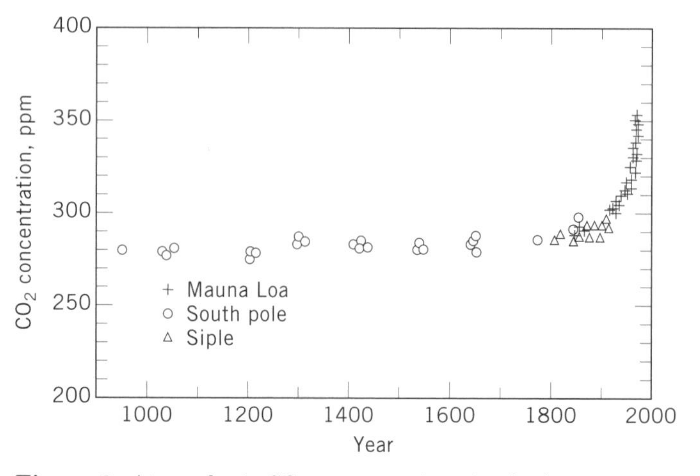

Figure 2. Atmospheric CO_2 concentrations for the last 1000 yrs determined from direct analysis of air at Mauna Loa, Hawaii (3), and from analysis of Antarctic ice cores from Siple station (4,5) and the South Pole (6).

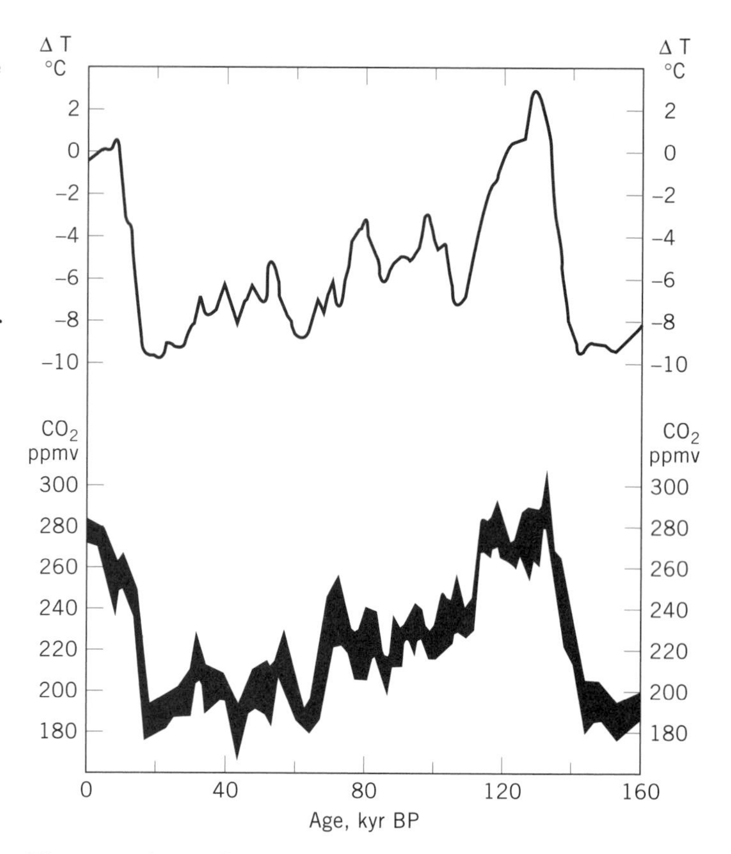

Figure 3. Atmospheric CO_2 concentrations and changes in surface temperatures over the last 160,000 yr determined from analysis of glacial ice at Vostok, Antarctica. Temperature changes were estimated based on measured deuterium concentrations. From Ref. 7.

as methane (CH_4) and as carbon monoxide (CO). Methane, another greenhouse gas, is at about 1/100 the concentration of CO_2 but is about 25 times more effective as a greenhouse gas than CO_2 per kilogram of emissions. The concentration of CH_4 was about 0.7 ppmv preindustrially. Methane, like CO_2, is produced as a result of respiration and decay, but under anaerobic conditions. Thus its major sources include wetlands and peatlands and the digestive processes of animals, notably ruminants. Unlike CO_2, CH_4 is chemically reactive, and a major sink for CH_4 is its destruction in the atmosphere by the hydroxyl radical. Carbon monoxide is not a greenhouse gas, but its reactivity with the hydroxyl radical means that it indirectly affects the concentrations of other greenhouse gases, principally CH_4.

Terrestrial ecosystems held about four times more carbon than the atmosphere before human disturbance. This terrestrial carbon is largely organic and exists in many forms, including living leaves and roots, animals, microbes, wood, decaying leaves, and humus. The turnover of these materials varies from less than 1 yr to more than 1000 yr. The living component of terrestrial ecosystems held approximately 800 Pg carbon, almost all of it in vegetation. Animals, including humans, account for less than 0.1% of the carbon in living organisms. About half the mass of dry plant material is carbon. The soils of the earth are estimated to have contained about 1500 Pg carbon, about twice the amount in the living biota. The carbon in soil is derived from the decomposition of plant material. Some reduction in the storage of terrestrial carbon had already occurred by 6000 yr ago; the values given in Figure 1 are approximate. Most of the carbon in terrestrial ecosystems was in forests. Forests once covered about 50% of the land surface and held more than 90% of the live organic carbon. When soils are included in the inventory, forests held about half of the terrestrial carbon. Grasslands, tundra, and wetlands store large amounts of organic carbon in their soils.

The amount of carbon in the surface ocean (ca. 1000 Pg) was also similar to the amount in the preindustrial atmosphere. The intermediate and deep waters of the ocean, not directly in contact with the atmosphere, held about 38 times this amount. At equilibrium, the concentration of CO_2 in the atmosphere is proportional to the concentration in the ocean, and hence, the oceans control the atmospheric concentration of CO_2. The equilibrium may require hundreds of years to reach; however, because the rate of circulation of oceanic waters is on the order of 1000 yr. The dramatic increase in atmospheric CO_2 over the last 30 yr demonstrates the extent of the present disequilibrium.

The form of carbon in the oceans, as in the atmosphere, is mostly inorganic carbon. The ocean's chemistry is such that less than 1% of this carbon is dissolved CO_2. Almost 90% exists in the form of the bicarbonate ion and the rest as carbonate ion. The three forms are in chemical equilibrium. About 1600 Pg carbon in the oceans (out of the total of more than 38,000 Pg carbon total) is dissolved organic carbon. Carbon in living organisms amounts to less than 2 Pg in the sea, compared with about 800 Pg on land. The mass of animal life in the oceans is almost the same as on land, however, pointing to the different trophic structures in the two environments. The ocean's plants are microscopic. They have a high productivity, but the production of plant material does not accumulate; it is either grazed or sinks to deeper layers where it decays. In contrast, terrestrial plants accumulate large amounts of carbon in long-lasting structures (ie, trees).

The amount of carbon stored in recoverable reserves of coal, oil, and gas is estimated to be about 5000 Pg, larger than any other reservoir except the deep sea, and about 10 times the carbon content of the atmosphere. The form of this carbon is organic, ie, it is the remains of terrestrial and marine vegetation once alive on the earth's surface. Until 100 to 150 yr ago, this reservoir of fossil carbon was not a significant part of the short-term cycle of carbon; the rate of burial of organic matter and its diagenesis to fossil deposits is several orders of magnitude lower than the cycling of carbon between land, sea, and air.

Each year the atmosphere exchanges about 100 Pg of carbon (about 15%) with terrestrial ecosystems and somewhat less with the surface ocean (Fig. 1). The exchanges with terrestrial systems are biotic flows of carbon. Terrestrial photosynthesis withdraws approximately 100 Pg carbon from the atmosphere annually in the production of organic matter. At least half of this photosynthetic production is respired by the plants, themselves, and the rest is consumed by animals or respired by decomposer organisms in the soil. This transfer of plant material to the consumer and decomposer communities is represented by the arrow from vegetation to soil in Figure 1 (50 Pg carbon/yr).

The most striking feature of CO_2 concentrations in the atmosphere is the regular sawtooth pattern (Fig. 4). This pattern repeats itself annually. The cause of the oscillation is the metabolism of temperate and boreal terrestrial ecosystems. Highest concentrations occur at the end of winter, following the season in which respiration has exceeded photosynthesis and caused a net release of CO_2 to the atmosphere. Lowest concentrations occur at the end of summer, following the season in which photosynthesis has exceeded respiration and drawn CO_2 out of the atmosphere. The latitudinal variability in the amplitude of this oscillation is consistent with the spatial and seasonal distribution of terrestrial metabolism: the highest amplitudes (up to about 16 ppmv) are in the Northern Hemisphere with the largest land area; the phase of the amplitude is reversed in the Southern Hemisphere, corresponding to seasonal metabolism there.

The annual rate of photosynthesis in the world's ocean is estimated to be about 50 Pg carbon. Marine photosynthesis does not affect the seasonal oscillation of atmospheric CO_2 greatly, because the chemical properties of CO_2 in seawater are such that little of the carbon involved in either photosynthesis or respiration in the sea is exchanged with the atmosphere. On the time scale of 1000 yr, however, marine photosynthesis is important in controlling the CO_2 concentration of the atmosphere. Marine photosynthesis and the sinking of organic matter out of the surface water (the so-called biological pump) are estimated to keep the concentration of CO_2 in air about 50% of what it would be in their absence.

The gross fluxes of carbon between the surface ocean and the atmosphere each year are largely the result of physicochemical processes. The rate of transfer of CO_2 across the sea surface is estimated to have been 74 Pg carbon annually before human disturbance. The flux is in both directions. The gross flows of carbon between the surface ocean and the intermediate and deep ocean are estimated to have been about 100 Pg per year, 10 Pg through the cycling of dissolved organic carbon (DOC) and the settling of particulate organic carbon (POC), and 90 Pg through physical processes of upwelling and downwelling.

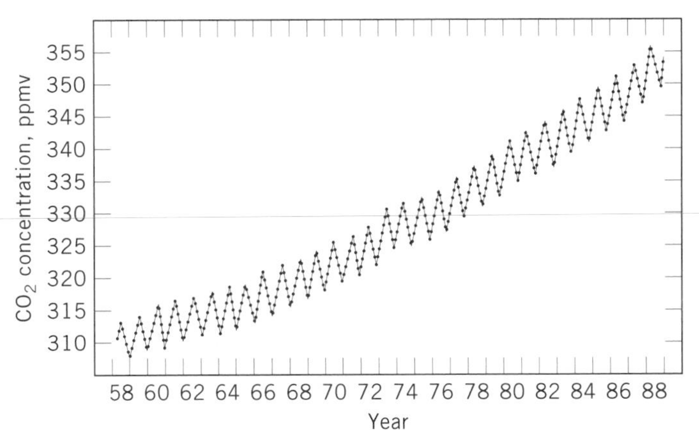

Figure 4. Atmospheric CO_2 concentrations as measured directly in air at Mauna Loa, Hawaii, since 1958. From Ref. 3.

CHANGES INDUCED BY HUMANS

Changes in the Atmosphere

For the 1000 years before 1800, or so, the concentration of CO_2 in the atmosphere remained between 275 and 285 ppmv (6). Since 1700, the concentration of CO_2 has increased from about 280 ppmv to about 353 ppmv (1990), an increase of more than 25% (Fig. 2). This increase is believed to have been due to the widespread replacement of forests with cleared land and emissions of CO_2 from the combustion of fossil fuels.

Numerous measurements of atmospheric CO_2 concentrations were made in the 19th century, but no reliable measure of the rate of increase was possible until after 1957 when the first continuous monitoring of CO_2 began at Mauna Loa, Hawaii, and at the South Pole (3). Measurements of the concentration of CO_2 in air made before 1957 had neither the continuous record nor the precision required to show the increase. In 1958, the average concentration of CO_2 in air at Mauna Loa was about 315 ppmv, and in 1990, it was 353 ppmv (Fig. 4); thus the average rate of increase was about 1 ppmv per year. The rate of increase has itself been increasing; the increase in recent years has been about 1.5 ppmv per year.

The atmosphere is completely mixed in about a year, so any monitoring station free of local contamination will show approximately the same year-to-year increase in CO_2. Today there are more than 30 monitoring stations, worldwide. They generally show the same year-to-year increase but vary with respect to absolute concentration, seasonal variability, and other characteristics useful for investigating the global circulation of carbon.

The concentration of methane (CH_4) in the atmosphere has increased by more than 100% in the past 100 yr, from background levels of less than 0.8 ppmv to a current value of about 1.7 ppmv. The pattern of the increase is similar to that of CO_2, with annual increases of 1.5 ppbv (parts per billion by volume) between 1700 and 1900, accelerating to 17 ppbv per year during the last decade. For the 1000 yr before 1700 there was no apparent trend in the concentration, but over the glacial–interglacial time frame, methane, like CO_2, was positively correlated with the mean global surface temperature.

Methane is released from anaerobic environments, environments with low oxygen concentrations, such as the sediments of marshes, peatlands, rice paddies and the guts of cattle and termites. The major sources of the increased methane are uncertain, but are thought to include the expansion of paddy rice, the increase in the world's population of cattle, and the warming of high latitude wetland ecosystems. Raising the temperature of peatlands stimulates anaerobic and aerobic respiration and may increase the release of both CH_4 and CO_2 during the last century. Atmospheric CH_4 budgets are more difficult to construct than CO_2 budgets because increased concentrations of CH_4 occur not only from increased sources from the earth's surface but from decreased destruction (by hydroxyl radicals) in the atmosphere itself. The increase in atmospheric CH_4 has been more significant for the greenhouse effect than it has for the carbon budget. The doubling of CH_4 since 1700 has contributed less than 1 ppmv, compared with the CO_2 increase of 70 ppmv. On the other hand, CH_4 is 25 times more effective than CO_2 as a greenhouse gas per kilogram released to the atmosphere.

Changes in Land Use

Changes in land use either release carbon to or withdraw it from the atmosphere. Releases occur when forests are replaced with agricultural crops, because the mass of carbon in vegetation per unit area is 20 to 50 times greater in forests than in croplands. The carbon in trees is released to the atmosphere when the wood is oxidized through burning or decay. Some of the organic matter of soils is also oxidized following disturbance and cultivation, and it contributes to the net release of carbon. Carbon is withdrawn from the atmosphere when agricultural land is abandoned and allowed to return to forest. Since about 6000 BP, the replacement of forests and grasslands with cultivated lands has released an estimated 250 Pg carbon to the atmosphere. The release is not known well because the amount of carbon held in vegetation and soils 6000 yr ago is uncertain. Over the last 150 yr the net loss of carbon to the atmosphere has been about 120 Pg as forests have been cleared and soils cultivated (9). This value is more reliable, but probably has an uncertainty of ± 30%. Estimates of the area and carbon content of major terrestrial ecosystems in 1980 are shown in Table 1.

The flux of carbon since 1850 is determined using historical records of land-use change and ecological data pertaining to the changes in vegetation and soil that accompany land-use change. The amount of carbon released to the atmosphere or accumulated on land depends not only on the magnitude and types of changes in land use but on the amounts of carbon held in different ecosystems. For example, the conversion of grasslands to pasture causes a smaller release of carbon to the atmosphere than the conversion of forests to pasture. The net release or accumulation also depends on the time lags introduced by the rates of decay of organic matter, the rates of oxidation of wood products, and the rates of regrowth of forests following harvest or following abandonment of agriculture land. The calculations include the accumulation of carbon in buildings, furniture, and land fills.

The long-term (1850–1990) flux of carbon to the atmosphere from global changes in land use is estimated to have been 120 Pg (Fig. 5). The annual flux of carbon increased from about 0.4 Pg/yr in 1850 to about 1.7 Pg/yr in 1990. The rate of the release has increased. It took 100 yr for the first increase of 0.6 Pg/yr (from 0.4 to 1.0 Pg/yr) and only 35 yr (1950–1985) for the next increase of 0.6 Pg/yr. Until about 1940, the region with the greatest flux was the temperate zone. Since 1950, the tropics have been increasingly important. Current emissions from northern temperate and boreal zones are thought to be close to zero, with releases from decaying wood products approximately balanced by annual accumulations of carbon in regrowing forests (10). Globally, the annual flux of carbon from changes in land use is estimated to have increased during the 1980s from about 1.5 Pg in 1980 to about 1.7 Pg in 1990. Tropical Asia, America, and Africa are estimated to

Table 1. Area, Total Carbon, and Mean Carbon Content of Vegetation and Soils in Major Ecosystems of the Earth in 1980

Ecosystem	Area (10^6 ha)	Carbon in Vegetation (10^{15} g)	Carbon in Soil (10^{15} g)	Mean Carbon Content	
				Vegetation (t/ha)	Soil (t/ha)
Tropical evergreen forest	562	107	62	190	111
Tropical seasonal forest	913	116	84	119	89
Temperate evergreen forest	508	81	68	161	134
Temperate deciduous forest	368	48	49	131	134
Boreal forest	1,168	105	241	90	206
Tropical fallows (shifting cultivation)	227	8	19	36	83
Tropical woodland	776	28	53	39	68
Tropical grass and pasture	1,021	17	49	16	48
Temperate woodland	264	7	18	27	69
Temperate grassland and pasture	1,235	9	233	7	189
Tundra and alpine meadow	800	2	163	3	204
Desert scrub	1,800	5	104	3	58
Rock, ice, and sand	2,400	0.2	4	0.1	2
Temperate cultivated land	751	3	96	4	128
Tropical cultivated land	655	4	35	7	53
Swamp and marsh	200	14	145	68	725
Total	*13,684*	*554*	*1,423*		

have contributed about 0.7, 0.6, and 0.3 Pg/yr, respectively.

About 66% of the total long-term flux, or 80 Pg carbon, was from oxidation of plant material, either burned or decayed. About 33%, or 40 Pg carbon, was from oxidation of soil carbon, largely from cultivation. Relative to the stocks of carbon in 1850, carbon in vegetation was reduced by about 12% over this 130-yr interval, and organic carbon in soil was reduced by about 4% worldwide.

The ratio of soil loss to biomass loss (1:2 in units of carbon) is somewhat surprising given the ratio of total stocks (1450:600, or approximately 2.4:1). The difference is explained by the fact that, on average, only 25 to 30% of the soil carbon is lost with cultivation, and a smaller fraction is usually lost when forests are converted to pastures or to shifting cultivation. In contrast, about 100% of the biomass is oxidized and released as CO_2 to the atmosphere with clearing. Assuming that, on average, 25% of

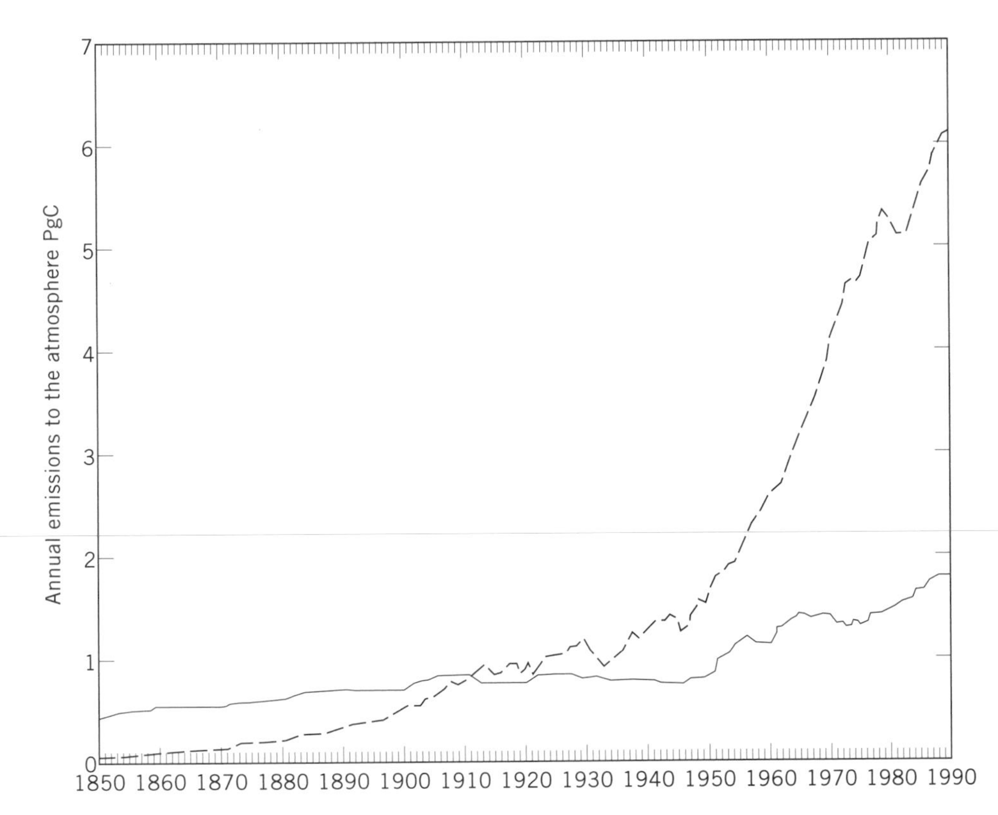

Figure 5. Annual sources of carbon from fossil fuels (dashed line) and land-use change (solid line).

soil carbon is lost, the effective ratio of soil carbon to biomass carbon becomes about 350:600, or approximately 1:2, in agreement with the ratio observed above.

Combustion of Fossil Fuels

The major fossil fuels are coal, oil, and natural gas. The CO_2 released annually from the combustion of these fuels is calculated from records of fuel production published by the United Nations since 1950 (11). Emissions before 1950 were also estimated (12). The rate of combustion has generally increased exponentially; the exceptions are interruptions caused by the two world wars and by the increase in oil prices during the 1970s. Despite the increase in price, the use of fossil fuels and the annual emission of CO_2 to the atmosphere increased yearly to 1980. From 1980 through 1983 the annual emissions decreased each year, but the downward trend was temporary. Starting in 1984, the annual emissions increased again; by 1985 they were similar to 1979. Since 1984, emissions have continued to increase. Annual emission of carbon from combustion of all fossil fuels was 6.0 Pg in 1991. Between 1850 and 1990, the total emission of carbon to the atmosphere from fossil fuels is estimated to have been about 225 Pg.

Total Emissions of Carbon from Human Activities

The net annual fluxes of carbon to the atmosphere from changes in land use (largely deforestation) and from fossil fuels are shown in Fig. 5. The net annual flux of carbon to the atmosphere from changes in land use before 1800 was probably less than 0.5 Pg and probably less than 1 Pg carbon per year until 1950. The combined emissions from both biotic and fossil fuel sources first exceeded 1 Pg at about the start of the 20th century. It was not until the middle of this century that the annual emission of carbon from combustion of fossil fuels exceeded the net biotic flux. Since then the fossil fuel contribution has predominated, although both fluxes have accelerated in recent decades with the intensification of industrial activity and the expansion of agricultural area. For the 1980s, the contemporary net flux averaged about 7.0 Pg carbon per year, 5.4 Pg from fossil fuels and 1.6 Pg from biotic sources (Fig. 6).

Although the industrial flows of carbon now dominate the net biotic flows, the gross natural flows of carbon still greatly exceed the anthropogenic flows. The natural flows of carbon in terrestrial photosynthesis and respiration are about 100 Pg annually, as are the gross flows between oceans and atmosphere (Figs. 1 and 6).

Changes in the Oceans

Direct measurement of changes in the amount of carbon contained in the world's oceans has not been possible for two reasons. First, the oceans are not mixed as rapidly as the atmosphere, so that spatial heterogeneity is large. Second, the background concentration of dissolved carbon in seawater is large relative to the change, so measurement of the change requires accurate methods that have not existed until the past few years. Most estimates of the uptake of terrestrial and fossil fuel CO_2 by the oceans have, therefore, been with models.

The world's oceans contain about 60 times more carbon than either the atmosphere or the world's terrestrial vegetation. Thus, at equilibrium, the ocean might be expected to have absorbed about 60 times more of the released carbon than the atmosphere, or 98% of total emissions. The capacity of the oceans to take up CO_2 decreases, however, as CO_2 is added to the atmosphere–ocean system. In preindustrial times, the distribution of CO_2 between atmosphere and ocean was 1.5 and 98.5%, respectively. If 150 Pg carbon are added to the atmosphere, that distribution

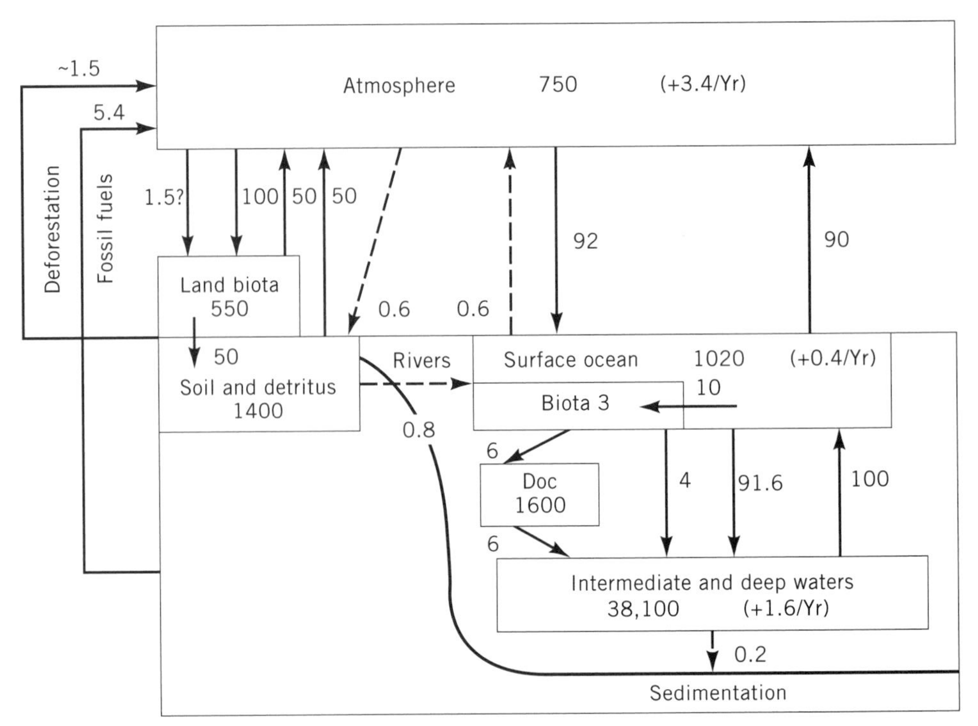

Figure 6. The global carbon cycle for the 1980s. Units are petagrams of carbon and petagrams of carbon per year. The atmosphere and oceans hold more carbon than they did before disturbance (Fig. 1), and the terrestrial biota and soils hold less. From Ref. 2.

becomes 16 and 84% (13). To date, the oceans have absorbed considerably less than 84%. As of 1980, the oceans are thought to have absorbed only about 40% of the emissions (20 to 47%, depending on the model used) (14), so there are clearly other mechanisms slowing the oceanic absorption of CO_2.

Two other mechanisms are the transfer of CO_2 across the air–sea interface and the mixing of water masses within the sea (15). The rate of transfer of CO_2 across the sea surface has been estimated by several different methods. One method is based on the fact that the transfer rate of naturally produced ^{14}C into the oceans should balance the decay of ^{14}C within the oceans. Both the production rate and the inventory of natural ^{14}C in the oceans are known, and the rate of transfer estimated by this method is about 100 Pg/yr (Fig. 1). A similar estimate is obtained by comparing the inventory of bomb ^{14}C in the oceans with the input of bomb ^{14}C from the atmosphere (16).

A third method for estimating the CO_2 gas transfer coefficient is based on measurement of radon gas in the surface ocean. Radon gas is generated by the decay of radon 226. The concentration of the parent radon 226 and its half-life allow calculation of the expected radon gas concentration in the surface water. The observed concentration is about 70% of expected, so 30% of the radon must be transferred to the atmosphere during its mean lifetime of 6 days. Correcting for differences in the diffusivity of radon and CO_2 allows an estimation of the transfer rate for CO_2. The transfer rates given by the ^{14}C methods and the radon method agree within about 10%. The rate of transfer of CO_2 across the air–sea interface is believed to have reduced the oceanic absorption of CO_2 by about 10% (15).

The most important process in slowing the oceanic uptake of CO_2 is the rate of vertical mixing. The surface layer of the ocean is approximately in equilibrium with atmospheric CO_2, but the bottleneck in oceanic uptake is in transporting surface waters to the intermediate and deeper layers.

The earliest attempts to model the uptake of carbon by oceans recognized that the ocean was not homogeneous but made up of at least two layers: a surface layer in contact with the atmosphere and a deep layer that can accumulate carbon only in exchange with the surface layer (17). Later versions of a one-dimensional ocean model consisted of a shallow surface ocean and a deep ocean composed of many boxes, stacked vertically, that exchanged carbon with those boxes immediately above or below (18). This type of model is commonly referred to as the box diffusion model.

More realistic models of the ocean are three-dimensional and include the mixing of tracers among water masses with similar densities (19). In addition to the surface and deep layers, these models include the thermocline, that portion of the ocean in which the temperature gradient is steepest. It is largely the steepness of this gradient in density that creates the bottleneck for deep-ocean absorption of CO_2.

All of the different ocean models are calibrated with tracers of some kind, either natural ^{14}C, bomb-produced ^{14}C, bomb-produced tritium, or other tracers. The profiles of these tracers were obtained during two extensive oceanographic surveys: geochemical ocean sections (GEOSECS), carried out between 1972 and 1978, and transient tracers in the ocean (TTO), carried out in the early 1980s. Both surveys measured profiles of carbon, radioisotopes, and other tracers along transects in the Atlantic and Pacific oceans. Current estimates of oceanic uptake of carbon based on all of these models range between 1.5 and 2.5 Pg/yr for the 1980s (2).

THE IMBALANCE

The global carbon cycle can be summarized by a simplified equation showing the atmospheric increase to be the sum of sources (combustion of fossil fuels and land-use change) and sinks (oceanic uptake) of carbon:

$$\text{Atmospheric increase} = \text{fossil source} + \text{land-use source} - \text{oceanic sink} - \text{residual sink}$$

For 1980 to 1989 the increase (in petagrams carbon per year) can be calculated as

$$3.2 \pm 0.2 = 5.4 \pm 0.5 + 1.6 \pm 1.0 - 2.0 \pm 0.5 - 1.8 \pm 1.2$$

and for 1850 to 1990:

$$140 = 225 + 120 - 100 - 105$$

As the equations show, an additional sink is needed to balance the equations for the last decade as well as for the longer term (Fig. 7). This unidentified sink (sometimes called the "missing carbon" sink) has received considerable attention over the last decade. The unidentified carbon sink makes projections of future atmospheric concentrations difficult. Should the sink be assumed to continue in the future? Will it increase in proportion to atmospheric concentrations of CO_2? Will it decrease? Will it saturate? When? What and where is the sink?

Two or three independent lines of geophysical evidence suggest that the unidentified sink is in terrestrial ecosystems. On the basis of a model of atmospheric transport, it was determined that the gradient in CO_2 concentrations between the Northern and Southern hemispheres was less steep than would be expected on the basis of the distribution of fossil fuel sources of CO_2 (20). Most of the CO_2 released from combustion of fossil fuels is released in the midlatitudes of the Northern Hemisphere and should cause a bulge in the concentrations at those latitudes. The bulge is there, but not as large as expected, indicating that CO_2 must be removed from the Northern Hemisphere before crossing the equator. Measurements of the partial pressure of CO_2 (P_{CO_2}) in surface waters suggest that the difference between atmospheric and surface ocean concentrations is not great enough to force CO_2 into the oceans. Thus it was concluded that the unidentified sink was on land in the northern midlatitudes (20).

Qualitatively, the inferred terrestrial sink is similar to that obtained with models of oceanic uptake, described earlier. However, the P_{CO_2} differences described above led the researchers to infer an oceanic uptake of carbon of only 0.3 to 0.8 Pg/yr, much less than the rate found by the more

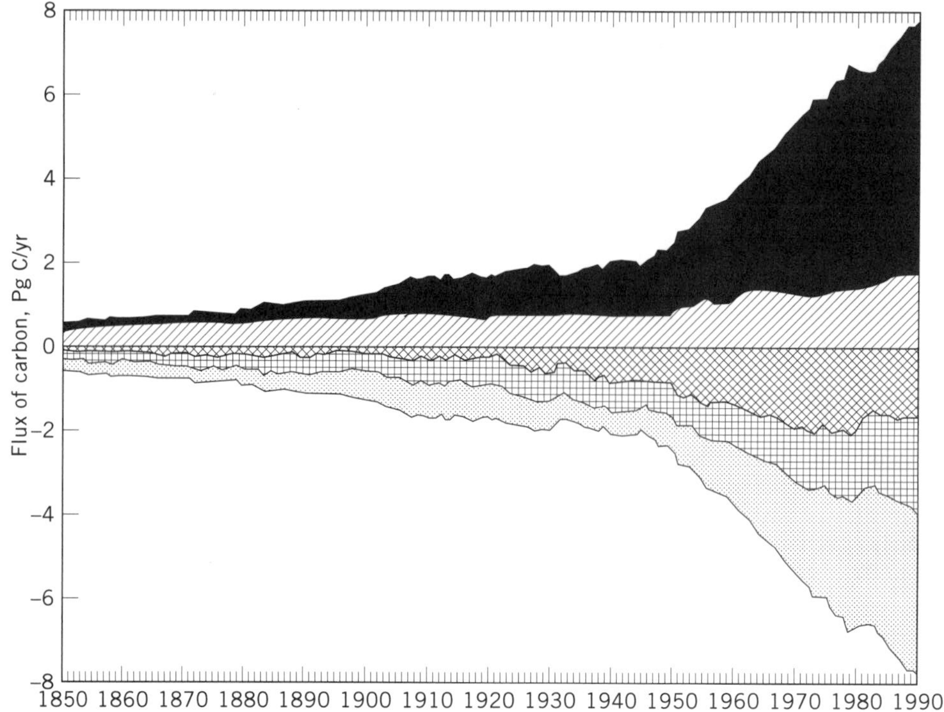

Figure 7. Annual sources and sinks of carbon in the global carbon cycle since 1850. In 1990, eg, the total source of 7.7 Pg (6.0 Pg from fossil fuels and 1.7 Pg from land-use change) must be equivalent to the total sink.
Emissions from fossil fuel
Net release from land-use change
Unidentified sink
Oceanic uptake
Atmospheric accumulation

traditional ocean models (2.0 ± 0.5 Pg/yr). Could the oceanic models be that wrong?

Three factors have been proposed to reconcile the difference: carbon in rivers, the skin temperature of the ocean, and carbon monoxide (21). The earlier estimate (20) concerned the air–sea exchange but not necessarily the net uptake of carbon by the oceans. As part of the natural carbon cycle, the oceans are estimated to release about 0.6 Pg/yr to the atmosphere to balance the input from rivers. Rivers discharge about 0.8 Pg/yr into the ocean, of which about 0.2 Pg of carbon is buried in ocean sediment (Figs. 1 and 6). Thus the background or steady-state flux is not 0 but an exchange of 0.6 Pg/yr of carbon from ocean to atmosphere.

The air–sea exchange is sensitive to temperature at the surface, and measurements of temperature (20) were the traditional bulk temperature measurements of waters near the surface. Because of evaporation, however, the skin temperature at the very surface of the ocean may be 0.1° to 0.2°C cooler than the surface waters. The correction is equivalent to about 0.4 Pg/yr of carbon.

Finally, there is a correction required to account for the fact that some of the carbon emitted from fossil fuel combustion is released as CO rather than as CO_2. The CO is eventually oxidized to CO_2 in the atmosphere, but not until some of it has been redistributed in the atmosphere. The net effect of the CO emissions and redistribution is to reduce the bulge in the north–south gradient of CO_2. Correction for the effect raises the estimate of oceanic uptake of carbon by about 0.2 Pg/yr.

Adding all three of these corrections to the earlier estimate (20) gives an oceanic uptake of about 1.8 Pg/yr, within the range determined independently by ocean models. If all of these corrections applied only to the northern oceans, the need for a large terrestrial sink in midlatitudes would vanish. As it is, the northern sink is only partially reduced. Other analyses, however, show the midlatitude terrestrial sink to be much less. The sink was inferred to be terrestrial on the basis of the difference in CO_2 concentrations between the surface ocean and air (20). The gradient in concentrations was not steep enough to make the oceans a significant sink. If, instead of using the air–sea gradient in CO_2 concentrations, one uses the observed ^{13}C:^{12}C ratio in atmospheric CO_2, the data support an oceanic sink rather than a terrestrial one (3). ^{13}C is a heavier isotope of carbon than the more common ^{12}C. The heavier isotope is discriminated against in photosynthesis, so plants and soils (and fossil fuels) are depleted in ^{13}C. Trees have a lighter isotopic ratio (−22 to −27 parts per thousand) than does air (−7 ppt); the ratios are expressed in relation to a standard. If there were a large terrestrial sink at midlatitudes, preferentially removing ^{12}C from the atmosphere, the ^{13}C:^{12}C ratios of atmospheric CO_2 would be enriched in ^{13}C. But the observed isotopic ratios of atmospheric CO_2 from northern midlatitudes do not show such an enrichment. On the contrary, they show a small depletion in ^{13}C, suggesting that terrestrial ecosystems at that latitude are a source rather than a sink.

The magnitude of a terrestrial midlatitude sink is in question. P_{CO_2} data from the oceans suggest a large sink; ^{13}C:^{12}C ratios in air do not. Nevertheless, these geophysical analyses, based on atmospheric and oceanic data and models as well as other analyses based on short-term records of ^{13}C:^{12}C in the ocean all require a terrestrial sink of approximately the same magnitude as the terrestrial source from tropical deforestation. The location of that sink, however, is unclear. Another inversion approach, similar to the one discussed earlier (20), found the same northern midlatitude sink but also found a small net source in the tropics. The tropical source is of the magnitude expected from the outgassing of CO_2 from the tropical oceans, indicating that the large source of carbon from tropical deforestation must be balanced by an equally large accumulation somewhere in the terrestrial ecosystems of the tropics.

Nowhere, either in the tropics or in northern midlatitude forests, has an accumulation of carbon been observed in trees or soil. Some scientists argue that the accumulation of 1 to 2 Pg/yr is so small relative to the background levels of carbon in these ecosystems and relative to the natural exchanges of carbon with the atmosphere, that it could not be detected. But is this the case? Recent analyses of forest growth in northern regions, indeed, show an accumulation of carbon, but the rate of accumulation is that expected from past histories of logging, ie, recovering forests accumulate carbon. What does this imply for the missing carbon? If additional carbon were accumulating in the northern temperate zone, could it be measured? The global carbon imbalance is approximately 1.5 Pg/yr. If all of that carbon were accumulating in the trees of the northern temperate zone (area of temperate zone forests is about 600×10^6 ha), the accumulation would be approximately 2.5 Mg/(ha · yr). In comparison, the observed accumulation of carbon in European forests was 0.4 to 0.6 Mg/(ha · yr) (22). Thus the accumulation of missing carbon would be approximately five times higher than observed. It would be difficult to miss in the data on growth rates. Including boreal forests in the calculation lowers the accumulation rate of carbon to 1.0 Mg/(ha · yr), but it is still high relative to observed growth rates.

If the unidentified carbon sink were accumulating throughout the temperate zone, not just in forests but in woodlands, grasslands, and cultivated lands, the average accumulation rate of carbon would be about 0.5 Mg/(ha · yr), about the rate observed in European forests. This rate of accumulation in the vegetation of grasslands and cultivated lands is high relative to the standing stocks of carbon there and would be obvious. Finally, if the missing carbon is accumulating in trees throughout the world, the rate, again, would average about 0.5 Mg/(ha · yr). However, this is the rate observed for regrowing European forests, which have a long history of intensive use. That an additional accumulation of this magnitude, representing the missing sink, could be in the forests of the world, many of which have not been logged and, hence, are not regrowing (eg, large regions of Canada, central and eastern Russia, and Brazil), seems unlikely. Regrowing forests are young; mortality is low, and as a result, little carbon is lost. Even if undisturbed forests around the world were growing larger in response to elevated concentrations of CO_2 or to increased deposition of nitrogen, they would be unlikely to be accumulating carbon at rates that approximate rates of regrowth immediately after logging. By these arguments, it seems unlikely that carbon is accumulating in vegetation anywhere at the rate required to balance the global carbon budget (23). The same is true for soils. Observed rates of carbon accumulation in soils of undisturbed ecosystems are too low to account for the unidentified sink (24).

In summary, most analyses of the contemporary carbon cycle based on geophysical models point to terrestrial ecosystems as the unidentified sink. Direct measurements of tree growth do not show such a sink, however. Uncertainties in these models leave the location of the sink ambiguous. Tropical as well as northern temperate zone lands are candidates, and the ocean itself cannot be ruled out as accounting for some portion of the unidentified sink. If the sink is spread throughout the terrestrial ecosystems of the earth, and in both vegetation and soil, it may be undetectable.

Further insights as to the mechanism or location of the unidentified sink may be obtained from a consideration of its change through time. The temporal pattern of the sink has been determined by comparing geophysical estimates of a terrestrial flux of carbon with estimates based on changes in land use (9). Models of oceanic uptake are traditionally used to estimate the concentration of CO_2 in the atmosphere that will result from different trends in emissions. The models can also be run in an inverse mode. In this mode, past concentrations of CO_2 are used as input to the models, and the models calculate what the sources and sinks of carbon must have been to yield the observed concentrations. In this way, the oceanic models have been used to infer the terrestrial flux of carbon over the last 150 yr or so. The historic record of CO_2 concentrations is available from ice cores. Bubbles of air trapped in glacial ice contain samples of air that was present at the time of ice formation. The ice is crushed under carefully controlled conditions, and the air is analyzed for CO_2. This record of CO_2 concentrations is then used, together with the direct atmospheric measurements since 1958, by models of oceanic uptake to calculate the annual sources of carbon necessary to reproduce the CO_2 record. Subtracting the sources of CO_2 from combustion of fossil fuels since about 1850 leaves a residual pattern of sources and sinks that is generally assumed to represent the net exchanges of carbon with terrestrial ecosystems. The assumption is consistent with the ratio of ^{13}C to ^{12}C in the air over this period.

The cumulative flux of this residual for the period 1850–1990 is calculated to have been a release to the atmosphere of between 25 and 50 Pg of carbon according to these inverse calculations (3,19,25). The temporal patterns of the estimates are almost identical. According to the most recent of these analyses, the residual flux (presumably terrestrial) was a net release of carbon of about 0.4 Pg/yr in 1800, rising gradually to 0.6 Pg/yr by 1900, remaining at 0.6 Pg/yr until about 1920, and then falling to 0 by the late 1930s (Fig. 8). The flux continued to fall, indicating a net accumulation of carbon in terrestrial ecosystems after 1930, reaching a maximum sink of 0.8 Pg/yr in the 1970s, and then abruptly returning to 0 by the early 1980s. It has remained near 0 since ca. 1982.

If this residual flux is from terrestrial ecosystems, it bears little resemblance to that calculated from changes in land use, which was about 120 Pg of carbon over the period since 1850 and increased almost steadily (Fig. 5). The disagreement, however, may not be the result of errors in one or the other of the approaches. The difference between these two estimates may represent a flux of carbon between terrestrial ecosystems and the atmosphere that is not related to changes in land use but to changes in the global environment.

Analyses of flux based on changes in land use assume that ecosystems not directly affected by land-use change are in a steady state with respect to carbon. This assumption is probably not valid. The amount of carbon stored in vegetation and soils is changed not only by deliberate human activity but also from inadvertent changes in climate, CO_2 concentrations, nutrient deposition, and pollution. Analyses of the flux of carbon due to land-use changes generally take into consideration deforestation (with the land turned into crops or pasture or degraded), reforestation (either purposefully or as a result of the abandonment of agriculture), and changes in the biomass within the forests (due to logging and regrowth, shifting cultivation and degration). Such studies tend to omit from consideration unmanaged, undisturbed areas (because the carbon cycle is assumed to be in a steady state); forest decline or enhancement; eutrophication (including CO_2 fertilization); frequency of fires (including fire surpression); climatic change, and desertification.

It is useful to distinguish between two causes or types of change: deliberate and inadvertent. Deliberate changes result directly from management, eg, cultivation. Inadvertent changes occur through changes in the metabolism of ecosystems, more specifically through a change in the ratio of the rates of primary production and respiration (respiration includes autotrophic and heterotrophic respiration, or plant and microbial respiration). Clearly, management practices affect metabolism, and the distinction between deliberate and inadvertent effects is blurred. Nevertheless, the distinction is a useful one because changes in carbon stocks that result from deliberate human activity are more easily documented than changes resulting from metabolic changes alone. Deliberate changes usually involve a well-defined area, eg, the area in agriculture or the area reforested. Furthermore, the changes in the density of carbon (t/ha) associated with land-use change are generally large: forests hold 20 to 50 times more carbon in their vegetation than the ecosystems that replace them following deforestation. In contrast to these large and well-defined changes associated with management, changes that occur as a result of metabolic changes are generally subtle. They occur slowly over poorly defined areas. They are difficult to measure against the large background levels of carbon in vegetation and soils and against the natural variability of diurnal, seasonal, and annual metabolic rates.

Assuming that the flux of carbon obtained from inversion of ocean models is a terrestrial flux, it includes both the flux from land-use change and a flux caused by other changes in terrestrial ecosystems. The flux of carbon from land-use change, on the other hand, includes only those lands directly and deliberately managed. Thus the difference between the two estimates of flux may define the flux of carbon to or from terrestrial ecosystems not directly modified by humans. This flux is the unidentified sink.

The difference is shown in Figure 8 (dashed line). The flux has always been negative and is interpreted to mean that some terrestrial ecosystems were accumulating carbon independent of land-use change. Positive differences, if they occurred, would indicate that some terrestrial ecosystems were releasing carbon to the atmosphere in addition to that released from changes in land use.

The long-term difference in these two estimates is a net sink of 70–95 Pg of carbon (a net release of 25 to 50 Pg minus the release of izo Pg from land use). As with the shorter-term sink, if this sink has been in mid- and high latitude forests, the increase in storage should be observed. Even if 66% of the sink has been in trees, and 33% in soil, the long-term sink in trees is equivalent to about a 20% increase in biomass or ring increment. Studies of tree rings do not consistently find an increased growth. It is important to recall here that the unidentified sink is over and above the sink attributable to regrowth of forests following abandonment of agricultural land or logging, ie, to changes in land use.

FEEDBACK BETWEEN THE CARBON CYCLE AND GLOBAL CLIMATE

One of the questions important to both climate modelers and ecologists is how the present-day dynamics of the carbon cycle will change if climate changes. If the earth

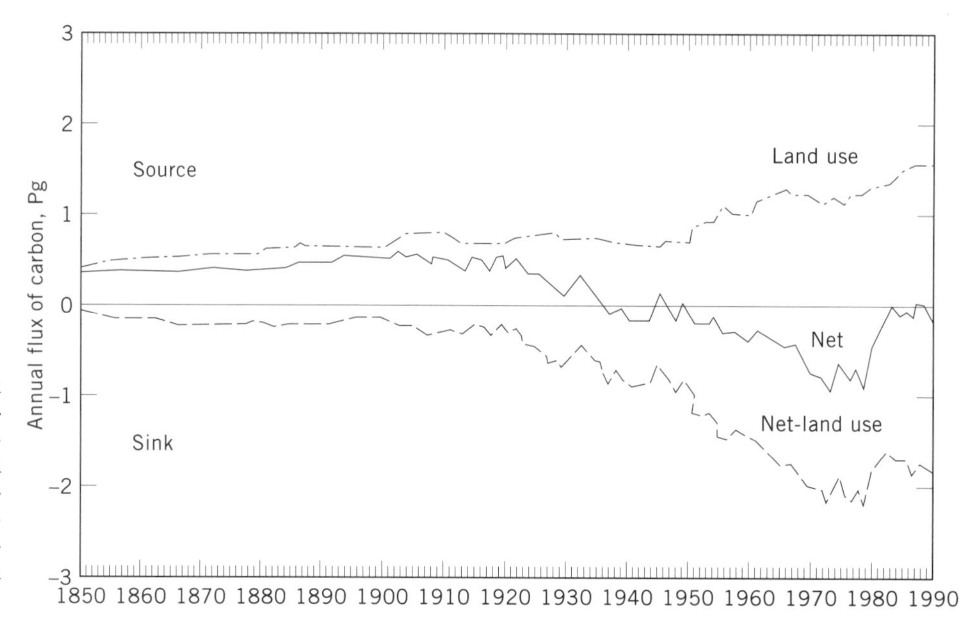

Figure 8. The residual flux of carbon determined by inverse methods (assumed to represent the total net terrestrial flux), the annual net flux of carbon from changes in land use, and the difference between them (assumed to represent changes in terrestrial ecosystems caused by factors other than changes in land use). From Ref. 19.

warms, will the oceans take up carbon more rapidly or more slowly? Will terrestrial ecosystems release additional carbon with an increase in temperature, or will rising concentrations of CO_2 enhance plant productivity and store more carbon on land? How will changes in precipitation affect this terrestrial storage? Answers to these questions about climatic change are probably related to the unidentified sink for carbon. The processes responsible for the unidentified sink, if they could be identified, might shed light on how the carbon cycle will respond to a warmer earth with higher CO_2 concentrations. Feedback between the global carbon and global climate systems is defined as a process that either enhances (positive feedback) or reduces (negative feedback) the warming. Feedback (26) and biotic feedback (27) have been reviewed.

Oceans

In general, the oceans tend to buffer changes in atmospheric CO_2 and thus serve as a negative feedback for any change. With respect to feedback between temperature and carbon, however, it is useful to consider physical, chemical, and biological processes separately. Physically, as the oceans warm, they hold less carbon because of the lower solubility of gases at higher temperatures. Furthermore, a warming may lead to increased stratification of the water column and, thereby, reduced circulation and reduced uptake of carbon. The greatest warming is expected to occur in higher latitudes; the least warming, at the equator. Such a pattern of warming would reduce the north–south thermal gradient and might, thereby, reduce the intensity of circulation. In both of these physical respects, the oceans will serve as a positive feedback.

Chemically, the ocean's buffering capacity tends to keep atmospheric concentrations of CO_2 from changing. As the oceans take up more carbon, they become more acidic and thus have a diminished capacity to absorb additional carbon. The net effect of the buffering is still a negative feedback, however. Furthermore, as additional carbon is taken up by the oceans, the increasing acidity of the waters will erode calcium carbonate sediments, thereby increasing alkalinity and decreasing atmospheric CO_2, again a negative feedback. This process is important over hundreds and thousands of years but is too slow to be significant over the next 100 yr.

There is another potential feedback between the oceanic carbon cycle and global warming that is more difficult to assess: the role of the marine biota. The *biological pump* refers to photosynthetic production of organic matter in the surface ocean, its sinking or export to depth, and its subsequent decomposition or decay, which releases nutrients and CO_2 back into solution at depth. In preindustrial times, the biological pump was responsible for maintaining the concentration of CO_2 in the atmosphere at about half of what it would have been in the absence of the biological pump. Might the effectiveness of the biological pump change with a warming? First, it is unlikely that marine phytoplankton will respond to an increased concentration of CO_2 in seawater because the supply from the other chemical forms of inorganic carbon (bicarbonate and carbonate ions) is almost inexhaustible. Not carbon but nutrients and light are the major factors the limit marine productivity. So the question becomes, How might a global warming affect the concentrations of nutrients in surface waters?

According to one argument, a global warming will reduce the quantity of nutrients delivered to the surface through a reduction in upwelling. Upwelling is related to the intensity of oceanic mixing, and that intensity might be diminished if the north–south gradient in temperature were reduced through global warming. But upwelling waters bring high CO_2 concentrations to the surface as well as nutrients. The two are related: they both result from decomposition of organic matter produced at the surface. Thus it is not clear that a reduction in the biological pump will result in higher atmospheric CO_2 concentrations. Reduced upwelling will bring fewer nutrients to the surface and thus reduce productivity, but it will also bring less high CO_2 water to the surface and thus reduce the release of CO_2 to the atmosphere.

Because of this link between nutrients and carbon, there are only two ways that the biological pump might change the net removal of carbon from the euphotic zone, according to this argument. One way is to change the total amount of nutrient in the oceans; the other way is to allow the carbon:nutrient ratios to change. There is little evidence that carbon–nutrient ratios are flexible, but a change in sea level might increase the nutrient inventory of the ocean through inundation of nutrient-rich soils. If a changing climate increased the intensity of winds or increased aridity, it might also increase the effectiveness of the biological pump through increased delivery of terrestrially eroded iron to the ocean. Iron is thought to limit production in many parts of today's ocean. Under this circumstance, the feedback would be negative.

The counterargument is that reduced upwelling will reduce the effectiveness of the biological pump and will reduce the rate of oceanic uptake of carbon, at least in the short term. The argument distinguishes between the preindustrial fluxes of carbon and the perturbational fluxes, or those due to human activity. Nutrients and CO_2 in upwelled water were presumably in balance preindustrially. Today, however, the atmosphere and deep ocean are not in equilibrium. Anything that enhances circulation or mixing thus enhances the equilibration of atmosphere and ocean. Although upwelled waters contain CO_2 at absolutely higher concentrations than the current atmospheric CO_2, the concentration in upwelled waters is low relative to an equilibrated ocean. Hence, decreased upwelling reduces the biological pump and thereby reduces oceanic uptake of carbon. The mechanism is largely a physical one (described above). In the oceans, positive feedback seems more likely to prevail with global warming.

Terrestrial Ecosystems

CO_2 Fertilization (Negative Feedback). There is also considerable uncertainty surrounding the role of terrestrial ecosystems in carbon–climate feedbacks. The two factors receiving most attention are the effects of temperature and CO_2 on carbon storage. The increase in the concentration of CO_2 over the last centuries is thought by many researchers to have increased the storage of carbon on land. Based on short-term experiments in greenhouses, it is clear that many crops, annual plants, and tree seedlings grow more rapidly under elevated levels of CO_2. It is less clear that such an increased growth will occur in natural ecosystems, that the growth response found in annual crops will apply to natural systems, that it will apply to periods longer than a growing season, or that any enhanced growth that occurs will lead to an increase in the amount of carbon held in ecosystems.

There have been two moderately long-term experiments in which natural ecosystems were exposed to elevated concentrations of CO_2, and the results were not consistent (28). The net uptake of CO_2 by a tundra community showed an increased storage of carbon, but the effect lasted only 3 to 5 yr, or until some other factor became limiting to the growth of plants and the storage of carbon in the system. The results suggested a limited response to elevated levels of CO_2. In a continuing study in a tidal marsh, the increased storage of carbon has continued for 6 yr. Rates of photosynthesis increased and rates of decomposition seemed to have decreased. In the 6th yr, the experimental systems still showed an annual accumulation of carbon greater than the control and as large a response as in the first 5 yr. Whether an increased storage will or has already occurred in forests is unclear. The expense and logistical problems of enclosing a forest in chambers have so far prevented such an experiment.

Tree rings might provide an answer. Do trees show an increased rate of growth over the last decades? If the answer were yes, then perhaps the unidentified sink would have been identified. The next question becomes whether the increase in growth can be attributed to elevated CO_2, to changes in climate, or to other factors. Studies of tree rings have been equivocal so far, with about as many showing decreasing rates as showing increasing rates of growth. An exception appears in trees selected for temperature sensitivity. A sytematic investigation of trees from high latitudes and high elevations in both hemispheres indicated increased rates of growth, apparently related to the warming over the last century (29). The authors of that study caution, however, that the trees selected were not necessarily representative of the earth's forested regions. Indeed, studies in midlatitude regions indicate that rates of forest growth there are consistent with the logging histories in those regions (23); there is no indication of an enhanced growth. Nevertheless, a systematic survey relating global changes in tree rings to carbon storage has not been carried out.

The increase in amplitude of the winter–summer oscillation of CO_2 concentrations is often interpreted to indicate a trend of increased photosynthesis. But increased photosynthesis is not the same as increased storage of carbon. An increase in photosynthesis might be balanced by an equivalent increase in respiration. Furthermore, the increasing amplitude is also consistent with an increased winter respiration associated with warmer winters (30). If the cause of the increase in the seasonal amplitude of CO_2 concentrations has been increased respiration in winter, it would indicate a decreased storage of carbon on land rather than an increased storage. The increasing amplitude is interesting, but its significance is not yet clear.

The difference (evaluated above) between the total net terrestrial flux of carbon (obtained from inverse models of oceanic uptake) and the flux from changes in land use implies an increased storage of carbon over the last 70 yr (Fig. 8). The difference exhibits three features: (1) a period before 1920 that showed a small terrestrial sink (not different from 0) with little variation, (2) a period between ca. 1920 and 1975 when some of the world's terrestrial ecosystems were apparently accumulating carbon at an increasing rate, and (3) a period since the mid-1970s in which the rate of accumulation has decreased. Over the period 1850 to 1990, the generally increasing sink was proportional to increasing industrial activity. This suggests that CO_2 fertilization or increased nitrogen deposition may have been responsible for an accumulation of carbon in undisturbed ecosystems. Over the shorter period since 1940, however, the difference is poorly correlated with increasing CO_2 concentrations. Temperature seems to explain more of the variation and suggests a positive feedback.

Temperature. Temperature is likely to cause a net release of carbon to the atmosphere in addition to that released from changes in land use. Warmer temperatures are known to increase the rates of respiration and decay, which convert organic matter to CO_2 and, under anaerobic conditions, to CH_4. Warmer temperatures may also increase rates of photosynthesis and growth, especially in cold regions, but respiration is generally more sensitive to temperature than photosynthesis is. Furthermore, the amount of carbon held in the world's soils is about three times greater than the amount held in plants and two times greater than that in the atmosphere. Soil carbon is thus a large reservoir of terrestrial carbon that might be released to the atmosphere. Only a fraction of the organic carbon in soils is labile, however. Cultivation, eg, causes a loss of carbon from soils, but typically about 25% of the amount held in the top meter of soil. The larger fraction of soil carbon is more refractory and less likely to be mobilized with an increase in temperature. Although a warming can be expected to enhance decomposition of organic matter in soil, decomposition of organic matter releases nutrients as well as CO_2. The release of additional nutrients may, in turn, allow productivity to increase and may lead to an increased storage of carbon in wood. Based solely on the difference in carbon:nitrogen ratios of soil (about 8) and wood (about 200), carbon released from soil might be more than compensated for by carbon accumulated in wood. The calculation assumes that the nitrogen mineralized will be incorporated into an enhanced production of wood. If little of the nitrogen is lost, the net effect of warming may be to store carbon.

The evidence for positive or negative feedback between carbon and temperature seems to show that positive feedback dominates at most time scales. Over tens of thousands of years, the CO_2 concentration of the atmosphere is related to mean global temperature. During the last glacial period, CO_2 concentrations were about 180 ppmv; during the interglacial, they were 280 ppmv. The temporal resolution of the Vostok core is not sufficient to determine whether the warming preceded the change in CO_2 or vice versa, but during cooling periods, changes in temperature preceded changes in CO_2. Whatever, the initiating factors, the correlation is interpreted as evidence for a positive feedback.

Over the last 30 yr, there is also a positive correlation between CO_2 concentration and mean global temperature (3). The change in CO_2 again followed changes in temperature, in this case with a lag of a few months. Based on atmospheric ^{13}C:^{12}C ratios in CO_2, it was suggested that the source of the carbon related to warming is terrestrial (3).

The unidentified flux of carbon shown in Figure 8 is also correlated with temperature over the last 50 yr but not in the preceding 100 yr (9). Over the shorter period since 1940, the difference is correlated more with temperature than with atmospheric concentrations of CO_2. Larger sinks were associated with cooler years; reduced sinks were associated with warmer years. The fact that a similar relationship was not apparent before 1940 suggests either that the relationship is accidental or that different mechanisms were dominating in the two periods. Perhaps before 1940, ecosystems were largely rebounding from the little Ice Age, while, after 1940, they were responding to a more rapid warming.

Although most evidence seems to support positive feedback, the magnitudes of the various mechanisms cannot be predicted at present. The effects may cancel each other or alternate in time. A warmer earth might support a larger area of forests with a greater storage of carbon. However, unless deliberate steps are taken to reduce emissions of greenhouse gases, the immediate prospect is not a new climate but a continuously changing one. The ability of forests to keep up with such changes through migration will be difficult. It is more likely that a continuously changing climate will result in increased areas affected by forest dieback or fires or both, releasing additional amounts of carbon to the atmosphere.

Other Factors. Finally, there are a number of factors that do not strictly feed back to the global climate system but that can be expected to affect the ability of other feedback systems to operate. If carbon has been accumulating in undisturbed forests, balancing the global carbon equation over the last decades, there are reasons to expect that continuation of this accumulation is limited. The area in forests has decreased by about 15%, globally, over the last 140 yr, and rates of deforestation in the tropics are higher now than ever before. If the world's population doubles in 30 yr, the need for additional agricultural land will only increase the pressure on the remaining forests. At the same time, decreasing concentrations of stratospheric ozone may increase the amount of ultraviolet radiation reaching the earth's surface. The effect is likely to reduce both marine and terrestrial productivity and, as a result, carbon storage. The effects of acid rain, which may have increased the productivity of some European forests through increased deposition of nitrate, may eventually saturate the system and decrease productivity. Most of these effects will not only enhance a global warming but reduce the capacity of the earth to support the human enterprise. The future need not be bleak, however. The steps for stabilizing concentrations of CO_2 and other greenhouse gases are clear. They include large reductions in emissions of greenhouse gases, ie, large reductions in the use of fossil fuels and in rates of deforestation. Such reductions would be good for a number of reasons besides slowing the rate of global warming.

BIBLIOGRAPHY

1. J. T. Houghton, G. J. Jenkins, and J. J. Ephraums, eds., *Climatic Change. The IPCC Scientific Assessment,* Cambridge University Press, Cambridge, UK, 1990.
2. U. Siegenthaler and J. L. Sarmiento, *Nature* **365,** 119–125 (1993).
3. C. D. Keeling, R. B. Bacastow, A. F. Carter, S. C. Piper, T. P. Whorf, M. Heimann, W. G. Mook, and H. Roeloffzen in D. H. Peterson, ed., *Aspects of Climate Variability in the Pacific and the Western Americas, Geophysical Monograph* **55,** American Geophysical Union, Washington, D.C., 1989, pp. 165–236.
4. A. Neftel, E. Moor, H. Oeschger, and B. Stauffer, *Nature* **315,** 45–47 (1985).
5. H. Friedli, H. Lotscher, H. Oeschger, U. Siegenthaler, and B. Stauffer, *Nature* **324,** 237–238 (1986).
6. U. Siegenthaler, H. Friedli, H. Loetscher, E. Moor, A. Neftel, H. Oeschger, and B. Stauffer, *Ann. Glaciol.* **10,** 151–156 (1988).
7. J. M. Barnola, D. Raynaud, Y. S. Korotkevich, and C. Lorius, *Nature* **329,** 408–414 (1987).
8. S. Arrhenius, *Philos. Mag.* **41,** 237 (1896).
9. R. A. Houghton in G. M. Woodwell and F. T. Mackenzie, eds., *Biotic Feedbacks in the Global Climate System: Will the Warming Feed the Warming?,* Oxford University Press, Oxford, UK, in press.
10. R. A. Houghton, R. D. Boone, J. R. Fruci, J. E. Hobbie, J. M. Melillo, C. A. Palm, B. J. Peterson, G. R. Shaver, G. M. Woodwell, B. Moore, D. L. Skole and N. Myers, *Tellus* **39B,** 122–139 (1987).
11. G. Marland and R. M. Rotty, *Tellus* **36B,** 232–261 (1984).
12. C. D. Keeling, *Tellus* **25,** 174–198 (1973).
13. J. L. Sarmiento, *Chem. Eng. News* **71,** 30–43 (1993).
14. B. Bolin in B. Bolin, B. R. Doos, J. Jager, and R. A. Warrick, eds., *The Greenhouse Effect, Climatic Change, and Ecosystems, SCOPE* **29,** John Wiley & Sons, Inc., New York, 1986, pp. 93–155.
15. W. S. Broecker, T. Takahashi, H. H. Simpson, and T.-H. Peng, *Science* **206,** 409–418 (1979).
16. W. S. Broecker, T.-H. Peng, G. Ostlund, and M. Stuiver, *J. Geophys. Res.* **90,** 6953–6970 (1988).
17. R. Bacastow and C. D. Keeling in G. M. Woodwell and E. V. Pecan, eds., *Carbon and the Biosphere, U.S. Atomic Energy Commission Symposium Series* **30,** National Technical Information Service, Springfield, Va., 1973, pp. 86–135.
18. H. Oeschger, U. Siegenthaler, U. Schotterer, and A. Gugelmann, *Tellus* **27,** 168–192 (1975).
19. J. L. Sarmiento, J. C. Orr, and U. Siegenthaler, *J. Geophys. Res.* **97,** 3621–3645 (1992).
20. P. P. Tans, I. Y. Fung, and T. Takahashi, *Science* **247,** 1431–1438 (1990).
21. J. L. Sarmiento and E. T. Sundquist, *Nature* **356,** 589–593 (1992).
22. P. E. Kauppi, K. Mielikainen, and K. Kuusela, *Science* **256,** 70–74 (1992).

23. R. A. Houghton, *Global Biogeochem. Cycles* **7,** 611–617 (1993).
24. W. H. Schlesinger, *Nature* **348,** 232–234 (1990).
25. U. Siegenthaler and H. Oeschger, *Tellus* **39B,** 140–154 (1987).
26. D. A. Lashof, *Climatic Change* **14,** 213–242 (1989).
27. G. M. Woodwell and F. T. Mackenzie, eds., *Biotic Feedbacks in the Global Climate System: Will the Warming Speed the Warming?* Oxford University Press, Oxford, UK, in press.
28. H. A. Mooney, B. G. Drake, R. J. Luxmoore, W. C. Oechel, and L. F. Pitelka, *BioScience* **41,** 96–104 (1991).
29. G. C. Jacoby and R. D. D'Arrigo in Ref. 27.
30. R. A. Houghton, *J. Geophys. Res.* **92,** 4223–4230 (1987).

CARBON DIOXIDE EMISSIONS, FOSSIL FUELS

Jae Edmonds
Pacific Northwest Laboratory, Washington, D.C.
Michael Grubb
Royal Institute of International Affairs, London
Patrick ten Brink
Michael Morrison
Caminus Energy Limited, Cambridge, UK

In the late 1980s, interest flourished in the issue of global climate change. Many studies focused on the options for limiting anthropogenic emissions of greenhouse-related gases and managing the consequences of global warming and climate change. Making appropriate policy choices requires information on both the costs and benefits, as they occur over time, of policy interventions, and an increasing number of studies have sought to quantify the costs especially of limiting CO_2 emissions, as the dominant anthropogenic source. Such analyses now form an important part of overall policy assessments and influence international negotiations on policy responses.

See also Coal combustion; Combustion modeling; Commercial availability of energy technology.

The majority of work in estimating the costs of reducing greenhouse gas emissions has occurred since 1988, but interest in the issue of costing emissions reductions began more than a decade earlier with the work of Nordhaus (1,2). Nordhaus's early work focused on the issue of reducing fossil-fuel CO_2 emissions, as did that of others (3–13). Only references 6, 11, and 13 contain data on non-CO_2 emissions, and these studies treated them separately and in an ad hoc manner; none of the studies took land-use change into account explicitly.

Although not the primary focus of their analysis, some of the studies conducted prior to 1988 analyzed the cost of emissions reductions. The results of these studies foreshadow the current debate. Reference 14 contains an assessment that explores the subsequent development, deepening, and broadening of research veins, focusing on the past five years of research on the costs of limiting CO_2 emissions. While this article focuses on the cost of reducing fossil fuel CO_2 emissions, it should be recalled that it is only one, albeit important, element in the full analysis of costs and benefits of policy intervention. A full analysis requires consideration of all greenhouse related gases from all related human activities, relevant biogeochemical processes, and damages to human and natural systems.

MODELING AND COSTING DEFINITIONS AND PARADIGMS

The cost of emissions reductions is always calculated as a difference in a given measure of performance between a reference scenario and a scenario that involves lower emissions. By far the most commonly used measures of performance are the net direct financial costs to the energy sector assessed at a specified discount rate, and the estimated impact on gross national product (GNP) or its close cousin GDP. GNP is the monetary value of new final goods and services produced in a given year, and it provides a measure of the scale of human activities that pass through markets, plus imputed values of some nonmarket activities. It is generally assumed that financial costs in the energy sector can be closely related to impacts on GNP, although this is not always the case.

Neither direct financial costs nor GNP provide direct measures of human welfare. One factor is that human welfare does not necessarily increase linearly with the degree of consumption; a given loss of income will likely matter far more to poor people, or poor countries, than to richer ones, for example. Some studies attempt to capture this through "equivalent welfare" measures, but these still rely centrally on a GNP basis. Some of the models discussed use the Hicks equivalent welfare variation, which assumes a national utility function to estimate the overall GNP increase that would be required to maintain the original level of utility.

GNP does not incorporate many nonmarket factors that affect welfare. Some studies have sought to examine explicitly the impact of abatement on various non-market costs, such as example for local air quality, and concluded that these can be very significant. However, in general, studies focus on financial costs or GNP impact. In the broader literature, other welfare indexes have been attempted (such as the United Nations Development Program's (UNDP) Human Development Index), but data are rarely adequate to quantify impacts in such terms in abatement costing studies. At present, for quantifying results there is little practical alternative to working with monetary cost and GNP impacts.

Economic studies use top-down models which analyze aggregated behavior based on economic indexes of prices and elasticities, and focus implicitly or explicitly on the use of carbon taxes to limit emissions. These studies have mostly concluded that relatively large carbon taxes (eg, that could much more than double the mine–mouth cost of coal) would be required to achieve goals such as the stabilization of fossil-fuel carbon emissions. Technology-oriented studies use bottom-up engineering models, which focus on the integration of technology cost and performance data. Many such studies have concluded conversely that emission reductions could be achieved with net cost savings. The division between the economic and engineering paradigm is closely related, but not identical, to the division between top-down and bottom-up models, as it has emerged in the literature.

There are also many modeling differences within each category. Most notably, since 1990 an important general division has become evident between the application of top-down models developed for long-run equilibrium analysis of energy and abatement costs (reflecting an idealized economy with optimal allocation of resources), and conventional macro-economic models designed for shorter-run analysis of the dynamic responses of economies which reflect many existing imperfections. Long-run equilibrium models generally estimate the costs of reducing emissions to be positive and high by the standards of most environmental measures implemented to date. Macro-economic models indicate a far more complex pattern of responses and cost indicators which may move in different directions and vary over time.

Technology Cost–Curve Results

Clearly a principal determinant of CO_2 abatement costs will be the costs and adequacy of technologies that can reduce emissions. Many studies of the technologies that could help to limit greenhouse gas emissions have been conducted; important reviews are given in Refs. 15–20. In addition, many international databases with information on energy technologies have now been established; the UNEP study (21) lists no less than 13 technology databases available.

Technology cost curves provide a useful way of summarizing the technical potential for limiting emissions as identified in such studies. The simplest approach is to stack up different technologies in order of the cost of emission reductions or energy displaced, though cost curves can be used to represent the output for almost any degree of sophistication in modeling. There are various ways of generating cost curves of successively greater sophistication and consistency, as discussed in UNEP (21).

Discrete technology cost curves for various developed countries are presented in references 22–25. An EPRI (26) study examined potential savings in the U.S. electricity sector and concluded that "if by the year 2000 the entire stock of electrical end-use stock were to be replaced with the most efficient end-use technologies (nearly all of them estimated to be cheaper than equivalent supply), the maximum savings could range from 24% to 44% of electricity supply." References 22 and 26 suggest much higher potential savings still.

The cost curves associated with the EPRI and references 22 and 26 are shown in Figure 1 as a percentage of system demand available by using various more efficient technologies, in order of increasing cost per unit saved. These costs compare with typical U.S. electricity prices of 6–8¢/kW·h; all costs below this thus involve net economic savings for the user at the discount rates employed for the analysis. The upper curve is an estimate by the U.S. Electric Power Research Institute; the lower curve is by the Rocky Mountain Institute (22,26). Figure 1 demonstrates that a good database does not necessarily ensure comparable results; the potential estimated in the upper curve is clearly much smaller than that estimated in the lower curve. Compared against typical U.S. electricity prices of at least ~7¢/kW·h, however, both illustrate that substantial emission reductions appear to be available at net cost savings.

Technology cost curves have by no means been confined to developed countries. A number are presented in studies of the TATA Energy Research Institute (27).

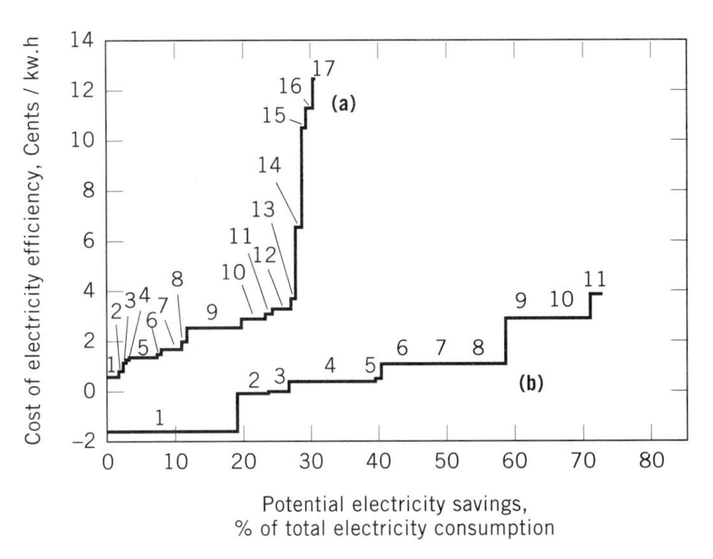

Figure 1. Discrete energy efficient technology cost curves for U.S. electricity (**a**) 1. Industrial process heating; 2. Residential lighting; 3. Residential water heating; 4. Commercial water heating; 5. Commercial lighting; 6. Commercial cooking; 7. Commercial cooking; 8. Commercial refrigeration; 9. Industrial motor drives; 10. Residential appliances; 11. Electrolytics; 12. Residential space heating; 13. Commercial and industrial space heating; 14. Commercial ventilation; 15. Commercial water heating (heat pump or solar); 16. Residential cooking; and 17. Residential water heating (heat pump or solar). (**b**) 1. Lighting; 2. Lighting's effect on heating and cooking; 3. Water heating; 4. Drive power; 5. Electronics; 6. Cooling; 7. Industrial process heat; 8. Electrolysis; 9. Residential process heat; 10. Space heating; 11. Water heating (solar).

Limitations and Interpretation of Technology Cost Curves

Abatement cost curves reflect the weaknesses and strengths of the procedures used to produce them. The simplest technology curves usefully summarize technical data, but may have substantial limitations as guides to actual abatement costs. In part this is because, unless they are developed iteratively using quite sophisticated system models, they may neglect interdependencies among abatement options and thus "double count" some emissions savings; eg, the CO_2 savings from reducing electricity demand may be much reduced if nonfossil sources are introduced later to displace coal power generation. They may also neglect interactions between various end uses, for example that between heat and lighting in widely diverse building environments. Frequently they also do not reflect adequately the timescales involved in bringing the technologies into place and the underlying growth in demand that may occur in the interim. Even after such issues are carefully incorporated, technology assessments and cost curves (particularly for end-use technologies) demonstrate a large potential for emission savings apparently at a negative cost, technologies that would both reduce emissions and yield net financial savings. Typically, the technologies suggest a potential to reduce emissions by well over 20% at net cost savings. The uncertainties, however, are very large, and to be meaningful the numbers have to be defined carefully in terms of scope and timescale. Based on an extensive review of technologies and related cost–curve studies, it is concluded that "substituting identified and cost-effective technologies on OECD countries could in principle increase the efficiency of electricity use by up to 50%, and of other applications by 15–40%, over the next two decades. Fully optimizing energy systems would yield larger savings, but it is far from clear how much of this potential can be tapped" (18).

Technology studies and cost curves show that a large energy efficiency gap exists between the apparent technical potential for cost-effective improvements and what is currently taken up in energy markets. In well-functioning markets, cost-effective options should be exploited anyway, because someone should profit by doing so. Some of the cost-effective potential may be taken up over time, but if such technologies are not being exploited, this may indicate that other important factors are not captured in technology analysis. For example, there may be hidden costs, people may be unaware of the options, or there may be other obstacles to uptake. The acceptability of different options may also vary, for least cost is by no means the only criterion that matters to people. This illustrates the fact that the apparent technical potential in fact comprises a number of different components. As shown in Figure 2, realizable gains consist of

- Those that are economically attractive in their own right and that will be installed without policy changes.
- Those that will not be obtained unless institutional constraints and barriers are removed, and/or other micro-economic policies are implemented to increase the take-up of cost-effective options.
- Those that are justified on the basis of nongreenhouse external benefits (eg reduced other environmental impacts, increased energy security).

In addition, some apparently cost-effective savings cannot be realized. The real economic potential can differ from apparent potential due to

- Take back or rebound of savings (improved efficiency reduces the cost of the associated energy service and therefore stimulates increased demand for the energy service).
- Unavoidal hidden costs (there may be costs associated with the use of a technology or policies to stimulate its uptake that are not revealed in a simplified analysis).
- Consumer preference (the technologies may not be a perfect substitute in the provision of the energy service, which may either increase or decrease consumer readiness to take up more efficient technologies, depending on their characteristics). For example, it would be technically possible and highly cost-effective to mandate a doubling of car efficiency in most countries, but it would probably make the vehicles available smaller and less powerful, which people may consider unacceptable (often this can also be considered an aspect of hidden costs).

These different components need to be understood before drawing conclusions about the scope for cost-free reductions. Cost curves may identify a technical potential, but there is no expectation that this can all be realized.

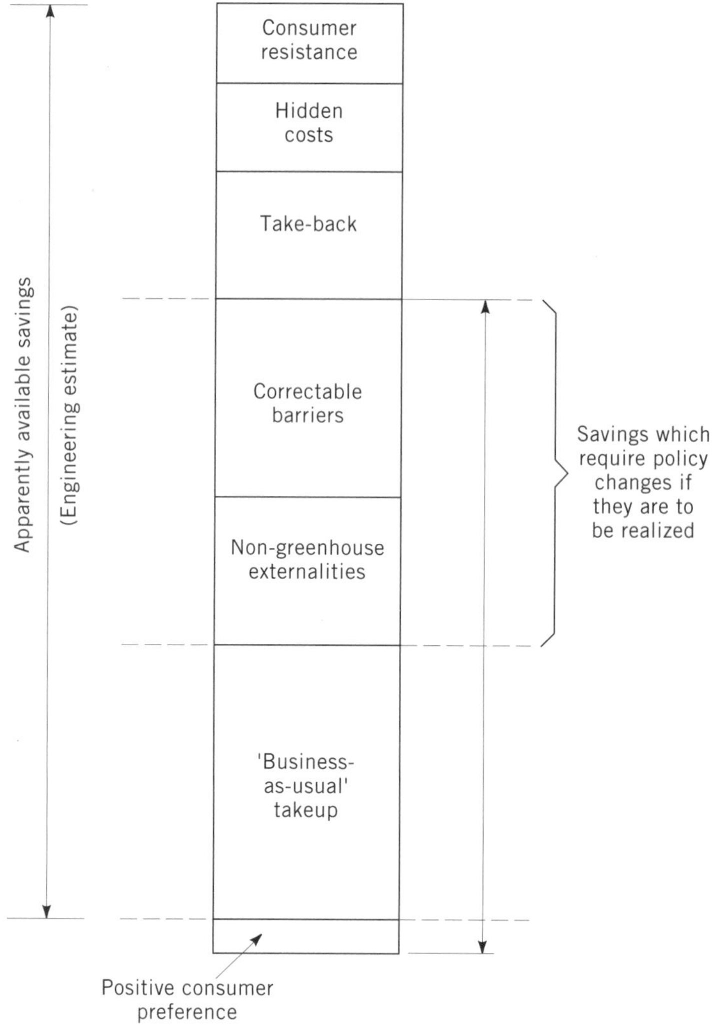

Figure 2. Energy efficiency: engineering potential and realizable gains, a classification. Hidden cost = unavoidable hidden costs and other barriers.

Part of the key is to consider specific policies for introducing better technologies set against explicit baseline scenarios that incorporate some "business as usual" uptake.

SYSTEM-WIDE ABATEMENT COST ESTIMATES

A wide variety of abatement costing studies at the global and national levels have been carried out. In addition, many local studies, which focus on policy-based studies for particular cities or utilities, have been conducted. These generally emphasize implementing the cost-free potential identified in bottom-up studies, but are too varied and specific to cover here.

Global Abatement

Table 1 reports to estimates of global costs of limiting CO_2 emissions. Four of the six models employed by these studies are fundamentally different. One uses an economic growth modeling framework oriented toward technology development, and one is a global bottom-up study. Many other global models have been developed. They have not

Table 1. Global CO_2 Abatement Cost Modeling Studies

Author (reference)	Key	CO_2 Reduction from Baseline % Reduction	CO_2 Reduction from Reference Year % Reduction[a]	GNP Impact/Cost from Baseline % Reduction
Anderson and Bird (28)	AB (2050)	68%	−17%	2.8%
Burnlaux and co-workers (29)	B (2020)	37%	17%	1.8%
Burnlaux and co-workers (30)	B (2050)	64%, 64%, 66%	−18%, −18%, −11%	2.1%, 1.0%, 0.3%[b]
Edmonds and Barns (31)	EB (2025)	14%, 36%, 47%, 70%		0.1%, 0.5%, 0.7%, 2.2%[c]
Edmonds and Barns (32)	EB (2020, 2050, 2095)	45%, 70%, 88%	22%, 41%, 53%	1.9%, 2.7%, 5.7%
Manne and Richels (33)	MR (2100)	75%	−16%	4.0%
Manne (34)	M (2020, 2050, 2100)	45%, 70%, 88%	13%, 25%, 21%	2.9%, 2.7%, 4.7%
Mintzer (13)	Mi (2075)	88%	67%	3.0%
Oliveira Martins and co-workers (35)	OM (2020, 2050)	45%, 70%	−2%, 2%	1.9%, 2.6%
Perroni and Rutherford (36)	PR (2010)	23%		1.0%
Rutherford (37)	R (2020, 2050, 2100)	45%, 70%, 88%	15%, 28%, 43%	1.5%, 2.4%, 3.6%
Whalley and Wigle (38)	WW (2030)	50%		4.4%, 4.4%, 4.2%[d]
Goldemberg and co-workers (19)	G (2020)	50%	0%	0%[e]

[a] Negative values imply an increase in CO_2.
[b] Toronto-type agreement in all three cases, with tradeable permits in the second and third cases and energy subsidy in the third case.
[c] Costs as estimated from consumer + producer surplus (see text).
[d] The three numbers refer to three different tax forms: a national producer tax, a national consumer tax, and a global tax with per capita redistribution of revenues.
[e] The cost value is indicative of the estimate by this study's authors that their scenario would incur no additional costs; it is not a modeling result.

been considered because they either are based on models already covered or they do not generate results in a relevant, comparable form.

The principal published results from these modeling studies are summarized in Figure 3; plotting the degree of abatement against the cost measured directly (for energy sector models) or as GNP loss, relative to the projected GNP. Figure 3**a** shows the results in terms of reductions from the baseline projection generated by the model; simple calculation shows that the average rate of abatement in almost all these studies, despite their apparent diversity, is 1.3–2.0% per year below the baseline projection. Figure 3**b** illustrates the same results, but in terms of the level relative to the base year (usually 1990).

Comparing the two curves in Figure 3 indicates that emission changes relative to a base year show less of a pattern, due primarily to the wide variation in the baseline emission projections. Therefore, one should focus on reductions relative to the reference projection without considering the base year.

In Figure 4**a** the relationship between the relative CO_2 reduction and required carbon tax (marginal abatement cost as reponed by the model in the target year) is shown. Figure 4**b**, shows the relationship between the reported carbon tax and the average GNP loss in the target year. For a fixed analysis the marginal cost must always be greater than the average cost, this plot shows no such relationship. For some models this may reflect anomalies in the way that marginal cost impacts are translated into GNP impacts; another source of the difference may be that in some models the CO_2 constraints are introduced to impose a much higher marginal cost over the first few decades than in the very long run. All these global studies impose a fixed path of emission constraint; only the simpler CETA (39) and DICE (40) models optimize the path of abatement to reach a given concentration level.

Even when normalized relative to the differing baseline projections, however, the results still show great variation. The Whalley-Wigle results define the high end of the cost spectrum. The relatively high costs probably reflect the limitations imposed by only having one generic carbon fuel, the lack of technology representation, and an extraordinarily high baseline projection, which scales up the global energy system 10-fold over the century because there is no allowance for autonomous efficiency improvements. The Goldemberg and co-workers (19) bottom-up results define a lower bound, with emissions reductions perhaps up to 50% estimated at no cost (7). However, some of the sensitivity runs of the global top-down models, when given more optimistic estimates of energy efficiency improvements and supply technology development, can also produce very low costs.

Excluding the Whalley-Wigle outlier, the spread of results roughly indicates that the costs of a long-run 50% reduction in global CO_2 emissions could range from negligible to a loss of about 3% in global GNP. Reductions of 80–90% depress GNP by 2–6%; at the other end of the curve, the global economic models, as well as the engineering models, suggest that emission reductions of at least 10–15% can be obtained at very low cost.

These global models are highly aggregated and capture technical issues, macro-economic issues, regional differences, and trade effects with varying degrees of detail. They inevitably sacrifice local and technological detail in order to represent important regional differences and (in some) incorporate trade.

Abatement Costing Studies for the United States

A wide range of national studies has been carried out; of these, those of the U.S. have received the most attention, as shown in Table 2. The range of estimates of the impact on GNP of CO_2 emissions reductions relative to the projected baseline emissions is shown in Figure 5.

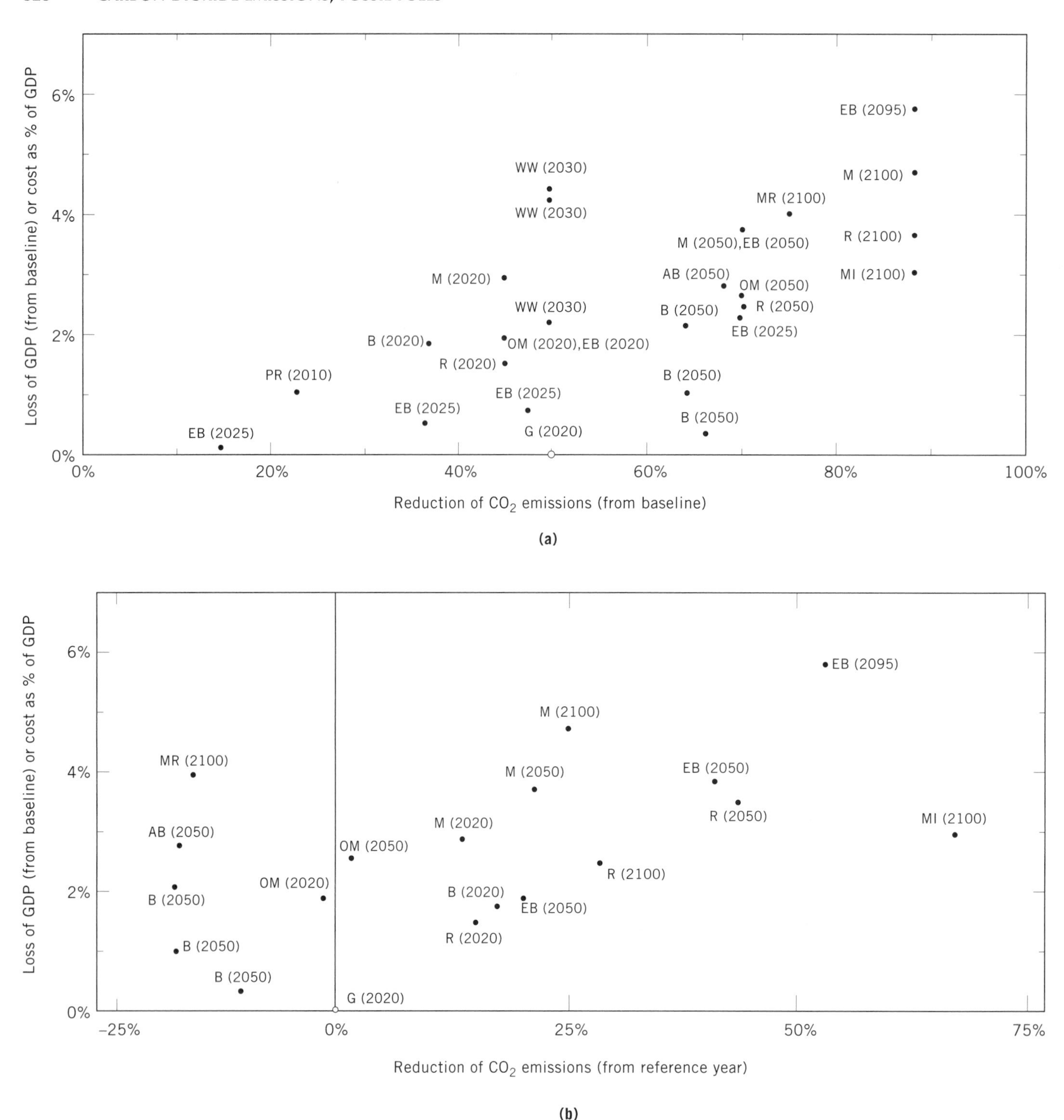

Figure 3. Global CO_2 emissions reductions cost estimates. (**a**) Reduction from baseline. Loss of GDP (from baseline) or cost as % of GDP. • Top-down; economic model, estimate, ○ Bottom-up, technology based estimate. For key see Table 1. (**b**) Reduction from reference year. Note that a negative reduction indicates an increase in emission.

U.S. studies are predominantly top-down economic studies. Many bottom-up/engineering studies have also been carried out, but most have been confined to cost–curve or subsectoral studies. In addition to the earlier examples, important recent studies include those of the National Academy of Sciences (NAS) (49) and the Office of Technology Assessment of the U.S. Congress (OTA) (50). These, like most engineering-based studies, maintain that main efficiency improvements (and hence CO_2 reductions) can be obtained at little or no cost. These estimates have been included in Figure 5 as indicative bottom-up cost estimates.

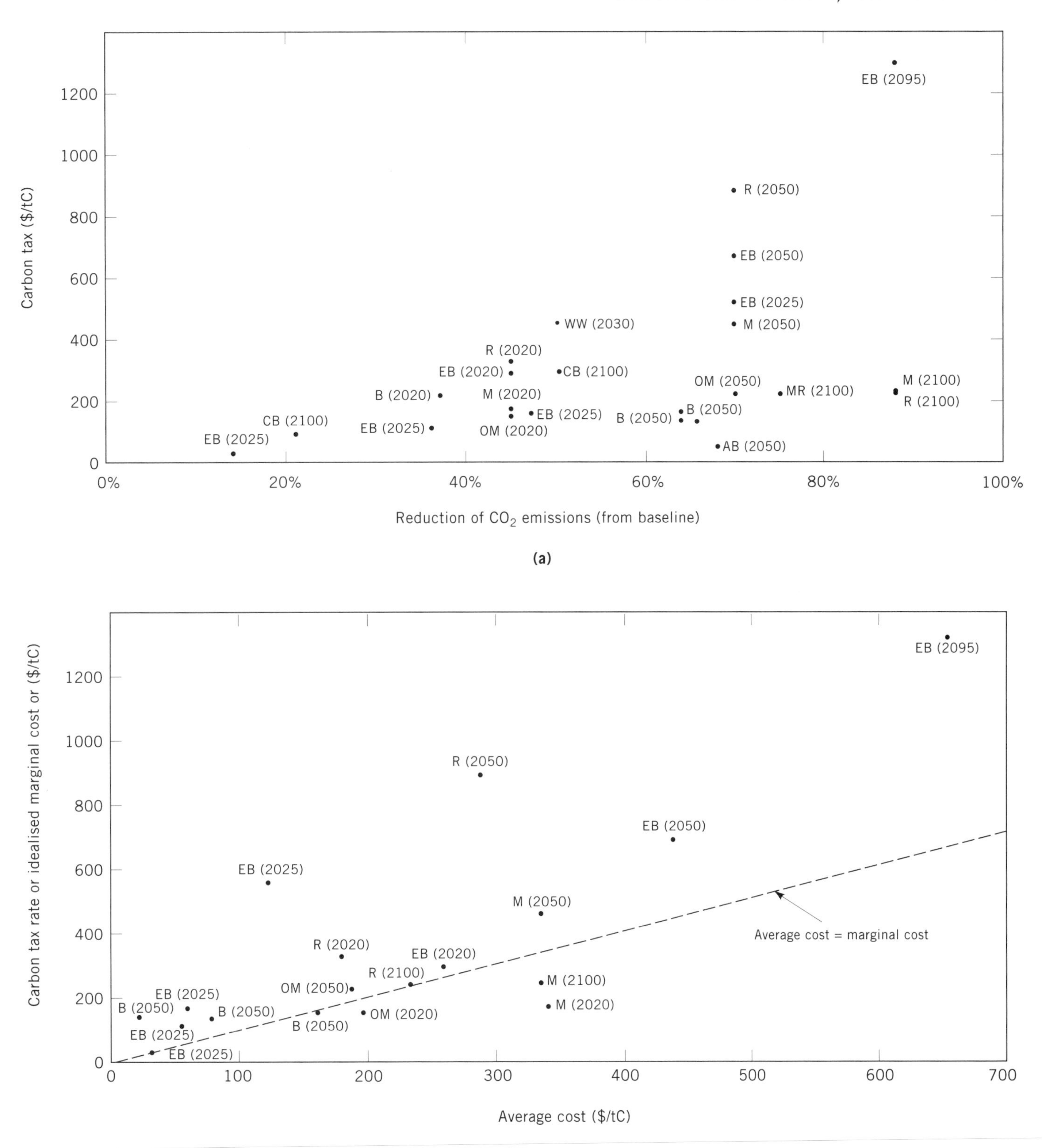

Figure 4. (**a**) Carbon taxes and CO_2 emissions reductions (from baseline), global studies. Some of the plotted carbon tax values are averages of different regional taxes. Where taxes vary over the study period, the end year tax is shown. For key see Table 1; (CB(2100), not shown in Table 1, refers to the CBO study). (**b**) Idealized marginal cost or carbon tax and the average cost, global studies. Some of the plotted carbon tax values are averages of different regional taxes. Where taxes vary over the study period, the end year tax is shown. For Key see Table 1.

Table 2. U.S. CO_2 Abatement Cost Modeling Studies

Author (reference)	Key[a]	CO_2 Reduction from Baseline % Reduction	GNP Impact/Cost from Baseline % Reduction
Barns and coworkers (41)	BG (2020)	26%, 45%, 60%	0.6%, 2.0%, 3.2%
Barns and co-workers (41)	BG (2050)	45%, 70%, 84%	1.9%, 4.9%, 7.5%
Barns and co-workers (41)	BG (2095)	67%, 88%, 96%	4.3%, 8%, 10.9%
Brinner and co-workers (42)	B (2020)	37%	1.8%
CBO-PCAEO, DRI (43	CB1, CB2 (2000)	8%, 16%	1.9%, 2%
CBO-DGEM (43)	CB3 (2000)	36%	0.6%
CBO-IEA-ORAU (43)	CBG (2100)	11%, 36%, 50%, 75%	1.1%, 2.2%, 0.9%, 3.0%[b]
Edmonds and Barns (44)	EB (2020), EB (2100)	35%, 59%	1.3%, 2.3%
Goulder (45)	G (2050)	13%, 18%, 27%	1.0%, 2.2%, 4.5%
Jackson (24)	J (2005)	34%, 40%, 46%	−0.2%, −0.1%, 0.1%
Jorgenson and Wilcoxen (45)	JW (2060)	20%, 36%	0.5%, 1.1%
Jorgenson and Wilcoxen (46)	JW (2100)	10%, 20%, 30%	0.2%, 0.5%, 1.1%
Jorgenson and Wilcoxen (47)	JW (2020)	8%, 14%, 32%	0.3%, 0.5%, 1.6%
Manne and Richels (48)	MR (2020)	45%	2.2%
Manne and Richels (48)	MR (2100)	50%, 77%, 88%	0.8%, 2.5%, 4.0%[c]
Manne (34)	MRG (2020)	26%, 45%, 60%	0.8%, 2.2%, 4.2%
Manne (34)	MRG (2050)	45%, 70%, 84%	1.4%, 2.7%, 3.3%
Manne (34)	MRG (2100)	67%, 88%, 96%	2.3%, 3.1%, 3.4%
Mills and co-workers (23)	M (2010)	21%	−1.2%[d]
NAS (49)	N	24%, 40%	0%, 0.8%
Oliveira Martins and co-workers (35)	OMG (2020)	26%, 45%, 60%	0.2%, 1.1%, 2.4%
Oliveira Martins and co-workers (35)	OMG (2050)	45%, 75%, 84%	0.4%, 1.3%, 2.4%
OTA (50)	O (2015)	23%, 53%, 53%	−0.2%[e], −0.2%[e], 1.8%
Rutherford (37)	RG (2020)	26%, 45%, 60%	0.5%, 1.3%, 2.5%
Rutherford (37)	RG (2050)	84%, 45%, 70%	2.4%, 1.2%, 2.5%
Rutherford (37)	RG (2100)	67%, 88%, 96%	1.8%, 2.6%, 2.8%
Shackleton and co-workers (51)	SJW (2010), SLINK (2010)	22%, 2%	−0.6%, 0.1%
Shackleton and co-workers (51)	SDR1 (2010), SG (2010)	5%, 28%	−0.4%, 0.2%

[a] If the model used is a global model, the Key includes the letter "G" before the date.
[b] The first two results use multilateral taxes, others use unilateral taxes; taxes are flat only in the 1st and 3rd estimates, rising in other estimates.
[c] Values represent different assumptions in technological developments: an optimistic, an intermediate, and a pessimistic view.
[d] Arising from 11 specified regulatory changes: estimated from claimed savings of $85 bn/yr.
[e] The benefit shown in the OTA cost estimates is only an indicative value; no explicit modeling value has been calculated.

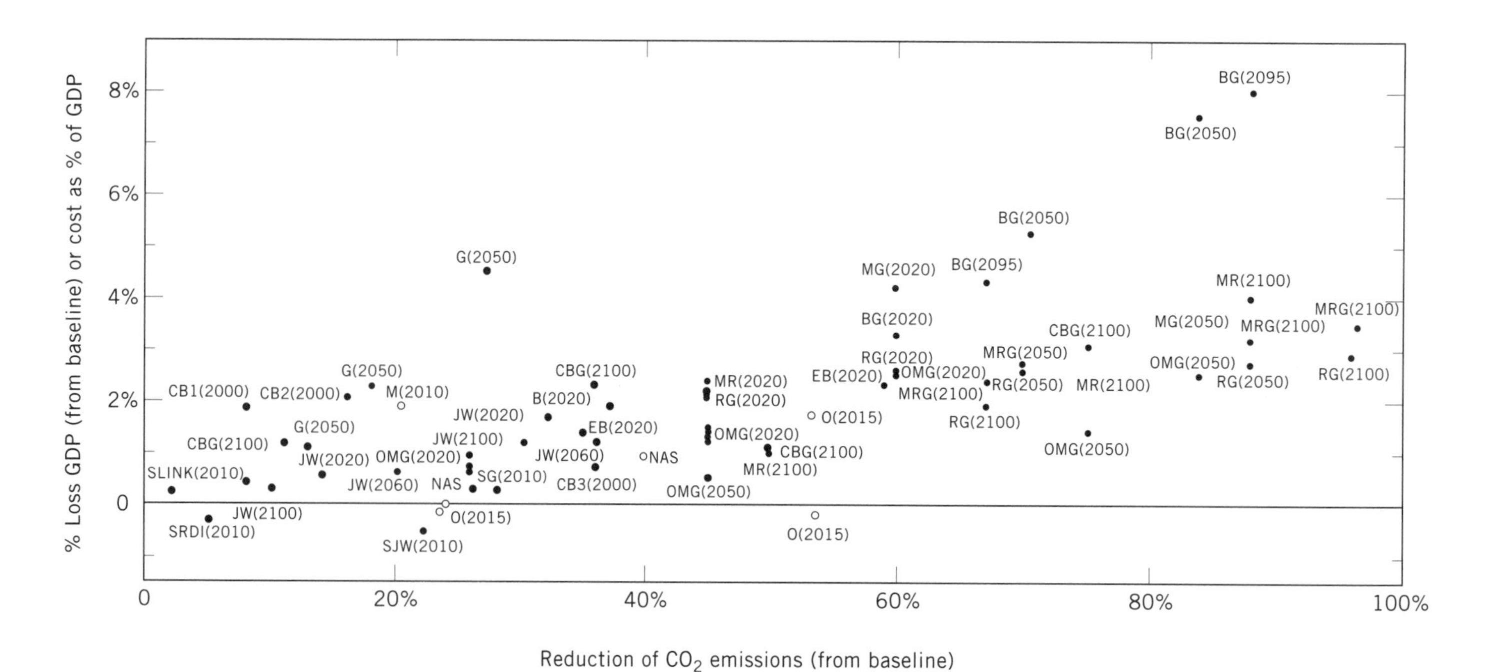

Figure 5. CO_2 emissions reductions cost estimates, United States, reduction from baseline. • Top-down; economic model estimate, ○ Bottom-up, technology based estimate. Labels are not attached to all points given danger of crowding. For key see Table 2.

As with the global studies, a wide range of cost estimates is observable. For example, for the same loss of GNP of about 2%, CO_2 emission reductions can range from 20 to 80% below baseline (52,53). One big reason for such differences is that the reductions are sought on different timescales; the CBO study seeks a 20% reduction by 2000, while the 80% reduction in the reference 53 study is achieved at the end of the next century.

Ignoring the rapid reductions imposed in the CBO studies for the year 2000, early studies form a high cost outlier (45). Bottom-up studies and those that examine the recycling of tax revenues have produced some very low and negative abatement cost estimates. With these exceptions, the spread of results is pretty consistent with the global results, with 50% emission reductions from baseline yielding losses up to a little more than 2% of GNP. The coincidence between U.S. and global results may partially reflect the importance of U.S. emissions and costs, but also the dominance of U.S.-based modeling approaches in global studies.

Non-U.S. OECD Studies

A number of emissions reduction studies of non-U.S. OECD countries are listed in Table 3. One striking feature is that nearly all these studies have a much shorter term focus than the global studies or most of the U.S. studies; most in fact are focused either on emissions stabilization by 2000 or on the Toronto target of 20% reduction by 2005. The results are displayed in Figure 6. Immense variation is once again apparent, even in terms of reductions from the baseline projection. No clear pattern emerges, other than the fact that the bottom-up studies again give much lower costs.

In general, the cost range is broader than in the global studies, perhaps reflecting the impact of transitional costs arising from the relatively rapid reductions required in some of these studies, captured by the short-run models, an issue discussed further below. The detailed EC studies of the macro-economic impacts of carbon taxation, using short-run macro-economic models, produce a wide variety of results; these results depend heavily on how the tax revenues are used, as they reflect rather the use of a tax to shift resources from one kind of economic activity to another. The Finnish study (54), with very high losses for modest reductions by 2010, is a similar outlier on the high side, and two of the Japanese studies also yield exceptionally high costs (55); this is because the carbon tax revenues are removed from the economy. The variations make it hard to discern any pattern, but even with these outliers excluded, the relative costs for this short-run national reductions are mostly larger than for equivalent long-run reductions in the U.S. and global studies, with several exceeding 3% GNP losses.

The Transitional Economics

Studies of abatement costs for the former centrally planned economies of Eastern Europe have also used both

Table 3. Non-U.S. OECD CO_2 Abatement Cost Modeling Studies

Country	Author (reference)	Key[a]	CO_2 Reduction from Baseline % Reduction	GNP Impact/Cost from Baseline % Reduction
Australia	Dixon and co-workers (56)	AD (2005)	47%	2.4%
Australia	Industry Commission (57)	AICG (2005)	40%	0.8%
Australia	Marks and co-workers (58)	AM (2005)	44%	1.5%[b]
Belgium	Proost and van Regemorter (59)	BP (2010)	28%	1.8%
EC	DRI (42)	ECDRI (2005)	12%	0.8%
Finland	Christenson (54)	FC (2010)	23%, 21%	6.9%, 4.8%[c]
France	Hermes-Midas (60)	FHM (2005)	11%	0.7%
Germany	Hermes-Midas (60)	GHM (2005)	13%	1.3%
Italy	Hermes-Midas (60)	IHM (2005)	13%	1.9%
Japan	Ban (61)	FB (2000)	18%, 18%	0.4%, 1.7%[d]
Japan	Goto (62)	JG (2000, 2010, 2030)	23%, 41%, 66%	0.2%, 0.8%, 1%
Japan	Nagata and co-workers (63)	JN (2005)	26%	4.9%
Japan	Yamaji and co-workers (55)	JYC (2005)	36%	6%[e]
Netherlands	NEPP (64)	NEN (2010)	25%, 25%	4.2%, 0.6%[f]
Norway	Bye, Bye, and Lorentson (65)	NB (2000)	16%	1.5%[g]
Norway	Glomsrod and co-workers (66)	NG (2010)	26%	2.7%
Sweden	Bergman (67)	SWB (2000)	10%, 20%, 30%, 40%, 51%	0%, 1.4%, 2.6%, 3.9%, 5.6%
Sweden	Mills and co-workers (23)	SWM (2010)	44%, 87%	−0.3%, −0.2%[h]
UK	Barker and Lewney (68)	UKBL (2005)	32%	0%
UK	Sondheimer (69)	UKS (2000)	4%	−0.5%
UK	Hermes-Midas (60)	UKHM (2005)	7%	1.9%

[a] The letters in the Key refer to the country and author; in cases where a global model is used a "G" is added before the date of the estimate.
[b] Study combines technology and macro-economic assessment of GDP impact.
[c] Unilateral and global action.
[d] Tax case and regulation case.
[e] Values of both 5 and 6% are given.
[f] National and global policy scenario.
[g] GDP costs for OECD range of 1–2%.
[h] Estimated from average value of saved emissions ($143/tC and $41/tC, respectively).

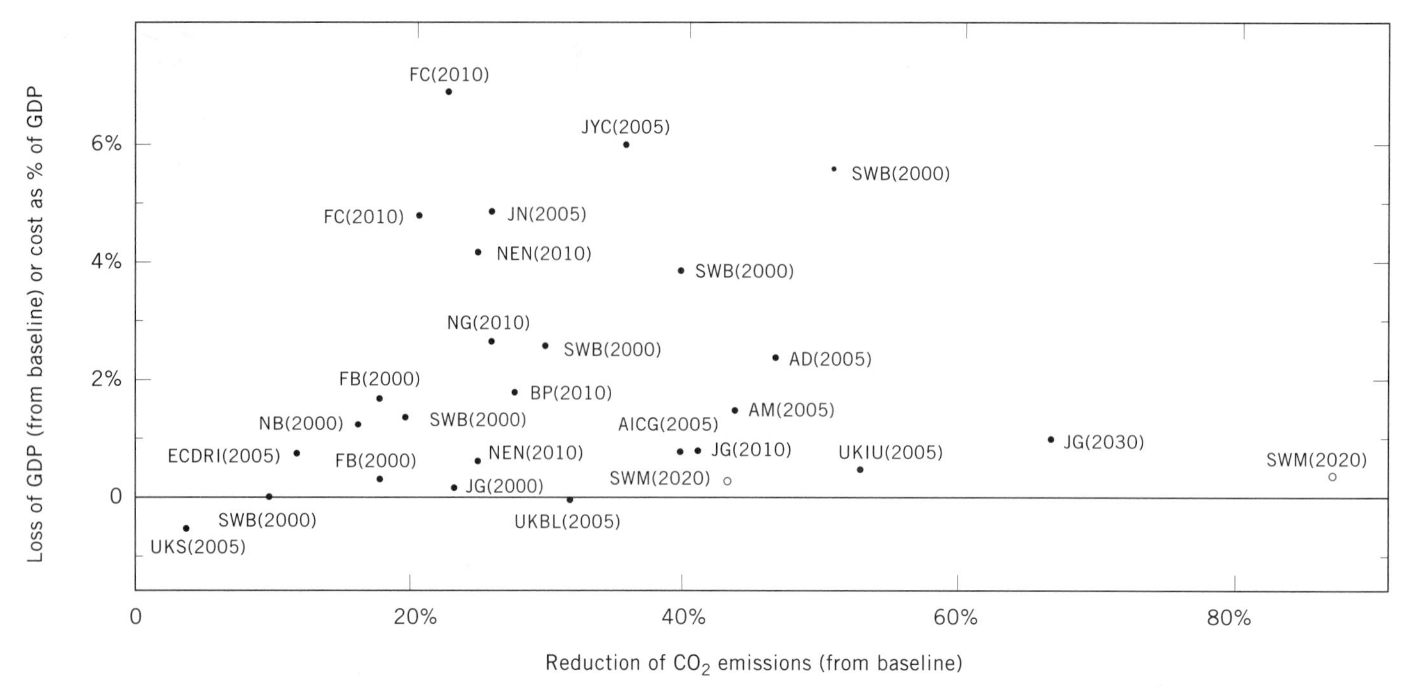

Figure 6. CO_2 emissions reductions cost estimates, industrialized countries, Non United States, reduction from baseline. • Top-down; economic model estimate, ○ Bottom-up, technology based estimate. For key see Table 3.

top-down and bottom-up approaches, but data are insufficient to summarize usefully as a scatter diagram. An energy technology cost curve estimated for Poland has been shown (70,71) (Fig. 7). All of the bottom-up studies indicate a large potential for reducing CO_2 emissions with net economic savings; specifically, the marginal cost of savings only rises above that of new supply for savings well in excess of 10 EJ, which is more than 20% of Soviet primary energy demand in 1989. This potential arises from the history of highly subsidized energy prices in these regions, and other cumulative inefficiencies in the structure of incentives. The economic savings potential is around 10% greater if the Soviet Union can access Western technologies.

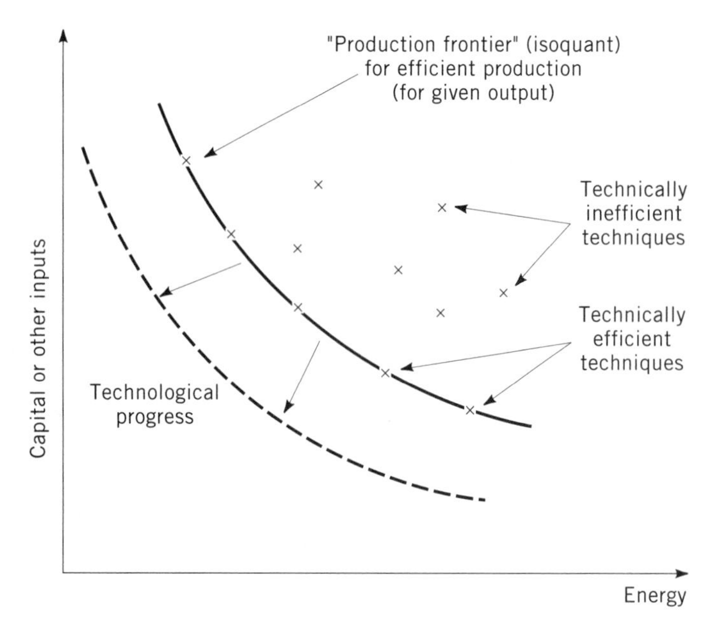

Figure 7. Technological efficiency and the "production frontier" UNEP, (21).

It was estimated in 1991 that increasing prices to world market levels in Poland, Hungary, and Czechoslovakia would reduce emissions by 30% (72). Of course, the realizable potential may be a very different matter and depends in part on the progress of economic restructuring. In fact, CO_2 emissions in the former East Germany have collapsed by at least 30% as uncompetitive heavy industry has more or less shut down in the process of unification. In Poland, provisional trend/technology results also indicate significant CO_2 reductions as a by-product of the economic restructuring process (73).

Manne and Richels include the Soviet Union as an independent region in their Global 2100 model (48), and calculate much higher GNP losses (5% in (33), reduced to 3% in (53)) there than for the rest of the world in the first half of the next century. This striking contrast with bottom-up studies reflects the difficulties top-down studies have with economies undergoing restructure. They rely on the existence of a market mechanism that in many instances is barely functioning, and such studies will likely miss many of the important features of these economies, such as current structural inefficiencies. It may take some time before a market-based modeling approach becomes appropriate. For the present, bottom-up engineering assessments appear much more relevant.

Developing Countries Including China. Some of these issues also apply to developing countries. Although concerns are frequently expressed about the limited data available concerning developing countries, considerable data exist concerning the situation in most principal developing countries, especially with respect to commercial energy supply. Data on detailed end uses, agriculture, and noncommercial energy sectors is sparser and less reliable, though usable estimates exist.

The past few years have also seen a number of studies of the potential for abating greenhouse gases in developing countries. Country reports to the IPCC Energy and Industries subgroup (74) include a number of studies from developing countries, some drawing on more extensive internal work. In addition, initial results from a series of nine developing-country studies coordinated by the U.S. Lawrence Berkeley Laboratory were reported (75). However, these studies focused on scenarios and did not attempt to estimate abatement costs. Some of the detailed studies for the Asian Energy Institute network reported in the UNEP study (21) attempt bottom-up estimates of short-term abatement costs or investment requirements.

Some of the global models do separate developing-country and oil-exporting regions. The global studies have also highlighted the growing importance of China in contributing to greenhouse gas emissions during the next century. The early results (33) suggested that Chinese GNP in the year 2050 could be depressed by up to 8% below the (greatly increased) reference level, if emissions are restricted to no more than double current levels, partly because of the apparent lack of alternatives to coal and modeling inability to import energy (though this was reduced to a 2% GNP loss, rising to 5% by 2100, in revised analysis (53)). The GREEN model suggests much lower abatement costs for China (76).

In estimates derived from global studies that model trade (eg, references 77 and 78), GNP impacts for some of the smaller developing countries especially can be dominated by trade and price effects arising from the action of other countries. These studies emphasize the large potential GNP loss for energy exporting developing countries; the loss could arise from abatement efforts elsewhere that depress the market for traded fuels. Conversely, energy-importing countries (which include the poorest countries) would gain from such effects.

Concerning domestic abatement efforts, a number of cost curves have been estimated which indicate substantial technical potential for savings with net economic benefits. However, the only integrated system-wide cost estimates found, excepting those from global models for China, are those of Blitzer and co-workers (79,80) for Egypt and an unpublished study of Zimbabwe by the UK consultants Touche Ross, reported in UNEP (21).

The Blitzer studies estimate a very large potential GNP impact from stabilizing CO_2 emissions in Egypt, with losses in some cases of more than 10% of GNP. This is based on a short-run macro-economic model of the Egyptian economy that is modeled with very limited capital mobility between sectors, and with oil and gas as the only future energy supply options. It recognizes none of the technical inefficiencies in the economy (ie, assumes zero scope for cost-free energy-efficiency improvements). Excepting the modest abatement available from switching from oil to gas, emissions savings can only be achieved by reducing energy consumption, which, given the constraints on capital mobility, can only be achieved by reducing economic activity or changing its structure. Consequently, the costs reported are clearly excessive. However, separate runs of the model that placed emission constraints on each sector of the economy individually did emphasize that GNP losses would be greater still if these additional restrictions were imposed.

The Touche Ross study of Zimbabwe reached precisely opposite conclusions. Using an engineering approach, widespread cost-effective options were identified that could both limit emissions growth and improve overall economic performance. However, these assessments neglect a variety of hidden costs and fundamental institutional obstacles; they also include some elements that are expected to be achieved anyway as part of current structural adjustments in the Zimbabwean economy. These and other limitations, which suggest that abatement costs in this case may be substantially underestimated, are summarized in the Zimbabwean case study of the UNEP report (21).

The limited range and appropriateness of studies for the transitional and developing economies make the use of scatter diagrams, as were used for presenting OECD results, not, in the authors' judgment, very meaningful in this case. One striking observation of note: the gap between top-down and bottom-up approaches is larger even than observed for OECD countries. Top-down models mostly report restricting developing-country emissions, even relative to projected increases, to be more expensive than equivalent relative constraints in OECD countries (37). It is not clear why this should be the case. Bottom-up studies, conversely, identify a potential for improving energy-efficiency in these regions at a net economic benefit that is even larger than that identified for OECD countries.

Despite this, it seems possible to draw two firm observations from the existing developing; country emissions abatement studies: many cost-effective technology options exist for improving energy-efficiency, but such potential will be swamped by the pressure for emissions growth in such rapidly expanding economies, so that actually stabilizing developing-country emissions at current levels is nevertheless likely to be very costly. More sophisticated and quantitative system-wide analysis of abatement impacts is, however, only just beginning, and as outlined in UNEP (21), the complexities are such that it may take many years to mature toward consensus even on very rough cost estimates and understanding of the key issues.

MODEL SCOPE AND TYPE

Cost estimates clearly vary greatly among models. It is difficult to disentangle the various reasons for these differences, due to the nonlinearity of the relationships involved, the diversity of the tools used to develop emissions reductions cost estimates, the many and varied assumptions employed, and the enormity of the task required to obtain all of the models, establish a protocol for analysis, and systematically unravel the relative contributions. Steps in that direction have been taken by teams at the Energy Modelling Forum (29,81) and OECD (45,82). In all these studies, participating modelers were asked to adopt standard assumptions to the degree possible and to provide standardized model results. Both sets of comparisons have focused on top-down models (the Energy Modelling Forum conducted a similar exercise with bottom-up models). It is clear from these activities that great variation in results can be generated through the use of different models. This variation is greatly reduced through the use of standardized assumptions.

All models share certain unavoidable limitations:

- Models are necessarily a simplified representation of reality, in terms of what the concerned researchers feel are important aspects that should be captured. A given model may not capture all important economic relationships.
- Despite their simplifications, all models are still rather complex and must necessarily rely on a large quantity of data and numerous parameter estimates. Robust estimation of these is itself significant research undertaking and serious doubts may arise about the validity of many of the actual numerical values employed. Studies of model sensitivities, and structured uncertainty studies in which the values of key parameters are varied over plausible ranges, are required to examine how much these uncertainties may affect model results. Such studies have not often been adequately performed.
- The timescales involved require assumptions to be made about changes in technology and lifestyles. Conjectures about such changes are inevitably uncertain and cannot be formally validated.

These limitations may be exacerbated by the fact that most studies employ models that were not initially designed to shed light on the cost of emissions reductions (the exceptions are the OECD's GREEN model (30,83) the ERB (3,4), and the Second Generation Model (84).

Consequently, all modeling results need to be treated with some caution, depending in part on the timescale of application, the care with which the model has been developed, the extent to which it is appropriate to the application used, and the care with which inputs have been formulated. The ultimate argument for such modeling efforts is not that they give precise and certain answers, but rather that they are the only consistent way of estimating abatement costs at all and of identifying the important factors that affect them.

Top-Down and Bottom-Up Models

The principal distinction between top-down economic models and bottom-up engineering/technology-based models have been noted. High positive abatement costs are frequently associated with top-down economic approaches and low and negative costs are frequently associated with bottom-up engineering approaches.

The underlying theoretical distinction lies between the economic and engineering paradigms. In economics, technology is features as the set of techniques by which inputs, such as capital, labor, and energy, can be transformed into useful outputs.

Figure 8 is a graph of energy versus other (eg, capital) inputs. Each cross represents an individual technique or technology. The best techniques define the production frontier, as illustrated. In principle, efficient markets should result in investment only in the technically efficient techniques on this frontier (after allowing for lags associated with old stock), because such investments can reduce all costs compared with other technologies.

Economic models all assume that markets work efficiently, therefore all new investments (after allowing for hidden costs) define the production frontier. This is assumed to be consistent with cost-minimizing (or utility maximizing) behavior in response to the observed price signals: various models can encompass other inefficiencies, such as externalities and fiscal imperfections arising from taxes and subsidies, but still share this assumption. Observed behavior (historical data) combined with the optimizing assumption defines an observed production frontier. The models assume that no investments are available that lie beyond this frontier, though future technical change may move it. To the extent that real-world inefficiencies exist, they are implicitly incorporated in the inferred frontier. Relative price changes move investments along this frontier (eg, substitute labor for energy) as defined by the estimated elasticities; a purely economic model has no explicit technologies, which are simply implicit in the elasticities used.

Studies using engineering models have often focused on identifying potential least-cost abatement opportunities by assessing directly the costs of all the technological options. Such an assessment is independent of observed market behavior and also defines a production frontier. If markets are technically efficient, the frontier revealed by market behavior should correspond to that calculated by engineering studies. However, this is not the case; engineering studies reveal widespread potential for investments beyond the limit of the production frontier suggested by market behavior and built into economic models. The explanation can be considered in terms of the contrasting limitations of the economic and engineering paradigms. Economic models are slave to the assumption of cost-minimizing behavior noted above. Limitations of purely engineering models include

- The cost concept is based on an idealized evaluation of technologies and options. The existence of hidden costs is typically ignored.
- The cost of implementation measures (eg, information campaigns, standard setting, and compliance processes) is not included.
- Market imperfections and other economic barriers mean that the technical potential can never be fully realized.
- Macro-economic relationships (multiplier effects, structural effects, price effects) and indicators (GNP, employment, etc) are not included in the models.

Top-down and bottom-up are very imprecise terms. Although models generally known as top-down, all determine energy demand through aggregate, economically driven indexes (GNP and/or productivity growth, and price elasticities), they can vary greatly in the modeling of energy supply. Some of the top-down models are purely economic, with supply changes being driven only by substitution elasticities. Others are primarily economic, but incorporate a "backstop technology"—a technology that can come in, in unlimited quantities, once a certain price threshold is reached. Yet in other top-down models (such as the ERB), supply is driven largely on an engineering basis of supply technology costs, chosen from a database of supply technology cost curves.

Nearly all bottom-up models contain extensive representation of supply technologies, but the key practical distinction, as it has emerged in the CO_2 costing literature, centers on the modeling of energy end use and the introduction of end-use technologies. Bottom-up end-use studies indicate a large potential for reducing both emissions and costs relative to a traditional top-down extrapolation of energy demand. In other words, they show that the top-down projections are not optimal interms of the technologies available, and the main savings come by contrasting this with a scenario that is an engineering optimum.

Why does this create such a large difference between bottom-up and top-down studies? The primary reason is to be found in Figure 2. Neglecting the segment referring to externalities, which may or may not be reflected in either top-down or bottom-up studies, top-down projections of energy demand incorporate only efficiency improvements corresponding to the bottom segment of the column—the "business as usual" takeup. Bottom-up models, on the other hand, include all the available technologies, without distinction as to under which category in the column they fall. Consequently, neglecting externalities:

- Top-down modeling studies tend to underestimate the potential for low-cost efficiency improvements (and overestimate abatement costs) because they ignore a whole category of gains that could be tapped by non-price policy changes; whereas
- Bottom-up end-use modeling studies overestimate the potential (and underestimate abatement costs) because they neglect various hidden costs and constraints that limit the uptake of apparently cost-effective technologies.

Which is more realistic depends on the relative size of different segments in Figure 2, something that cannot be determined without separate study of specific implementation policies. But the near-term potential for limiting CO_2 emissions at low or negative costs lies somewhere between the optimism of bottom-up studies, and the relative pessimism of many top-down studies.

The systematically differing results are largely a reflection of the non-optimality of the baseline implicit in such bottom-up studies, and the questions this raises about the assumptions built into top-down model baseline and abatement projections. This is perhaps the key difference between the modeling approaches. If the baseline in bottom-up studies used optimal technologies, the baseline emission projections would be much lower. This was illustrated clearly in a study in reference 85, which included end-use technologies in the MARKAL engineering model and found that obtaining base-case emissions anything near as high as official or macro-economic forecasts proved almost impossible; the model chose more energy-efficient end-use technologies, and more renewable energy technologies, irrespective of CO_2 constraints. Further reductions were, however, relatively expensive, relying more on supply substitution, as the stock of more efficient end-use technologies was largely selected already in the optimal baseline.

Thus there is no inherent reason why top-down studies should yield positive costs or bottom-up models should yield negative costs. The sign of the cost hinges critically on the approach applied to computing costs, in particular, assumptions regarding optimality of the baseline. For example, references 86 and 87 recognize a non-optimal baseline and illustrate negative abatement costs within a top-down energy-economy approach; whereas the study in reference 85 uses a wholly engineering model and obtains positive costs for any reductions beyond those captured in the (optimal, and much lower) baseline.

Some attempts have been made to integrate top-down and bottom-up models explicitly. Most notably, the Global 2100 model has been linked with the MARKAL engineering model (by replacing the energy technology submodelin Global 2100 that formed the energy component previously) in a bid to combine the best features of both into a single computational framework (88). However, this still does not resolve the dilemma about whether projections, for baseline or abatement scenarios, adopt the engineering optimum or econometric extrapolation of energy demand, and this linked model has been criticized on the grounds that one still dominates the other (81).

Time Horizons and Adjustment Processes: Short-term Transitional versus Long-term Equilibrium

Different models are designed for application over different timescales. There are no standard definitions, but in many relevant branches of economic analysis the short term is taken to be less than 5 years, the medium term is between 3 and 15 years, and the long term is more than 10 years. The timescale is a distinction of significant importance, particularly for economic models, because different economic processes are important on different timescales, and thus the timescale for which models are designed fundamentally affects their structures and objectives. Models for relatively long-run analysis may to a reasonable approximation assume an economic equilibrium in which resources are fully allocated. Short-run models focus on transitional and disequilibrium effects such as transitional market responses, capital constraints, unemployment, and inflation. This distinction parallels the structural distinction drawn by Boero and co-workers (89) between resource allocation models and macro-economic models.

MEDIUM TO LONG TERM: EQUILIBRIUM/RESOURCE ALLOCATION MODELS

These models focus on the allocation of available resources, within the energy sector or the broader economy. This category includes both optimizing bottom-up models (which seek to optimize resource allocation within the bounds set by available technology), and all the main long-run and global energy/CO_2 models. The latter are generally termed equilibrium models.

At one extreme of the long-term modeling dimension are models that can only consider the energy/investment mix for a "snapshot" year and compare this to another, without any information on the transition between them; these are comparative static models (29,77,78). Such models can enable detail in representing the system, but at the expense of modeling developments over time. In contrast, dynamic models cover medium- and long-term phenomena, extending across several time periods.

At the opposite of this extreme within the equilibrium models, some are designed to run in annual steps over a period of a few decades. These can include considerable detail on different sectors whose use of different resources in response to price changes is estimated econometrically from data over previous years. The main examples are the Jorgensen/Wilcoxen (90) and Goulder (45) models for the United States.

Equilibrium models such as Global 2100 and GREEN (in its more recent versions (30,83)) and the ERB model lie between these extremes. They are designed to operate in steps of 5–15 years, to look at the changing allocation of resources under different constraints over periods of many decades, and the way this may change under CO_2 constraints.

The treatment of capital stocks in these models can have important implications for costs. "Putty–putty" models represent capital stocks as perfectly interchangeable between sectors and over time. Nuclear power plants can be transformed into solar photovoltaic arrays instantaneously and without cost. "Putty–clay" models, on the other hand, allow no transfer of capital between applications. Once an investment has occurred, the technology cannot be altered. Resource allocation/equilibrium models cannot, however, model other aspects of transitional costs arising from disequilibria. In this respect, and in their assumption of optimal investments subject to constraints, they have been criticized for underestimating likely abatement costs (though the same caveats apply to the reference projection as well, which is similarly optimal within constraints and free from disequilibria).

SHORT TO MEDIUM RUN: MARKET SIMULATION/ MACROECONOMIC MODELS

Short-run models by contrast focus primarily or exclusively on the dynamics of transition, rather than the long-term equilibrium allocation of resources. One class of short-run models are sectoral market simulation models, such as detailed models of electricity or oil markets and pricing, or of industrial sector energy demand. The diversity of such models reflects the range of specific markets that they have attempted to model. Few such models as yet appear to have been applied to assessing abatement costs; one notable exception is the application of separate market models for the industrial, domestic, and transport sectors in the United Kingdom (82). However, such models may come to assume much greater importance as governments move closer to considering detailed policy measures tailored to specific sectors. Frequently they focus on sector responses to carbon taxes rather than costs (82).

Of more general interest for costing is the recent application of models known usually simply as macro-economic models. This is the name usually (if imprecisely) given to the models developed over many years for studying the short-run dynamic behavior of national economies. Typically these models contain explicit representation of investment and consumption in different sectors, and markets do not necessarily clear; there can be unemployment, idle production capacity, or capital shortage. Such models generally contain a strong Keynesian component, though many other aspects of macro-economic theory have also been brought to bear in them. Recent applications to CO_2 abatement are discussed below.

Such models can generate a wide range of macro-economic indexes such as GNP, inflation, employments, etc. For this reason they are of particular interest for assessing the short-term macro-economic impact of CO_2 abatement. However, such models may contain very limited representation of the energy sector, and some may not even model energy as a separate good within the economy. Few contain a representative set of energy-technology options. Such short-run macro-economic models are thus highly country and model-specific, and vary greatly in the extent to which they can be applied to assessing CO_2 abatement. The models are best at representing transitional costs; results may become unstable and questionable when the models are run too far ahead, because the economic feedbacks that keep economies from straying too far from economic equilibrium are generally not well represented.

No study has focused primarily on a comparison of short-run macroeconomic with general equilibrium (GE) models, but such a comparison is implicit in the studies of Shackleton and co-workers (51). This took two macroeconomic models (DRI and LINK) and two general equilibrium models (DGEM and Goulder), all of which the model authors considered appropriate to run over a period of 2–3 decades. Each was subject to the same carbon tax ($40/ton C). The models behaved in very different ways. Concerning the impact on CO_2, the macro models suggest a reduction of 0–8% depending on the way in which carbon tax revenues are used; even a CO_2 increase is observed from a tax recycling that boosts economic growth; the equilibrium models suggest 20–30% reductions depending only on which model is used. Even more striking, the macro models show a wide variety of GNP responses, varying greatly over time, and including substantial increases from some tax recycling options; the GE models show a much smoother response, with more modest impacts.

All this corresponds to the theoretical differences. The short-run macro models reflect the resistance of the economy to change, but also the cumulative impact of greater investment unconstrained by equilibrium requirements. The GE models allow capital to move easily across the economy to respond to the price changes, giving a much stronger CO_2 response. It seems reasonable to suggest that the longer-term results from the short-run macro-economic models are questionable (and highly sensitive to various assumed macro-economic investment and other responses), as are the short-term GE results; the most realistic outcome may be to assume a progression over time from the macro toward the GE results, but even this is speculative. Beyond this we cannot generalize, but we emphasize the importance of resolving such great variation based purely on model structure. McKibben and Wilcoxen (91) are the only researchers to have yet applied a model that can simultaneously address unemployment and GE class problems.

Finally we note that between the short- and long-run models, some progress has been made in developing medium-run models that combine elements from both short-run macro-economic and resource allocation/equilibrium models (see discussion in Reference 89). These may be seen as short-run models that are expanded to include adjustment to long-term equilibria. That is, medium-run models are mainly demand-determined and allow for market disequilibria; however, a central part of the models aims to describe the adjustment process from short-term market disequilibria to long-term market equilibria. Finally, there are some top-down modeling approaches that do not fit into these categories at all, such as the growth models employed in References 28 and 40.

Sectoral Coverage: Energy versus General Economy. Another important distinction is that between models that address only a limited part of the economy (in this case, the energy sector) and those that encompass the whole economy, which usually have a much more simplified representation of any particular sector. Among equilibrium models, the distinction is known as that between partial equilibrium and general equilibrium models; the parallel distinction for short-run models is that between energy market and macro-economic models; a more general terminology would be that of [energy] sector and general economy model.

Sectoral models focus heavily or exclusively on particular economic sectors (in this case, usually just the energy sector or parts thereof); insofar as the rest of the economy is represented, it is in a highly simplified way. They address the problem of describing behavior in a single area of the economy, for example energy, but ignore or treat summarily all other economic functions. An energy market model would have no description of the labor or capital markets.

Energy sector models come in many varieties. Bottom-up technology models are all sectoral; so are many equilibrium models, such as the ERB (3,4). Investment planning models, such as the electricity expansion models widely used for assessing power sector investments, are more focused sectoral models. Models of the international oil market—a very important but little analyzed issue in assessing abatement costs—are sectoral but global models.

Sectoral models yield an estimate of costs in a particular sector, but cannot take account of the macro-economic linkages of that sector with the rest of the economy. They cannot, for example, estimate how the labor or investment requirements in that sector may affect the resources available to other sectors.

By contrast, general economy models encompass all principal economic sectors simultaneously. They recognize feedbacks and interrelationships between sectors. In principle, this enables them to estimate the full long-run GNP impacts of restrictions (such as CO_2 constraints). In practice, this is obtained at the expense of considerable simplification of the energy sector.

General economy models are the only kind that can reflect important features in the rest of the economy. This may include, for example, economic distortions outside the energy sector. In particular, existing taxes impose varying burdens on economies. If the revenues from a carbon tax are used to reduce such taxes, the gains may in principle offset the losses, completely altering assumptions about the net impact on the GNP of such taxes. This seems to have attracted attention only recently (92); relatively early discussions of this are given by the CBO (43) and Grubb (93), and subsequent modeling studies are reviewed.

The past few years have seen the development of linked models that seek to integrate the detail of energy sector models with the economic consistency of general economy models. In Europe, the macro-economic HERMES model has been linked with the MIDAS energy supply model and applied to analyzing abatement strategies in the four largest EC countries, including interactions between these economies (60), to generate the results illustrated in Figure 6. The Manne and Richels GLOBAL 2100 model (33) contains considerably more energy supply detail than other general equilibrium models by integrating an energy technology model (ETA) into a simple general equilibrium model (MACRO, which clears markets for all goods and services treated as a single aggregate). The SGM (84), extends this approach. It is a general equilibrium model that was designed to address greenhouse-related issues, and thus desegregates economic activities on the basis of importance to the greenhouse issue.

The ERB model is an energy market rather than general economy model, but it contains a GNP feedback parameter, which was incorporated to ensure that the impact of large changes in energy sector costs on the scale of economic activity would be reflected in terms of its impact back on energy demand. The parameter is not intended to provide consistent estimates of the costs of emissions reductions, though many studies have used it as such. As noted above, costs must be developed using consumer plus producer surplus techniques (31,32). Other sectoral models, such as Fossil (94) require similar approaches to make consistent cost estimates.

Optimization and Simulation Techniques. Models adopt different approaches to optimization. Some models optimize energy investments over time by minimizing explicitly the total discounted costs (or per capita consumption), using linear or nonlinear techniques. Several engineering models use linear programming, most notably the EFOM model (31) used extensively within the EC, and MARKAL (32), promoted internationally by the ETSAP program of the International Energy Agency. The top-down Global 2100 model uses nonlinear dynamic optimization, as does its derivative CETA (39). Such optimization approaches in effect assume perfect investment (within the confines of the model), with perfect foresight. It is debatable whether this is a drawback or advantage.

Modeling partial foresight is, however, difficult, and the main alternative approach is to simulate investment decisions on the basis of static expectations, ie, static projection of conditions at the time of investment. This myopic assumption is used in ERB, GREEN, and the CRTM trading derivative of Global 2100. Moreover, no software yet exists for solving such large dynamic general equilibrium models under the assumption of perfect foresight. The SGM model incorporates a variety of options for determining investments on the basis of future expectations (including a formulation of partial foresight).

The mechanism for selecting investments is fundamentally different between investment simulation and optimization models, and this might be expected to have a significant impact on results. In fact, this does not appear to be the case. A recent comparison by the Energy Modelling Forum (unpublished) of results from Global 2100 (dynamic optimization) with those from the ERB model (investment simulation) shows that standardizing for key assumptions leads to remarkably similar energy and emission results. Assuming a competitive economy, variations in key input assumptions (such as those discussed below) thus appear far more important than the approach used for selecting investments in the model.

Short-run macro-economic models are concerned primarily with the simulation of aggregated investment responses in terms of labor, capital, etc, but do not seek to optimize investments; they do not represent specific technologies at all. Some other models are purely for simulation of system operation, without automated investment modeling, and they report on the implications of an investment strategy that is specified externally. This can enable much greater detail in representing the system, and avoids the limitations of linear optimization especially, though there are inevitably drawbacks from having to specify investments manually and check their consistency (21).

Level of Aggregation

Models differ greatly in their degree of disaggregation. To some extent this is the obverse of the model scope. Models that, for example, focus on household electricity demand can represent this and the options for improving household electricity efficiency in great detail. Global, economy-wide models have to be highly aggregated.

At one end of the spectrum are models such as LEAP (95) and the BRUS model (96) used in the Danish Energy 2000 study (97). Their demand sectors are generally disaggregated with respect to specific industrial subsectors and processes, residential and service categories, transport modes, etc, with the aim of achieving homogeneous entities whose long-term behavior can be defined through consistent scenario projections. Similarly, energy conversion and supply technologies are represented at the plant-type and device levels. This allows detailed modeling of the alternatives for technical innovation, fuel switching, etc.

With regard to emissions of pollutants and CO_2, this type of disaggregation into specific technologies makes it possible to take account of the different characteristics of energy technologies. A detailed analysis of abatement options can thus be carried out. This includes energy savings at the end-use level, changes in the conversion system, and fuel substitution. At the other end of the spectrum are models such as GREEN and Global 2100, which treat energy and the world in a highly aggregated manner.

In general, the level of aggregation is closely related to the other aspects; for example, a multiperiod, global, general-economy model by necessity has a highly aggregated representation of energy demand and supply, with little if any technological detail. Great detail in representing energy supply, conversion, and end-use markets and technologies is only possible in models that are specific to the energy sector, and focus on simulation rather than full system optimization. The benefits need to be weighed against these limitations.

Modeling Classifications

An understanding of the differences that are incorporated into models is required because each have different strengths and weaknesses. Models for studying CO_2 abatement costs have been developed in different ways, often by adapting existing models. They are able to handle some issues (or sectors) better than others; different models are thus suited to different purposes. Models cannot, or at least should not, be interpreted as giving complete and accurate answers, but rather used for the insights they offer when the results are combined with an understanding of the model structure and limitations.

There is no universal or accepted way of classifying models. Not all the attributes of model scope and type have been reviewed here. Attributes of particular importance are

- Time horizons and adjustment processes: short-term transitional versus long-term equilibrium.
- Sectoral coverage: energy versus general economy.
- Optimization and simulation techniques.
- Level of aggregation.
- Geographic coverage, trade, and "leakage" (20).

The division between top-down economic and bottom-up engineering is of great practical importance, despite its occasional ambiguity. Within top-down models, the distinction between long-run equilibrium and shorter-run macro-economic models is central, as is that between partial (sectoral) and general economy models. Neither of these latter distinctions is relevant to bottom-up models, for which important distinctions are those between partial models (generating a cost curve of savings related to a top-down or unspecified baseline), and full system representation, and within the latter, the choice between optimization and simulation. Table 4 classifies the primary models discussed in this paper according to this scheme, which has a pragmatic focus on the factors of greatest importance to abatement costing.

NUMERICAL ASSUMPTIONS AND SENSITIVITIES

Assumptions drive model results. Critical parameters can be usefully, though not exclusively, divided into those that govern the overall scale of the system and reference mis-

Table 4. Model Classification and Examples Applied to CO_2 Abatement Costing

	Bottom-up	Top-down						
		Equilibrium (resource allocation)			Growth	Energy market/macro-economic		
Sectoral division Models/authors	Partial		Linked	General		Partial	Linked	General
National	OTA NAS Mills (and many others)	MARKAL	ETA-MACRO MARKAL-MACRO	DGEM (Jorgensen and Wilcoxen) Goulder Bergman Glomsrod Proost and Van Regemorter		eg, electricity market models Ingham and Ulph	GDMEEM (N. Goto)	LINK MDM (Barker) ORANI (Marks)
Regional	EFOM	MIDAS					MIDAS-HERMES	QUEST HERMES G-CUBED (Mckubben and Wilcoxen) DRI
Global	Goldemburg	ERB	Global 2100 CRTM CETA	SGM GREEN Whalley and Wigle WEDGE	(Optimising) DICE (Nordhaus) (Non-optimising) Anderson and Bird	eg, international oil market models		

sions, and those that directly affect the relative cost of emissions reductions. GNP and population growth rates, income elasticities, and the rate of "autonomous end-use energy-intensity improvement" primarily affect the baseline scale; background fossil-fuel prices strongly affect both the baseline emissions and abatement costs; the cost of low carbon technologies and price elasticities largely drive abatement costs though they also affect baseline emissions.

Population and GNP Growth Rates

The demand for energy is driven by population and per capita energy demand, and all economic models at least assume that the latter is driven by per capita GNP. It is consequently much more difficult to restrict the growth in emissions for a developing country with a high population and economic growth rate, such as India, than for a more slowly growing developed economy, for example, Germany, which has a static or declining population.

Uncertainties in future global population are reflected in abatement studies; estimates for the year 2025 for example include 9.5 billion for the NAS (49) and 8.2 billion for Edmonds and Barns (31). However, the latter study found that there was little impact from a reduction in population growth for approximately 15 years, that is until labor force and therefore GNP was affected. Projections suggest that the increase in global GNP will be much greater and more uncertain than population growth, and so differences in GNP projections account for a greater part of variation in baseline emissions.

Different baseline GNP and energy demand assumptions across studies complicate comparisons of the GNP loss associated with a target CO_2 reduction. In models that derive GNP from labor productivity, this is correspondingly critical. The difference that baseline GNP makes to energy demand is more complex, as growth in any economy inevitably varies across sectors over time; greater GNP growth in reality would not necessarily imply that the growth within all sectors of the economy is increased proportionately.

Energy/GNP Relationships

Whereas GNP is a principal determinant of energy demand, many factors can affect the relationship between them. A few models incorporate explicitly a non-unitary energy-income elasticity, which implies a changing energy/GNP ratio as GNP grows. Most models, however, express such a change, if any, in terms of an exogenous parameter that defines the rate at which the energy/GNP ratio would change in the absence of price changes. This rate of exogenous (or autonomous) end-use energy-intensity improvement (AEEI) then becomes a determinant of baseline energy demand for long-term projection; the higher the rate of energy-intensity improvement, the lower the baseline CO_2 emissions, and the lower the costs of reducing relative to a given base year. The parameter has been widely described as a measure of technical progress, but it compounds many different elements including for example composition of energy using activities and the influence of non-price policy interventions.

Future Energy Prices, Resource Modeling, and Supply Elasticities

High background fossil-fuel prices lower energy demand and CO_2 emissions, and reduce the relative costs of moving to lower carbon fuels. Limited resources (eg, of oil) have the same effect, implicitly or explicitly raising the prices as resources are depleted. Resources are, however, uncertain, and the course of fuel prices is even more uncertain.

Various approaches may be taken toward estimating the future cost and availability of different fuels. National studies may define national production costs but define exogenous global prices with great variation.

Global models reflect resource/supply cost issues explicitly, through direct estimation of the resource base and supply elasticities. High supply elasticities mean lower fuel price rises as supply increases. GREEN assumes zero supply elasticity for oil outside OPEC (ie, volume set by fixed production constraints), but price is determined by OPEC supply elasticities varying from 1 to 3; supply elasticities are higher for gas and coal and much lower for nonfossil sources until backstop technologies become relevant.

Price/Substitution Elasticity of Demand for Energy

The impact of price changes on energy demand is determined by the price elasticity of energy demand, or in general economy models, the substitution elasticity between energy and other factors of production. The lower the relevant elasticity, the less energy demand is curtailed by higher energy prices, and the greater the tax that is required to reduce energy demand and consequently CO_2 emissions.

In the global models, assumed long-run elasticities range from −0.3 to −0.4 for the Manne and Richels U.S. and Global 2100 studies to −0.6 to −1.0 for the OECD's GREEN model; short-run elasticities are about one-tenth this value.

The notion of elasticities assumes a symmetric response to price changes; if prices rise and then fall back to previous levels, the energy-intensity (after allowing for lags) returns to former levels. This basic assumption, and the values assumed for modeling, are disputed and have been particularly called into question by recent trends and studies.

Technology Developments and Costs

Assumptions concerning the cost and rate of implementation of more efficient or lower carbon technologies affect both baseline emissions and relative abatement costs. The initial results for the United States (33) were strongly criticized (98) as being based on unreasonably pessimistic assumptions for efficiency improvements and the costs of alternative supply technologies. In response, Manne and Richels (99) examined three different background scenarios, which they termed technology optimistic, technology intermediate, and technology pessimistic. The last of these corresponds to their initial famous estimate that CO_2 abatement could cost the U.S. $3.6 trillion over the next century, and yields some of the higher cost points on Figures 3 and 5 above.

The assumptions for the technology optimistic scenario reduced these total costs by a factor of 20 for the (fixed) abatement target set, because of the combined impact on baseline emissions (primarily from the higher AEEI) and the halved costs for backstop low carbon supply technologies. As a result, Manne and Richels noted that "the direct economic losses are quite sensitive to assumptions about both demand and supply . . . for the losses (from carbon constraints) to approach zero; however, the most optimistic combination of supply and demand assumptions must be adopted." This confirms that results are sensitive to the assumptions concerning technology costs, a result again noted in the sensitivity study in reference 31. Almost all studies fix the costs of supply technologies as exogenous data; Reference 28 contains the only study in which technology costs decline with increasing investment.

Energy Sector Impact on GNP

The impact of changes in energy demand and energy sector costs on GNP is complex. General economy models capture the relationship consistently, but the resulting elasticity can still vary considerably. For example, the Global 2100 modeling approach (and by implication, the CRTM and CETA models also) contains a nested CES (constant elasticity of substitution) production function to relate energy input to economic output. (A CES function has the form $Q = (\sum_{i=1}^{N} a;x_i^{\rho})^{1/\rho}$ where Q is output, x_i for $i = 1. . . . N$ are inputs, and a; for $i = 1, . . . , N$ and ρ are parameters.) Cline (100) criticizes the parameters chosen, claiming that they yield an excessive impact of energy sector changes on GNP, a claim disputed in reference 99.

For sectoral or partial equilibrium models, the impact of energy sector costs or carbon taxes on GNP may be estimated, by a direct elasticity (GNP feedback) parameter, as available in the ERB model. This approach has been criticized for overestimating the feedback and consequently overestimating the cost of reducing CO_2; Reference 31 recognizes this limitation and suggests the use of consumer plus producer surplus changes as the best method of computing cost for the ERB model.

Interfuel Substitution Elasticities

Substantial CO_2 reduction can be achieved by switching toward less carbon-intensive fuels. In models having a purely engineering approach to supply this is captured by supply technology costs; for models with econometric supply modeling, it is governed by the interfuel substitution elasticity. GREEN assumes a long-run interfuel elasticity in production of 2.0 and a short-run value (reflecting existing supply infrastructure) of 0.5. Halving these values lowered global baseline emissions by 13% in 2050; the impact on abatement costs may be expected to be much larger unless the bulk of substitution is governed by backstop technologies.

KEY DETERMINANTS

The range of reported results on the costs of limiting CO_2, and the technical modeling and data issues that affect such estimates, have been examined in a recent review (20). The gulf between "technology-engineering" and "economic" perspectives, in terms of assumptions concerning energy demand and policy, structural, and technological assumptions as well as the abatement strategy and scope of analysis affect the reported results.

It would be easy to suppose that the differences between economic and engineering views are confined to a few modeling parameters. They reflect very different perspectives, almost paradigms, about driving forces in the energy economy. A report from a UNEP workshop that sought to bridge the divide (101) observed that:

> To economists, energy is a factor of production: it is an input into economic growth, and one which can substitute for labor or capital, depending on relative prices. While energy-efficiency may improve due to technical development, so does that of other factor inputs; and efficiency improvements lower the relative price of energy, increasing the extent to which it may be applied. Also, fossil fuels dominate because of demon-

strable convenience and low cost. Thus whilst recognizing the potential importance of technical improvements, and even market imperfections which prevent optimal energy use, to most economists there is every reason to expect energy consumption to grow with expanding GNP, and no particular reason to expect technical developments to reduce CO_2 emissions relative to the business-as-usual case without incurring substantial costs. It is a compelling case, with much long-term historical weight behind it.

To scientists and engineers, on the other hand, energy is not an abstract input but a physical means to particular ends. The applications which consume much energy are those of heating, heavy construction, metals, etc; basic infrastructure and comfort; and travel. In developed economies, most infrastructure needs have been met, travel may be approaching limits of congestion and time budgets, and much new economic activity is in areas which consume trivial amounts of energy, such as information technology, general entertainments, etc. Thus to most scientists and engineers economic growth is becoming less and less relevant to energy needs (in developed economies). In addition, very large technical improvements in efficiency, which need not incur much extra cost, are readily demonstrable; technology is powerful and adaptable to changing conditions (such as requirement for nonfossil sources); and it is hard to believe that human society is incapable of finding ways of putting such options into effect (this applies primarily to developed economies, but may also be of great relevance to developing ones if they can move directly to advanced technologies). This too appears to be a strong case, with at least partial support from recent trends, but it leads to a very different outlook from the economist's outlook sketched above.

This shows that the different perspectives lead to widely different assumptions concerning several factors: the relationship between energy demand and future economic growth in the absence of any abatement measures; the scope for exploiting more efficient technologies; and the scope for developing new technologies as needs (such as CO_2 reduction) require. The complex issues involved in each of these have been reviewed recently (20).

Baseline Energy Demand and the AEEI

Energy demand in the absence of abatement measures depends on GNP (population times per capita GNP) growth rates; energy prices, and the response of demand to these; and the parameter usually translated as the rate of autonomous end-use energy-intensity improvement (AEEI). The impact of GNP and energy price changes is recognized by all analysts, as are the uncertainties. The main contrast in views arises from the differing assumptions about the AEEI which has an enormous impact. Reference 83 states that "unfortunately there is relatively little backing in the economic literature for specific values of the AEEI . . . the inability to tie [it] down to a much narrower range . . . is a severe handicap, an uncertainty which needs to be recognized."

Among the principal global and U.S. studies, adopt the lowest values have been adopted for AEII (0 and around 0.5, respectively) (33,48,77,78). Williams (98) strongly criticized such values as too low; Manne and Richels (102) defend their value of AEEI on the grounds that it appears consistent with observed trends in the post world war two United States. However, it is difficult to separate the various factors in their analysis (eg, price, income distribution, and time sampling effects), and Wilson and Swisher (81) strongly dispute their interpretation, concluding that "one can produce an experiment that justifies whatever AEEI one likes within very broad ranges."

Whatever the value for AEEI, it is important to understand what it is. The parameter is badly misnamed: it is a measure of all nonprice-induced changes in gross energy-intensity, which may be neither autonomous, nor concern energy-efficiency alone. It is not simply a measure of technical progress, for it conflates at least three different factors: one is technical developments that increase energy productivity, but another is structural change, ie, shifts in the mix of economic activities (which may require widely different amounts of energy per unit value added). The third is policy-driven uptake of more efficient technologies, due to regulatory (as opposed to price) changes, greater than would occur without those changes.

Technical change is indeed hard to predict. Studies by (103,104) suggest that in manufacturing alone, technical change has increased energy productivity in OECD countries by about 2% per year for at least the past two decades; this includes price effects, but in fact there is no clear change in the trend that correlates with the energy price shocks (perhaps because of the lags in manufacturing equipment). Technical change appears more closely related to the price shocks in other sectors.

Structural change incorporates the phenomena of saturation in energy intensive activities (such as home heating and primary heavy industries), and shifts toward less energy-intensive activities. A range of studies have noted that structural change, both between sectors and within manufacturing industry, has played an important part in restraining energy demand in the OECD in the past 20 years (104). Williams and co-workers (105) provide considerable evidence for expecting the observed trend in manufacturing to continue. On the basis of this and other relevant literature, and various saturation effects, reference 93 argues that an autonomous structural trend toward lower energy-intensities (ie, rising AEEI) is to be expected as countries develop and as economic growth moves toward increasingly refined products and services.

Low values of AEEI in long-term studies (significantly below 1.0, especially for OECD countries) are dubious because of saturation and structural change effects. Low values make any fixed target much harder to reach, also increasing the scale and hence relative costs of reductions. This is, however, tentative; there are substantial uncertainties and a clear need for greater understanding of technical and structural trends, and integration of such studies in abatement cost modeling.

The AEEI is not the only controversial issue surrounding the relationship between GNP and energy demand. Considerable uncertainty also surrounds estimates of the price elasticity of energy demand. Most studies have sought to estimate elasticities from the response of demand to the energy price rises of the 1970s and early 1980s. More recent trends, if anything, increase the uncertainties. Although energy demand has started to rise again in OECD countries following the price falls since the mid-1980s, the response has not been nearly as great as predicted by the simple reversible statistics of elasticity. Engineers have long maintained that the efficiency gains would not be lost, because they are embodied in better knowledge and techniques that will not be abandoned even if energy prices fall.

Regulatory Instruments and the Energy-Efficiency Gap

Most top-down models assume the optimal operation of markets in response to observed price signals, in which case economic theory suggests taxes to be the optimal means of abatement. The observation of the large efficiency gap demonstrated by engineering studies calls this into question. Many economic discussions reject the relevance of this data.

The efficiency gap comprises many different components. If it can be wholly explained in terms of lags in take-up, unavoidable hidden costs, etc, then it is a phenomenon of little direct relevance to the actual costs of limiting CO_2 emissions. However, many barriers to the uptake of more efficient technologies have been identified; an extensive and clear analysis of the different kinds of barriers has been given (106).

Some things can be done to improve the uptake of efficient techniques, for example, with government campaigns to improve information and the awareness of consumers of energy-efficiency and conservation. Most top-down studies in effect assume such measures to be enacted irrespective of CO_2 abatement, and ignore or dispute the scope for other more direct cost-effective actions. However, experience and modeling studies of regulatory policies demonstrate that such measures can and have been effective.

Regulatory changes to encourage utility demand-side management programs have been widely advanced as a way of capturing the "free lunch" by getting utilities to invest in end-use efficiency. The hidden costs in such policies are debated. Joskow and Marron (107) conducted a survey of experience in U.S. programs and concluded that "reported costs exceed those of the technology potential analysis because program costs are higher and energy savings are lower than these studies assume . . . although many of the programs still appear cost effective."

The important fact remains that modeling studies of specific regulatory options frequently yield lower, rather than higher, estimates of abatement costs than those derived from carbon taxation designed to achieve similar abatement. Such studies of regulatory measures thus contrast with the common economic assumption that regulatory options are more expensive than using economic incentives. This is because such policies address areas of significant market failure.

So how large is the potential "free lunch" of zero-cost energy-efficiency improvements captured by regulatory change? End-use technology studies frequently suggest (104) a long-run technical potential to reduce energy demand without extra costs by 20–50%. Schipper and coworkers discuss the implementation and potential of energy-efficiency programs across the OECD in detail. They too conclude that there is large potential, but emphasize that the achievable potential will always be substantially smaller than the apparent technical potential, and that exploiting it will depend on more aggressive and sophisticated policies. Taking account of the various implementation issues points at best to the lower end of the 20–50% range available.

A different view is given by other comparisons of top-down and bottom-up results. A critique and comparison of data from economic modeling is presented (81), and indicates that bottom-up studies suggest cheaper abatement right across the range, up to abatement of 70% or more. These data suggest that the whole cost curve from the top-down studies reviewed by Nordhaus has to be shifted by 20–40% to reflect the technical opportunities identified by bottom-up studies, which would greatly alter the pattern shown in the graphs of Section 4. In effect this means that if the baseline in top-down studies is equated with a "business as usual" rather than a "least cost" path, all the points derived from top-down models in the scatter diagrams of Section 4 should be shifted perhaps 20 percentage points to the right so that emissions reductions of 20% are costless. This excepts the shorter-run studies run over 10–15 years, when a "free lunch" potential of perhaps 1% per year (ie, an increase in the AEEI by 1 percentage point) below baseline might be more appropriate.

The data from studies are too scattered to reach a definitive conclusion; the actual situation is likely to lie between these extremes, implying a need for some reduction in abatement costs from top-down models across the range to allow for the potential of regulatory-driven improvements in energy-efficiency. A further factor is that, with expanded markets for more energy-efficient technologies, these technologies might develop faster, thus permanently raising the AEEI.

Technology Assumptions

Ultimately, the costs of limiting CO_2 emissions will depend heavily on the technologies available, not only technologies for more efficient use of energy, but also for the production, conversion, and utilization of lower carbon energy sources. The importance of technology, as well as assumptions concerning its developments and costs, has been widely acknowledged: it forms the central element of Williams's (98) critique of the Manne and Richels (78) conclusions, and sensitivity studies by Manne and Richels (102) and Edmonds and Barns (31) have demonstrated that estimates of abatement costs depend crucially on technology assumptions.

Yet the care with which such assumptions have been developed varies widely, and the models employed to date have limited representation of technology issues. Models that do incorporate some explicit representation of technology include the Global 2100 model and its derivatives (CRTM and CETA), more recent variants of the OECD GREEN model, the SGM model and the ERB model, which has a fuller representation of technology in compensation for its weaker macro-economic linkages. Sensitivity studies with all these models illustrate the crucial importance of technological assumptions.

To establish reasonable assumptions, it is pertinent to start with current data, and visible trends and options. Supply-side options span technologies that are proven and largely mature (such as combined-cycle gas turbines (CCGTs) proven but still developing (such as wind energy and higher efficiency clean coal conversion), confidently predicted (such as integrated biomass gasification), to a wide variety of lesser or more speculative options. Most recently, an immense study of the prospects for modernized renewable energy technologies (108) argued that these could meet about half global energy demand by the middle of the next century at little if any additional cost; the studies of wind energy in this volume estimated that costs of modern wind turbines were already almost competitive against coal power stations for large-scale exploitation in countries such as the United States with extensive wind energy resources. None of the CO_2 abatement models contain explicit representation of wind energy technology, and for large countries such as the United States and the former Soviet Union, this alone might substantially lower abatement cost estimates.

No models can capture the full range of options available; by inevitably excluding some, there can be a builtin tendency to overestimate abatement costs (unless this is offset by using over-optimistic assumptions concerning those that are included). Some abatement studies use data already being rendered obsolete, as options already identified for cost reductions are exploited. Williams (98) provides one of the most detailed critiques of the technological assumptions used in the Manne and Richels base case assumptions and their derivatives, and argues that many of the assumed costs are higher than can already be predicted with confidence.

This naturally leads to the issue of technology development and cost reductions. This is uncertain terrain, but not a complete black box. Technologies do not arise, improve, and penetrate markets at random, especially for large and complex technologies such as those involved in energy provision. Technology development follows market demand, with the associated public and private R&D investment and learning processes as technologies develop toward market maturity. Yet despite this well-attested and understood fact, almost all the abatement costing studies to date model technology development as exogenous—the costs of abatement technologies are defined as input data and do not vary explicitly with the level of investment, incentives, and market penetration in the model. That alone must be considered as a severe limitation.

A different approach to the issue of endogenous technological development is that by Hogan and Jorgenson (109). This econometric study related changes in productivity trends (which are equated with technical progress), in different sectors of the U.S. economy, to price changes in the different inputs. Although energy productivity did improve when energy prices rose, this was more than offset by reduced productivity growth in other factor inputs at the same time. They found that overall, "technology change has been negatively correlated with energy prices . . . if energy prices increase, the rate of productivity growth will decrease." However, such results may be sensitive to the model specification, and as argued in Grubb (93), the transition from "has been" to "will" in this excerpt conceals the importance of innumerable extraneous factors in the years analyzed, most notably the macro-economic impact of the sudden and externally imposed oil price shocks. It is highly debatable whether the data reveal anything useful about the economy-wide and long-term technological impact of smoother price changes arising from domestic policies such as carbon taxes and other abatement policies.

The potential for substantial cost reductions associated with larger-scale deployment of low CO_2 technologies, combined with the observations above about possible irreversibilities in the impact of price changes, points to the possibility that there may be various choices of technological trajectories differing little in cost. One is to continue along a carbon-intensive path. Another is to invest enough to alter the course of new investments over the next decade or two toward more efficient and low carbon technologies. As investment patterns and institutions and infrastructures adapt to these new technologies, their costs will fall, perhaps until they become the naturally preferred options. The world would be on a different technological trajectory. Although the transition may be costly, especially if it is forced rapidly, given the nature of technology development and economies of scale it cannot be assumed that this would be a much more costly long-term path (110,111).

The prospects for technology development, production-scale economies, and exploitation of bifurcations to lower emissions, suggest that the cost estimates in many economic studies are implausibly high. Indeed, some use data that appear to indicate costs higher than those of some currently identified technologies, and make little or no allowance for future technical improvements, especially in nonfossil sources. However, there are considerable constraints on the rate at which such technologies could be developed and deployed, as modeled most clearly by Anderson and Bird (28), the deployment of significant new supply technologies will take many decades. On these ground, the higher long-term (beyond ca 2025) costs illustrated in Figures 3–7 may be considered relatively implausible.

Subsidies, Tax Forms, and the Use of Tax Revenues

Various forms of taxes and subsidies can be used to limit emissions. Different types of taxes lead to different reductions in CO_2 and to different impacts on the economy. Most economic models assume abatement to be achieved by a carbon tax, imposed on the carbon content of primary fuels. However, taxes could be applied to subsets of fuels, downstream on derived fuel products, or otherwise not in proportion to carbon content of fuels, eg, on gasoline only or on the energy content of fuels. This generally results in greater economic costs (lower tax efficiency). Thus a gasoline tax is less efficient than a carbon tax at reducing carbon emissions (112), and taxes on electricity production are much less efficient than a tax on input fuels; the latter do not encourage fuel switching, only reducing CO_2 by depressing demand for electricity (113).

Of greater relevance to the assessment of total abatement costs is that energy production in most countries is subject to a complex set of taxes and subsidies, and abatement costs inevitably depend on the existing tax structure. Where heavy taxes are already imposed (as with oil products in many OECD countries), the macro-economic impact of additional taxation is likely to be greater than in the absence of existing taxation; conversely, where energy is subsidized, removal of subsidies (or equivalent taxation) often yields macro-economic benefits. Many models ignore initial subsidies and taxes.

If the tax revenues are taken out of the economy (eg, unaccounted for or spent abroad), all impacts on the national economy are negative. For other uses of the tax revenues, different macro-economic indexes frequently move in different directions. For the carbon tax levels considered (up to about $80/t C), nearly all these studies find some ways in which the net effect is to boost GNP, as compared to a projection in which existing tax structures are unchanged.

A carbon tax raises money largely from consumption; this may be transferred to qualitatively different economic activities. If the revenues are used to stimulate investment directly, this increases savings and reduces consumption with a short-term negative impact (though obviously, other benefits are associated with deficit reduction). Conversely, if the revenues are used to boost investment, the resulting GNP growth in part reflects simply the transfer of resources from consumption to investment.

Carbon taxes at the levels considered may be genuinely less distortionary than the most distortionary existing taxes. There is still some debate as to how valid it is to count such changes as a credit for carbon taxation. As with the efficiency "free lunch" debate, should not governments make the current tax structures optimal irrespective of abatement efforts, with gains that should not be credited to carbon taxation? There are no easy answers to this. Certainly it is hoped that tax structures become more optimal over time. But there are also real constraints on taxation policy, and objectives other than just efficiency; Hourcade (114) notes growing political and trade constraints on traditional taxes and concludes that "it is timely to consider the taxation of 'bads' (such as pollution) as an answer to the general problem of raising revenues." In any case, if carbon taxes represent an efficient way of limiting emissions, it would be perverse to say that the revenues should not be used in the most desirable way, or alternatively, to say that economists should not model reality because the starting point is not optimal.

For longer-term assessment, current quirks of taxation systems may be less significant, but it is still important to recognize that governments need to raise revenues; carbon taxes will reduce the revenue required from other taxes and models need to reflect this reality. In general, the distortionary impact of a tax increases nonlinearly with its level, so it is efficient to spread the tax base as broadly as possible. Consequently, even in the absence of a CO_2 problem, the optimal level of energy/carbon taxation would not be zero.

Modeling studies that neglect such distortions in the rest of the tax system—and the consequent potential gains from carbon tax recycling—tend to overestimate the GNP impact of carbon taxation. This includes all the long-run (post 2010) points in Figures 3–7 and many of the shorter-term studies as well, where the practical GNP gains may be more significant.

Scope of Abatement

Climate change is a global problem, and the more countries that take part in abatement, the lower the costs of limiting global emissions. Studies emphasize that action by industrialized countries may have a significant short-term impact, but that the costs of restraining global emissions rise rapidly if developing countries are not soon included. Furthermore, even if all the principal countries participate, the costs are greater if each is bound to fixed emission targets, because some regions may then incur much higher costs than others as noted by, eg, Edmonds and co-workers (87).

Table 5 shows one estimate of the impact of allowing countries freedom to choose where to invest to limit emissions (ie, trade in emission commitments). Despite a no-trade case that involves similar rate of emissions reduction below baseline (2% per year), there are still significant differences in marginal abatement costs (reflected in the tax rate) between regions, and thus savings (10–45%) to be made from trading. Less uniform (relative to baseline) initial commitments would increase the benefits of trading accordingly; an extreme case was a special run of the Whalley-Wigle model (77,78) that showed the costs of obtaining the same degree of abatement, but with equal per capita emissions globally, would be roughly twice the costs of achieving the same reductions if emissions trading is allowed. Clearly models that do not allow such trade report costs greater than necessary.

Rate and Pattern of Emissions Abatement

Most studies, including those reviewed, have focused on measuring costs for a given degree of emissions abatement. It is clear that the costs must also depend on the rate of abatement. This is partly because relatively slow abatement can be achieved by introducing lower carbon technologies as older vintages are naturally retired. The modeling improvement has little effect on the costs of CO_2 abatement at a rate of 1% per year below the baseline

Table 5. Cost Savings from Emissions Trading[a]

		ERB		GREEN	
Target-Date	Trade	Tax ($/tC)[b]	GDP loss, %[c]	Tax ($/tC)[a]	GDP loss, %
2020	No trade	283	1.9	149	1.9
	Trade	238	1.6	106	1.0
2050	No trade	680	3.7	230	2.6
	Trade	498	3.3	182	1.9

[a] Ref. 83.
[b] Global tax or mean of taxes required to achieve CO_2 reduction at 2%/yr below baseline trend in each region.
[c] GDP loss reported by model, not net cost.

trend, but for abatement at a rate of 3% per year below the baseline trend, the costs are much higher when the stock effect is included.

In reality, many other factors also make more rapid abatement more costly. A variety of macro-economic disequilibria come into play; capital or labor cannot move rapidly from one sector to another, in response to relative price and other changes. Since resource allocation/equilibrium models do not capture these effects, they tend (other things being equal) to underestimate the costs of rapid abatement; unlike most other factors noted, this suggests that most of the points in Figures 3–5, most of which are derived from resource-allocation models, underestimate costs in this respect, especially for more rapid (shorter-term and more extreme) abatement.

The observation that short-run macro-economic models indicate a much smaller response of emissions to a given level of carbon tax is another indication that such models show the economy to be more resistant to change than do equilibrium models such as GREEN. A plot of the data from modeling studies against the rate of abatement (not reproduced here) does not, however, reveal a particularly strong relationship. This may be because the results are mostly from equilibrium models, which themselves have a wide variety of capital stock modeling (if any). Few studies with short-run macro-economic models have been carried out, and these focus on different ways of recycling carbon tax revenues; the issue of how costs vary with the rate of abatement does not seem to have been examined methodically with macro models.

A third issue relating directly to the rate of abatement is that of technology development and diffusion. This is inherently a process that takes time, particularly in energy supply. Marchetti and others at IIASA (115) proposed a general principle that it takes approximately 50 years for a new supply technology to become dominant. More detailed technology studies reveal a more complex picture, with the rates of diffusion dependent on the nature of the technology and existing infrastructure; Grubb & Walker (116) argue that electricity supply could change much more rapidly than the "50 years" rule implies, but that fundamental changes to noncarbon transport fuels could be still slower.

Clearly, long timescales are involved, and these estimates do not include timescales for basic research and development. Consequently, rapid reductions will have to use less developed technologies than will slower reductions. This observation has been interpreted as a way of saying that delayed reductions are cheaper, but the implication rather is that costs will be minimized by setting the process of transition in motion as early as possible, because that will give the maximum time in which to develop, deploy, and refine lower carbon technologies. For a given total emissions over the next century, for example, simply delaying initial abatement efforts would both delay the start of such a transition, and increase the rate and degree of abatement that would ultimately have to be pursued.

In this context, the relevance of the whole pattern of abatement imposed in abatement costing studies is noted. The ultimate objective is to reduce accumulation of CO_2 in the atmosphere. Studies that impose a set time path of emissions make no attempt to explore possible less costly emission paths to the same end. The original Manne-Richels studies, for example, show carbon taxes rising to more than $650/t C, settling back down to a level of $250/t C set by the backstop technology. Peck and Tiesberg (39) point out that this is clearly not efficient; greater long-run abatement can be achieved with less long-run GNP impact by a time path that involves steadily rising carbon taxes. A more efficient time path of abatement would thus lower the GNP impact of studies shown in Figures 3–6. By how much is difficult to say, indeed, there are so many issues inadequately addressed in this context (eg, concerning technology development and diffusion) that firm conclusions would be premature, but it is an important area for further exploration.

Externalities and Multicriteria Assessment

CO_2 is but one of many external impacts associated with energy. Other pollutants are produced, and other issues, ranging from energy security to road congestion, are important but not included in traditional economic studies. Attempts to quantify other externalities (eg, Hohmeyer (117); Ottinger and co-workers (118)) suggest these can be significant. A study by Glomsrod and co-workers for Norway (66) appears to be the only one that includes the reduction of these other impacts as

CONCLUSIONS

Our survey and analysis of the literature on fossil-fuel CO_2 abatement costing shows a wide spread of reported results, and there are many different features that explain these differences. The principal factors that affect estimates are as follows:

- The choice of modeling approach and focus, notably
  - top-down or bottom-up
  - associated regulatory modeling
  - short-run macro-economic or long-run equilibrium
  - the linkage between energy sector in impacts and GNP
  - modeling of technology development in response to incentives
- The numerical assumptions concerning
  - energy-GNP trends as governed by income elasticities and autonomous energy-intensity improvements
  - fossil-fuel reserves and associated future prices and supply elasticities
  - the degree (elasticity) and reversibility of energy demand responses to price changes
  - technology development and costs (including production scale economies), or equivalent substitution elasticities
- The nature of the abatement strategy and the scope of analysis:
  - the reflection of existing tax and subsidy distortions and associated use of carbon tax revenues

the scope of abatement and allowance for emissions trade among participants, the rate and pattern of emissions abatement

the extent to which external costs and benefits associated with energy supply are reflected

Bottom-up engineering models tend to underestimate costs by neglecting issues of implementation and other hidden costs. Top-down economic models tend to overestimate costs by neglecting the potential for enhancing structural change and energy-efficiency gains through regulatory policy. We suggest that real cost-free reductions of up to 20% below the baseline projection over a couple of decades is a reasonable estimate of this potential, but realizing such reductions requires extensive implementation of policy instruments to improve market efficiencies.

Within the range of abatement issues and options there are at least four kinds of options that reduce CO_2 emissions but may incur macro-economic gains: regulatory policies to reduce the efficiency gap as noted above; removal of subsidies for carbon fuels; use of carbon tax revenues to reduce existing tax distortions; and the reduction of high non-CO_2 externalities. In each case the possibilities raise similar questions about the extent to which the benefits of associated policy reform should be credited to CO_2 abatement. In each case it is concluded that some credit is justified, but not all the potential should be credited as benefits of CO_2 abatement.

For comparative cost purposes, it is useful to express abatement relative to a baseline projection of expected emissions in the absence of specific abatement efforts. More attention needs to be devoted to definition of the baseline case. However, in general, the following is concluded.

Short-run Abatement Costs. The results for regional and shorter-term studies of abatement costs vary widely; the 15 reported costs of emission reductions of 15 to 40% below baseline (not base year) over the next 15–20 years, for example, range from GNP gains of more than 1% to a several percent loss in GNP. However, the higher losses in this range reflect the fiscal effect of removing carbon tax revenues from the economy, and also ignore the potential for energy-efficiency programs, while the larger GNP gains either reflect the use of carbon taxes to transfer resources from consumption to investment or reflect pure technology potentials. These extremes are of questionable relevance as measures of abatement costs, which in reality may be expected to lie well within these outlier values.

Long-run Abatement Costs. The range in long-run cost estimates is not quite so broad. With few exceptions, long-run modeling results portray that reducing CO_2 emissions at a rate averaging up to 2% per year below baseline, leading to a halving of relative emissions by the middle of the next century, may reduce the associated long-run global GNP by up to 3%, with the lower bound of costs being almost negligible.

For most but not all of the issues listed above, using more realistic or plausible assumptions reduces abatement cost estimates as compared with the most widely reported top-down economic modeling studies. The evidence presented from sensitivity studies and other modeling studies suggests that reasonable allowance for technology development, removal of subsidies and recycling of carbon tax revenues, and the reflection of avoided externalities alone might halve the more pessimistic estimates of abatement costs. Consequently, the realistic range for the GNP loss from halving long-run CO_2 emissions (relative to the emissions in most baseline projections) is 0 to 1.5%.

The upper bound of this range in absolute terms (for example it represents $600 billion out of a projected global GNP of $40,000 billion), but it small compared with other uncertainties and influences on GNP; it is equivalent to reducing average GNP growth rates over 50 years from 3% per year to 2.97% per year. Efficient distribution of abatement efforts between regions and over time could further lower these.

Costs will depend heavily on the rate at which abatement is imposed. The conclusions cited above refer to moderate rates of abatement. Study of this issue is seriously inadequate, but the costs may start to rise sharply for abatement at rates exceeding 1–2% per year below the baseline trend.

Projections by the World Energy Council (119) and many others suggest that global CO_2 emissions, in the absence of CO_2 abatement policies or other policy changes that significantly restrain CO_2 emissions, may grow at a rate averaging around 1.6% per year (±0.4% per year) for many decades. Given all the uncertainties and possibilities for lower abatement costs identified in this paper, the survey and analysis therefore suggests that keeping long-term global CO_2 emissions to about current levels, which is much more severe than current proposals to stabilize emissions in industrialized countries, may if carried out in an efficient manner be expected to reduce global GNP toward the middle of the next century by no more than 1–1.5%, and average GNP growth rates over the period by less than 0.02–0.03% per year.

The real difficulties of abatement lie in the design and implementation of nationally and globally efficient policies, and the geographical, sectoral, and social distribution of abatement impacts. Many analytic issues remain to be resolved, but for constraining CO_2 emissions, the key problems appear not to be massive, macro-economic losses, but rather the politics of implementation, winners, and losers.

Acknowledgments

The work for this paper was supported in part by the Office of Energy Research of the U.S. Department of Energy under contract DE-AC06-76RLO 1830. The authors wish to express their appreciation to those who helped make the paper possible, including Ari Patrinos, John Houghton, Andrew Dean, Robert Shackleton, Wendy Nelson, and Liz Malone.

BIBLIOGRAPHY

1. W. D. Nordhaus, *Strategies for the Control of Carbon Diolide*, Cowles Foundation, Discussion Paper, No. 443, Yale University, New Haven, Conn., 1977.
2. W. D. Nordhaus, *The Efficient Use of Energy Resources*, Yale University Press, New Haven, Conn., 1979.
3. J. A. Edmonds and J. Reilly, *Energy J.* **4**(3), 21–47 (1983).

4. J. A. Edmonds and J. Reilly, *Global Energy: Assessing the Future*, Oxford University Press, New York, 1985.
5. R. F. Kosobud, T. A. Daly, and Y. I. Chang, "Long Run Energy Technology Choices," CO_2 abatement policies and CO_2 shadow, prepared for the *Atlantic Economic Society Meeting*, Paris, France, 1983.
6. S. Seidel and D. Keyes, *Can We Delay a Greenhouse Warming? I~S* Environmental Protection Agency, Washington, D.C., 1983.
7. D. J. Rose, M. M. Miller, and C. Agnew, *Global Energy Futures and CO_2-Induced Climate Change*, MITEL 83 015, MIT Energy Laboratory; also as MITNE-259, MIT Dept. of Nuclear Engineering Cambridge, Mass., 1983.
8. A. B. Lovins, L. H. Lovins, F. Krause, and W. Bach, "Energy Strategies for Low Climate Risk," (R&D No. 104 02 513), report for the German Federal Environmental Agency, International Project for Soh-Energy Paths, San Francisco, Calif., 1981.
9. R. H. Williams, J. Goldemberg, T. B. Johansson, A. K. N. Reddy, and E. Larson, "Overview of an End-Use-Oriented Global Energy Strategy," presented at the Hubert Humphrey Institute of Public Affairs Symposium Greenhouse Problems: Policy Options, May 29–31, 1984, University of Minnesota, Minneapolis, 1984.
10. A. S. Manne, ETA-MACRO projections, in *Global Carbon Dioxide Emissions—A Comparison of Two Models*, Electrical Power Research Institute, Palo Alto, Calif., 1984, Chapt. 2.
11. A. M. Perry, K. J. Araj, W. Fulkerson, D. J. Rose, M. M. Miller, and R. M. Rotty, *Energy* **7**(12), 991–1004 (1982).
12. W. D. Nordhaus and G. W. Yohe, in *Changing Climate*, National Academic Press, Washington, D.C., 1983, pp. 87–153.
13. I. Mintzer, *A Matter of Degrees: The Potential for Controlling the Greenhouse Effect*, research report #5, World Resource Institute, Washington, D.C, 1987.
14. J. Edmonds and J. Reilly, "Future Global Energy and Carbon Emissions," in J. Trabalka, ed., *Atmospheric Carbon Dioxide and the Global Carbon Cycle, DOE/ER-0239*, National Technical Information Service, U.S. Dept. of Commerce, Springfield, Va., 1985.
15. IPCC EIS (Energy and Industry Subgroup), report to the Intergovernmental Panel on Climate Change, UNEP/World Meteorological Organization (WMO), Geneva, 1990.
16. W. Fulkerson, S. I. Auerbach, A. T. Crane, D. E. Kash, A. M. Perry, D. B. Reister, *Energy Technology R&D: What Could Make a Difference?* ORNL-6541, Oak Ridge National Laboratory, Oak Ridge, Tenn., 1989.
17. International Energy Agency, *Energy Conservation in IEA Countries*, OECD, Paris, 1987.
18. M. J. Grubb, *Energy Policies and the Greenhouse Effect*, Vol. 11, Country Studies and Technical Options, Dartmouth, Brookfield, Vt., 1991.
19. J. Goldemberg, T. B. Johansson, A. Reddy, and R. Williams, *Energy for a Sustainable World*, World Resources Institute, Washington, D.C., Wiley-Eastern, New Delhi, India, 1987.
20. M. Grubb, J. Edmonds, P. ten Brink, and M. Morrison, *Ann. Rev. Energ. Environ.* **18,** 397–470 (1993).
21. United Nations Environmental Programme, *UNEP Greenhouse Gas Abatement Costing Studies*, Phase I report, UNEP Collab. Cent. Energy Environ. Riso Natl. Lab., Denmark, 1992.
22. A. B. Lovins and L. H. Lovins, *Ann. Rev. Energ. Environ.* **16,** 433–531 (1991).
23. E. Mills, D. Wilson and T. B. Johansson, *Energy Policy* **19**(6), 526–542 (1991).
24. T. Jackson, "Least-cost Greenhouse Planning: Supply Curves for Global Warming Abatement," *Energy Policy* (Jan/Feb. 1991).
25. F. Krause, W. Bach, and J. Kooney, *Energy Policies in the Greenhouse*, Int. Proj. Sustain. Energy Paths, Palo Alto, Calif., 1991.
26. Electronic Power Research Institute (EPRI), *Efficient Electricity Use: Estimates of Maximum Savings*, EPRI, CU-6746, Palo Alto, Calif., 1990.
27. Tata Energy Research Institute *UNEP National Greenhouse Costing Study—Indian Country Report* (Phase I Report), TATA Energy Research Institute (TERI), New Delhi, 1992.
28. D. Anderson and C. D. Bird, "Carbon Accumulations and Technical Progress—A Simulation Study of Costs," *Oxford Bull. Econ. Stat.* **42**(1) (1992).
29. J.-M. Burniaux, J. P. Martin, G. Nicoletti, and J. O. Martins, *The Costs of Policies to Reduce Global Emissions of CO_2: Initial Simulation Results with GREEN*, OECD Dept. Economic Statistics, Working Paper No. 103, OECD/GD, Resource Allocation Division, Paris, 1991, p. 115.
30. J.-M. Burniaux, G. Nicoletti, and J. O. Martins, "GREEN: A Global Model for Quantifying the Costs of Policies to Curb CO_2 Emissions," *OECD Economic Studies 19*, Winter 1992.
31. J. Edmonds and D. W. Barnes, *Estimating the Marginal Cost of Reducing Global Fossil Fuel CO_2 Emissions*, PNL-SA-18361, Washington, D.C., Pacific Northwest Laboratory, 1990.
32. J. A. Edmonds and D. W. Barns, *Int. J. Glob. Energy Issues* **4**(3), 140–166 (1992).
33. A. S. Manne and R. G. Richels, *Energy J.* **12**(1), 87–102 (1990).
34. A. S. Manne, *Global 2100: Alternative Scenarios for Reducing Emissions*, OECD Working Paper 111, OECD, Paris, 1992.
35. J. O. Martins, J.-M., Bumiaux, J. P. Martin, and G. Nicoletti, *The Cost of Reducing CO_2 Emissions: A Comparison of Carbon Tax Curves with Green*, OECD Economic Working Paper No. 118, OECD, Paris, 1992.
36. C. Perroni and T. Rutherford, *International Trade in Carbon Emissions Rights and Basic Materials: General Equilibrium Calculations for 2020*, Dept. of Economics, Wilfred Laurier University of Waterloo, Ontario and Dept. of Economics, University of Western Ontario, London, Ontario, 1991.
37. T. Rutherford, *The Welfare Effects of Fossil Carbon Reductions: Results from a Recursively Dynamic Trade Model*, Working Papers, No. 112, OECD/GD Paris, 1992, p. 89.
38. J. Whalley and R. Wigle, *Energy J.* **12**(1), 109–124 (1990).
39. S. C. Peck and T. J. Tiesberg, *Energy J.* **13**(1) (1992).
40. W. D. Nordhaus, *Energy Econ.* **15**(1), 27–50 (1993).
41. D. W. Barns, J. A. Edmonds, and J. M. Reilly, *The Use of the Edmonds-Reilly Models to Model Energy Related Greenhouse Gas Emissions*, Economic Department Working Paper No. 113, OECD, Paris, 1992.
42. Data Resources, Inc. (DRi), *The Economic Effects of Using Carbon Taxes to Reduce Carbon Dioxide Emissions in Major OECD Countries*, prepared for the U.S. Dept. Commerce DRI/McGraw-Hill, Lexington, Mass., 1993; structural comparison of the models in EMF-12; Special issue on policy modeling for climate change, *Energy Policy*, **21**(3) (1992).
43. The Congressional Budget Office, *Carbon Charges as a Response to Global Warming: The Effects of Taxing Fossil Fuel*, Congress of the United States, Washington, D.C., 1990.
44. J. A. Edmonds and D. W. Barns, *Use of the Edmonds-Reilly Model to Model Energy-Sector Impacts of Greenhouse Emissions Control Strategies*, prepared for the U.S. Dept. Energy

under Contract DE-AC06-76RLO 1830, Washington, D.C., 1991.

45. L. H. Goulder, *Effects of Carbon Taxes in an Economy with Prior Tax Distortion: An Intertmporal General Equilibrium Analysis for the U.S.* Working paper, Stanford University, Stanford, Calif., 1991.
46. D. W. Jorgenson and P. J. Wilcoxen, "The Cost of Controlling U.S. Carbon Dioxide Emissions," presented at the *Workshop on Econ.l Energy/Environ. Model. Clim. Policy Anal., Washington, D.C.*, Harvard University, Cambridge, Mass., 1990.
47. D. W. Jorgenson and P. J. Wilcoxen, *Reducing U.S. Carbon Dioxide Emissions: The Cost of Different Goals*, Harvard Institute Economic Research, Harvard University, Cambridge, Mass., 1990.
48. A. S. Manne and R. G. Richels, *Energy J.* **22**(2), 51–74 (1990); *Emissions from the Energy Sector: Cost and Policy Options*. Stanford Press, Stanford, Calif., 1990.
49. D. Gaskins and J. Weyant, eds. National Academy of Science (NAS), *Policy Implications of Greenhouse Warming*, National Academy Press, Washington, D.C., 1991.
50. U.S. Congress, Office of Technological Assessment (OTA), *Changing by Degrees: Steps to Reduce Greenhouse Gases: Summary*, Washington, D.C., 1991.
51. R. Shackleton and co-workers, *The efficiency value of carbon tax revenues*, 1993; *Proc. Semin. PRISTE-CNRS*, Paris, Oct. 1992.
52. Congressional Budget Office, *Carbon Charges as a Response to Global Warming: The Effects of Taxing Fossil Fuel*, U.S. Congress, Washington, D.C., 1990.
53. A. S. Manne and R. G. Richels, *Buying Greenhouse Insurance: The Economic Costs of CO_2 Emission Limits*, MIT Press, Cambridge, Mass., 1992. This book uses revised assumptions that lower costs, especially for the global studies.
54. A. Christensen, *Stabilization of CO_2 Emissions—Economic Effects for Finland*, Ministry of Finance Discussion Paper No. 29, Helsinki, Mar. 1991.
55. K. Yamaji, See W. U. Chandler, ed. *Carbon Emissions Control Strategies: Case Studies in International Cooperation*. World Wildlife Fund & The Conservation Foundation, Baltimore, 1990, pp. 155–172.
56. P. B. Dixon, R. E. Marks, P. McLennan, R. Schodde, and P. L. Swan, *The Feasibility and Implications for Australia of the Adoption of the Toronto Proposal for CO_2 Emissions*, report to Charles River & Assoc., Sydney, Australia, 1989.
57. Ind. Comm. *Costs and Benefits of Reducing Greenhouse Gas Emissions*, Vol. I, report (draft), Australia, 1991.
58. R. E. Marks and co-workers, *Energy J.* **12**(2), 135–152 (1990).
59. S. Proost and D. van Regemorter, *Economic Effects of a Carbon Tax—With a General Equilibrium Illustration for Belgium*, Public Economic Research Paper No. 11, Katholieke University Leuven, 1990.
60. P. Karadeloglou, Energy Tax versus Carbon Tax: A Quantitative Macro-economic Analysis with the Hermes/Midas models, *Eur. Econ.* Special Ed. No. 1, CEC DG-II, Brussels, 1992.
61. K. Ban, *Energy Conservation and Economic Performance in Japan—An Economic Approach to CO_2 Emissions*, Faculty Econ., Osaka University, Discussion Paper Series No. 112, Oct. 1991.
62. U. Goto, "A study of the impacts of carbon taxes using the Edmonds-Reilly model and related sub-models," in A. Amano, ed., *Global Warming and Economic Growth*, Cent. Glob. Environ. Res., 1991.
63. Y. Nagata, K. Yamaji, and N. Sakarai, *CO_2 Reduction by Carbon Taxation and its Economic Impact*, Economic Research Institute Report No. 491002, Central Research Institute of the Electronic Power Industry, 1991.
64. National Environment Policy Plan (NEPP), *To Choose or to Lose*, Second Chamber of the States General Gravenhage, the Netherlands, 1989.
65. B. Bye, T. Bye, and L. Lorentson, *SIMEN: Studies in Industry, Environment and Energy Towards 2000*, Cent. Bur. Stat. Discuss. Paper No. 44, Oslo, 1989.
66. S. Glomsrod, H. Vennemo, and T. Johnson, *Stabilization of Emissions of CO_2: A Computable General Equilibrium Assessment*, Cent. Bur. Stat., Discussion paper No. 48, Oslo, 1990.
67. L. Bergman, *General Equilibrium Effects of Environmental Policy: A CGE-Modelling Approach*, Research Paper No. 6415, Handelshgskolan, Stockholm, 1990.
68. T. Barker and R. Lewney, A green scenario for the UK economy. In *Green Futures for Economic Growth: Britain in 2010*, Cambridge Econometrics, Cambridge, UK, 1991, Chapt. 2.
69. J. Sondheimer, Energy policy and the environment: Energy policy and the Green 1990s—macroeconomic effect of a carbon tax, paper at *Semin. sponsored by Cambridge Econometrics and UKCEED: The Economy and the Green 1990s, 11–12 July, Ribonson College*, Cambridge, UK, 1990.
70. S. Sitnicki, K. Budzinski, J. Juda, J. Michna, and A. Szpilewica, Poland in Ref. 55, pp. 55–80.
71. S. Sitnicki, K. Budzinski, J. Juda, J. Michna, and A. Szpilewica, *Energy Policy* **19**(10) (1991).
72. E. Unterwurzacher and F. Wirl, Wirtschaftliche und politicsche Neuordnung in Mitteleuropa: Auswirkungen auf die Energiewirtschaft und Umweltsituation in Polen, der CSFR und Ungarn. *Z. Energie-wirtschaft* (in German) (1991).
73. G. Leach and Z. Nowak, *Energy Policy* **19**(10) (1991).
74. IPCC–EIS (Energy and Industry Subgroup), *Report to the Intergovernmental Panel on Climate Change*, UNEP/World Meteorological Organization (WMO), Geneva, 1990.
75. J. Sathaye, ed. *Energy Policy*, Special issue: Climate Change—Country Case Studies, **19**(10), 1991.
76. J. P. Martin, J. M. Burniaux, G. Nicoletti, and J. Oliveira-Martins, The costs of international agreements to reduce CO_2 emissions: Evidence from GREEN. *OECD Econ. Stud.* 19—Winter (formerly Dept. Econ. Stat. Working Paper No. 115).
77. J. Whalley and R. Wigle, *Energy J.* **12**(1), 109–124 (1990).
78. J. Whalley and R. Wigle, in R. Dornbusch and J. M. Poterba, eds., *Global Warming: Economic Policy Responses*, MIT Press, Cambridge, Mass., 1990.
79. C. R. Blitzer, R. S. Ekhaus, S. Lahiri, and A. Meeraus, 1990. A general equilibrium analysis of the effects of carbon emission restrictions on economic growth in developing country. In *Proc. Workshop Econ./Energy/Environ. Model. Clim. Policy Anal.* World Bank, Washington, D.C.
80. C. R. Blitzer, R. S. Ekhaus, S. Lahiri, and A. Meeraus, 1990. The potential for reducing carbon emissions from increased efficiency: a general equilibrium methodology. See Ref. 57.
81. D. Wilson and J. Swisher, *Energy Policy* **21**(3), 48 (1993). Exploring the gap: top-down versus bottom-up analyses of the cost of mitigating global warming.
82. A. Ingham, J. Maw, and A. Ulph, *Oxford Rev. Econ. Policy* **7**(2) (1991). Empirical measures of carbon taxes.
83. A. Dean and P. Hoeller, *OECD Econ. Stud.* **19** (Winter, 1992) (formerly published as OECD Econ. Dept. Working Papers No. 122). Costs of reducing CO_2 emissions: Evidence from six global models.

84. J. A. Edmonds, H. M. Pitcher, D. W. Bams, R. Baron, and M. A. Wise, in *Proc. United Nations Univ. Conf. "Global Change and Modelling,"* Tokyo, 1992. Modeling future greenhouse gas emissions: The second generation model description.
85. S. C. Morris, B. D. Solomon, D. Hill, J. Lee, and G. Goldstein, in J. W. Tester and N. Ferrari, eds., *Energy and Environment in the 21st Century*, 1990. A least cost energy analysis of U.S. CO_2 reduction options.
86. R. A. Bradley, E. C. Watts, and E. R. Williams, eds. *Limiting Net Greenhouse Gas Emissions in the United States*, DOE/pElolol. U.S. Dept. Energy, Washington, D.C. See especially Chapt. 7.
87. J. Edmonds, D. Barnes, M. Wise, and M. Ton, *Carbon Coalitions: The Cost and Effectiveness of Energy Agreements to Alter Trajectories of Atmospheric Carbon Dioxide Emissions*, prepared for the U.S. Office of Technological Assessment and the PNL Global Studies Program, 1992.
88. C.-O. Wene, Top down—bottom up: A system engineers's view. Paper presented to *Int. Workshop Costs, Impacts, and Possible Benefits of CO_2 Mitigation*, Sept. 28–30, IIASA, Laxenburg, Austria; Energy Systems Technology, Dept. Energy Conversion, Chalmers University of Technology, S-41296 Goteborg, Sweden, 1992.
89. G. Boero, R. Clarke, and L. A. Winters, *The Macroeconomic Consequences of Controlling Greenhouse Gases: A Survey*. Dept. Environ., Environ. Econ. Res. Ser., London, Her Majesty's Stationery Office, 1991.
90. D. W. Jorgenson and P. J. Wilcoxen, *Intertemporal General Equilibrium Modelling of U.S. Environmental Regulation*, Harvard Inst. Econ. Res. Cambridge, Mass., 1989.
91. W. J. McKibben and P. J. Wilcoxen, *The Global Costs of Policies to Reduce Greenhouse Gas Emissions*, Brookings Discussion papers no. 97, Brookings Institute, Washington, D.C., 1992.
92. R. C. Dower and M. B. Zimmemman, *The Right Climate for Carbon Taxes*, World Resources Institute, Washington, D.C., 1992.
93. M. J. Grubb, *Energy Policies and the Greenhouse Effect, Vol. I: Policy Appraisal*, Dartmouth Brookfield, Vt., 1990.
94. AES Corp., *An Overview of the Fossil2 Model*, prepared for the U.S. DOE Office Policy Eval., Arlington, Va., 1990.
95. SEI-B, LEAP: *Long-Range Energy Alternative Planning System, Vol. 1: Overview, Vol. 2: User Guide*, Stockholm Environmental Institute, Boston, 1992.
96. P. E. Morthorst, and J. Fenhann, *Brundtland Scenarienmodelien—BRUS (The Brundtland scenario model)*, Danish Ministry of Energy, Denmark, 1990.
97. Danish Ministry of Energy, *Energy 2000: A Plan of Action of Sustainable Development*, Danish Ministry of Energy, Denmark, 1990.
98. R. Williams, *Energy J.* **11**(4) (1990).
99. A. Manne and R. G. Richels, *Energy J.* **11**(4) (1990).
100. W. R. Cline, *The Economics of Global Warming*, Institute of International Economics, Washington, D.C., 1992.
101. *Report of UNEP Workshop on the Costs of Limiting Fossil Fuel CO_2 Emissions*, London, Jan. 7–8, 1991, UNEP Environmental Economics Series paper no. 5, UNEP, Nairobi, 1991.
102. A. S. Manne and R. G. Richels, *Buying Greenhouse Insurance: The Economic Costs of CO_2 Emission Limits*, MIT Press, Cambridge, Mass., 1992. This book uses revised assumptions that lower costs, especially for the global studies.
103. S. Meyers, and L. Schipper, *Annu. Rev. Energy Environ.* **17,** 463–505 (1992). World energy use in the 1970s and 1980s: Exploring the changes.
104. L. Schipper, S. Meyers, with R. Howarth and R. Steiner, *Energy Efficiency and Human Activity: Past Trends, Future Prospects*, Cambridge University Press, Cambridge, UK, 1992.
105. R. H. Williams, E. D. Larson, and M. Ross, *Annu. Rev. Energy* **12,** 99–144 (1987).
106. A. K. N. Reddy, *Energy Policy* **19**(10) (1991).
107. P. L. Joskow and D. B. Marron, *Science* **260.** (1993).
108. T. B. Johansson, H. Kelly, A. K. Reddy, R. H. Williams, and L. Bumham, *Renewable Energy: Sources for Fuels and Electricity*, Island Washington, D.C., 1993.
109. W. W. Hogan and D. W. Jorgenson, *Energy J.* **12**(1), 67–86 (1991). Productivity trends and the cost of reducing CO_2 emissions.
110. S. Boyle and co-workers, *A Fossil-free Energy Future*, technical appendices, Greenpeace International, London, 1993.
111. T. B. Johansson, H. Kelly, A. K. Reddy, R. H. Williams, and L. Burnham, *Renewable Energy: Sources for Fuels and Electricity*, Island, Washington, D.C., 1993.
112. W. U. Chandler and A. K. Nicholls, *Assessing Carbon Emissions Control Strategies: A Carbon Tax or a Gasoline Tax?* American Council on Energy Efficiency and Economics, 1990.
113. Capros and co-workers, *The Impact of Energy and Carbon Tax on CO_2 Emissions*, report to the CED, DGXII, Brussels, 1991.
114. J. C. Hourcade, *Energy Policy* **21**(3) (1993). Modeling long-run scenarios: methodology lessons from a prospective study on a low CO_2 intensive country.
115. J. Andere, W. Hafele, N. Nakicenovic, and A. McDonald, *Energy in a Finite World*, Ballinger, Cambridge, Mass., 1981.
116. M. J. Grubb and J. Walker, eds, *Emerging Energy Technologies: Impacts and Policy Implications*, Dartmouth, Brookfield, Vt., 1992. Especially Chapt. I.
117. O. Hohmeyer, *Social Costs of Energy Consumption*, Springer-Verlag, Berlin, 1988.
118. R. Ottinger, D. Wooley, N. Robinson, D. Hodas, and S. Babb, *The Environmental Costs of Electricity*, Oceana, New York, 1990.
119. World Energy Council, *Energy for Tomorrow's World*, London, 1992.

CARBON RESIDUE

Ramon Espino
Exxon Research and Engineering
Annandale, New Jersey

There is not a universally-accepted definition of the term carbon residue. Anyone involved in energy technology is bound to encounter as many variants of this term as there are applications. Generally, it refers to the fraction of a hydrocarbon-containing product or byproduct that is largely composed of carbon atoms. The residue in most cases is not pure carbon since it can contain measurable amounts of other elements, particularly hydrogen. Sulfur, nitrogen, oxygen and metals are also present in many carbon residues.

The fraction of a hydrocarbon that is converted to a "carbon residue" is strongly dependent on temperature. While very small amounts are produced at temperatures below 200°C, significantly greater amounts are generated at higher temperatures. Carbon hydrogen bonds begin to break, at a significant rate, at temperatures above 400°C, and most hydrocarbons are thermally unstable above 500°C, resulting in the evolution of hydrogen and the combination of the carbon atoms into larger molecular structures comprised mainly of carbon with small amounts of hydrogen. The conversion of hydrocarbons into a carbonaceous residue and hydrogen can be partially controlled by the presence of hydrogen at high pressure, and further suppressed if a catalyst that favors hydrogenation is present.

The thermal stability of the carbon-hydrogen bond varies. The hydrogens in methane and benzene are much more stable than hydrogens in paraffinic or olefinic hydrocarbons, particularly branched olefins and paraffins. In the latter, there are carbon atoms with only one hydrogen atom. These carbon-hydrogen bonds are the weakest and tend to break at lower temperatures. The thermal and catalytic chemistry of carbon-hydrogen molecules is well-advanced with a significant amount of theory and experimental data that allows the scientist and engineer to predict the yield of carbonaceous residue depending on the process conditions. Molecules containing carbon-oxygen, carbon-sulfur, and carbon-nitrogen can also produce carbonaceous residues, and the amount is also determined by the bond strength of these systems. The level of scientific understanding is not as advanced as for carbon-hydrogen systems but is still possible to estimate the amount of carbonaceous residue as a fraction of temperature, pressure, and reaction time.

The thermal decomposition of hydrocarbons at high temperatures (500°C and above) is the basis for the production of coke from petroleum residua and coal. Byproducts of these thermal reactions are volatile hydrocarbons that have a higher ratio of hydrogen to carbon than the original feed (residua or coal). Smaller amounts of molecular hydrogens, methane, water, carbon oxides, hydrogen sulfide, and ammonia are also produced if the feed contains oxygen, sulfur and nitrogen. These thermal processes are optimized for the production of maximum amounts of hydrogen-rich volatile compounds and a solid "residue" that is enriched in its carbon content. In these processes, the thermal decomposition behavior of hydrocarbons is used to produce a useful carbon residue in the form of coke.

However, in many other circumstances carbonaceous residues are not the preferred products, but rather undesirable byproducts. For example, during the combustion of hydrocarbons, the desired products are those of the complete combustion of carbon to carbon dioxide and carbon monoxide, and of hydrogen to water. However, in most combustion processes a fraction of the hydrocarbon is converted in a carbonaceous residue. This residue can be either small carbon particles called soot, or carbonaceous residues that deposit on the surfaces of the combustion chambers. A well-controlled combustion system produces very small amounts of these two byproducts, but their total elimination has not been achieved in practical applications. After all, at the high temperatures of most combustion processes the carbon-hydrogen bonds are unstable and both carbon–carbon and carbon–oxygen reactions are favored. The key to "clean combustion" is to maximize the latter at the expense of the former. In combustion processes carbon residues are undesirable. Soot particles are not only visually unattractive, but if inhaled can deposit in the lungs. Since soot particles can adsorb cancer-causing polynuclear aromatic hydrocarbons (PAH), they can also represent a health hazard. Solid carbonaceous residues can plug hydrocarbon feed nozzles (fuel injectors) and deposit on valves, combustion chambers and heat transfer equipment. Carbonaceous residues are carefully monitored in most combustion processes. For example, soot particles are analyzed for their content of soluble hydrocarbons (where PAHs are concentrated), ash (sulfates, nitrates and metals), and insoluble hydrocarbons (mainly coke-type molecules). The amount of carbonaceous residues is measured on fuel injector nozzles, inlet valves, and combustion chambers since these deposits can interfere with the operation of internal combustion engines.

Carbonaceous residues are observed in many refinery and chemical processes. These residues can deposit on reactor walls, flow systems, catalyst particles and many other types of process equipment. In many of these processes, the carbon residue is measured and monitored. For example, in the most widely used refining process, fluid catalytic cracking, the carbon residue that is deposited on the cracking catalyst rapidly kills its catalytic activity. In a fluid catalytic cracker, the catalyst flows through pipes and reactors like a fluid. The carbon containing catalyst is transferred to a combustion zone where the carbon is burned off the catalyst with air and thus regenerated. A large portion of the heat generated during the burning of carbon is retained by the hot catalyst particles which flow back to the catalytic reactor and provide the energy required for the cracking reaction to occur. The elegant simplicity of this two-step process is the reason for its worldwide applicability.

When hydrocarbons are used as lubricants, carbon residues often appear and interfere with the lubrication process. Therefore the tendency of lubricants to form carbon residues is evaluated in laboratory tests as well as in their actual application. There are a myriad of tests since the temperature and reaction time as well as the environment (catalyst, presence of hydrogen, oxygen, air, combustion gases) can vary. The carbon residue that is produced can also vary in composition, including carbonaceous deposits on surfaces, sludge and varnish on surfaces, and soot or carbonaceous particles dispersed in the liquid lubricant. The composition of these carbon residues can be highly variable, ranging from very carbon rich solids to solids containing significant amounts of metal salts and rich in nitrogen, oxygen and hydrogen.

BIBLIOGRAPHY

Reading List

Harry Marsh (Ed), *Introduction to Carbon Science*, Butterworth, Boston, 1989.

Martin A. Elliott, *Chemistry of Coal Utilization*, John Wiley and Sons, Inc. New York, 1981, 2nd suppl. vol.

E. F. Obert, *Internal Combustion Engines and Air Pollution*, Harper and Row, Publishers.

G. D. Hobson and W. Pohl, *Modern Petroleum Technology*, Applied Science Publishers, Ltd., Chapt. 17.

CARBON STORAGE IN U.S. FORESTS

Richard A. Birdsey
Linda S. Heath
U.S.D.A.—Forest Service
Radnor, Pennsylvania

Global concern about increasing atmospheric concentrations of greenhouse gases, particularly carbon dioxide (CO_2), and the possible consequences of future climate changes have generated interest in evaluating the role of terrestrial ecosystems in the global carbon cycle. Recent efforts to estimate changes in the global carbon budget by accounting for CO_2 emissions from burning fossil fuels, deforestation in the tropics and CO_2 uptake by the atmosphere and oceans have revealed an imbalance between sources and sinks of CO_2. Additional carbon sinks of approximately 1.8 Pg/yr are suspected in the terrestrial biosphere or the oceans (1). Identification of all sources and sinks, which are small relative to the magnitude of global carbon reservoirs, has been difficult because of inability to measure and monitor accurately the effects of natural and human disturbances on ecosystems over large regions.

See also Air quality modeling; Carbon balance modeling; Carbon cycle; Forest resources; Global climate change, mitigation.

Carbon compounds have a central role in plant processes that build structures and maintain physiological functions. Rates of carbon uptake through photosynthesis and release through respiration are a function of light, atmospheric CO_2, temperature, water, nutrients, and other factors such as air pollution. Loss of leaves and woody material and death of fine roots are sources of organic matter for soil processes that in turn support uptake of nutrients and plant growth. Forest ecosystems are capable of storing large quantities of carbon in biomass and soils. Forest disturbances such as fire and timber harvest may reduce biomass and cause oxidation of soil carbon, adding to the pool of CO_2 in the atmosphere. Growing forests may reduce atmospheric CO_2 through increases in biomass and soil carbon. Carbon in wood products may be effectively stored for long periods of time, depending on the end use of the wood. With a complete accounting for all of the forest changes and effects on carbon in each of the components of the system, it would be possible to determine whether a land area containing forests is a net source or sink of CO_2.

Increasing inventories of timber in European forests, and corresponding increases in biomass and soil carbon, could account for 5–9% of the "missing" carbon (2). Similar estimates for U.S. forests could account for 12–21% of the unexplained flux since 1952 (3). It is likely that Canadian forests have been a carbon sink in recent decades (4), and estimates for the territory of the former Soviet Union indicate a carbon sink there as well (5). Together, the world's temperate and boreal forests have been adding carbon in biomass and soil at a rate of 0.7–1.2 Pg/yr (6). This accounts for only some of the suspected imbalance in the global carbon cycle.

ESTIMATING CARBON IN FOREST ECOSYSTEMS

Carbon storage is usually estimated for different forest ecosystem components: biomass in trees, soil organic matter, litter on the forest floor, coarse woody debris, and understory vegetation. Live-tree biomass includes aboveground and below-ground portions of all trees: the merchantable stem, limbs, stump, foliage, bark and rootbark, and coarse tree roots. The soil component includes organic and inorganic carbon in mineral soil horizons to a specified depth (usually 1 m), excluding coarse tree roots. The forest floor includes all dead organic matter (litter and humus) above the mineral soil horizons except standing dead trees. Coarse woody debris includes standing and fallen dead trees and branches. Understory vegetation includes all live vegetation besides live trees.

Estimates of carbon storage and flux in trees over large regions are typically based on periodic forest inventories designed to provide statistically valid estimates of timber volume, growth, removals, and mortality (7). Timber volume includes only the merchantable portions of live trees that meet specified criteria for size and quality. Estimation of carbon from timber volume involves several steps to account for nonmerchantable tree biomass and to convert from volume units to weight units. Total tree biomass, if not estimated directly, may be estimated by multiplying timber volume times conversion factors derived from representative biomass inventories (8). Depending on species and location, total biomass in all trees averages from 1.7 to 2.6 times the biomass in merchantable timber (9). Approximately 50% of tree biomass is carbon.

Less direct methods are used to estimate carbon storage in other ecosystem components over large regions. Simple models can be devised to estimate carbon storage in the soil, forest floor, coarse woody debris, and understory vegetation, using data from compilations of intensive ecosystem studies (10). Soil carbon is related to mean annual temperature, precipitation, soil texture, and stand age and is sometimes estimated with a regression model (11).

Because carbon cannot be directly measured for any but the smallest of areas, estimates using the preceding methods include some uncertainty. Regional forest inventories are based on a statistical sample designed to represent the broad range of forest conditions actually present in the landscape. Therefore, estimates of carbon storage in forest trees are representative of the true average values, subject to sampling errors, estimation errors, and errors in converting data from one reporting unit to another. The estimated error for tree carbon is usually quite small if the forest inventories used to derive the estimates have small sampling errors over large areas. Accurate regional forest inventories are available for some countries with advanced forest inventory systems such as the United States.

Most regional estimates of carbon storage in the soil and forest floor are not based on statistical samples but on compilations of the results of many separate ecological studies of specific ecosystems. Published estimates for soil carbon show wide variation for terrestrial ecosystems (12).

CARBON STORAGE IN THE UNITED STATES

Forest ecosystems in the United States contain approximately 54.6 Pg organic carbon above and below the ground (Table 1). This is about 4% of all the carbon stored in the world's forests (13). The area of U.S. forests is 298 × 10^{10} m^2, or 5% of the world's forest area. The average forest in

Table 1. Area of Forest Land[a] and Carbon Storage by Region and Forest Ecosystem Component, 1992

Region	Area, m^2	Forest Ecosystem Component, pg				
		Soil	Forest Floor	Understory	Trees	Total
Northeast	$34,554 \times 10^4$	4.73	0.62	0.06	2.24	7.64
North Central	$33,634 \times 10^4$	3.50	0.51	0.05	1.68	5.73
Southeast	$35,645 \times 10^4$	2.52	0.21	0.11	2.09	4.92
South Central	$50,085 \times 10^4$	3.29	0.25	0.17	2.83	6.54
Rocky Mountains	$56,549 \times 10^4$	3.89	0.72	0.06	2.71	7.39
Pacific Coast	$35,408 \times 10^4$	3.41	0.63	0.15	2.24	6.41
Alaska	$52,259 \times 10^4$	12.01	1.63	0.20	2.16	16.00
United States	$298,134 \times 10^4$	33.33	4.56	0.80	15.94	54.64

[a] Land at least 10% stocked by forest trees of any size, including land that formerly had such tree cover and that will be naturally or artificially regenerated.

the United States contains 18.3 kg/m^2 of organic carbon. Trees, including tree roots, account for 29% of all forest ecosystem carbon (Fig. 1). Live and standing dead trees contain 15.9 Pg carbon, or an average of 5.3 kg/m^2. Of this total, 50% is in live tree sections classified as timber, 30% is in other solid wood above the ground, 6% is in standing dead trees, and 3% is in the foliage.

The largest proportion of carbon in the average U.S. forest is found in the soil, which contains 61% of the carbon in the forest ecosystem, or approximately 11.2 kg/m^2. Most of the soil carbon is found in the top 30 cm of the soil profile, the location of most fine roots and adjacent to the litter inputs to the soil. About 8% of all carbon is found in litter, humus, and coarse woody debris on the forest floor, and about 1% is found in the understory vegetation. By adding carbon in tree roots to the carbon in the soil, the average proportion of carbon below the ground in the United States is estimated to be 66%.

Carbon storage and accumulation rates in a particular region or forest are influenced by many factors such as climate, solar radiation, disturbance, land-use history, age of forest, species composition, and site and soil characteristics. Even though all trees have similar physiological processes, there are significant differences in growth rates and wood density between species and individual organisms. The combination of species and site differences produces a wide variety of carbon densities across a landscape. Historical land-use patterns and landscape attributes produce characteristic regional profiles of carbon storage.

The proportion of carbon in the different ecosystem components varies considerably between regions (Fig. 2). Alaska has the highest estimated amount of carbon in the soil, about 75% of the total carbon. The southeastern and south-central states each have about 50% of total carbon in the soil but have a higher percentage in the trees. Soil carbon is closely related to temperature and precipitation, with higher amounts of soil carbon being found in regions with cooler temperatures and higher precipitation. Cooler temperatures slow the oxidation of soil carbon, while higher rainfall tends to produce greater growth of vegetation, fine roots, and litter, which are the main sources of organic soil carbon. Carbon in the forest floor varies by region in a way similar to carbon in the soil. Western and northern states contain the most carbon on the forest floor, and southern states contain the least.

Significant differences in carbon storage are apparent for different forest types (Fig. 3). The differences are related both to biological and human influences on the landscape. For example, loblolly pine plantations are younger on average than natural loblolly pine, so there is less carbon in the trees. Likewise, because loblolly pine is located in the South, there is less carbon in the soil than in forest types such as Douglas fir or spruce–fir that grow in cooler,

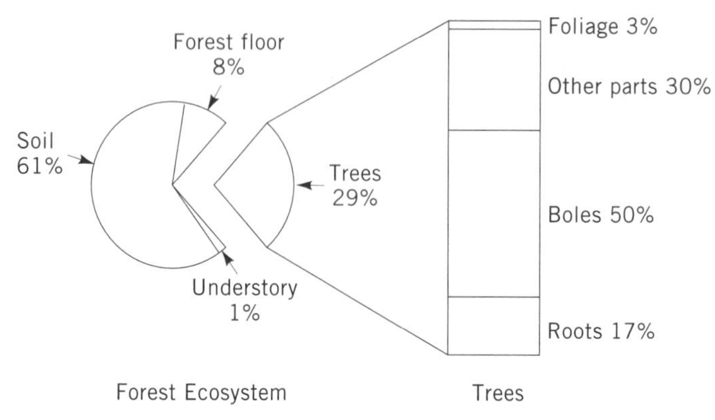

Figure 1. Allocation of carbon in forest ecosystems and trees for U.S. forests in 1992. Total storage in the United States is 54.6 billion t.

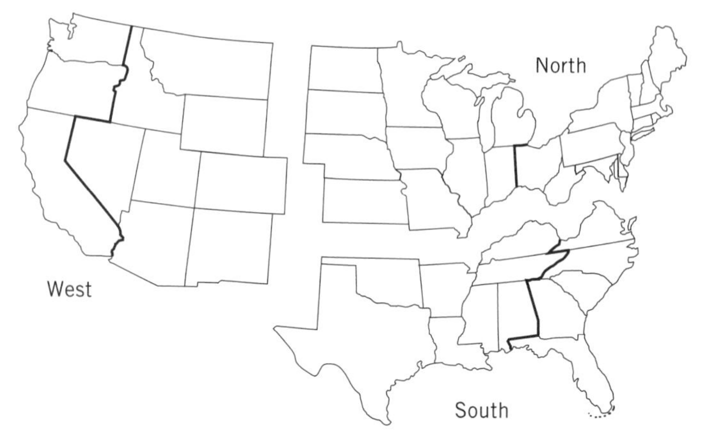

Figure 2. Regions of the United States used to estimate trends in carbon storage.

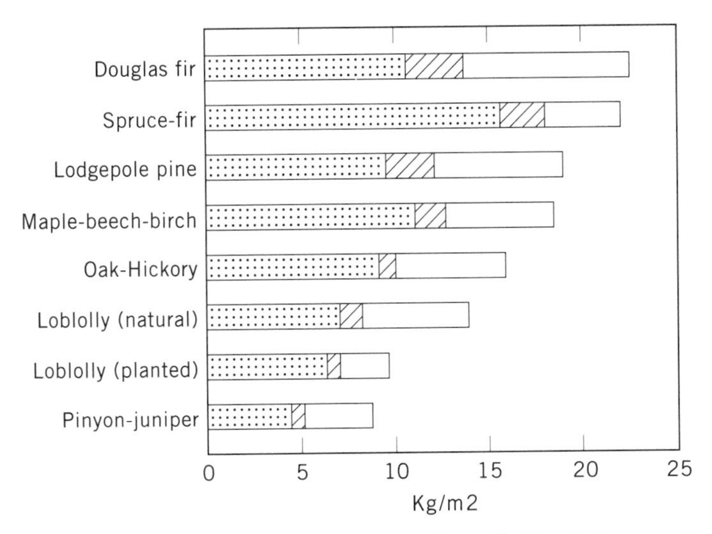

Figure 3. Average carbon storage in the soil, forest floor, and trees for selected forest types. ⊡, soil; ▨, forest floor; □, trees.

wetter climates. Of the examples shown, piñon–juniper has the lowest amount of carbon because it occurs in dry climates that support lower vegetation densities.

The age of the forest and degree of disturbance or management intensity have significant effects on carbon storage. Generally, older forests contain more carbon than younger forests, but younger forests accumulate carbon faster than older forests. These patterns are evident in plots of the relationship between carbon storage and age of forest for representative forest ecosystems (Fig. 4). For loblolly pine, a type in which trees represent a large proportion of total ecosystem carbon, a fourfold increase in total carbon is possible over an 80-yr period. Spruce–fir, which grows in the northeastern and north-central United States, shows less total variation over the same period of time, because the most dynamic ecosystem component, trees, represents a smaller proportion of the total carbon. The effects of management and natural disturbance on carbon are less well known than the effects of age but can be approximated by substituting "time since disturbance" for "stand age" and adjusting the initial estimates for each of the ecosystem components.

RECENT TRENDS IN CARBON STORAGE

U.S. forests are constantly changing. The total area of forest land declined by 1.6×10^{10} m^2 between 1977 and 1987 (7). Most of the net loss was removal of trees from forest clearing for urban and suburban development, highways, and other rights-of-way. A much larger area was cleared for agricultural use, but this loss was roughly balanced by agricultural land that was planted with trees or allowed to revert naturally to forest. In addition to land-use changes, each year about 1.6×10^{10} m^2 of timberland are harvested for timber products and regenerated to forests, 1.6×10^{10} m^2 are damaged by wildfire, and 1×10^{10} m^2 are damaged by insects and diseases. And of course, all forests change continually as individual trees and other vegetation germinate, grow, and die.

The sum effect of all these changes is evident in periodic carbon estimates for the last 40 yr. Between 1952 and 1992, carbon stored on forest land in the conterminous United States increased by an estimated 11.3 Pg (Table 2). This is an average of 0.281 Pg of carbon sequestered each year over the 40-yr period, an amount that has offset about 25% of U.S. emissions of CO_2 to the atmosphere (14). Most of the increase occurred in the eastern and central regions of the United States, offsetting a much smaller decline in the western region. Over the past 100 yrs or more, large areas of the East have reverted from agricultural use to forest. As these reverted forests have grown, biomass and soil organic matter have increased substantially. Recently in the South, increased harvesting and intensive forest management have slowed the rate of increase in carbon storage. Northeastern and north-central forests have continued to accumulate carbon at a rapid rate, because the mixed hardwood forests are less intensively used or

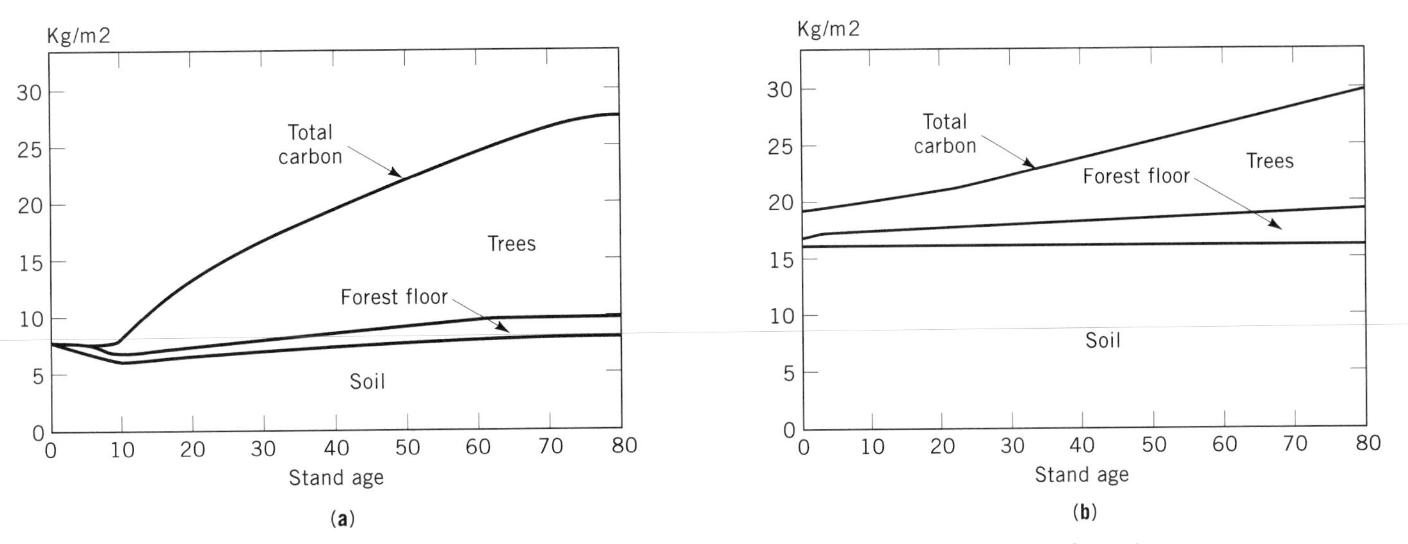

Figure 4. Carbon storage by forest ecosystem component for (**a**) a loblolly pine plantation in the Southeast and (**b**) a spruce–fir forest in the Northeast. Both forests have been clearcut and regenerated.

Table 2. Summary of Estimates (in petagrams) of Carbon Storage by Geographic Region, Conterminous U.S. Forest Land, 1952–1992

Region	1952	1962	1970	1977	1987	1992
Northeast	3.92	4.76	5.48	6.27	7.02	7.64
North Central	2.70	3.41	3.84	4.41	5.27	5.73
Southeast	3.10	3.51	3.96	4.50	4.91	4.92
South Central	3.76	4.47	4.99	5.60	6.21	6.54
Rocky Mountains	6.91	6.68	6.68	7.06	7.41	7.39
Pacific Coast	6.08	5.91	5.71	5.63	5.53	5.50
Totals	*26.47*	*28.74*	*30.67*	*33.47*	*36.35*	*37.72*

managed for wood products. Although the West has not had the major land-use shifts characteristic of the East, forest disturbance nonetheless has dominated the landscape as the original forests have been harvested and converted to second-growth forests. Declining carbon storage in the Pacific Coast region reflects the smaller amount of carbon that younger forests contain.

Past changes in forest carbon storage vary significantly by ownership group (Table 3). Most of the recent historical increase in carbon storage has been on privately owned land in the East, the regrowing forests that were cleared by the early 1900s. As these lands approach full stocking of relatively large trees with low rates of biomass accumulation or are more intensively used for timber products, accumulation of carbon may decline in the future. Carbon storage on forest industry lands has increased slightly in the past, but with intensive management for a steady supply of timber, carbon is expected to remain relatively constant. Carbon storage in national forests has declined in the recent past, a consequence of a reduction in the area of western old-growth forests, which contain large quantities of carbon, and an increase in the area of younger, faster-growing forests. With recent reductions in the amount of allowable harvest in western national forests, less carbon will be released, producing a net gain in carbon storage.

Projections show additional increases in carbon storage in U.S. forests, but at a slower rate than the past 50 yr. This projected trend reflects (*1*) a slow down in the rate of accumulation in the North as the average forest has reached an age of slower growth than in the past and increases in soil carbon on reverted land are less, (*2*) increasingly intensive use of forests for wood products in the South so that accumulation is balanced by removal, and (*3*) reduced harvest of public forests in the West coupled with a large area of younger, more vigorous and intensively managed forests on former old-growth forest land.

The effects of increasing atmospheric CO_2 and prospective climate change on productivity could have a significant impact on carbon storage in forests, but there is little agreement on the direction of such changes for temperate forests. Boreal forests, such as found in Alaska, are expected to become a net source of atmospheric CO_2 under global warming scenarios because respiration of soil carbon would likely increase more than biomass growth under higher temperatures.

CARBON IN HARVESTED WOOD

There is a substantial amount of carbon removed from forests for wood products. Some ends up in long-term storage in durable wood products and landfills, and some is burned for energy or decomposes and is released to the atmosphere (15). Wood products are goods manufactured or processed from wood, including lumber and plywood for housing and furniture and paper for packaging and newsprint. A large portion of discarded wood products is stored in landfills. This carbon eventually decomposes and is counted as emissions on release to the atmosphere. Emissions also include carbon from wood burned without generation of usable energy and from decomposing wood residues and byproducts not stored in landfills. Wood used for energy is usually considered as a separate category from emissions, because substitution of renewable for nonrenewable sources of energy has an effect on carbon flux.

Harvested wood can represent a substantial carbon sink. An estimated 0.04 Pg of additional carbon may be added to physical storage in wood products and landfills each year. This estimate is sensitive to assumptions about recycling, size of harvested timber, other factors that affect

Table 3. Summary of Estimates (in petagrams) of Carbon Storage by Ownership Group, U.S. Timberland,[a] 1952–1992

Owner	1952	1962	1970	1977	1987	1992
National forest	8.42	8.93	8.99	8.96	8.16	7.90
Other public	3.11	3.46	3.75	4.00	3.98	3.85
Forest industry	3.57	3.93	4.21	4.45	4.50	4.44
Other private	10.15	11.58	12.85	14.20	16.69	18.13
Totals	*25.25*	*27.90*	*29.80*	*31.61*	*33.33*	*34.32*

[a] Forest land that is producing or is capable of producing crops of industrial wood and not withdrawn from timber use by statute or administrative regulation.

the amount of wood entering the various pools and the average retention periods.

USE OF BIOMASS FOR ENERGY CONSERVATION

The use of biomass energy from wood or wood products and byproducts and the conservation of energy by using trees for shading, shelter belts, and windbreaks are considered important factors in reducing CO_2 emissions (16). Biomass is converted to energy using a variety of technologies that, depending on the efficiency of the process, can offset the combustion of a substantial amount of nonrenewable fossil carbon. Use of biomass can effectively reduce CO_2 emissions when substituted for fossil fuels at a favorable conversion efficiency. Because biomass is renewable, the amount of carbon released through combustion can be replaced with new biomass.

Living trees have a major role in energy conservation. Windbreaks, shelter belts, and properly placed urban trees can reduce the use of energy to heat and cool buildings and to farm affected croplands. Such trees also increase carbon storage in woody plants and soil. Globally, the potential savings are estimated to be substantial.

MITIGATION OPTIONS

As a consequence of expected increases in emissions of greenhouse gases and an acceleration of global change effects on terrestrial and oceanic systems, analysts have proposed various strategies to reduce emissions of CO_2 to the atmosphere or to offset emissions by storing additional carbon in forests or other terrestrial carbon sinks (17). Carbon sinks are a component of the U.S. strategy for limiting national contributions to greenhouse gas concentrations in the atmosphere. Of particular interest in the United States are options for increased tree planting, increased recycling, changes in harvesting and ecosystem management practices, and carbon offset programs.

One mitigation option favored for increasing carbon sinks is an increase in reforestation of marginal cropland and pasture. Investigation of several scenarios shows the possibility of medium- to long-term gains in carbon storage, on the order of 5–10 Tg/yr for a relatively low cost program treating about 8.094×10^{10} m^2 of timberland that would also produce an economic return on investment (9). A more ambitious program that treated more of the biologically suitable land in the United States could achieve substantially higher gains.

Increased recycling of paper and other wood products would keep more of the harvested carbon in use instead of decomposing and reduce the area of forest that needs to be harvested for wood products. Harvesting and forest management practices that are less disturbing to forest sites could increase carbon storage in forest areas used for timber production. Although the savings from increased recycling and different management practices are likely to be substantial, few reliable estimates of the potential carbon storage changes are available yet.

Biomass plantations, energy conservation measures, and energy offset programs have substantial potential for increasing carbon storage or offsetting carbon emissions. Short-rotation woody crops specifically established for energy production could be established on marginal cropland and pasture land unsuitable for economic production of other crops. Energy offset programs have been proposed as a way to increase the storage of carbon in the terrestrial biosphere by establishing tree plantations to offset emissions of CO_2.

All together, the various carbon sink options can be a significant part of the U.S. program to reduce net greenhouse gas emissions. With the participation of other nations in a global emissions-reduction program, it may be possible to have an impact on atmospheric chemistry that would lessen the possible effects of global change.

BIBLIOGRAPHY

1. E. T. Sundquist, *Science* **259,** 934–937 (1993).
2. P. E. Kauppi, K. Mielikäinen, and K. Kuusela, *Science* **256,** 70–74 (1992).
3. R. A. Birdsey, A. J. Plantinga, and L. S. Heath, *Forest Ecol. Manage.* **58,** 33–40 (1993).
4. M. J. Apps and W. A. Kurz, *World Resources Rev.* **3,** 333–344 (1991).
5. T. P. Kolchugina and T. S. Vinson, *Can. J. Forest Res.* **23,** 81–88 (1993).
6. R. N. Sampson and co-workers, *Water Air Soil Pollution* **70,** 3–15 (1993).
7. K. Waddell, D. D. Oswald, and D. S. Powell, *Forest Statistics of the United States*, Resource Bulletin PNW-RB-168, U.S. Department of Agriculture, Forest Service, Pacific Northwest Research Station, Portland, Ore., 1989.
8. N. D. Cost, J. Howard, B. Mead, W. H. McWilliams, W. B. Smith, D. D. Van Hooser, E. H. Wharton, *The Biomass Resource of the United States*, General Tech. Rep. WO-57, U.S. Department of Agriculture, Forest Service, Washington, D.C., 1990.
9. R. A. Birdsey in R. N. Sampson and D. Hair, eds., *Forests and Global Change, Vol. 1, Opportunities for Increasing Forest Cover, American Forests*, Washington, D.C., 1992.
10. K. A. Vogt, C. C. Grier, and D. J. Vogt, *Adv. Ecolog. Res.* **15,** 303–377 (1986).
11. W. M. Post and co-workers, *Nature* **298,** 156–159 (1982).
12. R. A. Houghton and co-workers in J. R. Trabalka, ed., *Atmospheric Carbon Dioxide and the Global Carbon Cycle*, DOE/ER-0239, U.S. Department of Energy, Washington, D.C., 1985.
13. L. L. Ajtay, P. Ketner, and P. Duvigneaud in B. Bolin and co-workers, eds., *The Global Carbon Cycle*, SCOPE Rep. No. 13, John Wiley & Sons, Inc., New York, 1979, pp. 129–181.
14. T. A. Boden, P. Kanciruk, M. P. Farrell, *Trends '90—A Compendium of Data on Global Change*, ORNL/CDIAC-36, Oak Ridge National Laboratory, Oak Ridge, Tenn., 1990.
15. C. Row and R. B. Phelps in *Agriculture in a World of Change, Proceedings of Outlook '91, 67th Annual Outlook Conference*, U.S. Department of Agriculture, Washington, D.C., 1991.
16. R. N. Sampson and co-workers, *Water Air Soil Pollution* **70,** 139–159 (1993).
17. Intergovernmental Panel on Climate Change, *Climate Change—The IPCC Response Strategies*, Island Press, Washington, D.C., 1991.

CHEMODYNAMICS

DANNY D. REIBLE
Louisiana State University
Baton Rouge, Louisiana
University of Sydney
Sydney, Australia

The assessment of the ambient concentration and thus the harm resulting from the release of chemicals is dependent on the dynamic transport, dilution and fate processes operative in the environment. Environmental chemodynamics is a term coined by Virgil Freed and popularized by Louis Thibodeaux (1) to describe the study of the dynamics of pollutants in the natural environment. The evaluation of the fate and transport of contaminants in the environment has often been limited to consideration of the equilibrium partitioning of these contaminants between environmental phases. This approach is one that is rarely likely to be valid and even then only locally. Instead the environment is in a state of disequilibrium and fate and transport cannot be separated from the concept of time. It is recognition of this dynamic that characterizes the study of environmental chemodynamics.

Environmental chemodynamics involves the application of fundamental energy and mass conservation principles to model and quantify pollutant concentrations and fluxes in the environment. Often, due to the relatively crude nature of our understanding of the natural environment and its inherent variability, the quantitative models are necessarily crude. Generally, these models attempt to capture only the key mechanisms of an environmental transport process and thus cannot hope to describe the full variance of field observations. The models are often lumped parameter models which assume limited spatially heterogeneity or quasi-steady conditions, while at other times, sufficient information may exist to allow spatially distributed and fully time dependent models of environmental transport processes that can be developed and applied.

Applications of environmental chemodynamics arise in any problem involving the assessment of the fate of chemicals in the environment and the resulting human or ecological exposure. The tools that are generated by chemodynamics, for example, allow one to estimate the mobility of contaminants and the exposure of potential receptors in the ecosystem. In addition, the formalism allows the evaluation of potential treatment processes by defining contaminant losses during in-situ treatment or during the collection and transportation of contaminants for conventional treatment processes. Environmental chemodynamics is thus the key to exposure assessment and the selection of remedial processes on the basis of minimum exposure. The Superfund Exposure Assessment Manual (2) illustrates the application of a variety of simple models to the hazardous waste site exposure assessment.

The purpose of the present discussion is to identify the fundamental basis for environmental chemodynamics and present some of the tools that allow estimation of the environmental dynamics of pollutants. The article is not intended to be comprehensive but instead designed to provide engineers and scientists new to the field, a starting point for the collection of information necessary to solve their own problems. In particular, the article will focus on the physico–chemical processes that influence the fate and transport of hydrophobic organic compounds. Metals and polar organic compounds are not considered although they often exhibit similar or analogous behavior. It should be emphasized that this discussion will include processes in all four phases commonly encountered in the environment: air, water, soil and nonaqueous liquid phases (eg, an oily phase). To avoid confusion, media specific parameters will be subscripted with a number to indicate air (phase 1), water (phase 2), soil (phase 3) and nonaqueous liquid phases (phase 4).

See also AIR QUALITY MODELING; WATER QUALITY ISSUES; WATER QUALITY MANAGEMENT; GLOBAL HEALTH INDEX.

THERMODYNAMICS AND TRANSPORT PHENOMENA

Fundamental Tools

The fundamental tools of the environmental chemodynamicist are no different than the tools used by the chemical process designer; that is, mass and energy conservation equations. This approach applies regardless of whether the mobile phase is air, water, a nonaqueous phase liquid or the pore fluid in soil. The mass conservation equation relates the mobile phase concentration of a pollutant A (C_A) to position (z) and time (t). One form of this equation can be written

$$\frac{\partial C_A^T}{\partial t} + U\frac{\partial C_A}{\partial z} = (D_A + D_{eff})\frac{\partial^2 C_A}{\partial z^2} - k_{fate} f(C_A) \qquad (1)$$

where C_A denotes the concentration of contaminant A in moles per unit volume of mobile phase, ρ_A will be used to denote mass concentration.

The first term on the left in equation 1 represents the time rate of change of total concentration of pollutant A (C_A^T). The total concentration is equal to the sum of the moles of contaminant in each phase that constitutes the system divided by the total volume of the system. The total concentration is equal to the mobile phase concentration, C_A, if the system does not contain any of the pollutant adsorbed onto immobile phases. Often the accumulation of contaminant occurs only in the mobile phase or is dependent only on the mobile phase concentration so that equation 1 can be written in terms of the single dependent variable C_A.

The second term in equation 1 represents the advection of contaminant as a result of the fluid velocity (U). The first term on the right-hand side represents transport by molecular diffusion (D_A) or by an effective diffusion process (D_{eff}). Molecular diffusion is the process of material transport by the random motion and collisions of molecules. D_A is dependent on the medium, the diffusing chemical, and the ambient temperature. The effective diffusion coefficient, however, is dependent primarily on the motion of the mobile phase (air or water). In the atmosphere or in surface bodies of water, the magnitude of D_{eff} is determined by the intensity of turbulence. In soil vapor or water transport, the D_{eff} is determined by the interrelationship of the mean flow with the soil heterogeneities. In some situations, however, other effective diffusion coefficients may be applicable. For example bioturbation, or the mixing of sediments by the normal activities of benthic organisms, gives rise to contaminant transport that is often modeled with an effective diffusion coefficient related to the density and activity of the benthic organisms.

The final term on the right represents chemical fate processes such as degradation reactions. The term is written in the equation as a loss term. Generally the magnitude of the loss term is a function of the local chemical concentration (ie, $f(C_A)$ as shown in equation 1). Often biological and hydrolysis reactions in soils, solution phase reactions in surface waters and reactions and deposition in the atmosphere are modeled as first order processes and the loss term is written as $k_{fate}C_A$. Irreversible fate processes will generally not be considered herein.

The relationship between the total system concentration, C_A^T, and the mobile phase concentration, C_A will be further examined. In the atmosphere or in bodies of water, C_A^T is generally taken equal to C_A. In groundwater systems, however, a contaminant can often accumulate on the immobile solid phase. Consider a linear partition coefficient, α_{32}, between the immobile solid (phase 3) and water (phase 2), that is, $\alpha_{32} = C_{A3}/C_{A2}$, where C_{A3} is the contaminant concentration on the solid phase (eg, in moles/kg) and C_{A2} is the contaminant concentration in the water phase (eg, in moles/L). Representing the water fraction or void volume by ε and the solid phase bulk density by ρ_s, then $C_A^T = C_{A2}(\varepsilon + \rho_s\alpha_{32})$. The ratio of the total concentration of the pollutant in the system to the concentration in the mobile phase (here, $\varepsilon + \rho_s\alpha_{32}$) is often termed a retardation factor, R_f, since it slows the response of the mobile phase concentration to transport via advection and diffusion. This relationship can be most readily seen by rewriting equation 1 with $C_A^T = C_AR_f$ and assuming R_f constant. Then equation 1 can be written:

$$\frac{\partial C_A}{\partial t} + \frac{U}{R_f}\frac{\partial C_A}{\partial z} = \left(\frac{D_A + D_{eff}}{R_f}\right)\frac{\partial^2 C_A}{\partial z^2} - \frac{k_{fate}}{R_f}f(C_A) \qquad (2)$$

Thus the ratio of the total concentration to the mobile phase concentration effectively produces a proportionate reduction in the contaminant velocity, diffusivity and reaction rate. Note that under steady state conditions, there is no transient accumulation in an immobile phase and therefore there is no retardation of the mobile phase processes.

A form of equation 1 (or 2) is widely used to predict dynamic contaminant concentrations and fluxes in the environment. In general, application of equation 1 requires generation of a numerical or analytical solution for concentration as a function of space and time. Carslaw and Jaeger (3) and Crank (4) present a wide variety of solutions to Equation 1 subject to various initial and boundary conditions. The vast majority of analytical solutions are limited to constant effective diffusion coefficient, linear or no reaction, and constant or zero velocity. These limitations have not hindered their usefulness in a variety of important situations when the available data does not support more sophisticated models.

Useful Solutions

Two analytical solutions have proven especially useful for environmental applications. In the first of these solutions,

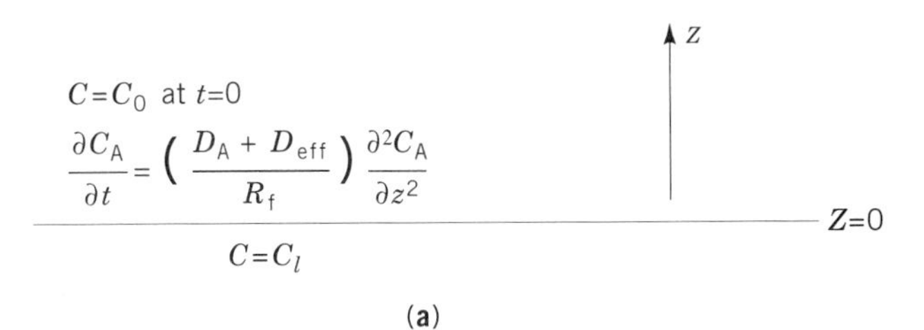

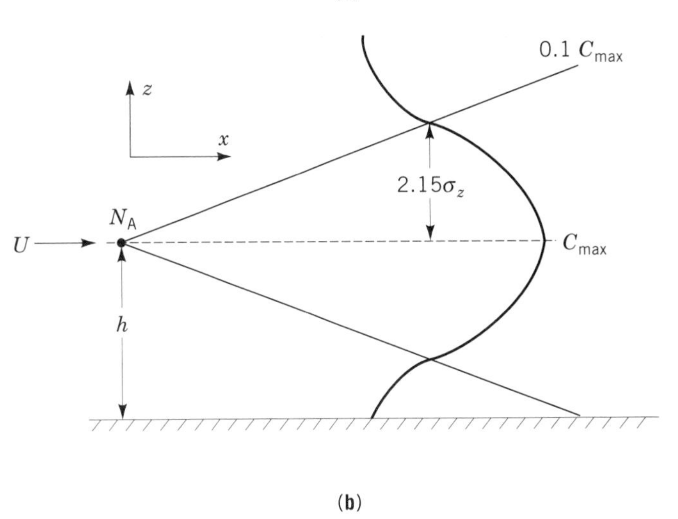

Figure 1. (**a**) Depiction of system described by equations 3 and 4. (**b**) Depiction of system described by equations 6 and 7.

depicted in Figure **1a**, it is assumed that a uniform and constant concentration, C_1, exists at a plane perpendicular to the direction of transport (designated the origin). That is, $C_A = C_1$ at $z = 0$. The environment above this plane is assumed quiescent ($U = 0$) and is initially uniform at a different concentration, C_0. Transport is assumed to occur only via a diffusive process (D_A or $D_{eff} \neq 0$) without reaction ($k_{fate} = 0$). The solution to Equation 1, subject to these conditions is given by the equation after substituting R_fC_A for C_A^T.

$$C_A = C_1 + (C_0 - C_1)\,erf\left(\sqrt{\frac{z^2R_f}{4t(D_A + D_{eff})}}\right) \qquad (3)$$

The erf (η) is the error function which is related to the cumulative normal distribution function and is tabulated (eg, CRC Standard Math Tables, 1974). Some values of the error function are included in Table 1. Note that the solution requires a constant ratio of total contaminant concentration to mobile phase concentration, ie, constant R_f. In the case of diffusion in soil groundwaters, this configuration is equivalent to assuming local equilibrium with linear partitioning between the soil and porewater. While this is a very common assumption, it can be quite restrictive.

The flux of contaminant through the interface ($z = 0$) by this model is given by

$$\dot{n}_A = -(D_A + D_{eff})\left.\frac{\partial C_A}{\partial z}\right|_0 = \left[\frac{(D_A + D_{eff})R_f}{\pi t}\right]^{1/2}(C_1 - C_0) \quad (4)$$

where use is made of the symbol convention of lower case letters representing a flux with the dot overstrike pre-

Table 1. Values of the Error Function

x	erf(x)
0	0.0
0.2	0.227
0.4	0.4284
0.6	0.6039
0.8	0.7421
1.0	0.8427
1.2	0.9103
1.4	0.9523
1.8	0.9891
2.2	0.9981
2.8	0.9999

senting a rate. An upper case symbol for mass or moles would represent an area integrated value. Assuming a constant total concentration in the contaminated region, C_1^T and recognizing $C_1 = C_1^T/R_f$, this model suggests that sorption reduces the flux by the factor $R_f^{1/2}$.

A second commonly encountered solution of equation 1, depicted in Figure 1b, is the concentration about a point source of strength $\dot{N}_A$ (eg, moles/s) released over the period δt. Since the emissions spread in all directions, the three-dimensional form of equation 1 is required. Advection ($U \neq 0$) is assumed in the x-direction and effective diffusion processes are assumed to occur in all three directions. If D_x, D_y, and D_z are the effective diffusion coefficients in the x, y and z directions, respectively, then the concentration field away from the source is given by

$$C_A = \frac{\dot{N}_A \delta t}{8(\pi\tau)^{3/2}\sqrt{D_x D_y D_z}} \exp\left[\frac{-(x-U_\tau)^2}{4D_x\tau}\right] \exp\left[-\frac{y^2}{4D_y\tau}\right]\left(\exp\left[-\frac{(z-h)^2}{4D_z\tau}\right] + \exp\left[\frac{-(z+h)^2}{4D_z\tau}\right]\right) \quad (5)$$

where τ is t/R_f. h is the location of the source relative to a plane of reflection $z = 0$. This model is only valid for $\delta t < \tau$. For $\delta t \gg \tau$, the source appears essentially continuous and the model can be integrated over time to give the steady solution. Under these conditions, transient accumulation in an immobile phase does not occur ($R_f \rightarrow 1$). Taking $\tau = x/U$, equation 5 becomes

$$C_A = \frac{\dot{N}_A}{4\pi\sqrt{D_y D_z}x} \exp\left[-\frac{y^2}{4D_y\frac{x}{U}}\right]\left(\exp\left[-\frac{(z-h)^2}{4D_z\frac{x}{U}}\right] + \exp\left[\frac{-(z+h)^2}{4D_z\frac{x}{U}}\right]\right) \quad (6)$$

This model is useful for estimating transport and dispersion from

1. an elevated emission source (stack) in the atmosphere at height h above the ground,
2. an outfall at depth h below the water surface of a lake, or,
3. a subsurface source at a height h above an impermeable layer in an aquifer.

In cases 1) and 2) the effective diffusion coefficients scale with the size of the turbulent eddies. In case 3) the effective diffusion coefficient scales with the size of the soil heterogeneities.

Equation 5 or 6 is often rewritten by setting $\sigma_i^2 = 2D_i\tau = 2D_i x/U$, where σ_i represents the standard deviation in concentration in direction i, a measure of the width of the concentration distribution. Equation 6, for example, becomes

$$C_A = \frac{\dot{N}_A}{2\pi\sigma_y\sigma_z U} \exp\left[-\frac{y^2}{2\sigma_y^2}\right]\left(\exp\left[-\frac{(z-h)^2}{2\sigma_z^2}\right] + \exp\left[\frac{-(z+h)^2}{2\sigma_z^2}\right]\right) \quad (7)$$

This form of the equation demonstrates that the model predicts profiles that are identical in shape to the Gaussian probability distribution function. The definition of the standard deviations in concentration suggest that they should grow with the square root of time or distance. Experiments in environmental media suggest, however, that linear rates of growth occur, at least initially. Empirical relationships relating σ_y and σ_z to turbulence and time or distance traveled are normally used to overcome this problem.

Note that the maximum concentration is directly downstream of the point source ($y = 0$) and at the same height or depth as the source ($z = h$). The edge of the dispersing plume can be arbitrarily selected as the location where the predicted concentration is 10% of this maximum. From the properties of the Gaussian probability distribution, this location is given in the y direction by $y = \pm 2.15\ \sigma_y$. In excess of 98% of the area in the Gaussian concentration distribution is located within these limits.

Interface Processes

The above solutions to the mass conservation equation (equation 1) describe intraphase transport processes. They assume simple forms of the concentrations and fluxes at the interface, either constant concentration or constant emission rate. Many attempts to assess transport and fate of chemicals in the environment, however, require prediction of the flux at the interface or more complicated models of its behavior. That is, in general, we must explicitly model the interface processes as well.

An environmental interface (eg, the air–water, air–soil or sediment–water) is typically assumed at equilibrium while the bulk phases are assumed to be in a state of disequilibrium. The interface is modeled as a two-dimensional surface which has no volume and in which accumulation (and thus disequilibrium) cannot occur. The rate of approach to equilibrium in the bulk phase as a result of transport across the interface is generally assumed to be proportional to the concentration difference between the interface and bulk phase, that is,

$$\dot{n}_A = k(C_A - C_{Ai}) \quad (8)$$

where $\dot{n}_A$ represents the chemical flux of A (eg, moles · area^{-1}·time^{-1}) and C_{Ai} represents the interfacial concentration of A in the same phase as the bulk phase. The coefficient of proportionality, or mass transfer coefficient, is a function of the processes operative in the particular phase for which the flux is being estimated. The concentration difference represents a deviation from equilibrium within the phase since rapid mass transfer and equilibration would result in the bulk phase attaining the concentration present at the interface.

A relationship equivalent to equation 8 could be written in an adjacent phase with a separate mass transfer coefficient describing processes within that phase and the concentration difference representing the concentration difference in that phase. As indicated above, the fluxes and concentrations are generally assumed in equilibrium at the interface, i, and thus the flux of contaminant, $\dot{n}_A$, out of phase j is the same as that entering phase k.

$$\dot{n}_A = k_j(C_A - C_{Ai})_j = -k_k(C_A - C_{Ai})_k \tag{9}$$

Here the subscript on the concentration differences refer to the phase (j or k).

The single phase mass transfer coefficients are primarily dependent on the flow conditions and are only weakly chemical dependent as a result of the influence of molecular diffusivity. The effect of molecular diffusivity can be approximated by

$$\frac{k_A}{k_B} = \left(\frac{D_A}{D_B}\right)^n \qquad \frac{1}{2} \le n \le 1 \tag{10}$$

where 1/2 is the power predicted by the penetration and surface renewal theories, 2/3 is predicted by boundary layer theory and 1 is predicted by film theory (5). There is very little variation in molecular diffusion coefficients between compounds in the same phase. Forty-six gases, organic and salt compounds displayed an average diffusivity in water of $1.1 \pm 0.48 \times 10^{-5}$ cm^2/s at 20–25°C (6). Similarly, the average molecular diffusivity in air at 20–25°C for forty-two compounds was 0.11 ± 0.07 cm^2/s. The specified ranges are standard deviations. This range suggests that the molecular diffusivity of most compounds in air or water at 25°C is well approximated by 0.1 and 10^{-5} cm^2/s, respectively. Thus the individual phase mass transfer coefficients are a strong function of only the flow conditions within the phase, and measurements for a particular compound are approximately applicable to other compounds either directly or through use of equation 10.

Use of equation 9, however, is hindered by the use of interfacial concentrations, C_{Ai}. Generally, only the bulk phase concentrations are of interest or are directly measurable. Thus it is often convenient to rewrite the flux relationship in terms of a global deviation from equilibrium,

$$\dot{n}_A = K_{jk}(C_{Aj} - C_{Aj}^*) = -K_{kj}(C_{Ak} - C_{Ak}^*) \tag{11}$$

Here K_{jk} represents the mass transfer coefficient for transport from phase j to phase k and C_{Aj}^* represents the concentration in phase j that would be in equilibrium with phase k. If α_{jk} represents the equilibrium ratio of the concentrations in phases j and k,

$$C_{Aj}^* = \alpha_{jk} C_{Ak} \tag{12}$$

Similarly, K_{kj} represents the mass transfer coefficient for transport from phase k to phase j and C_{Ak}^* represents the concentration in phase k that would be in equilibrium with phase j ($C_{Ak}^* = \alpha_{kj} C_{Aj} = C_{Aj}/\alpha_{jk}$). With equation 11, the chemical flux can be determined from a bulk concentration in each phase, thermodynamic equilibrium relationships and the "overall" mass transfer coefficients K_{jk} or K_{kj}. The overall mass transfer coefficients in equation 11 are dependent upon both the chemical (specifically the equilibrium partitioning of a chemical at the interface) and the flow conditions.

The more fundamental definition of the individual phase mass transfer coefficients (eq. 9) can be related to the lumped coefficients (eq. 11) for which the driving force is more easily measured. Equality of fluxes regardless of the form of the equation suggests that:

$$\frac{1}{K_{jk}} = \frac{1}{k_j} + \frac{1}{k_k}\left(\frac{C_{Aj}}{C_{Ak}}\right)_{eq} \tag{13}$$

The ratio of concentrations in the two phases (ie, the group in parentheses in eq. 13) is simply the equilibrium distribution coefficient, α_{jk}, since the interface is at equilibrium. Note that for large values of α_{jk}, the last term in eq. 13 dominates and the overall mass transfer coefficient is controlled by the mass transfer coefficient in phase k. Similarly for small values of α_{jk} (or large values of α_{kj}), the overall mass transfer coefficient is controlled by the mass transfer coefficient in phase j.

Quantitative determination of the mass transfer rate thus requires determination of the equilibrium partitioning of the contaminant between the two adjacent phases. Only linear partitioning will be considered here and is discussed below for each of the most common environmental interfaces.

Fluid–Fluid Interfaces. The distribution coefficient for a hydrophobic organic compound at the air–water interface is the ratio of the concentration in air to that in water,

$$\alpha_{12} = \frac{C_{A1}}{C_{A2}} = \frac{p_A^*}{RT\,C_{A2}^*} \tag{14}$$

where p_A^*/RT is the molar air concentration of the contaminant at its pure compound vapor pressure and C_{A2}^* is the molar solubility in water. This relationship reflects the fact that any compound exerts its pure compound vapor pressure, p_A^*, when present at its molar solubility. Thus a water body containing, for example, benzene at its water solubility of 1780 g/m^3 exerts the same benzene vapor pressure as pure benzene. For hydrophobic compounds (HOC) in water, this distribution ratio, α_{12}, remains essentially constant at any concentration below the solubility limit and is termed a Henry's Law constant. The Henry's Law constant is tabulated in a variety of different units which include the ratio of air and water concentrations or mole fractions (both dimensionless) and pressure over solubility units (eg, atm-m^3/mole). Note that Equation 13 requires use of an α_{jk} defined by the same units of concentration used in the definition of the fluxes, in this case the ratio of molar concentrations in the air and water phases. Because of their low solubility, α_{12} tends to be large for hydrophobic organic compounds even if the pure component vapor pressure is low. The PCB mixture Aroclor 1260, for example, exhibits a Henry's Law constant that is slightly higher than that of benzene despite having a vapor pressure that is more than a million times smaller (7). As a result, the evaporation of contaminants from water tends to be controlled by water-side mass transfer resistances since the water phase term dominates in equation 13. Vapor pressures, water solubilities and air–water distribution coefficients are given for selected compounds in Table 2.

The distribution coefficient for a hydrophobic organic compound at an air–nonaqueous phase liquid (NAPL) interface is governed by Raoult's Law if the compound is assumed to form an ideal solution in the NAPL phase. This can be written

$$\alpha_{14} = \frac{C_{A1}}{C_{A4}} = \frac{p_A^*}{RT\,C_4} \tag{15}$$

where C_4 is the molar density of the NAPL phase. Note that C_4 is, at high contaminant fractions (eg, >5%), a function of the concentration of the contaminant. At an air–oil interface, α_{14} tends to be small, especially for low vapor pressure compounds. The evaporation of an organic compound will then tend to be controlled by air side resistances.

At the interface between water and an oily phase, the distribution coefficient, α_{24}, is simply the ratio α_{14}/α_{12}.

$$\alpha_{24} = \frac{C_{A2}}{C_{A4}} = \frac{C_{A2}^*}{C_4} \tag{16}$$

Fluid–Solid Interfaces. At the sediment–water interface, the mobile fluid on both sides of the interface is water (ie, pore-water and overlying water) and the distribution coefficient between these "phases" is unity. Similarly, the thermodynamic distribution coefficient between soil, air and the atmosphere is unity. The partitioning between the soil or sediment phase and the adjacent pore fluid, however, is much more complicated. Partitioning between a solid and fluid phase is generally assumed to follow one of the following equilibrium relationships:

- Linear partitioning
- Langmuir or monolayer partitioning
- Fruendlich or empirical power-law partitioning
- Brunauer-Emmett-Teller or multiple layer partitioning

The latter three relationships are nonlinear and are necessary to describe partitioning in many systems. Space precludes formal presentation of the nonlinear partitioning relationships but the linear partitioning relationship is commonly used and directly analogous to Henry's Law for fluid phase equilibrium. Hydrophobic organic compounds in pore water tend to follow this linear partitioning relationship if present at low concentrations (concentrations small relative to the water solubility limit). Under such conditions, the soil or sediment–water partition coefficient is constant and can be written

$$\alpha_{32} = \frac{C_{A3}}{C_{A2}} \tag{17}$$

Since the concentration of pollutant in the solid phase is normally measured in units of mass or moles per mass of solid (eg, mg/kg) and that in the pore water is mass or moles per volume of fluid (eg, mg/L), the distribution co-

Table 2. Physical Properties of Selected Compounds[a]

Compound	MW	P_v mm Hg	Water Solubility mg/L	α_{12}	Log α_{oc}
Acenaphthene	154	0.005	3.47	0.011934	1.3
Anthracene	178	2×10^{-4}	0.045	0.042547	4.3
Benzene	78	95	1800	0.221397	1.9
Benzo[a]pyrene	252	5.5×10^{-9}	0.004	1.86E-05	6.0
Chlordane	410	1×10^{-5}	0.056	0.003938	5.2
p,p'-DDT	354	1.9×10^{-7}	0.003	0.001206	5.4
1,2-Dichloroethane	99	233	5060	0.24517	1.8
Dieldrin	381	3×10^{-6}	0.2	0.000307	4.6
Fluoranthene	202	5×10^{-6}	0.26	0.000209	4.6
Hexachlorobenzene	285	1×10^{-5}	0.005	0.030655	3.6
Methylene chloride	85	439	19000	0.105623	0.94
Naphthalene	128	0.233	30	0.053465	3.1
PCB-1242 (avg)	261	4×10^{-4}	0.24	0.023395	3.7
PCB-1254 (avg)	327	7.7×10^{-5}	0.057	0.023757	5.6
Pentachlorophenol	266	1.7×10^{-4}	20	0.000122	3.0
Phenanthrene	178	6.8×10^{-4}	1	0.00651	3.7
Pyrene	202	6.9×10^{-7}	0.135	5.55E-05	4.7
Tetrachloroethylene	166	20	150	1.190348	2.4
Trichloroethylene	131	74	1100	0.473955	2.0

[a] From Ref. 7. All data are estimates at 25°C.

efficient has units of volume fluid per mass of solid (eg, L/kg). For hydrophobic organics, the sorbed phase concentration also tends to be directly proportional to the organic carbon fraction of the solid phase, f_{oc}. The constant of proportionality is the organic carbon based-partition coefficient, α_{oc}, which is a tabulated chemical-specific property, as listed in Table 2. This quantity is a measure of the hydrophobicity of a compound and therefore has been found to correlate well with the octanol-water partition coefficient (8), given α_{oc}, $\alpha_{32} = f_{oc}\alpha_{oc}$. The ratio of the total contaminant concentration in the soil–water system to the concentration in the water phase is given by

$$R_f = (\varepsilon + \rho_s f_{oc}\alpha_{oc}) \tag{18}$$

Partitioning between soil and an adjacent vapor phase is often described by the Brunauer-Emmett-Taylor isotherm (9). Linear partitioning is sometimes assumed, however, for simplicity. In addition, the unsaturated soil zone generally contains significant amounts of water as well as air and soil. Thus accumulation in each of these phases must often be considered. Since the water tends to wet the soil surface more than the air, chemical partitioning in such a system seems to be well-represented by separate partitioning between the air and water phases and between the water and soil phases. Thus the retardation factor or the ratio of the total concentration of the contaminant in the soil–water–air system to its vapor phase concentration is given by

$$R_f = \left(\varepsilon_1 + \frac{\varepsilon_2}{\alpha_{12}} + \rho_s f_{oc}\frac{\alpha_{oc}}{\alpha_{12}}\right) \tag{19}$$

where ε_1 is the air-filled porosity and ε_2 is the water-filled porosity.

The picture of the air–soil equilibrium provided by equation 19 is valid as long as linear partitioning can be assumed and as long as the volumetric water content in the soil exceeds 1–3%. At lower volumetric water contents, portions of the solid surface may be exposed to the vapor space and direct sorption of vapors onto the surface may occur. This disposition can lead to significantly greater sorption of the vapors onto the solid phase and sorption that can rarely be modeled as a linear process. The process of sorbed vapor release, as a solid is wetted, has been explained as simply the reverse of this competitive sorption process. This behavior has important practical implications; for example, rapid release of pesticide or herbicide vapors can occur upon wetting an initially dry field surface.

The previous sections have introduced the dynamic, spatially distributed mass balance as the basic tool for the evaluation of contaminant transport and fate in the environment. These sections have also identified the equilibrium relationships at environmental interfaces that determine the boundary conditions for the intraphase transport processes as well as define the concentration differences that serve as driving forces for pollutant transport.

SUMMARY AND CONCLUSIONS

The key concepts of chemodynamics as applied to exposure assessment and the evaluation of in-situ remediation and containment technologies have been presented. The discussion of the concepts and the examples, emphasize the similarity of approaches to analyze contaminant migration problems in these media.

The key feature of many chemodynamic models is the capture of just the most fundamental physiochemical processes. By such an approach it is often not possible to describe all of the behavior of an environmental system, but the models presented herein do describe at least semi-quantitatively the systems for which they were designed. In some cases more sophisticated models are not yet supported by experiment despite the obvious limitations of the simpler models. In addition, the sparse dataset typically available in any particular field application often limits the ability to use more sophisticated models regardless of their level of validation by theory and laboratory experiments. The level of models presented herein often represent an excellent compromise between the ability to describe a physical system and the data input required to drive the model. The ultimate test of such models, of course, is a comparison to field observations. In this work, the field observations have been presented only in the form of empirical correlations for some of the model parameters. The reader is referred to the original works for the data and motivation for the form of the models.

Acknowledgments

The author would like to acknowledge partial support for this work by the U.S. Environmental Protection Agency Risk Reduction Engineering Laboratory (supporting studies of capping contaminated sediments); the U.S. Environmental Protection Agency Office of Exploratory Research and Louisiana State University Hazardous Substances Research Center (supporting studies of bioturbation); the Louisiana Board of Regents; and the United States National Science Foundation (supporting studies of non-aqueous phase liquid dynamics and colloidal facilitation of transport).

BIBLIOGRAPHY

1. L. J. Thibodeaux, *Chemodynamics Movement of Chemicals in Air, Water and Soil*, Wiley-Interscience, New York, 1979.
2. Environmental Protection Agency. *Superfund Exposure Assessment Manual*, 1988, EPA/540/188/001.
3. H. S. Carslaw, J. C. Jaeger, *Conduction of Heat in Solids,* 2nd ed., Oxford Science Publications, Oxford, UK, 1959.
4. J. Crank, *The Mathematics of Diffusion*, Clarendon Press, Oxford, UK, 1975.
5. Treybal, *Mass Transfer Operations*, McGraw-Hill, New York, 1974.
6. L. J. Thibodeaux, *Chemodynamics*, Wiley-Interscience, New York, 1979.
7. J.H. Montgomery and L. M. Welkom. *Groundwater Chemicals Desk Reference*, Lewis Publishers, Chelsea, Mich., 1990.
8. W. J. Lyman, W. F. Reehl, and D. H. Rosenblatt, *Handbook of Chemical Property Estimation Methods*, American Chemical Society, Washington, D.C., 1990.
9. K. T. Valsaraj and L. J. Thibodeaux. "Equilibrium Adsorption of Chemical Vapors on Surface Soils, Landfills and Landfarms—A review," *J. of Hazardous Materials,* **19,** 79–99 (1988).

THE CLEAN AIR ACT

SUSAN MAYER
Congressional Research Service
Washington, D.C.

The Clean Air Act (CAA), codified as 42 U.S.C. 7401 *et seq.*, seeks to protect human health and the environment from emissions that pollute ambient, or outdoor, air. To ensure that air quality in all areas of the United States meets certain federally mandated minimum standards, it assigns primary responsibility to States to assure adequate air quality. The Act deems areas not meeting standards as nonattainment areas, and requires them to implement specified air pollution controls. In addition to provisions relating to ozone nonattainment, the Act addresses mobile sources, air toxics, and the special problem of acid rain. It establishes a comprehensive permit system for all sources. Other provisions address ozone depleting substances, enforcement, clean air research, disadvantaged business concerns, and employment transition and assistance.

BACKGROUND

Prior to 1955, air pollution was controlled at the State and local level. Federal legislation controlling air pollution was first passed in 1955. The Federal role was strengthened in subsequent amendments and changed significantly with the passage of the Clean Air Act Amendments of 1990 (P.L. 101-549) which was signed into law on November 15, 1990.

Significant new provisions added to the Act by the 1990 amendments included (1) the acid rain control program, including the use of marketable allowances for introducing flexibility into the sulfur oxides reduction program; (2) a State-run program requiring permits for the operation of many sources of air pollutants, including a requirement that fees be imposed to cover administrative costs; and (3) the authorization for a 5-year, $250 million program for retraining and unemployment benefits for workers displaced by requirements of the CAA.

Changes to the Act by the 1990 amendments included, in part, provisions to (1) classify areas in nonattainment according to the extent to which they exceed the standard and to tailor deadlines, planning, and controls according to each area's status and problem; (2) tighten automobile emission standards and to require reformulated and alternative fuels in the most-polluted areas; (3) revise the air toxics section, establishing a new program of technology-based standards and addressing the problem of sudden, catastrophic releases of toxics; (4) change the stratospheric ozone protection provision to a phase-out of the most ozone-depleting chemicals; and (5) update the enforcement provisions so they parallel those in other pollution control acts, including authority for EPA to assess administrative penalties.

See also COAL COMBUSTION; COMBUSTION MODELING; ELECTRIC POWER GENERATION; ENERGY EFFICIENCY-ELECTRIC UTILITIES; ENVIRONMENTAL ECONOMICS.

The following sections describe each of these new and revised programs required by the Act.

NATIONAL AMBIENT AIR QUALITY STANDARDS (NAAQS) (SECTION 109)

The Act requires EPA to establish NAAQS for several types of air pollutants. The NAAQS must be designed to protect public health and welfare with an adequate margin of safety. The NAAQS must be attained as expeditiously as possible, and in no case more than five years after EPA determines that an area does not meet the standards, unless EPA provides an extension of the deadline. For areas not in attainment with NAAQS, the Clean Air Act Amendments of 1990 established special compliance schedules, staggered according to each area's air pollution problem and status, as described below.

EPA has promulgated NAAQS for six air pollutants: sulfur dioxide (SO_2), particulate matter (PM_{10}), nitrogen dioxide (NO_2), carbon monoxide (CO), ozone (O_8), and lead (Pb). The Act requires EPA to review the scientific data upon which the standards are based, and revise the standards, if necessary, every five years. More often than not,

Table 1. Clean Air Act and Principal Amendments[a]

Year	Act	Public Law Number
1955	Air Pollution Control Act	P.L. 84-159
1959	Reauthorization	P.L. 86-353
1960	Motor vehicle exhaust study	P.L. 86-493
1963	Clean Air Act Amendments	P.L. 88-206
1965	Motor Vehicle Air Pollution Control Act	P.L. 89-272, title I
1966	Clean Air Act Amendments of 1966	P.L. 89-675
1967	Air Quality Act of 1967	P.L. 90-148
	National Air Emission Standards Act	P.L. 90-148, title II
1970	Clean Air Act Amendments of 1970	P.L. 91-604
1973	Reauthorization	P.L. 93-13
1974	Energy Supply and Environmental Coordination Act of 1974	P.L. 93-319
1977	Clean Air Act Amendments of 1977	P.L. 95-95
1980	Acid Precipitation Act of 1980	P.L. 96-294, title VII
1981	Steel Industry Compliance Extension Act of 1981	P.L. 97-23
1987	Clean Air Act 8-month Extension	P.L. 100-202
1990	Clean Air Act Amendments of 1990	P.L. 101-549

[a] Codified generally as 42 U.S.C. 7401-7671.

however, EPA has taken more than five years in reviewing and revising the standards.

STATE IMPLEMENTATION PLANS (SIPS) (SECTION 110)

Although the Act authorizes the EPA to set NAAQS, the States are responsible for establishing procedures to attain and maintain the NAAQS. The States adopt plans, known as State Implementation Plans (SIPs), and submit them to EPA to ensure that they are adequate to meet the statutory requirements. Each State is responsible for achieving the NAAQS within its jurisdiction. Areas not meeting NAAQS are called "nonattainment" areas.

SIPs are based on emission inventories and computer models to determine whether air quality violations will occur. If these data show that standards would be exceeded, the State imposes necessary controls on existing sources to ensure that emissions do not cause "exceedences" of the standards. Proposed new and modified sources must obtain State construction permits in which the applicant shows how the anticipated emissions will not exceed allowable limits. In "nonattainment" areas, emissions from new or modified sources must be offset by reductions in emissions from existing sources.

The 1990 amendments *require* EPA to impose one of two sanctions in areas which fail to submit an SIP, fail to submit an adequate SIP, or fail to implement an SIP: either a 2-to-1 emissions offset for the construction of new polluting sources, or a ban on Federal highway grants (an additional ban on air quality grants is discretionary).

NONATTAINMENT REQUIREMENTS (PART D OF TITLE I)

In a great departure from the prior law, the 1990 Clean Air Act Amendments group nonattainment areas into classifications based on the extent to which the NAAQS is exceeded, and establish specific pollution controls and attainment dates for each classification.

Nonattainment areas are classified on the basis of a "design value," which is derived from the pollutant concentration (in parts per million) recorded by air quality monitoring devices. The design value for ozone is the fourth highest reading measured over a 3-year period. The Act creates 5 classes of ozone nonattainment. Only Los Angeles falls into the "extreme" class. A simpler classification system establishes moderate and serious nonattainment areas for carbon monoxide and particulate matter with correspondingly more stringent control requirements for the more polluted class.

For ozone nonattainment areas, the deadlines are established for each classification. For carbon monoxide, the attainment date for moderate areas is December 31, 1995, and for serious areas is December 31, 2000. For particulate matter, the deadline for areas currently designated moderate nonattainment areas is December 31, 1994; for those areas subsequently designated as moderate, the deadline is 6 years after designation. For serious areas, the respective deadlines are December 31, 2001 or 10 years after designation.

Although areas with more severe air pollution problems have a longer time to meet the standards, more stringent control requirements are imposed through the SIP. Each category of nonattainment must have specified air pollution control measures imposed. Marginal ozone nonattainment areas are required to have few controls imposed. Each area must meet the requirements of lower nonattainment categories in addition to the requirements of its own category. A summary of the primary ozone control requirements for each nonattainment category follows.

Marginal Areas. Inventory emissions sources to be updated every 3 years. Require 1.1 to 1 offsets, ie, industries must reduce emissions from existing facilities by 10 percent more than emissions of any new facility opened in the area. Impose reasonable available control technology (RACT) on all major sources emitting more than 100 tons per year for the nine industrial categories where EPA had already issued control technique guidelines describing RACT prior to 1990.

Moderate Areas. Meet all requirements for marginal areas. Impose a 15 percent reduction in volatile organic compounds (VOCs) in 6 years. Adopt a basic vehicle inspection and maintenance program. Impose RACT on all major sources emitting more than 100 tons per year for all additional industrial categories where EPA will issue control technique guidelines describing RACT. Require vapor recovery at gas stations selling more than 10,000 gallons per month. Require 1.15 to 1 offsets.

Serious Areas. Meet all requirements for moderate areas. Reduce definition of a major source of VOCs from emissions of 100 tons per year to 50 tons per year for the purpose of imposing RACT. Reduce VOCs 3 percent annually for years 7–9 after the 15 percent reduction already required by year 6. Improve monitoring. Adopt an enhanced vehicle inspection and maintenance program. Require fleet vehicles to use clean alternative fuels. Adopt transportation control measures if the number of vehicle miles travelled in the area is greater than expected. Require 1.2 to 1 offsets. Adopt contingency measures if the area does not meet required VOC reductions.

Table 2. Ozone Nonattainment Classifications

Class	Marginal	Moderate	Serious	Severe	Extreme
Deadline	3 years	6 years	9 years	15 years[a]	20 years
Areas	42 areas	31 areas	14 areas	9 areas	1 area
Design	0.121 ppm–	0.138 ppm–	0.160 ppm–	0.180 ppm–	>0280 ppm
Value	0.188 ppm	0.160 ppm	0.180 ppm	0.280 ppm	

[a] Areas with a 1988 design value between 0.190 and 0.280 ppm have 17 years to attain, rather than 15 years.

Severe Areas. Meet all requirements for serious areas. Reduce definition of a major source of VOCs from emissions of 50 tons per year to 25 tons per year for the purpose of imposing RACT. Adopt specified transportation control measures. Implement a reformulated gasoline program for all vehicles in the area. Require 1.3 to 1 offsets. Impose $5,000 per ton penalties on major sources if the area does not meet required reductions.

Extreme Areas. Meet all requirements for severe areas. Reduce definition of a major source of VOCs from emissions of 25 tons per year to 10 tons per year for the purpose of imposing RACT. Require clean fuels or advanced control technology for boilers emitting more than 25 tons per year of NO_x. Require 1.5 to 1 offsets.

As with ozone nonattainment areas, CO nonattainment areas are subjected to specified control requirements. A summary of the primary CO control requirements for each nonattainment category follows.

Moderate Areas. Conduct an inventory of emissions sources. Forecast total vehicle miles travelled in the area. Adopt an enhanced vehicle inspection and maintenance program, and demonstrate annual improvements sufficient to attain the standard.

Serious Areas. Adopt specified transportation control measures. Implement a reformulated gasoline program for all vehicles in the area, and reduce definition of a major source of CO from emissions of 100 tons per year to 50 tons per year if stationary sources contribute significantly to the CO problem.

Serious areas failing to attain the standard by the deadline have to revise their SIP and demonstrate reductions of 5 percent per year until the standard is attained.

EMISSION STANDARDS FOR MOBILE SOURCES (TITLE II)

The 1990 amendments tightened automobile emission standards. (For cars, the hydrocarbon standard is reduced by 40%, the nitrogen oxides (NO_x) standard by 50%). These standards would be phased in over the 1994–1996 model years. If a study due in 1997 shows both the technological feasibility and the need for air quality improvement, yet another round of tightened standards would be imposed in 2004.

The 1990 amendments also require that a cleaner, "reformulated" gasoline be sold, starting in 1995, in the nine worst ozone nonattainment areas (Los Angeles, San Diego, Houston, Baltimore, Philadelphia, New York, Hartford, Chicago, and Milwaukee). Other ozone nonattainment areas will be able to opt in to the reformulated gasoline program at a later date.

Use of alternative fuels other than gasoline will be stimulated by two programs. First, California is required to develop a program requiring low emission vehicles (LEVs) and ultralow emission vehicles (ULEVs) for use in the State to be introduced starting in 1996 at 150,000 vehicles per year and rising to 300,000 vehicles per year in 1999. Second, in all of the most seriously polluted ozone nonattainment areas and the worst CO nonattainment areas, centrally fueled fleets of 10 or more vehicles must purchase some "clean fuel vehicles" when they add new vehicles to existing fleets starting in 1998. The percentage of any new vehicles purchased that must be clean fuel vehicles is 30 percent in 1998, 50 percent in 1999, and 70 percent in 2000. The "clean fuel" vehicles must be the same as those sold in California.

The 1990 amendments also imposed tighter requirements on certification (an auto's useful life is defined as 100,000 miles instead of the earlier 50,000 miles), on emissions allowed during refueling, on low temperature CO emissions, on in-use performance over time, and on warranties for the most expensive emission control components (8 years/80,000 miles for the catalytic converter, electronic emissions control unit, and onboard emissions diagnostic unit). Regulations were also extended to include nonroad fuels and engines.

HAZARDOUS AIR POLLUTANTS (SECTION 112)

Completely rewritten by the Clean Air Act Amendments of 1990, section 112 of the Act establishes programs for protecting the public health and environment from exposure to toxic air pollutants. Section 112, as revised by the 1990 amendments, contains four major provisions including, Maximum Achievable Control Technology (MACT) requirements, health-based standards, standards for stationary area sources, and requirements for the prevention of catastrophic releases.

First, EPA is to establish technology-based emission standards, called MACT standards, for sources of 189 pollutants listed in the legislation, and to specify categories of sources subject to the emission standards. (Public Law 102-187, enacted on December 4, 1991, deleted hydrogen sulfide from the list of toxic pollutants leaving only 188.)

EPA is to revise the standards periodically (at least every eight years). EPA can, on its initiative or in response to a petition, add or delete substances or source categories from the lists. Section 112 establishes a presumption in favor of regulation for the designated chemicals; it requires regulation of a designated pollutant unless EPA or a petitioner is able to show "that there is adequate data on the health and environmental effects of the substance to determine that emissions, ambient concentrations, bioaccumulation or deposition of the substance may not reasonably be anticipated to cause any adverse effects to human health or adverse environmental effects."

Section 112 requires EPA to set standards for sources of the listed pollutants that achieve "the maximum degree of reduction in emissions" taking into account cost and other non-air-quality factors. The standards for new sources "shall not be less stringent than the most stringent emissions level that is achieved in practice by the best controlled similar source." The standards for existing sources may be less stringent than those for new sources, but must be more stringent than the emission limitations achieved by either the best performing 12 percent of existing sources (if there are more than 30 such sources in the category or subcategory) or the best performing 5 similar sources (if there are fewer than 30). Existing sources are given 3 years following promulgation of standards to achieve compliance, with a possible 1-year extension; additional extensions may be available for special circumstances or for certain categories of sources. Existing sources that achieve voluntary early emissions reductions will receive a 6-year extension for compliance with MACT.

For solid waste incinerators, EPA is to establish performance standards for compliance with both hazardous air pollutant provisions (section 112) and the new source performance requirements (section 111) of the Act. However, rules for both small and large incinerators are overdue.

The second major provision of section 112 sets health-based standards to address situations in which a significant residual risk of adverse health effects or a threat of adverse environmental effects remains after installation of MACT. This provision requires that EPA, after consultation with the Surgeon General of the United States, submit a report to Congress on the public health significance of residual risks, and make recommendations as to legislation regarding such risks. If Congress does not legislate in response to EPA's recommendations, then EPA is required to issue standards for categories of sources of hazardous air pollutants as necessary to protect the public health with an ample margin of safety or to prevent an adverse environmental effect. A residual risk standard is required for any source emitting a cancer-causing pollutant that poses an added risk to the most exposed person of more than 1-in-a-million. Residual risk standards would be due 8 years after promulgation of MACT for the affected source category. Existing sources have 90 days to comply with a residual risk standard, with a possible 2-year extension. In general, residual risk standards do not apply to area sources.

This provision also directs EPA to contract with the National Academy of Sciences for a study of risk assessment methodology, and creates a Risk Assessment and Management Commission to investigate policy implications and appropriate uses of risk assessment and risk management. These studies are designed to ensure that regulatory decisions are well based technically.

Third, in addition to the technology-based and health-based programs for major sources of hazardous air pollution, EPA is to establish standards for stationary "area sources" determined to present a threat of adverse effects to human health or the environment. (Area sources are numerous, small sources such as gas stations or dry cleaners that may cumulatively produce significant quantities of a pollutant.) The provision requires EPA to regulate the stationary area sources responsible for 90 percent of the emissions of the 30 hazardous air pollutants that present the greatest risk to public health in the largest number of urban areas. EPA is to list the sources and pollutants within 5 years of enactment, and promulgate regulations within 10 years. In setting the standard, EPA can impose less stringent "generally available" control technologies, rather than MACT.

Finally, section 112 addresses prevention of sudden, catastrophic releases of air toxics by establishing an independent Chemical Safety and Hazard Investigation Board. The Board will investigate accidents involving releases of hazardous substances, and conduct studies and prepare reports on the handling of toxic materials, as well as measures to reduce the risk of accidents.

EPA is also authorized to issue prevention, detection, and correction requirements for catastrophic releases of air toxics, which shall require owners and operators to prepare risk management plans for the listed chemicals, including a hazard assessment, measures to prevent releases, and a response program. EPA is to issue these regulations by November 1993, which become effective 3 years after promulgation (or after a substance is listed, whichever is later).

NEW SOURCE PERFORMANCE STANDARDS (NSPS) (SECTION 111)

NSPS establish nationally uniform, technology-based standards for categories of new industrial facilities. These standards accomplish two goals: first, they establish a consistent baseline for pollution control that competing firms must meet, and thereby remove any incentive for States or communities to weaken air pollution standards in order to attract polluting industry; and second, they preserve clean air to accommodate future growth, as well as for its own benefits.

NSPS establish maximum emission levels for new or extensively modified principal stationary sources—powerplants, steel mills, and smelters—with the emission levels determined by the best "adequately demonstrated" continuous control technology available, taking costs into account. EPA must regularly revise and update NSPS applicable to designated sources as new technology becomes available, since the goal is to prevent new pollution problems from developing and to force the installation of new control technology.

PREVENTION OF SIGNIFICANT DETERIORATION (PSD) (TITLE I, PART C)

Prevention of Significant Deterioration reflects the principle that areas where the air quality is better than required by NAAQS should be protected from significant new air pollution even if NAAQS would not be violated. The Act divides clean air areas in three classes, and specifies the increments of SO_2 and particulate pollution allowed in each. Class I areas include international and national parks, wilderness areas, or other such pristine areas; allowable increments of new pollution are very small. Class II areas include all attainment and not classifiable areas, not designated as Class I; allowable increments of new pollution are modest. Class III represents selected areas that States designate for development; allowable increments of new pollution are large (but not exceeding NAAQS).

Polluting sources in PSD areas must install best available control technology (BACT) that may be more strict than that required by NSPS. The justifications of the policy are that it protects air quality, provides an added margin of health protection, preserves clean air for future development, and prevents firms from gaining a competitive edge by "shopping" for "clean air" to pollute.

ACID DEPOSITION CONTROL (TITLE IV)

The Clean Air Act Amendments of 1990 added an acid deposition control program to the Act. It sets goals for the year 2000 of reducing annual SO_2 emissions by 10 million tons from 1980 levels and reducing annual NO_x emissions by 2 million tons, also from 1980 levels.

The SO_2 reductions are imposed in two steps. Under Phase 1, owners/operators of 111 facilities listed in the law that are larger than 100 megawatts must meet tonnage emission limitations by January 1, 1995. This would reduce SO_2 emission by about 3.5 million tons. Phase 2 would include facilities larger than 75 megawatts.

To introduce some flexibility in the distribution and timing of reductions, the Act creates a comprehensive permit and emissions allowance system. An allowance is a limited authorization to emit a ton of SO_2. Issued by EPA, the allowances would be allocated to Phase 1 and Phase 2 units in accordance with baseline emissions estimates. Powerplants which commence operation after November 15, 1990 would not receive any allowances. These new units would have to obtain allowances (offsets) from holders of existing allowances. Allowances may be traded nationally during either phase. The law also permits industrial sources and powerplants to sell allowances to utility systems under regulations to be developed by EPA. Allowances may be banked by a utility for future use or sale.

Beginning in 1993, the Act provides for two sales to improve the liquidity of the allowance system and ensure the availability of allowances for utilities and independent power producers who need them. A special reserve fund consisting of 2.8% of Phase 1 and Phase 2 allowance allocations is set aside for sale. Allowances from this fund (25,000 annually from 1993–1999 and 50,000 thereafter) are sold at a fixed price of $1,500 an allowance. Independent power producers have guaranteed rights to these allowances under certain conditions. An annual auction of allowances (150,000 from 1993–1995, and 250,000 from 1996–1999) is an open auction with no minimum price. Utilities with excess allowances may have them auctioned off at this auction, and any person may buy allowances.

The Act essentially caps SO_2 emissions at individual existing sources through a tonnage limitation, and at future plants through the allowance system. First, emissions from most existing sources are capped at a specified emission rate times an historic baseline level. Second, for plants commencing operation after November 15, 1990, emission must be completely offset with additional reductions at existing facilities beginning after Phase 2 compliance. However, as noted above, the law provides some allowances to future powerplants which meet certain criteria. The utility SO_2 emission cap is set at 8.9 million tons, with some exceptions.

The Act provides that if an affected unit does not have sufficient allowances to cover its emissions, it is subject to an excess emission penalty of $2,000 per ton of SO_2 and required to reduce an additional ton of SO_2 of pollution the next year for each ton of excess pollutant emitted.

The Act also requires EPA to inventory industrial emissions of SO_2 and to report every 5 years, beginning in 1995. If the inventory shows that industrial emissions may reach levels above 5.60 million tons per year, then EPA is to take action under the Act to ensure that the 5.60 million ton cap is not exceeded.

The Act requires EPA to set specific NO_x emission rate limitations—0.45 lb. per million Btu for tangentially fired boilers and 0.50 lb. per million Btu for wall-fired boilers—unless those rates can not be achieved by low-NO_x burner technology. Tangentially and wall-fired boilers affected by Phase 1 SO_2 controls must also meet NO_x requirements. EPA is to set emission limitations for other types of boilers by 1997 based on low-NO_x burner costs. In addition, EPA is to propose and promulgate a revised new source performance standard for NO_x from fossil fuel steam generating units by January 1, 1994.

PERMITS (TITLE V)

The Clean Air Act Amendments of 1990 added a Title V to the Act which requires States to administer a comprehensive permit program for the operation of sources emitting air pollutants. These requirements are modeled after similar provisions in the Clean Water Act. Previously, the Clean Air Act contained limited provision for permits, requiring only new or modified major stationary sources to obtain construction permits (Title I of the CAA).

Sources subject to the new permit requirements would generally include major sources that emit or have the potential to emit 100 tons per year of any regulated pollutant, plus stationary and area sources that emit or have potential to emit lesser specified amounts of hazardous air pollutants. However, in nonattainment areas, the permit requirements also include sources which emit as little as 10 tons per year of VOCs, depending on the severity of the region's nonattainment status (serious, severe, or extreme).

States are required to develop permit programs and to submit those programs for EPA approval by November 15, 1993. If a State does not have an approved plan within 5 years, EPA can impose and administer its own plan in the State. States are to collect annual fees from sources sufficient to cover the "reasonable costs" of administering the permit program, with revenues to be used to support the agency's air pollution control program. The fee would be not less than $25 per ton of regulated pollutants (excluding carbon monoxide). Permitting authorities will have discretion to not collect fees on emissions in excess of 4,000 tons per year and may collect other fee amounts, if appropriate.

The permit states which air pollutants a source is allowed to emit. As a part of the permit process, a source must prepare a compliance plan and certify compliance. The term of permits is limited to no more than 5 years; sources are required to renew permits at that time. State permit authorities must notify contiguous States of permit applications that may affect them; the application and any comments of contiguous States must be forwarded to EPA for review. EPA can veto a permit; however, this authority is essentially limited to major permit changes. EPA review need not include permits which simply codify elements of a State's overall clean air plan, and EPA has discretion to not review permits for small sources. Holding a permit to some extent shields a source from enforcement actions: the Act provides that a source cannot be held in violation if it is complying with explicit requirements addressed in a permit, or if the State finds that certain provisions do not apply to that source.

Table 3. Principal U.S. Code Sections of the Clean Air Act

42 U.S.C.	Section Title	Clean Air Act, as Amended
Subchapter I	Programs and Activities	
Part A	Air Quality Emissions and Limitations	
7401	Findings, Purpose	sec. 101[a]
7402	Cooperative activities	sec. 102[a]
7403	Research, investigation, training	sec. 103[a]
7404	Research relating to fuels and vehicles	sec. 104[b]
7405	Grants for support of air pollution and control programs	sec. 105[a]
7406	Interstate air quality agencies; program cost limitations	sec. 106
7407	Air quality control regions	sec. 107[a]
7408	Air quality criteria and control techniques	sec. 108[c]
7409	National primary and secondary air quality standards	sec. 109[c]
7410	SIPs for national primary and secondary air quality standards	sec. 110[c]
7411	Standards of performance for new stationary sources	sec. 111[c]
7412	National emission standards for hazardous air pollutants	sec. 112[c]
7413	Federal enforcement procedures	sec. 113[c]
7414	Recordkeeping inspections, monitoring and entry	sec. 114[c]
7415	International air pollution	sec. 115[a]
7416	Retention of State authority	sec. 116[c]
7417	Advisory committees	sec. 117[a]
7418	Control of pollution from Federal facilities	sec. 118[a]
7419	Primary nonferrous smelter orders	sec. 119[d]
7420	Noncompliance penalty	sec. 120[d]
7421	Consultation	sec. 121[d]
7422	Listing of certain unregulated pollutants	sec. 122[d]
7423	Stack heights	sec. 123[d]
7424	Assurance of adequacy of State plans	sec. 124[d]
7425	Measures to prevent economic disruption	sec. 125[d]
7426	Interstate pollution abatement	sec. 126[d]
7427	Public notification	sec. 127[d]
7428	State boards	sec. 128[d]
Part B	Ozone Protection (repealed)	
Part C	Prevention of Significant Deterioration of Air Quality	
Subpart I	Clean Air	
7470	Congressional declaration of purpose	sec. 160[d]
7471	Plan requirements	sec. 161[d]
7472	Initial classifications	sec. 162[d]
7473	Increments and ceilings	sec. 163[d]
7474	Area designations	sec. 164[d]
7475	Preconstruction requirements	sec. 165[d]
7476	Other pollutants	sec. 166[d]
7477	Enforcement	sec. 167[d]
7478	Period before plan approval	sec. 168[d]
7479	Definition	sec. 169[d]
Subpart II	Visibility Protection	
7491	Visibility protection of Federal class I areas	sec. 169a[d]
7492	Visibility	(New sec. 816)
Part D	Plan Requirements for Nonattainment Areas	
7501	Definitions	sec. 171[d]
7502	Nonattainment plan provisions	sec. 172[d]
7503	Permit requirements	sec. 173[d]
7504	Planning procedures	sec. 174[d]
7505	Environmental Protection Agency grants	sec. 175[d]
7506	Limitations on certain Federal assistance	sec. 176[d]
7507	New motor vehicle emission standards in nonattainment areas	sec. 177[d]
7508	Guidance documents	sec. 178[d]
7509	Sanctions (NEW)	(NEW, sec. 102(g)
7511	Additional Provisions for Ozone Nonattainment Areas (NEW)	(NEW, sec. 103, 819)
7512	Additional Provisions for Carbon Monoxide Areas (NEW)	(NEW, sec. 104)
7513	Particulate Matter (PM_{10})	(sec. 105(a))

Table 3. *(Continued)*

42 U.S.C.	Section Title	Clean Air Act, as Amended
7514	Additional Provisions for Areas Designated Nonattainment for Sulfur Oxides, Nitrogen Dioxide, and Lead (NEW)	(NEW, sec. 106)
7415	Savings Provisions (NEW)	(NEW, sec. 108(1))
Subchapter II	Emission Standards for Moving Sources	
Part A	Motor Vehicle Emission and Fuel Standards	
7521	Emission standards for new motor vehicles or engines	sec. 202[e]
7522	Prohibited acts	sec. 203[e]
7523	Actions to restrain violations	sec. 204[e]
7524	Penalties	sec. 205[e]
7525	Motor vehicle and engines testing and certification	sec. 206[e]
7541	Compliance by vehicles and engines in actual use	sec. 207[e]
7542	Records and reports	sec. 208[e]
7543	State standards	sec. 209[b]
7544	State grants	sec. 210[b]
7545	Regulation of fuels	sec. 211[b]
7546	Low emission vehicles (repealed)	sec. 212[c]
7547	Fuel economy improvement for new vehicles	sec. 213[c]
7548	Study of particulate emissions from motor vehicles	sec. 214[d]
7549	High altitude performance adjustments	sec. 215[d]
7550	Definitions	sec. 216[e]
7551	Study and report on fuel consumption	sec. [d]
7552	Compliance Program Fees (NEW)	(NEW, sec. 225)
7553	Prohibition on Production of Engines Requiring Leaded Gasoline (NEW)	(NEW, sec. 226)
7554	Urban Buses (NEW)	(NEW, sec. 227(a))
Part B	Aircraft Emissions Standards	
7571	Establishment of standards	sec. 231[c]
7572	Enforcement of standards	sec. 232[c]
7573	State standards and control	sec. 233[c]
7574	Definitions	sec. 234[c]
Part C	Clean Fuel Vehicles (NEW)	
7581	Definitions (NEW)	(NEW, sec. 229(a))
7582	Requirements Applicable to Clean Fuel Vehicles (NEW)	(NEW, sec. 229(a))
7583	Standards for Light-Duty Clean Fuel Vehicles (NEW)	(NEW, sec. 229(a))
7584	Administration and Enforcement as Per California (NEW)	(NEW, sec. 229(a))
7585	Standards for Heavy Duty Clean Fuel Vehicles (NEW)	(NEW, sec. 229(a))
7586	Centrally Fueled Fleets (NEW)	(NEW, sec. 229(a))
7587	Vehicle Conversions (NEW)	(NEW, sec. 229(a))
7588	Federal Agency Fleets	(NEW, sec. 229(a))
7589	California Pilot Test Program	(NEW, sec. 229(a))
7590	General Provisions	(NEW, sec. 229(a))
Subchapter III	General Provisions	
7601	Administration	sec. 301[a]
7602	Definitions	sec. 302[a]
7603	Emergency powers	sec. 303[c]
7604	Citizen suits	sec. 304[c]
7605	Representation in litigation	sec. 305[c]
7606	Federal procurement	sec. 306[c]
7607	Administrative proceedings and judicial review	sec. 307[c]
7608	Mandatory licensing	sec. 308[c]
7609	Policy review	sec. 309[c]
7610	Other authority	sec. 310[a]
7611	Records and audits	sec. 311[a]
7612	Cost studies	sec. 312[b]
7613	Additional reports to Congress (repealed)	sec. 313[b]
7614	Labor standards	sec. 314[b]
7615	Separability of provisions	sec. 315[a]
7616	Sewage treatment plants	sec. 316[d]
7617	Economic impact assessment	sec. 317[d]
7618	Financial disclosure (repealed)	sec. 318[d]

Table 3. *(Continued)*

42 U.S.C.	Section Title	Clean Air Act, as Amended
7619	Air quality monitoring	sec. 319[d]
7620	Standardized air quality modeling	sec. 320[d]
7621	Employment effects	sec. 321[d]
7622	Employee protection	sec. 322[d]
7623	Repealed	sec. 323[d]
7624	Cost of vapor recovery equipment	sec. 324[d]
7625	Vapor recovery for small business marketers of petroleum products	sec. 324[d]
7625a	Statutory construction	sec. 325[d]
7626	Authorization of Appropriations	sec. 326[d]
7651	Acid Disposition Control (NEW)	(NEW, sec. 401, 404, 406, 407, 408)
7661	Permits (NEW)	(NEW, sec. 501)
Title VI	Stratospheric Ozone Protection (New)	
7671	Stratospheric Ozone Protection	(NEW, sec. 602(a))
[29 U.S.C. 655]	Chemical Process Safety Management (NEW)	(NEW, sec. 304)
[29 U.S.C. 1502]	Grants for Clean Air Employment Transition Assistance	(NEW, sec. 1101(b)(2))

Codified generally as 42 U.S.C. 7401-7671. Shows original sections renumbered by P.L. 88-206, and indicates when added, although similar sections may have existed earlier. "Added" usually indicates a general amendment or complete revision. For a detailed history, see the notes in the U.S. [a] Code. P.L. 88-206, general amendment which renumbered sections. [b] added by P.L. 90-148, Nov. 21, 1967. [c] added by P.L. 91-604, Dec. 21, 1970. [d] added by P.L. 95-95, August 7, 1977; (New) added by P.L. 101-548. [e] added by P.L. 89-272, Oct. 20, 1965.

ENFORCEMENT (SECTION 113)

In general, this section establishes Federal authority to impose penalties for violations of the requirements of the Act. The 1990 Amendments elevated penalties for some knowing violations from misdemeanors to felonies; removed the ability of a source to avoid an enforcement order by ceasing a violation within 30 days of notice; gave authority to EPA to assess administrative penalties; and authorized $10,000 awards to persons supplying information leading to convictions under the Act. The Act authorizes EPA to require sources to submit reports, to monitor emissions, and to certify compliance with the Act's requirements.

STRATOSPHERIC OZONE PROTECTION (TITLE VI)

As amended in 1990, the Act establishes a national policy of phasing out production and use of "Class I" ozone-depleting substances (ie, all fully halogenated chlorine or bromine containing compounds and carbon tetrachloride) by the year 2000, and methyl chloroform by 2002. "Class II" substances (ie, other halogenated chlorine and bromine compounds, and others designated by EPA) will have their production frozen in 2015, and production ended by 2030. EPA can accelerate the phase-out schedules. Limited exemptions are allowed (eg, for medical, aviation, and fire-suppression uses). EPA is to issue regulations on use and disposal of these chemicals that will minimize production and foster recycling and reuse. This title's authorities are expressly linked to the Montreal Protocol (an international accord phasing out chlorofluorocarbons), specifying that the more stringent requirement of either is effective.

In the omnibus budget reconciliation act of 1989 (P.L. 101-239), Congress imposed a tax on CFCs and halons covered by the Montreal Protocol to accelerate the transition to substitutes. The tax starts at $1.37/lb. in 1990, escalates to $2.65/lb in 1993, and will increase an additional 45 cents/lb/year after 1994.

BIBLIOGRAPHY

Reading List

U.S. Office of Technology Assessment, *Catching our Breath: Next Steps for Reducing Urban Ozone, Summary,* U.S. Office of Technology Assessment, Washington, D.C. 1989, 24 p.

U.S. Environmental Protection Agency, *The Clean Air Act Amendments of 1990 Detailed Summary of Titles,* Environmental Protection Agency, Washington, D.C. 1990.

CLEAN AIR ACT AMENDMENT OF 1990—MOBILE SOURCES

MICHAEL WALSH
Arlington, Virginia

As the dominant source of hydrocarbons, carbon monoxide, and nitrogen oxides, motor vehicles were singled out for special attention in the 1970 Clean Air Act Amendments. Twenty years later, in spite of significant progress, this situation has not changed. Because more than 50 million additional cars are on U.S. highways than in 1970, motor vehicles remain the dominant emissions source and therefore received special attention and focus in the Clean Air Act Amendments of 1990. In addition to more stringent

Table 1. Emission Standards for Light-Duty Vehicles (Passenger Cars) and Light-Duty Trucks of up to 2721.6 kg (6000 lbs) GVWR

	Column A (5 yrs/80,450 km)				Column B (10 yrs/160,900 km)			
Vehicle Type	NMHC	CO	NO_x	Part.	NMHC	CO	NO_x	Part.
Non diesel								
LTDs 0–1701 kg (0–3, 750 lbs) LVW and light-duty vehicles	0.25	3.4	0.4		0.31	4.2	0.6	
LDTs 1701.4–2608.2 kg (3,751–5,750 lbs) LVW	0.32	4.4	0.7		0.40	5.5	0.97	
Diesel								
LTDs 0–1701 kg (0–3, 750 lbs) LVW and light-duty vehicles	0.25	3.4	1.0	0.08	0.31	4.2	1.25	0.10
LDTs 1701.4–2608.2 kg (3,751–5,750 lbs) LVW	0.32	4.4		0.08	0.40	5.0	0.97	0.10

standards for cars, trucks, and buses, the Amendments will require substantial modification to conventional fuels, provide greater opportunity for the introduction of alternative fuels (but without mandating them), and extend the manufacturers' responsibility for compliance with auto standards in use to 10 years or 160,900 km (100,000 miles).

See also EXHAUST CONTROL, AUTOMOTIVE; REFORMULATED GASOLINE; TRANSPORTATION FUELS—AUTOMOTIVE FUELS; METHANOL VEHICLES; AUTOMOBILE EMISSIONS; AUTOMOBILE EMISSIONS, CONTROL; ENERGY TAXATION.

Major provisions dealing with mobile sources are summarized as follows.

CONVENTIONAL VEHICLES

Light-Duty Vehicle Tailpipe Standards—Tier 1

Emissions Standards. The Tier 1 tailpipe standards for light-duty vehicles are summarized in Tables 1 and 2.

Intermediate In-Use Standards. For the first two years that passenger cars and light-duty trucks are subject to the Tier I certification standards shown in Table 1, less stringent, intermediate in-use emission standards apply for purposes of recall liability, and the useful life period is only five years or 80,450 km (50,000 miles), whichever occurs first.

The intermediate in-use standards for passenger cars and light-duty trucks up to (6000 lbs) GVWR are shown in Table 3 and for light-duty trucks above (6000 lbs) GVWR in Table 4.

Implementation Schedule. For purposes of certification, the standards in Table 1 will be phased in over a three-year period (applicable to 40% of MY 1994 vehicles, 80% of MY 1995 vehicles, and 100% of MY 1996 vehicles, as illustrated in Table 5). In use, starting in 1994, the standards will start to be phased in—40% the first year, 80% the second and 100% the third year (1996) for NO_x. For HC, in the first year, 40% will be required to meet the *intermediate* in use standard (0.32 grams per mile NMHC) with the remainder achieving the current standard. This will rise to 80% in the second year. By the third year only 60% will meet the intermediate standard with the other 40% meeting the *final* (0.25 NMHC) standard. In the fourth year, 80% will be required to meet the 0.25 level, rising to 100% by 1998.

Starting in 1996, new in-use standards start to be phased in for 160,900 kg (100,000 miles) (120,675 kg [75,000 miles] for Recall testing) that allow for 25% higher

Table 2. Emission Standards for Light-Duty Trucks of More Than 2721.6 kg (6000 lbs) GVWR[a]

	Column A (5 yrs/80,450 kg)			Column B (11 yrs/193,080 kg)			
LDT Test Weight	NMHC	CO	NO_x	NMHC	CO	NO_x	PM
1701.4–2608.2 kg (3,751–5,750 lbs)	0.32	4.4	0.7[b]	0.46	6.4	0.98	0.10
Over 2608.2 kg (5,750 lbs)	0.39	5.0	1.1[b]	0.56	7.3	1.53	0.12

[a] Standards are expressed in grams per mile (gpm).
[b] Not applicable to diesel-fueled LDTs.

Table 3. Intermediate In-use Standards for Passenger Cars and Light-duty Trucks up to 2721.6 kg (6000 lbs) GVWR

Vehicle Type	NMHC	CO	NO_x
Passenger cars	0.32	3.4	0.4[a]
LDTs 0–1701 kg (0–3750 lbs) LVW	0.32	5.2	0.4[a]
LDTs 1701.4–2608.2 kg (3751–5750 lbs) LVW	0.41	6.7	0.7[a]

[a] Not applicable to diesel-fueled vehicles.

emissions levels between 80,450 and 160,900 kg (50,000 and 100,000 miles) for HC and CO, and 50% higher for NO_x. Diesel vehicles are also allowed to comply with a relaxed 1.0 gpm NO_x standard. Light trucks under 1701 kg (3750 lbs) loaded vehicle weight (LVW) will be required to achieve the same standards as cars.

Recall Requirements. For the purpose of Recall, if a vehicle is tested in-use before five years or 80,450 kg (50,000 miles), the 80,450-kg (50,000-mile) certification standards apply. If the vehicle is tested after five years or 80,450 kg (50,000 miles), the 160,900-kg (100,000-mile) certification standards apply in the case of passenger cars and light-duty trucks up to 2721.4 kg (6000 lbs) GVWR and 193,080-kg (120,000-mile) certification standard for light-duty trucks over 2721.4 kg (6000 lbs) GVWR.

While passenger cars and light-duty trucks are expected to meet standards in-use for their full useful lives, EPA may not conduct recall testing on passenger cars and light-duty trucks up to 2721.4 kg (6000 lbs) that are over seven years old or have been driven more than 120,675 kg (75,000 miles) or on light-duty trucks over 2721 kg (6000 lbs) GVWR that are over seven years old or been driven more than 144,810 kg (90,000 miles).

Tier 2 Standards

EPA is required to study and report to Congress no later than June 1, 1997, on the technological feasibility, need for, and cost-effectiveness of the standards shown in Table 6 for passenger cars and light-duty trucks of 1701 kg (3750 lbs) LVW or less.

Within three years of reporting to Congress, but not later than December 31, 1999, EPA is required to (a) promulgate the Tier 2 standards for model years commencing not earlier than model year 2003 and not later than model year 2006, (b) promulgate standards different from those shown in Table 6 provided they are more stringent than

Table 5. Implementation Schedule for Tier 1 Standards for Light-duty Vehicles (Passenger Cars) and Light-duty Trucks of up to 2721.4 kg (6000 lbs) GVWR

Model Year Certification	in Use	Percentage of Manufacturers Sales Volume %
1994	1996	40
1995	1997	80
1996	1998	100

Implementation Schedule for Standards for Light-duty Trucks of More Than 2721.4 kg (6000 lbs) GVWR

Model Year Certification	in Use	Percentage of Manufacturers Sales Volume, %
1996	1998	50
1997	1999	100

the Tier 1 standards for model years commencing not earlier than January 1, 2003, or later than the 2006 model year, or (c) determine that standards more stringent than the Tier 1 levels are not technologically feasible, needed, or cost-effective. If the administrator fails to act, the Tier 2 standards become effective with the model years commencing after January 1, 2003.

Potential Revision of Standards

EPA retains the authority to revise emission standards for all classes of motor vehicles and engines based on the need to protect the public welfare *except that* the administrator may not revise the specific emission standards established in the Act for passenger cars, light-duty trucks, and heavy-duty trucks before the 2004 model year.

Cold Temperature CO Standards. EPA is required to establish a cold-temperature CO standard (−6.66°C) beginning with the 1994 model year of 10.0 gpm for passenger cars and a level comparable in stringency for light-duty trucks according to the phase-in schedule listed in Table 7.

If, as of June 1, 1997, six or more nonattainment areas (not including Stubenville and Winebago, where the CO problem is primarily caused by stationary sources) have a carbon monoxide design value of 9.5 ppm or greater, EPA is required to establish second-phase cold-temperature (−6.66°C) CO standards beginning with the 2002 model year of 3.4 gpm for passenger cars, 4.4 for light-duty trucks up to (6000 lbs) and a level comparable in stringency for light-duty trucks (6000 lbs) LVWR and above.

The useful life for the cold-temperature CO standards is 5 years/80,450 km (50,000 miles), but EPA may extend this period, if it determines such requirements to be technologically feasible.

Table 4. Intermediate In-use Standards for LDTs More than 2721.6 kg (6000 lbs) GVWR

Vehicle Type	NMHC	CO	NO_x
LDTs 1701.4–2608.2 kg (3751–5750 lbs) LVW	0.40	5.5	0.88[a]
LDTs 2608.2 kg (over 5750 lbs) LVW	0.49	6.2	1.38[a]

[a] Not applicable to diesel-fueled vehicles.

Table 6. Tier 2 Emission Standards for Gasoline and Diesel-Fueled Passenger Car and Light-duty Trucks 1701 kg (3750) lbs LVW or Less

Pollutant	Emission Level
NMHC	0.125 gpm
NO_x	0.2 gpm
CO	1.7 gpm

Table 7. Phase-in Schedule for Cold Start Standards

Model Year	Percentage of Manufacturer's Sales Volume, %
1994	40
1995	80
1996 and after	100

Trucks and Buses

New Heavy-Duty Vehicles. The statutory standards for HC, CO, and NO_x for heavy-duty vehicles and engines created by the 1977 Amendments are eliminated. In their place, a general requirement that standards applicable to emissions of HC, CO, NO_x, and particulate reflect "the greatest degree of emission reduction achievable through the application of technology which the Administrator determines will be available for the model year to which such standards apply, giving appropriate consideration to cost, energy, and safety factors associated with the application of such technology." Standards adopted by EPA that are currently in effect will remain in effect unless modified by EPA.

EPA may revise (relax) such standards, including standards already adopted, on the basis of information concerning the effects of air pollutants from heavy-duty vehicles and other mobile sources, on the public health and welfare, taking into consideration costs.

A 4.0 g/BHP-h NO_x standard is established for 1998 and later model year gasoline- and diesel-fueled heavy-duty trucks.

Standards adopted for heavy-duty vehicles must provide at least four years lead time before they go into effect and then must remain in effect for at least three years before they are changed.

Rebuilt Engines. The EPA is required to study the practice of rebuilding heavy-duty engines and the impact rebuilding has on engine emissions. EPA *may* establish requirements to control rebuilding practices, including emission standards. No deadlines are established.

Cold Temperature CO. EPA *may* establish cold temperature CO standards for heavy-duty vehicles and engines.

Urban Buses (New). The existing 1991 model year 0.1 g/BHP-h particulate standard is relaxed to 0.25 g/BHP-h for the 1991 and 1992 model years. For the 1993 model year, the particulate standard is 0.1 g/BHP-h.

Beginning with the 1994 model year, EPA is required to establish separate emission standards for urban buses including a particulate standard 50% more stringent than the 0.1 g/BHP-h (eg, 0.05 g/BHP-h) standard. If EPA determines the 50% level is technologically infeasible, EPA is required to increase the allowable level of particulate to no greater than 70% of the 0.1 g/BHP-h level (eg, 0.07 g/BHP-h).

EPA is required to conduct annual tests on a representative sample of operating urban buses to determine whether such buses meet the particulate standard over their full useful life. If EPA determines that buses are not meeting the particulate standard, EPA must require buses sold in areas with populations of 750,000 or more to operate on low-polluting fuels (methanol, ethanol, propane natural gas, or any comparably low polluting fuel). EPA may extend this requirement to buses sold in other areas if it determines there will be a significant benefit to the public health. The low-polluting fuel requirement would be phased in over five model years commencing three years after EPA's determination.

Retrofit Requirement for Buses. Not later than November 15, 1991, EPA is required to promulgate emission standards or emission-control technology requirements reflecting the best retrofit technology and maintenance practices reasonably achievable. Such standards are to apply to engines replaced or rebuilt after January 1, 1995, for buses operating in areas with populations of 750,000 or more.

Onboard Refueling Control Systems

By November 15, 1991, and after consultation with the Department of Transportation on safety issues, EPA is required to issue regulations mandating that passenger cars be equipped with vehicle-based ("onboard") systems to control 95% of the HC emissions emitted during refueling. The regulations take effect beginning in the fourth model year after the regulations are adopted and are to be phased in over three years (40% of total sales in the first year, 80% in the second, and 100% in the third and subsequent years).

Warranties

The warranty period for 1995 and later model year passenger cars and light-duty trucks for the catalytic converter, electronic emission control unit, onboard diagnostic device, and other emission-control equipment designated by EPA as a "specific major emission control component" will be eight years or 128,720 km (80,000 miles), whichever occurs first. To be designated a "specific major emission control component," the device or component could not have been in general use on vehicles and engines before the 1990 model year and must have a retail cost (exclusive of installation costs) that exceeds $200.

The Warranty for all remaining emission control components is two years or 38,616 km (24,000 miles), whichever occurs first.

EPA is given the authority to establish the warranty period for other classes of motor vehicles and engines.

Evaporative Controls

By June 15, 1991, EPA is required to establish evaporative hydrocarbon emission standards for all classes of gasoline-fueled motor vehicles. The standards, which are to take effect as expeditiously as possible, must require the greatest degree of reduction reasonably achievable of evaporative HC emissions during operation ("running losses") and over two or more days of nonuse under ozone-prone summertime conditions.

Toxics

EPA is required to complete a report by June 15, 1992, on the need for and feasibility of controlling unregulated motor vehicle toxic air pollutants including benzene, formaldehyde, and 1, 3-butadiene.

By June 15, 1995, EPA is required to issue regulations to control hazardous air pollutants to the greatest degree achievable through technology that will be available considering cost, noise, safety, and necessary lead time. These regulations must apply, at a minimum, to benzene and formaldehyde. No effective date for the regulations is specified in the Amendments.

Onboard Diagnostics

EPA must issue regulations requiring all passenger cars and light-duty trucks to be equipped with onboard diagnostic systems capable of

1. accurately identifying emission-related system deterioration or malfunctions including, at a minimum, the catalytic converter and oxygen sensor;
2. alerting vehicle owners of the need for emission-related component or system maintenance or repair; and
3. storing and retrieving diagnostic fault codes that are readily accessible.

The regulations are to be phased in starting in 1994 (covering 40% of sales the first year, 80% the second, and 100% the third) but EPA may delay them for up to two years for any class or category of motor vehicles based on technological feasibility considerations.

EPA *may* also establish onboard diagnostic control requirements for heavy-duty vehicles.

States are required to establish programs for inspecting onboard diagnostic systems as part of their periodic inspection and maintenance program requirements.

Compliance Program Fees

EPA may establish fees to recover the costs from manufacturers of (a) new vehicle or engine certification tests, (b) new vehicle or engine compliance monitoring (eg, assembly line testing), and (c) in-use vehicle or engine compliance monitoring (eg, recall). This could be extremely critical as the budget deficit squeezes harder.

Motor Vehicle Testing and Certification

By November 15, 1991, EPA must revise the certification test procedures to add a test to determine whether 1994 and later model year passenger cars and light-duty trucks are capable of passing state inspection emission (I/M) tests.

By June 15, 1992, EPA must review and revise, *as necessary,* the certification test procedures to ensure that motor vehicles are tested under conditions that reflect actual current driving conditions, including fuel, temperature, acceleration, and altitude.

Manufacturers with projected U.S. sales of vehicles or engines of no more than 300 are subject to less stringent durability demonstration requirements.

Information Collection

EPA is given authority to require manufacturers to maintain records and perform emission tests and to report testing results to the Agency. EPA is also given expanded authority to inspect manufacturers' facilities and records.

High-Altitude Testing

EPA is required to establish at least one high-altitude emission testing center to evaluate in-use emissions of vehicles at high altitudes. As part of the testing center, EPA must establish a research facility to evaluate the effects of high altitude on aftermarket emission-control components, dual-fueled vehicles, conversion kits, and alternative fuels. The center must also offer training courses designed to maximize the effectiveness of inspection and maintenance (I/M) programs at high altitude.

Antitampering

The Amendments extend the prohibition against tampering with emission controls to individuals. Also, the manufacture, sale, or installation of an emission-control defeat device is prohibited.

CONVENTIONAL FUELS

Reformulated Gasoline

Conventional gasoline will undergo significant modification during the 1990s as a result of the Amendments. There are two separate sets of requirements: one for ozone problems and one for CO problems.

Severe and Serious Ozone Nonattainment Areas. By November 15, 1991, EPA is required to establish requirements for reformulated gasoline requiring the greatest reduction of ozone-forming VOCs and toxic air pollutants achievable considering costs and technological feasibility.

Beginning January 1, 1995, cleaner, "reformulated" gasoline must be sold in the nine worst ozone nonattainment areas ("severe" and "serious") with populations over 250,000. Other ozone nonattainment areas ("marginal," "moderate," or "serious") are permitted to "opt-in," but EPA may delay on a limited basis requirements for reformulated gasoline in these areas if it determines the fuel will not be available in adequate quantities.

Emission Performance Requirements. At a minimum, reformulated gasoline must (a) not cause NO_x emissions to increase (EPA may modify other requirements discussed below if necessary to prevent an increase in NO_x emissions), (b) have an oxygen content of at least 2.0% by weight (EPA may waive this requirement if it would interfere with attaining an air quality standard), (c) have a benzene content no greater than 1.0% by volume, and (d) contain no heavy metals, including lead or manganese (EPA may waive the prohibition against heavy metals other than lead if it determines that the metal will not increase on an aggregate mass or cancer-risk basis, toxic air-pollution emissions from motor vehicles).

In addition, VOC and toxic emissions must be reduced by 15% over baseline levels beginning in 1995 and by 25% beginning in the year 2000. EPA may adjust the 25% requirement up or down based on technological feasibility and cost considerations, but in no event may the percent reduction beginning in the year 2000 be less than 20%. Toxic air pollutants are defined by the Amendments in terms of the aggregate emissions of benzene, 1,3 butadiene, polycyclic organic matter (POM), acetaldehyde, and formaldehyde.

Credits. Fuel refiners and suppliers may accumulate and trade credits if they produce reformulated gasoline that exceeds the minimum requirements specified by EPA.

Anti Dumping Provisions. By November 15, 1991, EPA must establish regulations that prohibit gasoline from being introduced into commerce that on average results in emissions of VOC, NO_x, or toxics greater than gasoline sold by that refiner, blender, or importer in 1990. These regulations must go into effect by January 1, 1995.

Carbon Monoxide Nonattainment Areas. Areas with a carbon monoxide design target of 9.5 ppm or above for 1988 and 1989 must have as part of their state implementation plan (SIP) a requirement that during that portion of the year in which the area is prone to high ambient concentrations of CO (winter months), all gasoline sold must contain not less than 2.7% oxygen by weight. Such requirements are to take effect no later than November 1, 1992 (or such other date in 1992 as determined by the Administrator). For areas that exceed the 9.5 CO design target for any two-year period after 1989, the 2.7% oxygen requirement must go into effect in that area no later than three years after the end of the two-year period.

Areas classified as serious CO nonattainment areas that have not achieved attainment by the date specified in the Act must require that the oxygen level in gasoline be 3.1% by weight.

Waivers of this requirement for up to two years can be granted by EPA on an area-by-area basis if it is petitioned and if it determines a) the use of oxygenated gasoline would present or interfere with the attainment by a given area with a federal or state ambient air quality standard; b) mobile sources do not contribute significantly to the CO levels in the area; or c) there is an inadequate domestic supply of, or distribution capacity for, oxygenated gasoline meeting the applicable requirements. EPA must act on any petition within six months of its filing.

Fuel Volatility

By June 15, 1991, the EPA must promulgate regulations limiting the volatility of gasoline to no greater than (9.0 pounds per square inch—(PSI) Reid Vapor Pressure (RVP) during the high ozone season. EPA may establish a lower RVP in an individual nonattainment area if it determines that a lower level is necessary to achieve comparable evaporative emissions (on a per-vehicle basis) in nonattainment areas. The fuel volatility requirements are to take effect not later than the high ozone season for 1992.

For fuel blends containing 10% ethanol, the applicable RVP limitation may be one pound per square inch greater than for conventional gasoline.

Detergent Requirements

Beginning January 1, 1995, any gasoline sold nationwide must contain additives to prevent the accumulation of deposits in the engine and fuel supply systems.

Lead Phasedown

After December 31, 1985, it will be unlawful to sell, supply, dispense, transport, introduce into commerce, or use gasoline that contains lead or lead additives for highway use.

Lead Substitute Gasoline Additives

EPA is required to develop a test procedure for evaluating the effectiveness of lead substitute gasoline additives and to arrange for independent testing and evaluation of each additive proposed to be registered as a lead substitute. EPA may impose a user fee to cover the cost of testing any lead-substitute fuel additive.

Prohibition on Engines Requiring Leaded Gasoline

EPA is required to prohibit the manufacture, sale, or introduction into commerce of any motor vehicle or nonroad engine that can only be operated on leaded gasoline and is manufactured after the 1992 model year.

Diesel Fuel Sulfur Content

Effective October 1, 1993, the sulfur content of motor vehicle diesel fuel may not exceed 0.05% by weight and must meet a cetane index minimum of 40.

EPA may exempt Alaska and Hawaii from the diesel fuel quality requirements.

The sulfur content of fuel used in certification of 1991–1993 model year heavy-duty vehicles and engines must be 0.10% by weight.

EPA may require diesel fuel manufacturers and importers to dye fuels to differentiate highway and nonhighway diesel fuels.

Ethanol Substitute for Diesel Fuel

EPA is required to commission a study and report to Congress within three years of enactment of the Amendments on the feasibility, engine performance, emissions, and production capability associated with an alternative diesel fuel composed of ethanol and high-erucic rapeseed oil.

Fuel and Fuel Additive Importers

The statutory requirements applicable to fuel and fuel additive manufacturers are expressly made applicable to importers.

Misfueling

The Amendments extend to individuals the prohibition against misfueling with leaded gasoline. After October 1,

1993, there will be a prohibition against fueling a diesel-powered vehicle with fuel containing a sulfur content greater than 0.05% by weight or that fails to meet a cetane index minimum of 40 (or such equivalent aromatic level as prescribed by the Administrator).

CLEAN ALTERNATIVE FUELS

The Amendments define "clean alternative fuel" as any fuel including methanol, ethanol, or other alcohols including any mixture thereof containing 85% or more by volume of such alcohol with gasoline or other fuels (reformulated gasoline, diesel, natural gas, liquified petroleum gas, and hydrogen) or power source (including electricity) used in a clean-fuel vehicle that complies with the Amendments' performance requirements.

Fleet Program

By November 15, 1992, EPA is required to issue regulations implementing the Clean Fuel Fleet Vehicle Program.

The fleet program applies in serious nonattainment areas to fleets of 10 or more vehicles that are centrally refueled or capable of being so (but not including vehicles garaged at personal residences each night under normal circumstances).

The program agreed upon will mandate California's Low Emission Vehicle (LEV) standards for light-duty vehicles (0.075 grams per mile nonmethane organic material, 3.4 CO and 0.2 NO_x) and light trucks below 2721.6 kg (6000 lbs) (0.1 grams per mile NMOG, 0.4 NO_x, 4.4 CO) by 1998, provided these vehicles are offered for sale in California. By 2001, these vehicles will be required without regard to availability in California.

EPA is mandated to establish an equivalent wrap-around standard (exhaust, evaporative, and refueling emissions combined) for LEVs below 3855.6 kg (8500 lbs) GVWR. This wraparound standard is to be based on LEV vehicles using reformulated gasoline meeting the reformulated gasoline standards for the applicable time period. It will be left to the manufacturers to decide which standard to use—the LEV tailpipe standards or the wrap around standards.

Congress followed California's lead in substituting a nonmethane organic gas (NMOG) standard in place of the total or nonmethane hydrocarbon standards currently used. A nonmethane organic gas is defined as the sum of nonoxygenated and oxygenated organic gases containing 5 or fewer carbon atoms and all known alkanes, alkenes, alkynes, and aromatics containing 12 or fewer carbon atoms. The test procedure for measuring NMOG is the recently adopted California NMOG Test Procedure. NMOG is a more appropriate substance to measure when evaluating the emission performance of alternative fueled vehicles.

Covered Fleets. ***Centrally Fueled Fleets.*** Centrally fueled fleets with 10 or more vehicles that are owned or operated by a single person and operate in a covered area are subject to the clean vehicle requirements. A number of vehicle fleets are exempted including rental fleets, emergency vehicles, enforcement vehicles, and nonroad vehicles. Also, vehicles garaged at personal residences each night under normal circumstances would not be covered.

Covered areas include any ozone nonattainment area with a population of 250,000 or more classified as Serious, Severe, or Extreme (approximately 26 areas), or any carbon monoxide nonattainment area with 250,000 or more population and a CO design value at or above 16.0 ppm.

States are required to implement clean-fuel vehicle phase-in as shown in Table 8.

In complying with this section, passenger cars and light-duty trucks up to 2721.6 kg (6000 lbs) GVWR must meet the Phase II emission standards shown in Table 9. Also, if passenger cars and light-duty trucks meeting the Phase II standards are sold in California in model years before 1998, then that model year becomes the applicable model year for phasing in clean vehicle fleets.

Federal Agency Fleets. The fleet requirements apply to federally owned fleets except vehicles exempted by the Secretary of Defense for national security reasons.

Conversions. The requirements for purchase of clean vehicles may be met through vehicle conversions of existing gasoline- or diesel-powered vehicles to clean fuel vehicles. EPA is required to establish regulations defining criteria for conversions.

Dedicated Clean-Fuel Light-Duty Vehicles. The emission standards applicable to clean-fuel vehicle passenger cars and light-duty (trucks weighing up to 2721.6 kg (6000 lbs) GWVR but not more than 1701 kg (3750 lbs) LVW are shown in Table 9.

The emission standards for light-duty trucks greater than 2721.6 kg (6000 lb) GVWR are shown in Table 10.

Flexible-Fueled Light-Duty Vehicles. The emission standards for flexible-fueled passenger cars and light-duty trucks up to 2721.6 kg (6000 lbs) GVWR are shown in Table 11 (when operating on clean alternative fuel) and Table 12 (when operating on conventional fuel).

Possible Modification to Requirement. The clean-vehicle and flexible-fueled vehicle standards in the Amendments are based on standards recently adopted by California as part of its Low Emission Vehicles and Clean Fuel Program. The Amendments provide that if California adopts less stringent standards applicable to clean fuel and dual-fueled passenger cars and light-duty trucks, the standards shown in Tables 11 and 12 are to be relaxed as well.

Table 8. Clean Fuel Vehicle Phase-in Requirements for Fleets

Vehicle Type	Applicable Model Year 1998	1999	2000
Light-duty trucks up to 2721.6 Kg (6000 lbs) GVWR and light-duty vehicles	30%	50%	70%
Heavy-duty trucks above 3855.6 Kg (8500 lbs) GVWR	50%	50%	50%

Table 9. Clean Fuel Vehicle Emission Standards for Light-Duty Vehicles and Light-duty Trucks of up to 6000 lb GVWR[a]

Pollutant	NMOG CO NO_x PM HCHO (formaldehyde)
Light-Duty Vehicles and Light-Duty Trucks of up to 1701 kg (3750 lbs) Loaded Vehicle Weight	
Phase I—1996 Model Year	
80,450-km (50,000-mile) standard	0.1253.40.40.015
160,900-km (100,000-mile) standard	0.1564.20.60.8[b]0.018
Phase II—2001 Model Year	
80,450-km (50,000-mile) standard	0.0753.40.20.015
160,900-km (100,000-mile) standard	0.904.20.30.080.018
Light-Duty Trucks From 1701.4 to 2608.2 kg (3751 to 5750 lb) Loaded Vehicle Weight	
Phase I—1996 Model Year	
80,450-km (50,000-mile) standard	0.1604.40.70.018
160,900-km (100,000-mile) standard	0.2005.50.90.08.0.023
Phase II—2001 Model Year	
80,450-km (50,000-mile) standard	0.1004.40.40.018
160,900-km (100,000-mile) standard	0.1305.50.50.080.023

[a] Standards are expressed in grams per mile.
[b] Standards for particulates (PM) shall apply only to diesel-fueled vehicles.

Enforcement of Standards. Where the numerical clean-fuel vehicle standards applying to vehicles of not more than 3855.6 kg (8500 lbs) GVWR are the same as the numerical standards applicable in California under the Low Emission Vehicle and Clean Fuels Regulations of the California Air Resources Board (CARB); the standards are to be administered and enforced by EPA in the same manner, and with the same flexibility, subject to the same requirements and using the same interpretations and policy judgments as applied by CARB including those applying for Certification, Assembly Line Testing, and In-Use Compliance.

Heavy-Duty Clean Fuel Vehicles. Model years 1998 and later heavy-duty vehicles and engines greater than 3855.6 kg (8500 lbs) GVWR and up to 11,793.6 kg (26,000 lbs) GVWR are required to meet a combined NO_x and non–methane hydrocarbon (NMHC) standard of 3.15 g/BHP-hr (equivalent to 50% of the combined HC and NO_x emission standards applicable to conventional 1994 model year heavy-duty diesel fueled vehicles or engines). EPA may relax this standard upon a finding that it is technologically infeasible for clean diesel-fueled vehicles. EPA must, however, require at least a 30% reduction from conventional-fueled vehicle and engine 1994 NO_x and HC standards (combined).

California Pilot Program

The Amendments also contain a clean fuels pilot program for California. Beginning in 1996, 150,000 Clean Fuel vehicles must be produced for sale in California; by 1999, this number must rise to 300,000. These vehicles must meet California's TLEV standards (0.125 NMOG, 0.4 NO_x, 3.4 CO, and 0.015 formaldehyde) until the year 2000 when they must meet the LEV requirements (0.075 NMOG, 0.2 NO_x, 3.4 CO, and 0.015 formaldehyde). California is required to develop an implementation plan revision within one year to ensure that sufficient clean fuels are produced, distributed, and made available so that all clean-fuel vehicles required can operate to the maximum extent practicable exclusively on such fuels in the covered area. If California fails to adopt a fuels program that meets the requirements, EPA is required to establish such a program within three years.

Other states with serious, severe, or extreme ozone non-attainment areas are authorized to "voluntarily" opt in to the program in whole or in part. This voluntary opt-in cannot include any production or sales mandate for vehicles or fuels but must rely on incentives to encourage their sale and use.

Urban Buses

The Amendments also set a performance criteria which mandates that beginning in 1994, buses operating more than 70% of the time in large urban areas using any fuel must reduce particulate by 50% compared to conventional heavy-duty vehicles (ie, 0.05 grams per brake horsepower-hour particulates). EPA is authorized to relax the control

Table 10. Clean Fuel Vehicle Emission Standards for Light-duty Trucks Greater Than 2721.6 kg (6000 lbs) GVWR[a]

Pollutant	NMOG CO NO_x PM[b] HCHO (formaldehyde)
Test Weight Category: Up to 1701 kg (3750 lb)	
80,450-km (50,000-mile) standard	0.1253.40.4[a]0.015
193,080-km (120,000-mile) standard	0.1805.00.60.080.022
Test Weight Category: Above 1701 (3750) but not above 2608.2 kg (5750 lb)	
80,450-km (50,000-mile) standard	0.1604.40.7[c]0.018
193,080-km (120,000-mile) standard	0.2306.41.00.100.027
Test Weight Category: Above 2608.2 (5750) but not above 3855.6 kg (8500 lb) GVWR	
80,450-kg (50,000-mile) standard	0.1955.01.1[c]0.022
54,432-kg (120,000-mile) standard	0.2807.31.50.120.032

[a] Standards are expressed in grams per mile.
[b] Standards for particulates (PM) shall apply only to diesel-fueled vehicles.
[c] Standard not applicable to diesel-fueled vehicles.

Table 11. NMOG Standards for Flexible- and Dual-fueled Vehicles When Operating on Clean Alternative Fuel

Vehicle Type	Column A 80,450-km (50,000-mi) Standard (gpm)	Column B 160,900 km (100,000-mi) Standard (gpm)
Light-duty Trucks up to 2721.6 kg (6000 lbs) GVWR and Light-duty Vehicles		
Beginning MY 1996: LTDs 0–1701 kg (0–3750 lb) LVW and light-duty vehicles	0.125	0.156
LTDs 1701.4–2608.2 kg (3751–5750 lbs) LVW	0.160	0.200
Beginning MY 2001: LTDs 0–1701 kg (0–3750 lb) LVW and light-duty vehicles	0.075	0.090
LTDs 1701.4–2608.2 kg (3751–5750 lb) LVW	0.100	0.130
Light-duty Trucks More than 2721.6 kg (6000 lb) GVWR		
	Column A 80,450-kg (50,000-mi)	Column B (120,000-mi)
Beginning MY 1998: LTDs 0–1701 kg (0–3750 lbs) tw.	0.125	0.180
LTDs 1701.4–2608.2 kg (3751–5750 lbs) tw.	0.160	0.230
LTDs above 2608.2 kg (5750 lbs) tw.	0.195	0.280

requirements to 30% on the basis of technological feasibility. Beginning in 1994, EPA is to do yearly testing to determine whether buses subject to the standard are meeting the standard *in use over their full useful life.* If EPA determines that 40% or more of the buses are not, it *must* establish a low pollution fuel requirement. Essentially, this provision allows the use of exhaust aftertreatment devices to reduce diesel particulates to a very low level provided they work in the field; if they fail, EPA will mandate alternative fuels.

OFF-HIGHWAY ENGINES

All Non-Road Engines/Vehicles Except Locomotives

By November 15, 1991, EPA must complete a study of the health and welfare effects of nonroad engines and vehicles (except locomotives). Within 12 months of completing the study, EPA must determine if HC, CO, or NO_x emissions from new or existing nonroad engines and vehicles significantly contribute to ozone or CO concentrations in more than one area in nonattainment for ozone or CO.

If EPA makes an affirmative finding, it must regulate those classes or categories of nonroad engines or vehicles by requiring the "greatest degree of emission reduction achievable considering technological feasibility cost, noise, energy, safety and lead time factors." EPA, in setting standards, must consider standards equivalent in stringency to onroad vehicle standards. No deadline for establishing standards has been set.

EPA *may* also regulate other pollutants from nonroad engines and vehicles (eg, diesel particulate) if it determines such standards are needed to protect the public health and welfare.

Locomotives

By November 15, 1995, EPA must establish standards for locomotive emissions that require the use of the best tech-

Table 12. NMOG Standards for Flexible- and Dual-fueled Vehicles When Operating on Conventional Fuel

Vehicle Type	Column A 80,450-km (50,000-mi) Standard (gpm)	Column B 160,900 km (100,000-mi) Standard (gpm)
Light-duty Trucks of up to 2721.6 kg (6000 lbs) GVWR and Light-duty Vehicles		
Beginning MY 1996: LTDs 0–1701 kg (0–3750 lbs) LVW and light-duty vehicles	0.25	0.31
LTDs 1701.4–2608.2 kg (3751–5750 lbs) LVW	0.32	0.40
Beginning MY 2001: LTDs 0–1701 kg (0–3750 lbs) LVW and light-duty vehicles	0.125	0.156
LTDs 1701.4–2608.2 kg (3751–5750 lbs) LVW	0.160	0.200
Light-duty Trucks More than 2721.6 kg (6000 lbs) GVWR		
Vehicle Type	Column A 80,450-km (50,000-mi) Standard	Column B 160,900-km (120,000-mi) Standard
Beginning MY 1998: LTDs 0–1701 kg (0–3750 lbs) tw.	0.25	0.36
LTDs 1701.4–2608.2 kg (3751–5750 lbs) tw.	0.32	0.46
LTDs (above 2608.2 kg (5750 lbs) tw.	0.39	0.56

nology that will be available considering cost, energy, and safety. The standards are to take effect at the earliest possible date considering the lead time needed to develop the control technology.

State Standards

States, including California, are prohibited from setting emission standards for a) new engines smaller than 175 horsepower used in construction equipment, vehicles, or farm equipment, and b) new locomotives or new engines used in locomotives.

SECTION 177 OPT IN TO THE CALIFORNIA STANDARDS

One of the most contentious issues, which reared its head very late in the process, was whether other states should continue to be allowed to adopt California emissions standards for motor vehicles, as is allowed by current law. As New York approached final adoption of the 1993 California requirements in the late summer of 1990, paving the way for other northeastern states to do the same, vehicle manufacturers persuaded some in Congress that such programs would create a patchwork of cars across the country and a production nightmare. Others raised concerns that the existing law did not provide sufficient flexibility for states to adopt California cars when these vehicles require new fuels because Section 211 C 4 could preempt state authority in this area. States were concerned that efforts to ensure that only vehicles meeting identical standards as applied in California and clarifying fuel authority were actually creating major new procedural hurdles that would make it more difficult for states to adopt these programs or make it easier for car manufacturers to challenge state enforcement decisions. Some proposals would have removed states entirely from the enforcement process and left it entirely to California and EPA.

The basic thrust of the provision is that states cannot carry out requirements that would force manufacturers to produce a third car.

COAL: AVAILABILITY, MINING, AND PREPARATION

James C. Hower
University of Kentucky
Lexington, Kentucky
Andrew C. Scott
Royal University of London
Halloway, Egham, Surrey
United Kingdom
Adrian C. Hutton
University of Wollongong
Wollongong, New South Wales
Australia
B. K. Parekh
University of Kentucky
Lexington, Kentucky
Bukin Dauley
University of Wollongong
Wollongong, New South Wales
Australia
(Co-Author on Indonesia Section)

RESOURCES AND RESERVES

Discussion of the quantity of coal in any coal field requires some understanding of the concept of resources and reserves. Generally, *resources* refers to the total amount of coal, discovered and undiscovered, given restrictions of coal thickness and depth. *Reserves* are that portion of the resources which are known through relatively tight exploration and which are considered to be economic and mineable within a certain time frame. For example, accessible reserves by the International Energy Agency definition is the amount of coal that could be mined to meet demand over the next 20 years using current mining technology (1). At the other end of the reporting scale are categories such as identified-hypothetical coal-in-place and speculative coal-in-place, which encompass undiscovered potential resources in formations which are coal bearing in explored areas. A good example would be the known, but relatively unexplored, extensions of the Appalachian coal field beneath the Cretaceous Gulf Coast cover in the southeastern United States.

World estimated recoverable reserves in tons is listed on Table 1. Production and consumption trends of high and low rank coals is listed on Table 2. The world coal flow (Table 3) provides an interesting corollary to the latter table. The United States and Australia contribute about half of the energy in the total coal flow. Coal involved in coal flow amounts to over 10% of the world consumption of coal-derived energy (2).

SOUTH AMERICA

Chile and Argentina

The Tertiary Magallanes coal field in Chile and the Río Turbio coalfield in Argentina are the most important coal-bearing basins in southern South America.

Brazil

Brazil has five coal-bearing regions: four in the Amazon basin and the commercial deposits in the Paraná Basin (Fig. 1).

Colombia

Colombia has a number of coal basins ranging in age from Maestrichtian–Paleocene in the Eastern Cordillera to middle Oligocene in the Western Cordillera (Fig. 2). Coal rank is generally high to medium volatile with some areas up to meta-anthracite where intruded by andesite.

Venezuela

Tertiary coals occur in several small coal fields (Fig. 3), three of which were identified as significant by IEA (2).

CENTRAL AMERICA

Owing to its relatively recent age and active tectonic setting, Central America does not have extensive coal deposits. The oldest coals are in the Triassic-Jurassic El Plan Formation of Honduras. Analyses published (4) for Mesozoic coals in Honduras and Guatemala show a wide range

Table 1. World Estimated Recoverable Reserves of Coal[a]

Region Country	Recoverable Anthracite and Bituminous		Recoverable Lignite		Total Recoverable Coal	
	World Energy Council[b]	British Petroleum[c]	World Energy Council[c]	British Petroleum	World Energy Council	British Petroleum
North America						
Canada	4,905	3,716	1,999	3,044	6,965	6,760
Mexico	1,885	1,222		634	1,885	1,856
United States[d]	208,600	112,670	31,963	127,894	240,920	
Total	215,450	117,608	33,963	131,572	249,413	249,182
Central and South America						
Argentina	130				130	
Brazil	1,245	1,933		2,323	1,245	4,256
Chile	1,180				1,180	
Colombia	4,663	9,612			9,663	9,611
Peru	960		100		1,060	
Venezuela	47	416			417	416
Other		983	23	1,403	23	2,389
Total	13,595	12,948	123	3,726	13,719	16,673
Western Europe						
Belgium	410				410	
France	258	177		43	258	220
Germany, West	23,913	23,694	34,233	34,825	59,053	58,523
Greece			2,999	2,862	2,999	2,862
Netherlands	497				497	
Spain	535	321	235	342	770	663
Turkey	175	158	5,928	5,778	6,102	5,936
United Kingdom	3,294	9,482	500		3,794	9,102
Yugoslavia	1,569		14,996	500	16,565	
Other	55	860	110	145	165	1,048
Total	30,710	33,859	59,909	44,405	90,619	78,355
Eastern Europe and U.S.S.R.						
Bulgaria	30		3,699		3,729	
Czechoslovakia	1,870		3,494		5,369	
Germany, East			20,994	20,139	20,994	20,139
Hungary	1,578		2,882		4,460	
Poland	28,692	28,182		11,487	40,390	39,670
U.S.S.R.	140,962	102,491	11,697	136,521	40,936	239,020
Other		2,221		26,792		29,013
Total	173,132	132,901	142,745	194,941	315,877	327,842
Middle East						
Iran	193	189			193	189
Total	193	189			193	189
Africa						
Botswana	3,499				3,499	
Mozambique	240				240	
Nigeria	190				190	
South Africa	55,318	54,812			55,318	54,812
Swaziland	1,820				1,820	
Tanzania	200				200	
Zaire	600				600	
Zimbabwe	734	722			734	722
Other	278	6,553		279	278	6,832
Total	62,878	62,087		279	62,878	62,366
Far East and Oceania						
Australia	49,027	44,894	41,890	45,461	90,916	90,355
China	610,537	152,833	119,965	13,292	730,505	166,125
India	60,632	60,099	1,900	1,874	62,531	61,973
Indonesia	1,400	986	1,599	2,000	2,999	2,986
Japan	855	826	17	17	873	843
Korea, North	600				600	
Korea, South	158	593			158	93
New Zealand	108	20	9	89	117	108
Pakistan	1,020				102	

Table 1. (*Continued*)

Region Country	Recoverable Anthracite and Bituminous		Recoverable Lignite		Total Recoverable Coal	
	World Energy Council[b]	British Petroleum[c]	World Energy Council[c]	British Petroleum	World Energy Council	British Petroleum
Taiwan	200	97		100	200	194
Thailand	14		890		914	
Vietnam	50				150	
Other	150	387	9	1,341	165	1,728
Total	723,939	260,234	166,290	64,175	890,230	324,410
World Total	1,219,900	619,827	403,030	439,189	1,622,930	105,902

[a] Amount in Place that can be recovered (extracted from the earth in raw form) under present and expected local economic conditions with existing available technology. British Petroleum definitions of "Proved Reserves": Proved reserves of coal are generally taken to be those quantities which geological and engineering information indicate with reasonable certainty can be recovered in the future from known deposits under existing economic and operating conditions. The Energy Information Administration does not certify the international reserves data but reproduces the information as a matter of convenience for the reader. Comparisons between sources for both the "Recoverable Anthracite and Bituminous" category and the "Recoverable Lignite" category require careful interpretation because of the different definitional groupings for subbituminous coal. Sum of components may not equal total due to independent rounding.

[b] Recoverable anthracite and bituminous data for the World Energy Council include subbituminous.

[c] Recoverable lignite data for British Petroleum includes subbituminous.

[d] U.S. data are calculated from ETA file information. Excludes certain resource data current under review: 7,315 million short tons of anthracite in 5 States: 1,407 short tons of subbituminous coal in Alaska, and a total of 164 million short tons of coal resources in noncoal-producing States. Data represent both measured and indicated tonnage, as of January 1, 1990. Those data have been combined prior to depletion adjustments and cannot be recaptured as "measured alone."

of coal quality, most of it poor with high ash contents, and a rank range from lignite to anthracite. Speculative Carboniferous deposits in Guatemala totaling 641 Mt have been noted (5).

NORTH AMERICA

Mexico

Economic coal deposits have been found in three regions of Mexico. General summaries are in Ref. 6.

Coal occurs in the Oaxaca region in the south in the early to middle Jurassic Rosario, Zorrillo, and Simón Formations.

Coal in Sonora in northwest Mexico occurs in the upper Triassic Barranca Formation (6,7).

The Sabinas basin of the Coahuila region in northeast Mexico has a 2-m thick split seam in the upper Cretaceous Olmos Formation which has supported mining since the 1880s (6,8–11).

Canada

Coal in Canada occurs in the western and Maritime provinces (Fig. 4) in Carboniferous through Tertiary formations (10).

Western Provinces. Coal reserves in western Canada occur from the Pacific islands in British Columbia; through to the Cordillera, foothills, and plains coal fields of Alberta; and to the lignite deposits of Saskatchewan.

Maritime Provinces. Pennsylvanian subbituminous to high volatile A bituminous, high ash, high sulfur reserves totaling 20 Mt occur in the Minto-Chipman area of central New Brunswick (Fig. 5). More important reserves occur in Nova Scotia in the Sydney coal field; in several coalfields on the west side of Cape Breton Island; in the Pictou, Stellarton, and several small mainland coal fields; and in the Springhill and Joggins coal fields in Cumberland County, western Nova Scotia.

Table 2. World Coal[a] Production and Consumption, 1989–1990, 10^6 t

Region Country	Production		Consumption	
	1989	1990[b]	1989	1990[b]
North America				
Canada	71	68	56.7	51.43
Mexico	10	11	10.1	10.56
United States[c]	890	934	807.9	811.54
Total	970	1,012	874.7	873.54
Central and South America				
Brazil	6	7	17.3	18.21
Chile	2	2	3.59	3.77

Table 2. (*Continued*)

Region Country	Production		Consumption	
	1989	1990[b]	1989	1990[b]
Colombia	20	20	8.31	8.72
Other	3	3	4.1	4.32
Total	30	32	33.35	35.02
Western Europe				
Austria	3	3	6.7	7.20
Belgium and Luxembourg	3	2	16.32	17.74
France	15	14	31.40	30.62
Germany, West	187	184	140.2	188.21
Greece	52	50	53.28	51.08
Italy	2	2	21.47	21.57
Norway	[d]	[d]	1.44	1.56
Spain	42	36	53.66	46.45
Turkey	40	39	41.77	43.56
United Kingdom	100	89	115.0	101.68
Yugoslavia	74	76	79.02	79.75
Other	[d]	[d]	21.17	18.67
Total	518	494	652.5	632.11
Eastern Europe and USSR				
Albania	3	3	2.65	2.79
Bulgaria	34	30	40.58	36.37
Czechoslovakia	118	107	117.93	107.70
Germany, East	301	256	306.93	257.74
Hungary	20	18	22.54	20.37
Poland	250	259	218.27	209.19
Romania	62	38	71.51	46.26
USSR	690	630	661.59	599.60
Total	1,477	1,343	1,442.99	1,300.00
Middle East				
Iran	1	1	1.56	1.64
Total	1	1	3.83	4.02
			5.39	5.66
Africa				
Morocco	1	1	1.87	1.97
South Africa	175	168	131.81	124.21
Zambia	[d]	[d]	0.39	0.38
Zimbabwe	5	5	5.17	5.43
Other	1	2	4.15	4.47
Total	183	177	143.39	136.47
Far East and Oceania				
Australia	196	210	95.28	107.04
China	1,054	1,107	1,040.77	1,092.80
India	209	219	212.49	223.12
Indonesia	4	4		
Japan	10	10	109.40	110.85
Korea, North	54	56	56.23	59.05
Korea, South	21	22	46.39	48.70
Mongolia	8	8	7.35	7.7
New Zealand	3	3	2.13	2.22
Pakistan	3	3	3.65	2.73
Philippines	2	2	2.48	2.61
Thailand	17	18	12.10	12.44
Vietnam	1	5	4.92	5.16
Other	1	1	32.49	3.41
Total	1,584	1,667	1,631.68	1,708.54
World Total	4,766	4,727	4,783.05	4691.33

[a] Coal includes anthracite, subanthracite, bituminous, subbituminous, lignite, and brown coal. Sum of components may not equal total due to independent rounding.
[b] Preliminary.
[c] Consumption data do not reflect net exports of coke.
[d] Denotes less than 450 thousand tons.

Table 3. World Coal Flow[a], 1989, P (10^{15}) J

	Exporting Region and Country															
	North, Central and South America			Western Europe				Eastern Europe			Africa	Far East				
Importing Region and Country	Canada	United States	Colombia	Belgium	Germany, West	Netherlands	United Kingdom	Czechoslovakia	Poland	USSR	South Africa	Australia	China	Japan	World Other	World Total
North America																
Canada	0	448	0	0	4	0	0	0	0	0	0	2	3	0	1	458
United States	37	0	28	2	73	1	1	0	3	0	0	24	3	24	0	137
Other	1	4	0	[b]	[b]	0	0	0	0	0	0	[b]	0	0	0	5
Central and South America																
Brazil	0	160	0	0	0	0	0	0	0	0	0	35	1	0	83	280
Chile	0	24	2	0	0	0	0	0	0	0	0	6	0	0	13	46
Other	0	11	2	0	79	0	[b]	33	0	5	0	8	2	0	0	81
Western Europe																
Austria	0	0	0	0	14	0	0	31	42	25	[b]	0	0	0	0	113
Belgium and Luxembourg	3	173	2	0	62	0	3	0	7	2	86	45	12	0	22	417
Denmark	16	76	61	0	[b]	0	12	[b]	21	19	0	74	4	0	0	282
Finland	0	24	4	0	[b]	0	5	0	57	64	0	0	1	0	0	156
France	19	163	38	8	44	0	8	0	0	64	18	79	30	0	40	493
Germany, West	[b]	55	8	2	0	[b]	2	5	44	22	80	9	4	0	0	233
Ireland	0	30	21	[b]	1	1	9	[b]	17	0	3	4	[b]	0	1	88
Italy	1	293	16	0	33	0	0	0	21	15	130	30	8	0	0	547
Netherlands	15	139	42	5	20	0	0	0	21	4	31	82	8	0	[b]	367
Portugal	0	34	0	0	0	0	6	0	2	44	0	0	0	1	8	96
Spain	0	90	13	1	9	0	3	0	0	7	130	26	1	0	0	280
Sweden	17	34	2	0	[b]	[b]	[b]	[b]	16	15	0	25	[b]	0	0	111
Turkey	0	57	0	0	0	0	0	0	0	13	0	30	0	0	0	99
United Kingdom	22	141	7	15	8	46	0	0	0	0	9	83	7	0	0	341
Yugoslavia	0	43	1	1	2	0	0	0	0	53	0	11	0	0	0	111
Other	1	27	2	15	14	1	6	4	17	[b]	1	3	1	1	0	95
Eastern Europe and U.S.S.R.																
Bulgaria	0	2	0	0	0	0	0	0	0	139	0	0	0	0	0	141
Czechoslovakia	0	0	0	0	0	0	0	0	35	67	0	0	0	0	16	119
Germany, East	0	0	0	0	0	0	0	22	22	83	0	1	1	0	14	145
Hungary	0	0	0	0	0	0	0	0	0	57	0	2	0	0	8	61
Romania	0	36	0	0	0	0	0	0	194	27	0	0	0	0	0	257
USSR	0	0	0	0	0	0	0	0	206	0	0	60	[b]	0	[b]	265
Other	0	12	0	0	0	0	0	2	3	5	0	5	0	0	0	26
Middle East																
Israel	0	12	12	0	0	0	0	0	0	0	60	15	0	0	0	97
Other	2	5	0	0	0	0	3	0	0	0	0	0	0	0	3	14
Africa																
Algeria	0	27	0	0	0	0	0	0	0	0	0	0	0	0	3	31
Egypt	0	14	0	0	0	0	0	0	0	12	0	0	0	0	0	26
Other	0	20	0	0	0	0	[b]	0	0	0	0	34	0	0	4	58
Far East and Oceania																
Hong Kong	0	0	9	0	0	0	0	0	0	0	103	100	44	0	0	257
India	0	5	0	0	0	0	0	0	0	0	0	95	0	0	15	118
Japan	527	341	3	0	2	0	0	0	0	72	287	1,443	115	0	0	2,792
Korea, South	264	140	0	0	0	0	0	0	0	40	0	220	42	0	4	708
Taiwan	26	131	0	0	0	0	0	0	0	0	137	158	5	1	0	406
Other	21	41	3	0	44	0	5	0	6	[b]	31	87	30	43	0	311
World Total	910	2,809	276	50	291	50	67	99	728	830	1,108	2,800	327	79	235	10,716

[a] Includes coke. Sum of components may not equal total due to statistical discrepancies, losses, unaccounted for coal and coal trade not in national accounts (eg, U.S. shipment to U.S. Armed Forces in Europe).
[b] Less that 527 billion kJ.

United States

The United States is one of the top three coal producers in the world and, indeed, the states of Wyoming, West Virginia, and Kentucky would individually rank with the top ten producing countries. The United States has an estimated 3.37 Tt of total resources, over 26% of the world total. Discussion of U.S. coal fields will proceed along the lines of the coal provinces: Appalachians, Eastern and Western Interior, Gulf Coast (including Cretaceous coals in Texas), Great Plains and Rocky Mountains, Pacific Northwest, and Alaska.

An overview of coal rank trends throughout the U.S. is presented in Ref. 13.

Figure 1. Coal fields of Brazil (14). I, Upper Amazonas region; II, Rio Fresco region; III, Tocantims-Araguaia region; IV, Western Piauí region; V, Southern Brazil region.

Appalachians. The U.S. Appalachian coal fields extend from Pennsylvania to Alabama (Fig. 5). With the exception of two breaks in Tennessee and Alabama, there is a contiguous region of coal producing counties from north-central Pennsylvania to Tuscaloosa County, Alabama, a distance of about 1300 km with a maximum width of about 340 km through Ohio and West Virginia.

Eastern Interior. The Eastern Interior, or Illinois Basin, includes the coal fields in Illinois, Indiana, and western Kentucky (Fig. 6).

Most of the principal coals are high sulfur throughout most of their extent. Notable exceptions are the "quality circle" of Herrin coal in central Illinois where sulfur is as low as 0.5% in contrast to >4% elsewhere.

Western Interior. The Western Interior coal fields include fields from Iowa to the Arkoma basin, Oklahoma and Arkansas, and in north central Texas. Coals range in age from Westphalian C/Atokan (Midcontinent series), which includes the Hartshorne coal bed in Oklahoma and Arkansas, through the Westphalian D/Desmoinesian, which includes the Croweburg and Bevier coal beds, to the Stephanian/Missourian and Virgilian.

Coal quality is variable but in general the northern coals are high sulfur while Oklahoma has some low-sulfur, metallurgical-grade coal, notably the Hartshorne coal bed. Texas coals are generally high sulfur with the Strawn Group Thurber coal bed having the highest quality. Coal rank through the various fields ranges from semianthracite/anthracite in Arkansas, to low volatile bituminous in portions of Arkansas and Oklahoma, medium volatile bituminous in Oklahoma, and high volatile bituminous in the remainder of the fields. Subbituminous coal is reported from Texas (14).

Gulf Coast. The Gulf Coast province stretches from Mexico north to Kentucky and east at least to Georgia. Commercial deposits are in Texas and Louisiana with Arkansas having some potential. Strippable low-rank coals occur in long belts along the strike of the Tertiary units.

The coals, where mined, are moderate ash with low sulfur content.

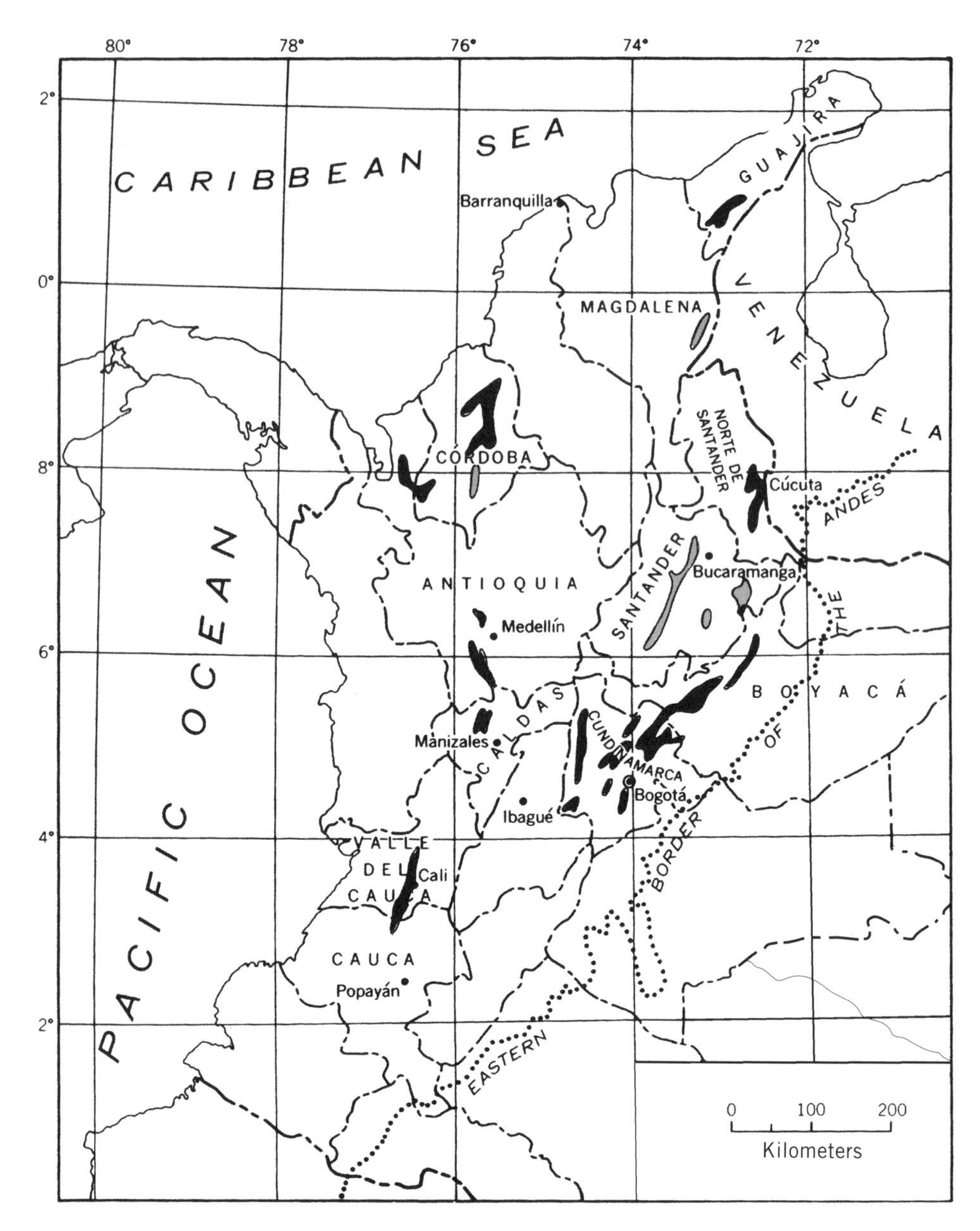

Figure 2. Coal fields of Colombia (3). Age of coal deposits ■, Tertiary; ▨ late Cretaceous.

Great Plains/Rocky Mountains. The coal fields of the Great Plains and Rocky Mountain provinces include some of the most important producing districts in the U.S., due to the continuum of Cretaceous-Tertiary coal-bearing strata.

The Cretaceous coal-bearing formations are thicker in the southern and central Rocky Mtn. region than in the north, up to 4.9 km vs 3 km. Important coals are early to middle Campanian age in the south and late Campanian to Maastrichtian in the central and northern Rockies.

Pacific Northwest. Washington state contains the most important reserves in the western states exclusive of Alaska. Discussion of other western fields may be found in Refs. 15–17.

Alaska. Alaska has an estimated 5.1 Tt of hypothetical resources compared to the 3.37 Tt of hypothetical resources for the conterminous United States. Much of the resource is north of the Arctic Circle. Summaries of Alaska coal deposits are from Refs. 15, 18, and 19. (Fig. 7). Development of much of the field will be hindered by the remote location, climate, and difficulty of mining in the tundra.

EUROPE

The coal industry in Europe has undergone significant decline over the past ten years. Coal production is likely to continue to be cut in many European countries and any shortfall made up by imports (20). A significant coal industry is likely to remain in Germany and Poland but the prospects for the British Coal Industry looks bleak.

Great Britain

The British coal industry has undergone severe decline over the past ten years and future prospects are uncertain. Currently there are three major coal basins (Midland Val-

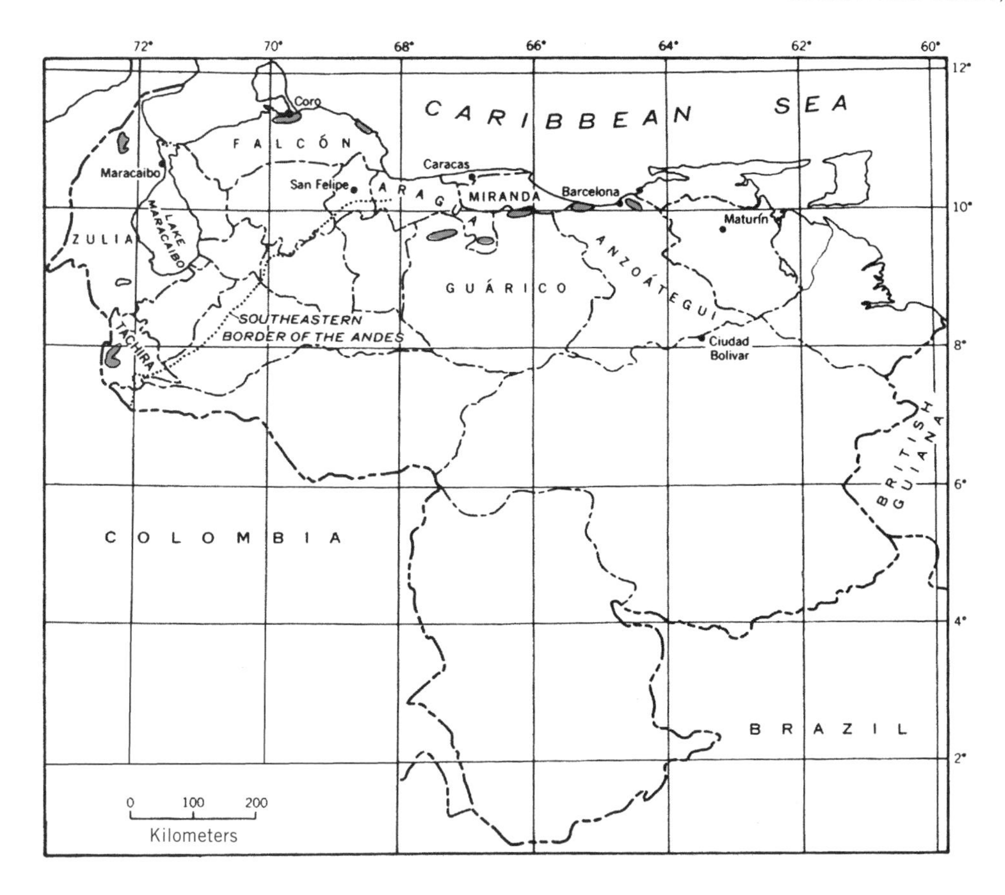

Figure 3. Coal fields of Venezuela (3). ■, Coal deposits of Tertiary age.

ley, South Wales, and Pennine) (Fig. 8) two of which now have very few operating mines.

Midland Valley Basin, Scotland. In Scotland, the Midland Valley Basin is a rift basin which underwent big subsidence in the late Palaeozoic. Within the basin there are several coalfields which have been structurally separated. Most of the deep mines have been closed.

Germany

Both the former East and West Germany contained significant quantities of lignites, bituminous coals and anthracites (Fig. 9). Proved recoverable reserves in the former Federal Republic (west) are 23.919 Gt with 60% coking but all underground mined (21).

Poland

Poland has significant reserves both of black bituminous coal and anthracite (28.7 Gt) and of lignite (11.67 Gt) of mainly Carboniferous and Tertiary age (21,22).

There are several large Carboniferous basins which were all formed as part of the tropical Euramerican coal province. The three main coal basins are those of Upper Silesia, Lower Silesia, and Lublin (Fig. 10**a**).

France

Of the principal western European countries with a long industrial record France has relatively few coal reserves and indeed its coal industry has been in decline for several years. It has 213 Mt reserves of Carboniferous bituminous coal and anthracite (21) and small reserves of about 30 Mt of Tertiary lignite in Provence (1) (Fig. 10**b**).

The small limnic basins contain minor reserves and many mines are now closed. Some mines still are working including at Monceau-les-Mines where low sulfur, low ash bituminous coal is mined.

Spain

Coals are found in several basins in Spain including Carboniferous bituminous coals and Cretaceous and Tertiary lignites (Fig. 10**c**). Total proved recoverable reserves of bituminous coal are 379 Mt, 155 Mt of sub-bituminous coal and 236 Mt of lignite (21). All of these tend to be high in ash and sulfur (21).

Greece

Coals in Greece are restricted to small Tertiary fault bounded basins in the north of the country where thick lignites occur (Fig. 10**d**). There is a total proved recoverable reserve of 3 Gt of lignite (21).

Turkey

Coals in Turkey comprise Carboniferous bituminous coals and Tertiary lignites. The Carboniferous sequence bears many similarities with other European basins. Total bituminous coal reserves are estimated at 162 Mt (1). In the Zonguldak coalfield (500 km^2) in the north coals are 0.7–10-m thick and are low ash, low sulfur bituminous coals which are used for coking. There is only 100 Mt of reserves. High gas levels are reported which may be of

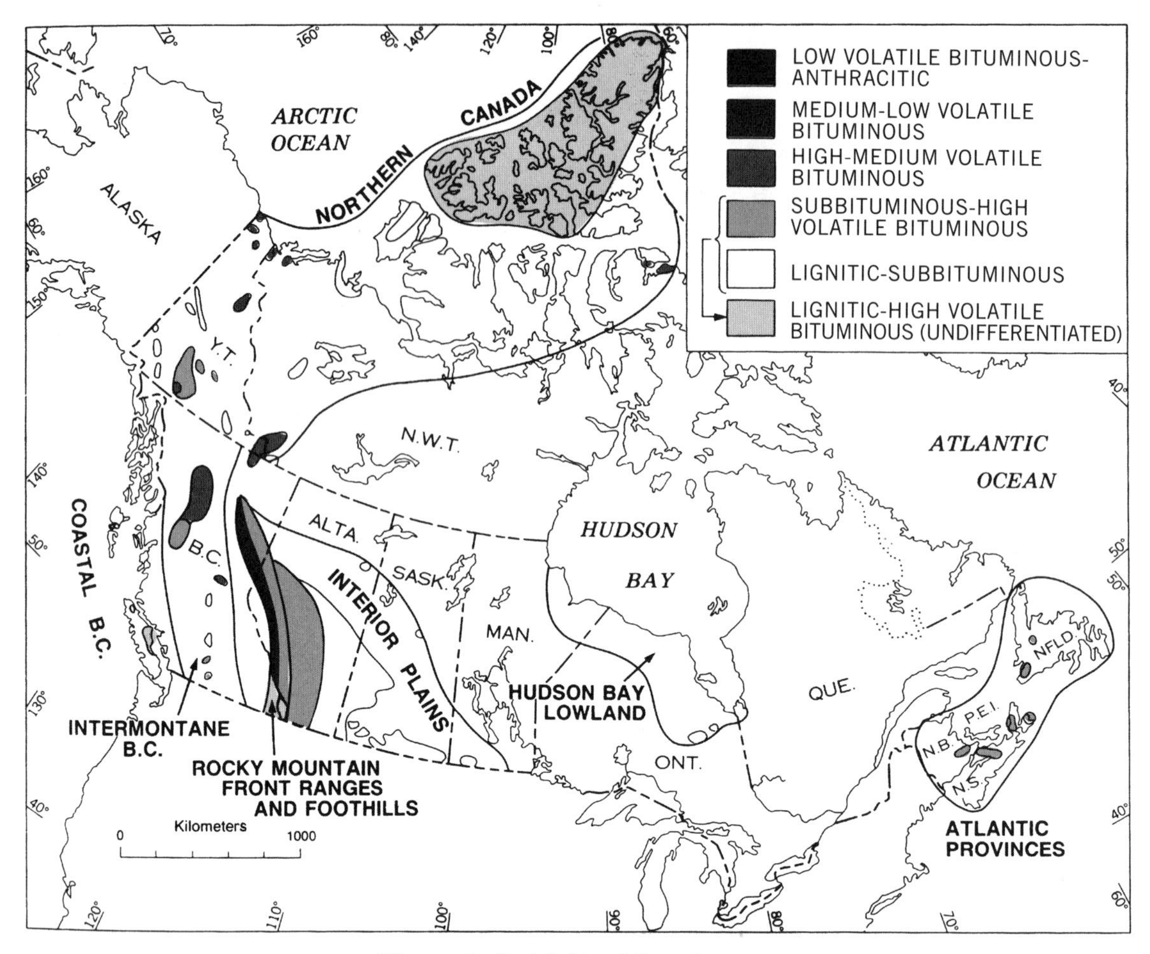

Figure 4. Coal fields of Canada (12).

significance for coal-bed methane exploitation in the future.

Czech Republic

Most of the coal in the former Czechoslovakia occurs in the Czech Republic (Fig. 11**a**). Most coals in the country are predominantly Tertiary low rank coals with high ash and sulfur contents.

OTHER EAST EUROPEAN COUNTRIES

Small coal deposits are also known in several East European countries such as Hungary, Romania, Bulgaria and the former Yugoslavia. Details are documented in Ref. 23.

Ukraine

The Ukraine (part of the former Soviet Union) possesses two major coalfields, one of bituminous coal, the Donetz Basin, and one of lignite, the Dnyepr Basin (Fig. 11**b**).

Kazakhstan

Kazakhstan (part of the former Soviet Union) possesses two principal coalfields, both of Carboniferous age, one of bituminous coal, the Karaganda Basin, and one of lignite and subbituminous coal, the Ekibastuz Basin.

Russia

The total Russian coal reserves are difficult to estimate as until now all figures published have been for the whole Soviet Union. In 1992 the total reserves for the former Soviet Union were 104 Gt of anthracite and bituminous coal and 137 Gt of lignite. Russia has vast reserves which range in age from the Carboniferous to the Tertiary but equally many of the largest basins are in very remote and inhospitable areas which together with problems of transport infrastructure makes their exploitation difficult.

EAST ASIA

North Korea

Coals are known in both onshore and offshore basins in North Korea (Fig. 12). They include seams of late Carboniferous/?Permian age, Jurassic, and Tertiary age. Some of these may also generate oil in the offshore West Korea Bay Basin (93). The total reserves of coal proved in place are 2 Gt and range from anthracites to lignites. WEC data (21)

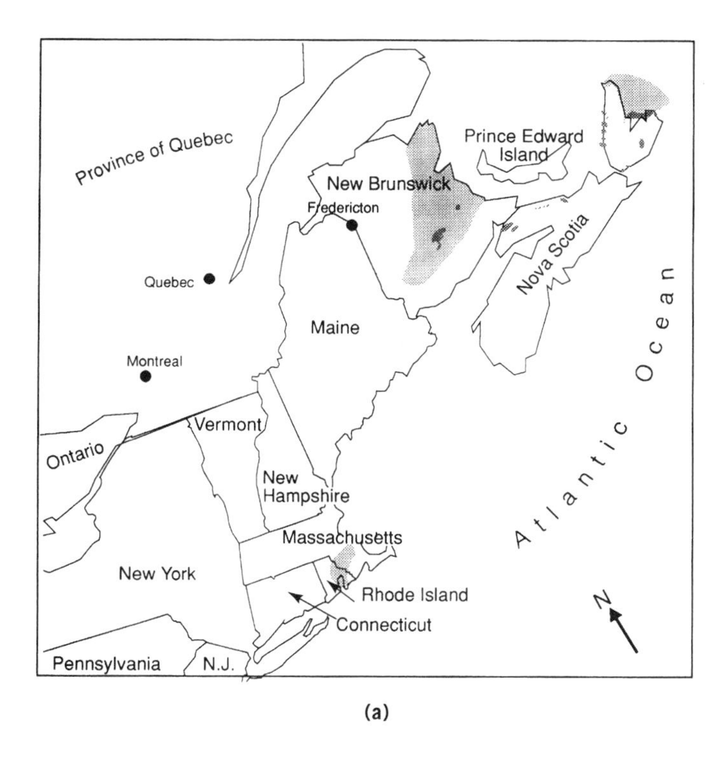

(a)

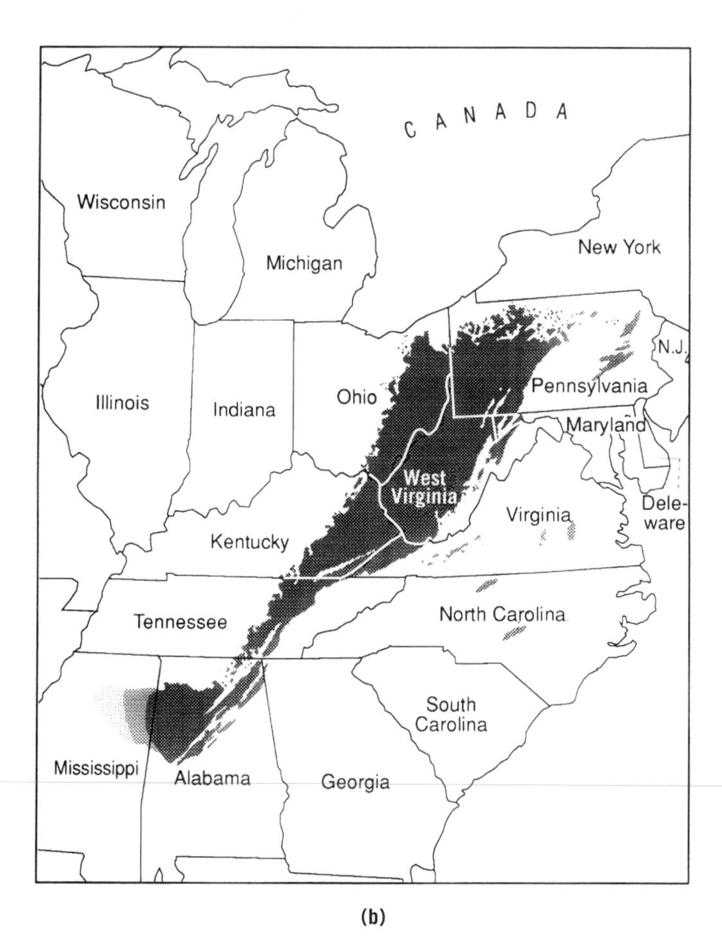

(b)

Figure 5. Coal fields in the (**a**) Maritime provinces, New England, and (**b**) Appalachian coal fields.

Figure 6. Eastern and Western Interior coal fields and the Michigan basin.

Figure 7. Coal fields in Alaska (19).

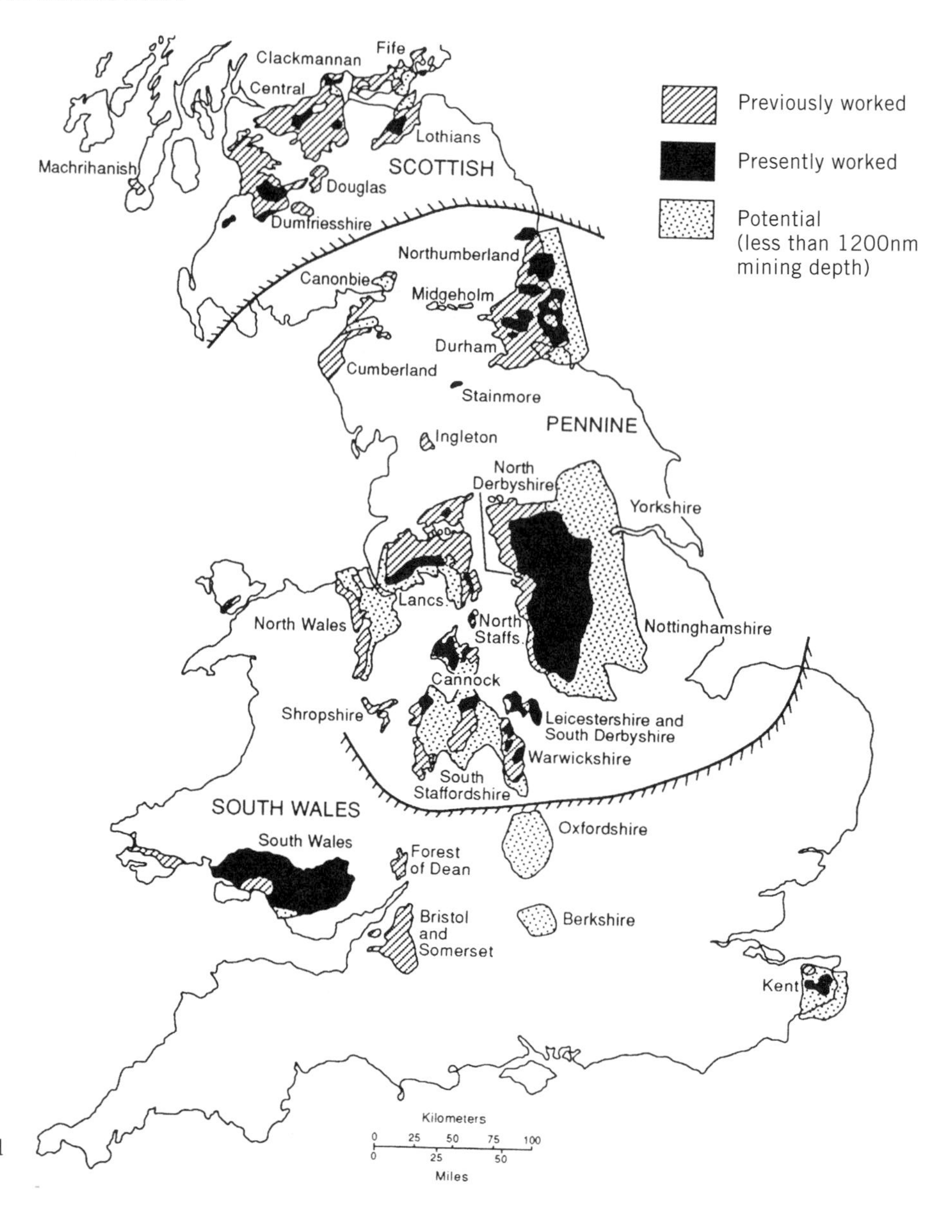

Figure 8. The coal fields of the United Kingdom.

indicates 300 Mt of bituminous coal and 300 Mt of lignite proved.

Palaeozoic coals were deposited on the Sino-Korean platform which was part of the Eurasian continent. Several phases of burial and tectonism have affected the coals so that all are anthracites (24). There are two main coalfields with significant economic potential: The Pyongyang-North Pyongyang coalfield and the Kowan-Muchon coalfield. Each of these have probable exploitable reserves of 300 Mt (1).

South Korea

As in the North, coals of late Palaeozoic, Jurassic, and Tertiary age occur including lignites to anthracites (Fig. 12**b**). Recent estimates indicate 158 Mt of recoverable reserves of bituminous coals and proved reserves of 238 Mt (21). However, other estimates indicate that anthracites have a total reserve of 1.6 Gt and a recoverable reserve of 754.3 Mt (26) and a proved reserve of 245 Mt (27). The anthracites are Carboniferous in age with low sulfur and seams 1–2-m thick but tectonically disturbed. The principal fields include Samchok (Samcheog) (26.7 Mt recoverable reserves), Chongson (Jeongseon) (45.2 Mt), Kangnung (Gangneung) (20.9 Mt), Danyang (15.8 Mt), and Mungyong (Munkyeong) (27.3 Mt) (29). Small amounts of Jurassic anthracites are known from the Mungyong and the Chunsan (Chungnam) coalfields (1).

Tertiary lignites are known in small areas bordering the south east coast.

Vietnam

The total amount of coal in Vietnam is a matter of conjecture but estimates of proven reserves range from 1 to 20 Gt with proven reserves of only 135 to 300 mt (25). Coals from Vietnam are mainly low volatile bituminous to anthracites mainly of Triassic age but also including Permo-

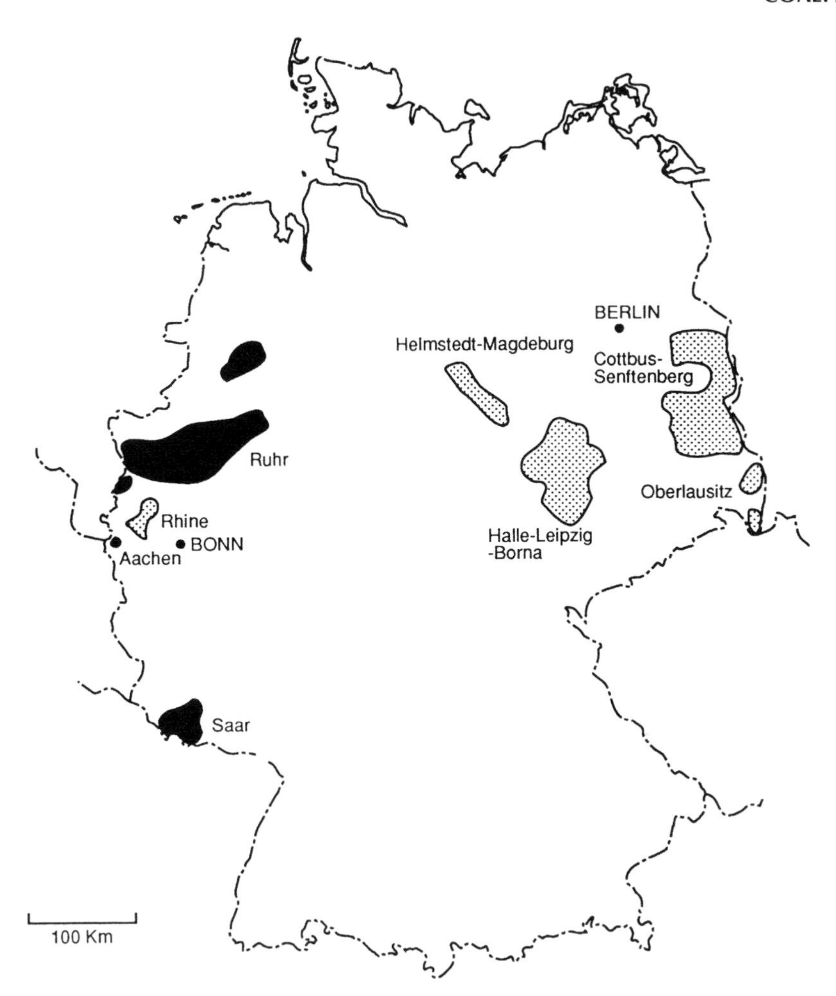

Figure 9. The coal fields of Germany ■, Bituminous coal and anthracite; ▩ brown coal.

Carboniferous which occur in a broad belt north of Hanoi (1,28) (Fig. 12**c**). Anthracites of Upper Triassic age in the Quang Ninh region have a total reserve of 6.5 Gt and a minable reserve of 557 Mt.

Thailand

Thailand is relatively poor in coal resources but has some semi-anthracite, bituminous coals, and lignites (Fig. 12**d**). Extensive petroleum exploration has shown the more widespread occurrence of Tertiary lignites in rift basins.

Taiwan

The coal reserves of Taiwan are relatively small.

In the northern province low-ash and low-sulfur high volatile bituminous and subbituminous coals occur but these are highly tectonized and suffer from the effects of igneous intrusions. In Chilung, in the north, semi-anthracites are mined.

Japan

Japan is a main coal importer, but although limited, it has over 8.6 Gt reserves, 3.2 Gt mineable (25), (which includes 17 Mt of lignite) ranging in age from Permian to Tertiary (Fig. 13**a**). The fact that Japan is within a major tectonic and volcanic belt has meant that extensive heat flows have created a wide range of coal ranks from anthracites to lignites (30). In 1989 9.6 Mt of coal were produced from 20 mines (32).

China

Calculations of Chinese coal reserves has been problematical but have almost certainly been underestimated. China has coals of every Phanerozoic Period and of every rank (Fig. 13**b**). Estimates by B. P. in 1992 give the reserves of bituminous coal as 62.2 Gt and lignite as 52.3 Gt. Recent estimates (31) indicate that in total known reserves are about 103.1 Gt and the predicted resources are 380 Gt. The literature on Chinese coals and basins is vast and a most useful guide is Ref. 33. China is divided into five coal accumulation regions (excluding Taiwan, dealt elsewhere) and the country is divided into 32 coal administrative regions. There are over 320 coal basins in China. China represents one of the most important areas for potential development.

AUSTRALIA

Although Australia is a big exporter of coal, most of the coal is taken from Permian deposits in the eastern states of New South Wales, the long-time principal producing state of thermal and coking coals, and Queensland, the

Figure 10. The coal fields of (**a**) Poland (23); (**b**) France (20); (**c**) Spain (20); (**d**) Greece (20).

state which has had the most rapid and greatest increase in production in the past 20 years. Of the other states, Victoria is a very large producer of soft brown coal for domestic markets and both Western Australia and South Australia produce significant volumes of coal. Historically, Tasmania has not been a large coal producer.

By world standards, particularly compared to China, United States, and USSR who together produce 80% of world coal, Australia is not a large coal producer with only 3.8% of world production (34). However, as an exporter, Australia is the world's largest with almost half total production shipped overseas (see Plate I).

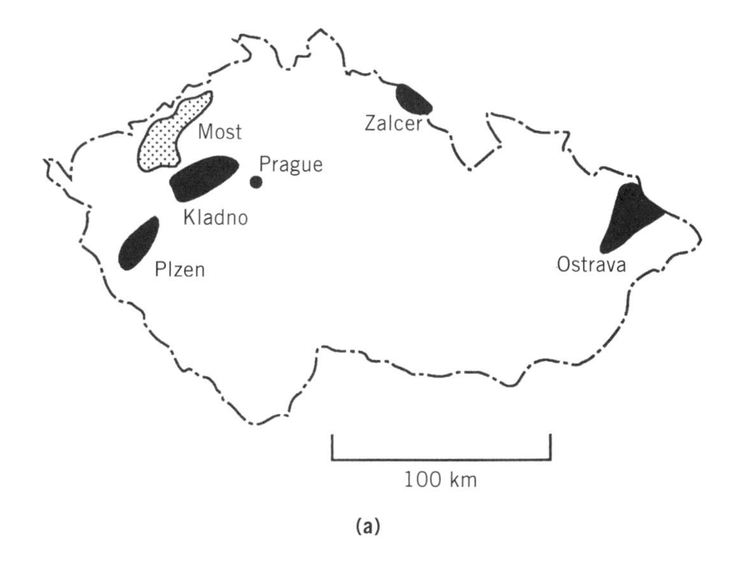

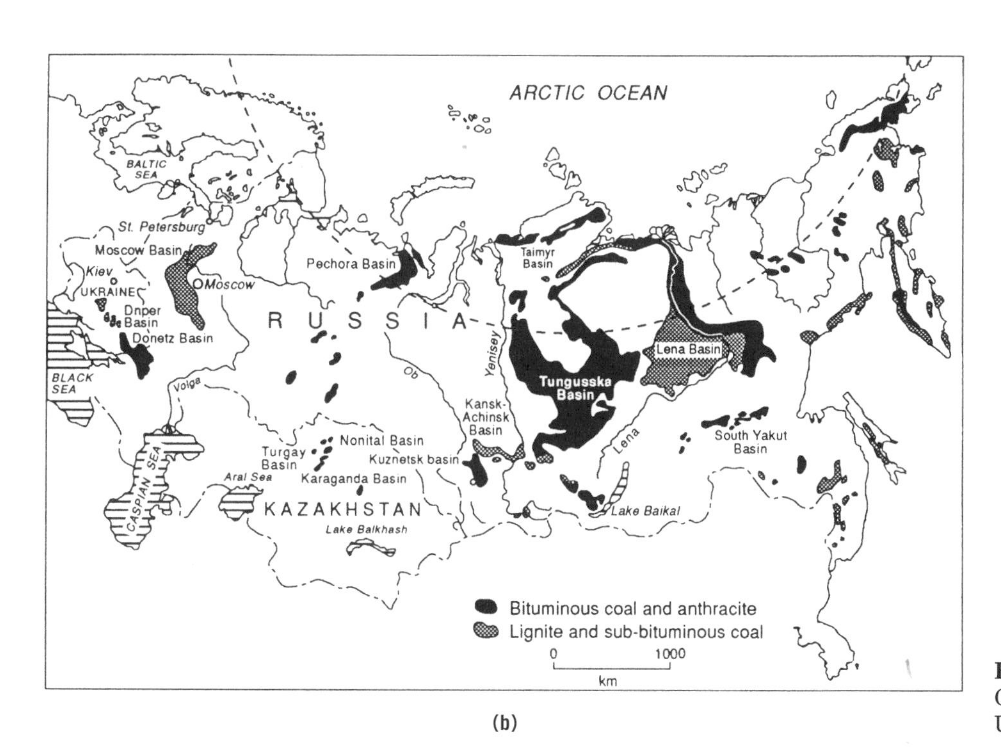

Figure 11. The coalfields of (**a**) the Czech Republic (23); (**b**) Russia, Ukraine, and Kazakhstan (22).

NEW ZEALAND

New Zealand is a relatively small producer of coal by world standards with most of its production used for domestic purposes and only small quantities exported. Most of the known coal resources (82% and most in the South Island) are of brown coal rank with approximately 14% subbituminous and 4% bituminous. Many areas of New Zealand have not been subjected to detailed feasibility studies and recovery rates are extremely variable ranging from as low as 18% in some underground mines to 95% in some open pit mines; typical underground recovery rates are approximately 30% (35).

Total coal resources in New Zealand are 15.7 Gt of in-ground coal of which 13.1 Gt are in the South Island and 2.6 Gt in the North Island. However the disparity in rank and quality of coal in the North and South Island is such that slightly more than 50% of production is from the North Island.

INDONESIA

Estimates of the total coal resources of Indonesia have increased rapidly over the past five years changing from 29 Gt (36) of which 12.96 Gt was regarded as demonstrated

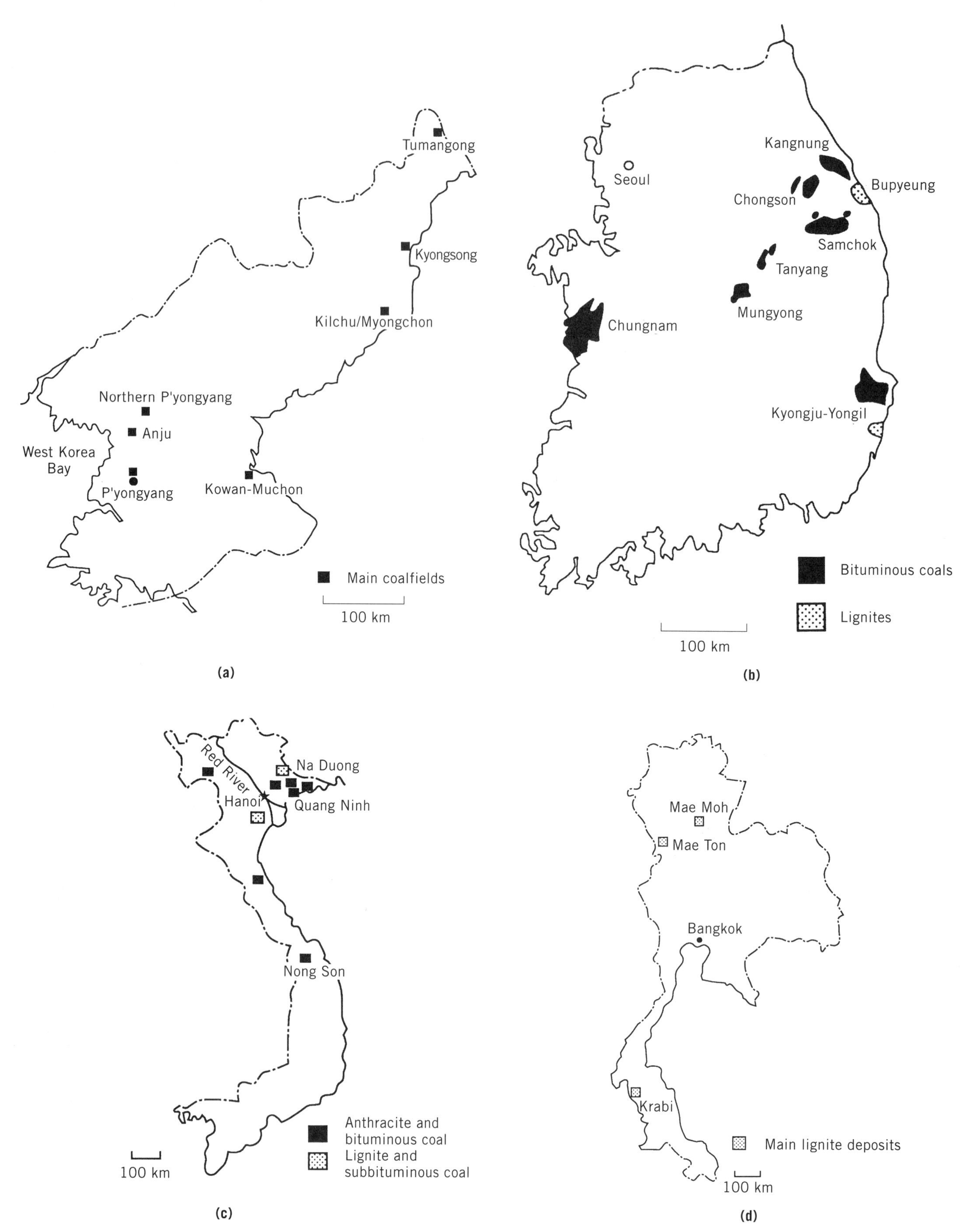

Figure 12. The coalfields of (**a**) North Korea (24,25); (**b**) South Korea (24,25–27); (**c**) Vietnam (28); (**d**) Thailand (25).

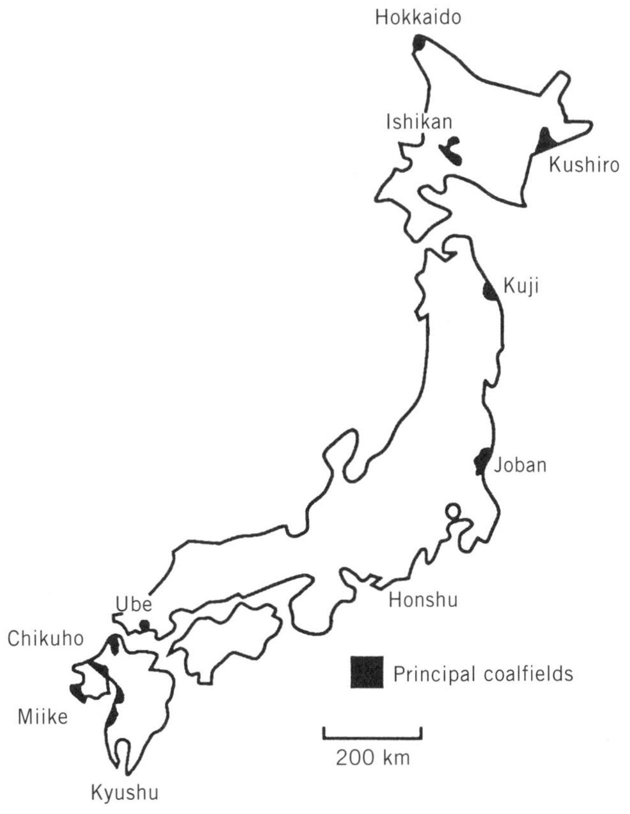

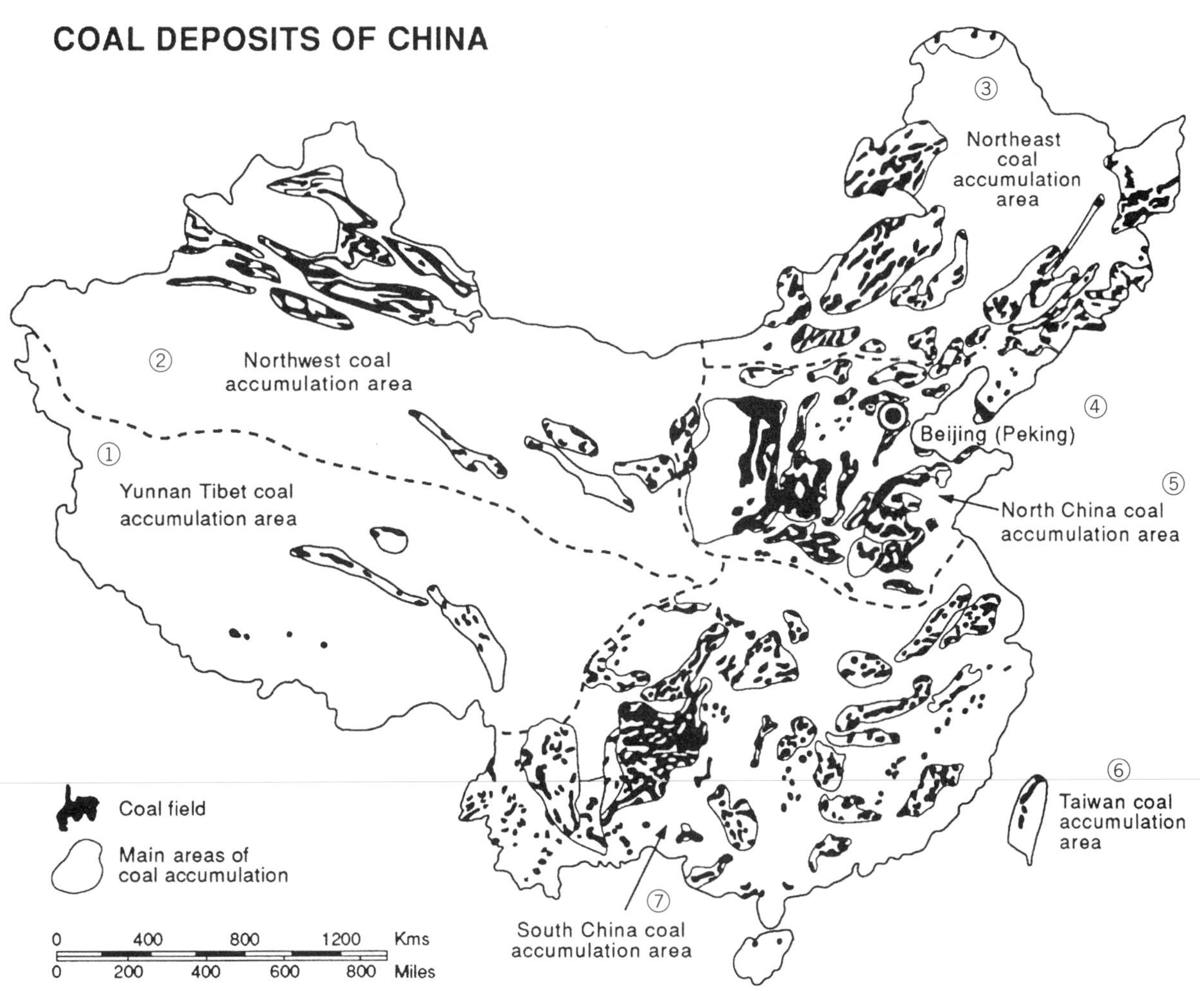

Figure 13. The coalfields of (**a**) Japan (30); (**b**) China (31).

Plate I. Australia's endowment of coal is enough to permit substantial exports. This stockpile, from the Moura field of Gladstone Harbor, Queensland, will be transported to Japan. Courtesy of Queensland State Government.

reserves with 2.39 Gt considered proven increasing to 36 Gt. Although coal deposits are widely distributed throughout the archipelago except in Nusatenggara, Maluka and northern Sulewesi, economic deposits are mostly in the Miocene to Eocene strata in Sumatera, Kalimantan and Java and Pliocene to Pleistocene strata in Sumatera (Fig. 14). Indonesia's coals span the spectrum of ranks with at least 14 Mt of anthracite, 760.2 Mt of bituminous coal, 4.4 Gt of sub-bituminous coal and 18 Gt of brown coal (37).

Coal has been known in Indonesia for many centuries but mining only commenced in 1849 in Kalimantan with exports of coal not commencing until the 1930s when coal was sent to neighboring countries.

INDIA

India has a long history of coal production covering at least 200 years although systematic exploration for coal did not commence until much later.

Ninety percent of India's major workable coal resources are of Permian age and are known as Gondwana coals (Fig. 15). Nearly 50 coalfields, ranging size from a few square kilometers to greater than 1500 km^2, have been mapped. Most occur in peninsular India in the quadrant bounded by 78°E longitude and 24°N latitude. Within the basins the number of coal seams ranges from as few as two or three up to 44 with total minable intervals are 0.5 to 160 m (Singrauli) with the rank of coals varying from sub-bituminous to bituminous rank; ash contents are generally high (20 to 35%) and phosphorus contents are low ($<$0.01%).

SOUTH AFRICA

Coal resources of the African continent are restricted to the Permo-Triassic Karoo Supergroup which is found in South Africa, Botswana, Zimbabwe, Zambia, South West Africa, and Mozambique (Fig. 16). In South Africa, the Karoo Supergroup is found in the central and eastern parts of the Cape Province, the southern part of the Orange Free State, Lesotho, Transvaal and Natal.

Coal was first discovered in Natal and Northwest Province and subsequently mined in the Molteno-Indwe Coal-

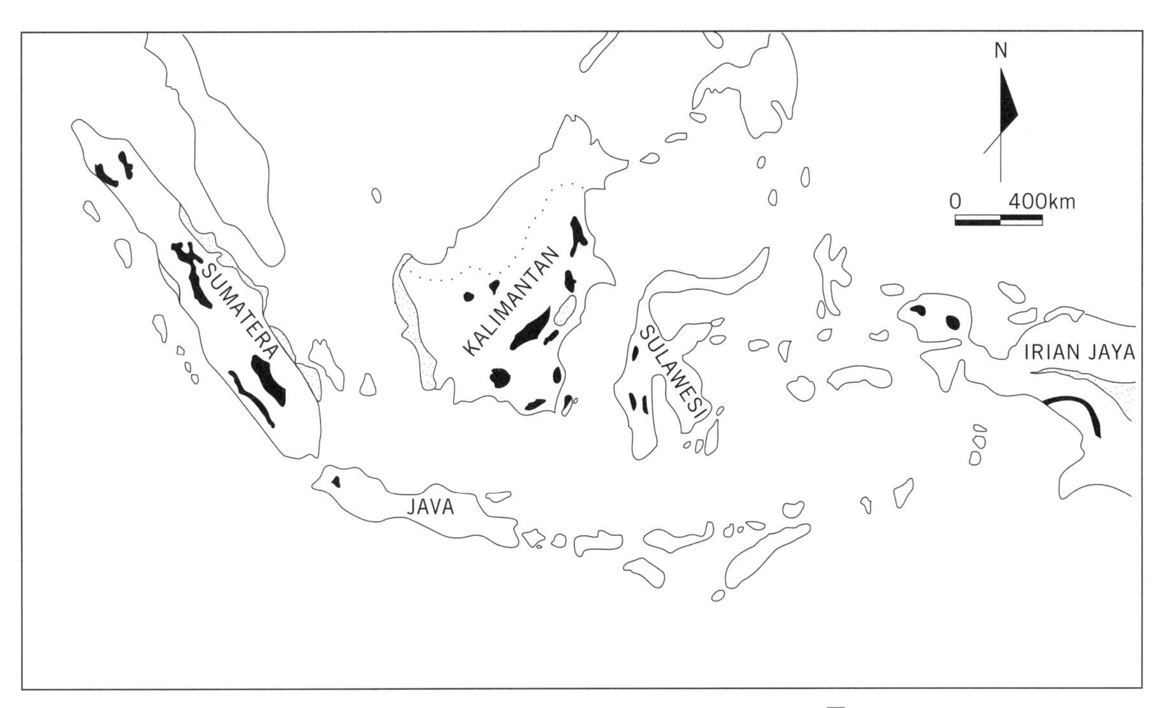

Figure 14. Main coal basins of Indonesia. ■, Coal; ▨ peat.

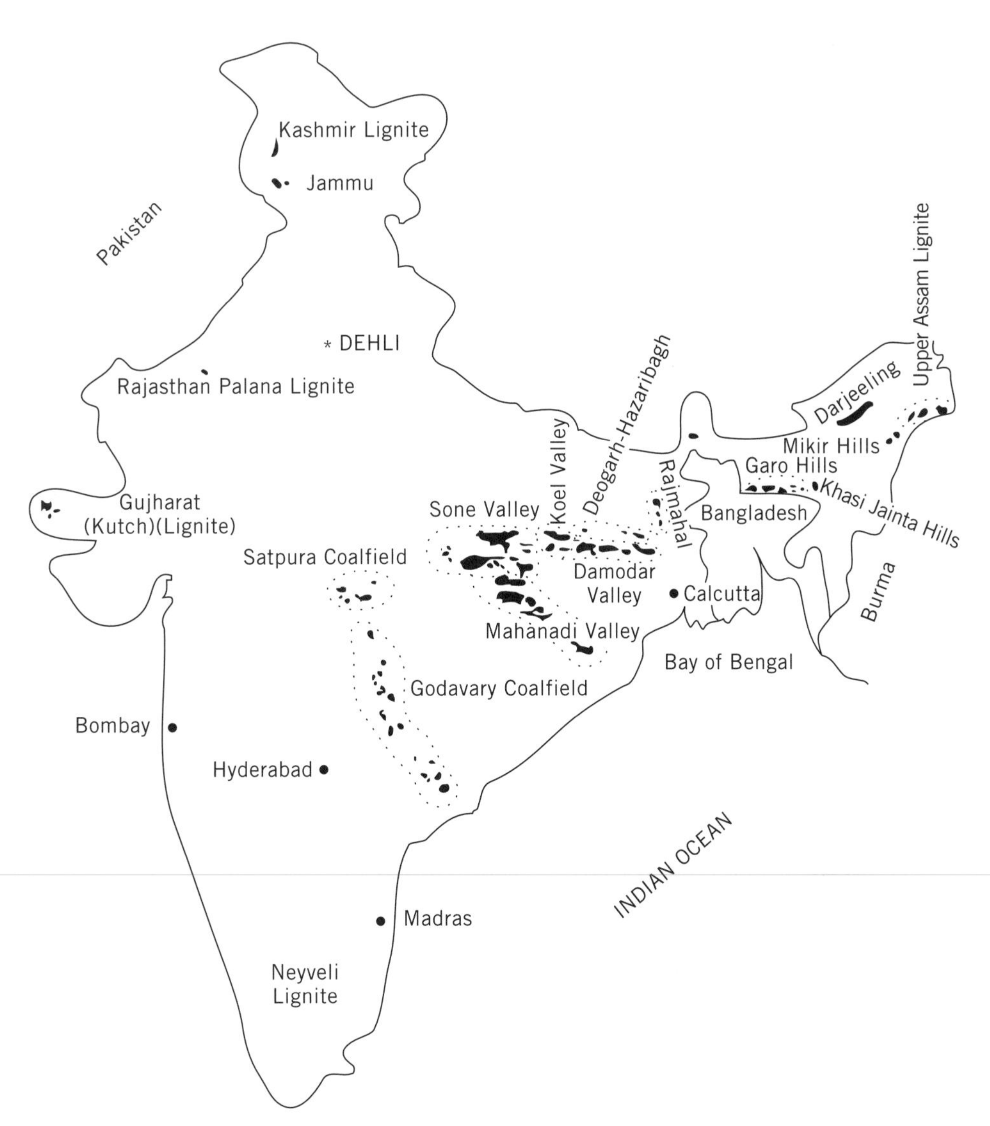

Figure 15. Coalfields of India.

field of North West Province from where it was transported by oxen-drawn wagons to supply the energy needs of diamond mining in the Kimberley region (38). Large scale mining commenced, in the East Rand area, soon after the discovery of diamonds with significant mining in the Witbank, Middleburg, and Vereeniging Coalfields.

A massive increase in coal production occurred after World War II, coinciding with rapid expansion of the South African economy. Production was further stimulated in the 1970s when the energy crisis was experienced. This rapid growth is reflected in the production figures which increased from a total of 34.2 Mt in 1957 to almost double this, 63.0 Mt in 1974.

The most recent available resources and production figures for South Africa are those given by the Commission of Inquiry into the Coal Resources of the Republic of South Africa in 1975. Minable *in situ* resources were given as 81.275 Gt of which 24.89 Gt was classified as high-grade

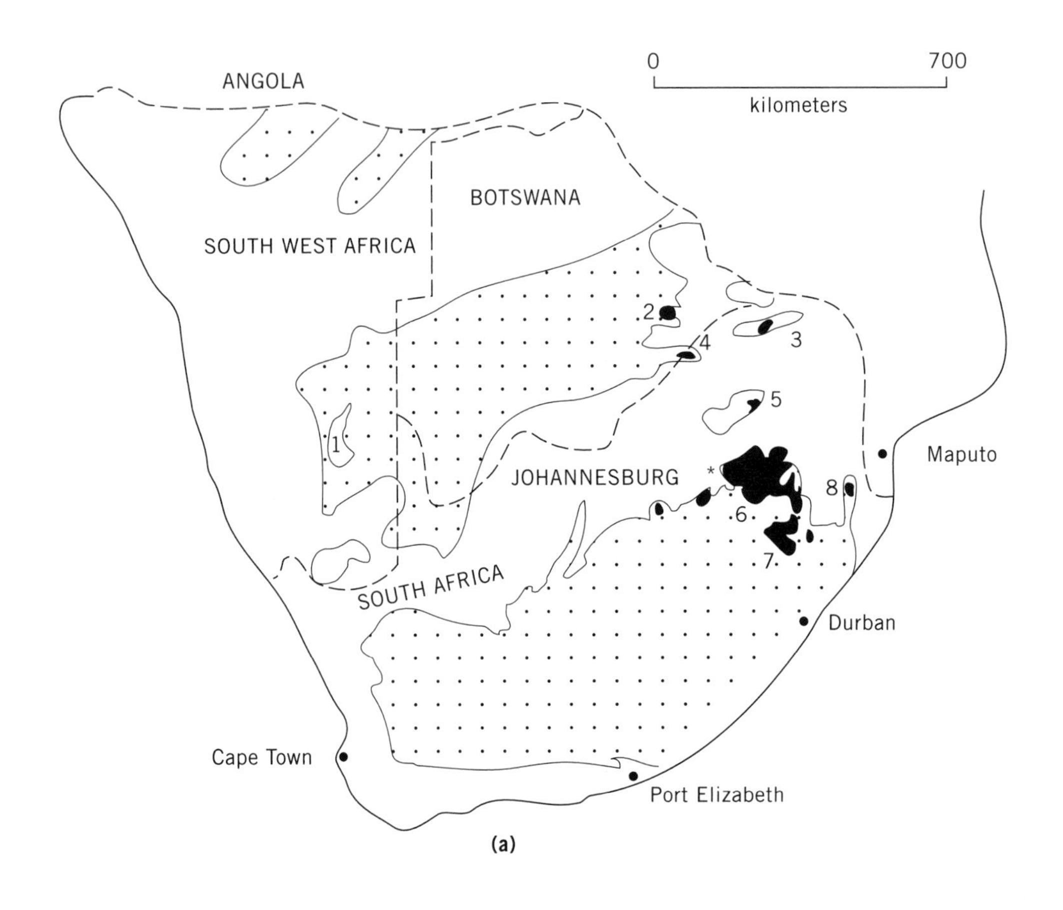

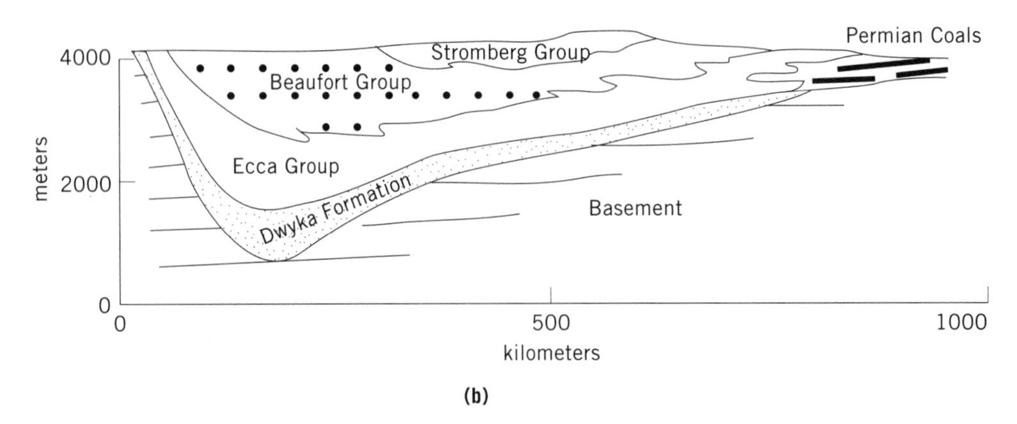

Figure 16. Coalfields of South Africa. (**a**) Location of the basins in South Africa; (**b**) stratigraphy of the Permian sequence. 1, Southwest Africa; 2, Botswana; 3, Limpopo-Southpansberg; 4, Watersberg; 5, Springbok flats; 6, Transvaal; 7, Natal; 8, Swaziland. ⊞ extent of the Ecca group.

steaming coal and 3.45 Mt was cited as washed metallurgical coal. Anthracite resources were estimated to 745 Mt. Of the total resources, excluding anthracite, 31% of the resources, or 24.91 Mt was minable by underground methods.

South Africa is an important producer and exporter of coal, being the fifth largest producer, 183 Mt in 1990, and the third largest exporter, 47 Mt in 1990 (39). This is a dramatic rise since the mid-1970s when total production was 65 Mt.

Domestically, coal has the dominant role in South Africa's economy with more than 80% of total primary energy requirements sourced from coal, possibly the highest proportion for any industrialized country.

COAL MINING AND PREPARATION

Coal mines fall into two general classifications: surface and underground.

Surface Mining

Surface mining techniques are used when the coal is present near the surface and the overburden is thin enough (see Plate II). These techniques include contour mining, strip mining, and auger mining.

Contour mining is used in hilly countryside areas where the slope of the surface will permit only a narrow bench to cut around the side of a hill; the excavation is backfilled immediately after the removal of coal. It is the only method that can be used on slopes of 15 degrees or higher.

Strip mining is used in flat or gently rolling lands of the Midwest and West where large and efficient equipment can be used. In this technique, the coal is exposed by removing the overlying strata, or overburden. Blast holes are drilled, and explosives are loaded into these holes to shatter the rock cover; earth-moving equipment is used to remove the soil and the shattered rock (see Plate III). The coal then is collected with power shovels or other coal-digging machines and loaded directly into trucks.

Auger mining is a supplementary method used to reach coal in stripped areas where the overburden has become too thick to be removed economically. Large augers are operated from the floor of the surface mines and bore horizontally into the coal face to produce some reserves not otherwise minable.

For a coal deposit where the seam is near the top of a hill, the entire hilltop can be removed to expose the coal.

Underground Mining

Underground mining techniques are somewhat more labor-intensive than surface mining and are used to remove coal located below too much overburden for surface mining; but here, too, machines are used in most instances to dig, load and haul the coal. Access to the coal seam is through a drift (horizontal passage), a slope, or a shaft (Fig. 17), depending on the location of the coal seam.

A *drift mine* is one that enters a coal seam exposed at the surface on the side of a hill or mountain. The mine follows the coal horizontally.

A *slope mine* is one where an inclined tunnel is driven through the rock to the coal, with the mined coal removed by conveyors or truck haulage.

Plate II. The typical strata above a layer of bituminous coal is shown in this surface mine excavation. Surface mining is used when the coal seam is not buried far underground. Courtesy Indiana Coal Producers Association.

A *shaft mine* is one where a vertical shaft is dug through the rock to reach the coal, which may be at great depth below the surface. The coal then is mined by horizontal entries into the seam, with the resulting coal hoisted to the surface through the vertical shaft.

Two general mining systems are used in underground mining, namely, room-and-pillar and longwall.

Room-and-pillar mining is an open-stopping method where mining progresses in a nearly horizontal or low-angle direction by opening multiple stopes or rooms, leaving solid material to act as pillars to support the vertical load. This system recovers about 50% of the coal and leaves an area much like a checkerboard. It is used in areas where the overlying rock or "roof" has geologic characteristics that offer the possibility of good roof support. This system was used in the old mines where the coal was hand-dug. Two current methods for extracting the coal from the seam are the conventional method, where the coal is undercut and blasted free (Fig. 18), and the continuous method, where a machine moves along the coal face to extract the coal instead of blasting it loose (Fig. 19, Plate IV). Roof control is the major problem of the room-and-pillar method of mining.

Longwall mining uses a machine which is pulled back and forth across the face of the coal seam in larger rooms. It is not a major mining method in the U.S. at this time.

Plate III. Sixty-feet of earth, shale and stone were removed from this site to recover a 40-inch seam of coal. In order to comply with reclamation laws, earth and shale must be replaced to a depth that will: prevent standing water, shape the material, and allow for seeding. Courtesy of the United States Department of Agriculture.

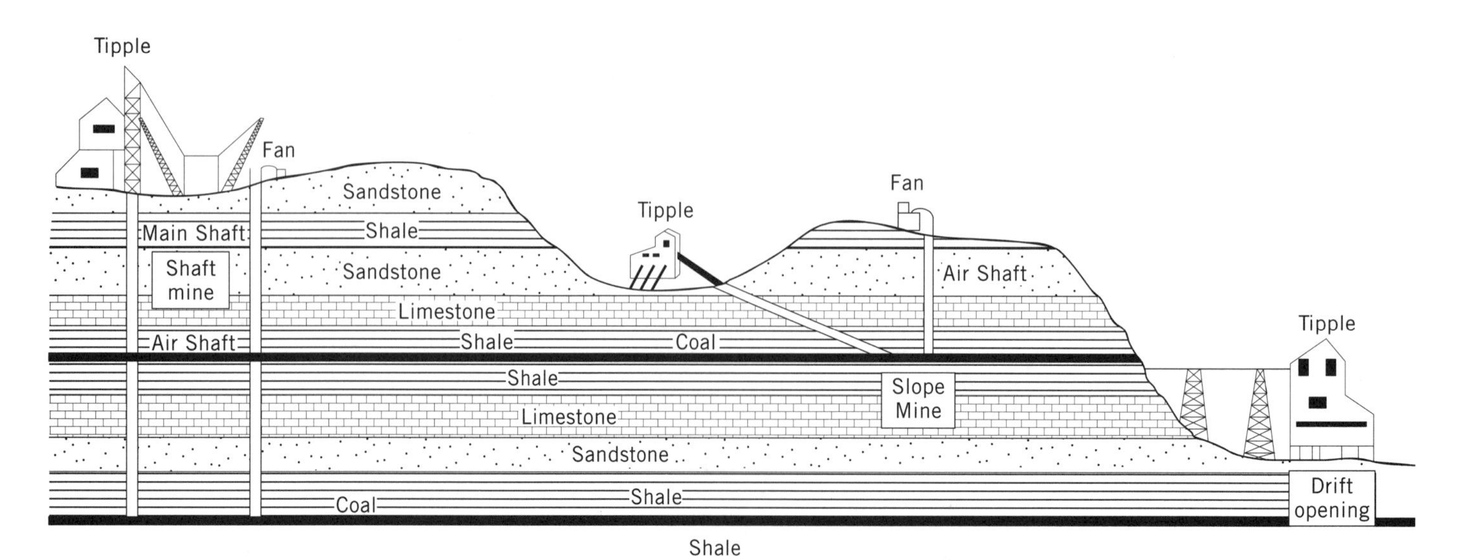

Figure 17. Three types of entrances to underground mines; shaft, slope, and drift (U.S. Bureau of Mines).

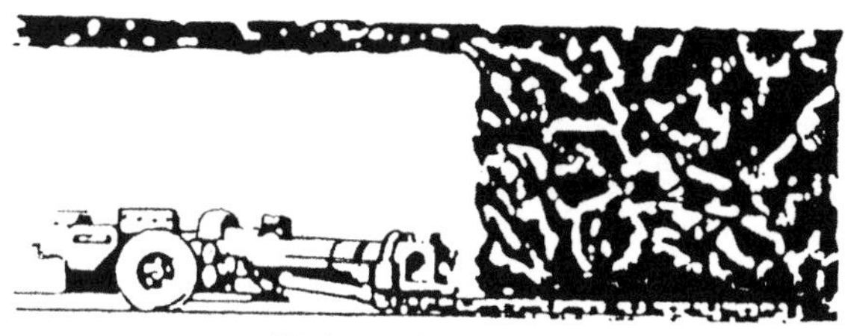

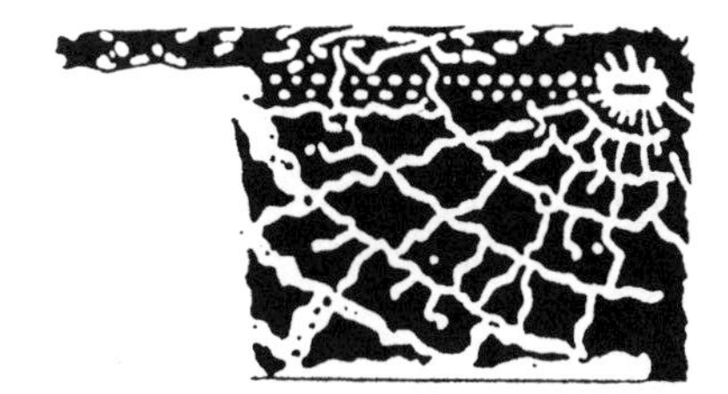

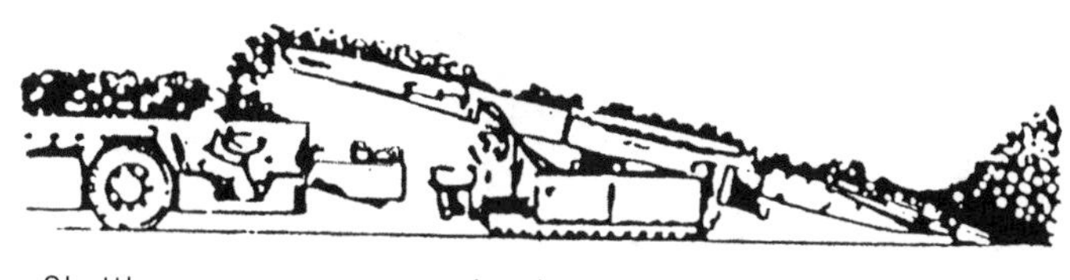

Figure 18. Basic steps in conventional mining (U.S. Bureau of Mines).

Coal recovery using this method is greater than in room-and-pillar mining and can be used where roof conditions are fair to poor. Strong roof rock, however, can be a problem. The seam should be over 42 inches (1.07 m) thick to accommodate this type of operation and the large coal cutter or plow that is used. A large reserve is necessary.

A modification of this method, using a continuous mining machine on faces up to 150 feet (45.7 m) long, is known as *shortwall mining*. It uses the room support system of self-advancing chocks developed for longwall operations.

Coal Preparation

"Coal Preparation or Beneficiation" is a term applied to upgrading of coal to make it suitable for a particular use. It includes blending and homogenization, size reduction and beneficiation or cleaning. It is this last aspect and the degree to which it is required, which most significantly governs the cost of coal preparation. Figure 20 shows levels of cleaning in terms of broad categories.

Only about one-third of the 3.3 Gt of coal produced every year is at present cleaned by breaking, crushing, screening, wet, and dry concentrating processes. This section will briefly describe the processes utilized for coal cleaning.

Coal is generally cleaned by physical methods to remove the mineral matter consisting of rock, slate, pyrite, and other impurities. In general, processes for removal of mineral matter utilize either the differences in density or surface chemical properties of coal and mineral matter. For convenience, coal cleaning can be divided into three parts: coarse-coal cleaning, medium-coal cleaning, and fine-coal cleaning. Figure 21 shows various coal cleaning methods and corresponding size ranges of coal treated.

The first step in most of the coal cleaning operations is size reduction, the main objective being to liberate mineral matter from coal. Size reduction equipment for coal application ranges from heavy-duty crushers and breakers, capable of crushing lumps of up to a meter cube in size, to coal pulverization equipment capable of milling coal to a fine powder. The hardness and the softness of the coal is defined in terms of Hardgrove Grindability Index (HGI). A high (>80) HGI indicates soft coal, easy to grind, and a low (~35) HGI indicates difficult-to-grind coal.

After the coal is crushed it is generally screened to separate the raw coal into various sizes for cleaning operations. Screening of coal above 12-mm size is usually is carried out dry. Double-deck vibrating screens mounted at 17–20° are commonly used for this purpose. For sizing below 12 mm, wet screening is done, using either high frequency vibrating screens or the Sieve Bend. Details on size reduction can be found in various books on coal preparation (40,41).

The specific gravity of coal ranges from 1.23–1.72 depending on rank, moisture, and ash content. The specific gravity of coal increases with rank. Shale, clay, and sandstone have a specific gravity of about 2.6 and mineral pyrite (FeS_2) has a specific gravity of about 5.0. These differences in specific gravities of coal and mineral matter suggest that their separation could be achieved by using a liquid with density in between coal and mineral matter. To determine the preparation method and to identify the best specific gravity for efficient separation of coal and mineral matter "washability" studies (or float-sink analysis) are conducted in the laboratory on a representative sample of coal. Coal is ground to a certain size and is mixed with organic liquids of various specific gravities

Figure 19. Continuous mining machine (U.S. Bureau of Mines).

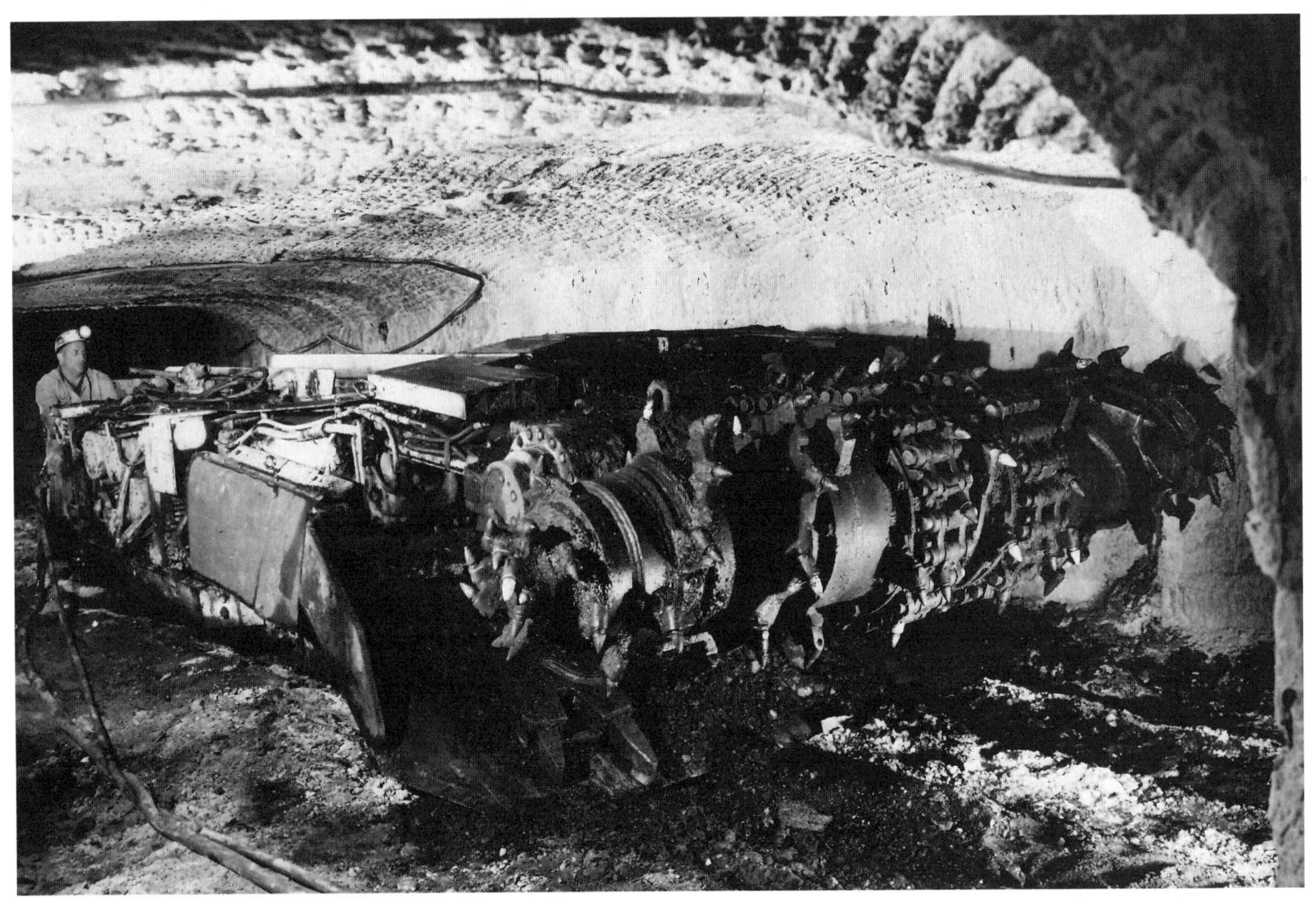

Plate IV. Continuous mining machines are used to extract coal in underground mines. Photo from Jeffrey Mfg. Co., Courtesy of the National Coal Association.

starting from 1.30 through 1.70. The float and sink fraction of each specific gravity is separated, dried, weighed, and analyzed for ash content. The data is mathematically combined on a weighted basis into cumulative float and cumulative sink and used to develop the washability curves that are characteristic for the coal. The scale value of near-gravity material which is determined from the washability curve is listed below. A 0 to 7% near gravity material indicates easy to clean coal, and above 25% represents impossible to clean coal. It can be seen that if a 5.8%

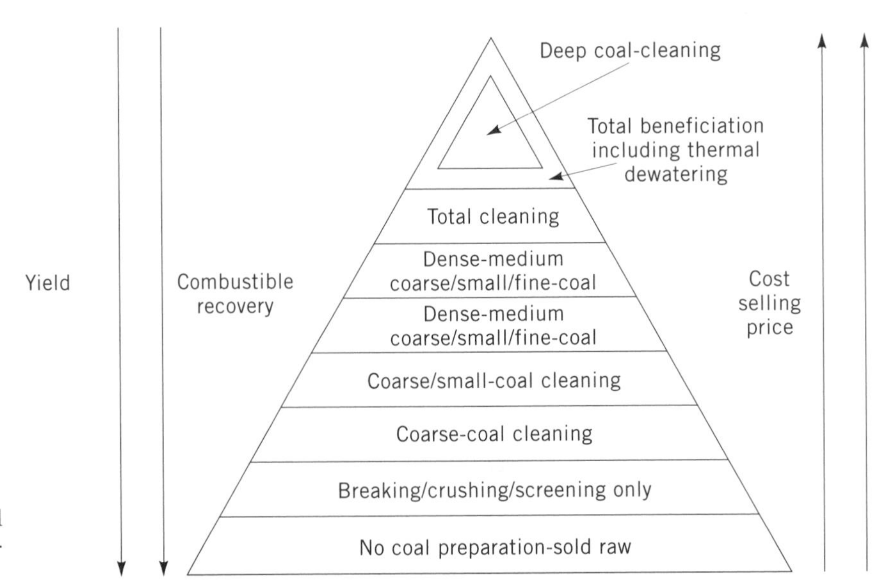

Figure 20. Different levels of coal cleaning and their effect on coal recovery and economics.

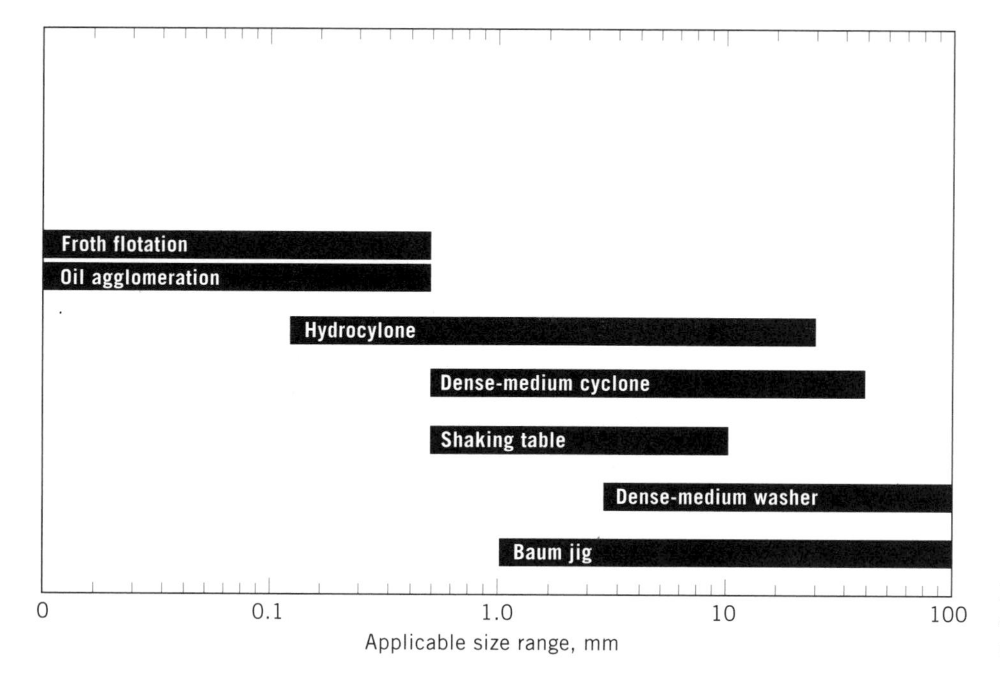

Figure 21. Coal cleaning equipment in common use in the coal industry with respect to the coal size cleaned.

ash clean coal is desired, then cleaning should be performed at a specific gravity of 1.48 and recovery of coal would be about 85.5%. The near specific gravity material will be about 10%, which means it is moderately difficult to clean coal. Scale of value of near-gravity material:

Quantity within +0.10 sp gr Range, %	Separation Problem
0–7	Simple
7–10	Moderately difficult
10–15	Difficult
15–20	Very difficult
20–25	Exceedingly difficult
Above 25	Formidable

Coarse-Coal Cleaning

Jigging and dense-medium separation are generally used for coarse coal cleaning.

Jigging. Jigging is among the oldest of mineral processing methods and one of the first adopted for coal cleaning. The separation of coal from mineral matter is accomplished via fluidized bed created by pulsing a column of water which produces a stratifying effect on the raw coal. Water is pulsated through the perforated bottom of an open top rectangular box. The lighter coal particles rise to the top, overflow the end of the jig and are removed as clean product. The denser mineral matter settles on a screen plate and is removed as refuse. Most jigs are of the Baum type (Fig. 22), in which the pulsations are generated by air pressure.

Dense-Medium Washing. Dense-medium separations include processes which clean raw coal by immersing it in a fluid medium having a density intermediate between clean coal and reject. Most dense medium washers use a suspension of fine magnetite in water to achieve the desired density. The process is very effective in providing a sharp separation and is relatively low in capital and operating costs.

Medium Coal Cleaning

Medium-size coal includes coal ranging in size from 3/4 in. (1.9 cm) to 28 mesh (0.5 mm). The principal techniques utilized are the wet concentrating tables, the dense-medium cyclone, the hydrocyclone and spirals. All of these equipment types are widely used in the coal industry.

Wet Concentrating Table. This is also known as a shaking table and works much like the classic miner's pan. The shape of the table is usually a parallelogram, and the deck of the table is at slope with rubber riffles on it. The upper side of the deck contains a feed box and a wash water box. The deck is vibrated longitudinally with a slow forward stroke and a rapid return. Normal decks are 5 m long along the longest edge and 2.5 m wide at right angles to the length. Such a machine can process up to 12 to 15 tons of coal per hour.

Dense-Medium Cyclone. The essential features of a cyclone are shown in Figure 23. A mixture of raw coal and dense medium (magnetite suspension) enters the cyclone tangentially near the top, producing free-vortex flow. The refuse moves along the wall and is discharged through the underflow orifice. The clean coal moves radially toward the cyclone axis, passes through the vortex finder to the overflow chamber and is discharged from a tangential outlet.

Cyclones are available in various diameters; a 500-mm diameter unit can process up to 50 tons of coal per hour, whereas one of 750-mm diameter can process up to 120 tons per hour.

Some other forms of the heavy-medium cyclone are the Vorsyl Separator, the Larcodems, the Dynawhirlpool, and the Tri-flo Separator. Each of these units utilizes cycloning principles.

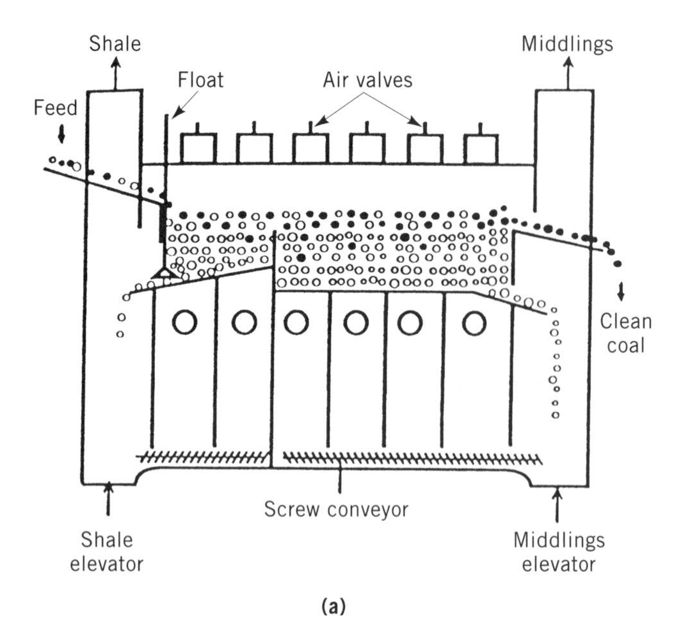

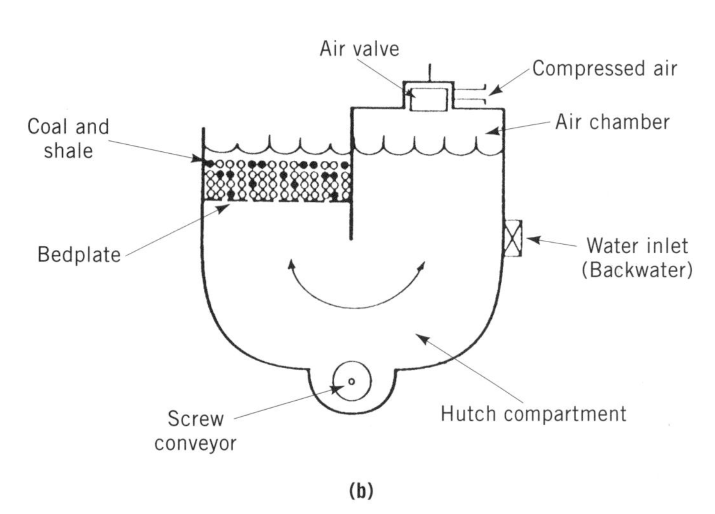

Figure 22. Sections of a typical Baum jig. (**a**) Longitudinal (**b**) Cross section.

Hydrocyclone. When only water is used in a cyclone for cleaning coal, it is called a hydrocyclone. Hydrocycloning has been applied to process coal finer than 0.5 mm. Its design differs from that of the dense-medium cyclone by having a greater cone angle and a longer vortex finder.

Spiral Separator. The spiral separator is usually 8 to 10 ft (2.44 to 3.05 m) in height and consists of a trough which goes downward in a spiral. The coal slurry is fed in at the top and as it follows the spiral down, centrifugal force separates the coal from the denser mineral matter. A splitter box at the bottom of the spiral can be adjusted to obtain the desired separation.

Fine Coal Cleaning

Coal below 0.5 mm size is classified as fine coal. This size coal is generally processed by surface chemical methods such as froth flotation. In the froth flotation process, the fine coal slurry, to which a small amount of flotation reagents such as fuel oil and a short chain alcohol (methyl isobutyl carbinol, MIBC) are added, is processed through a flotation cell. In the cell, fine bubbles are generated using either forced air or suction. The coal, being hydrophobic, attaches to the air bubbles and rises to the top where it is removed as froth. The refuse, being hydrophilic, remains in the water and is removed from the bottom. The process is very effective in recovering high-grade coal at moderate cost. A simplified diagram of a froth flotation cell is given in Figure 24.

The reagents used in froth flotation of coal are classified as collectors, frothers, and modifying agents. Collectors (kerosene or No. 2 fuel oil) provide a thin coating of oil on the coal and promote attachment of air bubbles to it. Frother (methyl isobutyl carbinol) or polypropylene glycols promotes a stable froth. Modifying agents have several functions such as depressing unwanted material, altering the surface of the particles to aid attachment of air bubbles, and providing acidity or alkalinity to the flotation pulp.

Recently, an advanced froth flotation technique called column flotation has been shown to be superior in obtaining a low-ash clean coal with high recovery of combustibles. It has been proven to be effective in recovering ultra-fine (minus 74 microns) coal (42,43). A schematic diagram of the column flotation process is shown in Figure 25. The basic difference between the conventional and the column flotation system is in the design, which provides quiescent flotation conditions in the column, thus effectively recovering fine coal. Also, gentle washing of froth at the top of the column removes entrained impurities from the froth, providing a low-ash clean coal product. The columns also have been effective in removing high amounts of liberated pyritic sulfur. In 1989 the University of Kentucky Center for Applied Energy Research designed four 2.4 m diameter columns which were installed at the Powell Mountain Coal Company, St. Charles, Virginia, for recovery of fine coal from refuse streams.

Recently, an air-sparged cyclone developed at the Uni-

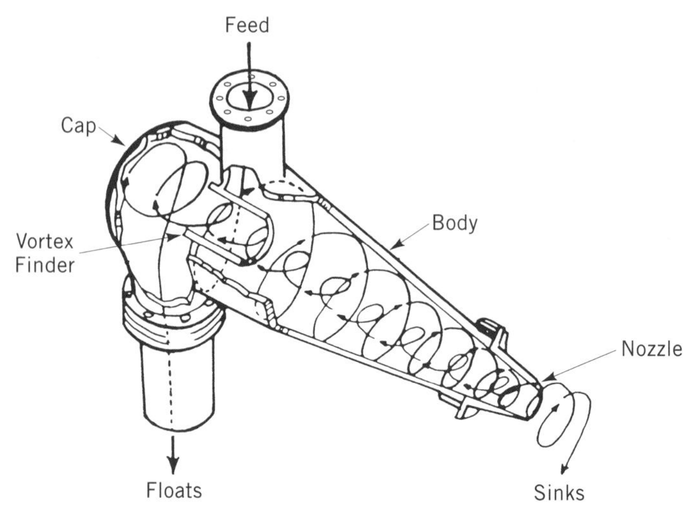

Figure 23. Heavy-medium cyclone separator.

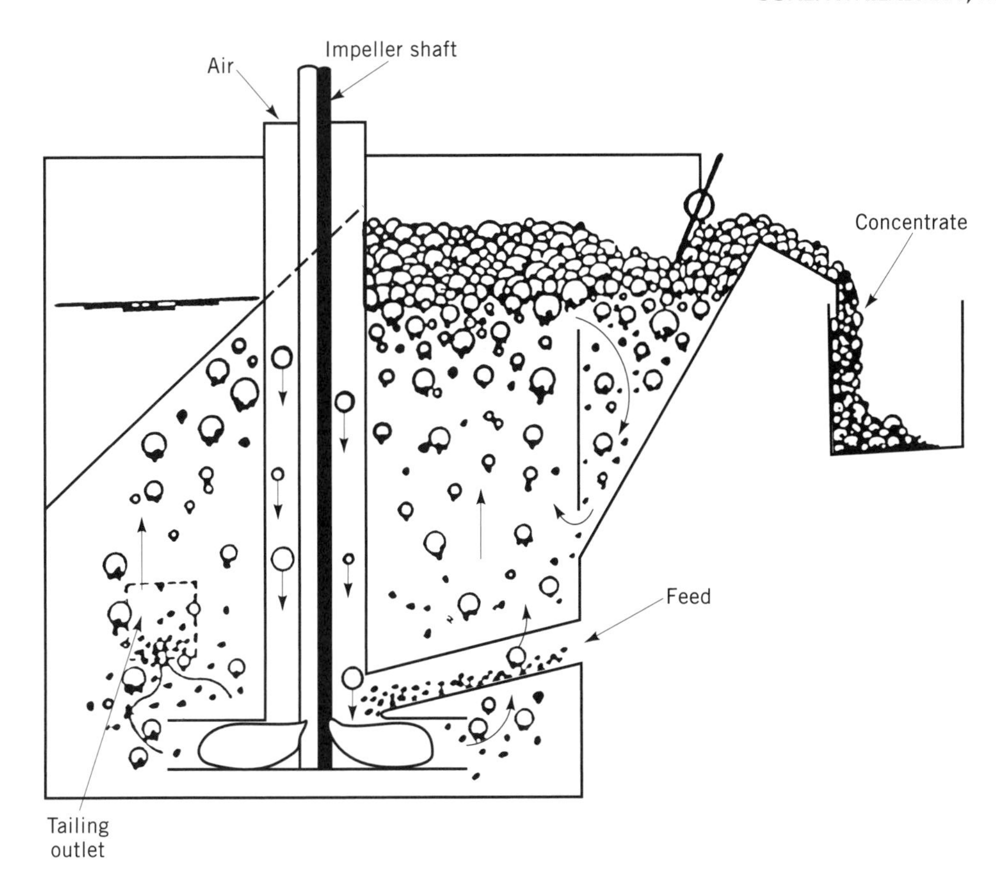

Figure 24. Froth flotation cell.

versity of Utah was tested on a pilot scale. In this technique, flotation is performed inside a cyclone that has porous walls through which air is injected.

Other Processes

An oil agglomeration process which utilizes oil or a hydrocarbon to agglomerate coal leaving mineral matter in aqueous suspension has been tested on pilot and commercial scale. The Otisca process, developed by Otisca Industries, Syracuse, New York, utilizes pentane as an agglomerant, which is recovered from clean coal and recycled back in the process. The National Research Council of Canada (NRCC) process utilizes fuel oil as an agglomerant. Another process, the high gradient magnetic and electrostatic cleaning process, utilizes differences in magnetic and electrical charge properties of the mineral matter present on coal. Both of these processes have achieved limited success.

Chemical cleaning processes utilize alkali or acid to leach the impurities present in the coal. TRW's Gravimelt process, which has been tested on a pilot scale, utilizes molten caustic to leach mineral matter. The biggest drawback of chemical cleaning process is the economics.

The chemical methods are not effective in removing major amounts of organic sulfur. Thus, removal of total sulfur to produce a clean burning fuel is very difficult. It is envisioned that the optimum desulfurization of coal will include physical, chemical, and microbial treatment. The raw coal will be ground fine to liberate most of the pyritic sulfur and then processed using the column flotation process to remove most of the liberated mineral matter. Then the clean coal will be treated chemically and finally by microbes to remove organic sulfur from the coal.

Fine Coal Drying. One of the biggest problems in processing fine-size coal is dewatering and drying. Conventional mechanical dewatering devices are effective in dewatering coarse and medium-size coal. For dewatering of fine-size coal, especially if a substantial amount is in the 74 micron range, the conventional techniques provide a dewatered product containing about 30% moisture (44).

Filtration devices that utilize high pressure forces, such as Andritz Hyperbaric and Ama's KDF filters, have the capability of reducing moisture in the filter cake to about 20%. However, the capital and operating costs are high for these devices. Other newly developed techniques which have been tested on a pilot scale include an ultra-high *g* centrifuge, which generates forces up to 4,000 *g* in the centrifuge, and an electro-acoustic technique which utilizes the synergistic effect of electric, ultrasonic, mechanical, and surface chemical forces to remove moisture from the cake.

Thermal drying is generally practiced on coarse and medium-sized coal. Fluidized bed (direct-heating) dryers are most popular in industry. Thermal drying of fine coal is generally avoided due to explosion possibilities and environmental problems.

Re-constitution of fine coal to form pellets and briquettes to improve handling of fines has not been practiced on large scale due to economic reasons. Research efforts are in progress to identify an economical binder, which can provide a low cost pellet or briquette of fine coal.

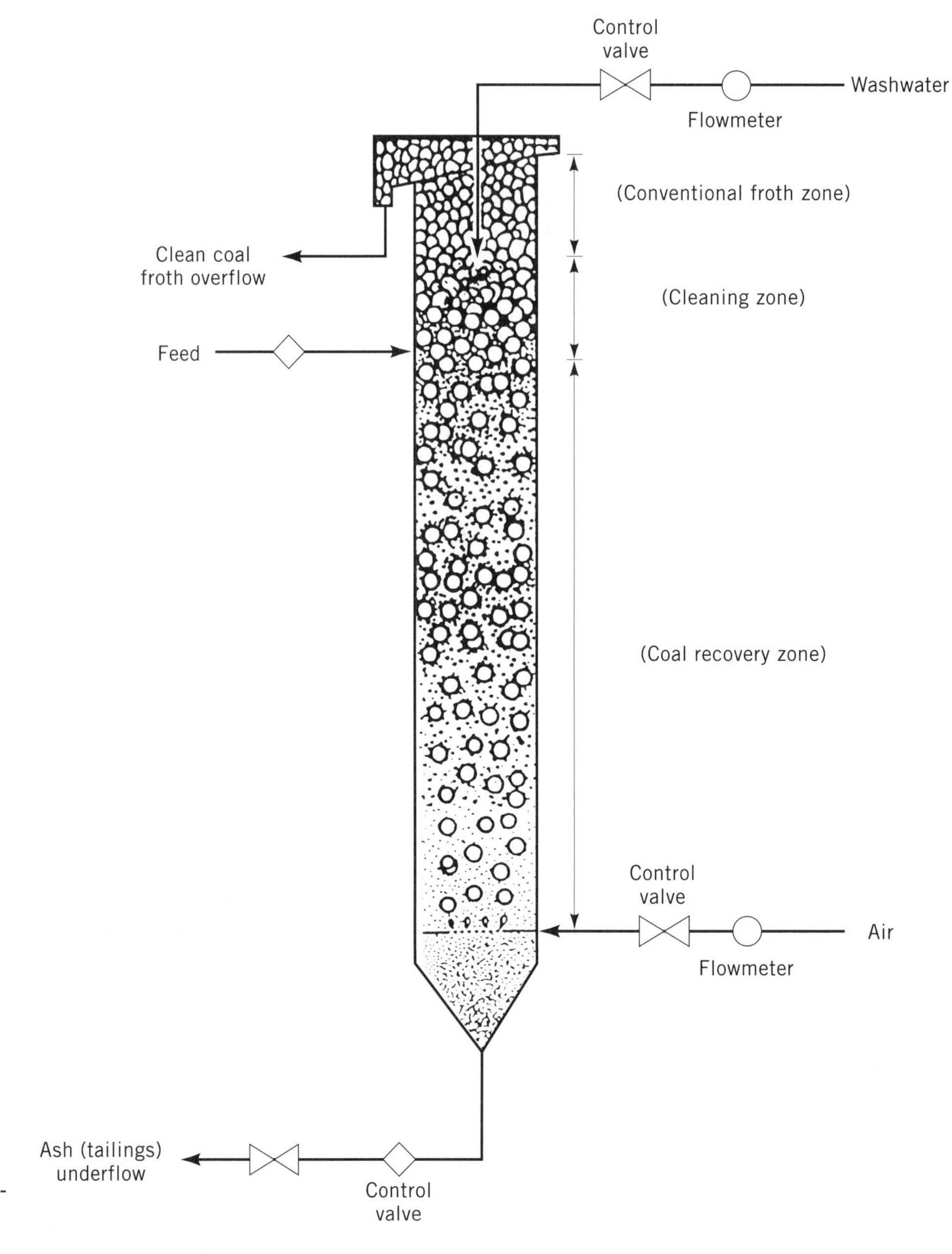

Figure 25. Schematic diagram of column flotation.

BIBLIOGRAPHY

1. International Energy Agency, *Concise Guide to World Coalfields,* IEA, London, 1983.
2. *International Energy Annual 1990,* Energy Information Administration, U.S. Department of Energy, 1992.
3. W. W. Olive, "Coal Deposits of Latin America," in F. E. Kottlowski and co-workers, *Coal Resources in the Americas,* Geological Society of America Special Paper 179, 1978.
4. O. H. Bohnenberger and G. Dengo, "Coal resources in Central America," in F. E. Kottlowski and co-workers, *Coal Resources in the Americas,* Geological Society of America Special Paper 179, 1978.
5. G. H. Wood, Jr., and W. V. Bour, III, *Coal Map of North America,* US Geological Survey Special Geologic Map, 1988, 44 p.
6. J. Ojeda-Rivera, "Main coal regions of Mexico," in F. E. Kottlowski and co-workers, *Coal Resources in the Americas,* Geological Society of America Special Paper 179, 1978.
7. E. T. Dumble, *Geol. Soc. America Bull.* **11,** 10–14 (1900).
8. E. J. Schmitz, *Trans. AIME* **13,** 388–405 (1885).
9. E. G. Tuttle, *Eng. and Mining J.* **58,** 390–392 (1894).
10. E. Ludlow, *Trans. AIME* **32,** 140–156 (1902).
11. J. G. Aguilera, *Eng. and Mining J.* **88,** 730–733 (1909).
12. G. G. Smith, *Coal Resources of Canada,* Geological Survey of Canada Paper 89-4, 1989.
13. H. H. Damberger, "Coalification in North American coal fields," in H. J. Gluskoter and co-workers, *The Geology of North America, vol. P-2, Economic Geology, U.S.,* Geological Society of America, 1991.
14. T. J. Evans, *Bituminous Coal in Texas,* Texas Bureau of Economic Geology, Handbook 4, 1974.
15. *Keystone Coal Industry Manual,* Maclean Hunter Publications, Chicago, Ill., 1992.
16. A. T. Cross, "Coals of far-western United States," in H. J. Gluskoter and co-workers, *The Geology of North America, vol. P-2, Economic Geology, U.S.,* Geological Society of America, 1991.

17. R. Choate and C. A. Johnson, "Geologic Overview, Coal and Coalbed Methane Resources of the Western Washington Region," TRW Energy Systems Group, Lakewood, Colorado.
18. R. D. Merritt, "Paleoenvironmental and Tectonic Controls in Major Coal Basins of Alaska," in P. C. Lyons and C. L. Rice, *Paleoenvironmental and Tectonic Controls in Coal-forming Basins of the United States,* Geological Society of America Special Paper 210, 1986.
19. G. D. Stricker, "Economic Alaskan coal deposits," in H. J. Gluskoter and co-workers, *The Geology of North America, vol. P-2, Economic Geology, U.S.,* Geological Society of America, 1991.
20. M. Daniel and E. Jamieson, *Coal Production Prospects in the European Community.* IEACR, 48, 1992, 78 pp.
21. World Energy Conference, *Survey of Energy Resources,* WEC, London, 1989, 190 pp.
22. G. Doyle, *Prospects for Polish and Soviet Coal Exports,* IEA Coal Research, **16,** 70 pp., 1989.
23. G. Couch, M. Hessling, A.-K. Hjalmarsson, E. Jamieson, and T. Jones, *Coal Prospects in Eastern Europe,* IEA Coal Research **31,** 1990, 117 pp.
24. Dai-Sung, Lee, *Geology of Korea,* Lyohak-Sa Publishing Co., 1988, 514 pp.
25. G. Doyle, *Coal supply prospects in Asia/Pacific region,* IEA Coal Research 26, 1990, 84pp.
26. Joo Heon Park, *Coal in Asia Pacific,* **3,** (3), 78–85 (1991).
27. Ministry of Energy and Resources, *Yearbook of Coal Statistics,* Korea Coal Industry Promotion Board, 1991, 283 pp.
28. Hiroaki Hirasawa, *Coal in Asia-Pacific,* **4**(1) 4–14 (1992).
29. M. S. Massoud, S. D. Killops, A. C. Scott, and D. P. Mattey, *J. of Petroleum Geol.* **14,** 365–386 (1991).
30. R. Takahashi, and A. Aihara, *Int. J. of Coal Geol.* (1989).
31. A. C. Scott and Mao Bangzhuo, *Geol. Today* **9,** 14–18 (1993).
32. M. Inone, *Coal in Asia-Pacific* **3**(1), 1–31 (1991).
33. G. Doyle, *China's Potential in International Coal Trade,* IEA Coal Research 2, 1987, 94 pp.
34. B. Lee, "Market trends and implication Bowen Basin coals," *Proceedings Bowen Basin Symposium 1990,* 54–73.
35. J. F. Anckhorn, M. P. Cave, J. Kenny, D. R. Lavill, and G. Carr, "The Coal Resources of New Zealand," *Coal Geology Report, 4,* Ministry of Energy, New Zealand, 1988, 43 pp.
36. Soehandojo, *Geologi Indonesia,* **12,** 279–325 (1989).
37. Hadiyanto and M. M. Faiz, *Indonesia's Coal Export Potential and its Prospects,* Second Asia/Pacific Mining Conference and Exhibition, March 14–17, 1990, Jakarta, 1990, pp. 648–663.
38. F. S. J. de Jager, "Coal," in C. B. Coetzee, *Mineral Resources of the Republic of South Africa. Handbook,* Geol. Survey, South Africa, Vol. 7, 1976, pp. 289–330.
39. R. G. Wadley, "A View on the South African Domestic Coal Market," *Abstracts Conference on South Africa's Coal Resources,* Witbank, Transvaal Branch Geological Society of South Africa, 6–9 November, 1991.
40. J. W. Leonard, ed., *Coal Preparation,* 5th ed., American Institute of Mining, Metallurgical and Petroleum Engineers, Inc., Littleton, Colorado, 1992.
41. D. G. Osborne, *Coal Preparation Technology,* Graham and Trotman, 1989.
42. R. P. Killmeyer, R. E. Hucko, and P. S. Jacobsen, "Interlaboratory Comparison of Advanced Froth Flotation Processes," Preprint No. 89-137, Society of Mining Engineers, Inc., Littleton, Colorado, 1989.
43. B. K. Parekh, J. G. Groppo, W. F. Stotts, and A. E. Bland, "Recovery of Fine Coal from Preparation Plant Refuse Using Column Flotation," K.V.S. Sastry, ed., Column Flotation '88, Society of Mining Engineers, Inc., Littleton, Colorado, 1988.
44. B. K. Parekh, and A. E. Bland, "Fine Coal and Refuse Dewatering–Present State and Future Consideration," B. M. Moudgil and B. J. Scheiner, eds., *Flocculation and Dewatering,* Engineering Foundation, New York, 1989.

COAL COMBUSTION

L. Douglas Smoot
Brigham Young University
Provo, Utah

GENERAL DESCRIPTION

Availability and Uses

Specific purposes for uses of direct coal combustion include (*1*) power generation, (*2*) industrial steam and heat production, (*3*) firing of kilns (cement, brick, etc), (*4*) coking for steel processing, and (*5*) space heating and domestic consumption. The various forms in which the coal is used also differ substantially. Coal particle diameters in different processes vary from micrometer size through centimeter size. Use is made of virginal coal, char, and coke. Pulverized coal is also slurried with water or other carriers, both for transporting and for direct combustion (1).

Coal is the world's most abundant fossil fuel. Recoverable world coal resources are estimated to be in excess of 0.9 trillion t (2). World coal production increased from 3.8 billion t in 1980 to 4.7 billion t in 1988. At this level of production, the recoverable world coal reserves should last for about 200 yr. Coal is also the most abundant fuel in the United States, where coal resources represent 90% of all known fossil energy resources (3). Recoverable reserves are estimated to be 264 billion t, with total resources far in excess of this amount. The U.S. coal production was 890 million t in 1989. At this production level, the U.S. recoverable coal reserves should last almost 300 yr (4). Coal represents a smaller fraction of the U.S. energy consumption than of its energy production, because of heavy reliance on imported oil for transportation. In 1989, coal accounted for 23% of total energy consumption, and its share is expected to increase in the future (4). Electric utilities are the largest coal-consuming sector by far and account for most of the growth in coal consumption. In 1989, 86% of coal consumption went into generating electricity, and about 55% of the electricity was produced from coal (3).

Most of the coal presently being consumed is by direct combustion of finely pulverized coal in large-scale utility furnaces for generation of electric power, and this is likely to remain the way through the end of this century. However, many other processes for the conversion of coal into other products or for the direct combustion of coal are being developed and demonstrated, including various coal combustion and gasification processes, magnetohydrodynamic generators, and fuel cells.

Increasing the use of coal presents many technical problems, particularly in protecting the environment while increasing combustion efficiency. To solve these problems and increase the use of coal, many countries in the world are supporting research and development of clean coal technologies. It is imperative for new coal technologies to reach the market in a timely manner with minimal environmental impact, and at a competitive cost. It was against this background that the International Energy Agency ministers decided in 1985 that the first area of emphasis for international collaboration in energy research, development, and demonstration should be the clean use of coal (5).

A key goal of the U.S. National Energy Strategy is to maintain coal as competitive and to establish it as clean fuel (3). The resulting U.S. Clean Coal Technology (CCT) Demonstration Program is a major effort in this direction. A multiyear effort consisting of five separate solicitations is under way. As a result of four solicitations through September 1991, the CCT program currently comprises 42 demonstration projects. Of these, 11 are in advanced electric power generation systems, 21 in high performance pollution control devices, 6 in coal processing for clean fuels, and 4 in industrial applications (6).

ENVIRONMENTAL CONTROL TECHNOLOGIES

Understanding the formation mechanisms and control methods of SO_x and NO_x pollutants is key to environmental concerns about coal combustion. These sulfur and nitrogen species, when emitted into the atmosphere, can cause acid rain. Burning of coal is a major source of these pollutants. Most coals contain about 1% nitrogen by weight, a part of which ends up as NO_x when combusted. Even low sulfur coals contain up to 1% sulfur, which forms SO_x when burned. The control of CO and hydrocarbon emissions is achieved by increasing the efficiency of combustion and, for the most part, has already been accomplished through boiler design and operating experience. The control of particulate matter has been successfully achieved using electrostatic precipitators and fabric filter bag houses (see PARTICULATE CONTROL) and thus is not often a major issue in pollution control research at this time. However, the impact of SO_2 and NO_x abatement technologies can have an adverse impact on the combustion efficiency and must be considered when applying the control applications. Control of trace metals is increasing in importance.

Nitrogen Oxides

Strategies. To comply with regulations for nitrogen oxides emissions, various abatement strategies have been developed. These strategies can be broken down into three categories: (*1*) modification of the combustion configuration, (*2*) injection of reduction agents into the flue gases, and (*3*) treatment of the flue gas by postcombustion denitrification processes. A compilation and discussion of current stationary combustion NO_x control systems, including those available by commercial manufacturers and those presently being developed by private and government research organizations, is available (7,8). A determination of the most effective and least cost abatement technology depends on specific boiler firing conditions and the mandated emission limits. A combination of technologies may be needed. Figure 1 (9) shows a composite drawing of an

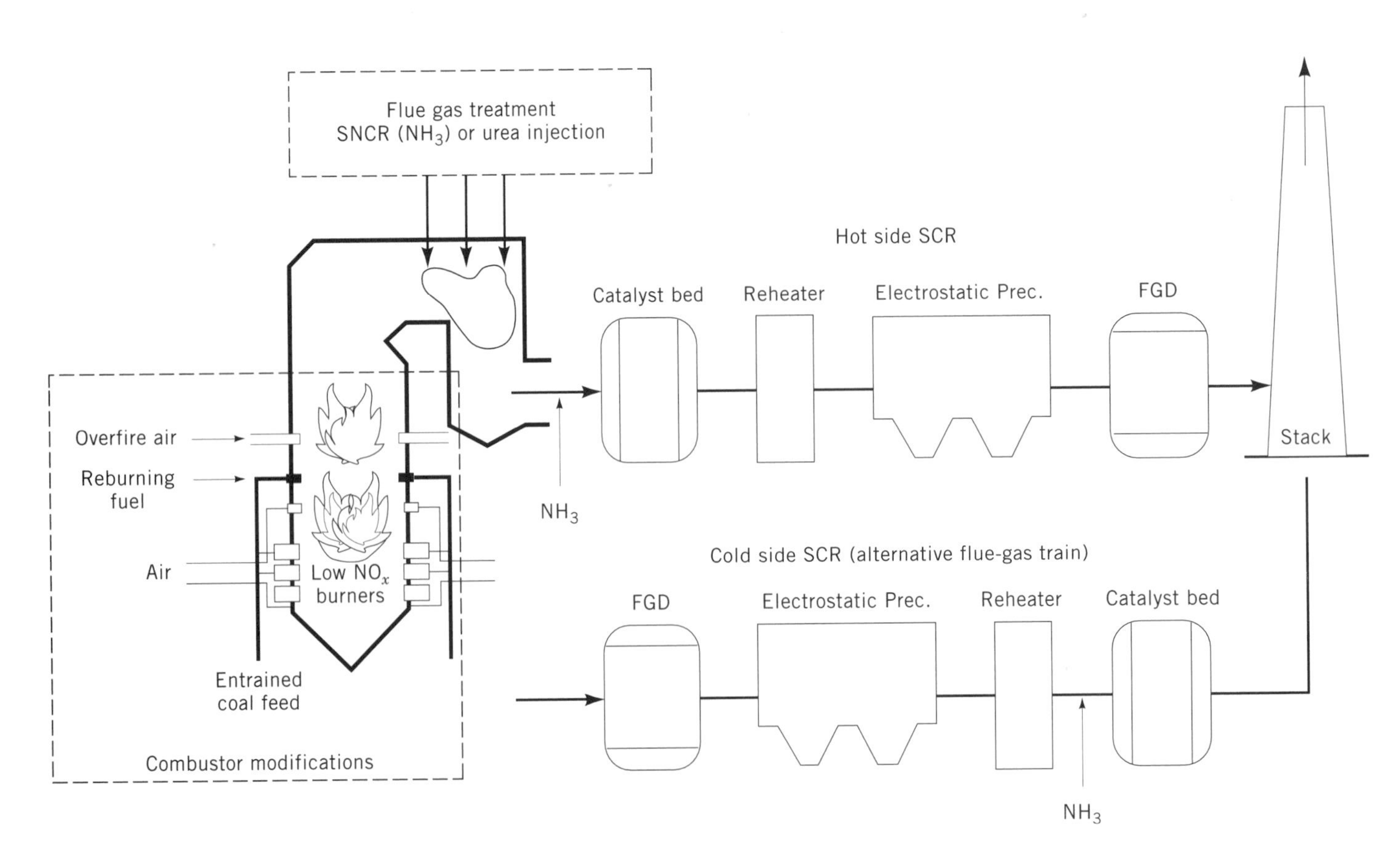

Figure 1. NO_x abatement control technologies and their point of application. From Ref. 9.

Table 1. Achievable Emissions Reduction for Selected NO_x Control Technologies for Coal Combustion[a]

Abatement Technology	Potential NO_x Reduction (%)	Advantages	Disadvantages
Low excess air	0–15	no significant capital cost, increased boiler efficiency	oxygen trim system requires maintenance
Flue gas recirculation	0–5	retrofits are usually possible, can easily be combined with other technologies, effective in reducing thermal NO	requires high temperature duct work and fans, higher capital cost than staged air or staged fuel to achieve equal or greater emissions reductions
Air staging	10–50 30 (typical)	low to moderate operating costs, can be adapted to existing boilers by taking selected burners out of service	may not be applicable to package boilers, access for secondary air ports may not be available
Fuel staging or low NO_x burners	30–75 50 (typical)	low capital costs, retrofits normally possible, low operating and maintenance costs	some increase in static pressure, may require package boilers to operate at a lower rate
Low NO_x burners and air staging or fuel staging	25–80 70 (typical)	highest reduction efficiency without resorting to flue-gas treatment technologies, retrofit of existing combustors normally possible	may require boiler to be taken offline for an extended period during retrofit, may require package boilers to operate at reduced rates
Ammonia injection	40–90 75 (typical)	lower capital costs than SCR	potential formation of fine particulates, potential NH_3 emissions, moderate to high costs, sensitive to temperature, catalyst poisoning possible
Urea of cyanuric acid injection	30–70	achieves lowest emission rate, can be retrofitted to most combustors	highest capital cost, potential for NH_3 emissions, increased pressure drop, potential for catalyst poisoning
Selective catalytic reduction (SCR)	60–95 90 (typical)		
Combined NO_x–SO_x	50–95 (NO_x) 50–95 (SO_x)	joint removal of NO_x and SO_x, often a salable byproduct is produced	high capital cost, most technologies only in development stage, solid or liquid waste disposal may be required

[a] From Ref. 9.

industrial, coal-fired boiler and illustrates the point of application of the various abatement techniques. A summary of the achievable emission rates for the various NO_x control technologies is presented in Table 1.

It is obvious that the selection of the pollution-control system must also include the impact on boiler efficiency because changes can affect the fuel oxidation process. The large variety of combustion configurations and different fuel types requires that the abatement strategy be applied on a case-by-case basis. Figure 2 illustrates potential NO_x reduction for selected technologies applied to pulverized coal furnaces (9).

Combustion Modification Approaches. The formation of nitrogen oxides strongly depends on temperature conditions and oxygen availability; both are influenced by the combustor design and firing configuration. Conventional boiler designs are classified as wall-fired, tangentially fired, cyclone, and stoker-fired boilers. Wall-fired boilers consist of rows of burner nozzles through which air and/or fuel are injected. Cell units employ a different type of wall-fired design with the burners arranged in clusters along the boiler wall. In a tangentially fired boiler, the burners are located in the corners and directed off-center to produce rotation of the bulk combustion zone gas. This en-

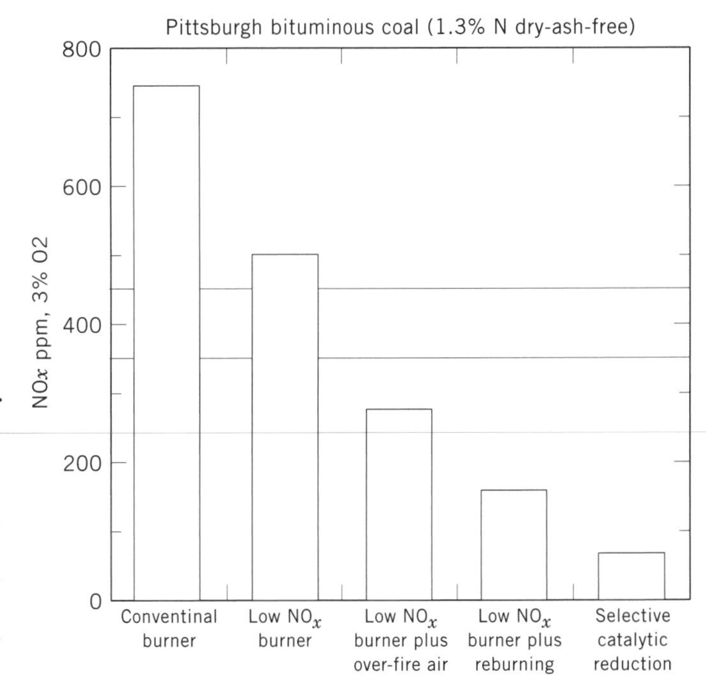

Figure 2. Potential reduction of NO_x by some low NO_x technologies for pulverized coal combustors. From Ref. 9.

sures a well-mixed condition throughout the boiler. Cyclone boilers are equipped with one or more smaller combustion chambers (cyclones) that burn the fuel at high turbulence and temperatures (1900–2150 K). They are designed to reduce fuel preparation costs (only coal crushing is required), furnace size, and fly ash in the flue gas. Stoker-fired boilers are not generally used for power generation but are used for heating and steam production by industry.

The chief objective of boiler operations is to achieve maximum conversion of the fuel to product to gain the highest possible thermodynamic efficiency. This objective can be accomplished by thoroughly mixing the oxidizing air and fuel and by increasing the excess air. In part, well-mixed conditions are achieved by the air that is used to pneumatically transport classified coal. Preheating the oxidizing air also is used in conventional designs to improve the thermal efficiency.

Unfortunately, the well-mixed high temperature and high excess air conditions that promote efficient oxidation of the fuel constituents also favor the production of nitric oxide (NO). The relative inertness and insolubility of nitric oxide makes flue gas treatment more difficult than sulfur oxide removal. Consequently, it is often easier to intervene right at the point of fuel combustion to avoid the formation of nitrogen oxides and their precursors. Modifications to conventional combustion configuration hardware can reduce nitrogen oxide emissions from 50 to 80%. They are the simplest and least expensive alternatives to apply. Modifications recognized as being technically viable in reducing nitrogen oxide emissions include low excess air, flue gas recirculation, staged air combustion, and staged fuel combustion (commonly called reburning).

Operating with lower excess air is achieved without any retrofit modifications and may reduce nitrogen oxide emissions up to 15% in some, but not all, combustors. Lowering the excess air often improves the thermal efficiency because higher excess air, while serving to reduce hydrocarbon emissions, also acts as a thermal diluent. The trade-offs in combustion efficiency and nitrogen oxide reductions must be monitored to achieve the best air: fuel ratio. Already most boilers, including pre-NSPS boilers, have implemented low excess air practices.

The purpose of flue gas recirculation is twofold: to dilute the inlet oxygen concentration and to lower the combustion zone temperature. Flue gas recirculation primarily affects thermal NO but has little influence on fuel NO. Furthermore, the high cost of equipment and operation (recirculation blowers) makes this modification generally unattractive. In fact, few facilities have retrofitted or designed boilers with flue-gas recirculation capability.

Air staging and fuel staging are effective in controlling fuel nitrogen oxide emissions. In concept, air and fuel staging can be accomplished by advanced burner design (aerodynamic staging) or by externally injecting the air or fuel at optimal locations throughout the furnace. Overfire air has been successfully implemented in wall-fired boilers to achieve nitrogen oxide reduction of up to 30%. Typically, 25% of the air is redirected from the primary combustion zone to an elevation above the fuel burners (Fig. 3) (10). Overfire air has the advantage of reducing nitrogen oxides without sacrificing boiler efficiency; however, thermal NO formation may result in the region of overfire air. Most applications of overfire air have been made to new facilities, although retrofitted applications are proposed for a number of existing facilities.

Air staging is accomplished in tangentially fired boilers by alternating fuel-rich and fuel-lean zones at different levels in the combustion zone. Another technique is to divert part of the combustion air along the boiler wall, thus establishing a concentric envelop of air rotating around a fuel-rich core, as in Figure 4. Staged fuel operation (reburning) involves injecting fuel into more than one combustion zone in the boiler. The objective is to inject part of the fuel with the bulk of the combustion air in the primary combustion zone. This produces a fuel-lean zone that reduces the formation of thermal NO by lowering the peak temperature; however, the potential for formation of fuel NO may be increased. In the secondary combustion zone where reburning fuel is added, the overall air:fuel ratio is fuel-rich. Fuel fragments are produced that react with NO, producing HCN. Subsequently, HCN favors reduction to N_2 under the prevailing fuel-rich conditions. Final air addition is then employed to burn out the unoxidized hydrocarbons at lower temperatures that do not favor thermal NO formation.

Fuel staging is particularly effective for abating fuel NO formation in oil and coal-fired boilers. The most effective reburning fuels are volatile, low nitrogen contain ing fuel oils and natural gas, although coal has been successfully applied in some pilot-scale tests. A 40–60% reduction in nitrogen oxide emissions can be achieved by reburning. One combustor that is a prime candidate for fuel staging is the cyclone combustor, because most modifications to the cyclone chambers are counterproductive.

Advanced low NO_x burners have been developed and tested by a number of boiler manufacturers and utilities (7). The general purpose of low NO_x burners is to aerodynamically stage the air or fuel. Variation in the momentum of the fuel and air streams, in the swirl of the streams, and in the angle of injection of the streams, affects the mixing and combustion pattern and retards the formation of both thermal and fuel NO. The majority of nitrogen oxides formed in coal flames occurs as the fuel nitrogen, which is released with the volatiles or contained in the char, is oxidized. Fortunately, the chemistry controlling nitrogen oxide formation is slower than the fuel oxidation rates. Thus the aim of low NO_x burners is to establish a fuel-rich zone of low temperature where the volatiles are oxidized but not the fuel nitrogen constituents. This zone is followed by gradual mixing of air into the outer fringes of the flame to burn out the char and soot. Figure 4 (11) illustrates the basic features of a low NO_x burner that is incorporated with some variations in commercially available burners.

Injection of Chemical Reductants. Under the right conditions, nitrogen oxides can be converted to nitrogen and water by injecting chemical reductants into the combustion flue gases. This technique involves selective noncatalytic homogeneous reactions with the decomposition products of the injected compounds. Ammonia is used in a well-established commercial process pioneered in the early 1970s (12). Other compounds that have been successfully tested

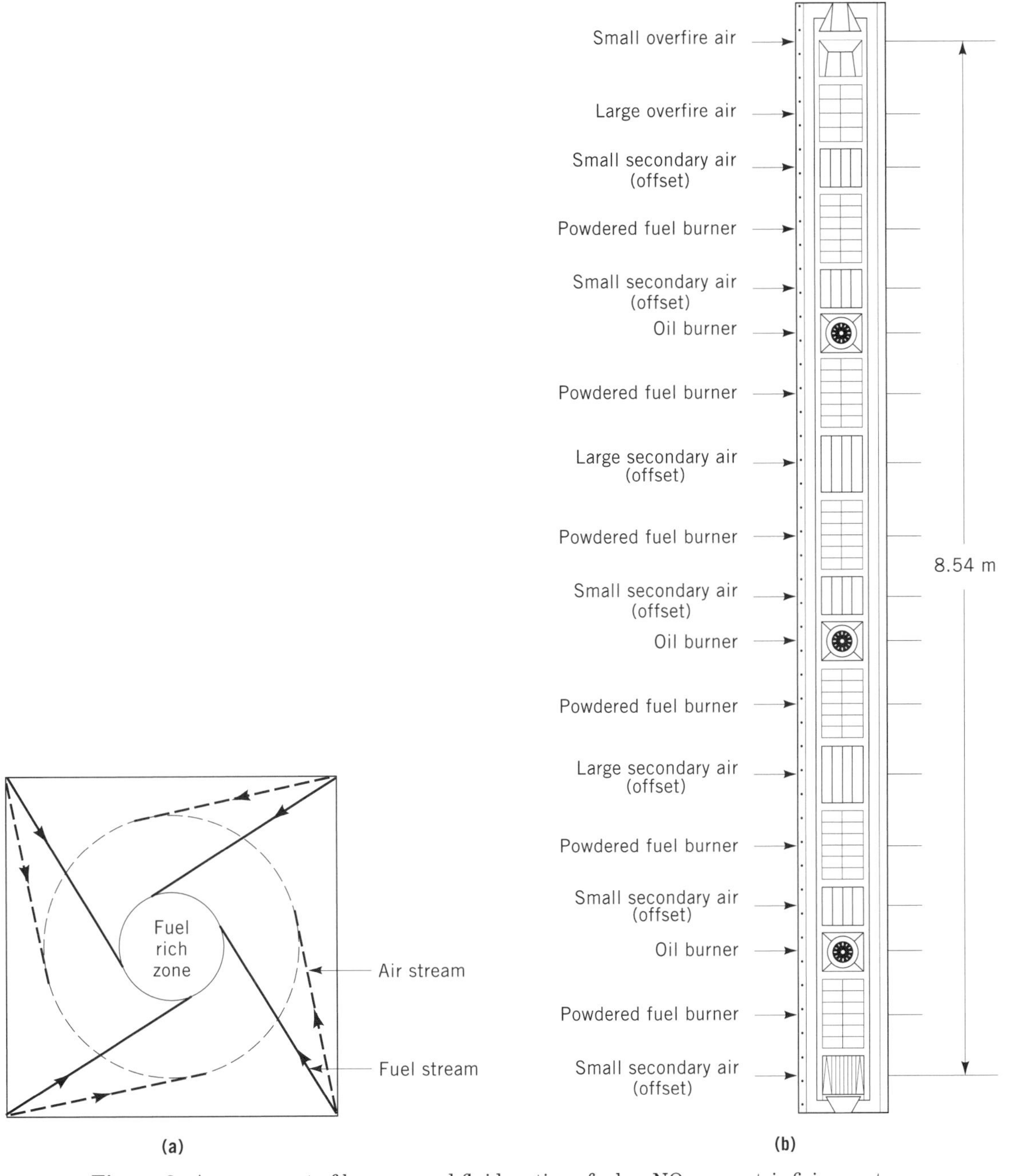

Figure 3. Arrangement of burners and fluid motion of a low NO_x concentric firing system. From Ref. 10.

include cyanuric acid, ($(HOCN)_3$), hydrazine hydrate (N_2H_4) solutions, urea ($CO(NH_2)_2$), and methanol (13,14).

Factors that determine the effectiveness of nitrogen oxide removal by reducing agents include the temperature of the flue gas, injection rate and mixing efficiency, and the composition of the flue gas. Injection jets are designed to achieve rapid and thorough dispersion of the reductant at the optimum temperature. Flue gas radicals initiate decomposition of the chemicals to nitrogen-containing intermediates that then selectively react with NO to form N_2. Thus the presence of moisture in the exhaust gas, or certain additives such as hydrogen and hydrogen peroxide, may enhance the initiation of reduction reactions. Normally, there is a narrow temperature range that is conducive to the reduction reactions. At excessive temperatures, the nitrogen intermediates will be oxidized and produce additional nitrogen oxides. At lower temperatures, undesired amounts of the injectant can slip through the process and be emitted with the nitrogen oxides. In some cases, the unreacted injectant can react with the nitrogen oxides to form undesirable particulate matter (eg, NH_4NO_3 and $(NH_4)_2SO_4$ aerosols).

Research efforts have succeeded in elucidating the chemical kinetics of ammonia decomposition and reaction with nitric oxide to the point at which the process is virtually predictable (12). This knowledge, coupled with practi-

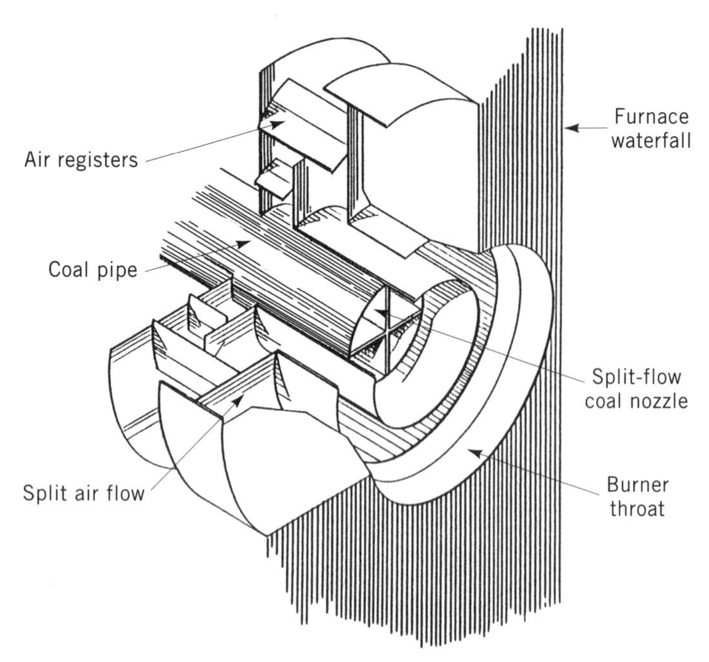

Figure 4. Schematic diagram of a basic low NO_x burner. From Ref. 11.

cal experience in a variety of commercial combustors, has allowed nitrogen oxide reduction efficiencies of around 75% to be obtained. Some retrofitted applications have achieved in excess of 90% removal of NO. The reaction sequence for NO_x reduction by ammonia is summarized in Figure 5 (12). The principal nitrogen intermediate, NH_2, either reduces NO or is competitively converted to NO. The $NH_2 + O$ reaction has a high activation energy; thus higher temperatures are required for this reaction pathway to be favorable. When the temperature is too low, the reaction of NH_3 with oxygen atoms and hydroxyl radicals is slow, allowing NH_3 to slip. The optimum temperature window for selective reduction of NO ranges from 1250 to 1400 K.

Ammonia reduction of nitric oxide is effective only when excess oxygen is available to produce oxygen atoms. In fact, the process can be inhibited by increasing the NH_3 concentration above the O_2 concentration or by increasing the moisture content of the flue gas. The presence of moisture can inhibit NO reduction at low temperatures by competing for the oxygen atom. At higher temperatures, moisture competes with the oxidation of NH_2 and inhibits NO formation. The combination of effects results in an increase of the optimum temperature. Additives such as hydrogen and hydrogen peroxide tend to lower the optimum temperature window.

The selection of cyanuric acid as a chemical reductant resulted from the recognition that HNCO could be involved in a series of reactions that would remove NO (15). When cyanuric acid is injected into the exhaust stream, it rapidly decomposes to form isocyanic acid (HNCO), which further decomposes in high temperatures to NH, NH_2, and N_2O. At lower temperatures, surface reactions are required to decompose isocyanic acid. The optimum temperature window ranges from 1000 to 1200 K (14). The detailed chemistry of urea with NO is not completely understood, but the reaction path depends on the thermal decomposition products. It is suggested that urea might decompose into NH_3 and HNCO and thus follow either or both the ammonia or isocyanic acid reaction paths.

One major disadvantage of flue-gas treatment by chemical reductants is the potential formation of byproduct nitrous oxide (N_2O). In view of the increased concern over N_2O emissions, this is an undesirable byproduct. An experimental and theoretical study recently conducted concluded that all chemical injectants, including NH_3, produce N_2O as a byproduct, depending on the amount of reductant used and the conditions at the point of injection (16). Ammonia produced the least concentrations of N_2O, while cyanuric acid produced the highest concentrations. Under some conditions, a molecule of N_2O may be formed for every molecule of NO removed by cyanuric acid (17). Nitrous oxide emissions due to urea injection were found to be between the levels produced by NH_3 and cyanuric acid.

Catalytic Reduction of Nitrogen Oxides. The most effective technology for removing nitrogen oxides from combustion gases is selective catalytic reduction (scr). Selective catalytic reduction involves introduction of ammonia before the catalyst bed (Fig. 2). The catalyst can be placed in different positions in the flue-gas flow as long as the gas temperature is appropriate for the type of catalyst used. Practical applications in Japan and Germany have obtained removal efficiencies approaching 90% (18,19). Nonselective catalytic reduction, in which hydrogen or hydrocarbons in the flue gas act as reducing agents, has also been researched but has not achieved the same level of success as scr.

The basic scr catalyst types that have been researched and are commercially available have been discussed in detail (7,19). These include noble metals (Pt, Pd, Ru/AL_2O_3, and Pt/Al_2O_3), metal oxides (Fe_2O_3/Cr_2O_3, V_2O_5/TiO_2, V_2O_5/

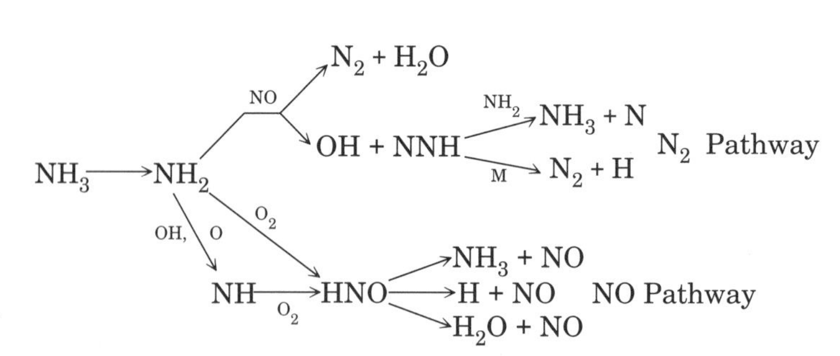

Figure 5. Mechanism of nitric oxide reduction in the flue gas by ammonia injection. From Ref. 12.

$MoO_3/WO_3/Al_2O_3$) and zeolite (synthetic mordenites). Each catalyst exhibits a unique optimum operating temperature and NH_3:NO_x ratio. The optimum operating temperature ranges from 500 to 700 K. At low temperatures, the reaction efficiency drops off, while the catalyst is prone to sintering at excessive temperatures. Typically, the catalyst bed will be placed upstream of the elctrostatic precipitator or bag house to take advantage of the hot flue gas. However, a problem exists for fouling and plugging in the catalyst bed by fly-ash particulates. Flue-gas products such as SO_3 can also combine with ammonia to form corrosive ammonium bisulfate $(NH_4)_2SO_4$.

Most metal-based catalysts are subject to a decrease in activity during operation. Catalyst poisons that deactivate the reaction sites include alkaline and alkaline-earth oxides or arsenic compounds. Sulfur oxides can also deactivate certain metal oxide catalysts and can reversibly inhibit noble metal catalysts. A catalyst activity loss of 20%/yr is considered normal for most combustion cleanup applications.

Selective catalytic reduction requires the highest capital and operating costs, despite many recent technological advances in catalyst support geometries and ammonia injection control. However, because of more stringent standards, scr may become the most viable technology for nitrogen oxide removal.

Sulfur Pollutants

Alternatives. Unlike nitrogen oxide pollutants, there are no benign gaseous species to which sulfur-containing pollutants can be converted. Limestone slurry scrubbers, although costly, have been effective in removing more than 90% of the sulfur oxides from the flue gas. A technical description on flue gas desulfurization systems has been published (20). Advanced combustion configurations, including the slagging combustor, fluidized-bed combustors, and the integrated gasification combined cycle can also achieve high sulfur pollutant reduction levels, while intermediate stream desulfurization of effluents is a promising technology for gasification processes. However, to reduce current sulfur oxide emissions, control technologies that can be retrofitted to existing facilities are being developed. The injection of sorbents directly into the post-combustion chamber is one option that can reduce SO_2 emissions from new and existing combustor configurations.

During the previous two decades, several desulfurization alternatives have been investigated in the laboratory and on a pilot plant scale. A few commercial applications have been demonstrated. New generation desulfurization can generally be categorized into four groups: (*1*) precombustion coal cleaning, (*2*) advanced combustion configurations, (*3*) *in situ* sulfur reduction, and (*4*) postcombustion cleaning. A fundamental discussion of many of the alternatives is available (21).

Precombustion Coal Cleaning to Remove Sulfur and Minerals. Environmental constraints, together with development of coal-based fuels for oil substitution, have increased interest in the cleaning of coal (22,23). Reduction in mineral and sulfur content of coals is of particular concern.

A certain amount of the sulfur can be removed from the coal before combustion by physically, chemically, or biologically cleaning the coal. Most coal receives some cleaning before it is burned. Physical cleaning typically separates inert matter from the coal by relying on differences in density or surface properties; however, it is not possible to remove physically the organically bound sulfur. Physical cleaning can remove 30–50% of the pyritic sulfur, which is 10–30% of the total sulfur. Advanced coal-cleaning methods are expected to remove even a greater amount of pyritic sulfur. Through chemical and biological cleaning, some of the sulfur that is chemically bound to the coal can also be removed. Chemical or biological cleaning is capable of removing 90% of the total organic and inorganic sulfur, but it is expensive. The predominate method of chemical coal cleaning is to leach both the organic sulfur and mineral matter from the coal using a hot sodium-based or potassium-based caustic. Biological cleaning involves exposing the coal to bacteria or fungi that have an affinity for sulfur. Often, sulfur-digesting enzymes are injected into the coal to stimulate sulfur consumption.

Ultrafine grinding (micronization) of the coal to an average particle diameter of about 10 μm separates much of the mineral matter from the organic coal. Subsequent separation of the coal and mineral matter by differences in specific gravity have led to coals with mineral matter levels of less than 1% and with sulfur levels about 50% the original value. Reduction in sulfur level by physical methods depends strongly on the relative proportions of sulfur in the mineral matter (pyritic and sulfates) and that bound in the organic structure. Work is continuing on several physical and chemical coal-cleaning processes (22,23) in an effort to demonstrate economically viable processes (1).

In Situ Sulfur Capture. Sulfur pollutants can be efficiently removed during the combustion process or in the flue-gas ducts by injecting sorbents into the postcombustion gases or by adding sorbents directly to the furnace or combustor. A variety of sorbents will absorb and react with sulfur pollutants, including calcium and magnesium oxides and metal oxides of zinc, iron, and titanium. The calcium content of coal ash provides some natural sulfur capture capability, which is limited by the Ca:S ratio of the coal (24), and the addition of other sorbents is required for substantial *in situ* sulfur capture. Calcium-based sorbents are particularly attractive due to their low cost and the inertness of the calcium sulfate or calcium sulfide byproduct. The reacted sorbents are typically collected with the fly ash by the bag house or electrostatic precipitator.

Sulfur oxide reduction by sorbents has been demonstrated in a number of pilot-scale facilities. Calcium-based sorbent–SO_2 capture has been characterized in a 380-kW, two-burner level, tangentially fired pilot-scale test (25). At the nominal Ca:S ratio of 2.1, the overall calcium use was 24%. Limestone injection tests conducted in a 440-kW pilot-scale combustor achieved SO_2 reductions of 40–60% for a range of test parameters. Conversions in excess of 50% appear possible for hydrated lime (26).

Fluidized-bed combustors commonly use crushed limestone for bed material. The particles, introduced into the bed, heat and calcine. The calcium oxide then reacts with sulfur dioxide to form calcium sulfate. Some coal, coal ash, and limestone particles are thrown out of the bed in a freeboard zone. The freeboard zone provides additional space and time to increase sulfur dioxide capture. The substantial reduction of SO_x during combustion (about 90%) reduces the need for postcombustion SO_x control processes (9).

Flue Gas Desulfurization. The most widely used technology for SO_x control is flue gas desulfurization (fgd). It is a highly efficient but expensive technology. More than 100 fgd technologies are commercially available or under development throughout the world (27). They are usually classified as wet throwaway, dry throwaway, and regenerative fgd. By far the most common fgd technology is the wet throwaway, or wet scrubbing. The largest user of fgd in terms of installed capacity is the United States, with more than 80,000 MW, followed by Japan, with about 45,000 MW, and Germany, with about 40,000 MW (27). The fgd systems are required on all new coal-fired power plants in the United States. The first wet scrubbers used lime as a sorbent and produced waste that required disposal. Most new wet scrubbers use limestone and produce gypsum, which can be marketed. In the wet scrubber, the flue gas passes through an absorber tower at low velocity while being sprayed by the limestone slurry.

Combined NO_x–SO_x Combustion Gas Treatment Technologies

Numerous processes also have been proposed in recent years for combined denitrification and desulfurization of combustion flue gases. Some are commercially available and have been implemented by full-scale combustion facilities (7,28). Most of the NO_x–SO_x removal processes are considered expensive and may only be considered as more stringent standards become mandatory for new and existing facilities. Combined NO_x–SO_x flue-gas treatment processes are generally classified as wet or dry. Wet processes normally use absorption with complexation, reduction, or oxidation chemistry; the dry processes include catalytic decomposition, absorption by activated carbon beds, spray drying, and electron-beam irradiation.

One innovative, NO_x–SO_x postcombustion concept involves streamer corona discharge to produce radical species in the flue gas. The chemistry proposed by various researchers is based on the interaction of electrons with molecular species, giving rise to atomic oxygen, nitrogen, and amine radicals. In the presence of ammonia and water, the radicals lead to the reduction of nitrogen oxides and SO_2, proceeding through reactions included in the ammonia–NO_x reduction mechanism and other ionic and free-radical reactions. However, in the absence of ammonia and water, NO_x emissions can be further exacerbated. The reactions leading to removal of SO_2 and NO by streamer corona discharge have been discussed in detail (29). Another scheme proposes denitrification of the flue gas using regenerable mixed oxide sorbents (eg, Na_2O, Co_3O_4, and Fe_2O_3). This process involves the sorption of nitrogen oxides onto porous sorbents that can be thermally regenerated on saturation. The concentrated gas stream is then treated by standard catalytic reduction methods. An in-depth description on the removal of nitric oxide from flue gas using mixed oxide sorbents has been published (30).

Demonstrations

Practical NO_x control experience has been obtained in pilot-scale and utility boilers over the past score of years. The majority of this experience has been reported in the *Proceedings of the Joint Symposium on Stationary Combustion NO_x Control* sponsored by the EPA and EPRI since 1979. Researchers from the international community have discussed the implementation of combustion modifications, injection of chemical reductants, and selective catalytic reduction. The bulk of scr experience has been provided by Japan. Other references include publications by the International Energy Agency (IEA), which is currently sponsored by 14 nations (Austria, Australia, Belgium, Canada, Denmark, Germany, Finland, Italy, Japan, The Netherlands, Spain, Sweden, the UK, and the United States). Most of these countries have strong governmental support for NO_x and SO_x abatement demonstration projects.

The U.S. CCT program has a goal to furnish showcase examples of advanced, more efficient, and environmentally viable coal-utilization alternatives. CCT demonstration projects include NO_x and SO_x control concepts that are applicable to conventional coal-fired power plants (eg, flue-gas desulfurization, limestone injection, gas reburning, and coal cleaning) and also nontraditional methods of coal combustion (eg, pressurized fluidized-bed combustion and integrated gasification-combined gas cycle (IGCC)).

Several CCT projects are being carried out to demonstrate sorbent capture of sulfur in entrained-flow and fluidized-bed combustors and also sorbent injection into the flue-gas duct (5). One project involves joint gas reburning and sorbent injection in a 71-MW, tangentially fired boiler burning a 3 wt % sulfur Illinois coal (31). The upper limit to SO_2 reduction using calcium-based sorbents appears to be 50–60%. With popular technology termed limestone injection multistage burner (LIMB) combined with lime injection in the flue-gas duct, reduction of 50–80% SO_2 is projected for a 104-MW wall-fired boiler. Other sorbent–SO_2-capture demonstration projects include injection into an integrated gasification combined cycle, a slagging combustor, and fluidized-bed combustors (32). The latter application is enhanced when the fluidized bed is pressurized (9).

Acknowledgment

Craig Eatough, postdoctoral associate in the Advanced Combustion Engineering Research Center, has contributed substantially to this section. Predrag Radulovic's and Richard Boardman's recently published work on coal combustion technologies and pollutant emissions and control, respectively, provided important contributions to this article. Support of the Advanced Combustion Engineering Research Center at Brigham Young University in preparing this manuscript is appreciated.

BIBLIOGRAPHY

1. L. D. Smoot, in W. Bartok and A. F. Sarofim, eds., *Fossil Fuel Combustion: A Science Source Book,* John Wiley & Sons, Inc., New York, 1991.

2. U.S. Department of Energy, *International Energy Annual 1988, Report* DOE/EIA-0219(88), Energy Information Administration, Washington, D.C., 1989.
3. U.S. Department of Energy, *"Coal," National Energy Strategy: Powerful Ideas for America, 1991–92,* Report DOE/S-0082P, U.S. Government Printing Office, Washington, D.C., 1991.
4. U.S. Department of Energy, *Annual Energy Outlook 1991 with Projections to 2010,* Report DOE/EIA-0383 (91), Energy Information Administration, Washington, D.C., 1991.
5. U.S. Department of Energy, *Clean Coal Technology Demonstration Program: Program Update 1990 (as of December 31, 1990),* Report DOE/FE-0219P, Washington, D.C., 1991.
6. P. T. Radulovic and L. D. Smoot, in L. D. Smoot, ed., *Fundamentals of Coal Combustion: For Clean and Efficient Use,* Elsevier, New York, 1993.
7. A.-K. Hjalmarsson and H. N. Soud, *Systems for Controlling NO_x from Coal Combustion,* IEA Coal Research, London, 1990.
8. A.-K. Hjalmarsson, *NO_x Control Technologies for Coal Combustion,* IEACR/24, IEA Coal Research, London, Dec. 1989.
9. R. D. Boardman and L. D. Smoot, in *Fundamentals of Coal Combustion for Clean and Efficient Use,* Elsevier, New York, 1993, Chapt. 6.
10. J. Redman, *Chem. Eng.,* **458,** 35 (Mar. 1989).
11. T. Moore, *EPRI J.* **26,** 26 (Nov. 1984).
12. R. K. Lyon, *Environ. Sci. Technol.* **21,** 231–235 (1987).
13. R. A. Perry and D. I. Siebers, *Nature* **324,** 657–658 (1986).
14. F. A. Caton and D. I. Siebers, *Combust. Sci. Technol.* **65,** 227 (1989).
15. J. A. Miller and G. A. Fisk, *C&E News* **22** (Aug. 1987).
16. L. J. Muzio, T. A. Montgomery, G. C. Quartucy, J. A. Cole, and J. C. Kramlich, paper presented at the EPRI/EPA Joint Symposium on Stationary Combustion NO_x Control, Washington, D.C., 1991.
17. J. A. Miller and C. T. Bowman, *Progr. Energy Combust. Sci.* **15,** 287–338 (1989).
18. K. M. Bently and Jelinek, *TAPPI J.* **72,** 123–130 (Apr. 1989).
19. F. P. Boer, L. L. Hegedus, T. R. Gouker, and K. P. Zak, *Chem. Tech.* **20,** 312 (1990).
20. J. C. Klingspor and D. Cope, *FGD Handbook: Flue Gas Desulphurization Systems,* ICEAS/B5, IEA Coal Research, London, May 1987.
21. Y. A. Attia, ed., *Proceedings of the First International Conference on Processing and Utilization of High Sulfur Coals,* Columbus, Ohio, 1985.
22. S. P. N. Singh, J. C. Moyers, and K. R. Carr, *Coal Beneficiation—The Cinderella Technology,* Coal Combustion Applications Working Group, Oak Ridge National Laboratory, Oak Ridge, Tenn., 1982.
23. Y. Liu, ed., *Physical Cleaning of Coal,* Marcel Dekker, Inc., New York, 1982.
24. M. Angleys and P. Lucat, paper presented at the 2nd International Symposium on Coal Combustion Science and Technology, Beijing, China, 1991.
25. W. Bartok, B. A. Folsom, T. M. Somer, J. C. Opatrny, E. Mecchia, R. T. Keen, T. J. May, and M. S. Krueger, paper presented at the Symposium on Stationary Combustion NO_x Control, Washington, D.C., 1991.
26. W. Bartok, B. A. Folsom, and F. R. Kurzynske, paper presented at the EPRI/EPA Joint Symposium on Stationary Combustion NO_x Control, New Orleans, La., 1987.
27. D. Thompson, T. D. Brown, and J. M. Beér in *Proceedings of the 14th Symposium (International) on Combustion,* Combustion Institute, Pittsburgh, Pa., 1973, pp. 787–799.
28. A. A. Siddigi and J. W. Tenini, *Hydrocarbon Proc* **60,** 115–124 (Oct. 1981).
29. J. S. Chang, *J. Aerosol Sci.* **20** (1989).
30. J. Gorst, D. D. Do., and N. J. Desai, *Fuel Process Technol.* **22,** 2 (1989).
31. R. J. Martin, J. T. Kelley, S. Ohmine, and E. K. Chu, *J. Eng. Gas Turbines Power* **110,** 111–116 (1988).
32. J. L. Vernon and H. N. Soud, *FGD Installations on Coal-Fired Plants,* IEACR/22, IEA Coal Research, London, Apr. 1990.

COALBED GAS

DUDLEY D. RICE
BEN E. LAW
JERRY L. CLAYTON
U.S. Geological Survey
Denver, Colorado

In addition to minable reserves, coal is a major source of hydrocarbons, particularly natural gas (see NATURAL GAS). The gas is a product of the coalification process. The presence of methane-rich gas in coal has been recognized for a long time because of dangerous explosions and outbursts associated with gassy underground mining. Only recently has coal been recognized as a reservoir rock in addition to being a source rock for gas, thus representing an enormous undeveloped "unconventional" energy resource.

Major resources of coalbed gas are associated with the immense amounts of coal. Worldwide estimates of coalbed gas are estimated to range from 85 to 262 $\times$ $10^{12}m^3$; the range of values indicates the scarcity of basic coal data in many coal-bearing areas of the world (1). Although the resource is widely distributed, most is concentrated in 12 countries where coal is commonly an important source of energy (1,2). The principal U.S. coal-bearing basins are estimated to have coalbed gas resources of about 11 $\times$ $10^{12}m^3$ of which 3 $\times$ $10^{12}m^3$ is expected to be recoverable (3,4). The largest part of the resource is in Rocky Mountain basins where the coal beds are of Cretaceous and Tertiary age; lesser amounts are in the eastern and central United States where the coal beds are of Paleozoic age.

Significant exploration for and production of coalbed gas in the United States began in the mid-1980s because of a federal tax credit given for the production of coalbed gas. The tax credit not only increased activity, but also was responsible for the development of new drilling and completion technology.

The knowledge and technology gained from exploration and production in the United States is currently being transferred to other countries where there has been no significant coalbed gas production because of economic, logistical, political, and geological reasons. However, considerable exploration activity is currently taking place in Australia, Canada, China, Czech Republic, the former Soviet Union, Poland, South Africa, Spain, and the United Kingdom as they become aware of the enormous resource potential.

GENERATION OF COALBED GAS

Generation of natural gas in coal beds takes place by two distinct processes: biogenic and thermogenic. Biogenic gas is composed primarily of methane and is the end result of the degradation of organic matter resulting from a complex series of processes by a diverse population of microorganisms. The major requirements for the generation of significant amounts of biogenic gas by methanogenic microorganisms, which are assigned to the Archaea domain (5), are: anoxic environment, low sulfate concentrations, low temperatures, abundant organic matter, high pH values, and adequate space (6,7). If these conditions are met, economically significant amounts of biogenic gas can be generated over a period of tens of thousands of years after burial. Although biogenic gas can be generated by two pathways—carbon dioxide reduction and methyl-type fermentation—most ancient accumulations have probably resulted from the former as indicated by molecular and isotopic composition (7).

Biogenic gas can be generated during two different stages in coal beds (8). Biogenic gas is formed early in the burial history of low rank coals (peat to subbituminous rank—vitrinite reflectance (R_O) values <0.5%). This biogenic gas is referred to as early stage and its generation and accumulation are favored by rapid sedimentation. In addition, biogenic gas can be generated in coal beds in recent geologic time (tens of thousands to a few million years ago) where there is groundwater flow in an aquifer creating a favorable environment for microbial activity. This type of biogenic gas is referred to as late stage and its generation can take place in coal beds of any rank provided that the requirements outlined previously have been met.

With increasing degree of coalification resulting from higher temperatures and pressures, coals become enriched in carbon as large amounts of volatile matter rich in hydrogen and oxygen are released. Methane and carbon dioxide (Fig. 1), and water are the most important products of this devolatilization (9). This methane is thermogenic and its generation begins at a rank of high-volatile bituminous (R_O values >0.6%). The quantities of these main volatile products can be estimated using the major elemental composition (C-H-O) of the coal which will vary for the different types of kerogen (10). Four types of kerogen (I-IV) are recognized using principal elemental analysis and a van Krevelen diagram (Fig. 2). These four types of kerogen generally correlate with the major maceral groups identified by petrography—liptinite, vitrinite, and inertinite. Liptinite-rich coals generally correspond to types I and II kerogen, vitrinite-rich coals to type III kerogen, and inertinite-rich coals to type IV kerogen.

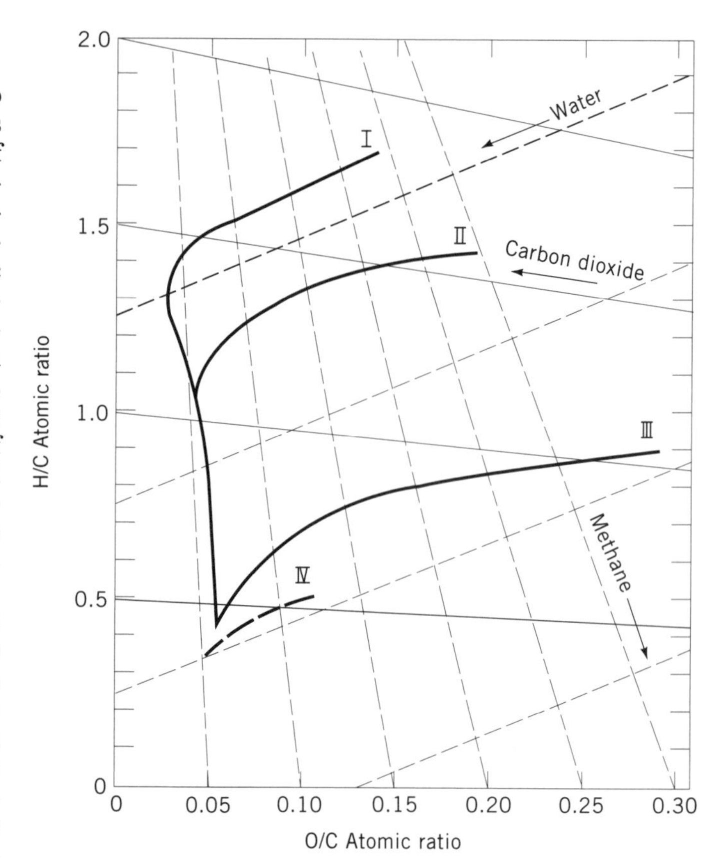

Figure 2. Van Krevelen diagram showing types of kerogen (I, II, III, IV) and pathways (solid and dashed lines) for generation of methane and elimination of carbon dioxide and water during coalification. Liptinite corresponds to types I and II kerogen, vitrinite to type III, and inertinite to type IV (12).

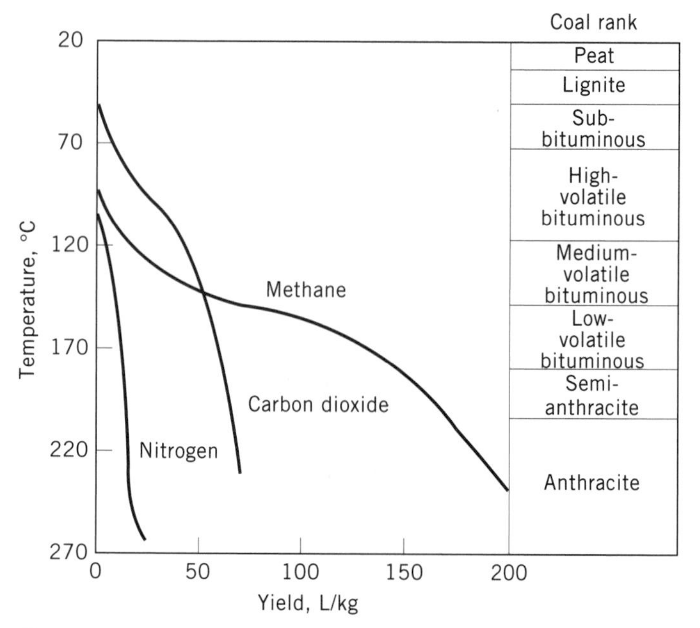

Figure 1. Calculated amounts of gases generated from coal during coalification (11).

In addition to methane, some coals are capable of generating wet gas and oil. Generation of these heavier hydrocarbons takes place in coals having significant amounts of hydrogen-rich components, such as liptinite or certain components of the vitrinite maceral, such as humic gel (13–16).

CHARACTERISTICS OF COALBED GAS

Compositional data for coalbed gas is obtained from desorption tests of coal samples from underground mines and exploration boreholes or from production of coalbed reservoirs. Compositional data presented in this article are for gas samples collected from coal beds. Analyses of coalbed gas samples are available from Australia, Canada, China, Germany, Poland, and the United States. The associated coals range in age from Pennsylvanian (Upper Carboniferous) to Tertiary and in rank from lignite to anthracite (R_O values of 0.3 to 4.9%) (8).

The molecular and isotopic composition of coalbed gases are highly variable (8). Methane is usually the major component with other hydrocarbon gases and carbon dioxide occurring in lesser amounts.

There is a tendency for all of the gases to be methane rich at low and high ranks, and for some to be wet at intermediate rank; however, many samples are methane rich at intermediate ranks. Methane $\partial^{13}C$ values generally are isotopically lighter at lower ranks and isotopically heavier at higher rank. However, gases from coal of a given rank display a wide range of methane $\partial^{13}C$ values.

Although wetness and methane $\partial^{13}C$ values of coalbed gases are scattered when plotted against rank of the associated coal, they illustrate a more systematic trend when plotted against present-day depth of burial (8). Shallow coalbed gases are composed of isotopically light methane, as compared with deeper coalbed gases, regardless of rank. The depth-related change in molecular and isotopic composition of coalbed gas can take place over a transition zone, can be abrupt, and can take place at varying depths, but usually in the range of a kilometer of the surface.

The reported coalbed gases are interpreted to be both biogenic and thermogenic based on molecular and isotopic composition (8). The primary controls on the composition of the gas are considered to be rank, composition, and depth/temperature of associated coal. Secondary controls are also reflected in the coalbed gas composition and they will be discussed later.

Biogenic gas generation is restricted to shallow depths and low temperatures, and can occur in coals of all ranks. Biogenic gas can generally be distinguished by its isotopic composition; methane $\partial^{13}C$ values are generally in the range of -55 to $-90‰$.

In comparison, thermogenic coalbed gases are characterized by (1) common presence of heavier hydrocarbons at intermediate ranks of high-volatile and medium-volatile bituminous coal, and (2) enrichment of heavy isotope ^{13}C in methane with increasing rank (methane $\partial^{13}C$ values commonly more positive than $-55‰$). At intermediate ranks, high-volatile and medium-volatile bituminous, sapropelic coals (liptinite and hydrogen-rich vitrinite) contain both wet gas and liquids, whereas humic coals (mostly vitrinite) contain a drier gas. These liquids contain high pristane : phytane ratios, a feature that is typical for coal-derived oils, and a slight predominance of odd-carbon numbered n-alkanes and in the C_{25+} fraction (17). However, the saturated fraction consists predominantly of low-molecular-weight components (C_4–C_{10}), which results in the liquids having high API gravities ($\approx 50°$). In addition, the coal-derived liquids are commonly characterized by high pour points because they contain abundant long-chain n-alkanes derived from terrestrial plant waxes. At high ranks, coalbed gases are mostly methane generated from residual kerogen and cracking of previously formed heavier hydrocarbons.

Thermogenic methane is isotopically more positive than biogenic methane because of the smaller kinetic isotope effects associated with thermal cracking (18,19). In general, methane $\partial^{13}C$ values for thermogenic methane become isotopically heavier with increasing rank because residual gas-producing carbon becomes enriched in ^{13}C due to $^{12}C—^{12}C$ bonds being broken more frequently by thermal processes than $^{12}C—^{13}C$ bonds. Also, coal composition, in addition to rank, influences the isotopic composition of the coalbed gas, whereby gases generated from coals with oxygen-rich kerogen (mostly vitrinite) (1) have more positive methane $\partial^{13}C$ values than those generated from hydrogen-rich kerogen (liptinite and hydrogen-rich vitrinite) at similar levels of thermal maturity, and (2) exhibit less spread in methane and ethane $\partial^{13}C$ values (19,20). These differences result from isotopically lighter methane being produced by thermal cracking of aliphatic-type structures, which are prevalent in hydrogen-rich kerogen, whereas isotopically heavier methane results from cracking of aromatic structures which are predominant in oxygen-rich kerogen.

Secondary processes also affect the composition of coalbed gases, particularly at shallow depths, such that isotopically light methane is dominant in shallow coalbed gases, regardless of rank (9). The depth interval where these secondary processes are active and affect the composition of the gases is referred to as the zone of alteration. Unaltered gas generally occurs in the deeper parts of basins and is referred to as original gas.

At shallow depths, coal beds are commonly aquifers where microorganisms can thrive. Microbial activity can affect the observed composition of coalbed gases in two ways. First, significant amounts of late-stage, isotopically light biogenic methane generated by anaerobic microorganisms can be mixed with previously generated thermogenic gas or fill degassed coal beds (8). Second, aerobic bacteria can preferentially destroy most of the wet gas components resulting in a methane-rich gas (21).

The zone of alteration, in which mixing and oxidation affect the coalbed gas composition, can extend from depths of a hundred meters to a kilometer below the surface. The zone is usually restricted to the margins of basins, however it can extend into and throughout a basin under certain conditions. The primary controls of the depth and lateral extent of the zone of alteration are the physical nature of the coal beds, burial history, and hydrology.

In addition to hydrocarbon gases, carbon dioxide is the other significant component of coalbed gases. The carbon dioxide presently in a coalbed gas may have an origin other than devolitilization. In addition, the concentration and isotopic composition ($\partial^{13}C$) of carbon dioxide in coalbed gases is commonly different from those in gases of adjacent reservoirs, suggesting different origins. Documented sources of large amounts of carbon dioxide in coalbed gas are (1) thermal destruction of carbonates, (2) bacterial degradation of organic matter, (3) bacterial hydrocarbon oxidation, and (4) migration from magma chambers or deep crust (8).

GAS STORAGE

One of the most important characteristics of coalbed gas reservoirs that set them apart from more conventional gas reservoirs is the amount of and manner in which gas is stored. In conventional reservoirs, such as sandstones or carbonates, gas occurs as either a free or dissolved phase. While some free and dissolved gas may occur in the coal, as much as 98% of the gas is sorbed in the coal (22). The free and dissolved gas in the coal occurs within the fractures (cleats) and pores, whereas the sorbed gas occurs as a monomolecular layer on the internal surfaces of coal (23,24).

In general, gas content increases with increasing rank, although there is a wide range of gas contents within each coal rank (Fig. 3). The large increase in gas content between the ranks of sub-bituminous A and high-volatile C bituminous is mainly due to the larger amounts of methane generated by thermal processes associated with the coalification of high-rank coals relative to the smaller volumes of methane generated by only microbial processes at lower ranks. At higher ranks (medium-volatile bituminous and higher), coals may have generated more methane than they can store resulting in possible expulsion into adjacent reservoirs (10,25).

Pressure and temperature also play an important role in the gas content of coal. For a given rank, sorbed gas content increases with increasing pressure and decreases with increasing temperature (10). Because high pressures and temperatures are more commonly associated with high-rank coals than low-rank coals, relatively high gas contents are generally expected in high rank coals. In some cases, however, natural desorption of gas facilitated by uplift and erosion in the subsurface environment may result in unexpected low gas contents, thereby accounting for some of the wide variation of gas content within a given coal rank (Fig. 3).

Regardless of rank, nearly all coals contain at least some water. For practical purposes, moisture content of coal is defined as the water present as H_2O and released at temperatures of 104° to 110°C (26). Water in coal occurs as free water, water of decomposition, and water of hydration.

The amount of inorganic material and the maceral composition of the coal are additional factors affecting the gas sorption capacity of coal. There is no affinity for gas to sorb onto the surfaces of inorganic matter. Inorganic matter occupies space that otherwise would be filled with organic matter, thereby reducing the surface area available for gas sorption (27). As a consequence, coals of equal rank with relatively high ash content will not contain as much gas as low-ash coals. The sorption capacity is increased again at higher ranks (low-volatile bituminous and higher) when these heavier hydrocarbons are thermally cracked.

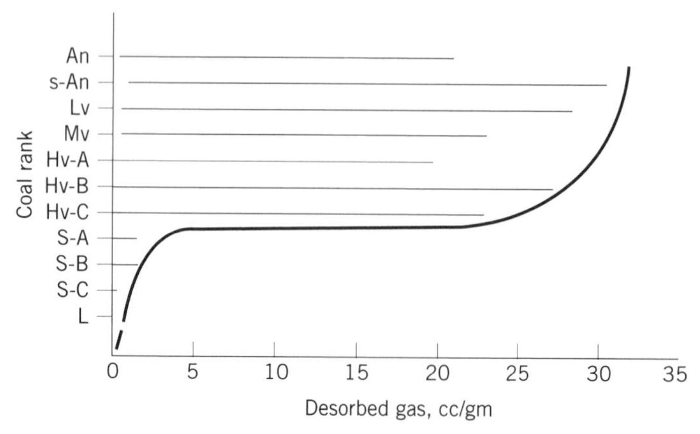

Figure 3. Cross-plot showing range of gas contents versus coal rank. Data from measurements of desorbed coalbed gas in the United States. From R. R. Charpentier and B. E. Law, 1992.

DETERMINATION OF GAS CONTENT

The most common method of determining the volume of gas contained in coal is by direct measurement from core or drill-cutting samples retrieved from a well during drilling. The direct method has three components of measurement—desorbed gas, residual gas, and lost gas.

The sorption capacity of a coal is commonly determined from a sorption isotherm, which provides the sorbed gas content as a function of the free gas pressure at a constant temperature (28). The sorption isotherm is a measure of the maximum amount of gas that a coal can sorb. However, in many cases, coals are undersaturated with respect to gas and the sorption isotherm, in conjunction with direct measurement of gas content, can be used to determine if a coal is gas saturated. Common reasons for the undersaturation of gas are miscalculation of gas content, laboratory errors in measuring sorption isotherm, degassing of coal beds resulting from uplift and erosion, and differential generation of coalbed hydrocarbons resulting from different coal composition.

GAS PRODUCTION

Economic quantities of gas can be produced from coalbed reservoirs. The diagrammatic sketch in Figure 4 shows the path that methane molecules, and other gases, must take in order to be produced.

The development of permeability is a critical aspect of gas production from coalbed reservoirs. Commonly, permeability in coalbed reservoirs is in the range of 1.0 to 10.0 md. For practical purposes, there is essentially no permeability within the coal matrix. Virtually all of the permeability in coal occurs within the fracture system, commonly referred to as the cleat system. Cleats constitute a roughly orthogonal set of fractures referred to as face and butt cleats. Face cleats are the dominant, more extensive set and butt cleats are the subordinate set. Due to the better development of face cleats, permeability in coal beds commonly exhibits varying degrees of anisotropy, with the better permeability developed parallel to the face cleat direction (30). The factors that affect the permeability of cleats include cleat frequency, connectivity, and aperture width. These factors are, in turn, affected by bed thickness, coal quality, rank, tectonic deformation, and stress (31–33). The preservation of cleats is largely dependent on the severity of the structural history. The economic production of gas in structurally complex regions may be severely hindered, even though the coals may contain large amounts of gas.

One of the largest obstacles to the economic recovery of coalbed gas is water. The presence of water in coal inhibits

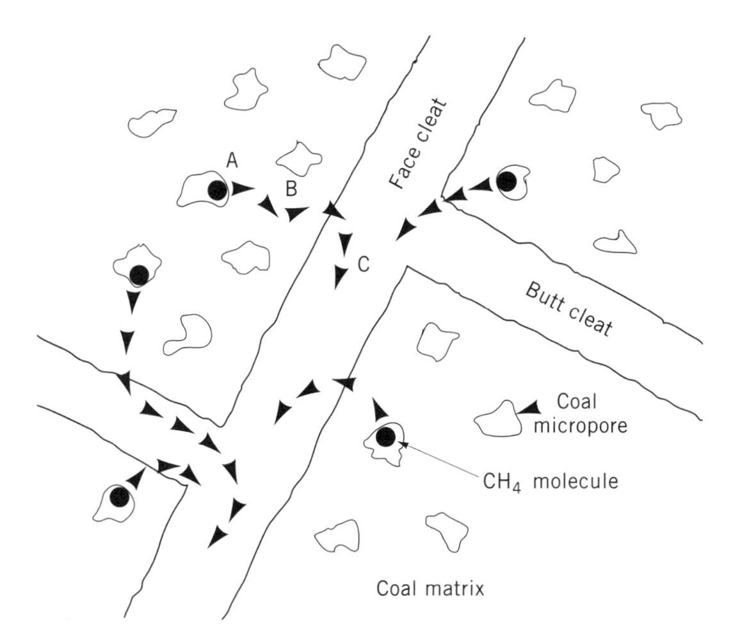

Figure 4. Diagram showing (A) desorption of methane from micropore in coal matrix, (B) diffusion path of methane through coal matrix, and (C) flow of methane in cleats (29).

desorption of gas from the coal matrix and flow to the wellbore, particularly in coals undersaturated with gas (Figure 4). Coal beds are commonly dewatered by drilling several wells and producing water by pumping to a point at which gas begins to desorb from the matrix.

Sources of water in coal may include inherent moisture, water of meteoric origin, and water from adjacent aquifers.

In areas where water is the pressuring phase, dewatering simultaneously reduces the reservoir pressure and the water content, thereby allowing gas to desorb from the coal. In areas where the coalbed reservoir is discontinuous and/or has low permeability and the supply of water is limited, the probability of conducting successful dewatering programs is favorable.

As a consequence of the gas storage characteristics of coal and the relation between water and gas production, coalbed gas wells exhibit a distinctive production history (Fig. 5).

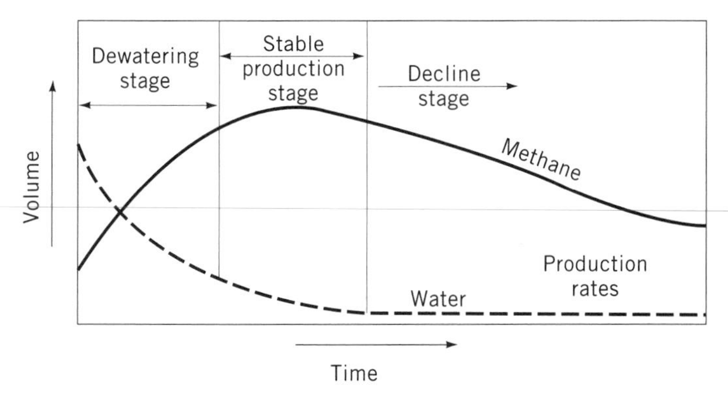

Figure 5. Generalized production history showing volumes of methane and water produced over time for a typical coalbed gas well (34).

In nearly all cases, coalbed gas wells require special drilling or completion techniques to effectively connect the reservoir to the wellbore. Both vertical and horizontal wells have been utilized, although vertically drilled wells are far more common than horizontally drilled wells, except in some of the coal mining areas. The completion techniques range from hydraulic fracturing of cased holes to open holes, some with enlarged cavities (35). The completion technique in each area must be tailored to the localized coalbed reservoir characteristics, such as rank, pressure, permeability, and gas content.

ENVIRONMENTAL CONCERNS

Methane Emissions From Coal Mining

Methane is a strong infrared absorber and important greenhouse gas. According to recent estimates, atmospheric methane accounts for approximately 15% of the "radiative forcing" added to the atmosphere during the past decade (36). Warming associated with methane emissions is realized in the first few decades after its emissions, whereas the warming from carbon dioxide takes place gradually over centuries. The short residence time indicates that reductions in methane emissions will have noticeable short-term impacts on atmospheric concentrations and that smaller reductions will be needed to stabilize atmospheric concentrations.

Currently, the concentration of methane in the atmosphere is about 1,700 parts per billion (ppb) (37). Evidence from analysis of gases trapped in ice cores indicates that atmospheric methane has more than doubled over the past 200–300 years and has increased at a rate of about 1% per year over the past 15 years.

This increase in atmospheric methane is correlative with human population growth and human activities. The principal anthropogenic sources of methane are rice paddies, domestic livestock, landfills, biomass burning, venting of natural gas, losses of natural gas during transmission of gas and petroleum, and coal mining (36). These human activities account for about 60% of the total global emissions that are estimated to be in the range of 440 to 640 Tg per year. The methane reductions required to stabilize atmospheric concentrations by the year 2000 are on the order of 25 to 50 Tg (36). These reductions are much lower than those required to stabilize other greenhouse gases, such as carbon dioxide and nitrous oxide.

Pressure reduction during coal mining results from the removal of overburden and dewatering. This pressure reduction leads to the desorption of methane, dominant in coalbed gas at shallow depths, and possible emissions to the atmosphere. In general, more methane is released from underground mines than surface mines because of the relation between methane storage, depth, and pressure. Methane emissions also continue during the transport, processing, and storage of the coal. This methane release is not only an environmental concern, but also a serious safety hazard because methane is highly explosive at relatively low concentrations of 5 to 15%. In the United States and some other countries, methane levels in the mines are kept at very low levels by installation of high-grade ventilation systems.

Annual emissions of methane from coal mining have been estimated to be about 30 to 50 Tg (36). However, currently available data are insufficient to satisfactorily assess the relative source strength of coalbed methane; current global estimates contain only "order of magnitude" levels of reliability.
Because of increasing energy demand, particularly from the developing countries and from the growth in the world's population, coal production is forecast to grow substantially over the next decade and coal mining could become an even greater source of atmospheric methane.

Coalbed gas can be recovered before, during, or after mining and the quality and quantity of the recovered methane will depend on the method used (38). Sometimes, methane emissions can be controlled using ventilation alone, but commonly this method has to be supplemented. Other recovery methods include: vertical wells; gob wells which remove gas released from rocks above and below the coal bed during and after mining; and horizontal and cross-measure boreholes (39). Because large amounts of air are mixed with methane during ventilation, this gas mixture is useful only for combustion air in a gas turbine or coal-fired boiler. Typical uses for gases recovered by wells and boreholes, which generally have higher percentages of methane, are pipeline injection and power generation.

Methane release related to coal mining can also cause safety hazards in nearby areas.

Water Disposal

Water, commonly in large quantities, is produced from coalbed gas wells, especially during the early stages of gas production. The disposal of this produced water cannot only affect the economics of the development of the coalbed gas, but also pose serious environmental concerns. In the United States, water disposal is controlled by Federal, State, and/or local agencies. The disposal of produced waters will also be an environmental and economic concern in other countries as they develop their coalbed gas resources.

The most economic methods of water disposal from coalbed gas reservoirs are discharge into streams, direct use for agricultural, wildlife, or industrial purposes, and evaporation. In addition, produced coalbed waters can be directly used in oil and gas fields as hydraulic fracture fluids and for enhanced oil recovery projects. More costly alternatives of disposal are underground injection and surface discharge after treatment, by methods such as distillation, dilution, ion-exchange, reverse osmosis, and electrodialysis. The specific method of disposal will depend on the quality and quantity of the water.

The quantity and quality of water in coal-bearing basins vary considerably and reflect the water's residence time and rock-water interactions in the subsurface and groundwater flow patterns. Total dissolved solids (TDS) of produced coalbed water range from less than 200 mg/L to more than 90,000 mg/L with most values less than 30,000 mg/L (40,41). TDS and salinity tend to increase with the water's residence time in the subsurface and its increased time for rock-water interactions. If enough chemical data are available, chlorinity maps and different water types can be used to map groundwater flow patterns (40,41).

Production rates of water from a well can be as high as 950 m^3 per day. However, most commercial coalbed gas wells produce less than 40 m^3 per day after the initial dewatering phase.

Aquifer Contamination by Coalbed Gas

In the San Juan basin, the upsurge of coalbed gas development and production has been partially contemporaneous with the contamination of an alluvial aquifer by natural gas in a populated area along a river valley (42). Based on water analyses (measurements of organic vapors), pressure testing of gas-producing wells, and molecular and isotopic composition of gas samples, some of the near-surface gas was determined to be coalbed gas which migrated vertically from a deeper interval (about 1000 meters). The coalbed gas has been interpreted to have been released by depressurization after the drilling of gas-producing wells, some completed in reservoirs below the coal beds, and the dewatering of coalbed-gas wells. The coalbed gas then migrated upward from the coal bed into the overlying strata via the uncemented portions of producing gas wells that penetrate the gas-bearing coal beds. The gas also escaped through leaks in the casing. Once inside the alluvial aquifer, the gas sometimes invaded the cathodic protection wells and the domestic water wells.

The State regulatory agency discovered that many gas wells were not cemented through the shallow alluvial aquifer and/or opposite the gas-bearing coal beds. To prevent further migration, the State and Federal regulatory agencies ordered operators to apply casing cement to all existing and new gas wells that penetrate the coal beds. In addition, they have continued to conduct pressure tests on producing wells in areas where gas wells or cathodic protection wells are blowing out. The San Juan basin is the first basin where large-scale development of coalbed gas has taken place subsequent to gas production from other types of reservoirs. Similar aquifer contamination problems might happen in other areas under similar circumstances.

Environmental Impact Studies

As demonstrated by experiences in the United States, mainly in the Black Warrior and San Juan basins, large-scale development of coalbed gas resources can strain a government and its regulatory agencies to adequately address the full range of environmental concerns. Multdisciplinary studies of all anticipated environmental concerns should be incorporated into a development plan prior to any development of the coalbed gas resource, and monitored and modified throughout the various stages of development of the resource.

ENVIRONMENTAL IMPACTS OF COAL MINING AND COMBUSTION VERSUS COALBED GAS DEVELOPMENT AND UTILIZATION

The environmental impacts of the mining and combustion of coal, as documented in Poland (43), appear to be more serious than those associated with the development and utilization of coalbed gas. First of all, coal mining results in large amounts of solid waste, only a small part of which is commonly utilized. In addition, the following water resource problems are associated with coal mining: (1) subsidence above and near mine areas can change local hydraulic gradients and drainage basin limits, and create numerous ponds; (2) drainage of fresh water aquifers can lower the water table and result in dry water wells; (3) leaching of coal waste piles can contribute to chlorides, sulfates, and heavy metals; and (4) highly mineralized mine waters can be discharged to streams. The combustion of coal commonly results in emissions of sulfur dioxide, nitrogen oxides, volatile organic compounds, such as benzene and toluene, carbon dioxide, particulate matter, and heavy metals.

In comparison, the main environmental concerns about the development of coalbed gas are the disposal of large amounts of production water. When burned, coalbed gas emits none of the sulfur dioxide or particulates associated with coal burning, and much less nitrogen oxides, carbon dioxide, and volatile organic compounds (44).

BIBLIOGRAPHY

1. V. A. Kuuskraa, C. M. Boyer II, and J. A. Kelafant, *Oil and Gas Journal* **90**(40), 49–54 (1992).
2. J. R. Kelafant, S. H. Stevens, and C. M. Boyer, II, *Oil and Gas Journal* **90**(44), 80–85 (1992).
3. C. T. Rightmire, G. E. Eddy, and J. N. Kirr, eds., *Coalbed Methane Resources of the United States*, American Association of Petroleum Geologists Studies in Geology No. 17, Tulsa, Okla.,1984.
4. ICF Resources, *Quarterly Review of Methane from Coal Seams Technology* **7,** 10–28 (1990).
5. C. R. Woese, O. Kandler, and M. L. Wheelis, *Proceedings of the National Academy of Science (U.S.)* **87,** 4576–4579 (1990).
6. D. D. Rice and G. E. Claypool, *American Association of Petroleum Geologists Bulletin* **65,** 5–25 (1981).
7. D. D. Rice, "Controls, Habitat, and Resource Potential of Ancient Biogenic Gas," in R. Vially, ed., *Bacterial Gas*, Editions Technip, Paris, 1992, pp. 91–118.
8. D. D. Rice, "Composition and Origins of Coalbed Gas," in B. E. Law and D. D. Rice, eds., *Hydrocarbons from Coal*, American Association of Petroleum Geologists *Studies in Geology*, no. 38, Tulsa, Okla., 1993, pp. 159–184.
9. E. Stach, M.-Th., Mackowsky, M. Teichmuller, G. H. Taylor, D. Chandra, and R. Teichmuller, *Stach's Textbook of Coal Petrology*, 3rd ed., Gebruder Borntraeger, Berlin, 1982.
10. H. Juntgen and J. Karwell, *Erdol Kohle, Erdgas, Petrochem* **19,** 251–258, 339–344 (1966).
11. J. M. Hunt, *Petroleum Geochemistry and Geology*, W. H. Freeman, San Francisco, Calif., 1979.
12. J. R. Levine, "Influence of Coal Composition on the Generation and Retention of Coalbed Natural Gas" in *The 1987 Coalbed Methane Symposium Proceedings*, 1987, pp. 15–18.
13. P. Bertrand, F. Behar, and B. Durand, *Organic Geochemistry* **1–3,** 601–608 (1986).
14. J. D. Saxby and M. Shibaoka, *Applied Geochemistry* **1,** 25–36 (1986).
15. G. K. Khorasani, "Oil-Prone Coals in the Walloon Coal Measures, Surat Basin, Australia," in A. C. Scott, ed., *Coal and Coal-Bearing Sequences-Recent Advances*, Geological Society Special Publication 32, London, 1987.
16. R. W. Stanton, D. D. Rice, J. L. Clayton, and R. M. Flores, "Matrix-Gel Vitrinite Types and Rock-Eval Analysis of Coal Samples, Cretaceous age, from the San Juan and Piceance Basins," in *Ninth Annual Meeting of the Society for Organic Petrography*, Abstracts and Agenda, 1992, pp. 57–58.
17. J. L. Clayton, "Composition of Crude Oils Generated and Expelled from Coal and Coaly Organic Matter Dispersed in Shales," in Ref. 8, pp. 185–201.
18. H. M. Chung, J. R. Gormly, and R. M. Squires, *Chemical Geology* **71,** 97–103 (1988).
19. C. Clayton, *Marine and Petroleum Geology* **8,** 232–240 (1991).
20. A. T. James, *American Association of Petroleum Geologists Bulletin* **74,** 1441–1458 (1990).
21. A. T. James and B. J. Burns, *American Association of Petroleum Geologists Bulletin* **68,** 957–960 (1984).
22. I. Gray, "Reservoir Engineering in Coal Seams," in *Society of Petroleum Engineers Preprints*, SPE Paper 12514, 1987.
23. R. F. Selden, *The Occurrence of Gases in Coal*, U.S. Bureau of Mines Report of Investigations 3233, 1934.
24. A. G. Kim, *Estimating Methane Content of Bituminous Coal from Adsorption Data*, U.S. Bureau of Mines Report of Investigations 8245, 1977.
25. F. F. Meissner, "Cretaceous and Lower Tertiary Coals as Sources for Gas Accumulations in the Rocky Mountain Area," in J. Woodward, F. F. Meissner, and J. L. Clayton, eds., *Hydrocarbon Source Rocks of the Greater Rocky Mountain Region*, Rocky Mountain Association of Geologists, Denver, 1984, pp. 401–431.
26. American Society for Testing Material, "Moisture in the Analysis Sample of Coal and Coke", *ASTM Standards*, Part 26, 1973.
27. J. Gunther, *Revue de l'Industrie Minerale* **47,** 693–708 (1965).
28. M. J. Mavor, L. B. Owen, and T. J. Pratt, "Measurement and Evaluation of Coal Sorption Isotherm Data," in *Society of Petroleum Engineers Preprints*, 1990, SPE Paper 20728.
29. W. Diamond in Ref. 8, pp. 237–267.
30. C. M. McCulloch, M. Deul, and P. W. Jeran, *Cleat in Bituminous Coal*, U.S. Bureau of Mines Report of Investigations 7910, 1974.
31. J. C. Macrae and W. Lawson, *Transactions of Leeds Geological Association*, **VI,** 1954, pp. 227–242.
32. I. I. Ammosov and I. V. Eremin, *Fracturing in Coal*, IZDAT Publishers, Moscow 1963. Translated from Russian by Israel Program for Scientific Translations, Jerusalem.
33. C. R. McKee, A. C. Bumb, and R. A. Koenig, "Stress-Dependent Permeability and Porosity of Coal," in *The 1987 Coalbed Methane Symposium Proceedings*, 1987, pp. 195–205.
34. V. A. Kruuskraa and C. F. Brandenberg, "Coalbed methane sparks a new energy industry," *Oil & Gas Journal,* **87**(41), fig. 5, 53 (1989).
35. I. D. Palmer, S. W. Lambert, and J. L. Spitler, "Coalbed Methane Well Completions and Stimulations," in Ref. 9, pp. 303–339.
36. Intergovernmental Panel on Climate Control, *Methane Emissions and Opportunities for Control*, EPA/400/9-90/007, Japan Environmental Agency and U.S. Environmental Protection Agency, Washington, D.C., 1990.
37. D. R. Blake and F. S. Rowland, *Science* **259,** 1129–1131 (1988).

38. M. Duel and A. G. Kim, eds., *Methane Control Research: Summary of Results, 1964–1980*, U.S. Bureau of Mines Bulletin 687, 1988.
39. C. M. Boyer II, J. R. Kelafant, V. A. Kuuskraa, K. C. Manger, and D. Kruger, *Methane Emissions from Coal Mining: Issues and Opportunities for Reduction*, U.S. Environmental Protection Agency, Washington, D.C., 1990, EPA/400/9-90/008.
40. J. C. Pashin, *Regional Analysis of the Black Creek-Cobb Coalbed-Methane Target Interval, Black Warrior Basin, Alabama*, Geological Survey of Alabama Bulletin 145, 1991.
41. W. R. Kaiser, T. E. Swartz, and G. J. Hawkins, "Hydrology of the Fruitland Formation, San Juan Basin," in W. B. Ayers, Jr. and co-workers, *Geologic and Hydrologic Controls on the Occurrence and Producibility of Coalbed Methane, Fruitland Formation, San Juan Basin,* Gas Research Institute, Chicago, Ill., 1991, Topical Report GRI-91/0072.
42. C. Shuey, "Policy and Regulatory Implications of Coal-Bed Methane Development in the San Juan Basin, New Mexico and Colorado," in *Proceedings of the First International Symposium on Oil and Gas Exploration and Production Waste Management Practices*, 1990, pp. 757–769.
43. R. C. Pilcher, C. J. Bibler, R. Glickert, L. Machesky, J. M. Williams, D. W. Kruger, and S. Schweitzer, *Assessment of the Potential for Economic Development and Utilization of Coalbed Methane in Poland*, U.S. Environmental Protection Agency, Washington, D.C., 1991, EPA/400/1-91/032.
44. *Oil & Gas Journal* **89**(29), 31–32(1991).

COLD FUSION

R. M. Shaubach
N. J. Gernert
Thermacore, Inc.
Lancaster, Pennsylvania

Two scientists at the University of Utah, Stanley Pons and Martin Fleischmann, announced on March 23, 1989, that they had carried out fusion through the electrolysis of water. The process was quickly dubbed "cold fusion" (1).

The announcement set off a flurry of activity directed at explaining and harnessing this apparent new source of energy. Over the next year, the world's scientific community expended significant funds trying to confirm the discovery at Utah. Cold fusion became the most controversial scientific event in decades. Work in the area has almost ceased, but a few groups continue to study the technology; at best, it remains a laboratory curiosity.

A theory consistent with the limited experimental observations is presented here. If the technology is proven to be feasible, the potential for generating large quantities of heat would exist. However, conversion of this heat to useful energy may present difficulties; its economic potential is not at all clear (ca. 1994).

See also Fusion energy; Magnetohydrodynamics.

THEORY

Many theories have been offered to explain the physics and chemistry behind the excess energy. Some, like Pons and Fleischmann's theory, include a nuclear fusion reaction, others include a chemical reaction. Experimental results to date show nuclear by-products orders of magnitude below those required to explain the excess energy. As a result, the authors believe a chemical rather than a nuclear explanation best fits existing observations in nature and experiments. A theory developed by Mills (2) provides a focal point for an explanation. However, this theory is controversial and is not widely accepted by the world's scientific community.

The theory starts with the hydrogen atom. Hydrogen is selected for ease of discussion; however, the theory applies to deuterium as well. The hydrogen atom has a single proton surrounded by an electron. The electron shell around the proton behaves in analogous terms much like an elastic balloon.

Energy is required to inflate a balloon; energy is stored in the expanded balloon. Similarly, the electron shell of a normal hydrogen atom can be made to expand by the absorption of energy. This is a common phenomenon known to scientists as photon absorption. When the stored energy of a balloon is released, the balloon contracts. Similarly, the electron of a normal hydrogen atom can be induced to contract below its normal ground state and release some of its stored energy. This is not a common phenomenon, and it runs counter to the usual model of the atom. The excess heat observed during electrolysis is derived from the energy released as hydrogen atoms contract.

The process has been stated by Mills as follows: "The predominant source of heat of the phenomenon denoted Cold Fusion is the electrocatalytically induced reaction whereby hydrogen atoms undergo transitions to quantized energy levels of lower energy than the conventional 'ground state.' These lower energy states correspond to fractional quantum numbers. The hydrogen electron transition requires the presence of an energy hole of approximately 27.21 eV provided by electrocatalytic reactant(s) (such as Pd^{2+}/Li^{+}, Ti^{2+}, or K^{+}/K^{+}), and results in 'shrunken atoms' analogous to muonic atoms. In the case of deuterium, fusion reactions of shrunken atoms yielding predominantly tritium are possible."

According to quantum mechanics, hydrogen can only have energies given by the following Rydberg formula:

$$E = \frac{-13.6\,eV}{n^2} \tag{1}$$

where n is an integer. Mills predicts that in addition to the energy states of hydrogen given by Eq. 1 with n as an integer, new lower energy states are possible given Eq. 1 with n as a fraction.

The hydrogen atoms that have achieved energy levels below the ground state are smaller in diameter than normal hydrogen. Mills has named these shrunken atoms hydrinos, Latin for baby hydrogen.

PRODUCT OF THE REACTION

Several methods are available to detect the presence of hydrinos. They appear to have been detected in interstellar space as well as in the laboratory.

Dihydrino and dideutrino molecules may have been identified by mass spectroscopy. For example, Miles (3–6)

Table 1. Mass Spectroscopy Intensity Cryofiltered DiHydrino and Hydrogen Molecules at Nominal Mass 2[a]

	Ionization Potential, eV		
Species	15	20	35
Hydrogen, H_2	0.000	0.005	0.005
Dihydrino, H_2^*	0.025	0.020	0.240

[a] These results show that H_2^* has an ionization potential above 20 eV; H_2 has an ionization potential below 20 eV.

and Bockris (7) report production of mass four as identified by mass spectroscopy of the cryofiltered gases evolved from an electrolysis cell comprising a palladium cathode and a $LiOD/D_2O$ electrolyte. According to Miles (5), the intensity of the mass four peak maintained an approximate correspondence to the amount of excess power or heat observed in electrochemical calorimetric cells. At the time, Miles believed the mass four peak to be helium as a result of fusion reactions.

The dideutrino molecule can also be identified by high-resolution mass spectroscopy. The dihydrino molecule (H_2^*) is predicted to be spin paired, to be smaller than the hydrogen molecule, to have a higher ionization energy than H_2, and to have a lower liquefaction temperature than H_2. Mass spectroscopy distinguishes a sample containing dihydrino molecules versus a sample containing H_2 by showing both a different ion production efficiency as a function of ionization potential and a different ion production efficiency at a given ionization potential.

Typical results for hydrogen and dihydrino gas samples are shown in Table 1.

BIBLIOGRAPHY

1. B. Stanley Pons and W. P. Pons, Mortin Press Conference, University of Utah, March 23, 1989. (As reported in Taubes, G., Bad Science, Random House, New York, 1993).
2. R. Mills, *Unification of Spacetime, the Forces, Matter, and Energy*, Science Press, Ephrata, Pa., 1992.
3. M. H. Miles, B. F. Bush, G. S. Ostrom, and J. J. Lagowski, *Electroanal. Chem.*, **304,** 271 (1991).
4. R. Daganl, *Chemical and Engineering News*, 31–33, (April 1, 1991).
5. M. H. Miles, B. F. Bush, G. Ostrom, and S. Lagowski, "The Science of Cold Fusion," *J. J. Conference Proceedings*, Vol. 33. T. Bressani, E. Del Giudice, and G. Preparata, eds., SIF, Bologna, 1992, pp. 363–372.
6. M. H. Miles, R. A. Hollins, B. F. Bush, J. J. Lagowski, and R. E. Miles, *J. Electroanal. Chem.*, **346,** 99 (1993).
7. C. C. Chien, D. Hodko, Z. Minevski, and J. Bockris, *J. J. Electroanal. Chem.* **338,** 189–212 (1992).

COMMERCIAL AVAILABILITY OF ENERGY TECHNOLOGY

SAM RASKIN
Research and Development Office MS 43
Sacramento, California

California's vulnerability during the energy crises of the 1970s demonstrated the importance of establishing energy programs and policies to provide the state with a secure mix of energy technologies. The California Energy Commission's *Energy Development Report (EDR)* has addressed this issue directly by establishing policy recommendations for California energy technology development based on evaluation criteria such as cost, rate impacts, diversity, environmental impacts, operating flexibility, planning flexibility and reliability. The *ETSR* is intended to serve as an important resource for this planning effort by supporting the *EDR* with a continually updated assessment of more than 230 energy technology options for electric generation and end-use efficiency.

Four levels of energy technology status information are provided: general conclusions, summary matrices, fact sheets and detailed technology evaluations. Each of these information levels covers the three evaluation components: commercial status, research and development goals, and deployment issues.

Conclusions provide an overview of evaluations for all fuel cycle, electric generation and end-use energy technologies and a list of possible state actions for supporting further energy technology development. This information provides a quick assessment of technology status and developments and also can be used to understand the broad range of options available to stimulate a preferred mix of technologies in California.

Summary Matrices (Fig. 1) provide the results of detailed analyses included in separate appendices on a single set of matrices for all *ETSR* technologies.

Fact Sheets are developed by compressing the most important information from the detailed evaluations into an easy-to-use graphic format.

Detailed Technology Evaluations include unabridged evaluations for electric generation and end-use technologies, respectively.

See also ELECTRIC POWER GENERATION; ENERGY CONSERVATION, ENERGY EFFICIENCY; FUELS FROM BIOMASS; FUELS FROM WASTE; FUELS, SYNTHETIC; NUCLEAR POWER; RECYCLING; RENEWABLE RESOURCES.

COMMERCIAL STATUS

Commercial status assessment for all *ETSR* technologies indicates energy options now available *in California,* or expected to be available within a 20-year planning period. The base year (first year of operation) for these assessments is 1997 for electric generation technologies and 1992 for end-use technologies. Three criteria are used for commercial status determinations: 1) technology maturity, 2) existence of supplier(s), and 3) competitive cost. All three criteria must be satisfied for a technology to be considered "commercially available" for operation (not order) in the base year. Technologies where any one criterion is not satisfied are automatically assessed as "not commercially available."

RESEARCH AND DEVELOPMENT GOALS

Energy technologies typically use mechanical equipment, electric devices and specialized materials that can be improved through research targeted at alternative materials,

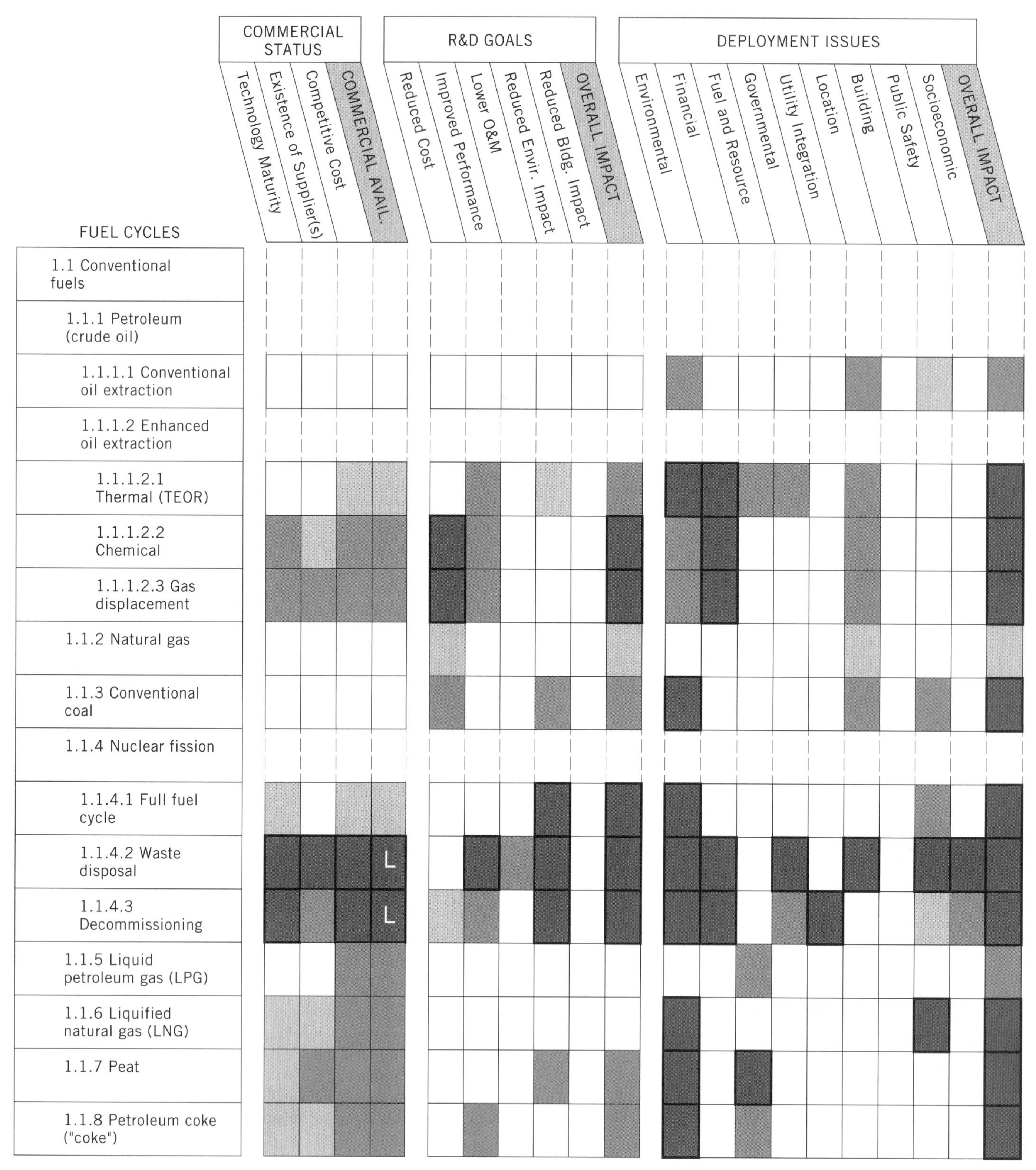

Figure 1. Technology Evaluation Matrix

designs, manufacturing techniques and volume production. The ultimate objective of research is to make energy technologies more acceptable to users and society. To achieve this objective, individual research and development activities frequently focus on a narrow aspect of technology development. Technological advancements usually are the result of an incremental process involving many small improvements that together lead to a commercially viable product or refinement.

Energy policymakers are most concerned with the end

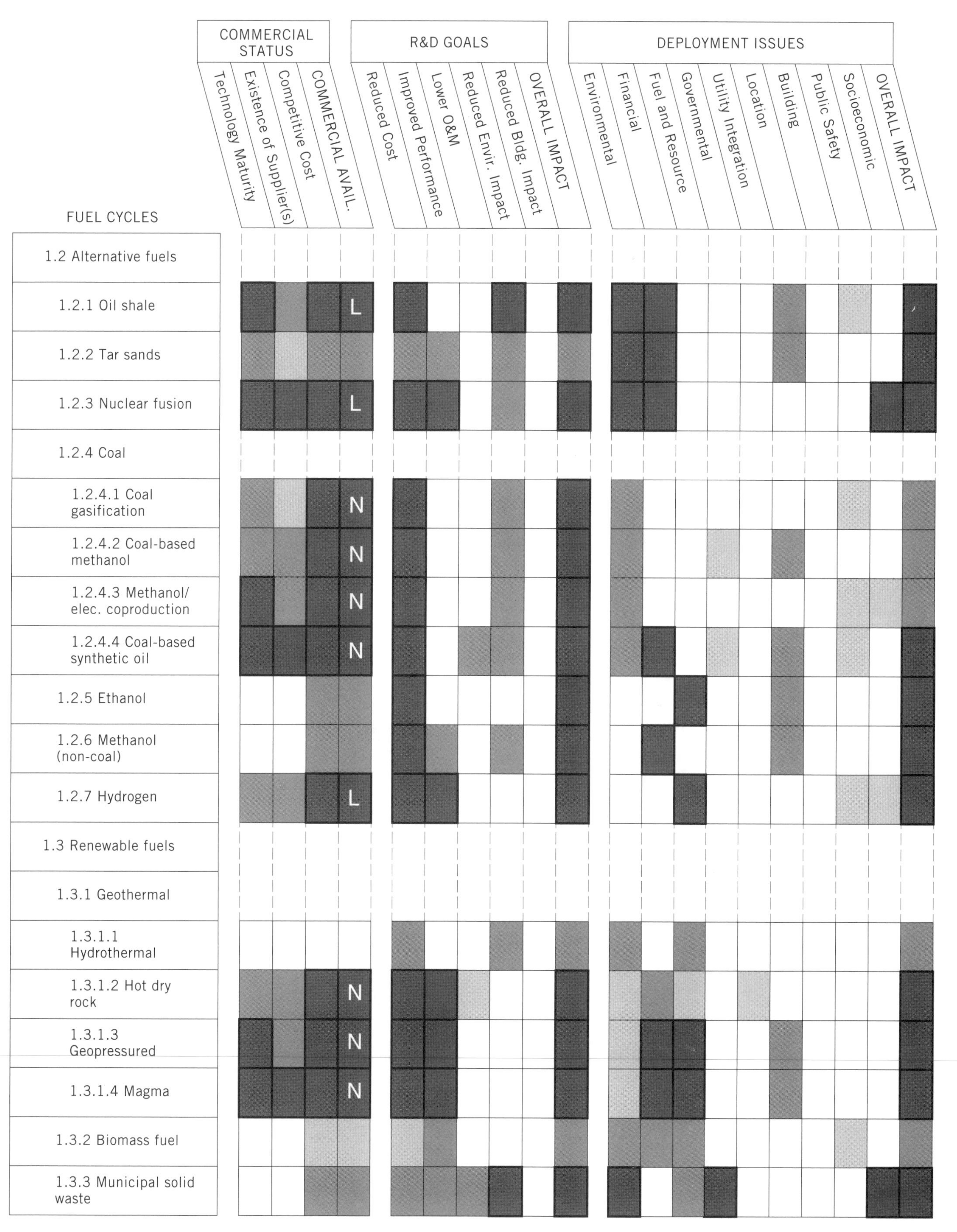

Figure 1. (Continued)

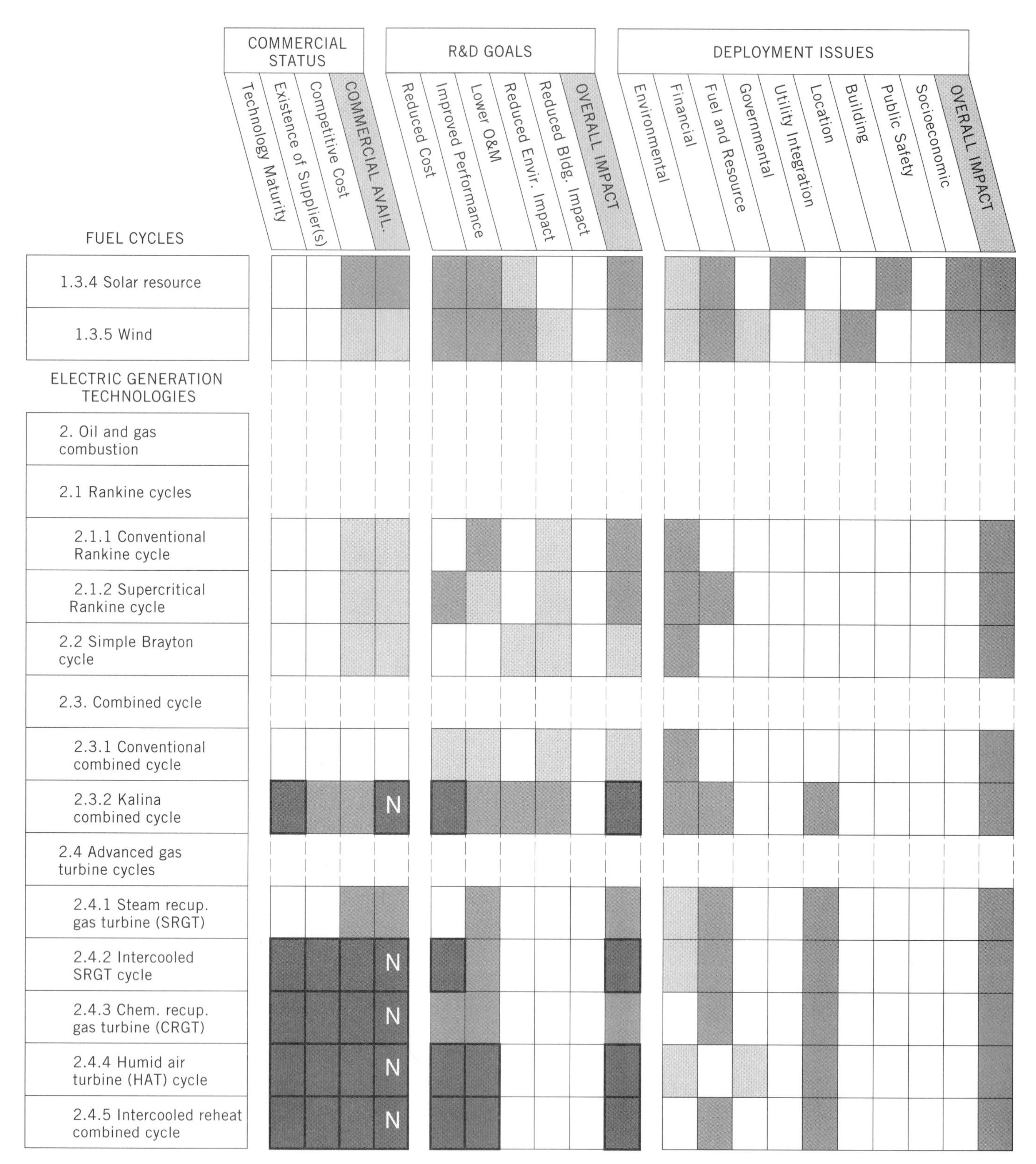

Figure 1. (Continued)

results of successful research and development rather than the details associated with highly specialized incremental developments. This "big picture" point of view allows policymakers to decide on broader issues such as the timeframe for commercial availability; necessity and suitability of research and development funding; and applicability of the technology in California's energy future.

To facilitate this type of policy analysis, each specific technology evaluation in this article categorizes research and development goals according to five identified generic

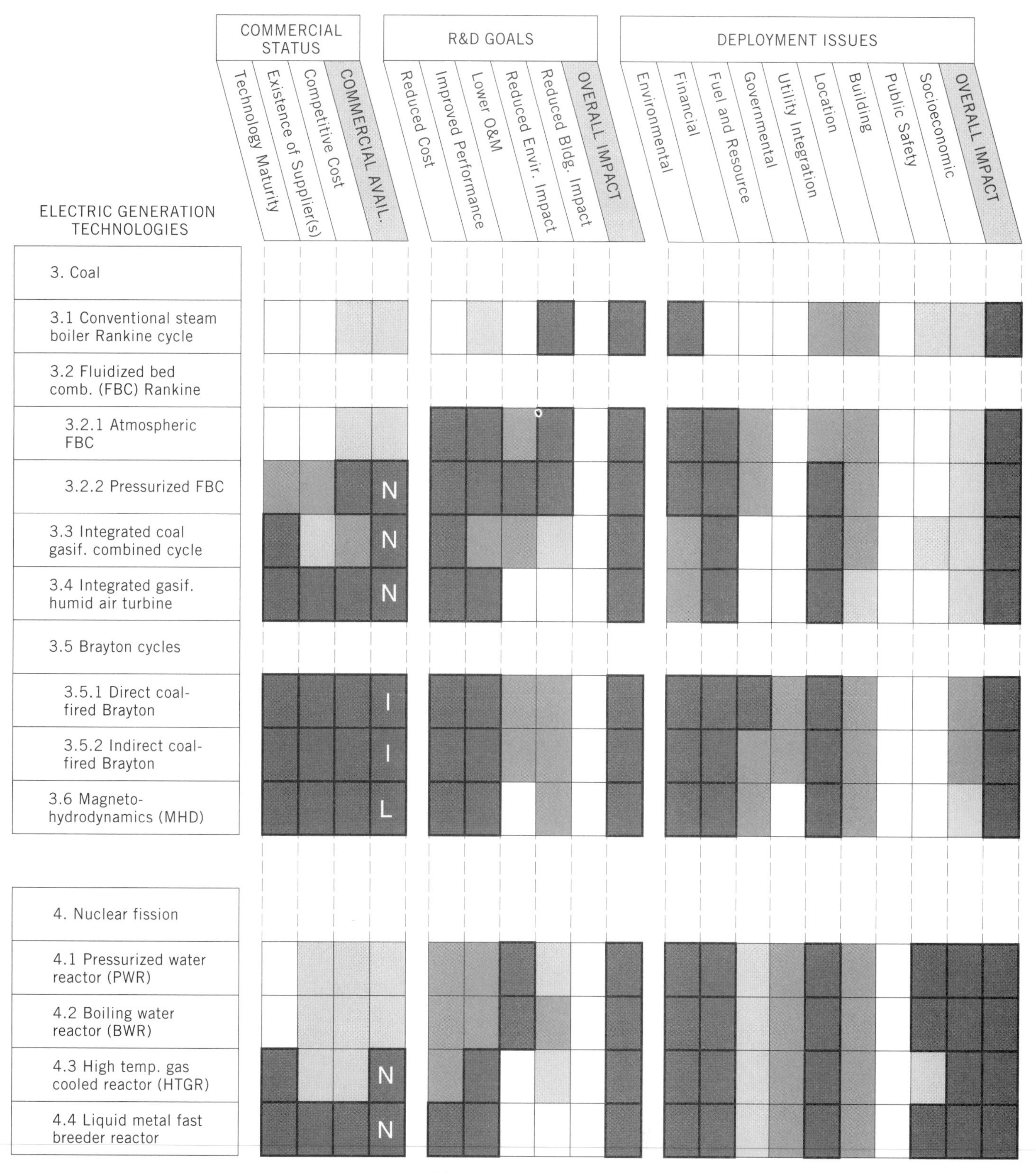

Figure 1. (Continued)

research goals based on research activities and programs currently underway at government and private sector research facilities: reduced cost, improved performance, lower operation and maintenance costs, reduced environmental impacts and reduced building impacts.

DEPLOYMENT ISSUES

A technology can become commercially available through research and development efforts yet require many years to achieve widespread market adoption due to constraints

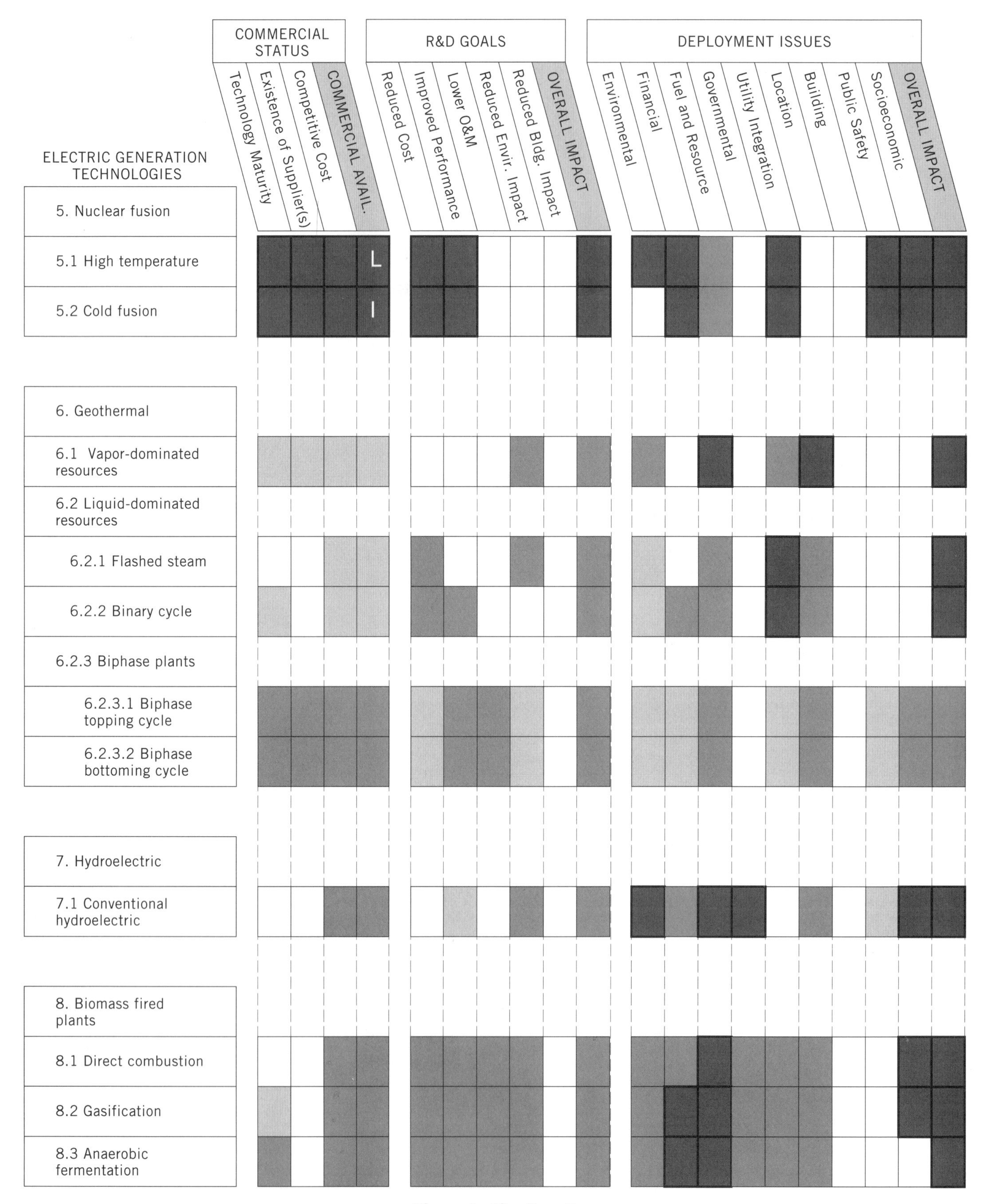

Figure 1. (Continued)

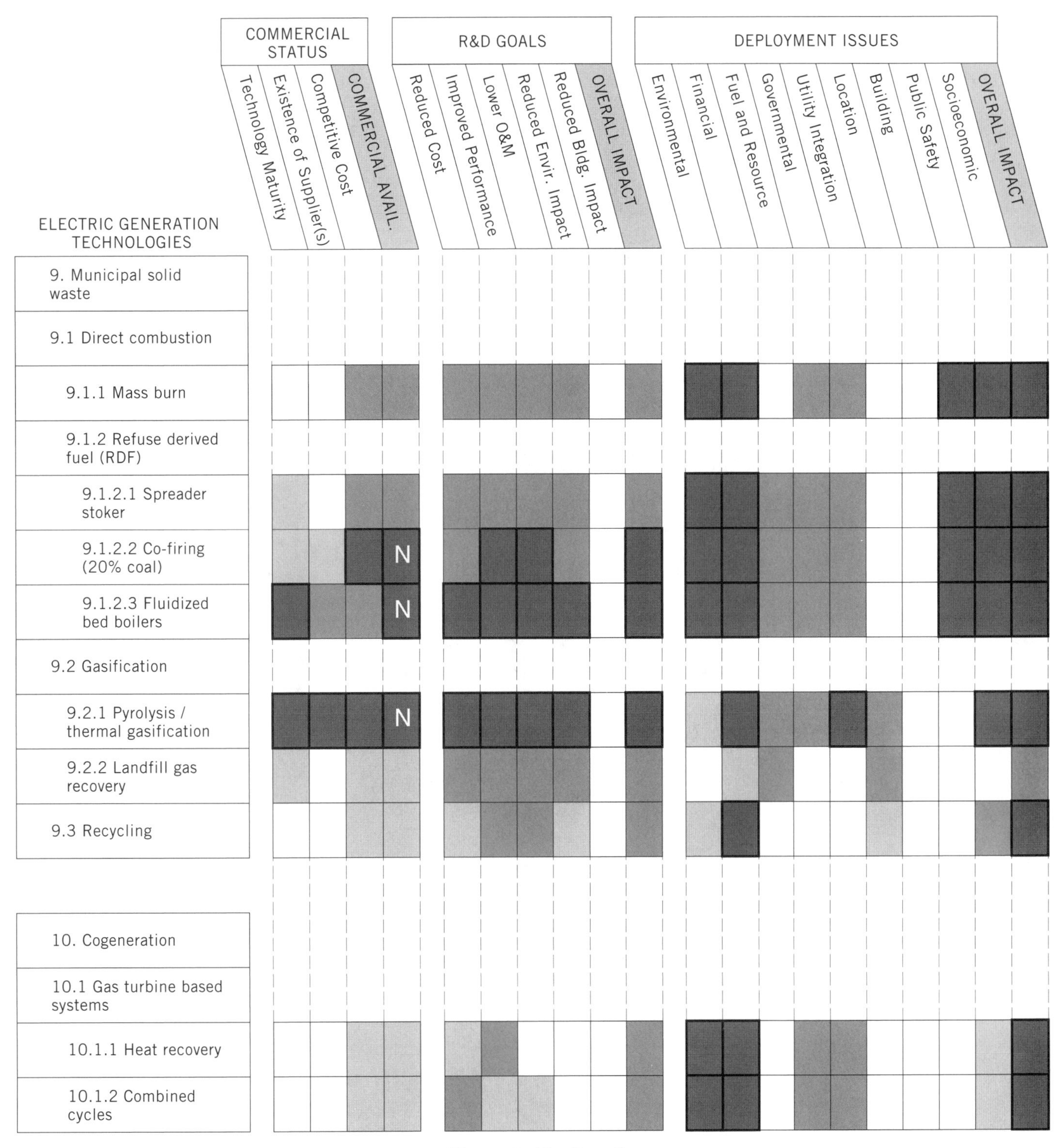

Figure 1. (Continued)

that preclude or limit a technology from consideration as a viable alternative. Energy policymakers must evaluate these issues when identifying programs and activities to support preferred energy options. To meet this need, a master list of generic deployment issues, included in the next paragraph, has been developed by identifying constraints to energy technology use. Each technology evaluation in this article details deployment issues evaluated according to these generic issues. These issues are most fully defined for commercially available technologies and less well defined for non-commercially available technologies. The absence of an identified deployment issue for a non-commercially available technology does not mean that a certain deployment issue will not exist. Some deployment

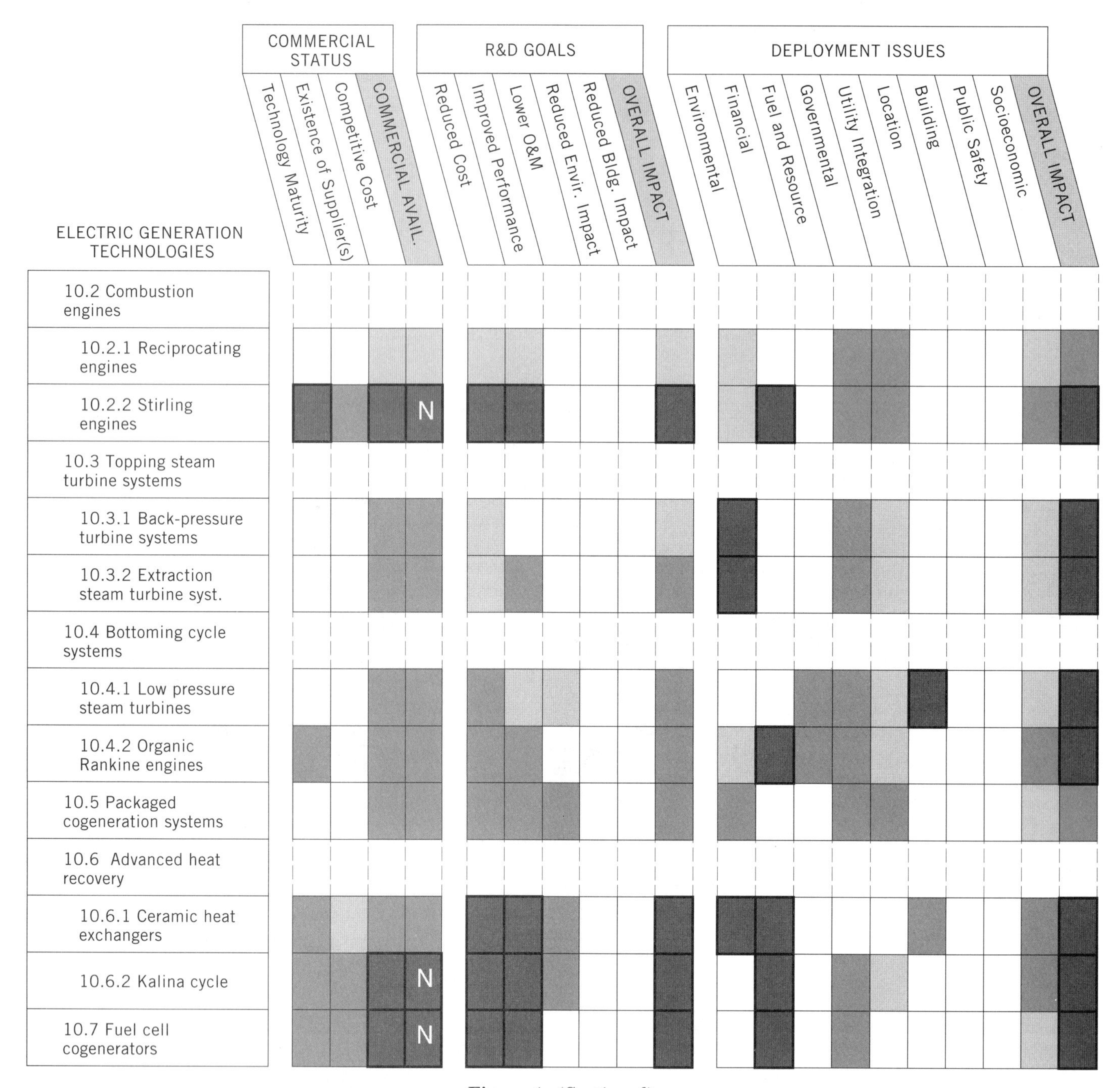

Figure 1. (Continued)

issues become important only when a technology has penetrated the marketplace above a certain level. All of these issues are discussed in more detail.

The following list includes nine issues constraining the deployment of energy technologies along with sub-issues that address more detailed aspects associated with each issue.

Environmental Constraints

- Air Pollution Impacts
- Water Pollution
- Waste Disposal
- Noise Pollution
- Radio/TV Signal Interference
- Thermal Discharge
- Destruction/Disturbance of Habitat
- Scenic Resource Impacts

Financial Constraints

- Availability of Financing
- High Capital Costs

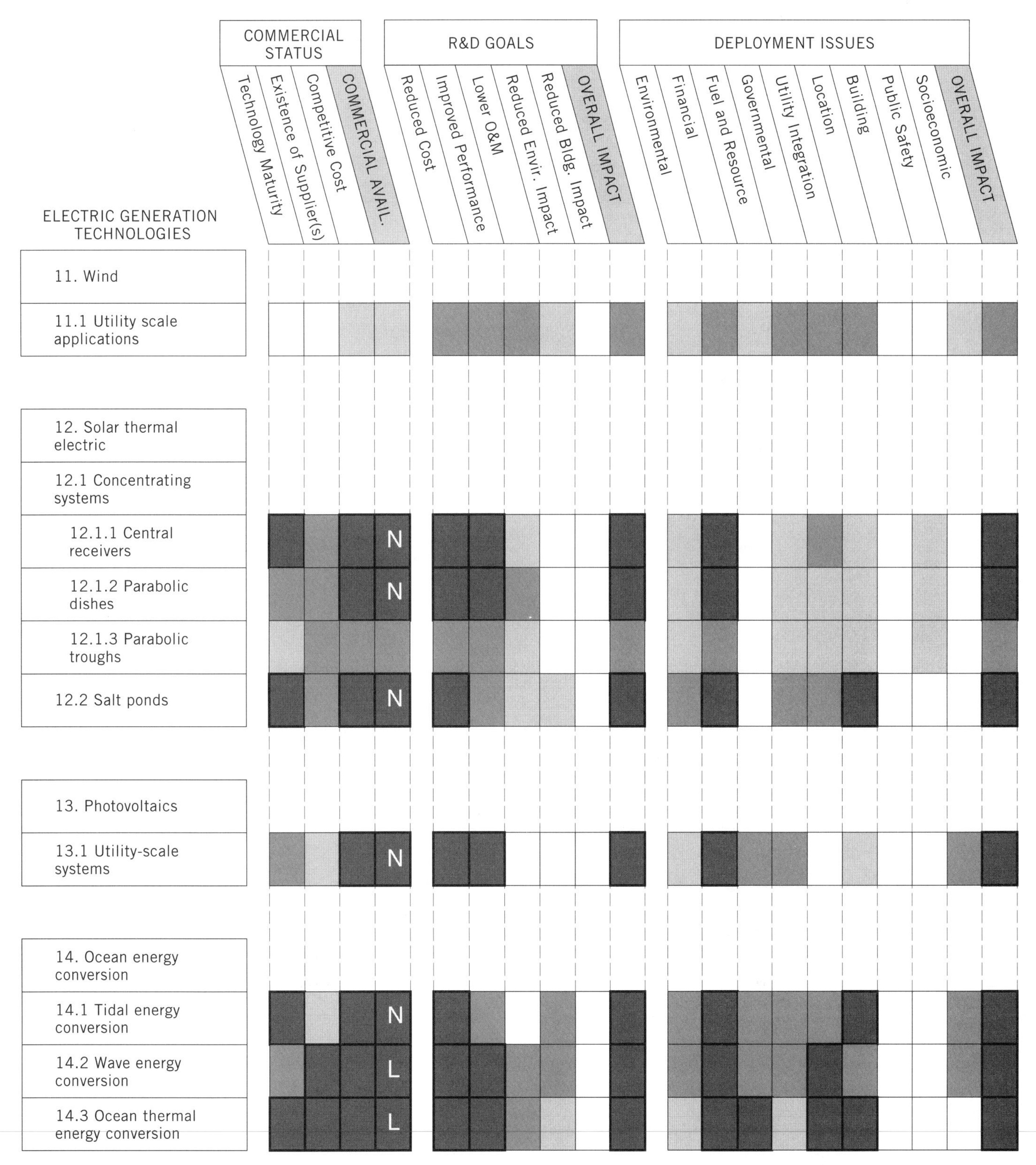

Figure 1. (Continued)

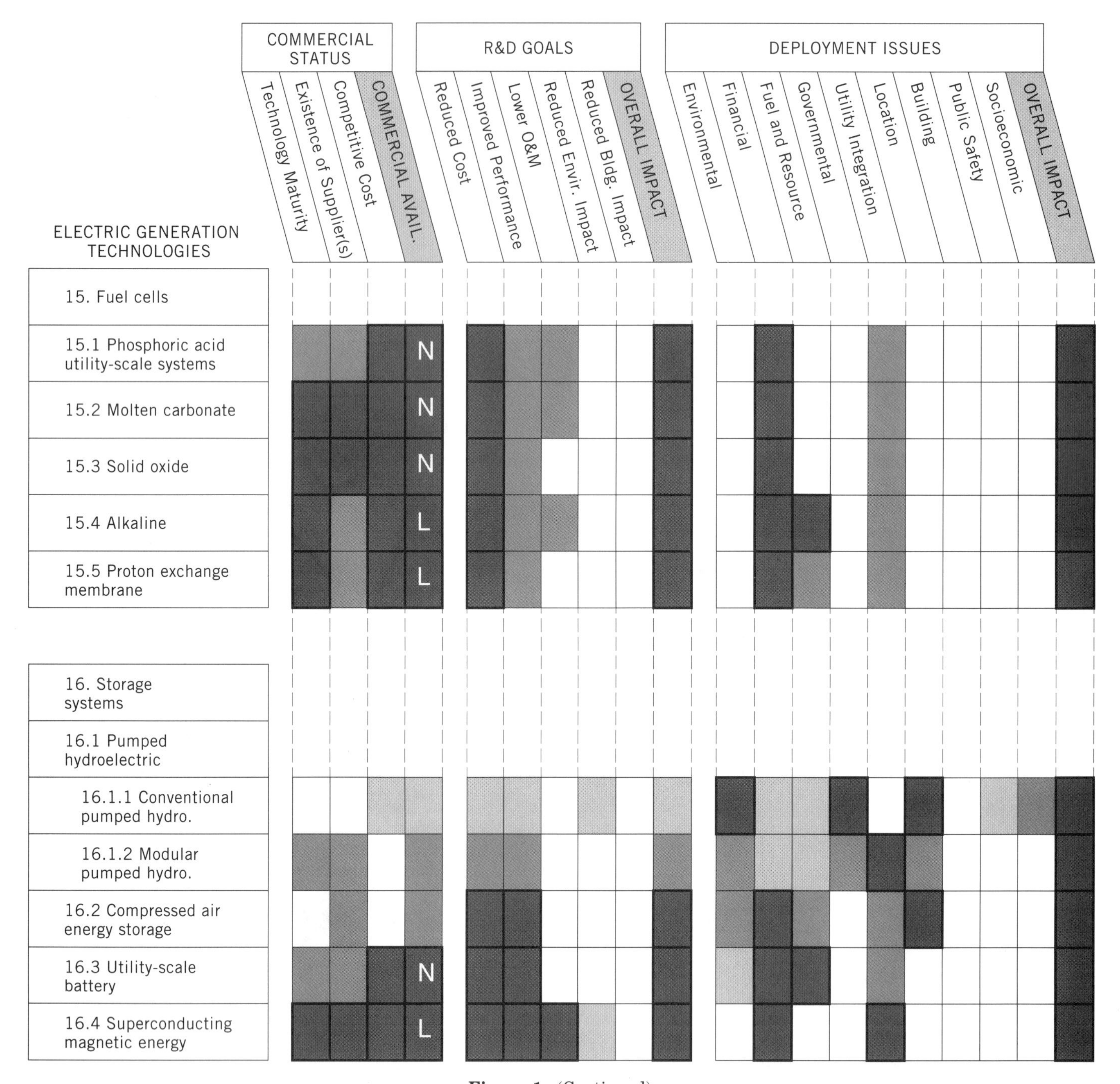

Figure 1. (Continued)

- High Operation and Maintenance Costs
- Availability of Tax Incentives

Fuel and Resource Constraints

- Availability of Fuel or Resource
- High Cost of Fuel
- Variation in Fuel or Resource Quality

Governmental Constraints

- Agency-Government Coordination
- Building Code/Planning Restrictions
- Undependable Avoided Cost Contracts
- Regulatory/Legislative Restrictions
- Permit Restrictions
- Intergovernmental Coordination

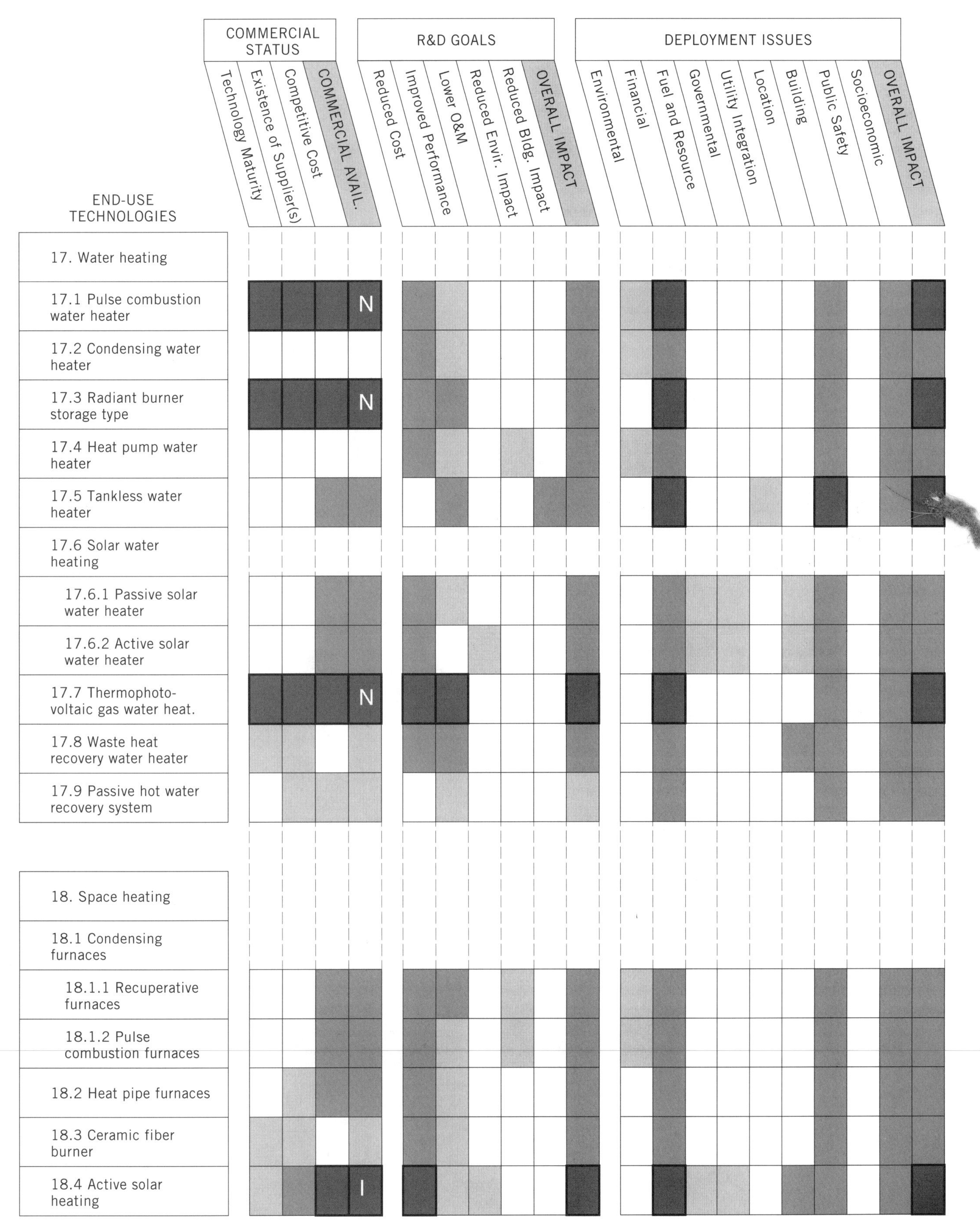

Figure 1. (Continued)

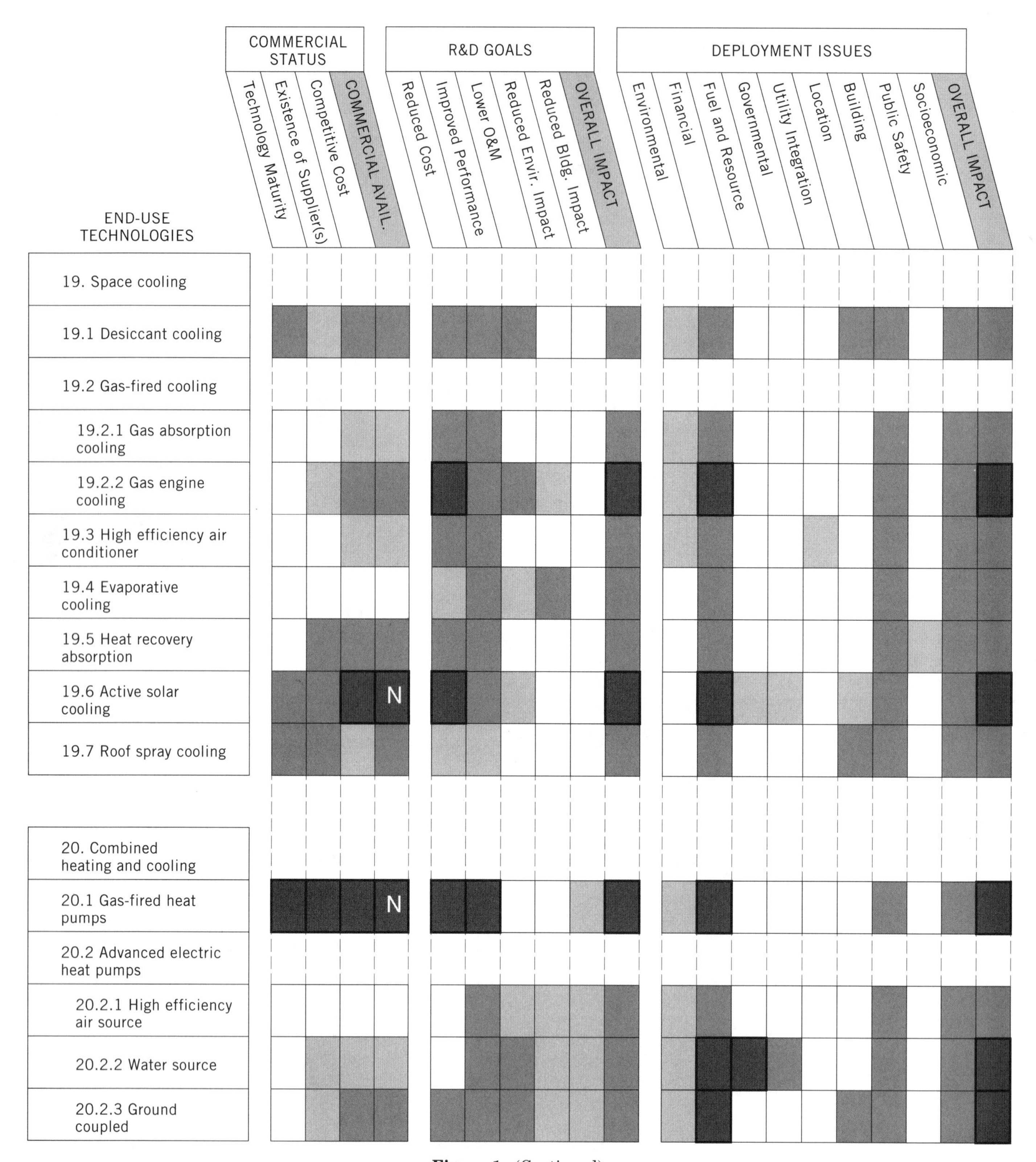

Figure 1. (Continued)

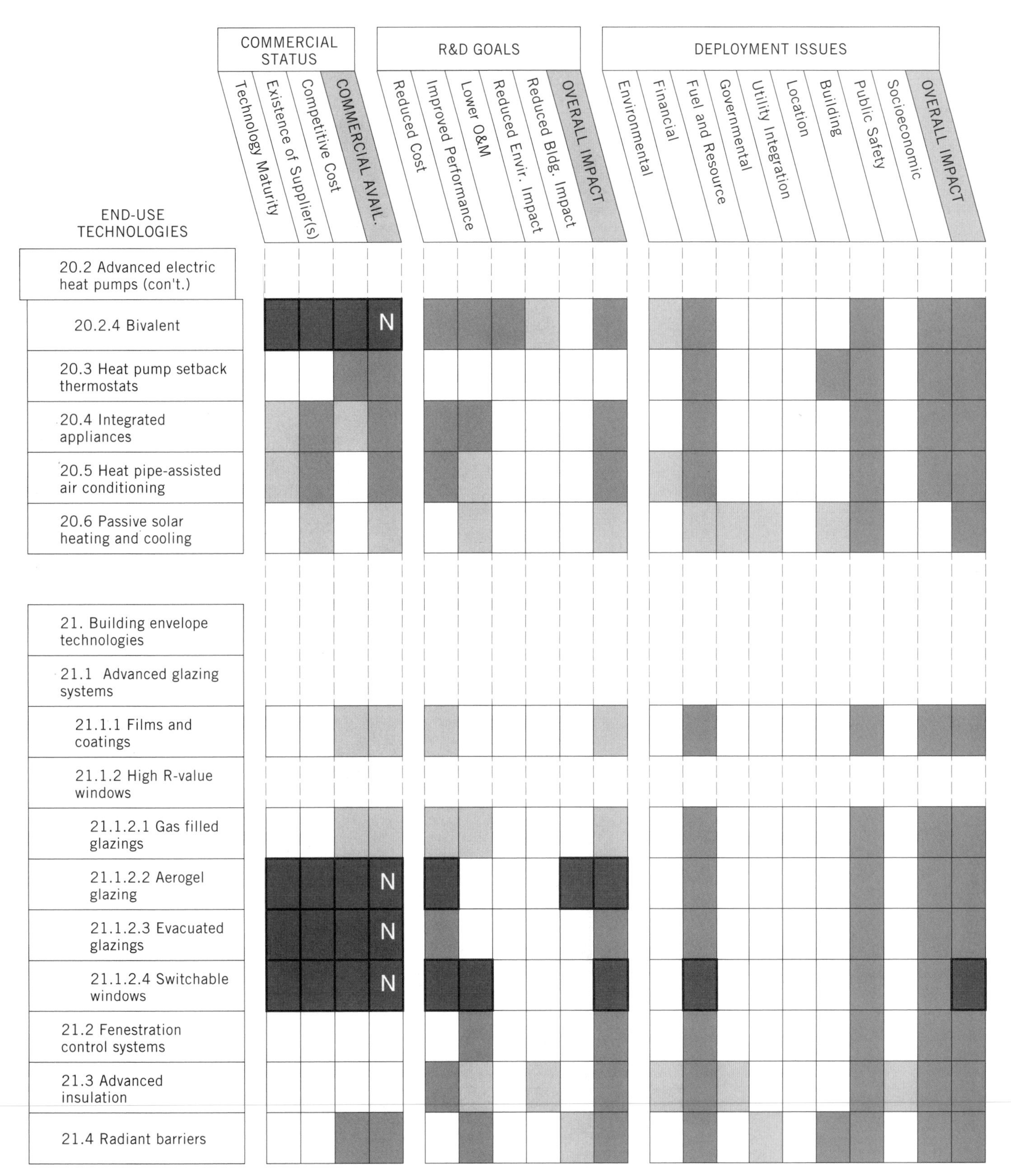

Figure 1. (Continued)

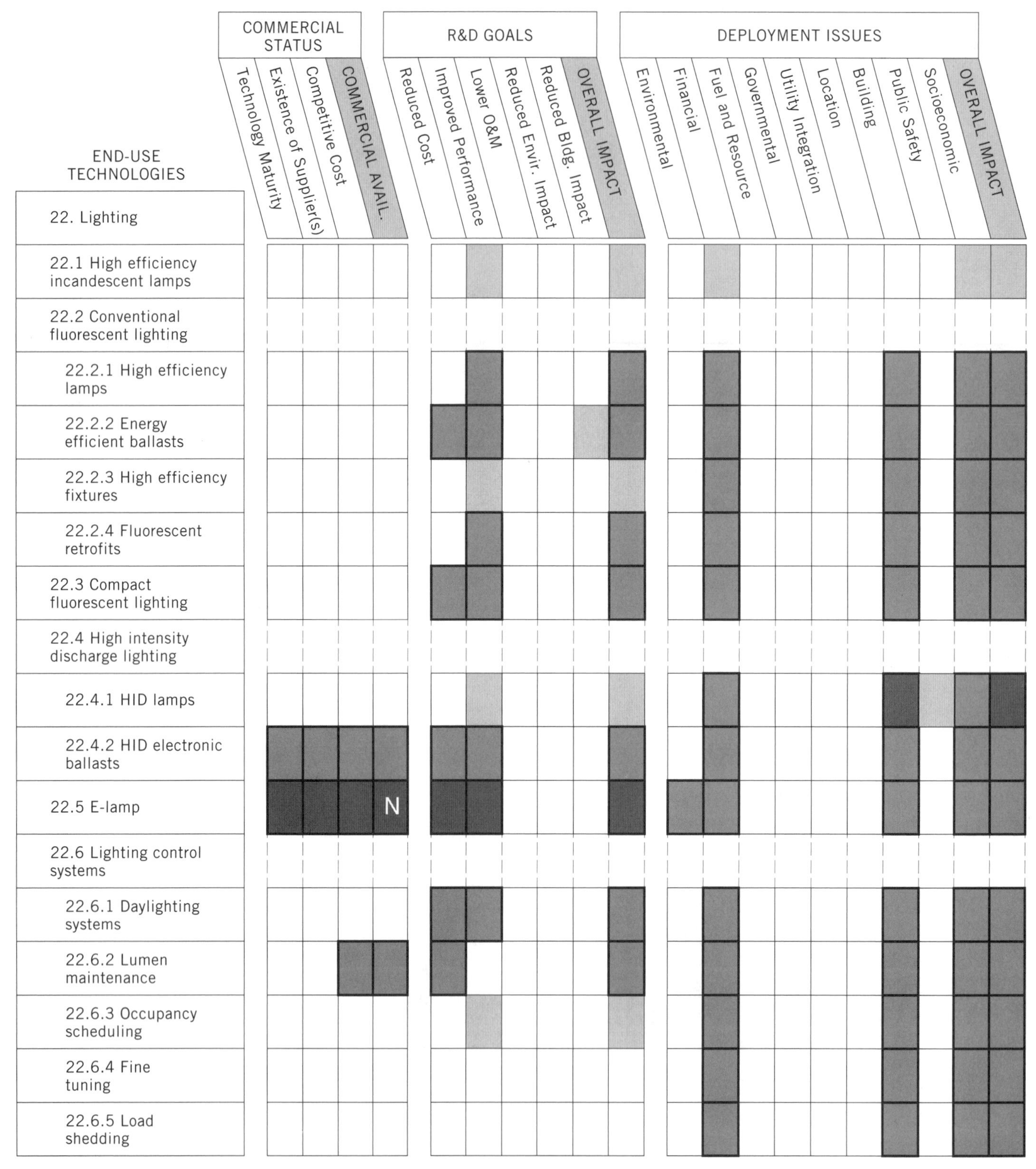

Figure 1. (Continued)

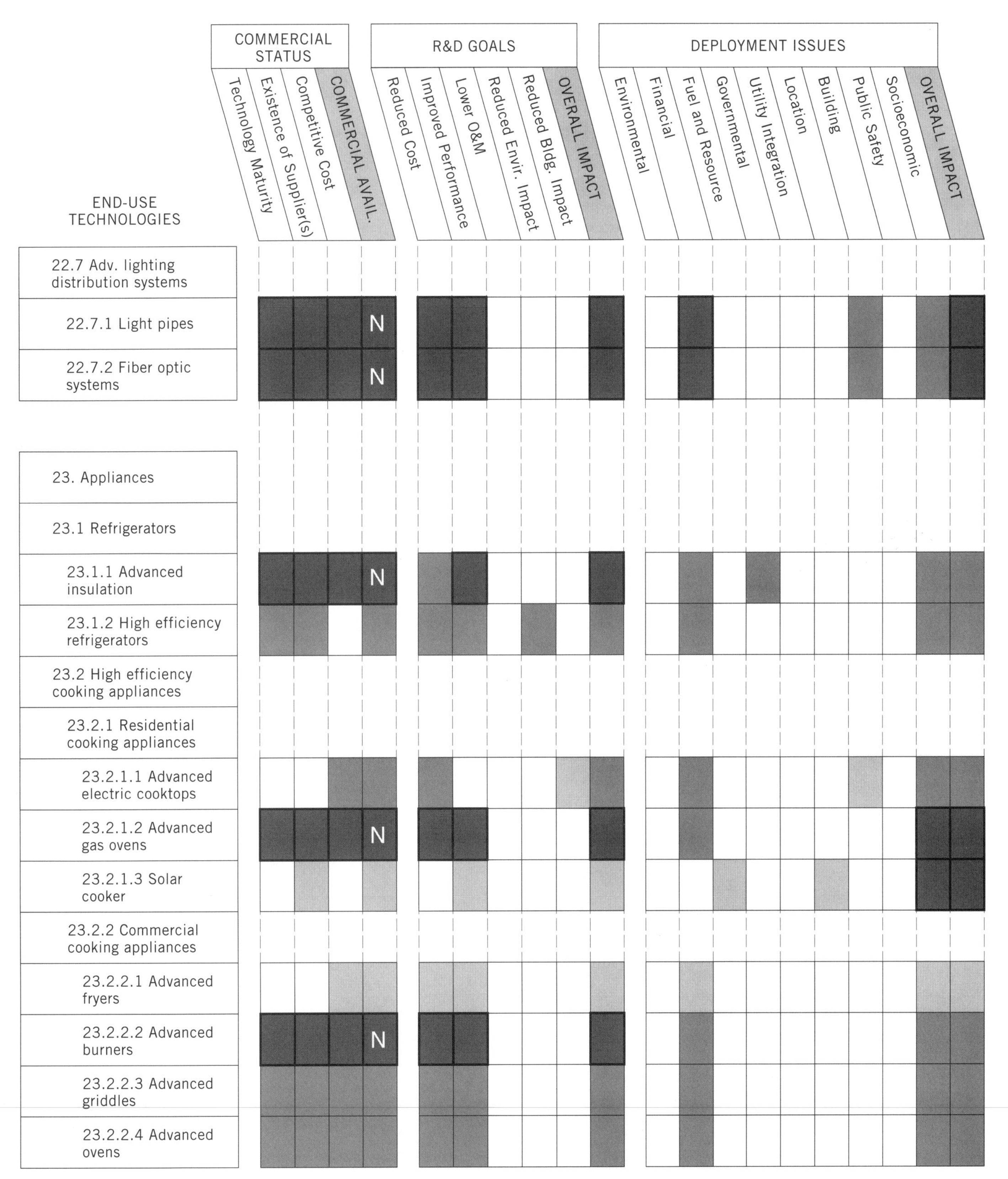

Figure 1. (Continued)

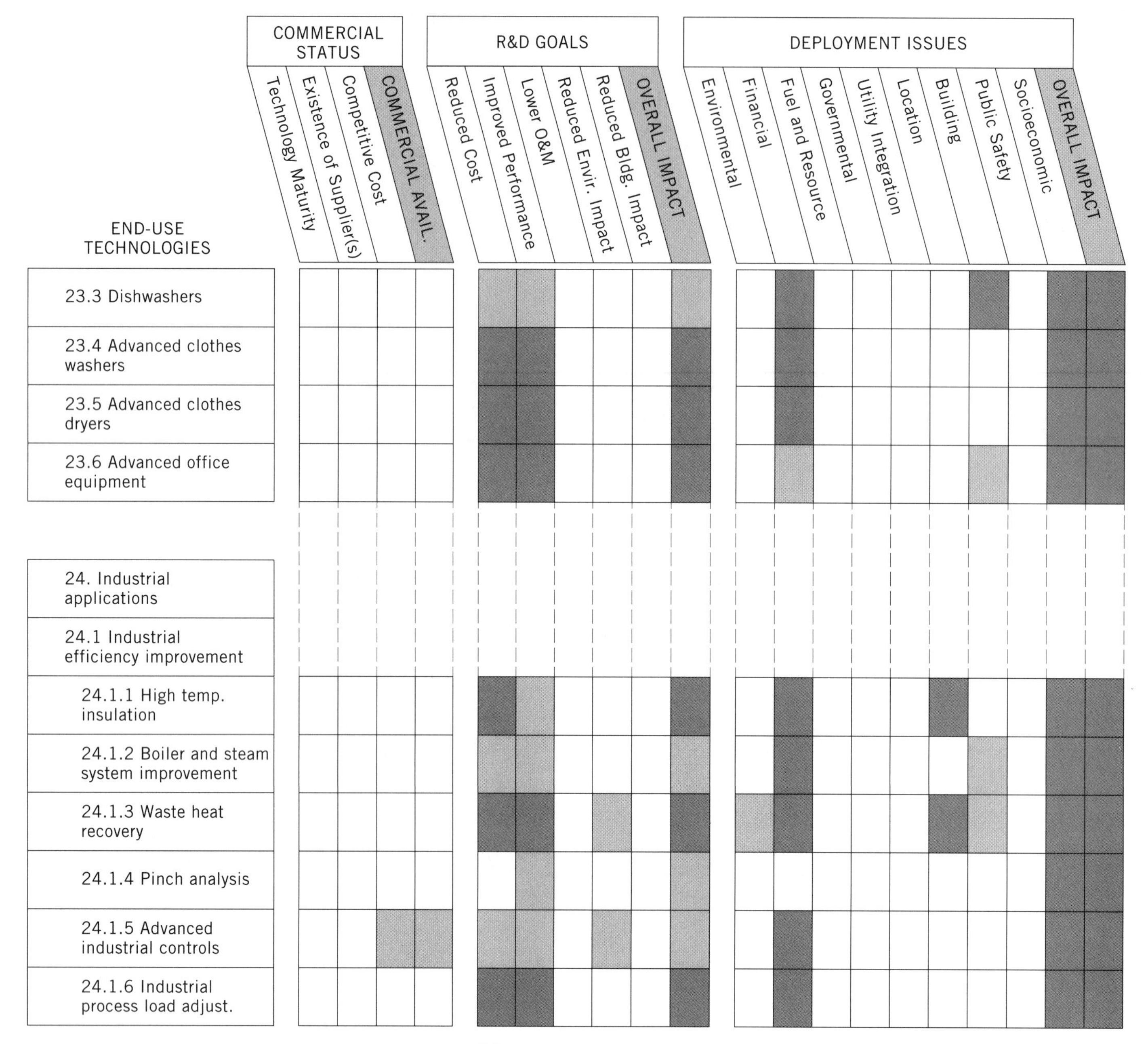

Figure 1. (Continued)

Utility Integration Constraints

- Control of Intermittent Sources
- Need Conformance
- Lack of Demonstrated Reliability/Performance
- Conformance with Interconnection Requirements
- Lack of Incentive for Utility Companies

Location Constraints

- Fuel Delivery Restriction/Cost
- Lack of Suitable Sites
- Adverse Subsidence Impacts
- Availability of Transmission Lines
- Availability of Water
- Risk of Seismic Damage

Building Constraints

- Adverse Structural Impacts
- Adverse Appearance Impacts
- Adverse Occupant Impacts
- Minimal Industry Acceptance
- Lack of Incentive for Building Owners/Developers

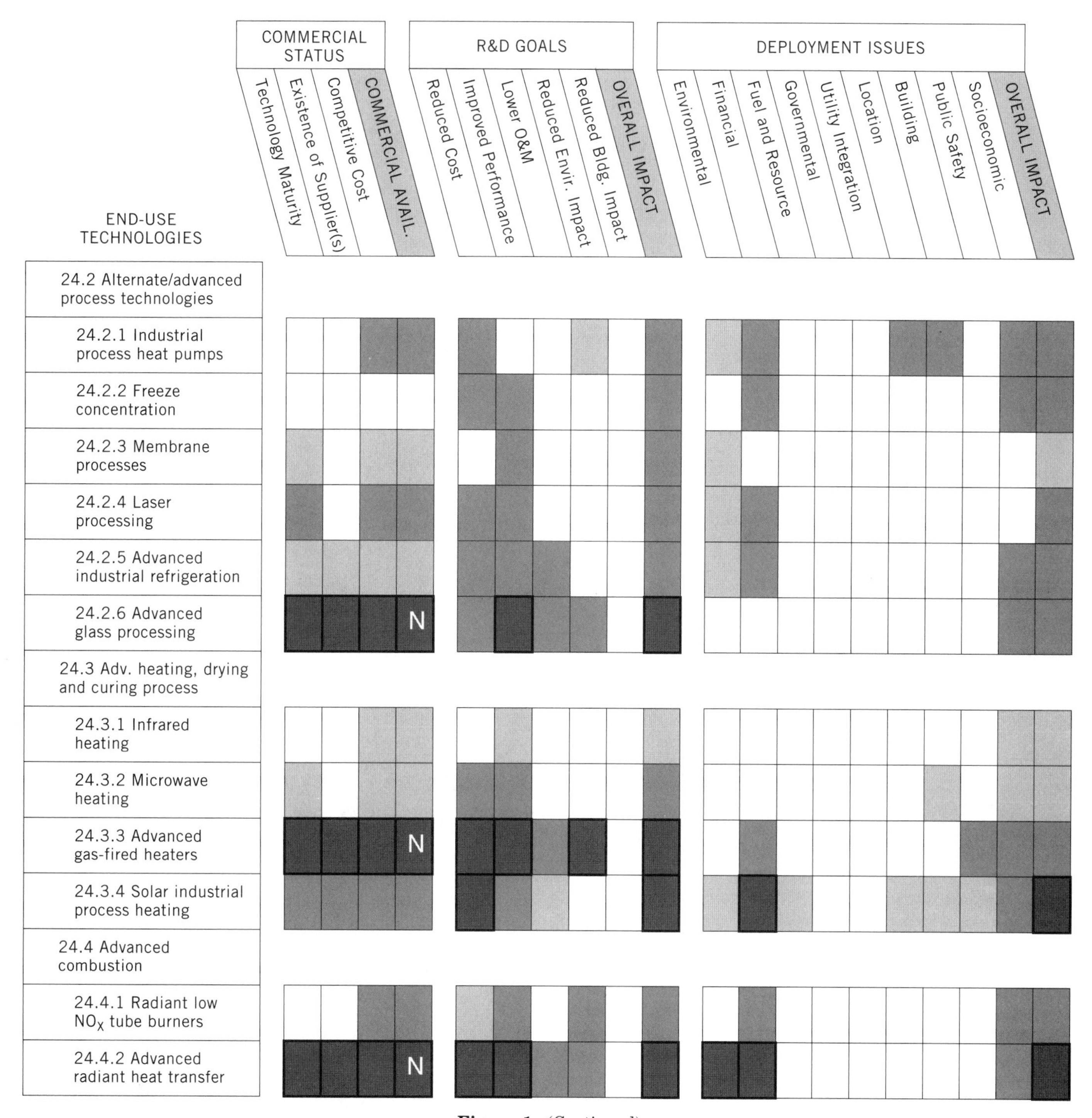

Figure 1. (Continued)

Public Safety Constraints

- Catastrophic Risks
- Fire Hazards
- Toxic Gas Hazards
- Health Risks

Socioeconomic Constraints

- Poor Public Opinion
- Low End User Awareness
- Complexity of Operation
- Adverse Agricultural Impacts

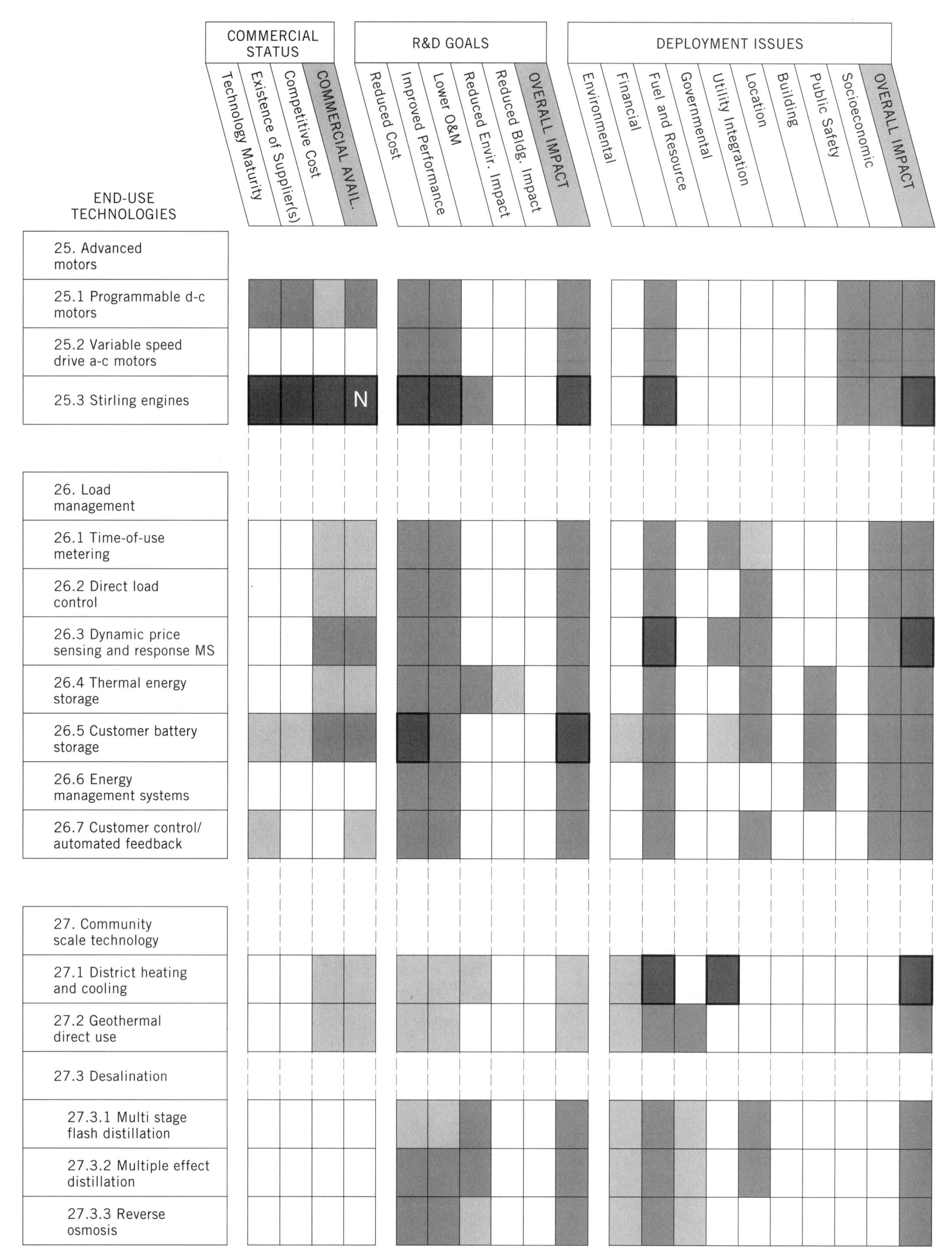

Figure 1. (Continued)

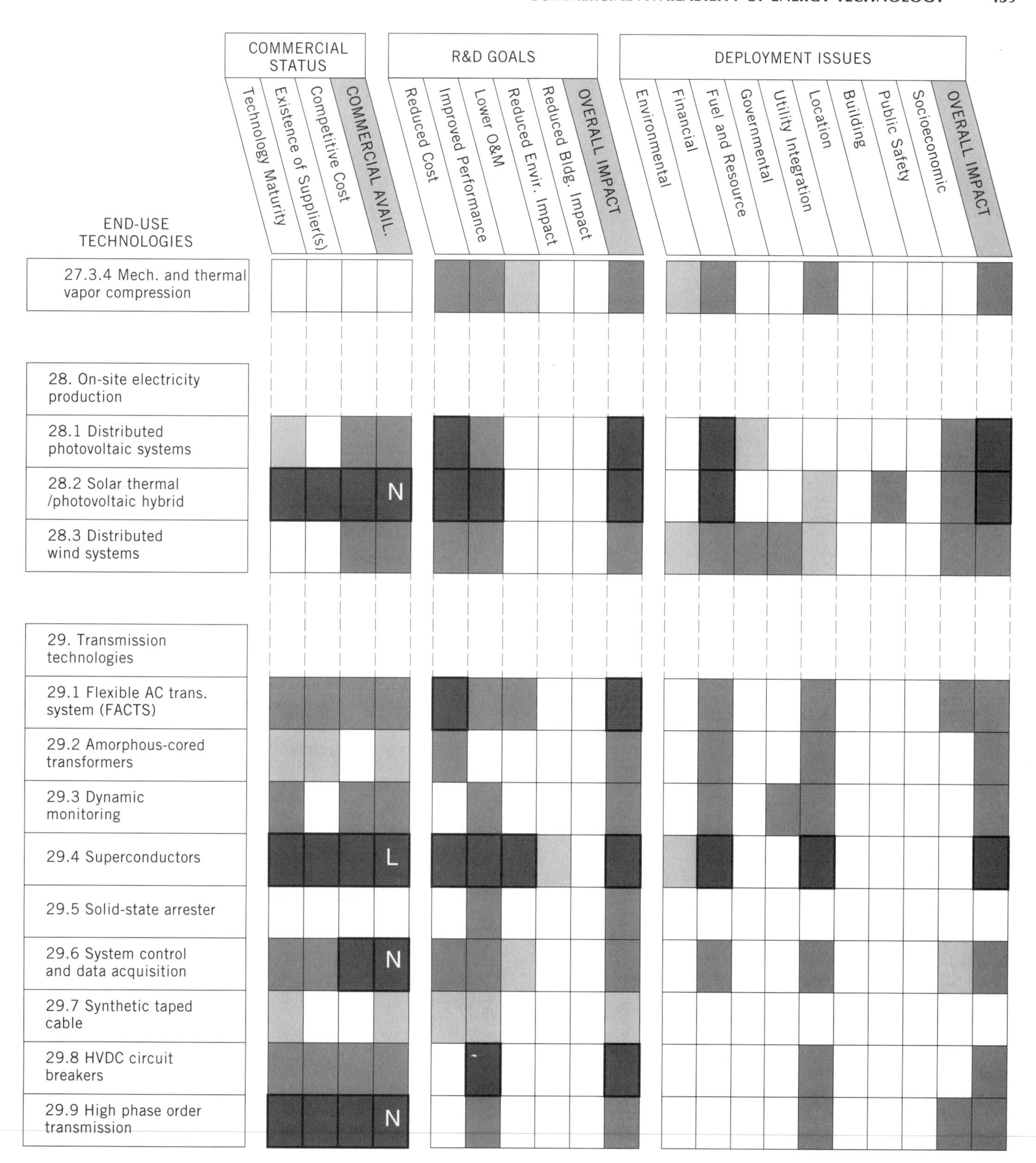

Figure 1. (Continued)

COMMERCIAL AVAILABILITY

This section provides an overview of all specific energy technologies assessed in terms of their commercial availability and unavailability and general conclusions concerning energy technology status, research and development goals and deployment issues. The conclusions are organized according to fuel cycle, electric generation and end-use energy technologies.

Fuel Cycle Technologies

Fuel Cycle Technologies Commercially Available. The following fuel cycle technologies have been determined to be commercially available. They are listed as "F" where all three criteria are "fully" satisfied, "MC" where one or more criteria are satisfied under "most conditions," and "LC" where one or more criteria are satisfied under rare or "limited conditions."

Conventional Fuels:
Conventional Oil Extraction (F)
Thermally Enhanced Oil Extraction (MC)
Chemical Enhanced Oil Extraction (LC)
Gas Displacement Enhanced Oil Extraction (LC)
Natural Gas (F)
Conventional Coal (F)
Full Fuel Nuclear Fission Cycle (MC)
Liquefied Petroleum Gas (LC)
Liquefied Natural Gas (LC)
Peat (LC)
Petroleum Coke (MC)

Alternative Fuels:
Tar Sands (LC)
Ethanol (LC)
Methanol (Non-Coal) (LC)

Renewable Fuels:
Hydrothermal Geothermal Resource (F)
Biomass Fuel (MC)
Municipal Solid Waste (LC)
Solar (LC)
Wind (MC)

Fuel Cycle Technologies Not Commercially Available. The following fuel cycle technologies have been determined to be not commercially available because they do not satisfy one or more of the commercial availability criteria. Dates of expected commercial availability are indicated in parentheses as either "near-term" (within 12 years), "long-term" (beyond 12 years) or "indeterminate" where there are a number of unresolved R&D issues or a low likelihood of commercialization in the foreseeable future.

Alternative Fuels:
Waste Disposal Nuclear Fission (long-term)
Decommissioning Nuclear Fission (long-term)
Oil Shale (long-term)
Nuclear Fusion (long-term)
Coal Gasification (near-term)
Coal-Based Methanol (near-term)
Methanol/Electricity Coproduction From Coal (near-term)
Coal-Based Synthetic Oil (near-term)
Hydrogen (long-term)

Renewable Fuels:
Hot Dry Rock Geothermal Resource (near-term)
Geopressured Geothermal Resource (near-term
Magma Geothermal Resource (long-term)

Electric Generation Technologies General Conclusions

- For electric generation technologies which are not commercially available, the research and development goals of Reduced Cost and Improved Performance and the Financial Deployment Issue are frequently listed as "Show Stoppers." The R&D Goals, the Deployment Issues and the Non-Commercial Status follow one from the other. High first cost and high operational cost (poor performance) are common reasons that a technology is not commercially available and has financial constraints on use. While a technology is in the R&D phase, production costs are high because of high engineering and fabrication costs. Once a technology has been successfully demonstrated, better and lower cost designs, equipment and materials can be chosen.
- For modular technologies, the high first cost results from high cost in tooling and labor. The development of a manufacturing infrastructure is an important R&D goal. Once production has started, cost reductions can be expected to follow a "learning curve" with perhaps a 13% reduction in production cost for every doubling of production volume.
- Although not always identified in this article, materials issues tend to impact the cost and performance of many technologies which must overcome problems of high materials costs, and corrosion, erosion, fatigue and thermal stress resistance.
- The Utility Integration (Lack of Demonstrated Reliability/Performance) and Socioeconomic (Poor End-User Awareness) Deployment Issues are frequently associated with Non-Commercially Available Technologies. The Deployment Issues follow directly from the commercial status: if a technology is not commercially available because it is immature (not demonstrated), utilities will not use it and they (or other end-users) may not be fully aware of the technology's potential.
- Deployment Issues associated with commercially available technologies are immediate opportunities for government or other organizations or individuals to affect the use of a technology, perhaps at a relatively low cost.

- Although gas technologies are substantially commercialized, research and development support to reduce costs is needed to design and demonstrate advanced gas turbine cycles. The advanced gas turbine cycles have the potential for low capital cost and high efficiency.
- Compared to other generation technologies, gas combustion technologies have fewer constraints–focused primarily on environmental and utility integration issues.
- Most coal technologies have significant research and development needs for cost reduction, improved performance, lower operations and maintenance costs, and environmental impact mitigation.
- As a group, coal technologies face multiple deployment issues associated with environmental, financial, resource, governmental, utility integration and location constraints. Environmental constraints must be considered because the resource extraction and some combustion processes are inherently dirty. Resource constraints become a factor for California because of limited indigenous resources and transportation costs for out-of-state resources to California.
- Among all coal technologies, integrated coal gasification combined cycle and integrated gasification humid air turbine may offer California the most environmentally acceptable options.
- Over the past few years, coal-fluidized bed technologies have developed rapidly. Applicability to utility operation remains uncertain.
- Nuclear energy technologies are faced with major research and development issues relative to cost reduction, improved performance, lower operations and maintenance costs, and reduced environmental impacts. Environmental impacts are particularly critical because future deployment of nuclear energy technologies in California depends on resolving difficult high-level waste disposal issues. The future of nuclear power may depend on the developments and acceptance of standardized designs.
- Many deployment issues confront nuclear technologies including environmental (high-level waste disposal), financial, public safety and socioeconomic (poor public opinion).
- Nuclear fusion technologies are long-term prospects that depend on meeting cost and performance research goals.
- Many advanced cogeneration technologies face cost and performance research goals before becoming commercially available.
- Most cogeneration technologies face varying degrees of environmental and financial constraints.
- Although fuel-cell technologies are not currently available, phosphoric acid, molten carbonate and solid oxide systems are expected in the near-term.
- All fuel-cell technologies must address substantial research goals for reduced cost and improved performance.
- All fuel-cell technologies must overcome financial and utility integration constraints before being effectively deployed.
- Storage technology options are lacking due to environmental, governmental and location constraints for pumped hydroelectric facilities, and the noncommercial status of compressed air, utility-scale battery and superconducting magnetic energy storage technologies.
- High capital costs and the need for improved performance are major research and development issues associated with most renewable energy technologies, including geothermal, biomass, municipal solid waste, wind, solar thermal electric, photovoltaics and ocean energy.
- Renewable and alternative energy technologies often are confronted with financial and utility integration constraints. Qualifying facility development is expected to be constrained because the low short-term avoided cost rate does not economically support technology development. In addition, short-term contracts inherently include too many uncertainties to risk intensive capital investments. A new long-term standard offer contract is available in locations that need baseload power, with the price set by competitive bidding.
- Resource constraints are a potential "show stopper" limiting the extensive use of many renewable energy technologies in California including vapor-dominated and advanced geothermal resources, hydroelectric, biomass and ocean energy conversion.
- Because utilities pass fuel costs on to ratepayers, they lack incentive to develop new technologies that could improve the energy efficiency of new and existing power plant facilities. The future bidding process for electric power generation resources may increase the incentive for utilities to improve energy efficiency.

End-Use Technologies General Conclusions

- The lack of incentive for building owners and developers to incorporate energy efficiency measures and minimal industry acceptance of cost-effective energy technologies limit the deployment of most end-use technologies provided as original equipment in new buildings. This situation is because building owners and developers have no incentive to improve energy efficiency where tenants or new owners pay their own energy bills. Until builders include more efficient technologies in new buildings, the only major market for energy efficient end-use technologies will be for replacing old equipment. Without government or utility actions, these constraints will be mitigated only when demands of tenants and building buyers create a market advantage to include cost-effective energy technologies.
- Low end-user awareness is a particularly critical deployment issue for new end-use technologies. Unlike electric generation technologies, end-users are widely

dispersed, have fewer resources and lack familiarity with technical concepts.

- Substantial energy savings can be derived from many commercially available end-use technologies that have very limited use though they are highly cost competitive. These technologies include: condensing and waste heat recovery commercial water heating, heat pump water heaters, ceramic fiber burner heating, evaporative cooling, passive solar heating and cooling, high efficiency lighting and control systems, high efficiency refrigerators, advanced clothes washers and dryers, advanced office equipment, advanced industrial technologies, variable speed drive AC motors, energy management systems, customer controlled automated feedback transmission systems, amorphous cored transformers, and synthetic taped transmission cables. Thus, government or utility programs targeted at the "lack of incentive for building owners and developers" and "low end-user awareness" issues identified in the previous two items could yield reduced energy consumption.
- Most advanced energy conservation technologies have favorable levelized costs compared to electric generation technologies.
- A broad range in the levelized cost of heating and cooling technologies exists because their economic viability is dependent on load requirements. Typically, high-cost heating and cooling technologies are not cost competitive in many of California's moderate climate areas.
- Most new lighting technologies are cost-effective energy options for new commercial buildings.

TECHNOLOGY EVALUATION MATRIX

The Technology Evaluation Matrix in Figure 1 summarizes the results of assessments of each energy technology for the three evaluation factors: 1) Commercial Status, 2) Research and Development Goals, and 3) Deployment Issues.

The matrix uses a graphic system to indicate the evaluation results. The darker the box under each criterion, the less a technology satisfies that criterion.

COMPUTER APPLICATIONS FOR ENERGY-EFFICIENT SYSTEMS

VIKAS R. DHOLE*
ROBIN SMITH
BODO LINNHOFF
University of Manchester Institute of Science and Technology

Energy is used whenever anything is made and traditional energy use means polluting emissions. This article deals with techniques that can, largely thanks to the development of computers, minimize that energy for industrial applications. The techniques, known collectively as process integration, were initially directed toward the design of chemical plants. But because they are based on fundamental principles, they may generally be applied to any process that uses networks of heating, cooling, and power to drive it. See also ENERGY EFFICIENCY; ENERGY MANAGEMENT.

Setting aside small items such as pumps and valves, there may be as many as 50 major plant items to consider in, eg, a modern petrochemical plant. If there are just three possible variations for each of these, that gives 3^{50} possible process designs. Of course, many of these alternatives will be too impractical for realistic consideration, but this simple calculation gives an idea of the immensity of the task of finding the best design (1,2).

Before the use of computers, designing a plant was an extremely laborious process. If any adjustment to completed designs was made, it took a lot of engineering time to recalculate the effect that this might have on energy consumption. So the advent of computer-aided engineering (CAE) (in this context building a mathematical model of the process inside the computer) allowed designers to get a step closer to the ultimate in energy efficiency by simulating the results of many possible changes to the process while it was still at the conceptual stage. This obviously made investigations much cheaper, quicker, and more wide ranging than when they were carried out by hand.

There is an important aspect of design, however, that this number crunching approach does not address. That is the problem of synthesis: knowing where exactly the process configuration should be changed and targeted and what the possible energy savings should be. Without proper synthesis and targeting, designers were effectively still working in the dark. While they might be able to find out the consequences of proposed changes more quickly, they still had to rely on common sense, rules of thumb, and experience to come up with suggestions as to which changes were the right ones. A lot of expensive computer time could be wasted trying to achieve the impossible, or the designer might (and this is the most likely outcome) end up with a process that uses more energy than it really needs to.

Toward the end of the 1970s, the initial steps were taken that would eventually lead to the family of techniques known today as process integration. The first big breakthrough was known as pinch technology and was centered around a graphic and thermodynamic approach. Figure 1 shows the effect that the use of pinch technology can have on the learning curve toward reduced energy consumption. Rather than building plants and thinking that the next one will be better, pinch technology allowed designers to take an entirely fresh and objective view of the energy demands of the process. The result was a "step change" in energy savings.

The use of pinch technology yielded typical energy savings of 20% to 30% at a sweep, coupled with capital cost savings related to optimizing the number and size of the heat exchangers needed to service the process (3). Pinch technology takes a two-stage approach to the problem. First, it defines realistic targets so that designers know where they are headed. Second, from the basic thermodynamic data generated in the targeting stage, it takes a logical, step-by-step approach to lead the designer to the right design to achieve the targets.

* Currently with Linnhoff March Ltd., UK.

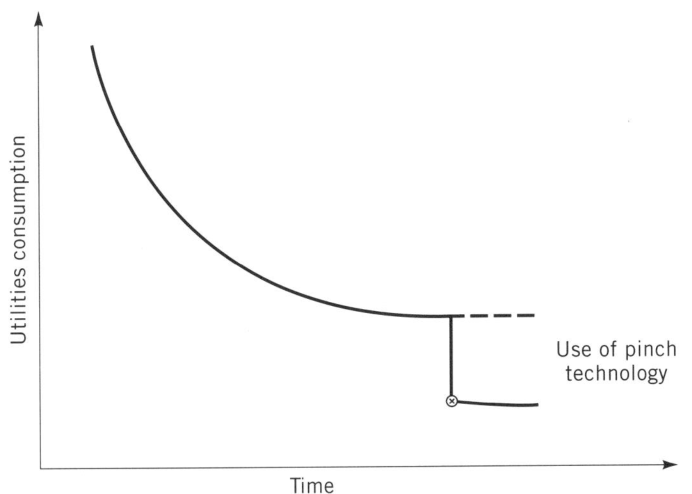

Figure 1. The use of pinch technology results in a step change in energy savings.

BASIC ENERGY TARGETING

Energy targeting using pinch technology starts from a basis of complete heat and material balance for the process. This initial stream information can be obtained from the plant data or from a converged computer simulation of the process.

Data Extraction

There may be several process streams, at a variety of different temperatures, and a majority of them must be heated or cooled before they can progress to the next stage of the process. But they can be simply divided into two groups: heat sources and heat sinks. Heat sources (also known as "hot streams," regardless of their temperature) need to be cooled down, whereas heat sinks (also known as "cold streams") need heat input.

Composite Curves

Each stream can be represented on a graph of heat content (enthalpy) versus temperature (2,5). Figure 2**a** shows two hot streams drawn in this way. The mass flow rates and heat capacities (CP) (how much energy is needed to heat the streams for each degree Celsius of temperature rise) of the streams are known. Stream 1 is cooled from 200° to 100°C. It has a total heat capacity (the mass flow rate times the specific heat capacity) of 1 kW/°C so it loses 100 kW of heat energy. Stream 2 is cooled from 150° to 50°C. It has a CP of 2 and loses 200 kW of heat.

The hot composite curve is produced by simply adding the heat content of the two streams over the different temperature ranges (Fig. 2**b**).

Figure 3 shows hot and cold composites plotted on the same temperature–enthalpy graph. The two curves overlap along the enthalpy axis for most of their length. In this region, it should be possible to heat the cold streams using the heat removed from the hot ones without the need to use utilities to provide external energy. But at each end of the curves there is an overhang where heat from within

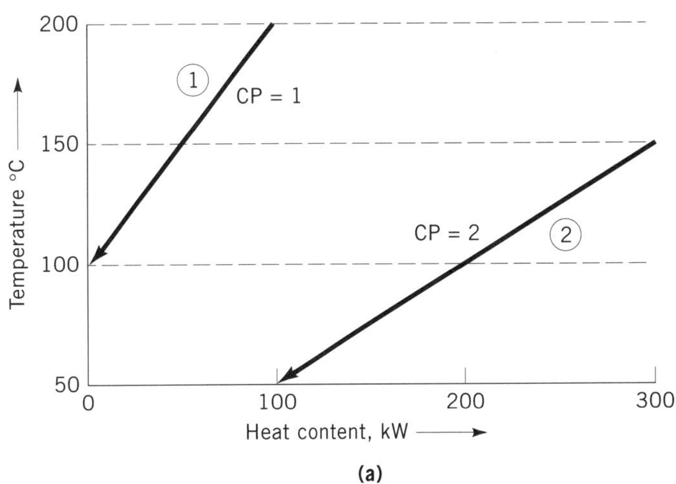

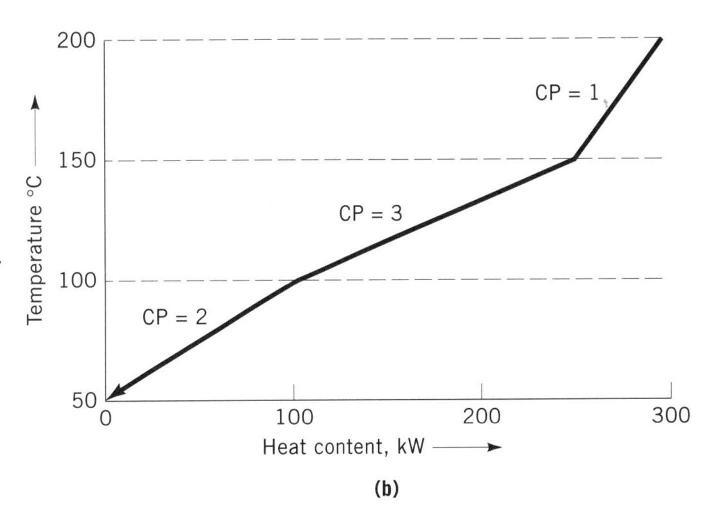

Figure 2. Composite curve construction.

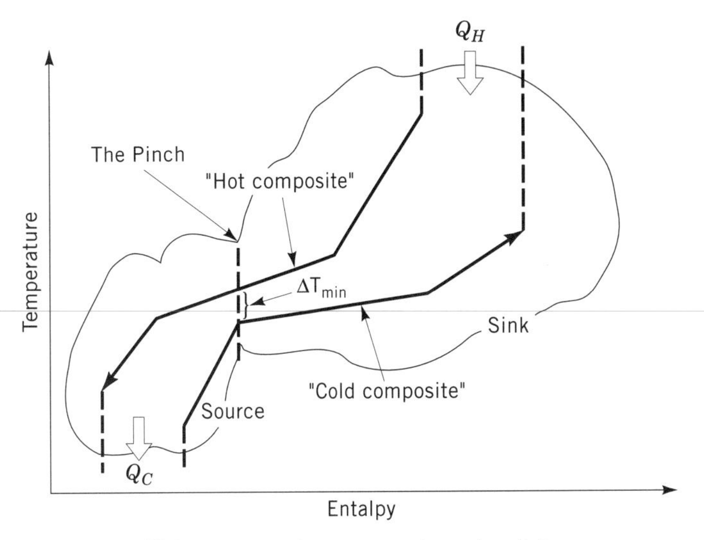

Figure 3. Using composite curves to set minimum energy targets.

the process itself cannot be used. The overhang at the top of the cold composite curve represents the minimum amount of external heating that will be needed to drive the process Q_H. Similarly, the overhang at the bottom of the hot composite represents the minimum amount of external cooling Q_C. The minimum hot and cold utility requirements are targets. They are set before design, purely on the basis of heat and material balance information.

At one point along their length, the composite curves reach their closest temperature approach. This is known as the "pinch," and recognizing its implications allows energy targets to be realized in practice. It will start designers on the road to designing a heat exchanger network that conserves energy.

Once the pinch has been identified, it is possible to consider the process as two systems: one above and one below the pinch. The system above the pinch requires a heat input and is, therefore, an overall heat sink. Below the pinch the system rejects heat and so is an overall heat source. This gives three golden rules that the designer must stick to in achieving the minimum energy targets (3):

1. Heat must not be transferred across the pinch.
2. There must be no outside cooling above the pinch.
3. There must be no outside heating below the pinch.

Targeting for Multiple Utilities: The Grand Composite Curve

The vast majority of industrial sites use steam to transfer heat to the process and cooling water for removing the excess heat. However, for high temperature heating hot oil circuits or furnace heating is usually used, while refrigeration is used for the low temperature cooling duties. Steam is normally provided by an on-site boiler or combined heat and power system using steam turbines. This design gives operators the flexibility to provide steam at several different temperatures and pressures.

It is usually preferable to use the cheapest utilities wherever possible, eg, steam at the lowest feasible temperature and pressures and cooling water instead of refrigeration. The composite curves give the designer overall energy targets but do not indicate how much energy needs to be supplied by different utilities. The tool that provides this information is called the grand composite curve (2,6).

The designer starts with the hot and cold composite curves. It is assumed that the temperature difference between the composite curves at the pinch is the minimum permissible temperature difference (ΔT_{min}). This also must be taken into account when transferring heat between the process streams and the utilities. The first step is to shift all the streams in the hot composite curve colder by $\frac{1}{2}\Delta T_{min}$. Similarly, all the streams in the cold composite are shifted to a temperature that is higher by $\frac{1}{2}\Delta T_{min}$. The reasons for this will become clear shortly. The grand composite curve can then be constructed from the enthalpy (horizontal) differences between the shifted composite curves at different temperatures. Because of the temperature shifts made to the composite curves, the curves will actually touch each other at the pinch. Thus the point on the grand composite curve that touches the temperature axis corresponds to the pinch, and the extremes of the curve correspond directly to the overall energy requirements. The section of the curve above the pinch is the overall heat sink of the process, and the heat source is represented below the pinch.

Capital Energy Trade-offs

Capital cost considerations are vital in determining whether an energy-saving project will go ahead, so they must be considered at an early stage. For heat-transfer networks, the overall capital cost is made up of two principal components: the minimum overall surface area for heat transfer (in effect the size of the heat exchangers) and the minimum number of units.

HEAT EXCHANGER NETWORK DESIGN

Targeting is only the first stage in creating a finished design. When it comes actually to deciding how to set up the network of heat exchangers, the grid diagram (2) representation of the network is much clearer.

The Grid Diagram

A basic grid diagram is shown in Figure 4. The hot streams are shown at the top of the figure, running left to right. Cold streams run across the bottom, from right to left. A heat exchanger transferring heat between the process streams is shown by a vertical line joining circles on the two matched streams. A heater is shown as a circle with an H and a cooler is a circle with a C.

The pinch is shown as a vertical line cutting the process into two parts. From the composite curves and optimum ΔT_{min}, it is known at what temperatures the hot and cold streams encounter the pinch. The heat sink section (above the pinch on the T–H plots) is shown to the left and the heat source section is to the right.

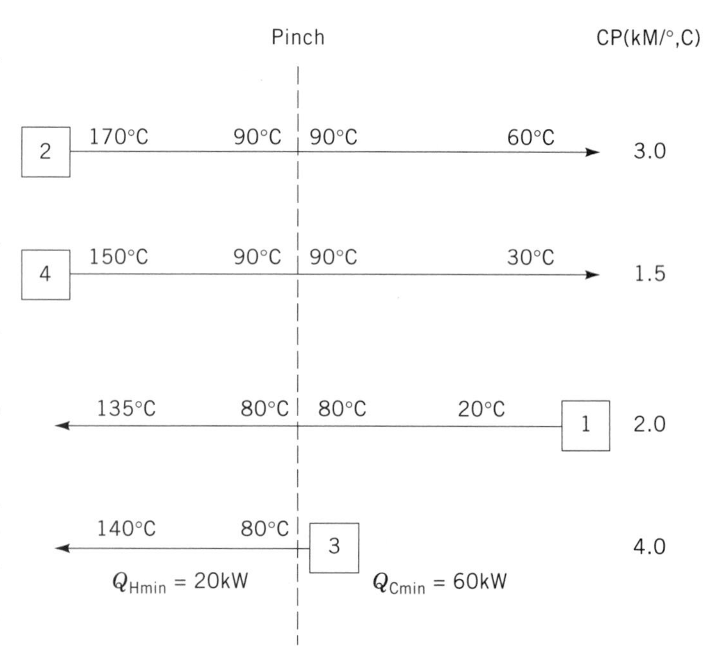

Figure 4. Grid diagram for the example problem.

Pinch Design Rules

The three golden rules are central to the way a designer goes about designing the network. There must be no external cooling used above the pinch (on the left) so hot streams on this side must be brought to pinch temperature by heat transfer with cold streams on the same side, ie, on the left. Similarly, cold streams on the right must be brought up to the pinch temperature using hot streams on the right rather than utility heating. The $\Delta T_{\min}$ puts another constraint on the design because it has been defined as the minimum temperature difference for heat transfer anywhere in the system.

Special Cases

For the constrained region around the pinch, there is a problem if there are more streams going into the pinch than there are coming out. According to the golden rules, each stream going into the pinch needs to be matched with an outgoing stream to bring it to the pinch temperature. Therefore, it is necessary to split the outgoing streams to equal the number of streams going into the pinch.

Network Optimization

After the network has been designed according to the pinch rules it can be further subjected to capital energy optimization. Remember that the network is subjected to optimization before design. By choosing the optimum $\Delta T_{\min}$ for the network design the designer addresses capital energy trade-offs. Optimization after network design, therefore, will only fine-tune the network costs. Network optimization is mainly carried out by using standard optimization routines. Optimization plays on extra degrees of freedom within the network.

RETROFIT MODIFICATION OF HEAT EXCHANGER NETWORKS

The designs examined so far started from scratch, but the majority of process integration projects are aimed at achieving energy savings on existing plants. Because of environmental problems, it is necessary to restrict emissions from industry; thus it could prove vital that energy consumption on existing installations is checked, helping industry to meet energy targets without having to cut production.

Before process integration, there were two approaches to retrofit design, both of which are still used to some extent. The first is known as design by inspection. The designer looks at the plant and chooses a project intuitively. The problem with this is that it will always leave remaining doubts. Because it does not necessarily involve a true understanding of all the factors at work, there might always be a better answer. The second approach uses computers purely as number crunchers. When simulation packages are available, some designers think that churning out many alternative network designs could be the answer. But, as with design by inspection, this approach lacks a true understanding of the process.

Targeting for Retrofits

Like grass-roots designs, proposed retrofit projects will always have a trade-off between energy and capital costs. In industry, this is usually expressed as a payback time on capital investment. It is common, for example, to expect to get the money that the company has invested at the start of a project back in energy savings over 2 years.

The basic difference between the cost considerations when designing for retrofits and designing from scratch is that a significant amount of hardware is already in place. The aim is to take up any slack in the system by making sure that existing units are used as effectively as possible. It then follows that any extra capital investment will also be minimized.

Designing for Retrofits

The targets indicate what the designer should be able to achieve. To get a clear picture of the situation, the existing heat-exchanger network is drawn up on a grid diagram. The position of the pinch and $\Delta T_{\min}$ were identified during the targeting stage. For this example, the target $\Delta T_{\min}$ of 19°C was identified. The corresponding grid representation is shown in Figure **5a**. This allows the designer to see which heat exchangers are working across the pinch (cross-pinch exchangers). These units must be removed from the design if heat transfer across the pinch is to be eliminated and hence the energy requirements of the process. This leaves the partial network shown in Figure **5b**.

It is then necessary to complete the network using the same procedures as used for new heat exchanger networks. In this case, however, the designer will obviously save on capital costs if he or she uses the existing units that were removed in the first step wherever possible. Figure 6 shows the resulting network for the example problem. A computerized optimization program can be used at this stage (eg, Supertarget). Loops and paths can be used to give added flexibility by helping to make old exchangers fit new duties.

PROCESS MODIFICATIONS

It has been accepted that heat-exchanger network design requires, as a starting point, the underlying process heat and material balance, and that much could be gained by identifying process changes to complement network design changes. However, it has also been accepted that opening up the question of process changes is like opening up Pandora's box. There are an infinite number of settings for reactor conversion, evaporator stages, distillation column pressures and reflux ratios, feed vaporization pressures, pump-around flow rates, etc. The number of choices is so large that it seems an impossible goal to predict confidently which three or four such parameters could be changed to advantage in the overall context. Against this background, the designer now has simple results that allow him or her to discuss with great confidence many different process parameters from the point of view of their ultimate impact on the overall design.

So far the discussion has concentrated on the effect of process modifications on energy costs. Evidently, if capital cost changes are not considered, the conclusions are likely to be a little tenuous. There are two relevant elements of capital cost. First, capital costs will change in the unit operations of the process (eg, the distillation columns and reactors). Second, capital costs will change in the heat exchanger network, which has as yet to be designed. Capital cost changes in the unit operation must be estimated by shortcut calculations. Capital cost changes in the network can be predicted by using network capital cost targets. With a little practice, the principles enable the engineer to screen quickly from among, say, 50 possible modifications 3 or 4 that will lead to beneficial overall cost effects.

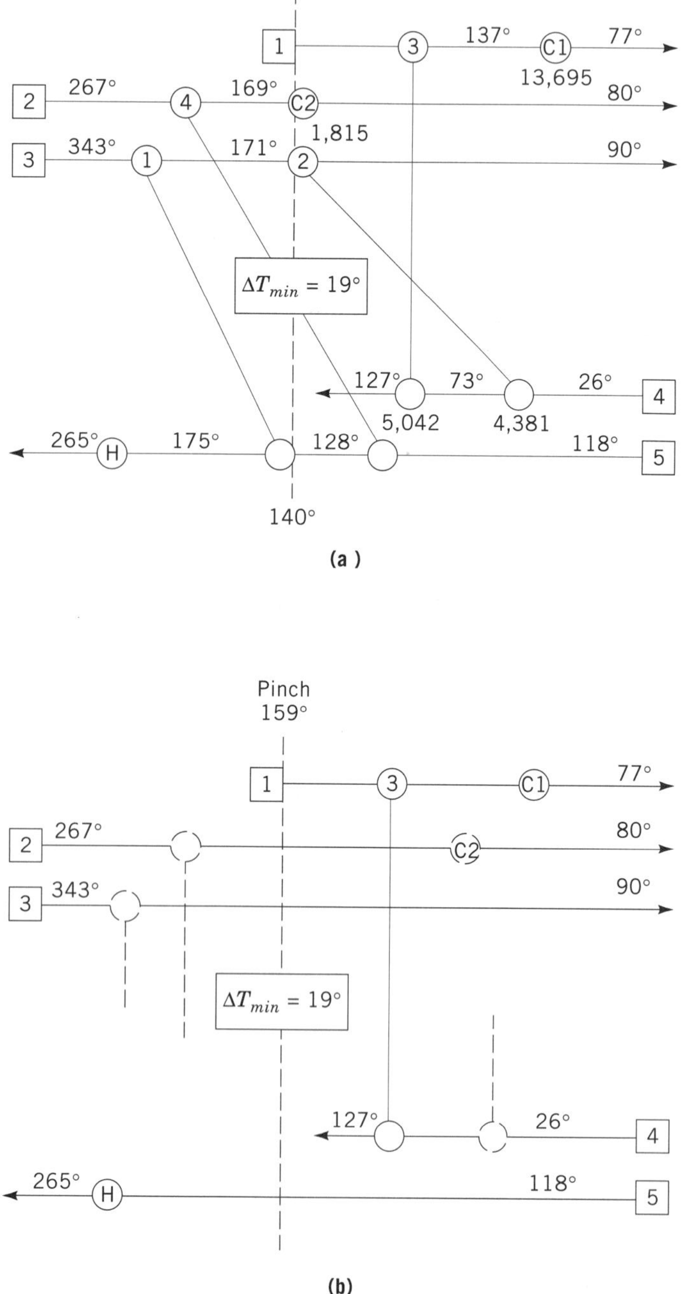

Figure 5. Designing the retrofit. (**a**) The network, initialized for retrofit, highlights exchangers that work across the pinch. (**b**) The cross-pinch exchangers must be eliminated before the network design is developed.

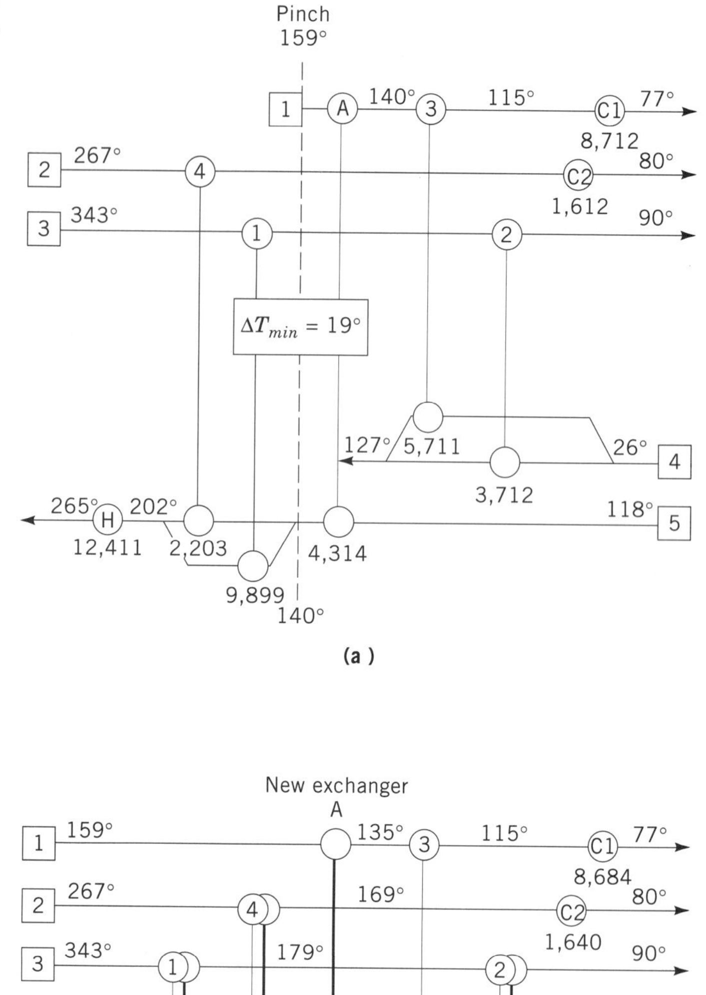

Figure 6. The completed network. (**a**) The preliminary design involves redeploying existing exchangers and adding new units. (**b**) The improved design employs all existing exchangers and offers a 1.9-yr payback.

PLACEMENT OF HEAT AND POWER UTILITY SYSTEMS

It would be a rare process that did not use mechanical energy, or power. Pumps, fans, process compressors, and refrigeration systems all use power, not heat, to drive them. Because mechanical energy is generated by heat in the first place, eg, by burning fossil fuels to produce the electricity to drive motors, and because no conversion of energy can ever take place without incurring losses, power is more expensive than heat. Typically, the price ratio between power and heat (or fuel) varies between 3 and 4.

So far, the discussion has only examined heat exchanger networks and their interaction with process conditions, but integrating power or combined heat and power systems correctly into the process is vital in regard to overall energy efficiency.

Heat Engines: Combined Heat and Power

A heat engine takes in heat from a higher temperature and rejects it at a lower temperature. The difference between the two heat loads goes into mechanical work or power. In practical terms, this is usually turbine systems. There are two main types of turbine systems: stream and gas.

Usually, processes use the heat engines to provide for the process heat as well as power demand. Steam turbine systems are used more often in the process industry than gas turbine systems. This is mainly because steam is a more convenient heating medium than gas.

Figure 7 shows three possibilities for the placement of heat engines with a process. The process is represented by two regions: one above and other below the pinch. If a heat engine falls across the pinch, it wastes energy (15). The use of the heat engine increases the hot and the cold utility requirements (Figure 7**a**). The energy is wasted because the heat engine would remove heat from the overall heat sink and transfer it to the overall heat source, therefore placing extra demands on both the hot and cold utilities.

If the heat engine is placed so that it rejects the heat into the process above the pinch temperature (Fig. 7**b**), it transfers that heat to the process heat sink, thereby reducing the hot utility demand. Due to the heat engine, the overall hot utility requirement is only increased by W (ie, shaft work). This implies a 100% efficient heat engine. The heat engine is, therefore, appropriately placed.

If the heat engine is placed so that it takes in energy from the process below the pinch temperature (Fig. 7**c**), it can take that energy from the overall process heat source and reject it directly to utilities. Here the engine effectively runs on process heat free of fuel cost and reduces the overall cold utility requirement by W. The heat engine is again appropriately placed.

When integrating the heat engines with the process, designers must consider the grand composite curve of the process. The grand composite curve determines the appropriate temperature levels and the heat loads for heat rejection from the steam and the gas turbines.

Placement of Distillation Columns as Separator Engines

Distillation is one of the most commonly used techniques for separation in the process industry. A distillation column separates a mixture into a low boiling fraction and a high boiling fraction. The column accepts heat at higher temperature (ie, reboiler temperature) and rejects heat at a lower temperature (ie, condenser temperature). Thus a distillation column can be treated as a separation engine.

Similar to the heat engines, a distillation column is inappropriately placed if it runs across the pinch. The designer must avoid the temperature overlap between the distillation column and the grand composite curve of the background process. The appropriate placement for a distillation column when the column lies entirely above the pinch or below the pinch and provided that the grand composite curve of the background process can accommodate the column. The integration possibilities between the column and the background process can be improved by introducing column modifications such as side condensing–reboiling and feed preheating–cooling. Recent developments in this area have been discussed (7).

Heat Pumps in the Process Context

A heat pump accepts heat at a lower temperature and, by using mechanical power, rejects the heat at a higher temperature. Heat pumps, therefore, operate in an opposite manner from heat engines. The best arrangement is to take in heat below the pinch and reject it at a temperature above the pinch. This provides savings in both hot

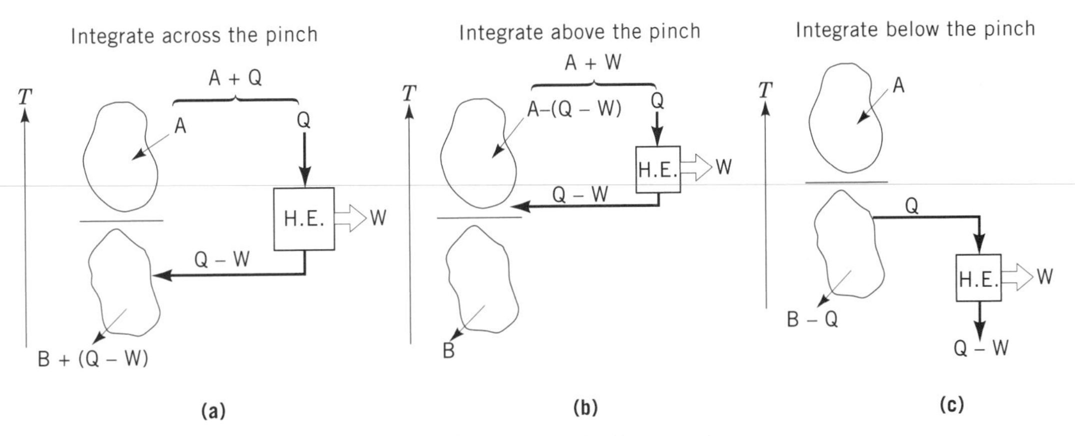

Figure 7. Placement of the heat engines.

and cold utilities because it is pumping heat from a source (below) to a sink (above) (2,8). So the best place to integrate heat pumps is across the pinch.

Shaft Work Targeting for Low-Temperature Processes

A low temperature process is serviced by a refrigeration system. A refrigeration system is in principle a heat pump that accepts heat at below ambient temperatures and rejects it to the ambient. In doing so it requires shaft work. In the process industry a refrigeration system usually operates through multiple levels of refrigeration and heat rejection. Using the grand composite curve for a low temperature process it is possible to determine appropriate heat loads for refrigeration and heat rejection. Based on these heat loads, the shaft work requirement for the refrigeration system can be determined through simulation.

A recent development in pinch technology provides a technique that bypasses the effort of refrigeration system simulation to predict its shaft work requirement (9). It directly sets shaft work targets for the refrigeration system. When the temperature axis of the grand composite curve is replaced by the Carnot factor ($\eta_c = (1 - T_o/T)$), the new grand composite curve is termed the exergy grand composite curve. The area between the exergy grand composite curve and the refrigeration levels is directly proportional to "exergy loss," or ideal shaft work loss in heat transfer. As the number of refrigeration levels is increased, the exergy loss will reduce. It can be shown that the change in the exergy loss is directly proportional to the change in the refrigeration shaft work (9). Therefore, targets can be set for the change in the refrigeration shaft work consumption for the various options by simply evaluating the change in the exergy loss. This concept can be further extended to minimize the overall refrigeration shaft work requirement of a low temperature process (10).

TOTAL SITE TARGETING

The discussion so far has concentrated on heat and power targeting for a single process. Typically refinery and petrochemical processes operate as parts of large sites or factories. These sites have several processes serviced by a centralized utility system involved in steam and power generation. Figure 8 shows a schematic of a total site, involving several processes and a central utility system. There is both consumption and recovery of process steam via the steam mains. The utility system consumes fuel (eg, gas and lignite), generates power, and supplies the necessary steam through several steam mains. The site imports or exports power to balance the power generated.

The process's steam heating and cooling demands dictate the sitewide fuel demand and co-generation potential via the utility system. The heating and cooling requirements of the individual processes are represented by the respective grand composite curves. The grand composite curve lays open the process–utility interface for a single process. However, for a site involving several processes, the grand composite curve of each process will suggest different steam levels and loads. How can the correct compromise in steam levels on a sitewide basis be identified?

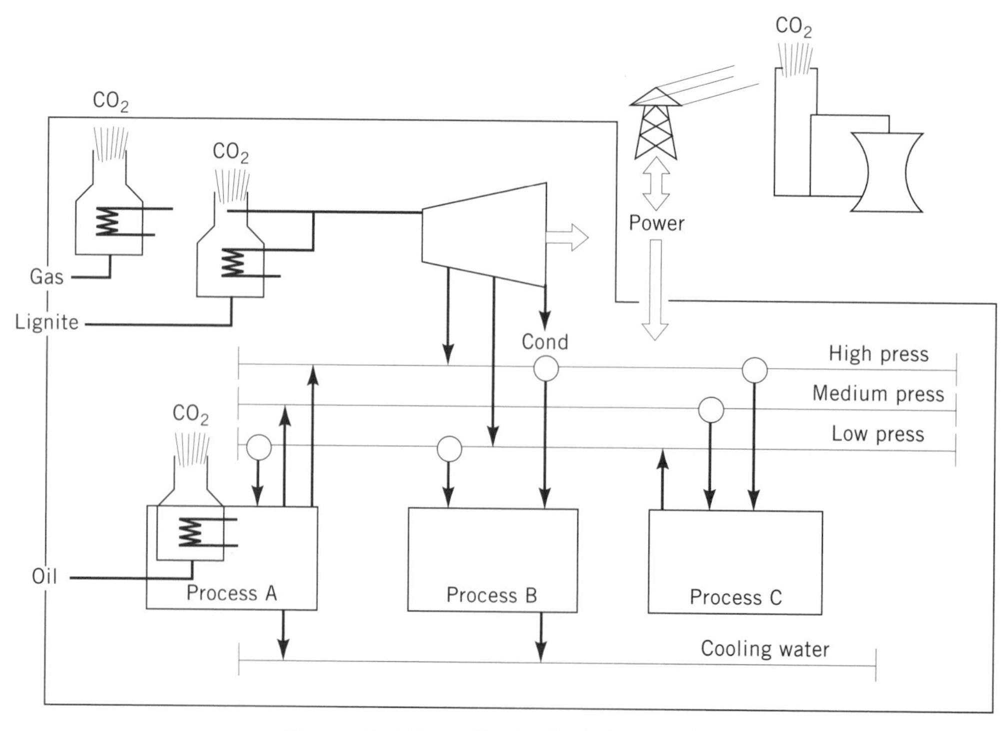

Figure 8. Schematic of a typical total site.

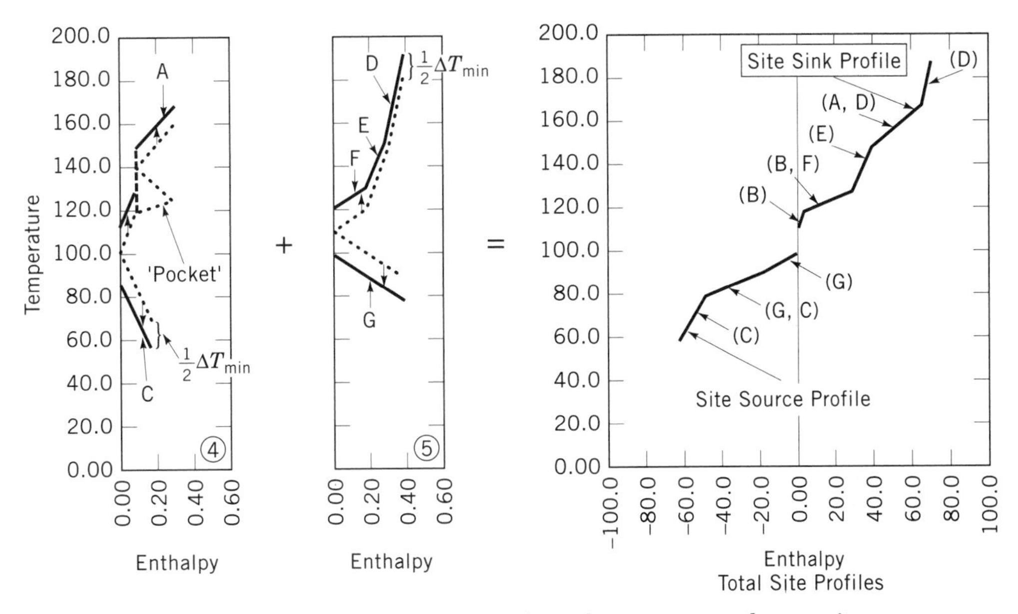

Figure 9. Constructing total site profiles from the process grand composite curves.

Total Site Profiles: Sitewide Utilities Targets

A recent development in pinch technology enables the designer to set utilities targets for total sites. This development is based on thermal profiles for the entire site called total site profiles (11). Total site profiles are constructed from the grand composite curves of the processes in the site. Figure 9 illustrates the construction of the total site profiles, using two processes. The construction starts with the grand composite curves of the individual processes. The grand composite curves are then modified in two steps. (*1*) The nonmonotonic parts, or the "pockets," in the individual grand composite curves are sealed off through vertical lines. (*2*) The source and sink elements of the resulting grand composite curves are shifted by $1/2\Delta T_{\min}$ (Fig. 9). Source element temperatures are reduced by $1/2\Delta T_{\min}$ while the sink element temperatures are increased by $1/2\Delta T_{\min}$.

Figure 10 illustrates an approach for setting the co-generation target using a plot of Carnot factor versus enthalpy.

Figure 11 illustrates sitewide utilities targeting based on total site profiles. The figure also shows the corresponding utilities schematic. Note that the vertical axis has been changed from temperature to the Carnot factor. This makes it possible to set direct targets for co-generation. Thus starting from individual processes grand composite curves the designer can set targets for site fuel, co-generation, site emissions, and cooling (11).

TARGETING FOR EMISSIONS

This article has summarized the techniques used to target the amount of energy a plant or a site needs. Remember that the principal aim is to use the energy-efficient designs

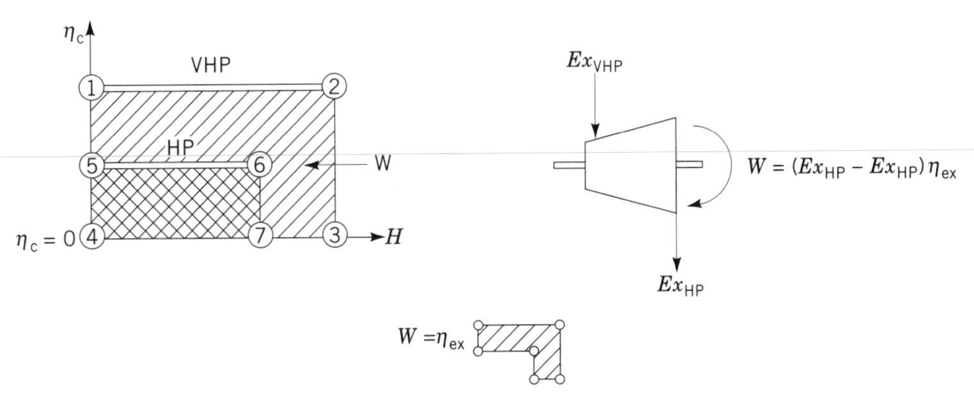

Figure 10. Shaft work targeting for co-generation.

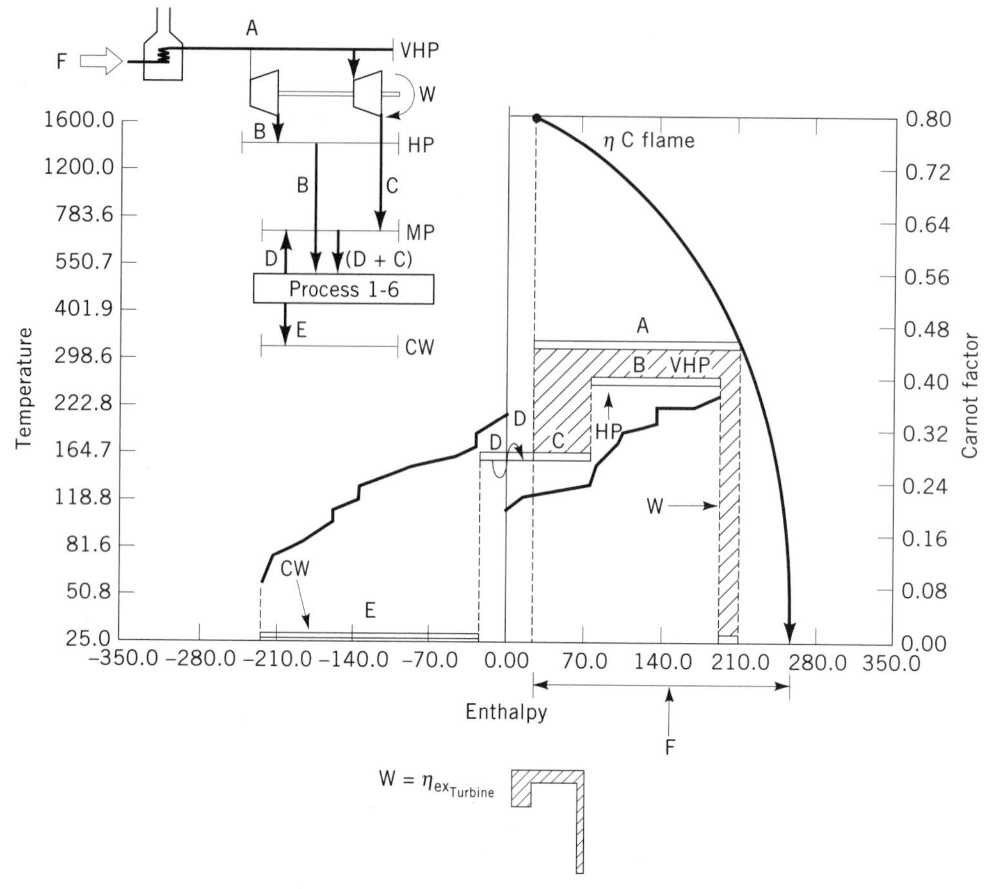

Figure 11. Total site targeting for fuel, co-generation, emissions, and cooling.

to improve the environmental performance of plants; thus how can the designer target for the specific pollutants that relate to energy consumption?

Defining Emissions Targets

Historically, air pollution control was established to tackle local problems like smog in the Los Angeles basin or in London. It is now recognized that problems such as global warming and acid rain must be tackled on a worldwide basis, with absolute limits placed on emissions. It is preferable to define the emissions of specific pollutants in terms of mass flow rate rather than concentration, which can be misleading (12). It is also important to look at the total energy-related emissions arising from the operations at a site on a global basis, ie, taking into account emissions from the power station supplying the site, rather than just emissions at the site itself.

Modeling for Specific Pollutants

Four pollutants will be considered:

1. Carbon dioxide (implicated as a major contributor to global warming).
2. Sulfur oxides (a major contributor to acid rain).
3. Nitrogen oxides (a factor responsible for acid rain).
4. Particulates (lead to smog and health hazards).

Carbon dioxide emissions are directly related to the fuel burned via a straightforward stoichiometric model:

$$C_nH_m + (n + m/4)O_2 \rightarrow n\ CO_2 + (m/2)H_2O$$

For predicting the amount of CO_2 it is usually accurate enough to assume that the combustion system is an efficient one with enough excess air added to the fuel to ensure that carbon monoxide emissions are negligible. In other words, all the carbon in the fuel results in CO_2.

The same type of model can be used for sulfur oxides:

$$C_nH_mS_p + (n + m/4 + p)O_2 \rightarrow n\ CO_2 + m/2\ H_2O + p\ SO_2$$

Some of the sulfur dioxide will also react to form SO_3, but because it forms roughly only 10% of the total sulfur oxide emissions, to predict the amount of SO_2, it is usually good enough to assume that SO_2 is the only product.

The situation with nitrogen oxides is more complex. Nitrogen oxides (NO_x) form via two main reaction paths (13). Thermal NO_x is formed from nitrogen in the combustion air, particularly at high temperatures:

$$N_2 + O_2 \rightleftharpoons 2\ NO$$

$$NO + \tfrac{1}{2}\ O_2 \rightleftharpoons NO_2$$

Fuel-bound NO_x is formed from nitrogen in the fuel at low as well as high temperatures. But part of the fuel nitrogen reacts directly to form N_2. To complicate things even more, N_2O and N_2O_4 are also formed in various reactions. The mixture of NO_x in stack emissions depends on kinetics, mixing, mass transfer, and thermodynamics.

It is virtually impossible to calculate a precise value for the NO_x for a real device. Modelers tend to use estimates based on experimental observations for different combustion technologies such as boilers, gas turbines, and fired heaters (14). At best, NO_x predictions are usually good enough only to indicate trends rather than precise figures.

Particulates derive from two sources. First, metals may be present in the fuel, which oxidize during combustion. Second, unburned hydrocarbons and carbon may exist. As with NO_x, kinetics, mixing, and thermodynamics will influence the amount of fuel that may remain unburned. It is usually sufficient to consider only the metals, assuming that all metals present in the fuel and inorganic ash are emitted after combustion.

Analyzing the Utility Options

It is possible to relate the emissions to the amount of fuel used. Thanks to pinch technology, it is also possible to calculate the amount of energy a utility system must supply. This still leaves a gap: it is necessary to model the utility systems, such as furnaces or turbines, to relate the amount of energy they product to the amount of fuel they consume. There are simple models available in the literature for furnaces and steam boilers and gas and steam turbines (2,12).

The method of relating utilities to the process using the grand composite curve and total site profiles has been discussed. By adding the models for utility fuel consumption and the emissions from the given fuel, the designer can use already established techniques to target directly for on-site emissions. Computers obviously provide the simplest means of putting this into practice.

A more global view can be taken if there is on-site power generation from the utility turbine system. While the turbine is creating emissions on site by placing extra demands on the furnace or boiler, it is reducing the emissions off site by cutting down on the amount of power that must be imported from the external supply grid (12,15–19). And because the big power generators may be operating less efficient plants (or using a different, dirtier fuel such as coal instead of gas) the global emissions from an on-site power device can even turn out to be negative. The use of total site profiles provides a direct interrelation between the process changes, fuel switching alternatives, and heat exchanger network modifications on local and global CO_2 (11,20).

THE ROLE OF SOFTWARE

The sheer size and complexity of most practical problems demands the use of software of some kind. In the implementation of an energy integration study it is possible to identify two broad activities that call for the use of software: targeting and network design.

Targeting

Targeting uses physical principles such as thermodynamics to define key parameters for the energy system before design. The parameters that can be targeted before design include the following:

- Minimum energy consumption.
- Minimum heat-exchanger surface area.
- Minimum number of heat exchangers.
- Minimum total cost of the heat-exchanger network (capital and energy).
- Minimum shaft work required by refrigeration systems.
- Maximum shaft work generated by co-generation systems.
- Minimum flue gas emissions (CO_2, SO_x, and NO_x).

The actual design task can be lengthy and it is usually impractical to study many design alternatives by carrying designs through in detail. The power of targeting is that it can be used to screen many design alternatives without having to resort to the design task itself.

Targeting clearly calls for software to explore these alternatives and sensitivities. However, because effective targeting also calls for creativity on the part of the designer to obtain the best results, it is important that the targeting software should be interactive with a graphical user interface.

Network Design

Once the major design parameters have been initialized during targeting, the task turns to network design. In network design, the topology of the heat-exchanger network and utility system is created. From the targeting, the designer knows what to expect in terms of the major design parameters such as energy consumption. Knowledge of the targets allows the validity of the design to be immediately assessed. More important, the targets allow the design to be initialized at the correct settings.

Two types of software are used at the network design stage. The designer can interact with a graphical presentation of the design, adding heat exchangers one at a time, while following certain rules and heuristics. The role of the software in this interactive mode is to carry out an evaluation of the design as pieces of equipment are added. Alternatively, the designer can use software that attempts to carry out an automatic network design. With large complex networks automated network design is attractive. However, the designer is removed from the decision-making processes, which can lead to designs that have some undesirable features.

Two different approaches can be used for targeting and network design.

Pinch Technology

Pinch technology as reviewed here has a philosophy of analysis and design of energy systems that is based entirely on physical principles. Targeting uses thermodynamics, graph theory, basic heat transfer theory, and economics to predict the key design parameters. Design relies on the decomposition of the problem into subproblems at the pinch (or pinches). In addition to the decomposition, thermodynamic rules are applied as the network is built up. A key feature of the approach is that it relies on the intervention of the designer. This has advantages and disadvantages. The principal advantage is that the designer can address the many intangibles, such as safety and layout, as the design evolves, maximizing the practicality of the final design. The major disadvantage of the intervention is that the designer must be an expert in pinch technology to obtain the best results.

Mathematical Programming

When mathematical programming is used, the intervention of the designer is replaced by an optimization algorithm that makes discrete decisions. This approach starts with the construction of a superstructure. The superstructure is a design that has all feasible alternatives embedded within it. The superstructure is then subjected to optimization. Should a structural parameter be optimized to zero, that feature is deleted from the design. The major disadvantage of this approach is that the designer is removed from the decision-making process, and so the many intangibles of process design cannot be accounted for.

The two types of software used in targeting and network design have advantages and disadvantages. It is likely, therefore, that each has a role to play.

BIBLIOGRAPHY

1. B. Linnhoff, *Chem. Eng. Res. Des.* **61** (July 1983); latest version in *Chem. Eng. Prog.* (*Aug.* 1994).
2. B. Linnhoff and co-workers, *A User Guide on Process Integration for the Efficient Use of Energy,* The Institution of Chemical Engineers, Rugby, UK, 1982.
3. B. Linnhoff and D. Vredeveld, *Chem. Eng. Prog.,* 33–40 (July 1984).
4. B. Linnhoff and V. Sahdev in *Ullmann's Encyclopedia of Industrial Chemistry,* 1987, Vol. B3, pp. 13-1–13-6.
5. B. Linnhoff, G. Polley and V. Sahdev, *Chem. Eng. Prog.,* 51–58 (July 1988).
6. B. Linnhoff and G. Polley, *Chem. Eng.* 25–32 (Feb. 1988).
7 V. Dhole and B. Linnhoff, *Comput. Chem. Eng.* **17**(5–6), 549–560 (1993).
8. R. Smith and B. Linnhoff, *Chem. Eng. Des. Res.* **66,** 195–228 (May 1988).
9. B. Linnhoff and V. Dhole, *Chem. Eng. Sci.* **47**(8), 2081–2091 (1992).
10. V. R. Dhole and B. Linnhoff, *Comput. Chem. Eng.* **18**(Suppl.), S105–S11 (1994).
11. V. Dhole and B. Linnhoff, *Comput. Chem. Eng.* **17**(Suppl.), S101–S109 (1993).
12. R. Smith and O. Delaby, *Trans IChemE* **69** (Part A), 495–505 (1991).
13. J. Glassman, *Combustion,* 2nd ed., Academic Press, Inc., Orlando, Fla., 1987.
14. Mills and co-workers, A Summary of Data on Air Pollution by Oxides of Nitrogen Vented from Stationary Sources. *Final Report, Report No. 4,* Emissions of Oxides of Nitrogen from Stationary Sources in Los Angeles County, California, 1961.
15. R. Smith and E. Petela, *Chem. Eng.* (Oct. 1991).
16. R. Smith and E. Petela, *Chem. Eng.* (Dec. 1991).
17. R. Smith and E. Petela, *Chem. Eng.* (Feb. 1992).
18. R. Smith and E. Petela, *Chem. Eng.* (Apr. 1992).
19. R. Smith and E. Petela, *Chem. Eng.* (July 1992).
20. B. Linnhoff and V. R. Dhole, *Chem. Eng. Tech.* **16,** 252–259 (1993).

D

DISTRICT HEATING

ATTILIO L. BISIO
Atro Associates
Mountainside, New Jersey

A range of technology is used to provide space heating in commercial buildings: residential-style oil and natural gas furnaces in smaller buildings, oil and natural gas boilers, heat pumps, and electric boilers. Gas boilers and furnaces are used to heat almost half of the commercial space in the United States (Table 1).

District heating involves the production of heat in the form of hot water or steam in a central plant. The steam or hot water is then distributed directly through underground pipes to buildings. District heating supplies the heat for over 10% of commercial building space (2).

Most district heating systems in the United States rely on dedicated steam generation facilities close to the buildings being served.

The efficiency of district heating depends on the method used to produce the heat. If a cogeneration system is used to produce both heat and electricity, the overall efficiency can be quite high. However, one must have a demand for hot water large enough to justify the cogeneration system. (See also AIR CONDITIONING; BUILDING SYSTEMS.)

Table 1. Space Heating Technologies in Commercial Buildings[a]

Technology	Percent[b]
Gas furnace/boiler	48
Oil furnace/boiler	25
Electric boiler	22
Electric heat pump	2
Other	3

[a] Ref. 1.
[b] The percent of all commercial square footage heated with the technology in 1988.

BIBLIOGRAPHY

1. *Baseline Projection Data Book,* Gas Research Institute, Washington, D.C., 1991, p. 127.
2. U.S. Department of Energy, Energy Information Administration, *Commercial Building Characteristics,* 1989, DOE/EIA-024689, Washington, D.C., June 1991, p. 128.
3. P. Kunjeer, "District Heating and Cooling: Solution for the Year 2000," *Proceedings from the International Symposium, Energy Options for the Year 2000,* Center for Energy and Urban Policy Research, University of Delaware, Newark, Del., 1988, pp. 1–109.

E

ELECTRIC HEATING

Karl S. Kunz
The Pennsylvania State University
University Park, Pennsylvania

Electric heating can occur via a number of processes and over a wide frequency range and may be intended or not. When intended, it may serve a host of useful purposes (Table 1). The electromagnetic phenomenon of current flow induced within a material body results in random particle motion in the material or heat, which may be used directly, ie, drying, curing, cooking, welding, etc., or indirectly, as in the production of light, as from an incandescent light bulb or the flow of electrons from a heated filament in a cathode ray tube (CRT). (See also Energy efficiency.)

DEFINITION OF ELECTRICAL HEATING

Current passing through a conductor obeys Ohm's law, $V = IR$, where V is the voltage across the conductor, I the current flowing through the conductor, and R a bulk or extensive property characterizing the resistance to the current flow. Here R is a macroscopic quantity based on atomic and electron interactions at the microscopic level that relates V and I in the conductor, which can be virtually any material. The physical process underlying the resistance R involves charge carriers, generally electrons, colliding with the atoms of the material. The atoms may be coupled to form a lattice structure within the material. Charge carrier motion and the attendant energy are gradually lost to motion of the material in the form of atomic or lattice vibrations. Energy imparted to the electrons, eg, from the electric field, is thereby converted via atomic collisions into mechanical motion of the atoms. This motion is random because of the random nature of the collisions and represents the electrical heating of the material.

Formally the electric power flow in the material P is given by $P = VI$. The heating is equal to this power flow for the case of a bulk material with voltage V across it and current I flowing through it when a lossless external circuit is assumed. Using Ohm's law, the power flow producing heating, P_{heat}, is either $P_{heat} = I^2R$ or $P_{heat} = V^2/R$. For a fixed current source, then, the heating rate increases as R does. Of course, as R increases for a fixed current source, the voltage $V = IR$ also increases and some practical limit to the permissible voltage will be reached. Similarly, for a fixed voltage source the heating rate increases as $1/R$. In this case, as R decreases, $1/R$ increases. A decreasing R for a fixed voltage implies an increasing current. Once again some practical limit is reached.

GOVERNING EQUATIONS

Electrical heating is a phenomenon that is subject to the Maxwell equations (1). The Maxwell equations may be expressed as

$$\nabla \times E = -\frac{\partial B}{\partial t}$$

$$\nabla \times H = \frac{\partial D}{\partial t} + J$$

with

$$B = \mu H \qquad D = \varepsilon E$$

These last two equations are the constitutive relations relating fields E and H to fluxes D and B. In addition

$$\nabla \cdot J + \frac{\partial \rho}{\partial t} = 0$$

This is the continuity equation relating current density flow J from an infinitesimal volume to the change in time of the charge ρ residing within the infinitesimal volume. The continuity equation is just the expression of the conservation of charge. Combined with the preceding two Maxwell equations, often referred to as the Maxwell curl equations, it yields the following two divergence equations:

$$\nabla \cdot D = \rho \qquad \nabla \cdot B = 0$$

These are ancillary equations contained in the curl and continuity equations that are often included in the set of Maxwell equations.

Maxwell's contribution was the inclusion of $\partial D/\partial t$ in the second curl equation. This term is referred to as the displacement current, and its inclusion results in coupled first-order differential equations that can be recast as wave equations for E and H that permit time harmonic, ie, oscillatory solutions. The J term in the second curl equation describes the motion of charge carriers as opposed to the time behavior of electric flux in the $\partial D/\partial t$ term.

The current density J is composed of two parts: a conduction current density $J_c = \sigma E$ and a convection current density $J_{conv} = \Sigma_i\, n_i q_i v_i\,(t)$, where $v_i(t)$ is not proportional to the electric field. The relationship given for the conduction current, $J_c = \sigma E$, is just a restatement of Ohm's law written as $I = (1/R)V$ when one accounts for density, electric field E as a voltage V over a path length d, and σ, the conductivity of the material, an intensive property characterizing the material and given by $\sigma = 1/\rho$, where ρ is the resistivity. The resistivity ρ is also an intensive property of the material. It can be related to the extensive property R, the resistance of a "slab" of material A in cross section and l in length by the formula

$$\rho = RA/l$$

The conduction current represents charge carriers moving in response to the applied electric field with a velocity

Table 1. Typical Electrical Heating Applications

Techniques	Frequency	Application	Advantage	Disadvantage
Resistive heating	50–60 Hz	Heating and lightning	Simple inexpensive	Only filaments heated to high temperatures
Induction heating	Kilo- to megahertz	Welding	Heat melts large metal objects	Heating over large area not for precision welding
Radio frequency and microwave	~100 kHz to gigahertz	Cooking, curing, drying	Efficient	Effects vary with material, especially depth of penetration
Infrared	$\sim 10^{-10}$ Hz	Similar to microwave	Robust, inexpensive	Broad area coverage only
Laser welding	1.06 and 10.6 μm	Welding	Precise, efficient	Cannot work Al and Cu or their alloys; expensive
Electron beam	—	Welding	Precise for vacuum implementation efficient	X-ray shielding required; expensive

proportional to the electric field. This situation pertains when the charge carrier is acted on by fields that change slowly with respect to a collision time τ_c. In this case many collisions occur while the field remains much the same. The multiple collisions ensure a randomization in the charge motion and through the collisions in the motion of the atoms or lattice of the material.

Imposed on the random motion of the charge carriers is a drift velocity μ that represents a net average velocity in the direction of the applied electric field of the charge carriers. This drift velocity arises from the collective motion of the charge carriers in the direction of E for the times between collisions. Thus a charge experiences a force along the direction of E for time τ_c, collides, and is sent into a new direction, but once again experiences the same force along the direction of E and hence, over time, while fluctuating in direction and velocity, maintains an average or drift velocity in the E direction. Thus we write

$$V_{\text{charge } i} = \mu_i E$$

where μ_i is the average or drift velocity of the ith charge species in the material where the collisions occur. Since current density is, by definition,

$$J_i = n_i q_i v_i$$

one can write

$$J_i = n_i q_i \mu_i E$$

or

$$J_i = \sigma_i E$$

where $\sigma_i = n_i q_i \mu_i$

The force equation for a charge carrier q_i is given by the Lorentz force law:

$$\mathbf{F} = q_i\,(\mathbf{E} + \mathbf{v} \times \mathbf{B})$$

Since $\mathbf{v} \times \mathbf{B}$ is always perpendicular to $\mathbf{v}$, no energy is added to the charge by the $\mathbf{B}$ field; it merely changes the direction of motion from that given by $\mathbf{E}$ alone. For a large enough $\mathbf{v} \times \mathbf{B}$ compared to $\mathbf{E}$ the motion is no longer along $\mathbf{E}$ and the above analysis would need modification. However, typically, $\mathbf{v}$ would need to have a magnitude approaching the speed of light c or B would have to be impractically large. In effect, for most practical situations the $\mathbf{v} \times \mathbf{B}$ term is ignored and the approximation $\mathbf{F} = q_i\mathbf{E}$ is made.

When a charge moves in free space, this approximation is often invalid when assessing motion over distances that are large compared to distances between collisions in a material and B must be considered. This motion is not the motion of charge associated with conduction current; rather it is the motion associated with convection current. More generally convection current is charge moving in response to the fields with no collisions occurring. Neglecting B, one can still write $\mathbf{F} = q_i\mathbf{E}$; however, without collisions there is no resulting drift velocity and in fact the approximate force equation can be written $\mathbf{a}_i\ (q_i/m_i)\mathbf{E}$. Thus the acceleration, not the velocity, is proportional to $\mathbf{E}$. This was the reason for the earlier distinction on $v_i(t)$ for convection current, namely, that it is not proportional to $\mathbf{E}$.

From the preceding discussion it is evident that only the conduction current involves collisions, and therefore only the conduction current produces electric heating.

The Maxwell equations can be solved directly in the time domain using finite differencing (2). Only the two curl equations are required. Conduction currents in the form of $J_c = \sigma E$ and the displacement current, $\partial D/\partial t$, are a part of these equations and their time behavior is part of the solution. The solution requires a specification of the material geometry and electrical properties, namely permittivity ε, permeability μ, and conductivity σ. A source current, voltage, or incident electromagnetic field must be prescribed.

All the heating problems discussed here can be treated in this very straightforward fashion with the advent of powerful computers. When convection currents are present, a particle distribution function describing the particles and subject to the Lorentz force law must be added.

SOURCES OF ELECTRICAL HEATING: TYPES

Electrical heating must concern itself with how the electron flow is generated and how collisions transfer energy of motion in the electron to random thermal motion of the material.

The generation of electron flow in a material, the source of the electrical heating, is a fundamental issue requiring its own treatment. The treatment presented here distinguishes between different sources or categories of electron or current flow.

The first distinction is between processes that require current flow for their operation, eg, the heating of a hot water heating coil, and processes for which current flow was not sought after or was unintended or a secondary effect, eg, heating of the walls of a waveguide.

The second distinction is between the forces that can produce charge carrier motion. A charge carrier (an electron for ease of discussion) can be set in motion by a number of forces including electromagnetic fields, short-range atomic forces in collisions with material particles, and gravitational effects. This last force has no technological utility for electrical heating and is mentioned only for completeness. Quantization effects, on the other hand, are not just a curiosity. High-energy photons in the form of gamma rays are an important form of ionizing radiation that can produce charge carriers from neutral materials and set these carriers in motion.

The third distinction that is useful to make is frequency regime. There are a number of distinctive frequency regimes in which electrical heating processes display unique characteristics.

Electrical heating effects can occur over a very broad range of frequencies. The frequencies can be as low as tens of hertz for low-frequency ionospheric wave effects, ELF signal generation and propagation effects, and power line heating effects, which can be of the lines themselves, of the nearby earth, or somewhat controversially, of biological material, such as human bodies, in close proximity to the power lines. The frequencies can be significantly higher at radio wave or microwave frequencies. For microwaves thermal motion can be efficiently produced by the internal vibrations of water molecules that result from the strong coupling of the microwave field energy to this mode of vibration. At even higher frequencies, such as infrared, visible, and ultraviolet, a different set of processes occur, such as the photoelectric effect. Finally for sufficiently high frequencies (eg, for gamma rays) any number of atomic scattering processes can take place, one example being Compton scattering of electrons.

As frequency increases, the simple picture of electron flow throughout the material characterized by a drift velocity gives way to a description in which individual scattering events and the details of the charge carrier motion and location are considered. When material size is very large compared to wavelength and conductivity is large, the effects can be mostly at the surface or more precisely over a depth of a few skin depths. Conversely, when the material size is small compared to a wavelength, as occurs in the structures making up a very large scale integrated (VLSI) chip, the effects are not noticeably attenuated. Thus material size and material properties, most importantly those related to the materials' conductivity, play an important role in how electrical heating may be produced at different frequencies.

A brief categorization of electrical conductivity for a limited set of different materials is given in Table 2.

Table 2. Representative Conductivity Values

Material	Conductivity
Good conductors	
Silver	6.17×10^7 mhos/m
Copper	5.8×10^7 mhos/m
Gold	4.1×10^7 mhos/m
Aluminum	3.82×10^7 mhos/m
Iron	1.0×10^7 mhos/m
Poor conductors	
Salt water	4 mhos/m
Average human tissue	0.95 mhos/m
Silicon	1.56×10^{-3} mho/m
Earth	10^{-3}–3×10^{-2} mho/m
Water	10^{-3} mho/m
Insulators	
Glass	10^{-10}–10^{-14} mho/m
Polystyrene	10^{-16} mho/m
Quartz	1.3×10^{-18} mho/m

INTENTIONAL PRODUCTION OF ELECTRICAL HEATING

Resistance Heating

The most commonly encountered form of intentional electrical heating occurs in poor conductors, metals treated for low conductivities, at low frequencies, namely power line frequencies. For a poorly conducting metal ring 1 cm in cross section and 20 cm in diameter a conductivity of 10^{-5} mho/m or equivalently a resistivity of 10^5 Ω-m yields a bulk resistance of around 10 Ω. The applied household voltage is 120 V peak so that the instantaneous peak power delivered to the coil and hence the heating is just $P_{\text{heat}} = V^2/R = 1440$ W. Integrating over time yields the average or root-mean-square (RMS) power $P_{<\text{heat}>}$, which is $\frac{1}{2} P_{\text{heat}}$, or about 700 W. Applications would include immersion water heating coils. The use of smaller wires produces high resistances even when the material has a higher conductivity. Such a configuration may be used for a toaster or, with higher conductivity and hence less resistance, for an electric blanket. Material properties of interest outside of conductivity are the ability to survive thermal cycling and, for the electric blanket, flexibility.

Other examples of desired heating are incandescent light filaments and electron emission from resistively heated cathodes in CRTs.

Induction Heating

Induction heating (3) works via magnetic induction. The Maxwell curl equation with a time-varying magnetic field is based on the Faraday law of induction. It can be converted to an integral equation via Stokes' theorem, which shows that a time-varying magnetic field linking a loop induces a voltage around the loop. A metal work piece placed in a time-varying field with less than perfect conductivity has electric fields induced on its surface and roughly over a depth given by the skin depth. These fields induce currents in the metals called eddy currents that produce heating via the same ohmic collision processes associated with direct resistance heating.

The advantage is that the work piece is not materially connected to the heat source. The connection is via the fields, and large pieces can be heated to very high temperatures without damaging the sources. Frequencies range from power line frequencies to radio frequencies with power levels as high as 100 MW at the lower frequencies to between 10 kW and 1 MW at radio frequencies. Applications include preheating and melting metals and surface hardening of steels to improve strength, wear, and fatigue properties. These applications are often in a continuous-line operation. Another application is sintering of carbide at high (2550°C) temperature. Induction heating provides on-demand heating because of its quick starting and quick heating. Induction heating does not lend itself to precision work, such as fine welds.

Microwave Heating

At higher frequencies in the radio frequency to microwave regime heating effects may also be desired, as in the case of microwave heating of food.

Microwave applications include Pasteurization, rendering, sterilization, and vulcanization. The important features here are uniformity of the field incident to the object to be heated and complete or at least adequate penetration of the field into the object being heated, which requires a skin depth comparable to the object's characteristic dimensions. This in turn may place limits on the frequency employed as the skin depth decreases with increasing frequency.

A medical application of microwave heating is the microwave generation of hyperthermia in tumors where as little as a 5°C increase in temperature exceeds certain tumor thermoregulatory capabilities and the tumor dies. Here field uniformity is not sought; rather the goal is a highly focused field on the tumor for rapid heating and minimal fields and minimal heating outside the tumor in the surrounding healthy tissue.

Determining the electrical heating of an object requires an understanding of not just the object's material properties, namely σ and the heating process, namely ohmic, but also the fields within the object. This determination of the fields is often a challenging task and can only be performed using relatively sophisticated computational tools.

Infrared Heating

Midway in frequency between microwave heating and visible and near visible light laser heating is infrared (IR) heating (4). Infrared can be produced from a bulb much like an incandescent bulb only with the filament heated to a lower temperature. This shifts the Planck blackbody radiation spectrum first presented for the incandescent bulb filament in the resistive heating section to a lower spectral peak. An industrial lamp with a filament at 2750 K has its spectrum centered at 10,500 Å. Alternatively, vapor arc lamps can be employed. Here the arc produces atomic collisions that excite one or more spectral lines in the atoms.

Finally, IR can be produced by heating in air filaments of various alloys such as Ni (80%) and Cr (20%) or Ni (60%) and Fe (25%) and Cr (15%), called 80/20 and 60/15 alloy, with resistivities of 106 and 110 $\mu\Omega/\text{cm}^3$, respectively. The temperature is in the range of 1000°C. This approach is ideally suited to space heaters as used in homes. Other IR applications include cooking, drying, and paint baking.

Laser Heating

At even higher frequencies there are fewer examples of intentional electrical heating. One example is laser metal cutting and welding (5). The requirement is to reach electric field strengths high enough to transfer enough energy into the metal's surface over a skin depth to heat that portion of the metal to melting, in fact beyond, so that it vaporizes and the heating process can continue at the layer below the one that was vaporized. Applications run from spot welding of pacemakers, CRT electron guns and automobile doors, as well as steel plates over 25 mm thick.

Electron Beam Welding

Electron beam welding (6) is in some respects similar to laser beam welding. It relies on a stream of electrons from a high-power electron gun to weld the metal instead of a laser beam. The advantage of an electron beam over a laser beam is that the electron beam can weld any metal or combination of dissimilar metals. The disadvantage is the need for a vacuum between the beam source and the work piece when the weld must be of narrow width. Nonvacuum electron welding produces a wider weld and more of the material is heated to where material properties may be adversely affected. Just as for laser welding, the weld can be very narrow.

UNINTENTIONAL ELECTRIC HEATING

The simplest example of unintentional electric heating is that occurring with overhead power lines. Even though the picture is one of very large currents flowing on very low resistance wires at very low frequencies, fields are still present.

The conservation of change requires a charge distribution on the line arising from the current. A radial E field about the wire as the result of this charge therefore arises. This radial E field is modified by the presence of a conducting earth beneath the wire in all real applications. The wire, its currents, and charges are imaged in the ground with electric field lines now bent between the wire and its image.

An exact solution includes the imaging and the attenuation in the fields arising from the finite conductivity of the earth.

Another loss mechanism is the ohmic heating of the wire itself.

Similar considerations hold at higher frequencies, namely in the microwave regime, for transmission lines supporting TEM waves predominately and for waveguides, generally either rectangular or coaxial supporting any number of modes, but principally TE_{10} for rectangular waveguides and TE_{11} for circular waveguides.

SOLID-STATE DEVICE HEATING

A final example of unintentional heating occurs in solids used for solid-state devices. In a solid at high frequencies a steady-state description based on drift diffusion solutions of the Boltzmann transport equation (BTE) may be needed (7). Such a solution is based on a probabilistic description of the collisions experienced by the electrons in the solid.

BIBLIOGRAPHY

1. J. D. Jackson, *Classical Electrodynamics,* 2nd ed., Wiley, New York, 1975.
2. K. S. Kunz and R. J. Luebbers, *The Finite Difference Time Domain Method for Electromagnetics,* CRC, Boca Raton, Fla., 1993.
3. S. Zinn and co-workers, *Elements of Induction Heating,* Electric Power Research Institute, 1988.
4. W. Summer, *Ultraviolet and Infrared Engineering,* Interscience, New York, 1962.
5. C. Dawes, *Laser Welding,* McGraw-Hill, New York, 1992.
6. M. M. Schwartz, *Source Book on Electron Beam and Laser Welding,* American Society for Metals, Engineering Bookshelf, 1981.
7. M. Lundstrom, *Modular Series on Solid State Devices,* Vol 10: *Fundamentals of Carrier Transport,* Addison-Wesley, Reading, Mass., 1990.

ELECTRIC POWER DISTRIBUTION

J. Duncan Glover
Failure Analysis Associates
Boston, Massachusetts

Distribution, including primary and secondary distribution, is that portion of a power system that runs from distribition substations to customers' service-entrance equipment. (See also Electromagnetic fields; lighting.) As an introduction to electric power distribution, the major components are shown in Figure 1.

PRIMARY DISTRIBUTION

Primary voltages in the "15-kV class" predominate among U.S. utilities. The 2.5- and 5-kV classes are older primary voltages that are gradually being replaced by 15-kV class primaries. In some cases, higher 25- to 50-kV classes are used in new, high-density-load areas (1–7).

The three-phase, four-wire primary system is the most widely used. Under balanced operating conditions, the voltage of each phase is equal in magnitude and 120° out of phase with each of the other two phases. The fourth wire in these Y-connected systems is used as a neutral for the primaries, or as a common neutral when both primaries and secondaries are present. Usually, the windings of transformers at distribution substations are Y-connected, with the neutral point grounded and connected to the common neutral wire. The common neutral is also grounded at frequent intervals along the primary, at distribution transformers, and at customers' service entrances. California is one state that uses three-phase, three-wire primary systems.

Rural areas with low-density loads are usually served by overhead primary lines, with distribution transformers, fuses, switches, and other equipment mounted on poles. Urban areas with high-density loads are served by underground cable systems, with distribution transformers and switchgear installed in underground vaults or in ground-level cabinets. There is also an increasing trend toward underground residential distribution (URD), particularly single-phase primaries serving residential areas. Underground cable systems are highly reliable and unaffected by weather. But the installation costs of underground distribution are significantly higher than overhead.

Primary distribution includes three basic types: radial, loop, and primary network systems.

Primary Radial Systems

The primary radial system, as illustrated in Figure 2, is a widely used, economical system often found in low-load-density areas (2,3,8).

To reduce the duration of interruptions, overhead feeders can be protected by automatic reclosing devices located at the substation or at the first overhead pole (9).

To further reduce the duration and extent of customer interruptions, sectionalizing fuses are installed at selected intervals along radial feeders.

Shunt capacitor banks including fixed and switched banks are used on primary feeders to reduce voltage drop, reduce power loss, and improve power factor.

One or more additional feeders along separate routes may be provided for critical loads, such as hospitals that cannot tolerate long interruptions. Switching from the normal feeder to an alternate feeder can be done manually or automatically with circuit breakers and electrical interlocks to prevent the connection of a good feeder to a faulted feeder.

Primary Loop Systems

The primary loop system, as illustrated in Figure 3 for overhead, is used where high service reliability is important (2,3,8). The feeder loops around a load area and returns to the distribution substation, especially providing two-way feed from the substation.

Figure 4 shows a typical primary loop for underground residential distribution (URD).

Primary Network Systems

Although the primary network system, as illustrated in Figure 5, provides higher service reliability and quality than a radial or loop system, only a few primary networks remain in operation today (2,3,8). They are typically found in downtown areas of large cities with high load densities. The primary network consists of a grid of interconnected feeders supplied from a number of substations.

SECONDARY DISTRIBUTION

Secondary distribution distributes energy at customer utilization voltages from distribution transformers up to me-

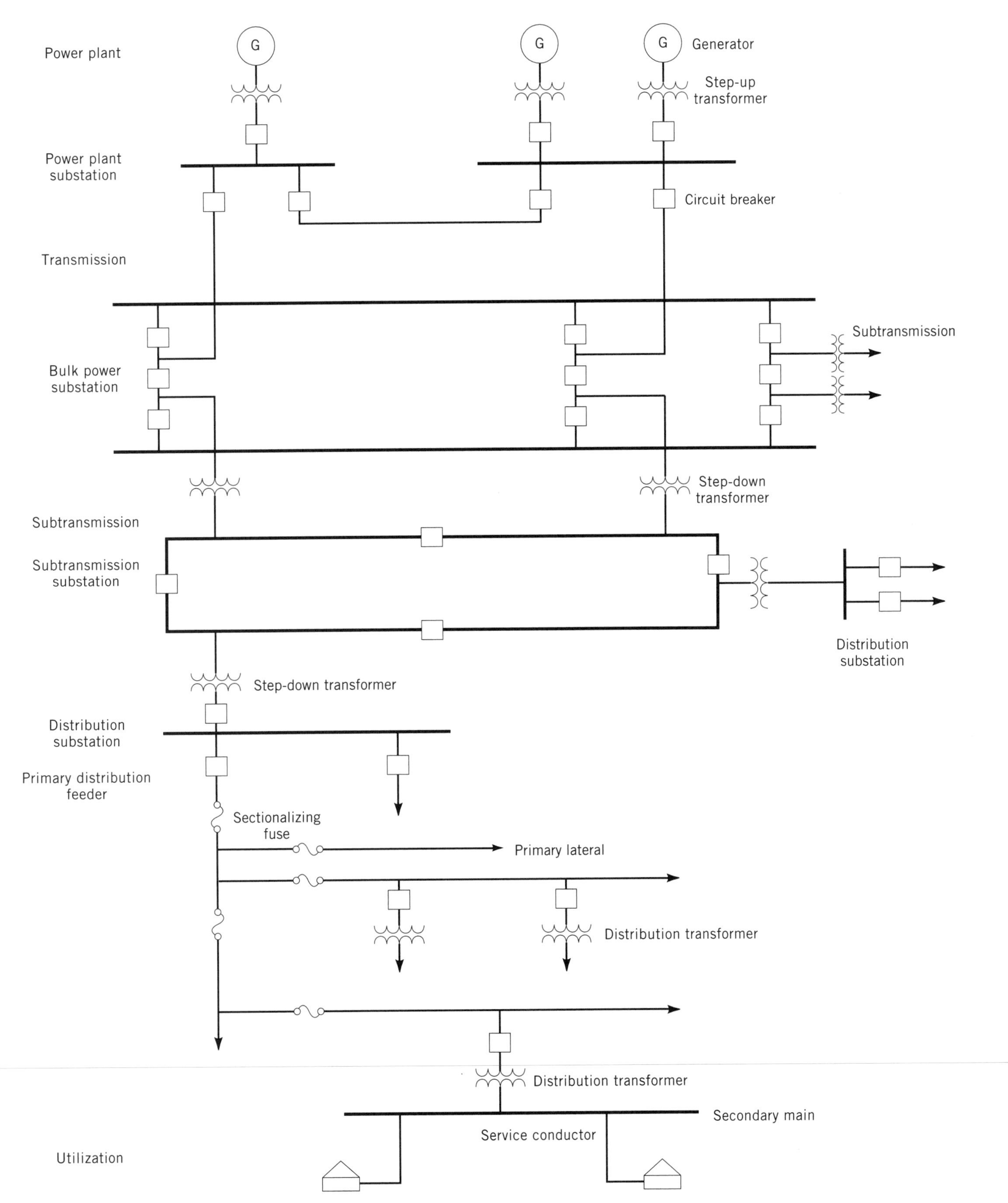

Figure 1. Principal components of an electric power system.

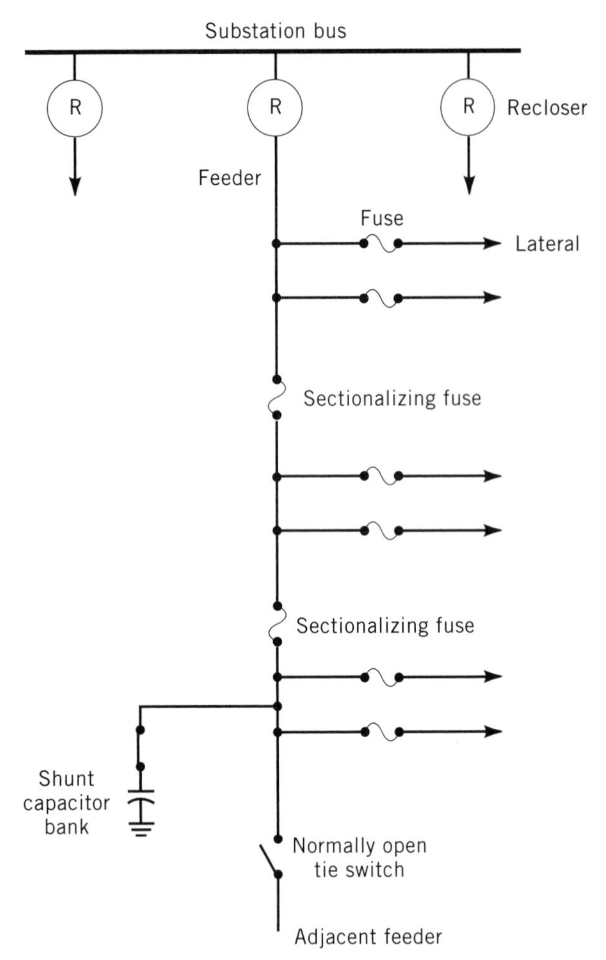

Figure 2. Primary radial system.

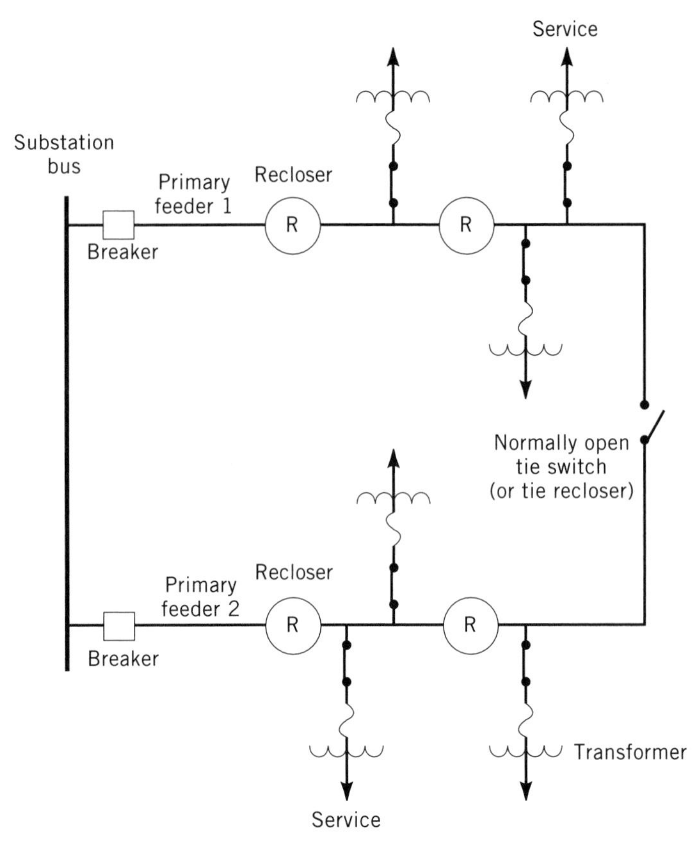

Figure 3. Overhead primary loop.

ters at customers' premises. Table 1 shows typical secondary voltages and applications in the United States (1–7).

There are four general types of secondary systems:

1. Individual distribution transformer per customer
2. Common secondary main
3. Secondary network
4. Spot network

Individual Distribution Transformer Per Customer

Figure 6 shows an individual distribution transformer with a single service supplying one customer, which is common in rural areas where distances between customers are large and long secondary mains are impractical (2,3). This type of system may also be used for a customer that has an unusually large load or for a customer that would otherwise have a low-voltage problem with a common secondary main.

Common Secondary Main

Figure 7 shows a primary feeder connected through one or more distribution transformers to a common secondary main with multiple services to a group of customers (2,3).

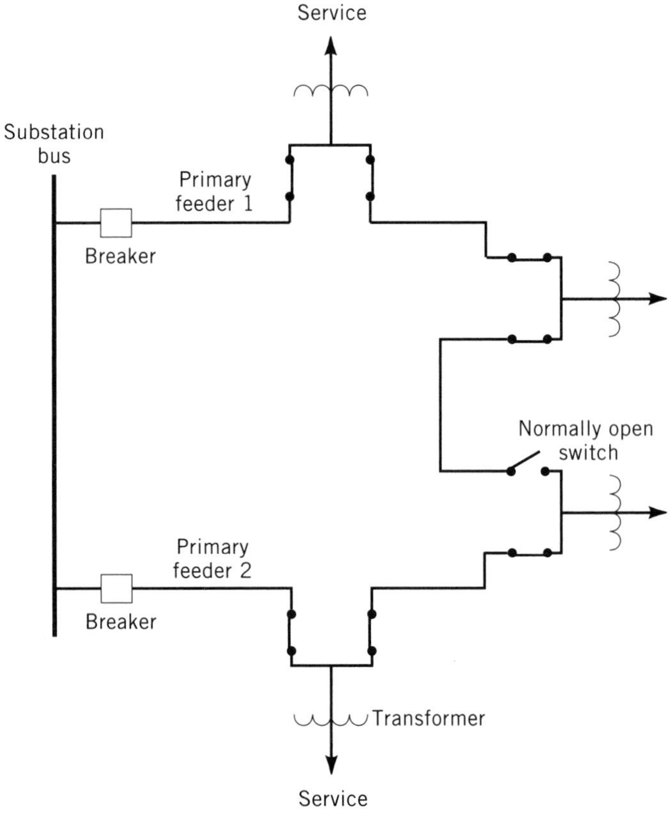

Figure 4. Underground primary loop.

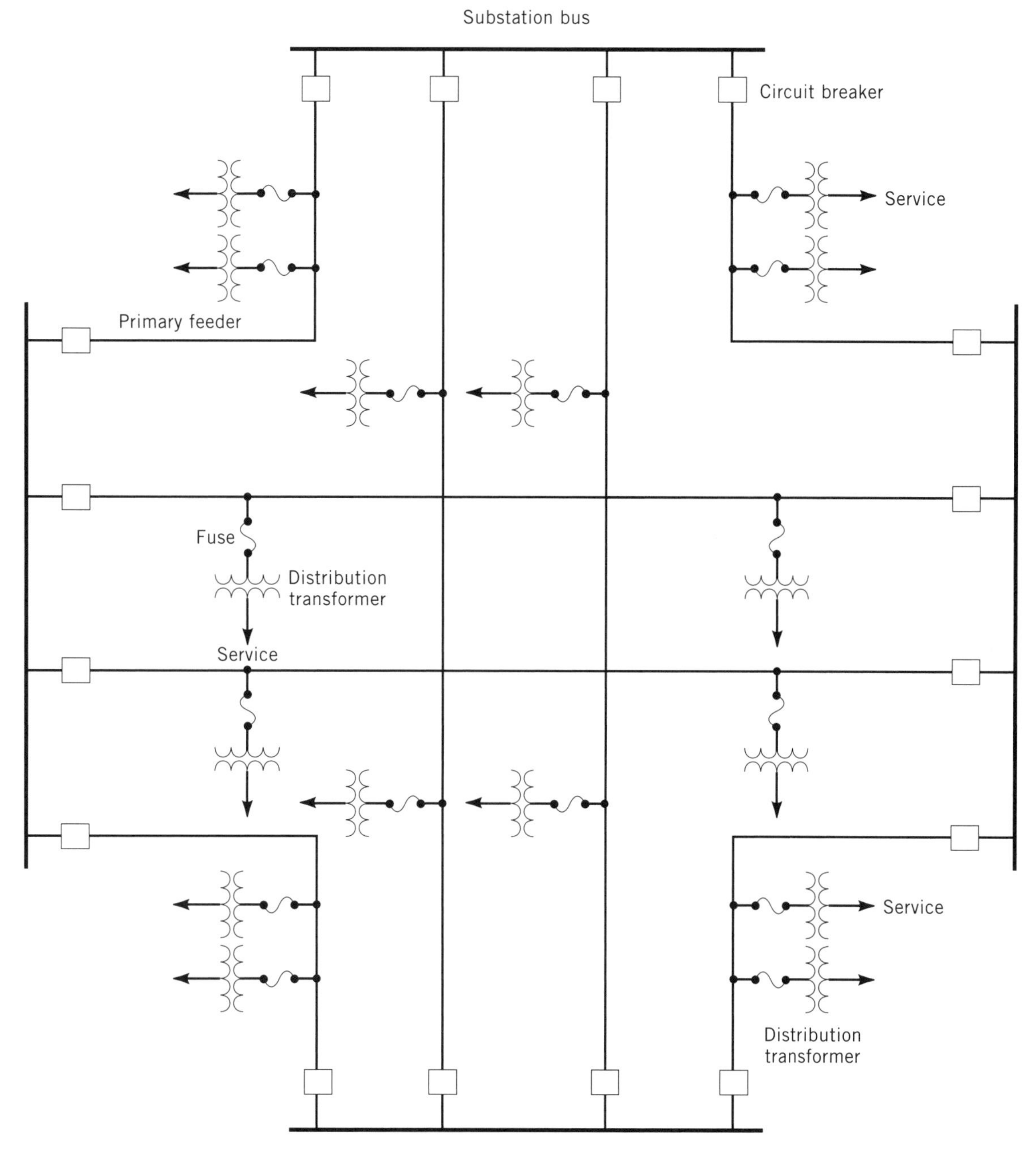

Figure 5. Primary network.

Secondary Network

Figure 8 shows a secondary network or secondary grid, which may be used to supply high-density load areas in downtown sections of cities, where the highest degree of reliability is required and revenues justify grid costs (2,3,8). The underground secondary network is supplied simultaneously by two or more primary feeders through network transformers.

Network transformers are protected by network protectors between the transformers and the secondary mains. A network protector is a circuit breaker with relays and auxiliary devices that opens to disconnect the transformer

Table 1. Typical Secondary Distribution Voltages in the United States

Voltage	# Phases	# Wires	Application
120/240 V	Single-phase	Three	Residential
208Y/120 V	Three-phase	Four	Residential/Commercial
480Y/277 V	Three-phase	Four	Commercial/Industrial/High Rise

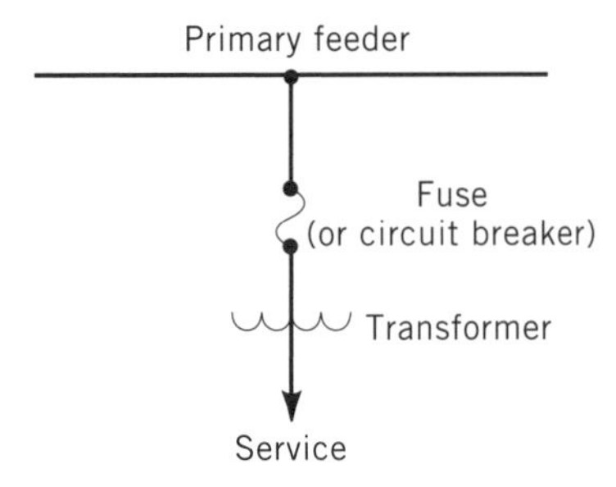

Figure 6. Individual distribution transformer supplying single-service secondary.

from the network when the transformer or the primary feeder is faulted, or when there is a power-flow reversal. Fuses may also be used for backup of network protectors. Also, special fuses called cable limiters are commonly used at tie points in the network to isolate faulted secondary cables.

Spot Network

Figure 9 shows a spot network consisting of a secondary network supplying a single, concentrated load such as a high-rise building or a shopping center, where a high degree of reliability is required (2,3,8). The secondary spot network bus is supplied simultaneously by two or more primary feeders through network transformers. Network protectors are used to disconnect transformers from the spot network bus for transformer/feeder faults or for power-flow reversal, and cable limiters are used to protect against overloads and faults on secondary cables.

DISTRIBUTION SOFTWARE

Computer programs are available for the planning, design, and operation of electric power distribution systems. Program functions include

1. Capacitor placement optimization
2. Circuit breaker duty
3. Conductor and conduit sizing—ampacity and temperature computations
4. Database management

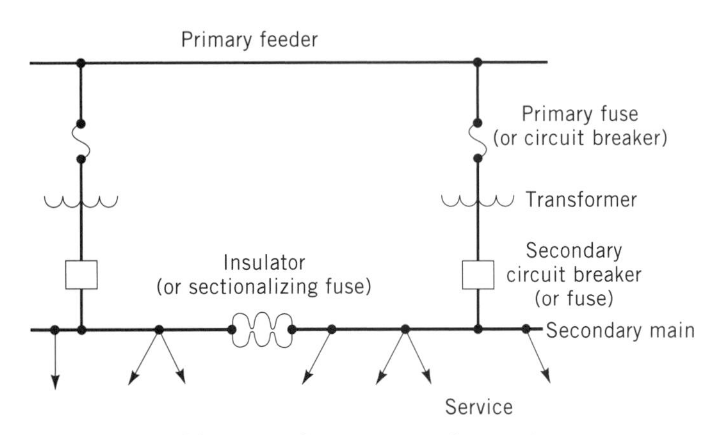

Figure 7. Common secondary main.

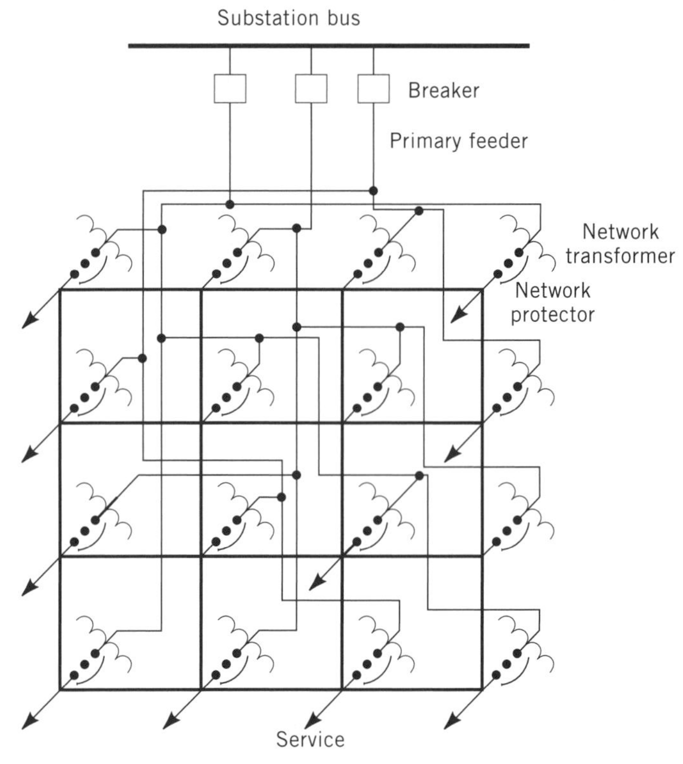

Figure 8. Secondary network.

5. Distribution reliability evaluation
6. Distribution short circuit computations
7. Graphics for single-line diagrams and mapping systems
8. Harmonics analysis
9. Motor starting
10. Power factor correction
11. Power flow/voltage drop computations
12. Power loss computations and costs of losses

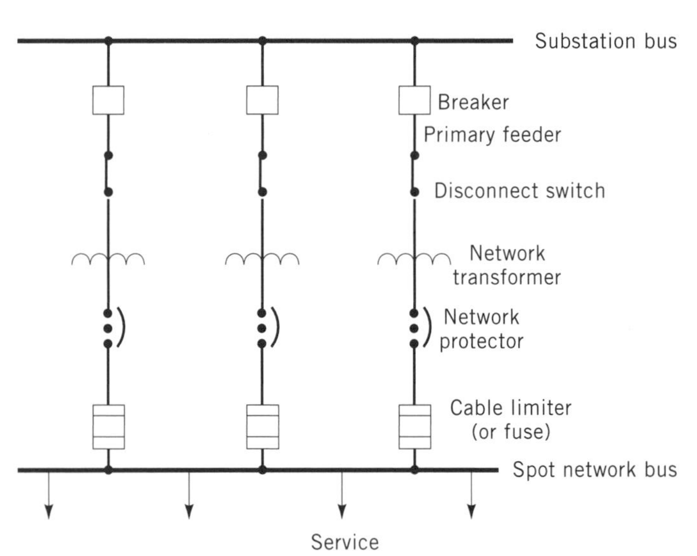

Figure 9. Secondary spot network.

13. Protective (overcurrent) device coordination
14. Tie capacity optimization
15. Transformer sizing—load profile and life expectancy.

Some of the vendors that offer distribution software packages are given as follows:

- ABB Network Control Ltd., Switzerland
- Cooper Power Systems, Pittsburgh, Pennsylvania
- Cyme International, Burlington, Massachusetts
- EDSA Corporation, Bloomfield, Michigan
- Electrocon International Inc., Ann Arbor, Michigan
- Power Technologies Inc., Schenectady, New York
- Operation Technology, Irvine, California

DISTRIBUTION RELIABILITY

Reliability in engineering applications, as defined in the *IEEE Standard Dictionary*, is the probability that a device will function without failure over a specified time period or amount of usage. In the case of electric power distribution, reliability concerns have come from customers who want uninterrupted continuous power supplied to their facilities at minimum cost (10–16).

A typical goal for an electric utility is to have an overall average of one interruption of no more than two hours' duration per customer per year. Given 8760 hours in a year, this goal corresponds to an Average Service Availability Index (ASAI) of:

$$\text{ASAI} \geq \frac{8758 \text{ service hours}}{8760} = 0.999772 = 99.9772\%$$

Table 2 lists generic and specific causes of outages, based on a U.S. Department of Energy study (15).

DISTRIBUTION AUTOMATION

Distribution automation is defined in a study by the Electric Power Research Institute (EPRI) as "an integrated systems concept for the digital automation of distribution substation, feeder and user functions. Distribution automation features include control, monitoring, and protection of distribution systems as well as load management and remote metering" (16).

Benefits of distribution automation include

- Improved distribution reliability
- Reduced customer outages and outage durations by automatically locating and isolating faulted sections of distribution circuits and automatically restoring service to unfaulted sections
- Reduced customer complaints
- Reduced power losses for substation transformers, distribution feeders, and distribution transformers
- More effective use of distribution through automatic voltage control, load management, load shedding, and other automatic control functions
- Improved methods for logging, storing, and displaying distribution data
- Improved engineering, planning, operating, and maintenance of distribution.

The installation and operating costs of distribution automation equipment are justified by increased revenues from faster service restoration and by lower operating costs from reduced power losses and corresponding gener-

Table 2. Generic and Specific Causes of Outages[a]

Weather	Miscellaneous	System Components	System Operation
Blizzard/snow	Airplane/helicopter	Electric and mechanical:	System conditions:
Cold	Animal/bird/snake	Fuel supply	Stability
Flood	Vehicle:	Generating unit failure	High/low voltage
Heat	Automobile/truck	Transformer failure	High/low frequency
Hurricane	Crane	Switchgear failure	Line overload/transformer overload
Ice	Dig-in	Conductor failure	Unbalanced load
Lightning	Fire/explosion	Tower, pole attachment	Neighboring power system
Rain	Sabotage/vandalism	Insulation failure:	Public appeal:
Tornado	Tree	Transmission line	Commercial and industrial
Wind	Unknown	Substation	All customers
Other	Other	Surge arrestor	Voltage reduction:
		Cable failure	0–2% voltage reduction
		Voltage control equipment:	Greater than 2–8% voltage reduction
		Voltage regulator	Rotating blackout
		Automatic tap changer	Utility personnel:
		Capacitor	System operator error
		Reactor	Power plant operator error
		Protection and control:	Field operator error
		Relay failure	Maintenance error
		Communication signal error	Other
		Supervisory control error	

[a] Ref. 15.

ator fuel savings. Lower operating costs are also achieved from reduced personnel time for repair and maintenance, meter reading, and handling customer complaints. In addition, investment savings come from deferral of additions or upgrades to generation, transmission, and substation facilities. Distribution automation can accomplish all these cost savings through more effective utilization of distribution (16,17).

Candidate distribution automation functions are listed below (16).

- Automatic Control
  - Switching and Sectionalizing
    - Overload detection
    - Fault location
    - Fault isolation
    - Service restoration
    - Circuit reconfiguration
  - Switching and Sectionalizing
    - Voltage regulators
    - Transformer load tap changers
    - Switched shunt capacitors
  - Substation Transformer Load Balancing
  - Cold Load Pickup
- Data Acquisition and Processing
  - Data Monitoring
    - Bus voltages
    - Equipment loadings
    - Circuit configurations
    - Sensors (CTs, VTs, equipment operating temperatures and pressures, liquid levels)
    - Alarms
    - Relay settings
  - Data Logging
    - Sequence of events
    - Relay targets
- Load Management
  - Load control
  - Remote service disconnect/reconnect
  - Load shedding
- Remote Metering
  - Load survey
  - Peak demand metering
  - Remote meter reading
  - Remote meter programming
  - Tamper detection
- Computer Databases and Graphics
  - Database management
  - Automated mapping/facilities management (AM/FM)
  - Global positioning system (GPS)
  - Reports
  - Single line diagrams

These functions involve actions such as opening and closing circuit switching devices, changing transformer taps, switching capacitor banks, monitoring voltages and equipment loadings, and reading customers' meters.

The main components of a distribution automation system are computer hardware and software, communications systems, remote terminal units, and distribution devices that are monitored or controlled. These components are briefly discussed as follows (16,17).

Distribution automation monitoring and control devices are listed below (16).

- Alarms
- Fault detectors
- Meters
- Reclosers
- Recorders
- Relays
  - Auxiliary relays
  - Circuit breaker relays
- Switches
  - Capacitor bank switches
  - Load break switches
  - Load control switches
  - Sectionalizer switches
  - Transformer load tap changer (LTC) switches
- Transducers
  - Ambient temperature transducers
  - Current transducers
  - Current transformers (CTs)
  - LTC position indicators
  - Transformer top oil temperature transducers
  - Var transducers
  - Voltage transducers
  - Voltage transformers (VTs)
  - Watt transducers

These devices include primary and secondary devices such as fault detectors, reclosers, relays, switches, transducers, and end-use customer devices such as meters and load-control switches.

BIBLIOGRAPHY

1. D. G. Fink and H. W. Beaty, *Standard Handbook for Electrical Engineers*, 11th ed., McGraw-Hill, New York, 1978, Sec. 18.
2. T. Gonen, *Electric Power Distribution System Engineering*, Wiley, New York, 1986.
3. A. J. Pansini, *Electrical Distribution Engineering*, 2nd ed., The Fairmont Press, Liburn, Ga., 1992.
4. Various co-workers, *Electric Distribution Systems Engineering Handbook, Ebasco Services Inc.*, 2nd ed., McGraw-Hill, New York, 1987.
5. Various co-workers, *Electrical Transmission & Distribution Reference Book*, Westinghouse Electric Corporation, Pittsburgh, 1964.
6. Various co-workers, *Distribution Systems Electric Utility Engineering Reference Book*, Vol. 3, Westinghouse Electric Corporation, Pittsburgh, 1965.
7. Various co-workers, *Underground Systems Reference Book*, Edison Electric Institute, New York, 1957.
8. R. Settembrini, R. Fisher, and N. Hudak, "Seven Distribution Systems: How Reliabilities Compare," *Electrical World* **206**(5), 41–45 (May 1992).
9. J. L. Blackburn, *Protective Relaying*, Marcel Dekker, New York, 1987.
10. R. Billinton, *Power System Reliability Evaluation*, Gordon and Breach, New York, 1988.

11. R. Billinton, R. N. Allan, and L. Salvaderi, *Applied Reliability Assessment in Electric Power Systems*, Institute of Electrical and Electronic Engineers, New York, 1991.
12. R. Billinton and J. E. Billinton, "Distribution Reliability Indices," *IEEE Transactions on Power Delivery*, **4**(1), 561–568 (January 1989).
13. D. O. Koval and R. Billinton, "Evaluation of Distribution Circuit Reliability," Paper F77 067-2, *IEEE Winter Power Meeting*, New York, NY, (January/February 1977).
14. IEEE Committee Report, "List of Transmission and Distribution Components for Use in Outage Reporting and Reliability Calculations," *IEEE Trans. on Power Apparatus and Systems* **PAS 95**(4), 1210–15 (July/August 1976).
15. U.S. Department of Energy, *The National Electric Reliability Study: Technical Study Reports*, DOE/EP-0003, April 1981.
16. J. B. Bunch and co-workers, *Guidelines for Evaluating Distribution Automation*, EPRI-EL-3728, Project 2021-1, Electric Power Research Institute, Palo Alto, CA, November, 1984.
17. T. Desmond, "Distribution Automation: What is it?, What does it do?," *Electrical World* **206**(2), 56–57 (February 1992).

ELECTRIC POWER GENERATION

SHEPPARD SALON
Rensselaer Polytechnic Institute
Troy, New York

When we switch on the lights at home we are the end users of electric power. This power was generated in a power plant by converting mechanical power from a turbine into electricity and then transmitted from the power plant to our homes. The electricity was generated by a synchronous machine which was driven by a turbine. On a global basis about 20% of the worlds electricity supply comes from turbines driven by falling water (hydro power), about 17% from steam turbines in which the steam is produced by nuclear fission (nuclear power), and almost all the rest from steam or gas turbines driven by heat produced by burning fossil fuels (coal, oil, and natural gas) (1).

In the United States it is estimated that 40% of all energy use goes into the production of electricity and this figure continues to rise. The widespread use of electricity for transportation, either automobiles or mass transit, could cause this percentage to rise dramatically.

See also COMMERCIAL AVAILABILITY OF ENERGY TECHNOLOGY; GEOTHERMAL ENERGY; HYDROPOWER; RENEWABLE RESOURCES.

SYNCHRONOUS GENERATORS

Essentially all of the world's electric power is generated by synchronous machines. The synchronous generator has proven to be a reliable and efficient device for converting mechanical power to electric power. Since the typical power system uses alternating current (60 Hz in the U.S.), the chief requirement of such a device is that it produces a constant and controllable voltage at a constant frequency.

Many configurations of a synchronous machine are possible but the most common synchronous machine consists of a rotor with a d-c winding and a stator with a three-phase a-c winding. The rotor has a d-c power supply and the stator is connected to the power system through a generator step up transformer.

To see how the machine operates, we first assume that the stator winding is open circuited. As the rotor turns, the magnetic field produced by the rotor induces a voltage in the stator windings. This voltage is approximately sinusoidal in time and due to the relative positions of the phases, the voltage in phase B lags the voltage in phase A by 120° and the voltage in phase C lags the voltage in phase A by 240°. This voltate, due to the rotor dc current alone, is called the internal voltage. Thus the turbine generator produces a sinusoidal three-phase voltage.

If the stator windings are connected to a load, current flows through the windings and the load. The magnetic field due to this load current produces a sinusoidal field in the air gap similar in spatial distribution to the rotor produced field. This magnetic field distribution rotates at the same speed as the rotor. For a given operating point (load and power factor), the stator-produced field and rotor-produced field are therefore always in the same relative position. If the machine is delivering load to the system (ie, normal operation) the magnetic field produced by the stator produdes a torque on the rotor to oppose the rotation. As the electrical load increases, the prime mover (turbine) must expend more mechanical energy to keep the rotor turning at a constant speed. Thus mechanical energy input by the turbine is being transformed into electrical energy.

Generators in hydro-plants are also synchronous machines, but due to the lower speed of the hydro-turbines these machines typically have a large radius and large number of poles. Mainly because of the low speed, the rotors of hydro-generators are salient pole instead of the round rotor construction. The salient pole is constructed of magnetic steel laminations. The d-c field winding is wound around the salient pole and is held in place by the overhang of the pole. The rotor is usually air cooled. The stator is similar in construction to the round rotor machines with a laminated steel core and three-phase winding. The stator is either air or water cooled.

Regulation of Power

The electrical power produced by a synchronous generator is almost equal to the mechanical power input, the efficiency being in the range of 98%. The division of electric load among a number of generators is determined by a number of factors, including economics. At a given operating point each turbine generator has an incremental fuel cost, which is the cost per kWh to generate an additional small amount of power. When this is included with other factors, such as the transmission losses, any increase in load to the system is allocated by an algorithm which minimizes the cost of producing the extra energy.

The control of the real power and regulation of the speed (which must be held constant to provide a constant frequency) is done with the speed governor. A simple configuration is shown in Figure 1. There are two basic classifications of governors: isochronous and droop. An isochronous governor holds the speed constant regardless of the load. In a droop governor the speed changes (slightly) as the load changes. Droop governors are used in electric power plants since loads are more easily shared among a number of generators with a droop characteristics. The simple Watt governor of Figure 1 operates as follows. Assume that the turbine is operating at steady state speed and load. If the speed drops, the fly-balls move inward. This causes point A to move upward. With the linkage arrangement shown, point B moves downward, opening the steam valve. This eventually increases the speed. A similar argument shows that as the speed increases the valve closes. A modern governor, usually electro-hydraulic, is more complicated, including hydraulic amplifiers, servo systems, damping, and possibly a number of auxiliary signal inputs for stability considerations.

Regulation of Reactive Power

Real and Reactive Power. There are two distinct types of power produced by the generator. Real power, measured in watts, is available to do work or produce heat and comes directly from the conversion of mechanical power into electrical power. Real power is sometimes called active power. Another form of power is the reactive power measured in volt-amperes. Although reactive power has no work equivalent, it is needed to maintain the voltage in the power system at desired levels and to run devices such as motors, which make up most of the power system load. The synchronous generator is capable of being a producer or an absorber of this reactive power.

Steady-State Generator Performance

The control of reactive power is accomplished by means of the rotor field winding current. For a given terminal voltage and real power output, the stator current can be represented by a set of phasors whose tips lie on the straight line perpendicular to the terminal voltage.

In the phasor diagram of the synchronous machine (see Fig. 2) using the angle of the terminal voltage of the machine as the reference angle, we see the current I_a lagging the voltage by the power factor angle θ. In the convention of generators, this is the case in which the generator is

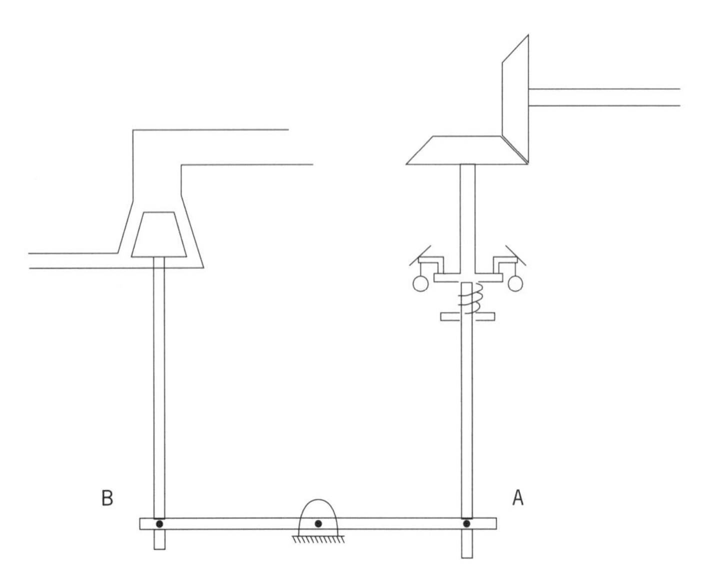

Figure 1. Simple watt governor.

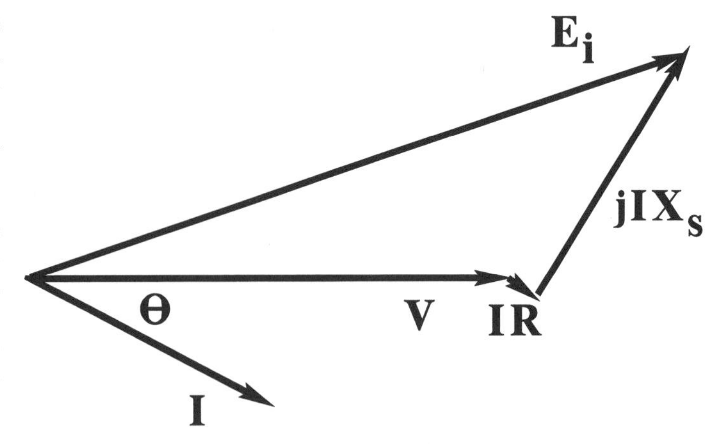

Figure 2. Phasor diagram of the synchronous machine.

delivering reactive power to the system. The internal voltage E_i is generated by the d-c field winding. This is the voltage which would appear at the generator terminals if the stator current ceased to flow and the field current remained the same. The voltage difference between the internal voltage and the terminal voltage is due to the load current. There is a resistance drop, I_aR, in phase with the stator current, and a reactance drop, I_aX_s, perpendicular to the stator current. The reactance X_s is called the synchronous reactance and is the equivalent inductive reactance of each phase of the generator when it is operating in steady state. As is shown, for constant terminal voltage, if the real power (input from the turbine) remains constant, the stator current magnitude and angle can be changed by adjusting the internal voltage, which is proportional to the field current.

Capability Curve. The limits within which the turbine generator can supply real and reactive power are illustrated in the capability curve. The curve is constructed for the case of constant terminal voltage. It is made of three different zones. The region where the generator is exporting reactive power is sometime called the overexcited or lagging power factor region. The reactive power is increased as the field current is increased. At some point the thermal limit of the rotor winding is reached. One section of the capability curve is a locus of constant field current, as the stator current and power factor change. The next section of points represents constant stator current. At some point on this curve all the power is real (unity power factor). In the region below the real axis the generator is absorbing reactive power. As the power factor changes to leading certain components of the end region, magnetic fields increase and stray losses in the end region thermally limit the operation.

STEAM GENERATION

The steam generation cycle begins when water is pumped into the boiler. The water absorbs heat from the furnace forming wet steam. Normally this liquid-vapor mixture flows through a series of tubes to increase the surface area for the heat exchange (watertube design). It is also possible to have the hot combustion gases flow through tubes to heat the vapor (firetube design). If this wet steam were used to drive the turbine, condensation would take place in the last stages. Further, the overall efficiency of the cycle would be rather low. Therefore, the steam is then heated again in a superheater which raises the temperature of the steam above it saturation point.

After the steam passes through the turbine it is changed back into water in a condenser and reused. The heat removed from the vapor is discharged to the atmosphere. To increase the overall cycle efficiency a feedwater heater is used to preheat the water before it goes back into the pump and the boiler.

STEAM TURBINES

There are two basic classifications of steam turbines, impulse turbines and reaction turbines. In an impulse turbine the steam flows through stationary nozzles against buckets mounted on the periphery of a rotating disk. When the nozzles are located on the rotating disk this is called a reaction turbine. In a modern power plant the turbines may be a combination of the two having impulse stages and reaction stages.

STEAM PLANTS

In a steam turbine generating plant, a fossil fuel is burned and the heat is used to produce steam. This steam drives a turbine which in turn drives a turbine generator. Steam turbines are heat engines and as such are limited in efficiency by the Carnot efficiency, which is the highest efficiency possible for a system such as this which turns heat energy into mechanical energy. The Carnot efficiency, E, is given by

$$E = \frac{T_{\text{in}} - T_{\text{out}}}{T_{\text{in}}}$$

where in our case T_{in} is the temperature of the steam input to the turbine in degrees Kelvin and T_{out} is the temperature of the exhaust. Taking typical numbers for a utility steam turbine, the steam entering the turbine may be 811 K (1000°F). The exhaust may be 311 K (100°F). This gives an efficiency of 61%. Although this is the theoretical limit, in practice the turbine efficiency is usually somewhat less than 50% (2). To find the overall efficiency of a power plant this efficiency is multiplied by the efficiency of the boiler and by the efficiency of the generator. Again using typical numbers a boiler efficiency of 88% and a generator efficiency of 98% are used to find an overall plant efficiently in the range of 40%. A range of between 30 and 35% has been given (3). It is common to classify these thermal plants in terms of heat rate. The heat rate is defined as the heat input from fuel divided by the electrical output of the plant. These are commonly expressed in units of Btu/KW·h (one kilowatt hour of energy is equivalent to 3412 Btu). Therefore high efficiency plants have lower heat rates than low efficiency plants. A typical number is 10,500 Btu/kW·h. In nuclear plants the steam temperature is somewhat lower than in a fossil fuel plant, eg, 623 K (2). The overall efficiency of the nuclear plant is in the range of 30%.

A steam power plant has the following basic structure which is illustrated in Figure 3.

The steam turbine used in power plants operates on a Rankine cycle. Liquid water at low pressure leaves the condenser where it flows through a pump where its pressure is raised to the turbine inlet pressure. The water is then heated at constant pressure in the boiler to its saturation temperature and then at constant temperature until it is a 100% saturated vapor. The steam can then either pass through the turbine or through a superheater and then through the turbine path. After expanding through the turbine the steam passes into a condenser and emerges as a saturated liquid. One problem that arises is that as the saturated vapor travels through the turbine, its quality is reduced (larger percentage of liquid). If the liquid exceeds more than a few per cent of the weight, severe erosion of the turbine blades can result. One solution is to extract the steam after it passes through the high pressure turbine and reheat it, then expand it in the low pressure stages of the turbine.

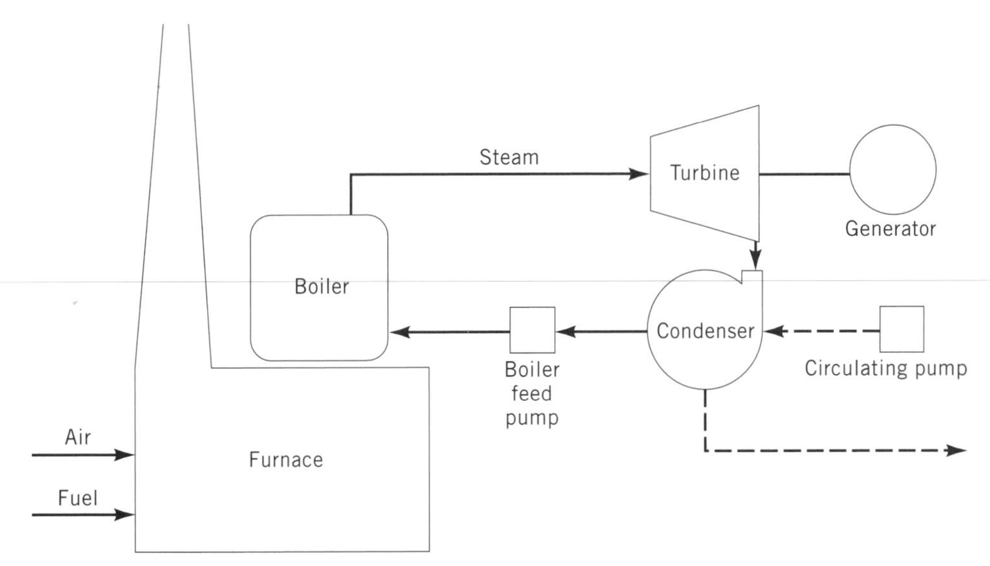

Figure 3. Basic configuration of a steam plant.

GAS TURBINES

An alternative and complement to steam turbines, gas turbines have enjoyed a resurgence in recent years (Fig. 4). Whereas a gas turbine normally has a higher cost than a steam turbine, a number of factors make them attractive.

Gas turbines are often used as peaking plants and have the following advantages:

1. The cost of installing a gas turbine plant is lower than installing a steam turbine plant.
2. The units have fast starting capability.
3. The units have a short delivery and installation time.
4. The units can pick up load very quickly.
5. The units have a black start capability.

The gas turbine operates on the Brayton cycle illustrated in Figure 5. Line 1–2 is the compression of air. Line 2–3 is the constant pressure combustion. Line 3–4 is the expansion of the gas through the turbine. Line 4–1 is the constant pressure discharge to atmospheric air. This is actually an open system.

Practical considerations such as noise abatement, air filtering, and exhaust heat recovery add pressure drops to the system. These result in a drop in the power output or, correspondingly, an increase in the heat rate.

The gas turbine is also used in a combined cycle plant along with a heat recovery system and a steam turbine. These plants have excellent heat rates. These gases still have a lot of heat energy and much of this energy can be recovered in the combined cycle plant.

Combined cycle plants use a simple cycle gas turbine. Instead of exhausting the high temperature gas output of the gas turbine to the atmosphere, the combined cycle plant uses the exhaust in heat recovery steam generators (HRSG) which drives a separate steam turbine generator. These plants can achieve an efficiency of approximately 50%.

NUCLEAR POWER

Nuclear power provides an alternative to the furnace described above under fossil fuel. In nuclear power plants the heat is generated by the nuclear reaction (fission) and used to heat water and produce steam. The fuel used is enriched uranium containing between 2–4% of U_{235}. In the

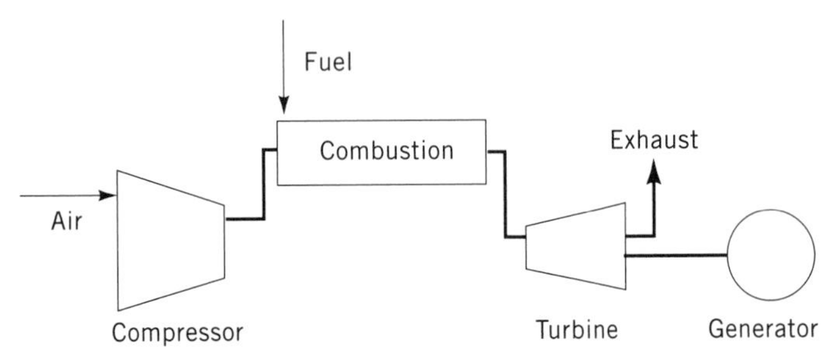

Figure 4. Simple cycle gas turbine.

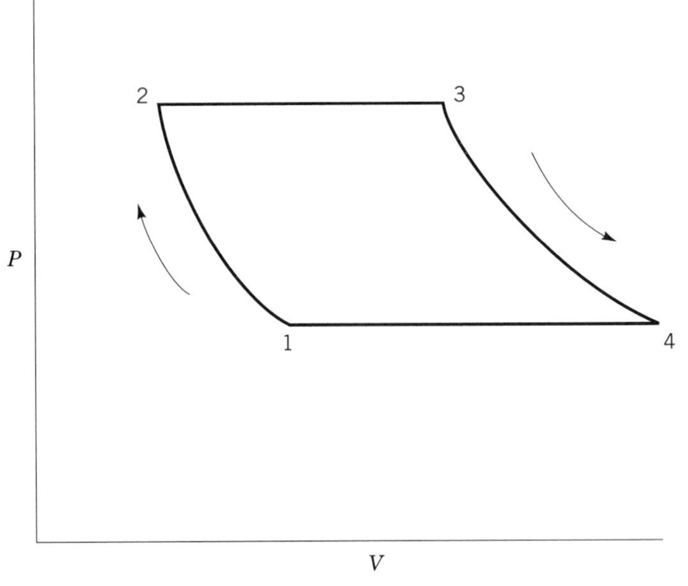

Figure 5. Brayton cycle.

United States the nuclear power plants use light (normal) water reactors.

There are two principal types of reactors used in the United States today: the pressurized water reactor and the boiling water reactor. Figure 6 gives a schematic of the pressurized water reactor.

In the pressurized water reactor the water is pumped through the reactor by a reactor coolant pump. This forms a closed loop referred to as the primary loop. In the secondary loop, the feedwater pump circulates water through a heat exchanger where it is turned to steam by heat from the primary loop. This steam then flows through the turbine driving a synchronous generator.

In the boiling water reactor, there is only one loop. The boiling water reactor generally has a higher overall efficiency then the pressurized water reactor, gained at the expense of having the turbine become radioactive.

HYDRO-GENERATORS

Hydro-generators use the potential energy stored in water to produce electricity. Water falls through a turbine and turns a synchronous generator. The arrangement of most hydro-generators is with vertical shafts as opposed to steam or gas turbine generators which have horizontal shafts. Some exceptions are so called low head hydro-generators usually defined as less than 66 feet of head (4). The energy in the water is proportional to the weight of the water through the turbine times the head or vertical distance the water travels to the turbine. The power is proportional to the head times the volumetric flow rate (m^3/s) through the turbine.

Pumped Storage

Pumped storage plants are designed so that during times of light load and inexpensive power from the rest of the system, water can be pumped back into a reservoir to be

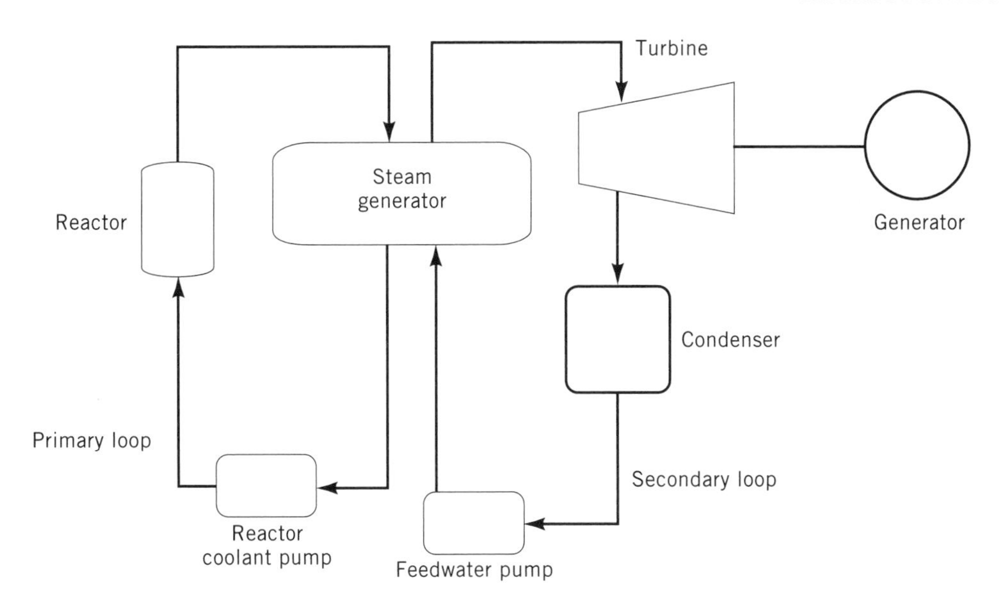

Figure 6. Pressurized water reactor.

later run through the turbine and produce power at times of greater load. In recent times reversible hydraulic pump turbines are used. Some of these are now equipped with adjustable speed drives used to increase the pumping efficiency.

WIND POWER

The interest in renewable energy sources has also spurred the development of wind energy generated electricity. The wind is used to drive a turbine which drives a generator. Because wind speed and direction are not controllable, substantial research has recently gone into turbine configurations of high efficiency and energy capture.

Due to the variable nature of the energy source, the speed of the wind turbine is not constant and machines other than synchronous generators have been applied. These include induction generators, d-c generators, and variable reluctance generators. The further development of power electronic controls will doubtless result in new configurations for wind power generating systems (5).

The U.S. Department of Energy has projected that the cost of wind-produced electricity will be competitive with coal energy by the year 2000 in sites with good wind resources (averaging 7.2 m/s at 10 m height) (7). In the United States, California leads the nation in wind energy production.

SOLAR POWER

Producing power directly from solar energy is not currently a significant factor in electricity generation. However, this technology has attracted the attention of electric utilities and industry as a viable alternative for future energy production. Photovoltaics are semiconductor devices that convert solar radiation (sunlight) directly into electricity. While the electricity is essentially free in that there are no fuel costs, photovoltaics have not been widely used even in attractive climates due to the high initial investment required for the devices. Collector types include thin-film technologies common in solar-powered calculators and crystalline silicon plates, the latter being somewhat more efficient (23%). Also of interest are concentrators or mirrors that focus the sunlight on a smaller area of cells thus reducing the number and cost of cells. The efficiency of these can be 35%. Also of interest is the solar–thermal concept in which solar energy is concentrated by a series of reflectors to boil water producing steam which is expanded through a turbine to produce electricity in the conventional manner. This technology is already in place in the California system.

BIBLIOGRAPHY

1. C. Starr, M. Searl, and S. Alpert, "Energy Source: A Realistic Outlook," *Science*, **256** (May 1992).
2. C. Summers, "The Conversion of Energy," *Sci. Am.*.
3. A. Wood and B. Wollenberg, *Power Generation Operation and Control*.
4. G. Friedlander, *Elect. World*, 57–63 (July 1980).
5. D. Torrey and S. Childs, "Development of Variable Reluctance Wind Turbine Generators," *Windpower '93 Conference Proceedings*, American Wind Energy Association, San Francisco, Calif., July 12–16, 1993, pp. 258–265.
6. R. Thresher, S. Hock, H. Dodd, and J. Cohen, "Advanced Technology for the Year 2000: Utility Applications," in Ref. 5, pp. 61–73.

ELECTRIC POWER SYSTEMS AND TRANSMISSION

J. A. Casazza
D. J. LeKang
CSA Energy Consultants
Arlington, Virginia

ELECTRIC POWER TRANSMISSION

The transmission system provides the means by which large amounts of power are transported from the generating stations that produce this power to other parts of the utility system or to other utilities (See Plate I). In terms of the three primary functions of the electric utility industry, transmission can be viewed as the link between the production and distribution functions (1). (see also COMMERCIAL AVAILABILITY OF TECHNOLOGY.)

The most common form of transmission in use in the United States today is three-phase, alternating current (ac) having a frequency of 60 Hz (cycles per second). In recent years, the use of direct current (dc) transmission, while a small portion of total transmission, has increased for a number of special applications.

The most common ac voltages in use in the United States today for transmission purposes are between 115,000 and 765,000 V. A standard voltage has been agreed upon for industry use.

Purpose

The purposes of the transmission system (2) are as follows:

- To deliver electric power from generating plants to large consumers and distribution systems.
- To interconnect systems and generating plants to reduce overall required generating capacity requirements by taking advantage of:

  The diversity of generator outages, ie, when outages of units occur in one plant, units in another plant can provide an alternative supply.

  The diversity of peak loads since peak loads occur at different times in different systems.
- To minimize fuel costs in the production of electricity by allowing its production at all times at the available sources having the lowest incremental production costs.
- To facilitate the location and use of the lowest cost additional generating units available.
- To make possible the buying and selling of electric energy and capacity in the marketplace.

In recognition of these functions, transmission systems have been designed to achieve acceptable reliability at minimum cost. This includes provision and maintenance of service without loss of load for "normal" generation and transmission contingencies under various load conditions. It also requires that transmission systems be able to withstand major contingencies, such as the unexpected loss of an entire power plant or a transmission right of way, without a total system shutdown.

Plate I. The pathway from the generating station to the utility system or utility is called the transmission system. Courtesy of Tennessee Valley Authority.

Components of Transmission System

Transmission Lines. The primary components of overhead transmission lines are conductors, insulators, support structures, and rights-of-way.

Conductors are the wires through which the electricity passes. They are usually made of aluminum with steel reinforcements. Insulators are made of materials (such as porcelain) that do not permit the flow of electricity. They are used to support the conductor while separating it from the supporting structure.

The most common form of support for transmission lines is a steel, lattice tower. Sometimes wood poles are used for lower transmission voltages. The primary purpose of the support structure is to maintain the electricity-carrying conductor at a safe distance above the ground.

Rights-of-way are strips of land along which the transmission line is built. In many cases rights-of-way are wide enough to accommodate two support towers with two transmission lines built on each tower.

In some areas of the country, population densities and environmental considerations preclude the possibility of obtaining a right-of-way for a transmission line. In such a case transmission is often built underground by using shielded cables located in large protective pipes. The pipes are filled with oil under pressure to provide insulation (3). This type of transmission is many times more costly to construct than overhead transmission and presents technical and accessibility problems.

Transmission Substations. Transmission substations occur wherever transmission lines terminate or connect to one another.

Busbars are used within substations to connect circuits of the same voltage level with one another and with transformers.

Transformers are used to transfer power between electrical circuits and buses operated at different voltages. Tap-changing equipment is used on transformers to vary the amount of voltage change in the transformer and to help compensate for voltage variations in the system.

Phase shifters, also called phase angle regulators, are devices similar in construction to transformers that induce a power flow into a circuit, in order to increase or decrease the power loading of that circuit by inserting a phase angle difference.

Switchgear is a general name given to equipment that can open connections and interrupt the flow of electricity between circuits, busbars, and transformers under normal and emergency conditions. The main types of switchgear are circuit breakers and disconnect switches.

Circuit breakers are designed to interrupt the flow of electricity in power system circuits and devices. When a fault or a short circuit occurs, the amount of current flowing through the transmission line and all of its substation equipment increases to many times the current under normal conditions. The circuit breaker is capable of interrupting the flow of electricity under these abnormal levels of current. In fact, the capability of these circuit breakers is given in terms of the maximum current they can interrupt and in terms of their current-carrying capability under normal continuous operation. An important characteristic of circuit breakers is the time that is required to interrupt the short circuit current, typically a very small fraction of a second.

Disconnect switches are used to keep circuits open once a circuit breaker had tripped. They generally cannot interrupt load or fault currents and are usually used in conjunction with circuit breakers.

Surge arresters are devices used in substations to protect equipment from overvoltages caused by lightning and circuit switching. They are designed to conduct high current to ground and away from substation equipment when voltages reach an undesirable level, thus limiting maximum voltages.

Protective relays are used, often in connection with communications equipment, to detect faults or abnormal conditions and signal circuit breakers or other equipment to take corrective action.

System Protection. Because faults and short circuits occur from time to time on electric transmission systems, system protection techniques have been used since the earliest days to disconnect the faulted equipment, thus allowing the remainder of the network to continue operation. This has been accomplished through the use of relays, which sense the existence of such faults and send an appropriate signal to circuit breakers, which disconnect the faulted line or apparatus. Relays have also been used to stabilize system conditions resulting from various types of perturbations.

Reactive Power Equipment Reactive power is required to supply the magnetic fields of electrical equipment (1, 3). It uses no energy and produces no useful work in itself.

An undersupply of reactive power in a part of the transmission system results in excessively low voltages in that area, and an oversupply produces excessively high voltages. In order to prevent these occurrences, reactive power must be supplied or absorbed as required.

The unit of reactive power is the VAR (volt-ampere reactive), but the practical unit used in transmission system operation and design is the megavar (MVAR), equal to 1 $\times 10^6$ VAR. Reactive power can be produced or absorbed by several types of equipment:

- Generators produce reactive power along with producing real power.
- Shunt capacitors are devices connected between a conductor and ground that produce reactive power.
- Shunt reactors absorb reactive power.
- Static VAR compensators (SVCs) utilize reactors, capacitors, and high-speed switching devices to provide a source of controlled reactive power.

Capacitors are sometimes also connected in series with lines. While a series capacitor produces some reactive power, its main function is to reduce the effective reactance of the line and cause it to deliver more current than other parallel lines. Similarly, series reactors can be used to increase the reactance of a line in order to reduce the load currents, but this application occurs infrequently.

Voltage Control. Voltage control on a transmission system is achieved by a combination of the settings of transformer fixed taps (which can only be changed when a transformer is out of service), transformer tap changing under load, and adjustment of the previously discussed reactive sources.

Voltage regulators monitor the voltage at the generator terminals and automatically adjust the excitation system for voltage fluctuations to obtain the voltage that the generating station operator has indicated or the voltage that is required under some emergency condition.

Direct Current Transmission

Direct current transmission, frequently referred to as high-voltage direct current (HVDC), is used in a growing number of instances where its characteristics are particularly beneficial (1,5).

HVDC Components. The dc transmission lines generally use two conductors, as compared with the three used in three-phase ac transmission. Both terminals of a dc line consist of dc conversion stations that transform ac power into dc (rectifiers) or dc into ac (inverters). The main components of a dc conversion station are

- Banks of thyristors (controllable solid-state diodes) arranged in "bridges" for the conversion of ac into dc or vice versa.
- Transformers.
- Capacitors used to filter out harmonics on the ac side and to provide reactive power to support the thyristor bridges.
- Switchgear on the ac side.
- Control systems to regulate the amount and direction of power flow.

Operating Characteristics of dc Transmission. By not being subject to the physical laws governing ac transmission, dc transmission offers some characteristics that are valuable in some situations.

- The amount and direction of the power converted from ac to dc and back is directly controllable and is independent of the characteristics and power flows on other ac lines in parallel that join the same areas.
- Reactive power, being inherently associated with ac transmission, is not carried by dc transmission. Conversion stations at both the sending and receiving terminals, however, require substantial reactive power support from their respective ac systems. Some of this support is generally supplied by capacitors at the conversion stations. These capacitors also help absorb some of the harmonics produced by the conversion stations.
- dc lines do not transmit short-circuit currents from adjoining ac systems and do not spread transient instability disturbances.

Application of dc Transmission. The physical characteristics of dc transmission make it particularly valuable for the following applications:

- Linking two large synchronous systems with relatively weak interties, which if linked by ac would be subject to instability or to an excessive inrush of power in case of severe disturbances.
- Providing links in parallel with ac transmission under conditions in which power would be transferred too much on some links and too little on others.
- Linking systems by submarine cables, which, if ac were used, would carry too much reactive power and severely limit the remaining capacity for real power.
- Linking areas that, if linked by ac circuits, would cause excessive currents to flow in the event of short circuits exceeding the capability of circuit breakers to interrupt them.
- Controlling the amount of power flowing from one area to another, allowing the link to be treated in a manner similar to a controllable generation source.

DEVELOPMENT AND OPERATION OF TRANSMISSION SYSTEMS

Economic Dispatch

The incremental cost of a generator is almost entirely the cost of its fuel and by how efficiently it converts this fuel into electric power. For a typical thermal unit, the incremental cost increases as production *increases* from the minimum to the maximum rated output. The incremental cost of a unit is not the same as its average cost, which generally decreases as its output increases (1–7).

Economic dispatch consists of supplying the system's total power requirements at any time by lading each available generating unit to the point where its incremental cost is the same as that of all other generations. Units whose highest incremental cost is lower than this common incremental cost are, of course, loaded to their maximum capability.

Today's electric power systems are being dispatched and operated based on the incremental costs of production. This general practice often results in one utility producing electricity for use by some other utility's customers when the first utility can produce the incremental power at a lower cost. *This leads to maximum economic efficiency in the use of the investment and in the cost of operating these systems.* No other industry operates in this way.

The only exception to this production based on incremental costs has resulted from the recent addition of cogenerators and qualifying facilities.

Transmission Losses

There are two basic types of power losses in transmission systems (3):

- *Core losses* are dissipated in the steel cores of transformers (8). These losses typically vary with at least the third, and often higher, powers of the voltage at their terminals. If transmission voltages are fairly constant, core losses are also constant. They depend on the transformer's design voltage tap settings and its rating. They are not affected by how much power is actually flowing through the transformer. An increase in the power carried by a transmission system does not substantially affect the core losses; a change in transmission voltage can.

- *Conductor losses* are dissipated in transmission lines and cables and transformer windings. As these losses vary with the square of the current, and therefore approximately with the square of the power carried by each component, they vary greatly between light-load and heavy-load times and are also affected by increases in the power carried by a transmission system (9). It should be noted that because of the "square" effect mentioned above, a given percent increase in power flow results in a much higher percent increase in power losses on the system.

Interconnection and Pooling

Benefits of Interconnections between Systems. With interconnections, a utility may be in a position to call upon its neighbors in the case of severe outages on its own system (2). Once neighboring utilities are interconnected, it is possible for them to exchange energy or buy and sell energy to each other in order to minimize their total expenses for energy. The customers of both utilities benefit through lower costs for electricity (10).

Electric utilities in the United States have been coordinating the development and use of their transmission systems and generating capacity for many years (11). As a result of this coordination, considerable savings have accrued. These savings have resulted from various causes:

- Reduced investment in installed generating capacity from sharing of installed generation reserves, load diversity and equipment outage diversity, optimum use of available generation sites, use of larger unit sizes, coordination of maintenance schedules, and long-term firm capacity purchases.
- Reduced operating costs from economic energy exchanges, use of regional economic dispatch and unit commitment, optimum use of hydro and pumped-hydro facilities, coordination of maintenance schedules, short-term capacity purchases and sales, and reduced spinning reserves.
- Reduced transmission investment from coordinated regional transmission planning, supply to other systems' loads, and back-up to other systems' substations.

Synchronous Operation. Because of the huge benefits, the addition of interconnections between systems and between regions in North America evolved rapidly, as they did elsewhere in the world. Figure 1 shows the five synchronous areas that currently exist in North America. All the systems within each area operate in synchronism. Four of these areas are interconnected by dc ties.

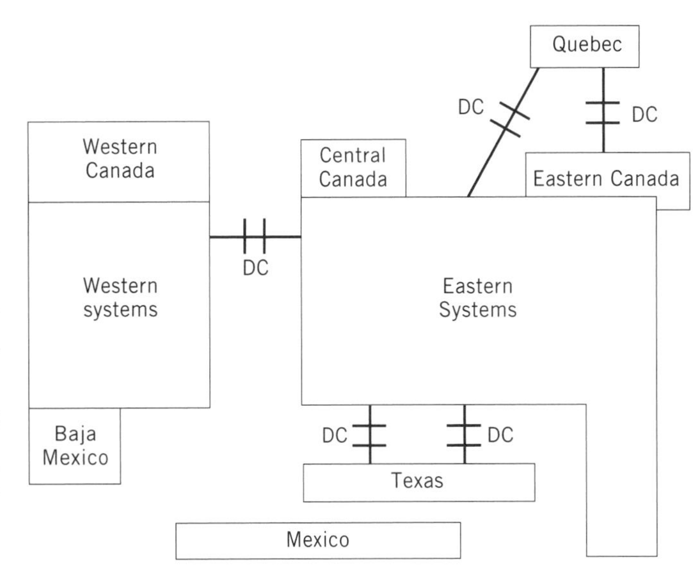

Figure 1. Fire synchronous systems of North America.

Transmission System Capacity

The capacity of an ac transmission system (12) to deliver power is also a complex matter and depends on the network arrangement, size and type of conductors, lengths of lines, location and size of generation sources and loads, etc. The network capacity may be limited by transient or steady-state stability conditions for longer lines, voltage instability, or thermal heating of conductors. The capability of the circuit breakers and protective systems must be carefully checked as new generators are added and system changes are made. These factors require that many highly technical analyses be made by skilled engineers to determine the capacity of a transmission network (1).

Equipment Ratings. When electricity flows through the conductors of a transmission line, the conductors and terminal equipment are heated. The amount of heat flowing through it increases. Each transmission line, transformer, and piece of switchgear is capable of withstanding a given amount of heat based on certain electrical and mechanical design factors. This amount of heat is usually defined in terms of the amount of current a line can carry. This current limit determines the maximum amount of power the equipment can deliver. The limiting factor for a line may be determined by conductor limitations, line sag limitations, current breaker limitations, wave trap limitations, etc., with the element having the lowest rating determining the overall line rating.

The rating depends upon the period for which the loading will exist and also what conditions (weather and line loading) existed prior to that period. Ratings are also dependent upon the loss of conductor life that the utility is willing to tolerate per high loading incident. Because heating effects also depend on ambient conditions, different line ratings are usually used by utilities for summer and winter conditions.

Voltage Drop. The passage of current through a wire results in a reduction in the voltage along that wire. Since utilization equipment is sensitive to voltage, there are limits beyond which the voltage cannot be permitted to decrease.

Power and Energy Losses. The passage of current through a line results in a loss of energy due to the internal heating of the conductor. At a certain point, the losses would render the operation uneconomic.

Stability of Operations. Stability refers to the property of the system to be able to maintain synchronism after a severe perturbation of some kind, such as a short circuit or the sudden outage of a transmission line. Such disturbances cause dynamic oscillations on the system with large power savings on various generators. This savings cause loss of synchronism and tripping of generators.

Voltage drop, power and energy losses, and stability limitations improve in proportion to the square of the voltage at which the line operates. Further, all three considerations are directly influenced by the distance over which power is to be transmitted. Thus, the longer is the transmission distance involved, the more likely is the voltage to be higher in order to avoid voltage drop, power and energy loss, and stability problems.

Division of Flow Among Transmission Routes

The flow of power over the specific circuits in an ac transmission network cannot be controlled. Kirchhoff's laws determine how the power will divide. Legislation, regulations, and contracts cannot specify the load to be earned by specific transmission circuits. This sometimes results in one transmission circuit being excessively loaded while other circuits in a network are lightly loaded.

Loop Flows. When various transmission systems are interconnected in a network, power from one system will flow over the lines of every other system even when each system is supplying its own load from its own generating source. This is caused by the physics of power flow. With this condition there is a circulating power flow around a closed loop. This flow is in addition to the flow that would exist on the lines of each system if there were no interconnections to other systems. Such flows are called "loop flows." They are the result of the network design and the distribution of generation and loads in each system. These loop flows can change considerably at various times and system loading levels. Loop flows are caused by the interconnection of systems to achieve generation capital cost savings and fuel cost reductions. These flows can be reduced or corrected through use of expensive phase-shifting transformers.

Parallel-Path Flow. When one utility sends power to another utility, a change in power flow occurs along all "parallel" electrical paths leading from the point (or points) of the seller's increased generation to the point(s) of the buyer's increased load or decreased generation. This occurs whether the seller or the buyer is a private entity or utility. Even if the two utilities are directly connected, some of the transacted power will flow over any and all available paths, including those through third-party utilities. In this case as well, however, some of the transacted power may flow in other utilities' systems. These flows through the systems of other utilities that are not involved in the transmission service transaction are generally referred to as "parallel-path flows." Such flows can be reduced through the use of phase-shifting transformers.

Control Areas

As the individual electric utility systems were being interconnected in the United States, it became quite apparent that new control technologies and procedures had to be developed so that operating decisions and problems in one system did not unnecessarily affect another (13–15). This led to the development of what is now known as control areas. A control area is a bounded geographical area within which all generation is controlled by a single dispatch center. All transmission lines entering or leaving this bounded geographic area are metered; the total flow in or out of the area is determined. The control center has a dispatch computer that controls generators to meet optimum economic dispatch and the scheduled amount of power flowing in or out on all of the tie lines (7,16,17).

In addition to the automatic computer control features of the control center, it monitors voltages. The control area concept has continued to evolve and will grow in the future.

Transmission System Reliability Criteria

Transmission system capacities sufficient to meet needs are determined on the basis of specified planning and operating criteria, such as those agreed to by the North American Electric Reliability Council (NERC) and members of the NERC Regional Electric Reliability Councils (18,19). Where justified by local requirements, individual companies often use criteria that exceed those specified by the reliability councils. The transmission capacities provided recognize both initial needs and future needs so the system can be expanded in the most economic manner.

The reliability criteria define the permissible range of deviations from the normal transmission conditions that may be tolerated for various contingencies. Reliability standards are generally based on both technical and economic considerations, often balancing increased costs against lower service reliability. The NERC sets these standards for North America. Regional Electric Reliability Councils may set more stringent standards for their specific regions, but the NERC standards must be adhered to by all regions.

Single-Contingency Criteria. One of the basic reliability criteria applied in system operation on most U.S. transmission systems consists of "single-contingency operation." It provides that the loadings on a transmission system must be limited to levels that will permit any single contingency (such as the loss of any single line or any generating unit) to occur without exceeding acceptable circuit or system limits.

Multiple-Outage Criteria. The possibility of a combination of disturbances and other conditions must also be considered, such as multiple transmission outages during lightning storms; transmission line outages occurring while other equipment such as generators, cables, or transformers are being maintained; or the outage of lines sharing a common right-of-way.

Transmission Reserves. One immediate effect of operation taking into account single or multiple contingencies is that a significant fraction of each system's transmission capacity must be held in reserve so that any likely contingency can be sustained.

Transmission Systems Operation

Operation of a transmission system is concerned with two principal goals:

- The reliability of supply to the customers.
- The production of power in the most economical way possible.

In pursuing these goals, the usual approach is for the operator to seek the most economic pattern of generation, while complying with all limitations imposed by the need for reliable operation (20).

The practice of limiting transfers to what can be safely transmitted even if the worst single contingency occurs is recommended by the NERC for general use by all systems. This limit is called the network transfer capability (21). It is based on what is sometimes called *preventive* operation. In some situations, operation based on *corrective* operation is used. In these situations, fast automatic controls are used to quickly reduce the transmission loading to safe limits if a contingency should occur. Corrective operation results in higher average transmission system loadings with some potential reliability risks.

The safe and reliable operation of the transmission system is based on three basic requirements:

- All components operating within their thermal ratings.
- Voltages at all points remaining within acceptable lower and upper limits. (This imposes both static and dynamic constraints.)
- Synchronous operation of all the generators connected to the system.

Transmission System Planning

Transmission system planning has been defined as the process of determining *what* facilities should be provided and *when* and *where* they should be provided to assure adequate transmission service at minimum present and future costs on a present-worth basis, consistent with environmental standards.

The design of the transmission system and the generating system must take into account the need to provide for customer loads both under normal conditions and under contingencies in the transmission system (transmission line failures) and in the generation system (generator failures).

The conditions that must be considered in system planning include any actions that the system may be required to take due to commercial, legal, contractual, and regulatory requirements.

Time Horizons. A transmission system must be planned many years in advance in order to satisfy a number of needs:

- To make decisions on the construction of transmission facilities far enough in advance that all the time needed to obtain the required governmental authorizations, obtain rights-of-way, make detailed designs, obtain construction bids, and construct the facilities be available before the facilities are needed.
- To determine whether planned generation facilities at given locations can be connected to the system and at what cost.
- To determine the best long-term transmission additions to meet near-term needs.
- To determine the additional facilities required, if any, to provide the transmission service requested by prospective transmission customers.

Since future expected circumstances, including load forecasts, available generation sources, fuel prices, and other conditions, change constantly, transmission planning is essentially a continuing process, requiring frequent reviews of the effect of changed circumstances and projections on the long-range plan.

Developing Transmission Technology

Research is in progress on new transmission technologies that may have long-range effects both on the types of transmission facilities installed and the capacity of the various networks. These technologies fall into the general classification of "network improvement" technologies and include devices such as solid-state phase shifters and fast control over series capacitators (22). They are typically "power electronic devices" that utilize thyristor and other solid-state devices to provide high-speed response.

TRANSMISSION SYSTEM COMPUTATIONAL METHODS

Types of Transmission Studies

Transmission studies are performed to determine if existing transmission lines will become overstressed under future loads, where and when new transmission facilities might be required, under what circumstances transmission facilities would be inadequate, and how existing and future transmission facilities will respond to emergency conditions.

In performing these studies, complex digital computer programs are used. These computer programs require detailed descriptions of the design and operating characteristics of the transmission and generation systems. Information such as transmission line ratings, circuit breaker ratings, impedances and length of each transmission line, impedances of transformers, turns ratios of transformers, estimated customer demand on each transmission and subtransmission substations, as well as estimated generating unit output levels on an existing and projected basis are typically recorded and maintained by the electric utility for use in these transmission studies.

Load Flow Studies

Load flow studies are used to simulate the operation of the electrical network under normal operation and various equipment outages. Reliable operation of the system depends on knowing the effects of various outages and new generation, new load, and new transmission lines before they are installed.

Short Circuit Studies

Every addition to a network increases the power that in the case of a short circuit is to be conducted or interrupted by switching equipment. Studies of potential fault currents are needed to ensure proper relaying and circuit breakers are installed.

Voltage Security Studies

These studies aim at detecting risks of voltage collapse due to severe contingencies, which may occur at high level of exchanges between the utilities.

Transient Stability Studies

All generators within an interconnected network should operate at a given speed accurately determined by the common network frequency. Sudden disturbances may create electrical and mechanical forces that may cause loss of synchronism for specific generators or portions of a system.

To analyze these different possibilities, stability studies have to be made that check the behavior of generators in different contingencies.

WHEELING AND TRANSMISSION ACCESS

Transmission Access

Interutility sales have long comprised a form of competition in generation as utilities with excess capacity competed to sell power to those that needed additional power. Typically these interutility transactions benefitted all parties and encouraged efficient operation.

The Energy Policy Act of 1992 provided expanded access to the transmission system. This was intended to increase competition by permitting (when possible) generating companies to deliver power to customers other than the local utility.

As the number of players in the electric power industry increases, demand for wheeling increased. Competing generators will want to sell to whomever will pay the most, whether that is the local utility or a distance customer, and some consumers will want to shop for supplies. In either case, they will require transmission services.

The subject of compulsory wheeling is a contentious one with many believing that it will result in overall increases in the use of electricity and adverse environmental effects because of increases in fuel consumption.

The technical and economic challenges of increased transmission access are significant. The available capacity on transmission systems is difficult to determine. It depends on the specific conditions at the time transmission is desired, the reliability and longevity levels selected by the utility, and the parallel-path flows that will result. Therefore disputes over the feasibility and cost of wheeling are difficult to resolve.

Wheeling Services

Wheeling services with many different conditions and arrangements are used: firm service and interruptible service.

Firm service, in almost all cases, refers to the firmness of the commitment to wheel only after the wheeling utility has agreed to provide the service and a rate schedule or service agreement has been signed stating the amount of power to be wheeled.

Interruptible service means that the wheeling service can be curtailed or interrupted for reasons other than those specified for firm service.

A very important characteristic of a transmission service is the priority of the particular transaction should curtailment be necessary. In general, firm transactions will have a much higher priority than nonfirm transmission.

Transmission Pricing

The provision of transmission service requires the use of facilities in generating plants and in the distribution systems. Costs for use of these facilities are being included in transmission costs. This becomes quite complicated since, in many cases, the generation and distribution companies are separately owned.

Also, transactions frequently result in the use of transmission systems other than that of the transmitting company. Procedures for handling this problem are proving difficult to develop.

SITING, ENVIRONMENTAL, AND HEALTH ISSUES

The continued need for additional transmission systems raises some public policy issues. Three of the most significant and contentious of these issues are transmission line siting, environmental impacts, and potential public health effects of electric and magnetic fields.

Siting

The siting of new electric transmission lines has become more difficult because of the obstacles encountered in the process of regulatory review and approval. The process of gaining approval for transmission line construction has become more formalized as opportunities have been provided for public involvement and greater scrutiny of potential environmental and social impacts of proposed projects. Competition for land to route transmission lines has become more intense and right-of-way costs are increasing (23).

Environmental Impacts

Overhead transmission lines are generally considered unsightly and undesirable. Transmission systems do provide, however, environmental benefits by reducing the total amount of generating capacity needed and by reducing the total amount of fuel burned to supply national electric energy needs.

Health Effects

Several recent epidemiologic studies have suggested an association between exposure to electric and magnetic fields and cancer. While these epidemiologic studies are controversial and incomplete, they do provide a basis for concern about effects from exposure.

BIBLIOGRAPHY

1. Rustebakke, *Electric Utility Systems & Practices*, 4th ed., Wiley, New York.
2. J. A. Casazza, *The Development of Electric Power Transmission*, The Institute of Electrical and Electronics Engineers, New York, 1993.
3. *Electrical Transmission and Distribution Reference Book*, Westinghouse Electric Corp., 1950.
4. L. V. Bewley, *Traveling Waves on Transmission Systems*, 2nd ed., Wiley, New York, 1961.
5. *Standard Handbook for Electrical Engineers*, McGraw-Hill, New York, 1993.
6. M. J. Steinberg, *Economic Loading of Steam Power Plants and Electric Systems*, 2nd ed., John Wiley & Sons, Inc., New York, 1947.
7. L. K. Kirchmayer, *Economic Control of Interconnected Systems*, Wiley, New York, 1959.
8. G. R. Slemon and A. Straughen, *Electric Machines*, Addison-Wesley, Reading, Mass., 1982.
9. L. K. Kirchmayer and G. W. Stagg, "Analysis of Total and Incremental Losses in Transmission Systems," *AIEE Transactions* **70,** 1197–1204 (1951).
10. J. A. Casazza, "Coordinated Regional EHV Planning in the Middle Atlantic States—U.S.A.," CIGRE Paper No. 315, Paris, France, 1964.
11. Federal Power Commission, *National Power Survey—A Report by the Federal Power Commission*, U.S. Government Printing Office, Washington D.C., October 1964.
12. National Electric Reliability Council (NERC), "Transfer Capability—A Reference Document," NERC, October 1980, pp. 6–7.
13. N. Cohn, *Control of Generation and Power Flow on Interconnected Systems*, Wiley, New York, 1966.
14. C. Concordia and L. K. Kirchmayer, "Tie-Line Power and Frequency Control of Electric Power Systems," *AIEE Transactions*, **72,** 562–572 (1953).
15. C. Concordia and L. K. Kirchmayer, "Tie-Line Power and Frequency Control of Electric Power Systems, Part II," *AIEE Transactions*, **73**(pt. III-A), 133–141 (1954).
16. R. H. Travers, "Load Control and Telemetering," *AIEE Transactions* **73**(pt. III-B), 522–527 (1954).
17. "Control Area Concepts and Obligations," North American Electric Reliability Council, 1992.
18. "Overview of Planning Reliability Criteria of the Regional Reliability Councils of NERC," North American Electric Reliability Council.
19. "Discussion of Regional Council Planning Reliability Criteria and Assessment Procedures," NERC.
20. P. J. Palermo, R. A. Bolley, and T. R. Woodward, *The Effects of Coordinated Operation on Energy Exchanges, System Operation and Data Exchange Requirements: A Comparison of Methods Used in the USA*, 1992.
21. "Transfer Capability—A Reference Document," NERC, 1980.
22. "The Future in High-Voltage Transmission: Flexible AC Transmission Systems (FACTS)," *Proceedings from the EPRI Workshop*, 1990.
23. *Non-Technical Impediments to Power Transfers*, National Regulatory Research Institute, 1987.

ELECTRIC AND MAGNETIC FIELDS IN THE ENVIRONMENT

MARKUS ZAHN
Massachusetts Institute of Technology
Cambridge, Massachusetts

We live in natural electromagnetic fields due to atmospheric electricity and the earth's magnetism. In addition, living organisms have internal electric fields due to electrolytic processes for cell and nerve functions, and magnetic fields from the magnetic moments of molecules. Superposed onto these natural fields are such man-made fields as from electric power lines, appliances and wiring. In order to better understand how electromagnetic fields can interact with living systems, we review the fundamentals of electric and magnetic field interactions with materials.

SOURCES OF ELECTRIC AND MAGNETIC FIELDS

All matter is composed of atoms that are in turn composed of negatively charged particles called electrons and positively charged particles called protons. Throughout this article, SI units are generally used for electrical quantities for which the base units are taken from the rationalized MKSA system of units. The unit of charge is a coulomb = 1 A·s. The charge magnitude on an electron and proton is 1.6×10^{-19} C so that 1 C of electricity contains about 6.24×10^{18} elementary charges. Each of these charges have associated with them an electric field **E**, which has SI units of volts per meter (V/m). The electric field emanates radially from a point charge and is proportional to the force that the charge exerts on other charges as given by Coulomb's law, where opposite charges attract and like charges repel. Usual matter is charge neutral with an equal amount of protons and electrons so that there is no external electric field. An electric field arises when there is an excess of protons or electrons on the material so that it is not charge neutral. See also ELECTRIC POWER DISTRIBUTION; ELECTRIC POWER SYSTEMS AND TRANSMISSION.

Most people have experienced nuisance frictional electrification caused by charge separation when walking across a carpet or pulling clothes out of a dryer. Using a plastic comb can cause hair to become charged and stand up, as the like-charged hairs repel one another. In a dry environment, in which large amounts of charge can significantly accumulate, these effects often result in small sparks because the electric forces from large amounts of charge actually pull electrons from air molecules. These sparks occur when the electric fields are larger than the electrical breakdown strength of air, $\sim 3 \times 10^6$ V/m.

When charges move, they constitute a current, and they give rise to the magnetic flux density **B**, which has SI units of teslas (T). Because a 1-T magnetic field is uncommonly large, a more common unit for **B** is the gauss (G); 1 T = 10,000 G. A related quantity, the magnetic field intensity **H**, which has units of amperes per meter (A/m), is also often used to describe magnetic fields; $\mathbf{B} = \mu_0 \mathbf{H}$ in free space. The quantity $\mu_0 = 4\pi \times 10^{-7}$ H/m is called the magnetic permeability of free space. The magnetic field from a small element of current is given by the Biot-Savart law, is in the direction perpendicular to the current element, and is proportional in magnitude to the force that the current element exerts on other current elements as given by the Lorentz force law where opposite flowing currents repel and like flowing currents attract. The magnetic Lorentz force on a current element is in the direction perpendicular to both the direction of current flow and the direction of the magnetic field acting on the current element. Currents generally flow in conductors, such as metallic wires, where electrons can easily flow. Insulators do not allow easy electron flow.

Net positive charge often accumulates on clouds with a negatively charged earth giving rise to a "fair weather" vertical dc electric field at ground level, typically of order 130 V/m. However, in thunderstorms these charges can increase, and when the electric field in air anywhere exceeds 3×10^6 V/m, air breaks down with luminous spark discharges as the charge passes from the cloud to another cloud or toward the ground, which is seen as lightning and heard as thunder (caused by the pressure wave of heated air). A typical lightning stroke passes 10,000–100,000 A of current for about 100 μs for a total charge of 1–100 C.

The earth's core consists of iron. An iron molecule behaves as if its nucleus of protons were spinning. This moving charge creates a dc magnetic field distribution around the earth, which at ground level has values of about 0.25–0.6 G. Because motion through a magnetic field creates an electric field, walking through the earth's magnetic field can induce body currents that add to biologically generated body currents.

Human bodies are composed of electrolytic materials, and bodily functions depend on ion concentrations and membrane potentials that are of order of the thermal voltage (kT/q), where $k = 1.38 \times 10^{-23}$ J/K is Boltzmann's constant, T is the absolute temperature in degrees Kelvin, and $q \approx 1.6 \times 10^{-19}$ C is the charge on an electron. At room temperature ($T \approx 300$ K), the thermal voltage is about 25 mV. The brain communicates with the rest of the body by electrical impulses. A typical nerve cell fires at $\sim$ 50–100 mV. It is possible to monitor the health of the body by measuring these electrical signals using electrocardiography (ECG) of the heart and electroencephalography (EEG) of the brain. Heart pacemaker electrodes in contact with heart muscle provide current pulses 0.1–10 mA a few milliseconds long to synchronize the firing of heart cells. Magnetic resonance imaging (mri) can image internal organs from the magnetic moments of molecules in the body using d-c magnetic fields up to 2T.

Generated electricity adds additional electric and magnetic fields to the environment, which can differ from natural electromagnetic fields in amplitude, direction, and frequency (1). In the United States, electric power uses 60 Hz ac, while Europe and some other parts of the world use 50 Hz. Table 1 lists representative values of generated electromagnetic fields.

The typical household background at 60 Hz is 1–10 V/m for the electric field and 0.1–10 mG for the magnetic field (3). People can generally sense 60 Hz electric fields above $\sim$ 20 kV/m through hair and skin sensations, although there is great variability among individuals (4). Some individuals can sense electric fields as low as 0.35 kV/m (4). People cannot generally sense magnetic fields even up to 20,000 G, although low frequency ($\sim$ 20 Hz) magnetic fields greater than 100 G produce luminous images known as "phosphenes," apparently due to induction of electric currents in the retina (3,4).

In the United States, maximum exposure standards for 60-Hz electromagnetic fields at transmission lines right of way are typically over the range of 1–11 kV/m rms for the electric field, and up to 250 mG in Florida for the magnetic field (3,5). Elsewhere there are no exposure standards for 50/60 Hz magnetic fields. Because these representative environmental fields are not greatly different from natural electromagnetic fields, it was generally thought that there were no health hazards (4–16).

ELECTROMAGNETIC FIELDS

The purpose of this article is to review fundamental electromagnetic field interactions with media so that people can try to minimize their exposure to fields and so that sound physical science can be used in the design and interpretation of continuing and future health studies to iden-

Table 1. Typical Steady-state Electric and Magnetic Power Frequency Fields[a]

Field	E (V/m) rms[b]	B (mG) rms
Home wiring	1–10	1–5
At electrical appliances	30–300	5–3000
Under distribution lines serving homes	10–60	1–10
Inside railroad cars on electrified lines		10–200
Under high voltage transmission lines	1000–7000	25–100

[a] From Ref. 2.

[b] rms is the root mean square; it is equal to the square root of the average of the square of periodic voltage or current over a period. For a sine wave with peak amplitude A, the rms value is $A/\sqrt{2} \approx 0.707$ A.

tify nonambiguously mechanisms of biological interaction between electromagnetic fields and living systems.

Quasistatics and Electrodynamics

There are three broad types of electromagnetic field interactions: (*1*) electroquasistatics, a low frequency range for which charges and voltages are the sources of the electric field coupled to dielectric and conducting media with negligible magnetic field; (*2*) magnetoquasistatics, a low frequency range for which currents are the source of the magnetic field coupled to magnetizable and conducting media with electromagnetic induction generating an electric field; and (*3*) electrodynamics, a high frequency range for which the electric and magnetic fields are of equal importance, resulting in radiating waves that travel at the speed of light ($c \approx 3 \times 10^8$ m/s) in free space. Figure 1 summarizes the frequency ranges for many common applications. Quasistatic fields occur when the wavelength of electromagnetic waves at frequency f ($\lambda = c/f$) is much larger than the size of the system. For power frequency of $f = 60$ Hz in the United States, $\lambda \approx 5 \times 10^6$ m, so that the usual small electrical appliances can be considered quasistatic. This power frequency regime is often called the extra low frequency (ELF) range. Quasistatic fields are confined to the immediate vicinity of the electrical device. Radio frequency and higher frequencies have propagating electromagnetic waves that travel at the speed of light away from a source.

The energy W in each photon, the fundamental equivalent particle of an electromagnetic field, is $W = hf$, where h is 6.6256×10^{-34} J·s is Planck's constant. Frequencies below the uv light region ($f < 10^{15}$ Hz) are called nonionizing radiation, because the energy W is too low to break chemical or molecular bonds. The energy of a 60-Hz photon ($W \approx 4 \times 10^{-32}$ J) is 11 orders of magnitude smaller than the thermal energy at room temperature ($T \approx 300$ K), $kT \approx 4 \times 10^{-21}$ J. Ultraviolet, x-ray, and γ-ray radiation have sufficiently high energy to break chemical and molecular bonds and can cause cancer and other health problems, depending on total exposure. Radio and microwaves do not have enough energy to break bonds, but they are absorbed by water in body tissues to cause heating. This is the principle of cooking with a microwave oven and of medical treatment by heating body tissue with diathermy machines.

Examples of electromagnetic field use in daily life are shown in Figure 2. Electroquasistatic applications include capacitors, vacuum tubes, and semiconductors in circuit applications; microphones; electrocardiography and electroencephalography; and electrostatic precipitators, printers, and copiers. Magnetoquasistatic applications include inductors and transformers in circuit and power system applications; audio speakers; magnetic resonance imaging devices; and motors and generators. The proposed new train system of magnetically levitated trains (Maglev) and perhaps electric cars provides further urgency to resolving adverse health issues in low level magnetic fields. Electrodynamic fields are present in radio, television, communications, optics, and microwave ovens.

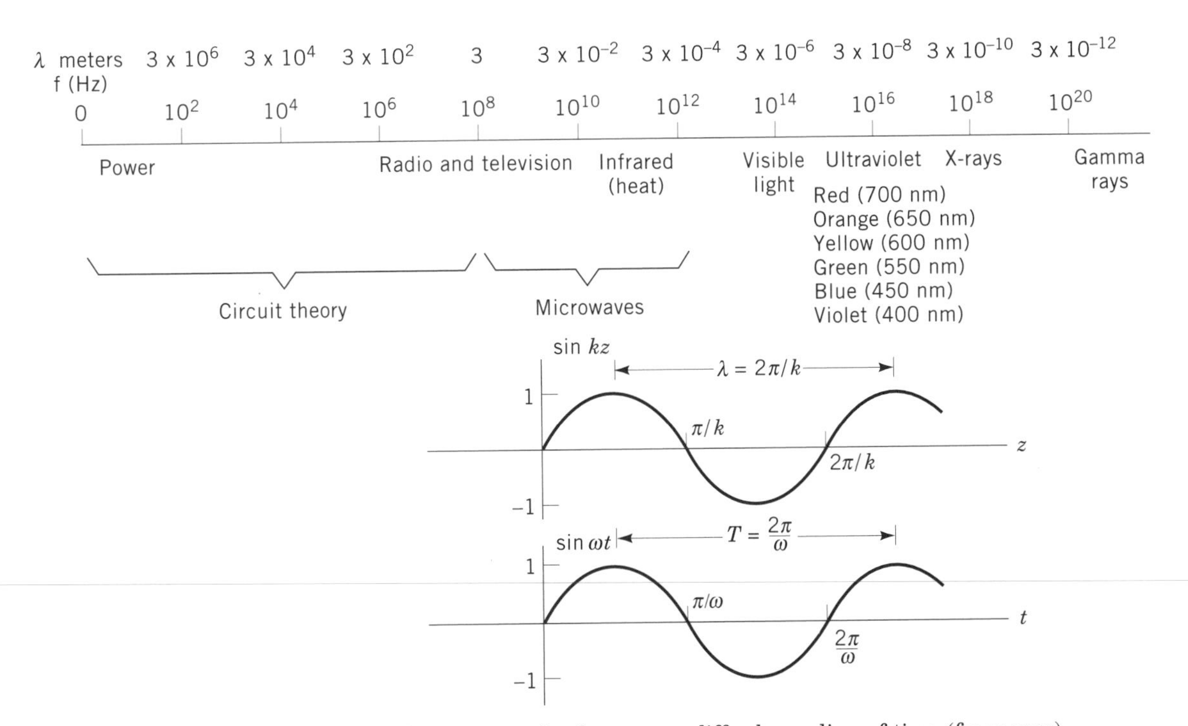

Figure 1. Time-varying electromagnetic phenomena differ by scaling of time (frequency) and size (wavelength). In free space, the frequency f in Hz (angular frequency $\omega = 2\pi f$ rad/s) and wavelength λ in m (wave number $k = 2\pi/\lambda$) are related as $f\lambda = c$ ($\omega = kc$) where $c = [\varepsilon_0\mu_0]^{-1/2} \approx 3 \times 10^8$ m/s is the speed of light in free space.

Figure 2. Examples of electromagnetic fields in daily life.

Imposed charges and voltages generate electric fields and imposed currents generate magnetic fields. However, these fields induce further charges and currents in dielectric, conducting, and magnetizable materials that then in turn create additional electric and magnetic fields. Often, the presence of material concentrates the fields so that the fields at sharp points at the surface can be much larger than when the materials are absent. Fields inside materials are also often much less than the fields just outside, being shielded by induced surface charges and currents. To further complicate analysis and understanding, time-varying magnetic fields create electric fields (Faraday's law) and time-varying electric fields create magnetic fields (Ampere's law). Electric fields in conducting materials cause current flow, voltage differences, and heating.

Quasistatic Fields from Electrical Devices

Small Appliances: Point Dipole Fields. The fundamental source for the electric field is a point charge q, while the source for a magnetic field is a current i of vector differential length $\mathbf{d\ell}$. Opposite polarity charges attract through the electric force, whereas like-charged particles repel. Currents flowing in the same direction attract through the magnetic force, while currents flowing in opposite directions repel. The electric field is the Coulomb force per unit charge on a charge due to all other charges. The magnetic field causes a Lorentz force on a moving charge at right angles to its motion due to all other moving charges or currents. The electric field from a single point charge q is given by Coulomb's law, and the magnetic field from a single current element $i\mathbf{d\ell}$ is given by the Biot-Savart law:

$$\mathbf{E} = \frac{q}{4\pi\varepsilon_0 r^2}\mathbf{i}_r; \mathbf{B} = \frac{\mu_0 i\mathbf{d\ell}\mathbf{x}\mathbf{i}_r}{4\pi r^2} \tag{1}$$

where $\varepsilon_0 \approx 10^{-9}/36\pi \approx 8.854 \times 10^{-12}$ F/m is the dielectric permittivity of free space. The quantities ε_0 and μ_0 are related by the speed of light in free space, $c = [\varepsilon_0\mu_0]^{-1/2} \approx$

3×10^8 m/s. Both fields at any point a distance r from the sources are inversely proportional to the square of the distance r. The unit vector $\mathbf{i}_r$ is in the direction from the source to the field point at distance r. The electric field $\mathbf{E}$ is in the direction of $\mathbf{i}_r$ while the magnetic field $\mathbf{B}$ is perpendicular to both the direction of the current $\mathbf{d}\ell$ and $\mathbf{i}_r$. If there is a distribution of charges and currents throughout space, the total electric and magnetic field is the vector superposition of equation 1 from all source charges and currents.

However, electrical devices must be charge neutral with at least two electrodes so that if one electrode has positive charge, the other electrode has an equal magnitude but negative charge. The charge on each electrode is proportional to the voltage difference between electrodes through the capacitance. The capacitance depends on electrode geometry and on the permittivity (dielectric constant) of the materials between the electrodes. From a distance away that is larger than the spacing d between opposite polarity charges ($r \gg d$), the system looks like a point electric dipole as in Figure 3a. Similarly, current must flow in closed loops as in the magnetic dipole in Figure 3b with small area S. For distances far from the dipoles compared with their size ($r \gg d$, $r \gg \sqrt{S}$), the electric and magnetic fields given by equation 1 approximately cancel, leaving weaker strength point dipole fields:

$$\mathbf{E} = \frac{qd}{4\pi\varepsilon_0 r^3}[2\cos\theta\mathbf{i}_r + \sin\theta\mathbf{i}_\theta];$$

$$\mathbf{B} = \frac{\mu_0 iS}{4\pi r^3}[2\cos\theta\mathbf{i}_r + \sin\theta\mathbf{i}_\theta] \quad (2)$$

These fields depend in magnitude and direction on angle θ, but most significant, they increase linearly with the source strength (charge or voltage for the electric field and current for the magnetic field), and source size (spacing or area), and decrease with distance as $1/r^3$. Figure 3c shows the dipole field lines that are tangent to the fields given in equation 2. The strength of the field is proportional to the density of drawn lines, which decreases as $1/r^3$. Electric field lines start on the positive charge and terminate on the negative charge, whereas magnetic field lines always form closed loops encircling a current loop.

Small appliances and motors have approximate point dipole fields at large distances compared with their size (17,18). Thus the near fields depend directly on the sources and size, but a doubling of the distance from the device center would approximately decrease the fields by a factor of eight. This indicates that by moving electric clocks, lights, motors, and other small electrical devices as far as possible from the body, the electric and magnetic field exposure would greatly decrease (16,19,20).

Within a typical home, appliances generally have the electrical and magnetic field values summarized in Table 2. Immediately next to these appliances, the maximum magnetic field is typically about 2 G and the maximum electric field is about 100 V/m (19,20,22–25). A 325-W soldering gun and hair dryer have localized magnetic fields from 10 to 25 G, whereas a clothes dryer, toaster, vacuum cleaner, dishwasher, clothes washer, electric iron, and refrigerator have magnetic fields less than 1 G (21). Electric blankets are thought to cause high exposure to power-frequency electric and magnetic fields because they are close to a body for a long period of time (3,25). Older-style designs have magnetic fields from 5 to 100 mG, with whole body–averged magnetic flux densities of ~22 mG and electric fields of 100–2000 V/m. Newer designs using partial magnetic field cancellation of parallel conductors with current flow in opposite directions have the magnetic field strength reduced by about 30 times while the electric field

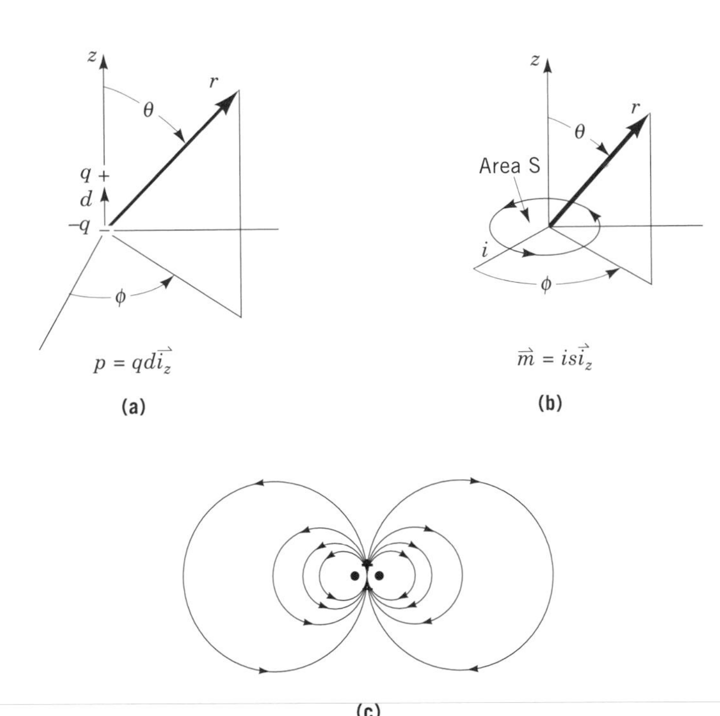

Figure 3. (**a**) Electrical devices must be charge neutral with positive and negative charges $\pm q$ a small distance d apart forming an electric dipole with dipole moment $p = qd$. (**b**) Current i must flow in closed loops with area S forming a magnetic dipole with moment $m = iS$. (**c**) The electric field from an electric dipole and the magnetic field from a magnetic dipole have the same shapes, which vary in magnitude and direction with radial distance r and angle θ. Shown are the field lines that are tangent to the electric and magnetic fields.

Table 2. Typical Residential Magnetic and Electric Field Exposures[a]

Field	B (mG rms)	E (V/m rms)
Power lines	5	
Appliances at 25 cm		
Electric range	20	4
Color television	12	30
Refrigerator	5	60
Clock radio (analogue)	24	15
Fluorescent light	21	

[a] From Refs. 20 and 21.

remains essentially unchanged (25). Twisting the two wires would further decrease the magnetic fields.

Home Wiring: Line Dipole Fields. The two wires connecting to an appliance are generally at an ac voltage difference of 120 V rms in the United States. At distances close to the wires compared with their length, the fields are essentially the same as if the wires were infinitely long.

The voltage on each line also imposes a line charge density $\pm\lambda$ C/m of opposite polarity on each wire. Each single cylindrical conductor alone with uniform line charge density λ and uniform current i have radial electric and encircling ϕ directed magnetic fields outside the conductor for radial distances r much less than the length:

$$\mathbf{E} = \frac{\lambda}{2\pi\varepsilon_0 r}\mathbf{i}_r; \mathbf{B} = \frac{\mu_0 i}{2\pi r}\mathbf{i}_\phi \quad (3)$$

The amount of charge $\pm\lambda$ on each conductor is proportional to the voltage difference between conductors through the capacitance, which only depends on the geometry and dielectric constant of the insulation. Because the pair of conducting wires at a distance d apart carry opposite polarity line charges $\pm\lambda$ and oppositely flowing currents $\pm i$, the fields in the vicinity of the wires tend approximately to cancel so that the fields die off quicker than the $1/r$ dependence in equation 3. These line dipole fields decay as $1/r^2$ with distance, and the field direction depends on angle ϕ:

$$\mathbf{E} = \frac{\lambda d}{2\pi\varepsilon_0 r^2}[\sin\phi\mathbf{i}_r - \cos\phi\mathbf{i}_\phi];$$

$$\mathbf{B} = \frac{\mu_0 i d}{2\pi r^2}[\sin\phi\mathbf{i}_\phi + \cos\phi\mathbf{i}_r] \quad (4)$$

The electric and magnetic fields can be even further decreased if the pair of wires is twisted, thereby greatly decreasing the effective spacing d. However, because for safety reasons additional grounds in the home wiring are often attached to water pipes, all the return current does not necessarily travel back via the second wire. Then the net current along the line pair is nonzero, and the magnetic field decays more slowly as $1/r$ as given by equation 3, greatly increasing a person's exposure to magnetic fields. Because the voltage difference across the line pair always imposes equal magnitude but opposite polarity line charges, the electric field dies off as $1/r^2$ as given by equation 4, even when the currents are unbalanced.

Europe and other parts of the world use 240 V residential power, which halves the current through the home wiring and to appliances. Then the resulting electric field is twice and magnetic field is half that in U.S. residences.

Fields from Power Lines

An electric power system consists of generating stations, transmission lines, and distribution systems (Fig. 4) (3,20,23,24). Transmission lines connect generators to the distribution systems, and the distribution systems connect to individual loads (see Plate I). A generator can typically produce on the order of 100–1000 MW power at 11–35 kV rms at typically 3,000–50,000 A rms. Because transmission line losses are primarily due to series resistance in the cables, to minimize these losses it is necessary to minimize the line current. Because the transmitted power is proportional to the product of line voltage and line current, the same power can be transmitted at reduced current by increasing the voltage. A voltage step-up transformer can increase the generator voltage at the primary winding up to the typical range of 69–765 kV rms at its secondary winding, thereby reducing the secondary current from the primary current by the inverse ratio of voltage step-up.

Typical power transmission systems use balanced four-wire, three-phase power, and the generator and voltage step-up transformer have voltages and currents of the fol-

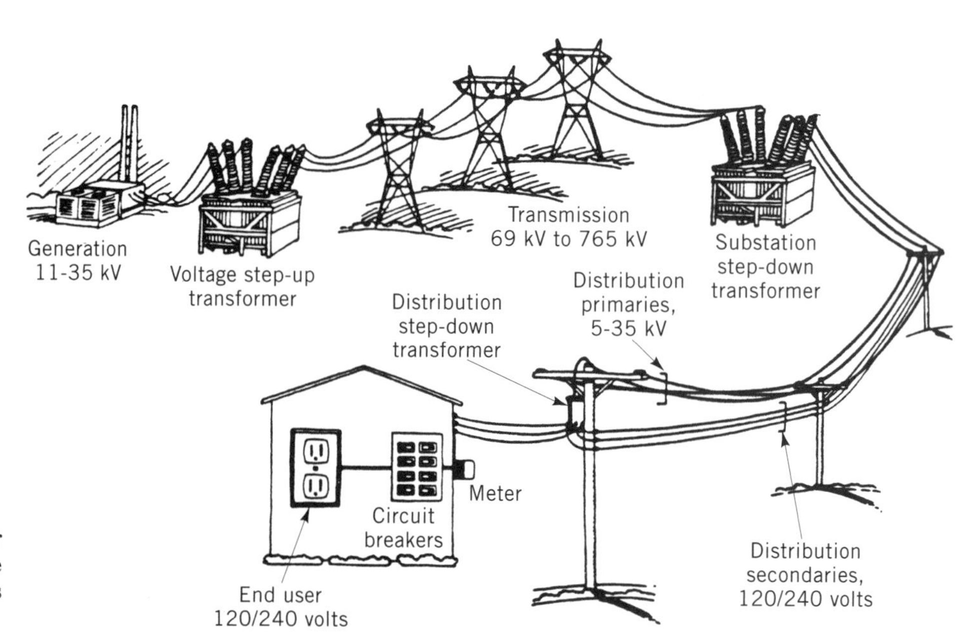

Figure 4. Three-phase electric power carried from the generator through the transmission and distribution systems to a house.

Plate I. Transformers allow for the connections between generators and transmission lines. Courtesy of Con Edison.

lowing form for each phase:

$$v_A = V \cos \omega t;$$
$$v_B = V \cos (\omega t - 2\pi/3);$$
$$v_C = V \cos (\omega t + 2\pi/3)$$

$$i_A = I \cos (\omega t + \phi);$$
$$i_B = I \cos (\omega t + \phi - 2\pi/3);$$
$$i_C = I \cos (\omega t + \phi + 2\pi/3) \qquad (5)$$

where ϕ is the phase angle between voltage and current, and for 60 Hz power, $\omega = 120\ \pi \approx 377$ rad/s. Each phase voltage is with respect to the neutral fourth wire, said to be at ground potential. Each phase has the same magnitude voltage and current, but each differs in phase angle from the other two phases by $\pm$ 120°. Three-phase power is used because the total power delivered by the generator

$$p = v_A i_A + v_B i_B + v_C i_C = \frac{3}{2} VI \cos \phi \qquad (6)$$

is constant in time and thus mechanically easier on the generator to deliver power at a constant rate.

Algebraically, the sum of voltages and the sum of currents for balanced three-phase lines are zero:

$$v_A + v_B + v_C = 0 \qquad (7)$$
$$i_A + i_B + i_C = 0$$

so that the electric and magnetic fields die off as line dipoles, given in equation 4. Because the neutral wire must carry the algebraic sum of currents in the three phases, a balanced system as in equation 7 has zero current in the neutral wire.

A three-phase ac 525 kV rms transmission line with cylindrical conductors about 3 cm in diameter, spaced about 10 m apart and 10.6 m above ground have a peak ground level electric field of about 8 kV/m rms. If this three-phase transmission line carries a representative current of 2000 A rms per phase, the peak ground level magnetic field is about 0.2 G rms.

Typically, each phase of a primary distribution line reaches a pole transformer to reduce the voltage to the secondary distribution line voltage of 120/240 V in the United States, which connects to residences by three wires. One wire is a neutral return and there are two hot wires at $\pm$ 120 V rms with respect to the neutral wire. Small appliances operate at 120 V between one hot line and neutral while large power appliances operate at 240 V between the two hot lines. The small pole transformer is like a small point dipole and so the fields die off as $1/r^3$ as given by equation 2. The power line to the house is designed to have no net current as all the current that flows in the two hot wires is supposed to return via the neutral. Such a balanced line again has fields that decrease as $1/r^2$ as given by equation 4. However, because for safety reasons electrical grounds are often attached to water pipes, cable television lines, telephone lines, and other connections to ground, all the return current does not necessarily travel back via the neutral wire. Also, each phase of the three-phase secondary distribution line is connected to different homes, and if the power requirements of each home differ, the current magnitudes flowing in each phase are not necessarily the same. Either of these cases leads to unbalanced line currents so that the net current along a set of distribution lines is not zero and the magnetic field decays more slowly as $1/r$ (eq. 3), greatly increasing a person's exposure to magnetic fields. However, because the line voltages in a three-phase system are always essentially balanced even if the currents are unbalanced, the power line electric fields die off as $1/r^2$, even when the magnetic field dies off as $1/r$ with unbalanced currents.

To lower electric and magnetic field exposure under power lines, the Electric Power Research Institute (EPRI) High Voltage Transmission Research Center (HVTRC) has been examining different conductor configurations such as multiple conductors per phase, twisted lines, shielding wires, and six-phase power (20,23). For a 115 kV rms line at 1000 A rms, at a person's waist level 50 feet from the line center, a conventional line configuration would give a magnetic field of 40 mG; the best experimental configuration would produce a much weaker magnetic field of 1.5 mG.

The step down of voltage from the transmission levels at a bulk-power substation typically reduces the voltage to the range of 35–138 kV rms. The next step down in voltage is at the distribution substation, reducing the voltage commonly to 5–35 kV rms. The first epidemiological study (10) noted a higher incidence of childhood leukemia for those homes near pole distribution transformers. When the voltage is stepped down from the 5–35 kV rms range to 120/240 V rms, about a factor of 40–300, the secondary distribution current from the transformer similarly increases by the same factor, thus increasing the magnetic field strength by the same factor. It was hypothesized that children living and sleeping near such a distribution transformer with high magnetic fields may be subject to adverse health effects (10,11).

Electric Railroads

Conventional electric railroads have typical steady-state magnetic fields within the cars of order 10–25 G (26,27), while the German Maglev System has a magnetic field mostly below 0.5 G.

Radiation Fields

Radio, television, communication, and radar systems operate at much higher frequencies f, 500 kHz (0.5×10^6 Hz) to 1000 GHz (10^{12} Hz). Measurements of FM radio and UHF and VHF broadcast signal field intensities indicate that typical population exposure is $\sim$ 50 μW/m^2, corresponding to $E \sim 0.14$ V/m and $B \sim 4.6$ μG (1). The simplest model for transmitting antennae is the point electric or magnetic dipole antennae (Fig. 5) excited by a sinusoidal current of the form $i(t) = I_0 \cos \omega t$. If the antenna length ℓ of the electric dipole antenna is small compared to the wavelength $\lambda = c/f$, ($\ell \ll \lambda$) the electric and magnetic fields are

$$\mathbf{E}(r,\theta,t) = \frac{I_0 \ell k^2}{4\pi}\sqrt{\mu_0/\varepsilon_0}\left[-\mathbf{i}_r\left[2\cos\theta\left(\frac{\cos(\omega t - kr)}{(kr)^2} + \frac{\sin(\omega t - kr)}{(kr)^3}\right)\right] + \mathbf{i}_\theta\left[\sin\theta\left(\frac{\sin(\omega t - kr)}{kr} - \frac{\cos(\omega t - kr)}{(kr)^2} - \frac{\sin(\omega t - kr)}{(kr)^3}\right)\right]\right] \tag{8}$$

$$\mathbf{H}(r,\theta,t) = -\mathbf{i}_\phi \frac{I_0 \ell k^2}{4\pi}\sin\theta\left[\frac{\sin(\omega t - kr)}{kr} - \frac{\cos(\omega t - kr)}{(kr)^2}\right] \tag{9}$$

where $k = 2\pi/\lambda = \omega/c$ is called the wave number. At distances close to the antenna, the fields die off as $1/r^3$ like a point electric dipole. There is an intermediate field region that varies as $1/r^2$, but the most important terms are the far radiation field terms that decrease only as $1/r$ and thus dominate at distances far from the antenna. In this far field, electric power flows radially and $E_\theta = [\mu_0/\varepsilon_0]^{1/2} H_\phi$ where $\eta = [\mu_0/\varepsilon_0]^{1/2} \approx 120\pi \approx 377\ \Omega$ is known as the radiation resistance of free space. In the far field, the electromagnetic power density in units of W/m^2 is $S_r = E_\theta H_\phi \approx \eta H_\phi^2 = E_\theta^2/\eta$. The magnetic dipole radiation fields are the dual to the electric dipole, where the electric and magnetic fields reverse roles ($E \rightarrow H$ and $H \rightarrow E$), replacing $I_0 \ell \cos \omega t/\omega\varepsilon_0$ by $-I_0 S \sin \omega t$, where S is the area of the small current loop.

If one uses transmitted radio or telephone communications, such as with CB radio or cellular phones, the electric and magnetic fields decrease by $1/r^3$ in the vicinity of the

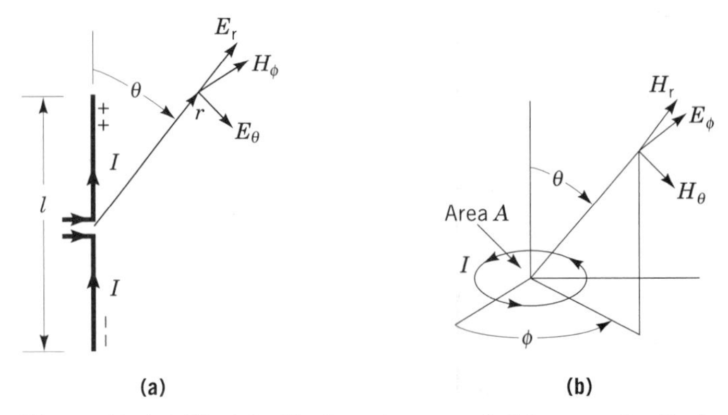

Figure 5. (**a**) Electric dipole antenna and (**b**) magnetic dipole antenna generate electromagnetic waves that radiate away from the antenna.

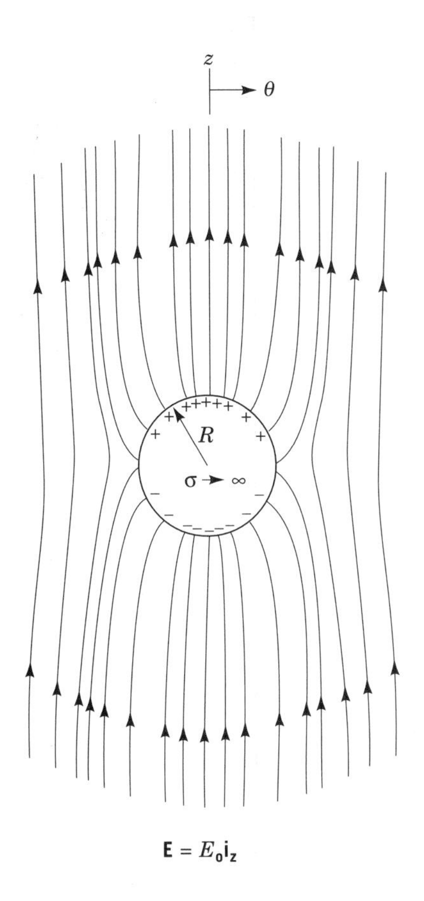

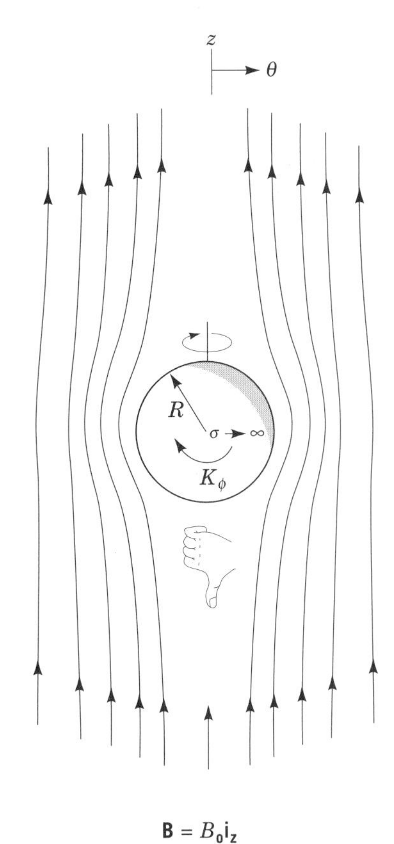

Figure 6. A perfectly conducting sphere is placed into uniform electric and magnetic fields. The induced surface charge causes the electric field to terminate perpendicularly and the induced surface current causes the magnetic field to pass tangentially to the sphere surface, with zero fields inside the sphere.

user. However, the electric and magnetic fields from large commercial transmitting and radar antennae die off as $1/r$ if one is more than a few wavelengths away. Low power radar is used for monitoring weather in aircraft, as navigational aids on small boats, and to determine vehicle speed by police. The maximum power density varies over the range of $S = 1 - 30$ W/m^2 so the fields are about $E = [S\eta]^{1/2} = 19 - 580$ V/m and $B = \mu_0[S/\eta]^{1/2} = 0.6 - 20$ mG (1).

A microwave oven is a microwave resonator at about 2.45 GHz with typical power levels of $\sim$ 500 W. The majority of microwave ovens have leakage below 50 W/m^2, which is the emission limit from ovens at a distance of 5 cm, as given by the U.S. Bureau of Radiological Health (1). This corresponds to microwave rms fields of $E \approx [S\eta]^{1/2} = 137$ V/m and $B = \mu_0[S/\eta]^{1/2} = 4.6$ mG. The 60 Hz magnetic field from a microwave oven at 25.4 cm is 1–120 mG (19).

Field Concentration

Dielectric and conducting objects placed into an electric field, change the electric field distribution in the vicinity of the object, often tending to enhance the electric field at selected positions along the medium surface. Similarly, conducting and magnetizable media tend to enhance magnetic fields at selected locations. For example, Figure 6 shows a perfectly conducting (ohmic conductivity $\sigma \rightarrow \infty$) sphere of radius R placed into uniform z directed electric and magnetic fields. The electric and magnetic fields inside the perfectly conducting sphere must be zero. To shield the interior of the sphere from the electric field, a dipolar surface charge forms on the sphere surface. A $-\phi$ directed surface current flows so that its self-field is in the opposite direction to the imposed magnetic field, as illustrated by the right-hand rule by which the thumb points in the direction of current-generated magnetic field when the fingers of the right hand are curled in the direction of current flow. The reaction fields due to induced surface charges and currents are point dipole fields so that the total electric and magnetic fields for $r > R$ are

$$\mathbf{E} = E_0\left[\left(1 + \frac{2R^3}{r^3}\right)\cos\theta\mathbf{i}_r - \left(1 - \frac{R^3}{r^3}\right)\sin\theta\mathbf{i}_\theta\right]$$
$$\mathbf{B} = B_0\left[\left(1 - \frac{R^3}{r^3}\right)\cos\theta\mathbf{i}_r - \left(1 + \frac{R^3}{2r^3}\right)\sin\theta\mathbf{i}_\theta\right] \quad (10)$$

Note that at the sphere surface, $r = R$, boundary conditions require that the electric field be purely perpendicular (radial) to the sphere and the magnetic field to be purely tangential (θ directed) to the sphere. The peak surface electric field is $3E_0$ at $\theta = 0$ and $\theta = \pi$, while the peak magnetic field is $3/2\, B_0$ at $\theta = \pi/2$. The sharper the geometry, the greater the field enhancements at the sharpest points.

Shielding

Electroquasistatic Fields. Most nonmagnetic materials can be described by their dielectric permittivity ε, or equivalently their relative dielectric constant $\varepsilon_r = \varepsilon/\varepsilon_0$, and their ohmic conductivity σ. When an electric field is first imposed in the vicinity of such a medium, the total electric field distribution depends only on the dielectric constants of all materials. However, the electric field in the conducting media causes a volume current to flow with current density $\mathbf{J} = \sigma\mathbf{E}$. As time progresses, the current carries charge to surfaces so that a new steady-state electric field distribution results. If such an object is surrounded by free space in a dc electric field, the resulting steady-state surface charge distribution is such that the dc electric field inside the conducting medium is zero. The time constant for the system to evolve to the dc steady-state is called the dielectric relaxation time $\tau = \varepsilon/\sigma$. Taking the relative dielectric constant of the body to be that of water, $\varepsilon_r = 80$, and a typical body conductivity to be $\sigma \approx 0.2$ Siemen/m, the dielectric relaxation time is $\tau = \varepsilon_r\varepsilon_0/\sigma \approx 3.5 \times 10^{-9}$ s. When a dc electric field is first turned on, the electric field penetrates into the conductor, but after a few multiples of τ, the electric field inside the isolated conductor is essentially zero. For low sinusoidal frequencies with $f \ll 1/(2\pi\tau)$ ($f \ll 45$ MHz), the body acts like a good conductor, and the electric field within the body is small, being 10^{-7}–10^{-4} times less inside the body than just outside for 60 Hz electric fields (8). Because the body is a good electrical conductor, it significantly perturbs the 60 Hz electric field from the distribution present when the body is absent. The external time varying electric fields must terminate essentially perpendicularly on the highly conducting body, which induces a surface charge that also varies with time. A time-varying surface charge causes surface currents. However the magnitude of these surface currents is typically much less than normal biologically induced currents. Measured body currents in grounded humans approximately obey the relation:

$$i = 15 \times 10^{-8} f W^{2/3} E_0 \tag{11}$$

where i has units of mA, frequency f is in Hz, body weight W is in g, and E_0 is the ambient electric field in V/m (8). Thus a 68-kg person in a 1000 V/m 60 Hz electric field would have ~ 15 mA current.

Magnetoquasistatic and Electrodynamic Fields. Time-varying magnetic fields generate electric fields as given by Faraday's law causing an ohmic current with density $\mathbf{J} = \sigma\mathbf{E}$. These induced "eddy" currents flow in the direction such as to generate self-magnetic fields in the opposite direction to the original magnetic field, tending to decrease the magnetic field strength in the conductor. The magnetic field then penetrates only about a skin depth distance, $\delta = [2/(\omega\mu_0\sigma)]^{1/2}$ into the conductor, where $\omega = 2\pi f$ is the angular frequency of the magnetic field. For a person, $\sigma \approx 0.2$ S/m, and at power frequency ($f = 60$ Hz) $\omega = 2\pi f = 120\pi \approx$

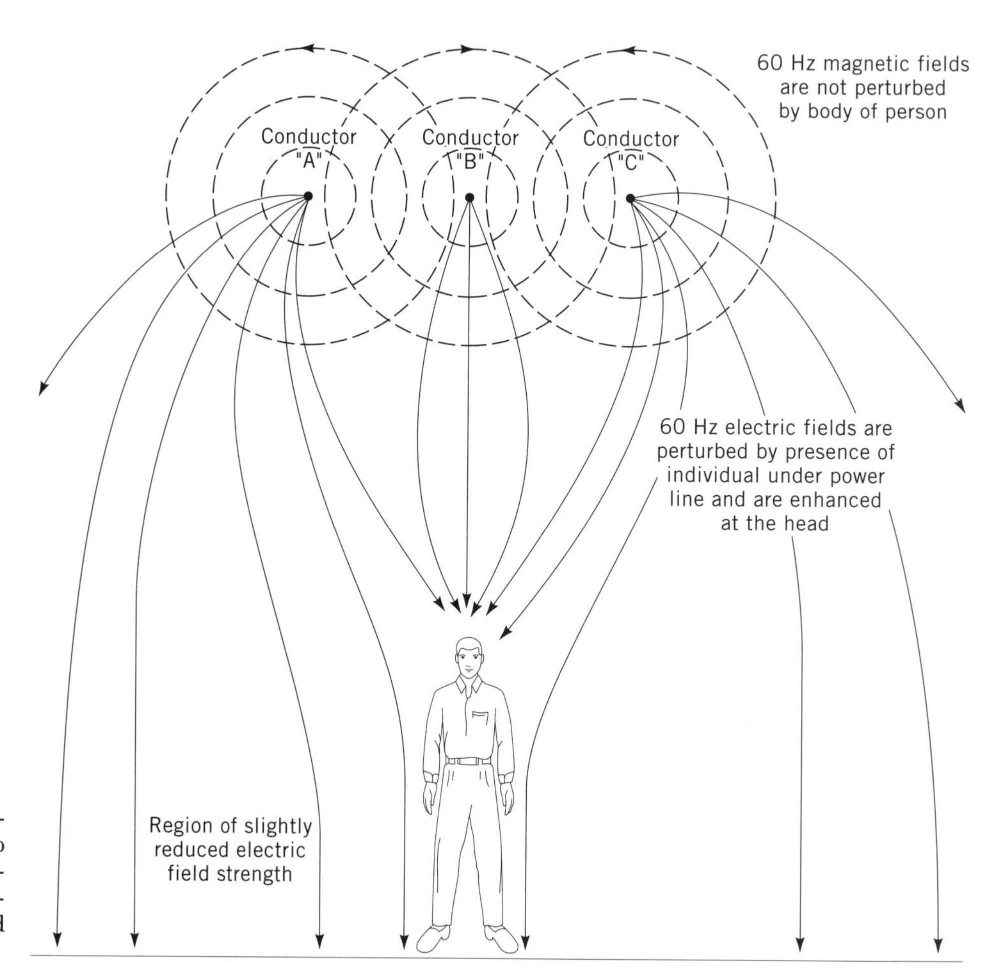

Figure 7. A person under a transmission line acts like a good conductor to the electric field, which terminates perpendicularly on the body but has negligible effect on the magnetic field (28,29).

377 rad/s, giving $\delta \approx 145$ m. Because this value is so much greater than the size of a person, the magnetic field essentially completely passes through the body with essentially no difference in distribution due to the presence of the body. At $f = 10$ MHz, the skin depth in the body is $\delta \sim 0.36$ m, somewhat comparable to a person's thickness. Thus for frequencies less than 10 MHz, the body has essentially no effect in perturbing a magnetic field distribution.

A house or vehicle also typically shields the interior from power line electric fields but has essentially no effect on power line magnetic fields. If one really wanted to shield their vehicle or house from power line magnetic fields, the good electrical conductor copper, which has conductivity $\sigma = 5.8 \times 10^7$ S/m, could be used. The skin depth of copper at 60 Hz is 8.5 mm, and to greatly attenuate magnetic field strength, copper sheet a few skin depths thick would be necessary, but with great weight and cost penalties. Of course, there would still be magnetic fields inside the home from home wiring and electricity use, so such an approach seems noneffective.

Power Frequency Fields and the Body. At 60 Hz the body is a good conductor for electric fields but a poor conductor for magnetic fields. The introduction of a body into an electric and magnetic field will locally change the electric field distribution but will have negligible effect on the magnetic field distribution. The heating rate in an ohmic conductor is $Q = \sigma E^2$ W/m^3. If the body is in an external electric field $E_{ext} = 10$ kV/m, the internal electric field is less than $E_{int} = 1$ V/m. With a body conductivity of $\sigma = 0.2$ S/m, the heating rate at $E = E_{int}$ is $Q = 0.2$ W/m^3. This power frequency heating is much less than body heating from normal metabolic processes.

Figure 7 shows a person standing under a three-phase power line (28,29). The electric field must terminate essentially perpendicularly on the body and on the ground with a surface charge distribution, with essentially no electric field inside the body or below ground. The magnetic field distribution is essentially the same whether the person is there or not and is not significantly perturbed by the presence of the ground. The time-varying electric field thus induces a time-varying surface charge density on the body and on the ground. A typical person standing on the ground in a 60 Hz, 10 kV/m rms electric field in the absence of the person has a body-averaged electric field enhanced to about 27 kV/m, with peak electric field of 180 kV/m rms at the head and a head current density of 60 nA/cm^2 (8). The current density through a person's foot to ground increases to about 2000 nA/cm^2.

Moving at velocity **v** through magnetic field **B** generates an electric field $\mathbf{E} = \mathbf{v} \times \mathbf{B}$, which is perpendicular to both **v** and **B** with magnitude $|E| = |vB \sin \theta|$ where θ is the angle between the vectors **v** and **B**. If a person walks at $v = 1$ m/s perpendicular ($\theta = 90°$) to the earth's dc magnetic field, $B \sim 0.5 \times 10^{-4}$ T, the induced dc electric field is $E \sim 0.5 \times 10^{-4}$ V/m. With the body's current density of $\sigma \sim 0.2$ S/m, the resulting dc current density is $J = \sigma E \sim 10^{-5}$ A/m^2 = 1 nA/cm^2.

Under a distribution line, the electric field is about 10 V/m, 1000 times less than the example of a 10 kV/m electric field, appropriate for standing under a transmission line. The induced current would similarly be 1000 times less or about 2 nA/cm^2 through a person's foot. Thus the induced dc current density of walking through the earth's magnetic field is much less than the induced 60 Hz currents of standing under a transmission line, but comparable to that of standing under a distribution line. Under most situations, 60 Hz induced currents in the body are comparable in magnitude to dc currents induced by walking through the earth's magnetic field.

Figure 8 illustrates the same principles as in Figure 7 for a video display terminal (VDT) operator (28,29). VDTs have 60 Hz fields from the power supply, dc fields with a 60 Hz modulation from the high voltage dc power supply applied to accelerate the electron beam in the cathode ray tube (CRT), and radio frequency (RF) fields (~10–300 kHz) caused by the circuitry associated with the deflection of the electron beam (28,29). At 300 kHz, the skin depth in the body is about 2 m, so that the magnetic field from the currents in the VDT are approximately that of a magnetic dipole and are negligibly perturbed by the presence of an operator or of ground and other nearby objects being about $H = 150$ mA/m ($B = 1.9$ mG) 10 cm in front of the screen. In the absence of an operator, the 300 kHz electric field would be essentially that of an electric dipole over a ground plane with electric field about 24 V/m rms just in

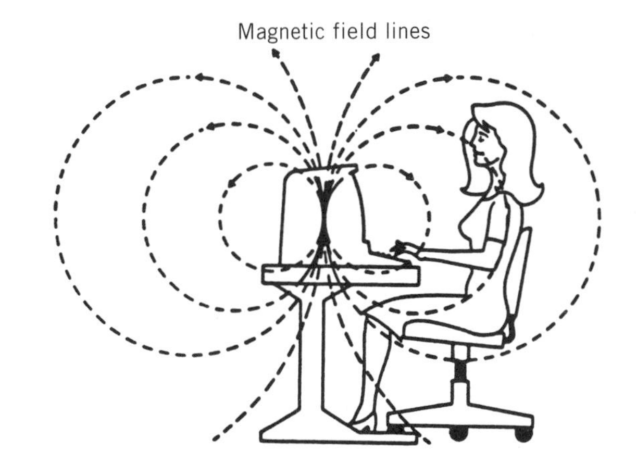

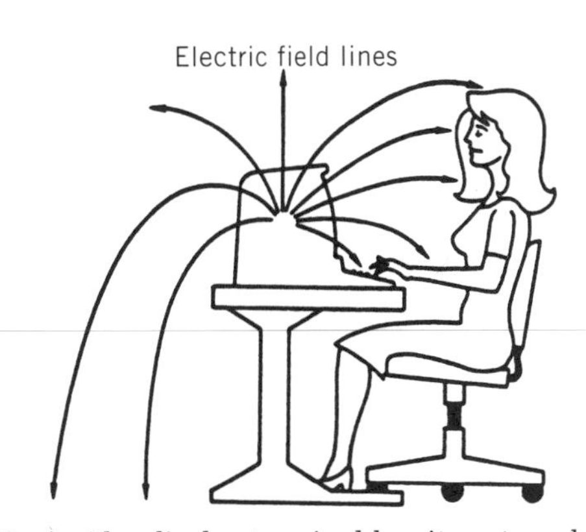

Figure 8. A video display terminal has its external electric field terminate perpendicularly on the operator, while the magnetic field is negligibly perturbed by the presence of the operator because the skin depth of all frequencies present are much larger than the person's thickness (29).

Table 3. IRPA/INIRC Limits of Exposure to 50/60 Hz Electric and Magnetic Fields[a]

Class	E (kV/m rms)	B (G rms)
Occupational workers		
Whole working day	10	5
Short-term	30	50
General public		
Whole day	5	1
Few hours per day	10	10

[a] From Refs. 31 and 32.

front of the screen. However, with the operator present, the electric field is perturbed to terminate essentially perpendicularly on the operator, with field strengths increased at sharper parts, such as the head.

REPRESENTATIVE STANDARDS

The first ANSI electromagnetic field standard established in 1966, limited exposure to 100 W/m^2 over the frequency range of 10 MHz to 100 GHz, which corresponds to an electric field limit of ~ 200 V/m rms and magnetic field limit of ~ 6.5 mG rms (30). In 1982, the ANSI standard was revised to make absorbed power dependent on frequency with a goal of limiting absorbed body average power density to 0.4 W/kg in the frequency range of 300 kHz–100 GHz. The International Radiation Protection Association (IRPA) formed the International Non-Ionizing Radiation Committee (INIRC) to formulate limits of exposure of 50/60 Hz and magnetic fields as given in Table 3 (31,32). The values in Table 3 were chosen so that the induced body current density is less than 10 mA/m^2, because this value does not exceed normal biological values.

IRPA and INIRC guidelines on radio frequency and microwaves (Table 4) assume no adverse health effects for energy deposition rates below 4W/kg. They further use a safety factor of 10 for long-term exposure and a further safety factor of 5 for general public limits.

PERSONAL MEASUREMENTS

Environmental electric and magnetic fields were measured in a house and community in Lexington, Massachusetts (33). A Holaday Industries, Inc. HI-3600-02 Power Frequency Field Strength Meter was used, which is designed to assist in the evaluation of electric and magnetic fields associated with 50/60 Hz electric power transmission and distribution lines and electrically operated equipment and appliances (28,29). A Field Star 1000 magnetic field meter with memory was also used to record magnetic fields as a function of time.

Electric fields are detected using a capacitive current sensor, which consists of two thinly separated conductive discs that are electrically connected together. When placed in an alternating electric field, charge is redistributed on the two parallel discs so that the electric field between the two discs at the same potential is zero. In a sinusoidally varying electric field, the redistribution of charge also changes sinusoidally with time giving a measured sinusoidal current in the connecting wire between the discs whose measured amplitude is proportional to the local electric field.

The magnetic field in the Holaday instrument is measured using a coil consisting of several hundred turns of fine gage wire. When placed in an alternating magnetic field, a current is induced in the coil whose amplitude is proportional to the magnetic field strength perpendicular to the coil. The Field Star 1000 instrument has three mutually perpendicular coils to measure the three vector components of magnetic field. The magnetic field magnitude is also computed by the instrument by taking the square root of the sum of squares of the three components.

The measured ambient electric field through the house varied over the range of 3–1600 V/m, the highest value being in front of an operating television. Far from an appliance the ambient electric field was of order 10 V/m. The magnetic field far from appliances was about H = 90 mA/m (B ~ 1 mG), with the highest values at a stove heating element at the high setting with 3330 mA/m (B ~ 42 mG). In the course of a 0.80-km distance from the house the outdoor electric field varied over the range of 0.35–17.2 V/m and the magnetic field varied from H = 5.2 to 620 mA/m (B = 0.065–7.8 mG). Within 3 m of a substation transformer, the electric field was 0.8 V/m and magnetic field was H = 1083 mA/m (B = 13.6 mG). For a typical high school student, the Field Star 1000 recorded a school-day magnetic field exposure of 0–2 mG; this increased to 16 mG in the school lunchroom (33).

Table 4. IRPA/INIRC Guidelines for Exposure Limits at Radio and Microwave Frequencies[a]

	Occupational Exposures		General Public Exposures	
f(MHz)	E (V/m rms)	B (mG rms)	E (V/m rms)	B (mG rms)
0.1–1	614	$20/f$	87	$2.9/f^{1/2}$
1–10	$614/f$	$20/f$	$87/f^{1/2}$	$2.9/f^{1/2}$
10–400	61	2	27.5	0.9
400–2000	$3f^{1/2}$	0.1	$1.37f^{1/2}$	$0.465f^{1/2}$
2000–300,000	137	4.5	61	2

[a] Refs. 31 and 32.

CONCLUSION

Despite epidemiological, laboratory, and human and animal studies that show possible biological effects from nonionizing electromagnetic fields, there are no well-established physical mechanisms or conclusive proof of adverse health effects. Nevertheless, it remains prudent that when possible, people should minimize their exposure to electromagnetic fields. This article has shown that typical exposures from power frequency and radio frequency fields result in voltages, currents, fields, and heating that are comparable or less in magnitude than those naturally occurring in the environment or from normal metabolic processes in the body.

Because at frequencies below ≈ 10 MHz the body is a

good conductor to electric fields but a poor conductor for magnetic fields, the presence of a body in an ambient electric and magnetic field will perturb the electric field distribution but will have essentially no effect on the magnetic field distribution. The electric field inside the body is greatly reduced from the electric field just outside the body, and the external electric field must terminate essentially perpendicularly to the body, resulting in a surface charge distribution and a surface current distribution if the external electric field is time varying. Those studies that show adverse health effects linked to the magnetic field but not the electric field may perhaps be due to the body conductivity that almost completely shields the body interior from the electric field but not the magnetic field.

Charges and voltages are the source of electric fields and currents are the source of magnetic fields. Appliances are charge neutral with current flow in a closed loop and thus approximately have electric fields like that of a point electric dipole and magnetic fields like that of a point magnetic dipole (fields that decrease with distances as $1/r^3$). Home wiring and balanced power lines have electric and magnetic fields like line dipoles that decrease with distances as $1/r^2$, but lines with unbalanced currents can have a larger magnetic field due to the net current flow on the lines that decreases with distance as $1/r$.

Because in the United States normal residential power is at 60 Hz, 120 V rms, whereas in Europe and elsewhere it is 50 Hz, 240 V, the normal household current is about half in Europe than in the United States. Health studies around the world need to recognize this difference in voltage and current magnitudes and perhaps frequency, although interpretation is difficult because health studies thus far have not shown the normal dose–response relationship that greater exposure is more harmful. While continuing health research tries to sort out physical mechanisms of possible health hazards, it is prudent for everyone to minimize unnecessary exposure to electric and magnetic fields.

Acknowledgments

Great appreciation is given to Donald L. Haes Jr., assistant radiation protection officer in the MIT Environmental Medical Service, and James C. Weaver of MIT for sharing of references on electromagnetic fields in the environment and for providing constructive comments on the manuscript. Appreciation is also given to Haes for loan of the Holaday Industries Power Frequency Field Strength Meter and the Field Star 1000.

BIBLIOGRAPHY

1. O. P. Gandhi, ed., *Biological Effects and Medical Applications of Electromagnetic Energy*, Prentice-Hall, Inc., Englewood Cliffs, N.J., 1990.
2. Oak Ridge Associated Universities Panel for the Committee on Interagency Radiation Research and Policy Coordination, *Health Effects of Low Frequency Electric and Magnetic Fields*, June 1992.
3. M. G. Morgan, *Electric and Magnetic Fields from 60 Hertz Power: What Do We Know About Possible Health Risks?*, Department of Engineering and Public Policy, Carnegie-Mellon University, Pittsburgh, Pa., 1989.
4. L. E. Anderson in B. W. Wilson, R. G. Stevens, and L. E. Anderson, eds., *Extremely Low Frequency Electromagnetic Fields: The Question of Cancer*, Battelle Press, Columbus, Ohio, 1990.
5. C. Polk and E. Postow, eds., *CRC Handbook of Biological Effects of Electromagnetic Fields*, CRC Press, Boca Raton, Fla., 1986.
6. M. Blank, ed., *Electricity and Magnetism in Biology and Medicine*, San Francisco Press, 1993.
7. E. L. Carstensen, *Biological Effects of Transmission Line Fields*, Elsevier, Amsterdam, The Netherlands, 1987.
8. W. T. Kaune and L. E. Anderson, Chap. 3 in Ref. 4.
9. T. S. Tenforde, Ch. 12 in Ref. 4.
10. N. Wertheimer and E. Leeper, *Am. J. Epidemiol.* **109,** 273–284 (1979).
11. P. Brodeur, *Currents of Death: Power Lines, Computer Terminals, and the Attempt to Cover Up Their Threat to Your Health*, Simon & Schuster, New York, 1989.
12. Electric Power Research Institute, *EMF Health Effects Research: A Selected Bibliography*, EPRI, Mar. 1992.
13. Electric Power Research Institute, *Exposure Assessment Fundamentals*, EPRI, Nov. 1992.
14. Electric Power Research Institute, *Fundamentals of Epidemiology: Parts I and II*, EPRI, Oct. 1993.
15. I. Nair, M. G. Morgan, and H. K. Florig, *Biological Effects of Power Frequency Electrical Fields*, NTIS PB89-209985, U.S. Congress Office of Technology Assessment, Washington, D.C., May 1989.
16. K. Fitzgerald, I. Nair, and M. G. Morgan, *IEEE Spectrum*, 22–35 (Aug. 1990).
17. D. L. Mader and S. B. Peralta, *Bioelectromagnetics* **13,** 287–301 (1992).
18. J. R. Gauger, *IEEE Trans. Power Apparatus Sys.* **PAS-104**(9), 2436–2444 (1985).
19. J. Douglas, *EPRI J.*, 18–25 (Apr.–May 1993).
20. J. Douglas, *EPRI J.*, 6–13 (July–Aug. 1993).
21. R. L. Loftness, *Energy Handbook*, Van Nostrand Reinhold Co., Inc., New York, 1978.
22. T. Moore, *EPRI J.*, 4–17 (Jan.–Feb. 1990).
23. T. Moore, *EPRI J.*, 4–19 (Oct.–Nov. 1990).
24. T. Moore, *EPRI J.*, 4–13 (Mar. 1992).
25. H. K. Florig and J. F. Hoburg, *Health Phys.* **58**(4), 493–502 (1990).
26. R. B. Goldberg, W. A. Creasey, and K. R. Foster, pp. 248–250 in Ref. 6.
27. B. W. Wilson, R. J. Reiter, and A. A. Pilla pp. 251–254 in Ref. 6.
28. *User Manual, HI-3600-2 Power Frequency Field Strength Meter*, #600040, Holaday Industries, Inc., Sept. 1989.
29. *User Manual, HI-3600 VDT Radiation Survey Meter*, #600031 B, Holaday Industries, Inc., Jan. 1988.
30. K. R. Foster and A. W. Guy, *Sci. Am.* **255**(3), 32–39 (1986).
31. A. S. Duchene, J. R. A. Lakey, and M. H. Repacholi, eds., *IRPA (International Radiation Protection Association) Guidelines on Protection against Non-Ionizing Radiation*, Pergamon Press, Oxford, UK, 1991.
32. M. Grandolfo and M. H. Repacholi pp. 77–80 in Ref. 6.
33. A. E. Zahn, *Environmental Electric and Magnetic Fields in the Lexington, MA Community,* 9th Grade Science Project, Lexington High School, Lexington, Mass., April, 1994.

ELECTROMAGNETIC FIELDS, HEALTH EFFECTS

Stephen F. Cleary
Medical College of Virginia
Virginia Commonwealth University
Richmond, Virginia

Life on earth has evolved in an electromagnetic environment comprised of natural terrestrial sources such as quasi-steady-state fields due to thunderstorms, lightening, and low frequency Schumann resonances (1). The sun and other extraterrestrial sources also contribute to our electromagnetic environment (1). With the exception of short duration transients from sun spots, giant radio bursts or lightening, the intensities of natural environmental low frequency electromagnetic fields, are of very low intensity and have remained relatively constant during past milleniums. This status is in distinct contrast to environmental levels of man-made electromagnetic fields which have continued to increase steadily since the turn of the century such that present levels are more than a million times higher than naturally occurring electromagnetic fields. See also Electric power distribution; Electric power systems and transportation.

The term electromagnetic indicates the coexistence in time and space of electric and magnetic field components, the interrelationship of which depends upon the nature of the source and the frequency (ν) of variation of the field amplitudes. The electric and magnetic field components are coupled since a time-varying electric field will induce a magnetic field at the same frequency and a time varying magnetic field will in turn induce an electric field in conducting objects such as the human body. The extent of electric–magnetic coupling is directly proportional to the rate-of-change, or frequency, of the field components. At low frequencies, such as in the extremely low frequency (ELF) range (0 to 10^3 Hertz (Hz)) coupling is minimal and thus either the electric or magnetic component will predominate depending upon whether the source is a current or voltage source. Under these minimal coupling conditions the components are referred to as electric or magnetic fields. At higher frequencies there is significantly greater coupling or energy exchange between the field components which results in propagation of the field through space as an electromagnetic wave or electromagnetic radiation. The basic relationship between the frequency (ν) and the wavelength (λ) of electromagnetic radiation is $\lambda\nu = c$, where c is the velocity of light (3×10^8 m/s). In the context of this article, electromagnetic fields (EMF) refer to fields having frequencies of 10^3 Hz or less, whereas higher frequency fields will be referred to as electromagnetic radiation (EMR). Predominant man-made environmental EMF sources are associated with the generation, transmission, and usage of 60 Hz power frequency fields. Common environmental sources of man-made EMR include radio frequency (RF) and microwave (MW) radiation in the frequency range of approximately 10^5 to 10^{11} Hz used for communications and heating. As the frequency increases, localized EMR energy coupling increases such that at frequencies of 10^{15} Hz or greater (corresponding to ultraviolet (UV) radiation) molecules or atoms are ionized. EMR in this range of frequencies, which includes in addition to UV, X- and gamma-rays, is referred to as ionizing radiation. Adverse health effects of ionizing EMR, including somatic and genetic mutations, have been well characterized as a result of numerous studies over the past 50 years (1). The remainder of this article will focus on the much less well-defined current health concerns resulting from exposure to nonionizing electromagnetic radiation, specifically ELF EMFs and RF and MW EMR.

Until relatively recently, recognized adverse health effects due to ELF EMFs were primarily shocks and/or burns resulting from inadvertent human contact with electrical conductors. Adverse effects of exposure to high frequency EMR, such as microwaves, were commonly attributed to radiation energy absorption in tissue resulting in excessive heating and consequent thermal damage. In either case, the acute nature of the exposure and the immediacy of the injury resulted in obvious cause–effect relationships (2).

Recently, attention has been focused on health effects of chronic or long-term low-intensity EMF and EMR exposure. The principal reason for concern are reports indicating elevated cancer incidence in human populations exposed to such fields in the home or in the workplace. In contrast to acute, high intensity exposure effects, it has been more difficult to establish cause–effect relationships resulting from chronic low-intensity exposure. This difficulty may be attributed to several factors including the limited amount of data presently available and the lack of a theoretical basis to explain how such low-intensity fields alter living systems. In contrast to effects of high intensity EMFs that involve significant field/tissue energy exchange, low-intensity field interactions in many instances occur at levels well below those associated with energies involved in classical physicochemical processes in living systems. Additional difficulty has been introduced by the dependence of such effects on field frequency or modulation, especially in the ELF range. It should be noted that in addition to potentially adverse exposure effects beneficial aspects of the biological activity of electromagnetic fields have also been demonstrated. For example, the capability of RF and MW EMR to induce localized heating has been exploited in tumor therapy and low intensity ELF pulsed magnetic fields have been used clinically to enhance rates of tissue healing (2,3).

Considering the ubiquitous and ever increasing levels of human exposure to EMFs and EMR in our environment, there is an obvious need to more precisely define the potential health effects. The remainder of this article will attempt to summarize the present state of knowledge regarding the health effects of the most common sources of human exposure; namely, ELF electric and/or magnetic fields and high frequency RF and microwave radiation.

ELF MAGNETIC FIELD EXPOSURE EFFECTS ON HUMANS

Cancer Incidence and Residential Exposure

In 1979, the results of an epidemiological study were published indicating a positive association between residential 60 Hz magnetic field exposure and childhood cancer (4). In this case-control study, general levels of 60 Hz magnetic field exposure of cancer victims and control subjects were estimated by observing the number and size of electric power supply lines and their distance from homes. This method of estimating magnetic field exposure levels is referred to as wire code configuration estimation. Children who lived closest to high electric current wiring configurations, and hence the highest potential environmental levels of 60 Hz magnetic fields, experienced a 2- to 3-fold increased risk of cancer compared to children who lived in lower magnetic field environments. The unexpected and ominous implications of this study generated significant controversy and criticism which was primarily related to the use of wire code configurations to categorize magnetic field exposures, an approach potentially subject to inaccuracies and bias.

Subsequent studies of 60 Hz magnetic field exposure and childhood cancer involved measurements of magnetic fields in the homes as well as exposure assessments based upon wire code configurations (5,6). The results of these studies presented an apparent inconsistency with respect to the previous study of Wertheimer and Leeper (4). In these studies, the association of measured residential magnetic field intensities and childhood cancer risk was not as strong as the risk indicated by wire codes. In a study (5), magnetic field intensities were not associated with childhood leukemia when measurements were made under conditions of high electric current use in the home. Under conditions of low average home power usage, there was a moderate association of magnetic field exposure with childhood leukemia risk, whereas there was a more obvious association when exposure was estimated by the use of wire code configurations, as in Ref. 4. In the study reported by London and co-workers (6), measured magnetic field intensities were not related to childhood leukemia risk, but again when risk was estimated using wire code configurations, the risk factor was approximately two, similar to that found in previous studies (4,5). In the studies in Refs. 5 and 6, there was a positive trend between the intensity of 60 Hz magnetic field exposure and leukemia risk.

Wire Code Configuration Estimation Versus Measurements

Questions regarding the apparent inconsistency between childhood leukemia risk and measured magnetic field intensities versus estimates based on wire code configuration remain unanswered. There are, however, a number of possible explanations. It is possible, for example, that an unidentified risk factor may be responsible for the positive association between leukemia and residential wire code configurations. Considering the designs of these studies and specific attempts to detect confounders and/or unknown risk factors, it does not appear likely that the outcomes resulted from an unknown risk factor. This conclusion is supported by the results of a recent Swedish epidemiological study of the association of magnetic field exposure from high voltage power lines and cancer, as discussed below (7). It has been suggested that social class may represent a confounding element in the association of residential magnetic field exposure and cancer. However, based on available information, it does not appear likely that social class *per se* is a significant confounder (8).

An alternative explanation for the inconsistencies in the risk estimates may be that the measured magnetic fields in the studies in Refs. 5 and 6 may not have characterized the long-term residential magnetic field exposure history as adequately as estimates based on wire code configuration. This possibility is supported by the results reported in Ref. 7.

Magnetic Field Interaction Mechanisms

As previously noted, there is limited theoretical understanding of basic interaction mechanisms responsible for effects of low intensity ELF magnetic fields on living systems. For example, it is uncertain whether effects are due to electric fields and/or currents induced in cells or tissue by time varying magnetic fields, or if the effects are due, in whole, or in part, to direct magnetic field interactions. Until quite recently, it was believed that magnetic fields did not interact directly with tissue, an assumption that was recently challenged by detection of trace levels of biogenic magnitite in human tissue and other living systems (9). As a consequence of uncertainties surrounding interaction mechanisms, the magnetic field exposure parameter(s) of greatest relevance to leukemia risk, or other effects, are presently undefined. It is possible, for instance, that instead of short term magnetic field intensities measured at one or a few locations (referred to as "spot measurements"), the most significant parameter could be one of the following: (a) instantaneous or peak magnetic or electric field intensity, (b) frequency or rate of occurrence of field transients, (c) rate of change of electric or magnetic fields, or (d) long-term average exposure intensity. Lacking knowledge of relevant magnetic field exposure parameters, it is not surprising that there are apparent inconsistencies in the results of epidemiological studies related to differences in exposure metrics.

Assuming that exposure to 60 Hz residential magnetic fields is, in fact, associated with increased cancer incidence in human populations, one may ask what effect inadequacies in exposure estimation, such as the use of wire code configurations, would have on risk estimation. In general, if exposure estimators such as wire code configurations are equally inadequate estimators for cancer cases as well as controls (non-cancer cases), the true risk would be underestimated rather than overestimated.

Estimates of cancer risk from residential exposure to 60 Hz magnetic fields differed significantly depending upon the selection of cutpoints distinguishing exposed from unexposed subjects (10). Using a higher magnetic field exposure cutpoint than used in an earlier study (5), Wartenburg and Savitz using the same data, obtained larger odd ratios, two of which achieved statistical significance (10). In addition they found that by using probability plots, the data showed greater consistency with measures of magnetic fields in both low- and high-power use situations in contrast to results previously reported (5). They also reported a lack of concordance with results based upon measures of electric fields, suggesting that increased leukemia risk was more directly associated with magnetic than electric field exposure. This study indicated statistically significant acute lymphocytic leukemia and lymphoma odds ratios for children exposed to residential 60 Hz magnetic field intensities of 0.3 micro Tesla (μT) [3 milliGauss (mG)] or greater (10). Magnetic field intensity units are Tesla (T) and Gauss (G) ($1T = 10^4G$).

A recent epidemiological study of exposure to 50 Hz magnetic fields and childhood leukemia involved both measured residential magnetic field intensities as well as long-term average intensities calculated on the basis of distance of homes from high voltage transmission lines and records of the electrical current carried by the lines during previous years of exposure (7). It is unfortunate that such records of line current are not available in the United States since they would permit studies of this type to be conducted in the U.S. Such studies could lead to a better understanding of the significance of long-term average magnetic field exposure as related to cancer incidence.

Children exposed in their homes to time-averaged magnetic fields of 0.3 μT (3mG) or more had a relative risk almost four times greater than children exposed to magnetic fields of less than 0.1 μT (1mG) (7). A significant aspect of this study was that cancer risk was directly proportional to average magnetic field exposure intensity. The detection of a statistically significant dose–response relationship is a criteria applied by epidemiologists to assess the internal consistency of study results. Children exposed to average magnetic field intensities of greater than 0.1 μT (1mG) had a doubling of leukemia incidence; exposure to more than 0.2 μT (2mG) was associated with a threefold increase and, as indicated above, exposure to more than 0.3 μT (3mG) resulted in an approximate four-fold increase in leukemia risk. The childhood leukemia risk ratios, which were controlled for potential confounders such as socioeconomic status and air pollution, were statistically significant (7). There was a 70% increased incidence of both acute myeloid leukemia (AML) and chronic myeloid leukemia (CML) in adults exposed in their residences to magnetic field intensities of greater than 0.2 μT (2mG). Although these results suggested an association of adult cancer with 50 Hz magnetic fields, as had the study in Ref. 10, the risk ratios were not statistically significant. It was also noted that statistically significant elevation in childhood leukemias occurred in children who lived in single- as opposed to multifamily dwellings (7).

This study involved the most detailed assessment of residential magnetic field exposure to date. The assessment involved reconstruction of magnetic field exposure levels in homes for periods of up to 10 years prior to diagnosis, using records of transmission line electric current flow. This strategy enabled the investigations to take into account short-term as well as seasonal changes in magnetic field intensities. Comparison of the long-term magnetic field exposure assessments with results of short term "spot" measurements made in the homes indicated a lack of correlation. "Spot" measurements tended to overestimate past exposures, which suggested that studies using measured values could be biased. Since the use of wire code configurations may provide a better estimate of long-term residential magnetic field exposure than spot measurements, the result of the study, in Ref. 7 provide a possible explanation for the consistent statistically significant association of childhood cancer with magnetic field exposure when wire code configurations were used to estimate exposure levels (4–6).

Sources of Magnetic Field Exposure

The apparent association of residential ELF magnetic fields and cancer focussed attention on the types and magnitudes of exposure, as well as methods of exposure assessment. Principal sources of ELF residential magnetic fields included: high voltage transmission lines, distribution lines, electrical service wiring, and household electric wiring. In this context, it should be noted that since the vast majority of residences in the United States and other industrialized countries are electrified, magnetic field exposure is essentially ubiquitous. Considering the results of epidemiological studies associating magnetic field exposure and cancer, it must be kept in mind that there are no truly unexposed persons to serve as controls for those exposed to higher intensity magnetic fields. It has been estimated, for example, that power lines and the grounding circuits in houses, result in average residential fields of approximately 0.1 μT (1mG) (12). The findings of Wartenberg and Savitz (10) and Feychting and Ahlbom (7) that exposures of 0.3 μT (3mG) or greater may significantly increase childhood cancer risk imply an unexpectedly high sensitivity for this magnetic field exposure effect.

The actual sensitivity of humans to ELF magnetic field health effects is difficult to assess since, as noted above, the exposure parameters of primary concern have not been adequately described and there are multiple sources of exposure having different spatial and temporal characteristics. Controversy exists, for instance, regarding the contribution of electric and magnetic fields from appliances compared to other sources of exposure (13). In contrast to other residential sources of magnetic fields that vary relatively little over time or space, appliance fields are intermittent and highly dependent upon distance from the appliance. The magnetic field intensity of appliances such as dishwashers, toasters, and irons varies from about 10 μT (100mG) at 3 cm from the appliance to approximately 0.01 μT (0.1mG) at 1 meter (11). Significantly higher magnetic field intensities exist in close proximity to appliances such as electric hair dryers that can produce fields of 2,000 μT (20G) at 3 cm (13). The volume- and time-averaged magnetic fields emitted by home appliances have been estimated (12). Calculations based on measurements for 98 appliances indicated that appliance generated magnetic fields were not a significant source of whole-body exposure but that body extremities could be exposed intermittently to peak fields of up to 100 μT (1G) or higher (12). Table 1 lists the spatial average magnetic field intensities in close proximity (3 to 30 cm) to appliances together with time averaged intensity. The time averaged intensity of magnetic field exposure of extremities is calculated by multiplying the spatial average intensity by the fraction of time during the day the appliance is in use.

Cancer Incidence and Occupational Exposure

Epidemiological studies have also suggested an association of occupational exposure to electric or magnetic fields and cancer incidence. A meta analysis combining the results of occupational studies conducted prior to 1987 led Savitz and Calle to conclude that there was a consistent pattern of increased leukemia risk in workers exposed to electric

Table 1. Spatial and Time Average Exposure of Extremities to 60 Hz Magnetic Fields from Appliances[a]

Source	Magnetic Field Spatial Average	(3–30 cm) (μT)	Magnetic Field Time Average (μT)	
Can opener	60.6	(606)[b]	0.47	(4.7)
Electric saw	32.5	(325)	0.11	(1.1)
Electric shaver	16.9	(169)	0.06	(0.6)
Mixer	16.2	(162)	0.18	(0.2)
Electric drill	12.7	(127)	0.06	(0.1)
Hair dryer	12.0	(120)	0.07	(0.1)
Blender	5.6	(56)	0.02	(0.2)
Iron	0.7	(7)	0.01	(0.1)

[a] Adapted from data of Mader and Peratta (12).
[b] Magnetic field intensity in milliGauss units.

or magnetic fields (15). Increased incidence of leukemia, acute leukemia (especially acute myeloid leukemia) were reported in post-1987 occupational studies as well (16–18). Workers at risk included powerline workers, electric utility workers, electricians and electronics workers (16).

A dose–response relationship for the association of cancer incidence and occupational exposure to ELF was reported (18). Based on measurements, workers were assigned to magnetic field exposure categories: Group 1, greater than 0.41 μT (4.1mG); Group 2, between 0.41 and 0.29 μT (4.1-2.9mG); Group 3, between 0.29 and 0.16 μT (2.9-1.6mG); and Group 4, less than 0.16 μT (1.6mG). There was a three-fold increase in chronic lymphocytic leukemia (CLL) and a 60% overall increase in leukemia in workers in Group 1. No association was found for acute myeloid leukemia. Among the most highly exposed Group 1 workers, there was a four-fold increased CLL risk. The study was based upon the occupational experiences of 850 cancer cases (approximately one-half leukemia cases and one-half brain tumors) and 1700 age-matched controls (non-cancer cases). Since the study group was not limited to any particular industries, the cases and controls were representative samples from the Swedish male working population. In addition to age, other potential confounders, including exposure to benzene, other organic solvents, and ionizing radiation, pesticides, and smoking, were taken into consideration (18).

The results of this study support the hypothesis that there is an association between ELF EMF occupational exposure and the development of CLL. Since the elevated risk was related to the job held longest during the 10 year period before diagnosis, the outcome was consistent with the possibility the EMF exposure acted as a cancer promoter.

The study also provided some evidence of elevated risk of brain tumors, but the dose–response relationship was less evident. Brain tumor relative risk was increased in the highest EMF exposure categories, based on median values for the job held longest during the decade before diagnosis. The EMF-brain tumor association was strongest for workers less than 40 years of age (18). Previous occupational studies suggested a positive association of brain cancer and EMF exposure (19,20).

Interaction Mechanisms

Insight regarding possible mechanisms involved in the elevation of cancer risk from low intensity EMF has been hindered by limited knowledge of cancer etiology in general. This limitation is particularly true in the case of specific types of cancer such as brain cancer and childhood leukemia. Some insight may, however, be provided by studies indicating that a rare form of cancer, male breast cancer, may be linked to occupational exposure to EMFs (21–23). If verified by future studies, the association of male breast cancer and EMFs suggests a possible mechanism related to the fact that breast cancer involves hormonal alterations.

Hormonal alterations have been reported to occur in laboratory animals exposed to ELF fields. Specifically, exposure affected biorhythms resulting in the suppression of the normal nocturnal increase in melatonin (24,25). This finding is potentially significant due to the interaction of melatonin with other hormones. Reduction in plasma melatonin concentrations causes increased levels of circulating steroid hormones such as estrogen and testosterone as well as increased prolactin release by the pituitary gland (26). Such hormonal alterations increase the rate of proliferation of breast tissue and suppress the immunological system, effects consistent with increased breast cancer risk (27). Animal experiments have indicated an association of melatonin levels with mammary tumorigenesis. Exposure of rats to 50 Hz magnetic fields of greater than 1 μT (10mG) resulted in statistically significant decrease in pineal and plasma melatonin levels (28). When the source of melatonin, the pineal gland, was removed, the incidence of breast tumors was increased significantly (29). The effect of ELF EMFs on melatonin levels in experimental animals thus suggests neuroendocrine involvement in the observed increase in breast cancer in occupationally exposed males. Obviously, this mechanism could be involved in exposure of females as well as males and could relate to other tumor types.

Epidemiological studies have also indicated an association of ELF EMF exposure and other physiological alterations including: altered biorhythms (30), behavioral and neurological disorders (31) and reproductive outcomes (32). The sensitivity of living systems to extremely weak electric fields is indicated by the alteration of normal biorhythms in humans exposed to fields as low 2.5 V/m, a field strength commonly encountered in the home or workplace (33). Exposure of human volunteers to combined electric and magnetic fields of higher intensity (9 kV/m, 20 μT (200mG) for 3 hr. periods resulted in statistically significant slowing of heart rate, changes in late components of event-related brain potentials, and decreased error rates on a choice reaction-time task (34). Functional changes in the central nervous system (CNS), including alterations in brain wave potentials and learning were also reported following 45 min. exposures of human volunteers to 1.25 mT (12.5G) 45 Hz magnetic fields (35). A 45 min. exposure of rats to 60 Hz magnetic fields at intensities of 0.75 mT (7.5G) altered brain neurochemistry (36). The results of this study indicated that magnetic fields altered endogenous brain opioids (36).

Exposure to residential magnetic fields from electric blankets, heated water beds and ceiling cable heat that produce fields in the range of 0.4 to 1.5 μT (4–15mG) have been associated with seasonal variations in fetal growth and fetal loss (36,37). The results of a recent epidemiological study indicated an association of residential 50 Hz magnetic fields and early pregnancy loss (EPL). Magnetic field measurements in residences indicated that exposure to fields of 0.63 μT (6.3mG) or more resulted in an EPL odds ratio of 5.1 (38). Animal studies have also indicated that pulsed low-intensity magnetic fields may, in some instances, increase the incidence of malformations (39).

ELF Magnetic and Electric Field Effects on Experimental Animals

In a general sense the results of studies of the effects of ELF magnetic and electric fields on laboratory animals are consistent with effects reported in humans. However, the animal studies database is limited due to the rather recent awareness of the potentially adverse effects of low intensity ELF magnetic fields and outcomes such as cancer. Prior to this time, attention was focussed on health effects of exposure to high intensity electric fields such as those encountered near high-voltage transmission lines. Few studies have investigated the association of long-term exposure of experimental animals to low-intensity magnetic fields and carcinogenesis. The results of one such study reported that 60 Hz magnetic fields had a tumor-promoting effect on mouse skin cancer (40). Considering the paucity of relevant experimental data and the complex nature of carcinogenesis, no conclusions are possible regarding the mechanisms of effects of 60 Hz magnetic fields on this endpoint. It may be concluded tentatively, however, that it is more likely that low-intensity ELF electric or magnetic fields act as tumor promoters rather than initiators. Biophysical considerations, as well as experimental cell studies indicating that such fields do not appear to induce direct chromosomal alterations, support this conclusion.

Although limited in extent and applicability to questions such as association with cancer incidence, effects of ELF electric and/or magnetic fields on cats, dogs, swine and nonhuman primates have been reported. Endpoints investigated include: behavioral changes, reproductive outcomes, brain neurochemistry, hormone levels, cardiovascular responses, and hematopoietic changes. The effects of ELF EMFs on laboratory animals have been the subject of detailed reviews (41,42). While these data do not reveal well-defined dose-responses, they do indicate general sensitivity of mammalian systems to ELF EMF exposure that are not explainable by known interaction mechanisms.

Dosimetric Considerations

Difficulties are encountered in establishing cause-effect relationships between exposure to ELF EMFs and health effects due to dosimetric complications. For example, the electric fields induced in laboratory animals or human beings exposed to ELF magnetic fields are complex and nonuniform. Nonuniformities are due in part to the fact that there are large differences in the electrical conductivity (or resistance) of tissues such as a muscle (high conductivity) versus fat (low conductivity). These differences cause significant differences in induced current densities in tissue even when the body is exposed to a spatially uniform magnetic field. Additional complications are introduced by the dependence of the magnitude of induced currents on the body dimensions. In an ELF magnetic field of constant intensity, tissue induced electric fields will be significantly greater in a human body as compared to a rodent, for example. Such factors introduce significant uncertainty into the extrapolation of data derived from animal studies to the prediction of human health effects.

ELF EMF Effects on Mammalian Cells

To minimize ambiguity introduced by uncertainty about the distribution and/or intensity of tissue induced, electric or magnetic fields studies have been conducted using isolated mammalian cells. Such *in vitro* studies afford an opportunity to directly relate cellular alterations to well characterized electric or magnetic fields, thus providing data of use in determining basic interaction mechanisms. Whereas it is obvious that *in vitro* results cannot be directly extrapolated to *in vivo* systems, knowledge of basic interaction mechanisms obtained from such studies will be an essential element in understanding health effects of nonionizing electromagnetic fields.

In vitro studies of the effects of ELF EMFs on normal as well as transformed (cancer) mammalian cells have identified a number of sensitive physiological endpoints including: a) cell proliferation, b) membrane signal transduction, c) biomolecular synthesis, d) ion fluxes and binding, e) immune responses, and f) energy metabolism. Generally, alterations were induced by short-term (ie, minutes or hours) exposure to low intensity electric and/or magnetic fields at frequencies of 100 Hz or less. The results, which provide extensive evidence that ELF EMFs are biologically active, have been reviewed in detail (43).

The results of *in vitro* studies provide much needed insight about EMF effects on living systems such as indicating that the cell membrane is a likely field interaction site. To date, the results of such studies have not provided the detailed data necessary for the development of interaction mechanisms. *In vitro* data have, however, led to the generation of a number of interesting hypothetical mechanisms that may advance understanding of the biological effects of ELF EMFs (44).

Electric Power Consumption and Cancer Mortality

It has been suggested that trends in electric power consumption should be considered in assessing the relationship of ELF EMFs (such as 60 Hz magnetic fields from electric power distribution) to adverse health effects. If there is a direct correlation between electric power consumption and time averaged level of human exposure to 60 Hz magnetic fields, and if exposure is directly related to cancer incidence, trends in power consumption should be reflected in trends in cancer incidence. Since electric power consumption has increased steadily during the past 40 years or so, this should be reflected in an increased cancer incidence during this period. The validity of this assumption is based upon the following premises: a) increased overall electric power consumption is directly correlated with *per capita* time-averaged 60 Hz magnetic field exposure levels, and b) cancer incidence has increased during the period of concern.

The relationship between increased power consumption and *per capita* exposure to 60 Hz magnetic fields is uncertain due to a number of factors. During the period of increased power consumption, demographic changes occurred such as large population migration from urban to suburban areas. In addition, there have been changes in the electrical wiring methods, materials, and building codes that may well have affected average exposure levels. There is evidence, in fact, that such factors may have acted to limit or reduce *per capita* exposure to 60 Hz EMFs (45,46).

Whereas overall cancer mortality has not increased during the past 20 years for persons under 45 years of age, there have been significant increases in the incidence of specific types of cancer (47,48). The fact that there has not been a substantial reduction in the overall cancer death rate in industrial societies may be due to the fact that new therapies have not kept pace with increased cancer incidence. For example, during the period 1950 to 1986, there was a 20% decrease in childhood leukemia mortality but leukemia incidence increased by 60% over this period (49). Changes in age-specific tumor incidence were also detected. In the United States during the period 1969–1986, there was a 20% decrease in brain tumor mortality in white males and females ages 0–44, but brain tumor mortality in the age group 65–84 increased by more than 80% and there was a 200% increase in persons aged 75 to 84 (50). Female breast cancer incidence and mortality have increased at a constant rate for the past 50 years or more for unknown reasons (51). Although overall mortality from multiple myeloma has remained essentially constant in industrialized countries, there has been persistent increased mortality in people over the age of 70 (52). This increase could be attributed to an ubiquitous environmental risk factor common to industrialized nations that increased during the latter quarter of the past century and the first quarter of the 20th century (52).

Trends in electric power consumption and cancer incidence thus are not inconsistent with the possibility of an age-dependent cause-effect relationship, especially for specific cancer types. It is obvious that there are multiple cancer risk factors in industrial societies and that cancers are most likely multi-factor diseases. Thus, the extent to which ELF magnetic field exposure contributes to cancer risk is difficult to determine from currently available information. The results of animal experiments as well as cell studies *in vitro* provide ample evidence of the biological activity of low intensity ELF electric and magnetic fields thus adding plausibility to concern about exposure-related health effects. There is an obvious need to conduct further investigations of the biological effects of ELF EMFs to determine interaction mechanisms and to better define the nature and extent of adverse health effects.

RADIOFREQUENCY AND MICROWAVE HEALTH EFFECTS

Radiofrequency (RF) and microwave radiation have physical characteristics, and hence interaction mechanisms, quite distinct from ELF electromagnetic fields such as 60 Hz fields associated with electric power transmission. The frequency of RF radiation extends from approximately 10 KHz (10^4 Hz) to 300 MHz (3×10^8 Hz). EMR in the frequency range of 300 MHz to 300 GHz (3×10^{11} Hz) is referred to as microwave radiation. Typical sources of RF radiation include AM, television, and FM broadcast signals, whereas the most common environmental sources of microwave radiation are microwave ovens in the home that operate at a frequency of 2450 MHz. The contrast between ELF and RF/microwave EMR is most dramatically illustrated by considering differences in wavelength. The wavelength of a 60 Hz electric field is measured in thousands of miles compared to RF/microwave radiation having wavelengths of yards to fractions of an inch. Based on the knowledge that the way EMR energy is coupled to a living system, such as the human body, relates to the wavelength and the size of the absorber, it would be logical to expect different biological effects of ELF versus RF/microwave radiation. Surprisingly, in spite of differences in physical characteristics, there are some similarities in biological responses resulting from exposure to ELF EMFs and EMR.

EMR and Cancer

A possible association of RF and MW EMR exposure and human cancer incidence emerged recently. The use of hand-held cellular telephones operating in the frequency range of 825–890 MHz has been linked with brain cancer. Exposure of police officers to higher frequency microwave radiation in the frequency range of 10 to 24 GHz, emanating from traffic speed detecting radar units, has been associated with increases in testicular, brain and eye cancer. Exposure to hand-held cellular telephone radiation and police radar both involve localized radiation absorption. Tumors reportedly occurred at or near the apparent site of maximum microwave energy absorption in either case. For example, parietal lobe brain tumors were reported to occur in the brain region closest to the location of the hand-held cellular telephone antenna. Localized microwave exposure of testicular tissue of police officers presumably resulted from resting the radar gun in the groin region while the unit was emitting radiation. Neither in the case of hand-held cellular telephones or police radar have epidemiological studies been conducted to establish the validity of the association with cancer incidence. To date, few studies have been conducted of the effects of these microwave frequencies on experimental animals or mammalian cells *in vitro*. Therefore, in the absence of a relevant scientific database, the possible association of microwave sources with cancer incidence must be viewed in terms of effects reported from exposure to other RF or microwave frequencies. The results of epidemiological studies, studies that involved exposure of laboratory animals, and investigations of the effects of RF or microwave radiation on mammalian cells, provide evidence for a possible association of such exposure with cancer incidence. The limited amount of relevant data precludes drawing firm conclusions. However, the results are most consistent with the hypothesis that under certain, presently not well-defined, exposure conditions, RF or microwave radiation may act as a cancer promoter, rather than as a cancer initiator. This hypothesis is consistent with the hypothesis discussed above that exposure to low intensity 60 Hz magnetic fields may also affect tumor promotion.

Epidemiological Studies

Until recently there has been limited concern about the relationship of RF or microwave exposure and cancer incidence. Consequently, few epidemiological studies have been conducted to investigate this possibility. This status is in distinct contrast to the number of epidemiological studies of the relationship of residential or occupational exposure to 60 Hz magnetic fields and cancer discussed previously in this chapter.

Milham (53) analyzed death certificates of 1,691 amateur radio operators from California and Washington states and found 24 leukemia cases. Compared to an expected number of such cases of 12.6, this was a statistically significant excess number of leukemias amongst amateur radio operators. In a subsequent study, Milham (54) used standardized mortality ratios (SMR) to characterize the cancer experience of amateur radio operators. The SMR for all cases of death of amateur radio operators was 71 and for combined cancers it was 89, indicating a better than average mortality history for the study group. This increase was expected based upon a comparison of this group with professional and academic cohorts. There was, however, significantly elevated leukemia mortality for amateur radio operators who had a SMR of 162 for cancers of lymphatic tissues including multiple myelomas and non-Hodgkins lymphomas. The SMR of 176 for chronic myeloid leukemia was statistically significant, whereas the elevated SMR of 124 for all leukemias was not significant. In addition to potential RF exposure from the operation of amateur radios, there was evidence of possible occupational RF exposure of some members of the study group from the state of Washington. It was noted that the elevated leukemia risk in the study group could have been due to other risk factors but that exposure to RF radiation should be considered as an etiological factor. In this study, it was not possible to determine the extent of exposure to RF radiation.

The association between EMR exposure and cancer including data from epidemiological as well as experimental studies has been investigated (55). An unexpectedly large number of cancer cases, including lymphomas, chronic myelocytic leukemia, acute myelocytic leukemia and pancreatic cancer was observed among military personnel exposed to microwave radiation. A retrospective epidemiological study of cancer morbidity in Polish military personnel aged 20–59 during the period 1971–1980 was reported (56). Cancer morbidity in personnel occupationally exposed to RF/microwave EMR was three times greater than the control group. Statistically significant increased cancer incidence was found for lymphatic and hematopoietic malignancies and for stomach, colorectal and skin neoplasms, including melanoma. The overall risk factor for lymphatic and hematopoietic malignancies in RF/microwave exposed workers was 6.7, which was statistically significant ($p < 0.01$) (56).

Evidence of an association of occupational EMR exposure and brain cancer was reported (57) in an epidemiological study of 951 cases of brain tumors among white male residents of Maryland during the period 1969–1982. Fifty cases of glioma and astrocytoma were observed among electricians, electrical engineers and high voltage transmission linemen. Compared to the expected number of 18 tumors, there was a statistically significant increase in brain tumors in EMR workers.

Epidemiological studies, although quite limited in number, provide evidence of an association of long term exposure to RF/microwave and lower frequency EMFs and cancer incidence. Although a number of different cancers have been reported to result from such exposure, leukemia and brain cancer appear to be the most prevalent.

Animal Studies

Effects of RF and microwave radiation on various psychophysiological responses in experimental animals such as rodents, dogs and non-human primates were reported. Effects on hematological and immunological systems, reproduction, nervous system, behavior, sensory systems and endocrine systems were the subject of extensive review articles (2,58). Numerous physiological alterations were reported to be induced by RF or microwave exposure, depending upon species and exposure conditions. Most studies involved acute or relatively short-term exposure to RF field intensities that induced varying degrees of temperature elevations in either the entire body or localized regions of the body. Whereas there were indications that certain physiological alterations, such as changes in behavior, might occur in the absence of tissue heating, it was generally assumed that such effects were of minor significance relative to assessments of potential effects of RF or microwave exposure on human health. Results of more recent studies involving long-term or chronic low-intensity exposure, studies of the relationship of microwave exposure to cancer promotion, and *in vitro* cell studies, have indicated the need to reconsider health effects issues from long-term low-intensity EMR exposure.

The potential tumor promoting effect of microwave exposure was investigated (56,60,61) and co-workers. Mice were exposed 2 h per day for 3 to 6 months to 2450 MHz microwave radiation at power densities of from 5 or 15 milliWatts per square centimeter (mW/cm^2) (specific absorption rate (SAR) 3–4 or 6–8 Watts per kilogram (W/kg), respectively). Microwave exposure accelerated the appearance and growth of skin neoplasms induced by ben carcinogens di-ethyl-nitrosoamine and methylchloran-trene. Szmigielski and co-workers (56) indicated that long-term exposure to low intensity (5 mW/cm^2) microwave radiation had an effect on tumor growth in mice similar to the effect of chronic stress.

Additional evidence of a possible association of chronic low-intensity microwave exposure and cancer in experimental animals was reported (62). Male rats were exposed, or sham-exposed, 21.5 h/day for 25 months to pulse-modulated 2450 MHz EMR at SARs that ranged from 0.15 to 0.4 W/kg. Physiological assays were routinely conducted throughout the experiment. Statistically significant alterations were reported in the following endpoints during the course of the experiment: behavior, serum corticosterone, lymphocyte (B and T cell) number, lymphocyte mitogenic stimulation, eosinophil and neutrophil counts, adrenal mass, O_2 consumption and CO_2 production, and benign adrenal neoplasia. Potentially of most significance with respect to the relationship of microwave exposure to cancer incidence, was the detection of 18 primary malignant neoplastic lesions in microwave exposed rats compared to 5 such lesions in the sham-exposed animals. This difference in malignant neoplasia was highly statistically significant ($p = 0.006$). There was also a statistically significant decrease in time of occurrence of primary tumors in microwave-exposed animals (62).

Results support the hypothesis that long-term or chronic low-intensity microwave exposure is associated with increased incidence and/or growth rate of cancer in laboratory animals (55,62). The significance of these observations with respect to human cancer incidence has yet to be determined. However, it should be noted that the microwave intensities used in these studies were in the range considered by certain regulatory and advisory bodies as being safe for long-term human exposure. It should also be noted that whole body RF or microwave absorption rates of 0.4 W/kg or less, as used in the study reported in Ref. 62, are assumed to be well below the level that causes tissue heating.

Animal studies have also provided evidence that low-intensity EMR alters brain neurochemistry under conditions not involving tissue heating. Effects of microwave exposure on: (*1*) actions of various psychoactive drugs, (*2*) the activity of the cholinergic systems in the brain, and (*3*) on neural mechanisms in the rat, were investigated by Lai (63). Neurological alterations were induced in specific parts of the rat brain by 45 minute exposures to 2450 MHz EMR at SARs of 0.6 W/kg. It was concluded that alteration of the levels of endogenous opiates were responsible for the observed EMR effects and that the effects depended upon the exposure parameters (63). The results of these studies are of interest since they provide evidence that low-intensity EMR can alter brain function which is consistent with numerous reports of behavioral, neurological and neuroendocrine alterations in humans due to EMR exposure.

In Vitro Studies

An impediment to assessing EMR effects on organisms is the fact that energy absorption in tissue is highly nonuniform and dependent upon the wavelength of the radiation and the size and orientation of the body and organs. Consequently, it is difficult to directly apply the results of studies using experimental animals to predict effects of RF or microwave exposure on human beings. In addition to dosimetric uncertainty, interspecies physiological differences and the interactive nature of various organ systems introduce additional complexities. In order to overcome such difficulties, studies have been conducted of the effects of EMR on mammalian cells *in vitro*. These studies can be conducted with precise knowledge and control of exposure conditions, thus providing data for the determination of cause-effect relationships. As in the case of ELF EMRs, effects of EMR on a variety of cellular endpoints have been investigated. Such effects include: a) membrane cation transport and binding, b) membrane structure, c) single ion channel kinetics, d) neuronal activity, d) energy metabolism, e) proliferation and activation and f) transformation. The results of these studies have been the subject of review articles (58,59,64). Due to their possible relevance to cancer incidence, effects of RF and microwave radiation on mammalian cell proliferation and transformation will be reviewed here.

To test the hypothesis that microwave radiation may interact synergistically with tumor promoters, mouse embryonic fibroblasts were exposed to low intensity 2450 MHz microwave radiation either in the presence or absence of the tumor promoter 12-0-tetradecanoylphorbol-13-acetate (TPA) (65). Cells were exposed to microwave radiation pulse modulated at 120 Hz at SARs of 0.1, 1, or 4.4 W/kg for 24 h. In the absence of TPA, EMR exposure did not affect fibroblast cell survival or the induction of neoplastic transformation. Neoplastic transformation was significantly enhanced, however, in cells treated with TPA. A synergistic TPA–EMR dose-response was detected for the frequency of neoplastic transformation in the SAR range of 0.1 to 4.4 W/kg. The neoplastic transformation frequency resulting from exposure at 4.4 W/kg, in the presence of TPA, was comparable to that induced by exposure to 1.5 Gy (1 Grey (Gy) = 100 rads) of X-radiation. It was noted that this dose of whole body X-radiation results in an approximate 6% risk of tumor induction. Comparison of the effects of X-radiation exposure, with or without 4.4 W/kg 2450 MHz microwave exposure, in the presence of TPA, indicated that the combined treatment induced two independent types of transformation damage, one due to microwave exposure and another to X rays, both promoted by TPA. This damage indicated separate sites of interaction for EMR versus X-radiation. Since genomic interactions of X-radiation are associated with neoplastic transformation, these data indicate the possibility of a non-genomic effect of microwave radiation leading to TPA promoted neoplastic transformation. Biophysical considerations, as well as results of studies of microwave radiation effects on other cell physiological endpoints, suggest the cell plasma membrane as a probable site of interaction leading to neoplastic transformation.

In addition to evidence that low-intensity EMR may affect neoplastic transformation of mammalian cells, *in vitro* studies have also revealed that the EMR may alter the rate of proliferation of cells that are already transformed or malignant. Cleary and co-workers (66,67,68) exposed human or rat glioma cells (LN71, RT2, respectively) for 2 h to 27 or 2450 MHz CW or pulse modulated RF radiation under isothermal ($37 \pm 0.2°C$) conditions. Cell proliferation was increased in a dose-rate dependent manner following exposures to either type of radiation at SARs of 0.5 to 25 W/kg. Maximum increased proliferation of 160% occurred following exposure to 25 W/kg 27- or 2450 MHz, 5 or 3 days after exposure, respectively. The effect on glioma proliferation was biphasic in that exposure at SARs of greater than 25 W/kg decreased the rate of cell proliferation. Statistically significant time-dependent effects were detected for up to 5 days postexposure which suggested a kinetic cellular response to EMR at these frequencies. The persistence of the RF or microwave radiation effect suggested the possibility of a cumulative effect on cell proliferation.

Evidence that EMR RF or microwave radiation may have a more general effect on cell proliferation was reported (66,68,69). Using the same experimental procedures used to study effects on glioma, human peripheral lymphocytes were exposed isothermally (37 ± 0.2°C) for 2 h to 27- or 2450 MHz radiation. Exposure to EMR at these frequencies had similar biphasic effects upon the rate of proliferation of mitogen (phytohemagglutinin (PHA)) stimulated lymphocytes. The maximum increase in proliferation, which occurred 3 days after exposure, was 40%, approximately one-fourth the magnitude of the proliferative effect of RF radiation on glioma (68). Cumulative effects were investigated by exposing lymphocyte cultures at the same SAR to two 1 h irradiations spaced 24 h apart. Such split-dose exposures caused similar effects but of reduced magnitude, indicating time dependent reversal of the exposure effect on lymphocyte proliferation with a time constant of somewhat greater than 24 h (69). Pulse modulation of 27- or 2450 MHz radiation caused generally similar effects on cell proliferation, at SARs of 5 W/kg or less, but there were differences in the dose responses. At higher SARs, pulse modulation resulted in significantly increased lymphocyte proliferation compared to CW exposures at the same SARs (69). The effects of 27- or 2450 MHz RF radiation on glioma or lymphocyte proliferation were attributed to a nonthermal, direct RF-induced alteration of the cell cycle (68).

In summary, *in vitro* studies of the effects of low-intensity RF and microwave radiation indicate dose-rate dependent increases in neoplastic transformation frequency and proliferation. In view of limitations on the extrapolation of *in vitro* results to *in vivo* responses, these results cannot be related directly to cancer incidence in human populations exposed to such radiation. However, these results are not inconsistent with the hypothesis that human exposure to RF- or microwave radiation, under presently not well-defined conditions, may affect cancer incidence.

CONCLUSIONS

There is increasing evidence of possible health effects of environmental exposures to EMFs and EMR in the home and in the work place. Epidemiological evidence indicates possible associations of long-term exposure and cancer incidence, adverse reproductive outcomes, and behavioral and neurological changes. Inherent limitations on exposure assessment, common to epidemiological studies, provide imprecise knowledge regarding time- or exposure intensity thresholds for these effects, thus making risk assessment difficult at this time. Whereas the results of animal experimentation and cellular studies of ELF EMFs and EMR effects are generally consistent with results of epidemiological studies, they provide insufficient data for meaningful risk assessment.

The greatest impediment to understanding the effects of EMFs and EMR on living systems is the limited knowledge of interaction mechanisms. One consequence is that research in this area has been treated with skepticism that has, together with other factors, resulted in serious limitations on research support. In view of the diverse nature of the physical properties of electromagnetic fields reviewed here, as well as the great variety of reported effects in living systems, the large gaps in our understanding are perhaps not surprising. The potential magnitude of exposure-related health effects in industrial societies indicates that these uncertainties must be resolved.

BIBLIOGRAPHY

1. *The Effects on Populations of Exposure to Low Levels of Ionizing Radiation,* Committee on Biological Effects of Ionizing Radiation (BEIR III), National Academy of Sciences, Washington, D.C., 1980.
2. S. F. Cleary, "Biological Effects of Nonionizing Electromagnetic Radiation," in J. G. Webster, ed. *Encyclopedia of Medical Devices and Instrumentation.* John Wiley & Sons, Inc., New York, 1988.
3. C. A. L. Bassett, S. N. Mitchell, and S. R. Gaston, *J. Am. Med. Assoc.* **247,** 623–628 (1982).
4. N. Wertheimer and E. Leeper, *Am. J. Epidemiol.* **109,** 273–284 (1979).
5. D. A. Savitz, H. Wachtel, F. A. Barnes, E. M. John, and J. G. Tvrdik, *Am. J. Epidemiol.* **128**(1), 21–38 (1988).
6. S. J. London, D. C. Thomas, J. D. Bowman, E. Sobel, L. Cheng, and J. M. Peters, *Am. J. Epidemiol.* **134,** 923–937 (1991).
7. M. Feychting, A. Ahlbom, "Magnetic fields and cancer in people residing near Swedish high voltage power lines," Report prepared for the *Swedish National Board for Industrial and Technical Development,* Karolinska Institute, Stockholm, Sweden, 1992.
8. M. R. Salzberg, S. J. Farish, and V. Delpizzo, *Bioelectromagnetics* **13,** 163–167 (1992).
9. J. L. Kirschvink, A. Kobayashi-Kirschvink, J. C. Diaz-Ricci, and S. J. Kirschvink, *Bioelectromagnetics* **Suppl 1,** 101–113 (1992).
10. D. Wartenberg and D. A. Savitz, *Bioelectromagnetics* **14,** 237–245 (1993).
11. N. Wertheimer and E. Leeper, *Int. J. Epidemiol.* **11,** 345–355 (1982).
12. D. L. Mader and S. B. Peratta, *Bioelectromagnetics* **13,** 287–301 (1992).
13. A. Leonard, R. Neutra, M. Yost, and G. Lee, "Electric and Magnetic Fields: Measurement and Possible Effects on Human Health," *Special Epidemiological Studies Program, California Dept. of Health Services,* Berkeley, Calif., 1990.
14. S. M. Harvey, *Evaluation of Residential Magnetic Field Sources,* 1988 Ontario Hydro Research Report No. E88-17-K.
15. D. A. Savitz, E. E. Calle, *J. Occup. Med.* **29,** 47–51 (1987).
16. G. Theriault, "Health Effects of Electromagnetic Radiation on Workers: Epidemiological Studies," in Peters Birenbaum eds., *Proceedings of the Scientific Workshop on the Health Effects of Electric and Magnetic Fields on Workers.* U.S. Dept. of Health and Human Services, Cincinnati, Ohio, 1991, Publ. No. 91-111.
17. G. Hutchinson, "The Epidemiology of EMF Exposures and Cancer Risks," presented at the *EPRI Conference on Future Epidemiologic Studies of Health Effects of EMF,* EPRI Workshop Agenda, Carmel, Calif., 1991.

18. B. Floderus and co-workers, "Occupational Exposure to Electromagnetic Fields in Relation to Leukemia and Brain Tumors. A Case-Control Study," *Report to the National Institute of Occupational Health,* Solna, Sweden, 1992.
19. M. A. Speers, J. G. Dobbins, and V. S. Miller, *Am. J. Ind. Med.* **13,** 629–38 (1988).
20. S. Preston-Martin, W. Mack, and B. E. Henderson, *Cancer Res.* **49,** 6137–6143 (1989).
21. G. M. Matanowski, P. N. Breysse, and E. A. Elliott, *Lancet* **1,** 737 (1991).
22. P. A. Demers and co-workers, *Am. J. Epidemiol.* **132,** 775–776 (1990).
23. T. Tynes and A. Andersen, *Lancet* **2,** 1596 (1990).
24. B. W. Wilson, L. E. Anderson, D. I. Hilton, and R. D. Phillips, *Bioelectromagnetics* **4,** 293 (1983).
25. B. W. Wilson, E. K. Chess, and L. E. Anderson, *Bioelectromagnetics* **7,** 239–242 (1986).
26. R. J. Reiter in B. W. Wilson, R. G. Stevens, L. E. Anderson, eds., *Extremely Low Frequency Electromagnetic Fields: The Question of Cancer,* Batelle Press, Columbus, Ohio, 87–107 (1990).
27. R. G. Stevens, *Am. J. Epidemiol.* **125,** 556–561 (1987).
28. L. Tamarkin, *Cancer Res.* **41,** 4432–4436 (1981).
29. M. Kato, K. Honma, T. Shigemitsu, and Y. Shiga, *Bioelectromagnetics* **14,** 97–106 (1993).
30. C. Martin, R. Moore-Ede, S. Scott, S. Campbell, R. J. Reiter, eds., *Electromagnetic Fields and Circadian Rhythmicity,* Birkhauser, Boston, Mass., 1992.
31. Z. Davanipour, *Neuroepidemiology* **10,** 308 (1991).
32. N. Wertheimer and E. Leeper, *Am. J. Epidemiol.* **129** (1), 220–224 (1989).
33. R. Wever, "ELF-Effects on Human Circadian Rhythms," in M. Persinger, ed. *ELF and VLF Electromagnetic Field Effects* Plenum Press, New York and London, 1974, pp. 101–144.
34. M. R. Cook, C. Graham, H. D. Cohen, and M. M. Gerkovich, *Bioelectromagnetics* **13,** 261–285 (1992).

ENERGY AUDITING

Naresh K. Khosla
Semra T. Love
Enviro-Management and Research, Inc.
Springfield, Virginia

An energy audit is the process that first identifies where, how, and why a building uses energy and then recommends energy conservation and savings opportunities that may exist. A general building audit examines building envelope, lighting, heating, air-conditioning, and ventilation requirements. A more detailed audit identifies specific system deficiencies and recommended solutions.

See also Energy efficiency; Energy management.

TYPES OF AUDITS

There are three types of energy audits: walk-through, intermediate, and comprehensive. These audits are different, but they are not defined by sharp boundaries. The type of audit performed determines the level of analysis accorded to the various systems. The following sections provide a more detailed description of each type of audit.

Walk-Through Energy Audit

A walk-through audit is the least expensive energy audit. It evaluates a building's energy costs and efficiency and identifies preliminary energy savings that can be obtained through low-cost capital investments. It typically begins with a review of a building's energy consumption over the previous year. A visual walk-through inspection of the building is then performed to detect obvious opportunities for energy savings in operations and maintenance. The data collected during this audit becomes the foundation for a more detailed analysis in which capital-intensive investments are considered.

A team that includes a facility manager, an energy management consultant, and sometimes a building owner's representative performs the walk-through audits. The team conducts two tours of the building to collect two specific types of data: one during occupied hours and another during unoccupied hours.

The first type of information collected identifies the various energy systems installed, the energy consumption of each system, and technical data from equipment nameplates. This information forms the basis for recommending areas where specific energy savings can be found. The second type of information evaluates the existing condition of systems and subsystems. This information determines opportunities for energy savings based on these conditions. A typical notation might read:

> *Exterior North Facade:* Window on the second floor does not close properly. Caulking around the window frame is cracked and some weatherstripping is missing.

Recommendations resulting from a walk-through audit include: reducing infiltration/exfiltration; improving the efficiency of heating, ventilating, and air-conditioning (HVAC) equipment and controls; modifying the lighting system; and optimizing tenant use practices.

Intermediate Energy Audit

An intermediate audit involves more detailed surveys and analyses of the building than a walk-through audit. Tests and measurements quantify energy uses and losses and estimate the economics of upgrades. The scope of this audit is determined by the building owner's concerns and economic criteria. Because of its adaptable format, an intermediate audit may provide detailed analysis of some systems but only limited information on others. This level of analysis is sufficient for most buildings and provides the following information:

Energy usage: identifies the areas of energy use within the building.

Savings and cost analysis: analyzes the savings and cost measures that would satisfy the owner's concerns and economic criteria.

Operations and maintenance procedures evaluation: examines the effect of operations and maintenance procedures on energy usage in the building.

Capital-intensive improvements list: lists potential capital-intensive improvements for consideration, recognizing that more thorough data collection and analysis is required. This list also includes an initial judgment of potential costs and savings.

Comprehensive Energy Audit

A comprehensive audit is a full-scale audit that evaluates the amount of energy used for each energy system function (such as lighting, HVAC, etc). It involves a comprehensive survey of all systems and identifies additional capital-intensive investments to provide maximum energy and cost savings. It requires highly detailed field data gathering, testing and measurements, engineering analysis, and model analysis (such as computer simulation). A comprehensive audit determines the energy usage of the building and its systems for an entire year and accurately predicts future energy usage. Unlike the walk-through and intermediate audits, a comprehensive audit provides detailed project cost and savings information for major capital investment modifications. This type of audit consists of the following elements:

Envelope Audit: surveys the building envelope for energy losses or gains due to leaks, building construction, doors, glass, lack of insulation, etc.

Functional Audit: determines the amount of energy required for a main system and identifies each system's energy savings potential. Main systems include HVAC, building envelope, lighting, domestic hot water, and air distribution.

Utility Audit: analyzes the daily, monthly, or yearly energy usage for each utility that services the building. This audit typically requires utility data from at least the three previous years.

The data collected from these elements is used to develop an energy-use profile that includes all end-use categories. When these energy-use profiles are analyzed, opportunities for energy savings are then developed and evaluated.

RECOMMENDED PRACTICES

To achieve the most accurate results from an energy audit, the building systems must be inspected thoroughly. Examining the end uses of a system as well as the system itself will provide more valuable information for developing opportunities for energy savings. For example, auditing a building with steam boilers would first observe all or most of the end uses of steam in the building before evaluating the efficiency of the boilers themselves. These end-use observations may uncover considerable quantities of steam being wasted by venting to the atmosphere, venting through defective steam traps, uninsulated lines, and unused heat exchangers. After end-use waste is eliminated, the boiler combustion efficiency can be measured and improvements recommended. However, discretion is required when evaluating end uses because tracking down every end use is not cost-effective.

It is also important to focus on operations and maintenance procedures during the audit. A judicious review of the operating and maintenance logs may suggest improvements to procedures that would provide energy savings in addition to the system savings.

It is always important to evaluate indoor air quality during an energy audit. Evaluating indoor air quality examines the ventilating system's ability to introduce and distribute air through the system, filter airborne contaminants, and maintain acceptable temperature and relative humidity levels. Manifestations of poor indoor air quality include an increasing number of health problems such as cough, eye irritation, headache, allergic reaction, and, in some extreme cases, life-threatening conditions (Legionnaire's disease, carbon monoxide poisoning); decreasing productivity due to discomfort; increasing absenteeism; and accelerating deterioration of furnishings and equipment. An indoor air quality profile should be developed to evaluate the effects of energy savings upgrades on the building's indoor air quality.

Four factors affect the development of indoor air quality problems:

Sources of contamination: There is a source of contamination or discomfort indoors, outdoors, or within the mechanical systems of the building.

HVAC systems: The HVAC system may not control existing air contaminants and ensure thermal comfort (temperature and relative humidity conditions comfortable for most occupants).

Pollutant pathways: One or more pollutant pathways connect the pollutant source to the occupants.

Building occupants: The building occupants are adversely affected by the pollutants.

It is important to ascertain how each of these factors affects the current indoor air quality conditions when investigating, resolving, and preventing indoor air quality problems.

PERFORMING THE ENERGY AUDIT

The exact parameters of an energy audit are based on the size of the building, the complexity of its systems, and budget constraints. The more comprehensive the audit, the more energy and cost saving opportunities it will identify. Guidelines for performing the audit are listed in the next section.

Assembling the Audit Team

Each member of the team will be responsible for completing different tasks in the audit. The team should include design professionals, qualified contractors, the building owner's representatives, operations and maintenance personnel, and utility representatives. Some, such as a representative of the electric utility, may provide off-site assistance.

Depending on the type and size of the audit, it may be necessary to include an architect on the team. For smaller buildings, an architect with mechanical and electrical engineering experience may be the only design professional required. If a significant percentage of the energy usage is from lighting, an illumination engineer should be on the team. If a significant part of the audit centers around the ventilation system, an indoor air quality professional should be included on the team or the consulting engineers should have sufficient experience in analyzing indoor air quality problems.

Design Professionals. For simple buildings, a qualified consultant or contractor with mechanical engineering expertise and electrical engineering experience (or vice-versa) will be the only design professional required. For larger, more complex buildings, a consultant(s) with expertise in both mechanical and electrical engineering will be required and should be designated as the team leader. The consulting engineer will conduct a comprehensive energy audit, develop a report indicating all savings opportunities, perform feasibility studies, provide recommendations and design guidelines for specific systems, and provide an economic analysis of the opportunities for energy savings.

Qualified Contractors. A reputable contractor with extensive experience in energy audits may substitute for the design professional. Contractors can provide information on the latest technology, products, and materials, including purchase and installation expenses, actual field performance evaluations, and the relative ease of operation and maintenance. Contractors should provide the same services as the consulting engineer.

Building Owner's Representatives. Building owner's representatives manage internal concerns. They help the data collection process by providing historical information on the building and its energy consumption. They are the critical liaisons that maintain cooperation between operations and maintenance personnel and provide access to the building for after-hours inspections. If required, the owner's representatives can elaborate on details about current lease agreements that affect entry into key areas, or that pertain to cost-sharing of modifications.

Operations and Maintenance Personnel. The facility manager or the chief operating engineer should escort members of the project team on most of their inspections. Staff members should be available to answer any questions. It is important to maintain amicable relations with members of the operations and maintenance staff.

Utility Representatives. Depending on the policies of the local electric utility, a utility representative may need to be involved with on-site inspections. All appropriate utility representatives should be contacted to obtain information on rate structures.

Obtaining Specific Information About the Building

The consultants performing the audit should have specific information about the building before actually visiting a building. Typically, this information includes the following:

Energy consumption data for previous years: Typically, energy consumption data for the three years prior to the base year is collected and available.

Weather data: Weather data should be obtained for each year that energy consumption data is available. Weather data usually are available from the local electric utility. It should include the daily minimum and maximum temperature and relative humidity readings.

Building data: The building owner/manger or the original consulting engineers or general contractor should provide original building plans and specifications, as-built plans, testing and balancing reports, and commissioning reports.

Operations and maintenance logs, manuals: Operations and maintenance (O&M) logs and equipment manuals provide information on equipment operation and maintenance and the manner in which equipment performance changes have been handled by modifications to O&M procedures. Complaint logs should also be included.

Utility rate schedules: Reviewing rate schedules will determine if modifications can result in better rates. The analysis should reflect future changes; eg, will utility policy changes concerning demand metering equipment or rate schedules affect restructuring? This information is important when formulating an energy savings strategy for the future.

Miscellaneous data: Other information available can assist in the auditing procedure. For example, details about energy conservation modifications that have already been completed should be obtained, along with plans for any future modifications. Of particular importance are any anticipated major tenant turnovers or scheduled physical modifications such as renovation, remodeling, or expansion. Other information pertains to codes that apply to the building and warranties and guarantees on equipment that may be modified.

Investigating On-Site Conditions

The main components of an on-site investigation are (1) testing and measurement and (2) inspection. These are addressed below.

Testing and Measurement. To accomplish the energy audit, qualify building energy uses and losses and perform extensive testing and measurements on each system. The instrumentation needed for this procedure is varied, and depends on the types of measurements required. It includes the following:

Cameras are used to take pictures of the condition of the equipment. The pictures highlight the importance of the modifications and are used in the report.

Infrared scanning devices detect the heat emitted by an object. The higher the temperature emitted by an object, the greater the infrared signature. These devices can locate building heating losses, inspect power transmission equipment, locate water leakage into building roof insulation, check for poor building insulation, inspect cooling coils for plugged tubes, inspect electronic circuits, detect plugged furnace tubes, and find leaks in buried steam lines.

Electrical Measurement Devices. These devices are used when measuring the electrical systems. These include the following:

Ammeters measure alternating current. For example, an ammeter determines the amount of current a motor is using under normal operating conditions. The amount of current used will establish if the motor is oversized or undersized.

Voltmeters measure the difference in electrical potential between two points in an electrical current. For example, a voltmeter would be used to determine the amount of incoming voltage a motor is receiving under normal operating conditions. If the voltage is too low or too high, the motor could fail prematurely.

Wattmeters measure electrical energy directly in watts. For example, a wattmeter uses a motor operating under normal operating conditions to determine how much energy it is using. This will verify if the motor should be replaced with a high efficiency motor that uses less energy.

Power factor meters measure the ratio of real power to apparent power. Utilities typically add a charge to the electric bills of buildings that have a low power factor. The power factor meter finds the sources of the low power factor, which can then be corrected.

Light intensity meters measure illumination in units of footcandles to determine lighting levels in office areas. This will establish where improvements in lighting can be made to reduce utility costs and occupant complaints.

Energy and demand meters measure electrical energy and electrical demand. These devices combine ammeters, voltmeters, and power factor meters into one meter to measure energy and demand. Measuring demand will locate equipment that incurs high demand during the peak demand period. This equipment can then become part of the peak-shaving or demand-limiting program.

Temperature Measurement Devices. These are an important part of any energy audit. These include the following:

Thermometers measure air, liquid, and surface temperatures. Glass-stem thermometers, electronic thermometers, and portable recording thermometers measure space temperature.

Surface pyrometers measure the temperature of various surfaces. They are typically used to measure heat loss through walls and to test steam traps.

Psychrometers measure relative humidity through a relationship between dry-bulb and wet-bulb temperatures. Sling psychrometers are typically used to measure the relative humidity in a space. Some electronic devices combine the thermometer and the psychrometer, taking both readings in the space with just one device.

Air Velocity Measurement Devices. These measure HVAC system performance. Smoke pellets, anemometers, pitot tubes, impact tubes, and heated thermocouples are all used to measure the air velocity on various types of systems and air movement in the space.

Pressure Measurement Devices. These measure pressure in HVAC systems. These include manometers, bourdon tubes, and draft gauges.

Flow Measurement Devices. These measure the flow of water, steam, oil, and gas. These instruments must be inserted into a pipe to take the measurements. Typically, three types of measurements are made: pressure differential, velocity, and positive displacement.

Combustion Efficiency Measurement Devices. They determine the composition of the flue-gas maximizing combustion efficiency. A boiler test kit that evaluates fireside boiler operation contains carbon dioxide, oxygen, and carbon monoxide gas analyzers. To measure stack conditions, an inclined manometer and/or a draft gauge measures pressure. A smoke detector measures the combustion completeness.

Portable PH Meters. PH meters measure the acidity or alkalinity of water treatment systems.

Tachometers and Stroboscopes. They measure the rotating speeds of fans and motors.

Indoor Air Quality Measurements. These measurements are made by thermometers and psychrometers or hygrothermographs to measure temperature and humidity; tracer gas or chemical smoke tubes to find pollutant pathways; a device to measure airborne particles; micromanometers to pressure differentials at fans, intakes, and ducts; flow hoods, anemometers, or velometers to measure the airflow at diffusers; and detector tubes or meters to measure carbon dioxide, carbon monoxide, and other contaminants.

Inspection. Because energy consumption in buildings is determined by the characteristics and interactions of all systems, information on these systems should aid in inspecting and monitoring operations. As equipment is inspected, information that identifies the specific type of equipment installed, the manufacturer, model number, nameplate ratings, current conditions, and other factors as required should be recorded.

Evaluations of maintenance operations and personnel are accomplished by examining equipment records and logs and inspecting the various fields of responsibility. If necessary, arrangements can be made for the team to witness a demonstration or explanation of how a given task is routinely performed by the personnel. Observing the office environment and general routine can provide insight into energy consumption. In the office environment, it is important to observe how people operate their equipment, lighting, appliances, drapes, blinds, etc. It is also important to observe the daily arrivals and departures, lunch breaks, and after-hours occupancy, including duration and equipment used to support the task performed.

The indoor air quality inspection consists of determining HVAC operating and maintenance schedules; inspecting the usage and storage of chemicals; looking for indoor air quality problem indicators, such as odors, unsanitary conditions, visible biological growth, poorly maintained filters, presence of hazardous substances, unusual noises from light fixtures or mechanical equipment, inadequate maintenance, signs of occupant discomfort and overcrowding, or blocked airflows; inspecting HVAC system condition and operation; performing an inventory of pollutant sources and pathways; collecting information on building occupancy; checking for indicators of adequate ventilation; determining whether the layout of air supplies, returns, and exhausts promotes efficient air distribution to all occupants and isolates or dilutes contaminants; and observing air movement direction.

Performing the Computer Simulation

In larger buildings, or those with more complex building systems, the consultant will use computer simulations or modeling to evaluate various opportunities for energy savings. Many computer programs are available and some, such as the DOE-II program, are more accurate than others.

The report from the simulation should provide an accurate estimate of monthly and annual energy consumption. If the monthly and annual data are not accurate, the cause must be determined. Once the program simulates the existing performance with a high degree of accuracy, several different modifications can be simulated to determine the energy savings impact of each modification or a combination of modifications.

Developing the Report

The end result of the audit is an extensive report detailing areas of energy consumption (problems) and listing the opportunities for energy savings (solutions) for each system. The report also contains information on how each energy saving opportunity affects other systems. It provides details on projected annual energy savings, annual dollar savings, the cost of modifications, payback periods, life-cycle costs, and return on investment. It establishes priorities specified by the owner's criteria, details proposed new procedures and types of equipment and controls to be added or modified, and may even provide specific guidelines for the modifications.

A feasibility study is provided whenever a modification involves a significant capital outlay or a main renovation that would result in closure of part of the building for a time (or an action of similar impact). This type of study provides the comprehensive information necessary for decision-making. The feasibility study is typically separate from the initial report and is prepared subsequently to review of the initial study, in full consideration of other alternatives that exist.

The report should also include an indoor air quality profile that can be referred to in planning for renovations, negotiating leases and contracts, or responding to future complaints. This profile should include information on how each opportunity for energy savings affects the indoor air quality.

BIBLIOGRAPHY

Reading List

Building Energy Audits, National Electrical Contractors Association, Washington, D.C., 1979.

Energy for the Year 2000: A Total Energy Management Handbook, National Electrical Contractors Association, Washington, D.C., 1993.

Energy Management and Control Systems, The Electrification Council, Washington, D.C., 1982.

ENERGY CONSERVATION

Dan Steinmeyer
Monsanto Co.
St. Louis, Missouri

The main driver for energy conservation is broadscale technological progress and continues in times of both rising and falling energy prices. It is responsible for the historical rise in energy efficiency of 1 to 3% per year achieved by the process industries. It includes a wide range of big and little steps such as:

- improved gas turbine efficiency
- structured packing in distillation
- computer control
- variable speed drives
- computer design tools
- improved catalysts and processes for a wide range of materials such as low density polyethylene, acrylonitrile, NH_3, and acetic acid.

The second component that has driven energy conservation increases is the trade of capital for energy. This is shown in Figures 1a,b,c,d,e. This trade is optimization within an existing technology and nets large increases when energy prices rise rapidly compared to capital price as it did in the 1975 to 1985 time period.

Many plants also offer opportunities for significant energy conservation by various cogeneration processes (see Energy efficiency; Thermodynamics).

ENERGY BALANCE

An energy balance has historically been prepared for components of a process, primarily to assure that heat exchangers and utility supply are adequate. Often, an overall process energy balance was not developed, but today, the energy balance for the overall process has become a document almost as important as the material balance.

The energy balance should analyze the energy flows by type and amount (ie, electricity, fuel gas, steam level, heat rejected to cooling water, etc). It should include realistic loss values for turbine inefficiencies and heat losses through insulation.

Exergy, Lost Work, and Second-Law Analysis

When energy is critically important to process economics, the simple energy balance is sometimes carried into an

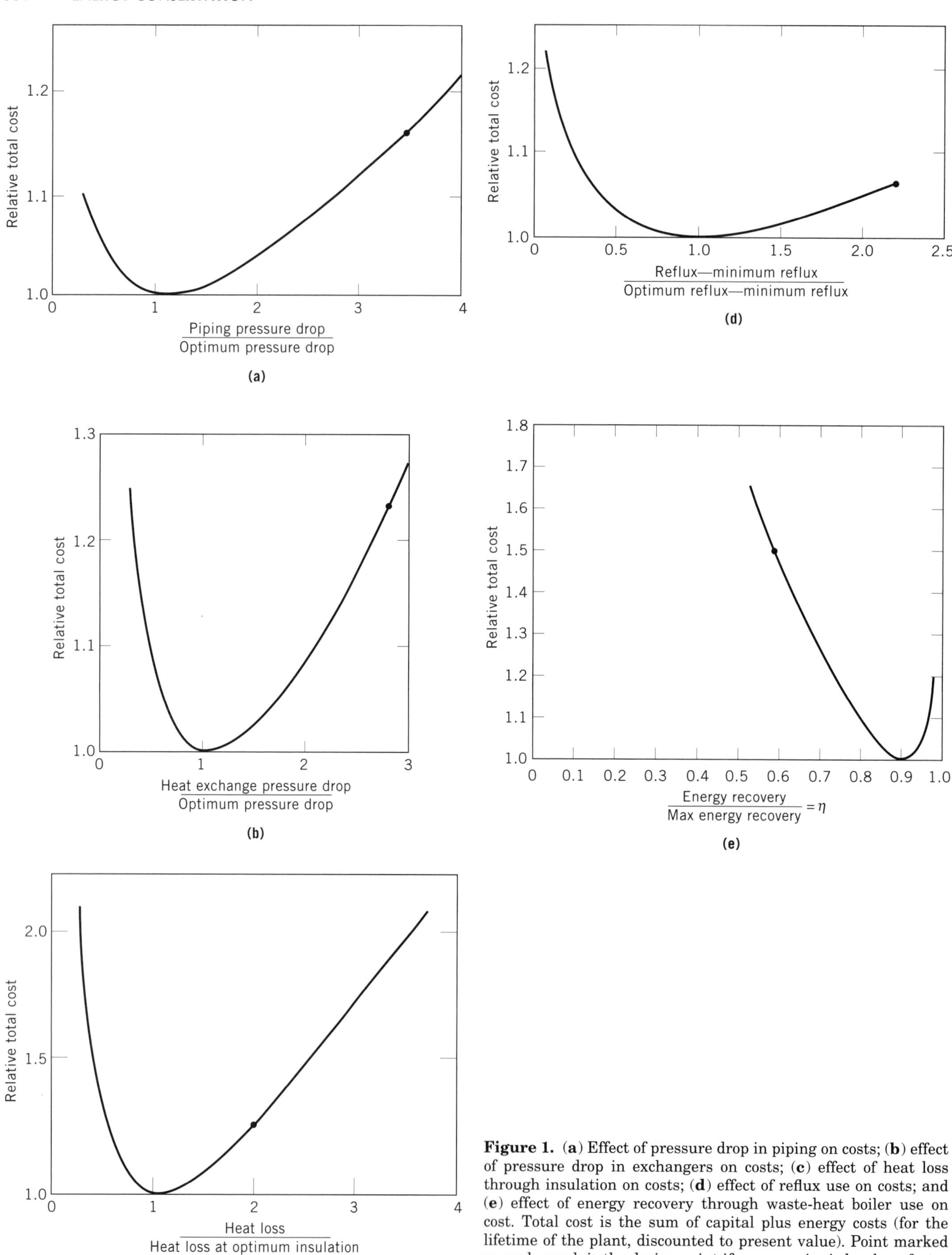

Figure 1. (**a**) Effect of pressure drop in piping on costs; (**b**) effect of pressure drop in exchangers on costs; (**c**) effect of heat loss through insulation on costs; (**d**) effect of reflux use on costs; and (**e**) effect of energy recovery through waste-heat boiler use on cost. Total cost is the sum of capital plus energy costs (for the lifetime of the plant, discounted to present value). Point marked on each graph is the design point if energy price is low by a factor of 4.

analysis of lost work. This compares the actual design against the theoretical ideal at each step and defines where the true energy use (or lost work) is occurring (see Thermodynamics).

In the following discussions of reaction, separation, heat exchange, compression, refrigeration, and steam systems, the importance of this concept is illustrated. A few terms are defined below.

Exergy (E) is the potential to do work. Thermodynamically, this is the maximum work a stream can deliver by coming into equilibrium with its surroundings.

$$E = (H - H_0) - T_0(S - S_0)$$

where E = exergy, the maximum theoretical work potential; H, S = enthalpy and entropy of the stream at its original conditions; H_0, S_0 = enthalpy and entropy of the same stream at equilibrium with the surroundings; and T_0 = temperature of the surroudings (sink).

Exergy is also sometimes called availability or work potential.

Free energy (G) is a related thermodynamic property. It is most commonly used to define the condition for equilibrium in a processing step.

$$\Delta G = \Delta H - T\Delta S$$

It is identical to ΔE if the processing step occurs at T_0.

Lost work (LW) is the irreversible loss in exergy that occurs because a process operates with driving forces or mixes material at different temperature or composition.

$$LW = E_{\text{in}} - E_{\text{out}}$$

Second-Law Analysis looks at the individual components of an overall process to define the causes of lost work. Sometimes it focuses on the efficiency of a step and ratios the theoretical work needed to accomplish a change (eg, a separation) to that actually used. Sometimes it is more cost effective simply to compare design against a second-law-violation checklist covering items such as (2): mixing streams at different temperatures and compositions; high pressure drops in control valves; reactions run far from equilibrium, high temperature differentials; and pump-discharge recirculation.

REACTOR DESIGN FOR ENERGY CONSERVATION

How close a designer comes to minimum energy is heavily determined by the raw materials and catalyst system chosen. However, if the designer only has the freedom to choose reaction temperature, residence time, and diluent, he or she still has a tremendous opportunity to influence energy use via their effect on yield. But even given none of these, there is still wide freedom to optimize the heat interchange system.

Maximizing Yield. Often the greatest single contribution to reduced energy cost is increased yield. High yield reduces the amount of material to be pumped, heated, and cooled while also simplifying downstream separation. This says nothing about the indirect energy reduction achieved through reduced raw material use. On average, the chemical industry uses almost as much energy in its raw materials as it does in direct purchases of fuel.

Minimizing Diluent. The case concerning diluent is less clear. A careful balance needs to be made of the benefits it gives in higher yield against the costs in mass handling and separation.

Optimizing Temperature. Temperature is usually dictated by yield considerations and often this opposes the simple dicta of energy: in an endothermic reaction, put the heat in at the lowest practical temperature; in an exothermic reaction, operate (and remove the heat) at the highest possible temperature.

In an exothermic reaction, there is an inevitable loss of work potential which is proportional to the free energy change (ΔG_{T_r}). With all reactants preheated to reaction temperature (T_r) and all heat recovered at T_r (3),

$$\text{lost work} = \Delta G^0_{T_r}\left(\frac{T_0}{T_r}\right)$$

For highly exothermic reactions, eg, oxidations, this is typically the dominant loss in the entire process.

Heat Recovery and Feed Preheat. The objective, to bring the reactants to and from reaction temperature with the least utility cost and to recover maximum waste heat at maximum temperature is generally achieved by the criteria given below under Heat Exchange. Sometimes, control and safety conditions prevent these criteria from being completely followed. The impact of feed preheating merits a particularly careful look. In an exothermic reaction, it permits the reactor to act as a heat pump, ie, "to buy low and sell high." The most common example is combustion-air preheating for a furnace.

Batch vs Continuous Reactors. Usually, continuous reactors yield much lower energy use because of increased opportunities for heat interchange. Sometimes the savings are even greater in downstream separation units than in the reaction step itself.

Especially on batch reactors, the designer should critically review any use of refrigeration to remove heat. Batch processes often have evolved little from the laboratory-scale glassware where refrigeration was a convenience.

SEPARATION

About one third of the chemical industry's energy is used for separation. A correlation exists between selling price and feed concentration (Fig. 2) (4). This in turn can be traced to the minimum work of separation.

$$W = RT_0\, \Sigma N_i \ln (x_i\gamma_i)$$

where T_0 is the sink temperature; N_i is the number of moles of a species present in the feed; $x_i = N_i/\Sigma N_i$; and γ_i is an activity coefficient.

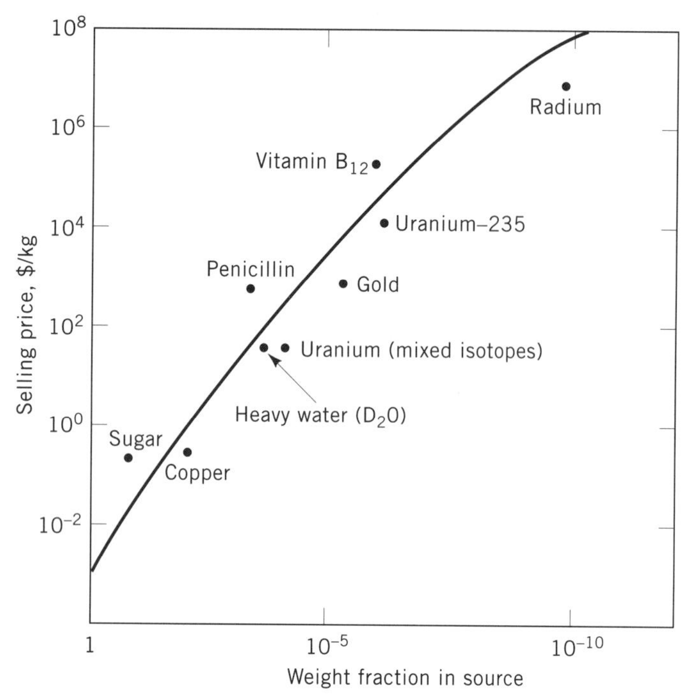

Figure 2. Commercial selling prices of some separated materials (6). Courtesy of McGraw-Hill Book Company.

This looks complicated, but actually, it provides a target that is easily calculated and approachable in practice. For example, work calculated from this expression closely approaches the performance of a real-world distillation after inefficiencies for driving forces are taken into account as illustrated below.

For ideal solutions ($\gamma_i = 1$) of a binary mixture, this simplifies to:

$$W = RT_0[x_1 \ln x_1 + (1 - x_1) \ln (1 - x_1)]$$

This applies whether the separation is by distillation or any other technique.

When a separation is not completed, less work is required. For x_1 equal to 0.5:

Product purity, %	*Relative work*
100	1
99.9	0.99
99.0	0.92
90.0	0.53

Note that it takes only a little work to move from 99% separation to 100%, but a great deal to move from 90% to 100%. This is important to recognize when comparing separation techniques. Some leave much of the work undone, as for example, in crystallization involving an unseparated eutectic mixture.

Distillation

Distillation is overwhelmingly the most common separation technique because of its inherent advantages: phase separation is clean; it is relatively easy to build a multistage, countercurrent device; and equilibrium is closely approached in each stage.

Minimum work for an ideal separation at first glance appears unrelated to the slender vertical vessel with a condenser at the top and a reboiler at the bottom. The connection becomes evident when one calculates the work embedded in the heat flow which enters the reboiler and leaves at the condenser. An ideal engine could extract work from this heat.

$$\text{work potential} = QT_0\left[\frac{1}{T_{\text{condenser}}} - \frac{1}{T_{\text{reboiler}}}\right]$$

Comparison of actual use of work potential against the minimum allows calculation of an efficiency relative to the best possible separation:

$$\eta = \frac{RT_0[x_1 \ln_1 + (1 - x_1)\ln(1 - x_1)]}{QT_0\left[\frac{1}{T_{\text{condenser}}} - \frac{1}{T_{\text{reboiler}}}\right]}$$

There is still no obvious reason to believe that the efficiency of separating a mixture with an α (relative volatility) of 1.1 will be related to that for an α of 2; however, it is known that when α is small, the required reflux and Q are large, but ($T_{\text{condenser}} - T_{\text{reboiler}}$) is small (see Distillation).

The two effects almost cancel to yield an approximation for the minimum thermal work used in a distillation (4–5).

$$\text{min thermal work} = RT_0(1 + [\alpha - 1]x_1)$$

When this is combined with the definition of minimum separation work for an ideal binary, an approximation for distillation efficiency can be obtained:

$$\eta = \frac{x_1 \ln x_1 + (1 - x_1)\ln(1 - x_1)}{1 + (\alpha - 1)x_1}$$

This efficiency is high and shows only minor dependence on α over a broad range of α. For $x_1 = 0.5$,

$\alpha = 1.1$	$\alpha = 1.5$	$\alpha = 2$
$\eta = 0.66$	$\eta = 0.55$	$\eta = 0.46$

The dependence on x_1 is greater:

	$\alpha = 1.05$	$\alpha = 2$
η for $x_1 = 0.1$	0.32	0.20
η for $x_1 = 0.01$	0.056	0.053

This suggests that distillation should be the preferred method for feed concentrations of 10–90% and that it is probably a poor choice for feed concentrations of less than 1%. This matches experience. Techniques such as adsorption, chemical reaction, and ion exchange are chiefly used to remove impurity concentrations $<1\%$.

The high η values above conflict with the common belief that distillation is always inherently inefficient. This belief arises mainly because past distillation practices utilized such high driving forces (pressure drop, reflux ratio, and temperature differentials in reboilers and condensers). A real example is instructive:

	C_2 *splitter* (*relative numbers*)
Theoretical work of separation	1.0
Min thermal work potential used	1.4

$$\eta = \frac{\text{Theoretical work}}{\text{Min thermal work}} = \frac{1.0}{1.4} = 0.7$$

Losses for driving forces:	0.1
Reflux above the minimum	0.1
Exchanger ΔT	2.1
ΔP in tower	0.5
ΔP in condenser and tower	0.8
Total losses	*3.5*

$$\eta_{\text{Including losses}} = \frac{1.0}{1.4 + 3.5} = 0.2$$

These numbers show first that the theoretical work can be closely approached by actual work after known efficiencies are identified, and second, that the dominant driving force losses are in pressure drop and temperature difference. This is a characteristic of towers with low relative volatilities.

What Does Optimum Design Look Like?

Condenser and Reboiler ΔT's. As shown by this example, the losses for ΔT typically are far greater than those for reflux beyond the minimum. As shown below under Heat Exchange, the economic optimum for temperature differential is typically under 15°C. This contrasts with the values over 75°C often used in the past. This is probably the biggest opportunity for improvement in the practice of distillation.

Adjusting the Process to Optimize ΔT. First glance may show only three or four utility levels (temperatures) to choose from. These might well be 100°C apart. Some ways to increase the options are to consider multieffect distillation, which spreads the ΔT across two or three towers; to use waste heat for reboil; and to recover energy from the condenser. Often to make some of these possible, the pressure in a column may have to be either raised or lowered.

Reflux Ratio. A number of studies have shown that the optimum reflux ratio is generally below 1.15 minimum. At this point, excess reflux is a minor contributor to column inefficiency. When designing to this tolerance, correct vapor–liquid equilibrium (VLE) and adequate controls are essential.

State-of-the-Art Control. Computer control with feed-forward capability can save 5–20% of a unit's utilities. It does this by reducing the margin of safety. This sounds hard to believe, particularly for systems designed to operate within 10% of minimum reflux; however, unless the discipline of a controller forces this to happen, operators typically opt for increased safety. They are probably right to do so unless the proper set of analyzers and controllers is provided and maintained.

Right Feed Enthalpy. Often it is possible to heat the feed with a utility considerably less costly than that used for bottom reboiling. Sometimes the preheating can be directly integrated into the column-heat balance by exchange against the condensing overhead or against the net bottoms from the column. Simulation and a careful look at the overall process are required to assess the value of feed preheating accurately.

A vapor feed is favored when most of the column feed leaves the tower as overhead product. The use of a vapor feed was a key component in the high efficiency cited above for the C_2 splitter where most of the feed goes overhead.

Low Column-Pressure Drop. The penalty for column-pressure drop is an increase in temperature differential

$$\Delta T = \left(\frac{dT}{dP}\right) \Delta P$$

$$\frac{dT}{dP} = \frac{R}{\Delta H}\frac{T^2}{P}$$

As this suggests, the penalty becomes very large for low vapor pressure materials, ie, for components that are distilled at or below atmospheric pressure. The work penalty associated with this ΔT is approximately defined by the ratio:

$$\frac{\Delta T \text{ for pressure drop}}{T_{\text{reboiler}} - T_{\text{condenser}}} = \text{fraction of work potential for } \Delta P$$

This penalty is severest for close-boiling mixtures. The most powerful technique for cutting ΔP is the use of packing. Conventional packings such as 5-cm (2-in.) pall rings can achieve a factor of four reduction over trays, and structured packing can give a factor of 10.

In applying this analysis, one must consider the overhead line and condenser pressure drop as well. (Note the high loss in the C_2 splitter example.)

Intermediate Condenser. As shown by Figure 3, an intermediate condenser forces the operating line closer to the equilibrium line, thus reducing the inherent inefficiencies in the tower. With the use of intermediate condensers and reboilers, it is possible to exceed the efficiency given above, particularly when the feed composition is far from 50:50 in a binary mixture.

	Max efficiency 50% of heavy component in feed	*Max efficiency 95% of heavy component in feed*
1 condenser, 1 reboiler	67	20
2 condensers, 1 reboiler	73	47
3 condensers, 1 reboiler	77	62

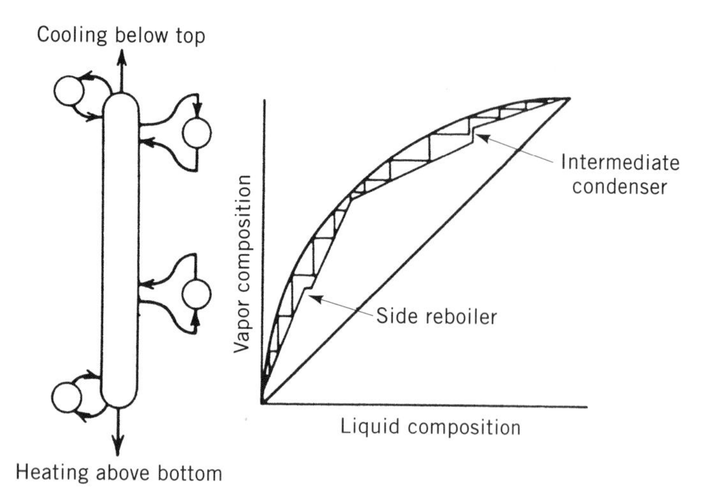

Figure 3. Intermediate condenser and reboiler.

The intermediate condenser is most effective when the feed leaves the column mainly as bottoms. This approach is economical if a less costly coolant can be substituted for refrigeration, or if it permits reuse of the heat of condensation.

Intermediate Reboiler. Inclusion of an intermediate reboiler moves the heat-input location up the column to a slightly colder point. It can permit use of waste heat for reboil when the bottoms temperature is too hot for the waste heat.

Heat Pumps. Because of added capital and complexity, heat pumps are rarely economical.

Lower Pressure. Usually, relative volatility increases as pressure drops. For some systems, a 1% drop in absolute pressure cuts required reflux by 0.5%. Again, if operating at reduced pressure looks promising, it can be evaluated by simulation (see Simulation and process design).

Steam Stripping. Steam (or other stripping gas) and vacuum are largely interchangeable. Steam stripping allows more tolerance for pressure drop, but this comes at the penalty of much higher energy use.

Other questions:

Is the separation necessary?

Is the purity necessary?

Are there any recycles that could be eliminated?

Can the products be sent directly to downstream units, thereby eliminating intermediate heating and cooling?

Other Separation Techniques

Under some circumstances, distillation is not the best method of separation. Among these instances are the following: when relative volatility is <1.05, as often happens with isomers; when ca 1% of a stream is removed, as in gas drying (adsorption or absorption) or C_2H_2 removal (reaction or absorption); when thermodynamic efficiency of distillation is <5% (for whatever reason); and when a high boiling point pushes thermal stability limits. A variety of other techniques may be more applicable in these cases.

Reaction. Purification by reaction is relatively common when concentrations are low (ppm) and a high energy but a low value molecule is present. Some examples are hydrogenation of acetylene and oxidation of waste hydrocarbons:

$$C_2H_2 + H_2 \rightarrow C_2H_4$$

$$\text{waste hydrocarbon} + O_2 \rightarrow H_2O + CO_2$$

Absorption (Extractive Distillation). As a separation technique, absorption starts with an energy deficit because it mixes in a pure material (solvent) and then separates it again. It is nevertheless quite common, because it shares most of the physical property advantages of distillation, and because it separates by molecular type it can be tailored to obtain a high α. The following ratios are suggested for equal costs (6):

$\alpha_{distillation}$	$\alpha_{extraction\ distillation}$	$\alpha_{extraction}$
1.2	1.4	2.5
1.4	1.9	5
1.6	2.3	8

In practice most of the applications have come where a small part (<5%) of the feed is removed. Examples include H_2S/CO_2 removal and gas drying with a glycol (see Azeotropic and extractive distillation).

Extraction. The advantage of extraction is that it purifies a liquid rather than a vapor, allowing operation at lower temperatures and removal of a series of similar molecules at the same time, even though they differ widely in boiling point. An example is the extraction of aromatics from hydrocarbon streams.

The disadvantage of extraction relative to extractive distillation is the greater difficulty of getting high efficiency countercurrent processing.

Adsorption. Adsorbents can achieve even more finely tuned selectivity than extraction. The most common application is the fixed bed with thermal regeneration, which is simple, attains essentially 100% removal, and carries little penalty for low feed concentration. An example is gas drying. A variant is pressure-swing adsorption. Here, regeneration is attained by a drop in pressure. By use of multiple stages, high impurity rejection can be achieved, but at the expense of also losing part of the desired product.

Another adsorption approach is the simulated moving-bed system. It has large-volume applications in normal-paraffin separation and *para*-xylene separation. Since its introduction in 1970, it has largely displaced crystallization in xylene separations. The unique feature of the system is that the bed is fixed but the feed point shifts to simulate a moving bed.

Melt Crystallization. Crystallization from a melt is inherently attractive in competition to distillation because

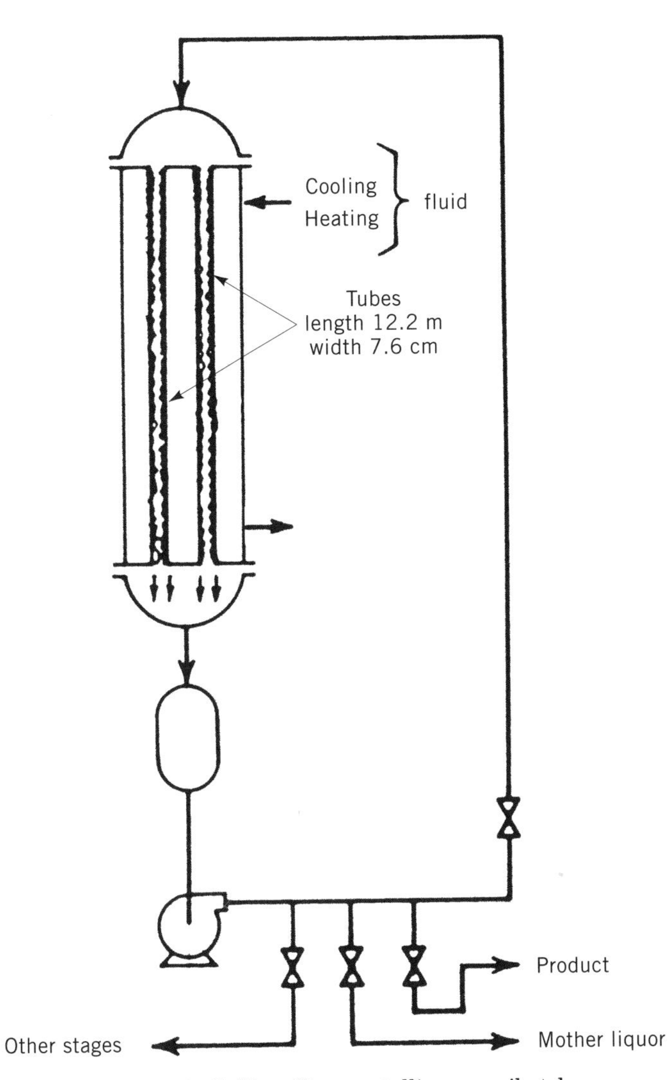

Figure 4. Falling-film crystallizer, semibatch.

the heat of fusion is much lower than that of evaporation. It also benefits from lower operating temperature. In addition, organic crystals are virtually insoluble in each other so that a pure product is possible in a one-stage operation.

However, crystallization has a unique set of disadvantages that generally outweigh the virtues and have sharply limited its application. Industry practice suggests the use of a workable alternative, if one exists. The disadvantages of melt crystallization are as follows:

Physical separation is difficult. Impure liquid is trapped as occlusions as the crystals grow, and liquid wets all surfaces.

It requires a second separating process to separate the eutectic mixture. In some senses, it is akin to formation of two liquid phases: little energy is required to get the two phases, but a great deal is required to finish the purification.

It is difficult to add or remove heat because of the thermal resistance of the crystal.

It is difficult to move the liquid countercurrent to the crystals.

Thermodynamic efficiency is hurt by the large ΔT between the temperatures of melting and freezing. In an analogy to distillation, the high α comes at the expense of a big spread in reboiler and condenser temperature. From a theoretical standpoint, this penalty is smallest when freezing a high concentration (ca 90%) material.

One process is shown in Figure 4. It is a semibatch operation in which liquid falls down the walls of long tubes. This permits both staged operation and sweating of crystals. Typically, the sweating and staged operation require melting 5 kg of material for each kilogram of product (7).

Membranes. Membrane separators are used in purification of hydrogen and air. The working principle is a membrane that is chemically tuned to pass molecular type (see Fig. 5).

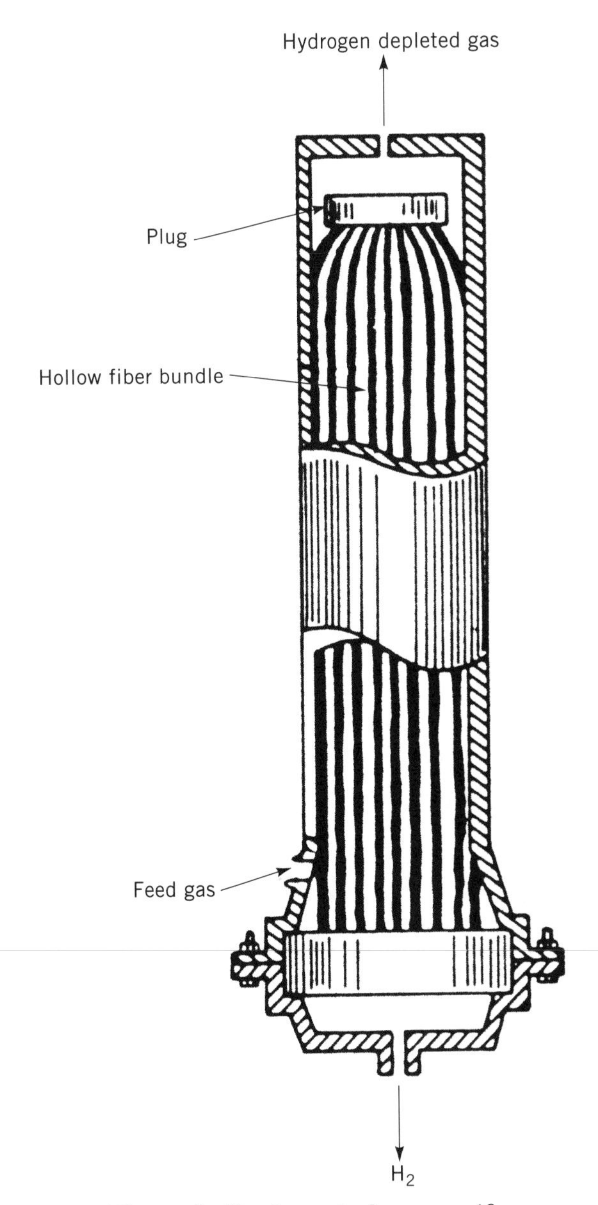

Figure 5. Membrane hydrogen purifier.

Liquid separation via membranes (reverse osmosis) is used in production of pure water from seawater. The chief limit to broader use of reverse osmosis is the high pressure required as the concentration of reject rises.

Mole fraction of reject	*Min* ΔP, MPa (psi)
0.05	7.6 (1100)
0.10	15.2 (2200)
0.20	31.8 (4600)

For most processes, the probability of finding a membrane with requisite strength, as well as selectivity and permeability, is low, and most systems are limited to achieving a mole fraction reject below 0.10.

HEAT EXCHANGE

Most processing is thermal. Reaction systems and separation systems are typically dominated by their associated heat exchange. Optimization of this heat exchange has a tremendous leverage on the ultimate process efficiency (see Heat-exchange technology).

Heat exchangers use energy two ways: as frictional pressure drop and as the loss in ability to do work when heat is degraded.

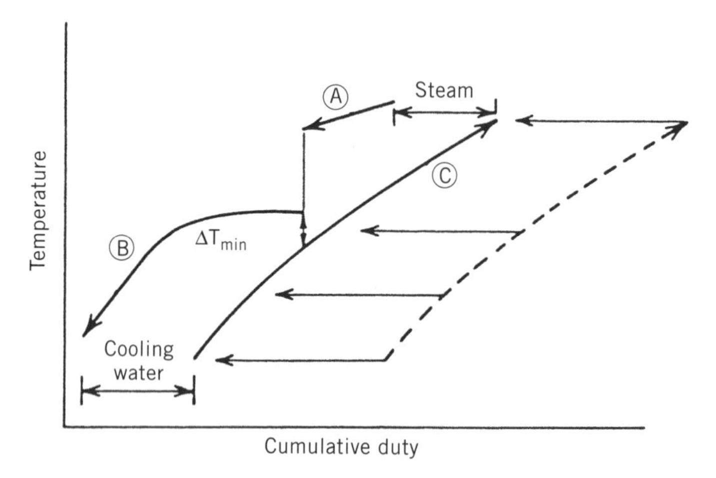

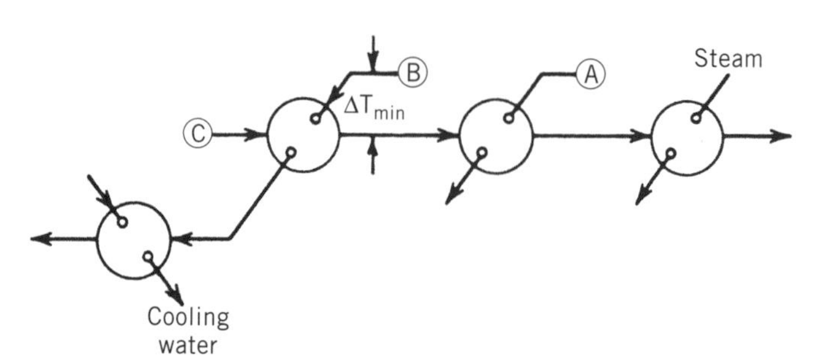

Figure 6. Simple heat-exchange network. Stream C is heated, and streams A and B are cooled.

$$\text{lost work} = QT_{\text{sink}}\left(\frac{1}{T_{\text{cold}}} - \frac{1}{T_{\text{hot}}}\right) + \text{frictional work}$$

(for ΔP)

In an optimized system the lifetime value of the lost work associated with Δt typically equals the cost of the heat exchanger. The lifetime value of the ΔP lost work in an optimized system is $\frac{1}{3}$ as great as the heat exchanger capital.

The selection of design numbers for ΔP and ΔT is frequently the most important decision the process designer makes, but often this goes unrecognized.

The designer commonly becomes lost in the detail of tube length and baffle cut in an effort to optimize the hardware to meet a target and spends far too little time on choosing that target.

Heat-Exchange Networks

A basic theme of energy conservation is to look at a process broadly, ie, to look at how best to combine process elements. The heat-exchange network analysis can be a useful part of this optimization Figure 6 illustrates a simple example of the basic concept of what network analysis does; it builds cumulative heating and cooling curves and merges them until a minimum ΔT is reached.

This example also illustrates one of the targets of network analysis, ie, obtaining the minimum number of shells:

$$\begin{aligned}\text{minimum shells} &= \text{no. of streams} - 1\\ &= 3 \text{ process streams} + 2 \text{ utilities} - 1 = 4\end{aligned}$$

Network analysis (or pinch technology) has become an increasingly powerful approach to process design that includes most of the virtues of 2nd law analysis (see ref. 8).

Overdesign

Overdesign also has a great impact on the cost of heat exchange and sometimes is confused with energy conservation. The best approach is to define clearly what the objective of overdesign is and then to explicitly specify it. If the main concern is a match to other units in the system, a multiplier is applied to flows. If the concern is with the heat balance or transfer correlation, the multiplier is applied to area. If the concern is fouling, a fouling factor is called for. But if low ΔT or ΔP is the principal concern, that should be specified. Adding extra surface saves energy only if the surface is configured to do so. Doubling the area may do nothing more than double the ΔP, unless it is configured properly.

ΔT and ΔP Optimization

Ideally, ΔT and ΔP are optimized by trying several values, making preliminary designs, and finding the point where savings in utility costs just balance incremental surface costs. Where the sums at stake are large, this should be

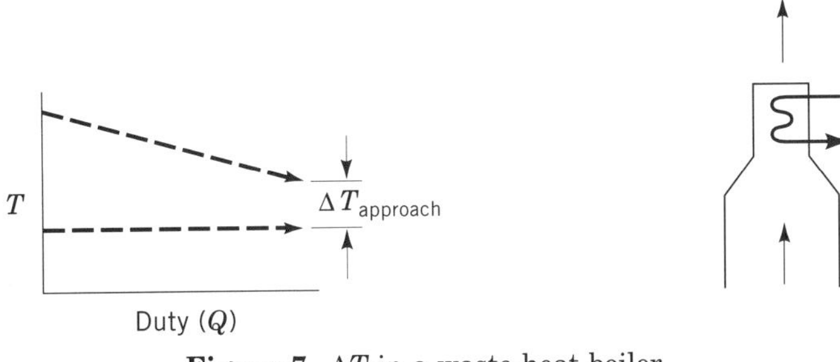

Figure 7. ΔT in a waste-heat boiler.

done. However, for many cases the simple guidelines given below are adequate. The primary focus is the impact of surface and utility prices; a secondary focus is the impact of fluid properties on heat-transfer coefficient (9).

Optimum ΔT. There are three general cases of high importance: the waste-heat boiler, in which only one fluid involves sensible heat transfer, ie, a temperature change; the feed–effluent exchanger, in which both fluids involve sensible heat transfer and are roughly balanced, ie, undergo essentially the same temperature change; and the reboiler, in which neither fluid involves a temperature change, ie, one fluid condenses and the other boils.

Waste-Heat Boiler

In a waste-heat boiler (Fig. 7), the optimum ΔT occurs when

$$\Delta T_{\text{approach}} = \frac{K_1}{K_v}\frac{1.33}{U}$$

where K_1 = annualized cost per unit of surface, \$/(m^2·yr); K_v = annualized cost per unit of utility saved, \$/(W·yr); and U = heat transfer coefficient, W/(m^2·K). The factor 1.33 includes the value of the pressure drop for the added surface.

For example, the optimum $\Delta T_{\text{approach}}$ is computed

$$K_1 = \frac{\$215/\text{m}^2}{2\ \text{yr}} = \frac{\$107.5}{(\text{m}^2\cdot\text{yr})}$$

$$K_v = \frac{0.017}{\text{kW}\cdot\text{h}}\cdot 8322\ \text{h/yr}$$

$$= \frac{\$142}{(\text{kW}\cdot\text{yr})} = \frac{\$0.142}{(\text{W}\cdot\text{yr})}$$

$$\Delta T_{\text{approach}} = \frac{107.5/(\text{m}^2\cdot\text{yr})}{0.142/(\text{W}\cdot\text{yr})}\ \frac{1.33}{56.8\ \text{W}/(\text{m}^2\cdot\text{K})} = 17.7\ \text{K}$$

where U = 56.8 W/(m^2·K) (10 Btu/(h·ft^2·°F)); surface cost = \$215/m^2 (\$20/ft^2); payout time = 2 yr; energy price \$0.017/kW·h (\$5/10^6 Btu); and onstream time = 8322 h/yr. This case underlines a dramatic change in process design. Note that $\Delta T_{\text{approach}}$ varies to the first power of the ratio of surface price to energy price. The most visible result has been a change in typical heater design efficiency from 65–75% to 92–94%. A secondary result has been appearance of waste-heat recovery units in many processes at the point where air coolers were once used.

Feed–Effluent Exchanger

The detailed solution for the optimum ΔT in a feed–effluent exchanger (Fig. 8) involves a quadratic equation for $\Delta T_{\text{approach}}$, but within the restrictions

$$0.8 < \frac{\Delta T_{\text{hot}}}{\Delta T_{\text{cold}}} < 1.25$$

$$\frac{T_{\text{hot}_{\text{in}}} - T_{\text{cold}_{\text{in}}}}{\Delta T_{\text{log mean}}} < 10$$

An excellent approximation is given by

$$\Delta T_{\text{log mean}} = \left[\frac{K_1}{K_v}\frac{1.33}{U}(T_{\text{hot}_{\text{in}}} - T_{\text{cold}_{\text{in}}})\right]^{1/2}$$

For example, the optimum $\Delta T_{\text{log mean}}$ for a feed–effluent exchanger is computed

$$K_1 = \frac{\$107.5/\text{m}^2}{2\ \text{yr}} = \$53.8/(\text{m}^2\cdot\text{yr})$$

$$K^{v} = \frac{\$0.027}{\text{kW}\cdot\text{h}}\cdot 8322\,\frac{\text{h}}{\text{yr}}\cdot 1000\,\frac{\text{kW}\cdot\text{h}}{\text{W}} = \frac{\$0.227}{\text{W}\cdot\text{yr}}$$

$$\Delta T_{\text{log mean}} = \left[\frac{53.8}{0.227}\ \frac{1.33}{284}(200 - 100)\right]^{1/2} = 10.5°\text{C}$$

where $T_{\text{hot}_{\text{in}}}$ = 200°C; $T_{\text{cold}_{\text{in}}}$ = 100°C; U = 284 W/(m^2·K) (50 Btu/(h·ft^2·°F)); surface cost = \$107.5/m^2 (\$10/ft^2); payout time = 2 yr; energy price = \$0.027/kW·h (\$8/10^6 Btu); onstream time = 8322 h/yr; and $\Delta T_{\text{hot}}/\Delta T_{\text{cold}}$ = 1.20.

The Reboiler

The case shown in Figure 9 is common for reboilers and condensers on distillation towers. Typically, this ΔT has a greater impact on the excess energy use in distillation

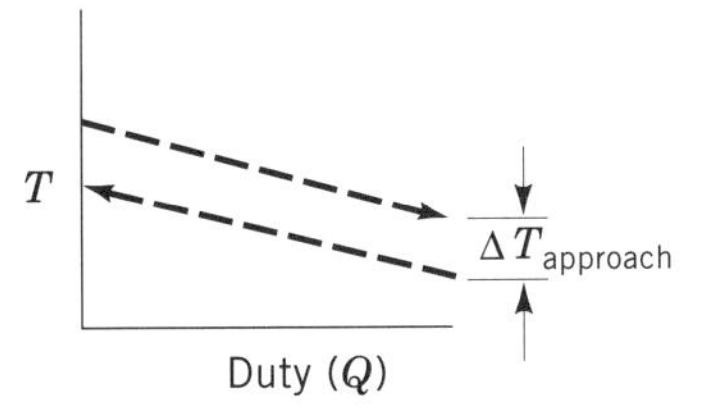

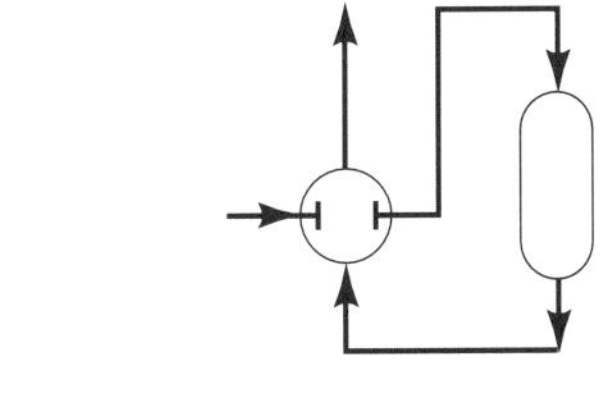

Figure 8. ΔT in a feed–effluent exchanger.

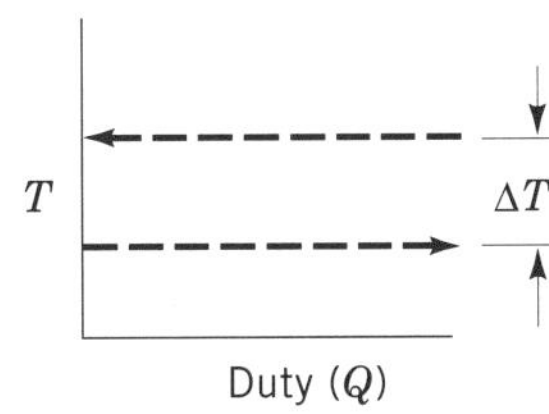

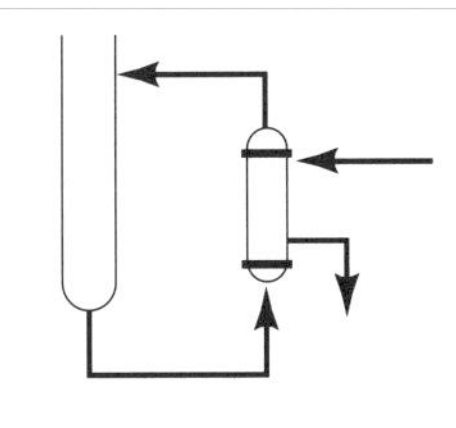

Figure 9. ΔT in the reboiler.

than does reflux beyond the minimum. The capital cost of the reboiler and condenser is often equivalent to the cost of the column they serve.

The concept of an optimum reboiler or condenser ΔT relates to the fact that the value of energy changes with temperature. As the gap between supply and rejection widens, the real work in a distillation increases. The optimum ΔT is found by balancing this work penalty against the capital costs of bigger heat exchangers.

If the Carnot cycle is used to calculate the work imbedded in the thermal flows with the assumption that the heat-transfer coefficient (U) is constant and that process temperature is much greater than ΔT,

$$\Delta T_{\text{optimum}} = T_{\text{p}} \left[\frac{K_1}{K_{\text{v}} U T_{\text{sink}}} \right]^{1/2}$$

where T_{sink} = temperature (absolute) at which heat is rejected; T_{p} = process temperature (absolute); K_1 = annualized cost per unit of surface; and K_{v} = annualized value of power.

For utilities above ambient temperature,

$$K_{\text{v}} = K_{\text{p}} \text{ (turbine efficiency)}$$

where K_{p} is the annualized cost of purchased power. The above relations will typically give ΔT values in the 10°C to 20°C range.

One strong caution is that the assumption of a constant U is usually inaccurate for boiling applications. Simulation is generally needed to fix ΔT accurately, particularly at ΔT values below 15°C.

Optimum Pressure Drop

For most heat exchangers, there is an optimum pressure drop. This results from the balance of capital costs against the pumping (or compression) costs. A common predjudice is that the power costs are trivial compared to the capital costs. The total cost curve is fairly flat within ± 50% of the optimum (see Fig. 1b), but the incremental costs of power are roughly one third of those for capital on an annualized basis. This simple relationship can be extremely useful in quick design checks.

The best approach is to have a computer program check a series of pressure drops and see how energy requirements decrease as surface increases. If this option is not available, the simple method shown below can be used to obtain specification sheet values.

Start with a pressure drop of 6.9 kPa (1 psi), and apply three correction factors ($F_{\Delta T}$, F_{cost}, F_{prop}):

$$\Delta P_{\text{opt}} = 6.9 \, (F_{\Delta T})(F_{\text{cost}})(F_{\text{prop}})$$

The correction for temperature difference is given by

$$F_{\Delta T} = \frac{T_{\text{in}} - T_{\text{out}}}{(T_{\text{hot}} - T_{\text{cold}})_{\text{mean}}}$$

This term is a measure of the unit's length. Sometimes it is referred to as the number of transfer units (see Mass Transfer).

The correction for costs is

$$F_{\text{cost}} = 0.017 \left(\frac{\$/(\text{m}^2 \cdot \text{yr})}{\$/\text{kW} \cdot \text{h}} \right)^{0.75}$$

The correction for physical properties is

$$F_{\text{prop}} = \left(\frac{c}{c_w} \right)^{0.6} \left(\frac{k_w}{k} \right)^{0.6} \left(\frac{\mu}{\mu_w} \right) \left(\frac{\rho}{\rho_w} \right)^{0.5}$$

where c = specific heat; k = thermal conductivity; μ = viscosity; ρ = density; and c_w, k_w, μ_w, and ρ_w are the same properties for water at 25°C.

From these equations, the optimum ΔP for a feed–effluent exchanger where the fluid has the physical properties of water and:

$$T_{\text{in}} - T_{\text{out}} = 20°\text{C}$$

$$\Delta T = (T_{\text{hot}} - T_{\text{cold}})_{\text{mean}} = 10°\text{C}$$

$$\left. \begin{array}{l} \text{surface cost} = \$215/\text{m}^2 \\ \text{payout time} = 2 \text{ yr} \end{array} \right\} \$107.5/(\text{m}^2 \cdot \text{yr})$$

$$\text{power cost} = \$0.03/\text{kW} \cdot \text{h}$$

is calculated:

$$F_{\Delta T} = \frac{20}{10} = 2$$

$$F_{\text{cost}} = 0.017 \left(\frac{107.5}{0.03} \right)^{0.75} = 7.8$$

$$F_{\text{prop}} = 1$$

$$\Delta P_{\text{opt}} = 6.9(2)(7.8)(1) = 107.67 \text{ kPa (15.6 psi)}$$

If all else remains the same as above except a gas is obtained with the following properties: $\mu = 0.02 \, \mu_w$; $\rho = 0.00081 \, \rho_w$; $c = 0.25 \, c_w$; and $k = 0.066 \, k_w$; then,

$$F_{\text{prop}} = (0.25)^{0.6} \left(\frac{1}{0.066} \right)^{0.6} (0.02)^{0.3} (0.00081)^{0.5} = 0.0194$$

$$\Delta P_{\text{opt}} = 6.9(2)(7.8)(0.0194) = 2.1 \text{ kPa (0.3 psi)}$$

The great impact of density in this example and in Table 1 should be noted. Probably the most common specification

Table 1. Impact of Fluid Density on Optimum ΔP

	$\dfrac{\rho}{\rho_w}$	$\left(\dfrac{\rho}{\rho_w}\right)^{0.5}$	=	Relative Optimum ΔP
Water	1	$(1)^{0.5}$		= 1
Oil	0.8	$(0.8)^{0.5}$		= 0.9
High pressure gas	0.05	$(0.05)^{0.5}$		= 0.22
Atm pressure gas	0.001	$(0.001)^{0.5}$		= 0.03
Vacuum	0.0002	$(0.0002)^{0.5}$		= 0.014

error is to use the large ΔPs characteristic of liquids in low density gas systems.

Fired Heaters

The fired heater is first a reactor and second a heat exchanger. Often, in reality, it is a network of heat exchangers (see Burner technology; Furnaces, fuel-fired).

The Fired Heater as a Reactor. When viewed as a reactor, the fired heater adds a unique set of energy considerations.

Can the heater be designed to operate with less air by O_2 and CO analyzers?

How does air preheat affect fuel use and efficiency?

How could a lower cost fuel (coal) be used?

Can the high energy potential of the fuel be used upstream in a gas turbine?

CO Control. Control of excess air by carbon monoxide is one of the most significant new energy technologies. The key is that CO is highly sensitive to excess air as shown by Figure 10.

Coal and Low Btu Gas. The much lower cost of coal has caused a rapid resurgence of coal-firing for steam generation. Its direct use in process heaters has been negligible because of historical prejudice and the space and capital demands of coal-handling facilities. Concern about attack on high alloy tubing by ash also precludes its direct use in applications such as ethylene furnaces.

Low Btu gas appears to be an attractive way of utilizing coal in process heaters for a number of reasons: Its technology is old and well established; it keeps the ash out of the process heater; and it requires only mild retrofitting, chiefly of burners. The flame temperature and total flue-gas generation for low Btu gas are both tolerably close to natural gas as shown in Table 2.

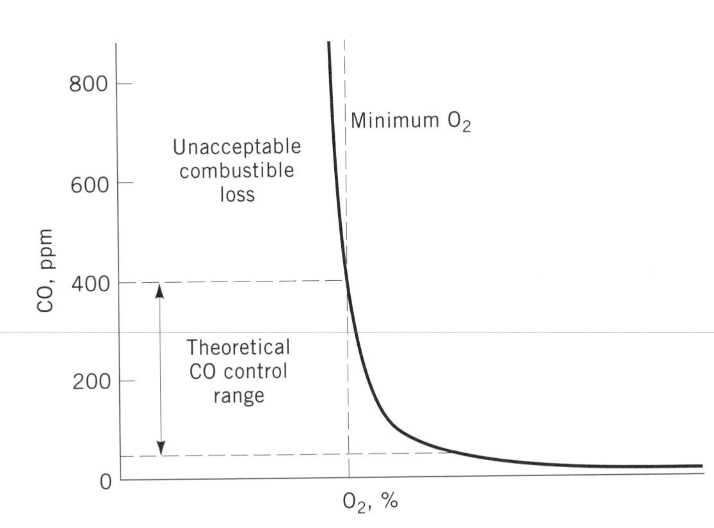

Figure 10. Relationship of CO concentration to O_2 concentration, in a fired heater.

Table 2. Comparison of Combustion Characteristics of Natural and Low Btu Gas

	Natural Gas	Low Btu Gas
Relative heating value per volume		
Fuel	1	0.17
Fuel and air mixture	1	0.78
Relative vol air:vol fuel	1	0.13
Relative flue-gas flow rate/unit of energy	1	1.11
Adiabatic flame temperature, °C	1927	1760

Air Preheating. Use of unpreheated air in the combustion step is probably the biggest waste of thermodynamic potential in industry (see Table 3). It is not practical to preheat to the flame burst temperature, the optimum thermodynamic situation, but some preheating is invariably profitable. Air preheating has the unique benefit of giving a direct cut in fuel consumed. It also can increase the heat-input capability of the firebox because of the hotter flame temperature.

The most common type of air preheater on new units is the rotating wheel. On retrofits, heat pipes or hot-water loops are often more cost-effective because of duct-work costs or space limits.

Upstream Firing of Fuel, Gas Turbines. Limitations in the material of construction make it difficult to use the high temperature potential of fuel fully. This restriction has led to insertion of gas turbines into power-generation steam cycles and even to use of gas turbines in preheating air for ethylene-cracking furnaces.

The Fired Heater as a Heat-Exchange System

Improved efficiency in fired heaters has tended to focus on heat lost with the stack gases. When stack temperatures exceed 149°C, that attention is proper, but other losses can be much bigger when viewed from a lost-work perspective. For example, a reformer lost-work analysis gave the breakdown shown in Table 3.

The losses for ΔT in the convection section are almost twice those for the very hot exit flue gas. Furnace optimization is the clearest illustration of the benefits of lost-work analysis. If losses from a stack are nearly transparent, the losses imbedded in an excessive ΔT in a convection section are even harder to identify. They do not show up even on the energy balance that highlights the hot stack. These losses can be cut by adding surface to the convection section and shifting load from the radiant section, as well as by looking at the overall process (including steam gen-

Table 3. Lost-Work Analysis for a Fired Heater

	Lost-Work Potential, %
Combustion step	54
Radiant section ΔT	7
Convection section ΔT	24
Stack losses (Exit temp 225°C)	13
Wall losses	2

eration) for streams to match the cooling curve of the flue gases.

Concern over corrosion from sulfuric acid when burning sulfur-bearing fuels often governs the temperature of the exit stack gas. However, the economics of heat recovery are so strong that flue gases are being designed into the condensing range of weak sulfuric acid. It is not a forbidden zone, but the designer should recognize that tube replacement is necessary.

Simple heat losses through the furnace walls are also significant. This follows from the high temperatures and large size of fired heaters, but these losses are not inevitable. In an optimized system, losses through insulation are roughly proportional to

$$\left(\frac{\text{refractory price}}{\text{energy price}}\right)^{1/2}$$

This means that if the price ratio has decreased by a factor of 9, then losses should be down by a factor of 3. If the optimum allowed a 3% loss in 1973, today's optimum would be closer to 1%.

Dryers

A drying (qv) operation needs to be viewed as both a separation and a heat-exchange step. When it is seen as a separation, the obvious perspective is to cut down the required work. This is accomplished by mechanically squeezing out the water. The objective is to cut the moisture in the feed to the thermal operation to less than 10%. In terms of hardware, this requires centrifuges and filters and may involve mechanical expression or a compressed air or superheated steam blow. In terms of process, it means big crystals.

When the dryer is seen as a heat exchanger, the obvious perspective is to cut down on the enthalpy of the air purged with the evaporated water. Minimum enthalpy is achieved by using the minimum amount of air and cooling as low as possible. A simple heat balance shows that for a given heat input, minimum air means a high inlet temperature. However, this often presents problems with heat-sensitive material and sometimes with materials of construction, heat source, or other process needs. All can be countered somewhat by exhaust-air recirculation.

Minimum exhaust-air enthalpy also means minimum temperature. If this cannot be attained by heat exchange within the dryer, preheating the inlet air is an option. The temperature differential guidelines of the feed–effluent interchange apply.

Like the fired heater, the dryer is physically large, and proper insulation of the dryer and its allied ductwork is critical. It is not uncommon to find 10% of the energy input lost through the walls in old systems.

OPTIMUM DESIGN OF PUMPING, COMPRESSION, AND VACUUM SYSTEMS

Pumping

Is piping pressure drop optimal? Many companies have optimum-pipe-sizing programs, but in the absence of one, a good rule of thumb is that in an optimized system the annualized cost for pumping power should be one seventh the annualized cost of piping (1).

Is exchanger pressure drop optimal? Similarly, for an optimized heat exchanger the annualized cost for pumping should be one third the annualized cost of the surface for the thermal resistance connected with that stream.

Is the pump specified for the right flow? As Figure 11 shows, a 50% overdesign factor will increase power by 35% in a combination of higher head and lower efficiency.

Can the allowance for control be reduced? One option is the use of a variable-speed drive. This eliminates the control valve and its pressure drop and piping. Its best application is where a large share of the head is required for friction and where process demands cause the required flow to vary.

What can be done to get a more efficient pump? Sometimes a higher available net positive suction head (NPSH) permits a more efficient machine.

Compression

The work of compession is typically compared against the isentropic-adiabatic case.

$$\eta_{\text{comp}} = \frac{W_{\min}}{E_{\text{out}} - E_{\text{in}}}$$

For an ideal gas, this can be expressed in terms of temperatures

$$\eta_{\text{comp}} = \frac{W_{\min}}{W_{\text{actual}}} = \frac{T_{\text{in}}\left[\left(\frac{P_{\text{out}}}{P_{\text{in}}}\right)^{R/cp} - 1\right]}{T_{\text{out}} - T_{\text{in}}}$$

where R/cp is the ratio of gas constant to molar specific heat. Minimum work is directly proportional to suction

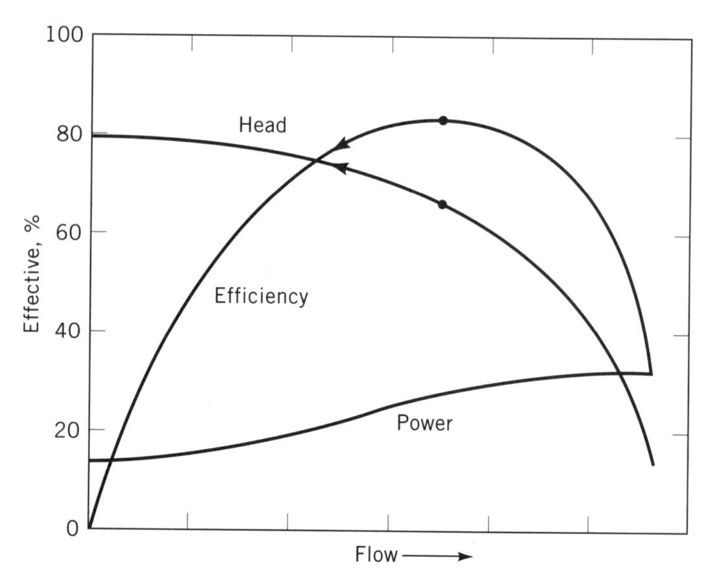

Figure 11. Impact of excess design capacity on pump energy use.

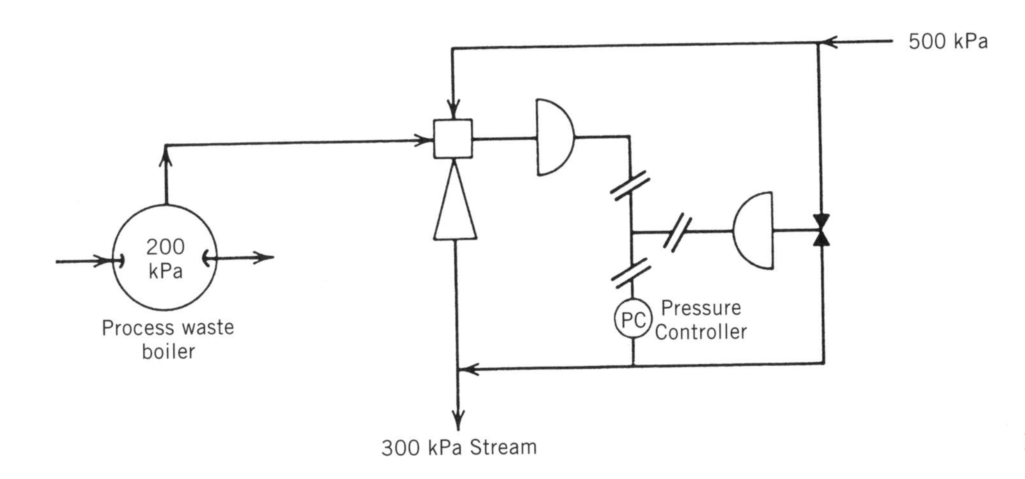

Figure 12. A thermocompressor.

temperature. This means that cooling-water systems should be run as cold as possible. Simply measuring temperature rise permits monitoring efficiency for a fixed pressure ratio and suction temperature.

Sometimes, W_{min} for compression is expressed for the isothermal case, in which it is always lower than for the adiabatic. The difference defines the maximum benefit from interstage cooling.

Efficiencies should always exceed 0.6, and 1.00 is approachable in reciprocating devices. Their better efficiency needs to be balanced against their greater cost, greater maintenance, and lower capacity.

The guidelines on pressure drop in piping and exchangers discussed above also apply here. The opportunity for variable-speed drive is possibly even greater, as is the importance of tight control of minimum flows.

Thermocompressors

A thermocompressor is a single-stage jet using a high pressure gas stream to supply the work of compression. The commonest application is in boosting waste-heat-generated steam to a useful level. An example is shown in Figure 12. Thermocompressors can also be used to boost a waste combustible gas into a fuel system by use of high pressure natural gas. The mixing of the high energy motive stream with the low energy suction stream inherently involves lost work, but as long as the pressures are fairly close, the net efficiency for the device can be respectable as the pressures are fairly close, the net efficiency for the device can be respectable (25–30%). Here, efficiency is defined as the ratio of isentropic work done on the suction gas to the isentropic work of expansion that could have been obtained from the motive gas. The thermocompressor has the big advantage of no moving parts and low capital cost.

Vacuum Systems

The most common vacuum system uses the vacuum jet. Because of the higher ratio of motive pressure to suction pressure, the efficiency of vacuum systems is lower than thermocompressors. Generally, it is 10–20%. The optimum system often employs several stages with intercondensers. Steam use in this range varies roughly as $(1/P)_{0.3}$, where P is absolute suction pressure (see Vacuum technology).

Because of the low efficiency of steam–ejector vacuum systems, there is a range of vacuum above 13 kPa (100 mm Hg) where mechanical vacuum pumps are usually more economic (13). The capital cost of the vacuum pump goes up roughly as (suction volume)$_{0.6}$ or $(1/P)_{0.6}$. This means that as pressure falls, the capital cost of the vacuum pump rises more swiftly than the energy cost of the steam ejector which increases as $(1/P)_{0.3}$. Usually below 1.3 kPa (10 mm Hg), the steam ejector is more cost-effective.

Other factors that favor the choice of the steam ejector are the presence of materials that could form solids and high alloy requirement. Factors that favor the vacuum pump are credits for pollution abatement and high cost steam. The mechanical systems require more maintenance and some form of backup vacuum system, but they can be designed with adequate reliability.

REFRIGERATION

Refrigeration (qv) is a very high value utility. The value of heat in a hot stream is the work it can surrender:

$$\frac{W}{Q} = \left(\frac{T - T_{\text{sink}}}{T}\right) \eta_{\text{turbine}}$$

And the value of refrigeration is the work required to heat pump it to the sink temperature:

$$\frac{W}{Q} = \left(\frac{T_{\text{sink}} - T}{T}\right) \frac{1}{\eta_{\text{compressor}}} \frac{1}{\eta_{\text{fluid}}}$$

The value of refrigeration is compared to heating in Figure 13 for $\eta_{\text{turbine}} = \eta_{\text{compressor}} = 0.7$ and for $\eta_{\text{fluid}} = 0.8$. Here, η_{fluid} accounts for cycle inefficiencies such as the letdown valve shown by Figure 14.

Because of its value, refrigeration justifies thicker insulation, lower ΔTs in heat exchange, and generally much more care in engineering (14). Some questions that the designer should ask are

Is refrigeration really necessary? Could river water or

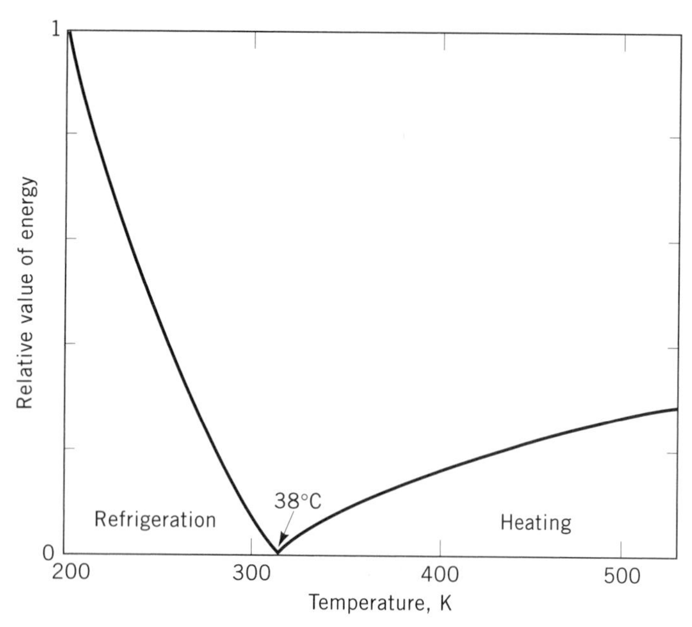

Figure 13. Relative value of energy at various temperatures.

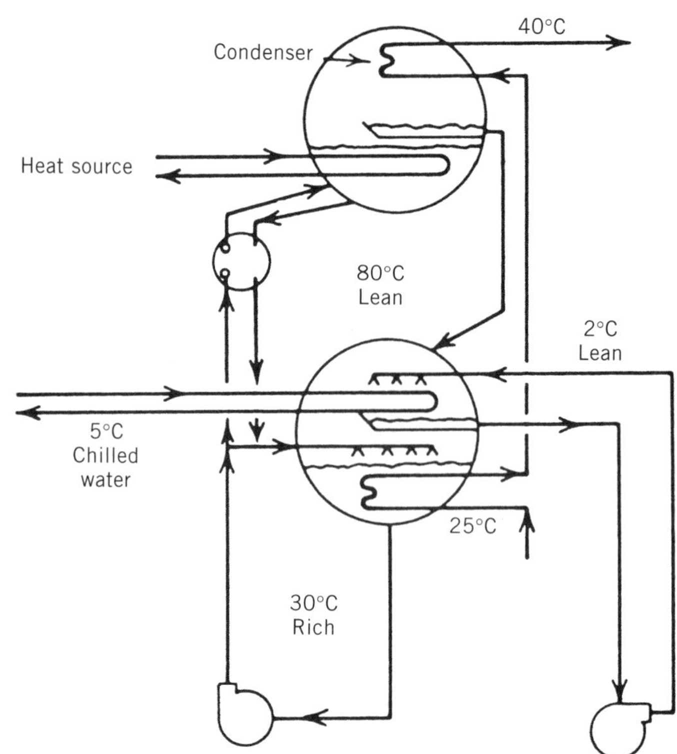

Figure 15. Absorption refrigeration.

cooling-tower water be used directly? Could they be used for part of the year? Could they replace part of the refrigeration?

Can the refrigerant-condensing temperature be reduced? Could it be reduced during part of the year?

Can the system be designed to operate without the compressor during the cold weather?

Is the heat-transfer surface the economic optimum?

Is a central system more efficient than scattered independent systems?

Does the control system cut required power for part-load operation? (Multiple units could yield increased reliability as well as efficiency.)

Are enough gauges and meters provided to monitor operation?

Is there an abundance of waste heat available from the plant (above 90°C)? If so, refrigeration could be supplied by an absorption system.

Absorption chiller units (Fig. 15) need 1.6–1.8 J of waste heat per joule of refrigeration. Commercially available LiBr absorption units are suitable for refrigeration down to 4.5°C.

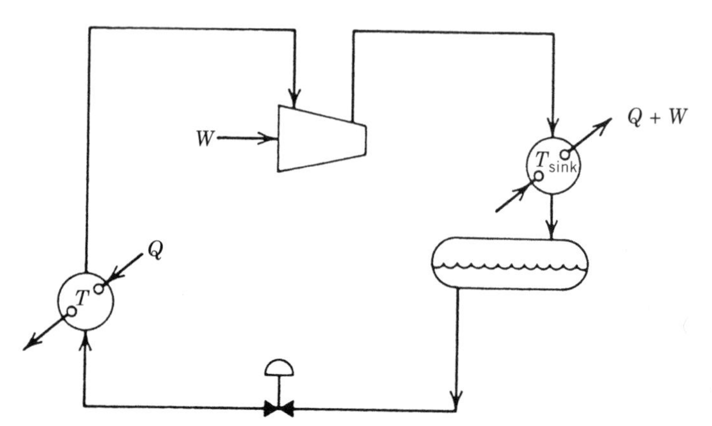

Figure 14. Compression refrigeration.

For low level waste heat (90–120°C), absorption chillers utilize waste heat as efficiently as steam turbines powering mechanical refrigeration units. Absorption refrigeration using 120°C saturated steam delivers 4.5°C refrigeration with an efficiency of 35% where efficiency is referred to the work potential in the steam.

STEAM AND CONDENSATE SYSTEMS

In the process industry, steam serves much the same role as money does in an economy, ie, it is the medium of exchange. If its pricing fails to follow common sense or thermodynamics, strange design practices are reinforced. For example, many process plants employ accounting systems where all steam carries the same price regardless of temperature or pressure. This may be appropriate in a polymer or textile unit where there is no special use for the high temperature; but it is clearly wrong in a petrochemical plant.

Some results of the constant-value pricing system are typically the following: generation in a central unit at relatively low pressure, <4.24 MPa (600 psig); tremendous economic pressure to use turbines rather than motors for drives; lack of incentive for high efficiency turbines; excessively high temperature differentials in steam users; tremendous incentive to recover waste heat as low pressure steam; and a large plume of excess low pressure steam vented to the atmosphere.

A number of alternative pricing systems have been proposed that hinge on turbine efficiency and the relative pricing of fuel and electricity (15–16). Perhaps the best system relates the value of steam to that at the generation pressure by its work potential (exergy content).

$$\frac{\text{value at pressure}}{\text{value at highest generation pressure}} = \frac{\text{exergy at pressure}}{\text{exergy at highest generation pressure}}$$

Design of a central power–steam system is beyond the scope of this discussion, but the interaction between the steam system and the process must be considered at all stages of design. There is a long list of factors to consider in designing a steam-using system:

Can steam-use pressure be lowered? (If ΔT in the heater is above 20°C, the steam pressure is probably above the economic optimum.)

Are there any turbines under 65% efficiency? (Today, turbines are being limited to large sizes above 500 kW, where good efficiencies can be obtained; they are used only for those small drives essential to the safe shutdown of the unit.)

Can a gas turbine be utilized for power generation upstream of the boiler?

Are there any waste streams with unutilized fuel value?

Is there a program to monitor turbine efficiency by checking temperatures in and out?

Is condensate recovered?

Is the flash steam from condensate recovered?

Is feedwater heating optimized?

Is there any pressure letdown without power recovery?

Has enough flexibility been built into the overall condensing–turbine system? (The balance changes over the history of a unit as a process evolves, generally in the direction of less condensing demand.)

Is steam superheat maintained at the maximum level permitted by mechanical design?

Can a thermocompressor be used to increase steam pressures from waste heat?

Are all users metered?

Is low level process heat used to preheat deaerator makeup?

Are ambient sensing valves used to turn off steam tracer systems?

Are steam traps appropriate to the service?

COOLING-WATER SYSTEMS

Cooling water is a surprisingly costly utility. On the basis of price per unit energy removed, it can cost one fifth as much as the primary fuel. Roughly half of this cost is in delivery (pump, piping, and power). This fact has several important implications for design.

Heat exchangers should be designed to use the available pressure drop. A heat exchanger that is designed for 10 kPa when 250 kPa is available will have five times the design flow.

If an exchanger cannot be economically designed to use available ΔP, orifices should be provided to balance the system. This can be done without compromising the guidelines that no unit should be designed for less than 0.8 m/s on tubeside or less than 0.3 m/s on shellside.

If temperature requirements permit, the system will cost less to operate with exchangers in series.

An installed measuring element is usually justified.

If only part of the system requires a high head, this should be supplied by a booster pump. The whole system should not be designed to use or waste the high head.

Other energy considerations for cooling towers include use of two-speed or variable-speed drives on cooling-tower fans, and proper cooling-water chemistry to prevent fouling and excess ΔT in users.

Air coolers can be a cost-effective alternative to cooling towers at 50–90°C (just below the level where heat recovery is economic).

SPECIAL TECHNIQUES

Heat Pumps and Temperature Boosting

A heat pump is a refrigeration system that raises heat to a useful level. The most common application is the vapor recompression system for evaporation (Fig. 16). Its application hinges primarily on low cost power relative to the alternative heating media. If electricity price per unit energy is less than 1.5 times the cost of the heating medium, it merits a close look. This tends to occur when electricity is generated from a cheaper fuel (coal) or when hydroelectric power is available.

Use in distillation systems is rare. The reason is a recognition that almost the same benefits can be achieved by integrating the reboiling–condensing via either steam system (above ambient) or refrigeration system (below ambient).

In an optimized system,

$$\frac{Q}{W} = \frac{T_{hot}\eta_{compressor}}{T_{hot} - T_{cold}}$$

where T_{hot} and T_{cold} are in absolute units, K.

This provides another criterion for testing whether a heat-pump system may be cost-effective. A power plant takes three units of Q to yield one unit of W; therefore, to provide any incentive for less overall energy use, Q/W must be far in excess of 3.

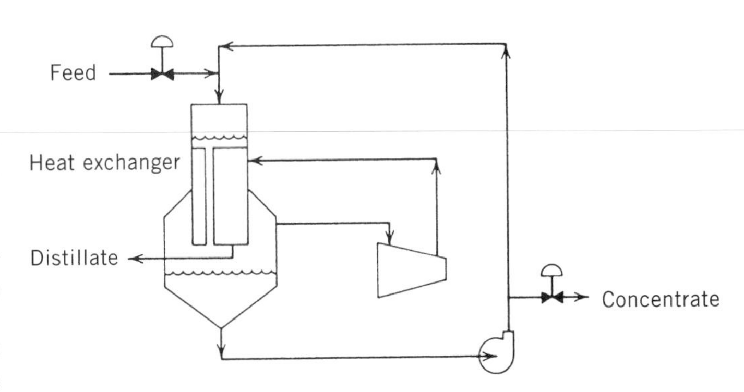

Figure 16. Vapor-recompression system.

Energy-Management Systems

The considerable reduction in computing costs has made it possible to do a wide range of routine monitoring and controlling. One can, for example, continuously monitor a distillation system and compare energy use against an optimum and display the cost-per-hour deviation from optimum setpoint. The computer can also test specific actions to achieve the optimum. A computer can monitor a steam system, advise how best to load a set of boilers and choose which turbines to run.

THE EXISTING PLANT

How do existing plants differ from new plants? Good ideas for new plants are also good ideas for existing plants, but there are three basic differences: *1.* Because a plant already exists, the capital vs operating cost curve differs; usually, this makes it more difficult to reduce utility costs to as low a level as in a new plant. *2.* The real economic justification for changes is more likely to be obscured by the plant accounting system and other nontechnical inputs. *3.* The real process needs are measurable and better defined.

An example in support of the first of these is the case of optimum insulation thickness. Suppose a tank was optimally insulated ten years ago. If the value of heat quadrupled in the interim, this change would justify twice the old insulation thickness on a new tank. However, the old tank may have to function with its old insulation. The reason is that there are large costs associated with preparations to insulate. This means that the cost of an added increment of insulation is much greater than assumed in the optimum insulation thickness formulas (Fig. 17). An example of the second difference is that many things appear to be strongly justified by savings in low pressure steam if the steam is valued artifically high. A designer of a new plant has the advantage of focusing attention on savings in the primary budget items, ie, fuel and electricity at the plant gate, rather than on cost-sheet items such as steam at battery limits. The third difference is that many process details are relatively uncertain when a plant is designed. For example, inert loading for vacuum jets is rarely known to within 50%. Although the first two differences are negative, the third provides a unique opportunity to measure the true need and revise the system accordingly.

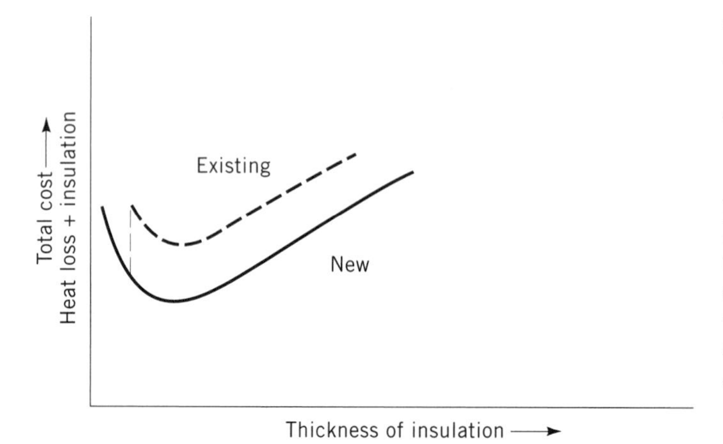

Figure 17. Tank-insulation costs, the existing vs the new.

Energy Survey. The energy survey has seven components: *"as-it-is" balance; field survey; equipment tests; check against optimum design; idea-generation meeting; evaluation* and *followup.*

As-It-Is Balance. This is a mandatory first step for the energy survey. It permits targeting of principal potentials; check of use against design; check of use against optimum (how a new plant would be designed today); definition of possible hot or cold interchanges; definition of the unexpected uses (eg, the large steam purge to process or high pressure drop exchanger); and contribution from specialists not familiar with the unit.

Field Survey. This is often done by a team of two: one who knows what to look for and one who knows the process. On field surveys, it is as important to talk to the operators ("What runs when the unit is down?"; "What happens when you cut reflux?"; "Where are the guidelines for steam–feed ratio? How close do you usually run?") as it is to record items like air leaks, high pressure drops across values, frost on piping, lights of the wrong type or at the wrong time, steam plumes (a reason to climb to the top of the unit), or minimum-flow bypasses in use. The field survey should develop detailed "fix" lists for leaking traps, uninsulated metal, lighting, and steam leaks.

Equipment Tests. Procedures for rigorous, very detailed, efficiency determination are available (ASME Test Codes) but are rarely used. For the objective of defining conservation potentials, relatively simple measurements are adequate. For fired heaters, stack temperature and excess O_2 in stack should be measured; for turbines, pressures (in and out) and temperatures (in and out) are needed.

Check Against Optimum Design. This attempts to answer the question, "Need a balance be as it is?"

In the large view, the first thing to compare against is the literature claims for chemicals such as NH_3, HNO_3, CH_3OH, and ethylene.

The second thing to do is look for the obvious: stack temperatures >149°C; process streams >121°C, cooled by air or water; process streams >65°C, heated by steam; $\eta_{turbine}$ <65%; reflux ratio >1.15 times minimum; and excess air >10% on clean fuels.

Idea-Generation Meeting. The idea-generation meeting is most productive if three important guidelines are followed: *1.* Get the right people. A good guiding principle is to get "two wise old Turks" for each corporate expert. It is extremely important to ask, "Given free choice, whom would you choose to attend?," and then get them. *2.* Discuss the as-it-is balance for each area and record all ideas. *3.* Assign follow-up responsibility.

Evaluation. The evaluation of each idea should include a technical description and its economic impact and technical risk. The ideas should be ranked for implementation. A report should provide a five-year framework for energy projects.

Follow-up. If no savings result, the effort has been wasted. The survey leader needs to be sure that the potential of the good ideas is recognized by management and the project-generation channels of his company.

What Do You Not Do?

Often, what looks like negligence can be a tried and proven practice, and one should be cautious about experts bearing lists and offering huge savings. The process has to work, and present utility saving may or may not compensate for future repair bills or lost products. Some examples are the following: an idling turbine may be necessary to permit a safe plant shutdown if a power failure occurs; a cooling-water flow that is throttled to below 0.6 m/s (in winter) will likely assure a heat-exchanger cleaning in late spring; a furnace that runs too low on excess air may run into after-burning; and a column run too close to the minimum reflux ratio without adequate controls runs a risk of off-specification product.

One also does not trust the plant accounting system unquestioningly. All energy is not created equal. The energy that is recovered from flashed steam or that is shaved off a reboiler's duty may not be worth its cost-sheet value. The meters that matter are the primary meters at the plant gate. Only if the recovered energy reduces the meter settings does it save the plant money.

One does not accept the first solution to an energy-waste problem without seeing how this problem fits into the overall plant-energy balance. The sudden rise in energy prices has lifted many options into the justifiable range. There might be better alternatives available.

Acknowledgment
Revised from the *Kirk-Othmer Encyclopedia of Chemical Technology,* 3rd ed., Suppl. Vol. John Wiley & Sons, Inc., New York, 1984.

BIBLIOGRAPHY

1. D. E. Steinmeyer, *Chemtech,* 188 (Mar. 1982).
2. W. F. Kenney, *Proceedings 1981 Industrial Energy Conservation Technology Conference,* Texas Industrial Commission, p. 247.
3. K. G. Denbigh, *Chem. Eng. Sci.* **6,** 11 (1956).
4. C. J. King, *Separation Processes,* 2nd ed., McGraw-Hill, Inc., New York, 1980.
5. C. S. Robinson and E. R. Gilliland, *Elements of Fractional Distillation,* 4th ed., McGraw-Hill, Inc., New York, 1950.
6. M. Souders, *Chem. Eng. Prog.* **60**(2), 75 (1964).
7. D. Carter, personal communication, Monsanto Corp., St. Louis, Mo., 1982.
8. B. Linnhoff, *Chem. Eng. Prog.,* **90**(8), 33 (1994).
9. D. Steinmeyer, *Hydrocarbon Process.,* **53** (April 1992).
10. W. F. Furgerson, *Conserving Energy in Refrigeration,* Manual 12 of *Industrial Energy–Conservation,* MIT Press, Cam-

ENERGY CONSUMPTION IN THE UNITED STATES

Paul Komor
Office of Technology Assessment, U.S. Congress
Washington, D.C.

Andrew Moyad
U.S. Environmental Protection Agency
Washington, D.C.

Energy is a crucial national and international concern for several reasons: it impacts the U.S. economy, it contributes to many environmental problems, and it is imported and therefore tied to issues of national security and international political stability. Further, its unequal global consumption raises questions of international resource equity and sustainability. The following facts highlight these points:

- The U.S. economy cannot function without energy, which is a principal component of Gross Domestic Product (GDP). In 1989, U.S. businesses, consumers, and government spent a total of $437 billon, about 8% of GDP, on energy (1).
- Energy consumption is closely tied to environmental concerns, notably global climate change and urban air quality. For example, virtually all U.S. carbon dioxide emissions stem from fossil fuel consumption; similarly, vehicles are responsible for much of the carbon monoxide and other harmful emissions found in urban areas (2).
- More than 40% of U.S. oil consumption is imported. This dependence is likely to climb in the future, contributing to the trade deficit and increasing U.S. vulnerability to economic shocks stemming from major shifts in international oil prices and availability.
- In 1992, total U.S. energy consumption was 82.4 quads of energy (3), more than any year in the past, and far more than any other country. In fact, the United States accounts for almost one-fourth of global energy use (4), despite having less than 5% of the total world population (1).

Given this U.S. dominance in world energy use, as well as the important role energy plays in economic, environmental, and international political issues, an understanding of U.S. energy consumption (how energy is used, how this use has changed over time, what factors influence this use, and how energy use may change in the future) is essential to understanding energy's potential impact on these various domestic and international issues. This discussion provides an overview of energy consumption from several perspectives: aggregate national energy use and trends, consumption by individual end-use sectors, and consumption by fuel.

See also Natural gas; Petroleum markets; Coal; Commercial availability of energy technology; Energy efficiency; Transportation fuels.

ENERGY USE: RECENT HISTORY

U.S. energy consumption increased steadily from 1950 to 1973. Since 1973, however, national energy consumption has moved erratically and has at times even decreased (see Fig. 1). A number of factors have contributed to these changes in U.S. energy use, including:

- Energy prices
- Population growth and migration
- Economic growth and structural economic shifts
- Technical advancements
- Lifestyle changes
- Major political events, notably the 1973 and 1979 oil crises
- Policy changes, at both the state and federal levels

The first four of these factors (price, population, economic growth and change, and technical advances) probably had the largest impact on national energy use, particularly since 1973. Energy prices in real terms, that is, adjusting for the effects of inflation, were generally stable until 1973 but shot up dramatically that year. In 1974 alone, energy prices increased 56%. As Figure 1 illustrates, real energy prices peaked in 1981 at more than 3.5 times the 1973 levels. Since 1981, though, real energy prices have generally decreased, dropping below the 1974 levels by 1991. Note that the prices discussed here are "composite," that is, a weighted average of the prices of the different types of energy. Trends in the prices of specific fuels are discussed below.

When confronted with rapidly rising prices, energy users generally act to reduce their energy costs (although the exact relationship between energy prices and energy consumption is not well-understood) by substituting fuels, choosing cheaper over more expensive fuels where possible; reducing consumption, through short-term behavioral changes (eg, lowering thermostats) and long-term structural changes (eg, shifting production away from energy-intensive goods); and increasing energy efficiency through behavioral and technical changes, such as adding insulation to houses, installing process controls in industry, and switching to energy-efficient fuel injection instead of carburetors in automobiles.

Prices clearly influence energy use, but other factors are also important. Energy provides energy services: moving goods and people, running industrial motors, space heating for buildings, and so on. Clearly, the more people requiring these services, the greater the consumption of energy. Population has increased steadily in the United States, from 151 million in 1950 to 252 million in 1992 (5); however, trends in energy use per capita have matched trends in total energy use, indicating the importance of other factors in influencing consumption (see Fig. 1).

Recently, the basic outputs of the U.S. economy have undergone a slow but steady shift, with attendant impacts on national energy use. The service sector, such as hotels, restaurants, and legal and medical services, account for an increasing fraction of total economic output. Industry, notably manufacturing, has shown a corresponding decrease. In 1970, goods accounted for 46% of GDP, whereas services accounted for 43%. (The remaining 11% was for structures.) By 1991, these numbers had shifted to 39% and 53%, respectively (1). These structural economic shifts have had major impacts on U.S. energy use, which grew more slowly than did the economy over the last two decades.

From 1950 to 1973, the growth in national energy use mirrored that of the GDP, suggesting that increased energy use was required for economic growth. Since 1973, however, these two variables have diverged. GDP continued to climb, whereas energy use remained relatively unchanged (see Fig. 2). One important reason for this decoupling of energy use and economic growth was structural change. Simply put, less energy is required to produce, for example, $1,000 of computers than $1,000 of steel. By one estimate, about one-third of the divergence between energy use and GDP growth is attributed to structural economic shifts that have occurred over the last two decades (6).

The other two-thirds of this divergence is attributed to technical improvements in energy efficiency. In all sectors of the economy, numerous technologies have been introduced that wring more useful energy service (eg, heat, light, and motor drive) from each unit of energy consumed. Refrigerators, electric motors, automobiles–all have shown considerable gains in energy efficiency over the past 20 years, and considerable potential for further improvements remain.

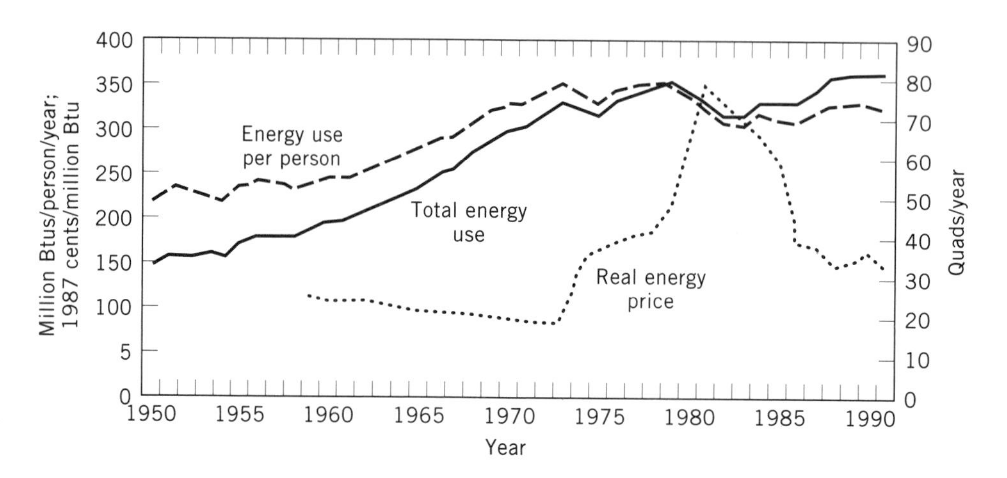

Figure 1. U.S. energy consumption, total and per person; and fossil fuel real energy price (1,3,5).

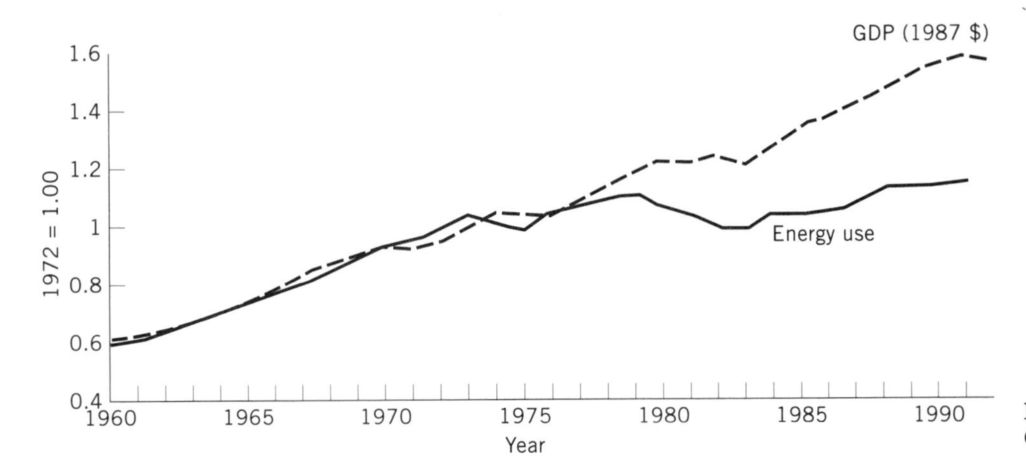

Figure 2. U.S. energy use and Gross Domestic Product (1,3,5).

Although debates continue over what caused these efficiency improvements, some argue that the energy price increases of the 1970s led to a search for ways to decrease energy use without reducing service. Others maintain that technical progress, including energy-efficiency improvements, has and will continue regardless of energy price changes. In any case, it is generally agreed that technical-efficiency improvements will continue, and that, all else being equal, higher prices lead to greater energy efficiency.

ENERGY USE BY SECTOR

Energy is used in three distinct end-use sectors: residential and commercial buildings, transportation, and industry.

Buildings

Energy use in buildings accounts for an increasing share of total U.S. energy consumption: from 27% in 1950, to 33% in 1970, to 36% in 1990. (See Reference 7 for a more complete discussion.) At present, more than 60% of national electricity consumption and almost 40% of national natural gas consumption occurs in the two basic building types: residential and commercial.

The Residential Sector. In 1989, residential buildings accounted for nearly 17 quads of energy (most as electricity, with significant contributions from natural gas and oil) at a total cost of $104 billion. As Figure 3 illustrates, space heating is responsible for almost half of this energy use, followed by water heating, refrigerators and freezers, space cooling, and lights. In the past 20 years, residential energy use increased at a modest average annual rate of about 1.2%.

Between 1970 and 1990, the combination of a growing population and a shrinking average household size (ie, fewer people per household) led to an almost 50% increase in the number of households. As each household requires space conditioning, hot water, and other energy services, these changes drove the growth in energy use in the residential sector. Increased demand for particular energy-intensive services also contributed to the growth in residential energy use, for example, central air conditioning is now routinely installed in over three-fourths of new single-family homes, and color televisions are found in almost all households.

Although total residential energy use increased from 1970 to 1989, annual energy consumption per household actually decreased by 15% in the same period. Whereas several factors contributed to this intensity drop, improved technology and better building practices were key: retrofits in older houses, greater use of energy-efficient building practices in new homes, and a dramatic improvement in

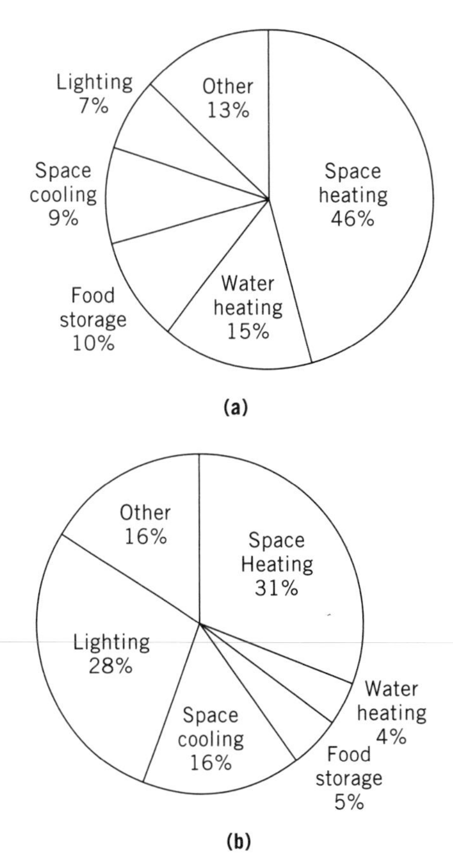

Figure 3. (**a**) Residential buildings and (**b**) commercial buildings; energy use (7).

the energy efficiency of household appliances and equipment.

Considerable efforts to improve the energy efficiency of U.S. buildings have been made. For example, by one estimate, about 26 million owner-occupied U.S. households added storm windows and/or storm doors, and 17 million added insulation, between 1983 and 1988. Careful evaluations of retrofits indicate that energy savings are often substantial. New houses have benefited from greater use of energy-efficient techniques; houses built in 1985, for example, were better insulated and included more energy-efficient windows than those built in 1973. Residential energy-using equipment is now more energy-efficient as well. The typical new gas furnace sold in 1975 had an efficiency of 63%, and by 1988, this increased to 75%. The efficiency gains in appliances were even greater; the typical new refrigerator sold in 1990 uses less than half as much electricity as a comparable unit sold in 1972.

The fuel mix of residential energy use also has changed. Whereas oil use for space heating has dropped sharply, electricity has become increasingly prevalent for both space and water heating. In 1970, electricity supplied 41% of residential energy; by 1988, this had climbed to 61%. Yet this 20-year trend toward greater residential electrification may be changing, as electric space heating in new single-family homes has decreased dramatically in recent years (from 49% in 1985 to 33% in 1990).

The Commercial Sector. In 1989, commercial buildings accounted for about 13 quads of energy use, at a cost of $68 billion. Electricity represented about two-thirds of this energy. Space heating, lighting, and space cooling were the principal end uses, as shown in Figure 4. Energy use in commercial buildings has increased rapidly since 1970, at an average annual rate of 2.3%, or about twice the rate of energy increase realized in residential buildings in the same period. A number of factors contributed to this growth, the most significant being the rapid growth in new commercial buildings. As measured by total square footage, the commercial building stock has increased more than 50% since 1970. Heating, cooling, and lighting these new buildings has considerably increased energy consumption. A greater demand for energy-intensive services, notably air conditioning and electronic office equipment, has further increased commercial building energy use. For example, in 1984, U.S. businesses used less than 2 million personal computers; their use in the commercial sector grew to 14 million, a sevenfold increase, only five years later.

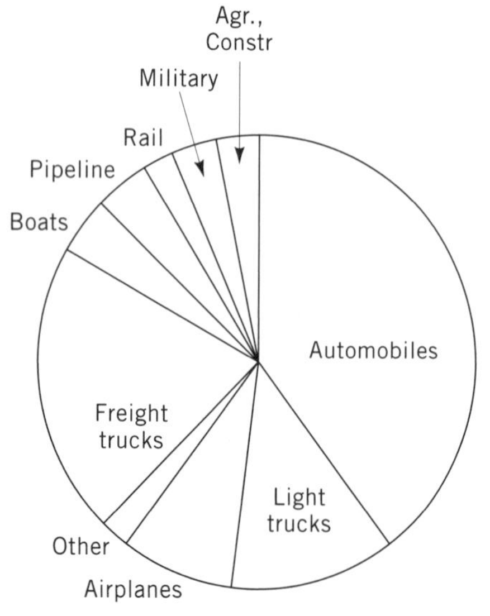

Figure 4. Commercial buildings energy use (7).

Despite increases in the use of air conditioning and electronic office equipment, annual energy use per commercial square foot actually stayed flat between 1970 and 1990. As in the residential sector, improved technology helped to dampen the growth in commercial building energy use. New commercial buildings use improved windows and shells, more efficient space-conditioning equipment, and better lighting systems. For example, commercial buildings constructed in the 1980s contained more ceiling and wall insulation, multipane and reflective windows, and shadings or awnings compared to buildings constructed in earlier years. Computer advances have permitted greater use of computer-aided building design and analysis methods, and retrofits have improved the energy efficiency of previously constructed commercial buildings.

Buildings—Summary and Outlook. Energy use in both residential and commercial buildings has grown in the past 20 years. Sheer increases in numbers underlie much of this growth: more people, more households, and more offices. Increased service demand (more air conditioning, more computers, larger houses) has contributed as well. However, the application of improved technology in the areas of building shells (windows and insulation), appliances (furnaces, air conditioners, refrigerators), and building design have helped to moderate the growth in building energy use.

Several factors will influence future energy use in buildings. First, studies indicate that greater use of commercially available technologies could reduce building energy use up to one-third. Second, the service sector offices, restaurants, and other energy-intensive buildings grew rapidly in the 1980s. If this growth resumes, then building energy use will also increase rapidly. Third, information technologies (eg, computers, printers, and copiers) are a small but rapidly growing energy user in commercial buildings. If their use continues to grow at current rates, these technologies will soon account for a significant share of commercial energy use. Fourth, electric and gas utility investments in energy efficiency, often called "demand-side management" (DSM), may substantially increase the use of energy-efficient technologies in buildings. Current utility DSM investments total about $2 billion per year. If these investments are as successful as planned, then significant energy savings will result.

Transportation

The movement of goods and people accounts for about one-fourth of U.S. energy use and almost two-thirds of U.S. oil.

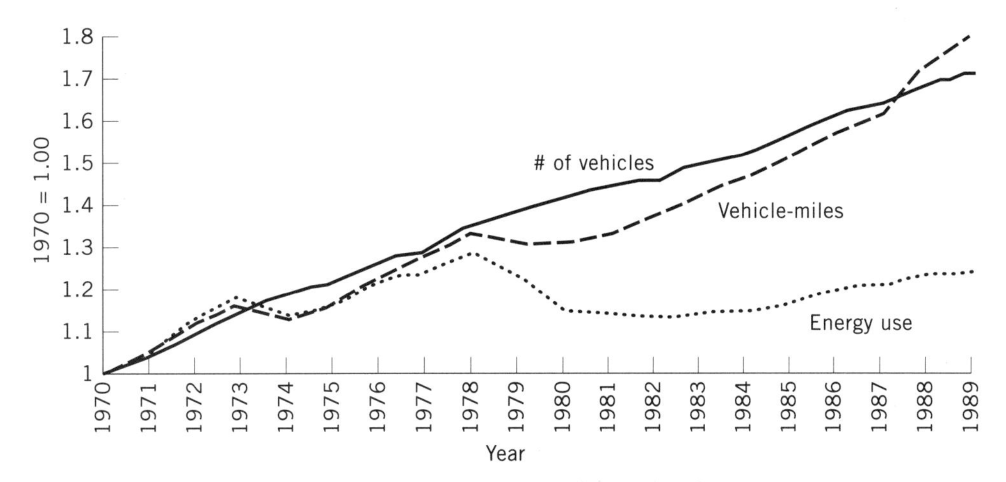

Figure 5. Automobile and passenger truck travel and energy use (8).

Although not the largest energy-consuming sector, transportation is often the most visible and controversial, due in part to its dependence on oil and to its impact on urban air quality. Transportation energy use is best understood by examining passenger and freight transport separately.

Passenger Transport Energy Use. The demand for passenger transport has grown rapidly in recent years, as the population has become more mobile and automobiles have become more widely available. Although vehicle energy efficiency has improved dramatically, this improvement has been outpaced by increased demand, resulting in significant increases in energy use. U.S. transport energy use grew faster between 1970 and 1990 than energy use in either the buildings or industrial sectors (8).

Private automobiles and light trucks used for personal transport account for over half of transport energy use use (see Fig. 5) and over 85% of passenger-miles (8). In the past 20 years, the number of automobile and passenger trucks, and the miles driven per year, have both increased, as illustrated in Figure 6. The combination of these two factors would have resulted in very rapid growth in energy use; however, energy use by these vehicles actually increased quite slowly in recent years, at an annual average rate of 0.3% from 1970 to 1989. This large increase in transportation services with only a small increase in energy use was made possible largely through efficiency improvements.

The efficiency of the private automobile fleet increased over 50% between 1975 and 1989, as measured by miles per gallon (8). The main impetus behind this improvement was corporate average fuel economy (CAFE) requirements. Federal legislation passed in 1975 set minimum fleet fuel economy standards, starting at 18 miles per gallon (mpg) in 1978 and increasing to 27.5 mpg by 1985 (see Fig. 7). This efficiency increase was achieved by several vehicle changes, including decreased weight, reduced engine size, and increased use of fuel-injection and other efficient technologies. Contrary to some popular perceptions, however, these efficiency improvements did not reduce vehicle performance. From 1977 to 1993, passenger car interior volume remained essentially constant and performance (as measured by 0–60 mph acceleration) improved (9).

Light trucks are playing an increasingly important role

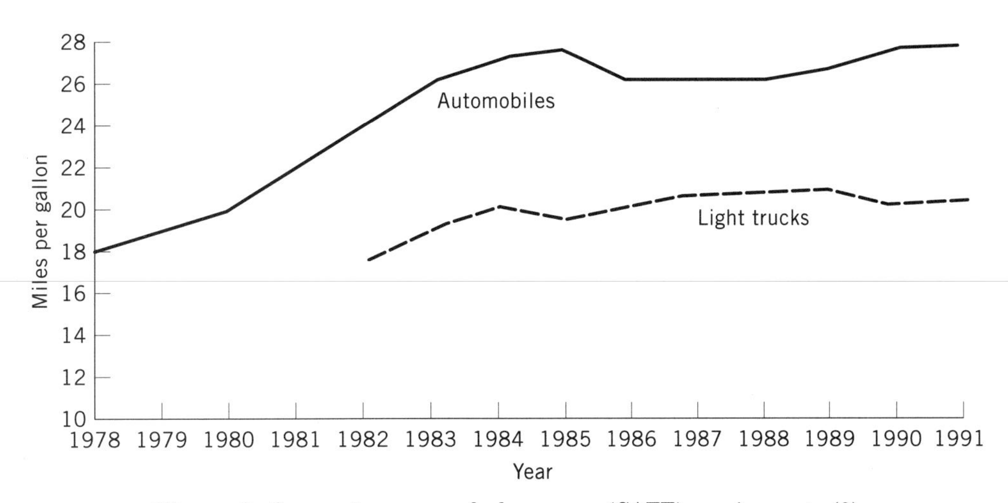

Figure 6. Corporate average fuel economy (CAFE) requirements (8).

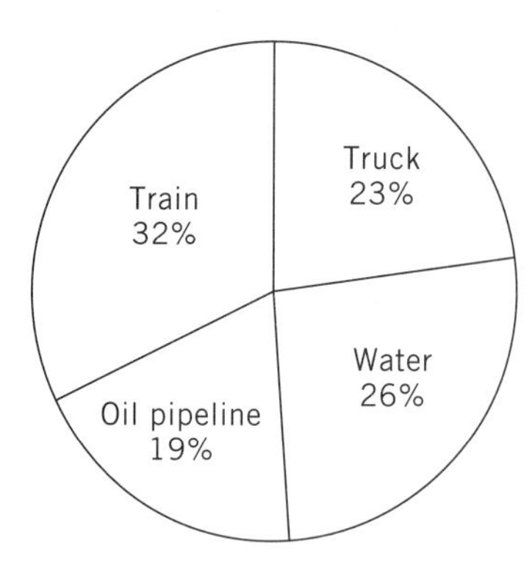

Figure 7. Freight ton-miles by mode, 1989 (8).

in passenger transport and energy use. These trucks are often used exclusively for passenger transport and now account for almost one-third of the new sales of light duty vehicles (ie, automobiles and light trucks) (8). Light trucks are generally far less energy-efficient than automobiles; the 1991 CAFE standard for automobiles was 27.5 mpg but only 20.2 mpg for light trucks (see Fig. 7). As a result, the growing population of light passenger trucks is dragging down the energy efficiency of the light duty fleet.

Airplanes account for 9% of transport energy use. Airplane travel, as measured by passenger-miles, has grown at an average annual rate of 6.6% from 1970 to 1989 (8). As with the other modes of travel, energy use increased more slowly than passenger-miles due to technical and operational improvements. More seats per aircraft, higher load factors (ie, the ratio of occupied to total seats), and improved engine efficiencies and aerodynamics all contributed to greater efficiency in passenger air transport.

Freight Transport Energy Use. Freight transport accounts for one-third of transport energy use (see Fig. 5). Trucks account for the bulk of this energy use, followed by barges, pipelines, and trains. However, as measured by ton-miles of travel, a common yardstick in freight, a different energy pattern emerges: trains and barges both exceed trucks (see Fig. 8).

In the past 20 years, a gradual but steady shift in the U.S. economy occurred, away from basic materials and toward greater consumption of services and higher value-added goods. Although production of raw materials (such as coal and minerals) has grown, production of manufactured goods has grown much faster. These economic shifts directly impacted the freight transport system, which has changed to accommodate the altering mix of industrial production. Over the past 20 years, movements by trains and barges, which typically carry basic commodities (coal, farm products, chemicals), grew slowly, whereas truck and air freight movements, which carry value-added goods, grew more rapidly. Truck and air movements generally require more energy per ton-mile than do trains and barges; therefore, these economic shifts have resulted in relatively rapid growth in freight transport energy use.

The volume of freight moved by trucks increased rapidly in the past 20 years. The energy intensity (Btu/ton-mile) of trucks, however, stayed roughly constant since 1970. (This discussion excludes light trucks used primarily for personal transport.) During this time, there were several technical improvements, notably electronic engine controls, demand-actuated cooling fans, aerodynamic improvements, and multiple trailers. The market penetration of these technologies varies, but some (eg, cab air deflectors) are now found in almost all trucks today. Despite these numerous technical improvements, however, truck energy intensity did not actually improve due to: (1) low turnover of the truck fleet, leading to slow adoption of new technologies, many of which cannot easily be retrofit to existing trucks; (2) increased highway speeds, which are less energy-efficient due to greater wind resistance; (3) changes in freight movements, eg, trucks may now be carrying less dense goods, causing trucks to fill up ("cube out") before they reach weight limits; and (4) other factors, including low load factors, poor driver training, and increased urban congestion.

For the train system, energy intensity has improved considerably. Between 1970 and 1990, train energy use ac-

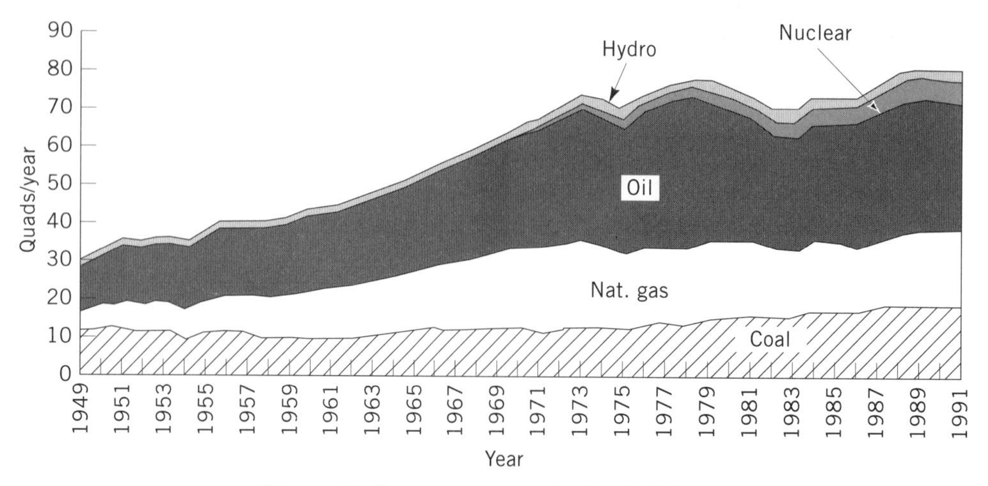

Figure 8. Energy consumption by fuel (3,5).

tually decreased, despite an increase of over one-third in ton-miles carried (10). Several factors contributed to these gains: increases in average trip lengths, operations and communications improvements, and technical improvements such as reduced idling speeds for locomotives and greater use of flange lubricators. Water-based freight transport showed a moderate improvement in energy intensity. Both technical and operational factors contributed to this improvement, including improved diesel engines, fuel management computer systems, and the use of larger barges and tugs.

The growth of intermodalism has strongly shaped the freight transport system. Intermodalism usually refers to the carriage of trailers and containers by trains, with delivery to and from the train terminals by truck, but the term can also refer to the use of barges or open ocean ships to move containers, which are then moved by train or truck. Several innovative technologies have been implemented, including sealed containers that can be moved by train, barge, or truck; roadrailers (truck trailers that can ride directly on train tracks); piggybacking (putting truck trailers onto railcars); and double-stack containers (putting two levels of containers onto railcars). Intermodal loadings on freight trains grew at an average annual rate of 4.9% from 1970 to 1990 (10). The energy implications of this shift are not well-documented, but it is thought that trains use about one-half the energy of trucks for long-distance movements of high density goods (11). Therefore, to the extent that it has led to a shift from trucks to trains for long-distance movements, intermodalism represents a significant improvement in energy efficiency.

Transport—Future Issues. Several key factors will influence transport energy use in the future. First, alternative fuels (including electricity, ethanol, hydrogen, methanol, and natural gas) are currently being evaluated for both passenger and freight transport. Although alternative fuels in transportation are not widely used at present, their use may increase in the near future, particularly with growing concerns about urban air quality. For example, beginning in the year 1998, 2% of new vehicles available for sale in California are required to be zero tailpipe-emission vehicles. Second, if oil prices remain low, then there will be little economic incentive to invest in efficiency, but if they increase, investment in efficient vehicles is likely to increase. Third, regulatory changes, notably increases in the CAFE requirements, have been proposed. If such changes are signed into law, significant improvements in the energy efficiency of the transportation system may occur. With maximal use of existing technologies, the efficiency of new U.S. automobiles could approach 40 mpg in the next decade (12).

Industry

Since at least 1950, industry has been the largest energy-consuming sector in the U.S. economy, although its share of total energy use has decreased from 47% in 1950, to 37% in 1990 (5). In 1990, the U.S. industrial sector consumed about 30 quads of energy, about twice as much as it had in 1950 (5). The industrial sector consumes significant portions of all major energy sources in the United States: 44% of all natural gas, 34% of electricity, 25% of petroleum, and 12% of coal (5).

Several heavy manufacturing industries consume about three-fourths of the energy in this sector: ceramics and glass, chemicals, food, petroleum refining, primary metals, and pulp and paper. Nonmanufacturing industries, such as agriculture, construction, and mining, account for another 15% of sectoral energy use. Finally, lighter manufacturing industries, such as automobile manufacturers and textiles, consume the remaining 11% of industrial energy, but these industries consume a larger share of industrial electricity, in part due to their extensive use of electric motors (13). Energy's share of production costs varies widely by industry. In most industries, energy represents 5% or less of production costs, but the portion exceeds 20% for several large industries, including aluminum, cement, and certain chemical manufacturers (13).

Industrial activities are even more diverse than their fuel supplies, ranging from agriculture and construction to manufacturing and mining. Despite this diversity, however, there are four basic uses for industrial energy: steam production, direct process heat, electric motors, and feedstocks. The main use of industrial energy, steam production, is fueled mostly from natural gas burned in conventional boilers and cogenerating equipment. Industrial steam is commonly used in steel and pulp and paper production. Direct process heat is used for curing, drying, melting, and smelting. As with steam, most process heat derives from natural gas burning. Electric motors account for most of the electricity used in industry. Feedstocks include natural gas for chemical and fertilizer manufacturing and coal for steel production (13).

One important trend associated with industrial energy use is cogeneration, which involves the combined production of heat (typically as steam) and electricity. With the passage of the Public Utility Regulatory Policies Act of 1978, cogeneration in the industrial sector has increased substantially; the legislation requires electric utilities to purchase cogenerated power produced by a qualifying non-utility generator. This legislation has encouraged many industries to install cogeneration capacity, both to help meet their own power needs and to generate salable electricity. Today, cogeneration provides about 12% of the electricity used by manufacturing industries, with major contributions from the chemicals, food, paper, petroleum, and steel industries (13).

To simplify comparisons between the relative energy use of different industrial activities, analysts often compare the amount of energy consumed per unit of economic output, a comparison termed "energy intensity." The existing range of industrial energy intensities, as the diversity of industrial activities may suggest, varies by a factor of about 200. Printing, for example, requires about 800 Btu per dollar of output, whereas manufacturing nitrogen fertilizers consumes about 160,000 Btu per dollar of output (13). Between 1970 and 1990, the energy intensity of U.S. industry declined more than 20%, from more than 10,000 Btu per dollar to less than 8,000 Btu (in constant 1982 dollars). For the economy as a whole, this intensity decline was due both to structural changes (eg, less steel and more computers) and to technical efficiency improvements. Over the next two decades, that intensity is projected to decline further, to about 6000 Btu per dollar (14). In that same period, industrial energy use is projected to increase slightly more than 1% annually (14), the lowest rate of growth projected among all three sectors.

The increasing international exchange of goods and services points to an important limitation of energy intensity measures: they fail to reflect the embodied energy of products. Many industries use materials with substantially differing amounts of embodied energy. Although analyzing energy intensities often provides the first indication of where to focus government and other efficiency programs, the measure is an imperfect way to determine where optimal efficiency savings are possible; it simply reflects the relative energy use at one point in an often extended production chain.

Because energy represents a relatively minor portion of production costs for most industries, corporate attention to its use and efficiency has historically been low (13). However, at least three recent trends have and are likely to continue to improve industrial energy efficiency, which may slow the growth of energy use in this sector. First, increasingly stringent federal environmental regulations, particularly those pursuant to the Clean Air Act, have prompted many firms to examine the potential of improved energy efficiency as a means to limit stack and other emissions. Second, a growing number of utility DSM programs are targeting industrial consumers as a means to offset the need for additional generating capacity; as more money and attention are channeled to DSM programs, industrial energy efficiency may increase considerably. Finally, process changes and materials substitutions are being increasingly used to improve outputs and reduce production costs; an additional benefit of such changes is often reduced energy use.

ENERGY USE BY FUEL

Greater insight can be gained by looking at consumption patterns of the various fuels consumed in the United States, as illustrated in Figure 9. Three fuels account for the bulk of direct consumption: oil, natural gas, and electricity (strictly speaking, not a fuel but a carrier). Coal, nuclear power, and hydropower are not discussed here, as they are used largely for electricity generation.

Oil

Oil is the single largest U.S. energy source, accounting for 41% of national energy use and 48% of national energy spending. This fuel is also the most problematic and politically volatile form of energy. Oil has many attractive features: it has a high energy density, making it less expensive to transport; it is a liquid and therefore relatively easy to produce and transport; and it can be used for a variety of purposes in different sectors, eg, transportation and large motor drive via internal combustion engines, space heating for buildings via oil-burning furnaces, industrial heat via oil-burning boilers, and so forth. Oil also has its problems. Its global distribution is uneven, leading to concentrations of wealth and power in those regions that have it and to disagreements and occasionally wars with those who do not. Further, like all fossil fuels, its production and use has detrimental environmental effects, including emissions of carbon dioxide (CO_2), volatile organic compounds (VOCs), and nitrogen oxides (NO_x), and occasionally widely publicized and locally disastrous oil spills.

Oil consumption in the United States increased rapidly from 1950 to 1973, due to growth in demand by all users—transportation, industry, buildings, and electric utilities. In 1973, however, Arab oil-producing countries, due in part to U.S. arms sales to Israel, briefly stopped selling oil to the United States. Oil prices climbed, and demand dipped slightly. In 1979, Iran briefly stopped exporting oil, and the market reacted much the same as it had in 1973, with large price increases and small demand decreases. Since then, both price and consumption have fluctuated (see Fig. 10).

At present, U.S. oil consumption is at about the same level it was in 1974, despite large increases in population, economic output, and the number of automobiles. This has been possible due in part to technical-efficiency improve-

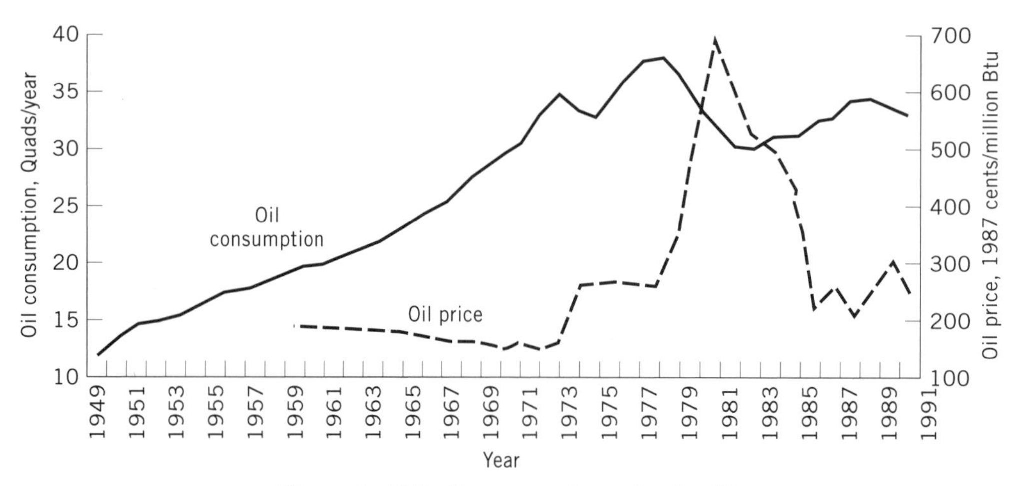

Figure 9. U.S. oil consumption and price (5).

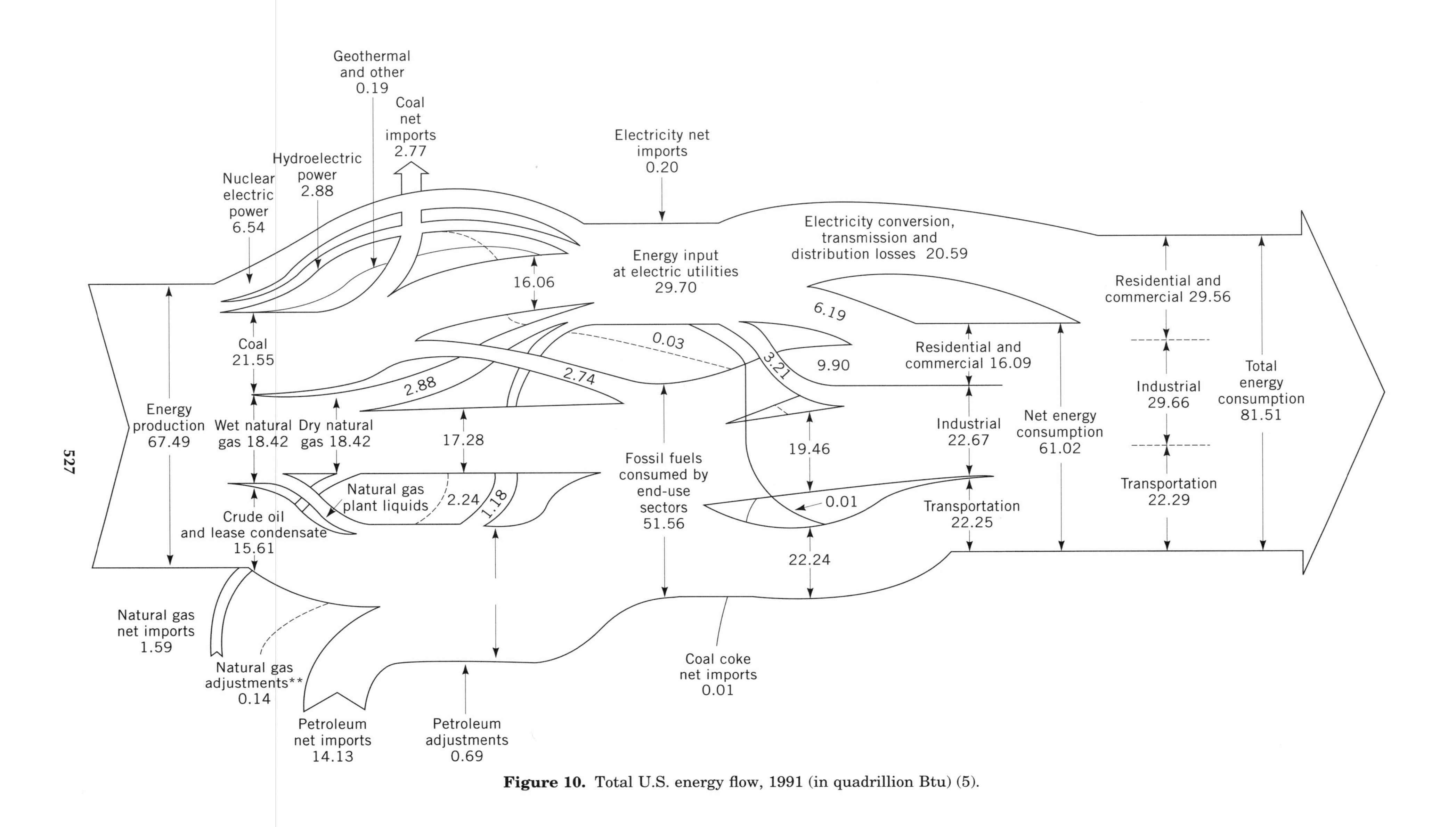

Figure 10. Total U.S. energy flow, 1991 (in quadrillion Btu) (5).

ments, and in part to shifts away from oil by those users who have other options. For example, industry and buildings both substituted other fuels, notably natural gas, when faced with increasing and volatile oil prices. The transport sector, however, has few nonoil technical alternatives; thus, the fraction of U.S. oil used for transport has increased from 53% in 1970, to 65% today. (Alternative fuels for transportation are being investigated and used in small numbers; but are currently used by only a small fraction of the total U.S. vehicle fleet.) Other oil users are industry (25%), buildings (7%), and electric utilities (3%) (5).

Natural Gas

Natural gas is an increasingly popular fuel, due to its reputation for environmental cleanliness, the relative abundance of domestic reserves (which makes it less susceptible to international price fluctuations), and to technical improvements that have allowed it to be used with impressive efficiency for electricity generation, space cooling, and other applications.

Historically, natural gas has generally been the second most-consumed U.S. energy source (behind oil) since 1958 (5). Low prices in the 1950s and 1960s (due in part to regulation) led to steady increases in demand for natural gas (5); demand peaked at 22 trillion cubic feet (tcf) in 1972. Since then, supply uncertainty, increasing prices, and restrictions in the Powerplant and Industrial Fuel Use Act of 1978 (repealed in 1987) slowed demand (5), although natural gas remains less expensive than oil or electricity on a per energy unit basis (5).

Industry consumes more natural gas than any other sector, accounting for more than 40% of total use. Over the last 15 years, greater use of cogeneration accounted for a significant share of the growth in industrial gas use (15). The relative use of natural gas in the residential and electric utility sectors has been fairly steady, representing about 25% and 15–20% of total demand since 1950, respectively. The commercial sector experienced the largest relative growth in natural gas demand over the last 40 years (from 7% to 14%), driven primarily by winter heating needs.

Over the next two decades, domestic natural gas use is expected to increase steadily, with growth averaging slightly more than 1% annually, in part due to its comparatively low cost and its environmental benefits relative to coal and oil. The greatest demand increases are expected from electric utilities and industry, where combined-cycle generation and cogeneration, respectively, are forecasted to increase quickly (14). By the year 2000, domestic natural gas consumption is projected to match its 1972 peak, and to increase to more than 24 trillion cubic feet by the year 2010 (14).

Electricity

U.S. electricity use has shown robust and steady growth since at least 1950. Over the past 40 years, annual electricity demand growth has averaged almost 6% (5), which translates to a doubling of total demand about every 12 years. However, demand growth during each decade has varied considerably. In the 1960s, annual electricity demand growth exceeded 7%, dropping to about 4% in the 1970s and less than 3% in the 1980s (14). Nonetheless, no other major energy source experienced as great of a demand growth in the past 40 years; since 1950, national electricity use has increased almost 10-fold (5). This dramatic growth in electricity use is largely explained by its convenience, end-use efficiency (which can approach or even exceed 100%), versatility, and cleanliness (for the end user), qualities important enough to overshadow the expense of this fuel (it is several times more expensive per unit of energy than other fuels).

In the 1950s, the industrial sector purchased about half of all U.S. electricity, but that relative share began dropping by the 1960s (5), as the residential and commercial sectors became more electrified, particularly with greater use of electric home appliances (notably space cooling) and electronic office equipment. Today, the residential, commercial, and industrial sectors each accounts for roughly one-third of electricity demand (5). Historically, coal has accounted for roughly half of all electricity generation; today, 55% of U.S. electricity is produced from coal, more than twice as much as the next largest source, nuclear power, at 22% (5).

In the next 20 years, national electricity demand is projected to increase between 1 and 2% per year, far lower than historic averages, in part due to the saturation of electric appliances and equipment, as well as to the proliferation of government and utility energy-efficiency programs. Most of the demand increase is expected in the commercial and industrial sectors, whereas demand growth in the residential sector will be more modest, due in part to low projected increases in U.S. population (14). If these projections are correct, annual U.S. electricity use will increase from about 2.7×10^{12} kWh in 1990 to 3.7×10^{12} by the year 2010 (14).

WHERE WE ARE TODAY

In 1991, the United States consumed about 81.5 quads of energy. One concise way of understanding where this energy comes from and where it goes is in the form of a "spaghetti chart." This chart shows all significant energy flows, starting with production on the far left, moving through transformation (eg, to electricity), and concluding with final consumption on the far right.

THE FUTURE OF ENERGY CONSUMPTION

Uncertainties over oil prices, technical advances, market penetration of energy-efficient technologies, economic growth, and other factors make it extremely difficult to forecast future energy consumption. It is possible, however, to identify likely trends and key uncertainties that will influence future levels of energy consumption.

Likely Future Trends

In the past 20 years, dramatic improvements in the energy efficiency of energy-using devices has occurred. New refrigerators, for example, use less than one-half the electricity used by units sold in the early 1970s. In addition, research by the Congressional Office of Technology Assessment and others suggests that dramatic further reductions in energy use are possible. Although the cost-effectiveness and market penetration rates of these technologies are unclear, technical advances will likely lead to continued improvements in energy efficiency.

There has been a gradual yet clear economic shift in the United States away from heavy manufacturing and toward greater demand for services. These trends, such as slow growth in the demand for transport of raw materials and minerals, rapid growth in computers and other information technologies, and increases in vehicle-miles per person per year, will likely continue in the near future.

Due largely to changes in state-level regulation, electric and gas utilities have invested considerable resources in energy efficiency. Current utility investments in energy efficiency exceeds \$2 billion/year. Although the energy savings resulting from these investments are uncertain, they will likely result in increased market penetration of energy-efficient technologies.

Key Uncertainties

Considerable research effort focuses on alternative fuels for transportation. Currently, no fuel is a clear successor to gasoline; some argue that in the future, the light-duty vehicle fleet will use several different fuels, for example, electric vehicles for short urban trips and gasoline for longer trips. In any case, a shift to nonoil-based fuels by a significant fraction of the vehicle fleet could dramatically reduce U.S. oil consumption.

Oil price fluctuations, often driven by political factors, have strongly influenced energy use in the past. Despite these fluctuations, the United States still heavily depends on imported oil; in the future, oil prices will continue to affect U.S. energy consumption.

The possible effects of global climate change, due largely to CO_2 emitted from fossil fuel burning, is attracting increased scientific and political attention. Although the effects of increased global temperatures are uncertain, concern over possible detrimental effects may result in efforts to reduce CO_2 emissions. This could significantly change energy use (eg, large increases in energy efficiency, and a shift toward renewable and nuclear energy).

Natural gas is seen by many as the fuel of the 1990s, due to its environmental benefits, low price, and high availability. However if demand increases, then existing pipeline capacity may be exceeded, requiring expensive and time-consuming construction of new pipelines. Furthermore, deregulation of natural gas prices will likey affect prices and therefore demand. In general, changes in natural gas prices may influence future energy consumption levels.

Finally, the market penetration of energy-efficient technologies significantly influences consumption levels. For example, by one estimate, full-market penetration of cost-effective technologies in the buildings sector would lead to a savings of 14 quads/year by the year 2015. These market penetration rates depend on energy and technology prices, perceptions of comfort and service effects, changes in consumer preferences, and other factors, and are a source of considerable uncertainty in future energy consumption.

Disclaimer

The opinions expressed in this article are those of the authors, and are not necessarily those of the Office of Technology Assessment, the United States Congress, or the United States Government.

BIBLIOGRAPHY

1. U.S. Department of Commerce, Bureau of the Census, *Statistical Abstract of the United States 1992,* Washington, D.C., 1992.
2. U.S. Congress, Office of Technology Assessment, *Changing By Degrees: Steps to Reduce Greenhouse Gases, OTA-O-482* U.S. Government Printing Office, Washington, D.C., Feb. 1991.
3. U.S. Department of Energy, Energy Information Administration, *Monthly Energy Review, DOE/EIA-0035(93/06),* Washington, D.C., June 1993.
4. U.S. Department of Energy, Energy Information Administration, *International Energy Annual 1991, DOE/EIA-0219(91),* Washington, D.C., Dec. 1992.
5. U.S. Department of Energy, Energy Information Administration, *Annual Energy Review 1991, DOE/EIA-0384(91),* Washington, D.C., June 1992.
6. U.S. Congress, Office of Technology Assessment, *Energy Use and the U.S. Economy, OTA-BP-E-57,* U.S. Government Printing Office, Washington, D.C., June 1990.
7. U.S. Congress, Office of Technology Assessment, *Building Energy Efficiency, OTA-E-518,* U.S. Government Printing Office, Washington, D.C., May 1992.
8. Oak Ridge National Laboratory, *Transportation Energy Data Book 12, ORNL-6710,* Oak Ridge, Tenn., Mar. 1992.
9. J. Murrell, K. Hellman, and R. Heavenrich, "Light-Duty Automotive Technology and Fuel Economy Trends Through 1993," *EPA/AA/TDG/93-01,* U.S. Environmental Protection Agency, Ann Arbor, Mich., May 1993.
10. *Railroad Facts 1992,* Association of American Railroads, Washington, D.C., Sept. 1992.
11. *Energy Use in Freight Transportation,* staff working paper, U.S. Congress, Congressional Budget Office, Feb. 1982.
12. U.S. Congress, Office of Technology Assessment, *Improving Automobile Fuel Economy: New Standards, New Approaches, OTA-E-504,* U.S. Government Printing Office, Washington, D.C., Oct. 1991.
13. U.S. Congress, Office of Technology Assessment, *Industrial Energy Efficiency, OTA-E-560,* U.S. Government Printing Office, Washington, D.C., Aug. 1993.
14. *Annual Energy Outlook 1993, DOE/EIA-0383(93),* U.S. Department of Energy, Energy Information Administration, Washington, D.C., Jan. 1993.
15. *Natural Gas Annual 1991, DOE/EIA-0131(91),* U.S. Department of Energy, Energy Information Administration, Washington, D.C., Oct. 1992.

ENERGY EFFICIENCY

Marc Ross
University of Michigan
Ann Arbor, Michigan

Energy use by people provides enormous benefits but causes substantial harm; and the harm will grow as energy use continues to grow. Improved energy efficiency is the centerpiece of action to limit the harm while enabling continued growth in the services provided by energy. Not only does improved efficiency reduce energy requirements and related environmental impacts, but the reduced spacial concentration or density of energy requirements eases the task of introducing more benign alternative energy supplies.

This article begins with needed background, including explanations of energy use and its efficiency. Next analytical tools and concepts are discussed. Opportunities for improving energy efficiency in industry and transportation are also examined. The discussion ranges from improving energy conversion devices such as engines and furnaces to that of whole processes such as steelmaking and automobiles to that of the whole system. An understanding of these different levels of organization and change is needed if one is intelligently to consider future developments. It is also needed for intelligent policymaking. For both industry and transportation, a wide range of policies is also considered from the perspective of their influence on energy efficiency. See also Thermodynamics; Energy conservation.

BACKGROUND

The Physics of Energy

The First Law. Energy is a quantitative property of all things (1). It has many forms, such as energy of bulk motion, gravitational energy, energy of light (or electromagnetic waves), chemical energy, and thermal energy. Transfer of energy from one form to another and from one place to another, or one set of matter to another, characterizes all happenings. The conversion of the chemical energy in gasoline to the energy of motion of a car plus considerable thermal energy as a by-product or waste, will be discussed.

An essential strength of energy as a concept is that energy is neither created nor destroyed. Its forms change, and it can move from one place to another, but for any region of space or system of matter the total energy is constant, after accounting for flows into and out of the system (the first law of thermodynamics). The constant numerical value for energy is a summary characteristic of any system. To determine it, all the different forms of energy must be expressed in the same units and added. This property of energy is similar to money accounting. Accounting for money flow and for changes in form makes it possible to conduct powerful analyses. There are many analogies between money and energy accounting, including the need to convert quantities of money into a common unit and to account for losses or diversions. Money can, however, be created and destroyed, whereas energy cannot.

The Second Law. Most energy forms are high quality, which means they can, in principle, be fully converted one into another. Thermal energy, the energy of random motion of atoms of matter, is of variable quality. Thermal energy can only be partly converted into a high quality form (the second law of thermodynamics).

The second law of thermodynamics is qualitatively shown in Figure 1. Two high quality forms such as chemical energy and electrical energy can, in principle, be wholly converted one into the other, and either of them can be converted into thermal energy, but the thermal energy cannot be wholly converted to a high quality form. Energy converted into a high quality form, or delivered from another place in a high quality form, is called work. Energy converted to the thermal form or delivered in that form is called heat.

Consider the hot gas in the cylinder of an engine, ie, the fuel–air mixture after combustion. As the gas pushes the piston, this thermal energy is converted to energy of bulk motion, a high quality form; so the hot gas does work. But this thermal energy cannot all be converted to work. Most of the energy is carried away as lower quality ther-

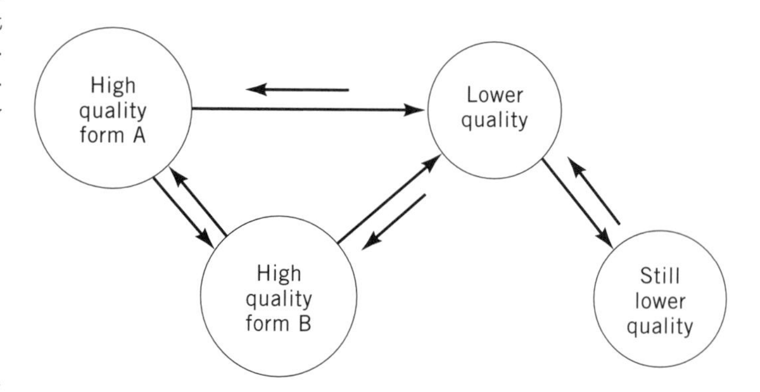

Although the total amount of energy does not change (first law), net transformations of energy increases the amount in low-quality forms (second law).

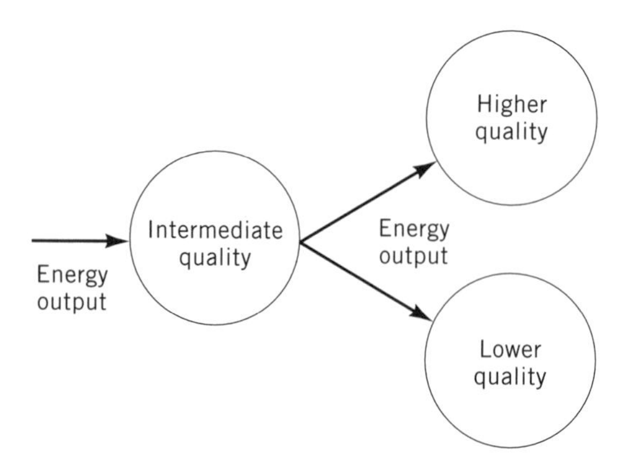

One cannot raise the quality of energy without either (1) reducing the quality of some of the energy shown, or (2) adding some higher quality energy (as in Fig. 2).

Figure 1. The second law of thermodynamics.

mal energy in the exhaust, and substantial "losses" of this kind cannot be avoided. Heat engines convert chemical fuel energy to thermal energy and then only partially into energy of bulk motion. They are fundamentally inefficient in this respect.

The quality of thermal energy depends on its temperature. High temperature thermal energy is of relatively high quality. At high temperatures most, but not all, of the thermal energy can, in principle, be converted to a high quality form. When the temperature is similar to the ambient temperature (that of the surroundings with which the system is in contact), then the thermal energy is of low quality. Little if any of it can be converted to a high quality form. The second law is illustrated in practical terms by heat pumps (or refrigerators) and heat engines: for a given amount of high quality energy input W, a heat pump delivers a larger amount of thermal energy Q_1 at a desired temperature higher than that of the ambient environment or the source from which thermal energy Q_0 is drawn (Fig. 2). The work primarily serves to increase the temperature of the thermal energy Q_0. The greater the quality (temperature) difference between the two forms of thermal energy,

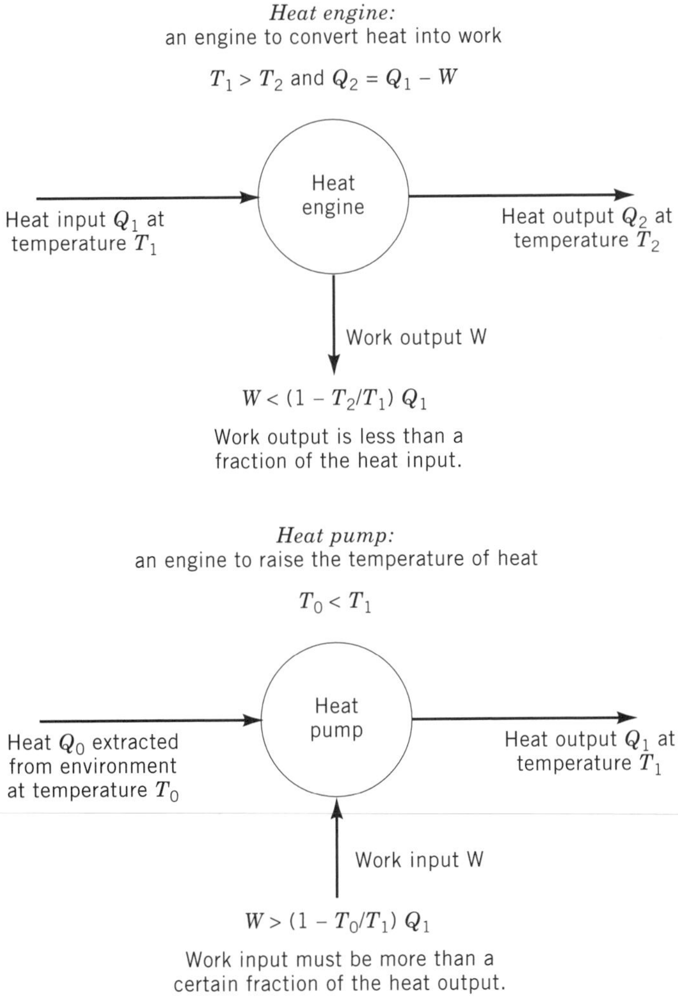

Figure 2. Second-law limitations on heat engines and heat pumps. Temperatures are absolute temperatures.

corresponding to Q_0 and Q_1, the more work must be done relative to the heat delivered. (A refrigerator operates in the same way except that the emphasis is on the thermal energy extracted at the lower temperature Q_0.) The first law is obeyed because the energy Q_0 is being extracted from an existing source, and $Q_1 = Q_0 + W$.

Turn this machine around, and it is a heat engine: with an input of thermal energy Q_1 at a high temperature T_1, work W can be done, but substantial thermal energy at lower temperature (ie, closer to ambient conditions) will have to be discharged. Thus $W < Q_1$. The greater the quality difference between the two forms of thermal energy, the greater the work output can be relative to the thermal energy input Q_1.

Consider the example of an automobile engine. Let 100 units of chemical (fuel) energy be introduced along with air into the cylinders. Almost all of this is converted into thermal energy by combustion. Some 40 units of work are done by the hot gases expanding against the pistons, while 60 units of lower quality energy are lost, primarily in the forms of exhaust and thermal energy transferred to the engine (eg, to the engine coolant). Thus $Q_1 = 100$, $W = 40$, and $Q_2 \lesssim 60$ energy units. Q_1 is slightly greater than $W + Q_2$ because some energy is lost in other forms, such as unburned fuel.

In contrast to converting one form of high quality energy into another using a heat engine, direct conversion between two high quality forms, such as chemical energy to energy of bulk motion, can be achieved without substantial losses of thermal energy. Efficient, direct conversion from chemical fuel to bulk motion (eg, using a fuel cell) has been technically difficult to achieve in practice, but it is a promising and active area for research and development.

Similar concepts apply, eg, to lighting. Light from a gas flame or incandescent electric bulbs is produced through the intermediary of thermal energy at high temperature. Only a modest fraction of the thermal energy can be converted to light, however. Fluorescent electric lights and other lamps not depending on the intermediate thermal stage are much more efficient.

Energy Use and Efficiency. What is meant then by energy use and efficiency of energy use? Energy use is the conversion of energy into a desired form at a desired time and place. For example, chemical fuel is converted into thermal energy at a temperature appropriate for industrial process heating, to operate an engine, or to keep a building warm. When energy is used it does not disappear, although much or all of it usually quickly degrades into thermal energy at near-ambient temperature, energy of low quality.

The thermal energy obtained by burning fuel in an automobile engine mostly goes out with the exhaust, where it quickly becomes useless, spreading into the surrounding air and moving toward ambient temperature. As mentioned, about 40% of the original energy remains as work done on the pistons. In average engine operations roughly 50% of this is lost to the thermal energy form as a result of frictional processes inside the engine, and only some 15% of the original fuel energy typically reaches the drive wheels in the form of rotating machinery. There it does what is desired; it moves the vehicle on the road. After doing so, this energy too is converted to thermal energy of low quality, via the frictional mechanisms of air resistance to the vehicle's motion, resistance of the tires to the deformation that occurs as the wheels roll, and braking friction. Thus all the fuel energy is soon converted to low temperature thermal energy; but on the way some 15% of it does the desired work.

As the automotive example shows, two effects mitigate against the desired transformation and movement of energy. First, the second law prevents thermal energy from being wholly converted into the high quality forms that may be desired. Losses of roughly 50% are implied for the best high temperature engines, and much more severe limitations apply near ambient temperatures. The second and independent effect is dissipation, of which the primary kind is friction. Friction in an internal-combustion engine, for instance, involves the rubbing of solid parts, the fluid friction in pulling air through small orifices into the cylinders and pushing exhaust out, and the fluid friction of pushing coolant and lubricant through yet smaller orifices.

There are other forms of dissipation. For example, it has proven difficult to reduce losses to near zero in chemical to electrical (and electrical to chemical) transformations. Electrical resistances in charging and discharging a good car battery result in loss (to heat, of course) of roughly 33% of the energy. Careful, inventive improvements over a long period have resulted in reducing the dissipative losses in converting aluminum oxide to metal in commercial electrolytic cells to about 50%. Meanwhile, success has been easier in conversions between bulk motion and electricity. Large motors and generators achieve these transformations with little loss.

One measure of the energy efficiency of a car is the ratio of the desired energy reaching the drive wheels to the fuel energy input, which is about 15% for average driving. The efficiency of the vehicle's engine, considering only the work done by the hot gases on the pistons and neglecting all friction in the engine, is typically about 40%. These are strictly defined thermodynamic efficiencies, ratios of energy outputs to inputs.

These thermodynamic efficiencies are useful, but a broader efficiency concept is also needed, because the fuel energy required to run a car is affected by how large the car is, how streamlined it is, and how hard its tires are. The fraction of fuel energy reaching the wheels is only part of the story. In principle, there are no limits to reducing the impediments to moving the vehicle (air, tires, and brakes) once the energy reaches the drive wheels. A car can be made more streamlined, the tires can be made harder, the use of frictional brakes can be minimized. In the extreme, it is possible to imagine vehicles running in evacuated tubes with almost frictionless magnetic cushions. Thus no simple ratio of energies can be defined as the efficiency for moving the vehicle as a whole.

It is useful, therefore, to define an energy intensity (EI), such as fuel use per passenger mile, and give the word *efficiency* an additional meaning. Efficiency improvement is the increase in the ratio of old to new energy intensity EI_0/EI. Thus the efficiency of a vehicle has been improved 10% when the energy use per passenger mile, for a certain kind of trip, is reduced some 9%, ie, 1.00/0.91 = 1.10.

Although the car has been used as an example, the same arguments apply to other energy uses, such as heating buildings and producing industrial products. For heating a building a thermodynamic efficiency can be defined, which expresses the ratio of desired thermal energy output from the heating system to the high quality input energy. (Actually, there are several definitions with different applications, such as nominal furnace efficiency, seasonal furnace efficiency, and efficiency compared with an ideal heat pump.) Analogous to the car, this approach can describe the furnace or heating system but not the effectiveness of the building as a whole from a heating standpoint. The building could be designed so that it needs no energy input as such, eg, by storing ambient warmth, or zero quality energy from the previous summer. Again the term *efficiency* can also be used to compare the energy intensities of the heating of buildings.

Comparing a heating system to an ideal heat pump was mentioned parenthetically as one way to define a strict thermodynamic efficiency. This is the second law efficiency approach (2,3). The first law efficiency is the ratio of desired heat output to actual energy input. The second law efficiency is the ratio of energy input to an ideal heat pump system to the input to the actual system (both systems delivering the same desired output). Household heating systems have first law efficiencies of 60% and higher, ie, 60% of the input energy is delivered as desired heat and 40% is lost from the system before delivery, eg, up the flue. Their second law efficiencies are roughly 5%, depending on how cold it is. From this perspective, low quality thermal energy from the outside air or water can, in principle, be extracted, upgraded, and delivered at the desired temperature with the work input only 5% of the energy delivered, because the energy delivered is of such low thermodynamic quality. From this second law perspective, any low temperature heating system is highly inefficient, unless it is based on a heat pump or makes use of heat that is waste from another process. In a much broader attempt using these techniques, it has been concluded that the second law efficiency of the entire U.S. economy is roughly 2.5% (4).

Stocks, Flows, and Units. It is often necessary to discuss energy numerically as a stock or quantity: barrels of oil, kilowatt hours of electricity, kilocalories of chemical energy. It is also necessary to discuss the time rate of energy flows: barrels of oil per day, kilowatts of electrical work being done, British thermal units (Btu) of heating per hour. Energy flows are called power by physicists and engineers. The basic power unit, the Watt (W), expresses simultaneously the quantity of energy and time of flow (analogous to the velocity unit knot). If, eg, a household uses an average of 1 kilowatt (kw) of electrical power per hour, its average power is 24 kw·h per day, or 8760 kw·h per year. The Watt unit includes time in the denominator and is converted to a stock or energy unit by multiplying by time, as in kilowatt hour (the energy corresponding to the flow of 1 kw for 1 h). This can again be converted to a power unit if one desires, as in kw·h/yr. The latter form of power unit is usually used to describe the average rate of flow, while the form kw can be used to describe the rate of flow at an instant, or an average rate.

Some common units for energy and power and conversion factors into Btu and Joules (the principal traditional and metric units, respectively) are given in the front of this volume. For example, 7.2 million barrels (bbl) of gasoline are used in the United States each day. Here are some conversions: $(7.2 \times 10^6)(5.25 \times 10^6)\,365 = 13.8$ quads/yr $= 13.8 \times 10^9 \times 33.4 = 461 \times 10^9$ W = 461 GWs (average rate of use). The conversion factors are 5.25×10^6 Btu = 1 bbl gasoline (using the higher heating value on combustion), 10^{15} Btu = 1 quad, 33.4 W = 10^6 Btu/yr, and 10^9 W = 1 GW. The energy values associated with fuels are, of course, the heat release on combustion. A significant part of the definition is whether it is higher heating value, in which the heat released converting any water from gas to liquid is counted, or lower heating value in which it is not counted.

Power units, like Watts, Btu/h, horsepower (hp), and bbl/day, commonly characterize energy equipment and facilities. The oil well, power plant, automobile, home appliance, transmission line, and pipeline all have a nominal maximum capacity or rate of energy conversion or energy carriage. An electric power plant may have a rating of 1 GW, and the capital cost of such plants is often expressed in dollars per kilowatt. In this case, Watts refer to the capability to produce electric power. The rate of fuel use is roughly three times higher.

The Earth's Energy Balance. One face of the earth is always bathed in sunlight, at least at the top of the atmosphere. Sunlight is absorbed by the earth at an enormous rate in terms of power units. (Some energy also reaches the surface of the earth from the interior and from radioactivity, and some is released by burning fossil fuels.) The energy arriving must be balanced by energy leaving, or the earth's average temperature would change. The balance is achieved by direct reflection and by the outflow of "earthshine" (Fig. 3). Earthshine is infrared light, similar to sunshine but of lower frequency and invisible to the human eye.

The flux of sunshine in space at the earth's orbit is $F_s = 1.4$ kw/m². The total solar energy absorbed by the earth is $F_s \times$ (area of earth facing the sun) $\times$ (1 − albedo), where the albedo (≈0.3) is the fraction of sunlight reflected. Thus

$$\text{sunlight absorbed (rate)} = F_s \cdot \pi R_e^2 \cdot 0.7 = 1.25 \times 10^{17}\ \text{W}$$

On the average, this flow is balanced by the outward flow of earthshine. The total earthshine emitted is

$$\text{earthshine emitted (rate)} = F_e \cdot 4\pi R_e^2$$

Setting these flows equal, the average flux of earthshine at the earth is $F_e = 0.7F_s/4 = 240$ W/m². (Much of the absorption and the emission into space is from the top of the atmosphere.) For comparison, the rate of energy use by people is about 1.2×10^{13} W, about 1 part in 10,000 of the average solar influx. However, energy use in U.S. metropolitan areas is roughly 2% of the average solar influx.

The propensity to emit earthshine depends on the temperatures of the surface and upper atmosphere and on the transparency of the atmosphere to infrared light. Greenhouse gases in the atmosphere decrease this transparency. If the concentration of greenhouse gases is increased, eg, by people's activities, then the surface will get warmer to emit the same amount of earthshine.

To summarize, energy, as used by people, is initially extracted from the natural environment, usually then converted into an energy carrier (a form convenient for use), transported to where it is needed, and finally used. The use consists of transforming the energy into a desired form at a desired place and time. None of the energy disappears in this use; however, inefficiencies in each step result in energy being lost to the system, almost always in the form of low temperature heat. In addition, the useful energy, such as chemical change, bulk motion, or light, is also usually transformed by natural processes into low temperature heat. All this heat is eventually radiated into space as infrared light as part of the earth's thermal balance.

The Commercial Energy System

The commercial energy system (Fig. 4) comprises extraction of energy supplies, conversion to energy carriers, and final use in the various sectors of the economy. The principal supplies are fossil fuels (primarily coal, oil, and natural gas), nuclear fuel (uranium for fission), and solar-related energy (especially hydropower, biomass, wind, and direct sunlight). Typically, the extracted energy forms are converted into energy carriers appropriate for fuel use (5). Fossil fuel is used to generate the carrier electricity via

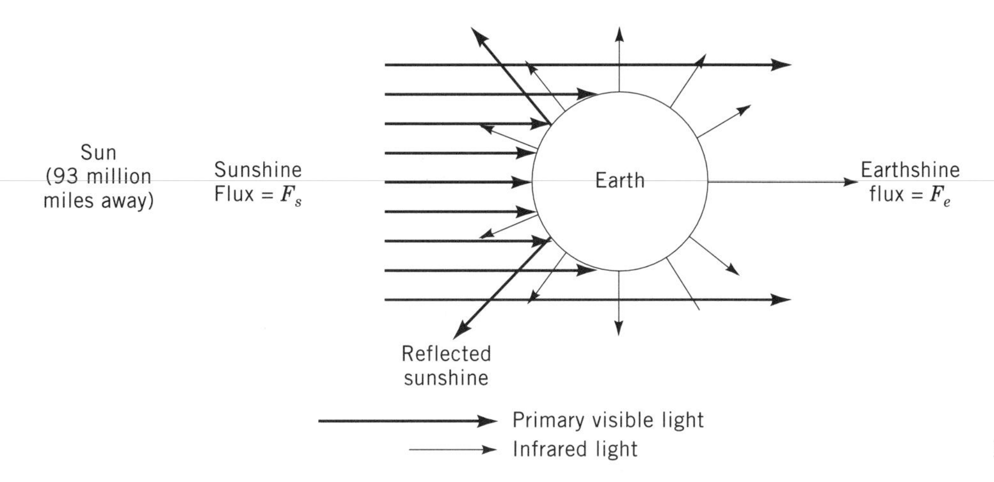

Figure 3. Energy balance of the earth.

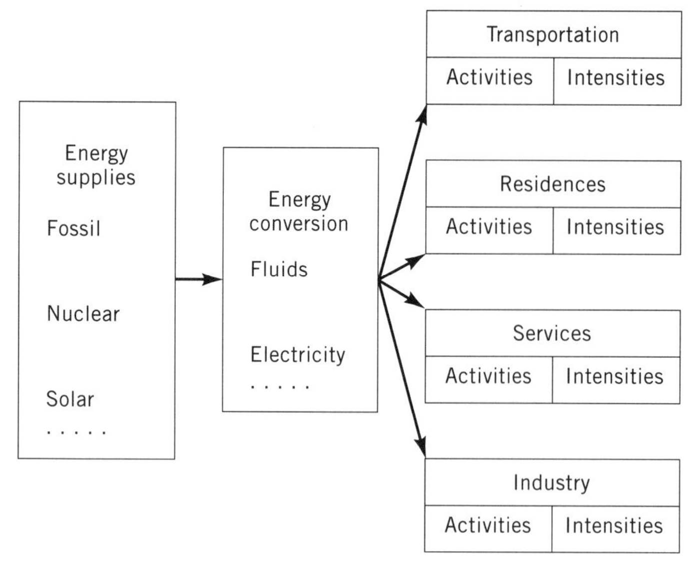

Figure 4. The commercial energy system.

the intermediary of thermal energy, with an average efficiency of about 32% (counting losses in transmission and distribution). That is, 32% of the fuel energy is delivered as electrical energy to customers, while 68% is lost as near ambient temperature heat. Crude oil is converted into products like gasoline with an overall efficiency of about 90%.

The principal energy carriers are electricity, dry natural gas, petroleum products, and washed and graded coal. In the future, gaseous hydrogen (eg, extracted from fossil fuel or from water) might become a major carrier. The properties of an energy carrier (especially its storability, its transportability, the ease of its conversion at the point of end use, and its environmental acceptability) establish its relative value for a particular end use. There is a wide range in the storage characteristics of different energy carriers. The storage characteristics of solid and liquid carriers (high energy density and low capital cost) are hard to beat.

As a general rule, the storability of a form of energy is closely linked to its transportability. Fuels that are easily stored (coal, oil, and uranium) can be economically shipped in batch mode via barge, railroad, truck, etc. Those forms that are expensive to store (electricity and natural gas) are most efficiently transported by line, or continuous, mode. The systems required for distribution to, and conversion at, the site of end use are also important determinants of the appropriateness of a carrier. In small-scale applications (buildings, transportation, and small industry) the use of solid fuels is inhibited.

The storability of energy affects not only the physical structure of the delivery system but also the institutional structure. Because coal and oil can be brought to the retail market in batch mode, they can be sold competitively by multiple distributors. Since natural gas and electricity, on the other hand, entail the installation of an immobile distribution system, they are most efficiently distributed by a local monopoly. If methods were discovered to store these carriers in an inexpensive, compact form, they might also come to market in batch mode and be sold competitively.

The efficiency with which raw energy supply is converted to carriers and the efficiency of transportation of energy are issues of major concern. In the interests of coherence and brevity, they will not be dealt with here.

Societal Concerns with Energy Use

Societal concerns are only partly associated with today's conditions (6). They are largely focused on predictable future developments. In the United States, energy consumption used to grow about 2% faster than the population. For the past 20 yr, it has been growing more slowly than the population, ie, per capita consumption has declined slightly. (The population has been growing 1.0%/yr.) Two principal uses are growing substantially faster: electricity in buildings and petroleum for transportation. Moreover, in the industrializing world these two kinds of use are growing extremely rapidly. Thus when considering the adequacy of resources, cost of facilities and environmental damage, it is important to consider the increasing pressure caused by population growth and growth in physical consumption per capita in these two areas of use.

Energy Resources. The energy resources in use are primarily fossil fuels; they account for 75% of the world's total consumption. On the scale of a few decades, conventional petroleum will become scarce, while natural gas will become more prominent, although it also will begin to be in short supply (7–11). Natural gas is a convenient fuel once pipelines and distribution are established, and the ultimate resources of natural gas are somewhat larger than those for petroleum; greenhouse gas emissions are also lower. Supplies of coal are even larger. Alternative fossil fuels are also available (heavy oils, tar sands, and shale oil), but the environmentally related costs of coal and the alternatives are likely to be high (12).

Solar-related energy flows are already being tapped for some 20% of current world energy use, primarily biomass and hydropower. New biofuels and wind and photovoltaic technologies to produce electricity are highly promising (13). Wind-generated electricity is beginning to become competitive in good locations without major subsidies. Biofuels grown for the purpose, rather than being created as by-products of other agricultural and forestry activities, are not important today, but with selected species in marginal croplands and using new methods of cultivation and conversion into carriers, they appear promising at today's prices. Photovoltaics are competitive today in niche markets, and reasonable projections of manufacturing costs at large scale suggest that photovoltaics could become competitive in favorable large markets, such as in the U.S. Southwest, in another decade or two.

Although some are promising from a cost standpoint, solar-related sources are diffuse. Biomass can take up a lot of land, for example. So there are capacity limitations. There is also powerful geographical variation, eg, in Michigan, there is not much sunlight in the winter and relatively little persistent wind. There is water, however, so biomass can be grown, but clearly there are limitations. In general, most forms of solar-related energy are constrained in most geographic areas. In desert country, however, photovoltaic electricity could be produced on a large scale.

Nuclear fission, at 5% of the world's total use, is faltering in competition with natural gas and coal and may be left behind in the foreseeable future by cost reductions in the solar-related energies. However, countries that are affluent and have high population density, like Japan and France, may continue to develop nuclear power.

It must be added that because energy purchases now make up a substantial 7.5% of all final expenditures in the United States society will not embrace much higher costs simply to have the energy. There are too many other things people want to do with their incomes and so many ways to do with less energy, that people will not accept large price increases, at least after a period for adjustment. In other words, only a fraction of energy consumption is "essential" and there are many strings to the supply bow in the long run. When expenditures on energy rose to 14% of GNP in 1980–1981, many things were done on both the supply and the demand sides to bring the costs down. Costs are back down. What was needed was some time and investment to make supply and demand adjustments. The main point here is that potentially huge new kinds of energy supply, which some people are enthusiastic about, will not be of wide interest if they are costly. Nuclear fusion may be an example.

Given this perspective, there is a mixed conclusion. There are limitations to the cheap fuels used now. These limitations will be felt strongly in a few decades. Certainly not all the proposed energy supply panaceas are practical, but there are a variety of promising new energy supplies that provide options. With these options and the critical option of using energy more efficiently, society has choices for developing a future energy system that can meet its goals.

Disruptions and National Security. The concentration of petroleum resources in the Middle East combined with the dependence of nations on petroleum as the principal transportation fuel has created concerns about vulnerability to interruptions in the petroleum supply. Twice in the 1970s there were serious interruptions: the Arab oil embargo of 1973–1974, and the interruption associated with the Iranian revolution of 1979. To diminish the impact of any future interruption, policies were established to help create alternative supplies, especially synthetic fuels from coal and shale oil, and to reduce petroleum demand (efficiency improvement and fuel switching). In addition, the United States has a strong military presence in the Middle East. In terms of the country's overall energy situation, the supply-side policies were not successful, and the demand-side policies were rather successful. However, the United States has returned to a strong dependence on imported oil. Although most other industrial countries rely even more heavily on imported oil, concern is stronger in the United States, because, as a world power it is thought that the country should not be sensitive to the influence of the principal oil-producing nations.

Capital Costs. The economic problem society faces with energy supply, conversion, and use is that it costs a great deal up front. The use of energy by people is a mammoth activity. In the United States, some 23 kg of fuels are used per day per person. A huge capital investment is required to accomplish this: wells, mines, pipelines, refineries, power plants, transmission lines, motors, furnaces, etc. Roughly 20% of all capital investment is energy related (14). This fraction rose to about 30% in the late 1970s to early 1980s. Energy investments strain the U.S. economy, and they severely constrain development in industrializing countries.

The constraint on development is one of the most important concerns about energy. Developing countries need energy, especially electricity and transportation fuels, for new industries and for improving lifestyles. The scale of conventional energy supply and conversion investment would soak up much of their capital, domestic and foreign. There are two ways to reduce this capital cost and yet provide the desired services: low cost energy efficiency improvement and energy supply systems with relatively low capital cost. Both these kinds of measures need to be strongly pursued. Energy efficiency opportunities are discussed here at some length. Biomass and natural gas systems are two supply options that tend to have relatively low capital costs.

The key conclusion is that capital for energy supply systems is scarcer than energy resources, in the limited meaning of the latter term as a natural endowment. Most of what consumers spend on energy goes for capital expenditures already made and to provide for their replacement.

In its simplest form, the price of energy pays for fixed (capital) and variable (operating) costs:

$$PQ = K \cdot CRF + VQ$$

where P is the energy price (eg, dollars per kilowatt hour), Q is the quantity (eg, kw·h) sold in a year, K is the capital cost of the plant and equipment (eg, of power plant, transmission, and distribution facilities), CRF is the capital recovery factor (fraction per year of the initial cost, primarily to repay the lenders or investors with interest), and V is the variable cost per unit of energy (primarily labor for operations and maintenance and fuel to operate the power plant). Considering an electric utility, a new power plant plus new transmission and distribution facilities might have the following cost parameters: CRF = 15% per year, capital cost = \$2000/kw of capacity, average use = 70% of rated capacity over the year, and variable costs = \$0.03/kw·h (of which \$0.02/kw·h is for fuel). With these numbers

$$P = \frac{2000 \times 0.15}{0.70 \times 8760} + 0.03 = \$0.08/\text{kw} \cdot \text{h}$$

where 8760 is the number of hours in a year. The average price of electricity in 1992 was \$0.68/kw·h. Of course, this calculation is crude and it neglects taxes and regulatory requirements, but it does illustrate that electricity from new plants can be somewhat more expensive than at present (though this opinion is controversial).

Environmental Impact of Energy Supply and Use. Energy supplies and intensive energy uses cause most air pollution and much other environmental damage (although there are other important anthropogenic sources of environmental degradation such as in agriculture and forestry) (15). Air pollution is the most important (16), and it is conveniently categorized in terms of a geographical scale.

Metropolitan smog associated with ozone at low altitude, especially in warm weather, and carbon monoxide in cold weather, are problems in many metropolitan areas around the world. Low altitude ozone is produced by sunlight acting on nitrogen oxides (NO_x) and hydrocarbons (HC). Fuel combustion is a significant source of these pollutants, especially from motor vehicles. Carbon monoxide is also largely created by motor vehicles. Power plants and heavy industry (such as refineries and steel mills) are also sources of air pollution in some cities. In China, eg, combustion of coal results in soot and sulfur oxide as well as NO_x pollution, which can be severe, especially in winter.

Low altitude ozone is also a regional problem, eg, northeast of Los Angeles and in the north central United States (17). More highly publicized is SO_x and NO_x pollution, primarily from power plants and heavy industry, which results in acid rain (18). This causes acidification of lakes, making them sterile, and is a factor in forest dieoffs, eg, in Germany's Black Forest. The sulfur is emitted in the combustion of high sulfur coal and high sulfur oil (although sulfur can be, and to a large extent is, removed at refineries). Most nitrogen oxide is a direct by-product of combustion in air, ie, although nitrogen is relatively unreactive, the high temperatures created when oxygen in the air combines with fuel lead to some reaction of the nitrogen in the air with oxygen.

Carbon dioxide from combustion of fossil fuels and, to a lesser extent, methane leaked from natural gas systems, are the greenhouse gases responsible for most of the global warming potential (19–21). The consequences of global warming are expected to be severe: summer temperatures that may be high enough to be dangerous to health, shifting rainfall patterns likely to dry up the U.S. grain belt, and eventual flooding of extensive low lying areas. In addition to these forms of air pollution, energy systems are also responsible for oil spills on the oceans and shorelines, severe degradation in many coal mining areas, and dams that displace people from valuable river areas and degrade important environmental values.

The environmental problems associated with energy have resulted in extensive regulations and new technology aimed at vehicular, industrial, and power plant air pollution as well as oil spills and damage caused by coal mining. (U.S. regulations do not as yet address global warming.) In the face of growth, especially in vehicular travel, these environmental policies have only been able to stop the increase in environmental damage at the metropolitan and regional levels in the United States (although absolute progress has been made in some areas). In the absence of greater progress on metropolitan and regional impacts and in the face of global climate change, environmental problems are probably the main challenge for energy policy. A gradual shift in the forms of energy supplies and substantial progress in energy efficiency are tools that must be brought into play to supplement cleanup technologies. Another option, cleanup technologies for capturing and sequestering greenhouse gases like carbon dioxide, may be too difficult but should be studied.

Policies to Improve Energy Efficiency, An Overview

A preliminary look at the social institutional opportunities to improve the efficiency of energy use is in order (22–24). Policies will be discussed in some detail later. The first point to be made is that the United States and most other governments have strenuously encouraged creation of energy supply. (Often these policies created extremely profitable opportunities to induce the desired investment.) Now that efficient energy use is a critical tool for managing energy-related economic, environmental, and security problems, it is timely for societies to take measures to encourage efficiency improvement. Fortunately, promising measures are available because efficiencies are typically low, and there are many opportunities in technology and practice to improve efficiencies cost effectively.

The policies are justified because society's interest in improving energy efficiency is much greater than the individual's. For most individuals and firms, energy costs are a minor part of total costs and potential savings do not warrant the priority needed to improve efficiency. Moreover, the individual cannot do much as an independent energy user to help manage goods held in common, like environmental quality.

Policies to encourage individuals and firms to deal with technical issues associated with efficiency include performance regulations for mass-produced energy-intensive products like light bulbs, refrigerators, and cars, and thermal standards for buildings. Policies to help energy users gain access to capital (a serious difficulty for many individuals and firms and a serious hassle for many more) include demand-side management by utilities based on financial assistance for efficiency projects. The creation of efficient new technologies is encouraged by policies to "push and pull" technology: government support of research and development, and directed government procurement, respectively.

ANALYZING THE EFFICIENCY OF ENERGY USE

Effective analysis of energy use is important because energy supply is capital intensive and needs must be anticipated and because energy efficiency calls for innovative public policies. In both areas good planning calls for good analysis. In this section the focus is on the language and methods of analysis. Later sections will deal with actual levels of energy use in different sectors and actual opportunities to improve efficiency.

Activity-Intensity Analysis

Effective analysis of energy use requires disaggregation. It is important to study lighting, driving, steelmaking, and other individual energy-intensive activities. Energy consumption is the sum over the energy uses in the various activities:

$$\text{Consumption Rate} = \Sigma_i A_i \, (EI)_i \qquad (1)$$

where A_i is the level of activity and EI_i is the energy intensity in subsector i. For example, when the use of lighting is involved, the activity could be area in square meters lit to a certain standard; thus the energy intensity is average electrical watts per square meter. The product of A and EI is then in Watts, a rate of energy use. Consider driving as another example. The activity could be vehicle miles per year; then the energy intensity is gallons of a standard fuel per mile of average driving. For steelmaking, the activity could be annual tons of steel produced, and the energy intensity is energy use per ton of an average mix of steel products.

It is critical for policy analysis to account separately for the amount of each activity and the intensity associated with it. It is also important to avoid the trap of thinking too narrowly about efficiency improvement. When analyzing furnace and process efficiencies, eg, it is important to avoid taking the activities as givens, independent of social and technological change and of energy-related policymaking. Passenger transportation is another example. It is provided by several modes, including walking. Planning related to changing the mix of modes is important to energy analysis, because the different modes have rather different energy implications.

Divisia Analysis. Change in aggregate energy consumption depends on three factors: the overall growth in activity, the change in the mix or composition of activities, and the change in the individual energy intensities. Overall activity can be represented by a summary index, depending on one's needs. It could be gross domestic product. One way to define summary measures for changes in the other two factors is Divisia analysis (25). The energy-weighted average energy intensity defines the real energy intensity. The energy-weighted average activity represents the combined effect of overall growth and relative changes in activities.

It is possible to express aggregate energy use as the product of three indices:

$$I_E = I(Q)I(S)I(EI)$$

or, equivalently, in terms of three growth rates:

$$G(E) = G(Q) + G(S) + G(EI) \tag{2}$$

where Q refers to the summary activity, such as GNP; S, to sectoral the mix of activities; and EI, to the real energy intensity. To obtain an expression like equation 2, define for the compound growth rate (in percent per year) of a variable X for the period t to $t + T$ years:

$$G(X) \equiv \frac{100}{T} \ln \left[\frac{X(t+T)}{X(t)} \right] \tag{3}$$

and the energy-weighted-average growth rate:

$$\langle G(X) \rangle = \Sigma_i W_i G(X_i) \tag{4}$$

where the sum is over the sectors or modes of equation 1 and the time-dependent energy weight for subsector, or mode, i is

$$W_i(t) = \frac{1}{2}\left(\frac{E_i(t+T)}{E(t+T)} + \frac{E_i(t)}{E(t)} \right) \tag{5}$$

which is the fraction of total energy used in the subsector. It is found that

$$G(E) \approx \langle G(A) \rangle + \langle G(EI) \rangle \tag{6}$$

To relate equation 6 to equation 2, $\langle G(A) \rangle = G(Q) + [\langle G(A) \rangle - G(Q)]$. The quantity in brackets is $G(S)$, the contribution of the change in the sectoral mix of activities. Equation 6 is not exactly what was wanted; it is an approximation because activity and intensity interact, but it is a good approximation.

Categorizing Energy Uses. One step in analyzing energy consumption is defining the categories of users. For example, should energy use by automobiles include the associated fuel consumption at petroleum refineries, the fuel consumption in maintaining and repairing automobiles, and that in manufacturing them? The user categories must be defined carefully or the same energy use will crop up in more than one place; there will be double counting. Economists have two ways of representing production (so that it can be summed without double counting) that provide good models for energy analysis. One is the final-product representation. In this picture, the values of all final products are added; these are products purchased by final consumers, not intermediate products purchased by businesses for further processing or resale. The categories that are summed are the different kinds of final products. The other method is the value-added representation. In this picture, the contributions of each category of production to the whole of production are added, eg, one term is the increase in the value of materials processed by the steel industry.

Theoretically, these two representations lead to quite different energy analyses. In the final product representation, automotive energy use is expressed as the combination of gasoline for driving and all the related energy uses for servicing cars, manufacturing cars, road building, manufacturing cement for the roads, and so on, in an endless sequence. This combination can be calculated with the help of input–output (IO) analysis, with its coefficients relating expenditures in one sector to expenditures in others (26). The advantage of the IO approach is that it expresses energy use in terms of more or less fundamental activities, tying the intermediate activities, like the part of steelmaking that is associated with cars, into the calculation of energy use for driving (27).

The value-added representation is simpler. It involves as separate items energy use in driving cars, servicing cars (and other equipment), assembling cars, steelmaking (for all purposes), etc. There is no doubt that, for a good energy analysis, analysts must in any case carefully consider developments in all such separate sectors that are energy intensive. That is not where the crux of the difference between the two representations lies. The weakness of the final-product analysis is that the time dependence of the

IO coefficients may be inaccurately specified, yet may be critical. The relative changes in demand for final products, like driving or housing, may be less important than changes in the input–output coefficients that, eg, relate driving to a certain amount of steelmaking. The complex final-product analysis is not easily used to analyze technical change, such as the changing materials content of final products. It can, however, be effective for analysis of a particular product, where a few coefficients instead of the n^2 coefficients of the full input–output matrix are involved. IO analysis is also ideal for analysis of the implications of imports and exports in terms of overall content of energy in products (28).

In this article, energy consumption will be discussed using the value-added representation. This does not mean that connections, eg, between driving and manufacturing, can be ignored. Analysts have the obligation to consider important connections whether or not their formulations are designed to internalize those connections.

Energy Accounting Issues

Commercial Energy. Commercial energy involves the conversion of natural fuels into convenient energy carriers and the delivery of those carriers to customers. These carriers, like electricity, gasoline, and dry natural gas are essential for services like high quality lighting, high speed transportation, communications, most building appliances, and most kinds of manufacturing. Total current annual consumption of primary commercial energy in the United States is 82 quadrillion Btu (quads); world consumption is about 340 quads (29). So, with 4.6% of the world's population, the United States consumes 24% of the commercial energy.

Noncommercial Energy. The noncommercial energy is mainly biomass, either wood harvested for the purpose (both trees and brush), or by-products of other activities (paper- and lumber-mill residues, crop residues, dung, household trash, etc). Total current annual consumption in the United States is about 3 quads, primarily industrial by-products (in the paper, wood products, forestry, and food industries), and firewood used in residences. World consumption of noncommercial energy is 40 to 50 quads (30).

Primary or Resource Energy. Primary energy is the energy from natural sources that is converted for use by people. In the United States, some 92% of the primary energy is used for heat and power. The remaining use of fuels is for materials in making products like plastics. This feedstock use is often counted in energy consumption totals, even though the energy is not converted into other energy forms. Furthermore, hydroelectricity and nuclear electricity output is often expressed as primary energy by multiplying by the fossil fuel to electricity conversion factor of about 3.3.

End Use Energy. What is called end use energy is usually the content of the carriers converted at the point of use. It is poor practice routinely to sum up the Btu values of all end use energies, even though it is often done. Electricity should normally be accounted for separately from fuels, because electricity typically costs about three times as much as fuels, and involves about three times as much primary energy use. (For petroleum-based carriers like gasoline the losses in conversion and delivery are much smaller, so for them the issue is much less important.) In addition, substitution between electricity and fuel is usually more difficult than among fuels.

Because a quantity of electricity expressed as electrical energy differs greatly from the primary energy involved, it is essential to know which accounting convention is being used for electricity and whether it is used consistently. Unfortunately, many authors are careless on this point, especially when reporting total energy that combines consumption of electricity with that of other forms of energy.

Cost of Conserved Energy and Conservation Supply Curves

Let K be the up-front or capital cost of equipment whose purpose is to save energy in some energy-using facility (31). The cost of conserved energy (CCE) is the annualized payment for the conservation equipment divided by the annual energy savings:

$$\mathrm{CCE} = \frac{K \cdot CRF}{S} \tag{7}$$

where S is energy savings per year when the equipment is installed in the facility and used normally (the units for S might be kw·h/yr) and CRF is the capital recovery factor. Consider that K has been obtained by borrowing at interest d (annual interest rate). Then $K \cdot CRF$ is the annual payment that covers both interest and capital, paying off the loan in n years.

$$CRF = d/(1 - (1 + d)^{-n}) \tag{8}$$

From a simple economic viewpoint, the behavior that should occur is that all investments will be made for which CCE is less than the price of energy (or the expected price of energy during the life of the equipment).

Example. Overhead cams are estimated to increase the retail cost of a typical automobile by ΔK = \$74 (in 1987 dollars). The fuel savings in average driving are estimated to be 6%. Consider applying this technology to a base car with 22 mpg and driven 10,000 mi/yr. The savings rate is then

$$S = 0.06 \times \frac{10{,}000}{22} = 27.3 \text{ gal/yr}$$

Assume the car has a useful life of 10 yr and the inflation-corrected, or real, loan rate to the buyer is 10% per year ($d = 0.1$). Then

$$CRF = 0.1/(1 - 1.1^{-10}) = 0.163/\text{yr}$$

Then the cost of conserved energy is

$$\mathrm{CCE} = 0.163 \times 74/27.3 = \$0.44/\text{gal}$$

If the price of gasoline is more than \$0.44/gal, as it of course is, this equipment should be adopted.

The interest rate in equation 8 is the decision maker's "discount rate." It is the overall cost to the decision maker of raising money (percent per year), including the costs of providing the down payment and other commitments that may be required. Proponents of energy conservation argue that a longer time horizon, ie, a lower discount rate, should be considered in designing energy efficiency policies. If, as they often suggest, a societal discount rate of 3% per year is adopted for the analysis, then $CRF = 0.117$ per year, and CCE = \$0.32/gal. Thus the use of this equipment is justified at a lower fuel price. Others, who oppose government intervention in new areas like energy conservation, argue that the correct calculation should only reflect the original buyer's perspective. If $n = 4$ yr is the original buyer's time of concern and is adopted for the analysis, and d is 10%, then $CRF = 0.315$ and CCE = \$0.86/gal. In this view, the equipment change is only justified at a higher fuel price.

Proponents of energy conservation also argue that for energy-policy analysis the price of energy to be compared with the CCE should include externalities or the net costs to society of the energy use beyond the costs included in the actual price. For example, energy suppliers and users do not pay for the air pollution damage caused by the energy system (which occurs even when all players satisfy environmental regulation). Unfortunately, the analysis of external costs is extremely uncertain.

The initial form of a conservation supply curve (CSC) to consider is shown in Figure 5. The y-axis, K/S, is the unit capital cost of each project in, eg, dollars per kilowatt saved. The x-axis shows the cumulative energy savings when the projects being analyzed are put in order of increasing unit capital cost. The CSC is analogous to an oil or gas supply curve: as money is spent to develop facilities, the capacity to deliver energy (eg, bbl/day) increases (32). With the CSC it is the capacity to save energy.

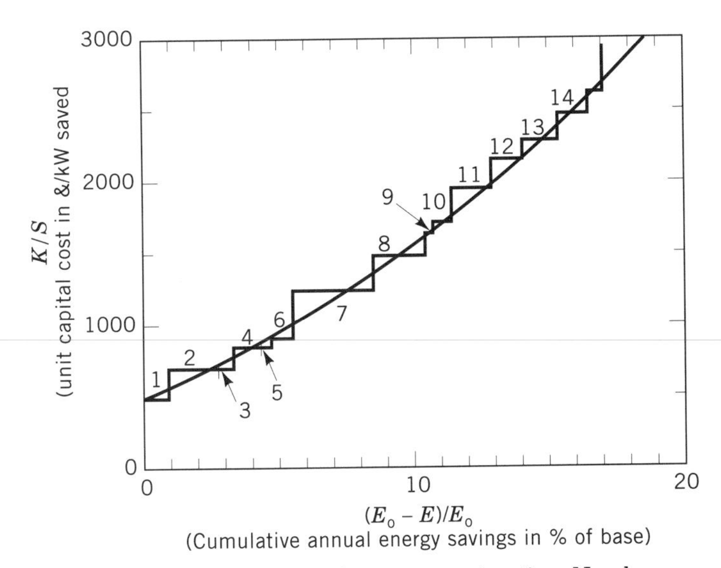

Figure 5. Conservation supply curve construction. Numbers on the curve correspond to hypothetical project numbers.

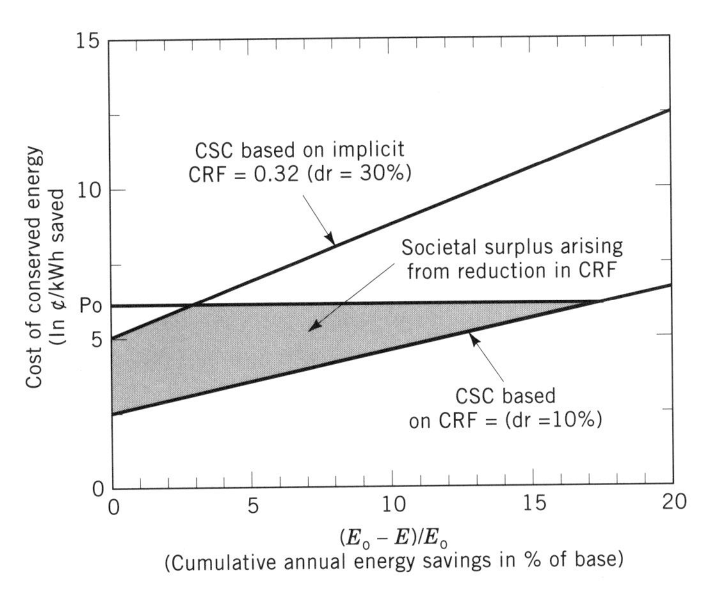

Figure 6. Dependence of conservation supply curve on capital recovery factor. CCE calculated assuming project life is equal to 10 yr.

Alternatively, to facilitate comparison with the purchase of energy, analysts often annualize project costs to obtain a CCE on the y-axis. Then the CSC takes on the form shown in Figure 6. Unlike Figure 5, it now depends on the CRF, or the discount rate.

These are normative views; they describe the effect of energy conservation investments that should occur according to a simple financial criterion. One change is often made in moving toward a description of actual behavior: the use of an implicit discount rate. After correcting for any investment inhibitions one may want explicitly to account for, one adjusts the discount rate d in equation 8 so that the actual investment behavior is described, or fit, by equation 7. Implicit discount rates are usually much higher than the apparent cost of raising money.

The Thermodynamic Potential for Improved Efficiency

The laws of physics, especially thermodynamics, determine the minimum conversion of energy that is needed to perform any task or "service." Many cases are straightforward: lifting a weight, creating a certain amount of light, raising the temperature of some material, and causing a certain chemical reaction in given input materials. For these cases, it is possible to compare the ideal energy use to that typically used in the economy.

Typically, this ratio, the thermodynamic efficiency, is in the range 0–50%. There is great variability. The more narrowly one specifies the service to be provided the higher the efficiency of today's equipment tends to be. For example, the process typically used to convert iron ore to molten iron, the blast furnace, is some 80% efficient. For the entire steelmaking process from iron ore to rolled products, the efficiency is about 25%.

As indicated by this example, if the minimum energy use as required by physical laws is relatively high, today's thermodynamic efficiencies tend to be high. However, if little or no energy is required or if, in principle, energy could be extracted in performing the service, then the efficiencies tend to be low. Thus fabrication and assembly operations in manufacturing, where ideally the minimum required energy is negligible, have thermodynamic efficiencies of essentially 0%. In transportation and building heating, as was mentioned, most energy conversion devices have substantial thermodynamic efficiencies; but the entire vehicle or building has a low, although essentially undefined, efficiency, because the heating or the transportation could, in principle, be achieved with little or no energy inputs.

In the following two sections, the methods described in this section and the policies alluded to earlier will be developed in connection with the energy efficiency of industry and transportation, respectively. Discussions of energy efficiency in residences and services, often collectively referred to as the buildings sector, are available (33,34).

INDUSTRIAL ENERGY EFFICIENCY

Around the globe, industry consumes almost 50% of all the commercial energy used and is responsible for roughly similar shares of greenhouse gases, the emissions responsible for regional air pollution, and the waste solids and liquids that contaminate land and water. In this section, industrial production and energy consumption are considered, focusing on their ratio, energy intensity. The aggregate energy intensity of the industrial system as a whole (E/A) depends on the composition of production and on the energy intensities of each sector. These dependencies can be separately summarized: the rate of change in aggregate intensity caused by the shift in the sectoral composition of production and the rate of change in the energy-weighted average of the several energy intensities, or real energy intensity (see above). The development of the real energy intensity is discussed from the perspectives of economics and technology. The section concludes with a discussion of policies. An up-to-date reference on the U.S. industry (35) and a global perspective (36) are available.

Historical Statistics

Production. Aggregate industrial production in the United States grew at an average rate of about 3.3% per year from the late 1950s to 1970. It grew about 2.4% per year during the 1970s and 1980s. Industrial production grew rapidly in the 1960s and early 1970s. Growth then slowed but has still been substantial. Heavy industry, however, including the energy-intensive materials sectors, grew quite slowly in the 1970s and 1980s, as will be discussed.

Unfortunately, these production growth rates are rather uncertain, because different measures of production disagree. One prominent measure is value added, industry's contribution to the gross national product, with dollar values deflated or adjusted for price changes. Another measure is deflated gross output, the value of shipments collected on a census form at the factory or establishment, deflated by the Bureau of Labor Statistics, and then added for all establishments. Another is an index of production created by the Federal Reserve Bank (FRB), which is based on quickly available measures of activity. These measures disagree, often at the 0.5 to 1.0% per year level. All these measures have serious flaws, as seen by examining each carefully and by comparing each with production tonnage in the bulk-materials sectors.

Value added is hard to deflate accurately because information about what materials a factory buys may not be available in sufficient detail. In addition, the quantitative connection between value added and production is weak because, in this time period, major changes have occurred in two aspects of value added; a declining return to capital from industrial establishments and a declining fraction of services done in-house.

A principal purpose of the FRB industrial production indices is quick publication. The indices are determined from a short list of quickly available indicators, often quite inappropriate for long-term energy analysis. Although widely used, eg, in modeling of the economy, FRB indices are poor for industrial analysis.

Gross output has the virtue that it is much closer to the primary data, the census questionnaire. And it involves deflation only of the value of output, which is relatively well defined. Unfortunately, several energy-intensive sectors are parts of vertically integrated firms, and both the value of shipments and price deflators are substantially in question because prices internal to a firm are not meaningful. The difficulty is apparent when comparing this index with tonnage in the homogeneous materials subsectors of primary aluminum and cement and with various measures of petroleum refining, another vertically integrated sector. A second difficulty is the extensive double counting that occurs in aggregating. All in all, however, this is the best general measure of growth of production in industrial sectors.

This digression on production data is worthwhile because the analysis of industrial efficiency can be no better than the data; and production data, surprisingly perhaps, are a weaker link than energy data, even though the latter are collected less frequently, and in less detail, and are less important in the general scheme of national accounts. Perhaps the key is that energy is a simpler concept than money value.

Aggregate Energy Intensity. Use of primary energy by industry, including the fuel used to generate the electricity consumed by industry, grew from about 20 quads/yr in the late 1950s to 31 quads/yr in the early 1970s. The 1960s were an extraordinary age of material growth. Then in the early 1970s came the natural gas shortages and the beginnings of real (ie, inflation-corrected) price increases for electricity, a reversal of the historical decline that blessed that energy form. In the fall of 1973 came the first oil shock and in 1979, the second, with major increases in the price of oil and gas. Industrial energy consumption hesitated in the mid and late 1970s, fell to a low of 26 quads in 1982–1983 and climbed back to 30 quads in the early 1990s, as oil and gas prices declined.

The aggregate energy intensity, ie, the ratio of the energy consumed by all of industry to total production was steady in the period from 1958 to 1972 and then fell by more than 33% to 1985. (The quantitative remarks about energy intensities here and below refer to energy use for heat and power in manufacturing, which accounted for

63% of total industrial energy use in 1985. The remainder is feedstock uses of fuel, and energy use in agriculture, mining, and construction.)

The different forms of energy had quite different histories, however. From 1958 to 1985, the aggregate coal and oil intensities fell 70 and 60%, respectively. (When specific fuels are referred to, it is consumption at the site, not including consumption at power plants.) During this period, coal was being eliminated, except at large facilities and in heavy industries. Coal is an inconvenient fuel to handle without special skills and special equipment; more recently coal has also been hampered by environmental regulation. During this period, petroleum was gradually losing share, and after the second oil shock, its use was drastically curtailed. Most consumption of petroleum is now as a feedstock, at remote sites such as forest products, mining, and agriculture and at refineries where by-products of the process supply much of the energy. Between 1958 and 1971, the aggregate natural gas intensity rose 30%, but by 1985, it dropped from its peak by more than 50%. Natural gas had been widely introduced through pipeline construction by the beginning of this period, and it was favored for its low price and convenience until the shortages associated with federal price controls occurred in 1971, shortly before the first oil shock.

The combined result of these developments was that the aggregate fossil fuel intensity fell 15% from 1958 to 1971 because of fuel efficiency improvement in energy-intensive sectors, even though real fuel prices were low and falling. Beginning in 1971, the decline quickened and, by 1985, the aggregate fossil fuel intensity declined by another 50%. Part of the reason for this accelerated decline was a relative shift in production away from energy-intensive production (see below).

The pattern for electricity is similar, except that continuing electrification (new uses of electricity) overlies the other developments (Fig. 7). Thus the aggregate electricity intensity grew rapidly between 1958 and 1970, even though the efficiency of electricity-intensive processes, such as electrolysis of brine to produce chlorine and smelting of aluminum, was being improved. Real electricity prices fell during this period. Since 1970, electricity prices have mostly been rising and the aggregate electricity intensity has gradually declined. The two forces for decline, efficiency improvement and the relative decline in electricity-intensive production, have slightly outweighed ongoing electrification in the aggregate.

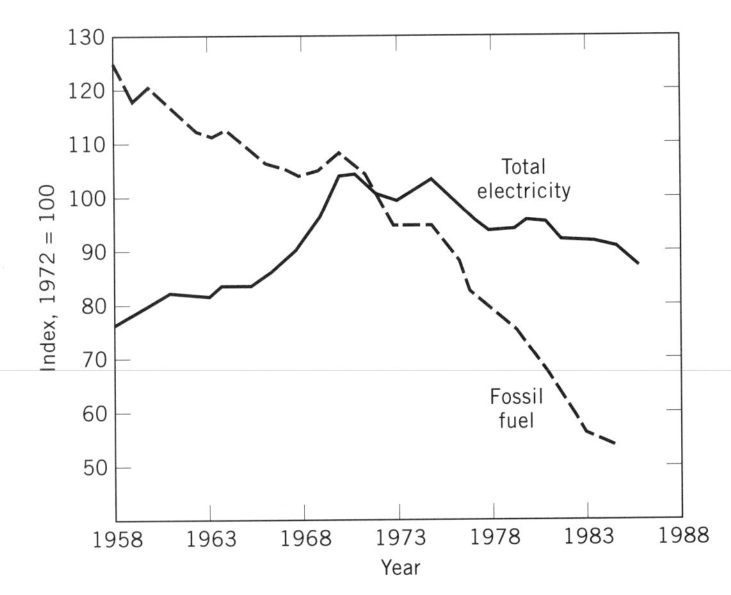

Figure 7. The aggregate electricity and fossil fuel intensities of U.S. manufacturing, relative to 1972.

In 1958, more than 20% of the total electricity used was generated and used on-site by manufacturers. On-site generation had been high in the early days of electricity but had long been falling. It dropped to less than 8% of the total by 1981. With the Public Utilities Regulatory Policy Act of 1978, on-site generation has begun a comeback, rising to 10% of the manufacturing total in 1986 and to 11.4% in 1990. These findings indicate that, up to 1981, electric utility sales to manufacturers were growing about 0.7% per year faster than total electricity use. Utility sales growth is now slower than the growth of total electricity use.

The Shifting Composition of Industrial Production

The relative roles of the different sectors in industrial production is important to energy efficiency because the energy intensities vary strongly from sector to sector (37) (Fig. 8). With a broad brush, Figure 8 shows roughly 10-times higher intensities in chemicals and primary metals than in general manufacturing. The ratio of the energy intensity of aluminum reduction to that of assembling computers or telephone equipment is about 300 to 1 (with value added as the denominator). Lumping these sectors together is similar to lumping cars with bicycles in the transportation sector.

The Maturing of Bulk Materials. The manufacture of bulk materials is much more energy intensive than manufacturing in general, and the use of materials in tons per year is declining relative to total production in all affluent societies. The consumption of materials used in large quantities (paper, fertilizers, synthetic fibers, plastics, other industrial chemicals, cement, glass, pottery, and metals) is declining relative to total industrial production. This phenomenon, called dematerialization (38), has contributed a 0.5–1.0% per year decline in the aggregate energy intensity of industry Organization for Economic Cooperation and Development (OECD) countries (25,39). Note that dematerialization is not associated with increasing imports of some materials by some nations.

Dematerialization is somewhat controversial. Consider two kinds of evidence for it: general statistics and a case study of a particular material, plastics. The conclusion from general statistics is controversial because (*1*) the long-term decline in older materials like steel (in kilograms per dollars of GNP) could be caused by materials substitution, (*2*) the decline in overall materials (in kilograms per dollars of GNP) dates only from the early 1970s and might be a temporary phenomenon associated with the oil shocks, and (*3*) there are disagreements about how to measure and aggregate materials consumption. Upon

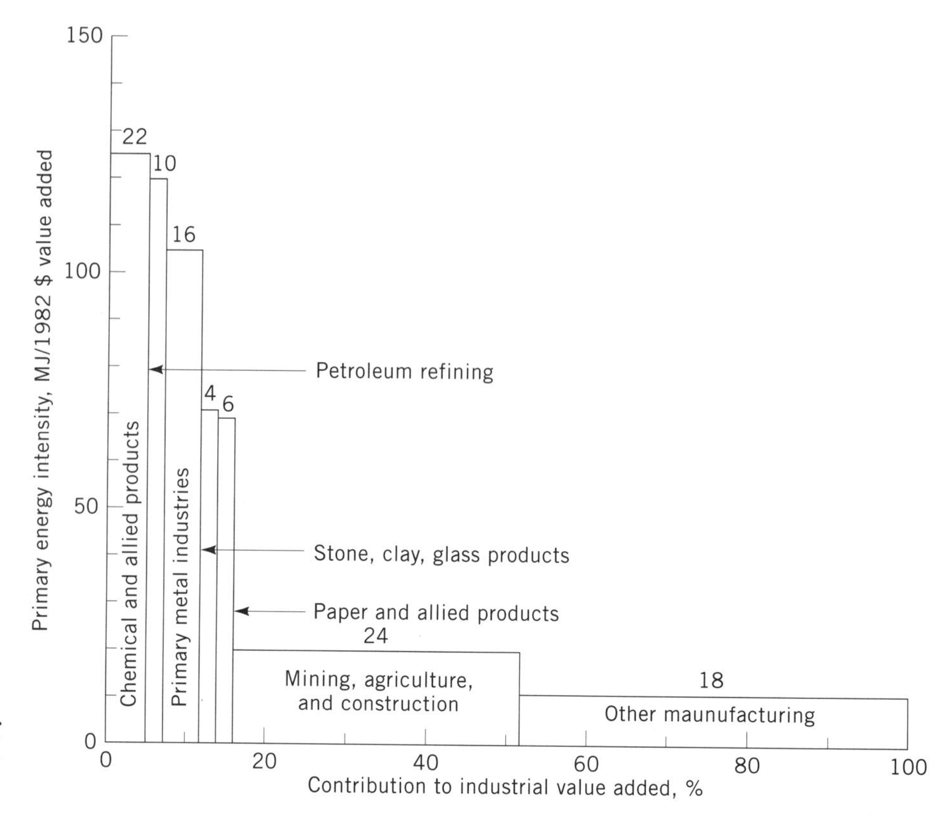

Figure 8. Primary energy intensities of U.S. industries versus value added by sector in 1981.

examining the developments for particular materials, these objections do not appear to be critical (40).

The only bulk materials whose consumption still grows as fast or faster than the economy in the United States are in the chemical products family: plastics and industrial gases. But growth rates for these materials as for all the bulk-chemical groups have, nevertheless, been falling dramatically. What is happening to plastics production? Saturation effects apply even to plastics. Markets for heavy consumer products are saturating. For example, while the application of plastics to motor vehicles is increasing, unit sales of vehicles are no longer growing with the economy. Innovative consumer products tend to have a low materials:cost ratio (kilograms per dollar). For example, although electronic equipment typically has a plastic structure or body, the ratio of weight to cost is low. Materials are being used more efficiently. For example, linear low density polyethylene, introduced in the late 1970s, allows the use of thinner films and is taking over low density polyethylene markets. This increasing efficiency is being driven, in part, by the competition among materials, eg, the competition for materials for grocery bags and beverage containers is fierce, putting a premium on efficient design and even beginning to bring in considerations of plastics recycling. Improved materials are increasing product durability. Plastic pipes, other uses of plastic in construction, and plastic auto parts that reduce rust and corrosion often contribute to longer product life.

In conclusion, materials manufacturing, which accounts for the lions' share of manufacturing fuel use and somewhat more than 50% of the electricity use, is likely to grow roughly with population rather than with the economy in industrialized countries.

The Fast-growing High Technology Sectors. Another structural development is taking place among some of the low energy-intensity sectors. Statistical information and case histories support the concept that high technology product sectors will play an important role in reducing the aggregate energy intensity of industry in the long term. During the 1970s and 1980s, the fastest growing sectors, especially electronic equipment, drugs, and instruments, were the main sources of growth in per capita industrial production. It seems likely that these and other research-intensive, innovative sectors will continue to propel growth in the U.S. economy as it matures.

In Figure 9, the growth rate in output is compared with energy intensity for the 71 manufacturing sectors in the National Energy Accounts (41). Total energy use in each sector is shown on the vertical axis (for 1985). The distribution is L-shaped with all the big energy users having low growth and high energy intensity (upper arm of the L), and all the high-growth sectors with real gross output growing roughly 4% per year and faster for 1971–1985 are furniture and fixtures (excluding household furniture), printing and publishing, drugs and toiletries, rubber and miscellaneous plastics products, computer and office equipment and miscellaneous industrial machinery, electronic equipment and components, communication equipment and miscellaneous electrical equipment, selected military equipment, and instruments.

The key energy-related behavior is the low and declining energy intensity of these high growth sectors. (Here, energy intensity can be measured as energy per unit of value added in a base year. Relative changes in energy intensity are, however, much better measured in terms of the ratio of energy to deflated gross output.) The decline

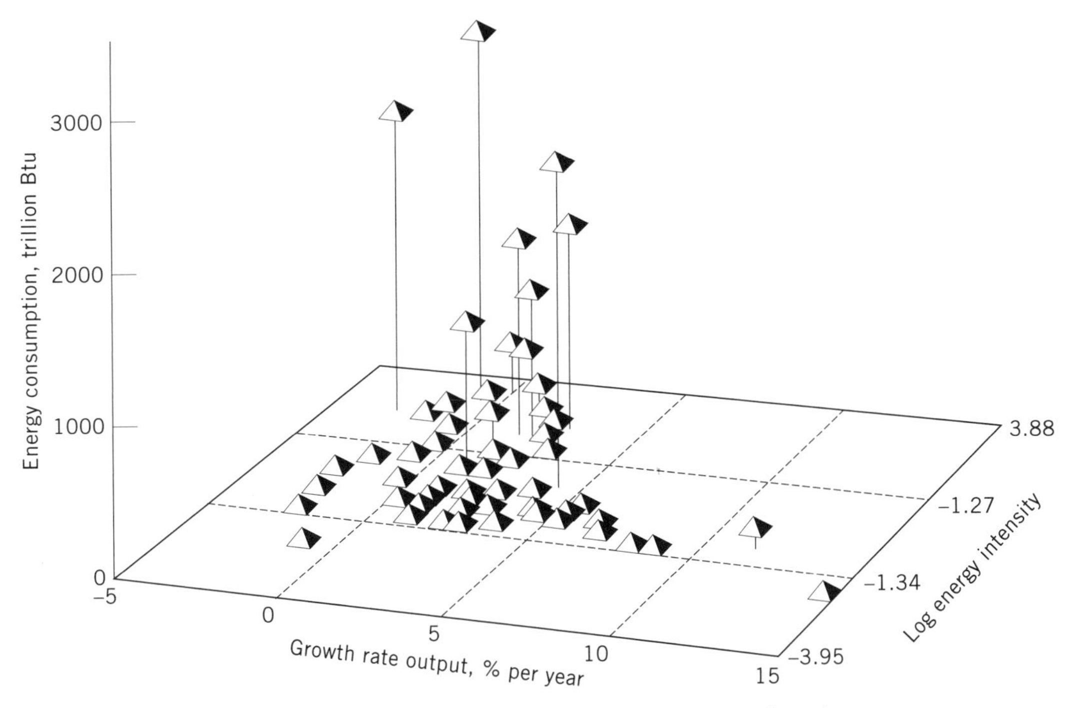

Figure 9. Growth rate of output (1971–1985), energy intensity, and total energy consumption (1985) for 71 sectors of U.S. manufacturing.

in the energy intensity, or equivalently, the increase in energy efficiency of these sectors is rapid. This decline is not due to efforts directed at improving the energy efficiency of production. Rather, it is due to continuing product innovation embodying design improvements that have principal impact on the product but relatively little on the materials and energy requirements. It is also due to the rapid creation of new, more modern, production facilities.

On the basis of statistics for 1975–1985, a judgment whether the rapidity of decline in energy intensity will match the rapid growth of these sectors is premature. During that decade there was an approximate balance, such that total energy use by these sectors grew with population rather than with the economy. Further experience and analysis are needed to form a judgment, but it is tempting to speculate that in the presence of substantial innovation, industrial energy consumption (as well as material tonnage consumption) will not grow with the economy, even in the absence of efforts to improve the real energy intensity.

Real Energy Intensities

In principle, most energy intensities could be reduced to zero or near zero (42). Thermodynamically, the difference between a collection of materials and the same materials shaped and assembled is nil. Substantial energy use is required for some endothermic chemical reactions, like reduction of metal ores (ie, removing the oxygen) certain organic reformations, and dissociating brines (such as separating the chlorine from salt). These constitute a modest part of all industry.

Actual and absolute minimum energy intensities are shown for a few materials in Table 1. Shown is energy consumption within the corresponding manufacturing sector. This and other accounting conventions affect the analysis. It is seen that in terms of fossil and electrical energy arriving at the site, aluminum smelting is about 50% efficient (thermodynamic minimum divided by carrier-energy consumption). The steel industry is about 25% efficient; it would appear more efficient if the shaping and finishing steps were not included. The industries that are less focused on a major endothermic chemical reaction are less efficient in these simple physical terms. For example, industries that use fuel for low temperature heating, as for drying paper, are relatively inefficient in these simple physical terms.

In practice, energy intensities are substantial even when no energy is needed in principle. But, historically, established energy-intensive processes, like smelting aluminum or making cement, have shown substantial ongoing reductions in energy intensity, even in the period 1958–1970 when energy prices were low and falling. Energy intensities, of course, fell more rapidly in the period 1971–1985.

There are three sources of improved energy efficiency in making a particular product: operational improvements; autonomous change, investment projects to improve the production process that are not associated with energy prices; and energy conservation investments associated with differences in the cost of energy.

Operational Improvements. Much of the efficiency improvement of the mid to late 1970s was achieved through better management without substantial investment in equipment. Basically, it is a question of giving energy a high priority in management goals. Employee involvement through training, inspection, suggestion programs, and contests are important, as are audits to identify no-cost

Table 1. Energy Intensities for Selected Basic Materials

Material	SIC	Energy Intensity (GJ/t)[a]: U.S. Industry, 1988: Primary[b]	Energy Intensity (GJ/t)[a]: U.S. Industry, 1988: Carrier[c]	Energy Intensity (GJ/t)[a]: Thermodynamic Minimum
Paper	26	24.1[d]	17.4[d]	—[e]
Petroleum refining[f]	2911	5.3	4.9	0.4
Cement	3241	5.4	4.4	0.8
Steel	3312	32.4[g]	28.1[g]	7.1
Primary aluminum	3334	207	69	29.3

[a] For millions of BTU per short ton, divide by 1.163.
[b] Electricity evaluated at 3.3 × electrical energy.
[c] Electricity evaluated at electrical energy.
[d] Excludes wood-derived fuels.
[e] The absolute value of the minimum is small, and its sign depends on accounting conventions and product.
[f] Per tonne of crude processed; thermodynamic minimum is nominal.
[g] Per tonne of rolled steel mill products.

and low cost opportunities. Metering and reporting procedures and accounting practices to provide feedback on energy use to groups within the plant are powerful tools. More thorough and frequent maintenance and use of sophisticated inspection and maintenance equipment are important. Often what works to improve operations depends on the culture at the plant. Modifications introduced by technical people may not be effective if production workers are not involved in the planning. Moreover, many of the best opportunities are known to production workers and not to management and technical staff.

The responses to the natural gas shortages of 1970–1971 and the price shock of 1973–1974 were rapid and resulted in a 10 to 20% reduction in energy intensities during the 1970s (relative to existing trends). Until the late 1970s there was little energy-efficiency investment, so the improvement was mainly the result of operational changes.

Autonomous Efficiency Improvement. The adoption of a fundamental production process usually has important energy efficiency implications but is usually autonomous. That is, the decision is essentially independent of energy price differences. For example, steel firms do not choose electric arc furnace technology with a primary concern for electricity prices (30). For minimills, the scrap-based electric-arc process involves much lower capital costs than the ore-based process: scrap is available in the region, the scale of production can be kept small, and product markets are promising. In this context, energy price differences are unimportant. This conclusion is even stronger for less energy-intensive sectors than steel. It may not be valid in some cases, however, eg, in the primary aluminum sector, where the cost of energy is an extreme 25 to 33% of total costs (and greater than value added within the sector).

General progress in production technology tends to have certain implications for energy consumption. (*1*) Total energy costs associated with established types of processes are reduced along with other costs (43). There is the evidence just mentioned from energy-intensive industries in the 1950s and 1960s when energy prices, including electricity prices, were low and, often, falling. (*2*) New applications of electricity continue to occur, with surface treatment and specific heating (as contrasted with general volume heating in a typical oven) being two currently important areas (44). In addition, some electricity consumption is associated with added environmental controls. Thus most current electrification is based on markedly superior production technology rather than energy price effects. (*3*) Learning or experience curve studies of particular processes show fairly steady and remarkably large cost reductions over long periods (45). The reductions correlate better with cumulative production than time. Energy costs decline along with total costs, although less rapidly.

An example of an experience curve is shown in Figure 10 (46). The x-axis is cumulative production of low density polyethylene, using the high pressure production process. The y-axis is manufacturing cost per pound, which is seen to decline to a mere 3% of its initial value in just under 40 yr. Energy intensity declined to 8% of its initial value dur-

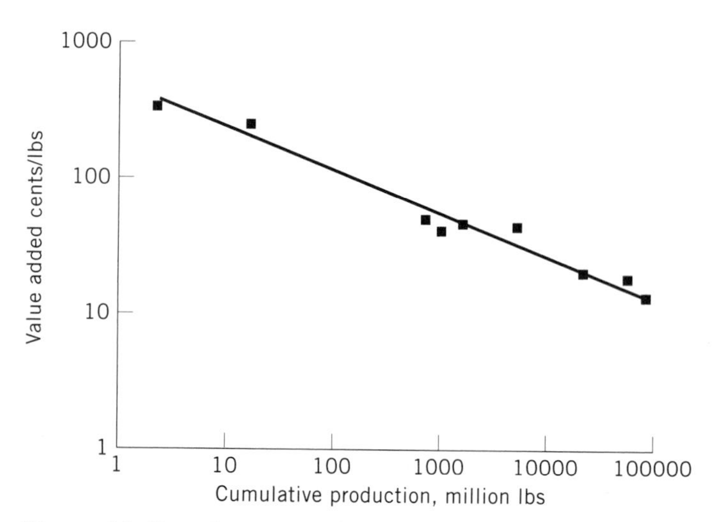

Figure 10. Experience curve for high pressure low density polyethylene. Based on constant dollar value added, obtained with 1981 factor prices. Energy use was reduced 88% (down by a factor of 0.125). Adapted from Ref. 46.

ing the same period. Moreover, just when it might be expected that opportunities for cost reduction were running out, a new low pressure process for making polyethylene was developed and is being adopted, enabling continuation of the downward trend in cost and energy intensity. With these developments, historical data indicate that there has been an autonomous trend for real fuel intensities to decline about 0.5% per year, and (excluding electricity-intensive sectors) for real electricity intensities to increase perhaps 2% per year (47).

Fundamental Process Technologies. Perhaps the most important recent examples of autonomous improvements, or fundamental process changes, are catalytic processes. In petroleum refining in the past half century, several processes to convert a larger fraction of the crude into gasoline have been created: catalytic cracking, catalytic reforming, alkylation, and isomerization. Molecular size, hydrogen:carbon ratio and molecular geometry are adjusted, and with enormous throughput and relatively modest cost. These technologies are now being turned to adjusting the chemical makeup to obtain reformulated gasoline that leads to reduced emissions, eg, to reduced evaporation.

Another industry that has been characterized by fundamental process change in recent decades is steel. While the initial step of ore reduction, or iron making in the blast furnace, has not fundamentally changed, steelmaking and shaping of products have. Refining of molten iron to make steel now takes place in a converting vessel (basic-oxygen furnace), in which oxygen is blown in, and the carbon already in solution with the molten iron is the fuel. This process greatly increases product throughput and saves energy compared with the previous standard, the open hearth furnace. Solidification, or casting, is now done continuously into a strand a few inches thick, instead of into thick ingots, which must then be heated and rough rolled to reach the stage achieved by the continuous caster. The continuous process greatly reduces the need to recycle unsatisfactory material from the beginning and end of each batch. In speeding up the process, these technologies have also helped the industry to bring the entire process under much better control, so that a much larger fraction of product meets specifications. Of course, in this way energy intensities are also substantially reduced.

The steel industry has also added a new subsector based on melting postconsumer scrap to make construction products like concrete reinforcing bars. The process technology is the electric arc furnace and continuous casting. Recent fundamental process changes in other industries and with major energy intensity implications are the float-glass process, induction heating and melting of metals, and surface heating of metals with electromagnetic beams.

Some examples of fundamental process change that may be achieved in the foreseeable future are listed below.

- Chemicals, paper, and food processes.
  - Improved separation processes based on membranes, adsorbing surfaces, critical solvents, freeze concentrations, etc.
  - Ethylene chemistry based on natural gas feedstocks.
  - Waste reduction using closed systems.
  - New and improved catalysts for chemical processing.
  - Recycling paper and plastics into new material and product areas.
  - Bioprocessing with genetically engineered organisms.
- Metals processes
  - Recycling postconsumer scrap into new product areas.
  - Near net shape casting, spray forming of products.
  - New surface treatments with electromagnetic beams.
  - Direct and continuous steelmaking.
  - Coal-based aluminum smelting, fuel-based scrap melting.

The improved separation techniques listed are all in areas where new processes have been developed and adopted in the last decade or two, and it is felt that further applications can be developed. Ethylene is the basic building block for most petrochemicals. It is now made from propane, butane, and naphtha feedstocks and might in the longer term have a more secure base in methane. The closing of systems by recycling, such as with water in bleaching of paper, is one way to avoid end-of-the-line treatment with discharge of harmful chemicals into the environment. The development of new catalysts, and new substrates on which to place the catalysts, is continuing to progress and will continue to lead to major innovations in manufacturing of chemicals, including creation of new and improved products. The challenge for recycling of paper and plastics is for manufacturers to make more valuable products from the postconsumer materials. This is an area of rapid change. Bioprocessing by organisms tailored to make the desired products and avoid undesirable by-products is a principal opportunity offered by genetic engineering. It is not known if these bioprocesses, which tend to have low throughput, can be competitive with petroleum-based processes, which have high throughput, in producing bulk products.

For the metals industry, as for plastics and paper, a principal opportunity is conversion of postconsumer scrap into more valuable products. The creation of sheet steel from scrap is one such area. High quality sheet requires low residual concentrations of copper and tin, so one approach would be to develop scrap processing techniques to separate these contaminants before melting the metal. Progress is being made in casting thinner slabs of metal at the steel mill, to reduce reheat and rolling requirements, and in casting closer to final shapes at foundries, to reduce machining requirements. Surface treatment, eg, with lasers and electron beams, is progressing as electromagnetic sources become stronger and more controlled. Direct and continuous steelmaking is under development in Germany, Sweden, Japan, and the United States. In the United States a joint government program with the old-line integrated steel firms was developed as the Steel Initiative in 1984. The concept is to replace four batch steps (agglomeration, coke ovens, blast furnace, and steel furnace) with a one- or two-step continuous process. Two motivations are the elimination of coke making with its air pollution and the reduction of capital cost per unit of production to help enable the industry to modernize. Replacing electricity-based aluminum smelting with fuel-based smelting would be partly motivated by the desire to retain the primary aluminum industry in industrialized regions. The primary aluminum industry is now in a gradual process of moving away to remote areas of the world where electricity is cheap and has few competing users.

If this were an analysis of the buildings sector, it would be necessary to discuss the percent energy savings offered by each technology; there are two reasons why this is not discussed here: (*1*) the technologies mentioned are only examples, and they represent only a few industrial sectors; and (*2*) these are fundamental process technologies. The decision to adopt such technologies, when and if they become available, will be based on broad business strategies, not on the price of energy or narrow energy-related considerations. In every example mentioned, issues involving product, environment, capital requirements, and/or location of production are more important than energy considerations as such. Still to be addressed are energy-efficiency technologies for which energy is a major component of the decision.

What can be concluded is that there are many opportunities to produce products differently. In terms of science and technology, this is a time of ferment in production process development, because of the new capabilities to design and evaluate systems through computation rather than trial and error, because of the new capabilities to operate systems accurately under demanding conditions using new sensors and controls, and because of the capabilities of new materials. There is every reason to believe that autonomous improvement in industrial processes, including energy efficiency, will continue at roughly the pace of the recent past, if investments in new facilities continue at roughly the same pace (see below).

Energy Efficiency Investments. Projects in industry that are largely motivated by energy cost reduction tend to be small or medium size (up to a few tens of million dollars but usually much smaller) and tend not to involve fundamental process change. In terms of technology, some of the projects are generic, like high efficiency motors and lighting systems. Others have names that are generic sounding, like variable-speed motor controls and automatic process controls, but these are mostly custom-designed systems, with a large cost for engineering the system (ie, in addition to the cost of equipment and installation as such). Other energy efficiency technologies, most of them in energy-intensive materials manufacturing, are process specific. For example, specific technologies from the grinding phase of cement manufacturing are use of roller mills instead of ball mills for grinding raw material, use of high efficiency classifiers for separating ground material that meets specifications from material that needs to be ground some more, and use of grinding media of increased hardness.

As these names suggest, energy efficiency investments are often not generic devices, like more insulation or more heat recovery. They are instead incremental production process improvements that, eg, reduce maintenance problems, improve product uniformity, increase throughput, and save energy. These multipurpose projects tend to be favored by decision makers over the single-purpose project whose only benefit is a simple reduction of costs. In other words, decision makers are attracted to projects that reduce uncertainty and hassle, and they prefer marketing advantages other than simple cost reduction. (Of course, everyone likes projects with really large cost savings.) The multipurpose project often is more easily sold to different players at the plant. It has potential advantages in a dynamic view of production, whereas the heat exchanger or insulation project carries the risk of having decreased value if the process changes. The multipurpose project can have the advantage of moving the state of the art at a plant, preparing the way for further technological advance.

In the Industrial Energy Efficiency Improvement Program (1978–1985), trade associations were asked to report examples of energy efficiency projects to the U.S. Department of Energy (DOE). Roughly 50 different kinds of projects, most of them process specific and few of them generic, can be found in each of these lists. Similarly, studies of five industries in the late 1980s (paper, steel, cement, glass, and textiles) list 50–100 different kinds of projects, mostly process specific, for each industry (48). The point is that each of the energy-intensive "process" industries is a world in itself, even if generic sounding terms like *controls* and *separation* are used. Energy conservation analysts have tended to select short lists of technologies, exaggerating their efficiency improvement potential, in the interest of simplification. Instead, achieving improved efficiency is not simple, even if some aspects of it, like efficient mass-produced devices for households may be. Achieving high efficiency in industry requires technological sophistication with specialized knowledge of the particular industry, its production processes, products, and markets. And then there is the hard work of implementing change. There is little simple efficiency magic.

That said, it is time to switch hats and consider some simplistic economics, while not entirely forgetting the above message. Assume that a good energy audit for a plant has been carried out. (For all the multipurpose projects for which energy is important, the energy-related part of the capital cost has been identified, eg, by estimating the benefit streams associated with the different purposes and allocating the capital costs according to the present values of these benefits, although normally such elaborate procedures are not carried out, except for large projects.) The projects can be ordered according to their cost of conserved energy, and a conservation supply curve can be created.

An estimate of the electricity conservation opportunity in fabrication and assembly sectors of industry is shown in Figure 11 for purposes of illustration. (The presentation of this conservation supply curve is illustrative. Such curves are not widely available for industry, nor are they likely to be accurate because of the heterogeneity of industry and the lack of experience at high energy prices.) Figure 11 shows the cumulative electricity savings that will

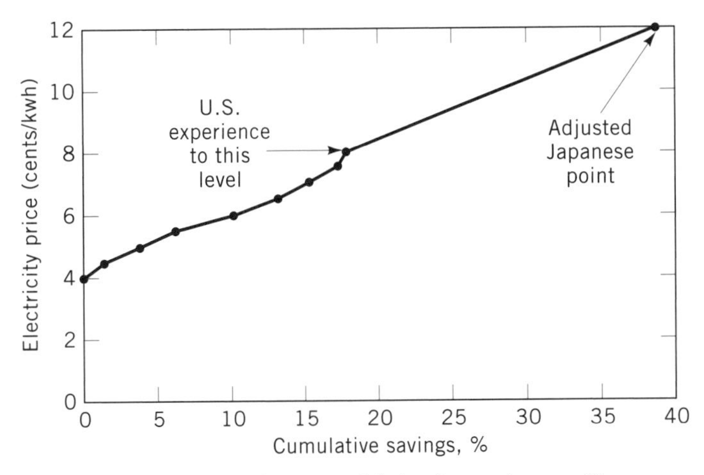

Figure 11. Supply curve: fabrication and assembly.

be achieved by conservation investments induced at a given electricity price. The curve represents typical behavior. The figure shows that with the financial criterion represented by a capital recovery factor (*CRF*) of 33%/yr, which roughly characterizes conservation investment by manufacturers, there is a relatively small opportunity for further conservation at the present price of about $0.05/kw·h. In particular, it shows that the electricity intensity is only about 5% higher than its equilibrium value, given the present price and investment behavior. (High implicit discount rates are discussed below.)

However, with a criterion more representative of capital markets, a *CRF* of 16%/yr, a 30–40% reduction of electricity intensity, would be justified at the present price and largely achieved after a delay. (With a 10% real discount rate and a 10-yr project life, the capital recovery factor is 16%/yr.) Figure 11 thus shows that there is substantial potential for efficiency improvement at negative overall cost, or a "free lunch" is to be had, if only there could be found an inexpensive way to change management practices and financing so that a longer time perspective would apply to conservation investments. There are schemes to accomplish just that. One that is beginning to be tried, but is highly controversial, is utility investment on the manufacturer's side of the meter, bringing into play the same interest rates that are used for financial decisions on energy supply projects. The barriers to investment in these projects and policies to reduce those barriers are discussed below.

Econometric Analyses of Energy Efficiency

The discussion continues to focus on real energy intensity, ie, the intensities of individual sectors, but shifts from a microeconomic view to statistical analysis of historical experience. In each sector of industry the energy intensity patterns for 1958–1985 were roughly as sketched in Figure 12 (47). The typical pattern for fuel intensity (Fig. 12**a**) shows a slow decline from (before) 1958 to the early 1970s, when prices were constant or falling, and a rapid decline from the early 1970s, when the natural gas shortage and first oil price shock occurred. (Data have not been analyzed beyond 1985.) The typical pattern for electricity intensity (Fig. 12b) shows a rapid increase from (before) 1958 to the early 1970s, when prices were falling. Electricity intensities have been roughly constant since the early 1970s.

What is desired is a simple analytical representation of these histories, with parameters that are easily interpreted. This is usually achieved through regression analysis (32): the researcher decides (ideally, after qualitative examination of the data, thought about what affects energy intensity, and many trials with different analytic expressions) on an expression, usually a linear mathematical formula, and uses a statistical package to find the parameter values that minimize the sum of the squares of the differences between the formula and historical data. There are many problems with selection of data, poor data, selection of the independent variables, and their mathematical form. There is considerable variation among the results. One good review is available (49).

Three kinds of parameters are determined: energy price elasticities, trends with production or time, and capacity–use dependence. The last of these will be briefly discussed, followed by the two more important kinds of parameters. The capacity–use effect is seen in the sketch for electricity intensity for the recession years of 1970, 1975, and 1982

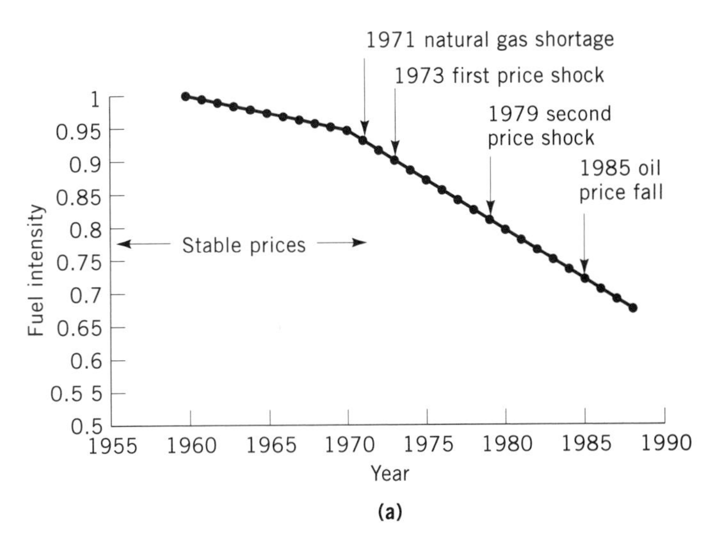

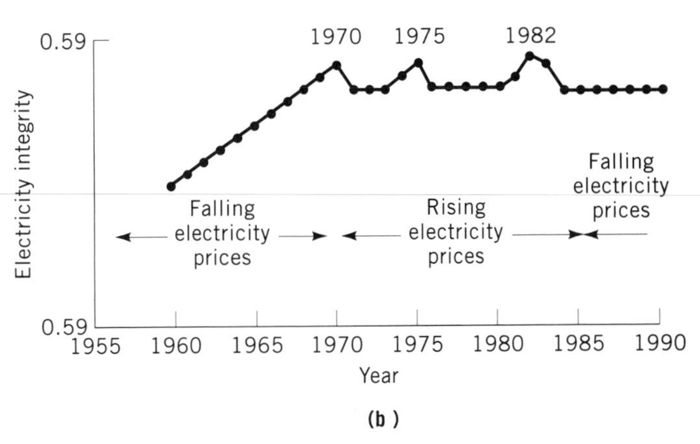

Figure 12. Fuel and electricity intensity histories (sketches).

(Fig. 12**b**). Roughly speaking, it is as if the motors were all left running where there is a downturn in production, so there would be a sizable increase in electricity intensity. Smaller but similar effects are seen for fuel in some sectors. It is interesting that simplistic economic ideas suggest that in a downturn the least efficient facilities would be shut down so that efficiency would increase instead of decrease, as observed. In fact, many other considerations than simple cost are involved in industrial decision making, such as the role of particular products and plant location in marketing.

Own-price electricities are the best known parameters in this kind of analysis (32): how much does an energy intensity decline due to decisions made as a result of a price increase in that form of energy? The assumed relationship may be:

$$EI(t) = EI(t_o)\left(\frac{p}{p_o}\right)^A e^{Bt}$$

where A is the own-price elasticity and B is a time trend (alternatively formulated as a production dependence). One recent analysis shows that the long-term elasticity is low ($A = -0.1$ to -0.3) in sectors where energy use is intensive and high (-0.5 to -1.0) in sectors where energy use is not intensive. For example, for cement, both fuel and electricity intensities are estimated to be -0.2 (47). This means that a 10% increase in price would lead to a factor of $1.1^{-0.2} = 0.98$, or an eventual 2% decline in energy intensity. On the other hand, the long-term elasticity for electricity in general manufacturing is -0.5. Then a 10% price increase would lead to a 5% decline in electricity intensity. (The exponential formula itself should be used for large price increases, eg, if $p/p_o = 2$ and $A = -0.5$ then the electricity intensity is modeled to decline by a factor of $2^{-0.5} = 0.71$, or 29%.) In truly light manufacturing, especially the fast-growth high technology sectors mentioned above, the elasticity for electricity is about -1.0. For this sector, a doubling of prices would thus essentially lead to a factor of $2^{-1} = 0.50$, or a 50% decline in electricity intensity. These results for general and light manufacturing, excluding the materials manufacturing industries, are roughly consistent with the conservation supply curve of Figure 11.

The relative sizes of these elasticities are qualitatively consistent with the underlying physical opportunities. In energy-intensive sectors an effort has been made over the decades to make the process efficient, so elasticities are near 0. In nonintensive sectors, energy efficiency has had little or no priority and thermodynamics says that no energy use is required in principle. For these sectors the elasticities are near -1. Moreover, because electricity is a newer form of energy than fuel in most sectors, the long run opportunity for improving the efficiency of electricity use in existing applications is good. The elasticities for electricity tend to be more negative than those for fuel, ie, the long-term response to price increases is greater for electricity.

The autonomous trends, represented by the parameters B, were discussed above. In summary, the trend in aggregate electricity intensity, shown in Figure 7, for 1970–1985 is a combination of four effects: (*1*) a decline associated with the shift away from production in electricity-intensive industries and movement into production in high tech industries, (*2*) a decline associated with increasing electricity prices, (*3*) a decline due to efficiency improvement associated with general productivity improvement, and (*4*) a substantial increase associated with continuing electrification. (The last two are autonomous changes.) The net overall result was a gradual decline in the aggregate intensity for that period.

Policy Analysis

Barriers to Adoption of Efficiency Improvements. Analysts have identified a variety of barriers to adoption of cost-effective conservation measures (50–53). The industrial sector is similar in behavior to the residential and commercial sector but has much more expertise, is more cost conscious, has a tendency to follow "rational" models of investment evaluation, and has more constraints on timing (54). Barriers relevant to the industrial sector in terms of effects are shown below:

- High implicit discount rate.
  - Capital rationing, internal investments restricted.
  - Capital rationing, market share or other strategic investments are favored.
  - Limited access to capital, financial market failure.
  - Split incentives, disconnect between investor and user.
  - Bias against small projects, poor evaluation techniques, or perceived riskiness of energy efficiency investments.
- Limited application (low penetration).
  - Unsuitably of some sites.
  - Information gap, slow diffusion of knowledge, lack of skills at plant.
  - New technology has not been demonstrated in a closely similar case.
- Delay in adoption.
  - Scheduling considerations.
  - Investor's financial position.

The three most important barriers are discussed more fully.

Capital Funds Are Limited for Conservation Projects. More generally, funds are limited for small- and medium-size cost-cutting projects. Available anecdotal information strongly suggests that in U.S. industry in the 1970s and 1980s implicit discount rates were high, approximately 30 to 100% per year compared with real interest rates of perhaps 5% and returns on stock of perhaps 15%. A study by the Alliance to Save Energy found that this behavior is primarily associated with capital availability problems, in particular the practice at many firms of rationing capital for all projects except the large projects specifically analyzed in detail by top management (55,56). Such capital rationing is widely practiced to enable top management easily to control capital requirements and thus the corporation's external financial arrangements (57,58). Usually external capital markets do not deny a firm access to capital, but they operate in a way that inhibits raising funds for discretionary purposes.

Firms tend to rely on internal cash flow to finance the smaller discretionary investments like energy conservation projects. In principle, top management could follow textbook rules, encouraging divisions and plants to propose as many good projects as they can, with top management raising the funds for all those that are likely to return more than the cost of capital. Instead, in most firms capital is severely rationed to divisions and plants for the medium and small projects for which they have primary responsibility, regardless of the investment opportunities that might exist. Under capital rationing, top management reserves the often arduous task of raising outside funds for its own strategic projects. (It is often arduous because lenders or investors tend to want some control over the firm and management wants to keep as much control as possible.) Capital rationing is widely practiced by marginal firms with poor cash flow. It is also often practiced by firms with good cash flow where top management has a financial strategy that strongly limits the availability of capital for internal investments, and by many firms that have highly centralized management.

Because of capital rationing, funds are limited for discretionary small and medium projects. Under this limitation, little effort is made to identify energy conservation projects. One should not misunderstand the rate-of-return criterion met by projects undertaken in this context. It is not that all good projects with a better rate of return are done but that the projects being done have a return at least that high. If more good opportunities were identified, the minimum return on investment would tend to increase, because the total capital allocation is fixed.

Many economists object as a matter of principle to the notion that U.S. markets, in this case capital markets for firms and investment allocation within firms, do not perform close to perfection. In their view, if management practices capital rationing, there must be real costs at the firm that management is more or less properly taking into account. Accepting the spirit of these concerns for the moment, consider two reasons for reluctance to make small cost-reduction investments, like conservation projects.

Some Sites for Conservation Technology Are Unsuitable. Although a technology is, in principle, applicable at a site, the application at that site may be disqualified by decision makers at the firm for several reasons: the product being made there has poor market prospects, the production process is expected to be changed, the plant is likely to be closed or sold in the next few years, the equipment that would be retrofitted is expected to be replaced, or management does not fully trust the people responsible for the project at the site. A substantial fraction of nominally applicable sites will be in this category at any time. This phenomenon does not, however, explain the high implicit discount rates that are common at sites that are suitable.

Smaller Projects Are Discounted Because They Tend to Return Less Than Projected. The perceived riskiness is more speculative as a reason for limited investment in conservation projects. There are two arguments. One involves the risk that the physical performance of smaller projects may tend to be poorer than project designs predict. This is not a large concern for industrial energy projects and is not included in the cost assessment, because most projects are straightforward and design is usually carried out by technical experts. The other argument is that engineering and management costs are likely to be relatively large for smaller projects but not included in the cost assessment. For small projects (eg, under $100,000), this can be an important issue. For medium-size projects (costing one or several million dollars), this argument does not explain the observed implicit rates of 30% per year and more.

This has not been a full-blown review of the controversy between energy analysts and engineers who find that there are many cost-effective opportunities to reduce costs and their economist critics who believe that opportunities not undertaken cannot be cost effective (59). The broad practice of capital rationing for all locally controlled investment projects (ie, plant- and branch-level investment decisions) is powerful evidence that high implicit discount rates are, primarily, a market failure. However, the dispute can only be settled by extensive empirical work.

Policies to Improve Industrial Energy Efficiency. Public policies aimed at improving industrial energy efficiency, or reduced CO_2 emissions, can be categorized by government action or by their impact. Generic policies available to governments are regulation, fiscal incentives, information, and technology policies (60,61). It is, perhaps, most useful to categorize policies by type of impact. Government policies may affect energy prices, financial criteria for equipment investment, rate of adoption of energy efficient equipment, rate of technical innovation, and criteria for investments other than financial. Consider some of these policies (62).

Energy Taxes Including Revenue-neutral Schemes. Relative to other sectors, industry responds strongly to energy prices. But this response is nevertheless inhibited by two effects: stiff financial criteria, like a 2-yr payback requirement for small conservation investments at many firms, and the low priority given energy cost savings at most firms in industries with low energy intensity, where energy costs are only 3% of value added or less. Nevertheless, the response to energy price increases was substantial in the 1970s and early 1980s. If energy prices are to be raised by taxation, an important question is what would happen to the revenues. There are three principal options: (*1*) return the revenues to the firms in proportion to their prior energy intensities, such that the tax is revenue neutral (discussed below), (*2*) use part of the revenues to fund incentives and technology policies (discussed below), and (*3*) use the revenues for general government purposes.

The political difficulties of substantially raising energy prices, especially their uneven impact, must be carefully considered. There is no escaping the major negative impact on the energy supply industries. There may be other heavily affected sectors. In particular, substantial energy price increases could seriously affect energy-intensive manufacturing. Unless the revenues were returned as suggested in option 1, their product prices would have to increase. In itself, this would probably not have an important impact on overall sales, because competition between quite different products is not likely to be affected by modest price increases. For example, plastics based on petroleum did not appear to be adversely affected in their competition with wood-based plastics during the recent oil price shocks. However, the competition between domestic and foreign producers of the same energy-intensive product would be altered if foreign competitors were not subject to similar energy tax increases. A possible mitigating policy would be refunds (as suggested above) or tariffs and refunds for energy-intensive products at the border. If these corrections were applied, say, only to products whose incremental energy costs due to the taxation were over 5% of sale price, relatively few sectors would be affected. Unfortunately, the administration of such tariffs and refunds might be complex.

Energy or carbon taxes could be part of a revenue-neutral fee-refund scheme, in which the tax revenues from each sector could be refunded to firms in that sector on a different basis. For example, the tax might be based on carbon emissions while the refund might be in proportion to production. This approach has two important advantages. While providing the economic inducement to reduce emissions it does so in a way that does not penalize any manufacturing sector and does not create inflationary pressures. There are, however, serious disadvantages in implementation. The outstanding difficulty is caused by the heterogeneity of products and product processes. This means that the administration of refunds would become complex or iniquitous as the level and mix of production by any firm changed. Another disadvantage is that little of the carbon tax might be passed on to customers, so they would not get the full price signal.

Utility Regulation. A specific policy tool for encouraging manufacturers to invest in conservation in spite of their often high discount rates is utility rebates to pay part of the investment cost. The concept is that many conservation investments are more attractive to society than energy supply investments. With their longer time horizon it may be reasonable to encourage utilities to share in the conservation investments. One approach for motivating this is adjustable returns to the utility on its rate base, determined by an evaluation of utility performance in minimizing its customers' costs. If most utilities were motivated to do this rather than simply selling kilowatt hours, as at present, innovative and effective programs might result (63). Many states have instituted programs of this form but without effective motivation. In a few states, like California, New York, and Massachusetts, motivation has been created through financial rewards to utilities. It will be interesting to evaluate program success by sector and kinds of technology (64–66).

Energy Performance–based Tax Credits. Narrowly focused performance-based taxes and rebates to influence equipment purchases by manufacturers are probably not appropriate for industry, but tax credits for broadly defined conservation projects, or groups of projects, may be appropriate. Tax credits for conservation projects were tried (1980–1985), with a 10% credit for qualifying equipment specified on a short list (Energy Tax Act of 1978 and Crude Oil Windfall Profits Tax Act of 1980). A study based on interviews and analysis of detailed data from 15 large energy-intensive firms showed that the 10% credit and short list did not influence firms significantly in their investment behavior (55). However, it is plausible that a much larger credit and more generous qualifying criteria might get a strong response. This incentive approach would have the flaw of generating windfalls for firms that would invest even without the credit.

General Financial Policies to Reduce Discount Rates. Conservation investments, process-change investments, and private R&D and demonstration programs are believed to be strongly inhibited in the United States by the short-time horizons of industrial managers and financial markets. Striking differences in plant and equipment investment and improvements in energy and labor productivity between Japanese and U.S. manufacturers can be conveniently described in terms of a difference in time horizons. Much more controversial is any analysis of causes or of policies to ameliorate the situation. One dramatic analysis, created in large part by a corporate leader in energy conservation equipment manufacture, associates the short-term horizon with low net national savings due to government policies such as income taxation of capital gains, social security as a pay-as-you-go system, and federal deficits (67). The purpose here is not, however, to propose specific policies but to point out the potential importance of this policy area. For example, if firms' de facto capital recovery rates for conservation projects could in some way be reduced from 33 to 20% (payback increased from 3 to 5 yr), then an energy price increase of 60% could motivate as much investment as a doubling of prices under present conditions.

Performance Standards for Equipment. The manufacturing sector is highly heterogeneous and most decisions are made by technically well-informed people (68). It is, therefore, not a sector where performance standards are generally appropriate. In the 1978 National Energy Conservation Policy Act energy-efficiency standards for motors and pumps were proposed and a study was mandated, but the concept was dropped. In the Energy Policy Act of 1992, efficiency standards for most general-purpose motors (1–200 hp) are included. While perhaps useful, such standards will not accomplish that much in the industrial sector. The real opportunity is in using motors more effectively in what they drive. This involves custom design. For lighting equipment, performance standards might, however, be effective. Lighting must be appropriate for production needs, but the energy efficiency of mass-produced lighting products (lamps, ballasts, fixtures) is of little or no concern to the plant engineer or manufacturer. Initial lighting-equipment performance standards are included in the Energy Policy Act of 1992.

Research and Development Programs. Many economists assume that because the U.S. R&D effort is large and generally successful it is basically sound, but there are many who criticize the design and priorities of the entire R&D system as it relates to "civilian" industries. The latter viewpoint can be summarized as follows: the basis of the problem is the underinvestment in research that characterizes the private sector, where a firm cannot expect to capture most of the benefits of research (because competitors can quickly copy successful ideas). The federal government's policy of leaving R&D to the private sector except in selected industries has meant massive research support in areas of military interest, and in certain other areas with historical precedent, but little research support in most civilian manufacturing sectors. Manufacturing processes for pulp and paper, inorganic chemicals, construction materials, and metals have not been fundamentally reexamined in light of the revolutions in basic science, modeling technology, and sensing and control technology that have occurred in recent decades. Engineering school education and research in these areas have largely withered.

In light of this analysis, a policy option is creation of strong R&D programs, including specialized research centers, in energy-intensive manufacturing processes and in other technology-based areas important to energy intensity reduction. This option differs from DOE conservation programs, which have selectively supported narrowly defined development efforts. The policy option suggested here involves creation of coherent programs of general research.

In creating such policies, one must be aware that R&D policies take a long time to have an impact. Many of the well-known fundamental process technologies were three or four decades or more in development and diffusion. However, the speed of creation and adoption may depend on the kind of technology. In particular, rapid development and adoption characterized many of the smaller-scale energy efficiency technologies of the 1970s–1980s. The timing of the introduction of technology for electricity conservation in automotive manufacturing (69) is listed below.

- New technology (first marketed after 1970).
  - Microprocessor-controlled energy management.
  - Variable speed drive based on inexpensive power semiconductors.
  - Die cushioning.
  - High efficiency lamps and lighting systems.
  - Large microprocessor-controlled heating and ventilation systems.
  - Aircraft-derivative gas turbine cogeneration.
- Old technology.
  - Isolation valving for segmenting compressed-air systems.
  - Small compressors for satellite compressed-air systems.
  - High efficiency motors.
  - Cog belts.

Demonstration Programs. The speed of diffusion of new technology depends on its demonstration, but the effectiveness of demonstration programs is controversial. The positive argument is that firms will delay use of a new technology, or new application of an existing technology, until they are assured by at least one demonstration, in the same manufacturing situation that concerns them, that production processes and product will be satisfactory. A typical Ford Motor Co. plant manager would like a technology to have been successfully demonstrated at another Ford plant. Demonstration in a GM plant might do, but the same technology in a different industry would not. Public subsidization of demonstrations and dissemination of technical information about them could accelerate the process. There are examples of successful programs (70). The negative argument is that the public sector tends to do a poor job in selecting technologies for demonstration and a poor job of managing its role in any demonstration. There are important examples of such failures (71).

Considerable experience has been gained about the design of demonstration programs in the past decade or so in the United States and Europe. Several characteristics of a successful program appear to be avoidance of large projects, a relatively small financial role for government, and a strong dissemination activity.

Goal Setting and Reporting of Performance. Goal setting (with the level of energy-intensity reduction being voluntary) data collection, and reporting of energy intensity improvement (through trade associations in each subsector of manufacturing) were mandated in the 1975 Energy Policy and Conservation Act. They were considered by many energy managers at firms to be remarkably effective, considering the mild nature of the legislation. The requirements expired in 1985. (The compilation and publication of the data by DOE was not considered effective.) The principal benefit from this policy was seen by energy managers as the creation of lines of responsibility within the firm for energy conservation and creation of energy consumption reporting systems within the firm. Similar requirements can be imagined for the area of recycled materials.

ENERGY EFFICIENCY OF TRANSPORTATION

Transportation is a critical sector for energy use because passenger travel is growing rapidly, because petroleum is the main energy resource used (requiring enormous imports in the United States and many other countries), and because it is the largest source for air pollution in most metropolitan areas. In this section the growth factors, especially in driving private motor vehicles, are considered. Then technical and economic aspects of improving the energy efficiency of motor vehicles are discussed. Some of the physical and technological issues will be explored in detail. Finally, a range of public policy issues is discussed. An up-to-date overview of U.S. transportation and energy issues is available (72).

Historical Statistics

The organization of the discussion is based on representing energy use as the sum of products of activity times intensity (eq. 1):

$$\text{Energy Consumption Rate} = \Sigma_i A_i\,(EI)_i$$

Transportation Activity. In the United States, passenger transportation dominates freight in importance from energy and environmental perspectives (Table 2). Passenger transportation averages 13,600 passenger mi/yr per capita or 37 mi/day (73). Of this, 85% is in motorized personal vehicles (cars as well as personal vans, utility vehicles like Jeeps, and pickup trucks when used like cars). This dominance of personal vehicles in per capita travel vehicles characterizes all industrialized countries, but the United States and Canada lead in the amount per capita of travel (36).

Total travel by personal motor vehicle has boomed in

the United States since World War II, growing 4.7% per year from 1950 to 1970, and 3.2% per year from 1970 to 1990. As with industrial production, the 1950s and 1960s were decades of especially intense growth in physical activity.

Energy Use and Intensities, 1990. Table 2 shows energy, activity, and energy intensity in some detail for all of transportation in 1990. The main characteristic is the dominance of personal passenger vehicles. Personal passenger vehicles account for 52% of all transportation energy use and for 85% of all passenger miles (PM). (These data may underestimate the use of light trucks as personal passenger vehicles).

The personal vehicle is more energy intensive than the other forms of passenger transportation, but not by as much as many think. The average car is shown in Table 2 to have an energy intensity of 5980 Btu per mile (an in-use fuel economy of 20.9 mpg). An urban transit bus has an energy-intensity of 3740 Btu/PM. So a car with two people, or a car with one person but twice the average fuel economy, not only goes where and when the driver wants but has lower energy intensity per passenger mile than an average urban bus. (The low energy intensity for buses in Table 2 is due to school buses and the shaky assumption that their average passenger load is 20. The average load of the urban transit bus is taken to be is 10.) The energy intensity of certificated air carriers is also not as great as one might, at first, think.

Freight energy use is also dominated by highway vehicles, but freight activity measured in ton miles (TM) is dominated by nonhighway modes. The nonhighway freight modes are much less energy intensive than heavy trucks. Note that gas pipelines are fairly energy intensive, however, because gas is much more difficult to pump than a liquid. There are at least three principal categories of freight: (*1*) light short-distance loads for which vehicle mile (VM) is the relevant measure of service; (*2*) long dis-

Table 2. Transportation Energy Use and Activity in the United States (1990)[a]

Use	Activity		Energy, quads[d]	Energy Intensity, 1000 Btu/activity unit
	Billions[b]	Units[c]		
Passenger				
automobiles	1515	VM	9.07	5.98
personal trucks	296	VM	2.68	9.06
motorcycles	10	VM	0.02	2.50
buses[e]	118	PM	0.16	1.38
railroad[f]	25	PM	0.08	3.13[g]
air (domestic and international)	359	PM	1.93	5.00
Subtotal			*13.94*	
Freight				
intercity truck[h]	735	TM[i]	2.47	3.36
other truck	182	VM	2.40	13.2
railroad[j]	1034	TM[i]	0.43	0.41
water	816	TM[i]	0.32	0.40
domestic				
foreign	1.04	T[k]	0.92	
pipeline	280[j]	TM[i]	0.72	2.6[g]
natural gas				
oil and products	583	TM[i]	0.16	0.27
other			0.05	
Subtotal			*7.47*	
Miscellaneous				
recreation boating			0.25	
general aviation			0.13	
military			0.76	
Total			*22.5*	

[a] From Refs. 73 and 80.
[b] For kilometers, multiply by 1.609.
[c] *VM*, vehicle mile; *TM*, short ton mile; *PM*, passenger mile; *T*, short ton.
[d] For exajoules, multiply by 1.055.
[e] Includes intercity, urban transit, and school.
[f] Includes intercity and urban transit.
[g] Electrical energy accounted for in terms of primary fuel use.
[h] Rural to rural and intracity deliveries excluded; see text.
[i] For tonne kilometers, multiply by 1.460.
[j] Average trip length for gas assumed to be 620 mi.
[k] For metric tonnes, multiply by 0.907.

tance loads that are packaged or finished goods; and (3) bulk cargoes, like coal, ores, grains, and semifinished goods. The ton mile is a good measure for the latter two services. The main point is that competition or substitutability across these categories is usually not practical, thus the much lower energy intensity of waterborne freight than intercity truck does not indicate much of an opportunity for intermodal substitution. The rail freight energy intensity is an average of somewhat lower intensity bulk cargo shipping and somewhat higher intensity finished goods shipping. In the latter area there is substantial opportunity for substituting rail for long-distance trucking, with transfer from and to trucks at the ends. This is now going on, using containers, putting truck trailers on flat cars, or putting a new kind of trailer directly on the rails. One requirement for this substitution is that the railroads achieve modern scheduling goals, eg, for transferring loads between trains.

Trends in Energy Use and Their Decomposition. Many of the transportation activities have been tracked in consistent or nearly consistent data series since 1970 and before. Energy consumption and average growth rates for activity during the period 1970–1990 are shown in Table 3. There are special weaknesses underlying some of the numbers in this table: the division of light truck VM into personal passenger and freight, the assignment of ton miles to all other trucks, and the assignment of ton miles to natural gas pipelines. In addition, several minor categories have been omitted. Table 3 reveals the rapid growth of the light truck as a personal passenger vehicle. It also shows the growing importance of air travel as well as the relatively slow growth of some freight activities. In the latter connection, there has been a relative decline in bulk materials transportation that parallels that in manufacture of bulk materials discussed earlier.

Table 3 shows strong improvements, in particular (modal) energy intensities, with amazing progress by airlines (cutting the intensity to less than half in 20 yr) and major progress by railroads and in light vehicles. The highly organized commercial transportation sectors, airlines and railroads, were able to do a great deal. The airline story, including the large potential for continued gains has been told (74). The light vehicle story, especially future possibilities, is the main subject of this section.

The data in Table 3 have been set up to enable a Divisia decomposition of the sources of change in aggregate energy consumption, an analysis that does not require activities in different subsectors to be measured in the same units (see eq. 6). The average growth rate in energy use (approximately) equals the average growth in activity plus the average growth in energy intensity (both energy weighted).

The results of the analysis are summarized in Table 4. The behavior for transportation as a whole is essentially the same as that for personal passenger vehicles alone: growth in activity at an average 4%/yr and a rapid decline in energy intensity, such that energy use grew only 1.5%/yr in this period. This leads to an important prediction: the decline in energy intensity is expected to end because the legislation that caused the improvement in personal passenger vehicles has now worked its way through the economy. Further decline would require further legislation, much higher energy prices, or some other event. So transportation energy use will soon begin to grow much faster than it has in the last 20 yr.

The separate results for passenger and freight activity show what is not surprising to any observer of the U.S. scene: travel is increasing rapidly, but so is the energy efficiency with which it is provided. Freight activity has been increasing less rapidly, a characteristic of an affluent and mature society. At the same time, it has proven more difficult to improve the energy efficiency of freight services. The truck freight data, excluding light trucks, involves a major inconsistency. "Intercity freight" grew 78% from 1970 to 1990, while truck vehicle miles grew 141% (and in addition the trucks became larger). By using the

Table 3. Transportation Energy Use and Activity, 1970–1990

Use	Energy, quads		Activity Unit	Growth Rates, %/yr	
	1970	1990		Activity	Energy Intensity
Passenger					
automobiles	8.48	9.05	VM	2.5	−2.2
personal trucks	0.61	2.68	VM	9.4	−2.0
(combined)	(9.09)	(11.73)		(3.2)	(−1.9)
buses	0.11	0.16	PM	1.7	0.4
railroad	0.04	0.06	PM	1.3	0.0
air	1.36	2.19	PM	6.4	−4.0
Subtotal	*10.60*	*14.14*			
Freight					
light truck	0.94	1.48	VM	3.9	−1.7
other truck	1.53	3.39	TM	4.9	−0.9
railroad	0.50	0.43	TM	1.5	−2.3
water (dom.)	0.33	0.33	TM	1.6	−1.6
pipeline energy	0.99	0.87	TM	0.9	−1.5
Subtotal	*4.28*	*6.50*			
Total	*14.88*	*20.64*			

Table 4. Divisia Analysis of Energy Used for Transportation 1970–1990, Growth Rates in Percent per Year

	Activity	Energy Intensity	Energy
Passenger	3.8	−2.3	1.5
Freight	3.4 (2.6)	−1.4 (−0.5)	2.1
Total	*3.7*	*−2.0*	*1.7*

intercity freight growth rate of 2.9% instead of the 4.9% growth rate based on vehicle miles and increased size, one obtains the results shown in parentheses in Table 4. These results conform even more strongly to the qualitative characteristics just mentioned.

If a Divisia decomposition is carried out for personal motor vehicles alone (cars and light trucks), because the units of activity are the same, $G(Q)$ and $G(S)$ can be calculated (see eq. 2). From Table 3, it is found that the growth rate for overall activity $G(Q) = 3.2$, the growth rate for the shift toward light trucks $G(S) = 0.4$, and the real intensity growth rate $G(EI) = -2.2$ (all in percent per year). (While these results are roughly correct, they depend on an educated guess for the 1970 use of light trucks as personal passenger vehicles.)

To understand these results, it is necessary to probe them in more detail. What is responsible for the rapid growth in travel? What is responsible for the decline in energy intensity? In the next two sections these questions will be explored with respect to personal passenger vehicles.

Growth in Driving per Driver

Vehicle travel continues to grow in spite of arguments that saturation is imminent. Total vehicle miles per adult (ie, total vehicle miles divided by the population aged 16 and over) is found from a regression analysis to be almost proportional to real disposable income per capita, corrected by a small fuel price elasticity effect (of −0.1), indicating that a 10% increase in the fuel price induces a 1% decline in consumption (75). This short-term response to increased gasoline prices is well determined, because it is a direct feature of the price shocks of 1974, 1978–1982, and 1990 (76).

The agreement between travel and income may not be more than the similarity of two increasing time trends. In any case, the causes can be expressed in more useful terms. Demographics show that much of the growth in driving since the late 1960s is associated with women moving into the labor force, those women becoming drivers, and those women driving distances that are moving toward men's distances. In 1969, 39% of women over 15 years old were employed; in 1990, 54% were. In 1969 women with driver's licenses drove an average of 5400 mi/yr; in 1990 they drove 9500 mi/yr (77).

There is, in addition, an upward trend in the annual miles per driver by employed men. This distance has been growing 1.3%/yr (1969–1990), and continues to grow. Employed men in the 25–45 age group average about 20,000 mi/yr. At an estimated average 35 mph, this is 1.6 h/day. This large fraction of time is the basis for believing that saturation must soon set in. But when and at what level?

A hint of the future might be obtained from cross-sectional comparisons (from place to place), rather than from longitudinal variations (from year to year). For example, it is possible to examine the cross-sectional effect of income on driving (and on that basis predict the future assuming that per capita incomes will increase). Of course, poor people drive less, but at issue here is whether upper-middle-class people drive more than lower-middle-class people. Unfortunately, this kind of question is difficult to resolve. Many things change with income, like household size, local population densities, and regional patterns. Depending on what else is explicitly accounted for, income is or is not associated with increased driving. Better research is needed.

A different analysis is under way that may shed light on the issue: the dependence of driving per driver on the kind of community in which he or she lives. The principal variable of interest is population density. (Other variables of interest may be density of services, pedestrian friendliness like sidewalks, frequency and density of public transportation, and distance to major centers. In addition, there are variables like income, and vehicle and fuel costs.)

Vehicle miles per capita for metropolitan areas around the world, with gasoline consumption as surrogate, were studied (78). Vehicle miles per capita for communities within metropolitan areas of California have been examined (79). These studies are not definitive because one should question the quality of the data and their interpretation, in part because effects of income have not yet been carefully addressed. Nevertheless, there is a strong population density dependence of driving:

$$VMPC \approx \text{const}\ \rho^{-\alpha} \tag{9}$$

where VMPC is vehicle miles per driver or vehicles miles per capita, and ρ is residential density (eg, people per square mile of developable land). One finds that α ranges from 0.5 to 1.0. This is an interesting line of research for two reasons. First, residential density should be expected to be a critical determinant of travel. People need access to work, services, social contacts, etc. Second, the choice of place to live, eg, urban versus rural, depends to some extent on public policies.

An interesting accompanying result of these studies is that people who drive less, travel much less. Each mile of travel on transit does not substitute for a mile of travel in a private vehicle or vice versa. Rather, if one lives in a denser community, one uses transit a little more and drives much less, with the result that 1 mi of travel on transit appears to substitute for, perhaps, 5 mi of travel by car or truck. This result is preliminary.

If one accepts, for the moment, this strong dependence on residential density in the United States, the important issue is: What motivates the growth of housing in low density communities? Is it primarily inherent human desires, as the real estate industry would suggest, or are there public policies that influence many people to live in rural or suburban areas, while nevertheless gaining a living in a city? Of course, some people strongly prefer to live in rural-like settings, but economic incentives are also powerful.

Public policies support investment in the land on which one's house stands, and newly developed areas are relatively cheap. Home mortgage interest is not taxed and infrastructure for real estate development is a prime focus of local government. Indeed, many U.S. local governments have been run by developers since they were founded. Such land investment has proven the best way to create a nestegg for those who are not in a business of their own or not specialists in speculation. Public policy also supports highway travel through diligent efforts to build and maintain safe high speed roads and to keep fuel prices low. Further exploration of the issue is beyond the scope of this article.

The description of vehicle miles traveled on the basis of population or of the number of drivers, as just presented, is in contrast to one based on the vehicles in use, an approach that has been widely used for analysis. The trouble with using miles per vehicle is that, in the United States, a fundamental shift in the use of private vehicles is now beginning to take place. The number of households with more vehicles than drivers is becoming large (77). This trend toward extra, probably special-use, vehicles may well continue as vehicles are kept in service longer and the adult population grows more slowly. For example, the median age of cars in use has increased 2 yr since the early 1970s (80). The growth in the number of vehicles and, especially, their use is thus difficult to analyze.

Choice of the variables on which the analysis is based is critical to the perspective created by the analysis as well as to the particular results. For example, the passenger transportation activity considered is usually passenger miles (mobility) or, as has been emphasized here, vehicle miles. With either of these variables there is a tendency to associate a substantial increase with progress. Certainly poorer Americans and people in poorer countries have less mobility. An alternative kind of activity variables is access, the number of trip destinations achieved. Because many people achieve destinations by walking in their neighborhoods or by short vehicular trips, there is a substantial difference between access and mobility or vehicle miles. In a course project at the University of Michigan (with a small sample size), the same amount of access per person-week for people in rural and urban samples was found, but twice as much travel was noted by the rural respondents. If data on access were available and the analysis of energy use for travel were carried out with access as the activity variable, one would probably find that fuel use per access has not been declining the way fuel use per vehicle mile has. This would create a different picture and, perhaps, different prognoses than the analysis presented here, with its emphasis on vehicles and only brief mention of community.

Fuel Economy Improvement: Technical Opportunities

Although this analysis will address a time period of about 15 yr, it is restricted to modifications of the present kind of gasoline-fueled vehicle (of today's size and power characteristics). This constraint is due to the need to be brief, not to lack of interest in alternative fuels, new propulsion systems, and alternative modes of transportation, all of which are important. Conventional vehicles are important because this country's gasoline-based personal transportation system involves an enormous investment in physical and human capital that will not be quickly replaced. Accordingly, moderate changes in light vehicles are an important means of addressing energy and environmental issues.

Fuel economy in miles per gallon is the number of miles of travel by a vehicle over a standardized sequence of speeds such that 1 gal of a standard fuel is consumed. Improvements in fuel economy are efficiency improvements as defined earlier (even though this generalized kind of efficiency does not include the concept of 100% efficiency).

Many believe that reducing vehicle size (and thus weight) is the best way to increase fuel economy. Reducing the maximum power per unit of vehicle weight can also make a major contribution. The subject of this analysis is yet a third kind of change: technological improvement without reducing vehicle size or performance. This change was behind most of the progress from 1975 to 1988. The industry is undergoing a remarkable period of increasing technical capabilities: electronic controls, new materials, and the capability, through computers, to design a car in detail without having to go through many stages of trial and error with real engines and vehicles are making it practical to do the things only dreamed of by earlier automotive engineers. This technological ferment can be sensed by reading papers of the Society of Automotive Engineers and attending its conferences.

The technologies for improving vehicle fuel economy fall into two categories. The first is load reduction, or decreasing the power required of the engine by reducing air drag, rolling resistance, weight, drivetrain friction, and vehicle accessory loads. The second is efficiency improvement of the engine, or increasing the effectiveness with which the energy in fuel is converted to useful work.

Load Reduction. In recent years the maximum power of new car engines has increased. The average new car power to weight ratio has risen from a low of 32 hp/1000 lb for the period of 1980–1982 to 43 hp/1000 lb in 1993, and acceleration times have fallen (81). This increase in power has been a useful marketing tool, as can be seen from the fact that many customers are sold the high power version of a model.

The power delivered to the drive wheels during typical patterns of driving is relatively low, however. High power is required only in unusual driving conditions, such as acceleration at speeds far over legal limits and climbing mountains at high speeds, situations that most drivers rarely encounter. Vehicles of average weight with modest engines, say 30 hp/1000 lb, can readily be used with good transmission management, to accelerate rapidly at moderate speeds, so acceleration at moderate speeds is not a rationale for high power.

In urban driving, a typical new U.S. car requires an average engine power output of 7 hp. This is low compared to engine capabilities that average 141 hp for 1993 models (81). At 7 hp, more fuel is being used merely to overcome the internal frictions of a typical engine than to provide the output power. The efficiency of an engine at part load is much lower than its best efficiency. This suggests the importance of engine downsizing coupled with load reduction and aggressive transmission management as key strategies for improving fuel economy.

For today's average new car, leaving the engine transmission unchanged, a 10% reduction in engine load results in a 4% reduction in fuel use in urban driving and a 5% reduction in highway driving (82). Technologies include reducing aerodynamic drag, reducing tire rolling resistance, and weight reduction at fixed vehicle size. Drivetrain efficiency can be improved through technologies such as torque converter lockup, electronically controlled standard gearing, and reducing transmission friction. Accessory loads (the largest of which is air-conditioning) can be cut by running accessories only when needed, improving component efficiencies, and reducing the need to run the accessories. Overall, there is a near-term potential for roughly a 15% reduction in weight and a 25% reduction in engine load, which would alone yield a 12% improvement in fuel economy (ie, without engine downsizing).

Types of Engines. There are two main kinds of internal combustion engines in use (83). In the spark ignition engine, combustion occurs by means of a flame front that proceeds from the spark through the mixture of vaporized fuel and the air. Typically the fuel:air ratio of the mixture is chemically correct, or stoichiometric, in that all the fuel present and all the oxygen in the air, could combine to form CO_2 and H_2O. (Ignition and smooth flame front propagation are challenging to achieve if the air:fuel ratio is more than about 1.5 times stoichiometric.) Variable (reduced) power output is achieved by reducing proportionately the fuel and the air admitted to the chamber. The amount of air is regulated by a throttle, which constricts the inlet to the intake manifold, thus creating a partial vacuum as each piston is pulled out during the intake stroke. The fuel is introduced through controlled injection. (Only about 0.017 as much fuel vapor as air is needed, by volume.) In cars sold in the United States, a catalytic converter oxidizes CO and hydrocarbons in the exhaust while simultaneously removing oxygen from (reducing) the NO_x. For this three-way balanced reaction to be achieved, the fuel:air mixture introduced to the engine must be stoichiometric.

The diesel engine uses fuel droplets and high compression. With the compression stroke, the temperature and pressure are high. After fuel injection, combustion spontaneously occurs on the surfaces of the droplets. The overall fuel:air ratio is typically lean (ie, excess air compared to stoichiometric). Variable power output is achieved by changing the amount of fuel injected, the amount of air admitted always being the same (no throttle). Very lean mixtures are satisfactory. With today's catalytic converters, NO_x in the diesel exhaust cannot be reduced, so it may be difficult to achieve low NO_x emissions in the regulatory test. In addition, soot or carbon can be emitted, especially when the fuel:air mixture approaches stoichiometric at high power output. New diesel truck engines are, however, meeting new emissions regulations. (Whether this means that soot will not be seen from new diesel trucks when they accelerate remains to be seen.) Finally, diesel engines are larger and heavier than spark ignition engines for the same power output. (They are also more robust.) These are not advantageous properties for most private passenger vehicles.

Thermal Efficiency. Engine efficiency is the product of two factors: thermal efficiency, expressing how much of the fuel energy is converted into work moving the pistons, and mechanical efficiency, the fraction of that work that is delivered by the engine to the vehicle (the rest going to overcome frictions in operating the engine). (Note: In many expositions the term *thermal efficiency* is used for the overall engine efficiency rather than the efficiency neglecting frictions as here.) Even the best practical combustion-based engines (boilers and steam turbines at electric power plants) have thermal efficiencies of only about 40%, and these engines are large, expensive, and stationary. About 50% efficiency is being achieved in new electric power plants with combined cycle technology, which involves energy recovery from the exhaust gases after the main energy conversion. (These efficiencies are based on the higher heating value of fuel.)

The thermal efficiency of typical internal combustion engines is about 40%, relative to the lower heating value of the fuel (or 37% relative to the higher heating value). This could, perhaps, be increased to near 50% through several changes: increased compression ratio, lean burn (increased air:fuel ratio), recovery of work from the exhaust, faster combustion, effective control of working characteristics such as the air:fuel ratio for each cylinder and each cycle, and control of valve timing and enhancement of breathing so that intake and exhaust are optimized at each engine speed (84).

One of the most interesting and promising of these is lean burn spark ignition engines. Lean burn is advantageous in terms of efficiency because a gas of simple molecules when heated increases its pressure more than a gas of complex molecules like vaporized gasoline; with complex molecules much of the thermal energy is diverted into motions internal to the molecule. Radically increasing the air:fuel ratio by a factor of 1.5 (above the chemically correct stoichiometric value), eg, would nominally increase efficiency by 11%. Moreover, if the air:fuel ratio could be widely varied while still obtaining satisfactory combustion with spark ignition engines, this method could be partly substituted for the throttle to regulate engine power output. Significant additional mechanical efficiency benefits would result. But lean operation prevents the emission control system from reducing nitrogen oxides, so NO_x emissions from the engine might not be as low as one would hope. Moreover, lean mixtures can fail to ignite (misfire) or lead to incomplete combustion. One way to address this problem is to make the mixture near the spark plug richer (stratified charge). Several engine manufacturers are working to overcome these drawbacks, and Honda, Toyota, and Mitsubishi have first-generation lean burn engines in production cars. Moreover, a more radical approach to lean burn is said to be achieving success: the two-stroke engine with modern fuel injection and controls.

In summary, improving thermal efficiency by a factor of as much as 1.25, from roughly 40 to 50%, is a potentially important but difficult goal. One way to achieve it is to solve the environmental problems of the diesel and adopt modern direct-injection diesels such as those now in use in several European cars. Another way to approach it is to develop successful lean burn spark ignition engines. Still another way is to switch to a fuel with much simpler molecules and high octane (eg, hydrogen or methane), designing a high efficiency engine for that fuel. Achieving still greater improvements in thermal efficiency in internal combustion engines is likely to be impractical.

Mechanical Efficiency. The mechanical efficiency of typical U.S. cars averaged over urban and highway driving is about 43%. It is lower for high powered cars and higher for low powered cars. The mechanical efficiency is 0 when the engine provides no power output (an idling engine). Near wide-open throttle, the mechanical efficiency is about 80%. Unlike thermal efficiency, for which it is not practical to achieve efficiencies more than about 50%, it is practical to achieve mechanical efficiencies approaching 100%.

The following analysis rests on a simple approximation for fuel use as a sum of two terms: one to overcome engine friction, the other to provide output power (85). The validity of this approximation is exemplified by Figure 13, in which fuel energy converted per revolution is shown on the y-axis and energy output per revolution on the x-axis, at various engine speeds. All the operating points lie essentially on one straight line. Thus the rate of fuel use has the form

$$P_f = aN + P_b / n_t \tag{10}$$

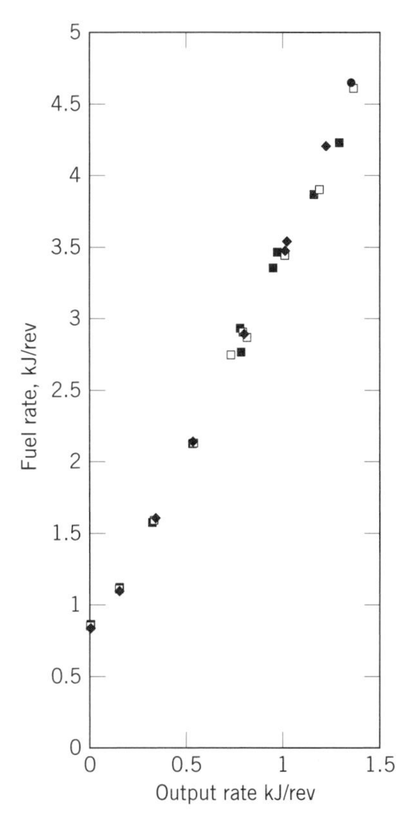

Figure 13. Fuel rate versus work output rate, Chevrolet V6 200CID 198 engine.

where N is the engine speed (revolutions per second) and P_b is the engine power output or load (in kw or hp). The constant n_t is the thermal efficiency discussed above, and a is the fuel energy per revolution needed to overcome engine frictions. There are three kinds of friction involved in operating the engine: the energy used for pumping, for overcoming rubbing friction, and for driving engine accessories. Pumping refers to moving the air and vaporized fuel into the cylinders and the combustion products out through the exhaust system. (For a given engine, a is essentially constant. Among engines, it is roughly proportional to engine displacement.)

The engine power output is the product of the mechanical and thermal efficiencies times the fuel energy input:

$$P_b = n_m \, n_t \, P_f \tag{11}$$

Figure 14 illustrates the two efficiencies. The slope of the line $\Delta P_f / \Delta P_b$ is a measure of the reciprocal of the thermal efficiency; here n_t is roughly 40%, independent of the load P_b. The mechanical efficiency varies strongly with the load. Solving equation 11 for n_m:

$$n_m = \frac{P_b}{n_t} / P_f = \frac{P_b / N}{a \, n_t + P_b / N} \tag{12}$$

At the point shown in Figure 14, $P_b / N = 0.34$ kJ/rev, and $P_f / N = 1.75$ kJ/rev, so that $n_m = 0.49$, or 49%. This is a typical highway driving situation.

In a typical urban driving situation, ($P_b / N = 0.20$, $P_f / N = 1.40$ kJ/rev in Figure 14), the mechanical efficiency is 36%, but in urban driving the car also spends some time idling, where $n_m = 0$, and some at high power, where n_m is high. At the top of the solid line in Figure 14 ($P_b / N = 1.3$, $P_f / N = 4.2$ kJ/rev), the mechanical efficiency is 77%.

Five general kinds of technology for improving average mechanical efficiency are

- Aggressive transmission management (ATM) to reduce average engine speeds.
- Reduced displacement, or engine size, at constant maximum power.
- Reduced rubbing friction and more efficient engine accessories.
- Stop–start (idle–off).
- Reduced pumping (elimination of throttling).

Technologies to progress in these five directions have been extensively discussed (76,86–89). It is of interest here to consider briefly the first two categories, ATM and engine downsizing. Not only are these important but they illustrate the rapid change in technology that is occurring and the possible conflict between some efficiency technologies and the product or service being offered. The tendency is to think of efficiency improvement as noninterfering. For example, additional wall insulation does not reduce the usefulness of a house, it may even expand the comfortable area. But some automotive efficiency improvements can reduce the amenities offered by a car.

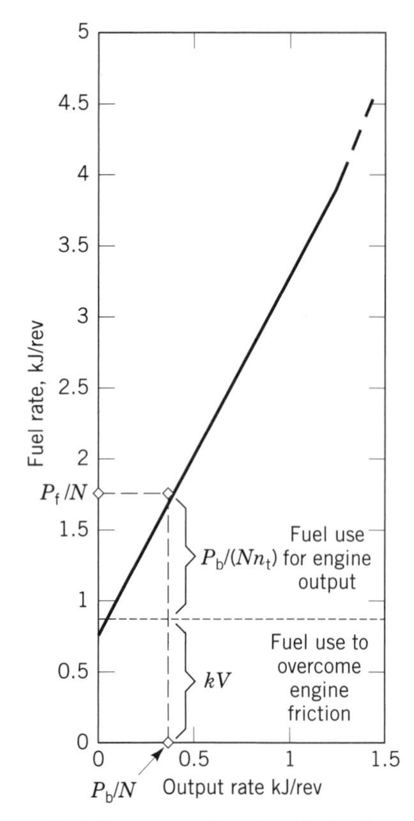

Figure 14. Thermal and mechanical efficiencies.

ATM. Modifying the transmission to reduce engine speed at a given power output is the best established method for improving average mechanical efficiency (90). To implement it, more gears and lower gear ratios can be built into the transmission and, in driving, gears are shifted up as soon as feasible. A feel for aggressive transmission management can be obtained by driving a 1990–1991 Honda CRX HF with a shift indicator light on the dashboard. As the car accelerates, the upshift light comes on very soon. If one follows the shift light's suggestion, one shifts up at the much lower engine speeds than is typical.

A critical consideration for fuel economy is the span, or the ratio of the highest to the lowest gear ratio. Consider a standard manual transmission. In the lowest gear (highest gear ratio), clutch slip is involved in getting the car moving, but the gear ratio must still be high enough to enable the engine to begin to move a stopped car up a grade (91). Good fuel economy performance in highway driving requires a low gear ratio in the highest gear. But the large span desired will not be feasible unless the ratios of adjacent gears are close enough to make shifting convenient. This requires many gears. For fuel economy, six forward gears are preferable to five in a manual transmission. For automatic transmissions with fluid coupling, fewer gears are needed because the lowest gear provides, roughly speaking, the function of the two lowest gears in manual transmissions. Four-speed automatics are now being widely used and five-speeds are being discussed.

Aside from the issue of creating the transmission technology, manufacturers may be reluctant to reduce engine speeds for two reasons. First, the engine may not run smoothly at low speed. Second, relatively high power is not immediately available at low engine speeds.

A Smaller High Tech Engine. The most rapid change in vehicle technology in recent years has been in specific power, the ratio of maximum engine power to engine size, or displacement. The average specific power increased 3.3% per year from 1976 to 1993 (81). This trend is expected to continue. The technologies are more valves per cylinder, high tech valve cams, higher compression ratios, advanced fuel injection, sophisticated controls (eg, of ignition timing), and tuning of the intake and exhaust manifolds. Variable valve timing is beginning to come into play. More sophisticated controls are in the offing, enabling management of cycle to cycle and cylinder to cylinder variations.

If engine displacement is reduced in proportion to the increase in specific power, then maximum power can be maintained, while the friction parameter a is reduced essentially in proportion to the displacement. With today's average new car, a 10% engine downsizing results in a 6.5% increase in efficiency (average over urban and highway driving cycles including the benefit of weight reduction).

Both engine downsizing and ATM reduce the available power at a typical engine speed. Consider the combined effect of these two technologies: In Figure 15 two overlapping engine maps are shown. Two variables fix the point at which any engine is operating at a given time. In Figure 15, engine speed and power output are the variables. A low engine speed (point A) and a higher engine speed (point A') with the same power level are shown. With an older design large engine, it is possible to move directly from point A to power at level B by opening the throttle (depressing the accelerator pedal). On the other hand,

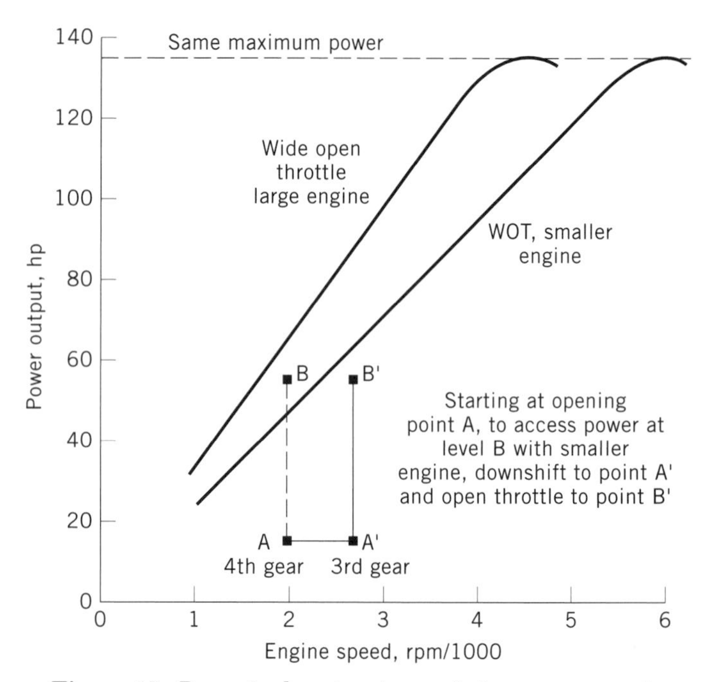

Figure 15. Downsized engine, transmission management.

with a small high tech engine, high power at point B' is immediately available if the driver starts from the higher engine speed at A', but downshifting is necessary if the driver starts from A.

This process is familiar to drivers of many cars with a four-cylinder engine and an automatic transmission, in which downshift and engine speedup occur when the accelerator pedal is floored. The action of declutching, engine speedup, and reclutching takes time; with a well-designed system, however, it takes half a second or less. Although this delay is a minor loss of amenity, it is a loss of amenity. In this respect, the ATM and engine downsizing technology are not analogous to many other technological changes that can be implemented without loss of amenity.

The Potential for Improving Automotive Efficiency. An ambitious goal for improving the average mechanical efficiency, with technology already in production or in development, would be a factor of 1.4, from about 43 to about 60%, still much less than the 80% achieved in the best operating region of today's engines. This involves a combination of reductions: (*1*) reducing the three kinds of engine friction, (*2*) reducing the size of the engine while maintaining power capabilities, and (*3*) reducing the average engine speed. This combination is exemplified by examining the aN term of equation 10, which is responsible for all the mechanical inefficiency. (Note that $1 - n_m = aN/P_f$.) Because a is essentially proportional to the engine displacement V, $aN = kVN$. The three reductions just listed are in the three factors, k, V, and N, respectively.

Consider the following technological scenario. With an engine of today's general type, thermal efficiency is increased 3%. With a vehicle of today's general type, weight is reduced 15% and overall load is reduced 27%. The average mechanical efficiency is increased 28% by the following steps: k is reduced 15%, mostly through reduced pumping and accessory losses. V is reduced 15% because of a corresponding vehicle weight reduction and an additional 40% through continuing specific-power increases. N is maintained at its base value, in spite of the major downsizing of the engine, through ATM (Table 5).

The accounting scheme of Table 5 may be unfamiliar. The allocation of fuel use in the table is based on five energy sinks, the fuel used to enable the work done at the five sinks shown. Two other accounting schemes are common: (*1*) only the work done at the final sinks (including the transmission) is shown. For the car called AVCAR corresponding to Table 5, the energies in this scheme are total engine losses 3480 (of which 478 is at idle), transmission 73, vehicle accessories 101, tires 231, air drag 220, and brakes 202, all in kJ/mi). (2) The engine frictions as well as the "lost work" associated with thermal inefficiency at the engine (as well as transmission loss) are allocated to the final sinks. For AVCAR in this scheme the energies are engine idle 478, vehicle accessories 467, tires 1190, air drag 1133, and brakes 1039.

In this scenario fuel economy is increased by 78% to 50 mpg. This goal has already been reached by two production cars, Geo Metro and Honda Civic VX. However, unlike the latter two cars, the car considered here is of average size for 1993 models and has an average power to weight ratio. The fuel use to overcome engine friction (top row of Table 5) is reduced a whopping 60%, but a good measure of the feasibility is the mechanical efficiency achieved (just under 60%), which is still low. The car described in Table 5 is hardly the ultimate that can be achieved with a fuel and internal-combustion (IC) based vehicle. The thermal efficiency was not pushed to a high value; neither a diesel engine or a lean burn spark ignition engine is considered. Nor was the load pushed to a really low value. In particular, radical new materials would enable weight reduction by perhaps 33%, instead of the 15% assumed. Moreover, a much narrower and smaller car for short trips and less carrying capacity, could enable even greater load reduction.

Table 5. Specific Fuel Energy Consumption for Two Cars, kJ/mi[a]

	AVCAR'93	High MPG
Engine friction		
under power	1835	787
idle	478	228
Subtotal	*2313*	*1015*
Loads		
tires	619	371
air drag	590	444
brakes	541	432
vehicle access	243	158
Subtotal	*1993*	*1404*
Total fuel energy	*4307*	*2191*
Fuel economy, mpg	28.0	49.9
Vehicle Characteristics		
inertial weight, lb[b]	3234	2749
displacement, L[b]	2.77	1.40
N/v, rpm/mph[b]	34.0	34.0
engine friction k, kJ/rev · l	0.25	0.212
thermal efficiency η_t, %	41.5	42.7
rolling resistance C_R	0.010	0.008
air resistance C_DA, m^2	0.663	0.530
transmission efficiency, ε	0.90	0.93

[a] EPA composite driving cycle, no adjustment.
[b] Sales-weighted averages from Ref. 81; the remainder of the AVCAR characteristics roughly describe modern cars.

Another option, still based on the internal combustion engine, is the hybrid car, using energy storage, such as battery, flywheel, and/or electric capacitor, to enable the IC engine to operate only in the zone of its high mechanical efficiency. Efficient storage technology, which is reliable and of moderate cost, is needed. Such a car could achieve 80 mpg or perhaps much better (92).

The Economics of Auto Efficiency Improvement

The analysis of technologies available for improving conventional vehicles and their costs has been updated to include recent technological advances (93). The central result of this analysis is that roughly a 70% improvement in average new car fuel economy at fixed performance would be cost effective and could be implemented in a decade or so. The incremental retail cost of a new car, associated with these measures, is $770 (in 1993 dollars), an amount similar to the estimated cost of improvements to improve fuel economy at fixed performance made since the mid 1970s. The criteria behind this cost-effectiveness result are 5%/yr real discount rate, similar to actual auto loan rates, and 12-yr life, corresponding to vehicle life (rather than the period of initial ownership).

Examples of conservation supply curves for new cars show that the cost-effective improvements in vehicular efficiency (ie, in mpg) range from about 25 (93) to 70% (94), depending on the list of technologies, the capital recovery factor, and the price of gasoline used for comparison. And these examples are from two sources with a somewhat similar set of performance and cost estimates for each technology.

However, while the incremental retail cost implied by implementing many of the particular technologies is rather uncertain, the cost of a major group of the technologies is limited on the high side. The reason is that most of them have been implemented in low priced production vehicles, without major impact on the vehicles' prices as illustrated by the Honda Civic VX (94). There simply is no room for these measures to add costs much greater than those estimated, if they are brought into production when models and components are being changed for other reasons.

Perhaps more important than these steady-state costs, are the up-front conversion costs. How much development and retooling is needed to incorporate these technologies in most cars, such that the manufacturing cost would be acceptable and the reliability, over millions of vehicles, would be excellent? Even though almost all the technologies mentioned are incorporated in some version of a production car, these up-front costs could be substantial for manufacturers who are not adept at changing the manufacturing technology or if the pace of change were rapid, not permitting the changes to be made when other large changes were made for other reasons.

These problems are particularly acute for engines, the site of most of the efficiency technologies. Engine manufacture is highly automated with the perhaps-surprising consequence that it is inflexible. The time between major changes in engines is long. Some Japanese manufacturers have, however, succeeded in moving to the next generation of engine manufacturing technology, where conversion of production lines for a new high tech engine is relatively easy. The U.S. manufacturers are catching up in this respect.

Policy Analysis

Selection of Policies for Discussion. The goal of a good transportation system with respect to daily travel is to provide convenient "access" at moderate cost, while maintaining or improving values such as safety and environmental quality. In a given situation, improving access may involve enabling people to travel farther, but it may instead involve reconfiguring land use patterns so that less travel is needed, eg, so the access can be achieved by walking.

Policy areas that bear directly on improved access are (95)

- Land use, including urban planning and community design.
- Public transport.
- Substitution of communication for transportation.
- Traffic and parking management, road controls, and road design.
- Driver behavior, including vehicle maintenance and driving style.
- Improvements in light vehicles.

Land use and public transportation policies are a principal focus of those interested in improved access (71,96,97). Their potential impact is suggested by the fact that per capita gasoline use in the Toronto metropolitan area is roughly half that in Houston, Phoenix, or Detroit (78). Yet Toronto is not that different. It is a relatively affluent, high quality of life North American metropolis. The key characteristics of Toronto that appear to be responsible for its low gasoline use are high density and a well-developed public transportation system, both associated with regional government.

Substitutes for transportation, such as telecommunications, which enable some people to work and shop at home, and satellite at places of work that rely heavily on telecommunications also have significant potential. These are not primarily issues for public policy, but technology policies and regulation of communication systems are important to their success.

Traffic management in the form of speed limits, high occupancy vehicle lanes, and car pooling assistance have had modest success in reducing travel demand and fuel consumption (98). A component to increase the success of many of these programs is charging full cost for parking privileges (99). Road charges in congested areas have long been considered in Europe and Asia and have been successful in Singapore, where they have been combined with provision of extensive modern public transport (100).

Highway controls such as sophisticated signal management to encourage smooth traffic flow in congested areas can be effective. "Metering" lights, eg, to control the entrance of vehicles onto freeways are effective in enabling the main body of traffic to move smoothly (101). A new generation of highway information and controls, with many potential forms, is the subject of widespread research: intelligent vehicle and highway systems (IVHS) could, if they do not go far, turn out to be no more than an improved means for advertising local businesses, or they could, if more successful, be a large step toward automating traffic flow on high speed arteries. In the latter case the traffic smoothing capability (such that average speed is close to the maximum sought) can yield significant fuel economy benefits (102).

Driving behavior is also potentially important but difficult to influence. Proper vehicle maintenance, such as regular engine tuning and maintaining correct tire pressure, and driving style, ie, smooth flow instead of rapid stops, can contribute perhaps 10% to the fuel economy of the average car. Public education may be useful in this area.

While all these areas are important to efforts to reduce the environmental impact and energy consumption of light vehicles, in the interest of brevity the focus of this section is on policies that encourage technological improvements in light vehicles. Improving light vehicle technology will not be a sufficient means of resolving all the environmental and energy problems created by light vehicle use, but it may be the most important, because merely modifying conventional vehicles could cut fuel use by almost 50%.

Several aspects of improving vehicle fuel economy are of interest to policymakers:

- Modifications to conventional vehicles.
- Alternative fuels.
- Radical vehicle technology, such as the fuel cell vehicle or the very light, small commuter car.
- Interactions with vehicle safety and with emissions of pollutants.

The emphasis here is on the first topic. Consider four kinds of public policies: fuel taxes, standards for fuel economy, fees and rebates related to their fuel economy, and assistance for developing improved technology. To evaluate these policies, it is necessary first to consider briefly why the present market does not create an efficient vehicle fleet.

Barriers to Improved Fuel Economy. A significant barrier to fuel economy improvement was removed by standardized testing. Before the current fuel economy tests, the various claims that were made could not be evaluated by the public. The tests available are not highly accurate in mimicking actual driving; they apparently underestimate fuel use by roughly 20% (103,104). But they are accurate enough to order most vehicles correctly according to their actual fuel consumption in typical driving.

The major remaining barrier is the apparent low priority given to operating costs by new vehicle buyers. Among the total costs of owning and operating a car, fuel plays a small role, about 10%, and the net annual savings from buying a higher fuel economy car are small (76,105). Surveys suggest that new car buyers find fuel economy to be a secondary consideration. Many other attributes have higher priority: brand, safety, interior volume, trunk size, handling, price, acceleration, reliability, etc (106). Indeed, manufacturers have decided that fuel economy is of so little interest to buyers at this time that they offer it only as part of a package in bottom-of-the-market vehicles (such as the Geo Metro), making it impossible for buyers simply to choose added fuel economy at extra cost while preserving the other vehicle attributes in which they are interested. The Honda Civic VX is the only car for which a buyer can pay more to obtain the same vehicle performance and amenities but with a better fuel economy; it is not a best-seller.

This barrier at the point of vehicle purchase should not be misinterpreted as a barrier to policymaking. Many buyers of a new car are not much interested in the fuel economy of their new car, since both the personal savings and the societal benefits associated with their purchase of a single high fuel economy car would be small. The same people as citizens may well be interested in vehicle fuel economy. In other words, when everyone participates in a move toward higher fuel economy, it becomes much more interesting.

Fuel Pricing. There are at least two issues to discuss in terms of fuel pricing: rationales for taxing fuels and the likely consequences of increased fuel prices. The starting point is, of course, the damage caused by use of vehicles and the costs borne to avoid or mitigate such damage. Many of these costs are directly associated with energy use. Many have not been internalized, ie, the costs are not paid by people in relationship to the use of vehicles or to the energy used in operating the vehicles (107). The advantage of fuel taxes is the potential breadth of response by individuals, based on their individual interests. (Economists refer to this as an efficiency.) Thus, in response to a fuel price increase, an individual may change behavior in any of a wide variety of ways, from land use to amount of driving to choice of vehicle. And those choices depend on the values and circumstances of each individual.

Unfortunately, the effect of a price increase on gasoline use is only known in terms of the small short-term response discussed above. The longer-term response, involving choice of vehicle and residential community, eg, is not known with any reliability for today's conditions, although there have been regression analyses (108).

Experience overseas suggests that a large fuel tax (of $1.00–3.00/gal as in Europe and Japan) would not by itself have a dramatic effect on vehicle choice. It might have more effect on driving (36). Such a large tax would certainly not be politically feasible in the United States at present.

Regulatory Standards for Fuel Economy. The Motor Vehicle Information and Cost Savings Act of 1975 set corporate average fuel economy (CAFE) standards that required the fuel economy of new cars to increase from about 14 mpg in the early 1970s to 27.5 mpg by 1985. The act provided flexibility for manufacturers by applying the standard to the sales-weighted average for each corporation, instead of to each vehicle. Moreover, the secretary of transportation was given the discretion to set a lower standard, as was done for 1986 through 1989, on appeal from manufacturers. The discretion to set standards for light trucks was also left to the secretary of transportation.

In hearings on the 1975 Act, the manufacturers stated that the technology to achieve 27.5 mpg was not available on the proposed time scale and that the only way to achieve the standard would be by making the average car much smaller. They said it would "outlaw full-size sedans and station wagons" (Chrysler), "require all sub-compact vehicles" (Ford), and "restrict availability of 5 and 6 passenger cars regardless of consumer needs" (General Motors) (109). Indeed, there was some reduction in the ratio of maximum power to weight, although almost none in the interior volume, in the early 1980s (81). By the late 1980s, however, the manufacturers were achieving the mandated standards with vehicles of interior volume and maximum power equal to and higher than those of the early 1970s. The CAFE standards were thus an important example of successful "technology forcing" by regulation.

Auto manufacturers have argued that higher gasoline taxes drove the fuel economy improvement of the late 1970s and early 1980s. That is, car buyers demanded higher fuel economy, and the manufacturers responded. A detailed look at the data does not support this argument; the CAFE regulations were the critical factor (110). However, synergism between performance regulation and fuel prices was helpful and is an important lesson for policymaking.

General Motors and Ford have also argued that the CAFE formulation placed them at a disadvantage because their mix of vehicles includes large cars while those of Asian manufacturers do not. As a consequence, they argue, it has been easier and less expensive for the Asian manufacturers to meet the standards. Most of the recent legislation proposed in the U.S. Congress to strengthen the fuel economy standards addresses this problem by changing the basis of the standards so that each manufacturer is required to improve its fuel economy by the same percentage above its base year fuel economy. Size-weighted average fuel economies could also serve the same purpose.

Because substantially higher fuel economies are practical and cost-effective and because society has a major interest in reducing petroleum demand, it is not surprising that stronger regulatory standards for fuel economy are actively being considered in Congress. Senator Bryan sponsored a bill that would have required each manufacturer to increase its average fuel economy 40% above its 1988 level by 2001. On average, the bill would have required new cars to reach 40 mpg. It was supported by a majority of the Senate but failed to overcome a filibuster in late 1990. The bill was reintroduced in 1992 and again came close, but failed.

Automobile manufacturers strongly oppose the legislation and claim, as they did in 1975 before the first CAFE standards were passed, that it is not practical to improve fuel economy substantially except by moving, on the average, to much smaller cars. Manufacturers are stonewalling on this point. Other, more compelling, reasons for their opposition are

- Large tooling investments would be needed to make the changes, especially if a moderately rapid time table is required as has been proposed.
- The required rate of improvement in fuel economy might prevent manufacturers from fully exploiting sales opportunities for low fuel economy models already in production.
- High fuel economy standards might somewhat restrict designers' options in developing new vehicles and markets.

It is important to address such concerns by creating a schedule of strengthened standards, allowing adequate time for manufacturers to adjust, and by enacting policy packages (with components discussed elsewhere in this section) such that the burden of compliance would not fall entirely on the manufacturer. Policies in addition to fuel economy standards should be enacted that motivate buyers to select high fuel economy vehicles.

Incentives at New Vehicle Purchase. The gas guzzler tax, enacted as part of the Energy Tax Act of 1978, has tended to be overlooked as an effective policy tool for improving fuel economy. There is strong evidence that the gas guzzler tax played an important role in improving new car fuel economy, especially between 1983 and 1986 (76). In 1978, Congress also considered rebates for new cars with high fuel economy, but rejected the concept. Such a program of fees and rebates can be designed to be revenue neutral, such that total rebates roughly equal total fees (111).

Fees and rebates on the purchase of a new vehicle could be important tools to improve fuel economy and emissions (112). Given U.S. society's sensitivity to first cost, it is probably easier and more effective to adjust for market imperfections and influence new car fuel economy and emission levels at the point of capital equipment purchase than it is to adjust for imperfections in the course of operations with, eg, a gasoline tax.

Technology Push and Pull Policies. The policies discussed above indirectly encourage the creation of new technology to meet the changed economic conditions or regulatory constraints. Experience shows, however, that a more direct policy focus on new technology can be highly effective (113). Before considering the policies, briefly consider the possibilities for new technology.

New technology here means vehicles, and their energy supply systems, which could radically reduce energy requirements and emissions but which are not close to being in mass production. (This technology would go beyond that considered in constructing Table 5.) There are three potential types of vehicle technologies:

1. Vehicles with much higher fuel economy but still based on gasoline or diesel fuel and still serving four or more passengers with, roughly, today's driving capabilities.
2. Special-purpose vehicles requiring much less energy at the drive wheels, such as small commuter cars.
3. Alternative-fuel vehicles.

In group 1 would be hybrid vehicles, mentioned above, which use energy storage to enable the fuel engine to achieve much higher mechanical efficiency that at present. Another possibility would incorporate an advanced, direct-injection, diesel engine. This engine, now entering production in Europe, is about 25% more efficient than comparable conventional gasoline-powered engines. High fuel economy prototype vehicles incorporating advanced diesels have been built or partially developed by Volvo, Volkswagen, Renault, Peugeot, and Toyota, with in-use fuel economies estimated to be almost 70 mpg and higher (114). Diesel emissions problems would have to be solved.

In group 2, there are vehicles such as the proposed Lean Machine and the demonstration electric vehicle called Impact, both developed by General Motors. The Lean Machine is a two seater with the passenger seat behind the driver. Both the Lean Machine and the Impact are small, have little air and tire drag, and require very little power to be delivered to the wheels in typical driving. (The fact that the Impact is an electric vehicle is incidental to this discussion.) Both of these prototype vehicles happen to have rather high acceleration performance. It is not clear if that is an important attribute for marketing such a vehicle. Safety is a critical issue for such small vehicles. It may be important to consider separate lanes on high speed roadways.

In group 3 there is an enormous range of possibilities. Only two of the most exciting are mentioned here: hybrid-electric and fuel-cell vehicles. The hybrid has an internal-combustion engine and one or two kinds of storage that can be used to drive an electric motor. An attractive combination uses the engine only in a high efficiency and low emissions operating region. One of the storage devices would provide high power, eg, to enable a good 0–60 mph acceleration time. The other would be energy storage that could be used to balance energy flows and might also be used for all-electric operation in restricted urban zones. The hybrid could overcome the severe disabilities of electric vehicles: their short daily range, long battery recharge period, short battery life under vehicular conditions, and high cost.

The fuel cell is a kind of battery, but different from the common electrochemical cell, with the advantage of relying on a stored fluid fuel like methanol or hydrogen. The fuel cell converts the chemical energy of the fuel to electricity without combustion. Extremely little, if any, emissions are associated with fuel-cell operation, with the exception of water and carbon dioxide. Much higher efficiencies of conversion are possible than with the present kind of engine, because the second law does not limit the efficiency.

Let us briefly consider some technology policies. The U.S. government has been highly effective in encouraging new technology in some sectors like agriculture, commercial aircraft, and semiconductors. The tools used are technology push and technology pull. Technology push concerns the creation of technology: research, invention, development, and demonstration. This is not a linear sequence of activities, in which one follows the next but a complex interaction in which new technologies are created. Technology push policies involve government support for research, development, and demonstration (R,D&D) and government encouragement of private sector R,D&D through tax incentives, patent law, etc.

Technology pull concerns the demand for new technology, ie, demand for it after it reaches initial commercial status. It cannot be overemphasized that the existence of a likely market for a new or improved process or product strongly motivates development and production of new technologies, and the apparent absence of a market strongly inhibits them. Government policies can provide technology pull through government purchases and by encouraging the private sector's propensity to purchase new technology (115).

An example of technology push policy is government-supported research and development on generic technologies that could form the basis for many new product developments. Modest government involvement in transportation R&D is proving beneficial in electrochemistry (new and improved batteries), combustion (understanding of knock and soot formation), and ceramic insulation (for the combustion chamber). It would be valuable to continue support in these areas and greatly expand the government's efforts in, eg, engine friction and control approaches for hybrid-electric vehicles.

Another kind of technology push policy is the cooperative government–industry venture, closely managed by committees and involving substantial industry investment. An example is the Advanced Battery Consortium for electric vehicles. As with the Steel Initiative mentioned above, there are serious difficulties with this kind of approach because the firms involved are often mature, and although willing to join the venture, they may, in fact, be lukewarm about radical innovation.

An attractive example of a technology pull policy would be providing extra fuel economy credit to manufacturers who produce automobiles or light trucks that attain exceptionally high levels of fuel economy. Such a provision would reward manufacturers for aggressively introducing new technology, providing an incentive for manufacturers to take a substantial leap forward with fuel economy technologies, instead of taking small incremental steps. The incentive could be made especially strong for improving the fuel economy of mid-size and large cars.

CONCLUSIONS

The opportunity to improve the energy efficiency of society is excellent. Researchers have a long way to go before bumping up against limitations imposed by physical laws. If cost-effective investments were widely adopted, and people followed efficient practices, the resulting energy efficiency improvement would substantially ease environment problems, reduce capital requirements for energy, remove the petroleum security problems of the United States, and enable renewable energy supplies to be more easily implemented. (What is cost effective depends, however, on who makes the evaluation, especially on their time horizon or discount rate.)

Both the mix of activities and the efficiencies of the narrowly defined activities are involved in the overall energy efficiency of society. (Here activity refers, eg, to annual travel by different modes and annual use of various manufactured materials.) And the activity mix and real energy intensity can both be improved through changes in technology, behavior, and social institutions. In addition, new technologies are easing the way.

In the examination of industry and transportation, it was found that the opportunity for efficiency improvement may have different characteristics. The mix of activities in industry is tending toward higher efficiency, while the mix in transportation is tending toward lower energy efficiency. Meanwhile, the sectoral, or modal, energy intensities show great room for technological improvement. Moreover, large corporations (which predominate in industry, and as producers, and often operators, of transportation equipment) have proven highly capable of improving their own energy efficiency when motivated. Individuals and smaller firms have less technical capability and may be harder to motivate.

Many feel that market-based decision making is the only proper means for society's energy efficiency to be determined. This is, however, simplistic. One can ask, What market? Today's? There are three problems with today's energy-use markets. (*1*) Energy prices do not reflect many of the costs involved in providing energy, such as unregulated environmental damage, petroleum security expenditures, and supply subsidies. (*2*) Market failures are common in the design of products (relative to their energy performance), for which the cost of energy is a small part of the total. When the motivation is strong, striking design improvements can be made that are cost-effective for general application. For example, new high performance portable computers are designed to have much smaller energy consumption than personal computers plugged into the grid. With some government encouragement, these improvements, especially with respect to stand-by operation, are now beginning to be used in standard PCs. (*3*) Market failures are also common in consumer decisions relating to equipment purchase and its energy efficiency. There appears to be a threshold in relative cost below which individuals and firms do not give energy any attention, aside from complaining about the bills. The problems are especially acute where the energy-related decisions are made by a series of players.

Public policies can be appropriate tools for achieving the benefits of increased energy efficiency. Until recently, energy policies focused on expanding the supply of energy and encouraging the exploitation of natural resources more generally (5). The same general kinds of policies are gradually being adopted to encourage efficiency improvement. While this represents a significant shift in philosophy and does adversely impact particular sectors, such as energy supply, the policies need not be draconian or intervene in the details of everyday life.

Having said that, it must be added that while some kinds of energy efficiency improvement are proving relatively easy to achieve, other kinds of improvement are difficult to achieve. Progress will be difficult in several areas.

An important example of improvement that should continue to be relatively easy is enhanced performance of mass-produced equipment, such as automobiles, household appliances, lighting equipment and commercial building space conditioning equipment. That is not to say that these improvements are implemented (in the absence of high energy prices) without intervention. The gains have been and will be made with the help of regulatory performance standards; government support of research, development, and demonstration; and economic incentives such as rebate programs by regulated utilities. In many cases, effective public policy interventions have yet to be made.

Another area of relatively rapid improvement is in the mix of industrial activities. Consumption has gradually been shifting from massive products to high technology products, which are less energy intensive to produce and often less energy intensive to use. This development is a fundamental historic change, not the result of public policy intervention, although several policies nudge it along, especially policies to encourage high tech industries and policies to reduce the environmental impacts of industries that extract and process raw materials.

Other aspects of energy use are less encouraging: improving the efficiencies of custom processes and facilities, like most housing and industrial production processes, and moving the mix of travel activities toward more efficiency. These aspects of energy use will be hard to improve without the commitment of extraordinary skills and resources or large energy price increases.

The difficulty with custom-designed processes and facilities is that policies aimed at overcoming market failures of the kinds just mentioned tend to be cumbersome and ineffective. For example, government and utilities can and do offer energy audits to identify efficiency improvement opportunities. But the customer often does not know if the auditor and the vendor who would install the improvement really do good work. More important, energy has a low priority for most customers. Their attention is focused elsewhere. For example, the factories provided with recommendations by a (Department of Energy sponsored) Energy Audit and Diagnostic Center will usually only undertake projects that pay back in a year or less.

There are, nevertheless, possibilities for progress in this difficult area. The keys are found in new technologies and in changes that have multiple benefits rather than in narrowly focused energy efficiency improvement. For example, industrial processes can be changed with sensors and controls, so that the product is of higher quality, there are fewer rejects, maintenance requirements are anticipated, production becomes more flexible, and energy consumption is reduced. In contrast, pushing insulation and heat recovery to their simple cost-effective limits may prove to be inflexible and, not, in the end, be a sound way to use limited capital and engineering resources.

The pattern of increasing per capita auto and light truck travel is encouraged by land-use policies favoring this growth, and those policies will be difficult to change. Personal values, established self-government by towns, and tax and other long-established policies favoring low density land development in rural areas will resist change. Improvements in land-use patterns and associated travel needs require, at least, the political will to remove the incentives for development of rural lands. There are some growing nonenergy reasons for eliminating those incentives, such as environmental concerns and saturation in the devotion of land to the automobile. Because energy in itself is seldom a governing consideration, such mutual benefits are essential if energy efficiency is to be increased rapidly. To be improved rapidly, energy efficiency must be seen as part of economic efficiency and part of environmental quality.

BIBLIOGRAPHY

1. J. Priest, *Energy Principles, Problems, Alternatives,* 4th ed., Addison-Wesley Publishing Co., Inc., Reading, Mass., 1991.
2. American Institute of Physics, *Efficient Use of Energy, Part 1–A Physics Perspective, Conference Proceedings,* vol. 25, New York, 1975.
3. J. H. Keenan, *Thermodynamics,* John Wiley & Sons, Inc., New York, 1941.
4. R. U. Ayres, *Energy Efficiency in the U.S. Economy: A New Case for Conservation,* International Institute for Applied Systems Analysis, Laxenburg, Austria, 1989.
5. M. Ross and R. Williams, *Our Energy: Regaining Control,* McGraw-Hill Book Co., Inc. New York, 1981.

6. Scientific American, *Energy for Planet Earth,* W. H. Freeman Publishers, San Francisco, 1990.
7. C. D. Masters, D. H. Root, and E. D. Attanasi, *Ann. Rev. Energy* **15,** 23–51 (1990).
8. National Research Council, *Undiscovered Oil and Gas Resources: An Evaluation of the Department of Interior's 1989 Assessment Procedures,* National Academy Press, Washington, D.C., 1991.
9. World Resources Institute, *World Resources 1992–93,* Washington, D.C., 1992.
10. U.S. Congress, *U.S. Natural Gas Availability: Gas Supply Through the Year 2000,* Office of Technology Assessment, Washington, D.C., 1985.
11. R. Nehring, *Ann. Rev. Energy* **7,** 175–200 (1982).
12. G. W. Hinman in R. Howes and A. Fainberg, eds., *The Energy Sourcebook: A Guide to Technology Resources and Policy,* American Institute of Physics, New York, 1991, pp. 99–126.
13. T. B. Johansson, H. Kelly, A. K. N. Reddy, and R. H. Williams, *Renewable Energy: Sources for Fuels and Electricity,* Island Press, Washington, D.C., 1993.
14. U.S. Department of Commerce, *Survey of Current Business,* Washington, D.C. (monthly).
15. J. M. Hollander, ed., *The Energy Environment Connection,* Island Press, Washington, D.C., 1992.
16. T. E. Graedal and P. J. Crutzen, *Sci. Amer.* **261,** 58–68 (Sept. 1989).
17. S. Sillman, *Ann. Rev. Energy Environ.* **18,** 31–56 (1993).
18. V. A. Mohnen, *Sci. Amer.* **259,** 30–38 (Aug. 1988).
19. S. M. Schneider, *Sci. Amer.* **263,** 312–320 (1990).
20. A. Revkin, *Global Warming: Understanding the Forecast,* Abbeville Press, New York, 1992.
21. J. P. Peixoto and A. H. Oort, *Physics of Climate,* American Institute of Physics, New York, 1992.
22. U.S. Congress, *Changing by Degrees: Steps to Reduce Greenhouse Gases,* Office of Technology Assessment, Washington, D.C., 1991.
23. M. Grubb and co-workers, *Energy Policies and the Greenhouse Effect,* 2 vols., Dartmouth Publishing, Aldershot, UK, 1991.
24. V. Anderson, *Energy Efficiency Policies,* Routledge, London, 1993.
25. G. Boyd, J. McDonald, M. Ross, and D. Hanson, *Energy J.* **8**(2), 77–97 (1987).
26. W. Leontief, *Sci. Amer.,* **185** 15 (Oct. 1951).
27. U.S. Congress, *Energy Use and the U.S. Economy,* Office of Technology Assessment, Washington, D.C., 1990.
28. J. M. Roop, *The Trade Effects of Energy Use in the U.S. Economy: An Input-Output Analysis,* paper presented at the North American meeting of the International Association of Energy Economists, 1986.
29. Energy Information Administration, *International Energy Annual 1991,* Washington, D.C., 1992.
30. J. Goldemberg, T. B. Johansson, A. K. N. Reddy, and R. H. Williams, *Energy for a Sustainable World,* World Resources Institute, Washington, D.C., 1987.
31. A. J. Meier, J. Wright, and A. H. Rosenfeld, *Supplying Energy Through Greater Efficiency,* University of California Press, Berkeley, 1983.
32. P. G. LeBel, *Energy Economics and Technology,* Johns Hopkins University Press, Baltimore, Md., 1982.
33. U.S. Congress, *Building Energy Efficiency,* Office of Technology Assessment, Washington, D.C., 1992.
34. A. H. Rosenfeld and E. Ward, in Ref. 15, pp. 233–257.
35. U.S. Congress, *Industrial Energy Efficiency,* Office of Technology Assessment, Washington, D.C., 1993.
36. L. Schipper and S. Meyers, *Energy Efficiency and Human Activity,* Cambridge University Press, Cambridge, UK, 1992.
37. M. Ross, *Proc. Nat. Acad. Sci. U. S. A.* **89,** 827–831 (1992).
38. R. Herman, S. A. Ardekani, and J. H. Ausubel, in J. H. Ausubel, ed., *Technology and Environment,* National Academy Press, Washington, D.C., 1989, pp. 50–69.
39. R. C. Marlay, *Science* **226,** 1277–1283 (1984).
40. R. H. Williams, E. D. Larson, and M. H. Ross, *Ann. Rev. Energy* **12,** 19–144 (1987).
41. Jack Faucett Assoc., *National Energy Accounts 1958–1985,* Bethesda Md., 1989.
42. D. Steinmeyer in Ref. 15, pp. 319–343.
43. C. Berg, *Science* **199,** 608–614 (1978).
44. P. S. Schmidt, *Electricity and Industrial Productivity: A Technical and Economic Perspective,* Pergamon Press, Oxford, UK, 1984.
45. L. Argote and D. Epple, *Science* **247,** 920–924 (1990).
46. W. H. Joyce, in J. Tester, D. Wood, and N. Ferrari, eds., *Energy and the Environment in the 21st Century,* MIT Press, Cambridge, Mass., pp. 427–435, 1991.
47. M. H. Ross, P. Thimmapuram, R. E. Fisher, and W. Maciorowski, *Long-Term Industrial Energy Forecasting (LIEF) Model,* ANL/EAIS/TM-95, Argonne National Laboratory, Argonne, Ill., 1992.
48. S. R. Venkateswaran and H. E. Lowitt, *The U.S. Cement Industry: An Energy Perspective,* a report by Energetics Inc., Columbia, Md., 1988.
49. M. J. King, *Guide to the INDEPTH Level 1 Econometric Models: Final Report,* Electric Power Research Inst., Palo Alto, Calif., 1990.
50. R. Carlsmith, W. Chandler, J. McMahon, and D. Santini, *Energy Efficiency: How Far Can We Go?* ORNLTM-11441, Oak Ridge National Laboratory, Oak Ridge, Tenn., 1990.
51. J. G. Koomey, *Energy Efficiency Choices in New Office Buildings: An Investigation of Market Failures and Corrective Policies,* Ph.D. dissertation, University of California, Berkeley, 1990.
52. A. C. Fisher and M. H. Rothkopf, *Energy Policy* **17,** 397–406 (1989).
53. C. Blumstein, B. Krieg, L. Schipper, and C. York, *Energy* **5,** 355–371 (1980).
54. J. S. Peters, ACEEE, *Summer Study, American Council for an Energy-Efficient Economy,* Washington, D.C., 1988.
55. Alliance to Save Energy, *Industrial Investment in Energy Efficiency: Opportunities, Management Practices and Tax Incentives,* Washington, D.C., 1983.
56. M. Ross, *Financial Manage.,* 15–22 (Winter: 1986).
57. J. C. van Horne, *Financial Management and Policy,* Prentice-Hall, Inc., Englewood Cliffs, N.J., 1980.
58. R. Pike and B. Neale, *Corporate Finance and Investment: Decisions and Strategies,* Prentice-Hall, Inc., Englewood, N.J., 1993.
59. W. D. Montgomery in M. A. Kuliasha, A. Zucker, and K. J. Ballew, eds., *Technologies for a Greenhouse-Constrained Society,* Lewis Publishers, Boca Raton, Fla., 1992.
60. D. A. Lashof and D. Tirpak, *Policy Options for Stabilizing Global Climate,* Office of Policy, Planning and Evaluation, U.S. Environmental Protection Agency, Washington, D.C., 1989.

61. U.S. Department of Energy, *A Compendium of Options for Government Policy to Encourage Private Sector Responses to Potential Climate Change,* Washington, D.C., 1989.
62. W. H. Chandler, H. Geller, and M. Ledbetter, *Energy Efficiency: A New Agenda,* American Council for an Energy-Efficient Economy, Washington, D.C., 1988.
63. D. H. Moskovitz, *Ann. Rev. Energy* **15,** 399–421 (1990).
64. E. Hirst, *A Good Integrated Resource Plan: Guidelines for Electric Utilities and Regulators,* ORNL/CON-354, Oak Ridge National Laboratory, Oak Ridge, Tenn., 1992.
65. E. Hirst, *Electric-Utility DSM Program Costs and Effects: 1991 to 2001,* ORNL/CON-364, Oak Ridge National Laboratory, Oak Ridge, Tenn., 1993.
66. S. Nadel, *Ann. Rev. Energy Environ.* **17,** 507–536 (1992).
67. G. Hatsopoulos, P. Krugman, and L. Summers, *Science* **241,** 299–307 (1988).
68. H. S. Geller and S. M. Nadel, *Consensus National Efficiency Standards for Lamps, Motors, Showerheads, and Faucets, and Commercial HVAC Equipment,* American Council for an Energy-Efficient Economy, Washington, D.C., 1992.
69. A Price and M. Ross, *Electricity J.* **2,** 40–52 (July 1989).
70. H. Geller, J. Harris, M. D. Levine, and A. H. Rosenfeld, *Ann. Rev. Energy* **12,** 357–396 (1987).
71. J. Ahearne, *Why Federal Research Fails,* discussion paper EM 88-02, Resources for the Future, Washington, D.C., 1988.
72. D. Gordon, *Steering a New Course: Transportation, Energy and the Environment,* Union of Concerned Scientists, Boston, 1991.
73. S. C. Davis and S. G. Strang, *Transportation Energy Data Book: 13,* ORNL-6743, Oak Ridge National Laboratory, Oak Ridge, Tenn., 1993.
74. D. L. Greene, *Ann. Rev. Energy Environ.* **17,** 537–573 (1992).
75. M. Ross, *Ann. Rev. Energy* **14,** 131–171 (1989).
76. M. Ross and M. Ledbetter, in Ref. 15, pp. 258–318.
77. Federal Highway Administration, *1990 Nationwide Personal Transportation Survey, Summary of Travel Trends,* Washington, D.C., 1992.
78. P. G. Newman and J. R. Kenworthy, *Cities and Automobile Dependence: A Sourcebook,* Gower Technical, Aldershot, UK, 1988.
79. J. Holtzclaw, *Explaining Urban Density and Transit Impacts on Auto Use,* Testimony before the California Energy Commission, Docket No. 89-CR-90, Sacramento, April 23, 1990.
80. American Automobile Manufacturers Association, *Facts and Figures '93,* Washington, D.C., 1993.
81. J. D. Murrell, K. H. Hellman and R. M. Heavenrich, *Light Duty Automotive Technology and Fuel Economy Trends Through 1993,* U.S. Environmental Protection Agency, Office of Mobile Sources, Ann Arbor, Mich., 1993.
82. F. An and M. Ross, *A Model of Fuel Economy and Driving Patterns,* 930328, Society of Automotive Engineers, Warrendale, Pa., 1993.
83. R. Stone, *Introduction to Internal Combustion Engines,* Macmillan Publishers, London, 1985.
84. C. A. Amann, *The Automotive Engine–a Future Perspective,* 891666, Society of Automotive Engineers, Warrandale, Pa., 1989.
85. M. Ross and F. An, *The Use of Fuel by Spark Ignition Engines,* 930329, Society of Automotive Engineers, Warrendale, Pa., 1993.
86. Energy and Environmental Analysis Inc., *Analysis of the Capabilities of Domestic Auto-Manufacturers to Improve Corporate Average Fuel Economy,* U.S. Department of Energy, Arlington, Va., 1986.
87. U. Seiffert and P. Walzer, *Automobile Technology of the Future,* Society of Automotive Engineers, Warrendale Pa., 1991.
88. U.S. Congress, *Improving Automobile Fuel Economy: New Standards, New Approaches,* Report OTA-E-504, Office of Technology Assessment, Washington, D.C., 1991.
89. National Research Council, *Automotive Fuel Economy: How Far Should We Go?* National Academy Press, Washington, D.C., 1992.
90. K. C. Ludema in J. C. Hilliard and G. S. Springer, eds., *Fuel Economy in Road Vehicles Powered by Spark Ignition Engines,* Plenum Press, New York, 1984.
91. R. Stone, *Automobile Fuel Economy,* Macmillan Education, London, 1989.
92. A. B. Lovins, J. W. Barnett, and L. H. Lovins, *Supercars: the Coming Light Vehicle Revolution,* Rocky Mountain Institute, Snowmass, Colo., 1993.
93. J. DeCicco and M. Ross, *An Updated Assessment of the Near-Term Potential for Improving Automotive Fuel Economy,* American Council for an Energy-Efficient Economy, Washington, D.C., 1993.
94. J. G. Koomey, D. E. Schechter, and D. Gordon, *Cost Effectiveness of Fuel Economy Improvements in 1992 Honda Civic Hatchbacks,* Transportation Research Record 1416, National Research Council, Washington, D.C., 1993.
95. M. Ledbetter and M. Ross in J. Byrne and D. Rich, eds., *Energy and Environment, Energy Policy Studies,* Vol. 6, Transaction Publishers, New Brunswick, N.J., 1992, pp. 187–233.
96. B. S. Pushkarev and J. M. Zupan, *Public Transportation and Land Use Policy,* Indiana University Press, Bloomington, 1977.
97. R. W. Burchell and D. Listokin, eds., *Energy and Land Use,* Center for Urban Policy Research, Rutgers University, Piscataway, N.J., 1982.
98. M. Burke, *High Occupancy Vehicle Facilities: General Characteristics and Potential Fuel Savings,* American Council for an Energy-Efficient Economy, Washington, D.C., 1990.
99. M. Replogle, *U.S. Transportation Policy: Let's Make it Sustainable,* Institute for Transportation and Development Policy, Washington, D.C., 1990.
100. B. W. Ang in D. L. Bleviss and M. L. Birk, eds., *Driving New Directions: Transportation Experiences and Options in Developing Countries,* International Institute for Energy Conservation, Washington, D.C., 1991, pp. 41–51.
101. Institute of Transportation Engineers, *A Toolbox for Alleviating Traffic Congestion,* Washington, D.C., 1989.
102. F. An and M. Ross, *A Model of Fuel Economy with Applications to Driving Cycles and Traffic Management,* Transportation Research Record 1416, National Research Council, Washington, D.C., 1993.
103. K. Hellman and J. D. Murrell, *Development of Adjustment Factors for the EPA City and Highway MPG Values,* 8400496, Society of Automotive Engineers, Warrendale, Pa., 1984.
104. M. Mintz, A. Vyas, and L. A. Conley, *Differences Between EPA-test and In-use Fuel Economy: Are the Correction Factors Correct?* Transportation Research, Record 1416, Washington, D.C., 1993.
105. F. von Hippel and B. G. Levi, *Resources Conservation,* **10,** 103–124 (1987).
106. P. S. McCarthy and R. Tay, *Transportation Res.* **23A,** 367–375 (1989).

107. J. J. MacKenzie, R. C. Dower, and D. Chen, *The Going Rate: What it Really Costs to Drive,* World Resources Institute, Washington, D.C., 1992.
108. D. R. Bohi and M. B. Zimmerman, *Ann. Rev. Energy* **9,** 105–154 (1984).
109. Energy Conservation Coalition, *The Auto Industry on Fuel Efficiency: Yesterday and Today,* Washington, D.C., 1989.
110. D. L. Greene, *Energy J.* **11**(3), 37–57 (1990).
111. D. Gordon and L. Levenson, *J. Policy Anal. Manage.* **9,** 409–415 (1990).

ENERGY EFFICIENCY CALCULATIONS

WILLIAM KENNEY
Exxon Research and Engineering
Florham Park, New Jersey

Equipment and processes with high energy efficiency are generally valued. Indeed, consumer advocates have arranged for practical measures of efficiency (eg, cost/year to operate) to be attached to a number of household appliances and automobiles. Yet, the basis for such efficiency projections often remains vague. Invariably, the tests that provide the data for these efficiency claims follow some government protocol intended to match the patterns in which an "average user" would employ the appliance or automobile. Although these protocols remain as obscure as the habits of the "average user" to most consumers, the labels do provide some guidance to the purchaser of relatively standard devices about the energy consumption of the alternatives that are being considered. See also THERMODYNAMICS; ENERGY AUDITING; ENERGY CONVERSION FACTORS.

However, when specialized industrial equipment and processes are considered, many bases for efficiency calculations are possible. Not only is there a choice between a first law (of thermodynamics) and a second law basis, but assumptions are also made about ambient conditions, the boundaries of the system to be included, coolant temperatures (winter vs summer vs average), the purity of working fluids and lubricants, fouling, maintenance status, and so forth.

During the energy crisis of the late 1970s, Senator Kennedy argued that all processes and equipment should be evaluated by their second law efficiency, which is generally quite low relative to the more conventional first law approach. The implication was that because most industrial processes had low second law efficiencies, they could be markedly improved if only the industry would attend to the problem. There was, perhaps, even the inference that the government might not permit certain low efficiency processes or equipment to continue to operate. In that time frame, much work was done in the second law analysis of processes and equipment, and the government funded a number of research projects aimed at improving analytical techniques and actual processes and hardware in pursuit of higher efficiencies.

Some rejected the second law approach because in many cases, high second law efficiencies could only be achieved theoretically; practical processes and equipment to carry out the desired operations did not exist. For example, there were no high temperature fuel cells that would permit electricity generation directly from a combustion reaction, and no membranes existed that could separate the components of the flue gas so that they could be expanded in an engine to their partial pressure in the atmosphere and generate power in the process.

Despite the fog that surrounds the concept, analysts have found the concept of efficiency useful over the years. The comparison of the performance of a practical device or process to that of a well-defined standard has been of value in many applications. A low efficiency does highlight steps or systems that may be suitable for upgrade (1–3). In some cases, improvement may be inherently impossible, but these are often easily identified. For example, in a Carnot cycle, the efficiency of the engine depends only on the temperature difference between the source of heat and the sink to which waste heat is rejected. The best possible efficiency for the process is given by (4):

$$E = \frac{T - T_0}{T}$$

where all temperatures are absolute. For a heat engine operating between 1,000°F (1460°R) and 100°F (560°R) the maximum efficiency possible is

$$E = \frac{1460 - 560}{1460} = 0.614$$

Aspiring to an efficiency of 65% is futile. However, the best steam-power plant cycles have an efficiency of 35–40%, so some modest improvement may be possible.

Low temperature power cycles, such as those that operate on waste heat or on the temperature differences between the surface and the depths of the ocean, have much lower ideal efficiencies. Suppose the ocean surface is 60°F, with 30°F at depth. The maximum possible efficiency is

$$E = \frac{520 - 490}{520} = 0.057 \text{ or} < 6\%$$

Similarly, an engine working with waste heat at 200°F and rejecting heat at 70°F can aspire to an ideal efficiency of

$$E = \frac{660 - 530}{660} = 0.197 \text{ or} < 20\%$$

Thus, low temperature heat engines clearly have less potential to be efficient. If a practical system is to be developed, it must have other attributes that would compensate for the inherent efficiency limitations of the cycle (5).

In cases where lower efficiencies are not inherent, technology improvements might have significant impact. For example, improvements in the materials for and design of rotors in gas turbines have allowed the machines to run reliably at higher temperatures, resulting in higher efficiencies, ie, closer to the theoretical limitation set by combustion temperature.

CALCULATING EFFICIENCIES

First Law

The first law efficiency of a process can be captured by an enthalpy balance, that is, by simply calculating the fraction of the input energy which is captured in the process. For the boiler in the steam-power cycle shown in Figure 1, the fraction of the combustion heat absorbed in the water equals the first law efficiency of the boiler. For modern boilers, this is typically 85–90% and is controlled by reducing the stack temperature of the flue gas as low as practical (in regard to economics and corrosion). In the system shown, 12.549 MBtu/h are fired in the boiler; when the flue gas exits to the atmosphere at 200°F, its enthalpy has been reduced to 1.631 MBtu/h by exchange with the water and combustion air in the boiler. Thus, 87% of the fuel heat has been captured in the steam, ie, the first law efficiency of the boiler is 87%.

The first law efficiency of a heat engine, like the steam turbine, is virtually 100%. The enthalpy in the steam is reduced only by the amount of work extracted (and some very small amount for the friction in the machine). Similarly, the efficiency of a heat exchanger, like the condenser in the steam-power plant, is 100% in that all of the heat contained in the hot stream is captured in the cooling medium. Of course, this represents the heat-rejection step so that unless warming the cooling medium is a desired objective, this energy is lost.

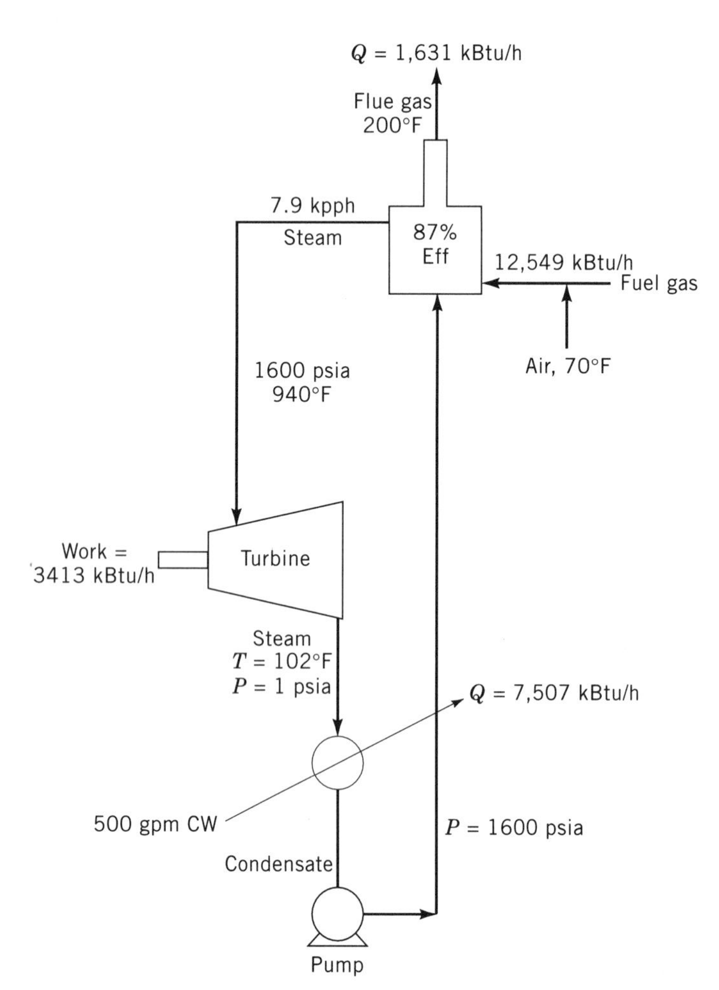

Figure 1. Heat balance for typical high pressure steam-power cycle showing fuel fired and heat rejected per megawatt hour power produced.

If the overall efficiency of the steam-power process is considered to be based on only the amount of work produced, the first law efficiency is calculated simply by dividing the work produced by the heating value of the fuel burned. If the boiler first law efficiency is 87%, as previously discussed, this is calculated by

$$\mathrm{EFF} = \frac{W}{Q_\mathrm{F}} = \frac{3413}{12549} = 27.2\%$$

This equation ignores the work needed to pump the process water and the cooling medium, and the forced air fans for the boiler (if any). These factors could be significant.

If both power and process heat are desired, the potential for higher efficiency is presented. One approach to providing a combined load is shown in Figure 2. Two separate operations are shown: the steam-power cycle of Figure 1 and a separate boiler to provide steam for a process heat load. The useful products from this arrangement are the 3.413 MBtu/h of work and 17.645 MBtu/h of process heat (Q_P) the fuel fired increased to 32.830 MBtu/h. The first law efficiency is then given by

$$\mathrm{EFF} = \frac{W + Q_\mathrm{p}}{Q_\mathrm{F}} = \frac{3.413 + 17.645}{32.830} = 64.1\%$$

As before, the heat rejected in the condenser and that lost in the flue gas is not counted as useful product from the process.

In a cogeneration configuration (see Fig. 3), both power and useful heat are produced in a single cycle. In many cases, the turbine exhaust steam is used to provide heat to some process or to commercial or residential buildings. Thus, a second component to the useful energy output is present and must be added to the numerator of the efficiency equation, resulting in a higher efficiency. This higher efficiency is, of course, the reason that legal incentives exist to use cogeneration.

From a practical standpoint, the base process conditions of the steam-power cycle must be changed to accommodate the dual role of providing both the power and the required process steam at 125 psia. Fundamentally, the exhaust pressure of the turbine must be raised to a level higher than 125 psia to provide a driving force to transfer heat to the process. Thus, 200 psia was chosen, and it was assumed that an additional heat exchanger would be provided so that the very pure water needed to produce the high pressure steam would not be contaminated by potential leakage in a process heat exchanger. This change in exhaust pressure means that the total steam flow through the turbine must be increased from 7.9 to 19.2 × 10^3 lb/hr (Kpph) because less work is extracted per pound of flow. However, only one boiler is required and total fuel fired is reduced to 24.21 MBtu/h because the heat rejected at the condenser is totally eliminated. Thus, the numerator in the efficiency equation remains the same, but the denomi-

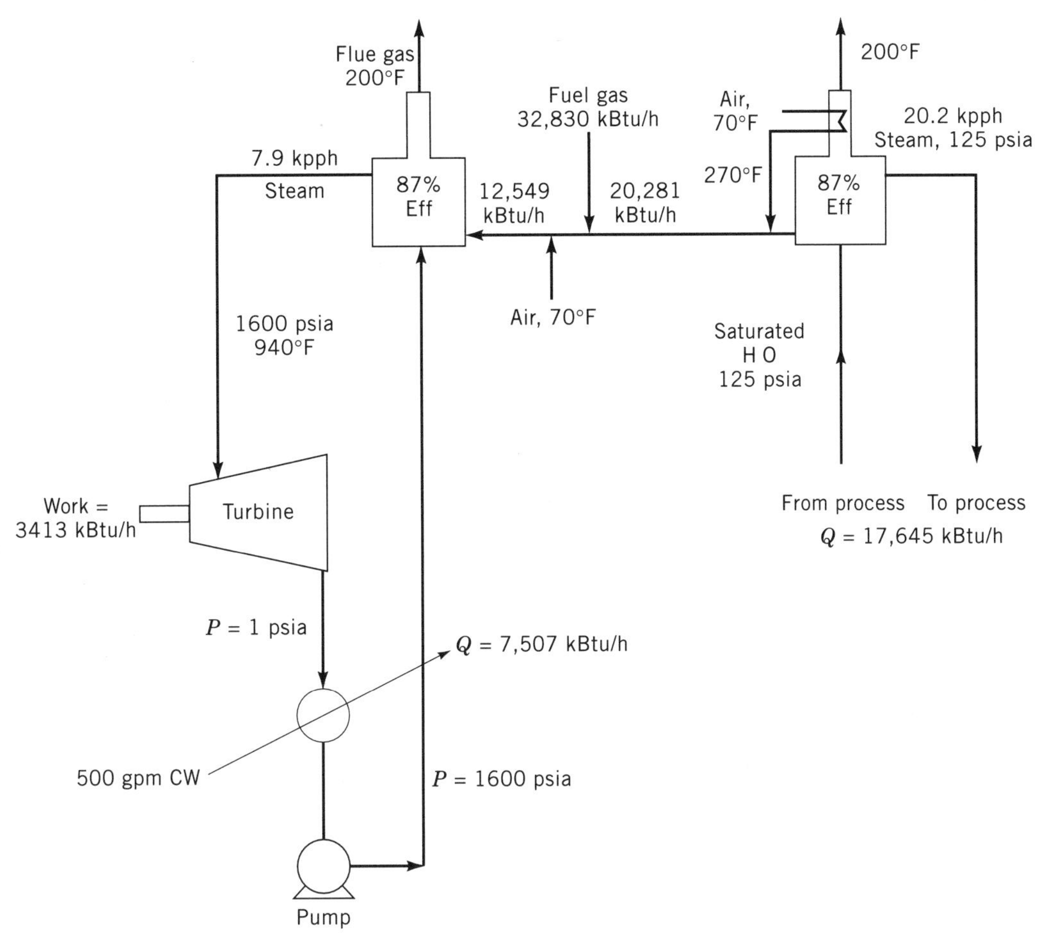

Figure 2. Heat balance for two separate systems producing 1 megawatt hour of power and a fixed amount of process heat. Heating requirements are additive.

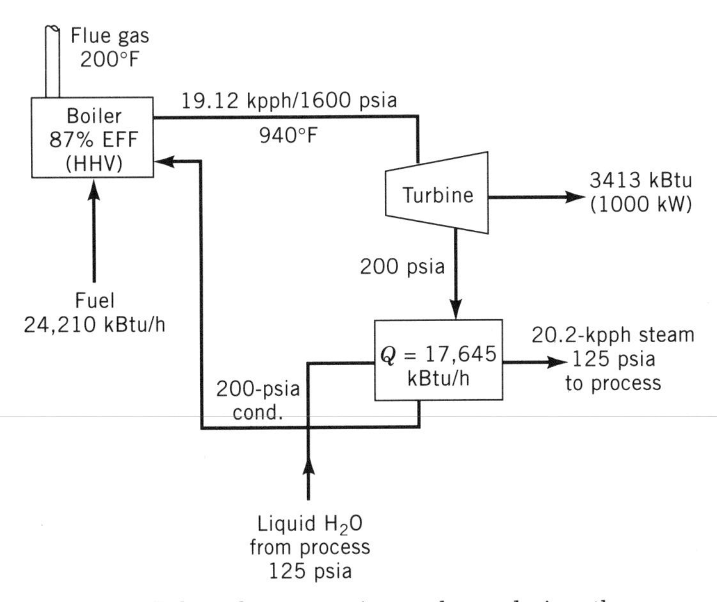

Figure 3. Balanced cogeneration cycle producing the same power and process heat demand as in Figure 2. Fuel fired is reduced because waste at condenser is eliminated, and turbine exhaust conditions and steam flow rate changed significantly.

nator is lower, yielding yet a higher first law efficiency:

$$\text{EFF} = \frac{W + Q_p}{Q_F} = \frac{3.413 + 17.645}{24.21} = 87\%$$

Not surprisingly, this equals the boiler efficiency, as there are no other losses taken into account.

Second Law

Calculating the second law efficiency of a process is another matter. Fundamentally, this efficiency is calculated from the availability (exergy) balance of the process. One approach is to take the ratio of the availability (exergy) in the products of the process to that input to the process. The appropriate boundaries need to be set up so that products and availability input are clearly defined. Thus, the efficiency is calculated as follows:

$$\text{EFF} = \frac{\text{Availability products}}{\text{Availability input}}$$

A second approach is to compare the availability (exergy, work) required or produced in the process to that of a reversible process producing the same product from the same inputs. For example, the Carnot cycle is a reversible process for a heat engine. Therefore, the amount of work produced from a given amount of heat at T_1 with rejection at T_0 in a real system might be compared to that which is theoretically available from a Carnot cycle operating between the same temperatures. In a process that requires work (availability) input, such as refrigeration, the reversible work required would be compared to that actually required to measure the efficiency. Thus, the efficiency for a work-producing process is

$$\mathrm{EFF} = \frac{W_A}{W_I} = \frac{\Delta A_A}{\Delta A_I}$$

For a process that requires work, the ratio is inverted, ie,

$$\mathrm{EFF} = \frac{W_I}{W_A} = \frac{\Delta A_I}{\Delta A_A}$$

Sign conventions must also be considered in these calculations. Work performed on the surroundings is considered positive and that done by the surroundings to the process is negative. Conversely, heat transferred into the process is positive and that rejected to the surroundings is negative.

In parallel with the development of availability balances, the work or availability lost or degraded in a process can be calculated directly from the total entropy change $(\Delta S)_T$ of the system:

$$W_L = \Delta A_L = T_0(\Delta S)_T$$

Even in reversible processes, an entropy change of the system will occur, causing some loss or degradation of availability. In a reversible process, the entropy change will be minimum, and the process will require minimal resources to implement; any real process will require more. As stated earlier, practical ways to approach reversible process steps may not exist.

To illustrate second law efficiencies, consider the processes in Figures 1–3. In Figure 1, the only product of value is considered to be the work generated. As work is 100% availability, both the first law and second law efficiencies are the same (neglecting the small difference between the availability and the fuel value for the fuel fired), ie,

$$\mathrm{EFF} = \frac{3{,}413}{12{,}549} = 27.1\%$$

However, when analyzing a process that produces a product other than work (Figs. 2 and 3), a different result emerges. The availability of the steam heat produced is not the same as that of the product produced directly as work. (From a conceptual standpoint, its capacity to cause change is clearly less than that of the electricity.) Thus, the second law efficiency of the heat plus power process is significantly lower than the corresponding first law efficiency.

Table 1. Lost Availability Comparison of Conventional and Cogeneration Steam-Power Systems

Parameter	Conventional System[a]	Cogeneration System[b]
Availability input, kBtu		
Fuel at 21,370 Btu/lb	32,830[c]	24,210[c]
Boiler feed water at 344°F and 125 psia	1,103	1,103
Total input	33,933	25,313
Availability output, kBtu		
Electricity	3,413	3,413
Steam	7,130	7,130
Total output	10,543	10,543
Second law efficiency, %	31.1[c]	41.7[c]
Total lost availability, kBtu	23,390	14,770

[a] See Figure 2.
[b] See Figure 3.
[c] Approximate figures; it has been demonstrated that the correct data should be 29.362 MBtu vs 32.830 and 21.640 MBtu/hr vs 24.210 for the conventional and cogeneration configurations, respectively; the corresponding efficiencies will be 34.6% and 46.3% (6).

Table 1 lists the availability balances for the two processes and the resultant second law efficiency calculated by the ratio of product availability to that input. For simplicity, it continues with the assumption that the fuel value corresponds to the input availability for the fuel and calculates the water and steam values from Keenen and Keyes, "Steam Tables." This simplified analysis is presented to reinforce the relationship between fuel fired and availability losses.

The results are clear; the useful output of both systems is identical, as is the input from the 125-psia boiler feed water. The difference between the two cycles rests with the fuel requirement. Because combustion is a particularly irreversible process, more lost work is associated with the conventional process, and the thermodynamic efficiency is lower, thus providing quantitative support for cogeneration in its simplest form.

The simple analysis of the two processes can be broken down to identify further the sources of lost work in each process. A more detailed analysis is given in Table 2,

Table 2. Sources of Availability Losses in Conventional and Cogeneration Steam-Power Heating Systems

	Lost Availability, kBtu	
Source	Conventional System[a]	Cogeneration System[b]
High pressure boiler		
Combustion	3686	7155
Steam generation	2604	3872
Low pressure boiler		
Combustion	5958	
Steam generation	6122	
Turbine (net)	1072	615
Heat exchange	419	606

[a] See Figure 2.
[b] See Figure 3.

which was calculated by the more rigorous approach of Sussman. (6) Combustion losses are proportional to the amount of fuel consumed. Steam-generation losses are really a function of the temperature-driving forces used in the boilers. The lower the steam pressure produced in a fired boiler, the greater the loss of available energy. Note that the work lost in heat exchange for the cogeneration cycle occurs in the 200-psia/125-psia boiler, and its contribution to the total lost work is only one-tenth that of steam generation in the fired 125-psia boiler. Turbine losses are lower because the exhaust pressure is higher in the cogeneration cycle.

In this simplified preview of the insight provided by thermodynamics, the first fundamental guidelines for the engineer dedicated to reducing a plant's energy bill emerge:

1. Minimize combustion.
2. Generate steam at maximum pressure and generate power from all low pressure steam demand (ie, maximize cogeneration).
3. Use steam for process heat at minimum temperature.

Note also the differences between the efficiency values and locations of losses between a first law viewpoint and the one advanced here. First law boiler efficiencies are listed at 87%, with the main waste occurring at the condenser. From the second law viewpoint, there is little lost in the condenser. Thus, a different energy conservation strategy is indicated by available energy analysis compared to simple energy accounting. The value of the second law approach lies not in the efficiency calculated, but rather in the analysis required to calculate the efficiency and in the basis the analysis provides for further identification of the process steps that offer opportunity for improvement.

As the total availability of the fuel in calculating the second law efficiencies of these processes has been considered, any differences that might have existed between the two approaches to calculating efficiencies mentioned previously are virtually eliminated. In an ideal combustion process, all of the availability liberated in the combustion reaction in the products would be recovered. In addition, the flue gases would be cooled to ambient temperatures (T_0) before leaving the system, and the reaction would take place in a reversible fuel cell.

However, this ideal process is far-removed from the process considered in the analysis. Yet, by considering all of the availability of the fuel as the basis for efficiency calculations, essentially the same result is obtained for efficiency. If only the availability of the flue gas in the boilers would be considered as input, a different (higher) result would be obtained because the availability loss of the combustion reaction would have been ignored. Thus, careful attention to the boundaries of the system being analyzed is important; identifying the original source of the availability driving the process is essential to correct analysis.

For the sake of comparison, the efficiency ranges of other processes that are of industrial and commercial interest are shown in Table 3. Those with high combustion components have the lowest second law efficiencies, whereas those driven by other sources of availability (eg, oxygen separation) give higher results. However, the data sources used in Table 3 are disparate, so that no direct comparison is possible.

Table 3. Typical Efficiencies

Process	Second Law Efficiency	First Law Efficiency
Ethylene manufacture (from ethane)	22–25%	~80%
Phthalic anhydride manufacture	22–28%	70–80%
Steam/power generation	33–38%	33–38%
Petroleum refining	~15%	~90%
Oxygen separation from air	27%	70–80%
High pressure steamboiler	50%	85–90%
Domestic water heater (fuel fired)	2–3%	40%
Residential heater (fuel fired)	10%	60%

A BROADER VIEW OF EFFICIENCY

Two views of efficiency have been explored so far: the first law (Btu accounting approach) and the second law (energy quality/usefulness overlay). A third, more selective, approach to measuring efficiency appears to be growing. Yet unnamed, it will be referred here as "molecular efficiency" and defined as follows:

> "The optimum management of all the molecules involved in a process system so that the desired product is produced with the minimum overall consumption of nature's resources."

The molecular considerations began with recycling scrap metal and has recently spread to recycling paper and container materials and the elimination of nondegradable packaging by some companies in the food-provider chain. These actions were driven by the growing problem in solid-waste disposal and by the realization that these materials, having served a useful purpose for society, had not yet exhausted that usefulness and could be used again.

A second force in molecular management has been SARA III, the Superfund reauthorization act in the United States, which requires manufacturing companies to report certain materials rejected to the environment in any form. These companies were thus forced to measure emissions, explore their sources, and understand how to manage better the materials in process to minimize the public outcry at the magnitude of the emissions reported. In addition, concepts such as "just-in-time" raw material supply offered economic advantages not only in working capital, but also in space utilization, work flow, and worker attitudes.

In many industries, the optimum use of feed and energy has become essential in a competitive world market to cut processing costs and to reduce waste handling costs, the administrative burdens of permits, tracking wastes, ensuring disposal methods and sites meet all present (and future) regulations, and dealing with public outrage. In addition, exposure to potential future liabilities is reduced.

Society has taken various initiatives to force a new definition of efficient operations upon those who would remain a viable enterprise. Thus, "pollution prevention," rather than clean-up at the end of the pipe, or no clean-up at all, has become a new measure of efficient operation. The evolution of automobile manufacturing in the United States provides an example of this process. Under the pressures for improved fuel economy, automobiles became smaller, lighter, driven by less powerful engines, and festooned with less chromeplated trim. All of this improved fuel economy, but reduced user functionality. In addition, less waste was produced and some wastes were eliminated altogether, eg, chromeplating bath chemicals.

As benefits from this "defunctionizing" phase approached the point of diminishing returns, newer technologies emerged. (That is, work began on the front end of the process to solve problems and restore functionality.) Better, longer-lasting lubricating oils were introduced to reduce waste; unleaded gasoline reduced emissions of a toxic substance and made possible the use of catalysts to reduce exhaust emissions further; and reliable, more economic fuel injection was developed to enhance fuel efficiency, thus restoring some of the lost performance encountered in the first phases.

Today, an even greater fundamental understanding of the manufacture and operation of an automobile is being used to maintain the car owner's independence while addressing environmental issues. Much of this comes from the availability of new materials, but some comes from further refinements in understanding combustion. New materials for bumpers and other car parts have restored some of the lost safety functionality formerly provided by larger cars, and have even enhanced the tolerance of some automobile parts. Anti-lock brakes and air bags are but a few of the newer technologies aimed at producing a safer vehicle; reformulated gasolines containing oxygenated compounds will further reduce emissions.

The common theme of these improvements is the application of new molecules to the problem and the better management and control of all molecules utilized in building and operating the product. This may well be the viewpoint required to be both competitive and acceptable to society in the future.

EFFICIENCY AND COST

The operating cost benefits of fundamental efficiency improvements are usually fairly obvious: lower energy cost, lower waste handling costs, lower administration costs, and lower exposures to future potential environmental liabilities. However, many believe that these benefits can only be achieved at higher investment. There are appreciable data (1) that demonstrate that truly fundamental improvements actually save investment as well as operating costs.

Often, cost-benefit analyses of what appear to be efficiency/investment trade-offs are flawed by either a too-narrow focus or inaccurate data. Utility cost figures do not necessarily represent the real cost of the availability supplied; a prime example of a mispriced utility is steam. Steam is usually priced on the basis of its enthalpy content, regardless of its pressure, and is often burdened with high fixed costs related to ensuring reliability of supply (such as 100% spare boiler capacity). This may have been a good decision when fuel cost $0.20/mBtu, but today, numerous implant steam pricing systems actually value low pressure steam more highly than high pressure steam, in spite of the fact that low pressure steam's capability to cause useful change in the plant is much lower than that of the high pressure steam.

Thermodynamicists will state that pricing steam (or any other utility) according to its availability (exergy) content is the only way to obtain correct analysis. However, pricing systems should be tested against their impact on decisions ie, do the prices established provide acceptable returns on energy efficiency investments? For example, if low pressure steam condenses at high enough temperatures to provide the needed process heat, but high pressure steam can do the same job in a smaller heat exchanger, then the price of low pressure steam must be sufficiently lower than that of the high pressure steam to give a good return on the larger investment for the heat exchanger. Of course, the low pressure steam cannot be produced by simply expanding high pressure steam through a valve for these values to be real; it must be produced by using the higher pressure material to produce work or do some other task that cannot be performed by the lower pressure steam. Otherwise, its capacity to cause change, or its availability content, is simply wasted.

A brief comparison of steam prices based on enthalpy content and availability content is often instructive. Assume that 1,500-psia steam is available at a cost of $10/1,000 lb, and that the plant also uses 140-psia and 15-psia steam. The prices developed for these three pressure levels by ratios of enthalpy and availability are shown in Table 4. This range of potential values would make it difficult to decide on any particular investment.

In many cases, using availability-based prices makes it difficult to justify using back-pressure turbines to produce the lower pressure steam. The decision will be based on the turbine efficiency, investment, and cost of supplying the required power by other means, eg, purchased electricity. Using enthalpy ratio prices, the credit for the exhaust steam would be much higher and turbines are thus easily justified.

The third test of correct steam pricing is whether the recovery of waste heat can be justified. Again, high values for low pressure steam make this easy to do. Unfortunately, many plants witnessed an overabundance of low pressure steam produced because of such pricing, steam that then had to be wasted somewhere in the system because it was in excess of plant needs.

Table 4. A Comparison of Steam Prices

	Prices	
Pressure	Enthalpy Ratio	Availability Ratio
1,500 psia	$10/1,000 lb	$10/1,000 lb
140 psia	8.34	4.77
15 psia	7.88	3.35

To price utilities accurately, how the plant energy balances impact on the prime inputs of availability (usually fuel and electricity) must be known. If recovery of waste heat can be traced back to a reduction in fuel somewhere in the plant, its true value can be calculated. Similarly, if generating work can be traced back to an electricity or fuel saving, its true value can also be seen. The benefit part of the cost-benefit ratio can now be correctly measured.

Cost is often obscured by a too-narrow focus. Investments required outside the plant premises, to provide the increased demand (which must ultimately be paid for in higher prices or taxes), are often not factored into the equation. This became clear in global studies provided for the federal government in the late 1970s. If the circle of analysis is drawn wide enough, the true cost to society of providing new supply or recovering energy can be calculated. The barrier to this more global approach is the challenge of accounting for the benefits to a single enterprise. Consortia have been organized to clear these barriers, but this is often not a trivial problem.

Yet, there is hope for those who take a systems approach. Data has been presented for cases where both capital and operating costs were saved in new investments. For example, one case compared the investment and operating costs of using the overhead vapors from a higher temperature tower to reboil two lower temperature towers versus simply rejecting this heat to cooling water and reboiling the two towers with steam. Energy cost savings (reduced steam, slight increase in fuel consumption) were $1.8 million/year. Increased investment costs for heat exchangers (pumps and piping) were $2.3 million higher, but these were balanced by savings in cooling water system and incremental boiler capacity of $3.5 million. (This analysis did not include any credits for reduced waste handling costs.) It is often true that investment in utility system capacity is greater than that in the process equipment to render it unnecessary, if the utility system investment can legitimately be drawn into the decision envelope.

Both costs and benefits must be assessed on the basis of the impact of any improvement on the entire plant energy system. The net savings on fuel and power usage, and the full impact on plant investment, should be considered in calculating cost-benefit ratios. At least for new investments, fundamental improvements in efficiency can offer savings in both investment and operating costs.

BIBLIOGRAPHY

1. W. F. Kenney, *Energy Conservation in the Process Industries,* Academic Press, Inc., Orlando, Fla., 1984.
2. E. P. Gyftopolous and T. F. Wiomer, *Potential Fuel Effectiveness in Industry,* Ballanger, Cambridge, Mass., 1974.
3. G. M. Reisad, "Available Energy Conversion and Utilization in the U.S.," *Journal of Engineering for Power* **97,** 429–434 (1975).
4. J. B. Fenn, *Engines, Energy and Entropy,* W. H. Freeman & Co., New York, 1982.
5. W. J. O'Brien, "Low Level Heat Recovery Technology," *Proceedings of the Fifth Industrial Energy Conservation Technology Conference,* Houston, Tex., 1982.
6. M. Sussman, *Availability (Energy) Analysis,* Milliken, Boston, 1980.

ENERGY MANAGEMENT, PRINCIPLES

DOUGLAS A. DECKER
Johnson Control, Inc.
Milwaukee, Wisconsin

HISTORICAL PERSPECTIVE

Viewed from a historical perspective, the statement "there is no time like the present" overwhelmingly applies to energy management: the systematic, ongoing science of favorably balancing an organization's energy needs and goals with the capabilities of people, technologies, and energy sources. At the risk of oversimplifying the concept, it is important to manage energy successfully because energy ultimately drives, in a modern, mechanized world, production aimed at meeting all three basic human needs: food, clothing, and shelter. The quality of human lives is thus defined by how, and how well, energy is used.

For those managing energy as it relates to buildings, the universe of energy management now presents a multitude of cost-effective and productive results. These results include improved management efficiencies, energy savings, operations and maintenance labor reduction, and the increased comfort and productivity of those using buildings. International events; national, state, and local governmental policies; public awareness; and levels of technology today offer those concerned with energy management the mix of incentives and resources that make results both reliable and rewarding. The incentives and resources range from utility rebates to enhanced understanding by those outside the energy management realm, to the greater societal benefits such as reduced greenhouse gases and less impact on the environment.

At one time energy managers faced a swinging pendulum of concern. As energy prices rose and fell, as supply and demand vacillated, and as governmental administrations came and went, energy managers faced varying and frustrating standards and levels of concern from officials, influentials, and the public at large. In fact, many of the 1970's energy czars in public and private institutions have long since been forgotten.

In addition to a historical perspective, this article discusses the organizational levels of energy management, the facilities management systems and services that help attain energy management goals, the economics of energy management, and the future of energy management. See also ENERGY AUDITING; ENERGY CONSERVATION.

Events and Trends

A quick review of the events and trends of the last few years reveals why the pendulum of concern may now have stopped swinging and may reflect a position that is best for continued advances in energy management. Among the top international events that have promoted energy management is the end of the Cold War. With fewer financial, governmental, and technological resources dedicated to strategic military needs, more resources are being devoted to tightening up the way strategic economic and domestic needs are managed. Indeed, the shift from military to domestic needs is complementary in terms of what constitutes national security, the capacity of the United States to protect and advance the interests of its people. The end of the Cold War cleared the national agenda for energy management. But the frequent and sudden shocks to world oil prices and supplies, emanating from the Middle East, should be given credit for putting energy management on the agenda in the first place. The oil crisis of 1973, the Iranian revolution of 1980, and the Persian Gulf War of 1991, typify events that heightened U.S. awareness of the need to have an energy strategy.

Sudden, dramatic changes in world oil prices harm the United States and other nations much more than a persistent but gradual rise in price. Even if the average price over the long term is the same, shocks do more harm. This stems not from how much oil is imported but from how oil dependent the U.S. economy is, its capacity for switching to alternative fuels, oil reserve stocks around the world, and the additional international production capacity that can be brought on line.

Leaders perceive achieving greater energy stability to be a principal part of enhancing America's security. They recognize that the United States must insulate itself from the shocks that emanate from the Persian Gulf, whose oil fields provide 25% of the oil the world presently consumes and nearly 66% of the world's proved oil reserves (1).

Data also show that conservation does not necessarily mean a decline in living standards, a position once held by many leaders. Since the oil shocks of the early 1970s, Americans have enjoyed a 35% rise in the gross national product without increasing their energy consumption (Fig. 1). The main reason is that the living standard–related services energy provides, such as comfort and mobility, are generated much more efficiently today than they were during the early 1970s. Similarly, by slowing the growth in demand for new energy capacity, it has been estimated that conservation could liberate 10% of U.S. industrial investment capital for other uses (2).

At national, state, local, and corporate levels the United States has faced disruptions, economic conditions, and budget constraints that have caused leaders to concentrate on the efficient use of funds, which equates to support of energy conservation and efficiency standards. In recent years a steady stream of congressional hearings and public forums involving leaders from the private and public sectors have led to several federal initiatives.

The National Energy Strategy, established by President George Bush in 1989, set the stage for Executive Order 12759, signed by Bush in 1991. Both advanced a commitment to improving energy use in federal buildings, facilities, and vehicles as part of a strategy to counter disruptions, economic conditions, and budget constraints. As part of this campaign to renew a federal energy conservation ethic, an industry coalition of trade organizations representing 4.7 million members collectively and individually supported signing of the Executive Order.

Executive Order 12759 led to the Energy Policy Act of 1992. Broadly stated, the purpose of the Energy Policy Act is to foster greater energy conservation in the traditional sectors of energy consumption: industrial, transportation, residential, and commercial. It requires states to incorporate energy efficiency standards in their commercial building codes that meet or exceed the 1992 voluntary standards established by the American Society of Heating, Refrigerating and Air-Conditioning Engineers (ASHRAE). The Energy Policy Act also requires the Department of Energy (DOE) to issue a new building code that contains minimum energy efficiency standards for new buildings.

Recent events and initiatives support the notion of sustained, politically bipartisan energy management awareness. In passing the Energy Policy Act, Congress has displayed leadership in promoting energy management. President Bill Clinton and Vice President Al Gore have track records of commitment to energy management and conservation. In his 1993 Earth Day address, Clinton announced initiatives to stimulate the energy technology industry, energy conservation, and a clean environment.

Two other Clinton initiatives indicate a move to a sustained energy policy for the country. The Climate Change Action Plan has a goal of reducing harmful greenhouse gases to 1990 levels by the year 2000 and creating jobs by promoting energy efficiency through public–private partnerships. Rather than relying exclusively on command-and-control mandates that tend to stifle innovation, the

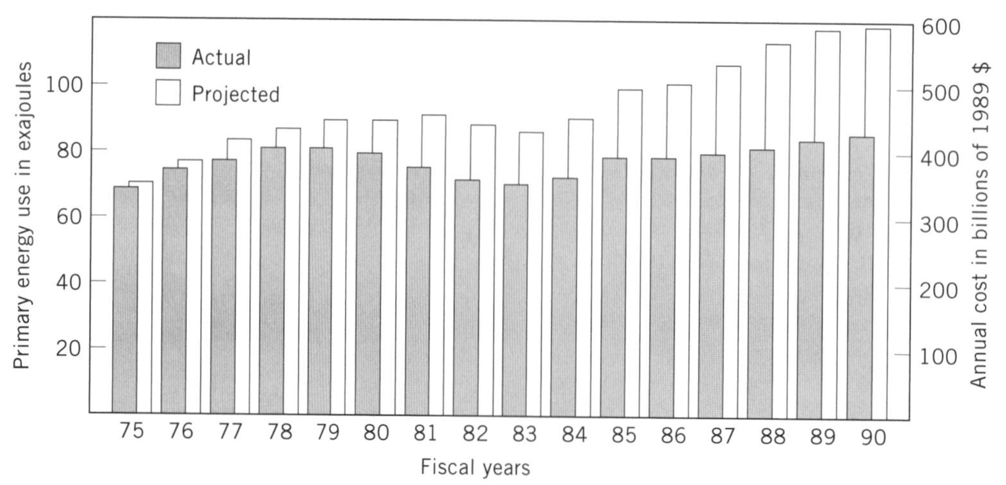

Figure 1. GNP projected energy use and GNP actual energy use. Data from the DOE and the Lawrence Berkeley Laboratory. ☐; Actual; ☐, projected.

partnerships reflect the mutual responsibility of both the private sector and the government.

These investments are estimated to save over $60 billion in reduced costs between 1994 and 2000, with continued benefits of over $200 billion in energy savings between 2001 and 2010. The voluntary nature of the programs is important because the threat of additional regulations and the success of existing voluntary programs will induce companies to participate.

The EPA's Green Lights program is one of the successful voluntary initiatives promoted in the Climate Change Action Plan. Almost 1200 firms currently participate by upgrading lighting systems and reducing electricity use by nearly 200 million kW annually, at a savings of $17.7 million. Approximately 1 in every 20 commercial buildings in the United States is now involved in Green Lights.

The costs will fall and the benefits will rise as participants become more experienced with the implementation process of voluntary programs. In addition, businesses can show they are good citizens by saving energy, helping the environment, and contributing to the local economy by hiring local workers to implement voluntary programs.

While the Climate Change Action Plan addresses private industry, another Clinton initiative directs the federal government to reduce energy more than ever before. The Executive Order on Energy Efficiency and Water Conservation at Federal Facilities supercedes the Bush Executive Order. The Clinton mandate calls for a 30% reduction in energy use in federal buildings by 2005 from 1985 levels. The order includes increased uses of innovative financing, such as performance contracting and utility demand side management, to implement the measures. It calls for agencies and employees to be rewarded for excellence in energy efficiency, and it holds agencies and employees responsible for meeting the goals.

By signing this new Executive Order and announcing his intentions to hold the agencies accountable, Clinton showed his personal commitment to the issue. In addition, the very nature of the document will break down barriers and encourage participation by federal agencies. It will result in the federal government acting as a role model for the rest of the country and proving that energy efficiency is the best economic and environmental course of action an organization can take.

Sustained, active leadership as demonstrated by the Bush and Clinton administrations is essential to making energy management work. Without active leadership and funding, laws such as the Energy Policy Act become totally irrelevant. Sustained industry leadership, such as that which led to the Energy Policy Act, will ensure that legislation reflects the will and ability of the marketplace to implement energy strategies, avoiding controversy.

Incentives to Implement

There is growing recognition that investment in energy-efficient technologies creates jobs. For example, a study conducted for the DOE reported that for each $1 million of energy efficiency improvements made within local hospitals, employment would increase by 56.1 jobs over a 20-yr period (3).

Perhaps one of the strongest incentives to implement energy management programs and technologies will come from efforts by private industry to improve productivity, while also reducing energy costs. A joint study by the Center for Architectural Research and the Center for Services Research and Education at Rensselaer Polytechnic Institute studied the effect of environmentally responsive workstations (ERWs) on the productivity of office workers at the new West Bend (Wisconsin) Mutual Insurance Company headquarters (4). The study reported the combined effect of the new building with the ERWs was a 16% increase in employee productivity. The ability of employees to control their internal environment was credited with having a significant impact on this increase. The integrated design of ERWs allow people to control temperature, airflow, noise, and light within their work spaces. Equally important, energy-efficient building design, energy management control systems (EMCS), and the ERWs contributed to a reduction in utility costs per square foot of building space, from $0.18 per square foot to $0.11 per square foot.

ERWs represent one of the many off-the-shelf technologies available to help energy managers. For federal energy managers, the General Services Administration and Department of Defense have procurement schedules that include proven energy-efficient and energy management products. Overall, energy technology improves with each year, making efficiency easier and providing better paybacks. Corporations are joining in strategic partnerships to advance the compatibility, and hence cost-effectiveness, of their equipment. And creative financing methods allow public officials, corporations and commercial building owners to defer large capital outlays of funds for energy-efficient equipment.

Times have indeed changed, and for those concerned with energy management, there is no time like the present. An indepth understanding of this new and improved universe of energy management can lead the way to opportunity.

But has the pendulum of concern stopped swinging permanently? Will the mix of events, policies, incentives, and resources favor strong energy management in the future as much as it does now? The forces that influence energy management, including environmental issues, utility supply and demand, industry cooperation, ease of use, economic benefit, consumer focus, and political will, all will play a role in determining whether the pendulum has stopped and will forever favor a commitment to energy management.

THE LEVELS OF ENERGY MANAGEMENT

To manage energy at any level, with an enhanced likelihood of success, it is necessary to have knowledge of the major regulations, agencies, issues, resources, trends, and technologies as they affect each level of the energy management model or pyramid (Fig. 2). Such knowledge also permits energy managers to implement winning strategies that will satisfy criteria set by entities outside as well as inside their organizations.

This section considers the levels of energy management

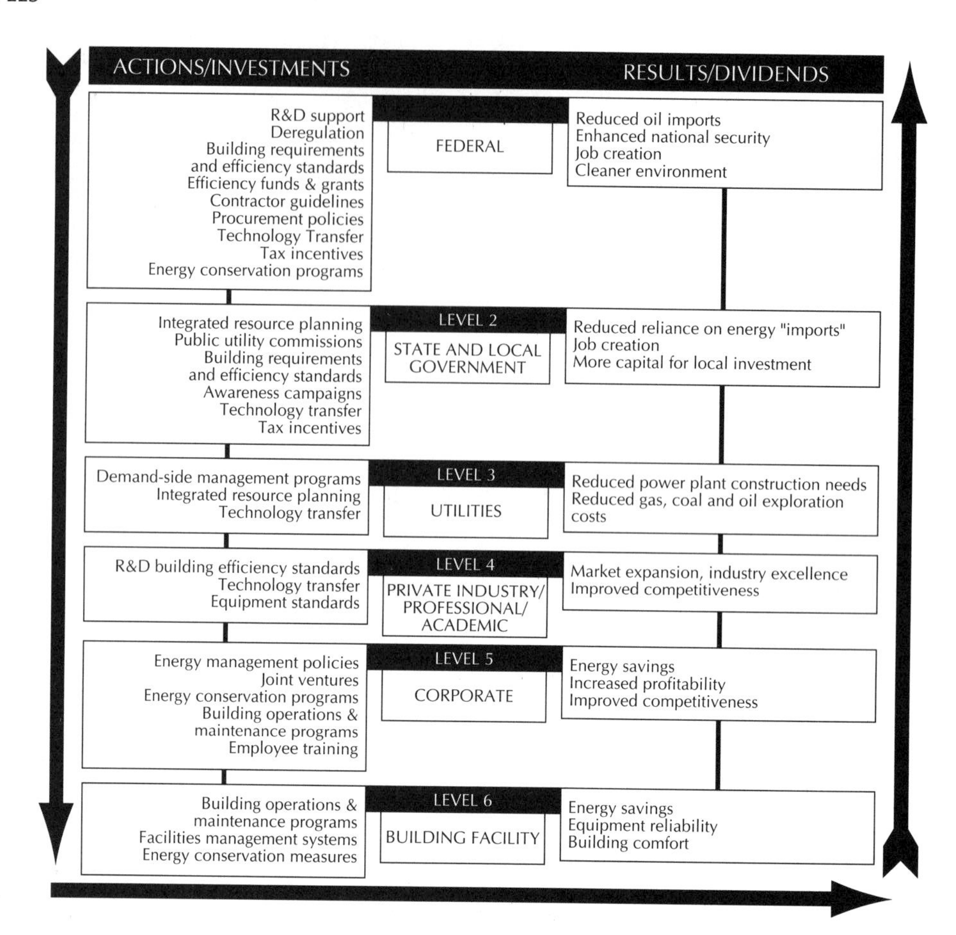

Figure 2. Levels of energy management.

as they relate to public (government), commercial, and industrial buildings to be

1. Federal, including the executive and congressional branches of the U.S. government.
2. State and local government.
3. Utilities.
4. Private industry, professional, and academic.
5. Corporate.
6. Building and facility.

In a country based on a free market economy and individual rights, the role of the federal government in promoting energy management and efficiency will always be subject to debate. The traditional U.S. policy has been one of market reliance, allowing markets to determine prices, quantities, and technology choices. In certain cases and periods of history, markets cannot or do not work efficiently. During these periods government action must be taken to remove or overcome barriers to efficient market operation.

However, no one who seeks to improve this nation's energy efficiency for reasons of national security, international industrial competitiveness, or reducing federal energy outlays can doubt that federal, state, and local governments can play a positive role in fostering energy efficiency. There is also the compelling need to protect the environment.

The extraction, conversion, and consumption of energy creates more global environmental damage than almost any other human activity (5). The damage inflicted includes deforestation, nuclear waste and CO_2-induced climate change. Because research indicates that pollution has economic as well as social costs, there is a need for those who manage the economy and society to take the lead in protecting the environment. Energy efficiency initiatives are a required part of leadership at the government level as well as all other levels of energy management. The challenge for government is to develop programs, laws, and policies for promoting better energy management. Government initiatives can provide significant stimuli to the free market. Help is needed to overcome barriers such as short payback times required by investors, lack of investment capital, lack of information, and limited supply of and demand for energy-efficient products.

The case for government-supported research and development of energy-efficient technologies is based on evidence that projected returns on investments make support worthwhile. One study showed that increased use of renewable technologies and energy-efficient products and procedures can save American consumers and businesses hundreds of billions of dollars over the next 40 yr (Fig. 3). Depending on the mix of energy-efficient technologies, market penetration rates, and pollution emission objectives, the net savings estimated for a 40-yr period range from \$1.9 trillion to \$2.3 trillion (6).

Noting that residential and commercial buildings account for about 33% of U.S. energy consumption, at an annual cost of \$170 billion, a separate government-author-

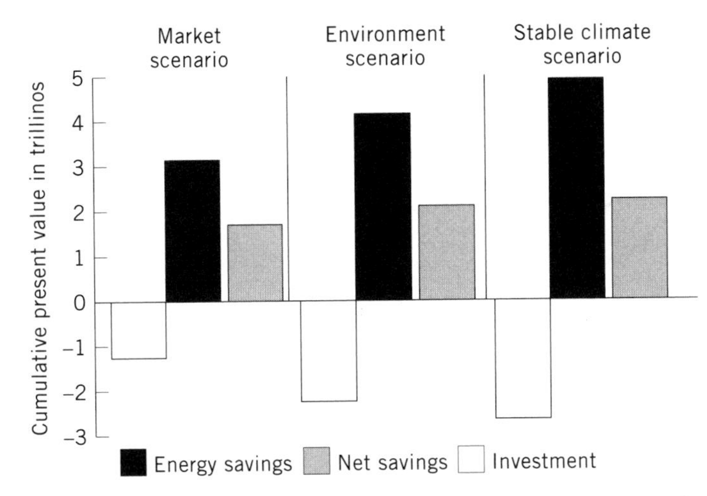

Figure 3. Analysis of energy savings and costs for three levels of investment in energy-efficient and renewable energy technologies, 1988–2030. The market scenario assumes moderate market penetration rates for the technologies. The environment scenario assumes more rapid market penetration rates and additional use of energy-efficient and renewable energy technologies. The stable climate scenario is based on use of a low and high cost mix of technologies. All scenarios lead to reductions in primary energy requirements and pollutant emissions from a reference scenario. The reference scenario was adapted from DOE projections, reflecting a current reliance on fossil fuels, nuclear power, and limited energy efficiency improvements. For additional discussion, see Ref. 6.

ized study projected significant savings using technologies that are already available. The 1992 study estimated that using commercially available, cost-effective technologies, building energy consumption could be reduced up to 33% by 2015, compared with a business-as-usual projection that did not employ these technologies (7).

Federal Government

In response to overwhelming evidence, the trend is toward the federal government taking a leadership role in promoting energy management and stimulating the private sector. Just as the Clean Air Act saw government leading the way toward a cleaner environment by setting regulations for pollution control, the Energy Policy Act of 1992 positions the government in a leadership role for energy conservation and management. The Energy Policy Act affects every level of energy management, and energy managers in the public as well as private sector.

From a building technology perspective, the Energy Policy Act provides leadership by supporting the research and development of energy-efficient technologies. The act sets the stage for energy efficiency standards for buildings, lighting products, motors, and other equipment. Efficiency will be possible by allowing more competition and permitting the federal government to act as a consumer of new technologies. Reflecting increasingly higher federal energy costs, the lag in funding for energy conservation in federal agencies (Figs. 4 and 5), and a desire to lead by example, the Energy Policy Act makes the previously voluntary goal of reducing energy consumption per gross square foot in federal buildings by 10% before 1995 into a requirement and extends it to require a reduction of 20% by 2000.

The act also impacts the Federal Building Code. Under the new law the DOE must issue a new building code that contains minimum energy efficiency standards for new federal buildings. The standards must be at least equivalent to those of the 1992 Council of American Building Officials (CABO) and ASHRAE.

Increased competition will come in part from a deregulated electric utility industry. The Energy Policy Act opens up the electric utility industry to competition, which should bode well in the long term for price stability and energy efficiency incentives. At the heart of this movement toward increased competition are provisions for exempt

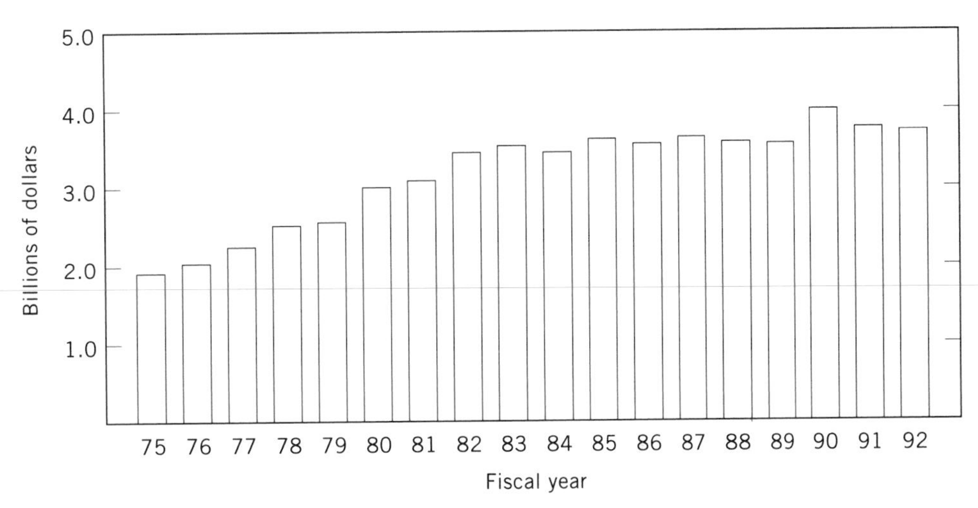

Figure 4. Federal energy costs for buildings and facilities, 1975–1992. Data from the DOE.

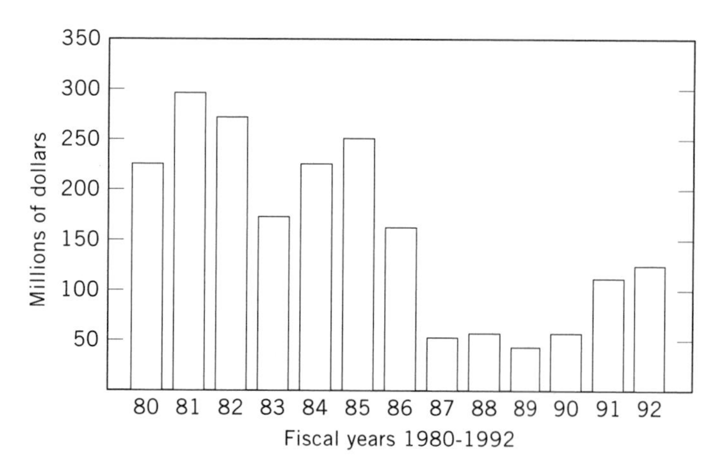

Figure 5. Federal building energy efficiency funding during fiscal years 1980–1992. Data from the DOE.

wholesale generators (EWGs). EWGs will be able to sell electricity to utilities and, according to the act, must have fair access to transmission lines.

Just as deregulation of the telephone industry led to restructuring along smaller, regional, and functional business lines, many experts believe this attempt to open up competition in the utility industry will spur development of separate power generation, transmission, and distribution companies.

The Energy Policy Act also focuses on commercial buildings in the private sector. The Energy Policy Act requires states "to incorporate energy efficiency standards in their commercial building codes that meet or exceed the 1992 voluntary standards established by ASHRAE." The standards are supported by the requirement that the DOE "provide incentive funding to states for implementation of this provision and to improve and implement their codes."

The purpose of the standard, officially known as ASHRAE/IES Standard 90.1, is to set minimum requirements for the energy-efficient design of new buildings and construction, to provide criteria for energy-efficient design and methodologies for measuring projects against these criteria, and to provide guidance in designing energy-efficient buildings and building systems (8). Although Standard 90.1 is not a code, its requirements can and have been adopted by governmental agencies empowered to enact codes through legislative or regulatory processes.

Standard 90.1 has a comparative standard in the federal government. The full reference for this standard is Code of Federal Regulations (CFR), Title 10, Part 435, Subpart A, *Performance Standards for New Commercial and Multi-Family High Rise Residential Buildings,* U.S. Department of Energy (DOE), January 1989. The DOE standard is intended to deal with private construction of federal buildings; Standard 90.1 encompasses all new construction, except low-rise residential, in all climates across the United States.

The act aims to enhance energy efficiency in process-oriented industries. The act requires the DOE to promote the use of energy-efficient technologies in certain manufacturing industries, such as food and food products, lumber and wood products, and petroleum and coal products. There is a tax break for commercial and industrial buildings that comply. The Energy Policy Act excludes from gross income 40% the value of any subsidy provided by a public utility for the purchase and installation of an energy conservation measure in 1994, 50% in 1995, and 65% in 1996 and thereafter. This incentive only applies to amounts received after December 31, 1993.

For the energy manager looking to finance energy efficiency initiatives, today's market offers a number of options. The most widely used in recent history have been the incentives offered by utilities as part of demand side management programs. These incentives typically take the form of rebates tied to purchases of energy-efficient equipment. Energy managers of federal buildings should note that the Energy Policy Act of 1992 authorized federal agencies to participate in utility incentive programs. A total of 50% of the rebates or payments received from these programs are to be used for the purchase of additional energy efficiency measures.

Investment tax credits also have been used to offset the initial costs of energy-efficient equipment. Offered by states but not currently by the federal government, these tax credits complement utility-sponsored demand side management programs and help industries lower their tax liability as well as their future energy costs (9). States also have authorized bond issues to cover the cost of energy efficiency projects at state and municipal buildings.

Relatively new is the emergence of performance-based contracting as a financing method. In performance-based contracting, one or two suppliers plan and implement a building renovation in its entirety, acting as general contractors. Financing options are tied to performance guarantees, with the contractor taking as payments future energy savings. Performance-based contracting offers energy managers single-source convenience, control over expenditures, predictable costs, supplier accountability, guaranteed service delivery, and unlimited options. Usually these contracts include operations and maintenance services as a means of ensuring favorable results.

The federal government recognizes the importance of performance-based contracting. The Energy Policy Act of 1992 simplified the procedures federal agencies need to follow to enter into energy-saving performance contracts. Congress intended to simplify this form of financing as a way for the private sector to invest in federal energy projects (10). The Energy Policy Act offers an additional financing opportunity to federal energy managers via the Federal Energy Efficiency Fund. Under the Energy Policy Act, the secretary of energy may award monies from this fund to any federal agency for energy conservation investments. The fund is capitalized with $10 million in fiscal year 1994, $50 million in fiscal year 1995, and will be capitalized with "such sums as may be necessary" in the fiscal years that follow.

The Energy Policy Act also opens up the possibility for an energy conservation bank with deeper pockets than the Federal Energy Efficiency Fund. The Energy Policy Act directs top-level federal treasury, budget, and energy officials to conduct a detailed study of financing options for investments in energy conservation measures by federal agencies. The study must include a review of energy banks and revolving funds as long-term options.

It should be noted that funding as provided for in the Energy Policy Act is but a small piece of the total funding available to federal energy managers. The Department of Energy's annual report for fiscal year 1991 (11) lists 13 federal agencies that will spend a projected $124.5 million in energy conservation measures in fiscal year 1992, with the trend moving toward increased funding. These agencies, such as the General Services Administration, each allocate a portion of their operational budgets to energy conservation projects.

Without continued economic incentives to reduce energy consumption, energy efficiency goals would be tigers without teeth. Public policy and societal goals are not enough to initiate and sustain energy efficiency campaigns, especially in a free market economy. To advance and protect their efforts, energy managers must incorporate into their programs strong economic underpinnings in the form of realistic savings estimates, periodic review of results, appropriate tools and services, and a favorable financing package.

Refinements to the Energy Policy Act will be made as subsequent presidents and Congresses advance their interpretations of the nation's energy priorities. The federal government's ability to impact the energy manager with legislation and programs is shared by state and local governments.

State and Local Governments

State governments have a proven track record of energy management. In fact, the Energy Policy Act reflects legislation that has been successful at a state energy management level. Many states have effective programs in integrated resource planning (IRP), building energy codes, and equipment standards. All states and most U.S. territories, including American Samoa, Guam, and Puerto Rico, have energy offices or departments of state government responsible for coordinating energy management efforts. A state's public utility commission (PUC) regulates the supply of energy in its various forms to businesses and consumers. The PUCs and their federal counterpart, the FERC, also regulate access to transmission lines, an issue brought to the fore by the Energy Policy Act.

The approach that an individual state takes to energy efficiency is affected in part by the degree of dependence on energy "imported" into the state. Iowa, for example, is an agricultural state with relatively few resources of its own for energy production. More than 98% of its total energy is imported. The state thus wisely chooses to take a strong leadership position regarding energy management. Iowa is known for energy programs that apply to all traditional sectors of energy consumption, including industrial, transportation, residential, and commercial (12).

Individual states also vary on the emphasis they place on energy management within these sectors, a result of different levels of consumption and vulnerability arising from that sector (Fig. 6). In Nevada, the industrial and commercial sectors are largely untouched by state energy programs. This can probably be attributed to perceived need: Nevada's industrial sector accounts for 24% of state energy use compared with the national average of 36%. Nevada's commercial sector uses approximately 18% of the total energy consumed, in line with a national average of 16%. Demand by casinos for electricity is considerable, but casinos have been the target of past energy-saving programs and are considered efficient (14). By contrast, the transportation sector comprises 37% of total state energy consumption compared with a national average of 28%. Nevada is geographically the seventh largest state, and state officials assert that tourists driving from California to Nevada contribute to the state's poor ranking in transportation energy efficiency (15). Tourists notwithstanding, Nevada officials place considerable emphasis on leadership in the transportation area, running car-care clinics and designating October as car-care month to educate consumers about the environmental, safety and energy-savings benefits of a well-tuned, well-maintained car. Nevada employers may soon be required to participate in ride sharing and mass transportation efforts.

New York State considers all sectors and end uses and is the most energy-efficient state. The lowest energy use per capita can be attributed to the fact that a large percentage of the state's population lives in New York City where mass transit and multifamily dwellings are the norm.

Wisconsin has implemented an ambitious $50 million

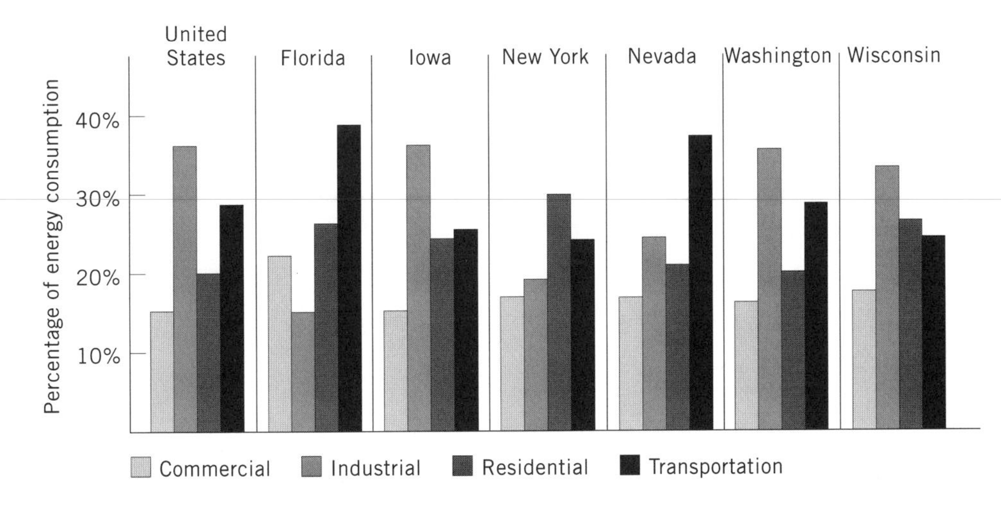

Figure 6. Six-state comparison of energy use by commercial, industrial, residential, and transportation users. Data from the DOE and Ref. 13.

six-year energy conservation program that will significantly reduce energy consumption in state facilities. The program is designed to result in cost savings to taxpayers and to have a positive effect on the environment. Wisconsin is also responding to energy and environmental challenges by implementing a program called "2000 by 2000" (16). This program will place 2000 alternative-fueled vehicles on the road by the year 2000. This commitment is 33% of the state's fleet and represents three times the requirements of the federal government's Energy Policy Act. Wisconsin involves the private sector in its programs through coalitions and task forces formed to address specific issues.

To have an effective energy management plan on a state level, a majority of states use their authority to regulate utilities and thus manage energy supplies and prices. Integrated resource planning (IRP) is a cornerstone of most state energy plans (17). IRP requires utility companies to give as much consideration to conservation programs as they do to building new power plants. IRP simultaneously examines all energy-saving and energy-producing options to optimize resources and minimize total consumer costs, while considering environmental and health concerns (18).

In early 1994, all but seven states have either initiated or established IRP. The popularity does not translate to commonality, however. Nearly every state government, public utility commission, or utility adds its own perspective to the process. IRP generally takes the following seven steps:

1. Development of a load forecast.
2. Inventory of existing resources.
3. Identification of future electricity needs that will not be met by existing resources.
4. Identification of potential resource options, including demand-side reduction programs.
5. Screening of options to identify those that are feasible and economic.
6. Performance of some form of uncertainty analysis.
7. Selection of a preferred mix of resources.

Figure 7 shows a model of the IRP process. State planners use the model as a tool to evaluate demand-side options such as conservation and load management, to compare these to supply-side options, and to structure an energy policy that incorporates environmental and social costs. The goal is to develop a long-term energy policy that acquires the most inexpensive resources first and internalizes social costs in the rate structure.

Most of the people involved in state utility issues who were interviewed for a study said that quantifying the environmental costs of electricity is likely to be the primary issue for utility planners in the next decade (19). Fostering energy efficiency is thus conducted with varying levels of intensity from state to state. States generally recognize there are spinoff benefits from increased energy efficiency and conservation. Reducing expenditures to import energy means more money can be spent on goods and services to

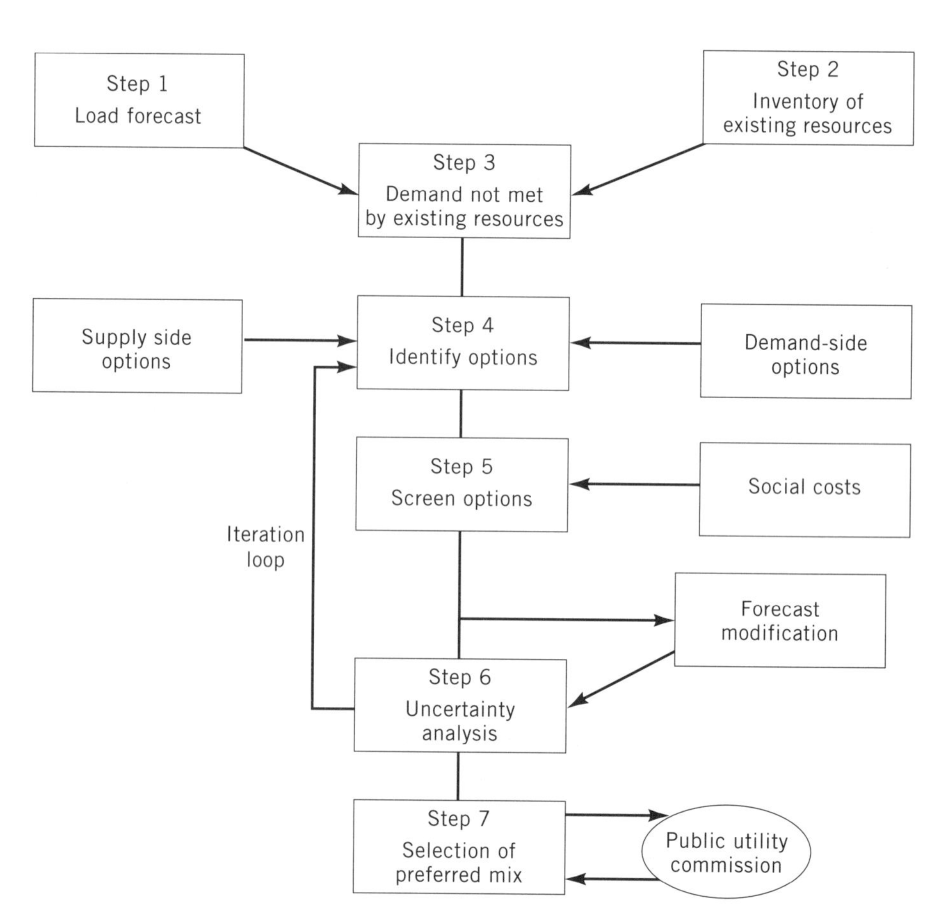

Figure 7. Schematic diagram of IRP process. Data from Ref. 12.

increase the state's sales revenues and employment levels. Savings put into financial institutions increase the supply of loanable funds and stimulate new business investments in the states. Decreasing dependence on imported energy is another spinoff benefit, as are improved air and water quality.

Utilities

The challenges that utilities and their customers have faced in the past stem, in part, from the Public Utility Holding Company Act of 1935 (PUHCA). Competition among wholesale electricity suppliers was not a practical possibility in 1935 (20). The utility structure imposed by PUHCA assumes that electricity will be generated by the local utility. Electric utilities vary widely in their ability to minimize power plant construction costs. PUHCA is thus a major obstacle to relying on the most efficient firms to build and operate the power plants that will be needed in the future.

Over the years the U.S. electricity supply system has been integrated into large regional networks. The Public Utility Regulatory Policies Act of 1978 led to limited competition among a small group of wholesale suppliers. The experience has led public policymakers to conclude that greater competition among wholesale suppliers is both feasible and likely to be beneficial, hence the Energy Policy Act's attempt to open up the nation's utilities to competition.

Demand side management (DSM) is a process used by utilities to plan, implement, and evaluate programs designed to influence the amount or timing of customer energy use. DSM programs affect the system energy and total capacity that a utility provides to meet demand. DSM uses four basic techniques to influence energy use. Peak clipping reduces system peak loads, valley filling builds off-peak loads through methods such as thermal energy storage, load shifting shifts load from on-peak to off-peak periods, and strategic conservation reduces end-use consumption. Gas utilities have recently begun to use DSM strategies as natural gas becomes more recognized as a bridging fuel to replace imported oil.

All utilities implementing DSM programs must consider the effect these programs have on sales and earnings. Investor-owned utilities in particular recognize that reduced sales mean reduced earnings. Pulling utilities in the other direction is the realization that without DSM programs, load growth could ensue that would necessitate the need for constructing new and expensive generating stations.

There are existing mechanisms that utilities and their regulators can use to deal with the conflict between shareholders and customers. These include command and control regulation; frequent rate cases, alternative rate designs that set energy and demand charges equal to short-term costs; compensation to the utility for the revenues lost because of its energy efficiency programs; and decoupling of electric revenues from sales (21).

A new method called statistical recoupling could help answer criticisms of other methods. Utilities would construct statistical models to estimate electricity use. The difference between estimated and actual electricity use would be used to adjust utility revenues. Refunds to customers or increases to retail electricity prices would take place in the following year.

Private Industry, Professional, and Academic

Professional associations, private industry groups, and academic organizations represent another level of energy management. One of the most influential is ASHRAE. ASHRAE Standard 90.1 represents the model used by many states and the federal government to develop regulations. The Council of American Building Officials (CABO), often works with ASHRAE. The Energy Policy Act of 1992 directs these two organizations to develop voluntary energy codes for buildings.

Other important groups are the Advanced Building Systems Integration Consortium (ABSIC), the Alliance to Save Energy (ASE), the American Institute of Architects (AIA), the Association of Energy Engineers (AEE), the Building Owners and Managers Association (BOMA), the Energy Efficient Building Association (EEBA), Intelligent Buildings Institute (IBI), the International Facility Management Association (IFMA), and the U.S. Energy Association (USEA). The University of California at Berkeley and its Lawrence Berkeley Laboratory is extremely active in researching energy-efficient technologies and transferring information to the public and private sectors. These organizations collect data on energy use, sponsor seminars and conferences, produce targeted publications, and track policy changes and technologies that affect their constituencies.

In many cases strategic partnerships are formed between industry associations or companies from private industry to develop codes, standards, and equipment solutions. Industry excellence, societal goals, and economic opportunity all drive these efforts.

Corporate

The corporate level of energy management represents the place where many people direct their energy efficiency initiatives. Together, the industrial processes and office buildings of the corporate world are big consumers of energy: the industrial and commercial building sectors combine to use 52% of the nation's energy annually (22).

Corporations should have an energy management policy as a matter of self-interest. Those that aggressively manage their energy use also reduce their exposure to potential environmental risks, cost increases, and future energy shortages. The energy management policies developed by companies range from formal policies that establish broad objectives to less formal policies that set reduction goals, and the time frames and means of achieving them (23).

Companies may employ design standards, operating standards and maintenance standards into their plans. Individual company cultures and management styles affect the nature of these plans. Energy managers in these organizations face moving targets that include process innovation, long-term quality planning, energy assessments of building and equipment purchases, employee or tenant awareness, and waste minimization and recovery.

A common denominator for the most successful programs appears to be assigning direct responsibility for implementing a company's energy policy to an energy manager. These energy managers should report to top management and also have the authority to make change. They work with division managers to develop and implement company policy in individual facilities. At their best, information and responsibility in these efforts flows up and down. Goals are set at the management level with input from the facility or building level, and accountability is at the lowest possible level.

Some of the strategies these energy managers use are monitoring energy usage and cost and establishing building performance criteria, purchasing reliable energy supplies, performing regular energy surveys of facilities and buildings, developing energy-efficient building design criteria, and providing technical assistance to facility and corporate staff. The use of facility management systems, another important strategy, will be discussed in detail below.

One of the most important aspects of energy management at this level is flexibility. A dramatic upswing in business may increase energy use, even though efficiency goals are being met. A drop in production may make it harder for some facilities to achieve goals because the plant is no longer operating 24 hours a day. The opportunity to reuse energy by-products is thus reduced.

Implementation of energy management policies pays off through even simple actions. At a General Motors plant, employees were given a monthly report that showed through computer-generated graphics the cost of neglecting to turn off lights and equipment. An awareness campaign resulted in 50% less energy wasted on lighting and 86% less energy wasted on major equipment, saving more than $309,500 a year (22).

Building Level

At a building level, energy managers, plant managers, and facility engineers implement policies developed in coordination with the corporate level. New building construction plans can be scrutinized with the goal of providing a comfortable environment while using energy cost effectively. Energy audits or detailed engineering studies can be done on existing buildings to determine if energy conservation measures, such as equipment retrofits, make operational and economic sense.

Building owners and managers need to recognize however, that interrelations and tradeoffs exist between energy and building environmental goals. For example, during nonoccupied periods opportunities exist for reductions of air flow that conserve energy and maintain air quality. During occupied periods, variable air volume systems provide a way to balance energy conservation and air quality (24).

Companies often put purchasing decisions on plant and equipment through energy assessments, requiring those making the decisions not only to consider the cost of the equipment but also to review how things can be done differently to save energy or reduce waste. Bringing employee talents and corporate resources together to focus on energy management can be as effective as purchasing expensive technologies. For commercial buildings, educating tenants about energy conservation measures can help keep building operating costs, and rents, down. Lighting in particular is an important area of concern.

Lighting typically comprises 30–50% of a building's electrical load. In the absence of computerized occupancy sensors and lighting controls, people should be encouraged to turn off lights that are not needed; lighting options such as retrofitting fixtures with optical reflectors and energy-efficient electronic ballasts can further reduce electrical load. By combining various lighting strategies, energy consumption savings can reach 40–60%.

Regulations, agencies, issues, resources, trends, and technologies, as they affect the above levels of energy management, will change as society's desire and ability to manipulate energy likewise changes. Strong technologies now exist, as well as an understanding of the long-term benefits of energy efficiency. At every level of energy management, there also is the will to improve our use of energy.

FACILITIES MANAGEMENT SYSTEMS AND SERVICES

Facilities management systems (FMS), also called energy management control systems (ECMS), represent technologies that act on all the goals of efficient energy management. Facilities management systems and related services, however, deliver benefits beyond energy savings. Broadly stated, facilities management systems, and related equipment and services create a comfortable, safe building environment; control energy and operating costs; and provide vital business information management to help optimize resource use. Equally important, FMS provide vital communications that are used to operate buildings and manage building activities more effectively.

One of the biggest challenges in the past has been how to optimize use of the many subsystems that must achieve building management objectives (25). These subsystems include heating, ventilation, and air-conditioning (HVAC); lighting; security; fire management; and power distribution.

Automation solutions previously involved master control rooms. The master control room attempted functionally to integrate building subsystems by placing all operator terminals in one control room. The building operator exercised the concept of "management by exception," seeking to manage variables so conditions stayed within predetermined parameters. The master control room was the place where optimizing strategies and the orchestration of schedules were conceived, tested, and tuned.

However, only a few high technology buildings have had the luxury of the master control room. More common has been the use of a nonintegrated control scheme. In most buildings, subsystems were managed passively or reactively. Analogue, digital, and quite often pneumatic displays were monitored by time-consuming visits to remote corners, rather than conveyed to a central point. Temperatures and other variables relating to specific setpoints were regulated as opposed to optimized. Important operational data on systems were observed and forgotten instead of being tracked as trends and constantly processed for management-by-exception reporting.

More recently, facility managers have used direct digital controllers to enhance their control capabilities.

Mounted on or near the equipment they control, these programmable devices can stand alone or send and receive data as part of a network. They do not depend on the network for basic control nor do they encumber the network by sending unwanted information. When asked or polled, they can respond fully.

Noncompatibility as a Hindrance

As various mechanical and electrical subsystems grew into the electronic age, each subsystem panel acquired information, but seldom would it capitalize on opportunities to distribute information to other panels. The existence of a direct communications port on both panels has not been enough to have one device talking to another across a network. The protocol or programming language used by a vendor for a given device has typically not been compatible with central system software or devices from other subsystems. There has thus been two levels of noncompatibility: peer-to-peer noncompatibility between individual manufacturers and noncompatibility between manufacturers. This incompatibility formed an impediment to centralized reporting and a more coordinated control scheme.

Fortunately, recent developments in building system integration and advances in personal computer technology have made possible the seamless integration of subsystems. An evolution occurred toward total integration by a single vendor who is able to supply compatible equipment for all subsystems. These network-based FMS are designed to be integrated and interactive on every level. Peer-to-peer networks and data highways provide infrastructure while personal computers and their graphic displays provide system operators with an improved means of knowing how their systems are operating. Operators have a number of ways of responding to conditions.

This desired flexibility is possible because current FMS are fully distributed systems built to achieve stand-alone control, supervisory control, and information management. Distributed systems offer many benefits. They ensure a more efficient system, because bottlenecks are eliminated and operators can make better use of processor power. Distributed systems offer a more reliable system through improved fault-tolerance; no single point of failure can bring down an entire system. The FMS are designed as scaleable systems, with building owners buying only what they need now and having the option of expanding later without the loss of original investments.

Connectivity at Work in FMS

Energy managers increasingly rely on the ability of devices employed in facilities management systems to exchange useful information, displaying connectivity with other devices and building management equipment. The basic architecture of these FMS consists of programmable control panels, called network control units (NCUs), and operator workstations that communicate with each other over a high-speed communication network or data highway (Fig. 8). NCUs directly control central plant equipment, while the management of smaller air handlers, heat pumps, lighting circuits, and other building subsystems is delegated to application specific controllers (ASCs).

With connectivity, vendors work together on equipment and protocol standards to produce an open control system for equipment deployed in buildings. This results in devices being able to communicate with other devices across

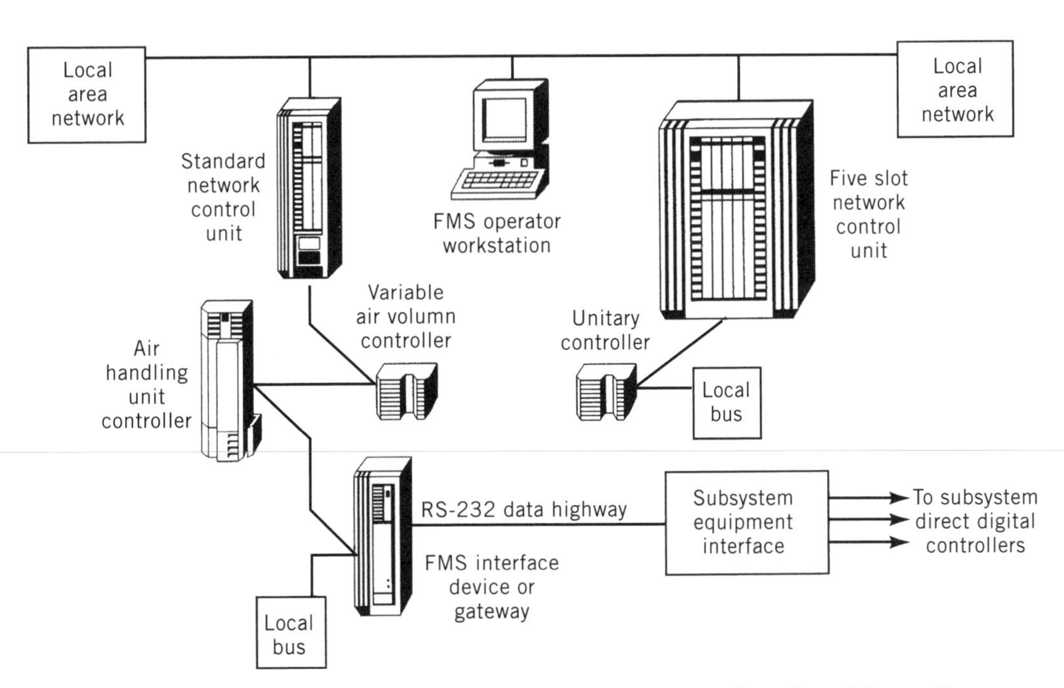

Figure 8. Basic architecture of a facilities management system. Data from Johnson Controls, Inc.

"transparent" barriers. Connectivity works because it understands the protocols of individual devices. Under connectivity, integration can be the function of interface units or gateways positioned between the controlled devices on a network's data highway. Alternatively, interface modules can be built into the hardware of central control equipment. Through short lead times for development of a common language or protocol, connectivity shortens technology curves, or the time it takes to introduce improvements to technology.

Connectivity also strives to use existing equipment and staff expertise. Existing HVAC equipment, for example, can be modified with sensors to make interactive communications with a broader control scheme possible. Because adaptations such as these do not represent wholesale changes, the current level of operator knowledge remains viable, and substantial retraining is not required.

Fostered by partnership, connectivity allows for even greater partnership by suppliers. Vendors of the different principal subsystems can come together to propose and provide packaged solutions that offer the best equipment and best technology, without the user being tied to a solution offered by a single supplier.

Guiding strategies for retrofits, renovations, and upgrades to FMS and new construction projects would include getting suppliers involved early in a team approach. With connectivity and technologies advancing at such a rapid rate, it is possible that building management personnel may not be aware of all the control synergies possible, including those linked with energy conservation measures. Power monitoring, lighting control and building access, for example, can easily be brought into the control system. Equipment suppliers are more likely to have at their fingertips information on "holistic" system possibilities, the possibilities for subsystems to interact optimally to form a well-tuned whole-building system.

Also, it is important to view building system needs from the perspectives of different types of building users. Office space has needs different from those of manufacturing areas, computer rooms, and controlled environments such as "clean rooms." Users or tenants that occupy multiple buildings may have an additional layer of needs. To bring suppliers and building users into the planning picture may well require a formal approach such as a building audit or survey. Factors such as the age of the existing FMS and how recently a study was conducted determine the level of complexity required by the audit or survey. The formal assessment will contain a technical proposal that lists options including energy conservation measures as well as the cost and benefits of those options.

To move beyond this stage toward implementation requires that the building owner or manager choose certain vendors and options over others. Adding connectivity potential to traditional vendor and project qualification criteria will ensure that a new or retrofitted FMS makes the most of current open system technologies.

A fully interactive building system would include physical and functional access to building operations data from any system or subsystem within the building, effectively interfacing all controllers, processors, or data acquisition paths (26). The most frequently integrated, useful, and owner-demanded controls are listed below (see also Fig. 9).

Boiler controls
Fume hood controls
HVAC controls
Lighting control systems
Maintenance management control systems
Motor drive controls
Fire management systems
Power monitoring systems
Programmable logic controllers
Security management and access control systems
Uninterruptible power supply (UPS) systems

	HVAC and controls	Fire	Lighting	Maintenance management
Fire	Integrate HVAC with fire alarm system for smoke control.			
Lighting	Use light switch to determine occupancy and start fan system.			Scheduled O & M program for maximum energy efficiency.
Security	Use occupancy sensor to start fan system.	Lights on in intrusion zone.	Lights on in intrusion zone.	
Electrical equipment and power monitoring	Optimize and monitor demand limit	Doors open for evacuation.	Schedule lighting	Monitor alarms from on-site or off-site. Monitor run time for lead/lag and maintenance purposes.
Chillers and boilers	Optimize	Shut down in case of fire.		
HVAC and controls				

Figure 9. Subsystem connectivity matrix. Data from Johnson Controls, Inc.

Energy management becomes much easier with an integrated control scheme. The integration of power monitoring and other subsystems with a centralized facilities management control system provides an example of the interprocess interaction that is possible from a single-seat user interface.

FMS as a Management Tool

Given a peak demand or time-of-day billing situation, it could become necessary to reduce or shift electrical use, because it saves operating dollars. An FMS can accomplish load reduction without a significant impact on occupant comfort. The lighting load could be decreased by using light sensors to monitor daylight levels, and adjust lighting conditions, provided sufficient daylight is available. Occupancy schedules, either programmed or monitored by the access control system, would automatically alter the load reduction strategy.

One of the strategies that can be employed is the use of ice storage for air-conditioning. This strategy uses electrical energy to prechill water for use during peak demand periods. This trend will continue because of financial incentives offered by utilities to adopt this form of load reduction. The FMS will be called on to optimize the start and stop time in accomplishing the above mission and providing documentation of these strategies.

Another off-shoot of these incentives will result in owners installing generators for off-peak demand and the use of load levelers. The generators will either be self-contained or use energy from a power plant's waste heat. In using waste heat, steam generated for heating purposes could be slightly increased and used to power steam turbines before channeling the energy for warming buildings.

Building owners will see value in generating their own peak energy and use alternative forms of energy production to minimize their peak demand. Included in these strategies may be operating an on-site peak demand generator and battery load-leveling.

Achieving effective energy management through use of integrated FMS requires properly trained operators working with the right equipment and over the long term, strong operations and maintenance (O&M) programs. These O&M programs ensure that equipment runs at peak energy efficiency. Scheduled servicing of equipment will increase the efficiency of machinery.

The FMS can monitor the operating efficiency of major energy-consuming equipment and alert management of the need for scheduling maintenance. Other advanced technologies that are employed include vibration analysis. This technique can predict failures before their occurrence. The procedures noted, all made possible by an FMS, can provide the building owner with the knowledge and power to perform predictive maintenance to save time, repair costs, and energy.

Planning as an Inclusive Process

Building experts acknowledge that one of the greatest challenges to greater efficiency is the fragmentation of the building industry (27). One consideration in the design of an FMS is to gather the views and needs of those who will operate the system once it is installed. The best systems have included facility managers in the system design process.

One trend that is occurring as a result of looking at the building holistically is a concept called the "smart building." According to the Intelligent Buildings Institute, a smart or intelligent building is one that provides a productive and cost-effective environment through optimization of its four basic components: structure, systems, services, and management as well as the interrelationships between them (28).

To develop a smart building, experts advocate an approach called participatory design. Specialists (including architects, engineers, facility managers, telecommunications experts, construction managers, owners, and even end-users) meet before a building is designed. The result of this meeting is a building plan, complete with facilities management systems, that optimizes the four basic building components. Smart buildings incorporate flexibility into building design and subsystems to meet the changing needs of occupants and to adjust automatically to internal and external conditions.

To counter static, inflexible building environments, smart buildings may employ environmentally responsive workstations (ERWs). Specified in open office environments, ERWs allow workstation occupants to control, modulate, and maintain the environmental qualities at the workstation (4). ERWs are at their best if they are integrated with an environmentally responsive architecture. ERWs and their benefits will be discussed below.

Yet while the energy management related advantages of smart buildings will chiefly be available to those with a new building, significant benefits are within the grasp of energy managers overseeing any existing building. A participatory design approach can certainly apply to renovations and retrofits of existing buildings to achieve the optimal building environment. If the building is designed properly, an integrated FMS exists, and hence effective energy management is in place, the optimal building environment is attainable. Facilities management, HVAC, and lighting and power monitoring systems as well as building material technologies offer every energy manager a host of energy management tools.

ECONOMICS OF ENERGY CONSERVATION

Public policy, societal goals, and the information management capabilities of an FMS notwithstanding, the most compelling reason for many organizations to change their use of energy is the opportunity to make or save money. Organizations typically ask energy managers to develop a winning formula by which the economic benefits derived from energy management efforts outweigh their costs.

Organizations soon learn that a progressive energy management program can be a profit center that yields bottom-line results, and can be a safety net to protect an organization from future energy shocks (23). In some organizations, it is believed that to show results one must start by buying one's way to energy efficiency. In reality, economic benefits can begin by managing the way to savings. Organizations can start with a simple campaign to save energy through turning off unneeded lights and equipment. The energy savings that accrue can boost profits. For example, a company implements a plan that nets \$10,000/yr. in energy savings. If that company has a modest pretax profit margin of 10%, the company would have needed to sell \$100,000 in products to earn the same \$10,000.

Pre-tax profits from reductions in energy use can also be demonstrated when considering energy cost as a percentage of revenue. For example, consider a company whose pretax profit is equal to 10% of revenues. If that company's energy bill is 15% of revenues and the company reduces energy use by 4%, the company's pretax profit would increase by 6% (Fig. 10).

Generally speaking, the lower the profit margin, the greater the impact energy savings will have on the bottom line. Energy managers at government agencies and not-for-profit organizations, while not concerned with profit margins, can appreciate the increased dollars available for organizational purposes, especially if those entities depend on outside sources of funding.

To achieve greater savings, organizations inside and outside of the private sector find they must, in effect, spend money to make money. Historically, this has been difficult for organizations for a number of reasons. Some organizations find it difficult to invest in energy-efficient technology when they are operating with tight budgets or close to their profit margins. In more secure companies, spending for energy efficiency competes with other capital investments, with research and development, and with sales and marketing needs. It may also be hard for organizations to justify investments in energy technologies at times when energy costs are not rising and the costs of new equipment are. Many organizations typically require equipment to pay for itself in 2–4 years. Choices for a given equipment upgrade or specification may include energy-efficient equipment that has initial costs higher than conventional, less efficient technologies. A comparison of initial or first costs does not normally put energy-friendly technologies in more favorable and realistic contexts.

Evaluating Cost-effectiveness

There are several mathematical methods for evaluating the cost-effectiveness of equipment purchases or energy efficiency alternatives (7). Payback, the simplest method, divides the initial cost by the annual savings in dollars. The payback method is easy to understand but ignores the time value of money, the value of savings after the payback period, and the limited life of some measures.

Energy costs as a percentage of revenue	Percent reduction in energy use: −2%	−4%	−6%	−8%	−10%
30%	+6%	+12%	+18%	+24%	+30%
15%	+3%	+6%	+9%	+12%	+15%
10%	+2%	+4%	+6%	+9%	+10%
5%	+1%	+2%	+3%	+4%	+5%

Figure 10. Pretax profit from energy savings based on pretax profit equal to 10% of revenues. Data from the DOE and National Association of Manufacturers.

Life cycle costing analyzes the aggregate cost of a capital outlay. Total aggregate costs would include the costs of energy consumed, maintenance, and impact on other systems. It demonstrates the real cost of a piece of equipment or building and includes not only the initial cost but also the total expense of operation over a useful life. In an energy efficiency context, life cycle costing enables energy managers to compare different procurement or specification alternatives to show the savings accrued over time.

There are two competing calculations used to attain life cycle savings estimates. Net present value translates all future costs and savings into their equivalent in today's dollars. If a piece of energy-efficient equipment is projected to have a useful life of 20 yr, the total savings would be

$$\text{20 yr} \times \text{annual savings in dollars} = \text{total savings}$$
$$\text{total savings} - \text{initial costs} = \text{net present value of savings}$$

A second life cycle calculation that is considered more realistic recognizes that a dollar received a year from now is worth less than a dollar received today. The theory is that a dollar received today can be put into an interest-earning account and thus grow in the year ahead. Future savings can be discounted by using percentages to reflect the time value of money:

$$\text{total savings with discounting} - \text{initial costs} = \text{total savings in discounted dollars}$$

Another method is the cost of conserved energy (CCE). This method first measures how much one pays for each unit of energy saved, independent of fuel price. The CCE is then compared with the cost of the supplied energy it displaces to determine the merits of an energy conservation measure. The CCE is defined as the initial cost times the capital recovery factor (CRF), which converts an initial investment into an equivalent series of annual payments incorporating a discount rate, divided by the annual energy savings in energy units. The equation for Capital Recovery Factor is

$$i(1 + i)^n/(1 + i)^{n-1}$$

where i is the discount rate and n is the number of years.

$$\text{initial cost} \times \text{CRF/annual savings} = \text{CCE}$$

To complete the energy manager's understanding of these mathematical tools, a discussion of electricity conversion ratios is required. Analyses of energy use often require that different forms of energy such as natural gas, oil, and electricity be combined into one common measure, typically Btus. Converting electricity into Btus presents a problem because there is no one correct conversion method. The site conversion ratio converts 1 kW·h of electricity directly into heat or 3412 Btu. This ratio ignores that energy is used to produce that 1 kW·h of electricity. An alternative to the site conversion ratio is the primary conversion ratio of 10,240 Btu/kW·h, which includes the energy used to produce the electricity.

While the primary conversion ratio has drawbacks, among them not accounting for energy losses during transportation, it does allow for a more accurate comparison of the true energy savings resulting from increases in the efficiency of electricity use. Cents per kilowatt·hour calculated using the site conversion ratio are significantly higher than the price calculated using the primary conversion ratio, reflecting the omission of the true costs of producing electricity.

When reviewing savings estimates, especially those provided by outside vendors bidding on possible projects, energy managers should check to see that computations are based on consistent methods and should have the vendors make adjustments. Energy managers may want to use the organization's own internal methods and principles, possibly those used in its financial accounting, as benchmarks for computing potential energy savings. Benchmark methods can be stipulated up front as part of the project specification and design process.

When a commercial building is being designed, attention must be paid to the way in which it will function as a whole. Because initial costs of construction are comparable to energy costs over a building's life span, an increase of only a few percent in efficiency still translates into a considerable sum of money (27).

Computer Programs as Tools

Energy managers have in computers and software programs powerful tools for analyzing building performance, including a building's response to ECMs. One example of these computer programs is *DOE-2*, a building energy analysis program. *DOE-2* was originally developed by the DOE's Office of Building Technologies as an objective, public-domain tool to assist in evaluating building performance (29).

Used worldwide, the *DOE-2* model is an analytical tool, a design tool, a tool to evaluate individual technologies by climate zone, and a cost–benefit evaluation tool. *DOE-2* has been used in the development of building standards and as an educational tool for new engineers, designers and architects. An upgrade of *DOE-2* and an improved version called *DOE-3* will expand the simulation tools available to energy managers. *DOE-3* will be easier to use, provide built-in guidance, and allow models for new HVAC technologies to be quickly built up from component modules (30).

Building energy analysis programs typically perform the calculations such as life cycle costing. Many of these programs allow users quickly and completely to examine the impact of building standards including ASHRAE Standard 90.1. For the energy manager juggling a host of possible discount rates, energy rates, and financing levels, these computer programs represent a valuable tool.

A Dynamic Environment

A few cautionary notes on estimates of energy savings. Estimates of potential savings are important for planning, but the aim of energy management programs is to realize actual reductions in energy use and overall costs (31). Actual savings may be less than expected. This can be because the estimates were based on idealized engineering analyses that may be different from conditions in practice.

Production levels can effect actual savings, especially if production in an industrial setting is increased, thereby requiring more energy. Weather abnormalities and unforeseen occupancy changes can also effect actual savings. Building use in general is dynamic, with new equipment and load factors changing constantly.

Case Study: Improved Productivity

Relatively new to the list of benefits attributed to energy-efficient buildings and technologies, but with potentially significant impact on the justification of FMS, is improved productivity. As earlier noted, a 1991 study of the new corporate headquarters of the West Bend (Wisconsin) Mutual Insurance Company reported an increase in productivity of 16% over productivity in the old building vacated during the course of the research (4). The new building used smart building techniques, including participatory design and energy-efficient technologies. But the focal point of the research was ERWs. ERWs provide integrated heating, cooling, lighting, ventilation, and other environmental qualities directly to the occupants of workstations.

To measure the impact of ERWs themselves, several units were randomly turned off during a 2-week period. Researchers found that productivity, measured as the number of claim files processed by individual employees, decreased 13%. Overall, researchers estimated that the ERWs were responsible for an increase in productivity of approximately 2%. One of the insurance company's executives estimated that the productivity increases resulted in an annual savings of $260,000, with the system of ERWs paying for itself in less than 1 yr (32).

When reviewing ERWs, managers should keep in mind that increased productivity could be used to justify costs. An increase in productivity of a few percentage points could have a big impact on their payback, because the cost of labor is generally the single largest cost factor in the production of goods and services. An energy management program with strong economic underpinnings, when joined with high quality technologies and management practices, forms a comprehensive approach to meeting the challenges of conserving energy.

Case Study: Performance Contracting

There is perhaps no better example of an energy management program with strong economic underpinnings than one based on performance contracting. As noted earlier, under performance contracting a supplier will finance the project for the organization, with the contractor taking as payments future energy savings.

An example is the Pendleton Memorial Methodist Hospital in New Orleans. To help control climbing utility bills, Pendleton initiated a long-range, multimillion-dollar energy conservation program. Under the energy services agreement with the 317-bed hospital, the supplier, Johnson Controls, Inc., provided both the financing and retrofitting of the principal improvements to the HVAC and lighting systems. The monitoring and control of the HVAC and lighting systems, duties that once consumed physical plant staff time, are now aided by a facilities management system. Plant personnel use computer terminals to view dynamic graphs of system schematics and change equipment schedules from a central, remote location.

The improvements are projected to save approximately $2 million over the next 10 yr. The supplier will share in those savings as compensation for financing the program and monitoring the program's progress. Meanwhile, even with the scheduled addition of a 39,624 m^2, six-story building, the plant operations staff will be able to manage the total 152,400 m^2 complex without additional personnel.

Case Study: Connectivity

At a commercial building in Oak Ridge, Tennessee, a facilities management system equipped with an interface device extends the capabilities of the building's old and new HVAC equipment. A zone-control scheme and lack of control power once led to 8° to 10° temperature swings from the north to the south side of the 17,069 m^2 building.

Two separate suppliers teamed-up in cooperative planning and specifying efforts that incorporated a 9144 m^2 addition to the existing building. They offered the building owner off-the-shelf connectivity between factory-mounted smart devices on HVAC equipment and a facilities management system, with the FMS recognizing all the control points in the building.

The new, connectivity-enhanced equipment and control scheme allows staff to operate the building at maximum efficiency and to monitor or adjust equipment at a central location. Building tenants have enhanced comfort and no temperature swings, with the control system using 140 thermostats instead of the old system's 12. Overall, the building owner credits the cooperative efforts with increasing the value of the building by saving time and by creating a happy tenant.

THE FUTURE OF ENERGY MANAGEMENT

Priorities are changing. Governments and businesses know they must make better use of limited natural and economic resources (33). Will the pendulum of concern swing back toward apathy, or has it stopped permanently in favor of a more responsible use of energy? For energy managers who favor the latter, a discussion of the future of energy management is equal parts products and programs, obstacles, and an agenda for progress.

The degrees to which products and programs will be successful depend on ease of use or application. A product's application as a building management tool requires that facility management personnel and building users be readily able to use it and can readily equate its use with the target benefits of increased efficiency and comfort.

An initiative such as a program designed to achieve increased energy efficiency in an organization's buildings must also be readily understood by its participants. Users of federal buildings need to be shown that while giving them a more comfortable and productive place to work or learn, increased energy efficiency conserves the economic resources of the nation's taxpayers and the funds available to federal agencies. It must be understood by all that strong energy management enhances national security by making the United States more self-sufficient and less susceptible to international disruptions of price and supply.

Energy Efficiency and Labor Efficiency

Much of the above rationale translates to users of commercial and industrial buildings. Increased energy efficiency can help them be more comfortable, happy and productive. In any building, the costs of the occupants, ie, their wages and benefits, far outweigh any of the operating costs in a building. Worker productivity, therefore, should be a prime consideration as the future unfolds. Because labor costs comprise such a large component of production expense, productivity increases of even 1–2% provide quick returns on investment for the latest building technologies (34). Energy managers should seek to apply technologies that improve the productivity of workers. This could include such equipment as environmentally responsive workstations.

Increased energy efficiency, from a traditional perspective, conserves the organization's limited financial resources, allowing more to be returned to the organization through product research and development, employee compensation and benefits, additional jobs, and job security. The organization itself will be less susceptible to disruptions in energy prices and supply that could affect product cost and hence market competitiveness.

Future developments also will depend on successful measurement and management. Energy conservation programs and measures must have efficiency and payback targets; technologies and products must have efficiency standards. Equally important is the provision for evaluating the performance of programs and products. If the performance of initiatives cannot be measured in some fashion, they cannot be managed. Facility management and control systems provide this measurement and management function on an interactive, continuous process basis. Energy audits study and report on energy use in a building. Energy managers and their organizations can use the information provided by these systems to further optimize their strategies.

The participatory design of intelligent buildings has been discussed as a means of delivering results to new buildings. Owners of existing buildings should do more in the future to incorporate elements of the participatory design process into retrofit and renovation projects.

Money to Save Money

Product technologies hold the future of energy management. Goals in developing them should be to help employees work smarter, not harder. Facility management systems have been discussed as proven technologies; they need to be more universally applied and their capabilities more completely tapped.

Simple solutions, such as "lights off" campaigns, do not depend on sophisticated technologies or large-scale infusions of capital. But the biggest obstacle to energy efficiency is that it can take money to save money. Where possible national, state, and local government leaders must expand the monetary resources made available to energy managers. Loans, grants, appropriations, and municipal bonds form just a part of the picture.

Legislation establishing self-financing energy efficiency banks should be promulgated to provide additional sources of funds. Legislation and procedures that enable energy managers in the government sector to easily form public–private partnerships must likewise be advanced.

The trend is toward developing for utilities (and hence their customers) the type of marketplace that encourages competition and efficiency. This will conserve natural and economic resources, and as it has done in other deregulated industries, should stabilize energy prices. Regulations such as those promoting the use of statistical recoupling of revenues would give investor-owned utilities an incentive to run energy efficiency programs, avoiding the disincentive that an energy management program causes when it reduces earnings. Public utility commissions promote this concept.

Cultivating Quality

In the future, energy-conservation experience and management capabilities should count for more in the vendor selection process as should the vendor's willingness to submit proposals based on cooperative or progressive financing methods. Qualified energy management personnel should also be cultivated. Energy managers will require a career path, adequate compensation, and a recurrent training program. Incentives in the form of recognition and enhanced compensation should be expanded as a means of rewarding and retaining qualified personnel.

Organizations, particularly government agencies, must continue to expand their energy-consumption measurement, record-keeping, and analysis activities. It will be difficult to meet future goals without good statistics. If energy managers can appropriately measure energy use, energy managers will be able to develop strategies and purchase systems designed to meet their requirements. Energy efficiency standards for buildings and products will continue to be reviewed and refined. Energy efficiency standards must be reinforced by operating procedures that require documentation of building and energy management activities and a commitment by top management to meet these goals.

In the past, proprietary equipment protocols and computer languages have impeded interprocess interaction between computerized equipment. Attempts have been made by international and domestic organizations to develop standard protocols. In this way, users can take advantage of the best technologies available, regardless of the manufacturer, and be assured they are compatible with equipment already in place.

Compatibility is already achievable, however, using the various protocols developed by the private sector. Interface devices and strategic partnerships forged by forward-thinking suppliers already satisfy the requirements of many public, commercial and industrial buildings. More suppliers, especially those that refuse to open their equipment to a compatible or connective scheme, need to join this movement.

To enhance energy management in the future, maintenance procedures need to be improved. Managing the maintenance schedules of the building's systems is a consuming task, but a task that is well-suited to computerized facility management systems. These systems can help ensure that correct equipment maintenance is performed and documented. This will result in longer equipment life and minimal down-time.

Creating Jobs Through Energy Efficiency

In the years to come, energy efficiency improvements will lower energy bills and reduce the cost of energy services, cut oil imports, reduce pollutant emissions, and provide other benefits. The creation of jobs is an equally compelling reason to improve energy efficiency. By shifting economic activity away from energy supply and by saving consumers and businesses money that will be re-invested throughout the economy, energy efficiency improvements will result in a net increase in jobs and personal income. It also will make America more globally competitive.

The American Council for an Energy-Efficient Economy estimates that efficiency improvements consistent with an attainable 2.4% annual reduction in national energy intensity could create nearly 500,000 new jobs by the year 2000 and nearly 1.1 million new jobs by 2010 on a net basis (35). The ultimate goal must be to make energy efficiency a top priority in the minds of individual leaders, executives, operations and maintenance personnel, and building users. Progress has been made, but the momentum must be sustained. Certain obstacles remain, including inadequate funding, inconsistent commitment from top management, limited vendor qualifications, personnel limitations, deficient recordkeeping and maintenance procedures, and weak enforcement of building standards.

The movement toward the universal adoption of energy management philosophies must continue if the benefits of reduced energy costs and increased building comfort are to be obtained on a building or facility level. The more universal adoption of energy management philosophies also will ensure that the goals of enhanced organizational efficiency, national security, reduced pollution, and economic vitality are realized over the long term.

In this era of increased energy awareness, is it possible to say the pendulum of concern has stopped swinging? Given the factors that have affected the will and shortened the attention spans of participants at each level of energy management in the past, it must be concluded that the jury is still out. Perceived benefit and self-interest, as always, will determine the position of the pendulum at any period in history.

BIBLIOGRAPHY

1. U.S. Department of Commerce, *National Energy Strategy,* Springfield, Va., 1991.
2. A. H. Rosenfeld and D. Hafemeister, *Sci. Am.,* 78, (Apr. 1988).
3. S. Laitner, *Using Input-Output Analysis in Energy Policy Review,* Economic Research Associates, Eugene, Ore., 1993, p. 27.
4. W. Kroner, J. A. Stark-Martin, and T. Willemain, *Rensselaer's West Bend Mutual Study: Using Advanced Office Technology to Increase Productivity,* The Center for Architectural Research, Troy, N.Y., 1992.

5. Energy and Environment Division, Lawrence Berkeley Laboratory, *Center for Building Science,* FY 1992 Annual Report, MS 90-3058, University of California, Berkeley, 1993.
6. Alliance to Save Energy, American Council for an Energy-Efficient Economy, Natural Resources Defense Council, Union of Concerned Scientists, *America's Energy Choices,* the Union of Concerned Scientists, Cambridge, Mass., 1991.
7. U.S. Congress, Office of Technology Assessment, *Building Energy Efficiency,* OTA-E-518, U.S. Government Printing Office, Washington, D.C., 1992.
8. American Society of Heating, Refrigerating and Air-Conditioning Engineers, Inc., *ASHRAE/IES Standard 90.1-1989 User's Manual,* 1992.
9. Commonwealth of Massachusetts, Massachusetts Division of Energy Resources, *The Massachusetts Energy Plan,* 1993.
10. M. Ginsberg, *Energy User News,* 24, (June 1993).
11. U.S. Department of Energy, *Annual Report to Congress on Federal Government Energy Management and Conservation Programs, Fiscal Year 1991,* 1992.
12. National Conference of State Legislatures, *Energy Manage. Conservation,* 161, 1993.
13. U.S. Department of Energy, *Energy Manage. Conservation* (1988).
14. Ref. 12, p. 191.
15. Ref. 12, p. 190.
16. *State of Wisconsin, Governor Thompson's Alternative Fuels Task Force,* 1993.
17. National Conference of State Legislatures, *State Legislative Rep.* **17**(23), (1992).
18. Ref. 12, p. 37.
19. Ref. 12, p. 130.
20. U.S. Department of Commerce, National Technical Information Service, *National Energy Strategy: Powerful Ideas for America,* Springfield, Va., 1991.
21. *Energy Conservation Digest* **16**(8), 73 (1993).
22. Ref. 12, p. 138.
23. U.S. Department of Energy, National Association of Manufacturers, *Energy Efficiency: The Competitive Edge,* Washington D.C., 1991.
24. N. Camejo, paper presented at the 14th World Energy Engineering Congress, 1991.
25. J. Enright, *New for the Nineties: Connectivity in Building Systems,* 1993.
26. T. R. Weaver in *The Intelligent Building Sourcebook,* Fairmont Press, Lilburn, Ga., 1988, p. 16.
27. R. Bevington and A. H. Rosenfeld, *Sci. Am.,* 77 (Sept. 1990).
28. N. M. Post, *Eng. News Rec.* (May 17, 1993).
29. Alliance to Save Energy, American Council for an Energy-Efficient Economy, *Achieving Greater Energy Efficiency in Buildings: The Role of DOE's Office of Building Technologies,* Washington, D.C., 1992.
30. Energy and Environment Division, Lawrence Berkeley Laboratory, *Building Technologies Program 1991 Annual Report,* University of California, Berkeley, 1991.
31. U.S. Congress, Office of Technology Assessment, *Energy Efficiency in the Federal Government: Government by Good Example?,* OTA-E-492, U.S. Government Printing Office, Washington, D.C., 1991, p. 46.
32. P. E. Beck, *Consulting-Specifying Eng.,* 35 (Jan. 1993).
33. K. Rospond, *Service Reporter,* 27 (June 1993).
34. D. A. Decker, *Testimony before the U.S. Senate Committee on Governmental Affairs,* Feb. 18, 1991.
35. H. Geller, J. DeCicco, S. Laitner, *Energy Efficiency and Job Creation: The Employment and Income Benefits from Investing in Energy Conserving Technologies,* American Council for an Energy-Efficient Economy, Washington D.C., 1992.

ENERGY MANAGEMENT, PROCESS

Dan Steinmeyer
Monsanto Company
St. Louis, Missouri

In the chemical industry, which is inherently energy intensive, energy costs, including feedstock, average approximately 8% of the value added. For large-volume chemicals these costs represent a much higher fraction. For example, for nitrogen-based fertilizer the energy costs are approximately 70% of the value added.

Energy management includes energy conservation, but also encompasses utility system reliability; the intermesh of process design with utility systems; purchasing, including plant location for minimum energy cost; environmental impacts of energy use; tracking energy performance; and the optimization of energy against capital in equipment selection.

ENERGY AND THE CHEMICAL INDUSTRY

Table 1 is an estimate of energy usage by United States industry for 1988 (1). The chemical industry used 21% of the energy consumed by the U.S. industrial sector, and the other three related process industries, paper, petroleum, and primary metals, combined for an additional 50% of the industrial consumption.

Feedstocks

A separate breakdown between fuels and feedstocks for the chemical industry (2) shows that the quantity of hydrocarbons used directly for feedstock is about as great as that used for fuel (see Fuels, synthetic). Much of this feedstock is oxidized accompanied by the release of heat, and in many processes, by-product energy from feedstock oxidation dominates purchased fuel and electricity. A classic example is the manufacture of nitric acid (qv) [*7697-37-2*], HNO_3. Ammonia [*7664-41-7*], NH_3, is burned in air on a catalyst at a pressure around 1 MPa (10 atm) and a temperature above 900°C (3). As shown in Figure 1, the process is built around the heat and power recovery from this reaction, including the integration of the power recovery turbine and the air compressor. Another example is the reaction of propylene [*9003-07-0*], C_3H_6, and ammonia to make acrylonitrile [*107-13-1*], C_3H_3N. Here, two-thirds of the hydrogen is combusted just to satisfy stoichiometry. In addition, CO and CO_2 formation consumes about 15% of the feed propylene.

Fuels

Two-thirds of the fuel used by the United States chemical industry in 1988 was natural gas [*8006-14-2*], which is clean and easy to combust (see Gas, natural). Although relatively inexpensive at the wellhead, natural gas is costly to transport. Hence the chemical industry is concen-

Table 1. Energy Usage by United States Industry in 1988[a]

	Consumption, EJ[b,c]				
Energy Source	Chemicals	Petroleum Refining	Paper	Primary Metals	Other
Electricity					
Direct use	0.44	0.11	0.20	0.54	1.24
Loss[d]	0.88	0.22	0.40	1.08	2.48
Oil	0.15	0.11	0.20	0.06	0.34
Gas and LPG	3.11	0.78	0.45	0.79	2.10
Coal and coke	0.34		0.34	1.59	0.68
Other	0.58	5.65[e]	1.31[f]	0.05	0.51
Total	*5.50*	*6.87*	*2.90*	*4.11*	*7.35*

[a] Values include feedstocks.
[b] One EJ = 1×10^{18} Joules.
[c] To convert Joules to Btu, multiply by 0.95×10^{-3}.
[d] Assumes delivered electricity has one-third of the energy content of the fuel used to produce it; ie, 1 Joule of electrical supply to an industrial user represents 3 Joules of fuel consumed.
[e] For petroleum, this value represents primarily feedstock used in nonenergy products such as asphalt, as well as some feedstock used for petrochemicals.
[f] For paper, this value represents the fuel value of the biomass.

trated in regions where natural gas is produced, keeping the average price paid by the U.S. chemical industry for natural gas in 1988 to only 80% of the average U.S. industrial price (1). Similarly the movement of chemical commodity production to the Middle East is driven by the desire to obtain low cost natural gas.

Waste Fuel Utilization

It is always preferable to minimize or upgrade by-products for sale as chemicals, however when this is not feasible the fuel value can still be recovered. Increased combustion of by-product gases and liquids was one of the principal components in the improvement in energy efficiency that occurred in the industry in the 1980s. An example of waste fuel utilization is the incineration of the off-gases from an acrylonitrile reactor followed by generation of high pressure steam, shown in Figure 2.

Electricity

Electricity, including the losses associated with production, represents 24% of the total energy used by the chemical industry (Table 1). On a cost basis, electricity represents a higher share at 29% of the energy bill including feedstocks. Increases in electrical costs have provided the

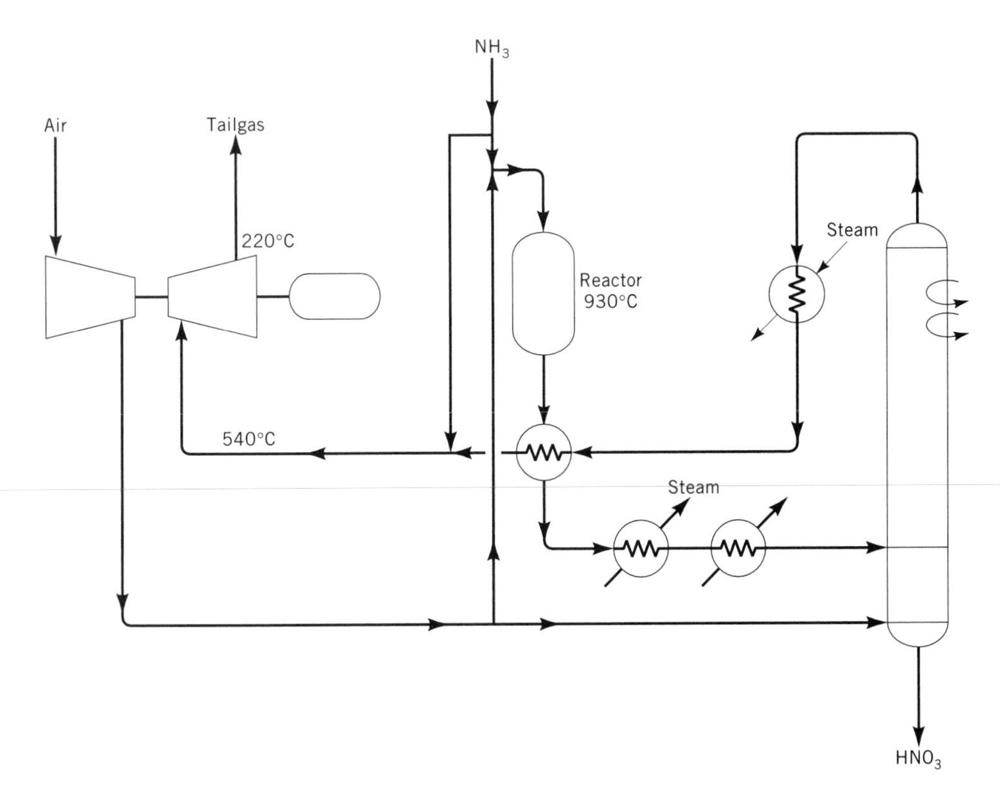

Figure 1. Schematic of nitric acid from ammonia showing integration of reactor heat recovery, power recovery from tailgas, and air compresssion (3).

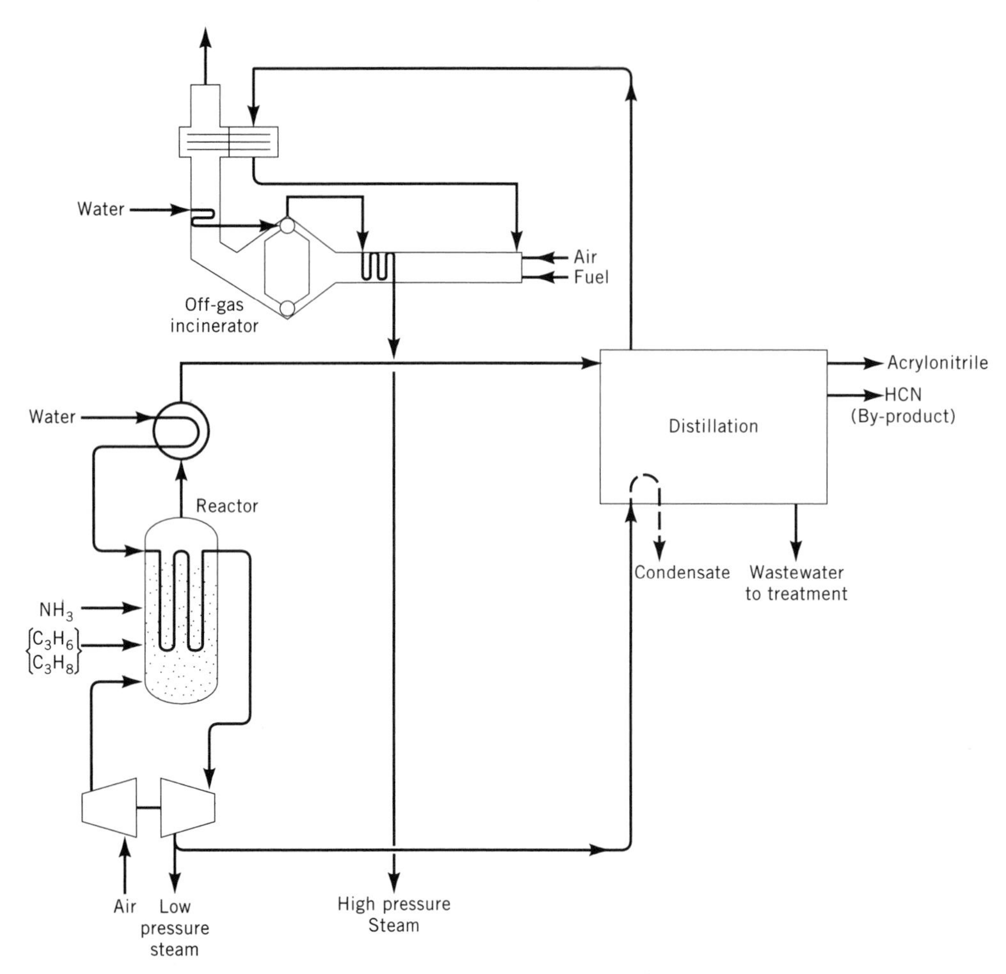

Figure 2. Acrylonitrile process showing integration of waste heat from reactor and off-gas incinerator. The reaction in the reactor is $C_3H_6 + NH_3 + 1.5\ O_2 \rightarrow C_3H_3N + 3\ H_2O$ + heat; in the off-gas incinerator, $C_3H_8 + 5\ O_2 \rightarrow 3\ CO_2 + 4\ H_2O$ + heat.

driving force for increased cogeneration, ie, the recovery of power as a by-product of other process plant operations (see also POWER GENERATION). The historic cogeneration example is the steam turbine associated with the boiler plant. The relatively high cost of electricity has also led designers to focus on the efficiency of rotating equipment, and has motivated a closer look at how processes can be controlled to reduce power, using such innovations as variable frequency motor drives.

Energy Efficiency Improvements

Energy management is basically a game of economics played with a special set of technical rules. A saving of millions of kilowatt hours or gigaJoules is only communicable when converted into dollars, or into a ratio of dollars saved per dollar of incremental investment. The question of where in a process to focus on improvement is often answered by determining the biggest energy cost items in a plant.

Efficiency improvement is driven by two distinct forces: technological progress which is the long-term trend, and cost optimization which is the short-term response to price swings. The baseline, long-term trend has been in the range of 2 to 3% per year improvement. The forecast for the 1990s is 1 to 2% per year, reflecting a more mature industry. A 1 to 2% per year energy reduction still yields a large saving over time.

The U.S. chemical industry achieved an annual reduction of 4.2% in energy input per unit of output for the period 1975–1985 (2). This higher reduction resulted from cost optimization, the tradeoff of increased capital for reduced energy use, that was driven by energy prices (4). In contrast, from 1985 to 1990, the energy input per unit of output has been almost flat (2) as a consequence of falling prices. The average price the U.S. chemical industry paid for natural gas fell by one-third between 1985 and 1988 (1,5).

Whereas energy conservation is an important component of cost reduction for the chemical industry, conservation is rarely the only driving force for technological change. Much of the increased energy efficiency comes as a by-product of changes made for other reasons such as higher quality, increased product yield, lower pollution, increased safety, and lower capital. For example, process energy integration in design, enabled by computer simulation, saves energy as well as capital; substitution of variable speed drives on motors for control valves saves energy as well as capital in the supply of power. One of the roles of energy management is to be sure energy use reduction is considered whenever processes are changed.

The refinement of processes such as occurred for production of low density polyethylene, where a process requiring over 100 MPa (1000 atm) was replaced by one taking place at 2 MPa (20 atm), may continue, but such refinement is expected to be less than in the 1970s and 1980s. The introduction of biotechnology-derived processes is expected to cause a shift to lower temperature and lower pressure processing in the chemical industry, but as of this writing the timing is unclear.

Energy and the Environment

The impact of energy usage on gaseous emissions has emerged as a primary environmental issue, and regulatory action has required emission reductions in NO_x and SO_2 (see AIR POLLUTION; AIR POLLUTION CONTROL METHODS). Control of NO_x is achieved by limiting the temperature of combustion and limiting excess oxygen. Control of SO_2 either requires changing fuel or flue gas scrubbing. Because the preferred fuel of the chemical industry is already predominantly low sulfur natural gas, the primary impact on the chemical industry of SO_2 regulation is expected to be to raise the price of electricity derived from coal.

Issues related to gases such as CO_2, which contribute to the global greenhouse effect, are also rising in importance (see ATMOSPHERIC MODELS). Energy conservation directly reduces CO_2 emissions. The elimination of fugitive hydrocarbon emissions as a result of improved maintenance procedures is also a tangible step that the industry is taking.

ENERGY TECHNOLOGY

Energy management requires the merging of such technologies as thermodynamics, process synthesis, heat transfer, combustion chemistry, and mechanical engineering.

Thermodynamics

The first law of thermodynamics, which states that energy can neither be created nor destroyed, dictates that the total energy entering an industrial plant equals the total of all of the energy that exits. Feedstock, fuel, and electricity count equally, and a plant should always be able to close its energy balance to within 10%. If the energy balance does not close, there probably is a big opportunity for saving.

The second law of thermodynamics focuses on the quality, or value, of energy. The measure of quality is the fraction of a given quantity of energy that can be converted to work. What is valued in energy purchased is the ability to do work. Electricity, for example, can be totally converted to work, whereas only a small fraction of the heat rejected to a cooling tower can make this transition. As a result, electricity is a much more valuable and more costly commodity.

Unlike the conservation guaranteed by the first law, the second law states that every operation involves some loss of work potential, or energy. The second law is a very powerful tool for process analysis, because this law tells what is theoretically possible, and pinpoints the quantitative loss in work potential at different points in a process.

Typically, the biggest loss that occurs in chemical processes is in the combustion step (6). One-third of the work potential of natural gas is lost when it is burned with unpreheated air. Figure 3 shows a conventional and a second law heat balance. The conventional analysis only points to recovery of heat from the stack as an energy improvement. Second law analysis shows that other losses are much greater.

The second law can also suggest appropriate corrective action. For example, in combustion, preheating the air or firing at high pressure in a gas turbine, as is done for an ethylene cracking furnace, improves energy efficiency by reducing the lost work of combustion (Fig. 4).

Converting Heat to Work

There has been a historic bias in the chemical industry to think of energy use in terms of fuel and steam systems. A more fundamental approach is to minimize the input of work potential embedded in the fuel and feedstock, as well as work purchased directly as electricity. Steam is really just a medium of exchange, like money in an economy.

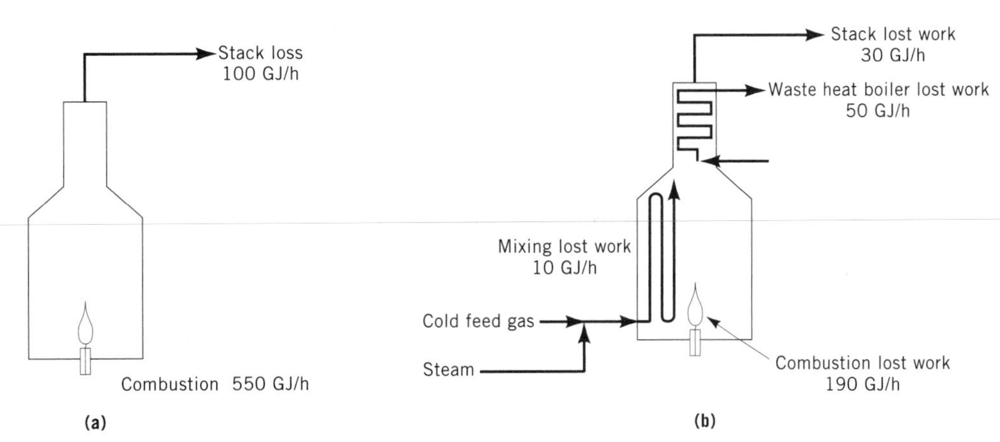

Figure 3. (**a**) A conventional heat balance, and (**b**) a balance employing the second law of thermodynamics. To convert J/h to Btu/h, multiply by 0.95×10^{-3}.

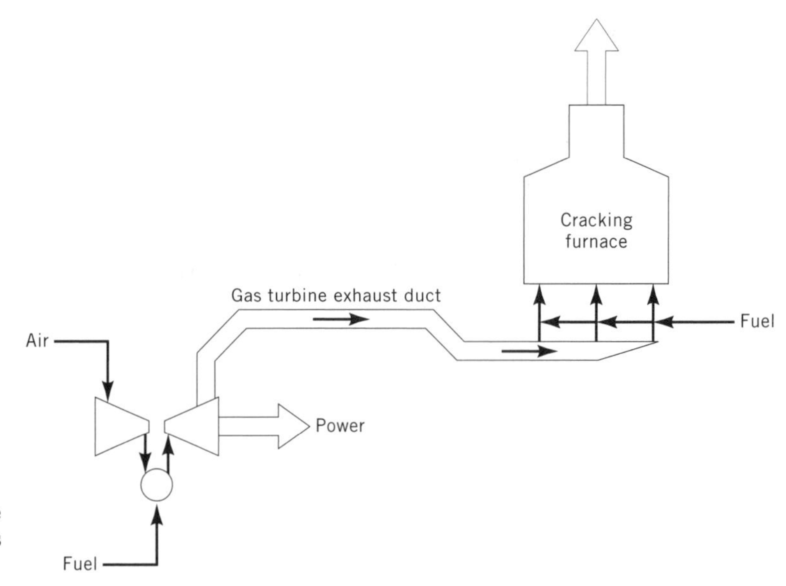

Figure 4. Use of gas turbine air preheat for ethylene cracking furnace. The gas turbine exhaust duct contains 17% oxygen at 400°C.

Waste heat tends to be very visible. It can be seen directly in steam plumes and easily measured by determining the temperature of the discharges. The loss of work potential or excess use of work is much harder to spot, but often is larger and more easily corrected. Small inefficient turbines, oversized pumps run on minimum bypasses, rewound motors having efficiency far below design, higher than optimum temperature differences that lead to below optimum power recovery, organics discharged in wastewater, and pressure drops taken across control valves are all examples of loss of work potential.

Economics of Energy Levels and Power Recovery

Cogeneration in a Steam System. The value of energy in a process stream can always be estimated from the theoretical work potential, ie, the determination of how much power can be obtained by running an ideal cycle between the actual temperature and the rejection temperature. However, in a steam system a more tangible approach is possible, because steam at high pressure can be let down through a turbine for power. The shaft work developed by the turbine is sometimes referred to as by-product power, and the process is referred to as cogeneration.

The by-product power takes only 40% as much energy to produce as on-purpose firing for power only. A comparison of Figure 5**b** and 5**d** illustrates this point. By-product power also takes much less equipment, as can be seen in Figure 6. By-product power also avoids the losses and capital associated with use of electricity such as power lines, transformers, and motors. As a result, by-product power is much cheaper than power purchased as electricity.

There is, however, only a limited quantity of by-product power available, and for large process operations the demand for power is usually far greater than the simple steam cycle can produce. Many steam system design decisions fall back to the question of how to raise the ratio of by-product power to process heat. One simple approach is to limit the turbines that are used to extract power to large sizes, where high efficiency can be obtained.

Another way to raise the power/heat ratio is by raising the pressure of the steam system. An increase in pressure from 4.2 to 10.1 MPa (600 to 1500 psi) almost doubles the power associated with a given steam load. (Power/heat ratio increases from 0.12 to 0.20). This, however, comes at appreciable capital cost for alloy materials of construction in the boiler, piping, and turbines. It also requires a substantially higher cost treatment system for the boiler feedwater, and mandates a relatively high recovery factor for condensate.

By-product power does not give enough power to match the demand for many processes such as ammonia synthesis, and designs have historically incorporated condensing turbines for incremental power with heat rejection to cooling water. A more effective response is use of the gas turbine combined cycle shown by Figures 5**c** and 6**c**.

Gas Turbines and the Combined Cycle. The combined cycle first fires fuel into a gas turbine and greatly increases the power extracted per unit of steam produced. As the numbers in Figure 5 show, the simple steam turbine gives by-product power at the lowest incremental energy use. The combined cycle shown in Figure 5**c** is intermediate in energy per unit of power between Figure 5**b** and Figure 5**d**. The big advantage of the gas turbine in cogeneration is that it permits a much higher ratio of power to heat. This ratio, which is routinely >0.8, gets bigger as the unit size of the turbine increases (7). The ratio of power to heat is also larger for aero-derivative systems which are basically jet engines exhausting into power recovery turbines. Because the original design assumed only a gas cycle, ie, no waste heat boiler, the aero-derivative was designed for a higher compression ratio, resulting in a cooler (500°C) gas turbine exhaust. Cooler exhaust means less heat to surrender, and hence a higher power/heat ratio. The heavy-duty machines of comparable vintage run discharge temperatures about 100°C higher. The aero-derivatives are also lighter in weight, giving a second advantage for a process operation because the lightweight permits very

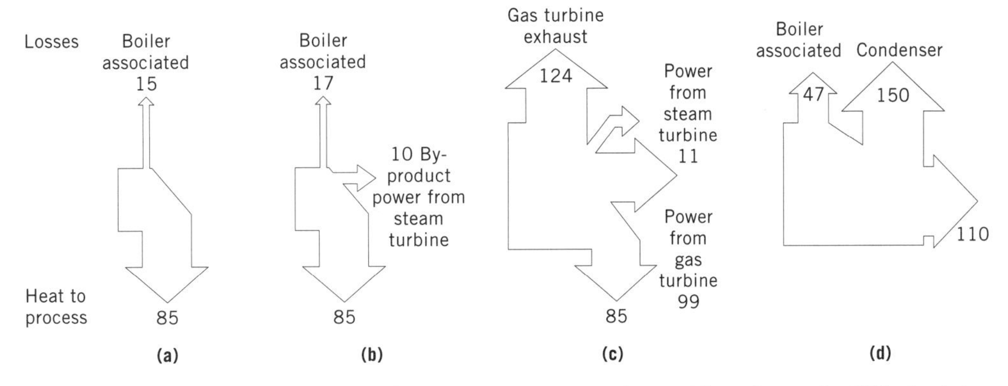

Figure 5. Relative energy flows showing power generation and heat losses in GJ/h for (**a**), boiler only; (**b**), boiler + steam turbine; (**c**), combined cycle employing gas turbine; and (**d**), condensing steam for power only. To convert J/h to Btu/h, multiply by 0.95×10^{-3}.

Parameter	(a)	(b)	(c)	(d)
fuel	100	112	319	307
incremental fuel	0	12	219	307
power	0	10	110	110
power/incremental fuel		0.85	0.5	0.36
power/heat to process	0	0.12	1.29	

fast plug-out/plug-in maintenance using spare rotating assemblies.

In contrast, the heavy-duty gas turbines, designed with capital cost per kilowatt as a key criterion, cost less for the same power output and come in much larger sizes than the aero-derivatives. This means that very large (>100 MW) installations invariably use the heavy-duty type. Heavy-duty turbines have benefited from such technology developments of the aero-derivatives as increased firing temperatures permitted by use of high temperature alloys (qv) and internal cooling of the blading. These developments have been a factor in continuing increased efficiency and increased output from a given frame size.

Gas turbine cogeneration is inherently relatively low in capital expenditures. The absence of heat exchange surface in the gas turbine part of the cycle provides the basic capital advantage, and standardized equipment, prepackaged as skid mounted components, adds to the capital advantage. When these factors are coupled with low priced natural gas, a situation results in which petrochemical plants have become exporters of cogenerated power to utilities. The gas turbine also has advantages that are firmly rooted in thermodynamics. It utilizes energy directly at a high temperature level, without large driving forces for pressure drop and temperature difference.

Most gas turbine applications in the chemical industry are tied to the steam cycle, but the turbines can be integrated anywhere in the process where there is a large requirement for fired fuel. An example is the use of the heat in the gas turbine exhaust as preheated air for ethylene cracking furnaces as shown in Figure 4 (8).

The combined cycle is also applicable to dedicated power production. When the steam from the waste heat boiler is fed to a condensing turbine, overall conversion efficiencies of fuel to electricity in excess of 50% can be achieved. A few public utility power plants use this cycle, but in general utilities have been slow to convert to gas turbines. Most electricity is generated by the cycle shown in Figures 5**d** and 6**d**.

Power Recovery in Other Systems. Steam is by far the biggest opportunity for power recovery from pressure letdown, but others such as tailgas expanders in nitric acid plants (Fig. 1) and on catalytic crackers, also exist. An example of power recovery in liquid systems, is the letdown of the high pressure, rich absorbent used for H_2S/CO_2 removal in NH_3 plants. Letdown can occur in a turbine directly coupled to the pump used to boost the lean absorbent back to the absorber pressure.

Heat Recovery, Energy Balances, and Heat-Exchange Networks

The goal of heat recovery is to be sure that energy does the maximum useful work as it cascades to ambient. An energy balance is a summary of all of the energy sources and all of the energy sinks for a unit operation, a process unit, or an entire manufacturing plant. Table 2 gives an energy balance for a simple propane-fired dryer. The energy balance is almost as important to understanding how a process works as the material balance. The energy balance is the basic tool for analyzing an operation for energy conservation opportunities. When incorporated into a computer program, the energy balance becomes the base for a model of the process. Operational changes, system configuration, and equipment alterations can be evaluated via the model.

The heat exchanger network analysis, sometimes called pinch technology, is a special kind of model that has been developed into a sophisticated way of attacking heat recovery problems. This type of analysis defines the optimum interchange between all the heat sinks and heat sources.

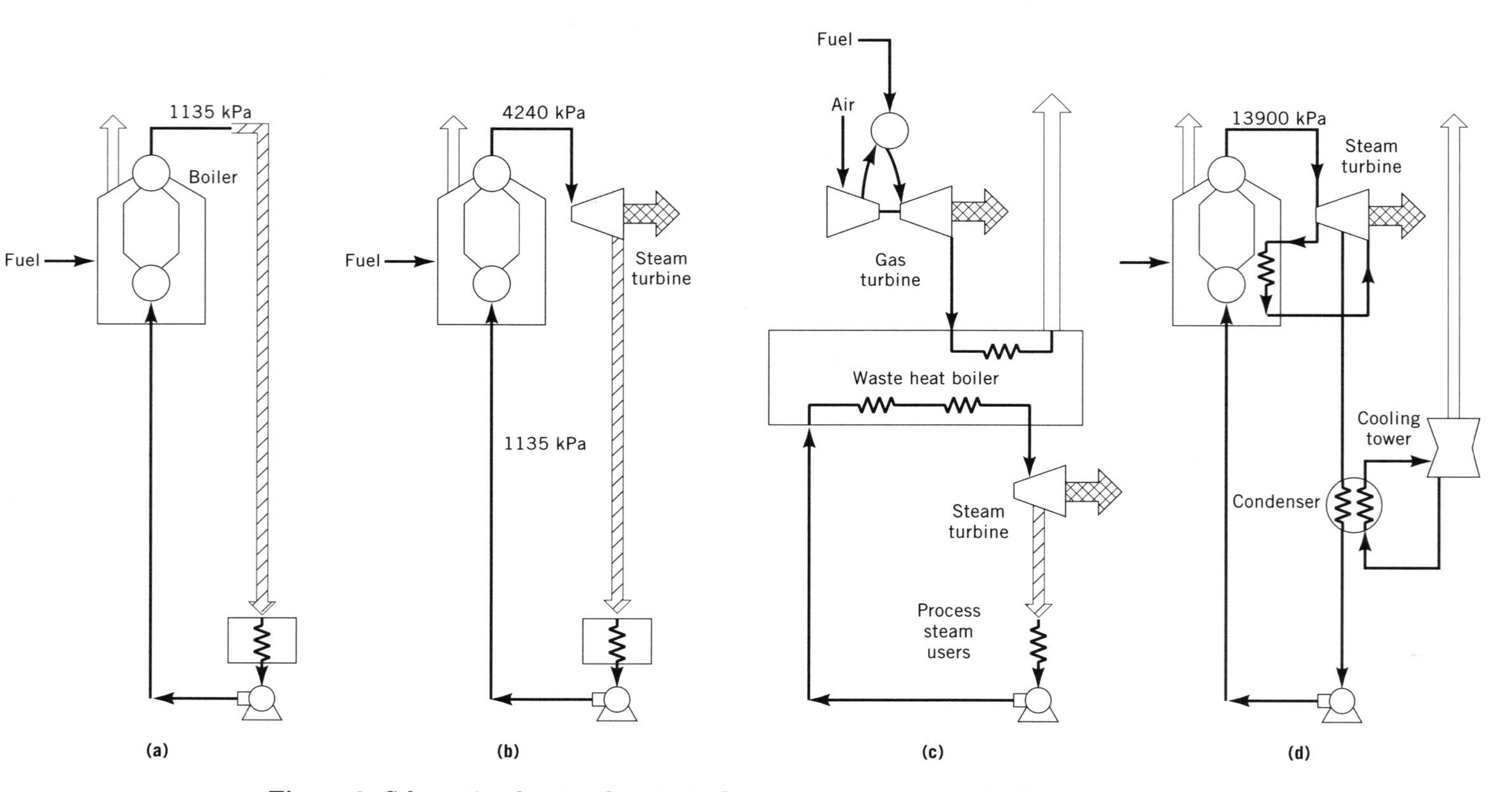

Figure 6. Schematics showing the principal equipment components for the energy systems shown in Figure 5. (**a**) through (**d**) correspond to Figure 5**a** through 5**d**, respectively. To convert kPa to psi, multiply by 0.145.

Table 2. Product Dryer Analysis Heat Balance

Material	Mass, kg/h	Energy, MJ/h[a]
Inputs		
Fuel, C_3H_8	130	6,553
Air		
Combustion	6,817	106
Secondary	14,846	232
In-leakage	4,289	67
Water with product	1,731	354
Dry product solids	4,478	249
Totals in	*32,291*	*7,561*
Outputs		
Water vapor		
From product	1,445	3,808
From combustion of H_2	212	560
Air and combustion products	25,870	1,933
Dry products solids	4,478	458
Water with product out	286	146
Heat losses		656
Totals out	*32,291*	*7,561*

[a] To convert J/h to Btu/h, multiply by 0.95×10^{-3}.

It is similar to the concept used by furnace designers for many years for matching various heat sinks against the flue gas source. Using a temperature vs enthalpy plot, the analysis can be done manually or via computer. It can even be set up to adjust distillation sequence or operating pressures to improve energy efficiency and minimize capital (9).

Waste-Heat Boilers. In most chemical process plants, the steam system is the integrating energy system. Recovering waste heat by generating steam makes the heat usable in any part of the plant served by the steam system. Many waste-heat recovery boilers are unique and adapted to fit a particular process. There is a long history of process waste-heat boiler failure resulting from inadequate attention to detail in design and the failure to maintain water quality (10). The high heat-transfer coefficients of boiling water are dependent upon clean surfaces. Designers should match the hardware as closely as possible to demonstrated designs, and operators should be sure that water treatment is monitored. Incinerators and gas turbines also involve heat recovery boilers. Here, a number of fairly standard designs have evolved (11).

Product-to-Feed Heat Interchange. Heat exchange is commonly used to cool the product of a thermal process by preheating the feed to that process, thus providing a natural stabilizing, feed forward type of process integration. Product-to-feed interchange is common on reactors as well as distillation trains.

Combustion Air Preheat. Flue gas to air exchange, a type of product-to-feed heat exchange, is extremely important because of the large loss associated with the combustion of unpreheated air. This exchange process has generated fairly unique types of hardware such as the Ljungstrom or rotary wheel regenerator shown in Figure 7**a**; the brick checkerwork regenerators used in metallurgical furnaces, hot oil, or hot water belts (also called "liquid runarounds") (Fig. 7**b**); and heat pipes (Fig. 8). Liquid runaround systems make it practical to use finned surface on both gas exchange surfaces. These are particularly useful for retrofit because of the ability to move the heat to physically separated units.

The heat pipe exchanger is a variant on the liquid runaround system where each tube (pipe) is sealed on both ends and filled with a vaporizing–condensing heat-transfer medium. At the hot end of the pipe, liquid is vaporized and moves to the cold end. At the cold end, vapor condenses and returns to the heat intake end. The flows are driven by gravity and capillary wicking. The heat pipe is particularly useful because it permits very compact, side by side ducting arrangements with countercurrent flow, as shown in Figure 8.

Boiler Economizers. Heat exchangers that use boiler flue gases to preheat the boiler feedwater are termed boiler economizers.

Heat Pumps. The use of heat pumps adds a compressor to boost the temperature level of rejected heat. It can be very effective in small plants having few opportunities for heat interchange. However, in large facilities a closer look usually shows an alternative for use of waste heat. The fuel/steam focus of energy use has led to application of heat pumps in applications where a broader examination might suggest a simpler system of heat recovery.

Heat Recovery Equipment

Factors that limit heat recovery applications are corrosion, fouling, safety, and cost of heat-exchange surface. Most heat interchange utilizes shell and tube-type units because of the rugged construction, ease of mechanical cleaning, and ease of fabrication in a variety of materials. However, there is a rich assortment of other heat exchangers. Examples found in chemical plants in special applications include the following.

Plate heat exchangers are made by sandwiching thin sheets of metal that have a corrugated pattern pressed into them. The corrugations provide mechanical support where the sheets contact each other, permitting compact, low cost construction and generating high turbulence and high heat-transfer coefficients. These plates are normally separated by gaskets, and are particularly useful in food processing, where cleaning is facilitated. However, the gaskets are the mechanical weak link, and limit the application in chemical plants to nontoxic, noninflammable fluids.

Brazed-fin aluminum cores are made of aluminum sheets having corrugated, cut layers sandwiched between flat sheets. The package is brazed together to form a very compact unit. The cut corrugated layers act as fins for both sides of the exchanger. The brazed aluminum construction needs to be protected from fire by special insulation or a coldbox filled with perlite. These units are used almost exclusively for very clean, cryogenic services such as air separation or hydrogen purification.

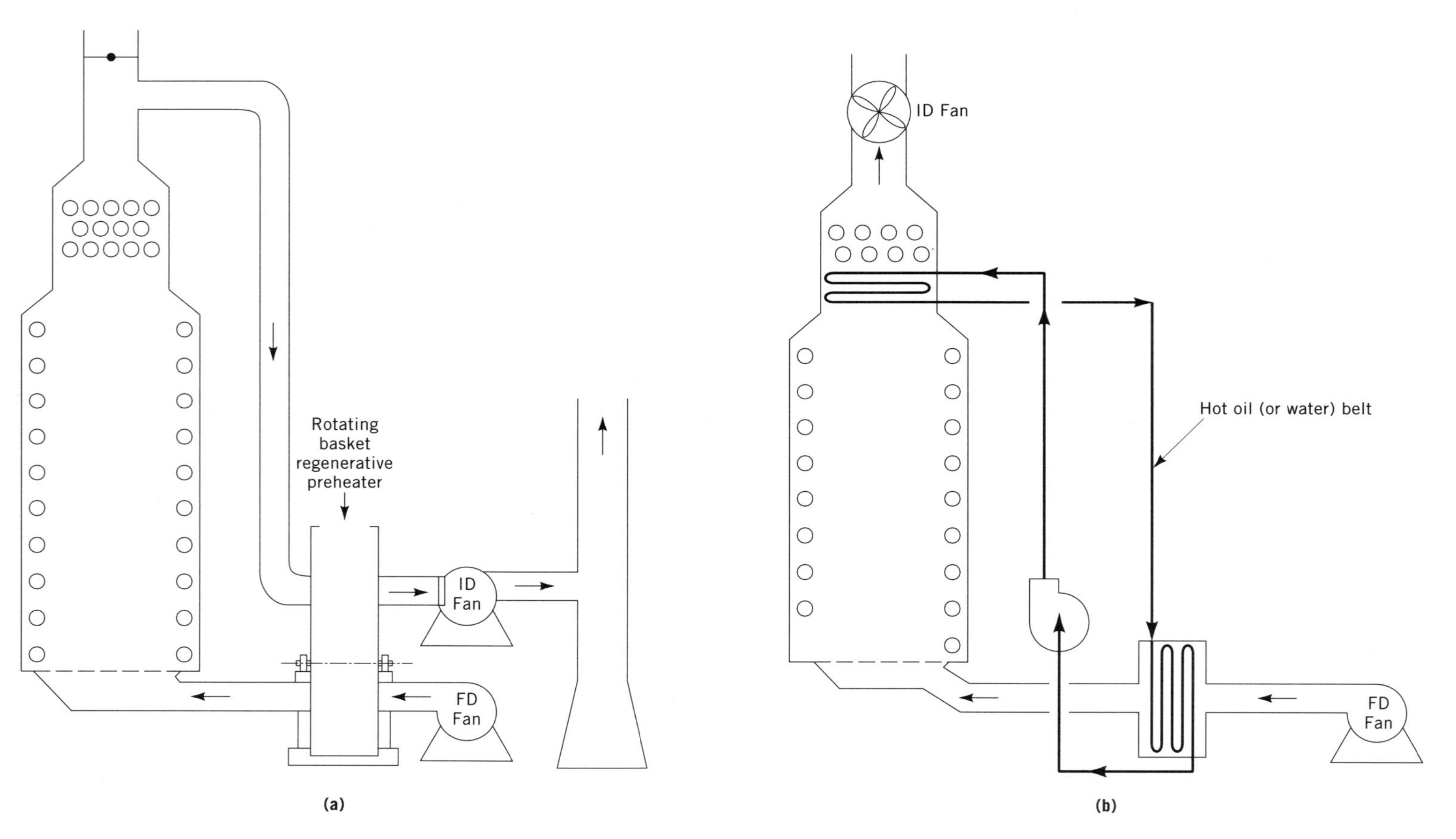

Figure 7. Air preheaters where ID = induced draft and FD = forced draft fan. (**a**) Rotating metal basket or Lungstrom regenerative preheater; and (**b**) hot oil or water belt (liquid runaround) used to move convection section heat to air preheater in furnace retrofit.

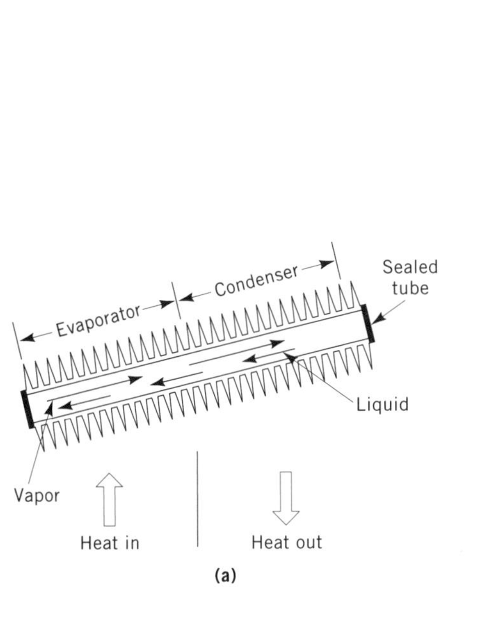

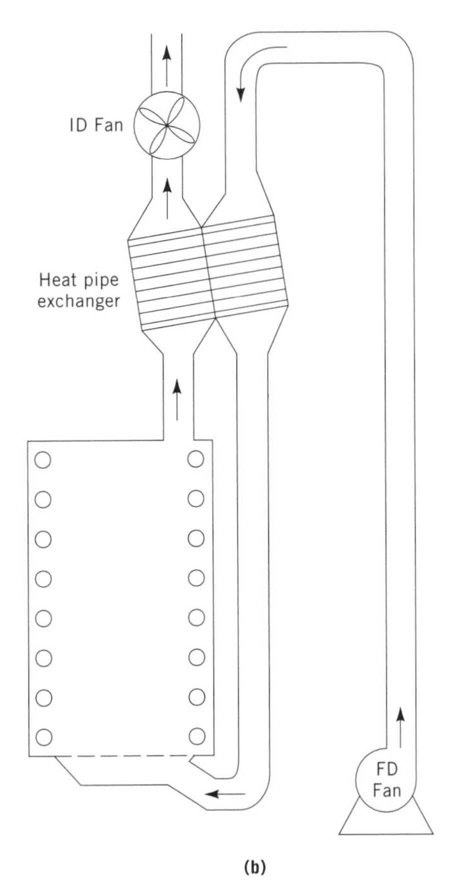

Figure 8. (**a**) Heat pipe showing the use of finned tubing for both heating and cooling; and (**b**), heat pipe exchanger air heater system. ID = induced draft and FD = forced draft fan.

In spiral plate construction, two plates are welded together and rolled into a jelly-roll shape. The prime advantage is that there is a single flow passage. Any pluggage generates a high local pressure drop and tends to erode the deposit. Thus the unit is less subject to pluggage than in the other constructions having parallel flow paths.

UTILITY SYSTEMS

Steam

The steam system serves as the integrating energy system in most chemical process plants. Steam holds this unique position because it is an excellent heat-transfer medium over a wide range of temperatures. Water gives high heat-transfer coefficients whether in liquid phase, boiling, or in condensation. In addition, water is safe, nonpolluting, and if proper water treatment is maintained, noncorrosive to carbon steel.

Steam Balances. The steam balance is usually the most important plantwide energy balance. It shows each service requirement, including the use of steam as a working fluid to develop power.

There are, however, some limitations. A steam balance depicts the steam flows at a given point in time or as an average over some period. The steam flow on a cold weekday morning in winter is quite different from that on a Sunday in summer, and neither matches the annual average steam balance. Startup flows are also usually far different and merit their own special balance. It is also wise to prepare a balance for the beginning and the end of the cycle between unit shutdowns. For example, the power required by a turbine driving a compressor rises as the compressor efficiency falls, and process heating requirements rise as interchangers foul. Use of a computer-based model permits easy adaptation to the calendar and onstream cycle.

There are two ways of presenting steam balance data, schematically or tabularly. For both presentation types, a balance is made at each pressure level. In a schematic balance, such as that shown in Figure 9, horizontal lines are drawn for each pressure. The steam-using equipment is shown between the lines, and individual flows are shown vertically. Table 3 contains the same data as shown in Figure 9. In both cases the steam balance has been simplified to show only mass flows. A separate balance should be developed that identifies energy flows, including heat losses and power extraction from the turbines.

Steam Turbines. Historically, back-pressure steam turbines were used as drives throughout processes to increase reliability to cover electrical power failures. A typical turbine would be a single-stage 375 kW machine having throttle steam at 4240 kPa (600 psig) and exhaust at 1135 kPa (150 psig). The turbine would be controlled by a centrifugal fly-weight governor operating a single throttle valve. The efficiency would be about 40% when operated at rated conditions; ie, for the amount of steam passing through the turbine, it would develop 40% of the power

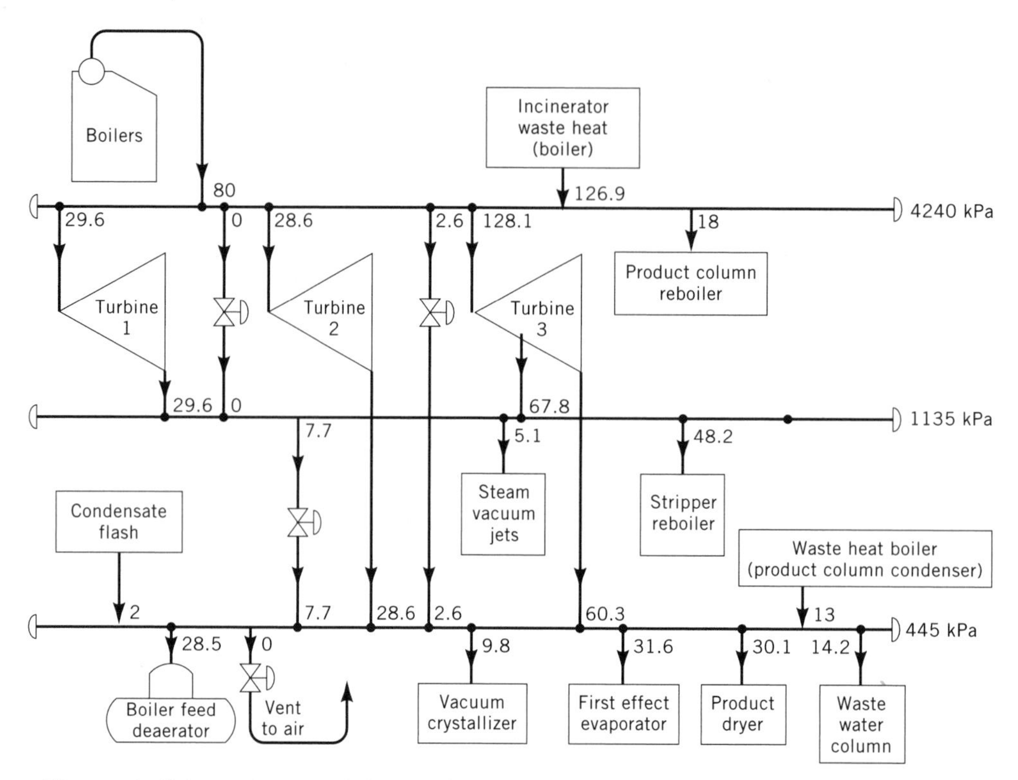

Figure 9. Schematic steam balance where the numbers represent steam flows in metric tons per hour. See Table 3.

that could be developed by an ideal turbine, expanding the steam isentropically. The efficiency was substantially lower when the machine was operated at part load, because a large portion of the pressure drop at part load was lost across the throttling valve, producing no work.

Because of increased emphasis on maximizing cogenerated power, newer plants are trying to utilize back-pressure turbines only in applications where efficiencies above 70% can be attained. This typically means limiting the applications to the large (>1000 kW) drives, and using small

Table 3. Steam Balance, Flows in Metric Tons per Hour[a]

	4240 kPa[b]		1135 kPa[b]		445 kPa[c]	
Equipment	Supply	Use	Supply	Use	Supply	Use
Boilers	80.0					
Incinerator waste heat	126.9					
Turbine 1		29.6	29.6			
Turbine 2		28.6			28.6	
Turbine 3		128.1	67.8		60.3	
Pressure reducing valve		0	0			
Pressure reducing valve		2.6			2.6	
Pressure reducing valve				7.7	7.7	
Product column reboiler		18.0				
Steam vacuum jets				5.1		
Stripper reboiler				48.2		
Finishing column reboiler				36.4		
Condensate flash					2.0	
Product column condenser						
Waste-heat boiler					13.0	
Vacuum crystallizer						9.8
1st effect evaporator						31.6
Product dryer						30.1
Wastewater column						14.2
Boiler feed deaerator						28.5
Total	*206.9*	*206.9*	*97.4*	*97.4*	*114.2*	*114.2*

[a] See Figure 9.
[b] To convert kPa to psi, multiply by 0.145.

machines only where they are necessary for the safe shutdown of the unit. Multistage turbines are used even on the smaller loads.

Most large plants also have some condensing turbines to handle process and seasonal swings and provide some flexibility to the steam balance. For large (>15,000 kW) applications, condensing turbines can compete with purchased electricity. For small applications, power can be provided at much lower cost by motors. Condensing turbines generally have high reliability, and are also used where the costs of electrical power failure, in process downtime, are high. Public utility plants usually have condensing turbines at the bottom of a power cycle as shown by Figure **6d**.

Condensate Return Systems. In a process plant, steam traps are used to drain and return condensate. Given proper application and continuous maintenance, these can operate with minimal steam leakage. Correct installation is also important (12).

For draining principal items of process equipment, level-controlled condensate chambers provide much better performance and reliability than steam traps. Usage is generally justified when condensate flow is greater than 4500 kg/h.

Electrical

The plant electrical system is sometimes more important than the steam system. The electrical system consists of the utility company's entry substation, any in-plant generating equipment, primary distribution feeders, secondary substations and transformers, final distribution cables, and various items of switch-gear, protective relays, motors starters, motors, lighting control panels, and capacitors to adjust power factor.

Electric Motors. Except for electrolytic, eg, chlorine production electric furnace processes, eg, phosphorus, typically 95% of the electricity used in a chemical plant is for electric motor drives. Induction electric motors in general use in chemical process plants range in efficiency from 90 to 95% depending primarily on size. The larger motors are more efficient. For any size, a range of efficiencies is available and high efficiency motors are somewhat more expensive than standard ones as shown in Figure 10. This price increment is normally justified in chemical process plants because of the high number of annual operating hours (13). For heating and ventilationg operations, the lower operating hours sometimes make the lower efficiency units the cost-effective choice.

Variable Frequency Drives. An important energy byproduct of solid-state electronics is the relatively low cost variable speed drive. These electronic devices adjust the frequency of current to control motor speed such that a pump can be controlled directly to deliver the right flow without the need for a control valve and its inherent pressure drop. Figure 11 shows that at rated load the variable speed drive uses only about 70% as much power as a standard throttle control valve system, and at half load, it uses only about 25% as much power.

In addition to energy conservation, the variable speed drives offer better control because of a faster response, ie, reduced dead band. They are also sometimes chosen for safety reasons because of elimination of the control station and accompanying valving. The capital saved by use of a smaller motor and elimination of the control valve partially compensates for the cost of the drive.

Lighting. High pressure sodium or metal halide lights offer significant savings over mercury vapor for process areas. Fluorescent lights have an even greater savings over incandescent lights for areas having low (mounting height <4 m) ceilings. Some other features that minimize electrical use in lighting are circuit arrangement so that unneeded lights can be turned off during non-use hours, and efficient reflectors and ballasts.

Other Energy Systems

Chemical plants usually require cooling water, compressed air, and fuel distribution systems. Sometimes also included are refrigeration, pressurized hot water, or specialized heat-transfer fluids such as Therminol liquid or condensing vapor. Each of these systems serves the process and reliability is the most important characteristic. Thus a project in any of them that achieves a 10% reduction in energy cost at the expense of a 1% loss of reliability loses money for the operation.

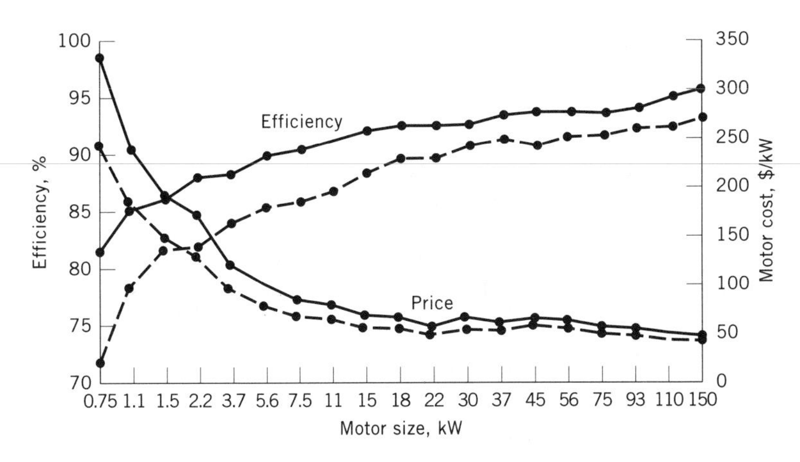

Figure 10. Full-load efficiencies and wholesale prices (in 1988 U.S. dollars) of (---) standard and (—) energy efficient (EEM) three-phase motors (13).

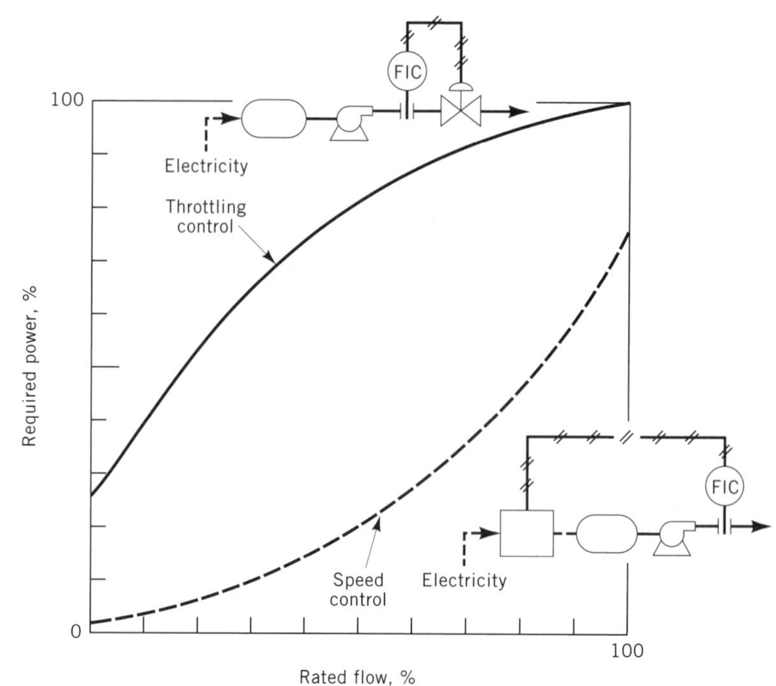

Figure 11. Power saving for variable speed drives. Power input for variable speed adjusts with flow to naturally match the frictional losses. FIC = flow indicating controller.

Cooling Water. The primary reliability concern is that water chemistry must be maintained in a low fouling, noncorroding regime. In addition, water flow velocity must be maintained above a certain threshold (ca 0.5 m/s in tubeside flow) to avoid fouling and corrosion.

The principal energy cost concern is avoidance of flows greatly in excess of need. Some cooling water systems are operated at such low temperature rises and high flows that pumping power cost equals the cost of the original heat input that it rejects. Care should be taken in design to assure that the system is balanced, and all heat exchangers utilize available pressure drop with the design flow. Every user should have a thermowell in the cooling water outlet to monitor temperature rise, and enough flow and temperature measuring elements should be provided to check overall heat balances (see TEMPERATURE MEASUREMENT).

Compressed Air. Enough flow elements should be provided to permit accurate assessment of use by different operating departments within a plant. Compressed air is often unmetered; thus there is little motivation to reduce use. A large fraction is often lost through leakage at fittings.

A key question for energy reduction is whether a lower discharge pressure can be used. Dropping pressure from 790 kPa (100 psig) to 650 kPa (80 psig) reduces the required power 12%, and reduces the driving force for air leakage losses by 20%. Controls such as inlet guidevanes on the air compressor can be provided to trim pressure to the required level. Another action that lowers energy use is lowering inlet and interstage temperatures. A drop from 30 to 20°C typically reduces power by 3%.

Refrigeration. In processes such as olefin separations, the economic importance of refrigeration exceeds that of the steam system. Refrigeration is an extremely valuable utility because of the work required to raise the energy to ambient temperature. Its value goes up directly as the temperature gap between ambient and use level goes up. For example, whereas refrigeration at −25°C is worth approximately as much as heat at 250°C, refrigeration at −75°C is worth twice as much.

A series of energy audit questions asking what can be done to reduce the work of raising this heat to ambient, or whether refrigeration is really necessary, should be addressed. Energy costs are cut if cooling water can be used directly or used for part of the year or for part of the load. An energy savings can also be realized if a higher refrigeration temperature can be used. If the refrigeration need is above 5°C, and there is waste heat available above 90°C, an absorption refrigeration system can be used to supply refrigeration.

The optimization of heat-transfer surfaces also plays a role. At the optimum, the lifetime cost of a surface is approximately equal in value to the lifetime cost of power used to overcome the temperature differential in the condenser and evaporator. Additionally, condensation on insulation is a sign of questionable insulation (see INSULATION, THERMAL). Frost is a certain signal that insulation can be improved.

Condensing Organic Vapor. The eutectic mixture of diphenyl and diphenyl oxide is an excellent vapor medium for precise temperature control at temperatures higher than those practical using steam. This mixture can achieve 315°C while holding pressure at 304 kPa (3 atm) absolute. In contrast, steam would require 10.6 MPa (105 atm) pressure.

These systems, commercially known as Therminol VP or Dowtherm A, differ from steam in some key areas which can result in operating problems unless handled properly in design (14). The low pressure–high temperature operation means that the $\Delta T/\Delta P$ ratio at saturation is quite high; for example, at 315°C the ratio is 25 times that of steam. This means that a pressure drop that would be nominal in a steam system (10 kpa (0.1 atm)), can not be tolerated if precise temperature control is needed.

Another difference is that molecular weight is much higher than that of the common noncondensables, and hence the noncondensables are harder to purge. In contrast, in a steam system almost all noncondensables are heavier than steam and tend to flush out with the condensate.

CAPITAL AND EQUIPMENT AREAS

Virtually all chemical processing is energy driven, but in separations such as distillation, drying, and evaporation, this is particularly clear. All three of these processes are simple thermal operations that involve separation through vaporization, and only a minor change in the chemical energy of the products. The capital related costs of those operations are typically three to five times as large on a lifetime basis as the energy costs. In almost all cases there is a balance between capital and energy costs, and typically one is traded against the other to achieve the lowest overall cost. Much can be said about this energy/capital trade based on first principles and engineering correlations (15).

Insulation

A surprisingly important capital element of energy management is insulation. On large projects the capital cost of insulation is in the same range as that for heat exchangers or distillation towers and trays. At the optimum insulation thickness, the lifetime value of the insulation approximates the lifetime value of the heat loss; ie, insulation is as costly as the heat loss that it prevents. Uninsulated flanges, when they exist, are a particularly severe loss point, and when flanges need to be opened periodically, insulation via removable blankets is usually justified.

Insulation provides other functions in addition to energy conservation. A key role for insulation is safety. It protects personnel from burns and minimizes hot surfaces that could ignite inflammables. It also protects equipment, piping, and contents in event of fire. Thus materials such as mineral wool are sometimes used despite relatively poor thermal qualities.

Corrosion under insulation is also a concern, particularly in refrigeration systems. The specification of the insulation system needs to include painting, vapor barriers, and external metal jackets (16).

Compressors

Compression equipment accounts for a large fraction of power use as well as a large fraction of installed capital. Usually the energy bill for a compressor is large enough to warrant a very visible monitor of the driver, such as a control room electric meter. Testing programs to ensure operation of the compressor and its driver at peak energy efficiency are also justified. Temperature rise across a compressor is a simple and effective way to monitor efficiency.

Compressors are relatively fragile, precision machines, and require more care in specification and maintenance than any other part of the process. The largest process compressors are axial flow machiness, for example the air compressors for gas turbines. Centrifugal machines can also handle large volumes, for example the cracked gas compressors in ethylene plants. Centrifugal machines are a bit less efficient, but are more rugged and tolerant of fouling service. Smaller volume compressors are reciprocating or rotary designs.

Energy consumption is but one of the selection criteria. For example, a reciprocating compressor usually delivers 5 to 25% higher efficiency than a centrifugal machine. In the size range where a single unit compressor can handle the flow, this usually pays for increased maintenance, but it rarely justifies the increased capital of parallel units in competition with a larger single train centrifugal.

A compressor is typically a specially designed device, and comes with far less surplus capacity than other process components. As a result compressors merit great care in specification of flow, inlet pressure, and discharge pressure. Similarly, the control system and equipment need to be carefully matched to provide turndown with maximum efficiency.

Because of the large volumetric flow inherent in gases, the cost of power for incremental pressure drop is high. To a first approximation, incremental power is proportional to the volumetric flow of suction or discharge, multiplied by incremental pressure drop; hence the high importance of pressure drop otpimization for the associated piping and heat exchangers. Volumetric flow also varies with the absolute value of temperature; hence suction cooling is another area for optimization. The drive for lower temperatures also provides the incentive for adding compression stages, with intercooling between stages.

Pumps

Energy use for pumps can best be controlled by design for the proper flow and discharge pressure. Constant speed electric motor driven pumps, having a large margin of safety on flow, are particularly wasteful. One solution is use of the variable speed drive. Another solution in an operating unit is trimming of oversized impellers.

Vacuum Systems

The basic question in vacuum systems is what can be done to cut design inert loading. Historically, inert flows were whimsically overspecified, and the vacuum systems were run until mechanical deterioration brought the system capacity down to the actual inert loading. One factor driving toward greater use of vacuum pumps is the large reduction they achieve in effluent to be treated.

Boilers and Process Furnaces

Boilers and process fired heaters are the entry point for the energy released from burning fuel. Fuel combustion is irreversible, and fired heaters are typically the principal loss point for work potential (6). The high irreversibility results from taking the chemical energy of fuel and degrading it to heat. Air preheat cuts energy losses by cutting fuel firing and increasing the flame burst temperature.

A more obvious energy loss is the heat to the stack flue gases. The sensible heat losses can be minimized by reduced total air flow, ie, low excess air operation. Flue gas losses are also minimized by lowering the discharge temperature via increased heat recovery in economizers, air preheaters, etc. When fuels containing sulfur are burned, the final exit flue gas temperature is usually not permitted to go below about 100°C because of several problems relating to sulfuric acid corrosion. Special economizers having Teflon-coated tubes permit lower temperatures but are not commonly used.

Inadequate mixing of air and fuel can result in unburned combustibles in the flue gases. This results in energy losses, environmental problems, and damage from afterburning in the convection section. Heat leakage through refractories constitutes still another type of energy loss from combustion equipment. Because of the high temperature, the heat leakage is higher than on most equipment and can represent as much as 3% of fired fuel. On newer designs this value is typically closer to 1%.

Distillation

Reflux Rate. The optimum reflux rate for a distillation column depends on the value of energy, but is generally between 1.05 times and 1.25 times the reflux rate, which could be used with infinite trays. At this level, excess reflux is a secondary contributor to column inefficiency. However, when designing to this tolerance, correct vapor–liquid equilibrium data and adequate controls are essential.

Control. Energy savings for improved control are surprisingly high. A 2 to 20% savings from a series of control projects has been reported (17).

A reflux reduction of 15% is typical. Improved control achieves this by permitting a reduction in the margin of safety that the operators use to handle changes in feed conditions. The key element is the addition of feed-forward capability, which automatically handles changes in feed flow and composition. One of the reasons for increased use of features such as feed-forward control is the reduced cost of computers and online analyzers.

Reboilers and Condensers. The real work used in a distillation varies with the temperature difference between the heating medium and the cooling medium. Part of this differential is the difference in boiling points between the overhead stream and the bottom stream. However, a large portion often results from the temperature difference used in the reboiler and condenser. The optimum differential is generally under 20°C, and if refrigeration is used, the optimum can be as small as 3°C. A signal that an excessive temperature difference may exist is a condenser or reboiler having a shell diameter less than one-third the diameter of the column.

There are a number of ways to provide the heating or cooling medium at temperatures closer to the optimum level. One is by use of double-effect distillation, which uses the overhead vapor from one column as the heat source for another column such that the second column's reboiler becomes the first column's condenser. This basically cuts the temperature differential in half, and shows up as an energy saving because external heat is supplied to only one of the units.

Column Pressure Drop. Another element that sets the temperature differential across the distillation is the pressure drop in the column and its auxiliaries. One of the more recent changes is the introduction of special structured packings (see DISTILLATION) which give extremely low (10% of an equivalent column with trays) pressure drop. This energy benefit can show up in an overhead temperature high enough to permit generation of by-product steam. It can also show up in a variety of other ways including lower bottoms temperature, yielding less fouling and product degradation to by-products, as in the styrene-ethylbenzene separation.

The Overall Process. Each distillation column should be examined in context with the rest of the process as well as by itself. For example, an opportunity for energy saving may be a reduction of process recycles via a purer overhead or bottoms stream. Lowering or raising column pressure to facilitate interchange with other parts of the process is another possible opportunity.

Drying

A typical dryer mass and energy balance (Fig. 9, Table 3) shows that the heat loss is 10% of the fuel input. Improving insulation is one of the simplest ways to reduce energy input. Another simple way to reduce energy input is improving the dewatering of the feed. There is a great difference in energy input for dewatering as compared to the subsequent drying step, as shown by Figure 12.

Some of the other energy conservation approaches applicable to dryers are interchange between the stack vapor and the incoming dryer air; recovering sensible heat from the product; use of waste heat from another operation for air preheat; and using less, but hotter drying air. This last is limited to nonheat-sensitive materials.

Evaporation. In most chemical industry evaporation systems, the objective is product recovery, although occasionally the objective is concentration of an organic waste from an aqueous solution, to facilitate incineration. Similar equipment is used extensively for desalination of salt or brackish water.

A single-effect evaporator produces slightly less than a kilogram of water vapor per kilogram of steam. By using the vapor produced by the first-effect as the heat source for a second-effect evaporator, steam use can be essentially halved. The performance can be improved almost in proportion to the number of effects employed. Six- and seven-effect evaporators are common in the wood pulp industry for concentration of black liquor. However, as the number

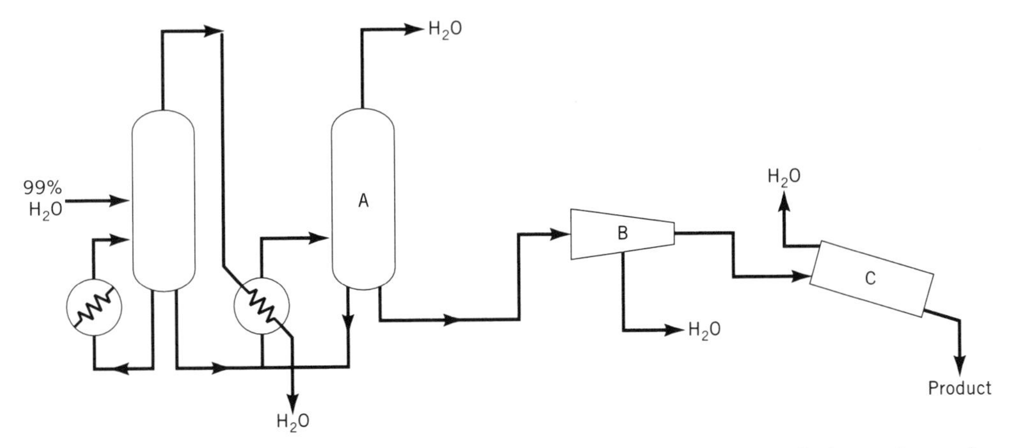

Figure 12. The relative energy input for removal of 1 kg of water relative to heat of vaporization is 0.7 in A, the double-effect crystallizer; 0.015 in B, the dewatering centrifuge; and 5 in C, the rotary dryer.

of effects goes up, the temperature driving force is spread over the additional units and the capital cost increases almost in direct proportion to the number of effects. As a consequence, high alloy systems are often limited to single- or double-effect.

Much as reverse osmosis can compete with evaporation in desalination applications, osmosis should also be considered as an alternative for process evaporation. Reverse osmosis is particularly attractive where the inlet stream is greater than 99% water.

ENERGY ACCOUNTING AND IMPROVEMENT

Energy Accounting

Long-term costs at plant gate meters are costs that should be considered in project evaluations. The real benefit of a technical action can be quite different from the savings allocated by the accounting system. The benefit also depends on other changes that happen at the plant site. For example, the incremental value of steam may be near zero because of production from waste heat boilers and utility boilers operating at minimum load. Prudent planning credits a steam saving project based on the probable plant energy balance during the project's operation rather than on current allocated cost.

The incremental cost of electricity is usually dictated by contract. Incremental cost is normally less than the average cost because most utilities have descending rate scales for high volume industrial customers and also because there is a large fixed charged associated with past peak demand. An important factor in the control of operations is recognizing that the cost of electricity is typically one-third for peak demand and two-thirds for the actual incremental use. The demand charge is typically set by the peak usage in a month or 12 month period. The demand charge recognizes that a large component in the cost of electricity is the capital used by the public utility to generate and transmit power. Many chemical plants have load shedding systems to avoid setting a higher peak and increasing demand charges. Usually the shed load is for non-process applications such as battery charging of forklift trucks, but sometimes large users such as electric furnaces need to be curtailed as well.

Because it is so readily measurable, energy conservation offers a unique opportunity for continuous improvement. Energy conservation is something that individuals want to do, and given the opportunity to see ideas converted into measurable progress, personnel pursue conservation with enthusiasm, generating larger savings than deemed possible (18).

Measurements and Audits

The enabling element of continuous improvement is measurement. An old rule of thumb says that increased accuracy in measuring an energy use ultimately yields a reduction in use equal to 10% of the increased closure of the balance. A basic principle of economics is that any thing that is free is used in excess, ie, an unmetered electrical use is bigger than expected by at least 10%. Metering of the cost elements at each unit in a chemical plant provides effective accountability. Measurements should be linked via computer software to production as well as to weather to result in maximum feedback.

Reprinted from the *Kirk-Othmer Encyclopedia of Chemical Technology,* 4th ed., Vol. 9, John Wiley & Sons, New York, 1994.

BIBLIOGRAPHY

"Energy Management" in *ECT* 3rd ed., Vol. 9, pp. 21–45, by A. F. Waterland, Waterland, Viar & Associates, Inc.

1. *Manufacturing Energy Consumption Survey: 1988*, U.S. Department of Energy, Washington, D.C., 1991.
2. *U.S. Chemical Industry Statistical Handbook*, Chemical Manufacturers Association, Washington, D.C., 1991.
3. W. M. Weiss, personal communication, Monsanto Enviro-Chem Systems Inc., St. Louis, Mo., 1991.
4. M. H. Ross and D. E. Steinmeyer, *Sci. Am.* **163,** 88 (1990).
5. *Manufacturing Energy Consumption Survey: 1985*, U.S. Department of Energy, Washington, D.C., 1988.

6. W. F. Kenney, *Energy Conservation in the Process Industries*, Academic Press, Inc., New York, 1984.
7. J. M. Kovacek, *Cogeneration Application Considerations*, General Electric Co., Schenectady, N.Y., GER-3430B, 1991.
8. W. F. Kenney, "Combustion Air Preheat on Steam Cracker Furnaces," *Proceedings, 1983 Industrial Energy Conservation Technology Conference*, Texas Industrial Commission, p. 595.
9. R. Smith and B. Linnhoff, *Chem. Eng. Res. Des.* **66**, 195 (1988).
10. P. S. Gupton and A. S. Krisher, *Chem. Eng. Prog.* **69**(1), 47 (1973).
11. V. Ganapathy, *Waste Heat Boiler Deskbook*, Prentice Hall, Inc., Englewood Cliffs, N.J., 1991.
12. F. S. Pychewicz, "Steam Traps—The Oft Forgotten Energy Conservation Treasure," *Proceedings, 1985 Industrial Energy Conservation Technology Conference*, Texas Industrial Commission, Houston, Tex., p. 392.
13. E. D. Larson and L. J. Nilsson, *ASHRAE Trans.* **97**(2), 363 (1991).
14. D. R. Frikken, K. S. Rosenberg, and D. E. Steinmeyer, *Chem. Eng.*, 86 (June 9, 1975).
15. D. E. Steinmeyer, *CHEMTECH* **188** (Mar. 1982).
16. L. G. Britton and H. G. Clem, *Chem. Eng. Prog.* **87**(11), 87 (1991).
17. T. L. Tolliver, *Chem. Eng.*, 99 (Nov. 24, 1986).
18. K. Nelson, *Chem. Proc.*, 77 (Jan. 1989).

ENERGY POLICY PLANNING (UNITED STATES)

Amy R. Maron
William Lanouette
Gary R. Boss
General Accounting Office

For decades, the federal government has attempted to develop a national energy policy. However, until 1977 when the Department of Energy (DOE) was established, the federal government did not have a framework for consolidating its efforts on energy matters. By the end of the decade, the executive branch's authority over energy and natural resources had expanded, and analytic and forecasting capabilities had been introduced. In addition, the first national energy plan had been developed and a formalized biennial process for planning energy policy had been established under title VIII of the Department of Energy Organization Act (1). See also Energy taxation; Environmental economics.

OBJECTIVES OF THE NATIONAL ENERGY POLICY PLAN

The Department of Energy Organization Act, which created the Department of Energy, also provided a mechanism through which a coordinated national energy policy could be formulated and implemented. This mechanism, in title VIII of the act, requires that the President submit a National Energy Policy Plan (NEPP) to the Congress by April 1 every 2 years, beginning in 1979. Title VIII called for the use of more than 50 kinds of information in developing the plan. The 5 most prominent elements, which incorporate many of these 50 provisions, are

- Holding public hearings to seek the views of a range of interests, including regional, state, and local governments and the private sector.
- Establishing 5- and 10-year energy production, utilization, and conservation objectives that pay particular attention to the need for full employment, price stability, energy security, economic growth, environmental protection, and nuclear nonproliferation as well as special regional needs and the efficient utilization of public and private resources.
- Identifying strategies to be followed to achieve the objectives and outlining the appropriate policies and actions of the federal government to maximize private production and investment in each significant energy supply sector.
- Estimating the domestic and foreign energy supplies on which the United States will be expected to rely and evaluating current and foreseeable trends in the price, quality, management, and utilization of energy resources and the effects of those trends on the social, economic, environmental, and other requirements of the nation; and
- Submitting whatever data and analyses are necessary to support the objectives, resource needs, and policy recommendations of the plan.

By calling on administrations to set objectives; identify the strategies needed to achieve the objectives; project energy supply, demand, and prices; provide the data and analyses to support goals and strategies; and invite public input throughout, the Congress sought a comprehensive approach that uses the most timely and relevant information available. At the time DOE was created, the federal government was not required by law to develop a comprehensive energy policy, nor was much of the information needed to achieve such a task available. In total, there have been six submissions under title VIII: one by the Carter administration (in 1979), four by the Reagan administration (in 1981, 1983, 1985, and 1987), and one by the Bush administration in 1991.

In late 1992, the Congress enacted the Energy Policy Act of 1992 (P.L. 102-486). The act modified title VIII to require that future plans also include a "least-cost energy strategy." This new provision entails developing, among other things, inventories and estimates of the costs of energy resources, preparing a program to ensure adequate supplies of energy, and estimating the life-cycle costs of existing energy production facilities. The act also requires the Secretary of Energy to prepare a report that assesses, among other things, the feasibility of stabilizing and/or reducing greenhouse gases; the extent to which the United States is responding, compared with other countries, to recommendations made by the National Academy of Sciences in its 1991 report *Policy Implications of Greenhouse Warming*; the feasibility of complying with recent international agreements on reducing greenhouse gases; and the potential impacts of policies needed to comply with these agreements. As part of the least-cost energy analysis, the Secretary is required to consider this new report in connection with the plan submitted under title VIII.

ENERGY PLANNING BEFORE 1977

Comprehensive national energy planning had been considered by the federal government for 40 years. However, it was the oil price and supply shocks of the 1970s that prompted the most concerted planning and government reorganization efforts regarding energy in this century—efforts designed largely to reduce the nation's dependence on imported oil through consideration of all energy sources and through conservation. These events eventually led to the title VIII provisions in the 1977 act.

As early as 1939, a report by an Energy Resources Committee, formed as part of the Temporary National Economic Committee, noted that all energy resources are closely interrelated and require the systematic attention of government (2). The report recommended that a "national energy resources policy" be prepared that would address these interrelationships rather than focusing on a fuel-by-fuel approach. Other comprehensive efforts followed, including those by the Department of the Interior and the National Security Resources Board from 1947 to 1949, the 1950–1952 Materials Policy Commission (known as the Paley Commission), the 1955 Cabinet Committee on Energy Supplies and Resources Policy, the U.S. Senate's 1961 National Fuels and Energy Study, and the Johnson administration's 1964 *Resource Policies for a Great Society: Report to the President by the Task Force on Natural Resources*.

Those who interacted with the federal government on energy matters found a lack of cohesion in the federal structure. For example, an oil executive said in 1972 that "the present dispersion of effort among some 61 different government agencies creates delays and confusion and will inevitably tend to accentuate whatever energy shortages may lie ahead." (3) While complaining about fragmentation and inefficiency, this executive acknowledged a strong energy policy role for the executive branch. In response to growing concerns about the dispersion of energy functions, President Nixon outlined a "comprehensive, integrated national energy policy" in April 1973 and in June established an Energy Policy Office to consolidate energy policy planning in the White House.

By the early 1970s, energy policy had crossed a watershed when, for the first time, supplies were inadequate to meet demand at prevailing energy prices. The role of the federal government shifted from support for energy-production industries to the regulation of increasingly (and temporarily) scarce and costly resources, particularly oil. By the spring of 1974, the Arab oil embargo that followed the November 1973 Arab-Israeli war had caused oil prices to climb to four times their level at the beginning of 1973. In early 1974, the Nixon administration launched "Project Independence," a research and development plan aimed at greater energy self-sufficiency for the nation.

During the same decade, increased concern about whether the nation would have adequate energy supplies at affordable prices to meet growing demand also prompted interest in employing economic and engineering analysis to forecast future energy needs and assist with policy decisions. In 1974, the Federal Energy Administration was created. Economic modelers at the Federal Energy Administration developed an 800-page draft *Project Independence Report*, using an analytical approach that "marked a turning point in the assessment of policy options" (4). Later that year, President Ford created an Energy Resources Council charged with developing a "single national energy policy and program." In November 1974, the final *Project Independence Report* was released.

In an April 1977 speech, President Carter declared the policies necessary for increasing energy self-sufficiency to be the "moral equivalent of war." Within three months of taking office, his administration had prepared the "National Energy Plan," (often referred to as "NEP I") (5). The plan was developed by a presidential team of economists, public administrators, and attorneys and was accompanied by the legislation to implement it.

In August of 1977, the Congress enacted the Department of Energy Organization Act, creating a single cabinet-level department responsible for overall coordination of energy programs. The act also established, in title VIII, a governmental planning function as a permanent mechanism for developing and implementing a national energy policy.

Under title VIII, the biennial energy plan is submitted by the President, while the Secretary of Energy is the principal adviser on formulating the plan and the principal agency head who defends the plan before the Congress. The Senate report explaining title VIII's planning provisions said that the plan should integrate the many disparate viewpoints within the executive branch. For all title VIII submissions, the Department of Energy has played a leading role as the integrator of views within the executive branch.

The 1977 National Energy Plan served as a kind of model for title VIII because it incorporated many of the criteria and exercises later codified in the statute. For example, it set the following goals for 1985: to reduce the annual growth rate of energy demand to below 2%, reduce gasoline consumption by 10%, cut U.S. oil imports from 16 to 6 million barrels a day, establish a 1-billion barrel Strategic Petroleum Reserve in the United States to protect against supply disruptions, increase coal production by two-thirds, insulate to minimum energy efficiency standards 90% of American homes and all new buildings, and use solar energy in more than 2.5 million homes.

This strategy reflected a time when many in the administration and the Congress believed that energy policy needed greater federal planning and control. Domestic oil production had been steadily declining from 1972 to 1976, while imports as a percent of consumption had risen from almost 29% to almost 42% during the same period. Two weeks before the 1977 plan was submitted to the Congress, the President had declared that the "energy crisis" was the "greatest challenge that our country will face during our lifetime." The administration's centralized approach to energy problems set the tone for the title VIII provisions that were subsequently enacted.

NATIONAL ENERGY PLANS HAVE VARIED SIGNIFICANTLY

The extent to which past energy plans conformed with the provisions of title VIII varied significantly from administration to administration. None of the energy plans submitted fully addressed the statute's provisions; the 1979 and 1991 plans represented the most thorough efforts to incorporate title VIII's provisions. Most plans contained strategies for achieving general goals, and all discussed energy trends, although none set 5- and 10-year objectives specified in the act. Furthermore, few plans offered analysis supporting their statements. Public hearings to solicit a range of input were held for all but one of the plans.

The primary factors that influenced how the administrations approached planning were (*1*) their differing views on the role of government in energy planning, supply, and price regulation and (*2*) developments in the nation's energy situation. Table 1 illustrates the wide variation in how the plans conformed with title VIII's provisions.

NO PLANS SET SPECIFIC 5- AND 10-YEAR OBJECTIVES

Although title VIII calls for 5- and 10-year energy production, utilization, and conservation objectives, no plans followed this approach. The only time such objectives were established was before the law was enacted: in the 1977 National Energy Plan. As noted earlier, this plan set numerous objectives to be achieved for 1985. Each subsequent plan instead developed its own unique approach to objective-setting.

1979: National Energy Plan II

The first plan submitted under title VIII's provisions, in 1979, did not specifically set 5- and 10-year objectives for energy production, utilization, and conservation. Instead, the plan, known as NEP II, reiterated the President's 1977 vision of a long-term energy transition, declaring broad objectives for the three periods of the transition: near-term (1979–1985), mid-term (1985-2000), and long-term (2000 and beyond). The plan's near-term objective was to reduce the nation's dependence on foreign oil and vulnerability to supply interruptions. The mid-term objectives were to (*1*) seek to keep imports sufficiently low to protect national security and extend the time it took before world oil demand reached the limits of production capacity and (*2*) develop the capability to use new, higher-priced technologies as world oil prices rose. The long-term objective was to use renewable and essentially inexhaustible sources of energy to sustain the economy. The plan mentioned specific times for achieving only a few objectives, such as reducing oil import demand by 5 percent of the consumption in International Energy Agency countries by the end of 1979 (the result of an agreement that year by the agency's governing board) and curbing the federal government's energy use by 5% in the year ending March 31, 1980.

By the time the 1979 plan was completed, major elements of the legislative package that accompanied the 1977 National Energy Plan had been enacted. As a result, few new strategies were announced in the 1979 plan; almost all of the specific strategies in the plan represented implementation of the newly authorized programs. The plan included a few new strategies. One strategy called for developing a 10-year energy conservation plan for federal buildings. Another strategy envisioned new legislation to phase out controls on domestic crude oil prices and estab-

Table 1. National Energy Policy Plans, 1979–1991

Title of Plan, Date Submitted, and Administration Responsible	Conformity with Title VII's Provisions				
	5- and 10-Year Energy Objectives	Strategies to Achieve Objectives	Projections of Energy Prices and Supplies	Supporting Analyses	Public Hearings
National Energy Plan II (NEP II), May 1979: Carter administration	No	Strategies to achieve general objectives only	Yes	Partial	Yes
Securing America's Energy Future: The National Energy Policy Plan (NEP III), July 1981; Reagan administration	No	Strategies to achieve general objectives only	Yes	None	Yes
The National Energy Policy Plan, Oct. 1983; Reagan administration	No	Strategies to achieve general objectives only	Yes	None	Yes
The National Energy Policy Plan, Mar. 1986;[a] Reagan administration	No	Strategies to achieve general objectives only	Yes	Partial	Yes
Energy Security: A Report to the President of the United States, Mar. 1987; Reagan administration	No	Weighed pros and cons of alternative strategies	Yes	Partial	No
The National Energy Strategy, Feb. 1991; Bush administration	No	Strategies to achieve general objectives only	Yes	Partial	Yes

[a] The plan that was due in 1985 was not submitted until 1986. However, in this article this plan is referred to as the 1985 plan.

Note: No plan was submitted for the 1989 requirement. DOE officials told us that a summary of the public hearings prepared for the 1991 plan was used to meet the 1989 requirement and issued in April 1990.

lish a tax on the so-called windfall profits resulting from the decontrol of oil prices. The proceeds of this tax were to be assigned, in part, to subsidize mass transit and alternative energy technologies.

1981: Securing America's Energy Future: The National Energy Policy Plan

The 1981 plan, known as NEP III, did not include 5- and 10-year energy production, utilization, or conservation objectives. Instead, it offered "guiding principles" for energy policy that included both broad objectives and specific strategies such as

- Recognizing that even though efficient displacement of imported oil is an important objective, achieving a low level of U.S. oil imports at any cost is not a major criterion for the nation's energy security and economic health.
- Cooperating with international oil partners.
- Increasing oil stockpiles against potential disruptions in world markets.
- Eliminating controls or other impediments that could discourage the private sector from dealing with disruptions efficiently should they recur.
- Implementing the President's Economic Recovery Program.
- Refocusing federal spending for energy-related purposes to cases in which the private sector is unlikely to invest.

The 1981 plan also briefly discussed other forthcoming actions, chief of which was the implementation of the President's Economic Recovery Program. To address the special needs of the poor that result from higher energy prices, the plan counted on the Economic Recovery Program to deal directly with the problems of inflation and unemployment. The plan also described as forthcoming actions a review of regulations affecting the coal and electric utility industries and improvements to the licensing process for nuclear power plants.

1983: The National Energy Policy Plan

The 1983 plan also lacked 5- and 10-year energy objectives. Instead, it contained a single objective, two strategies for pursuing that objective, and a discussion of specific federal programs and actions determined by those strategies. The plan's objective was broadly to "foster an adequate supply of energy at reasonable costs." The strategies to meet these objectives were, according to the plan, to minimize federal control and involvement while maintaining public health and safety and environmental quality, and to promote a balanced, mixed energy resource system responsive to both domestic and international market forces that would also protect U.S. national security interests.

Describing the 1983 plan's approach to title VIII's provisions on 5- and 10-year objectives, a former DOE official who worked on the 1983 plan explained that the Department made a deliberate decision not to develop a "prescriptive set of domestic regulations," but instead structured the plan as a statement of policy, rather than as a plan. The administration, the official stated, believed that market forces rather than mandates would lead to socially desirable results for energy. As an example, he pointed to the administration's early opposition to setting minimum efficiency standards for appliances. Although appliance standards were eventually set by the Congress, the administration's approach of not setting general energy standards in the plan was not challenged and continued to be used in successive plans.

1985: The National Energy Policy Plan

The 1985 plan, like the three plans before it, did not set 5- and 10-year objectives and therefore did not set specific strategies to achieve objectives. However, continuing the trend begun in 1981, the plan discussed strategies to achieve the general objective of adequate supplies of energy at reasonable costs. These strategies were similar to those described in the 1983 plan: opening up additional offshore oil and gas fields; eliminating federal price controls on domestic natural gas production; redirecting research dollars to basic, rather than applied, science; and introducing legislation to reform licensing and regulation of nuclear power plants.

1987: Energy Security

The 1987 plan, entitled *Energy Security: A Report to the President of the United States*, also did not establish 5- and 10-year objectives. According to a former DOE official, because the report was prepared with the purpose of examining the domestic and international dimensions of energy security, it was not intended as a prescriptive document containing specific 5- and 10-year energy goals or strategies. The report did not purport to make any choices but rather to show what choices could be made, the former official said.

These choices were the strategies that could be used to reduce vulnerability to supply disruptions and enhance the domestic oil industry. For each strategy, the report weighed the pros and cons. However, no specific changes to existing law were proposed along with the report. The report raised, in general terms, the administration's belief that a revised regulatory framework would allow natural gas to compete more freely with other fuels, but it discussed few specific actions for making such revisions. The report also noted that DOE would pay increasing attention to electricity policy, although no specific steps were enumerated. The report included a general review of potential changes in federal requirements and regulations to increase development and improve the transportation of coal. More specific was a discussion of a four-part initiative for nuclear power, including licensing reform, federal-industry cooperation on reactor design, rate regulation, and creation of a repository for high-level waste disposal.

1991: National Energy Strategy

The Department of Energy had intended to develop a plan to meet the 1989 requirement. But the report submitted in April 1990 was an interim report on the 1991 plan, containing only a summary of public opinion but no proposed objectives and strategies. However, the 1991 title VIII submission, known as the National Energy Strategy (NES), had an overall objective that had been set by the President in July 1989: "achieving balance among our increasing need for energy at reasonable prices, our commitment to a safer, healthier environment, our determination to maintain an economy second to none, and our goal to reduce dependence by ourselves and our friends and allies on potentially unreliable energy suppliers." The plan also had more distinct objectives and approaches for energy production, utilization, and conservation than any plan before it. These goals included diversifying sources of oil supply outside the Persian Gulf, increasing energy efficiency, and enhancing environmental quality. Although few goals were identified with a specific date for their achievement, one objective was to have an operating nuclear fusion demonstration plant by about 2025 and an operating commercial plant by about 2040.

Although no 5- and 10-year timetables were included for most of its objectives, the NES gave more attention than any previous plan to identifying specific strategies that might achieve the plan's many distinct objectives. The plan included approaches for improving electricity generation and use in commercial and residential sectors, allowing private access to oil and gas on federal lands, and increasing the ability of natural gas to compete with other fuels. The plan also listed administrative actions and legislative proposals for a number of areas, and the administration submitted a major legislative package to the Congress shortly after the NES was released.

ALL PLANS DESCRIBED ENERGY TRENDS

As set forth in title VIII, all of the plans provided estimates of energy supply and demand for the major energy sources (oil, gas, coal, nuclear, and renewable energy) and descriptions of current and foreseeable trends in world oil prices. Most plans discussed, in varying depth, the effects of these trends on the economy. Only the 1991 plan discussed the impact of energy trends on the environment.

The 1979 plan's evaluation of energy trends was contained in three statistical appendixes published during the year after the plan was released. These appendixes predicted energy supply, demand, and prices to the year 2000. The appendixes also analyzed likely macroeconomic impacts of the President's April 1979 oil price decontrol and tax proposals (on inflation, employment, the trade balance, and household expenditures on energy in 10 regions of the country and for nine income-group levels). The appendixes also included detailed input-output model calculations showing the likely effects of the new proposals on 157 sectors of the economy.

A separate series of reports in support of the 1981 plan evaluated current and foreseeable energy trends and their effects on the economic health of the nation. These reports presented the macroeconomic impacts of energy prices for the period 1980–90; described energy and economic interaction in 1973–1980; analyzed the effects of energy price changes on various income groups; and provided estimates of energy prices, consumption, and supply. All projections assumed "the continuation of existing programs and policies." The reports stated that "*since the projections take into account only policies or programs that were in effect as of June 1981, the projections should not be viewed as a statement of Administration energy goals* [italics in original]. The plan projections are a starting point, or 'base case' for evaluating the potential impacts of new energy initiatives or developments."

The 1983 plan summarized projections of oil prices, energy consumption, energy production, and primary electricity inputs. These projections were presented in a technical report, *Energy Projections to the Year 2010*. The plan stated, however, that these projections "do not necessarily represent Administration policy or the beliefs of the President or the Secretary of Energy." A second technical report in support of the plan, *Energy Activity and Its Impact upon the Economy*, analyzed the macroeconomic effects of projected high and low energy prices using two different economic models.

The 1985 plan also contained forecasts for oil prices and energy consumption and production. A technical report, *National Energy Policy Plan Projections to 2010*, summarized expected world oil prices and U.S. energy supply and demand. Like the 1979 plan, the 1985 plan showed energy trends that might result if certain elements in the administration's energy policy were accepted. These elements included comprehensive decontrol of the natural gas market, increased rates of leasing federal lands for energy development, and a focus of federal research and development on long-term development rather than on subsidized commercialization.

The 1987 plan also provided an analysis of future energy trends, but it did not describe the potential effects of proposed administrative or legislative actions. Instead, it used two general scenarios. (1) A higher world oil price, indicating less dependence on imported oil, and (2) a lower world oil price, indicating greater dependence. These scenarios were used to generate projections of energy consumption, production, and imports for 1985–1995, by both world region and end-use sector. The 1987 report contained the most comprehensive worldwide projections to date; the trends were developed through an interagency process that included analysts from DOE and other federal agencies.

The 1990 *Interim Report* on the National Energy Strategy compared future energy production, consumption, and utilization trends from the Energy Information Administration and academic and interest groups. Then, in 1991, the NES presented DOE's projections of energy prices, supply, and demand for 1990–2030. This plan also described trends in environmental emissions from energy sources. In addition, it provided possible scenarios with and without many of the proposed strategies. A technical annex published in July 1991 explained some methodologies and assumptions used for these projections.

PLANS DID NOT ALWAYS PROVIDE DATA AND ANALYSIS

Title VIII also states that the plan should be accompanied by "whatever data and analysis are necessary to support the [plan's] objectives, resource needs, and policy recommendations." Title VIII did not specify what types of data and analysis should be submitted or at what time, but the

Senate report on title VIII suggested that "the relevant data and analysis necessary to demonstrate the feasibility of achieving the plan's objectives and to support the estimates made in the plan" be included in a report "accompanying the plan." The types and timing of data and analysis included varied from plan to plan.

One approach to providing data and analysis to support the plan's objectives was to project future outcomes—such as impacts on the economy, energy supply and demand, and the environment—should the proposals in the plans be adopted. This approach was used in the 1979, 1985, and 1991 plans.

Neither the 1981 nor the 1983 plan included analysis to support its purposes. But as described earlier, by projecting possible results of the administration's proposals, the 1985 plan provided the data and analyses to support some of the plan's objectives (decontrolling natural gas, leasing of federal lands, refocusing federal research and development).

Although the 1987 submission discussed a wide range of theoretical strategies, an appendix to the report presented the data and analyses used to support only one proposal under consideration: a hypothetical oil import fee. The administration rejected this proposal because its analysis showed adverse macroeconomic impacts.

In 1991, the projections in the NES that reflected proposed administration strategies concerned (*1*) changes in the mix of fuels used in generating electricity or in transportation, (*2*) changes in oil import levels, (*3*) the impacts on electricity prices of clean coal technologies, and (*4*) changes in the levels of emissions and effluent from various pollutants. The plan presented more data and analyses to support its strategies than had any previous plan. Nevertheless, the NES lacked analytical support for several of its strategies. DOE has subsequently published four technical annexes containing analytical support for the proposals. In July 1991 testimony on the process DOE used in developing this plan (6), the U.S. General Accounting Office (GAO) stated that the administration had not published alternative analyses that it had examined in developing the plan's policy options, such as those examined at the request of the cabinet-level Economic Policy Council. At the time, the GAO stated that disclosure of all the relevant analyses conducted for the plan would have provided the Congress with better information to judge the relative merits of various energy policy proposals.

PUBLIC HEARINGS WERE USUALLY HELD

Another title VIII objective is active participation through public hearings. Except for the 1987 plan, each administration held public hearings to solicit input for its national energy plan. The public-hearings process for the 1991 National Energy Strategy, conducted over an 18-month period, was the most extensive effort in this regard.

For the 1979 plan, DOE conducted six seminars in Washington, D.C., to obtain the views of several principal constituency groups concerned with energy policy. Represented were energy producers and consumers, state and local government agencies, large and small businesses, large industrial energy users, environmental groups, and labor. These seminars were followed by public hearings in six cities: Boston, Dallas, Denver, Omaha, San Francisco, and Washington, D.C. Members of Congress, representatives of state and local governments, and environmental and other interest groups were also consulted. An appendix to the plan summarized the public participation.

For the 1981 plan, DOE held public hearings to solicit views of minority groups in Atlanta, Boston, Dallas, Denver, Kansas City, and San Francisco. The plan provided a supplemental report on these hearings. DOE sought comments for preparation of the 1983 plan in an invitation to hearings published in the *Federal Register*. Hearings were held in seven U.S. cities. The comments of the 136 persons who testified and the 111 who submitted written responses were summarized by general topic in the final plan. The plan reported that "energy security and emergency preparedness" were "mentioned frequently," while "some believe" in "free-market forces" to ensure fair and equitable distribution of energy supplies during an emergency. Similarly, the plan reported that while "many respondents expressed concern" about the role of nuclear energy in the nation's future and "questioned the continued funding" in light of "the Administration's commitment to free-market forces, others supported the development of nuclear power."

For the 1985 plan, DOE received 275 letters in response to a notice published in the *Federal Register* soliciting comments on energy policy. Public hearings were held in seven U.S. cities, and 124 people testified. In a two-page summary, DOE grouped public comments into five broad categories. Of these five categories, conservation was the issue most often cited, specifically, in support of continued federal funding of residential conservation and energy-efficiency programs and extension of energy tax credits. The summary stated that some speakers encouraged a "more market-based approach." Comments on fossil fuel and environmental issues were said to include support for offshore oil and gas leasing, full deregulation of natural gas, and increased research into nuclear waste disposal and acid rain. Nuclear energy prompted a range of suggestions as well, from streamlining licensing procedures to lessening federal promotion of nuclear power plants.

For the 1987 plan, DOE made a limited effort to obtain public participation. A *Federal Register* notice soliciting public comments resulted in 50 submissions from state and local government officials, trade associations and industry representatives, public interest groups, university and research organizations, and private citizens. A second *Federal Register* request for comments on a "Study of Crude Oil Production and Refining Capacity in the United States" produced 28 written comments. But no public hearings were held to provide input to the report.

Public hearings were a major feature of the process for developing the 1991 NES. In April 1990, a year after the 1989 plan was due, DOE released the *Interim Report* on the NES (a summary of these hearings). As noted earlier, the *Interim Report* contained no proposed objectives, strategies, or data and analysis to support such objectives.

For the NES, DOE held 18 sessions in 14 cities, more hearings than had been held for any previous plan. Over 499 witnesses from 43 states, two U.S. territories, and two Canadian provinces appeared, and DOE received more than 2,000 written submissions. When DOE released its *Interim Report* on April 2, 1990, it said the report was not intended as a first draft of the NES but as a step in building a national consensus.

ADMINISTRATIONS' VIEWS AND ENERGY CONDITIONS SHAPED PLANNING APPROACHES

The different approaches taken in the title VIII submissions prepared through 1991 largely reflected the specific views of each administration regarding the proper role for the federal government in energy policy. Changes in the nation's energy situation itself during the last 15 years have also influenced approaches to energy planning.

Stronger Federal Role Was Sought During the 1970s

A belief that a federal government plan could reduce dependence on foreign oil supplies and develop domestic alternatives shaped the national energy plans produced in 1977 and 1979. According to an energy analyst who witnessed much of the debate at the time, a prevailing view was that "if the government sat down and thought ahead, then we would have a better energy policy for the nation." Developing models which portrayed future scenarios and seeking public opinion, according to this expert, were also believed to contribute to a better energy policy.

The 1979 plan was issued about a month after the end of the Iranian oil embargo that had led to rapidly increasing world oil prices. Reducing dependence on foreign oil and the vulnerability to higher world oil prices were the principal energy concerns of the administration, and the 1979 plan described the economic, political, and strategic risks posed by continued dependence on foreign oil. At the same time, the plan predicted that U.S. dependence on imported oil would endure, that world oil prices would continue to rise, and that the nation would in turn continue to be vulnerable to oil price shocks or supply shortages.

During the 1970s, as a result of two big oil price shocks and a natural gas shortage, as well as a belief that energy supplies would be inadequate to meet growing demand, a crisis atmosphere shaped U.S. energy policies. These policies, which envisioned an activist government and national energy management, were designed primarily to insulate consumers from high world oil prices and increase energy self-sufficiency. As a result, by the end of the decade a number of demand-reducing (conservation) and supply-enhancing policies were put in place. Conservation measures included oil price controls, natural gas regulation, a tax on windfall profits, taxes on fuel-inefficient vehicles, tax credits for purchases of energy-saving equipment, and weatherization grants for low-income households, schools, and hospitals. To increase supplies of alternatives to imported oil, policies encouraged funding for renewable energy research and development, tax incentives for domestic production, and the development of synthetic fuels. The Department of Energy was created to manage many of these new programs and to fund research and development of renewable resources and conservation.

However, by mid-1979 some policies favoring more government involvement in energy began to be reversed. While the 1979 plan, published in May of that year, proposed continued federal support for alternatives to imported oil, it also proposed less federal intervention in some areas. The removal of federal price controls on domestic crude oil and implementation of the Natural Gas Policy Act of 1978, which deregulated some natural gas, were major features of the 1979 plan.

More Market-Based Approach Was Pursued in the 1980s

During the 1980s, the administration's views about the appropriate role of government in energy matters shifted further, and these views were reflected in each of the plans produced during that decade. The plans revealed the administration's opposition to setting energy goals with specific timetables, intervention in energy markets, and planning for different supply and demand conditions.

The 1981 plan departed from energy policies that had been in place since the 1973–1974 energy crises. The plan stated that "increased reliance on market decisions" rather than a "stubborn reliance on government dictates" was likely to lead to "the most appropriate energy policy." The plan presented a general set of guiding principles that were consistent with the administration's Economic Recovery Program—a plan for federal spending reductions, tax cuts, and regulatory reform. The American economy, not the government, according to these principles, would choose the appropriate energy consumption level for a strong, productive, and secure society in the year 2000. According to the plan, the best guarantee of maintaining a wholesome balance among competing interests in regard to energy lay in reversing policies that insert the government into the energy market and "allow[ing] the American people themselves to make free and fully informed choices."

Furthermore, on December 17, 1981, the President announced his intention to propose to the Congress a reorganization of federal energy programs. Federal efforts to finance the commercialization of energy technologies were to be greatly reduced. Concluding that a cabinet-level department was no longer necessary for managing energy matters, and that other highly critical energy functions (including DOE's nuclear weapons and basic research activities) could be more effectively carried out elsewhere in the government, the President proposed abolishing DOE and shifting many of its principal functions to other federal departments.

Beginning in early 1981 and continuing throughout most of the 1980s, a number of other developments had taken place in energy systems and markets as well as in government policies toward them. One of the most far-reaching of the government energy reforms was the suspension of oil price controls with deregulation of the oil industry in 1981. In addition, during this decade, federal funding for conservation programs and alternative energy resources was reduced and funds were redirected toward basic research. The government-sponsored synthetic fuels and breeder reactor projects were canceled, efficiency standards for automobiles were softened, natural gas prices were further decontrolled, and the windfall profits tax was eventually rescinded.

Federal policies to reduce government intervention in energy markets were further bolstered by simultaneous developments not necessarily related to the administration's energy policies. Oil prices had peaked in 1981, were fully decontrolled by September 30, 1981, and continued to fall until 1987. During the period 1981–86, the world price of oil had dropped from about $50 to about $17 a barrel, adjusted for inflation. From 1981 through 1985, electric utilities cut back on their petroleum consumption. Falling world oil prices between 1981 and 1986 also precipitated a drop in the value of U.S. production. As a result, domestic producers were receiving less value for each barrel of oil.

The 1983 plan attributed several of these developments to an improved national energy situation after 1981. As had the 1981 plan, it rejected the notion of government planning but did indicate that "protecting the environment, maintaining health and safety standards, and improving energy security are appropriate government responsibilities" and that "limited control and intervention may be required to reflect nonmonetary costs to society as a whole of energy production and use."

The 1985 plan, stating that "heavyhanded government planning has been abandoned," noted the success of the "*energy policy planning*" [italics in original] of recent years as "distinct from the earlier efforts at micromanaged energy planning." Repeating the goal of the administration's 1983 plan—that Americans should have an adequate supply of energy at a reasonable cost—the 1985 plan stated that progress had been made since 1981 through a climate of reduced government regulation, fewer controls, lower tax rates, and freer international trade in energy. According to the plan, "Market-oriented policies that build upon America's vast production and conservation resources and its technological genius, free of arbitrary regulation, provide the best hope of maintaining the momentum of energy progress of the past five years."

After the 1985 plan was issued, the U.S. energy situation took another turn, marked by steady increases of oil imports from countries in the politically unstable Middle East, continued plummeting of world oil prices, and a declining domestic oil industry. Moreover, in the spring of 1987, the United States and several of its allies had begun using the military to protect the safety of Kuwaiti oil tankers in the Persian Gulf in Operation Earnest Will.

The 1987 report *Energy Security* was issued in the wake of these events. In presenting the report, the Secretary of Energy stated that it had been written at the request of the President in response to his concern over declining domestic oil production and rising oil imports. When the report was published, imported oil accounted for over 40% of U.S. consumption. According to the report, worldwide oil price reductions, coupled with lower inflation and lower interest rates, had deeply affected the domestic oil industry. The net income of the 22 largest oil companies had been cut in half between 1985 and 1986, and exploration expenditures and active drilling by oil companies had declined by more than 40%. The most pressing question raised by the oil price collapse, according to the report, was what would happen if the United States and its principal allies and trading partners became much more dependent on oil supplies from the Persian Gulf region and from other member countries of the Organization of Petroleum Exporting Countries (OPEC). Because the administration was primarily concerned about rising oil imports and not about developing a national energy policy plan, the report made no specific proposals but rather weighed the costs and benefits of a range of policy options for meeting the nation's energy security objectives.

1991 Plan Responded to Concerns About Oil Vulnerability, the Economy, and the Environment

Concerns about the vulnerability of the United States to "sudden, dramatic changes in world oil prices" and about maintaining a "safer, healthier environment" and "an economy second to none" shaped the NES issued in 1991. Among the steps toward achieving these goals, the plan emphasized diversifying world oil supplies, increasing funding for renewable energy programs, extending renewable energy tax credits, and implementing the Clean Air Act Amendments of 1990.

Although oil imports as a percentage of U.S. consumption continued to rise in the latter half of the 1980s, concerns since then have been less about the level of these imports than about the vulnerability of the United States economy to oil price shocks and developing ways to mitigate the effects of such shocks. The NES stated that the United States would continue to be heavily dependent on oil imports into the next century, but that alternatives to Persian Gulf supplies in particular, such as oil from Western Hemisphere sources, domestic oil supplies, and alternative fuels, should be developed.

Several energy programs that had been scaled back during the 1980s were revived somewhat during 1991 and 1992. Between fiscal years 1980 and 1990, appropriations for DOE's renewable, fossil, nuclear, and conservation research and development programs fell by 83, 50, 68, and 34%, respectively. The administration's budgets in fiscal years 1991, 1992, and 1993 all proposed increases above the previous year's levels in these programs.

Despite these changes, the NES largely repeated a pattern of opposition to government energy planning, goal setting, and intervention in energy markets. In fact, as noted earlier, this document was not called a National Energy Policy Plan but rather a National Energy Strategy, although DOE officials told us it was submitted in response to title VIII. A DOE official in the Office of the Assistant Secretary for Domestic and International Energy Policy who worked on the plan stated that setting objectives or goals represented a kind of "command and control" approach to energy policy that the administration opposed. The NES reflected the administration's philosophy that, wherever possible, markets should determine energy prices, quantities, and technology choices, and that when markets fail to do so, government actions should be aimed at removing or overcoming barriers to efficient market operation. For example, the plan proposed removing regulations that prohibited energy exploration in the Arctic National Wildlife Refuge and the Outer Continental Shelf, further decontrolling natural gas, and accelerating the licensing of commercial nuclear power plants.

ENERGY PLANNING IS A CHALLENGING PROCESS

Developing consensus on energy policy can be a contentious activity, as administrations have learned in 15 years of responding to title VIII. Addressing title VIII's many provisions requires coming to grips with many complex issues, such as conflicting national goals, differing agency missions, and regional disparities in energy supplies. The difficulty of resolving these and other issues does not fully explain why plans have not addressed all of title VIII's provisions, but it does illustrate the challenge of completing the planning exercise within a short time.

Balancing Conflicting Goals and Values Is Difficult

Preparing energy plans requires coping with multiple and often conflicting goals and social values. Under title VIII, energy plans are to consider the "needs for full employment, price stability, energy security, economic growth, environmental protection, nuclear non-proliferation, special regional needs, and the efficient utilization of public and private resources." In the 1979 plan, the administration characterized these conflicts as "a complex tangle of sometimes competing national goals: market efficiency and greater production, equity among income classes and regions, environmental protection, national security, economic growth, and inflationary restraint. It will be difficult, and sometimes impossible, to reconcile all these goals." In 1983, the administration's commitment to securing an "adequate supply of energy at reasonable costs" signified a belief that Americans had abundant, affordable energy but that financial assistance to low-income Americans might be necessary. During the first of many hearings in 1991 on comprehensive energy legislation, the president of Cambridge Energy Research Associates and author, Daniel Yergin, observed that over the last 20 years energy policy has pursued three sometimes contradictory objectives—"cheap energy, secure energy, and clean energy" (7).

Conflicting goals also occur in other aspects of energy policy. Fundamental differences between the interests of consumers and producers blend with geographic disparities to make agreement on energy policy contentious. For example, most domestic oil is produced in the Pacific and West South Central states or offshore (in the Gulf of Mexico, in Alaska, or on the West Coast). But the Middle Atlantic, South Atlantic, and East North Central states, which produce almost no oil, account for 41% of U.S. consumption, giving rise to questions of fairness.

Energy Policy Involves Multiple Federal Agencies

Interagency conflicts arise from the multiple goals that need to be addressed by a comprehensive energy policy. For example, a Department of Energy or Department of the Interior program to encourage energy production may conflict with the Environmental Protection Agency's missions to protect the air, land, and water.

Furthermore, despite the consolidation of administrative and information-gathering functions within the new Department of Energy in 1977, important policy activities continue to take place elsewhere within the executive branch. At the time DOE was created, energy functions were spread among 20 executive departments and agencies, and more than 100 energy data programs were run by four separate entities. While some of this fragmentation has been corrected, jurisdiction over energy exploration on federal lands and the Outer Continental Shelf remains with the Department of the Interior, fuel-efficiency standards for motor vehicles are the responsibility of the Department of Transportation, and tax policy affecting both domestic and imported energy sources is initiated and regulated by the Department of the Treasury.

Cognizant of these diverse authorities, the drafters of title VIII placed the responsibility for developing plans not with the Secretary of Energy but with the President. Traditionally, however, the Secretary of Energy has managed the process of developing a plan.

Other National Issues Compete With Energy for Attention

A final reason for the difficulty of creating an energy plan is that energy policy is an adjunct to so many other policy matters. Energy serves as a medium or a catalyst to social, political, and economic activities and, as such, is inextricably linked to the complexity of the nation's heterogeneous society. Yet, unless policymakers perceive a crisis, such as after the first oil shock in 1973–1974, or when increased reliance on imported oil threatened the domestic industry in 1986–1987, or when some oil supplies were lost during the Persian Gulf War in 1990–1991, energy does not always receive consistent and focused attention.

A NATIONAL ENERGY PLANNING PROCESS IS USEFUL

Despite the inconsistencies in the approaches taken in past energy plans, experts we interviewed believed that the process of preparing energy plans is beneficial. Energy planning gives the federal government an opportunity to periodically assess long-term energy needs and develop a "base case" against which to weigh future decisions. It also provides a forum for competing interests to express their views and have them challenged.

GAO believes that title VIII provides a useful framework for achieving these planning benefits. However, the frequency with which plans must be submitted and the timing of those submissions under title VIII have not contributed to more effective planning. Once an administration prepares a comprehensive plan setting forth its goals and strategies, little is gained by repeating the full planning exercise two years later unless a new administration or energy developments warrant a new plan. It will be particularly difficult for incoming administrations to fully address all of title VIII's provisions, including additional provisions in the Energy Policy Act of 1992, by April of the first year in office.

EXPERTS SUPPORT THE NEED FOR ENERGY PLANNING

After 15 years of plans with varying approaches, there is no clear agreement among energy experts, agency officials, and congressional staff about the type of plan that may be the most useful to the Congress. Many believe the process itself is the most useful aspect of energy planning. A few including congressional staff, questioned whether the planning process should be continued, given that title VIII's objectives have been disregarded in the past and the Congress does not always find the plans useful.

Despite the inconsistencies of past plans, most of the experts and officials generally believed that a periodic evaluation of energy trends such as prices, supplies, and consumption is an important executive branch function, serving to focus attention and debate on key issues. Most who had an opinion on the question believed that requiring such as evaluation as often as every two years was probably not necessary. In addition, most favored the approach of setting some kind of energy goal or goals. But those interviewed did not all agree that the administration should set goals with specific dates for their achievement and accompanying strategies, as currently specified in title VIII but not included in the plans that have been submitted to date. This wide range of opinion may help explain why title VIII's provision for 5- and 10-year objectives has not been uniformly observed.

Opinions Differ on the Value of Objective Analysis Versus Policy-Oriented Plans

Some experts, including congressional staff, when asked what kind of national energy plan or report they believed would be most useful, stated that a "state of energy" report that also contained projections but that did not serve as a justification for current policy or make additional recommendations might be the most valuable. Such a report, they believed, would present a vision of the future under various energy price, supply, and demand scenarios. The information in such a report would be useful to a wide variety of interests, and those who chose to do so could challenge the data and assumptions. Although the 1987 report *Energy Security* did not address all of title VIII's provisions, it was regarded by some experts we interviewed as highly valuable because it objectively examined the future of the nation's energy needs.

On the other hand, some DOE officials and other experts believed that plans that contained policy statements and specific strategies, such as the 1991 plan, were more useful than those that contained only relatively neutral analyses of future scenarios. One congressional staff member believed that a plan should lay out all possible paths to reach the same goal. A few saw no value in continuing the planning exercise.

Because of the important role energy plays in the economy and national security, and its impact on the environment, the need to develop a national energy policy plan or energy "strategy" (as DOE called the 1991 title VIII submission) was a principal objective of the Department of Energy Organization Act in 1977. In a June 1990 report, GAO stated that DOE's effort to develop a national energy strategy in 1989 was a step in the right direction toward addressing the nation's future energy needs and the environmental and budgetary implications that should be considered when developing energy policies (8).

Experts Believe Goals Are Important to Energy Planning

From the first plan, administrations have chosen not to set 5- and 10-year objectives for energy production, utilization, and conservation. However, most plans did contain one or more goals without dates attached to them. Goals with specific target dates were only set in 1977, before title VIII was enacted. When asked whether objectives with target dates or some other kinds of quantifiable goals should be part of plans, those we interviewed did not agree on any single approach. But they all agreed that the nation needs at least "general energy goals." On the question of the level of specificity that such goals should include (eg, the percentage of oil imports in five years), concerns were raised that (*1*) the choice of specific goals runs the risk of choosing the wrong numbers, (*2*) failing to meet the goals does not necessarily indicate a lack of progress, (*3*) choosing specific goals sets up a debate over how much the nation would be willing to spend to reach them, and (*4*) the goals themselves become more important than the policies employed to achieve them. One argument in favor of specific goals was that since the administration currently has specific goals for education, nutrition, and health, why not for energy? Some of those we interviewed believed that certain environmental goals already imposed or that may be imposed, such as meeting fuel efficiency standards or national targets for carbon dioxide emissions by specific dates, could influence energy policy.

The Process Itself May Be the Most Valuable Part

Some experts noted that the real value of a plan is in the process that the administration must go through to prepare it. Soliciting a wide range of opinions, developing models, weighing alternative policies, setting goals, and choosing strategies may be more important than the final product itself, they argued. The exercise itself stimulates debate on the issues and serves as a forum for discussing alternatives and dealing with competing interests. "The point was that everything was heavily debated; that is exactly what you want in putting together an energy plan," stated a former DOE official in describing the approach taken to develop the 1987 plan.

FREQUENCY AND TIMING REQUIREMENTS DO NOT CONTRIBUTE TO BETTER PLANS

The requirement that plans be submitted biennially has contributed to the differing approaches and inconsistent adherence to title VIII's provisions. Incoming administrations have had difficulty meeting title VIII's provisions in their first year of office and have, in fact, chosen not to address all of title VIII's provisions every two years. In 1981, a new administration provided only a limited response to title VIII in a plan that set forth no specific energy policies or strategies but offered only "guiding principles." In addition, in responding to title VIII's provision for projections of energy prices and supplies, a supplement to the plan, *Energy Projections to the Year 2000*, pointed out that "since the projections take into account only policies or programs that were in effect as of June 1981, the projections should not be viewed as a statement of Administration energy goals. The plan's projections are a starting point, or base case, for evaluating the potential impacts of new energy initiatives or developments."

In response to the requirement that a plan be submitted in 1989, another new administration prepared only a summary of its public hearings and issued it a year late. The administration chose instead to work toward a comprehensive plan, the National Energy Strategy (NES), in time for the 1991 deadline. To explain the approach taken by this administration, one DOE official who worked on the NES said that in January 1989 when the new administration was still choosing staff, the Department was not ready to publish even a general energy policy statement. Yet the deadline for submission of the plan fell three months later. This official and others also acknowledged the time-consuming process of developing the NES. Because of the level of effort involved, DOE officials said that the process for meeting title VIII's requirement in 1993 would be a much more limited effort.

Plans submitted by the same administration every one to two years are not likely to vary much from one year to the next, according to some current and former DOE staff. Another DOE official stated that starting to prepare a new plan immediately after completing one leaves no time for retrospection. In addition, requiring an administration to submit a new plan every two years if no significant changes in the energy situation or in administration policies warrant a substantially revised plan could result in plans that are generally identical.

By the spring of 1993, with a National Energy Policy Plan due on April 1st, the incoming Clinton administration requested a one-year extension of the deadline. Since then, the Energy Department planned to issue a National Energy Strategy in 1995 that is based on the theme of "sustainable development."

CONCLUSIONS

Title VIII was developed at a time when the Congress believed that a biennial, step-by-step planning process in the executive branch would lead to a more efficient national energy policy. But administrations have not always closely followed the process of setting objectives, developing strategies, and projecting energy supply and demand from plan to plan. Each plan has reflected the current administration's philosophy toward energy and toward planning itself as well as the status of energy prices and markets. No plan has fully addressed all of title VIII's provisions.

Developing the required plans is difficult and sometimes contentious because of the time needed to address title VIII's provisions, the fact that energy policy cuts across many issues, and the large number of competing interests that must be considered. Thus, the current requirement that plans be submitted biennially, with the additional analysis the 1992 energy legislation requires, is unlikely to result in the comprehensive planning exercise the Congress intended in title VIII. Changing the frequency and timing to require that plans be submitted every four years, with the first deadline falling during the administration's second year in office, would allow each administration time to develop, and the Congress to review, a thorough energy policy statement.

GAO and most experts agree that there is value in periodically evaluating the nation's energy needs and developing a strategy to address these needs. The process of developing an energy plan serves as a forum for debating and discussing alternatives and for dealing with competing interests.

BIBLIOGRAPHY

1. *Department of Energy Organization Act, P.L. 95-91*, August 4, 1977.
2. C. D. Goodwin, "The Truman Administration: Toward a National Energy Policy," in C. D. Goodwin, ed., *Energy Policy in Perspective: Today's Problems, Yesterday's Solutions*, The Brookings Institution, Washington, D.C., 1981, pp. 6–7.
3. R. H. K. Vietor, Energy Policy in America Since 1945: A Study of Business-Government Relations, Cambridge University Press, Cambridge, UK, 1984, p. 320.
4. N. de Marchi, "Energy Policy Under Nixon: Mainly Putting Out Fires," in Ref. 2, p. 395.
5. *The National Energy Plan*, Executive Office of the President, Office of Energy Policy and Planning, Washington, D.C. April 29, 1977.
6. *Full Disclosure of National Energy Strategy Analyses Needed to Enhance Strategy's Credibility*, GAO/T-RCED-91-76, GAO, Washington, D.C., July 8, 1991.
7. *National Energy Strategy: A New Start*, statement of D. Yergin before the Subcommittee on Energy and Power, House Committee on Energy and Commerce, Feb. 20, 1991.
8. *Energy Policy: Developing Strategies for Energy Policies in the 1990s*, GAO/RCED-90-85, GAO, Washington, D.C., June 9, 1990.

Reading List

H. H. Landsberg, Ed., *Making National Energy Policy*, Resources for the Future, Washington, D.C., 1993.

A. R. Maron, W. J. Lanouette, G. R. Boss, *Energy Policy: Changes Needed to Make National Energy Planning More Useful*, U.S. General Accounting Office, Washington, D.C., GAO/RCED-93-29, April, 1993.

ENERGY TAXATION: AUTOMOBILE FUELS

DAVID GUSHEE
SALVATORE LAZZARI
Congressional Research Service
Washington, D.C.

Over the past few years, increasing emphasis has been placed on the role that alternative motor vehicle fuels might play in reducing oil imports, improving urban air quality, and providing domestic jobs. During the past five years Congress has passed three bills designed to stimulate the technology, economics, and infrastructure needed for these fuels to compete with the incumbent fuels, gasoline and diesel.

The latest of these bills, the Energy Policy Act of 1992, includes tax incentives applied to alternative fuel vehicles and to refueling facilities for alternative fuels. However, no action has been taken to rationalize the disparate highway taxes applied to the different fuels that result from a series of independent actions taken over many decades.

Similarly, at the state level, motor vehicle fuel taxation has been driven by the twin needs of revenue generation and highway infrastructure development. Motor fuels are taxed primarily on a gallonage basis without regard to energy content. Until recently, and even then only in a few states, conscious stimulation of alternative motor fuels has not been a policy goal. In most of the states where stimulation has been a goal, ethanol has been the primary beneficiary, with its use mainly as an additive to gasoline rather than as an alternative fuel.

As a result, the different fuels have widely disparate tax rates. Thus, estimated prices at the retail service station pump or equivalent, particularly when corrected to a common energy content, are affected very differently by highway taxes. Compressed natural gas (CNG) and electricity are favored, electricity by not being viewed as a highway fuel and thus not carrying a highway tax burden at either the Federal or state level, and CNG, (a) by virtue of being taxed at the Federal level only at the newly-imposed additional deficit reduction rate, and (b) by being treated on an equivalent energy content basis by the states. Three others, propane, liquefied natural gas (LNG), and methanol, are significantly disadvantaged, because they have lower energy densities than gasoline (and the energy-equated CNG) and thus pay higher tax rates per unit of energy. Methanol is disadvantaged the most of any alternative fuel.

FEDERAL HIGHWAY TAXES ON MOTOR FUELS

Structure

The Internal Revenue Code (IRC) imposes excise taxes on a variety of motor fuels used in highway transportation, with the tax rates, varying by type of fuel and its use. The traditional fuels, gasoline and diesel, used in highway motor vehicles, are taxed at 18.4¢ per gallon and 24.4¢ per gallon, respectively (Internal §4081 and 4091).

These tax rates have several components: The 18.4¢ rate comprises an 11.5¢ highway trust fund rate (which finances the Federal Highway Trust Fund), a 6.8¢ deficit reduction rate (which is designated to the general fund for deficit reduction), and a 0.1¢ Leaking Underground Storage Tank (LUST) trust fund rate. (The LUST trust fund is a Federal program that finances the cost of cleaning up spills from underground fuel tanks). The 24.4¢ diesel tax rate comprises the 17.5¢ highway trust fund rate, the 6.8¢ deficit reduction rate, and the 0.1¢ LUST trust fund rate (§4091, and §4041(a)).

Many non-highway uses of motor fuels, such as farm uses or commercial uses in stationary motors and some highway uses of motor fuels such as uses by school districts or state and local governments, are tax-exempt. Some non-highway uses of motor fuels are, however, taxed at varying rates also. The 18.4¢ tax rate on gasoline and special motor fuels generally applies to fuels used in noncommercial motorboats. Fuel used for transportation on island waterways by commercial cargo vessels is taxed at 21.4¢ per gallon, rising to 24.4¢ by 1995 (§4042). Fuels used in noncommercial aviation are taxed at either 19.4¢ per gallon (in the case of gasoline, §4041(c), and §4081) or 21.9¢ per gallon for jet fuel (§4041(c) and §4091). Jet fuel used in commercial aviation only pays the 0.1¢ LUST fund tax. However, diesel used in trains is taxed at 6.9¢ per gallon, comprising the deficit reduction rate and the LUST fund rate only.

Liquid special motor fuels such as naphtha, benzene, benzol, casinghead gasoline, and natural gasoline–these are also known as gasoline substitutes–are subject to the 18.4¢ rate if the fuel is used in a highway vehicle for purposes that are not specifically tax-exempt (§4041(a)(2)). Liquefied petroleum gas (LPG) is taxed at 18.3¢ per gallon since it is the only gasoline substitute not subject to the 0.1¢ LUST tax. Where LPG is sold by weight, the equivalent of a gallon for purposes of computing the motor fuels tax is 4.25 pounds per gallon (see Rev. Rul. 71-464, 1971-2 CB 357).

Mixtures of motor fuels and biomass-derived alcohols (either methanol or ethanol) are partially tax-exempt, with the amount of the exemption depending upon the fraction of alcohol that is in the mixture and the type of alcohol. Gasohol mixtures–blends of gasoline and ethanol that are 10% ethanol–are taxed at 13.0¢ per gallon (the exemption is 5.4¢). Under the IRC, blends of gasoline with biomass-derived methanol would also qualify, but such blends are disqualified under the Clean Air Act because of the associated increase of emissions of ozone-forming pollutants. Mixtures that are 7.7% ethanol are taxed at 14.24¢ per gallon (the exemption is 4.16¢). And finally, mixtures that are 5.7% ethanol are taxed at 15.32¢ per gallon (the exemption is 3.08¢ per gallon). In all these cases, the exemption equates to 54 cents per gallon of ethanol used.

The exemption for alcohol fuels mixtures also applies to blends of diesel and biomass-derived alcohol and blends of a special motor fuel and biomass-derived alcohol. Alcohol blended with diesel–sometimes called "dieselhol"–is taxed at the rate of 19.0¢ (the exemption is 5.4¢, the same as for gasoline). Alcohol blended with one of the special motor fuels is assessed a 13.0¢ tax rate (the exemption is again 5.4¢).

It should be noted that in all these cases of alcohol blends, the alcohol must be at least 190 proof (95% pure alcohol) and the alcohol cannot be derived from petroleum, natural gas, or coal (including peat). In the case of the blended fuels, methanol produced from coal or natural gas did not originally qualify for the tax exemptions. This dispensation is because when the ethanol exemption was first introduced in 1977, the Congress believed that the cost of producing methanol from coal and natural gas was low enough without a Federal tax subsidy; whereas, the cost of producing methanol from wood and ethanol from grain was costly and did warrant a subsidy.

Exempt alcohols can currently only be derived from biomass. Thus, while both methanol and ethanol qualify for the exemption, methanol is in effect disqualified because it currently is mostly produced from natural gas. However, in the case of special motor fuels that are 85% alcohol (ethanol or methanol) derived from natural gas, there is a separate exemption of 7.0¢ per gallon (the tax rate is 11.4¢ per gallon (§4041(m)). This partial exemption has the effect for methanol of equalizing the Federal tax rate to that of gasoline on an energy-content basis.

Alcohol fuels mixtures that contain 85% alcohol made from biomass are taxed at varying rates. In the case of 85% ethanol blends, the tax rate is 12.35¢ consisting of a 5.5¢ highway trust fund rate (a 6.0¢ exemption), the 6.8¢ deficit reduction rate, and a 0.05¢ LUST fund rate. In the case of 85% methanol blends, the tax rate is 12.95¢ per gallon, consisting of a 6.1¢ highway trust fund rate (a 5.4¢ exemption), the 6.8¢ deficit reduction rate, and the 0.05¢ LUST fund rate. Note that the LUST fund rate on these 85% mixtures is one-half the rate that applies to all other taxable fuels (4041(b)(2)).

Finally, it is important to underscore the point that the highway motor fuels excise taxes historically applied to liquid fuels only; electricity and gaseous fuels, such as compressed natural gas, were not taxed. This situation is still true of electricity. Compressed natural gas, as a result of the Omnibus Budget Reconciliation Act of Intermodal Surface Transportation Efficiency Act of 1991 (P.L. 102-240). It should be noted, however, that the various motor fuels have always had expiration dates, which have always been extended prior to the actual expiration. The excise tax exemptions for alcohol fuels expire on September 30, 2000.

A summary of the tax rates on the various highway motor fuels is given in Table 1. The first row of numbers in Table 1 presents the tax rates on a per gallon basis, as discussed in the text. As these data clearly show, motor fuel tax rates display wide variation among different fuels.

The second row of numbers in Table 1 shows the energy content per gallon of each of the fuels in British thermal units (Btu). All of the alternative fuels listed have less energy per gallon than gasoline or diesel fuel. The amount of energy in a "gallon" of CNG is about 20 percent of the energy in a gallon of gasoline. The exact amount depends upon the temperature and pressure, which will vary depending on the fuel fill rate and the ambient temperature. Therefore, industry practice is to measure the amount of gas fed into the fuel tank, to calculate its energy content, and to divide the energy content by a factor to convert the energy to billable quantity (eg, therms, which equals 100,000 Btus). The number of therms equivalent to the energy content of a gallon of gasoline is defined by each State. The intent is to equate the energy content of a "gallon" of CNG to the energy content of a gallon of gasoline. Table 1, follows that intent. The higher energy density of gasoline and diesel fuel is one of their advantages compared to the alternatives, in that a given storage capacity for fuel leads to a longer driving range per tankful.

The third row of numbers in Table 1 shows the tax rates adjusted for the energy content compared to that of a gallon of gasoline. For example, since propane contains 85,000 Btu per gallon compared to gasoline's 115,000 Btu, 35% more of it is required to deliver the same quantity of energy as a gallon of gasoline. Thus, the "effective" excise tax rate is 24.9¢ per gallon, 35% more than the 18.4¢ per gallon tax. In Table 1, we have used lower heat values than those used by the Energy Information Administration and by some authors. The difference is whether water vapor is calculated as vapor (which results in lower heat values) or as liquid. The choice of one or the other does not affect the conclusions. Another cause of differences in heat values used in various sources is that all of these fuels are mixtures whose compositions vary from place to place and time to time.

Evolution of Federal Highway Excise Taxes on Motor Fuels

The present structure of Federal excise taxes on motor fuels evolved from three public policy concerns: (1) revenue generation for budget deficit reduction; (2) revenue generation for highway infrastructure financing; and (3) energy policy considerations.

Deficit Reduction. Revenue for purposes of deficit reduction was the rationale for enacting the gasoline tax in 1932 and for several of the many subsequent increases in tax rates as well as extensions of the expiration dates. The Federal excise tax on gasoline was first enacted as part of the Revenue Act of 1932 (P.L. 154), although the first known Federal proposal to tax gasoline goes back to the Wilson Administration (1). The tax, which was initially 1¢ per gallon, was enacted as part of a program of tax increases designed to generate additional revenue to reduce budget deficits, which were looming due to the deepest and longest economic recession in U.S. history. Revenues from the tax were allocated to the general fund for deficit reduction.

Revenue generation for deficit reduction was also the underlying rationale for the gasoline tax rate increases of 1940, 1941, 1951, and 1954. The increases of 1951, which were part of the Revenue Act of 1951, raised the gasoline tax from 1.5¢ to 2¢ per gallon and introduced the tax on diesel fuel, also at the rate of 2¢ per gallon. Revenue generation to help finance the Korean War was an additional reason for these tax rate increases. Revenue generation for deficit reduction was part of the rationale for the tax rate increases of 1990. The Omnibus Budget and Reconciliation Act of 1990 (P.L. 101-508) raised the gasoline and diesel fuel taxes, which had increased to 9.1¢ and 15.1¢ per gallon respectively during the 1980s, by another 5¢ per gallon and authorized that revenues from 2.5¢ of the 5.0¢ in-

Table 1. Federal Highway Motor Fuels Taxes Adjusted for Energy Content

Gasoline	Diesel	Methanol (M100)	Ethanol (E100)	Natural Gas (CNG)	Natural Gas (LNG)	Propane (LPG)	Electricity
Tax per gallon, cents							
18.4[a]	24.4	11.4	13.0	5.6	18.4	18.3	None
Energy content per gallon (Btu)							
115,000	135,000	57,000	76,000	115,000	75,000	85,000	
Tax adjusted for energy content (cents per gallon of gasoline equivalent)[b]							
18.4	20.8	23.0	19.7	5.6	28.2	24.9	None

[a] In addition to the highway tax (18.3 cents/gal) and the LUST Fund tax (0.10 cents/gal), there are two additional taxes on imported petroleum products, including gasoline: the Hazardous Substance Trust Fund (0.23 cents/gal), and Oil Spill Liability Trust Fund (0.12 cents/gal). Domestically produced gasoline carries equivalent taxes, but they are imposed on the crude oil used in its manufacture.

[b] Obtained by multiplying a fuel's "unadjusted" tax by the ratio of per gallon energy content of gasoline to that of the particular fuel.

crease would go toward deficit reduction rather than the highway trust fund. The 1990 law also allowed diesel fuel used in trains to be taxed at 2.6¢ per gallon, with 2.5¢ for deficit reduction and 0.1¢ for the LUST fund. Prior to the 1990 law, diesel used in trains was tax-exempt because it was a non-highway use.

Highway Finance. In 1956, the Highway Trust Fund was created under the Federal-Aid Highway Act of 1956 (P.L. 84-627). This act marked a fundamental change in Federal highway financing–from general revenues to motor fuels taxes. All gasoline tax revenues and most other highway user revenues went into that fund for highway construction finance. The purpose of the trust fund was to finance the cost of the interstate highway system. Thus, revenues to finance highway infrastructure rather than revenue generation for deficit reduction became the primary rationale underlying most of the increases in tax rates and expansion of the tax bases since then. From 1956 to 1982, there were two increases in tax rates, seveal extensions of expiration dates, and repeals of scheduled declines in tax rates. Each of these amendments was made to generate more money for the Highway Trust Fund programs.

Beginning in late 1982, another objective was added to the list. Rather than fiscal deficits or energy security, attention began to focus on the large portion of the roads and highways in this country that had fallen into disrepair, and on the unemployment rate, which had risen steeply as a result of the 1981–1982 economic recession. Between 1982 and 1990 there were four increases in the motor fuels excise taxes. Title I of the Surface Transportation Assistance Act of 1982 (P.L. 97-424) boosted the motor fuel excise taxes by 5 cents per gallon (to 9 cents). The 1982 law also provided that 1¢ of the 5¢ increase would be allocated to a special mass transit account. The Tax Reform Act of 1984 (P.L. 98-369) increased the diesel fuel tax another 6¢ per gallon in return for a repeal of a scheduled boost in truck taxes based on vehicle weights. This made the tax on diesel fuel 15¢ per gallon. A 0.1¢ per gallon tax was added by the Superfund Amendments and Reauthorization Act of 1986 (P.L. 99-499) to pay for the cleaning up of leaking underground storage tanks.

Energy Policy Considerations. Beginning in the 1970s, energy policy considerations began to influence both the level of motor fuel taxation and, more importantly, the structure of tax rates. Reducing petroleum consumption and importation made it easier to support motor fuels excise tax increase proposals, and was the rationale for reducing the tax rates on alternative fuels, particularly alcohol fuels. Proposals to increase the Federal excise tax on gasoline became common during and after the 1973–1974 Arab oil embargo and subsequent rises in crude oil prices. Coming in the aftermath of the 1973–1974 oil shock, such proposals were intended largely to reduce consumption of motor fuels (by raising their prices), and thereby reduce oil imports. Perhaps the most ambitious of these proposals was that of Senator Henry Jackson, proposing to increase the tax by $1.00 per gallon.

The concept of taxing alternative fuels at lower rates, which began in the middle 1970s in response to the first oil shock, was actually realized in 1978 with the enactment of the Energy Tax Act (P.L. 95-618). The Energy Tax Act (ETA) also provided for the gas-guzzler tax, incentives for van pooling, and miscellaneous energy tax provisions. The underlying rationale for the ETA was the perceived failures in the energy markets in allocating resources efficiently and fairly, in coping with the 1973 oil embargo, and in adjusting to the sharp increases in energy prices, the shortages, and the associated adverse economic and social problems. Prior to this law, there were no special exemptions for highway use of alternative motor fuels.

The Federal exemption for alcohol fuels under the 1978 law was for the full amount of the gasoline tax: 4¢ per gallon. The Crude Oil Windfall Profits Tax (P.L. 96-223) extended the 4¢ exemption from October 1, 1984, to December 31, 1992. The Surface Transportation Assistance Act of 1982 (P.L. 97-424) raised the gasoline tax from 4¢ to 9¢ per gallon and also changed the exemption for gasohol from the complete 4¢ exemption to a partial 5¢ exemption (gasohol would be taxed at 4¢ per gallon instead of 9¢ per gallon). The Deficit Reduction Act of 1984 (P.L. 98-369) raised the diesel fuel tax from 9¢ to 15¢ per gallon as part of a compromise that also lowered the highway use taxes on trucks. The 1984 tax law also raised the gasohol exemption from 5¢ to 6¢ (i.e., it reduced the tax rate for gasohol from 4¢ to 3¢), and retained the 9¢ exemption for "neat" alcohol fuels, and provided that alcohol produced from natural gas would also qualify for the exemption.

The Tax Reform Act of 1986 (P.L. 99-514) reduced the excise tax exemption for 85 percent alcohol from 9¢ to 6¢ per gallon (for sales made beginning in 1987). The Technical and Miscellaneous Revenue Act of 1988 (P.L. 100-647) made minor liberalizations to the excise tax rules. Finally, the OBRA of 1990 reduced the alcohol fuels exemption to 5.4¢ per gallon. The 1990 OBRA also introduced a new tax credit for small ethanol producers (less than 15 millon gallons per year).

The Energy Policy Act of 1992 (P.L. 102-486) extended the gasohol excise tax exemption to gasohol that contains less than 10 percent alcohol. Two categories of gasohol mixtures were prescribed: mixtures containing 7.7-percent alcohol; and mixtures containing 5.7-percent alcohol. The exemption for 7.7 percent mixtures is 4.16¢ per gallon; the exemption for 5.5 percent mixtures is 3.08¢ per gallon.

The Omnibus Budget Reconciliation Act of 1993, in addition to imposing a new tax on CNG, raised the tax rate on motor fuels used in highway transportation by 4.3¢ per gallon.

STATE HIGHWAY TAXES ON MOTOR FUELS

Structure

States tax motor fuels through a combination of excise taxes and other fees and taxes. State excise taxes are tabulated in Table 2. Some impose sales taxes on top of the Federal and State excise taxes; others use the pre-tax gasoline price as the base. Many states impose specific-purpose fees such as their own versions of the Leaking Underground Storage Tank tax, inspection fees, and the like. Fifteen states have authorized cities and counties to add their own charges. Communities in 11 of those states have chosen to do so.

Many states have tax policies for ethanol used in motor fuels and propane used as a motor fuel. With respect to ethanol, the special provisions take the form of a waiver of part of the highway tax or a credit per gallon of ethanol produced in the state and used in the motor fuel (as in gasohol, for example) or, in some cases, both. In most of these cases, however, the state does not have a specified policy for taxing ethanol as the primary component of the motor fuel as in, for example, E100 (100% ethanol) or E85 (85% ethanol and 15% gasoline).

With respect to propane, a number of states substitute a fee on vehicles (mandatory in some states, optional in others) based on vehicle type in place of the fuel tax. The fee in most cases comes out close to what the tax would have cost for a vehicle of average mileage. High–mileage vehicles end up paying a somewhat lower tax rate per gallon. These states also have a gallonage tax for out-of-state or nonfee-paying vehicles or, in some cases, require an LPG-fueled vehicle to buy a permit to operate in the state. Two states (Colorado and Texas), now have fee systems for natural gas-fueled vehicles.

Most states, however, do not have special policies for ethanol (as discussed above), methanol, natural gas (either compressed or liquefied), or electricity, since they are too

Table 2. State Tax Rates on Motor Fuels—As of October 1, 1993 (Cents per Gallon)

State	Gasoline	Diesel	Methanol	Ethanol	CNG	LNG	Propane	Electricity
Alabama	18	19	18	18	[c]	18	[c]	None
Alaska	8	8	None	None	[b]	[b]	None	None
Arizona	18	18	18	18	1	18	18	None
Arkansas	18.7	18.7	18.5	18.5	5	5	16.5[c]	None
California	17	17	8.5	8.5	7	6	[d]	None
Colorado	22	20.5	20.5	20.5	20.5[d]	20.5[c]	20.5[d]	None
Connecticut[a]	29	18	28	28	29	29	18	None
Delaware	22	19	19	19	19	19	19	None
Dist. of Columbia	20	20	20	20	20	20	20	None
Florida[a]	11.8	21	11.8	11.8	[c]	[c]	[c]	None
Georgia[a]	7.5	7.5	7.5	7.5	7.5	7.5	7.5	None
Hawaii[a]	16	16	16	16	[b]	[b]	11	None
Idaho	21	21	21	21	21[d]	21[d]	15.2[d]	None
Illinois[a]	19	21.5	19	19	19	19	19	None
Indiana[a]	15	16	15	15	[c]	[c]	[c]	None
Iowa[a]	20	22.5	20	20	18.4	20	20	None
Kansas[a]	18	20	20	None	17	17	17[d]	None
Kentucky[a]	15.4	12.4	15	15.4	15	15	15	None
Louisiana[a]	20	20	20	20	20[d]	20	20[d]	None
Maine	19	20	[b]	[b]	[b]	[b]	18	None
Maryland	23.5	24.75	23.5	23.5	23.5	23.5	23.5	None
Massachusetts[a]	21	21	21	21	21	21	9.6	None
Michigan	15	15	15	15	None	None	15	None
Minnesota[a]	20	20	20	18	20	20	20	None
Mississippi	18.2	18.2	18.2	18.2	18.2	18.2	17	None
Missouri	13.03	13	13	13.03	[c]	13	[c]	[c]
Montana[a]	24	24	24	24	7.49	7.49	[c]	None
Nebraska[a]	24.4	24.4	24.4	24.4	24.4	24.4	23.8	None
Nevada[a]	23	27	23	23	23	23	17	None
New Hampshire[a]	18.7	18.7	[c]	[c]	[c]	[c]	18	None
New Jersey[a]	10.5	13.5	10.5	10.5	5.25	5.25	5.25	None
New Mexico[a]	23	19	23	22	23[d]	[b]	18[d]	None
New York[a]	23.03	25.03	22.84	22.84	22.84	22.84	8	None
North Carolina[a]	22.3	22.3	22.3	22.3	25.6	22.3	22.3	None
North Dakota[a]	17	17	17	17	17	17	17	None
Ohio[a]	22	22	22	22	22	22	22	None
Oklahoma	17	14	17	17	[c]	[c]	[c]	None
Oregon[a]	24	24	24	24	24	24	24	None
Pennsylvania[a]	22.4	22.4	12	22.4	12	12	22.4	None
Rhode Island[a]	28	28	28	28	28	28	28	None
South Carolina	16	16	16	16	16	16	16	None
South Dakota[a]	18	18	16	16	18	18	16	None
Tennessee	21.4	17	21.4	21.4	13	14	14[d]	None
Texas	20	20	20	20	[c]	[c]	[c]	None

Table 2. *(Continued)*

State	Gasoline	Diesel	Methanol	Ethanol	CNG	LNG	Propane	Electricity
Utah	19	19	19	19	19	19	19[d]	None
Vermont[a]	16	17	None	None	None	None	[c]	None
Virginia[a]	17.5	16	17.5	17.5	10	10	10	10
Washington[a]	23	23	23	3.45	[c]	[c]	[c]	None
West Virginia[a]	25.35	25.35	25.35	25.35	25.35	25.35	25.35	None
Wisconsin[a]	23.2	23.2	23.2	23.2	23.2	23.2	23.2	None
Wyoming[a]	9	9	None	None	None	None	None	None

[a] The following states have special provisions:
Arkansas: Natural gas tax rate will increase as number of vehicles using it increases. Breakpoints are 1000, 1500, 2000, 2500, and 3000. At 3000, tax rate will be 16.5 cents per equivalent gallon.
California: Gasoline and diesel tax rates rise to 18 cents per gallon on 1/1/94.
Connecticut: Gasoline tax will go up one cent per gallon on 1/1/94. Propane was given the diesel tax rate on 10/1/93.
Florida: Tax rates are adjusted annually. For gasoline and gasohol, there is also a State Comprehensive Enhanced Transportation System Tax (SCETS) that varies by county from 0 to 4.2 cents per gallon. CNG, LNG, and LPG pay on a decal system; out of state vehicles pay 4 cents per gallon.
Hawaii: Ethanol is exempt from the State's general excise tax of 4%.
Illinois: Motor carriers pay an additional 5.9 cents per gallon on diesel.
Iowa: Ethanol produced in the State receives 20 cents per gallon subsidy.
Kansas: Ethanol is taxed as a motor fuel only when blended as gasohol. Trucks can buy pertrip permits in lieu of highway tax.
Kentucky: Tax rates are adjusted quarterly. A 2 percent surtax is imposed on gasoline and 4.7 percent on special fuels for any vehicle with 3 or more axles. There is an additional 2 cents per gallon on vehicles over 50,000 lbs.
Louisiana: There is a producer credit of $1.40 per gallon of ethanol.
Massachusetts: Tax rates are adjusted quarterly.
Minnesota: There is a credit of 20 cents per gallon of ethanol used to make gasohol. The LPG decal system has been repealed. Taxation by energy equivalence is likely in 1995. A temporary reduced tax rate for E85, CNG, LNG, and LPG is possible.
Missouri: The decal fee increases with vehicle weight. Three State agencies are considering highway tax changes to encourage alternative fuels.
Montana: There is an alcohol distillers' credit of 30 cents/gallon of ethanol produced in qualified facilities in the State.
Nebraska: Rates are adjusted quarterly. There is a producer incentive credit of 20 cents per gallon of ethanol produced in qualified facilities in the State.
Nevada: 125 cubic feet of natural gas or LPG is defined as equal to one gallon of gasoline and subject to the gasoline tax rate.
New Hampshire: Alternative fuel vehicles pay twice the usual registration fee in lieu of the gallonage tax.
New Jersey: There is a gross receipts tax of 4 cents per gallon on gasoline, diesel, and on-road propane. There is also a proposal to reduce the price of natural gas sold for motor vehicle use.
New York: Tax rates are adjusted annually. The rates include a State Petroleum Business Tax not applied to propane. There is a State sales tax of 4% and local sales taxes ranging from 0 to 4.5%.
North Carolina: Tax rates are adjusted semiannually. CNG is taxed at 22.3 cents per 100,000 Btu. The State is considering changing that to a higher energy content gallon equivalency, such as 115,000 or 120,000.
North Dakota: A special excise tax of 2 percent is imposed on all sales of LPG and diesel that are exempted from the gallonage tax if the fuel is sold for use in the State. There is a producer credit of 40 cents per gallon of agriculturally derived alcohol produced in the State and used to make gasohol.
Ohio: Dealers are refunded 15 cents per gallon of each qualified fuel (methanol or ethanol) blended with unleaded gasoline.
Oklahoma: The decal fee increases with increasing vehicle weight.
Oregon: 50% of local property taxes are waived for five years for an ethanol production plant.
Pennsylvania: Motor carriers pay an additional 6 cents per gallon.
Rhode Island: Tax rates are adjusted quarterly.
South Dakota: There is a credit at the rate of the gasoline tax to distributors blending ethanol with gasoline to produce ethanol. There is also a producer incentive payment of 20 cents per gallon of ethanol used in gasohol.
Vermont: Diesel vehicles over 10,000 lbs. pay 26 cents/gallon. Incentive proposals for AFVs have not **South Carolina:** Legislature to consider committee proposals to tax on energy equivalence, with reduced tax for clean-burning fuels, the reduction proportional to the air pollution benefit passed.
Virginia: Favorable tax rates for CNG, LNG, LPG, hydrogen, hythane, and electricity begin on 1/1/94 and run until 7/1/98.
Washington: Decal free for CNG, LNG, and LPG increases with vehicle weight. There is a credit of 60 percent of the gasoline tax for every gallon of alcohol used in gasohol, year-round for small producers (<10 million gallons per year) and for all but the 4 winter months for others.
West Virgina: Tax rates are adjusted annually.
Wisconsin: Tax rates are adjusted annually. The legislature has discussed waiving taxes on alternative fuels, but specific legislation has not moved forward.
Wyoming: All alternative fuels are subject to sales and use taxes. WDoT is studying alternative fuels tax policy.
[b] No tax because State has not addressed tax policy for this fuel.
[c] Registered vehicles using this fuel must pay an annual fee in lieu of the gallonage tax.
[d] Registered vehicles may pay an annual fee in lieu of the gallonage tax.
Sources: Federal Highway Administration. Office of Highway Information Management. Telephone calls to State energy and taxation departments.

new as motor fuels to have received specific attention. In most states, these new alternative fuels have been swept up under a general approach; if liquid, they are taxed on a volume-equivalent basis while, for compressed natural gas, they are taxed on some definition of energy equivalence to gasoline. In a few states, some alternative fuels are not taxed at all, either through lack of attention or because of special policy (particularly the case where a fuel-producing state seeks to develop motor fuel markets for its indigenous fuel). There has not been a standard definition of Btu equivalency be-tween CNG and gasoline. The equivalency depends on the gas composition, the pressure to which it is compressed, and the gasoline to which it is compared. The National Conference on Weights and

Measures had recommended a standard Btu equivalency to gasoline for CNG of 1.14 therms (100 cubic feet at standard atmospheric conditions). The energy content of a therm varies over time by a few percent but averages about 100,000 Btu. This recommendation was recently changed, after a review of the issues, to a standard weight of 4.6601 lbs. of gas as the equivalent of one gallon of gasoline. The recommendation has not yet been adopted by states.

As the impact of the Energy Policy Act of 1992 is felt in the states (which must soon start to buy alternative fuel vehicles for their fleets), administrative and legislative attention to tax policy is increasing rapidly, with a number of states already having taken actions such as substituting up-front registration fees for natural gas (and sometimes propane) for the usual highway tax. Arkansas is the only state so far to introduce a reduced tax on an alternative fuel (compressed natural gas), phasing the tax back toward equivalence with gasoline as the number of natural gas vehicles increases. Several other states are considering similar approaches but are finding resistance from those sectors dependent on highway tax revenues.

Thirty-three states have sales taxes on motor fuels. They range from 2% to over 6%. They are applied differently from state to state–in some cases to the sales price of gasoline less all highway taxes, in other cases to the sales price including all highway taxes. Details are shown in Table 3.

Evolution of State Highway Excise Taxes on Motor Fuels

The present structure of state motor fuels taxes evolved predominantly from the need for revenues to finance highway construction, and secondarily for general revenue purposes, reasons generally true for each state. Energy policy considerations were not a factor in the evolution of the

Table 3. State Sales Taxes on Motor Fuels

State	Percent	Remarks
Alabama	4	Applies to fuel not taxable under gallonage tax laws
Arizona	5	Applies to fuel not taxed under the motor fuel or use fuel taxes
Arkansas	4.5	Special fuel for municipal buses and gasoline are exempt
California	6	Applies to sales price including Federal and State motor fuel taxes
District of Columbia	6	Applies to fuel not taxable under gallonage tax laws
Georgia	4	Applies to sales price including Federal motor fuel tax
Hawaii	4	Applies to sales price excluding Federal and State motor fuel taxes. Alcohol fuels are exempt
Illinois	6.25	Applies to sales price excluding Federal and State motor fuel taxes. For gasohol, only 70 percent of the price is subject to sales tax
Indiana	5	Applies to sales price excluding Federal and State motor fuel taxes
Iowa	5	Fuel on which the gallonage tax was paid and not refunded is exempt. Gasohol is exempt
Kansas	4.9	Applies to fuels not taxable under the gallonage tax laws
Kentucky	6	Applies to sales price, exclusive of Federal tax, of fuels not taxable under gallonage tax laws
Louisiana	4	Fuels subject to gallonage tax are exempt. Gasohol is exempt if alcohol is produced in State
Maine	6	Applies to motor fuel not taxed at the maximum rate for highway use under the gallonage tax laws
Massachusetts	5	Applies to fuels not taxable under the gallonage tax laws
Michigan	4	Applies to sales price including Federal motor fuel tax except for certain multi passenger, for hire vehicles on scheduled routes
Minnesota	6	Applies to fuels not taxable under the gallonage tax laws
New Mexico	5	Applies to fuels not taxable under gallonage tax laws. Ethanol blends deductible under the gasoline tax laws are exempt
New York	4	Applies to sales price including Federal motor fuel tax
North Dakota	5	Applies to fuels not taxable under gallonage tax laws
Ohio	6	Applies to fuels not taxable under gallonage tax laws
Oklahoma	4.5	Applies to fuels not taxable under gallonage tax laws
Pennsylvania	6	Applies to fuels not taxable under gallonage tax laws
Rhode Island	7	Applies to sales price. Gasoline is exempt
South Carolina	5	Applies to aviation gasoline only
South Dakota	4	Applies to fuels not taxable under gallonage tax laws
Tennessee	4.5	Applies to aviation fuel only
Texas	6.25	Applies to fuels not taxed or exempted under other laws
Utah	5	Applies to fuels not taxable under gallonage tax laws
Virginia	2	Applies to retail sales within counties and cities with subway or bus systems owned and operated by transportation agencies
Washington	6.5	Applies to sales price excluding Federal and State gallonage taxes. Alcohol for use as a motor vehicle fuel is exempt
Wisconsin	6	Applies to fuels not taxable under gallonage tax laws
Wyoming	3	Applies to sales price of LPG. Gasoline and diesel subject to gallonage tax are exempt

Source: Federal Highway Administration. Office of Highway Information Management.

structure of motor fuels taxes in the states, although in recent years some states introduced special provisions for ethanol and propane, as was discussed above.

New York was the first state to require the licensing of automobiles. By 1917, every state had similar rules. Oregon enacted the first gasoline tax in 1919 at the rate of 1¢ per gallon thus initiating the policy of user financing of highway spending (2). By 1929, every state in the Union had a gasoline tax for highway finance, at rates ranging from 3¢ to 7¢ per gallon. Prior to 1919, all highway construction was financed from general tax revenues, generated primarily from the property tax.

Between 1919 and 1980, state gasoline taxes were changed infrequently, and usually .01¢ at a time. But, through most of this period, state gasoline tax collections grew enormously due to the growth in the demand for highway travel and in the number of automobiles. For instance, gasoline tax collections grew from $1 million in 1919 to $1,124 million in 1947. By 1948, motor fuels taxes yielded more revenue than any other state excise tax.

Lagging gasoline tax revenues due to energy conservation in response to the energy price shocks of the 1970s, combined with rising highway repair costs and demand for additional highways, created pressure to raise gasoline tax rates. As a result, during the 1980s most states increased taxes frequently and by relatively large amounts. Between 1980 and 1988, for example, there were 107 increases in gasoline taxes at the state level (3,4).

IMPACT OF FEDERAL AND STATE MOTOR FUEL TAXES

The combination of Federal and state motor fuel and sales taxes on gasoline adds up to 47 cents or more to the pretax price at the service station. For alternative fuels with lower energy densities than gasoline, the taxes, when applied per gallon of fuel unadjusted for energy content add an additional bite–up to 25 to 30 cents for methanol and up to 10 cents for propane. The precise amount extra depends on the state's gallonage tax and sales tax (if any), the fuel's energy content relative to gasoline, and the vehicle efficiency of fuel use compared to that for gasoline.

Electricity is such a special case that highway taxes, were they to be applied, would pay only a minor role compared to vehicle purchase price and battery cost and short life.

Since the Federal highway tax on compressed natural gas is only about 30% of that applied to gasoline, since state highway taxes on CNG are imposed in most cases on a Btu equivalency basis, and since the state highway taxes on the other alternative fuels are, in most cases, on a gallon basis rather than on energy equivalency, the net effect of the combined Federal and state taxes on the prices at the pump of alternative fuels compared to each other and compared to gasoline is very significant. Table 4 summarizes the estimated service station pump prices per gallon and per gasoline-equivalent gallon by state. This table shows clearly that the way highway taxes are applied narrows the difference between diesel fuel and gasoline, while among the alternative fuels it heavily favors CNG, disfavors LPG somewhat, disfavors LNG and ethanol considerably, and just about wipes out methanol in most states as a viable economic competitor.

The key factors driving these outcomes are the disparate Federal highway taxes and the strong tendency in the states to tax on a gallonage basis unadjusted for the fuels' energy contents (except for CNG and, in a few cases, LPG).

State-by-State Fuel Price Assumptions Underlying Table 4

To construct Table 4, Relative State-by-State Fuel Prices, the pretax prices at the service station pump must be estimated. For gasoline, an average price at the refinery rack was assumed to be $0.607 per gallon. An average combined distribution cost and service station markup of $0.209 per gallon was assumed. The pump pretax price would thus be $0.816. With a Federal gasoline tax of $0.184 per gallon, the final pump price, not counting any sales taxes or other local taxes, is $1.00 plus the state gasoline tax. Clearly, the assumptions include the goal of making the calculations as simple as possible, but, in addition, they are based on industry experience. The estimate is consistent with data over the past several years from the Energy Information Administration.

This gasoline price is a composite of regular and premium gasolines. Average premium pump price is about .15¢ higher; regular is about .05¢ lower.

For diesel fuel, the starting point is a recent estimate by the *Lundberg Letter* that the national average pump price was $1.22 per gallon. Subtracting the Federal tax of $0.244 and the median State tax of $0.19 per gallon (taken from the Federal Highway Administration report of October 1, 1993) gives an average pretax pump price of $0.786, or $1.03 plus the state tax.

Methanol price at the plant was assumed to be $0.45 per gallon which, with a distribution cost of $0.209 and a Federal tax of $0.114, brings the pump price to $0.773 plus the State tax. Methanol price is currently varying between $0.35 and $0.45 per gallon; a number of studies estimate that a price nearer to $0.45 per gallon is needed to attract project-based investment capital.

An average "city gate" price for natural gas of $2.75 per thousand cubic feet was assumed. This estimate is "soft" in that it is highly dependent on how far down the gas pipeline the take-off point is and whether the gas will be priced on an interruptible or noninterruptible (higher price) basis. The closer to the wellhead, the lower this price is likely to be and vice versa. The greater the volume of gas going to vehicles, the greater the likelihood that it will be priced on a noninterruptible basis.

The assumed "city gate" price for a natural gas equates to $0.315 per energy-equivalent gallon of gasoline. Adding $0.209 for local distribution and service station markup, plus an additional $0.11 per gallon for compression costs brings the pretax pump price to $0.634 per gallon. The Federal highway tax of 48.54 cents per million Btu (5.6 cents per gallon equivalent) brings the price at the pump to $0.69 per gallon of gasoline equivalent before State tax is added.

For liquefied natural gas, Gas Research Institute studies (5) on the economics of LNG as a vehicle fuel, considered a number of different potential ways to make and deliver LNG, and estimated delivered cost to be anywhere from $0.48 per gallon to $1.03 per gallon, depending on the scale of operation, the location, and the method of liquefaction. For this exercise, the case where LNG is im-

Table 4. Relative State-by-State Fuel Prices at the Pump (October 1, 1993) (Cents per Gallon of Gasoline and Alternative Fuel Price Relative to Gasoline)[a]

State	Gasoline[b] $/Gal.(Index)	Diesel/Gasoline		Methanol/Gasoline		CNG/Gasoline[e]	LNG/Gasoline[f]		LPG/Gasoline[g]	
		Gal.	GGE[c]	Gal.	GGE	GGE	Gal.	GGE	Gal.	GGE
Alabama	$1.18(1.00)	1.05	0.90	0.81	1.41	0.78	0.87	1.33	0.82	1.12
Alaska	$1.08(1.00)	1.05	0.89	0.72	1.25	[h]	[h]	[h]	0.73	0.99
Arizona	$1.18(1.00)	1.04	0.89	0.81	1.41	0.59	0.87	1.33	0.82	1.12
Arkansas	$1.19(1.00)	1.04	0.89	0.81	1.41	0.62	0.75	1.15	0.81	1.09
California	$1.17(1.00)	1.05	0.89	0.73	1.28	0.65	0.77	1.18	0.73	0.99
Colorado	$1.22(1.00)	1.03	0.88	0.80	1.40	0.73	0.86	1.32	0.82	1.11
Connecticut	$1.29(1.00)	0.96	0.81	0.82	1.43	0.76	0.88	1.35	0.75	1.02
Delaware	$1.22(1.00)	1.02	0.87	0.79	1.38	0.72	0.85	1.30	0.81	1.09
Dist of Columbia	$1.20(1.00)	1.04	0.89	0.81	1.42	0.74	0.87	1.33	0.83	1.12
Florida	$1.12(1.00)	1.13	0.96	0.80	1.39	0.88	0.86	1.32	0.81	1.10
Georgia	$1.08(1.00)	1.05	0.89	0.79	1.38	0.71	0.85	1.31	0.81	1.09
Hawaii	$1.16(1.00)	1.05	0.89	0.80	1.41	[h]	[h]	[h]	0.78	1.05
Idaho	$1.21(1.00)	1.04	0.89	0.81	1.42	0.74	0.87	1.33	0.78	1.06
Illinois	$1.19(1.00)	1.07	0.91	0.81	1.42	0.80	0.87	1.33	0.83	1.12
Indiana	$1.15(1.00)	1.05	0.90	0.80	1.40	0.91	0.86	1.32	0.82	1.11
Iowa	$1.20(1.00)	1.07	0.91	0.81	1.42	0.97	0.87	1.33	0.83	1.12
Kansas	$1.18(1.00)	1.06	0.90	0.82	1.44	0.73	0.86	1.32	0.82	1.10
Kentucky	$1.15(1.00)	1.02	0.87	0.80	1.40	0.90	0.86	1.32	0.82	1.11
Louisiana	$1.20(1.00)	1.04	0.89	0.81	1.42	0.84	0.87	1.33	0.83	1.12
Maine	$1.19(1.00)	1.05	0.90	[h]	[h]	[h]	[h]	[h]	0.82	1.11
Maryland	$1.24(1.00)	1.05	0.90	0.82	1.43	0.75	0.87	1.34	0.83	1.13
Massachusetts	$1.21(1.00)	1.04	0.89	0.81	1.42	0.92	0.87	1.33	0.73	0.99
Michigan	$1.15(1.00)	1.05	0.89	0.80	1.40	0.77	0.73	1.12	0.82	1.11
Minnesota	$1.20(1.00)	1.04	0.89	0.81	1.42	0.87	0.87	1.33	0.83	1.12
Mississippi	$1.18(1.00)	1.04	0.89	0.81	1.41	0.74	0.87	1.33	0.81	1.10
Missouri	$1.13(1.00)	1.05	0.89	0.80	1.40	0.88	0.86	1.32	0.82	1.10
Montana	$1.24(1.00)	1.04	0.89	0.82	1.43	0.74	0.74	1.14	0.83	1.13
Nebraska	$1.24(1.00)	1.04	0.89	0.82	1.43	0.91	0.87	1.34	0.83	1.12
Nevada	$1.23(1.00)	1.08	0.92	0.82	1.43	0.75	0.87	1.34	0.78	1.06
New Hampshire	$1.19(1.00)	1.04	0.89	0.81	1.42	0.94	0.87	1.33	0.82	1.11
New Jersey	$1.11(1.00)	1.08	0.92	0.79	1.39	0.86	0.81	1.24	0.77	1.04
New Mexico	$1.23(1.00)	1.01	0.86	0.82	1.43	0.75	[h]	[h]	0.79	1.07
New York	$1.23(1.00)	1.06	0.90	0.81	1.42	0.75	0.87	1.34	0.71	0.96
North Carolina	$1.22(1.00)	1.04	0.89	0.81	1.43	0.92	0.87	1.34	0.83	1.12
North Dakota	$1.17(1.00)	1.05	0.89	0.81	1.41	0.85	0.87	1.33	0.82	1.11
Ohio	$1.21(1.00)	1.04	0.89	0.81	1.42	0.81	0.87	1.34	0.83	1.12
Oklahoma	$1.17(1.00)	1.02	0.87	0.81	1.41	0.73	0.87	1.33	0.82	1.11
Oregon	$1.24(1.00)	1.04	0.89	0.82	1.43	0.94	0.87	1.34	0.83	1.13
Pennsylvania	$1.22(1.00)	1.04	0.89	0.73	1.28	0.81	0.79	1.21	0.83	1.12
Rhode Island	$1.26(1.00)	1.04	0.89	0.82	1.44	0.80	0.88	1.35	0.84	1.13
South Carolina	$1.16(1.00)	1.05	0.89	0.80	1.41	0.73	0.86	1.33	0.82	1.11
South Dakota	$1.18(1.00)	1.04	0.89	0.79	1.38	0.93	0.87	1.33	0.81	1.09
Tennessee	$1.21(1.00)	1.01	0.86	0.81	1.42	0.89	0.81	1.24	0.77	1.04
Texas	$1.20(1.00)	1.04	0.89	0.81	1.42	0.88	0.87	1.33	0.83	1.12
Utah	$1.19(1.00)	1.04	0.89	0.81	1.42	0.74	0.87	1.33	0.83	1.12
Vermont	$1.16(1.00)	1.05	0.90	0.67	1.17	0.74	0.73	1.11	0.82	1.11
Virginia	$1.18(1.00)	1.03	0.88	0.81	1.41	0.88	0.80	1.23	0.76	1.03
Washington	$1.23(1.00)	1.04	0.89	0.82	1.43	0.85	0.87	1.34	0.83	1.13
West Virginia	$1.25(1.00)	1.04	0.89	0.82	1.43	0.75	0.87	1.34	0.83	1.13
Wisconsin	$1.23(1.00)	1.04	0.89	0.82	1.43	0.88	0.87	1.34	0.83	1.13
Wyoming	$1.09(1.00)	1.05	0.89	0.71	1.24	0.80	0.77	1.19	0.73	0.98

[a] Ref. 3. Estimated gasoline prices are presented in $/gallon. For each State, the estimated gasoline price becomes the base point against which the other fuels are compared. All other prices are relative to that State's gasoline price. For Alabama, for example, the diesel price of $1.24 is divided by the gasoline price of $1.18 to give the index of 1.05, or 105% of the gasoline price per gallon. The diesel price on an energy basis is only 90% of the gasoline price per Btu.

[b] Pretax gasoline pump price assumed to be $0.816 per gallon.

[c] GGE: Gallons of gasoline equivalent (adjusts price for energy content of the fuel). Pretax pump price for diesel assumed to be $0.809/gallon.

[d] Pretax methanol pump price assumed to be $0.659 per gallon. Methanol FFV assumed to be 12 percent more efficient on methanol than on gasoline, based on California data.

[e] Pretax CNG pump price assumed to be $0.69 per gallon at 8.7 gallons per 1000 cubic feet.

[f] Pretax LNG pump price assumed to be $0.659 per gallon.

[g] Pretax LPG pump price assumed to be $0.61 per gallon. In States with fees in lieu of gallonage taxes, an average effective tax rate has been estimated.

[h] Alaska and Hawaii have not addressed their tax policy toward CNG and LNG.

ported and trucked 250 miles to fuel a medium duty fleet; has been picked; estimated delivered cost is \$0.66 per gallon.

An LPG price at the refinery or gas separation plant of \$0.40 per gallon was assumed. Adding the standard assumed \$0.209 (adjusted to \$0.21 for simplicity) for distribution and \$0.183 for Federal highway tax brings the price to \$0.79 plus the State tax. However, retail markups for LPG vary widely; the markup assumed here assumes dealer commitment to a vehicle fuel market.

Except for the methanol flexible fuel vehicle (FFV), no adjustment has been made for the potential improvements in energy efficiency which might be available from use of engines designed specifically to take advantage of the alternative fuels's characteristics. For the methanol FFV, an assumption of a 12% better efficiency on methanol than on gasoline has been made, based on operating experience gained in California on such vehicles over a decade. As a result, the multiplier from gallons of gasoline used is 1.76, instead of 2, the energy content ratio. Potential design efficiencies might reduce the energy equivalent pump price for alternative fuels, including a dedicated methanol vehicle on M100 (100% methanol), by as much as 20%. Such adjustments were not made for the other fuels because engines designed to take advantage of each fuel's characteristics are not yet available commercially.

The pump prices for each alternative fuel in each state calculated on the basis of these assumptions were then divided by the gasoline price to get a ratio of the alternative fuel price to the gasoline price, on both the volume and energy bases. The ratios show clearly, for example, that lower pump prices for methanol and LNG per gallon (ratios less than 1.0) translate into higher pump prices per unit of energy (ratios higher than 1.0).

States charging a fee for alternative fuel vehicles instead of a tax collected at the pump have in general set their fee at a level designed to generate about the same amount of revenue as a vehicle driven an average number of miles. Thus, in those states, the higher the mileage actually driven, the lower the imputed tax rate per gallon. For LPG, for example, in states where the ratio in energy terms is close to but higher than 1.0, vehicles driven more than the average 15,000 miles per year would become economic and, the more miles driven, the more economic they become.

CONCLUSIONS

All of these assumptions are challengeable for any specific location or specific set of circumstances with unique characteristics. Pump prices vary for a number of reasons not related to taxes. Nonetheless, the assumptions used generate a reasonable starting point from which to compare the impacts of the fuel energy density and other alternative fuel characteristics in combination with the highly variable Federal and state taxes.

An obvious conclusion is that compressed natural gas, which is taxed at an energy-equivalent rate, benefits the most among the alternative fuels from highway excise taxes, most of which are imposed on a gallonage basis. Methanol, as the contending liquid fuel with the lowest energy content, is the fuel subject to the greatest disadvantage.

The survey from which these numbers were generated showed also that states are beginning to think more explicitly about tax policies for the alternative fuels, particularly in light of the new mandates for their fleet vehicles in the Energy Policy Act and the mandates in the Clean Air Act Amendments of 1990 for cleaner vehicles and fuels.

However, states also report that the need for the revenue generated by highway taxes is just about as strong an imperative as is the need to provide incentives for alternative fuels. So far, the states are not of one mind on how to respond and will be actively considering their options over the next several years.

BIBLIOGRAPHY

1. Much of this historical information is taken from: U.S. Library of Congress, Congressional Research Service, *Federal Excise Tax on Gasoline and the Highway Trust Fund: A Short History*, Report No. 94-354-174 E, Washington, D.C., April 22, 1994.
2. A. M. Sharp and B. F. Sharp, *Public Finance: An Introduction to the Study of the Public Economy*, Business Publications, Inc., Austin, Texas, 1970, p. 377.
3. The Road Information Program, *1989 State Highway Funding Methods*, p. 20.
4. J. H. Bowman and J. L. Mikesell, "Recent Changes in State Gasoline Taxation: An Analysis of Structure and Rates," *National Tax J.* **36** (June 1983).
5. *Preliminary Assessment of LNG Vehicle Technology, Economics, and Safety Issues*, Revision 1, GRI 91/0347, Prepared for GRI by Acurex Environmental Systems Division, Jan. 10, 1992.

ENERGY TAXATION: SUBSIDIES FOR BIOMASS

SALVATORE LAZZARI
Congressional Research Service
Washington, D.C.

Biomass is broadly defined as any organic material or substance other than oil, natural gas, coal, or any product or byproduct of oil, natural gas, and coal. Biomass includes plants, wood, crops, plant and animal wastes, solid wastes including municipal and industrial waste, sewage, and sludge. Historically, the Federal tax system favored oil and gas, which qualified for two major tax subsidies, over energy from biomass and other alternative energy resources, which received no tax subsidies. Oil and gas companies could use percentage depletion (instead of cost depletion) for the recovery of the investment in a well, and could expense (deduct currently, rather than capitalize) intangible drilling costs (IDCs). This preferential treatment was curtailed during the 1970s and 1980s. Restrictions were imposed upon the two oil and gas tax subsidies, and several new oil taxes were introduced, all of which raised the industry's effective tax rates and narrowed its preferential treatment relative to investments in other industries. And a variety of tax incentives were introduced targeted specifically for the development of alternative energy resources. While these new tax incentives have been provided to a wide spectrum of renewable and nonconventional forms of energy resources (solar, wind, geothermal, synfuels, shale oil, coalbed methane and many others), biofuels (such as ethanol) and other types of biomass energy became the focus of virtually every energy tax legislation.

The Federal tax system has shifted away from oil and gas toward energy conservation and the development of alternative energy resources such as biomass. Combined with the two expired biomass tax incentives, the general investment tax incentives, and the spending programs and Federal research and development programs in effect at that time, these biomass tax incentives contributed to the enormous growth in the biomass industry over the last 20 years and they have helped to sustain it today.

This chapter examines the provisions of the Federal tax code that promote biomass energy and biofuels. The first section describes each incentive in detail, along with its limitations, legislative history and any expansion or liberalization adopted by the recently enacted Energy Policy Act of 1992. The second section describes the new biomass tax incentives enacted as part of the 1992 act. The third section compares the estimated tax benefits for biomass energy and other energy alternatives in relation to oil and gas, the benchmark energy resource. Current Federal energy tax policy clearly favors biomass and other alternatives to oil and gas.

This study examines only those tax provisions (incentives) specifically targeted to biomass. Oil and gas tax incentives and penalties will be discussed only to the extent that they are useful in placing the biomass tax incentives in perspective. General tax provisions of the corporate or individual income tax systems (such as accelerated depreciation, or tax rate structure) are not discussed because they are not likely to produce significant differential effects. However, under certain restrictive conditions, some types of biomass equipment may qualify for a slightly accelerated depreciation (1). Two biomass tax incentives–the 1978 business energy investment tax credit and the 1980 tax-exempt bond provisions for biomass facilities are not discussed since they have expired.

CURRENT FEDERAL TAX INCENTIVES FOR BIOMASS ENERGY

The current Federal tax code contains four incentives for biomass energy: the partial exemptions from the various motor fuels excise taxes, particularly the 5.4¢ per-gallon exemption for gasohol; the $5.35 per barrel alternative fuels production tax credit; the tax credits for blended and pure ethanol fuels; and the new 10¢ per gallon small ethanol-producer credit. Each of these tax incentives is provided for the conversion of biomass, broadly defined, into either a liquid or gaseous fuel or for the use of the biofuel; no incentives are currently provided for the direct combustion of biomass, such as wood. However, the Energy Policy Act of 1992 contains a tax incentive for biomass used directly to generate electricity. Moreover, in the case of biomass conversion, the tax incentives are for the conversion of biomass into an alcohol fuel, primarily ethanol.

Excise Tax Exemptions for Alcohol Fuels

Current Law. The most important tax incentives for biofuels are the exemptions for blends of ethanol and motor fuels, such as gasoline and diesel, from the various Federal motor fuels excise tax. The current Internal Revenue Code (IRC) imposes excise taxes on a variety of motor fuels used in highway transportation; the tax rates depend on the type of fuel and its use. Gasoline and special motor fuels (gasoline substitutes and additives such as liquified petroleum gas, naphtha, and benzene) are taxed at 18.4¢ per gallon (IRC §4041(a)(2), and §4081). Diesel fuel is taxed at 24.4¢ per gallon (§4091, and §4041(a)). Fuels used in noncommercial aviation are taxed at either 19.4¢ per gallon (in the case of gasoline, §4041(c), and §4081) or 21.9¢ per gallon for jet fuel (§4041(c), and §4091). Fuel used for transportation on inland waterways by commercial cargo vessels is taxed at 21.4¢ per gallon, rising to 24.4¢ by 1995 (§4042). The 18.4¢ tax also applies to gasoline and special motor fuels used in noncommercial motorboats.

In the first three of the above four excise taxes, mixtures of ethyl alcohol and the otherwise taxable fuel are partially tax-exempt. For example, ethanol blended either with gasoline ("gasohol") or one of the other qualifying special motor fuels is exempt from 5.4¢ of the 18.4¢ per gallon tax (ie, the tax rate is 13.0¢. Ethanol blended with diesel—sometimes called "dieselhol"—is exempt from 5.4¢ of the 24.4¢ tax (the tax rate is 19.0¢). Ethanol blended with either aviation gasoline or jet fuel also qualifies for a 5.4¢ exemption (making the taxes on the ethanol mixtures 14.0¢ and 16.5¢, respectively).

These tax exemptions apply to mixtures that are at least 10% ethanol. If the fuel contains at least 85% pure alcohol the exemption is 6.05¢. This tax structure would make the tax rates for mixtures with 85% alcohol as follows: 12.35¢ for gasoline and alcohol; 18.35¢ for diesel and alcohol; 13.35¢ for gasoline and alcohol used in noncommercial aviation; and 15.55¢ for jet fuel and alcohol.

The excise tax exemptions for alcohol fuels mixtures, particularly the 5.4¢ gasohol exemption, has been the single most important tax incentive for biofuels, causing (along with high oil prices and state tax exemptions) fuel ethanol production (produced mostly from corn) to increase from about 40 million gallons in 1978 to over 1 billion gallons in 1991.

The 5.4¢ per gallon exemption for gasohol represents 29% of the 18.4¢ gasoline tax. When the gasoline tax was 9.1¢ per gallon the gasohol tax exemption was raised to 6¢ per gallon (66% of the tax). Under the original 1978 statute, the exemption was 4¢ per gallon (100% of the tax). Thus in absolute terms, the gasohol tax exemption first increased from 4¢ to 6¢, then decreased to 5.4¢. In relative terms, however, the gasohol tax exemption has steadily decreased. The dieselhol tax exemption has decreased from 100% to 40%, and to 22% of the excise tax.

Limitations. It should be noted that, in the case of alcohol blends, the alcohol must be at least 190 proof (95% pure alcohol), the mixture must be at least 10% alcohol, and the alcohol cannot be derived from petroleum, natural gas, or coal (including peat). The latter limitation means that exempt alcohol is, generally, derived from biomass.

The only exception to the latter rule is for special motor fuels that are 85% alcohol (ethanol or methanol) derived from natural gas, which is exempt for 7.0¢ of the 18.4¢ tax (the tax rate is 7.1¢ per gallon (§4041(m)).

In the case of the blended fuels, methanol produced from coal or natural gas does not qualify for the tax exemptions because in 1977, when the ethanol exemption was first introduced, Congress believed that the cost of producing methanol from coal and natural gas was low at that time and did not warrant a Federal tax subsidy; whereas, the cost of producing methanol from wood and ethanol gram grain was costly and did warrant a subsidy.

Finally, it is important to underscore the point that the highway motor fuels excise taxes historically applied to liquid fuels only; electricity and gaseous fuels, such as compressed natural gas, were tax-exempt. Compressed natural gas, as a result of the Omnibus Budget Reconciliation Act of 1993 (P.L. 93-66), is now taxed at 48.54¢ per million Btu, the equivalent of 5.6¢ per gallon of gasoline.

Legislative History. The excise tax exemptions for alcohol fuels originated with the Senate's version of the Energy Tax Act of 1978 (P.L. 95-618), which introduced the special tax provisions for biomass and other alternative energy resources, and signaled the new shift in Federal energy tax policy.

The other four components of Carter's National Energy Plan were: the Public Utilities and Regulatories Policies Act (P.L. 95-617); the National Energy Conservation Policy Act (P.L. 95-619); the Powerplant and Industrial Fuel Use Act (P.L. 95-620); and the Natural Gas Policy Act (P.L. 95-621). For biomass energy, the Energy Tax Act introduced the excise tax exemptions for alcohol fuels, the business energy investment tax credits, and the tax-exempt bond provisions to provide financing incentives. The Energy Tax Act also provided for the gas-guzzler tax, incentives for van pooling, and miscellaneous energy tax provisions.

The Tax Act was part of the Carter administration's National Energy Plan. However, tax provisions for conversion to alternative transportation fuels (gasohol) were not part of the original House bill (H.R. 8444, 95th Congress), which embodied the tax provisions of the National Energy Plan. The version of H.R. 8444 reported out of the Senate Finance Committee included the excise tax exemptions for gasohol fuels. The exemption was intended to induce a substitution of ethanol and methanol for gasoline used in transportation. The underlying rationale for the ETA was the perceived failures in the energy markets in allocating resources efficiently and fairly, in coping with the 1973 oil embargo, and in adjusting to the sharp increases in energy prices, the shortages, and the associated adverse economic and social problems. The belief was that the Federal Government had to influence resource allocations through tax incentives and other financial incentives. Each of these tax instruments was intended to contribute to the general goal of conserving conventional energy resources (primarily oil and gas), stimulate production of alternatives to oil and gas, reduce oil imports, and achieve energy security.

The original exemption was for the full amount of the gasoline tax: 4¢ per gallon. The Crude Oil Windfall Profits Tax (P.L. 96-223) extended the 4¢ exemption from October 1, 1984 to December 31, 1992. The Surface Transportation Assistance Act of 1982 (P.L. 97-424) raised the gasoline tax from 4¢ to 9¢ per gallon and also changed the exemption for gasohol from the complete 4¢ exemption to a partial 5¢ exemption (gasohol would be taxed at 4¢ per gallon instead of 9¢ per gallon). The Tax Reform Act of 1984 (P.L. 98-369) raised the diesel fuel tax from 9 to 15¢ per gallon as part of a compromise that also lowered the highway use taxes on trucks. The 1984 tax law also raised the gasohol exemption from 5 to 6¢ (ie, it reduced the tax rate for gasohol from 4 to 3¢), and retained the 9¢ exemption for "neat" alcohol fuels, and provided that alcohol produced from natural gas would also qualify for the exemption. The Tax Reform Act of 1986 (P.L. 99-514) reduced the excise tax exemption for 85% alcohol from 9¢ to 6¢ per gallon (for sales made beginning on 1987). The Technical and Miscellaneous Revenue Act of 1988 (P.L. 100-647) made minor liberalizations to the excise tax rules. Finally, the OBRA of 1990 reduced the exemption to 5.4¢ per gallon.

Amendments in the Energy Policy Act of 1992. P.L. 102-486 extends the gasohol excise tax exemption to gasohol that contains less than 10% alcohol. Two categories of gasohol mixtures are prescribed: mixtures containing 7.7% alcohol; and mixtures containing 5.7% alcohol. The exemption for 7.7% mixtures is 4.16¢ per gallon (the tax rate is 9.94¢); the exemption for 5.5% mixtures is 3.08¢ per gallon (the tax rate is 11.02¢).

The Alternative Fuels Production Tax Credit

The second major tax break for biomass fuels is the alternative fuels production tax credit, also known as the "§29 tax credit," named after the section of the IRC in which it resides.

Current Law. The alternative fuels production tax credit is a credit against the producer's income tax for the *production and sale* of fuels derived from a wide variety of alternative energy resources. Qualifying fuels are grouped into three categories: oil from shale or tar sands; liquid or gaseous synthetic fuels from coal, gas from coal seams (coalbed methane), tight sands, Devonian shale and geopressurized brine; and liquid or gaseous fuels from wood, agricultural byproducts, and other biomass. Certain types of alcohol fuels qualify for this credit including both ethanol and methanol, and alcohol produced from coal and lignite. Moreover, alcohol fuels produced from coal or lignite may be used as feedstocks, unlike other fuels, without invalidating the tax credit.

Two other types of biomass have qualified for the production tax credit under the provisions of the original 1980 law: processed wood fuel treated to increase the BTU content of the wood by at least 40%, and steam from agricultural products. These two types of biomass no longer qualify for the tax credit under original expiration dates. The credit for processed wood fuel expired on October 1, 1983, and the credit for steam from solid agricultural products expired on December 31, 1985. Solid agricultural products included only solid byproducts of farming or agriculture and excluded timber products or other forms of biomass.

Thus waste wood used directly as a fuel did not qualify for the credit.

The production tax credit is $3.00 per barrel, in real terms, in barrels of oil or oil equivalent. The credit is linked with the BTU (British Thermal Unit) content of oil. Each 5.8 million BTUs of fuel, the energy content of one barrel of oil, qualifies for the $3.00 credit. The $3.00 amount is adjusted for inflation (using 1979 as the base year and the GNP deflator as the price index), which makes the current credit about $5.75 per barrel of oil equivalent (for qualifying gases the credit is about $1.00 per thousand cubic feet).

The availability of the credit is linked to the average wellhead price of domestic crude oil (called the reference price). When the reference price of oil is below $23.50 (in real terms), the tax credit becomes available; when the price of oil is between $23.50 and $29.50, the credit is phased out proportionately; when the price of oil is above $29.50, no credit is available. These trigger or threshold prices are also adjusted for inflation so that a comparison may be made with the reference price in nominal terms. At this writing, the phase-out range in current dollars is between $40 and $50 per barrel. With the market price of West Texas Intermediate crude oil (the reference price in nominal terms) at about $22 per barrel, well below the $40 ($23.50 times the inflation adjustment factor of 1.70), the credit is available. The Congress has believed that when oil prices are high, market incentives should suffice to stimulate production of alternative fuels.

Until recently, most of the current credit has gone for coalbed methane gas and very little for biomass. Coalbed methane is eligible for the production tax credit, and it has generated much of the revenue losses from the credit. Coalbed methane is a colorless and odorless natural gas that permeates coal seams, and is virtually identical to conventional natural gas. Under IRC§29, coalbed methane is treated as an unconventional gas because it resides in an unconventional location–coal beds–as opposed to conventional gas reservoirs. The combined effect of the $1.00 per MCF tax credit (which was at times 100% of natural gas prices) and declining production costs (due to technological advances in coalbed methane drilling and production techniques) was sufficient to offset the decline in oil and natural gas prices, and the resulting decline in domestic conventional natural gas production. Data show that production of coalbed methane has increased from 0.1 billion cubic feet in 1980 to over 300 billion cubic feet in 1991, a large part of it in response to the production tax credits, and virtually all of it at the expense of conventional gas production. The credit for coalbed methane benefits largely oil and gas producers, both independent producers and major integrated oil companies, and complicates calculation of the special tax benefits to biomass and other alternative energy resources in relationship to the net tax burdens on oil and gas.

Limitations. The production tax credit is available for fuels produced through December 31, 2002. However, the facilities must be placed-in-service (or from wells drilled) after 1979 and before 1993. To prevent "double dipping" the credit is reduced by any subsidized financing (grants, loans, tax-exempt financing) energy tax credits, including the enhanced oil recovery tax credit enacted as part of the OBRA 90. The credit is nonrefundable and is limited to the excess of a taxpayer's regular tax over several tax credits and the tentative minimum tax.

Legislative History. The production tax credit was introduced by the 1980 windfall profit tax (WPT) and has been amended several times. Most of the amendments have concerned the placed-in-service deadlines. The original 1980 WPT established a placed-in-service deadline of December 31, 1989 and was extended to December 31,1990 by the Technical and Miscellaneous Revenue Act of 1988, and to December 31, 1991 by OBRA of 1990, which also liberalized the treatment of tight-sands gas.

Amendments in the Energy Policy Act of 1992. H.R. 776 provides for a three-year extension of the placed-in-service rule, and an extension of the credit, for biomass and synthetic fuels only. Under these new rules, producers of gas from biomass or liquid, gaseous or solid synthetic fuels from coal or lignite, pursuant to a binding contract signed before January 1, 1996, would have until December 31, 1996 to build a facility and begin production. If these conditions are met, then production tax credits for these fuels would be available for another five years through December 31, 2007 (instead of December 31, 2002). These amendments mean that production of biomass and synfuels from new facilities or new wells will no longer qualify for the §29 credit if the new facilities are placed in service after 1997. For all other alternative fuels, production from facilities or wells placed in service after 1992, including coalbed methane wells, no longer qualify for the credit.

The Two Alcohol Fuels Tax Credits

In place of an exemption, the Federal tax code provides an alcohol fuels mixtures tax credit (the blender's credit), and a tax credit for straight alcohol fuel. Both credits are part of IRC §40.

Current Law. The blender's tax credit is 54¢ per gallon for alcohol that is at least 190 proof, and 40¢ per gallon for alcohol that is at least 150 proof but less than 190 proof. No credit is available for alcohol that is less than 150 proof. This credit is available to the blender only for use as a motor fuel in a trade or business whether produced and used by him or produced and sold by him. The straight alcohol fuels credit is 60¢ per gallon for alcohol that is at least 190 proof, and 40¢ per gallon for alcohol between 150 and 190 proof. This credit is typically available to the retail seller that dispenses it in the fuel tank of the buyer's vehicle. These credits have been available since October 1, 1980, and will; under current law, continue to be available through December 31, 2000.

The alcohol fuels tax credits apply to most types of ethanol (ie, alcohol derived from renewable energy resources such as vegetative matter, crops, and other biomass) and to methanol derived from wood. The alcohol cannot be derived from petroleum, natural gas, or coal (including peat). This rule effectively limits the tax credits to ethanol since most economically feasible methanol is derived primarily from natural gas and does not qualify for the credits. Currently, about 95% of current ethanol production is derived from corn. A 1990 IRS ruling has allowed mixtures of gasoline and ETBE—Ethyl Tertiary Butyl Ether–to qualify for the 54¢ blender's credit. ETBE is a compound that results from a chemical reaction between ethanol (which may be produced from renewable energy) and isobutylene. ETBE, which is no longer an alcohol after its chemical reaction, is being considered for use as a substitute for MTBE–Methyl Tertiary Butyl Ether—as the oxygenate in reformulated gasoline mentioned under the Clean Air Act of 1990 for use in designated ozone nonattainment areas. Allowing ETBE to qualify for the blender's tax credit is designed to stimulate the production of ethanol for use in reformulated gasoline and to reduce the production of MTBE, an alternative oxygenate made from natural gas. This strategy would increase the share of the U.S. corn crop allocated to ethanol production above the current 4–5%. It would also significantly increase Federal revenue losses from the alcohol fuels credits, which heretofore have been negligible due to blender's use of the exemption over the credit.

Limitations. There are several limitations to the two alcohol fuels tax credits: First, the alcohol cannot be derived from petroleum, coal, or natural gas. Thus, the alcohol must be derived from renewable energy resources such as vegetative matter, crops, wood, and other biomass to qualify for the credits. This limitation means that the credits are currently available only to ethanol, since most economically feasible methanol is made from natural gas, which does not qualify for the credits. Second, the alcohol fuels tax credits are a component of the general business credit under IRC §38, which includes the jobs tax credit, research and development tax credit, low-income housing tax credit, and other credits. The general business tax credit is limited to the taxpayer's net income tax over the larger of either 25% net regular tax liability above $25,000 or the tentative minimum tax. Any unused general business credits may generally be carried forward 15 years or carried back three years.

Any taxpayer who claims the credit must also report it as gross income for the tax year in which the credit is earned (IRC §87). Fourth, the credits are not refundable; they may be used only against a positive tax liability; they are of no value of the producer has no tax liability. Finally, and more importantly, the credits are offset by the excise tax exemptions claimed on the same fuel. Typically, blenders prefer the excise tax exemption over the credit because the exemption is an immediate "up-front" tax benefit, which increases cash-flow, while the benefits from the tax credit must await the preparation of the tax returns. The limitations under IRC §38 also could reduce the value of the alcohol fuels tax credits.

Legislative History. The alcohol fuels tax credits were enacted as part of the 1980 WPT at the initial amounts of 40¢ and 30¢ per gallon for proofs greater than 190 and between 150 and 190 respectively, and raised by the Surface Transportation Assistance Act of 1982 to 50¢ and 37.5¢, respectively. The Tax Reform Act of 1984 raised the credits to 60¢ and 45¢. This law also introduced the provision that alcohol produced from peat, like coal, would not qualify for the credits.

Ethanol Tax Credit for Small Producers

An additional tax incentive for biofuels production is the small ethanol-producer credit introduced as part of OBRA 90.

Current Law. IRC §40 provides for an income tax credit of 10¢ per gallon ($4.20 per barrel) for up to 15 million gallons of annual ethanol production by a small ethanol producer. A "small producer" is defined as one with ethanol production capacity of less than 30 million gallons per year (about 2,000 barrels per day). This credit is available for alcohol produced by a small producer and sold to another person for blending into a qualified mixture in the buyer's trade or business, for use as a fuel in the buyer's trade or business or for sale at retail where such fuel is placed in the fuel tank of the retail customer. Casual off-farm production of ethanol does not qualify for this credit.

Limitations. The new ethanol tax credit is limited to small producers, and aggregation rules are provided to prevent the credit from benefiting producers wth a capacity in excess of the 30 million gallons per year limit, or from going to production in excess of the 15 million gallon limit.

Legislative History. The small-producers credit was added by the Omnibus Budget Reconciliation Act of 1990 (P.L. 101-508); it has not been amended.

New Biomass Tax Incentives in The Energy Policy Act of 1992

The end of 1992 has witnessed the enactment of broadly-based energy legislation–The Energy Policy Act of 1992–designed to resuscitate a Federal energy policy that many argued was missing. The law, P.L. 102-486, which some call the crowning achievement of the 102nd Congress, includes numerous provisions intended to reduce dependence on petroleum imports: a restructuring of the electric utility industry to enhance competition, promotion of alternative transportation fuels, provisions that mandate new and more stringent energy efficiency standards, and numerous other provisions. Title XIX of the law contains a variety of energy taxes and tax incentives. In addition to the liberalization of existing biomass tax incentives, as discussed above, H.R. 776 creates two new tax incentives for biomass: an income tax deduction for the costs of a vehicle that burns alcohol or some other clean-burning fuel; and an income tax credit for electricity generated from "closed-loop" biomass facilities.

Income Tax Deduction for Alcohol Fuel Vehicles and Alcohol Storage Facilities

Beginning on July 1, 1993, the purchaser of a new vehicle that burns either ethanol, methanol, ether, any combination of these, or some other clean-burning fuel, has qualified for a limited income tax deduction for the costs of the vehicle allocable to the engine and any collateral equipment used to store or deliver the fuel. The maximum deduction is $2,000 for cars, $5,000 for light trucks, and $50,000 for heavy trucks or buses. This deduction is also provided for the costs of retrofitting used non-qualifying vehicles to clean-fuel burning vehicles, and for the costs of equipment used to store or dispense the alcohol fuels and other clean-burning fuels into the tank of the clean-burning vehicle. Qualifying storage and dispensing equipment includes equipment used to compress natural gas into a usable fuel, provided that the equipment is located on the site that dispenses the fuel into the vehicle. The storage equipment deduction is limited to $100,000 per year. Alcohol fuels must contain 85% alcohol. Other clean burning fuels that qualify for the deduction are compressed natural gas, liquefied petroleum gas, and hydrogen.

Income Tax Credit for Closed-Loop Biomass Facilities

The second new tax incentive in P.L. 102-486 that would benefit biomass energy is the income tax credit for production of electricity from "closed-loop" biomass systems defined as systems that use renewable plant matter exclusively as an energy source to generate electricity. The credit, which also applies to electricity generated from wind energy systems, equals 1.5¢ per kilowatt hour of electricity and is phased out, proportionately, as the national average price of electricity produced from renewables rises from 8.0¢ to 11¢ per kilowatt hour. Both the credit and the phase-out limit are adjusted for inflation. It is important to note that biomass is narrowly defined for purposes of this new credit. It does not apply to municipal or agricultural waste or to scrap wood and other wastes. This credit will be part of the general business credit, and its limitations, which were discussed above. It will also be offset by any type of grant or subsidized financing, including tax-exempt bond financing.

THE MAGNITUDE OF BIOMASS ENERGY TAX BENEFITS IN RELATION TO OIL AND GAS

This section on the study estimates the tax benefits to biomass and other energy alternatives that result from the various tax incentives peculiar to that industry. These are compared to any tax benefits that might accrue to oil and gas, the benchmark energy resource, as a result of Federal tax incentives and taxes peculiar to that industry. For each industry, first we estimate aggregate tax benefits, then we estimate tax benefits per unit of output.

Aggregate tax benefits to each industry are measured as the sum of the losses in Federal tax revenue–also referred to as "tax expenditures"–that result from each of the various tax incentives peculiar to that industry. The source for these estimates is the annual tax expenditure study published by Congress's Joint Committee on Taxation, which shows these losses for each of the industry-specific tax incentives. It is assumed that these losses in Federal tax revenues are equivalent to the reduction in industry tax liabilities. Tax benefits per unit of output are measured as the total industry tax benefits divided by estimated output. The latter is an indicator of the effect of the tax benefits on relative prices among the fuels, which forms the basis of economic decisions. This particular framework of analysis abstracts from the general provisions of the income tax laws; it considers only those provisions peculiar or special to each industry. The comparison will show that biomass energy receives a fairly sizeable net tax subsidy compared with oil and gas, which is subject to a small, but positve, net tax.

Aggregate Tax Benefits

Table 1 shows the four tax incentives for biomass and other alternative energy resources and the corresponding annual revenue loss–reductions in industry tax liabilities–resulting solely from those incentives as calculated by the Joint Committee on Taxation. The table includes only the incentives prior to the enactment of P.L. 102-486.

House of Representatives Bill 776 contains tax relief for both the oil and gas industry as well as the biomass industry. In absolute dollar terms, the tax relief for the two industries over five years is about the same, reducing taxes

Table 1. Special Tax Incentives for Biomass and Alternative Energy Industry and Corresponding Reductions in Industry Tax Liabilities, Fiscal Year 1993[a]

Description of Tax Incentive	Principal Limitation	Expiration Date	Reduction in Tax Liabilities ($ millions)
	Special Tax Incentives (−)		
5.4¢ Excise tax exemption	Must be at least 10% alcohol	90-30-2000	−400
$5.35 Per barrel alternative fuels production tax credit	Market oil price below $50 per barrel	12-31-2000	−800
54¢ Per gallon alcohol fuels tax credits	Applies primarily to ethanol	12-31-2000	−50
10¢ Per gallon new ethanol producer credit	Up to 15 million gallons of output per year	12-31-2000	[b]
	Special Taxes (+)		
None			0
			Net Tax = −1,250

[a] Ref. 3

[b] The revenue loss for this provision is part of the revenue loss corresponding to the alcohol fuels tax credits, as shown in the third row of the table

by about $200 million annually. But, the tax relief for biomass and alternative energy resources is larger, in relation to industry size, than for oil and gas. Moreover, oil and gas would still be a net special taxpayer, while the net tax subsidy to alternative energy resources would increase. Thus after H.R. 776, the posture of energy tax policy is even more slanted in favor of biomass and other energy alternatives. These tax incentives resulting from H.R. 776 are excluded from our analysis because they are negligible for FY93, the period of analysis, and because we were unable to separate the tax decreases for biomass from the tax decreases for other qualifying fuels (2). The biomass and alternative energy industry tax liabilities will be approximately $1,250 million lower in FY93 as a result of the four tax incentives peculiar to that industry. In other words, biomass and other alternative energy resources will receive a net tax subsidy, ie, above and beyond the general provisions of the income tax laws. Note that, unlike oil and gas, there are no industry-specific Federal taxes targeted for this industry. It is difficult to determine the share of this $1,250 million that accrues specifically to biomass energy because the sum revenue loss corresponding to the alternative fuels tax credit is not disaggregated. Much of this benefit undoubtedly accrues to biomass, but a portion of it also accrues to such fuels as coalbed methane, which some consider not to be an alternative fuel. If half of this tax benefit accrues to coalbed methane production, then the total tax benefit to alternative energy resources is reduced to $850 million for FY 1993.

In order to put these estimates in perspective, Table 2 shows Joint Committee of Taxation calculations of the various industry-specific Federal tax incentives and penalties imposed on oil and gas. Oil and gas is helped from four tax subsidies (including the two discussed in the introduction) and is penalized by three special taxes. The tax subsidies are: percentage depletion allowance, which is available only to independent producers, and only for limited quantities of output; expensing of domestic IDCs, which is available to all who drill for oil and gas, subject to limitations for corporate producers; exemption from the material participation requirements of the passive loss limitation rules; and a new 15% tax credit for enhanced oil recovery techniques. The tax penalties are: the alternative minimum tax imposed on oil and gas producers; a 9.7¢ tax per barrel on domestic and imported crude oil to finance Superfund; and a 5¢ tax per barrel on domestic and imported crude oil to finance the Oil Spill Liability Trust Fund.

The Comprehensive Environmental Response, Compensation, and Liability Act of 1980 (P.L. 96-510)–commonly known as Superfund–originally imposed a 0.79¢ excise tax on oil received by a domestic refinery. Under the Superfund Amendments and Reauthorization Act of 1986 (P.L. 99-499), this tax was increased to 8.2¢ per barrel of domestic oil and 11.7¢ per barrel of imported oil. This tax differential was ruled to be in violation of GATT–the Gen-

Table 2. Special Oil and Gas Tax Incentives and Taxes for and the Corresponding Reduction (−) and Increase (+) in Tax Liabilities[a]

Description of Tax Provision	Principal Limitation	Expiration Date	Revenue Effect, ($ Millions)
	Special Tax Subsidies (−)		
Expensing of intangible drilling costs Corporations must amortize 30% of costs	None	None	−200
Percentage depletion allowance[b]	Available only for domestic properties Availabe to independents on 1,000 barrels per day	None	−100
Exempt from passive loss rules	None	None	[c]
15% Credit for enhanced oil recovery	Oil prices < $28	None	−50
	Special Taxes (+)		
9.7¢ Per barrel Superfund tax	None	12-31-95	+553
5.0¢ Per barrel oil spill tax	None	12-31-94	+285
Alternative minimum tax	None	None	[c]
			Net Tax = +488

[a] Refs. 3, 5.
[b] Includes special tax incentives for stripper oil and heavy oil enacted as part of OBRA 900.
[c] The revenue effects from this provision is part of the estimated revenue loss from the expensing and percentage depletion tax incentives.

eral Agreement on Tariffs and Trade–to which the United States is a signatory and a 9.7¢ tax was imposed equally on domestic and imported oil as part of the Steel Trade Liberalization Program Implementation Act (P.L. 101-221). The 5¢ per barrel tax on crude oil for the ol spill liability trust fund was enacted under Omnibus Budget Reconciliation Act of 1989 (P.L. 101-239). This tax, also part of the 1980 Superfund law, was originally 1.3¢ per barrel, but no revenues were ever collected because the enabling legislation required to activate the tax under the 1908 law was not enacted. Another excise tax on domestic crude oil was introduced in the 1908 windfall profit tax but this tax was repealed in 1988 by the Omnibus Trade and Competitiveness Act (P.L. 100-418).

Table 2 shows that the four special tax subsidies for oil and gas development are projected to reduce industry tax liabilities by about $350 million in FY 1993 (net of the alternative minimum tax). If the estimated $400 million production tax credit for coalbed methane is scored as a tax subsidy for conventional fuels, then the oil and gas subsidy increases to $750 million for FY 1993. Table 2 also shows that there are two special excise taxes on oil and gas that increase the tax burdens on the oil and gas industry. These special taxes are projected to increase the tax burden on oil and gas by an estmated $838 million in FY 1993. The net effect of the special tax preferences and special taxes is a net special tax burden of +$488 million for FY 1993 ($838–350) without coalbed methane and +$88 million with coalbed methane.

The revenue loss from the expensing of IDCs shown in Table 2 has two important shortcomings: First, it excludes expensing of losses from abandoned properties because the Joint Tax Committee, which determines tax expenditure items and estimates the corresponding revenue losses, does not consider it a tax expenditure. Second, a revenue loss estimate for one year shows only a cash-flow effect, which is not a good indicator of the value of IDCs that are based on timing. Calculations that take these two shortcomings into account lower the next tax burden on the oil industry somewhat, but it is still a positive tax burden. These calculations do not affect the net tax subsidy to biomass and other energy alternatives, therefore do not affect the major conclusions of the report.

Tax Benefits Per Unit of Output

Estimates of aggregate tax benefits do not take into account the fact that economic decisions are based on relative prices per unit of output. To estimate the effect of the various tax incentives and taxes (in the case of oil and gas) on the relative prices of the two fuels we merely have to correct for the size difference between the two industries, which is significant. The U.S. oil and gas industry is ostensibly very large in comparison with the alternative energy industry, which was basically nonexistent until the 1970s. In 1990, for example, oil and gas output accounted for 53% of total domestic energy production, as compared with 9% for total renewable energy and 4.5% for biomass, broadly defined (4).

Size differences between the two industries are accounted for by computing the net tax subsidy per unit of output for both the biomass and alternative energy sources industry and the oil and gas industry. This requires an estimate of each industry's output, which is then projected for 1993. In the case of oil and gas, this is relatively easy undertaking because the data are readily available. The absence of output data makes it more difficult for biomass and other alternative fuels. As a result, output data must be imputed from the tax expenditure data presented in table 1. At $5.35 per barrel, the production tax credit is estimated to lose $800 million in FY 1993, implying industry output of 149.5 million barrels of oil equivalent of alternative fuels ($800 ÷ $5.35). The same approach is applied to each tax incentive item in Table 1, and the results are summed to determine total energy output for 1993. Each industry's total tax subsidy (for the alternative energy industry) or net penalty (for the oil and gas industry) is divided by each industry's total energy output. These estimates are made under two scenarios, depending upon whether coalbed methane is treated as an alternative or conventional fuel.

If coalbed methane is treated as an alternative fuel, the estimated Federal subsidy for alternative energy resources as a whole is about 72¢ per million BTUs or about $4.25 per barrel of oil equivalent. In contrast, oil and gas experiences a net tax burden estimated to be about 1.3¢ per million BTUs, which is equivalent to about 7.5¢ per barrel of oil. When coalbed methane is treated as a conventional fuel, the tax subsidy for alternative energy sources declines to about $3.00 per barrel and the net tax burden on oil and gas increases to about 2¢ per barrel. Thus, the net special tax burden on the oil and gas industry is small per unit of output, while the net tax subsidy for alternative energy resources is large per unit of output.

These estimates suggest that Federal tax incentives lower the market price of biomass energy by an estimated $3.00 to $4.25 per barrel of oil equivalent ($0.50 to $0.75 per MCF of gas) below what the market price would be without the incentives. Clearly, the Federal tax incentives help to narrow, but probably not completely eliminate the competitive disadvantage of biomass and other energy alternatives relative to conventional oil and gas. The estimates also indicate that Federal energy tax policy is, at this writing, slanted in favor of alternative energy resources, a policy posture that constitutes a reversal of the historical posture in favor of oil and gas.

BIBLIOGRAPHY

1. G. L. Middleton, Jr., Impact of Future Tax Incentive Legislation on the Development of Biomass Energy. *Biolog*, **7**, 7–11 (Feb./March 1990).
2. U.S. Congress House, Committee On Ways and Means, *Comprehensive National Energy Policy Act*, House Report No. 102-474, Part 6, U.S. Government Printing Office, Washington, D.C., May 5, 1992, p. 94.
3. U.S. Congress, Joint Tax Committee, *Estimates of Federal Tax Expenditures for Fiscal Years 1993–1997*, Joint Committee Print, Washington, D.C., April 24, 1992, p. 24.
4. U.S. Department of Energy, *Annual Energy Review*, DOE/EIA-0384(91), Washington, D.C., June, 1992, pp. 241–257.
5. U.S. Congress, Joint Committee on Taxation, *Schedule of Present Federal Excise Taxes (As of January 1, 1992)*. Joint Committee Print No. JCS-7-92, 102d Cong., 2d sess. U.S. Government Printing Office, Washington, D.C., March 27, 1992, 35p.

ENERGY TAXATION: BIOMASS

Salvatore Lazzari
Congressional Research Service
Washington, D.C.

In general, the outlook for Federal energy tax policy is quite favorable for biomass. Biomass energy enjoys considerable support in the Congress, in the environmental community, and among farming interests. Biomass is a renewable energy resource with a huge resource base. Wood, an important energy resource in the underdeveloped world, is still used as a fuel in the United States. A variety of liquid and gaseous fuels can be produced from crops, crop wastes, residues, solid waste, municipal waste, and other plant and animal matter. Direct combustion of biomass can generate steam, heat, and electricity, in addition to ethanol, methanol, synthetic natural gas, and synthetic gasoline.

In the short-run, Federal energy tax policy is likely to witness frequent amendments of the type enacted in OBRA 1990, or recently in the Energy Policy Act of 1992. Federal energy tax policy will be activist and interventionist, as tax policy in general is likely to be. While both biomass and oil and gas are likely to be the focus of the energy tax laws, the thrust of this short-term policy will likely be to favor alternatives to oil and gas, especially biomass. The trend in energy tax policy that began in the 1970s is likely to continue. For example, we are likely to witness additional tax incentives for oil and gas, in the spirit of those recently enacted in OBRA 1990 (which introduced a new tax credit for enhanced oil recovery costs). These incentives proposals are motivated by the concern over declining domestic oil production, which peaked in 1970 and has since declined by an average of 4 percent per year, and rising oil imports which reached new records recently. These incentives are likely to focus on marginal, high cost oil such as from small wells or from enhanced oil recovery projects. However, any future tax breaks for oil are likely to be small due to the large deficits in the Federal budgets, and they are unlikely to have important implications for the development of biomass and other alternative energy resources, which is affected relatively more by the price of oil.

In the longer run, the outlook for biomass energy tax incentives will be shaped by three main issues: oil import dependence and energy security, which has driven energy policy for the last 20 years; air pollution problems (greenhouse effects and global warming, ozone); and Federal budget deficits. The first two issues suggest tax policy initiatives that promote alternative transportation fuels and alternative forms of energy such as biomass, which can improve the environment and reduce energy consumption and petroleum imports.

Large budget deficits, however, make it unlikely that significant tax incentives or subsidies will be enacted for either oil and gas or biomass. Rather, large budget deficits suggest that energy taxes on fossil fuels (particularly oil and coal) will be the preferred policy instrument, rather than tax subsidies, to achieve energy, environmental and fiscal policy objectives. Three types of energy tax options are likely to be seriously considered by Federal policymakers: gasoline taxes, oil import taxes, and broadly-based energy taxes such as carbon taxes. Each of these proposals would have important implications for the future of biomass.

HIGHER GASOLINE TAXES

The Federal excise tax on gasoline and special motor fuels is currently 18.4¢ per gallon; the rate on diesel is 24.4¢ per gallon. As recently as 1982 the tax rate on motor fuels was 4¢ per gallon. Increases in rates were enacted in 1982, 1984, and 1990. Including State and local gas taxes, the current tax rate on gasoline averages over 30¢ per gallon in the U.S. (about 30% of retail price), considerably lower than Western industrialized countries where gas taxes average about $3.50 per gallon (about 70% of price), and approach $4.00 per gallon (Italy). Low excise taxes and falling oil prices recently have reduced the real price of gasoline to the lowest level in about forty years. Proposals to increase gasoline excise taxes were common after the 1973–1974 oil embargo and subsequent increases in oil prices. Such proposals were motivated primarily as an effective method to conserve fuel and reduce imported oil. Gasoline tax hike proposals are perennial favorites in Washington as a relatively painless revenue raising and deficit reducing option. More recently proponents have mentioned gas tax hikes as an environmental policy, and as a way to finance public infrastructure investments. One popular, recent proposal would have hiked the gas tax by a total of 50¢ per gallon to be phased-in at 10¢ per year.

Higher gas taxes can simultaneously generate large revenues, reduce petroleum consumption and imports, and reduce vehicle emissions. It is estimated that revenues increase by more than $1 billion per year for every penny increase gas tax increase. An additional $200 million per year can be collected from each penny increment to the diesel fuel tax. A sizeable tax could be imposed and still leave gasoline prices relatively low. Nearly 2/3 of the 17 million barrels per day of petroleum is used in transportation, and nearly 2/3 of that—approximately the level of petroleum imports—is used in passenger cars. Emissions from mobile sources are a significant source of urban air pollution due to the large numbers of vehicles and the total miles driven.

In addition to the potential revenue, energy security, and environmental benefits, higher taxes on gasoline would be an important economic incentive for further development of biofuels, such as ethanol, methanol, and other alternative transportation fuels. A higher gas tax could render methanol from natural gas much more competitive, especially if the excise tax exemption limitations are repealed. If the increases are large enough ethanol from biomass might also be rendered competitive with gasoline and diesel. For a more comprehensive discussion of the pros and cons of a gas tax increase, see Ref. 1.

OIL IMPORT DEPENDENCE AND OIL IMPORT TAXES

Another energy tax option that would have significant economic benefits for biomass energy is the oil import tax or fee. As was mentioned above, there are presently two Federal excise taxes on imported and domestic oil that finance (in part) the Superfund and the oil spill trust fund. From 1981 to 1986, various oil import fee proposals were introduced primarily for the purpose of raising revenues and reducing large Federal budget deficits. The collapse of oil prices in 1986 generated much interest in an oil import tax as a way of helping the distressed U.S. oil industry. More recently, the oil import tax is also viewed as a method of lessening U.S. dependence upon foreign oil, thereby enhancing energy security and improving the balance of trade.

Federal policymakers are concerned that excessive dependence upon OPEC (Organization of Petroleum Exporting Countries) and other foreign oil producers entails a vulnerability to either supply disruptions or a sudden price spike. U.S. oil imports have been rising from about 8% of demand in 1947 to a record 54 percent of total oil demand, a total of 9.1 million barrels per day, in January 1990. Over 50% of oil imports come from OPEC. This is in sharp contrast to the 1930s, when the U.S. was the world's leading oil producer and exporter, accounting for about 2/3 of world oil output. Oil import dependence and vulnerability and energy security have been the single most important issue driving U.S. energy policy and energy tax policy over the last 20 years. This concern goes back to the two energy crises of the 1970s and was accentuated by the oil price shock of 1990–1991 caused by the conflict with Iraq.

Although many policies have been proposed to address the concern of energy security and rising dependence upon foreign oil, including tax incentives to expand oil and gas production and an increase in the gasoline excise tax, one prominent proposal is the imposition of a sizeable tax, usually of $5 or $10 per barrel, on imported crude oil and petroleum products. Taxes of these magnitudes would raise oil prices and promote the production of not only oil and gas and other fossil fuels, but also stimulate the commercialization of alternative energy technologies and the production of renewable and unconventional energy resources. In terms of its effects on the posture of energy tax policy, an oil import tax would, initially tend to favor oil and gas because the tax would raise oil prices and stimulate increased investment in oil and gas. In the long run, higher oil prices would lead to substitution of biomass and alternative energy resources for oil and gas. A sizeable oil import tax would, thus, be very beneficial in stimulating the development of biomass and other energy alternatives (2).

The oil import fee would also raise revenues, although its potential is significantly less than the gasoline tax. This is an important feature when $400 billion deficits are looming. The environmental consequences, however, are not all positive, which renders the oil import fee a less desirable option than the gasoline tax. Opponents of the fee also argue that it would increase inflation and unemployment in the economy, hurt lower income groups more than higher income groups, and discriminate against oil consuming States in the Northeast.

CARBON TAXES, BTU TAXES, AND OTHER ENERGY TAXES

A variety of broad-based energy taxes proposed in recent years to address the United States' energy, environmental, and fiscal problems would have significant economic effects on biomass energy. These include carbon taxes (excise taxes on fossil fuels based on their carbon content), BTU taxes (taxes based on the heat content of various fuels), ad-valorem taxes on energy (excises on the value of energy resources), and other broadly-based energy taxes. While the precise impacts on biomass and other energy alternatives varies, these proposals would, like the oil import tax, increase the price of oil (the benchmark energy resource) and promote production and use of energy from biomass (3).

BIBLIOGRAPHY

1. S. Lazzari, *Gasoline Excise Tax: Economic Impacts of an Increase*, Issue Brief No. 93028, Washington, Congressional Research Service, 1992, 13p.
2. S. Lazzari, *Taxation of Imported Oil*, Issue Brief No. 87189, Washington, Congressional Research Service, 1988. 14p.
3. S. Lazzari, *Btu Taxes, Carbon Taxes, and Other Energy Tax Options for Deficit Reduction*, CRS Report #90-384 E, 1990, 19 p.

ENVIRONMENTAL ANALYSIS

The results given by analytical methods are critical to making sound decisions about the changes in the environment and the sources of those changes. Sometimes what is required is only a qualitative determination of the elemental and molecular components of a selected specimen such as trace elements in soil. Other times the goal is the quantitative measurement of the fractional distribution of constituents in a gas such as the emissions in the tailpipe from an automobile. Sometimes it is also necessary to monitor a stream over an extended period of time. Information concerning analytical methods useful for specific purposes may be found in many articles dispersed alphabetically throughout the *Encyclopedia*. The follow articles are introductions to specific analytical methods of considerable importance for environmental analysis.

ENVIRONMENTAL ANALYSIS (CHEMICAL SENSOR APPLICATIONS)

BARRY M. WISE
JIRI JANATA
Pacific Northwest Laboratory
Richland, Washington

The latter part of the 20th Century has been dubbed the "information age." In fact, the information age really began with an explosion of data which resulted from the use of computers and data acquisition hardware. The increased availability of data has driven the need for advanced data processing algorithms that turn data into information. As data algorithms and computer hardware have advanced, the need for new sources of data has become apparent. The result is the self-stoking cycle of data and information that is occurring now. Nowhere is this more apparent than in the field of chemical sensors.

Chemical sensors provide data about the chemical state of our surroundings, such as the atmosphere of our habitats, about the course of dynamic chemical processes taking place in chemical manufacturing plants, etc. However, in most complex sensing situations, the simple functional relationship between the concentration of some species in the system and the raw electrical output from the sensor for that species is rarely adequate. Interferences from other species present in the system often result in biased measurements. The response of analytical chemists has been to collect more data on the system under observation and attempt to deconvolute the measurements.

Modern statistical techniques popularly known as chemometrics have been developed to overcome this measurement deconvolution problem. It has been amply demonstrated that multivariate measurements, combined with modern data and system modeling methods, can produce information well beyond the limits achievable with individual sensors (1). Thus, the advanced data processing techniques have become an integral part of the sensing process.

The driving force behind environmental science is the need to obtain a rational and quantitative assessment of the ecological impact of past and present human activities on the future prospects for our continued existence. The development of a dynamic ecological model is the subject of this emerging scientific discipline. Because many of the interactions between humans and environment are chemical in nature, chemical sensors are expected to play an important role in environmental science. Predictably, they will be called "environmental sensors" very much like "biosensors" became a poorly defined but popular catchword for a group of sensors that had anything to do with "bio-" several years ago.

There are close to 1000 papers dealing with chemical sensors published annually, and the trend indicates further increase in this number (2–4). There are actually more sensor papers than species that are sensed. This paradox is partly due to the fact that most applications require a "customized" sensor with different attributes dictated by the application. A classic example of this situation is the sensing of hydrogen ion. Although the principle of operation of a pH sensor is the same, its implementation for pH measurement in different situations, for example soils, chemical process streams or *in vivo* measurement of pH in whole blood, are very different.

There have been several reviews on the general subject of environmental sensing published already (5–11) and the monitoring of effluent gases has been covered as a separate topic (12–15). In spite of the apparently large number of environmental sensors, there have been few successful implementations of chemical sensors in environmental applications. This is due largely to the problems created by non-specific sensors and the great number of potentially interfering species in most real world scenarios.

SCOPE OF ENVIRONMENTAL SENSING

Environmental problems range from those that make life unpleasant or difficult to those that could lead to a large alteration of the present ecosystem and ability of humans to survive. Most of the well-publicized environmental issues such as the effect of fluorocarbons on the ozone layer in the upper atmosphere, the effect of phosphates on natural water, adverse health effect of pesticides, etc, fall nearer the "make-life-difficult" end of the spectrum (16).

A good example for an environmental sensor in a make-life-difficult application is the monitoring of stack emissions from industrial processes. Large chemical and power plants can afford fairly expensive equipment for monitoring emissions (such as on-line spectrometers), but smaller plants will have to rely on less expensive sensing alternatives. It has also recently been found that in many urban areas, such as Los Angeles, the main contributors to air pollution are nonpoint sources such as dry cleaners and bakeries. Small businesses such as these cannot support expensive equipment for monitoring emissions, yet it is clear that if progress is to be made in improving air quality, these small sources must be monitored.

Another potentially huge market for environmental sensors is the detection of leaks of volatile organic compounds from valves and fittings in chemical plants. Current regulations require frequent testing of these systems, but it is entirely possible that continuous monitoring will eventually be required. Related areas include the monitoring of chemical storage tanks for fugitive emissions and the monitoring of oil rigs.

An example of a potentially more serious environmental issue arises from the prolonged and concentrated production of military materials, namely transuranium fuel for nuclear weapons. The United States Department of Energy production site at Hanford, Washington, is the largest U.S. environmental problem of this type. Specific examples of environmental problems at Hanford will be referred to throughout this article. Unlike other synthetic materials, such as synthetic polymers for consumer products, transuranics and fission products have been produced by nuclear reactions and are not subject to chemical degradation processes. The lifetimes of some transuranium elements (such as Pu^{239} and Am^{241}) and fission products (such at Tc^{99} and I^{129}) are very long, as measured on the historical time scale of the human species, and many of these materials are known to be chemically and genetically toxic. Release of large quantities of these materials into the biosphere could make life "more than difficult."

The entire spectrum of chemicals has been used during the large-scale production of nuclear materials at Hanford (17). These chemicals include halogenated hydrocarbons, heavy metals, nitrates, nitrites, ferrocyanides, toxic and flammable gases, a variety of complexing agents, etc. Any of those materials would feature prominently on a list of environmental contaminants. What makes them particularly serious, in the context of the nuclear materials production site, are the quantities involved and the fact that they are mixed in unknown ratios with each other and the radioactive waste. In addition, soil and groundwater contamination, both chemical and radiochemical, is a problem at Hanford. Several thousand wells at the Hanford site must be monitored for groundwater contaminants. These wells are checked at least yearly, with many of them requiring testing more often than that. Sensing technology that would eliminate sampling of the wells and transport of the samples to testing laboratories would greatly improve the efficiency of the monitoring effort. There are many other site specific needs within the DOE complex.

CHEMICAL INTELLIGENCE

Chemical sensors are only one component of *characterization*, which is defined as follows:

Characterization is the act of developing a model or understanding of the chemical, physical, and spatial properties of a system.

In this context, characterization can be seen as chemical intelligence, which provides operational information about the chemical state of the system under consideration. From the analytical point of view data is obtained from: (*1*) off-line, *discontinuous*, batch chemical analysis; (*2*) on-line, *continuous*, *in situ* chemical sensors and *ex situ* sensor systems; and (*3*) transformed into information through application of advanced data processing. This scheme is diagrammed in Figure 1.

The validity of the information obtained by these two intertwined routes is critically dependent on the *sampling*. There is a problem with the meaning of the term "characterization" in systems as complicated as a waste tank or a human body in which several phases (or body compartments) coexist in a state of quasi-equilibrium. The reason is that a sample obtained from one phase is in a complex and often undefined relationship with respect to the other phases. The perturbation of such a system during any operation, such as the sampling event itself, may perturb any such relationship and render the result of the analysis invalid. The critical step in such a case is the separation of the phases and/or homogenization of the sample. Only then can a rational characterization of the original content of the object be contemplated. The situation is substantially similar in the health effects and in soils and groundwater areas. In all these cases, *the sampling space needs to be precisely delineated.*

On the other hand, local sampling or local sensing may have merit in obtaining chemical intelligence about critical localities within a complex system. For example, measurement of hydrogen-ion activity at the boundary between the organic liquid and aqueous phase adjacent to the container wall may be critically important in assessing the corrosion risk of that metal container. Similarly, accumulation of a heavy metal in certain organ tissues may be highly important for assessing the associated health risks.

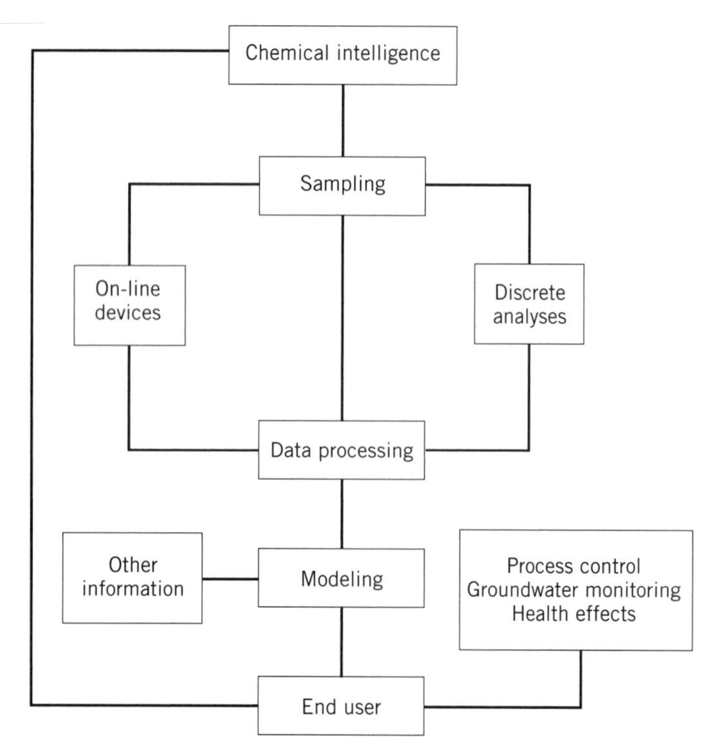

Figure 1. Chemical intelligence scheme.

The characterization sequence defined in chemical intelligence terms typically begins with:

1. Noninvasive physical diagnosis
2. Separation of macroscopic phases
3. Batch analysis and continuous sensing of the individual phases
4. Development and updating of a model of the system

The last step yields more detailed information about the chemical content of the sample. This information node represents the boundary condition on which the further treatment during the processing stage depends. From this information the strategy for development of a *model*, decisions about the raw data treatment, and deployment of additional chemical sensors/techniques must evolve.

Advantages and Limitations of Chemical Sensors

Chemical sensors are *complementary* to batch chemical analysis. Ordinary chemical sensors are generally not suitable for sustained, long-term quantitative monitoring because the baseline stability is rarely adequate. On the other hand they can economically provide valuable information about rapidly changing systems, something that batch techniques cannot do. If the required accuracy is traded for stability and selectivity, it is possible to design a sensor that performs according to its specification for a very long (months) period of time. Such sensors are often called *threshold detectors or alarms*. Many scenarios can be envisaged in which such devices would be useful in the framework of environmental monitoring. A better approach to solving the baseline stability problem may be design of higher order chemical sensors and sensor arrays, as will be shown later.

Single chemical sensors generally do not perform well in situations where many different chemical species are present, which is typically the case in environmental monitoring applications. This is the issue of chemical selectivity. Univariate sensors cannot reach the selectivity of modern analytical instruments (eg, of the hyphenated techniques such as gc-ms) in which the analytical separation step (eg, gc or hplc) precedes the actual quantification step. The trade off, *vis-a-vis* the sensor approach, is the speed of information acquisition, portability, remote operation, and economy. Again, it will be shown that higher performance can be expected from higher order sensors and sensor arrays.

A very unique aspect of chemical sensors is that some of them can be designed to respond to the change of *activity*, rather than concentration, of a given species. This is critically important in obtaining information about processes that are driven by thermodynamic, rather than by kinetic considerations. Thus, for example, interphasial partitioning equilibrium occurring in soil transport are driven by the activities rather than concentrations of the species involved.

Given the dangerous nature of some environmental remediation operations, remote sensing may be required. Another driving factor is the economy of obtaining information through chemical sensing as opposed to a discrete chemical analysis. These special considerations may affect the relative importance of the two routes of information acquisition shown in Figure 1. Multiple data inputs are then processed in order to formulate and refine the model of any given operation.

It is important first to identify the distinguishing features of environmental sensors by identifying the special requirements encountered in environmental sensing. In detection and determination of various species in the environment gaseous (eg, air), solid (eg, soils) and liquid (eg, groundwater) samples are encountered. The volume of the sample and the availability of the sampling space are not expected to be limiting factors. Therefore, the size of the individual chemical sensor is not a dominating issue. However, design and fabrication of multisensor arrays is important from the point of view of enhancement of chemical selectivity. Thus, microfabrication is expected to play an important role in development of environmental sensors. For environmental processing (ie, remediation) the speed of response may be the most important factor. On the other hand the detection limit may be most important in mapping the dynamics of contaminant spreading in soils, air, and groundwater. The attributes of chemical sensors required for different types of sensing situations are below.

Environmental	Health	Processing
Detection limit	Size	Speed
Selectivity	Safety	Safety
Stability	Selectivity	Robustness
Dynamic range	Detection limit	Stability

DEFINITION OF TERMS

In view of the existing apparent misconceptions about chemical sensors, it is pertinent to review the terminology and some essential features of these devices (18). Chemical sensors translate information about the concentration of chemical species in the sample into an electrical signal. Chemical sensors can be categorized by the four basic mechanisms by which this is done: thermal, mass, electrochemical, and optical. These devices operate in a *continuous* mode, in contrast to *chemical sensor systems*, such as flow-injection analysis or chromatography that requires manipulation of the sample and provides information in a *discrete* manner. Sensor systems can be automated and are expected to play an important role in environmental characterization. There is a low limit of concentration below which the sensor does not respond to a change of concentration of the detected species. This is the so-called *detection limit*. At the high end of concentration, the sensor again does not respond to the concentration change. This is called the *saturation limit*. The difference between these two limits is the *dynamic range*, which from the practical point of view is the usable concentration range of the sensor. Different sensors have different dynamic ranges. The slope of the signal vs. concentration dependence within the dynamic range is called the *sensitivity* of the sensor. Another important parameter is the *selectivity*, which is defined as the ability of the sensor to respond to the species of interest while being inert to other species. *Robustness* is a qualitative term that describes the ability of the sensor to maintain its essential performance characteristics under adverse operating conditions. Robustness is related to the operational lifetime of the sensor. Speed of response of the sensor is characterized by its time-constant, which for exponential processes is given as the time required for the signal to reach 63% of its final value. For sensors that are governed by more complicated response kinetics the response times of 95% or 90% are often used.

All of these parameters that characterize sensor performance are critically dependent on the conditions under which the sensor is used. This is why there are so many seemingly identical reports of applications of seemingly identical sensors. The key in a rational use of any sensor is to design it, package it, and evaluate its performance under the intended conditions of its application.

CHEMICAL SENSORS IN ENVIRONMENTAL REMEDIATION

It is useful to divide chemical environmental sensors according to their application. At the Hanford site, environmental remediation includes retrieval and processing of tank wastes and contaminated soils and requires monitoring of process input streams and process unit operations for control purposes and effluent monitoring. Also sensors are used for the monitoring of contaminants already released into the biosphere. Environmental monitoring would include mapping out of the movement of already released contaminants through soil and groundwater and the monitoring of wastes stored in underground tanks. The monitoring and remediation of contaminated industrial sites, such as toxic waste dumps, will require similar sensors.

Dynamic Models

A dynamic model relates the response of a system to a stimulus as a function of time. Dynamic models can be obtained directly from theoretical consideration of the governing physical laws of the system (such as material and energy balances) and other known or estimated system properties (mass and heat transfer coefficients, equilibrium constants, reaction mechanism, etc). Often, however, many of the fundamental properties of the system are unknown or not known accurately enough to allow development of an accurate dynamic model. In these cases it may be necessary to develop a model based entirely on data from controlled experiments (more desirable) or from observations of the system (less desirable). There is an extensive body of literature on the development of dynamic models. The foremost publication is probably *System Identification* (19). Recently, more work has been done on nonparametric model methods which require less *a priori* knowledge of the process dynamics (see, eg, ref. 20,21).

Advanced Process Control

The waste tanks at Hanford and many contaminated industrial sites contain an incredibly complex, heterogeneous mixtures. A rough idea about the number of possible sampling spaces can be obtained from the Gibbs phase rule which, loosely interpreted, says that number of phases will be comparable to the number of components (ie, chemicals). Even though some species are linked by equilibrium relationships and others may be accumulated only at interphases, the number of coexisting phases is expected to be large. This conclusion is supported by photographs of Hanford waste tank interiors. An example is shown in Figure 2. From the sampling point of view, it can be immediately concluded that the condensed phases must be first separated before any meaningful information can be obtained. On the other hand, the composition of the head space in a waste tank or the vapors over contaminated soil can be obtained both by batch sampling and analysis and by direct sensing techniques.

Sensors used in environmental remediation will be required to meet rigorous safety and robustness requirements. For instance, sensors used in nuclear waste process could be exposed to very high radiation fields (~1000 rad/h). In other applications, potentially explosive gases exist in high concentrations and impose restrictions on the operational safety of electrical devices. It is expected that optical sensors will play a prominent role in these processing situations (22).

Even the simplest processing operation such as stirring or physical separation of the condensed phases (eg, filtration) must be done under controlled conditions in order to ensure the safety of the operation. Sensors for this application do not need to be particularly accurate, but they must be fast. Processing operations that result in some conversion of the raw waste materials can be very complex. A diagram of a more sophisticated operation such as the so called "clean option" process for the separation of Cs^+ from Hanford tank waste is shown in Figure 3. The nodes for possible deployment of sensors and/or discrete automatic assay have been added to probable locations in this diagram. They include sensors and sensing systems for ions and for neutral species in both liquid and gaseous phase.

Figure 2. Photograph of the inside a double-shell tanks taken with a remotely controlled camera.

The raw data obtained from this distributed sensor network are processed through the control system, which incorporates a dynamic model and is linked by the actuators back to the process. This feedback loop constitutes *advanced process control* (Fig. 4). Any operation in which macroscopic quantities of materials are manipulated must have advanced process control to operate the system safely and efficiently.

A dynamic model is an essential part of the control of a dynamic system. All control schemes use either an implicit or explicit model of the system. In classical PID control (proportional, integral, derivative), the implicit model is the response time and approximate order of the system. In more advanced control systems, an explicit model of the system is incorporated directly in the control loop (see Fig. 4a). In the figure the system dynamic model is placed in parallel with the system and given identical inputs (u). Thus the difference between the output of the model ($y^\wedge$) and the measured output of the system (y) form an estimate of any unmeasured disturbance to the system (v). The controller bases its actions on this disturbance estimate and the desired output of the system, the setpoint. Modern control models consist, in part, of a mathematical inverse of the dynamic system model. In this sense the controller cancels the dynamics of the system and replaces them with the desired system dynamics. The more accurately the model represents the system, the more accurately the system can be controlled to conform to the setpoint.

The dynamic model is also used to decrease the uncertainty in the measurements. This problem was first solved by Kalman and the approach of optimally combining information on system inputs, system outputs and the dynamic system model to improve measurements is known as Kalman filtering (23). This procedure can also be used to monitor the health of the process sensors and determine if sensor failures have occurred. Thus, it can be seen that system models can improve the effective accuracy and robustness of the process sensors.

CHEMICAL SENSORS IN ENVIRONMENTAL MONITORING

The contamination of soils and groundwater has resulted from the discharge of the hazardous liquid wastes over the preceding half century. This was an accepted industrial practice over this period. At Hanford, soil and groundwater contamination was the result of nuclear materials production. For example, "inverse wells," where liquid waste was injected into the subterranean zone, were used for liquid waste disposal until 1980. More typical industrial processes would be the production of petrochemicals such as pesticides and plastics. Chemical sensors used in agriculture should be directly applicable to many of these sensing needs. However, there are special requirements, such as sensing of the organic liquids and vapors in the subterra-

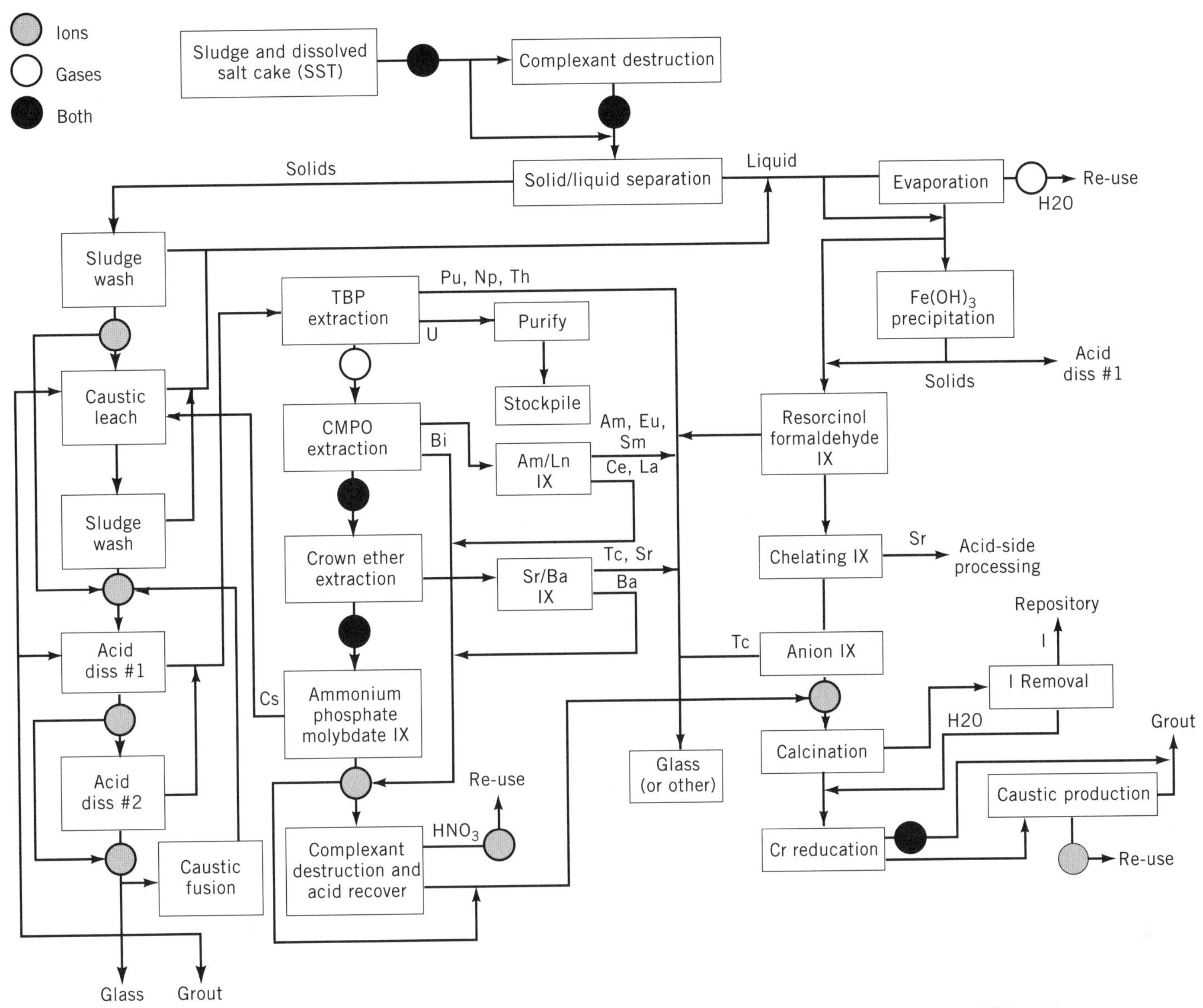

Figure 3. Flow-diagram of the proposed "clean option" process for extraction of Cs^+ with inserted "chemical intelligence" points.

nean (vadose) zone, that require design of special class of sensors. Evaluation of the soil and groundwater contamination at Hanford has been prepared (24). Similar studies will be required, and in many cases have been prepared, for industrial sites.

In contrast to processing, changes of contaminant concentrations in soils and groundwater are slow. Moreover, the concentrations of the species to be monitored are often low and the important parameter is likely to be the *activity* rather than concentration. This consideration virtually eliminates optical sensing of ionic species as a viable possibility (25). The monitoring of the movement of contaminants through the soils and groundwater at Hanford is done primarily through a system of sampling wells from which samples of the liquid or suspension of soils can be obtained from various depths and at various time intervals (Figs. 5, 6). The samples are usually brought to the laboratory for analysis. Deployment in *in situ* chemical sensors in this situation is a principal economic incentive. Because the information must be obtained over long periods of time the calibration stability is of premium importance. This and the requirement of low detection limits makes the deployment of self-calibrating sensor systems and higher order sensors particularly attractive.

The pH of the soil, and its buffer and redox capacity play a dominating role in the dynamics of subterranean plumes of heavy metals and transuranium metals. Sensors and sensing systems for these applications should be at the top of the priority list.

Dynamic modeling and Kalman filtering may be applied to the problem of sensing changes in soils and groundwater in much the same way that it is applied in processing, with the exception that the feedback part of the loop no longer exists. This is shown in Figure 7b. Here the output from the model and raw measurement on the system are combined in an optimal fashion to achieve a better esti-

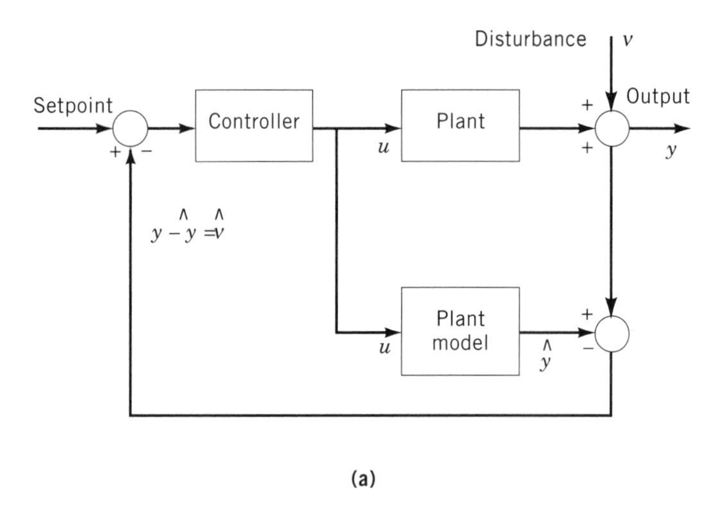

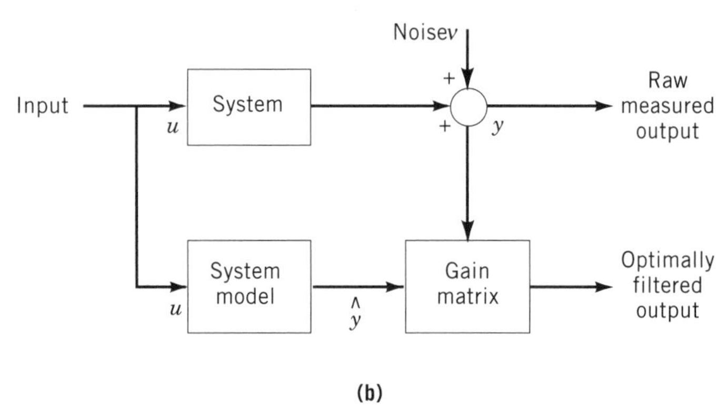

Figure 4. Schematic diagram of dynamic model used in (**a**) advanced process control loop and (**b**) for soil monitoring.

mate. The general approach has been demonstrated by several researchers in analytical chemistry (23,26).

STRATEGIES FOR DEVELOPMENT OF ADVANCED CHEMICAL SENSORS

It is possible to classify sensing systems according to the dimensionality of the data they produce. The dimensionality of the data defines the *order* of the system. For instance, a pH electrode, which takes one datum per measurement, is a zero-order instrument since a single point has dimension zero. A spectroscopic instrument, which measures absorbance as a function of wavelength, is a first-order instrument. Each measurement produces a vector of data. A vector, of course, is one-dimensional (a first-order tensor). There is a growing trend towards instruments that produce a matrix (second-order tensor) of data with each measurement. Typical examples are the hyphenated techniques, such as gc-ms, and many types of time resolved spectroscopy. In these techniques, measurements are made as a function of two variables to produce a matrix of measurements. For example, in gc-ms the quantity of ions produced is measured as a function of their atomic mass and the retention time on the gc.

It is impossible to detect the presence of an interferent with a zero-order instrument. There are no data with which to cross reference the measurement. In a first-order instrument, it is generally possible to detect the presence of interferents; however, it is not generally possible to correct for the interferents and quantitate the species of interest correctly in their presence. With a second-order instrument, both detection of interferents and quantitation in their presence is possible.

The concept of order is relatively new to chemical sensors and sensing systems. Traditional chemical sensors provide output that is uniquely related to the concentration (or activity) of one species. They are economical and simple to operate. However, they require frequent calibration and any deviation from the calibration status between calibration steps is not detected. Another drawback of a zero-order sensor is that it relies solely on the specificity of the interaction of the detected species with the selective layer for the selectivity of its response. It is a unique and fortunate coincidence that a glass pH electrode has such an extraordinary selectivity and dynamic range for the most ubiquitous chemical species, the hydrogen ion. Ion selective electrodes, which represent an entire class of most developed chemical sensors, do not match the performance of a glass pH electrode for detection of other ions.

In the last decade individual zero-order sensors have been grouped to form first-order instruments. The output of these systems have been processed through advanced statistical and mathematical algorithms collectively known as chemometrics.

Chemometrics

Chemometrics can be broadly defined as *the science of relating measurements made on a chemical system to the state of the system via application of mathematical or statistical methods*. It is the name given to a collection of methods that have been found useful in chemical applications.

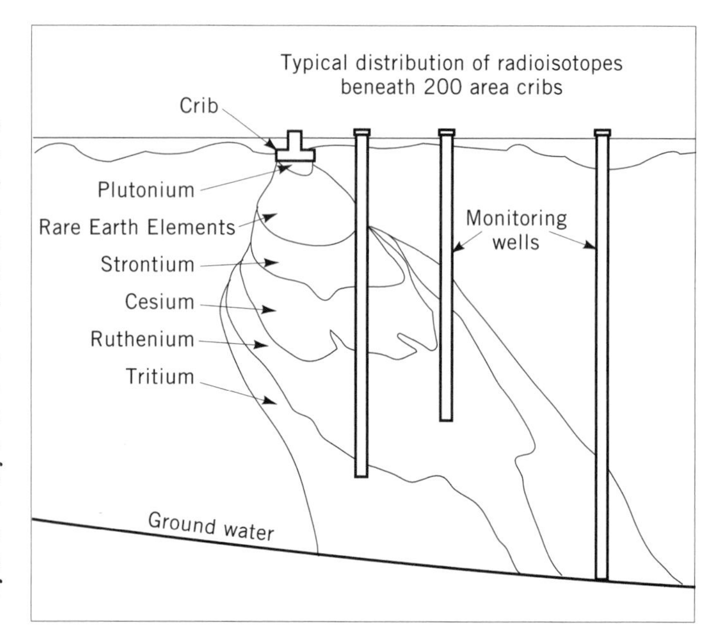

Figure 5. Discharge of the waste into a crib and ensuing contamination.

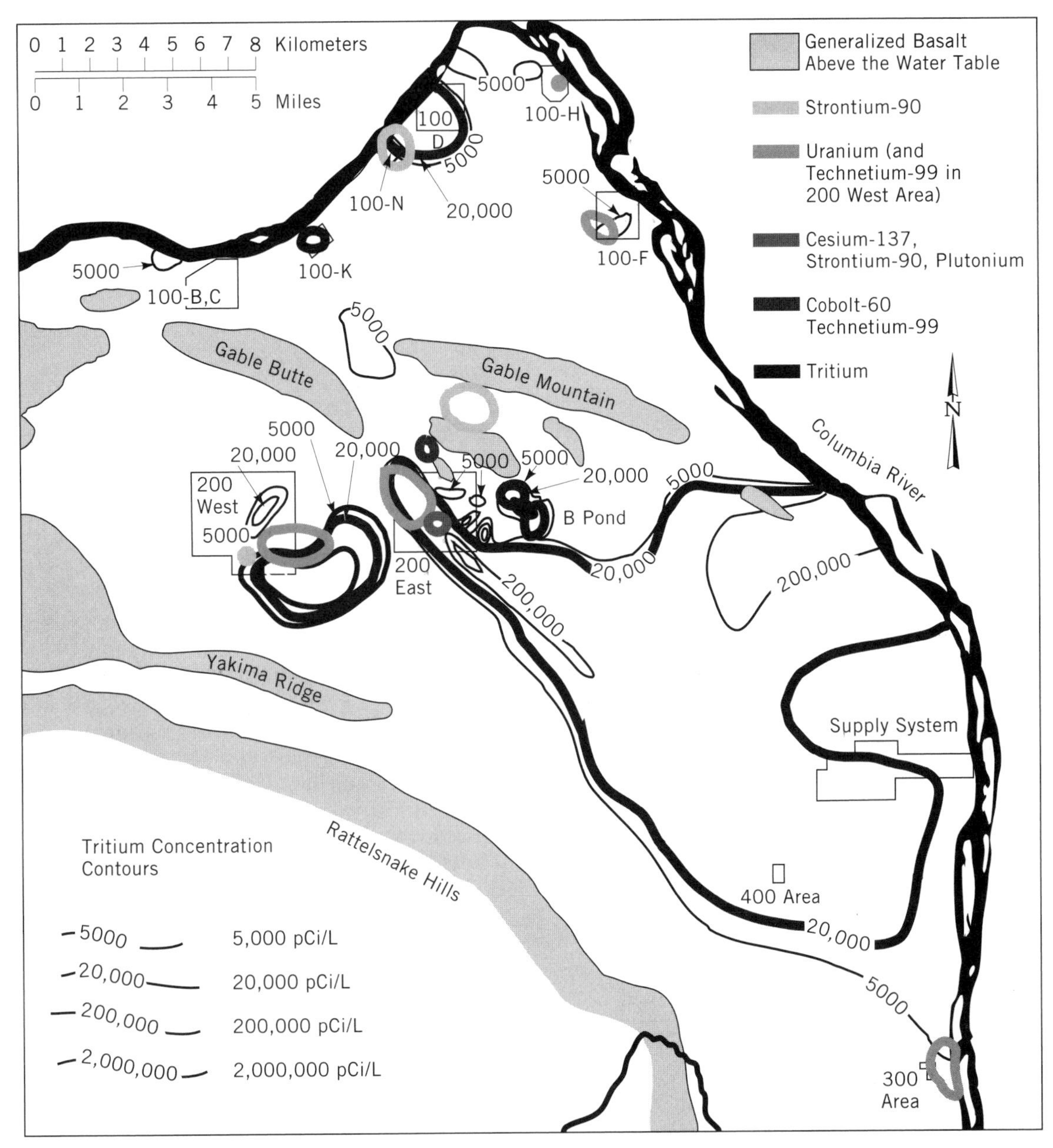

Figure 6. Distribution of various contaminants at the Hanford site and comparison to the Drinking Water Standard.

A principal area of interest in chemometrics is the general model identification problem. This problem can be divided into calibration of analytical instruments and sensors and identification of dynamic models of chemical systems.

Methods for calibration of zero-order instruments have existed since the time of Newton and Gauss. Typically, measurements are made with the instrument on several known samples. A simple bivariate model is then developed relating the instrument response to the property or concentration of interest. The model is often a simple least squares fit of the response to the system property, though often a linearizing transformation must be used first. An example of this is the calibration of a pH electrode. The response of the instrument (in millivolts) is a nonlinear function of the concentration (activity) of hydrogen ion. Taking the log of the hydrogen ion concentration (ie, pH) linearizes the problem since the relationship between response and log(activity of H^+) == pH is linear. The final model consists of a slope, relating the sensitivity of the measurement to changes in pH and temperature and an intercept that gives the baseline offset, due to factors such as the reference electrode and the liquid junction that connects it to the sample.

In some cases the instrument response is a nonlinear function of the property of interest and, unlike pH measurement, it is not possible to obtain a linearizing relationship explicitly. In these cases a general nonlinear regression technique must be used to calibrate the instrument.

Once the sensor is calibrated, the model is used to relate instrument response to the property or concentration of interest in an unknown sample. This system works well

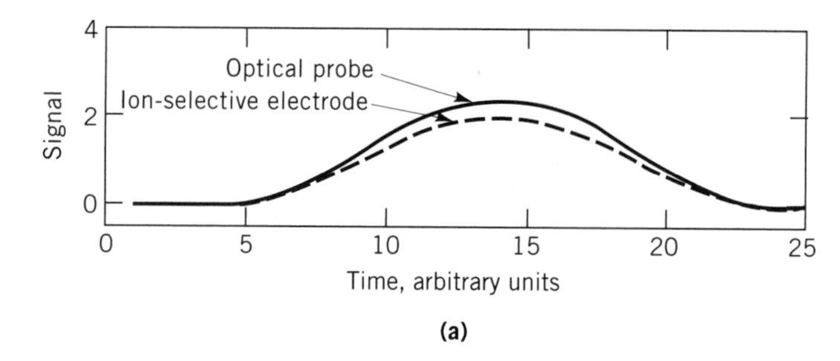

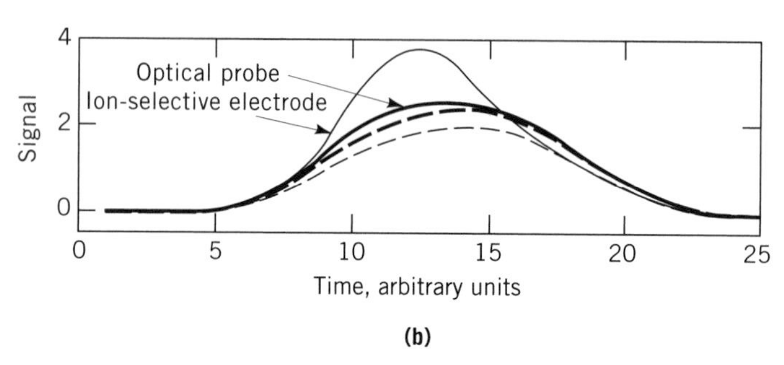

Figure 7. Response of second-order probe consisting of low resolution separation step followed by ion-selective electrode (light gray) and optic probe (solid black). (**a**) Response of the system to pure analyte (dashed lines). (**b**) Response of the system with interferent.

until an unknown interferent that changes the sensitivity of the instrument to the property is introduced into the system. Unfortunately, it is impossible to detect the presence of such an interferent without some knowledge about the system other than the instrument responses.

Often, it is necessary to make measurements in systems where the concentration of more than one analyte or system property varies. This would not be a problem if it were possible to build sensors that were totally specific for the analyte or property of interest. In general, however, it is not possible to do this and instead we must rely upon collections of sensors that are only partially selective for the analyte of interest. Models are then developed (using any of the methods discussed below) that relate the responses on all of the sensor elements to the analyte or property of interest.

There are many possible calibration methods for these linear first order systems. Perhaps the oldest is the classical least squares (CLS) approach (also known as the K-matrix method) (1,26), which requires knowledge of the pure component response of each analyte to each sensor. In some cases, however, it is not possible to obtain pure component responses and another method must be employed. In these instances, techniques such as multiple linear regression (MLR), principal components regression (PCR) (27), and partial least squares (PLS) regression (28,29), which can all be unified under the technique continuum regression (CR) (20,30), and ridge regression (RR) (31) may be employed. In each case data is collected on the response of the sensors to known mixtures of analytes. It is critical that the mixtures "span the space" of expected concentrations during use of the measurement system. This includes species that may not be of interest analytically but whose concentration may vary during the course of using the sensing system.

The MLR approach (also known as ordinary least squares or OLS) fits a model to the data that minimizes the sum of squares error of the fit of the model. In this sense it provides an unbiased estimate of the property of interest. In systems where there are more sensors than independently varying analytes or if the response of the instrument to several analytes is identical, the MLR problem becomes "ill-conditioned." This "co-linearity" problem is due to the fact that if there are fewer independent analytes than measurements, some of the measurements are necessarily linear combinations of other measurements. In effect, there are more parameters in the model than independent variations in the data. In these cases the MLR model obtained may be a strong function of small changes in the calibration data caused by "noise" and may fit the calibration set quite well but not provide accurate estimates for new samples.

The PCR, PLS, and CR approaches solve the colinearity problem by determining linear combinations of variables (sensor elements) or latent variables that are independent. In general, these methods will produce a smaller number of latent variables than original variables. This "reduced set" of latent variables are then regressed against the properties of interest. A cross-validation procedure is used to determine the optimum number of latent variables to retain for prediction of new samples.

Ridge regression solves the co-linearity problem in a different way, essentially by restricting the magnitude of the coefficients that relate the sensor response to the analyte concentration. Experience has shown that this method can be very effective in practice. However, ridge regression does not provide the useful diagnostics for detecting of outlier samples, i.e., samples that contain analytes not in the original calibration matrix.

An example of a multivariate calibration problem follows. Suppose that you have two sensors, one which measures the activity of Mg^{2+} in solution by ion-selective electrode and the other which measures concentration by a fiber optic probe. It is desirable to use both of these measurements to calculate $[Mg^{2+}]$, but in the example calibration set, only $[Mg^{2+}]$ was allowed to change.

If the calibration done for this system was attempted with MLR, it would be found that the response of the two sensors was co-linear (assuming that the log of the ion-selective electrode was used and that no nonlinear effects were observed) and a good solution would not be obtained. Using a CR method, it would be found that one linear combination of the response of both instruments would be used to form the calibration model. The CR model is essentially two models in one. The first model relates the response of the two probes to each other and the second model relates their collective response to $[Mg^{2+}]$. This is illustrated graphically in Figure 8, where the response of the ion-selective electrode and the fiber optic probe are plotted against each other. Note that a straight line fits through the data quite well. Any nonsystematic deviations

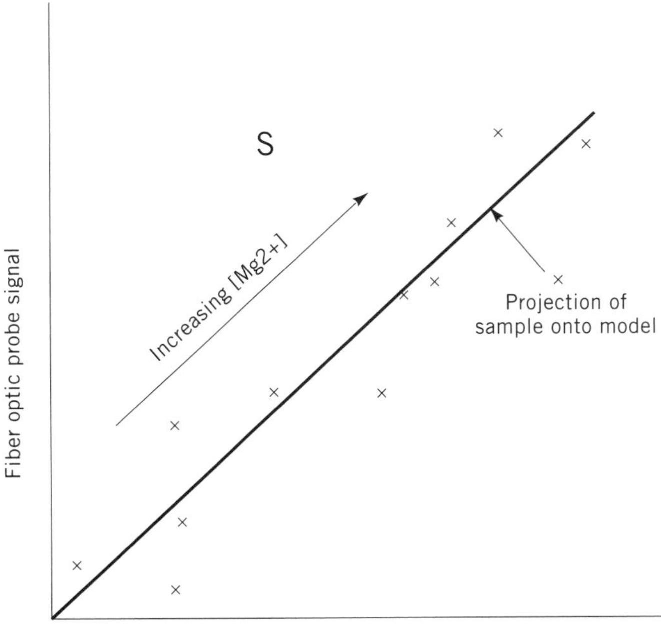

Figure 8. Relationship of linearized ion selective electrode signal to fiber optic probe signal under calibration conditions (xs) with an unusual sample (**S**).

from the line can be modeled as noise (random fluctuations of the output of the sensors). Systematic deviations from a line with a slope of 1 would be due to the fact that the response of the optical probe does not depend upon the activity coefficient of [Mg^{2+}] while the ion-selective electrode does. The concentration of Mg^{2+} is estimated from the projection of the measured data onto the line. This is commonly known as a "score" value.

The CR techniques offer the advantage that they provide a model (the latent variables) that describes the normal variation in the response of the sensors (in linear algebra terms, the subspace spanned by the calibration set responses). When a new sample is analyzed, the difference between the new sample response and the normal subspace, or residuals, can be calculated (these are the deviations from the line in Figure 8). Statistical tests can be used to determine if the residuals are significant or within normal variation. It is in this way that interferents are detected. When the instrument responses do not "span the same space" as the calibration set, something new must have been added to the system, or the properties of the instrument itself must have changed.

Now suppose that an unknown change was made in the system (say, addition of another ion for which the ion selective electrode has some response), which affected the apparent activity of the Mg^{2+} relative to its concentration. When the CR model was applied, it would be found that the relationship between the measurements had changed. This is shown graphically in Figure 8, where the new data point is plotted. Note that the residual is very large, much larger than that associated with random fluctuations in the system. In this instance, it would be known that the model was no longer valid and that the estimates of [Mg^{2+}] were no longer reliable.

Unfortunately, even though it is possible to detect the presence of an interferent, it is not possible to correct for it and do accurate quantitation on new samples. This is due to the fact that the exact nature of the change is impossible to quantify since it is not known how orthogonal the interferent is to normal changes in the system. In our example, deviation from the line relating the response of the two sensors could just as easily be caused by changes to either sensor. For instance, change of ionic strength could also account for such a shift.

Second order techniques can be classed as "bilinear" or "non-bilinear." In a bilinear technique, the response of the instrument to a single pure analyte can be modeled as the (linear) outer product of two vectors (32–34). The resulting matrix of data will have a rank of one (rank = number of linearly independent columns or rows in the matrix). Many common hyphenated instruments involving a separation step are bilinear, including lc/uv, gc/ms, gc/ftir, and emission-excitation fluorescence. A nonbilinear instrument produces data matrices that have rank greater than one for pure analytes. Examples of this are ms/ms and 2d nmr, where the rank of pure analyte matrices may be as high as 20–30 (35,36).

As might be expected, different techniques have been developed to deal with calibration of second order instruments depending upon whether the data produced is bilinear, weakly nonbilinear or strongly nonbilinear. The Generalized Rank Annihilation Method (GRAM) works well for bilinear data. Several approaches have been suggested for nonbilinear data, including Nonbilinear Rank Annihilation (37) and Residual Bilinearization (38,39). Most recently, techniques such as Restricted Tucker Models have been developed to deal with cases where the data is only mildly nonbilinear (40).

GRAM can be used with bilinear data to quantitate for the species of interest in spite of the presence of interferents. The basic idea behind GRAM is that multiples of the pure analyte response are subtracted from the sample response and the rank of the response matrix is observed. When the multiple of the pure analyte response is correct, the rank decreases by 1. In the presence of interferents, the response matrix rank will increase, but GRAM will work in the same way since it looks for changes in the rank of the response matrix.

Moving back to our example, assume for a moment that we have combined our sensors for measurement of [Mg^{2+}] with a low resolution separation step preceding the measurement. This would result in a "bilinear" second-order instrument. Suppose that the interferent is sufficiently different from the analyte of interest so that the separation step results in some separation (the interferent and analyte do not completely overlap). The response of the instrument to the pure analyte and to the mixture is now shown in Figure 7. The rank of the pure analyte matrix is 1, while that matrix with addition of interferent has a rank of 2. By subtracting multiples of the pure analyte response from the response of the mixture, a point can be found that reduces the matrix rank back to one. This point represents the correct amount of analyte present in the system.

As mentioned above, it is not always possible to identify a linearizing transformation that will allow one to use the linear techniques discussed above for calibration. There are a variety of approaches being used for such nonlinear problems. In the last several years, artificial neural networks (ANNs) have been applied to sensor calibration problems (41). Additional techniques include locally weighted regression (42), multivariate adaptive regression splines (43), and PLS techniques with nonlinear inner relationships (44,45).

In remote systems that may be typical of the monitoring of hazardous waste sites, the recalibration of an instrument is likely to become an issue. It would be desirable to transfer a calibration (mathematical model) from another instrument in a nonhazardous environment. This problem has been studied by several researchers and some methods are beginning to gain acceptance. Some spectroscopic instrument manufacturers are developing the concept of an *absolute virtual instrument* and relating the response of any particular instrument to this virtual response.

The basic idea behind calibration transfer is that it is possible to develop a "transform" between instruments based on the results of testing a few "well chosen" samples on both instruments. A full calibration model is then developed on one instrument, usually the laboratory (rather than field) instrument that is generally both easier to work with (get to) and of higher quality. The calibration for the second instrument is the transform that relates the instruments times to the calibration for the first instrument.

There are several methods for developing a transform between first order instruments. The piece-wise direct standardization (PDS) method has been shown to be quite effective for first-order instruments such as spectroscopic instruments and arrays of temperature sensors where it is expected that the correlation between responses on specific channels decreases with "distance" (46,47). With temperature measurement, this distance is the physical distance between sensors while for spectroscopic measurements it is the difference between wavelengths.

As might be expected, less work has been done to date on transforms for second-order instruments. The second-order problem is considerably more complex than the first-order problem. However, simultaneous standardization of both instrument dimensions has been demonstrated for simulated lc-uv data (48).

Microfabrication

Almost every type of chemical sensor can be made by using silicon-based microfabrication technology (49,50). In the context of the needs for environmental sensing the fabrication of multisensor arrays is particularly important. Multisensor chemiresistor arrays or multiple chemically sensitive field-effect transistors (CHEMFET) chips are the obvious example. Microfabrication can be extended to development of miniature sensing systems such as a potentiostatic electrochemical microcell (51). The requirements on lateral resolution of the fabrication process rarely exceed 5 μm; however, ability to deposit and geometrically pattern unusual combinations of materials is important. Because, during the course of its lifetime, the solid state device comes in an intimate contact with quite harsh environments new encapsulation procedures had to be developed (52). The microfabrication of environmental sensors is driven by the need for multisensor capability rather than by the size of the available sampling space.

Synergism Between Sensor and Remediation Process Development

As a general rule it can be stated that the interphasial interaction on which a macroscopic advanced process, eg, for environmental remediation, is based should also be the basis for development of an optimal chemical sensor. In other words, the development of advanced processes must go hand-in-hand with the development of chemical sensors and sensor systems for advanced process control. Thorough knowledge of interphasial physical chemistry is the mandatory common base for these two elements of environmental restoration.

For example, a chemical sensor for Cs^+ based on a conventional polymeric membrane/ionophore combination may not be usable in the high radiation fields encountered during the separation of the Cs^{137} isotope. Instead a new sensor based on radiation resistant Cs ion exchanger may have to be developed. Such a sensor may have a lower selectivity than the conventional polymer-based Cs ion-selective electrode but in terms of overall usability may be more appropriate. On the other hand, development of a new material for a Cs^+ sensor may lead to a new or improved material for the macroscopic conversion process. In any case, it is apparent that concurrent development of remediation processes and the required process sensors will result faster development and a higher likelihood of success.

CONCLUSIONS

A superficial examination of environmental sensing needs may lead to an erroneous conclusion that any of the existing, even commercially available, chemical sensors may be used and that no further sensor development is required. In exceptional circumstances this may be true; however, the state of characterization at Hanford and other DOE and industrial sites clearly shows that this is not the general case. The problem lies in the fact that the specifics of each sensing situation can degrade the performance characteristics of a commercially available sensor and in the extreme case make it unusable. This is often the result of unknown interferents in environmental samples that lead to biased sensor response. A careful examination and understanding of the operating principle of a given sensor under the application conditions is the least requirement for its rational deployment.

It is clear that there are many applications in environmental monitoring and remediation where no suitable sensor or sensor system exists and new devices need to be developed. It is the author's contention that first order sensors will become the norm for environmental applications and that the potential advantages of second order sensors will make them common. This will require continued development of both sensors, producing ever more data, and calibration and diagnostic methods, for extracting the required information. Thus, the field of environmental sensors will experience the data and information cycle characteristic of the "information age."

Acknowledgment
Help of Roy E. Gephart in compiling the Hanford inventory information is gratefully acknowledged.

BIBLIOGRAPHY

1. M. A. Sharaf, D. L. Illman, and B. R. Kowalski, *Chemometrics*, John Wiley & Sons, Inc., New York, 1986.
2. J. Janata, and A. Bezegh, *Anal. Chem.* **60,** 62R–74R (1988).
3. J. Janata, *Anal. Chem.* **62,** 33R–44R (1990).
4. J. Janata, *Anal. Chem.* **64,** 196R–219R (1992).
5. T. Hori, T.; ed., International Symposium on New Sensors and Methods for Environmental Characterization 1986, Kyoto, Japan *Pure Appl. Chem.* **59**(4) (1987).
6. R. B. Smart, *Hazard Assess. Chem.* **5,** 1–27 (1987).
7. J. D. R. Thomas, *Int. J. Environ. Anal. Chem.* **38**(2), 157–169 (1990).
8. R. Kalvoda, *Electroanalysis (N.Y.)*, **2**(5), 341–346 (1990).
9. S. P. Banerjee, *J. Mines, Met. Fuels*, **37**(5), 178–185 (1989).
10. A. A. Tumanov and E. A. Korostyleva, *Zh. Anal. Khim.* **45**(7), 1304–1311 (1990).
11. I. Giannini, *Fis. Tecnol. (Bologna)* **12**(1), 23–34 (1989).
12. Chen, Aifan, Luo, Ruixian, Tan, Thiam Chye, Liu, Chung Chiun, *Sens. Actuators*, **19**(3), 237–248 (1989).
13. L. L. Altpeter, Jr., T. A. Williams, M. Rupich and H. Wise, *Oper. Sect. Proc. Am. Gas Assoc.*, Volume Date 1989, p. 144–147.
14. J. Riegel and K. H. Haerdtl, *Sens. Actuators, B*, **B1**(1–6), 54–57 (1990).
15. A. Accorsi, G. Delapierre, C. Vauchier, and D. Charlot, *Sens. Actuators, B*, **B4**(3–4), 539–543 (1991).
16. A. Gore, *Earth in the Balance*, Plume Publishers, 1993.
17. D. L. Illman, *Chemical and Engineering News*, 9–21 (June 21, 1993).
18. J. Janata, *Principles of Chemical Sensors*, Plenum, New York, 1989.
19. L. Ljung, *System Identification—Theory for the User*, Prentice-Hall International, London, 1989.
20. B. M. Wise and N. L. Ricker, *J. Chemometrics* **7,** 1–14 (1993).
21. B. M. Wise and N. L. Ricker, *Process Control and Quality*, **4,** 77–86 (1992).
22. M. A. Arnold, *Anal. Chem.* **64,** 1015A–1025A (1992).
23. R. K. Woodruff, R. W. Hanf, and R. E. Lundgren, *Hanford Site Environmental Report for Calendar Year 1991.*
24. J. Janata, *Anal. Chem.* **64,** 921A–927A (1992).
25. S. D. Brown, *Chemom. Intell. Lab. Syst.* **10,** 87–105 (1991).
26. C. L. Erickson, M. J. Lysaght, and J. B. Callis, *Analytical Chemistry*, **64,** 1155A–1163A (1992).
27. T. Naes and H. Martens, *J. Chemometrics* **2,** 155–167 (1988).
28. P. Geladi and B. R. Kowalski, *Anal. Chim. Acta* **185,** 1–18 (1986).
29. A. Lorber, L. E. Wangen, and B. R. Kowalski, *J. Chemometrics* **1,** 19–31 (1987).
30. M. Stone and R. J. Brooks, *J. R. Statist. Soc. B*, **52,** 337–369 (1990).
31. H. R. Draper and H. Smith, *Applied Regression Analysis*, 2nd ed. John Wiley & Sons, Inc., New York, 1981.
32. E. Sanchez and B. R. Kowalski, *J. Chemometrics* **2,** 247–263 (1988).
33. E. Sanchez and B. R. Kowalski, *J. Chemometrics* **2,** 265–280 (1988).
34. T. Hirschfeld, *Anal. Chem.* **52,** 297A–312A (1980).
35. B. E. Wilson, W. Lindberg, and B. R. Kowalski, *J. Am. Chem. Soc.* **111,** 3797–3804 (1989).
36. Y. Wang, O. Borgen, B. R. Kowalski, M. Gu, and F. Turecek, *J. Chemometrics*, in press.
37. B. E. Wilson, W. Lindberg, and B. R. Kowalski, *J. Am. Chem. Soc.* **111,** 3797–3804 (1989).
38. J. Ohman, P. Geladi, and S. Wold, *J. Chemometrics* **4,** 79–91 (1990).
39. J. Ohman, P. Geladi, and S. Wold, *J. Chemometrics* **4,** 135–148 (1990).
40. A. K. Smilde, Y. Wang, and B. R. Kowalski, *J. Chemometrics*, (Dec. 1992).
41. R. Long, H. T. Mayfield, M. V. Henley, and P. R. Kromann, *Analytical Chemistry* **63,** 1256–1261 (1991).
42. T. Naes, T. Isaksson, and B. Kowalski, *Analytical Chemistry*, **62,** 664–673 (1990).
43. J. H. Friedman, *Annals of Statistics*, **19,** 1–141 (1991).
44. S. Wold, N. Kettaneh-Wold, and B. Skagerberg, *Chemometrics and Intelligent Laboratory Systems* **7,** 53–65 (1989).
45. S. Wold, *Chemometrics and Intelligent Laboratory Systems* **14,** 71–84 (1992).
46. Y. Wang, D. J. Veltkamp, and B. R. Kowalski, *Analytical Chemistry*, **63,** 2750–2756 (1991).
47. Y. Wang, M. J. Lysaght, and B. R. Kowalski, *Analytical Chemistry*, **64,** 562–565 (1992).
48. Y. Wang and B. R. Kowalski, *Analytical Chemistry*, in press.
49. J. N. Zemel, *Rev. Sci. Instrum.* **61,** 1579–1606 (1990).
50. A. Manz, J. C. James, E. Verpoorte, H. Luedi, H. M. Widmer, and D. J. Harrison, *Trends Anal. Chem.* **10**(5), 144–149 (1991).
51. M. Koudelka, F. Rohner-Jeanrenaud, J. Terrattaz, E. Bobbioni-Harsch, N. F. DeRooij, B. Jeanrenaud, *Biosens. Bioelectron.* **6,** 31–36 (1991).
52. K. Domansky, J. Janata, M. Josowicz, and D. Petelenz, *Analyst*, **118,** 335–340 (1993).

ENVIRONMENTAL ANALYSIS (MASS SPECTROMETRY)

JOAN BURSEY
Radian Corporation
Research Triangle Park, North Carolina

Mass spectrometry pertains to the separation of charged particles (ions) according to their mass-to-charge ratio. The separated ions and their relative abundances are then indicative of the structure of the original molecule. The most common route to formation of the charged particles is by interaction of vaporized molecules with energetic electrons emitted from a filament, usually at an energy of 70 electron volts (eV), a process called electron ionization. Positively charged ions, as well as negatively charged ions and excited neutral species, are formed in the ion source of the mass spectrometer, and then electrically accelerated into a device to perform mass analysis. More than 99% of the sample molecules are removed from the ion source continuously by vacuum pumps, which maintain the mass spectrometer under vacuum. The separation of the charged particles is accomplished in numerous ways:

- Magnetic Sector Mass Analyzer. In a magnetic sector mass spectrometer, a magnetic field acts as a mass analyzer. The mass of the ions which can pass through the magnetic field at any given time depends upon the radius of ion path in the magnetic field, the strength of the magnetic field, and the potential with which the ion is accelerated out of the ion source in which it is formed.
- Quadrupole Mass Filter. In a quadrupole mass filter, ions from the ion source are electrically accelerated into a combination of radiofrequency (rf) and electric (dc) fields on four precisely-machined stainless steel rods, with opposite polarities on pairs of opposing rods. Ion paths oscillate according to a changing rf field. At any given time, only ions of a particular mass can achieve a stable orbit and arrive at the detector; all other ions are in unstable orbits and are discharged by impact with the rods.
- High Resolution Mass Spectrometer. The magnetic sector mass spectrometer and the quadrupole mass filter typically separate unit masses. Separating mass spectrometric peaks with a small mass difference (very much less than one mass unit difference) requires the use of high resolution mass spectrometry. Ions formed in the ion source of a mass spectrometer exhibit a wide range of energies. This spread in energies causes divergence in the orbits of these ions through a mass analyzer and, consequently, limits the ability of the instrumentation to achieve resolution of masses. If the ions which are formed are focused according to energies before they enter a magnetic field, a much higher level of resolution is attainable in the mass analysis. One common arrangement of double focusing instrumentation places an energy focus (electrostatic sector) prior to the magnetic sector. It is also possible to achieve high resolution in other configurations, such as a magnetic sector preceding an electrostatic sector.
- Time-of-Flight Mass Analyzer. In a time-of-flight mass analyzer, ions are alternately formed and accelerated; mass values of the ions are differentiated on the basis of the time of their arrival at the detector.
- Ion Trap Mass Spectrometer. In an ion trap mass spectrometer, ions are typically formed by interaction with a filament that emits electrons. The electron beam is gated so that packets of electrons are produced. When a packet of ions is formed, the ions and ionic decomposition products are trapped by a cyclotron radiofrequency voltage. When this trapping rf voltage is swept, ejection of ions from the trapping cell occurs, and the ejected ions are detected.

When vaporized molecules are bombarded with energetic electrons, the excess energy transferred to the vaporized molecules results not only in their ionization but also in their fragmentation: charged fragments of the original molecule are deflected through the appropriate field and separated according to their mass-to-charge ratios, to produce a characteristic pattern of mass and abundance. Extensive scholarly efforts have been devoted to the interpretation of the mechanism of the fragmentation which occurs. Common fragmentation features have been elucidated for families of molecules, and pattern recognition techniques have been developed to allow computerized identification of the molecule from its fragmentation pattern. The utility of the mass spectrometer in environmental applications stems from its ability to produce these characteristic fragmentation patterns, which allow the qualitative identification of a molecule as well as quantitative analysis as a function of the signal which is produced.

ENVIRONMENTAL ANALYSIS: APPLICATIONS

The mass spectrometer, as a scientific instrument, has been in use since the historic experiments of J. J. Thomson in 1910, when the isotopes of neon (^{20}Ne and ^{22}Ne) were discovered. F. W. Aston built the first mass spectrometer in 1919 to use velocity focusing. However, only since the coupling of the gas chromatograph and computer with the mass spectrometer in approximately the last twenty years has the application of the mass spectrometer to environmental analysis been extensive. The crux of environmental analysis is the ability to identify and perform quantitative analysis of trace quantities of a substance present in a complex matrix. Separation of the analyte of interest from the matrix is usually the most serious problem facing the environmental analyst. The mass spectrometer itself provides no separation of analytes from the matrix: analyte and matrix constituents are ionized together in the ion source of the mass spectrometer, and differentiation of the resulting mass of signals at a large number of masses is a challenge to the most powerful mass spectrometric system. The ability to perform separations, to differentiate the analyte from the environmental matrix in which it is found, is crucial to the ability to perform environmental analysis.

Gas Chromatograph/Mass Spectrometer (Gc/Ms)

The gas chromatograph is central to the utilization of the analytical capabilities of the mass spectrometer. The commercial availability of computerized gas chromatograph/mass spectrometer systems in the late 1970s led to the designation of mass spectrometry as the primary monitoring tool for trace organic compounds by the United States Environmental Protection Agency.

The gas chromatograph, first employed in 1905, is used to perform separations of organic compounds from complex matrices due to their differential solubilities in a liquid phase coated on particles in a column or chemically bonded in a thin film with the walls of a column through which a stream of inert gas, usually helium, is flowing. The basis for the separation is the partitioning of the sample in and out of the liquid phase. The gas chromatograph operates at a positive pressure of helium; most mass spectrometers require a vacuum of 10^{-5} to 10^{-6} torr for successful operation. The development of a molecular separator, a device fabricated usually of glass, has been required for the coupling of the two apparently disparate techniques: the pressurized gas stream exiting from the gas chromatograph enters the separator and, since the rapidly diffusing helium particles are pumped away at a greater rate than the heavier molecules of the separated sample components, an enrichment of the chromatographic effluent is achieved and the flow of helium is reduced to a level at which the vacuum system of the mass spectrometer can operate. In many cases, if a chromatographic column of very narrow internal diameter (0.25 mm to 0.32 mm) is used, the flow of helium required to operate the column is sufficiently low (1–2 mL/min) so that the gas chromatograph can be coupled directly to the mass spectrometer without the use of a separator; the pumping system of the mass spectrometer will accommodate this low flow of helium. Columns of these narrow diameters, usually made of fused silica capillary tubing, are a development of the last ten to fifteen years. Fused silica capillary columns are now in very common use in laboratories around the world.

With the routine coupling of the gas chromatograph and the mass spectrometer and with the computerization of the mass spectrometer, the application of the coupled technique of gas chromatography/mass spectrometry (gc/ms) to environmental analysis, as well as any other areas of analysis requiring separation of complex mixtures with qualitative and quantitative characterization of the components, could proceed. The use of a prior separation technique prior to the application of mass spectrometric techniques is essential because environmental matrices are typically very complex: the compounds of environmental interest are present as minor constituents in a medium such as sewage sludge or process waste which may contain hundreds of other components. The gas chromatograph can be coupled successfully to all types of mass analyzers.

Gc/Ms Applications

One of the extensive applications of gc/ms to environmental analysis is in the area of execution of numbered methods promulgated by the United States Environmental Protection Agency and other government agencies and scientific groups such as the American Society for Testing and Materials. In a numbered method, the entire procedure of taking a sample, preparing the sample for analysis, performing the analysis, interpreting and reporting the results, and ensuring the quality of the analysis is addressed in detail. In areas of concern such as the remediation of Superfund chemical dumpsites, for example, the ability of mass spectrometry to perform qualitative and quantitative analysis is essential in establishing the presence of toxic volatile and semivolatile organic compounds prior to remediation, and the absence of these compounds after remediation of the site has taken place. A chromatographic method, such as gas chromatography, is usually a very sensitive mode of analysis, but the result is nonspecific; that is, the gas chromatographic analyst can never be absolutely sure of the identification of the compounds which are observed, since retention time and detector response are the only parameters which the gas chromatograph can report. In addition, it is impossible to say with certainty in a chromatographic analysis, that no other compound is eluting at the same time as the compound of interest. When the unique ability of the mass spectrometer to produce a characteristic fragmentation pattern is coupled with the ability of the gas chromatograph to provide retention time information and resolve the compound of interest from its matrix, very sensitive (typically in the nanogram or picogram range) and highly specific analytical information is produced. Gc/ms, as applied in numbered methods, is used for qualitative and quantitative analysis of organic compounds in matrices such as drinking water, wastewater, groundwater, sewage sludge, soil, sediment, gaseous stationary source emissions, ambient air, indoor air, and ash. The numbered methods are modified and adapted to address matrices as diverse as still bottoms, incinerator flyash, process feeds, barrel coatings, printed circuit boards, wire coatings, cement, wood, wood smoke, upholstery, food, building materials, and a wide variety of complex liquid, solid, and gaseous matrices.

The challenge in the area of development of techniques and technology is always to make the analysis more sensitive and more specific. Computer models have been formulated to allow the assessment of the risk posed to populations by the presence of trace levels of organic constituents in media such as air, groundwater and drinking water. With levels of acceptable risk proceeding ever lower, it is essential that the analytical technology detect the presence of the compounds of interest at ever lower levels in the various media of environmental interest.

Toxic Dioxins: High Resolution Gc/Ms

Polychlorinated dibenzodioxins (PCDD) and dibenzofurans (PCDF) are among the most toxic synthetic organic chemicals known to humans. The entire group of dioxins and furans comprises 210 compounds, of which approximately one dozen compounds are considered to be very toxic and one compound, 2,3,7,8-tetrachloridibenzo-*p*-dioxin, is extremely toxic. Public attention has been focused on this group of compounds because of prominent incidents of environmental contamination which have occurred at locations such as Seveso, Italy; Times Beach, Missouri; and Love Canal, New York. PCDD and PCDF are formed as byproducts in the synthesis of herbicides and are formed in combustion processes (for example, in a municipal solid waste incinerator) by unknown mechanisms during the combustion of halogenated organic compounds such as polychlorinated biphenyls and chlorinated phenols. PCDD and PCDF have also been reported in vegetation, human and animal tissue, milk, blood, in discharges from pulp and paper mills, and in paper products.

Because of the toxicity of the polychlorinated dibenzodioxins and dibenzofurans and the high cost of dealing with their presence in the environment (the evacuation of Times Beach, for example), it is absolutely essential that identification of these compounds be correct beyond a reasonable doubt and that quantitative data be accurate. For these reasons, gas chromatography/mass spectrometry is used as the method of choice in the analysis of samples containing these compounds. However, the mass spectrometer found in the average laboratory is not adequate for the performance of this level of sensitive and specific analysis. A specialized mass spectrometer capable of high resolution of masses is required to perform the analysis at the required level of sensitivity and specificity.

High resolution gas chromatographic techniques are required for the separation of the 210 chemically similar compounds which make up this family. Additionally, sophisticated laboratory techniques for the extraction of these compounds from a complex environmental matrix such as flyash, with extensive compound purification techniques, are required for a successful analysis. With the coupling of the sensitivity and mass resolution of the instrumentation with the accurate masses of the molecule and its fragment ions and using complex computerized algorithms to relate accurate mass to the information yielded by the high resolution mass spectrometer, it is possible to obtain accurate mass measurement to allow for the qualitative identification and quantitative analysis of polychlorinated dibenzodioxins and dibenzofurans at the parts per trillion (10^{-12} grams/gram) level. The sensitivity is critical to be able to provide an accurate assessment of the risk posed by the presence of these chemicals. Accurate identification, beyond any reasonable doubt, is required as the basis of regulations or legal actions.

Applications of High Resolution Gc/Ms

The single most valuable piece of information which can be obtained from the mass spectrum is the molecular weight of an unknown compound. When the analyst has established the molecular weight of an unknown compound, potential atomic compositions can be suggested for the molecule and the fragmentation pattern observed in the mass spectrum can be interpreted to determine the arrangement of atoms in the original molecule. In many cases, the molecular weight information, coupled with the fragmentations which occur in the molecule under electron ionization, is sufficient to specify a structure for the molecule.

Frequently, the mass spectrum of a compound will identify that compound as a member of a family but cannot specify the particular isomer. Often the coupling of chromatographic retention times with the mass spectrum can specify the atomic composition, but a complete identification of the compound, including the exact position within the molecule of all of the functional groups, cannot be made. Sometimes the ionized molecular species formed under conditions of electron ionization is sufficiently energetic that extensive fragmentation occurs with formation of very stable fragments and no ion with a mass characteristic of the molecular weight is observed. High resolution mass spectrometry is extensively employed in environmental analysis for accurate mass identifications of unknown compounds, even when these compounds are present in complex matrices.

A mass spectrometric technique which can be quite useful in obtaining molecular weight information is chemical ionization. In chemical ionization mass spectrometry, the vaporized compound of interest enters the ion source as a minor constituent of a reagent gas consisting mostly of methane, isobutane, water, or ammonia, for example. The major constituent of the mixture (the reagent gas) is ionized and the compound of interest is ionized by ionic chemical reaction with the reagent gas. Reactions which typically occur, with judicious selection of the appropriate reagent gas, involve transfer of a proton or characteristic addition of some group such as a methyl group to the compound of interest. Because the ionization process occurs by chemical reaction, high levels of energy are not transferred to the compound of interest so minimal fragmentation occurs. The species which is detected in the mass spectrometer is typically the (M + H) ion, ie, the protonated molecule. Since the analyst is aware of the common reactions which can occur with the use of a given reagent gas, the mass of the charged adduct species can be related to the molecular weight of the original molecule to aid in the identification of an unknown compound.

If the chemical ionization reactions are performed in a high resolution mass spectrometer, accurate mass measurements can be used to determine the composition of an ion at a given mass. If chemical ionization techniques are combined with standard electron ionization techniques (not necessarily in the same analysis), both molecular weight information and fragmentation patterns can be used for the characterization of unknown compounds. Instrumentation which is commercially available can apply both electron ionization and chemical ionization techniques, nearly simultaneously.

Research/Methodologies: Lc/Ms Applications

Not all compounds of environmental interest are directly amenable to gas chromatographic techniques for separation of the compounds of interest from the environmental matrix in which they occur. Many compounds are thermally unstable and decompose upon vaporization or application of heat, while other compounds are insufficiently volatile for analysis by gas chromatography because they cannot be vaporized. In these situations, high performance liquid chromatography (hplc), a liquid, ambient temperature analog to gas chromatography, can be employed as a separation technique without the requirement of vaporization of the compound. Technological advances over the last 5–10 years have been applied to the coupling of liquid chromatography with mass spectrometry. Although several commercial lc/ms systems are available, many different interfacing methodologies are currently being used for coupling with the mass spectrometer, with variable levels of reproducibility and reliability. No single interface for lc/ms is universally accepted, and operating procedures are not standard.

The research field of lc/ms is moving in the direction of standardization of instrumentation and operating parameters. No numbered methods are presently available for application of lc/ms to the solution of environmental problems, but numbered methods in this area are being written and will be evaluated and adopted within the next few years. The lc/ms technique is essential in environmental analysis for qualitative and quantitative analysis of dyes and dyestuffs in media such as groundwater and wastewater. Dyes are typically large and very polar organic molecules. Their molecular weight and polar nature, for the most part, preclude their analysis by gas chromatography. The polar groups in the molecule in many cases make the dye molecules biologically active and either toxic or carcinogenic or both. The magnitude of the environmental problem afforded by the presence of aqueous waste containing dyes cannot be assessed accurately until successful analytical methods are available. Lc/ms techniques are being developed to address this area. Hplc/ms techniques have also been applied to the qualitative and quantitative analysis of pesticides and herbicides, especially the highly polar pesticides such as carbamates, which are not amenable to gas chromatographic techniques.

The problem of separating compounds of interest from complex environmental matrices so that qualitative and quantitative analytical mass spectrometric techniques can be applied extends into chromatographic research. It has been more than twenty years since the first reports on supercritical fluid chromatography (sfc), but coupling of supercritical chromatographic techniques with mass spectrometry is a development of the last ten years. The coupled technique, sfc/ms, is still far from a routine laboratory technique but the technology and areas of application are developing rapidly. The applications of lc/ms and sfc/ms overlap in many areas, but there are also many areas of application in which the two techniques complement each other.

Supercritical fluids exist at temperatures and pressures above the supercritical point of a compound. By controlling pressure and temperature, the physical properties of a supercritical fluid are variable between normal gas and nearly liquid. At each density, the solvent characteristics of the supercritical fluid are different. By judicious selection of the supercritical fluid and various materials which can be added to this fluid as modifiers, it is possible to perform selective extractions with a minimum of solvent. With the appropriate chromatographic equipment, it is possible to use supercritical fluid to perform chromatographic separations. Lower viscosities and higher diffusion coefficients observed in supercritical fluids (as compared to liquids) produce far higher chromatographic efficiency compared to hplc techniques.

Interfacing Options: Sfc/Ms

Since supercritical fluids are pressurized dense gases, there are numerous options for interfacing SFC to mass spectrometry. A direct fluid injection interface allows the supercritical fluid to expand into a region where ionization can occur. Since a gas is produced upon expansion of a supercritical fluid, either electron ionization or chemical ionization can occur upon expansion and ionization of the fluid. Various techniques which involve cooling and condensation of the supercritical fluid can be used, so that the interfaces which are effective for lc/ms are effective for sfc/ms. The development of the instrumentation is a prominent area of research. Some of the areas of application of sfc/ms include characterization of polymeric mixtures such as surfactants, and labile and nonvolatile compounds such as pesticides, herbicides, rodenticides, mycotoxins, polar substituted biphenyls, and alkaloids. Surfactants are components of laundry detergents and are widely dispersed in the environment. Analytical techniques which could be applied to their characterization and quantitative analysis have only recently become available. Azo dyes are widely used polar compounds which are difficult to separate and analyze by gas chromatographic techniques. Because of their highly polar nature, the compounds do not chromatograph well, if at all. Hplc/ms is successful in characterizing some of the members of the class, and sfc/ms is proving to have a broad application in this area as well.

Analysis of Semivolatile Organic Compounds: Gc/Ftir/Ms

Another combined technique with important advantages for the analysis of semivolatile or even relatively nonvolatile organic compounds is the combination of gas chromatography with Fourier transform infrared spectroscopy with mass spectrometry (gc/ftir/ms). At least one combined system is commercially available to allow the collection of ftir data simultaneously with MS information to provide complementary spectral information. The mass spectrometer used in this combination is a quadrupole mass filter. The ms provides molecular weight data, a characteristic fragmentation pattern, and isotopic cluster information. The ftir can distinguish isomers, provide frequency data for organic functional groups, and can provide absorption information for quantitative analysis, as a confirmation of the quantitative information available from the mass spectrometric measurements. The two techniques are complementary in that the ftir can be used to confirm the identifications made by the mass spectrometer. In some cases, when the mass spectrometer is able to characterize a compound as a member of a particular class, the ftir can even pinpoint the location of organic functional groups within the molecule. The characterization of specific isomers is frequently impossible for the mass spectrometer alone. In many cases, the ability of the ftir to characterize organic functional groups may yield sufficient information to assess biological hazard of constituents of the environment.

The application of the gc/ftir/ms systems to nonvolatile organic compounds has been slow to develop, although these compounds constitute the greater portion of the extractable portion of an environmental sample, and this nonvolatile fraction of an environmental sample frequently exhibits mutagenic properties. This nonvolatile material is difficult to analyze because nonvolatile materials do not vaporize readily and do not chromatograph using standard techniques. However, some nonvolatile materials can often be made amenable to chromatographic analysis by preparation of an appropriate derivative which will enhance the volatility of a nonvolatile compound. Compounds which are nonvolatile because of polar groups in the molecule can frequently be converted to an analyzable form by chemical reaction. Common examples of derivatization include preparation of esters of carboxylic acids and preparation of 2,4-dinitrophenylhydrazone derivatives of aldehydes and ketones, as well as ethers of alcohols.

Ms/Ms Technique

A technique known as mass spectrometry/mass spectrometry (ms/ms) often utilizes three quadrupole mass filters in tandem. The first quadrupole is used to perform a mass separation of a vaporized species. The second quadrupole serves as a reaction chamber, where a gas is introduced to collide with the energetic ions emerging from the first quadrupole. Highly specific reactions occur in this collision cell, and the products of these reactions are mass-analyzed in the final quadrupole. By utilizing characteristic reactions which can occur for a particular species, it is possible to characterize trace quantities of a given organic compound in a complex matrix without extensive purification of the compound or removal of the matrix. The ms/ms technique for mass spectrometric analysis can be combined with gas chromatographic separation, with liquid chromatographic separation, or with supercritical fluid chromatography. The hplc/ms/ms technique has been applied to the analysis of dioxins and furans without prior cleanup, to the analysis of carbamate and organophosphorus pesticides, to analysis of dyes, dye wastes, and dye-manufacturing intermediates and byproducts, and to manufacturing wastes. A mobile-van-based ms/ms system has been used for the direct analysis of gaseous mixtures in the field. The net effect of utilization of highly specific ionic reactions to identify compounds can be achieved by other combinations of mass spectrometric technique than three quadrupole analyzers; for example, the ion trap detector can trap ions and the occurrence of ionic chemical reactions can be observed, so the ion trap functions as an ms/ms system.

Icp/Ms Applications

The quadrupole mass filter has also been coupled to inductively coupled argon plasma as an ion source. The primary application of the coupled inductively coupled plasma/mass spectrometry (icp/ms) technique is to inorganic analysis, such as the determination of trace metals in various types of water such as groundwater, sea water, lake water, and drinking water. The technique is also applicable to the characterization of trace metals in soils and sediments, for as many as 22 elements. The combined technique of icp/ms can also be used to determine accurate elemental isotopic ratios. These accurate isotopic ratios may be used to pursue the source of metallic environmental contaminants so that regulatory action can be taken for remediation of the contamination. Different trace elements may be used to define regional sources for atmospheric aerosol pollution. Because toxicological assessment requires speciated information for trace metals rather than a value for total metal concentration, the next development in the area of icp/ms will probably be interfacing this technique with an amenable chromatographic separation technique such as ion chromatography or hplc.

Api: Advantages

It has been possible to form ions at atmospheric pressure and analyze these ions using mass spectrometric techniques for more than 30 years. The use of atmospheric pressure ionization (api) offers definite advantages for coupling of mass spectrometry with chromatographic techniques, since chromatographic techniques have a gaseous or liquid effluent that is at atmospheric pressure or above. Use of conventional mass spectrometric techniques requires that this pressurized flow be attenuated to accommodate the vacuum system of the mass spectrometer. In api, the ion source region is separated from the high-vacuum mass analyzer by a very small orifice. This orifice must be large enough to allow the entrance of sufficient ions to perform analysis, while keeping the pressure sufficiently low to allow efficient operation of the mass analyzer. Historically, quadrupole mass filters have been used for apims, although some use of magnetic sector mass spectrometers and ion trap mass spectrometers has been reported. Ions can be formed by interaction with a thin foil of radioactive material which emits energetic particles, or by interaction with a corona discharge needle (an electrical discharge at the tip of a needle held at high voltage). When the ions are formed, they undergo reaction with the air that is present as well as with other gaseous molecules. Thus, the ionic products that are ultimately observed are the products of chemical ionization reactions, and the ion chemistry that occurs can be very complex. Both positive and negative ions can be formed under api conditions. Formation of positive or negative species can be enhanced by judicious choice of reagent gases. Strongly electronegative compounds such as pesticides are more sensitive as negative ions. Since the ionization is mild, little fragmentation of the ions occurs and molecular weight information is readily available. However, to obtain structural information, the instrumentation can be modified for ms/ms capability. Chromatographic introduction systems such as gas chromatography, high performance liquid chromatography, and supercritical fluid chromatography can be utilized as sample introduction systems. Capillary electrophoresis, a simple high-efficiency separation for mixtures containing organic ions in solution, can also be used as a sample introduction system for apims, for environmental applications such as the characterization of sulfonylurea herbicides, sulfonated azo dyes, and phenoxyacetic acids.

FTICRMS: Technique for Mass Measurement

Fourier transform ion cyclotron resonance mass spectrometry (fticrms) is a technique for conversion of the mass-to-charge ratio of an ion to a cyclotron orbital frequency which is experimentally measurable. Since frequency can be measured more accurately than any other physical property, fticrms offers the potential for ultrahigh accuracy in mass measurement. Extensive instrument development research is presently being performed, with the coupling of numerous ionization techniques to produce ionized species of nonvolatile and high molecular weight compounds, which can then be subjected to accurate mass measurement of compounds of environmental interest using fticrms.

Surface Mass Spectrometry: Applications

Surface phenomena are an area of very high interest in environmental applications. Organic compounds can condense into solids from the gas phase, and this condensing organic material can interact with particulate matter, being deposited on the surface of the particle or becoming entrained within a porous particle. The environmental impact of the organic material is governed by the molecular structure of the uppermost molecular layers of the surface of a solid. Analytical elemental information, which can be obtained by digestion or extraction of the particulate matter with subsequent analysis for organic or inorganic analytes, provides characterization of the bulk composition of the particulate and is not sufficient to characterize the particulate surface. An analytical instrument with high sensitivity (these organic materials are present on the surface in a monolayer) and sufficient spatial resolution to provide detailed molecular information about the surface structures, is required.

Surface mass spectrometry shows high sensitivity and can provide detailed and specific molecular information at the surface of a particle: identification and quantitative analysis of elements, isotopes, and organic molecular species (nonvolatile and thermally labile compounds) can be performed. Two mass spectrometric techniques have been applied to the solution of this analytical problem: static secondary ion ms (sims) and laser secondary neutral ms (laser snms). In sims, surface species are desorbed by particle bombardment at an energy of thousands of electron volts. Adsorbed particles are ionized during their bombardment, and time-of-flight (tof) mass spectrometry offers a high transmission rate for these ions with detection of ions of all masses in parallel, although initial development of the field employed both magnetic sector mass spectrometers and quadrupole mass filters. With an unlimited mass range, high resolution and accurate determination of masses, and short time for analysis, tof–sims is a very powerful technique for the analysis of surfaces. Mass spectrometric surface characterization techniques have been applied to the characterization of polymers, and to identification and localization of contaminants on semiconductor surfaces (tof–sims can be used to determine the step in a manufacturing process where surface contamination occurs). Failure analysis, where failure in function can be characterized by a change in the molecular structure of the outer layers, is a prominent application of the tof–sims technique.

Laser Snms: Applications

Formation of positive ions from a surface using laser excitation (laser snms), in combination with the use of an ion microprobe, yields quantitative elemental mapping coupled with high sensitivity. Laser snms may be applied to elemental analysis as well as the characterization of molecular surfaces. Surfaces such as metals and ceramics are amenable to analysis, with flat or rough surfaces. Environmental applications are currently being reported in the literature, and technologies and techniques are under development in laboratories around the world.

Ims: Laboratory Techniques

Ion mobility spectrometry (ims) is a technique which has been in existence since the sixties, primarily for the study of ion-molecule reactions in the research laboratory and for qualitative trace analysis. Like many other laboratory techniques, ims is finding environmental applications in the continuous monitoring of gases such as ammonia, hydrogen chloride, hydrogen fluoride, chlorine, and chlorine dioxide. These gases are all toxic and corrosive, and are difficult to monitor by other techniques. In an ion mobility spectrometer, a diluted sample (diluted through a dilution probe from stationary source emissions or from chemical processes) is forced over a semi-permeable membrane. Purified air sweeps the sample into a reaction region, where the sample is ionized by a weak plasma formed by a small radioactive source (^{63}Ni). Ionized sample molecules drift through the cell under an electric field, and are periodically introduced into a drift tube by an electronic shutter grid. In the drift tube, the ions separate on the basis of charge, mass, and shape. The current at the detector is measured as a function of time, so the operation is equivalent to a time-of-flight mass spectrometer. Development of this type of system for use as a continuous emissions monitor is in its very early stages; its ultimate application in this area will depend upon its cost, ease of use, transportability, and maintenance.

Fab Analysis Technique

Fast atom bombardment analysis (fab) is a mass spectrometric ionization technique which improves the response of the ms system to solid nonvolatile materials. From its beginnings, mass spectrometry has focused on the analysis of samples in the vapor phase, an approach which usually requires heating samples in order to vaporize them. There has, therefore, always been some level of concern about the contribution of thermal processes to the mass spectral fragmentation which occurs. In the fast atom bombardment technique, a beam of fast argon atoms is directed to the surface of a target which is carrying the sample. The fast argon atoms are produced in a fast atom "gun": argon ions with an energy between 5 and 10 kV are generated by a gaseous discharge. The ions formed in this discharge pass through a collision chamber which is filled with neutral argon atoms. Collisions occur between the charged and neutral atoms. Charge exchange occurs, usually without loss of significant amounts of kinetic energy, and a beam of fast argon atoms is produced. When these fast argon atoms hit the surface of a sample, ions are generated and focussed into the analytical portion of the instrument. At present, the major applications of the fab technique have been large biomolecules, but use of the fab technique coupled with higher performance liquid chromatography will broaden the area of application to large thermally-labile molecules of environmental interest such as dyes and toxins for which there has been no viable analytical technique.

CONCLUSION

In most cases, the transition from research laboratory technique to commercial availability and hence to extensive application is quite short—sometimes a period of one to two years. In other cases, particularly coupled high performance liquid chromatography/mass spectrometry techniques, the sheer complexity of the coupled technologies makes widespread routine application slower. The instrumentation is costly, and a high level of expertise is required for effective utilization of the technique. The areas of application for hplc/ms have been far less urgent than the areas which required the development of gc/ms techniques and their nearly universal application. The current trends are to make the instrumentation simpler, more compact (and, therefore, less costly) and to make the instrumentation accessible for more laboratories in order to broaden the areas of application.

BIBLIOGRAPHY

J. J. Thomson, *Rays of Positive Electricity*, Longmans, Green, 1913.

D. H. Smith, ed., *Computer-Assisted Structure Elucidation*, American Chemical Society, Washington, D.C, 1977, p. 151.

W. H. McFadden, *Techniques of Combined Gas Chromatography / Mass Spectrometry*, Wiley, New York, 1973, p. 463.

P. H. Dawson, *Quadrupole Mass Spectrometry and its Applications*, Elsevier, Amsterdam, The Netherlands, 1976.

W. W. Lowrance, ed., *Public Health Risks of the Dioxins*, William Kaufmann, Inc., Los Altos, Calif., 1984.

M. Gough, *Dioxin, Agent Orange—the Facts*, Plenum Press, New York, 1986.

J. S. Stanley, T. M. Sack, *Protocol for the Analysis of 2,3,7,8-Tetrachlorodiobenzo-*p*-Dioxin by High Resolution Gas Chromatography / High Resolution Mass Spectrometry*, Jan. 1986, EPA 600/4-86-004.

J. M. L. Penninger, M. Radosz, M. A. McHugh, V. J. Krukonis, eds., *Supercritical Fluid Technology*, Elsevier, Amsterdam, The Netherlands, 1985.

F. W. McLafferty, ed., *Tandem Mass Spectrometry*, John Wiley & Sons, Inc., New York, 1983.

M. V. Buchanan, ed., *Fourier Transform Mass Spectrometry: Evolution, Innovation, and Applications*, American Chemical Society, Washington, D.C., 1987.

D. M. Lubman, ed., *Lasers in Mass Spectrometry*, Oxford University Press, New York, 1990.

D. P. Woodruff, T. A. Delchar, *Modern Techniques of Surface Science*, Cambridge University Press, Cambridge, UK, 1986.

A. S. Czanderna, D. M. Hercules, *Ion Spectroscopies for Surface Analysis*, Plenum Press, New York, 1991.

G. W. A. Milne, ed., *Mass Spectrometry: Techniques and Applications*, Wiley-Interscience, New York, 1971.

ENVIRONMENTAL ANALYSIS—MASS SPECTROMETRY, ADVANCES

RONALD A. HITES
School of Public and Environmental Affairs
and Department of Chemistry
Indiana University
Bloomington, Indiana

Innovations in mass spectrometry are being applied increasingly to environmental issues. The primary features of environmental mass spectrometry can be usefully compared with those of biomedical mass spectrometry, because these are the two largest areas of application for this analytical technique.

Environmental mass spectrometry tends to deal with anthropogenic compounds with molecular weights less than about 1000 daltons. On occasion, petroleum and the combustion of organic compounds are studied; but in both cases, there is an anthropogenic component. On the other hand, biomedical mass spectrometry tends to deal with natural products. Some of these are of modest molecular weight, but most have very high molecular weights and include biopolymers such as proteins and carbohydrates.

Environmental mass spectrometry tends to be highly quantitative. The utmost sensitivity is sought, and many laboratories routinely measure a few picograms of many compounds. Biomedical mass spectrometry, however, tends to be largely qualitative, and the materials studied have a high mass. For example, the sequence of amino acids in a protein might be the experimental goal; analysis of proteins of molecular weights of 150,000 or more is now routine (1).

Environmental mass spectrometry often uses official methods promulgated or approved by regulatory agencies, such as the Environmental Protection Agency (EPA) in the United States (2). Biomedical methods tend to use *ad hoc* procedures designed to solve the problem at hand. There is however, one area of commonality: environmental mass spectrometry of pesticide metabolites is similar to biomedical drug metabolite studies.

FIELDABLE INSTRUMENTS

A growing trend is the use of fieldable instruments that can be taken to the site of environmental contamination, for example, a hazardous waste landfill that is being cleaned up. Here, fieldable mass spectrometers are used to measure the quantity of specific compounds as the cleanup proceeds. Such an approach reduces the analytical costs associated with the cleanup through rapid turnaround and high throughput of samples. While instruments have been taken to the site in a truck, van, or recreational vehicle, an instrument that can be carried by a worker (Figure 1) has been developed by Urban and co-workers (3).

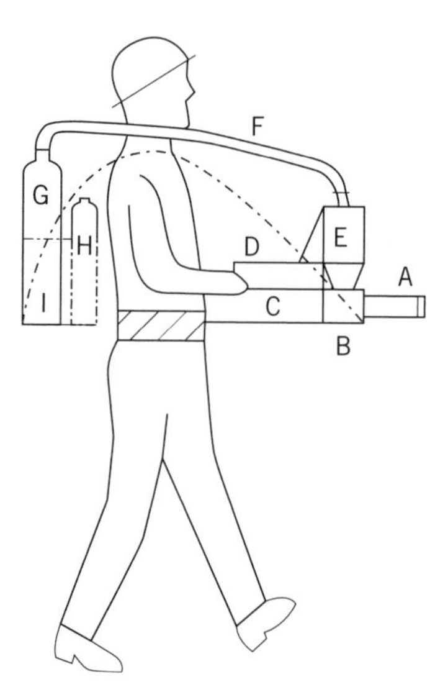

Figure 1. A fieldable mass spectrometer that can be carried by a single worker; from reference 3. Its total weight is about 35 kg. A, Inlet and gc column; B, mass spectrometer; C, electronics; D, computer; E, molecular drag pump; F, vacuum hose; G, vacuum reservoir; H, carrier gas; I, battery (24 VDC).

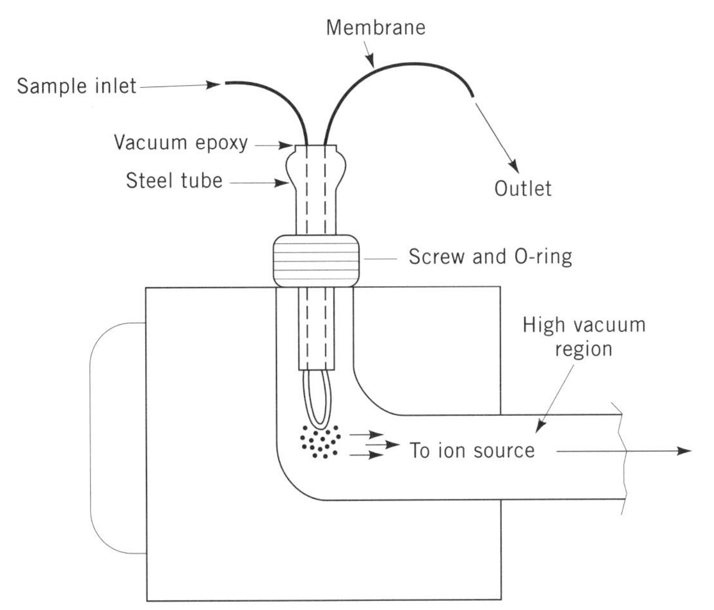

Figure 2. Principle of operation of membrane mass spectrometry; from reference 4.

MEMBRANE MASS SPECTROMETRY

"Membrane mass spectrometry" is illustrated by the apparatus shown in Figure 2. The sample (water or air) passes through a tube made from a semipermeable polymer that is in the vacuum system of the ion source. The analytes of interest pass through the membrane and into the ion source. By using a triple-quadrupole mass spectrometer, considerable qualitative and quantitative information can be obtained without subjecting the sample to chromatography. This work has been popularized in the last several years in Cooks's laboratory (4), but important contributions have also been made by Enke and his students (5).

LIQUID CHROMATOGRAPHY MASS SPECTROMETRY

Considerable attention has also been given to liquid chromatographic mass spectrometry (LC/MS). Work by Behymer and colleagues (6), for example, has used particle beam LC/MS to develop an "official" EPA method for the analysis of benzidine. Unfortunately, developmental issues still exist for some environmental applications of LC/MS. For example, chromatographic resolution and sensitivity are often inadequate for some samples.

LC/MS is ideal for thermally unstable or nonvolatile compounds because these compounds cannot be analyzed by GC/MS. Unfortunately, many LC/MS methods heat the sample in the interface. This reduces the suitability of the techniques for just those compounds for which they should be most suited.

One exception, continuous flow fast-atom bombardment (CF-FAB), takes compounds from a flowing liquid stream directly into the ionic phase. Azo dyes have been analyzed by LC/CF-FAB using tandem mass spectrometry in our laboratory. Figure 3 (top) shows the liquid chromatogram of a mixture of five azo dyes, all of which are aromatic sulfonates. Note that the peak widths increase with time as a result of isocratic elution. Figure 3 (bottom) is the product ion spectra of peaks *B* and *C*, which are the dyes Acid Orange 7 and 8, respectively. These compounds differ only by the presence of a methyl group *ortho* to the azo linkage; therefore, the spectra are simply offset by 14 daltons. Abundant ions at m/z 171 and 185 are due to cleavage of the azo group with a hydrogen rearrangement. The ions at m/z 93 and 107 are due to further rearrangement of the sulfonate group.

FLOW INJECTION MASS SPECTROMETRY

The cleanness and simplicity of the spectra in Figure 3 imply that in some cases, chromatographic separation before mass spectrometric analysis may not be necessary. If the chromatographic step could be avoided, the throughput would be increased by a factor of at least 5 to 10. This concept has been demonstrated for linear alkylbenzene sulfonates (LAS), widely used surfactants (7).

A typical LAS structure is shown in the upper left of Figure 4. The compound shown has 12 carbons in the side chain, commercial mixtures would have homologues with 10 to 13 carbons in the side chain. All LAS homologues show an abundant product ion at m/z 183, due to cleavage *beta* to the aromatic ring. Therefore, scanning the precursor ions of m/z 183 analyzes all the homologues in an LAS mixture. Branched alkylbenzene sulfonates can be distin-

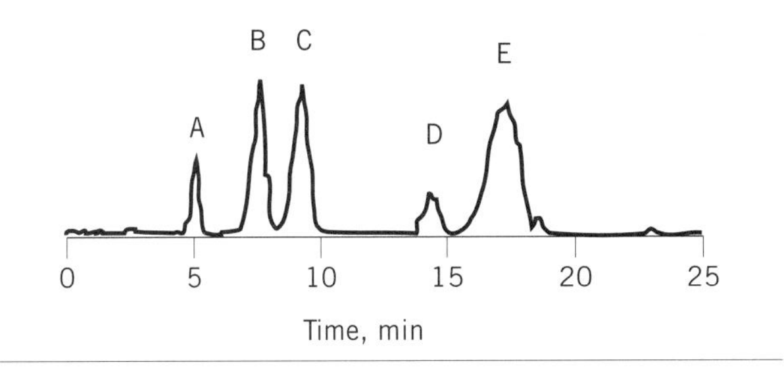

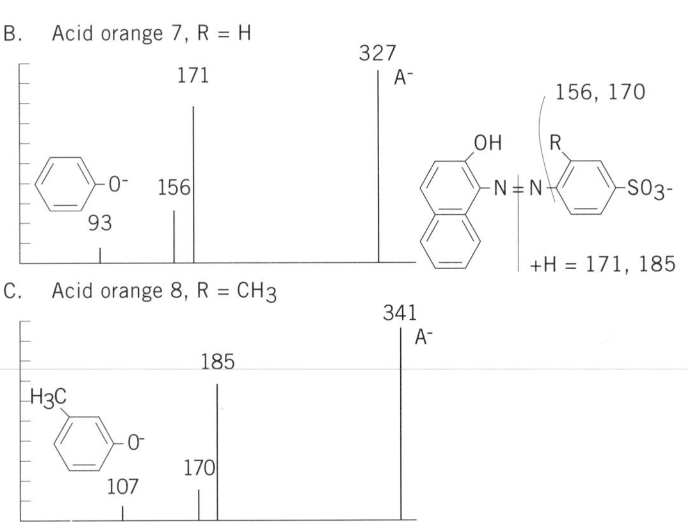

Figure 3. Liquid chromatogram of a standard mixture of 5 azo dyes detected by continuous-flow fast-atom bombardment mass spectrometry (top) and daughter ion mass spectra of A⁻ from peaks *B* and *C* (bottom).

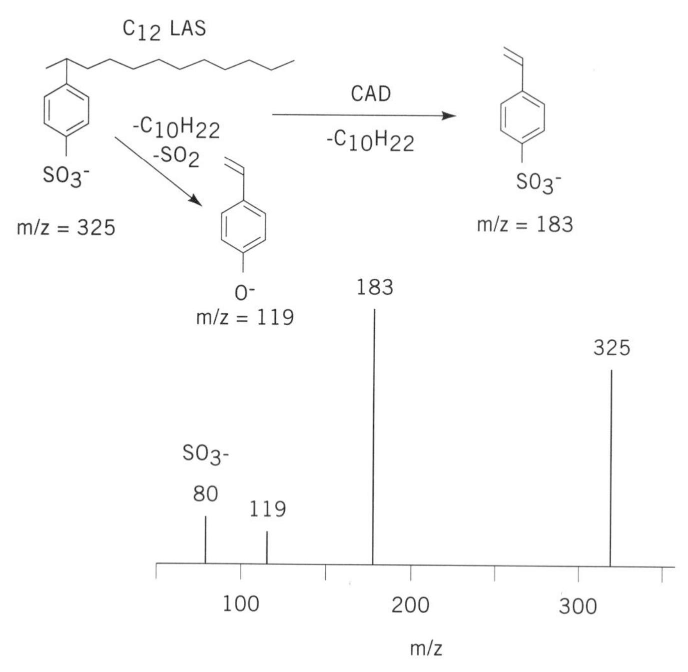

Figure 4. Product ion mass spectrum of a C_{12} linear alkyl benzene sulfonate and scheme showing the sources of the various ions.

guished from their linear cousins because the product ion at m/z 183 is shifted to m/z 197.

Because the different homologues give equal molar responses, the CF-FAB technique can be used for the quantitative analysis of LAS with only a simple calibration (7). For a sample containing 10 nanograms of LAS, the signal-to-noise ratio is about 5:1; this gives a sensitivity of about 6 picomoles.

The data obtained from an analysis of LAS samples in the input and primary treatment stages of a wastewater treatment plant and in the river 100 m downstream from the discharge point are shown in Figure 5. Concentrations of the 10- to 13-carbon LAS homologues decrease by factors of 10 to 80 between the influent and the river. This entire analysis, including calibration, took about one hour (the LASs were isolated by solid phase extraction), less time than going to the wastewater treatment plant to obtain the samples.

NEGATIVE IONIZATION MASS SPECTROMETRY

Electron capture (EC) mass spectrometry is a very important tool for environmental analysis (8). The sensitivity and selectivity of this technique make it particularly useful for analytes that contain electronegative functional groups or atoms. For example, many organochlorides, such as PCBs and chlordane, can be analyzed by electron capture mass spectrometry without extensive sample cleanup and with exquisite sensitivity. Figure 6 shows the electron impact (EI) mass spectrum of endosulfan (top) and the electron capture, mass spectrum of endosulfan (bottom) (8). This is probably the world's ugliest EI mass spectrum; there is no molecular ion, and very few structurally specific fragment ions are present. Furthermore, because of the large number of fragment ions, the *absolute* abundance of each ion is smaller than if most of the ions were at only a few m/z values. On the other hand, the EC mass spectrum shows an abundant molecular anion at m/z 404 and a relatively abundant ion because of the loss of a chlorine and the addition of a hydrogen at m/z 370. The electron capture mass spectrum is clearly more useful for the quantitation of this compound.

EC mass spectrometry is being used for the routine analysis of endosulfan in ambient air. An example of the data that can be obtained with this technique is shown in Figure 7, a plot of the natural logarithm of the concentration (in torr) versus the reciprocal of the absolute temperature. The slope of this line gives the heat a vaporization of endosulfan, 110 kJ/mole. Each air sample takes less than one hour to analyze becasue there is no sample cleanup.

DIOXINS

2,3,7,8-Tetrachlorodibenzo-*p*-dioxin alone has been responsible for the sale of more mass spectrometers than any other compound. Even today, 15 to 20 years after the first warnings about this compound were made public, debate about its toxic effects on humans continues. The literature on the use of mass spectrometry for the analysis of dioxins is so vast that one example suffices.

There are two major sources of dioxins and the related dibenzofurans: by-products in chlorinated aromatic compounds (9) and the combustion of municipal and chemical wastes (10). The latter was an important discovery. No longer could the simple presence of dioxin in a sample indict a chemical production facility. Indeed, it had been suggested that "dioxins have been with us since the advent of fire" (11).

Chlorinated dioxins and furans formed during combustion are emitted into the atmosphere. Depending on the ambient temperature, some of these compounds are adsorbed onto particles and some are in the vapor state. In either case, these compounds can travel through the atmosphere for considerable distances. While they are in the atmosphere, several things can happen to them.

The compounds can reequilibrate between the particle and vapor phases; this is a temperature-dependent pro-

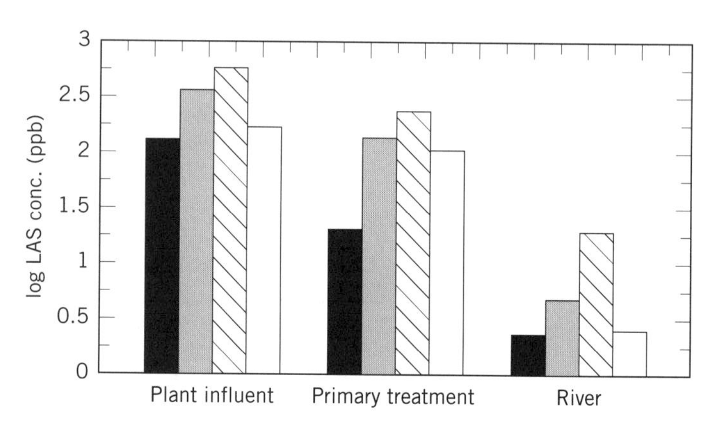

Figure 5. Concentrations (log scale) of C_{10} to C_{13} linear benzene sulfonates (LAS) in wastewater, in the treatment plant, and in the receiving river as measured by parent ion scans of m/z 183.

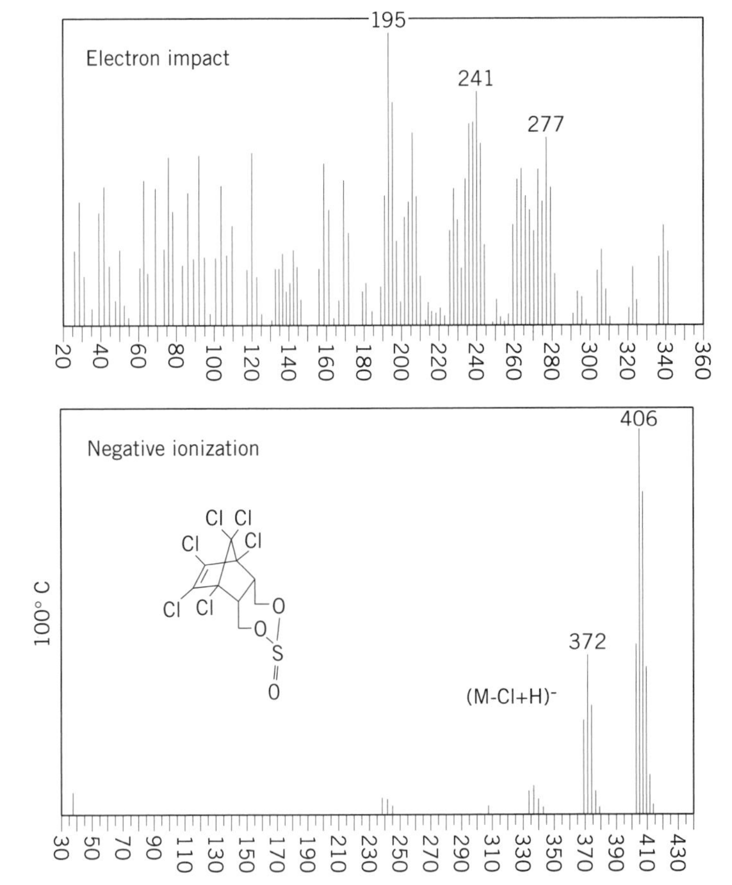

Figure 6. Electron impact mass spectrum (top) and electron capture, negative ion, mass spectrum (bottom) of endosulfan.

cess. They may also degrade by photo-oxidative or other chemical processes; the extent of this degradation depends on the physical state of the reactant. Eventually, the dioxins and furans leave the atmosphere by a number of routes. Particles with their load of absorbed compounds settle out of the air, scavenging both particle-bound and vapor-phase compounds. Dioxins and furans from industrial sources also enter the atmosphere; however, except for sporadic and localized events, these sources are minor.

Lake sediment preserves a record of atmospheric deposition because there is a rapid transport of material deposited on the top of a lake to its bottom and the regular accumulation of sediment at the bottom. Cores of sediment from the bottom of a lake, sliced into 0.5- to 1-cm layers, have been analyzed for the tetrachloro- through the octachlorodioxins and furans with isotope dilution and EC GC/MS at sensitivities of less than 100 attomoles (12). Radioisotopic analysis establishes when the sediment core was in contact with the atmosphere.

Figure 8 gives the concentrations of the dioxins and furans in a sediment core taken from Siskiwit Lake (on Isle Royale in Southern Lake Superior [13]) as a function of year of deposition. These data are typical for all the sediment cores that have been studied (12–15). Octachlorodioxin is always the most abundant of the compounds; the heptachlorodioxins and -furans are next in abundance. Although other chlorinated dioxins and furans are present, their concentrations are very small. Moreover, the concen-

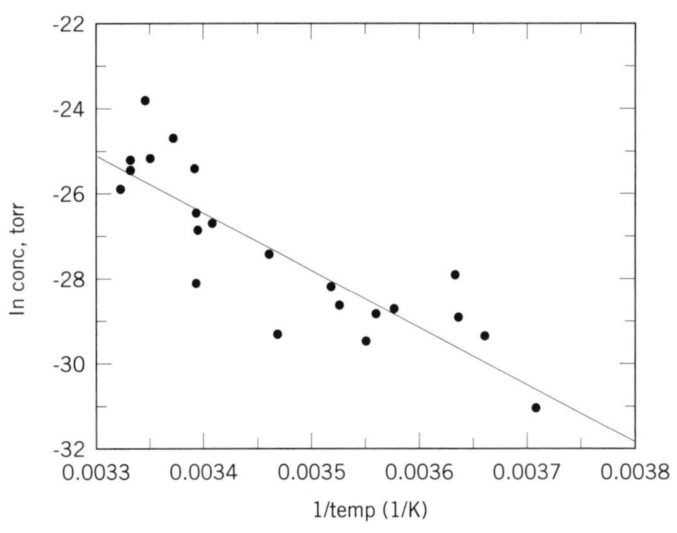

Figure 7. Ambient atmospheric concentrations of endosulfan measured in Bloomington, IN, plotted as a function of atmospheric temperature.

trations of dioxins and furans have not been constant over the last century. The concentrations maximized about 1970, and they were at unmeasurable levels before 1930. This finding suggests that atmospheric dioxin and furan levels increased slowly starting about 1935 and have decreased since about 1970.

Sediment cores from the other Great Lakes and from three high-altitude lakes in Switzerland have also been analyzed (14,15). In every case, dioxins and furans were not present in the sediments before about 1935, and by implication, they were not present in the atmosphere then. This is true despite large differences in both the rates of sediment accumulation and the locations of the lakes. The overall average horizon date is 1938, a date well after the "advent of fire."

What happened in the mid- to late 1930s that led to the emission of dioxins? The products produced by the chemical industry changed about then. Before World War II (1939 to 1945), the chemical industry was commodity

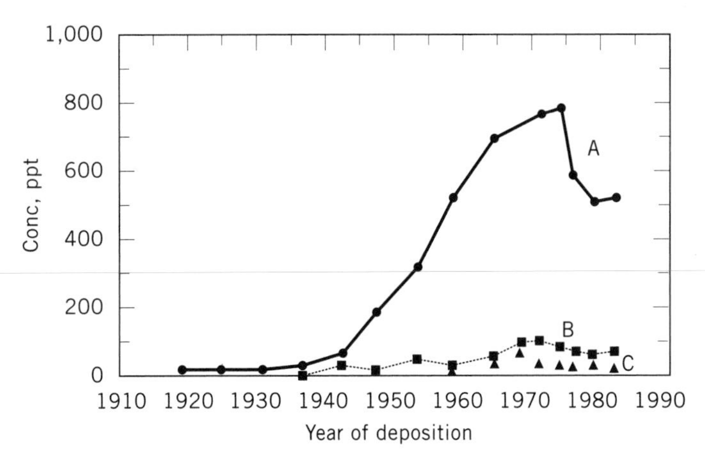

Figure 8. Concentrations of octa- (A) and heptachlorodioxins (B) and of heptachlorofurans (C) (in parts per trillion) versus year of deposition into Siskiwit Lake on Isle Royale.

based, selling large amounts of inorganic products. During the war, organic products were introduced, and plastics became an important part of the chemical industry. Some of these products were organochlorine based; polyvinylchloride is but one example. As waste materials containing these chemicals were burned, dioxins and furans were produced and released into the atmosphere. These compounds eventually ended up in lake sediments.

Incidentally, coal combustion cannot account for the increase in dioxin and furan levels. Coal combustion has been almost constant since 1910; there has been no major shift either in amount burned or in combustion technology since 1935 (12,15). The 1970 maximum in concentration in most sediment cores suggests that emission-control devices, which were beginning to be widely installed at about this time, have been effective in removing dioxins, furans, and the more conventional air pollutants.

BIBLIOGRAPHY

1. M. M. Siegel, I. J. Hollander, M. Karas, A. Ingendoh, and F. Hillenkamp, *Proc. 38th ASMS Conf. on Mass Spectrom.*, 1990, 158.
2. R. A. Hites and W. L. Budde, *Environ. Sci. Technol.* **25,** 998 (1991).
3. D. T. Urban, N. S. Arnold, and H. L. C. Meuzelaar, *Proc. 38th ASMS Conf. on Mass Spectrom.*, 1990, 615.
4. J. S. Brodbelt and R. G. Cooks, *Anal. Chem.* **57,** 1153 (1985).
5. M. A. LaPack, J. C. Tou, and C. G. Enke, *Anal. Chem.* **62,** 1265 (1990).
6. T. D. Behymer, T. A. Bellar, and W. L. Budde, *Anal. Chem.*, **62,** 1686 (1990).
7. A. J. Borgerding and R. A. Hites, *Anal. Chem.* **64,** 1449 (1992).
8. T. W. Burgoyne and R. A. Hites, *Environ. Sci. Technol.* **27,** 910 (1993).
9. T. Pollock, *Dioxins and Furans: Questions and Answers*, Academy of Natural Sciences, Philadelphia, 1989.
10. K. Olie, P. L. Vermuelen, and O. Hutzinger, *Chemosphere* **6,** 455 (1977).
11. R. L. Rawls, *Chem. Engin. News* (Feb. 12, 1979), p. 23.
12. J. M. Czuczwa and R. A. Hites, *Environ. Sci. Technol.* **18,** 444 (1984).
13. J. M. Czuczwa, B. D. McVeety, and R. A. Hites, *Science* **226,** 568 (1984).
14. J. M. Czuczwa, F. Niessen, and R. A. Hites, *Chemosphere* **14,** 1175 (1985).
15. J. M. Czuczwa and R. A. Hites, *Environ. Sci. Technol.* **20,** 195 (1986).

ENVIRONMENTAL ANALYSIS (OPTICAL SPECTROSCOPY)

JAMES F. KELLY
ROBIN S. MCDOWELL
Pacific Northwest Laboratory
Richland, Washington

Optical spectroscopy, broadly defined as spectroscopic measurements and instrumentation using the electromagnetic spectrum from the ultraviolet to the infrared and microwave regions, is a versatile and powerful analytical technique with many environmental applications. This article covers atmospheric analysis and pollution monitoring, in which the sample consists of an *in situ* volume of air in the atmosphere, and laboratory analysis, where samples of environmental interest are either removed to or constrained within a laboratory environment and analyzed there. The distinction is not rigid, eg, laser detection and ranging (lidar) involves sounding the atmosphere by both remote optical interrogation methods and mobile air samplers that employ laboratory instrumentation to effect *in situ* point-source analyses.

As in any analytical process, the object is to determine the composition of the sample (speciation) and to measure the amounts of different species present (quantification). Spectroscopic analysis can meet both of these demands and may offer advantages over other analytical methods. Spectroscopic techniques can identify and quantify species in a single measurement; they can detect a wide range of compounds and at the same time are highly specific, allowing the molecular identification of each species in multicomponent mixtures. They are quantitatively accurate, and they are noninvasive, may not require sample collection or pretreatment, do not introduce sample contamination, and can be remotely situated, requiring only optical access. They are capable of giving rapid results in real time, and they are easily adaptable to continuous long-term monitoring, including data logging and analysis. Environmental conditions such as temperature and pressure can be recovered from the data.

In spectroscopic analysis, species are identified by the positions and shapes of their absorption, scattering, or emission features, and quantified by the intensities of these features. Numerous applications of optical methods have been made to pollution and atmospheric monitoring; these rely on a few basic mechanisms of light–matter interaction. Absorption spectroscopy in the infrared (ir) or ultraviolet–visible (uv–vis) regions records energy depletion of transmitted radiation at characteristic resonance frequencies involving various rotational, vibrational, and/or electronic energy levels in molecules or atoms.

Scattering techniques monitor the change of a probe signal resulting from scattering by atmospheric species of interest. This can involve either elastic (energy-conserving) interactions, such as Rayleigh or Mie scattering, with the photons undergoing only a change in momentum, or the inelastic (energy-changing) Raman effect, in which a strong monochromatic probe beam is scattered with discrete changes in frequency. A spectroscopic analysis of the scattered Raman light reveals spectral shifts characteristic of the different species. Rayleigh scattering is much stronger (ca 10^3 times) than Raman scattering and occurs for all atoms, molecules, and small particles with diameters smaller than the wavelength of the probe light (eg, $\lesssim 0.05\ \lambda$), while even stronger Mie scattering occurs when particulate sizes are $> 0.05\ \lambda$. These elastic scattering processes are a fundamental source of background scatter (noise) but can also serve as the source of a return signal for lidar sounding.

Fluorescence detection measures the emission of light from atoms or molecules that have been selectively excited to higher electronic levels by a spectrally intense light source and then decay to lower lying energy states. Such laser-induced fluorescence (lif) is commonly used to study transparent gases, liquids, and solids, while x-ray fluorescence is used extensively to analyze opaque or heterogeneous samples such as rock specimens.

Radiometry is the direct detection and spectroscopic analysis of radiation emitted naturally by the target molecules (ie, emission spectroscopy). This can involve thermal emissions of the atmosphere and of opaque surfaces or fluorescence from naturally induced excitation phenomena, such as airglow from the upper atmosphere due to solar wind and cosmic-ray bombardment.

Many novel interrogation schemes for each of these physical processes have been developed around the experimental details of available light sources, detectors, and spectral analyzers. The advent of powerful and broadly tunable laser sources has led to the ascendancy of the first three processes for active remote sensing of effluents and for ultrasensitive detection of trace samples under laboratory conditions. The laser's extremely high spectral intensity (photons per unit bandwidth) and spatial coherence (low divergence, allowing tight focusing) makes even weak scattering processes like the Raman effect useful over large distances (kilometers). These attributes have also been exploited to induce a variety of new nonlinear responses in media. New laser spectroscopies like multiphoton absorption (leading to fluorescence or ionization) have permitted the detection of single atoms in background gases at atmospheric pressure. Multiphoton scattering processes can be induced by powerful lasers, causing trace effluents to undergo stimulated emission (induced lasing) in highly preferred directions, thus enhancing the detection limits over spontaneous processes that scatter into many directions.

An extensive review of all such techniques is beyond the scope of this article. The conceptual issues of important analytical methodologies and instrumentation will be presented in the following sections. The reader is referred to more detailed treatments of the basic principles of spectroscopy, radiative transfer, and active laser-based remote sensing (1–3), passive remote atmospheric sounding and radiometry (4–6), conventional spectroscopic methodologies and basic atomic physics (7,8), and molecular spectroscopy and structure (9,10). Special treatments of infrared spectroscopy (11,12), laser principles and laser spectroscopy (13,14), and laboratory-based ultrasensitive spectroscopies (15,16) may be useful. Specific methodologies of optical applications to analytical chemistry are treated in the monograph series *Chemical Analysis* and new developments in lasers and their applications are reviewed periodically in the *Springer Series on Optical Sciences.*

SPECTROSCOPIC BACKGROUND

Electromagnetic radiation is characterized by its wavelength λ, frequency ν, or wave number $\overline{\nu}$, which are related by

$$\nu = c / \lambda, \overline{\nu} = 1/\lambda = \nu/c \quad (1)$$

where c is the speed of light. The photon energy is $E = h\nu = hc\overline{\nu}$, where h is Planck's constant and hence is proportional to frequency (expressed in some multiple of cycles per second or Hertz) and wave number (usually given in cm^{-1}). The units for wavelength are commonly nm or μm (1 nm = 10 Å = 10^{-9}m; 1 μm = 10^{-6} m). The conversion of wave numbers to frequency is given directly as $\nu = c\overline{\nu}$; ie, 1 cm^{-1} = 3 × 10^{10} Hz = 30 GHz.

For convenience, the electromagnetic spectrum is divided into several energy regions characterized by the differing experimental techniques employed and the various nuclear, atomic, and molecular processes that can be studied. These are, in order of increasing energy (decreasing wavelength), with their approximate wavelength limits: radio waves (>30 cm), microwaves (1 mm–30 cm), far infrared (50–1000 μm), mid infrared (2.5–50 μm), near infrared (0.8–2.5 μm), visible (0.4–0.8 μm or 400–800 nm), near ultraviolet (180–400 nm), vacuum ultraviolet (10–180 nm), x rays (0.01–10 nm), and γ rays (<0.01 nm, or energy >0.1 MeV). The regions most useful for atmospheric and pollution monitoring are the mid infrared to near ultraviolet.

Molecular spectroscopy can be divided into the three broad areas: rotational, vibrational, and electronic transitions, named in order of increasing energy required for excitation. Rotational transitions occur in the far infrared and microwave regions (wavelengths longer than about 100 μm, except for a few light molecules such as H_2O that extend to shorter wavelengths). From about 15 μm (670 cm^{-1}) out to the extreme far infrared–microwave region near 1000 μm (1 mm or 10 cm^{-1}), there is strong and near-continuous absorption due to the rotational spectrum of water vapor, and atmospheric studies are difficult or impossible here. Still longer wavelengths (> 1 mm) comprise the various microwave and radar bands; these are useful for the detection and ranging of extended objects, ranging in size from raindrops to aircraft to large weather systems. Spectroscopy in the region covering 1 mm to 1 m has been applied successfully to the study of molecules in astronomical sources and interstellar dust clouds (radioastronomy), but this is not a useful region for general spectroscopic monitoring of the earth's atmosphere, because microwave absorption by molecules is extremely weak. However, spectroscopy in the region 260–280 GHz (~ 1.1 mm) has been used to quantify trace polar molecules such as ClO, O_3, HO_2, HCN, and N_2O at the ppt level in the stratosphere, using a ground-based millimeter-wave superheterodyne receiver (17). In such cases, the large fluence available from microwave sources, used in conjunction with coherent (heterodyne) detection, can overcome weak absorption.

Molecules vibrate at fundamental frequencies corresponding to electromagnetic radiation with wavelengths of approximately 2.5 μm (wavenumber 4000 cm^{-1}) to 100 μm (100 cm^{-1}) (the mid infrared), with some overtone and combination transitions at shorter wavelengths. Because this provides enough energy to excite rotational motions also, the infrared spectrum of a gas consists of rovibrational bands in which the basic vibrational transition is broadened by the superposition of numerous lower energy rotational transitions that occur simultaneously with the vibration. The vibrational frequencies depend on the masses of the atoms involved in the various motions and on the force constants and geometry of the bonds connecting them. At the same time, the shapes of the bands are determined by the rotational structure and hence by the molecular symmetry and moments of inertia. The rovibrational spectrum thus provides direct molecular structural information, resulting in high specificity: the vibrational spectrum of any molecule is unique (except those of optical isomers). Every molecule, except homonuclear diatomics such as O_2, N_2, and the halogens, has at least one vibrational absorption in the infrared.

Still shorter wavelength (higher energy) radiation will promote transitions between electronic orbitals in atoms and molecules. Outer bonding electrons can usually be excited by near ultraviolet or visible radiation. At higher energies, in the vacuum ultraviolet, innershell transitions occur, but this region is not useful for the present purposes because of strong absorption by atmospheric O_2 for $\lambda <$ 180 nm. Deep innershell electronic transitions can be induced via x-ray excitation, and this provides a technique for laboratory elemental analysis. Electronic transitions in molecules will also include structure due to associated vibrational and rotational transitions (rovibronic band). Electronic transitions tend to have larger absorption cross-sections than rotational or vibrational transitions and hence higher sensitivity for analytical purposes, but spectral overlap and interferences are more likely to be problems.

LINEWIDTHS, LINE STRENGTHS, AND DETECTIVITY

The detectivity for a given species in optical spectroscopy is a product figure-of-merit of spectral resolution and sensitivity. Resolution, the ability to distinguish between transitions of nearly equal wavelength, is determined practically by the degree to which neighboring spectral features overlap. All spectral features have a narrow natural width, but line-broadening effects and instrumental limitations cause additional line broadening.

The sensitivity of an optical process is determined by the strength of the photon–molecule interaction, expressed usually as a frequency-dependent cross-section $\sigma(\nu)$ (cm^2) for absorption, or as a differential cross-section $d\sigma(\nu)/d\Omega$ (cm^2/steradian) for scattering (1). The latter measures the likelihood that active species scatter some portion of the incident laser fluence $\Phi(\nu)$ (photons/cm^2) into a viewing solid angle $\Delta\Omega$ (steradians) (Fig. 1). In most cases of practical significance for environmental monitoring, the frequency-dependent cross-sections can be written

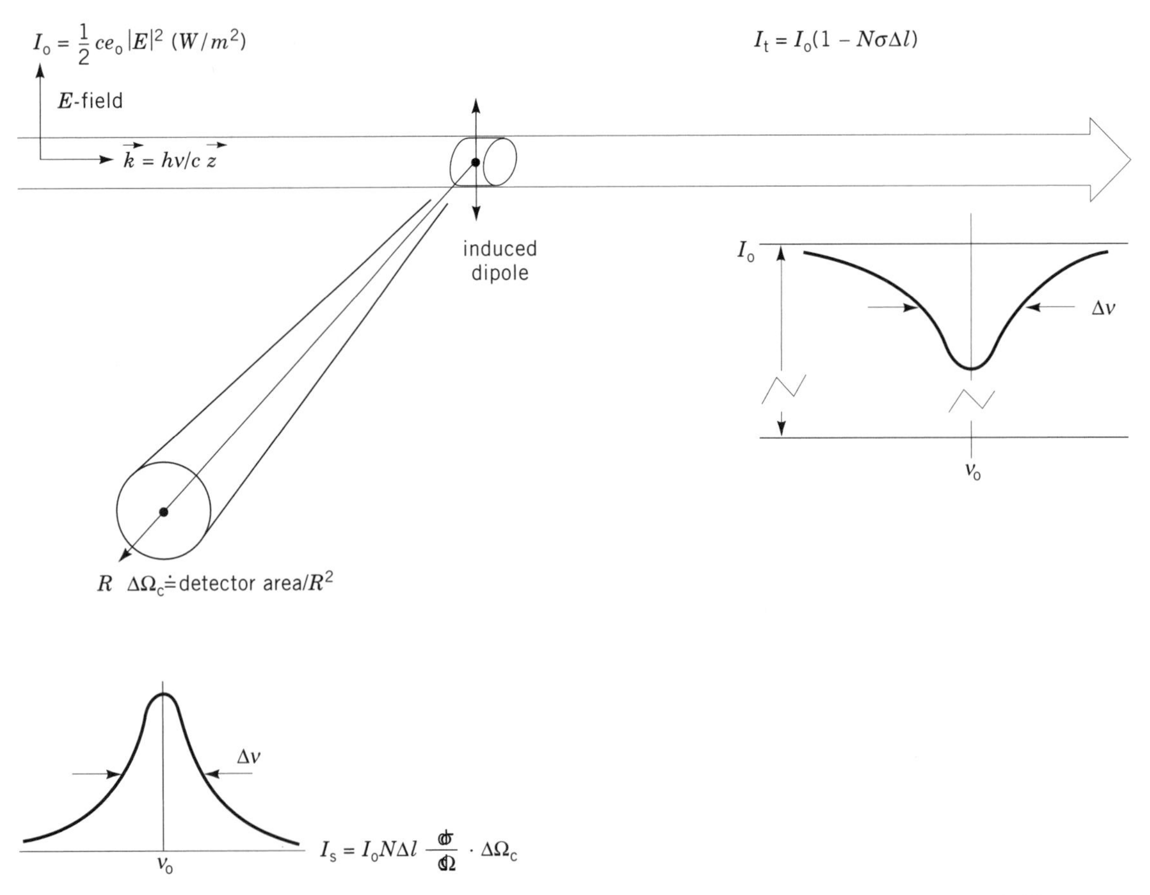

Figure 1. Schematic of an absorption–scattering process. An incident electromagnetic field of intensity I_0, with an associated electric field $\vec{E}$, induces dipole oscillation in the absorbers, resulting in light I_s scattered into a collection solid angle $\Delta\Omega_c$, and corresponding absorption resulting in transmitted intensity I_t.

separately in the form

$$\sigma(\nu) = \sigma_0 \cdot \mathfrak{L}(\nu - \nu_0, \Delta\nu), \quad (2)$$

where σ_0 (or $d\sigma/d\Omega$) is the peak value for absorption (scattering) and $\mathfrak{L}(\nu - \nu_0, \Delta\nu)$ is a symmetrical lineshape function that is parameterized by a center frequency ν_0 and linewidth $\Delta\nu$.

An important "semiclassical" measure of the frequency-integrated absorption cross-section (Ladenburg's formula) is

$$\int \sigma(\nu)dv = \pi r_c cf \equiv S \text{ (cm}^2\text{·frequency)} \quad (3)$$

where r_c is the classical electron radius (2.8×10^{-13} cm) and f is the oscillator strength for the radiative transition (12,18). Similar expressions can be written for the differential fluorescence or scattering cross-sections, since they are interrelated processes (1). The f-values or line strengths S can be calculated from first principles of quantum mechanics but most often are obtained empirically. S is related to a matrix element d_{ij} connecting the initial and final states of a transition, $S \propto (d_{ij})^2$, where d_{ij} has typical dimensions of the atom ($\sim10^{-8}$ cm). The maximum value of the line shape function $\mathfrak{L}(\nu - \nu_0, \Delta\nu)$ is inversely proportional to the line width, so the peak absorption (or angle-integrated scattering) cross-sections can be approximated as

$$\sigma_0 \doteq S / \Delta\nu \quad (4)$$

The peak absorption (scattering) cross-sections are thus a useful comparative measure of detectivity, because the latter is a product of the line strength and the practical line resolution. Table 1 lists typical peak values for absorption, fluorescence, and Rayleigh and Raman scattering. Four causes of line broadening are of importance in environmental analysis:

1. Natural broadening is due to the intrinsic lifetimes of the states involved in the transition. The line width (full width at half-maximum, fwhm) $\Delta\nu_N$ is $(2\pi\tau)^{-1}$, where τ is the natural lifetime for a transition to undergo spontaneous decay in the absence of external perturbations. $\Delta\nu_N$ is ≤ 80 MHz for strong uv–visible transitions and ~ 1 MHz for ir vibrational transitions. These line widths can be resolved by modern laser spectroscopies under laboratory conditions.

2. Collisional broadening results when an active absorber or scatterer experiences perturbations by other atoms or molecules. These collisions effectively reduce the natural lifetime of the optically active system, so the net broadening depends on a characteristic impact time τ_c:

$$\Delta\nu_C = \frac{1}{2\pi}\left(\frac{1}{\tau} + \frac{1}{\tau_c}\right) \quad (5)$$

3. Doppler broadening arises from the random thermal agitation of the active radiators, each of which, in its own rest frame, see the applied light field with a different frequency. When averaged over a Maxwellian velocity distribution, this yields a line width (fwhm)

$$\Delta\bar{\nu}(\text{cm}^{-1}) = 7.16 \times 10^{-7}\bar{\nu}_0\sqrt{T/A}, \quad (6)$$

where T is the sample temperature (K), A the molecular weight of the species (amu), and $\bar{\nu}_0$ is the transition energy (cm^{-1}). Typical Doppler line widths are $\sim$0.05 cm^{-1} for visible transitions ($\sim$500 nm) at room temperature, and are correspondingly larger (smaller) for uv (ir) transitions. Unless special nonlinear spectroscopies are exploited, the Doppler broadening of a transition represents the fundamental lower limit for practical resolution.

Natural and collisional broadening are considered homogeneous processes, because all radiators experience the same local effects. Such lines have a Lorentzian shape, with $\Delta\nu$(fwhm) $= (2\pi\tau')^{-1}$, where τ' is the effective "lifetime" of a radiator's uninterrupted oscillation period. The homogeneous line profile is distorted by inhomogeneous broadening effects, of which the most important is the Doppler broadening, which has a Gaussian distribution. For gaseous samples at STP, absorptions and emissions actually have the Voigt profile, which is the mathematical combination (convolution) of the Lorentzian and Gaussian shapes.

Table 1. Typical Lineshape Parameters and Relationships

Parameter	Rotational Transitions	Vibrational Transitions	Electronic Transitions
Line frequency, $\bar{\nu}$	1–100 cm^{-1}	100–4,000 cm^{-1}	up to 50,000 cm^{-1}
Natural line width, $\Delta\bar{\nu}_N$	$<10^{-11}$ cm^{-1}	$<10^{-7}$ cm^{-1}	$<3 \times 10^{-3}$ cm^{-1}
Doppler width at 300 K, $\Delta\bar{\nu}_D$	<0.0003 cm^{-1}	<0.01 cm^{-1}	0.01–0.2 cm^{-1}
Natural radiative lifetime, τ_N	$>10^{-1}$ s	$>2 \times 10^{-4}$ s	$>2 \times 10^{-9}$ s
Peak Doppler broadened absorption cross-section, σ_A	$<10^{-20}$ cm^2	$\leq10^{-18}$ cm^2	10^{-11}–10^{-16} cm^2 (atoms) $<10^{-17}$ cm^2 (molecules)
Peak Rayleigh differential scattering cross-section, $d\sigma/d\Omega$	Negligible	Negligible	$<2 \times 10^{-13}$ cm^2 sr^{-1} (atoms)[a] $<10^{-26}$ cm^2 sr^{-1} (molecules)[b]
Peak Raman differential scattering cross-section, $d\sigma/d\Omega$	$<10^{-27}$ cm^2 sr^{-1}	$<10^{-28}$ cm^2 sr^{-1}	$\sim10^{-24}$ cm^2 sr^{-1} (atoms)
Peak fluorescence differential scattering cross-section, $d\sigma/d\Omega$	Negligible	Negligible	$<5 \times 10^{-16}$ cm^2 sr^{-1} (atoms)[c] 10^{-20}–10^{-25} cm^2 sr^{-1} (molecules)

[a] Values are for resonant scattering by atomic vapors.
[b] Values are for nonresonant scattering of visible–near uv radiation by atmospheric gases; resonant Raman scattering approaches 10^{-24} cm^2 sr^{-1}.
[c] Values are for STP atmospheric conditions.

4. Instrumental broadening occurs when the spectral resolution of the probe light or spectral analyzer are larger than the intrinsic broadening of the radiating species. The observed line shapes are the convolution of the instrumental line shape and the environmentally perturbed Voigt profile (Fig. 2) (8,13).

With inadequate resolution much of the information in a complex vapor-phase spectrum can be blended or lost, with consequent degradation of specificity and loss of quantitative accuracy. To make full use of the spectral information available, the instrumental resolution should be comparable with, or smaller than, the spectral width of the transitions observed. Under atmospheric conditions, collisional broadening is the dominant mechanism. There is usually a linear relationship between pressure and line width, with pressure-broadening coefficients typically in the range 0.01–0.5 cm^{-1} atm^{-1} for infrared rovibrational transitions. Ir spectrometers used in remote sensing will typically have 0.1-cm^{-1} resolution for scanning the troposphere (lower atmosphere), while analysis of the stratosphere (heights > 25 km) will often require < 0.01 cm^{-1}. Usually somewhat lower resolution will suffice for the uv–visible region because of the combined effects of collisional and Doppler broadening. For general analysis, 0.5–1 cm^{-1} may be adequate. Studies in liquid-phase solutions generally require even lower spectral resolution, because solvent effects significantly broaden molecular transitions to well over 1 cm^{-1}.

ATMOSPHERIC ANALYSIS

Two limiting cases define the range of analytical situations that might be encountered: (*1*) characterization of multicomponent mixtures, which might be a portion of the atmosphere itself or the emission plume from a stack or incinerator, for which numerous species are of concern and for which the nature of the pollutants might change with time, and (*2*) monitoring of emissions from particular sources, for which relatively few species, whose identity may already be known, must be quantified continuously, but there is less need for qualitative analysis.

Infrared Absorption Spectroscopy

The strong fundamental vibrational transitions of molecules fall in the mid infrared (2.5–100 μm) range and provide two types of specificity (12). At the shorter of these wavelengths ($\lambda \lesssim 8$ μm) many chemical functional groups exhibit characteristic "group frequencies" that are relatively independent of the molecular environment and provide useful information on the chemical nature of the absorber. At longer wavelengths the frequencies are influenced more by the skeletal vibrations of the molecule: this is the "fingerprint region," where even similar molecules may have quite different spectra, allowing individual identification. Overtones and combinations of some of these vibrational modes appear at shorter wavelengths in the near infrared (nir) or visible but with intensities reduced by ca 100 for the first overtone and an additional factor of 10 for each successive higher overtone. Nir is useful for some industrial analyses (19), for which the potential loss of sensitivity may be compensated by the ability to analyze concentrated process streams (including aqueous systems) that would totally absorb mid-ir radiation, but most atmospheric monitoring occurs in the mid infrared.

Infrared methods are well established for qualitative and quantitative analyses of molecules and are noted for their high specificity and reasonably good quantitative accuracy (20). The pattern of the infrared absorption frequencies for vapor-phase molecules reflects the nature of the rotational and vibrational structure of a given molecule and so is unique for each species. While the rotational structure enhances specificity, there are some disadvantages to the resulting band broadening. The band intensity is spread over all the component rotational transitions, resulting in lower peak cross-sections for absorption (the analytical parameter that determines sensitivity), and the broadened bands are more likely to overlap and interfere. However, if the rotational structure can be resolved, valuable information about the temperature distributions of remote samples can be recovered. The high specificity of ir spectroscopy reduces difficulties caused by interferences, as when one is seeking a minor species present in a complex mixture. Atmospheric H_2O and CO_2 can interfere at some wavelengths (21), but for most molecules at least one absorption feature can be found that falls within any of several broad transmission windows. In process monitoring, commercial methods have been developed to remove water vapor from stack emissions for analytical purposes. Particulates smaller than the wavelength of the radiation (< 1 μm) have little effect on an infrared beam, but heavy loadings of larger particulates may present a problem.

Quantitative accuracy of most infrared methods can be characterized as adequate but not outstanding; typically, one might expect uncertainties of about 5%. In assessing quantitative accuracy it is assumed that the optical path length through the sample is known. This is true for laboratory analyses and can be achieved in stack configurations, but in the open atmosphere one obtains a column density averaged over the total path, which may not reflect the concentration of a "slug" of pollutant that occupies only a small volume in the sampled region. This may be acceptable when the location of the analyte is already known, as when sampling over a stack or a highway.

Detection limits depend strongly on the intrinsic band strengths and on the optical configuration employed (especially the path length) but are typically in the subparts-per-million range. For optically thin, scatter-free conditions, the intensity of radiation transmitted I_T is related to the incident intensity I_0 by the Beer-Bouguer-Lambert law,

$$I_T(\nu) = I_0(\nu)e^{-\sigma(\nu)Nl} \tag{7}$$

where $\sigma(\nu)$ is the absorption cross-section and N is the density of absorbing species along the path length l. The fractional absorption is

$$\frac{I_0 - I_T}{I_0} = 1 - e^{-\sigma Nl} \approx \sigma Nl \tag{8}$$

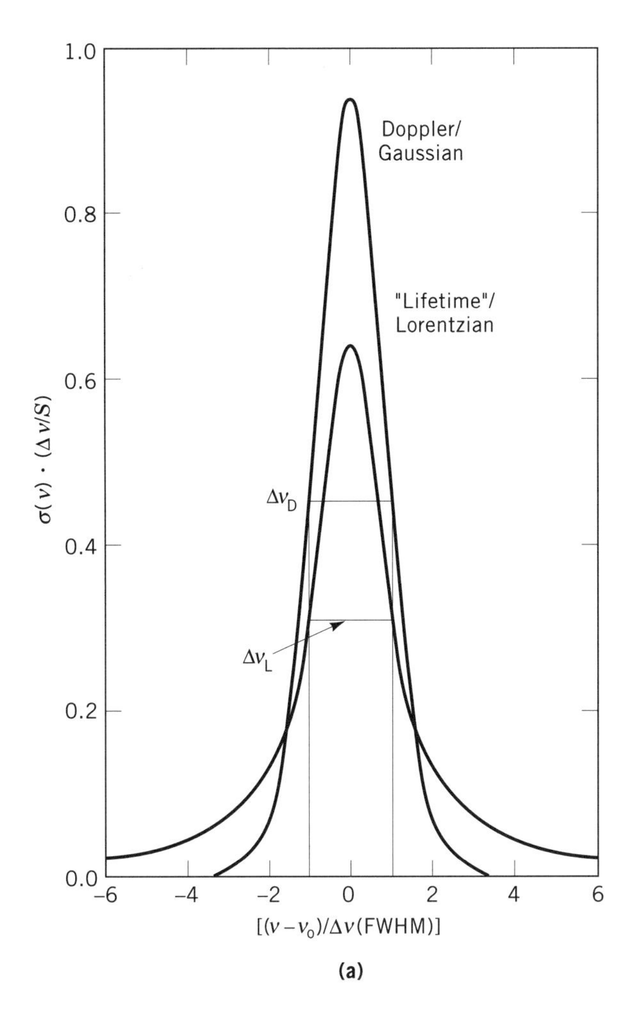

$$\exp(-4(\nu-\nu_0)^2 \ln 2/\Delta\nu_D)$$

$$L_L(\nu-\nu_0) = \frac{1}{\pi} \cdot \frac{(\Delta\nu_L/2)}{(\nu-\nu_0)^2 + (\Delta\nu_L/2)^2}$$

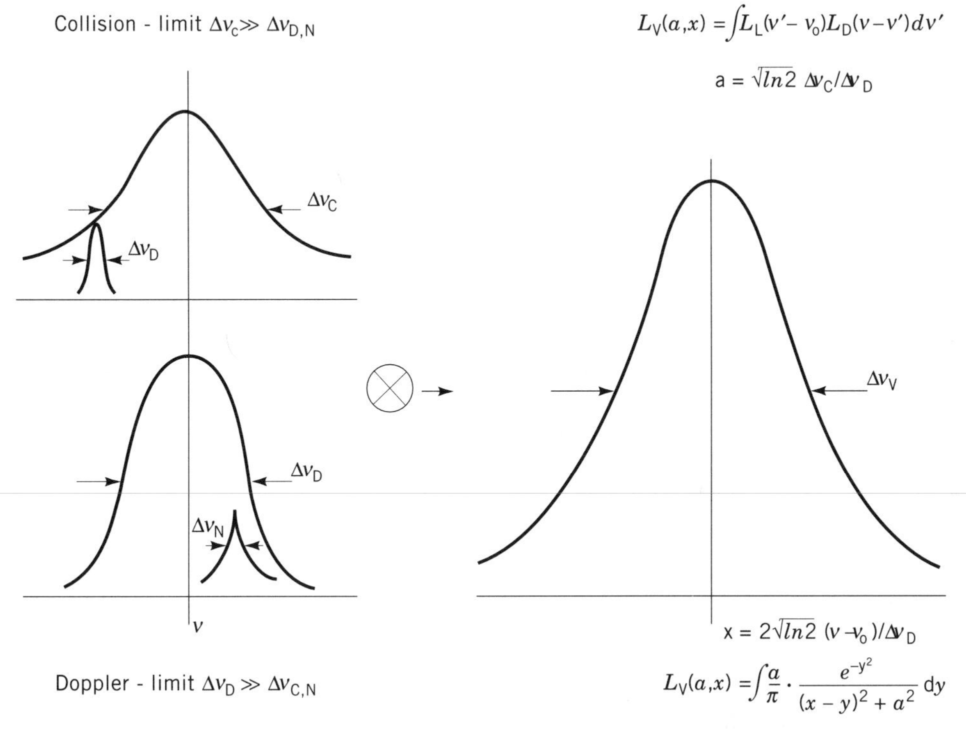

Figure 2. Comparison of line shape functions $\mathfrak{L}(\nu-\nu_0,\Delta\nu)$. (**a**) Normalized (integrated areas equal 1) absorption cross-sections are shown for the two natural limits of pure Doppler and lifetime broadening; under typical conditions, the line shape is a combination of two or more effects. (**b**) The convolution of a Lorentzian with a Gaussian line shape leads to the Voigt profile.

where the latter limit of weak absorption usually holds for atmospheric samples.

A typical peak absorption cross-section for an infrared transition is $\sigma = 10^{-18}$ cm^2. If the minimum detectable fractional absorption is assumed to be 0.01 and $l = 10$ m, then the minimum detectable density is $N = 10^{13}$ cm^{-3} or (10^{13} mol cm^{-3})/$L \approx 400$ ppb, where L is the Loschmidt number (2.69×10^{19} mol cm^{-3} at STP). The relatively low sensitivity of conventional absorption spectroscopy stems from the fact that concentration measurements are made as the difference of two large numbers, I_0 and I_T. Novel techniques have been developed to improve sensitivity by increasing the effective integrated sample path length; improving the practical resolution of the absorbing features, so the measured peak absorption is as large as theoretically possible; and/or eliminating the background subtraction by "null-background" techniques.

Sampling. The two basic optical arrangements for real-time monitoring by absorption are point to point, for which the source and detector are separated and the atmospheric volume to be sampled lies between them, and single-ended remote sensing, in which a probe beam is returned from a retroreflecting target (a mirror or topographic feature) and traverses the sampled volume twice. Because sensitivity is proportional to the path length, mirrored multipass sample cells (White or Herriot cells) (22) are often used in the laboratory to obtain optical paths of 200 m or more with a small volume.

It has been assumed here that the optical path from source to sample to detector will be through the atmosphere. This is the usual arrangement, but it may be desirable in some cases to transmit the beam in part through fiber optics, which are cylindrical wave guides that channel electromagnetic radiation (23,24). This allows one safely to access a small sample region in remote or harsh environments far removed from the spectrometer. Silica fibers are available that transmit from 220 nm in the uv to 2.3 μm in the infrared with low losses; fibers of fluoride and chalcogenide glasses extend further into the ir, to about 5 μm, but have not found wide application because of their brittleness. An elaboration of fiberoptic sensors has given rise to optrodes, a specialized type of analytical instrumentation. Optrodes (25) have optical transducers mounted at the distal end of a fiber to monitor chemically selective changes in transmission or fluorescence, often employing an immobilized reagent on the surface of the fiber. Conventional ir absorption can be measured with a transmission-based optrode; the output of a low power, light-emitting diode is directed over a silica fiber, through the sample, and returned over another fiber for analysis. Such systems have been demonstrated for monitoring explosive gases (methane, propane) and NH_3, using near and mid ir absorption.

Numerous specialized sampling techniques are available (26). Planar wave guide cells have been developed that employ frustrated multiple total internal reflection, where the infrared beam is transmitted inside a thin, highly refractive crystal (ZnSe or Ge) and is absorbed by a sample in optical contact with the surface. Matrix isolation is a cryogenic sampling technique that freezes analytes onto windows, internal-reflection substrates, or mirrors for spectroscopic analysis. Cold metal mirrors offer the easiest sampling strategy, by which a thin film is condensed onto the mirror, from which the probe beam is reflected, traversing the sample twice (27). For atmospheric samples, CO_2 naturally present will condense as a matrix in which other minor species are trapped in individual lattice sites, so that Doppler and collisional broadening are eliminated, and because there is no molecular rotation in a solid, the resulting collapse of the rotational structure results in all the intensity of the band being concentrated in one sharp absorption line (ca 1 cm^{-1} wide), increasing sensitivity and reducing the likelihood of interferences. Detection limits are on the order of 10 μg for moderate absorbers, such as NO_x or CO_2, to 100 ng or less for a strong absorber such as a metal carbonyl. This approach involves some time delay, both for the deposition of the sample and for subsequent removal of the film from the substrate in preparation for the next analysis.

Nondispersive ir Gas Analyzers. A simple method of infrared analysis is to monitor the absorption of a specific band of a species by using two narrow bandpass filters to isolate the analytical frequency and a nearby reference frequency where there is no absorption (28). Such nondispersive infrared (ndir) spectroscopy can be an effective low cost method for gas analysis. Typically, radiation from a source is collimated, passed through the filters and sample cell, and focused on a detector. The resulting on- and off-resonance signals are processed. An analyzer can be tailored with the proper filter set for any species with a strong ir absorption that is reasonably free from interferences, and multichannel units can accommodate several filters to monitor different absorption frequencies and a background. Sensitivities are of the order of 0.1% to low ppm levels, depending on the species, and response is fast, providing continuous real-time analysis. Such ndir analyzers are currently marketed for CO, CO_2, CH_4, C_2H_6, freons, SF_6, and total hydrocarbons, the last monitoring a wavelength region around 3.4 μm, where most hydrocarbons exhibit absorption due to the C—H stretching vibration.

Correlation Spectrometers. In a correlation spectrometer, the beam, after traversing the sample, is alternated between two cells (6,29). One contains the analyte gas, so radiation passed by this cell is unaffected by concentration changes in the sample; the other cell is empty (or contains an interfering gas, if one is present). The fractional change between the two signals provides an output sensitive to the species sought and relatively independent of other disturbances. Correlation analyzers can be sensitive and specific and have an advantage over filter systems in that they are readily adaptable to different gases as conditions warrant, it being necessary only to fill the appropriate cell with the target gas. They lack flexibility because they can analyze for just one specific species, and interfering gases can sometimes cause difficulties.

Ir Spectrometers. An outstanding characteristic of rovibrational spectroscopy is its ability to analyze for a broad range of compounds, but this requires obtaining a significant portion of the infrared spectrum, say from 2 to 25 μm, with a recording spectrometer. From such a spectrum one can identify and quantify nearly all species present.

There are two basic methods for recording broad-band infrared spectra: dispersive and interferometric. In dispersive spectrometers a narrow beam defined by a mechanical entrance slit is dispersed by a prism, or (for better resolution) a diffraction grating, into its constituent wavelengths. Because photographic emulsions are not useful at wavelengths $\gtrsim 1.3$ μm, the dispersed radiation is swept slowly across an exit slit by the rotation of the prism or grating, and the signal is photoelectrically detected, amplified, and recorded to furnish a spectral plot of intensity versus wavelength. The optical path between the entrance and the exit slits serves to isolate a single narrow frequency or wavelength range out of the spectrum, and accordingly this portion of the instrument is called a monochromator; when the source and detector optics and amplifier and recording electronics are added, it becomes a recording spectrophotometer.

Commercial ir spectrophotometers usually have a double-beam configuration, in which a system of rotating mirrors switches the source beam alternately through the sample and through an equivalent reference path many times per second; these signals are compared at the detector, and their ratio provides a spectrum free from the influences of unwanted atmospheric absorption, variations in source intensity, etc. High performance dispersive spectrophotometers have for the most part been replaced by interferometric Fourier-transform spectrometers (discussed below), which offer many advantages in performance and flexibility. But inexpensive grating instruments will still play a large role in pollution monitoring, because multielement detector arrays can allow multiplexed detection of several absorption features simultaneously. Such polychromators have a solid-state optoelectronic array detector in the focal plane of a monochromator. These devices are discussed in more detail below; most are silicon based and usable only for $\lambda <$ 1.1 μm, but commercially available Pt-Si devices now cover 1.5–5.0 μm, and research continues to extend their response to longer wavelengths, which will strongly impact ir instrumentation. InSb and HgCdTe arrays are also becoming available for still longer wavelengths (5.5 and 11 μm, respectively), but are very expensive. An array-equipped ir polychromator might provide spectral readouts for both speciation and quantification with much greater flexibility than a tunable laser and at much less expense than a Fourier-transform spectrometer.

Fourier-Transform Spectrometers. A Fourier-transform spectrometer (FTS) is actually a Michelson interferometer that records the interference produced between two alternate light paths reaching the same detector, as the optical delay (or phase retardation) of one of these paths is changed by the linear motion of a mirror (4,11). The resulting signal strength as a function of mirror travel is an interferogram, from which the desired spectrum (intensity as a function of wave number) can be obtained by the mathematical procedure known as a Fourier transform. Such instruments incorporate a dedicated computer to perform the transform using the FFT (fast Fourier transform, or Cooley-Tukey) algorithm. They are expensive and sophisticated devices, but they offer a high level of performance that makes them attractive for optical monitoring.

There are several reasons for the high performance of Fourier instruments compared with dispersive spectrometers. While a scanning spectrometer must record the spectrum sequentially, one spectral slit width at a time, an interferometer processes information from all frequencies simultaneously: the multiplex (or Fellgett) advantage. In this it has some of the characteristics of a spectrometer with a continuously recording array detector, but it offers much better resolution and broader spectral coverage. Interferometers also have a throughput (or Jacquinot) advantage, in that they accept a large solid angle of radiation and hence pass a much greater light flux than do spectrometers, which are limited by the necessity of defining the beam with narrow slits. These two advantages can be converted into orders-of-magnitude improvements in resolution, scan time, and/or signal:noise (S:N) ratio, the three mutually related instrumental parameters that are of the greatest practical interest to the analyst. The resolution of a FTS is approximately the reciprocal of the maximum optical path difference or of twice the maximum mirror travel. A resolution of 0.1 cm^{-1} thus requires a mirror travel of only 5 cm, and many commercial instruments offer at least this capability.

Many types of FTIR spectrometers are marketed, ranging from simple, rugged, low resolution instruments intended for basic analytical applications to expensive high performance instruments, offering high resolution and broad flexibility for research spectroscopy. A typical system for recording mid infrared spectra in the 2–25 μm region might include a silicon–carbide globar source, a germanium-coated KBr beam splitter, and a pyroelectric or photoconductive detector. Different spectral regions can be accessed with appropriate substitutions of source, beam splitter, and detector. The sample compartments of commercial instruments are designed for standard 10-cm gas absorption cells, but they can also accommodate multiple-reflection cells, and transfer optics are available or can be designed for other configurations that might be needed for pollution and atmospheric monitoring. FTIR spectrometers have now become an effective tool for quantifying pollutants in remote sensing spectroscopy and have allowed temperature determinations to within ± 2 K from rotational analysis of spectra, as shown in Fig. 3. FTIR provides near-real-time results. The scan time depends to some extent on the mirror travel (ie, resolution) required, but its main determinant is the number of individual scans that are co-added to provide a final spectrum with an acceptable S:N level. A typical commercial FTIR can produce a spectrum from 10 co-added interferograms at a resolution of 1 cm^{-1} in ca 10 s.

Compact, robust FTIR instrumentation is currently marketed specifically for industrial analysis and on-line process monitoring. Permanently aligned, industrially hardened instruments are vibration and shock resistant and sealed against dust and moisture. Using sources that need no external cooling and room temperature detectors, they require no utilities other than a 120-V line. Units are available that can monitor a dozen or more selected gases with sensitivities of better than 1 ppm for most pollutants and with recording and automatic alarm capabilities.

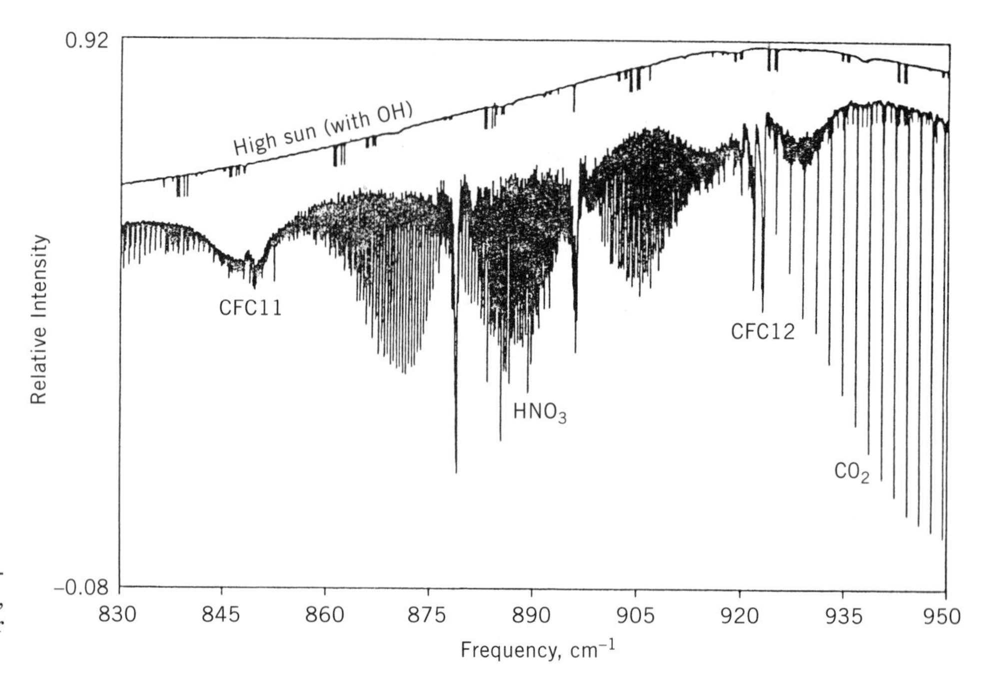

Figure 3. Real spectrum of the atmosphere at a tangent height of 17 km, taken with a remote sensing Fourier transform spectrometer (4).

Laser Sources. A (fixed-frequency) laser that happens to emit at a pollutant absorption frequency can provide a sensitive method of analysis (13,14). The most useful for this purpose is the CO_2 laser, which provides high power monochromatic emission in the region 8.7–11.8 μm; this falls in an atmospheric window and also corresponds to the "fingerprint" region in which many organic molecules have characteristic absorptions. The emission is not, unfortunately, continuously tunable, but $^{12}C^{16}O_2$ will lase on any of some 100 rovibrational lines, and its various isotopomers ($^{13}C^{16}O_2$, $^{16}O^{12}C^{18}O$, etc) provide many hundreds of additional lines. Analysis with line-tunable lasers offers little flexibility because the frequency must be selected for a given species, but it can provide high sensitivity in some cases. The detection limits depend on the offset or detuning between the analyte absorption feature and the laser emission line. Sensitivity varies from 3 ppb/km for ethylene (a strong absorber of a specific CO_2 line) to 3 ppm/km for SO_2. The CO_2 laser is particularly suited for long-range remote sensing because of its high efficiency and power.

Tunable ir Laser Sources. Fully tunable lasers have the advantage that the output frequency can be adjusted to correspond exactly to that needed for a specific analyte (30–32). The development of nearly monochromatic lasers that can be continuously frequency tuned throughout much of the ir region has revolutionized vibrational spectroscopy. Such a source provides in effect a compact high resolution spectrometer without the need for a bulky monochromator or for the computational complexities of Fourier-transform instrumentation. Several such systems have been developed, but semiconductor diode lasers are by far the most widely used. Tunable lead salt semiconductor diode lasers (TDLs) can be fabricated to cover any portion of the mid infrared from 3.3 to 28 μm. They can be tuned, by varying the temperature and/or injection current, quasi-continuously over frequency ranges of 50–100 cm^{-1} in continuous scans of over 1 cm^{-1}.

Commercial TDL systems (33) are modular, consisting of a cryogenic laser source assembly, a power and control module, collimating optics, a simple mode-selecting monochromator, and a Ge:Cu detector with lock-in amplifier. While the laser source assembly has usually included a relatively expensive closed-cycle helium refrigerator, recent improvements in manufacturing technology have produced high performance diodes that operate reliably above 77 K, allowing the use of simple liquid nitrogen Dewars with substantial reduction in cost and complexity.

The limited tuning ranges of TDLs make them unsuitable for general survey spectroscopy, but once a specific interference-free molecular absorption feature has been identified, their high resolution (ca 0.0003 cm^{-1}) and easy tunability are useful for monitoring such a feature. In addition to continuous tuning, they can be operated at discrete frequencies, tuning on and off a resonance many times per second to provide instant quantification. Industrial analytical applications to TDLs include stack-gas monitors for H_2SO_4 and other pollutants.

Null-Background Techniques. Conventional absorption spectroscopies measure small fractional changes of transmitted light (eq. 8), to determine concentrations. The uncertainties in obtaining small differences between large values of the incident and transmitted light limits measurable absorption values $N\sigma l$ to $\geq 10^{-3}$. A significant improvement can be achieved with null-background absorption methods that exploit the high spectral sensitivity and frequency stability of laser sources. These techniques have allowed sensitive determinations of $N\sigma l \sim 10^{-7}$ under laboratory conditions.

One of these techniques is harmonic or derivative spectroscopy (30), which involves modulating a relatively narrow-band light source, passing the modulated radiation through the sample, and then using notch filters and synchronously tuned amplification of the signal at the modulation frequency (or at one of its nth-order harmonics).

Scanning the wavelength of the light source gives a signal that is proportional to the first (or *n*th-order) derivative of the absorption profile. Greater sensitivity results from elimination of the background and reduction of low frequency drifts in the light source, and the method discriminates against spurious signals that do not have a sharp wavelength dependence at the modulated frequency. The modulation can be effected by varying either the amplitude or the frequency of the source; the latter (fm modulation) offers the highest sensitivity, while amplitude modulation (am) adds additional frequency components that are harder to filter during postdetection processing. TDLs can be readily frequency modulated with high fidelity (little residual am) by adding a small ac modulation to their dc bias current. Balloon-borne TDL spectrometers have measured absorbances of $\sim 10^{-5}$ with modulation frequencies of $\leq$ 10 kHz (34), and such systems have been proposed as *in situ* probes of planetary atmospheres (35).

Another method of eliminating background radiation is to measure secondary signals that are directly indicative of the energy depletion of the light. Examples include photoacoustic (pas) and photothermal spectroscopies, which measure the change of a thermodynamic parameter during absorption (15). When a molecular gas, liquid, or solid absorbs radiation, some of the energy is converted into kinetic motion and the sample is heated; this nearly adiabatic transfer induces pressure pulses that can be synchronously detected by a sensitive microphone (spectrophone). The principal applications of pas have been in solid sampling and in the analysis of gases at low concentrations. A closed gas cell is required (for maximum sensitivity, an acoustically resonant one), and the microphones used are very small, so pas is better suited for the analysis of small volumes of gas rather than for probing broad regions of the atmosphere. Gas cells that can accommodate flowing samples have been designed for atmospheric and pollution studies. An alternative technique, useful for localized gas concentrations such as might arise from a leak, is photoacoustic detection and ranging (padar) (36): a laser pulse tuned to an absorption line of the gas generates an acoustic signal that is detected with a parabolic microphone, with a range resolution of 1 cm out to 100 m. In photoacoustic deflection spectroscopy, the pressure changes are observed remotely by tracking the deflection of a probe laser; this technique has been exploited in studies of flame chemistry, where a spectrophone is not feasible because of background acoustic noise and thermal dissipation.

Examples of Infrared Analysis. Volatile organic compounds all have sufficiently distinctive ir spectra to be easily identified and quantified. Detection limits depend on the intrinsic strengths of the absorption bands used, but as a rough guide, detectability (ppm) is given approximately by (0.1 to 1)/(path length (m)). Ground-based solar ir spectra obtained at the South Pole have been used to monitor the column density of freon-12 and confirmed its secular increase (37). Ethylene (C_2H_4) has been analyzed at the 20-ppt level in air using pas at 10 μm (38); the system could detect the production of this gas, a plant growth regulator, by a single orchid flower. Ethylene emissions from a petrochemical plant have been followed at concentrations of some ppb and ranges up to 500 m with ir absorption using two CO_2 lasers and retroreflecting mirrors to return the signal (39).

Ultraviolet Absorption Spectroscopy

Ultraviolet absorption spectroscopy is similar to infrared, but it detects electronic rather than vibrational transitions, using shorter wavelength (higher energy) photons (40). The region of interest for the present purposes is 200–400 nm, extending for some molecules into the visible. Uv spectroscopy is a mature technique that has long been used for both qualitative and quantitative analysis.

The photon energies in this region can promote electronic transitions between outer (bonding) orbitals. This is also enough energy to excite vibrational and rotational transitions, so the electronic spectra of vapor phase molecules consist of highly structured rovibronic bands, offering the potential for high specificity. One potential problem with uv atmospheric monitoring is that this shorter wavelength radiation is much more susceptible to the effects of Rayleigh and particulate scattering and of turbulence than is the infrared. Also, spectroscopic congestion is a problem, with many strong transitions occurring in a relatively narrow wavelength band. Spectral overlap limits the use of uv spectroscopy for identifying complex mixtures of pollutants; most of the compounds sought will absorb in about the same spectral region, and so will tend to interfere. This limits the flexibility of the method, making uv less suitable than ir for general survey spectroscopy and long-range identification of unknown constituents. On the other hand, uv offers high sensitivity and excellent quantitative accuracy, for the greater uv absorption cross-sections (Table 1) and more efficient uv sources and detectors result in detection limits several orders of magnitude better than in the ir. Sampling methods for uv absorption spectroscopy are essentially the same as in the infrared; window materials are less of a problem, for quartz and glass are suitable. Uv atmospheric monitoring, like ir, provides an average column density of absorbing species.

Dispersive Spectrometers. Packaged uv spectrophotometers are manufactured similar to those widely used in the ir and typically operate in the uv–visible –near ir regions (say, 190 nm to 3 μm), with a tungsten–halogen or deuterium lamp source, holographic gratings, and for a detector either a photomultiplier tube (PMT) or (in the near infrared) a photoconductor such as PbS. Typical resolution may be 0.05 nm in the uv–vis and 0.5 cm^{-1} in the near infrared for a 0.2-m monochromator; it is proportionally better with larger instruments.

Simple monochromators are also widely used with solid-state imaging arrays (41), called optoelectronic imaging devices (OIDs) or optical multichannel analyzers (OMAs), which can be placed in the focal plane of a monochromator to record the spectrum nearly instantaneously. These are in effect the modern equivalent of the traditional photographic plate, with the advantage over an emulsion of rapid response with real-time results, instead of requiring long exposure times and subsequent development of the image. They also have a multiplex advantage in that for a given observation time, the signal:noise ratio is increased by a factor of $\sqrt{N}$ when N detectors are used or, for a given S:N, recording time is reduced by $1/N$. An array detector consists of a set of photodiodes that respond to incident electromagnetic radiation, together with an integral electronic readout scheme that serially accesses the signal from each photodetector. The usual spectral response extends from ca 170 nm to ca 1 μm, and they are available in lengths of > 5000 pixels. Arrays with time-gated windows as short as 5 ns have also become useful in the spectroscopy of short-lived species and unstable chemical systems and in kinetics studies, for which they have obvious advantages over mechanically scanned spectrophotometers with single-channel detectors and are often superior to Fourier-transform spectroscopy. This speed is less of a necessity for pollution monitoring, for which time resolutions of the order of some seconds are perfectly adequate. Two-dimensional arrays, designed for image recording, also have spectroscopic applications in kinetics and chemical reaction monitoring, for which a temporally changing spectrum can be rastered across the second dimension of the arrray for time-resolved spectroscopy.

Fourier-Transform Spectrometers. Interferometry is difficult in the uv region because of much higher demands on optical alignment and mechanical stability imposed on the instrumentation by the shorter wavelength of the radiation. Most commercial Fourier-transform spectrometers have been designed for visible and (especially) infrared use. While in principle any interferometer can be operated in the uv with the proper choice of source, beam splitter, and detector, in practice it has proved difficult to obtain reliability from commercial instruments at wavelengths much shorter than the visible. Recently, some manufacturers have claimed satisfactory performance out to 55,000 cm^{-1} (182 nm), but this was achieved only with difficulty. This covers the region of potential usefulness for atmospheric monitoring, but there may be problems in operating such delicate instrumentation outside the laboratory in the open atmosphere and/or under conditions of vibrational and thermal stress.

Tunable Laser Sources. Tunable uv–vis lasers are better developed than comparable infrared systems (14). Dye lasers provide especially useful continuously tunable sources of near uv–visible–near ir light. The lasing action takes place in certain large organic dye molecules in solution, which are optically pumped to excited electronic states by more intense sources and then fluoresce with high quantum efficiency down to the ground state, emitting a broad band of longer wavelengths. The output frequency is selected with a dispersive optical cavity. Pulsed dye lasers (pumped with fixed-frequency Nd:YAG, excimer, or Cu-vapor lasers or by flash lamps) can provide high peak powers at repetition rates to 100 Hz and line widths of 0.1–1 cm^{-1}, which can be improved to < 0.01 cm^{-1} with an intracavity etalon. A given dye–solvent combination can typically be tuned continuously over a 40–80 nm range, and overall coverage with available dyes is from about 320 nm in the near uv to 1.2 μm in the near ir. With Ar^+ or Kr^+ pump lasers, cw operation is possible with output powers of 0.1–1 W. Multimode cw cavities can achieve line widths of a few GHz, while commercial ring-laser geometries improve the resolution to < 0.5 MHz over a 1-cm^{-1} continuous scan. Recently Ti:sapphire has been used as the gain medium for near-infrared cw ring lasers that can provide > 1-W output over much of the range 700–900 nm. Tunable solid-state optical parametric oscillators (OPOs) are being developed that tune continuously from 400 nm to 2 μm with a resolution of 0.1 cm^{-1}.

Tunable dye lasers are widely used in laboratory research on spectroscopy and photochemistry. For routine monitoring of pollutants they have serious drawbacks: they are inherently complex devices needing a separate pump laser (or at least a flash lamp), with demanding optical and alignment requirements (especially the synchronization of the tuning elements), and usually requiring a flowing dye system (for cw operation, a liquid dye jet) to dissipate heat generated by the pump. One could not reasonably consider the remote installation of a dye laser in an industrial environment, such as could be done with a tunable diode laser. But their high output power and tunability make them attractive as sources for lidar, Raman, and fluorescence analysis.

Examples of Ultraviolet Analysis. Uv absorption may be a quite satisfactory analytical technique when there are only a few species present. Volatile organics all have distinctive rovibronic bands in the near uv, which in the vapor phase exhibit detailed rovibronic structure and are in principle suitable for identification and quantification if interferences can be avoided. Perhaps the most suitable molecule for uv monitoring is ozone, which has strong bands in the near ultraviolet: the Huggins (300–370 nm) and Hartley (210–300 nm) bands. This pollutant has been monitored near 283 nm by differential optical absorption spectroscopy (doas), using a correlation technique in which the absorption spectrum is compared with that of a known amount of O_3 (42). In this case light from a Xe lamp was returned from a retroreflector 750 m distant; ozone in an urban environment was continuously monitored over 1 yr with this apparatus, with a sensitivity of about 3 $\mu g/m^3$ (2.5 ppb). In cases for which detecting O_3 is of paramount importance and few interfering species are present, simple uv systems like these are appropriate. Various pollutants have been recorded in the Black Forest at the ppb level over paths of ca 9 km, using retroreflectors to return the beam (43), including the OH radical using a dye laser source at 308 nm and O_3, NO_2, SO_2, and CH_2O with a Xe arc source and spectrometer. Other examples of uv–visible analyses of organics include formaldehyde in the ppb range by pas, using a pulsed laser at 303 nm (44) and formic and acetic acids by pas at 220 nm, with detection limits of ca 120 ppb (45).

Scattering Techniques

In lidar, a laser pulse is propagated coaxially along a telescope's field of view, and the return signal from atmospheric scattering is collected by the telescope for detection and in some cases spectral analysis (1). The azimuth and elevation of the scatterers is determined from the orientation of the telescope and the range from the time delay of the return signal. Several scattering processes can contribute to the signal: Rayleigh, elastic scattering (no change in frequency) from particles much smaller than the wavelength of the probe beam (atoms, molecules, fine dust), Mie (or Tyndall) elastic scattering from larger particulates or aerosols, and Raman inelastic scattering from molecules, with a frequency shift characteristic of the molecule. Fluorescence can be considered a natural limit of resonant Rayleigh or Raman scattering from excited electronic states of atoms and molecules, but this technique is sufficiently unique to justify its separate treatment below. An important modification of lidar is differential absorption lidar (dial), which uses two laser frequencies, one on and one off resonance, to monitor a specific molecular absorption feature; the backscatter can be either from a retroreflecting target or from aerosol and dust returns. These methods will all be referred to as lidar techniques, with the term *simple lidar ranging* used to distinguish, when necessary, range and velocity determination by elastic scattering from the more sophisticated Raman, dial, and fluorescence methods.

The accurate and rapid three-dimensional spatial information provided about target species by laser scattering techniques makes these ideal for monitoring air mass movements and plume transport and for tracking aerosol and pollutant species (46). Such data are useful in meteorological site characterization and are vital for directing and evaluating emergency response to an inadvertent hazardous release. In addition to information on air mass position, Raman and fluorescence scattering both contain molecular spectral information that can identify the target species much as can absorption spectroscopy. These latter techniques, and dial, can also be used for quantification.

Lidar techniques can be very sensitive in some circumstances. Ground-based sensing of Na and Li atoms in the stratosphere (range 30–90 km) has been reported for concentrations of only a few atoms/cm^3. More typical detection ranges in the lower atmosphere are of the order of hundreds of meters and concentrations of ppm to ppb. Quantitative accuracy is difficult to assess, especially considering the variety of molecular processes subsumed under the term *lidar*, but generally is comparable with absorption spectroscopy: uncertainties of ca 5% can be expected for typical atmospheric conditions and perhaps 1% under optimum conditions. The sample in lidar and other scattering techniques is the column of air interrogated by the probe beam and is determined by the orientation of that beam. The return signal is proportional to the column density of the scattering molecules rather than to concentration at a given point. However, point concentrations can be estimated from signal return times, which can yield the range of the scatterers with resolutions on the order of a meter. Some further characteristics of the individual scattering methodologies are described below.

Elastic Scattering: Lidar. Lidar is a ranging technique that detects an elastically scattered return signal, predominantly from particulates and aerosols. The backscattered signal contains information on the distance of the scatterers (from the temporal delay of the return), their abundance (from the intensity), and in Doppler lidar (below), their velocity components along the line of sight (from the Doppler shift). No spectral information is present, so the method does not provide the identification or quantification of specific compounds. It is included here because of the ready adaptability of lidar instrumentation to such techniques as dial, Raman, and fluorescence and also because it is useful in problems closely related to pollution monitoring, such as tracking plume transport and dispersal.

Lidar range resolution depends on the laser pulse width and the detector gate width. As an example, temporal pulse lengths of 10 ns are readily available from excimer and solid-state lasers, corresponding to a spatial length of some 0.3 m. The beam spread for a laser of divergence of 0.5 mrad is 5 m at a range of 10 km, so spatial resolution of the order of a few meters is readily achievable at kilometer distances. Quantification is difficult, even for a single known species, because not all of the factors that affect the return signal intensity (laser power, fraction of signal returned, extinction coefficient, optical efficiency of the system, etc) are adequately known.

Aerosols and particulates are strong scatterers and are present in most plant and stack emission sources. Accordingly, the use of lidar for air mass and plume tracking is well established and has been proven in many field applications. Dense aerosols can be successfully located with a fast lidar search pattern and tracked over daylight ranges approaching 150 km under favorable conditions. Even faint plumes, orders of magnitude fainter than visible contrails, can be observed at 40 km or more in clear air, though a range limit of 10 km is more typical under usual atmospheric conditions near sea level. With lidar one can use time-dependent studies of aerosol transport to obtain wind velocity vectors that characterize actual wind fields and plume transport paths under various meteorological conditions and to study plume dispersal, stack effluent behavior near buildings, flow characteristics over complex terrain and the effects of terrain features, the influence of inversion and boundary layers, vortex and eddy structure, molecular flux gradients, etc. Deriving such information from meteorological sampling would require an expensive three-dimensional array of many sensors covering an extensive ground area to heights of hundreds of meters.

Dial. Two-frequency differential absorption lidar (47,48) combines lidar and absorption spectroscopy by using two probe frequencies, one at a molecular or atomic absorption feature and one at a nearby frequency where there is no absorption. The frequencies are alternated, generating a small differential absorption signal from the much stronger Raleigh or Mie scattering returns, which are relatively insensitive to small changes in frequency. Because one of the two probe frequencies provide essentially a reference signal, some of the quantification problems that arise with simple single-frequency lidar are overcome. One can thus obtain absolute range-resolved concentrations of specific molecules.

Dial has the disadvantage of requiring considerably greater experimental complexity than simple lidar systems. Either two different lasers must be used as the sources or a single laser must be tuned on and off resonance. Usually a tunable laser is chosen, for economy, but there are obviously stringent requirements on the tunability and monochromaticity of the laser selected, because it must be able to cycle repeatedly and rapidly to precise frequencies; missing the peak of an absorption feature by any significant fraction of a line width will greatly degrade quantitative accuracy. But with an appropriate tunable source, dial can provide species-specific concentration and range information that simple lidar cannot. Because the frequencies must be chosen carefully for any given species, dial does not offer the qualitative analytical capability and flexibility of spectroscopic methods.

If one is willing to sacrifice range information for sensitivity, dial can be performed with the signal returned by a reflecting target such as a mirror, topographic feature, or cloud. Because molecular scattering is not involved in the signal return, this is now essentially a single-ended remote sensing infrared absorption technique, giving path-averaged or column densities. Sensitivities for some molecules of the order of 1 to 10 ppb-km can be achieved, ie, a few parts per billion over ranges of several kilometers. In ranging experiments, concentration-path products of a few ppm-m can be measured out to ranges of 1–5 km.

Range-resolved dial measurements can provide three-dimensional maps of the measured species, but here the return signal is much smaller than that from a retroreflecting target. This requires either higher power lasers or sophisticated detection techniques such as photon counting or heterodyne detection. Most such work has been limited to relatively abundant pollutants such as O_3, SO_2, and NO_2 in urban and industrial sites. Detection limits in most cases are of the order of ppm. It should be noted that some innocuous compounds that have high sensitivity to dial, such as ethylene, simple freons, isopropanol, and SF_6, might be deliberately released as plume tracers.

Raman Scattering. Molecular Raman scattering consists of the inelastic scattering of an incident photon $\bar{\nu}_L$ to a new frequency $\bar{\nu}_R = \bar{\nu}_L \pm \bar{\nu}_{\text{mol}}$, where $\bar{\nu}_{\text{mol}}$ is the energy taken ($-$) or given up ($+$) by the molecule as a vibrational and/or a rotational transition (1,49,50). Typically the uv–vis laser frequency $\bar{\nu}_L$ is far detuned from any electronic or vibrational transition energy (ie, is nonresonant), in which case the differential scattering cross-section is

$$\frac{d\sigma}{d\Omega} \cong \left(\frac{2\pi}{\lambda}\right)^4 A_{if}^2 \sin^2\theta \tag{9}$$

where λ is the wavelength of the incident light and A_{if} (the polarizability matrix element connecting the initial and final molecular states) has values of $< 10^{-24}$ cm^3 (comparable with the "volume" d^3 of the molecule). Typical nonresonant angle-integrated Raman cross-sections are of the order 10^{-27} cm^2 for uv–vis radiation and correspondingly smaller (by λ^{-4}) for longer wavelengths. If the frequency of the laser is within a few absorption line widths of an electronic transition, the resonant enhancements can be considerable ($\sim 10^3$). The selection rules for Raman scattering differ from those of absorption; in particular, rotational and vibrational transitions in homonuclear diatomics such as N_2 and O_2 appear strongly in the Raman effect. Because all molecules are Raman active, Raman lidar can provide a complete range-resolved spectroscopic fingerprint of each species in real time.

In Raman lidar, spectrometric detection is necessary to resolve the wavelength-shifted signal from the elastically scattered background. The Raman spectrum consists of a strong line at the exciting frequency, due to Rayleigh scattering, and much weaker (by factors of 10^3 or more) red-shifted lines at lower frequencies (Stokes lines) caused by inelastic scattering in which the molecules have gained energy from the photons, plus corresponding but weaker blue-shifted anti-Stokes lines at higher frequencies, indicating collisions in which the molecules have lost rovibrational energy to the photons. Usually a reasonably broad spectral region is detected that may include several Raman-shifted lines. The resulting molecular rovibrational spectra permit species identification and quantification. The relative strengths of the Stokes and anti-Stokes lines can provide valuable additional information about the temperature of the sample.

Raman scattering has one major disadvantage, its inherent weakness. The process must, therefore, be excited by high energy pulsed lasers; it is most useful in the visible–uv region, where sensitive photomultiplier detectors can be used, and it is most applicable to abundant species or to short-range analysis. Atmospheric N_2 can be detected at ranges of tens of km, but most pollutants are limited to a hundred meters or so. Sensitivity is no better than 100 ppm, or even into the parts-per-thousand range, for most species. Some of these drawbacks can be overcome with nonlinear and coherent techniques, but these are most often used in laboratory bench studies. On the other hand, Raman is broadly applicable and highly specific and selective, because the return signal from every laser pulse contains spectral information on each molecular constituent present, and several components can be monitored simultaneously. Atmospheric water causes less interference than with infrared absorption spectroscopy, because water has a small Raman scattering cross-section. In contrast, many organic compounds are strong Raman scatterers. The laser wavelength used is often not critical because it does not have to be resonant with a molecular transition.

Laser Raman spectroscopy has been widely used for probing flame and combustion chemistry with a spatial resolution of better than 1 mm^3. In the case of atmospheric monitoring, an intense and dependable laser source is required, and there are advantages in using wavelengths of 200–370 nm, where sunlight is absorbed by stratospheric ozone, but oxygen absorption is not significant. Such solar-blind operation eliminates background solar and sky radiation, allowing the reliable detection of small Raman signals at any time of day or night. Typically, the returning Raman signal is detected as a function of time, and the signal intensity yields concentration versus range. Concentrations can be calibrated by ratioing the Raman signal from the monitored species against that from N_2, with separate filtered photomultiplier tubes providing the analyte and N_2 signals.

Special Raman Spectroscopies. Because of the weakness of the Raman effect, even with the advantages of laser excitation typically no more than 10^{-8} of the incident laser photons are converted to a usable Raman signal. In addition, collision-induced off-resonant fluorescence is often a problem, overwhelming the much weaker Raman scattering. Thus the sensitivity of conventional spontaneous Raman spectroscopy does not compare with that of absorption spectroscopy. This drawback can often be overcome with some clever alternative approaches.

In resonance Raman (RR) scattering, the probe frequency is tuned near an electronic absorption of the molecule under investigation, which increases the intensities of both fluorescence and Raman scattering by several orders of magnitude. This permits the highly selective enhancement of a particular species in a mixture, and the sensitivity can be increased to the point at which trace constituents can be routinely monitored. To date RR has been applied to a relatively few molecules with accessible electronic transitions such as O_2 and the halogens.

For the intense exciting beams provided by laser sources, the dipole moment induced in the molecule may vary nonlinearly with the electric field, giving rise to new nonlinear spectroscopic phenomena. As the irradiance of the source increases above the threshold for nonlinear interaction, multiphoton scattering and absorption processes arise. Because the scattered radiation is coherently driven by the pump laser, it emerges from the sample in a specific direction with a small solid angle of divergence. In contrast, noncoherent Raman signals are relatively isotropic and are emitted into a solid angle of 4π steradians. A coherent, directional signal can be collected efficiently and is much easier to separate from incoherent emissions such as fluorescence and Rayleigh scattering.

All scattering processes can be induced to stimulated emission at sufficiently high threshold intensities, but the substantial laser flux required makes these schemes impractical except for laboratory use. Stimulated Raman scattering (SRS) by atmospheric N_2 has the lowest threshold, by virtue of its high concentration in the atmosphere, with a threshold for gain of $\sim$ 2 MW/cm^2.

Doppler lidar. If the signal return in any of these lidar processes is detected with sufficient spectral resolution to allow a measurement of the Doppler shift, then obviously one can determine the velocity component of the probed species along the line of sight. Such shifts are small ($\Delta\nu/\nu \approx 10^{-8}$ at 10 μm for a velocity of 1 m/s), but can be measured by optical heterodyne methods, using high resolution laser sources (bandwidth < 1 MHz) (51). Optical heterodyne detection involves the superposition of two signals in the same detector (photomixer) to produce a beat frequency at the difference between them. In the case of Doppler lidar, the return signal usually has a well-defined frequency offset and high spatial and temporal coherence that allows high quality heterodyne detection with a reference beam from the incident source laser. Spatial distortion and frequency broadening of the return signal by atmospheric inhomogeneities and turbulence, both in the transit path and by aerosol backscatter, will determine the ultimate sensitivities of heterodyne detection. Unless there are heavy winds, the line width of the return signal is typically less than 1 MHz for 10-μm radiation, which sets the limits for the laser stability needed. Effects of spatial distortion can be minimized by reducing the field of view of the receiving optics. Most such work has been carried out with CO_2 lasers at 10 μm. Doppler velocimetry has been done in conjunction with simple lidar ranging, dial, Raman, and fluorescence measurements. It can supply useful additional information on airmass movements but is not an analytical technique for monitoring pollutant species.

Instrumentation. All lidar methods share the basic requirement of an intense, nearly monochromatic exciting source, usually a laser (1–3). Given this source, lidar, dial, Raman, and lif can all be employed, depending on how the source is tuned and whether or not the return signal is spectroscopically analyzed. Because much of the instrumentation, including the expensive laser, monochromator, and detector system, can be shared by the various techniques, it makes sense to treat them together for this purpose.

The source is the most important feature of lidar instrumentation and must have sufficient output power at the appropriate wavelength to provide an adequate return signal; for dial and fluorescence lidar, tunability is also a requirement. In the infrared, laser pulse energies of 1–10 mJ are needed for detection of the return from a hard target at a range of a few km, and about 1 J for range-resolved returns from atmospheric aerosols. Heterodyne detection can reduce these requirements by factors of 10 to 100. Lower energy is also acceptable in the visible and uv, where better detectors are available and there is more atmospheric backscatter. These considerations have limited most experiments to certain lasers: discretely tunable systems such as rare gas–halide excimer (190–355 nm), ruby (694 nm), Nd:YAG (0.95–1.8 μm), CO (4.9–8.2 μm), and CO_2 (8.7–11.8 μm), and continuously tunable dye lasers (340–1150 nm) OPOs (400–2000 nm), and alexandrite lasers (700–800 nm).

Lidar systems are not commercially available as such and must be assembled from components. Many such systems have been described in the literature; typical is a portable lidar developed for spatial and temporal atmospheric monitoring (52). It consists of a laser source, a beam-steering theodolite, a 16-in receiving telescope coaxial with the laser, and a grating spectrometer with microchannel-plate signal intensifiers and a 1024-element silicon OMA detector, all mounted on a truck that carries its own 12-kW diesel generator and necessary utilities. The usual Raman-scatter laser is a high-power KrF excimer emitting at 248 nm; the detectors can be made solar blind so the system can operate day or night. To demonstrate the usefulness of remote sensing data for agricultural measurements, this system has been used to study water vapor concentrations and transport properties over a cotton field with a resolution of 1.5 m at ranges of 100–500 m, revealing fine-scale details of distribution, eddy structure, and vertical flux gradients and velocities. The same equipment can monitor a pollutant such as CCl_4 at the 0.1% level in air at distances over 1 km. Other portable lidar systems have been described in the literature (53,54).

Examples of Analysis by Scattering Methods. Most pollutants thus far investigated by lidar techniques have been relatively simple species such as NO, O_3, SO_2, NO_2, and H_2S. There has, however, been some interesting work reported on organics. Methane leaks from underground pipelines and landfills were detected with a two-frequency HeNe laser at 2948 cm^{-1} using topographic-backscatter dial (55); detection limits were 3 ppm-m out to 90-m ranges. Ethylene (C_2H_4) has been measured over a refinery using a CO_2-laser dial system, providing a 3-D concentration map for a 70 × 400-m area with a sensitivity of about 10 ppb (56). Other pollutant analyses that have been carried out with lidar-related techniques include ammonia by dial (detection limit 5 ppb at ranges to 2.7 km) (57), SO_2 at high concentrations (0.1–20%) by cars (58), and hydrazine (N_2H_4) and methyl hydrazines by dial (40 ppb at ranges to 5 km) (59).

Laser-induced Fluorescence

The emission of uv, visible, or ir light as a result of a radiative transition from an excited to a lower state is broadly termed luminescence (1,46,60,61). Such processes are conventionally divided into short-lived (usually spin-allowed) emission or fluorescence, and long-lived (usually spin-forbidden) emission or phosphorescence. "Short-" and "long-lived" are defined relative to the natural lifetimes of allowed radiative transitions, which are of the order of nanoseconds for visible light. The resonant Raman scattering discussed above can be considered a variety of fluorescence, but there are differences that justify treating fluorescence as a separate technique and limiting scattering to only those processses that do not involve electronic excitation. The redistribution of radiation by fluorescence and scattering (both Rayleigh and resonant Raman) becomes subtle under near-resonant conditions, especially at high incident laser intensities and high pressures. Generally, fluorescence is considered to be the portion of incident laser radiation that undergoes spectral broadening as a result of the natural lifetime and collisional broadening of the excited state, whereas resonant Raman and Rayleigh scattering are considered to be two-photon scattering processes that are not so broadened.

Fluorescence (and resonant Raman) lidars require tunable pump lasers but offer greater sensitivity than spontaneous Raman lidar. In principle, any strong broad-band light source, such as xenon discharge lamps, can serve as an optical pumping source, but lasers are preferable because their high spectral intensity ensures that a greater fraction of absorbers can be selectively pumped into a desired excited state. Usually laser excitation provides sufficient discrimination of an analyte, but Rayleigh scattering by other species could mask the presence of a trace species. This is especially true if the temporal decay of the fluorescence cannot be studied. In such cases, a branch fluorescence to other states of the absorber can serve as an additional spectral discrimination of its presence. If there is just one well-isolated fluorescence transition, then it is economical to use interference band-pass filters to reject other emission frequencies. The analysis of complex effluent streams may be more effective with a polychromator.

The strength of a fluorescence signal is directly related to the absorption cross section σ_A:

$$\frac{d\delta}{d\Omega} = \frac{1}{4\pi}\,\sigma_A F. \tag{10}$$

The geometrical factor ($1/4\pi$) accounts for the isotropic reemission in all possible directions, and F represents a fluorescence yield factor. F seldom approaches unity because excited atoms often fluoresce to several different energy states, but the greatest yield reduction usually results from nonradiative quenching of the excited states before fluorescence ever occurs. At atmospheric pressures, the most important quenching occurs by collisional transfer to the ground or other states, which competes directly with the main emission. Quenching causes fluorescence intensities to depend strongly on pressure, temperature, and composition, and the resulting signal may be difficult to interpret quantitatively. Fluorescence is typically more intense in the uv than in the ir, due to both the greater uv absorption cross-sections and the longer ir radiative lifetimes, which permit more collisional deexcitation. Typically, $F < 10^{-3}$ for atomic and molecular transitions in the uv and is progressively smaller for longer-wavelength transitions.

Other types of luminescence than fluorescence might be useful for analysis in some situations. In chemiluminescence, eg, part of the energy of a chemical reaction is emitted as light. Methods of sensitive nitric oxide analysis (62–64) have been based on the detection of chemiluminescence generated by the reaction of NO with ozone, and ozone can be detected by the emission at 585 nm from its reaction with ethylene. Such techniques can be extremely sensitive, reaching ppb detection limits with relatively simple instrumentation. Chemiluminescent analysis tends to be highly specific, well suited for detecting a certain species but not for general monitoring of more than a few substances.

Photoluminescence has long been applied to both qualitative and quantitative analysis. Lif provides, much as does infrared absorption spectroscopy, "fingerprints" of different organic molecules, and these can be quantified by measuring fluorescence intensities. Emission methods such as lif frequently can offer much greater sensitivity than absorption analysis, because one is measuring small signals on a near-zero background rather than the small difference between two large signals (I and I_0), which is much more difficult. Selectivity can also be quite high in lif, because both the pump frequency and the fluorescence frequency can be individually chosen for optimum performance. Additional selectivity may be possible with measurements of fluorescence lifetimes and polarization behavior. There are, however, several factors that limit the applications of lif as a monitoring technique in the troposphere. Quenching of the excited states of species at atmospheric pressure has already been mentioned. Operation in the infrared beyond 1 μm is limited by the lack of sensitive photomultiplier detectors. And solar background radiation can interfere. Most applications reported to date have been to atomic species (Na, K, Li, Ca, Ca^+) and to radicals such as OH, CN, NH, and CH (63). Lif detection of trace contaminants in the stratosphere has been especially effective, because the collisional redistribution at pressures less than 30 Torr is small.

Some variations of lif should be mentioned that are applicable in special situations. Because molecular fluorescence extends over a broad wavelength region, it is of limited use for analyzing multicomponent systems. But by scanning both the excitation wavelength and the fluorescence detection wavelength synchronously with a fixed wavelength separation, the fluorescence can be reduced to a narrow signal only at the region of overlap between the excitation and fluorescence spectra. This technique, synchronous detection by lif (sdlif), allows the components in complex mixtures to be distinguished. A tunable laser is normally used, but similar results have been reported (at lower sensitivity) with a white-light source followed by a monochromator. Another technique is multiple-photon lif, in which the fluorescing state is pumped by a two-photon or higher process, using a source of longer wavelength than would be required for conventional lif. This results in a considerably weakened fluorescence signal but certain advantages accrue: because the fluorescence is blue shifted relative to the pump, noise sources can be discriminated against by using long-wavelength blocking filters and solar-blind PMTs; and the longer pump wavelength reduces scattering problems and may excite fewer potentially interfering species.

Lif requires an intense exciting laser and equipment for spectroscopically and temporally analyzing the return fluorescence signal. It thus can be combined with the lidar group of techniques (simple lidar ranging, dial, and Raman scattering), and in fact most lidar installations for atmospheric monitoring are equipped also for laser fluorescence measurements. As with other scattering techniques, in lif the "sample" is the column of air interrogated by the probe beam and is determined by the orientation of that beam. The integrated return signal is proportional to the column density of the scattering molecules, but point concentrations can often be estimated by the analysis of signal return times.

Fluorescence optrodes have been developed using reagents immobilized on porous glass and polymer films. Here, as with ir absorption optrodes, the effective sample space is a small volume right at the tip of the optrode. Such devices have been demonstrated for the environmental monitoring of such gases as CO_2, H_2S, NH_3, I_2, and Cl_2.

Examples of Analysis by Lif. Despite its disadvantages, lif may be appropriate for certain molecules that are strong fluorescers even at atmospheric pressure. An important group of such compounds are the polycyclic aromatic hydrocarbons (PAHs) such as anthracene, pyrene, chrysene, and benzo-[α]-pyrene. These are potent carcinogens that are produced by combustion and appear especially in the exhaust of diesel engines; they may also occur in incinerator stack gases as products of incomplete combustion. Many of them are near-ideal fluorophors, with large fluorescence quantum yields and long fluorescence lifetimes. Detection limits of a few ppt in water solution have been reported using lif, and comparable sensitivities should be achievable in the vapor phase (66).

Another strong fluorescer is the tryptophan amino acid component in some bacteria, and this has been made the basis of an effective tracer technique for following air mass movements. Live spores of a harmless species such as *Bacillus globigii* are dispersed into an airflow and their fluorescence monitored by lidar. The spore plume can easily be detected at distances of several kilometers in minute releases.

Radiometry

Radiometry is the detection and measurement of radiant electromagnetic energy (4–6). The term properly includes spectroscopy, but it is convenient to consider separately under this heading the direct detection of molecular emissions, as distinguished from absorption and scattering techniques in which the sample is actively illuminated. Because this detection will include the spectroscopic analysis of the emitted radiation, radiometry as used here is synonymous with emission spectroscopy. Such detection of the weak emissions of remote thermal sources is the ultimate passive and noninvasive technique, requiring not even an optical probe of the sampled volume. At any temperature above absolute zero, some fraction of a sample of atoms or molecules will be in excited levels that have been populated by thermal energy; transitions from these states to the ground state will radiate this energy as thermal emission. The emission maximum for molecules at $T = 300$ K is near 10 μm, in the mid infrared. This radiation will be strongest at just those molecular frequencies at which absorption occurs, ie, at the energies of rovibrational transitions. Thermal emission, therefore, carries essentially the same qualitative and quantitative information as does an infrared absorption spectrum.

Thermal radiation is intrinsically weak, and in the presence of intense background radiation from sunlight, its direct spectroscopic detection and analysis is difficult. There is, however, a sensitive technique for detecting these weak thermal signals that has applications in pollution monitoring and atmospheric studies: laser heterodyne radiometry, in which one measures not the emitted frequency itself but the beat frequency between this and another, accurately known, frequency. The incident radiation is combined with the output of a coherent local oscillator (lo), usually a fixed-frequency laser, in a high speed photomixer, thus generating a difference frequency called the intermediate frequency (if), which is synchronously detected and amplified. The if preserves the spectral characteristics of the source but shifts this information into the radiofrequency region where sensitive radio detection techniques can be used, thus providing significantly higher sensitivities than can be obtained with background-limited incoherent detection. With a fixed-frequency lo, the spectral analysis is performed by a multichannel rf filter bank, with the channel width determining the resolution, which is typically better than 0.05 cm^{-1}. The tuning range is limited by the if bandwidth of the mixer; for the HgCdTe photodiodes used in the infrared this yields a spectrum in the region ± 0.08 cm^{-1} (2.4 GHz) around the lo frequency. The requirement of finding a molecular transition this close to a gas laser frequency is highly restrictive; the technique is not suitable for speciation but rather for monitoring one or a few specific molecular features. Tunable diode lasers now have sufficient power to serve as local oscillators, permitting continuous tunability out to 30 μm, but one is still limited with any single TDL to a tuning range of some 100 cm^{-1}.

Heterodyne radiometry provides excellent sensitivity, illustrated by the fact that it has been used (with appropriate receiving telescopes) for detection of various constituents of stellar and planetary atmospheres. Quantitative accuracy is generally good, but the signal strength depends on the total emissivity of all molecules radiating in the field of view at the detected frequency, and thus it yields an integrated column abundance.

For direct (nonheterodyne) detection of ir emission, any

dispersive spectrometer or Fourier-transform interferometer can in principle be used, but in pollution monitoring, where the emitting molecules are generally at 300 K or colder, a dispersive spectrometer is unlikely to provide satisfactory S:N ratios for these weak signals. The multiplex and throughput advantages of FTIR instruments make them much better suited for this purpose. Recognizing this potential use, many commercial FTIR spectrometers have an input port that can accept the beam from a collecting telescope without disturbing the usual source optics. The *Voyager* interferometric spectrometer (iris) spectra from the outer planets and Titan illustrate that direct FTIR emission spectroscopy can offer effective analytical capabilities. In contrast, instrumentation for laser heterodyne radiometry is fairly complex and must be assembled from individual components (67,68). The optical system normally includes a collecting telescope, a beam combiner, and a monochromator for coarse calibraton. Detecting and signal-processing electronics are necessarily elaborate: the signal and lo are combined in a wide-bandwidth photomixer–detector to generate the if, which is processed with a lock-in rf amplifier and signal averager. As an example of what can be accomplished in industrial monitoring, smokestack effluents have been analyzed by FTIR (range 74 m), using H_2O:CO_2 concentration ratios to discriminate between gas and oil combustion (69). Field-deployable spectrometers are available that achieve sensitivities of the order of ppm-m out to 1 km using retroreflectors.

LABORATORY LASER ANALYSIS OF ENVIRONMENTAL SAMPLES

The principles of optical detection in the laboratory are similar to those outlined for atmospheric analysis (14,16). However, several laser spectroscopies best exploit the unique resolution and/or high focused intensities that are achieved only in the laboratory. This section describes some very high resolution, or ultrasensitive, laser spectroscopies used primarily for laboratory analysis. Some of these techniques are now being adapted for field applications or as *in situ* sensors, but most are still fairly complex or cumbersome because of the lasers or sampling methods involved.

Absorption Methods

Absorption detectivity can be improved by increasing the spectral resolution, by using null-background techniques, and by increasing the integrated path density Nl of the sample by multipass or other methods, such as using long hollow-core wave guides.

Atomic and Molecular Beam Spectroscopy. In the absence of collisions, Doppler broadening $\Delta\nu_D$ usually overwhelms the radiative absorption line widths $\Delta\nu_N$ of optical transitions, unless the sample is extremely "cold" (70). Strong uv–vis transitions have natural widths of 10–50 MHz, while mid ir transitions have natural line widths of <1 MHz. Doppler broadening of room temperature uv–vis transitions is of order 1–3 GHz, so the effective peak absorption cross-sections measured by narrow-line lasers are $\sim(\Delta\nu_N/\Delta\nu_D)$ smaller than their theoretical maximum. Put differently, the absorbances measured with high resolution lasers (line widths $<\Delta\nu_N$) can be enhanced by $(\Delta\nu_D/\Delta\nu_N)$ if the Doppler broadening is suppressed. This can be accomplished in molecular beams by probing the beam transversely to its direction of flow; the frequency shift of the absorption becomes negligible when $\vec{k} \cdot \vec{u} \cong 0$, where $\vec{k}$ is the momentum of the radiation field ($\propto \bar{\nu}/c$) and $\vec{\mu}$ is the forward velocity of the molecular beam. In practice, slight divergences produce residual Doppler broadening, but the peak absorption cross-sections can be improved dramatically with this refinement.

For molecular analytes there is another reason to employ beam techniques. Under conditions of high pressure expansion into vacuum, the transverse motions of the effusive particles can be strongly ordered by collisional effects near the orifice of the jet (the regime of supersonic expansion). The rapid expansion also results in further cooling the thermal motions of the molecules, leading to a collapse of much of the rotational and vibrational structure into transitions originating from a few lowest lying states, and predominantly the ground state. The effective rovibrational energies of the molecules can be cooled to <10 K, although their kinetic energy of translation along the direction of the jet is quite high. The conditions downstream are also nearly collisionless. A supersonic molecular beam thus has the threefold benefit of greatly reducing environmental perturbations that degrade specificity; concentrating the molecules into a few quantum states, which enhances both specificity and sensitivity; and increasing the flux (particles passing through a unit area per second) of analyte probed by the laser beam(s). The disadvantage is that supersonic beams require large and expensive vacuum pumps. While much beam work has been primarily concerned with the basic science of molecular structure and aggregation, the methods developed may eventually find widespread use in environmental analysis. In addition to detecting light absorption by its direct attenuation, the absorption process has also been monitored in molecular beams by directly observing the increased energy of the beam with sensitive pyrometers or other calorimeters. Laboratory detection limits for strongly absorbing species are at the <10 ppb level using 1-mW lead salt diode lasers and can theoretically be improved by at least two orders of magnitude. Detection of atoms or molecules with uv–vis lasers can achieve higher sensitivities, but typically more sensitive fluorescence techniques are used. Fluorescence by ir transitions is usually undetectable for $\lambda > 3$ μm.

Nonlinear Saturated Absorption. Early laser researchers observed that gas-phase lasers exhibited a slight dip in their emitted gain at certain well-defined wavelengths within the Doppler-broadened gain profile of the laser's working medium (16). These "Lamb dips," named after their discoverer, result from intense nonlinear laser-matter interactions. A velocity subset of the randomly moving gas particles can undergo strong resonance with the narrow-line laser light, which suppresses their availability as emitters. This phenomenon, saturated absorption, permitted the first sub-Doppler line shapes to be resolved in static cells and is a useful method of studying absorption features at the highest possible resolution, without the need to collimate molecular beams. Saturated absorption offers high selectivity of detection, at higher sample densities, than in molecular beams. Gas pressures < pressures $<10^2$ Pa will typically produce collisional broadening comparable with or smaller than the natural radiative line widths in the visible range. The signal strength is the result of higher order nonlinear absorption cross-sections, which make the absolute quantification of species densities difficult to obtain.

Saturated absorption is used mainly as a sensitive null-background spectroscopy to measure precise line positions. Saturated absorption profiles are observed when a strong pump beam (intensity 10–100 mW/mm^2) and weaker probe beam (from the same laser) are made to overlap and counterpropagate through the sample. As the laser frequency is tuned, the probe beam measures pump-induced changes of the absorbing medium's response (the nonlinear susceptibility). The advantage of saturated absorption over many other nonlinear optical techniques is that it measures resonant absorption positions precisely with modest cw laser powers (<100 mW), which minimizes light-field–induced level shifts and broadening. The technique is used most in the uv–vis, because of the availability of suitable lasers. Also, ir transitions are substantially weaker, and more powerful lasers that can saturate ir transitions, such as pulse lasers, often do not afford significant improvements in resolution. Furthermore, higher order multiphoton absorption or scattering processes are often induced with higher power lasers that offer new possibilities for sensitive or high resolution detection.

Doppler-free Multiphoton Absorption. A monochromatic light source of frequency ν_0 will be observed by a particle in thermal motion at a new apparent frequency, $\nu_0(1 + v_z/c)$, where v_z is the velocity component of the absorber along the direction of the photon (16). If the absorber could interact with two photons, one each from two counterpropagating beams from the same laser, then the transition frequency observed by the particle would be

$$\nu = \nu_0\left(1 - \frac{v_z}{c}\right) + \nu_0\left(1 + \frac{v_v}{c}\right) = 2\nu_0 \tag{11}$$

The frequency of this sum is independent of the velocity of any one absorber. When simultaneous two-photon absorption is induced in this way, the Doppler width of an entire thermal collection of atoms can be suppressed. There is the possibility that two photons can be absorbed from one direction, but only a much smaller velocity subset of molecules participates, so the Doppler-free process results in a strong narrow absorption by all species, with peak amplitude $\sim(\Delta\nu_D/\Delta\nu_N)$ times larger than the broader Doppler-shifted two-photon absorptions from each beam. For weak, nearly forbidden transitions this can represent a 1000-fold improvement in selectivity and sensitivity. Although this is an important high resolution spectroscopic technique, the sensitivity is especially low for cw lasers, unless there is a near resonant intermediate state that can enhance the cross-section for a two-photon absorption. A two-photon absorption is analogous to the Raman process, discussed above, and distinct from a sequential two-step transition, because the former is not lifetime broadened by the intermediate state, while the latter experiences greater line broadening due to each resonant interaction. Multiple resonance experiments trade off some spectral selectivity but gain significant enhancement of sensitivity ($>10^3$).

Multiphoton Ionization. Photoionization of atoms or molecules can permit sensitive detection by absorption because the ionic signal is a clear signature, without background (14–16,71–73). The first demonstrations of multiphoton ionization (mpi) in the 1960s used the focused output of powerful lasers to produce plasma sparks in air; this is nonresonant mpi from the simultaneous absorption of several light quanta to energies much greater than that carried by a single photon. Mpi has been used in both gases and liquids and at interfaces, but it is most useful as a probe of low pressure gases or molecular beams, for which the charged photofragments can be extracted with high efficiency and analyzed by electron spectrometers or mass spectrometers. Mpi and its many variants are categorized here as null-background absorption techniques, but they are often considered as a distinct category of nonlinear optical spectroscopies, requiring initiation by an intense laser irradiance (typically >100 MW/cm^2).

Resonant Ionization Spectroscopy. Multiphoton absorption through bound states is a potentially important spectroscopic probe of high lying levels and/or levels normally forbidden by the selection rules of one-photon absorption. The development of powerful, broadly tunable dye lasers made resonantly enhanced multiphoton absorption practical; the ionization effected by this process is an especially sensitive spectroscopic technique for detecting trace absorbers, because charged photofragments can easily be collected with nearly unit probability (unlike photons). Resonant ionization spectroscopy (ris) is sensitive enough to detect single atoms of an analyte like Cs, even in the presence of an atmosphere of noble gas (72). In this demonstration, the background buffer gas was used in a gas proportional counter (similar to a Geiger counter); the resonantly enhanced ionization of the trace Cs induced a strong electrical breakdown that greatly amplified the laser-induced signal.

Resonant Ionization Mass Spectrometry. The use of resonantly enhanced multiphoton ionization (rempi) for analytical spectroscopic detection of atoms and molecules became practical when electron–ion multipliers and mass spectrometers were used to detect the charged photofragments. Resonant ionization mass spectrometry (rims) is now a mature analytical methodology. The optimization of the laser ionization process has evolved to the point at which this is usually the most efficient method to induce ionization of trace analytes. Selective ionization by rempi is now used routinely to eliminate isobaric interferences and enhance isotopic selection by mass spectrometry. There are many technical trade-offs that must be considered in detail, but generally laser-induced ionization of the analyte can provide as much as 10^7–10^8 isotopic selectivity and accomplish efficient (~10%) ionization of the desired sample. The selectivity can be enhanced by factors of $\sim 10^4$ by subsequent mass spectrometry. Under laboratory conditions, cascading rempi with high resolution mass spectrometers has demonstrated isotopic selection greater than 10^{11} with sufficient throughput to analyze microgram samples in <10 h. Detection limits of 10^{-17} g have been achieved with moderately long-lived isotopes, representing the reliable analytic detection of less than 50,000 trace atoms. Isotopic detection of several hundred trace atoms like Pb, Cd, and Tl is theoretically possible (73).

Achieving these levels of trace detection depends on improving details of the ionization and detection of the analyte, but three basic strategies can be identified that are crucial to any ultrasensitive rims experiment: (*1*) availability of narrow line width lasers capable of resolving the natural line widths of the transitions being probed, (*2*) ionization induced by Doppler-free multiphoton absorption or multiple resonance schemes that permit the natural line width to be nearly resolved, and (*3*) the use of atomic or molecular beam techniques to reduce the effective Doppler line widths of neighboring lines so that undesirable off-resonant absorption pathways are suppressed.

The high isotopic sensitivities noted above for elemental analyses circumvented another fundamental issue of resolution by using cw lasers. In the case of resonant transitions in atoms, the atomic transitions are strong, so spectrally intense lasers can often induce rapid stimulated absorption and stimulated emission of a transition, which leads to light-field–induced shifts and broadening of transitions. The rapid excitation–deexcitation (known as Rabi cycling) further shortens the effective lifetime of the excited atoms, so the "lifetime-limited" width is increased with greater light intensity, which degrades spectral resolution and reduces peak sensitivity. A transition is said to be saturated when the rate of Rabi cycling equals the natural transition rate for spontaneous emission by the radiator. Because the peak absorption cross-sections are greatest for transitions with natural line widths, it is always most effective to induce multiphoton transitions with high resolution lasers at powers just sufficient to begin saturating the desired transitions. Besides optimizing the desired excitation pathway for highest selectivity, this also reduces the off-resonant ionization paths, which are usually nonlinear in laser intensity I (eg, if n laser photons are needed to ionize an atom, then the nonresonant ionization signal scales as I^n). Detectivity, which is related to the ratio of resonant to nonresonant ionization, is thus optimized. Atomic transitions require a laser irradiance of ~ 100 mW/mm^2 to achieve saturation if their natural width can be resolved. Cw lasers can easily achieve this irradiance and spectral resolution, and they provide the added advantage of unity-duty-cycle interrogation of the sample. Cw-double resonance ionization mass spectrometry (drims) (73) has achieved detection limits approaching 10^{-18} g for ^{90}Sr and offers similar isotopic detection limits for other species like ^{210}Pb.

Rims is a particularly sensitive null-background absorption technique, but it is sometimes cumbersome to implement without pretreating samples for use in vacuum systems, and it suffers from limited throughput. Ris and Rims techniques require fairly low analyte densities, because a dense discharge tends to experience electron–ion recombination. This and other space-charge perturbations limit efficient extraction of the ions from the laser focal volume when the intensity of photofragments is $>10^8$ cm^{-3}.

Photoacoustic Spectroscopy. A number of other null-background absorption techniques have been developed to detect small absorptions in gases, liquids, and solids, or at their interfaces, as well as in flame or plasma discharges (14,74,75). Many of these techniques are simpler than rims, and can achieve impressive detection limits with higher sampling throughput. These techniques can be categorized as calorimetric (the absorbed radiation induces thermal changes in the medium) or direct. Important examples of the former are photoacoustic and photothermal spectroscopy. Under optimal conditions, both these techniques can measure absorbances as low as 10^{-7}.

Photoacoustic methods rely on the efficient transformation of absorbed radiation into heat by various radiationless processes of relaxation, and the efficient conversion of local heating into pressure waves. Gas-phase samples are the least effective media for conversion, but the technology of gas microphonic cavities (spectrophones) is fairly well developed, so absorbances of $<10^{-6}$ in 1-cm path lengths can be achieved. Because the acoustic signal is proportional to the product of the source intensity and the energy absorption in the sample, high sensitivity can be achieved by increasing the power. Liquid or solid samples can be placed in direct contact with piezoelectric or pyroelectric transducers that convert pressure waves or temperature changes to an electrical signal. Properly coupling the sample to the transducer is important for obtaining high sensitivity. Usually liquids and solids offer good elastic coupling to piezoelectric crystals, and frequency responses to >1 GHz have been obtained. Absorbances in both aqueous and solid samples of ca 10^{-7} or slightly less have been measured.

The principal applications of pas have been in solid sampling and in the analysis of gases at low concentrations, but because the tuning range of the source lasers must be selected for the specific analytes wanted, it has limited flexibility. Within its limits, pas can provide high specificity and sensitivity. Pas detection of various volatile organic compounds at the ppb and sub-ppb levels has been demonstrated using line-tunable CO and CO_2 lasers. Such compounds as vinyl chloride and acrolein have been detected with limits of 1 to 3 ppb. The ultimate potential of pas may be in trace absorption studies of interfacial systems, such as thin films and coatings. In work on gas–solid absorption, pas has demonstrated a detection limit of 0.002 monolayers of species like SF_6 and NH_3 on silver substrates deposited on pyroelectric detectors (76).

Photothermal Methods. Photothermal methods include two related approaches, thermal lensing and photothermal deflection, which in principle can be as sensitive as photoacoustic detection and which are truly noncontact optical methods of interrogation. In both methods, the probe beam can pass through, be reflected from, or make glancing passes over a sample. Thermal lensing is primarily a low speed technique that monitors the thermal focusing or defocusing of a laser. The change of focusing of a pump laser beam can be studied directly, but the most sensitive methods use an amplified modulated cw laser periodically to heat a sample, while the transmission of a coaxial probe laser is synchronously monitored with a pinhole-detector combination. This configuration has been used to measure absorbances of $<10^{-7}$, but in general this method is limited by its geometry to the study of flat, homogeneous, and reasonably transparent samples and substrates. Photothermal deflection is more adaptable and is especially suited to analyzing thin films. It is useful for studying solid–liquid interfaces, with the flexibility to change quickly the position of measurement by adjusting the probe-pump beam overlap. Quantification by photothermal deflection is difficult, because one is studying sharp thermal gradients at the edge of the pump beam.

Optogalvanic Effect. Laser absorption in high temperature flames or gaseous discharges can change the steady-state rate of ionization; the resulting impedance (conductance) changes are sensed as changes in the steady-state discharge voltage established between two electrodes. When cw lasers are used, their amplitude is modulated and the ac response of the discharge is monitored synchronously, while transients generated by pulsed lasers are detected using gated electronics. The optogalvanic effect (oge) is becoming a useful method of diagnosing combustion and gaseous discharges, with excellent potential for trace element detection and chemical analysis. Ir lasers have been used to induce the oge in small molecules like NO, N_2O, CO_2, NH_3, SO_2, H_2S, and H_2O_2 (77), while an assortment of atomic and small molecular species have been studied with visible dye lasers. Molecular detection with visible laser sources is still effective, despite the fact that weak overtone transitions are probed. Uv and multiphoton absorption, which promotes transitions to high lying levels near the first ionization limit, appear to be the most desirable methods to effect strong oge signals. Uv absorption with the oge in aspirated flames and sputter discharge sources have shown potential for ultrasensitive detection of trace elements down to 10^{-15} g limits (78).

Frequency-modulated (fm) Spectroscopy. Fm absorption spectroscopy is analogous to the modulation detection techniques discussed earlier but was developed for use with well-stabilized single-mode lasers. It has emerged as an effective laboratory method to measure very weak absorption (and dispersive) spectral features. The technique combines high sensitivity and acquisition speed to achieve absorbance sensitivities $N\sigma l < 10^{-7}$ in a 1-s integration time, usng 1 mW of cw laser power; this corresponds to shot-noise–limited detection for laser light undergoing natural fluctuations of phase. Stabilized laser radiation exhibits well-defined statistical properties for a stable single-mode laser producing n photons the variance is given by the Poisson distribution of $\overline{(\Delta n)^2} = \overline{n}$. For a 1-mW single-mode laser light source, $n > 10^{15}$, so the fluctuations of a well-stabilized laser will produce shot noise of $\sqrt{n}/n \leq 10^{-7}$. Sensitivity can be improved by increasing the probe intensity, for it scales as $(I_0)^{1/2}$. Shot-noise–limited detection has been demonstrated with visible cw dye lasers (79) and lead salt laser diodes in the mid infrared (80). Fms has also been demonstrated with single-line sources, most notably with a CO_2 laser, as well as pulse lasers, and it has been successfully implemented after second harmonic conversion to the uv. Standard nanosecond pulse lasers of 5 ns duration should yield a single-shot absorption sensitivity of 2×10^{-3}, and this has been nearly achieved experimentally.

The technique involves frequency modulating a laser source at ω_0 to produce a carrier frequency with sidebands at $\omega_0 \pm m\omega_m$, where $m\omega_m$ is an integer multiple of the modulation frequency. For laser diodes, this can be effected by modulating the drive current with standard rf sources, whereas dye lasers and many other single-line sources can be frequency modulated using an external electrooptic modulator (eom). The technique of generating side bands on a laser field is based on the equivalence of phase modulation and frequency modulation given by

$$E(t) = E_0 \sin(\omega_0 t + \eta \sin\omega_m t) = E_0 \sum J_n(\eta) \sin(\omega_0 \pm n\omega_m)t \tag{12}$$

Applying a sinusoidal change of phase $\eta\sin\omega_m t$ at modulation frequency ω_m on a monochromatic light field produces side bands. A variety of eom materials are available to frequency-modulate lasers at 0.42–5.2 μm ($LiTaO_3$ and $LiNbO_3$) and in the deeper ir (GaAs, 2.0–11 μm; CdTe, 1.0–25 μm).

The pure fm laser light is passed through a sample, where near-resonant absorption and/or dispersion leads to differential absorption (scattering) of the side bands. This produces a resultant am beat frequency, which can be coherently detected by standard phase-sensitive mixing technology. In principle, fm spectroscopy produces zero-background signal in the absence of an absorption process, so signal calibration is straightforward while yielding a high S:N ratio with low absorption. If the side band modulation ω_m is small compared with the spectral line width, the observed signature is a derivative of the line shape when the carrier frequency is tuned through a resonance (as in modulation spectroscopy). Greater S:N is achieved when the side band spacing is much larger than the spectral line width. This imposes a technical difficulty of requiring high speed detectors with bandwidths exceeding 3 GHz if Doppler- or collision-broadened lines are to be detected at the greatest sensitivity. This drawback can be eliminated by using a two-tone fm scheme, in which a pair of closely spaced side bands (with large average modulation frequency) are demodulated at their smaller relative separation. Two-tone demodulation yields second-derivative line shapes. The demodulation frequency is still kept fairly high (50–100 MHz), but it is now within limits imposed by standard detector technology. The advantage of using these high radio frequencies is that the signal is recovered in a higher frequency band where the noise of the laser amplitude fluctuations is usually negligible, so that the detection sensitivity is now limited by amplifier noise and/or quantum fluctuations. A balloon-borne *in situ* laser sensor (bliss) using fm absorption has demonstrated ~0.5 ppb sensitivities over 500-m path lengths for soundings made in the stratosphere above 20 km (32). Although impressive, this is still more than two orders of magnitude less sensitive than theoretical limits for fms. The bliss experiments were limited by using a small phase modulation index η and quite low modulation frequency (<2 kHz).

With a low power lead salt laser diode, one can typically detect molecular species at atmospheric pressures to the 0.5 ppb level over a single-pass 10-cm path length. The spectral resolution could be enhanced significantly (10- to 100-fold) by reducing the ambient pressure to eliminate collisional line broadening. Trace sensitivity would remain relatively invariant because the peak line strength of a molecular absorption remains fixed until either the Doppler or laser line width is achieved. This improvement in selectivity is especially important when heavier molecules are being probed. Fm detection has also been used with saturation absorption to achieve high resolution and sensitivity. Currently there is extensive work to develop "high temperature" ir laser diodes for thermoelectrically cooled operation near 250 K. Such devices in conjunction with fms detection could provide sub-ppb molecular analysis in small *in situ* sensor packages.

Laser-induced Fluorescence

Fluorescence detection is one of the most sensitive schemes for detecting uv–vis emission when environmental perturbations can be minimized so that scattering and quenching are negligible. Lif has demonstrated ultrasensitive elemental detection sensitivity (eg, <1 atom/mm^3) with excellent isotopic discrimination using atomic beam sources (66). These levels of sensitivity are generally obtained using resonance fluorescence of atomic systems having a dominant transition to the ground state with little branching to other levels. The Group IA and IIA metals are especially suited to ultrasensitive resonance lif, because their principal absorption lines are strong and have branching ratios to the ground state of >99%. Lif used in conjunction with beam sources is useful for spectroscopic analysis of atomic and molecular structure but less so for analytic quantification, because many collisional and radiative redistribution effects complicate the analysis. Under laboratory conditions, these secondary effects can be eliminated by design or by careful temporal analysis. In such cases, lif has provided useful information about collisional transfer rate coefficients, with applications to kinetic modeling of atmospheric chemistry and gaseous discharges.

Coherent Scattering Spectroscopies

As the irradiance of the source increases above the threshold for nonlinear interaction, multiphoton processes arise, in which two or more incident photons are effectively scattered (14,16,81). Figure 4 shows some examples of stimulated scattering processes that are induced by a strong pump laser of frequency ν_L, detuned near a resonance condition. The very intense light fields shift the atomic or molecular level positions slightly (ac Stark shifts) and induce new resonant conditions (denoted by the dashed lines). With sufficiently high intensities, two, three, or more photon scatterings can be induced to emit strongly in the direction of the incident pump beam. Stimulated Rayleigh scattering occurs at $\nu_R = \nu_L$, stimulated Raman scattering ν_S is induced to lower lying levels, and stimulated three- and four-photon scatterings are induced at frequencies ν_3

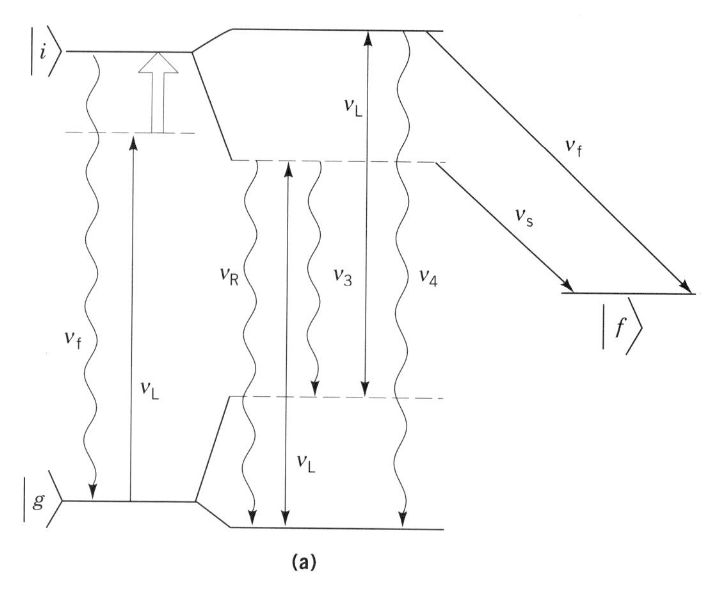

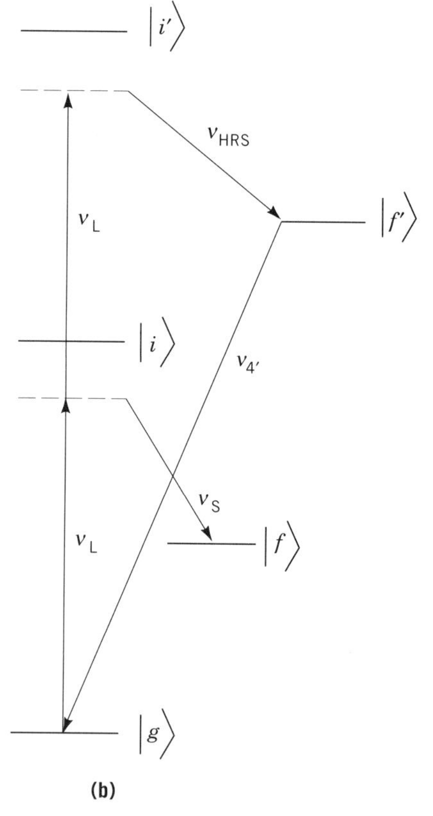

Figure 4. Various scattering–fluorescence processes can be induced in the presence of strong light fields. (**a**) A two-level radiator being pumped by a strong laser source at ν_L. The positions of the ground and intermediate levels are shifted and additional "dressed" levels are induced (denoted by dashes). These lead to Rayleigh scattering at ν_R, induced scattering at ν_3 and ν_4, and possible Stokes Raman emission at ν_S to lower lying levels. If collisions occur (denoted by the outline arrow), collision-induced fluorescence occurs at ν_F and ν_F'. At sufficiently high intensities all scattering–fluorescence processes can be induced to create stimulated emission. (**b**) Possible two-photon scattering via higher excited states and possible hyper-Raman scattering to a different final state $|f'\rangle$. In some cases a radiative transition can occur between $|f'\rangle$ and $|g\rangle$, leading to four-wave mixing and efficient generation of ν_4'.

and ν_4 close to the pump laser frequency. Two laser photons can also create scattering at υ_{HRS} from higher excited states, known as hyper-Raman scattering, and stimulated emissions via these higher excited states can induce cascade emissions back to lower levels. Certain combinations of scatterings (four-wave mixing at ν_4 (ν_4')) can return the scatterers back to the ground state, to undergo continued cycling. These latter processes have the potential for significant amplification because the available scatterers are never depleted. The scattered radiation from these processes is coherent and directional: it emerges from the sample in a small solid angle and can be efficiently collected and separated from incoherent fluorescence and Raman scattering. Some of these, and other, coherent emission phenomena have laboratory applications for air monitoring.

Even for large detunings in molecules, with high excitation power coherent pumping can occur, leading to stimulated Raman scattering (SRS) at $\nu = \nu_L \pm \nu_{\text{vib}}$ (Fig. 5a). The amplified frequency occurs at the strongest Raman transition in the molecule, and the amplification process can increase the scattered signal by as much as 10^{13} over the spontaneous Raman effect. SRS is characterized by high conversion efficiency, resulting in high intensity and strong coherent forward scattering; at high enough intensities the SRS signals can also experience SRS to create new frequencies at $\nu = \nu_L \pm n\nu_{\text{vib}}$, where n is an integer.

While SRS is of limited use in analysis, some variations of it have more promise. Stimulated Raman gain spectroscopy (SRGS) can be considered as a stimulated gain or induced emission process at a Stokes frequency ν_S (see Fig. 5b). Two lasers are used, a pump ν_L and a tunable probe. The intensity of either of these beams is monitored as the probe laser is scanned. When the probe ν_P is scanned, the Raman (Stokes) frequency $\nu = \nu_L - \nu_{\text{vib}}$ is driven and ν_P gains in intensity at the expense of ν_L. The inverse of this process (Fig. 5c) might be called stimulated Raman loss spectroscopy but for historical reasons is generally termed inverse Raman scattering (IRS). Here there is stimulated loss at an anti-Stokes–shifted frequency, with gain on the pump beam ν_L: the probe is attenuated at frequencies corresponding to Raman-active vibrations.

Finally, perhaps the most useful stimulated technique is coherent anti-Stokes Raman spectroscopy (CARS). This employs two pump beams ν_1 and ν_2, interacting through the third-order nonlinear susceptibility of the medium to generate coherent anti-Stokes emission at $\nu_3 = 2\nu_1 - \nu_2$ in a four-photon mixing process (Fig. 5d). The laserlike beam of υ_3 is greatly enhanced when the frequency interval $\nu_1 - \nu_2$ is made equal to a Raman-active molecular infrared frequency by tuning the pump line ν_2. In addition to the high intensity of this scattering (signals stronger by factors of 10^5 to 10^{10} than spontaneous Raman), the anti-Stokes output is shifted to higher frequencies, so CARS spectra are free from Rayleigh interference from both lasers. (Coherent Stokes Raman scattering also occurs, but does not have this Rayleigh rejection advantage, offers no advantages over CARS, and is little used.) The laserlike CARS output beam emerges at a slightly different angle than the pump beams, and so it can be easily separated from interfering incoherent light and collected and detected with high efficiency. Because a single frequency is determined by the resonance conditions of the lasers, no monochromator is necessary, and the resolution depends only on the spectral purity of the exciting laser, which can reach 10^{-3} cm^{-1}. A significant disadvantage to CARS is the presence of a nonresonant background signal that can mask the weaker resonant signal; CARS detection limits are, therefore, 0.1% to as high as 1%. But like linear Raman scattering, CARS signals can be resonantly enhanced by tuning υ_1 near an electronic transition, improving detection limits by many orders of magnitude.

CARS has been widely used in combustion and flame diagnostics, for the identification and quantification of major species with high spatial resolution. Rovibrational bands of the molecule of interest can be scanned to obtain

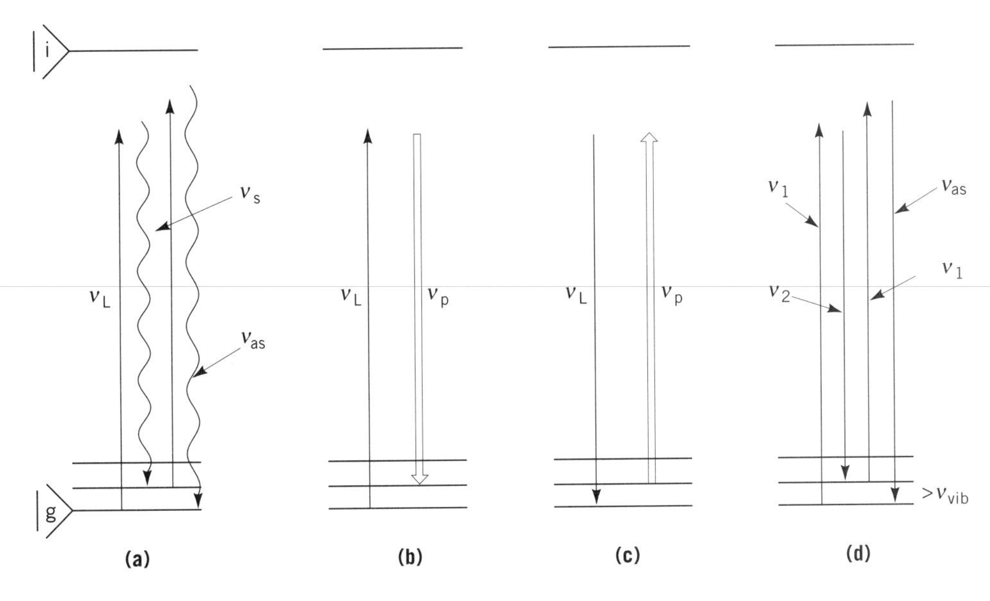

Figure 5. Possible molecular Raman scattering processes. (**a**) Spontaneous Raman scattering can occur from a ground or excited state, producing a Stokes or anti-Stokes emission at $\nu_L - \nu_{\text{vib}}$ and $\nu_L + \nu_{\text{vib}}$, respectively. If a second probe laser is tuned to $\nu_S = \nu_L - \nu_{\text{vib}}$, then (**b**) gain can occur at ν_P (SRGS) or, (**c**) if ν_P is strong, on ν_L (inverse Raman scattering). (**d**) If the intensity *and* momenta of the pump (ν_1) and probe (ν_2) lasers are matched, then strong gain can occur at the anti-Stokes transition (CARS).

temperature information with an accuracy of $\pm 10°$ or better. Because the laser sources in CARS must be tuned so their frequency difference corresponds to a resonant frequency in a single molecule, this is not a technique that can monitor many species simultaneously. Beam alignment is critical for optimum results, for the phase-matching angle must be realized to within 0.1°. The theory of the process is complex, and not all factors contributing to signal intensity are fully understood; especially troublesome is the nonlinear dependence of signal intensity on species density. But adequate quantitative accuracy can be achieved by ratioing the scattered signal against that from atmospheric N_2. Other potential problems include signal degradation by scattering and turbulence and the experimental complexity inherent in using two lasers.

Degenerate Four-wave Mixing

Degenerate four-wave mixing (dfwm), a variation of the coherent Raman scattering process, has novel characteristics that offer great promise as a sensitive analytical technique. Dfwm uses a strong nonresonant pump and weaker probe beam from the same laser to induce a large backscattered coherent signal at the same laser frequency (82). All coherent scattering spectroscopies involve a nonlinear response by the scattering medium that is related to a third-order susceptibility, which is often small compared with the linear susceptibility for incoherent scattering or absorption. As such, intense laser irradiance (>1 MW/cm^2) is typically needed to induce most scatterings, even near resonance. Dfwm is like CARS, except that all four legs of excitation–scattering are near resonance, so the induced signal is quite strong and requires irradiance 10^3-fold smaller than CARS. Also, dfwm has the unusual property of being self-corrective, unlike CARS, which requires rigorous alignments to achieve proper spatial overlap (phase matching). Dfwm is simple to align and has the proven unique capability of producing a highly directional return signal that can faithfully "unmap" the spatial optical distortions of inhomogeneous or turbulent media. This offers two benefits: it ensures excellent phase matching over longer lengths (and, therefore, greater signal enhancement) and the return signal exhibits negligible optical distortion. Finally, the process can employ a Doppler-free pump-probe scheme to induce the response, which makes it especially useful for studying high temperature combustion or plasma process streams. The sub-Doppler dfwm response probes a smaller velocity subset of the medium, which reduces sensitivity. Nevertheless, dfwm is almost the perfect noncontact sampling method, from simplicity of execution, imaging, and spectral sensitivity. It does have the drawback of nonlinear signal strength, but progress has been made toward using it to quantify species concentration. Currently, dfwm has been used for high fidelity spatial mapping of analytes in flames and has shown excellent detection sensitivities for molecular species (for example, OH at concentrations of 10^{10} cm^{-3}), and it is even more sensitive to strong atomic transitions. Detection of excited-state Na atoms in the range of 10^8 cm^{-3} has been reported with an irradiance of ~1 kW/cm^2 (83). This corresponds to a total sensitivity of ~10^6 atoms in a 1-mm diameter by 1-cm long interaction volume.

FUTURE TRENDS IN LASER ANALYTICAL METHODS

An update of ultrasensitive laser spectroscopies is published every odd year as part of a series on laser-based ultrasensitive spectroscopy and detection, in the *Proceedings of the SPIE—The International Society for Optical Engineering*. Typically, three categories of ultrasensitive detection processes are reviewed: rims, single atom–molecule fluorescence detection, and advanced concepts in absorption measurements. Currently, the methodologies and principles used in laser-based analysis are those discussed above, but new techniques of laser atom or molecule deflection, cooling, and trapping are quickly emerging as potential new analytical techniques. Many of the laser spectroscopies that are now being exploited for analysis first appeared in basic research to make precise physical measurements. The applications to ultrasensitive analytical chemistry were often obvious, but generally occurred as secondary developments 10–15 yr later.

How the latest developments in laser spectroscopy may prove out as analytical methods can be guessed by realizing two important consequences of Ladenburg's formula (eq. 3): the peak absorption cross-section for any radiative process varies inversely with the resolved line width of the transition, and when the limit of the natural line width is achieved this peak absorption cross-section scales simply as $\sigma_0 \doteq \lambda_0^2$, where λ_0 is the transition wavelength. In principle, then, *any* radiative transition could be induced to absorb strongly if the laser line width could be made to match the natural width, *and* if the environmental perturbations of the absorber can be reduced below the natural line width. Transitions with natural line widths of ≥ 1 MHz can be resolved with commercial lasers by Doppler-free techniques of laser and/or molecular beam spectroscopies. Much weaker transitions could ultimately be resolved if the atoms or molecules can be cooled and trapped so that residual Dopper and collisional broadening were reduced to limits approaching a few hundreds of Hertz. The straightforward detection of weak or nearly forbidden transitions is already under way in several laboratories, where ions or neutral atoms are trapped and cooled to ~1 μK; despite such low temperatures this still represents a Doppler line width in excess of 40 kHz in the visible. Currently tunable cw dye lasers can be stabilized to line widths ~1 Hz, while tunable laser diodes are now capable of stabilities ≤ 10 kHz. It can be anticipated that these ultrastable laser sources, used with atom or molecule laser cooling–trapping concepts, could become the absorption spectrometers for future chemical and elemental analysis.

BIBLIOGRAPHY

1. R. M. Measures, *Laser Remote Sensing*, John Wiley & Sons, Inc., New York, 1984.
2. R. M. Measures, ed., *Laser Remote Chemical Analysis*, John Wiley & Sons, Inc., New York, 1988.
3. E. D. Hinkley, ed., *Laser Monitoring of the Atmosphere*, Springer-Verlag, Berlin, 1976.
4. R. Beer, *Remote Sensing by Fourier Transform Spectrometry*, John Wiley & Sons, Inc., New York, 1992.

5. J. T. Houghton, F. W. Taylor, and C. D. Rodgers, *Remote Sounding of Atmospheres*, Cambridge University Press, Cambridge, 1984.
6. H. S. Chen, *Space Remote Sensing Systems*, Academic Press, Inc., Orlando, Fla., 1985.
7. A. Corney, *Atomic and Laser Spectroscopy*, Oxford University Press, 1977.
8. A. P. Thorne, *Spectrophysics*, 2nd ed., Chapman and Hall, London, 1988.
9. J. I. Steinfeld, *Molecules and Radiation*, 2nd ed., MIT Press, Cambridge, Mass., 1985.
10. C. H. Townes and A. L. Schawlow, *Microwave Spectroscopy*, Dover, New York, 1985.
11. P. R. Griffiths and J. A. de Haseth, *Fourier Transform Infrared Spectroscopy*, John Wiley & Sons, Inc., New York, 1986.
12. J. T. Houghton and S. D. Smith, *Infra-Red Physics*, Oxford University Press, Oxford, 1966.
13. P. W. Milonni and J. H. Eberly, *Lasers*, John Wiley & Sons, Inc., New York, 1988.
14. W. Demtröder, *Laser Spectroscopy*, Springer-Verlag, Berlin, 1982.
15. D. S. Kliger, ed., *Ultrasensitive Laser Spectroscopy*, Academic Press, Inc., New York, 1983.
16. M. D. Levenson and S. S. Kano, *Introduction to Nonlinear Laser Spectroscopy*, Academic Press, Inc., Boston, 1988.
17. A. Parrish, R. L. deZafra, P. M. Solomon, and J. W. Barrett, *Radio Sci.* **23,** 106–118 (1988).
18. A. C. G. Mitchell and M. W. Zemansky, *Resonance Radiation and Excited Atoms*, Cambridge University Press, Cambridge, 1961.
19. D. A. Burns and E. W. Ciurczak, eds., *Handbook of Near-Infrared Analysis*, Marcel Dekker, Inc., New York, 1992.
20. K. Narahari Rao and A. Weber, eds., *Spectroscopy of the Earth's Atmosphere and Interstellar Medium*, Academic Press, Inc., New York, 1992.
21. R. Beer in Ref. 2, pp. 85–162.
22. J. Altmann, R. Baumgart, and C. Weitkamp, *Appl. Optics* **20,** 995 (1981), and references therein.
23. M. J. Webb, *Spectroscopy* **4**(6), 26–34 (July–Aug. 1989).
24. U. Krull and R. S. Brown in Ref. 2, pp. 505–532.
25. S. M. Angel, *Spectroscopy* **2**(4), 38–48 (Apr. 1987).
26. P. B. Coleman, ed., *Practical Sampling Techniques for Infrared Analysis*, CRC Press, Boca Raton, Fla., 1993.
27. D. W. T. Griffith and G. Schuster, *J. Atmos. Chem.* **5,** 59–81 (1987).
28. D. W. Hill and T. Powell, *Nondispersive Infrared Gas Analysis in Science, Medicine, and Industry*, Hilger, London, 1968.
29. R. H. Wiens and H. H. Zwick in J. S. Mattson, H. B. Mack, Jr., and H. C. MacDonald Jr., eds., *Infrared, Correlation, and Fourier Transform Spectroscopy*, Marcel Dekker, Inc., New York, 1977.
30. C. Webster, R. Menzies, and E. D. Hinkley in Ref. 2, pp. 163–272.
31. R. Grisar, H. Preier, G. Schmidtke, and G. Restelli, eds., *Monitoring of Gaseous Pollutants by Tunable Diode Lasers*, D. Reidel, Dordrecht, The Netherlands, 1987.
32. R. S. McDowell, *Vibrational Spectra Struct.* **10,** 1–151 (1981).
33. W. Lo, ed., *Tunable Diode Laser Development and Spectroscopy Applications, Proc. SPIE* **438** (1983).
34. C. R. Webster and R. D. May, *J. Geophys. Res.* **92,** 11931 (1987).
35. C. R. Webster and co-workers, *Appl. Opt.* **29,** 907–917 (1990).
36. D. J. Brassington, *J. Phys.* **D15,** 219–228 (1982).
37. C. P. Rinsland, A. Goldman, F. J. Murcray, F. H. Murcray, D. G. Murcray, and J. S. Levine, *Appl. Opt.* **27,** 627–630 (1988).
38. F. J. M. Harren, J. Reuss, E. J. Woltering, and D. D. Bicanic, *Appl. Spectrosc.* **44,** 1360–1368 (1990).
39. H. Ahlberg, S. Lundqvist, and B. Olsson, *Appl. Opt.* **24,** 3924–3928 (1985).
40. R. E. Huffman, *Atmospheric Ultraviolet Remote Sensing*, Academic Press, Inc., San Diego, 1992.
41. J. V. Sweedler, R. D. Jalkian, and M. B. Denton, *Appl. Spectrosc.* **43,** 953–962 (1989).
42. H. Axelsson, H. Edner, B. Galle, P. Ragnarson, and M. Rudin, *Appl. Spectrosc.* **44,** 1654–1658 (1990).
43. U. Platt, M. Rateike, W. Junkermann, J. Rudolph, and D. H. Ehhalt, *J. Geophys. Res.* **93,** 5159 (1988).
44. M. Boutonnat, D. A. Gilmore, K. A. Keilbach, N. Oliphant, and C. H. Atkinson, *Appl. Spectrosc.* **42,** 1520–1524 (1988).
45. P. V. Cvijin, D. A. Gilmore, and G. H. Atkinson, *Appl. Spectrosc.* **42,** 770–774 (1988).
46. D. K. Killinger and N. Menyuk, *Science* **235,** 37–45 (1987).
47. K. Fredriksson, in Ref. 2, pp. 273–332.
48. K. A. Fredriksson, *Appl. Opt.* **24,** 3297–3304 (1985).
49. H. Inaba in Ref. 3, pp. 153–236.
50. J. G. Grasselli and B. J. Bulkin, eds., *Analytical Raman Spectroscopy*, John Wiley & Sons, Inc., New York, 1991.
51. R. T. Menzies in Ref. 3, pp. 297–353.
52. F. J. Barnes, R. R. Karl, K. E. Kunkel, and G. L. Stone, *Remote Sens. Environ.* **32,** 81–90 (1990).
53. J. D. Houston, S. Sizgoric, A. Ulitsky, and J. Banic, *Appl. Opt.* **25,** 2115–2121 (1986).
54. H. Edner, K. Fredriksson, A. Sunesson, S. Svanberg, L. Unéus, and W. Wendt, *Appl. Opt.* **26,** 4330–4338 (1987).
55. W. B. Grant, *Appl. Opt.* **25,** 709–719 (1986).
56. K. W. Rothe, *Radio Electron. Eng.* **50,** 567–574 (1980).
57. A. P. Force, D. K. Killinger, W. E. DeFeo, and N. Menyuk, *Appl. Opt.* **24,** 2837–2841 (1985).
58. M. Aldén and W. Wendt, *Appl. Spectrosc.* **42,** 1421–1427 (1988).
59. N. Menyuk, D. K. Killinger, and W. E. DeFeo, *Appl. Opt.* **21,** 2275–2286 (1982).
60. J. R. Lakowicz, *Principles of Fluorescence Spectroscopy*, Plenum Press, New York, 1983.
61. E. L. Wehry, ed., *Modern Fluorescence Spectroscopy*, Vol. 1 of *Modern Analytical Chemistry Series*; Plenum Press, New York, 1976.
62. B. A. Ridley and L. C. Howlett, *Rev. Sci. Instrum.* **45,** 742–746 (1974).
63. H. I. Schiff, D. Pepper, and B. D. Ridley, *J. Geophys. Res.* **84,** 7895–7897 (1979).
64. A. Torres, *J. Geophys. Res.* **90,** 12875–12880 (1985).
65. A. E. S. Green, ed., *The Middle Ultraviolet: Its Science and Technology*, John Wiley & Sons, Inc., New York, 1966.
66. J. H. Richardson, *Mod. Fluorescence Spectrosc.* **4,** 1–24 (1981).
67. R. T. Ku and D. L. Spears, *Opt. Lett.* **1,** 84–86 (1977).
68. D. Glenar, T. Kostiuk, D. E. Jennings, D. Buhl, and M. J. Mumma, *Appl. Opt.* **21,** 253–259 (1982).
69. R. C. Carlson, A. F. Hayden, and W. B. Telfair, *Appl. Opt.* **27,** 4952–4959 (1988).
70. G. Scoles, *Atomic and Molecular Beam Methods*, vol. 1, Oxford University Press, New York, 1988.

71. V. S. Letokhov, *Laser Photoionization Spectroscopy*, Academic Press, Inc., Orlando, Fla., 1987.
72. G. S. Hurst and co-workers, *Rev. Mod. Phys.* **51,** 767 (1979).
73. B. A. Bushaw, *Prog. Analyt. Spectrosc.* **12,** 247–276 (1989).
74. P. Hess and J. Pelzl, eds., *Photoacoustic and Photothermal Phenomena*, Springer-Verlag, Berlin, 1987.
75. V. P. Zharov and V. S. Letokhov, *Laser Optoacoustic Spectroscopy*, Springer-Verlag, Berlin, 1986.
76. H. Coufal, F. Trager, T. J. Chuang, and A. C. Tam, *Surface Sci.* **145,** L504 (1984).
77. R. E. Muenchausen et al., *Opt. Commun.* **48,** 317 (1984), and references therein.
78. B. W. Smith, G. A. Petrucci, R. G. Badini, and J. D. Winefordner, *Anal. Chem.* **65,** 118–122 (1993).
79. M. G. Gehrtz, G. C. Bjorklund, and E. A. Whittaker, *J. Opt. Soc. Am.* B **2,** 1510 (1985).
80. C. B. Carlisle, D. E. Cooper, and H. Preier, *Appl. Optics* **28,** 2567 (1990).
81. A. B. Harvey, *Chemical Applications of Nonlinear Raman Spectroscopy*, Academic Press, New York, 1981.
82. R. L. Farrow and D. J. Rakestraw, *Science* **257,** 1894–1900 (1992), and references therein.
83. P. Ewart and S. V. O'Leary, *J. Phys.* B **15,** 3669 (1982); **17,** 4609 (1984).

ENVIRONMENTAL ECONOMICS, SURVEY

J. Lon Carlson
Illinois State University
Normal, Illinois

The foundation for the study of environmental economics can be traced to the work of Pigou (1) and, in particular, his work on the theory of externalities. Systematic economic analyses of environmental problems started to appear in the 1960s (2,3). Since that time, a considerable amount of research has been devoted to numerous issues that arise within the context of environmental policymaking. An excellent review of the important advances that have occurred in the field over the last 25 yr is available (4). In addition to a large number of journal articles, monographs, reports, numerous books have also been written on the subject (5–11).

The purpose of this article is to highlight the principal features of the theory of environmental economics and the practical implications of that theory for the development of environmental policy. (See also Energy efficiency; Energy policy planning; Energy taxation.)

ECONOMIC THEORY

The Standard Market Model

One of the fundamental premises of microeconomic analysis is that any action should be undertaken up to the point at which the marginal benefits gained equal the marginal costs incurred. Thus, eg, in the case of a marketed good, additional units of the good should be produced so long as consumers' willingness to pay for the last unit produced exceeds the marginal costs of production. This concept is illustrated in Figure 1, which depicts the demand for and supply of good X. The market demand curve is the horizontal sum of the demand curves of the individual consumers of X. The demand curve for X, which is downward sloping to reflect the declining marginal benefits that are derived from the consumption of X, can also be thought of as a willingness-to-pay schedule. The market supply curve is the horizontal sum of the marginal cost curves of the firms that produce X. The supply curve is upward sloping reflecting the increasing opportunity costs of producing additional units of X. Opportunity costs are increasing as a result of diminishing returns in production. Discussion of the theory of supply and demand are available (12–15).

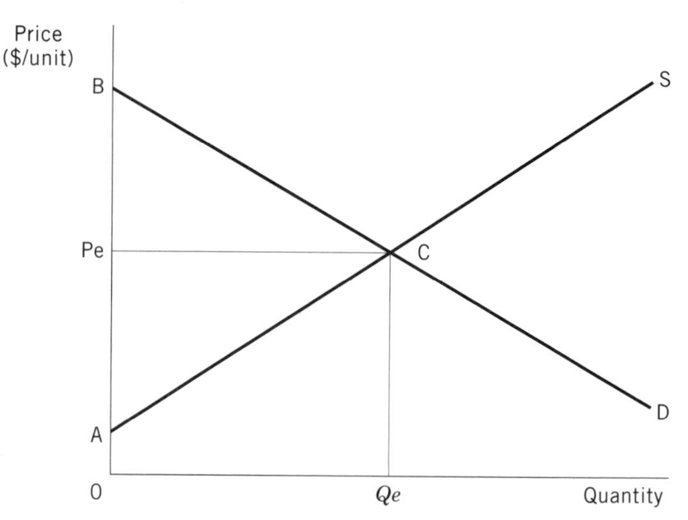

Figure 1. Supply and demand for a normal good. The equilibrium price and quantity of good X are P_e and Q_e. Total willingness to pay for Q_e units is $OBCO_e$. Total expenditures equal OP_eCQ_e, and total consumer surplus is, therefore, P_eBC. Producer surplus is AP_eC. Thus the combined net benefits from the production and consumption of Q_e units are equal to the area ABC.

According to Figure 1, the equilibrium price and quantity of X are P_e and Q_e. Both consumers and producers benefit from the production of X. Consumers realize net benefits to the extent that total willingness to pay exceeds the total amount of money spent on X. In Figure 1, total willingness to pay for Q_e units of X is equal to the area $0BCQ_e$. However, because the equilibrium price is P_e per unit, consumers only spend $P_e \times Q_e$, or the area $0P_eCQ_e$ for the Q_e units. The difference between total willingness to pay and total expenditures, the area P_eBC, is called consumer surplus, and represents the net benefits to consumers. Producer surplus, which measures the difference between total revenues and total variable costs, represents the net benefit to the producers of X. In Figure 1, producer surplus is equal to the area AP_eC. Thus the combined benefits from the production and consumption of Q_e units are equal to the area ABC. Net benefits are maximized by producing Q_e. For output levels less than Q_e, net benefits would increase with additional production because marginal benefits exceed marginal costs. Production in excess of Q_e would cause net benefits to decline since marginal costs exceed marginal benefits. The concepts of consumer and producer surplus are developed within the framework of welfare economics. A number of excellent treatments of the theory of welfare economics are available (16–18).

Assuming that all of the costs and benefits associated with the production and consumption of X are incurred by either the producers or consumers of X, the equilibrium in Figure 1 is socially efficient. Moreover, in the course of determining the socially efficient level of output, the market has also determined the socially efficient allocation of inputs to the production of X. This is especially important in light of the constraint imposed by scarcity of resources. A number of conditions must be met for a privately determined equilibrium to be economically efficient from society's perspective, including full information, a high degree of competition on the part of both sellers and buyers, and a well-defined system of property rights. As is discussed below, in many situations one or more of these conditions is violated. The result is a socially inefficient equilibrium.

External Costs

In the preceding example it was assumed that all costs and benefits associated with the production and consumption of X were incurred by the producers and consumers of X. However, in many situations this assumption does not hold. Instead, some costs are also incurred by individuals who are neither producers nor consumers of the good in question. In the usual market setting, the producer does not have an incentive to consider such costs. Hence, they are external to the decision-making process. Such costs are referred to as negative externalities. External benefits are also possible. In this case, the marginal benefits curve, ie, the demand curve, understates the benefits derived from the consumption of each unit of the good.

A considerable amount of attention has been devoted to the concept of externalities and how to define them (analysts have identified a number of different types of externalities). For the purposes of this discussion, an externality can be defined as "the case where an action of one economic agent affects the utility or production possibilities of another in a way that is not reflected in the market place" (19). Pollution is an excellent example of a negative externality. The adverse effects of pollution, such as acid deposition or the risks associated with the land disposal of hazardous wastes, constitute costs to third parties.

The failure to consider external costs in the decision-making process results in the overallocation of resources to the production of certain goods. Figure 2 illustrates the situation in which the production of X also results in the production of a pollutant, eg, air emissions. Assume that the adverse effects of the air emissions impose costs on society that are not accounted for in the firm's decision-making process. The curve labeled *MPC* represents the firms' private, or internal, production costs. These costs include expenditures on labor and raw materials and any other out-of-pocket expenses. The curve labeled *MSC* represents the marginal social costs of production, which are equal to the sum of internal and external costs. Thus the vertical difference between *MPC* and *MSC* represents the marginal external costs resulting from the production of each successive unit of X. For simplicity it has been assumed that the marginal external costs are constant. However, an assumption of increasing marginal external costs is also quite plausible. In this case, the *MSC* curve would be steeper than the *MPC* curve. In addition, it is assumed that there are no external benefits associated with the consumption of X. The socially efficient equilibrium is found by equating marginal benefits and marginal social costs. In Figure 2, the socially efficient price and level of output are P_s and Q_s. Comparing this equilibrium with the privately determined equilibrium (P_p,Q_p) confirms the earlier observation; failure to incorporate external costs into the decision-making process results in the overallocation of resources to the good in question. Failure to account for external benefits results in an underallocation of resources to the good in question; in this case, the privately determined equilibrium quantity would be too low. The net benefits of producing Q_s units are equal to the area *DBE*. For units of output to the right of Q_s, marginal social costs exceed the marginal benefit from each additional unit of output. Production of the privately determined equilibrium output would result in a loss of benefits equal to the area *EFC*.

Further consideration of Figure 2 reveals another important observation: in many cases, the optimal amount of pollution is not zero. Production of Q_s units results in external costs equal to the area *ADEG*; pollution is still being produced. However, according to the demand curve for X, the benefits from production of X exceed the costs, both internal and external, incurred. Thus external costs are still being borne by some segment of the population. Because the gainers (those benefitting from the production of X) could compensate the losers (those incurring external costs) and still be better off, the outcome is considered to be economically efficient. This condition is sometimes referred to as the Kaldor-Hicks compensation test (20). The fact that those experiencing the external costs are not compensated is immaterial to the question of efficiency. In fact, it has been shown that compensation of those experiencing external costs would result in inefficient behavior

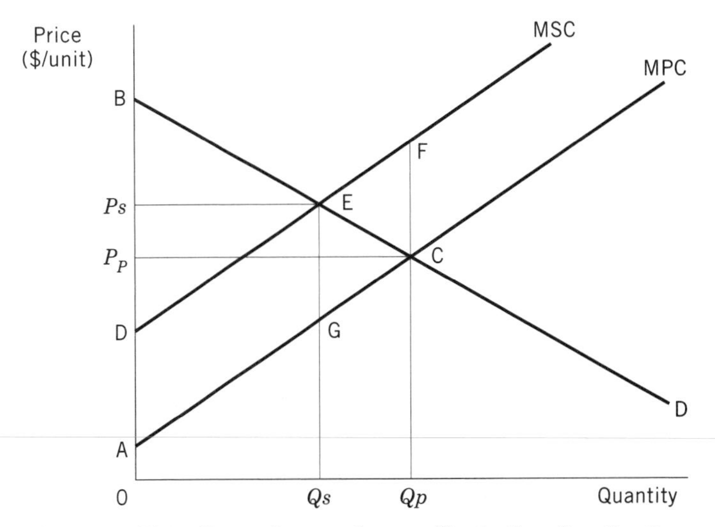

Figure 2. The effects of external costs, illustrating the situation in which production also results in external costs. The curve labeled *MPC* represents the firms' private, or internal, production costs; the curve labeled *MSC* represents the marginal social costs of production (the sum of internal and external costs). The socially efficient price and level of output are P_s and Q_s. The net benefits of producing Q_s units are equal to the area *DBE*.

on their part unless the compensation is in the form of a lump-sum payment (21).

Internalizing Negative Externalities

The failure of firms and individuals to incorporate external costs into their decision-making processes results from a lack of incentives to do otherwise. In a market economy, the assumed primary objective of firms is the maximization of profits. A necessary condition for profit maximization is cost minimization. Thus firms have an incentive to manage their wastes and any resulting pollution at the lowest possible cost; as used here, pollution refers to any substance, energy form, or action that, when introduced into the natural environment, results in a lowering of the ambient quality level (22). To the extent that there are no restrictions, eg, a positive price, on using the atmosphere, waterways or land as a waste sink these options will be exploited. Failure to price the use of environmental media as waste sinks on the basis of the costs such uses impose on society results in the overallocation of these resources to such uses. Individual consumers are confronted with a similar situation. While the objective is different (utility maximization as opposed to profit maximization) the outcome is the same. Consumption results in, among other things, the generation of wastes (packaging and so forth). To the extent that waste disposal services are underpriced, ie, price does not reflect both the internal and external costs of waste management, too much consumer waste will be generated and disposed of.

Environmental media (air, water, and land) will only be priced and used efficiently if property rights for these media are well defined. In fact, property rights to the air and water historically have been poorly defined or nonexistent. When property rights are not completely defined, the resulting market equilibrium is likely to be socially inefficient. However, the question of whether poorly defined property rights require some form of market intervention depends on the level of transactions costs incurred in resolving disputes involving external costs. As has been demonstrated (23), so long as transactions costs, ie, the costs of bringing together the affected parties and negotiating a settlement, are low, private negotiation will result in the efficient allocation of property rights. Regardless of how the property right in question is initially allocated, it will eventually go to the party that values it most highly. This conclusion has come to be known as the Coase Theorem. However, an equally important point made by Coase is that in the majority of cases involving disputes over property rights, transactions costs are *not* low. In such cases, private negotiation is not possible. Thus the initial allocation of the property right will have important implications for the resulting allocation of resources. In those cases where the resulting allocation is inefficient as a result of, say, failure to consider external costs, it may be necessary for the government to intervene in the market in an effort to move the affected parties toward the efficient solution.

There are a number of approaches that can be employed to address the problem of negative externalities, including reliance on legal remedies, direct regulation, and the use of market-based incentives (6,8,11). Legal remedies include the use of civil and criminal law to address disputes over property rights (4,8,11). Civil law, such as the use of a negligence standard, allows the plaintiff the opportunity to recover damages incurred as a result of some action by the defendant. Actions brought under a criminal standard provide for punitive measures in the event that the defendant is found to be at fault. Direct regulation involves the use of various standards to control the behavior of individuals. Direct regulation has been the chief means employed by the U.S. Environmental Protection Agency (EPA) to control the amount of pollution generated in the United States. Market-based incentives, including pollution charges (taxes) and permit systems, are also used to control the generation of pollution. As the name implies, these approaches rely on market forces, rather than direct government intervention, to determine who will ultimately be responsible for pollution control.

Public Goods

Thus far, the discussion has focused on the effects of externalities, such as pollution, on the allocation of resources. An equally important issue concerns the benefits that are derived from pollution control. To be specific, once pollution is controlled, everyone who was adversely affected by the pollution benefits. Moreover, the amount of benefit one person derives from pollution control does not affect the amount of benefits available to others. A single unit of pollution control can be "consumed" simultaneously by any number of individuals. In addition, individuals cannot be excluded from enjoyment of the benefits of pollution control. Goods that possess these properties, referred to respectively as indivisibility and nonexcludability, are called public goods. Public goods stand in contrast to private goods, which are both divisible and excludable. Examples of public goods include national defense, police and fire protection, national parks, and pollution control.

The fact that pollution control is a public good has important implications for the measurement of the benefits of pollution control and the manner in which pollution control will be achieved. With respect to benefits, because a public good is indivisible, the aggregate benefits from, or market demand for, each unit of the public good are calculated by summing the willingness to pay of all affected individuals *for each unit* of the good produced. This amounts to vertically summing the individual demand curves to derive the market demand curve for the public good. In the case of private goods, the market demand curve is calculated by horizontally summing the individual demand curves. This approach yields the total number of units that will be purchased at each price (ie, at each level of willingness to pay). Discussions of the derivation of the demand curves for private and public goods are available (8,12).

The practical importance of the distinction between the market demand for public goods and private goods can be illustrated with a simple example. Assume that five individuals each wish to purchase a unit of good X, and that the willingness to pay of each of the five individuals for one unit of X is $10. If X is a private good, five units of X must be produced to satisfy the demands of all five individuals. In addition, the marginal cost of the fifth unit of X cannot exceed $10 if the demand of all five individuals is to be met in an efficient manner. If the marginal cost of the fifth unit (or the first through fourth unit for that matter) exceeds $10, marginal cost exceeds marginal benefit ($10); production of the marginal unit is inefficient. In contrast, if X is a public good, as a result of the property of indivisibility a single unit of X provides marginal benefits of $10 to each individual. Collectively, they are willing to pay $50 for one unit. Thus marginal cost could be as much as $50 and production of the good would still be efficient.

Because public goods are indivisible, the marginal benefit from each unit of the good produced is potentially much greater than it would be if the good were a private good. However, the property of indivisibility, as well as nonexcludability, also prevents public goods from being produced in a market setting. Because public goods cannot be divided into discrete units for sale and individuals cannot be excluded from consuming the good unless they pay the market price, private entrepreneurs have no incentive to engage in the production of public goods. Instead, production of public goods must be undertaken by the government or, alternatively, the government must take actions to ensure that the public good is produced in the private sector. Thus the government has assumed the responsibility of providing for the nation's defense as well as police and fire protection. In the case of pollution control, the government has taken steps, primarily legislative and regulatory, designed to force firms to internalize the external costs of pollution and, in so doing, produce pollution control.

ANALYTICAL APPROACHES TO POLLUTION CONTROL POLICY

Depending on the needs and objectives of the decision maker, various analytical approaches can be employed to assess the effects of alternative policies designed to improve environmental quality. The most basic approach is referred to as impact analysis. The fact that it is the most basic approach should not be interpreted to imply that impact analysis is a simple, straightforward procedure. In many instances, as is described below, impact analyses can be quite complex and subject to considerable uncertainty.

Cost-effectiveness analysis considers the costs of alternative policy options and provides the decision maker with insights regarding the value of the resources that will be required to achieve a particular objective, such as a specified reduction in a particular type of air emissions. Benefit–cost analysis is intended to shed light on the relative value of the benefits and costs of policy alternatives. Depending on the decision rule that is used, benefit–cost analysis can assist in the identification of the economically efficient solution or those options for which net benefits are positive. Risk assessment is applicable to those situations in which policy is intended to influence the risks of certain adverse effects on humans or the environment.

Impact Analysis

The purpose of an impact analysis is to summarize the various impacts a proposed action is predicted to have on the affected environment. The affected environment can include both the natural environment and human activities. Thus impacts may include effects on natural resources, environmental quality, and economic activities. Environmental and economic impact assessments often are required at the federal and state levels. The National Environmental Policy Act (NEPA) of 1969 requires that an impact analysis be conducted for all major federal actions. The result of many of these analyses is an Environmental Impact Statement (EIS), which summarizes the predicted impacts of the proposed action and alternatives to the proposed action on the affected environment. The content of an EIS is dictated by regulations promulgated by the Council on Environmental quality (CEQ), which was created by NEPA and is responsible for, among other things, overseeing the EIS process (24). At the state level, similar requirements are often imposed on regulatory agencies. For example, in Illinois an Economic Impact Statement (Ecis) is required as part of the process of promulgating many environmental regulations. The determination of whether an Ecis is required in a particular situation is the responsibility of the Illinois Pollution Control Board.

Impact analyses, such as an EIS, provide decision makers with a large amount of information on the proposed action and its alternatives. However, because the impacts vary across the different affected resources, and some impacts are positive while others are negative, it is often difficult to determine, on net, whether the proposed action will be beneficial when viewed from society's perspective. As has been pointed out, "Current environmental impact statements are more sophisticated than their early predecessors. . . . Historically, however, the tendency had been to issue huge environmental impact statements that are virtually impossible to comprehend in their entirety" (25).

The EIS process has become more sophisticated over time. For example, the U.S. Bureau of Reclamation (BOR) has completed a draft of an EIS that examines the impacts of the operations of Glen Canyon Dam on the affected environment (26). This EIS employs state-of-the-art analytical techniques to analyze the impacts of a change in the operations of Glen Canyon Dam on such variables as water quality, sedimentation and river bed characteristics, and beaches and backwaters; biological resources including vegetation, wildlife and habitat potential, and fisheries; recreational resources including fishing, day rafting, white-water rafting, and camping; hydropower resources; sociocultural resources; economic resources; and aesthetic resources. With respect to hydropower and economic resources, sophisticated models were developed to analyze the impacts of a change in dam operations on the mix of resources (hydropower and fossil-fuel generation) that would be required to meet peak and off-peak energy demands and the implications for electrical power supplier in the region. In addition, these models were used to estimate the change in retail rates charged to various power customers in the affected region. The recreation analysis provided, among other things, estimates of the change in the value of recreational activities (angling and white-water boating) resulting from a change in streamflows below the dam. A separate component of the analysis is attempting to estimate the effects of a change in ambient environmental conditions, resulting from a change in dam operations, on the so-called nonuse values individuals attach to the affected environment. The term nonuse value refers to the amount an individual would be willing to pay for a change in environmental quality at a specific location, even though they do not intend to visit or otherwise use the site personally. As this brief discussion illustrates, impact analysis can be extremely complex, time-consuming, and costly. Thus far, the Glen Canyon EIS has required several years and tens of millions of dollars to complete.

Cost-effectiveness Analysis

Cost-effectiveness analysis involves comparison of the costs of alternative approaches to achieving a specific objective, eg, a reduction in pollution (27). Unlike benefit–cost analysis, which requires monetization of both the benefits and costs of the proposed action, cost-effectiveness analysis does not require that benefits be expressed in monetary terms. Instead, benefits are expressed as the reduction in different types of damages or sources of damages, eg, decreased risk of mortality or decrease in the concentration of a pollutant in a body of water. Cost-effectiveness can be assessed by choosing a target level of effectiveness (or goal) for which the least-cost alternative is identified. Conversely, a fixed budget can be applied to alternative strategies to identify the alternative that yields the maximum effectiveness for the funds available. In either case, performing a cost-effectiveness analysis requires the implementation of three steps: (*1*) identification of the alternatives to be evaluated, (*2*) selection of the appropriate measure(s) of effectiveness, and (*3*) measurement of the costs incurred and the effectiveness of each alternative.

Identification of the preferred policy option depends on whether the effects of the proposed actions vary across the policy alternatives. If each of the policy alternatives has the same effect, ie, each alternative reduces pollution, and therefore damages, by the same amount, the decision rule is straightforward: select the policy alternative that achieves the desired reduction in pollution at least cost. The least-cost solution is determined by calculating the cost-effectiveness ratio, which is simply the total cost of an alternative divided by its total effectiveness, for each alternative. The costs of implementing the alternative actions are then compared to determine the least-cost strategy. While this approach is rather straightforward in the case in which each alternative yields the same level of effectiveness, it can produce ambiguous results when alternatives yield differing levels of effectiveness.

When the effectiveness of the alternatives varies, selection of the appropriate decision rule is more complex. For example, one could calculate the average cost per unit of effectiveness as is described above. However, the appropriate decision rule is not obvious. Consider, for example, the situation in which one option results in both a higher average cost and total amount of effectiveness than another option. In this case, it is not clear which option should be preferred. As an alternative, one could compare the marginal costs of pollution control across the various options. A frequently cited decision criterion is the equivalent marginal cost approach (28). The marginal cost of effectiveness is measured as the change in costs associated with a one unit change in the effectiveness of the action. In this approach alternatives that have the same marginal cost effectiveness are grouped together, and a predetermined selection process is then used to maximize the stated objective. For the marginal cost effectiveness criteria to be applicable, a number of conditions must be met, including continuity among alternatives, costs, and effectiveness measures.

As an alternative to the marginal cost approach, incremental cost-effectiveness is measured as the difference between the cost of an alternative and the cost of the next most stringent (or effective) alternative divided by the difference in effectiveness between the two alternatives (28). For example, assume two alternatives, A and B, are contiguous and A is more stringent than B, ie, A yields a greater level of effectiveness. In this case, incremental cost-effectiveness (IC-E) is measured as:

$$IC\text{-}E = \frac{C_A - C_B}{E_A - E_B}$$

The incremental cost effectiveness approach is most useful in the case of discrete alternatives and subalternatives. The alternatives are grouped according to equality of incremental cost-effectiveness. Once grouped in this way, a choice among alternatives can be made.

Finally, the decision rule could be based on minimization of control costs subject to some minimum threshold level of control. Alternatively, the threshold could be defined as the maximum cost that could be accepted (to be economic). In either case, the threshold value could be used as a decision point for choosing alternatives. Any alternative that results in a level of effectiveness or cost that falls short of or exceeds the corresponding threshold value would be eliminated from consideration.

Benefit-Cost Analysis

Simply put, the objective of a benefit-cost analysis is to compare the benefits and costs of a proposed policy action. More thorough treatments of the subject are available (29–31). Depending on the objectives of the decision maker, benefit–cost analysis can be more or less rigorous. For example, if the objective is simply to ensure that the net benefits of a proposed policy action are positive, the net present value criterion is sufficient (32). Under this decision rule, so long as the present value of the total benefits exceeds the present value of total costs, the proposed action is acceptable. This is equivalent to requiring that the ratio of total benefits to total costs (appropriately discounted) be greater than 1.

In the case in which the objective of the analysis is identification of the economically efficient policy action the decision rule is much more rigorous. Identification of the economically efficient solution requires knowledge of the marginal benefits and marginal costs of pollution control. This concept is illustrated in Figure 3, which depicts the marginal benefits and marginal costs associated with increasing levels of pollution control. The external costs of a particular pollutant, such as the sulfur emissions from a coal-fired generating unit are measured by the resulting

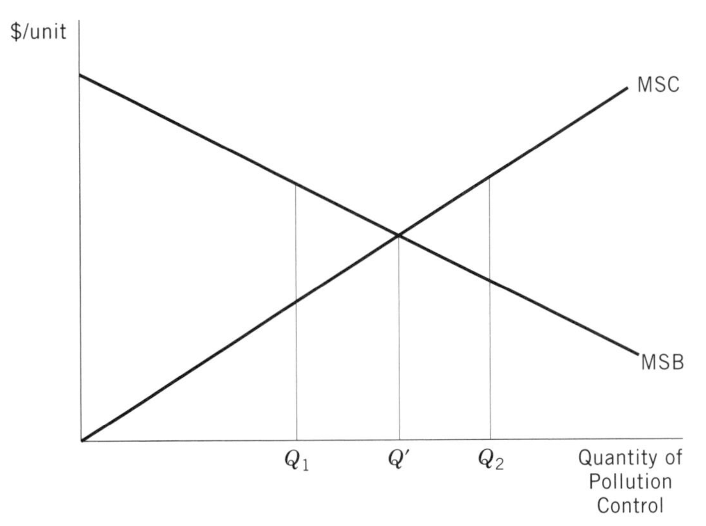

Figure 3. Determining the optimal level of pollution control, depicting the marginal benefits and marginal costs associated with increasing levels of pollution control. The marginal social benefit (MSB) curve represents society's willingness to pay for each additional unit of pollution control. The marginal social cost (MSC) curve represents the incremental cost of each additional unit of pollution control. The optimal level of pollution control is Q'. Any point other than Q' is inefficient. At levels such as Q_1, MSB > MSC: marginal net benefits are positive. At levels such as Q_2, MSB < MSC, and marginal net benefits are negative.

damages imposed on society. As such, the benefits of successive reductions in the quantity of sulfur emissions are measured by society's valuation of the reduction in damages attributable to each unit of emissions reduction. Technically, this benefit is measured as the value of the reduction in the risk of environmental damage that results from the reduction in pollution. In Figure 3, pollution reduction is measured on the horizontal axis. The dollar value of the marginal costs and benefits associated with successive increments of pollution reduction is measured on the vertical axis.

The marginal social benefit (MSB) curve represents society's willingness to pay for each additional unit of pollution control. The area under the MSB curve to the left of any level of pollution reduction represents the total benefits of that level of pollution control. As illustrated in Figure 3, the marginal benefits of pollution control are assumed to decrease as the total amount of pollution control increases. The marginal social cost (MSC) curve represents the incremental cost of each additional unit of pollution control. The area under the MSC curve to the left of any particular level of control represents the total cost of pollution control. As the total amount of pollution control increases, the marginal cost of each additional unit of pollution control is assumed to increase. This feature of pollution control costs, which has its origins in the law of diminishing returns, is regularly observed in actual situations involving pollution control (33).

The optimal level of pollution control is that level at which marginal benefits and marginal costs are equal (Q' in Fig. 3). Any point other than Q' is inefficient. At levels of control less than Q', such as Q_1, the marginal benefits of additional control exceed the marginal costs incurred. As such, the net benefits of additional pollution control would be positive. At levels of control in excess of Q', such as to Q_2, the marginal costs of pollution control exceed marginal benefits: marginal net benefits are negative. Another important point to emphasize is that the optimal level of pollution control is not the level at which total benefits equal total costs (8,11).

As an alternative to the model presented in Figure 3, the question of how much pollution to control can be approached from the perspective of cost minimization. Figure 4 depicts the marginal costs of pollution control and marginal damages resulting from each increment of pollution. The amount of pollution generated is measured on the horizontal axis, this contrasts with Figure 3, in which the amount of pollution *controlled* is measured on the horizontal axis. Marginal damage costs, depicted by the curve labeled MDC, are assumed to increase as the amount of pollution increases. (Marginal damage costs could also be assumed to be constant without changing the results of the analysis that follows.) Marginal control costs, depicted by the curve labeled MCC, decrease as the level of pollution increases, reflecting reduced levels of control. (Q_0 represents the amount of pollution that would be generated in the absence of market intervention.)

For the situation illustrated in Figure 4, the total costs associated with the pollutant in question would be minimized by allowing Q^* units to be generated. The total costs attributable to Q^* units of pollution, ie, the sum of damage costs and control costs, is equal to the area $a + b + c + d$, where $a + b$ represents total damage costs and $c + d$ represents total control costs. Restricting the level of pollution to something less than Q^*, such as Q_1, would result in

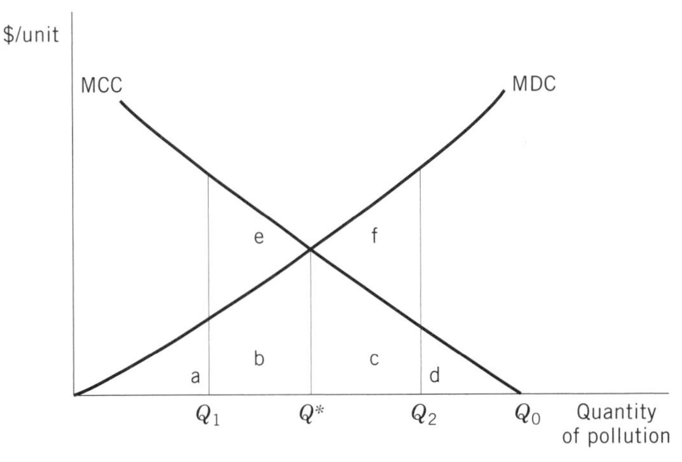

Figure 4. Marginal control costs versus marginal damage costs of pollution, depicting the marginal costs of pollution control (MCC) and marginal damage costs (MDC) resulting from each increment of pollution. In this example, the total costs of pollution are minimized by allowing Q^* units to be generated. The total costs attributable to Q^* units of pollution are equal to the area $a + b + c + d$, where $a + b$ represents total damage costs and $c + d$ represents total control costs. Restricting the level of pollution to something like Q_1 would result in MCC > MDC; total costs would increase by the area e. In a similar manner allowing the level of pollution to exceed Q^*, eg, Q_2, MDC > MCC and total costs would increase by the area f.

marginal control costs that exceed marginal damage costs avoided. Thus relative to the cost minimizing solution, total costs would increase by the area *e*. In a similar manner if the level of pollution were to exceed Q^*, eg, Q_2, marginal damage costs would exceed marginal control costs and total costs would once again exceed the minimum, this time by the area *f*.

In fact, Figures 3 and 4 convey the same information with respect to the optimal amount of pollution, albeit from different perspectives. This is because the marginal benefits from pollution control are measured by the reduction in damages attributable to the pollution that is being controlled. Thus Q^* in Figure 4 is equal to the initial amount of pollution minus Q' in Figure 3, the optimal amount of pollution control.

A number of important points need to be borne in mind when discussing benefit–cost analysis. First, in many instances the costs and benefits of pollution control will be dispersed over many individuals and entities. Aggregating the gains and losses of all affected parties and then comparing these aggregate measures ignores the question of distribution. However, recall the earlier discussion of the Kaldor-Hicks compensation test. According to this criterion, if the gainers can compensate the losers and still be better off, efficiency is increased. This approach allows decision makers to consider a wider range of alternatives than would be available if the rule were that no one could be made worse off. However, it does ignore questions of equity and fairness. To be specific, the compensation test does not require that compensation of the losers actually occur. Another important point concerns the various difficulties that are likely to be encountered when conducting a benefit–cost analysis.

Risk Analysis

Risk analysis (11) has been developed in recent years as an alternative approach to addressing certain environmental issues. Risk analysis is composed of two basic components: risk assessment and risk management. The concept of risk analysis can be illustrated by reference to the case of superfund sites. Superfund sites are locations at which hazardous materials or wastes have been deposited, and that have been designated for inclusion on the National Priorities List by the EPA. Superfund sites are of concern due to the risks they pose to the ambient environment. These risks include the potential for contamination of ground or surface waters, as well as the soil and atmosphere around the site. Contamination can then lead to adverse effects on ecosystems and human health and welfare.

Risk assessment is used to determine the types of adverse effects that could occur as a result of contamination or migration of materials away from the site and the probability of such occurrences. Thus, in the case of a superfund site such as an abandoned hazardous waste landfill, the risk assessment would entail identification of the types of wastes that were buried at the site, the possible pathways of migration of materials away from the site, the adverse effects that could result from exposure to such materials, and the probabilities of migration, exposure, and occurrence of adverse effects. Risk management focuses on the question of what constitutes the appropriate policy response to the risks that have been identified in the risk assessment. For example, it might be determined that disturbing the site would result in a greater probability of exposure than simply leaving the site intact and monitoring the area around the site. In addition, available evidence suggests that the values that society places on different types of risks can vary considerably. Thus willingness to pay, and therefore the benefits from risk mitigation, is likely to vary depending on the specifics of the situation. Studies of the public's attitudes toward different types of risks suggests that responses are influenced by a number of factors including whether the risk is voluntarily or involuntarily assumed and the specific types of adverse effects that could occur (34–36).

Comparison–Contrast of the Approaches

Going from impact analysis to cost-effectiveness analysis to benefit–cost analysis and risk analysis, the amount of information that can be made available to decision makers increases, at least in principle. However, the information and data requirements also increase. In effect, the outputs of an impact assessment serve as inputs to a cost-effectiveness analysis. In a similar fashion, the results of a cost-effectiveness analysis constitute information required to complete a benefit–cost analysis. In the case of impact analysis, the primary focus is on the effects the policy alternatives will have on the affected environment. Cost-effectiveness analysis goes beyond an impact analysis by estimating the opportunity costs, measured in dollars, of each of the policy alternatives. To complete a benefit–cost analysis it is also necessary to express, to the extent possible, the effects (benefits) of each alternative in monetary terms. Consequently, some discussion of cost and benefits estimation is in order. Brief discussions of the components of total value and the role of discounting are also presented.

Cost Estimation Techniques. The costs of a particular pollution control policy are measured by the value of the resources that must be used to achieve the policy's objectives and any net welfare losses in the affected markets (37). Economists refer to this concept as opportunity cost. Opportunity cost is measured as the value of the next best use of the resources in question. Consider, for example, a proposed regulation that would require sewage treatment plants to install an additional filtration device that would reduce the concentration of a particular chemical in the public's drinking water supply. In this case, the cost of the proposed regulation would depend on the cost of the filtration device, installation costs, and any operation and maintenance costs that would be incurred over the service life of the device as well as price and quantity adjustments in the affected markets. This example raises a number of important issues, including the methods used to obtain cost data and the use of market prices to approximate opportunity cost.

In practice, the most common approaches to cost estimation include the use of surveys, construction of an engineering process model (or series of models), or some combination of the two (38). Surveys are used to compile estimates of compliance costs from those entities, often firms in an industry, that would be subject to the proposed policy. Engineering process models entail construction of a mathematical model of a representative firm or set of firms in the affected industry. The model relates inputs and outputs and enables the analyst to estimate the costs of production for a given set of input prices and level of demand. Both approaches have their respective strengths and weaknesses. Surveys rely on input from those entities who are presumably in the best position to estimate compliance costs, ie, the affected firms. However, there is an incentive for affected firms to overstate expected compliance costs in an effort to influence the stringency of the final policy. Process models are able to circumvent this problem, but are limited since the models are for a "representative" firm (or firms) in the industry. Actual firms may differ significantly from the hypothetical average and therefore incur costs that differ significantly from those estimated using the process model. A detailed discussion of alternative cost estimation methods including various types of process models and econometric estimation techniques is presented in (39).

The extent to which pollution control expenditures based on market prices reflect the actual opportunity costs of a proposed policy is another important issue in cost estimation. If all of the markets for inputs and outputs in the economy were perfectly competitive, market prices would be an accurate measure of opportunity cost. However, in many of the markets in the U.S. economy, competition is limited. The extreme case is monopoly, in which there is a single seller of the good or input in question. When competition is constrained, for whatever reason, a likely outcome is that price will exceed marginal cost, which is the correct measure of opportunity cost, assuming marginal cost is determined in competitive markets). Hence, expenditures will tend to overstate the actual costs, ie, opportunity costs, of the proposed regulation. The difference between price and marginal cost, although paid by the purchaser of the good or input, represents a transfer of income and does not reflect the value of resources required to comply with the proposed policy. Hence, it is not properly considered as part of the costs of the proposed policy from society's perspective.

Benefits Estimation. Estimation of the benefits of a particular policy option is complicated by the fact that many of the effects of pollution control are not valued in a market setting. Consider, for example, the situation in which a pollutant adversely affects agriculture and water-related recreation. To the extent that the proposed policy reduces adverse effects on agriculture, such benefits could be measured by the market value of the increased production of the affected outputs. In contrast, water-related recreation is not "purchased" in the usual sense. Instead, recreators purchase a set of inputs, eg, boats, swimming gear, fishing tackle, and so forth, and combine these with a specific site to produce an output, water-related recreation. Thus what is of interest is the value attached to the quality of the recreation site. Because the price paid for access to a site is generally low or nonexistent, this price is not a good indicator of an individual's willingness to pay for access. Thus the value of the site is unknown. However, economists have developed a number of techniques, which can be classified as either indirect or direct, to assess such values. Indirect methods include the hedonic estimation technique and the travel cost method. The direct method consists of so-called contingent valuation.

Hedonic Estimation Technique. The hedonic estimation technique, first developed in the early 1970s (40,41), is based on the observation that in many situations, willingness to pay for a good depends on a number of characteristics of the good (4,9,42). Consider, for example, the case of housing. The price paid for a house is likely to depend on a number of factors, including overall size (square footage), the number rooms (baths, bedrooms, etc), and the location of the house (including quality of the neighborhood and schools). It is also reasonable to expect that the level of environmental quality could influence willingness to pay for a specific house. For example, two houses may be identical in all respects with the exception that one house is located next to a municipal waste landfill and the other is located in a quiet suburban neighborhood. In this case, it is reasonable to expect that the price of the latter would be higher than the price of the former. The objective of the hedonic estimation technique is to estimate that portion of the price that is attributable to the environmental characteristic of interest, eg, distance from a landfill. A good example of the application of the hedonic technique is presented in (43).

The hedonic technique has also been applied to the problem of risk valuation. Studies (44,45) have attempted to estimate that portion of the wage rate paid for certain jobs that represents compensation for an increased risk of death. The results of these "wage-risk premium" analyses have been used by some analysts to estimate the value of policies that would mitigate certain environmental risks. While this approach does shed some light on the potential magnitude of the benefits of such policies, it raises a number of issues including the difference between voluntarily and involuntarily assumed risks and the resulting value of a change in each type of risk.

In cases in which property values or wage estimates are available, the hedonic technique can prove useful in estimating willingness to pay for environmental improvements. However, this is also a main source of its limitations (46). To be specific, many instances arise in which the good to be valued is not related to property values or wages. Thus willingness to pay for environmental improvements must be estimated by alternative means.

Travel Cost Method. One particular case in which the hedonic technique often is not particularly useful is that of recreation values. Pollution can adversely affect the quality, and therefore value of recreation, especially water-based recreation. However, in many cases there is no fee for access to a recreation site, and in other cases, eg, national parks, there is a nominal fee for access; this price is far less than actual value of the experience to the recreator. Consequently, the hedonic technique cannot be used directly. An alternative approach (which was first proposed by Harold Hotelling in a letter to the director of the National Park Service in 1947 and since described formally by (47)), uses travel costs as a proxy for the price paid for access to a particular recreation site. The travel cost method has been used in a number of different studies to estimate the value of recreation (eg, angling) at numerous locations. Briefly, the demand for access to a recreational resource is estimated on the basis of the travel costs (both direct and implicit) incurred by individuals using the site. The area under the demand curve is then taken as a measure of the benefits derived from the site. Extended discussions of the travel cost method are available (4,9,42). Applications and assessments of the travel cost method can also be found (48–50).

Contingent Valuation. Contingent valuation entails the use of a survey instrument to generate individuals' estimates of their willingness to pay for the item in question, eg, an improvement in visibility at the Grand Canyon. The resulting estimates are then statistically analyzed to generate estimates of the affected population's aggregate willingness to pay for the proposed policy. Detailed discussions of the contingent valuation method have been presented (5,52). In many situations, contingent valuation constitutes the only method for estimating willingness to pay for a change in environmental quality. This is especially true in such cases as a change in the characteristics of a specific recreation site or the estimation of so-called nonuse values (which are discussed below).

The practice of contingent valuation has evolved considerably over the past 15 yr. Many of the changes have occurred in direct response to the numerous criticisms that have been leveled at previous studies. These criticisms include, but are not limited to, potential biases that may enter into the process (53). For example, starting point bias, or anchoring, occurs when the respondent's expressed willingness to pay is influenced by an initial value of willingness to pay suggested in the survey. To the extent that the respondent sees the potential to gain from a particular response, even though the response does not reflect his or her true willingness to pay, strategic bias can occur. Hypothetical bias refers to the possibility that the respondent may react to the hypothetical nature of the survey and in so doing respond in a manner that does not reflect his or her true willingness to pay for the change in question.

Each of the issues noted above, as well as a variety of others, has been the subject of careful analysis. And in many cases, significant progress has been made in overcoming such obstacles. However, in spite of the considerable advances that have made, the validity of the contingent valuation method is still a matter of some dispute. For example, the results of one recent study (54) suggest that survey respondents may have considerable difficulty isolating the specific good or environmental attribute that is being valued in the survey. While such results have in turn been challenged (55), other analysts have argued for significant revisions to the approach to enhance the validity of the resulting estimates of willingness to pay for changes in environmental quality (56).

Total Valuation: Use Values Versus Nonuse Values. In many cases, pollution control results in reduced damages, eg, adverse health effects and ecosystem damage, as well as an improvement in the quality of certain resources. To ensure that all of the benefits of a proposed policy are included, it is necessary to consider the change in the total value of the affected resources. The total value of an environmental attribute or natural resource is composed of two components: use value and nonuse value. In addition, the concept of option value has been developed to further distinguish the separate components that make up the total value of a resource.

Use value refers to the value of the direct (and sometimes indirect) uses to which a resource can be put. The concept of option value was first developed by Weisbrod (57). In its initial conception, option value referred to the value an individual attaches to the option to use a resource he or she is not currently using but may decide to use at some future point in time. Subsequent theoretical work has led to the rejection of this interpretation. According to Smith (58), option value is the difference between two different measures of use value in the presence of uncertainty and, as such, is not a separate component of the total value of a resource. Nonuse values include what have been termed existence value and bequest value. Existence value refers to the value an individual attaches to the existence or quality of a resource, even though he or she does not intend to ever use the resource personally. Bequest value refers to the value individuals may attach to the preservation of resources for future generations.

There is, in fact, some disagreement among economists regarding the legitimacy of nonuse value as a value distinct from use value. Some critics have argued that there is little or no reason to believe that existence values (or bequest values) are not already reflected in use values. It has been suggested that although existence values may be distinct from use values, they should nonetheless be excluded from benefit–cost analyses, because they reflect considerations other than the efficiency motive (59). More recently, it has been argued that nonuse values present a number of problems that raise serious questions about the legitimacy of their inclusion in the decision-making process (60). It is argued that the range of possible existence values may be limitless and that accurate estimation of existence values is extremely difficult, if not impossible. Other arguments against the consideration of nonuse values are available (61).

Proponents of the case for nonuse values have offered strong rebuttals to the arguments described above. For example, there is a well-developed theoretical basis for the consideration of non-use values (62,63). In addition, the fact that nonuse values are difficult to estimate is not, by itself, sufficient justification to exclude them from policy analyses (64). On the empirical side of the debate, studies suggest that nonuse values may account for a substantial portion of the total value of a resource. For example, a number of studies were reviewed that estimated use and nonuse values of particular resources (65). The authors concluded that "non-use benefits generally are at least half as great as recreational use benefits" (66). This suggests that, all else constant, failure to consider nonuse values could result in serious underestimation of the benefits of many proposed environmental policies.

The Question of Discounting. Because many of the costs and benefits of pollution control policies will occur in the future, discounting is also an important issue. To be specific, costs and benefits that occur at various times in the future must be adjusted so that all costs and benefits can be compared on an equal basis. The usual approach to this problem is to calculate the present value of all future costs and benefits and then aggregate the respective values for purposes of comparison. The immediate question that arises is what value of the discount rate to use. Not surprisingly, this issue has been the subject of considerable debate. Excellent discussions can be found in a number of references (31,67,68). It is sufficient here to note that this issue is far from settled. Discount rates used by various federal agencies in recent years have ranged from 10% by the Office of Management and Budget to 2% by the Congressional Budget Office (69).

ECONOMIC EFFICIENCY VERSUS COST EFFECTIVENESS: TRADE-OFFS IN POLICY MAKING

Standard Setting

As the discussion of benefit–cost analysis suggested, identification of the economically efficient level of pollution control is often difficult, if not impossible, to achieve in practice. This stems largely from the fact that the benefits of pollution control are often difficult to measure, let alone quantify in dollars. In addition, benefit–cost analysis is concerned only with the issue of efficiency. No consideration is given to such factors as equity, eg, income redistribution, or the feasibility of the efficient solution. However, these are important concerns for policymakers. In light of these constraints, environmental policy in the United States has tended to rely on the use of specific standards to achieve improvements in environmental quality. The Clean Air Act, for example, requires that primary ambient air quality standards be set at a level that ensures protection of human health, without consideration of the costs that might be incurred in achieving the proposed standard. Although establishment of a standard circumvents the problem of identifying the efficient solution, ie, the optimal amount of pollution reduction, the question of how to achieve the standard remains. Any one (or combination) of a number of different policy options could be employed. However, from the perspective of efficiency, the objective is to identify the policy option that achieves the standard at least cost.

For a particular standard to be met at least cost, the marginal costs of compliance must be equal across all of the affected firms. To see this, assume that two different firms (A and B) are the only generators of a particular pollutant and that between them they currently generate a total of 50 units of pollution per time period. Assume also that the control authority (eg, EPA) has determined that the aggregate amount of the pollutant generated per time period should be limited to 34 units, ie, the total amount of pollution must be reduced by 16 units. Figure 5 depicts the marginal costs of pollution control incurred by each of the two firms. Marginal cost is measured on the vertical axis and pollution control is measured on the horizontal axis. The amount of pollution control undertaken by firm A is measured moving from left to right. The amount of pollution control by firm B is measured moving from right to left.

According to Figure 5, requiring firm A to control 5 units of pollution and firm B to control 11 units would result in the control of 16 units (the reduction necessary to meet the standard). With this allocation of pollution control marginal costs of control are equal across the two firms and total costs of control equal the area $a + b + c$. A deviation in either direction (eg, to an equal reduction of 8 units by each firm) would increase the marginal costs incurred by one firm by more than the reduction in marginal costs experienced by other firm; total costs increase. Allocating control responsibility equally across the two firms would result in total control costs of $a + b + c + d$, or a net increase of d. Thus the total costs of pollution control are minimized by allocating pollution control such that the marginal control costs of the affected firms are equal. The equal marginal cost rule for cost minimization extends to any number of firms.

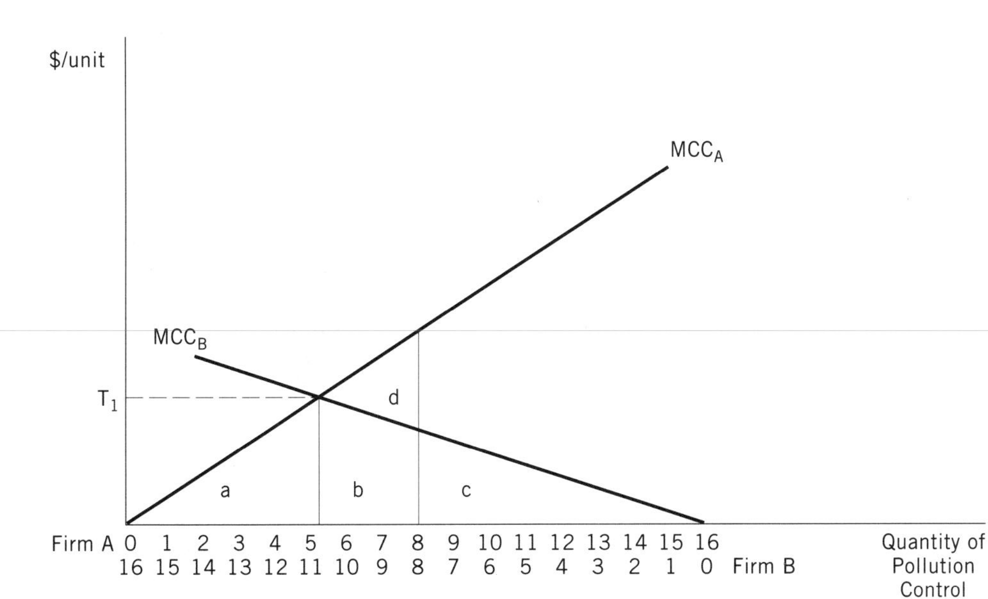

Figure 5. Determining the cost-effective allocation of pollution control. The marginal costs of pollution control incurred by each of two firms are shown. The amount of pollution control undertaken by firm A is measured moving from left to right. The amount of control by firm B is measured moving from right to left. Assume that 16 units of pollution must be eliminated. Requiring firm A to control 5 units of pollution and firm B to control 11 units would result in the control of 16 units. With this allocation of pollution control marginal costs of control are equal across the two firms and total costs of control equal the area $a + b + c$. A deviation in either direction, eg, to an equal reduction of 8 units by each firm, would increase the marginal costs incurred by one firm by more than the reduction in marginal costs experienced by the other firm; total costs increase.

Policy Options

Following one approach (70), it is useful to think of policy in terms of distinct components: goals, objectives, and instruments. In the case of pollution, the goal of the different policies that could be implemented might be described as "an improvement in the overall level of environmental quality." To this end, policies can be designed to achieve different objectives such as a reduction in the quantity or harmful characteristics of a particular water pollutant, or an increase in air quality. Instruments refer to the specific methods that are used to achieve the policy's objectives and, ultimately, the policy goals. In the case of pollution control policies, the primary instruments include direct regulation, charges, and permits. Legal remedies can also be used (4,8,11).

Direct Regulation. Historically, direct regulation has constituted the dominant approach to achieving improvements in environmental quality. Also referred to as command and control (CAC), direct regulation may, eg, require firms to use a particular production process or one that achieves a specified level of performance, reduce emissions or effluents to specified levels, or otherwise meet specific criteria. In addition, direct regulation generally requires that all affected firms meet the same criteria, eg, restrict emissions of a particular air pollutant below some threshold level.

One of the advantages of direct regulation is its relative simplicity. Once a standard has been established, each affected firm is responsible for meeting it. However, if the standard is applied uniformly across affected firms, it is almost certain that it will not be cost effective. The exception is the case in which all of the affected firms have identical marginal control costs. In this special case, a uniformly applied standard would result in marginal costs that are equal across the affected firms and the outcome would be cost effective. The upshot is that, in most cases, for direct regulation to be cost effective it is necessary to require the affected firms to control pollution by varying amounts. To be specific, policymakers would have to ascertain the marginal costs of pollution control of each of the affected firms and then impose a standard on each firm that achieves the cost-effective outcome. In many, if not virtually all, cases this approach would require a substantial amount of data that are in all likelihood not available to regulators. It is also reasonable to expect that this would involve the expenditure of significant resources by the regulatory authority and be time-consuming as well.

Two additional points regarding the CAC approach are worth noting. The first concerns the distinction between design criteria and performance criteria. Design criteria specify technologies, inputs, chemical processes, and so forth that must be employed by affected firms. In contrast, performance criteria specify a necessary end result that can be achieved by any means, presently available or yet to be developed, available to the firm. Performance standards possess significant advantages relative to design criteria, especially when dealing with complex production processes. In particular, performance standards allow, and encourage, firms to innovate in the effort to achieve a particular standard at least cost.

A second point is that, relative to the incentives-based approaches discussed below, attempting to achieve a standard via direct regulation is likely to result in a greater amount of pollution control than attempts to achieve the same standard using an incentives-based mechanisms. Thus, *ceteris paribus*, the benefits associated with the use of CAC are likely to be larger as well. This fact must be considered when comparing the cost-effectiveness of alternative policy strategies.

Charges. Economists have identified alternatives to direct regulation that rely on market-based incentives (ie, prices) to control pollution. One approach forces firms to account for the external costs they generate by imposing a per unit charge (tax) on the firm equal to the external cost of the firm's production activities (1); more recent discussions of the use of charges to control pollution are available (7,8,71). In particular, a charge could be levied on each unit of pollution (emissions) produced by the firm. Such a charge affects the firm's decision-making process by altering the relative prices of the inputs to the production process. Referring to the previous example, by imposing a charge equal to T_1 (the level at which the firms' marginal control costs are equal) on each of the affected firms, pollution would be reduced by 16 units. To ensure that the standard is met in a cost-effective manner, the charge levied on each unit of pollution generated by a particular source should be a function of whether the pollutant is uniformly or nonuniformly mixed. A discussion of the difference between uniformly and nonuniformly mixed pollutants and how this issue affects the determination of the appropriate charge is available (72).

Firm A would reduce its pollution by 5 units because the marginal cost of control is less than the tax on the first 5 units of waste. However, paying the tax is cheaper than reducing waste beyond 5 units. Using the same logic, firm B would choose to reduce the amount of pollution it generates by 11 units. Thus the charge leads firms to undertake the cost-effective allocation of pollution control.

In order for the use of charges to result in the desired level of pollution control, the control authority must identify the appropriate level of the charge. To accomplish this, policymakers must have information on the control costs incurred by the affected firms to be able to predict the amount of pollution reduction a charge of a given magnitude is likely to induce. A charge that is set too high will result in overcontrol of pollution relative to the standard. If the charge is set too low, the opposite outcome will occur. An iterative process that entails adjusting the charge until the standard is met could be used to overcome this problem. However, this approach is politically unattractive and may have adverse incentives effects on firms who must periodically modify their pollution control strategies in response to changes in charge rates.

Permits. A second incentives-based approach to pollution control policy entails the creation of a market for permits to pollute. Description of this approach are available (3,7,8,73). A system of transferable permits entails the creation of a market in which a predetermined number of permits to emit a specific pollutant within a well-defined geographic region can be bought and sold. Like charges, the use of transferable permits results in a cost-effective allocation of pollution control across affected firms. Referring once again to the example illustrated in Figure 5, the regulatory authority would issue a total of 34 permits, assuming each permit represents one unit of pollution. Thus the two firms must reduce their emissions by a combined total of 16 units for the standard to be met. Regardless of the initial allocation of the permits, firms would find it in their interest to trade permits until the per unit price of the permits was equal to the marginal cost of control for both firms. The firms would in turn engage in the same amount of pollution reduction that they did under the charge scheme. As in the case of charges, to ensure that the standard is met in a cost-effective manner, the amount of pollution that each permit allows the holder to emit should be a function of whether the pollutant is uniformly or nonuniformly mixed (74).

Comparison of Alternative Instruments. The use of permits offers a number of attractive features relative to both direct regulation and charges. Compared with direct regulation, permits allow firms to achieve the target level of control in a cost-effective manner. In addition, permits (as well as charges) have the added advantage of offering firms a greater degree of flexibility with respect to how they will comply with the standard. Compared with charges, permits offer a greater degree of certainty with respect the amount of pollution control that will occur. In addition, in the case in which an industry is growing, permits ensure that the amount of pollution that is generated will remain fixed at the target level. In the case of charges, total pollution generated is likely to increase with an increase in the number of firms in the affected market. A potential drawback of permits is that there may be less incentive to reduce control costs than there would be under a charge system. Additional discussion of these issues can be found (4,75,76).

The potential cost savings from the use of incentives-based approaches has been the subject of numerous studies (73). The results of many early studies suggested that the use of charges or permits to achieve a specific standard could result in substantial cost savings relative to the use of direct regulation. For example, studies of various air pollution control policies have estimated that the use of direct regulation could result in total costs as much as 22 times greater than the costs that would be incurred under a system of permits (77). However, more recent work suggests that in many instances the predicted cost savings attributable to incentives-based systems, and in particular permits, may greatly exceed actual savings. As has been described (78), this outcome is the result of differences between the manner in which permit trading is assumed to occur (multilateral and simultaneously) and the actual trading process (which is more likely to be bilateral and sequential). Nonetheless, market-based incentives continue to be preferred to direct regulation in most settings on efficiency grounds.

CURRENT POLICY

The majority of pollution control laws that have been passed over time are directed at specific environmental media. The Clean Air Act and Water Pollution Control Act (and their respective amendments) govern emissions to the air and water ways. The Resource Conservation and Recovery Act, as amended, governs the land disposal of wastes and, in particular, hazardous wastes. Other laws, such as the Toxic Substances Control Act; the Comprehensive Environmental Response, Compensation, and Liability Act; the Federal Insecticide, Fungicide, and Rodenticide Act; the Safe Water Drinking Act; and their respective amendments are directed at specific sources of pollution. A complete discussion of the various aspects of this body of policy is obviously well beyond the scope of this article. The discussion that follows is intended only to highlight some of the major features of these laws and provide references to more complete discussions of environmental policy in the United States.

Air Pollution Control

At the federal level, air pollution control policy (8,10) dates back to the Air Pollution Control Act passed in 1955. Since that time, the federal approach to air pollution control responsibility has evolved considerably. Before 1970, policy relied heavily on the individual states to undertake steps to reduce the amount of air emissions. The role of the federal government was restricted primarily to funding research on the effects of air pollution and air pollution control strategies, and assistance to the individual states in support of their air pollution control efforts. However, once it became apparent that economic incentives and other factors were preventing states from assuming an active role, the federal government took the lead in developing policies designed to reduce the amount of air pollution in the United States.

The passage of the Clean Air Act amendments of 1970 marked a significant change in air pollution control policy, with the federal government assuming much more responsibility than it had previously. Three features of the 1970 amendments are particularly noteworthy. First, the amendments called on the administrator of the EPA to establish the National Ambient Air Quality Standards for a number of specific air pollutants. These standards were to be set so as to provide "an adequate margin of safety" without regard to the costs of achieving the standard. Second, uniform emissions standards were established for mobile sources of air emission, ie, cars. Third, in the case of new stationary sources of air emissions (eg, factories and electric utilities), standards were to be based on the best available technology that was deemed to be affordable. Thus Congress did allow for some consideration of costs in the setting of new source performance standard (NSPS). However, referring to the previous discussion of the conditions necessary for efficiency or cost-effectiveness, it is clear the requirements just described fail on both counts.

In spite of the limitations posed by the language in the Clean Air Act, the EPA has modified its air pollution control program over time in a number of innovative ways. Three innovations, which are parts of the EPA's emissions trading program and reflect a move toward greater use of economic incentives include the offset program, the bubble policy, and emission banking.

The offset program allows new air pollution sources in

a particular geographic region to pay existing sources to reduce their emissions below the level required by the standard in lieu of installing control technology at the new plant so long as the total level of the pollutant does not increase (and, in fact decreases). Thus new sources are able to exploit the least-cost approach to pollution reduction in the geographic region so long as the air quality in the region is better after the new plant is established. The bubble policy allows a generator (or group of generators) to treat a number of closely situated sources, usually within a given production facility, as if they were encased in a giant bubble. The standard then applies to emissions coming out of the bubble. Firms are able to reduce emissions from those sources within the production facility that are the cheapest to control and to relax controls on the more costly sources, thus reducing total control costs. The emission banking program gives firms credits for reducing emissions below the emissions level allowed by the standard. These credits can then be saved for use at a later time or sold to another firm. This approach gives existing firms an incentive to adopt new, cost-saving pollution-control technologies.

The Clean Air Act Amendments of 1990. The Clean Air Act Amendments (CAAA) of 1990 contain five main sections that deal with the following topics: motor vehicles and fuels, urban air quality, air toxics, acid deposition, and the problems resulting from stratospheric ozone. In many respects, the CAAA are similar to previous legislation. For example, the use of CAC is still the principle policy instrument, and control of criteria pollutants is a major issue (as was the case with legislation passed in both 1970 and 1977). However, Title IV, which is directed at the control of SO_2 and NO_x emissions from coal-burning electric utilities, provides for the use of allowances (permits) to control SO_2 emissions. In addition, the CAAA identify 189 specific toxic air emissions for which the EPA is to set emissions standards. Discussions of the major features of the CAAA of 1990 and their possible impacts can be found (79,80).

Title IV: Allowances for SO_2 Emissions. The objective of Title IV of the CAAA is to reduce SO_2 emissions to roughly 50% of their current level, or 8.9 million tons per year. This objective is to be achieved, in large part, by restricting the total amount of SO_2 emissions from electric utilities. The requirements of Title IV will be implemented in two phases. Phase I, which runs from January 1, 1995, to December 31, 1999, will affect 110 utility plants. Phase II, which begins on January 1, 2000, will affect all of the remaining fossil-fuel burning plants. In each phase allowances, or permits, for SO_2 emissions will be allocated to each of the affected firms. Each allowance represents the right to emit one ton of SO_2 in a specified calendar year. Allowances can be used to offset current emissions, sold to other entities, or saved for use at a later date.

According to a recent report (80), many of the utilities affected during Phase I plan to comply with the new restrictions on SO_2 by using a number of approaches. The most common response appears to be the use of coal blending and fuel switching. Other approaches include the installation of flue gas desulfurization, reduced utilization, altered dispatch ordering, and purchase of allowances. Somewhat surprisingly, it appears that the use of allowances purchased from other plants will be relatively limited. Possible explanations from this include uncertainty with respect to the regulatory treatment of allowances and future regulatory actions regarding hazardous air pollutants and CO_2 emissions.

Water Pollution Control

Water pollution control policy at the federal level dates back to the Water Pollution Control Act of 1948. A concise summary of the history of water pollution control policy has been presented (81). Additional discussions of the economic aspects of water pollution can be found (4,8,11). Water pollution is attributable to a variety of sources. Rivers and lakes have been used as natural conduits for many types of commercial and industrial wastes. Agricultural practices have contributed to water pollution through runoff containing pesticides, herbicides, and fertilizers as well as groundwater contamination as water leaches down through the soil. Leachate contamination of groundwater has also been linked to various waste management practices, eg, landfills, and deep well injection. Sewage treatment plants can also contribute to water pollution depending on the level of biological oxygen demand present in the wastes that are ultimately released from the plant.

With respect to energy-related issues, water pollution problems and impacts on the quality of water resources are most obvious in the cases of acid deposition, nuclear power generation, and hydro power. One of the most common illustrations of the potential effects of acid deposition is the status of many of the lakes in the Adirondacks. Acid rain has been credited with rendering many of the lakes in this region and elsewhere unable to support aquatic life. While acid deposition has been traced to SO_2 emissions and is therefore addressed via air pollution policy, it nonetheless has significant adverse effects on water quality. The cooling requirements of nuclear-power and coal-fired plants can also have water-related impacts. For example, as a result of a significant difference in temperature, the discharge water from plants built next to a lake or river may affect the ecosystem of the receiving water body. Such concerns have had significant impacts on the siting and licensing process for power plants. In the case of hydropower generation, the construction of a dam and the resulting reservoir has various irreversible effects on the surrounding environment. While some of these impacts are viewed as costs (eg, loss of scenic vistas, range land, or other uses of the flooded area) many benefits also arise, such as the increased opportunities for water-based recreation.

Land Pollution

Compared with air and water pollution, land-based pollution received relatively little attention until the passage of the Resource Conservation and Recovery Act (RCRA) in 1976. The primary thrust of RCRA is to control the generation and disposal of hazardous wastes from "cradle to grave." In the time since RCRA was passed (and amended in 1984) the EPA and the states have developed a large body of regulations designed to reduce the risks posed by hazardous wastes.

Energy production facilities such as coal-fired and nuclear-powered electric plants pose special problems with respect to land-based waste management. In 1993, the

EPA published its final regulatory determination for certain coal combustion wastes (eg, fly ash, bottom ash, slag, and FGD wastes) exempting the wastes from RCRA Subtitle C hazardous waste regulations. However, EPA retained the option of considering these wastes in its ongoing assessment of industrial nonhazardous wastes under Subtitle D. Consequently, these wastes could be subject to requirements that are more stringent than existing regulations under Subtitle D as a result of future actions by the agency. In addition, the determination did not address low volume, potentially hazardous "clean coal technology" wastes. Additional information must be acquired before a determination can be made regarding these wastes. One possible outcome is that these wastes could be classified as hazardous in the future.

High level radioactive wastes, which are the by-product of nuclear-powered generation of electricity, are governed by the Nuclear Waste Policy Act (NWPA) of 1982. Although the NWPA requires the U.S. Department of Energy to develop an underground depository for high level wastes, the United States is still without a single repository for such wastes. Attempts to site such a facility, most recently in Nevada near Yucca Mountain, have met considerable opposition from the public and many local government officials. Low level radioactive wastes are also posing serious problems. At present, there are only three repositories for such wastes in the United States and these facilities are filling rapidly. Moreover, the prospects for developing additional sites look bleak (82).

Impacts of Pollution Control Policy

Viewed from an economic perspective, the primary interest is in the relative costs and benefits that have been realized over time. According to a study by the EPA (83) annualized environmental compliance costs in 1990 (measured in 1986 dollars) were approximately \$28 billion for air pollution, \$42.4 billion for water pollution, and \$26.5 billion for land pollution. Other environmental statutes added another \$1.6 billion to the total. However, these figures only reflect out-of-pocket expenditures and as such are likely to understate the true economic costs of pollution control. While estimation of the benefits that have accrued over time is even more difficult, the evidence that has been collected suggests that total air pollution control benefits exceeded total costs through 1981 (84). However, water pollution is another matter; available estimates suggest that total costs have exceeded total benefits (85). In both instances, there is considerable room for improvement vis-à-vis cost effectiveness.

CASE STUDY: THE USE OF ADDERS IN LEAST-COST UTILITY PLANNING

Over the past decade the concepts of least-cost utility planning and integrated resource planning (IRP) have become important components in the regulation of electric utilities. The objective of IRP is to identify the mix of energy sources and demand-side management (DSM) techniques that satisfies an expected level of demand and reliability requirements at a lower cost than any other possible combination of energy sources and DSM. As a rule, in those jurisdictions where IRP is used, each utility is required to formulate an integrated resource plan for meeting its obligation to serve.

As was discussed earlier, efficiency requires that the price of a good (in this case, electrical service) reflect all of the marginal costs, both internal and external, associated with the production of that good. Thus electricity produced by a coal-fired unit should be priced to include both the internal costs, eg, fuel and labor expenses, and the external costs resulting from, among other things, the SO, NO_x, CO, CO_2, and particulate emissions from the unit. Assuming that output is priced on the basis of marginal costs, failure to include any external costs will result in a price for coal-fired electricity that is too low. To be specific, price will be less than full marginal costs of production. Moreover, the relative price of electricity produced by a coal-fired unit (ie, its price relative to the price of electricity produced by other types of generating units) may be understated, depending on how externalities attributable to other sources are treated. The result is that this source of electricity could make up too large a proportion of the utility's energy portfolio, ceteris paribus.

Numerous states have attempted to address the problem of externalities resulting from the production of electricity by requiring utilities to explicitly consider external costs in the process of planning capacity expansions. As of 1991, 19 states had implemented such requirements (86). Such costs have come to be referred to as "externality adders." In some cases, estimates of the external costs of different types of emissions, measured, eg, in dollars per kilowatt hour, are added to production costs to arrive at a total cost figure. In other cases, the cost of additional pollution control equipment is added to production costs to account for external costs. However, such efforts have been criticized by different agencies, including the Department of Energy (DOE) and the Federal Energy Regulatory Commission (FERC) (87), and numerous analysts (88,89). Three issues are addressed briefly here: (*1*) the validity of existing estimates of external costs of pollution and the methods used to estimate such costs, (*2*) the potential distortionary effects of adders, and (*3*) the implications of existing pollution-control policies for determining the correct value of adders.

Validity of Available Estimates. Questions regarding the validity of available estimates are certainly justified. Reviews of the literature on externalities associated with electricity production have pointed out that many of the existing estimates of external costs are relatively crude approximations that, in many cases, have been derived via flawed research methods (86,90,91). However, the fact that there are questions about how well existing estimates approximate actual external costs is not, in and of itself, sufficient support for the argument that until better estimates are forthcoming, public utility commissions (PUCs) should simply ignore external costs in the planning process. Rather, it emphasizes the need for the use of extreme care in how such estimates are used.

The question of whether existing estimates are a good approximation of actual costs notwithstanding, the different methods that have been used to generate estimates of external costs merit further consideration. In practice, three approaches have been used: mitigation or averting costs, control costs, and the damage function approach (86,90). Of these, mitigation costs and the damage function approach have the most to recommend them from a theoretical perspective. In addition, the damage function approach is the only one that attempts to relate costs directly to the damages attributable to the pollutant in question.

In some cases, the costs that would be incurred to mitigate or avert the potential damages attributable to a pollutant have been used to measure damage costs. For example, potential damages to agricultural crops might be averted by increased use of fertilizer, water, or some other input (4). The value of the additional resources required to offset the potential damage constitutes the damage costs attributable to the pollutant, ie, the additional production costs constitute the damages incurred by farmers. However, in many cases, opportunities to offset fully the adverse effects of a pollutant are unavailable, and hence this approach cannot be used.

As a second approach, some analysts have suggested that pollution control costs already incurred provide insights to the marginal damage costs of certain pollutants (92). This assertion is based on the assumption that because policymakers have decided to require the existing level of control, it is reasonable to assume that the perceived marginal benefits of the existing level of control (ie, marginal damages avoided) are equal to or exceed the marginal costs of control, ie, policymakers are behaving in an efficiency enhancing manner. However, this argument has no sound basis on which to stand. In many, if not most, instances, pollution standards have been based on considerations other than economic efficiency. For example, according to the Clean Air Act, the standards for criteria pollutants are to be set so as to ensure the health and safety of the affected population. The act specifically states that control costs are not to be considered in setting these standards. A number of other environmental statutes also explicitly exclude the consideration of costs or benefits in the decision-making process. Thus there is no reason to believe that marginal control costs are in any way related to marginal damage costs.

The theoretically valid approach to estimation of the external costs of pollution is to rely on the use of damage functions and measures of the corresponding willingness to pay for a reduction in damage. Admittedly, this approach is anthropocentric, ie, assumes that all values should be based on the relationship between a specific damage and its effects on the welfare of humans. As such, pollution is only considered a problem to the extent that any damages from the pollution have economic value. Under this approach, the amount of damage and resulting value of that damage attributable to a particular pollutant are determined by the interaction of a number of factors.

As a practical matter, the damage function approach entails several complex steps, the first of which is to identify the specific types of damages that could be attributed to the pollutant in question. The second step is to estimate the relationship between the quantity of the pollutant and the resulting level of damages. In the third step, the relationship estimated in step two is used to estimate the level of damages associated with a given level of the pollutant. The fourth step entails assigning a monetary value to the damages calculated in step three. Each of these steps is likely to be data intensive. In addition, there is likely to be considerable uncertainty regarding the functional form of the damage function and the estimates of the monetary value of damages.

Data limitations (with respect to both quantity and quality) have the potential to affect each of the steps outlined above. For example, to be able to relate a pollutant to various damages, all of the potential damages that could be linked to the pollutant must be identified. However, in many cases, latency is a problem, ie, many adverse effects do not manifest themselves for many years. As such, there may be a considerable time lag between exposure to a pollutant and the occurrence of adverse effects. Limited data will also affect estimation of the damage function and, therefore, estimates of the amount of damage attributable to a given quantity, eg, concentration, of a pollutant. Finally, data limitations with respect to willingness to pay to avoid damages constrain the estimation of the monetary value of the damages.

Data limitations and other uncertainties notwithstanding, the damage function approach focuses directly on the link between pollutants and damages, measured in both physical and monetary terms. Thus, of the three approaches that have been discussed here, it is the most defensible from a theoretical perspective. In addition, it is the only method that is applicable, at least in theory, to all of the types of damages that might be linked to electricity production. It is for this reason that agencies such as the DOE and FERC have decided to rely on the damage function approach to estimate the external costs of electricity production (87).

Potential Distortionary Effects. A second issue concerns the distortionary effects that the use of externality adders might introduce in the IRP process (88). To be specific, external costs currently are considered only in the case of new capacity brought on line by independently owned utilities (IOUs), ie, utilities whose rates are governed by PUCs. Although the production activities of other sources of electrical power, such as cogeneration facilities and utilities located outside a particular PUC's jurisdiction, may also result in external costs, these sources are not subject to the same requirements. In the case of utilities located outside of the PUC's jurisdiction, the exception would be those instances in which the state where the "outside" utility is located also requires consideration of external costs. However, external costs per unit of output would have to be measured using the same method to ensure that no distortion in relative costs occurs.

Forcing IOUs to include external costs when determining the marginal cost of power could put them at a competitive disadvantage relative to alternative suppliers solely because the alternative suppliers do not have to factor external costs into their marginal cost calculations. The consequent distortion of relative costs could result in alternative suppliers producing a larger share of the total amount of electricity consumed in the region. If these alternative suppliers also produce greater amounts of pollution per unit of output than the regulated IOU, the result could be an increase in the total amount of air pollution in the region. Obviously, this is not the intended result of the use of externality adders.

Impact of the Current Regulatory Structure. The fact that electric utilities have already undertaken efforts to internalize at least some of the external costs resulting from the generation of electricity also has important implications for the socially efficient quantities of electricity and corresponding pollution. Over the past two decades electric utilities have undertaken a number of steps, including the installation of flue gas desulfurization (FGD) units, coal cleaning, and the adoption of newer "clean coal" or "green field" technologies that result in reduced emissions of SO_2, NO_x, particulates, and other air-borne pollutants. The effect of these efforts has been to internalize at least some amount of what were previously external costs. In the context of Figure 2, such efforts have shifted the MPC curve to the left, toward the MSC curve.

Pollution control efforts also have affected the marginal external costs attributable to electricity production. Referring again to Figure 2, the value of marginal external costs is based on preregulatory conditions; internal costs and a specific relationship between the firm's production function and the amount of pollution that is produced per unit of output in the absence of pollution control policies. The amount of pollution produced per unit of output and the value of the damages attributable to each unit of pollution combine to determine the value of external costs at each level of output. Depending on the actions taken by a utility in response to legislative and regulatory mandates directed at pollution reduction, marginal external costs per unit output are likely to have decreased (93). This would be the case if, eg, the types of pollutants produced are unchanged, but there is a reduction in the quantity of one or more of the pollutants produced per unit of output.

The preceding discussion suggests that marginal external costs and the amount of inefficiency attributable to a failure to account for such costs in the decision-making process in the electricity sector have both decreased as a result of regulations implemented to date. However, without knowing the actual value of the marginal external costs associated with electricity production and the marginal costs of pollution control it is not possible to determine whether the resulting level of pollution is efficient. Three outcomes are possible; the efficient level of pollution, undercontrol relative to the efficient level, and overcontrol relative to the efficient level. As has been demonstrated (89), and summarized below, the correct value of any adder that should be included in the calculation of the marginal social costs of electricity will depend on the type of policy instrument that has been used.

In the first case, assuming pollution control responsibility has been allocated across firms in a cost-effective manner (marginal control costs are equal across the affected firms) and the amount of control has been set at the efficient level, firms are producing the socially efficient output level. However, depending on the policy that was used to achieve this outcome, any remaining external costs may or may not be internalized, ie, they may or may not be included in the firm's decision-making process. If pollution control has been achieved via direct regulation, any remaining external costs should be added to private costs to ensure that total social costs are accounted for. However, if pollution control has been achieved via either charges or permits, the correct adder is zero. In the case of charges, the efficient output level will only occur if the charge is set equal to marginal external cost at the efficient level of output. Thus, if as has been assumed, the efficient solution has been achieved, it follows that the charge is an accurate measure of marginal damage costs. Because the charge is paid for each unit of pollution emitted by the firm, external costs have been internalized. In the case of permits, so long as the rules governing the trading of permits reflect the characteristics of the pollutant being regulated, all external costs have been internalized as well; determination of the appropriate trading rules depends primarily on the characteristics of the pollutant in question (8). Incorporating adders in either of these latter two cases would result in double counting of external costs.

A second possibility is that the current level of pollution control is less than the efficient level. In this case, if current policy consists of direct regulation, the external cost attributable to any remaining pollution would once again constitute the correct value of the adder. In the case in which charges are being used to control pollution, the adder should be set equal to the difference between the current charge and actual external costs. Under a system of permits, and assuming the correct rules for trading are in place, the correct value for the adder is once again zero. In this latter case, the appropriate response to the undercontrol of pollution would be to decrease the number of permits to a level consistent with the economically efficient level of pollution control.

The third possibility is that there is excess pollution control. In this case, and assuming policy consists of direct regulation, the external cost attributable to any remaining pollution would once again constitute the correct value for the adder. While one might be tempted to argue for a negative adder to compensate for the overcontrol attributable to excessive restrictions on pollution, this would only introduce further distortions into the market. In the case in which charges are being used to control pollution, the adder should be set equal to the difference between the current charge and actual external costs; in the case of overcontrol this amount will be negative. Under a system of permits, and assuming the correct rules for trading, the correct value for the adder is once again zero. In this latter case, the appropriate response to the overcontrol of pollution would be to increase the number of permits to a level consistent with the economically efficient level of pollution control.

Presumably, the purpose of adders is to improve the overall efficiency of the electricity sector by accounting for external costs in the decision-making process. However, as the preceding discussion illustrates there are a number of important issues that must be addressed for adders to achieve that purpose. Failure to do so could result in even greater inefficiencies.

BIBLIOGRAPHY

1. A. C. Pigou, *The Economics of Welfare*, 4th ed., Macmillan, London, 1932.
2. A. V. Kneese and B. T. Bower, *Managing Water Quality: Economics, Technology, Institutions*, Johns Hopkins University Press for Resources for the Future, Inc., Baltimore, Md., 1968.

3. J. H. Dales, *Pollution, Property, and Prices*, University of Toronto Press, Toronto, Ont., 1968.
4. M. L. Cropper and W. E. Oates, *J. Econ. Lit.*, **30**(2), 675–740 (1992).
5. K. G. Maler, *Environmental Economics: A Theoretical Inquiry*, Johns Hopkins University Press for Resources for the Future, Inc., Baltimore, Md., 1974.
6. W. J. Baumol and W. E. Oates, *Economics, Environmental Policy, and Quality of Life*, Prentice-Hall, Inc., New York, 1979.
7. W. J. Baumol and W. E. Oates, *The Theory of Environmental Policy*, 2nd ed., Cambridge University Press, Cambridge, UK, 1988.
8. T. Tietenberg, *Environmental and Natural Resource Economics*, 3rd ed., HarperCollins, New York, 1992.
9. A. M. Freeman, *Measurement of Environmental and Resource Values: Theories and Methods*, Resources for the Future, Washington, D.C., 1993.
10. P. R. Portney, ed., *Public Policies for Environmental Protection*, Resources for the Future, Washington, D.C., 1990.
11. B. C. Field, *Environmental Economics: An Introduction*, McGraw-Hill Book Co., Inc., New York, 1994.
12. A. E. Dillingham, N. T. Skaggs, and J. L. Carlson, *Economics: Individual Choice and its Consequences*, Allyn & Bacon, Needham Heights, Mass., 1992.
13. P. A. Samuelson and W. D. Nordhaus, *Economics*, 14th ed., McGraw-Hill Book Co., Inc., New York, 1992.
14. W. Nicholson, *Macroeconomic Theory: Basic Principles and Extensions*, 2nd. ed., Dryden Press, Hinsdale, Ill., 1978.
15. E. Silberberg, *The Structure of Economics: A Mathematical Analysis*, McGraw-Hill Book Co., Inc., New York, 1978.
16. P.-O. Johansson, *An Introduction to Modern Welfare Economics*, Cambridge University Press, New York, 1991.
17. R. Boadway and N. Bruce, *Welfare Economics*, Basil Blackwell, Inc., Cambridge, Mass., 1984.
18. R. E. Just, D. L. Hueth, and A. Schmitz, *Applied Welfare Economics and Public Policy*, Prentice-Hall, Inc., Englewood Cliffs, N.J., 1982.
19. Ref. 18, p. 269.
20. Ref. 18, chapt. 3.
21. Ref. 7, chapt. 4.
22. Ref. 11, p. 28.
23. R. H. Coase, *J. Law Econ.* **3,** 1–44 (1960).
24. *Environmental Quality 1984*, Council on Environmental Quality, Washington, D.C., 1985.
25. Ref. 8, p. 95.
26. U.S. Bureau of Reclamation, *Operation of Glen Canyon Dam Colorado River Storage Project, Arizona*, Draft Environmental Impact Statement, Salt Lake City, 1993.
27. J. L. Carlson and co-workers, *An Economic Study on Proposed IPCB Regulation R86-9: Hazardous Waste Prohibitions*, ILENR/RE-EA-90/10, Illinois Department of Energy and Natural Resources, Springfield, 1990.
28. T. G. Walton and A. C. Basala, *Cost-Effectiveness Analysis and Environmental Quality Management*, paper presented at the 74th meeting of the Air Pollution Control Association, Philadelphia, 1981.
29. E. J. Mishan, *Cost-Benefit Analysis*, 2nd. ed., Praeger, New York, 1976.
30. R. Sugden and A. Williams, *The Principles of Practical Cost-Benefit Analysis*, Oxford University Press, Oxford, UK, 1986.
31. E. M. Gramlich, *Benefit-Cost Analysis of Government Programs*, Prentice-Hall Inc., Englewood Cliffs, N.J., 1981.
32. Ref. 8, chapt. 4.
33. Ref. 8, p. 365.
34. C. Starr, *Science* **165,** 1232–1238 (1969).
35. D. Kahneman and A. Tversky, *Econometrica* **47**(2), 263–291 (1979).
36. M. C. Weinstein and R. J. Quinn, *Natural Resource J.* **23,** 659–673 (1983).
37. Ref. 4, pp. 721–722.
38. Ref. 8, pp. 84–86.
39. J. B. Braden and co-workers, *Pollution Control Cost Analysis*, U.S. Environmental Protection Agency, Washington, D.C., 1985.
40. Z. Griliches, ed., *Price Indexes and Quality Change*, Harvard University Press, Cambridge, Mass., 1971.
41. S. Rosen, *J. Political Econ.* **82,** 34–55 (1974).
42. P.-O. Johansson, *The Economic Theory and Measurement of Environmental Benefits*, Cambridge University Press, Cambridge, UK, 1987.
43. D. S. Brookshire and co-workers in V. K. Smith, ed., *Advances in Applied Microeconomics*, Vol. 1, JAI Press, Greenwich, Conn., 1981.
44. R. Arnould and L. Nichols, *J. Political Econ.* **91**(2), 332–340 (1983).
45. W. K. Viscusi, *Rev. Econ. Statistics* **60**(3), 408–416 (1978).
46. Ref. 42, p. 111.
47. M. Clawson and J. L. Knetsch, *Economics of Outdoor Recreation*, Johns Hopkins University Press for Resources for the Future, Baltimore, Md., 1966.
48. W. J. Vaughn and C. S. Russell, *Land Econ.* **58,** 450–463 (1982).
49. N. E. Bockstael, K. E. McConnell, and I. E. Strand in J. B. Braden and C. D. Kolstad, eds., *Measuring the Demand for Environmental Quality*, North-Holland, Amsterdam, 1991.
50. V. K. Smith and Y. Kaoru, *Am. J. Agri. Econ.* **72,** 419–433 (1990).
51. R. C. Mitchell and R. T. Carson, *Using Surveys to Value Public Goods: The Contingent Valuation Method*, Resources for the Future, Washington, D.C., 1989.
52. R. G. Cummings, D. S. Brookshire, and W. D. Schultze, *Valuing Public Goods: The Contingent Valuation Method*, Rowman & Allanheld, Totwa, N.J., 1986.
53. Ref. 51, chapts. 5 and 11.
54. D. Kahneman and J. L. Knetsch, *J. Environ. Econ. Manage.* **22**(1), 57–70 (1992).
55. V. K. Smith, *J. Environ. Econ. Manage.* **22**(1), 71–89 (1992).
56. R. Gregory, S. Lichtenstein, and P. Slovic, *J. Risk Uncertainty* **7,** 177–197 (1993).
57. B. A. Weisbrod, *Q. J. Econ.* **78,** 471–477 (1964).
58. V. K. Smith, *South. Econ. J.* **54,** 19–26 (1987).
59. D. S. Brookshire, L. S. Eubanks, and C. S. Sorg, *Water Resources Res.* **22,** 1509–1518 (1986).
60. D. H. Rosenthal, D. H. and R. H. Nelson, *J. Policy Anal. Manage.* **11**(1), 116–122 (1992).
61. J. Quiggin, *J. Policy Anal. Manage.* **12**(1), 195–199 (1993).
62. A. Randall in Ref. 49.
63. R. C. Bishop and M. P. Welsh, *Land Econ.* **68**(4), 405–417 (1992).
64. R. Kopp, *J. Policy Anal. Manage.* **11**(1), 123–130 (1992).
65. A. Fisher, and R. Raucher in V. K. Smith, ed., *Advances in Applied Micro-Economics*, Vol. 3, JAI Press, Greewich, Conn., 1984, pp. 37–66.
66. Ref. 65, p. 60.

67. R. C. Lind, *J. Environ. Econ. Manage.* **18**(2), S-29–S-50 (1990).
68. W. R. Cline, *The Economics of Global Warming*, Institute for International Economics, Washington, D.C., 1992.
69. Ref. 11, p. 122.
70. A. Breton, *The Economic Theory of Representative Government*, Aldine, Chicago, 1974.
71. F. R. Anderson and co-workers, *Environmental Improvement Through Economic Incentives*, Johns Hopkins University Press, Baltimore, Md., 1977.
72. Ref. 8, pp. 372–380.
73. T. H. Tietenberg, *Emissions Trading: An Exercise in Reforming Pollution Policy*, Resources for the Future, Washington, D.C., 1985.
74. Ref. 8, pp. 375–382.
75. M. L. Weitzman, *Rev. Econ. Studies* **41**(4), 477–491 (1974).
76. P. Bohm and C. F. Russell, in A. V. Kneese and J. L. Sweeney, eds., *Handbook of Natural Resource and Energy and Economics*, Vol. 1, North-Holland, Amsterdam, 1985.
77. Ref. 10, pp. 70–74.
78. S. Atkinson and T. H. Tietenberg, *J. Environ. Econ. Manage.* **21**(1), 17–31 (1991).
79. Ref. 11, pp. 295–298.
80. Bailey and co-workers, *Examination of Utility Phase I Compliance Choices and State Reactions to Title IV of the Clean Air Act Amendments of 1990*, ANL/DIS/TM-2, Argonne National Laboratory, Argonne, Ill., 1993.
81. Ref. 10, chapt. 4.
82. Ref. 11, pp. 342–343.
83. Ref. 4, p. 712.
84. Ref. 10, p. 69.
85. Ref. 10, pp. 122–127.
86. R. L. Ottinger and co-workers, *Environmental Costs of Electricity*, Oceana Publications, New York, 1991.
87. Federal Energy Regulatory Commission Staff, *Report on Section 808, Renewable Energy and Energy Conservation Incentives of the Clean Air Act Amendments of 1990*, FERC, Washington, D.C., 1992.
88. P. L. Joskow, *Elec. J.* **5,** 53–67 (1992).
89. M. A. Freeman and co-workers, *Elec. J.* **5,** 18–25 (1992).
90. C. B. Szpunar and J. L. Gillette, *Environmental Externalities: Applying the Concept to Asian Coal-Based Power Generation*, ANL/EAIS/TM-90, Argonne National Laboratory, Argonne, Ill, 1992.
91. D. Pearce, C. Bann, and S. Georgiou, *The Social Cost of Fuel Cycles*, UK Department of Trade and Industry by the Centre for Social and Economic Research on the Global Environment, London, 1992.
92. P. Chernick and E. Caverhill, *Elec. J.* **4**(2), 46–53 (1991).
93. Ref. 18, pp. 274–275.

ENVIRONMENTAL CONSERVATION ORGANIZATIONS

JIMMY LANGMAN and HEATHER AUYANG
Earth Island Institute
San Francisco, California

Many consider environmental issues to be the most pressing problems confronting humanity in the 1990's. As public concern for the environment increases, so does the need for a comprehensive guide to organizations addressing ecological challenges.

Activists, research, scientific, and legal groups, all devoted to addressing and providing innovative solutions to various facets of the global environmental crisis are listed. Issue areas cover a wide range: from energy to marine protection to land conservation and restoration to population growth.

Fax numbers for many organizations, and in some cases computer mailbox addresses have been included. The Institute for Global Communications, 18 De Boom St., San Francisco, CA, 94107, (415) 442-0220; econet e-mail address igcoffice@igc.apc.org offers more information on computer networking.

Abalone Alliance
2940 16th Street, Room 310
San Francisco, CA 94103
P(415) 861-0592
E-Mail(EcoNet): abalone
Through raising public awareness about the dangers of nuclear power, the ultimate goal of the Alliance is a nuclear-free world.

Acid Rain Foundation, Inc.
1410 Varsity Dr.
Raleigh, NC 27606
P(919) 828-9443
Created to foster a greater understanding of air quality issues, including acid rain, air pollutants, and global climate change, and to help bring about solutions.

African Wildlife Foundation
1717 Massachusetts Avenue, NW
Washington, DC 20036
P(202) 265-8393
F(202) 265-2361
Finances and operates wildlife conservation projects in Africa.

Alaska Conservation Foundation
430 West 7th Ave., Suite 215
Anchorage, AK 99501
P(907) 276-1917
F(907) 274-4145
Provides supplemental financial support to Alaskan groups emphasizing activist projects with an immediate environmental focus.

Alaska Environmental Lobby, Inc.
P.O. Box 22151
Juneau, AK 99802
P(907) 463-3366
A parent organization for 19 Alaskan environmental groups which lobby the Alaskan legislature for better environmental laws.

Alliance for the Wild Rockies
Box 8731
Missoula, MT 59807
Dedicated to preserving wilderness in the Wild Rockies Bioregion; now has 39 member organizations which work at local, state, and national levels to educate the public.

Alternative Energy Resources Organization
44 N. Last Chance Gulch #9
Helena, MT 59874
P(406) 443-7272
Fosters exchange of information between U.S. and Canadian farmers and farm organizations concerning sustainable farming and agricultural management.

American Cetacean Society
P.O. Box 2639
San Pedro, CA 90731-0943
P(213) 548-6279
F(213) 548-6950
Dedicated to the conservation and protection of whales, dolphins, porpoises and the oceans they live in.

American Conservation Association, Inc.
30 Rockefeller Plaza, Rm. 5402
New York, NY 10112
P(212) 649-5600
Preserves and develops natural resources for public use.

American Farmland Trust
1920 N St., NW, Suite 400
Washington, DC 20036
P(202) 659-5170
Dedicated to preserving America's farmland, this organization purchases land and/or its development rights.

American Friends Service Committee
2160 Lake Street
San Francisco, CA 94121
P(415) 752-7766
This independent Quaker committee works through nonviolent means on environmental and social justice issues.

American Forestry Association
1516 P St., NW
Washington, DC 20005
P(202) 667-3300
Educates the public of the best ways to use natural resources.

American Littoral Society
Sandy Hook
Highlands, NJ 07732
P(201) 291-0055
A national organization dedicated to the study and conservation of coastal habitats and a publisher of scientific and popular materials.

American Oceans Campaign
725 Arizona Ave. #102
Santa Monica, CA 90401
P(301) 452-2206
The primary focus of this organization is to establish a National Oceans Policy which will end toxic and sewage dumping in the ocean and stop offshore drilling in sensitive areas.

American Rivers
801 Pennsylvania Ave., SE
Suite 303
Washington, DC 20003-2167
P(202) 547-6900
F(202) 543-6142
E-Mail(EcoNet): amrivers
Dedicated to preserving the rivers and landscapes of the United States.

American Water Resources Association
5410 Grosvenor Ln., Suite 220
Bethesda, MD 20814
P(301) 493-8600
This group works to advance the research, planning, and development of water resources in the United States.

American Wildlands
7500 E. Arapahoe Rd., #355
Englewood, CO 80112
P(303) 771-0380
The primary focus of this group is to preserve the timberlands, wildlands, and wildlife of the west.

Anglers for Clean Water, Inc.
P.O. Box 17900
Montgomery, AL 36141
P(205) 272-9530
Educates the public on the conditions of pollution nationwide and promotes clean streams, lakes, and rivers.

AT Work
300 Broadway, Suite 28
San Francisco, CA 94133
P(415) 788-3666
F(415) 788-7324
E-Mail(EcoNet): atwork
AT Work supports appropriate technology projects in the Third World with materials and skilled volunteers, focusing on the promotion of socially sustainable communities.

Atlantic Center for the Environment
39 S. Main St.
Ipswich, MA 01938
P(508) 356-0038
This organization promotes sustainable development in the Atlantic Region by developing leadership opportunities and providing technical assistance.

Atlantic Salmon Federation
P.O. Box 684
Ipswich, MA 01938
P(506) 529-8889
F(506) 529-4438
Dedicated to bringing the salmon back through stream restoration, public education, and hatcheries.

Better World Society
1100 Seventeenth St., NW Suite 502
Washington, DC 20036

P(202) 659-1833
F(202) 331-3779
Dedicated to developing programming which will foster awareness of global issues of sustainability and promote the benefits of preserving the world's rainforests.

California Action Network
P.O. Box 464
Davis, CA 95617
P(916) 756-8518
This organization focuses on the water uses of the agricultural industry in order to redefine the water allocation policy in California.

California Energy Commission
1516 9th Street, MS-25
Sacramento, CA 95814-5512
P(916) 324-3000
Forecasts energy uses, promotes energy conservation, develops renewable energy resources and alternative energy-generating technologies.

California Energy Extension Service
1400 Tenth St., Rm. 209
Sacramento, CA 95814
P(916) 323-4388
Works to lower energy use for small-scale consumers by energy management projects and demonstrations.

California Marine Mammal Center
Marin Headlands
Golden Gate National Recreation Area
Sausalito, CA 94965
P(415) 331-0161
This organization rescues injured and orphaned marine mammals from the California coast in hopes of returning them to the wild, and conducts research in marine mammal science and medicine.

California Trout
870 Market St., Suite 859
San Francisco, CA 94102
P(415) 392-8887
Dedicated to preserving trout fisheries and habitats in California.

California Wilderness Coalition
2655 Portage Bay East, Suite 5
Davis, CA 95616
P(916) 758-0380
E-mail(EcoNet): jeaton
This group promotes preservation of wildlands throughout California.

Californians Against Waste
909 12th St., Suite 201
Sacramento, CA 95814
P(916) 443-5422
This organization focuses on recycling programs and legislation across the country and publishes a quarterly newsletter, *Waste Watch*.

CARE
660 First Ave.
New York, NY 10016
P(212) 686-3110
F(212) 696-4005
CARE works to improve the livelihoods of the rural poor through sustainable resource management.

CEIP Fund
68 Harrison Ave.
Boston, MA 02111
P(617) 426-4375
F(617) 423-0998
An environmental career organization offering job and intern opportunities at local, state and federal government organizations, corporations and consulting firms, and non-profit organizations.

Catalina Conservancy
P.O. Box 2739
Avalon, CA 90704
P(310) 510-1421
Works to preserve the many native plants and animals of Catalina Island.

Cause for Concern
RD 1, Box 570
Stewartsville, NJ 08886
P(201) 479-6778
Cause for Concern works to increase consumer awareness and promote informed choices of environmentally sound products.

Center for Conservation Biology
Department of Biological Sciences
Stanford University
Stanford, CA 94305
P(415) 723-5924
F(415) 723-5920
E-Mail(EcoNet): conbio
Works on the development of conservation biology and its application to environmental problems.

Center for Economic Conversion
222 View St., Suite C
Mountain View, CA 94041
P(415) 968-8798
F(415) 968-1126
E-Mail(EcoNet): cec
CEC works for the conversion of military production towards projects that encourage environmental restoration.

Center for Environment, Commerce, and Energy
733 6th St., SE, Suite 1
Washington, DC 20003
P(202) 543-3939
The Center focuses on air quality and pollution, water resources and pollution, land use, and toxic substances.

Center for Environmental Education, Inc.
1725 DeSales St., NW
Washington, DC 20036
P(202) 429-5609
Dedicated to the conservation of marine species and their habitats.

Center for Marine Conservation, Inc.
1725 DeSales St., NW, Suite 500
Washington, DC 20036
P(202) 429-5609
F(202) 872-0619
Determines who are major manufacturers of pollution in marine habitats and seeks to bring action against them.

Center for Resource Economics
1718 Connecticut Ave., NW
Suite 300
Washington, DC 20009
P(202) 232-7933
This organization develops, publishes, and markets books concerning global environmental problems.

Central States Resource Center
809 S. Fifth
Champaign, IL 61820
P(217) 344-2371
Grass-roots organization researches hazardous waste, water and transportation issues.

Cetacean Society International
Box 9145
Wethersfield, CT 06109
P(203) 563-6444
Works to protect the oceans and marine inhabitants.

Children & The Environment: UNEP Youth Forum
United Nations
New York, NY 10017
P(212) 963-4931
United Nations committee which works to bring together school children active in environmental issues at an annual conference in New York.

Citizen Action
1300 Connecticut Ave., NW
Washington, DC 20036
P(202) 857-5153
Encourages citizens to increase involvement in environmental, social and economic decisions.

Citizens Against Nuclear Power & Weapons
53 W. Jackson, Suite 1306
Chicago, IL 60604
P(312) 786-9041
Opposed to nuclear power and weapons, this organization educates the public on alternatives.

Citizens Clearinghouse for Hazardous Waste
119 Rowell Court
Falls Church, VA 22046
P(703) 237-2249
Works to halt illegal dumping of hazardous wastes and promote recycling and other alternatives to landfills and incinerators.

Citizens Environmental Task Force
321 Calle Loma Norte
Santa Fe, NM 87501
P(505) 983-2894
F(505) 982-6412
Through education and lobbying this citizens group deals with the management of natural resources.

Citizens for a Better Environment
33 East Congress Parkway
Suite 523
Chicago, IL 60605
P(312) 939-1530
This organization fights environmental health threats.

Citizens for Alternatives to Chemical Contamination
9496 School St.
Lake, MI 48632
P(517) 544-3318
As well as functioning as a clearinghouse to alert citizens to potentially harmful substances, this organization also focuses on education.

Clean Sites, Inc.
1199 North Fairfax St., Suite 400
Alexandria, VA 22314
P(703) 683-8522
F(703) 548-8773
Works to accelerate cleanup of hazardous wastes by assisting cleanup and developing effective public policy.

Clean Water Action
1320 18th St., NW
Washington, DC 20036
P(202) 457-1286
This national organization works for safe water as well as protection of our nation's resources.

Climate Change & Energy Program
1616 P Street, NW
Washington, DC 20036
P(202) 332-0900
F(202) 332-0905
E-Mail(EcoNet)
This organization presses for reduction of carbon dioxide emissions from fossil fuels by working on better energy and transportation policies.

Coalition for New Budget Priorities
43 Samoset St.
Dorchestor, MA 02124
P(617) 727-4596
Demands to cut military spending and to use the money towards social and environmental projects.

Coast Alliance
235 Pennsylvania Ave., SE
Washington, DC 20003
P(202) 546-9554
Increase public awareness of the value of the coast and works to strengthen policies and programs to protect coastal ecosystems.

Coastal Conservation Association, Inc.
4801 Woodway, Suite 220 West
Houston, TX 77056
P(713) 626-4222
This association promotes and conserves marine and plant life along coastal areas of the United States.

Commonweal
P.O. Box 316, 451 Mesa Rd.
Bolinas, CA 94924
P(415) 868-0970
This nonprofit institute fosters projects serving humanity and the Earth.

Community Environmental Council
930 Miramonte Dr.
Santa Barbara, CA 93109
P(805) 963-0583
F(805) 962-9080
Through research and education, the Council promotes recycling of resources and sustainable development.

Concern, Inc.
1794 Columbia Rd., NW
Washington, DC 20009
P(202) 328-8160
F(202) 328-8161
Concern provides environmental information to communities in order to find solutions to problems that threaten public health.

Conservation Foundation
1250 24th St., NW
Washington, DC 20037
P(202) 293-4800
Promotes wise use of the earth's resources through research and public education.

Conservation Fund
1800 North Kent St., Suite 1120
Arlington, VA 22209
P(703) 525-6300
A national nonprofit organization searching for innovative ways to advance land and water conservation.

Conservation Law Foundation, Inc.
3 Joy St.
Boston, MA 02108
P(617) 742-2540
An environmental law firm that takes polluters to court, and works to pass legislation and to protect important natural resources.

Co-op America
2100 M St., NW, Suite 403
Washington, DC 20063
P(202) 872-5307 or (800) 424-2667
Provides alternatives for consumers to buy and invest in businesses that are concerned about the environment.

Council on Economic Priorities
30 Irving Place
New York, NY 10003
P(212) 420-1133 or (800) 822-6435
F(212) 420-0988
Promotes solutions to environmental problems and fosters arms control by encouraging corporate social responsibility.

Council on Ocean Law
1709 New York Ave., NW, Suite 700
Washington, DC 20006
P(202) 347-3766
F(202) 638-0036
Furthers the laws governing ocean uses and preserves the ocean's abundance.

Cousteau Society, Inc., The
930 W. 21st St.
Norfolk, VA 23517
P(804) 627-1144
F(804) 627-7547
Telex: 6974570 COUSTEAUNFK
Dedicated to the study and exploration of the earth's oceans and to the education of the public on the pollution and development of the oceans.

Craighead Environmental Research Institute
Box 156
Moose, WY 83012
P(307) 733-3387
The Craigheads have been active in wildlife research from the migratory habits of caribou to the effects of environmental contamination on the raptor population.

Critical Mass Energy Project of Public Citizen
215 Pennsylvania Ave., SE
Washington, DC 20003
P(202) 546-4996
Opposes nuclear power and promotes safe alternatives through lobbying and litigation.

J.N. (Ding) Darling Foundation, Inc.
c/o J.M. Redman, Treasurer
P.O. Box 703
Des Moines, IA 50303-0703
P(515) 281-0812
Committed to conservation education by providing grants to students and initiating educational projects.

Declaration of Earth Ethics
700 East Daisy Lane
Milwaukee, WI 53217-3632
P(414) 351-2737
An urgent statement to establish world harmony in conjunction with the ecosystem.

Defenders of Wildlife
1244 19th St., NW
Washington, DC 20036
P(202) 659-9510
F(202) 833-3349
This group opposes any practice that harms wildlife diversity by advocating governmental, citizen, and legal action.

Duck Unlimited, Inc.
One Waterfowl Way
Long Grove, IL 60047
P(708) 438-4300
Works to increase waterfowl and other wildlife population on the North American continent by restoration and management of wetland areas.

Earth First!
P.O. Box 5871
Tucson, AZ 85703
P(602) 622-1371
E-Mail(EcoNet): earthfirst
A nonviolent movement that works to preserve natural diversity as typified in wildness.

Earth Island Institute
300 Broadway, Suite 28
San Francisco, CA 94133
P(415) 788-3666
F(415) 788-7324
E-Mail(EcoNet): earthisland
Telex: 6502829302 MCI UW
Develops innovative projects for the conservation, preservation, and restoration of the global environment.

Earth Right Institute
Gates-Briggs Building
Room 322
White River Junction, VT 05001
P(802) 295-7734
An educational and action center for environmental concerns.

EarthDance
P.O. Box 2155
Asheville, NC 28802
P(704) 252-8188
An educational project of the Youth Environmental Service Network that links young people from all over the world.

Earthsave Foundation
706 Frederick St.
Santa Cruz, CA 95062
P(408) 423-4069
F(408) 425-0255
EarthSave focuses on an ecologically sustainable future by providing education and developing leadership.

Earthwatch
P.O. Box 403
Mt. Auburn St.
Watertown, MA 02272
P(617) 926-8200
F(617) 926-8532
Telex: 5106006452
One of the largest private sponsors of research expeditions in the world, they provide funds to assist environmental scholars and scientists internationally.

Eco-Cycle
P.O. Box 4193
Boulder, CO 80306
P(303) 444-6634
Collects and recycles a portion of Boulder's solid waste to financial self-sufficiency in order to demonstrate the potential of recycling.

Eco-Home Network
4344 Russell Ave.
Los Angeles, CA 90027
P(213) 662-5207
Promotes and demonstrates urban ecological living.

Ecological Society of America
Center for Environmental Society
Arizona State University
Tempe, AZ 85287
P(602) 965-3000
Through published research the Center facilitates the dissemination of ecological principles.

EcoNet
3226 Sacramento St.
San Francisco, CA 94115
P(415) 923-0900
E-Mail(EcoNet)
An environmental electronic conference network.

Educational Communications, Inc.
P.O. Box 35473
Los Angeles, CA 90035
P(213) 559-9160
Works on media productions of all environmental issues to foster a better quality of life on earth.

Elmwood Institute
P.O. Box 5765
Berkeley, CA 94705
P(510) 845-4595
F(510) 845-1439
Founded by noted physicist Fritjof Capra, the Institute is known for its application of systemic thinking to a wide variety of contemporary issues.

Endangered Species UPDATE
School of Natural Resources

University of Michigan
Ann Arbor, MI 48109-1115
P(313) 763-3243
Magazine of data and articles that reports on endangered species.

Energy Conservation Coalition
1525 New Hampshire Ave., NW
Washington, DC 20036
P(202) 745-4874
This citizens' group is dedicated to finding and improving the efficiency of the nation's energy use.

Environmental Action Foundation, Inc.
46930 Carroll Ave
Tacoma Park, MD 20912
P(301) 891-1100
F(301) 891-2218
A combination of Environmental Action and the Environmental Task Force, the foundation lobbies the courts to make positive environmental changes.

Environmental Defense Center
906 Garden Street, Suite 2
Santa Barbara, CA 93101
P(805) 963-1622
F(809) 962-3162
E-mail(EcoNet): edc
A legal defense organization that counsels citizens on how to apply environmental laws.

Environmental Defense Fund, Inc.
257 Park Avenue South
New York, NY 10010
P(212) 505-2100
F(212) 505-2375
E-Mail(EcoNet): edf
This organization works to enforce laws preserving our natural resources, pursuing projects that fight against acid rain and the destruction of tropical rain forests.

Environmental Law Institute
1616 P St., NW, Suite 200
Washington, DC 20036
P(202) 328-5150
Works on research and education of environmental law and policy.

Environmental Protection Information Center, Inc.
P.O. Box 397
Garberville, CA 95440
P(707) 923-2931
Focuses on efforts to stop the clear-cutting of old growth forests.

Food First!/Institute for Food and Development Policy
145 Ninth Street
San Francisco, CA 94103
P(415) 864-8555
F(415) 864-3909
Encourages citizens to participate in finding solutions to critical social problems locally, nationally, and globally.

Forest Trust
P.O. Box 9238
Santa Fe, NM 87504-9238
P(505) 983-8992
Dedicated to improving forest ecosystems and resources through innovative land management and resource protection.

Fossil Fuel Policy Action Institute
P.O. Drawer 8558
Fredericksburg, VA 22404
P(703) 371-0222
Works to connect the U.S. internationally in a stronger environmental movement for projects such as tracking global warming.

Freshwater Foundation
2500 Shadywood Rd., Box 90
Navarre, MN 55392
P(612) 471-8407
F(612) 471-8142
Focuses on education and research of usable water to help people understand water issues and their environment.

Friends of the Earth
218 D St., SE
Washington, DC 20003
P(202) 544-2600
F(202) 543-4710
E-Mail(EcoNet): foedc
Telex: 650-192-5483
Fights to protect the earth from environmental disaster such as ozone depletion, global warming, and rainforest destruction.

Friends of the River, Inc.
Friends of the River Foundation
Bldg. C, Fort Mason Ctr.
San Francisco, CA 94123
P(415) 771-0400
This premier river preservation organization is dedicated to preserving over 100 natural river segments in California.

Fundamental Action to Conserve Energy
75 Day St.
Fitchburg, MA 01420
P(508) 345-5385
Through education, action, and a co-op store, this organization is dedicated to energy conservation, water conservation, and recycling.

Future Resources Associates, Inc.
2000 Center St., Suite 418
Berkeley, CA 94704
P(510) 644-2700
A consulting firm in energy and development issues.

Garden Club of America
598 Madison Ave.
New York, NY 10022
P(212) 753-8287
Dedicated to conservation and restoration of the environment, advocating wise land use and pollution control.

Global Action Plan
449 A Rt. 28A
West Hurley, NY 12491
P(914) 331-1312
F(914) 331-3241
Global Action Plan works on environmental restoration through prompting individual action.

Global Conservation, Protection, Restoration (CPR) Service
300 Broadway, Suite 28
San Francisco, CA 94133
P(415) 788-3666
F(415) 788-7324
E-mail(EcoNet): earthisland
Catalyses environmental restoration through education and advocacy, and by providing people with the opportunity to volunteer.

Global ReLeaf
1516 P St., NW
Washington, DC 20009
P(202) 667-3300
F(202) 667-7751
Organizes tree planting, providing technical assistance for new urban forestry programs.

Global Tomorrow Coalition, Inc.
1325 G St., NW
Suite 915
Washington, DC 20005-3104
P(202) 628-4016
F(202) 628-4018
E-Mail(EcoNet): gtc
Educates the public on social, environmental and economic issues to foster sustainable development in the U.S. and international communities.

Great Swamp Research Institute
Office of the College Dean
College of Natural Sciences and Math
Indiana University of Pennsylvania
305 Weyandt Hall
Indiana, PA 15705
P(412) 357-2609
Seeks innovative solutions to the increasing number of environmental problems.

Greenbelt Alliance
116 Montgomery, Suite 640
San Francisco, CA 94105
P(415) 543-4291
Works to protect the Greenbelt and promote better land development.

Greenhouse Crisis Foundation
1130 17th St., NW
Suite 630
Washington, DC 20036
P(202) 466-2823
F(202) 429-9602
E-Mail(EcoNet): gcf
Telex: 904059-WAS
Unites organizations to find solutions to the problem of global warming.

Greenpeace
1436 U St., NW
Washington, DC 20009
P(202) 462-1177
F(202) 462-4507
Telex: 89-2359
The largest environmental group in the world, Greenpeace was formed in 1971 and is dedicated to "protect all forms of life and to obstruct wrongs, without committing violence."

Human Environment Center
1001 Connecticut Ave., NW
Suite 827
Washington, DC 20036
P(202) 331-8387
The Human Environment Center encourages environmental organizations to work together and serves as a clearinghouse for youth conservation and service corps.

Humane Society of the United States
2100 L St., NW
Washington, DC 20037
P(202) 452-1100 or (800) 223-5400
F(202) 778-6132
Committed to the welfare of animals, this organization works to protect both domestic and wild species.

Inform
381 Park Ave. South
New York, NY 10016
P(212) 689-4040
F(212) 447-0689
A research, educational, and action organization that focuses on critical environmental issues.

Institute for Alternative Agriculture
9200 Edmonston Rd., Suite 117
Greenbelt, MD 20770
P(301) 441-8777
F(301) 220-0164
Encourages and facilitates environmentally sound and low cost farming methods.

Institute for Environmental Negotiation
Campbell Hall
Univ of Virginia
Charlottesville, VA 22903
P(804) 924-1970

Serves as a mediator between governments, businesses, and citizens to settle issues dealing with land use and other environmental issues.

Institute of Environmental Sciences
940 E Northwest Hwy.
Mount Prospect, IL 60056
P(312) 255-1561
Focuses on trying to understand the relationship between nature and the impact of humanity in order to promote reliable and safe operations avoiding further contamination of the ecosystem.

International Ecology Society
1471 Barclay St.
St. Paul, MN 55106-1405
P(612) 774-4971
Committed to the correct treatment of animals and the protection of natural wildlands.

International Fund for Animal Welfare
P.O. Box 193
Yarmouth Port, MA 02675
P(508) 362-4944
Works to preserve wildlife and secure humane treatment of domestic animals.

International Rivers Network
1847 Berkeley Way
Berkeley, CA 94703
P(510) 848-1155
F(510) 848-1008
E-Mail(EcoNet): irn
Telex: 6503532706
Works on reclamation and restoration of rivers and informs the public concerning issues such as development.

Izzak Walton League of America, Inc., The
1401 Wilson Blvd., Level B
Arlington, VA 22209
P(703) 528-1818
Committed to educating the public in the conservation and restoration of soil, water and other natural resources.

Jackson Hole Alliance for Responsible Planning
Box 2728
Jackson, WY 83001
P(307) 733-9417
Publishes the *Alliance Newsletter* periodically and focuses on maintaining high quality land usage in NW Wyoming.

John Muir Institute for Environmental Studies, Inc.
743 Wilson St.
Napa, CA 94559
P(707) 252-8333
Works to identify and study environmental problems.

Keep America Beautiful, Inc.
9 West Broad St.
Stamford, CT 06902
P(203) 323-8987
Through community level participation, this organization is dedicated to improving waste handling and recycling practices.

Land and Water Fund of the Rockies
1405 Arapahoe Ave.
Suite 200
Boulder, CO 80302
P(303) 444-1188
Develops legal strategies with grassroot groups working on pollution issues.

Land Institute
2440 E Water Well Rd.
Salina, KN 67401
P(913) 823-5376
Works on agricultural issues in order to sustain the health of the earth.

League for Coastal Protection
P.O. Box 421698
San Francisco, CA 94142-1698
P(415) 777-0221
Dedicated to protecting California's coastal waters, beaches, wetlands and wildlife.

League of Conservation Voters
1150 Connecticut Ave., NW
Suite 201
Washington, DC 20036
P(202) 785-8683
Works to help elect pro-environmental candidates to the U.S. House of Representatives and the Senate.

League To Save Lake Tahoe
P.O. Box 10110
S. Lake Tahoe, CA 95731
P(916) 541-5388
Non-profit organization committed to the preservation of the Lake Tahoe Basin.

Legal Environmental Assistance Foundation
203 North Gadsden St.
Suite 200
Tallahassee, FL 32301
P(904) 681-2591
This law foundation works to provide technical assistance in environmental and civil rights issues to grassroot organizations and citizens.

Local Environmental Action Group
717 1/2 Pujo St.
P.O. Box 3244
Lake Charles, LA 70601-4368
P(318) 474-6133
E-Mail(Environet)
This action group works to restore the ecosystem, from toxics and waste reduction to air quality issues.

Low Input Sustainable Agriculture
237 Hatchville Rd.
East Falmouth, MA 02536
P(508) 564-6301
Works on developing environmentally sound methods for food production, energy use and waste recycling.

Maine Organic Farmers and Gardeners Association
P.O. Box 2176
Augusta, ME 04338-2176
P(207) 622-3118
Committed to aiding farmers and gardeners grow organic food as well as helping to increase yield and public awareness about the benefits of organic food.

Marine Mammal Fund
Fort Mason, Center, Bldg. E
San Francisco, CA 94123
P(415) 921-3140
Projects and research to protect sea mammals including many films that show the tragic death of marine life.

Max McGraw Wildlife Foundation
P.O. Box 9
Dundee, IL 60118
P(312) 741-8000
The Foundation conducts research, management, and conservation education in wildlife and fisheries.

Midwest Consortium on Groundwater and Farm Chemicals
c/o Minnesota Project
2222 Elm St., SE
Minneapolis, MN 55414
P(612) 378-2142
Focuses on projects that protect groundwater from agricultural pesticides.

Mono Lake Committee
P.O. Box 29
Lee Vining, CA 93541
P(619) 647-6386 or 647-6596
This organization is dedicated to the preservation of Mono Lake.

National Academy of Sciences
2101 Constitution Ave., NW
Washington, DC 20037
P(202) 334-2644
F(202) 334-2614
The National Academy of Sciences studies issues such as environmental exposure, epidemiology and toxicology.

National Association for Environmental Education
P.O. Box 400
Troy, OH 45373
P(513) 649-3000
Works to aid individuals who work in environmental education, research, and service.

National Association for State River Conservation Programs
801 Pennsylvania Ave., SE
Suite 302
Washington, DC 20003
P(202) 543-2682
Encourages conservation and restoration of rivers and their shore land environments as well as fostering a forum to discuss river conservation.

National Association of Conservation Districts
509 Capitol Ct., NE
Washington, DC 20002
P(202) 547-6223
Works to promote conservation of natural resources.

National Audubon Society
950 Third Ave.
New York, NY 10022
P(212) 832-3200
F(212) 593-6254
A lobbying, litigation and citizens' action organization that works to protect the world's ecosystem.

National Coalition Against the Misuse of Pesticides
530 7th St., SE
Washington, DC 20003
P(202) 543-5450
Focuses on the pesticide poisoning problem and promotes better alternative pest management solutions.

National Coalition for Marine Conservation
P.O. Box 23298
Savannah, GA 31403
P(912) 234-8062
This nonprofit, privately supported organization focuses on the conservation of ocean fish and their environment by increasing the public's awareness of its responsibility for the natural world.

National Environmental Health Association
720 S. Colorado Blvd.
South Tower, 970
Denver, CO 80222
P(303) 756-9090
F(303) 691-9490
Supports people interested in environmental issues and is the largest society of environmental health practioners in the nation.

National Geographic Society
Colorado and M Sts., NW
Washington, DC 20036
P(202) 857-7000 or (800) 638-4077
F(202) 775-6141
Through exploration and research projects, the National Geographic Society has increased the knowledge of earth, sea, sky, and space.

National Parks and Conservation Association
1015 31st St., NW
Washington, DC 20007

P(202) 944-8530
This organization founded in 1919 acts to preserve our national parks using research, wilderness preservation, and direct action.

National Recycling Coalition
1101 30th St., NW
Suite 305
Washington, DC 20007
P(202) 625-6406
Works to maximize recycling and conservation.

National Toxics Campaign
1168 Commonwealth Ave.
Boston, MA 02134
P(617) 232-0327
A coalition of citizens, consumer organizations, environmental groups, family farmers, and others working for solutions to the nation's toxic and environmental problems.

National Water Well Association
6375 Riverside Dr.
Dublin, OH 43017
P(614) 761-1711
F(614) 761-3446
Committed to the protection of ground water through education and the publication of such water issues as toxic substances and solid wastes.

National Wetlands Technical Council
1616 P St., NW Suite 200
Washington, DC 20036
P(202) 328-5150
Advises on wetlands policies and research priorities to provide assistance to the nation's wetland conservation efforts.

National Wildflower Research Center
2600 FM 973 North
Austin, TX 78725
P(512) 929-3600
Dedicated to the restoration and conservation of native plants, providing information through publications, seminars, programs, and tours.

National Wildlife Federation
1400 Sixteenth St., NW
Washington, DC 20036-2266
P(202) 797-6800
E-Mail(EcoNet): nwfdc
With over 5 million members, this organization promotes the wise use and proper management of natural resources upon which humanity depends.

National Wildlife Refuge Association
10824 Fox Hunt Ln.
Potomac, MD 20854
P(303) 249-8717
Dedicated to the protection of the National Wildlife Refuge System by increasing public awareness and appreciation of the system.

National Woodland Owners Association
374 Maple Ave., E., Suite 204
Vienna, VA 22180
P(703) 255-2700
A nationwide association of woodland owners that work together toward wise management of nonindustrial private forest lands.

Native Americans for a Clean Environment
P.O. Box 1671
Tahlequah, Oklahoma 74465
P(918) 458-4322
Focuses on nuclear industry issues as well as environmental issues in order to raise public awareness.

Native Forest Council
P.O. Box 2171
Eugene, OR 97402
P(503) 688-2600
F(503) 461-2156
Dedicated to preservation and protection of all remaining native forests in the United States.

Natural Areas Association
320 S. Third St.
Rockford, IL 61104
P(815) 964-6666
Works to identify, preserve and manage natural areas and diversity.

Natural Guard
125 W. 44th St., Suite 11E
New York, NY 10036
P(212) 704-0346
F(212) 869-3045
Acts to encourage young people, especially in inner cities, to address environmental issues.

Natural Land Institute
302 S Third St.
Rockford, IL 61104
P(815) 964-6666
Purchases or receives natural areas for preservation.

Natural Lands Trust
Hildacy Farm, 1031 Palmers Mill Road
Media, PA 19063
This organization concentrates on the protection and preservation of natural lands in the mid-Atlantic region.

Natural Resources Council of America
801 Pennsylvania Ave., SE
Suite 410
Washington, DC 20003
P(202) 547-7553
Concerned with proper management of natural resources in the public interest by providing citizens with policy information on conservation issues.

Natural Resources Defense Council
40 West 20th St.

New York, NY 10011
P(212) 727-2700
E-Mail(EcoNet): nrdc
This lobbying group works to protect endangered natural resources, combining legal action and a scientific approach.

Nature Conservancy
1815 North Lynn St.
Arlington, VA 22209
P(703) 841-5300
F(703) 841-1283
E-Mail(EcoNet): natconsv
In order to maintain genetic diversity among species, the Conservancy is committed to finding, maintaining, and protecting natural lands and ecosystems.

Negative Population Growth, Inc.
16 East 42nd St., Suite 1042
New York, NY 10017
P(201) 837-3555
As population growth is becoming a serious issue, this organization educates on the need to limit human population for sustainable development.

New Alchemy Institute
237 Hatchville Rd.
East Falmouth, MA 02536
P(508) 564-6301
The goal of this organization is to work towards a world which meets all basic human needs of food, water, and shelter.

New York Zoological Society
The Zoological Park
Bronx, NY 10460
P(212) 220-5100
Funds research and development of wildlife management, establishes parks and reserves, and works to increase the public understanding of zoology and the environment.

North American Lake Management Society
1000 Connecticut Ave., NW
Suite 300
Washington, DC 20036
P(202) 466-8550
Works to protect and restore lakes, reservoirs and their watersheds. This organization also sponsors annual conferences, and regional workshops as well as publications.

North American Wildlife Foundation
102 Wilmot Rd., Suite 410
Deerfield, IL 60015
P(708) 940-7776
F(708) 940-3739
Operates research programs concerned with wetland ecology and improving marsh management.

Northwest Coalition for Alternative Pesticides
P.O. Box 1393
Eugene, OR 97440
P(503) 344-5044
The main goal of the Northwest Coalition is to eliminate the use of pesticides by education, watchdogging, and direct action.

Northwest Renewable Resources Center
1133 Dexter Horton Bldg.
710 Second Ave.
Seattle, WA 98104
P(206) 623-7361
Tries to offer alternatives on disputes dealing with natural resources issues.

Ocean Alliance
Building E
Fort Mason Center
San Francisco, CA 94123
P(415) 441-5970
Dedicated to preserve and protect the ocean through conservation and marine education programs.

Oceanic Society
Executive Offices
1536 16th St., NW
Washington, DC 20036
P(202) 329-0098
Promotes understanding and stewardship of marine and coastal environments as well as working to protect the oceans.

Oregon Natural Resources Council
Yeon Bldg., Suite 1050
522 Southwest Fifth Avenue
Portland, OR 97204
P(503) 223-9001
Concentrates on preserving wild forest lands by lobbying for better forest management.

Organization of Wildlife Planners
Box 7921
Madison, WI 53707
P(307) 777-7461
Focuses on constructing effective management systems of resources.

Pacific Energy and Resource Center
Building 1055
Fort Cronkite
Sausalito, CA 94965
P(415) 332-8200
Provides policy research, professional consulting services and natural resource education programs such as exhibits and lectures.

Pacific Whale Foundation
Kealia Beach Plaza, Ste. 25
101 N. Kihei Rd.
Kihei, HI 96753
P(808) 879-8811

Works to conserve and research all marine mammals, focusing primarily on Hawaiian humpback whales.

Partners for Livable Places
1429 21st Street, NW
Washington, D.C. 20036
P(202) 887-5990
Works to improve economic resources and the quality of life of certain communities by preservation, conservation and cultural resources management.

Peregrine Fund, Inc., The
5666 West Flying Hawk Ln.
Boise, ID 83709
P(208) 362-3716
The Fund works on the preservation and restoration of peregrine falcons.

Pesticide Action Network
P.O. Box 610
San Francisco, CA 94101
P(415) 771-2763
F(415) 541-9253
E-Mail(EcoNet): panna
Telex: 15683472 PANNA
Works to replace pesticides with sustainable agriculture practices and technologies.

Planning and Conservation League Foundation
909 12th St., Suite 203
Sacramento, CA 95814
P(916) 444-8726
This organization dedicated to environmental education and research works on solving problems concerning energy, environmental research, water quality, and coastal use.

Population-Environmental Balance
1325 G St., NW Suite 1003
Washington, DC 20005
P(202) 879-3000
F(202) 879-3019
Works to educate the public on the relationship between population growth and the well-being of the United States.

Prevention of Global Warming Project
26 Church St.
Cambridge, MA 02238
P(617) 547-5552
Seeks to involve scientists in the need to develop a national energy policy based on renewable energy.

Public Media Center
466 Green Street
San Francisco, CA 94133
P(415) 434-1403
PMC offers a service for environmental organizations to create and run effective ads on TV, radio, and in the written media.

Rainforest Action Network
301 Broadway, Suite A
San Francisco, CA 94133
P(415) 398-4404
F(415) 398-2732
E-Mail(EcoNet): rainforest
Telex: 151276475
This organization works to preserve the rainforests by drawing attention to the short term use of rainforests and by educating the public as to citizen actions in pressuring businesses to change.

Rainforest Alliance
270 Lafayette St., Suite 512
New York, NY 10012
P(212) 941-1900
F(212) 941-4986
This organization's goals are to preserve the rainforests and to create a way of utilizing the rainforests without further destruction.

Redwood Alliance
P.O. Box 293
Arcata, CA 95521
P(707) 822-7884
This organization promotes safe energy alternatives to nuclear power through education and lobbying.

Redwood Community Action Agency
904 G St.
Eureka, CA 95502
P(707) 445-0881
Among their many projects, this agency promotes the preservation of redwood forests, and the belief that improving the environment enhances the local economy.

Rene Dubos Center for Human Environments
100 East 85th St.
NY 10028
P(212) 249-7745
Works to create new environmental values and resolution of environmental conflicts through education, research, and publications.

Renew America
1400 16th St., NW Suite 710
Washington, DC 20036
P(202) 232-2252
Works to build a sustainable society by encouraging effective public policy.

Renewable Natural Resources Foundation
5430 Grosvenor Ln.
Bethesda, MD 20814
P(301) 493-9101
Renewable natural resources and public policy alternatives are the focuses of this organization.

Resource Renewal Institute
Ft. Cronkite, #1055
Sausalito, CA 94965
P(415) 332-8082

Creates long-term solutions to renewable resource problems by training professionals to implement changes and by setting up data banks.

Resources for the Future
1616 P St., NW
Washington, DC 20036
P(202) 328-5000
Develops research projects and education in the use of natural resources, quality of the environment, and conservation.

Restoring the Earth
1713C MLK Jr. Way
Berkeley, CA 94709
P(510) 843-2645
Develops creative solutions to environmental problems by means of ecological restoration.

Rocky Mountain Institute
1739 Snowmass Creek Road
Snowmass, CO 81654-9199
P(303) 927-3128
F(303) 927-4178
E-Mail(EcoNet): rmi
Promotes sustainable agriculture activities and ecologically sound and economically viable communities.

Sacramento River Preservation Trust
P.O. Box 5366
Chico, CA 95927
P(916) 345-4050
The purpose of this organization is to preserve, protect and improve the Sacramento River by educating the public

Save the Dolphin Project
Earth Island Institute
300 Broadway, Suite 28
San Francisco, CA 94133
P(415) 788-3666/800 3-DOLFIN
E-Mail(EcoNet): earthisland
This project is determined to eliminate the unnecessary slaughter of dolphins by promoting legal, political, and economic reforms.

Save the Redwoods League
114 Sansome St., Rm. 605
San Francisco, CA 94104
P(415) 362-2352
Hopes to rescue the redwoods by creating Redwood Parks and educating the public of the value of the Sequoias.

Sea Shepherd Conversation Society
P.O. Box 7000-S
Redondo Beach CA 90277
P(213) 394-3198
This is a direct, nonviolent action organization that focuses on illegal whaling, the dolphin slaughter, and drift netting.

Sea Turtle Restoration Project
300 Broadway, Suite 28
San Francisco, CA 94133
P(415) 788-3666
F(415) 788-7324
E-Mail(EcoNet): earthisland
Acts to preserve, restore and investigate threats to the world's endangered sea turtles

Sierra Club
730 Polk St.
San Francisco, CA 94109
P(415) 776-2211
F(415) 776-0350
With 57 chapters and 386 groups in North America, this organization's work includes legislation, litigation, public information, publishing, wilderness outings, and conferences.

Sierra Club Legal Defense Fund, Inc.
180 Montgomery St., Suite 1400
San Francisco, CA 94104
P(415) 627-6700
F(415) 627-6740
This public interest law firm brings lawsuits on behalf of environmentalists and citizens organizations to protect the environment.

Soil and Water Conservation Society
7515 NE Ankeny Rd.
Ankeny, IA 50021-9764
P(515) 289-2331
F(515) 289-1227
Through education this organization promotes good land and water use, and natural resources conservation.

Spill Control
400 Renaissance Center
10th Floor
Detroit, MI 48243-1895
P(313) 567-0500
Organized to combat pollution incidents by minimizing their effects.

Student Conservation Association, Inc.
P.O. Box 550
Charlestown, NH 03603
P(603) 826-4301
F(603) 826-7755
Publishes *Job Scan,* a monthly environmental job listing, and supports volunteers in conservation of public lands.

Student Environmental Action Coalition
P.O. Box 1168
Chapel Hill, NC 27514
P(919) 967-4600
F(919) 962-5604
E-Mail: SEAC UNC BITNET
The largest student environmental organization in the United States, SEAC was founded in 1989 to provide network services for campus environmental groups.

Surfrider Foundation
P.O. Box 230754
Encinitas, CA 92023
P(619) 792-9940
The Surfrider Foundation works towards enhancing the quality of water and beaches.

Thorne Ecological Institute
5398 Manhattan Cir.
Boulder, CO 80303
P(303) 499-3647
Brings together adults in order to educate them on the application of ecological principles to enhance the human and natural environment.

Threshold, Inc.
International Center for Environmental Renewal
Drawer CU
Bisbee, AZ 85603
P(602) 432-7353
Established in 1972, this organization works to develop ecologically sound alternatives in order to improve upon human understanding of the ecosystem.

Treepeople
12601 Mulholland Dr.
Beverly Hills, CA 90210
P(818) 753-4600
F(818) 753-4625
E-Mail(EcoNet): treepeople
The goal of this organization is to end global deforestation by taking surplus trees and sending them to needy areas for planting, and by encouraging citizens to plant trees.

Trout Unlimited
800 Follin Lane, Suite 250
Vienna, VA 22180-4906
P(703) 281-1100
F(703) 281-1825
Works to maintain trout fisheries by protecting and preserving rivers and streams.

Trust for Public Land
116 New Montgomery St.
4th Floor
San Francisco, CA 94105
P(415) 495-4014
F(415) 495-4103
A land conservation organization preserving land for present and future generations.

Union of Concerned Scientists
26 Church St.
Cambridge, MA 02238
P(617) 547-5552
F(617) 664-9405
Scientists and citizens concerned about the effects of technology on society, analyzing issues such as nuclear power and weapons.

United Nations Environment Programme
United Nations
Rm. DC2-0803
New York, NY 10017
P(212) 963-8138
Serves as an advocate for environmental management among U.N. agencies, monitoring and assessing issues of water, land and atmospheric areas.

US Public Interest Research Group
215 Pennsylvania Ave., SE
Washington, DC 20003
P(202) 546-9707
US PIRG engages in many issues including the environment, energy and government reform, and consumer protection.

Urban Creeks Council
2530 San Pablo Avenue
Berkeley, CA 94702
P(510) 540-6669
The Urban Creeks Council encourages the preservation and restoration of natural streams in urban or human-made environments.

Urban Ecology
P.O. Box 10144
Berkeley, CA 94709
P(510) 549-1724
Builds and rebuilds cities in a manner that makes them environmentally healthy and culturally viable.

Urban Habitat Program
300 Broadway, Suite 28
San Francisco, CA 94133-3312
P(415) 788-3666
F(415) 788-7324
EcoNet: earthisland
Develops multi-cultural, urban environmental leadership in the Bay Area.

Water Pollution Control Federation
601 Wythe St.
Alexandria, VA 22314-1994
P(703) 684-2400
F(703) 684-2492
Concerned with the collection and disposal of domestic and industrial wastewater.

Wetlands for Wildlife, Inc.
P.O. Box 344
West Bend, WI 53095
P(414) 628-0103
Promotes and participates in the preservation of wetlands and wildlife habitat in the United States.

Whale Center
3929 Piedmont Ave.
Oakland, CA 94611
P(415) 654-6621
Works to protect whales and their ocean habitat.

Wilderness Society
900 17th St., NW
Washington, DC 20006-2596
P(202) 833-2300
F(202) 429-3958
Dedicated to preserving the US's forests, parks, rivers and shorelands.

Wilderness Watch
P.O. Box 782
Sturgeon Bay, WI 54235
P(414) 743-1238
Based on scientific research, this organization is dedicated to sustained use of the United States's woodlands and waters.

Wildlife Conservation International
New York Zoological Society
185th St., and So. Blvd., Bldg. A
Bronx, NY 10460
P(212) 220-5155
Strives to understand the ecosystem in order to apply that research to conservation efforts.

Wildlife Forever
12301 Whitewater Dr., Suite 210
Minnetonka, MN 55343
P(612) 936-0605
Promotes the need for scientific wilderness management and preservation.

Wildlife Habitat Enhancement Council
1010 Wayne Ave., Suite 1240
Silver Springs, MD 20910
P(301) 588-8994
Works to bring together conservation and corporate communities to benefit environmental legislation.

Wildlife Information Center, Inc.
629 Green St.
Allentown, PA 18102
P(215) 434-1637
Researches wildlife conservation and educates the public through outreach programs.

Wildlife Management Institute
Suite 725
1101 14th St., NW
Washington, DC 20005
P(202) 371-1808
Works to improve professional management of natural resources to benefit all of society.

Wildlife Society
5410 Grosvenor Ln.
Bethesda, MD 20814
P(301) 897-9770
F(301) 530-2471
Organized in 1937, this organization engages in wildlife research, management, education, and administration to increase awareness and appreciation of wildlife values.

Windstar Foundation
2317 Snowmass Creek Rd.
Snowmass, CO 81654
P(303) 927-4777
F(303) 927-4779
Serves to inspire citizens towards direct action in achieving an environmentally sustainable future.

World Environment Center, Inc.
419 Park Ave., Suite 1404
New York, NY 10016
P(212) 683-4700
F(212) 683-5053
Serves to strengthen the bridge between industry and government to achieve environmentally beneficial goals.

World Forestry Center
4033 SW Canyon Rd.
Portland, OR 97221
P(503) 228-1367
A nonprofit organization dedicated to preserving the forests as well as other natural resources through school and community education and publications.

World Nature Association, Inc.
P.O. Box 673
Silver Spring, MD 20901
P(301) 593-2522
Funds small projects and scholarships for conservation and education projects in various parts of the world.

World Resources Institute
1709 New York Ave., NW
Washington, DC 20006
P(202) 638-6300
World Resources Institute provides environmental information to facilitate debate among people with differing perspectives.

World Wildlife Fund
1250 24th St., NW
Washington, DC 20037
P(202) 293-4800
F(202) 293-9211
Telex: 64505 PANDA
Works to protect the world's wildlife and biological diversity.

Worldwatch Institute
1776 Massachusetts Ave., NW
Washington, DC 20036
P(202) 452-1999
A research organization elucidating global problems and issues such as energy, population growth and migration, global economy, and the environment.

Zero Population Growth, Inc.
1400 16th St., NW Suite 320
Washington, DC 20036
P(202) 332-2200
Through educational material and lobbying this organization encourages public support for population stabilization.

EXHAUST GAS RECIRCULATION

RAMON ESPINO
Exxon Research and Engineering
Annandale, New Jersey

One of the first deliberate steps taken in engine design to reduce atmospheric pollution consisted of recirculating a fraction of the exhaust gas. The dilution of the air–fuel mixture with a relatively inert exhaust gas reduced the maximum combustion temperature significantly. Since the formation of NO in an internal combustion engine is strongly influenced by temperature, Exhaust Gas Recirculation (EGR) allowed NO_x emissions to decrease below the 4 g/mile level by the mid-1970s. (See also AIR POLLUTION: AUTOMOBILE; AUTOMOBILE EMISSIONS—CONTROL.

The first techniques used to recirculate exhaust gases to the combustion chamber accomplished the task, but at a price in terms of performance. One method was to increase the time overlap when both the exhaust valve and the inlet valve are open. This leads to a certain amount of exhaust combustion gas remaining in the combustion chamber. This method of lowering the combustion temperature and NO emissions should be called Exhaust Gas Retention instead of Exhaust Gas Recirculation. A true recirculation method was implemented in parallel consisting of a calibrated tube that bled a constant volume fraction of the high pressure exhaust gas into the lower pressure air–fuel mixture in the inlet manifold of the engine.

These two methods suffered from very poor idling and poor acceleration response at low speeds (when the amount of exhaust gas was relatively large compared with the fresh air–fuel mixture). Moreover, these techniques significantly reduced NO emissions for a level of about 6 g/mile prior to 1968 to almost 4 g/mile by 1973. However, the standards for NO emissions in the USA were below that level by 1974.

To meet the more stringent standards, most engine manufacturers developed variable-flow Exhaust Gas Recirculation (EGR) systems. They also went back to the normal valve overlap. The variable-flow EGR system substituted the fixed diameter pipe connecting the exhaust to the inlet manifold with a spring-loaded, vacuum-controlled, temperature compensated metering valve. The range of flow of the exhaust gas varies between zero when the car is idling and reaching a peak in flow when the engine is cruising in the 30–70 mph range. The valve also closes when the engine is operated at wide open throttle (WOT). The temperature control is used to shut off recirculation when the ambient temperature is below 14°C. With zero EGR during idling and wide open throttle, the problems of poor idling and slow acceleration were reduced. The new design increased the amount of exhaust gas at the conditions for peak NO formation and this enabled auto manufacturers to meet the 3g/mile standard for 1974/1975.

Exhaust gas recirculation has a negative impact on engine performance and fuel economy. Every percent of exhaust gas that is recirculated yields a corresponding reduction in fuel economy and acceleration times. When the variable flow EGR valves were introduced, there were also many concerns about their operability and driveability. The government required that these EGR systems be inspected every 12,000 miles. However, the design proved rugged and inspection schedules have been extended to 30–40,000 mile intervals. While a number of European cars still use EGR, most American and Japanese OEMs have moved from EGR to catalytic exhaust gas converters to reduce NO. Further restrictions in NO emissions standards, as well as better vehicle performance were the main drivers for this change.

BIBLIOGRAPHY

K. Owen and T. Coley, *Automotive Fuels Handbook,* Society of Automotive Engineers, Inc.

E. F. Obert, *Internal Combustion Engines an Air Pollution,* Harper and Row, Publishers.

Effect of Automotive Emissions Requirements on Gasoline Characteristics, ASTM Publication, STP 487.

EXHAUST CONTROL, AUTOMOTIVE

JOHN J. MOONEY
Engelhard Corporation
Iselin, New Jersey

Automobiles and trucks consume large amounts of gasoline, producing commensurately large amounts of gaseous exhaust consisting primarily of carbon dioxide [*124-38-9*], water, unburned hydrocarbons (qv), carbon monoxide [*630-08-0*] (qv), and oxides of nitrogen, NO_x. The latter three atmospheric pollutants have been regulated since the 1970s by the U.S. government and more stringently by the State of California (see AIR POLLUTION). Automobile companies have developed fuel metering and exhaust systems using the catalytic converter to meet emission regulations. Carbon dioxide emissions are indirectly controlled by corporate average fuel economy (CAFE) standards for passenger cars and small trucks.

Prior to emission control, passenger car and truck emissions together were the largest contributors to atmospheric pollution in the United States (1). By 1994, when the Tier I exhaust emission standards mandated by the Clean Air Act Amendments of 1990 take effect in the United States, the degree of cleanup for automobiles and small trucks must be 97.4% for hydrocarbon emissions, 96.0% for carbon monoxide, and 90% for oxides of nitrogen as compared to pre-control of exhaust emissions. For areas in the United States that do not reach minimum ambient air standards by the end of the 1990s, the U.S. Congress has set conditional Tier II standards for vehicles to take effect as early as the year 2003. Also, California, because of unusually severe atmospheric pollution conditions, has established the most stringent automobile exhaust regulations in the world to control exhaust emissions to the absolute minimum levels. These California standards, called Low Emission Vehicle (LEV) Standards, start to take effect in 1996. Other states are expected to adopt the California standards.

Key to achieving the degree of exhaust emission control by the early 1990s were the monolithic catalytic converter, the three-way catalyst, and the closed loop system based on the oxygen (qv) sensor (see SENSORS). Initially, the cata-

Table 1. Federal Exhaust Emission Standards for Conventionally Fueled Passenger Cars and Light Trucks[a], g/km

Vehicle and Year	Fleet, %	NMHC[b]	CO	NO_x	Particulates
		Light-duty (0–2,700 kg gross vehicle weight rating) vehicles			
Passenger cars					
1991–1993	100	0.25[c]	2.1	0.6	0.12
1994	40	0.16[d]	2.1	0.2	0.05
1994[e]	40	0.19	2.6	0.4	0.06
1995	80	0.10[d]	2.1	0.2	0.05
1995[e]	80	0.19	2.6	0.4	0.06
1996–2003	100	0.10[d]	2.1	0.2	0.05
1996–2003[e]	100	0.19	2.6	0.4	0.06
2004–2006[e,f]		0.08	1.1	0.1	0.06
Trucks, loaded vehicle weight <1700 kg					
1991–1993	100	0.50[c]	6.2	0.7	0.08
1994	40	0.16[d]	2.1	0.2	0.08
1994[e]	40	0.19	2.6	0.4	
1995	80	0.16[d]	2.1	0.2	0.05[g]
1995[e]	80	0.19	2.6	0.4	0.06[g]
1996–2003	100	0.16[d]	2.1	0.2	0.05[h,i]
1996—2003[e]	100	0.19	2.6	0.4	0.06[h,i]
2004—2006[e,f]		0.08	1.1	0.1	0.6
Trucks, loaded vehicle weight 1700–2600 kg					
1991–1993[j]	100	0.50[c]	6.2	1.1	0.08
1994	40	0.20[d]	2.7	0.4	
1994[e]	40	0.25	3.4	0.60	0.08
1995	80	0.20[d]	2.7	0.4	0.05[g]
1995[e]	80	0.25	3.4	0.60	0.06[g,h]
1996–2003	100	0.20[d]	2.7	0.4	0.05[h,i]
1996–2003[e]	100	0.25	3.4	0.60	0.06
		Light-duty (2,700–3,900 kg gross vehicle weight rating) trucks			
Trucks, 1,700–2,600 test weight[k]					
1991–1993[j]	100	0.50[c]	6.2	1.1	0.08
1994[j]	100	0.50[c]	6.2	1.1	0.08
1995[j]	100	0.50[c]	6.2	1.1	0.08
1996	50	0.20	2.7	0.4	
1996[j]	50	0.28[d]	4.0	0.61	0.06
1997–2003	100	0.20[d]	2.7	0.4	
1997–2003[e]	100	0.28	4.0	0.61	0.06
Trucks, >2,600 kg test weight[k]					
1991–1993[j]	100	0.50[c]	6.2	1.1	0.08
1994[j]	100	0.50[c]	6.2	1.1	0.08
1995	100	0.50[c]	6.2	1.1	0.08
1996[j]	50	0.24[d]	3.1	0.7	
1996[j]	50	0.35	4.5	0.95	0.07
1997–2003	100	0.24[d]	3.1	0.7	
1997–2003[e]	100	0.35	4.5	0.95	0.07

[a] The useful life of the emissions control system is expected to be five years or 80,000 km, unless otherwise noted.
[b] NMHC = nonmethane-reactive hydrocarbons.
[c] Total hydrocarbons.
[d] A set of intermediate in-use standards also applies during the phase-in period, 1994–1997 for passenger cars and small light-duty trucks, and 1996–1998 for larger light-duty trucks.
[e] Useful life is 10 years or 160,000 km; in-use compliance is seven years or 120,000 km.
[f] After an EPA/OTA study due June 1, 1997, the EPA shall by December 31, 1999, set standards more stringent than 1996 standards for passenger cars and certain light trucks, effective after January 1, 2003, but not later than model year 2006.
[g] Corresponds to 40% of fleet for particulate.
[h] Corresponds to 80% of fleet for particulate.
[i] 100% of fleet in 1997 and thereafter.
[j] Useful life is 11 years or 192,000 km; in-use compliance is seven years or 144,000 km starting 1994.
[k] Test weight = (gross vehicle weight rating + curb weight)/2.

lytic converter contained an oxidation catalyst to control hydrocarbon (HC) and carbon monoxide emissions. Exhaust gas recirculation was used to control NO_x emissions. Subsequently, the three-way catalyst was developed to control HC, CO, and NO_x in a single catalyst. Over 200 million vehicles have been equipped with the catalytic converter which has been rated among the top 10 engineering breakthroughs of the twentieth century (2). Emission control is achieved without negatively affecting fuel economy or performance. The shift to alternative transportation fuels such as methanol, ethanol, natural gas, liquefied petroleum gas (LPG), and reformulated gasoline in accordance with the Clean Air Act and the National Energy Policy Act of 1992 (GAS, NATURAL) is expected to produce further modifications to the catalytic converter.

Diesel engine emission control technology is not discussed herein. As of early 1993 only one passenger car manufacturer was marketing diesel fuel cars. Emission control technology for diesel engines, used in some light-duty trucks and in medium- and heavy-duty trucks, is evolving at a rapid pace. This technology includes engine modifications such as high pressure fuel injection, variable valve timing, and intercooled turbochargers. Catalytic aftertreatment that is quite different from that discussed herein has also been developed.

See also AIR POLLUTION: AUTOMOBILE; AUTOMOTIVE EMISSIONS; CLEAN AIR ACT OF 1990, MOBILE SOURCES.

EMISSION REGULATIONS AND TESTING

In the United States, federal regulations require automobile manufacturers to certify that vehicles are in compliance with exhaust emission standards when tested under specific test procedures.

Clean Air Act Amendments

Exhaust emission standards were set by the Clean Air Act Amendments of 1970 requiring automobile manufacturers to control the amount of hydrocarbons, carbon monoxide, and oxides of nitrogen emitted from vehicles. The act was amended in 1990. Table 1 summarizes the emission standards for passenger cars and light trucks between the years 1991 and 2006. In 1994, emissions from vehicles come under more stringent control, and the useful life of the emission control system is extended from five years or 80,000 km to 10 years or 160,000 km. The regulations are set up so that after 80,000 km or five years of actual use, a slightly more relaxed standard is applicable for the next 80,000 km.

Test Procedure

To comply with emission standards, representative vehicles must be run for 80,000 km (Appendix IV of the Federal Test Procedure (FTP)) (3). The first 6,400 km are considered a break-in portion. Exhaust emissions are measured each 8,000 km between approximately 6,400 and 80,000 km of accumulation and a deterioration factor (DF) of emissions is calculated. A DF of 1.15 for HC indicates that HC emissions increased by 15% between 6,400 and 80,000 km, and were within the 80,000 km standard. This DF is applied to the 6,400 km emission test data points for all other model variations of the family of vehicles represented by the 80,000 km durability car.

The test for evaluating individual vehicle emissions, the FTP (4), specifies that a test vehicle be stored in an area where the ambient temperature is between 20 and 29°C for at least 12 hours immediately prior to the emission test. Then, the vehicle is placed on a chassis dynamometer which is calibrated for the vehicle weight and road load. The vehicle is started and driven for 41 min on a prescribed cycle of accelerations, cruises, decelerations, idles, a 10-min shutdown (called the hot soak), and a period of rerun.

During the FTP, all exhaust passes through a constant volume sampling system (CVS) which dilutes the exhaust with air so that the total flow of air plus exhaust is con-

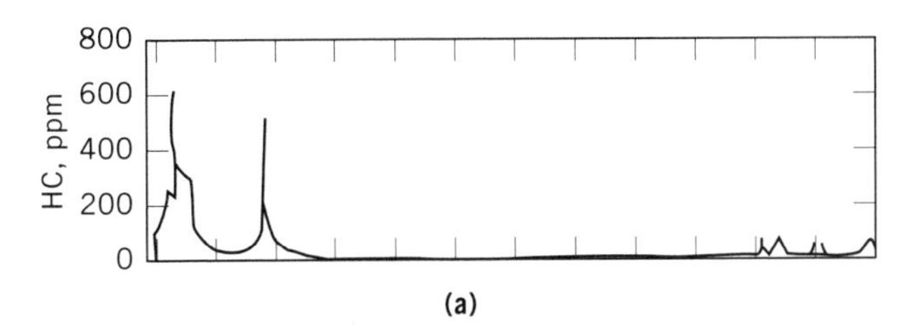

(a)

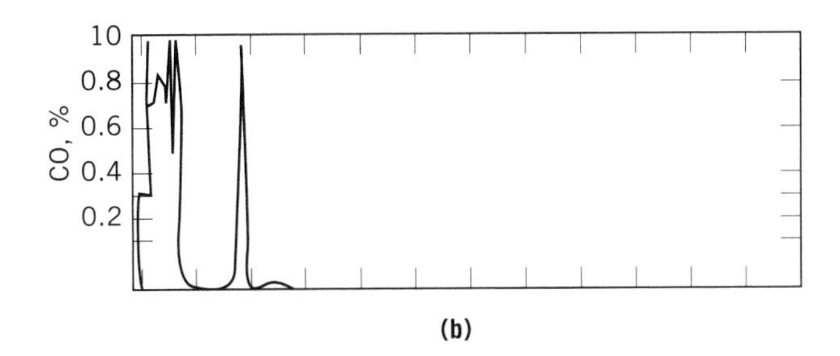

(b)

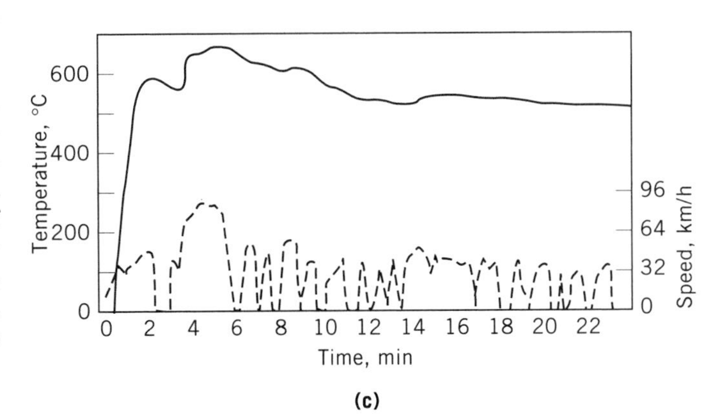

(c)

Figure 1. CO and hydrocarbon tailpipe emissions. Data from a test vehicle during a test cycle where the catalyst was mounted ~1.2 m from the exhaust part of the engine: (**a**) hydrocarbon and (**b**) CO tailpipe emissions; and (**c**), (—) catalyst temperature and (---) speed. As can be seen, the principal CO and hydrocarbon emissions occur during the cold start for this vehicle, ie, during the catalyst warmup period. When hot, the catalyst is very effective. In practice, one can expect between 60 and 90% of the engine CO and hydrocarbon emissions, as measured over the whole test cycle, to be removed by the catalyst after 50,000 miles of use (6).

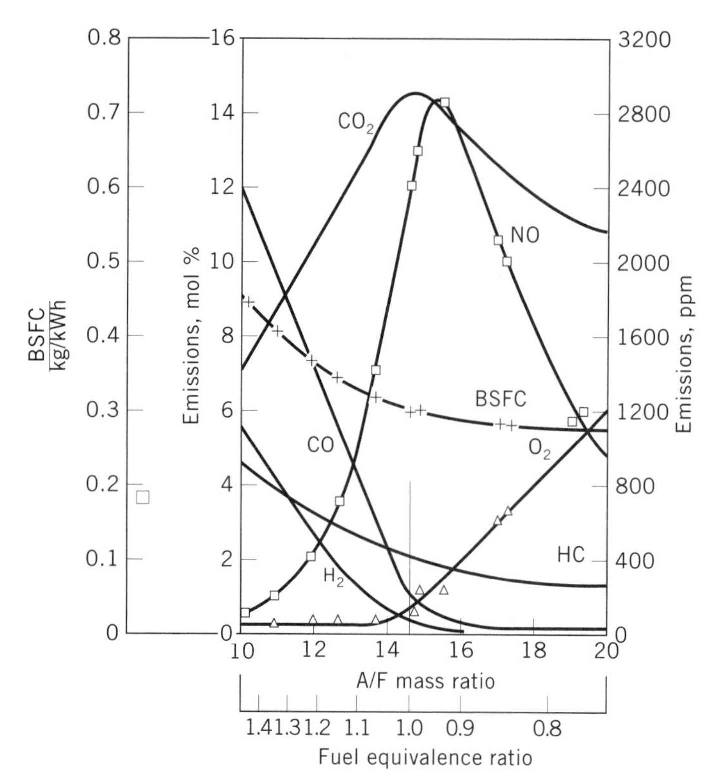

Figure 2. Effect of mixture strength on exhaust gas composition (dry basis) and brake specific fuel consumption (BSFC) for an unsupercharged automotive-type engine using indolene fuel, H/C = 1.86, where the ignition is tuned to achieve maximum best torque (MBT), the brake mean effective pressure (BNEP) is 386 kPa at 1200 rpm (7,8).

stant. A sample of the diluted gas is metered to three gas sample collection bags sequentially. The first, or cold start bag, contains the gas sample from the first 505 seconds of the test. The second bag is the gas sample of the hot transient portion of the test, between 506 and 1372 seconds, after which time the ignition is turned off for 10 minutes. The third bag gas sample is taken after the 10 min hot soak, from the point of ignition of the hot restart, for 505 seconds. Each bag is then measured for hydrocarbons, carbon monoxide, and oxides of nitrogen using, respectively, a flame-ionization detector (FID) HC analyzer, an infrared CO analyzer, and a chemiluminescent NO_x analyzer. A CO_2 analyzer (NDIR) is used to calculate dilution. The concentration measurements are converted into grams of each emission per unit of distance. The FTP prescribes weighting factors for each phase to give a composite value for the total test.

Typical HC and CO emissions from a vehicle undergoing the FTP are shown in Figure 1**a**. Figure 1**b** shows the inlet gas temperature to a typical underfloor catalytic converter. The HC and CO are very high during the first 100 seconds after the engine is started, and drop off precipitously after the engine is up to running temperature. From 75 to 85% of the HC and CO emissions pass through the catalytic converter during the time the converter is being heated by the hot exhaust gases (5). Very little of these gases pass unconverted through a hot catalyst of a properly calibrated system.

EXHAUST GAS COMPOSITION

The exhaust composition from gasoline/air combustion is dependent on many factors. Total combustion in the engine is not possible even when excess oxygen is present (6). Formation of the air/fuel mixture as well as the design of the combustion chamber influence the combustion process, as do engine power and ignition system timing. However, for emission control, the main factor affecting the composition of the exhaust gas is the air/fuel mixture or ratio. The composition of exhaust varies according to the air/fuel ratio as shown in Figure 2 (7) for a standard gasoline fuel where the hydrogen to carbon ratio (H/C) is 1.86. The exhaust gas composition changes as the H/C ratio changes (9), as shown in Figure 3 (8,10,11).

Unburned hydrocarbons in the exhaust originate primarily from crevices in the combustion chamber, such as gaps between the piston and cylinder wall, where the combustion flame cannot burn. The composition of unburned hydrocarbons is dictated primarily by the composition of the fuel (12). Carbon monoxide results from areas of insufficient oxygen. Oxides of nitrogen are produced in the high temperature zones during combustion by the reaction of

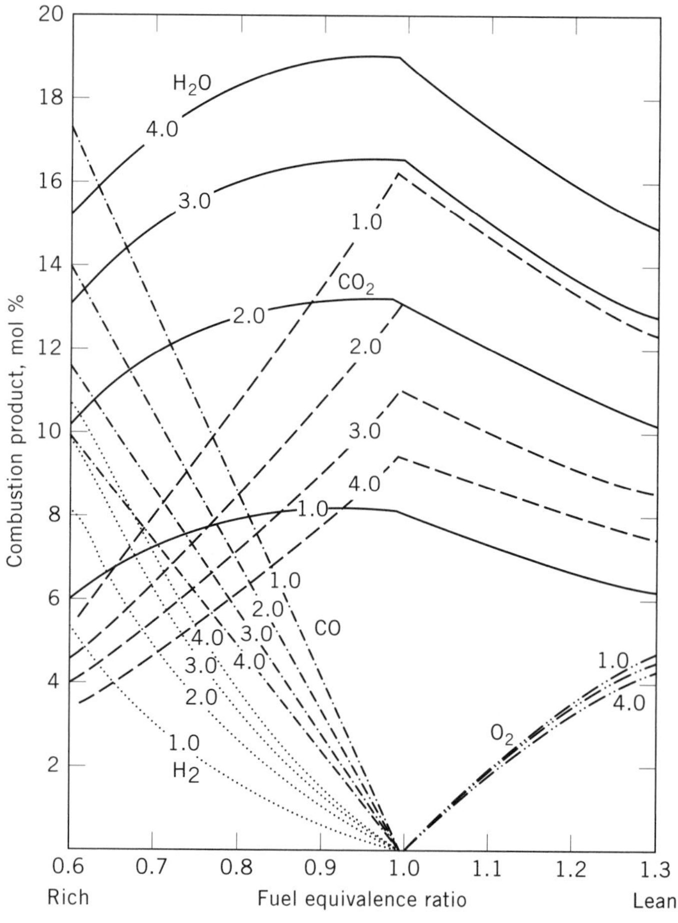

Figure 3. Theoretical mole percent of the principal combustion products of hydrocarbon fuels for fuel hydrogen:carbon ratios from 1, eg, C_6H_6, to 4, eg, CH_4, wet basis, where (—) represents H_2O; (– – –) CO_2; (–·–·–) CO; (–··–··–) O_2; and (·········) H_2 (8,10,11).

nitrogen molecules and oxygen atoms thermally produced from oxygen and oxygen-containing species, according to the Zeldovich mechanism (6,13).

Hydrocarbons and carbon monoxide emissions can be minimized by lean air/fuel mixtures (Fig. 2), but lean air/fuel mixtures maximize NO_x emissions. Very lean mixtures (> 20 air/fuel) result in reduced CO and NO_x, but in increased HC emissions owing to unstable combustion. The turning point is known as the lean limit. Improvements in lean-burn engines extend the lean limit. Rich mixtures, which contain excess fuel and insufficient air, produce high HC and CO concentrations in the exhaust. Very rich mixtures are typically used for small air-cooled engines, needed because of the cooling effect of the gasoline as it vaporizes in the cylinder, where CO exhaust concentrations are 4 to 5% or more.

The best power is achieved slightly rich of stoichiometric air/fuel; the best fuel economy is achieved slightly lean of stoichiometric mixture.

Over 150 hydrogen and carbon species are present in the exhaust of a gasoline fueled engine (14–18) including methane, various paraffins, olefins, aldehydes, aromatics, and polycyclic hydrocarbons as well as unburned gasoline. Exhaust gas also contains sulfur dioxide from the combustion of sulfur contained in the gasoline. The average U.S. gasoline sulfur content is about 300 ppm, which results in about 20 ppm SO_2 in the exhaust. Gasoline sulfur has ranged between 50 and >1000 ppm in the United States. Reformulated gasolines are expected to have lower sulfur because of the greater processing by the petroleum (qv) refinery (11,12). Additionally, the State of California has specified a Phase II gasoline having a sulfur content between 30–40 ppm for the LEV program.

Exhaust gas also contains small amounts of hydrogen cyanide and ammonia depending on the air/fuel ratio.

EMISSION CONTROL SYSTEM

A typical 1993 model year automobile emission control system containing a multipoint fuel injection fuel metering system, which meters the fuel in response to a measured amount of air, is shown in Figure 4. Incorporated into the exhaust stream prior to the catalytic converter is an oxygen sensor which indicates whether the exhaust air/fuel mixture is rich or lean of the stoichiometric point (defined as neither air nor fuel in excess). The catalytic converter is located in the exhaust line leading from the exhaust manifold, upstream of the acoustic muffler. The signal generated by the oxygen sensor is sent to the computer controller as is a signal from an air flow measurement device. The computer controller regulates the fuel metered by the fuel injection system in response to the air measurement signal. Air flow varies in response to the throttle position and load (inlet vacuum). The oxygen sensor quickly detects any change in oxygen concentration in the exhaust and the controller adjusts for this change. Thus the air/fuel ratio is constantly being adjusted back and forth slightly rich and slightly lean of the stoichiometric mixture. The three-way conversion (TWC) catalyst therefore receives exhaust gas that reflects this constant change back and forth in air/fuel mixture and is designed to operate under those conditions to convert NO_x by reduction and HC and CO by oxidation of at least 80 to 90%.

Catalytic Converter

The converter consists of a catalytic unit contained in a metal canister which surrounds the fragile ceramic catalytic unit with a steel shell. The converter shell assembly is usually made from an iron/chrome Series 409 muffler-grade stainless steel that is resistant to internal and external oxidative corrosion. In between the steel shell and the exterior of the catalytic unit is a compliant layer that grips the catalytic unit with sufficient force to prevent movement of the catalytic unit within the canister, and which compensates for the differences in thermal expansion between the catalyst and the metal shell. Several types of compliant layers are used, and all have spring-like properties under compression that provide the necessary gripping force at all exhaust temperatures. Corrugated knitted wire mesh was the first successful compliant layer. As of this writing, a material based on vermiculite, which expands upon application of heat (about 300°C), is used, as is a wire mesh material wrapped several times around the catalytic unit. The compliant layer mounting system has proved to be durable for the life of the vehicle. The converter design, flow, and pressure drop characteristics are described in the literature (19–23).

The catalytic unit is designed to provide enough surface area so that all exhaust gases contact the catalyst surfaces as they pass from the engine to the tailpipe (24–27). In order to function quickly after the engine is started, the catalytic unit must rapidly heat up to operating temperature. It therefore must possess good heat-exchange properties to extract the necessary heat from the exhaust gas. Once the minimum catalytic operation temperature is reached, the catalytic unit is designed to maximize transfer of the pollutants from the exhaust gas to the surface of the catalytic unit. Heat transfer and mass transfer (qv) are driven by temperature difference and concentration difference, respectively. At operating temperatures above 300 or 350°C, the catalytic reactions are so fast that only the exterior surfaces of the catalyst are utilized for the catalytic function (28).

Automobile exhaust catalysts have been developed that maximize the catalyst surface area available to the flowing exhaust gas without incurring excessive pressure drop. Two types have been extensively studied: the monolithic honeycomb type and the pellet type.

Use of the pelletted converter, developed and used by General Motors starting in 1975, has declined since 1980. The advantage of the pelletted converter, which consists of a packed bed of small spherical beads about 3 mm in diameter, is that the pellets were less costly to manufacture than the monolithic honeycomb. Disadvantages were the pelletted converter had 2 to 3 times more weight and volume, took longer to heat up, and was more susceptible to attrition and loss of catalyst in use. The monolithic honeycomb can be mounted in any orientation, whereas the pelletted converter had to be downflow. Additionally, the pressure drop of the monolithic honeycomb is one-half to one-quarter that of a similar function pelletted converter.

Pelletted converters are used by General Motors for

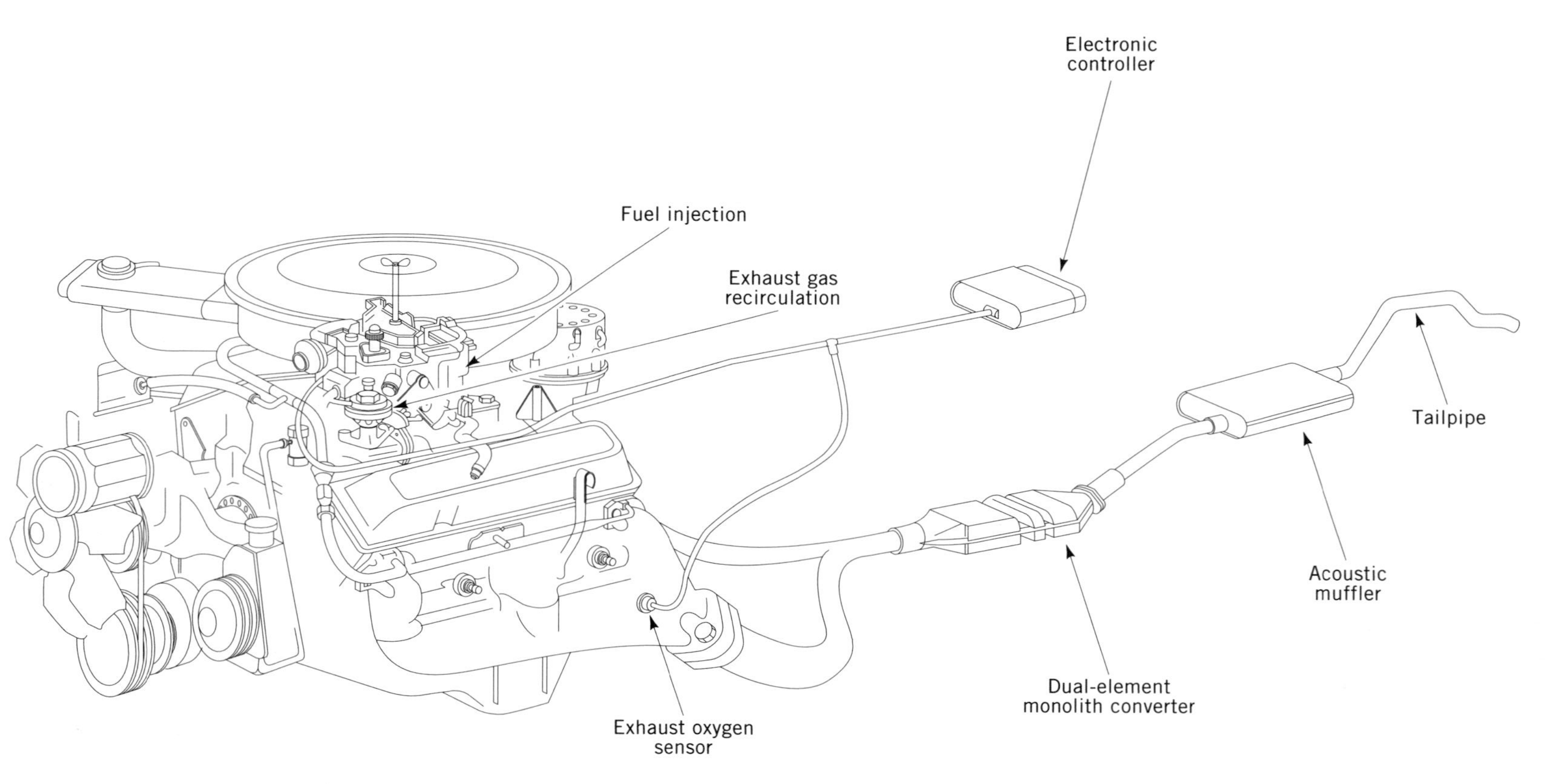

Figure 4. Closed-loop dual-catalyst system for emissions control using dual element monolith converter, which is three-way and oxidizing.

heavy-duty gasoline fueled trucks because the pellets have better high temperature physical characteristics and the converter has been redesigned to minimize pressure drop for this application (19). However, these are expected to be replaced by monolithic converters. In Japan, taxicabs have used pelletted converters and LPG (propane) fuel since the 1960s. The catalyst in these converters is changed once per year. It is thought easier to change pellets than to change a monolithic honeycomb catalyst. Although this practice of changing catalyst remains, the monolithic honeycomb converter remains active and fully functional for the life of the vehicle, especially when operated on a clean fuel such as LPG.

Catalytic Unit

The catalytic unit consists of an activated coating layer spread uniformly on a monolithic substrate. The catalyst predominantly used in the United States and Canada is known as the three-way conversion (TWC) catalyst, because it destroys all three types of regulated pollutants: HC, CO, and NO_x. Between 1975 and the early 1980s, an oxidation catalyst was used. Its use declined with the development of the TWC catalyst. The TWC catalytic efficiency is shown in Figure 5. At temperatures of >300°C a TWC destroys HC, CO, and NO_x effectively when the air/fuel mixture is close to stoichiometric, as shown. Other conventions for describing the air/fuel mixture are use of the Greek letter lambda (λ) where $\lambda = 1.0$ is the stoichiometric mixture, REDOX ratio, and equivalence ratio. When the air/fuel mixture is near the stoichiometric point, an optimum value of conversion for all three components is achieved. The principal chemical reactions of a TWC catalyst are

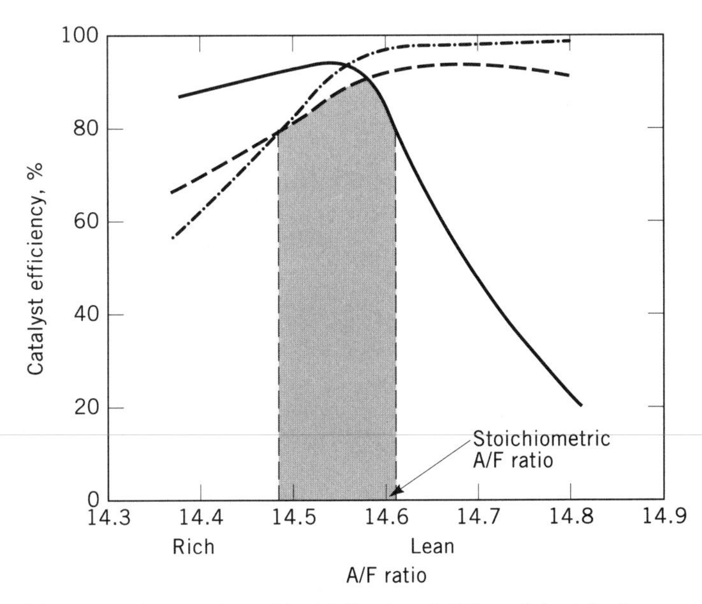

Figure 5. Conversion of (—) NO_x, (–·–) CO, and (– –) hydrocarbons for a TWC as a function of the air/fuel (A/F) ratio. The shaded area shows the A/F ratio window where the TWC catalyst functions at 80% efficiency at 400°C.

Oxidation reactions

$$H_2 + \tfrac{1}{2} O_2 \rightarrow H_2O$$

$$CO + \tfrac{1}{2} O_2 \rightarrow CO_2$$

$$C_mH_n + (m + n/4)\, O_2 \rightarrow m\, CO_2 + n/2\, H_2O$$

TWC reactions at stoichiometric A/F mixture

$$CO + NO \rightarrow \tfrac{1}{2} N_2 + CO_2$$

$$C_mH_n + 2(m + n/4)\, NO \rightarrow (m + n/4)\, N_2 + n/2\, H_2O + m\, CO_2$$

$$H_2 + NO \rightarrow \tfrac{1}{2} N_2 + H_2O$$

$$CO + H_2O \rightarrow CO_2 + H_2 \quad \text{water gas shift}$$

$$C_mH_n + m\, H_2O \rightarrow m\, CO + \left(m + \frac{n}{2}\right) H_2 \quad \text{steam reforming}$$

Other NO-reduction reactions

$$2\, NO + 5\, CO + 3\, H_2O \rightarrow 2\, NH_3 + 5\, CO_2$$

$$2\, NO + 5\, H_2 \rightarrow 2\, NH_3 + 2\, H_2O$$

$$NO + \text{hydrocarbons} \rightarrow N_2 + H_2O + CO_2 + CO + NH_3$$

$$\left.\begin{array}{l} 2\, NO + CO \rightarrow N_2O + CO_2 \\ 2\, NO + H_2 \rightarrow N_2O + H_2O \end{array}\right\} \text{at 200°C (below exhaust gas temperature)}$$

Fuel sulfur reactions

$$S + O_2 \rightarrow SO_2$$

$$SO_2 + \tfrac{1}{2} O_2 \xrightarrow[Pt]{} SO_3$$

$$3\, SO_3 + Al_2O_3 \overset{>600°C}{\rightleftharpoons} Al_2(SO_4)_3$$

$$Al_2(SO_4)_3 + 12\, H_2 \xrightarrow[\text{PM catalyst}]{} 3\, H_2S + Al_2O_3 + 9\, H_2O$$

$$SO_2 + 3\, H_2 \xrightarrow[\text{PM catalyst}]{} H_2S + 2\, H_2O$$

where PM is precious metal (6,29).

Activated Coating. The activated coating layer of a TWC is applied to the high geometric surface area monolithic honeycomb body or substrate. Precious metals are dispersed within this catalytic layer, which contains particles as small as 10 μm. The thickness of the layer is between 25 and 100 μm. The honeycomb substrate typically has a geometric surface area of 27 cm^2/cm^3 so that a typical vehicle catalyst volume of 1.6 L would contain 4.6 m^2 of geometric surface area. The activated coating layer increases the total surface to 7000 m^2, ie, more than three orders of magnitude. The total surface area consists of the combined internal and external surface of all the minute particles contained in the activated coating layer. Total surface area is measured by the classic Brunauer-Emmet-Teller (BET) technique, which involves N_2 adsorption at low temperature (30).

The small (10 μm) coating particles are typically aluminum oxide [*1344-28-1*], Al_2O_3. These particles can have

BET surface areas of 100 to 300 m^2/g. The thermal and physical properties of alumina crystalline phases vary according to the starting phase (aluminum hydroxide or hydrate) and thermal treatment.

Alumina is used because it is relatively inert and provides the high surface area needed to efficiently disperse the expensive active catalytic components. However, no one alumina phase possesses the thermal, physical, and chemical properties ideal for the perfect activated coating layer. A great deal of research has been carried out in search of modifications that can make one or more of the alumina crystalline phases more suitable. For instance, components such as ceria, baria, lanthana, or zirconia are added to enhance the thermal characteristics of the alumina. Figure 6 shows the thermal performance of an alumina-activated coating material.

An unstabilized high surface area alumina sinters severely upon exposure to temperatures over 900°C. Sintering is a process by which the small internal pores in the particles coalesce and lose large fractions of the total surface area. This process is to be avoided because it occludes some of the precious metal catalyst sites. The network of small pores and passages for gas transfer collapses and restricts free gas exchange into and out of the activated catalyst layer resulting in thermal deactivation of the catalyst.

The activated coating layer must possess two additional properties. It must adhere tenaciously to the monolithic honeycomb surface under conditions of rapid thermal changes, high flow, and moisture condensation, evaporation, or freezing. It must have an open porous structure to permit easy gas passage into the coating layer and back into the main exhaust stream. It must maintain this porous structure even after exposure to temperatures exceeding 900°C.

Precious Metal Catalysts. Precious metals are deposited throughout the TWC-activated coating layer. Rhodium plays an important role in the reduction of NO_x, and is combined with platinum and/or palladium for the oxidation of HC and CO. Only a small amount of these expensive materials is used (31) The metals are dispersed on the high surface area particles as precious metal solutions, and then reduced to small metal crystals by various techniques. Catalytic reactions occur on the precious metal surfaces. Whereas metal within the crystal cannot directly participate in the catalytic process, it can play a role when surface metal oxides are influenced through strong metal to support reactions (SMSI) (32,33). Some exhaust gas reactions, for instance the oxidation of alkanes, require larger Pt crystals than other reactions, such as the oxidation of CO (34).

The small precious metal crystals can exist as metal crystallites or as metal oxides, both of which are catalytic (31). Rodium oxide has a tendency to react with alumina to form a solid solution (35). To minimize this reaction, zirconia is used with the alumina (36). Publications regarding the TWC function of precious metals abound (37–42).

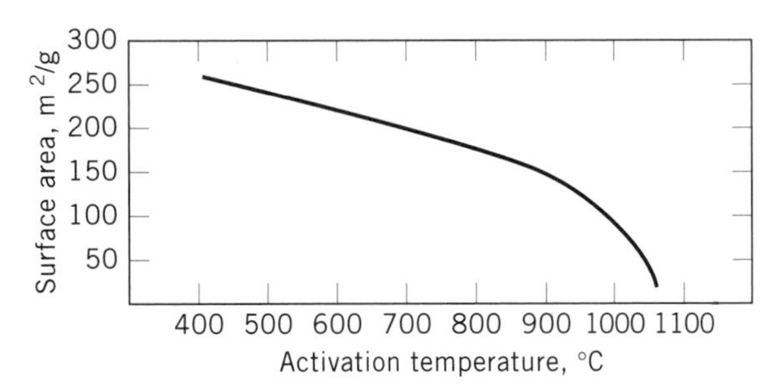

Figure 6. Surface area stability of Pural SB. The activation time was 3 h.

Catalytic Support Body Monolithic Honeycomb Unit

The terms substrate and brick are also used to describe the high geometric surface area material upon which the active coating material is placed. Monolithic honeycomb catalytic support material comes in both ceramic and metallic form. Both are used in automobile catalysts and each possesses unique properties. A common property is a high geometric surface area which is inert and does not react with the catalytic layer.

Ceramic. The ceramic substrate is made from a mixture of silicon dioxide, talc, and kaolin to make the compound cordierite [*12182-53-5*]. Cordierite possesses a very low coefficient of thermal expansion and is thermal-shock resistant. The manufacturing process involves extruding the starting mixture (which is mixed with water and kneaded into a sort of dough) through a complex die to form the honeycomb structure. The extruded piece is dried and fired in a kiln to form the cordierite. The outside or circumferential dimension is formed by the die, and the length is cut later with a ceramic saw.

The extrusion of the honeycomb was a significant advance aiding the adoption of the monolithic honeycomb catalyst by the automobile industry as the preferred means of emission control (43). Coupled with this was the development of the mounting system (44) securing the fragile ceramic substrate-based catalytic unit in a protective canister. Some of the physical properties considered when selecting a substrate for a catalytic converter are given in Table 2. The most common cell structure used is 62 $cells/cm^2$ with a 0.152 mm thick wall. This cell density is about optimal with respect to pressure drop, geometric surface area, physical strength, and general ruggedness (24–27).

Metallic. Metallic substrates are also used as a support for the activated coating layer. A class of metal alloys containing Fe, Cr, and Al, when stabilized with Y or Ce, have excellent oxidation resistance at the extreme temperatures found in automobile exhaust (45). Melting temperatures are about in the same range as that of the ceramic cordierite. A feature of these alloys is that upon heating in air, an aluminum oxide surface layer is formed. The yttria or ceria is present to stabilize the aluminum oxide surface and resist the spalling of the alumina which would otherwise occur. The ceria-stabilized alloy develops a whisker-like surface. The catalytic activated coating layer has been found to adhere very well to these surfaces.

Table 2. Physical Properties of Catalytic Converter Substrates

Parameter	Cordierite Value		
Cell structure			
Wall thickness, mm	0.152	0.203	0.304
cells/cm^2	62	46	46
Pressure drop, %	100	90	120
Geometric surface area, cm^2/cm^3	27	24	22
Isostatic strength,[a] N/m^2	>103	>103	>206
Bulk density, g/cm^3	0.4	0.4	0.5
Cross-sectional shape			

[a] To convert N/m^2 (=Pa) to kgf/cm^2, multiply by 1.02×10^{-5}.

The metallic substrate is designed to provide a cell density similar to that of the ceramic counterpart, and can be constructed in versions of 46, 62, or 93 cells/cm^2. However, the wall thickness can be thinner, ie, 0.05 to 0.15 mm thick (46). Additionally, 0.05 mm thick wall material can be constructed in versions of 124 to 248 cells/cm^2. The unique advantage that a metal substrate has over its ceramic counterpart is that either the same geometric surface can be made into a smaller volume, or that for the same volume and geometric surface area, there is a lower pressure drop (47). The 124 to 248 cells/m^2 metal versions have an additional potential performance advantage over ceramic because of large increases in geometric surface area. Metallic substrate catalysts are used by Porsche for high performance vehicles and by Chrysler for the Viper model (45,48). In both of these cases, the lower exhaust gas pressure drop results in increased horsepower and performance. In the future, metallic supports may be used for small catalysts located close to the exhaust manifold in order to heat up faster in emission control systems calibrated to meet the strict California Low Emission Standards.

Mass Transfer

Exhaust gas catalytic treatment depends on the efficient contact of the exhaust gas and the catalyst. During the initial seconds after start of the engine, hot gases from the exhaust valve of the engine pass through the exhaust manifold and encounter the catalytic converter. Turbulent flow conditions (Reynolds numbers above 2000) exist in response to the exhaust stroke of each cylinder (about 6 to 25 times per second) times the number of cylinders. However, laminar flow conditions are reached a short (~ 0.6 cm) distance after entering the cell passages of the honeycomb (5,49–52).

The process of catalyst heating and initiation of the catalytic function is shown in Figure 7, where there are three or four distinct regions. Depending on the location of the catalytic unit in the exhaust system and the thermal mass present prior to the catalytic unit, it can take from 30 to 120 seconds for the catalyst unit to reach the catalyst ignition temperature of approximately 250–300°C (Region I). At temperatures above this ignition point, the activity of the catalyst increases rapidly with temperature (Region II). Some catalyst reaches a point where the sharp rise with temperature abruptly takes on a mild positive slope (Region III). Then, a point is reached at which catalytic performance improves only slightly to an increase in temperature (Region IV).

In Region I (below the ignition point) there is no catalytic activity. Within Region II, the specific activity of the catalyst is the rate-limiting step; this is called the kinetically controlled region. Highly active catalysts have a lower ignition point and exhibit large increases in catalytic performance associated with small increases in temperature. The behavior of a catalyst in Region II is important to selecting an auto emission catalyst because catalyst light-off is a prime factor in achieving adequate emission control early in the FTP test.

Region III is not present in all catalysts. It depends on the porous structure design of the catalyst and is often only found in a used catalyst. If present, the rate-limiting step is called pore diffusion control. The volume of the catalytic unit dictates the catalyst performance in Regions III and IV. Catalysts clogged with masking poisons such as lube oil ash would exhibit Region III behavior. Region IV is a mass-transfer limited region, ie, catalytic reactions occur so fast that the rate-limiting factor is getting the

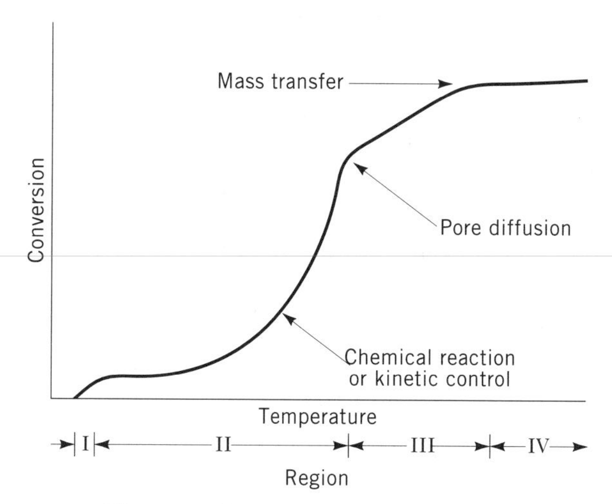

Figure 7. Conversion vs temperature.

Table 3. Specific Reaction Rates

	$CO + O_2$[a]		$C_2H_4 + O_2$[b]	
Catalyst	Temperature, °C	R^c	Temperature, °C	R^c
Pd (wire)	300	500	300	100
Pt (wire)	300	100	300	10
Au (wire)	300	15	300	0.03
Co_3O_4	200	25 + 5	300	0.33
CuO	200	11	300	0.6
$CuCr_2O_4$	200	5	300	0.8
$LaCoO_3$	200	2.3	400	0.53
$BaCoO_3$	300	5.3	400	0.1
SnO_2	300	5.2	400	0.4
MnO_2	300	3.4	300	0.04
$LaMnO_3$	300	2	400	0.3
$La_{0.5}Sr_{0.5}MnO_3$	300	1.2	400	0.26
$La_{0.5}Pb_{0.3}MnO_3$	300	0.5	400	<0.05
Fe_2O_3	300	0.4	400	0.06
$FeCr_2O_4$	300	0.33		
Cr_2O_3	300	0.03	300	0.006
NiO	300	1.5		
ZrO_2	300	0.013	400	0.002

[a] 1% O_2, 1% CO, 0% H_2O.
[b] 1% O_2, 0.1% C_2H_2, ~0.1% H_2O.
[c] Units in terms of CO_2 formation mL/(min · m²).

reactants to the surface of the catalyst. Mathematical models describing the entire process are found in several references (5,49–52).

Catalyst Function

Automobile exhaust catalysts are perfect examples of materials that accelerate a chemical reaction but are not consumed. Reactions are completed on the catalyst surface and the products leave. Thus the catalyst performs its function over and over again. The catalyst also permits reactions to occur at considerably lower temperatures. For instance, CO reacts with oxygen above 700°C at a substantial rate. An automobile exhaust catalyst enables the reaction to occur at a temperature of about 250°C and at a much faster rate and in a smaller reactor volume. This is also the case for the combustion of hydrocarbons.

Concerning the reduction of NO_x, automobile three-way catalysts exhibit a property called selectivity. Catalyst selectivity occurs when several reactions are thermodynamically possible but one reaction proceeds at a faster rate than another. In the case of a TWC catalyst, CO, HC, and H_2 are all potential reductants of NO. On the other hand, O_2 is present, which oxidizes the CO, HC, and H_2. If these oxidation reactions are too rapid, no reductant is available to convert NO. Using modern TWC catalysts, however, NO reduction is fast enough that it is substantially completed before the reductants are consumed by O_2.

Two classes of metals have been examined for potential use as catalytic materials for automobile exhaust control. These consist of some of the transitional base metal series, for instance, cobalt, copper, chromium, nickel, manganese, and vanadium; and the precious metal series consisting of platinum [*7440-06-4*], Pt; palladium [*7440-05-3*], Pd; rhodium [*7440-16-6*], Rh; iridium, [*7439-88-5*], Ir; and ruthenium [*7440-18-8*], Ru. Specific catalyst activities are shown in Table 3.

The precious metals possess much higher specific catalytic activity than do the base metals. In addition, base metal catalysts sinter upon exposure to the exhaust gas temperatures found in engine exhaust, thereby losing the catalytic performance needed for low temperature operation. Also, the base metals deactivate because of reactions with sulfur compounds at the low temperature end of auto exhaust. As a result, a base metal automobile exhaust catalyst would need to be considerably larger than a precious metal one and, even if a large bed were used, it would not heat up quickly enough to achieve the catalytic performance demanded of the emission control systems (6).

Catalyst function in an exhaust gas stream can be understood by examining the catalyst performance curves shown in Figures 5 and 8. Figure 8 shows catalyst function at a specific set of flow conditions as the catalyst inlet tem-

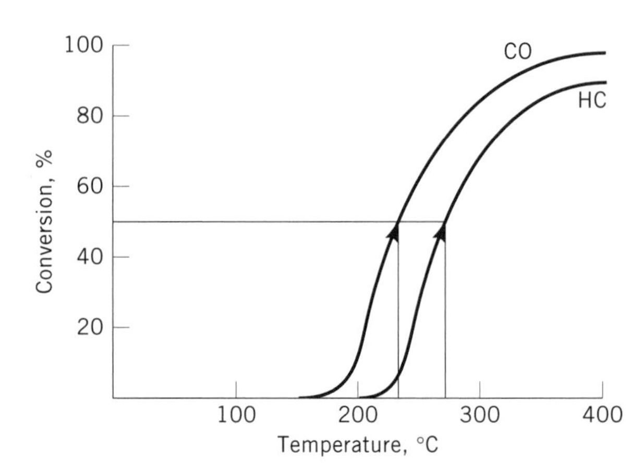

Figure 8. Effect of temperature on catalyst performance using synthetic exhaust: 0.45% CO, 12% CO_2, 500 ppm NO, 0.15% H_2, 3% O_2, 200 ppm C_3H_6, 10% H_2O, and remainder N_2. The T_{50} for CO and HC are 230°C and 270°C, respectively. See text.

perature is slowly increased. This test, conducted in a laboratory reactor, uses a gaseous mix to simulate the composition of automobile exhaust. As the temperature increases from ambient to about 200°C, there is no apparent action on the part of the catalyst to consume any of the reactants. The carbon monoxide present strongly chemisorbs on the surface of the catalyst, and prevents oxygen access. As the inlet gas temperature approaches 200°C, the CO bonds to the metal surface are relaxed. Oxygen molecules are now able to chemisorb, and catalytic ignition occurs.

Beyond the catalytic ignition point there is a rapid increase in catalytic performance with small increases in temperature. A measure of catalyst performance has been the temperature at which 50% conversion of reactant is achieved. For carbon monoxide this is often referred to as T_{50} CO. The catalyst light-off property is important for exhaust emission control because the catalyst light-off must occur reliably every time the engine is started, even after extreme in-use engine operating conditions.

A representation of a catalytic reaction is shown in Figure 9. A gaseous pollutant such as CO diffuses from the main gas stream into the porous catalyst matrix. There is a net CO flow from the main flow stream to the catalytic surface because CO is at higher concentration in the exhaust stream, and it is attracted to a precious metal site that is temporarily unoccupied. For the same reasons, oxygen molecules are attracted to the site. The oxygen molecule is chemisorbed onto the precious metal surface where oxygen molecule bond stretching occurs, leading to dissociation to oxygen atoms that bond to the catalyst surface. The rate at which the reaction proceeds depends strongly on the temperature at the catalyst surface. The reaction of carbon monoxide and oxygen proceeds to form CO_2, which desorbs and reenters the exhaust. There is less CO_2 in the flowing exhaust stream and more at the surface so there is a net flow of CO_2 away from the catalyst surface.

The oxidation of HC and CO must proceed in balance with the reduction of NO_x by CO, HC, or H_2. For the NO_x removal reaction, a reductant is required. First NO is adsorbed on the catalyst surface and dissociates forming N_2 which leaves the surface, but the O atoms remain. CO is required to remove the O atoms to complete the reaction cycle (53).

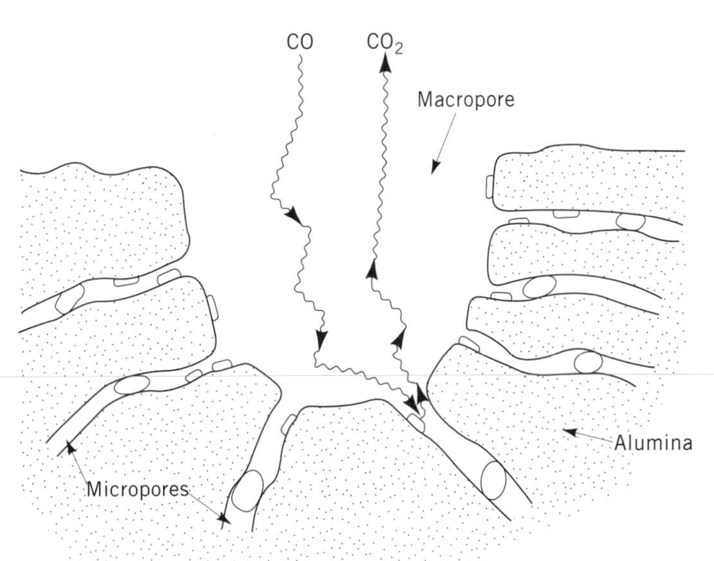

Figure 9. Catalyst pore and reaction. The CO diffuses into a precious metal site ; reacts with O_2; and leaves as CO_2.

A TWC catalyst must be able to partition enough CO present in the exhaust for each of these reactions and provide a surface that has preference for NO adsorption. Rhodium has a slight preference for NO adsorption rather than O_2 adsorption; Pt prefers O_2. Rh also does not catalyze the unwanted NH_3 reaction as does Pt, and Rh is more sinter-resistant than Pt (6). However, the concentrations of O_2 and NO have to be balanced for the preferred maximum reduction of NO and oxidation of CO. This occurs at approximately the stoichiometric point with just enough oxidants (O_2 and NO_x) and reductants (CO, HC, and H_2). If the mixture is too rich there is not enough O_2 and no matter how active the catalyst, some CO and HC is not converted. If the mixture is too lean, there is too much O_2 and the NO cannot effectively compete for the catalyst sites (53–58).

In an actual exhaust system controlled by the signal of the oxygen sensor, stoichiometry is never maintained, rather, it cycles periodically rich and lean one to three times per second, ie, one-half of the time there is too much oxygen and one-half of the time there is too little. Incorporation of cerium oxide or other oxygen storage components solves this problem. The ceria adsorbs O_2 that would otherwise escape during the lean half cycle, and during the rich half cycle the CO reacts with the adsorbed O_2 (32,44,59–63). The TWC catalyst effectiveness is dependent on the use of Rh to reduce NO_x and upon the proper balance of reactants; the oxygen sensor plays the key role in maintaining this proper balance.

Catalyst Durability

Automobile catalysts last for the life of the vehicle and still function well at the time the vehicle is scrapped. However, there is potential for decline in total catalytic performance from exposure to very high temperatures, accumulation of catalyst poisons, or loss of the active layer (29,64–68).

High Temperatures. Exposure to temperatures above 1000°C can cause the active layer surfaces to sinter, collapse, agglomerate, shrink, and possibly exfoliate. Collapse of the surfaces occludes precious metal sites, preventing free passage to these sites. Also, the small precious metal crystals can unite and grow into larger crystals, thereby diminishing the total precious metal surface and thus the total number of active surface sites. If catalyst temperatures exceed 1200°C, the substrate softens and perhaps shrinks. At about 1465°C, the substrate melts and is destroyed. A thermally damaged pore is shown in Figure 10**a**.

High catalyst temperatures are caused by two types of factors. The first are engine related and can be corrected by calibration. That is, ignition timing and the air/fuel mixture, as well as the engine load, govern the exhaust gas temperature and thus the catalyst inlet gas temperature. The catalyst consumes unburned exhaust components, and most of the corresponding reactions are exo-

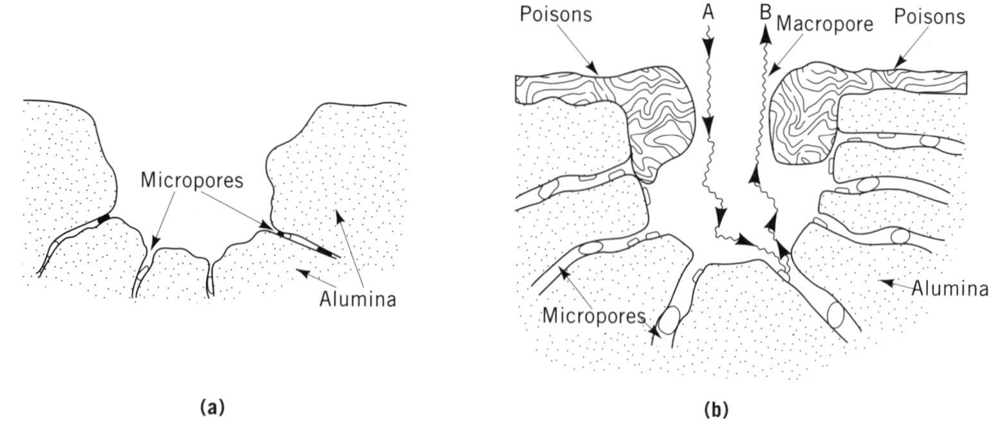

Figure 10. Catalyst macropores showing □ noble metal sites and (**a**) narrowed micropores after exposure to high temperatures where □ represents thermally damaged noble metal sites; and (**b**) pore mouth plugging from poisons where A, if allowed, diffuses in to be converted to B.

thermic, resulting in increased catalyst and exhaust gas temperatures (33). For instance, the combustion of 1% CO would result in a 87°C adiabatic temperature rise of the exhaust gas (33); the combustion of 1000 ppm C_6H_6 would yield a 103°C temperature rise.

The second type is the quality of combustion. A misfire or partial misfire of the air/fuel charge in the combustion chamber causes the unburned air and fuel mixture to pass to the catalyst, where combustion occurs. The temperature rise from combustion in the catalyst is considerable. Two misfiring cylinders in a four-cylinder engine could cause the catalyst to approach the melting temperature of the substrate (1465°C) (69).

There are many reasons for misfire. The principal causes are interruption of ignition energy caused by faulty or wet ignition wiring and either lean or rich noncombustible air/fuel mixtures. Electronic ignition, and the development of very durable ignition wiring, as well as highly improved fuel metering systems, have greatly lessened the occurrence of misfire. Unleaded fuel does not clog, corrode, or wear spark plugs, extending spark plug life. Thus the ignition and fuel metering systems are vastly improved over pre-1975, pre-catalyst equipped vehicles. Catalysts have been designed to resist thermal degradation by the incorporation of base metal oxide stabilizers (70).

Poisons and Inhibitors. Catalyst poisons and inhibitors can come from the fuel, the lube oil, from engine wear and corrosion products, and from air ingestion. There are two types of catalyst poisons: one poisons active sites, the other is a masking agent.

The main catalyst site poison for many years was tetraethyllead [*78-00-2*], $C_4H_{12}Pb$, even after use of unleaded gasoline. Not only is lead a catalyst poison, but automotive source lead is also a health hazard (66). The source of this lead came from manufacture and distribution of leaded and unleaded gasoline in common transport equipment and storage facilities (67). In the early 1990s, so little leaded gasoline was being distributed that Pb contamination was approaching zero ($<$0.26 ppm/L).

The mechanism of poisoning automobile exhaust catalysts has been identified (71). Upon combustion in the cylinder tetraethyllead (TEL) produces lead oxide which would accumulate in the combustion chamber except that ethylene dibromide [*106-93-4*] or other similar halide compounds were added to the gasoline along with TEL to form volatile lead halide compounds. Thus lead deposits in the cylinder and on the spark plugs are minimized. Volatile lead halides (bromides or chlorides) would then exit the combustion chamber, and such volatile compounds would diffuse to catalyst surfaces by the same mechanisms as do carbon monoxide compounds. When adsorbed on the precious metal catalyst site, lead halide renders the catalytic site inactive.

Lead compounds were not found on the surrounding activated coating layer, rather only associated with the precious metal. The Pt sites are less poisoned by lead than are Pd or Rh sites because the Pt sites are protected by the sulfur in the fuel. Fuel sulfur is converted to SO_2 in the combustion process, and Pt easily oxidizes SO_2 to SO_3 on the catalyst site. The SO_3 reacts with the lead compounds to form $PbSO_4$, which then moves off the catalyst site so that lead sulfate is not a severe catalyst poison. Neither Pd nor Rh is as active for the SO_2 to SO_3 reaction, and therefore do not enjoy the same protection as Pt.

Sulfur oxides resulting from fuel sulfur combustion often inhibit catalyst performance in Regions II, III, and a portion of Region IV (see Fig. 7) depending on the precious metals employed in the catalyst and on the air/fuel ratio. Monolithic catalysts generally recover performance when lower sulfur gasoline is used so the inhibition is temporary. Pd is more susceptible than Rh or Pt. The last is the most resistant. Pd-containing catalysts located in hotter exhaust stream locations, ie, close to the exhaust manifold, function with little sulfur inhibition (72–74).

Fuel sulfur is also responsible for a phenomena known as storage and release of sulfur compounds. Sulfur oxides (SO_2,SO_3) easily react with ceria, an oxygen storage compound incorporated into most TWC catalysts, and also with alumina. When the air/fuel mixture temporarily goes rich and the catalyst temperature is in a certain range, the stored sulfur is released as H_2S yielding a rotten egg odor to the exhaust. A small amount of nickel oxide incorporated into the TWC removes the H_2S and releases it later as SO_2 (75–79).

Masking agents also deactivate catalysts and are loosely called poisons. Small droplets of lubricating oil, for instance, are emitted from the engine and deposit on the surfaces of the catalyst. Lubricating oil contains a small amount of inorganic elements such as zinc, phosphorus, calcium, and barium (see LUBRICATION AND LUBRICANTS). When the organic fractions of lube oil are combusted on the surface of the catalyst, these inorganic fractions can remain on the catalyst, usually as oxides. Zinc oxide and phosphorus pentoxide have been found in prodigious amounts on catalyst surfaces. A zinc pyrophosphate glaze has been found to form on the catalyst surface after exposure to certain temperatures (80–82). This glaze completely masks large areas of the catalyst surfaces, preventing passage of the gases into the catalyst porous structure. Materials such as calcium inhibit the formation of this glaze.

Silicone residue introduced to gasoline with toluene plugged catalysts on vehicles (83). Also a manganese-based octane improver known as MMT has been shown to clog catalyst surfaces (84).

The accumulation of matter on the surface of the catalyst restricts gas passage into the catalyst by a mechanism known as pore mouth plugging (see Fig. 10**b**). It takes only a small amount of material on the surface of the catalyst to restrict the free passage of gases into and out of the active porous catalyst layer.

Activated Layer Loss. Loss of the catalytic layer is the third method of deactivation. Attrition, erosion, or loss of adhesion and exfoliation of the active catalytic layer all result in loss of catalyst performance. The monolithic honeycomb catalyst is designed to be resistant to all of these mechanisms. There is some erosion of the inlet edge of the cells at the entrance to the monolithic honeycomb, but this loss is minor. The pelletted catalyst is more susceptible to attrition losses because the pellets in the catalytic bed rub against each other. Improvements in the design of the pelletted converter, the surface hardness of the pellets, and the depth of the active layer of the pellets also minimize loss of catalyst performance from attrition in that converter.

OXYGEN SENSOR AND THE CLOSED LOOP FUEL METERING SYSTEM

The first commercial application of the TWC catalyst and closed loop fuel metering system using an oxygen sensor came with the introduction of the 1977 Volvo for the California market (85–88). This catalyst was developed by Engelhard. Other car companies introduced the system the following year, and in the 1990s almost 100% of U.S. and Canada passenger cars and light-duty vehicles utilize it (89). The fully developed system is described in the literature (90).

The function of the oxygen sensor and the closed loop fuel metering system is to maintain the air and fuel mixture at the stoichiometric condition as it passes into the engine for combustion; ie, there should be no excess air or excess fuel. The main purpose is to permit the TWC catalyst to operate effectively to control HC, CO, and NO_x emissions. The oxygen sensor is located in the exhaust system ahead of the catalyst so that it is exposed to the exhaust of all cylinders (see Fig. 4). The sensor analyzes the combustion event after it happens. Therefore, the system is sometimes called a closed loop feedback system. There is an inherent time delay in such a system and thus the system is constantly correcting the air/fuel mixture cycles around the stoichiometric control point rather than maintaining a desired air/fuel mixture.

Oxygen Sensor

The oxygen sensor is also known as the lambda sonde or lambda sensor, from the greek letter used to denote the air/fuel ratio, and as an exhaust gas oxygen sensor (EGO). The lambda sensor responds in about 3 ms to a change in exhaust gas composition, yielding a voltage signal that is monitored by the computer control unit. The signal variation is from 50 mV at $\lambda = 1.05$ (lean) to 900 mV at $\lambda = 0.99$ (rich) as shown in Figure 11. The voltage changes sharply in the immediate vicinity of the stoichiometric ratio. A schematic of the lambda sensor is shown in Figure 12**a**. The sensor consists of a ceramic porous solid electrolyte that is ionically conductive at operating temperatures. The outside of the ceramic is coated with platinum electrodes: one electrode exposed to the exhaust gas, the second to outside air. Between these electrodes is a zirconia solid electrolyte. The voltage generated depends on the oxygen concentration difference between each electrode. The exhaust platinum electrode is a catalyst at the exhaust gas surface and equilibrates the mixture, consuming, by catalytic reaction, any unreacted oxygen and CO, HC, and H_2, yielding a net amount of oxygen present. Near the stoichiometric point, ie, $\lambda = 1.0$, the difference between atmospheric O_2 and exhaust O_2, and thus the voltage generated,

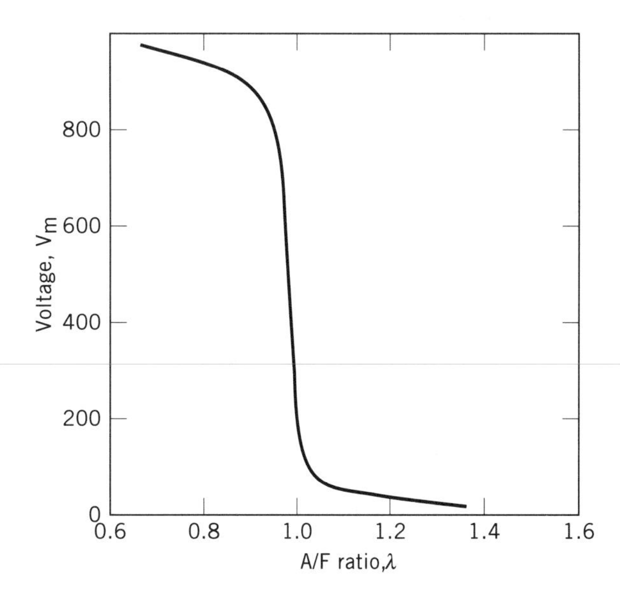

Figure 11. The A/F ratio and the lambda sensor's voltage signal. Courtesy of Robert Bosch.

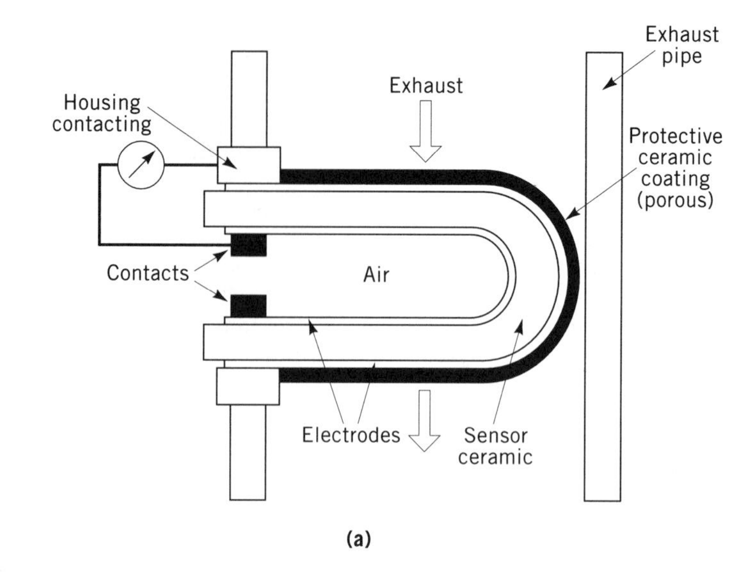

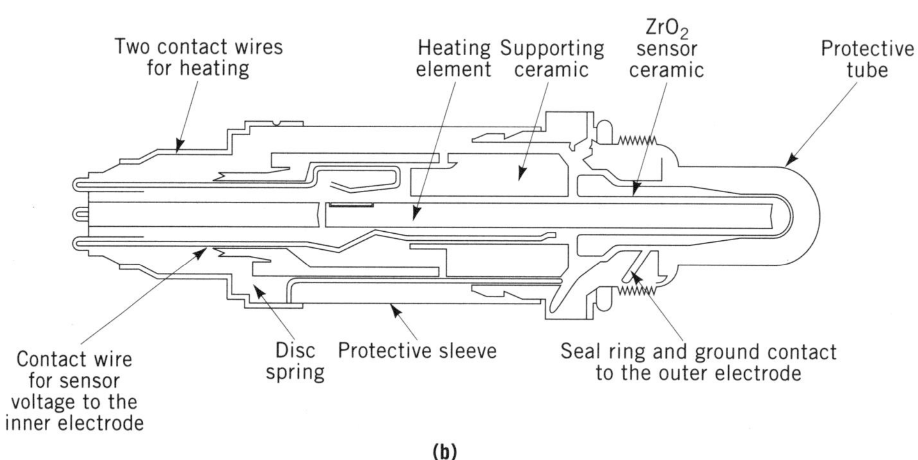

Figure 12. Schematic of lambda sensor (**a**) in exhaust pipe and (**b**) internal components. Courtesy of Robert Bosch.

approach a maximum. Thus the sensor signal is used to detect the stoichiometric point (91–95). In order for the sensor to function properly, it needs to be heated to above 350°C. To accommodate the need for quick catalyst heatup, a heated version of the oxygen sensor is used (96–99) (see Fig. 12**b**).

Computer Controller

The computer controller takes inputs of speed, load, and temperature to assist the engine in cold starting and to select the optimal air/fuel trim adjustment for optimal power, fuel economy, and emissions control. The control system uses the sensor signal as a switching device using the logic "rich! go lean!" or "lean! go rich!" Sophisticated adjustment logic is incorporated into the control unit to correct the air/fuel mixture in anticipation of change.

The oxygen sensor closed loop system automatically compensates for changes in fuel content or air density. For instance, the stoichiometric air/fuel mixture is maintained even when the vehicle climbs from sea level to high altitudes where the air density is lower.

The oxygen sensor is, however, fooled by certain gases. For instance, methanol combustion results in higher H_2 content in the exhaust, and because H_2 has a much higher rate of diffusion than other exhaust gases, a higher concentration is registered at the exhaust side electrode of the sensor than is actually present in the exhaust, causing a richer signal. The computer controller would then adjust the air/fuel to a leaner setting. This factor has to be compensated for when using fuels such as methanol.

Closed loop control has been designed for both carburetors and fuel injection metering systems. The latter are used in almost all 1990 models. Two types of fuel metering exist: a single fuel injector to serve all cylinders, called single-point fuel injection; and fuel injectors for each cylinder, called multipoint fuel injection. The multipoint fuel injection systems may be continuous or individual electrically activated. An electronic fuel injection valve is located in the inlet manifold just ahead of each inlet valve. All valves are connected in parallel and open for a calibrated time as called for by the computer controller. The injected fuel is swept into the cylinder along with the air. The latest development is sequential fuel injection that meters fuel to each cylinder according to the firing order.

All closed loop control systems must measure the amount of air needed under all conditions of engine demand. Air measurement is most often done using a hot wire anemometer, usually referred to as a mass air meter (99,100).

The performance of the catalytic converter is affected by the conditions of air/fuel control provided by the fuel metering system. A slowly responding fuel metering system can dramatically decrease the conversion efficiency of the converter compared to a fast response multipoint fuel injection system.

On-Board Diagnostics

State of California regulations require that vehicle engines and exhaust emission control systems be monitored by an on-board system to assure continued functional performance. The program is called OBD-II, and requires that engine misfire, the catalytic converter, and the evaporative emission control system be monitored (101). The U.S. EPA is expected to adopt a similar regulation.

One system for measuring catalyst failure is based on two oxygen sensors, one located in the normal control location, the other downstream of the catalyst (102,103). The second O_2 sensor indicates relative catalyst performance by measuring the ability to respond to a change in air/fuel mixture. Other techniques using temperatures sensors have also been described (104–107). Whereas the dual O_2 sensor method is likely to be used initially, a criticism of the two O_2 sensors system has been reported (44) showing that properly functioning catalysts would be detected as a failure by the method.

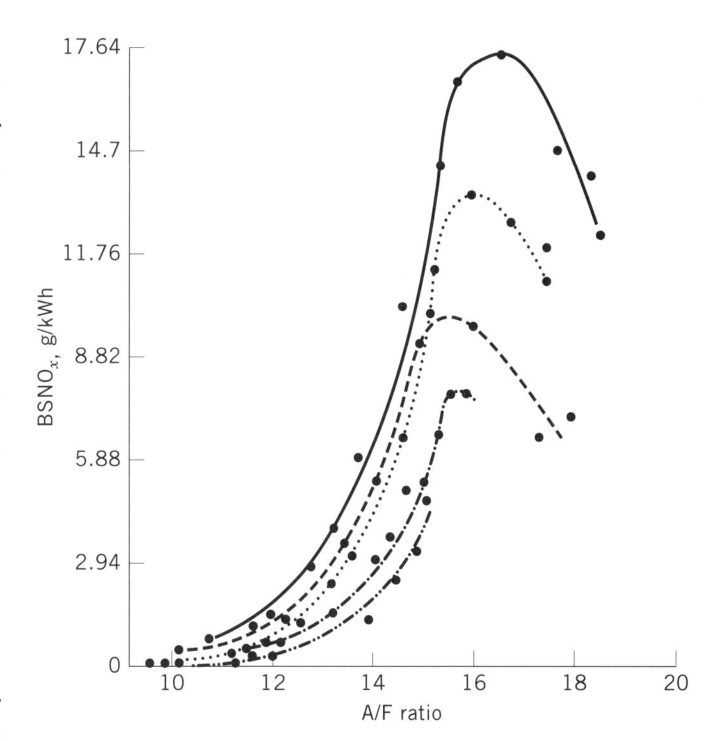

Figure 13. Effect of EGR on brake specific NO_x ($BSNO_x$) production where (—) represents no EGR, (– – –) no EGR and a 20° retard, (⋯⋯) 5% EGR, (·–·–·) 10% EGR, and (–··–··) 15% EGR.

Oxidation Catalyst

An oxidation catalyst requires air to oxidize unburned hydrocarbons and carbon monoxide. Air is provided with an engine driven air pump or with a pulse air device. Oxidation catalysts were used in 1975 through 1981 models but thereafter declined in popularity. Oxidation catalysts may be used in the future for lean burn engines and two-stroke engines.

The oxidation catalyst (OC) operates according to the same principles described for a TWC catalyst except that the catalyst only oxides HC, CO, and H_2. It does not reduce NO_x emissions because it operates in excess O_2 environments. One concern regarding oxidation catalysts was the ability to oxidize sulfur dioxide to sulfur trioxide, because the latter then reacts with water to form a sulfuric acid mist which is emitted from the tailpipe. The SO_2 emitted has the same ultimate fate in that SO_2 is oxidized in the atmosphere to SO_3 which then dissolves in water droplets as sulfuric acid.

Some OCs were of the monolithic honeycomb type, but all General Motors cars used pelletted OCs. For a period in the late 1970s and throughout the 1980s, both TWC and OC were used in a dual-bed catalyst. Oxygen needed for OC performance was provided by an engine driven air pump or a reed valve (pulse air valve) positioned in the exhaust pipe or manifold. In the case of dual-bed catalysts, air was injected between the TWC and OC. OCs utilize Pt and/or Pd but not Rh as a rule. The OC is subjected to the same exhaust environments as the TWC, except the air/fuel ratio is different (108,109).

Exhaust Gas Recirculation

In one method of NO_x emission control, exhaust gas is fed back into the inlet manifold and mixed with the fuel and inlet air. The resultant mixture upon combustion in the cylinder results in lower peak combustion temperature and lower NO_x formation because the reaction of $N_2 + O_2 \rightarrow NO_x$ is strongly dependent on the combustion flame temperature (99,109–112). The degree of NO_x depression is dependent on the amount of exhaust gas recirculation (EGR) as shown in Figure 13. EGR provides a diluent gas having high molecular weight and CO_2 which absorbs heat. Also, EGR affects the flame speed of the mixture, and thus provides a certain antiknock quality to the combustion process. The impact of EGR on engine parameters has been detailed (113).

EGR can seriously degrade engine performance, especially at idle, under load at low speed, and during cold start. Control of the amount of EGR during these phases can be accomplished by the same electronic computer controller used in the closed loop oxygen sensor TWC system. Thus the desired NO_x reduction is achieved while at the same time retaining good driveability.

OTHER EMISSIONS CONTROL

Evaporative Emission

Fumes emitted from stored fuel or fuel left in the fuel delivery system are also regulated by U.S. EPA standards. Gasoline consists of a variety of hydrocarbons ranging from high volatility butane (C-4) to lower volatility C-8 to C-10 hydrocarbons. The high volatility HCs are necessary for cold start, and are especially necessary for temperatures below which choking is needed to start the engine. Stored fuel and fuel left in the fuel system evaporates into the atmosphere.

Evaporative emissions sources are as follows. (*1*) Diurnal evaporative emissions are those that occur as ambient temperature fluctuates between daily high and low. The actual quantities of loss depend on the composition of the fuel and the daily temperatures. High concentrations of high volatility C-4 result in high vapor pressure and high evaporative emissions. (*2*) When the engine is turned off the engine and exhaust system heat up the fuel lines. This is known as hot soak, and fuel left in the fuel metering system or inlet manifold evaporates. (*3*) Evaporation also results from running loss, which occurs during operation at unusual conditions such as at low speeds on a hot day when the rate of vapor generation exceeds the rate that the engine can consume the vapor. (*4*) Evaporative loss source occurs during fueling as the liquid fuel displaces the fuel tank vapor.

Fuel vapor pressure, or Reid vapor pressure (RVP) is predominantly influenced by the amount of C-4 in the fuel. California, other of the United States, and the EPA have or are restricting fuel RVP for both seasonal and areas (latitude). These restrictions greatly reduce all gasoline evaporative losses.

The EPA regulation prohibits more than 2 grams per test to escape into the atmosphere (114). The test consists of a diurnal cycle of 1 hour where the temperature of the fuel is raised from 15.6 to 28.9°C during a 17 mile run on a chassis dynamometer. An immediate hot soak in a shed enclosure follows the dynamometer run.

The common method of controlling evaporative emission is an activated charcoal canister that connects to the intake manifold. When the engine is shut off, the valve permits hydrocarbon fumes to be absorbed and stored by the activated charcoal. When the engine is operating, the stored hydrocarbons are purged back into the manifold where they enter the combustion process. A new California regulation also controls running loss emissions of hydrocarbons to 0.03 g/km. There is discussion as to whether control of fueling emissions should be on-board the vehicle or incorporated into the fueling station system (115–118). As of 1993 the control is in the fueling station system.

Crankcase Emissions

Exhaust gases are also found in the crankcase. The principal source of these exhaust gases results from what is known as blow-by or gases from the combustion chamber, past the piston rings to the crankcase, or from the intake and exhaust valve mechanisms. As the engine wears, blow-by increases. Control systems are required to feed these gases back into the inlet manifold so that the hydrocarbons and carbon monoxide contained are consumed in the combustion process. The control device used is called a positive crankcase ventilation (PCV) valve. This device or one of similar function is required by law (119–121). Formerly, these gases were ventilated to the atmosphere through the engine breather tube. The test to measure crankcase emissions is defined in Reference 121.

ALTERNATIVE FUELS

Under the National Energy Policy Act of 1992 nonpetroleum-based transportation fuels are to be introduced in the United States. Such fuels include natural gas (see GAS, NATURAL), liquefied petroleum gas (LPG), methanol, ethanol, and hydrogen, although hydrogen fuels are not expected to be a factor until after the year 2000 (see also ALCOHOL FUELS; HYDROGEN ENERGY).

Natural Gas

Natural gas, an abundant fuel resource in the United States, has sufficient reserves to fuel over 10×10^6 U.S. vehicles per year for the next 50 years (122). Natural gas is used in two forms as a transportation fuel compressed or liquefied at low temperatures. Tanks for the storage of compressed natural gas are heavy and larger in volume than for liquid fuels. However, the added cost is offset by an expected lower pump price compared to gasoline (123). Whereas the lack of public natural gas fueling stations and other factors make natural gas more attractive for fleet vehicles in the United States, small compressors for overnight natural gas refueling at household sites have been developed. Natural gas has a high antiknock quality, and engines designed for this fuel can have a higher compression ratio, yielding higher power and fuel economy. A disadvantage of compressed natural gas is that the driving range for a fuel load is usually 50% of that of gasoline. Work on an adsorptive system of natural gas storage at reduced storage pressures has been announced that should reduce compressed tank costs and may be able to improve range per fueling (124). Dual fuel systems have been developed in which the engine can run on either natural gas or gasoline.

Emission control for natural gas fueled engines consists of either the same principal components as used for gasoline or those designed for lean combustion. No evaporative emission control is required.

The catalysts for natural gas run engines are designed differently, however, because a primary portion of the exhaust from natural gas combustion is methane, and the catalysts are based on Pd rather than Pt. Spark ignited engines calibrated for natural gas usually are calibrated at stoichiometric conditions. Diesel engines re-engineered for natural gas are usually spark ignited and calibrated very lean. Pd/Rh is used for TWC catalysts for the stoichiometric engine; Pd oxidation catalysts are used for the lean re-engineered diesel engine. Both the U.S. EPA and the California standards exclude methane in the calculation of hydrocarbon emissions because methane has low reactivity in the atmosphere and has low ozone (qv) formation potential (125), and natural sources produce large quantities of methane. Methane, however, is a strong global warming gas and is regulated by a total hydrocarbon standard. The symmetrical methane molecule is the most refractory of the hydrocarbon family, and resists reactions on traditional auto exhaust catalysts at low temperatures. Pd catalysts need 450°C to oxidize methane but are nevertheless much better than all other precious metal catalysts. Methane is a poor reductant of NO_x (126,127). More recently, a Pd-based catalyst having poor dispersion (low metal surface area) was found to be superior to highly dispersed Pd for these reactions (128). Pd suffers temporary deactivation from SO_2 present in the exhaust. Natural gas contains very little sulfur. A small amount of mercaptan is added to give a characteristic odor to the fuel so that leaks in the distribution system can be detected. Pd-based catalysts therefore maintain performance much better than when used with gasoline (129–131).

LPG

LPG could be a principal alternative transportation fuel if its other uses were displaced by natural gas. A significant number of LPG fueling stations are located throughout the United States. LPG is a liquid fuel and does not suffer the same driving range problem as natural gas. Because LPG vapor pressure is high, the storage tank has to withstand 2800 kPa (400 psi).

The emission control system for LPG is the same as is used for gasoline fueled engines with the exception of the fuel metering system. No evaporative emission system is required. Both Pt–Rh and Pd–Rh catalysts are good for emission control of LPG fuel exhaust. Pt provides the lowest light off temperature for C_3H_8. The sulfur content of LPG is also very low so that Pd catalysts perform very well.

Methanol

Methanol is a liquid fuel made from natural gas, but can also be made from coal (qv), and the U.S. has huge coal resources. Engines designed for 100% methanol usually take advantage of the properties of methanol to provide high power. Methanol has high antiknock properties and can be used in high compression engines. Although methanol has a lower energy content, it has a higher latent heat of vaporization than gasoline. Evaporation in the inlet manifold cools and densifies the air/fuel charge to the engine, thus increasing the energy charge to the engine. Theoretically, therefore, the engine could be smaller and still provide the same power as a gasoline fueled engine. The disadvantage of 100% methanol engines is cold starting of the engine. Methanol has very low vapor pressure at cold temperatures, so there is insufficient vapor phase to provide a combustible mixture with a choked amount of air to start the engine.

Addition of 15% gasoline to methanol to produce M85 fuel is an alternative. At temperatures above −6.7°C, reliable ignition of M85 fuel occurs because the gasoline provides the vapor phase necessary for ignition under choked condition.

Engines are also designed to use either gasoline or methanol and any mixture thereof (132–136). Such a system utilizes the same fuel storage system, and is called a flexible fueled vehicle (FFV). The closed loop oxygen sensor and TWC catalyst system is perfect for the flexible fueled vehicle. Optimal emissions control requires a fuel sensor to detect the ratio of each fuel being metered at any time and to correct total fuel flow.

The principal hydrocarbon emissions from a methanol fueled car are methanol and formaldehyde. Formaldehyde is a carcinogen and regulated to very low levels by California and the U.S. EPA (0.015 g/m for light-duty vehicles (LDV)) and standards going to 0.008 g/m for California ultra low emission vehicle (ULEV) standards. Uncontrolled formaldehyde emissions from a methanol fueled light-duty vehicle are on the order of 0.60 g/m. Most TWC catalysts destroy formaldehyde effectively after operating temperature is reached, ie, about 1 to 2 minutes after the engine is started. However, during the 1 to 2 minutes that the catalyst is heating, about 0.05 to 0.1 g/m of formaldehyde is passed unreacted through the catalyst. The emission control system for either 100% methanol or for M85 must have catalysts located as close as possible to the manifold to heat up faster. Specially designed Pd–Rh TWC catalysts and close coupled manifold catalysts are being utilized for FFV to minimize formaldehyde emissions as well as to meet the other emission standards (136). An electrically heated catalyst system obtained very low formaldehyde emissions from a methanol fueled vehicle (137).

FUTURE ENGINES AND EMISSION CONTROL SYSTEMS

Two engines are under development as of this writing: the two-stroke engine and the lean burn engine. The driving forces behind this development are fuel economy and global warming.

The National Highway and Safety Administration regulates vehicle fuel consumption through the Corporate Average Fuel Economy (CAFE). When vehicle manufacturers submit vehicles for certification, fuel consumption values must be given for a specified city and highway driving cycle (138). The average must be equal to or more than 11.6 km/L (27.5 mpg). Carbon dioxide is a principal global warming gas and conservation actions, such as increased CAFE, are thought to be a measure to stave off the buildup of CO_2 in the atmosphere.

Whereas automobile companies continue to improve the four-stroke engine, making it more efficient and more powerful to run vehicles that are smaller and lighter and have sufficient space, the two-stroke and the lean burn engines are also being studied.

Two-Stroke Engine

The two-stroke engine has the potential of delivering the same power as the four-stroke engine, but has fewer moving parts, lower weight, and a smaller engine volume, ie, it would be ideal for smaller, lighter, more fuel efficient vehicles having smaller engine compartments. Baseline emissions of two-stroke engines are very high in hydrocarbons. For example, a four-stroke engine may have 250 ppm hydrocarbons in its exhaust; the two-stroke engine could have 5000 ppm or greater. Designs that use inlet valves, fuel injection, and air compression have reduced baseline emissions, but have added complexity to the original simple two-stroke engine (139–147).

The operating air/fuel mixture of the two-stroke engine designs range from 1.3 to 2.0 stoichiometric. This lean mixture plus the characteristic internal exhaust gas recirculation lowers the peak combustion temperatures and results in low NO_x formation.

Emission control systems for two-stroke engines depend heavily on an efficient oxidation catalyst. These may be based on Pt and/or Pd. Higher lube oil consumption characteristics of two-stroke engines may result in modification to the lube oil or require the development of oxidation catalysts more resistant to lube oil ash compounds.

Lean Burn Engine

An engine calibrated at >18/1 A/F ratio, ie, very lean, encounters what is known as the lean limit (148). At this point, ignition of the air/fuel mixture becomes more difficult to achieve and subsequent propagation of the flame more difficult to sustain. Engineering solutions to both problems are diametrically opposed. At A/F ratios even more lean, combustion within the cylinder deteriorates and hydrocarbon emissions increase. Another result is loss of power. However, the production of NO_x is decreased at leaner A/F mixtures, and more importantly, thermal efficiency is improved and specific fuel consumption decreases. If engine designs could push back the lean limit, a practical lean burn engine may be a reality (149). To overcome the poor power and performance of a lean burn engine, a partial lean burn system was designed. The engine operated lean during idle and low cruise engine operating conditions. When acceleration was required, the engine adjusted to the stoichiometric air/fuel ratio. Toyota marketed a lean burn engine in Japan in the late 1980s and for a period in Europe without much customer acceptance (150). More recently, since 1990, several automobile manufacturers, such as Honda, Mitsubishi, Nissan, and Hyundai, have announced the development of the lean burn engine (151,152). All of these later developments used the Ford partial lean and partial stoichiometric calibration. Emission control for the lean burn/stoichiometric engine requires a TWC catalyst that can operate as a TWC catalyst at stoichiometric air/fuel mixtures and as an oxidation catalyst when the engine is operated lean.

Work on the development of the full time lean burn engine continues with efforts to push back the lean burn limit. A problem recognized by the developers is that although low basic engine NO_x emissions are possible, it is not yet possible to meet the NO_x emissions required by California and the Tier II emission levels of the U.S. Clean Air Act of 1990 (see Table 1).

There has been a growing demand for a lean NO_x catalyst in order to decrease the relatively low NO_x emission of the lean burn engine sufficiently to meet the future standards. Lean NO_x catalysts have been developed based on zeolites. Cu-promoted ZSM-5 zeolite has shown ability to reduce NO_x in an exhaust having excess oxygen at an efficiency of 30 to 50% (153). Durability is not proven. Research has revealed that certain hydrocarbons are preferred for the reduction of NO_x, and that CO and H_2 apparently do not reduce NO_x over such lean NO_x catalysts (154). Considerable effort is being expended to develop a practical lean NO_x catalyst system (155–159).

Emission Control Technologies

The California low emission vehicle (LEV) standards has spawned investigations into new technologies and methods for further reducing automobile exhaust emissions. The target is to reduce emissions, especially HC emissions, which occur during the two minutes after a vehicle has been started (53). It is estimated that 70 to 80% of nonmethane HCs that escape conversion by the catalytic converter do so during this time before the catalyst is fully functional.

Technologies being investigated include improved catalysts that can be located closer to the exhaust manifold (high temperature resistant), reduced thermal mass in the exhaust system, use of alternative fuels, use of air pumps to assist catalyst light off, and engine adjustments that yield higher exhaust temperatures for this period. Three new technologies are undergoing intense development.

Electrically Heated Catalyst. An electrically heated catalyst (EHC) is an electrical resistive metallic foil or matrix which, like a toaster, will heat upon the application of voltage. Two basic types are being developed. One is based on a metal foil; the other is based on sintered metal powder. Figure 14**a** shows a cross-sectional view of a metal foil type, and Figure 14**b** shows a cross-sectional view of the sintered metal powder type. A voltage, typically 12 V or 24 V supplied from the automobile battery, is applied across the electrodes for about 15 to 20 seconds, in order to heat the foil or metal matrix from ambient temperature to 300–600°C in this time period. The metal foil or metal matrix surfaces are covered with a catalytic layer as described for the monolithic converter. Thus, when the metal surface achieves 300°C, catalyst light-off occurs. The system usually employs an air pump so that there is sufficient oxygen present to consume significant quantities of CO and HCs, resulting in an additional temperature rise.

The main converter, which is located downstream of the EHC, heats to functional temperature much more quickly because of catalytic combustion of exhaust gases that would otherwise pass unconverted through the catalyst during the cold start period. The EHC theoretical power required for a reference case (161) was 1600 watts to heat an EHC to 400°C in 15 s in order to initiate the catalytic reactions and obtain the resultant exotherm of the chemical energy contained in the exhaust. Demonstrations have been made of energy requirements of 15–20 Wh and 2 to 3 kW of power (160,161). Such systems have achieved nonmethane HC emissions below the California ULEV standard of 0.025 g/km. The principal issues of the EHC are system durability, battery life, system complexity, and cost (137,162–168).

Hydrocarbon Trap System. The concept of a hydrocarbon trap or adsorber system is based on molecular sieve hydrocarbon adsorber systems. The temperatures at which hydrocarbon adsorption takes place exist in the auto engine exhaust system during the period of cold start of an automobile when the catalytic control system has not yet reached functional temperature. Zeolites have been reportedly useful for hydrocarbon adsorption (53,169). Zeolites desorb hydrocarbons at temperatures of 400°C, ie, once the catalytic control system is functional. Therefore, hydrocarbons adsorbed by the zeolite can also be desorbed then oxidized by a catalyst. Methods to accomplish cold start hydrocarbon adsorption, heatup of the main catalyst, and desorption have been identified. Some of these systems use exhaust pipe valves to divert the exhaust gases to the hydrocarbon trap for the low temperature portion, and bypass the gases around the trap after the main catalyst has heated up. One device that uses a heat exchanger is shown in Figure 15 (44). The Si–Al ratio in the zeolite is important, and by lowering the alumina content, the zeolite is rendered more hydrophobic and more able to adsorb hy-

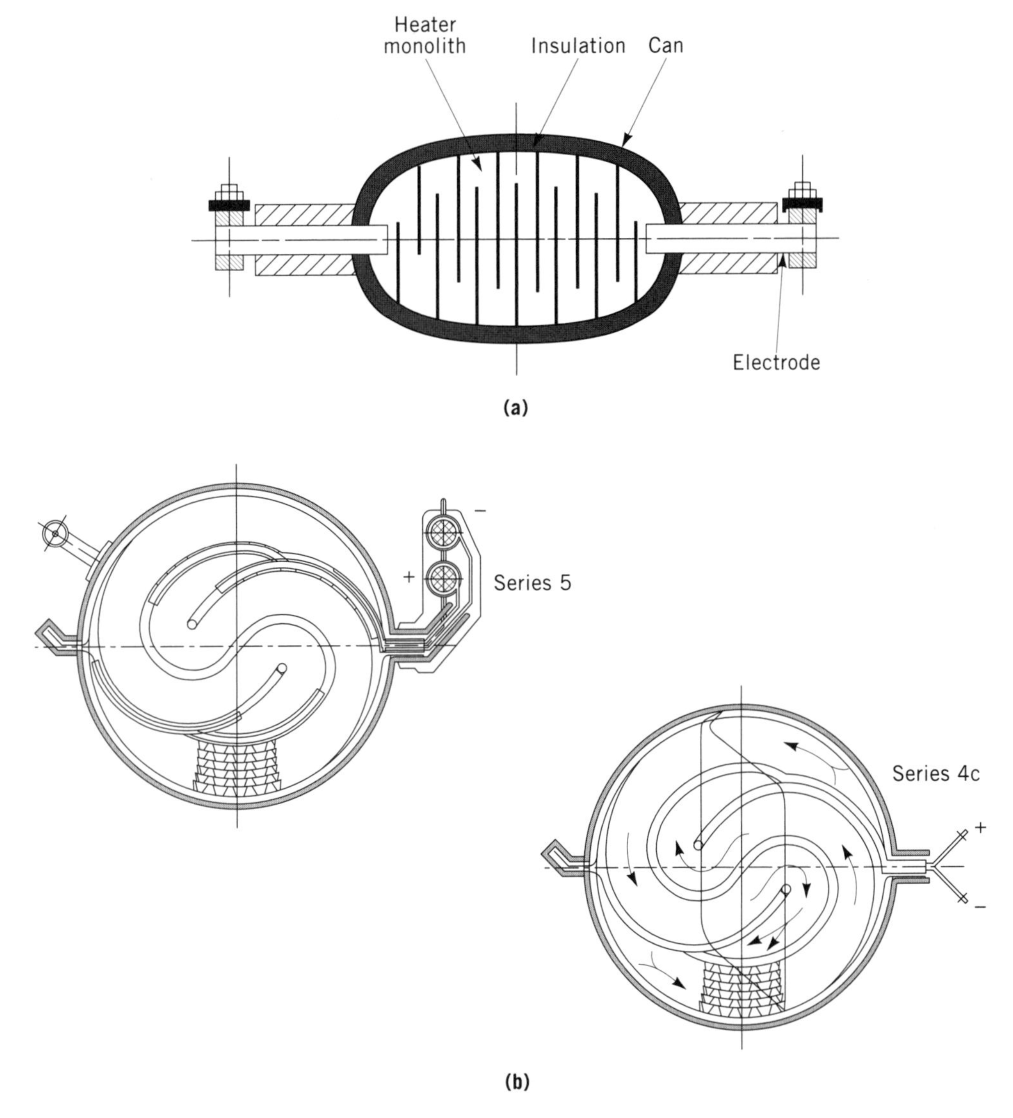

Figure 14. Cross-sectional schematics of electrically heated catalyst (EHC) for emission control (**a**) extruded sintered metal powder EHC (160); (**b**) two sintered metal foil EHCs. Courtesy of Emitec GmbH.

drocarbons. Adsorption of 50 to 60% of the cold start hydrocarbons have been demonstrated (44,169–171). Zeolites do not efficiently adsorb C-2 to C-4 hydrocarbons.

Exhaust Gas Igniter. Exhaust gases during cold start have high concentrations of CO and HCs because of air restriction caused by choking. Using a severe choke strategy and addition of air by a pump the exhaust gas/air mixture is flammable. The Ford Exhaust Gas Igniter incorporates a spark igniter in the exhaust line to ignite and sustain the flame. The flame heats up the main catalyst in 15 to 20 seconds, and in the process consumes those

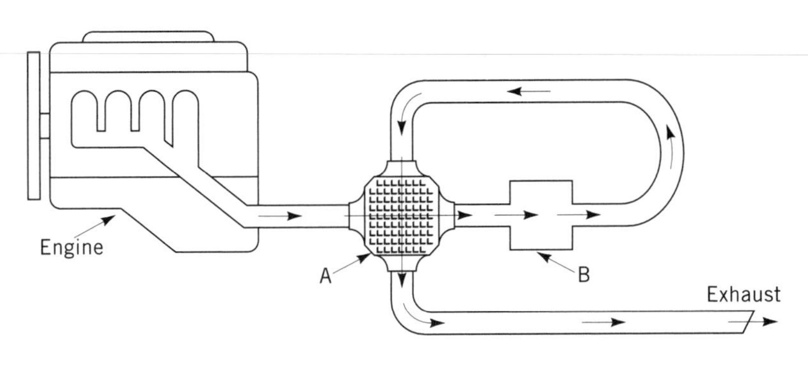

Figure 15. Low hydrocarbon emission control system utilizing a cross-flow heat exchanger TWC catalyst, A, and a zeolite-based hydrocarbon absorber system. Cold start HCs are absorbed by the hydrocarbon trap, B, until the cross-flow heat exchanger catalyst is hot enough to destroy the HCs that subsequently are eluted from the HC trap.

hydrocarbons and carbon monoxide that would otherwise pass through the cold catalyst (172,173). The system uses energy available in the exhaust gases during the initial seconds of engine starting. On the other hand, the choked condition is extreme, and there are associated engine concerns.

Reprinted from *Kirk-Othmer Encyclopedia of Chemical Technology,* 4th ed., Vol. 9, John Wiley & Sons, Inc., New York, 1994.

BIBLIOGRAPHY

1. M. P. Walsh, *Plat. Met. Rev.* **30**(3), 106–115 (1986).
2. C. Csere, *Car Driver,* 60–64 (Jan. 1988).
3. *Code of Federal Regulations,* Title 40, Part 86, Appendix IV, AMA Mileage Accumulation Route, Washington, D.C., July 1992, p. 951.
4. *1991 SAE Handbook,* Vol. 3. Society of Automotive Engineers, Warrendale, Pa., 1991, Chapt. 25.
5. A. L. Boehman, S. Niksa, and R. J. Moffatt, *A Comparison of Rate Laws for CO Oxidation Over Pt on Alumina,* SAE 930252, Society of Automotive Engineers, Warrendale, Pa., 1993.
6. J. T. Kummer, *Prog. Ener. Combus. Sci.* **6,** 177–199 (1981).
7. B. A. D'Alleva, *Procedure and Charts for Estimating Exhaust Gas Quantities and Composition,* General Motors Laboratory Report 372, Warren, Mich., 1960.
8. N. A. Heinen and D. J. Patterson, *Combustion Engine Economy, Emissions, and Controls,* University of Michigan, Ann Arbor, Mich., Chapt. 5, 1992.
9. L. Eltinge, *Fuel–Air Ratio and Distribution from Exhaust Gas Composition,* SAE 680114, Society of Automotive Engineers, Warrendale, Pa., 1968.
10. J. B. Edwards, *Combustion: The Formation and Emissions of Trace Species,* Ann Arbor Science Publishers, Ann Arbor, Mich., 1974, p. 16.
11. J. D. Benson and co-workers, *Effects of Gasoline Sulfur on Mass Exhaust Emissions–Auto/Oil Air Quality Improvement Research Program,* SAE 912323, Society of Automotive Engineers, Warrendale, Pa., 1991.
12. *Auto/Oil Air Quality Improvement Research Program,* SAE SP-920, Society of Automotive Engineers, Warrendale, Pa., 1992, 16 pp.
13. P. Blumberg and J. T. Kummer, *Combust. Sci. Technol.* **4,** 73 (1971).
14. N. Pelz and co-workers, *The Composition of Gasoline Engine Hydrocarbon Emissions–An Evaluation of Catalytic and Fuel Effects,* SAE 902074, Society of Automotive Engineers, Warrendale, Pa., 1990.
15. P. R. Shore, D. T. Humphries, and O. Hadded, *Speciated Hydrocarbon Emissions from Aromatic, Olefinic and Parafinic Model Fuels,* SAE 930373, Society of Automotive Engineers, Warrendale, Pa., 1993.
16. G. J. den Otter, R. E. Malpas, and T. D. B. Morgan, *Effect of Gasoline Reformulation on Exhaust Emissions in Current European Vehicles,* SAE 930372, Society of Automotive Engineers, Warrendale, Pa., 1993.
17. J. Laurikko and N-O. Nyland, *Regulated and Unregulated Emissions from Catalyst Vehicles at Low Ambient Temperatures,* SAE 930946, Society of Automotive Engineers, Warrendale, Pa., 1993. Good reference for low ambient temperature emissions.
18. P. R. Shore and R. S. de Vries, *On-Line Hydrocarbon Speciation Using FTIR and CI-MS,* SAE 922246, Society of Automotive Engineers, Warrendale, Pa., 1992.
19. D. W. Wendland and W. R. Matthes, *Visualization of Automotive Catalytic Converter Internal Flows,* SAE 861554, 1986; D. W. Wendland, P. L. Sorrell, and J. E. Kreucher, *Sources of Monolithic Catalytic Converter Pressure Loss,* SAE 912372, Society of Automotive Engineers, Warrendale, Pa., 1991.
20. H. Weltans, H. Bressler, and P. Krause, *Influence of Catalytic Converters on Acoustics of Exhaust Systems for European Cars,* SAE 910836, Society of Automotive Engineers, Warrendale, Pa., 1991.
21. D. W. Wendland, W. R. Matthes, and P. L. Sorrell, *Effect of Header Truncation on Monolith Converter Emission Control Performance,* SAE 922340, Society of Automotive Engineers, Warrendale, Pa., 1992.
22. H. Weltans and co-workers, *Optimization of Catalytic Converter Gas Flow Distribution by CFD Prediction,* SAE 930780, Society of Automotive Engineers, Warrendale, Pa., 1993.
23. I. Sword and co-workers, SAE 930943, Society of Automotive Engineers, Warrendale, Pa., 1993.
24. J. Howitt, *Thin Wall Ceramics as Monolithic Catalyst Supports,* SAE 910611, Society of Automotive Engineers, Warrendale, Pa., 1991.
25. L. S. Socha, J. P. Day, and E. M. Barnett, *Impact of Catalytic Support Design Parameters on FTP Emissions,* SAE 892041, Society of Automotive Engineers, Warrendale, Pa., 1989.
26. H. Yamamoto and co-workers, *Warm-Up Characteristics of Thin Wall Honeycomb Catalysts,* SAE 910611, Society of Automotive Engineers, Warrendale, Pa., 1991.
27. J. P. Day and L. S. Socha, Jr., *The Design of Automotive Catalyst Supports for Improved Pressure Drop and Conversion Efficiency,* SAE 910371, Society of Automotive Engineers, Warrendale, Pa., 1991.
28. M. Luoma, P. Lappi, and R. Lylykangas, *Evaluation of High Cell Density Z-Flow Catalyst,* SAE 930940, Society of Automotive Engineers, Warrendale, Pa., 1993. Good reference for mass-transfer limited model reactions.
29. K. Otto, W. B. Williamson, and H. Gandhi, *Ceram. Eng. Sci. Proc.* **2**(6), (May/June, 1981). Good review of various catalyst deactivation processes.
30. R. J. Farrauto and M. C. Hodson, "Catalyst Characterization," *Encyclopedia of Physical Science and Technology,* Vol. 2, Academic Press, Inc., New York, 1987.
31. J. T. Kummer, "The Use of Noble Metals in Automobile Exhaust Catalyst," *Chicago Symposium,* 1985.
32. J. E. Kubsh, J. S. Rieck, and N. D. Spencer, in A. Crucq, Ed., *Catalysis and Automotive Pollution Control II,* Elsevier, New York, 1991, p. 125.
33. J. K. Hockmuth and co-workers, *Hydrocarbon Traps for Controlling Cold Start Emissions,* SAE 930739, Society of Automotive Engineers, Warrendale, Pa., 1993.
34. Y. T. Yu Yao, *Indust. Eng. Chem. Prod. Res. Dev.* **19,** 293 (1980).
35. H. C. Yao, H. K. Stephen, and H. S. Gandhi, *J. Catal.* **61,** 547 (1980).
36. H. C. Yao, S. Japar, and M. Shelef, *J. Catal.* **50,** 407 (1977).
37. S. H. Oh, P. J. Mitchel, and R. M. Siewert, *J. Catal.* **132,** 287 (1991).
38. R. F. Hicks, C. Rigano, and B. Pang, *Catal. Lett.* **6,** 271 (1990).
39. T. Yamada, K. Kayano, and M. Funibike, *The Effectiveness*

of Pd for Converting Hydrocarbons in TWC Catalysts, SAE 930253, Society of Automotive Engineers, Warrendale, Pa., 1993.

40. J. C. Dettling and W. K. Lui, *A Non-Rhodium Three-Way Catalyst for Automotive Applications,* SAE 920094, Society of Automotive Engineers, Warrendale, Pa., 1992.
41. C. N. Montreiul, S. D. Williams, and A. A. Adanczyk, *Modeling Current Generation Catalytic Converters: Laboratory Experiments and Kinetic Parameter Optimization–Steady State Kinetics,* SAE 920096, Society of Automotive Engineers, Warrendale, Pa., 1992.
42. M. A. Harkonen and P. Talvitie, *Optimization of Metallic TWC Behavior and Precious Metal Costs,* SAE 920395, Society of Automotive Engineers, Warrendale, Pa., 1992.
43. J. T. Kummer, *J. Phys. Chem.* **90,** 4747 (1986).
44. G. B. Fisher and co-workers, *The Role of Ceria in Automotive Exhaust Catalysis and OBD-II Catalyst Monitoring,* SAE 931034, Society of Automotive Engineers, Warrendale, Pa., 1993.
45. S. Pelters, F. W. Kaiser, and W. Maus, *The Development and Application of Metal Supported Catalyst for Porsche's 911 Carrera 4,* SAE 890488, Society of Automotive Engineers, Warrendale, Pa., 1989.
46. F. W. Kaiser and S. Pelters, *Comparison of Metal Supported Catalysts with Different Cell Geometries,* SAE 910837, Society of Automotive Engineers, Warrendale, Pa., 1991.
47. P. Oser, *Novel AutoCatalyst Concepts and Strategies for the Future with Emphasis on Metal Supports,* SAE 880319, Society of Automotive Engineers, Warrendale, Pa., 1988.
48. K. Nishizawa and co-workers, *Metal Supported Automotive Catalysts for Use in Europe,* SAE 880188, Society of Automotive Engineers, Warrendale, Pa., 1988.
49. S. H. Oh and D. L. Van Ostrom, *A Three Dimensional Model for the Analysis of Transient Thermal and Conversional Characteristics of Monolithic Catalytic Converters,* SAE 880282, Society of Automotive Engineers, Warrendale, Pa., 1988.
50. J. C. W. Kuo, C. R. Morgan, and H. G. Lasson, *Mathematical Modelling of CO and HC Catalytic Converter Systems,* SAE 710289, Society of Automotive Engineers, Warrendale, Pa., 1971.
51. C. N. Montreuil, S. D. Williams, and A. A. Adanczyk, *Modeling Current Generation Catalytic Converters; Laboratory Experiments and Kinetic Parameter Optimization–Steady State Kinetics,* SAE 920096, Society of Automotive Engineers, Warrendale, Pa., 1992.
52. K. S. Creamer and J. H. Sanders, *Evaluation of a Catalytic Converter for a 3.73 kW Natural Gas Engine,* SAE 930221, Society of Automotive Engineers, Warrendale, Pa., 1993.
53. S. Tauster, Engelhard Corp., Iselin, N.J., 1992.
54. C. M. Friend, *Sci. Am.* **268**(4), 74–79 (Apr. 1993).
55. M. Golze and co-workers, *Phys. Rev. Lett.* **53**(8), 850–853 (1993).
56. R. J. Maddix, *Science* **V233,** 1159–1166 (Sept. 12, 1986).
57. G. Fisher and co-workers, "Mechanism of the Nitric Oxide–Carbon Monoxide–Oxygen Reaction Over a Single Crystal Rhodium Catalyst," in M. J. Philips and M. Ternan, eds., *Proceedings of the 9th International Congress on Catalysis,* Vol. 3, *Characterization and Metal Catalysts,* Chemical Institute of Canada, Ottawa, 1988.
58. J. T. Yates, Jr., *Chem. Eng. News.* **70**(13), 22–35 (Mar. 30, 1992).
59. J. C. Summers and S. A. Ansen, *J. Catal.* **58,** 131 (1979).
60. H. C. Yao and Y. F. Yu Yao, *J. Catal.* **86,** 254 (1984).
61. K. Ihara, H. Marakami, and K. Ohkubo, *Improvement of Three-Way Catalyst Performance by Optimizing Ceria Impregnation,* SAE 902168, Society of Automotive Engineers, Warrendale, Pa., 1990.
62. A. F. Duvell, R. R. Rojaram, H. A. Shaw, and T. J. Truex, in A. Crucq, ed., in Ref. 32, p. 139.
63. R. Hicks and co-workers, *J. Catal.* **122,** 296–306 (1990).
64. H. S. Gandhi, W. B. Williamson, and J. L. Bomback, *Appl. Catal.* **3,** 79–88 (1982).
65. E. Jobson and co-workers, *Deterioration of Three-Way Automotive Catalysts, Part I–Steady State and Transient Emission of Aged Catalyst,* SAE 930937, Society of Automotive Engineers, Warrendale, Pa., 1993.
66. S. Lundgren and E. Jobson, *Deterioration of Three-Way Automotive Catalysts, Part II–Oxygen Storage Capacity at Exhaust Conditions,* SAE 930944, Society of Automotive Engineers, Warrendale, Pa., 1993.
67. L. A. Carol, N. E. Newman, and G. S. Mann, *High Temperature Deactivation of Three-Way Catalyst,* SAE 892040, Society of Automotive Engineers, Warrendale, Pa., 1989.
68. N. Miyoshi and co-workers, *Development of Thermal Resistant Three-Way Catalyst,* SAE 891970, Society of Automotive Engineers, Warrendale, Pa., 1989.
69. C. D. Tyree, *Emission Level in Catalyst Temperature as a Function of Ignition-Induced Misfire,* SAE 920298, Society of Automotive Engineers, Warrendale, Pa., 1992.
70. N. Miyoshi and co-workers, *Development of Thermal Resistant Three-Way Catalysts,* SAE 891970, Society of Automotive Engineers, Warrendale, Pa., 1989.
71. M. P. Walsh, *The Advantages of Removing Lead from Gasoline and Using Catalytic Converters to Control Vehicle Exhaust Pollution,* sponsored by Corning Glass Works, 1983.
72. Needleman and co-workers, *New Engl. J. Med.* **300**(13), (Mar. 29, 1979).
73. A. M. Hochhauser and co-workers, *Effects of Gasoline Sulfur Level on Mass Exhaust Emissions–Auto/Oil Air Quality Improvement Research Program,* SAE 912323, Society of Automotive Engineers, Warrendale, Pa., 1991.
74. J. C. Summers and co-workers, SAE 920558, Society of Automotive Engineers, Warrendale, Pa., 1992.
75. U.S. Pat. 4,552,733 (Nov. 12, 1985), C. Thompson, J. Mooney, C. D. Keith, and W. Mannion (to Engelhard Corp.)
76. I. Gottberg, E. Hogberg, and K. Weber, *Sulfur Storage and Hydrogen Sulphide Release From a Three-Way Catalyst Equipped Car,* SAE 890491, Society of Automotive Engineers, Warrendale, Pa., 1989.
77. J. C. Dettling and co-workers, *Control of H2S Emissions from High-Tech TWC Converters,* SAE 900506, Society of Automotive Engineers, Warrendale, Pa., 1990.
78. S. E. Golunski and S. A. Roth, *Catal. Today* **9,** 109 (1991).
79. U.S. Pat. 5196390 (Mar. 23, 1993), S. Tauster, J. Dettling, and J. Mooney (to Engelhard Corp).
80. W. B. Williamson and co-workers, *Catalyst Deactivation Due to Glaze Formation from Oil-Derived Phosphorus and Zinc,* SAE 841406, Society of Automotive Engineers, Warrendale, Pa., 1984.
81. K. Inoue, T. Kurahashi, and T. Negishi, SAE 92064, Society of Automotive Engineers, Warrendale, Pa., 1992.
82. W. B. Williamson and co-workers, *Appl. Catal.* **15,** 277–292 (1985).
83. H. S. Gandhi and co-workers, "Affinity of Lead for Moble Metals on Different Supports," *Surface Interface Analy.* **6**(4), (1984).

84. W. B. Williamson, H. S. Gandhi, and E. E. Weaver, *Effects of Fuel Additive MMT on Contaminent Retention and Catalyst Performance,* SAE 821193, Society of Automotive Engineers, Warrendale, Pa., 1982.
85. G. T. Engh and S. Wallman, *Development of the Volvo Lambda-Sond System,* SAE 770295, Society of Automotive Engineers, Warrendale, Pa., 1977.
86. J. J. Mooney, C. E. Thompson and J. C. Dettling, *Three-Way Conversion Catalysts–Part of the New Emission Control System,* SAE 770365, Society of Automotive Engineers, Warrendale, Pa., 1977.
87. E. Koberstein, *Characterization of Multifunctional Catalysts for Automotive Exhaust Purifications,* SAE 770366, Society of Automotive Engineers, Warrendale, Pa., 1977.
88. P. Oser, *Catalyst Systems with and Emphasis on Three-Way Conversion and Novel Concepts,* SAE 790306, Society of Automotive Engineers, Warrendale, Pa., 1979.
89. K. Oblander, J. Abthoff, and H. D. Schuster. *Der Driewegkatalysator–eine Abgasreinigungstechologie fur Kraftfahrzenge mit Ottomotoren (Three-Way Catalyst–An Exhaust Cleaning Technology for Automobiles with Internal Combustion Engines),* VDI-Berichte 630, VDI-Verlag, Dusseldorf, Germany, 1984.
90. J. Abthoff and co-workers, *MTZ Motortechnische Zeitschrift* **53,** 11 (1992).
91. E. Logothetis, *Sci. Technol. Zirconia,* **3,** 388 (1981); R. E. Hetrich, W. A. Fate, and W. C. Vassell, *Oxygen Sensing by Electrochemical Pumping,* SAE 8104333, Society of Automotive Engineers, Warrendale, Pa., 1981.
92. W. J. Fleming, *Zirconia Oxygen Sensor—An Equivalent Circuit Model,* SAE 800020, Society of Automotive Engineers, Warrendale, Pa., 1980.
93. E. Hamann, H. Manger, and L. Stencke, *Lambda Sensor with Y203–Stabilized ZrO2–Ceramic for Application in Automotive Emission Control Systems,* SAE 770401, Society of Automotive Engineers, Warrendale, Pa., 1977.
94. J. W. Butler and co-workers, *Fast Response Zirconia Sensor–Based Instrument for Measurement of the Air/Fuel Ratio of Combustion Exhaust,* SAE 840061, Society of Automotive Engineers, Warrendale, Pa., 1984.
95. I. Gorille, N. Rittmansberger, and P. Werner, *Bosch Electronic Fuel Injection with Closed Loop Control,* SAE 750368, Society of Automotive Engineers, Warrendale, Pa., 1975.
96. E. Hendricks, T. Vesterholm, and S. C. Sorenson, *Non Linear Closed Loop SI Engine Control Observers,* SAE 920237, Society of Automotive Engineers, Warrendale, Pa., 1992.
97. A. J. Beumont, A. D. Noble, and A. Scariobrick, "Adaptive Transient Air Fuel Ratio Control to Minimize Gasoline Engine Emissions," *Fisita Congress,* London, 1992.
98. M. J. Anderson, *A Feedback A/F Control System for Low Emission Vehicles,* SAE 930388, Society of Automotive Engineers, Warrendale, Pa., 1993.
99. *Automotive Handbook,* 2nd ed., Robert Bosch GmbH, Stuttgart, Germany, 1986, p. 442.
100. C. O. Probst, *Bosch Fuel Injection & Engine Management,* Robert Bentley, Cambridge, Mass., 1989.
101. Technical Support Document *Revisions to the Malfunction and Diagnostic System Requirements Applicable to 1994 and Later New California Passenger Cars, Light-Duty Trucks, and Medium Duty Vehicles with Feedback Fuel Control Systems (OBD-II),* California Air Resources Board, Sacramento, Sept. 14, 1989.
102. J. W. Koupal, M. A. Sabourin, and W. V. Clemmens, *Detection of Catalyst Failure On-Vehicle Using the Dual Oxygen Sensor Method,* SAE 91061, Society of Automotive Engineers, Warrendale, Pa., 1991.
103. W. Clemmens, M. Sabourin, and T. Rao, *Detection of Catalyst Failure Using On-Board Diagnostics,* SAE 900062, Society of Automotive Engineers, Warrendale, Pa., 1992.
104. S. H. Oh, *Thermal Response of a Monolith Catalyst Converter During Sustained Misfiring: A Computational Study,* SAE 881591, Society of Automotive Engineers, Warrendale, Pa., 1988.
105. W. Cai, Novel Sensors for On Vehicle Measurements of Emissions, Ph.D. dissertation University of Cambridge, Mass., 1992.
106. W. Cai and N. Collings, *A Catalytic Oxidation Sensor for the On-Board Detection of Misfire and Catalyst Efficiency,* SAE 922248, Society of Automotive Engineers, Warrendale, Pa., 1992.
107. N. Collings and co-workers, *A Linear Catalyst Temperature Sensor for Exhaust Gas Ignition (EGI) and On Board Diagnostics of Misfire and Catalyst Efficiency,* SAE 930938, Society of Automotive Engineers, Warrendale, Pa., 1993.
108. D. D. Bech and co-workers, *The Performance of Pd, Pt and Pd–Pt Catalysts in Lean Exhaust,* SAE 930084, Society of Automotive Engineers, Warrendale, Pa., 1993.
109. H. E. Jaaskelainen and J. S. Wallace, *Performance and Emissions of a Natural Gas-Fueled 16-Valve DOHC Four-Cylinder Engine,* SAE 930380, Society of Automotive Engineers, Warrendale, Pa., 1993.
110. A. A. Quader, *Why Intake Charge Dilution Decreases Nitric Oxide Emissions from Spark Ignited Engines,* SAE 710009, Society of Automotive Engineers, Warrendale, Pa., 1971.
111. J. A. Harrington and R. C. Shishu, *Zirconia Oxygen Sensor–An Equivalent Circuit Model,* SAE 730476, Society of Automotive Engineers, Warrendale, Pa., 1973.
112. E. A. Mayer, *Electro-Pneumatic Control Valve for EGR/ATC Actuation,* SAE 810464, Society of Automotive Engineers, Warrendale, Pa., 1981.
113. J. J. Gumbleton and co-workers, *Optimizing Engine Parameters with Exhaust Gas Recirculation,* SAE 740104, Society of Automotive Engineers, Warrendale, Pa., 1974.
114. Ref. 4, Chapt. 25.89, SAE J171.
115. H. M. Haskey, W. R. Cadman, and T. F. Liberty, *The Development of a Real-Time Evaporative Emission Test,* SAE 901110, Society of Automotive Engineers, Warrendale, Pa., 1990.
116. L. L. Lave, W. E. Wecher, W. S. Reis, and D. A. Ross, *Environ. Sci. Technol.* **24,** 8 (1990).
117. *The Effects of Temperature and Fuel Volatility on Vehicle Evaporative Emissions,* Report No 90/1, Concawe, Brussels, Belgium, 1990.
118. T. Cam, K. Cullen, and S. L. Baldus, *Running Loss Temperature Profile,* SAE 930078, Society of Automotive Engineers, Warrendale, Pa., 1993; H. M. Haskew, W. R. Cadman, and T. F. Liberty, *Evaporative Emissions Under Real-Time Conditions,* SAE 891121, Society of Automotive Engineers, Warrendale, Pa., 1989.
119. C. B. Tracy and W. W. Frank, *Fuels, Lubricants, and Positive Crankcase Ventilation System,* SAE PT112, 451, Society of Automotive Engineers, Warrendale, Pa., 1963–1966.
120. R. J. Templin, *Discussion of Reference 2,* SAE TP-6, 249, Society of Automotive Engineers, Warrendale, Pa., 1964.
121. Ref. 4, Chapt. 25.68, SAE J900.
122. A. Unich, R. M. Bada, and K. W. Lyons, *Natural Gas: A Promising Fuel for I.C. Engines,* SAE 930929, Society of Au-

tomotive Engineers, Warrendale, Pa., 1993. An extensive reference source.

123. American Gas Association (AGA), "Projected Natural Gas Demand From Vehicles Under the Mobile Sources Provisions of the Clean Air Act Ammendments," *AGA Energy Analysis EA 1990–1991,* Chicago, 1991.
124. M. Samsa, *Potential for Compressed Natural Gas Vehicles in Centrally-Fueled Automobile, Truck and Bus Fleet Application,* Gas Research Institute, Chicago, 1991, pp. 44–61.
125. Technical staff report on Reactivity Adjustment Factors (RAF), California Air Resources Board, 1993.
126. J. J. White, *Low Emission Catalysts for Natural Gas Engines,* GRI Report 91/0214, Gas Research Institute, Chicago, Southwest Research Institute, San Antonio, SwRI 3178–22, 1991.
127. J. Klimstra, *Catalytic Converters for Natural Gas Engines–A Measurement and Control Problem,* SAE 872165, Society of Automotive Engineers, Warrendale, Pa., 1987.
128. R. Hicks and co-workers, *Structure Sensitivity of Methane Oxidation over Platinum and Palladium; J. Catal.,* 280–306 (1990).
129. W. M. Burkmyre, W. E. Liss, and M. Church, *Natural Gas Converter Performance and Durability,* SAE 930222, Society of Automotive Engineers, Warrendale, Pa., 1993.
130. S. Subramanian, R. J. Kudla, and M. S. Chattha, *Treatment of Natural Gas Vehicle Exhaust,* SAE 930223, Society of Automotive Engineers, Warrendale, Pa., 1993.
131. J. E. Sinor and B. K. Bailey, *Current and Potential Future Performance of Ethanol Fuels,* SAE 930376, Society of Automotive Engineers, Warrendale, Pa., 1993.
132. J. J. Mooney, J. G. Hansel, and K. R. Burns, *Three-Way Conversion Catalysts on Vehicles Fueled with Ethanol–Gasoline Mexuture,* SAE 790428, Society of Automotive Engineers, Warrendale, Pa., 1979.
133. R. K. Pfefly, University of Santa Clara, personal communication, 1979.
134. C. L. Mynng and co-workers, *Research and Development of Hyundai Flexible Fuel Vehicles (FFVs),* SAE 930330, Society of Automotive Engineers, Warrendale, Pa., 1993.
135. T. Suga and Y. Hamazaki, *Development of Honda Flexible Fuel Vehicle,* SAE 922276, Society of Automotive Engineers, Warrendale, Pa., 1992.
136. J. K. Hochmuth and J. J. Mooney, *Catalytic Control of Emissions from M-85 Fueled Vehicles,* SAE 930219, Society of Automotive Engineers, Warrendale, Pa., 1993.
137. K. H. Hellman, G. K. Piotrowski, and R. M. Schaefer, *Start Catalyst Systems Employing Heated Catalyst Technology for Control of Emissions from Methanol-Fueled Vehicles,* SAE 93082, Society of Automotive Engineers, Warrendale, Pa., 1993.
138. Ref. 4, Chapt. 24.373.
139. R. Douglas and G. P. Blair, *Fuel Injection of a Two-Stroke Cycle Spark Ignition Engine,* SAE 820952, Society of Automotive Engineers, Warrendale, Pa., 1982.
140. D. Plohberger and co-workers, *Development of a Fuel-Injected Two-Stroke Gasoline Engine,* SAE 880170, Society of Automotive Engineers, Warrendale, Pa., 1988.
141. M. Nuti, *A Variable Timing Electronically Controlled High Pressure Injection System for 2S S.I. Engines,* SAE 9009799, Society of Automotive Engineers, Warrendale, Pa., 1990.
142. G. E. Hundleby, *Development of a Poppet-Valved Two-Stroke Engine–The Flagship Concept,* SAE 900802, Society of Automotive Engineers, Warrendale, Pa., 1990.
143. G. P. Blair and co-workers, "The Reduction of Emissions and Fuel Consumption by Direct Air-Assisted Fuel Injection into a Two-Stroke Engine," *The 4th Graz Two-Wheeler Symposium,* Technical University, Graz, Austria, 1991.
144. J. Stokes and co-workers, *Development Experience of a Popped-Valved Two-Stroke Flagship Engine,* SAE 920778, Society of Automotive Engineers, Warrendale, Pa., 1992.
145. H. H. Huang and co-workers, *Improvement of Irregular Combustion of Two-Stroke Engine by Skip Injection Control,* SAE 922310, Society of Automotive Engineers, Warrendale, Pa., 1992.
146. C. Csere, *Car Driver,* 87–95, (June 1991).
147. *Auto. Eng.* **99**(7), 11–14 (July 1991).
148. F. Markus, *Car Driver,* 72–75 (Feb. 1992).
149. K. Oblander, J. Abthoff, and L. Fricker, "From Engine Testbench to Vehicle–An Approach to Lean Burn by Dual Ignition," *I. Mech E. C80/79,* (1979).
150. S. Matsushita and co-workers, *Development of the Toyota Lean Combustion System,* SAE 850044, Society of Automotive Engineers, Warrendale, Pa., 1985.
151. I-Y Ohm and co-workers, *Development of HMC Axially Stratified Lean Combustion Engine,* SAE 930879, Society of Automotive Engineers, Warrendale, Pa., 1993.
152. T. Inoue and co-workers, *Toyota Lean Combustion System–the Third Generation System,* SAE 930873, Society of Automotive Engineers, Warrendale, Pa., 1993.
153. M. Iwamoto and H. Hamada, *Catal. Today* **10,** 57 (1991).
154. B. H. Engler and co-workers, *Catalytic Reaction of NO_x with Hydrocarbons under Lean Diesel Exhaust Gas Conditions,* SAE 930735, Society of Automotive Engineers, Warrendale, Pa., 1993.
155. W. Held and co-workers, *Catalytic NO_x Reduction in Net Oxidizing Exhaust Gas,* SAE 900496, Society of Automotive Engineers, Warrendale, Pa., 1990.
156. T. J. Truex, R. A. Searles, and D. C. Sun. *Plat. Met. Rev.* **36**(1), 2–11 (Jan. 1992). A good NO_x reduction background reference.
157. G. Muramatsu and co-workers, *Catalytic Reduction of NO_x in Diesel Exhaust,* SAE 930135, Society of Automotive Engineers, Warrendale, Pa., 1993.
158. M. J. Heimrich and M. L. DeViney, *Lean NO_x Catalyst Evaluation and Characterization,* SAE 930736, Society of Automotive Engineers, Warrendale, Pa., 1993.
159. D. R. Monroe and co-workers, *Evaluation of a Cu–Zeolite Catalyst to Remove NO_x from Lean Exhaust,* SAE 930737, Society of Automotive Engineers, Warrendale, Pa., 1993.
160. L. S. Socha, D. F. Thompson, and P. S. Weber. *Reduced Energy and Power Consumption for Electrically Heated Extruded Metal Converters,* SAE 930383, Society of Automotive Engineers, Warrendale, Pa., 1993.
161. F. W. Kaiser and co-workers, *Optimization of an Electrically-Heated Catalytic Converter System–Calculations and Application,* SAE 930384, Society of Automotive Engineers, Warrendale, Pa., 1993.
162. M. J. Heimrich, S. Albu, and J. Osborne, *Electrically Heated System Conversion on Two Current Technology Vehicles,* SAE 910612, Society of Automotive Engineers, Warrendale, Pa., 1991.
163. I. Gottberg and co-workers, *New Potential Exhaust Gas Aftertreatment Technologies for Clean Car Legislation,* SAE 910849, Society of Automotive Engineers, Warrendale, Pa., 1991.

164. K. H. Hellman, C. K. Piotroski, and R. M. Shaefer, *Evaluation of Different Resistively Heated Technologies,* SAE 912382, Society of Automotive Engineers, Warrendale, Pa., 1991.
165. R. J. Hurley and co-workers, *Experiences with Electrically Heated Catalysts,* SAE 912384, Society of Automotive Engineers, Warrendale, Pa., 1991.
166. J. E. Kubsh and P. W. Lissuik, *Vehicle Emission Performance with an Electrically Heated Converter System,* SAE 912385, Society of Automotive Engineers, Warrendale, Pa., 1991.
167. L. S. Socha and D. F. Thompson, *Electrically Heated Extruded Metal Converter for Low Emission Vehicles,* SAE 920093, Society of Automotive Engineers, Warrendale, Pa., 1992.
168. O. Haddad and co-workers. *The Achievement of ULEV Emission Standards for Large High Performance Vehicles,* SAE 930389, Society of Automotive Engineers, Warrendale, Pa., 1993.
169. EP 424966 (1990), T. Minami and T. Nagase.
170. B. H. Engler and co-workers, *Reduction of Exhaust Gas Emission by Using Hydrocarbon Adsorber Systems,* SAE 930738, Society of Automotive Engineers, Warrendale, Pa., 1993.
171. M. J. Heimrich, L. R. Smith, and J. Kitowski, *Cold-Start Hydrocarbon Collection for Advanced Exhaust Emission Control,* SAE 920847, Society of Automotive Engineers, Warrendale, Pa., 1992.
172. T. Ma, N. Collings, and T. Hands, *Exhaust Gas Ignition (EGI)–A New Concept for Rapid Light-Off of Automotive Exhaust Catalyst,* SAE 920400, Society of Automotive Engineers, Warrendale, Pa., 1992.
173. N. Collings and co-workers, *A Linear Catalyst Temperature Sensor for Exhaust Gas Ignition (EGI) and In-Board Diagnostics of Misfire and Catalyst Efficiency,* SAE 930938, Society of Automotive Engineers, Warrendale, Pa., 1993.

General References

Alcoa Alumina Handbook

R. B. Bird and co-workers, *Transport Phenomena,* John Wiley & Sons, Inc., New York, 1965.

Code of Federal Regulations, Title 40, Part 86, Washington, D.C., July 1, 1992.

K. C. Taylor, in J. R. Anderson and J. R. Boudart, eds., *Catalysis Science and Technology,* Berlin, 1984, Chapt. 2, pp. 119–170.

G. C. Bond and G. Webb, eds, "Catalysis," *Royal Soc. Chem., London,* **6,** (1984).

D. D. Eley, H. Pines, and P. B. Wiez, eds., *Advances in Catalysis,* Vol. 33, Academic Press, Inc., New York, 1985.

H. Heinman and J. J. Carberry, eds., *Catalysis Reviews–Science and Engineering,* Vol. 26, Marcel-Dekker, New York, 1984.

C. N. Satterfield, *Heterogeous Catalysis in Practice,* 1st ed., McGraw-Hill Book Co., Inc., New York, 1980.

J. R. Mondt, *J. Eng. Gas Turbines Power* **109**(2), 200–206 (1987).

C. D. Falk and J. J. Mooney, *Three-Way Conversion Catalysts–Effect of Closed Loop Feedback Control and Other Parameters on Catalyst Efficiency,* SAE 800462, Society of Automotive Engineers, Warrendale, PA., 1980.

M. Luomo, P. Loppi, and R. Lylykaugas, *Evolution of High Cell Density Z-Flow Catalyst,* SAE 930940, Society of Automotive Engineers, Warrendale, Pa., 1993.

EXHAUST CONTROL, INDUSTRIAL

RONALD L. BERGLUND
The M. W. Kellogg Company
Houston, Texas

Limits for exhaust emissions from industry, transportation, power generation (qv), and other sources are increasingly legislated (see also EXHAUST CONTROL, AUTOMOTIVE) (1,2). One of the principal factors driving research and development in the petroleum and chemical processing industries in the 1990s is control of industrial exhaust releases. Much of the growth of environmental control technology is expected to come from new or improved products that reduce such air pollutants as carbon monoxide [*630-08-0*], (CO), volatile organic compounds (VOCs), nitrogen oxides (NO_x), or other hazardous air pollutants (see AIR POLLUTION). The mandates set forth in the 1990 amendments to the Clean Air Act (CAAA) (see CLEAN AIR ACT OF 1990) push pollution control methodology well beyond what, as of this writing, is in general practice, stimulating research in many areas associated with exhaust system control (see AIR POLLUTION CONTROL METHODS). In all, these amendments set specific limits for 189 air toxics, as well as control limits for VOCs, nitrogen oxides, and the so-called criteria pollutants. An estimated 40,000 facilities, including establishments as diverse as bakeries and chemical plants are affected by the CAAA (3).

There are 10 potential sources of industrial exhaust pollutants which may be generated in a production facility (4): (*1*) unreacted raw materials; (*2*) impurities in the reactants; (*3*) undesirable by-products; (*4*) spent auxiliary materials such as catalysts, oils, solvents, etc; (*5*) off-spec product; (*6*) maintenance, ie, wastes and materials; (*7*) exhausts generated during start-up and shutdown; (*8*) exhausts generated from process upsets and spills; (*9*) exhausts generated from product and waste handling, sampling, storage, and treatment; and (*10*) fugitive sources.

Exhaust streams generally fall into two general categories, intrinsic and extrinsic. The intrinsic wastes represent impurities present in the reactants, by-products, co-products, and residues as well as spent materials used as part of the process, ie, sources (*1*)–(*5*). These materials must be removed from the system if the process is to continue to operate safely. Extrinsic wastes are generated during operation of the unit, but are more functional in nature. These are generic to the process industries overall and not necessarily inherent to a specific process configuration, ie, sources (*6*)–(*10*). Waste generation may occur as a result of unit upsets, selection of auxiliary equipment, fugitive leaks, process shutdown, sample collection and handling, solvent selection, or waste handling practices.

CONTROL STRATEGY EVALUATION

There are two broad strategies for reducing volatile organic compound (VOC) emissions from a production facility: (*1*) altering the design, operation, maintenance, or manufacturing strategy so as to reduce the quantity or toxicity of air emissions produced, or (*2*) installing aftertreatment controls to destroy the pollutants in the generated air emission stream (5). Whether the exhaust stream contains a specific hazardous air pollutant, a VOC, a nitrogen oxide, or carbon monoxide, the best way to control the pollutant is to prevent its formation in the first place. Many technologies are being developed that seek to minimize the generation of undesirable by-products by modifying specific process materials or operating conditions. Whereas process economics or product quality may restrict the general applicability of these approaches, an increased understanding of the mechanisms and conditions by which a pollutant is created is leading to significant breakthroughs in burner design and operation (for nitrogen oxide control), equipment design, maintenance, and operation for fugitive and vent VOC emission control, and product and waste storage and handling design and operation (for VOC emission control).

Once an undesirable material is created, the most widely used approach to exhaust emission control is the application of add-on control devices (6). For organic vapors, these devices can be one of two types, combustion or capture. Applicable combustion devices include thermal incinerators (qv), ie, rotary kilns, liquid injection combusters, fixed hearths, and fluidized-bed combustors; catalytic oxidation devices; flares; or boilers/process heaters. Primary applicable capture devices include condensers, adsorbers, and absorbers, although such techniques as precipitation and membrane filtration are finding increased application. A comparison of the primary control alternatives is shown in Table 1.

The most desirable of the control alternatives is capture of the emitted materials followed by recycle back into the process. However, the removal efficiencies of the capture techniques generally depend strongly on the physical and chemical characteristics of the exhaust gas and the pollutants considered. Combustion devices are the more commonly applied control devices, because these are capable of a high level of removal efficiencies, ie, destruction for a variety of chemical compounds under a range of conditions. Although installation of emission control devices requires capital expenditures, these may generate useful materials and be net consumers or producers of energy. The selection of an emission control technology is affected by nine interrelated parameters: (*1*) temperature, T, of the inlet stream to be treated; (*2*) residence time; (*3*) process exhaust flow rate; (*4*) auxiliary fuel needs; (*5*) optimum energy use; (*6*) primary chemical composition of exhaust stream; (*7*) regulations governing destruction requirements; (*8*) the gas stream's explosive properties or heat of combustion; and (*9*) impurities in the gas stream. A process flow diagram for the consideration of these parameters is shown in Figure 1. Given the many factors involved, an economic analysis is often needed to select the best control option for a given application.

Capture devices are discussed extensively elsewhere (see AIR POLLUTION CONTROL METHOD). Oxidation devices are either thermal units that use heat alone or catalytic units in which the exhaust gas is passed over a catalyst usually at an elevated temperature. These latter speed oxidation and are able to operate at temperatures well below those of thermal systems.

OXIDIZATION DEVICES

Thermal Oxidation

Thermal oxidation is one of the best known methods for industrial waste gas disposal. Unlike capture methods, eg, carbon adsorption, thermal oxidation is an ultimate disposal method destroying the objectionable combustible compounds in the waste gas rather than collecting them. There is no solvent or adsorbent of which to dispose or regenerate. On the other hand, there is no product to recover. A primary advantage of thermal oxidation is that virtually any gaseous organic stream can be safety and cleanly incinerated, provided proper engineering design is used (7).

A thermal oxidizer is a chemical reactor in which the reaction is activated by heat and is characterized by a specific rate of reactant consumption. There are at least two chemical reactants, an oxidizing agent and a reducing agent. The rate of reaction is related both to the nature and to the concentration of reactants, and to the conditions of activation, ie, the temperature (activation), turbulence (mixing of reactants), and time of interaction.

Thermal oxidation relies on a homogenous gas-phase reaction condition. Exhaust emissions from industrial sources usually contain organic compounds (the reducing agents) well mixed with oxygen (the oxidizing agent). Imparting the necessary, uniform temperature for reaction within this mixture is of primary importance in the design of the oxidizer and related equipment. Proper activation requires establishing the minimum required temperature (650–800°C) for an adequate time (0.1–0.3 s). General design consideration is given to minimizing heat input and reactor size under the constraints of time, turbulence, and temperature.

Thermal oxidization devices are widely used, and generally provide a high degree of assurance that the process oxidizes the material in the exhaust gas. The high temperature operation causes other problems, however, especially compared to alternatives such as catalytic oxidation. The thermal oxidation devices often incur higher fuel costs because of the higher temperatures necessary, and require exotic high temperature materials, because the high temperatures entailed can bring about serious mechanical design problems as a result of operating in the temperature range in which metal creep takes place (1,8). In addition, equipment durability is reduced by the extent of thermal cycling. Thermal oxidation systems are susceptible to thermal stress effects which result in distortion of the ductwork and heat-exchange surfaces, creating the potential for cracks and leaks. A further consequence of these high temperatures is that a thermal oxidizer may produce nitrogen oxides (NO_x), and sometimes yield undesirable byproducts such as dioxins from chlorinated materials (5).

Some of the problems associated with thermal oxidizers have been attributed to the necessary coupling of the mixing, the reaction chemistry, and the heat release in the burning zone of the system. These limitations can reportedly be avoided by using a packed-bed flameless thermal oxidizer which is under development. This system relies on radiant heat from a large heat sink to raise the temper-

Table 1. Emission Control Technologies

Technology	Reduction Effectiveness	Recovery	Waste Generation	Advantages	Disadvantages
Activated carbon adsorption	90–98%	Chemical recovery possible with regeneration	Spent carbon or regenerant	Good for wide variety of VOCs	Carbon replacement, regeneration costs, potential for bed fires
Adsorption in wet scrubbers	75–90%+	Chemical recovery possible through decanting/ distillation	Spent solvent or regenerant	Simple operation	Not efficient at low concentration
Vapor condensation	50–80%	Chemical recovery possible through decanting/ treatment	Liquid wastes, needs off-gas treatment	Simple operation, effective for high VOC concentration	Low removals applicability limits to some VOCs, high power costs
Thermal oxidation	99%	Heat recovery	NO_x generation, CO_2 generation	Handles any VOC concentration	High operating costs, capital costs, temperatures, and maintenance
Catalytic oxidation	95–99%	Heat recovery	Spent catalyst regeneration acids and alkalines	Simple systems, lower T than thermal economical operation	Fouling of catalysts, temperature limits

ature of the exhaust gas to its ignition temperature. The heat sink, a ceramic matrix, is preheated radiatively using an electric preheating element, or a natural gas preheater, prior to introducing the exhaust gas. Because the system temperature is reasonably constant, NO_x generation within the flame is minimized. High (99.99%) VOC reductions at low contact times were reported (9). However, as with all thermal oxidation systems, this system is most effective for higher concentration exhaust streams, and requires significant auxiliary heat to treat low concentration streams.

Catalytic Oxidization

A principal technology for control of exhaust gas pollutants is the catalyzed conversion of these substances into innocuous chemical species, such as water and carbon dioxide. This is typically a thermally activated process commonly called catalytic oxidation, and is a proven method for reducing VOC concentrations to the levels mandated by the CAAA. Catalytic oxidation is also used for treatment of industrial exhausts containing halogenated compounds.

As an exhaust control technology, catalytic oxidation enjoys some significant advantages over thermal oxidation. The former often occurs at temperatures that are less than half those required for the latter, consequently saving fuel and maintenance costs. Lower temperatures allow use of exhaust stream heat exchangers of a low grade stainless steel rather than the expensive high temperature alloy steels. Furthermore, these lower temperatures tend to avoid the emissions problems arising from the thermal oxidation processes (10,11).

Critical factors that need to be considered when selecting an oxidation system include (12) (*1*) waste stream heating value and explosive properties. Low heating values resulting from low VOC concentration make catalytic systems more attractive, because low concentrations increase fuel usage in thermal systems; (*2*) waste gas components that might affect catalyst performance. Catalyst formulations have overcome many problems owing to contaminants, and a guard bed can be used in catalytic systems to protect the catalyst; (*3*) the type of fuel available and optimum energy use. Natural gas and no. 2 fuel oil can work well in catalytic systems, although sulfur in the fuel oil may be a problem in some applications (13). Other fuels should be evaluated on a case-by-case basis; and (*4*) space and weight limitations on the control technology. Catalysts are favored for small, light systems.

There are situations where thermal oxidation may be preferred over catalytic oxidation: for exhaust streams that contain significant amounts of catalyst poisons and/or fouling agents, thermal oxidation may be the only technically feasible control; where extremely high VOC destruction efficiencies of difficult to control VOC species are required, thermal oxidation may attain higher performance; and for relatively rich VOC waste gas streams, ie, having $\geq$20–25% lower explosive limit (LEL), the gas stream's explosive properties and the potential for catalyst overheating may require the addition of dilution air to the waste gas stream (12).

Whereas the catalytic converter has been used in automobiles to control air pollutants only since 1975 (5), catalytic oxidation of industrial exhaust emissions began in the late 1940s, and is a reasonably mature technology (14). Initially it was used only in circumstances where an extremely serious odor problem was associated with an industrial system, or where the concentration of organic solvents in the gases to be discharged to the air was high enough that these could be burned and the heat utilized in the process (8). By the mid-1950s there were several dozen catalytic incinerators in California, primarily in Los

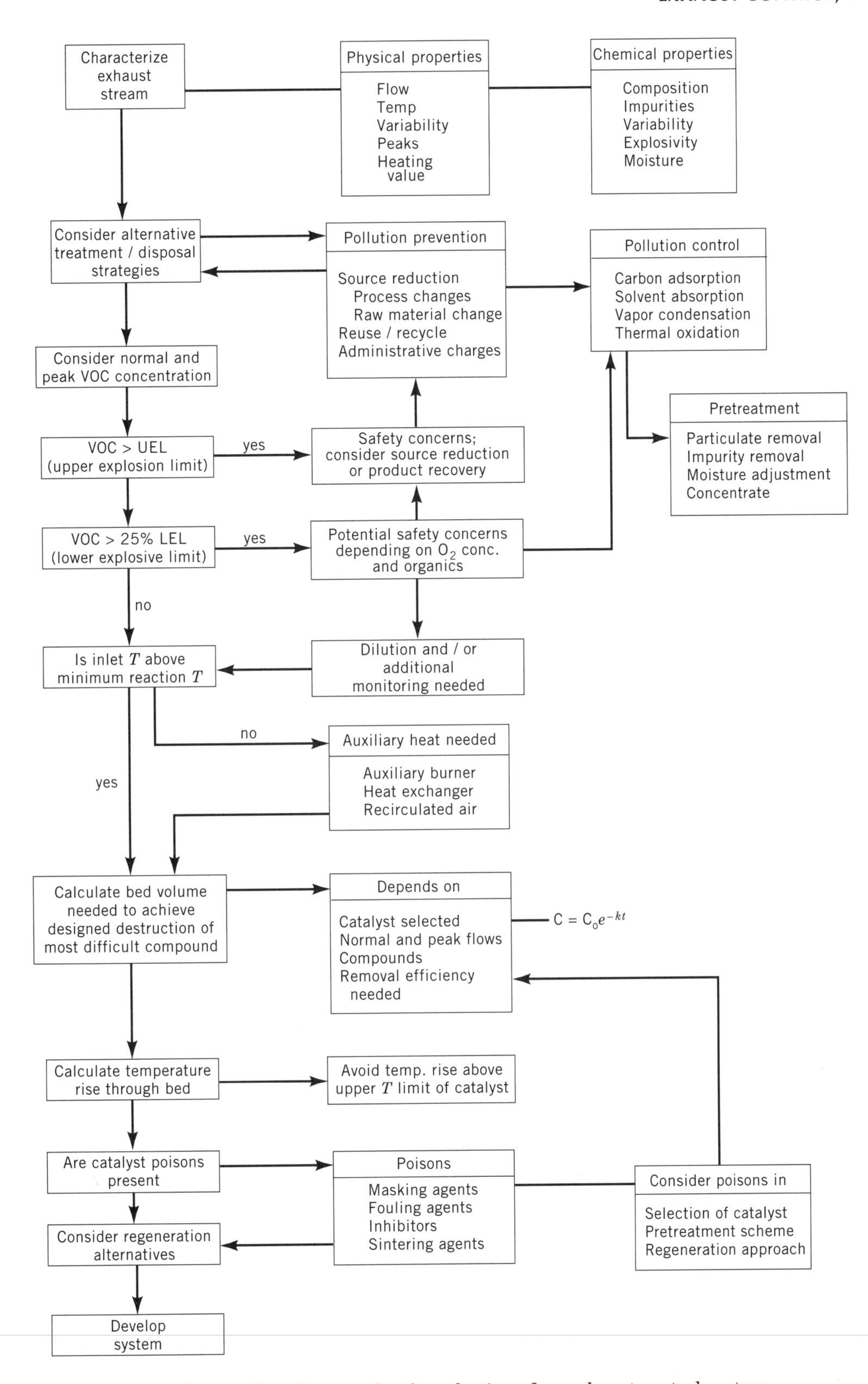

Figure 1. Process flow diagram for the selection of an exhaust control system.

Angeles county, the first sizable area within the United States to experience a serious air pollution problem. Early applications of this technology involved some serious odor, eye irritation, or visible organic emission problems resulting from halogen poisoning and catalyst fouling (15).

The chemical industry was the first to utilize catalytic oxidation extensively for emission control, building units capable of treating up to 50 m³/s (100,000 scfm) of exhaust gas containing VOCs. Catalytic systems accounted for roughly one-fourth of the $200 million market for VOC control systems in 1992, and over one thousand catalytic oxidation devices were in place by the end of that year (5).

Catalysts. A catalyst has been defined as a substance that increases the rate at which a chemical reaction approaches equilibrium without becoming permanently involved in the reaction (16). Thus a catalyst accelerates the kinetics of the reaction by lowering the reaction's activation energy (5), ie, by introducing a less difficult path for the reactants to follow. For VOC oxidation, a catalyst decreases the temperature, or time required for oxidation, and hence also decreases the capital, maintenance, and operating costs of the system.

A key feature of a catalyst is that the catalytic material is not consumed by the chemical oxidation reactions, rather it remains unaltered by the reactions which occur generally on its surface and thus remains available for an infinite number of successive oxidation reactions.

Many chemical elements exhibit catalytic activity (5) which, within limits, is inversely related to the strength of chemisorption of the VOCs and oxygen, provided that adsorption is sufficiently strong to achieve a high surface coverage (17). If the chemisorption is too strong, the catalyst is quickly deactivated as the active sites become irreversibly covered. If the chemisorption is too weak, only a small fraction of the surface is covered and the activity is very low (17) (Fig. 2).

Catalysts vary both in terms of compositional material and physical structure (18). The catalyst basically consists of the catalyst itself, which is a finely divided metal (14,17,19); a high surface area carrier; and a support structure. Three types of conventional metal catalysts are used for oxidation reactions: single- or mixed-metal oxides, noble (precious) metals, or a combination of the two (19).

The precious metal or metal oxide imparts high intrinsic activity, the carrier provides a stable, high surface area for catalyst dispersion, and the mechanical support gives a high geometric surface area for physical support and engineering design features (20). Only the correct combination of these components provides suitable performance and long catalyst life of a properly designed catalytic system (21).

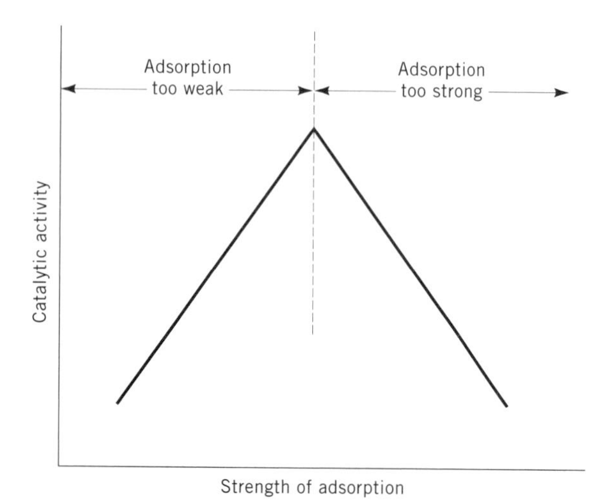

Figure 2. Catalytic activity as a function of adsorption strength (17). Courtesy of Oxford University Press.

Metal Oxides. The metal oxides are defined as oxides of the metals occurring in Groups 3–12 (IIIB to IIB) of the Periodic Table. These oxides, characterized by high electron mobility and the positive oxidation state of the metal, are generally less active as catalysts than are the supported nobel metals, but the oxides are somewhat more resistant to poisoning. The most active single-metal oxide catalysts for complete oxidation of a variety of oxidation reactions are usually found to be the oxides of the first-row transition metals, V, Cr, Mn, Fe, Co, Ni, and Cu.

Noble Metals. Noble or precious metals, ie, Pt, Pd, Ag, and Au, are frequently alloyed with the closely related metals, Ru, Rh, Os, and Ir. These are usually supported on a metal oxide such as α-alumina, α-Al_2O_3, or silica, SiO_2. The most frequently used precious metal components are platinum [*7440-06-4*], Pt, palladium [*7440-05-3*], Pd, and rhodium [*7440-16-6*], Rh. The precious metals are more commonly used because of the ability to operate at lower temperatures. As a general rule, platinum is more active for the oxidation of paraffinic hydrocarbons; palladium is more active for the oxidation of unsaturated hydrocarbons and CO (19).

Each precious metal or base metal oxide has unique characteristics, and the correct metal or combination of metals must be selected for each exhaust control application. The metal loading of the supported metal oxide catalysts is typically much greater than for nobel metals, because of the lower inherent activity per exposed atom of catalyst. This higher overall metal loading, however, can make the system more tolerant of catalyst poisons. Some compounds can quickly poison the limited sites available on the noble metal catalysts (19).

Carrier. The metal catalyst is generally dispersed on a high surface area carrier, ie, the carrier is given a washcoat of catalyst, such that very small (2–3 nm dia) precious metal crystallites are widely dispersed over the surface area, serving two basic functions. It maximizes the use of the costly precious metal, and provides a large surface area thereby increasing gas contact and associated catalytic reactions (18).

Proper selection of the carrier is critical (20). For example, in most cases, the carrier of a precious metal catalyst is a high surface area alumina having an effective surface area in the order of 120 m^2/g of material. Alumina is often used because of its unique phase transformation properties. Various phases of the aluminum hydroxides exist as a function of temperature and starting phase. For a catalyst to be stable, the correct starting phase of alumina must be selected for the projected commercial operating temperature. Otherwise the alumina may undergo a transition during operation resulting in a carrier of less surface area, and hence less catalytic activity (1).

Efforts to redesign catalyst formulations involve both catalyst and washcoat. One thrust of this research is to manipulate the catalytic surface so that it can handle larger quantities of catalyst poisons and to incorporate more catalytic sites, redistributed within the washcoat to

Table 2. Conventional Catalyst Bed Geometries[a]

Geometry	Advantages	Disadvantages
Metal ribbons	Low pressure drop; high surface-to-volume	Less active than ceramic-supported catalysts
Spherical pellets	Can be used in both fixed and fluidized beds	High pressure drop; attrition problem
Ceramic rods	Low pressure drop	Low surface-to-volume ratio
Ceramic honeycomb	Low pressure drop; high surface-to-volume ratio	May have nonuniform catalyst coating
Metal honeycomb	Low pressure drop; high surface-to-volume ratio; high mechanical strength	Less active than ceramic-supported catalysts

[a] Ref. 14.

make them more accessible to exhaust molecules. Altering the composition of the alumina washcoat by including various nonprecious metal oxides, such as oxides of barium, cerium, and lanthanum as stabilizers, is being looked at to promote catalyst activity before sintering and stabilize precious metal dispersion (18). Reformulation efforts are aided by use of computer controls and cleaner reactants and continuous monitoring, all of which help make exhaust composition more predictable (1).

The Support Structure. After the catalytic element is placed on the high surface area carrier, it is deposited on a mechanical support structure which determines the form of the catalyst. The support structure may have many forms, such as spheres, pellets, woven mesh, screen, honeycomb, or other ceramic matrix structures designed to maximize catalyst surface area (6). Some of the advantages of these different supports are listed in Table 2.

The pelleted and honeycomb support structures are most widely used; honeycombs are the most commonly employed. Pelleted structures are generally spherical beads or cylinders having diameters ranging from 0.16–0.64 cm. The pellets are assembled into a packed bed containing large numbers of these pellets through which the exhaust passes. The honeycomb supports are monolithic structures having numerous parallel channels through which the exhaust passes, the channel sizes ranging from about 8 to 50 cells/cm^2 of catalyst frontal area. Each cell has a width opening ranging from 0.29 to 0.13 cm, respectively. Some commercial honeycombs are available from 1.6 to 100 cells/cm^2 (20). The shape in the individual honeycomb channel is unlimited, eg, circle, square, triangle, etc.

Although more expensive to fabricate than the pelleted catalyst, and usually more difficult to replace or regenerate, the honeycomb catalyst is more widely used because it affords lower pressure losses from gas flow; it is less likely to collect particulates (fixed-bed) or has no losses of catalyst through attrition, compared to fluidized-bed; and it allows a more versatile catalyst bed design (18), having a well-defined flow pattern (no channeling) and a reactor that can be oriented in any direction.

The honeycomb structure is either fabricated of ceramic or stainless steel. The high surface area carrier and catalytic precious metal crystallites are coated onto the walls of the channels in the honeycomb. The honeycomb catalyst blocks generally range from 15 to 30 cm square at depths from 5 to 10 cm. These blocks are packed into larger modules containing many catalyst blocks. Flow through a honeycomb catalyst structure is shown in Figure 3 (18,22). A typical 30 cells/cm^2 honeycomb structure has about 4600 m^2 of geometric wall area per cubic meter of catalyst volume (18). The actual shape of the individual honeycomb channel is unlimited, eg, it may be circular, square, triangular, etc. In addition, the channel density can be varied. Commercial honeycombs are available that range from 1.5 to 100 cells/cm^2 (10 to 600 cells/in.^{2} (20).

Only by using the carrier can the catalyst be sufficiently active because the majority of applications require 10 to

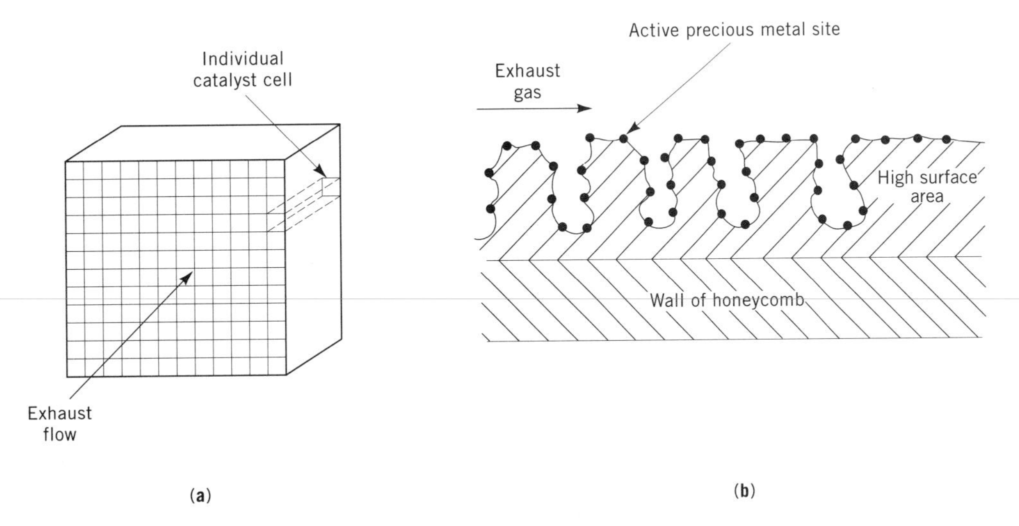

Figure 3. Schematic of the flow through (**a**) a honeycomb catalyst structure and (**b**) a cross section of a honeycomb channel.

100 m^2/g of surface area (20). Surface areas for a typical monolith support structure and a carrier are given (20).

Catalyst	BET surface area, m^2/g
geometric smooth monolith wall	0.003
natural porosity of monolith wall	0.3
carrier on monolith wall	20

Mechanistic Models. A general theory of the mechanism for the complete heterogeneous catalytic oxidation of low molecular weight vapors at trace concentrations in air does not exist. As with many catalytic reactions, however, certain observations have led to a general hypothesis (17).

The overall process of any catalytic reaction is a combination of mass-transfer (qv), describing transport of reactants and products to and from the interior of a solid catalyst, and chemical reaction kinetics, describing chemical reaction sequences on the catalyst surface. The most cost-effective catalytic oxidation systems require use of a solid catalyst material having a high specific surface area, ie, high surface area per net weight of catalyst. The presence of many small pores necessarily introduces pore transport diffusion resistance as a factor in the overall, or global, kinetics. The overall process consists of (17,23): (*1*) transport of reactants from the bulk fluid through the gas film boundary layer to the surface of the catalytic particle; (*2*) transport of reactants into the catalyst particle by diffusion through the catalyst pores; (*3*) chemisorption of at least one reactant on the catalyst surface; (*4*) chemical reaction between chemisorbed species or between a chemisorbed species and a physisorbed or fluid-phase reactant; (*5*) desorption of reaction products from the catalyst surface; (*6*) diffusive transport of products through the catalyst pores to the surface of the catalyst particle; and (*7*) mass transfer of products through the exterior gas film to the bulk fluid.

In principle, any of these steps or some combination can be rate controlling. In practice, temperature plays a primary role in determining the rate-controlling stage. Any comprehensive analysis of actual catalytic oxidation systems of practical interest must include a quantitative understanding of the relative effects of mass transfer (steps *1,2,6,7*) and surface reaction (steps *3,4,5*). The temperature relationship of these two mechanisms is shown in Figure 4 (17,18,20). As a catalyst is heated, conversion of the pollutant is negligible until a critical temperature is reached, then the rate of conversion increases rapidly with rising temperature. This is referred to as the kinetically limited region. Conversion increases in this region because catalytic reaction rates increase with temperature, until the catalyst's normal operating temperature is achieved. Then the conversion rate increases only slightly with further temperature rise in the mass-transfer limited region. At some advanced temperature, the conditions reach a point where thermal oxidation begins to play a role, and the rate of conversion again increases rapidly.

In the mass-transfer limited region, conversion is most commonly increased by using more catalyst volume or by increasing cell density, which increases the catalytic wall area per volume of catalyst. When the temperature reaches a point where thermal oxidation begins to play a role, catalyst deactivation may become a concern.

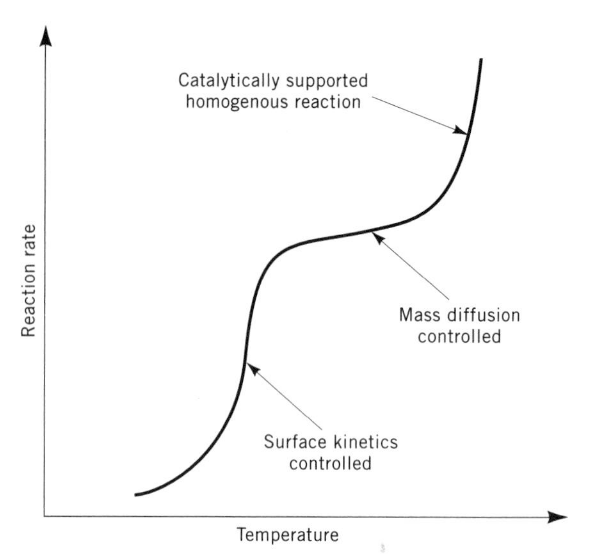

Figure 4. Reaction rate profile as a function of temperature (20).

Reaction Rate. The kinetics for a single catalytic reaction can be modeled as

$$-r_m = k(T)f(C)n$$

where $-r_m$ is the rate of the main reaction; $k(T)$ is the rate constant, a function of temperature, T; $f(C)$ is the function of reactant and product concentration, C; and n is the effectiveness factor, which accounts for pore-diffusional resistance (24). The form of the terms $k(T)$ and $f(C)$ depends on the kinetic model for the system. Kinetic models for the catalytic oxidation can either be empirical or mechanistic.

Empirical Models. In the case of an empirical equation, the model is a power law rate equation that expresses the rate as a product of a rate constant and the reactant concentrations raised to a power (17), such as

$$r_m = kC_1^aC_2^b$$

where r_m is the reaction rate; k is the rate constant; C_1 is the concentration of reactant 1; C_2 is the concentration of reactant 2; and a and b are empirically determined reaction orders.

For combustion of simple hydrocarbons, the oxidation reactions appear to follow classical first-order reaction kinetics sufficiently closely that practical designs can be established by application of the empirical theory (8). For example, the general reaction for a hydrocarbon:

$$C_xH_y + (x + y/2)\ O_2 \rightarrow x\ CO_2 + (y/2)\ H_2O$$

can be represented by the rate equation

$$r_m = (dC/dt) = -kC$$

where C = hydrocarbon concentration, r_m = rate of change of contaminant concentration, t = time, and k = reaction rate constant, which must be determined experimentally from the burning of various organic materials. The pattern of variation with t is predictable from kinetic theory and follows the Arrhenius equation,

$$k = A \exp(-\Delta E/RT)$$

where A is the Arrhenius collision constant, ΔE the activation energy, and R is the universal gas constant. A catalyst increases the rate of reaction by adsorbing gas molecules on catalytically active sites. The catalyst may function simply to bring about a higher concentration of reactive materials at the surface than is present in the bulk gas phase, which has the effect of increasing the collision constant, A, or the catalyst may modify a molecule of adsorbed gas by adding or removing an electron or by physically opening a bond. This has the effect of decreasing the activation energy, ΔE, in the Arrhenius equation. In either circumstance, it is necessary for the reactive materials to reach the active catalyst surface by diffusion through the gas phase, and for the reaction products to leave the surface. For the conditions encountered in most hydrocarbon emission control applications, the oxygen partial pressure is much larger than the organic reactant partial pressure, and can be treated as a constant.

Mechanistic kinetic expressions are often used to represent the rate data obtained in laboratory studies, and to explain quantitatively the effects observed in the field. Several types of mechanisms have been proposed. These differ primarily in complexity, and on whether the mechanism assumes that one compound that is adsorbed on the catalyst surface reacts with the other compound in the gas phase, eg, the Eley-Rideal mechanism (23); or that both compounds are adsorbed on the catalyst surface before they react, eg, the Langmuir-Hinshelwood mechanism (25).

The volatile organic compounds on the list of hazardous air pollutants under the CAAA have been classified into four main categories: (*1*) pure hydrocarbons (qv), (*2*) halogenated hydrocarbons, (*3*) nitrogenated hydrocarbons, and (*4*) oxygenated hydrocarbons. The compounds in these groups are characterized by the following oxidation reactions (26):

Hydrocarbons

$$C_6H_6 + 9\ O_2 \rightarrow 6\ CO_2 + 6\ H_2O$$

Halogenated hydrocarbons

$$CCl_4 + O_2 + 2\ H_2O \rightarrow CO_2 + 4\ HCl + O_2$$

Nitrogenated hydrocarbons

$$2\ HCN + 3\ O_2 \rightarrow 2\ CO_2 + 2\ H_2O + 2\ N_2$$

Oxygenated hydrocarbons

$$C_4H_8O + 5\tfrac{1}{2}\ O_2 \rightarrow 4\ CO_2 + 4\ H_2O$$

Temperature reaction rate profiles for representative compounds are available (21,26). Particularly important are the operating temperatures required before destruction is initiated. Chemical reactivity by compound class from high to low is (27) alcohols > cellosolves/dioxane > aldehydes > aromatics > ketones > acetates > alkanes > chlorinated hydrocarbons. In general, within a class the higher the molecular weight, the higher the relative destructibility. All of these compound classes, except chlorinated hydrocarbons, can be destroyed with 98–99% efficiency at sufficiently low space velocities and/or high enough inlet temperatures (28). Table 3 (22) presents oxidation temperatures for a number of hydrocarbons.

Historically, the destruction efficiency for chlorinated hydrocarbons is quite low. In addition, tests conducted after the chlorinated hydrocarbon is treated show that the catalyst is partially deactivated. More recent advancements in catalyst technology have resulted in the development of a number of catalysts and catalytic systems capable of handling most chlorinated hydrocarbons under a variety of conditions (19).

Mixture Effects. Care must be taken in determining the oxidation kinetics for a mixture of chemicals (29). In principle, given one set of conditions and a two-component mixture, the overall conversion of one component A may be controlled by mass transfer to the catalyst surface and the conversion of another component B by surface-reaction kinetics. Of course, the controlling regime (mass transfer or reaction) can change with temperature. Thus for two independent parallel reactions the combined effect of diffusional and reaction rate resistances can have a considerable influence on the relative rate of the two reactions. Additionally, a third, fourth, or nth component can conceivably affect the other components by, for instance, competing more successfully for active surface sites than B while simultaneously influencing the mass transfer of A. Thus even for a simple two- or three-component mixture, interpretation of observed results can be difficult. Extrapolation of mixture behavior from single-component data is ill-advised.

In a mixture of n-hexane and benzene (29), the deep catalytic oxidation rates of benzene and n-hexane in the binary mixture are lower than when these compounds are singly present. The kinetics of the individual compounds

Table 3. Ignition Temperatures for 90% Conversion[a]

Component	Temp, °C
Hydrogen	93
Acetylene	177
Carbon monoxide	218
Cyclohexanone	218
Propylene	232
Toluene	232
2-Propanol	260
Ethylene	260
Benzene	260
Xylene	260
Ethanol	260
Methyl ethyl ketone	274
Ethyl acetate	288
Cyclohexane	288
n-Hexane	316
Methyl isobutyl ketone	316
Propane	399
Methane	427

[a] Ref. 22.

can be adequately represented by the Mars-VanKrevelen mechanism. This model needs refinements to predict the kinetics for the mixture.

One important consideration in any catalyst oxidation process for a complex mixture in the exhaust stream is the possible formation of hazardous incomplete oxidation products. Whereas the concentration in the effluent may be reduced to acceptable levels by mild basic aqueous scrubbing or additional vent gas treatment, studying the kinetics of the mixture and optimizing the destruction cycle can drastically reduce the potential for such emissions.

Design and Operation. The destruction efficiency of a catalytic oxidation system is determined by the system design. It is impossible to predict *a priori* the temperature and residence time needed to obtain a given level of conversion of a mixture in a catalytic oxidation system. Control efficiency is determined by process characteristics such as concentration of VOCs emitted, flow rate, process fluctuations that may occur in flow rate, temperature, concentrations of other materials in the process stream, and the governing permit regulation, such as the mass-emission limit. Design and operational characteristics that can affect the destruction efficiency include inlet temperature to the catalyst bed, volume of catalyst, and quantity and type of noble metal or metal oxide used.

Catalytic oxidation systems are normally designed for destruction efficiencies that range from 90 to 98% (27). In the early 1980s, typical design requirements were for 90% or higher VOC conversions. More recently, however, an increasing number of applications require 95 to 98% conversions to meet the more stringent emission standards (20).

Operational Considerations. The performance of catalytic incinerators (28) is affected by catalyst inlet temperature, space velocity, superficial gas velocity (at the catalyst inlet), bed geometry, species present and concentration, mixture composition, and waste contaminants. Catalyst inlet temperatures strongly affect destruction efficiency. Mixture compositions, air-to-gas (fuel) ratio, space velocity, and inlet concentration all show marginal or statistically insignificant effects (30).

Operating Temperature. The operating temperature needed to achieve a particular VOC destruction efficiency depends primarily on the species of pollutants contained in the waste stream, the concentration of the pollutants, and the catalyst type (14). One of the most important factors is the hydrocarbon species. Each has a catalytic initiation temperature which is also dependent on the type of catalyst used (14).

For a given inlet temperature, the quantity of supplied heat may be provided by (6) the heat supplied from the combustion of supplemental fuel, the sensible heat contained in the emission stream as it enters the catalytic system, and the sensible heat gained by the emission stream through heat exchange with hot flue gas (6). Three types of systems for catalytic oxidation of VOCs are shown in Figure 5 (11). The simplest (Fig. **5a**) uses a direct contact open flame to preheat the gas stream upstream of the catalyst. The second (Fig. **5b**) involves only a catalyst bed over which the gas stream passes, usually after some indirect preheating. The third (Fig. **5c**) involves more extensive indirect preheating and heat exchange. The difference in the three configurations is the method for preheating the gas.

There are two general temperature policies: increasing the temperature over time to compensate for loss of catalyst activity, or operating at the maximum allowable temperature. These temperature approaches tend to maximize destruction, yet may also lead to loss of product selectivity. Selectivity typically decreases with increasing temperature; faster deactivation; and increased costs for reactor materials, fabrication, and temperature controls.

Reactor Design. The catalytic reactor is designed to be operated in the mass-transfer controlled catalytic region. The prime design parameter is the geometric surface area. The honeycomb catalyst shows substantial advantage over other forms because of the high geometric surface areas obtainable with low pressure drop (20).

Catalyst Selection. The choice of catalyst is one of the most important design decisions. Selection is usually based on activity, selectivity, stability, mechanical strength, and cost (31). Stability and mechanical strength, which make for steady, long-term performance, are the key characteristics. The basic strategy in process design is to minimize catalyst deactivation, while optimizing pollutant destruction.

Both catalyst space velocity and bed geometry play a role. The gas hourly space velocity (GHSV) is used to relate the volumetric flow rate to the catalyst volume. GHSV has units of inverse hour and is defined as the volume flow rate per catalyst volume.

The size of the catalyst bed depends mainly on the degree of VOC reduction required (14). VOC destruction efficiencies up to 95% can usually be attained using reasonable space velocities (14). However, the low GHSVs, and subsequently high catalyst volumes required to achieve extremely high (eg, 99%) conversions, can sometimes make catalytic oxidation uneconomical. Conventional bed geometries may be found in the literature (14).

Process Conditions. To effectively design a catalytic control system, the Manufacturers of Emissions Controls Association recommends the following data be obtained (5): list of all VOCs present and range of concentration of each, flow rate of exhaust and expected variability, oxygen concentration in exhaust and expected variability, temperature of exhaust and expected variability, static pressure, potential uses for heat recovery, particular performance criteria and/or regulations to be met, capture efficiency, ie, fraction of all organic vapors generated by the processes that are directed to the control device, presence of hydrocarbon aerosols in the effluent exhaust, identity and quantity of all inorganic and organic particulate, amount of noncombustibles, presence of possible catalyst deactivators, and anticipated start-up/shutdown frequency of the system.

Pilot Studies. Applications requiring the reduction of VOC emissions have increased dramatically. On-site pilot tests are beneficial in providing useful information regarding VOC emission reduction applications. Information that can be obtained includes optimum catalyst operating con-

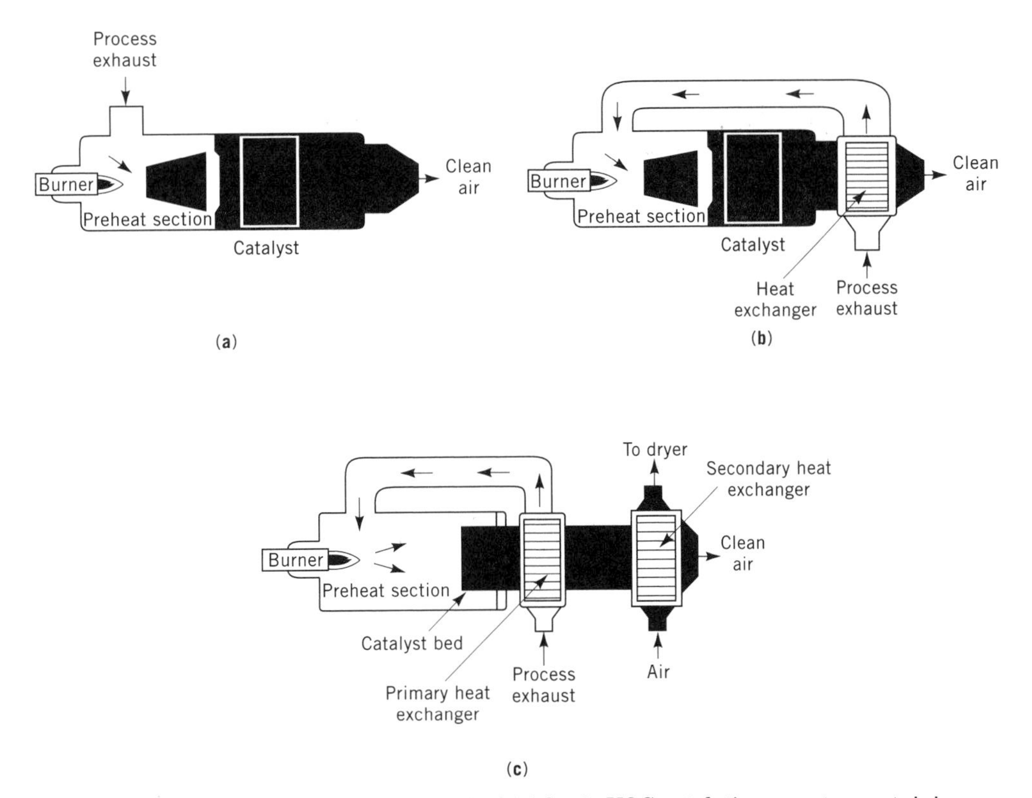

Figure 5. Catalytic system designs (11) of (**a**) basic VOC catalytic converter containing a preheater section, a reactor housing the catalyst, and essential controls, ducting, instrumentation, and other elements; (**b**) a heat exchanger using the cleaned air exiting the reactor to raise the temperature of the incoming process exhaust; and (**c**) extracting additional heat from the exit gases by a secondary heat exchanger.

ditions, the presence of contaminants in the gas stream, and the effects of these contaminants.

Catalyst Inhibition. A number of potential applications for catalytic oxidation of organic materials have resulted in serious odor or eye irritation, or visible emission problems (8). Some of these failures are a result of fouling of the catalyst surface. Others occur because materials such as halogens in the gas stream interfere with or suppress the activity of the catalyst, or because the substances react with the precious metals, rendering them permanently inactive. Finally, all catalysts eventually deteriorate by aging or thermal processes (8).

Many of the exhaust streams that must be purified contain significant amounts of halogenated organics, such as polychlorinated ethanes and ethylenes vented in the manufacture of vinyl chloride monomer or released in usage solvents (32). However, the catalysts used in the conventional catalytic oxidation are severely inhibited by the halogen atoms in these compounds (32). Other trace contaminants of concern in air streams may include phosphorus, nitrogen, and sulfur-containing compounds. Whereas gases containing chlorine, sulfur, and other atoms can deactivate supported noble metal catalysts such as platinum, chlorinated VOC can be treated by certain supported metal oxide catalysts (7).

The four basic mechanisms of catalyst decay are shown in Figure 6 (5,18,24). These are fouling or masking, poisoning, thermal degradation through aging or sintering, and loss of catalyst material through formation and escape of vapors. Poisoning and vapor transport are basically chemical phenomena, whereas fouling is mechanical. Table 4 lists substances that inhibit catalyst activity (5).

Masking or Fouling. Masking or fouling is a physical deposition of species from the fluid phase onto the catalyst surface (Fig. 6**a**) that results in blockage of reaction sites or pores (24). Masking or fouling is caused by a gradual accumulation of noncombusted, solid material that mechanically coats the catalyst's surface and prevents or slows down the diffusion of reactants to the catalyst.

Typical masking or fouling agents include (5,8,21,24) airborne dust or dirt; metal oxides formed from materials in the process, such as silicon dioxide ash remaining when silicone compounds are oxidized; aggregate compound formation on the catalyst surface, ie, phosphorus for lubricating oils; corrosion products from the duct system; and organic char or tars formation from incomplete combustion products, often caused by too low a reactor operating temperature.

Low levels of particulates or potential poisons can sometimes be tolerated without a dramatic decrease in performance. Generally it has been recommended that the maximum particulate concentration not exceed 115 mg/m^2 and that the maximum poison concentration not exceed 25 ppm (14). In addition, every effort should be made to avoid flow over a cold catalyst bed for any extended period of time, as a process stream containing volatile organics may condense on a cold catalyst bed (5).

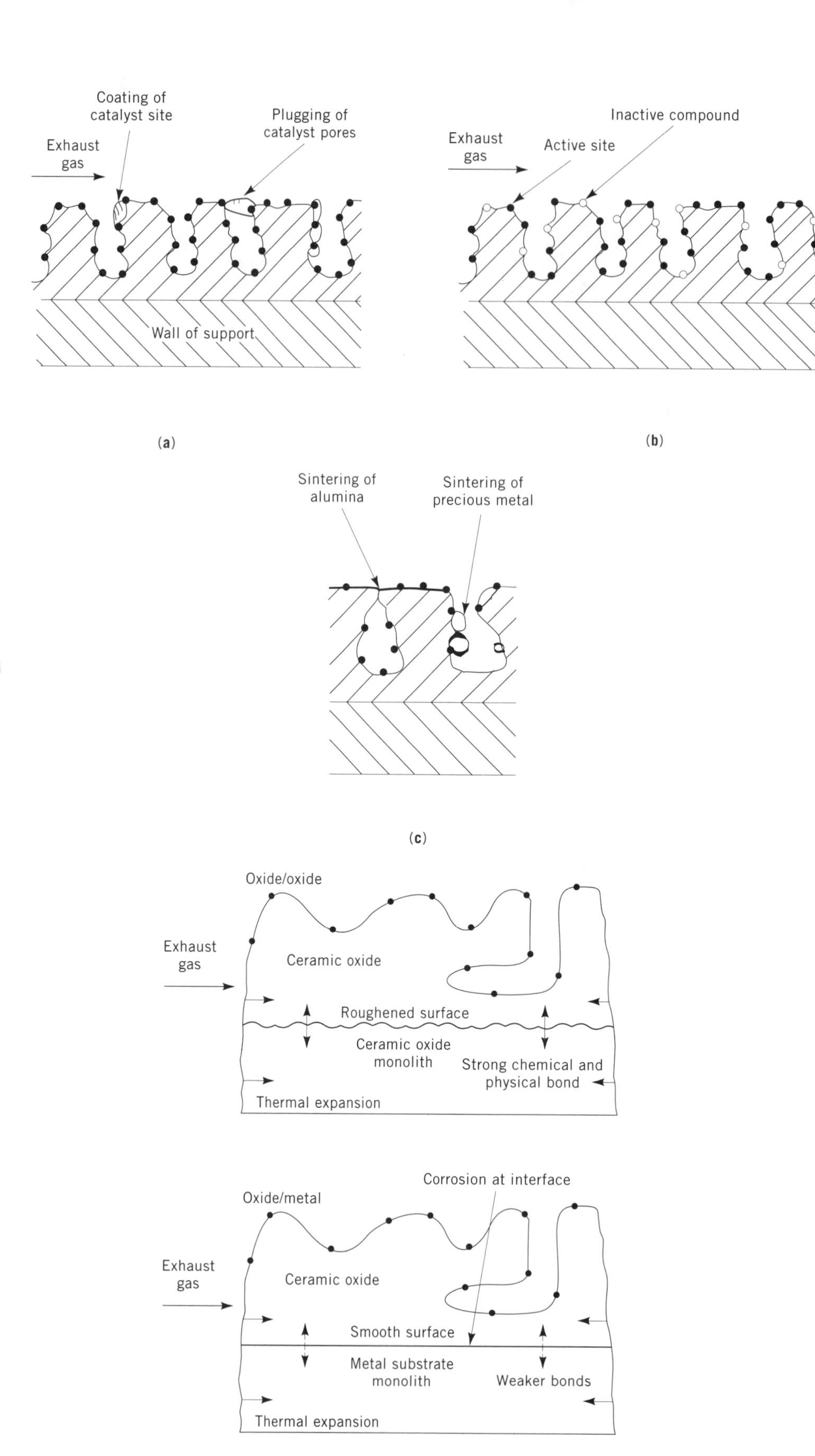

Figure 6. Catalyst inhibition mechanisms where (●) are active catalyst sites; ▨, the catalyst carrier; and ▧ the catalytic support: (**a**) masking of catalyst; (**b**) poisoning of catalyst; (**c**) thermal aging of catalyst; and (**d**) attrition of ceramic oxide metal substrate monolith system, which causes the loss of active catalytic material resulting in less catalyst in the reactor unit and eventual loss in performance.

Table 4. Substances that Inhibit Catalyst Activity[a]

Type of Inhibitor	Effect	Examples	Regeneration
Fast-acting inhibitors	Reduction of catalyst activity at rate depending on concentration and temperature	Phosphorus, bismuth, lead, arsenic, antimony, mercury	Catalyst regeneration is sometimes difficult or impossible
Slow-acting inhibitors	Reduction of catalyst activity; higher concentrations than those of fast-acting catalyst inhibitors may be tolerated	Iron, tin, silicon	Catalyst regeneration remains difficult or impossible
Reversible inhibitors/maskers	Surface coating of catalyst active area; rate also dependent on concentration and temperature	Sulfur, halogens, silicon, zinc, phosphorus	Regeneration is possible
Surface maskers	Surface coating of catalyst active areas	Organic solids	Removed by increasing catalyst temperature or by acid and alkaline washing
Surface eroders and maskers	Surface coating of catalyst active area, or erosion of catalyst surface; both result in loss of catalyst activity; rate dependent on particle size, grain loading, and gas stream velocity	Inert particulates	Surface coating is easily removed by washing
Thermal degradation and sintering	Loss of catalyst surface area because of catalyst dispersion and crystal growth, or catalyst support collapse through sintering	Higher temperatures for extended time, temperature excursions, hot spots in bed	Regeneration generally very difficult; best avoided by operating in optimum temperature range and avoiding temperature excursions
Vapor transport and attrition	Loss of catalytic material through formation of metal carbonyl oxides, sulfides, and halides, or surface shear effects resulting from exhaust gas velocity, particulates, or thermal shock	CO, NO, hydrogen sulfide, halogens, and particulates	Must replace lost catalytic material; vaporization generally not a factor; attrition particularly important in fluidized beds

[a] Refs. 5 and 22.

Combustible masking materials such as organic char may be partially or completely removed by periodic elevations of the catalyst bed temperature. Noncombustible masking materials may be removed by air lancing or aqueous washing generally with a leaching solution (20,21).

Poisons. Halogens, sulfur dioxide [*7446-09-5*], SO_2, nitrogen dioxide [*10102-44-0*], NO_2, and numerous other materials act as catalyst suppressants for precious-metal oxidation catalysts. These compounds tend to adsorb strongly on the catalytic surface, preventing the reactants from doing so. The strength of adsorption is ordinarily such that the suppressant materials can gradually be stripped off after there is no longer a concentration of suppressant materials in the gas stream passing through the catalyst (8). In other cases, the adsorption is irreversible. A poison blocks the catalytic sites, and may also induce changes in the surface to result in formation of compounds (24). Active precious-metal sites become inactive, reducing catalyst performance (see Fig. 6**b**).

At low (>450°C) temperatures, the presence of these materials, particularly the oxides, leads to simple masking or fouling. In some cases, a catalyst that shows reduced activity believed to be from poisoning may simply be masked, and activity can be rejuvenated by cleaning with aqueous leaching solutions (21).

Poisoning is operationally defined. Often catalysts believed to be permanently poisoned can be regenerated (5). A species may be a poison in some reactions, but not in others, depending on its adsorption strength relative to that of other species competing for catalytic sites (24), and the temperature of the system. Catalysis poisons have been classified according to chemical species, types of reactions poisoned, and selectivity for active catalyst sites (24).

Group 14 and 15 (VA and VIA) elements act as poisons. The interaction depends on the poison's oxidation state and chemical structure. For sulfur, the order of decreasing poisoning activity is $H_2S > SO_2 > SO_4^{2-}$. Adsorption studies indicate that H_2S adsorbs strongly and dissociates on nickel surfaces. The sulfur adsorbs essentially irreversibly and over most of the catalyst–metal surface. It has been observed that SO_2 and SO_3 also poison catalysts differently (13); SO_2 selectively adsorbs on Pt or Pd in an oxidation catalyst, whereas SO_3 reacts with the Al_2O_3 carrier, forming $Al_2(SO_4)_3$, which destroys the structure of the catalyst. The latter can be prevented by using a more inert support such as SiO_2 or TiO_2. The former requires a change in operating conditions such as a higher temperature. When

No. 2 fuel oil is used in some gas turbines, the sulfur compound in the fuel oil can be converted to SO_2 at levels of 40 to 150 ppm in the exhaust. For such applications, the presence of 100 to 200 ppm SO_2 can require 150 to 200°C higher temperature for the catalyst to give the same CO conversion as without SO_2.

Toxic heavy metals and ions, eg, Pb, Hg, Bi, Sn, Zn, Cd, Cu, and Fe, may form alloys with catalytic metals (24). Materials such as metallic lead, zinc, and arsenic react irreversibly with precious metals and make the surface unavailable for catalytic reactions. Poisoning by heavy metals ordinarily destroys the activity of a precious-metal catalyst (8).

Molecules having unsaturated bonds, eg, CO, NO, HCN, and benzene, may chemisorb through multiple bonds (24).

Catalysts having improved poison resistance have been developed. Catalysts are available that can destroy chlorine-, fluorine-, or bromine-containing organic compounds (5).

Thermal Degradation and Sintering. Thermally induced deactivation of catalysts may result from redispersion, ie, loss of catalytic surface area because of crystal growth in the catalyst phase (21,24,33) or from sintering, ie, loss of catalyst-support area because of support collapse (18). Sintering processes generally take place at high (>500°C) temperatures and are generally accelerated by the presence of water vapor (see Fig. 6**c**). Another thermal effect is the transformation of catalytic phases to noncatalytic ones, eg, the reaction of nickel and alumina to form nickel aluminate (24). Each catalyst has a recommended temperature window of operation. At temperatures above this window (usually ≥760°C), sintering can occur.

Loss of Catalyst by Vapor Transport. The direct volatilization of catalytic metals is generally not a factor in catalytic processes, but catalytic metal can be lost through formation of metal carbonyl oxides, sulfides, and halides in environments containing CO, NO, O_2 and H_2S, and halogens (24).

The ceramic oxide carrier is bonded to the monolith by both chemical and physical means. The bonding differs for a ceramic monolith and a metallic monolith. Attrition is a physical loss of the carrier from the monolith from the surface shear effects caused by the exhaust gas, a sudden start-up or shutdown causing a thermal shock as a result of different coefficients of thermal expansion at the boundary between the carrier and the monolith, physical vibration of the catalyzed honeycomb, or abrasion from particulates in the exhaust air (21) (see Fig. 6**d**).

Avoiding Catalyst Deactivation. Catalyst deactivation is more easily prevented than cured. Poisoning by impurities may be prevented by removing impurities from the reactants. Carbon deposition and coking may be prevented by minimizing formation of precursors and manipulating mass-transfer regimes so as to minimize the carbon's or coke's effect on activity. Most sintering is irreversible, or reversible only with great difficulty, so it is important to choose reaction, ie, lower temperatures, that do not sinter the catalyst. Additionally, when process upsets that could release inhibitors or cause small fluctuations in the heating value of the oxidizer are highly probable a thermal system is favored over a catalytic one (7).

Except for No. 2, fuel oil should not be considered as auxiliary fuel when using a catalytic system because of the sulfur and vanadium the fuel oil may contain (7). In some cases even the sulfur in No. 2 fuel oil can present a problem. Galvanized metal should not be used in process ovens or ductwork because zinc is a catalyst poison.

Proper system design and catalyst maintenance are key to minimizing deactivation and providing long-term catalyst service. For example, control of air dilution, use of temperature control loops, and use of catalysts having high intrinsic thermal stability can provide necessary protection against high temperature damage caused by reaction exotherms and from operational upsets (20).

Experimental Evaluation. Often the deactivation kinetics for a catalytic oxidation system can be evaluated in a series of laboratory studies (24). Reactors should be gradientless with respect to reactant poison concentration and temperature. Heat- and mass-transfer effects should be avoided because these disguise the intrinsic kinetics. Experiments should be designed to study one deactivation process at a time, and accelerated targets must be representative of the process. Deactivation can be accelerated by using smaller amounts of catalyst, operating at higher temperatures or different pressures, at greater residence times, or at different gas compositions.

Whereas changing catalyst volume or residence time rarely yields complications, changing temperature or pressure could introduce sintering. The properties of the catalyst should be measured both before and after deactivation and inlet and outlet streams should be analyzed by chromatography (qv) or spectrometry.

Around 1972, it was reasoned that the problem of catalyst deactivation could not always be entirely eliminated, but that continuous replacement of a portion of the catalyst bed during normal operation would allow continuing operation at high efficiency even in the presence of poisoning agents. Hence the fluidized bed was born (8). In some applications, fluidized-bed oxidation processes overcame poisoning, masking, and thermal aging. A process in which performance depends on the continuous attrition of the external surface of the catalyst particles, however, has many unattractive features, including the effort required for trapping, collecting, and disposing of the fine particulate released from the reactor (32).

At least one printer using catalytic oxidation has experienced relatively rapid catalyst deactivation, requiring replacement after useful lives as short as 3–6 months, when producing high quality printed matter using some of the most desirable lithographic printing plates and materials (34). It was observed that the use of phosphorus additives caused rapid deactivation of the conventional catalyst used to destroy the hydrocarbons in the solvent-laden air (SLA) discharged from the press dryer. A precious-metal catalyst containing platinum and palladium was being used in this application. It had replaced an earlier base metal catalyst, which showed rapid deactivation as a result of sulfur in the SLA, presumably introduced in the fuel used to fire the heatset dryer. It was found that the P concentration in the SLA might be as high as 0.16 ppm. The phosphorus concentration on the deactivated catalyst was found to be

1.4% of the catalyst. The printer was urged by the supplier to find and eliminate the cause of the phosphorus contamination of the waste gas entering the catalyst bed. However, after it was determined that use of phosphorus-containing additives was crucial to many of the high quality printing jobs, attention was directed at the catalyst bed itself.

A catalyst with a substantially improved resistance to poisoning by phosphorus in catalytic oxidation applications was developed. In part, the catalyst in this program permitted printers to use lithographic technology without paying an unreasonable cost in terms of frequent replacement of oxidation catalysts.

Catalyst Reactivation. Some catalytic systems are reported to have operated continuously for more than 10 years with little or no loss in control efficiency (5). In most processes catalysts inevitably lose activity, and when the activity has declined to a critical level, the catalyst needs to be discarded or regenerated (24). Regeneration is only possible when the deactivation is reversible by chemical washing or heat treatment or oxidation (20,24).

Thermal Treatment. A thermal treatment for catalyst regeneration is usually effective when deactivation is a result of coking or masking of the catalyst surface. Thermal treatment can usually be done on-site, by elevating the temperature of the catalyst bed by 50 to 100°C above the normal operating point and running at this oxidizing condition for a specified limited period of time (20). The elevated temperature vaporizes or oxidizes the organic compounds or char that may be masking the catalyst surface.

Physical Treatment. If inspection of the catalyst indicates deposits again, or if an excessive pressure drop across the catalyst is noted, then the catalytic bed may be lanced, on-site, using compressed air or water until the deposits are removed. Abrasion by contact with excessively high pressure from the compressed air should be avoided (20). If this treatment is combined with heating, hot spots or overtemperatures that could further deactivate the catalyst should be avoided (24). In many cases, periodic maintenance, removing the catalyst bed and blowing or washing off residues, has restored catalyst to original or near-original activity levels (5).

Chemical Treatment. The most involved regeneration technique is chemical treatment (20) which often follows thermal or physical treatment, after the char and particulate matter has been removed. Acid solution soaks, glacial acetic acid, and oxalic acid are often used. The bed is then rinsed with water, lanced with air, and dried in air. More involved is use of an alkaline solution such as potassium hydroxide, or the combination of acid washes and alkaline washes. The most complex treatment is a combination of water, alkaline, and acid washes followed by air lancing and drying. The catalyst should not be appreciably degraded by the particular chemical treatment used.

Analyses of a catalyst used in a process involving cleaning products and pigments and achieving a hydrocarbon destruction capacity of only 13% showed deposition of P, Sn, Pb, and Na contaminants (20). Initial acid treatment increased the hydrocarbon destruction capacity from 13 to 63%. Alkaline treatment increased the capacity to 90% of that new.

EXHAUST CONTROL TECHNOLOGIES

In addition to VOCs, specific industrial exhaust control technologies are available for nitrogen oxides, NO_x, carbon monoxide, CO, halogenated hydrocarbon, and sulfur and sulfur oxides, SO_x.

Nitrogen Oxides

Annual releases of nitrogen oxides (NO_x) into the atmosphere amounted to ca 550 × 10^6 t in the early 1990s. A number of states, in addition to California, regulate NO_x emissions (35). The production of nitrogen oxides can be controlled to some degree by reducing formation in the combustion system. The rate of NO_x formation for any given fuel and combuster design are controlled by the local oxygen concentration, temperature, and time history of the combustion products. Techniques employed to reduce NO_x formation are collectively referred to as combustion controls and U.S. power plants have shown that furnace modifications can be a cost-effective approach to reducing NO_x emissions. Combustion control technologies include operational modifications, such as low excess air, biased firing, and burners-out-of-service, which can achieve 20–30% NO_x reduction; and equipment modifications such as low NO_x burners, overfire air, and reburning, which can achieve 40–60% reduction (36). As of this writing, approximately 600 boilers having 10,000 MW of capacity use combustion modifications to comply with the New Source Performance Standards (NSPS) for NO_x emissions (37).

When NO_x destruction efficiencies approaching 90% are required, some form of post-combustion technology applied downstream of the combustion zone is needed to reduce the NO_x formed during the combustion process. Three post-combustion NO_x control technologies are utilized: selective catalytic reduction (SCR); nonselective catalytic reduction (NSCR); and selective noncatalytic reduction (SNCR).

Selective Catalytic Reduction. Selective catalytic reduction (SCR) is widely used in Japan and Europe to control NO_x emissions (1). SCR converts the NO_x in an oxygen-containing exhaust stream to molecular N_2 and H_2O using ammonia as the reducing agent in the presence of a catalyst. NO_x removals of 90% are achievable. The primary variable is temperature, which depends on catalyst type (38). The principal components of an SCR system include the catalyst, the SCR reactor, the ammonia injection grid (AIG), the ammonia–air dilution system, the ammonia storage–vaporization system, the ammonia addition control system, and a continuous emissions monitoring system (39).

The AIG is used to uniformly inject diluted ammonia into the reactor. Uniform mixing of the ammonia into the flue gas is necessary to maintain catalyst performance at its highest level and to minimize ammonia leakage (ammonia slip) past the catalyst.

The ammonia–air dilution system dilutes the vaporized ammonia by a factor of 20 to 25 with air for better admixing through the AIG and to prevent explosive ammonia–air mixtures. Once the catalyst volume is selected, the NO_x removal is set by the NH_3/NO_x mole ratio at the inlet of the SCR system (39).

SCR was first developed in the United States in the late 1950s, targeted at nitric acid tail-gas exhausts, using precious-metal catalysts. In the mid-1970s, SCR entered widespread commercial use in Japan using base metal catalysts. The first SCR systems in Germany started up in 1986 (39), and German utilities are committed to installing SCR systems on the majority of oil- and coal-fired boilers to achieve between 60 and 90% NO_x reductions (39). By the end of 1992, there were about 120 SCR plants in service in Germany alone (40). A limited amount of experience has been documented in the United States (41,42), although commercial service began in 1985 and is expected to increase (43).

Performance criteria for SCR are analogous to those for other catalytic oxidation systems: NO_x conversion, pressure drop, catalyst/system life, cost, and minimum SO_2 oxidations to SO_3. An optimum SCR catalyst is one that meets both the pressure drop and NO_x conversion targets with the minimum catalyst volume. Because of the interrelationship between cell density, pressure drop, and catalyst volume, a wide range of optional catalyst cell densities are needed for optimizing SCR system performance.

Reactions. The SCR process is termed selective because the ammonia reacts selectively with NO_x at temperatures $> 232°C$ in the presence of excess oxygen (44). The optimum temperature range for the SCR catalyst is determined by balancing the needs of the redox reactions.

SCR reactions

$$4\,NO + 4\,NH_3 + O_2 \rightarrow 4\,N_2 + 6\,H_2O$$

$$2\,NO_2 + 4\,NH_3 + O_2 \rightarrow 3\,N_2 + 6\,H_2O$$

The NO reduction is the most important because NO_2 accounts for only 5–10% of the NO_x in most exhaust gases.

Ammonia oxidation reactions

$$4\,NH_3 + 5\,O_2 \rightarrow 4\,NO + 6\,H_2O$$

$$4\,NH_3 + 3\,O_2 \rightarrow 4\,N_2 + 6\,H_2O$$

When sulfur dioxide is also present there are important side reactions in which SO_2 is oxidized to SO_3. The main side reaction in the SCR catalyst is the conversion of SO_2 to SO_3, thus facilitating the reaction above. The SO_3 in turn reacts with ammonia to form ammonium sulfates.

$$NH_3 + SO_3 + H_2O \rightarrow NH_4HSO_4$$

$$2\,NH_3 + SO_3 + H_2O \rightarrow (NH_4)_2SO_4$$

The formation of ammonium bisulfate is strongly temperature dependent. Formation is favored at the lower temperatures. The temperature at which ammonium bisulfate is not formed depends strongly on the SO_3 concentration in the exhaust gas. The temperature needed to minimize bisulfate formation has been reported to increase by about 15°C (around about 350°C) when the SO_3 concentration increases from 5 to 15 ppm (23). The formation of the bisulfate is reversible, ie, if the temperature is raised to 20°C above the minimum temperature, the reaction is shifted to result in the decomposition of the bisulfate formed. When chlorides are present, ammonium chlorides can be formed:

$$NH_3 + HCl \rightarrow NH_4Cl$$

When sulfuric acid is present, ammonium bisulfate can be formed:

$$NH_3 + H_2SO_4 \rightarrow NH_4HSO_4$$

These various reactions should be minimized to avoid plugging the catalyst and to prevent fouling of the downstream air preheaters, when these components condense from the gas at the lower temperatures.

The SCR Process. The first step in the SCR reaction is the adsorption of the ammonia on the catalyst. SCR catalysts can adsorb considerable amounts of ammonia (45). However, the adsorption must be selective and high enough to yield reasonable cycle times for typical industrial catalyst loadings, ie, uptakes in excess of 0.1% by weight. The rate of adsorption must be comparable to the rate of reaction to ensure that suitable fronts are formed. The rate of desorption must be slow. Ideally the adsorption isotherm is rectangular. For optimum performance, the reaction must be irreversible and free of side reactions.

It has been suggested that the first step, weak coadsorption of NO and O_2 on a reduced vanadium site, may represent the slow step in the mechanism. Subsequent formation of a N_2O_3-like intermediate could be a fast step, because it is known that in the gas phase the equilibrium of NO–NO_2 and N_2O_3 is established within microseconds (35).

At low temperatures the SCR reactions dominate and nitrogen oxide conversion increases with increasing temperature. But as temperature increases, the ammonia oxidation reactions become relatively more important. As the temperature increases further, the destruction of ammonia and generation of nitrogen oxides by the oxidation reactions causes overall nitrogen oxide conversion to reach a plateau then decreases with increasing temperatures. Examples are shown in Figure 7 (44).

In the SCR process, ammonia, usually diluted with air or steam, is injected through a grid system into the flue/exhaust stream upstream of a catalyst bed (37). The effectiveness of the SCR process is also dependent on the NH_3 to NO_x ratio. The ammonia injection rate and distribution must be controlled to yield an approximately 1:1 molar ratio. At a given temperature and space velocity, as the molar ratio increases to approximately 1:1, the NO_x reduction increases. At operations above 1:1, however, the amount of ammonia passing through the system increases (38). This ammonia slip can be caused by catalyst deterioration, by poor velocity distribution, or inhomogeneous ammonia distribution in the bed.

Types of SCR Catalysts. The catalysts used in the SCR were initially formed into spherical shapes that were placed either in fixed-bed reactors for clean gas applications or moving-bed reactors where dust was present. The moving-bed reactors added complexity to the design and in some applications resulted in unacceptable catalyst

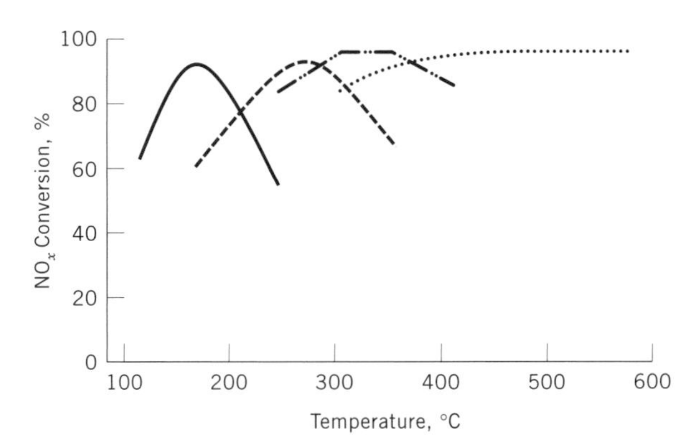

Figure 7. NO_x reduction vs temperature for SCR catalysts: (—), (–––), precious metal; (–··–), base metal; and (·········) zeolites (41).

abrasion. As of 1993 most SCR catalysts are either supported on a ceramic or metallic honeycomb or are directly extruded as a honeycomb (1). A typical honeycomb block has face dimensions of 150 by 150 mm and can be as long as one meter. The number of cells per block varies from 20 by 20 up to 45 by 45 (39).

No SCR catalyst can operate economically over the whole temperature range possible for combustion systems. As a result, three general classes of catalysts have evolved for commercial SCR systems (44): precious-metal catalysts for operation at low temperatures, base metals for operation at medium temperatures, and zeolites for operation at higher temperatures.

The precious-metal platinum catalysts were primarily developed in the 1960s for operation at temperatures between about 200 and 300°C (1,38,44). However, because of sensitivity to poisons, these catalysts are unsuitable for many combustion applications. Variations in sulfur levels of as little as 0.4 ppm can shift the catalyst required temperature window completely out of a system's operating temperature range (44). Additionally, operation with liquid fuels is further complicated by the potential for deposition of ammonium sulfate salts within the pores of the catalyst (44). These low temperature catalysts exhibit NO_x conversion that rises with increasing temperature, then rapidly drops off, as oxidation of ammonia to nitrogen oxides begins to dominate the reaction (see Fig. 7).

The most popular SCR catalyst formulations are those that were developed in Japan in the late 1970s comprised of base metal oxides such as vanadium pentoxide [*1314-62-1*], V_2O_5, supported on titanium dioxide [*13463-67-7*], TiO_2 (1). As for low temperature catalysts, NO_x conversion rises with increasing temperatures to a plateau and then falls as ammonia oxidation begins to dominate the SCR reaction. However, peak conversion occurs in the temperature range between 300 and 450°C, and the fall-off in NO_x conversion is more gradual than for low temperature catalysis (44).

A family of zeolite catalysts has been developed, and is being increasingly used in the United States in SCR applications. Zeolites which can function at higher temperatures than the conventional catalysts, are claimed to be effective over the range of 300 to 600°C, having an optimum temperature range from 360 to 580°C (37,38). However, ammonia oxidation to NO_x begins around 450°C and is predominant at temperatures in excess of 500°C. Zeolites suffer the same performance and potential damage problems as conventional catalysts when used outside the optimum temperature range. In particular, at around 550°C the zeolite structure may be irreversibly degraded because of loss of pore density. Zeolite catalysts have not been continuously operated commercially at temperatures above 500° C (37).

Using zeolite catalysts, the NO_x reduction takes place inside a molecular sieve ceramic body rather than on the surface of a metallic catalyst. This difference is reported to reduce the effect of particulates, soot, SO_2/SO_3 conversions, heavy metals, etc, which poison, plug, and mask metal catalysts. Zeolites have been in use in Europe since the mid-1980s and there are approximately 100 installations on stream. Process applications range from use of natural gas to coal as fuel. Typically, nitrogen oxide levels are reduced 80 to 90% (37).

Catalyst Selection. For an SCR application, catalyst selection depends largely on the temperature of the flue gas being treated. A given catalyst exhibits optimum performance within a temperature range of about 30 to 50°C. Below this optimum temperature range, the catalyst activity is greatly reduced, allowing unreacted ammonia to slip through. Above this range, ammonia begins to be oxidized to form additional NO_x. Operations having adequate temperature controls are important, as are uniform flue gas temperatures (37,38).

Problems. A number of difficulties in utilizing SCR operations have been identified. Problems in European installations are of particular interest because SCR systems are subjected to conditions not experienced in Japan, but also encountered in the United States. Difficulties include matching the NH_3 injection pattern to the nonuniform flow of NO_x in the ductwork ahead of the SCR rector; inability to optimize the NH_3 injection rate by feedback control of slip ammonia for lack of a reliable NH_3 monitor; erosion and plugging on units retrofitted to boilers that fire high ash coal; catalyst deactivation caused by arsenic poisoning on wet-bottom units that recycle flyash; process control under load swings; and spent catalyst disposal. For medium to high sulfur coals, the potential exists for accelerated catalyst deactivation caused by sulfur poisoning and contamination by trace metals in flyash, and deposit buildup and corrosion of the air heater.

For boilers, SCR $DENO_x$ plants can be installed at the exhaust exit just before the air preheater. These are called high dust plants (17 g dust per cubic meter) (23) because the flue gas still contains volatile trace elements as well as flyash particles. When the SCR system is installed after the flue gas desulfurization (FGD) system, it is called a low dust or tail-end plant. The high dust plant has the advantages of not requiring any additional energy because of the high temperatures present. The low dust plant requires a regenerative heat exchanger, but requires less catalyst because (40): there is less dust to contaminate the bed, potential catalyst poisons are removed from the flue gas by the FGD system, and a high activity catalyst can be used when only low concentrations of poisons such as SO_2 remain after the flue gas system.

Other problems that can be associated with the high

dust plant can include alkali deterioration from sodium or potassium in the stack gas deposition on the bed, calcium deposition, when calcium in the flue gas reacts with sulfur trioxide, or formation and deposition of ammonium bisulfate. In addition, plugging of the air preheater as well as contamination of flyash and FGD wastewater discharges by ammonia are avoided if the SCR system is located after the FGD (23).

A significant problem area for initial SCR systems has been the continuous emission monitoring (CEM) systems. In power plants, all sites equipped with CEM systems report the highest failure frequency. The CEM systems are the most labor intensive component, requiring as much as full-time attention from one technician. At one power plant CEM systems were responsible for 100% of 73 reported SCR system shutdowns (38). As CEM systems improve, these concerns may disappear.

Nonselective Catalytic Reduction. Hydrocarbons, hydrogen, or carbon monoxide can be used as reducing agents for NO_x in applications where the exhaust oxygen concentration is low, as it is in fuel rich-burn reciprocating engines, where it is less than 1%, and in nitric acid plants, when it is from 2 to 3%. This approach is called nonselective catalytic reduction (NSCR). In some applications, the oxygen must be removed from the feed stream prior to the catalyst (35). An oxygen sensor in the exhaust stream signals the air–fuel delivery system to adjust the air–fuel ratio so it is just slightly fuel-rich, having enough reducing agent present to react with all the oxygen and nitrogen oxides (1).

Nonselective catalytic reduction systems are often referred to as three-way conversions. These systems reduce NO_x, unburned hydrocarbon, and CO simultaneously. In the presence of the catalyst, the NO_x are reduced by the CO resulting in N_2 and CO_2 (37). A mixture of platinum and rhodium has been generally used to promote this reaction (37). It has also been reported that a catalyst using palladium has been used in this application (1). The catalyst operation temperature limits are 350 to 800°C, and 425 to 650°C are the most desirable. Temperatures above 800°C result in catalyst sintering (37). Automotive exhaust control systems are generally NSCR systems, often shortened to NCR.

Typically NO_x conversion ranges from 80 to 95% and there are corresponding decreases in CO and hydrocarbon concentrations. Potential problems associated with NSCR applications include catalyst poisoning by oil additives, such as phosphorus and zinc, and inadequate control systems (37).

Carbon Monoxide

Carbon monoxide is emitted by gas turbine power plants, reciprocating engines, and coal-fired boilers and heaters. CO can be controlled by a precious-metal oxidation catalyst on a ceramic or metal honeycomb. The catalyst promotes reaction of the gas with oxygen to form CO_2 at efficiencies that can exceed 95%. CO oxidation catalyst technology is broadening to applications requiring better catalyst durability, such as the combustion of heavy oil, coal (qv), municipal solid waste (qv), and wood (qv). Research is underway to help cope with particulates and contaminants, such as flyash and lubricating oil, in gases generated by these fuels (1).

CO conversion is a function of both temperature and catalyst volume, and increases rapidly beginning at just under 100°C until it reaches a plateau at about 150°C. But, unlike NO_x catalysts, above 150°C there is little benefit to further increasing the temperature (44). Above 150°C, the CO conversion is controlled by the bulk phase gas mass transfer of CO to the honeycomb surface. That is, the catalyst is highly active, and its intrinsic CO removal rate is exceedingly greater than the actual gas transport rate (21). When the activity falls to such an extent that the conversion is no longer controlled by gas mass transfer, a decline of CO conversion occurs, and a suitable regeneration technique is needed (21).

It has been reported that below about 370°C, sulfur oxides reversibly inhibit CO conversion activity. This inhibition is greater at lower temperatures. CO conversion activity returns to normal shortly after removal of the sulfur from the exhaust (44). Above about 315°C, sulfur oxides react with the high surface area oxides to disperse the precious-metal catalytic agents and irreversibly poison CO conversion activity.

Catalyst contamination from sources such as turbine lubricant and boiler feed water additives is usually much more severe than deactivation by sulfur compounds in the turbine exhaust. Catalyst formulation can be adjusted to improve poison tolerance, but no catalyst is immune to a contaminant that coats its surface and prevents access of CO to the active sites. Between 1986 and 1990 over 25 commercial CO oxidation catalyst systems operated on gas turbine cogeneration systems, meeting both CO conversion (40 to 90%) and pressure drop requirements.

Halogenated Hydrocarbons

Destruction of halogenated hydrocarbons presents unique challenges to a catalytic oxidation system (45–51). The first step in any control strategy for halogenated hydrocarbons is recovery and recycling (45). However, even upon full implementation of economic recovery steps, significant halocarbon emissions can remain. In other cases, halogenated hydrocarbons are present as impurities in exhaust streams (45). Impurity sources are often intermittent and dispersed.

The principal advantage of a catalytic oxidation system for halogenated hydrocarbons is in operating cost savings. Catalytically stabilized combusters improve the incineration conditions, but still must employ very high temperatures as compared to VOC combustors; eg, carbon tetrachloride [*56-23-5*], CCl_4, has a 40-fold lower heat of combustion than a typical organic vapor such as toluene [*108-88-3*], thus CCl_4 requires much more supplemental fuel to burn than do typical organics (45). Alternatively, the low temperature catalytic oxidation process is typically designed for a maximum adiabatic temperature rise of only 200°C. This would correspond to only about 1500 ppm of an organic compound in the exhaust stream. But, with the lower heat of combustion, up to 40,000 ppm of carbon tetrachloride could be treated in the same temperature rise, or with less dilution air.

By-Product Formation. The presence of halogenated hydrocarbons dramatically increases the yield of aldehydes from the oxidation process (45). For example, in the partial oxidation of methane on a PdO sponge catalyst (19), methylene chloride, CH_2Cl_2, was added in pulses to the inlet gas, which also contained the methane. Methane oxidation was strongly inhibited and formaldehyde was formed. The formaldehyde production continued even after the methylene chloride addition was stopped, suggesting a strong interaction of chlorine with the catalyst. However, pulses of pure CH_4 plus oxygen gradually restored the original activity to the catalyst, indicating that the effect of this interaction was reversible.

Catalyst Deactivation. Catalyst deactivation (45) by halogen degradation is a very difficult problem particularly for platinum (PGM) catalysts, which make up about 75% of the catalysts used for VOC destruction (10). The problem may well lie with the catalyst carrier or washcoat. Alumina, for example, a common washcoat, can react with a chlorinated hydrocarbon in a gas stream to form aluminum chloride which can then interact with the metal. Fluid-bed reactors have been used to offset catalyst deactivation but these are large and costly (45).

Catalytic Reaction. The desired reaction of the chlorine group on a chlorinated hydrocarbon is

$$RCl + O_2 \rightarrow CO_2 + HCl + R'$$

It is important to produce HCl rather than elemental chlorine, Cl_2, because HCl can be easily scrubbed out of the exhaust stream, whereas Cl_2 is very difficult to scrub from the reactor off-gas. If the halogenated hydrocarbon is deficient in hydrogen relative to that needed to produce HCl, low levels of water vapor may be needed in the entering stream (45) and an optional water injector may be utilized. For example, trichloroethylene [*79-01-6*], C_2HCl_3, and carbon tetrachloride require some water vapor as a source of hydrogen (45).

$$C_2HCl_3 + H_2O + 1.5\ O_2 \rightarrow 2\ CO_2 + 3\ HCl$$

Groundwater contaminated with chlorinated hydrocarbons is being remediated by a conventional air stripper or a rotary stripper, producing an air stream containing the halogenated hydrocarbon vapors and saturated with water vapor (45), which is then passed through a catalyst bed.

At least two catalytic processes have been used to purify halogenated streams. Both utilize fluidized beds of probably nonnoble metal catalyst particles. One has been estimated to oxidize > 9000 t/yr of chlorinated wastes from a vinyl chloride monomer plant (45). Several companies have commercialized catalysts which are reported to resist deactivation from a wider range of halogens. These newer catalysts may allow the required operating temperatures to be reduced, and still convert over 95% of the halocarbon, such as trichlorethylene, from an exhaust stream. Conversions of C-1 chlorocarbons utilizing an Englehardt HDC catalyst are shown in Figure 8. For this system, as the number of chlorine atoms increases, the temperatures required for destruction decreases.

USES

Catalytic oxidation of exhaust streams is increasingly used in those industries involved in the following (13,15,17):

Surface Coatings

Aerospace
Automobile
Auto refinishing
Can coating
Coil coating
Fabric coating
Large appliances
Marine vessels
Metal furniture
Paper coating
Plastic parts coating
Wire coating and enameling
Wood furniture

Printing Inks

Flexographic
Lithographic
Rotogravure
Screen printing

Solvent Usage

Adhesives
Disk manufacture
Dry cleaning
Fiber glass manufacture
Food tobacco manufacture
Metal cleaning
Pharmaceutical
Photo finishing labs
Semiconductor manufacture

Chemical and Petroleum Processes

Cumene manufacture
Ethylene oxide manufacture
Acrylonitrile manufacture
Caprolactam manufacture
Maleic anhydride manufacture
Monomer venting
Phthalic anhydride manufacture
Paint and ink manufacture
Petroleum product refining
Petroleum marketing
Resin manufacture
Textile processing

Industrial/Commercial Processes

Aircraft manufacture
Asceptic packaging
Asphalt blowing
Automotive parts manufacture
Breweries/wineries
Carbon fiber manufacture
Catalyst regeneration
Coffee roasting
Commercial charbroiling
Electronics manufacture
Film coating
Filter paper processing
Food deep frying
Gas purification
Glove manufacture
Hospital sterilizers
Peanut and coffee roasting
Plywood manufacture
Rubber processing
Spray painting
Tire manufacture
Wood treating

Engines

Diesel engines
Lean burn internal combustion
Natural gas compressors
Oil field steam generation
Rich burn internal combustion
Gas turbine power generation

Cross Media Transfer

Air stripping
Groundwater cleanup
Soil remediation (landfills)
Hazardous waste treatment
Odor removal from sewage gases

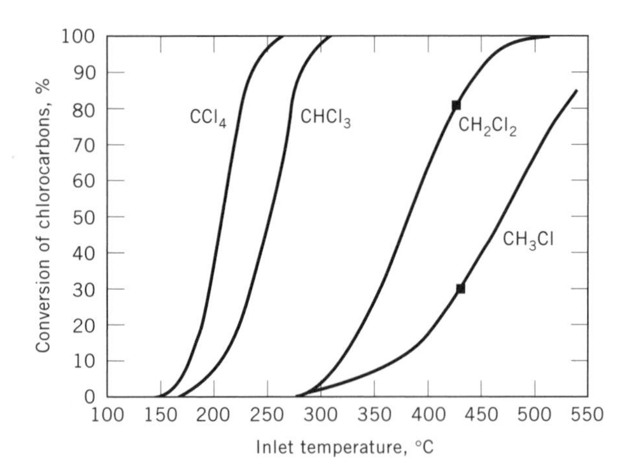

Figure 8. Destruction of C-1 chlorocarbons over HDC in the presence of 1.5% H_2O in air at 15,000 h^{-1} GHSV at STP. Chlorocarbon concentrations are CCl_4, 900 ppmv; $CHCl_3$, 500 ppmv; CH_2Cl_2, 800 ppmv; and CH_3Cl, 600 ppmv.

The most important factors affecting performance are operating temperature, surface velocity, contaminant concentration and composition, catalyst properties, and the presence or absence of poisons or inhibitors.

Air Stripping of Groundwater

Treatment of exhaust streams from the air stripping of contaminated groundwater is a particular challenge, because the emissions from air stripping units may consist of a complex mixture of both fuel and solvent fractions (6). The catalytic oxidation of any given compound is generally negatively impacted by the presence of others in mixtures, and higher catalyst bed operating temperatures are required to achieve adequate destruction.

Some catalysts exposed to air stripping off-gas were subject to deactivation. However, using a catalytic oxidizer at a U.S. Coast Guard facility (Traverse City, Mich.) for the destruction of benzene, toluene, and xylene stripped from the groundwater, the catalytic oxidization unit operated at 260 to 315°C, and was able to achieve 90% destruction efficiency.

Printing and Graphic Arts

In the graphic arts industry, the catalyst in the oxidizer needs to be monitored regularly because it is susceptible to contamination by phosphorus from fountain solutions, silica from silicone gloss enhancer sprays, and chlorides from chlorinated solvents or blanket wash solutions. Phosphorus and silica accumulate most rapidly on the leading edge of the catalyst bed, deactivating the catalyst by masking the precious metals. In a fluidized-bed configuration, the catalyst surface is continually renewed by abrasion and the problem of masking the catalyst surface with silicones is avoided.

Chemical Processing

Terephthalic Acid Production. The control of exhaust from production of pure terephthalic acid (PTA) has been a challenge (see CARBOXYLIC ACIDS) (52). Eight million metric tons of PTA are produced annually worldwide for use primarily in high grade polyester fiber production. Based on a *para*-xylene feedstock, vent gases from the process contain such by-products as methyl acetate, organic acids, and often methyl bromide. These exhausts have been estimated to total 34,000 m^3 worldwide. Historically, the presence of the methyl bromide limited the use of fixed-bed catalytic oxidation as a control technology using precious-metal catalysts. Thus base metal catalysts in fluidized-bed reactors have been the primary catalytic technology of choice. In this application, the continuous abrasion of the outer layer of the catalyst particle exposes a fresh surface of unpoisoned material to the reactants, allowing the catalyst to effectively treat the exhaust stream.

In the late 1980s, however, the discovery of a noble metal catalyst that could tolerate and destroy halogenated hydrocarbons such as methyl bromide in a fixed-bed system was reported (52,53). The products of the reaction were water, carbon dioxide, hydrogen bromide, and bromine. Generally, a scrubber would be needed to prevent downstream equipment corrosion. However, if the focus of the control is the VOCs and the CO rather than the methyl bromide, a modified catalyst formulation can be used that is able to tolerate the methyl bromide, but not destroy it. In this case the methyl bromide passes through the bed unaffected, and designing the system to avoid downstream effects is not necessary. Destruction efficiencies of hydrocarbons and CO of better than 95% have been reported, and methyl bromide destructions between 0 and 85% (52).

Latex Monomer Production. ARI Technologies, Inc. has introduced a catalyst system which, it is claimed, can operate at an average bed temperature of 370°C while achieving conversion efficiency in excess of 99.99% on exhaust streams from latex monomer production (see LATEX TECHNOLOGY).

Acrylonitrile Manufacture. In the manufacture of acrylonitrile (qv), off-gases containing from 1–3% of CO plus various hydrocarbons are emitted. Catalytic beds of platinum-group metals are used to reduce the regulated compounds to acceptable levels. Close attention to bed design is required to prevent the formation of appreciable quantities of NO_x caused by the fixation of combustion–air nitrogen. Some NO_x also is produced from fuel nitrogen by oxidation. Because of the high thermal energy content of the off-gases, considerable heat recovery is possible in abating acrylonitrile plant emissions.

Vinyl Monomer Manufacturing. Process vent gases containing small quantities of halogenated hydrocarbons and substantial quantities of nonhalogenated hydrocarbons have been successfully reduced to comply with regulatory objectives in large-scale laboratory–pilot catalytic fume abaters having satisfactory long-term catalyst performance. The design freedoms offered by precious metals on ceramic honeycomb support catalysts have been demonstrated in equipment that utilizes the heat energy resulting from the substantial exotherm of the nonhalogenated hydrocarbon oxidation to preheat the exhaust gases. Fuel consumption is thereby minimized.

Coatings Industries

Surface coating processes (qv) produce similar air pollution problems in a number of different industries.

Can Manufacturing. An internal coating is necessary to protect the purity and flavor of can contents for beverages or any edible product that might react with the container metal. Both the exterior decorative and interior sanitary coatings are applied to the metal surface by rolls or spray guns using a solvent vehicle. Catalytic oxidation systems are used by the principal can manufacturers to treat coatings exhaust streams. The can manufacturers' industry is estimated to utilize more catalysts than any of the other surface coating industries.

A large number of diverse solvents are used in exterior and interior coatings in plants for manufacturing both three- and two-piece cans. Most of the organic solvents are found in the cure-oven exhausts at concentrations of 2–16% of the lower explosive limit (LEL). The oven exhaust volumes are usually 1–35 m^3/s. When burned, these concentrations of combustibles provide an exotherm of 30 to 220°C. The heat that is released is used for preheating the incoming effluent and/or heating the cure oven by recycling the hot, cleaned gases to the supply blowers or by heating makeup air by heat exchange. A few plants use the heat of the cleaned exhaust to produce hot water for the two-piece can line washers, hot air for dry-off ovens, or building space heating. For example, one large can company utilizes the heat energy contained in the stream, leaving some of their catalytic fume abaters to supply all the heat energy required by the oven's heating zones, which have no burners. The fuel energy supplied to the catalytic fume abater is less than would be needed to heat the oven if the solvent fumes were exhausted directly to the atmosphere without use of the fume abater. The exhaust rate of the oven is adjusted to maintain a solvent concentration of at least 8% of the LEL, equivalent to a 110°C temperature differential.

The various reaction rate properties of the different solvents influence the design of a catalytic reactor. For example, for a specific catalyst bed design, an effluent stream containing a preponderance of monohydric alcohols, aromatic hydrocarbons, or propylene requires a lower catalyst operating temperature than that required for solvents such as isophorone and short-chain acetates.

Design considerations and costs of the catalyst, hardware, and a fume control system are directly proportional to the oven exhaust volume. The size of the catalyst bed often ranges from 1.0 m^3 at 0°C and 101 kPa per 1000 m^3/min of exhaust, to 2 m^3 for 1000 m^3/min of exhaust. Catalyst performance at a number of can plant installations has been enhanced by proper maintenance. Annual analytical measurements show reduction of solvent hydrocarbons to be in excess of 90% for 3–6 years, the equivalent of 12,000 to 30,000 operating hours. When propane was the only available fuel, the catalyst cost was recovered by fuel savings (vs thermal incineration prior to the catalyst retrofit) in two to three months. In numerous cases the fuel savings paid for the catalyst in 6 to 12 months.

Can manufacturers often regenerate the catalyst beds on an annual or biannual basis during a weekend downtime. Both air lancing and an aqueous bath are utilized to remove noncombustible particulates that mask the active sites. Frequently, condensed organic material on the catalyst is removed by short-term (4–6 h) heating excursions to 370 or 430°C; the organic matter is removed much like a self-cleaning oven. The gaseous and organic smoke, which is usually evolved from the first few cm of catalyst bed depth, is oxidized in the latter part of the bed. If allowed to operate too long at temperatures that promote condensation, high boiling organic compounds, the subsequent carbon char that is formed, may require temperatures of 480 to 540°C to convert the carbon to carbon monoxide for subsequent oxidation. The higher temperatures required for burn-offs should be approached in small (0–30°C) increments to bring about slow evolution and partial oxidation; this prevents autogenous combustion of local high concentrations of combustible material.

Day-to-day operating techniques that are employed by one large can manufacturer and are intended to prevent organic condensation are dictated by the use of a low cost, well-established, sanitary coating for beer and beverage three-piece cans. Polybutadiene and other sanitary coatings may have volatile resin monomers entrained in the oven atmosphere as a result of rapid evaporation of solvent before polymerization takes place. A short (4 to 6 h) heating excursion up to a catalyst inlet temperature of 400°C after use of the coating usually burns off any condensed organic materials. It has become standard practice in some plants to turn the catalytic afterburner up to 370°C for these coatings vs the normal 315°C operating temperature for vinyls, acrylics, etc.

Coil Coating. Coil coating is the prefinishing of many sheet metal items with protective and decorative coatings that are applied by roll coating on one or both sides of a fast-moving metal strip. The metal strip (from 13 mm to 1.7 m in width) unwinding from a coil travels at rates of 30–150 m/min through the coating applicator rolls and bake ovens. It is rewound into a coil for transport to a forming operation for products that are to be used in cans, appliances, industrial and residential siding, shelving, cars, gutters, downspouts, etc. The source of hydrocarbon emissions in coil coating is the coating application area and the cure-oven exhaust. The coatings include primers, finishes, and metal protective (5 μm) films or backers.

The increasing use of siliconized coatings for weather durability caused severe masking problems for the all-metal, filter mesh-like catalyst elements available in the 1970s. Interest in catalytic afterburners increased when dispersed-phase precious metal–alumina-on-ceramic honeycomb catalysts offered economically attractive results.

The hot (260–370°C) oven exhaust can be oxidized catalytically without preheat, but when the coater-area exhaust (at room temperature) is combined with the oven exhaust, a preheat burner becomes necessary. The greatest energy savings potential having the least capital investment is obtained by recycling a portion of the hot, cleaned exhaust to the oven. This principle has been demonstrated at a number of can manufacturing plants and at least four coil coating facilities. One operator preheats oven make-up air which has been taken from the oven cooler section by means of a heat exchanger; whereas others recycle directly to the oven. In one case, the heat energy for the dry-off oven is supplied by the catalytic incinerator exhaust (482°C) remaining after supplying most of the heat energy to operate the four zones in the oven. The concentration of solvents in the exhaust is about 12% LEL (167°C temperature differential). The net fuel energy consumption is about 20% of that required to fire the paint, bake, and dry-off ovens without fume control.

Coil coaters operate equipment continuously and, in most cases, operate catalytic fume abaters 6000–7000 h/yr. Under these conditions the anticipated catalyst life is years, with an annual aqueous solution cleaning. However, the catalyst may last no more than two years if frequent maintenance is needed, such as in-place air lancing every 60 to 90 days to remove noncombustible particulates. Frequent maintenance may be needed if coatings such as siliconized polyester (15–40% silicones) comprise 30% of the coatings put through the system.

Filter Paper Processing. In the fabrication of fuel oil and air filters for vehicles such as motorcycles and diesel locomotives, heat processing of the filter paper is required to cure the resin (usually phenolic) with which the paper (qv) is impregnated. The cure-oven exhaust, which contains water vapor, alcohols, and dimers and trimers of phenol, produces a typical blue haze aerosol having a pungent odor. The concentration of organic substances in the exhaust is usually rather low.

The paper-impregnation drying oven exhausts contain high concentrations (10–20% LEL) of alcohols and some resin monomer. Vinyl resins and melamine resins, which sometimes also contain organic phosphate fire retardants, may be used for air filters. The organic phosphates could shorten catalyst life depending on the mechanism of reduction of catalyst activity. Mild acid leaching removes iron and phosphorus from partially deactivated catalyst and has restored activity in at least one known case.

Catalysis is utilized in the majority of new paper filter cure ovens as part of the oven recirculation/burner system which is designed to keep the oven interior free of condensed resins and provide an exhaust without opacity or odor. The application of catalytic fume control to the exhaust of paper-impregnation dryers permits a net fuel saving by oxidation of easy-to-burn methyl or isopropyl alcohol, or both, at adequate concentrations to achieve a 110–220°C exotherm.

BIBLIOGRAPHY

1. R. J. Farrauto, R. M. Heck, and B. K. Speronnelo, *C & EN,* 34–44 (Sept. 7, 1992).
2. C. C. Lewis, *CPI Purch.* 29–33 (Aug. 1991).
3. J. C. Summers, J. E. Sawyer, and A. C. Frost "The 1990 Clean Air Act and Catalytic Emission Control Technology for Stationary Sources," in R. G. Silver, J. E. Sawyer, and J. C. Summers, eds., *Catalytic Control of Air Pollution: Mobile and Stationary Sources,* ACS Symposium Series 495, 1992.
4. R. L. Berglund and C. T. Lawson, "Pollution Prevention in the Chemical Process Industries," *Chem. Eng.* (Sept. 1991).
5. *Catalytic Control of VOC Emissions, A Cost Effective Viable Technology for Industrial, Commercial and Waste Processing Facilities,* Manufacturers of Emission Controls Association, Washington, D.C., 1992.
6. M. Kosusko and C. M. Nunez, "Destruction of Volatile Organic Compounds Using Catalytic Oxidation," *JAWMA* **40**(2) (Feb. 1990).
7. D. R. van der Vaart, W. M. Vatvuk, and A. H. Wehe *JAWMA* **41**(1), 92–98 (Jan. 1991).
8. L. C. Hardison and E. J. Dowd, *Chem Eng Prog.* 31–35 (Aug. 1977).
9. R. J. Martin, R. E. Smyth, and J. T. Schofield, "Elimination of Petroleum Industry Air Toxic Emissions with a Flameless Thermal Oxidizer," *Petro-Safe '93,* Houston, Tex., Jan. 29, 1993.
10. G. Parkinson, *Chem Eng.* 37–43 (July, 1991).
11. A. F. Hodel, *Chem Proc.,* 88–90 (June 1992).
12. R. Yarrington and L. Morris, "The VOC-Incinerator Option," *Indust. Finish.* (Mar. 1992).
13. J. Chen, R. M. Heck, and R. J. Farraoto *Catal. Today* **11,** 517–545 (1992).
14. M. S. Jennings, N. E. Krohn, and R. S. Berry, *Control of Industrial VOC Emissions by Catalytic Incineration,* Vol. 1, U.S. Environmental Protection Agency, Research Triangle Park, N.C., July 1984.
15. E. J. Dowd, W. M. Sheffer, and G. E. Addison, "A Historical Perspective on the Future of Catalytic Oxidation of VOCs," paper 92-109.03, *85th Annual Meeting of Air and Waste Management Association,* Kansas City, Mo., June 21–26, 1993.
16. D. M. VanBenshchoten, "On-Site Pilot Testing Demonstrates Catalytic Emission Control Technology for New VOC Applications," paper 92-109.01, presented at *85th Annual AWMA Meeting & Exhibition,* Kansas City, Mo., June 21–26, 1992.
17. J. J. Spivey, *Ind. Eng. Chem. Res.* **26,** 2165–2180 (1987).
18. K. R. Bruns, "Use of Catalysts for VOC Control," presented at the *New England Environmental Expo,* Boston, Mass., Apr. 10–12, 1990.
19. J. J. Spivey and J. B. Butt, *Catal. Today* **11,** 465–500 (1992).
20. R. M. Heck, M. Durilla, A. G. Bouney, and J. M. Chen "Air Pollution Control–Ten Years Operating Experience with Commercial Catalyst Regeneration," paper presented at the *81st APCA Annual Meeting & Exhibition,* Dallas, Tex., June 19, 1988.
21. R. M. Heck, J. M. Chen, and M. F. Collins "Oxidation Catalyst for Cogeneration Applications–Regeneration of Commercial Catalyst," paper 90-105.1, presented at the *83rd Annual AWMA Meeting & Exhibition,* Pittsburgh, Pa., June 24–29, 1990.
22. R. E. Kenson, "Control of Volatile Organic Emissions," Bulletin 1015, Series 1000, Met-Pro Corp., 1981.
23. W. L. Prins and Z. L. Nuninga, *Catal. Today* **16,** 187–105 (1993).
24. C. H. Bartholomew, *Chem. Eng.,* 96–119 (Nov. 12, 1984).
25. A. C. Frost and co-workers, *Environ. Sci. Technol.* **25**(12), 2065–2070 (1991).
26. D. Ciccilella and B. Holt *Environ. Protec.,* 41–47 (Sept. 1992).
27. K. J. Herbert, "Catalysts for Volatile Organic Control in the 1990's," presented at the *1990 Incineration Conference,* San Diego, Calif., May 14–18, 1990.
28. M. A. Palazzolo, J. I. Steinmetz, D. L. Lewis, and J. F. Beltz "Parametric Evaluation of VOC/HAP Destruction via Catalytic Incineration," U.S. Environmental Protection Agency, Report no. EPA/600/S2-85/041, Research Triangle Park, N.C., July 1985.

29. S. K. Gangwal, M. E. Mulling, J. J. Spivey, and P. R. Caffrey, *Appl. Catal.* **36,** 231–247 (1988).

30. M. A. Palazzolo, and C. L. Jamgonhian "Destruction of Chlorinated Hydrocarbons by Catalytic Oxidation," U.S. Environmental Protection Agency, Report EPA/600/82-86/079, Research Triangle Park, N.C., Jan. 1987.

31. Difford and Spensor, *Chem. Eng. Prog.* **71,** 31 (1975).

32. G. R. Lester, "Catalytic Destruction of Hazardous Halogenated Organic Chemicals," presented at the *82nd Annual AWMA Meeting & Exhibition,* Anaheim, Calif., June 25–30, 1992.

33. S. E. Wanke and P. C. Flynn, *Cat. Rev. Sci. Eng.* **12,** 93 (1975).

34. G. R. Lester, and J. C. Summers "Poison-Resistant Catalyst for Purification of Web Offset Press Exhaust," presented at the *Air Pollution Control Association, 81st Annual Meeting,* Dallas, Tex., June 19–24, 1988.

35. F. Luck and J. Roiron, *Catal. Today* **4,** 205–218 (1989).

36. E. Cichanowicz, *Power Eng.,* 36–38 (Aug. 1988).

37. L. M. Campbell, D. K. Stone, and G. S. Shareef, *Sourcebook: NO_x Control Technology Data,* EPA Report NO. EPA600/S2-91/029., Washington, D.C., Aug. 1991.

38. G. S. Shareef, D. K. Stone, K. R. Ferry, K. L. Johnson, and K. S. Locke "Selective Catalytic Reduction NO_x Control for Small Natural Gas-Fired Prime Movers," paper 92-136.06, in Ref. 16.

39. J. R. Donnelly and B. Brown "Joy/Kawasaki Selective Reduction DE-NOX Technology," paper No 89-96B.6, in Ref. 32.

40. H. Gutberlet and B. Schallert, *Catal. Today* **16,** 207–236 (1993).

41. R. D. Walloch, *Oil Gas J.,* 39–41 (June 13, 1988).

42. R. Craig, G. Robinson, and P. Hatfield, "Performance of High Temperature SCR Catalyst System at Unocal Science and Technology Division," paper 92-109-08, in Ref. 15.

43. D. L. Champagne, *Gas Turb. World,* 20–23 (Nov.–Dec. 1987).

44. B. K. Speronello, J. M. Chen, and R. M. Heck "Family of Versatile Catalyst Technologies for NO_x and CO Removal in Co-Generation," paper 92-109.06, in Ref. 16.

45. D. W. Agar and W. Ruppel, "Extended Reactor Concept for Dynamic $DeNO_x$ Design," *Chem. Eng. Sci.* **43**(8), (1988); J. R. Kittrell, C. W. Quinian, and J. W. Eldridge *J. Air Waste Manage. Assoc.* **41**(8), 1129–1133 (Aug. 1991).

46. T. D. Hylton, *Environ. Prog.* **11**(1), 54–57 (Feb. 1992).

47. Y. Wang, H. Shaw, and R. J. Farrauto, "Catalytic Oxidation of Trace Concentrations of Trichloroethylene over 1.5% Platinum on alpha-Alumina," in Ref. 3.

48. T. C. Yu, H. Shaw, and R. J. Farrauto, "Catalytic Oxidation of Trichloroethylene over PdO Catalyst on Alpha Al_2O_3," in Ref. 3.

49. J. L. Lin and B. E. Bent, "Thermal Decomposition of Halogenated Hydrocarbons on a Cu (111) Surface," in Ref. 3.

50. S. L. Hung and L. D. Pfefferie *Environ. Sci. Technol.* **23**(9), 1085–1091 (1989).

51. M. Mirghanbari, D. J. Muno, and J. A. Bacchetti, *Chem. Proc.* **45,** 14 (Dec. 1982).

52. T. G. Otchy and K. J. Herbert "First Large Scale Catalytic Oxidation System for PTA Plant CO and VOC Abatement," paper presented at the *85th Annual Meeting & Exhibition, AWMA,* June 1992, p. 2026.

53. K. J. Herbert, "Catalytic Oxidation of the Vent Gas from a PTA Plant," 1992.

EXTRA HEAVY OILS

JAMES SPEIGHT
Western Research Institute
Laramie, Wyoming

DEFINITIONS

When petroleum occurs in a reservoir that allows the crude material to be recovered by pumping operations as a free-flowing dark to light colored liquid, it is often referred to as "conventional" petroleum.

Heavy oils are the other "types" of petroleum that are different from conventional petroleum insofar as they are much more difficult to recover from the subsurface reservoir. The definition of heavy oils is usually based on the API gravity or viscosity, and the definition is quite arbitrary.

For many years, petroleum and heavy oils were very generally defined in terms of physical properties. This has led to the development of a more formal method of classification which depends upon gravity and viscosity (Table 1). This system affords a better classification of petroleum, heavy oils and bitumen; the scale can also be used for residua or other heavy feedstocks.

Extra heavy oils are materials that occur in the near-solid state and are almost incapable of free flow under ambient conditions. Therefore, it is more appropriate that native asphalt (often referred to as "bitumen") that occurs in various locations throughout the world (1) be also included in the extra heavy oil definition. Bitumen includes a wide variety of reddish brown to black materials of semisolid, viscous to brittle character that can exist in nature with no mineral impurity or with mineral matter contents that exceed 50% by weight. The bitumen is frequently found filling pores and crevices of sandstones, limestones, or argillaceous sediments, in which case the organic and associated mineral matrix is known as rock asphalt.

Alternative names, such as bituminous sand or oil sand, are gradually finding usage, with the former name (bitu-

Table 1. Crude Oil Classification Using Specific Gravity, API Gravity, and Viscosity

Type of Crude	Characteristics
1. Conventional or "light" crude oil	Density–gravity range less than 934 kg/g^3 (>20° API)
2. "Heavy" crude oil	Density–gravity range from 1000 kg/m^3 to more than 934 kg/m^3 (10° API to <20° API) Maximum viscosity of 10,000 mPa.s (cp)
3. "Extra-heavy" crude oil; may also include atmospheric residue (bp >340°C	Density–gravity greater than 1000 kg/m^3 (<10° API) Maximum viscosity of 10,000 mPa.s (cp)
4. Tar sand bitumen or natural asphalt; may also include vacuum residua (bp >510°C)	Viscosity greater than 10,000 mPa.s (cp) Density–gravity greater than 1000 kg/m^3 (<10° API)

minous sand) more technically correct. The term "oil sand" is also used in the same way as the term "tar sand" and these terms are used interchangeably throughout this text.

Other materials, albeit not naturally-occurring but not a chemically-altered material, that fit into the category of extra heavy oils are the residua (singular: residuum; often shortened to "resid"). A residuum is the nonvolatile material obtained from petroleum after nondestructive distillation has removed all the volatile materials. Residua are black, viscous materials which may be liquid at room temperature (generally atmospheric residua) or almost solid (generally vacuum residua), depending upon the nature of the crude oil. The differences between a parent petroleum and the residua are due to the relative amounts of various constituents present, which are removed or remain by virtue of their relative volatility.

When a residuum is obtained from a crude oil and thermal decomposition has commenced, it is more usual to refer to this product as pitch. Being a thermally altered material, pitch is not classed as an extra heavy oil.

When the asphalt is produced simply by distillation of an asphaltic crude, the product can be referred to as residual, or straight run, petroleum asphalt. It is actually a residuum. If the asphalt is prepared by solvent extraction of residua or by light hydrocarbon (propane) precipitation, or if blown or otherwise treated, the term should be modified accordingly to qualify the product (eg, propane asphalt).

The chemical composition of extra heavy oil is complex. Physical methods of fractionation indicate high proportions of asphaltenes and resins, even in amounts up to 50% (or higher). In addition, the presence of ash-forming metallic constituents, including such organometallic compounds as those of vanadium and nickel, is also a distinguishing feature of the extra heavy oils.

The properties of extra heavy oils can be summarized quite conveniently. These feedstocks (*1*) are usually nonvolatile below 200°C, (*2*) have an API gravity less than 10° and a high viscosity, (*3*) contain high proportions of asphaltenes and resins, (*4*) contain high proportions, often more than 2% w/w, of sulfur as organically-bound sulfur, and (*5*) contain high proportions, several thousand parts per million, of metallic ash-forming constituents.

BIBLIOGRAPHY

1. J. G. Speight, *In Fuel Science and Technology Handbook,* Marcel Dekker Inc., New York, 1990.

F

FOREST RESOURCES

JOHN S. SPENCER, JR.
USDA Forest Service
St. Paul, Minnesota

The world's forests have often presented a dilemma to those who lived nearby. Forests have provided food, shelter, and raw material, while hindering agriculture, national growth, and expansion. Shifting agriculturists cleared the forest to grow crops, and forest clearing expanded as sedentary agriculture became a way of life. Area of this old-growth forest continues to diminish today as previously inaccessible areas are opened to utilization.

Today's global forests, which may cover from one-half to two-thirds of the forested area of preagricultural times, provide a broad array of commodities, amenities, and environmental services (see Plate I). Fuelwood, wildlife habitat, pasture for livestock, industrial forest products, recreation, soil moisture retention, production of atmospheric oxygen, climate regulation, a source of new agricultural or grazing land, and spiritual renewal are a few examples. Currently, a growing world population, an increasingly industrialized world economy with accelerating energy needs, and rising material expectations of the people combine to generate unparalleled pressures on the global forest resource. Today, the clash between the philosophies of the environmentalist and those of the forest commodity user is more pronounced than ever as people are becoming more aware of the disparate benefits that flow from the forest, as well as the consequences of realizing those benefits. Society's goal, which is both elusive and difficult to achieve, is to find ways to use the forest bounty that are environmentally sound, socially acceptable, and economically rewarding.

See also ACID RAIN; CARBON CYCLE; CARBON STORAGE IN FORESTS; FUELS FROM BIOMASS.

HISTORICAL PERSPECTIVE

A large accessible supply of wood, the principal building material and fuel of past societies, was essential to the flowering of civilizations. History abounds with examples of societies that have grown rapidly because of an abundance of wood and that have collapsed after exhausting their forests. In 2700 BC the Sumerians in Mesopotamia thrived in the lower reaches of the Tigris and Euphrates Rivers of present-day Iraq. But by 2000 BC the Sumerian empire had collapsed largely because of the progressive decline of barley yields caused by salinization of soils, triggered by the clearing of the forests in the watersheds of the two rivers, exposing salt-rich sedimentary rock on the denuded slopes (1).

The extensive forests surrounding Athens probably provided the material needed to build the Athenian fleet that defeated the Persian navy in 480 BC, thrusting Athens into its Golden Age. By the time of the defeat in 404 BC of Athens by Sparta in the Peloponnesian War, the forests had been denuded and much of the topsoil had washed away, prompting Aristotle to recommend that laws be adopted to protect forests and to regulate their use.

The Macedonians translated their wealth in forests to economic, political, and military power, and in the process conquered much of what was then the civilized world under Alexander the Great, from 334 BC to 323 BC.

The Romans responded to their wood scarcity by seizing new forests in Iberia, Gaul, Britain, and North Africa, transferring Rome's problems to the provinces. Rome financed the growth of its empire largely from silver mined and smelted in Spain, using local wood for fuel. During the 400 years they operated, the furnaces consumed an estimated 500 million trees, deforesting more than 1.8 million hectares (7,000 square miles) of the Spanish landscape (1). When silver production declined, not because the ore supply was exhausted but because fuel was inaccessible, emperors were forced to debase the coinage progressively, until by the end of the third century AD the public and the government had little confidence in the almost-silverless metal.

The scarcity of wood in England from centuries of iron- and glass-making, ship-building, and domestic heating and cooking generated much of England's interest in timber-rich North America in the seventeenth and eighteenth centuries, especially in the tall white pines of New England prized for masts on Royal Navy ships.

As in centuries past, the distribution and condition of the world's forests continue to be altered by human activities, generated by increasing demands for space, food, fiber, and energy, and fueled by a population that grows by nearly 88 million people annually (2).

DISTRIBUTION AND CLASSIFICATION OF GLOBAL FORESTS

Distribution

The United Nations Food and Agriculture Organization (FAO) estimates the world's land area as of 1989 to be 13,076 million hectares (32,311 million acres; hectares may be converted to acres by multiplying by 2.471) (3), of which forest and woodland account for 31% (Fig. 1).

Forests of the world range in composition and structure from the closed, old-growth Douglas-fir (*Pseudotsuga menziesii*) forests of the mountains of Oregon and Washington to the open, dry woodland of the plains of Africa. Closed forests are those in which the canopy allows little light to fall to the ground, and open forests are those in which the canopy has openings that permit the ground to receive some light. The FAO defines forest and woodland as land under natural or planted stands of trees, whether productive or not, including land from which forests have been cleared but which will be reforested in the foreseeable future. The FAO estimates the world's area of forest and woodland as of 1989 to be 4,087 million hectares (10,098 million acres) (Table 1); approximately the same size as

Plate I. Global forests are an important resource that provide industrial, and environmental services. Ways to use the forests that are environmentally sound, socially acceptable and economically rewarding are being sought. Courtesy of the United States Forest Service.

the total combined areas of North, Central, and South America (3).

The global area of forest and woodland continues to shrink as it is used for other purposes. Area declined from 4,224 to 4,087 million hectares from 1974 to 1989 (Fig. 2), a loss of 137 million hectares (3.3%). In general, however,

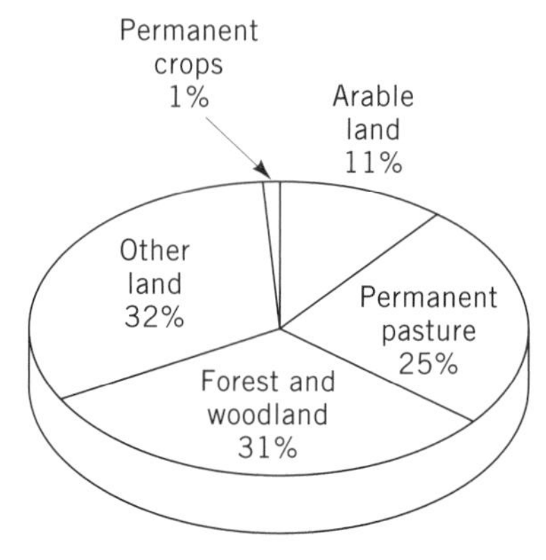

Figure 1. The world's land area by type of land use (3).

temperate forest area increased, and tropical forest area declined. Temperate forests are generally located on the globe between the treeless tundra region near the poles, and the Tropics of Cancer and Capricorn; tropical forests are generally located within or near the Tropics of Cancer and Capricorn. A much larger area of temperate forest is planted annually than tropical forest, and tropical forests are being converted at a much faster rate than temperate forests. Also, some nonforest lands, such as cropland or pasture, may revert naturally to forest if abandoned, a situation that occurs more frequently in the temperate forest region, because there is less competition there for agricultural land than in the tropics.

Between 1974 and 1989, the rate of decline of forest and woodland area in Oceania was greater than that of any other region (16.8%). Oceania is made up of a wide group of islands in the Pacific Ocean, including Australia, New Zealand, American Samoa, French Polynesia, Cook Islands, Fiji, Guam, Midway Islands, Papua New Guinea, Solomon Islands, Wake Island, and others. The rate of decline was next largest in Asia (7.2%), followed by South America (6.6%), and Africa (5.4%). However, the largest area of forest and woodland converted to other uses was in South America (63 million hectares), led by Brazil (35 million hectares).

The former Soviet Union ranked first in 1989 among individual countries in area of forest and woodland (946 million hectares), 71% greater than second-place Brazil (Table 2).

Closed forests, defined by FAO as land with a forest cover and with tree crowns covering more than 20% of the land area, make up between 2,500 and 3,000 million hectares, or about two-thirds of the total forest and woodland area (4). The remainder is open woodland, discontinuous forest stands with tree-crown cover of 10–20%, perhaps interspersed with grass or brush cover.

Classification of Forests

Because of the diverse nature of forests, several systems for classifying them have been developed. One system, devised by the FAO (5), includes the five groups discussed below.

Cool Coniferous Forests. These boreal forests are found in the northern latitudes of North America, Scandinavia, and the former Soviet Union, as well as at high elevations in temperate zones. Principal tree species, ie, spruce (*Picea* spp), fir (*Abies* spp), larch (*Larix* spp), aspen (*Populus* spp), and birch (*Betula* spp), are small and growth is slow because of the short growing season and cool temperatures.

Temperate Mixed Forests. This mix of forest ecosystems is found between the cool coniferous forests and the tropical forests of the northern hemisphere, and south of the tropical forests of the southern hemisphere. The range includes southern Canada, the United States, Europe (except Scandinavia), much of the former Soviet Union, China, Japan, parts of East Asia, Chile, Argentina, New Zealand, and Australia.

Table 1. Area of Forest and Woodland in the World by Region, 1974 and 1989[a]

Region	Area of Forest and Woodland[b] 1974	1989	Difference
Former Soviet Union	920,000	946,000	+ 26,000
South America	954,520	891,338	(−) 63,182
North and Central America	707,609	716,175	+ 8,566
Africa	722,661	683,574	(−) 39,087
Asia	576,939	535,398	(−) 41,541
Oceania	189,005	157,245	(−) 31,760
Europe	153,639	156,964	+ 3,325
Total	*4,224,373*	*4,086,694*	*(−)137,679*

[a] Reference 3.
[b] In thousand hectares.

Tropical Moist Evergreen Forests. These tropical rainforests are found where annual precipitation is more than 2,000 mm and evenly distributed throughout the year. They are generally located in the Amazon Basin, northern South America, eastern coastal Central America, southern Mexico, equatorial Africa, western coastal India, Southeast Asia, and northeast coastal Australia. Tropical rainforests include the most biologically diverse ecosystems on earth, containing nearly half the world's known plant and animal species (6). As elevation increases, rainforests grade into cloud forests, where trees are shorter and less diverse.

Tropical Moist Deciduous Forests. These tropical forests are located where annual precipitation is 1,000 to 2,000 mm, and where a dry season occurs for at least one month each year. They are generally found in central South America, western coastal Central America, south-central Africa, India, and Southeast Asia. Commercially valuable teak (*Tectona* spp) and Philippine mahogany (*Shorea* spp) are found in these forests in Asia.

Dry Forests. These forests are located in both temperate and tropical zones where annual precipitation is less than 1,000 mm. They range from closed forests to open woodland, thornlands, shrublands, savannahs, and other sparse woody vegetation. These forests generally occur in the southwestern United States, Mexico, South and Central America, the Mediterranean Basin, India, Australia, and over much of sub-Saharan Africa. In the arid western United States, pinyon pine (*Pinus edulis*)–juniper (*Juniperus* spp) stands are representative of dry forests.

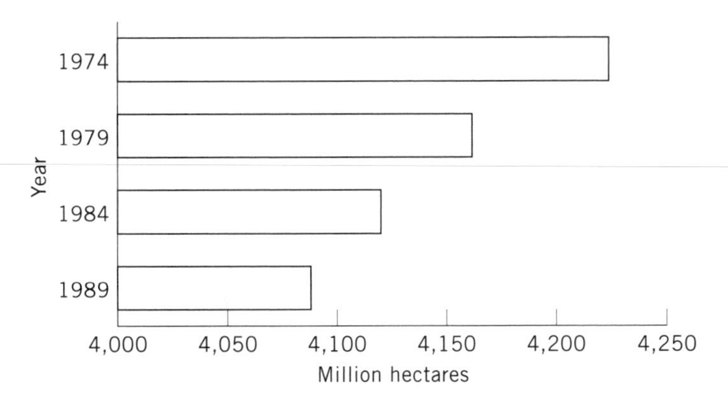

Figure 2. Area of global forest and woodland, 1974, 1979, 1984, and 1989 (3).

Limiting Factors of Forests

Climate and geographical location determine distribution and type of forests. Temperature is the limiting factor at higher latitudes and elevations. Generally speaking, forests do not grow poleward beyond the isotherm of 10°C (50°F) average temperature, eg, northern Norway and Siberia, during the growing season (7). However, in areas close to the sea they may extend from this isotherm toward the pole, and in continental areas they may contract

Table 2. Area of Forest and Woodland by Major Forested Country, 1989

Country	Area[a]
Former Soviet Union	946
Brazil	553
Canada	358
United States	294
Zaire	175
China	126
Indonesia	113
Australia	106
Peru	69
India	67
Argentina	59
Bolivia	56
Angola	53
Colombia	51
Sudan	45
Mexico	43
Tanzania	41
Papua New Guinea	38
Central African Republic	36
Myanmar	32
Venezuela	30
Zambia	29
Sweden	28

[a] In million hectares.

toward the equator. The amount and seasonal distribution of precipitation determine the nature of forests at midlatitudes. Tropical rainforests exist where rainfall is greater than 2,000 mm (about 80 in.) annually, eg, Brazil's Amazon Basin, while open woodland, scrub, and savannah exist where rainfall is less than 1,000 mm annually, eg, plains of New Mexico. Little woodland is found where rainfall amounts to less than 400 mm/yr, eg, Sahara Desert. Other factors such as soil type and drainage, slope, and aspect modify the local situation further to produce the mosaic of diverse forests found throughout the world.

IMPORTANCE OF FORESTS

Ecological Significance of Forests

Much more remains to be learned about the many complex interactions in forests between trees, other plants, animals, soil, and the environment, as well as about the role of forests in maintaining the health of the planet. However, we do know that forest ecosystems provide a host of environmental services including maintaining biological diversity, providing wildlife habitat, cycling nutrients, producing oxygen, affecting regional rainfall patterns, and sequestering carbon in the global carbon pool. They also regulate streamflow, reduce flooding, store water, moderate wind erosion, and reclaim degraded land.

Disturbance of forest ecosystems in a particular location may result in important changes in other ecosystems that may be separated by great distances. For example, removal of Central and South American tropical forests used as wintering grounds by many species of neotropical migratory songbirds, along with fragmentation of the forest habitat in their summer range and along their migratory routes, are thought to be the primary reasons for the decline in the numbers of these birds (see Plate II).

Plate II. Forest habitats provide for birds. As forest ecosystems are disturbed corresponding declines are seen in the populations of many bird species. Courtesy of Whittlesey Birnie.

Forest decline, ie, needle loss, tree growth reduction, and tree death, has been observed over broad areas in Europe and North America, and is thought to be caused by acid rain or airborne pollutants. However, at this time ongoing research has documented only that ozone has caused decline in the forest health of ponderosa pine (*Pinus ponderosa*) in southern California, but can not confirm pollutant-caused decline elsewhere (8). Ozone is produced in the atmosphere when nitric oxide (NO) is created by the combustion of fossil fuels and emitted as exhaust, such as from automobiles. The nitric oxide combines with oxidants in the atmosphere to form nitrogen dioxide (NO_2), which, in the presence of sunlight, breaks down into free oxygen atoms (O) and nitric oxide (NO). The oxygen atoms then combine with oxygen molecules (O_2) in the atmosphere to produce ozone (O_3).

Many pollutants may simply stress trees, predisposing them to natural causes of death such as drought, insects, and disease. However, continuing research may reveal a causal effect between other pollutants and forest decline. Already there is a growing body of evidence to link the decline of red spruce (*Picea rubens*) in the northern Appalachian Mountains of the United States to acid deposition. Over the past three decades, red spruce in the higher elevations of the region have exhibited reduced rates of growth and poor crown condition. In some areas, more than 50% of the red spruce trees have died. This area of forest decline is unique because of its high degree of exposure to acidic depositions due to the frequent immersion in clouds, which are more polluted than precipitation. Acidic deposition has been linked to increased frequency and severity of winter injury to red spruce foilage, and to reduction of nutrients in the soil that are essential to tree growth (9).

Scientists at the U.S. Department of Agriculture (USDA) Forest Service's Northeastern Forest Experiment Station in Durham, New Hampshire, have found that calcium is being depleted from mountain forest soils in parts of the eastern United States by leaching (probably induced by acid deposition) and by frequent whole-tree clearcutting of forests (10). At the present rate of loss, 50% of the total soil and biomass calcium could be removed within 120 yr, a level that would result in serious calcium deficiency and subsequent reduced tree growth. As a result of these findings, timber harvest practices in the eastern United States now leave more large, woody residue on the site so the calcium and other nutrients within it will be returned to the soil as the residue decomposes.

Industrial emissions can spread far beyond the point of origin and deposit pollutants which build up in forest ecosystems with no present visible adverse effects. USDA Forest Service scientists at the North Central Forest Experiment Station Laboratory in Grand Rapids, Minnesota, demonstrated a relationship between emissions from fossil fuel combustion and the acidity of precipitation, as evidenced by near background levels of acidic deposition in northwestern Minnesota, increasing to high levels in southeastern Michigan (11). These emissions probably originated in places like the industrialized Ohio River valley, and were carried by winds to the Lake States. Studies of lakes along this deposition gradient found evidence of water chemistry changes related to acidic deposition. Investigations of soil and tree tissue along the gradient showed that the amount of sulfate deposited in acid deposition is related to the amount of sulfur in soil and trees, suggesting that effects of acid deposition may be cumulative and that continued monitoring is advisable. No apparent effect of elevated sulfur levels on tree growth has been observed to date, regardless of location on the gradient. However, long-term effects of this and other buildups of potential pollutants are unknown.

Volume of Timber in Forests

The diversity of forests translates into widely different volumes of standing timber. It is estimated (12) that average volumes per hectare range from 50 cubic meters in dry savannah woodlands to 150 cubic meters in temperate forests, and to 350 cubic meters in tropical rainforests. These averages translate into an estimate of total standing volume for global closed forests of 396 billion cubic meters, half of which is in tropical rainforests. By comparison, there were an estimated 23 billion cubic meters (816 billion cubic feet) in live trees on timberland in the United States in 1987 (13). Some of the global volume is in tree species considered unsuitable for industrial forest products, is in forests legally protected from logging, or is located where exploitation is precluded by inaccessibility. The area of exploitable closed forests is estimated to be about 2,150 million hectares with around 300 billion cubic meters of standing volume (14). Exploitable forests are those on which there are no legal, economic, or technical restrictions on timber harvesting.

Production of Wood from Forests

The FAO estimates that total production (removals) of roundwood from global forests in 1991 was 3,429 million cubic meters (Table 3), or about 1.2% of the exploitable inventory (15). Roundwood is defined as wood in the rough, from logs, bolts, or other round sections cut from trees. Asia produced the largest share of the world's output (32%). The United States led all individual countries with 495.8 million cubic meters, followed by the former Soviet Union (355.4 million), China (282.3 million), India (279.8 million), Brazil (264.6 million), Canada (178.0 million), and Indonesia (173.0 million).

The 1991 production of 3,429 million cubic meters was 80% higher than the 1960 production of 1,901 million cubic meters (Fig. 3). Coniferous (softwood) species accounted for about 39% of the total production, and deciduous (hardwood) species accounted for 61%. Fifty-three percent of the volume produced was used for fuelwood and charcoal, and 47% was industrial roundwood such as saw logs, veneer logs, and pulpwood. Developed countries produced 76% of the industrial roundwood; developing countries produced 87% of the fuelwood and charcoal, still the principal source of energy for three-fourths of the population of the developing world.

Table 3. Volume of Global Roundwood Production by Region, 1991[a]

Region	Roundwood Production[b]	Total, %
Asia	1,086.1	31.7
North and Central America	737.4	21.5
Africa	527.2	15.4
Former Soviet Union	355.4	10.3
South America	345.4	10.1
Europe	335.5	9.8
Oceania	42.4	1.2
Total	*3,429.4*	*100.0*

[a] Reference 15.
[b] Million cu. meters.

Wood: A Vital but Dwindling Energy Source in Developing Countries

The silent energy crisis, unknown by most people in industrialized countries, and unaffected by oil price fluctuations, is the daily struggle by more than one third of the world's population to find wood to cook their meals. An Indian official enunciated the looming fuelwood crisis in many countries when he said, "Even if we somehow grow enough food for our people in the year 2000, how in the world will they cook it?"

About half the world's roundwood production in 1991 was used to produce energy (15). Wood is the primary source of energy for poorer urban households and for the vast majority of rural households in developing countries, where other fuels are economically beyond the reach of most people. The developed countries consume wood primarily for industrial purposes (lumber, plywood, paper, etc), but the developing world consumes wood primarily for energy production. Eighty percent of the total roundwood consumption in developing countries in 1986 was for

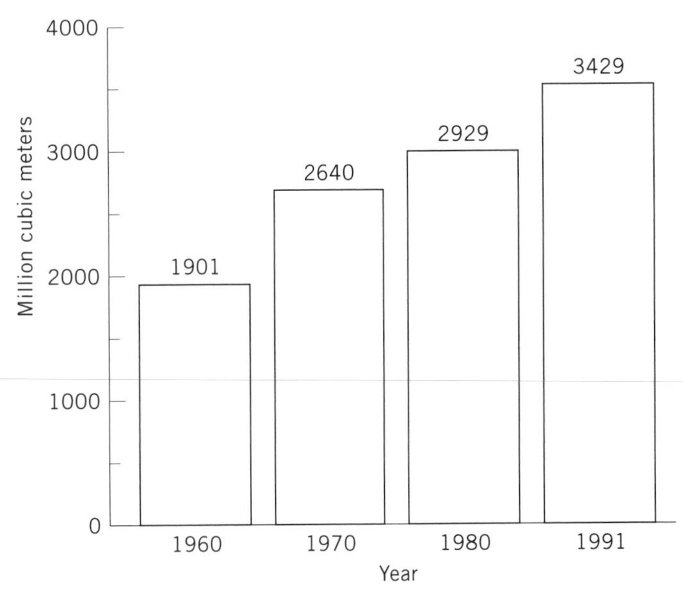

Figure 3. Volume of global roundwood production 1960, 1970, 1980, and 1991 (15).

fuelwood (16). Fuelwood, which may be converted to charcoal, is usually consumed in-country for domestic heating and cooking, and to a lesser extent, for small industrial enterprises, such as baking, pottery making, and coffee and tobacco drying. Urban and rural fuelwood needs differ, as illustrated in Africa (south of the Sahara) where 67 to 75% of urban domestic energy use is from fuelwood, compared to 90 to 98% of rural energy use (17).

An estimated 1,688.7 million cubic meters of fuelwood were produced from global forests in 1991 (15), as shown in Table 4. The 1991 volume was 23% greater than the 1980 volume (1,373.3 million cubic meters). Eighty-seven percent of the volume was from hardwood species, and the remaining 13% was from softwood species. Asia produced 48% of the fuelwood, followed by Africa (24%) and South America (12%). India led all individual countries in fuelwood production with 243 million cubic meters (14% of the world's production), followed by China (192 million), Brazil (154 million), Indonesia (143 million), and Nigeria (94 million).

Dead trees and branches have traditionally supplied much of the fuelwood requirement in many countries, but increasingly demand must be satisfied by cutting live trees. The pressure to use livewood is exacerbated by rapid population growth, urbanization, or restriction of access for wood gathering, as parts of the forest that were formerly common property become privatized or nationalized (18). The consequences of such pressure on a limited forest resource is often degradation of the resource and fuelwood scarcity. There are now shortages of fuelwood in 57 developing countries. Of the 2,000 million people there who depended on fuelwood in 1980, approximately 100 million experienced acute fuelwood scarcity, and 1,050 million people lived in areas of increasing fuelwood scarcity and met their fuelwood needs at the expense of depleting existing wood resources. By the year 2000, the number of people in situations of acute shortages will increase to 2,400 million unless action is taken (19).

In many cases, local populations do not perceive an impending fuelwood shortage, and are unaware that by depleting their forest capital, they are destroying their future source of fuelwood and other forest resources. However, trees represent a capital stock that appreciates through growth, and can regenerate and maintain itself as long as timber harvesting and other removals do not exceed the level of regeneration and net growth.

This potential fuelwood crisis takes an increasing human toll; in parts of Tanzania, villagers had to walk 10 km for firewood in 1983, compared to only 1 km twenty years earlier (20). In parts of East Africa, a family may spend up to 40% of its income to purchase fuelwood (21). In parts of West Africa, people now have only one instead of the customary two cooked meals a day (6). The fuelwood shortage also has a profound effect on soils and crop yields because people are forced to burn animal dung and crop residues that once were plowed under to increase soil fertility. However, Dewees (22) argues that the projected shortage may be overstated because it does not take into account practices that could moderate future consumption, such as fuel sharing, shared cooking, and increased labor use for fuel collection.

National energy planners in many developing countries neglected fuelwood in energy development programs because of their presumption that family incomes would increase as a result of national development programs, allowing households to substitute oil, liquid petroleum gas, kerosene, electricity, or coal for wood. This has not occurred because of the still prohibitive costs of nonwood energy sources for most people, and the outlook is for fuelwood to continue to be a principal energy source in the foreseeable future for most developing countries (17). Improvement of the fuelwood situation requires a two-pronged approach: (*1*) an increase in fuelwood production, and (*2*) better fuelwood conservation and conversion efficiency. It has been suggested (23) that the integration of trees into agricultural systems is the most worthwhile method of increasing fuelwood production (see Social Forestry). Widespread use of more efficient cookstoves, better use of waste wood, and improved methods of charcoal production will stretch existing fuelwood stocks.

Nontimber and Amenity Uses of the Forest

Besides timber products, forests provide nuts, fruits, herbs, medicinal plants, pharmaceuticals, resins, gums, oils, forage, commercial flowers, spices, syrups, thatch, and rattan. Additionally, forests protect soil and watersheds; provide areas for ecosystem research; furnish opportuni-

Table 4. Volume of Global Fuelwood Production by Region and by Softwood and Hardwood Production, 1991[a,b]

Region	Softwood Production	Hardwood Production	Total Fuelwood Production
Asia	99.0	713.7	812.7
Africa	6.5	395.4	401.9
South America	16.3	180.4	196.7
North and Central America	29.9	108.7	138.6
Former Soviet Union	52.7	28.4	81.1
Europe	15.3	33.9	49.2
Oceania	0.6	7.9	8.5
Total	*220.3*	*1,468.4*	*1,688.7*

[a] Reference 15.
[b] In million cubic meters.

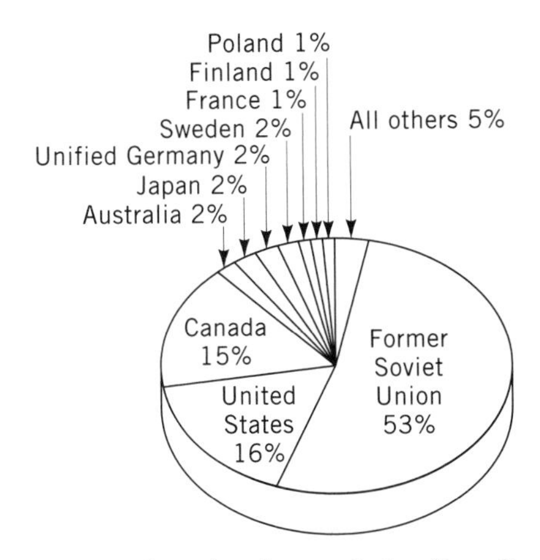

Figure 4. Percent of total volume of standing, live (growing-stock) trees in global temperate forests, by country, 1980s (24).

ties for recreation and spiritual renewal; and inspire literature, music, religion, and art.

TEMPERATE FORESTS

Depending on their location on the globe, forests can be classed as either temperate or tropical. The temperate forest zone generally occurs between the treeless tundra region and the frost-free tropical forest region in both the northern (north temperate zone) and southern (south temperate zone) hemispheres. The north temperate zone includes the boreal forest, often called "taiga," located closest to the tundra in the northern hemisphere, and composed of stands of spruces, firs, larches, birches, and aspens. The southern hemisphere contains nothing comparable to the boreal forest. The remainder of the two forest zones contains stands of other frost-hardy coniferous, or deciduous, or mixed coniferous–deciduous trees, such as the longleaf (*Pinus palustris*)–slash pine (*Pinus ellottii*) stands of the southeastern United States, or the maple (*Acer* spp)–beech (*Fagus grandifolia*)–birch (*Betula* spp) stands of the Lake States and northeastern United States. Tree species found naturally in the northern hemisphere are related to but different from those in the southern hemisphere. Temperate zone forests occur largely in developed countries, where they are intensively used and generally receive a higher level of forest management than tropical forests.

In 1990, the area of forest land (land with tree crown cover of more than 20% of the area) in the developed temperate zone was estimated to be 1,432 million hectares (24), or 27% of the land area of the zone. The developed temperate zone includes all the countries in Europe (including Cyprus, Israel, Turkey, and the former Soviet Union), Canada, the United States, Japan, Australia, and New Zealand. The proportion of forest to total land area ranges from 66% in Japan and Finland to 5% in Israel and Australia. No estimates of recent change in area of temperate forest land are available. However, the FAO estimates that the area of forest and other wooded land (includes forest, open woodland and scrub, and shrub and brushland) increased by 2.0 million hectares between 1980 and 1990, probably from new plantations and from other land reverting naturally back to forest. An average of 63% of the temperate forest area is exploitable. The exploitable share by country ranges from 100% for Belgium, Czeck

Table 5. Characteristics of Tropical Forests by Ecological Zone[a]

Ecological Zone	Type of Forest	Elevation, m[b]	Mean Annual Rainfall, mm	Region
Tropical rainforest	Wet and very moist evergreen and semievergreen forests	<800	>2,000	Brazil, Ecuador, Colombia, Central Africa, Indonesia, Malaysia
Moist deciduous forest	Moist semideciduous and deciduous forest, woodlands, and tree savanna	<800	1,000–2,000	Brazil, Bolivia, Paraguay, Central and West Africa, India, Bangladesh
Dry deciduous forest	Drought-deciduous and evergreen forest, woodlands, and tree savanna	<800	500–1,000	Brazil, Bolivia, Paraguay, eastern and tropical southern and Sub-sahelian Africa, Sudan, India, Thailand, Laos, Myanmar
Very dry forest	Discontinuous thickets, tree/shrub savanna	<800	200–500	Namibia, Zimbabwe, Kenya, Sudan, Ethiopia, Madagascar, Sub-sahelian Africa, northwestern India, Pakistan
Desert	Tree/shrub steppe	<800	<200	Sub-sahelian Africa, Namibia, Pakistan
Hill and montane forest	Premontane to alpine forest	>800 (upland)	Includes wet, moist, and dry conditions	Bolivia, Peru, Ethiopia, northern Pakistan, Nepal, Bhutan, Mexico

[a] Reference 26.
[b] Lowland unless indicated.

and Slovak Federal Republic, Denmark, and the United Kingdom, to 28% for New Zealand. Eighty-one percent of the temperate forest is publicly owned, ranging from 100% in the former Soviet Union to an average of 62% in North America and 49% in Europe.

The volume of standing, live (growing-stock) trees in global temperate forests in the 1980s amounted to 159 billion cubic meters, about three-quarters of which is in softwood species and one-quarter of which is in hardwood species (24). More than half of the standing volume is in the former Soviet Union (Fig. 4).

TROPICAL FORESTS

The global tropical forest situation is of particular importance because of the alarming rate at which forests are being cleared. Between 1980 and 1990, global tropical forest area declined from 1,910 to 1,756 million hectares, an 8% loss (25). This amounts to about 29 hectares or 72 acres each minute. This rapid rate of increase has generated increasing concerns about the environmental consequences of the loss of these forests; the economic impact of the forest resource, especially to developing countries; and the social effects of this deforestation on indigenous persons. The term tropical forest may be misleading in that it implies a forest in a moist tropical environment. In fact, tropical forests grow on both humid and semiarid lands within or near the Tropics of Cancer and Capricorn in Asia, the Pacific region, Africa, Latin America, and the Caribbean region. The FAO classifies tropical forests into six primary ecological zones, as shown in Table 5.

The 1990 area of tropical forests (1,756 million hectares) roughly equals the total area of South America. The largest tropical forest area is in Latin America and the Caribbean region (918 million hectares), followed by Africa (528 million hectares), and Asia and the Pacific region (310 million hectares), as shown in Table 6.

The greatest area loss was in the Latin America and Caribbean region where 74.1 million hectares (7.5% of the region's total) were deforested between 1980 and 1990. The FAO defines deforestation as a change of land use with depletion of tree crown cover to less than 10%. However, the largest proportional loss occurred in the Asia and Pacific region where the 39.0 million hectares deforested amounted to 11.2% of the region's total. The average of 15.4 million hectares (an area the size of the state of Georgia in the United States) of tropical forest cleared annually during 1981–1990, is 36% higher than the average of 11.3 million hectares cleared annually during 1976–1980 (27).

In addition to this deforestation, a large but unestimated area of tropical forest is degraded, changed from closed to open forest by human activity or by natural phenomena (insects, disease, wind, fire, or flooding), but with enough live tree crown cover left per hectare to be classified as forest. Degraded forests contain a reduced volume of timber and biomass, may contain fewer commercially desirable tree species, exhibit a loss of biodiversity, and may reflect damage to soils. Many of these degraded lands will be further cleared and converted to other uses in the future. Clearly, there is cause for concern over the declining area of tropical forests, and the ecological impact of their conversion (28).

Although efforts to manage tropical forests are increasing, the results are not encouraging. The FAO defines forest management as inventorying the site and stand, preparing a management plan, and applying silvicultural treatments and logging controls in a forest area set aside

Table 6. Area of Global Tropical Forests by Region, 1980 and 1990[a]

Region	Tropical Forest Area, 1980[b]	Tropical Forest Area, 1990[b]	Area Deforested Annually, 1981–1990[b]	Annual Rate of Deforestation, 1981–1990, %
Latin America and Caribbean	992.2	918.1	7.4	0.8
Central America and Mexico	79.2	68.1	1.1	1.5
Caribbean	48.3	47.1	0.1	0.3
Tropical South America	864.6	802.9	6.2	0.7
Asia and Pacific	349.6	310.6	3.9	1.2
South Asia	69.4	63.9	0.6	0.8
Continental Southeast Asia	88.4	75.2	1.3	1.6
Insular Southeast Asia	154.7	135.4	1.9	1.3
Pacific	37.1	36.0	0.1	0.3
Africa	568.6	527.6	4.1	0.7
West Sahelian Africa	43.7	40.8	0.3	0.7
East Sahelian Africa	71.4	65.5	0.6	0.9
West Africa	61.5	55.6	0.6	1.0
Central Africa	215.5	204.1	1.1	0.5
Tropical Southern Africa	159.3	145.9	1.3	0.9
Insular Africa	17.1	15.8	0.1	0.8
Total[c]	*1,910.4*	*1,756.3*	*15.4*	*0.8*

[a] Reference 25.
[b] Million hectares.
[c] Totals may not add due to rounding.

to be managed for future production of forestry goods and services. The FAO and International Timber Trade Organization (ITTO) studied the status of forest management in the tropics and reported that South Asia, principally India, accounted for the largest area under management in 1980: 59 million hectares covered by working plans, or 3.1% of the 1980 global tropical forest area (25). However, much of that area is being degraded slowly by intense grazing of livestock and unauthorized fuelwood removals by local people. Outside of South Asia, the tropical forest area under sustained yield management was less than 1 million hectares.

The FAO estimates that about three-fourths of the tropical forest area in 1990 was in two zones, the tropical rainforest and moist deciduous forest zones (25). The largest area deforested annually between 1981 and 1990 was in the moist deciduous forest (6.1 million hectares) (Table 7). Annual deforestation rates are highest in the hill and montane forest (1.1%). The hill and montane forest is perhaps most threatened because of the relatively small area of forest remaining, a population density that is higher than the average for tropical forests, and the high deforestation rate.

DEFORESTATION

Causes of Deforestation

The forest area in some developed regions has expanded (Table 1) as economic development has encouraged the reversion of agricultural lands to forest. However, in developing countries, the trend is toward deforestation, particularly in tropical forests. Although broad issues of poverty, rapidly increasing population pressures, unequal political power, lack of opportunities to make a living, landlessness, and inadequate knowledge and means to exploit the tropical forest without destroying it are at the root of deforestation, there are more specific causes.

Shifting cultivation (also called slash and burn agriculture), practiced by landless indigenous people who clear trees to grow subsistence crops, is the principal cause of deforestation in the tropics, accounting for 70, 50, and 35%, respectively, in Africa, Asia, and tropical America (29) (see Plate III). Farmers must move to new sites after several years because most tropical forest soils are of low productivity; the already limited nutrients in these soils are lost to leaching and erosion, rather than being quickly recycled into plants as they had been under a forest cover. The abandoned patches, called forest fallows, are being created at the rate of more than 5 million hectares per year, but may revert back to forest if left undisturbed. However, because rising populations and the ensuing competition for land are forcing farmers to return to these fallows at increasingly shorter intervals, little of this land is allowed to revert to forest. About 500 million people (nearly 10% of the world population) and 240 million hectares of closed forest are involved in shifting cultivation, which is increasing at an average annual rate of 1.25% (30). Improvement of the efficiency of agricultural production in many developing countries is a prerequisite for controlling tropical deforestation (31).

Deforestation also occurs because conversion of forest to pasture for domestic animals is widely practiced, stimulated by a growing network of roads, and, in some places, by government incentives. Fuelwood gathering may be an important deforestation agent in dry forests.

Commercial logging may not be a primary cause of deforestation in the tropics (except in parts of West Africa) because the number of trees left after logging may be sufficient to classify the site as forested. However, it is often a secondary cause because new logging roads permit shifting cultivators and fuelwood gatherers to gain access to logged areas and fell the remaining trees. When logging is performed poorly, it results in a degraded forest.

Finally, expansion of agribusinesses that grow oil palm, rubber, and fruit trees, and ornamental plants has resulted in forests being cleared, and government-sponsored programs that resettle landless farmers on forested sites have contributed to deforestation around the world.

Consequences of Deforestation

A large effect of deforestation, although incompletely understood, is its contributions to global warming by releasing carbon dioxide and other greenhouse gases into the atmosphere from the cutting and burning of trees.

Other consequences of deforestation are better known, including loss of industrial timber and nontimber prod-

Table 7. Area of Global Tropical Forest in 1990, Area Deforested Annually Between 1981 and 1990, and Annual Rates of Deforestation, by Ecological Zone[a]

Ecological Zone	Area, 1990[b]	Area Deforested Annually, 1981–1990[b]	Annual Rate of Deforestation, 1981–1990, %
Tropical rainforest	718.3	4.6	0.6
Moist deciduous forest	587.3	6.1	1.0
Dry and very dry deciduous forest	238.3	2.2	0.9
Hill and montane forest	204.3	2.5	1.1
Desert and alpine zone	8.1	0.1	1.0
Total[c]	*1,756.3*	*15.4*	*0.8*

[a] Reference 25.
[b] Million hectares.
[c] Totals may not add due to rounding.

Plate III. This region has been subject to shifting cultivation or slash and burn agriculture. Shifting cultivation is a primary cause of deforestation in the tropics. Courtesy of United States Forest Service.

ucts, and loss of long-term forest productivity on the site. In many places the lack of fuelwood following deforestation challenges local people, especially where fuelwood had already been scarce. Forest fragmentation, the reduction of a large block of forest to many smaller tracts, promotes loss of biodiversity because some species of plants and animals require large continuous areas of similar habitat to survive. Agriculture may be negatively impacted if deforestation causes soil loss or compaction, or sedimentation of irrigation systems. Human life and downstream structures may be endangered by floods that may be intensified by clearing forests on upstream watersheds. Species of plants and animals, which may occupy narrow ecological niches and whose potential value to humans is unknown, may be eliminated: it is estimated (32) that two-thirds (3.3 million) of all species of organisms on earth live in the tropics, and that two-thirds (2.2 million) of these are unique to tropical forests. Indigenous people may be forced into a new way of life for which they are unprepared. Encroaching deserts may follow deforestation in dry forest areas.

REVERSING DEFORESTATION

Forest Plantations

Estimates of the area of global forest plantations (temperate and tropical) differ, but center around a total of 100 million hectares as of 1985, or about 3.5% of the closed forest area. It is estimated (33) that the four countries or regions with the largest plantation areas (the former Soviet Union, western Europe, China, and the United States) account for 62% of the total (Table 8).

An estimated 60% of the area of global plantations is devoted to industrial timber production, and the remaining 40% is devoted to nonindustrial uses, such as fuelwood production and environmental protection (29). Most of the plantation area is outside the tropical zone, yet most of the deforestation occurs in the tropics. Although the area planted is increasing at an accelerating rate, global deforestation surpasses it by about 5 million hectares annually. Most plantations contain only one or several tree species, and cannot replace a natural forest in terms of biological diversity. Nevertheless, plantations play a crucial role in slowing down the loss of natural forests by supplying large quantities of industrial timber from small, highly productive areas. For example, in Latin America, industrial plantations made up less than 1% of the forest area, but accounted for 30% of industrial wood production during the 1980s. This proportion is expected to rise to 50% by the year 2,000 (34).

Tropical forest plantations in 90 countries as of 1990 amounted to 43.9 million hectares (25). Based on an average tree survival rate of 70% from 56 plantation inventories in 18 tropical countries, the FAO estimates the net area of successful plantations to be 30.7 million hectares. Between 1981 and 1990, the annual deforested area (15.4 million hectares) was about 8.5 times greater than the annual net planted area (1.8 million hectares). Five countries accounted for 85% of the gross plantation area: India (18.9

Table 8. Area of the World's Forest Plantations by Countries or Regions, ca 1985[a]

Country or Region	Plantation Area[b]	Percent
Developed countries		
Former Soviet Union	21,900	22.8
Western Europe	13,000	13.6
United States	12,100	12.6
Japan	9,600	10.0
Canada	1,500	1.6
New Zealand	1,100	1.2
Australia	800	0.8
Subtotal	*60,000*	*62.6*
Developing countries		
China	12,700	13.2
Brazil	6,100	6.4
India	3,100	3.2
Indonesia	2,600	2.7
Republic of Korea	2,000	2.1
Chile	1,200	1.3
Argentina	800	0.8
Others	7,400	7.7
Subtotal	*35,900*	*37.4*
Total	*95,900*	*100.0*

[a] Reference 33.
[b] In thousand hectares.

million hectares), Indonesia (8.8 million hectares), Brazil (7.0 million hectares), Vietnam (2.1 million hectares), and Thailand (0.8 million hectares). About 60% of the total reported plantation area was established during the period 1981 to 1990. The principal tree species planted were pines (*Pinus* spp.) and teak (*Tectona* spp.), planted mostly for industrial timber production; eucalypts (*Eucalyptus* spp.), planted for both industrial and nonindustrial purposes; and acacias (*Acacia* spp.), used primarily for fuelwood and gum arabic production.

Agroforestry: A Positive Tool

Agroforestry, the deliberate integration of trees, shrubs, and other woody perennials on the same land with agricultural crops, livestock, or both, recognizes the symbiotic relationship between trees and agriculture that has existed in most countries for centuries. It offers a practical way to intensify land use and to reduce the economic and social pressures for clearing additional closed forests. Growing annual crops in rows between nitrogen-fixing trees or shrubs, and growing canopy trees for plantation crops, eg, coffee or cacao, are examples of this strategy, which represents part of a solution to the growing fuelwood shortage.

Conservation agroforestry, the practice of working trees into agricultural and community ecosystems for human, as well as economic and environmental, benefits, is a variation of traditional agroforestry that is gaining acceptance in many parts of the semiarid world. Examples of these practices include planting windbreaks, living fences, livestock havens, and riparian buffer strips, and establishing wildlife habitat.

Social Forestry

A modification of agroforestry is social forestry, a broad range of tree- or forest-related activities that rural people and community groups in developing countries undertake to provide products for their own use and to generate local income (17). Examples are: farmers growing wood to sell or use for firewood; individuals or communities earning income from the gathering, processing, and sale of minor forest products (fruits, nuts, herbs, basketry materials, etc); or governments or other groups planting trees on public lands to meet local village needs. A basic premise is to change land use in such ways that people get what they need on a sustainable basis from a relatively fixed, or even shrinking, land base. Another premise is the critical importance of widespread local participation, possible only when people have the ability to cooperate, the knowledge of what is required, the proper mix of incentives to stimulate them, and the institutions to support and sustain them. Solutions to many environmental problems ultimately reside with each land user and his/her land use practices (35).

Social forestry programs are helping to ameliorate some of the important issues facing developing countries, such as energy shortages, food shortages, and unemployment. Fuelwood production is one of the key objectives of social forestry because of seriously declining fuelwood availability and the closely linked environmental instability. Other objectives are to reverse declining agricultural productivity associated with poor land use, deforestation, erosion, and declining water supplies, and to provide employment opportunities and income from forestry and related processing activities. To stimulate the practice of social forestry in developing countries, the World Bank made loans totaling $1.3 billion during the decade 1977 to 1986 (17).

SUMMARY

Forests are a renewable resource, and are part of a complex network of natural systems that affect life on the globe. When forests are misused, the consequences may be felt in places far removed from the offending act. Increasingly, international cooperation, public education, political will, personal incentives, and institutional structures are needed to manage the world's forests in harmony with the environment and with economic development.

BIBLIOGRAPHY

1. J. Perlin, *A Forest Journey: The Role of Wood in the Development of Civilization*, W. W. Norton and Co., New York, 1989.
2. World Resources Institute, *World Resources 1990–91*, Oxford University Press, New York, 1990.
3. Food and Agriculture Organization of the United Nations (FAO), *Production Yearbook, 1990*, Vol. 44, FAO Statistics Series No. 99, Rome, 1991.
4. A. Mather, *Global Forest Resources*, Belhaven Press, London, 1990.
5. A. Sommer, *Unasylva* **28**(112/113), 5–25 (1976).

6. J. Spears and E. Ayensu, in R. Repetto, ed., *The Global Possible*, Yale University Press, New Haven, 1985, pp. 299–335.
7. H. Gleason and A. Cronquist, *The Natural Geography of Plants*, Columbia University Press, New York, 1964.
8. P. Miller, in R. Olson, D. Binkley, and M. Bohm, eds., *The Response of Western Forests to Air Pollution*, Springer-Verlag, New York, 1992; *Ecological Studies* **97,** 461–500 (1992).
9. C. Eagar and M. Adams, eds., *Ecology and Decline of Red Spruce in the Eastern United States*, Springer-Verlag, New York, 1992.
10. C. Federer and co-workers, *Env. Manage.* **13,** 593–601 (1989).
11. E. Verry and A. Harris, *Water Resources Res.* **24,** 481–492 (1988).
12. R. Persson, *World Forest Resources: Review of the World's Forest Resources in the Early 1970's*, Research Note 17, Royal College of Forestry, Stockholm, 1974.
13. K. Waddell, D. Oswald, and D. Powell, *Forest Statistics of the United States, 1987*, U.S. Department of Agriculture, Forest Service, Resource Bulletin PNW-RB-168, Portland, Oreg., 1989.
14. C. Binkley and D. Dykstra, in M. Kallin, D. Dykstra, and C. Binkley, eds., *The Global Forest Sector: An Analytical Perspective*, John Wiley & Sons, Inc., New York, 1987, pp. 508–533.
15. Food and Agriculture Organization of the United Nations (FAO), *Yearbook of Forest Products, 1991,* FAO Forestry Series No. 26, FAO Statistics Series No. 110, Rome, 1993.
16. Food and Agriculture Organization of the United Nations (FAO), *Forest Products: World Outlook Projections*, FAO Forestry Paper 84, Rome, 1988.
17. H. Gregersen, S. Draper, and D. Elz, eds., *People and Trees—The Role of Social Forestry in Sustainable Development*, Economic Development Institute of the World Bank, Washington, D.C., 1989.
18. G. Goodman, *Ambio* **16,** 111–119 (1987).
19. Food and Agriculture Organization of the United Nations (FAO), *Fuelwood Supplies in Developing Countries*, FAO, Rome, 1983.
20. M. Fergus, *Geographical J.* **142,** 29–38 (1983).
21. E. Mnzava, *Unasylva* **33**(131), 24–29 (1981).
22. P. Dewees, *World Dev.* **17**(8), 1159–1172 (1989).
23. C. Bailly and co-workers, *Bois For Trop.* **197,** 23–43 (1982).
24. Food and Agriculture Organization of the United Nations (FAO), *The Forest Resources of the Temperate Zones*, Vol. 1, The UN-ECE/FAO 1990 Forest Resource Assessment, ECE/TIM/62, New York, 1992.
25. Food and Agriculture Organization of the United Nations (FAO), *Summary of the Final Report of the Forest Resources Assessment 1990 for the Tropical World*, (Report prepared for the United Nations Eleventh Session of the Committee on Forestry), Rome, 1993.
26. Personal communication, Dr. K. D. Singh, Project Coordinator, Forest Resources Assessment 1990 Project, FAO, Rome.
27. Food and Agriculture Organization of the United Nations (FAO), *Second Interim Report of the State of Tropical Forests*, FAO, Forest Resources Assessment 1990 Project, Hand-out at 10th World Forestry Congress, Paris, Sept. 1991, Rome, 1991.
28. B. Freezailah, "The Evolving Role of International Organizations," in *Proceedings of Society of American Foresters 1992 National Convention*, Oct. 25–28, 1992, Richmond, Va., 1992, pp. 399–403.
29. J. Lanly, *Tropical Forest Resources*, Food and Agriculture Organization of the United Nations, FAO Forestry Paper 30, Rome, 1982.
30. J. Lanly, *Unasylva* **37**(147), 17–21 (1985).
31. A. Lugo, "Tropical Forest Management—Time to Do Something About It," in Ref. 28, pp. 382–391.
32. N. Myers, in R. Hoage, ed., *Animal Extinctions: What Everyone Should Know*, Smithsonian Institution Press, Washington, D.C., 1985.
33. J. Laarman and R. Sedjo, *Global Forests: Issues for Six Billion People*, McGraw-Hill, Inc., New York, 1992.
34. S. McGaughey and H. Gregersen, *Forest-Based Development in Latin America*, Inter-American Development Bank, Washington, D.C., 1983.
35. L. Brown, *The Woodlands Forum* **3**(2), 1–4 (1986).

FORESTRY, SUSTAINABLE

DAVID N. WEAR
USDA Forest Service
Research Triangle Park, North Carolina

Sustainability has become the primary criterion for evaluating human endeavors that involve natural resources and influence the environment. Defined in such a way, and given the ecological maxim that all things are connected, sustainability applies to nearly every human endeavor. Because forestry represents a globally important, perhaps the greatest, direct interface between human endeavor and natural process, its practice has an especially important influence on the function of ecological systems and the quality of the environment. Sustainable forestry is now seen by many as the ultimate objective for the practice and profession of forestry.

See also AGRICULTURE AND ENERGY; BIODIVERSITY MAINTENANCE; CARBON CYCLE; CARBON STORAGE IN FORESTS.

The word sustainable implies a stasis. That is, it suggests a process that can be continued into perpetuity. Beyond some consensus on the definition of sustainability at this broad level, that activities undertaken today should not compromise the opportunities left for tomorrow, there is very little agreement on how sustainability should be measured and implemented in the practice of forestry.

There are several reasons for disparate definitions. One relates to differences in basic values that define "land ethics" (1). For example, deep differences in land ethics may be rooted in the difference between modern rural and urban cultures. Clearly, not all people value the same quantities and qualities in forests and nature. However, a great deal of dissonance can perhaps be traced to contradictions in the root concept of sustainability. Stasis is perhaps not an especially useful concept for guiding the management of essentially dynamic ecological systems in a world of expanding populations and changing climate.

Consequently, practical guiding principles for forest management have yet to result from the discourse on sustainability. This, in fact, may be an unreasonable expectation. At this time it is perhaps most accurate to view sustainability as a platform upon which many public concerns regarding the practice of resource management and its consequences for ecosystem health are debated and, one may hope, eventually sorted out.

Resource management issues have, of course, been individually debated in several forums. By and large, however, the debates have been undertaken in the context of protest, advocacy, and litigation. In contrast, sustainability seems to be emerging as a focal point for discussions between researchers in several disciplines, policy makers, and advocates as well, on how to construct management objectives and approaches that reconcile the direct use of resources with ecological health and environmental quality (2–4). The task is enormous. Viewed in this way, defining sustainability emerges from the fray of resource management and environmental debates as an attempt at synthesis; as an attempt to recognize the tradeoffs and interactions between human endeavors and consumption and resulting environmental health. Defining sustainability therefore holds promise as a positive rather than a reactionary endeavor that has gathered considerable attention both in the United States and globally.

For sustainability to move from a platform for discussion and synthesis to a means to guiding resource management towards large-system goals, several challenges must be surmounted. These range from defining the structure and operation of forest-ecological systems to fashioning instruments and institutions that can effectively address resource and environmental goals at appropriate scales. Of course, these issues hinge on being able to settle on some notion of what it is that should be sustained.

FORESTRY AND CONCEPTS OF PRODUCTIVITY

Forestry throughout the world has historically involved some element of social crusade. Not surprisingly, these social concerns have typically related to protecting and enhancing the productivity of forest lands. Modern concerns, especially those brought under the discussion of sustainability, can be easily seen as extensions of historical movements to conserve or protect natural resources.

In the United States (of which the forestry history serves as the vehicle for discussing the evolution of resource management in this article), forestry started in the late 1800s as a public response to the private exploitation of resources. The rhetoric of the times saw severe timber shortages resulting from the expansive harvesting that moved from region to region in the U.S. The expanding country would be crippled by a shortage of its most important raw material.

Conservation, broadly defined as wise use of renewable resources, was the public sector's response to rapidly increasing timber consumption and deforestation at the start of industrial expansion in the United States. Between 1850 and 1909, per capita consumption of wood products increased four-fold (5). This expansion derived from vast demands for wood to construct the railroads, to build cities and ships, and to smelt iron (6).

The strong demand for wood products was fed by a lumber industry in the United States that moved from the Northeast to the lake states and eventually to the Pacific Northwest and the South (see Plate I). The scale of forest clearing and lumber production was enormous and impressive. However, it was not the scale of production alone that suggested impending timber famine. It was also the prevalence of fraud and theft in the procurement of timber from public lands that fueled the perception of resource use gone out of control (7).

The setting in which forestry developed in the U.S. strongly parallels conditions now motivating conservation activities in developing parts of the world. In both cases, rapidly increasing populations coupled with uncertain frontier land ownership have resulted in rapid deforestation. Typically, other pressures for land use or improved transportation have resulted in forests becoming more accessible and therefore more marketable. Also, the demand for agricultural land has typically fueled deforestation. Both then and now, public intervention to control forest exploitation has emerged as the solution to resource degradation.

In the late nineteenth century, the private timber industry was seen as failing to account for the long-term economic interests of the developing U.S. The response was a conservation movement based on wise use and reforestation. As a result of these efforts, a system of, first, federal forest reserves and then, national forests was established. Presidential authority to establish forest reserves from the public domain was set out by the Creative Act of 1891. The national forests were established with the Organic Act of 1897 and the Weeks Law of 1911 allowed for the purchase of land and the establishment of national forests in the eastern United States. At roughly the same time, forestry attained credibility as a profession and the first forestry schools and the two principal professional forestry associations were established in the United States (8).

The young profession of U.S. forestry sought to bring the forests of the country under a scientific model of management, and its earliest textbooks reflect this emphasis on forest management as an engineering exercise. At root, however, forestry was based on a social vision. Indicative of this view, the Society of American Foresters in 1917 defined forestry as "the science and art of managing forests in *continuity* for forest purposes" (9). This and other textbooks of the day clearly establish perpetual production as the goal of forestry practice. In the same text, forest management is described as being motivated by ". . . the ultimate aim of securing a sustained yield" (10).

Sustained yield, perhaps a primitive form of sustainability, has provided the motivating concept as well as the guiding ideology for forestry throughout its history. Its value in the latter role has only recently been challenged. The context of the rapid deforestation experienced in the 1890s and the early part of the twentieth century, easily justified a public forestry that sought maximum output to ward off national timber shortages. That is, the government intervened in private resource use on the grounds that private production could not provide for the future (11).

Sustained yield provided, then, the initial impetus for the development of forestry as a profession in the U.S. Social movements in general are most effectively motivated by policy capsules that summarize the good qualities of

Plate I. Fresh-cut logs are marked so that their source can be identified. This is one method of preventing fraud and theft in the lumber industry. The hillside background shows a lumber field that was clear-cut 10 years prior and is again fully stocked. Restocking restores resources and enables their future use. Courtesy of the United States Forest Service.

highly complex social phenomena (12). These encapsulations often take on a life of their own, forming the ideology of a profession. As such, they are discarded only with great trepidation and debate.

The shortcomings of a forest policy and practice guided solely by the goal of maximum perpetual yields of timber were evident early on. They related to two arguments. One was that many goods and services in addition to timber were valued outputs from forests. These included water and forage as well as recreation and wildlife habitats. The other was that society was not necessarily served best by producing more timber at any cost. Because it focused on a single measure of production, the sustained yield model of forest management could not accommodate value tradeoffs.

The history of policy and legislation directed at the management of national forests can be seen as gradually expanding the spectrum of forest products and services. The progression of events that expanded the agenda of public forestry from one focused on timber and watershed protection commenced early in the century and led up to the Multiple Use Sustained Yield Act (MUSYA) of 1960 (13). The MUSYA clearly defines a broad agenda for forestry in the United States: "It is the policy of the Congress that the national forests are established and shall be administered for outdoor recreation, range, timber, watershed, and wildlife and fish purposes."

While the MUSYA came only in 1960, the concerns for the nontimber benefits derived from forests were clear from the start. The enabling legislation for the National Forests emphasized the protection of waterways in addition to sustained timber production. Wildlife, recreation and wilderness considerations developed in the U.S. Forest Service beginning in the 1920s (13).

Public forests have been seen as progressively more than just trees. Since 1960, a progression of forest management and environmental protection laws have codified the multiple concerns and procedures for managing the government's forest assets. Public lands do not, however, constitute the entirety of forests in the United States. It is important to remember that, from the Multiple-Use Sustained-Yield Act on, public and private visions of forestry have essentially diverged. This is not necessarily a distinct dichotomy. Some public holdings, for example State trust lands in the western states, are managed for maximum long-run profit, though constrained to a sustained yield style of management. In addition, environmental and species protection laws clearly have important influence on private land management.

As the complement of objectives has expanded, sustained yield has become a broader and more ambiguous encapsulation of the goals of private and public forestry. Because it encapsulates so much, it can easily be interpreted in many different ways. The private sector may view sustained yield as the physical timber production model emphasized by the forestry profession in its early years. The public sector is perhaps more likely to view it as a part of multiple-use sustained yield as defined by MUSY. It is quite possible and perhaps likely that debates over what constitutes the difference between the tradition of sustained yield and the innovation of sustainability is based on these different, though rarely discussed, definitions of the root concept.

Multiple Use did not, however, supplant sustained yield as the guiding principle for public forestry. Rather as the name Multiple-Use Sustained-Yield indicates, the concept was simply extended to the full range of uses listed in the Act. While this may appear egalitarian, a simple extension to all relevant resources is simply not possible. This is because once more than one resource service is addressed, it becomes impossible to unambiguously maximize their production. That is, as long as the multiple resources are not perfectly complementary (and they clearly are not), producing more of one often implies producing less of another. Multiple use therefore raises the issue of tradeoffs.

However, the MUSY did not explicitly define how tradeoffs should be resolved. At the time, it can be argued that the greater availability of forests (especially on a per capita basis) limited the extent of conflicts over the use of public forests. That is, the tradeoffs were not, initially, severe because there was more than enough room to accommodate most uses. However, the most modern phase of public forestry is marked by the process of sorting out tradeoffs that have become progressively more severe. Driven primarily by the National Forest Management Act of 1976, National Forest managers sought to define resource tradeoffs on public lands through the 1980s. The platform for this exercise was a massive planning exercise guided by the NFMA with public involvement and impact analysis guided by the National Environmental Policy Act of 1969 (NEPA).

Sorting out tradeoffs between dissimilar goods requires boiling the arguments down to a common currency. The analyses undertaken in the 1980s were expansive and involved scientific management tools in the form of linear programming models and cost benefit analysis, as well as massive efforts to collect and utilize public opinion to define the relative values of resource uses. The result was a complex public forest management process that matched an equally complex set of resource management concerns.

The outcome of public forest planning and management in the United States, rather than improving consensus over appropriate forest usage, fomented an intense level of debate. Hundreds of administrative appeals and law suits resulted. Some were not satisfied with the outcome of the process; their relative values did not reflect the perceived majority view. Others took issue with the structure of the process, that is, the casting of the problem as a scientific exercise left out important elements that could not be modeled. It is also likely that the practice of placing dollar values on all resources, for example wilderness preservation, offended many at philosophical levels. Another argument directly relevant to the discussion of sustainability is that the process focused on many parts of resource management issues but failed to address the integrity of overarching ecological systems.

In sum, public forestry in the United States has become an increasingly complex and an increasingly divisive issue. As forests and their services become more scarce the issues become increasingly difficult. Not surprisingly, public forestry is where much of the discussions of sustainable forestry has focused. Sustainable forestry can be viewed as an outgrowth of the public discourse over national forests that started in the 1970s, reached a crescendo in the 1980s, and continues to the present.

MODERN ORIGINS OF SUSTAINABILITY

Concerns for sustainability are, however, much broader than forest management issues. Rather, "sustainability" or "sustainable development" has become an important criterion for judging all human activities that impinge on ecological systems. The modern focus on sustainability is easily traced to the Brundtland Commission Report (World Commission on Environment and Development 1987), which emphasizes that environmental health is and will continue to be a necessary condition for economic prosperity. It also makes an explicit link between poverty and disparate distributions of wealth and the state of this ecological-economic system. Gro Harlem Brundtland, the commission's chair and the report's namesake, views "alleviating poverty (as) priority number one . . ." in successfully achieving sustainability (14).

Sustainability is therefore an extremely broad concept, encapsulating ecological, physical, economic, and social aspects of resource use and management. It can indeed be argued that the concept is too broad to be productive, especially if it spawns endless debates on defining what sustainability means. Perhaps, however, debate over what is and who affects sustainability is productive in itself. If the linkages between ecological health, economic prosperity and the human condition are critical, then discourse between ecologists, economists, anthropologists, policy makers, etc, is a necessary first step in formulating these ideas and then organizing human activities towards desired ends. Sustainability has begun to stimulate discussions between disparate disciplines and spawn interdisciplinary investigations (2,4).

While sustainability may not enjoy ubiquitous definition, it does command near-ubiquitous attention. For example, sustainable development was the general theme of the 1992 Earth Summit. It encapsulates a number of concepts regarding the connection between humankind and its environment and serves as a focal point in a world of disparate ideas, models, and policies. Perhaps the concepts of sustainable development can focus disparate scientific disciplines on these clearly important issues in the way that George Marshall seemed to organize disparate viewpoints to address the complex problems of post-war Europe. It can be argued that, in both cases, simplified statements catalyzed action where careful analysis of complicated issues could (and can) not. It may therefore be more useful to view sustainability as useful for organizing ideas about the impacts of human activity and growth on the functioning of the biosphere, than as any operable set of physical standards or guidelines at this point.

If sustainability is about the interactions between resource use, environmental health, and the human condition, then forestry is clearly a complete microcosm of the sustainability debate. Forestry is the most land-extensive human endeavor so that forestry practice may have enormous influence on the function of ecological systems. For example, commercial tree planting can replace old and mixed species forests with vast monocultures of a single species and, in the process, completely alter the matrix of sub-canopy flora and fauna as well. Impacts can range from small local to large regional scales.

Several other elements of sustainable development are also found in the context of forestry issues. The time required to regrow forests to commercial or old-growth conditions may be an impediment to regenerating land and has been raised as a point of concern for centuries. That is, popular sentiment has held that the very nature of the exercise of growing trees precludes investment in sustainable forestry. Accordingly forestry decisions are often seen as favoring the present over future generations. Intergenerational equity is at the crux of sustainability issues (15).

The pattern of forest allocations within a particular generation also define a set of issues within forestry. As a primary resource, wood production has a larger than average variation. Resulting boom and bust patterns of resource use typically lead to expanding and contracting economies and considerable dislocation in rural areas. Public forest policy is often motivated by a desire to stabilize the rural communities built upon wood products (16). In this sense forest policy is often concerned with equity and distribution of income in poorer rural areas (17).

Another area where forests (but not necessarily forestry) are central to sustainable development is forest clearing and land conversions. Clearing aboriginal forests is often the first step in transforming a place dominated by nature and ecological structure to one that is dominated by humankind and its cultural structures. While, in an increasingly global and crowded community, it may be counter productive to separate humankind from "natural" systems (2), forest harvesting remains a symbol of man's conquest of nature (see Plate II).

As with most symbolic meaning, the issues are actually much more subtle. Not all forest harvesting is an act of deforestation. Rather, it may also be the first step in reforestation, either active tree planting or a more passive natural regeneration. Furthermore, the practice of forestry (forest growing as well as harvesting) may in fact be the only practical alternative to less desirable alternatives such as an erosive agricultural use. In comparison, the active practice of forestry may prevent soil erosion and other ancillary problems found in the agricultural uses of marginal lands. In such a case, forestry plays a positive role in sustaining soil and land productivity.

Forestry seems to capture or is central to several elements of the broader sustainable development debate. It defines a direct interaction between human consumption and ecological function and directly involves intergenerational tradeoffs. Sustainable forestry is, accordingly, a complex issue, effectively encapsulating ecological, intergenerational, equity, and productivity concerns. Structuring these multiple issues into coherent resource management strategies and policies remains the essential challenge in developing an operational definition of sustainable forestry.

ESTABLISHING SUSTAINABILITY GOALS

The first step in establishing a practice of sustainable forestry is defining a set of objectives. Concerns for environmental quality as well as human welfare constantly raise questions about tradeoffs. It is clear that sustainability is not an absolute concept. That is, there appears to be no single "ecologically correct" approach to managing ecosystems. Rather, and especially due to an expanding human population, priorities and objectives must be carefully articulated. Without objectives, sustainable policy and management decisions are essentially arbitrary and their effectiveness cannot be measured. However, while crucial, defining these goals is far from straightforward.

There are many different perspectives in the discussion of sustainable forestry. For example, there is the obvious split between an industrial or agricultural view and the natural system view of forests. In policy discussions, there is also a parallel dichotomy between how the social sciences and the natural sciences view sustainability issues. Both perspectives recognize that ecological health is important and that sustainable forestry requires a long-term view. Differences arise over how to view human use of ecological assets, the role of ecosystem health in human welfare, and the emphasis placed on human endeavors in global systems.

The differences between economic and ecological perspectives define the salient features of defining sustainability goals. While generally aimed at the same issues it is important to recognize that the tension between economic and ecological perspectives may often result from confusing the ends and means of sustainable forestry.

The social sciences, quite naturally, put the human condition at the center of the discourse over resource use and ecological health. While anthropology, economics, and sociology may measure welfare in decidedly different terms, ultimately they measure the health of ecosystems in terms of observed and expected effects on human welfare. In addition, the social sciences place great emphasis on the ability of humanity to adapt to changing conditions and to improve its lot in the face of impending resource scarcity.

Economics in particular views the human community as an adaptive organism and actively engages human progress and adaptability as important elements in global systems. Notably, the twin engines of substitution and technology have guided humanity away from resource scarcity and has, at least through the modern age, vastly increased the productivity of the earth's reources in terms of their consumable products. Assuming that degraded ecological systems define a kind of scarcity that can be signaled through resource markets, these signals might

Plate II. Areas of once-forested-land are being leveled to provide space for human development. This area in the Amazon Basin of northern Brazil is naturally a tropical forest, it has been burned to build settlements. Courtesy of the Food and Agricultural Organization.

similarly motivate innovations that would mitigate ecological damage and improve the human condition. These signals take the form of changes in price and the correct signal is a price that reflects the true cost of a resource, including the costs of ecological damages.

At the root of this line of argument is the assumption that shifting ecological conditions can be recognized and signaled in the process of economic enterprise. There are several reasons why this may be a weak assumption. One is simply that many environmental qualities are not traded in markets. Examples include classical externalities of production, such as air and water pollution. Another reason is that a considerable lag may occur between human actions and ecological consequences. This is often the case with species extinction arising from habitat destruction, where populations may fragment and gradually decline before being recognized as threatened. In such cases, impacts are cumulative, with effects being felt only several years or decades after initial causes.

The argument that technological innovation and substitution will always alleviate problems is a somewhat polar case. Economics, in the main, recognizes several factors that may justify a society's intervention in the free operation of markets. All of these relate to the welfare of citizens and include the fair pricing of goods, protection from crime, and providing a quality environment. In fact, recent government initiatives in the United States have created markets for environmental goods such as air quality. By placing these goods in the care of markets (i.e., internalizing them), these policies attempt to allow scarcity to be recognized by producers and consumers and thereby be alleviated by substitution and technological innovation.

There are perhaps more fundamental objections to the hope that economic man's present and future ingenuity can alleviate ecological problems. One is that the world of the future will be a new and wholly unprecedented condition due mainly to expanding human populations (18). Simply put, the problems of the world are becoming progressively larger and they may not be effectively anticipated by the marketplace. Another argument is that human ingenuity necessarily approaches a point of diminishing returns. That is, while the processing of materials through industry and agriculture has become more and more efficient, all processes are eventually bounded by the laws of thermodynamics (19). Of course, it is not clear just how close the laws of thermodynamics are to becoming binding constraints on human enterprize.

It may also be argued that ecological health is too broad

and intricate to be understood by the firms and consumers who interact in markets. That is, the integrity, quality, and interactions of complex ecosystems is what matters, not their mere abundance. If this is the case, then it follows that market mechanisms that have historically recognized scarcity and averted resource famine are simply not designed to respond to changes to ecological systems. Furthermore, and at a most fundamental level, ecological health may therefore depend on a concept that is incongruous with the competitive market place. It may depend on the orchestration of resource management among different and otherwise competing landowners and firms.

This notion of an essential complexity that cannot be recognized or protected by a collection of freely-acting economic agents, is a central element of the natural science perspective on sustainability. This perspective places the condition of ecological systems rather than human condition at the center of the discourse on sustainability. This focus leads to rules for managing ecosystems in ways that preserve their natural complexity and information, generally without consideration of tradeoffs or cost. Typically, the perspective defines desired ecosystem conditions without explicit consideration of the role of humankind in ecosystems (2).

Of course how to measure ecosystem health remains an unanswered question. Because these systems represent complex and dynamic assemblages of organisms and processes, defining essential linkages is at present impossible. In light of this uncertainty, various rules for managing ecosystems have been proposed. For example, Franklin (20) suggests that two principles guide ecosystem management. One reflects a concern for the essential genetic information contained in an ecosystem: "prevent . . . the degradation of the productive capacity of our lands and waters—no net loss of productivity." The other reflects concerns for process: "prevent . . . the loss of genetic diversity, including species—no loss of genetic potential" (20).

Concern for preserving biological information or biodiversity is an ubiquitous element of contemporary conservation. The modern rate of species loss is estimated to be several times higher than any recorded by history or fossils. The cause of species loss is easily traced to the impacts of development and other human activities which, as a rule, reduce the complexity of ecosystems.

The erosion of biodiversity may have severe consequences. One consequence is related to the loss of biological information (in the form of survival strategies encoded in DNA) that has evolved over the millennia. Extinction erases the biochemical recipes of species which may have human benefit in future application.

However, more severe consequences may relate, not to the specific information encoded within any particular species, but to the interactions among different species. Organisms do not interact with their physical environment in isolation. Rather organisms are connected, for example, in the transfer of energy from one trophic level to another, or as vectors in the reproduction of other organisms. Commensal as well as parasitic relationships define the overall organization of an ecosystem and its cycling of materials.

These interactions and a complex coevolution of species suggests that, while species have a range of adaptability, their survival and persistence are inextricably tied to other species. The biological parable for this type of dependence is the interacting population dynamics of predator and prey, for example, between foxes and rabbits. Fluctuations in predator or prey populations are constantly moderated by starvation or procreation. Loss of the predator or the prey eliminates a critical regulatory process, thereby severely affecting the remaining species and its habitat. However, many important ecological interactions are decidedly more subtle, involving multiple interactions between and among flora and fauna. Damage to the specific complexes of interacting species that defines habitat is the most common cause of extinction. Productive intervention in the demise of endangered species is, accordingly, very difficult.

Followed to conclusion, this line of argument also suggests that loss of a species may eventually lead to collapse of larger systems of interacting species. System disfunction clearly depends, however, on the amplitude of other species adaptability. The human population, while capable of massive system intervention and modification is, in the end, tied to other organisms as a part of ecological systems. Accordingly, species persistence may have direct bearing on human prospects.

Because of these important relationships, biodiversity has been proposed as the key index for measuring the sustainability of human endeavors. Concern for extinction in particular, has become a critical element in checking development in the U.S. as well as in other countries. The U.S. Endangered Species Act of 1973 lays out strict regulations for the protection of species that become threatened by development or other causes. In fact, some argue that the Endangered Species Act places a nearly absolute value on the preservation of individual species once they are declared endangered. Evidence for this valuation can be found in recent actions taken to protect several species, including the Northern Spotted Owl in the Pacific Northwest. The human cost of protecting owl habitat on commercial timber lands has been very high.

However, the process of monitoring and protecting individual species that are on the brink of extinction may have only limited effectiveness. Because there is often a lag between the loss of important habitat and eventual species endangerment, the ESA may signal a problem only after it is insurmountable (21). Furthermore, a general demise of an ecosystem's function and health, for example, severe forest decline observed in the Blue Mountains of Oregon, may not be signaled by an endangered species. In such cases, the consequences of changes may be very severe at a local scale while not immediately threatening the persistence of species at a larger scale.

In addition, species persistence can only be monitored for those species that have been identified. While a large share of the vertebrates and vascular plants have been cataloged, these groups make up a very small share of the world's biota, and the proportion of species actually cataloged is placed somewhere between 2 and 15 percent (22). Accordingly, the extinction of many species may simply go unnoticed.

An index of biodiversity, based on the number of species that occur in an area, can be an essentially arbitrary and even misleading indicator of ecological function when temporal species dynamics are considered (23). That is, species may flurish or decline depending on the stage of a forest's development. Furthermore, common biodiversity indices are not weighted to place emphasis on critical or scarce species. Instead all species are treated equally. At present the information required to understand the role of species and therein identify critical species is not known.

One way of summarizing the differences between ecological and economic perspectives on sustainability then is the relative emphasis placed on ecological health and human welfare. Economics views sustainable development in general as an issue of intergenerational fairness. That is, it is a matter of arranging natural resource use and investment so that future living standards are protected (24). If markets work well and therefore prices reflect the true quality and costs of the resources we use, then sustainability can readily be achieved in a market economy. Substitutions and technical progress will steer humanity away from resource famine, and investment will transfer the necessary wealth to posterity. Of course, markets don't always work well and the major role of sustainability policy is to correct for these problems and to adjust prices so that they reflect true scarcities.

In countries where markets are not the primary mechanism for resource allocation or where markets are not mature, the issues are more basic. In such cases, sustainability policy would still aim to get prices right, but creating and stabilizing the market foundations must necessarily come first. Chief among these foundations is secure ownership of property, including land. Without a reasonable assurance of long-term tenure, there can be no incentive for long-term resource management. As a result the dual effects of over-exploitation and under-investment short change future generations. Regardless of the setting, the economic version of sustainability goals is establishing the setting in which well-informed decisions about resource use can be made.

The ecological perspective can be summarized as a concern for the maintenance of ecological systems. Its tacit assumptions are that human progress has steadily eroded the diversity of life and the integrity of ecosystem processes and that the implication of this erosion is a potentially catastrophic eventual shift in the structure of ecosystems at local and global scales. While admittedly uncertain, the implication of a rapidly unravelling system of fundamental energy transfer and resource production is a greatly diminished human prospect.

The ecological perspective on sustainability takes issue with the economic perspective at two basic levels. One is that the erosion of biological diversity (biological capital) has progressed at an increasingly rapid rate over the last century and that markets have not recognized the phenomenon, mainly because there is no linear connection between this scarcity and human welfare. Rather, the connection is subtle and ultimately cumulative. If the effects of an erosion of biodiversity are irreversible, and felt only after several decades, then we cannot rely on markets to account for these values.

The economist might reply that the market mechanism should not be abandoned because of incomplete information. Instead, resource prices should be adjusted to reflect these important values. However, it might be countered that, in fact, the market mechanism is fundamentally flawed for protecting ecological functions. Due to the spatial nature of important ecological processes, individual landowners and resource managers cannot effect sustainable forest ecosystems while pursuing independent ends. As discussed above, this suggests that ecosystem maintenance in the presence of resource management may require an orchestration of management efforts across owners.

At the same time, the ecological perspective is generally not set within a social context. Protecting ecosystem health and ecological functions is only one of several important social goals and cannot be accomplished without cost. Without comparing the costs of achieving these goals with the derived benefits as well as with other social objectives, calls to protect ecosystems imply either arbitrary valuation or an absolute value for ecosystem health. One implication is that ecosystems hold intrinsic value that overwhelms all other social concerns.

However, the outcome of undisclosed value comparisons does not define the motivations behind them. There are at least two ways to interpret values in this context. One is what Toman (15) calls "ecological centrisism," where nature is accorded values and rights. The other is that an undisclosed tradeoff analysis places high priority on ecological functions. It may be that the consequences of ecological failure are so severe that they overwhelm all foreseeable costs of ecological protection. This latter perspective might be labeled ecological utilitarianism.

Are there ways to reconcile these disparate views on the need for and the structure of sustainability goals? Perhaps there is, if one takes the perspective of ecological utilitarianism, rather than ecological centrisism. With the latter, the prospects for consensus are limited simply because a conflict with absolute values cannot be resolved. However, if the perspective is utilitarian, then differences in opinions regarding sustainability goals are the result of different perspectives on physical outcomes and risk. These differences can be ameliorated over time with better information on outcomes. That is, science can play a strong role in the latter situation, but not the former.

Taking this perspective allows us to reconcile these perspectives in a hierarchial approach to defining sustainability goals. At one level, the goals of ecological sustainability need to be prioritized with other important social goals such as improving education, fighting crime, and reducing poverty. At another level, science needs to define the connections between "states of nature" and different levels of social welfare. That is, science can inform the process of social goal setting with information on feasible states of nature and their anticipated outcomes. Clearly, this area of knowledge is fraught with uncertainty, so that relationships between nature and welfare need to be couched in the language of risk analysis. Communicating the degree of uncertainty is as important as communicating the state of knowledge.

The process of sorting out sustainability goals in a social context begins with defining the relative values social systems place on natural systems as well as an understanding (albeit necessarily limited) of the connections between natural productivity and human prosperity. This stage of the process seeks an understanding of how much society is willing to pay for ecological health, in direct terms of forgone consumption and in terms of transfers of natural wealth to future generations.

Ultimately this stage of the process can only be conducted as a political exercise where the discourse is played out in the context of an economy and other social concerns. Clearly the context is crucial and the ranking of ecological goals will necessarily differ between countries. For example, France, Pakistan, Ukraine, and Brazil will likely place different relative values on alleviating poverty, improving transportation infrastructure, and ecological sustainability.

EFFECTING SUSTAINABLE FORESTRY

The previous section focused on formulating and articulating sustainable forestry goals. This section focuses on the subsequent issue of designing plausible approaches to moving the practice of forestry towards these goals. The following discusses three aspects of implementing action to achieve sustainability. One is the collection of physical management strategies that might be applied. These necessarily depend on the scale of application and on site specific features of the forest. Another aspect is the influence of human institutions on sustainability. The practicality of any management strategy depends crucially on the institutional setting within which it is applied. For example, the institutions that define land ownership and taxation can strongly influence resource management decisions. The final aspect of implementing sustainable forestry is a program of research that addresses ecological uncertainty and monitors the effectiveness of different forest management strategies.

Management Approaches

Regardless of their form, goals for sustainable forestry can only be implemented through modifications in the practice of forestry. Forest management will need to address two kinds of questions. One is where will forestry be practiced and, conversely, where will natural processes be left to operate without significant human intervention. The other is how will forestry (the practice of harvesting and regenerating trees) be practiced.

Allocating forest lands to various management emphases has long been a part of public forestry in the United States. While, historically, public forestry has emphasized producing the largest number of multiple use services from each forest stand, it is clear that not all of these services are complimentary. Accordingly, in most cases multiple use can only be accomplished by zoning forest lands to different uses (25). For example, different areas may be zoned to emphasize timber, forage, wildlife, or recreation values.

Another important forest land allocation mechanism in the United States is wilderness designation. Since passage of the Wilderness Act in 1973 (and the subsequent Eastern Wilderness Act of 1975), the setting aside of large blocks of large, essentially undisturbed, natural areas has been a major focus of conservation efforts. The Acts set aside several wilderness areas but also set out a mechanism for adding to this initial constellation. The process requires Congressional action to designate an area as wilderness and fully engages public debate over land use and wilderness values.

While the processes of forest zoning on National Forests and Wilderness allocation have long histories, the process has been motivated by a set of goals that are distinct from ecological sustainability. As discussed above, there are important dissimilarities between a multiple-use agenda and one that focuses on biodiversity and ecological processes. In addition, wilderness designation has been motivated not by relative ecologic merits but primarily by the lack of human disturbance or presence and the potential for wilderness recreation. Accordingly, the present system of wilderness areas are generally remote and mountainous. These physical barriers to access are, in fact, what prevented their development.

The present collection of preserved areas in the U.S., wilderness areas along with national parks and wildlife preserves, does not fully represent the spectrum of ecological conditions. Scarce and therefore important ecological conditions are significantly missing from this network of preserves. These and other problems of designing reserve systems to effectively protect diversity and ecological function have been the focus of the relatively new field of conservation biology (26).

Among these concerns is how to define the minimum size of an effective reserve. In effect, there are certain economies of scale associated with ecosystems. Species persistence relates not only to the quantity of available habitat but to a complement of different habitats and their interconnections. In addition, the shape of the reserve is critical because of the phenomenon known as edge effect (27). At the interface between a forest and another kind of land cover, for example, pasture, environmental conditions are different from conditions in the interior of the forest, and therefore support a different complement of species. Therefore, the shape and size of the reserve interact to define the relative extent of edge and interior habitat conditions. Relatively large reserve areas may be shaped in ways that cause disturbance-related edge habitat to dominate.

Edge effects and the importance of connections between various habitats suggests that the condition of lands immediately adjacent to reserves can play a crucial role in their function. Because the condition of adjacent areas defines edge effects as well as potential barriers or corridors, for example, for seed dispersal and animal migration, their management is a critical element in defining the effectiveness of reserves. Coordinating management on adjacent lands is a challenge especially in a multiple owner landscape.

The phenomenon of spatial juxtaposition raises a complex set of issues for sustainable forestry. Ecological functions are rarely determined by conditions found within a small area. Rather they are connected across the landscape among areas. One example of these spatial connections is a large mammal that requires foraging habitat as well as hiding and resting cover. Another is the gravitational definition of water courses that connects upstream activities to downstream impacts.

These spatial connections define an "economy of configuration" (28) in addition to an economy of scale in the function of ecological reserves. That is, the ability of reserves to protect biodiversity or ecological function depends crucially on where they are located in absolute terms and in reference to other landscape conditions. In addition, these conditions are dynamic—forests may be harvested and then regrown—so that controlling for ecological function may become very complex.

One element of the problem is simply defining where high-valued (ie, scarce) ecological conditions exist. One way of prioritizing areas is to order ecosystems by their species diversity. An example of an ecosystem type with an exceptionally high diversity of species is wetlands, especially riparian corridors. These zones of gradation between water courses and upland habitats are extremely dynamic and diverse (29). They define portions of the landscape that, because of their flood-related disturbance regimes, natural variability, and connections across land forms, serve as species reservoirs in their respective ecosystems. Accordingly they may deserve special management emphasis.

At a larger scale, areas of ecological importance can be identified by mapping the distribution of endangered species. Areas that have relatively large concentrations of threatened and endangered species may indicate where ecological functions are imperilled. Accordingly, the ecological returns to preserves may be highest in these areas.

In general, however, ecology has yet to provide a comprehensive scheme for classifying ecosystems at any scale (30). Instead the focus of classification and systematics has been on the identification and classification of individual species. Without a classification of ecosystem types, however, it is difficult to judge indeed where ecological conditions are scarce or threatened.

While one aspect of a physical strategy for sustainable forestry focuses on where to undertake forestry, the other focuses on how to manage forests. The application of silviculture ultimately defines the condition of forests under management. Because managed forests represent the largest share of forested lands, at least in the temperate zones, the practice of silviculture plays a critical role in defining the ecological condition and biodiversity provided by the entire landscape.

Managed forest lands differ from unmanaged lands in several ways. In general, however, forest management tends to simplify the spatial distribution and the internal structure of forest stands (31). This reduction in complexity results from "streamlining" stands to efficiently produce high yields of commercial timber products.

At temperate latitudes, the typical transition to a managed forest involves harvesting a mixed-species and mixed-age forest and replacing it with a single, fast growing species. This species is generally adapted to early seral conditions and quickly occupies the site and grows to maturity, in a financial rather than a biological sense. Timber harvesting is then associated with a regeneration method that establishes a new forest. The regeneration method is the crucial activity defining the structure of the subsequent forest.

Timber harvesting combined with regeneration methods defines what Smith (32) calls a silvicultural system. These systems lead to either an even-aged or an uneven-aged forest stand. Even aged stands can be established by cutting the entire stand (clearcutting) and then either planting seedlings or spreading seed across the site. Two other approaches allow for natural regeneration in connection with a harvest. One leaves a few scattered trees to naturally seed the site (seed-tree methods). The other removes trees in two or three stages so that natural regeneration can take place under the shelter of partial forest cover (shelterwood methods). An uneven-aged forest can be maintained by constantly removing a small proportion of the forest in a way that promotes regeneration (33). Another type of forest stand is generated by vegetative reproduction or coppice regeneration. These forests are regrown from root stock and are generally used to produce fuelwood or other forms of biomass fuel stocks.

The choice of silvicultural methods, therefore, has direct implications for the age and species composition of forest stands. Even-aged approaches yield forest stands of the same age and species. However, and depending on the growth period and the amount of effort expended on controlling vegetation, an understory of competing tree species will also emerge. These are the natural successors of the early seral species emphasized by even-aged management. In contrast, uneven-aged management through selection silvicultural systems, generally yields a highly diverse stand in terms of both species and age composition.

The majority of forest management utilizes even-aged silvilculture. The result in some regions is a great reduction in the species composition of managed forests. Furthermore, the genetic diversity of the regenerated species may often be quite limited. This is because tree nurseries and seed orchards breed trees to emphasize certain productivities and disease resistance. Quite often this can result in a small number of superior trees parenting vast areas of forest. As a result, the adaptive amplitude of a commercial species can be greatly reduced from its natural range.

Because commercial forests are harvested much earlier than natural disturbance would ordinarily replace forests, the average age of managed forests is much lower than typically found in a natural setting. This obviously skews habitat conditions to favor species complexes in early seral stages and away from later seral stages. The endangerment of the Red-Cockaded woodpecker and Northern spotted Owl for example, reflect substantial reductions of old-forest habitats in the southern and Pacific Northwestern parts of the United States, respectively. Preserving what remains of old-growth conditions has been a focus of efforts to preserve regional biodiversity.

Active forest management is not, however, the only major vector of change in forested systems. In addition to forestry, other uses of land may have both direct and indirect effects. For example, in some rural areas—notably the Southern Appalachians—low density residential development continues to spread from central urban areas and across large areas of forest. Agricultural demands for land have also played an historically important role in modifying the landscape and clearing forests for crop and pasture uses continues to be behind the majority of global deforestation activities.

These activities reduce habitat by removing forests. However, they also have indirect impacts on remaining forests. One effect is that the remaining forest patches are smaller and increasingly isolated from one another. In addition, the interspersion of land uses also increases the occurrence of edge habitat within forest patches. This may completely change the character of remaining forests in a developed area. Accordingly, measuring the amount of forest disturbance may greatly understate the area of forest that is actually impacted by development.

Another vector of change in forest systems is human effects on natural disturbance regimes. The most effective example of this is the campaign against forest fires waged over the last century. This campaign has been enormously successful and has greatly reduced the area of forest burned by natural and human-caused fires (34). Protecting large areas of forests in the temperate zones from fire has also resulted in eliminating the chief natural agent of material cycling in forests. As a result, the composition of forests has shifted away from fire adapted species. In addition, some larger-scale effects have also occurred. Among these are an increased severity and extent of insect epidemics in the U.S. (eg, southern pine beetles and mountain pine beetles, gypsy moth, etc). Also, because fuel loads in forest stands are no longer regulated by frequent and low intensity natural fires, fuel builds up in forests over time and eventually promotes large, high intensity fires (35). The fires in and around Yellowstone National Park during the summer of 1989 is one example of this metaphenomenon.

All of these factors change and, in one way or another, decrease the complexity of a forested landscape. Accordingly, the choice of silvicultural systems and forest management strategies can have a great impact on the ecological function and biodiversity of forests. There are ways to balance the production of valuable services and products from forests while maintaining more of their ecological integrity.

A critical factor in determining ecosystem level sustainability is not just the quantity of certain kinds of habitat but their spatial juxtaposition. While not a traditional focus of forest management, spatial arrangement of harvest units is increasingly emphasized in public forestry (36). Careful selection of harvest patches through time allows important habitat connections to be maintained while providing a continuous flow of timber products (37,38). In addition, the size of cutting units is variable and can be adjusted to an appropriate scale for a given ecosystem type and set of objectives. For example, smaller units spread disturbance more widely across a landscape. However, depending on the intensity of harvesting, smaller cutting units may also lead to smaller forest patches, greatly increasing edge effects.

Streams and riparian corridors define especially important connectors in a landscape (29), and the effects of forestry activities, such as sedimentation, water flow, and debris in streams, may have critical impacts on watershed integrity (29). Because of their central role in ecological health, riparian corridors and wetlands have been a focus of regulation. Some argue that maintaining ecosystem health will require even more careful management of these critical landscape elements.

The techniques of harvesting and regeneration (ie, silvicultural systems) within cutting units may have the greatest impact on the species and age structure of forest patches. The key to managing for these conditions depends on understanding the dynamic processes of regeneration, growth, and species succession particular to individual ecosystem types. New silvicultural systems may be designed to mimic natural disturbance regimes (39) thereby maintaining more ecological structure and biodiversity.

Shifting from even-aged to uneven-aged silvicultural systems is a direct means of increasing species diversity within a forest patch. However, this less intensive harvest approach requires accessing forest stands much more frequently and accordingly, increasing a different kind of human intervention. The more frequent presence of people and machinery may render areas ineffective as habitat for certain animal species, especially mammals. In effect, while less intensive, uneven-aged management spreads annual disturbance regimes out across a much larger area.

Other choices in the design of forest management strategies may also hold crucial influence over the structure and function of large forested landscapes. For example, techniques used to prepare a site for regeneration will have a critical influence on the nutrients left on a site following harvest. Rather than clear harvested sites—often an aesthetically desirable approach—more stems and other waste can be left to decompose on the site. Decisions regarding site preparation need to be balanced against resulting fuel loads and fire hazards as well, but they afford a means to increasing forest material cycling and could play a role in protecting and extending soil productivity, a critical element of sustainable forestry.

Certain engineering aspects of forest management plans may also have a critical role in defining the composition and health of regenerated forests. These include decisions regarding the type and the location of forest roads. Road design is perhaps the most important variable in determining stream sedimentation (40) and roads can serve as corridors for the spread of exotic weed species. It may be possible to apply low-impact temporary roads where highly engineered permanent roads have been constructed in the past. In addition, the equipment used to harvest trees, move logs to loading areas, and then transport logs to market will also have residual impacts of ecosystems through, for example, soil compaction. Alternative harvesting machinery and skidders may be applied to mitigate these impacts where important.

Institutions

The preceding discussion emphasizes a combination of site and system level management approaches to sustainable forestry. Implementing these types of approaches may be feasible in the setting of public land management where large areas are under a common management agency, though this may not be as straightforward as it appears. However, system-level planning is extraordinarily difficult in the mixed private ownership setting of most forest areas. The institutional setting of forest management, including land ownership and regulations, defines the crucial context for designing forest policies to accomplish sustainable forestry.

An institution is any rule or organization which governs the behavior of humans. In the context of forest management, relevant institutions include the structures of land ownership (public and private), the size and distribution of land holdings, and policy instruments, such as forest management laws and property taxes, which influence land management decisions. Because human behavior, expressed through land use decisions, is the dominant cause of landscape change, institutions define crucial control mechanisms for achieving goals otherwise unaddressed by human activities.

There are two broad institutional contexts for land management in the United States. One is private ownership with extensive property rights held by autonomous landowners. Under the right conditions, enlightened self-interest should guide landowners to allocate land to highest-valued uses and, in the process, to effectively produce marketable goods. However, there are goods and services which do not transact in markets but which may be of considerable value to society, including the ecological services discussed above. The production of nonmarket goods is a primary rationale for public ownership of forest land, the other major institutional context of land management (11). In theory, public land management aims to provide all important goods and services including those left unaddressed by markets.

Of course, public ownership is not the only mechanism for providing nonmarket goods. The actions of individual land owners might be directed towards producing other benefits by altering their incentives and selectively restricting property rights. An example of the former would be a severance tax on forest products, and an example of the latter would be forest practices regulations. These types of policy tools have a cost side, including the costs of administration, but also the cost of foregone market benefits. Balancing these regulatory costs against public benefits is a critical part of policy design. However, using these types of policy tools to address economies of configuration may be very costly.

The costs of public land management and of central planning in general may also be high, and this has been highlighted by recent international events. The private or market model of resource allocation seems to have enjoyed near global vindication and a great run of international favor. The fall of Soviet-style socialism, the Reagan Revolution, and the Sagebrush Rebellion all involve passing the control of resources from the public to the private sector. Even environmental advocates have adopted market arguments for regulating air quality with tradeable pollution permits and managing wilderness through private vendors, in effect, creating markets for environmental goods.

The collapse of centrally planned economies and the development of environmental markets reflect, on the one hand, the cumbersome and costly approach of central planning, and on the other, the efficiency of market allocation through competition. Where markets are well-structured or where they can be effectively created, there seems to be no more efficient mechanism for allocating resources. However, not all problems can be solved with market tools and economies of configuration seem to define problems which may not be shaped into a well-structured market. There seems to be little hope of organizing the actions of neighboring landowners through regulated competition alone. Rather, organizing landscape patterns and spatial processes seems to call for directly orchestrating the decisions of landowners.

Standard policy tools are applied without spatial discrimination across land owners and landscapes. They apply to all lands. However, in many cases critical habitat or corridors may lie on only a small portion of the relevant landscape. In effect, the policy may regulate and therefore penalize land management on all acres to influence production on a few, imposing considerable costs. Furthermore, it is unclear that these policy tools could be designed to influence these types of issues at all. A severance tax might, for example, encourage forest owners to grow trees longer so that, on average, more old-forest habitat is available but if harvesting occurs in large blocks then this may not address the configuration of habitats at the right scale.

One approach to directly shaping landscapes might call for land management zoning. Of course, severe land use regulations amount to de facto public ownership. That is, regulation can remove nearly all of the relevant property rights, often without compensation. In this kind of regulatory environment uncertainty might be high, encouraging accelerated resource extraction while policies and regulations are being developed. It may well be better to condemn and purchase the critical properties outright. As far as orchestrated action is concerned, public control over designated critical habitat seems to have an advantage over other regulatory mechanisms because it allows specific important areas to be targeted.

Sustainability seems to require an approach which can orchestrate land conditions across landscapes. The policy question, as usual, is whether the medicine is worse than the illness. While some type of land management coordination seems called for, central planning of any sort has its own kinds of costs, related mainly to the burdens of collecting and processing massive amounts of information and enforcing complicated plans. On the other hand, marginal regulation may be costly in two ways. It may be ineffective for targeting highly important habitats and it imposes costs on the management of all lands in the process. Marginal regulation may be more benign in an ideological sense because it works through markets but at the same time it may be very expensive.

At present, efforts to implement ecologically sustainable forestry have focused on the public lands, especially the national forests (41). These entities have a comparative advantage in this regard because they control large, though not necessarily contiguous, areas of forested land. Even so, managing for ecological conditions will require expanding the focus of national forest management to consider the context of neighboring forest lands and connections at a larger scale (42). In effect, public lands become control mechanisms in a larger dynamic landscape.

Shifting management of the national forests to focus on ecological health may be a cost-effective approach to sustainable forestry. Allowing public lands to specialize in this area, might reduce the need to impose onerous regulation on private forests which efficiently produce commercial timber products.

However, this kind of specialization (ecological services on public lands, commercial products on private lands) raises a crucial question for sustainability. Is the present arrangement of public lands, defined in another time with different objectives, sufficient to achieve sustainability goals? If not, then what lands should be public and what lands should be private? An inventory of critical habitat would be a first step towards evaluating the potential impacts of this kind of approach.

The preceding discussion has focused on difficult system-level approaches to sustainable forestry. At the level of forest management practices implementation may be much more straightforward. However, these approaches necessarily address different aspects of sustainable forestry. There is a long history of requiring "Best Management Practices" (BMP's) and other regulatory stipulations for the practice of forestry (43). These regulations have been applied at state and local levels and address everything from adequate forest regeneration to environmental protection.

BMP's could obviously be extended to provide for certain kinds of management in special areas. For example, they could, as they have in the past, target wetlands and riparian corridors. In addition, they could be targeted to other scarce and important types of habitat. These types of regulatory tools hold the same kinds of enforcement costs, but not the same extent as the management orchestration discussed above. Rather, enforcement can easily be implemented through, for example, a harvest permitting process.

Research

The third element of implementing sustainable forestry is a research program aimed at generating new ecological and economic insights into forest management. In addition, research will need to focus on monitoring the impacts of forest management efforts. The preceding discussion mentioned several areas where knowledge deficits exist. These include the following:

1. Classification of ecosystem types and definition of relatively scarce types.
2. Understanding the relative valuation of ecological systems.
3. Establishing the relationships between ecological conditions and human welfare.
4. Defining the effects of various silvicultural practices on forest stand dynamics and resulting biodiversity.
5. Designing institutional structures to affect sustainable forestry in a cost effective manner.

BIBLIOGRAPHY

1. A. Leopold, *A Sand County Almanac*, Oxford University Press, New York, 1949, 226 pp.
2. J. Lubchenco and co-workers, *Ecology* **72**(2), 371–412 (1991).
3. M. A. Harwell and co-workers, "Ecological Sustainability and Human Institutions: Case Studies of Three Biosphere Reserves," *Research Proposal submitted to the U.S. Man and the Biosphere Program*, May 1991.
4. R. G. Lee and co-workers, "Land Use Patterns in the Olympic and Southern Appalachian Biosphere Reserves: Implications for Long-Term Sustainable Development and Environmental Vitality," *Research Proposal submitted to U.S. Man and the Biosphere Program*, May 15, 1990.
5. T. R. Cox, R. S. Maxwell, P. D. Thomas, and J. J. Malone, *This Well-Wooded Land: Americans and Their Forests from Colonial Times to the Present*, University of Nebraska Press, Lincoln, 1985, 325 p.
6. Ref. 5, pp. 111–132.
7. G. Pinchot, *Breaking New Ground*, reprinted by Island Press, Washington, D.C., 1947, pp. 79–86.
8. Ref. 7, pp. 147–153.
9. A. B. Recknagel and J. Bentley, *Forest Management*. John Wiley & Sons, Inc., London, 1919, 269 pp.
10. Ref. 9, p. 124.
11. J. V. Krutilla and J. A. Haigh, *Environmental Law* **8,** 373–415 (1978).
12. A. Downs, *Inside Bureaucracy*, Little, Brown, and Company, Boston, 1967, 292 pp.
13. C. F. Wilkenson and H. M. Anderson, *Land and Resource Planning in the National Forests*, Island Press, Washington, D.C., 1987, 396 p.
14. Technology Review, "The Road from Rio: An interview with Gro Harlem Brundtland," *MIT Technology Review* **96**(3), 60–65 (1993).
15. M. A. Towman, *Resources* 3–6, (Winter 1992).
16. C. H. Schallou and R. M. Alston, *Environmental Law* **17**(3), 429–482 (1987).
17. D. N. Wear and W. F. Hyde, *J. of Business Administration* **21,** 297–314 (1992).
18. P. R. Ehrlich and A. H. Ehrlich, "Humanity at the crossroads" in H. E. Daly, ed., *Economics, Ecology, and Ethics: Essays Toward a Steady-State Economy*, W. H. Freeman and Co., New York, 1973, pp. 38–43.
19. H. E. Daly, *J. of Pol. Econ.* **76**(3), 392–406 (1968).
20. J. F. Franklin, "The fundamentals of ecosystem management with applications in the Pacific Northwest," in G. H. Aplet, ed., *Defining Sustainable Forestry*, Island Press, 1993, pp. 127–144.
21. D. J. Rohlf, *Conservation Biology* **5**(3), 273–282 (1991).
22. P. H. Raven and E. O. Wilson, *Science* **258,** 1099–1100 (Nov. 13, 1992).
23. J. F. Franklin, *Ecol. Applications* **3**(2), 202–205 (1993).
24. R. Solow, *An Almost Practical Step Toward Sustainability*, Resources for the Future, 1992, 22 pp.
25. S. E. Daniels, *Environ. Law* **17**(3), 483–506 (1987).
26. J. E. Rodiek and E. G. Bolen, eds., *Wildlife and Habitats in Maanaged Landscapes*, Island Press, Washington, D.C., 1991, 220 pp.
27. D. A. Saunders, R. J. Hobbs, and C. R. Margules, *Conserv. Biol.* **5**(1), 18–32 (1991).
28. D. N. Wear, "Forest Management, Institutions, and Ecological Sustainability," *Proceedings of the Appalachian Society of American Foresters Meeting*, Asheville, N.C., 1992, pp. 12–16.
29. R. J. Naiman, H. DeCamps, and M. Pollock, *Ecol. Applications* **3**(2), 209–212 (1993).
30. G. H. Orians, *Ecol. Applications* **3**(2), 209–212 (1993).
31. A. J. Hansen, T. A. Spies, F. J. Swamson, and J. L. Ohmann, *BioScience* **41**(6), 382–392 (1991).
32. D. M. Smith, *The Practice of Silviculture*, John Wiley & Sons, Inc., New York, 1962, 578 pp.
33. Ref. 31, pp. 353–358 for extensive discussion of these techniques.
34. S. F. Arno, *J. Forestry* **78**(8), 460–465 (1980).

35. M. A. Marsden, "Modeling the Effect of Wildfire Frequency on Forest Structure and Succession in the Northern Rocky Mountains," *J. Environ. Manag.* **16,** 45–62 (1983).
36. J. P. Roise, *Forest Science* **36,** 487–501 (1990).
37. J. G. Hof and L. A. Joyce, *Forest Science* **38,** 489–508 (1992).
38. J. Sessions, *Forest Science* **38**(1), 203–207 (1992).
39. J. F. Franklin and R. T. T. Forman, *Landscape Ecology* **1,** 5–18 (1987).
40. W. F. Megahan and W. J. Kidd, *J. Forestry* **70,** 136–141 (1972).
41. D. J. Brooks and G. E. Grant, *J. Forestry* **90**(1), 25–28 (1992).
42. S. K. Swallow, and D. N. Wear, *J. Environ. Econ. Manag.* **25**(2), 103–120 (1993).
43. F. W. Cubbage, J. O'Laughlin, and C. S. Bullock III, *Forest Resource Policy*, John Wiley & Sons, New York, 1993, 562 pp.

Reading List

R. W. Fri, "Sustainable Development: Can We Put These Principals into Practice?" *J. Forestry* **89**(7), 24–26 (1991).

R. P. Gale and S. M. Cordray, "What Should Forests Sustain? Eight Answers." *J. Forestry* **89**(5), 31–36 (1991).

R. J. Kopp, "The Role of Natural Assets in Economic Development," *Resources* 7–10 (Winter 1992).

R. G. Lee, R. Flamm, M. G. Turner, C. Bledsoe, P. Chandler, D. De Faerrari, R. Gottfried, R. J. Naiman, N. Schumaker, and D. Wear, "Integrating Sustainable Development and Environmental Vitality," in R. J. Naiman, ed., *Watershed Management: Balancing Sustainability and Environmental Change*, Springer-Verlag, New York, 1992, pp. 497–518.

R. G. Lee, "Ecologically Effective Social Organization as a Requirement for Sustaining Watershed Ecosystems," in R. J. Naiman, ed., *Watershed Management: Balancing Sustainability and Environmental Change*, Springer-Verlag, New York, 1992, pp. 73–90.

D. Ludwig, R. Hilborn, and C. Walters, "Uncertainty, Resource Exploitation, and Conservation: Lessons from History," *Ecolog. Applications* **3**(4), 547–549 (1993).

G. H. Orians, "Ecological Concepts of Sustainability," *Environment* **32**(9), 10–39 (1990).

J. R. Probst and T. R. Crow, "Integrating Biological Diversity and Resource Management," *J. Forestry* **89**(2), 12–17 (1991).

G. C. Ray and J. F. Grassle, "Marine Biological Diversity." *BioScience* **41**(7), 453–457 (1991).

H. Salwasser, "Sustainability as a Conservation Paradigm." *Conserv. Biol.* **4**(3), 213–216 (1990).

M. G. Turner, V. H. Dale, and R. H. Gardner, "Predicting Across Scales: Theory Development and Testing," *Landscape Ecology* **3**(3/4), 245–252 (1989).

FUEL CELLS

JOHN APPLEBY
Texas A&M University
College Station, Texas

DIRECT ENERGY CONVERSION

Fuel cells (1) are devices which directly convert the Gibbs energy ("free energy") of fuel oxidation into work in the form of direct current electricity. They are devices operating at constant temperature with a controlled combination of reactants, unlike the uncontrolled process which takes place when the fuel is burned to produce heat. Work is a directed flow of energy in the form of particles in motion, whereas heat is a random movement of particles in all directions.

A number of different devices are said to offer direct energy conversion. These include thermoelectric and thermionic converters and photovoltaic cells. The first two convert heat, and the third photons, into direct current electricity. Thermoelectric and thermionic converters are not strictly direct energy conversion devices. They are simply thermal engines which convert heat into electrical work using no moving parts, compared with devices which first produce mechanical work using moving parts, followed by electricity using a rotating or linear generator.

A photovoltaic cell accepts Gibbs energy in the form of directed beams of photons and converts it to electrical work in the form of a directed stream of electrons under isothermal conditions. It consumes no heat which is converted into work. A fuel cell, a specialized form of a battery, directly converts the Gibbs energy of a chemical reaction into a directed stream of electrons, again under isothermal conditions. Again, it consumes no heat which is converted into work.

See also BATTERIES; HYBRID VEHICLES; ELECTRIC POWER.

THERMODYNAMICS OF ENERGY CONVERSION

The first law of thermodynamics states that energy is conserved in a system of constant mass. If the energy of the system changes, it must do so by a combination of exchanging heat with its surroundings and performing work on its surroundings or having work performed on it. If dE is an infinitesimally small change in internal energy, q is the amount of heat absorbed by the system, and w is the amount of work performed by the system. If we consider systems at constant pressure P, then $w = P\,dV$, where dV is the change in volume of the system, doing work by moving against the pressure. Thus, from the first law,

$$dE = q - P\,dV \tag{1}$$

The second law of thermodynamics connects the change in the amount of heat and the absolute temperature of the system (T in kelvins, K) by the change in state of disorder or entropy, dS, such that $dS = q/T$. Thus, combining the first and second laws,

$$dE = T\,dS - P\,dV \tag{2}$$

If we define the heat content of the system, H, as the sum of the energy E and the energy required to occupy the volume of the system at constant pressure, PV, then

$$dH = dE + P\,dV + V\,dP = q + V\,dP \tag{3}$$

Thus, for a system at constant pressure, $dH = q$. The "bound" energy of the system at constant pressure, the part associated with its finite temperature, is equal to $T\,dS$. Thus the change in "free" energy at constant pressure, which may be converted into work, is

$$dG = dH - T\,dS = q - T\,dS \tag{4}$$

We can change the infinitesimal changes to gross changes at constant pressure and temperature by integration, writing Δ instead of d. Thus

$$\Delta G = \Delta H - T\,\Delta S = Q - T\,\Delta S \tag{5}$$

where Q is the gross change in heat absorbed by the system. In any system at constant temperature and pressure, a change in ΔH, or a change in Q, can produce a maximum amount of useful work equal to ΔG, the Gibbs energy of reaction.

When a constant-pressure system consists of chemical equivalents of two substances which interact in an exothermic chemical reaction, the system acquires a quantity of energy equal to the heat or enthalpy of reaction per equivalent, ΔH. For exothermic reactions, ΔH is a negative quantity representing the loss of energy from the interacting atoms. If we assume that the process is spontaneous, so that all of the energy appears as heat and none as work, i.e., the system remains at constant volume V_1, then the temperature of the system will be raised from its initial value T_1 to T_2, and its pressure from P_1 to P_2. If the system is then allowed to expand isothermally to pressure P_1 and volume V_2, it will do work on its surroundings. It is then allowed to cool down at constant volume to its original temperature T_1, rejecting heat at that temperature to its surroundings. It will then be at a pressure P_3, where $P_3 < P_1$, Following this, it contracts to its original volume V_1 and pressure P_1 at constant temperature T_1. During this part of the cycle, the surroundings do work on the system. The net work w performed by the system is given by the heat in at T_2 minus the heat rejected at T_1, i.e., the difference in area between the isotherms at T_2 and T_1, respectively. These are given by $\int V\,dP$ integrated between the appropriate limits. From the ideal-gas law, for all states, $PV = nRT$, where n is the number of equivalents per mole and R is the gas constant. Thus, w and q' are given by the expressions

$$w = nR(T_2 - T_1)\ln(V_2/V_1) \qquad q' = nR(T_2)\ln(V_2/V_1) \tag{6}$$

Thus ε, the theoretical thermal efficiency for converting heat into work for this particular cycle, is given by

$$\varepsilon = (T_2 - T_1)/T_2 \tag{7}$$

ie, Carnot's theorem. The cycle considered above is the Stirling cycle (2). Carnot's theorem was devised for a cycle in which the system accepts heat in a compression at constant entropy from T_1 to T_2. It then expands isothermally at T_2, then expands at constant entropy while its temperature changes from T_2 to T_1, during which it rejects heat to the sink. This is followed by isothermal compression and then constant-entropy compression at constant entropy while it receives heat to increase its temperature from T_1 to T_2, thus completing the cycle. In principle, an ideal cycle has two isothermal stages and two at constant volume (the Stirling cycle), with two either at constant entropy (the Carnot cycle) or at constant pressure (the Ericsson cycle). All show the same relationship for the maximum efficiency (2).

For an ideal gaseous chemical fuel and oxidant, we can arrive at the maximum Carnot efficiency as follows. We assume that the combination acts as its own working fluid for one complete thermodynamic cycle. We also assume that the specific heat at constant pressure is independent of temperature in the ideal system; then it follows that ΔH and ΔS are temperature independent. Since no work is done during the initial reaction, the change in ΔG is zero, and from Eq. (5), $T_2\,\Delta S = \Delta H$, where ΔS is the entropy of reaction. At T_1, the Gibbs energy of reaction is

$$\Delta G_{T_1} = \Delta H - T_1\,\Delta S \tag{8}$$

Thus, since $T_2 = \Delta H/\Delta S$ and $T_1 = (\Delta H - \Delta G_{T_1})/\Delta S$, the maximum efficiency of a thermal machine which uses the heat from this chemical process, rejecting it at temperature T_1, is

$$(T_2 - T_1)/T_2 = \Delta G_{T_1}/\Delta H \tag{9}$$

Thus, a thermal engine operating on the heat from a chemical reaction having a heat source at the maximum spontaneous temperature that the system can reach and rejecting energy at a temperature T_1 would have a maximum efficiency equal to $\Delta G_{T_1}/\Delta H$, where ΔG_{T_1} and ΔH are the Gibbs energy and enthalpy of reaction at T_1. Operation of a heat engine using hydrogen as the fuel and oxygen as the oxidant, producing liquid water as the product at 25°C (298 K) would require a source temperature T_2 approaching 2000°C (2273 K). Materials considerations make this impossible, to which must be added the limitations of heat transfer to the working fluid and the nonideality of both the materials used and of the thermal cycle itself. Practical machines, e.g., gas turbines, are restricted to source temperatures of 1250°C (1523 K). In a combined steam cycle application, their sink temperatures are about 40°C (313 K). Thus, such a system has a theoretical maximum efficiency of about 79%. The maximum practical efficiency resulting from the nonideality of the thermodynamic cycle and practical losses is about two-thirds of the theoretical value, or 52%.

FUEL CELLS

Primary batteries contain electrode materials which are consumable. As the materials are oxidized at the anode and reduced at the cathode, the electrode structure degrades and finally collapses, and the battery will cease to operate. A fuel cell is a primary battery with invariant electrode structures and reaction sites, arranged in such a way that an oxidizable material (a fuel) can be fed continuously as required to the anode. Similarly, a reducible material (an oxidant) can be continuously fed to the cathode. Finally, provision is needed to remove the reaction product as it is formed. Thus, a fuel cell is a primary battery which is capable of continued operation as long as an oxidizable fuel and an oxidant are fed to it.

Fuel cells are generally required to operate on common hydrogen-containing fuels. For reasons to be explained, the fuel which is actually electrochemically consumed in high-performance fuel cells is gaseous hydrogen, although there are certain exceptions. Because of its availability, the oxidant is almost invariably molecular oxygen. For terrestrial purposes, this is obtained from air without separation, whereas for use in space cryogenic oxygen is normally used.

FUEL CELL ELECTROLYTES

In a hydrogen–oxygen fuel cell, possible electrode reactions might be

$$2\,H_2 \rightarrow 4\,H^+ + 4\,e^- \quad \text{(anode)} \qquad (10)$$

$$O_2 + 4\,H_2O + 4\,e^- \rightarrow 4\,OH^- \quad \text{(cathode)} \qquad (11)$$

$$2\,H_2 + O_2 \rightarrow 2\,H_2O \quad \text{(overall)} \qquad (12)$$

A fuel cell must perform all work in the external electronic circuit, and hence it must minimize the voltage losses in the ionic circuit. Above all, it must avoid any concentration gradients for ions formed (or consumed) in the electrode processes, since these result in an irreversible change in the effective Gibbs energy of reaction equal to $RT\ \ln(C_1/C_2)$, where C_1/C_2 is the concentration ratio of the H^+ or OH^- ions (3). In electrical terms, this represents an opposing voltage equal to $RT/nF\ \ln(C_1/C_2)$ per mole, or $RT/F\ \ln(C_1/C_2)$ per equivalent, where n is the change on the ion and F is the faraday (96,485 in coulombs per equivalent or in joules per electron-volt, J/eV). To avoid this loss, the fuel cell must use an electrolyte whose majority conductor is the ion produced in the reaction at one electrode and consumed at the other. An electrolyte for a high-power-density aqueous fuel cell must therefore be a concentrated solution containing this ion. A practical aqueous-electrolyte fuel cell operating at economically effective current densities requires a strongly acid or strongly alkaline solution to function well. Under these conditions, the cell reactions will be

$$2\,H_2 \rightarrow 4\,H^+ + 4\,e^- \quad \text{(anode, acid electrolyte)} \qquad (13)$$

$$O_2 + 4\,H^+ + 4\,e^- \rightarrow 2\,H_2O \quad \text{(cathode, acid electrolyte)} \qquad (14)$$

$$2\,H_2 + 4\,OH^- \rightarrow 4\,H_2O + 4e^- \quad \text{(anode, alkaline electrolyte)} \qquad (15)$$

$$O_2 + 2\,H_2O + 4\,e^- \rightarrow 4\,OH^- \quad \text{(cathode, alkaline electrolyte)} \qquad (16)$$

In molten salt electrolytes, the rule that the majority charge carrier should be an ion produced at one electrode and consumed at the other requires the ions to be H^+ (in a molten proton conductor), OH^- (in a molten hydroxide), or O^{2-} (in a molten oxide). The latter would be produced in a simple four-electron charge transfer process in the reduction of molecular oxygen. The only molten salt electrolyte which has found application is a mixture of molten alkali carbonates, usually Li–Na or Li–K. A mixture is used to lower the melting point and to give improved ion transport and other physicochemical properties. In the molten carbonate fuel cell (MCFC), oxide ion (O^{2-}) carries the current from anode to cathode in the form of a carrier ion, CO_3^{2-}. This is formed by supplying CO_2 along with O_2 as a reactant at the cathode:

$$O_2 + 2\,CO_2 + 4\,e^- \rightarrow 2\,CO_3^{2-} \qquad (17)$$

The reaction with hydrogen at the anode is

$$2\,H_2 + 2\,CO_3^{2-} \rightarrow 2\,H_2O + 2\,CO_2 + 4\,e^- \qquad (18)$$

The CO_2 in the anode exit gas stream is then collected and recycled to the cathode gas. In this way, the overall cell process is reaction 12. The use of CO_2 as a depolarizer in this manner eliminates the possibility of concentration gradients.

While no molten oxides are feasible as electrolytes, at sufficiently high temperatures certain ionically doped oxides become effective O^{2-} ion conductors. Thus, they may be used as electrolytes in the solid oxide fuel cell (SOFC) operating on hydrogen-containing fuel and oxygen.

FUEL CELL THERMODYNAMICS

Irreversible Effects

A fuel cell is intended to convert a fuel at a net rate, not produce reversible work at zero rate of working. Under conditions of net reaction rate, the measured values of the cell potential are always less than the reversible value; ie, some of the available Gibbs energy is converted into $T\ \Delta S_{\text{irrev}}$.

FUEL CELL KINETICS

Overpotential

The deviations from the reversible thermodynamic potential values due to the efffect of reaction rate are usually called overpotentials or overvoltages. They are conceptually explained, at least in outline, by the theory of irreversible thermodynamics of Prigogine (4), using the so-called Marcellin-DeDonder equation (5), as applied to electrochemical processes by Van Rysselburgh (6). In short, this states that to induce a net rate, a reaction must be driven by part of the (reversible) Gibbs energy of reaction, known as the (irreversible) reaction affinity ($-\Delta A$). This appears as irreversible (ie, irrecoverable) enthalpy, $T\ \Delta S_{\text{irrev}}$. At least for relatively small displacements from thermodynamic equilibrium, ie, for low net rates of reaction.

CARBON-CONTAINING FUELS

The high reactivity, ie, exchange current, of hydrogen is unfortunately the exception, rather than the rule. Other hydrogen-containing fuels without exception show much lower rates of reaction (7). For the series gaseous methane (eight electrons), liquid methanol (six electrons), gaseous formaldehyde (four electrons), and solid formic acid (two electrons), the enthalpy of combustion per equivalent (HHV) is in the order 1.0, 1.09, 1.26, and 1.21. With allowance for the latent heat of the initial states, it can be seen that the available enthalpy of combustion of successive hydrogens increases as oxidation becomes more extensive. However, the enthalpies of combustion per unit mass fall dramatically in the ratio 1:0.41:0.34:0.11. For this and toxicological reasons, only methanol has been seriously considered as a fuel.

The electrochemical oxidation of methanol and other oxidized carbon species in aqueous media has been extensively studied (7–9). The most effective electrocatalysts are platinum alloys, e.g., those with tin and ruthenium (8,9). Direct methanol cells are still of some interest, but their low cell potential (ie, low thermal efficiency), generally low current density, requirement for high catalyst loading, and the general tendency for unreacted methanol to diffuse through the electrolyte and affect the performance of the oxygen cathode have limited their application.

HYDROGEN–OXYGEN FUEL CELL

The only practical high-performance fuel for use in fuel cells is hydrogen, which shows oxidation behavior on suitable catalysts which is sufficiently rapid to be close to being thermodynamically reversible, even at ambient temperatures. If common fuels, eg, natural gas, other hydrocarbons, or even coal, are to be employed with fuel cells, they must first be converted into hydrogen or hydrogen-rich gases.

At ambient temperatures, the highest rates for electrochemical hydrogen oxidation are on platinum and palladium (10). These catalysts may be supported on high-surface-area carbon blacks (11,12), which may be graphitized for the highest corrosion resistance (13,14). Typical catalyst powders have 10% by weight of platinum on carbons with a specific surface area of about 250 m^2/g. The corresponding specific surface area of the platinum catalyst itself is about 100 m^2/g. A typical loading at the hydrogen anode is 0.25 mg/cm^{-2}, resulting in an effect catalyst area of 250 cm^2 per geometric square centimeter (14).

Platinum catalysts of this type are generally used in aqueous media and in the phosphoric acid fuel cell operating at up to approximately 200°C. As temperature rises, the nature of the electrocatalyst becomes less critical, since reaction rates rise with temperature. The usable specific surface areas of suitable catalysts are also lower as the operating temperature increases, because of the increased rate of sintering.

So far, we have said nothing about the reactivity of oxygen in fuel cells. Both hydrogen and oxygen molecules (dihydrogen and dioxygen) have two atoms. However, the two hydrogen atoms are joined by a single two-electron bond. When this is adsorbed on a suitable catalyst (preferably platinum or palladium, somewhat less effectively nickel) for which the Gibbs energy of adsorption is near zero or slightly negative, it will adsorb with electron transfer to two adjacent catalyst surface atoms. This results in dissociation of the bond, which can be followed by rapid electron transfer to give protons. Three possible rate-determining steps can occur in the anodic direction (with corresponding steps for hydrogen evolution):

$$H_2 \rightarrow 2H_{ads} \tag{19}$$

$$H_{ads} \rightarrow H^+ \tag{20}$$

$$H_2 \rightarrow H^+ + H_2 \tag{21}$$

Reaction (19) in the anodic direction is often called the combination or Tafel reaction, (20) the discharge or Volmer reaction, and (21) the ion plus atom or Heyrovsky reaction, after their proposers. The first is normally rate-determining near-equilibrium on platinum-group metals, whereas the others are important at high overpotentials and on less reactive surfaces (10,15).

Other single-bonded molecules, eg, difluorine or dichlorine, also have high electrode reactivity, for the same reasons as hydrogen. However, dioxygen is much less reactive. Since electrons must be transferred one at a time in a reaction sequence, dihydrogen only requires two electronic steps for oxidation, giving two protons. However, dioxygen requires four separate electronic steps to complete its reduction. If water is to be the final product, this requires four protons also, either from an acid electrolyte or from water molecules. This leads to a complex and less probable (ie, slower) reaction sequence for dioxygen reduction, in which many parallel pathways are possible (12,18).

Dioxygen reduction on effective catalysts, e.g., platinum, does not have excessively high activation energies, but it seems to require specialized or favorable sites which must occur only in small numbers on the electrocatalyst surface (16). Since electrocatalysts such as platinum show reaction with water, ie, coverage with adsorbed OH, which favors the oxidation of hydrocarbon residues, sites showing the most ready adsorption are occupied (17), reducing the chances of dioxygen adsorption and reduction.

As a result, dioxygen reduction is a slow process close to ambient temperatures, with complex kinetics, and often shows two Tafel slopes corresponding to a change of mechanism or a change in oxygen radical adsorption conditions. Near ambient temperatures in acid electrolyte, the low-overpotential slope is equal to RT/F (about 60 mV/pdecade, $\alpha_c = 1.0$), with j_0 equal to approximately 10^{-10} A/cm^{-2} (18). Crossover to a higher slope ($2RT/F$, about 120 mV/decade, $\alpha_c = 0.5$) occurs at about 0.8 V vs. hydrogen, i.e., about −0.43 V from the hypothetical reversible oxygen potential, with an extrapolated j_0 equal to about 5×10^{-10} A/cm^{-2}. The reversible oxygen potential was called "hypothetical" above because it is never attained. At about 0.95 V vs. hydrogen, the cathodic process encounters equal and opposite anodic processes such as oxidation of the electrocatalyst or its substrate (19) and/or the oxidation of impurities (20). In consequence, in aqueous medium the open-circuit potential of practical oxygen electrodes normally lies about 0.25 V below the theoretical value under these conditions.

In the absence of other limitations, the kinetic rates of both the hydrogen and oxygen electrode processes are proportional to the concentrations of reacting molecules, i.e., to the gas partial pressures. The electrodes used in practical fuel cells are termed *three-phase-boundary* or *gaseous diffusion* electrodes. They have porous structures to maximize the area of contact between the gas, electronically conducting catalyst, and ionic phases and to maximize gaseous diffusion. Even so, local depletion of a reactant may occur at high current densities due to the inability of gaseous diffusion to keep up with reaction rate. Product accumulation (eg, water vapor at the oxygen cathode in acid media) will also dilute the reactant. This results in deviations from the Tafel line at high current densities. These are more marked for oxygen than for hydrogen, because of the lower diffusivity of the heavier molecule.

In general, the low rates of dioxygen reduction require heavier catalyst loadings than those at the anode to obtain effective performance. Acid fuel cells for terrestrial applications normally use cathodes with loadings in the range 0.5–0.75 mg/cm^{-2}, although efforts are being made to reduce these values to facilitate a wider range of applications. The carbon-supported platinum catalysts are similar to those used at the anode, generally using graphitized carbons because of the aggressive conditions at the cathode, especially in phosphoric acid fuel cells operating in the 200°C range (14). It has been discovered that a number of platinum alloys with a base metal as a minority component are more active than pure platinum (14,21–23) by up to about 50 mV at constant current density, implying a three-fold increase in j_0 on these materials compared with the value on pure platinum. It is assumed that these are now being employed by developers.

Dioxygen becomes very reactive at high temperatures, at red heat and beyond. In high-temperature fuel cells, the oxygen electrode may show very rapid kinetics, similar to those for hydrogen at ambient temperatures. Unfortunately, the consequent reduction in overpotential is offset by the change in ΔG with temperature. As a result of this effect, the voltage developed by a fuel cell at high temperatures is not very different from that in aqueous electrolytes, in spite of the vastly improved kinetics for dioxygen reduction. However, noble metal catalysts are not required in high-temperature cells, which in any case possess other advantages.

So far, we have assumed that gaseous hydrogen and oxygen reactants and product water can be made to react, virtually to completion, under standard thermodynamic conditions (with regard to the phase and the pressure, if not necessarily to the temperature). This is indeed possible if the fuel cell operates below the boiling point of water, so that liquid water is the product, which can be readily separated from the reactants. However, terrestrial fuel cells normally operate on air rather than on pure oxygen. In this case, air is allowed to enter into the cathode chamber and flows toward the cathode exit, oxygen being consumed along the way. The oxygen therefore becomes progressively more dilute, and the current density becomes lower as the air traverses the cell. Increasing the utilization of the oxygen in the air results in diminishing returns, but clearly, the mean partial pressure of oxygen is considerably less than that in the standard state at 1 atm.

In an acid fuel cell, conduction is via H^+ ions produced at the anode, which react to produce water at the cathode. If the cell operates at temperatures sufficient to produce water vapor, this further dilutes the cathode air, more so toward the cathode chamber exit, further reducing the partial pressure of oxygen. In general, a maximum oxygen utilization from air is usually 85%, which also determines the partial pressure of water vapor in the cell, and hence the electrolyte concentrations, which may have important implications for the ionic conductivity of the electrolyte. The cathode air is almost always run on a one-pass basis without feedback to economize on the work of circulation. In an alkaline fuel cell (AFC), ionic conduction is via OH^- ions produced at the cathode, which react with hydrogen, producing electrons and water at the anode. Hence, if pure hydrogen is used as fuel, it is diluted by water vapor as it passes through the anode chamber. It is not possible to consume the hydrogen, since if this were to be so, both the current density and the reversible hydrogen potential vs. that of the oxygen electrode (ie, the cell potential) would go to zero. In consequence, the hydrogen is generally operated in a feedback loop, in which a condenser continuously removes water vapor, discharging the latent heat of evaporation. In such cases, the hydrogen fuel may be operated at 30% utilization per pass, depending on the temperature. Aerospace AFCs, such as the United Technologies 12-kW PC-17C units in the Space Shuttle *Orbiter*, operate on this principle using pure oxygen as oxidant at 4 atm pressure to increase performance and thus render the system more compact (24).

For terrestrial applications, pure hydrogen is generally not available as the fuel. Fuel cells are supplied with manufactured fuel, consisting of various mixtures of H_2 and CO and/or CO_2. Pure hydrogen (and indeed chemically scrubbed air to eliminate CO_2) is required for use with terrestrial AFCs. In spite of their potential advantages, these considerations have severely limited their ordinary use. Mixtures of H_2 and CO_2, with CO reduced by the water–gas shift reaction to levels which avoid anode poisoning (see above), are the norm for acid fuel cells. As these mixtures react, the hydrogen becomes progressively more dilute toward the anode exit. All of the hydrogen cannot be used, since the current density, as well as the cell voltage, would then go to zero. The practical utilization represents what is economically feasible. Since the anode exit gas, even at 80% utilization, has some heating value which can be used for other purposes, such as to provide heat for steam reforming, reaction (27) in an external reactor. Thus a figure of 80% is often used as a practical utilization level.

In contrast to acid–electrolyte fuel cells, the HTFCs can handle CO, since it is spontaneously converted inside the anode chamber of the cell to H_2 and CO_2, via the water–gas shift process, reaction (32). The HTFCs (MCFC and SOFC) rely on O^{2-} transport from the cathode to the anode, either alone or via CO_3^{2-} as a carrier ion. Thus, water vapor is produced at the anode as H_2 is oxidized, which drives the water–gas shift reaction, consuming any CO which is present until equilibrium is reached. As hydrogen is consumed across the face of the fuel cell anode, the fuel gas becomes more and more dilute, resulting in a fall in current density. Under these conditions, it is not possible to drive the reaction to zero hydrogen content in the fuel gas stream. As before the hydrogen utilization is determined by economic considerations. In the SOFC, it is usually limited to 85%, and it may be lower in the MCFC. If internal steam reforming of the fuel, eg, methane (natural gas), is used with fuel cell stack waste heat, then the heating value of the gas in the anode exit stream is wasted, although it might serve for cogeneration purposes.

HISTORY OF FUEL CELL DEVELOPMENT

The battery was invented by Volta in 1800 using the zinc–silver oxide couple. It was immediately used to show that electricity could decompose many compounds by electrolysis, which, eg, led to the discovery of sodium and potassium by Davy in 1807. The realization that water electrolysis might be reversible was first realized by Grove, a British jurist and gentleman scientist, in 1839, with the invention of what he called the "gaseous voltaic battery." Each cell consisted of two platinum wires platinized with platinum black dipping into sulfuric acid electrolyte. One had a meniscus in contact with hydrogen gas, the other with oxygen. He could draw a small current from the cell at a voltage of about 0.5–0.6 V. Four such cells in series could be seen to decompose water. A series of papers describing Grove's fuel cells appeared in the *Philosophical Magazine* and the *Proceedings of the Royal Society* over the period 1839–1845 (25). Grove realized that an effective fuel cell would require what he described as a "notable surface of action," ie, the largest possible interface per unit area between the gas phase, the electrolyte, and the electrically conducting active surface, which would later be called the electrocatalyst. In later editions (26) of his book *The Correlation of Physical Forces*, first published in 1846, he described fuel cells operating fuels other than hydrogen. In 1889, the German-born scientists Mond and Langer constructed a fuel cell prototype in England which resembled many modern designs (27). It consisted of two sheets of perforated gold leaf covered with platinum black, both in contact with a thin stable porous diaphragm or matrix containing the electrolyte. When one electrode contacted hydrogen and the other air at ordinary temperatures, they were able to obtain 0.73 V at a current density of 3.5 A/cm^2. They were the first to realize the high efficiency of direct electrochemical energy conversion, stating that "this gives a useful effect of nearly 50% of the total energy contained in the hydrogen absorbed in the battery." This was before the work of Gibbs was generally known, and Mond and Langer meant 50% of the enthalpy (HHV) of combustion of hydrogen. The standard enthalpy of combustion of hydrogen to liquid water is 286.0 kJ/mol, or 143 kJ/equivalent. In voltage terms, this is equal to 143/F or 1.482 eV. Thus, the HHV thermal efficiency of Mond and Langer's cell (the ratio of the work done to the enthalpy of combustion in electron-volts) was 49.3%.

In his discourse on the foundation of the Bunsengesellschaft in 1894 (28), Ostwald looked forward to a future of electrochemical, rather than thermal, combustion, which would be more efficient and much cleaner than the coal-fired steam engine of his day. However, the competition was the internal combustion engine operating on oil distillates, which took over the world for all but very large steam plants. Even so, work was going on at about the same time on the direct use of coal in high-temperature fuel cells, eg, that of Jacques and of Baur and co-workers. In later work, Baur was the first to make use of molten carbonate as a high-temperature electrolyte.

Grove had remarked on the difficulty of maintaining what he called "a notable surface of action" in his electrodes. Those of Mond and Langer became wetted by the electrolyte an the product water formed as time progress, so that the surface available for gaseous reaction was reduced. Wetting of high-temperature electrodes by molten carbonate electrolyte was controlled by Baur and co-workers by using capillary action. Their matrix contained a finer mean porosity than the electrodes, so that the latter did not flood (29).

Starting in 1933, Bacon in England used concentrated KOH electrolyte in pressurized cells at 200°C. This work continued (with a break during World War II) until the late 1950s (30). Schmid had previously used a fine-pore platinum black electrode on a coarse-pore graphite layer in 1923 in an attempt to control wetting (31). The electrodes eventually used by Bacon consisted of fine-pore sintered nickel toward the gas side, with a coarse-pore support toward the electrolyte, which was circulated to remove heat. Pure hydrogen and oxygen were used as fuel and oxidant at 45 bars pressure. Bacon had demonstrated 1.0 A/cm^2 at 0.8 V under these conditions by the late 1950s (30). Justi and Winsel developed dual-porosity nickel electrodes in parallel withh Bacon's later work, but their structures were intended for use in more dilute aqueous KOH at low temperatures and pressure. The lower reactivity of the gases under these conditions was offset by the use of high-surface-area Raney nickel in the active layer (32). Another approach to maintaining a high-surface-area three-dimensional microscale meniscus in the electrode for aqueous electrolytes is the incorporation of nonwetting materials in the electrodes for meniscus control. Heise and Schumacher (33) attempted to use paraffin wax for this purpose in 1932. After polytetrafluoroethylene (PTFE, Teflon, E.I. DuPont de Nemours) became available in the 1950s, it began to be used for this purpose, apparently first by workers at the laboratories of General Electric, Schenectady, NY, and at Union Carbide Corporation, Parma, OH (7). It is now used in virtually all fuel cell electrodes for use in aqueous media to control internal wetting of electrodes and to maximize the total area of triple contact between gas, electrolyte, and electronically conducting catalyst per unit of geometrical area (the "three-phase boundary").

Acid fuel cells demonstrated during the later 1950s, eg, at the General Electric Research and Development Laboratory, used high-surface-area Teflon-bonded platinum black electrodes and sulfuric acid as the electrolyte (7). The aim was to develop systems which would operate directly on hydrocarbon fuels, which could not be used in alkaline fuel cells because of the problem of carbonate formation in the electrolyte. It soon became apparent that high temperatures in excess of 150°C would be required for effective operation, even with electrodes containing 40 mg/cm^2 of pure platinum black. Sulfuric acid could only be operated up to about 80–90°C, since it was reduced (at least by hydrogen) at the anode. A search for other possible acids showed that only phosphoric acid had both the required stability and lack of volatility to operate at these temperatures. Orthophosphoric acid is not normally considered to be a strong acid, but at temperatures beyond the boiling point of water it becomes successively dehydrated, forming pyrophosphate polymer chains. Sulfuric acid (a relatively strong acid) shows the same property, though in a less degree, since it can dimerize to form pyrosulfuric acid. Their structures are as follows:

$$[(OH)_3P{=}O]_n \rightarrow —O—PO(OH)—O—PO(OH)—O—PO(OH)— \quad (22)$$

$$(OH)_2S({=}O)_2 \rightarrow HO—S({=}O)_2—O—S({=}O)_2—OH \quad (23)$$

The corresponding Group IV acid, silicic acid, is not only weaker than phosphoric acid, but its valence state means that corresponding polymers are totally deprotonated. The protons can ionize in both the Group V and Group VI pyroacids, but the fact that phosphoric acid is a relatively weak acid with a proton on each phosphate group gives it unique properties. Not only can it form stable polyanions of the type $—P({=}O)_2^-—O—$, but each of these groups also has the capability of forming cations with the $—P(OH)_2^+—O—$ structure. Pyrosulfuric acid is not only unstable at the hydrogen anode, but also has a much lower probability of forming corresponding cations. That the cation, anion, and neutral groups in pyrophosphoric acid have an approximately equal probability of formation and that protons can transfer down the chain via a hopping mechanism from one phosphate group to the next give pyrophosphoric acid an ionic conductivity which is apparently unique. Other aqueous acids, including sulfuric and pyrosulfuric, have protons which become immobile as water is lost, so that their monohydrates (in which the proton is present as H_3O^+) form ion pairs with their anions, giving very low conductivity. When phosphoric acid was first used as a high-temperature acidic electrolyte, its unique proton conductivity properties were accepted, but the fact that other potentially stable acids might not show similar properties was overlooked. At constant water activity, the conductivity of phosphoric acid improved with temperature, whereas that of true aqueous acids decreases. In effect, the phosphoric acid fuel cell (PAFC) operating in the temperature range in which it shows advantageous properties and performance (ie, 150–210°C) is not strictly an aqueous fuel cell, but a molten acid salt system (34).

Even at temperatures up to the maximum available, the performance of the PAFC as a direct converter of hydrocarbons was shown to be disappointingly low. It showed uneconomic performance even with methanol fuel. However, it could operate at temperatures well beyond the boiling point of water, and therefore its waste heat could be used to raise steam for a reformer and shift converter, which would increase the efficiency of steam reforming of natural gas and light hydrocarbons. In addition, the platinum anode of an acid fuel cell operating at 90°C is poisoned by the presence of more than a few ppm of CO, whereas at 200°C, a PAFC anode will tolerate up to 1–2% of CO in the anode inlet gas from a shift converter. These two features attracted the engineers at Pratt and Whitney Aircraft Division of United Aircraft to the PAFC system for use with reformed hydrocarbon fuels in the 1960s (1,14). Even though the Bacon-type AFC would operate at similar temperatures, it required total purification of both hydrogen and air streams to eliminate CO_2. The first in particular was considered to be economically and energetically impossible.

The challenge presented by the PAFC was that of the high cost of materials. Pourbaix's Atlas of Electrochemical Equilibria at 25°C (35), the standard compilation on the thermodynamic stability of the elements in noncomplexing aqueous solutions at 25°C as a function of potential and pH, shows that only niobium, tantalum, gold, and platinum and certain of its relatives were thermodynamically stable under PAFC cathode operating conditions, even at ordinary temperatures. By the early 1970s, it had been shown that various carbons were *kinetically*, if not *thermodynamically*, stable at the cathode and could show reasonable lifetimes (13,14). After graphitization, stabilities were even better. Careful work showed that corrosion was dependent on the partial pressure of water vapor and temperature, which allowed definitions of the temperature and pressure ranges in which stable operation could be expected. This led to the development of low-loading platinum electrodes supported on carbon and graphite structural elements, which could be fabricated at low cost in mass production. Since acid electrolyte fuel cells require platinum metal catalysts, an early goal was to obtain a performance equal to that of the best high-loading platinum systems with low-loading electrodes. During the early 1970s, a great deal of effort went into the development of carbon-supported catalysts, which allowed stable surface surface areas (in square meters per gram) about five times higher than those of pure platinum black (about 20 m^2/g). At the same time, it was clear that platinum was not being effectively used in high-loading electrodes, so effforts where made to increase the amount in contact with the gas phase and the electrolyte in thinner electrodes. As a result, platinum loadings were lowered by more than one order of magnitude and platinum utilizations were increased to over 50% (14). Performance increased to as much as 325 mA/cm^2 and 0.62 V at atmospheric pressure in the most recent PAFC systems (14).

In parallel to work on the direct-hydrocarbon PAFC system, the General Electric Company developed another acid system for application as a power source for the *Gemini* space capsule, in which its product water could also be used for drinking. The system had to be simple and light, because of the weight constraints of the small space capsule. The system was called the Solid Polymer Electrolyte system (SPE) by its developer and used a solid sulfonic acid electrolyte based on a nonfluorinated polystyrene divinyl benzene copolymer backbone ion-exchange membrane (7,36). Water dissolved in the material, but it was not itself soluble in water, so that product water was rejected automatically in pure form. It was completed by platinum black electrodes applied to the membrane and bonded under pressure. An early demonstrator made a suborbital flight on October 30, 1960, as the first fuel cell in space. The *Gemini* contract was awarded in early 1962, and the result was a-1 kW (nominal) pure hydrogen–oxygen sytem operating at 2 atm and 50°C at 0.78 V, but only 37 mA/cm^2. The system weighed 31 kg/kW, corresponding to 27 L/kW, and equal to 9 kg/m^2 of active area. In contrast, the same statistics for the Pratt & Whitney 1.5-kW, 115-kg PC-3A Apollo AFC prototype of the same year were 77 kg/kW, 100 L/kW, and 92 kg/m^2 of active area (24). The United Technologies 12-kW PC-17C Space Shuttle *Orbiter* system of 1974 represented a great improvement at 5.3 kg/kW (of which 55% were the cell stacks), 12.8 L/kW, and 26.0 kg/m^2 (24). In spite of its poor performance, the *Gemini* system remains one of the lightest ever constructed in terms of kilograms per square meter of active area.

The early material used in the *Gemini* fuel cell was much less stable than the later fluorinated materials typified by Nafion, a product of E. I. DuPont de Nemours and Company developed as a sodium-conducting membrane to produce pure NaOH for the chlor-alkali industry, replacing the mercury cell, which posed environmental problems. Fluorination made Nafion not only more stable than sulfonated polystyrene divinyl benzene sulfonic acid, but it also had much lower ionic resistance and higher performance (36). The fluorinated backbone of Nafion contains the same elements as a combination of the Teflons TFE (polytetrafluoroethylene), FEP (perfluoroethylene-perfluoropropylene copolymer), and PFA (perfluoroalkoxy), ie,

$$\text{—}(CF_2CF_2)_n\text{—}CF(CF_2\text{—})\text{—}O\text{—}[CF_2\text{—}CF(CF_3)\text{—}O\text{—}]_m \quad (CF_2)_xSO_3H \qquad (24)$$

where n is normally 2–3, m = 1, and x = 2. The material is a relatively dilute acid, with about the same conductivity as 1 M sulfuric acid (36). It behaves as a typical aqueous acid from the viewpoint of conductivity (34), so that its use is limited to temperatures below the boiling point of water. Work at GE showed cells using Nafion to be capable of attaining 10 times the current density of the *Gemini* fuel cell under comparable conditions, provided that the cathodes were improved by adding Teflon to provide a surface to avoid flooding by liquid product water (36). When GE sold its Nafion-based technology to the Hamilton Standard Division of United Technologies Corporation in 1984, the successor company retained all rights to the SPE trademark. In consequence, the technology is now known elsewhere as proton exchange membrane (or polymer electrolyte membrane), PEM.

Inspired by work by Greger (37) and Gorin (38), which was founded on experiments by Baur in the early 1920s (29), development of today's MCFC was started in the Netherlands by Broers and Ketelaar in the 1950s (39). The laboratory cells made by Broers and Ketelaar used sintered nickel anodes similar to those of Bacon's AFC, though with only one sintered layer. Broers and co-workers used various alkaline carbonate eutectics as electrolytes, usually retained by a magnesia matrix. Different cathodes were used, including silver, copper, and lithium-doped nickel oxide prepared in situ from sintered nickel structures similar to the anode. Only nickel oxide was shown to be sufficiently stable. Finally, lithium aluminate was adopted as the matrix material in late work by Broers and by the Institute of Gas Technology (IGT) in Chicago in the late 1960s. Early work at IGT showed that stainless steel was sufficiently stable at the cathode side as a structural component, although it required cladding with nickel to avoid corrosion in the humid atmosphere at the cathode outlet. It also required protection from corrosion at the edges because of the presence of an oxidizing atmosphere outside and a reducing atmosphere inside, in the present of electrolyte. This required the use of aluminizing. The electrolyte was standardized as 62 mol % Li_2CO_3–38 mol % K_2CO_3 eutectic operating at a mean temperature of 650°C (40).

In work started in the late 1950s at Westinghouse, Weissbart and Ruka (41) were inspired by the operation of the zirconia-based oxide-ion-conducting high-temperature light source, the "Nernst Glower," invented in 1900 (42). The geometry of the system has seen changes to improve assembly (1,43), but the basis is a thin layer of yttria-doped zirconia oxide-ion-conducting electrolyte operating at 1000°C, which is also used in the oxygen sensors of spark ignition engines with catalytic converters. The anode is a nickel cermet, and stable conducting oxides with compatible coefficients of thermal expansion are used for the cathode.

FUEL CELL CLASSIFICATION

Today's fuel cells are descended from the above systems. They may be classified in various ways, ie, by their temperature of operation (high-, medium-, and low-temperature fuel cells) or by the type of electrolyte they use. The latter classification is perhaps the most convenient and logical.

AFCs are confined to using KOH electrolytes. The main reason for this choice is the higher solubility of potassium carbonate than sodium carbonate. Any CO_2 entering an AFC is converted to carbonate, so that both fuel and oxidant supplies must be as free as possible from this contaminant. The fuel must be pure hydrogen and the oxidant pure oxygen or CO_2-scrubbed air. Hydrogen manufactured from natural gas for aerospace applications always contains traces of CO_2, so some accumulation of carbonate in the electrolyte is inevitable. The use of KOH instead of NaOH reduces the possibility of precipitation of carbonate deposits in the electrodes. Even though lithiated sintered nickel oxide cathodes had a marginal lifetime for the 1968 and later Apollo missions (ie, approximately 15 days at 260°C), other materials considerations have now reduced the operating temperature of the AFC to about 115°C. Systems using practical materials are obliged to operate at 80°C or less. The lower limit of operating temperature is determined by the freezing point of the electrolyte. A cell designed to operate at 65°C will generally use 34 wt % or approximately 8 N KOH as the electrolyte, and it will be able to operate down to perhaps −30°C (34 wt %), although its low-temperature performance is limited by increasing electrolyte viscosity and reduced ionic conductivity.

The only true acid fuel cell electrolytes in practical use today are the perfluorinated proton exchange membrane materials such as Nafion, a somewhat similar material from the Dow Chemical Company in which m = 0 and x = 2 in formula (50), and the material known as Aciplex-S from the Asahi Chemical Industry Company, in which m = 0–2 and x = 2–5. Typical electrolyte film thicknesses are in the 100–125-μm range, resulting in cells with resistances of about 0.1 Ω-cm^{-2} when operating at temperatures below the boiling point of water under the particular imposed pressure conditions. Such cells operate at about 65–70°C under atmospheric pressure conditions and up to slightly over 100°C under pressurized conditions.

We have already seen that the PAFC uses a self-ionizing oxyacid as its electrolyte. This does not require water for conduction, so it can operate at temperatures which exceed the boiling point of water. Systems designed for operation at 190–200°C use an electrolyte concentration corresponding to 97 wt % orthophosphoric acid, which is in equilibrium with the typically partial pressure of product water vapor in this temperature range. This electrolyte will freeze at about +40°C. The lower limit of operating temperature is determined by performance considerations, and it is usually about 150°C for use with H_2/CO_2 fuel gas mixtures which have been water–gas shifted to give acceptable CO contents. At lower temperatures, its CO tolerance is reduced. The upper limit of operating temperature is determined by materials considerations at the cathode, involving corrosion of the graphitized carbon cathode support and slow sintering and area loss of the platinum catalyst. The latter shows a voltage decay which is proportional to log time (14). Corrosion of the graphite intercell separator plate (bipolar plate) can also be a problem, especially if there is an ionic contact so that a galvanic corrosion cell can be set up driven by the potential difference between sites which cathodically reduce oxygen and anodic corrosion sites (14).

Like the PAFC, both the MCFC and SOFC have upper operating temperature which is limited by materials considerations. In the case of the MCFC, the maximum temperature point in the cell (about 675°C) is determined by corrosion of components in the presence of molten carbonate electrolyte. The lowest temperature point (perhaps 625°C) is determined by performance considerations. In the SOFC, the upper temperature limit of the cell (about 1050°C) is determined by interdiffusion of solid-state components, particularly manganese into zirconia, which result in a loss in desired properties. The lower temperature limit is determined by performance considerations, which are largely due to increased electronic and/or ionic resistance of cell and intercell components. Today's minimum operating temperature with standard components is about 925°C. Attempts are being made to reduce this to allow the use of steel peripheral components (heat exchangers) instead of those constructed from superalloys of ceramics.

Thus, fuel cells can operate from below ambient temperature to about 115°C, which is followed by a gap to another operating range from 150 to 200°C. Followed by other gaps, there are two other operating ranges from 625 to 675°C and from 925 to 1050°C. These ranges are determined by the nature and properties of the only available electrolytes.

ENERGY CONVERSION EFFICIENCY OF FUEL CELLS

It is clear from the discussion given earlier that the maximum nominal thermal efficiency of a fuel cell operating at temperature T will be given by $\Delta G_{T1}/\Delta H^\circ$, where ΔG_{T1} is the Gibbs energy of reaction at T_1 and where $-\Delta H^\circ$ is the enthalpy of combustion of the hydrogen used in the fuel cell, which is normally defined under standard conditions. In practice, this can only be achieved in an ideal fuel cell operating under conditions which are truly thermodynamically reversible, ie, at zero net forward rate. A real fuel cell must consume fuel at a positive rate, and since fuel is valuable, it must consume as much of it as possible. A hypothetical cell with infinitely fast anode and cathode kinetics producing a combustion product which may be continuous separated and removed from the reacting anode and cathode gas streams would indeed have an efficiency given by the above expression. In practice, all fuel cells operating under temperature conditions where liquid water is a product 1 atm pressure are limited by kinetic considerations, so the example is only of theoretical interest.

Practical fuel cells are required to operate at the highest possible power densities, which means at the highest temperatures their materials allow. This almost invariably means that product water is eliminated in vapor form, although certain low-temperature alkaline fuel cell and acid (PEM systems) do produce product water. In most cell designs, water vapor will progressively dilute either the anode or the cathode reactant gas. This depends on whether the cell operates by conduction of a proton carrier ion (H_3O^+) or an oixde carrier ion (eg, OH^-, O^{2-}, or CO_3^{2-}). In the former case, product water vapor is formed on the cathode side and, in the latter case, at the anode side. In aqueous systems, the electrolyte has a composition which is not far from being in equilibrium with the product water vapor, so both anode and cathode side become diluted. The molten salt (ie, molten carbonate) and solid oxide systems produce all their water vapor at the anode side.

In principle, the appropriate reactant gas or gases can be bled from the operating cell and cooled, and the water can be condensed as a liquid after a single pass of a volume of reactant gas which is much larger than the stoichiometric requirement. The reactant gases may then be recycled to the cell after heat exchange. This will ensure that the reactant gas remains at a high concentration everywhere in the cell. In real systems, especially those operating on practical fuel and oxidant mixtures, a line must be drawn between efficiency and capital cost. As a result, one or both of the reactant gases will normally be allowed to become progressively more dilute as it traverses the cell from inlet to exit, as the concentration of product water vapor rises.

If pure hydrogen and oxygen are used in an acid fuel cell, eg, a PEM system, water is produced on the cathode side. The water vapor will be in equilibrium with the water activity in the PEM electrolyte, which will in turn humidify the hydrogen at the anode. The system must include a provision to remove product water. In principle, water can be collected at either the anode or the cathode, but collection is more efficient at the cathode, where the product is formed. Hence, the hydrogen anode may be operated dead headed, and the oxygen at the cathode may be converted to water until its partial pressure becomes too low to give satisfactory performance, after which water is removed in a condenser or by pressurizing, and the oxygen is recycled. Pure hydrogen–oxygen systems are generally for space applications and are normally operated pressurized to give maximum performance. Under these conditions, pressurization at temperatures of about 90°C will result in the idealized situation of formation of liquid water at the cathode. This must be removed from the cathode itself, to avoid blockage of the reaction surface by a liquid film. In a recent development, United Technologies uses a microporous membrane in the cathode chamber opposite and parallel to the plane of the electrode. Pure water is circulated on the opposite side of this membrane to cool the cell, and the system is designed so that the cathode gas is at a higher pressure than the circulating water, so that product water is forced through the membrane and eliminated from the cell. Hence, both pure hydrogen and oxygen can be used dead headed. In practice, the reactant gases are never absolutely pure, and inert impurities, generally nitrogen and argon, will build up in the anode and/or cathode chambers. Hence, a purge is required from time to time to remove impurities or a small amount of gas may be bled off at the dead-headed end of a large cell.

In alkaline space cells, typified by the United Technologies PC-17C used in the Space Shuttle, water vapor is produced at the anode side in a matrix cell. This dilutes the KOH electrolyte, which must be provided with an expansion volume in the form of an electrolyte reservoir plate (ERP) which is an electronically conducting sintered nickel structure to carry the bipolar current flow. It is in contact with the anode (Fig. 1), arranged so that it does not interfere with the flow of hydrogen. The hydrogen is partially converted in the cell, to about 50% utilization, and the mixture of hydrogen and water vapor is recycled in a feedback loop which contains a condenser for product water removal. The oxygen supply is operated dead headed.

Whereas space cells must carry and use all of their reactant gases, for terrestrial use, the oxidant is normally air, which is available in any required quantity subject to the energy required to circulate and/or to compress it, if applicable. In acid electrolyte cells, eg, the PEM, product water vapor is formed in the air as it traverses the cell, as oxygen is consumed. The major design consideration is how much air must be circulated, ie, the oxygen utilization in the cell. 50% utilization is normally the aim, which will result in exit oxygen and water vapor partial pressures of 0.091 and 0.182 atm, respectively. The situation at an acid fuel cell anode in which hydrogen is supplied as part of a gas mixture with inert components, eg, from a reformer and shift converter, is similar. Using natural gas as the fuel with a steam-to-carbon ratio of 3, followed by two-stage water–gas shifting will result in a mixture with the composition $4H_2 + H_2O + CO_2$, along with small amounts of CO (about 1.5% for feedstock for the PAFC). Initially, the mixture contains about 67% of hydrogen. At 80% utilization, the hydrogen content is reduced to 13.3%, ie, a H_2–CO ratio of about 9:1. Under these conditions, the operation of the hydrogen anode begins to show a reduction in performance due to the start of the effects of poisoning by CO, so this is about the practical limit for the anode exit gas composition.

Since the cathodes of aqueous cells are entirely limited by kinetics, the low concentration of oxygen in air near the exit is the controlling factor, since the presence of product water has no effect on reaction rate in a kinetically controlled situation. However, in an alkaline cell in which water is produced at the anode, the system may operate close to thermodynamic equilibrium due to the rapid hydrogen oxidation kinetics. Hence, as the reaction is driven toward higher hydrogen utilizations and rising product water partial pressures, from the Nernst equation the reversible cell potential will be effectively reduced by the amount.

$$\Delta V = RT/2F(\ln p_{H_2^\circ}/p_{H_2O^\circ} - \ln p_{H_2}/p_{H_2O} \quad (24)$$

as the reactant traverses the cell. In this expression, $p_{H_2^\circ}$ and $p_{H_2O^\circ}$ are the partial pressures of hydrogen and water vapor at the anode inlet and p_{H_2} and p_{H_2O} are the corresponding values at a given point in the cell. It is clear from material balance considerations that

$$p_{H_2} = p_{H_2^\circ} - (p_{H_2O} - p_{H_2O^\circ}) \quad \text{and} \quad p_{H_2O} = p_{H_2O^\circ} + (p_{H_2^\circ} - p_{H_2}) \quad (25)$$

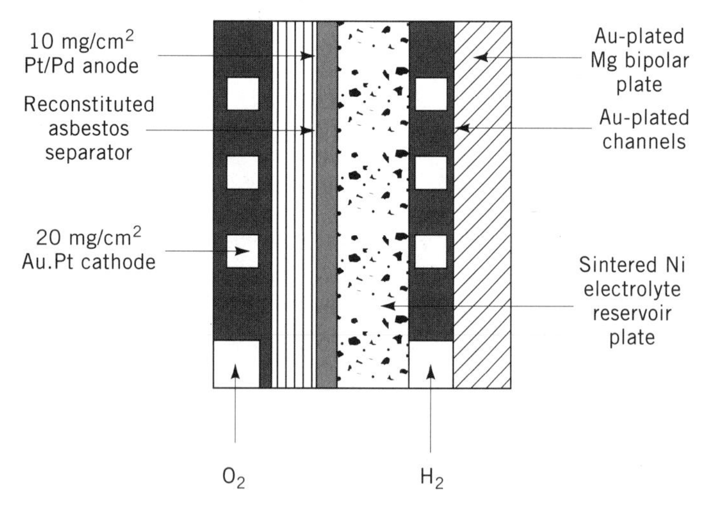

Figure 1. Schematic of United Technologies (now IFC) PC-17C pure hydrogen–oxygen (4 atma) AFC used in the Space Shuttle, in which water vapor is produced at the anode side in a crossflow matrix cell. The KOH electrolyte is provided with an expansion volume in contact with the anode in the form of an electrolyte reservoir plate (ERP), an electronically conducting sintered nickel structure to carry the bipolar current flow. It is arranged so that it does not interfere with the flow of hydrogen, which passes through a feedback loop for product water and heat removal. The oxygen cathode is operated dead headed. The schematic does not include two-piece dielectric liquid cooling plates.

In high-temperature fuel cells (MCFCs and SOFCs) both electrodes show the reversible potential under zero-current conditions. The revesible or Nernst potential at the cell anode exit then takes on a particular significance which is easiest to appreciate for parallel-flow (coflow or counterflow) systems. It is the highest potential the cell can attain under the particular conditions of fuel utilization assumed at the exit, when no driving force for reaction will be present and the current density will have fallen to zero. Neglecting any small Nernst effects at the cathode, the effect of utilizations of 50% ($p_{H_2} = p_{H_2O}$ = 0.5 atm), 75% (p_{H_2} = 0.25 atm, p_{H_2O} = 0.75 atm), 85% (p_{H_2} = 0.15 atm, p_{H_2O} = 0.85 atm), and 90% (p_{H_2} = 0.1 atm, p_{H_2O} = 0.9 atm) at an anode operating on pure hydrogen as a function of temperature is the following: The maximum operating voltage of the cell shows significant losses as operating temperatures increase. This is a major practical problem of the SOFC, in which the anode and cathode polarizations may be negligible and the IR reduced to acceptable values yet may operate at lower cell voltages than the MCFC, in which the anode and cathode polarizations are certainly not negligible. Thus, the MCFC is the most efficient fuel cell to date.

DESIGN CONSIDERATIONS

Fuels and Fuel Clean-up

Fuels Available. The anodic reactant in a fuel cell should be hydrogen. However, common fuels are hydrocarbons and their derivatives, which must therefore be converted to either a hydrogen-rich gas or pure hydrogen. The most important fuel considered is as-delivered natural gas, which is generally of lower cost than other hydrocarbons and derivatives. For certain applications, the use of methanol, ethanol, kerosene, jet fuel, diesel fuel, military combined jet–diesel fuel (JP-8), gasoline, liquid propane gas (LPG), naphtha, or refinery off-gas may be desirable. The use of coal-derived synthesis gas is a special case, which is discussed separately. Finally, biomass gasified by fermentation, by steam reforming, or by oxidative gasification is possible.

Integrated Reforming. A PAFC operating at 190–200°C can supply the excess steam required for efficient reforming via waste heat from the cell stack. In general, steam-to-carbon ratios of 2.5–3.5 are used, compared with a theoretical value of 1, to drive the reaction as far as possible to equilibrium and improve kinetics. The anode off-gas is used in the reformer burner to provide the enthalpy of reforming. The availability of free excess steam reduces the energy requirement for reforming by 11–14%, giving a substantial gain in overall system efficiency. The reforming process is followed by two-stage shift conversion at high and low temperatures to reduce CO levels to acceptable values for the PAFC operating at 190–200°C. The overall HHV efficiency of a large pressurized 11-MW PAFC unit was estimated at 42% overall (net ac), or 44.3% (gross d-c power output).

Hydrogen Fuel. The HHV efficiency of the fuel cell operating on hydrogen at 0.73 V is 49.3%. AFCs require pure hydrogen as a fuel. Recent work has shown that PEMs may not operate effectively on reformate containing a few ppm of CO produced by partial oxidation following low-temperature methanol reforming or two-stage water–gas shift after natural gas reforming. State-of-the-art hydrogen production from natural gas might use an advanced reformer, perhaps based on the integrated concept developed by Rolls-Royce and Associates (44). Operating as an autothermal reformer, this has a natural gas to reformate HHV efficiency of 91% (85.7% LHV). The latest data for the work requirements to separate CO_2 using pressure-swing adsorption indicate a maximum of 0.35 kW·h/H-m^3 (45). If this electric power is produced from natural gas at a modest 38% LHV efficiency, this will produce hydrogen from natural gas at 78.7% LHV efficiency. Assuming 5% compression loss for hydrogen storage or transportation, a hydrogen-fueled AFC or PEM operating at 0.73 V could operate from natural gas at 43.5% LHV efficiency (39.2% HHV), based on natural gas input with advanced processing and hydrogen separation. This may provide an effective way of reducing greenhouse gas emissions in the transportation field in the future.

Coal Gas. This is a special case. As produced in a gasifier, a partial oxidation device with steam injection to perform what is in effect noncatalytic high-temperature adiabatic reforming, the resulting gas contains variable amounts of H_2, CO, and steam, with some CO_2 and CH_4, together with N_2 if air is the reactant instead of oxygen separated via a liquid air plant. The exact gas composition depends on the gasifier type, i.e., on the reaction temperature and chemistry, which depends on whether air of oxygen is used and on the oxygen–carbon and steam-to-carbon ratio. Depending on the coal used, the resulting gas may contain considerable amounts of H_2S and COS, which may be removed via a number of chemical or physical processes. Most require that the synthesis gas should be cooled to ambient temperature, e.g., the chemical Stretford process, which uses sodium carbonate as a stripping agent, with a regenerative step to produce H_2S, 50% of which is then combusted with air to SO_2. The resulting SO_2 and the remaining H_2S are then reacted in the gas-phases Claus process to produce sulfur and water. In the Selexol process, the synthesis gas is chilled to near 0°C, and the sulfur products are removed via an organic solvent, which also removes some of the CO_2. These processes require multiple heat exchangers, which add to the capital cost of the plant. In consequence, high-temperature desulfurization processes, typically using metal oxides which may be regenerated, are being sought if the gas is to be used in a high-temperature fuel cell.

In general, systems producing clean coal gas from coal may be up to about 75% efficient, depending on the gasifier used.

If the hydrogen derived from coal gas is used in an (unspecified) fuel cell system operating at 0.73 V at 100% utilization, in which no steam recovery for gasification occurs, then the overall system HHV efficiency will be given by multiplying the gasification efficiency by 0.73/1.48, ie, an unimpressive 33–37%. If clean synthesis gas produced at 80% efficiency can be directly used in the fuel cell at 85% utilization, the corresponding efficiency at the same cell voltage will be 0.8 × 0.85 × 0.73/1.48, i.e., 33.5%. However, if the anode exit gas stream and fuel cell waste heat can be used in the process, e.g., for the production of steam and for preheating, then the efficiency may rise to 42%. Efficient integration of the total system can yield even higher values.

The objective of fuel-cell-combined cycle-gasifier processes is to recover as much waste heat as possible from each element of the system. An oxygen-blown gasifier which yields a medium-Btu gas containing no nitrogen is most advantageous. It will generally be pressurized and will yield a clean low-sulfur pressurized gas stream after clean-up. Such a gas stream will most advantageously be used in a pressurized fuel cell, preferably operating at high temperature. Such a system, operating at 0.8 V at

85% utilization, may have an HHV efficiency of 50%. However, its initial capital cost and the cost of replacing fuel cell stacks at 5-year intervals may be such that it would be more economical to regard a smaller MCFC as a "topping cycle" for an integrated gasifier gas turbine combined cycle (IGGCC) plant, in which a larger turbine burns some of the synthesis gas and the anode exit fuel. The efficiency of such a plant will be lower (about 45% based on the coal HHV), but lower capital and O&M cost may mean less expensive electricity. Their low emissions, combined with their relatively high efficiency, may make them attractive in the next century, either with MCFC or SOFC systems.

State-of-the-Art Fuel Cell Generators

On-Site PAFC Systems. The PAFC represents the most developed integrated fuel cell generation technology, with the start of system design for operating on natural gas fuel using an integrated fuel processing dating from the American Gas Association TARGET (Team to Advance Research in Gas Energy Technology) Program, conducted at the Pratt and Whitney Division of United Technologies Corporation, in 1967. This was intended to supply individual households with 12.5-kW combined heat and power units, making them independent of the electricity distribution network. The cost of delivered gas was then one-sixth of the cost of delivered electricity. Even at the cost ratio of gas to electricity in the late 1960s, it became evident that the small-atmospheric-pressure 12.5-kW TARGET unit would not be economical, at least under the then-prevalent materials and economics situation, so that scale-up would be desirable. A number of units were however given proof-of-principle testing during the 1970s, and they showed successful operation, although their materials were high cost, e.g., pure platinum electrodes in high loadings per square centimeter of electrode area were used. They showed the principle of integrating PAFC cell stacks to the fuel processor so that steam for reforming could be raised, increasing overall system efficiency.

It was expected that, by 1995, Japanese electric utilities would introduce about 30 PAFC units in the 50–200-kW class, as a preliminary to building up about 1000 MW of capacity by the year 2000. Japanese gas utilities were expected to introduce similar amounts. Demonstrations of PAFCs in Japan (whether of Japanese or overseas technology) were to be one-third financed by the Japanese government during the early 1990s as an aid to commercialization. In the decade from 1983 to 1993, over 100 PAFC units of different sizes had been operated in Japan.

PAFC Electric Utility Units. During the late 1970s and early 1980s, the Power Systems Division of United Technologies Corporation (UTC) actively pursued the development of electric utility power generation equipment, which should cost less than on-site cogeneration units and make a profit of electric sales alone.

During the 1980s, a new design with 0.93-m^2 cells was proposed by UTC as a generator for urban sites with negligible emissions, no cooling water requirements, and an imperceptible noise level of 55 dBA at the fence. This would have been an 11-MW unit with a lower power footprint than that of the 4.5-MW unit, to ease maintenance problems. The unit was designated PC-23 and would have used eighteen 650-kW (dc) stacks. After certain utility interest in the United States, it failed to attract orders because of low energy costs, its high capital cost, lack of utility financing, and not least, utilities that had specific requirements which could not be met by a standardized design.

PAFC Issues. Since PAFCs use an acid electrolyte, it is inevitable that platinum catalysts must be used. The key to reducing platinum loadings to economic levels is that as much as possible of the total platinum surface area can be used effectively. If the platinum particles on graphitized high-surface-area carbon have an initial specific area of about 100 m^2/g, which will eventually result in 50 m^2/g as aging occurs according to a logarithmic time scale, then 5.5 g platinum/cm^2 resulted in satisfactory performance. Of this two-thirds was at the cathode (0.5 mg/cm^2) and one-third at the anode (0.25 mg/cm^2). More recently, 0.1 mg/cm^2 has been used at the anode (46). Some developers may use somewhat higher cathode loadings to increase the system efficiency and stack cost effectiveness. Certain base-metal platinum catalysts, eg, Pt–30% Cr, appear to result in stable materials with little or no base-metal corrosion and 50% greater activity at constant cathode potential. Since cathodes generally cannot double current density beyond the state-of-the-art levels available today because of diffusion limitations, the increase in performance of the alloy catalysts is better taken in the form of an increased cell voltage, eg, by 50 mV. This underscores the fact that today's electrodes are essentially current density limited for any given set of reactant partial pressures. Under present operating conditions, the remaining components of the PAFC cells are sufficiently stable to allow 40,000 h operation. Some corrosion of the graphitized high-surface-area carbon catalyst support occurs at the cathode, which contributes to platinum catalyst area loss. The matrix material which immobilizes the electrolyte between the electrodes by absorption and serves as a bubble-pressure barrier is a layer of fine-particle-size silicon carbide about 0.25 mm thick bonded with some PTFE. It generally shows high stability. The graphite parts show good stability provided that they are not exposed to conditions where galvanic corrosion might occur.

There are three economic issues with the electric utility PAFC, namely efficiency, durability, and capital cost. All of the above are involved in determining the cost of electricity, the ultimate economic criterion.

The PAFC today is expensive. The unpressurized PC-25 200-kW cogeneration unit in limited-series production costs approximately $3500/kW. The next prototype multimegawatt electric utility unit is likely to cost $10,000/kW. In mass production, costs will certainly fall on a learning curve (14). In the early to mid-1980s it was expected that by 1993 eight 10-MW units per year would be installed at a cost (adjusted for inflation, 1993 dollars) of about $2500/kW. This would be close to the break-even point for a cogeneration unit but twice the affordable cost of a stand-alone electric utility generator (14). These figures still stand today. At $2500/kW, the PC-25 should start to break even in some cogeneration applications. At $1500/kW, it will be widely used. Similarly, a multimegawatt plant may break even in a stand-alone electric utility application if its installed cost is $1200/kW. In niche markets, the allowable cost might be higher.

Today, we are a long way from these costs, and little or no learning is taking place on how to further reduce costs and improve performance of pressurized systems because no large units are being bought.

The lifetime of the stack is another issue. Its design life is 40,000 operating hours for decay corresponding to the nameplate performance. While this is 5 years, the life of the balance of the plant will be 30 years, so six sets of stacks will be required during the system lifetime. This will clearly further increase operating costs compared with conventional generating equipment.

It is difficult to see how larger electric utility PAFCs are going to find a niche in the marketplace. However, the on-site cogeneration PAFC unit seems assured of a place.

MCFCs. During the 1980s, the MCFC went from laboratory single cells to small stacks (40). Much of the development work during this period was conducted at United Technologies (International Fuel Cells), the Institute of Gas Technology (IGT) in Chicago, and Energy Research Corporation, Danbury, CT (ERC). Scale-up of UTC stacks with external manifolding containing a porous gasket material the height of the stack showed eventual failure (at about 1000–4000 h) due to loss of electrolyte via the gasket. This was traced to an electroosmotic mechanism, which was alleviated by modifications to the gasket or, in another approach, by the use of internal manifolding in the stack (1). During the same period, nickel anode creep was reduced to acceptable limits by the use of nickel alloys, first with chromium, then with aluminum (1).

Other problems which required solution included corrosion of stainless steel cathode current collectors and the cathode sides of stainless steel bipolar plates, which were rendered immune from corrosion by wet anode exit gas by nickel cladding on the anode side. To avoid galvanic corrosion around the edges associated with omnipresent carbonate electrolyte films, the sides of the plates were aluminized. This places an electronically insulating layer in the galvanic pathway, eliminating corrosion of this type. A life-limiting factor involves slow dissolution of the lithiated nickel oxide cathode, which allows migration of Ni^{2+} dissolved in the electrolyte and precipitated as nickel nodules close to the anode. This may eventually cause shorting of the cell (40). The rate of nickel oxide dissolution was proportional to CO_2 partial pressure, so that it was estimated that the lifetime of pressurized cells operating on a cathode gas mixture containing 30% CO_2 at 5 atma pressure would be less then 10,000 h, although 40,000 h might be attained at atmospheric pressure (1). The major aim of the U.S. Department of Energy Development Program managed from Morgantown, WV, was the development of pressurized cells to operate on coal gas in large integrated gasifier combined-cycle systems. An active effort was therefore started at Argonne National Laboratory and elsewhere to find a stable substitute cathode (1).

Active support in the United States included emphasis on simple, inexpensive natural gas systems supported by the Electric Power Research Institute on behalf of the electric utility industry. System designs in the mid-1980s by ERC using internal reforming showed the possibility of attaining 50% HHV efficiency in atmospheric pressure systems using natural gas fuel. These used a minimum of heat exchangers with internal reforming (1).

In the United States, the major developers supported by the U.S. DOE are (1993) MC-Power Corporation, a company using IGT technology, and ERC. The emphasis of the first is on pressurized (3-atma) 1-m^2 technology with internal manifolding, using external reforming of natural gas with integrated flat-plate reformers. Their system contains four heat exchangers, which may be used in part to supply cogenerated steam equal to 40% of electrical output. ERC concentrates on externally manifolded systems with internal reforming of natural gas using indirect internal reforming (IIR) plates within the stack (one per six cells) which remove sensible heat, with cells scaled up to 0.56 m^2. The program at IFC uses externally manifolded 0.74-m^2 cells in natural gas systems incorporating sensible heat reforming in the anode feedback loop. the large stacks are designed to be capable of thermal cycling, including cool-down to ambient temperature.

Emphasis in all cases is on simplification.

SOFC. The most successful SOFC technology to date has been the tubular technology at Westinghouse. This was developed as a system with a porous nonconducting calcia-stabilized zirconia-supported tube 1.56 cm in diameter on which was deposited a layer of strontium-doped lanthanum manganite, which was then sintered to become the cathode. This was followed by a masking operation, after which a strip of magnesium-doped lanthanum chromite was deposited via a process called electrochemical vapor deposition (EVD). In this, the tube is exposed on the outside to the mixed chlorides in the vapor phase, and hydrogen and steam are supplied on the inside. Since the material is electronically conductive with slight ionic conductivity, it continues to grow even when densified, via hydrogen oxidation and chloride reduction with conductivity via the dense ceramic layer. This layer is called the interconnect and is stable under both oxidizing and reducing conditions. The interconnect is then masked, and a 250-μm layer of dense yttria-stabilized zirconia electrolyte is deposited, again by EVD. The anode (nickel powder) is then deposited as a slurry. It is dried and then turned into a cermet impregnated with yttria-stabilized zirconia electrolyte, again by EVD. The final step is nickel plating of the interconnect. The complete tubes are up to 1 m in length with an active area of about 450 cm^2. There are plans for further scale-up. The tubes are connected from the interconnect of one of the external nickel cermet anode of the next by nickel felts which sinter in place in situ at the operating temperature of 1000°C. Preheated air (four to six times stoichiometric for the cathode process and cooling) is supplied to the interior of the tubular cells, and reformate is supplied to the bottom of the tubes at the outside. The system is typically operated at 85% fuel utilization. Spent fuel and air are allowed to mix and burn in a ceramic plenum chamber at the top of the cells, which acts as an air preheater. Air enters the plenum chamber at 600°C from a stainless steel heat exchanger. Reforming of fuel is accomplished in a reforming preheater.

Since 1991, the calcia-stabilized zirconia support tube has been replaced by a tube made of air electrode material, which serves as the cathode. This has greatly improved the peripheral electronic conductivity of the system, which is required for current collection between tubes. The performance is now excellent, although it is naturally limited by the thermodynamic considerations. The waste heat from the system may be used for cogeneration or in a bottoming cycle. In 1994, Westinghouse was scaling demonstration 25-kW units up to 100 kW using 1-m-length tubes. Two 25-kW units with five hundred seventy-six 50-cm cells arranged in 18 cell bundles, each in strings of 192 cells, had been tested in Japan by gas industry and electric industry consortia. Good performance was obtained, and individual bundles had performed well to 6000–10,000 h. Emissions were measured on smaller 3-kW Westinghouse systems at Osaka Gas and Tokyo Gas in the early 1990s. They varied from 0.3 to 1.3 ppmv NO_2 at six times stoichiometric air flow, or about 7–30 g/MW·h.

Worldwide interest in the SOFC has grown enormously since 1990, since it is perceived as a simple system which will use what are expected to be inexpensive ceramic materials. It requires no electrolyte management techniques and should show good durability. Emphasis is on small planar systems, operating at temperatures less than 1000°C to avoid the use of ceramic heat exchangers (cf. the Westinghouse system). These may also allow the use of metal bipolar plates between cells. Such approaches may require alternative ceramics (47).

AFCs. Modifications of the Bacon cell with sintered nickel electrodes operating at up to 260°C, as in the Project Apollo fuel cell, have been abandoned due to the short lifetime of nickel cathodes at high temperatures, which were required so that performance would not be significant by operating at less than 4 bars pressure rather than 45 atma (24). This change was necessary to reduce the mass of the pressure containment. Stable alternative materials, eg, rhodium, are not cost-effective. For simplicity, a stationary (although free) electrolyte was used in the Apollo fuel cell, rather than the circulating system favored by Bacon. In the Space Shuttle fuel cell system, whose design was frozen in the 1970s, an immobilized stationary electrolyte was used. This contained a purified crysotile mat as a matrix in each cell. This set a limit to the cell operating temperature at about 80°C. In consequence, Teflon-bonded noble metal electrodes were used (gold with 20% platinum at the cathode in 20 mg/cm^2 total loading on a gold-plated nickel screen, platinum with 20% palladium at the anode in 10 mg/cm^2 loading on a silver-plated nickel screen). The system operated at 4 atma on pure hydrogen and oxygen and could give a nominal performance of 0.86 V at 470 mA/cm^2 and 0.8 V at 750 mA/cm^2. Stacks consisted of 32 cells each with 465 cm^2 active area. Each Space Shuttle unit in its final form had three parallel-connected stacks, each capable of 6 kW at nominal power, derated to give the normal power requirement of 12 kW. The Space Shuttle contained three such units, weighing about 120 kg. The AFC uses PC-17C Space Shuttle fuel cell cathodes consisting of 10 mg/cm^2 of 10 wt % platinum and 90 wt % gold black on a gold-plated nickel screen. Modifications of these cells using advanced materials operating at high temperatures and pressures with thin components will offer up to 6 A/cm^2 at 0.8 V. They were developed for applications under the Strategic Defense Initiative.

Clearly, electrodes of the type used in the Space Shuttle fuel cell are unaffordable for common applications, and they are normally replaced by low-loading carbon-supported platinum electrodes. However, little recent work has been carried out to optimize these for common applications, because of the requirement for pure hydrogen fuel and CO_2-purified air for the AFC.

The PEMFC. For specialized military or aerospace applications, this may be operated under pressure on pure hydrogen and oxygen, with 4 mg/cm^2 platinum black electrodes. The PEMFC system is also attractive for terrestrial applications, particularly for electric vehicles. During the late 1980s, the system did not show much promise for this application because of the poor interface formed between typical electrodes, which required high platinum loadings. Since 1986, remarkable progress in platinum utilization has occurred. The kinetics of the oxygen electrode in the PEM are more than an order of magnitude more rapid in the PEM at 80°C than in the PAFC at 190°C. Today, the best results observed so far are about a factor of 2 higher than those for the PAFC (about 800 mA/cm^2 at 0.6 V, compared with 350 mA/cm^2, with hydrogen and air at atmospheric pressure), but these results can be obtained with platinum loadings which are 10 times lower (0.05 mg/cm^2 Pt at the cathode, vs. 0.5 mg/cm^2). With an anode with only 0.025 mg/cm^2 of Pt, this indicates a total platinum catalyst requirement of only 0.15 g per peak kilowatt. A small car can therefore be powered by 3 g of platinum costing $40, similar to the amount of noble metal in a catalytic converter.

A small 1000-kg subcompact commuter car requires about 0.15 kW/km for 95 kph cruise, or 0.1 kW·h/kg for urban driving. Based on an averaged figure, a 300-km range will require 37.5 kW·h. This could be achieved with 1000 mol of hydrogen, or slightly more than 100 L of compressed gas at 200 atma. How this hydrogen might be stored (as compressed gas, metal hydride, liquid hydrogen, or some other form) may be the basis of the future transportation economy.

BIBLIOGRAPHY

1. A. J. Appleby and F. R. Foulkes, *Fuel Cell Handbook*, Van Nostrand Reinhold, New York, 1989.
2. K. Wark, *Thermodynamics*, McGraw-Hill, New York, 1983; A. W. Culp, *Principles of Energy Conversion*, McGraw-Hill, New York, 1979.
3. E. A. Moelwyn-Hughes, *Physical Chemistry*, Pergamon, New York, 1961.
4. I. Prigogine, *Introduction àla Thermodynamique des Processus Irreversibles*, Gallimard, Paris, 1971.
5. P. Van Rysselberghe, *J. Chem. Phys.* **29,** 640–642 (1958).
6. P. van Rysselberghe, in J. O'M. Bockris, ed., *Modern Aspects of Electrochemistry*, vol. 4, Plenum, New York, 1966, pp. 1–46.
7. H. A. Leibhavsky and E. J. Cairns, *Fuel Cells and Fuel Batteries*, Wiley, New York, 1968.
8. B. D. McNicol, *Specialist Reports on Catalysis*, vol. 2, The Chemical Society, London, 1979.
9. B. D. McNicol, R. T. Short, and A. G. Chapman, *Journal of the Chemical Society, Faraday Transcripts 1* **72,** 2735–2743 (1976); *Journal of Applied Electrochemistry* **6,** 221–227 (1976); M. R. Andrew, J. S. Drury, B. D. McNichol, C. Pin-

nington, and R. T. Short, *Journal of Applied Electrochemistry* **6,** 99–106 (1976).

10. A. J. Appleby, "Electrocatalysis," in B. E. Conway and J. O'M. Bockris, eds., *Modern Aspects of Electrochemistry*, vol. 9, Plenum, New York, 1974, pp. 369–478.
11. U.S. Pats. 3,992,331 and 3,992,512 (November 16, 1976), 4,044,193 (August 23, 1977), 4,059,541 (November 22, 1977), and 4,082,695 (April 4, 1978), H. G. Petrow and R. J. Allen (to Prototech., Inc.).
12. U.S. Pats. 4,136,056 and 4,137,373 (January 23, 1979), V. L. Jalan and C. L. Bushnell (to United Technologies, Inc.).
13. A. J. Appleby, in S. Sarangapani, J. R. Akridge, and B. Schumm, eds., *The Electrochemistry of Carbon*, The Electrochemical Society, Princeton, N.J., 1964, pp. 251–272.
14. A. J. Appleby, *Energy* **11,** 13–94 (1986).
15. R. Parsons, *Transactions of the Faraday Society* **54,** 1053–1063 (1958).
16. A. J. Appleby, *Journal of Electroanalytical Chemistry* **357,** 117–179 (1993).
17. H. Dahms and J. O'M. Bockris, *Journal of the Electrochemical Society* **111,** 728–736 (1964); A. T. Kuhn, H. Wroblowa, and J. O'M. Bockris, *Transactions of the Faraday Society* **63,** 1458–1467 (1967).
18. A. Damjanovic and V. Brusic, *Electrochim. Acta* **12,** 615–628 (1967).
19. A. J. Appleby, *Journal of Electroanalytical Interfacial Electrochemistry* **35,** 193–207 (1972).
20. J. O'M. Bockris and A. K. S. Huq, *Proceedings of the Royal Society of London* **A237,** 277–296 (1956).
21. U.S. Pats. 4,186,410 (January 29, 1980), V. M. Jalan and D. A. Landsman; 4,192,907 (March 11, 1980), V. M. Jalan, D. A. Landsman, and D. M. Lee; 4,202,934 (May 13, 1980), V. M. Jalan (all to United Technologies, Inc.).
22. P. N. Ross, "Oxygen Reduction on Supported Pt Alloys and Intermetallic Compounds in Phosphoric Acid," EPRI-EM-1553, Electric Power Research Institute, Palo Alto, Ca., 1980.
23. V. M. Jalan and E. J. Taylor, *Journal of the Electrochemical Society* **130,** 2299–2302 (1983).
24. J. O'M Bockris and A. J. Appleby, *Energy* **11,** 95–135 (1986).
25. W. R. Grove, *Philosophical Magazine, Series 3* **14,** 127–130 (1839); **21,** 417–420 (1843); *Proceedings of the Royal Society of London* **4,** 463 (1843); **5,** 557–559 (1845).
26. W. R. Grove, *The Correlation of Physical Forces*, 6th. ed., Longmans Green, London, 1874.
27. L. Mond and C. Langer, *Proceedings of the Royal Society of London* **46,** 296–304 (1889).
28. W. Ostwald, *Zeitschrift fuer Elektrochemie* **1,** 122–125 (1894).
29. E. Baur, W. D. Treadwell, and G. Trumpler, *Zeitschrift fuer Electrochemie* **27,** 199–208 (1921).
30. A. M. Adams, F. T. Bacon, and R. G. H. Watson, in W. Mitchell, Jr., ed., *Fuel Cells*, Academic, New York, 1963, pp. 129–192.
31. A. Schmid, *Die Diffusionsgaselektrode*, Enke, Stuttgart, 1923.
32. E. W. Justi and A. W. Winsel, *Cold Combustion Fuel Cells*, Steiner, Wiesbaden, 1962.
33. G. W. Heise and E. A. Schumacher, *Transactions of the Electrochemical Society* **52,** 383–391 (1932).
34. A. J. Appleby, O. A. Velev, J-G. LeHelloco, A. Parthasarthy, S. Srinvasan, D. D. DesMarteau, M. S. Gillette, and J. K. Ghosh, *Journal of the Electrochemical Society* **140,** 109–111 (1993).
35. M. Pourbaix, *Atlas of Electrochemical Equilibria at 25°C*, National Association of Corrosion Engineers, Houston, Tex., 1966.
36. A. J. Appleby and E. B. Yeager, *Energy* **11,** 137–152 (1986).
37. U.S. Pat. 2,175,523 (October 10, 1939), H. H. Greger (to H. H. Greger).
38. U.S. Pat. 2,581,651 (January 8, 1952), 2,654,661 (October 6, 1953), 2,654,662 (October 6, 1953), E. Gorin (to Pittsburgh Consolidated Coal Company).
39. G. H. J. Broers and J. A. A. Ketelaar, in G. H. Young, eds., *Fuel Cells*, vol. 1, Rheinhold, New York, 1960, pp. 78–93.
40. R. Selman, *Energy* **11,** 153–208 (1986).
41. J. Weissbart and R. Ruka, *Journal of the Electrochemical Society* **109,** 723–726 (1962).
42. W. Nernst and W. Wild, *Zeitschrift fuer Elektrochemie* **7,** 373–380 (1900).
43. J. T. Brown, *Energy* **11,** 209–229 (1986).
44. J. P. Shoesmith, R. D. Collins, M. K. Oakley, and R. D. Stevenson, *Journal of Power Sources* **49,** 129–142 (1994).
45. A. J. Appleby, *Energy* **19** (in press).
46. J. H. Hirschenhofer, D. B. Stauffer, and R. R. Engleman, *Fuel Cells, A Handbook (Revision 3)*, U.S. Department of Energy, Office of Fossil Energy, Morgantown Energy Technology Center, Morgantown, W. Va., 1994.
47. B. H. C. Steele, *Journal of Power Sources* **49,** 1–14 (1994).

FUEL RESOURCES

DAVID A. TILLMAN
JEFFREY WARSHAUER
DAVID E. PRINZING
Foster Wheeler Environmental Coop.
Sacramento, California

The wheel is considered to be the greatest invention and fire the greatest discovery of all time. Together, the invention of the wheel and the discovery of fire as a useful force have led to the application of energy. From the invention of the wheel has come such innovations as steam and combustion turbines, rotors and stators used in electricity generation, diesel and Otto-cycle engines for transportation systems, and windmills, water wheels, and hydroelectric turbines. Similarly, the harnessing of fire has led to the use of various materials as fuels: coal, lignite, petroleum, natural gas (see GAS, NATURAL), tar sands, oil shale (qv), peat, wood (qv), and the biofuels (see FUELS FROM BIOMASS), organic wastes (see FUELS FROM WASTE), uranium and nuclear power, wind, falling water (for hydroelectric power), geothermal steam and hot water (see GEOTHERMAL ENERGY), sunlight, ocean thermal gradients, and the range of conversion products including both electricity and synthesis gas from coal (see FUELS, SYNTHETIC) have been used. These fuels are used both to power the wheel-related inventions and to supply energy for process applications: iron- and steelmaking, nonferrous metal smelting and refining, process heat and steam for pulp and paper operations, process energy for chemicals manufacture, etc. Harnessed fuels supply the needs of commercial and residential users as well.

Evaluations of fuel resources or total fuel supply focus on critical economic and environmental issues as well as existence. These issues include availability, utilization

Table 1. U.S. Energy Consumption by Source from 1870–1990, Exajoules (EJ)[a,b]

Year	Wood and Biomass	Coal	Petroleum	Natural Gas	Hydroelectric	Nuclear[c]	Other[d]	Total
1870	3.1	1.1						4.1
1880	3.1	2.1	0.1			qc		5.2
1890	2.6	4.3	0.2	0.3				7.5
1900	2.1	7.2	0.2	0.3	0.3			10.1
1910	2.0	13.4	1.1	0.5	0.5			17.5
1920	1.7	16.4	2.7	0.8	0.8			22.4
1930	1.6	14.4	5.7	2.1	0.8			24.5
1940	1.5	13.2	7.9	2.9	1.0			26.3
1950	1.3	13.6	14.2	6.5	1.5			37.1
1960	0.3	10.7	21.2	13.4	1.8			47.4
1970	1.1	13.4	28.9	23.2	2.9	0.2	0.01	69.6
1980	2.53	16.27	36.08	21.51	3.29	2.89	0.08	80.13
1985	2.6	18.44	32.62	18.81	3.54	4.38	0.21	78.01
1986		18.21	33.97	17.63	3.58	4.72	0.23	78.33
1987		19.00	34.68	18.72	3.24	5.18	0.26	81.08[e]
1988		19.89	36.10	19.57	2.79	5.97	0.30	84.61
1989		19.98	36.09	20.45	3.04	5.99	0.26	85.81
1990	3.3	20.17	35.40	20.36	3.11	6.50	0.22	85.76

[a] References 1–6.
[b] To convert EJ to Btu, multiply by 9.48×10^{14}.
[c] Nuclear energy is that generated by electric utilities.
[d] Other includes net imports of coal coke and electricity produced from wood, waste, wind, photovoltaic, and solar thermal sources connected to electric utility distribution systems. It does not include consumption of wood energy (other than consumed by electric utility industry).
[e] An estimated additional 2.5 EJ of wood energy was consumed for residential heating and light industry.

patterns, environmental consequences, and related economic considerations (See also ENERGY CONSUMPTION IN THE UNITED STATES; RENEWABLE RESOURCES).

HISTORICAL PATTERNS IN FUEL UTILIZATION

Preindustrial society relied primarily on wood, other biomass, and falling water for energy. These energy sources provided carbon for steelmaking, heat for domestic and commercial purposes, energy for modest shaft power applications, eg, grinding of grain, and fuel for transportation on riverboats and early railroads. These fuels were readily available and could be gathered up or otherwise harnessed with little capital investment and scant attention to technology. U.S. energy consumption by fuel source from 1870 to 1990 is shown in Table 1.

Industrialization in the United States and northern Europe demanded significant sources of carbon for steelmaking, fuels for pumping water from mines, and energy for manufacturing processes. As the process of industrialization gained momentum manufacturing shifted away from optimal sites along rivers and connected regional economies with transcontinental railroads. Industrialization created a national economy, along with strong regional economies, through the use of energy for manufacturing and transportation systems, and coal was the fuel of choice (Table 1). With the advent of industrialization also came the shift in agriculture toward development of mechanized equipment and chemical fertilizers, and petroleum became the dominant fuel. The emergence of pipelines (qv), has enabled natural gas, then complemented oil, to be the desired form of energy. Fuel selection factors, in all cases, include availability, energy density (J/kg or J/m^3), energy transportability, fuel cost, and fuel reliability.

Fuel Production and Consumption Since 1970

In recent hisotry, both technological and political forces have influenced fuel consumption in the United States and throughout the world. Events such as the oil embargo of 1973, the political unrest in the Middle East, and the collapse of the Union of the Soviet Socialist Republics, have caused disruptions and shifts in petroleum supply systems. The emergence of the North Sea oil field, the construction of the Alaska Pipeline bringing North Slope, Alaska crude to refineries, and other technical developments have also occurred. Most recently, environmental concerns have influenced the selection of fuels, such as the potential to form air pollutants such as particulates NO_x, SO_2, and most recently air toxics (3,7–10) (see AIR POLLUTION; AIR POLLUTION CONTROL METHODS). Moreover, there has been the passage of numerous energy and environmental laws within the United States. Legislation has included the Clean Air Act amendments of 1990, and the National Energy Policy Act of 1992. These laws complement the move toward energy conservation, and the emphasis on materials recycling (qv) for resource management. Further, actions by local and state regulatory agencies in the 1990s, including Public Utility Commissions, have further increased the complexity of fuel supply in the United States.

Trends in commercial fuel, eg, fossil fuel, hydroelectric power, nuclear power, production and consumption in the United States and in the Organization of Economic Cooperation and Development (OECD) countries, are shown in Tables 2 and 3. These trends indicate (6,13); (*1*) a significant resurgence in the production and use of coal throughout the U.S. economy; (*2*) a continued decline in the domestic U.S. production of crude oil and natural gas leading to increased imports of these hydrocarbons (qv); and (*3*) a

Table 2. U.S. Energy Production and Consumption, 1982–1992, EJ[a,b]

Fuel Source	1982	1984	1986	1988	1990	1992
			Consumption			
Petroleum	31.90	32.76	33.97	36.11	35.40	35.31
Dry Natural gas	19.52	19.53	17.63	19.57	20.36	21.44
Coal	16.17	18.01	18.21	19.88	20.15	19.96
Hydroelectric	3.77	4.01	3.64	2.81	3.11	2.94
Nuclear	3.30	3.75	4.72	5.97	6.50	7.02
Total	*74.66*	*78.06*	*78.17*	*84.34*	*85.52*	*86.67*
			Production			
Crude oil	19.32	19.89	19.39	18.23	16.43	16.02
Natural gas liquids	2.31	2.40	2.27	2.38	2.39	2.49
Dry natural gas	19.26	18.92	17.38	18.48	19.37	19.27
Coal	19.66	20.8	20.58	21.88	23.70	22.75
Hydroelectric	3.45	3.57	3.24	2.46	3.09	2.65
Nuclear	3.30	3.75	4.72	5.97	6.50	7.02
Total	*67.30*	*69.33*	*67.58*	*69.40*	*71.38*	*70.20*

[a] References 11, 12.
[b] To convert EJ to Btu, multiply by 9.48×10^{14}.

Table 3. Total Final Consumption per Gross Domestic Product OECD Countries, 1973–1989[a,b]

Country	1973	1979	1987	1988	1989
OECD North America	0.45	0.41	0.32	0.32	0.31
Canada	0.55	0.52	0.41	0.41	0.41
United States	0.44	0.40	0.31	0.31	0.31
OECD Pacific	0.30	0.26	0.20	0.20	0.20
Australia	0.36	0.35	0.31	0.31	0.31
Japan	0.30	0.25	0.18	0.19	0.18
New Zealand	0.33	0.35	0.39	0.40	0.41
OECD Europe	0.38	0.35	0.30	0.29	0.28
Austria	0.35	0.34	0.30	0.29	0.28
Belgium	0.58	0.51	0.40	0.39	0.38
Denmark	0.36	0.33	0.23	0.23	0.22
Finland	0.50	0.44	0.38	0.36	0.36
France	0.36	0.31	0.26	0.25	0.24
Germany	0.39	0.36	0.30	0.29	0.28
Greece	0.38	0.38	0.40	0.40	0.41
Iceland	0.48	0.35	0.29	0.31	0.33
Ireland	0.45	0.43	0.38	0.36	0.35
Italy	0.34	0.29	0.25	0.24	0.24
Luxembourg	1.43	1.18	0.78	0.76	0.78
Netherlands	0.49	0.47	0.41	0.38	0.37
Norway	0.39	0.36	0.31	0.30	0.29
Portugal	0.39	0.43	0.46	0.49	0.49
Spain	0.31	0.35	0.29	0.30	0.29
Sweden	0.44	0.41	0.32	0.32	0.30
Switzerland	0.21	0.21	0.20	0.19	0.19
Turkey	0.65	0.62	0.61	0.60	0.61
United Kingdom	0.40	0.36	0.30	0.29	0.28
Total OECD	*0.41*	*0.37*	*0.29*	*0.29*	*0.28*

[a] Reference 13.
[b] Ratio of total final consumption of energy to gross domestic product (GDP). Measured in metric tons of oil equivalent per $1000 of GDP at 1985 prices and exchange rates; changes in ratios over time reflect the combined effects of efficiency improvements, structural changes, and fuel substitution.

continued trend of energy conservation, expressed in terms of energy consumed per dollar of gross domestic product.

U.S. ENERGY PRODUCTION, CONSUMPTION, AND AVAILABILITY

Production and consumption of commercially available fossil fuel, nuclear power, and hydroelectric power in the United States for the year 1992 is shown in Table 2 (12). Coal production is most significant followed by natural gas and petroleum. Electricity generation and utilization patterns are shown in Table 4. Coal is overwhelmingly the most significant energy source used to generate electricity.

The data presented in Tables 2 and 4 focus on commercially traded sources of energy. During the period 1970–1990, increased emphasis was placed on renewable energy resources (qv), including wood and wood waste; municipal solid waste and refuse-derived fuel; other sources of biomass and waste, eg, agricultural crop wastes, tire-derived fuels, and selected hazardous wastes burned as fuel substitutes in cement kilns; wind and solar energy; geothermal steam and hot water; andd other unconventional energy sources. Estimates of the contribution of these energy sources vary. As of this writing biofuel utilization in the United States runs about 3.7 EJ/yr (3.5×10^{15} Btu/yr) in support of process energy needs for industry, cogeneration facilities, and small stand-alone power plants (5), and geothermal energy is about 0.21 EJ/yr (0.2×10^{15} Btu/yr) (6).

Coal Availability and Utilization

There are vast reserves of coal (qv) and lignite (see LIGNITE AND BROWN COAL) in the United States (see Table 5). The total reserve base exceeds 425 billion metric tons equivalent to 11,200 EJ (10.6×10^{18} Btu) and is distributed throughout 32 states. This reserve base has increased by 8.3% since the 1970s despite the high levels of fuel produc-

Table 4. Electricity Supply and Disposition, 1990[a]

Supply and Disposition	Quantity, kW·h × 10^9	Percent of Total
Fuel type for electric utilities generation		
Coal	1560	55.6
Petroleum	117	4.2
Natural gas	264	9.4
Nuclear power	577	20.5
Pumped storage hydroelectric	−2	
Renewable sources/other[b]	293	10.4
Total	*2808*	*100*
imports	2	
Fuel type for nonutilities[c] generation		
Coal	33	15
Petroleum	5	2.3
Natural gas	100	45.9
Renewable sources/other[b,d]	80	36.7
Total	*218*	*100*
Sales to utilities	106	
Generation for own use	111	
Electricity sales by sector		
Residential	924	34.1
Commercial/other	843	31.0
Industrial	946	34.9
Total	*2713*	*100*

[a] Reference 14.
[b] Renewable sources/other includes hydroelectric, geothermal, wood, wood waste, municipal solid waste, other biomass, and solar and wind power.
[c] Nonutilities includes cogenerators, small power producers, and all other sources, except electric utilities which produce electricity for self-use or for delivery to the grid. The generation values for nonutilities represent gross generation rather than net generation (net of station use).
[d] Includes waste heat, blast furnace gas, and coke oven gas.

Table 5. U.S. Coal Reserves by State, 1990, EJ[a,b,c,d]

State	Reserves	State	Reserves
Alabama	114	Montana	2,848
Alaska	146	New Mexico	106
Arizona	6		
Arkansas	10	North Dakota	229
Colorado	403	Ohio	438
		Oklahoma	38
Illinois	1,857	Pennsylvania	691
Indiana	241	South Dakota	9
Iowa	52	Tennessee	20
Kansas	23	Texas	316
Kentucky	697	Utah	146
Louisiana	12	Virginia	62
Maryland	18	Washington	34
Michigan	3	West Virginia	880
Missouri	143	Wyoming	1,614
Total	*11,155*		

[a] References 6 and 15.
[b] Reserve data is based on demonstrated reserve base. Minable reserves differ from these figures.
[c] Georgia, Idaho, North Carolina, and Oregon also have some reserves.
[d] To convert EJ to Btu, multiply 9.48 × 10^{14}.

Table 6. Largest Coal-Producing States in 1990, EJ[a,b]

State	1990 Production	Rank
Wyoming	4.37	1
Kentucky	4.11	2
West Virginia	4.02	3
Pennsylvania	1.67	4
Illinois	1.43	5
Texas	1.32	6
Virginia	1.11	7
Montana	0.89	8
Indiana	0.85	9
Ohio	0.84	10
Total	*20.63*	

[a] Reference 6.
[b] To convert EJ to Btu, multiply by 9.48 × 10^{14}.

tion (6). Total U.S. recoverable reserves exceed 240 billion metric tons or 6100 EJ (5.8 × 10^{18} Btu) and are distributed among three geographic areas: the Appalachian, Interior, and Western coal producing regions. Of these, the western region contains 53.6% of the recoverable reserves, the interior region 25.8%, and the Appalachian region 20.6%. Reserves can also be evaluated in terms of the sulfur (qv) content of the coal. The sulfur is important owing to environmental considerations. Of the recoverable reserves, 34.3% contains $<$0.6% sulfur, 33.9% contains 0.61–1.67% sulfur, and 31.8% contains $>$1.68% sulfur.

Coal production and consumption in the 1990s reflects the shift toward the use of western, lower sulfur coal. In 1970, West Virginia, Kentucky, and Pennsylvania ranked 1–3 in coal production, respectively. In 1990, Wyoming, Kentucky, and West Virginia held those ranks, and Texas and Montana entered the top 10 coal producers. Whereas Appalachia remained the most significant energy production region, the western coal producing states surpassed Emissionsthe interior states in solid fossil fuel production (see Table 6). The average coal heating value reflected the shift from Appalachia and the interior to the west, declining from 25.8 × 10^6 J/kg (11.1 × 10^3 Btu/lb) in 1973 to 24.8 × 10^6 J/kg (10.7 × 10^3 Btu/lb) in 1980 and 24.3 × 10^6 J/kg (10.4 × 10^3 Btu/lb) in 1990. The shift in coal production toward western coal deposits also reflects the shift in coal utilization patterns (see Table 7). Electric utilities are increasing coal consumption on both absolute and percentage bases, whereas coke plants, other industrial operations, and residential and commercial coal users are decreasing use of this solid fossil fuel.

Environmental considerations also were reflected in coal production and consumption statistics, including regional production patterns and economic sector utilization characteristics. Average coal sulfur content, as produced, declined from 2.3% in 1973 to 1.6% in 1980 and 1.3% in 1990. Coal ash content declined similarly, from 13.1% in 1973 to 11.1% in 1980 and 9.9% in 1990. These numbers clearly reflect a trend toward utilization of coal that produces less SO_2 and less flyash to capture. Emissions from coal in 1990s were 14 × 10^6 t/yr of SO_2 and 450 × 10^3 t/yr of particulates generated by coal combustion at electric utilities. The total coal combustion emissions from all

Table 7. U.S. Coal Consumption by Sector, 1970–1990, EJ[a,b]

Year	Electric Utilities	Industrial		Residential and Commercial	Total Consumption
		Coke Plants	Other		
1970	7.60	2.29	2.15	0.38	11.42
1975	9.64	1.98	1.51	0.22	13.35
1980	13.51	1.58	1.43	0.15	16.67
1985	16.47	0.98	1.79	0.19	19.43
1990	18.36	0.94	1.81	0.16	21.27

[a] Reference 6.
[b] To convert EJ to Btu, multiply by 948 × 10^{14}.

sources were only slightly higher than the emissions from electric utility coal utilization (6).

Oil and Natural Gas Availability and Utilization

U.S. resources and reserves of petroleum (qv) and natural gas (see GAS, NATURAL), including natural gas liquids (NGL) are limited. As of January 1, 1992, U.S. reserves of petroleum were some 151 EJ (24.7 × 10^9 bbl) and U.S. reserves of natural gas were 182 EJ (17.3 × 10^{16} Btu) (11). Since 1976, the United States has experienced a significant decline in oil reserves. In 1976, proven petroleum reserves totaled 205 EJ (33.5 × 10^9 bbl). Between 1976 and 1993, some 210 EJ (3.4 × 10^{10} bbl) were added to the reserves, and 263.5 EJ (4.31 × 10^{10} bbl) were produced, yielding a net reserve loss of 53.8 EJ (8.8 × 10^9 bbl) (14). Similarly from 1976 to 1992, there was a net reserve loss of 44.5 EJ (4.22 × 10^{16} Btu) of dry natural gas (16).

As shown in Table 8, U.S. distribution of oil and natural gas reserves is centered in Alaska, California, Texas, Oklahoma, Louisiana, and the U.S. outer-continental shelf. Alaska reserves include both the Prudhoe Bay deposits and the Cook Inlet fields. California deposits include those in Santa Barbara, the Wilmington Field, the Elk Hills Naval Petroleum Reserve No. 1 at Bakersfield, and other offshore oil deposits. The Yates Field, Austin Chalk formation, and Permian Basin are among the producing sources of petroleum and natural gas in Texas.

The decrease in petroleum and natural gas reserves has encouraged interest in and discovery and development of unconventional sources of these hydrocarbons. Principal alternatives to conventional petroleum reserves include oil shale (qv) and tar sands (qv). Oil shale reserves in the United States are estimated at 20,000 EJ (19.4 × 10^{18} Btu) and estimates of tar sands and oil sands reserves are on the order of 11 EJ (10 × 10^{15} Btu) (see TAR SANDS). Of particular interest are the McKittrick, Fellows, and Taft quadrangles of California, the Asphalt Ridge area of Utah, the Asphalt, Kentucky area, and related geographic regions.

The unconventional reserves of natural gas occur principally in the form of recoverable methane from coal beds. As of 1991, reserves of coal bed methane totaled 8.6 EJ (8.2 × 10^{15} Btu), principally in the states of Alabama, Colorado, and New Mexico (16).

Domestic petroleum, natural gas, and natural gas liquids production has declined at a rate commensurate with the decrease in reserves (see Table 2). Consequently, the reserves/production ratio, expressed in years, remained relatively constant from about 1970 through 1992, at 9–11 years (16). Much of the production in the early 1990s is the result of enhanced oil recovery techniques: water flooding, steam flooding, CO_2 injection, and natural gas reinjection.

Whereas the use of petroleum and natural gas is significant in the electricity generating sector, this usage declined from 1970 to 1990, in part owing to the 1977 Fuel Use Act (see Table 9). The legislation of the 1990s and the growth of independent power producers (IPP) generating electricity for utilities in combined cycle combustion turbine (CCCT) facilities, may mean a reversal in the trend for oil and natural gas utilization for power generation (qv). In any event, total U.S. oil and gas consumption (Table 1) remains high, and these are the fuels of choice for residential, commercial, industrial, and transportation applications.

Table 8. Crude Oil and Natural Gas Proved Reserves, EJ[a,b,c]

State	Oil Proved Reserves	Gas Proved Reserve[d]
Alaska	37.22	10.22
California	25.80	3.19
Louisiana	4.15	11.79
Oklahoma	4.28	15.83
Texas	41.59	39.87
Wyoming	4.63	11.02
Federal offshore	16.03	31.02
Total	*133.72*	*122.94*

[a] Reference 16.
[b] To convert EJ to Btu, multiply by 9.48 × 10^{14}.
[c] As of Dec. 31, 1991.
[d] Gas reserves equal dry natural gas plus natural gas liquids.

Table 9. Fuel for Electric Utility Generation of Electricity 1970–1990, kW·h × 10^9[a]

Year	Petroleum	Gas-Fired	Internal Combustion and Gas Turbine	Total
1970	174	361	22	557
1975	273	288	28	589
1980	238	326	28	592
1985	97	279	16	392
1990	113	246	22	381

[a] Reference 6.

Table 10. Municipal Waste to Energy Projects[a]

Location	Capacity, t/d
Mass burn	
Hillsborough County, Fla.	1100
Pinnelas County, Fla.	2700
Tampa, Fla.	910
Baltimore, Md.	2000
North Andover, Mass.	1350
Saugus, Mass.	1350
Peekskill, N.Y.	2000
Tulsa, Okla.	1000
Marion County, Oreg.	500
Nashville, Tenn.	1000
Refuse-derived fuel	
Akron, Ohio	1000
Duluth, Minn.	360
Niagara Falls, N.Y.	1800
Dade County, Fla.	2700
Columbus, Ohio	1800
Hartford, Conn.	1800

[a] References 18, 19.

Other Fuel Availability and Utilization

As shown in Table 2, nuclear, hydroelectric, and geothermal resources now contribute some 9.8 EJ (9.3×10^{15} Btu) annually to the U.S. economy. Of these energy sources, nuclear power is the dominant force having over 70% of the total. U.S. nuclear power production continued to increase through 1990, but nuclear electricity generation may have peaked at the 6.5 EJ for political and social reasons. Hydroelectric power generation remains relatively stable. There are annual variations in supply which depend on local weather, eg, rainfall, snowpack, and regional economic conditions. Geothermal energy (qv) has been developed to only a modest extent.

Biomass and waste fuels contributed some 3.7 EJ to the economy (see FUELS FROM BIOMASS; FUELS FROM WASTE). These fuels include wood and wood waste; spent pulping liquor at pulp and paper mills; agricultural materials such as rice hulls, bagasse, cotton gin trash, coffee grounds, and a variety of manures. When wood waste and numerous other forms of biomass are added to municipal solid waste (MSW), refuse-derived fuel (RDF), methane recovered from landfills and sewage treatment plants, and special industrial and municipal wastes such as tire-derived fuel, these together contribute about 5 EJ (4.7×10^{15} Btu) to the U.S. economy (17). Of these fuels, wood and the biofuels are typically employed in industrial settings either to generate process steam (qv) or to cogenerate electricity and process steam. Some condensing power plants have been built by such utilities as Washington Water Power, Burlington Electric, and several IPP firms. There are some 1500 MW_e of electricity generating capacity based on wood and the biofuels in existence as of 1993.

MSW incinerators (qv) are typically designed to reduce the volume of solid waste and to generate electricity in condensing power stations. Incineration of unprocessed municipal waste alone recovers energy from about 34,500 t/d or 109 million metric tons of MSW annually in some 74 incinerators throughout the United States. This represents 1.1 EJ (1.05×10^{15} Btu) of energy recovered annually (18). Additionally there are some 20 RDF facilities processing from 200 to 2000 t/d of MSW into a more refined fuel (19). Representative projects are shown in Table 10.

Other sources of energy worth noting are the extensive wind farms, solar projects, and related engineering unconventional technologies. These renewable resources provide only small quantities of energy to the U.S. economy as of this writing.

Trends in Energy Technology and Future Fuel Consumption

Increased economic activity usually means an increase in energy consumption particularly for generation of electricity, manufacturing of products, transportation, and residential and commercial applications. Regulatory and political requirements associated with energy supply and utilization require increasing attention to environmental concerns in order to ensure reliable energy availability without undue environmental degradation. Thus attention is being paid to increasing the efficiency of fuel utilization as well as to reducing the formation of airborne emissions ranging from particulates NO_x and SO_2 to the management of air toxics such as HCl and trace metals.

Coal. Technologies traditionally deployed for coal utilization include using pulverized coal (PC), cyclone, and stoker-fired boilers. For PC boilers, technologies being deployed or developed include the use of micrometer-sized coal, staged fuel-staged air low NO_x burners, limestone injection multistage burners (LIMB), reburning for NO_x control, and advanced techniques for overfire air management. Cyclone-fired boilers also are capitalizing on reburning technologies and air management techniques. Further, both PC and cyclone-fired boilers are utilizing cofiring techniques, blending nitrogen and sulfur-free biomass fuels, and low sulfur tire-derived fuels with the coal for both cost control and emissions reduction (17,20).

Advanced coal utilization technologies include the development of bubbling, circulating, and pressurized fluidized-bed combustion for electricity generation and process energy production (see COAL CONVERSION PROCESSES; FLUIDIZATION). Since the early 1980s, over 250 fluidized beds have been installed that have capacities ranging from 25 GJ/h to 1 TJ/h. Whereas fluidized beds are fired using every solid fuel available, these beds are predominantly used for coal combustion. Projects include the Shawnee #10 boiler of TVA, the Black Dog project of Northern States Power, and the Colorado-Ute circulating fluidized-bed project. Advanced coal utilization technologies also include integrated gasification combined cycle (IGCC) systems where coal is gasified and the low or medium heat-value gas is then utilized in a combustion turbine and the exhaust is ducted to a heat recovery steam generator (HRSG). The initial demonstration of this technology was at the Cool Water project near Barstow, California. More recent projects are under development in Polk and Martin counties in Florida, and at the Tracy Station of Sierra Pacific Power Co. in Nevada. Post combustion technologies including alternative acid gas scrubbing technologies, urea or ammonia injection for NO_x control, and combinations of air pollution control systems are also emerging to support coal utilization.

Petroleum and Natural Gas. The dominant technologies under development for oil and gas include advances in combustion turbine design. These include applying aircraft engine technology to power generation (qv), and placing emphasis on higher temperatures and increased efficiencies in the combustion turbine. Other technologies of significance include dry low NO_x combustion systems employing principles of staged fuel-staged air, and various catalytic and noncatalytic ammonia injection systems in the post-combustion environment.

Other Fuels. The emerging technologies for unconventional and renewable fuels somewhat mirror those associated with coal: cofiring of biofuels and coal in PC and cyclone boilers, fluidized-bed combustion systems fed with biofuel or a blend of various solid fuels, combustion air management, and gasification–combustion systems. Additionally, technologies associated with post-combustion pollution control for the alternative solid fuels are similar to those developed for coal utilization. At the same time, significant advances have been made in such alternative technologies as fuel cells (qv), photovoltaic cells (qv), wind energy, and the broad range of renewable resources. All of these technologies, combined with those for the dominant

Table 11. World Fossil Fuel Reserves, EJ[a,b]

Region	Coal Reserves 1991[c]		Petroleum Reserves 1992[d]		Natural Gas Reserves 1992[d]	
North America	6,570	(6,565)	499	(500)	346	(344)
Central and South America	254	(254)	419	(447)	172	(193)
Western Europe	2,964	(2,578)	90	(136)	187	(222)
Eastern Europe and former USSR	7,818	(8,252)	358	(378)	1,818	(1,929)
Middle East	5	(5)	4,048	(3,651)	1,360	(1,387)
Africa	1,623	(1,623)	370	(462)	320	(344)
Far East and Ocenia	7,950	(7,942)	270	(345)	309	(401)
Total	*27,185*	*(27,221)*	*6,054*	*(5,918)*	*4,512*	*(4,820)*

[a] Refs. 11 and 12.
[b] To convert EJ to Btu, multiply by 9.48×10^{14}.
[c] Data from the World Energy Council. Values in parentheses are from British Petroleum.
[d] Data from *Oil and Gas Journal*. Data in parentheses are from *World Oil*.

Table 12. World Energy Production 1991, EJ[a,b]

Region	Crude Oil	Natural Gas Liquids	Dry Natural Gas	Coal	Hydroelectric	Nuclear	Total
North America	26.22	3.75	24.64	24.73	6.64	7.94	102.28
Central and South America	10.56	0.31	2.56	0.82	3.73	0.11	14.90
Western Europe	9.52	0.46	8.21	9.60	5.06	8.10	68.64
Eastern Europe and former USSR	22.86	0.89	29.52	20.97	2.67	2.93	74.12
Middle East	36.83	1.64	4.59	0.03	0.14	0	12.53
Africa	15.04	0.43	2.81	4.64	0.48	0.10	10.49
Far East and Oceania	14.85	0.33	6.29	36.38	4.81	3.21	82.97
Total	*135.88*	*7.81*	*78.61*	*97.18*	*23.51*	*22.40*	*365.40*

[a] Ref. 11.
[b] To convert EJ to Btu, multiply by 9.48×10^{14}.

Table 13. World Energy Consumption, 1991, EJ[a,b]

Region	Petroleum	Natural Gas	Coal	Hydroelectric	Nuclear	Total
North America	41.95	24.58	21.17	6.64	7.94	102.28
Central and South America	7.88	2.46	0.72	3.73	0.11	14.90
Western Europe	29.67	12.29	13.70	4.98	8.00	68.64
Eastern Europe and former USSR	20.44	27.63	20.27	2.75	3.03	74.12
Middle East	7.78	4.45	0.16	0.14	0.00	12.53
Africa	4.76	1.62	3.53	0.48	0.10	10.49
Far East and Oceania	31.19	6.42	37.34	4.81	3.21	82.97
Total	*143.67*	*79.44*	*96.91*	*23.51*	*22.40*	*365.93*

[a] Ref. 11.
[b] To convert EJ to Btu, multiply by 9.48×10^{14}.

Table 14. World Net Electricity Consumption, 1982–1991, kW·h × 10⁹[a]

Region	1982	1983	1984	1985	1986	1987	1988	1989	1990	1991
North America[b]	390.1	406.3	435.9	458.1	477	499.1	524.2	541.6	543.2	552.5
United States	2,086.4	2,151.0	2,285.8	2,324.0	2,368.8	2,457.3	2,578.1	2,646.8	2,712.6	2,759.3
Central and South America	309.9	333.4	355.6	371.1	391.1	403.7	421.7	427.8	437.0	446.9
Western Europe	1,725.1	1,783.4	1,872.2	1,962.6	2,001.7	2,064.9	2,069.4	2,150.8	2,188.9	2,217.1
Eastern Europe and former USSR	1,554.4	1,610.8	1,694.8	1,748.9	1,727.0	1,787.5	1,825.7	1,853.9	1,838.8	1,751.0
Middle East	111.4	121.7	131.3	143.5	149.4	158.3	169.7	180.2	195.6	185.6
Africa	182.9	202.0	206.5	217.7	248.6	254.7	270.5	280.1	282.0	281.0
Far East and Ocenia	1,300.3	1,387.0	1,477.8	1,568.9	1,654.4	1,783.0	1,904.8	2,047.9	2,199.1	2,288.5
World total	*7,661.6*	*7,995.6*	*8,461.0*	*8,794.9*	*9,018.1*	*9,408.5*	*9,764.1*	*10,128.4*	*10,397.2*	*10,481.9*

[a] Ref. 11.
[b] Excluding the United States.

fossil fuels, are designed to promote fuel supply and utilization within the framework of an environmentally conscious society.

WORLD FUEL RESERVES, PRODUCTION, AND CONSUMPTION

Energy reserves, production, and consumption in the world economy are given in Tables 11, 12, and 13, respectively. As these tables indicate the overwhelming sources of petroleum reserves, and supoply are in the Middle East. Other significant sources of reserves include Russia, the North Sea, North American countries, and parts of southeast Asia. There are also significant concentrations of coal reserves in Russia and China. The dominant coal-producing countries include China and the United States, plus Poland, South Africa, Australia, India, Germany, and the United Kingdom. China is the single largest coal producer and consumer, utilizing over 22 EJ/yr (21×10^{15} Btu/yr) of this solid fossil fuel.

In addition to the significant consumption of coal and lignite, petroleum, and natural gas, several countries utilize modest quantities of alternative fossil fuels. Canada obtains some of its energy from the Athabasca tar sands development (the Great Canadian Oil Sands Project). Oil shale is burned at two 1600 MW power plants in Estonia for electricity generation. World reserves of tar sands total some 6400 EJ (6.1×10^{18} Btu), and world reserves of oil shale total some 20,400 EJ (19.3×10^{18} Btu).

Renewable and unconventional energy sources are used more extensively in other parts of the world than in the United States. Tables 12 and 13 document the significance of hydroelectric power throughout industrialized and developing economies. Biofuels are also a significant contributor to certain economies, with proportional contributions as follows: Kenya, 75%; India, 50%; China, 33%; Brazil, 25%; and Scandanavia, 10% (5,21). Peat is a significant source of energy for Russia, Finland, and Ireland.

World electricity generation and consumption, shown in Table 14 increased from 1982 to 1991, and is expected to continue to do so. Further, the industrialized economies are focusing on issues of energy conservation, materials conservation through recycling, and environmental protec-

Table 15. Projections of World and U.S. Energy Consumption, 1995–2005, EJ[a,b]

	1995			2000			2005		
Energy Source	Base	Low	High	Base	Low	High	Base	Low	High
				World projection[c]					
Oil	152	141	157	159	147	170	171	152	190
Gas	82	80	83	92	89	94	107	100	112
Coal	106	102	108	112	107	116	129	118	136
Nuclear	24	23	24	26	25	26	31	27	32
Other	31	31	32	35	34	36	42	39	46
Total	*395*	*377*	*404*	*424*	*402*	*442*	*480*	*436*	*516*
				U.S. projection					
Oil	38	37	38	40	39	41	42	40	44
Gas	23	23	23	25	24	25	26	25	27
Coal	21	21	21	21	21	22	22	22	23
Nuclear	7	7	7	7	7	7	7	7	8
Other	8	8	8	9	9	10	10	10	10
Total	*97*	*96*	*97*	*102*	*100*	*105*	*107*	*104*	*112*

[a] Ref. 14.
[b] To convert EJ to Btu, multiply by 9.48×10^{14}.
[c] World consumption totals also include the United States.

tion. Given the world trends in fuels availability and consumption, projections of energy production and consumption have been made, as shown in Table 15. These projections, to the year 2005, reflect the emphases on fuel availability, energy economics, and environmental awareness of a world community. Further, they reflect the trend toward increased technology development, leading to economically and environmentally sound energy utilization.

Reprinted from *Kirk-Othmer Encyclopedia of Chemical Technology*, 4th ed., vol. 11, John Wiley & Sons, Inc., New York, 1994.

BIBLIOGRAPHY

1. H. Enzer, W. Dupree, and S. Miller, *Energy Perspectives: A Presentation of Major Energy Related Data*, U.S. Department of the Interior, Washington, D.C., 1975.
2. H. H. Lansberg, L. L. Fishman, and J. L. Fisher, *Resources in America's Future*, Johns Hopkins University Press, Baltimore, Md., 1963.
3. H. C. Hottel, and J. B. Howard, *New Energy Technology: Some Facts and Assessments*, MIT Press, Cambridge, Mass., 1971.
4. *A National Plan for Energy Research, Development, and Demonstration: Creating Energy Choices for the Future*, Energy Research and Development Administration, Washington, D.C., 1976.
5. D. A. Tillman, *The Combustion of Solid Fuels and Wastes*, Academic Press, Inc., San Diego, Calif., 1991.
6. *The U.S. Coal Industry, 1970–1990: Two Decades of Change*, Energy Information Agency, Washington, D.C., 1992.
7. F. W. Brownell, *Clean Air Handbook*, Government Institutes, Inc., Rockville, Md., 1993.
8. S. Bruchey, *Growth of the Modern American Economy*, Dodd, Mead, New York, 1975.
9. N. Rosenberg, *Technology and American Economic Growth*, Harper and Row, New York, 1972.
10. H. M. Jones, *The Age of Energy*, Viking Press, New York, 1970.
11. *International Energy Annual 1991*, Energy Information Agency, Washington, D.C., 1992.
12. *EIA's Annual Energy Review 1992*, Energy Information Agency, Washington, D.C., 1993.
13. *Energy Policies of IEA Countries, 1990 Review*, International Energy Agency, Washington, D.C., 1991.
14. *Annual Energy Outlook, 1993*, Energy Information Agency, Washington, D.C., 1993.
15. *Keystone Coal Industry Manual*, Maclean Hunter Publishing Co., Chicago, 1992.
16. *U.S. Crude Oil, Natural Gas, and Natural Gas Liquids Reserves*, Energy Information Agency, Washington, D.C., 1992.
17. D. Tillman, E. Hughes, and B. Gold, "Cofiring of Biofuels in Coal Fired Boilers: Results of Case Study Analysis," *Proceedings First Biomass Conference of the Americas: Energy, Environment, Agriculture, and Industry*, Burlington, Vt., 1993.
18. D. Tillman, A. Rossi, and K. Vick, *Incineration of Municipal and Hazardous Solid Wastes*, Academic Press, Inc., San Diego, Calif., 1989.
19. J. L. Smith, *Early and Current Systems Utilizing Refuse Derived Fuels*, Combustion Engineering Co., Windsor, Conn., 1986.
20. V. Nast, G. Eirschele, and W. Hutchinson, "TDF Co-firing Experience in a Cyclone Boiler," *Proceedings, Strategic Benefits of Biomass and Waste Fuels Conference*, EPRI, Washington, D.C., 1993.
21. D. O. Hall and R. P. Overend, eds., *Biomass, Regenerable Energy*, John Wiley & Sons, Inc., New York, 1987.

FUELS FROM BIOMASS

DONALD L. KLASS
Entech International, Inc.
Barrington, Illinois

The contribution of biomass energy to U.S. energy consumption in the late 1970s was over 1.8×10^{15} kJ/yr [850,000 barrels of oil equivalent per day (BOE/d)] or more than 2% total energy consumption (1). By 1987, biomass energy had increased to approximately 3.1×10^{15} kJ/yr (1,400,000 BOE/d, 3.0×10^{15} Btu/yr), or 3.7% of U.S. primary consumption (2). Projections indicate that by the year 2000, the biomass energy contribution will increase to about 4.2×10^{5} kJ/yr (1,900,000 BOE/d), ie, over 4% of total U.S. primary energy consumption (2). Land- and water-based vegetation, organic wastes, and photosynthetic organisms are categorized as biomass and are nonfossil, renewable carbon resources from which energy, eg, heat, steam, and electric power, and solid, liquid, and gaseous fuels, ie, biofuels, can be produced and utilized as fossil fuel substitutes (see Plate I).

See also ALCOHOL FUELS; COMMERCIAL AVAILABILITY OF ENERGY TECHNOLOGY; FOREST RESOURCES; FORESTRY, SUSTAINABLE; ENERGY TAXATION, BIOMASS.

Renewable carbon resources is a misnomer because the earth's carbon is in a perpetual state of flux. Carbon is not consumed such that it is no longer available in any form. Reversible and irreversible chemical reactions occur in such a manner that the carbon cycle makes all forms of carbon, including fossil resources, renewable. It is simply a matter of time that makes one carbon form more renewable than another. If it is presumed that replacement does in fact occur, natural processes eventually will replenish depleted petroleum or natural gas deposits in several million years. Fixed carbon-containing materials that renew themselves often enough to make them continuously available in large quantities are needed to maintain and supplement energy supplies; biomass is a principal source of such carbon.

The capture of solar energy as fixed carbon in biomass via photosynthesis is the initial step in the growth of biomass. It is depicted by the equation

$$CO_2 + H_2O + \text{light} \xrightarrow{\text{chlorophyll}} (CH_2O) + O_2$$

Carbohydrate, represented by the building block CH_2O, is the primary organic product. For each gram mole of carbon fixed, about 470 kJ (112 kcal) is absorbed. Oxygen liberated in the process comes exclusively from the water, according to radioactive tracer experiments. There are many unanswered questions regarding the detailed molecular mechanisms of photosynthesis, but the prerequisites for plant biomass production are well established; ie, car-

Plate I. This photo depicts a typical backyard biogas digester under construction for a community outside Beijing, China. Waste is deposited down the stone chute (right) and gas is extracted from the open hole (left). An average digester produces 2–3 cubic meters of gas each day. Biogas is being explored in the development of alternative energy sources. Courtesy of the United Nations.

bon dioxide, light in the visible region of the electromagnetic spectrum, the sensitizing catalyst chlorophyll, and a living plant are essential. The upper limit of the capture efficiency of incident solar radiation in biomass is estimated to range from about 8% to as high as 15%; in most situations it is generally in the range 1% or less (3).

The primary features of biomass-to-energy technology as a source of synthetic fuels are illustrated in Figure 1. Conventionally, biomass is harvested for feed, food, and materials-of-construction applications, or is left in the growth areas where natural decomposition occurs. Decomposing biomass, and waste products from the harvesting and processing of biomass, disposed of on or in land, in theory can be partially recovered after a long period of time as fossil fuels. This is indicated by the dashed lines in Figure 1. Alternatively, biomass, and any wastes that result from its processing or consumption, can be converted directly into synthetic organic fuels if suitable conversion processes are available. The energy content of biomass also can be diverted to direct heating applications by combustion. Certain species of biomass can be grown, eg, the rubber tree (*Hevea braziliensis*), in which high energy hydrocarbons are formed within the biomass by natural biochemical mechanisms. The biomass serves the dual role of a carbon-fixing mechanism and a continuous source of hydrocarbons without being consumed in the process. Other plants, such as the guayule bush, also produce hydrocarbons but must be harvested to recover them. Thus,

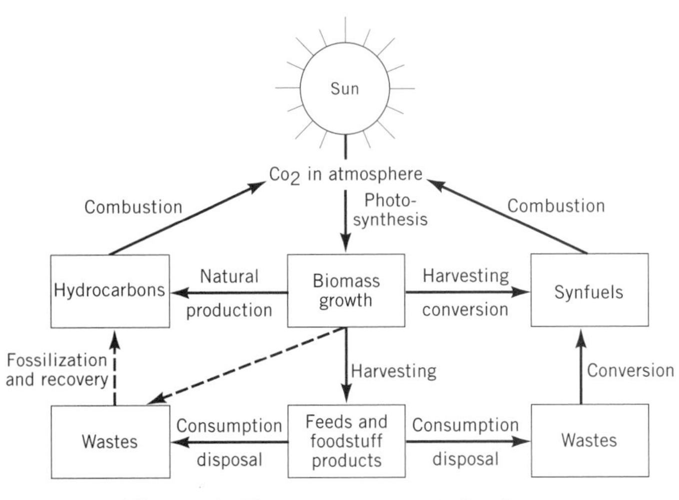

Figure 1. Biomass-to-energy technology.

conceptually, there are several different pathways by which energy products and synthetic fuels might be manufactured (Fig. 1).

Fixed carbon supplies also can be developed from renewable carbon sources by the conversion of carbon dioxide outside the biomass species into synthetic fuels and organic intermediates. The ambient air, which in 1992 contained an average of 350 ppm by volume of carbon dioxide, the dissolved carbon dioxide and carbonates in the oceans, and the earth's large terrestrial carbonate deposits serve as renewable carbon sources. However, because carbon dioxide is the final oxidation state of fixed carbon, it contains no chemical energy and energy must be supplied in a reduction step. A convenient method of supplying the required energy and reducing the oxidation state is to reduce carbon dioxide with elemental hydrogen. The end product can be, eg, methane (CH_4), the dominant component of natural gas.

$$CO_2 + 4\ H_2 \rightarrow CH_4 + 2\ H_2O$$

With all components in the ideal gas state, the standard enthalpy of the process is exothermic by −165 kJ (−39.4 kcal) per gram mole of methane formed. Biomass can serve as the original source of hydrogen, which then effectively acts as an energy carrier from the biomass to the carbon dioxide, to produce substitute (or synthetic) natural gas (SNG).

Distribution of Carbon

Estimation of the amount of biomass carbon on the earth's surface is a problem in global statistical analysis. Although reasonable projections have been made using the best available data, maps, surveys, and a host of assumptions, the validity of the results is impossible to support with hard data because of the nature of the problem. Nevertheless, such analyses must be performed to assess the feasibility of biomass energy systems and the gross types of biomass available for energy applications.

The results of one such study are summarized in Table 1 (4). Each ecosystem on the earth is considered in terms of area, mean net carbon production per year, and standing biomass carbon; ie, carbon contained in biomass on the earth's surface and not including carbon stored in biomass underground. Forest biomass, produced on only 9.5% of the earth's surface, contributes more than any other source to the total net carbon fixed on earth. Marine sources of net fixed carbon also are high because of the large area of earth occupied by water. However, the high turnover rates of carbon in the marine environment result in relatively small steady-state quantities of standing carbon. The low turnover rates of forest biomass make it the largest contributor to standing carbon reserves. Forests produce about 43% of the net carbon fixed each year and contain over 89% of the standing biomass carbon of the earth; tropical forests are the largest sources of these car-

Table 1. Estimate of Net Photosynthetic Production of Dry Biomass Carbon and Standing Biomass Carbon for World Biosphere[a,b]

		Carbon Production Mean Net		Standing Biomass Carbon	
Ecosystem	Area, 10^6 km^2 [c]	t/(hm^2·yr) [c]	10^9 t/yr [c]	t/hm^2	10^9 t
Tropical rain forest	17.0	9.90	16.83	202.5	344
Boreal forest	12.0	3.60	4.32	90.0	108
Tropical seasonal forest	7.5	7.20	5.40	157.5	118
Temperate deciduous forest	7.0	5.40	3.78	135.0	95
Temperate evergreen forest	5.0	5.85	2.93	157.5	79
Total	*48.5*		*33.26*		*744*
Extreme desert-rock, sand, ice	24.0	0.01	0.02	0.1	0.2
Desert and semidesert scrub	18.0	0.41	0.74	3.2	5.8
Savanna	15.0	4.05	6.08	18.0	27.0
Cultivated land	14.0	2.93	4.10	4.5	6.3
Temperate grassland	9.0	2.70	2.43	7.2	6.5
Woodland and shrubland	8.5	3.15	2.68	27.0	23.0
Tundra and alpine	8.0	0.63	0.50	2.7	2.2
Swamp and marsh	2.0	13.50	2.70	67.5	14.0
Lake and stream	2.0	1.80	0.36	0.1	0.02
Total	*100.5*		*19.61*		*85*
Total continental	*149.0*		*52.87*		*829*
Open ocean	332.0	0.56	18.59	0.1	3.3
Continental shelf	36.6	1.62	4.31	0.004	0.1
Estuaries excluding marsh	1.4	6.75	0.95	4.5	0.6
Algae beds and reefs	0.6	11.25	0.68	9.0	0.5
Upwelling zones	0.4	2.25	0.09	0.9	0.04
Total marine	*361.0*		*24.62*		*4.5*
Grand total	*510.0*		*77.49*		*833.5*

[a] Ref. 4.
[b] Dry biomass is assumed to contain 45% carbon.
[c] 1 km^2 = 1 × 10^6 m^2 (0.3861 sq. mi); to convert t/(hm^2·yr) to short ton/(acre · yr), divide by 2.24.

bon reserves. Temperate deciduous and evergreen forests also are large sources of biomass carbon, followed by the savanna and grasslands. Cultivated land is one of the smaller producers of fixed carbon and is only about 9% of the total terrestrial area of the earth.

Human activity, particularly in the developing world, continues to make it more difficult to sustain the world's biomass growth areas. It has been estimated that tropical forests are disappearing at a rate of tens of thousands of hm^2 per year. Satellite imaging and field surveys show that Brazil alone has a deforestation rate of approximately 8×10^6 hm^2/yr (5). At a mean net carbon yield for tropical rain forests of 9.90 $t/hm^2 \cdot yr$ (4) (4.42 short ton/acre·yr), this rate of deforestation corresponds to a loss of 79.2×10^6 t/yr of net biomass carbon productivity.

The remaining carbon transport mechanisms on earth are primarily physical mechanisms, such as the solution of carbonate sediments in the sea and the release of dissolved carbon dioxide to the atmosphere by the hydrosphere (6). The great bulk of carbon, however, is contained in the lithosphere as carbonates in rock. These carbon deposits contain little or no stored chemical energy, although some high temperature deposits could provide considerable thermal energy, and all of the energy for a synfuel system must be supplied by a second raw material, such as elemental hydrogen. These carbon deposits consist of lithospheric sediments and atmospheric and hydrospheric carbon dioxide. Together, these carbon sources comprise 99.9% of the total carbon estimated to exist on the earth. Fossil fuel deposits are only about 0.05% of the total, and the nonfossil energy-containing deposits make up the remainder, about 0.02%.

Biomass carbon is thus a very small, but important, fraction of the total carbon inventory on earth. It helps maintain the delicate balance among the atmosphere, hydrosphere, and biosphere necessary to support all life forms, and serves as a perpetual source of food and materials. Biomass carbon also has served as a primary energy source for the industrialized nations of the world; it continues to do so for developing countries. Biomass carbon may again become a dominant source of energy products throughout the world because of fossil fuel depletion and environmental problems, eg, the effect that large-scale fossil fuel combustion is believed by many to have on atmospheric carbon dioxide build-up (7). The utilization of biomass carbon as a primary energy source does not add any new carbon dioxide to the atmosphere; it is simply recycled between the surface of the earth and the air over a period of time that is extremely short compared to the recycling time of fossil-derived carbon dioxide.

ENERGY POTENTIAL

The percentage of energy demand that could be satisfied by particular nonfossil energy resources can be estimated by examination of the potential amounts of energy and biofuels that can be produced from renewable carbon resources and comparison of these amounts with fossil fuel demands.

The average daily incident solar radiation, or insolation, that strikes the earth's surface worldwide is about 220 W/m^2 (1675 Btu/ft^2). The annual insolation on 0.01% of the earth's surface is approximately equal to all energy consumed (ca 1992) by humans in one year, ie, 321×10^{18} J (305×10^{15} Btu). In the United States, the world's largest energy consumer, annual energy consumption is equivalent (1992) to the insolation on about 0.1 to 0.2% of its total surface.

Based on the state of technology in the early 1990s, the most widespread and practical mechanism for capture of this energy is biomass formation. The energy content of standing biomass carbon, ie, the above-ground biomass reservoir that in theory could be harvested and used as an energy resource (Table 1) is about 110 times the world's annual energy consumption (8). Using a nominal biomass heating value of 16×10^9 J/dry t (13.8×10^6 Btu/short ton), the solar energy trapped in 17.9×10^9 t of biomass, or about 8×10^9 t of biomass carbon, would be equivalent to the world's fossil fuel consumption in 1990 of 286×10^{18} J. It is estimated that 77×10^9 t of carbon, or 171×10^9 t of biomass equivalent, most of it wild and not controlled, is fixed on the earth each year. Biomass should therefore be considered as a raw material for conversion to large supplies of renewable substitute fossil fuels. Under controlled conditions dedicated biomass crops could be grown specifically for energy applications.

A realistic assessment of biomass as an energy resource is made by calculating average surface areas needed to produce sufficient biomass at different annual yields to meet certain percentages of fuel demand for a particular country (Table 2). These required areas are then compared with surface areas available. The conditions of biomass production and conversion used in Table 2 are either within the range of 1993 technology and agricultural practice, or are believed to be attainable in the future.

Figure 2 shows the three yield levels in Table 2 together with the percentage of the U.S. area needed to supply SNG from biomass for any selected gas demand. Although relatively large areas are required, the use of land- or freshwater-based biomass for energy applications is still practical. The area distribution pattern of the United States (Table 3) shows selected areas or combinations of areas that might be utilized for biomass energy applications (9), ie,

Table 2. Potential Substitute Natural Gas in United States from Biomass at Different Crop Yields

	Average Area Required, km^2[b,c,d]		
Demand, %[a]	25 $t/(hm^2 \cdot yr)$	50 $t/(hm^2 \cdot yr)$	100 $t/(hm^2 \cdot yr)$
1.58	20,400	10,200	5,100
10	129,000	64,500	32,300
50	645,500	323,000	161,000
100	1,291,000	645,500	323,000

[a] United States demand estimated to be 244×10^8 GJ or 653×10^9 m^3 (231×10^{11} standard cubic feet at 15.5°C, 101.5 kPa (60°F, 30.00 in Hg) dry (SCF)). A percentage of 1.58 is equal to a daily production of 28.3×10^6 m^3 at normal conditions (1×10^9 SCF) of SNG.
[b] Biomass, whether trees, plants, grasses, algae, or water plants, has a heating value of 15.1×10^9 J/dry t, and is converted in integrated biomass planting, harvesting, and conversion systems to SNG at an overall thermal efficiency of 50%.
[c] 1 km^2 = 0.3861 sq. mi.
[d] Yields expressed as dry t.

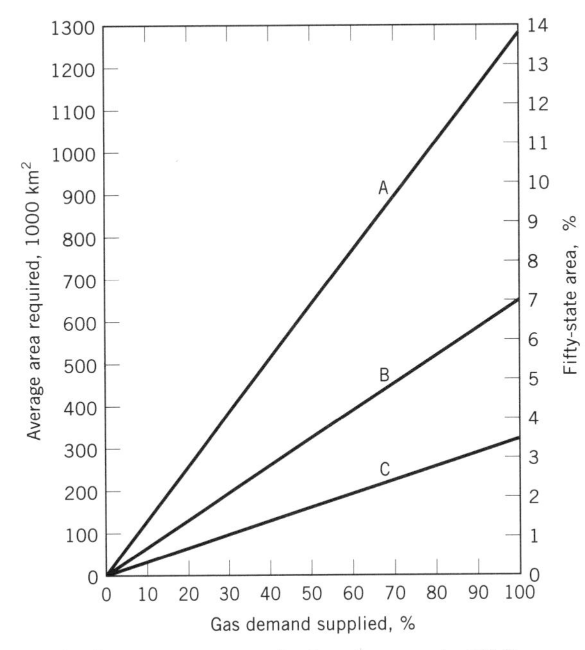

Figure 2. Average area required and percent of U.S. area vs gas demand, %, for a total gas demand of 24.4 EJ/yr. A, 25 $t/hm^2 \cdot yr$; B, 50 $t/hm^2 \cdot yr$; C, 100 $t/hm^2 \cdot yr$ (Table 2). To convert $t/hm^2 \cdot yr$ to short ton/acre·yr, divide by 2.24.

areas not used for productive purposes. It is possible that biomass for both energy and foodstuffs, or energy and forest products applications, can be grown simultaneously or sequentially in ways that would benefit both. Relatively small portions of the bordering oceans also might supply needed biomass growth areas, ie, marine plants would be

Table 3. Land and Water Areas of United States[a]

Area Classification	10^6 hm^2 [b]	%
Nonfederal land		
Forest[c]	179.41	18.8
Rangeland[d]	178.66	18.7
Other land[e]	279.09	29.2
Transition land[f]	14.41	1.5
	651.57	*68.2*
Federal land		
Forest[c]	102.14	10.7
Rangeland[d]	133.10	13.9
Other land[e]	25.70	2.7
	260.94	*27.3*
Water		
Inland water	24.75	2.6
Other water	19.28	2.0
	44.03	*4.6*
Total land and water	*956.55*	

[a] Ref. 9. Data for forest, rangeland, and other land as of 1982; data on inland water as of 1980; data on other water as of 1970.
[b] 1 hm^2 = 2.471 acre.
[c] At least 10% stocked by trees of any size, or formerly having such tree cover and not currently developed for nonforest use.
[d] Climax vegetation is predominantly grasses, grass-like plants, forbs, and shrubs suitable for grazing and browsing.
[e] Includes crop and pasture land and farmsteads, strip mines, permanent snow and ice, and land that does not fit into any other land cover.
[f] Forest land that carries grasses or forage plants used for grazing as the predominant vegetation.

Table 4. Potential U.S. Biomass Energy Available in 2000, EJ[a,b]

Energy Source	Estimated Recoverable	Theoretical Maximum
Wood and wood wastes	11.0	26.4
Municipal solid wastes		
Combustion	1.9	2.1
Landfill methane	0.2	1.1
Herbaceous biomass and agricultural residues	1.1	15.8
Aquatic biomass	0.8	8.1
Industrial solid wastes	0.2	2.2
Sewage methane	0.1	0.2
Manure methane	0.05	0.9
Miscellaneous wastes	0.05	1.1
Total	*15.4*	*57.9*

[a] Ref. 10. 1 EJ = 0.9488 × 10^{15} Btu.
[b] Gross heating value of biomass or methane. Conversion of biomass or methane to another biofuel requires that the process conversion efficiency be used to reduce the potential energy available. These figures do not include additional biomass from dedicated energy plantations.

grown and harvested. The steady-state carbon supplies in marine ecosystems can conceivably be increased under controlled conditions over current low levels by means of marine biomass energy plantations in areas of the ocean dedicated to this objective.

Waste biomass is another large renewable carbon resource. It consists of a wide range of materials and includes municipal solid wastes (MSW), municipal sewage, industrial wastes, animal manures, agricultural crop and forestry residues, landscaping and tree clippings and trash, and dead biomass that results from nature's life cycles. Several of these wastes can cause serious health or environmental problems if not disposed of properly. Some wastes, such as MSW, can be considered a source of recyclables such as metals and glass in addition to energy. Waste biomass is a potential energy resource in the same manner as virgin biomass.

To assess the potential availability and impact of energy from wastes on energy demand, the energy contents and availabilities of different types of wastes generated must be considered. For example, in the United States an average of 2.3 kg of MSW/d is discarded per person. From an energy standpoint, one t of MSW has an as-received energy content of about 10.5 × 10^9 joule (10 million Btu), so about 2.2 EJ/yr of energy potential resides in the MSW discarded in the United States.

The amount of energy that can actually be recovered from a given waste and utilized depends on the waste type. The amount of available MSW is larger than the total amount of available agricultural wastes even though much larger quantities of agricultural wastes are generated. A larger percentage of MSW is collected for centralized disposal than the corresponding amounts of agricultural wastes, most of which are left in the fields where generated; the collection costs would be prohibitive for most agricultural wastes.

Several studies estimate the potential of available virgin and waste biomass as energy resources (Table 4) (10). In Table 4, the projected potential of the recoverable materials is about 25% of the theoretical maximum; woody biomass is about 70% of the total recoverable potential. These estimates of biomass energy potential are based on existing, sustainable biomass production and do not include new, dedicated biomass energy plantations that might be developed.

U.S. Market Penetration

Table 5 shows U.S. consumption of biomass energy in 1990 and projected consumption for 2000 (10,11). The projected consumption for 2000 is about 50% greater than the consumption of biomass energy in 1990.

A projection of biomass energy consumption in the United States for the years 2000, 2010, 2020, and 2030 is shown in Table 6 by end-use sector (12). This analysis is based on a National Premiums Scenario which assumes that specific market incentives are applied to all new renewable energy technology deployment. The scenario depends on the enactment of federal legislation equivalent to a fossil fuel consumption tax. Any incentives over and above those in place (ca 1992) for use of renewable energy will have a significant impact on biomass energy consumption.

The market penetration of synthetic fuels from biomass and wastes in the United States depends on several basic factors, eg, demand, price, performance, competitive feedstock uses, government incentives, whether established fuel is replaced by a chemically identical fuel or a different product, and cost and availability of other fuels such as oil and natural gas. Detailed analyses have been performed to predict the market penetration of biomass energy well into the twenty-first century. A range of from 3 to about 21 EJ seems to characterize the results of most of these studies.

Table 5. U.S. Consumption of Biomass Energy, EJ[a]

Resource	1990,[b] EJ	2000,[c] EJ
Wood and wood wastes		
Industrial sector	1.646	2.2
Residential sector	0.828	1.1
Commercial sector	0.023	0.04
Utilities	0.013	0.01
Subtotal (wood)	*2.510*	3.35
Municipal solid wastes	0.304	0.63
Agricultural and industrial wastes	0.040	0.08
Methane		
Landfill gas	0.033	0.100
Digester gas	0.003	0.004
Thermal gasification	0.001	0.002
Subtotal (methane)	*0.037*	*0.106*
Transportation fuels		
Ethanol	0.063[d]	0.1[d]
Other biofuels	0	0.1
Subtotal (transportation fuels)	*0.063*	*0.2*
Total all resources	*2.954*	4.37
Primary energy consumption, %	3.3	4.8

[a] 1 EJ = 0.9488×10^{15} Btu. To convert from EJ to barrels of oil equivalent per day (BOE/d), multiply by 448,200.
[b] Refs. 10 and 11. Total energy consumption including biomass energy estimated to be 88.426 EJ in 1990.
[c] Ref. 10. Assumes noncrisis conditions, tax incentives and PURPA in place continued to 2000, no legislative mandates to embark on an off-oil campaign, and total consumption of 91.7 EJ in 2000.
[d] Domestic consumption only.

Table 6. Projected Biomass Energy Consumption in the United States from 2000 to 2030,[a] EJ[b]

End Use Sector	2000	2010	2020	2030
Industry[c]	2.85	3.53	4.00	4.48
Electricity[d]	3.18	4.41	4.95	5.48
Buildings[e]	1.05	1.53	1.90	2.28
Liquid fuels[f]	0.33	1.00	1.58	2.95
Total	*7.41*	*10.47*	*12.43*	*15.19*

[a] Ref. 12.
[b] 1 EJ = 0.9488×10^{15} Btu. Assumes market incentives of 2 ¢/kWh on fossil fuel-based electricity generation, \$2.00/10^6 Btu on direct coal and petroleum consumption, and \$1.00/10^6 Btu on direct natural gas consumption.
[c] Combustion of wood and wood wastes.
[d] Electric power derived from present (ca 1992) technology via the combustion of wood and wood wastes, MSW, agricultural wastes, landfill and digester gas, and advanced digestion and turbine technology.
[e] Biomass combustion in wood stoves.
[f] Ethanol from grains, and ethanol, methanol, and gasoline from energy crops.

U.S. capacity for producing biofuels manufactured by biological or thermal conversion of biomass must be dramatically increased to approach the potential contributions based on biomass availability. For example, an incremental EJ per year of methane requires about 210 times the biological methane production capacity that now exists, and an incremental EJ per year of fuel ethanol requires about 14 times existing ethanol fermentation plant capacity. The long lead times necessary to design and construct large biomass conversion plants makes it unlikely that sufficient capacity can be placed on-line before 2000 to satisfy EJ blocks of energy demand. However, plant capacities can be rapidly increased if a concerted effort is made by government and private sectors.

Projections of market penetrations and contributions to primary consumption of energy from biomass are subject to much criticism and contain significant errors. However, even though these projections may be incorrect, they are necessary to assess the future role and impact of renewable energy resources, and to help in deciding whether a potential renewable energy resource should be developed.

Global Market Penetration. The consumption of all energy resources worldwide in 1990, according to the United Nations, is presented in Table 7 (8). Detailed and time-consuming analysis is necessary to assure validity of the results for an energy resource that is as widespread, dispersed, and disaggregated as biomass, and many nations of the world do not require the archiving of historical energy production and consumption data. Table 7 indicates that biomass energy is a significant source of energy in the developing regions of the world; Africa, 36.8% of total energy consumed; South America, 23.7%; and Asia, 10.9%. It is a small energy resource in the industrialized areas relative to fossil fuels. The markets for biomass energy as replacements and substitutes for fossil fuels are large and have only been developed to a limited extent. As fossil fuels are either phased out because of environmental issues or become less available because of depletion, biomass energy is expected to acquire an increasingly larger share of the organic fuels market.

Table 7. Global Energy Consumption in 1990, EJ[a]

Region	Fossil Fuels[b] Solids	Liquids	Gases	Electricity[c]	Biomass[d]	Total[e]
Africa	2.96	3.36	1.55	0.18	4.68	12.73
N. America	21.55	38.48	22.13	4.69	1.77[f]	88.62
S. America	0.68	4.66	2.09	1.29	2.71	11.43
Asia	35.52	27.58	8.38	2.57	8.89	82.94
Europe[g]	35.18	40.90	37.16	6.25	1.29	120.85
Oceania	1.64	1.70	0.85	0.14	0.19	4.53
Total world	*97.52*	*116.68*	*72.18*	*15.13*	*19.53*	*321.10*

[a] Ref. 8.
[b] Solids are hard coal, lignite, peat, and oil shale. Liquids are crude petroleum and natural gas liquids. Gases are natural gas.
[c] Includes hydro, nuclear, and geothermal sources, but not fossil fuel-based electricity, which is included in fossil fuels.
[d] Includes fuelwood, charcoal bagasse, and animal, crop, pulp, paper, and municipal solid wastes, but does not include derived biofuels.
[e] Sums of individual figures may not equal totals because of rounding.
[f] Less than Table 5 value of 2.954 EJ for the United States because does not include biofuels from biomass.
[g] Includes former Soviet Union.

CHEMICAL CHARACTERISTICS OF BIOMASS

The chemical characteristics of biomass vary over a broad range because of the many different types of species. Table 8 compares the typical analyses and energy contents of land- and water-based biomass, ie, wood, grass, kelp, and water hyacinth, and waste biomass, ie, manure, urban refuse, and primary sewage sludge, with those of cellulose, peat, and bituminous coal. Pure cellulose, a representative primary photosynthetic product, has a carbon content of 44.4%. Most of the renewable carbon sources listed in Table 8 have carbon contents near this value. When adjusted for moisture and ash contents, it is seen that with the exception of the sludge sample, the carbon contents are slightly higher than that of cellulose, but span a relatively narrow range.

The organic components that make up biomass depend on the species. Alpha-cellulose [*9004-34-6*], or cellulose as it is more generally known, is the chief structural element and a principal constituent of many biomass types. In trees, eg, the concentration of cellulose is about 40 to 50% of the dry weight; materials such as lignin and compounds related to cellulose, such as hemicelluloses, comprise most of the remaining organic components. However, cellulose is not always the dominant component in the carbohydrate fraction of biomass. For example, it is a minor component in giant brown kelp; mannitol [*87-78-5*], a hexahydric alcohol that can be formed by reduction of the aldehyde group of D-glucose to a methylol group, and alginic acid [*9005-32-7*], a polymer of mannuronic and glucuronic acids, are the primary carbohydrates.

Fat and protein content of plant biomass are much less on a percentage basis than the carbohydrate components. Fatty constituents are usually present at the lowest concentration; the protein fraction is much higher in concentration but lower than that of carbohydrate. Crude protein

Table 8. Composition and Heating Value of Biomass, Wastes, Peat, and Coal

Analysis	Pure Cellulose	Pine Wood	Kentucky Bluegrass	Giant Brown Kelp[a]	Feedlot Manure	Urban Refuse[b]	Primary Sewage Sludge	Reed Sedge Peat	Illinois Bituminous Coal
Elemental, wt %									
C	44.44	51.8	45.8	27.65	35.1	41.2	43.75	52.8	69.0
H	6.22	6.3	5.9	3.73	5.3	5.5	6.24	5.45	5.4
O	49.34	41.3	29.6	28.16	33.2	38.7	19.35	31.24	14.3
N		0.1	4.8	1.22	2.5	0.5	3.16	2.54	1.6
S		0.0	0.4	0.34	0.4	0.2	0.97	0.23	1.0
C (MAF)[c]	44.44	52.1	52.9	45.3	45.9	47.9	59.5	57.2	75.6
Proximate, wt %									
Moisture		5–50	10–70	85–95	20–70	18.4	90–97	84.0	7.3
Volatile matter		99.5	86.5	61.1	76.5	86.1	73.47	92.26	91.3
Ash		0.5	13.5	38.9	23.5	13.9	26.53	7.74	8.7
High heating value, MJ/kg[d]									
Dry	17.51	21.24	18.73	10.01	13.37	12.67[e]	19.86	20.79	28.28
MAF[c]	17.51	21.35	21.65	16.38	17.48		27.03	22.53	30.97
Carbon	39.40	41.00	40.90	36.20	38.09		45.39	39.38	40.99

[a] *Macrocystis pyrifera.*
[b] Combustible fraction.
[c] Moisture and ash free.
[d] To convert MJ/kg to Btu/lb, multiply by 430.
[e] As received with metals.

Table 9. Fuel Values of Biomass Components[a,b]

Component	Carbon, %	MJ/kg[c,d]
Monosaccharides	40	15.6
Disaccharides	42	16.7
Polysaccharides	44	17.5
Lignin	63	25.1
Crude protein	53	24.0
Fat	75	39.8
Carbohydrate	41–44	16.7–17.7
Crude fiber[e]	47–50	18.8–19.8

[a] Ref. 13. Approximate values.
[b] Product water in liquid state.
[c] Dry.
[d] To convert MJ/kg to Btu/lb, multiply by 430.
[e] Contains ca 15–30% lignin.

values can be approximated by multiplying the organic nitrogen analyses by 6.25. The average weight percentage of nitrogen in pure dry protein is about 16%, although the protein content of each biomass species can best be determined by amino acid assay. The calculated crude protein values of the biomass species listed in Table 8 range from a low of about 0% for pine wood, to a high of about 30% for Kentucky Blue Grass. For grasses, the protein content is strongly dependent on growing procedures used before harvest, particularly fertilization methods. However, some biomass species, such as legumes, fix nitrogen from the ambient atmosphere and often contain high protein concentrations.

The energy content of biomass is a very important parameter from the standpoint of conversion to energy products and synfuels. The different components in biomass have different heats of combustion because of different chemical structures and carbon content. Table 9 lists heating values for each of the classes of organic compounds. The more reduced the state of carbon in each class, the higher the heating value. As carbon content increases and degree of oxygenation is reduced, the structures become more hydrocarbon-like and heating value increases. Fatty components thus have the highest heating values of the components in Table 9. Cellulose, the dominant component in most biomass, has a high heating value of 17.51 MJ/kg (7533 Btu/lb).

Typical low heating values of selected biomass are given in Table 10. The water-based algae, *Chlorella*, has a higher energy content value than woody and fibrous materials because of its higher lipid or protein contents. Oils derived form plant seeds are much higher in energy content and approach the heating value of paraffinic hydrocarbons. High concentrations of inorganic components in a given biomass species greatly affect its heating value because inorganic materials generally do not contribute to heat of combustion, eg, giant brown kelp, which leaves an ash residue equivalent to about 40 wt % of the dry weight, has a high heating value on a dry basis of about 10 MJ/kg, and on a dry, ash-free basis, the heating value is about 16 MJ/kg (Table 8).

Table 10. Low Heating Values[a] of Biomass and Fossil Materials[b]

Material	MJ/kg[c]
Wood	
Pine	21.03
Beech	20.07
Birch	20.03
Oak	19.20
Oak bark	20.36
Bamboo	19.23
Fiber	
Coconut shells	20.21
Buckwheat hulls	19.63
Bagasse	19.25
Green algae	
Chlorella	26.98
Seed oils	
Cottonseed	39.77
Rapeseed	39.77
Linseed	39.50
Amorphous carbon	33.8
Paraffinic hydrocarbon	43.3
Crude oil	48.2

[a] Product water in vapor state.
[b] Refs. 14 and 15.
[c] Dry; to convert MJ/kg to Btu/lb, multiply by 430.

BIOMASS CONVERSION

Various processes can be used to produce energy or gaseous, liquid, and solid fuels from biomass and wastes. In addition, chemicals can be produced by a wide range of processing techniques. The following list summarizes the principal feed, process, and product variables considered in developing a synfuel-from-biomass process.

Feeds	Conversion Processes	Products
Land-based trees plants grasses	Separation, combustion, pyrolysis, hydrogenation, anaerobic fermentation, aerobic fermentation, biophotolysis, partial oxidation, steam reforming, chemical hydrolysis, enzyme hydrolysis, other chemical conversions, natural processes	Energy thermal, steam, electric
Water-based single-cell algae multicell algae water plants		Solid fuels char, combustibles
		Gaseous fuels methane (SNG), hydrogen, low and medium thermal-value gas, light hydrocarbons
		Liquid fuels methanol, ethanol, higher hydrocarbons, oils
		Chemicals

There are many interacting parameters and possible feedstock–process–product combinations, but all are not feasible from a practical standpoint; eg, the separation of small amounts of metals present in biomass and the direct combustion of high moisture content algae are technically possible, but energetically unfavorable.

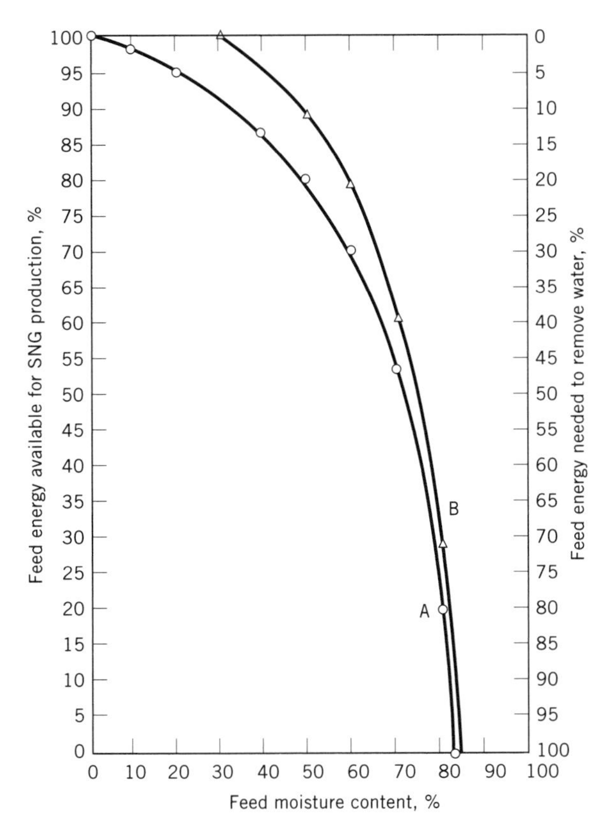

Figure 3. Effect of feed moisture content on energy available for synfuel production. Assumes feed has a heating value of 11.63 MJ/kg (5000 Btu/lb) dry. A, 0% moisture in dried feed; B, 30% moisture in dried feed. For example, reduction of an initial moisture content of 70 wt % by thermal drying to 30% moisture content requires the equivalent of 37% feed energy content and leaves 63% feed energy available for SNG production.

Moisture content of the biomass chosen is especially important in the selection of a suitable conversion process. The giant brown kelp, *Macrocystis pyrifera*, contains as high as 95% intracellular water, so thermal gasification techniques such as pyrolysis and hydrogasification cannot be used directly without first drying the algae. Anaerobic digestion methods are preferred because the water does not need to be removed. Wood, on the other hand, can often be processed by several different thermal conversion techniques without drying. Figure 3 illustrates the effects of thermal drying on biomass used for synfuel production as SNG. A large portion of a feed's equivalent energy content can be expended for drying, so the properties of the feed must be considered carefully in relation to the conversion process.

Table 11 lists the important feed characteristics to be examined when developing a successful conversion process for a specific biomass feedstock. A particular process also may have specific requirements within a given process type; eg, anaerobic digestion and alcoholic fermentation are both biological conversion processes, but animal manure, which has a relatively high biodegradability, is not equally applicable as a feedstock for both processes. The degree of complexity of the process design also affects practical utility of the conversion process. Some processes, such as combustion, are simple in design, whereas others, such as alcoholic fermentation, consist of several different unit operations and are complex. Capital and operating costs are dictated by the particular process design, the logistics of raw material supply, plant size, and operating conditions. Generally, the more complex the process, the higher the costs.

The need to meet environmental regulations can affect processing costs. Undesirable air emissions may have to be eliminated and liquid effluents and solid residues treated and disposed of by incineration or/and landfilling. It is possible for biomass conversion processes that utilize waste feedstocks to combine waste disposal and treatment with energy and/or biofuel production so that credits can be taken for negative feedstock costs and tipping or receiving fees.

The primary types of conversion processes for biomass can be divided into four groups, ie, physical, biological–biochemical, thermal, and chemical.

Physical Processes

Particle Size Reduction. Changes in the physical characteristics of a biomass feedstock often are required before it can be used as a fuel. Particle size reduction (qv) is performed to prepare the material for direct fuel use, for fabrication into fuel pellets, or for a conversion process. Particle size of the biomass also is reduced to reduce its storage volume, to transport the material as a slurry or pneumatically, or to facilitate separation of the components.

The ultimate particle size required depends on the conversion process used, eg, for thermal gasification processes the particle size of the material converted can influence

Table 11. Biomass Feedstock Characteristics that Affect Suitability of Conversion Process

	Process Type			
Feedstock Characteristic	Physical	Thermal	Biological	Chemical
Water content	+	+	+	+
Energy content	+	+		+
Chemical composition		+	+	+
Bulk component analysis	+	+	+	+
Size distribution	+	+	+	+
Noncombustibles	+	+	+	
Biodegradability			+	
Carbon reactivity		+		+
Organism content/type			+	

the rate at which the gasification process occurs. Biological processes, such as anaerobic digestion to produce methane, also are affected by the size of the particle; the smaller the particle, the higher the reaction rate because more surface area is exposed to the organisms. Particle size reduction consists of one or more unit operations that make up the front end of the total processing system. Two basic types of machines are in commercial use (ca 1992) for particle size reduction, ie, wet shredders and dry shredders. Wet shredders utilize a hydropulping mechanism in which a high speed cutting blade pulverizes a water slurry of the feed over a perforated plate. The pulped material passes through the plate and the nonpulping materials are ejected. The two most common types of dry shredders in commercial use are the vertical and horizontal shaft hammermills. Rotating metal hammers on a shaft reduce the particle size of the feed material until the particles are small enough to drop through the grate openings. Particle size reduction units such as agricultural choppers and tree chippers are usually hammermills or are equipped with knife blades that reduce particle size by a cutting or shearing action. Maintenance costs for dry shredders generally are higher than those for wet shredders.

Separation. It may be desirable to separate the feedstock into two or more components for different applications. Examples include separation of agricultural biomass into foodstuffs and residues that may serve as fuel or as a raw material for synfuel manufacture, separation of forest biomass into the darker bark-containing fractions and the pulpable components, separation of marine biomass to isolate various chemicals, and separation of urban refuse into the combustible fraction, ie, refuse-derived fuel (RDF), and metals and glass for recycling. Common operations such as screening, air classification, magnetic separation, extraction, distillation, filtration, and crystallization often are used as well as industry-specific methods characteristic of farming, forest products, and specialized industries.

Drying. Drying refers to the vaporization of all or part of the water in the feedstock. In cases where the biomass or waste is thermally processed directly for energy recovery, it may be necessary to partially dry the raw feed before conversion; otherwise, more energy might be consumed to operate the process than that produced in the form of recovered energy or fuels. Open-air solar drying is perhaps the cheapest drying method if it can be used. Raw materials that are not sufficiently stable to be dried by solar methods can be dried more rapidly using spray driers, drum driers, and convection ovens. For large-scale applications, forced-air-type furnaces and driers designed to use stack gases are more efficient. Special driers such as those that use powdered feeds, and hot metal balls separated from the feed for reheating, also have been successful.

Fabrication. Processes for fabricating solid fuel pellets from a variety of feedstocks, particularly RDF, wood, and wood and agricultural residues, have been developed. The pellets are manufactured by extrusion and other techniques and, in some cases, a binding agent such as a thermoplastic resin is incorporated during fabrication. The fabricated products are reported to be more uniform in combustion characteristics than the raw biomass. Depending on the composition of the additives in the pelletized fuel, the heat of combustion can be higher or lower than that of the unpelletized material.

Biological–Biochemical Processes

Fermentation is a biological process in which a water slurry or solution of raw material interacts with microorganisms and is enzymatically converted to other products. Biomass can be subjected to fermentation conditions to form a variety of products. Two of the most common fermentation processes yield methane and ethanol. Biochemical processes include those that occur naturally within the biomass.

Anaerobic Digestion. Methane can be produced from water slurries of biomass by anaerobic digestion in the presence of mixed populations of anaerobes. This process has been used for many years to stabilize municipal sewage sludges for purposes of disposal. Presuming the biomass is all cellulose, the chemistry can be represented in simplified form as follows:

$$(C_6H_{10}H_6)_x + x\,H_2O \xrightarrow{\text{hydrolysis}} x\,C_6H_{12}O_6$$

$$x\,C_6H_{12}O_6 \xrightarrow{\text{acidification}} 3x\,CH_3COOH$$

$$3x\,CH_3COOH \xrightarrow{\text{methanation}} 3x\,CH_4 + 3x\,CO_2$$

Complex organic compounds are first converted in the water slurry to lower molecular weight soluble products, primarily carboxylic acids, by the acidogenic bacteria present in the digester. Methanogenic bacteria then convert these intermediates to a medium heat value (MHV) gas which has heating values ranging from about 19.6 to 29.4 MJ/m^3 at normal conditions [500 to 750 Btu/SCF, dry at 60°F, 30 in. Hg (15.5°C, 101.5 kPa)]. The principal components in the gas are methane and carbon dioxide (Fig. 4). Residual ungasified solids which contain more nitrogen, phosphorus, and potassium than the feed solids also are formed. In some systems, these solids have application as animal feeds and fertilizers. The conventional high-rate digestion process is conducted under nonsterile conditions in large, mixed, anaerobic fermentation vessels at near-ambient pressures, temperatures of about 35°C (mesophilic range) or 55°C (thermophilic range), and reactor residence times of 10 to 20 days. The pH is maintained in the range 6.8 to 7.2. The raw digester gas has been used for many years as a fuel to heat the digesters and for steam and electric power production. It also can be upgraded to SNG by removing the carbon dioxide by means of adsorption or acid-gas scrubbing processes.

The overall thermal efficiency of the anaerobic digestion of biomass to methane is a function of the process design and the raw material characteristics. Without pretreatment of the feed or the recycled ungasified solids to increase biodegradability, the feed can generally be gasified at overall thermal efficiencies ranging from about 30 to 60%. Yields of methane can range up to 0.30 m^3 at normal

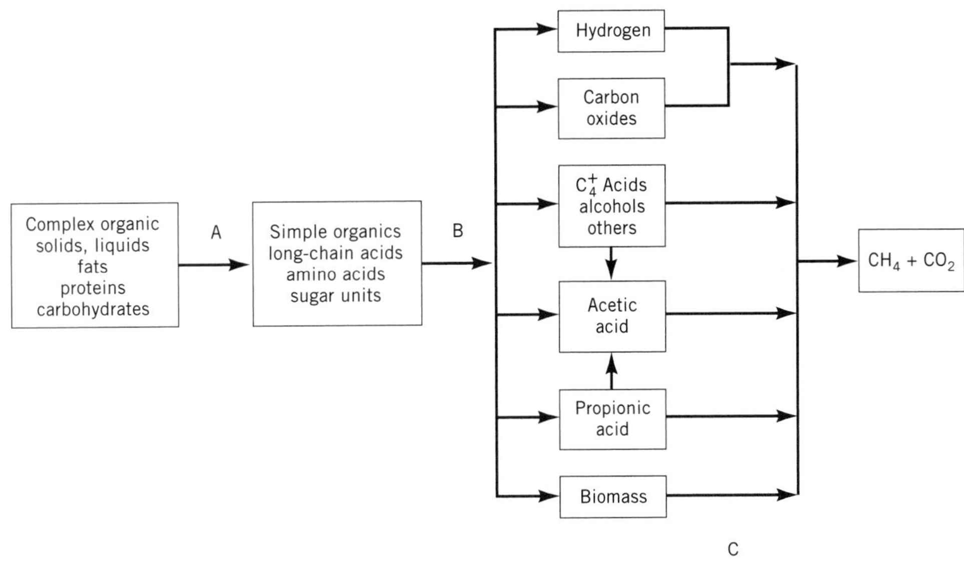

Figure 4. Microbial phases in anaerobic digestion: A, Hydrolysis; B, acidification; C, methane fermentation.

conditions/kg (5 SCF/lb) of volatile solids (VS) added to the digesters. A typical biological gasification plant can contain hydrolysis units, anaerobic digesters, gas cleanup and dehydration units, and liquid effluent treatment units (Fig. 5). Advanced digester designs under development include two-phase, plug-flow, packed-bed, fluidized-bed, and sludge blanket digesters (16).

Typical methane yields and volatile solids reductions observed under standard high-rate conditions are shown in Table 12. Longer detention times will increase the values of these parameters, eg, a methane yield of 0.284 m^3 at normal conditions/kg VS added (4.79 SCF/lb VS added) and volatile solids reduction of 53.9% for giant brown kelp at a detention time of 18 days instead of the corresponding values of 0.229 and 43.7 at 12 days under standard high rate conditions. However, improvements might be desirable in the reverse direction, ie, at shorter detention times.

Alcoholic Fermentation. Certain types of starchy biomass such as corn and high sugar crops are readily converted to ethanol under anaerobic fermentation conditions in the presence of specific yeasts (*Saccharomyces cerevisiae*) and other organisms (Fig. 6). However, alcoholic fermentation of other types of biomass, such as wood and municipal wastes that contain high concentrations of cellulose, can be performed in high yield only after the cellu-

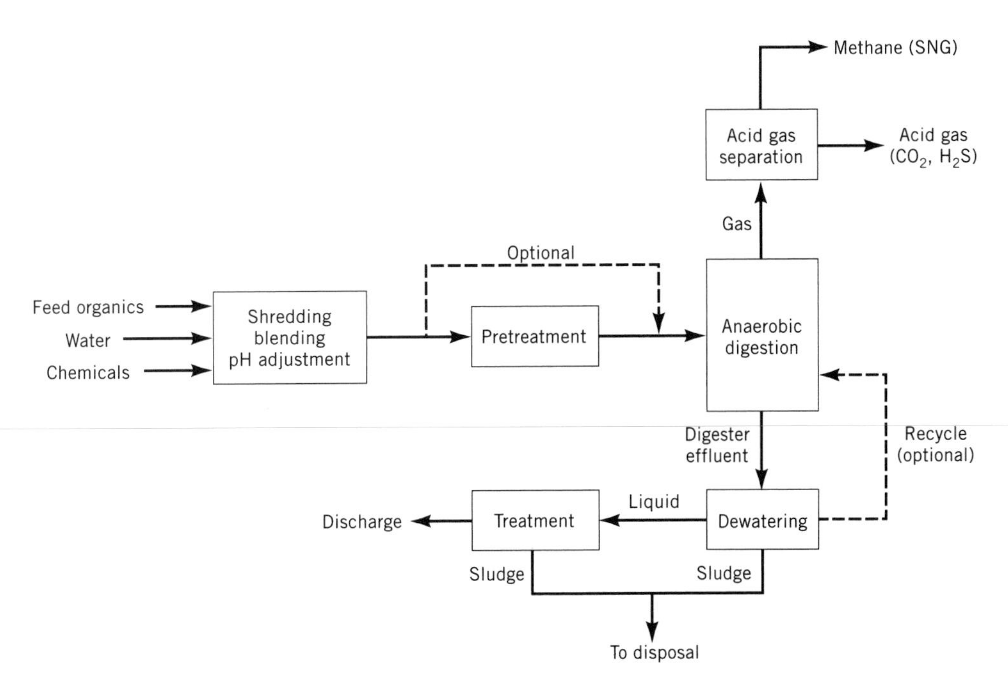

Figure 5. Methane production by anaerobic digestion of biomass.

Table 12. Comparison of Methane Fermentation Performance Under High Rate Mesophillic Conditions[a,b]

Component or Measure of Performance	Primary Sewage Sludge	Primary Activated Sludge	RDF–Sludge Blend	Biomass–Waste Blend	Coastal Bermuda Grass	Kentucky Bluegrass	Giant Brown Kelp	Water Hyacinth
Carbon, wt % (dry)	43.7	41.8	42.1	43.1	47.1	46.2	26.0	41.0
Nitrogen, wt % (dry)	4.02	4.32	1.91	1.64	1.96	4.3	2.55	1.96
Phosphorus, wt % (dry)	0.59	1.30	0.81	0.43	0.24		0.48	0.46
Ash, wt % of total solids	26.5	23.5	8.4	17.2	5.05	10.5	45.8	22.7
Volatile matter, wt % of total solids	73.5	76.5	91.6	82.8	95.0	89.8	54.2	77.3
Heating value, MJ/kg (dry)	19.86	18.31	17.20	20.92	19.04	19.19	10.26	16.02
C/N ratio	10.9	9.7	22.0	26.3	24.0	10.7	10.2	20.9
C/P ratio	74.1	32.2	52.0	100	196		54.2	89.1
Gas production rate, volume(n)/liquid volume-day	0.74	0.84	0.59	0.52	0.56	0.52	0.62	0.47
Methane in gas, mol %	68.5	65.5	60.0	62.0	55.9	60.4	58.4	62.8
Methane yield, m^3(n)/kg VS added	0.313	0.327	0.210	0.201	0.208	0.150	0.229	0.185
Volatile solids reduction, %	41.5	49.0	36.7	33.3	37.5	25.1	43.7	29.8
Substrate energy in gas, %	46.2	54.4	39.7	38.3	41.2	27.6	49.1	35.7

[a] Ref. 16.
[b] Daily feeding, continuous mixing, 35°C, pH 6.7 to 7.2, 12-day hydraulic retention time, 1.6 kg volatile solids/m^3·d except for kelp, which was 2.1 kg VS/m^3·d. All biomass substrates 1.2 mm or less in size.

losics are converted to sugar concentrates by acid- or enzyme-catalyzed hydrolysis:

$$(C_6H_{10}O_5) + x\,H_2O \xrightarrow{\text{hydrolysis}} x\,C_6H_{12}O_6$$

$$x\,C_6H_{12}O_6 \xrightarrow{\text{fermentation}} 2x\,C_2H_5OH + 2x\,CO_2$$

Advanced processes for conversion of cellulosics are being developed (17). A commercial alcohol fermentation plant for biomass would include units to shred and separate the combustible fermentable organic fraction from the nonfermentable components, hydrolysis units to produce glucose concentrates if the feed were high in low degradability cellulose components, fermenters, distillation towers, and dehydration units (Fig. 7). Under conventional conditions, the degradable organics are converted in the fermenters, at high efficiencies and residence times of 1 to 2 days, to a beer that contains about 10% ethanol. This broth is heated to remove the product alcohol as overhead by distillation, and the resulting 50 to 55% alcohol distillate is distilled again to yield 95% alcohol and by-product aldehydes and fusel oil. Bottoms from the beer still contain low volatility components from the fermentation called stillage, which is often processed further to yield high protein animal feeds.

The thermal efficiency of ethanol production from fermentable sugars is high, but the overall thermal efficiency of the process is low because of the many energy-consuming steps, the nonfermentable fraction in biomass, and the by-products formed. Alcohol yields are about 40 to 50 wt % of the weight of the fermentable fraction in the feed. Substantial improvements in the overall thermal efficiency of alcoholic fermentation are possible by improving

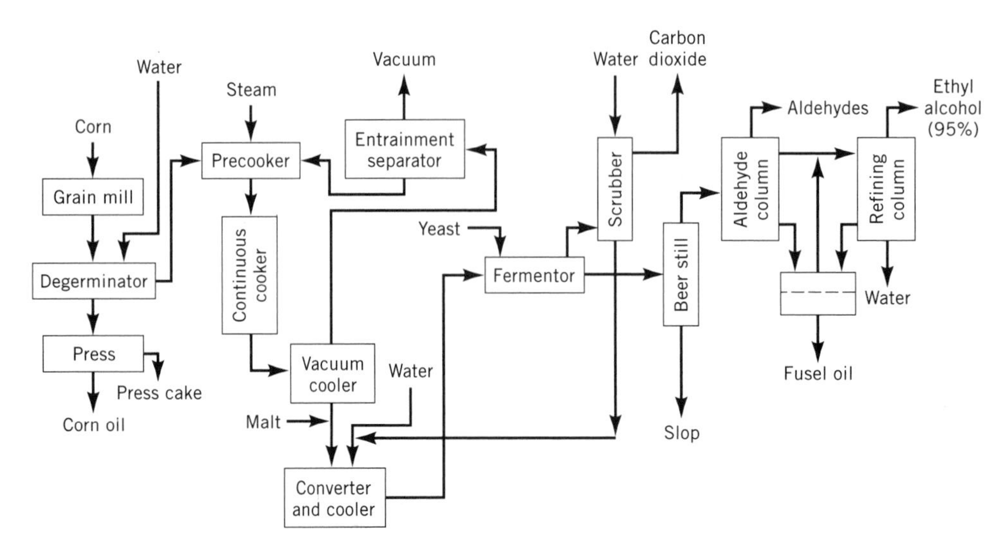

Figure 6. Flow scheme for manufacture of ethyl alcohol from corn.

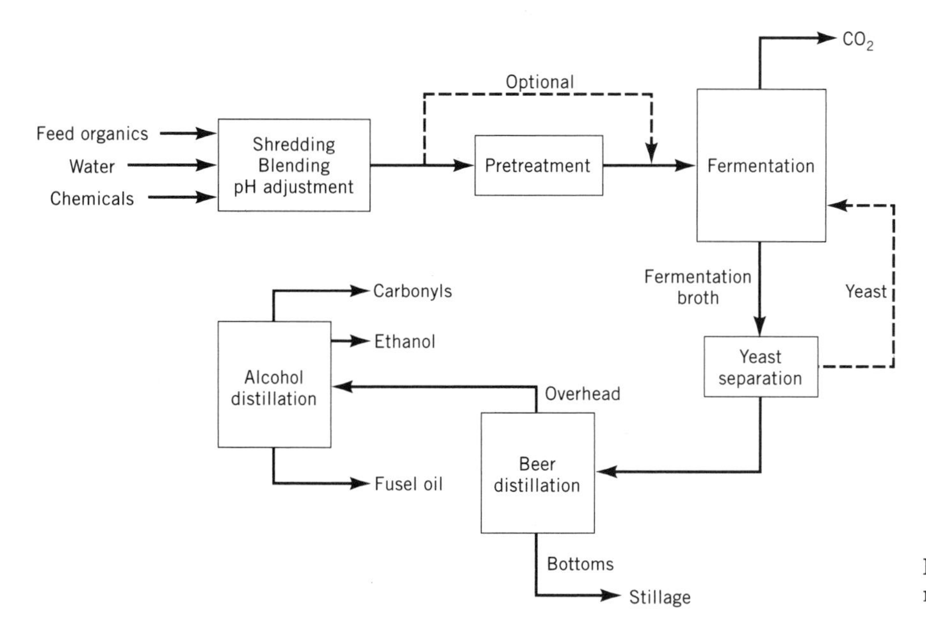

Figure 7. Ethanol production by alcoholic fermentation.

the thermal efficiencies of the auxiliary unit operations. Alcoholic fermentation under reduced pressure, systems in which polysaccharide hydrolysis and fermentation occur together, and improved heat exchange and alcohol-drying processes are under development to provide improved process performance.

Biophotolysis. The decomposition of water (splitting) to hydrogen and oxygen using the radiant energy of visible light and the photosynthetic apparatus of green plants, certain bacteria, and blue-green algae, is called biophotolysis. The concept has been studied in the laboratory, but has not been developed to the point where a practical process exists. Basically, biophotolysis involves the oxidation of water to liberate molecular oxygen and electrons which are raised from the level of the water-oxygen couple, eg, +0.8 V, to 0.0 V by Photosystem II.

$$H_2O \xrightarrow{\text{light}} \tfrac{1}{2}\,O_2 + 2\,H^+ + 2\,e^-$$

The electrons undergo the equivalent of a partial oxidation process in a dark reaction to a positive potential of +0.4 V, and Photosystem I then raises the potential of the electrons to as high as −0.7 V. Under normal photosynthesis conditions, these electrons reduce tryphosphopyridinenucleotide (TPN) to TPNH, which reduces carbon dioxide to organic plant material. In the biophotolysis of water, these electrons are diverted from carbon dioxide to a microbial hydrogenase for reduction of protons to hydrogen:

$$2\,H^+ + 2\,e^- \rightarrow H_2$$

Thus, the overall chemistry is simply the photolysis of water to hydrogen and oxygen.

Synthetic water-splitting membranes that contain the biochemical and other catalysts necessary to form hydrogen also are under development. These membranes mimic natural photosynthesis except that the electrons are directed to form hydrogen. Several sensitizers and catalysts are needed to complete the cycle, but progress is being made. Various single-stage schemes, in which hydrogen and oxygen are produced separately, have been studied, and the thermodynamic feasibility of the chemistry has been experimentally demonstrated.

The upper limit of efficiency of the biophotolysis of water has been projected to be 3% for well-controlled systems. This limits the capital cost of useful systems to low cost materials and designs. But the concept of water biophotolysis to afford a continuous, renewable source of hydrogen is quite attractive, and may one day lead to practical hydrogen-generating systems.

Natural Processes. Hydrocarbon production in land-based biomass by natural chemical mechanisms is a well-known phenomenon. Commercial production of natural rubber, the highly stereospecific polymer *cis*-1,4-polyisoprene [*9003-31-0*], is an established technology. Natural rubber has a mol wt range between about 500,000 and 2,000,000 and is tapped as a latex from the hevea rubber tree (*Hevea braziliensis*). The desert shrub guayule, which grows in the southwestern United States and in northern Mexico, is another biomass species studied as a source of natural rubber almost identical to hevea rubber (18). The idea of growing guayule and extracting the rubber latex from the whole plant was tested in full-scale plantations during the rubber shortage in World War II and found technically feasible. Terpene extraction from pine trees and other biomass species is also established technology.

Many plants native to North America, or that can be grown there, have been tested as sources of oils (triglycerides) and hydrocarbons (19,20). The objectives of this work have been to identify those biomass species that produce hydrocarbons, especially those of lower molecular weight than natural rubber, so that they would be more amenable to standard petroleum refining methods; to characterize hydrocarbon yields and those of other organic compounds; and to learn what controls the structure and molecular

Table 13. Oil- and Hydrocarbon-Producing Biomass Species Potentially Suitable for North America[a]

Family	Genus and Species	Common Name
Aceraceae	*Acer saccharinum*	Silver maple
Anacardiaceae	*Rhus glabra*	Smooth sumac
Asclepiadiaceae	*Asclepias incarnata*	Swamp milkweed
	sublata	Desert milkweed
	syriaca	Common milkweed
	Cryptostegia grandiflora	Madagascar rubber vine
Buxaceae	*Simmondsia chinensis*	Jojoba
Caesalpiniaceae	*Copaifera langsdorfi*	Copaiba
	multijuga	
Caprifoliaceae	*Lonicera tartarica*	Red tarterium honeysuckle
	Sambucus canadensis	Common elder
	Symphoricarpos orbiculatus	Corral berry
Companulaceae	*Companula americana*	Tall bellflower
Compositae	*Ambrosia trifida*	Giant ragweed
	Cacalia atriplicifolia	Pale Indian plantain
	Chrysathamnus nauseosus	Rabbitbrush
	Circsium discolor	Field thistle
	Eupathorium altissimum	Tall boneset
	Parthenium argentatum	Guayule
	Silphium integrifolium	Rosin weed
	laciniatum	Compass plant
	terbinthinaceum	Prairie dock
	Solidago graminifolia	Grass-leaved goldenrod
	leavenworthii	Edison's goldenrod
	rigida	Stiff goldenrod
	Sonchus arvensis	Sow thistle
	Vernonia fasciculata	Ironweed
Curcurbitaceae	*Cucurbita foetidissima*	Buffalo gourd
Euphorbiaceae	*Euphorbia denta*	
	lathyris	Mole plant, gopher plant
	pulcherima	Poinsetta
	tirucalli	African milk bush
Gramineae	*Agropyron repens*	Quack grass
	Elymus canadensis	Wild rye
	Phalaris canariensis	Canary grass
Labiatae	*Pycnanthemum incanum*	Western mountain mint
	Teucrium canadensis	American germander
Lauraceae	*Sassafras albidium*	Sassafras
Rhamnaceae	*Ceanothus americanus*	New Jersey tea
Rosaceae	*Prunus americanus*	Wild plum
Phytolaccaceae	*Phytolacea americana*	Pokeweed

[a] Ref. 20.

weight of hydrocarbons within the plant so that genetic manipulation or other biomass modifications can be applied to control hydrocarbon structures. Some efforts have concentrated on desert plants that might be grown in arid or semiarid areas without competition from biomass grown for foodstuffs. Other work has been aimed at perennial species adapted to wide areas of North America. Several biomass species have been found to contain oils and/or hydrocarbons (Table 13). It is apparent that oil or hydrocarbon formation is not limited to any one family or type of biomass. Interestingly, some species in the Euphorbiaceae family, which includes *Hevea braziliensis*, form hydrocarbons having molecular weights considerably less than that of natural rubber at yields as high as 10 wt % of the plant. This corresponds to hydrocarbon yields of about 3.97 $m^3/hm^2 \cdot yr$ (25 $bbl/hm^2 \cdot yr$).

Figure 8 illustrates one of the processing schemes used for separating various components in a hydrocarbon-containing plant. Acetone extraction removes the polyphenols, glycerides, and sterols, and benzene extraction removes the hydrocarbons. If the biomass species in question contain low concentrations of the nonhydrocarbon components, exclusive of the carbohydrate and protein fractions, direct extraction of the hydrocarbons with benzene or a similar solvent might be preferred.

The principal steps in the mechanism of polyisoprene formation in plants are known and should help to improve the natural production of hydrocarbons. Mevalonic acid, a key intermediate derived from plant carbohydrate via acetylcoenzyme A, is transformed into isopentenyl pyrophosphate (IPP) via phosphorylation, dehydration, and decarboxylation (see ALKALOIDS). IPP then rearranges to di-

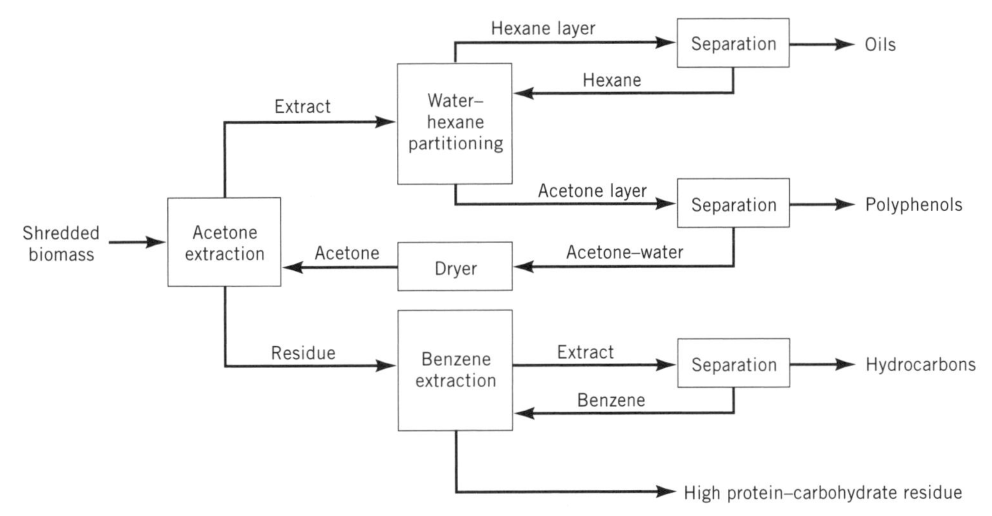

Figure 8. Operations for processing hydrocarbon-containing biomass.

methylallyl pyrophosphate (DMAPP). DMAPP and IPP react with each other, releasing pyrophosphate to form another allyl pyrophosphate containing 10 carbon atoms. The chain can successively build up by five-carbon units to yield polyisoprenes by head-to-tail condensations; alternatively, tail-to-tail condensations of two C_{15} units can yield squalene, a precursor of sterols. Similar condensation of two C_{20} units yields phytoene, a precursor of carotenoids. This information is expected to help in the development of genetic methods to control the hydrocarbon structures and yields.

Other sources of natural oils are the oilseed crops, many of which have been used in nonfuel applications, and microalgae (21). Oils from these types of biomass are largely triglycerides and typically contain three long-chain primary fatty acids, each bound to one of the carbon atoms of glycerol via an ester linkage. The viscosity and other properties of these oils vary with the degree of saturation of the fatty acid; the more paraffinic oils have higher viscosities and melting points. The oils can be upgraded to diesel fuels or gasoline plus diesel fuels by transesterification or by catalytic cracking or hydrocracking. Transesterification with methanol or ethanol yields monoesters in the C_{15-20} range depending on the oil source (22). The monoesters have viscosity and volatility characteristics similar to conventional diesel fuels. Carbon build-up and crankcase oil contamination in diesel engines vary with the degree of saturation and with the service characteristics of the diesel engine. Catalytic cracking of the oils over shape-selective zeolites gives substantial yields of aromatic-rich gasoline-range liquids (23). Catalytic hydrocracking of similar vegetable oils is reported to yield diesel fuel additives (24) or high quality gasolines (25).

Average and potential yields for seed oils are shown in Table 14 (26). If the seed oil is the only product for which revenue is realized, even the high potential yields are insufficient to justify seed oil use as fuel at 1992 petroleum prices, ie, about $20–22 per 159 L (42-gal bbl) of crude oil. Some studies indicate that small-scale transesterification facilities operated as farm cooperatives in the United States can produce biodiesel fuels from seed oils at a profit provided advantage is taken of the Minor Oilseed Provision of the 1990 Farm Bill (27). In this option, the farmer grows rapeseed, for example, on land that is removed from production of a crop such as corn, wheat, cotton, and soybeans. The 1990 Farm Bill permits a farmer to harvest and sell minor oilseed crops grown on this set-aside land without losing his program participation payment. In effect, the farmer is paid land rent by the Government, but can still produce minor oilseed crops.

Work on microalgae has focused on the growth of these organisms under conditions that promote lipid, ie, algal oil, formation. This eliminates the high cost of cell harvest because the lipids often can be separated by simple flotation or extraction (28). Lipid yields greater than 50% of the cell dry weight have been reported when the organisms are grown under nitrogen-limited conditions (29). The oils are high in triglycerides and can be transesterified to form biodiesel fuels in the same manner as seed oils (30). However, the estimated production costs of these fuels still appear to be too high to compete with petroleum-based diesel fuels (28,30).

Thermal Processes

Thermal processes for the production of energy and fuels from biomass and wastes usually involve irreversible chemical reactions, heat, and the transfer of chemical energy from reactants to products. The two largest classes of thermal processes are combustion and pyrolysis. A third class of processes can be described either as a combination of combustion and pyrolysis reactions, or as a thermochemical process in which conversion of the feed is facilitated by a reactant such as water or hydrogen. For convenience, these processes are grouped together as miscellaneous processes. Gaseous, liquid, and solid fuels can be produced by pyrolysis processes and several processes in the miscellaneous category.

Combustion. Complete combustion, eg, incineration, direct firing, burning, is the rapid chemical reaction of the feed and oxygen to form carbon dioxide, water, and heat. The heat released is a function of the enthalpy of combustion of the biomass. Agricultural products, such as bagasse

Table 14. Commercial Yields of Oilseeds and Seed Oils in the United States[a,b]

	Seed Yield, kg/hm²		Seed Oil Yield			
			Average		Potential	
Species	Average	Potential	kg/hm²	L/hm²	kg/hm²	L/hm²
Castorbean[c] (*Ricinus communis*)	950	3,810	428	449	1,504	1,590
Chinese tallow tree (*Sapium sebiferum*)	(12,553)[d]		(5,548)[d]			(6,270)[d]
Cotton[c] (*Gossypium hirsutum*)	887	1,910	142	150	343	370
Crambe[c] (*Crambe abssinica*)	1,121	2,350	392	421	824	940
Corn (high oil) (*Zea mays*)		5,940			596	650
Flax (*Linum usitatis-simum*)	795	1,790	284	309	758	840
Peanut (*Arachis hypogaea*)	2,378	5,160	754	814	1,634	1,780
Safflower (*Carthamus tinctorius*)	1,676	2,470	553	599	888	940
Soybean (*Glycine max*)	1,980	3,360	354	383	591	650
Sunflower (*Helianthus annuus*)	1,325	2,470	530	571	986	1,030
Winter rape (*Brassica napus*)		2,690			1,074	1,220

[a] Ref. 26.
[b] Growth conditions are dryland unless otherwise noted.
[c] Irrigated.
[d] Not an average yield from several sources; it is one reported yield equivalent to 6,270 L/hm² of oil plus tallow. It is believed that yield would be substantially less than this in a managed dense stand, but still higher than that of conventional oilseed crops.

generated in sugarcane plantations, forestry residues, wood chips, RDF, and even raw garbage, have been used as fuels in combustion systems for many years. The recovered heat has been used for steam production, electric power production in a steam-electric plant, and drying.

Many types of combustion equipment are available commercially. The basic differences in various units reside mainly in the design of the combustion chambers, the operating temperature, and the heat transfer mechanism. Refractory-lined furnaces operating at about 1000°C were standard until the introduction in early 1990s of water-wall incinerators. Ash buildup occurs rapidly in refractory-lined furnaces, and excess air must be introduced to limit the wall temperature. The water-wall incinerator has combustion chamber walls containing banks of tubes through which water is circulated, thereby reducing the amount of cooling air needed. Heat is transferred directly to the tubes to produce steam.

Another type of combustion unit operates at about 1600°C to produce a molten slag which forms a granular frit on quenching rather than the usual ash. The higher operating temperature is obtained by preheating the combustion air or by burning auxiliary fuel.

Fluidized-bed combustion represents still another approach. In these systems, air is dispersed through an orifice plate at the bottom of the combustion unit. The dispersed air passes through a bed of sand or residual inorganic particles recovered from combustion causing the effective volume of the bed to increase and the bed to become fluidized. The feed is fed to this rapidly mixed bed, where flameless combustion occurs at about 650°C. This temperature is substantially below flame temperature and because of the lower heat input requirements, many high moisture feeds can be combusted without supplemental fuel. Many other furnace variations, such as stationary and rotating shaft furnaces, suspension firing systems, and stationary and moving grates, are in commercial use or available for biomass and waste combustion applications.

The specific design most appropriate for biomass, waste combustion, and energy recovery depends on the kinds, amounts, and characteristics of the feed; the ultimate energy form desired, eg, heat, steam, electric; the relationship of the system to other units in the plant, independent or integrated; whether recycling or co-combustion is practiced; the disposal method for residues; and environmental factors.

Pyrolysis. Pyrolysis, eg, retorting, destructive distillation, carbonization, is the thermal decomposition of an organic material in the absence of oxygen. For biomass and wastes, pyrolysis generally starts at temperatures near 300 to 375°C. Chars, organic liquids, gases, and water are formed in varying amounts, depending particularly on the feed composition, heating rate, pyrolysis temperature, and residence time in the pyrolysis reactor. Higher temperatures and longer residence times promote gas production, while higher liquid and char yields result from lower tem-

Table 15. Product Yields from Pyrolysis of Municipal Solid Waste Organics[a]

	Products, wt %			Gas Yield	
Temperature, °C	Gases	Liquids	Char	Combustibles,[b] m^3/kg	Combustibles, MJ/kg
500	12.3	61.1	24.7	0.114	1.39
650	18.6	59.2	21.8	0.166	2.63
800	23.7	59.7	17.2	0.216	3.33
900	24.4	58.7	17.7	0.202	3.05

[a] Ref. 31.
[b] At normal conditions.

peratures and shorter residence times. No matter what the pyrolysis conditions are, with the exception of extremely high temperatures, the product mixture has a complex composition and selectivity for specific products is low even with a single feed component.

Depending on the pyrolysis temperature, the char fraction contains inorganic materials ashed to varying degrees, any unconverted organic solids, and fixed carbon residues produced on thermal decomposition of the organics. The liquid fraction contains a complex mixture of organic chemicals having much lower average molecular weights than the feed. For highly cellulosic feeds, the liquid fraction will usually contain acids, alcohols, aldehydes, ketones, esters, heterocyclic derivatives, and phenolic compounds. The pyrolysis gas is a low heat value (LHV) gas having a heating value of 3.9 to 15.7 MJ/m^3 at normal conditions (100 to 400 Btu/SCF). It contains carbon dioxide, carbon monoxide, methane, hydrogen, ethane, ethylene, minor amounts of higher gaseous organics, and water vapor. It is immediately apparent that if pure pyrolysis products are desired, product separation is a significant problem or further processing to refine the products is necessary.

Pyrolysis processes may be endothermic or exothermic, depending on the temperature of the reacting system. For most biomass feeds containing highly oxygenated cellulosic fractions as the principal components, pyrolysis is endothermic at low temperatures and exothermic at high temperatures. Energy to drive the process often is obtained from a portion of the feed or the pyrolysis products such as the char. At low temperatures, pyrolysis generally is reaction-rate controlled; at high temperatures, the process becomes mass-transfer controlled. The experimental data in Tables 15 and 16 show how temperature affects product yields and gas and char compositions with the combustible fraction of municipal solid waste (31). Gas yield increases as the temperature is increased from 500 to 900°C. Although the heating value of the product gas remains about the same, significant increases in hydrogen concentration and energy yield in the gas occur with increasing temperature. Substantial decreases occur in the carbon dioxide concentration over the same temperature range. Also, as the temperature increases from 500 to 900°C, the char yields decrease along with the volatile matter concentration, but the energy value of the char does not undergo similar changes.

Pyrolysis reactor designs are as varied as combustion unit designs. They include fixed beds, moving beds, suspended beds, fluidized beds, entrained feed solids reactors, stationary vertical shaft reactors, inclined rotating kilns, horizontal shaft kilns, high temperature (1000 to 3000°C) electrically heated reactors with gas-blanketed walls, single and multihearth reactors, and a host of other designs. One of the more innovative pyrolysis processes in development for gas production is a fluidized two-bed system (32–34). This system uses two fluidized-bed reactors containing sand as a heat transfer medium. Combustion of char from the pyrolysis reactor takes place within the combustion reactor. The heat released supplies the energy for pyrolysis of the combustible fraction in the pyrolysis reactor. Heat transfer is accomplished by sand flow from the combustion reactor at 950°C to the pyrolysis reactor at 800°C and return of the sand to the combustion reactor (Fig. 9). This configuration separates the combustion and pyrolysis reactions and keeps the nitrogen in air separated from the pyrolysis gas. It yields a pyrolysis gas that can be readily upgraded to SNG by shifting, scrubbing, and methanating without regard to nitrogen separation. The initial pyrolysis gas from high cellulose feeds contains about 37 mol % hydrogen, 35 mol % carbon monoxide, 16 mol % carbon dioxide, and 11 mol % methane. This is an LHV gas having a high heating value of about 13.5 MJ/m^3 at normal conditions (344 Btu/SCF). The projected gas yields are about 667 m^3 at normal conditions (17,000 SCF) of pyrolysis gas and about 196 m^3 at normal conditions (5000 SCF) of methane per dry ton of feed (32).

Table 16. Char and Gas Composition from Pyrolysis of Municipal Solid Waste Organics[a]

	Temperature, °C			
Component	500	650	800	900
Gas				
Carbon dioxide, mol %	44.8	31.8	20.6	18.3
Carbon monoxide, mol %	33.5	30.5	34.1	35.3
Methane, mol %	12.4	15.9	13.7	10.5
Hydrogen, mol %	5.56	16.6	28.6	32.5
Ethane, mol %	3.03	3.06	0.77	1.07
Ethylene, mol %	0.45	2.18	2.24	2.43
High heat value (HHV), MJ/m^3[b]	12.3	15.8	15.4	15.1
Char				
Volatile matter, %	21.8	15.1	8.13	8.30
Fixed carbon, %	70.5	70.7	79.1	77.2
Ash, %	7.71	14.3	12.8	14.5
High heat value (HHV), MJ/kg[c]	28.1	28.6	26.7	26.5

[a] Ref. 31.
[b] At normal conditions. To convert MJ/m^3 to Btu/SCF, multiply by 25.45.
[c] To convert MJ/kg to Btu/lb, multiply by 430.

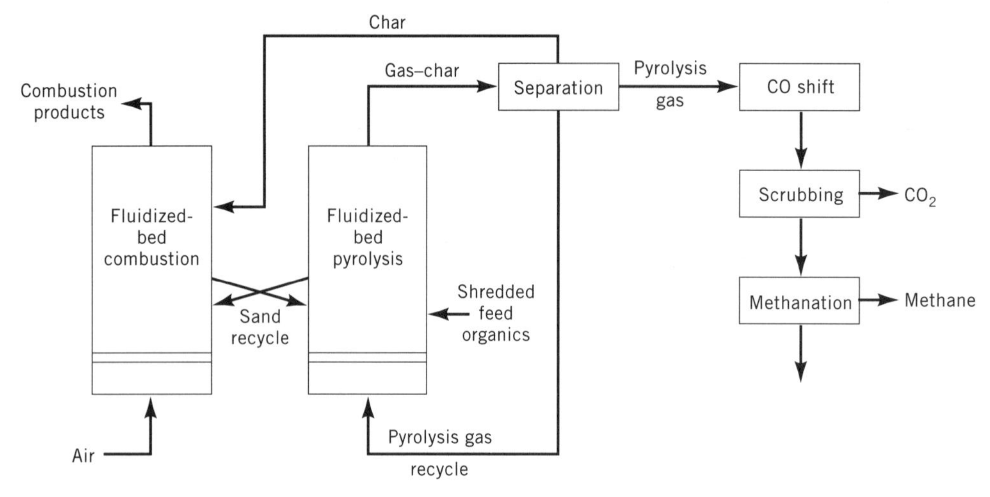

Figure 9. Methane production by pyrolysis using sand and char recycle in fluidized two-bed system.

An example of a liquid fuel production system under development in a short-residence time pyrolysis process is shown in Figure 10 (35). In this process, high cellulose RDF is dried in a rotary kiln to about 4 wt % moisture content, and finely divided to a particle size of which 80% is smaller than 14 mesh (1200 μm). The feed, about 0.23 kg of recycled char preheated to 760°C per kilogram of this finely divided material, is rapidly passed through the pyrolysis reactor. The raw product mixture, which consists of product gas, the char fed to the reactor, and new char formed on pyrolysis, leaves the reactor at about 510°C. Separation of the gas from the char and rapid quenching to about 80°C yields the liquid fuel. The remaining gas goes through a series of cleanup steps for in-plant use. Part of the gas is used as an oxygen-free solids transport medium and part of it as fuel. The raw product yields are about 10 wt % water, 20 wt % char, 30 wt % gas, and 40 wt % liquid fuel. The product char has a heating value of about 20.9 MJ/kg (9000 Btu/lb), contains about 30 wt % ash, and is produced at an overall yield of about 7.5 wt % of the dry feed. The corresponding values of the liquid fuel are about 24.4 MJ/kg (10,500 Btu/lb), 0.2 to 0.4% ash, and 22.5 wt % of dry feed as received (approximately 1 bbl/short ton of raw refuse). This product has been proposed for use as a heating oil; its properties are compared with a typical No. 6 fuel oil in Table 17. It is apparent that some differences exist, but successful combustion trials in a utility boiler with the liquid fuel have been performed.

A report on the continuous flash pyrolysis of biomass at atmospheric pressure to produce liquids indicates that

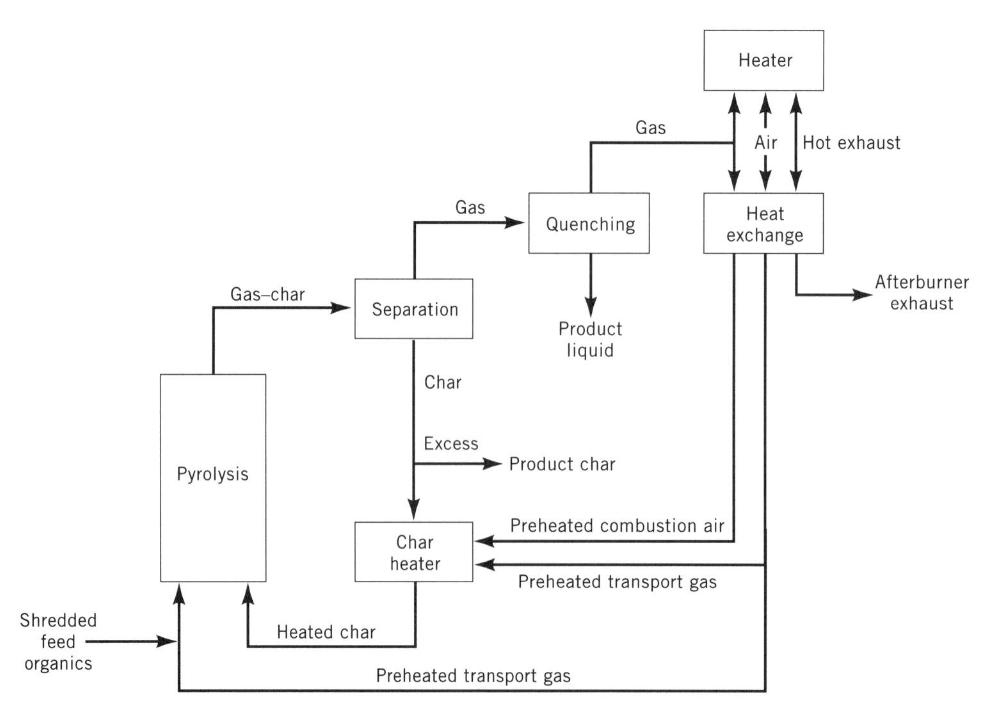

Figure 10. Liquid-fuel production by flash pyrolysis using char recycle.

Table 17. Properties and Analysis of Liquid Fuel and No. 6 Fuel Oil[a]

Properties	Liquid Fuel	No. 6 Fuel Oil
Heating value, MJ/kg[b]	24.6	42.3
Density, g/cm^3	1.3	0.98
Pour point, °C	32[c]	15–30
Flash point, °C	56[c]	65
Viscosity at 87.8°C (SUs)	1150[c]	90–250
Pumping temperature, °C	71	46
Atomization temperature, °C	116	104
Analysis,[d] wt %		
Carbon	57.5	85.7
Hydrogen	7.6	10.5
Sulfur	0.1–0.3	0.5–3.5
Chlorine	0.3	
Nitrogen	0.9	2.0
Oxygen	33.4	
Ash	0.2–0.4	0.5

[a] Ref. 35. Liquid fuel produced by flash pyrolysis using char recycle (Fig. 10).
[b] To convert MJ/kg to Btu/lb, multiply by 430.
[c] Containing 14% water as produced.
[d] Dry basis.

pyrolysis temperatures must be optimized to maximize liquid yields (36). It has been found that a sharp maximum in the liquid yields vs temperature curves exist and that the yields drop off sharply on both sides of this maximum. Pure cellulose has been found to have an optimum temperature for liquids at 500°C, while the wheat straw and wood species tested have optimum temperatures at 600°C and 500°C, respectively. Organic liquid yields were of the order of 65 wt % of the dry biomass fed, but contained relatively large quantities of organic acids.

Miscellaneous Thermal Processes. Many thermal conversion processes can be classified as partial oxidation processes in which the biomass or waste is supplied with less than the stoichiometric amount of oxygen needed for complete combustion. Under these conditions, LHV gases similar to pyrolysis gases are formed that can contain high concentrations of hydrogen and carbon monoxide. Such gaseous mixtures are termed synthesis gases and can be converted to a large number of chemicals and synthetic fuels by established processes (Fig. 11). In some partial oxidation processes, the various chemical reactions may occur simultaneously in the same reactor zone. In others, the reactor may be divided into a combustion zone, which supplies the heat to promote pyrolysis in a second zone, and perhaps a third zone for drying, the overall result of which is partial oxidation. Both air and pure oxygen have been utilized for such systems.

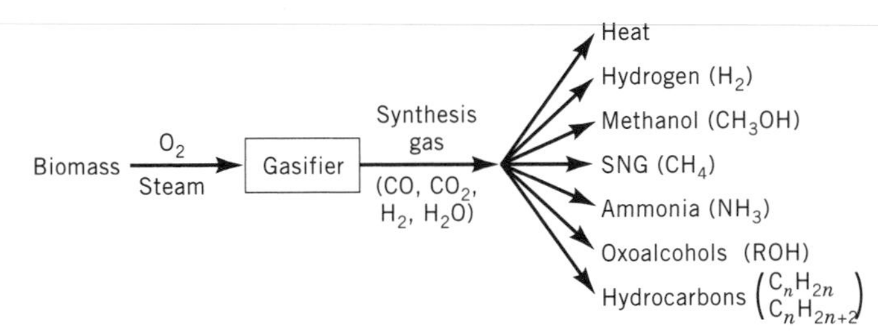

Figure 11. Applications of synthesis gas from biomass (37).

In one system, the three-zoned vertical shaft reactor furnace (Fig. 12), coarsely shredded feed is fed to the top of the furnace. As it descends through the first zone, the charge is dried by the ascending hot gases, which are also partially cleaned by the feed. The gas is reduced in temperature from about 315°C to the range of 40 to 200°C. The dried feed then enters the pyrolysis zone in which the temperature ranges from 315 to 1000°C. The resulting char and ash then descend to the hearth zone, where the char is partially oxidized with pure oxygen. Slagging temperatures near 1650°C occur in this zone and the resulting molten slag of metal oxides forms a liquid pool at the bottom of the hearth. Continuous withdrawal of the pool and quenching forms a sterile granular frit. The product gas is processed to remove fly ash and liquids, which are recycled to the reactor. A typical gas analysis is 40 mol % carbon monoxide, 20 mol % hydrogen, 23 mol % carbon dioxide, 5 mol % methane, and 5 mol % C_2. This gas has an HHV of about 14.5 MJ/m^3 at normal conditions (370 Btu/SCF).

An example of partial oxidation in which air is supplied without zone separation in the gasifier is the molten salt process (39). In this process, the shredded biomass or waste and air are continuously introduced beneath the surface of a sodium carbonate-containing melt which is maintained at about 1000°C. The resulting gas passes through the melt. Acid gases are absorbed by the alkaline salts and the ash is also retained in the melt. The melt is continuously withdrawn for processing to remove the ash and returned to the gasifier. No tars or liquid products are formed in this rather simple process. The heating value of the gases produced depends on the amount of air supplied, and is essentially independent of the type of feed organics. The greater the deficiency of air needed to achieve complete combustion, the higher the fuel value of the product gas. Thus with about 20, 50, and 75% of the theoretical air, the respective high heating values (HHVs) of the gas are about 9.0, 4.3, and 2.2 MJ/m^3 at normal conditions (230, 110, and 55 Btu/SCF).

Steam also is blended with air in some gasification units to promote the overall process via the endothermic steam–carbon reaction to form carbon monoxide and hydrogen. This was common practice at the turn of the nineteenth century, when so-called producer gasifiers were employed to manufacture LHV gas from different types of biomass and wastes. The producer gas from biomass and wastes had heating values around 5.9 MJ/m^3 at normal conditions (150 Btu/SCF), and the energy yields as gas ranged up to about 70% of the energy contained in the feed. Many gasifier designs were offered for the manufacture of producer gas from biomass and wastes; several types of units are still available for purchase (ca 1992). Thousands of producer gasifiers operating on air and wood were used during World War II, particularly in Sweden, to power automobiles, trucks, and buses. The engines needed only slight modification to operate on LHV producer gas.

Hydrogenation. Another approach to the production of energy products from biomass and wastes is based on hydrogenation. Hydrogen, which can be either generated

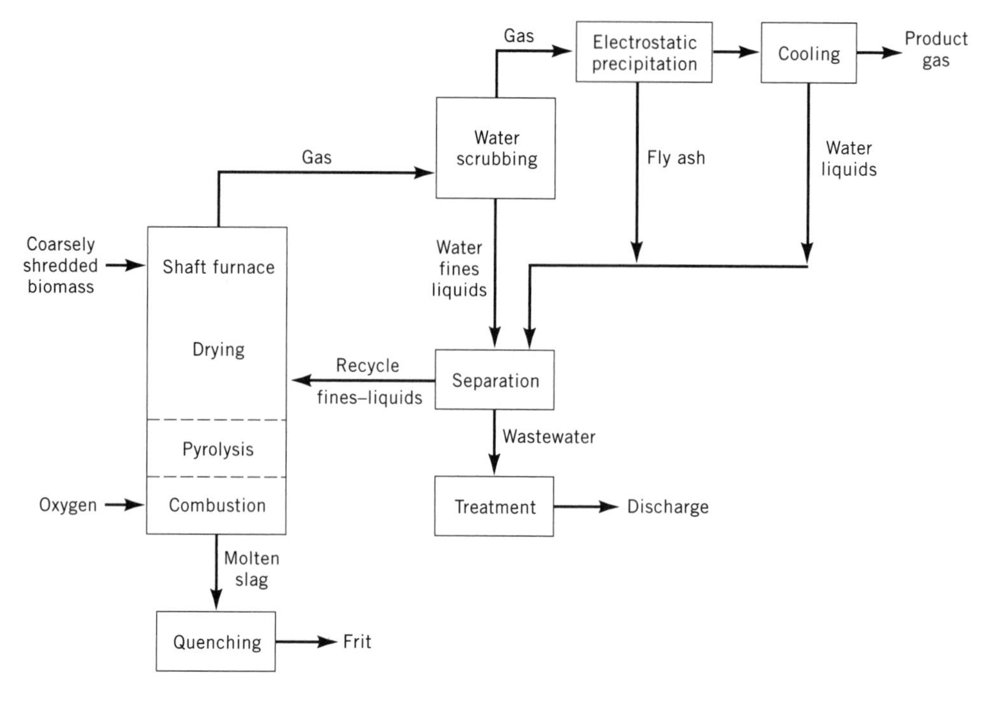

Figure 12. Production of synthesis gas in three-zone shaft reactor furnace (38).

from the feed or the conversion products, or obtained from an independent source, reacts directly with the feed organics or intermediate process streams at elevated pressures and temperatures to yield substitute fuels. In theory, highly oxygenated feeds should be capable of reduction to liquid and gaseous fuels at any level between the initial oxidation state of the feed and methane.

$$R(OH)_x + y\,H_2 \rightarrow RH_y(OH)_{x-y} + y\,H_2O$$

$$R - R' + H_2 \rightarrow RH + R'H$$

For a cellulosic material containing hydroxyl groups, the reactions might consist of dehydroxylation and depolymerization by hydrogenolysis, during which there is a transition from solid to liquid to gas.

Most of the work on hydrogenation has been concentrated on hydrogasification to produce methane as the final product. One route to methane involves the sequential production of synthesis gas and then methanation of the carbon monoxide with hydrogen to yield methane. The routes shown in Figure 13 involve the direct reaction of the feed with hydrogen (40). In this process, shredded feed

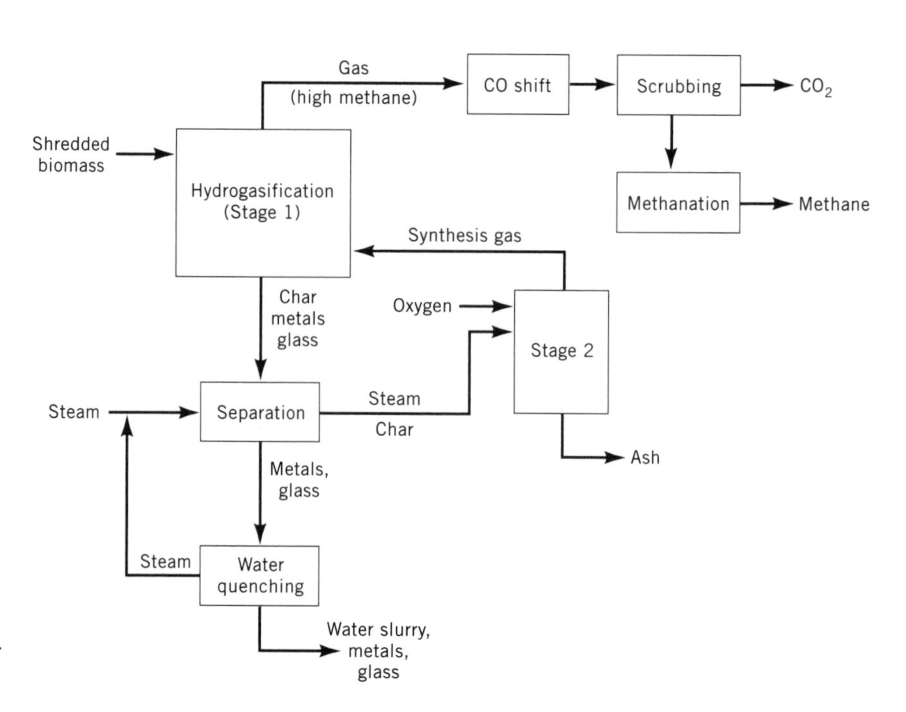

Figure 13. Methane production by hydrogasification.

Table 18. Gas Composition and Yield from Integrated Hydrogasification Process at Stage 1[a]

Product	Free Fall	Moving Bed
Composition, mol %		
Carbon monoxide	45.9	51.9
Hydrogen	31.9	13.3
Methane	10.4	17.2
Carbon dioxide	10.1	16.1
Ethane	1.2	1.1
Benzene	0.5	0.4
Yield, m^3/kg[b] dry feed	1.1	0.95
Fraction of total methane produced in stage 1 after methanation	0.26	0.52

[a] Fig. 13, estimated.
[b] To convert m^3/kg to ft^3/lb, multiply by 16.0.

is converted with hydrogen-containing gas to a gas containing relatively high methane concentrations in the first-stage reactor. The product char from the first stage is used in a second-stage reactor to generate the hydrogen-rich synthesis gas for the first stage. From experimental results with the first-stage hydrogasifier operated in the free-fall and moving-bed modes at 1.72 MPa and 870°C with pure hydrogen, calculations shown in Table 18 were made to estimate the composition and yield of the high methane gas produced when the first stage is integrated with an entrained char gasifier as the second stage. Although the methane content of the raw product gas is projected to be higher in the moving-bed reactor than in the falling-bed reactor, gas from the first stage must still be adjusted in H_2/CO ratio in a shift converter, scrubbed to remove carbon dioxide, and methanated to obtain SNG.

Another hydrogenation process utilizes internally generated hydrogen for hydroconversion in a single-stage, noncatalytic, fluidized-bed reactor (41). Biomass is converted in the reactor, which is operated at about 2.1 kPa, 800°C, and residence times of a few minutes with steam-oxygen injection. About 95% carbon conversion is anticipated to produce a medium heat value (MHV) gas which is subjected to the shift reaction, scrubbing, and methanation to form SNG. The cold gas thermal efficiencies are estimated to be about 60%.

Another advancement involves low temperature catalytic gasification of 2 to 10% aqueous biomass slurries or solutions that range from dilute organics in wastewater to waste sludges from food processing (42). The estimated residence time in the metallic catalyst bed is less than 10 min at 360°C and 20,635 kPa (3000 psi) at liquid hourly space velocities of 1.8 to 4.6 L of feedstock/L of catalyst/hr depending on the feedstock. The product fuel gas contains 45–70 vol % methane, 25–50 vol % carbon dioxide, and less than 5% hydrogen with as much as 2% ethane. The by-product water stream carries residual organics from 40 to 500 ppm COD. The product gas is MHV gas produced directly in contrast to MHV gas-phase processes that require either oxygen in place of air or a two-bed reaction system to keep the nitrogen in air separated from the fuel gas product.

Studies on the gasification of wood in the presence of steam and hydrogen show that steam gasification proceeds at a much higher rate than hydrogasification (43). Carbon conversions 30 to 40% higher than those achieved with hydrogen can be achieved with steam at comparable residence times. Steam/wood weight ratios up to 0.45 promoted increased carbon conversion, but had little effect on methane concentration. Other experiments show that potassium carbonate-catalyzed steam gasification of wood in combination with commercial methanation and cracking catalysts can yield gas mixtures containing essentially equal volumes of methane and carbon dioxide at steam/wood weight ratios below 0.25, with atmospheric pressure and temperatures near 700°C (44). Other catalyst combinations produced high yields of product gas containing about 2:1 hydrogen/carbon monoxide and little methane at steam/wood weight ratios of about 0.75 and a temperature of 750°C. Typical results for both of these studies are shown in Table 19. The steam/wood ratios and the catalysts used can have significant effects on the product gas compositions. The composition of the product gas also can be manipulated depending on whether a synthesis gas or a fuel gas is desired.

Direct hydroliquefaction of biomass or wastes can be achieved by direct hydrogenation of wood chips on treatment at 10,132 kPa and 340 to 350°C with water and Raney nickel catalyst (45). The wood is completely converted to an oily liquid, methane, and other hydrocarbon gases. Batch reaction times of 4 hours give oil yields of about 35 wt % of the feed; the oil contains about 12 wt % oxygen and has a heating value of about 37.2 MJ/kg (16,000 Btu/lb). Distillation yields a significant fraction that boils in the same range as diesel fuel and is completely miscible with it.

A catalytic liquefaction process for heavy liquids production reacts biomass in a water solution of sodium carbonate and carbon monoxide gas at elevated temperature and pressure to form a heavy liquid fuel (46). Biomass and the combustible fraction of wastes have been converted at weight yields of 40 to 60% at temperatures of 250 to 425°C and pressures of 10 to 28 MPa. Lower viscosity products are generally obtained at higher reaction temperatures

Table 19. Product Gases from Steam Gasification of Wood

Gas or Parameter	Value			
Gas composition, mol %				
H_2	0	53	29	50
CH_4	52	4	15	17
CO_2	48	12	17	11
CO	0	30	34	17
Reactor temperature, °C	740	750	696	762
Pressure, kPa (gauge)	0	0	129	159
Primary catalyst	K_2CO_3	K_2CO_3	none	wood ash
Secondary catalyst	Ni:SiAl	SiAl	none	none
Steam/wood weight ratio	0.25	0.75	0.24	0.56
Carbon conversion to gas, %	64	77	68	52
Feed energy in gas, %	76	78		
Heating value of gas,[a] MJ/m^3	20.6	12.1	16.6	17.7
Reference	44[b]	44[b]	45[c]	45[c]

[a] At normal conditions.
[b] Laboratory results with unspecified wood.
[c] Process development unit (PDU) results with unspecified hardwood.

and solid or semisolid products are obtained when the reaction temperature is below 300°C. However, the high nitrogen and oxygen contents and the boiling characteristics and high viscosity range of the liquid products make it difficult to classify them as synthetic crude oils. Conventional refining methods could not be used to upgrade this kind of material to standard petroleum derivatives. The original process consisted of a sequence of steps: drying and grinding wood chips to a fine powder, mixing the powder with recycled product oil (30% powder to 70% oil), blending the mixture with water containing sodium carbonate, and treatment of the slurry with synthesis gas at about 27,579 kPa (4000 psi) and 370°C. The modified process consists of partially hydrolyzing the wood in slightly acid water and treating the water slurry containing dissolved sugars and about 20% solids with synthesis gas and sodium carbonate at 27,579 kPa and 370°C on a once-through basis. The resulting oil product yield is about 1 bbl/400 kg (158.9 L/400 kg) of chips and is approximately equivalent to No. 6 grade boiler fuel. It contains about 50% phenolics, 18% high boiling alcohols, 18% hydrocarbons, and 10% water.

Study of the mechanism of this complex reduction-liquefaction suggests that part of the mechanism involves formate production from carbonate, dehydration of the vicinal hydroxyl groups in the cellulosic feed to carbonyl compounds via enols, reduction of the carbonyl group to an alcohol by formate and water, and regeneration of formate (46). In view of the complex nature of the reactants and products, it is likely that a complete understanding of all of the chemical reactions that occur will not be developed. However, the liquefaction mechanism probably involves catalytic hydrogenation because carbon monoxide would be expected to form at least some hydrogen by the water-gas shift reaction.

Table 20. Product Yields from Wood Pyrolysis[a,b]

Product	Yield per t Dry Wood
Gas,[c] m^3	156
Charcoal, kg	300
Ethyl acetate, L	61.1
Creosote oil, L	13.6
Methanol, L	13.0
Ethyl formate, L	5.3
Methyl acetate, L	3.9
Methyl ethyl ketone, L	2.7
Other ketones, L	0.9
Allyl alcohol, L	0.2
Soluble tar, L	91.8
Pitch, kg	33.0

[a] Ref. 47.
[b] Feed: 70% maple, 25% birch, 5% ash, elm, and oak; av temperature, 515°C.
[c] CO_2, 37.9 mol %; CO, 23.4 mol %; CH_4, 16.8 mol %; N_2, 16.0 mol %; O_2, 2.4 mol %; H_2, 2.2 mol %; hydrocarbons, 1.2 mol %. To convert m^3 to ft^3, multiply by 35.3.

Chemical Processes

Biological–biochemical and thermal conversion processes are chemical processes, too, but a few specific chemical processes are mentioned separately because they are directed more to conventional chemical processing and production. These processes have been grouped together as chemical processes.

Chemicals have long been manufactured from biomass, especially wood (silvichemicals), by many different fermentation and thermochemical methods. For example, continuous pyrolysis of wood was used by the Ford Motor Co. in 1929 for the manufacture of various chemicals (Table 20) (47). Wood alcohol (methanol) was manufactured on a large scale by destructive distillation of wood for many years until the 1930s and early 1940s, when the economics became more favorable for methanol manufacture from fossil fuel-derived synthesis gas.

In the production of chemicals from biomass, wood is still the raw material of choice for the manufacture of certain chemicals, although many of them cannot compete with fossil-based products. The chemistry of silvichemical production is related directly to the chemical composition of trees, ie, 50% cellulose, 25% hemicelluloses, and 25% lignins. However, specialty chemicals are often manufactured from nonwoody biomass because they occur naturally in certain plant species or can easily be derived from these plants. Examples are the alginic acids from *Macrocystis pyrifera*, ie, giant brown kelp, and physiologically active alkaloids from particular plants. Ethanol has been manufactured for chemical applications from starchy biomass by fermentation for many years. Figure 14 lists some of the more important primary biomass-derived chemicals, the principal intermediates, and the dominant processing methods used. All chemicals listed in Figure 14 are either manufactured commercially in 1992 by the indicated routes or were manufactured in the past. Secondary processing of these primary chemicals would appear to make it possible to manufacture almost all heavy and fine organics produced from fossil raw materials, eg, ethanol (qv) can be converted to ethylene, acetaldehyde, and acetic acid, which can be converted to other organic chemicals by established routes.

The availability of C_6 sugars such as glucose is an important factor in the development of a biomass chemicals industry as alluded to in Figure 14. Unfortunately, most biomass is higher in cellulose than C_6 sugars. Cellulose, because of its relatively low biodegradability, cannot easily be converted to fermentation alcohol and other products without first liberating the monosaccharides. Trees and fibrous biomass, such as plant stalks and reedy plants, contain cellulose in partially crystalline form sometimes complexed with other materials. The low degradability of wood cellulose, which can exist as lignin–cellulose complexes, is attributed to these factors. Such forms of cellulose can be degraded to glucose concentrates by several hydrolytic methods, the most common of which is hydrolysis with dilute sulfuric acid. At the temperatures necessary to form glucose, a large portion of the product is ordinarily converted to by-products such as hydroxymethylfurfural. Glucose yields are usually near 50% of theory because of by-product formation. Enzyme-catalyzed hydrolysis of cellulose has afforded much higher yields of glucose, but the particle size of the cellulosic material subjected to hydrolysis must be reduced to facilitate

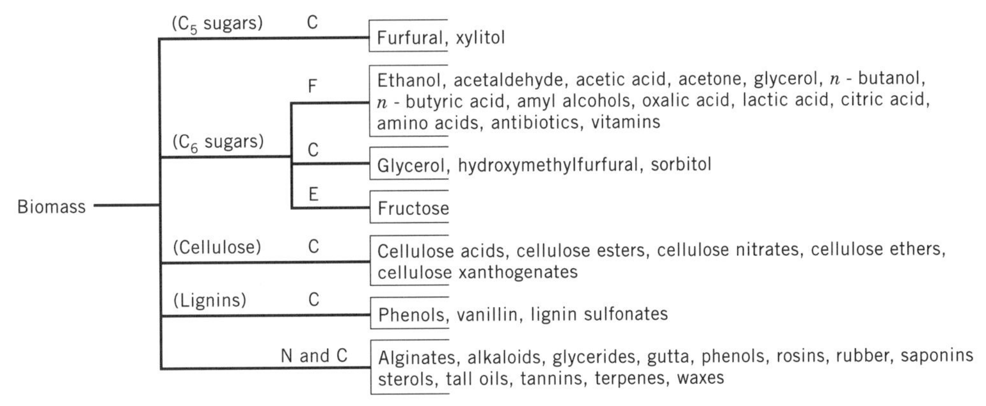

Figure 14. Primary biomass-derived chemicals. Dominant processing methods are chemical (C), fermentation (F), enzymic (E), and natural (N) processes; products in parentheses represents intermediates.

depolymerization. The cost of particle size reduction tends to outweigh the advantages of higher glucose yields. One approach to improving yield of monosaccharides is first to dissolve the cellulose in a solvent, separate the insolubles from the cellulose solution, and then subject the solution to hydrolysis conditions or precipitate the cellulose before hydrolysis. This method destroys crystallinity and any cellulose complex that may be present, and makes it possible to achieve high yields of glucose even with highly fibrous biomass. Special solvents such as Cadoxen [*14874-24-9*] and 65% sulfuric acid have been suggested for this application (48). Another approach involves the use of a blocking agent, acetone, which temporarily forms a cyclic ketal (acetonide) with vicinal diols to protect the sugars from degradation under hydrolysis conditions (49). The process has been reported to afford very high yields of sugars and to be equally applicable to high fiber biomass residues.

As mentioned in the biological–biochemical section, another approach to improve alcoholic fermentation combines saccharification and fermentation, ie, simultaneous saccharification and fermentation (SSF). Enzyme-catalyzed cellulose hydrolysis and fermentation to alcohol takes place in the same vessel in the presence of enzyme and yeast (50). Reduced fermenter pressures and enzyme and yeast recycling result in 70 to 80% ethanol yields. These process modifications, coupled with more energy-efficient distillation and heat exchanger improvements, are projected to make fermentation ethanol from low value biomass competitive with industrial ethanol (51).

Multiproduct processes for biomass chemical plants in which the operating conditions can be manipulated to vary the product distribution as a function of demand or other factors, or in which an optimum mix of products is chosen based on feedstock characteristics, appear to have some merit even though no full-scale systems have yet been commercialized. The process depicted in Figure 15 illustrates how such a plant might function. Mild acid hydrolysis of the hemicelluloses in wood affords either a predominantly xylose or mannose solution, depending on the type of wood feed, and a cellulose–lignin residue. Strong acid treatment of this residue yields a glucose solution, which can be combined with mannose for alcoholic fermentation, and a lignin residue. Phenols can be made from this residue by hydrogenation, and furfural can be made from xylose by strong acid treatment. The products are thus ethanol, furfural, and phenols.

There are many different routes to organic chemicals from biomass because of its high polysaccharide content and reactivity. The practical value of the conversion processes selected for commercial use with biomass will depend strongly on the availability and price of the same chemicals produced from petroleum and natural gas.

BIOMASS PRODUCTION

The manufacture of synfuels and energy products from biomass requires that suitable quantities of biomass be grown, harvested, and transported to the conversion plant site. Many variables must be considered when selecting the proper species or mixture of species for operation of a system: growth cycle; fertilization; insolation; temperature; precipitation; propagation and planting procedures; soil and water needs; harvesting methods; disease resistance; growth area competition from biomass for food, feed, and fiber; growth area availability; possibilities for

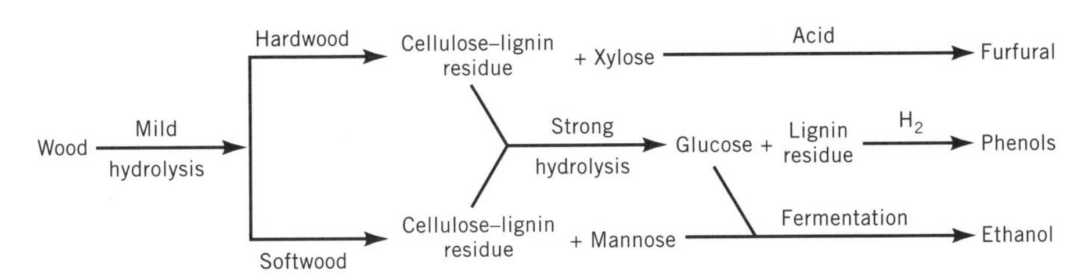

Figure 15. Furfural, ethanol, and phenols production from wood in a multiproduct process biomass chemical plant (52). Wood is a ca 50% cellulose, 25% lignin, and 25% hemicellulose.

simultaneous or sequential growth of biomass for synfuels and foodstuff or other applications; and nutrient depletion. At least 250,000 botanical species, of which only about 300 are cash crops, are known in the world. A relatively small number are, and will be, used as biomass feedstocks for the manufacture of synfuels and energy products.

In the ideal case, biomass chosen for energy applications should be high yield, low cash-value species that have short growth cycles and grow well in the area and climate chosen for the biomass energy system. Fertilization requirements should be low and possibly nil if the species selected fix ambient nitrogen, thereby minimizing the amount of external nutrients that must be supplied to the growth areas. In areas having low annual rainfall, the species grown should have low water needs and be able to efficiently utilize available precipitation. For land-based biomass, the requirements should be such that the crops can grow well on low grade soils and do not need the best classes of agricultural land. After harvesting, growth should commence again without the need for replanting. Surprisingly, several biomass species meet many of these idealized characteristics and appear to be quite suitable for energy applications. There are a number of important factors that relate to biomass production for energy applications.

Photosynthesis

The basic biochemical pathways in ambient carbon dioxide fixation involve decomposition of water to form oxygen, protons, and electrons; transport of these electrons to a higher energy level via Photosystems I, II, and several electron transfer agents; concomitant generation of reduced nicotinamide adenine dinucleotide [*53-57-6*] (NADPH) and adenosine triphosphate [*56-65-5*] (ATP); and reductive assimilation of carbon dioxide to carbohydrate. The initial process is believed to be the absorption of light by chlorophyll [*1406-65-1*], which promotes decomposition of water. The ejected electrons are accepted by ferredoxin [*9080-02-8*] (Fd), a nonheme iron protein. The reduced Fd initiates a series of electron transfers to generate ATP from adenosine diphosphate [*58-64-0*] (ADP), inorganic phosphate, and NADPH. For each of the two light reactions, one photon is required to transfer each electron; a total of eight photons is thus required to fix one molecule of carbon dioxide. Assuming that the carbon dioxide is in the gaseous phase and that the initial product is glucose, the standard Gibbs free energy change at 25°C is +0.48 MJ(+114 kcal) per mole of carbon dioxide assimilated and the corresponding enthalpy change is +0.47 MJ (+112 kcal).

The maximum efficiency with which photosynthesis can occur has been estimated by several methods. The upper limit has been projected to range from about 8 to 15%, depending on the assumptions made; ie, the maximum amount of solar energy trapped as chemical energy in the biomass is 8 to 15% of the energy of the incident solar radiation. The rationale in support of this efficiency limitation helps to point out some aspects of biomass production as they relate to energy applications.

The relationship of the energy and wavelength of a photon is energy $= \hbar_c/\lambda$ where $\hbar$ is Planck's constant, 6.624×10^{-34}; c is velocity of light; and λ is wavelength. Assume that the wavelength of the light absorbed is 575×10^{-9} m and is equivalent to the light absorbed between the blue (400×10^{-9} m) and red (700×10^{-9} m) ends of the visible spectrum. This assumption has been made by several investigators for green plants to calculate the upper limit of photosynthesis efficiency. The energy absorbed in the fixation of one mole of carbon dioxide, which requires 8 photons/molecule, is then given by

$$\text{energy absorbed} = [(6.624 \times 10^{-34})(3.00 \times 10^{8})/(575 \times 10^{-9})] \times 8 \times 6.024 \times 10^{23} = 1.67\ \text{MJ}$$

Since 0.47 MJ of solar energy is trapped as chemical energy in this process, the maximum efficiency for total white-light absorption is 28.1%. Further adjustments are usually made to account for the percentages of photosynthetically active radiation in white light, the light that can actually be absorbed, and respiration. The amount of photosynthetically active radiation in solar radiation that reaches the earth is estimated to be about 43%. The fraction of the incident light absorbed is a function of many factors, such as leaf size, canopy shape, and reflectance of the plant; it is estimated to have an upper limit of 80%. This effectively corresponds to the utilization of eight photons out of every 10 in the active incident radiation. The third factor results from biomass respiration. A portion of the stored energy is used by the plant, the amount of which depends on the properties of the biomass species and the environment. For purposes of calculation, assume that about 25% of the trapped solar energy is used by the plant, thereby resulting in an upper limit for retention of the nonrespired energy of 75%. The upper limit for the efficiency of photosynthetic fixation of biomass can now be estimated to be 7.2%, ie, $0.281 \times 0.43 \times 0.80 \times 0.75$. For the case where little or no energy is lost by respiration, the upper limit is estimated to be 9.7%, ie, $0.281 \times 0.43 \times 0.80$. The low efficiency limit might correspond to land-based biomass, while the higher efficiency limit might be closer to water-based biomass such as unicellular algae. These figures can be transformed into dry biomass yields by assuming that all of the fixed carbon dioxide is contained in the biomass as cellulose, $(C_6H_{10}O_5)_x$, from the equation

$$Y = \frac{CIE}{F}$$

where Y is yield of dry biomass, t/ha·yr; C is a constant, 3.1536; I is average insolation, W/m^2; E is solar energy capture efficiency, %; and F is energy content of dry biomass, MJ/kg. Thus for high cellulose dry biomass, an average insolation of 184 W/m^2 (1404 Btu/ft^2·d), which is the average insolation for the continental United States, a solar energy capture efficiency of 7.2%, and a high heat of combustion of 17.51 MJ/kg for cellulose, the yield of dry biomass is 239 t/hm^2·yr (107 short tons/acre-yr). The corresponding value for an energy capture efficiency of 9.7% is 321 t/hm^2·yr (143 short t/acre-yr). These yields of organic matter can be viewed as an approximation of the theoretical yield limits for land- and water-based biomass. Some estimates of maximum yield are higher and some are lower than these figures, depending on the values used for *I*, *E*, and *F*, but they serve as a guideline to indicate the highest yields that a biomass production system could be expected to achieve under normal environmental conditions. Unfortunately, real biomass yields rarely approach these limits. Sugarcane, for example, which is one of the high yielding species of biomass, typically produces total dry plant matter at yields of about 80 t/hm^2·yr (36 short tons/acre·yr).

Yield is plotted against solar energy capture efficiency in Figure 16 for insolation values of 150 and 250 W/m^2, which spans the range commonly encountered in the United States, and for biomass energy values of 12 and 19 MJ/kg (dry). The higher the efficiency of photosynthesis, the higher the biomass yield. For a given solar energy capture efficiency and incident solar radiation, the yield is projected to be lower at the higher biomass energy values, ie, curves A and C, curves B and D. From an energy production standpoint, this means that a higher energy content biomass could be harvested at lower yield levels and still compete with higher yielding but lower energy content biomass species. It is also apparent that for a given solar energy capture efficiency, yields similar to those obtained with higher energy content species should be possible with a lower energy content species even when it is grown at a lower insolation, eg, curves B and C. Finally, at the solar energy capture efficiency usually encountered in the field, about 1% or less, the spread in yields is much less than at the higher energy capture efficiencies. It is important to emphasize that this interpretation of biomass yield as functions of insolation, energy content, and energy capture, although based on sound principles, is still a theoretical analysis of living systems. Because of the many uncontrollable factors in the field, such as the changes that occur in climate and seasonal changes in biomass composition, departures from the norm can be expected.

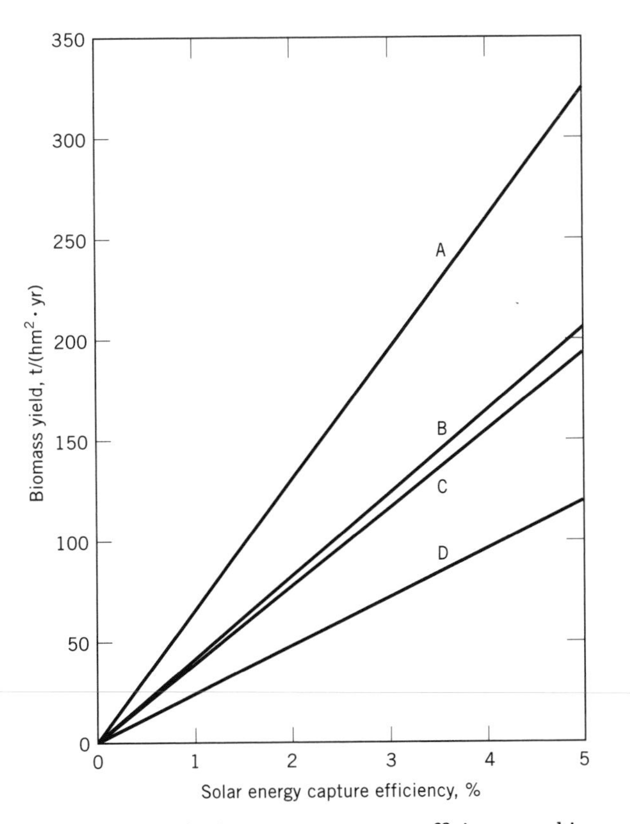

Figure 16. Effect of solar energy capture efficiency on biomass yield. A, insolation value (I) of 250 W/m^2, and biomass energy value (F) of 12 MJ/kg (dry); B, I of 250 W/m^2, F of 19 MJ/kg; C, I of 150 W/m^2, F of 12 MJ/kg; D, I of 150 W/m^2, F of 19 MJ/kg. To convert W/m^2 to Btu/(ft^2·d), multiply by 7.616; to convert MJ/kg to Btu/lb, multiply by 430.

The previous discussion of photosynthesis has concentrated on the gross features of ambient carbon dioxide fixation in biomass. However, the biochemical pathways involved in the conversion of carbon dioxide to carbohydrate play an important role in understanding the molecular events of biomass growth. Three different biochemical energy transfer pathways occur during carbon dioxide fixation (53–55). One pathway, the Calvin three-carbon cycle, involves an initial three-carbon intermediate of phosphoglyceric acid [*82-11-1*] (PGA). This cycle, often referred to as the reductive pentose phosphate cycle, is used by autotrophic photochemolithotrophic bacteria, algae, and green plants. For every three molecules of carbon dioxide converted to glucose in a dark reaction, nine ATP, six NADPH, and 12 Fd molecules are required. Six molecules of PGA are formed from three molecules each of carbon dioxide and ribulose-1,5-diphosphate (RuDP) in the chloroplasts. After these carboxylation reactions, a reductive phase occurs in which six molecules of PGA are successively transformed into six molecules of diphosphoglyceric acid (DPGA) and then six molecules of 3-phosphoglyceraldehyde, a triose phosphate. Five molecules of the triose phosphate are then used to regenerate three molecules of RuDP, which initiates the cycle again. The other triose phosphate molecule is used to generate glucose-6-phosphate and energy and electron carriers. Plant biomass species that use the Calvin cycle are called C_3 plants; it is common in many fruits, legumes, grains, and vegetables. C_3 plants usually exhibit low rates of photosynthesis at light saturation, low light saturation points, sensitivity to oxygen concentration, rapid photorespiration, and high carbon dioxide compensation points, ie, about 50 ppm. The carbon dioxide compensation point is the carbon dioxide concentration in the surrounding environment below which more carbon dioxide is respired by the plant than is photosynthetically fixed. Typical C_3 biomass species are peas, sugar beet, spinach, alfalfa, *Chlorella*, *Eucalyptus*, potato, soybean, tobacco, oats, barley, wheat, tall fescue, sunflower, rice, and cotton.

The second pathway is called the C_4 cycle because the carbon dioxide is initially fixed as the four-carbon dicarboxylic acids, malic or aspartic acids. Phosphoenol pyruvate (PEP) reacts with one molecule of carbon dioxide to form oxaloacetate (OAA) in the mesophyll of the biomass, and then malate or aspartate is formed. The C_4 acid is transported to the bundle sheath cells, where decarboxylation occurs to regenerate pyruvate, which is returned to the mesophyll cells to initiate another cycle. The carbon dioxide liberated in the bundle sheath cells enters the C_3 cycle in the usual manner. Thus no net carbon dioxide is fixed in the portion of the C_4 cycle, and it is the combination with the C_3 cycle which ultimately results in carbon dioxide fixation. The subtle differences between the C_4 and C_3 cycles are believed responsible for the wide variations in biomass properties. In contrast to C_3 biomass, C_4 biomass is usually produced at higher yields and has higher rates of photosynthesis, high light saturation points, insensitivity to oxygen concentrations below 21 mol %, low levels of respiration, low carbon dioxide compensation points, and greater efficiency of water usage. C_4 biomass often occurs in areas of high insolation, hot daytime temperatures, and seasonal dry periods. Typical C_4 biomass includes important crops such as corn, sugarcane, and sorghum, and forage species and tropical grasses such as Bermuda Grass. Even crabgrass is a C_4 biomass. At least 100 genera in 10 plant families are known to exhibit the C_4 cycle.

The third pathway is called crassulacean acid metabolism (CAM). CAM refers to the capacity of chloroplast-containing tissues to fix carbon dioxide in the dark via phosphoenolpyruvate carboxylase leading to the synthesis of free malic acid. The mechanism involves the β-carboxylation of PEP by this enzyme and the subsequent reduction of OAA by malate dehydrogenase. CAM has been documented in at least 18 families, including the family Crassulaceae, and 109 genera of the Angiospermae. Biomass species in the CAM category are typically adapted to arid environments, have low photosynthesis rates, and have high water usage efficiencies. Examples are cactus plants and the succulents, such as pineapple. The information developed to date on CAM biomass indicates that CAM has evolved so that initial carbon dioxide fixation can take place in the dark with much less water loss than the fully light-dependent C_3 and C_4 pathways. CAM biomass also conserves carbon by recycling endogenously formed carbon dioxide. Several CAM species show temperature optima in the range 12 to 17°C for carbon dioxide fixation in the dark. The stomates in CAM plants open at night to allow entry of carbon dioxide and then close by day to minimize water loss. The carboxylic acids formed in the dark are converted to carbohydrates when the radiant energy is available during the day. Relatively few CAM plants have been exploited commercially.

Significant differences in net photosynthetic assimilation of carbon dioxide are apparent between C_3, C_4, and CAM biomass species. One of the principal reasons for the generally lower yields of C_3 biomass is its higher rate of photorespiration; if the photorespiration rate could be reduced, the net yield of biomass would increase. Considerable research is in progress (ca 1992) to achieve this rate reduction by chemical and genetic methods, but as yet, only limited yield improvements have been made. Such an achievement with C_3 biomass would be expected to be very beneficial for foodstuff production and biomass energy applications.

The specific carbon dioxide-fixing mechanism used by a plant will affect the efficiency of photosynthesis, so from an energy utilization standpoint, it is desirable to choose plants that exhibit high photosynthesis rates to maximize the yields of biomass in the shortest possible time. There are numerous factors that affect the efficiency of photosynthesis other than the carbon dioxide-fixing mechanism, eg, insolation; amounts of available water, nutrients, and carbon dioxide; temperature; and transmission, reflection, and biochemical energy losses within or near the plant. For lower plants such as the green algae, many of these parameters are under human control. For conventional biomass growth subjected to the natural elements, it is not feasible to control all of them.

Climate and Environmental Factors

The biomass species selected for energy applications and the climate must be compatible to facilitate operation of fuel farms. The three primary climatic parameters that have the most influence on the productivity of an indigenous or transplanted species are insolation, rainfall, and temperature. Natural fluctuations in these factors remove them from human control, but the information compiled over the years in meteorological records and from agricultural practice supplies a valuable data bank from which to develop biomass energy applications. Ambient carbon dioxide concentration and the availability of nutrients are also important factors in biomass production.

Insolation. The intensity of the incident solar radiation at the earth's surface is a key factor in photosynthesis; natural biomass growth will not take place without solar energy. Insolation varies with location and is high in the tropics and near the equator. The approximate changes with latitude are illustrated in Table 21. At a given latitude, the incident energy is not constant and often exhibits large changes over relatively short distances. A more quantitative summary of insolation values over the continental United States is shown in Table 22. To place the amount of energy that strikes the earth in the proper perspective, the annual insolation on about 0.1 to 0.2% of the surface of the continental United States is equivalent to all the energy consumed by the United States in one year. The production figures shown in Table 23 represent annual yields obtained under good growth conditions (13,14,53,59,60). The estimated solar energy capture efficiencies (SECD) for the biomass listed assumes that all organic matter is cellulose. These are only rough approximations and most are probably too high, but they indicate that C_4 plants are usually better photosynthesizers than C_3 plants, and that high insolation alone does not always correlate with high biomass yield and capture efficiency. Although there are a few exceptions in Table 23, there appears to be a trend in this direction.

Table 21. Insolation at Various Latitudes for Clear Atmospheres[a]

		Insolation, W/m^2[b]		
Location	Latitude	Maximum	Minimum	Average[c]
Equator	0°	315	236	263
Tropics	23.5°	341	171	263
Mid-earth	45°	355	70.9	210
Polar circle	66.5°	328	0	158

[a] Ref. 56.
[b] To convert W/m^2 to Btu/ft(^{2}·d), multiply by 7.616.
[c] Yearly total divided by 365.

Table 22. Daily Solar Radiation in the United States[a]

Location	Total Daily Insolation, W/m^2 [b,c]				
	January	April	July	October	Annual[c]
Tucson, Arizona	146	289	288	208	229
Fresno, California	93.2	290	338	187	229
Lakeland, Florida	135	260	247	189	210
Indianapolis, Indiana	90.0	188	242	120	157
Lake Charles, Louisiana	109	215	236	175	191
Saint Cloud, Minnesota	75.9	178	275	104	157
Glasgow, Montana	72.0	190	299	118	175
Ely, Nevada	108	257	288	176	210
Oklahoma City, Oklahoma	80.8	212	264	155	183
San Antonio, Texas	113	198	286	182	199
Burlington, Vermont	76.3	182	208	99.7	146
Sterling, Virginia	90.9	173	233	113	159
Seattle–Tacoma, Washington	36.5	179	276	98.1	151

[a] Ref. 57.
[b] To convert W/m^2 to $Btu/(ft^2 \cdot d)$, multiply by 7.616.
[c] Average.

Precipitation. Precipitation as rain, snow, sleet, or hail is governed by movement of air and is generally abundant wherever air currents are predominately upward. The greatest precipitation should therefore occur near the equator. The average annual rainfall in the United States is about 79 cm.

The moisture needs of aquatic biomass presumably are met in full because growth occurs in liquid water, but the growth of land biomass often can be water-limited. Requirements for good growth of many biomass species have been found to be in the range of 50 to 76 cm of annual rainfall (63). Some crops, such as wheat, exhibit good growth with much less water, but they are in the minority. Without irrigation, water is supplied during the growing season by the water in the soil at the beginning of the season and by rainfall. Figure 17 depicts the normal pre-

Table 23. Annual Production of Dry Matter and Solar Energy Capture Efficiency (SECD)[a]

Location	Biomass Community	Productivity, $t/(hm^2 \cdot yr)$ [b]	Insolation, W/m^2 [c]	SECD,[d] %
Sweden	Enthrophic lake angiosperm	7.2	106	0.38
Denmark	Phytoplankton	8.6	133	0.36
Mississippi	Water hyacinth	11.0–33.0	194	0.31–0.94
Minnesota	Maize	24.0	169	0.79
New Zealand	Temperate grassland	29.1	159	1.02
West Indies	Tropical marine angiosperm	30.3	212	0.79
Nova Scotia	Sublittoral seaweed	32.1	133	1.34
Georgia	Subtropical saltmarsh	32.1	194	0.92
England	Coniferous forest, 0–21 yr	34.1	106	1.79
Israel	Maize	34.1	239	0.79
New South Wales	Rice	35.0	186	1.04
Congo	Tree plantation	36.1	212	0.95
Holland	Maize, rye, two harvests	37.0	106	1.94
Marshall Islands	Green algae	39.0	212	1.02
FRG	Temperate reedswamp	46.0	133	1.92
Puerto Rico	*Panicum maximum*	48.9	212	1.28
California	Algae, sewage pond	49.3–74.2	218	1.26–1.89
Colombia	Pangola grass	50.2	186	1.50
West Indies	Tropical forest, mixed ages	59.0	212	1.55
Hawaii	Sugarcane	74.9	186	2.24
Puerto Rico	*Pennisetum purpurcum*	84.5	212	2.21
Java	Sugarcane	86.8	186	2.59
Puerto Rico	Napier grass	106	212	2.78
Thailand	Green Algae	164	186	4.90

[a] Refs. 13, 14, 53, 59, 60.
[b] To convert $t/(hm^2 \cdot yr)$ to short ton/(acre·yr), divide by 2.24.
[c] Average. To convert W/m^2 to $Btu/(ft^2 \cdot d)$, multiply by 7.616.
[d] Approximate estimates of solar energy capture efficiencies; probably too high.

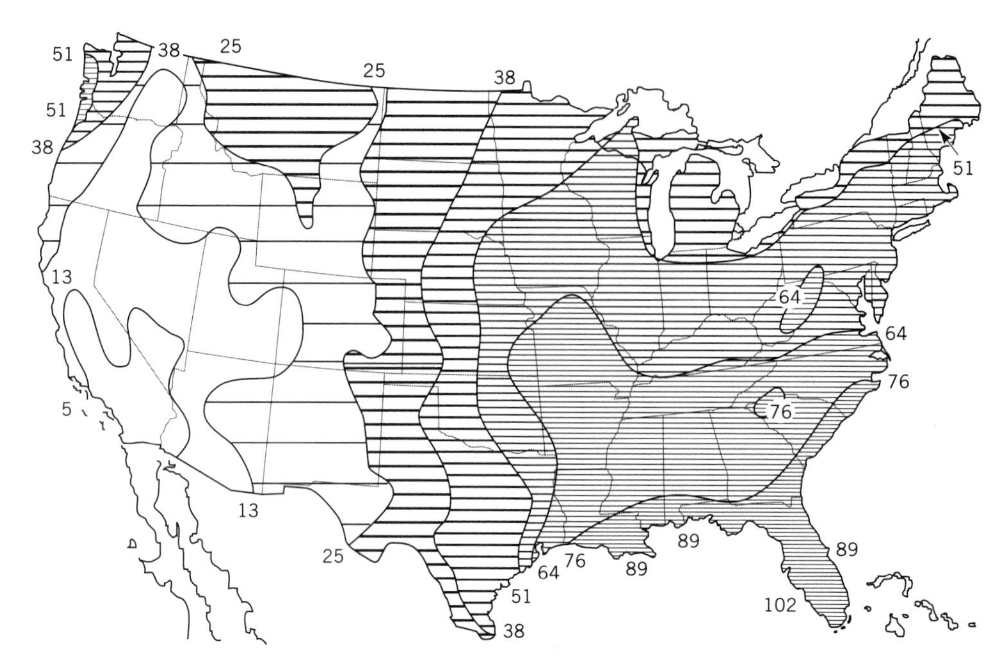

Figure 17. Normal precipitation in cm during the growing season, April to September, in the United States (61).

cipitation recorded in the 48-state area during the normal growing season, April to September. This type of information and the established requirements for the growth of land-based biomass can be used to divide the United States into precipitation regions. Regions more productive for biomass generally correlate with precipitation regions. It should be realized, however, that rainfall alone is not quantitatively related to productivity of land biomass because of the differences in soil characteristics, water evaporation rates, and infiltration. Also, certain areas that have low rainfall can be made productive through irrigation. Finally, some areas of the country that vary widely in precipitation as a function of time, such as many Western states, will produce moderate biomass yields, and often sufficient yields of cash crops without irrigation, to justify commercial production.

Temperature. Most biomass species grow well in the United States at temperatures between 15.6 and 32.3°C. Typical examples are corn, kenaf, and napier grass. Tropical grasses and certain warm-season biomass have optimum growth temperatures in the range 35 to 40°C, but the minimum growth temperature is still near 15°C (64). Cool-weather biomass such as wheat may show favorable growth below 15°C, and certain marine biomass such as the giant brown kelp only survive in water at temperatures below 20 to 22°C (65). The growing season is longer in the southern portion of the United States; in some areas such as Hawaii, the Gulf states, southern California, and the southeastern Atlantic states, the temperature is usually conducive to biomass growth for most of the year (61,62).

The effect of temperature fluctuations on net carbon dioxide uptake is illustrated by the curves in Figure 18. As the temperature increases, net photosynthesis increases for cotton and sorghum to a maximum value and then rapidly declines. Ideally, the biomass species grown in an area should have a maximum rate of net photosynthesis as close as possible to the average temperature during the growing season in that area.

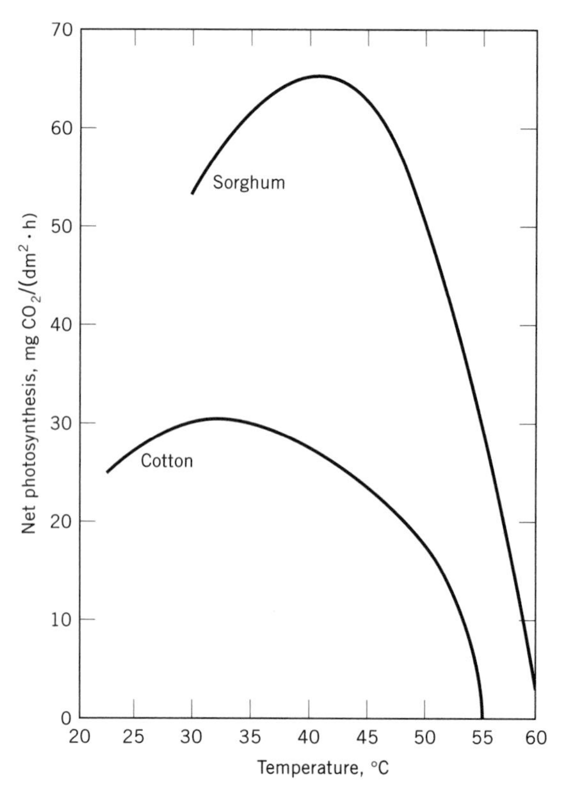

Figure 18. Effect of temperature on net photosynthesis for sorghum and cotton leaves. To convert mg/(dm^2·h) to lb/(ft^2·h), multiply by 2.373×10^{-3}.

Ambient Carbon Dioxide Concentration. Many studies have been performed which show that higher concentrations of carbon dioxide than are normally present in air will promote more carbon fixation and increase biomass yields. In confined, environmentally controlled enclosures such as hothouses, carbon dioxide-enriched air can be used to stimulate growth. This is not practical in large-scale open systems such as those envisaged for biomass energy farms. For aquatic biomass production, carbon dioxide enrichment of the water phase may be an attractive method of promoting biomass growth if carbon dioxide concentration is a limiting factor; the growth of biomass often occurs by uptake of carbon dioxide from both the air and liquid phase near the surface.

For some high growth-rate biomass species, the carbon dioxide concentration in the air among the leaves of the plant often is considerably less than that in the surrounding atmosphere. Photosynthesis may be limited by the carbon dioxide concentrations under these conditions when wind velocities are low and insolation is high.

Nutrients. All living biomass requires nutrients other than carbon, hydrogen, and oxygen to synthesize cellular material. Principal nutrients are nitrogen, phosphorus, and potassium; other nutrients required in lesser amounts are sulfur, sodium, magnesium, calcium, iron, manganese, cobalt, copper, zinc, and molybdenum. The last five nutrients, as well as a few others not listed, are sometimes referred to as micronutrients because only trace quantities are needed to stimulate growth. For land-based biomass, these elements are usually supplied by the soil, so the nutrients are depleted if they are not replaced through fertilization. Some biomass, such as the legumes, are able to meet all or part of their nitrogen requirements through fixation of ambient nitrogen. Water-based biomass such as marine kelp use the natural nutrients in ocean waters. Freshwater biomass such as water hyacinth is often grown in water enriched with nutrients in the form of wastewater. The growth of the plant is stimulated, and at the same time the influent waste is stabilized because its components are taken up by the plant as nutrients. So-called luxuriant growth of water hyacinth on sewage, in which more than the needed nutrients are removed from the waste, can be used as a substitute wastewater treatment method (28).

Whole plants typically contain 2 wt % N, 1 wt % K, and 0.5 wt % P, so at a yield of 20 t/hm^2·yr (8.9 short ton/acre·yr), harvesting of the whole plant without return of any of the plant parts to the soil corresponds to the annual removal of 400 kg N, 200 kg K, and 100 kg P per hm^2. This illustrates the importance of fertilization, especially of these macronutrients, to maintain fertility of the soil. Biomass growth is often nutrient-limited and yield correlates with fertilizer dose rates. Average nitrogen fertilizer applications for production of wheat, rice, potato, and brussel sprouts are about 73, 134, 148, and 180 kg/hm^2, respectively, in the United States (67). Estimates of balanced fertilizers needed to produce various land biomass species are shown in Table 24. Note that alfalfa does not require added nitrogen because of its nitrogen-fixing ability. It is estimated that this legume can fix from about 130 to 600 kg of elemental nitrogen per hm^2 annually (68).

Table 24. Fertilizer Requirements of Biomass[a,b], kg

Biomass	N	P_2O_5	K_2O	CaO
Alfalfa	0.0	12.3	34.0	20.7
Corn	11.8	5.7	10.0	0.0
Kenaf	13.9	5.0	10.0	16.1
Napier grass	9.6	9.3	15.8	8.5
Slash pine[c]	3.8	0.9	1.6	2.3
Potato	16.8	5.3	28.3	0.0
Sugar beet	18.0	5.4	31.2	6.1
Sycamore	7.3	2.8	4.7	0.0
Wheat	12.9	5.3	8.4	0.0

[a] Ref. 62.
[b] Estimated per whole dry plant.
[c] Five years.

Normal weathering processes that occur in nutritious soils release nutrients, but they are not available at rates that promote maximum biomass yields. Fertilization is usually necessary to maximize yields. Since nitrogenous fertilizers are largely manufactured from fossil fuels, mainly natural gas, and since fertilizer needs are usually the most energy intensive of all inputs in a biomass production system, careful analysis of the integrated biomass production-conversion system is needed to ensure that net energy production is positive. Trade-offs between synfuel outputs, nonsolar energy inputs, and biomass yields are required to operate a system that produces only energy products.

Land Availability

The availability of sufficient land suitable for production of land-based biomass can be estimated for the United States by several techniques. One method relies on the land capabilities classification scheme developed by the U.S. Department of Agriculture (69), in which land is divided into eight classes. Classes I to III are suited for cultivation of many kinds of crops; Class IV is suited only for limited production; and Classes V to VIII are useful only for permanent vegetation such as grasses and trees. The U.S. Department of Agriculture surveyed nonfederal land usage for 1987 in terms of these classifications (70). Out of 568 million hm^2, which corresponds to 60% of the 50-state area, 43% of the land (246.3×10^6 hm^2) was in Classes I to III, 13% (75.59×10^6 hm^2) in Class IV; and 43% (246.3×10^6 hm^2) in Classes V to VIII. The actual usage of this land at the time of the survey is shown in Table 25 (70). Table 25 illustrates that of all the land judged suitable for cultivation in Classes I to III, only about one-half of it is actually used as cropland (70), and that the combined areas of pasture, range, and forest lands is about 66% of the total nonfederal lands. There is ample opportunity to produce biomass for energy applications on nonfederal land not used for foodstuffs production. Large areas of land in Classes V to VIII not suited for cultivation, and sizable areas in Classes I to IV not being used for crop production, also would appear to be available for biomass energy applications; land used for crop production could be considered for simultaneous or sequential growth of biomass for foodstuffs and energy. Portions of

Table 25. Nonfederal Rural Land Use in United States by Type, 1987,[a] 10^6 hm^2

Land Class	Cropland[b]	Pastureland[c]	Rangeland[d]	Forest[e]	Minor	Total
I	11.58	0.82	0.17	0.66	0.23	13.47
II	77.31	12.81	6.58	18.00	2.66	117.36
III	54.28	16.00	18.76	24.26	2.23	115.53
IV	18.60	10.30	21.62	23.56	1.51	75.59
V	1.16	1.86	1.99	7.52	1.03	13.55
VI	6.56	6.84	53.34	37.34	2.63	106.71
VII	1.59	3.91	58.40	46.88	4.08	114.86
VIII	0.035	0.067	1.68	1.40	7.81	11.00
other[f]	0	0	0	0	2.08	2.08
Total	*171.12*	*52.60*	*162.56*	*159.62*	*24.25*	*570.15*
%	30.01	9.22	28.51	28.00	4.25	

[a] Ref. 70. Totals may not be precise summations due to rounding.
[b] Land used for production of crops for harvest alone or in rotation with grasses and legumes.
[c] Land used for the production of adapted, introduced, or native species in a pure stand, grass mixture, or a grass–legume mixture.
[d] Land on which the vegetation is predominantly grasses, grass-like plants, forbs, or shrubs suitable for grazing or browsing.
[e] Land that is at least 10% stocked by forest trees of any size or formerly having had such tree cover and not currently developed for nonforest use.
[f] Land, such as farmsteads, stripmines, quarries, and other lands, that do not fit into any other land class category.

federally owned lands, which are not included in the survey, might also be dedicated to biomass energy applications. Careful design and management of land-based biomass production areas could result in improvement or upgrading of lands to higher land capability classifications.

Water Availability

The production of marine biomass in the ocean, even on the largest scale envisaged for energy applications, would require only a very small fraction of the available ocean areas. The U.S. Navy has estimated that a square area 753 km on each edge off the coast of California may be sufficient to produce enough giant brown kelp for conversion to methane to supply all of the nation's natural gas needs. This large area is very small when compared with the total area of the Pacific Ocean. Also, the benefits to other marine life from a large kelp plantation have been well documented (65). Any conflicts that might arise would be concerned primarily with ocean traffic. With the proper plantation design for marine biomass and precautionary measures to warn approaching ships, it is expected that marine biomass growth could be sustained over long periods.

Freshwater biomass in theory can be grown on the 20 million hm^2 of fresh water in the United States. However, several difficulties mitigate against large-scale freshwater biomass energy systems. About 80% of the fresh water in the United States is located in the northern states, while several of the freshwater biomass species considered for energy applications require a warm climate such as that found in Gulf states. The freshwater areas suitable for biomass production in the southern states, however, are much smaller than those in the North, and the density of usage is higher in southern inland waters. Overall, these characteristics make small-scale aquatic biomass production systems more feasible for energy applications. It may be advisable in the future to examine the possibility of constructing large artificial lakes for this purpose; this does not seem practical in the early 1990s.

Land-Based Biomass

Much effort to evaluate land-based biomass energy applications has been expended. This work aims at selecting high yield biomass species, characterizing physical and chemical properties, defining growth requirements, and rating energy use potential. Several species have been proposed specifically for energy usage, while others have been recommended for multiple uses, one of which is as an energy resource. The latter case is exemplified by sugarcane; bagasse, the fibrous material remaining after sugar extraction, is used in several sugar factories as a boiler fuel. Most land-based biomass plantations operated for energy production or synfuel manufacture also will yield products for nonenergy markets. Land-based biomass for energy production can be divided into forest biomass, grasses, and cultivated plants.

Forest Biomass. About one-third of the world's land area is forestland. Broadleaved evergreen trees are a dominant species in tropical rain forests near the equator (71). In the northern hemisphere, stands of coniferous softwood trees such as spruce, fir, and larch dominate in the boreal forests at the higher latitudes, while both the broadleaved deciduous hardwoods such as oak, beach, and maple, and the conifers such as pine and fir, are found in the middle latitudes. Silviculture, ie, the growth of trees, is practiced by five basic methods: exploitative, conventional extensive, conventional intensive, naturalistic, and short-rotation (71). The exploitative method harvests trees without regard to regeneration. The conventional extensive method harvests mature trees so that natural regeneration is encouraged. Conventional intensive silviculture grows and harvests commercial tree species in essentially pure stands such as Douglas fir and pine on tree farms. The naturalistic method has been defined as the growth of selected mixed tree species, including hardwoods, in which the species are selected to match the ecology of the site. The last method, short-rotation silviculture, ie, short-rotation intensive culture (SRIC) or short-rotation woody crops (SRWC), has been suggested as the most suitable method for energy applications. In this technique, trees that grow quickly are harvested every few years, in contrast to once every 20 or more years. Fast-growing trees such as cottonwood, red alder, and aspen are intensively cultivated and mechanically harvested every 3 to 6 years when they are 3 to 6 m high and only a few centimeters in diameter. The young trees are converted into chips for further processing or direct fuel use and the small remaining stems or stumps form new sprouts by coppice growth and are intensively cultivated again. SRWC production affords dry yields of several tons of biomass per hm^2 annually without large energy inputs for fertilization, irrigation, cultivation, and harvesting.

Historically, trees have been important resources and still serve as significant energy resources in many developing countries. Several studies of temperate forests indicate productivities from about 9 to 28 $t/hm^2 \cdot yr$, while the corresponding yields of tropical forests are higher, ranging from about 20 to 50 $t/hm^2 \cdot yr$ (72). These yields are obtained using conventional forestry methods over long periods of time, ie, 20 to 50 years or more. Productivity is initially low in a new forest, slowly increases for about the first 20 years, and then begins to decline. Coniferous forests will grow even in the winter months if the temperatures are not too low; they do not exhibit the yield fluctuations characteristic of deciduous forests.

One of the tree species studied in great detail as a renewable energy source is the *Eucalyptus* (73), an evergreen tree which belongs to the myrtle family, *Myrtaceae*. There are approximately 450 to 700 identifiable species of *Eucalyptus*. The *Eucalyptus* is a rapidly growing tree native to Australia and is a prime candidate for energy use because it reaches a size suitable for harvesting in about seven years. Several species have the ability to coppice, ie, resprout, after harvesting; as many as four harvests can be obtained from a single stump before replanting is necessary. In several South American countries, *Eucalyptus* trees are converted to charcoal and used as fuel. *Eucalyptus* wood has also been used to power integrated sawmill, wood distillation, and charcoal–iron plants in Western Australia. Several large areas of marginal land in the United States may be suitable for establishing *Eucalyptus* energy farms. These areas are in the western and central regions of California and the southeastern United States.

Various species and hybrids of the genus *Populus* are some of the more promising candidates for SRWC growth and harvesting as an energy resource (74). The group has long been cultivated in Europe and more recently in North America. Poplar hybrids are easily developed and the resulting progeny are propagated vegetatively using stem cuttings. Short-rotation growth of poplar hybrids has been reported by several investigators to afford yields of biomass that range as high as 112–202 green $t/hm^2 \cdot yr$ (50–90 green short ton/acre·yr) (75). These results were reported with very high density plantings and selected clones; this type of tree growth has been termed woodgrass in which the tree crop is harvested several times each growing season in the same manner as perennial grasses. However, there is some dispute regarding the benefits of woodgrass growth vs SRWC growth (76).

It can be concluded from other studies that deciduous trees are preferred over conifers for the production of biofuels (77). Several species can be started readily from clones, resprout copiously and vigorously from their stumps at least five or six times without loss of vigor, and exhibit rapid initial growth. They also can be grown on sites with slopes as steep as 25%, where precipitation is 50 cm or more per year. It has been estimated that yields between about 18 and 22 dry $t/hm^2 \cdot yr$ are possible on a sustained basis almost anywhere in the Eastern and Central time zones in the United States from deciduous trees grown in dense plantings. A representative list of deciduous trees judged to have desirable growth characteristics for methane plantations, and shown to grow satisfactorily at high planting densities on short and repeated harvest cycles, is available (77).

Grasses. Grasses are very abundant forms of biomass (78). About 400 genera and 6000 species are distributed all over the world and grow in all land habitats capable of supporting higher forms of plant life. Grasslands cover over one-half the continental United States; about two-thirds of this land is privately owned. Grass, as a family Gramineae, includes the great fruit crops, wheat, rice, corn, sugarcane, sorghum, millet, barley, and oats. Grass also includes the many species of sod crops that provide forage or pasturage for all types of farm animals. In the concept of grassland agriculture, grass also includes grass-related species such as the legumes family, ie, the clovers, alfalfas, and many others. Grasses are grown as farm crops, for decorative purposes, for preserving the balance of productive capacity of lands by crop rotation, for controlling erosion on sloping lands, for the protection of water sheds, and for the stabilization of arid areas. Many advances in grassland agriculture have been made since the 1940s through breeding and the use of improved species of grass, alone or in seeding mixtures; cultural practices, including amending the soil to promote herbage growth best suited for specific purposes; and the adoption of better harvesting and storage techniques. Until the mid-1980s, very little of this effort had been directed to energy applications. A few examples of energy applications of grasses can be found, ie, as the combustion of bagasse for steam and electric power, but many other opportunities exist that have not been developed.

Perennial grasses have been suggested as candidate raw materials for conversion to synfuels (77). Most perennial grasses can be grown vegetatively, and they reestablish themselves rapidly after harvesting. Also, more than one harvest can usually be obtained per year. Warm-season grasses are preferred over cool-season grasses because their growth increases rather than declines as the temperature rises to its maximum in the summer months. In certain areas, rainfall is adequate to permit harvesting every 3 to 4 weeks from late February into November, and yields between about 18 and 24 $t/hm^2 \cdot yr$ of dry grasses may be obtainable in managed grasslands. Table 26 lists promising warm-season grasses proposed for conversion to synfuels.

Experimental work has shown that cool-season grasses such as Kentucky bluegrass and warm-season grasses such as Coastal Bermuda grass (*Cynodon dactylon*) can be converted to methane by conventional anaerobic digestion techniques (79,80). The compositions of some grasses indi-

Table 26. Warm-Season Grass Species for Methane Plantations[a,b]

Species	Localities[c]	Comments
Perennial sorghums and their hybrids	Plains, South	Sudan grasses, Johnson grasses, and other warm-season hybrids are promising for localities with alkaline soils; several harvests per year
Bermuda grasses[d]	South and South Central states	Most promising of all warm-season grasses, especially for localities with acid soils; can be harvested several times per year
Related to sugarcane[e]	Louisiana and Florida	Limited suitable sites
Related to bamboo[f]	South Central United States	
Bahia grasses	Florida and southern coastal plains	Competes with Bermuda grasses when fertilized; effect on overall yield is in dispute

[a] Ref. 77.
[b] High annual yields in the range of 18–22 dry t/(hm^2·yr) unless otherwise noted.
[c] Regions in which species grow naturally, have been successfully introduced, or have been tested extensively.
[d] Coastal, midland, and Suwanne grasses.
[e] Very high annual yield up to 45 dry (t/(hm^2·yr) in specially suitable sites.
[f] Untested annual yield.

cate that fertilizatioin procedures can incorporate certain nutrients into the harvested grass so that they can be converted by biological means without the use of excessive chemical additions to the conversion units (80).

Sugarcane is used commercially as a combination food and fuel crop. A great deal of information has been compiled about sugarcane, and it might well be used as a model for other biomass energy systems. It grows rapidly, produces high yields, the fibrous bagasse is used as boiler fuel, and cane-derived ethanol is used as a motor fuel in gasoline blends, ie, gasohol. Sugarcane plantations and the associated sugar processing and ethanol plants are in reality biomass fuel farms. About one-half of the organic material in sugarcane is sugar and the other half is fiber. Dry cane yields per year have been reported to range as high as 80 to 85 t/hm^2 (36 to 38 short ton/acre). Normal cultivation of sugarcane provides dry annual sugarcane yields of about 50 to 59 t/hm^2 (22 to 26 short ton/acre) (13).

Other productive grasses given serious consideration as raw materials for the production of energy and synfuels include sorghum and their highbreds. Tropical grasses are very productive and normally yield 50 to 60 t of organic matter per hm^2 annually on good sites (13). The tropical fodder grass *Digitaria decumbens* has been grown at yields of organic matter as high as 85 t/hm^2·yr (38 short ton/acre·yr) (13).

There are many grasses and related plants that can be considered for energy applications because they have the desirable characteristics needed for land-based biomass energy systems.

Other Cultivated Crops. Other high yielding land biomass species have been proposed as renewable energy sources (81). Promising species are kenaf, *Hibiscus cannabinus*, an annual plant reproducing by seed only; sunflower, *Helianthus annuus L.*, which is an annual oil seed crop grown in several parts of North America; and a few others, such as the polyisoprene-containing plant species described previously. Kenaf is highly fibrous and exhibits rapid growth, high yields, and high cellulose content. It is a potential pulp crop and is several times more productive than the traditional pulpwood trees. Maximum economic growth usually occurs in less than 6 months, and consequently two croppings may be possible in certain regions of the United States. Without irrigation, heights of 4 to 5 m are average in Florida and Louisiana; 6-m plants have been observed under near-optimum growing conditions. Yields as high as 45 t/hm^2·yr have been observed on experimental test plots in Florida, and it has been suggested that similar yields could be achieved in the Southwest with irrigation.

The sunflower is a prime candidate for biomass energy applications because of its rapid growth, wide adaptability, drought tolerance, short growing season, massive vegetative production, and adaptability to root harvesting. Dry yields have been projected to be as high as 34 t/hm^2 per growing season.

Water-Based Biomass

The average net annual productivities of dry organic matter on good growth sites for land- and water-based biomass are shown in Table 27. With the exception of phytoplankton, which generally has lower net productivities, aquatic biomass seems to exhibit higher net organic yields than land biomass. Water-based biomass considered to be the most suitable for energy applications include the unicellular and multicellular algae and water plants.

Algae. Unicellular algae, eg, the species *Chlorella* and *Scenedesmus*, have been produced by continuous processes in outdoor light at high photosynthesis efficiencies. *Chlorella*, for example, has been produced at a rate as high as 1.1 dry t/hm^2·day (82); this corresponds to an annual rate of 401 t/hm^2·yr. These figures are probably in error, but there is no theoretical reason why yields cannot achieve these high values because the process of producing algae can be almost totally controlled.

Algae production is not composed only of surface growth. Algae are produced as slurries in lakes and ponds, so the depth of the biomass-producing area as well as plant yield per unit volume of water are important parameters. The nutrients for algae production can be supplied

Table 27. Annual Biomass Yields on Fertile Sites[a,b]

Dry Organics, t/(hm^2·yr)	Climate	Ecosystem Type	Remarks
1	Arid	Desert	Better yield if hot and irrigated
2		Ocean phytoplankton	
2	Temperate	Lake phytoplankton	Little influence by humans
3		Coastal phytoplankton	Probably higher in some polluted estuaries
6	Temperate	Polluted lake phytoplankton	In agricultural and sewage runoffs
6	Temperate	Freshwater submerged macrophytes	
12	Temperate	Deciduous forest	
17	Tropical	Freshwater submerged macrophytes	
20	Temperate	Terrestrial herbs	Possibly higher yields if grazed
22	Temperate	Agriculture, annuals	
28	Temperate	Coniferous forests	
29	Temperate	Marine submerged macrophytes	
30	Temperate	Agriculture, perennials	
30		Salt marsh	
30	Tropical	Agriculture, annuals	Including perennials in continental climates
35	Tropical	Marine submerged macrophytes	Including coral reefs
38	Temperate	Reedswamp	
40	Subtropical	Cultivated algae	Better yield if CO_2 supplied
50	Tropical	Rain forest	
75	Tropical	Agriculture, perennials, reedswamp	

[a] Ref. 13.
[b] Average net values.

by sewage and other wastewaters. It should be pointed out that most unicellular algae are grown in fresh water, which limits their energy applications to small-scale algae farms. The high water content of unicellular algae also limits the conversion processes to biological methods.

Macroscopic multicellular algae, or seaweeds, have been considered as renewable energy resources. Candidates include the giant brown kelp *Macrocystis pyrifera* (83–85), the red algae *Rhodophyta*, and the floating algae *Sargassum*. Giant brown kelp has been studied in detail and harvested commercially off the California coast for many years (65). Because of its high potassium content, it was used as a commercial source of potash during World War I; in the 1990s, organic gums and thickening agents and alginic acid derivatives are manufactured from it. Laminaria seaweed is harvested off the East Coast for the manufacture of alginic acid derivatives. In tropical seas not cooled by upwelled water, species of the *Sargassum* variety of algae may be suitable as renewable energy resources. Several species of *Sargassum* grow naturally around reefs surrounding the Hawaiian Islands. However, only a small amount of research has been done on *Sargassum* and little detailed information is available about this algae. A considerable amount of data on yields and growth requirements is available, however, on the *Macrocystis* and *Laminaria* varieties. The high water content of macroscopic algae suggests that biological conversion processes, rather than thermochemical conversion processes, should be used for synfuel manufacture. The manufacture of coproducts from macroscopic algae, such as polysaccharide derivatives, along with synfuel may make it feasible to use thermochemical processing techniques on intermediate process streams.

Water Plants. The productivity of some salt marshes is similar to that of seaweeds. *Spartina alterniflora* has been grown at net annual dry yields of about 33 t/hm^2·yr (14.7 short ton/acre·yr), including underground material, on optimum sites (13). Other emergent communities in brackish water, including mangrove swamps, appear to have annual organic productivities of up to 35 t/hm^2·yr (15.6 short ton/acre·yr) (13); insufficient information is available to judge their value in biomass energy systems. Freshwater swamps may be highly productive and offer opportunities for energy production. Both the reed *Arundo donax* and bulrush *Scirpus lacustris* appear to produce 57 to 59 t/hm^2·yr (25.4–26.3 short ton/acre·yr) yields (13); if these could be sustained, they will be suitable candidates for biomass energy applications.

A strong candidate for energy applications is the water hyacinth, *Eichhornia crassipes* (3,86). This species of aquatic biomass is highly productive, grows in warm climates, and has submerged roots and aerial leaves like reedswamp plants. It is estimated that water hyacinth could be produced at rates up to about 150 t/hm^2·yr (67 short ton/acre·yr) if the plants are grown in a good climate, the young plants always predominated, and the water surface was always completely covered (13). Some evidence has been obtained to support this growth rate (87,88). If it can be sustained on a steady-state basis, water hyacinth may be one of the best candidates as a nonfossil carbon source for synfuels manufacture. It has no competitive uses (ca 1992) and is considered to be an undesirable species on inland waterways. Many attempts have been made to rid navigable streams in Florida of water hyacinth without success; the plant is a very hardy, disease-resistant species (89).

SYSTEMS ANALYSIS

The overall design of an integrated biomass-to-synfuel system is very important to its successful operation. The system is large and requires coordination of many different operations, such as planting, growing, harvesting, transporting, and converting biomass to gaseous and liquid synfuels. The detailed design of a biomass-to-synfuel system depends on several parameters, such as the type, size, number, and location of the biomass growth and processing areas. In the ideal case, synfuel production plants are located in or near the biomass growth areas to minimize cost of transporting the harvested biomass to the plant. All nonfuel effluents are recycled to the growth areas as shown in Figure 19. This type of synfuel plantation, if developed, would be equivalent to an isolated system with inputs of solar radiation, air, carbon dioxide, and minimal water, and one output, synfuel. The nutrients are kept within the ideal system so that the addition of external fertilizers and chemicals is not necessary. Also, environmental and disposal problems are minimized.

Various modifications of the idealized design in Figure 19 can be conceived for large-scale usage. One modification might consist of the addition of wastewater influent into the biomass growth area and the growth of water hyacinth in the Southern United States for two purposes, ie, the treatment of wastewater by luxuriant uptake of nutrients and the conversion of water hyacinth to synfuels. In this case, inorganic material would build up in the biomass growth area so that the residual material from the conversion plant could be partially removed or bled from the system as the synfuel is produced. This product might be considered to be a coproduct along with the synfuel.

Alternatively, short-rotation hybrid poplar and selected grasses can be multicropped on an energy plantation in the U.S. Northwest and harvested for conversion to liquid transportation fuels and cogenerated power for on-site use in a centrally located conversion plant. The salable products are liquid biofuels and surplus steam and electric power. This type of design may be especially useful for larger land-based systems.

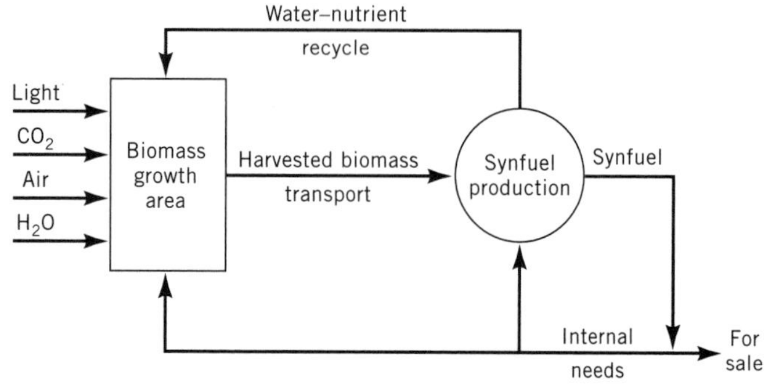

Figure 19. Idealized biomass-to-synfuel plantation system.

Another possibility, especially for small-scale farm use, is integration of agricultural crop, farm animal, and biofuel production into one system, eg, a farmer in the Midwest United States might grow corn as feedstock for a farm cooperative fuel ethanol plant. The residual distillers dried grains from the plant is used as hog feed, and the hog manure is used to generate medium heating value (MHV) fuel gas by anaerobic digestion. The fuel gas is used as plant fuel and the residual ungasified solids, which are high in nitrogen, potassium, and phosphorus, are recycled to the fields as fertilizer to grow more corn. The salable products are ethanol and hogs; the residuals are kept within the system.

Still another possibility is a marine biomass plantation such as that envisaged for giant brown kelp grown off the California Coast and conversion of the kelp to methane in a system similar to that shown in Figure 19. The location of the SNG plant could be either on a floating platform near the kelp growth area or located on shore, in which case the biomass or fuel transport requirements would be different.

Many different biomass energy system configurations are possible. As the technology is refined and developed to the point where commercialization activities are well under way, optimum designs will evolve. A great deal of attention has been given to the cost factors in the operation of biomass energy systems for the production of energy and biofuels. Of equal importance is the net energy production efficiency of the total system.

Economics

The practical value of biomass energy ultimately depends on the end-user costs of salable energy and biofuels. Consequently, many economic analyses have been performed on biomass production, conversion, and integrated biofuels systems. Conflicts abound when attempts are made to compare results developed by two or more groups for the same biofuels because methodologies are not the same. Technical assumptions made by each group are sometimes so different that valid comparisons cannot be made even when the same economic ground rules are employed. Comparative analyses, especially for hypothetical processes conducted by an individual or group of individuals working together, should be more indicative of the economic performance and ranking of groups of biofuels systems.

Several important generalizations can be made. The first is that fossil fuel prices are primary competition for biomass energy. Table 28 summarizes 1990 U.S. tabulations of average, consumption-weighted, delivered fossil fuel prices by end-use sector (90). The delivered price of a given fossil fuel is not the same to each end user; ie, the residential sector normally pays more for fuels than the other sectors, and large end users pay less.

In the context of biomass energy costs, dry, woody, and fibrous biomass species have an energy content of approximately 20 MJ/kg (8600 Btu/lb) or 20 GJ/t (17.2 MBtu/short ton). If such types of biomass were available at delivered costs of $1.00/GJ ($1.054/MBtu), or $20.00/dry t ($18.14/dry short ton), biomass on a strict energy content basis without conversion to biofuels would cost less than

Table 28. U.S. Delivered Fossil Fuel Prices to End Users, 1990, $/GJ[a]

Fossil Fuel	Residential	Commercial	Industrial	Transportation[b]	Utility Electricity[c]
Coal	2.87	1.52	1.60		1.38
Natural gas	5.34	4.45	2.79		2.20
Petroleum	8.43	5.65	5.32	7.90	3.24
LPG	10.38	8.17	5.12	8.03	
Kerosine	8.41	6.40	6.25		
Distillate fuel	7.60	5.79	5.39	8.03	
Motor gasoline		8.68	8.68	8.65	
Residual fuel		3.25	2.94	2.83	

[a] Ref. 90. All figures are consumption-weighted averages for all states in nominal dollars and include taxes.
[b] Aviation gasoline is delivered at $8.84/GJ; jet fuel at $5.39/GJ.
[c] Heavy oil, ie, grade nos. 4, 5, and 6, and residual fuel oils; light oils, ie, no. 2 heating oil, kerosine, and jet fuel; and petroleum coke are delivered at $3.13/GJ, $5.33/GJ, and $0.79/GJ, respectively.

most of the delivered fossil fuels listed in Table 28. The U.S. Department of Energy has set cost goals of delivered biomass energy drops at $1.90–2.13/GJ ($2.00–2.25/MBtu) (91) and $0.18/L ($0.67/gal) for fuel ethanol from biomass without subsidies in 2000, $0.22 to $0.26/L ($0.85 to $1.00/gal) by the year 2007 for biocrude-derived gasoline, $3.32/GJ ($3.50/MBtu) for methane from the anaerobic digestion of biomass by the year 2000, and 4.5 cents/kWh for electricity from biomass by the late 1990s (92).

An economic analysis of the delivered costs of biomass energy in 1990 dollars has been performed (ca 1992) for herbaceous and woody biomass for different regions of the United States (91). The analysis was done for each decade from 1990 to 2030 for Class I and II lands; results for biomass grown on Class II lands for the years 1990 and 2030 are shown in Table 29. Estimates of the total production costs for biomass were calculated with discounted cash flow models, one for the herbaceous crops switch-grass, napier grass, and sorghum, and one for the short-rotation production of sycamore and hybrid poplar trees. The delivered costs are shown in Table 29 in 1990 $/dry t and 1990 $/MJ and are tabulated by region and biomass species. The yield figures for 1990 were obtained from the literature and the projected yields for 2030 were assumed achievable from continued research. The annual, dry biomass yields per unit area have a great influence on the final estimated costs. This analysis indicates that the lowest cost energy crop of those chosen may be different for different regions of the country. A few of the biomass-region combinations appear to come close to providing delivered biomass energy near the U.S. Department of Energy cost goal. Realizing that there are many differences in the methodologies and assumptions used to compile the 1990 costs for delivered fossil fuels in Table 28 and deliv-

Table 29. Projected Delivered Costs for Candidate Biomass Energy Crops in 1990 and 2030[a,b]

	1990			2030		
		Delivered Cost			Delivered Cost	
Region and Species	Assumed Yield[c]	$/t	$/GJ	Assumed Yield[c]	$/t	$/GJ
Great Lakes						
Switchgrass	7.6	104.07	5.26	15.5	61.32	3.60
Energy sorghum	15.5	62.56	3.17	30.9	36.79	2.16
Hybrid poplar	10.1	113.79	5.76	15.9	72.82	4.29
Southeast						
Switchgrass	7.6	105.89	5.36	17.3	52.91	3.11
Napier grass	13.9	63.72	3.22	30.9	33.31	1.96
Sycamore	8.1	88.61	4.49	14.3	53.19	3.13
Great Plains						
Switchgrass	5.4	74.32	3.77	10.3	44.05	2.59
Energy sorghum	6.3	91.73	4.65	13.7	48.07	2.83
Northeast						
Hybrid poplar	8.1	105.26	5.33	11.9	71.69	4.26
Pacific Northwest						
Hybrid poplar	15.5	66.69	3.56	23.8	44.73	2.63

[a] Ref. 91. Discounted cash-flow models account for use of capital, working capital, income taxes, time value of money, and operating expenses. Real after-tax return assumed to be 12.0%. Short-rotation model used for sycamore and poplar. Herbaceous model used for other species. Costs in 1990 dollars. Dry tons.
[b] Yields in 1990 obtained from literature on Class II lands. Average total field yields are for entire region on prime to good soil, less harvesting and storage losses. Yields in 2030 assumed to be attained through research and genetic improvements. Short-rotation woody crops (hybrid poplar and sycamore) grown on six-year rotations on six independent plots. Net income is negative for first five years for each plot.
[c] Yield in $t/hm^2 \cdot hr$.

ered biomass energy in Table 29, it is evident that many of the biomass energy costs are competitive with those of fossil fuels in several end-use sectors, even without incorporating yield improvements that are expected to evolve from continued research on biomass energy crops.

It is essential to recognize several other factors, in addition to the cost of virgin biomass and its conversion to biofuels, when considering whether the costs of biomass energy are competitive with the costs of other energy resources and fuels. Some potential biomass energy feedstocks have negative values; ie, waste biomass of several types such as municipal biosolids, municipal solid wastes, and certain industrial and commercial wastes must be disposed of at additional cost by environmentally acceptable methods. Many generators of these wastes will pay a service company for removing and disposing of the wastes, and many of the generators will undertake the task on their own. These kinds of feedstocks often provide an additional economic benefit and revenue stream that can help support commercial use of biomass energy.

Another factor is the potential economic benefit that may be realized due to possible future environmental regulations from utilizing both waste and virgin biomass as energy resources. Carbon taxes imposed on the use of fossil fuels in the United States to help reduce undesirable automobile and power plant emissions to the atmosphere would provide additional economic incentives to stimulate development of new biomass energy systems. Certain tax credits and subsidies are already available for commercial use of specific types of biomass energy systems (93).

Energetics

The net energy production efficiency of an integrated biomass energy system is extremely important to its development and practical use. The ultimate goal is to design and operate environmentally acceptable systems to produce new supplies of salable energy whether they be low heat value gas, substitute natural gas, substitute gasolines or diesel fuels, methanol, ethanol, hydrogen, or electric power from biomass at the lowest possible cost and energy consumption. It is necessary to quantify how much energy is expended and how much salable energy is produced in each fully integrated system. An energy budget similar to an economic budget should be prepared because the capital, operating, and salable energy cost projections and the conversion process efficiency are insufficient to choose and design the best systems. These values do not necessarily correlate with net energy production. Also, the capital energy investment consumed during construction of the system should be recovered during its operation. Comparative analyses of similar systems for production of synthetic liquid and gaseous fuels from the same feedstock or of different systems that yield the same fuels from different biomass should be performed by consideration of the economics and the net energetics.

One method of analyzing net energetics of a biomass energy system is to let E_f, E_x, and E_p represent the energy content of the dry biomass feed, E_f, the sum of the external nonsolar energy inputs into the total system, E_x, and the energy content of the salable fuel products, E_p. The ratio $(E_p - E_x)/E_x$, which can be termed the net energy production ratio, indicates how much more, or less, salable fuel energy is produced than that consumed in the integrated system if the external energy consumed is replaced and it is assumed that the biomass feed energy is zero. This is a reasonable assumption because the energy value of biomass is derived essentially 100% from solar radiation. Net energy production ratios greater than zero indicate that an amount of energy equivalent to the sum of the external nonsolar energy inputs and an additional energy increment of salable fuel are produced; the larger the ratio, the larger the increment. The ratio $100E_p/(E_f + E_x)$ is the overall fuel production efficiency of the system or simply the energy output divided by the gross energy input. Finally, another useful ratio is the value of net energy output divided by the gross energy out, $100(E_p - E_x)/E_p$. This ratio expresses the percentage of the energy output that is new energy added to the economy.

A simple model for biomass energy production, excluding conversion to synfuels, is illustrated in Table 30 (94) which presents the results of an analysis of a short-rotation tree plantation that produces dry biomass at a yield of 11.2 t/hm^2·yr. The principal sources of energy consumption are fertilization, which includes the energy cost of ammonia production, and biomass transportation. The important result of this study is that the net energy production efficiency is high and that about 14 times more energy is produced in the form of woody biomass than the external nonsolar energy inputs needed to operate the system.

A mixed biomass plantation, including species of short-rotation hardwoods, sunflower, and kenaf, is projected to produce dry biomass at a yield of 67.3 t/hm^2·yr (95). Again, fertilization consumes the largest amount of external energy, but the energy cost for irrigation is also high. The

Table 30. Net Energy Analysis of Short-Rotation Wood Biomass Production[a,b]

Operation	Energy Consumed per Dry Wood Produced, MJ/t[c]	Total Consumption, %
Cultivation and planting	9.3	0.8
Fertilization	604	52.1
Harvesting	87	7.5
Transport to trucks	115	9.9
Load trucks	55	4.7
80.5-km transport to user	221	19.0
Unload	nil	
Auxiliary	69	5.9
Total	*1160*	*99.9*
Biomass energy produced, GJ/(hm^2·yr)[d]	196[e]	
Energy production efficiency, %	93.4[f]	
Net energy production ratio	14.1[g]	

[a] Ref. 94.
[b] Yield is 11.2 t/(hm^2·yr) dry; 20-yr planting cycle; fertilization is 224 kg N/(hm^2·yr).
[c] To convert MJ/t to Btu/1000 lb, multiply by 430.
[d] To convert GJ to 10^6 Btu, divide by 1.054.
[e] Assumes heating value of biomass is 17.5 MJ/dry kg.
[f] (Net energy output ÷ gross energy output) × 100, ie, 100 $(E_p - E_x)/E_p$.
[g] Net energy output ÷ energy consumed, ie, $(E_p - E_x)/E_x$.

Table 31. Net Energy Production Ratios for Ethanol Production from Corn in Integrated System[a]

Salable Energy Products, E_p	Nonfeed Energy Inputs, E_x	N^b
Alcohol	Corn production, fermentation, bottoms drying	−0.65
Alcohol	Corn production, fermentation	−0.51
Alcohol, chemicals	Corn production, fermentation	−0.50
Alcohol, chemicals, cattle feed	Corn production, fermentation, bottoms drying	−0.44
Alcohol, chemicals, cattle feed	Corn production, fermentation, bottoms drying, 50% residuals[c]	−0.10
Alcohol, chemicals, cattle feed	Corn production, fermentation, bottoms drying, 75% residuals[c]	0.29
Alcohol	Corn production, fermentation, 75% residuals[c]	1.43
Alcohol, chemicals	Corn production, fermentation, 75% residuals[c]	1.47

[a] Ref. 96.
[b] $N = (E_p - E_x)/E_x$.
[c] Percent of cobs and stalks collected and used as fuel within system to replace fossil fuel inputs.

net energy production efficiency is high; about 20 times more energy is produced as biomass energy than the external nonsolar energy inputs.

These systems analyses suggest that biomass plantations can be designed to operate at high net energy efficiencies, and that further improvements might best be incorporated in the systems by concentrating on fertilization. The use of nitrogen-fixing biomass and the recycling of nutrients from the conversion facilities may offer additional benefits.

For the total integrated biomass production–conversion system, the arithmetic product of the efficiencies of biomass production and conversion is the efficiency of the overall system. An overall conversion efficiency near 45% would thus be produced by integrating the biomass plantation illustrated in Table 30 with a conversion process that operated at an overall efficiency of 50%. Every operation in the series is thus equally important.

These simplified treatments correspond to net energy analyses using the First Law of Thermodynamics. Some energy analysts feel that only an analysis based on the Second Law can provide the ultimate answers in terms of where more available energy, in the thermodynamic sense, can be found to permit efficiency maximization. Others believe that the conventional energy balance is optimal because it is more realistic and easier to use. Indeed, for integrated synfuel production systems, entropic losses may not always be definable for all segments of the system, and a rigorous Second Law analysis may not be possible.

Location of the system boundaries also is important in the net energy analysis of integrated biomass energy systems. Thus tractors may be used to plant and harvest biomass. The fuel requirements of the tractors are certainly part of E_x, but is the energy expended in manufacturing the tractors also part of E_x? Some analysts believe that a complete study should trace all materials of construction and fossil fuels used back to their original locations in the ground.

Another important factor in net energy analysis concerns energy credits taken for by-products; they have an important effect and can determine whether the net energy production efficiency is positive or negative. An example of this effect is shown in Table 31 (96), which was derived from the integrated alcohol-from-corn production system illustrated in Figure 20. The boundary of the system depicted in Figure 20 circumscribes all the operations

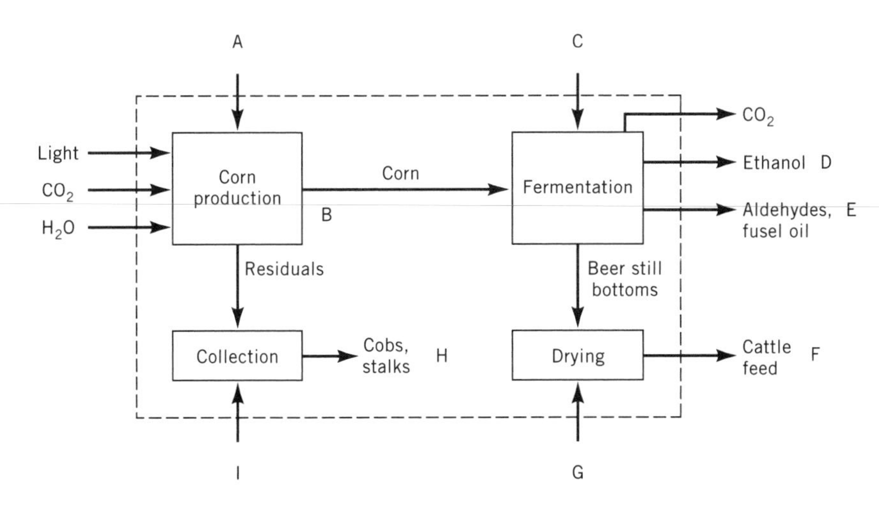

Figure 20. Energy inputs and outputs to manufacture 3.785 L of anhydrous ethanol from corn. (---) denotes system boundary. All KJ figures are lower heating value (LHV). A, 48,469 KJ; B, 128,314 KJ; C, 113,748 KJ; D, 79,682 KJ; E, 1,138 KJ; F, 47,483 KJ; G, 66,634 KJ; H, 174,768 KJ; I, 1,644 KJ. To convert KJ to Btu, divide by 1.054.

necessary to grow and harvest the corn, to collect residual cobs and stalks if they are used as fuel, to operate the fermentation plant for the production of anhydrous alcohol and by-product chemicals, and to dry the stillage to produce distillers' dried grains plus solubles for sale as cattle feed. The capital energy investment in the system is not included within the boundary.

Various net energy production ratios were calculated as shown in Table 31. It is apparent that the ratio can be either positive or negative, depending on whether credit is taken for the by-product chemicals and cattle feed, whether the energy for drying the stillage is included as an input to the system, and whether a portion of the residual corn cobs and stalks is collected and used as fuel within the system to replace fossil fuel inputs. To permit comparisons, net energy analyses must be clearly specified as to all details. On the whole, net energy production in a modern corn-to-ethanol plant would appear to be borderline if petroleum fuels comprise a significant part of the nonfeed energy inputs. However, to improve net energy production, these fuels can be replaced by fuels generated within the system or by renewable fuels, and credit can be taken for the by-products. Another route is to use more of the corn plant or the stillage as feedstock to the fermentation plant. The use of the grain alone as feedstock places a severe limitation on net energy production. Some of the information available on designs of corn wet-milling plants in which corn oil and other products are produced indicate that integration with an alcohol plant may be more efficient than a conventional corn alcohol plant.

COMMERCIAL USE OF BIOMASS ENERGY IN THE UNITED STATES

Relatively few biomass energy technologies are in commercial use in 1992. With two possible exceptions, ie, a few small tree plantations, and fuel ethanol from corn, sugarcane, and sugar beet, there is no biomass species grown in the United States specifically for conversion to biofuels. But conversion processes in commercial use in the early 1990s still span the basic technologies of combustion, gasification, and liquefaction. These include combustion of wood, wood wastes, and forestry residues; combustion of agricultural residues such as rice husks and bagasse; combustion of MSW and RDF for simultaneous disposal and energy recovery; biological gasification of animal manures in farm- and feedlot-scale anaerobic digestion systems for simultaneous waste disposal and production of biogas as well as upgraded solids for feed, fertilizer, and animal bedding; biological gasification of municipal wastewater sewage by anaerobic digestion for simultaneous waste disposal and biogas production; biological gasification of MSW in sanitary landfills and recovery of biogas for fuel, which also mitigates environmental and safety problems caused by gas migration in the landfill; thermal gasification of biomass for LHV gas production for on-site use; and alcoholic fermentation of starchy and sugar crops for fuel ethanol for use as a fuel extender and octane enhancer in motor gasoline blends.

Most of these commercial processes have been in use for many years. Some were greatly improved in the early 1990s, such as alcoholic fermentation that has undergone large process steam requirement reductions, thereby increasing net energy production efficiencies of fuel ethanol. Other commercial processes are relatively new, such as two-phase anaerobic digestion that permits higher plant capacities at low capital costs and the production of higher methane-content biogas at higher rates.

Inventories of commercial usage in the United States are available (97,98) Table 32 offers a summary (2). Although most of the data available does not refer to a specific time or year, wood use as a fuel in the industrial and residential sectors is responsible for the largest portion of biofuels consumption in the United States. Those states that have large forest products industries are the principal wood energy users. Similarly, states in the Corn Belt are the largest fuel ethanol producers. With few exceptions, those states having the most populated cities tend to process more municipal solid wastes by simultaneous disposal-energy recovery technologies. The biomass energy industry covers the entire nation; not one state is devoid of commercial biofuels production or utilization. The practical limitations to transport distance of some biomass

Table 32. Biofuels Utilization and Production and Biomass-Fueled Electric Power Plant Capacities in the United States[a]

Utilization	United States, Total	Energy Equivalents,[b] 10^{12} kJ/yr
Wood fuel utilized		
Commercial, 10^6 t/yr	83	1,553
Residential, 10^3 m^3/yr	198,496	1,154
Agricultural wastes utilized as fuels, 10^6 t/yr	9.5	142
Fuel ethanol produced, 10^6 L/yr	2,542	53.5
Biogas produced, 10^6 m^3/yr	9.3	0.2
MSW converted to energy,[c] 10^6 t/yr	14.0	113
Biomass-fueled electric plant capacity, MW	5,154	

[a] Ref. 2. The indicated biofuels consumption and capacity figures are the estimated values for various time periods since 1985 and do not refer to a specific year.

[b] The energy equivalents were calculated by the author using the following assumptions for Btu values: HHV of commercial wood fuel is 16×10^6 Btu/t, HHV of residential wood fuel is 20×10^6 Btu/cord, HHV of agricultural waste is 12.9×10^6 Btu/t, HHV of ethanol is 75,500 Btu/gal, HHV of biogas is 500 Btu/cf, HHV of MSW is 7×10^6 Btu/t, HHV of 1.00 barrel of oil equivalent (BOE) is 5.8×10^6 Btu.

[c] Ref. 98. These values are from 1987 and 1988. Energy products are steam and/or electric power.

such as wood, and the requirement for nearby or local processing, correlates with the concentration of biomass energy-processing facilities by state.

Combustion

The Public Utility Regulatory Policies Act of 1978 (PURPA), which provides benefits to cogenerators and small power producers, has stimulated commercialization of biomass combustion for electric power production. To be eligible for benefits under PURPA, small power production systems are limited to 80 MW and must receive 75% or more of their total energy input from renewables. Cogeneration systems do not have these limitations. Emissions limitations must be taken into consideration in developing commercial biofuels combustion projects. The regulations apply in a rather complex fashion for boilers burning either wood or municipal solid wastes alone or in combination with fossil fuels (99); even wood-burning stoves must meet national standards. With few exceptions, all stoves built after July 1988 must comply with a strict set of environmental regulations established by the U.S. Environmental Protection Agency and some states (100).

Wood Fuels

Wood fuels are the largest contributor to biomass energy usage. Approximately 88% of the total is attributed to wood energy consumption. Distribution of the annual consumption of wood fuel by region and sector from 1980 to 1984 is available (101); the South and the industrial sector are the largest wood energy consumers. Table 33 presents the composition of wood fuel by resource type from 1972 to 1987 (102). The largest concentration of wood fuel usage is in the lumber and wood products industry and the pulpwood and paper industry. It has been reported that the pulp and paper industry meets about 70% of its own fuel needs with wood energy (103). Over the period covered in Table 33, the pulp and paper industry utilized about one-half of total wood energy consumption, primarily as black liquor. This lignin-containing material is a by-product of the pulping process and is not wood as such. It is noteworthy that commercial, utility, and other industrial usage of wood fuels is a very small part of total consumption, and that residential usage more than doubled over the time period 1972–1984, whereas total wood energy usage increased about 61%.

The increase in residential fuelwood consumption over this period parallels the sharp increase in costs of oil, natural gas, and electricity and can be tracked by the number of wood-burning stoves in homes. Between 1950 and 1973, the estimated number of stoves dropped from 7.3 million to 2.6 million, but grew to an estimated 11 to 14 million in 1981 (104). The trend in the wood-burning stove inventory suggests that a four-fold increase in residential fuelwood use may have occurred during the 1970s and 1980s, which can be explained in part by heavy rural wood burning (105). However, since attaining about a one EJ level of consumption in 1984, there has been little gain in total residential wood energy use (106). Steady gains in industrial use have been counterbalanced by a decrease in residential use (106).

The data in Table 33 indicate that there is a small but relatively steady increase in total wood energy usage; that the industries having captive sources of wood or wood-derived products consume the bulk of wood energy, although at a generally flat rate; that there appears to be much opportunity for growth in wood energy consumption in the utility and commercial sectors; and that residential fuelwood usage has shown the largest incremental growth until the late 1980s.

Municipal Solid Waste

In the early 1990s, the need to dispose of municipal solid waste (MSW) in U.S. cities has created a biofuels industry because there is little or no other recourse (107). Landfills and garbage dumps are being phased out in many communities. Combustion of MSW, ie, mass-burn systems, and RDF, ie, refuse-derived fuel, has become an established waste disposal–energy recovery industry.

In May 1988 the United States had 105 operating MSW-to-energy plants, 29 plants under construction, 61 plants in the advanced planning stage, and 5 plants temporarily shutdown (98). About one-half of the 105 operating plants had been placed in operation since 1985, and about 80% of all operating plants use mass-burn, modular technology. Mass-burn, waterwall designs predominated for plants under construction or in the advanced planning stage. The sum total design capacity of these

Table 33. Wood Energy Use, 10^{12} kJ

Sector	1972	1978	1982	1986	1987
Lumber and wood products industry	373.1	444.7	362.5	474.3	484.8
Pulp and paper industry					
As hog fuel	473.2	104.0	178.1	268.7	271.9
As bark	94.5	99.3	116.9	139.1	142.3
As black liquor	737.8	854.8	811.5	949.6	1001.3
Other industry	38.9	47.4	45.3	51.6	51.6
Residential	420.5	689.3	985.4	927.5	885.3
Commercial	9.5	15.8	23.2	23.1	23.2
Utilities	3.2	1.1	3.2	9.4	9.5
Total	*2150.7*	*2256.7*	*2526.7*	*2843.3*	*2869.9*

[a] Refs. 2 and 102.

MSW-to-energy plants and corresponded to about 30% of the total MSW generated in the United States. A 1988 inventory of 25 MSW-to-energy plants that have been permanently shutdown shows the technologies for these plants consist of both mass-burn and RDF combustion systems, as well as a few pyrolysis plants. They have a combined total design capacity of 13,500 t/d, and most were built in the 1970s (98). The majority of these plants had operating difficulties caused by equipment and environmental concerns as well as cost factors. Problems that can cause permanent shutdown do not appear to be prevalent in more modern designs, presumably because combustion and waste-processing equipment have been improved.

According to the Solid Waste Association of North America, by 1992 there tense 137 municipal solid waste-to-energy plants operating in 36 states. They process about 16% of the 185 million tons of solid waste generated in the United States, and produce the equivalent of more than 2,300 MW of electricity. Nearly 100 other waste-to-energy projects are in various stages of planning or implementation. The U.S. Environmental Protection Agency has estimated that there will be more than 300 waste-to-energy plants in operation in the United States by 2000; these will process about 25% of U.S. municipal solid waste.

Power Production

PURPA has stimulated commercial use of biomass combustion for electric power production. It has prompted many companies to build and operate small power plants fueled with fossil and nonfossil energy resources. The power is used on-site in many cases, and the surplus is injected into the grid for sale to the utility at avoided cost. The number of U.S. filings as of December 31, 1987, submitted to the Federal Energy Regulatory Commission (FERC) for biomass-fueled and all cogeneration and small power production facilities illustrates the phenomenal growth of this industry from fiscal 1980 to 1988 (Table 34) (108,109). Of the total number of 1730 cogeneration filings that have been qualified by FERC as eligible for PURPA benefits, 138 (8.0% of total) are biomass-fueled and have a capacity of about 2699 MW, ie, 6.0% of total. Similarly, of the 1987 small power production filings that have been qualified by FERC, 468 (23.6% of total) are biomass-fueled and have a capacity of about 6260 MW, ie, 36.8% of total. The number of filings qualified by FERC and the total electric capacity for each biomass type are: 231 and 3703.8 MW of wood wastes, 140 and 654.951 MW for biogas, 131 and 2546.186 MW for MSW, and 108 and 1719.01 MW for agricultural wastes (109).

Data on the installed electric generating capacity and generation in 1986 by nonutility and utility power producers as of December 31, 1986 are available (110,111). Nonutility generators accounted for 3.45% of total U.S. capacity (25.3×10^3 MW) (73.3×10^4 MW) in 1986 and generated 4.31% (112×10^6 MWh) of the power produced in the same year. Biomass-fueled electric capacity and generation was 19.2% (4.9×10^3 MW) and 21.2% (23.7×10^6 MWh) respectively, of total nonutility capacity and generation. Biomass-fueled capacity experienced a 16% increase in 1986 over 1985, the same as natural gas, but it was not possible to determine the percentage of the total power production that was sold to the electric utilities and used on-site. Total production should be substantially more than the excess sold to the electric utilities. Overall, the chemical, paper, and lumber industries accounted for over one-half of the total nonutility capacity in 1986, and three states accounted for 45% of total nonutility generation, ie, Texas, 26% of total; California, 12% of total; and Louisiana, 7% of total. There were 2449 nonutility producers with operating facilities in 1986, a 15.8% increase over 1985, 75% of whose capacity was interconnected to electric utility systems.

Installed nonutility electric generation capacity and generation for each biofuel is presented in Table 35 for 1986. Wood was the largest contributor in 1986 to total capacity and generation, followed in decreasing order by MSW, agricultural wastes, landfill gas, and digester gas.

Table 34. Biomass-fueled Cogeneration and Small Power-production Capacities and Facilities, kW

Facility	1980	1985	1988	Biomass Total	Qualified Total All[c]
			Cogeneration		
New	400(1)	383,003 (23)	13,685 (3)	1,617,390 (122)	41,947,273 (1,705)
Existing		115,000 (2)		570,497 (18)	3,341,629 (78)
Both[d]	161,000 (1)	88,000 (1)		615,000 (0)	1,784,543 (33)
Total	161,400 (2)	586,003 (26)	13,685 (3)	2,802,887 (149)	47,073,445 (1,816)
Qualified[c]	161,400 (2)	585,503 (24)	725 (2)	2,698,802 (138)	44,943,616 (1,730)
			Small power production		
New	50 (1)	1,078,690 (112)	339,611 (19)	6,219,125 (487)	16,869,289 (1,964)
Existing		−800 (2)		116,925 (9)	319,099 (62)
Both[d]				11,500 (2)	136,485 (15)
Total	50 (1)	1,077,890 (114)	339,611 (19)	6,347,550 (498)	17,324,873 (2,041)
Qualified[c]	50 (1)	1,075,490 (110)	339,611 (19)	6,260,309 (468)	17,006,761 (1,987)

[a] Refs. 108 and 109. Filings under Public Utilities Regulation Policies Act of 1978. Totals are for years 1980–1988.
[b] Number of facilities in parentheses.
[c] Qualiied for PURPA benefits, ie, only owners or operators of facilities who claim qualifying status for PURPA benefits would make filings, some filings are submitted after the facilities begin operation, FERC does not review notices of qualifying status and has not completed review of all listed applications for certification, and the data provided in the filings are not verified by FERC inspection of the facilities.
[d] Combination of existing and new incremental capacity.

Table 35. Installed Nonutility Electricity Generation Capacity and Generation by Biofuel, 1986[a,b]

	Cogeneration		Small Power Producers		Total	
Biofuel	Capacity, MW	Generation, MWh	Capacity, MW	Generation, MWh	Capacity, MW	Generation, MWh
Wood	3,119.8	16,650,778	624.7	2,403,718	3,744.5	19,054,496
Agricultural wastes	252.0	1,022,573	68.2	310,387	320.2	1,322,960
Municipal solid waste	75.4	217,599	463.9	2,198,941	539.3	2,416,540
Landfill gas	0	0	184.6	622,031	184.6	622,031
Digester gas	21.2	117,146	49.6	186,750	70.8	303,896
Total	*3,468.4*	*18,008,096*	*1,391.0*	*5,721,827*	*4,859.4*	*23,729,923*

[a] Ref. 110.
[b] Total number of facilities reported in early 1985 was 111 in operation, 50 under construction, and 72 in the planning stages (112, 113).

The incremental changes in capacity and generation between 1985 and 1986 were 11.5% and 1.7% for wood, 7.8% and 15.9% for agricultural wastes, 30.4% and 9.2% for MSW, 184% and 207% for landfill gas, and 7.4% and 27.0% for digester gas. All incremental changes were positive.

Utility production of biomass-fueled electric power is much less than nonutility production. In early 1985, there were only 18 facilities having a total capacity of 245 MW, ie, nine fueled with wood (180.7 MW), five fueled with MSW (33.8 MW), two fueled with agricultural residues (22.5 MW), and two fueled with digester gas (8 MW) (112,113). The largest was a 50-MW plant in Burlington, Vermont (114).

Gasification

Conversion of biomass to gaseous fuels can be accomplished by several methods; only two are used by the biomass energy industry (ca 1992). One is thermal gasification in which LHV gas, ie, producer gas, is produced. The other process is anaerobic digestion, which yields an MHV gas.

Thermal Gasification. A survey of commercial gasifiers in use and under construction in 1983 indicates that the commercialization rate was low, but that several process developers and vendors have installed 30 to 35 operating systems (115). Feed rates of biomass range from 0.1 to 13.6 t/h, LHV gas output ranges from 1.1 to 211 billion J/h (1.0 to 200 million Btu/h) for a wide range of reactor configurations, and the gas is used for several different applications. Several large U.S. plants have been shutdown because of operating difficulties, eg, plants in Baltimore, Maryland; Orlando, Florida; and El Cajon, California.

By 1993, many U.S. gasifier vendors had gone out of business or were focusing marketing activities overseas or on other conversion technologies, particularly combustion for power generation, in states where combined federal and state incentives make economic factors attractive. Some existing gasification installations also have been shutdown and placed in a stand-by mode until natural gas prices make biomass gasification competitive again.

A survey of commercial thermal gasification in the United States shows that few gasifiers have been installed since 1984 (115). Most units in use are retrofitted to small boilers, dryers, and kilns. The majority of existing units operate at 0.14 to 1.0 t/h of wood wastes on updraft moving grates. The results of this survey are summarized in Table 36. Assuming all 35 of these units are operated continuously, extremely unlikely, the maximum amount of LHV gas that can be produced is about 0.003 to 0.006 EJ/yr (222–445 m^3/d).

Table 36. Commercial Thermochemical Gasifiers, June 1988[a]

Size, 10^6 kJ/h[b]	Number of Gasifiers	Feedstock	Gas Use
	Updraft reactor		
6.0	2	Corn cobs	Corn dryer
4.9–25.0	4	Wood	Space or process heat
0.94–25.9	14	Wood	Dry kiln, space heat
12.0–69.9	3	Wood	Brick kilns
	Downdraft reactor		
0.2–6.0	5	Wood, peach pits	Greenhouse
10.0	2	Wood	Power boiler
	Fluid-bed reactor		
25.0	2	Rice hulls	Process heat
82.2	1	Wood	Power boiler
124.4	2	Wood	Clay dryers

[a] Ref. 116. All units are LHV gasifiers.
[b] To convert 10^6 kJ/h to 10^6 Btu/h, divide by 1.054.

Biological Gasification. Several surveys have been performed on the use of anaerobic digestion for biogas production and waste stabilization. In 1984, 84 farm-scale and industrial anaerobic digestion facilities, exclusive of municipal wastewater treatment plants, were identified in 32 states and were estimated to have a total reactor volume of 90,600 m^3 (117). Individual digesters were found to range in size up to 13,000 m^3, but the majority had volumes in the range 249 to 750 m^3. A comprehensive inventory was conducted in 1985 (118). Exclusive of municipal wastewater treatment, 96 farm-scale and industrial anaerobic digestion systems were found to have been built between 1972 and mid-1985 in 30 states; in Puerto Rico, 2 units. Eighty-seven of the systems had digestion volumes of 100 m^3 or more; 60 were operational, 7 were temporarily shutdown, and others were in various stages of design or development. Forty-four of the operational or shutdown systems, ie, 43.900 m^3, were used for digesting animal manures, 35 for dairy or beef cattle manure, and the remainder for swine or poultry manure. Fourteen of the facilities, ie, 107,900 m^3, provided wastewater clean-up services to agricultural product processing plants, breweries, and related food production facilities. The designs that were used extensively for the farm-scale digesters were plug flow and stirred tank configurations and unheated lagoons. Only a few additional commercial digestion systems have been installed since this inventory was completed.

An estimate of the potential methane production possible from existing (ca 1992) municipal wastewater treatment plants that produce and use biogas as a fuel and from the farm-scale and industrial anaerobic digestion plants identified in the inventory is presented in Table 37. The maximum amount of methane that could be produced from this commercial anaerobic digestion capacity under conventional operating conditions is about 0.005 EJ/yr (2,400 BOE/d).

Gas production is considerably greater from commercial landfill methane recovery systems. In 1987, 94 plants (50 operational, 44 scheduled) had an estimated design production of 1.2×10^6 m^3/d and an estimated actual production of 0.314×10^6 m^3/d, or 114×10^6 m^3/yr; estimated electric capacity was 231.2 MW (120–123).

The initial biogas recovered is an MHV gas and is often upgraded to high heat value (HHV) gas when used for blending with natural gas supplies. The annual production of HHV gas in 1987, produced by 11 HHV gasification facilities, was 116×10^6 m^3 of pipeline-quality gas, ie, 0.004 EJ (121). This is an increase from the 1980 production of 11.3×10^6 m^3. Another 38 landfill gas recovery plants produced an estimated 218×10^6 m^3 of MHV gas, ie, 0.005 EJ. Additions to production can be expected because of landfill recovery sites that have been identified as suitable for methane recovery. In 1988, there were 51 sites in preliminary evaluation and 42 sites were proposed as potential sites (121).

Liquefaction

Since the 1970s attempts have been made to commercialize biomass pyrolysis for combined waste disposal–liquid fuels production. None of these plants were in use in 1992 because of operating difficulties and economic factors; only one type of biomass liquefaction process, alcoholic fermentation for ethanol, is used commercially for the production of liquid fuels.

Fermentation ethanol, primarily from corn, but also from sugarcane, sugar beet, or derivatives, has shown extraordinarily high production rate increases since 1979 when it was reintroduced in the United States as a blending component in motor fuels. In 1979, 24 operating plants, with a design capacity of 151×10^6 L/yr, produced 75.7×10^6 L/yr; 35 additional plants were planned. By 1988, the number of plants had increased to 55, with a design capacity of 3743×10^6 L/yr and an actual production of 3160×10^6 L/yr; 70 additional plants were shut down, with total unused design capacity of 1400×10^6 L/yr (124). Tax incentives provided by the federal and state governments coupled with generally high gasoline prices, low corn prices, and the phase-out of leaded fuels, have helped establish the fuel ethanol industry.

Since most fuel ethanol manufactured in the United States is made from corn, its price plays a crucial role in determining the competitive position of ethanol in an open market. With corn priced at about $2.50/bu, the embedded feedstock cost of product ethanol is about $0.14–0.23/L ($0.52–0.87 gal), depending on overall yield and by-products ignored (125). Fuel ethanol plants may have contingency plans to close if corn prices rise to a certain level, eg, $3.50/bu or above (126).

A listing of fuel ethanol plants in operation, with total anhydrous capacity of 3743.2×10^6 L/yr, is available (124). Leading producers include Archer Daniels Midland, with over one-half of all domestically produced fuel ethanol from four locations: Pekin Energy Co., Pekin, Illinois; South Point Ethanol, South Point, Ohio; and New Energy Company of Indiana, South Bend. Several plants terminated operations even when the price of corn was in the $2.00/bu range. Continuous operation at a profit is difficult to sustain when the selling price of fuel ethanol must remain competitive with gasoline prices and alternative octane-enhancing methods. Without tax incentives, it is doubtful that fuel ethanol producers, particularly those who operate smaller plants and use older technologies, will be able to survive during times of high corn prices and low crude oil prices.

Biomass Production

In 1992, there was no biomass species grown and harvested in the United States specifically for conversion to biofuels, with the possible exceptions of feedstocks for fuel

Table 37. Potential Methane Production from Commercial U.S. Anaerobic Digestion Systems

Number of Plants	Feedstock	Estimated Digester Volume, 10^6 m^3	Estimated Methane Production Potential[a] 10^6 m^3/d	EJ/yr
209[b]	Municipal wastewater	0.213	0.208	0.0028
44	Animal manures	0.044	0.043	0.0006
14	Industrial wastes	0.108	0.105	0.0015
Total			*0.356*	*0.0049*

[a] Calculated assuming 65 vol % of methane in product gas and 1.5 vol gas/culture vol · d.
[b] Ref. 119. These are treatment plant unit processes, not individual digesters, that produce and use digester gas; the flow capacity is 14.2×10^3 m^3/d.

ethanol and a few tree plantations. This is understandable from an economic standpoint. For example, the average natural gas price in the United States in 1991 at the point of production, not end use cost, was estimated to be \$1.51/BJ (\$1.59/MBtu) (U.S. Energy Information Agency, Washington, D.C.). For biomass to compete on an equivalent basis, it must be grown, harvested, and gasified to produce methane at an average cost of \$1.51/GJ (\$1.59/MBtu). Assuming an unrealistic gasification cost of zero, the maximum biomass cost that is acceptable under this condition is \$29.73/dry ton. At an optimistic yield of 4.45 dry t/hm^2·yr (10 dry short ton/acre·yr) a biomass energy crop producer for a gasification plant will realize not more than \$667.60/hm^2·yr (\$270.30/acre·yr), a marginal amount to permit a net return on an energy crop without other incentives. This simplistic calculation emphasizes the effect of depressed fossil fuel prices on biomass energy crops. Negative feedstock costs, ie, wastes, substantial by-product credits, captive uses, and/or tax incentives, are needed to justify energy crop production on strict economic grounds.

Most of the commercial tree plantations that produce wood for captive use as a raw material in manufacturing operations use a portion as fuel. Examples of short-rotation plantations are listed in Table 38 (127). Paper companies in the southeastern United States are reported to have short-rotation plantings also, eg, Weyerhaeuser, James River Corp., Buckeye Cellulose, and Lykes Brothers, but the intensity of maintenance is not known (127).

The advances in biomass growth technologies developed in the United States for agricultural crops, trees, and aquatic species, and that are commercial and being improved further through research, are available for growth of biomass energy crops. Multicropping designs and multiple-use crops will be the most likely candidates for biomass energy when conditions warrant commercial plantations.

ECONOMIC AND LEGISLATIVE IMPACTS

An interminable number of studies have been performed to predict future energy consumption patterns, resources, imports, and prices. If the predictions of higher oil prices had been accurate in the late 1970s, or if the oil price had stabilized at its peak in 1981, the biomass energy industry would have exhibited much greater growth than it has (128).

Biofuels usage has slowly increased since the mid-1980s because of environmental problems, eg, MSW disposal; favorable legislation, eg, tax incentives and PURPA; and combinations of both, eg, oxygenated transportation fuels. Although environmental problems continue to increase, many tax incentives for alternative renewable energy resources have been reduced or eliminated (129). Commercialization of biomass energy in 1992 is driven by waste disposal, alternative fuels and environmental issues, and the available incentives for PURPA power plants.

Capacity Limitations and Biofuels Markets. Large biofuels markets exist (130–133), eg, production of fermentation ethanol for use as a gasoline extender (see ALCOHOL FUELS). Even with existing (1987) and planned additions to ethanol plant capacities, less than 10% of gasoline sales could be satisfied with ethanol–gasoline blends of 10 vol % ethanol; the maximum volumetric displacement of gasoline possible is about 1%. The same condition applies to methanol and alcohol derivatives, ie, methyl *t*-butyl ether [*1634-04-4*] and ethyl-*t*-butyl ether.

In 1987, taxable motor gasoline sales were 415.89×10^9 L (109.88×10^9 gal) (131). In the same year, the methanol nameplate capacity was 5.30×10^9 L (1.40×10^9 gal) and actual production was 4.12×10^9 L (1.09×10^9 gal); synthetic ethanol capacity was 0.80×10^9 L (0.21×10^9 gal) and 0.30×10^9 L (0.08×10^9) was actually produced (133,134); fermentation ethanol capacity was 3.62×10^9 L (0.957×10^9 gal) and actual production for blending with gasoline was 2.84×10^9 L (0.750×10^9 gal) (124).

Only a small portion of motor fuel needs could be satisfied if truly large-scale alcohol–gasoline blending or fuel switching occurred via transition to fuel-flexible vehicles and ultimately to neat alcohol-fueled vehicles (132).

Capacities for producing virtually all biofuels manufactured by biological or thermal conversion of biomass must be dramatically increased to approach their potential contribution to primary energy demand (Table 4). As already pointed out, an incremental EJ per year of biogas requires about 210 times the existing digestion capacity, including wastewater treatment plants, whereas an incremental EJ per year of ethanol requires about 14 times the existing

Table 38. Commercial Tree Production for Energy Use,[a] 1988

Company	Area, hm^2	Species	Rotation, yr	Comments
Simpson Timber Co.[b]	283	Eucalyptus		
West Vaco[c]	6475	Cottonwood, Sycamore	10	Primarily for pulp with some to fuel paper mills
Packaging Corp. of America[d]	1214	Hybrid poplar		
Hagerstown[e]	202	Hybrid poplar		Wastewater disposal site, energy use of wood planned
Reynolds Metals Co.[f]	91	Hybrid poplar	6	Captive energy use of wood planned
Union Corp. of North Carolina	8903	Sweetgum	10	Captive for pulp with some to fuel paper mills
James River Corp. of Nevada	2975	Hybrid poplar	6	Captive for fiber and fuel for paper mills, larger plantings are planned

[a] Ref. 127. [b] California. [c] Kentucky.
[d] Michigan [e] Maryland. [f] New York.

fermentation plant capacity. Thus, biofuels cannot be expected to satisfy large EJ markets in the short- to midterm. Since most nonwaste-derived biofuels are not economically competitive with fossil fuels in the early 1990s, large additions to plant capacity will not occur except in those cases where environmental concerns or legislative incentives are governing factors.

Investment Opportunities and Capital Requirements. Despite some of the temporary economic problems that confront the biomass energy industry in the early 1990s, several business opportunities are being developed at rapid rates. These projects are distributed across the nation and include landfill gas recovery plants, MSW-to-energy systems, and nonutility power generation that qualifies under PURPA. Conventional combustion technology is utilized in the majority of plants; gasification seems to have been largely ignored and should offer several advantages (112). A production tax credit equivalent to $0.48/m^3 ($3.00/BOE) indexed to inflation and linked to the price of oil is available; it amounted to about $0.71/GJ ($0.75/MBtu) of product gas in mid-1985 (112,129), and can have a significant beneficial impact on the profitability of a biofuels project. The lower the cost of oil, the greater the credit. Taking the most optimistic view of the language in the law, wastes are included in the definition of biomass, so it appears the production tax credit is applicable to all of the above projects, not just those based on wood and other nonwaste biomass.

The Tax Reform Act of 1986 has resulted in a transition away from capital supplied by individuals for the financing of biofuels projects toward conventional financing and greater use of institutional capital sources (128,134). The capital requirements can be large, eg, $20.7 billion in 10 years to complete 240 early and advanced-planned MSW-to-energy plant projects (98), and $35 million per plant for 300–400 small plants (112). In the United States, however, there are also biofuels opportunities that do not involve such large capital needs. Numerous landfill gas recovery plants have been installed for well under $30 million each; most have a capital cost of $5 million or less, depending on scale and end use, although the capital cost of one of the largest landfill gas recovery projects is in the $20 million range (115).

The Energy Policy Act of 1992 (H.R. 776) has liberalized the rules concerning biofuels and provided tax incentives for increased usage. Many states also have gasohol fuel tax exemptions in place, and some have enacted legislation that requires use of oxygenated fuels under certain conditions. Most of these laws impact favorably on biofuels usage.

Many energy analysts believe that it is only a matter of time before petroleum prices, the economic parameter that influences almost all other energy prices, begin to return to market prices of at least $3.97/m^3 ($25/bbl). It is widely believed that the gas bubble, which provided excess gas deliverability in the 1980s, will decline in the 1990s. Thus, energy prices are expected to rise again under any scenario. If petroleum prices stabilize at or continue to increase to levels over $3.97/m^3 ($25/bbl) it is expected that this, along with environmental issues, will provide the market forces that will increase biomass energy usage.

RESEARCH

A large variety of biomass feedstock developments and advanced conversion processes for the production of energy, fuels, and chemicals are in the research stage in the United States. Many other countries are also developing biomass energy in the laboratory and in the field. The research is aimed at reducing the cost of biomass and increasing the efficiency of production of the final products, eg, new fuels, substitute fossil fuels, and energy, so that biofuels can compete with other energy resources, especially fossil fuels.

Feedstock Development

Most of the research in process in the United States in the early 1990s on the selection of suitable biomass species for energy applications is limited to laboratory studies and small-scale test plots. Many of the research programs on

Table 39. Reported Maximum Productivities for Recommended Herbaceous Plants[a]

U.S. Region[b]	Species	Yield, dry t/hm^2·yr
Southeastern prairie delta and coast	Kenaf	29.1
	Napier grass	28.5
	Bermuda grass	26.9
	Forage sorghum	26.9
General farm and North Atlantic	Kenaf	18.6
	Sorghum hybrid	18.4
	Bermuda grass	15.9
	Smooth bromegrass	13.9
Central	Forage sorghum	25.6
	Hybrid sorghum	19.1
	Reed canary grass	17.0
	Tall fescue	15.7
Lake states and Northeast	Jerusalem artichoke	32.1
	Sunflower	20.0
	Reed canary grass	13.7
	Common milkweed	12.3
Central and southwestern plains and plateaus	Kenaf	33.0
	Colorado River hemp	25.1
	Switchgrass	22.4
	Sunn hemp	21.3
Northern and western great plains	Jerusalem artichoke	32.1
	Sunchoke	28.5
	Sunflower	19.7
	Milkvetch	16.1
Western range	Alfalfa	17.9
	Blue panic grass	17.9
	Cane bluestem	10.8
	Buffalo gourd	10.1
Northwestern/ Rocky Mountain	Milkvetch	12.1
	Kochia	11.0
	Russian thistle	10.1
	Alfalfa	8.1
California subtropical	Sudan grass	35.9
	Sudan–sorghum hybrid	31.6
	Forage sorghum	28.9
	Alfalfa	19.1

[a] Ref. 135.
[b] As defined by U.S. Dept. of Agriculture, Agriculture Handbook 296, Mar. 1972; excludes Alaska and Hawaii.

Table 40. Production Costs for Annual Herbaceous Plants[a]

Plant Groups	Model Crop Used	Whole Plant Yield, dry t/(hm^2 · yr)	Cost, \$/t
Tall grasses	Corn	17.3	19.1
Short grasses	Wheat	9.9	17.2
Tall broadleaves	Sunflower	15.0	12.7
Short broadleaves	Sugar beet	13.9	77.1
Legumes	Alfalfa[b]	13.7	20.9
Tubers	Potatoes	9.2	136

[a] Ref. 135. Average cost.
[b] Is a perennial.

feedstock development were started in the 1970s or early 1980s.

Herbaceous Biomass. Considerable research has been conducted to screen and select nonwoody herbaceous plants as candidates for biomass plants that are unexplored in the continental United States; other research has concentrated on cash crops such as sugarcane and sweet sorghum; and still other research has emphasized tropical grasses. In the late 1970s, a comprehensive screening study of the United States generated a list of 280 promising candidates from which up to 20 species were recommended for field experiments in each region of the country (135). The four highest-yielding species recommended for further tests in each region are listed in Table 39 (135). Since many of the plants in the original list of 280 species had not been grown for commercial use, the production costs were estimated as shown in Table 40 for the various classes of herbaceous species and used in conjunction with yield and other data to develop the recommendations in Table 39.

A large number of small-scale field tests on potential herbaceous energy crops have been carried out. The productivity ranges for some of the most important species for the midwestern and southeastern United States are shown in Table 41 (136). The results of this research helped to establish a strategy that these crops should be primarily grasses and legumes produced using management systems similar to those used for conventional forage crops. It was concluded from this work that the ideal selection of herbaceous energy crops for these areas would consist of at least one annual species, one warm-season perennial species, one cool-season perennial species, and one legume. Production rates, cost estimates, and environmental considerations indicate that perennial species will be preferred to annual species on many sites, but annuals may be more important in crop rotations.

In greenhouse, small-plot, and field-scale tests conducted to screen tropical grasses as energy crops, three categories have emerged, based on the time required to maximize dry-matter yields: short-rotation species (2–3 months), intermediate-rotation species (4–6 months), and long-rotation species (12–18 months) (137). A sorghum–sudan grass hybrid (Sordan 70A), the forage grass napier grass, and sugarcane are outstanding candidates in these categories. Minimum-tillage grasses that produce moderate yields with little attention are wild *Saccharum* clones, and Johnson grass in a fourth category. The maximum yield observed was 61.6 dry t/hm^2·yr for sugarcane propagated at narrow row centers over 12 months. The estimated maximum yield is of the order of 112 dry t/hm^2·yr using new generations of sugarcane and the propagation of ratoon, ie, regrowth, plants for several years after a given crop is planted.

Overall, research on the development of herbaceous energy crops shows that a broad range of plant species may ultimately be prime energy crops.

Short-Rotation Woody Crops. Research to develop trees as energy crops via short-rotation intensive culture (SRIC) made significant progress in the 1980s. Projections indicate that yields of organic matter can be substantially increased by coppicing techniques and genetic improvements. Advanced designs of whole-tree harvesters, logging residue collection and chipping units, and rapid planting machinery have progressed to the point where prototype units are being evaluated in the field. It is expected that

Table 41. Productivity Rates for Reproductive Herbaceous Biomass species in Southeast and Midwest, dry t/hm^2 · yr[a]

Biomass Type and Species	Southeast	Midwest
Annuals		
Warm-season		
Sorghums[b]	0.2–19.0	1.9–29.1
Cool-season		
Winter rye[c]	0.0–7.2	2.4–6.1
Perennials[d]		
Warm-season		
Switchgrass[c]	2.9–14.0	2.5–13.4
Weeping lovegrass[c]	5.4–13.7	
Napier grass–energycane[b]	20.4–28.3	
Cool-season		2.7–10.8
Reed canary grass[c]		
Legumes		
Alfalfa		1.6–17.4
Flatpea	2.1–12.9	3.9–10.2
Sericea lespedeza	1.8–11.1	

[a] Ref. 136. Figures are average annual productivities.
[b] Thick-stemmed grass.
[c] Thin-stemmed grass.
[d] Productivity rates after 1–2 yr establishment period.

several of these devices will be manufactured for commercial use. Some tree species being targeted for research are red alder, black cottonwood, Douglas fir, and ponderosa pine in the Northwest; *Eucalyptus*, mesquite, Chinese tallow, and the leucaena in the West and Southwest; sycamore, eastern cottonwood, black locust, catalpa, sugar maple, poplar, and conifers in the Midwest; sycamore, sweetgum, European black alder, and loblolly pine in the Southwest; and sycamore, poplar, willow, and sugar maple in the East. Generally, tree growth in test plots is studied in terms of soil type and the requirements for site preparation, planting density, irrigation, fertilization, weed control, disease control, and nutrients. Harvesting methods are also important, especially in the case of coppice growth for SRIC hardwoods. Although tree species native to the region are usually included in the experimental design, non-native and hybrid species are often tested too. Advanced biotechnological methods and techniques, such as tissue culture propagation, genetic transformation, and somaclonal variation, are being used in research to clonally propagate individual genotypes and to regenerate genetically modified species.

After an intensive 10-year research effort, short-rotation woody crop yields in the United States, based on data accumulated to 1992, were projected to be 9, 9, 11, 17, and 17 dry t/hm^2·yr in the Northeast, South/Southeast, Midwest/Lake, Northwest, and Subtropics, respectively (138). The corresponding research goals are 15, 18, 20, 30, and 30 dry t/hm^2·yr. Hybrid poplar, which can grow in many parts of the United States, and *Eucalyptus*, which is limited to Hawaii, Florida, southern Texas, and part of California, have shown the greatest potential thus far for attaining exceptionally fast growth rates (138). Both have achieved yields in the range of 20 to 43 t/hm^2·yr in experimental trials with selected clones. Research indicates other promising species to be black locust, sycamore, sweetgum, and silver maple.

Hybridizing techniques seem to be leading to super trees that have short growth cycles and yield larger quantities of woody biomass. Fast-growing clones are being developed for energy farms in which the trees are ready for harvest in as little as 10 years and yield up to 30 m^3/hm^2·yr. Genetic and environmental manipulation has also led to valuable techniques for the fast growth of saplings in artificial light and with controlled atmospheres, humidity, and nutrition. The growth of infant trees in a few months is equivalent to what can be obtained in several years by conventional techniques.

Chemical injections into pine trees have been reported to have stimulatory effects on the natural production of resins and terpenes and may result in high yields of these valuable chemicals. Combined oleoresin–timber production in mixed stands of pine and timber trees is under development, and it appears that when short-rotation forestry is used, the yields of energy products and timber can be substantially higher than the yields from separate operations.

One of the largest research projects on SRIC trees in the Western World, the Large European Bioenergy Project (LEBEN), was reported to be scheduled for initiation in the Abruzzo Region of Italy in the mid-1980s and to be established near the end of that decade (139,140). This project integrates SRIC tree production, agricultural energy crops and residues, and biomass conversion to fuels and energy. About 400,000 t/yr of biomass consisting of 260,000 t/yr of woody biomass from 700 hm^2, and 120,000 t/yr of agricultural residues from 700 hm^2 of vineyards and olive and fruit orchards, will be used. Later, 110,000 t/yr of energy crops from 1050 hm^2 will be utilized. The energy products include liquid fuels, ie, biomass-derived oil and charcoal, 200 million kWh/yr of electric power, and waste heat for injection into the regional agro-forestry and industrial sectors. This project is still in the start-up stage.

In the Amazon jungle of Brazil, perhaps the largest SRIC tree plantation in the world is being integrated with energy, pulp, and chemical production facilities (139). Fast-growing Caribbean pine, *Eucalyptus*, and *Gmelina arborea* grown on 51,400 hm^2 are converted to these products. Although gmelina failed on some of the planted sites, it is doing well on about one-third of the planted sites that have the best soil conditions. The other tree species are apparently being grown successfully on the other two-thirds of the sites with sandy and transition soils. It has been reported that the Brazilian Government and industry have taken control of the project from non-Brazilian interests.

Aquatic Biomass. Aquatic biomass, particularly micro- and macroalgae, are more efficient at converting incident solar radiation to chemical energy than are most other biomass species. For this reason, and the fact that most aquatic plants do not have commercial markets, research was performed in the late 1970s and 1980s to evaluate several species as energy crops. The overall goals of the research have generally been directed to either biomass production, often with simultaneous waste treatment, for subsequent conversion to fuels by fermentation, or to species that contain valuable products. The aquatics studied and their main applications are microalgae for liquid fuels, the macrophyte water hyacinth for wastewater treatment and conversion to methane, and marine macroalgae for specialty chemicals or conversion to methane.

Research in the United States on microalgae focuses on the growth of these organisms under conditions that promote lipid formation. This eliminates the high cost of cell harvest because the lipids often can be separated by simple flotation or extraction. The United States Department of Energy research program on microalgae in the 1980s was one of the largest of its kind. It consisted of several projects and emphasized the isolation and characterization of the organisms and the development of microalgae that afford high oil yields. The research included projects on siting studies; collection, screening, and characterization of microalgae; growth of certain species in laboratory and small-scale production systems; exploration of innovative approaches to microalgae production; and innovative methods for increasing oil formation. Some microalgae, such as *Botryococcus braunii*, have been reported to produce lipid yields that are 40–50% of the dry cell weight under nitrogen-limited conditions (29). However, in other research, *B. braunii* has been reported to yield 20–52% of the dry cell weight as liquid hydrocarbons (139).

Conversion

Combustion. Biomass combustion accounted for about 4% of total U.S. energy consumption in 1992, primarily in the industrial, residential, and utility sectors. Electric power capacity fueled by biomass grew from 200 MW in the early 1980s to about 6000 MW in 1992. The direct combustion of biomass for heat, steam, and power has been, and is expected to continue to be, the principal end use of biomass energy. Conventional biomass-fired technology uses a variety of combustion equipment designs that are usually capable of burning a wet, nonhomogeneous fuel with large variations in moisture content and particle size (141). Spreader stoker-fired boilers have evolved from the designs of the past to systems which include several designs for controlled fuel distribution and automatic ash removal. Research on biomass combustion has focused on improvements of existing systems with respect to ease of operation, increased efficiency, and lower capital and operating costs; emission controls and abatement; and development of new technologies to permit utilization of solid biomass fuels in a wider range of applications (142). Some of the biomass combustion research developments since the early 1930s include whole-tree burning technologies (143), cyclonic incineration of waste biomass (144), direct wood-fired gas turbines (145), improved combustion cycles for biomass (141), fluid-bed biomass combustion (146), pulverized biomass combustion (147), catalytic wood-burning stoves (148), cofiring of biomass and fossil fuels to reduce emissions (149), and control of biomass combustion to reduce emissions (150). Even though the burning of biomass is one of the oldest energy-producing methods used, research continues to make significant advancements in the art and science of biomass combustion. Recent U.S. legislation concerned with air quality and waste biomass disposal has a significant impact on the direction of ongoing research to develop advanced biomass combustion systems (151).

Anaerobic Digestion (Methane Fermentation). A large amount of research was performed on the anaerobic digestion of biomass in the 1970s and 1980s to develop biological gasification processes that are capable of producing methane (16). Basic research on methane fermentation provides a better understanding of the kinetics and mechanisms of biomass conversion under anaerobic conditions; improvements in digestion efficiencies in terms of methane yield and volatile solids reduction have been slow to evolve from this knowledge. A large number of agricultural residues, animal wastes, and biomass species have been evaluted as potential feedstocks for methane production in laboratory digesters and small-scale digestion facilities. Considerable laboratory work has also been done to develop pre- and post-digestion treatments that improve biodegradability. A plateau of about 50–60% volatile solids destruction efficiencies and energy recoveries in the product gas seems to exist for most methane fermentation systems.

Research in the early 1990s has addressed several potentially beneficial methods of improving the process, eg, two-phase digestion in which the acetogenic and methanogenic phases are physically separated. Practical implementation of two-phase digestion is achieved by control of the hydraulic retention times in the acid and methane reactors or reaction zones. This process configuration provides several advantages over conventional high rate digestion such as enhanced stability, an optimum environment for acetogenic and methanogenic bacteria, substantial increases in throughput rates for given size reactors, increased gas and methane production rates, and higher methane content in the product gas (16,152). The process has been scaled-up for treatment of municipal biosolids, ie, the Acimet Process (153), and has been applied to industrial wastes (16). From a practical standpoint, two-phase anaerobic digestion of biomass is capable of retrofit to existing digestion systems of any design and is projected to be capable of doubling plant capacity at about 50% of the capital cost of a grassroots plant. A significant market is anticipated for this advanced technology, particularly for wastewater treatment.

Some of the other research studies have addressed topics such as high solids biomass digestion (154), the utilization of superthermophilic organisms (155), advanced reactor designs (156), landfill gas enhancement (157), and microbiology of the mixed cultures involved in methane fermentation (158).

Thermochemical Gasification. Extensive research and pilot studies have been carried out since 1970 to develop thermochemical processes for biomass conversion to energy and fuels. Basic studies on the effects of various operating conditions and reactor configurations have been performed in the laboratory and at the process development unit (PDU) and pilot scales on steam, steam-air, air-blown, and oxygen-blown gasification, and on hydrogasification. Other research has also been done on the rapid pyrolysis of biomass which, in addition to gaseous products, yields coproduct liquids and solids.

Over one million air-blown gasifiers were built during World War II to manufacture LHV gas to power vehicles and to generate steam and electric power. Units are available in a variety of designs, some of which have been retrofitted to gas-fired furnaces. Although some research is in progress to refine air-blown wood gasifiers in North America, particularly portable units, most of the research has been conducted in Europe. The Swedish automobile manufacturers, Volvo and Saab, have ongoing programs to develop a standard gasifier design suitable for mass production.

Research on thermochemical biomass gasification in North America has tended to concentrate on MHV gas production, scale-up of the advanced process concepts that have been evaluated at the PDU scale, and the problems that need to be solved to permit large-scale thermochemical biomass gasifiers to be operated in a reliable fashion for power production, especially advanced power cycles. Many different reactor designs have been evaluated under a wide range of operating conditions. Exemplary advanced gasifiers and gasification systems, some of which are in the scale-up stage, include IGT's single-stage, pressurized, fluid-bed gasifier; the National Renewable Energy Laboratory's (NREL) pressurized, fixed-bed, down-draft gasifier;

the fire tube-heated, fluid-bed system of the University of Missouri-Rolla; the indirectly heated, fluid-bed, dual reactor system of Battelle Columbus Laboratory; the pulse-enhanced, indirectly heated, fluid-bed gasifier of M.T.C.I.; and the catalytic, pressurized, gasification system for wet biomass of Pacific Northwest Laboratory.

An example of a scale-up project is the pressurized, fluid-bed Renugas™ plant in Hawaii (159). The gasifier is designed for 63.5 t/d of sugarcane bagasse. In addition to bagasse, other feedstocks such as wood, waste biomass, and RDF may be evaluated. The demonstration provides process informaton for both air- and oxygen-blown gasification at low and high pressures. Renugas will be evaluated for both fuel gas and synthesis gas production, and for electric power production with advanced power generation schemes.

Although only limited research has been carried out on small-scale, LHV, producer gasifiers for biomass, significant design advancements have been made even though they have been used for over 100 years. One development is the open-top, stratified, downdraft gasifier in which air is drawn in through successive reaction strata (160). The unit is simple to operate, inexpensive, and can be close-coupled to an engine–generator set without complex gas-cleaning equipment. The gasifier dimensions are sized to deliver gas to the engine based on its fuel-rate requirements, and no controls are needed.

Emissions from the thermochemical gasification of biomass can affect the operation of advanced power cycles and include particulates, alkalies, oils and tars, and heavy metals. One of the high priority research efforts is to develop hot-gas cleanup methods that will permit biomass gasification to supply suitable fuel gas for these cycles (161). Some of the other research needs that have been identified include versatile feed-handling systems for a wide variety of biomass feedstocks; biomass feeding systems for high pressure gasifiers; determination of the effects of additives including catalysts for minimizing tar production and capturing contaminants; and suitable ash disposal and wastewater treatment technologies (162).

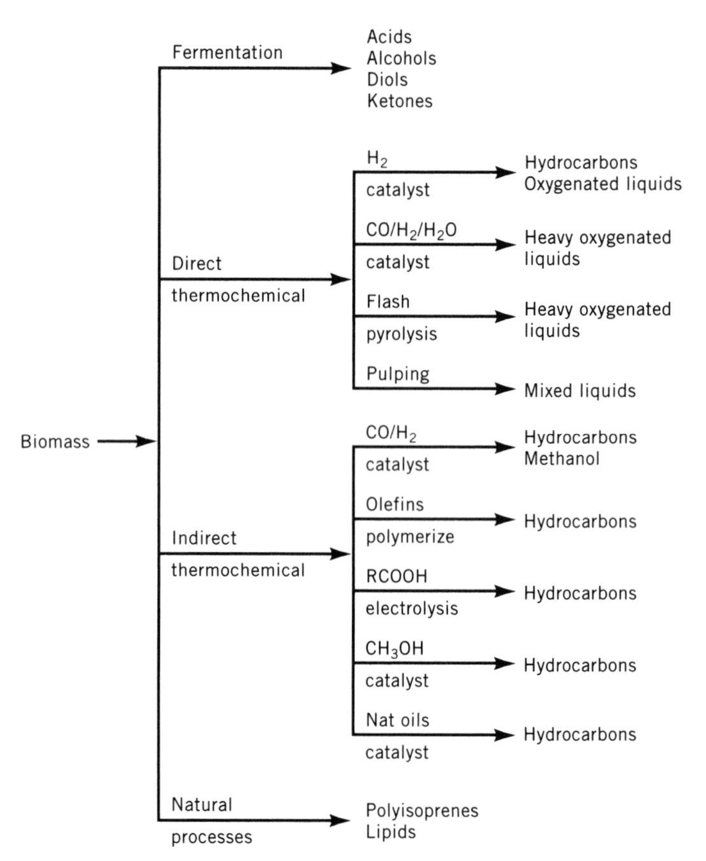

Figure 21. Biomass liquefaction routes under development (ca 1992).

Liquefaction. Figure 21 outlines most of the biomass liquefaction methods under development. There are essentially three basic types of biomass liquefaction technologies, ie, fermentation, natural, and thermochemical processes.

Much research has been conducted since the early 1970s to improve the alcohol fermentation process, ie, the technology by which biomass is converted to fuel ethanol (96,112,125). A large portion of this research has been focused on development of the process so that it will be suitable for conversion of low grade lignocellulosics. This type of biomass generally contains about 50 wt % celluloses, 25 wt % hemicelluloses, and 25 wt % lignins. The hexose sugars obtained on hydrolysis of celluloses are converted by conventional alcohol-forming yeasts to ethanol. The pentose sugars require other specialized organisms for conversion. The lignins are essentially inert.

Research has focused on minimizing the energy inputs for the distillation steps needed to produce 190 and 200 proof ethanol; on the fermentation process itself to increase ethanol yields and reduce fermentation times; on the pretreatment and hydrolysis processes needed to afford high yields of sugars and to make low grade cellulosics suitable feedstocks for fermentation; on development of organisms that ferment pentoses separately, and pentoses and hexoses together; on development of advanced processes that permit simultaneous saccharification and fermentation (SSF); and on development of immobilized organisms and genetically engineered bacteria that afford much shorter fermentation times and avoid yeast recycling.

Several advancements have been developed and incorporated into grain fermentation processes to improve operating efficiencies and reduce energy consumption. Ethanol yields and production rates have improved slightly, but not significantly. In contrast, research on the development of low grade biomass fermentation processes is on the verge of process demonstrations with a variety of feedstocks such as waste paper, agricultural residues, wood biomass, and refuse-derived fuel.

Much of the research to apply ethanol fermentation to low grade biomass is funded by the U.S. Department of Energy. The goal is to produce fuel ethanol at a cost of \$0.145/L, or \$6.90/GJ, by 1995 (166).

Oilseed and Vegetable Oil Fuels. Limited research has continued on the utilization of seed and vegetable oils as motor fuels, particularly as substitute diesel fuels and diesel fuel extenders (164). Work has focused on studies of the yields and properties of oils from oilseed and vegetable oil crops, the performance of neat oils and oil–diesel fuel blends as fuels for compression ignition engines, improvement of the transesterification process and the fuel characteristics of the resulting esters as diesel fuels, upgrading vegetable oils to gasolines and diesel fuels by hydrocracking processes, and field tests of the liquid fuels made from seed and vegetable oils in trucks and buses. Although several operating problems have been observed, such as lubricating oil deterioration and crystal formation in cooler weather even with some of the lower viscosity seed oil esters, significant advances have been made. Esters of selected vegetable oils are very promising candidates for both indirect and direct fuel-injected engines.

The cost of seed and vegetable oil fuels is still not competitive with petroleum-based diesel fuels. The cost of esterification alone can add up to 50% to the cost of the fuel, depending on the size of the processing operation and the market value of by-product glycerol. In 1992 the cost of oilseed and vegetable oil diesel fuels was about twice that of conventional diesel fuels. One approach to elimination of the cost differential is to use waste vegetable oils from large-scale restaurant operations as the feedstock for transesterification plants (165). This offers the possibility of taking credit for waste oil disposal, ie, the analogue of a tipping fee in the solid waste disposal field, and of recycling of oils as fuel. Another approach is to utilize tax credit legislation available in the United States to farmers who produce minor oilseed crops in place of principal commodity crops.

Thermochemical Liquefaction. Most of the research done since 1970 on the direct thermochemical liquefaction of biomass has been concentrated on the use of various pyrolytic techniques for the production of liquid fuels and fuel components (96,112,125,166,167). Some of the techniques investigated are entrained-flow pyrolysis, vacuum pyrolysis, rapid and flash pyrolysis, ultrafast pyrolysis in vortex reactors, fluid-bed pyrolysis, low temperature pyrolysis at long reaction times, and updraft fixed-bed pyrolysis. Other research has been done to develop low cost, upgrading methods to convert the complex mixtures formed on pyrolysis of biomass to high quality transportation fuels, and to study liquefaction at high pressures via solvolysis, steam–water treatment, catalytic hydrotreatment, and noncatalytic and catalytic treatment in aqueous systems.

Essentially all of these conversion processes are technically feasible and can be used to convert biomass to a wide range of liquid products. Unfortunately, because of the complex composition of biomass and the chemistry of direct thermal cracking of biomass, complex product mixtures are always formed. Selectivities for individual products are low; this is sometimes advantageous. Because of the oxygenated nature of biomass, higher yields of certain oxygenated products can be obtained. This offers the possibility of producing specific organic liquids that have higher intrinsic value as chemicals rather than as fuels. A few research efforts are cited here to illustrate the versatility of direct liquefaction processes for biomass.

The principal products in conventional, long-term pyrolysis of biomass at about 400°C or lower are char, gas, organic liquids, and water, as shown in Table 20. Research has shown that fast pyrolysis of biomass at 475–525°C and vapor residence times of a few seconds or less can maximize organic liquid yields. Wood and grasses yield 55–65 wt % and 40–65 wt % of the dry biomass as organic liquids, respectively (168). Products from the fast pyrolysis of wood, for example, contain significant amounts of low molecular weight oxygenated compounds such as hydroxyacetaldehyde, acetaldehyde, formic acid, acetic acid, and glyoxal. Fast pyrolysis of waste lignocellulosics such as newsprint or pulp mill sludges also affords similar liquid products. Pretreatment of the wood before pyrolysis gives dramatic changes in product selectivity.

Ultrafast pyrolysis in the vortex reactor is capable of pyrolyzing biomass at high heat-transfer rates on the reactor wall by ablation and has been found to be useful for a variety of biomass and waste feedstocks. In this reactor, biomass particles are entrained tangentially at high velocities by a carrier gas into the vortex reactor, which causes the biomass particles to be preferentially heated relative to the carrier gas and the pyrolysis vapors (169). Products recovered from this innovative reactor have been demonstrated to be about 55 wt % organic liquids, 14 wt % gases, 13 wt % char, and 12 wt % water (94% closure). The pyrolysis vapors can be condensed to a low viscosity liquid, thought to be suitable for combustion in furnaces and turbines, or can be cracked to form about 15 wt % C_2+ hydrocarbons. Zeolites have been found to catalyze the conversion of the pyrolysis vapors to gasoline range hydrocarbons in yields that approach the theoretical upper limit as determined from stoichiometry (170). The energy conversion efficiency was about 45% for C_2–C_8 range hydrocarbons and 55% for C_2+ hydrocarbons. The ablative fast pyrolysis process has been scaled up to a unit that converts 11,400 t/yr of waste sawdust into 5.3 million L/yr of fuel oil and 1,720 t/yr of charcoal (171). This relatively small plant is expected to have a pretax revenue of $263,000 presuming the feedstock cost is zero, and the fuel oil and char can be sold for $4.74/GJ and $72.60/t, respectively. The capital cost of the plant was $850,000.

Prospects

Despite the slow development of renewable biomass as a primary source of energy, the large research effort in progress on feedstock production and conversion is expected to lead to greater commercialization of advanced energy and organic chemical processes based on biomass. Small-scale systems for the individual farmer are being designed and marketed to make it possible to install and operate complete on-site total energy packages that will supply all of the farm's energy requirements. These systems will be fueled with captive sources of biomass and wastes generated on the farm. It is likely that large building complexes such as schools, apartments, shopping malls, and theme parks in urban areas will be able to incorporate similar systems using captive wastes and delivered biomass. Individual, small-scale, farmers' cooperative and industrial-scale fuel ethanol plants will continue to be built and operated as long as government tax incentives are provided. Tax subsidies for fuel ethanol are expected to become unnecessary as the technologies for use of low grade lignocellulosic feedstocks for fuel ethanol are perfected. New, larger scale, biomass-fueled and waste-to-energy power plants, especially those that incorporate cogeneration, will continue to show modest growth as the technology advances and the disposal of waste biomass in an environmentally acceptable manner is implemented. The development of improved methane fermentation processes for waste biomass such as municipal biosolids, industrial wastes from food-processing and beverage alcohol plants, and refuse-derived fuel, is expected to result in more efficient waste treatment and disposal and increased methane recovery and utilization.

Fossil fuels are still sufficiently low in cost to make the economics of large-scale production of substitute transportation fuels, fuel gases, and fuel oils from biomass borderline or unattractive if the biomass systems are used only to produce energy. Large-scale integrated biomass energy plantations are therefore not expected to be constructed and operated until some time during the first or second quarter of the twenty-first century. Biomass grown strictly as profitable energy crops is expected to occur in that time frame as fossil fuels are phased out or their prices increase because of shortages or additional taxes.

Growing environmental concerns and federal and state environmental regulations are expected to be the driving force behind increased usage of biomass energy. Carbon taxes applied to fossil fuel usage, especially for vehicles and utility power plants, are expected to provide very strong incentives to convert to renewable biomass energy resources for both mobile and stationary applications.

Reprinted from *Kirk-Othmer Encyclopedia of Chemical Technology*, 4th ed., vol. 12, John Wiley & Sons, Inc., New York, 1995.

BIBLIOGRAPHY

1. D. L. Klass, in D. L. Klass and J. W. Weatherly III, eds., *Energy from Biomass and Wastes IV*, IGT, Chicago, 1980, pp. 1–41.
2. D. L. Klass, in D. L. Klass, ed., *Energy from Biomass and Wastes XIII*, IGT, Chicago, 1990, pp. 1–46.
3. D. L. Klass, *Chemtech* **4**(3), 161 (1974).
4. R. H. Whittaker and G. E. Likens, in H. Leith and R. H. Whittaker, eds., *Primary Productivity of the Biosphere*, Springer Verlag, New York, 1975.
5. R. Repetto, *Sci. Am.* **262**(4), 36 (1990).
6. B. Bolin, "The Carbon Cycle," in *The Biosphere*, W. H. Freeman and Co., San Francisco, Calif., 1970.
7. J. T. Houghton, G. J. Jenkins, and J. J. Ephraums, eds., *Climate Change: The IPCC Scientific Assessment*, Cambridge University Press, Cambridge, 1990, 365 pp; D. L. Klass, *Energy & Environment* **3**(2), 109 (1992); D. L. Klass, *Energy Policy* **21**(11), 1076 (1993).
8. United Nations, *1990 Energy Statistics Yearbook*, Department of Economic and Social Development, New York, 1992.
9. *RPA Assessment of the Forest and Rangeland Situation in the U.S., 1989*, No. 26, USDA, Forest Service Washington, D.C., Oct. 1989.
10. D. L. Klass, *Chemtech* **20**(12), 720 (1990).
11. U.S. Department of Energy, *Estimates of U.S. Biofuels Consumption 1990*, DOE/EIA-0548(90), Energy Information Administration, Washington, D.C., Oct. 1991.
12. U.S. Department of Energy, *The Potential of Renewable Energy, An Interlaboratory White Paper*, SERI/TP-260-3674, DE90000322, Office of Policy, Planning and Analysis, Washington, D.C., Mar. 1990.
13. D. F. Westlake, *Biol. Rev.* **38,** 385 (1963).
14. J. S. Burlew, ed., *Algae Culture From Laboratory to Pilot Plant*, Publication 600, Carnegie Institute of Washington, Washington, D.C., 1953, pp. 55–62.
15. C. D. Hodgman, ed., *Handbook of Chemistry and Physics*, 31st ed., Chemical Rubber Publishing Co., Cleveland, Ohio, 1949, p. 1537.
16. D. L. Klass, *Science* **223,** 1021 (1984).
17. L. R. Lynd and co-workers, *Science* **251,** 1318 (1991).
18. *Guayule: An Alternative Source of Natural Rubber*, National Academy of Sciences, Washington, D.C., 1977.
19. R. A. Buchannan and F. O. Otey, "Multi-Use Oil- and Hydrocarbon-Producing Crops in Adaptive Systems for Food, Material, and Energy Production," paper presented at *19th Annual Meeting, Society for Economic Botany*, St. Louis, Mo., June 11–14, 1978.
20. M. Calvin, *Chemtech* **7**(6), 353 (1977); *Bioscience* **29,** 533 (1979); *Die Naturwissenschaften* **67,** 525 (1980); E. K. Nemethy, J. W. Otvos, and M. Calvin, in D. L. Klass and G. H. Emert, eds., *Fuels from Biomass and Wastes*, Ann Arbor Science Publishers, Ann Arbor, Mich., 1981; J. D. Johnson and C. W. Hinman, *Science* **208,** 460 (1980).
21. D. K. Schmalzer and co-workers, *Biocrude Suitabilities for Petroleum Refineries*, ANL/CNSV-69, Argonne National Laboratory, Argonne, Ill., June 1988.
22. K. R. Kaufman, in E. D. Schultz and R. P. Morgan, eds., *Fuels and Chemicals from Oil Seeds: Technology and Policy Options*, Westview Press, Boulder, Colo., 1982, pp. 143–174.
23. P. B. Weisz and J. F. Marshall, *Science* **206,** 257 (1979); R. M. Furrer and N. N. Bakshi, in Ref. 2, pp. 897–914.
24. M. Stumborg and co-workers, in D. L. Klass, ed., *Energy from Biomass and Wastes XVI*, IGT, Chicago, 1993, pp. 721–738.
25. E. S. Olson and R. K. Sharma, in Ref. 24, pp. 739–751.
26. E. S. Lipinsky and co-workers, in Ref. 22, pp. 205–223.
27. E. E. Gavett and D. VanDyne, in Ref. 24, pp. 709–719.
28. D. L. Klass, in D. L. Klass, ed., *Energy from Biomass and Wastes IX*, IGT, Chicago, 1985, pp. 1–83.
29. D. M. Tillett and J. R. Benemann, in D. L. Klass, ed., *Energy from Biomass and Wastes XI*, IGT, Chicago, 1988, pp. 771–786.
30. A. M. Hill and D. A. Feinberg, *Fuel Products from Microalgae*, SERI/TP-231-2348, Solar Energy Research Institute, Golden, Colo., 1984.
31. D. A. Hoffman and R. A. Fitz, *Environ. Sci. Technol.* **2**(11), 1023 (1968).
32. S. B. Alpert and co-workers, *Pyrolysis of Solid Wastes: A Technical and Economic Assessment*, NTIS PB 218-231, SRI, Menlo Park, Calif., Sept. 1972.
33. M. A. Paisley, H. F. Feldmann, and H. R. Appelbaum, in D. L. Klass and H. H. Elliott, eds., *Energy from Biomass and Wastes VIII*, IGT, Chicago, 1984, pp. 675–696.
34. R. C. Bailie, "Results from Commercial-Demonstration Pyrolysis Facilities (35–45 tons/day refuse) Extended to Producing Synfuels from Biomass," in D. L. Klass and J. W. Weatherly III, eds., *Energy from Biomass and Wastes V*, IGT, Chicago, 1981, pp. 549–569.
35. G. T. Preston, "Resource Recovery and Flash Pyrolysis of Municipal Refuse," in F. Ekman, ed., *Clean Fuels from Biomass, Sewage, Urban Refuse, Agricultural Wastes*, IGT, Chicago, 1976, pp. 89–114.

36. D. S. Scott and J. Piskorz, "Continuous Flash Pyrolysis of Wood for Production of Liquid Fuels," in D. L. Klass and H. H. Elliott, eds., *Energy from Biomass and Wastes VII*, IGT, Chicago, 1983, pp. 1123–1146.
37. D. L. Klass, "Wastes and Biomass as Energy Resources: An Overview," in ref. 35, pp. 21–58.
38. T. F. Fisher, M. L. Kasbohm, and J. R. Rivero, "The Purox system," in ref. 35, pp. 447–459.
39. S. J. Yosim and K. M. Barclay, *Preprints of Papers, 171st National Meeting ACS, Div. of Fuel Chem. 21* (1), 73 (Apr. 5–9, 1976).
40. H. F. Feldmann and co-workers, *Hydrocarbon process. 55* (11), 201 (1976).
41. S. P. Babu, D. Q. Tran, and S. P. Singh, "Noncatalytic Fluidized-Bed Hydroconversion of Biomass to Substitute Natural Gas," in ref. 1, pp. 369–385.
42. D. C. Elliott and co-workers, "Low-Temperature, Catalytic Gasification of Wastes for Simultaneous Disposal and Energy Recovery," in D. L. Klass, ed., *Energy from Biomass and Wastes XV*, IGT, Chicago, 1991, pp. 1013–1021.
43. H. F. Feldmann and co-workers, "Gasification of Forest Residues," in D. L. Klass, ed., *Biomass as a Nonfossil Fuel Source*, ACS Symposium Series 144, American Chemical Society, Washington, D.C., 1980, pp. 351–375.
44. L. K. Mudge and co-workers, "Catalytic Gasification of Biomass," in *3rd Annual Biomass Energy Systems Conference Proceedings The National Biomass Program,* SERI/TP-33-285, Solar Energy Research Institute, Golden, Colo., 1979, pp. 351–357.
45. D. G. B. Boocock and D. Mackay, "The Production of Liquid Hydrocarbons by Wood Hydrogenation," in ref. 1, pp. 765–777.
46. H. R. Appel and co-workers, "Conversion of Cellulosic Wastes to Oil," U.S. Bur. of Mines Rep. Invest. *8013* (1975).
47. E. R. Riegel, *Industrial Chemistry,* 2nd ed., The Chemical Catalog Co., New York, 1933, Chapt. 16, p. 257.
48. M. R. Ladisch, C. M. Ladisch, and G. T. Tsao, *Science 201* 743, (1978).
49. L. Paszner and co-workers, "Two-Stage, Continuous Hydrolysis of Wood by the Acid Catalyzed Organosolv Saccharification (ACOS) Process," in ref. 24, in press.
50. W. H. Hoge, U.S. Patent 4,009,075, assigned to BioIndustries, Inc., Feb. 22, 1977.
51. L. R. Lynd and co-workers, *Science 251,* 1318 (1991).
52. R. Katzen Associates, *Chemicals from Wood Wastes*, U.S. Department of Agriculture Forest Products Laboratory, Madison, Wisc., Dec. 14, 1975.
53. R. S. Loomis, W. A. Williams, and A. E. Hall, *Ann. Rev. Plant Physiol. 22* 431 (1971).
54. E. I. Rabinowitch, *Photosynthesis*, Vols. 1–2 Interscience, New York, 1956.
55. C. B. Osmond, *Ann. Rev. Plant Physiol. 29* 379 (1978).
56. B. J. Brinkworth, *Solar Energy for Man,* John Wiley & Sons, Inc., New York, 1973.
57. U.S. Dept. of Commerce, *Climatological Data, National Summary*, Vol. 21, Nos. 1–12, U.S. Government Printing Office, Washington, D.C., 1970.
58. H. J. Critchfield, *General Climatology*, 3rd ed., Prentice-Hall, Inc., Englewood Cliffs, N.J., 1974, p. 22.
59. R. S. Loomis and W. A. Williams, *Crop Sci. 3,* 67 (1963).
60. T. R. Schneider, *Energy Convers. 13*, 77 (1973).
61. S. S. Visher, *Climatic Atlas of the United States*, Harvard University Press, Cambridge, Mass., 1954.
62. *Statistical Abstracts of the United States*, U.S. Department of Commerce, U.S. Government Printing Office, Washington, D.C., 1976.
63. W. L. Roller and co-workers, *Grown Organic Matter as a Fuel Raw Material Source, NASA Report CR-2608*, Ohio Agricultural Research and Development Center, Washington, D.C., Oct. 1975.
64. M. M. Ludlow and G. L. Wilson, *J. Aust. Inst. Agric. Sci. 36*, 43 (Mar. 1970).
65. W. J. North, ed., *The Biology of Giant Kelp Beds (Macrocystis) in California*, Cramer, Lehre, FRG, 1971, p. 12.
66. T. A. El-Sharkawy and J. D. Hesketh, *Crop Sci.* **4,** 514 (1964).
67. J. Krummel, in Ref. 35, pp. 359–370.
68. H. J. Evans and L. E. Barber, *Science* **197,** 332 (1977).
69. *Land Capability Classification, Agricultural Handbook 210*, U.S. Department of Agriculture, Soil Conservation Service, Washington, D.C., 1966, 21 pp.
70. *Summary Report 1987 National Resources Inventory*, No. 790, U.S. Department of Agriculture, Soil Conservation Service, Washington, D.C., Dec. 1989, 37 pp.
71. S. H. Spurr, *Sci. Am.* **240,** 76 (1979).
72. A. A. Nichiporovich, *Photosynthesis of Productive Systems*, Israel Program for Scientific Translations, Jerusalem, Isreal, 1967.
73. E. O. Mariani, in D. L. Klass and W. W. Waterman, eds., *Energy from Biomass and Wastes*, IGT, Chicago, 1978, pp. 29–38.
74. R. L. Sajdak and co-workers, in Ref. 43, pp. 21–48.
75. J. C. Dula, in Ref. 33, pp. 193–207.
76. L. L. Wright and co-workers, in D. L. Klass, ed., *Energy from Biomass and Waste XII*, IGT, Chicago, 1989, pp. 261–274.
77. *Solar SNG, Final Report American Gas Association Project IU-114-1*, Prepared by InterTechnology Corp., American Gas Association, Washington, D.C., Oct. 1975.
78. *Grass: The Yearbook of Agriculture 1948*, U.S. Department of Agriculture, U.S. Government Printing Office, Washington, D.C., 1948.
79. D. L. Klass, S. Ghosh, and J. R. Conrad, in Ref. 35, pp. 229–252.
80. D. L. Klass and S. Ghosh, in Ref. 43, pp. 229–249.
81. J. A. Alich, Jr., and R. E. Inman, *Effective Utilization of Solar Energy to Produce Clean Fuel, Grant No. GI 38723*, Final Report for National Science Foundation, Stanford Research Institute, Palo Alto, Calif., June 1974.
82. R. Retovsky, *Continuous Cultifation of Algae, Theoretical and Methodological Bases of Continuous Culture of Microorganisms*, Academic Press, Inc., New York, 1966.
83. D. L. Klass and S. Ghosh, in W. W. Waterman, ed., *Clean Fuels from Biomass and Wastes*, IGT, Chicago, 1977, pp. 323–351.
84. D. L. Klass, S. Ghosh, and D. P. Chynoweth, *Process Biochem.* **14,** 18 (1979).
85. D. P. Chynoweth, D. L. Klass, and S. Ghosh, in Ref. 73, pp. 229–251.
86. D. L. Klass and S. Ghosh, in D. L. Klass and G. H. Emert, eds., *Fuels from Biomass and Wastes*, Ann Arbor Science Publishers, Ann Arbor, Mich., 1981, pp. 129–149.
87. M. G. McGarry, *Process Biochem.* **6,** 50 (1971).
88. J. L. Yount and R. A. Grossman, *J. Water Pollut. Control Fed.* **42,** 173 (1970).
89. E. S. Dell Fosse, in Ref. 83, pp. 73–99.

90. *State Energy Price and Expenditure Report 1990*, DOE/EIA-0376(90), U.S. Department of Energy, Energy Information Administration, Washington, D.C., Sept. 1992.
91. M. D. Fraser, in Ref. 24, pp. 295–330.
92. R. F. Moorer, D. K. Walter, and S. Gronich, in Ref. 24, pp. 139–153.
93. S. Lazzari, in Ref. 24, pp. 275–294.
94. N. Smith and T. J. Corcoran, *Preprints of Papers Presented at 171st National Meeting, ACS, Fuel Chemistry Division, Symposium on Net Energetics of Integrated Synfuel Systems* **21**(2), 9 (Apr. 1976).
95. R. E. Inman, in Ref. 94, pp. 21–27.
96. D. L. Klass, *Energy Topics*, 1 (Apr. 14, 1980).
97. National Wood Energy Association, *NWEA State Biomass Statistical Directory*, Arlington, Va., 1988.
98. E. Berenyi and R. Gould, *1988–1989 Resource Recovery Yearbook*, Governmental Advisory Associates, Inc., New York, 1988, 718 pp.
99. R. M. Dykes, in Ref. 76, pp. 379–397.
100. *Stove and Fireplace Catalog IX*, Consolidated Dutchwest, Plymouth, Mass., 1988, 67 pp.
101. *Estimates of U.S. Wood Energy Consumption 1980–1983*, DOE/EIA-0341(83), U.S. Department of Energy, Energy Information Administration, Washington, D.C., Nov. 1984.
102. J. W. Koning, Jr. and K. E. Skog, in D. L. Klass, ed., *Energy from Biomass and Wastes X*, IGT, Chicago, 1986, pp. 1309–1322; J. C. Nicolello, *U.S. Pulp and Paper Industry's Energy Use-Calendar Year 1986*, New York, Apr. 20, 1987; J. C. Nicello, *U.S. Pulp and Paper Industry's Energy Use-Calendar Year 1987*, New York, May 17, 1988; *Annual Energy Review 1987*, DOE/EIA-0384(87). U.S. Department of Energy, Energy Information Addministration, Washington, D.C., 1988.
103. National Wood Energy Association, *Wood Energy, America's Renewable Resource*, Arlington, Va., 1988, 2 pp.
104. *Past, Present, and Future Trends in the U.S. Forest Sector: 1952–2040, Review Draft*, U.S. Department of Agriculture, Forest Service, Washington, D.C., June 1988.
105. J. I. Zerbe, *Forum for Applied Research and Public Policy*, 38–47 Winter (1988).
106. Table 1, in Ref. 11.
107. D. L. Klass and C. T. Sen, *Chem. Eng. Prog.* **83**(7), 46 (1987).
108. *The Qualifying Facilities Report*, Federal Energy Regulatory Commission, Washington, D.C., Jan. 11, 1988.
109. J. L. Easterly, personal communication, Meridian Corp., Alexandria, Va., Aug. 16, 1988.
110. D. A. Flint and C. Norris, *1986 Capacity and Generation of Non-Utility Sources of Energy*, Edison Electric Institute, Washington, D.C., July 1988.
111. B. DeCampo, D. A. Flint, and C. Norris, *Electric Perspectives*, 22 (Summer 1988).
112. D. L. Klass, *Resources and Conservation* **15,** 7 (1987).
113. J. L. Easterly, S. Lees, and B. Detwiler, *Electric Power from Biofuels: Planned and Existing Projects in the U.S.*, rev. Jan. 1985, DOE/CE/307841/1, U.S. Department of Energy, Washington, D.C., Aug. 1985.
114. C. Tewksbury, in D. L. Klass, ed., *Energy from Biomass and Wastes X*, IGT, Chicago, 1987, pp. 555–578.
115. D. L. Klass, *Resources and Conservation* **11,** 157 (1985).
116. T. R. Miles and T. R. Miles, Jr., *Biomass* **18,** 163 (1989).
117. R. L. Wentworth, "Anaerobic Digestion in North America," paper presented at *Symposium Anaerobic Digestion and Carbohydrate Hydrolysis of Waste, sponsored by Commission of the European Communities*, Luxembourg, May 8–10, 1984.
118. J. H. Ashworth, Y. M. Bihun, and M. Lazarus, *Universe of U.S. Commercial-Scale Anaerobic Digesters: Results of SERI/ARD Data Collection*, Solar Energy Research Institute, Golden, Colo., May 30, 1985; J. H. Ashworth, *Problems With Installed Commercial Anaerobic Digesters in the United States: Results of Site Visits*, Rev. ed., Solar Energy Research Institute, Golden, Colo., Nov. 6, 1985.
119. *1984 Needs Survey Report to Congress: Planned and Existing Projects in the U.S.*, rev. Jan. 1985, DOE/CE/30784/1, U.S. Department of Energy, Washington, D.C., Aug. 1985.
120. Table 19, in Ref. 28.
121. S. Doelph, *Gas Energy Review* **16**(1), 14 (1988).
122. "Landfill Gas Summary Update," *Waste Age* **19**(3), 167 (1988).
123. "Resource Recovery Activities," *City Currents* **6**(4), 1 (1987).
124. F. L. Potter, personal communication, Information Resources, Inc., Washington, D.C., Aug. 11, 1988.
125. D. L. Klass, *Energy Topics*, 1 (Aug. 1, 1983).
126. *Alc. Update*, Aug. 8, 1988; *Alc. Wk.* **9**(32) (Aug. 8, 1988).
127. L. L. Wright, personal communication, Oak Ridge National Laboratory, Oak Ridge, Tenn., July 1988.
128. S. Fenn, *Institutional Investment in Renewable Energy Technologies*, Renewable Energy Institute, Washington, D.C., Feb.. 1987, 50 pp.
129. S. Lazzari, *A History of Federal Tax Policy: Conventional as Compared to Renewable and Nonconventional Energy Resources*, 88-455E, The Library of Congress, Congressional Research Service, Washington, D.C., June 7, 1988.
130. W. A. Rains, paper presented at *1982 Annual Meeting, National Petroleum Refiners Association*, San Antonio, Tex., Mar. 21–23, 1983.
131. W. R. Keene, Lundberg Survey, Inc., N. Hollywood, Calif., Aug. 16, 1988.
132. *Chem. & Eng. News* **66**(25), 40 (June 20, 1988).
133. *Assessment of Costs and Benefits of Flexible and Alternative Fuel Use in the U.S. Transportation Sector, Progress Report One*, DOE/PE-0080, U.S. Department of Energy, Washington, D.C., Jan. 1988.
134. B. Paul, *WSJ LXIX* (217), Sec. 2, 19 (Aug. 19, 1988).
135. K. A. Saterson and M. W. Luppold, *3rd Annual Biomass Energy Systems Conference Proceedings*, SERI/TP-33-285, U.S. Department of Energy, Golden, Colo., June 5–7, 1979, pp. 245–254.
136. J. H. Cushman and A. F. Turhollow, in D. L. Klass, ed., *Energy from Biomass and Wastes XIV*, IGT, Chicago, 1991, pp. 465–480.
137. A. G. Alexander, in Ref. 136, pp. 367–374.
138. L. L. Wright, in L. L. Wright and W. G. Hohenstein, eds., *Biomass Energy Production in the United States: Situation and Outlook*, Oak Ridge National Laboratory, Oak Ridge, Tenn., Aug. 1992, Chapt. 2.
139. D. L. Klass, in Ref. 114, pp. 13–113.
140. G. Grassi, in Ref. 114, pp. 1545–1562.
141. A. Ismail and R. Quick, in Ref. 42, pp. 1063–1100.
142. J. E. Robert and E. N. Hogan, in Ref. 42, pp. 1245–1265.
143. L. D. Ostlie and T. E. Drennen, in Ref. 76, pp. 621–650.
144. A. Rehmat and M. Khinkis, in Ref. 42, pp. 1111–1139.
145. J. T. Hamrick, in Ref. 114, pp. 517–528.
146. M. L. Murphy, in Ref. 29, pp. 371–380; in Ref. 42, pp. 1167–1179; S. C. Bhattacharya and W. Wu, in Ref. 76, pp. 591–601.
147. J. F. L. Lincoln and T. C. Litchney, in Ref. 29, pp. 357–369.
148. S. G. Barnett and S. J. Morgan, in Ref. 136, pp. 191–236.

149. G. A. Norton and A. D. Levine, in Ref. 2, pp. 513–527.
150. C. Tewksbury, in Ref. 42, pp. 95–127.
151. S. M. Turner and D. A. Rowley, in Ref. 42, pp. 43–63; D. R. Patrick, in Ref. 42, pp. 65–72; R. N. Sampson, in Ref. 42, pp. 159–186.
152. U.S. Pat. 4,022,665 (May 10, 1977), S. Ghosh and D. L. Klass (IGT); U.S. Pat. 4,318,993 (Mar. 9, 1982), S. Ghosh and D. L. Klass (IGT); T. L. Miller and S. Ghosh, in Ref. 136, pp. 869–876.
153. *Acimet Technical Briefing*, Illinois Department of Energy and Natural Resources and the DuPage Group, Woodridge, Ill, Oct. 7, 1992.
154. W. J. Jewell and co-workers, in Ref. 28, pp. 669–693; R. Legrand and W. J. Jewell, in Ref. 114, pp. 1077–1095; B. De Wilde and L. De Baere, in Ref. 136, pp. 915–929; C. J. Rivard, in Ref. 24, pp. 1025–1041.
155. J. W. Deming, in Ref. 114, pp. 1097–1111.
156. R. W. Meyer and W. R. Guthrie, in Ref. 28, pp. 857–872; S. R. Harper, G. E. Valentine, and C. C. Ross, in Ref. 29, pp. 637–664; L. M. Safley and P. D. Lusk, in Ref. 136, pp. 955–980; R. R. Dague and S. Sung, in Ref. 24, pp. 1001–1023.
157. J. J. Walsh and co-workers, in Ref. 114, pp. 1115–1125; P. Fletcher, in Ref. 76, pp. 1001–1027.
158. A. J. L. Macario and co-workers, in Ref. 114, pp. 1009–1020; S. J. Schropp and co-workers, in Ref. 114, pp. 1035–1043; D. P. Chynoweth and co-workers, in Ref. 76, pp. 965–981; V. Chitra and K. Ramasamy, in Ref. 24, pp. 1043–1060.
159. A. R. Trenka and co-workers, in Ref. 42, pp. 1051–1061.
160. H. LaFontaine, in Ref. 29, pp. 561–575.
161. *Summary Report for Hot-Gas Cleanup, Compiled by Institute of Gas Technology*, for International Energy Agency, IGT, Chicago, Dec. 1991, 50 pp.
162. *Research Needs for Thermal Gasification of Biomass, Compiled by Studsvik AB Thermal Processes*, for International Energy Agency, IGT, Chicago, Mar. 1992, 4 pp.
163. *Conservation and Renewable Energy Technologies for Transportation*, DOE/CH 10093-84, U.S. Department of Energy, Washington, D.C., Nov. 1990, 20 pp.
164. G. R. Quick, in G. Robbelen, R. K. Downey, and A. Ashri, eds., *Oil Crops of the World*, McGraw-Hill Publishing Co., New York, 1989, pp. 118–131.
165. T. B. Reed, M. S. Graboski, and S. Gaver, in Ref. 42, pp. 907–914.
166. A. V. Bridgwater and G. Grassi, eds., *Biomass Pyrolysis Liquids Upgrading and Utilization*, Elsevier Applied Science, New York, 1991, 377 pp.
167. D. C. Elliott and co-workers, *Energy & Fuels* **5,** 399 (1991).
168. D. S. Scott, J. Piskorz and D. Radlein, in Ref. 24, pp. 797–809.
169. J. Diebold, in Ref. 166, pp. 341–350.
170. J. Diebold and co-workers, in E. Hogan and co-workers, eds., *Biomass Thermal Processing*, The Chameleon Press Ltd., London, 1992, pp. 101–108.
171. D. A. Johnson, G. R. Tomberlin, and W. A. Ayres, in Ref. 42, pp. 915–925.

General References

D. L. Klass, ed., *Energy from Biomass and Wastes*, Vols. 4–16, IGT, Chicago, 1980–1993.

D. L. Klass, ed., *Biomass as a Nonfossil Fuel Source*, ACS Symposium Series 144, American Chemical Society, Washington, D.C., 1981, 564 pp.

D. L. Klass and G. H. Emert, eds., *Fuels from Biomass and Wastes*, Ann Arbor Science Publishers, Inc., Ann Arbor, Mich., 1981, 592 pp.

J. L. Jones and S. B. Radding, eds., *Thermal Conversion of Solid Wastes and Biomass*, ACS Symposium Series 130, American Chemical Society, Washington, D.C., 1980, 747 pp.

S. S. Sofer and O. R. Zaborsky, eds., *Biomass Conversion Processes for Energy and Fuels*, Plenum Press, New York, 1981, 420 pp.

E. Hogan and co-workers, eds., *Biomass Thermal Processing*, The Chameleon Press Limited, London, 1992, 255 pp.

A. V. Bridgwater and G. Grassi, eds., *Biomass Pyrolysis Liquids Upgrading and Utilization*, Elsevier Applied Science, London, 1991, 377 pp.

A. V. Bridgwater, ed., *Thermochemical Processing of Biomass*, Butterworths, London, 1984, 344 pp.

Biomass, International Directory of Companies, Processes & Equipment, Macmillan Publishers, Ltd., New York, 1986, 243 pp.

D. L. Klass, ed., *A Directory of U.S. Renewable Energy Technology Vendors, Biomass, Photovoltaics, Solar Thermal, Wind*, Biomass Energy Research Association, Washington, D.C., 1990, for U.S. Agency for International Development, 74 pp.

P. F. Bente, Jr., ed., *Bio-Energy Directory*, The Bio-Energy Council, Washington, D.C., 1980, 768 pp.

Biofuels Technical Information Guide, SERI/SP-220-3366, Solar Energy Research Institute, Golden, Colo., Apr. 1989, 198 pp.

A Guide to Federal Programs in Biomass Energy, Meridian Corporation and Science Applications International Corp., Washington, D.C., Sept. 1984, 157 pp.

W. H. Smith, ed., *Biomass Energy Development*, Plenum Press, New York, 1986, 668 pp.

First Biomass Conference of the Americas: Energy, Environment, Agriculture, and Industry, NREL/CP-200-5768, DE930/0050, Proceedings Vols. I–III, National Renewable Energy Laboratory, Golden, Colo., 1993, 1942 pp.

FUELS FROM WASTE

DAVID A. TILLMAN
Foster Wheeler Environmental Corporation
Sacramento, California

A significant number and variety of organic wastes are combusted in energy recovery systems including municipal solid waste (MSW), various forms of refuse-derived fuel (RDF) produced from MSW, and municipal sewage sludge; bark and other wood wastes from sawmills and other forest industry operations; spent pulping liquor from chemical pulp mills such as kraft and sulfite mills; wastewater treatment solids (WTS) or sludges from pulp and paper operations; agribusiness wastes including bagasse from sugar-refining operations, rice hulls, orchard and vineyard prunings, cotton gin trash, and a host of other food and fiber-producing operations; manure from feedlots and dairy cattle, chickens, and other agricultural animals; methane-rich gases generated from municipal-waste landfills; industrial trash and specific wastes such as demolition debris, tire derived fuel (TDF), broken pallets, unrecyclable paper wastes, and related materials; off-gases from pulp mills and chemical manufacturers; incinerable hazardous wastes generated regularly as a function of production processes, eg, spent solvents, or found on Superfund sites targeted for clean-up; and a broad range of other specific specialty wastes. The practice of incinerating these materials has become increasingly prevalent (ca 1990) in order to accomplish disposal in a cost-effective, environmentally sensitive manner. The combustion of such wastes already contributes some 5 EJ (5×10^{15} Btu) to the U.S. economy and over 15 EJ ($>14 \times 10^{15}$ Btu) to the economies of the industrialized world (1). Combustion of such wastes reduces the volume of material which must be disposed of in a landfill, reduces the airborne emissions resulting from plant operations and landfill operations, and permits some economic benefit through energy recovery.

The technologies used to combust wastes depend on the form and location of components to be burned. Typically solid wastes are burned, alone or in combination, and both with and without supplementary fossil fuels. Solid wastes can be burned in mass-burn or pile-burning systems such as hearth furnaces, spreader–stokers, ashing and slagging rotary kilns, or fluidized beds. The choice of combustion technology depends on the degree of waste preparation which is practical; the availability of existing combustion systems, eg, a spreader–stoker for hog fuel utilization, adapted to the cofiring of hog fuel and WTS; and the type of energy recovery contemplated. Energy recovery from the solid wastes can be accomplished in the form of medium or high pressure steam, eg, 4.5–8.6 MPa/672–783 K) suitable for cogeneration or condensing power generation purposes; low pressure steam, eg, 314–1030 kPa, saturated, suitable for process purposes; or the direct production of process heat in the form of heated air or hot combustion products. Energy recovery from gaseous wastes can be accomplished through electricity generation from gas-fired boilers, combustion turbines, or internal combustion engines. Alternatively, these gaseous fuels can be used to generate process heat in conventional fashion.

The success of waste-to-energy programs using municipal wastes and biowastes reduces the volume of material being interred in the ground in landfills. This action also changes the character of materials being landfilled, reducing the organic content with its associated generation of methane gas, and leachates with their significant concentrations of organic compounds. Waste-to-energy, applied to municipal and biomass wastes, can simultaneously provide renewable energy while addressing environmental issues.

Critical concerns associated with energy generation from wastes include fuel composition characteristics; combustion characteristics; formation and control of airborne emissions including both criteria pollutants and air toxics, eg, trace metals; and the characteristics of bottom and flyash generated from waste combustion. These issues are particularly important given the U.S. Clean Air Act Amendments of 1990, the Resource Recovery and Conservation Act (RRCA), and related state and regional regulations. Further, these issues are of critical importance given the capital intensive nature of organic waste-to-energy systems.

See also COMMERCIAL AVAILABILITY OF ENERGY TECHNOLOGY; ELECTRIC POWER GENERATION; EXHAUST CONTROL, INDUSTRIAL; INCINERATION; WASTE-TO-ENERGY.

FUEL CHARACTERISTICS OF ORGANIC WASTES

Fuel characteristics of organic wastes include physical characteristics such as state, specific gravity, bulk density, porosity, and void volume, and related thermal properties; traditional chemical analyses such as proximate and ultimate analyses, including chlorine; calorific content; elemental analyses of the ash, including trace metal contents, base/acid, slagging, and fouling ratios of the various ash products; and certain chemical structural analyses such as aromaticity. These characteristics are governed by the sources of waste-based fuels. They determine the performance of materials in fuel preparation systems such as particle size reduction and drying systems, and also govern the combustion characteristics of the various wastes being burned.

Sources of Waste-Based Fuels

The general architecture of waste-based fuels is a function of waste origination. MSW characteristics are governed by the product composition of the waste stream, as shown in Table 1. The composition of RDF is governed by the processing technologies used to generate the fuel. RDF production technologies involve, at a minimum, coarse shredding of the MSW stream, followed by magnetic separation of ferrous metals. Primary separation techniques for concentration of combustibles involve trommels, air classifiers, or eccentric screens. Trommels have become the most popular separation systems; their overall separation efficiency can be as high as 98.5% (Table 2). Process flow sheets using trommel separation of MSW follow the pattern shown in Figure 1. The composition of MSW, and RDF, ultimately is a function not only of the general composition of the waste stream and the RDF production technology, but also of community and industrial recycling programs. Such programs are accelerating and will influence the amount and relative concentration of paper, plastic, aluminum, and other commodities in the waste stream.

The basic architecture of wood-waste fuels is governed by sawmill or plywood mill configuration, and the consequent blend of bark, trim ends, sawdust, planer shavings, and related residuals. All chippable wastes typically are

Table 1. Product Composition for Municipal Solid Waste,[a] Wt %

Product	1990[b]	2000[c]
Paper and paperboard	38.3	41.0
Yard waste	17.0	15.3
Food waste	7.7	6.8
Plastics	8.3	9.8
Wood	3.7	3.8
Textiles	2.2	2.2
Rubber and leather	2.5	2.4
Glass	8.8	7.6
Metals	9.4	9.0
Miscellaneous	2.1	2.1

[a] Ref. 2.
[b] Approximate.
[c] Estimated.

Table 2. MSW Separation Efficiencies for Trommels as a Function of Waste Component,[a] Wt %

Waste component	Separation efficiency
Paper, plastic	61.1–69.4
Other combustibles[b]	74.6–86.8
Ferrous metal	61.6–80.1
Aluminum	76.7–93.6
Glass, stones, and other	96.6–100
Fines	97.0–98.0
Overall efficiency	*81.0–98.5*

[a] Ref. 3.
[b] For example, wood.

directed to pulp chips. Planer shavings and some sawdust may be directed to alternative products including oriented strand board (OSB), particleboard, animal bedding, a range of other materials applications, and fuel. The characteristics of pulp-mill wastes, eg, bark, WTS, and spent pulping liquor, also are determined by the production processes. The characteristics of wastes from food processing, eg, bagasse, rice hulls, peach pits, cotton gin trash, etc, also are governed by the basic product manufacturing technology and its efficiency of separation.

Physical Properties

Physical properties of waste as fuels are defined in accordance with the specific materials under consideration. The greatest degree of definition exists for wood and related biofuels. The least degree of definition exists for MSW, related RDF products, and the broad array of hazardous wastes. Table 3 compares the physical property data of some representative combustible wastes with the traditional fossil fuel bituminous coal. The solid organic wastes typically have specific gravities or bulk densities much lower than those associated with coal and lignite.

Specific gravity is the most critical of the characteristics in Table 3. It is governed by ash content of the material, is the primary determinant of bulk density, along with particle size and shape, and is related to specific heat and other thermal properties. Specific gravity governs the porosity or fractional void volume of the waste material, ie,

$$\text{FVV} = (1 - \text{SG})/1.5 \quad (1)$$

where FVV is fractional void volume, SG is specific gravity, and the value, 1.5, is the approximate specific gravity of the cell wall in wood fiber (5). Specific gravity and moisture content (MC) together determine thermal conductivity characteristics, k, of cellulosic waste-based fuels:

$$k_{\text{MC}<30\%} = \text{SG}\,(1.39 + 0.028 \times \text{MC}) + 0.165 \quad (2)$$

and

$$k_{\text{MC}>30\%} = \text{SG}\,(1.39 + 0.028 \times \text{MC}) + 0.165 \quad (3)$$

Specific gravity is directly related to the bulk density of waste fuels prepared in a variety of ways. Solid oven-dry (OD) wood, for example, has a typical bulk density of 48.1 kg/m^3 (30 lb/ft^3). In coarse hogged form, eg, <1.9-cm minor dimension, this bulk density declines to about 24 kg/m^3 (15 lb/ft^3). In pulverized form, at a particle size <0.16 cm, this bulk density declines to 16–19 kg/m^3 (10–2 lb/ft^3). Similar relationships hold for municipal waste, agricultural wastes, and related fuels.

Chemical Composition

Chemical compositional data include proximate and ultimate analyses, measures of aromaticity and reactivity, elemental composition of ash, and trace metal compositions of fuel and ash. All of these characteristics impact the combustion processes associated with wastes as fuels. Table 4 presents an analysis of a variety of wood-waste fuels; these energy sources have modest energy contents.

The analysis of solid fuels (Table 4) contains the bases for calculating reactivities, ie, volatile:fixed carbon ratios, volatile carbon:total carbon ratios, hydrogen:carbon and oxygen:carbon ratios, and aromaticity, which is estimated from the chemical components of the waste stream. The

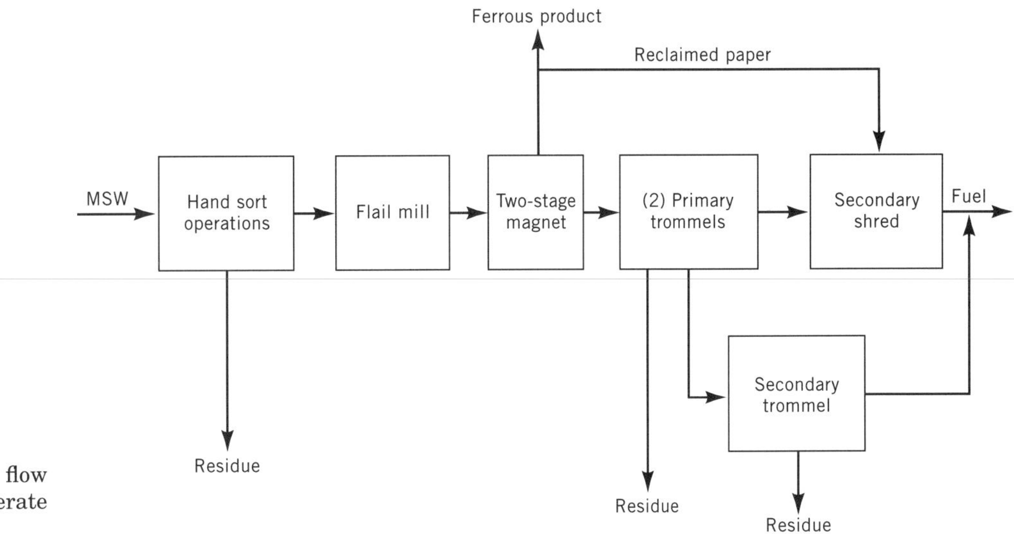

Figure 1. Simplified schematic flow sheet for the production of a moderate RDF, using trommel separation.

Table 3. Physical Properties of Waste-Based Fuels[a]

Fuel	Specific Gravity[b]	Bulk Density,[c] kg/m^3	Moisture Content, wt %
Municipal waste		160–320	25–35
Waste paper	1.2–1.4	80–160	15–25
Waste wood	0.37–0.65	100–320	5–15[d]
			40–65[e]
Bagasse			50–55
Rice hulls			7–10
Orchard and vineyard prunings	0.45–0.55		20–40
Bituminous coal	1.12–1.35	672–1393	3.5–5.0

[a] Refs. 1 and 4.
[b] Ovendry.
[c] To convert kg/m^2 to lb/ft^3, divide by 16.01.
[d] Dry waste.
[e] Wet waste.

typical source of aromatic carbon in waste fuels, such as municipal solid waste and wood waste, is lignin (qv) found in either groundwood-based papers or wood products. Lignin [*9005-53-2*] has a typical empirical formula (1) of $C_9H_{10}(OCH_3)_{0.9-1.7}$, and a higher heating value (HHV) of 26.7 MJ/kg (11,500 Btu/lb), resulting from its basic building blocks of phenyl propane units. These basic building blocks contain aromatic structures. Other sources of aromatic structures in waste fuels include plastic polymers in the waste stream. The aromaticity of a solid fuel can be estimated as a function of H:C atomic ratio (6):

H:C atomic ratio	Carbons as aromatic carbons, %
1.5	0.0
1.2	3.0
0.7	9.0
0.5	16.0

Typically, 40–50% of the carbon atoms in lignite are in aromatic structures while 60–70% of the carbon atoms in Illinois bituminous coal are in aromatic structures (7,8). By all of these measures, waste fuels are significantly more reactive than coal, peat, and other combustible solids.

The chemical analysis of waste fuels also demonstrates that the wood-based fuels contain virtually no sulfur and little nitrogen. Unless the hog fuel contains bark from logs previously stored in salt-water, the chlorine content is very modest to nonmeasurable.

Municipal waste contains, nominally, about 0.5% nitrogen and 0.5% chlorine, the latter coming largely from plastics. Municipal waste also contains moderate amounts of sulfur. The actual composition of MSW, or RDF generated from MSW, is a function of the relative percentages of various components in the waste stream as shown in Table 5. Use of these wastes provides a means for reducing acid gas emissions from energy generation.

Table 4. Analysis of Wood-Based Fuels,[a] **Wt %**

Wood material	Volatile Matter	Fixed Carbon	Ash	Element					HHV,[b] MJ/kg
				C	H	O	N	S	
Big leaf maple	87.9	11.5	0.6	49.9	6.1	43.3	0.14	0.03	16.9
Douglas fir	87.3	12.6	0.1	50.6	6.2	43.0	0.06	0.02	18.3
Douglas fir bark	73.6	25.9	0.5	54.1	6.1	38.8	0.17	[c]	19.7
Oak									
Black[d]	85.6	13.0	1.4	49.0	6.0	43.5	0.15	0.02	16.8
Tan[e]	87.1	12.4	0.5	48.3	6.1	45.0	0.03	0.03	17.2
Oak bark			5.3	49.7	5.4	39.3	0.2	0.1	17.5
Pine bark			2.9	53.4	5.6	37.9	0.1	0.1	18.4
Pitch pine			1.1	59.0	7.2	32.7			21.8
Popular			0.7	51.6	6.3	41.5			17.2
Red alder	87.1	12.5	0.4	49.6	6.1	43.8	0.13	0.07	17.3
Red alder bark	77.3	19.7	3.0	50.9	5.5	40.7	0.39		17.2
Western hemlock[f]	87.0	12.7	0.3	50.4	5.8	41.4	0.1	0.1	19.8

[a] Ref. 4.
[b] Higher heating value (OD basis); to convert MJ/g to Btu/lb, multiply by 430.3.
[c] Trace amounts.
[d] Black oak bark has 81.0 wt % volatile matter, 16.9 wt % fixed carbon, and 2.1 wt % ash.
[e] Tan oak bark has 76.3 wt % voltatile matter, 20.8 wt % fixed carbon, and 2.9 wt % ash.
[f] Western hemlock bark has 73.9 wt % volatile matter, 24.3 % fixed carbon, and 0.8 wt % ash.

Table 5. Components of Municipal Solid Waste[a]

Material	Components, wt %								HHV, MJ/kg[b]
	Carbon	Hydrogen	Oxygen	Nitrogen	Chlorine	Sulfur	Moisture	Ash	
Corrugated paper	36.79	5.08	35.41	0.11	0.12	0.23	20.0	2.26	13.0
Newsprint	36.62	4.66	31.76	0.11	0.11	0.19	25.0	1.55	13.0
Magazine stock	32.93	4.64	32.85	0.11	0.13	0.21	16.0	13.13	11.4
Other paper	32.41	4.51	29.91	0.31	0.61	0.19	23.0	9.06	11.5
Plastics	56.43	7.79	8.05	0.85	3.00	0.29	15.0	8.59	24.2
Rubber and leather	43.09	5.37	11.57	1.34	4.97	1.17	10.0	22.49	17.6
Wood	41.20	5.03	34.55	0.24	0.09	0.07	16.0	2.82	14.5
Textiles	37.23	5.02	27.11	3.11	0.27	0.28	25.0	1.98	13.8
Yard waste	23.29	2.93	17.54	0.89	0.13	0.15	45.0	10.07	8.37
Food waste	17.93	2.55	12.85	1.13	0.38	0.06	60.0	5.10	6.82
Fines[c]	15.03	1.91	12.15	0.50	0.36	0.15	25.0	44.90	5.41

[a] Ref. 9.
[b] To convert MJ/kg to Btu/lb, multiply by 430.3.
[c] Smaller than 2.54 cm (1 in.).

Ash Characteristics. The elemental ash composition of biomass waste and municipal solid waste differs dramatically from that of coal (qv). Wood wastes have ash compositions that are quite alkaline (Table 6) and that have consequent low ash fusion temperatures (Table 7). When firing solid wastes with coal or lignite, the potential exists to have eutectic mixtures formed by the two ash products.

The Clean Air Act of 1990 has made trace metal content in fuels and wastes the final ash-related compositional characteristic of significance. Considerable attention is paid (ca 1993) to emissions of such metals as arsenic, cadmium, chromium, lead, mercury, silver, and zinc. The concentration of these metals in both grate ash and flyash is of significance as a result of federal and state requirements; of particular importance is the mobility of metals. This mobility, and the consequent toxicity of the ash product, is determined by the Toxic Characteristic Leaching Procedure (tclp) test. Tables 8–10 present trace metal contents for wood wastes and agricultural wastes, municipal waste, and refuse-derived fuel, respectively. In Table 8, the specific concentration of various components in the RDF governs the expected average concentration of trace metals.

Biofuels, ie, wood and agricultural waste, are relatively low in metal contents, and typically have a lower metals content when compared to coals being burned for energy generation. However, municipal waste and its derivative fuels (RDF) can be quite high in trace metals. The RDF production process removes approximately 67% of the incoming metals content, but significant quantities of components such as lead and cadmium remain in some compositions of RDF. These metals must be controlled for safe energy generation from such combustible materials. The wood waste in RDF is typically not the same as the wood waste from forest products manufacture. Commonly the wood in RDF is treated with compounds, eg, copper chromium arsenate (CCA), which make it more suitable in outdoor service, such as in deck construction. Such wood treating adds trace metals to the fuel feed (13).

COMBUSTION OF SOLID WASTE-BASED FUELS

It is useful to examine the combustion process applied to solid wastes as fuels and sources of energy. All solid wastes are quite variable in composition, moisture con-

Table 6. Elemental Analysis of Wood Waste Ash[a]

Compound	CAS Registry Number	Source, wt%		
		Pine Bark	Oak Bark	Spruce Bark
SiO_2	*[14808-60-7]*	39.0	11.1	32.0
Fe_2O_3	*[1309-37-1]*	3.0	3.3	6.4
TiO_2	*[13463-67-7]*	0.2	0.1	0.8
Al_2O_3	*[1344-28-1]*	14.0	0.1	11.0
MnO_4	*[12502-70-4]*	[b]	[b]	1.5
CaO	*[1305-78-8]*	25.5	64.5	25.3
MgO	*[1309-48-4]*	6.5	1.2	4.1
Na_2O	*[12401-86-4]*	1.3	8.9	8.0
K_2O	*[12136-45-7]*	6.0	0.2	2.4
SO_3	*[7446-11-9]*	0.3	2.0	2.1
Cl	*[7782-50-5]*	[b]	[b]	[b]

[a] Ref. 10.
[b] Trace amounts.

Table 7. Ash Fusion Temperatures for Some Wood Waste Ash,[a] K

Wood Species	Initial		Softening		Fluid	
	Oxidizing	Reducing	Oxidizing	Reducing	Oxidizing	Reducing
Tan oak	1663	1650	1713	1711	1730	1727
Pine bark	1483	1467	1522	1500	1561	1540
Oak bark	1744	1750	1772	1766	1783	1777

[a] Refs. 4 and 10.

tent, and heating value. Consequently, they typically are burned in systems such as grate-fired furnaces or fluidized-bed boilers where significant fuel variability can be tolerated.

Combustion characteristics of consequence include the overall mechanism of solid waste combustion, factors governing rates of waste fuels combustion, temperatures associated with waste oxidation, and pollution-formation mechanisms.

Mechanisms and Rates of Combustion

All solid fuels and wastes burn according to a general global mechanism (Fig. 2). The solid particle is first heated. Following heating, the particle dries as the moisture bound in the pore structure and on the surface of the particle evaporates. Only after moisture evolution does pyrolysis initiate to any great extent. The pyrolysis process is followed by char oxidation, which completes the process.

The rate of solid waste combustion is controlled by diffusion, rather than by reaction kinetics. In general, the time required for combustion of a single particle of waste (1) can be expressed as:

$$T_b = T_h + T_d + T_p + T_{co} \tag{4}$$

where T_b is time for complete particle burnout, T_h is time for initial particle heatup, T_d is time required for particle drying, T_p is time required to pyrolyze the particle into volatiles and char, and T_{co} is time for char oxidation. The first two terms, initial heating plus drying, $T_h + T_d$, can be taken as the drying step. This time component is governed by the temperature of the environment, the particle size, the moisture content of the particle, and the porosity of the particle. The term T_p is strictly governed by heat transfer through the particle (14,15). The rate of pyrolysis is governed by the heat capacity of the solid waste, its porosity, and its thermal conductivity. The T_{co} term is mass-transfer-limited, with diffusion of oxygen to the surface of the char particle being rate-limiting. Of these steps, either drying or char oxidation may be rate-limiting, as shown in Figure 3, depending on the moisture content of the solid waste.

Temperatures Associated with Combustion

The temperatures achieved by solid waste combustion are typically lower than those associated with fossil fuel oxidation, and are governed by the following general equation (1):

$$T_{f,\,\text{solid waste}} = [695 - 10.1\,\text{MC}_t + 1734(1/\text{SR}) + 0.61(A - 298)]\ \text{K} \tag{5}$$

Where T_f is flame temperature is K; MC_t is moisture content of the waste, expressed on a total weight basis; SR is defined as stoichiometric ratio or moles O_2 available/moles O_2 required for complete oxidation of the carbon, hydrogen,

Table 8. Trace Metal Concentrations[a] in Ash from Agricultural Biofuels and Wood-Fired Boilers, mg/kg

Metal	Agricultural Biofuel			Wood-Fired Boilers[b]
	Cotton Gin Trash	Orchard Prunings	Vineyard Prunings	
Barium	120	220	41	130
Silver	<0.08	<0.08	<0.08	<0.08
Arsenic	12	5.5	3.4	3.0–6.3
Beryllium	0.1	0.1	0.06	0.1
Cadmium	1.1	0.36	0.39	1.5–16
Cobalt	14	9.0	2.8	8.5–20
Chromium	20	12	11	16.8–25
Copper	23	14	31	40–76.9
Mercury	<0.05	<0.05	<0.05	<0.05–<0.5
Molybdenum	16	2	2	3.0–14
Nickel	4.6	5.8	4.4	11–50
Lead	21	22	55	38–70
Antimony	10	10	10	10
Selenium	<0.2	<0.2	<0.2	5.0
Vanadium	20	12	11	26–27
Zinc	87	190	40	130–560
Thallium	15	10	2	6.5

[a] Ref. 1.
[b] Range of concentrations from various locations.

Table 9. Trace Metals in Municipal Solid Waste and Solid Waste Ash,[a] ppmw

Metal	Solid Waste[b,c]	Solid Waste Ash
Arsenic	0.73–12.5	2.9–50
Barium	19.8–675	79–2,700
Beryllium	ND–0.6	ND–2.4
Boron	6–43.5	24–174
Cadmium	0.05–25	0.18–100
Chromium	3.0–375	12–1,500
Cobalt	0.43–22.75	1.7–91
Copper	10–1,475	40–5,900
Lead	7.75–9.15	31–36,600
Magnesium	175–4,000	700–16,000
Molybdenum	0.6–72.5	2.4–290
Manganese	3.5–782.5	14–3,130
Mercury	ND–4.38	0.05–17.5
Nickel	3.25–3,228	13–12,910
Selenium	0.03–12.5	0.10–50
Strontium	3.0–160	12–640
Zinc	23–11,500	92–46,000

[a] Range of concentration. Ref. 11.
[b] Based on ash measurements. Imputed to waste.
[c] ND = nondetectable.

and sulfur in the fuel, ie, 1/SR = equivalence ratio; and A is temperature of the combustion air, expressed in K. In English units, this equation is as follows:

$$T_{f,\,\text{solid waste}} = [3870 - 15.6\,\text{MC}_t - 130.4\,\text{EO}_2 + 0.59(A - 77)]\,^\circ\text{F} \quad (6)$$

where EO_2 is excess oxygen in the stack gas, ie, total, not dry, basis.

Whereas solid wastes can achieve significant flame temperatures, they are substantially below those associated with fossil fuels. These differences are largely caused by the lower calorific value; the chemical composition, eg, oxygen content of the waste; and the higher moisture and ash contents commonly associated with various solid wastes. Typically for grate-fired systems the use of wastes as fuel requires maintaining temperatures in excess of 1256 K (1800°F) and for residence times exceeding 2 s in order to ensure complete combustion and minimize dioxin and furan formation. As shown from equation 5, such temperatures are readily achieved under most conditions.

Given the mechanisms and temperatures, waste combustion systems typically employ higher percentages of excess air, and typically also have lower cross-sectional and volumetric heat release rates than those associated with fossil fuels. Representative combustion conditions are shown in Table 11 for wet wood waste, with 50–60% moisture total basis, municipal solid waste, and RDF.

Formation of Airborne Emissions

Airborne emissions are formed from combustion of waste fuels as a function of certain physical and chemical reactions and mechanisms. In grate-fired systems, particulate emissions result from particles being swept through the furnace and boiler in the gaseous combustion products, and from incomplete oxidation of the solid particles, with consequent char carryover. If pile burning is used, eg, the

Table 10. Analysis of Refuse-Derived Fuel[a]

Parameter	Glossy Paper	Nonglossy Paper	Cardboard	Film Plastics	Rubber, Leather, and Hard Plastics	Wood and Textiles	Other Organics	Total RDF
Trace metals, mg / kg fuel								
Arsenic	3.1	3.3	3.5	2.7	2.5	5.2	4.6	4.0
Barium	285.1	78.9	48.7	186.5	724.3	96.7	210.0	173.2
Beryllium	1.1	1.3	1.2	0.5	0.4	1.5	1.5	1.3
Cadmium	1.1	1.3	3.8	7.7	17.3	3.0	3.1	3.4
Chromium	23.8	37.3	23.2	69.4	95.9	34.8	44.5	42.7
Copper	74.8	40.3	27.0	2740.7	12.1	202.3	61.4	220.1
Lead	88.4	621.2	66.2	836.6	668.1	747.6	475.1	495.5
Manganese	61.2	137.6	101.1	311.8	83.1	183.9	367.3	260.4
Mercury	0.3	0.7	0.4	1.0	0.4	0.9	1.2	0.9
Nickel	10.4	15.5	25.5	45.6	170.4	27.4	17.7	24.4
Selenium	3.1	2.9	3.3	2.1	2.0	3.3	3.0	2.9
Strontium	62.4	73.2	47.8	88.5	88.6	198.9	474.8	283.2
Zinc	164.5	227.6	161.4	482.2	2494.5	449.4	360.0	380.2
Ultimate analysis, wt %								
Carbon	43.4	47.3	49.6	59.8	53.8	50.1	34.6	41.1
Hydrogen	5.3	6.1	6.4	8.2	8.9	6.0	4.3	5.3
Oxygen	27.5	32.0	35.7	13.8	23.3	31.5	41.1	35.2
Nitrogen	0.62	1.58	0.72	1.01	0.83	1.07	1.07	1.12
Sulfur	0.25	0.25	0.24	0.56	0.57	0.28	0.38	0.34
Chlorine	0.04	0.04	0.05	0.10	0.05	0.05	0.01	0.07
Ash	23.0	12.7	7.4	16.6	12.5	11.0	18.3	16.5
Higher heating value, MJ/kg[b]	14.7	19.7	18.5	31.0	25.4	21.0	16.5	18.7

[a] Fuel produced in Tacoma, Wash. Values on ovendry (OD) fuel basis. Ref. 12.
[b] To convert MJ/kg to Btu/lb, multiply by 430.3.

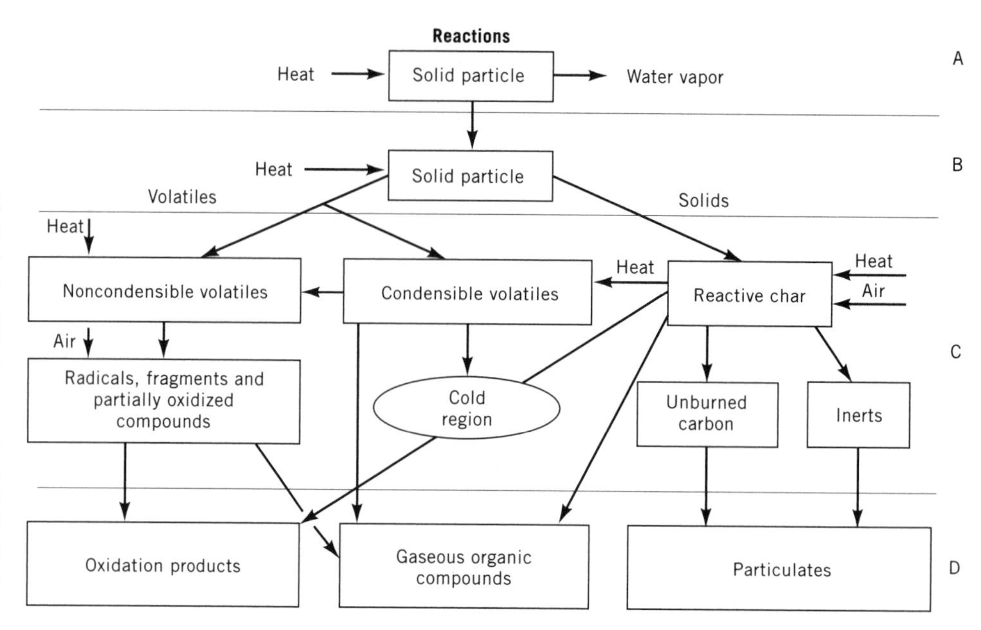

Figure 2. Overall schematic of solid fuel combustion (1). Reaction sequence is A, heating and drying; B, solid particle pyrolysis; C, oxidation; and D, postcombustion. In the oxidation sequence, left and center comprise the gas-phase region, right is the gas–solids region. Noncondensible volatiles include CO, CO_2, CH_4, NH_3, H_2O; condensible volatiles are C-6–C-20 compounds; oxidation products are CO_2, H_2O, O_2, N_2, NO_1; gaseous organic compounds are CO, hydrocarbons, and polyaromatic hydrocarbons (PAHs); and particulates are inerts, condensation products, and solid carbon products.

mass burn units employed for unprocessed MSW, typically only 20–25% of the unburned solids and inerts exit the combustion system as flyash. If spreader–stoker technologies are employed, between 75 and 90% of the unburned solids and inerts may exit the combustion system in the form of flyash.

Sulfur dioxide [*7446-09-5*] is formed as a result of sulfur oxidation, and hydrogen chloride is formed when chlorides from plastics compete with oxygen as an oxidant for hydrogen. Typically the sulfur is considered to react completely to form SO_2, and the chlorine is treated as the preferred oxidant for hydrogen. In practice, however, significant fractions of sulfur do not oxidize completely, and at high temperatures some of the chlorine atoms may not form HCl.

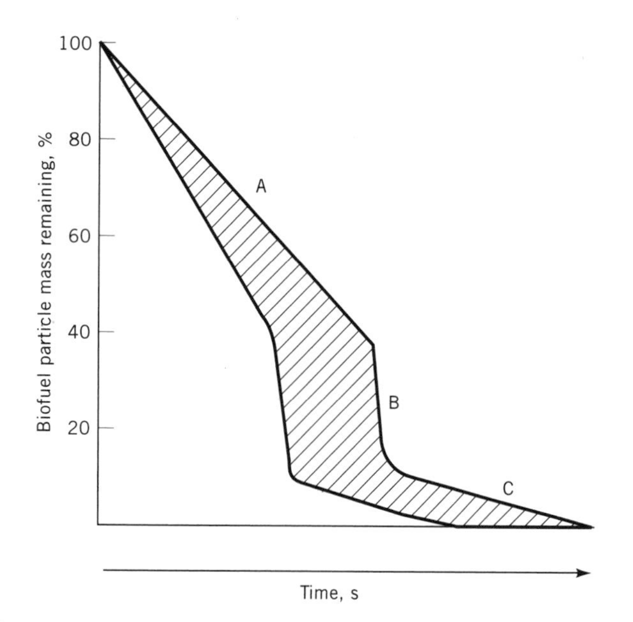

Figure 3. Schematic of the sequential nature of waste fuel combustion (1). A, particle heating and drying; B, solid particle pyrolysis; and C, char oxidation. A and C may be rate-limiting.

Nitrogen oxide, NO_x, formation results from conversion of nitrogen in the fuel to NO, since combustion temperatures are below those typically associated with thermal NO_x formation, eg, 1483°C as the threshold for thermal NO_x has been documented (17,18). The conversion of fuel nitrogen to NO_x typically proceeds along the pathways of nitrogen volatilization followed by oxidation of the nitrogen volatiles in the presence of excess oxygen. In the absence of available oxygen, the nitrogen volatiles react with each other to form N_2. Conversion of nitrogen from waste fuels into NO_x is typically 15–25% of the fuel nitrogen converted, depending on combustion technology and firing conditions.

Formation of emissions from fluidized-bed combustion is considerably different from that associated with grate-fired systems. Flyash generation is a design parameter, and typically >90% of all solids are removed from the system as flyash. SO_2 and HCl are controlled by reactions with calcium in the bed, where the limestone fed to the

Table 11. Combustion Conditions for Conventional Waste Fuel Boilers[a]

Parameter	Wood[b]	MSW	RDF
Maximum fuel moisture, %	55–60	30–40	20–35
Grate fuel feed rate, kg/m^2h[c]	1000–1500		
Grate heat release, GJ/m^2h[d]	8.5–11.35	3.4	5.7–8.5
Volumetric heat release, MJ/m^3h[e]	480–560	335–410	450–480
Stoichiometric ratio	1.25–1.5	1.8–2.0	1.6–1.8
Excess air, %	25–50	80–100	60–80
Representative overfire air, %	20–40	30–40	30–40

[a] Refs. 1 and 16.
[b] Wet wood waste, 50–60% moisture total basis.
[c] To convert $kg/(m^2h)$ to $lb/(ft^2h)$, multiply by 0.204.
[d] To convert $GJ/(m^2h)$ to $Btu/(ft^2h)$, multiply by 8.8×10^4.
[e] To convert $MJ/(m^3h)$ to $Btu/(ft^3h)$, multiply by 26.9.

bed first calcines to CaO and CO_2, and then the lime reacts with sulfur dioxide and oxygen, or with hydrogen chloride, to form calcium sulfate and calcium chloride, respectively. SO_2 and HCl capture rates of 70–90% are readily achieved with fluidized beds. The limestone in the bed plus the very low combustion temperatures inhibit conversion of fuel N to NO_x.

Trace metal emissions from waste combustion are a function of metal content in the feed, combustion temperatures, and the percentage of ash exiting the combustion chamber as flyash. They are also a function of the temperatures in the air pollution control system, eg, the precipitator, baghouse, or scrubber, and the consequent degree to which these metals undergo homogeneous nucleation and become a fine flume, eg, submicron particles in the flyash, or undergo heterogeneous condensation on existing flyash particles. For some metals, such as arsenic and lead, emissions are also a function of the combustion system and the presence of lime. In fluidized beds, it has been shown that the arsenic and lead are captured and stabilized by the presence of reactive lime (1,19).

Dioxin [*828-00-2*] and furan [*110-00-9*] emissions are the final pollutants of consideration and are of most concern for combustors using MSW, RDF, or hazardous waste. Dioxins are formed, at some concentration level ranging from inconsequential to problematical, whenever aromatic compounds and trace quantities of chlorine are present in the boiler feed. Several mechanisms have been postulated for dioxin and furan emission formation including the passage of such molecules, unreacted, through the furnace; the formation of dioxins and furans from such precursers as lignins and trace concentrations of chlorine; and the formation of dioxins in post-combustion reactions in the economizer section at temperatures of about 550–700 K (2,20,21). Of these, the post-combustion mechanism is shown to dominate. However, the impact of this mechanism can be minimized by maintaining temperatures in excess of 983°C for 2 s, while ensuring complete mixing of fuels and oxidants and ensuring the absence of cool zones in the furnace. A general equation for approximating dioxin and furan emissions is as follows (22):

$$D + F_{\mu g/Nm^3} = 0.0376(EA) - 3.305 \quad (7)$$

where $D + F$ is dioxins plus furans in the gaseous combustion products, corrected to 12% CO_2, and EA is the percentage of excess air above about 70%. Dioxin emissions are a problem more for MSW burners than other types of waste fuel systems, largely as a result of chlorine in the waste feed. Dioxin emissions typically are minimized in fluidized-bed combustion as a consequence of the solids mixing and solids turbulence in the bed.

APPLICATIONS OF FUELS FROM WASTE

Because fuels from combustible organic wastes have long been economic in specific industries such as pulp mills, sawmills, sugar mills or factories, and other biomass processing operations, and because municipal waste-to-energy is becoming increasingly cost effective, these systems are continuing to be installed. The typical industrial system is used either to generate process steam or to generate both electricity and steam in a cogeneration application. Typically these applications involve power boilers which are the essential source of process energy in pulp mills and food processing operations. Typical larger installations generate some 200–300 t/h of steam used to generate 25–35 MW plus process heat. As stand-alone electricity generating stations, these units are capable of 50–60 MW, with a typical thermal efficiency of 65–75%. Thermal efficiency depends on moisture and ash content of the feed waste, consequent firing conditions employed, and extent of heat-transfer surface available for combustion air heating as well as steam generation.

Since the early 1980s, there have been several stand-alone power plants built to fire biomass wastes including such materials as wood waste, rice hulls, and vineyard prunings; these facilities typically generate 20–50 MW_e for sale to electric utilities. MSW and RDF are typically consumed in condensing power plants generating 15–50 MW_e while reducing the volume of solids to be landfilled. These units have thermal efficiencies comparable to the large power boilers of the pulp and paper industry, depending again on waste fuel condition and firing strategy.

There has been increased interest in firing wood waste as a supplement to coal in either pulverized coal (PC) or cyclone boilers at 1–5% of heat input. This application has been demonstrated by such electric utilities as Santee-Cooper, Tennessee Valley Authority, Georgia Power, Delmarva, and Northern States Power. Cofiring wood waste with coal in higher percentages, eg, 10–15% of heat input, in PC and cyclone boilers is being carefully considered by the Electric Power Research Institute (EPRI) and Tennessee Valley Authority (TVA). This practice may have the potential to maximize the thermal efficiency of waste fuel combustion. If this practice becomes widespread, it will offer another avenue for use of fuels from waste.

Reprinted from *Kirk-Othmer Encyclopedia of Chemical Technology*, 4th ed., vol. 12, John Wiley & Sons, Inc., New York, 1995.

BIBLIOGRAPHY

1. D. A. Tillman, *The Combustion of Solid Fuels and Wastes*, Academic Press, San Diego, Calif., 1991.
2. W. R. Seeker, W. S. Lanier, and M. P. Heap, *Municipal Waste Combustion Study: Combustion Control of MSW Combustors to Minimize Emissions of Trace Organics*, EER Corporation, Irvine, Calif., 1987.
3. J. Barton, *Evaluation of Trommels for Waste to Energy Plants, Phase I*, National Center for Resource Recovery, Washington, D.C., 1982.
4. A. J. Rossi, in D. A. Tillman and E. Jahn, eds., *Progress in Biomass Conversion*, Vol. 5, Academic Press, New York, 1984, pp. 69–99.
5. U.S. Forest Service, *Wood Handbook: Wood as an Engineering Material*, U.S. Government Printing Office, Washington, D.C., 1974.
6. F. Shafizadeh and Y. Sekuguchi, *Carbon* **21**(5), 511–516 (1983).

7. K. E. Chung and I. B. Goldberg, in *Proceedings of the 12th Annual EPRI Contractors' Conference on Fuel Science and Conversion*, EPRI, Palo Alto, Calif., 1988.
8. K. E. Chung, I. B. Goldberg, and J. J. Ratto, *Chemical Structure and Liquefaction Reactivity of Coal*, EPRI, Palo Alto, Calif., 1987.
9. R. E. Kaiser, "Physical-Chemical Character of Municipal Refuse," *Proceedings of the 1975 International Symposium on Energy Recovery from Refuse*, University of Louisville, Louisville, Ky., 1975.
10. *Steam: Its Generation and Use*, 40th ed., Babcock & Wilcox, Barberton, Ohio, 1991.
11. A. .M. Ujihara and M. Gough, *Managing Ash From Municipal Waste Incinerators*, Resources for the Future, Washington, D.C., 1989.
12. D. A. Tillman and C. Leone, "Control of Trace Metals in Flyash at the Tacoma, Washington Multifuels Incinerator," *Proceedings of the American Flame Research Committee Fall International Symposium*, San Francisco, 1990.
13. D. A. Tillman, "The Fate of Arsenic at the Tacoma Steam Plant #2," paper presented at *1992 Fall International Symposium*, American Flame Research Committee, Boston, Mass., 1992.
14. M. Hertzberg, I. A. Zlochower, and J. Edwards, *Coal Particle Pyrolysis Temperatures and Mechanisms*, RI 9169, U.S. Department of the Interior, Bureau of Mines, Washington, D.C., 1988.
15. M. Hertzberg, I. Zlochower, R. Conti, and K. Cashdollar, *Am. Chem. Soc.* **32**(3), 24–41 (1987).
16. D. A. Tillman, A. J. Rossi, and K. M. Vick, *Incineration of Municipal and Hazardous Solid Wastes*, Academic Press, San Diego, Calif., 1989.
17. D. W. Pershing and J. Wendt, *Proceedings of the 16th International Symposium*, The Combustion Institute, Pittsburgh, Pa., 1976.
18. D. W. Pershing and co-workers, *Proceedings of the 17th International Symposium*, The Combustion Institute, Pittsburgh, Pa., 1978.
19. T. C. Ho and co-workers, "Metal Capture During Fluidized Bed Incineration of Wastes Contaminated with Lead Chloride," presented at the *Second International Congress on Toxic Combustion By-Products: Formation and Control*, Salt Lake City, Utah, Mar. 26–29, 1991.
20. R. G. Barton, W. O. Clark, W. S. Lanier, and W. R. Seeker, "Dioxin Emissions During Waste Incineration," presented at *Spring Meeting, Western States Section of the Combustion Institute*, Salt Lake City, Utah, 1988.
21. *National Incinerator Testing and Evaluation Program: Mass Burning Incinerator Technology*, Vol. II, Lavalin, Inc. Quebec City, Quebec, Canada, 1987.
22. M. Beychok, *Atmos. Envir.* **21**(1), 29–36 (1987).

FUELS, SYNTHETIC, GASEOUS

JAMES G. SPEIGHT
Western Reserve Institute
Laramie, Wyoming

Substitute or synthetic natural gas (SNG) has been known for several centuries. When SNG was first discovered, natural gas was largely unknown as a fuel and was more a religious phenomenon (see GAS, NATURAL) (1). Coal was the first significant source of substitute natural gas and in the early stages of SNG production the product was more commonly known under variations of the name coal gas (2,3). Whereas coal continues to be a principal source of substitute natural gas (4) a more recently recognized source is petroleum (5). See also NATURAL GAS.

GAS FROM COAL

Coal can be converted to gas by several routes (2,6–11), but often a particular process is a combination of options chosen on the basis of the product desired, ie, low, medium, or high heat-value gas. In a very general sense, coal gas is the term applied to the mixture of gaseous constituents that are produced during the thermal decomposition of coal at temperatures in excess of 500°C (>930°F), often in the absence of oxygen (air) (3). A solid residue (coke, char), tars, and other liquids are also produced in the process:

$$C_{coal} + \text{heat} \rightarrow C_{char} + \text{tar/liquid} + CO + CO_2 + H_2$$

The tars and other liquids (liquor) are removed by condensation leaving principally hydrogen, carbon monoxide, and carbon dioxide in the gas phase. This gaseous product also contains low boiling hydrocarbons (qv), sulfur-containing gases, and nitrogen-containing gases including ammonia (qv) and hydrogen cyanide. The solid residue is then treated under a variety of conditions to produce other fuels which vary from a purified char to different types of gaseous mixtures. The amounts of gas, coke, tar, and other liquid products vary according to the method used for the carbonization (especially the retort configuration), and process temperature, as well as the nature (rank) of the coal (3,11).

The recorded chronology of the coal-to-gas conversion technology began in 1670 when a cleryman, John Clayton, in Wakefield, Yorkshire, produced in the laboratory a luminous gas by destructive distillation of coal (12). At the same time, experiments were also underway elsewhere to carbonize coal to produce coke, but the process was not practical on any significant scale until 1730 (12). In 1792, coal was distilled in an iron retort by a Scottish engineer, who used the by-product gas to illuminate his home (13).

The conversion of coal to gas on an industrial scale dates to the early nineteenth century (14). The gas, often referred to as manufactured gas, was produced in coke ovens or similar types of retorts by simply heating coal to vaporize the volatile constituents. Estimates based on modern data indicate that the gas mixture probably contained hydrogen (qv) (ca 50%), methane (ca 30%), carbon monoxide and carbon dioxide (ca 15%), and some inert material, such as nitrogen (qv), from which a heating value of approximately 20.5 MJ/m^3 (550 Btu/ft^3) can be estimated (6).

Blue gas, or blue-water gas, so-called because of the color of the flame upon burning (10), was discovered in 1780 when steam was passed over incandescent carbon (qv), and the blue-water gas process was developed over the period 1859–1875. Successful commercial application of the process came about in 1875 with the introduction of the carburetted gas jet. The heating value of the gas was low, ca 10.2 MJ/m^3 (275 Btu/ft^3), and on occasion oil was added to the gas to enhance the heating value. The new product was given the name carburetted water gas and the technique satisfied part of the original aim by adding luminosity to gas lights (10).

Coke-oven gas is a by-product fuel gas derived from coking coals by the process of carbonization. The first by-product coke ovens were constructed in France in 1856. Since then they have gradually replaced the old and primitive method of beehive coking for the production of metallurgical coke. Coke-oven gas is produced in an analogous manner to retort coal gas, with operating conditions, mainly temperature, set for maximum carbon yield. The resulting gas is, consequently, poor in illuminants, but excellent as a fuel. Typical analyses and heat content of common fuel gases vary (Table 1) and depend on the source as well as the method of production.

In Germany, large-scale production of synthetic fuels from coal began in 1910 and necessitated the conversion of coal to carbon monoxide and hydrogen.

Water gas reaction

$$C_{coal} + H_2O \rightarrow CO + H_2$$

The mixture of carbon monoxide and hydrogen is enriched with hydrogen from the water gas catalytic (Bosch) process, ie, water gas shift reaction, and passed over a cobalt–thoria catalyst to form straight-chain, ie, linear, paraffins, olefins, and alcohols in what is known as the Fisher-Tropsch synthesis.

$$n\,CO + (2n + 1)H_2 \xrightarrow[\text{catalyst}]{\text{cobalt}} C_nH_{2n+2} + n\,H_2O$$

$$2n\,CO + (n + 1)H_2 \xrightarrow[\text{catalyst}]{\text{iron}} C_nH_{2n+2} + n\,CO_2$$

$$n\,CO + 2n\,H_2 \xrightarrow[\text{catalyst}]{\text{cobalt}} C_nH_{2n} + n\,H_2O$$

$$2n\,CO + n\,H_2 \xrightarrow[\text{catalyst}]{\text{iron}} C_nH_{2n} + n\,CO_2$$

$$n\,CO + 2n\,H_2 \xrightarrow[\text{catalyst}]{\text{cobalt}} C_nH_{2n+1}OH + (n - 1)H_2O$$

$$(2n - 1)CO + (n + 1)H_2 \xrightarrow[\text{catalyst}]{\text{iron}} C_nH_{2n+1}OH + (n - 1)CO_2$$

In Sasolburg, South Africa, a commercial plant using the Fischer-Tropsch process was completed in 1950 and began producing a variety of liquid fuels and chemicals. The facility has been expanded to produce a considerable portion of South Africa's energy requirements (15,16).

In 1948, the first demonstration of suspension gasification was successfully completed by Koppers, Inc. The product gas was of 11.2 MJ/m^3 (300 Btu/ft^3) calorific value and consisted primarily of a mixture of hydrogen and oxides of carbon. In the United States, so-called second-generation coal gasification processes came into being as a result of the recognized need to develop reliable, domestic energy sources to replace the rapidly diminishing supply of conventional fuels (3,9). More recently, the biological conversion of coal and synthesis gas (carbon monoxide–hydrogen mixtures) into liquid fuels by methanogenic bacteria has received some attention (17–19).

Gas Products

The originally designated names of the gaseous mixtures are used herein, with the understanding that since their introduction there may be differences in means of production and in the make-up of the gaseous products. Properties of fuel gases are available (20). There are standard tests to determine properties and character of gaseous mixtures. These tests are accepted by the American Society for Testing and Materials (ASTM), by the British Standards Institution (BSI), by the Institute of Petroleum (IP), and by the International Standards Organization (ISO) (1,10,20).

Low Heat-Value Gas. Low heat-value (low Btu) gas (7) consists of a mixture of carbon monoxide and hydrogen

Table 1. Analyses of Fuel Gases

	Gas Composition, vol %							
Type of Fuel Gas	CO	CO_2	H_2	N_2	O_2	CH_4	Illuminants	Heat Value,[a] MJ/m^3
Blast-furnace	27.5	10.0	3.0	58.0	1.0	0.5		3.8
Producer (bituminous)	27.0	4.5	14.0	50.9	0.6	3.0		5.6
Blue-water	42.8	3.0	49.9	3.3	0.5	0.5		11.5
Carburetted water	33.4	3.9	34.6	7.9	0.9	10.4	8.9[b]	20.0
Retort coal	8.6	1.5	52.5	3.5	0.3	31.4	2.2[c]	21.5
Coke-oven	6.3	1.8	53.0	3.4	0.2	1.6	3.7[d]	21.9
Natural								
Mid-continent		0.8		3.2		96.0		36.1
Pennsylvania				1.1		67.6	31.3[e]	46.0

[a] To convert MJ/m^3 to Btu/ft^3, multiply by 26.86.
[b] 6.7 vol % C_2H_4 plus 2.2 vol % C_6H_6.
[c] 1.1 vol % each of C_2H_4 and C_6H_6.
[d] 2.7 vol % C_2H_4 plus 1.0 vol % C_6H_6.
[e] 31.3 vol % C_2H_6.

and has a heating value of less than 11 MJ/m^3 (300 Btu/ft^3), but more often in the range 3.3–5.6 MJ/m^3 (90–150 Btu/ft^3). The gas is formed by partial combustion of coal with air, usually in the presence of steam (7).

$$2\ C_{coal} + O_2 \rightarrow 2\ CO$$

$$C_{coal} + H_2O \rightarrow CO + H_2$$

$$CO + H_2O \rightarrow CO_2 + H_2$$

This gas is of interest to industry as a fuel gas or even, on occasion, as a raw material from which ammonia and other compounds may be synthesized.

The first gas producer making low heat-value gas was built in 1832. (The product was a combustible carbon monoxide–hydrogen mixture containing ca 50 vol % nitrogen). The open-hearth or Siemens-Martin process, built in 1861 for pig iron refining, increased low heat-value gas use. The use of producer gas as a fuel for heating furnaces continued to increase until the turn of the century when natural gas began to supplant manufactured fuel gas.

The combustible components of the gas are carbon monoxide and hydrogen, but combustion (heat) value varies because of dilution with carbon dioxide and with nitrogen. The gas has a low flame temperature unless the combustion air is strongly preheated. Its use has been limited essentially to steel (qv) mills, where it is produced as a byproduct of blast furnaces. A common choice of equipment for the smaller gas producers is the Wellman-Galusha unit because of its long history of successful operation (21).

Medium Heat-Value Gas. Medium heat-value (medium Btu) gas (6,7) has a heating value between 9 and 26 MJ/m^3 (250 and 700 Btu/ft^3). At the lower end of this range, the gas is produced like low heat-value gas, with the notable exception that an air separation plant is added and relatively pure oxygen (qv) is used instead of air to partially oxidize the coal. This eliminates the potential for nitrogen in the product and increases the heating value of the product to 10.6 MJ/m^3 (285 Btu/ft^3). Medium heat-value gas consists of a mixture of methane, carbon monoxide, hydrogen, and various other gases and is suitable as a fuel for industrial consumers.

High Heat-Value Gas. High heat-value (high Btu) gas (7) has a heating value usually in excess of 33.5 MJ/m^3 (900 Btu/ft^3). This is the gaseous fuel that is often referred to as substitute or synthetic natural gas (SNG), or pipeline-quality gas. It consists predominantly of methane and is compatible with natural gas insofar as it may be mixed with, or substituted for, natural gas.

Any of the medium heat-value gases that consist of carbon monoxide and hydrogen (often called synthesis gas) can be converted to high heat-value gas by methanation (22), a low temperature catalytic process that combines carbon monoxide and hydrogen to form methane and water.

$$CO + 3\ H_2 \rightarrow CH_4 + H_2O$$

Prior to methanation, the gas product from the gasifier must be thoroughly purified, especially from sulfur compounds the precursors of which are widespread throughout coal (23). Moreover, the composition of the gas must be adjusted, if required, to contain three parts hydrogen to one part carbon monoxide to fit the stoichiometry of methane production. This is accomplished by application of a catalytic water gas shift reaction.

$$CO + H_2O \rightleftarrows CO_2 + H_2$$

The ratio of hydrogen to carbon monoxide is controlled by shifting only part of the gas stream. After the shift, the carbon dioxide, which is formed in the gasifier and in the water gas reaction, and the sulfur compounds formed during gasification, are removed from the gas.

The processes that have been developed for the production of synthetic natural gas are often configured to produce as much methane in the gasification step as possible thereby minimizing the need for a methanation step. In addition, methane formation is highly exothermic which contributes to process efficiency by the production of heat in the gasifier, where the heat can be used for the endothermic steam–carbon reaction to produce carbon monoxide and hydrogen.

$$C + H_2O \rightarrow CO + H_2$$

Carbonization

Next to combustion, carbonization represents one of the largest uses of coal (2,24–26). Carbonization is essentially a process for the production of a carbonaceous residue by thermal decomposition, accompanied by simultaneous removal of distillate, of organic substances.

$$C_{organic} \rightarrow C_{coke/char/carbon} + \text{liquids} + \text{gases}$$

This process may also be referred to as destructive distillation. It has been applied to a whole range of organic materials, more particularly to natural products such as wood (qv), sugar (qv), and vegetable matter to produce charcoal (see FUELS FROM BIOMASS). However, in the present context, coal usually yields coke, which is physically dissimilar from charcoal and appears with the more familiar honeycomb-type structure (27).

The original process of heating coal in rounded heaps, the hearth process, remained the principal method of coke production for over a century, although an improved oven in the form of a beehive was developed in the Durham-Newcastle area of England in about 1759 (2,26,28). These processes lacked the capability to collect the volatile products, both liquids and gases. It was not until the mid-nineteenth century, with the introduction of indirectly heated slot ovens, that it became possible to collect the liquid and gaseous products for further use.

Coal carbonization processes are generally defined according to process operating temperature. Terms are defined in Table 2.

Low Temperature Carbonization. Low temperature carbonization, when the process does not exceed 700°C, was mainly developed as a process to supply town gas for lighting purposes as well as to provide a smokeless (devolatilized) solid fuel for domestic consumption (30). However,

Table 2. Coal Carbonization Methods[a]

Carbonization Process	Final Temperature, °C	Products	Processes
Low temperature	500–700	Reactive coke and high tar yield	Rexco (700°C) made in cylindrical vertical retorts; coalite (650°C) made in vertical tubes
Medium temperature	700–900	Reactive coke with high gas yield, or domestic briquettes	Town gas and gas coke (obsolete); phurnacite, low volatile steam coal, pitch-bound briquettes carbonized at 800°C
High temperature	900–1050	Hard, unreactive coal for metallurgical use	Foundry coke (900°C); blast-furnace coke (950–1050°)

[a] Ref. 29.

the process by-products (tars) were also found to be valuable insofar as they served as feedstocks for an emerging chemical industry and were also converted to gasolines, heating oils, and lubricants (31).

Coals preferred for the low temperature carbonization were usually lignites or subbituminous, as well as high volatile bituminous, coals (see LIGNITE AND BROWN COAL). These yield porous solid products over the temperature range 600–700°C. Certain of the higher rank (caking) coals were less suitable for the process, unless steps were taken to destroy the caking properties, because of the tendency of these higher rank coals to adhere to the walls of the carbonization chamber.

The options for efficient low temperature carbonization of coal include vertical and horizontal retorts which have been used for batch and continuous processes. In addition, stationary and revolving horizontal retorts have also been operated successfully, and there are also several process options employing fluidized or gas-entrained coal. Coke production from batch-type carbonization of coal has been supplanted by a variety of continuous retorting processes which allow much greater throughput rates than were previously possible. These processes employ rectangular or cylindrical vessels of sufficient height to carbonize the coal while it travels from the top of the vessel to the bottom and usually employ the principle of heating the coal by means of a countercurrent flow of hot combustion gas. Most notable of these types of carbonizers are the Lurgi-Spulgas retort and the Koppers continuous steaming oven (2).

High Temperature Carbonization. When heated at temperatures in excess of 700°C (1290°F), low temperature chars lose their reactivity through devolatilization and also suffer a decrease in porosity. High temperature carbonization, at temperatures >900°C, is, therefore, employed for the production of coke (27). As for the low temperature processes, the tars produced in high temperature ovens are also sources of chemicals and chemical intermediates (32).

A newer concept has been developed that is given the name mild gasification (33). It is not a gasification process in the true sense of the word. The process temperature is some several hundred degrees lower, hence the term mild, than the usual gasification process temperature and the objective is not to produce a gaseous fuel but to produce a high value char (carbon) and liquid products. Gas is produced, but to a lesser extent.

Documented efforts at cokemaking date from 1584 (34), and have seen various adaptations of conventional wood-charring methods to the production of coke including the eventual evolution of the beehive oven, which by the mid-nineteenth century had become the most common vessel for the coking of coal (2). The heat for the process was supplied by burning the volatile matter released from the coal and, consequently, the carbonization would progress from the top of the bed to the base of the bed and the coke was retrieved from the side of the oven at process completion.

Some beehive ovens, having various improvements and additions of waste heat boilers, thereby allowing heat recovery from the combustion products, may still be in operation. Generally, however, the beehive oven has been replaced by wall-heated, horizontal chamber, ie, slot, ovens in which higher temperatures can be achieved as well as a better control over the quality of the coke. Modern slot-type coke ovens are approximately 15 m long, approximately 6 m high, and the width is chosen to suit the carbonization behavior of the coal to be processed. For example, the most common widths are ca 0.5 m, but some ovens may be as narrow as 0.3 m, or as wide as 0.6 m.

Several (usually 20 or more, alternating with similar cells that contain heating ducts) of these chambers are constructed in the form of a battery over a common firing system through which the hot combustion gas is conveyed to the ducts. The flat roof of the battery acts as the surface for a mobile charging car from which the coal (25–40 t) enters each oven through three openings along the top. The coke product is pushed from the rear of the oven through the opened front section onto a quenching platform or into rail cars that then move the coke through water sprays. The gas and tar by-products of the process are collected for further processing or for on-site use as fuel.

Most modern coke ovens operate on a regenerative heating cycle in order to obtain as much surplus gas as possible for use on the works, or for sale. If coke-oven gas is used for heating the ovens, the majority of the gas is surplus to requirements. If producer gas is used for heating, much of the coke-oven gas is surplus.

The main difference between gas works and coke oven practice is that, in a gas works, maximum gas yield is a primary consideration whereas in the coke works the quality of the coke is the first consideration. These effects are obtained by choice of a coal feedstock that is suitable to the task. For example, use of lower volatile coals in coke ovens, compared to coals used in gas works, produces lower yields of gas when operating at the same temperatures. In addition, the choice of heating (carbonizing) conditions and the type of retort also play a principal role (10,35).

Gasification

The gasification of coal is essentially the conversion of coal by any one of a variety of processes to produce combustible gases (7,8,11,22,36–38). Primary gasification is the thermal decomposition of coal to produce mixtures containing various proportions of carbon monoxide, carbon dioxide, hydrogen, water, methane, hydrogen sulfide, and nitrogen, as well as products such as tar, oils, and phenols. A solid char product may also be produced, and often represents the bulk of the weight of the original coal.

Secondary gasification involves gasification of the char from the primary gasifier, usually by reaction of the hot char and water vapor to produce carbon monoxide and hydrogen.

$$C_{char} + H_2O \rightarrow CO + H_2$$

The gaseous product from a gasifier generally contains large amounts of carbon monoxide and hydrogen, plus lesser amounts of other gases and may be of low, medium, or high heat value depending on the defined use (Table 3) (39,40).

The importance of coal gasification as a means of producing fuel gas(es) for industrial use cannot be underplayed. But coal gasification systems also have undesirable features. A range of undesirable products are also produced which must be removed before the products are used to provide fuel and/or to generate electric power (22,41).

Chemistry. Coal gasification involves the thermal decomposition of coal and the reaction of the carbon in the coal, and other pyrolysis products with oxygen, water, and hydrogen to produce fuel gases such as methane by internal hydrogen shifts

$$C_{coal} + H_{coal} \rightarrow CH_4$$

or through the agency of added (external) hydrogen

$$C_{coal} + 2\,H_2 \rightarrow CH_4$$

although the reactions are more numerous and more complex as can be seen in Table 4.

If air is used as a combustant, the product gas contains nitrogen and, depending on design characteristics, has a heating value of approximately 5.6–11.2 $\times$ 10^3 MJ/m^3 (150–300 Btu/ft^3). The use of pure oxygen, although expensive, results in a product gas having a heating value of 11–15 MJ/m^3 (300–400 Btu/ft^3) and carbon dioxide and hydrogen sulfide as by-products.

If a high heat-value gas (33.5–37.3 MJ/m^3 (900–1000 Btu/ft^3)) ie, SNG, is the desired product, efforts must be made to increase the methane content.

Shift conversion reaction

$$CO + H_2O \rightarrow CO_2 + H_2$$
$$CO + 3\,H_2 \rightarrow CH_4 + H_2O$$
$$2\,CO + 2\,H_2 \rightarrow CH_4 + CO_2$$
$$CO + 4\,H_2 \rightarrow CH_4 + 2\,H_2O$$

The gasification is performed using oxygen and steam (qv), usually at elevated pressures. The steam–oxygen ratio along with reaction temperature and pressure determine the equilibrium gas composition. The reaction rates for these reactions are relatively slow and heats of formation are negative. Catalysts may be necessary for complete reaction (2,3,24,42,43).

Process Parameters. The most notable effects in gasifiers are those of pressure (Fig. 1) and coal character. Some ini-

Table 3. Gas Composition Requirements for Substitute Natural Gas and for Power Generation

Characteristic	Product Requirement	
	Synthetic Natural Gas	Power Generation
Methane content	High, less synthesis required	Low, probably means no tars
H_2/CO ratio	High, less shifting required	Low, CO more efficient fuel
Moisture content	High, steam required for shift	Low, lower condensate treatment costs
Outlet temperature	Low, maximizes methane minimizes sensible heat loss leads to high cold gas efficiency	High, precludes tar formation provides for steam generation reduces cold gas efficiency
Gasifier oxidant	O_2 only, cost of N_2 removal excessive	Air or O_2, low heating gas value acceptable fuel

Table 4. Gasification Reactions

Reaction	ΔH, kJ[a]
Gasification zone, 595–1205°C	
$C + CO_2 \rightarrow 2\,CO$	9.65[b]
$CO + H_2O \rightarrow CO_2 + H_2$	−1.93[c]
$C + H_2O \rightarrow CO + H_2$	7.76[d]
$C + 2\,H_2 \rightarrow CH_4$	−5.21
$C + 2\,H_2O \rightarrow 2\,H_2 + CO_2$	5.95
Combustion zone, >1205°C	
$2\,C + O_2 \rightarrow 2\,CO$	−22.6
$C + O_2 \rightarrow CO_2$	−6.74
Methanation reactions	
$CO + 3\,H_2 \rightarrow CH_4 + H_2O$	−12.9
$CO_2 + 4\,H_2 \rightarrow CH_4 + 2\,H_2O$	−11.0
$2\,C + 2\,H_2O \rightarrow CH_4 + CO_2$	0.62
Other reactions	
$2\,H_2 + O_2 \rightarrow 2\,H_2O$	
$CO + 2\,H_2 \rightarrow CH_3OH$	
$2\,C + H_2 \rightarrow C_2H_2$	
$CH_4 + 2\,H_2O \rightarrow CO_2 + 4\,H_2$	

[a] To convert kJ to kcal, divide by 4.184.
[b] Boudouard reaction.
[c] Water gas shift reaction.
[d] Water gas reaction.

tial processing of the coal feedstock may be required. The type and degree of pretreatment is a function of the process and/or the type of coal.

Depending on the type of coal being processed and the analysis of the gas product desired, some or all of the following processing steps may be required: (*1*) pretreatment of the coal (if caking is a problem); (*2*) primary gasification of the coal; (*3*) secondary gasification of the carbonaceous residue from the primary gasifier; (*4*) removal of carbon dioxide, hydrogen sulfide, and other acid gases; (*5*) shift conversion for adjustment of the carbon monoxide–hydrogen mole ratio to the desired ratio; and (*6*) catalytic methanation of the carbon monoxide–hydrogen mixture to form methane. If high heat-value gas is desired, all of these processing steps are required because coal gasifiers do not yield methane in the concentrations required.

An example of application of a pretreatment option occurs when the coal displays caking or agglomerating characteristics. Such coals are usually not amenable to gasification processes employing fluidized-bed or moving-bed reactors. The pretreatment involves a mild oxidation treatment, usually consisting of low temperature heating of the coal in the presence of air or oxygen. This destroys the caking characteristics of coals.

Gasification technologies for the production of high heat-value gas do not all depend entirely on catalytic methanation, that is, the direct addition of hydrogen to coal under pressure to form methane.

The hydrogen-rich gas for hydrogasification can be manufactured from steam by using the char that leaves the hydrogasifier. Appreciable quantities of methane are formed directly in the primary gasifier and the heat released by methane formation is at a sufficiently high temperature to be used in the steam–carbon reaction to produce hydrogen so that less oxygen is used to produce heat for the steam–carbon reaction. Hence, less heat is lost in the low temperature methanation step, thereby leading to a higher overall process efficiency.

There are three fundamental reactor types for gasification processes: (*1*) a gasifier reactor, (*2*) a devolatilizer, and (*3*) a hydrogasifier (Fig. 2). The choice of a particular design is available for each type, eg, whether or not two stages should be involved depending on the ultimate product gas desired. Reactors may also be designed to operate over a range of pressure from atmospheric to high pressure.

Gasification processes have been classified on the basis of the heat-value of the produced gas. It is also possible to classify gasification processes according to the type of reactor and whether or not the system reacts under pressure. Additionally, gasification processes can be segregated according to the bed types, which differ in the ability to accept and use caking coals. Thus gasification processes can be divided into four categories based on reactor (bed) configuration: (*1*) fixed bed, (*2*) moving bed, (*3*) fluidized bed, and (*4*) entrained bed.

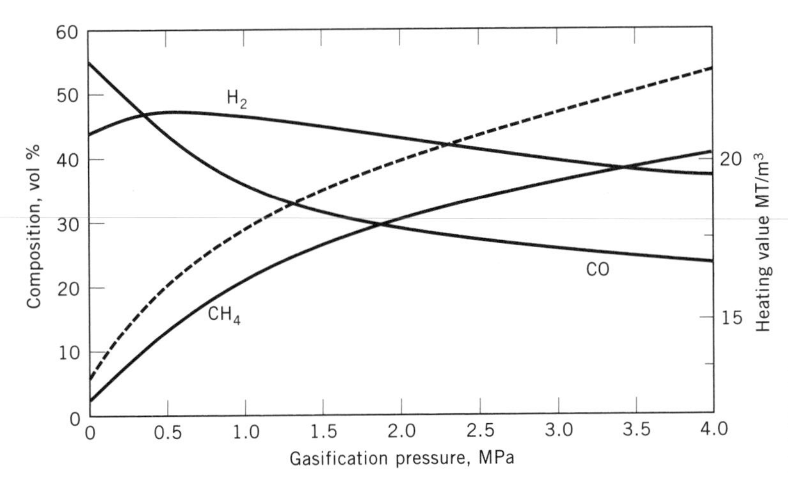

Figure 1. Variation of (—) gas composition and (---) heating value with gasifier pressure. To convert MJ/m³ to Btu/ft³, multiply by 26.86. To convert MPa to psi, multiply by 145.

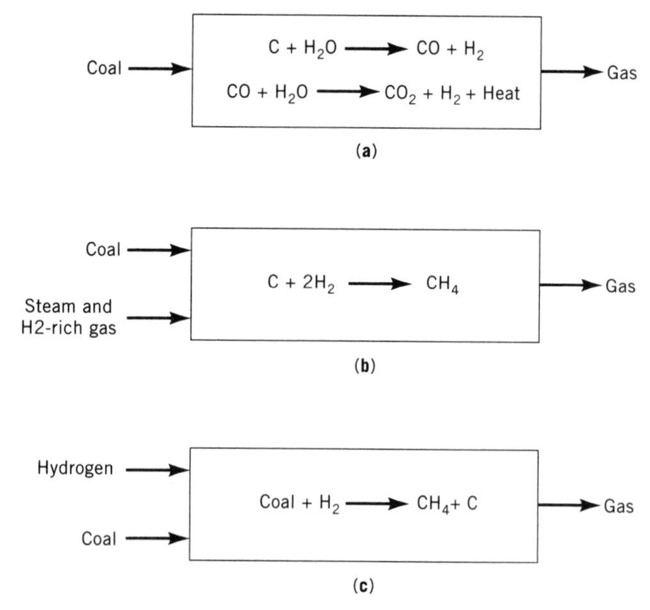

Figure 2. Chemistry of (**a**) gasifier; (**b**) hydrogasifier; and (**c**) devolatization processes. The gaseous product of (**a**) is of low heating value; that of (**b**) and (**c**) is of intermediate heating value.

In a fixed-bed process the coal is supported by a grate and combustion gases, ie, steam, air, oxygen, etc, pass through the supported coal whereupon the hot produced gases exit from the top of the reactor. Heat is supplied internally or from an outside source, but caking coals cannot be used in an unmodified fixed-bed reactor. In the moving-bed system (Fig. 3), coal is fed to the top of the bed and ash leaves the bottom with the product gases being produced in the hot zone just prior to being released from the bed.

The fluidized-bed system (Fig. 3) uses finely sized coal particles and the bed exhibits liquid-like characteristics when a gas flows upward through the bed. Gas flowing through the coal produces turbulent lifting and separation of particles and the result is an expanded bed having greater coal surface area to promote the chemical reaction. These systems, however, have only a limited ability to handle caking coals.

An entrainment system (Fig. 3) uses finely sized coal particles blown into the gas steam prior to entry into the reactor and combustion occurs with the coal particles suspended in the gas phase; the entrained system is suitable for both caking and noncaking coals.

Gasifier Options. The standard Wellman-Galusha unit, used for noncaking coals, is an atmospheric pressure, air-blown gasifier (Fig. 4) fed by a two-compartment lockhopper system. The upper coal-storage compartment feeds coal intermittently into the feeding compartment, which then feeds coal into the gasifier section almost continuously except for brief periods when the feeding compartment is being loaded from storage. In small units up to 1.5 m internal diameter ash is removed by shaker grates. In larger units, ash is removed continuously from the bottom of the gasifier into an ash-hopper section by a revolving grate.

The grate is constructed of flat, circular steel plates set one above the other with edges overlapping. The grate, revolving eccentrically within the gasifier, causes ash to fall from the coal bed as the space between the grate and the shell increases, and then pushes the ash down into the ash-hopper as the space decreases. The smaller size units are brick-lined, although the larger sizes are unlined and water-jacketed. Combustion air, provided by a fan, passes over the warm (82°C) jacket water causing the water to vaporize to provide the necessary steam for gasification. Gas leaves from above the fuel bed at 428–538°C for bituminous coal.

On the other hand, the agitator type of Wellman-Galusha unit, used for gasification of any type of coal, uses a slowly revolving horizontal arm which also spirals vertically below the surface of the fuel bed to retard channeling and maintain a uniform bed. Use of the agitator not only

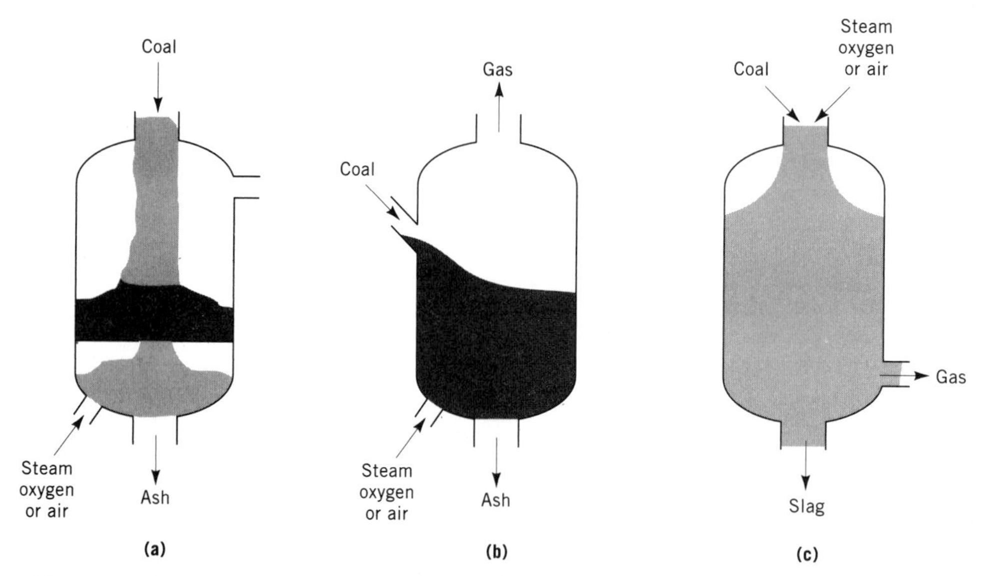

Figure 3. Gasifier systems: (**a**), moving bed (dry ash); (**b**), fluidized bed; and (**c**), entrained flow.

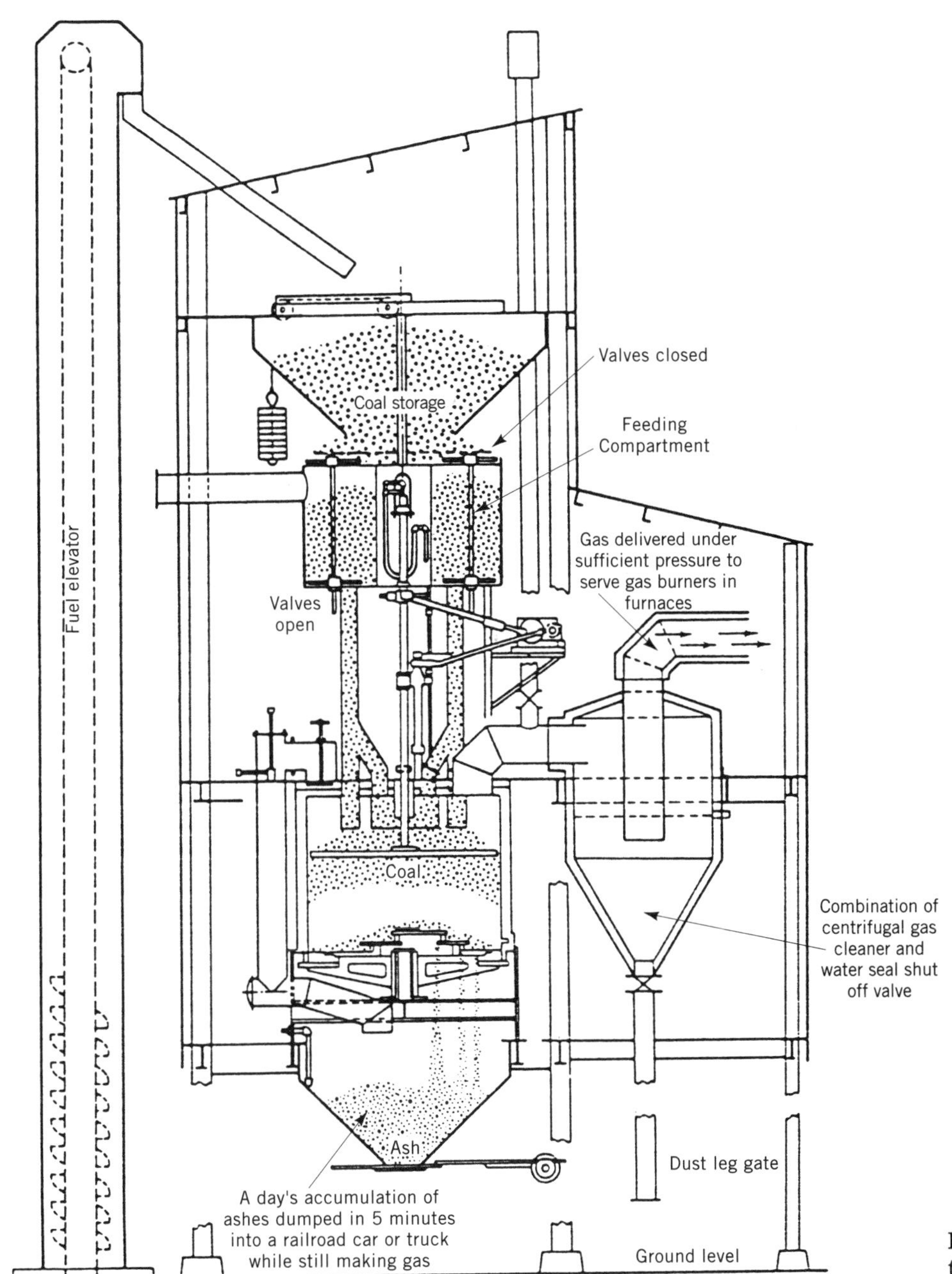

Figure 4. A Wellman-Galusha agitator-type gas producer.

allows operation with caking coals but also can increase the capacity of the gasifier by about 25% for use with other coals.

The Winkler gasifier (Fig. 5) is an example of a medium heat-value gas producer which, when oxygen is employed, yields a gas product composed mainly of carbon monoxide and hydrogen (43).

In the process, finely crushed coal is gasified at atmospheric pressure in a fluidized state; oxygen and steam are introduced at the base of the gasifier. The coal is fed by lockhoppers and a screw feeder into the bottom of the fuel bed. Sintered ash particles settle on a grate, where they are cooled by the incoming oxygen and steam; a rotating, cooled rabble moves the ash toward a discharge port. The ash is then conveyed pneumatically to a disposal hopper.

The gas, along with entrained ash and char particles, which are subjected to further gasification in the large space above the fluid bed, exit the gasifier at 954–1010°C. The hot gas is passed through a waste-heat boiler to recover the sensible heat, and then through a dry cyclone. Solid particles are removed in both units. The gas is further cooled and cleaned by wet scrubbing, and if required, an electrostatic precipitator is included in the gas-treatment stream.

The Koppers-Totzek process is a second example of a process for the production of a medium heat-value gas (44,45). Whereas the Winkler process employs a fluidized bed, the Koppers-Totzek process uses an entrained flow system. In the Koppers-Totzek process, dried and pulverized coal is conveyed continuously by a screw into a mixing

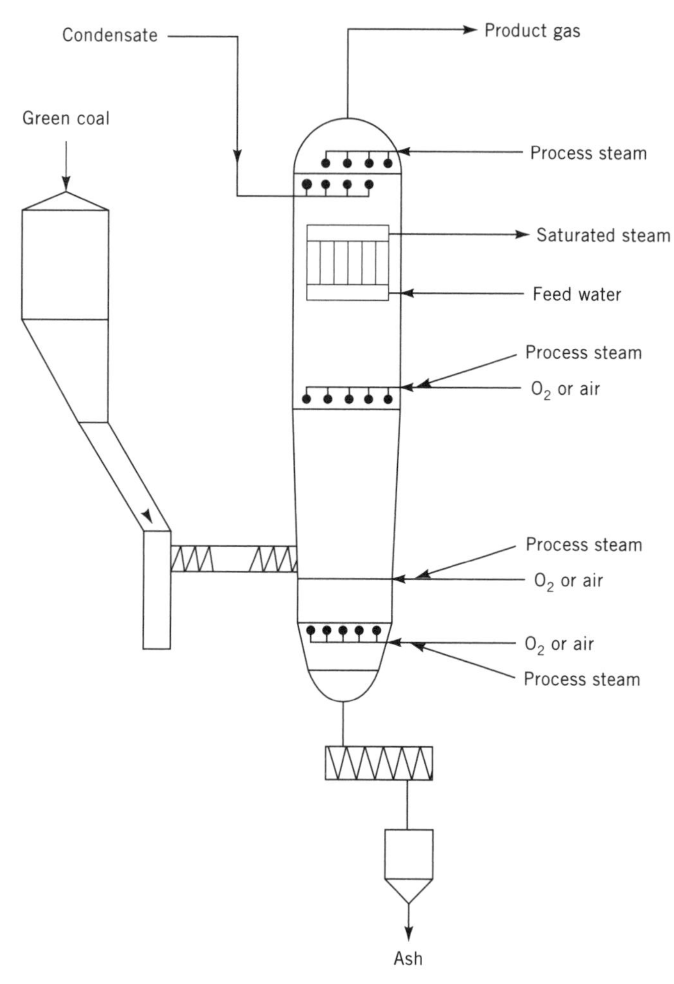

Figure 5. The Winkler gas producer.

nozzle. From there a high velocity stream of steam and oxygen entrains the coal into a gasifier. The gasifier (Fig. 6) is a cylindrical vesel with a refractory-lining that is designed to conduct a selected amount of heat to a surrounding water jacket in which low pressure process steam is generated. The lining is thin (about 5 cm) and made of a high alumina cast material. In a two-headed gasifier two burner heads are placed 180° apart at either end of the vessel. Four burner heads, 90° apart, are used in a four-headed gasifier. The largest gasifiers are 3–4 m diameter at the middle, tapering to 2–3 m at the burner ends and are about 19 m long. The reactor volume is about 30 m^3 for the two-headed design, and 64 m^3 for the larger, four-headed models.

The process is carried at moderate (slightly above atmospheric) pressures, but at very high temperatures that reach a maximum of 1900°C. Even though the reaction time is short (0.6–0.8 s) the high temperature prevents the occurrence of any condensable hydrocarbons, phenols, and/or tar in the product gas. The absence of liquid simplifies the subsequent gas clean-up steps.

Normally ca 50% of the coal ash is removed from the bottom of the gasifier as a quenched slag. The balance is carried overhead in the gas as droplets which are solidified when the gas is cooled with a water spray. A fluxing agent is added, if required, to the coal to lower the ash fusion temperature and increase the molten slag viscosity.

Conversion of carbon in the coal to gas is very high. With low rank coal, such as lignite and subbituminous coal, conversion may border on 100%, and for highly volatile A coals, it is on the order of 90–95%. Unconverted carbon appears mainly in the overhead material. Sulfur removal is facilitated in the process because typically 90% of it appears in the gas as hydrogen sulfide, H_2S, and 10% as carbonyl sulfide, COS; carbon disulfide, CS_2, and/or methyl thiol, CH_3SH, are not usually formed.

The production of synthetic natural gas can be achieved by use of the Lurgi gasifier (Fig. 7), which is similar in principle to the Wellman-Galusha unit and is designed for operation at pressures up to 3.1 MPa (450 psi) (46). Three distinct reaction zones are identifiable in a pressurized (1.9–2.9 MPa (280–425 psi)) Lurgi reactor: (*1*) the drying/devolatilization and pyrolysis zone, 370–595°C, nearest the coal feed end, commences the process by converting the coal to a reactive char; (*2*) the gasification zone (595–1205°C); and (*3*) the combustion zone (>1205°C), nearest the discharge end, which provides the heat requirements for the endothermic steam–carbon reaction. Equations and reaction enthalpies for the last two zones are given in Table 4.

The operating conditions in the gasifier (temperature and pressure) and the reaction kinetics (residence time, concentration of the constituents, and rate constants) determine the extent of conversion or approach to equilibrium.

The coal is fed through a lockhopper mounted on the top of the gasifier where a rotating distributor provides uniform coal feed across the bed. When processing caking coals, blades attached to the distributor rotate within the bed to break up agglomerates. A revolving grate supports the bed at the bottom and serves as a distributor for steam and oxygen. Solid residue is removed at the bottom of the gasifier through an ash lockhopper. The entire gasifier ves-

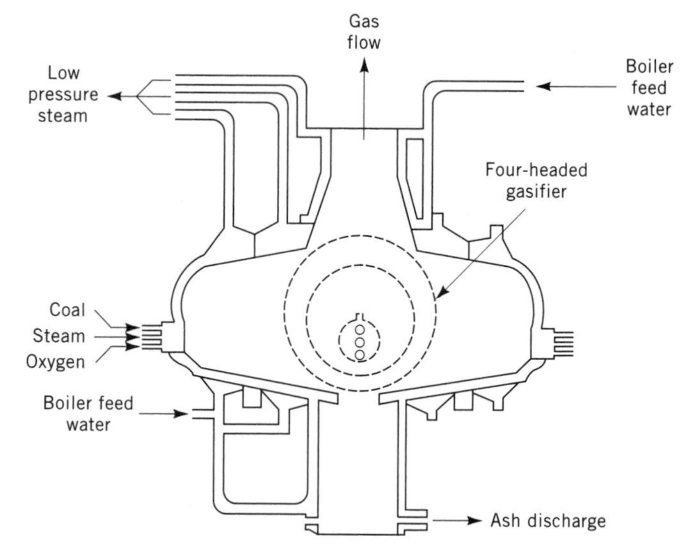

Figure 6. The Koppers-Totzek gas producer.

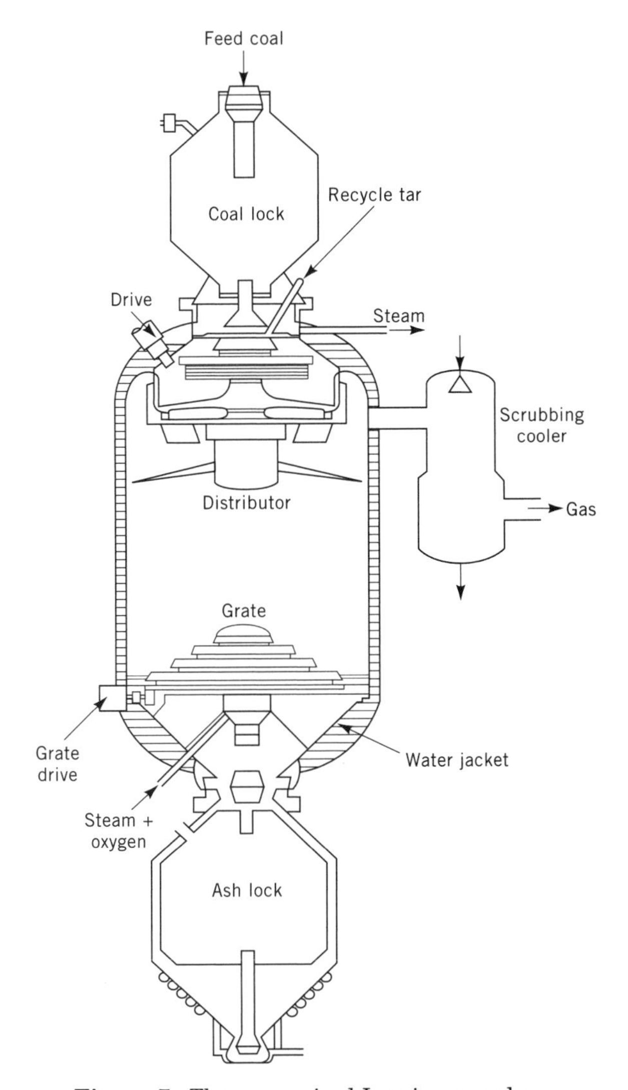

Figure 7. The pressurized Lurgi gas producer.

sel is surrounded by a water jacket in which high pressure steam is generated.

Crude gas leaves from the top of the gasifier at 288–593°C depending on the type of coal used. The composition of gas also depends on the type of coal and is notable for the relatively high methane content when contrasted to gases produced at lower pressures or higher temperatures. These gas products can be used as produced for electric power production or can be treated to remove carbon dioxide and hydrocarbons to provide synthesis gas for ammonia, methanol, and synthetic oil production. The gas is made suitable for methanation, to produce synthetic natural gas, by a partial shift and carbon dioxide and sulfur removal.

As in most of the high heat-value processes, the raw gas is in the medium heat-value gas range and can be employed directly in that form. Removing the carbon dioxide raises the heating value, but not enough to render the product worthwhile as a high heat-value gas without methanation.

Methanation of the clean desulfurized main gas (less than 1 ppm total sulfur) is accomplished in the presence of a nickel catalyst at temperatures from 260–400°C and pressure range of 2–2.8 MPa (300–400 psi). Equations and reaction enthalpies are given in Table 4.

Hydrogenation of the oxides of carbon to methane according to the above reactions is sometimes referred to as the Sabatier reactions. Because of the high exothermicity of the methanization reactions, adequate and precise cooling is necessary in order to avoid catalyst deactivation, sintering, and carbon deposition by thermal cracking.

Catalytic methanation processes include (*1*) fixed or fluidized catalyst-bed reactors where temperature rise is controlled by heat exchange or by direct cooling using product gas recycle; (*2*) through wall-cooled reactor where temperature is controlled by heat removal through the walls of catalyst-filled tubes; (*3*) tube-wall reactors where a nickel–aluminum alloy is flame-sprayed and treated to form a Raney-nickel catalyst bonded to the reactor tube heat-exchange surface; and (*4*) slurry or liquid-phase (oil) methanation.

To enable interchangeability of the SNG with natural gas, on a calorific, flame, and toxicity basis, the synthetically produced gas consists of a minimum of 89 vol % methane, a maximum of 0.1% carbon monoxide, and up to 10% hydrogen. The specified minimum acceptable gross heating value is approximately 34.6 MJ/m^3 (930 Btu/ft^3).

In a combined power cycle operation, clean (sulfur- and particulate-free) gas is burned with air in the combustor at elevated pressure. The gas is either low or medium heat-value, depending on the method of gasification.

The hot gases from the combustor, temperature controlled to 980°C by excess air, are expanded through the gas turbine, driving the air compressor and generating electricity. Sensible heat in the gas turbine exhaust is recovered in a waste heat boiler by generating steam for additional electrical power production.

The use of hot gas clean-up methods to remove the sulfur and particulates from the gasified fuel increases turbine performance by a few percentage points over the cold clean-up systems. Hot gas clean-up permits use of the sensible heat and enables retention of the carbon dioxide and water vapor in the gasified fuel, thus enhancing turbine performance. Further, additional power may be generated, prior to combustion with air, in an expansion turbine as the hot product fuel gas is expanded to optimum pressure level for the combined cycle. Future advances in gas turbine technology (turbine inlet temperature above 1650°C) are expected to boost the overall combined cycle efficiency substantially.

More recently, advanced generation gasifiers have been under development, and commercialization of some of the systems has become a reality (36,41). In these newer developments, the emphasis has shifted to a greater throughput, relevant to the older gasifiers, and also to high carbon conversion levels and, thus, higher efficiency units.

For example, the Texaco entrained system features coal–water slurry feeding a pressurized oxygen-blown gasifier with a quench zone for slag cooling (Fig. 8). In fact, the coal is partially oxidized to provide the heat for the gasification reactions. The Dow gasifier also utilizes a coal–water slurry fed system whereas the Shell gasifier

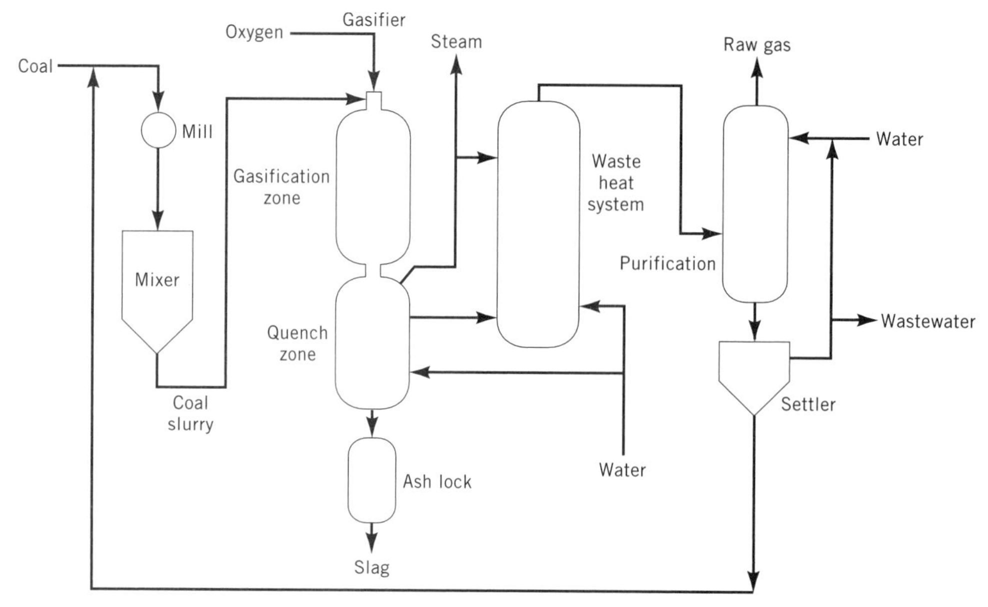

Figure 8. The Texaco gasification process.

features a dry-feed entrained gasification system which operates at elevated temperature and pressure. The Kellogg Rust Westinghouse system and the Institute of Gas Technology U-Gas system are representative of ash agglomerating fluidized-bed systems.

In response to the disadvantage that the dry ash Lurgi gasifier requires that temperatures have to be below the ash melting point to prevent clinkering, improvements have been sought in the unit; as a result the British Gas-Lurgi GmbH gasifier came into being. This unit is basically similar to the dry ash Lurgi unit insofar as the top of the unit is identical but the bottom has been modified to include a slag quench vessel (Fig. 9). Thus the ash melts at the high temperatures in the combustion zone (up to 2000°C) and forms a slag that runs into the quench chamber, which is in reality a water bath where the slag forms granules of solid ash. Temperatures and reaction rates are high in the gasification zone so that coal residence time is markedly educed over that of the dry ash unit.

In summary, these second-generation gasifiers offer promise for the future in terms of increased efficiency as well as for use of other feedstocks, such as biomass. The older, first-generation gasifiers, however, continue to be used.

Combustion

Coal combustion, not being in the strictest sense a process for the generation of gaseous synfuels, is nevertheless an important use of coal as a source of gaseous fuels. Coal combustion, an old art and probably the oldest known use of this fossil fuel, is an accumulation of complex chemical and physical phenomena. The complexity of coal itself and the variable process parameters all contribute to the overall process (8,10,47–50).

There are two principal methods of coal combustion: fixed-bed combustion and combustion in suspension. The first fixed beds, eg, open fires, fireplaces, and domestic stoves, were simple in principle. Suspension burning of coal began in the early 1900s with the development of pulverized coal-fired systems, and by the 1920s these systems were in widespread use. Spreader stokers, which were developed in the 1930s, combined both principles by providing for the smaller particles of coal to be burned in suspension and larger particles to be burned on a grate (10).

A significant issue in combustors in the mid-1990s is the performance of the process in an environmentally acceptable manner through the use of either low sulfur coal or post-combustion clean-up of the flue gases. Thus there is a marked trend to more efficient methods of coal combustion and, in fact, a combustion system that is able to accept coal without the necessity of a post-combustion treatment or without emitting objectionable amounts of sulfur oxides, nitrogen oxides, and particulates is very desirable (51,52).

The parameters of rank and moisture content are regarded as determining factors in combustibility as it relates to both heating value and ease of reaction as well as to the generation of pollutants (48). Thus, whereas the lower rank coals may appear to be more reactive than higher rank coals, though exhibiting a lower heat-value and thereby implying that rank does not affect combustibility, environmental constraints arise through the occurrence of heteroatoms, ie, noncarbon atoms such as nitrogen and sulfur, in the coal. At the same time, anthracites, which have a low volatile matter content, are generally more difficult to burn than bituminous coals.

Chemistry. In direct combustion coal is burned to convert the chemical energy of the coal into thermal energy, ie, the carbon and hydrogen in the coal are oxidized into carbon dioxide and water.

$$2\ H_{coal} + O_2 \rightarrow H_2O$$

After burning, the sensible heat in the products of combustion can then be converted into steam that can be used for external work or can be converted directly into energy to drive a shaft, eg, in a gas turbine. In fact, the combustion process actually represents a means of achieving the complete oxidation of coal.

The combustion of coal may be simply represented as the staged oxidation of coal carbon to carbon dioxide.

$$2\,C_{coal} + O_2 \rightarrow 2\,CO$$
$$2\,CO + O_2 \rightarrow 2\,CO_2$$

with any reactions of the hydrogen in the coal being considered to be of secondary importance. Other types of combustion systems may be rate-controlled as a result of the onset of the Boudouard reaction (see Table 4).

The complex nature of coal as a molecular entity (2,3,24,25,35,37,53) has resulted in the chemical explanations of coal combustion being confined to the carbon in the system. The hydrogen and other elements have received much less attention but the system is extremely complex and the heteroatoms, eg, nitrogen, oxygen, and sulfur, exert an influence on the combustion. It is this latter that influences environmental aspects.

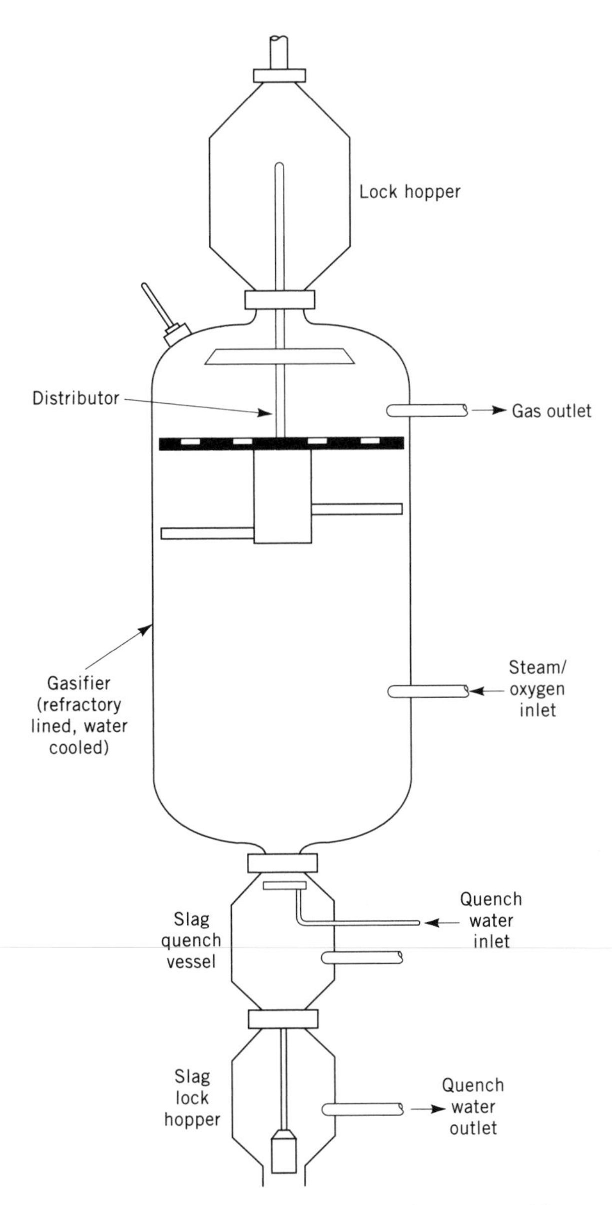

Figure 9. The British Gas-Lurgi slagging gasifier.

For example, the conversion of nitrogen and sulfur, during coal combustion, to the respective oxides during combustion cannot be ignored:

$$S_{coal} + O_2 \rightarrow SO_2$$
$$2\,SO_2 + O_2 \rightarrow 2\,SO_3$$
$$2\,N_{coal} + O_2 \rightarrow 2\,NO$$
$$2\,NO + O_2 \rightarrow 2\,NO_2$$
$$N_{coal} + O_2 \rightarrow NO_2$$

The sulfur and nitrogen oxides that escape into the atmosphere can be converted to acids by reaction with moisture in the atmosphere (see also AIR POLLUTION; AIR QUALITY MODELING.)

Sulfurous acid	$SO_2 + H_2O \rightarrow H_2SO_3$
Sulfuric acid	$2\,SO_2 + O_2 \rightarrow 2\,SO_3$
	$SO_3 + H_2O \rightarrow H_2SO_4$
Nitrous acid	$NO + H_2O \rightarrow H_2NO_2$
Nitric acid	$2\,NO + O_2 \rightarrow 2\,NO_2$
	$NO_2 + H_2O \rightarrow HNO_3$

Combustion Systems. Combustion systems vary in nature depending on the nature of the feedstock and the air needed for the combustion process (54). However, the two principal types of coal-burning systems are usually referred to as layer and chambered. The former refers to fixed beds; the latter is more specifically for pulverized fuel.

Fixed or Slowly Moving Beds. For fuel-bed burning on a grate, a distillation effect occurs. The result is that liquid components which are formed volatilize before combustion temperatures are reached; cracking may also occur. The ignition of coal in a bed is almost entirely by radiation from hot refractory arches and from the flame burning of volatiles. In fixed beds, the radiant heat above the bed can only penetrate a short distance into the bed.

Consequently, convective heat transfer determines the intensity of warming up and ignition. In addition, convective heat transfer also plays an important part in the overall flame-to-surface transmission. The reaction of gases is greatly accelerated by contact with hot surfaces and, whereas the reaction away from the walls may proceed slowly, reaction at the surface proceeds much more rapidly.

Fluidized Beds. Fluidized-bed combustion occurs in expanded beds (Fig. 3) and at relatively lower (925°C) temperatures; high convective transfer rates exist resulting from the bed motion. Fluidized systems can operate under substantial pressures thereby allowing more efficient gas clean-up. Fluidized-bed combustion is a means for providing high heat-transfer rates, controlling sulfur, and reducing nitrogen oxide emissions from the low temperatures in the combustion zone.

There are, however, problems associated with pollution control. Whereas the sulfur may be removed downstream using suitable ancillary controls, the sulfur may also be captured in the bed, thereby adding to the separations and recycle problems. Capture during combustion, however, is recognized as the ideal and is a source of optimism for fluidized combustion.

A fluidized bed is an excellent medium for contacting gases with solids, and this can be exploited in a combustor because sulfur dioxide emissions can be reduced by adding limestone, $CaCO_3$, or dolomite, $CaCO_3{\cdot}MgCO_3$, to the bed.

$$2\,SO_2 + O_2 \rightarrow 2\,SO_3$$

$$SO_3 + CaCO_3 \rightarrow CaSO_4 + CO_2$$

or

$$2\,SO_2 + O_2 + 2\,CaCO_3 \rightarrow 2\,CaSO_4 + 2\,CO_2$$

The spent sorbent from fluidized-bed combustion may be taken directly to disposal and is much easier than the disposal of salts produced by wet limestone scrubbing. Alternatively, the spent sorbent may be regenerated using synthesis gas, CO/H_2.

$$CaSO_4 + H_2 \rightarrow CaO + H_2O + SO_2$$

$$CaSO_4 + CO \rightarrow CaO + CO_2 + SO_2$$

The calcium oxide product is supplemented with fresh limestone and returned to the fluidized bed. Two undesirable side reactions can occur in the regeneration of spent lime leading to the production of calcium sulfide:

$$CaSO_4 + 4\,H_2 \rightarrow CaS + 4\,H_2O$$

$$CaSO_4 + 4\,CO \rightarrow CaS + 4\,CO_2$$

which results in the recirculation of sulfur to the bed.

Entrained Systems. In entrained systems, fine grinding and increased retention times intensify combustion but the temperature of the carrier and degree of dispersion are also important. In practice, the coal is introduced at high velocities which may be greater than 30 m/s and involve expansion from a jet to the combustion chamber.

Types of entrained systems include cyclone furnaces, which have been used for various coals. Other systems have been developed and utilized for the injection of coal–oil slurries into blast furnaces or for the burning of coal–water slurries. The cyclone furnace, developed in the 1940s to burn coal having low ash-fusion temperatures, is a horizontally inclined, water-cooled, tubular furnace in which crushed coal is burned with air entering the furnace tangentially. Temperatures may be on the order of 1700°C and the ash in the coal is converted to a molten slag that is removed from the base of the unit. Coal fines burn in suspension; the larger pieces are captured by the molten slag and burn rapidly.

GAS FROM OTHER FOSSIL FUELS

As of this writing natural gas is a plentiful resource, and there has been a marked tendency not to use other fossil fuels as SNG sources. However, petroleum and oil shale (qv) have been the subject of extensive research efforts. These represent other sources of gaseous fuels and are worthy of mention here.

Petroleum

Thermal cracking (pyrolysis) of petroleum or fractions thereof was an important method for producing gas in the years following its use for increasing the heat content of water gas. Many water gas sets operations were converted into oil-gasification units (55). Some of these have been used for baseload city gas supply, but most find use for peak-load situations in the winter.

In the 1940s, the hydrogasification of oil was investigated as a follow-up to the work on the hydrogasification of coal (56,57). In the ensuing years, further work was carried out as a supplement to the work on thermal cracking (58,59), and during the early 1960s it became evident that light distillates having end boiling points $<182°C$ and containing no sulfur could be catalytically reformed by an autothermic process to pure methane (60). This method was extensively used in England until natural (North Sea) gas came into use.

Prior to the discovery of plentiful supplies of natural gas, and depending on the definition of the resources (1), there were plans to accommodate any shortfalls in gas supply from solid fossil fuels and from gaseous resources by the conversion of hydrocarbon (petroleum) liquids to lower molecular weight gaseous products.

$$CH_{petroleum} \rightarrow CH_4$$

Thermal Cracking. In addition to the gases obtained by distillation of crude petroleum, further highly volatile products result from the subsequent processing of naphtha and middle distillate to produce gasoline, as well as from hydrodesulfurization processes involving treatment of naphthas, distillates, and residual fuels (5,61) and from the coking or similar thermal treatment of vacuum gas oils and residual fuel oils (5).

The chemistry of the oil-to-gas conversion has been established for several decades and can be described in general terms although the primary and secondary reactions can be truly complex (5). The composition of the gases produced from a wide variety of feedstocks depends not only on the severity of cracking but often to an equal or lesser extent on the feedstock type (5,62,63). In general terms, gas heating values are on the order of 30–50 MJ/m^3 (950–1350 Btu/ft^3).

Catalytic Processes. A second group of refining operations which contribute to gas production are the catalytic cracking processes, such as fluid-bed catalytic cracking, and other variants, in which heavy gas oils are converted into gas, naphthas, fuel oil, and coke (5).

The catalysts promote steam reforming reactions that lead to a product gas containing more hydrogen and carbon monoxide and fewer unsaturated hydrocarbon products than the gas product from a noncatalytic process (5). Cracking severities are higher than those from thermal cracking, and the resulting gas is more suitable for use as a medium heat-value gas than the rich gas produced by straight thermal cracking. The catalyst also influences the reactions rates in the thermal cracking reactions, which can lead to higher gas yields and lower tar and carbon yields (5).

The basic chemical premise involved in making synthetic natural gas from heavier feedstocks is the addition of hydrogen to the oil:

$$CH_3(CH_2)_nCH_3 + (n + 1)H_2 \rightarrow (n + 2)CH_4$$

In general terms, as the molecular weight of the feedstock is increased, similar operating conditions of hydrogasification lead to decreasing hydrocarbon gas yields, increasing yields of aromatic liquids, with carbon also appearing as a product.

The principal secondary variable that influences yields of gaseous products from petroleum feedstocks of various types is the aromatic content of the feedstock. For example, a feedstock of a given H/C (C/H) ratio that contains a large proportion of aromatic species is more likely to produce a larger proportion of liquid products and elemental carbon than a feedstock that is predominantly paraffinic (5).

Another option for processing crude oils which are too heavy to be hydrogasified directly involves first hydrocracking the crude oil to yield a low boiling product suitable for gas production and a high boiling product suitable for hydrogen production by partial oxidation (57). Alternatively, it may be acceptable for carbon deposition to occur during hydrogasification which can then be used for heat or for hydrogen production (64,65).

Partial Oxidation. It is often desirable to augment the supply of naturally occurring or by-product gaseous fuels or to produce gaseous fuels of well-defined composition and combustion characteristics (5). This is particularly true in areas where the refinery fuel (natural gas) is in poor supply and/or where the manufacture of fuel gases, originally from coal and more recently from petroleum, has become well established.

Almost all petroleum fractions can be converted into gaseous fuels, although conversion processes for the heavier fractions require more elaborate technology to achieve the necessary purity and uniformity of the manufactured gas stream (5). In addition, the thermal yield from the gasification of heavier feedstocks is invariably lower than that of gasifying light naphtha or liquefied petroleum gas(es) because, in addition to the production of hydrogen, carbon monoxide, and gaseous hydrocarbons, heavy feedstocks also yield some tar and coke.

As in the case of coal, synthetic natural gas can be produced from heavy oil by partially oxidizing the oil to a mixture of carbon monoxide and hydrogen

$$2\ CH_{petroleum} + O_2 \rightarrow 2\ CO + H_2$$

which is methanated catalytically to produce methane of any required purity. The initial partial oxidation step consists of the reaction of the feedstock with a quantity of oxygen insufficient to burn it completely, making a mixture consisting of carbon monoxide, carbon dioxide, hydrogen, and steam. Success in partially oxidizing heavy feedstocks depends mainly on details of the burner design (66). The ratio of hydrogen to carbon monoxide in the product gas is a function of reaction temperature and stoichiometry and can be adjusted, if desired, by varying the ratio of carrier steam to oil fed to the unit.

To make synthetic natural gas by partial oxidation, virtually all of the methane in the product gas must be produced by catalytic methanation of carbon monoxide and hydrogen. The feed is mixed with recycled carbon and fed, together with steam and oxygen, to a reactor in which partial combustion takes place. Heat from the reaction gasifies the rest of the feed, and by-product coke is formed. The heavier feedstocks tend to produce more carbon than can be consumed through recycling; thus some must be withdrawn. After carbon separation by water scrubbing, the synthesis gas that is available can be converted into hydrogen or into synthetic natural gas by methanation.

Steam Reforming. When relatively light feedstocks, eg, naphthas having ca 180°C end boiling point and limited aromatic content, are available, high nickel content catalysts can be used to simultaneously conduct a variety of near-autothermic reactions. This results in the essentially complete conversions of the feedstocks to methane:

$$CH_3(CH_2)_3CH_3 + 2\ H_2O \rightarrow 4\ CH_4 + CO_2$$

Because of limitations on the activity of practical catalysts, this reaction must be carried out in stages, the first of which is carried out at 425–480°C and 1.5–2.9 MPa (200–400 psi) and amounts approximately to the following reaction:

$$C_5H_{12} + 3\ H_2O \rightarrow 3\ CH_4 + CO_2 + 3\ H_2 + CO$$

In ensuing catalytic stages, usually termed hydrogasification and methanation (not to be confused with the noncatalytic, direct hydrogasification processes described above), the remaining carbon monoxide and hydrogen are reacted to produce additional methane.

Oil Shale

Oil shale (qv) is a sedimentary rock that contains organic matter, referred to as kerogen, and another natural resource of some consequence that could be exploited as a source of synthetic natural gas (67–69). However, as of this writing, oil shale has found little use as a source of substitute natural gas.

Biomass

Biomass is simply defined for these purposes as any organic waste material, such as agricultural residues, animal manure, forestry residues, municipal waste, and sewage, which originated from a living organism (70–74).

Biomass is another material that can produce a mixture of carbonaceous solid and liquid products as well as gas:

$$C_{organic} \rightarrow C_{coke/char/carbon} + \text{liquids} + \text{gases}$$

Whereas biomass has not received the same attention as coal as a source of gaseous fuels, questions about the security of fossil energy supplies related to the availability of natural and substitute gas have led to a search for more reliable and less expensive energy sources (75). Biomass resources are variable, but it has been estimated that substantial amounts (up to 20×10^6 mJ (20×10^{15} Btu)) of energy, representing ca 19% of the annual energy consumption in the United States. In addition, environmental issues associated with the use of coal have led some energy producers to question the use of large central energy generating plants. However, biomass may be a gaseous fuel source whose time is approaching (see FUELS FROM BIOMASS).

The means by which synthetic gaseous fuels could be produced from a variety of biomass sources are variable and many of the known gasification technologies can be applied to the problem (70,71,76–82). For example, the Lurgi circulatory fluidized-bed gasifier is available for the production of gaseous products from biomass feedstocks as well as from coal (83,84).

GAS TREATING

The reducing conditions in gasification reactors effect the conversion of the sulfur and nitrogen in the feed coal to hydrogen sulfide, H_2S, and ammonia, NH_3. Some carbonyl sulfide, COS, carbon disulfide, CS_2, mercaptans, RSH, and hydrogen cyanide, HCN, are also formed in the gasifier. These compounds, along with carbon dioxide, are removed simultaneously, either selectively or nonselectively, from the gas stream in the clean-up stages of the process using commercially available physical or chemical solvents and scrubbing agents (1,5,85–88).

Solvents used for hydrogen sulfide absorption include aqueous solutions of ethanolamine (monoethanolamine, MEA), diethanolamine (DEA), and diisopropanolamine (DIPA) among others:

$$2\,RNH_2 + H_2S \rightarrow (RNH_3)_2S$$

$$(RNH_3)_2S + H_2S \rightarrow 2\,RNH_3HS$$

$$2\,RNH_2 + CO_2 + H_2O \rightarrow (RNH_3)_2CO_3$$

$$(RNH_3)_2CO_3 + CO_2 + H_2O \rightarrow 2\,RNH_3HCO_3$$

$$2\,RNH_2 + CO_2 \rightarrow RNHCOONH_3R$$

These solvents differ in volatility and selectivity for the removal of H_2S, mercaptans, and CO_2 from gases of different composition. Other alkaline solvents used for the absorption of acidic components in gases include potassium carbonate, K_2CO_3, solutions combined with a variety of activators and solubilizers to improve gas–liquid contacting.

Whereas most alkaline solvent absorption processes result in gases of acceptable purity for most purposes, it is often essential to remove the last traces of residual sulfur compounds from gas streams. This is in addition to ensuring product purity such as the removal of water, higher hydrocarbons, and dissolved elemental sulfur from liquefied petroleum gas. Removal can be accomplished by passing the gas over a bed of molecular sieves, synthetic zeolites commercially available in several proprietary forms. Impurities are retained by the packed bed, and when the latter is saturated it can be regenerated by passing hot clean gas or hot nitrogen, generally in a reverse direction.

By-product water formed in the methanation reactions is condensed by either refrigeration or compression and cooling. The remaining product gas, principally methane, is compressed to desired pipeline pressures of 3.4–6.9 MPa (500–1000 psi). Final traces of water are absorbed on solica gel or molecular sieves, or removed by a drying agent such as sulfuric acid, H_2SO_4. Other desiccants may be used, such as activated alumina, diethylene glycol, or concentrated solutions of calcium chloride.

Reprinted from *Kirk-Othmer Encyclopedia of Technology,* 4th ed. Vol. 12, John Wiley & Sons, Inc. New York, 1995

BIBLIOGRAPHY

1. J. G. Speight, ed., *Fuel Science and Technology Handbook,* Marcel Dekker, Inc., New York, 1990.
2. J. G. Speight, *The Chemistry and Technology of Coal,* Marcel Dekker, Inc., New York, 1983.
3. R. A. Hessley, in Ref. 1, pp. 645–734.
4. E. J. Parente and A. Thumann, eds., *The Emerging Synthetic Fuel Industry,* Fairmont Press, Atlanta, Ga., 1981.
5. J. G. Speight, *The Chemistry and Technology of Petroleum,* 2nd ed., Marcel Dekker, Inc., New York, 1991.
6. A. Kasem, *Three Clean Fuels from Coal: Technology and Economics,* Marcel Dekker, Inc., New York, 1979.
7. L. L. Anderson and D. A. Tillman, *Synthetic Fuels from Coal,* John Wiley & Sons, Inc., New York, 1979.
8. A. D. Dainton, in G. J. Pitt and G. R. Millward, eds., *Coal and Modern Coal Processing: An Introduction,* Academic Press, Inc., New York, 1979.
9. D. M. Considine, *Energy Technology Handbook,* McGraw-Hill Book Co., Inc., New York, 1977.
10. A. Francis and M. C. Peters, *Fuels and Fuel Technology,* Pergamon Press, Inc., New York, 1980.
11. J. L. Johnson, in M. A. Elliott, ed., *Chemistry of Coal Utilization,* 2nd suppl. vol. John Wiley & Sons, Inc., New York, 1981, Chapt. 23.
12. A. Elton, in C. Singer and co-workers, eds., *A History of Technology,* Vol. 4, Oxford University Press, Oxford, U.K., 1958, Chapt. 9.
13. L. Shnidman, in H. H. Lowry, ed., *Chemistry of Coal Utilization,* John Wiley & Sons, Inc., New York, 1945, Chapt. 30.
14. C. M. Jarvis, in Ref. 12, Vol. 5, Chapt. 10.
15. P. F. Mako and W. A. Samuel, in R. A. Meyers, ed., *Handbook of Synfuels Technology,* McGraw-Hill Book Co., Inc., New York, 1984, Chapt. 2–1.
16. *Oil Gas J.* **90**(3), 53 (1992).
17. K. T. Klasson and co-workers, C. Akin and J. Smith, eds., *Gas, Oil, Coal, and Environmental Biotechnology II,* Institute of Gas Technology, Chicago, 1990, p. 408.
18. B. D. Faison, *Crit. Revs. Biotechnol.* **11,** 347 (1991).
19. S. R. Bull, *Energy Sources* **13,** 433 (1991).
20. R. C. Reid, J. M. Prausnitz, and T. K. Sherwood, *The Properties of Gases and Liquids,* McGraw-Hill Book Co., Inc., New York, 1977.

21. *Wellman Galusha Gas Producers,* research bulletin no. 576A, McDowell-Wellman Engineering Co., Cleveland, Ohio, May 1976.
22. R. F. Probstein and R. E. Hicks, *Synthetic Fuels,* pH Press, Cambridge, Mass., 1990.
23. J. S. Sinninghe Damste and J. W. de Leeuw, *Fuel Processing Technol.* **30,** 109 (1992).
24. N. Berkowitz, *Introduction to Coal Technology,* Academic Press, Inc., New York, 1979.
25. R. K. Hessley, J. W. Reasoner, and J. T. Riley, *Coal Science,* John Wiley & Sons, Inc., New York, 1986.
26. M. O. Holowaty and co-workers, in R. A. Meyers, ed., *Coal Handbook,* Marcel Dekker, Inc., New York, 1981, Chapt. 9.
27. W. Eisenhut, in Ref. 11, Chapt. 14.
28. F. W. Gibbs, in Ref. 12, Vol. 3, Chapt. 25.
29. G. J. Pitt and G. R. Millward, eds., in Ref. 8, p. 52.
30. L. Seglin and S. A. Bresler, in Ref. 11, Chapt. 13.
31. E. Aristoff, R. W. Rieve, and H. Shalit, in Ref. 11, Chapt. 16.
32. D. McNeil, in Ref. 11, Chapt. 17.
33. C. Y. Cha and co-workers, *Report No. DOE / MC / 24268-2700 (DE89000967),* United States Department of Energy, Washington, D.C., 1988.
34. F. M. Fess, *History of Coke Oven Technology,* Gluckauf Verlag, Essen, Germany, 1957.
35. M. A. Elliott, ed., in Ref. 11.
36. D. Hebden and H. J. F. Stroud, in Ref. 11, Chapt. 24.
37. R. A. Meyers, ed., in Ref. 15.
38. D. R. Simbeck, R. L. Dickenson, and A. J. Moll, *Energy Prog.* **2,** 42 (1982).
39. W. W. Bodle and J. Huebler, in Ref. 26, Chapt. 10.
40. L. K. Rath and J. R. Longanbach, *Energy Sources* **13,** 443 (1991).
41. S. Alpert and M. J. Gluckman, *Ann. Rev. Energy* **11,** 315 (1986).
42. R. A. Meyers, ed., in Ref. 26.
43. J. H. Martin, I. N. Banchik, and T. K. Suhramaniam, *Report of the Committee on Production of Manufactured Gases,* report no. 1GU/B-76, London, 1976.
44. J. F. Farnsworth, *Ind. Heat.* **41**(11), 38 (1974).
45. J. F. Farnsworth, *Proceedings Coal Gas Fundamentals Symposium,* Institute of Gas Technology, Chicago, 1979.
46. J. C. Hoogendoorn, *Proceedings of the Ninth Pipeline Gas Symposium,* Chicago, Oct. 31–Nov. 2, 1977.
47. R. Essenhigh, in Ref. 11, Chapt. 19.
48. M. A. Field and co-workers, *Combustion of Pulverized Coal,* British Coal Utilization Research Association, Leatherhead, Surrey, U.K., 1967.
49. A. Levy and co-workers, in Ref. 26, Chapt. 8.
50. N. Chigier, *Combustion Measurements,* Hemisphere Publishing Corp., New York, 1991.
51. United States Congress, *Public Law 101-549, An Act to Amend the Clean Air Act to Provide for Attainment and Maintenance of Health Protective National Ambient Air Quality Standards, and for Other Purposes,* Nov. 15, 1990.
52. United States Department of Energy, *Clean Coal Technology Demonstration Program,* DOE/FE-0219P, U.S. Dept. of Energy, Washington, D.C., Feb. 1991.
53. J. E. Funk, in J. A. Kent, ed., *Riegel's Handbook of Industrial Chemistry,* Van Nostrand Reinhold Co., New York, 1983, Chapt. 3.
54. F. J. Ceely and E. L. Daman, in Ref. 26, Chapt. 20.
55. J. M. Reid, *Proceedings SNG Symposium 1,* Institute of Gas Technology, Chicago, May 12–16, 1973.
56. F. J. Dent, *Gas J.* **288**(12), 600, 606, 610 (1956).
57. F. J. Dent, *Gas World,* **144**(11), 1078, 1080 (1956).
58. H. R. Linden and E. S. Pettyjohn, *Research Bulletin No. 12,* Institute of Gas Technology, Chicago, 1952.
59. G. B. Schultz and H. R. Linden, *Research Bulletin No. 29,* Institute of Gas Technology, Chicago, 1960.
60. F. J. Dent, *Proceedings of the Ninth International Gas Conference,* The Hague (Scheveningen), the Netherlands, Sept. 1–4, 1964.
61. J. G. Speight, *The Desulfurization of Heavy Oils and Residua,* Marcel Dekker, Inc., New York, 1981.
62. B. B. Bennett, *J. Inst. Fuel* **35**(8), 338 (1962).
63. H. R. Linden and M. A. Elliot, *Am. Gas J.* **186**(2), 22 (1959).
64. *Oil Gas J.* **71**(7), 36, 37 (1973).
65. *Oil Gas J.* **71**(15), 32, 33 (1973).
66. C. J. Kuhre and C. J. Shearer, *Oil Gas J.* **71**(36), 85 (1971).
67. P. Nowacki, *Oil Shale Technical Data Handbook,* Noyes Data Corp., Park Ridge, N.J., 1981.
68. V. D. Allred, ed., *Oil Shale Processing Technology,* Center for Professional Advancement, East Brunswick, N.J., 1982.
69. C. S. Scouten, in Ref. 1.
70. J. S. Robinson, *Fuels from Biomass: Technology and Feasibility,* Noyes Data Corp., Park Ridge, N.J., 1980.
71. J. L. Jones and S. B. Radding, eds., *Thermal Conversion of Solid Wastes and Biomass, Symposium Series No. 130,* American Chemical Society, Washington, D.C., 1980.
72. M. P. Kannan and G. N. Richard, *Fuel* **69,** 747 (1990).
73. L. Jimenez, J. L. Bonilla, and J. L. Ferrer, *Fuel* **70,** 223 (1991).
74. L. Jimenez and F. Gonzalez, *Fuel* **70,** 947 (1991).
75. *Energy World* **145**(3), 11 (1987).
76. M. P. Sharma and B. Prasad, *Energy Management (New Delhi)* **10**(4), 297 (1986).
77. A. A. C. M. Beenackers and W. P. M. van Swaaij, *Thermochemical Processing of Biomass,* Butterworths, London, 1984, p. 91.
78. G. J. Esplin, D. P. C. Fung, and C. C. Hsu, *Can. J. Chem. Eng.* **64,** 651 (1986).
79. S. Gaur and co-workers, *Fuel Sci. Technol. Int.* **10,** 1461 (1992).
80. M. A. McMahon and co-workers, *Preprints, Div. Fuel Chem.* **36**(4), 1670 (1991).
81. K. Dura-Swamy and co-workers, *Preprints, Div. Fuel Chem.* **36**(4), 1677 (1991).
82. J. T. Hamrick, *Preprints, Div. Fuel Chem.* **36**(4), 1986 (1991).
83. R. Reimert and co-workers, *Bioenergy 84,* Vol. 3, Elsevier Applied Science Publishers, London, p. 102.

FUELS, SYNTHETIC, LIQUID

SCOTT HAN
CLARENCE D. CHANG
Mobil Research and Development Corporation
Princeton, New Jersey

The creation of liquids to be used as fuels from sources other than natural crude petroleum (qv) broadly defines synthetic liquid fuels. Hence, fuel liquids prepared from naturally occurring bitumen deposits qualify as synthetics, even though these sources are natural liquids. Synthetic liquid fuels have characteristics approaching those of the liquid fuels in commerce, specifically gasoline, kerosene, jet fuel, and fuel oil (see AIRCRAFT FUELS; MIDDLE DISTILLATE; TRANSPORTATION FUELS, AUTOMOTIVE). For much of the twentieth century, the synthetic fuels emphasis was on liquid products derived from coal upgrading or by extraction or hydrogenation of organic matter in coke liquids, coal tars, tar sands, or bitumen deposits. More recently, however, much of the direction involving synthetic fuels technology has changed. There are two reasons.

The potential of natural gas, which typically has 85–95% methane, has been recognized as a plentiful and clean alternative feedstock to crude oil (see GAS, NATURAL). Estimates (1–3) place worldwide natural gas reserves at ca 1×10^{14} m^3 (3.5×10^{15} ft^3) corresponding to the energy equivalent of ca 1×10^{11} m^3 (637×10^9 bbl) of oil. As of this writing, the rate of discovery of proven natural gas reserves is increasing faster than the rate of natural gas production. Many of the large natural gas deposits are located in areas where abundant crude oil resources lie such as in the Middle East and Russia. However, huge reserves of natural gas are also found in many other regions of the world, providing oil-deficient countries access to a plentiful energy source. The gas is frequently located in remote areas far from centers of consumption, and pipeline costs can account for as much as one-third of the total natural gas cost (1) (see PIPELINES). Thus tremendous strategic and economic incentives exist for on-site gas conversion to liquids.

In general, the proven technology to upgrade methane is via steam reforming to produce synthesis gas, $CO + H_2$. Such a gas mixture is clean and when converted to liquids produces fuels substantially free of heteroatoms such as sulfur and nitrogen. Two commercial units utilizing the synthesis gas from natural gas technology in combination with novel downstream conversion processes have been commercialized.

The direct methane conversion technology, which has received the most research attention, involves the oxidative coupling of methane to produce higher hydrocarbons such as ethylene. These olefinic products may be upgraded to liquid fuels via catalytic oligomerization processes.

A second trend in synthetic fuels is increased attention to oxygenates as alternative fuels (4) as a result of the growing environmental concern about burning fossil-based fuels. The environmental impact of the oxygenates, such as methanol (qv), ethanol (qv), and methyl *tert*-butyl ether (MTBE) is still under debate, but these alternative liquid fuels are gaining new prominence. The U.S. Alternative Fuels Act of 1988, and the endorsement of oxygenate fuels that act contains, clearly underscore the idea that economics is no longer the sole consideration with regard to alternative fuels production (5).

Despite reduced prominence, coal technology is well positioned to provide synthetic fuels for the future. World petroleum and natural gas production are expected ultimately to level off and then decline. Coal gasification to synthesis gas is utilized to synthesize liquid fuels in much the same manner as natural gas steam reforming technology. Although as of this writing world activity in coal liquefaction technology is minimal, the extensive development and detailed demonstration of processes for converting coal to liquid fuels should serve as solid foundation for the synthetic fuel needs of the future.

Coal, tar, and heavy oil fuel reserves are widely distributed throughout the world. In the Western hemisphere, Canada has large tar sand, bitumen (very heavy crude oil), and coal deposits. The United States has very large reserves of coal and shale. Coal comprises ca 85% of the U.S. recoverable fossil energy reserves (6). Venezuela has an enormous bitumen deposit and Brazil has significant oil shale (qv) reserves. Coal is also found in Brazil, Colombia, Mexico, and Peru. Worldwide, the total resource base of these reserves is immense and may constitute >90% of the hydrocarbon resources in place.

The driving force behind the production of combustible liquids before 1900 was the search for low cost lighting. Gas produced during coal distillation was used to light homes at the end of the eighteenth century (7). Large-scale use of coal, which began in England in the nineteenth century, led to significant reductions in the costs of hydrocarbon liquids. The production of coal tar, and the separation therefrom of various coal liquids concomitant to the production of illuminating gas, probably predates production from the coking operations associated with iron (qv) production. The coal tars produced in gas works may have been the first synthetic liquids turned to fuel use in quantity.

Proof of the existence of benzene in the light oil derived from coal tar (8) first established coal tar and coal as chemical raw materials. Soon thereafter the separation of coal-tar light oil into substantially pure fractions produced a number of the aromatic components now known to be present in significant quantities in petroleum-derived liquid fuels. Indeed, these separation procedures were for the recovery of benzene–toluene–xylene and related substances, ie, benzol or motor benzol, from coke-oven operations (8).

By the middle of the nineteenth century it was realized, both in England and in the United States, that kerosene, or coal oil, distilled from coal, could produce a luminous combustion flame. Commercialization was rapid. By the time of the U.S. Civil War, ~87 m^3/yr (23,000 gal/yr) of lamp oil was produced in the United States from the distillation not only of coal, but also of oil shale and natural bitumen. In 1859, high gravity, low sulfur crude oil was discovered in the United States. This produced high quality kerosene with minimal processing, and the world's first oil boom erupted. Until the end of the nineteenth century, kerosene was the only substance of value extracted from natural crude oil. It cost too much for heating purposes, but was used widely for lighting until replaced by electricity. Refiners slowly learned to use the residues from kerosene production, and as the market for lamp oil collapsed, gasoline increased in value. The widespread use of liquids as fuels dates from that time.

Liquid fuels possess inherent advantageous characteristics in terms of being more readily stored, transported, and metered than gases, solids, or tars. Liquid fuels also are generally easy to process or clean by chemical and catalytic means. The energy densities of clean hydrocarbon liquids may be very high relative to gas, solid, or semisolid fuel substances. Moreover, liquid fuels are the most compatible with the twentieth century world fuel infrastructure because most fuel-powered conveyances are designed to function only with relatively clean, low viscosity liquids. In general, liquid hydrocarbon fuels possess an intermediate hydrogen-to-carbon content. Production of synthetic fuels from alternative feedstocks to natural petroleum crude oil is based on adjusting the hydrogen-to-carbon ratio to the desired intermediate level.

There is an inherent economic penalty associated with producing liquids from either natural gas or solid coal feedstock. Synthetic liquid fuels technologies are generally not economically competitive with crude oil processing in the absence of extraneous influences such as price supports or regulations.

INDIRECT LIQUEFACTION/CONVERSION TO LIQUID FUELS

Indirect liquefaction of coal and conversion of natural gas to synthetic liquid fuels is defined by technology that involves an intermediate step to generate synthesis gas, CO + H_2. The main reactions involved in the generation of synthesis gas are the coal gasification reactions:

Combustion

$$C + O_2 \rightleftharpoons CO_2 \quad \Delta H_{298\,K} = -394 \text{ kJ/mol } (-94.2 \text{ kcal/mol}) \tag{1}$$

Gasification

$$C + 1/2\ O_2 \rightleftharpoons CO \quad \Delta H_{298\,K} = -111 \text{ kJ/mol } (-26.5 \text{ kcal/mol}) \tag{2}$$

$$C + H_2O\ (g) \rightleftharpoons CO + H_2 \quad \Delta H_{298\,K} = +131 \text{ kJ/mol } (31.3 \text{ kcal/mol}) \tag{3}$$

$$C + CO_2 \rightleftharpoons 2\ CO \quad \Delta H_{298\,K} = +172 \text{ kJ/mol } (41.1 \text{ kcal/mol}) \tag{4}$$

Water gas shift

$$CO + H_2O\ (g) \rightleftharpoons CO_2 + H_2 \quad \Delta H_{298\,K} = -41 \text{ kJ/mol } (-9.8 \text{ kcal/mol}) \tag{5}$$

the methane steam reforming reactions:

Partial oxidation

$$CH_4 + 1/2\ O_2 \rightleftharpoons CO + 2\ H_2 \quad \Delta H_{298\,K} = -36 \text{ kJ/mol } (-8.6 \text{ kcal/mol}) \tag{6}$$

Reforming

$$CH_4 + H_2O\ (g) \rightleftharpoons CO + 3\ H_2 \quad \Delta H_{298\,K} = 206 \text{ kJ/mol } (49.2 \text{ kcal/mol}) \tag{7}$$

and the water gas shift reaction (eq. 5), used to increase the H_2/CO ratio of the product synthesis gas.

Coal gasification technology dates to the early nineteenth century but has been largely replaced by natural gas and oil. A more hydrogen-rich synthesis gas is produced at a lower capital investment. Steam reforming of natural gas is applied widely on an industrial scale (9,10) and in particular for the production of hydrogen.

The conversion of coal and natural gas to liquid fuels via indirect technology can be achieved by the routes shown in Figure 1. Two pathways from synthesis gas can be taken. Both have been commercialized. One pathway involves coupling with Fischer-Tropsch technology to produce fuel range hydrocarbons directly or upon further processing. Using coal feedstock, this route has been commercialized in South Africa since the 1950s and a process using natural gas was commercialized in Malaysia by Shell Oil Co. in 1993. An alternative route relies on the production of methanol from synthesis gas and subsequent transformation of the methanol to fuels using zeolite catalyst technology introduced by Mobil Oil Corp. This route was commercialized using indigenous natural gas in New Zealand in 1985.

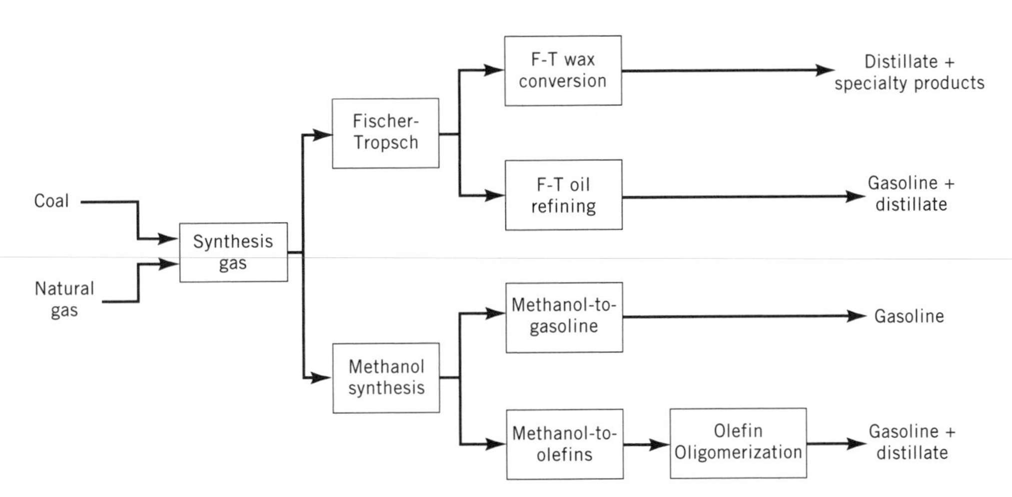

Figure 1. Routes to liquid fuels from natural gas and coal via synthesis gas. F-T is the Fischer-Tropsch process.

Coal Upgrading via Fischer-Tropsch

The synthesis of methane by the catalytic reduction of carbon monoxide and hydrogen over nickel and cobalt catalysts at atmospheric pressure was reported in 1902 (11).

$$CO + 3\,H_2 \rightarrow CH_4 + H_2O\,(l) \qquad \Delta H_{298\,K} = -250\ \text{kJ/mol}\,(-59.8\ \text{kcal/mol}) \quad (8)$$

$$2\,CO + 2\,H_2 \rightarrow CH_4 + CO_2 \qquad \Delta H_{298\,K} = -247\ \text{kJ/mol}\,(59.0\ \text{kcal/mol}) \quad (9)$$

In the early 1920s Badische Anilin- und Soda-Fabrik announced the specific catalytic conversion of carbon monoxide and hydrogen at 20–30 MPa (200–300 atm) and 300–400°C to methanol (12,13), a process subsequently widely industrialized. At the same time Fischer and Tropsch announced the Synthine process (14,15), in which an iron catalyst effects the reaction of carbon monoxide and hydrogen to produce a mixture of alcohols, aldehydes (qv), ketones (qv), and fatty acids at atmospheric pressure.

In the classical normal pressure synthesis (16), higher hydrocarbons are produced by net reactions similar to those observed in the early 1900s, but at temperatures below the level at which methane is formed:

$$n\ CO + 2n\ H_2 \rightarrow C_nH_{2n} + n\ H_2O + \text{heat} \quad (10)$$

$$2n\ CO + n\ H_2 \rightarrow C_nH_{2n} + n\ CO_2 + \text{heat} \quad (11)$$

In the mid-1930s improvements in catalysts and techniques (17–19) culminated in the licensing of the process to Ruhrchemie to produce liquid hydrocarbons and paraffin waxes using precipitated cobalt-on-kieselguhr catalysts. Subsequently, a medium pressure synthesis was developed (20) at 0.5–2 MPa (5–20 atm) using dispersed cobalt catalysts which improved hydrocarbon yields by 10–15%. The yields of paraffin wax, in particular, could be increased to 45% of the total liquid product. Hydrotreating of catalyst (required in the normal pressure process) was avoided, and catalyst life was extended (21–23). There is a marked influence of pressure on product yields. Beyond the optimum pressure of about 2 MPa (20 atm), paraffin yield decreases. Little change is found in the gasoline and gas oil yields, however.

Furthermore it was discovered that reasonable yields could be obtained using precipitated iron catalysts at 1–3 MPa (10–30 atm), and that very high melting waxes could be synthesized at 10–100 MPa (100–1000 atm) over ruthenium catalysts. At the same time a related process, the oxo synthesis, was announced (24). Early in World War II, the iso-synthesis process was developed for the production of low molecular weight isoparaffins at high temperatures and pressures over thoria and mixtures of alumina and zinc oxide (25–28). In the early 1960s polymethylenes were synthesized using activated ruthenium catalysts at high pressures.

Industrial operation of the Fischer-Tropsch synthesis involved five steps: (*1*) synthesis gas manufacture; (*2*) gas purification by removal of water and dust, and hydrogen sulfide and organic sulfur compounds; (*3*) synthesis of hydrocarbons; (*4*) condensation of liquid products and recovery of gasoline from product gas; and (*5*) fractionation of synthetic products. Only the synthesis reactor and its method of operation were unique to the process. For low pressure synthesis the reactor incorporated elaborate bundles of cooling tubes immersed in the catalyst, whereby circulating water removed the heat of reaction, limiting the conversion to methane which produced high temperatures. In the pressure process, bundles of concentric tubes, with catalyst arranged in the annuli, through and around which cooling water flowed, served as conversion units. In both systems, the conversion units each contained about 10 m^3 (ca 350 ft^3) of catalyst, and were rated at a capacity of ca 4.8 m^3 (30 bbl) of liquid product per day.

During World War II, nine commercial plants were operated in Germany, five using the normal pressure synthesis, two the medium pressure process, and two having converters of both types. The largest plants had capacities of ca 400 m^3/d (2500 bbl/d) of liquid products. Cobalt catalysts were used exclusively.

Development Outside Germany. In the late 1930s experimental work in England (29–31) led to the erection of large pilot facilities for Fischer-Tropsch studies (32). In France, a commercial facility near Calais produced ca 150 m^3 (940 bbl) of liquid hydrocarbons per day. In Japan, two full-scale plants were also operated under Ruhrchemie license. Combined capacity was ca 400 m^3 (2500 bbl) of liquids per day.

In the mid-1930s Universal Oil Products reported (33,34) that gasoline of improved quality could be produced by cracking the high boiling fractions of Fischer liquids, and a consortium, the Hydrocarbon Synthesis, Inc., entered into an agreement with Ruhrchemie to license the Fischer synthesis outside Germany.

In 1955 the South African Coal, Oil, and Gas Corp. (Sasol) commercialized the production of liquid fuels utilizing Fischer-Tropsch technology (35). This Sasol One complex has evolved into the streaming of second-generation plants, known as Sasol Two and Three. The Sasol One process, shown in Figure **2a** (36), combines fixed-bed Ruhrchemie-Lurgi Arge reactor units with fluidized-bed Synthol process technology (37). For Sasol One, 16,000 t/d of coal is crushed and gasified with steam and oxygen. After a number of gas purification steps in which by-products and gas impurities are removed, the pure gas is processed in both fixed- and fluidized-bed units simultaneously. Table 1 gives product selectivity comparisons of fixed-bed and

Table 1. Product Selectivities for Commercial Fixed-Bed and Synthol Units[a]

Product	Fixed-bed	Synthol
Methane	2.0	10
Ethylene	0.1	4
Ethane	1.8	4
Propylene	2.7	12
Propane	1.7	2
Butenes	3.1	9
Butanes	1.9	2
C_5 and higher	83.5	51
Soluble chemicals	3.0	5
Water-soluble acids	0.2	1

[a] Ref. 36.

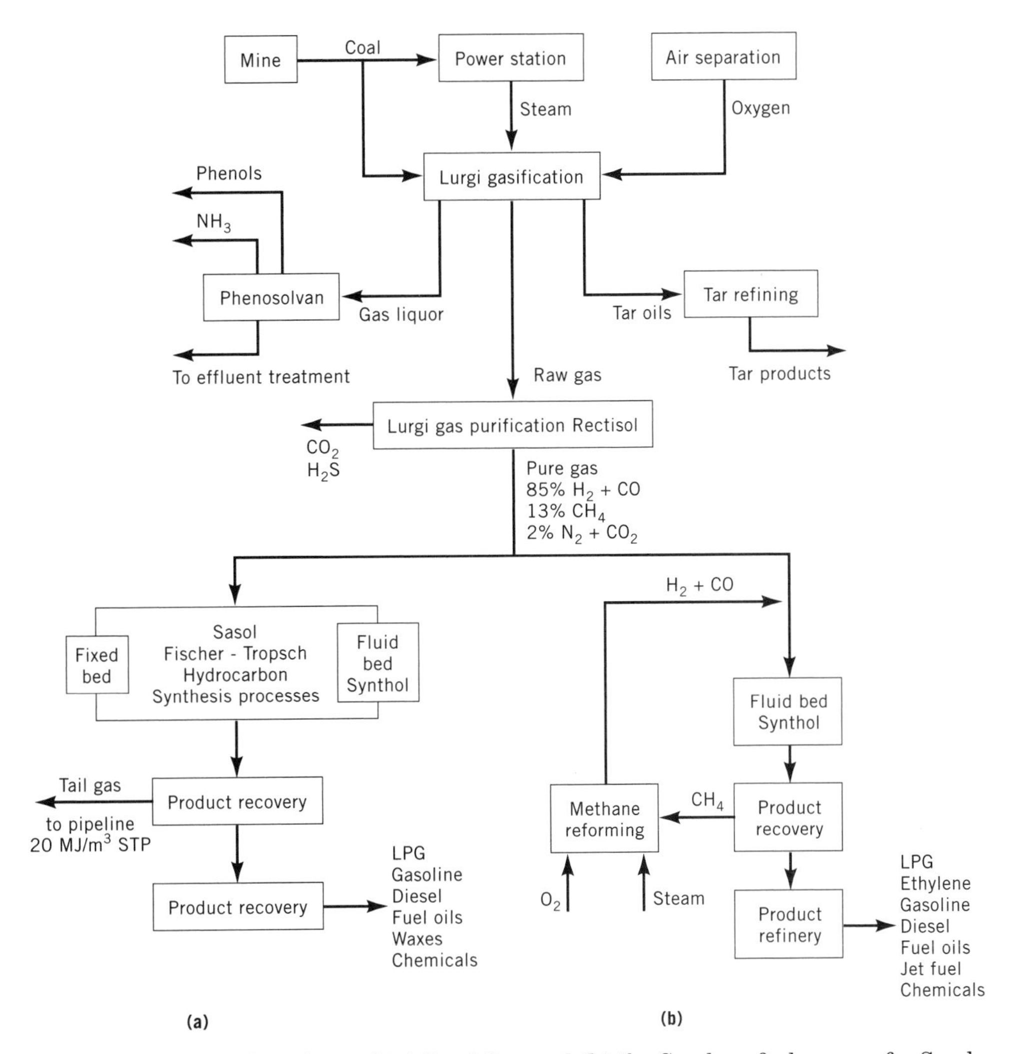

Figure 2. Process flow sheet of (**a**) Sasol One and (**b**) the Sasol synfuel process for Sasol Two and Three (36). LPG is liquefied petroleum gas; other terms are defined in the text. To convert J to cal, divide by 4.184.

Synthol operations. Conversion to hydrocarbons is higher in the Synthol unit and the H_2/CO ratio is also higher. Because the fixed-bed Arge reactor favors the formation of straight-chain paraffins, there is greater production of diesel and wax fractions than the Synthol unit. The Arge reactor products have lower gasoline octane number but higher diesel cetane number relative to Synthol. The high wax production using the Arge reactor was disadvantageous at the time owing to market limitations of wax fuels. Sasol One produced a vast array of chemical and fuel products, including gasoline at 1.5×10^6 t/yr.

The 1973 oil crisis resulted in the Sasol Two unit which started up in early 1980 followed by the nearly identical Sasol Three unit two years later. Figure 2**b** gives the schematic flow diagrams for the Sasol Two and Three processes. Sasol Two uses 36 Lurgi gasifiers in parallel to process ca 31,000 t/d of sized coal. By-product effluents and gas impurities are removed in Rectisol (sulfur compounds and CO_2 removal), Phenosolvan (oxygen compounds and ammonia removal), and tar separation units. Synthol fluid-bed units were employed because of the product distribution and ease of design scale-up. Approximately 80,000 t/d of coal are needed for the two plants. Composition and manufacturing information for Sasol Fischer-Tropsch catalysts are trade secrets, but the catalyst is widely accepted as being an alkali-metal promoted iron-based material.

More recently, Sasol commercialized a new type of fluidized-bed reactor and was also operating a higher pressure commercial fixed-bed reactor (38). In 1989, a commercial scale fixed fluid-bed reactor was commissioned having a capacity similar to existing commercial reactors at Sasol One (39). This effort is aimed at expanded production of higher value chemicals, in particular waxes (qv) and linear olefins.

Properties. Fischer-Tropsch liquid obtained using cobalt catalysts is roughly equivalent to a very paraffinic natural petroleum oil but is not so complex a mixture. Straight-chain, saturated aliphatic molecules predominate but monoolefins may be present in an appreciable concentration. Alcohols, fatty acids, and other oxygenated compounds may represent less than 1% of the total liquid product. The normal pressure synthesis yields ca 60% gasoline, 30% gas

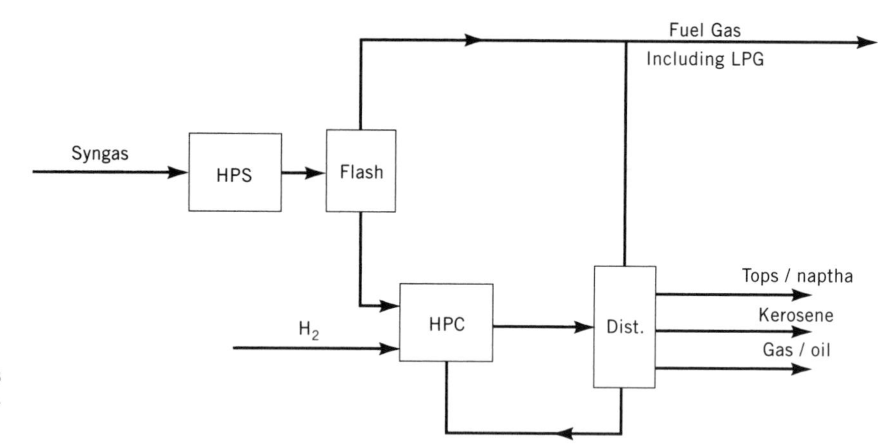

Figure 3. The Shell middle distillate synthesis (SMDS) process. HPS = heavy paraffin synthesis. HPC = heavy paraffin conversion.

oil, and 10% paraffin (mp 20–100°C). The medium pressure synthesis yields 35% gasoline, 35% gas oil, and 30% paraffin. The octane rating of the gasoline is too low for direct use as motor fuel (40).

Most of the German gasoline production was blended into motor fuels using benzene derived from coking. The gas oil could be used directly as a superior diesel fuel; some was also used in soap manufacture. The paraffin (referred to as gatsch) was used primarily for the synthesis of fatty acids and hard soaps. The propane and butane gases were also used as motor fuels. Some propylene and butylenes were polymerized over phosphoric acid to high octane gasoline, and some olefins to lubricating oils. Typical values for the composition of the technical-scale reaction products of the normal and medium pressure synthesis are available (41).

Natural Gas Upgrading via Fischer-Tropsch

In the United States, as in other countries, scarcities from World War II revived interest in the synthesis of fuel substances. A study of the economics of Fischer synthesis led to the conclusion that the large-scale production of gasoline from natural gas offered hope for commercial utility. In the Hydrocol process (Hydrocarbon Research, Inc.) natural gas was treated with high purity oxygen to produce the synthesis gas which was converted in fluidized beds of iron catalysts (42).

Shell Middle Distillate Synthesis. The Shell middle distillate synthesis (SMDS) process developed by Shell Oil Co., uses remote natural gas as the feedstock (43–45). A simplified flow scheme is given in Figure 3. This two-step process involves Fischer-Tropsch synthesis of paraffinic wax called the heavy paraffin synthesis (HPS). The wax is subsequently hydrocracked and hydroisomerized to yield a middle distillate boiling range product in the heavy paraffin conversion (HPC). In the HPS stage, wax is maximized by using a proprietary catalyst having high selectivity toward heavier products and by the use of a tubular, fixed-bed Arge-type reactor. The HPC stage employs a commercial hydrocracking catalyst in a trickle flow reactor. The effect of hydrocracking light paraffins is shown in Figure 4. The HPC step allows for production of narrow range hydrocarbons not possible with conventional Fischer-Tropsch technology.

After years of bench-scale and pilot-plant studies, construction was begun on a gca 1600 m^3/d (10,000 bbl/d) unit in Sarawak, Malaysia, by Shell in a joint venture with Mitsubishi and the Malaysian government. Plant commissioning was in early 1993 at a capital investment of ca

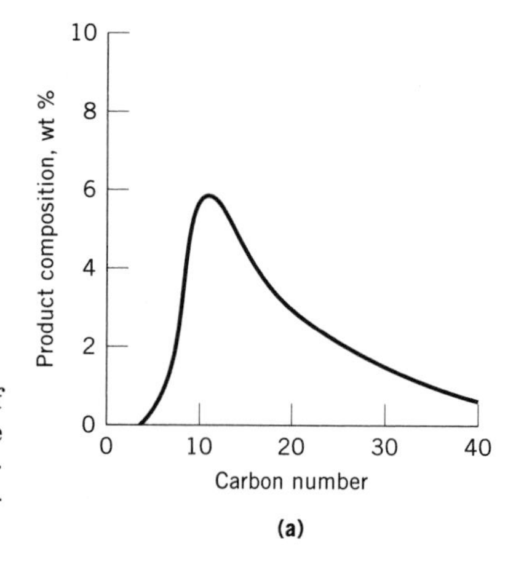

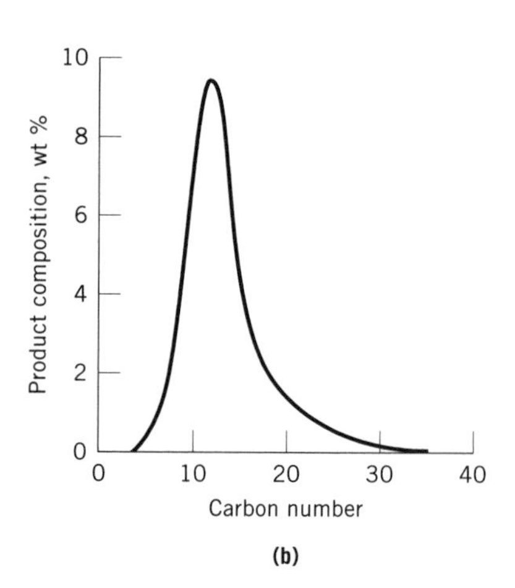

Figure 4. Product compositions as a function of carbon number for the Shell middle distillate synthesis process: (**a**) the Fischer-Tropsch product following HPS, and (**b**) the final hydrocracking product following HPC. See text (45).

Table 2. SMDS Product Quality[a,b]

Parameter	SMDS	Specification
	Gas oil	
Cetane number	70	40 to 50
Cloud point, °C	−10	−10 to + 20
	Kerosene	
Smoke point, mm	45	19
Freezing point, °C	−47	−47

[a] Ref. 44.
[b] The tops/naphtha fraction is similar to straight-run material.

$600–700 × 10^6. The plant uses natural gas from offshore fields and is located adjacent to the existing Malaysian Liquefied Natural Gas (LNG) plant. The production of liquid transportation fuels via SMDS cannot compete economically with fuels derived from crude oil. However, economics vary greatly with site location, and subsidies from the Malaysian government, eg, reduced natural gas cost, brought this plant to commercialization. In addition, premium selling prices for the high quality products made from SMDS are a primary influence on commercialization potential (44).

A similar process to SMDS using an improved catalyst is under development by Norway's state oil company, den norske state olijeselskap AS (Statoil) (46). High synthesis gas conversion per pass and high selectivity to wax are claimed. The process has been studied in bubble columns and a demonstration plant is planned.

Properties. Shell's two-step SMDS technology allows for process flexibility and varied product slates. The liquid product obtained consists of naphtha, kerosene, and gas oil in ratios from 15:25:60 to 25:50:25, depending on process conditions. Of particular note are the high quality gas oil and kerosene. Table 2 gives SMDS product qualities for these fractions.

The products manufactured are predominantly paraffinic, free from sulfur, nitrogen, and other impurities, and have excellent combustion properties. The very high cetane number and smoke point indicate clean-burning hydrocarbon liquids having reduced harmful exhaust emissions. SMDS has also been proposed to produce chemical intermediates, paraffinic solvents, and extra high viscosity index (XHVI) lubeoils (see LUBRICATION AND LUBRICANTS) (44).

Liquid Fuels via Methanol Synthesis and Conversion

Methanol is produced catalytically from synthesis gas. By-products such as ethers, formates, and higher hydrocarbons are formed in side reactions and are found in the crude methanol product. Whereas for many years methanol was produced from coal, after World War II low cost natural gas and light petroleum fractions replaced coal as the feedstock.

Methanol-to-Gasoline. The most significant development in synthetic fuels technology since the discovery of the Fischer-Tropsch process is the Mobil methanol-to-gasoline (MTG) process (47–49). Methanol is efficiently transformed into C_2–C_{10} hydrocarbons in a reaction catalyzed by the synthetic zeolite ZSM-5 (50–52). The MTG reaction path is presented in Figure 5 (47). The reaction sequence can be summarized as

$$n/2\ [2\ CH_3OH \rightleftarrows CH_3OCH_3 + H_2O] \rightarrow C_nH_{2n} \rightarrow n[CH_2] \qquad (12)$$

where $[CH_2]$ represents an average paraffin–aromatic mixture.

How the initial C—C bond is formed from the C_1 progenitor is unknown and much debated (48,53–55). Light olefins are key intermediates in the reaction sequence. These undergo further transformation, ultimately forming aromatics and light paraffins. Table 3 lists a typical MTG product distribution. A unique characteristic of these products is an abrupt termination in carbon number at around C_{10}. This is a consequence of molecular shape-selectivity (56–58), a property of ZSM-5. The composition and proper-

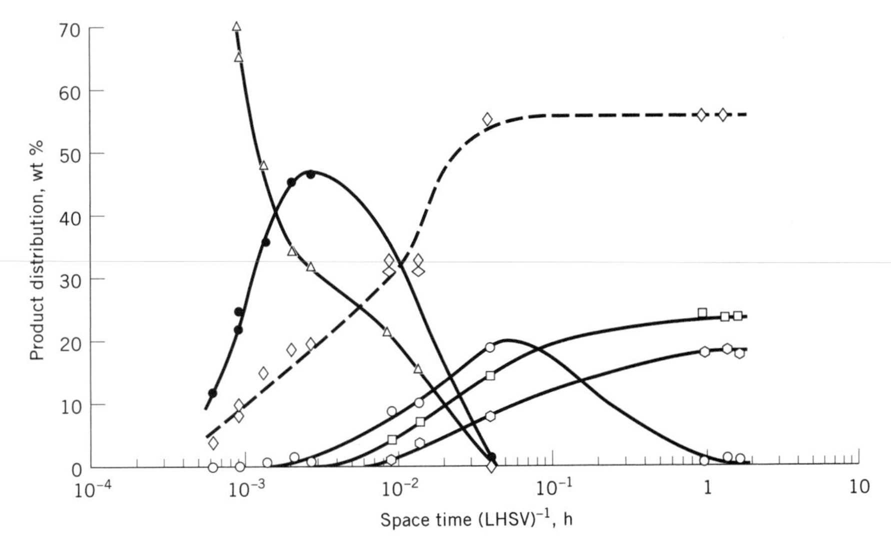

Figure 5. Methanol-to-hydrocarbons reaction path at 371°C △—△— methanol; ●—●— = dimethyl ether; where (◇—◇—) is water; (□—□—) paraffins (and C_6+ olefins); (○—○—) aromatics; and (○—○—) C_2—C_5 olefins. LHSV = Liquid hourly space velocity.

Table 3. MTG Product Distribution[a,b]

Hydrocarbon	Distribution, wt %
Methane	1.0
Ethane	0.6
Ethylene	0.5
Propane	16.2
Propylene	1.0
i-Butane	18.7
n-Butane	5.6
Butenes	1.3
i-Pentane	7.8
n-Pentane	1.3
Pentenes	0.5
C_6+ Aliphatics	4.3
Benzene	1.7
Toluene	10.5
Ethylbenzene	0.8
Xylenes	17.2
C_9 Aromatics	7.5
C_{10} Aromatics	3.3
C_{11}+ Aromatics	0.2

[a] Reaction conditions of 371°C and LHSV of 1.0 h^{-1}.
[b] 100% conversion.

ties of the C_5+ fraction are those of a typical premium aromatic gasoline. Interestingly, C_{10} also is the end point of conventional gasoline.

The MTG process was developed for synfuel production in response to the 1973 oil crisis and the steep rise in crude prices that followed. Because methanol can be made from any gasifiable carbonaceous source, including coal, natural gas, and biomass, the MTG process provided a new alternative to petroleum for liquid fuels production. New Zealand, heavily dependent on foreign oil imports, utilizes the MTG process to convert vast offshore reserves of natural gas to gasoline (59).

Two versions of the MTG process, one using a fixed bed, the other a fluid bed, have been developed. The fixed-bed process was selected for installation in the New Zealand gas-to-gasoline (GTG) complex, situated on the North Island between the villages of Waitara and Motonui on the Tasman seacoast (60). A simplified block flow diagram of the complex is shown in Figure 6 (61). The plant processes over 3.7 × 10^6 m^3/d (130 × 10^6 SCF/d) of gas from the offshore Maui field supplemented by gas from the Kapuni field, first to methanol, and thence to 2.3 × 10^3 m^3/d (14,500 bbl/d) of gasoline. Methanol feed to the MTG section is synthesized using the ICI low pressure process (62) in two trains, each with a capacity of 2200 t/d.

A flow diagram of the MTG section is shown in Figure 7. Methanol feed, vaporized by heat exchange with reactor effluent gases, is converted in a first-stage reactor containing an alumina catalyst to an equilibrium mixture of methanol, dimethyl ether (DME), and water. This is combined with recycle light gas, which serves to remove reaction heat from the highly exothermic MTG reaction, and enters the reactors containing ZSM-5 catalyst. As indicated in Figure 7, five parallel swing reactors are used. Four reactors are on feed while the fifth is under regeneration. The multiple-bed configuration is used to minimize pressure drop as well as to control product selectivity. Reaction conditions are 360–415°C, 2.17 × 10^3 kPa (315 psia), and 9/1 recycle/fresh feed ratio. The overall thermal efficiency of the plant is ca 53%.

A fluid-bed version of the MTG process has been developed (60,63–65) and demonstrated in semiworks scale of 15.9 m^3/d (100 bbl/d), but has not been commercialized to date (ca 1993). Heat management of the exothermic MTG reaction is greatly facilitated by use of fluid-bed reactors. The turbulent bed, with its excellent heat-transfer characteristics, ensures isothermality through the reaction zone and permits steam generation by direct exchange with steam coils in the bed. A schematic diagram appears in Figure 8. The reactor system consists of three principal parts: the reactor, the catalyst regenerator, and an external catalyst cooler. The reactor is also equipped with internal heat-exchanger tubes. Methanol is converted in a single pass at 380–430°C, 276–414 kPa (40–60 psia). The methanol feed rate is 500–1050 kg/h. The fluid-bed demonstration was carried out in 1982–1983 (66).

Properties. Table 4 contains typical gasoline quality data from the New Zealand plant (67). MTG gasoline typically contains 60 vol % saturates, ie, paraffins and naphthenes; 10 vol % olefins; and 30 vol % aromatics. Sulfur and nitrogen levels in the gasoline are virtually nil. The MTG process produces ca 3–7 wt % durene *[95-93-2]* (1,2,4,5-tetra-methylbenzene) but the level is reduced to ca 2 wt % in the finished gasoline product by hydrodealkylation of the durene in a separate catalytic reactor.

Methanol-to-Olefins and Olefins-to-Gasoline-and-Distillate. Because the MTG process produces primarily gasoline, a variation of that process has been developed which allows for production of gasoline and distillate fuel (68). This process integrates two known technologies, methanol-to-olefins (MTO) and Mobil olefins-to-gasoline-and-distillate (MOGD). The MTO/MOGD process schematic is shown in Figure 9. The combined process produces gasoline and distillate in various proportions and, if needed, olefinic by-products.

In the MTO process, methanol is converted over ZSM-5 giving high (up to ca 80 wt % hydrocarbons) olefin yields and low ethylene and light saturate yields. The low ethyl-

Table 4. MTG Gasoline Quality[a]

Parameter	Average	Range
Density at 15°C, kg/m^3	730	728–733
Reid vapor pressure, kPa[b]	86.2	83.4–91.0
Octane number		
Research	92.2	92.0–92.5
Motor	82.6	82.2–83.0
Durene content, wt %	2.0	1.74–2.29
Induction period, min	325	260–370
Distillation, % evaporation		
at 70°C	31.5	29.5–34.5
at 100°C	53.2	51.5–55.5
at 180°C	94.9	94.0–96.5
Distillation end point, °C	240.5	196–209

[a] Ref. 67
[b] To convert kPa to psia, multiply by 0.145.

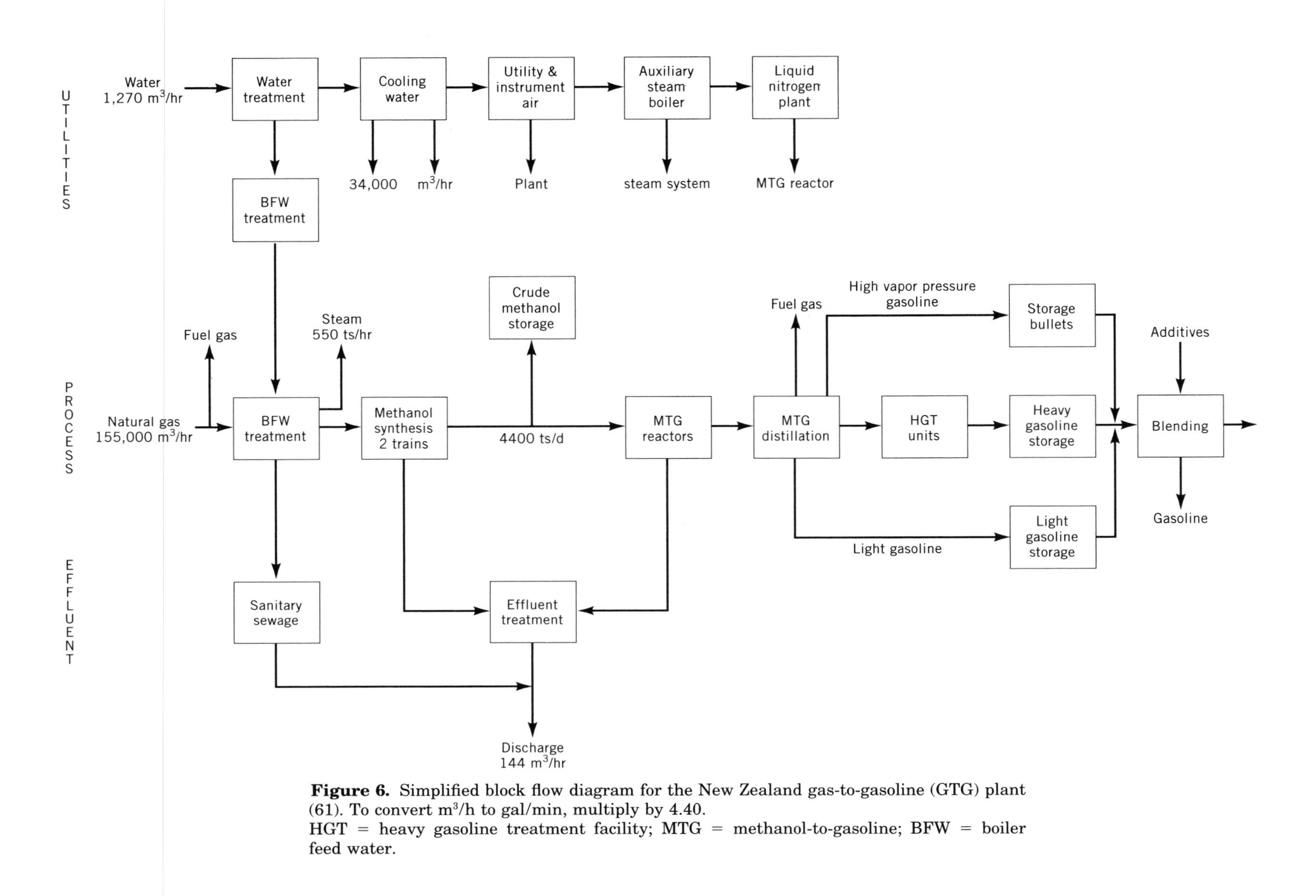

Figure 6. Simplified block flow diagram for the New Zealand gas-to-gasoline (GTG) plant (61). To convert m^3/h to gal/min, multiply by 4.40.
HGT = heavy gasoline treatment facility; MTG = methanol-to-gasoline; BFW = boiler feed water.

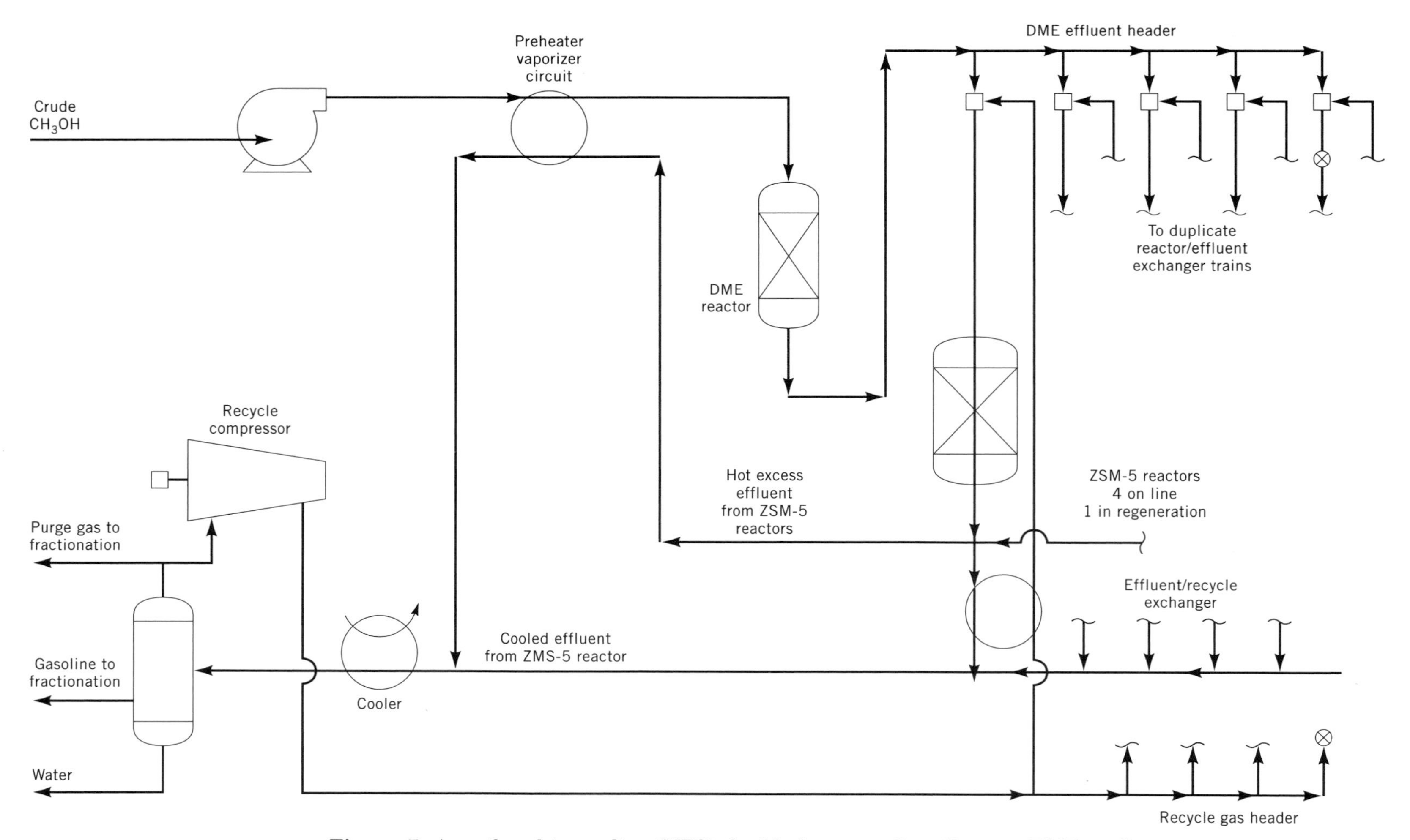

Figure 7. A methanol-to-gasoline (MTG) fixed-bed process flow diagram. DME = di-methyl ether.

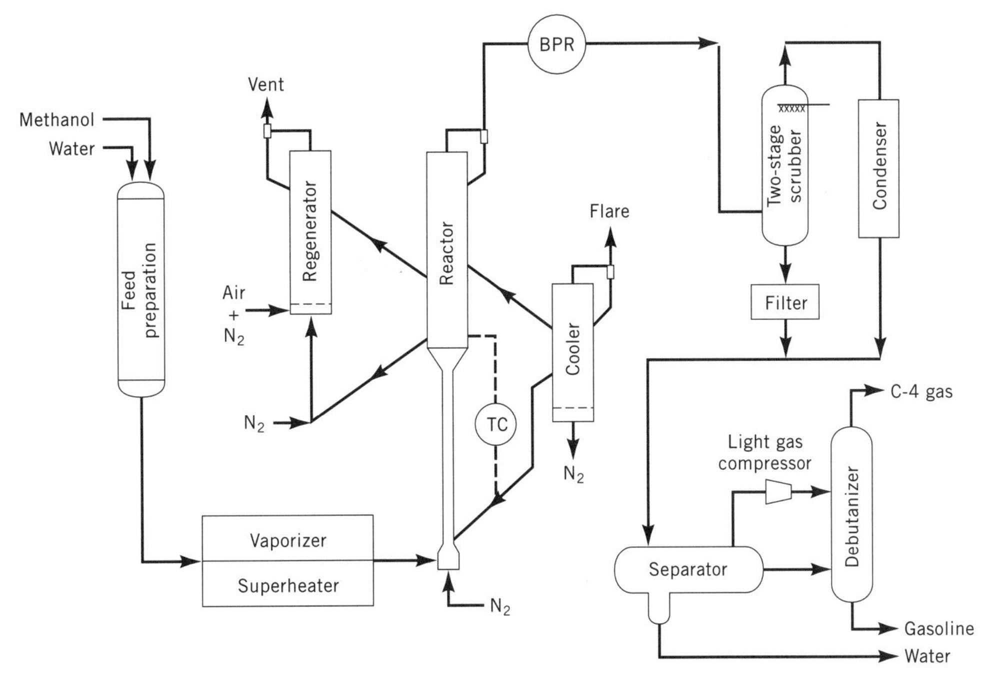

Figure 8. Fluid-bed MTG demonstration plant schematic diagram. BPR = Back pressure regulator; TC = temperature controller.

ene yields are desirable in achieving high distillate yields using MOGD. Figure 5 shows that the production of olefins rather than gasoline from methanol is governed by the kinetics of methanol conversion over ZSM-5 catalyst (69). Generally, catalyst and process variables which increase methanol conversion decrease olefins yield.

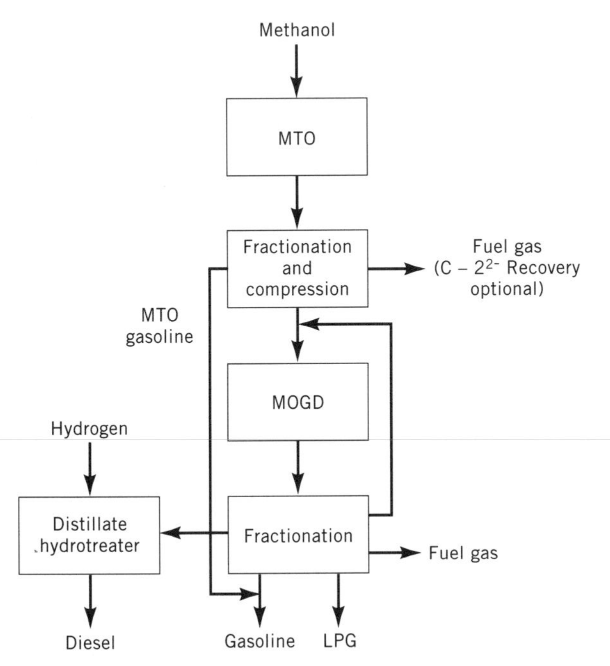

Figure 9. Methanol-to-olefins (MTO) and Mobil olefins-to-gasoline (MOGD) and distillate process schematic.

The MTO process employs a turbulent fluid-bed reactor system and typical conversions exceed 99.9%. The coked catalyst is continuously withdrawn from the reactor and burned in a regenerator. Coke yield and catalyst circulation are an order of magnitude lower than in fluid catalytic cracking (FCC). The MTO process was first scaled up in a 0.64 m^3/d (4 bbl/d) pilot plant and a successful 15.9 m^3/d (100 bbl/d) demonstration plant was operated in Germany with U.S. and German government support.

The MOGD process oligomerizes light olefins to gasoline and distillate products over ZSM-5 zeolite catalyst. Gasoline and distillate selectivity is >95% of the feed olefins and gasoline/distillate product ratios can vary, depending on process conditions, from 0.2 to >100. High octane MTO gasoline is separated before the MOGD section and blended with MOGD gasoline. Some MOGD gasoline may be recycled. The distillate product requires hydrofinishing. Generally, the process scheme uses four fixed-bed reactors, three on-line and one in regeneration. A large-scale MOGD refinery test run was conducted by Mobil in 1981.

Properties. The gasoline product from the integrated MTO/MOGD process is predominately olefinic and aromatic. The gasoline quality (ca 89 octane) is comparable to FCC gasoline. Typical distillate product properties are given in Table 5. After hydrofinishing, the distillate product is mostly isoparaffinic and has high cetane index, low pour point, and negligible sulfur content. MOGD diesel fuel has somewhat lower density than typical conventional fuels (0.8 vs 0.86). Low aromatics levels contribute to a stable jet fuel with very little smoke emission during combustion. MOGD diesel and jet fuels meet or exceed all conventional specifications.

Table 5. MTO/MOGD Distillate Properties

Parameter	Total Distillate	Jet Fuel	Diesel Fuel
Quantity, vol %	100	30	70
Density, g/mL	0.792	0.774	0.800
Pour point, °C	−50		−30
Freeze point, °C	−60	−60	
Flash point, °C	60	50	100
Cetane number	50		52
Smoke point, mm	25	25	
Aromatics, vol %	4	5	
Sulfur, ppm	50		

DIRECT CONVERSION OF NATURAL GAS TO LIQUID FUELS

The capital costs associated with indirect natural gas upgrading technology are high, thus research and development has focused on direct conversion of natural gas to liquid fuels. Direct conversion is defined as upgrading methane to the desired liquid fuels products while bypassing the synthesis gas step, ie, direct transformation to oxygenates or higher hydrocarbons. Direct upgrading routes which have been extensively studied include direct partial oxidation to oxygenates, oxidative coupling to higher hydrocarbons, and pyrolysis to higher hydrocarbons. Owing to the inert nature of methane, the technology is limited by the yields of desired products which in turn affects the process economics. Only one direct oxidative methane conversion process has been commercialized. A plant at Copsa Mica (Romania) in the 1940s (70) produced formaldehyde directly from methane and air by partial oxidation. This plant is no longer in operation. Plants to produce acetylene from methane by high temperature pyrolysis routes have been commercialized.

Generally, the most developed processes involve oxidative coupling of methane to higher hydrocarbons. Oxidative coupling converts methane to ethane and ethylene by

$$2\,CH_4 + 1/2\,O_2 \rightarrow H_3CCH_3 + H_2O \quad (13)$$

$$H_3CCH_3 + 1/2\,O_2 \rightarrow H_2C{=}H_2 + H_2O \quad (14)$$

The process can be operated in two modes: co-fed and redox. The co-fed mode employs addition of O_2 to the methane/natural gas feed and subsequent conversion over a metal oxide catalyst. The redox mode requires the oxidant to be from the lattice oxygen of a reducible metal oxide in the reactor bed. After methane oxidation has consumed nearly all the lattice oxygen, the reduced metal oxide is reoxidized using an air stream. Both methods have processing advantages and disadvantages. In all cases, however, the process is run to maximize production of the more desired ethylene product.

Direct conversion of natural gas to liquids has been actively researched. Process economics are highly variable and it is unclear whether direct natural gas conversion technologies are competitive with the established indirect processes. Some emerging technologies in this area are presented herein.

ARCO Gas-to-Gasoline Process

A two-step process using oxidative coupling to upgrade natural gas to liquid fuels has been proposed by ARCO (Atlantic Richfield Co.) (71,72). A simplified process scheme is given in Figure 10. Methane is passed through a redox-mode oxidative coupling reactor which generates C_2 + hydrocarbons such as ethylene. The olefinic products are then oligomerized over a zeolite catalyst in a second reactor to produce gasoline and distillate. Unreacted methane is recycled. ARCO claims 25% conversion of methane with 75% C_2 + selectivity (ethylene to ethane ratio up to 10) in the oxidative coupling first stage and 95% ethylene conversion with 70% selectivity to gasoline and distillate in the olefin oligomerization second stage. This technology has been developed through bench-scale and pilot-plant stages.

OXCO Process

The OXCO process for upgrading natural gas has been proposed by the Commonwealth for Scientific and Industrial Research Organization (CSIRO) in Australia (73,74). This process involves C_2+ pyrolysis and oxidative coupling of natural gas in a two-stage reactor; the entire concept is shown schematically in Figure 11. The methane in natural

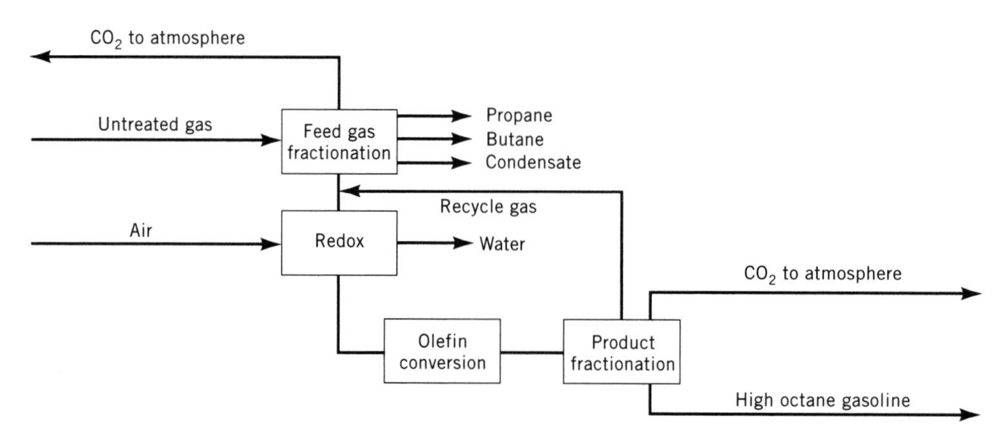

Figure 10. Simplified flow diagram depicting the ARCO gas-to-gasoline process for a conceptual gasoline production plant (72).

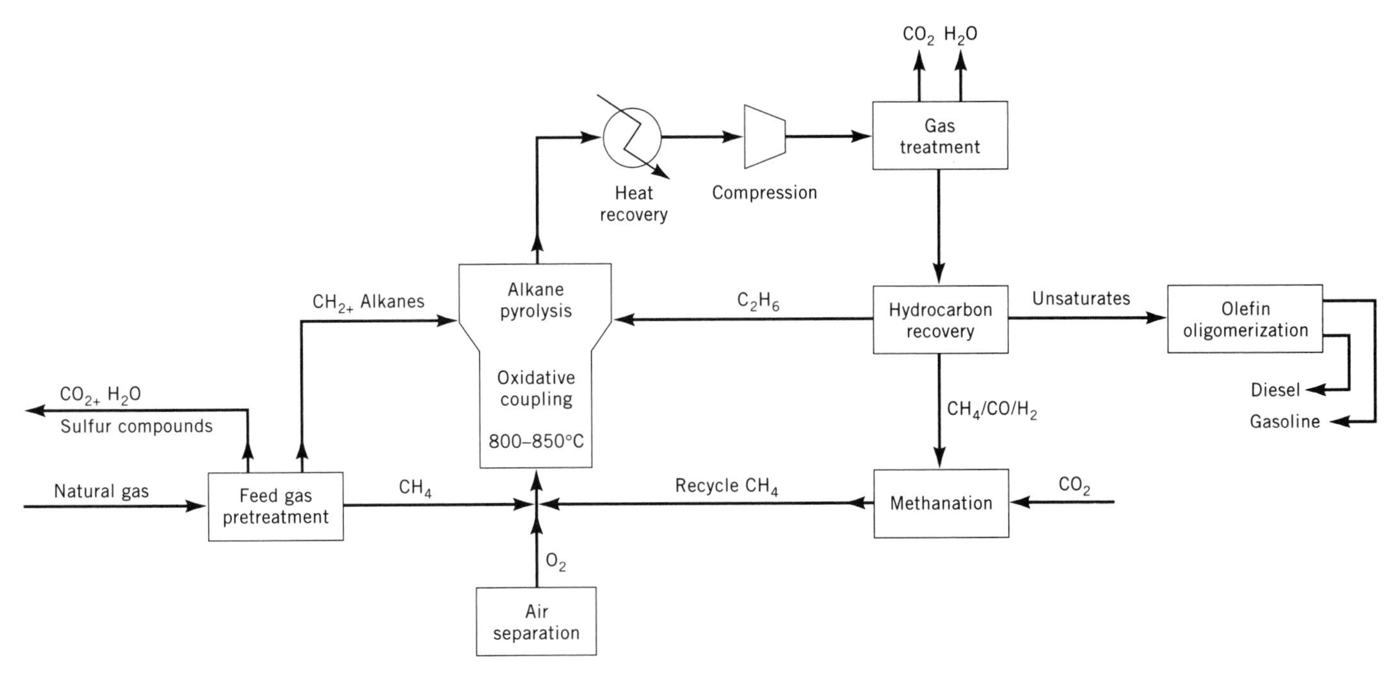

Figure 11. The OXCO process for natural gas conversion (74).

gas is separated and oxidatively coupled in a fluidized-bed reactor operating in co-fed mode to produce ethylene and ethane. The higher alkanes from the natural gas as well as the product ethane from the first stage are injected into an oxygen-free pyrolysis stage to make additional ethylene. The heat from the coupling reactor is used in the pyrolysis reaction. The overall carbon conversion to unsaturates plus CO_2 per pass is 30% with an overall carbon selectivity to unsaturates of 86%. The ethylene may be subsequently upgraded by oligomerization to liquid fuels. This process, which produces higher yields of ethylene and has a more favorable heat balance than conventional oxidative coupling technology, has been demonstrated in 30- and 60-mm fluidized-bed reactors.

Properties. Liquid fuels derived from oxidative coupling/olefin oligomerization processes would be expected to have properties similar to those derived from olefin oligomerization pathways such as MTO/MOGD.

OXYGENATE FUELS

Alcohols and ethers, especially methanol, ethanol, and methyl *tert*-butyl ether [*1634-04-4*] (MTBE), have been widely used separately or in blends with gasolines (reformulated gasoline) and other hydrocarbons to fuel internal combustion engines. Fuel properties of key oxygenates are presented in Table 6 (5). These compounds, as a class, may be considered to be partially oxidized, ie, each has a mole of oxidized hydrogen. They differ from the hydrocarbons that make up gasoline principally in lower heating values and in higher vaporization heat requirements. This constitutes a serious disadvantage to the substitution of

Table 6. Fuel Oxygenates Properties[a]

Oxygenate	Blending Octane, 1/2(RON + MON)[b]	Heat of Combustion, MJ/L[c]	Specific Gravity	Boiling Point, °C
Methanol	101	18.0	0.79	64.6
Ethanol	101	21.3	0.79	78.5
2-Propanol	106	26.4	0.79	82.4
2-Butanol	99	28.3	0.80	99.5
tert-Butyl alcohol	100	28.1	0.80	82.6
MTBE[d]	108	30.2	0.75	55.4
ETBE[e]	111	32.5	0.74	72.8
TAME[f]	102	31.2	0.77	86.3
Gasoline	87	34.8	0.74	

[a] Ref. 5.
[b] RON = research octane number; MON = motor octane number.
[c] To convert MJ/L to Btu/gal, multiply by 3589.
[d] MTBE = methyl *tert*-butyl ether.
[e] ETBE = ethyl *tert*-butyl ether.
[f] TAME = *tert*-amyl methyl ether.

oxygenates, especially lower alcohols, for motor gasoline. For example, the heating value of methanol is about half that of gasoline on an equivalent volume basis. Other properties which greatly influence the potential of oxygenates as fuels include octane performance, solubility in gasoline, effect on gasoline vapor pressure, sensitivity to water, and evaporative/exhaust emissions. Oxygenate fuels tests are often debated because the tests employed were developed for conventional gasolines.

The addition of small percentages of oxygenates to gasoline can produce large gains in octane. Thus, as blending components in gasoline, oxygenates improve octane quality. As neat fuels for spark-ignition engines, octane values for oxygenates are not useful in determining knock-limited compression ratios for vehicles because of the lean carburetor settings relative to gasoline. Neither do these values represent the octane performance of oxygenates when blended with gasoline.

In part because neat alcohols are insufficiently volatile to enable a cold engine to start, even at moderate temperatures, the use of neat alcohols for automotive motor fuel is problematic (see ALCOHOL FUELS). Manufacturers have, however, reported that alcohol-powered cars, after being started and warmed up, can have the same or better driveability as gasoline cars (75–77). As for gasoline vehicles, port fuel injector fouling has occurred in some methanol vehicles and has affected driveability and emissions. Other problems related to high alcohol content gasoline in conventional engines include vapor lock and corrosion. Flexible-fuel vehicles (FFV), which can operate on either neat methanol or gasoline, or mixtures thereof, are being evaluated.

Gasoline blends containing oxygenates change the emissions characteristics of a motor vehicle designed for gasoline. Oxygenates and oxygenate-blends approved for use by the U.S. government are expected to have desirable emissions features as automotive fuels, and governmental environmental mandates and regulations have necessitated increased examination and implementation of oxygenates as fuels. As of this writing, however, no process can produce alcohols or ethers at equivalent or lower cost per volume than gasolines derived from natural petroleum.

Methanol

Methanol production in the 1990s is dominated by reaction of synthesis gas produced from natural gas. The economics of producing methanol as fuel are highly variable and site-specific. The natural gas feedstock has a broad range of values depending on location. Delivered costs for methanol are probably double that for gasoline from petroleum. Impurities, including water, vary according to the synthesis process employed. The term methyl fuel describes certain methanol products that may also contain significant quantities of water and higher alcohols as well as other oxygenated compounds. Water removal from such mixtures by distillative processes is generally complicated because of the formation of azeotropes. Methanol is used to produce MTBE, another oxygenate fuel.

Methanol as a fuel has been proposed in various ratios with gasoline. In gasoline formulations having relatively low methanol content, eg, M3 (3% methanol) and M15 (15% methanol), solubilizers are used and stringently dry conditions must be maintained. High methanol content fuels, M85 (85% methanol) and M100 (neat methanol), have special engine requirements. The use of high methanol content fuels is limited by methanol cost plus poor compatibility with the existing gasoline infrastructure.

Methanol is more soluble in aromatic than paraffinic hydrocarbons. Thus varying gasoline compositions can affect fuel blends. At room temperature, the solubility of methanol in gasoline is very limited in the presence of water. Generally, cosolvents are added to methanol–gasoline blends to enhance water tolerance. Methanol is practically insoluble in diesel fuel.

Concerns about using methanol–gasoline blend fuels include problems with vapor lock, cold start, and warmup. Oxygenate–gasoline blends, and in particular those containing methanol, have unusual volatility characteristics and cannot be accurately characterized using test methods developed for gasoline. Vapor pressure is an important volatility parameter that is adversely affected by the addition of methanol. Gasoline blends containing oxygenates form nonideal solutions with varying characteristics. In general, methanol–gasoline blend fuels exhibit increases in Reid vapor pressure (RVP) over that of gasoline itself. This effect contributes to vapor lock and evaporative emissions. The data available for assessing methanol's impact on exhaust emissions, and consequently on air quality, are limited. Formaldehyde has been reported in the exhausts of cars fueled with straight methanol or methanol–gasoline blends (see also EXHAUST CONTROL, AUTOMOTIVE).

The use of methanol as a motor fuel has been discussed since the 1920s. Straight methanol has long been a preferred fuel for racing engines because of the much higher compression ratios at which methanol may be combusted relative to gasolines. This is translatable for racing purposes at equivalent power outputs to engines of considerably reduced weight. However, fuel consumptions are roughly three times that of gasoline on a km/L (mi/gal) basis, and extremely high emissions of unburned fuel and carbon monoxide can result (78).

In Germany in the early 1950s, a 50:50 mixture of methanol and 2-propanol was blended with gasoline, first at a level of 7.5% and later at 1.5% (79). Complaints about stalling, power loss, and phase separation caused the ratio to be changed to 60:40::methanol:2-propanol but this apparently aggravated the problems. The practice was discontinued in 1970 when a tax was placed on alcohol.

In the United States, the Clean Air Act of 1970 imposed limitations on composition of new fuels, and as such methanol-containing fuels were required to obtain Environmental Protection Agency (EPA) waivers. Upon enactment of the Clean Air Act Amendments of 1977, EPA set for waiver unleaded fuels containing 2 wt % maximum oxygenates excluding methanol (0.3 vol % maximum). Questions regarding methanol's influence on emissions, water separation, and fuel system components were raised (80).

In 1979 Sun Oil Co. was granted a waiver for a gasoline blend containing 2.75 vol % methanol and an equal volume of *tert*-butyl alcohol (TBA) (2 wt % total oxygen). Cosolvents such as TBA were shown to reduce adverse effects of methanol on volatility and water tolerance. ARCO obtained EPA waiver in 1981 for a 3.5 wt % oxygen fuel blend containing Oxinol, also comprising equal parts of methanol and TBA. In 1985 a waiver was granted to Du Pont, Inc. for a gasoline blend containing 5 vol % maximum methanol with at least 2.5 vol % higher alcohol cosolvents. The waiver incorporated a water tolerance or phase separation requirement (81).

The most extensive worldwide program on methanol blend gasoline was in Italy where from 1982 to 1987 a 1.9×10^4 m^3/yr (5×10^6 gal/yr) plant produced a mixture containing 69% methanol. The balance contained higher alcohols. This mixture was blended into gasoline at the 4.3% level and marketed successfully as a premium gasoline known as Super E (82).

Methanol, a clean burning fuel relative to conventional industrial fuels other than natural gas, can be used advantageously in stationary turbines and boilers because of its low flame luminosity and combustion temperature. Low NO_x emissions and virtually no sulfur or particulate emissions have been observed (83). Methanol is also considered for dual fuel (methanol plus oil or natural gas) combustion power boilers (84) as well as to fuel gas turbines in combined methanol/electric power production plants using coal gasification (85).

Owing to its properties, methanol is not recommended for aircraft or marine fuel uses. Methanol cannot be used in conventional diesel-powered vehicles without modifications to the fuel system and engine. Simple methanol–diesel blends are not possible because of insolubility. Heavy-duty diesel engines have been adapted to use neat methanol by many U.S. manufacturers, and several are being used in field demonstrations (82) (see ALCOHOL FUELS).

Ethanol

Ethanol is produced both from ethylene derived from the cracking of petroleum fractions and by the fermentation of sugars derived from grains or other biomass. Many of its relevant properties are similar to those of methanol. Although ethanol may be a more desirable fuel or fuel component than methanol, its significantly higher cost (volume basis) may outweigh these advantages. Broad implementation of ethanol-containing fuels would require government action, eg, in the form of subsidies to farmers and fuel waivers.

The term gasohol has come into wide usage to identify, generally, a blend of gasoline and ethanol, with the latter derived from grain. The term may also be applied to blends of methanol or other alcohols in gasolines or other hydrocarbons, without regard to sources of components.

Brazil's Alcohol Program.

In Brazil, the enactment of legislation in 1931 made ethanol addition to gasoline compulsory at a level of 5% (86). Excess molasses and sugar were converted to alcohol in distilleries attached to sugar mills as a means to stabilize sugar prices. Production of fuel ethanol in the 1990s is mostly from biomass.

Starting in the city of Sao Paulo in 1977, and extending to the entire state of Sao Paulo in 1978, a gasohol incorporating 20% ethanol was mandated. Brazil's National Alcohol Program (Proalcool) set an initial goal of providing the 20% fuel mixture nationwide by 1980–1981 and a system of special tax, warranty, and price considerations were enacted to advance the aims of Proalcool.

For a considerable period, >90% of the new cars in Brazil operated on E96 fuel, or a mixture of 96% ethanol and 4% water (82). The engines have high compression ratios (ca 12:1) to utilize the high knock resistance of ethanol and deliver optimum fuel economy. In 1989 more than one-third of Brazil's 10 million automobiles operated on 96% ethanol/4% water fuel. The remainder ran on gasoline blends containing up to 20% ethanol (5).

Gasohol in the United States. Over 90% of the fuel ethanol in the United States is produced from corn. Typically, 0.035 m^3 (1 bushel) of corn yields 9.5 L (2.5 gal) of ethanol. Ethanol is produced by either dry or wet milling (87). Selection of the process depends on market demand for the by-products of the two processes. More than two-thirds of the ethanol in the United States is produced by wet milling. Depending on the process used, the full cost of ethanol after by-product credits has been estimated to be between \$0.25–0.53/L (\$1–2/gal) for new plants (88). Feedstock costs are a significant factor in the production of fuel ethanol. A change in corn price of \$0.29/m^3 (\$1.00/bushel) affects the costs of ethanol by \$0.08/L (\$0.30/gal).

Ethanol can also be produced from cellulose or biomass such as wood (qv), corn stover, and municipal solid wastes (see ENERGY FROM BIOMASS; FUELS FROM WASTE). Each of these resources has inherent technical or economic problems. The Tennessee Valley Authority (TVA) is operating a 2 t/d pilot plant on converting cellulose to ethanol.

After the oil embargo in 1973, gasohol use was stimulated by tax incentives. An application for EPA waiver of gasohol fuels (up to 10 vol % ethanol) was granted in 1979. From 1981 to 1983 the California Energy Commission field tested alcohol-powered cars equipped with a gasoline-assist starting system, ie, having an onboard auxiliary supply of volatile fuel for cold start tests. In 1989 about 8% of U.S. gasoline contained 10% ethanol plus a corrosion inhibitor (82). As of this writing, government waivers of RVP standards for gasohol fuels are being considered (89).

Methyl *t*-Butyl Ether

MTBE is produced by reaction of isobutene and methanol on acid ion-exchange resins. The supply of isobutene, obtained from hydrocarbon cracking units or by dehydration of *tert*-butyl alcohol, is limited relative to that of methanol. The cost to produce MTBE from by-product isobutene has been estimated to be between \$0.13 to \$0.16/L (\$0.50–0.60/gal) (90). Direct production of isobutene by dehydrogenation of isobutane or isomerization of mixed butenes are expensive processes that have seen less commercial use in the United States.

More than 95% of MTBE produced worldwide is used to blend with gasoline. In 1987 U.S. production of MTBE exceeded 3.8×10^6 m^3/yr (1×10^9 gal/yr) (82). The worldwide capacity for MTBE is increasing, especially in the United States and Europe, and has been projected to exceed production for years to come.

MTBE's gain in prominence as a fuel-blend component is a result of inherent technical advantages over other oxygenates, especially the lower alcohols. MTBE has a high blending octane number (Table 6) although this number varies somewhat with gasoline composition. The low vapor pressure relative to the lower alcohols results in no increase in RVP for MTBE-gasoline blends and consequently better evaporative emission and vapor lock characteristics. No phase separation occurs in blends with other fuels. MTBE, in blends of <20 vol % with gasoline, does not deleteriously affect other fuel or driving characteristics such as cold start, fuel consumption, and engine materials compatibility.

MTBE has been used in motor fuels in Europe since the early 1970s and is undergoing rapid growth, particularly in the United States. MTBE-blended gasoline containing up to 11 vol % MTBE received EPA waiver in 1981. Later legislation increased the MTBE waiver up to 15 vol %. In 1987–1988 Colorado began mandating use of winter oxygenate-based fuels in the Denver region. About 90% of the fuel in this period used a gasoline blend containing 8 vol % MTBE and in 1988–1989 the fuel was required to contain at least 2% oxygen (11 vol % MTBE). Based on the success of this program and EPA assessments that CO reductions of 10–20% over the next decade were possible with oxygenate-blend fuels, numerous state governments enacted legislation requiring the use of these fuels in winter and in cities having high ozone (smog) concentrations. The Clean Air Act Amendments of 1990 have mandated the use of reformulated gasolines, especially in serious ozone problem areas, by 1995.

The effectiveness of MTBE, however, is under discussion (91). Based on Denver, Colorado vehicle emissions data from 1981 to 1991 and theoretical models, Colorado scientists have claimed that the use of MTBE-blended fuels had no statistically significant effect on atmospheric CO levels, but increased pollutants such as formaldehyde. A drop in CO levels in Denver during this time period was attributed to fleet turnover of older, more polluting cars being replaced by newer cars having cleaner burning engines. In addition, health problems associated with direct exposure to MTBE in Fairbanks, Alaska has resulted in EPA exemption of the oxygenated fuel requirement in that area (91).

DIRECT LIQUEFACTION OF COAL

Direct liquefaction, the production of liquids from feed coal in a single processing scheme without a synthesis gas intermediate step, includes two routes for the upgrading of coal: hydrogenation and pyrolysis. In hydrogenation, the conversion of coal to liquids having higher hydrogen-to-carbon ratio involves the addition of hydrogen. Generally, the additional hydrogen required is added either from molecular hydrogen or from a hydrogen-donor solvent such as tetralin. Processes classified under pyrolysis are those which produce liquids by removal of carbon. This occurs when coal is thermally processed under inert or reducing atmospheric conditions. The use of hydrogen in a pyrolytic process to increase yields of distillate products is known as hydropyrolysis. Coal carbonization to produce metallurgical coke involves much the same chemistry as pyrolysis.

Coal and Coal-Tar Hydrogenation

If paraffinic and olefinic liquids are extracted from solid fuel substances, the hydrogen content of the residual material is reduced even further, and the residues become more refractory. The yields of liquids so derivable are generally low, even when a significant fraction of the hydrogen is extractable. Thus production of fuel liquids from nonliquid fuel substances such as coal and coal tars may be enhanced only by the introduction of additional hydrogen in a synthesis process. The principal differences in the processes are from the modes in which hydrogen is introduced and the catalysts used.

Hydrogenation of coal and other carbonaceous matter using high pressure hydrogen has been patented (92), and subsequently the Nobel Prize in chemistry was won for this accomplishment. By 1992, a 1 t/d plant was operating and using hydrogen at 10 MPa (100 atm) and 400°C to treat brown coal tar to give a liquid that comprised 25 wt % gasoline boiling at 75–210°C and 40 wt % middle oil, 210–300°C (see LIGNITE AND BROWN COAL). The pitch residue had a specific gravity of 1.04, and a solidification point of 15°C. The degree of liquefaction was shown to increase with decrease in oil rank (93). Liquid products were of low quality, being high in oxygen, nitrogen, and sulfur content, owing to low hydrogenation rates and polymerization of primary products (94).

In 1935 an ICI coal hydrogenation plant at Billingham, U.K., produced ca 136,000 t/yr motor fuel from bituminous coal and coal tar. By 1936, 272,000 t/yr of motor fuel were produced by improved hydrogenation of brown coal and coal tar at a facility constructed at Leuna and some 363,000 t/yr was being produced in three other German plants (95). Two years later the total German output from these facilities was ca 1.4×10^6 t/yr (96). The number of coal hydrogenation plants in Germany increased during World War II to 12, with total capacity of about 4×10^6 t/yr (100,000 bbl/d) of aviation and motor gasolines.

Experimental plants for hydrogenating coal or coal tar were operated in Japan, France, Canada, and in the United States before or during World War II. Much of that technology has remained proprietary. In general, coal-in-oil slurries containing iodine or stannous oxalate catalyst were subjected to liquid-phase hydrogenation at pressures of 25–70 MPa (250–700 atm). Liquids produced were fractionated, and the middle oils were then subjected to vapor-phase hydrogenation over molybdenum-, cobalt-, or tungsten sulfide-on-alumina catalysts (97). About 1 t of crude motor fuel was recovered from 4.5 t of coal, from which all necessary hydrogen and power requirements for the production were also obtained.

Developments in the United States. A large number of proprietary coal hydrogenation process variants have been proposed. Much of the technology originally directed to the catalytic hydrogenation of coals and coal tars in Germany has been applied to the hydrorefining of petroleum fractions, but U.S. commercial interest in coal hydrogenation was offset by the relative abundance of domestic petroleum up to World War II.

The huge demand for liquid fuels during World War II prompted the passage of the Synthetic Liquid Fuels Act of 1944. There were various programs relating to demonstration plants to produce liquid fuels from coal, oil shale, and other substances, including agricultural and forestry products. The Bureau of Mines had begun work on coal liquefaction in 1936, at which time a 45 kg/d experimental coal hydrogenation unit was constructed (98). The expanded program, after 1944, culminated in the construction and operation of a 45 t/d coal hydrogenation demonstration plant at Louisiana, Missouri, in 1949 (99), where a variety of problems and processing variations were investigated (100,101). Cost studies (102) showed that production of gasoline from coal hydrogenation could not compete with using natural petroleum as a gasoline source. The demonstration plant operations were terminated in 1953.

Work on coal hydrogenation continued by the Bureau of Mines on a laboratory scale (103–105). In one of these variants (106) coal-oil pastes admixed with catalyst in tubular reactors were hydrogenated at high pressure and low residence times to give improved yields of liquid products. The original thrust of the work was to hydrodesulfurize coal economically to produce environmentally acceptable boiler fuel (107). In the mid-1970s, a process sponsored by the Bureau of Mines named Synthoil (108) was developed, but the efforts were terminated by 1978 owing to limited catalyst lifetimes.

H-Coal Process. The H-coal process (Hydrocarbon Research, Inc., HRI, subsidiary of Dynalectron Corp.), for the conversion of coal to liquid products (109), is an application of HRI's ebullated-bed technology for the conversion of heavy oil residues into lighter fractions. Coal is dried, pulverized, and slurried with coal-derived oil (110). The coal-oil slurry is charged continuously with hydrogen to a reactor of unique design (111) containing a bed of ebullated catalyst, where the coal is hydrogenated and converted to liquid and gaseous products. The liquid product is a synthetic crude oil that can be converted to gasoline or heating oil by conventional refining processes. Alternatively, under milder operating conditions, a clean fuel gas and low sulfur fuel oils may be produced. The relative yields of these products depend on the desired sulfur level in the heavy fuel oil. In general, reaction products are separated by fractionation and absorption (qv). Unreacted coal may be fed into a fluid coker that produces gas, gas oil, and dry char. The coker gas oil, along with gas oils separated from the main reactor effluent, may be subjected to hydrocracking for conversion to lighter products.

In 1976, Ashland Oil (Ashland Synthetic Fuels, Inc.) was awarded the prime contract to construct a 540 t/d H-coal pilot plant adjacent to its refinery at Catlettsburg, Kentucky, by an industry–government underwriting consortium. Construction was completed in 1980 (112). The pilot-plant operation ended in early 1983.

Properties. The properties of naphtha, gas oil, and H-oil products from an H-coal operation are given in Table 7. These analyses are for liquids produced from the syncrude operating mode. Whereas these liquids are very low in sulfur compared with typical petroleum fractions, they are high in oxygen and nitrogen levels. No residual oil products (bp >540°C) are formed.

Solvent-Refined Coal Process. In the 1920s the anthracene oil fraction recovered from pyrolysis, or coking, of coal was utilized to extract 35–40% of bituminous coals at low pressures for the purpose of manufacturing low cost newspaper inks (113). Tetralin was found to have higher solvent power for coals, and the I. G. Farben Pott-Broche process (114) was developed, wherein a mixture of cresol and tetralin was used to dissolve ca 75% of brown coals at 13.8 MPa (2000 psi) and 427°C. The extract was filtered, and the filtrate vacuum distilled. The overhead was distilled a second time at atmospheric pressure to separate solvent, which was recycled to extraction, and a heavier liquid, which was sent to hydrogenation. The bottoms product from vacuum distillation, or solvent-extracted coal, was carbonized to produce electrode carbon. Filter cake from the filters was coked in rotary kilns for tar and oil recovery. A variety of liquid products were obtained from the solvent extraction-hydrogenation system (113). A similar process was employed in Japan during World War II to produce electrode coke, asphalt (qv), and carbonized fuel briquettes (115).

In the United States there was little interest in solvent processing of coals. A method to reduce the sulfur content of coal extracts by heating with sodium hydroxide and zinc oxide was, however, patented in 1940 (116). In the 1960s the technical feasibility of a coal deashing process was studied (117), and a pilot plant able to process ca 45 t/d was completed in late 1974 (118).

A flow diagram of the solvent-refined coal or SRC process is shown in Figure 12. Coal is pulverized and mixed with a solvent to form a slurry containing 25–35 wt %

Table 7. Properties of Syncrude from H-Coal Process[a]

Property	Boiling Range			Total
	Initial to 190°C	190–343°C	343–524°C	
Specific gravity, (°API)[b]	0.767 (53.0)	0.915 (23.2)	1.05 (3.5)	0.863 (32.4)
Vol % on total	40.0	54.2	5.8	100.0
Analysis, wt %				
Carbon	84.5	88.8	89.4	87.3
Hydrogen	13.6	11.0	10.2	11.9
Oxygen	1.7			0.6
Nitrogen	0.1	1.0	0.1	0.1
Sulfur	0.1	0.1	0.3	0.1
Total	*100*	*100*	*100*	*100*

[a] Ref. 111.

[b] $°API = \frac{141.5}{\rho} - 131.5$ where ρ is specific gravity.

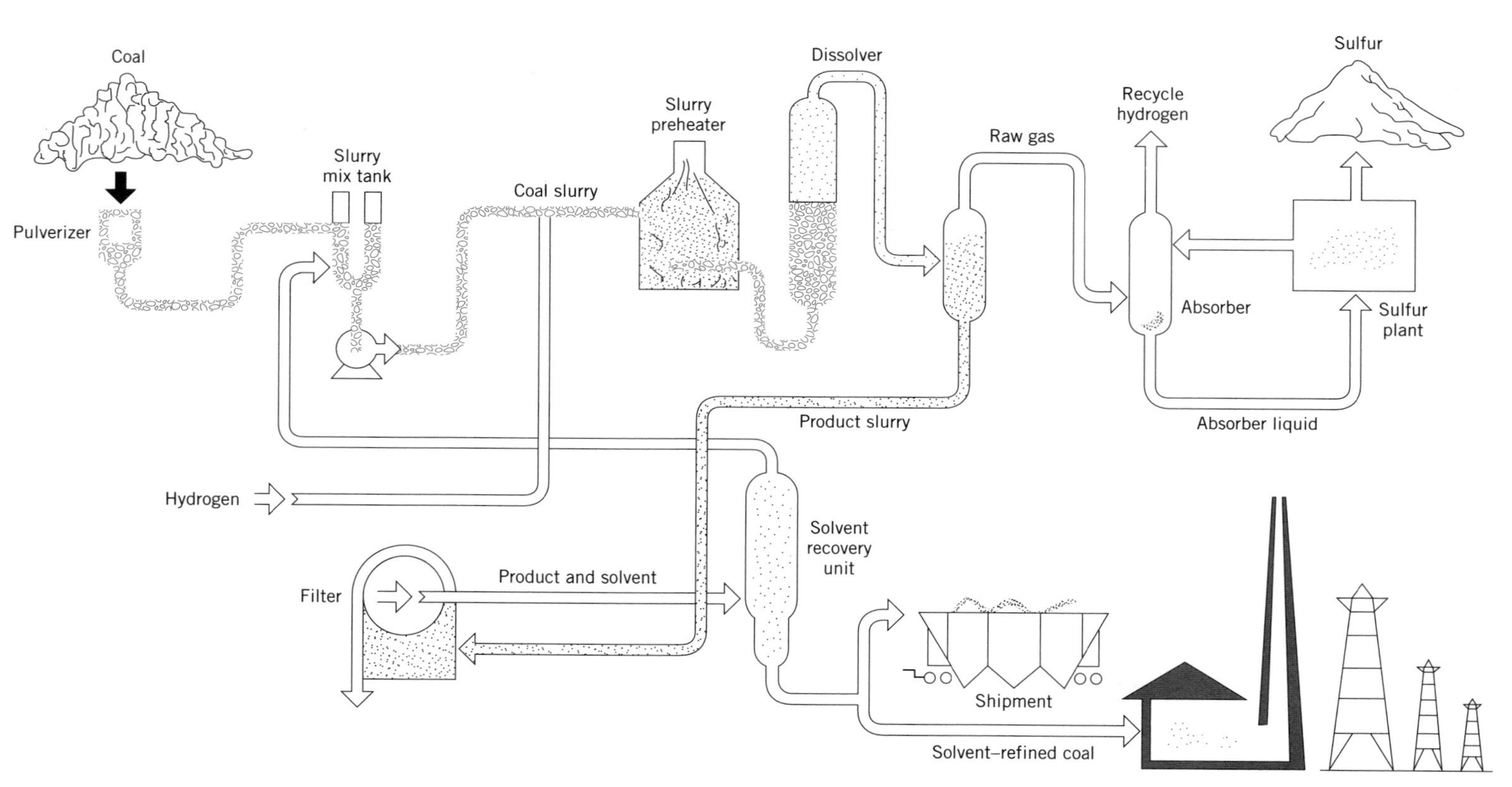

Figure 12. Solvent-refined coal process (119).

coal. The slurry is pressurized to ca 7 MPa (1000 psig), mixed with hydrogen, and heated to ca 425°C. The solution reactions are completed in ca 20 min and the reaction product flashed to separate gases. The liquid is filtered to remove the mineral residue (ash and undissolved coal) and fractionated to recover the solvent, which is recycled.

The liquid remaining after the solvent has been recovered is a heavy residual fuel called solvent-refined coal, containing less than 0.8 wt % sulfur and 0.1 wt % ash. It melts at ca 177°C and has a heating value of ca 37 MJ/kg (16,000 Btu/lb), regardless of the quality of the coal feedstock. The activity of the solvent is apparently more important than the action of gaseous hydrogen in this type of uncatalyzed hydrogenation. Research has been directed to the use of petroleum-derived aromatic oils as start-up solvents (118).

In the early 1970s production of low sulfur, ashless (solid) boiler fuel was the preferred commercial application (119). This basic process (SRC-I) yielded small amounts of liquid oil products with additional processing. Liquid output was significantly increased by the coal-oil-gas (COG) refinery concept (120–122) which incorporated high degrees of hydroconversion and hydrotreating. A SCR-II process has been developed, in which hydrocracking occurs in the solution (hydrogenation) vessels (123). A low viscosity fuel oil is the primary distillate product in this case, although naphtha and LPG are also recovered.

Two pilot plants have been built and operated to demonstrate the feasibility of the SRC process. These included a 6 t/d plant at Wilsonville, Alabama (*vide infra*) and a 50 t/d plant at Ft. Lewis, Washington which was operated from 1974 to 1981.

In an effort to obtain higher value products from SRC processes, a hydrocracking step was added to convert resid to distillate liquids. The addition of a hydrocracker to the SRC-I process was called nonintegrated two-stage liquefaction (NTSL). The NTSL process was essentially two separate processes in series: coal liquefaction and resid upgrading. NTSL processes were inefficient owing to the inherent limitations of the SRC-I process and the high hydrocracker severities required.

Properties. The properties of the liquid fuel oil produced by the SRC-II process are influenced by the particular processing configuration. However, in general, it is an oil boiling between 177 and 487°C, having a specific gravity of 0.99–1.00, and a viscosity at 38°C of 40 SUs (123). Pipeline gas, propane and butane (LPG), and naphtha are also recovered from an SRC-II complex.

Exxon Donor Solvent-Coal Liquefaction Process. The EDS process from Exxon is a hydrogenation process using a donor solvent for the direct conversion of a broad range of coals to liquid hydrocarbons (124). In the process sequence, shown in Figure 13 the feed coal is crushed, dried, and slurried with hydrogenated recycle solvent (the donor solvent) and fed to the reactor with hydrogen. The reactor is an upward plug-flow design operating at 430–480°C and at ca 14 MPa (2000 psi) total pressure.

The reactor effluent is separated by conventional distillation into recycle solvent, light gases, C_4 to 537°C bp distillate, and a heavy vacuum bottoms stream containing unconverted coal and ash. The recycle solvent is hydrogenated in a separate reactor and sent back to the liquefaction reactor.

The heavy vacuum bottoms stream is fed to a Flexicoking unit. This is a commercial (125,126) petroleum process that employs circulating fluidized beds at low (0.3 MPa (50 psi)) pressures and intermediate temperatures, ie, 480–650°C in the coker and 815–980°C in the gasifier, to produce high yields of liquids or gases from organic material present in the feed. Residual carbon is rejected with the ash from the gasifier fluidized bed. The total liquid product is a blend of streams from liquefaction and the Flexicoker.

The EDS process was developed starting from 1976 in a 10-year joint undertaking between DOE and private in-

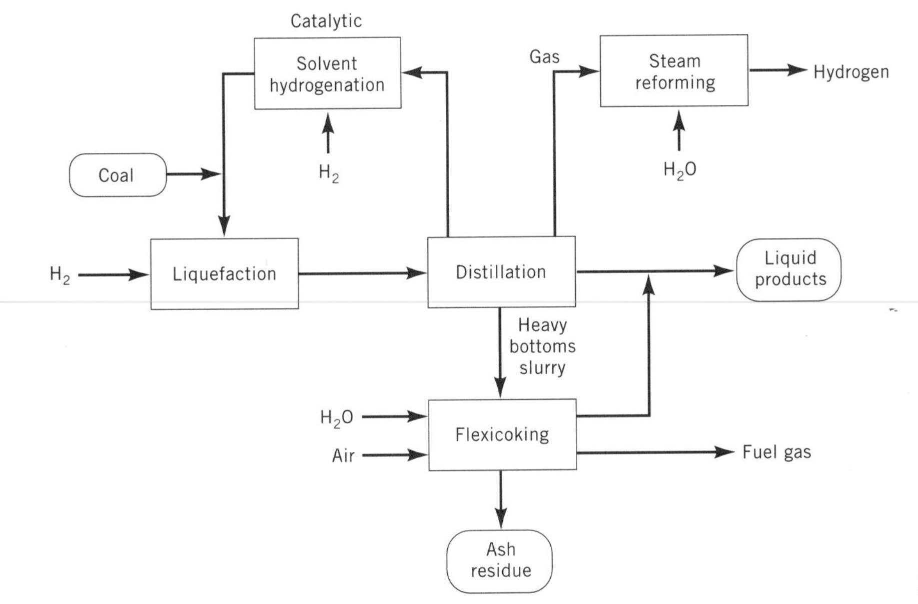

Figure 13. Exxon donor solvent process (124).

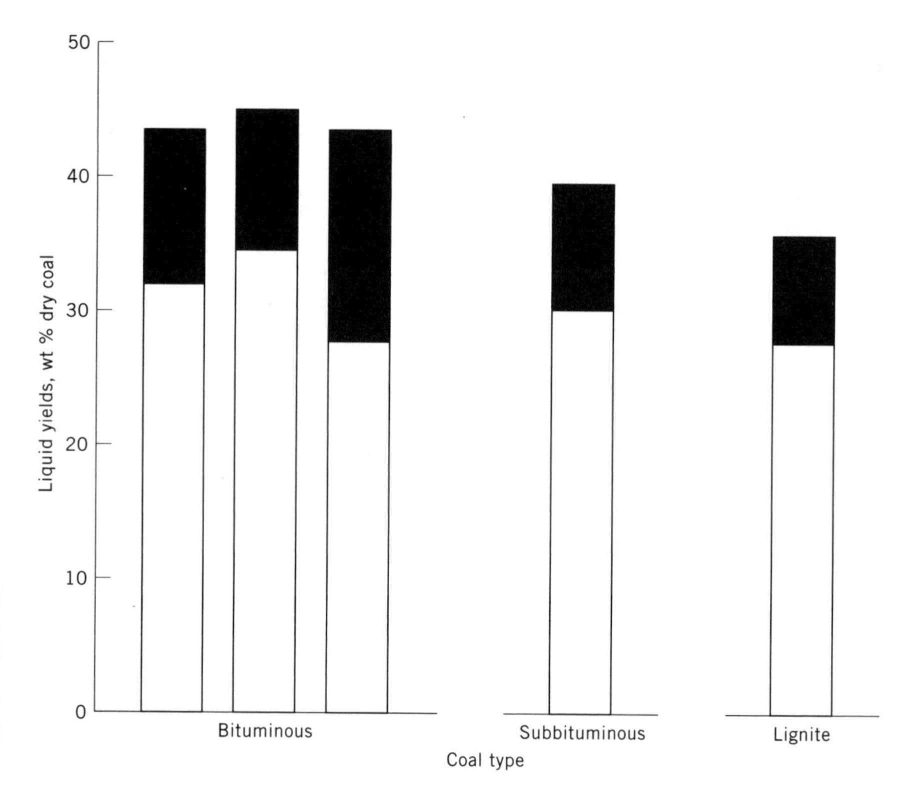

Figure 14. Preferred liquefaction-coking liquid yields in the EDS process for various coals where ■ represents Flexicoking liquids and, □ liquefaction liquids (124). A, Ireland (West Virginia); B, Monterey (Illinois); C, Burning Star (Illinois); D, Wyodak (Wyoming); and E, Big Brown (Texas).

dustry (127). Under the direction of Exxon Co. USA, a 250 t/d pilot plant was operated at Baytown, Texas. Operation of this unit began in 1980 and was completed by late 1982.

Properties. Pilot-unit data indicate the EDS process may accommodate a wide variety of coal types. Overall process yields from bituminous, subbituminous, and lignite coals, which include liquids from both liquefaction and Flexicoking, are shown in Figure 14. The liquids produced have higher nitrogen contents than are found in similar petroleum fractions. Sulfur contents reflect the sulfur levels of the starting coals: ca 4.0 wt % sulfur in the dry bituminous coal; 0.5 wt % in the subbituminous; and 1.2 wt % sulfur in the dry lignite.

Table 8 shows that the naphthas produced by the EDS process have higher concentrations of cycloparaffins and phenols than do petroleum-derived naphthas, whereas the normal paraffins are present in much lower concentrations. The sulfur and nitrogen concentrations in coal naphthas are high compared to those in petroleum naphthas.

Table 8. Composition of Naphthas[a] from Various Sources[b]

	EDS Coal Liquefaction		Petroleum Naphthas	
Component, wt %	Illinois Coal	Wyodak Coal	Cycloparaffinic[c]	Paraffinic[d]
Saturated compounds	69.9	60.8	77.9	81.3
Paraffins	13.4	19.4	36.0	58.2
Cycloparaffins	56.5	41.4	41.9	23.1
Olefins[e]	0.5	3.7		
Aromatics	17.0	28.6	22.1	18.7
Benzenes	11.7	25.1	20.8	17.2
Indanes, tetralins	5.1	3.5	1.0	1.2
Indenes	0.2	0.0		
Naphthalenes	0.01	0.0	0.3	0.3
Phenols	12.6	6.9	Traces	Traces
Total	*100.0*	*100.0*	*100.0*	*100.0*
Sulfur	0.57	0.10	0.035	0.049
Nitrogen	0.15	0.18	0.0001	0.0001
Oxygen	1.60	2.49		

[a] C_5 to 204°C bp.
[b] Ref. 128.
[c] Prudhoe Bay.
[d] Arab Light.
[e] Values are approximate.

Gas oil fractions (204–565°C) from coal liquefaction show even greater differences in composition compared to petroleum-derived counterparts than do the naphtha fractions (128). The coal-gas oils consist mostly of aromatics (60%), polar heteroaromatics (25%), asphaltenes (8–15%), and saturated compounds (<10%). Petroleum-gas oils, on the other hand, contain more than 50% saturated compounds, less than 5% polar heteroaromatics, and no asphaltenes. Furthermore, the aromatics of petroleum-gas oils have longer side chains.

Coal Liquefaction at Wilsonville. Starting in 1974 the Advanced Coal Liquefaction R&D Facility at Wilsonville, Alabama operated a 6 t/d pilot plant and studied various coal liquefaction processing schemes. The facility, cosponsored by the DOE, the Electric Power Research Institute (EPRI) and Amoco Oil Co, was shut down in early 1992.

Initial operation at the Wilsonville pilot plant was in SRC-I mode and later evolved into a two-stage process (129) by operation in NTSL mode. NTSL limitations described previously combined with high hydrogen consumptions resulted in subsequent focus on a staged integrated approach, which was to be the basis for all further studies at Wilsonville.

The integrated two-stage process (ITSL) combined short contact time liquefaction in one reactor with ebullated-bed hydrocracking in a second stage (130). The short contact time conditions permitted better hydrogen transfer from the solvent rather than from the gas phase. The hydrocracking step operated at lower severity resulting in lowered gas make and improved hydrogen efficiency. Recycle solvent was generated from the hydrocracked distillates and coupled the two reaction stages. Results of ITSL processing of Illinois No. 6 coal at Wilsonville are given in Table 9. Distillate yields and coal throughput for ITSL were higher than those obtained by NTSL.

Further developments of the ITSL process resulted in incremental gains in distillate yields (131). Reconfigured integrated two-stage liquefaction (RITSL) involved placing the solvent deasher after the hydrocracker thus producing a recycle solvent consisting of deashed resid and distillate. This resulted in reduction of feed to the deasher and reduced organic rejection. Close coupled integrated two-stage liquefaction (CC-ITSL) linked the two reactors and removed several operations between the two stages. A deleterious effect of these two processing modes was increased hydrogen consumption over ITSL.

From 1985 to 1992, development activity at Wilsonville

Table 9. Wilsonville Plant Operating Conditions and Yields for ITSL and CTSL Modes[a]

Parameter	Mode of Operation[b] ITSL	CTSL
Operating conditions		
Run number	7242BC; 243JK/244B	253A
Catalyst	Shell 32M	Shell 317
First stage[c]		
Average reactor temperature, °C	460; 432	432
Space velocity	690; 450[d]	4.8[e]
Pressure, MPa[f]	17; 10–17	17.9
Second stage		
Average reactor temperature, °C	382	404
Space velocity, feed/catalyst[e]	1.0	4.3
Catalyst age, resid/catalyst	278–441; 380–850	100–250
Yields[g]		
C_1–C_3 gas	4; 6	6
C_4 + distillate	54; 59	70
Resid	8; 6	−1
Hydrogen consumption	4.9; 5.1	6.8
Other		
Hydrogen efficiency, C_4+ distillate/H_2 consumed	11, 11.5	10.3
Distillate Selectivity, C_1–C_3/C_4 + distillate	0.07; 0.10	0.08
Energy Content of feed coal reject to ash concentrate, %	24; 20–23	20

[a] Feed is Illinois No. 6 coal.
[b] CTSL = catalytic two-stage liquefaction; ITSL = integrated two stage liquefaction.
[c] First stage is thermal for ITSL.
[d] Value given is coal space velocity at temp >371°C in kg/m^3.
[e] Value given is in h^{-1}.
[f] To convert MPa to psia, multiply by 145.
[g] Wt % on a moisture- and ash-free (MAF) coal basis.

was on catalytic two-stage liquefaction (CTSL). CTSL, initiated by HRI (132), consists of catalytic processing in two ebullated-bed reactors which lower reaction temperatures and increase distillate yields, up to 78% yield. CTSL results from Wilsonville for Illinois No. 6 coal are also given in Table 9. Distillate yields were shown to be significantly higher for CTSL over ITSL; however, hydrogen consumption was somewhat increased.

Properties. CTSL distillates have qualities comparable to or better than No. 2 fuel oil and have good hydrogen content and low heteroatom contents. Distillates having a higher boiling point distribution from Wilsonville CTSL operation (131) showed 26.8°API gravity with heteroatom levels of 0.11 wt % sulfur, <1 wt % oxygen, and 0.16 wt % nitrogen.

Coal Pyrolysis

Pyrolysis is the destructive distillation of coal in the absence of oxygen typically at temperatures between 400 and 500°C (133). As the temperature of carbonaceous matter is increased, decomposition ultimately occurs. Melting and dehydration may also occur. Coals exhibit more or less definite decomposition temperatures, as indicated by melting and rapid evolution of volatile components, including potential fuel liquids, during destructive distillation (134). Table 10 summarizes an extensive survey of North American coals subjected to laboratory pyrolysis. The yields of light oils so derived average no more than ca 8.3 L/t (2 gal/short ton), and tar yields of ca 125 L/t (30 gal/short ton) are optimum for high volatile bituminous coals (135).

Coal pyrolysis has been studied at both reduced and elevated pressures (136), and in the presence of a variety of agents and atmospheres (137). Although important to the study of coal structure and reactions, coal pyrolysis, as a means to generate liquids, has proved to have limited commercial value.

COED Process. Sponsored by the Office of Coal Research of the U.S. Department of the Interior, the COED process was developed by FMC Corp. as Project Char-Oil-Energy Development (COED) through 1975 (138–140). Bench-scale experiments led the way to construction in 1965 of a process development unit employing multistage, fluidized-bed pyrolysis to process 45 kg/h (141). Correlated studies included hydrotreating of COED oil (142), high temperature hydrodesulfurization of COED char, and investigations of char-oil and char-water slurry pipe-lining economics (143). A pilot plant capable of processing up to 33 t/d and hydrotreating 4.7 m^3/d (30 bbl/d) was started up in 1970 (144), and was operated successfully for a number of years (145).

The COED concept (139), designed to recover liquid, gaseous, and solid fuel components, consists of four stages. Heat is generated by the reaction of oxygen with a portion of the char in the last pyrolysis stage and is also introduced by the air combustion of gas to dry feed coal. The number of stages in the pyrolysis, and the operating temperatures in each, may be varied to accommodate high volatile bituminous and subbituminous feed coals with widely ranging caking or agglomerating properties.

Oil condensed from the released volatiles from the second stage is filtered and catalytically hydrotreated at high pressure to produce a synthetic crude oil. Medium heat-content gas produced after the removal of H_2S and CO_2 is suitable as clean fuel. The pyrolysis gas produced, however, is insufficient to provide the fuel requirement for the total plant. Residual char, 50–60% of the feed coal, has a heating value and sulfur content about the same as feed coal, and its utilization may thus largely dictate process utility.

Properties. The properties of char products from two possible coal feeds, a low sulfur Western coal, and a high sulfur Midwestern coal, are shown in Table 11. The char derived from the low sulfur Western coal may be directly suitable as plant fuel, with only minor addition of clean process gas to stabilize its combustion. Flue gas desulfurization may not be required. Flue gas from the combustion of the char derived from the high sulfur Illinois coal, however, requires desulfurization before it may be discharged into the atmosphere.

Typical COED syncrude properties are shown in Table

Table 10. Average Yields and Range of Yields of Fischer Assay of Various Coals[a,b]

Rank of coal	Coke, %		Tar, L/t[c]		Light oil, L/t[c]		Gas, m^3/t[d]		Water, %	
	Average	Range	Average	Range	Average	Range	Average	Range	Average	Range
Semianthracite			3.2		0.14					
Low volatile bituminous	89.7	85.8–93.3	39.6	29.0–58.4	4.69	3.36–7.41	59.8	54.4–66.6	3.2	1.1–6.6
Medium voltaile bituminous	83.3	77.4–90.4	86.9	44.6–117.8	7.68	4.92–10.58	66.0	47.3–76.2	4.1	2.8–7.0
High volatile A bituminous	75.5	68.8–81.4	142.1	105.3–187.2	10.53	6.81–15.09	67.0	57.5–80.2	6.0	3.0–9.2
High voltatile B bituminous	70.4	66.0–73.2	139.4	111.8–198.3	10.03	7.13–15.82	68.3	56.4–82.3	11.1	10.2–13.1
High volatile C bituminous	67.1	65.4–68.6	124.2	85.1–178.5	8.65	5.93–12.47	61.2	53.0–70.4	15.9	12.0–19.1
High volatile C bituminous or subbituminous A	59.1		94.3	84.6–112.2	7.59	6.26–8.88	90.4		23.4	
Subbituminous A			81.9	81.0–82.8	6.21	6.12–6.26				
Subbituminous B	57.6	54.8–59.9	70.8	60.7–76.8	6.12	5.24–7.13	90.4	62.2–93.8	27.8	23.3–30.4
Lignite	36.5		69.9	30.8–124.2	5.47	2.90–8.69	71.4		44.0	
Cannel	58.8	44.1–69.0	338.1	247.0–498.2	23.28	16.84–34.13	61.5	51.0–72.1	3.7	2.0–4.8

[a] Ref. 135.
[b] As-received basis; maximum temperature, 500°C.
[c] To convert L/t to gal/short ton, divide by 4.6
[d] To convert m^3/t to ft^3/short ton, multiply by 29.4.

Table 11. Properties of COED Char Product[a]

Property	Utah Coal	Illinois No.6
Volatile matter, wt %	6.1	2.7
Fixed carbon, wt %	80.2	77.0
Ash, wt %	13.7	20.3
Higher heating value, MJ/kg,[b] dry	28.6	25.6
Elemental analysis, wt %, dry		
Carbon	81.5	73.4
Hydrogen	1.3	0.8
Nitrogen	1.5	1.0
Sulfur	0.5	3.4
Oxygen	1.5	1.0
Chlorine	0.006	0.1
Iron[c]	0.28	

[a] Ref. 139.
[b] To convert MJ/kg to Btu/lb, multiply by 430.
[c] Included in ash above.

12. The properties of the oil products depend heavily on the severity of hydroprocessing. The degree of severity also markedly affects costs associated with hydrogen production and compression. Syncrudes derived from Western coals have much higher paraffin and lower aromatic content than those produced from Illinois coal. In general, properties of COED products have been found compatible with expected industrial requirements.

Occidental Petroleum Coal Conversion Process. Garrett R&D Co. (now the Occidental Research Co.) developed the Oxy Coal Conversion process based on mathematical simulation for heating coal particles in the pyrolysis unit. It was estimated that coal particles of 100-mm diameter could be heated throughout their volumes to decomposition temperature (450–540°C) within 0.1 s. A large pilot facility was constructed at LaVerne, California, in 1971. This unit was reported to operate successfully at feed rates up to 136 kg/h (3.2 t/d).

Hot product char carries heat into the entrained bed to obtain the high heat-transfer rates required. Feed coal must be dried and pulverized. A portion of the char recovered from the reactor product stream is cooled and discharged as product. The remainder is reheated to 650–870°C in a char heater blown with air. Gases from the reactor are cooled and scrubbed free of product tar. Hydrogen sulfide is removed from the gas, and a portion is recycled to serve as the entrainment medium.

Properties. A high volatile western Kentucky bituminous coal, the tar yield of which by Fischer assay was ca 16%, gave a tar yield of ca 26% at a pyrolysis temperature of 537°C (146–148). Tar yield peaked at ca 35% at 577°C and dropped off to 22% at 617°C. The char heating value is essentially equal to that of the starting coal, and the tar has a lower hydrogen content than other pyrolysis tars. The product char is not suitable for direct combustion because of its 2.6% sulfur content.

The TOSCOAL Process. The Oil Shale Corp. (TOSCO) piloted the low temperature carbonization of Wyoming subbituminous coals over a two-year period in its 23 t/d pilot plant at Rocky Falls, Colorado (149). The principal objective was the upgrading of the heating value in order to reduce transportation costs on a heating value basis. Hence, the solid char product from the process represented 50 wt % of the starting coal but had 80% of its heating value.

Furthermore, 60–100 L (14–24 gal) oil, having sulfur content below 0.4 wt %, could be recovered per metric ton coal from pyrolysis at 427–517°C. The recovered oil was suitable as low sulfur fuel. Figure 15 is a flow sheet of the Rocky Flats pilot plant. Coal is fed from hoppers to a dilute-phase, fluid-bed preheater and transported to a pyrolysis drum, where it is contacted by hot ceramic balls. Pyrolysis drum effluent is passed over a trommel screen that permits char product to fall through. Product char is thereafter cooled and sent to storage. The ceramic balls are recycled and pyrolysis vapors are condensed and fractionated.

Properties. Results for the operation using subbituminous coal from the Wyodad mine near Gillette, Wyoming, are shown in Table 13. Char yields decreased with increasing temperature, and oil yields increased. The Fischer assay laboratory method closely approximated the yields and product assays that were obtained with the TOSCOAL process.

Table 12. Typical COED Syncrude Properties[a]

Property	Utah A-Seam	Illinois No. 6 Seam
Specific gravity, (°API)[b]	0.934 (20)	0.929 (22)
Pour point, °C	16	−18
Flash point, closed cup, °C	24	16
Viscosity, at 38°C, mm²/s (= cSt)	8	5
Ash, wt %	0.01	0.01
Moisture, wt %	0.1	0.1
Metals, ppm	10	10
Elemental analysis, wt %		
C	87.2	87.1
H	11.0	10.9
N	0.2	0.3
O	1.4	1.6
S	0.1	0.1
ASTM distillation initial bp, °C	138	88
10%	221	134
30%	277	199
50%	349	270
70%	416	316
90%	493	362
End point (95%)	510	397
Hydrocarbon type analysis, liquid vol %		
Paraffins		10.4
Olefins	23.7	0
Naphthenes	0	41.4
Aromatics	42.2	48.2
	34.1	

[a] Ref. 139.
[b] $°API = \frac{141.5}{\rho} - 131.5$ where ρ is specific gravity.

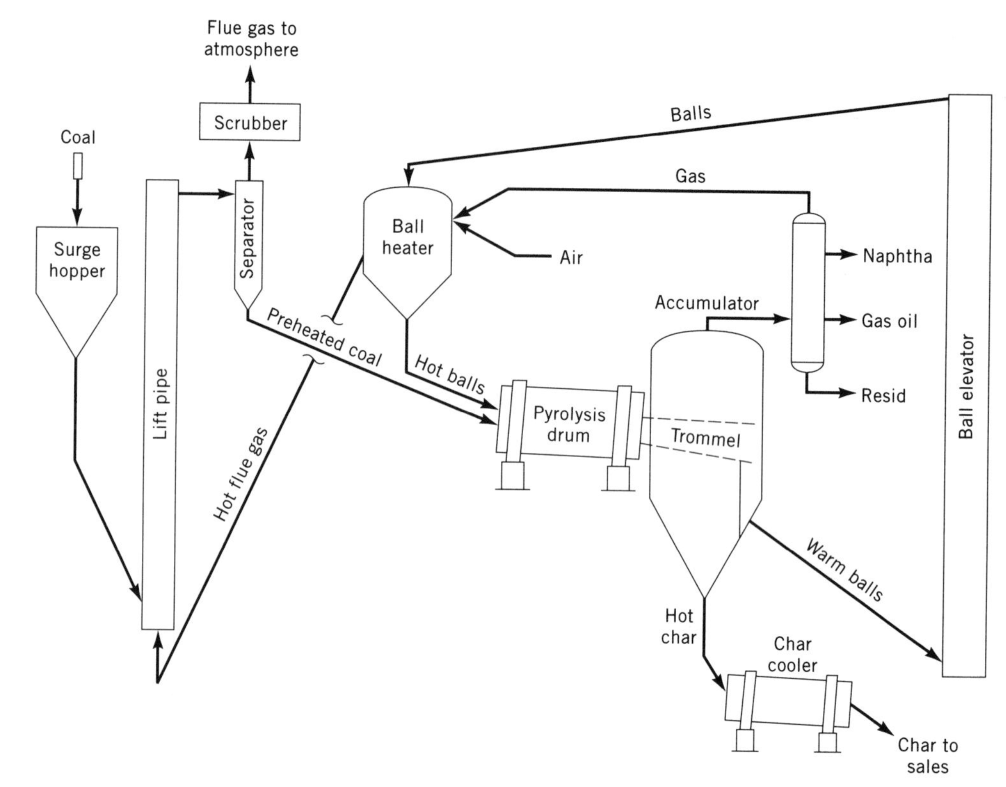

Figure 15. TOSCOAL process (149).

The volatiles contents of product chars decreased from ca 25–16% with temperature. Char (lower) heating values, on the other hand, increased from ca 26.75 MJ/kg (11,500 Btu/lb) to 29.5 MJ/kg (12,700 Btu/lb) with temperature. Chars in this range of heating values are suitable for boiler fuel application and the low sulfur content (about equal to that of the starting coal) permits direct combustion. These char products, however, are pyrophoric and require special handling in storage and transportation systems.

Properties of the tar oil products are given in Table 14. The oils change only slightly with change in the retorting temperature; sulfur levels are low. The fraction boiling up to 230°C contains 65 wt % of phenols, cresols, and cresylic acids.

TOSCO tar oils have high viscosity and may not be transported by conventional pipelines. Heating values of product gas on a dry, acid gas-free basis are in the natural gas range if butanes and heavier components are included.

Table 13. TOSCOAL Retorting of Wyodak Coal[a,b]

	Retort Temperature		
Component	427°C	482°C	521°C
Char	524.5	505.8	484.4
Gas, C_3 and lighter	59.5	78.4	63.0
Oil, C_4 and heavier	57.0	71.5	93.1
Water[c]	351.0	351.0	351.0
Total	*992.0*	*1006.7*	*991.5*
Recovery, %	*99.2*	*100.7*	*99.1*

[a] Ref. 150.
[b] Product yields, kg/t, of as-mined coal.
[c] Value assumed from Fischer assay and moisture content. The addition of steam to the process prevented accurate measurement of water produced in retorting.

Coal Carbonization. In the by-product recovery of a modern coke oven, coal tar is removed first by cooling the gases emanating, and light oil is removed last by scrubbing the gas with solvents. Other products, including ammonia, phenols, pyridine, or naphthalene, may be recovered between these operations. The constituents of coal tar, light oil, and gas usually overlap considerably, ie, the fractional condensation does not effectively separate individual components. Assuming the lowest boiling coal tar constituent to be benzene (bp 80.09°C), and the highest boiling to be naphthalene, the overlapping compositions of gas, light oil, and tar may be as shown graphically in Figure 16. Many chemical compounds have been identified (8) in these substances. Included are most of the significant constituents of petroleum-derived fuel liquids, although only a few components are present in sufficient quantity to make commercial recovery feasible.

The precise compositions of the light oil and coal tar recovered from coke-oven gas is a distinct function of the design of the recovery system, as well as of the properties of the starting coal. In general, 12.5–16.7 L/t (3–4 gal/m light oil per short ton) of coal carbonized is recovered from high temperature coke-oven operations. Light oil may contain 55–70% benzene, 12–20% toluene, and 4–7% xylene. Unrecovered light oil appearing in the effluent coal gas may comprise ca 1 vol % and contribute ca 5% of the gas's heating value. Refining of light oil consists mainly of sulfuric acid washing, followed by fractional distillation.

Large-scale recovery of light oil was commercialized in

Table 14. TOSCOAL Oil Properties[a]

Properties	Retort temperature		
	427°C[b]	482°C	521°C
Analysis, wt %			
Carbon	81.4	80.7	80.9
Hydrogen	9.3	9.1	8.7
Oxygen	8.3	9.4	9.3
Nitrogen	0.48	0.7	0.7
Sulfur	0.43	0.2	0.2
Chlorine	0.0	0.0	0.0
Ash	0.0	0.2	0.1
Total	*99.91*	*100.3*	*99.9*
Heating values			
Gross, MJ/kg[c]	38.59	37.72	37.13
Net, MJ/kg[c]	36.61	35.75	35.26
Specific gravity, (°API)[d]		1.040 (4.5)	1.061 (1.9)
Primary oil	1.015 (7.9)	0.985 (12.1)	1.027 (6.2)
Calculated, with C_4 and heavier components of gas added	0.978 (13.2)		
Pour point, °C	32	38	35
Conradson carbon, wt %	7.6	9.9	11.4
Distillation,[e] °C			199
2.5 vol %	212	216	235
20 vol %	302	288	385
50 vol %	407	413	
Viscosity, SUs			
at 82°C	122	123	128
at 90°C	63	66	69

[a] Ref. 150.
[b] Feed coal was different from that used for 482 and 521°C.
[c] To convert MJ/kg to Btu/lb, multiply by 430.
[d] $°API = \frac{141.5}{\rho} - 131.5$ where ρ is specific gravity.
[e] Combination of true boiling point and D1160 distillations.

England, Germany, and the United States toward the end of the nineteenth century (151). Industrial coal-tar production dates from the earliest operation of coal-gas facilities. The principal bulk commodities derived from coal tar are wood-preserving oils, road tars, industrial pitches, and coke. Naphthalene is obtained from tar oils by crystallization, tar acids are derived by extraction of tar oils with caustic, and tar bases by extraction with sulfuric acid. Coal tars generally contain less than 1% benzene and toluene, and may contain up to 1% xylene. The total U.S. production of BTX from coke-oven operations is insignificant compared to petroleum product consumptions.

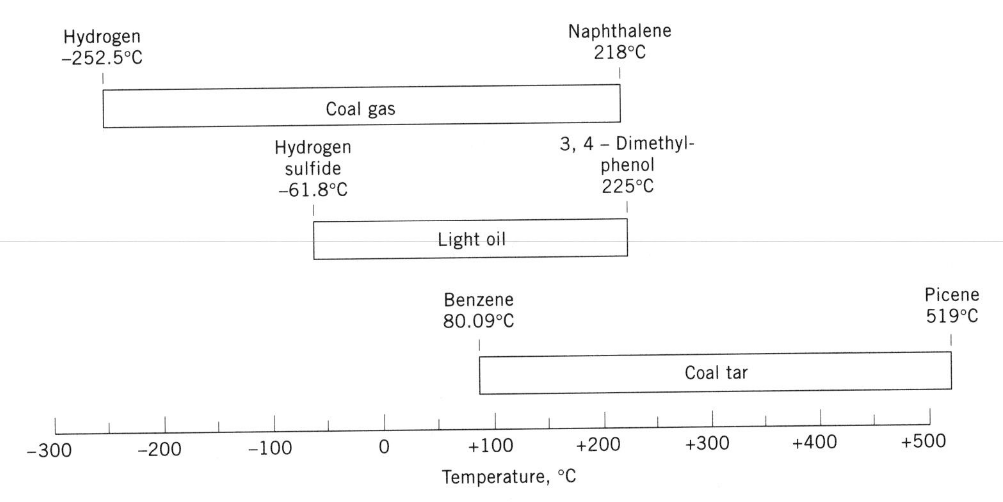

Figure 16. Boiling ranges of carbonization products (8).

OTHER PROCESSES

Shale Oil

In the United States, shale oil, or oil derivable from oil shale, represents the largest potential source of liquid hydrocarbons that can be readily processed to fuel liquids similar to those derived from natural petroleum. Some countries produce liquid fuels from oil shale. There is no such industry in the United States although more than 50 companies were producing oil from coal and shale in the United States in 1860 (152,153), and after the oil embargo of 1973 several companies reactivated shale-oil process development programs (154,155). Petroleum supply and price stability has since severely curtailed shale oil development. In addition, complex environmental issues (156) further prohibit demonstration of commercial designs.

Heavy Oil

The definitions used to distinguish among naturally occurring heavy petroleum oils, bitumens, asphalts, and tars, are subject to broad variations. More than 10% of the world's current crude oil production has an API gravity below 20°, or a specific gravity greater than 0.934^{15}_{15}. Oils having sp $gr^{15}_{15} > 0.904–0.934$ (20–25°API) are considered heavy oils in most classifications. However, Safaniyah crude oil produced in Saudi Arabia having sp gr^{15}_{15} of 0.893 (27°API) carries the designation Arabian heavy, and in petroleum parlance is generally referred to as heavy crude oil. Yet its production method does not differ from that of Arabian medium or Arabian light crude oils.

Energy in the form of injected water or CO_2 may be supplied to increase the rate of production of light crude oils. Application of heat to the reservoirs, eg, using hot water, steam, heated CO_2, fireflood, or *in situ* combustion, however, is generally associated with the production of heavier, viscid crudes.

Heavy crude oil is widely distributed, and it is difficult to estimate reserves separate from normal crude oil reserves or from tar sands deposits. Estimates of petroleum reserves frequently include a large heavy oil component, which can only be produced at significantly higher cost than light oil.

Most heavy oil production is concentrated in California, Canada, and Venezuela. There is significant production of heavy oil in California from the Kern River field near Bakersfield and in Canada from the Cold Lake deposit in Alberta. Production generally involves steam drives, or the injection of steam into reservoirs through special wells in prescribed sequences. Oil–water mixtures are recovered, and often separated water is treated and reinjected.

Heavy oil may be upgraded through two main routes: coking and hydroprocessing. Virtually all established upgrading schemes involve some variant of those two routes. The challenges in upgrading and refining are from the low hydrogen content and specific gravity and high sulfur, nitrogen, and metals content of the heavy oil.

Tar Sands

Tar sands are considered to be sedimentary rocks having natural porosity where the pore volume is occupied by viscous, petroleum-like hydrocarbons. The terms oil sands, rock asphalts, asphaltic sandstones, and malthas or malthites have all been applied to the same resource. The hydrocarbon component of tar sands is properly termed bitumen.

Distinctions between tar sands' bitumens and heavy oils are based largely on differences in viscosities. The bitumen in oil sand has a specific gravity of less than 0.986 g/mL (12°API), and thus oil sands may be regarded as a source of extremely heavy crude oil. Whereas heavy oils might be produced by the same techniques used for the lighter crude oils, the bitumens in tar sands are too viscous for these techniques. Consequently these oil-bearing stones have to be mined and specially processed to recover contained hydrocarbon.

Tar sands have been reported on every continent except Australia and Antarctica. The best known deposits are the Athabasca of Canada, where almost 60,000 km^2 in northeastern Alberta is underlain with an estimated 138×10^9 m^3 (870×10^9 bbl) of recoverable bitumen (157). The Alberta deposits may contain up to 215×10^9 m^3 (1350×10^9 bbl) of bitumen reserves. Venezuela may have the largest accumulations in the world; the Orinoco heavy-oil belt has been estimated by some (157) to contain as high as 636×10^9 m^3 (4000×10^9 bbl). The Olenek reserves in the former USSR may contain ca 95×10^9 m^3 (600×10^9 bbl). The United States is estimated to have deposits of about 4.5×10^9 m^3 (28×10^9 bbl).

The Great Canadian Oil Sands, Ltd. (GCO) (Sun Oil Co.) has been operating a plant at Fort McMurray, Alberta, Canada, since 1967. Initially, some 8050 t/d (55,000 bbl/d) of synthetic crude oil was produced from coking (158) with the project expanding to 9220 t/d (63,000 bbl/d). Since 1978, Syncrude Canada has been producing ca 22,000 m^3/d (140,000 bbl) synthetic crude oil by fluid coking from their plant at Cold Lake, Alberta, Canada (159) with expansion planned for ca 35,000 m^3/d (225,000 bbl/d).

ECONOMIC ASPECTS OF SYNTHETIC FUELS

As of this writing, processes for production of synthetic liquid fuels by upgrading natural gas, coal, or heavy oil are generally not directly competitive with crude oil upgrading (160). The key controlling factors in the economics are crude oil price and availability. Many economic analyses for synthetic liquid fuels give a crude oil price target whereupon the alternative technology becomes attractive, but these studies sometimes neglect the fact that the natural gas, coal, and heavy oil prices often track those of crude oil. In addition, conversion of a refractory gas (methane) or solid (coal) to liquids is a greater technical challenge than that of processing crude oil. Thus there are processing cost penalties which inevitably exist even after considerable technological development. Nevertheless, synthetic fuels technology is projected to play a primary role in providing liquid fuels once crude oil depletion is of concern. Economic competitiveness plays a reduced role in commercialization only when environmental legislation mandates the use of certain fuels such as oxygenates.

The commercialization potential of synthetic fuels technology relies on site-specific economic and political factors. This complex network of factors may include capital costs, crude oil price, product yields and value, government subsidies, strategic impact, alternative uses for the feed, and environmental and geographical constraints. Whereas no direct coal liquefaction process has gone to commercial stage, technologies involving indirect conversion of natural gas or coal have been commercialized. In all cases, special conditions allowed the technology to progress. In the Sasol project at South Africa, coal upgrading was possible due to factors such as no indigenous petroleum, minimal environmental standards, and cheap labor (160). The New Zealand GTG process became economically feasible owing to high oil prices, abundance of indigenous natural gas, and government commitment to energy self-sufficiency (59). Government support and long-term strategic benefits were also keys to Shell's SMDS project in Malaysia.

At 1994 crude oil prices of ca $\$94–125/m^3$ (ca \$15–20/bbl), conversion of natural gas to liquid fuels exists only in unique situations. For natural gas upgrading via New Zealand-type technology, economics by Mobil for a 1987 plant start-up on the U.S. Gulf Coast (161) indicated an investment of $\$895 \times 10^6$ was required for a $2.3 \times 10^3\ m^3/d$ (14,600 bbl/d) gasoline production unit. Thus this and other natural gas-to-fuels processes are highly capital-intensive and capital recovery remains the dominant factor even with incremental advances in conventional technology. This is especially the case using indirect upgrading of natural gas because the cost of synthesis gas manufacture may account for more than 50% of the total process capital cost (44). An analysis by Shell of the SMDS process published in 1988 showed capital expense for a $1600\ m^3/d$ (10,000 bbl/d) plant to be $\$300 \times 10^6$ for a developed site in an industrialized country and $\$600 \times 10^6$ for a developing site in a developing country (44). Direct upgrading of natural gas to gasoline and distillate by oxidative coupling plus olefin oligomerization has been evaluated to be ca 10% costlier in capital than upgrading via indirect technologies (162).

Natural gas upgrading economics may be affected by additional factors. The increasing use of compressed natural gas (CNG) directly as fuel in vehicles provides an alternative market which affects both gas price and values (see GAS, NATURAL). The hostility of the remote site environment where the natural gas is located may contribute to additional costs, eg, offshore sites require platforms and submarine pipelines.

The economic feasibility of coal upgrading, and in particular direct coal liquefaction, are closely tied to crude oil price and capital expense. H-coal technology was evaluated as a base case in 1981 and the results showed economic feasibility was possible only at crude oil price of $\$630/m^3$ (\$100/bbl) or greater (160). A more recent analysis by HRI on coal/oil coprocessing technology indicated the required light crude price was $\$138–182/m^3$ (\$22–29/bbl) for economic feasibility (163). The cost of capital could add over $\$60/m^3$ (\$10/bbl) to the cost of products. A study of EDS, H-coal, ITSL, and two-stage Wilsonville systems showed capital costs for a 30,000 t/d plant processing Illinois No. 6 coal to run between $\$4100–\$4700 \times 10^6$ (131). Required break-even selling prices for products from these technologies ranged from $\$226–314/m^3$ (\$36–50/bbl). The two-stage system was the most economical at $\$229.94/m^3$ (\$36.56/bbl). An evaluation of coprocessing Cold Lake vacuum bottoms using Alberta subbituminous coal in a 3200 m^3/d (20,000 bbl/d) synthetic crude oil production unit indicated a selling price of $\$189–220/m^3$ (\$30–35/bbl) was necessary for the process to be competitive (164). In general, the economics of direct coal liquefaction depend more on the high cost of liquefaction rather than the cost for upgrading product coal liquids (165).

Factors which may affect the cost of coal upgrading are environmental considerations such as toxicity, hazardous waste disposal, and carcinogenic properties (131). These and other environmental problems from process streams, untreated wastewaters, and raw products would figure significantly into the cost of commercialization.

Reprinted from *Kirk-Othmer Encyclopedia of Chemical Technology,* 4th ed., Vol. 12, John Wiley & Sons, Inc., New York, 1995.

BIBLIOGRAPHY

1. H. Mimoun, *New J. Chem.* **11,** 4 (1987).
2. U. Preuss and M. Baerns, *Chem. Eng. Technol.* **10,** 297 (1987).
3. "Liquid Fuels From Natural Gas," *Petrole Informations,* API 34-5250, 96 (May 1987).
4. *Chem. Eng. News,* 25 (Aug. 14, 1989).
5. E. E. Ecklund and G. A. Mills, *Chemtech,* 549 (Sept. 1989).
6. L. Haar, in W. P. Earley and J. W. Weatherly, eds., *Advances in Coal Utilization Technology IV,* Institutes of Gas Technology, Chicago, 1981, pp. 787–952.
7. L. Shnidman, in H. H. Lowry, ed., *Chemistry of Coal Utilization,* Vol. 2, John Wiley & Sons, Inc., New York, 1945, pp. 1252–1286.
8. W. L. Glowacki, in Ref. 7, pp. 1136–1231; E. O. Rhodes, in Ref. 7, pp. 1136–1231.
9. J. R. Rostrup-Nielsen, *Steam Reforming Catalysts,* Teknisk Forlag A/S, Copenhagen, 1975.
10. *Catalyst Handbook,* Wolfe Scientific Texts, London, 1970.
11. P. Sabatier, *Catalysis, Then and Now,* Part II, Franklin Publishing Co., Englewood, N.J., 1965.
12. Fr. Pat. 571,356 (May 16, 1924), (to Badische Anilin- und Soda-Fabrik).
13. Fr. Pat. 580,905 (Nov. 19, 1924), (to Badische Anilin- und Soda-Fabrik).
14. F. Fischer and H. Tropsch, *Ber.* **56,** 2428 (1923).
15. F. Fischer, *Die Umwandlung der Kohle in Ole,* Borntraeger, Berlin, 1923, p. 320.
16. F. Fischer and H. Tropsch, *Ber.* **59,** 830, 832, 923 (1926).
17. F. Fischer, *Brennstoff-Chem.* **11,** 492 (1930).
18. F. Fischer and H. Koch, *Brennstoff-Chem.* **13,** 61 (1932).
19. F. Fischer and K. Meyer, *Brennstoff-Chem.* **12,** 225 (1931).
20. F. Fischer and H. Pichler, *Brennstoff-Chem.* **20,** 41, 221 (1939).
21. K. Fischer, *Comparison of I. G. Work on Fischer Synthesis, Technical Oil Mission Report, Reel 13,* Library of Congress, Washington, D.C., July 1941.
22. H. Pichler, *Medium Pressure Synthesis on Iron Catalyst, (Pat. Appl), Technical Oil Mission Report, Reel 100,* Library of Congress, Washington, D.C., 1937–1943.

23. H. Pichler, *Medium Pressure Synthesis on Iron Catalyst, Technical Oil Mission Report, Reel 101,* Library of Congress, Washington, D.C., June 1940.
24. U.S. Pat. 2,327,066 (Aug. 17, 1943). O. Roelen.
25. F. Fischer, *Ole Kohle* **39,** 517 (1943).
26. H. H. Storch, N. Golumbic, and R. B. Anderson, *The Fischer-Tropsch and Related Synthesis,* John Wiley & Sons, Inc., New York, 1951.
27. W. G. Frankenburg, V. I. Komarewsky, and E. D. Rideal, *Advances in Catalysis,* Vol I., Academic Press, Inc., New York, 1948, pp. 115–156.
28. H. H. Storch, in Ref. 7, p. 1797.
29. O. C. Elvins and A. W. Nash, *Fuel* **5,** 263 (1926).
30. O. C. Elvins, *J. Soc. Chem. Ind. (London)* **56,** 473T (1927).
31. A. Erdeley and A. W. Nash, *J. Soc. Chem. Ind. (London)* **47,** 219T (1928).
32. W. W. Myddleton, *Chim. Ind.* **37,** 863 (1937); *J. Inst. Fuel* **11,** 477 (1938); *Colliery Guardian* **157,** 286 (1938).
33. G. Egloff, *Brennst.-Chem.* **18,** 11 (1937).
34. G. Egloff, E. F. Nelson, and J. C. Morrell, *Ind. Eng. Chem.* **29,** 555 (1937).
35. F. Mako and W. A. Samuel, in R. A. Meyers, ed., *Handbook of Synfuels Technology,* McGraw-Hill, Inc., New York, 1984, pp. 2-5–2-43.
36. J. C. Hoogendoorn, *Phil. Trans. R. Soc. Lond. A* **300,** 99 (1981).
37. W. B. Johnson, *Pet. Ref.* **35** (Dec. 1956).
38. M. E. Dry, "Fischer-Tropsch Synthesis Over Iron Catalysts," paper presented at *1990 Spring AIChE National Meeting,* Orlando, Fla., Mar. 18–22, 1990.
39. B. Jager and co-workers, in Ref. 38.
40. A. E. Sands, H. W. Wainwright, and L. D. Schmidt, *Ind. Eng. Chem.* **40,** 607 (1948); A. E. Sands and L. D. Schmidt, *Ind. Eng. Chem.* **42,** 2277 (1950).
41. H. Pichler, *Technical Oil Mission Report, Reel 259,* Library of Congress, Washington, D.C., 1947, frames 467–654.
42. P. C. Keith, *Oil Gas J.* **345**(6), 102 (1946).
43. *Oil Gas J.,* 74 (Feb. 17, 1986).
44. M. J. v. d. Burgt and co-workers, in D. M. Bibby and co-workers, eds., *Methane Conversion,* Elsevier Science, Inc., New York, 1988, pp. 473–482.
45. I. E. Maxwell and J. E. Naber, *Catal. Lett.* **12,** 105 (1992).
46. P. T. Roterud and co-workers, in Ref. 38.
47. C. D. Chang and A. J. Silvestri, *J. Catal.* **47,** 249 (1977).
48. C. D. Chang, *Catal. Rev.-Sci. Eng.* **25,** 1 (1983).
49. C. D. Chang and A. J. Silvestri, *Chemtech* **17,** 624 (1987).
50. U.S. Pat. 3,702,886 (1972), R. J. Argauer and G. R. Landolt (to Mobil Oil Corp.).
51. G. T. Kokotailo and co-workers, *Nature* **272,** 437 (1978).
52. D. H. Olson, G. T. Kokotailo, and S. L. Lawton, *J. Phys. Chem.* **85,** 2238 (1981).
53. C. D. Chang, in Ref. 44, pp. 127–143.
54. G. J. Hutchings and R. Hunter, *Catal. Today* **6,** 279 (1990).
55. F. Bauer, *Zfl-Mitt.* **156,** 31 (1990).
56. P. B. Weisz and V. J. Frilette, *J. Phys. Chem.* **64,** 382 (1960).
57. S. M. Csicsery, *ACS Monograph* **171,** 680 (1976).
58. N. Y. Chen, W. E. Garwood, and F. G. Dwyer, *Shape Selective Catalysis in Industrial Applications,* Marcel Dekker, Inc., New York, 1989.
59. C. J. Maiden, in Ref. 44, pp. 1–16.
60. J. E. Penick, W. Lee, and J. Maziuk, *ACS Symp. Ser.* **226,** 19 (1983).
61. J. Z. Bem, in Ref. 44, pp. 663–678.
62. P. L. Rogerson, in Ref. 35, pp. 2-45–2-73.
63. A. Y. Kam, M. Schreiner, and S. Yurchak, in Ref. 35, pp. 2-75–2-111.
64. D. Liederman and co-workers, *Ind. Eng. Chem. Proc. Des. Devel.* **17,** 340 (1978).
65. H. R. Grimmer, N. Thiagarajian, and E. Nitschke, in Ref. 44, pp. 273–291.
66. K. H. Keim and co-workers, *Erdol. Erdgas, Kohle* **103,** 82 (1987).
67. K. G. Allum and A. R. Williams, in Ref. 44, pp. 691–711.
68. A. A. Avidan, in Ref. 44, pp. 307–323.
69. C. D. Chang, *Catal. Rev.-Sci. Eng.* **26** (3&4), 323 (1984).
70. M. M. Holm and E. H. Reichl, *Fiat Report No. 1085,* Office of Military Government for Germany (U.S.), Mar. 31, 1947.
71. J. A. Sofranko, "Gas to Gasoline: The ARCO GTG Process," paper presented at *Bicentenary Catalysis Meeting,* Sydney, Australia, Sept. 1988.
72. J. A. Sofranko and J. C. Jubin, "Natural Gas to Gasoline: The ARCO GTG Process," paper presented at *International Chemical Congress of Pacific Basin Societies,* Honolulu, Hawaii, Dec. 1989.
73. J. H. Edwards, K. T. Do, and R. J. Tyler, in Ref. 72.
74. J. H. Edwards, K. T. Do, and R. J. Tyler, in E. E. Wolf, ed., *Methane Conversion by Oxidative Processes,* Van Nostrand Reinhold, New York, 1992, pp. 429–462.
75. R. J. Nichols, "Applications of Alternative Fuels," *SAE Special Publication SP-531,* Society of Automotive Engineers, Warrendale, Pa., Nov. 1982.
76. R. A. Potter, "Neat Methanol Fuel Injection Fleet Alternative Fuels Study," paper presented at *Fourth Washington Conference on Alcohol,* Washington, D.C., Nov. 1984.
77. N. D. Brinkman, *Ener. Res.* **3,** 243 (1979).
78. T. Powell, "Racing Experiences with Methanol and Ethanol Based Motor-Fuel Blends," paper 750124, *Automotive Engineering Congress and Exposition,* Society of Automotive Engineers, Detroit, Mich., Feb. 1975.
79. American Petroleum Institute, Task Force EF-18 of the Committee on Mobile Source Emissions, *Alcohols–A Technical Assessment of Their Application as Fuels, Publication No. 4261,* API, New York, July 1976.
80. U.S. Environmental Protection Agency, *Fed. Reg.* **46**(144) (July 28, 1981).
81. U.S. Environmental Protection Agency, *Fed. Reg.* **50**(12), 2615 (Jan. 17, 1985).
82. G. A. Mills and E. E. Ecklund, *Chemtech,* 626 (Oct. 1989).
83. KVB, Inc., *KVB Report Number 72-804830-1998,* Vol. 2, California Energy Commission, Irvine, Calif., Mar. 1985, pp. 1–2.
84. A. J. Weir and co-workers, "Methanol Dual-Fuel Combustion," paper presented at *1987 Joint Symposium on Stationary Combustion NO_x Control,* New Orleans, La., Mar. 23–26, 1987.
85. S. B. Alpert and D. F. Spencer, *Methanol and Liquid Fuels from Coal–Recent Advances,* Electric Power Research Institute, Palo Alto, Calif., 1987.
86. V. Yand and S. C. Trindade, *Chem. Eng. Prog.,* 11 (Apr. 1979).
87. *Alcohols: Economics and Future U.S. Gasoline Markets,* Information Resources, Inc., Washington, D.C., 1984.

88. H. L. Muller and S. P. Ho, "Economics and Energy Balance of Ethanol as Motor Fuel," paper presented at *1986 Spring AIChE National Meeting*, New Orleans, La., Apr. 1986.
89. *Chem. Eng. News,* 7 (Nov. 2, 1992).
90. ARCO Chemical Co., *Testimony to the Colorado Air Quality Control Commission on Proposed Regulation No. 13 (Oxygenate Mandate Program),* Denver, Colo., June 4, 1987.
91. *Chem. Eng. News,* 28 (Apr. 12, 1993).
92. U.S. Pat. 1,251,954 (Jan. 1, 1918), F. Bergius and J. Billwiller.
93. F. Fischer and H. Tropsch, *Ges. Abhandl. Kenntis Kohle* **2,** 154 (1918).
94. F. Bergius, *Pet. Z.* **22,** 1275 (1926).
95. *Gas World* **104,** 421 (1936).
96. *Pet. Times* **42,** 641 (1939).
97. H. H. Storch and co-workers, *U.S. Bur. Mines. Tech. Pap.* **622** (1941).
98. A. C. Fieldner and co-workers, *U.S. Bur Mines Tech. Pap.* **666** (1944).
99. M. L. Kastens and co-workers, *Ind. Eng. Chem.* **41,** 870 (1949).
100. J. L. Wiley and H. C. Anderson, *U.S. Bur. Mines Bull.* **485,** I (1950), II (1951), III (1952).
101. C. C. Chaffee and L. L. Hirst, *Ind. Eng. Chem.* **45,** 822 (1953).
102. Bituminous Coal Staff, *U.S. Bur. Mines Rep. Invest.* **5506** (1959).
103. E. L. Clark and co-workers, *Ind. Eng. Chem.* **42,** 861 (1950).
104. L. L. Newman and A. P. Pipilen, *Gas Age* **119**(10), 16 (1957); **119**(11), 18 (1957).
105. U.S. Pat. 2,860,101 (Nov. 11, 1958), M. G. Pelipetz (to the United States of America).
106. S. Akhtar, S. Friedman, and P. M. Yavorsky, *U.S. Bur. Mines Tech. Prog. Rep.* **35** (1971).
107. S. Akhtar and co-workers, "Process for Hydrodesulfurization of Coal," paper presented at *71st National AIChE Meeting,* Dallas, Tex., Feb. 20, 1972.
108. B. Linville and J. D. Spencer, *U.S. Bur. Mines Inf. Cir.* **8612** (1973).
109. U.S. Pat. 3,321,393 (May 23, 1967), S. C. Schuman, R. H. Wolk, and M. C. Chervenak (to Hydrocarbon Research, Inc.).
110. *Coal Age,* 101 (May 1976).
111. G. A. Johnson and co-workers, "Present Status of the H-Coal Process," paper 30, *IGT Coal Symposium,* Chicago, 1973.
112. J. E. Papso, in Ref. 35, pp. 1-47–1-63.
113. "High Pressure Hydrogenation at Ludwigshafen-Heidelberg," Vol. IA, *FIAT Final Report No. 1317, ATI No. 92,762,* Central Air Documents Office, Dayton, Ohio, 1951.
114. H. H. Lowry and H. J. Rose, *U.S. Bur. Mines Inf. Cir.* **7420** (1947).
115. A. Baba and co-workers, *Rep. Resources Res. Inst. Jpn.,* (22) (1955).
116. U.S. Pat. 2,221,866 (Nov. 19, 1940), H. Dreyfus.
117. D. L. Kloepper and co-workers, *Solvent Processing of Coal to Produce a De-ashed Product,* Contract 14-01-0001-275, OCR Report No. 9, U.S. Government Printing Office, Washington, D.C., 1965.
118. V. L. Brant and B. K. Schmid, *Chem. Eng. Prog.* **68**(12), 55 (1969).
119. B. K. Schmid and W. C. Bull, "Production of Ashless, Low-Sulfur Boiler Fuels From Coal," paper presented at *ACS Division of Fuel Chem. Symposium on Pollution Control,* New York, Sept. 12, 1971.
120. *Demonstration Plant, Clean Boiler Fuels From Coal,* OCR R&D report no. 82, Interim report no. 1, Vols, 1–3, Ralph M. Parsons Co., Los Angeles, Calif., 1973–1975.
121. M. E. Frank and B. K. Schmid, "Economic Evaluation and Process Design of a Coal–Oil–Gas (COG) Refinery," paper presented at *Symposium on Conceptual Plants for the Production of Synthetic Fuels From Coal, AIChE 65th Annual Meeting,* New York, Nov. 26, 1972.
122. U.S. Pat. 3,341,447 (Sept. 12, 1967), W. C. Bull and co-workers (to the United States of America and Gulf Oil Corp.).
123. B. K. Schmid and D. M. Jackson, "The SRC-II Process," paper presented at *Third Annual International Conference on Coal Gasification and Liquefaction,* University of Pittsburgh, Aug. 3–5, 1976; D. M. Jackson and B. K. Schmid, "Production of Distillate Fuels by SRC-II," paper presented at *ACS Div. of Ind. and Eng. Chem. Symposium,* Colorado Springs, Col., Feb. 12, 1979.
124. W. R. Epperly and J. W. Taunton, "Exxon Donor Solvent Coal Liquefaction Process Development," paper presented at *Coal Dilemma II ACS Meeting,* Colorado Springs, Colo., Feb. 12, 1979.
125. D. E. Blaser and A. M. Edelman, "Flexicoking for Improved Utilization of Hydrocarbon Resources," paper presented at *API 43rd Mid-Year Meeting,* Toronto, Canada, May 8, 1978.
126. S. F. Massenzio, in Ref. 35, pp. 6-3–6-18.
127. T. A. Cavanaugh, W. R. Epperly and D. T. Wade, in Ref. 35, pp. 1-3–1-46.
128. L. E. Swabb, Jr., G. K. Vick, and T. Aczel, "The Liquefaction of Solid Carbonaceous Materials," paper presented at *The World Conference on Future Sources of Organic Raw Materials,* Toronto, Can., July 10, 1978.
129. E. L. Huffman, *Proceedings of the Third Annual International Conference on Coal Gasification and Liquefaction,* Pittsburgh, Pa., 1976.
130. H. D. Schindler, J. M. Chen, and J. D. Potts, *Final Technical Report on DOE Contract No. DE-AC22-79ET14804,* Department of Energy, Washington, D.C., 1983.
131. H. D. Schindler, *Final Technical Report on DOE Contract No. D-AC01-87ER30110,* Vol. 2, Department of Energy, Washington, D.C., 1989.
132. A. G. Comolli and co-workers, *Proceedings of the DOE Direct Liquefaction Contractors' Review Meeting,* Pittsburgh, Pa., 1986.
133. M. G. Thomas, in B. R. Cooper and W. A. Ellingson, eds., *The Science and Technology of Coal and Coal Liquefaction,* Plenum Press, New York, 1984, pp. 231–261.
134. M. J. Burges and R. V. Wheeler, *Fuel* **5,** 65 (1926).
135. W. A. Selvig and W. H. Ode, *U.S. Bur. Mines. Bull.* **571,** (1957).
136. H. C. Howard, in Ref. 7, Vol. 1, pp. 761–773.
137. *Ibid.,* Suppl. Vol., pp. 340–394.
138. J. F. Jones and co-workers, *Chem. Eng. Prog.* **62**(2), 73 (1966).
139. J. A. Hamshar, H. D. Terzian, and L. J. Scotti, "Clean Fuels From Coal by the COED Process," paper presented at *EPA Symposium on Environmental Aspects of Fuel Conversion Technology,* St. Louis, Mo., May 1974.
140. C. D. Kalfadelis and E. M. Magee, *Evaluation of Pollution Control in Fossil Fuel Conversion Processes, Liquefaction, Section 1, COED Process,* EPA-650/2-74-009e, Environmental Protection Agency, Washington, D.C., 1975.

141. R. T. Eddinger and co-workers, *Char Oil Energy Development, Office of Coal Research R&D Report No. 11,* Vol. 1 (PB 169,562) and Vol. 2 (PB 169,563), U.S. Government Printing Office, Washington, D.C., Mar. 1966.
142. J. F. Jones and co-workers, *Char Oil Energy Development, Office of Coal Research R&D Report No. 11,* Vol. 1 (PB 173,916) and Vol. 2 (PB 173,917), U.S. Government Printing Office, Washington, D.C., Feb. 1967.
143. M. E. Sacks and co-workers, *Char Oil Energy Development, Office of Coal Research Report 56, Interim Report No. 2,* GPO Cat. No. 163.10:56/Int.2, U.S. Government Printing Office, Washington, D.C., Jan. 1971.
144. J. F. Jones and co-workers, *Char Oil Energy Development, Office of Coal Research R&D Report No. 56, Final Report,* GPO Cat. No. 163.10:56, U.S. Government Printing Office, Washington, D.C., May 1972.
145. J. F. Jones and co-workers, *Char Oil Energy Development, Office of Coal Research R&D Report No. 73, Interim Report No. 1,* GPO Cat. No. 163.10:73/Int 1, U.S. Government Printing Office, Washington, D.C., Dec. 1972.
146. A. Sass, "The Garrett Research and Development Company Process for the Conversion of Coal into Liquid Fuels," paper presented at *65th Annual AIChE Meeting,* New York, Nov. 29, 1972.
147. *Oil Gas J.,* 78 (Aug. 26, 1974).
148. A. Sass, *Chem. Eng. Prog.* **70**(1), 72 (1974).
149. F. B. Carlson and co-workers, *Chem. Eng. Prog.* **69**(3), 50 (1973).
150. F. B. Carlson, L. H. Yardumian, and M. T. Atwood, "The TOSCOAL Process for Low Temperature Pyrolysis of Coal," paper presented at *American Institute of Mining, Metallugical, and Petroleum Engineers,* San Francisco, Calif., Feb. 22, 1972, and to *American Institute of Chemical Engineers,* New York, Nov. 29, 1972.
151. W. Tiddy and M. J. Miller, *Am. Gas J.* **153**(3), 7 (1940).
152. M. J. Gavin, *U.S. Bur. Mines Bull.* **210,** (1922).
153. M. J. Gavin and J. S. Desmond, *U.S. Bur. Mines Bull.* **315** (1930).
154. H. Shaw, C. D. Kalfadelis, and C. E. Jahnig, *Evaluation of Methods to Produce Aviation Turbine Fuels From Synthetic Crude Oils-Phase I, Technical Report AFAPL-TR-75-10,* Vol. I, Air Force Aero Propulsion Laboratory, Wright-Patterson Air Force Base, Dayton, Ohio, Mar. 1975.
155. C. D. Kalfadelis, *Evaluation of Methods to Produce Aviation Turbine Fuels From Synthetic Crude Oils-Phase II, Technical Report AFAPL-TR-75-10,* Vol. II, Air Force Aero Propulsion Laboratory, Wright-Patterson Air Force Base, Dayton, Ohio, May 1976.
156. Colony Development Operation, *An Environmental Impact Analysis for a Shale Oil Complex at Parachute Creek, Colorado,* Vols. 1–3, Denver, Colo., 1974.
157. H. L. Erskine, in Ref. 35, pp. 5-3–5-32.
158. R. D. Hynphreys, F. K. Spragins, and D. R. Craig, "Oil Sands–Canada's First Answer to the Energy Shortage," *Proceedings 9th World Petroleum Congress,* Tokyo, Japan, May 11, 1975, Vol. 5, p. 17.
159. C. W. Bowman, R. S. Phillips, and L. R. Turner, in Ref. 35, pp. 5-33–5-79.
160. M. Crow and co-workers, *Synthetic Fuel Technology Development in the United States–A Retrospective Assessment,* Praeger Publishing, New York, 1988.
161. S. Yurchak and S. S. Wong, "Mobil Methanol Conversion Technology," *Proceedings IGT Asian Natural Gas Seminar,* Singapore, 1992, pp. 593–618.
162. J. L. Matherne and G. L. Culp, in Ref. 74, pp. 463–482.
163. J. E. Duddy, S. B. Panvelker, and G. A. Popper, "Commercial Economics of HRI Coal/Oil Co-Processing Technology," paper presented at *1990 SummerAIChE National Meeting,* San Diego, Ca., 1990.
164. M. Ikura and J. F. Kelly, "A Techno-Economic Evaluation of CANMET Coprocessing Technology," *Proceedings Annual International Pittsburgh Coal Conference,* 1990, pp. 719–728.
165. J. G. Sikonia, B. R. Shah, and M. A. Ulowetz, "Technical and Economic Assessment of Petroleum, Heavy Oil, Shale Oil and Coal Liquid Refining," paper presented at *Synfuels' 3rd Worldwide Symposium,* Washington, D.C., Nov. 1–3, 1983.

General Reference

American Petroleum Institute, *Alcohols and Ethers-A Technical Assessment of Their Application as Fuels and Fuel Components,* API Publication 4261, American Petroleum Institute, Wash-

FUSION ENERGY

William R. Ellis
Raytheon Engineers & Constructors
New York, New York

As far as is known, nuclear fusion, which drives the stars, including the sun, is the primary source of energy in the universe. The process of nuclear fusion releases enormous amounts of energy. It occurs when the nuclei of lighter elements, such as hydrogen, are fused together at extremely high temperatures and pressure to form heavier elements, such as helium. Whereas practical methods for harnessing fusion reactions and realizing the potential of this energy source have been sought since the 1950s, achieving the benefits of power from fusion has proved to be a difficult, long-term challenge. See also Electric power generation; Cold fusion; Nuclear power.

Fusion is widely held to be the ultimate resource for the world's long-term energy needs. The fuel reserves for fusion are virtually limitless and available to all countries. Fusion fuels can be extracted from water. Additionally, fusion promises to be an energy source which is potentially safe and environmentally benign. Radiological and proliferation hazards are much smaller than for fission power plants. The atmospheric impact is negligible compared to fossil fuels, and adverse impacts on the Earth's ecological and geophysical processes are smaller than for large-scale renewable energy sources (see also Fuels, Synthetic; Gas, Natural). The economics and costs of fusion power plants are still being studied, but appear comparable to those for other medium- and long-term energy sources. The tantalizing promise of affordable essentially unlimited supplies of clean, safe energy, free of political boundaries, has motivated a worldwide research effort to develop this energy resource.

The nuclear burning mechanism of the Sun was elucidated in the 1930s (1). In a complex sequence of reactions starting with hydrogen, atomic nuclei are fused to form heavier species. Because of a mass deficit, Δm, exhibited by the reaction products, large amounts of energy, E, are released, as dictated by the well-known Einstein equivalence $E = \Delta mc^2$ where c is the speed of light. Large-scale fusion energy production was demonstrated dramatically on earth in the early 1950s with the explosion of thermonuclear fusion, ie, hydrogen, bombs. These weapons used the heat of nuclear fission (atomic bombs) to cause the fusion of deuterium [*16873-17-9*], D, and tritium [*15086-10-9*], ^{3}H or T. Subsequently, an international research effort was undertaken to harness this awesome power on a controllable scale for peaceful purposes. Several impressive advances in the 1980s and early 1990s have led to a well-founded feeling of optimism that fusion energy should become a practical energy source during the early twenty-first century.

In order to effect a fusion reaction between two atomic nuclei, it is necessary that these nuclei be brought together closely enough to experience an attractive nuclear force. All nuclei are positively charged and repel one another via Coulomb's law, the electrostatic law of the repulsion of like charges. This electrostatic barrier can be overcome by imparting sufficient kinetic energy to the reacting species so that the nuclei can approach closely enough together that quantum mechanical tunneling can occur. The repulsive forces increase rapidly with the magnitude of the nuclear charge; therefore, nuclear fusion research has concentrated on the lightest elements and the isotopes having the lowest atomic numbers.

The reactions of deuterium, tritium, and helium-3 [*14762-55-1*], ^{3}He, having nuclear charge of 1, 1, and 2, respectively, are the easiest to initiate. These have the highest fusion reaction cross-sections and the lowest reactant energies.

DEUTERIUM–TRITIUM FUSION

The D–T reaction involving the two heavy isotopes of hydrogen

$$\mathrm{D} + \mathrm{T} \rightarrow {}^4\mathrm{He} + n$$

is especially attractive to fusion scientists because of its relative ease of ignition. The products of this reaction are an alpha particle, ie, the helium nucleus, ^{4}He, and a free neutron, n, carrying kinetic energies of 3.5 and 14.1 MeV, respectively. In an electric power-generating facility the neutrons would be absorbed in a blanket surrounding the fusion region, and the kinetic energy converted into heat. Conventional power conversion systems could then be used to transform this heat into electrical energy. Fusion reactions are extremely energetic, and yields are measured in units of millions of electron volts, MeV (1 MeV = 1.6×10^{-13} J).

Another set of reactions of practical interest involves only deuterons. The D–D reaction can proceed along either of two pathways with roughly equal probabilities:

$$\mathrm{D} + \mathrm{D} \rightarrow \mathrm{T} + p + 4.0\ \mathrm{MeV}$$

or

$$\mathrm{D} + \mathrm{D} \rightarrow {}^3\mathrm{He} + n + 3.2\ \mathrm{MeV}$$

Finally, the D–^{3}He reaction

$$\mathrm{D} + {}^3\mathrm{He} \rightarrow {}^4\mathrm{He} + p$$

is noteworthy not only because of its high (18.3 MeV) energy release, but also because the reaction products are both charged particles, offering the possibility of high efficiency, direct energy conversion. Direct energy conversion would involve the extraction of the positively charged ions and negatively charged electrons from the reaction region directly onto collection electrodes having a potential difference of the same order of magnitude as the mean kinetic energy of the charged particles. A variation of these above reaction schemes is the catalyzed D–D reaction wherein the external feedstock is deuterium but the ^{3}He and T produced in the D–D reactions are recycled and burned *in situ* to enhance the net energy yield.

Because of its relatively high reactivity (2), the D–T fusion-fuel cycle is very likely to be employed in the first generation of fusion reactors. This implies the use of a neutron absorbing blanket and thermal (Carnot) conversion efficiencies. Deuterium, also known as heavy hydrogen, occurs naturally in the ratio of 1:6700 relative to ordinary hydrogen; 30,000 kg water contains one kilogram of deuterium. The separation of deuterium from water is a relatively simple and inexpensive process. Tritium, on the other hand, is a radioactive isotope of hydrogen found in nature only in trace amounts and has a half-life of only 12.36 yr. The initial inventory of tritium for a power-producing D–T fusion power plant is a few kilograms and could be supplied, for example, from heavy-water fission reactors where it is produced as a by-product. Further tritium needs can be met by breeding additional tritium in the fusion reactor itself, by absorbing the fusion-produced neutrons in a blanket of lithium and exploiting the reaction

$$n + {}^6\mathrm{Li} \rightarrow \mathrm{T} + {}^4\mathrm{He} + 4.8\ \mathrm{MeV}$$

^{6}Li accounts for about 7.5% of natural lithium and is abundantly available in the earth's crust and oceans. Detailed fusion-blanket designs incorporate additional isotopes, such as ^{7}Li and ^{9}Be, which provide neutron-multiplying reactions, to compensate for the leakage and absorption losses of neutrons. A D–T fusion reactor, then, is in reality a consumer of deuterium and lithium. The estimated reserves of lithium should prove sufficient for at least several hundred years of D–T fusion power plant operation, even allowing for a significant increase in the world demand for energy (3). A common fusion evolution scenario relies on D–T fusion to fulfill the energy needs until the more difficult fuel cycle involving pure deuterium can be implemented. Then deuterium alone would be the fuel for the fusion energy economy. Because each liter of seawater contains enough deuterium to supply the energy equivalent of 300 L of gasoline, long-term energy needs would be assured.

Although the D–T reaction is the easiest route to fusion power production, it is no easy task to meet the conditions required to produce net fusion energy. Relative kinetic energies between the deuterons and tritons of 10 keV or more are necessary for practical energy generation, corresponding to relative particle velocities on the order of 10^6 m/s. Fusion-produced neutrons have, in fact, been created by impinging a beam of accelerated deuterons onto a solid target containing tritium. Unfortunately, a fusion reactor cannot be built around this concept because most of the incident beam energy is dissipated nonproductively through scattering and collision events in the target, and only a relatively small number of energy-producing fusion reactions occur. Other approaches, involving colliding beams of particles, have been proposed, but such schemes are inherently of very low power density and are not likely to yield practical energy sources.

PLASMA CONDITIONS REQUIRED FOR NET ENERGY RELEASE

The most promising approach to attaining significant reaction rates is to heat the reacting species to a high temperature, thereby imparting large kinetic energies to the nuclei in the form of thermal motions. By doing so, the particles, eg, deuterons and tritons, may scatter among themselves many times before undergoing fusion reactions, without losing significant energy from the system. At any given temperature, a system of particles in thermal equilibrium is characterized by a Maxwellian distribution of kinetic energies. The particles at the high energy end of this distribution account for most of the fusion reactions in fusion experiments.

The fusion fuel, when undergoing thermonuclear reactions, exists as an ionized gas called a plasma. In physics, the plasma state usually means a high temperature gas of net electrical neutrality consisting of free electrons and ions exhibiting collective behavior. The collection of charged particles exhibits characteristics of an electrically conducting fluid that can interact with electromagnetic fields. As such, its physical behavior is much more complex than that of an ordinary gas, and plasma confinement can be disrupted or reduced by many different kinds of plasma instabilities and other loss mechanisms. There exists a large literature and a number of outstanding books on plasma physics, such as Reference 4.

In a plasma undergoing fusion reactions, the reactivity, and thus the fusion-power output rate, increases with increasing temperature. However, over a wide range of temperatures, as the temperature of the plasma is raised, the radiation losses are also increased, primarily because of bremsstrahlung, or continuum, ie, braking, radiation from the electrons. For any fusion-fuel system there exists a unique temperature at which the fusion power production is precisely balanced by the radiation losses. This temperature is called the ideal ignition temperature, and equals about 50 million K (5 keV) for a D–T plasma (1 keV = 11.6×10^6 K). For a D–D plasma, this temperature is considerably higher, about 400×10^6 K (40 keV), a fact which considerably increases the difficulty of using pure deuterium fuel. Furthermore, a fusion system must be operated above the ideal ignition temperature for net power production, typically by a factor of 2–5.

Besides having to satisfy a minimum temperature requirement, the plasma must be sufficiently dense and contained for a long enough time to yield net power. If the plasma burns above the ideal ignition temperature for some time period, τ, the fusion energy released must at least equal the energy required to heat the plasma to that temperature plus the energy radiated during that period. It can be shown that this condition is met by requiring that the product of the plasma density, n, and confinement time, τ, exceed a characteristic value which depends only on the temperature. The minimum value of the product $n\tau$ represents the least stringent condition for the plasma to be a net producer of fusion energy. For D–T plasmas, this minimum occurs at a temperature of about 100×10^6 K, for which $n\tau \sim 10^{20}$ s/m^3. This minimum $n\tau$ product is called the Lawson criterion (5). For D–D, the minimum $n\tau$ product is about 10^{22} s/m^3 at a higher temperature, again indicating that a pure deuterium system requires a higher quality of confinement. A commonly used measure of the quality of plasma confinement is given by the triple product of the plasma density, n, ion temperature T_i, and energy confinement time, τ, usually expressed in units of keV·s/m^3. A primary goal of fusion research is to achieve $n\tau T_i$ values of $\sim 10^{22}$ keV·s/m^3, as required for a D–T reactor. Experiments as of this writing (1993) have reached a value of 1.1×10^{21} keV·s/m^3 in the JT-60 tokamak in Japan (6).

Plasmas at fusion temperatures cannot be kept in ordinary containers because the energetic ions and electrons would rapidly collide with the walls and dissipate their energy. A significant loss mechanism results from enhanced radiation by the electrons in the presence of impurity ions sputtered off the container walls by the plasma. Therefore, some method must be found to contain the plasma at elevated temperatures without using material containers.

Once a fusion reaction has begun in a confined plasma, it is planned to sustain it by using the hot, charged-particle reaction products, eg, alpha particles in the case of D–T fusion, to heat other, colder fuel particles to the reaction temperature. If no additional external heat input is required to sustain the reaction, the plasma is said to have reached the ignition condition. Achieving ignition is another primary goal of fusion research.

PATHS TO FUSION POWER

Two diverse technical approaches to fusion power, magnetic confinement fusion, also known as magnetic fusion energy (MFE) and inertial confinement fusion, also known as inertial fusion energy (IFE) are being pursued worldwide. These form the basis of a large number of fusion research programs. Magnetic confinement techniques, studied since the 1950s, are based on the principle that charged particles such as electrons and ions, ie, deuterons and tritons, tend to be bound to magnetic lines of force. Thus the essence of the magnetic confinement approach is to trap a hot plasma in a suitably chosen magnetic field configuration for a long enough time to achieve a net energy release, which typically requires an energy confinement time of about one second. In the alternative IFE approach, fusion conditions are achieved by heating and compressing small amounts of fuel ions, contained in capsules, to the ignition condition by means of tightly focused energetic beams of charged particles or photons. In this case the confinement time can be much shorter, typically less than a millionth of a second.

Magnetic Confinement

In magnetic confinement, strong magnetic fields are used to confine the plasma. Electrons and ions in magnetic fields spiral in circles around the field lines but translate freely along the direction of the magnetic field. Thus the magnetic field of a long solenoid, for example, confines the plasma in two directions but does not prevent the particles from streaming from either end of the system. Furthermore, collisions between particles displace them from one field line to another, producing a net diffusion of plasma across the field toward the walls of the container. By employing more complex magnetic field configurations, fusion researchers have made significant progress toward solving the problem of magnetic confinement of plasmas by substantially reducing plasma losses.

One of the earliest configurations studied was the simple magnetic mirror. A simple mirror system is depicted in Figure 1. Particles gyrating about the field lines move freely along these lines until they enter regions of increased field strength at either end of the device. Conservation of angular momentum considerations dictate that, as the particles approach the end regions, they gyrate more energetically about the field lines and slow down in the direction of motion along the lines. Ultimately, their kinetic energy is completely converted into gyration energy, at which point the particles are reflected from these mirror points and return to the central, weaker field region. Particles having motion exactly along the axis of the device are not reflected and are lost through the ends. Although ingenious attempts have been made to reduce end losses from mirror machines and to make them stable against magnetohydrodynamic (MHD) and other instabilities, all single-cell mirror reactor designs have suffered from a high recirculating power fraction, ie, a large fraction of the output power has to be used to operate the reactor itself. In single-cell mirror machines these losses are fundamentally too high. The machines are referred to as too lossy, and the amount of injected power required to maintain the plasma, usually in the form of high energy neutral beams, has been too large to be practical.

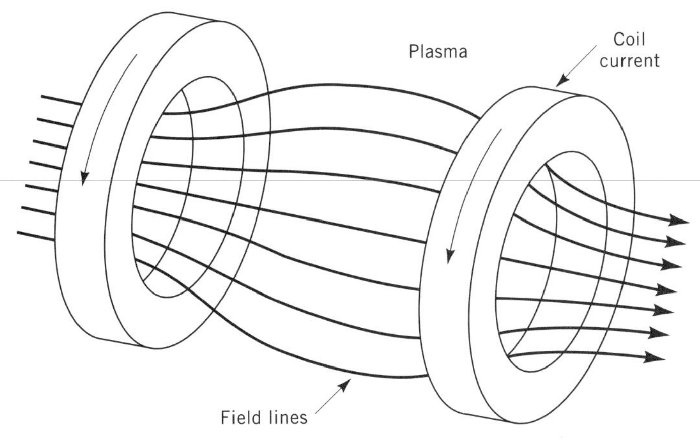

Figure 1. Simple magnetic mirror open configuration.

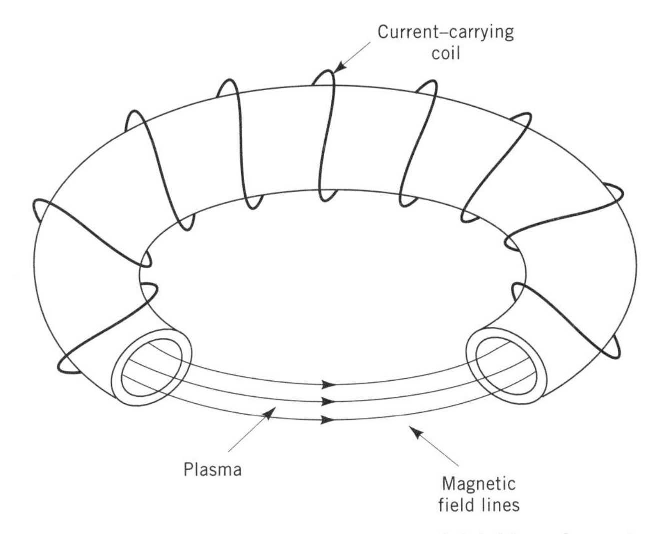

Figure 2. Cutaway view of a simple toroidal field configuration.

A more advanced mirror approach involving multicells, called the tandem mirror, has been studied as a means to overcome the leakage problem. One way to view the tandem mirror is as a long uniform magnetic solenoid with two single-cell mirrors installed at the ends to electrostatically plug the device. Plasma end losses are impeded by electrostatic potentials developed by the plasma as the electrons and ions attempt to leave the device at different rates.

Another mirror variation is the field-reversed mirror configuration, in which the diamagnetic nature of the plasma is exploited to cause the interior magnetic field lines within the central region of a single-cell mirror to reverse and close on themselves. The plasma current responsible for this field modification is at right angles to the original field lines.

The problem with all the mirror approaches is that none has achieved the degree of confinement quality that the closed systems have. Closed systems are characterized by magnetic field lines that close on themselves so that charged particles following the field lines remain confined within the system.

The simplest way of producing a closed configuration is to employ a torus or doughnut-shaped container having current-carrying coils wrapped around the minor diameter as shown in Figure 2. In this geometry, the magnetic lines of force are circles that traverse the torus and provide endless paths for the plasma ions and electrons to spiral about. Unfortunately, such a simple toroidal configuration is well known to have very poor confinement properties, because the magnetic field strength is not constant across the plasma. Instead, it is stronger at the inner wall and weaker at the outer wall of the toroidal chamber. As a result, the positive ions and electrons drift in opposite directions across the field lines and establish an electric field within the plasma. This electric field, coupled with the applied magnetic field, then causes the plasma as a whole to move to the container wall and dissipate its energy (2).

This deleterious effect can be obviated by introducing

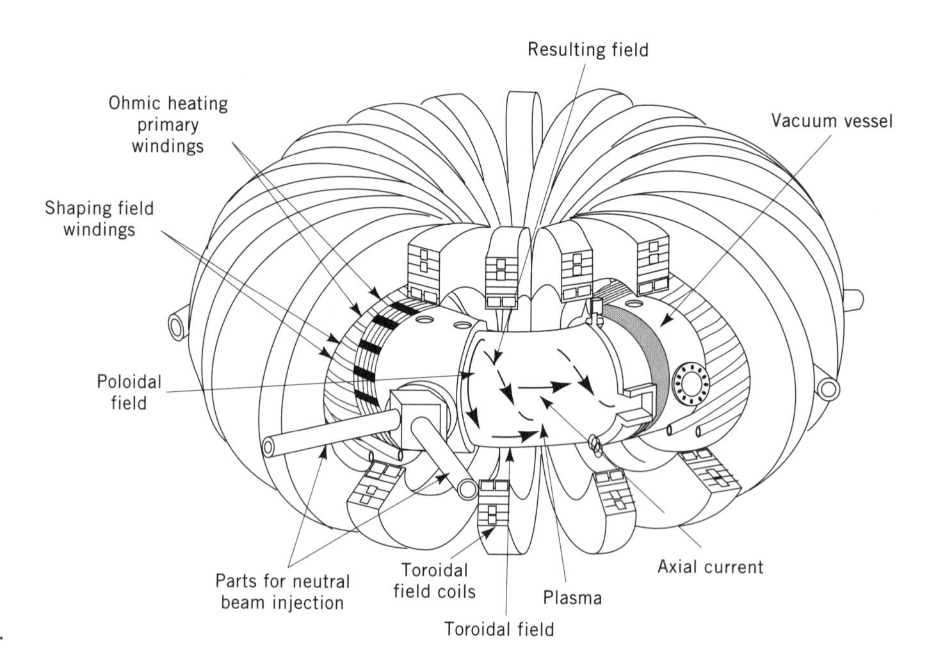

Figure 3. The tokamak fusion approach.

additional components of magnetic field, causing the field lines to circumscribe the torus without ever closing on themselves. The net magnetic field is then composed of a major, or toroidal, field component produced by the current coils, plus a smaller poloidal component which gives the desired twist to the lines. Particle drifts weaken or nullify the harmful electrical field and the plasma no longer tends to move to the walls.

Several geometries for producing the required poloidal magnetic field component have been studied. The class of plasma devices called tokamaks generates the poloidal (ie, around the minor circumference) field component from a toroidal (ie, around the major circumference) current in the plasma itself, either induced by a pulse from an external transformer or driven by external current-drive systems. The basic components of a tokamak are shown in Figure 3. External current-drive systems, such as high energy neutral beam injection or radio-frequency (r-f) current drive, impart a net toroidal momentum to one of the charged species, ions or electrons. A toroidal system related to the tokamak, which has a higher plasma energy density, is called the reversed field pinch (RFP). The RFP is an inherently pulsed, or batch-burn device.

The poloidal field component can also be created externally by using current-carrying coils that wind helically around the outside surface of the torus. Such devices, called stellerators or torsatrons, have the advantage of not requiring a net toroidal current. The helical field windings, however, make these machines mechanically and magnetically more complex than the tokamak.

Tokamak. The design concept that has come the closest by far to achieving energy breakeven conditions is the tokamak. Invented in the 1950s by the Russian physicists Andrei Sakharov and Igor E. Tamm (7), the tokamak derives its name from the Russian acronym for toroidal magnetic chamber. Technical progress in tokamaks was dramatic in the late 1980s and early 1990s (8,9). Central ion temperatures of 400×10^6 K have been reached, and energy confinement times have increased from 0.02 to about 1.4 seconds for strongly heated plasmas. The result has been $n\tau T_i$ triple products of about 10^{21} keV·s/m^3, compared to a value of $\sim 10^{22}$ keV·s/m^3 required for a steady state D–T power plant.

Other important parameters have also shown dramatic gains. The normalized plasma pressure which is usually

Table 1. Tokamak Plasma Parameters

Parameter	Achieved 1971	Achieved 1981	Achieved 1991	Required for Steady-State D–T power plant
Central ion temperature, T_i, keV	0.5	7	35	30
Central electron temperature, T_e, keV	1.5	3.5	15	30
Energy confinement time, τ, s	0.007	0.02	1.4	3
Triple product, $n\tau T_i$, keV·s/m^3	1.5×10^{17}	5.5×10^{18}	9×10^{20}	7×10^{21}
Normalized plasma pressure (β), %	0.1	3	11	5
Fusion reactivity, reactions per second				
D–D		3×10^{14}	1×10^{17}	
D–T			6×10^{17}	10^{21}

called beta, the plasma pressure divided by the confining magnetic field pressure, has been increased fourfold, to about 10%. This value is actually higher than that needed in a reactor. Bootstrap currents have been measured for the first time in several experiments. Bootstrap current is the name given to a toroidal current, theoretically predicted to arise spontaneously in tokamaks under near-reactor conditions. These in principle can eliminate much of the need for external current drive. Bootstrap currents open the possibility of a self-sustained steady-state tokamak power plant. Results from the large Japanese tokamak JT-60 are particularly interesting in this regard, where up to 80% of the 500,000 A of total plasma current is attributed to the bootstrap effect (8). Table 1 summarizes some of the progress made in the parameters of interest for magnetic fusion since 1971.

Some of the tokamaks in operation around the world, on which the data in Table 1 were obtained are

Designation	Tokamak	Location
ALC-A	Alcator-A	Plasma Fusion Center, Massachusetts Institute of Technology (MIT), Cambridge, Mass.
ALC-C	Alcator-C	Plasma Fusion Center, MIT
ASDEX	Axially Symmetric Divertor Experiment	Max Planck Institute for Plasma Physics, Garching, Germany
ATC	Adiabatic Toroidal Compressor	Princeton Plasma Physics Laboratory (PPPL), Princeton, N.J.
C-MOD	ALC-C Modified	MIT
DIII	Doublet III	General Atomics, San Diego, Calif.
DIII-D	Doublet III-D	General Atomics, San Diego, Calif.
ISX-B	Impurity Studies Experiment B	Oak Ridge National Laboratory (ORNL), Oak Ridge, Tenn.
JET	Joint European Torus	Abingdon, England
JFT-2M		Japan Atomic Energy Research Institute, Tokai, Japan
JT-60		Japan Atomic Energy Research Institute, Naka, Japan
ORMAK	Oak Ridge Tokamak	ORNL
PDX	Princeton Divertor Experiment	PPPL
PLT	Princeton Large Torus	PPPL
T-3	Tokamak-3	Kurchatov Institute, Moscow
T-10	Tokamak-10	Kurchatov Institute, Moscow
T-15	Tokamak-15	Kurchatov Institute, Moscow
TFR	Tokamak Fontenay-aux-Roses	Centre d'Etudes Nucleaire, Fontenay-aux-Roses, France
TFTR	Tokamak Fusion Test Reactor	PPPL

Additionally, two other tokamaks, the International Thermonuclear Experimental Reactor (ITER) for which the location is under negotiation, and the Tokamak Physics Experiment at PPPL, Princeton, New Jersey, are proposed. The most impressive advances have been obtained on the three biggest tokamaks, TFTR, JET, and JT-60, which are located in the United States, Europe, and Japan, respectively. As of this writing fusion energy development in the United States is dependent on federal funding (10–12).

Until 1992, tokamak experiments were performed using deuterium or hydrogen only. The use of radioactive tritium greatly complicates the operation of experimental facilities, impeding the pace of research. Certain experiments, however, such as those directly involving D–T fusion, cannot be done without the use of tritium. A European research team in 1992 produced nearly 2 million watts of fusion power for about one second in the JET device, and opened the modern frontier of D–T fusion experiments (13). Only about half of the JET fusion energy release came from fusion in the thermal plasma, at temperatures of 15–20 keV. The other half came from fusion of the injected tritium beams striking the deuterium in the plasma. The ratio of tritium to deuterium was about 2% in JET. If a 50:50 mixture of tritium and deuterium had been used instead, an amount of fusion energy would have been released roughly equal to the energy required to heat and sustain the plasma, giving an energy gain, Q, of about unity. In December 1993, scientists at the Princeton Plasma Physics Laboratory initiated a series of experiments on the Tokamak Fusion Test Reactor (TFTR), introducing D–T fuel into the machine and producing over 6 MW of fusion power. For the first time in a tokamak experiment an approximately 50:50 mixture of deuterium and tritium was used as the fusion fuel. Preliminary analysis of the first 100 experimental runs indicated that the confinement in a D–T fuel mixture was better than in a pure deuterium plasma, the ion and electron temperatures were higher, and the plasma stored energy longer. No enhanced loss of alpha particles (the product of D–T fusion reactions) was observed as the fusion power was increased. These results are encouraging for tokamak-based power generation.

International Thermonuclear Experimental Reactor. One of the largest obstacles to the development of fusion power has been that high powered, and correspondingly expensive, research facilities are needed at each step of the reactor development path. ITER (pronounced "eater") is a project supported by the United States, Japan, the European community, and Russia, wherein each party contributes equally to the effort and shares equally in the results (9). The main reason for making the ITER an international effort is cost sharing. The project is managed under the auspices of the International Atomic Energy Agency (IAEA), and the design is based on the tokamak concept. The central purpose of the ITER is to demonstrate the scientific and technological feasibility of fusion power by achieving, for the first time, controlled ignition and extended burn in a D–T plasma. ITER is expected to accomplish this by demonstrating technologies essential to a reactor in an integrated system, and by integrated steady-state testing of the high heat-flux and nuclear components (9).

A conceptual design of ITER, done in 1988–1990 by an international team (14), utilizes superconducting magnets. The heating and current drive are provided by a combination of 1.3 MeV negative-ion neutral beams, lower-hybrid frequency rf, and electron-cyclotron frequency rf. The negatively charged beams of deuterons or tritons are to be accelerated to 1.3 MeV, neutralized, and then injected, unperturbed by the confining magnetic field, into the plasma. The design is based on a conservative assessment of physics knowledge and allows for operational and experimental flexibility. The conceptual design calls for a plasma major radius of 6 m, plasma minor radius of 2.1 m, plasma current of 22 MA, magnetic field of 4.85 T, average neutron wall loading of about 1 MW/m^2, and fusion power of about 1 GW thermal.

The second phase of ITER, the engineering design activity (EDA), was begun in 1992 and is scheduled to be completed in 1998. The ITER engineering design is being conducted at three co-centers: San Diego, California; Garching, Germany; and Naka, Japan. At these co-centers, multinational teams focus on developing a mature design in sufficient detail to allow the construction of the machine, with industrial vendors able to bid on the fabrication and installation of ITER systems. The first ITER plasma could be made as early as 2005. D–T operation could begin a few years later.

Inertial Confinement

Because the maximum plasma density that can be confined is determined by the field strength of available magnets, MFE plasmas at power plant conditions are very diffuse. Typical plasma densities are on the order of one hundred-thousandth that of air at STP. The Lawson criterion is met by confining the plasma energy for periods of about one second. A totally different approach to controlled fusion attempts to create a much denser reacting plasma which, therefore, needs to be confined for a correspondingly shorter time. This is the basis of inertial fusion energy (IFE). In the IFE approach, small capsules or pellets containing fusion fuel are compressed to extremely high densities by intense, focused beams of photons or energetic charged particles as shown in Figure 4 and Plate I. Because of the substantially higher densities involved, the confinement times for IFE can be much shorter. In fact, no external means are required to effect the confinement; the inertia of the fuel mass is sufficient for net energy release to occur before the fuel flies apart. Typical burn times and fuel densities are 10^{-10} s and 10^{31}–10^{32} ions/m^3,

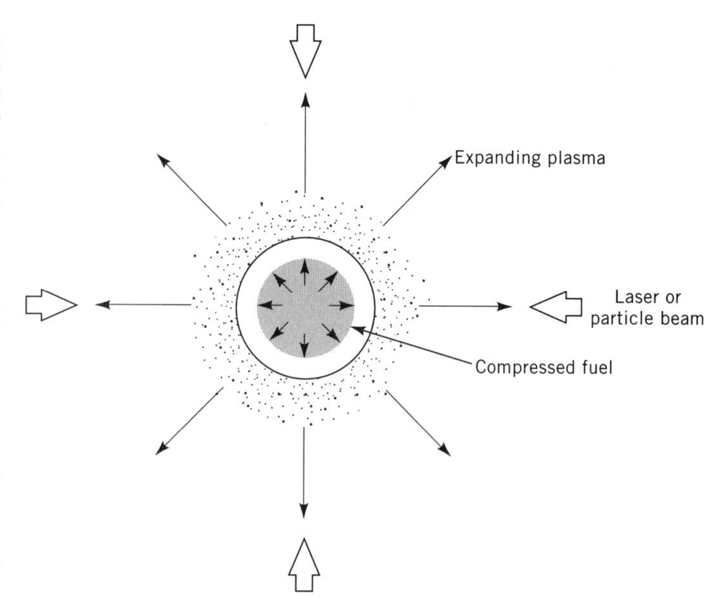

Figure 4. IFE capsule compression.

respectively. These densities correspond to a few hundred to a few thousand times that of ordinary condensed solids. IFE fusion produces the equivalent of small thermonuclear explosions in the target chamber. An IFE power plant design, therefore, must deal with very different physics and technology issues than an MFE power plant, although some requirements, such as tritium breeding, are common to both. Some of the challenges facing IFE power plants include the highly pulsed nature of the burn, the high rate at which the targets must be made and transported to the beam focus, and the interface between the driver beams and the reactor chamber (15).

Drivers. In inertial fusion the fuel is compressed and heated using driver beams. Achieving ignition requires a large amount of energy to be precisely controlled and delivered to the fuel target in a very short time, and the target must be capable of absorbing this energy efficiently. To produce net energy, the IFE system must have gain, ie, more energy output than was used to make, compress, and heat the fuel. Driver efficiency and capsule design and fabrication are therefore important issues for an IFE reactor (16).

The necessary energy can be delivered to the fuel by a variety of possible drivers. The four types of drivers receiving the most research attention are solid state lasers, KrF lasers, light-ion accelerators, and heavy ion accelerators. The leading driver for target physics experiments worldwide is the solid-state laser, and in particular the Nd:glass laser. The reason is that the irradiances required for IFE are in the 10^{18}–10^{19} W/m^2 range (17). The Nd:glass laser was the first driver which could produce these large power densities on target and it has remained in the forefront because of its high performance, reliable technology, and relative ease of maintenance. Low efficiencies and pulse rates have traditionally eliminated Nd:glass lasers from serious consideration in IFE reactor designs. However, new Nd:glass technology, replacing flash lamp pumping

Plate I. This inertial confinement fusion target chamber divides a pulse of light into ten beams which deliver 100-trillion watts of power; the beams simultaneously strike a small pellet of teh hydrogen isotopes deuterium and tritium. The impact heats the pellet to 100-million degrees Clesius and compresses it to a density 20 times that of lead, causing the isotopes to fuse, form helium, and to release energy. Courtesy of the United States Department of Energy.

with higher efficiency diode pumping and utilizing crystalline disks and gas cooling, could change this view. Higher driver efficiencies are achievable in KrF lasers and particle beam accelerators. Particle beams have thus far had difficulty in achieving the low divergences and small focal spots required for IFE experiments, a technical area where lasers have a natural advantage. In IFE power plants, however, focal spots as large as 1 cm are permitted, and it appears that both light ion and heavy ion drivers could meet this requirement.

Targets. Two types of IFE targets have been investigated, known as direct and indirect drive targets. Direct-drive targets absorb the energy of the driver directly onto the fuel capsule, whereas indirect-drive targets use a cavity, called a hohlraum, to convert the driver energy to x-rays which are then absorbed by the fuel capsule. This latter method can tolerate greater inhomogeneities in driver illumination, albeit at the expense of the efficient delivery of energy to the capsule.

The extremely high peak power densities available in particle beams and lasers can heat the small amounts of matter in the fuel capsules to the temperatures required for fusion. In order to attain such temperatures, however, the mass of the fuel capsules must be kept quite low. As a result, the capsules are quite small. Typical dimensions are less than 1 mm. Fuel capsules in power plants could be larger (up to 1 cm) because of the increased driver energies available.

Laser Fusion. The largest and most powerful operating laser in the world is the NOVA 10-beam Nd:glass laser facility at the Lawrence Livermore National Laboratory in California. NOVA can deliver up to 40 kJ of 351-nm light in a 1-ns pulse onto the target. NOVA is primarily used for indirect-drive experiments. Other large Nd:glass laser facilities include the GEKKO XII laser at Osaka University in Japan, and the OMEGA laser at the Laboratory for Laser Energetics at the University of Rochester (Rochester, New York). The latter is used primarily for direct-drive experiments.

The krypton-fluoride (KrF) laser, which uses a gaseous lasing medium, can in principle operate at much higher pulse repetition rates and efficiencies than solid-state Nd:glass lasers. Moreover, the shorter (250 nm) wavelength and broad bandwidth, both of which improve coupling to the target, provide additional advantages. However, the use of KrF lasers is complicated by the long pulse length, which, for the 1 ns time scales of IFE, has to be shortened by a factor of about 100. At least two schemes to do this have been proposed and demonstrated (15). In one method, angular multiplexing, many short, low power pulses are sent sequentially through the laser power amplifier stage for the entire duration of the pumping pulse, each at a different angle. After traversing paths of different optical length, these pulses are recombined at the target into a single high amplitude short pulse. In the second method, a long pulse is extracted and subsequently shortened in a Raman scattering cell filled with, for example, SF_6 gas (see INFRARED TECHNOLOGY AND RAMAN SPECTROSCOPY). Through Raman backscattering, the pulse can be shortened by the desired factor of 100. KrF laser technology is not as well developed as the technology for Nd:glass lasers, however, and no KrF lasers have been constructed which are as powerful as NOVA. The efficiency of KrF lasers may also fall a little short of that needed for a power plant.

Particle Beam Fusion. Advances in pulsed power technology have enabled large quantities of electrical energy to be generated in short pulses with relatively high efficiency and low cost. In a light-ion particle accelerator, an initial electrical pulse of the required energy is progressively shortened through a series of pulse forming steps to be delivered with an amplitude of several tens of megavolts to a diode which emits and accelerates the selected ions, eg, lithium, across a short gap to converge on the fuel capsule. The light-ion Particle Beam Fusion Accelerator II (PBFA II) at Sandia National Laboratory in New Mexico is the most energetic particle beam driver, delivering up to 1 MJ on target. However, obtaining good beam divergence has been a challenge.

To survive the effects of the target explosion, the diode must be located at least several meters away from the target. The diode on PBFA II is only about 15 cm from the target. Long-lived, reliable diodes having 10 Hz repetition rates and beam-transport systems several orders of magnitude longer than those in use as of this writing, are required to make a light-ion beam power plant feasible (15).

The Fusion Policy Advisory Committee of the Department of Energy has identified the heavy-ion accelerator as the leading candidate for an IFE power plant driver (16). The reasons include ruggedness, reliability, high pulse-rate capabilities, and potential for high efficiency. There are two different technologies being developed for heavy-ion accelerators: induction acceleration and radio frequency (rf) acceleration. The induction accelerator approach is pursued mainly in the United States, at the Lawrence Berkeley Laboratory. The rf accelerator approach is pursued primarily in Europe and Japan (15). The same types of heavy ions can be utilized in both approaches; typically cesium, bismuth, or xenon are chosen. To obtain the required 10^{18}–10^{19} W/m^2 on target for a power plant, using targets of 1 cm^2 size and accelerator energies limited to 5 GeV (to provide the requisite stopping distance inside the target fuel), particle beam currents of around 100,000 A are required. These currents are quite large compared to traditional high energy physics accelerators, and in experiments where high currents have been achieved, the beam divergence has been unsatisfactorily large.

ENVIRONMENTAL AND SAFETY ASPECTS

Fusion power plants are expected to be relatively benign environmentally when compared to other sources of power. A 1989 National Research Council report cites the environmental issue as a persuasive reason for pursuing the fusion energy option (10). A general environmental advantage of nuclear power plants whether fission or fusion, compared to fossil fuel plants, is the minimization of mining requirements and no emission of noxious effluents. A further advantage of fusion, relative to fission, is the absence of meltdown dangers and avoidance of long-lived radioactive wastes (see NUCLEAR REACTORS). An accidental runaway reaction cannot occur in a fusion reactor for two reasons. First, the amount of deuterium and tritium in the reactor at any given time is small, and any uncontrolled burning quickly consumes all the available fuel and extinguishes itself. Second, a neutron chain reaction of the fission-reactor type is impossible in fusion, because fusion reaction rates are not sustained by neutrons.

The fusion of deuterium and tritium produces only energetic neutrons and alpha particles (helium nuclei), which are not themselves radioactive. The 14-MeV neutrons are absorbed in a blanket surrounding the reacting plasma, and the only unavoidable ash of the D–T reaction is ordinary helium gas. The main concern about radiation comes from a secondary process, namely activation of the reactor components by the fusion neutrons. The secondary nuclear reactions which result from the energetic neutrons depend on the materials selected for the reactor blanket and support structure (18). The materials aspects of fusion reactors have been reviewed (19), and the calculated decay of radioactivity following shutdown of D–T fusion reactors constructed of various materials is shown in Figure 5, together with that of a fission reactor (8,18). If advanced structural materials such as silicon carbide, SiC, can be used, fusion power plants are expected to reduce the amount of radioactive waste by six orders of magnitude or more.

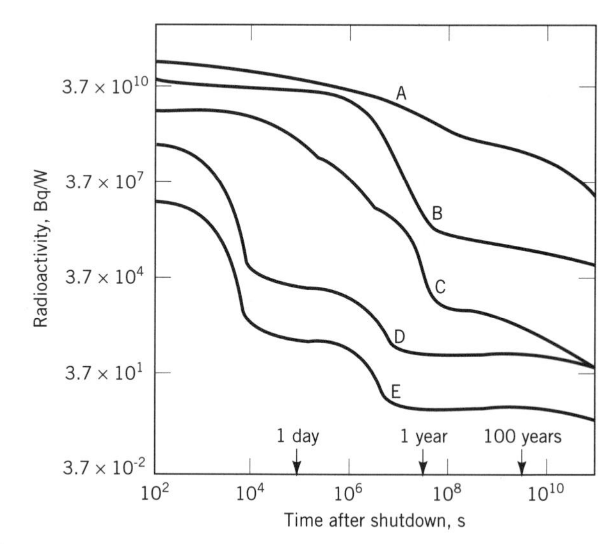

Figure 5. Radioactivity after shutdown per watt of thermal power for A, a liquid-metal fast breeder reactor, and for a D–T fusion reactor made of various structural materials: B, HT-9 ferritic steel; C, V-15Cr-5Ti vanadium–chromium–titanium alloy; and D, silicon carbide, SiC. There is the million-fold advantage of SiC over steel a day after shutdown. The radioactivity level after shutdown is also given for E, a SiC fusion reactor using the neutron reduced D–^{3}He fuel cycle.

A D–T fusion power plant is expected to have a tritium inventory of a few kilograms. Tritium is a relatively short-lived (12.36 year half-life) and benign (beta emitter) radioactive material, and represents a radiological hazard many orders of magnitude less than does the fuel inventory in a fission reactor. Clearly, however, fusion reactors must be designed to preclude the accidental release of tritium or any other volatile radioactive material. There is no need to have fissile materials present in a fusion reactor, and relatively simple inspection techniques should suffice to prevent any clandestine breeding of fissile materials, eg, for potential weapons diversion.

FUTURE DEVELOPMENTS AND APPLICATIONS

The goal of fusion development is central station electrical power generation. Using the D–T fuel cycle, power would be extracted from the thermalization of the neutron kinetic energy deposited in the blanket. Pulsed systems such as inertial fusion require storage techniques to provide a continuous output of electrical power. In some cases, this storage medium may be simply the thermal blanket surrounding the reaction chamber. In MFE, significant technological challenges include the development of large superconducting magnets, efficient current drive systems, and adequate diverter plates and plasma facing components to handle the high particle and radiation heat loads. Provisions must also be made for the replacement and maintenance of components by remote handling techniques.

Potential fusion applications other than electricity production have received some study. For example, radiation and high temperature heat from a fusion power plant could be used to produce hydrogen by the electrolysis or radiolysis of water, which could be employed in the synthesis of portable chemical fuels for transportation or industrial use. The transmutation of radioactive actinide wastes from fission reactors may also be feasible. This idea would utilize the neutrons from a fusion power plant to convert hazardous isotopes into more benign and easier-to-handle species. The practicality of these concepts requires further analysis.

Fusion energy research is also the primary avenue for the development of plasma physics as a scientific discipline. The technologies and the science of plasmas developed en route to fusion power are already important in other applications and fields of science.

COLD FUSION

In the spring of 1989, it was announced that electrochemists at the University of Utah had produced a sustained nuclear fusion reaction at room temperature, using simple equipment available in any high school laboratory. The process, referred to as cold fusion, consists of loading deuterium into pieces of palladium metal by electrolysis of heavy water, D_2O, thereby developing a sufficiently large density of deuterium nuclei in the metal lattice to cause fusion between these nuclei to occur. These results have proven extremely difficult to confirm (20,21). Neutrons usually have not been detected in cold fusion experiments, so that the D–D fusion reaction familiar to nuclear physicists does not seem to be the explanation for the experimental results, which typically involve the release of heat and sometimes gamma rays.

Room temperature fusion reactions, albeit low probability ones, are not a new concept, having been postulated in 1948 and verified experimentally in 1956 (22), in a form of fusion known as muon-catalized fusion. Since the 1989 announcement, however, international scientific skepticism has grown to the point that cold fusion is not considered a serious subject by most scientists. Follow-on experiments, conducted in many prestigious laboratories, have failed to confirm the claims, and although some unexplained and intellectually interesting phenomena have been recorded, the results have remained irreproducable and, thus far, not accepted by the scientific community.

Reprinted from *Kirk-Othmer Encyclopedia of Chemical Technology,* 4th ed., Vol. 12, John Wiley & Sons, Inc., New York, 1995.

BIBLIOGRAPHY

1. H. A. Bethe, *Phys. Rev.* **55,** 103 (1939).
2. S. Glasstone and R. H. Lovberg, *Controlled Thermonuclear Reactions,* D. Van Nostrand, New York, 1960.
3. R. A. Gross, *Fusion Energy,* John Wiley & Sons, Inc., New York, 1984.
4. L. Spitzer, Jr., *Physics and Fully Ionized Gases,* 2nd rev. ed., John Wiley & Sons, Inc., New York, 1962.
5. J. D. Lawson, *Proc. Phys. Soc.* **B70,** 6(1957).
6. O. Naito and co-workers, *Plasma Phys. Control. Fusion* **35,** B215–B222 (1993); T. Kondo and co-workers, "High Performance and Current Drive Experiments in JA-ERI Tokamak-60 Upgrade," *Phys. Plasmas* (in print) (1994); H. Ninomiya and co-workers, "Recent Progress and Future Prospect of the JT-60 Program," in the *Proceedings of the 15th Symposium on Fusion Engineering,* Hyannis, Mass., 1993.
7. I. E. Tamm and A. D. Sakharov, in M. A. Leontovich, ed., *Plasma Physics and the Problem of Controlled Thermonuclear Reactions,* Vol. 1, Pergamon Press, New York, 1961.
8. H. P. Furth, *Science* **249,** 1522 (Sept. 1990); J. G. Gordey, R. J. Goldston, and R. R. Parker, *Phys. Today,* 22 (Jan. 1992).
9. R. W. Conn and co-workers, *Sci. Am.* **266,** 103 (Apr. 1992).
10. *Pacing the U.S. Magnetic Fusion Program,* National Academy Press, Washington, D.C., 1989.
11. *Fusion Policy Advisory Committee Final Report, DOE / S-0081,* Department of Energy, Washington, D.C., Sept. 1990.
12. *National Energy Strategy, First Edition 1991 / 1992,* Department of Energy, Washington, D.C., Feb. 1991.
13. The JET Team, *Nuc. Fusion* **32,** 187 (1992).
14. International Atomic Energy Agency, *ITER Conceptual Design Report,* IAEA, Vienna, 1991.
15. W. J. Hogan, R. Bangerter, and G. L. Kulcinski, *Phys. Today,* **42** (Sept. 1992).
16. *Review of the Department of Energy's Inertial Confinement Fusion Programs,* National Academy Press, Washington, D.C., Sept. 1990.
17. J. D. Lindl, R. L. McCrory, and E. M. Campbell, *Phys. Today,* **32** (Sept. 1992).

18. R. W. Conn and co-workers, *Nucl. Fusion* **30,** 1919 (1990).
19. J. P. Holdren and co-workers, *Report of the Senior Committee on Environmental, Safety, and Economic Aspects of Magnetic Fusion Energy,* report UCRL-53766, Lawrence Livermore National Laboratory, Livermore, Calif., Sept. 25, 1989.
20. *Cold Fusion Research, DOE Report S-0073,* U.S. Dept. of Energy, Washington, D.C., Nov. 1989.
21. F. Close, *Too Hot to Handle: The Role for Cold Fusion,* Princeton University Press, Princeton, N.J., 1991.
22. B. V. Lewenstein and co-workers, *Forum for Applied Research and Public Policy,* **7**(4), 67–107 (Winter 1992).

REFERENCE BOOK
DOES NOT CIRCULATE